Molecular Biology of
THE CELL

Sixth Edition

Molecular Biology of
THE CELL

Sixth Edition

Bruce Alberts

Alexander Johnson

Julian Lewis

David Morgan

Martin Raff

Keith Roberts

Peter Walter

With problems by

John Wilson

Tim Hunt

W. W. NORTON & COMPANY
Independent Publishers Since 1923

W. W. Norton & Company has been independent since its founding in 1923, when William Warder Norton and Mary D. Herter Norton first published lectures delivered at the People's Institute, the adult education division of New York City's Cooper Union. The firm soon expanded its program beyond the Institute, publishing books by celebrated academics from America and abroad. By midcentury, the two major pillars of Norton's publishing program—trade books and college texts—were firmly established. In the 1950s, the Norton family transferred control of the company to its employees, and today—with a staff of five hundred and hundreds of trade, college, and professional titles published each year—W. W. Norton & Company stands as the largest and oldest publishing house owned wholly by its employees.

For Garland Science
Vice President: Denise Schanck
Associate Editor: Allie Bochicchio
Production Editor and Layout: EJ Publishing Services
Senior Production Editor: Georgina Lucas
Text Editors: Sherry Granum Lewis and Elizabeth Zayatz
Illustrator: Nigel Orme
Structures: Tiago Barros
Designer: Matthew McClements, Blink Studio, Ltd.
Copyeditor: Jo Clayton
Proofreader: Sally Huish
Indexer: Bill Johncocks
Permissions Coordinator: Sheri Gilbert
Back Cover Photograph: Photography, Christophe Carlinet; Design, Nigel Orme

Molecular Biology of the Cell Interactive Media:
Artistic and Scientific Direction: Peter Walter
Narration: Julie Theriot
Director of Digital Publishing: Michael Morales
Editorial Assistant: Leah Christians
Production Editor: Natasha Wolfe

Permission to use copyrighted material is included alongside the appropriate content.

Library of Congress Cataloging-in-Publication Data
Alberts, Bruce, author.
 Molecular biology of the cell / Bruce Alberts, Alexander Johnson, Julian Lewis, David Morgan, Martin Raff, Keith Roberts, Peter Walter ; with problems by John Wilson, Tim Hunt. -- Sixth edition.
      p. ; cm.
 Preceded by Molecular biology of the cell / Bruce Alberts ... [et al.]. 5th ed. c2008.
 Includes bibliographical references and index.
  ISBN 978-0-8153-4432-2 (hardcover) -- ISBN 978-0-8153-4464-3 (paperback)
 I. Title.
 [DNLM: 1. Cells. 2. Molecular Biology. QU 300]
 QH581.2
 572.8--dc23
                    2014031818

W. W. Norton & Company, Inc., 500 Fifth Avenue, New York, NY 10110
wwnorton.com
W. W. Norton & Company Ltd., 15 Carlisle Street, London W1D 3BS

15 14 13 12 11 10 9 8 7 6

About the Authors

Bruce Alberts received his PhD from Harvard University and is the Chancellor's Leadership Chair in Biochemistry and Biophysics for Science and Education, University of California, San Francisco. He was the editor-in-chief of Science magazine from 2008 until 2013, and for twelve years he served as President of the U.S. National Academy of Sciences (1993–2005). **Alexander Johnson** received his PhD from Harvard University and is Professor of Microbiology and Immunology at the University of California, San Francisco. **Julian Lewis** (1946–2014) received his DPhil from the University of Oxford and was an Emeritus Scientist at the London Research Institute of Cancer Research UK. **David Morgan** received his PhD from the University of California, San Francisco, and is Professor of the Department of Physiology there as well as the Director of the Biochemistry, Cell Biology, Genetics, and Developmental Biology Graduate Program. **Martin Raff** received his MD from McGill University and is Emeritus Professor of Biology at the Medical Research Council Laboratory for Molecular Cell Biology at University College London. **Keith Roberts** received his PhD from the University of Cambridge and was Deputy Director of the John Innes Centre, Norwich. He is Emeritus Professor at the University of East Anglia. **Peter Walter** received his PhD from the Rockefeller University in New York and is Professor of the Department of Biochemistry and Biophysics at the University of California, San Francisco, and an Investigator at the Howard Hughes Medical Institute. **John Wilson** received his PhD from the California Institute of Technology and pursued his postdoctoral work at Stanford University. He is Distinguished Service Professor of Biochemistry and Molecular Biology at Baylor College of Medicine in Houston. **Tim Hunt** received his PhD from the University of Cambridge where he taught biochemistry and cell biology for more than 20 years. He worked at Cancer Research UK until his retirement in 2010. He shared the 2001 Nobel Prize in Physiology or Medicine with Lee Hartwell and Paul Nurse.

Cover design: Cell biology is not only about the structure and function of the myriad molecules that comprise a cell, but also about how this complex chemistry is controlled. Understanding the cell's elaborate regulatory feedback networks will require quantitative approaches.

Julian Hart Lewis

August 12, 1946—April 30, 2014

Preface

Since the last edition of this book appeared, more than five million scientific papers have been published. There has been a parallel increase in the quantity of digital information: new data on genome sequences, protein interactions, molecular structures, and gene expression—all stored in vast databases. The challenge, for both scientists and textbook writers, is to convert this overwhelming amount of information into an accessible and up-to-date understanding of how cells work.

Help comes from a large increase in the number of review articles that attempt to make raw material easier to digest, although the vast majority of these reviews are still quite narrowly focused. Meanwhile, a rapidly growing collection of online resources tries to convince us that understanding is only a few mouse-clicks away. In some areas this change in the way we access knowledge has been highly successful—in discovering the latest information about our own medical problems, for example. But to understand something of the beauty and complexity of how living cells work, one needs more than just a wiki- this or wiki- that; it is enormously hard to identify the valuable and enduring gems from so much confusing landfill. Much more effective is a carefully wrought narrative that leads logically and progressively through the key ideas, components, and experiments in such a way that readers can build for themselves a memorable, conceptual framework for cell biology— a framework that will allow them to critically evaluate all of the new science and, more importantly, to understand it. That is what we have tried to do in *Molecular Biology of the Cell*.

In preparing this new edition, we have inevitably had to make some difficult decisions. In order to incorporate exciting new discoveries, while at the same time keeping the book portable, much has had to be excised. We have added new sections, such as those on new RNA functions, advances in stem cell biology, new methods for studying proteins and genes and for imaging cells, advances in the genetics and treatment of cancer, and timing, growth control, and morphogenesis in development.

The chemistry of cells is extremely complex, and any list of cell parts and their interactions—no matter how complete—will leave huge gaps in our understanding. We now realize that to produce convincing explanations of cell behavior will require quantitative information about cells that is coupled to sophisticated mathematical/computational approaches—some not yet invented. As a consequence, an emerging goal for cell biologists is to shift their studies more toward quantitative description and mathematical deduction. We highlight this approach and some of its methods in a new section at the end of Chapter 8.

Faced with the immensity of what we have learned about cell biology, it might be tempting for a student to imagine that there is little left to discover. In fact, the more we find out about cells, the more new questions emerge. To emphasize that our understanding of cell biology is incomplete, we have highlighted some of the major gaps in our knowledge by including *What We Don't Know* at the end of each chapter. These brief lists include only a tiny sample of the critical unanswered questions and challenges for the next generation of scientists. We derive great pleasure from the knowledge that some of our readers will provide future answers.

The more than 1500 illustrations have been designed to create a parallel narrative, closely interwoven with the text. We have increased their consistency between chapters, particularly in the use of color and of common icons; membrane pumps and channels are a good example. To avoid interruptions to the text, some material has been moved into new, readily accessible panels. Most of the important protein structures depicted have now been redrawn and consistently colored. In each case, we now provide the corresponding Protein Data Bank (PDB) code for the protein, which can be used to access online tools that provide more information about it, such as those on the RCSB PDB website (www.rcsb.org). These connections allow readers of the book to explore more fully the proteins that lie at the core of cell biology.

John Wilson and Tim Hunt have again contributed their distinctive and imaginative problems to help students gain a more active understanding of the text. The problems emphasize quantitative approaches and encourage critical thinking about published experiments; they are now present at the end of all chapters. The answers to these problems, plus more than 1800 additional problems and solutions, all appear in the companion volume that John and Tim have written, *Molecular Biology of the Cell, Sixth Edition: The Problems Book.*

We live in a world that presents us with many complex issues related to cell biology: biodiversity, climate change, food security, environmental degradation, resource depletion, and human disease. We hope that our textbook will help the reader better understand and possibly contribute to meeting these challenges. Knowledge and understanding bring the power to intervene.

We are indebted to a large number of scientists whose generous help we mention separately in the detailed acknowledgments. Here we must mention some particularly significant contributors. For Chapter 8, Hana El-Samad provided the core of the section on Mathematical Analysis of Cell Functions, and Karen Hopkin made valuable contributions to the section on Studying Gene Expression and Function. Werner Kuhlbrandt helped to reorganize and rewrite Chapter 14 (Energy Conversion: Mitochondria and Chloroplasts). Rebecca Heald did the same for Chapter 16 (The Cytoskeleton), as did Alexander Schier for Chapter 21 (Development of Multicellular Organisms), and Matt Welch for Chapter 23 (Pathogens and Infection). Lewis Lanier aided in the writing of Chapter 24 (The Innate and Adaptive Immune Systems). Hossein Amiri generated the enormous online instructor's question bank.

Before starting out on the revision cycle for this edition, we asked a number of scientists who had used the last edition to teach cell biology students to meet with us and suggest improvements. They gave us useful feedback that has helped inform the new edition. We also benefited from the valuable input of groups of students who read most of the chapters in page proofs.

Many people and much effort are needed to convert a long manuscript and a large pile of sketches into a finished textbook. The publishing team that managed this conversion was outstanding. Denise Schanck, directing operations, displayed forbearance, insight, tact, and energy throughout the journey; she guided us all unerringly, ably assisted by Allie Bochicchio and Janette Scobie. Nigel Orme oversaw our revamped illustration program, put all the artwork into its final form, and again enhanced the back cover with his graphics skills. Tiago Barros helped us refresh our presentation of protein structures. Matthew McClements designed the book and its front cover. Emma Jeffcock again laid out the final pages, managing endless rounds of proofs and last-minute changes with remarkable skill and patience; Georgina Lucas provided her with help. Michael Morales, assisted by Leah Christians, produced and assembled the complex web of videos, animations, and other materials that form the core of the online resources that accompany the book. Adam Sendroff provided us with the valuable feedback from book users around the world that informed our revision cycle. Casting expert eyes over the manuscript, Elizabeth Zayatz and Sherry Granum Lewis acted as development editors, Jo Clayton as copyeditor, and Sally Huish as proofreader. Bill Johncocks compiled the index. In London, Emily Preece fed us, while the publishing team's professional help, skills, and energy, together with their friendship, nourished us in every other way throughout the revision, making the whole process a pleasure. The authors are extremely fortunate to be supported so generously.

We thank our spouses, families, friends, and colleagues for their continuing support, which has once again made the writing of this book possible.

Just as we were completing this edition, Julian Lewis, our coauthor, friend, and colleague, finally succumbed to the cancer that he had fought so heroically for ten years. Starting in 1979, Julian made major contributions to all six editions, and, as our most elegant wordsmith, he elevated and enhanced both the style and tone of all the many chapters he touched. Noted for his careful scholarly approach, clarity and simplicity were at the core of his writing. Julian is irreplaceable, and we will all deeply miss his friendship and collaboration. We dedicate this Sixth Edition to his memory.

Note to the Reader

Structure of the Book

Although the chapters of this book can be read independently of one another, they are arranged in a logical sequence of five parts. The first three chapters of Part I cover elementary principles and basic biochemistry. They can serve either as an introduction for those who have not studied biochemistry or as a refresher course for those who have. Part II deals with the storage, expression, and transmission of genetic information. Part III presents the principles of the main experimental methods for investigating and analyzing cells; here, a new section entitled "Mathematical Analysis of Cell Functions" in Chapter 8 provides an extra dimension in our understanding of cell regulation and function. Part IV describes the internal organization of the cell. Part V follows the behavior of cells in multicellular systems, starting with development of multicellular organisms and concluding with chapters on pathogens and infection and on the innate and adaptive immune systems.

End-of-Chapter Problems

A selection of problems, written by John Wilson and Tim Hunt, appears in the text at the end of each chapter. New to this edition are problems for the last four chapters on multicellular organisms. The complete solutions to all of these problems can be found in *Molecular Biology of the Cell, Sixth Edition: The Problems Book*.

References

A concise list of selected references is included at the end of each chapter. These are arranged in alphabetical order under the main chapter section headings. These references sometimes include the original papers in which important discoveries were first reported.

Glossary Terms

Throughout the book, boldface type has been used to highlight key terms at the point in a chapter where the main discussion occurs. Italic type is used to set off important terms with a lesser degree of emphasis. At the end of the book is an expanded glossary, covering technical terms that are part of the common currency of cell biology; it should be the first resort for a reader who encounters an unfamiliar term. The complete glossary as well as a set of flashcards is available on the Student Website.

Nomenclature for Genes and Proteins

Each species has its own conventions for naming genes; the only common feature is that they are always set in italics. In some species (such as humans), gene names are spelled out all in capital letters; in other species (such as zebrafish), all in lowercase; in yet others (most mouse genes), with the first letter in uppercase and rest in lowercase; or (as in *Drosophila*) with different combinations of uppercase and lowercase, according to whether the first mutant allele to be discovered produced a dominant or recessive phenotype. Conventions for naming protein products are equally varied.

This typographical chaos drives everyone crazy. It is not just tiresome and absurd; it is also unsustainable. We cannot independently define a fresh convention for each of the next few million species whose genes we may wish to study.

Moreover, there are many occasions, especially in a book such as this, where we need to refer to a gene generically—without specifying the mouse version, the human version, the chick version, or the hippopotamus version—because they are all equivalent for the purposes of our discussion. What convention then should we use?

We have decided in this book to cast aside the different conventions that are used in individual species and follow a uniform rule: we write all gene names, like the names of people and places, with the first letter in uppercase and the rest in lower-case, but all in italics, thus: *Apc, Bazooka, Cdc2, Dishevelled, Egl1*. The corresponding protein, where it is named after the gene, will be written in the same way, but in roman rather than italic letters: Apc, Bazooka, Cdc2, Dishevelled, Egl1. When it is necessary to specify the organism, this can be done with a prefix to the gene name.

For completeness, we list a few further details of naming rules that we shall follow. In some instances, an added letter in the gene name is traditionally used to distinguish between genes that are related by function or evolution; for those genes, we put that letter in uppercase if it is usual to do so (*LacZ, RecA, HoxA4*). We use no hyphen to separate added letters or numbers from the rest of the name. Proteins are more of a problem. Many of them have names in their own right, assigned to them before the gene was named. Such protein names take many forms, although most of them traditionally begin with a lowercase letter (actin, hemoglobin, catalase), like the names of ordinary substances (cheese, nylon), unless they are acronyms (such as GFP, for Green Fluorescent Protein, or BMP4, for Bone Morphogenetic Protein #4). To force all such protein names into a uniform style would do too much violence to established usages, and we shall simply write them in the traditional way (actin, GFP, and so on). For the corresponding gene names in all these cases, we shall nevertheless follow our standard rule: *Actin, Hemoglobin, Catalase, Bmp4, Gfp*. Occasionally in our book we need to highlight a protein name by setting it in italics for emphasis; the intention will generally be clear from the context.

For those who wish to know them, the table below shows some of the official conventions for individual species—conventions that we shall mostly violate in this book, in the manner shown.

Organism	Species-Specific Convention		Unified Convention Used in This Book	
	Gene	Protein	Gene	Protein
Mouse	*Hoxa4*	Hoxa4	*HoxA4*	HoxA4
	Bmp4	BMP4	*Bmp4*	BMP4
	integrin α-1, Itgα1	integrin α1	*Integrin α1, Itgα1*	integrin α1
Human	*HOXA4*	HOXA4	*HoxA4*	HoxA4
Zebrafish	*cyclops, cyc*	Cyclops, Cyc	*Cyclops, Cyc*	Cyclops, Cyc
Caenorhabditis	*unc-6*	UNC-6	*Unc6*	Unc6
Drosophila	*sevenless, sev* (named after recessive phenotype)	Sevenless, SEV	*Sevenless, Sev*	Sevenless, Sev
	Deformed, Dfd (named after dominant mutant phenotype)	Deformed, DFD	*Deformed, Dfd*	Deformed, Dfd
Yeast				
Saccharomyces cerevisiae (budding yeast)	*CDC28*	Cdc28, Cdc28p	*Cdc28*	Cdc28
Schizosaccharomyces pombe (fission yeast)	*Cdc2*	Cdc2, Cdc2p	*Cdc2*	Cdc2
Arabidopsis	*GAI*	GAI	*Gai*	GAI
E. coli	*uvrA*	UvrA	*UvrA*	UvrA

Molecular Biology of the Cell, Sixth Edition: The Problems Book
by John Wilson and Tim Hunt (ISBN: 978-0-8153-4453-7)
The Problems Book is designed to help students appreciate the ways in which experiments and simple calculations can lead to an understanding of how cells work. It provides problems to accompany Chapters 1–20 of *Molecular Biology of the Cell*. Each chapter of problems is divided into sections that correspond to those of the main textbook and review key terms, test for understanding basic concepts, pose research-based problems, and now include MCAT-style questions which help students to prepare for standardized medical school admission tests. *Molecular Biology of the Cell, Sixth Edition: The Problems Book* should be useful for homework assignments and as a basis for class discussion. It could even provide ideas for exam questions. Solutions for all of the problems are provided in the book. Solutions for the end-of-chapter problems for Chapters 1–24 in the main textbook are also found in *The Problems Book*.

RESOURCES FOR INSTRUCTORS AND STUDENTS

The teaching and learning resources for instructors and students are available online. The instructor's resources are password-protected and available only to adopting instructors. The student resources are available to everyone. We hope these resources will enhance student learning and make it easier for instructors to prepare dynamic lectures and activities for the classroom.

Instructor Resources

Instructor Resources are available at wwnorton.com/instructors. Adopting instructors can obtain access to the site from their sales representative. You can identify your representative by visiting this site: wwnorton.com/educator, and clicking the "Find My Rep" button.

Art of Molecular Biology of the Cell, Sixth Edition
The images from the book are available in two convenient formats: PowerPoint® and JPEG. They have been optimized for display on a computer.

Figure-Integrated Lecture Outlines
The section headings, concept headings, and figures from the text have been integrated into PowerPoint presentations. These will be useful for instructors who would like a head start creating lectures for their course. Like all of our PowerPoint presentations, the lecture outlines can be customized. For example, the content of these presentations can be combined with videos and questions from the book or Question Bank, in order to create unique lectures that facilitate interactive learning.

Animations and Videos
174 animations and videos are available to students and instructors at the Digital Landing Page for the book: digital.wwnorton.com/mboc6. The movies are correlated to each chapter of the book and are available at no additional charge.

Question Bank
Written by Hossein Amiri, University of California, Santa Cruz, this greatly expanded question bank includes a variety of question formats: multiple choice, short answer, fill-in-the-blank, true-false, and matching. There are 35–60 questions per chapter, and a large number of the multiple-choice questions will be suitable for use with personal response systems (that is, clickers). The Question Bank was created with the philosophy that a good exam should do much more than simply test students' ability to memorize information; it should require them to reflect upon and integrate information as a part of a sound understanding. This resource provides a comprehensive sampling of questions that can be used either directly or as inspiration for instructors to write their own test questions.

Medical Topics Guide
This document highlights medically relevant topics covered throughout *Molecular Biology of the Cell* and *The Problems Book*. It will be particularly useful

for instructors with a large number of premedical, health science, or nursing students.

Resources for Students

The resources for students are available on the *Molecular Biology of the Cell*, Sixth Edition, Digital Landing Page located at digital.wwnorton.com/mboc6.

Animations and Videos
There are 174 movies, covering a wide range of cell biology topics, which review key concepts in the book and illuminate subcellular processes. The movies are correlated to each chapter.

Cell Explorer Slides
This application teaches cell morphology through interactive micrographs that highlight important cellular structures.

Flashcards
Each chapter contains a set of flashcards, built into the student site, that allow students to review key terms from the text.

Glossary
The complete glossary from the book is available on the student site.

Acknowledgments

In writing this book we have benefited greatly from the advice of many biologists and biochemists. We would like to thank the following for their suggestions in preparing this edition, as well as those who helped in preparing the first, second, third, fourth, and fifth editions. (Those who helped on this edition are listed first, those who helped with the first, second, third, fourth, and fifth editions follow.)

General:
Steven Cook (Imperial College London), Jose A. Costoya (Universidade de Santiago de Compostela), Arshad Desai (University of California, San Diego), Susan K. Dutcher (Washington University, St. Louis), Michael Elowitz (California Institute of Technology), Benjamin S. Glick (University of Chicago), Gregory Hannon (Cold Spring Harbor Laboratories), Rebecca Heald (University of California, Berkeley), Stefan Kanzok (Loyola University Chicago), Doug Kellogg (University of California, Santa Cruz), David Kimelman (University of Washington, Seattle), Maria Krasilnikova (Pennsylvania State University), Werner Kühlbrandt (Max Planck Institute of Biophysics), Lewis Lanier (University of California, San Francisco), Annette Müller-Taubenberger (Ludwig Maximilians University), Sandra Schmid (University of Texas Southwestern), Ronald D. Vale (University of California, San Francisco), D. Eric Walters (Chicago Medical School), Karsten Weis (Swiss Federal Institute of Technology)

Chapter 2: H. Lill (VU University)

Chapter 3: David S. Eisenberg (University of California, Los Angeles), F. Ulrich Hartl (Max Planck Institute of Biochemistry), Louise Johnson (University of Oxford), H. Lill (VU University), Jonathan Weissman (University of California, San Francisco)

Chapter 4: Bradley E. Bernstein (Harvard Medical School), Wendy Bickmore (MRC Human Genetics Unit, Edinburgh), Jason Brickner (Northwestern University), Gary Felsenfeld (NIH), Susan M. Gasser (University of Basel), Shiv Grewal (National Cancer Institute), Gary Karpen (University of California, Berkeley), Eugene V. Koonin, (NCBI, NLM, NIH), Hiten Madhani (University of California, San Francisco), Tom Misteli (National Cancer Institute), Geeta Narlikar (University of California, San Francisco), Maynard Olson (University of Washington, Seattle), Stephen Scherer (University of Toronto), Rolf Sternglanz (Stony Brook University), Chris L. Woodcock (University of Massachusetts, Amherst), Johanna Wysocka and lab members (Stanford School of Medicine)

Chapter 5: Oscar Aparicio (University of Southern California), Julie P. Cooper (National Cancer Institute), Neil Hunter (Howard Hughes Medical Institute), Karim Labib (University of Manchester), Joachim Li (University of California, San Francisco), Stephen West (Cancer Research UK), Richard D. Wood (University of Pittsburgh Cancer Institute)

Chapter 6: Briana Burton (Harvard University), Richard H. Ebright (Rutgers University), Daniel Finley (Harvard Medical School), Michael R. Green (University of Massachusetts Medical School), Christine Guthrie (University of California, San Francisco), Art Horwich (Yale School of Medicine), Harry Noller (University of California, Santa Cruz), David Tollervey (University of Edinburgh), Alexander J. Varshavsky (California Institute of Technology)

Chapter 7: Adrian Bird (The Wellcome Trust Centre, UK), Neil Brockdorff (University of Oxford), Christine Guthrie (University of California, San Francisco), Jeannie Lee (Harvard Medical School), Michael Levine (University of California, Berkeley), Hiten Madhani (University of California, San Francisco), Duncan Odom (Cancer Research UK), Kevin Struhl (Harvard Medical School), Jesper Svejstrup (Cancer Research UK)

Chapter 8: Hana El-Samad [major contribution] (University of California, San Francisco), Karen Hopkin [major contribution], Donita Brady (Duke University), David Kashatus (University of Virginia), Melanie McGill (University of Toronto), Alex Mogilner (University of California, Davis), Richard Morris (John Innes Centre, UK), Prasanth Potluri (The Children's Hospital of Philadelphia Research Institute), Danielle Vidaurre (University of Toronto), Carmen Warren (University of California, Los Angeles), Ian Woods (Ithaca College)

Chapter 9: Douglas J. Briant (University of Victoria), Werner Kühlbrandt (Max Planck Institute of Biophysics), Jeffrey Lichtman (Harvard University), Jennifer Lippincott-Schwartz (NIH), Albert Pan (Georgia Regents University), Peter Shaw (John Innes Centre, UK), Robert H. Singer (Albert Einstein School of Medicine), Kurt Thorn (University of California, San Francisco)

Chapter 10: Ari Helenius (Swiss Federal Institute of Technology), Werner Kühlbrandt (Max Planck Institute of Biophysics), H. Lill (VU University), Satyajit Mayor (National Centre for Biological Sciences, India), Kai Simons (Max Planck Institute of Molecular Cell Biology and Genetics), Gunnar von Heijne (Stockholm University), Tobias Walther (Harvard University)

Chapter 11: Graeme Davis (University of California, San Francisco), Robert Edwards (University of California, San

Francisco), Bertil Hille (University of Washington, Seattle), Lindsay Hinck (University of California, Santa Cruz), Werner Kühlbrandt (Max Planck Institute of Biophysics), H. Lill (VU University), Roger Nicoll (University of California, San Francisco), Poul Nissen (Aarhus University), Robert Stroud (University of California, San Francisco), Karel Svoboda (Howard Hughes Medical Institute), Robert Tampé (Goethe-University Frankfurt)

Chapter 12: John Aitchison (Institute for System Biology, Seattle), Amber English (University of Colorado at Boulder), Ralf Erdmann (Ruhr University of Bochum), Larry Gerace (The Scripps Research Institute, La Jolla), Ramanujan Hegde (MRC Laboratory of Molecular Biology, Cambridge, UK), Martin W. Hetzer (The Salk Institute), Lindsay Hinck (University of California, Santa Cruz), James A. McNew (Rice University), Nikolaus Pfanner (University of Freiberg), Peter Rehling (University of Göttingen), Michael Rout (The Rockefeller University), Danny J. Schnell (University of Massachusetts, Amherst), Sebastian Schuck (University of Heidelberg), Suresh Subramani (University of California, San Diego), Gia Voeltz (University of Colorado, Boulder), Susan R. Wente (Vanderbilt University School of Medicine)

Chapter 13: Douglas J. Briant (University of Victoria, Canada), Scott D. Emr (Cornell University), Susan Ferro-Novick (University of California, San Diego), Benjamin S. Glick (University of Chicago), Ari Helenius (Swiss Federal Institute of Technology), Lindsay Hinck (University of California, Santa Cruz), Reinhard Jahn (Max Planck Institute for Biophysical Chemistry), Ira Mellman (Genentech), Peter Novick (University of California, San Diego), Hugh Pelham (MRC Laboratory of Molecular Biology, Cambridge, UK), Graham Warren (Max F. Perutz Laboratories, Vienna), Marino Zerial (Max Planck Institute of Molecular Cell Biology and Genetics)

Chapter 14: Werner Kühlbrandt [major contribution] (Max Planck Institute of Biophysics), Thomas D. Fox (Cornell University), Cynthia Kenyon (University of California, San Francisco), Nils-Göran Larsson (Max Planck Institute for Biology of Aging), Jodi Nunnari (University of California, Davis), Patrick O'Farrell (University of California, San Francisco), Alastair Stewart (The Victor Chang Cardiac Research Institute, Australia), Daniela Stock (The Victor Chang Cardiac Research Institute, Australia), Michael P. Yaffe (California Institute for Regenerative Medicine)

Chapter 15: Henry R. Bourne (University of California, San Francisco), Dennis Bray (University of Cambridge), Douglas J. Briant (University of Victoria, Canada), James Briscoe (MRC National Institute for Medical Research, UK), James Ferrell (Stanford University), Matthew Freeman (MRC Laboratory of Molecular Biology, Cambridge, UK), Alan Hall (Memorial Sloan Kettering Cancer Center), Carl-Henrik Heldin (Uppsala University), James A. McNew (Rice University), Roel Nusse (Stanford University), Julie Pitcher (University College London)

Chapter 16: Rebecca Heald [major contribution] (University of California, Berkeley), Anna Akhmanova (Utrecht University), Arshad Desai (University of California, San Diego), Velia Fowler (The Scripps Research Institute, La Jolla), Vladimir Gelfand (Northwestern University), Robert Goldman (Northwestern University), Alan Rick Horwitz (University of Virginia), Wallace Marshall (University of California, San Francisco), J. Richard McIntosh

(University of Colorado, Boulder), Maxence Nachury (Stanford School of Medicine), Eva Nogales (University of California, Berkeley), Samara Reck-Peterson (Harvard Medical School), Ronald D. Vale (University of California, San Francisco), Richard B. Vallee (Columbia University), Michael Way (Cancer Research UK), Orion Weiner (University of California, San Francisco), Matthew Welch (University of California, Berkeley)

Chapter 17: Douglas J. Briant (University of Victoria, Canada), Lindsay Hinck (University of California, Santa Cruz), James A. McNew (Rice University)

Chapter 18: Emily D. Crawford (University of California, San Francisco), James A. McNew (Rice University), Shigekazu Nagata (Kyoto University), Jim Wells (University of California, San Francisco)

Chapter 19: Jeffrey Axelrod (Stanford University School of Medicine), John Couchman (University of Copenhagen), Johan de Rooij (The Hubrecht Institute, Utrecht), Benjamin Geiger (Weizmann Institute of Science, Israel), Andrew P. Gilmore (University of Manchester), Tony Harris (University of Toronto), Martin Humphries (University of Manchester), Andreas Prokop (University of Manchester), Charles Streuli (University of Manchester), Masatoshi Takeichi (RIKEN Center for Developmental Biology, Japan), Barry Thompson (Cancer Research UK), Kenneth M. Yamada (NIH), Alpha Yap (The University of Queensland, Australia)

Chapter 20: Anton Berns (Netherlands Cancer Institute), J. Michael Bishop (University of California, San Francisco), Trever Bivona (University of California, San Francisco), Fred Bunz (Johns Hopkins University), Paul Edwards (University of Cambridge), Ira Mellman (Genentech), Caetano Reis e Sousa (Cancer Research UK), Marc Shuman (University of California, San Francisco), Mike Stratton (Wellcome Trust Sanger Institute, UK), Ian Tomlinson (Cancer Research UK)

Chapter 21: Alex Schier [major contribution] (Harvard University), Markus Affolter (University of Basel), Victor Ambros (University of Massachusetts, Worcester), James Briscoe (MRC National Institute for Medical Research, UK), Donald Brown (Carnegie Institution for Science, Baltimore), Steven Burden (New York University School of Medicine), Moses Chao (New York University School of Medicine), Caroline Dean (John Innes Centre, UK), Chris Doe (University of Oregon, Eugene), Uwe Drescher (King's College London), Gordon Fishell (New York University School of Medicine), Brigid Hogan (Duke University), Phil Ingham (Institute of Molecular and Cell Biology, Singapore), Laura Johnston (Columbia University), David Kingsley (Stanford University), Tom Kornberg (University of California, San Francisco), Richard Mann (Columbia University), Andy McMahon (University of Southern California), Marek Mlodzik (Mount Sinai Hospital, New York), Patrick O'Farrell (University of California, San Francisco), Duojia Pan (Johns Hopkins Medical School), Olivier Pourquie (Harvard Medical School), Erez Raz (University of Muenster), Chris Rushlow (New York University), Stephen Small (New York University), Marc Tessier-Lavigne (Rockefeller University)

Chapter 22: Simon Hughes (King's College London), Rudolf Jaenisch (Massachusetts Institute of Technology), Arnold Kriegstein (University of California, San Francisco), Doug Melton (Harvard University), Stuart Orkin (Harvard

University), Thomas A. Reh (University of Washington, Seattle), Amy Wagers (Harvard University), Fiona M. Watt (Wellcome Trust Centre for Stem Cell Research, UK), Douglas J. Winton (Cancer Research UK), Shinya Yamanaka (Kyoto University)

Chapter 23: Matthew Welch [major contribution] (University of California, Berkeley), Ari Helenius (Swiss Federal Institute of Technology), Dan Portnoy (University of California, Berkeley), David Sibley (Washington University, St. Louis), Michael Way (Cancer Research UK)

Chapter 24: Lewis Lanier (University of California, San Francisco).

Readers: Najla Arshad (Indian Institute of Science), Venice Chiueh (University of California, Berkeley), Quyen Huynh (University of Toronto), Rachel Kooistra (Loyola University, Chicago), Wes Lewis (University of Alabama), Eric Nam (University of Toronto), Vladislav Ryvkin (Stony Brook University), Laasya Samhita (Indian Institute of Science), John Senderak (Jefferson Medical College), Phillipa Simons (Imperial College, UK), Anna Constance Vind (University of Copenhagen), Steve Wellard (Pennsylvania State University), Evan Whitehead (University of California, Berkeley), Carrie Wilczewski (Loyola University, Chicago), Anna Wing (Pennsylvania State University), John Wright (University of Alabama)

First, second, third, fourth, and fifth editions:
Jerry Adams (The Walter and Eliza Hall Institute of Medical Research, Australia), Ralf Adams (London Research Institute), David Agard (University of California, San Francisco), Julie Ahringer (The Gurdon Institute, UK), Michael Akam (University of Cambridge), David Allis (The Rockefeller University), Wolfhard Almers (Oregon Health and Science University), Fred Alt (CBR Institute for Biomedical Research, Boston), Linda Amos (MRC Laboratory of Molecular Biology, Cambridge), Raul Andino (University of California, San Francisco), Clay Armstrong (University of Pennsylvania), Martha Arnaud (University of California, San Francisco), Spyros Artavanis-Tsakonas (Harvard Medical School), Michael Ashburner (University of Cambridge), Jonathan Ashmore (University College London), Laura Attardi (Stanford University), Tayna Awabdy (University of California, San Francisco), Jeffrey Axelrod (Stanford University Medical Center), Peter Baker (deceased), David Baldwin (Stanford University), Michael Banda (University of California, San Francisco), Cornelia Bargmann (The Rockefeller University), Ben Barres (Stanford University), David Bartel (Massachusetts Institute of Technology), Konrad Basler (University of Zurich), Wolfgang Baumeister (Max Planck Institute of Biochemistry), Michael Bennett (Albert Einstein College of Medicine), Darwin Berg (University of California, San Diego), Anton Berns (Netherlands Cancer Institute), Merton Bernfield (Harvard Medical School), Michael Berridge (The Babraham Institute, Cambridge, UK), Walter Birchmeier (Max Delbrück Center for Molecular Medicine, Germany), Adrian Bird (Wellcome Trust Centre, UK), David Birk (UMDNJ—Robert Wood Johnson Medical School), Michael Bishop (University of California, San Francisco), Elizabeth Blackburn (University of California, San Francisco), Tim Bliss (National Institute for Medical Research, London), Hans Bode (University of California, Irvine), Piet Borst (Jan Swammerdam Institute, University

of Amsterdam), Henry Bourne (University of California, San Francisco), Alan Boyde (University College London), Martin Brand (University of Cambridge), Carl Branden (deceased), Andre Brandli (Swiss Federal Institute of Technology, Zurich), Dennis Bray (University of Cambridge), Mark Bretscher (MRC Laboratory of Molecular Biology, Cambridge), James Briscoe (National Institute for Medical Research, UK), Marianne Bronner-Fraser (California Institute of Technology), Robert Brooks (King's College London), Barry Brown (King's College London), Michael Brown (University of Oxford), Michael Bulger (University of Rochester Medical Center), Fred Bunz (Johns Hopkins University), Steve Burden (New York University School of Medicine), Max Burger (University of Basel), Stephen Burley (SGX Pharmaceuticals), Keith Burridge (University of North Carolina, Chapel Hill), John Cairns (Radcliffe Infirmary, Oxford), Patricia Calarco (University of California, San Francisco), Zacheus Cande (University of California, Berkeley), Lewis Cantley (Harvard Medical School), Charles Cantor (Columbia University), Roderick Capaldi (University of Oregon), Mario Capecchi (University of Utah), Michael Carey (University of California, Los Angeles), Adelaide Carpenter (University of California, San Diego), John Carroll (University College London), Tom Cavalier-Smith (King's College London), Pierre Chambon (University of Strasbourg), Hans Clevers (Hubrecht Institute, The Netherlands), Enrico Coen (John Innes Institute, Norwich, UK), Philip Cohen (University of Dundee, Scotland), Robert Cohen (University of California, San Francisco), Stephen Cohen (EMBL Heidelberg, Germany), Roger Cooke (University of California, San Francisco), John Cooper (Washington University School of Medicine, St. Louis), Michael Cox (University of Wisconsin, Madison), Nancy Craig (Johns Hopkins University), James Crow (University of Wisconsin, Madison), Stuart Cull-Candy (University College London), Leslie Dale (University College London), Caroline Damsky (University of California, San Francisco), Johann De Bono (The Institute of Cancer Research, UK), Anthony DeFranco (University of California, San Francisco), Abby Dernburg (University of California, Berkeley), Arshad Desai (University of California, San Diego), Michael Dexter (The Wellcome Trust, UK), John Dick (University of Toronto, Canada), Christopher Dobson (University of Cambridge), Russell Doolittle (University of California, San Diego), W. Ford Doolittle (Dalhousie University, Canada), Julian Downward (Cancer Research UK), Keith Dudley (King's College London), Graham Dunn (MRC Cell Biophysics Unit, London), Jim Dunwell (John Innes Institute, Norwich, UK), Bruce Edgar (Fred Hutchinson Cancer Research Center, Seattle), Paul Edwards (University of Cambridge), Robert Edwards (University of California, San Francisco), David Eisenberg (University of California, Los Angeles), Sarah Elgin (Washington University, St. Louis), Ruth Ellman (Institute of Cancer Research, Sutton, UK), Beverly Emerson (The Salk Institute), Charles Emerson (University of Virginia), Scott D. Emr (Cornell University), Sharyn Endow (Duke University), Lynn Enquist (Princeton University), Tariq Enver (Institute of Cancer Research, London), David Epel (Stanford University), Gerard Evan (University of California, Comprehensive Cancer Center), Ray Evert (University of Wisconsin, Madison), Matthias Falk (Lehigh University), Stanley Falkow (Stanford

University), Douglas Fearon (University of Cambridge), Gary Felsenfeld (NIH), Stuart Ferguson (University of Oxford), James Ferrell (Stanford University), Christine Field (Harvard Medical School), Daniel Finley (Harvard University), Gary Firestone (University of California, Berkeley), Gerald Fischbach (Columbia University), Robert Fletterick (University of California, San Francisco), Harvey Florman (Tufts University), Judah Folkman (Harvard Medical School), Larry Fowke (University of Saskatchewan, Canada), Jennifer Frazier (Exploratorium®, San Francisco), Matthew Freeman (Laboratory of Molecular Biology, UK), Daniel Friend (University of California, San Francisco), Elaine Fuchs (University of Chicago), Joseph Gall (Carnegie Institution of Washington), Richard Gardner (University of Oxford), Anthony Gardner-Medwin (University College London), Peter Garland (Institute of Cancer Research, London), David Garrod (University of Manchester, UK), Susan M. Gasser (University of Basel), Walter Gehring (Biozentrum, University of Basel), Benny Geiger (Weizmann Institute of Science, Rehovot, Israel), Larry Gerace (The Scripps Research Institute), Holger Gerhardt (London Research Institute), John Gerhart (University of California, Berkeley), Günther Gerisch (Max Planck Institute of Biochemistry), Frank Gertler (Massachusetts Institute of Technology), Sankar Ghosh (Yale University School of Medicine), Alfred Gilman (The University of Texas Southwestern Medical Center), Reid Gilmore (University of Massachusetts, Amherst), Bernie Gilula (deceased), Charles Gilvarg (Princeton University), Benjamin S. Glick (University of Chicago), Michael Glotzer (University of Chicago), Larry Goldstein (University of California, San Diego), Bastien Gomperts (University College Hospital Medical School, London), Daniel Goodenough (Harvard Medical School), Jim Goodrich (University of Colorado, Boulder), Jeffrey Gordon (Washington University, St. Louis), Peter Gould (Middlesex Hospital Medical School, London), Alan Grafen (University of Oxford), Walter Gratzer (King's College London), Michael Gray (Dalhousie University), Douglas Green (St. Jude Children's Hospital), Howard Green (Harvard University), Michael Green (University of Massachusetts, Amherst), Leslie Grivell (University of Amsterdam), Carol Gross (University of California, San Francisco), Frank Grosveld (Erasmus Universiteit, The Netherlands), Michael Grunstein (University of California, Los Angeles), Barry Gumbiner (Memorial Sloan Kettering Cancer Center), Brian Gunning (Australian National University, Canberra), Christine Guthrie (University of California, San Francisco), James Haber (Brandeis University), Ernst Hafen (Universitat Zurich), David Haig (Harvard University), Andrew Halestrap (University of Bristol, UK), Alan Hall (Memorial Sloan Kettering Cancer Center), Jeffrey Hall (Brandeis University), John Hall (University of Southampton, UK), Zach Hall (University of California, San Francisco), Douglas Hanahan (University of California, San Francisco), David Hanke (University of Cambridge), Nicholas Harberd (University of Oxford), Graham Hardie (University of Dundee, Scotland), Richard Harland (University of California, Berkeley), Adrian Harris (Cancer Research UK), John Harris (University of Otago, New Zealand), Stephen Harrison (Harvard University), Leland Hartwell (University of Washington, Seattle), Adrian Harwood (MRC Laboratory for Molecular Cell Biology and Cell Biology Unit, London),

Scott Hawley (Stowers Institute for Medical Research, Kansas City), Rebecca Heald (University of California, Berkeley), John Heath (University of Birmingham, UK), Ramanujan Hegde (NIH), Carl-Henrik Heldin (Uppsala University), Ari Helenius (Swiss Federal Institute of Technology), Richard Henderson (MRC Laboratory of Molecular Biology, Cambridge, UK), Glenn Herrick (University of Utah), Ira Herskowitz (deceased), Bertil Hille (University of Washington, Seattle), Alan Hinnebusch (NIH, Bethesda), Brigid Hogan (Duke University), Nancy Hollingsworth (State University of New York, Stony Brook), Frank Holstege (University Medical Center, The Netherlands), Leroy Hood (Institute for Systems Biology, Seattle), John Hopfield (Princeton University), Robert Horvitz (Massachusetts Institute of Technology), Art Horwich (Yale University School of Medicine), David Housman (Massachusetts Institute of Technology), Joe Howard (Max Planck Institute of Molecular Cell Biology and Genetics), Jonathan Howard (University of Washington, Seattle), James Hudspeth (The Rockefeller University), Simon Hughes (King's College London), Martin Humphries (University of Manchester, UK), Tim Hunt (Cancer Research UK), Neil Hunter (University of California, Davis), Laurence Hurst (University of Bath, UK), Jeremy Hyams (University College London), Tony Hyman (Max Planck Institute of Molecular Cell Biology and Genetics), Richard Hynes (Massachusetts Institute of Technology), Philip Ingham (University of Sheffield, UK), Kenneth Irvine (Rutgers University), Robin Irvine (University of Cambridge), Norman Iscove (Ontario Cancer Institute, Toronto), David Ish-Horowicz (Cancer Research UK), Lily Jan (University of California, San Francisco), Charles Janeway (deceased), Tom Jessell (Columbia University), Arthur Johnson (Texas A&M University), Louise Johnson (deceased), Andy Johnston (John Innes Institute, Norwich, UK), E.G. Jordan (Queen Elizabeth College, London), Ron Kaback (University of California, Los Angeles), Michael Karin (University of California, San Diego), Eric Karsenti (European Molecular Biology Laboratory, Germany), Ken Keegstra (Michigan State University), Ray Keller (University of California, Berkeley), Douglas Kellogg (University of California, Santa Cruz), Regis Kelly (University of California, San Francisco), John Kendrick-Jones (MRC Laboratory of Molecular Biology, Cambridge), Cynthia Kenyon (University of California, San Francisco), Roger Keynes (University of Cambridge), Judith Kimble (University of Wisconsin, Madison), Robert Kingston (Massachusetts General Hospital), Marc Kirschner (Harvard University), Richard Klausner (NIH), Nancy Kleckner (Harvard University), Mike Klymkowsky (University of Colorado, Boulder), Kelly Komachi (University of California, San Francisco), Eugene Koonin (NIH), Juan Korenbrot (University of California, San Francisco), Roger Kornberg (Stanford University), Tom Kornberg (University of California, San Francisco), Stuart Kornfeld (Washington University, St. Louis), Daniel Koshland (University of California, Berkeley), Douglas Koshland (Carnegie Institution of Washington, Baltimore), Marilyn Kozak (University of Pittsburgh), Mark Krasnow (Stanford University), Werner Kühlbrandt (Max Planck Institute for Biophysics), John Kuriyan (University of California, Berkeley), Robert Kypta (MRC Laboratory for Molecular Cell Biology, London), Peter Lachmann

(MRC Centre, Cambridge), Ulrich Laemmli (University of Geneva, Switzerland), Trevor Lamb (University of Cambridge), Hartmut Land (Cancer Research UK), David Lane (University of Dundee, Scotland), Jane Langdale (University of Oxford), Lewis Lanier (University of California, San Francisco), Jay Lash (University of Pennsylvania), Peter Lawrence (MRC Laboratory of Molecular Biology, Cambridge), Paul Lazarow (Mount Sinai School of Medicine), Robert J. Lefkowitz (Duke University), Michael Levine (University of California, Berkeley), Warren Levinson (University of California, San Francisco), Alex Levitzki (Hebrew University, Israel), Ottoline Leyser (University of York, UK), Joachim Li (University of California, San Francisco), Tomas Lindahl (Cancer Research UK), Vishu Lingappa (University of California, San Francisco), Jennifer Lippincott-Schwartz (NIH), Joseph Lipsick (Stanford University School of Medicine), Dan Littman (New York University School of Medicine), Clive Lloyd (John Innes Institute, Norwich, UK), Richard Locksley (University of California, San Francisco), Richard Losick (Harvard University), Daniel Louvard (Institut Curie, France), Robin Lovell-Badge (National Institute for Medical Research, London), Scott Lowe (Cold Spring Harbor Laboratory), Shirley Lowe (University of California, San Francisco), Reinhard Lührman (Max Planck Institute of Biophysical Chemistry), Michael Lynch (Indiana University), Laura Machesky (University of Birmingham, UK), Hiten Madhani (University of California, San Francisco), James Maller (University of Colorado Medical School), Tom Maniatis (Harvard University), Colin Manoil (Harvard Medical School), Elliott Margulies (NIH), Philippa Marrack (National Jewish Medical and Research Center, Denver), Mark Marsh (Institute of Cancer Research, London), Wallace Marshall (University of California, San Francisco), Gail Martin (University of California, San Francisco), Paul Martin (University College London), Joan Massagué (Memorial Sloan Kettering Cancer Center), Christopher Mathews (Oregon State University), Brian McCarthy (University of California, Irvine), Richard McCarty (Cornell University), William McGinnis (University of California, San Diego), Anne McLaren (Wellcome/Cancer Research Campaign Institute, Cambridge), Frank McNally (University of California, Davis), Freiderick Meins (Freiderich Miescher Institut, Basel), Stephanie Mel (University of California, San Diego), Ira Mellman (Genentech), Barbara Meyer (University of California, Berkeley), Elliot Meyerowitz (California Institute of Technology), Chris Miller (Brandeis University), Robert Mishell (University of Birmingham, UK), Avrion Mitchison (University College London), N.A. Mitchison (University College London), Timothy Mitchison (Harvard Medical School), Quinn Mitrovich (University of California, San Francisco), Peter Mombaerts (The Rockefeller University), Mark Mooseker (Yale University), David Morgan (University of California, San Francisco), Michelle Moritz (University of California, San Francisco), Montrose Moses (Duke University), Keith Mostov (University of California, San Francisco), Anne Mudge (University College London), Hans Müller-Eberhard (Scripps Clinic and Research Institute), Alan Munro (University of Cambridge), J. Murdoch Mitchison (Harvard University), Richard Myers (Stanford University), Diana Myles (University of California, Davis), Andrew Murray (Harvard University), Shigekazu Nagata (Kyoto University, Japan), Geeta Narlikar (University of California, San Francisco), Kim Nasmyth (University of Oxford), Mark E. Nelson (University of Illinois, Urbana-Champaign), Michael Neuberger (deceased), Walter Neupert (University of Munich, Germany), David Nicholls (University of Dundee, Scotland), Roger Nicoll (University of California, San Francisco), Suzanne Noble (University of California, San Francisco), Harry Noller (University of California, Santa Cruz), Jodi Nunnari (University of California, Davis), Paul Nurse (Francis Crick Institute), Roel Nusse (Stanford University), Michael Nussenzweig (Rockefeller University), Duncan O'Dell (deceased), Patrick O'Farrell (University of California, San Francisco), Bjorn Olsen (Harvard Medical School), Maynard Olson (University of Washington, Seattle), Stuart Orkin (Harvard University), Terry Orr-Weaver (Massachusetts Institute of Technology), Erin O'Shea (Harvard University), Dieter Osterhelt (Max Planck Institute of Biochemistry), William Otto (Cancer Research UK), John Owen (University of Birmingham, UK), Dale Oxender (University of Michigan), George Palade (deceased), Barbara Panning (University of California, San Francisco), Roy Parker (University of Arizona, Tucson), William W. Parson (University of Washington, Seattle), Terence Partridge (MRC Clinical Sciences Centre, London), William E. Paul (NIH), Tony Pawson (deceased), Hugh Pelham (MRC, UK), Robert Perry (Institute of Cancer Research, Philadelphia), Gordon Peters (Cancer Research UK), Greg Petsko (Brandeis University), Nikolaus Pfanner (University of Freiburg, Germany), David Phillips (The Rockefeller University), Jeremy Pickett-Heaps (The University of Melbourne, Australia), Jonathan Pines (Gurdon Institute, Cambridge), Julie Pitcher (University College London), Jeffrey Pollard (Albert Einstein College of Medicine), Tom Pollard (Yale University), Bruce Ponder (University of Cambridge), Daniel Portnoy (University of California, Berkeley), James Priess (University of Washington, Seattle), Darwin Prockop (Tulane University), Mark Ptashne (Memorial Sloan Kettering Cancer Center), Dale Purves (Duke University), Efraim Racker (Cornell University), Jordan Raff (University of Oxford), Klaus Rajewsky (Max Delbrück Center for Molecular Medicine, Germany), George Ratcliffe (University of Oxford), Elio Raviola (Harvard Medical School), Martin Rechsteiner (University of Utah, Salt Lake City), David Rees (National Institute for Medical Research, London), Thomas A. Reh (University of Washington, Seattle), Louis Reichardt (University of California, San Francisco), Renee Reijo (University of California, San Francisco), Caetano Reis e Sousa (Cancer Research UK), Fred Richards (Yale University), Conly Rieder (Wadsworth Center, Albany), Phillips Robbins (Massachusetts Institute of Technology), Elizabeth Robertson (The Wellcome Trust Centre for Human Genetics, UK), Elaine Robson (University of Reading, UK), Robert Roeder (The Rockefeller University), Joel Rosenbaum (Yale University), Janet Rossant (Mount Sinai Hospital, Toronto), Jesse Roth (NIH), Jim Rothman (Memorial Sloan Kettering Cancer Center), Rodney Rothstein (Columbia University), Erkki Ruoslahti (La Jolla Cancer Research Foundation), Gary Ruvkun (Massachusetts General Hospital), David Sabatini (New York University), Alan Sachs (University of California, Berkeley), Edward Salmon (University of North Carolina,

Chapel Hill), Aziz Sancar (University of North Carolina, Chapel Hill), Joshua Sanes (Harvard University), Peter Sarnow (Stanford University), Lisa Satterwhite (Duke University Medical School), Robert Sauer (Massachusetts Institute of Technology), Ken Sawin (The Wellcome Trust Centre for Cell Biology, UK), Howard Schachman (University of California, Berkeley), Gerald Schatten (Pittsburgh Development Center), Gottfried Schatz (Biozentrum, University of Basel), Randy Schekman (University of California, Berkeley), Richard Scheller (Stanford University), Giampietro Schiavo (Cancer Research UK), Ueli Schibler (University of Geneva, Switzerland), Joseph Schlessinger (New York University Medical Center), Danny J. Schnell (University of Massachusetts, Amherst), Michael Schramm (Hebrew University, Israel), Robert Schreiber (Washington University School of Medicine), James Schwartz (Columbia University), Ronald Schwartz (NIH), François Schweisguth (Institut Pasteur, France), John Scott (University of Manchester, UK), John Sedat (University of California, San Francisco), Peter Selby (Cancer Research UK), Zvi Sellinger (Hebrew University, Israel), Gregg Semenza (Johns Hopkins University), Philippe Sengel (University of Grenoble, France), Peter Shaw (John Innes Institute, Norwich, UK), Michael Sheetz (Columbia University), Morgan Sheng (Massachusetts Institute of Technology), Charles Sherr (St. Jude Children's Hospital), David Shima (Cancer Research UK), Samuel Silverstein (Columbia University), Melvin I. Simon (California Institute of Technology), Kai Simons (Max Planck Institute of Molecular Cell Biology and Genetics), Jonathan Slack (Cancer Research UK), Alison Smith (John Innes Institute, Norfolk, UK), Austin Smith (University of Edinburgh, UK), Jim Smith (The Gurdon Institute, UK), John Maynard Smith (University of Sussex, UK), Mitchell Sogin (Woods Hole Institute), Frank Solomon (Massachusetts Institute of Technology), Michael Solursh (University of Iowa), Bruce Spiegelman (Harvard Medical School), Timothy Springer (Harvard Medical School), Mathias Sprinzl (University of Bayreuth, Germany), Scott Stachel (University of California, Berkeley), Andrew Staehelin (University of Colorado, Boulder), David Standring (University of California, San Francisco), Margaret Stanley (University of Cambridge), Martha Stark (University of California, San Francisco), Wilfred Stein (Hebrew University, Israel), Malcolm Steinberg (Princeton University), Ralph Steinman (deceased), Len Stephens (The Babraham Institute, UK), Paul Sternberg (California Institute of Technology), Chuck Stevens (The Salk Institute), Murray Stewart (MRC Laboratory of Molecular Biology, Cambridge), Bruce Stillman (Cold Spring Harbor Laboratory), Charles Streuli (University of Manchester, UK), Monroe Strickberger (University of Missouri, St. Louis), Robert Stroud (University of California, San Francisco), Michael Stryker (University of California, San Francisco), William Sullivan (University of California, Santa Cruz), Azim Surani (The Gurdon Institute, University of Cambridge), Daniel Szollosi (Institut National de la Recherche Agronomique, France), Jack Szostak (Harvard Medical School), Clifford Tabin (Harvard Medical School), Masatoshi Takeichi (RIKEN Center for Developmental Biology, Japan), Nicolas Tapon (London Research Institute), Diethard Tautz (University of Cologne, Germany), Julie Theriot (Stanford University),

Roger Thomas (University of Bristol, UK), Craig Thompson (Memorial Sloan Kettering Cancer Center), Janet Thornton (European Bioinformatics Institute, UK), Vernon Thornton (King's College London), Cheryll Tickle (University of Dundee, Scotland), Jim Till (Ontario Cancer Institute, Toronto), Lewis Tilney (University of Pennsylvania), David Tollervey (University of Edinburgh, UK), Ian Tomlinson (Cancer Research UK), Nick Tonks (Cold Spring Harbor Laboratory), Alain Townsend (Institute of Molecular Medicine, John Radcliffe Hospital, Oxford), Paul Travers (Scottish Institute for Regeneration Medicine), Robert Trelstad (UMDNJ—Robert Wood Johnson Medical School), Anthony Trewavas (Edinburgh University, Scotland), Nigel Unwin (MRC Laboratory of Molecular Biology, Cambridge), Victor Vacquier (University of California, San Diego), Ronald D. Vale (University of California, San Francisco), Tom Vanaman (University of Kentucky), Harry van der Westen (Wageningen, The Netherlands), Harold Varmus (National Cancer Institute, United States), Alexander J. Varshavsky (California Institute of Technology), Donald Voet (University of Pennsylvania), Harald von Boehmer (Harvard Medical School), Madhu Wahi (University of California, San Francisco), Virginia Walbot (Stanford University), Frank Walsh (GlaxoSmithKline, UK), Trevor Wang (John Innes Institute, Norwich, UK), Xiaodong Wang (The University of Texas Southwestern Medical School), Yu-Lie Wang (Worcester Foundation for Biomedical Research, MA), Gary Ward (University of Vermont), Anne Warner (University College London), Graham Warren (Yale University School of Medicine), Paul Wassarman (Mount Sinai School of Medicine), Clare Waterman-Storer (The Scripps Research Institute), Fiona Watt (Cancer Research UK), John Watts (John Innes Institute, Norwich, UK), Klaus Weber (Max Planck Institute for Biophysical Chemistry), Martin Weigert (Institute of Cancer Research, Philadelphia), Robert Weinberg (Massachusetts Institute of Technology), Harold Weintraub (deceased), Karsten Weis (Swiss Federal Institute of Technology), Irving Weissman (Stanford University), Jonathan Weissman (University of California, San Francisco), Susan R. Wente (Vanderbilt University School of Medicine), Norman Wessells (University of Oregon, Eugene), Stephen West (Cancer Research UK), Judy White (University of Virginia), William Wickner (Dartmouth College), Michael Wilcox (deceased), Lewis T. Williams (Chiron Corporation), Patrick Williamson (University of Massachusetts, Amherst), Keith Willison (Chester Beatty Laboratories, London), John Wilson (Baylor University), Alan Wolffe (deceased), Richard Wolfenden (University of North Carolina, Chapel Hill), Sandra Wolin (Yale University School of Medicine), Lewis Wolpert (University College London), Richard D. Wood (University of Pittsburgh Cancer Institute), Abraham Worcel (University of Rochester), Nick Wright (Cancer Research UK), John Wyke (Beatson Institute for Cancer Research, Glasgow), Michael P. Yaffe (California Institute for Regenerative Medicine), Kenneth M. Yamada (NIH), Keith Yamamoto (University of California, San Francisco), Charles Yocum (University of Michigan, Ann Arbor), Peter Yurchenco (UMDNJ—Robert Wood Johnson Medical School), Rosalind Zalin (University College London), Patricia Zambryski (University of California, Berkeley), Marino Zerial (Max Planck Institute of Molecular Cell Biology and Genetics).

Contents

Special Features

Detailed Contents

Chapter 14 Energy Conversion: Mitochondria and Chloroplasts 753

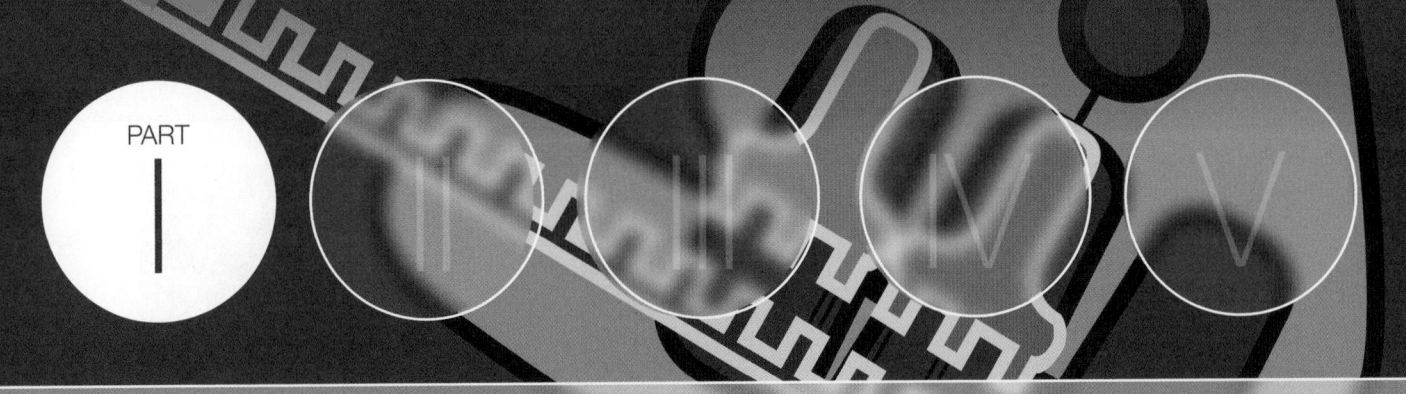

INTRODUCTION TO THE CELL

Cells and Genomes

The surface of our planet is populated by living things—curious, intricately organized chemical factories that take in matter from their surroundings and use these raw materials to generate copies of themselves. These living organisms appear extraordinarily diverse. What could be more different than a tiger and a piece of seaweed, or a bacterium and a tree? Yet our ancestors, knowing nothing of cells or DNA, saw that all these things had something in common. They called that something "life," marveled at it, struggled to define it, and despaired of explaining what it was or how it worked in terms that relate to nonliving matter.

The discoveries of the past century have not diminished the marvel—quite the contrary. But they have removed the central mystery regarding the nature of life. We can now see that all living things are made of cells: small, membrane-enclosed units filled with a concentrated aqueous solution of chemicals and endowed with the extraordinary ability to create copies of themselves by growing and then dividing in two.

Because cells are the fundamental units of life, it is to *cell biology*—the study of the structure, function, and behavior of cells—that we must look for answers to the questions of what life is and how it works. With a deeper understanding of cells and their evolution, we can begin to tackle the grand historical problems of life on Earth: its mysterious origins, its stunning diversity, and its invasion of every conceivable habitat. Indeed, as emphasized long ago by the pioneering cell biologist E. B. Wilson, "the key to every biological problem must finally be sought in the cell; for every living organism is, or at some time has been, a cell."

Despite their apparent diversity, living things are fundamentally similar inside. The whole of biology is thus a counterpoint between two themes: astonishing variety in individual particulars; astonishing constancy in fundamental mechanisms. In this first chapter, we begin by outlining the universal features common to all life on our planet. We then survey, briefly, the diversity of cells. And we see how, thanks to the common molecular code in which the specifications for all living organisms are written, it is possible to read, measure, and decipher these specifications to help us achieve a coherent understanding of all the forms of life, from the smallest to the greatest.

IN THIS CHAPTER

THE UNIVERSAL FEATURES OF CELLS ON EARTH

THE DIVERSITY OF GENOMES AND THE TREE OF LIFE

GENETIC INFORMATION IN EUKARYOTES

Figure 1–1 **The hereditary information in the fertilized egg cell determines the nature of the whole multicellular organism.** Although their starting cells look superficially similar, as indicated: a sea urchin egg gives rise to a sea urchin (A and B). A mouse egg gives rise to a mouse (C and D). An egg of the seaweed *Fucus* gives rise to a *Fucus* seaweed (E and F). (A, courtesy of David McClay; B, WaterFrame/Alamy Stock Photo; C, courtesy of Patricia Calarco, from G. Martin, *Science* 209:768–776, 1980. With permission from AAAS; D, Rudmer Zwerver/Alamy Stock Photo; E and F, courtesy of Colin Brownlee.)

THE UNIVERSAL FEATURES OF CELLS ON EARTH

It is estimated that there are more than 10 million—perhaps 100 million—living species on Earth today. Each species is different, and each reproduces itself faithfully, yielding progeny that belong to the same species: the parent organism hands down information specifying, in extraordinary detail, the characteristics that the offspring shall have. This phenomenon of *heredity* is central to the definition of life: it distinguishes life from other processes, such as the growth of a crystal, or the burning of a candle, or the formation of waves on water, in which orderly structures are generated but without the same type of link between the peculiarities of parents and the peculiarities of offspring. Like the candle flame, the living organism must consume free energy to create and maintain its organization. But life employs the free energy to drive a hugely complex system of chemical processes that are specified by hereditary information.

Most living organisms are single cells. Others, such as ourselves, are vast multicellular cities in which groups of cells perform specialized functions linked by intricate systems of communication. But even for the aggregate of more than 10^{13} cells that form a human body, the whole organism has been generated by cell divisions from a single cell. The single cell, therefore, is the vehicle for all of the hereditary information that defines each species (**Figure 1–1**). This cell includes the machinery to gather raw materials from the environment and to construct from them a new cell in its own image, complete with a new copy of its hereditary information. Each and every cell is truly amazing.

All Cells Store Their Hereditary Information in the Same Linear Chemical Code: DNA

Computers have made us familiar with the concept of information as a measurable quantity—a million bytes (to record a few hundred pages of text or an image from a digital camera), 600 million bytes for the music on a CD, and so on. Computers have also made us well aware that the same information can be recorded in many different physical forms: the discs and tapes that we used 20 years ago for our electronic archives have become unreadable on present-day machines. Living

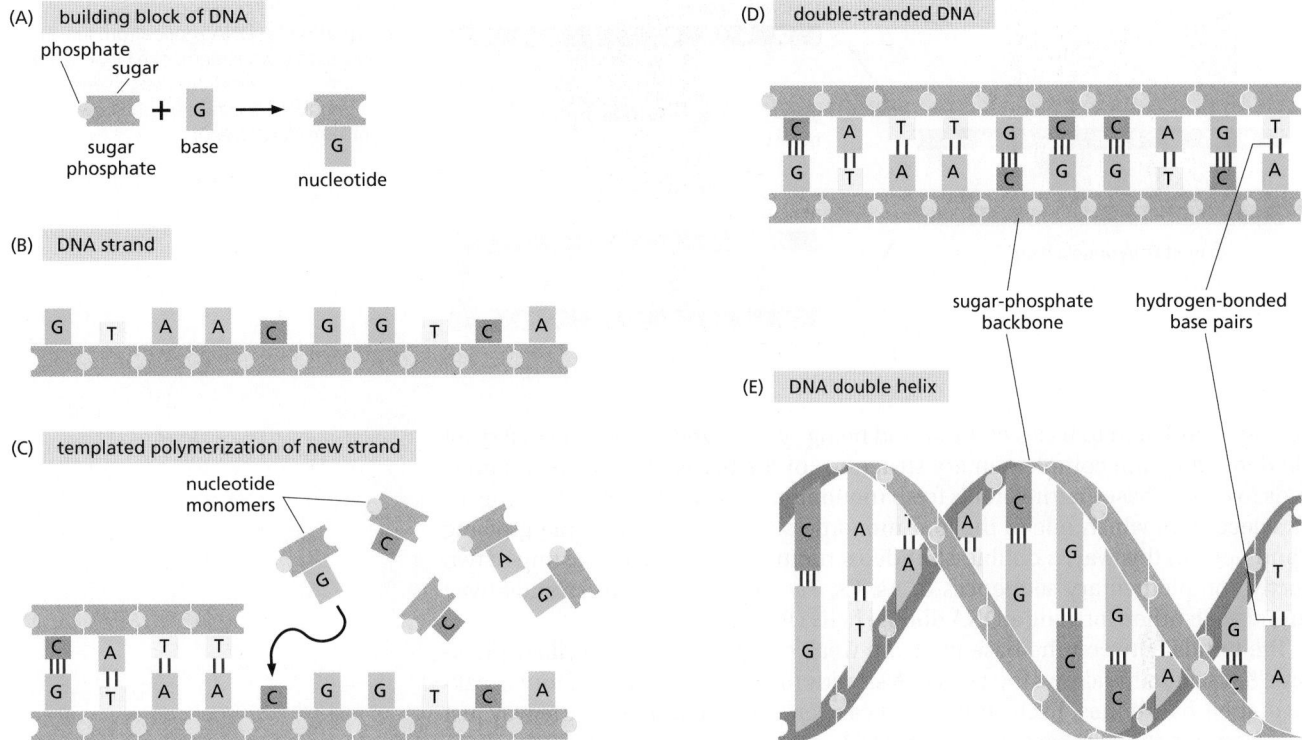

cells, like computers, store information, and it is estimated that they have been evolving and diversifying for over 3.5 billion years. It is scarcely to be expected that they would all store their information in the same form, or that the archives of one type of cell should be readable by the information-handling machinery of another. And yet it is so. All living cells on Earth store their hereditary information in the form of double-stranded molecules of DNA—long, unbranched, paired *polymer* chains, formed always of the same four types of *monomers*. These monomers, chemical compounds known as nucleotides, have nicknames drawn from a four-letter alphabet—A, T, C, G—and they are strung together in a long linear sequence that encodes the genetic information, just as the sequence of 1s and 0s encodes the information in a computer file. We can take a piece of DNA from a human cell and insert it into a bacterium, or a piece of bacterial DNA and insert it into a human cell, and the information will be successfully read, interpreted, and copied. Using chemical methods, scientists have learned how to read out the complete sequence of monomers in any DNA molecule—extending for many millions of nucleotides—and thereby decipher all of the hereditary information that each organism contains.

All Cells Replicate Their Hereditary Information by Templated Polymerization

The mechanisms that make life possible depend on the structure of the double-stranded DNA molecule. Each monomer in a single DNA strand—that is, each **nucleotide**—consists of two parts: a sugar (deoxyribose) with a phosphate group attached to it, and a *base*, which may be either adenine (A), guanine (G), cytosine (C), or thymine (T) (**Figure 1–2**). Each sugar is linked to the next via the phosphate group, creating a polymer chain composed of a repetitive sugar-phosphate backbone with a series of bases protruding from it. The DNA polymer is extended by adding monomers at one end. For a single isolated strand, these monomers can, in principle, be added in any order, because each one links to the next in the same way, through the part of the molecule that is the same for all of them. In the living cell, however, DNA is not synthesized as a free strand in isolation, but on a template formed by a preexisting DNA strand. The bases protruding from the

Figure 1–2 DNA and its building blocks. (A) DNA is made from simple subunits, called nucleotides, each consisting of a sugar-phosphate molecule with a nitrogen-containing side group, or base, attached to it. The bases are of four types (adenine, guanine, cytosine, and thymine), corresponding to four distinct nucleotides, labeled A, G, C, and T. (B) A single strand of DNA consists of nucleotides joined together by sugar-phosphate linkages. Note that the individual sugar-phosphate units are asymmetric, giving the backbone of the strand a definite directionality, or polarity. This directionality guides the molecular processes by which the information in DNA is interpreted and copied in cells: the information is always "read" in a consistent order, just as written English text is read from left to right. (C) Through templated polymerization, the sequence of nucleotides in an existing DNA strand controls the sequence in which nucleotides are joined together in a new DNA strand; T in one strand pairs with A in the other, and G in one strand with C in the other. The new strand has a nucleotide sequence *complementary* to that of the old strand, and a backbone with opposite directionality: corresponding to the GTAA... of the original strand, it has ...TTAC. (D) A normal DNA molecule consists of two such complementary strands. The nucleotides within each strand are linked by strong (covalent) chemical bonds; the complementary nucleotides on opposite strands are held together more weakly, by hydrogen bonds. (E) The two strands twist around each other to form a double helix—a robust structure that can accommodate any sequence of nucleotides without altering its basic structure (see Movie 4.1).

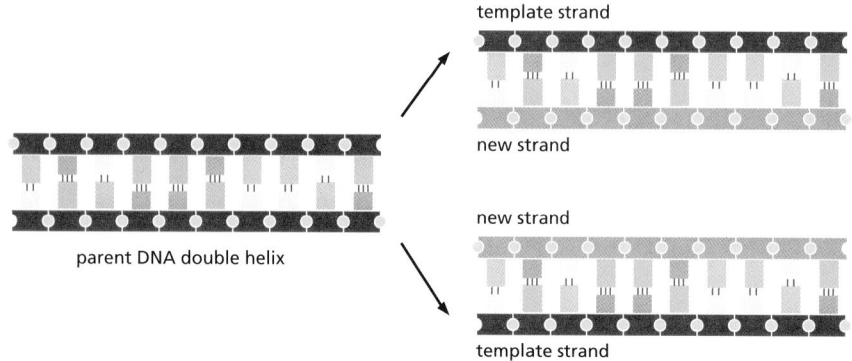

Figure 1–3 The copying of genetic information by DNA replication. In this process, the two strands of a DNA double helix are pulled apart, and each serves as a template for synthesis of a new complementary strand.

template strand

new strand

new strand

parent DNA double helix

template strand

existing strand bind to bases of the strand being synthesized, according to a strict rule defined by the complementary structures of the bases: A binds to T, and C binds to G. This base-pairing holds fresh monomers in place and thereby controls the selection of which one of the four monomers shall be added to the growing strand next. In this way, a double-stranded structure is created, consisting of two exactly complementary sequences of As, Cs, Ts, and Gs. The two strands twist around each other, forming a DNA double helix (Figure 1–2E).

The bonds between the base pairs are weak compared with the sugar-phosphate links, and this allows the two DNA strands to be pulled apart without breakage of their backbones. Each strand then can serve as a template, in the way just described, for the synthesis of a fresh DNA strand complementary to itself—a fresh copy, that is, of the hereditary information (**Figure 1–3**). In different types of cells, this process of **DNA replication** occurs at different rates, with different controls to start it or stop it, and different auxiliary molecules to help it along. But the basics are universal: DNA is the information store for heredity, and *templated polymerization* is the way in which this information is copied throughout the living world.

All Cells Transcribe Portions of Their Hereditary Information into the Same Intermediary Form: RNA

To carry out its information-bearing function, DNA must do more than copy itself. It must also *express* its information, by letting the information guide the synthesis of other molecules in the cell. This expression occurs by a mechanism that is the same in all living organisms, leading first and foremost to the production of two other key classes of polymers: RNAs and proteins. The process (discussed in detail in Chapters 6 and 7) begins with a templated polymerization called **transcription**, in which segments of the DNA sequence are used as templates for the synthesis of shorter molecules of the closely related polymer **ribonucleic acid**, or **RNA**. Later, in the more complex process of **translation**, many of these RNA molecules direct the synthesis of polymers of a radically different chemical class—the *proteins* (**Figure 1–4**).

In RNA, the backbone is formed of a slightly different sugar from that of DNA—ribose instead of deoxyribose—and one of the four bases is slightly different—uracil (U) in place of thymine (T). But the other three bases—A, C, and G—are the same, and all four bases pair with their complementary counterparts in DNA—the A, U, C, and G of RNA with the T, A, G, and C of DNA. During transcription, the RNA monomers are lined up and selected for polymerization on a template strand of DNA, just as DNA monomers are selected during replication. The outcome is a polymer molecule whose sequence of nucleotides faithfully represents a portion of the cell's genetic information, even though it is written in a slightly different alphabet—consisting of RNA monomers instead of DNA monomers.

The same segment of DNA can be used repeatedly to guide the synthesis of many identical RNA molecules. Thus, whereas the cell's archive of genetic information in the form of DNA is fixed and sacrosanct, these *RNA transcripts* are

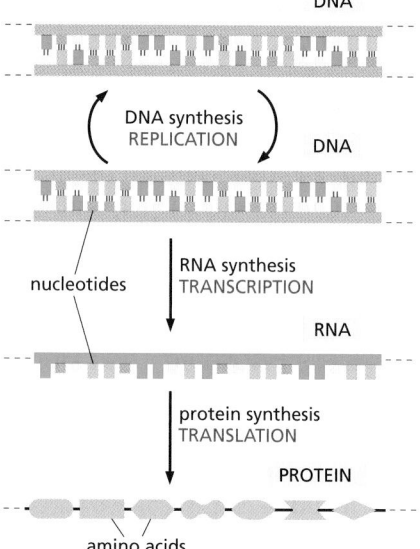

DNA

DNA synthesis
REPLICATION

DNA

nucleotides

RNA synthesis
TRANSCRIPTION

RNA

protein synthesis
TRANSLATION

PROTEIN

amino acids

Figure 1–4 From DNA to protein. Genetic information is read out and put to use through a two-step process. First, in *transcription*, segments of the DNA sequence are used to guide the synthesis of molecules of RNA. Then, in *translation*, the RNA molecules are used to guide the synthesis of molecules of protein.

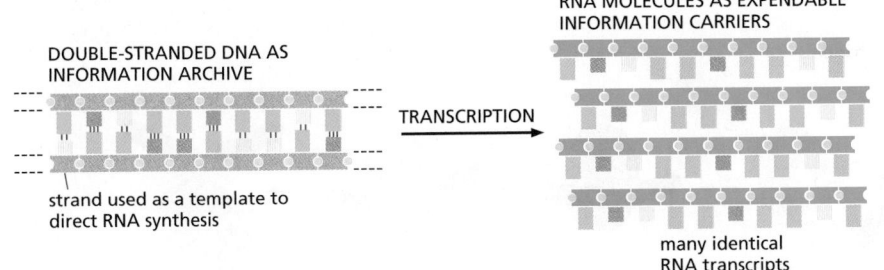

DOUBLE-STRANDED DNA AS
INFORMATION ARCHIVE

RNA MOLECULES AS EXPENDABLE
INFORMATION CARRIERS

TRANSCRIPTION

strand used as a template to
direct RNA synthesis

many identical
RNA transcripts

Figure 1–5 **How genetic information is broadcast for use inside the cell.** Each cell contains a fixed set of DNA molecules—its archive of genetic information. A given segment of this DNA guides the synthesis of many identical RNA transcripts, which serve as working copies of the information stored in the archive. Many different sets of RNA molecules can be made by transcribing different parts of a cell's DNA sequences, allowing different types of cells to use the same information store differently.

mass-produced and disposable (**Figure 1–5**). As we shall see, these transcripts function as intermediates in the transfer of genetic information. Most notably, they serve as **messenger RNA** (**mRNA**) molecules that guide the synthesis of proteins according to the genetic instructions stored in the DNA.

RNA molecules have distinctive structures that can also give them other specialized chemical capabilities. Being single-stranded, their backbone is flexible, so that the polymer chain can bend back on itself to allow one part of the molecule to form weak bonds with another part of the same molecule. This occurs when segments of the sequence are locally complementary: a ...GGGG... segment, for example, will tend to associate with a ...CCCC... segment. These types of internal associations can cause an RNA chain to fold up into a specific shape that is dictated by its sequence (**Figure 1–6**). The shape of the RNA molecule, in turn, may enable it to recognize other molecules by binding to them selectively—and even, in certain cases, to catalyze chemical changes in the molecules that are bound. In fact, some chemical reactions catalyzed by RNA molecules are crucial for several of the most ancient and fundamental processes in living cells, and it has been suggested that an extensive catalysis by RNA played a central part in the early evolution of life (discussed in Chapter 6).

All Cells Use Proteins as Catalysts

Protein molecules, like DNA and RNA molecules, are long unbranched polymer chains, formed by stringing together monomeric building blocks drawn from a standard repertoire that is the same for all living cells. Like DNA and RNA, proteins carry information in the form of a linear sequence of symbols, in the same way as a human message written in an alphabetic script. There are many different protein molecules in each cell, and—leaving out the water—they form most of the cell's mass.

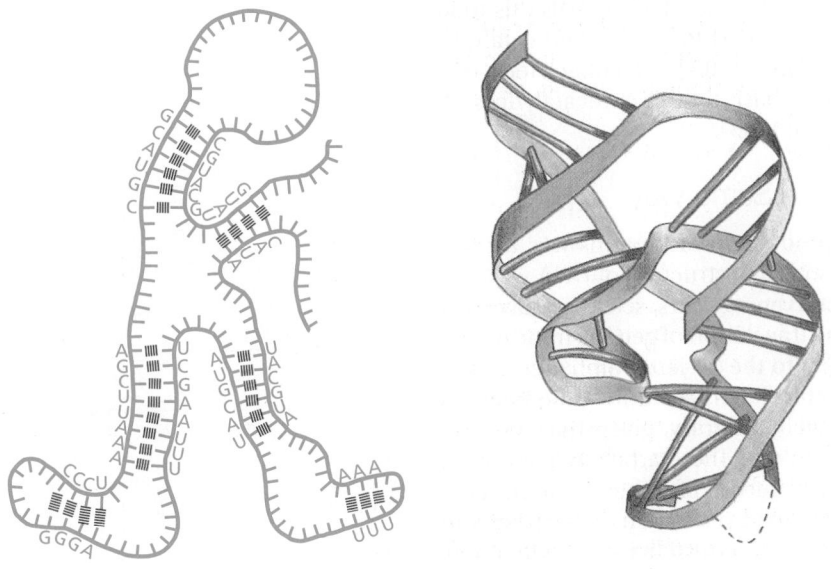

(A)   (B)

Figure 1–6 **The conformation of an RNA molecule.** (A) Nucleotide pairing between different regions of the same RNA polymer chain causes the molecule to adopt a distinctive shape. (B) The three-dimensional structure of an actual RNA molecule produced by hepatitis delta virus; this RNA can catalyze RNA strand cleavage. The *blue* ribbon represents the sugar-phosphate backbone and the bars represent base pairs (see Movie 6.1). (B, based on A.R. Ferré-D'Amaré, K. Zhou, and J.A. Doudna, *Nature* 395:567–574, 1998.)

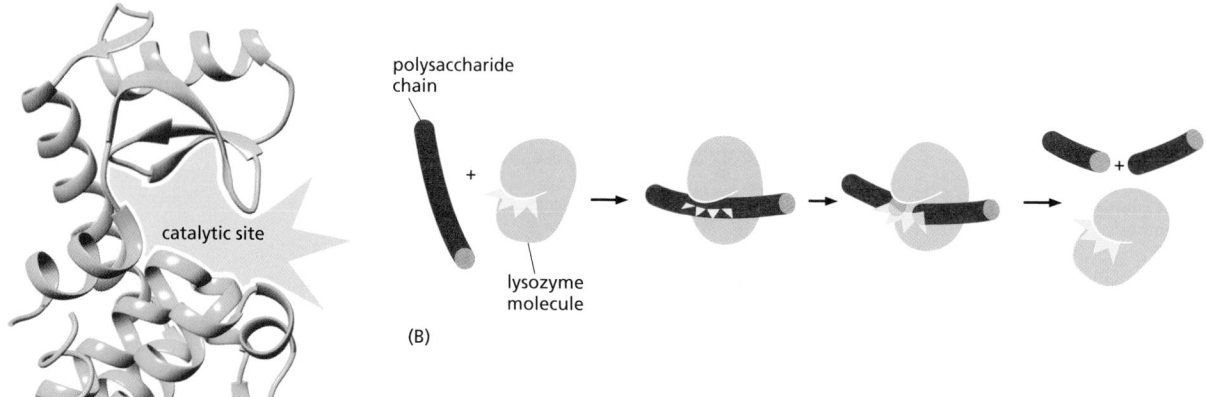

Figure 1–7 **How a protein molecule acts as a catalyst for a chemical reaction.** (A) In a protein molecule, the polymer chain folds up into a specific shape defined by its amino acid sequence. A groove in the surface of this particular folded molecule, the enzyme lysozyme, forms a catalytic site. (B) A polysaccharide molecule *(red)*—a polymer chain of sugar monomers—binds to the catalytic site of lysozyme and is broken apart, as a result of a covalent bond-breaking reaction catalyzed by the amino acids lining the groove (see Movie 3.9). (PDB code: 1LYD.)

The monomers of protein, the **amino acids**, are quite different from those of DNA and RNA, and there are 20 types instead of 4. Each amino acid is built around the same core structure through which it can be linked in a standard way to any other amino acid in the set; attached to this core is a side group that gives each amino acid a distinctive chemical character. Each of the protein molecules is a **polypeptide**, created by joining its amino acids in a particular sequence. Through billions of years of evolution, this sequence has been selected to give the protein a useful function. Thus, by folding into a precise three-dimensional form with reactive sites on its surface (**Figure 1–7A**), these amino-acid polymers can bind with high specificity to other molecules and can act as **enzymes** to catalyze reactions that make or break covalent bonds. In this way they direct the vast majority of chemical processes in the cell (Figure 1–7B).

Proteins have many other functions as well—maintaining structures, generating movements, sensing signals, and so on—each protein molecule performing a specific function according to its own genetically specified sequence of amino acids. Proteins, above all, are the main molecules that put the cell's genetic information into action.

Thus, polynucleotides specify the amino acid sequences of proteins. Proteins, in turn, catalyze many chemical reactions, including those by which new DNA molecules are synthesized. From the most fundamental point of view, a living cell is a self-replicating collection of catalysts that takes in food, processes this food to derive both the building blocks and energy needed to make more catalysts, and discards the materials left over as waste (**Figure 1–8A**). A feedback loop that connects proteins and polynucleotides forms the basis for this autocatalytic, self-reproducing behavior of living organisms (Figure 1–8B).

All Cells Translate RNA into Protein in the Same Way

How the information in DNA specifies the production of proteins was a complete mystery in the 1950s when the double-stranded structure of DNA was first revealed as the basis of heredity. But in the intervening years, scientists have discovered the elegant mechanisms involved. The translation of genetic information from the 4-letter alphabet of polynucleotides into the 20-letter alphabet of proteins is a complex process. The rules of this translation seem in some respects neat and rational but in other respects strangely arbitrary, given that they are (with minor exceptions) identical in all living things. These arbitrary features, it is thought, reflect frozen accidents in the early history of life. They stem from the chance properties of the earliest organisms that were passed on by heredity and have become so deeply embedded in the constitution of all living cells that they cannot be changed without disastrous effects.

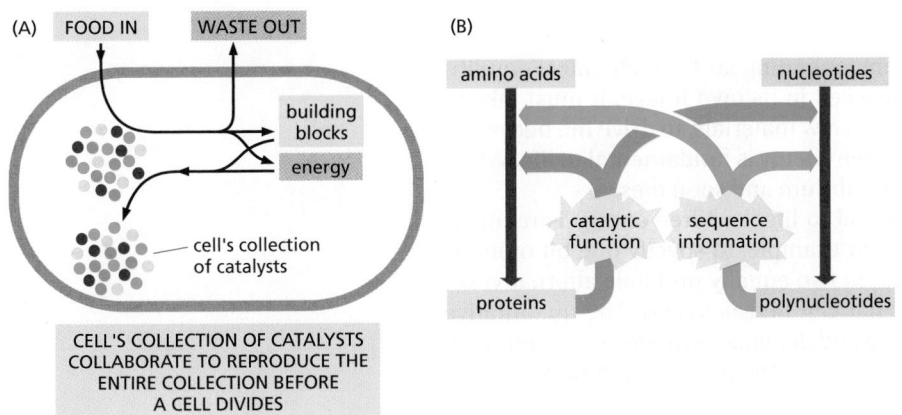

(A) FOOD IN   WASTE OUT

building blocks

energy

cell's collection of catalysts

CELL'S COLLECTION OF CATALYSTS COLLABORATE TO REPRODUCE THE ENTIRE COLLECTION BEFORE A CELL DIVIDES

(B)

amino acids   nucleotides

catalytic function   sequence information

proteins   polynucleotides

Figure 1–8 **Life as an autocatalytic process.** (A) The cell as a self-replicating collection of catalysts. (B) Polynucleotides (the nucleic acids DNA and RNA, which are nucleotide polymers) provide the sequence information, while proteins (amino acid polymers) provide most of the catalytic functions that serve—through a complex set of chemical reactions—to bring about the synthesis of more polynucleotides and proteins of the same types.

It turns out that the information in the sequence of a messenger RNA molecule is read out in groups of three nucleotides at a time: each triplet of nucleotides, or *codon*, specifies (codes for) a single amino acid in a corresponding protein. Since the number of distinct triplets that can be formed from four nucleotides is 4^3, there are 64 possible codons, all of which occur in nature. However, there are only 20 naturally occurring amino acids. That means there are necessarily many cases in which several codons correspond to the same amino acid. This *genetic code* is read out by a special class of small RNA molecules, the *transfer RNAs* (*tRNAs*). Each type of tRNA becomes attached at one end to a specific amino acid, and displays at its other end a specific sequence of three nucleotides—an *anticodon*—that enables it to recognize, through base-pairing, a particular codon or subset of codons in mRNA. The intricate chemistry that enables these tRNAs to translate a specific sequence of A, C, G, and U nucleotides in an mRNA molecule into a specific sequence of amino acids in a protein molecule occurs on the *ribosome*, a large multimolecular machine composed of both protein and *ribosomal RNA*. All of these processes are described in detail in Chapter 6.

Each Protein Is Encoded by a Specific Gene

DNA molecules as a rule are very large, containing the specifications for thousands of proteins. Special sequences in the DNA serve as punctuation, defining where the information for each protein begins and ends. And individual segments of the long DNA sequence are transcribed into separate mRNA molecules, coding for different proteins. Each such DNA segment represents one **gene**. A complication is that RNA molecules transcribed from the same DNA segment can often be processed in more than one way, so as to give rise to a set of alternative versions of a protein, especially in more complex cells such as those of plants and animals. In addition, some DNA segments—a smaller number—are transcribed into RNA molecules that are not translated but have catalytic, regulatory, or structural functions; such DNA segments also count as genes. A gene therefore is defined as the segment of DNA sequence corresponding to a single protein or set of alternative protein variants or to a single catalytic, regulatory, or structural RNA molecule.

In all cells, the *expression* of individual genes is regulated: instead of manufacturing its full repertoire of possible proteins at full tilt all the time, the cell adjusts the rate of transcription and translation of different genes independently, according to need. Stretches of *regulatory DNA* are interspersed among the segments that code for protein, and these noncoding regions bind to special protein molecules that control the local rate of transcription. The quantity and organization of the regulatory DNA vary widely from one class of organisms to another, but the basic strategy is universal. In this way, the **genome** of the cell—that is, the totality of its genetic information as embodied in its complete DNA sequence—dictates not only the nature of the cell's proteins, but also when and where they are to be made.

Life Requires Free Energy

A living cell is a dynamic chemical system, operating far from chemical equilibrium. For a cell to grow or to make a new cell in its own image, it must take in free energy from the environment, as well as raw materials, to drive the necessary synthetic reactions. This consumption of free energy is fundamental to life. When it stops, a cell decays toward chemical equilibrium and soon dies.

Genetic information is also fundamental to life, and free energy is required for the propagation of this information. For example, to specify one bit of information—that is, one yes/no choice between two equally probable alternatives—costs a defined amount of free energy that can be calculated. The quantitative relationship involves some deep reasoning and depends on a precise definition of the term "free energy," as explained in Chapter 2. The basic idea, however, is not difficult to understand intuitively.

Picture the molecules in a cell as a swarm of objects endowed with thermal energy, moving around violently at random, buffeted by collisions with one another. To specify genetic information—in the form of a DNA sequence, for example—molecules from this wild crowd must be captured, arranged in a specific order defined by some preexisting template, and linked together in a fixed relationship. The bonds that hold the molecules in their proper places on the template and join them together must be strong enough to resist the disordering effect of thermal motion. The process is driven forward by consumption of free energy, which is needed to ensure that the correct bonds are made, and made robustly. In the simplest case, the molecules can be compared with spring-loaded traps, ready to snap into a more stable, lower-energy attached state when they meet their proper partners; as they snap together into the bonded arrangement, their available stored energy—their free energy—like the energy of the spring in the trap, is released and dissipated as heat. In a cell, the chemical processes underlying information transfer are more complex, but the same basic principle applies: free energy has to be spent on the creation of order.

To replicate its genetic information faithfully, and indeed to make all its complex molecules according to the correct specifications, the cell therefore requires free energy, which has to be imported somehow from the surroundings. As we shall see in Chapter 2, the free energy required by animal cells is derived from chemical bonds in food molecules that the animals eat, while plants get their free energy from sunlight.

All Cells Function as Biochemical Factories Dealing with the Same Basic Molecular Building Blocks

Because all cells make DNA, RNA, and protein, all cells have to contain and manipulate a similar collection of small molecules, including simple sugars, nucleotides, and amino acids, as well as other substances that are universally required. All cells, for example, require the phosphorylated nucleotide ATP (*adenosine triphosphate*), not only as a building block for the synthesis of DNA and RNA, but also as a carrier of the free energy that is needed to drive a huge number of chemical reactions in the cell.

Although all cells function as biochemical factories of a broadly similar type, many of the details of their small-molecule transactions differ. Some organisms, such as plants, require only the simplest of nutrients and harness the energy of sunlight to make all their own small organic molecules. Other organisms, such as animals, feed on living things and must obtain many of their organic molecules ready-made. We return to this point later.

All Cells Are Enclosed in a Plasma Membrane Across Which Nutrients and Waste Materials Must Pass

Another universal feature is that each cell is enclosed by a membrane—the **plasma membrane**. This container acts as a selective barrier that enables the cell to concentrate nutrients gathered from its environment and retain the products it

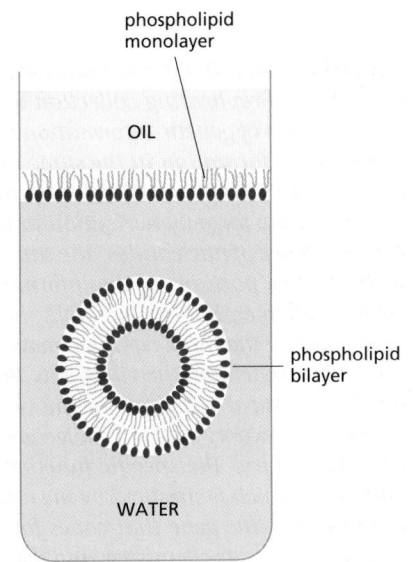

Figure 1–9 **Formation of a membrane by amphiphilic phospholipid molecules.** Phospholipids have a hydrophilic (water-loving, phosphate) head group and a hydrophobic (water-avoiding, hydrocarbon) tail. At an interface between oil and water, they arrange themselves as a single sheet with their head groups facing the water and their tail groups facing the oil. But when immersed in water, they aggregate to form bilayers enclosing aqueous compartments, as indicated.

synthesizes for its own use, while excreting its waste products. Without a plasma membrane, the cell could not maintain its integrity as a coordinated chemical system.

The molecules that form a membrane have the simple physicochemical property of being *amphiphilic*—that is, consisting of one part that is hydrophobic (water-insoluble) and another part that is hydrophilic (water-soluble). Such molecules placed in water aggregate spontaneously, arranging their hydrophobic portions to be as much in contact with one another as possible to hide them from the water, while keeping their hydrophilic portions exposed. Amphiphilic molecules of appropriate shape, such as the phospholipid molecules that comprise most of the plasma membrane, spontaneously aggregate in water to create a *bilayer* that forms small closed vesicles (**Figure 1–9**). The phenomenon can be demonstrated in a test tube by simply mixing phospholipids and water together; under appropriate conditions, small vesicles form whose aqueous contents are isolated from the external medium.

Although the chemical details vary, the hydrophobic tails of the predominant membrane molecules in all cells are hydrocarbon polymers ($-CH_2-CH_2-CH_2-$), and their spontaneous assembly into a bilayered vesicle is but one of many examples of an important general principle: cells produce molecules whose chemical properties cause them to *self-assemble* into the structures that a cell needs.

The cell boundary cannot be totally impermeable. If a cell is to grow and reproduce, it must be able to import raw materials and export waste across its plasma membrane. All cells therefore have specialized proteins embedded in their membrane that transport specific molecules from one side to the other. Some of these *membrane transport proteins*, like some of the proteins that catalyze the fundamental small-molecule reactions inside the cell, have been so well preserved over the course of evolution that we can recognize the family resemblances between them in comparisons of even the most distantly related groups of living organisms.

The transport proteins in the membrane largely determine which molecules enter the cell, and the catalytic proteins inside the cell determine the reactions that those molecules undergo. Thus, by specifying the proteins that the cell is to manufacture, the genetic information recorded in the DNA sequence dictates the entire chemistry of the cell; and not only its chemistry, but also its form and its behavior, for these too are chiefly constructed and controlled by the cell's proteins.

A Living Cell Can Exist with Fewer Than 500 Genes

The basic principles of biological information transfer are simple enough, but how complex are real living cells? In particular, what are the minimum requirements? We can get a rough indication by considering a species that has one of the smallest known genomes—the bacterium *Mycoplasma genitalium* (**Figure 1–10**). This organism lives as a parasite in mammals, and its environment provides it with many of its small molecules ready-made. Nevertheless, it still has to make all the large molecules—DNA, RNAs, and proteins—required for the basic processes of heredity. It has about 530 genes, about 400 of which are essential. Its genome of 580,070 nucleotide pairs represents 145,018 bytes of information—about as much as it takes to record the text of one chapter of this book. Cell biology may be complicated, but it is not impossibly so.

The minimum number of genes for a viable cell in today's environments is probably not less than 300, although there are only about 60 genes in the core set that is shared by all living species.

Summary

The individual cell is the minimal self-reproducing unit of living matter, and it consists of a self-replicating collection of catalysts. Central to this reproduction is the transmission of genetic information to progeny cells. Every cell on our planet stores its genetic information in the same chemical form—as double-stranded DNA. The cell replicates its information by separating the paired DNA strands and using each as a template for polymerization to make a new DNA strand with a complementary sequence of nucleotides. The same strategy of templated polymerization is used to transcribe portions of the information from DNA into molecules of the closely related polymer, RNA. These RNA molecules in turn guide the synthesis of protein molecules by the more complex machinery of translation, involving a large multimolecular machine, the ribosome. Proteins are the principal catalysts for almost all the chemical reactions in the cell; their other functions include the selective import and export of small molecules across the plasma membrane that forms the cell's boundary. The specific function of each protein depends on its amino acid sequence, which is specified by the nucleotide sequence of a corresponding segment of the DNA—the gene that codes for that protein. In this way, the genome of the cell determines its chemistry; and the chemistry of every living cell is fundamentally similar, because it must provide for the synthesis of DNA, RNA, and protein. The simplest known cells can survive with about 400 genes.

THE DIVERSITY OF GENOMES AND THE TREE OF LIFE

The success of living organisms based on DNA, RNA, and protein has been spectacular. Life has populated the oceans, covered the land, infiltrated the Earth's crust, and molded the surface of our planet. Our oxygen-rich atmosphere, the deposits of coal and oil, the layers of iron ores, the cliffs of chalk and limestone and marble—all these are products, directly or indirectly, of past biological activity on Earth.

Living things are not confined to the familiar temperate realm of land, water, and sunlight inhabited by plants and plant-eating animals. They can be found in the darkest depths of the ocean, in hot volcanic mud, in pools beneath the frozen surface of the Antarctic, and buried kilometers deep in the Earth's crust. The creatures that live in these extreme environments are generally unfamiliar, not only because they are inaccessible, but also because they are mostly microscopic. In more homely habitats, too, most organisms are too small for us to see without special equipment: they tend to go unnoticed, unless they cause a disease or rot the timbers of our houses. Yet microorganisms make up most of the total mass of living matter on our planet. Only recently, through new methods of molecular analysis and specifically through the analysis of DNA sequences, have we begun to get a picture of life on Earth that is not grossly distorted by our biased perspective as large animals living on dry land.

In this section, we consider the diversity of organisms and the relationships among them. Because the genetic information for every organism is written in the universal language of DNA sequences, and the DNA sequence of any given organism can be readily obtained by standard biochemical techniques, it is now possible to characterize, catalog, and compare any set of living organisms with reference to these sequences. From such comparisons we can estimate the place of each organism in the family tree of living species—the "tree of life." But before describing what this approach reveals, we need first to consider the routes by which cells in different environments obtain the matter and energy they require to survive and proliferate, and the ways in which some classes of organisms depend on others for their basic chemical needs.

Cells Can Be Powered by a Variety of Free-Energy Sources

Living organisms obtain their free energy in different ways. Some, such as animals, fungi, and the many different bacteria that live in the human gut, get it by feeding on other living things or the organic chemicals they produce; such organisms

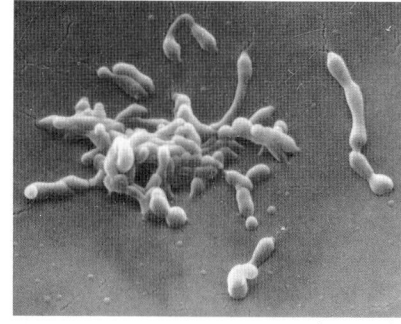

(A)
5 μm

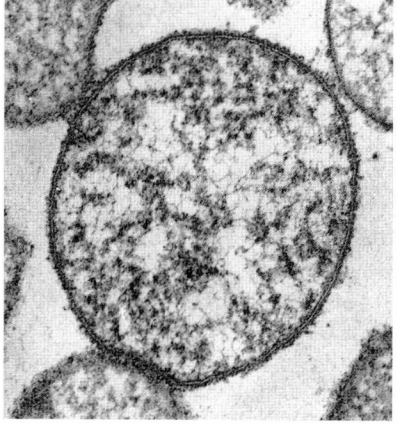

(B)
0.2 μm

Figure 1–10 *Mycoplasma genitalium*.
(A) Scanning electron micrograph showing the irregular shape of this small bacterium, reflecting the lack of any rigid cell wall. (B) Cross section (transmission electron micrograph) of a *Mycoplasma* cell. Of the 525 genes of *Mycoplasma genitalium*, 43 code for transfer, ribosomal, and other non-messenger RNAs. Functions are known, or can be guessed at, for most of the 482 protein-coding genes. Of these, 154 are involved in replication, transcription, translation, and related processes involving DNA, RNA, and protein; 98 in the membrane and surface structures of the cell; 46 in the transport of nutrients and other molecules across the membrane; 71 in energy conversion and the synthesis and degradation of small molecules; and 12 in the regulation of cell division and other processes. Note that these categories are partly overlapping, so that some genes feature twice. (A, from S. Razin et al., *Infect. Immun.* 30:538–546, 1980. With permission from the American Society for Microbiology; B, courtesy of Roger Cole, in Medical Microbiology, 4th ed. [S. Baron ed.]. Galveston: University of Texas Medical Branch, 1996.)

are called *organotrophic* (from the Greek word *trophe,* meaning "food"). Others derive their energy directly from the nonliving world. These primary energy converters fall into two classes: those that harvest the energy of sunlight, and those that capture their energy from energy-rich systems of inorganic chemicals in the environment (chemical systems that are far from chemical equilibrium). Organisms of the former class are called *phototrophic* (feeding on sunlight); those of the latter are called *lithotrophic* (feeding on rock). Organotrophic organisms could not exist without these primary energy converters, which are the most plentiful form of life.

Phototrophic organisms include many types of bacteria, as well as algae and plants, on which we—and virtually all the living things that we ordinarily see around us—depend. Phototrophic organisms have changed the whole chemistry of our environment: the oxygen in the Earth's atmosphere is a by-product of their biosynthetic activities.

Lithotrophic organisms are not such an obvious feature of our world, because they are microscopic and mostly live in habitats that humans do not frequent—deep in the ocean, buried in the Earth's crust, or in various other inhospitable environments. But they are a major part of the living world, and they are especially important in any consideration of the history of life on Earth.

Some lithotrophs get energy from *aerobic* reactions, which use molecular oxygen from the environment; since atmospheric O_2 is ultimately the product of living organisms, these aerobic lithotrophs are, in a sense, feeding on the products of past life. There are, however, other lithotrophs that live anaerobically, in places where little or no molecular oxygen is present. These are circumstances similar to those that existed in the early days of life on Earth, before oxygen had accumulated.

The most dramatic of these sites are the hot *hydrothermal vents* on the floor of the Pacific and Atlantic Oceans. They are located where the ocean floor is spreading as new portions of the Earth's crust form by a gradual upwelling of material from the Earth's interior (**Figure 1–11**). Downward-percolating seawater is heated and driven back upward as a submarine geyser, carrying with it a current of chemicals from the hot rocks below. A typical cocktail might include H_2S, H_2, CO, Mn^{2+}, Fe^{2+}, Ni^{2+}, CH_4, NH_4^+, and phosphorus-containing compounds. A dense

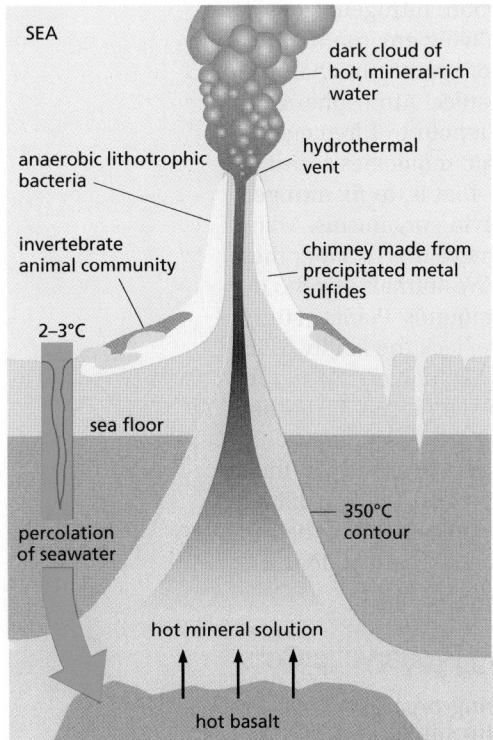

Figure 1–11 **The geology of a hot hydrothermal vent in the ocean floor.** As indicated, water percolates down toward the hot molten rock upwelling from the Earth's interior and is heated and driven back upward, carrying minerals leached from the hot rock. A temperature gradient is set up, from more than 350°C near the core of the vent, down to 2–3°C in the surrounding ocean. Minerals precipitate from the water as it cools, forming a chimney. Different classes of organisms, thriving at different temperatures, live in different neighborhoods of the chimney. A typical chimney might be a few meters tall, spewing out hot, mineral-rich water at a flow rate of 1–2 m/sec.

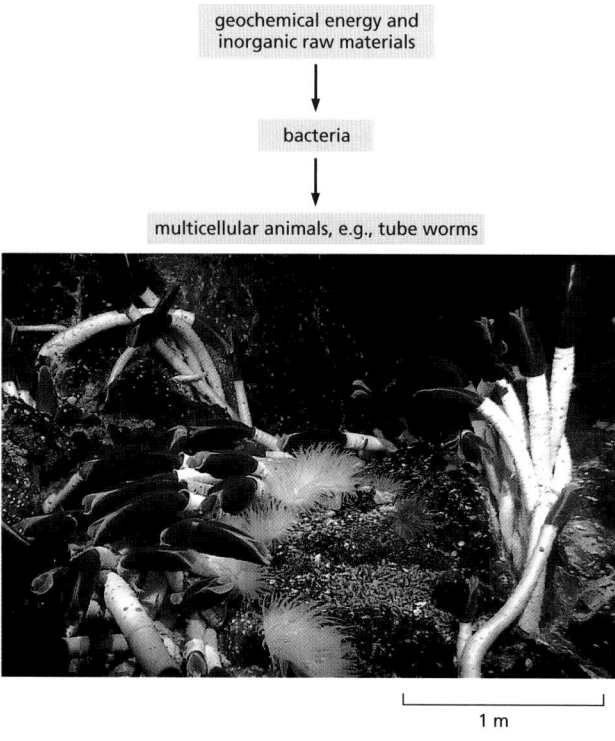

geochemical energy and
inorganic raw materials

↓

bacteria

↓

multicellular animals, e.g., tube worms

1 m

Figure 1–12 Organisms living at a depth of 2500 meters near a vent in the ocean floor. Close to the vent, at temperatures up to about 120°C, various lithotrophic species of bacteria and archaea (archaebacteria) live, directly fueled by geochemical energy. A little further away, where the temperature is lower, various invertebrate animals live by feeding on these microorganisms. Most remarkable are these giant (2 meter) tube worms, *Riftia pachyptila,* which, rather than feed on the lithotrophic cells, live in symbiosis with them: specialized organs in the worms harbor huge numbers of symbiotic sulfur-oxidizing bacteria. These bacteria harness geochemical energy and supply nourishment to their hosts, which have no mouth, gut, or anus. The tube worms are thought to have evolved from more conventional animals, and to have become secondarily adapted to life at hydrothermal vents. (Science History Images/Alamy Stock Photo.)

population of microbes lives in the neighborhood of the vent, thriving on this austere diet and harvesting free energy from reactions between the available chemicals. Other organisms—clams, mussels, and giant marine worms—in turn live off the microbes at the vent, forming an entire ecosystem analogous to the world of plants and animals that we belong to, but powered by geochemical energy instead of light (**Figure 1–12**).

Some Cells Fix Nitrogen and Carbon Dioxide for Others

To make a living cell requires matter, as well as free energy. DNA, RNA, and protein are composed of just six elements: hydrogen, carbon, nitrogen, oxygen, sulfur, and phosphorus. These are all plentiful in the nonliving environment, in the Earth's rocks, water, and atmosphere. But they are not present in chemical forms that allow easy incorporation into biological molecules. Atmospheric N_2 and CO_2, in particular, are extremely unreactive. A large amount of free energy is required to drive the reactions that use these inorganic molecules to make the organic compounds needed for further biosynthesis—that is, to *fix* nitrogen and carbon dioxide, so as to make N and C available to living organisms. Many types of living cells lack the biochemical machinery to achieve this fixation; they instead rely on other classes of cells to do the job for them. We animals depend on plants for our supplies of organic carbon and nitrogen compounds. Plants in turn, although they can fix carbon dioxide from the atmosphere, lack the ability to fix atmospheric nitrogen; they depend in part on nitrogen-fixing bacteria to supply their need for nitrogen compounds. Plants of the pea family, for example, harbor symbiotic nitrogen-fixing bacteria in nodules in their roots.

Living cells therefore differ widely in some of the most basic aspects of their biochemistry. Not surprisingly, cells with complementary needs and capabilities have developed close associations. Some of these associations, as we see below, have evolved to the point where the partners have lost their separate identities altogether: they have joined forces to form a single composite cell.

The Greatest Biochemical Diversity Exists Among Prokaryotic Cells

From simple microscopy, it has long been clear that living organisms can be classified on the basis of cell structure into two groups: the **eukaryotes** and the

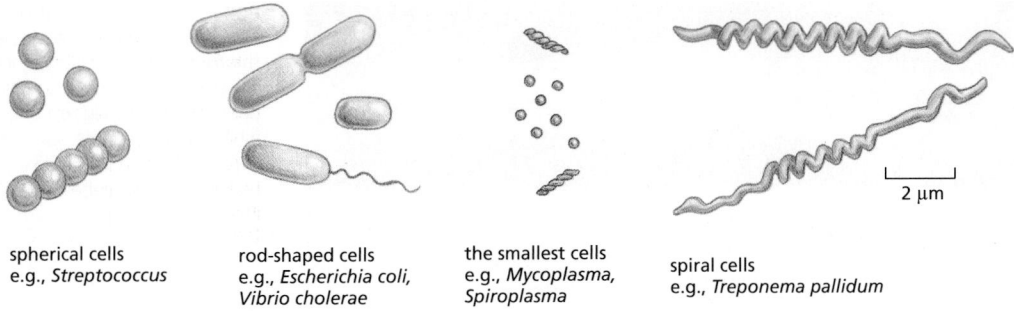

spherical cells
e.g., *Streptococcus*

rod-shaped cells
e.g., *Escherichia coli*,
Vibrio cholerae

the smallest cells
e.g., *Mycoplasma*,
Spiroplasma

spiral cells
e.g., *Treponema pallidum*

2 μm

prokaryotes. Eukaryotes keep their DNA in a distinct membrane-enclosed intracellular compartment called the nucleus. (The name is from the Greek, meaning "truly nucleated," from the words *eu*, "well" or "truly," and *karyon*, "kernel" or "nucleus.") Prokaryotes have no distinct nuclear compartment to house their DNA. Plants, fungi, and animals are eukaryotes; bacteria are prokaryotes, as are archaea—a separate class of prokaryotic cells, discussed below.

Most prokaryotic cells are small and simple in outward appearance (**Figure 1–13**), and they live mostly as independent individuals or in loosely organized communities, rather than as multicellular organisms. They are typically spherical or rod-shaped and measure a few micrometers in linear dimension. They often have a tough protective coat, called a *cell wall*, beneath which a plasma membrane encloses a single cytoplasmic compartment containing DNA, RNA, proteins, and the many small molecules needed for life. In the electron microscope, this cell interior appears as a matrix of varying texture without any discernible organized internal structure (**Figure 1–14**).

Prokaryotic cells live in an enormous variety of ecological niches, and they are astonishingly varied in their biochemical capabilities—far more so than eukaryotic cells. Organotrophic species can utilize virtually any type of organic molecule as food, from sugars and amino acids to hydrocarbons and methane gas. Phototrophic species (**Figure 1–15**) harvest light energy in a variety of ways, some of them generating oxygen as a by-product, others not. Lithotrophic species can feed on a plain diet of inorganic nutrients, getting their carbon from CO_2, and relying on H_2S to fuel their energy needs (**Figure 1–16**)—or on H_2, or Fe^{2+}, or elemental sulfur, or any of a host of other chemicals that occur in the environment.

Figure 1–13 **Shapes and sizes of some bacteria.** Although most are small, as shown, measuring a few micrometers in linear dimension, there are also some giant species. An extreme example (not shown) is the cigar-shaped bacterium *Epulopiscium fishelsoni*, which lives in the gut of a surgeonfish and can be up to 600 μm long.

Figure 1–14 **The structure of a bacterium.** (A) The bacterium *Vibrio cholerae*, showing its simple internal organization. Like many other species, *Vibrio* has a helical appendage at one end—a flagellum—that rotates as a propeller to drive the cell forward. It can infect the human small intestine to cause cholera; the severe diarrhea that accompanies this disease kills more than 100,000 people a year. (B) An electron micrograph of a longitudinal section through the widely studied bacterium *Escherichia coli (E. coli)*. The cell's DNA is concentrated in the lightly stained region. Part of our normal intestinal flora, *E. coli* is related to *Vibrio*, and it has many flagella distributed over its surface that are not visible in this section. (B, courtesy of E. Kellenberger.)

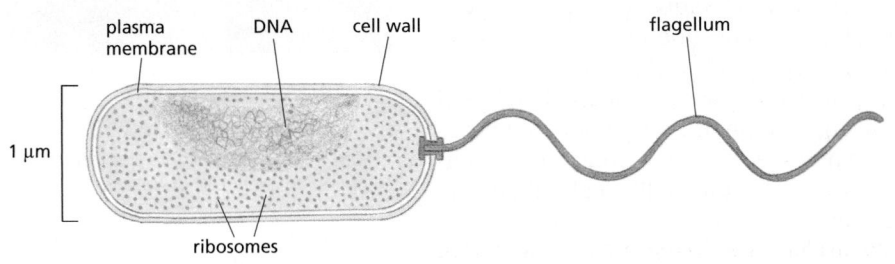

plasma membrane   DNA   cell wall   flagellum

1 μm

ribosomes

(A)

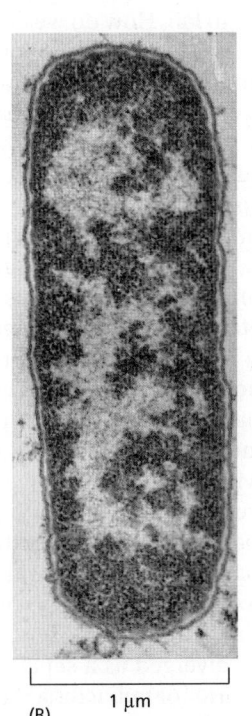

1 μm

(B)

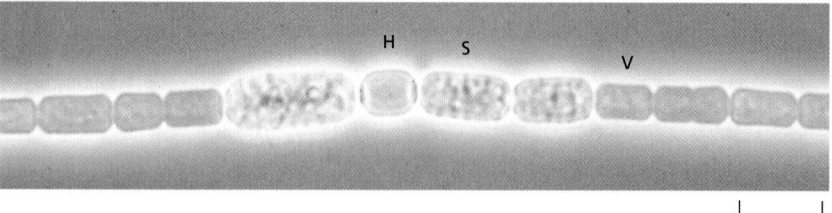

Figure 1–15 The phototrophic bacterium *Anabaena cylindrica* viewed in the light microscope. The cells of this species form long, multicellular filaments. Most of the cells (labeled V) perform photosynthesis, while others become specialized for nitrogen fixation (labeled H) or develop into resistant spores (labeled S). (Courtesy of Dave G. Adams.)

10 μm

Much of this world of microscopic organisms is virtually unexplored. Traditional methods of bacteriology have given us an acquaintance with those species that can be isolated and cultured in the laboratory. But DNA sequence analysis of the populations of bacteria and archaea in samples from natural habitats—such as soil or ocean water, or even the human mouth—has opened our eyes to the fact that most species cannot be cultured by standard laboratory techniques. According to one estimate, at least 99% of prokaryotic species remain to be characterized. Detected only by their DNA, it has not yet been possible to grow the vast majority of them in laboratories.

The Tree of Life Has Three Primary Branches: Bacteria, Archaea, and Eukaryotes

The classification of living things has traditionally depended on comparisons of their outward appearances: we can see that a fish has eyes, jaws, backbone, brain, and so on, just as we do, and that a worm does not; that a rosebush is cousin to an apple tree, but is less similar to a grass. As Darwin showed, we can readily interpret such close family resemblances in terms of evolution from common ancestors, and we can find the remains of many of these ancestors preserved in the fossil record. In this way, it has been possible to begin to draw a family tree of living organisms, showing the various lines of descent, as well as branch points in the history, where the ancestors of one group of species became different from those of another.

When the disparities between organisms become very great, however, these methods begin to fail. How do we decide whether a fungus is closer kin to a plant or to an animal? When it comes to prokaryotes, the task becomes harder still: one microscopic rod or sphere looks much like another. Microbiologists have therefore sought to classify prokaryotes in terms of their biochemistry and nutritional requirements. But this approach also has its pitfalls. Amid the bewildering variety of biochemical behaviors, it is difficult to know which differences truly reflect differences of evolutionary history.

Genome analysis has now given us a simpler, more direct, and much more powerful way to determine evolutionary relationships. The complete DNA sequence of an organism defines its nature with almost perfect precision and in exhaustive detail. Moreover, this specification is in a digital form—a string of letters—that can be entered straightforwardly into a computer and compared with the corresponding information for any other living thing. Because DNA is subject to random changes that accumulate over long periods of time (as we shall see shortly), the number of differences between the DNA sequences of two organisms can provide a direct, objective, quantitative indication of the evolutionary distance between them.

This approach has shown that the organisms that were traditionally classed together as "bacteria" can be as widely divergent in their evolutionary origins as is any prokaryote from any eukaryote. It is now clear that the prokaryotes comprise two distinct groups that diverged early in the history of life on Earth, before the eukaryotes diverged as a separate group. The two groups of prokaryotes are called the **bacteria** (or eubacteria) and the **archaea** (or archaebacteria). Detailed genome analyses have recently revealed that the first eukayotic cell formed after a

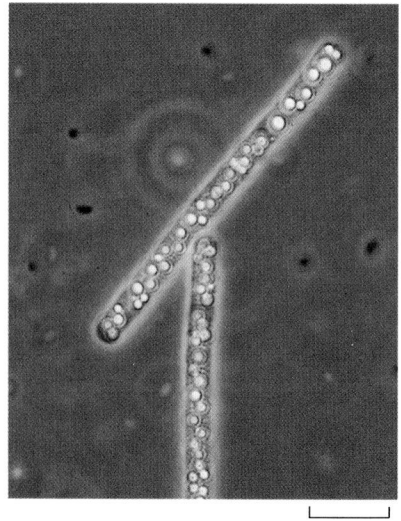

6 μm

Figure 1–16 A lithotrophic bacterium. *Beggiatoa*, which lives in sulfurous environments, gets its energy by oxidizing H_2S and can fix carbon even in the dark. Note the yellow deposits of sulfur inside the cells. (Courtesy of Ralph S. Wolfe.)

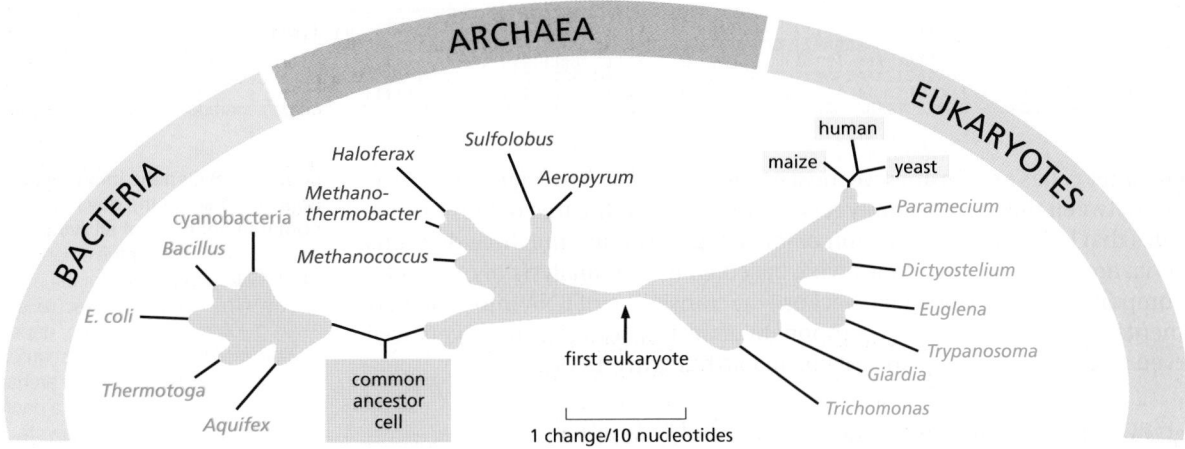

Figure 1–17 **The three major divisions (domains) of the living world.** Note that the word *bacteria* was originally used to refer to prokaryotes in general, but more recently has been redefined to refer to eubacteria specifically. The tree shown here is based on comparisons of the nucleotide sequence of a ribosomal RNA (rRNA) subunit in the different species, and the distances in the diagram represent estimates of the numbers of evolutionary changes that have occurred in this molecule in each lineage (see Figure 1–18). The parts of the tree shrouded in *gray cloud* represent uncertainties about details of the true pattern of species divergence in the course of evolution: comparisons of nucleotide or amino acid sequences of molecules other than rRNA, as well as other arguments, can lead to somewhat different trees. As indicated, the nucleus of the eukaryotic cell is now thought to have emerged from a sub-branch within the archaea, so that in the beginning the tree of life had only two branches—bacteria and archaea.

particular type of ancient archaeal cell engulfed an ancient bacterium (see Figure 12–3). Thus, the living world today is considered to consist of three major divisions or *domains*: bacteria, archaea, and eukaryotes (**Figure 1–17**).

Archaea are often found inhabiting environments that we humans avoid, such as bogs, sewage treatment plants, ocean depths, salt brines, and hot acid springs, although they are also widespread in less extreme and more homely environments, from soils and lakes to the stomachs of cattle. In outward appearance they are not easily distinguished from bacteria. At a molecular level, archaea seem to resemble eukaryotes more closely in their machinery for handling genetic information (replication, transcription, and translation), but bacteria more closely in their apparatus for metabolism and energy conversion. We discuss below how this might be explained.

Some Genes Evolve Rapidly; Others Are Highly Conserved

Both in the storage and in the copying of genetic information, random accidents and errors occur, altering the nucleotide sequence—that is, creating **mutations**. Therefore, when a cell divides, its two daughters are often not quite identical to one another or to their parent. On rare occasions, the error may represent a change for the better; more probably, it will cause no significant difference in the cell's prospects. But in many cases, the error will cause serious damage—for example, by disrupting the coding sequence for a key protein. Changes due to mistakes of the first type will tend to be perpetuated, because the altered cell has an increased likelihood of reproducing itself. Changes due to mistakes of the second type—*selectively neutral* changes—may be perpetuated or not: in the competition for limited resources, it is a matter of chance whether the altered cell or its cousins will succeed. But changes that cause serious damage lead nowhere: the cell that suffers them dies, leaving no progeny. Through endless repetition of this cycle of error and trial—of *mutation* and *natural selection*—organisms evolve: their genetic specifications change, giving them new ways to exploit the environment more effectively, to survive in competition with others, and to reproduce successfully.

Some parts of the genome will change more easily than others in the course of evolution. A segment of DNA that does not code for protein and has no significant regulatory role is free to change at a rate limited only by the frequency of random errors. In contrast, a gene that codes for a highly optimized essential protein or RNA molecule cannot alter so easily: when mistakes occur, the faulty cells are almost always eliminated. Genes of this latter sort are therefore *highly conserved*. Through 3.5 billion years or more of evolutionary history, many features of the genome have changed beyond all recognition, but the most highly conserved genes remain perfectly recognizable in all living species.

```
GTTCCGGGGGGAGTATGGTTGCAAAGCTGAAACTTAAAGGAATTGACGGAAGGGCACCACCAGGAGTGGAGCCTGCGGCTTAATTTGACTCAACACGGGAAACCTCACCC    human
 I   I IIIIIIII I III IIIIIII    IIIIIIIIIIIII III I IIIII III IIIIII IIIIIII III II I IIIII  II IIIII  I III IIII
GCCGCCTGGGGAGTACGGTCGCAAGACTGAAACTTAAAGGAATTGGCGGGGGAGCACTACAACGGGTGGAGCCTGCGGTTTAATTGGATTCAACGCCGGGCATCTTACCA    Methanococcus
  IIIIIIIIIIIIIIIIIIIIII IIIIII    I IIIIII III IIIIII IIIII    I  I IIII  IIIIIIII II IIIIIIII III IIIIII   I IIIIII
ACCGCCTGGGGAGTACGGCCGCAAGGTTAAAACTCAAATGAATTGACGGGGGCCCGC.ACAAGCGGTGGAGCATGTGGTTTAATTCGATGCAACGCGAAGAACCTTACCT    E. coli
  II IIIIIIII IC    IIII I I IIIIIII   I IIIII   I  II IIIIII     I  II IIIIIIIII II IIIIIIII III IIIIII  IIIII III
GTTCCGGGGGGAGTATGGTTGCAAAGCTGAAACTTAAAGGAATTGACGGAAGGGCACCACCAGGAGTGGAGCCTGCGGCTTAATTTGACTCAACACGGGAAACCTCACCC    human
```

Figure 1–18 Genetic information conserved since the days of the last common ancestor of all living things. A part of the gene for the smaller of the two main rRNA components of the ribosome is shown. (The complete molecule is about 1500–1900 nucleotides long, depending on species.) Corresponding segments of nucleotide sequence from an archaean (*Methanococcus jannaschii*), a bacterium (*Escherichia coli*), and a eukaryote (*Homo sapiens*) are aligned. Sites where the nucleotides are identical between species are indicated by a vertical line; the human sequence is repeated at the bottom of the alignment so that all three two-way comparisons can be seen. A dot halfway along the *E. coli* sequence denotes a site where a nucleotide has been either deleted from the bacterial lineage in the course of evolution or inserted in the other two lineages. Note that the sequences from these three organisms, representative of the three domains of the living world, still retain unmistakable similarities.

These latter genes are the ones we must examine if we wish to trace family relationships between the most distantly related organisms in the tree of life. The initial studies that led to the classification of the living world into the three domains of bacteria, archaea, and eukaryotes were based chiefly on analysis of one of the rRNA components of the ribosome. Because the translation of RNA into protein is fundamental to all living cells, this component of the ribosome has been very well conserved since early in the history of life on Earth (**Figure 1–18**).

Most Bacteria and Archaea Have 1000–6000 Genes

Natural selection has generally favored those prokaryotic cells that can reproduce the fastest by taking up raw materials from their environment and replicating themselves most efficiently, at the maximal rate permitted by the available food supplies. Small size implies a large ratio of surface area to volume, thereby helping to maximize the uptake of nutrients across the plasma membrane and boosting a cell's reproductive rate.

Presumably for these reasons, most prokaryotic cells carry very little superfluous baggage; their genomes are small, with genes packed closely together and minimal quantities of regulatory DNA between them. The small genome size has made it easy to use modern DNA sequencing techniques to determine complete genome sequences. We now have this information for thousands of species of bacteria and archaea, as well as for hundreds of species of eukaryotes. Most bacterial and archaeal genomes contain between 10^6 and 10^7 nucleotide pairs, encoding 1000–6000 genes.

A complete DNA sequence reveals both the genes an organism possesses and the genes it lacks. When we compare the three domains of the living world, we can begin to see which genes are common to all of them and must therefore have been present in the cell that was ancestral to all present-day living things, and which genes are peculiar to a single branch in the tree of life. To explain the findings, however, we need to consider a little more closely how new genes arise and genomes evolve.

New Genes Are Generated from Preexisting Genes

The raw material of evolution is the DNA sequence that already exists: there is no natural mechanism for making long stretches of new random sequence. In this sense, no gene is ever entirely new. Innovation can, however, occur in several ways (**Figure 1–19**):

1. *Intragenic mutation*: an existing gene can be randomly modified by changes in its DNA sequence, through various types of error that occur mainly in the process of DNA replication.

2. *Gene duplication*: an existing gene can be accidentally duplicated so as to create a pair of initially identical genes within a single cell; these two genes may then diverge in the course of evolution.

3. *DNA segment shuffling*: two or more existing genes can break and rejoin to make a hybrid gene consisting of DNA segments that originally belonged to separate genes.

4. *Horizontal (intercellular) transfer*: a piece of DNA can be transferred from the genome of one cell to that of another—even to that of another species. This process is in contrast with the usual *vertical transfer* of genetic information from parent to progeny.

Each of these types of change leaves a characteristic trace in the DNA sequence of the organism, and there is clear evidence that all four processes have frequently

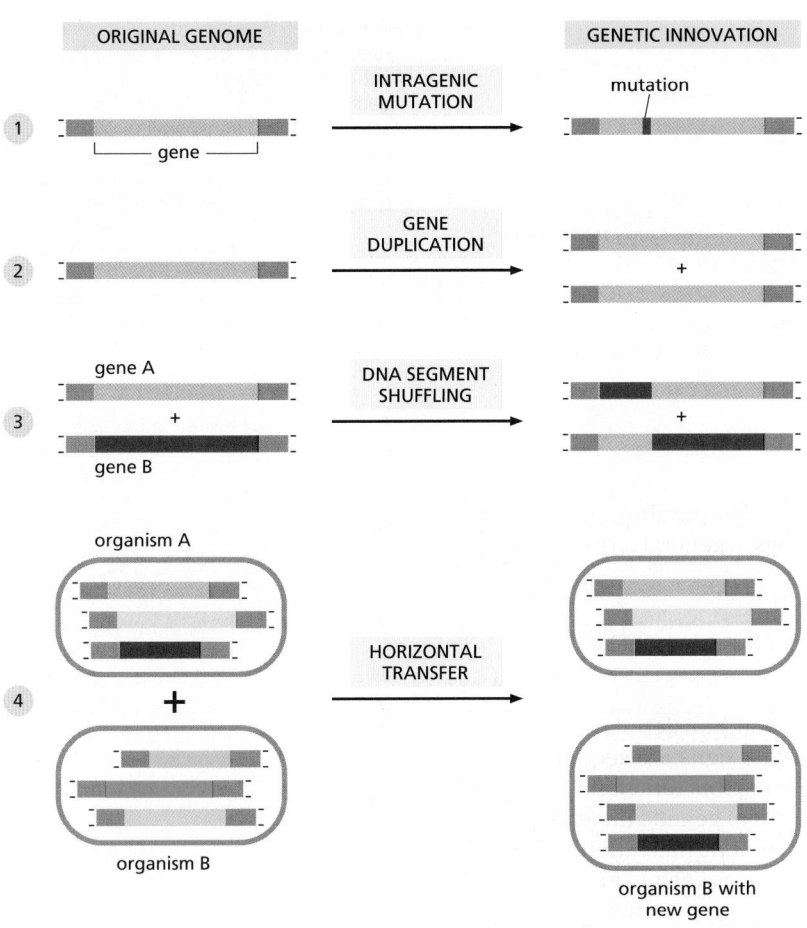

Figure 1–19 **Four modes of genetic innovation and their effects on the DNA sequence of an organism.** A special form of horizontal transfer occurs when two different types of cells enter into a permanent symbiotic association. Genes from one of the cells then may be transferred to the genome of the other, as we shall see below when we discuss mitochondria and chloroplasts.

occurred. In later chapters, we discuss the underlying mechanisms, but for the present we focus on the consequences.

Gene Duplications Give Rise to Families of Related Genes Within a Single Cell

A cell duplicates its entire genome each time it divides into two daughter cells. However, accidents occasionally result in the inappropriate duplication of just part of the genome, with retention of original and duplicate segments in a single cell. Once a gene has been duplicated in this way, one of the two gene copies is free to mutate and become specialized to perform a different function within the same cell. Repeated rounds of this process of duplication and divergence, over many millions of years, have enabled one gene to give rise to a family of genes that may all be found within a single genome. Analysis of the DNA sequence of prokaryotic genomes reveals many examples of such **gene families**: in the bacterium *Bacillus subtilis*, for example, 47% of the genes have one or more obvious relatives (**Figure 1–20**).

When genes duplicate and diverge in this way, the individuals of one species become endowed with multiple variants of a primordial gene. This evolutionary process has to be distinguished from the genetic divergence that occurs when one species of organism splits into two separate lines of descent at a branch point in the family tree—when the human line of descent became separate from that of chimpanzees, for example. There, the genes gradually become different in the course of evolution, but they are likely to continue to have corresponding functions in the two sister species. Genes that are related by descent in this way—that is, genes in two separate species that derive from the same ancestral gene in the last common ancestor of those two species—are called **orthologs**. Related genes that have resulted from a gene duplication event within a single genome—and

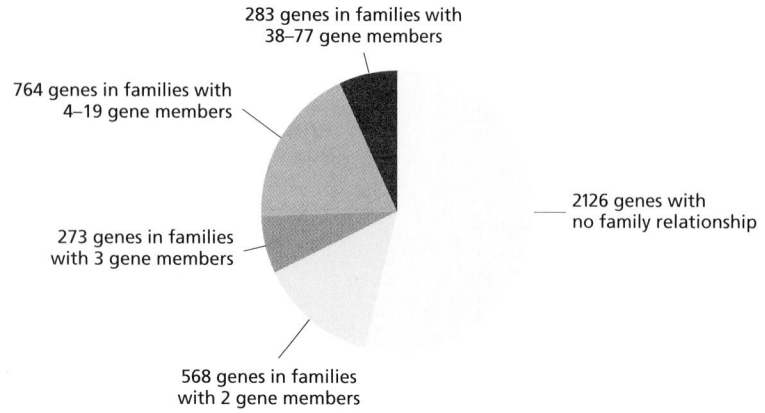

283 genes in families with 38–77 gene members

764 genes in families with 4–19 gene members

273 genes in families with 3 gene members

568 genes in families with 2 gene members

2126 genes with no family relationship

Figure 1–20 Families of evolutionarily related genes in the genome of *Bacillus subtilis*. The largest gene family in this bacterium consists of 77 genes coding for varieties of ABC transporters—a class of membrane transport proteins found in all three domains of the living world. (Adapted from F. Kunst et al., *Nature* 390:249–256, 1997.)

are likely to have diverged in their function—are called **paralogs**. Genes that are related by descent in either way are called **homologs**, a general term used to cover both types of relationship (**Figure 1–21**).

Genes Can Be Transferred Between Organisms, Both in the Laboratory and in Nature

Prokaryotes provide good examples of the horizontal transfer of genes from one species of cell to another. The most obvious tell-tale signs are sequences recognizable as being derived from viruses, those infecting bacteria being called *bacteriophages* (**Figure 1–22**). **Viruses** are small packets of genetic material that have evolved as parasites on the reproductive and biosynthetic machinery of host cells. Although not themselves living cells, they often serve as vectors for gene transfer. A virus will replicate in one cell, emerge from it with a protective wrapping, and then enter and infect another cell, which may be of the same or a different species. Often, the infected cell will be killed by the massive proliferation of virus particles inside it; but sometimes, the viral DNA, instead of directly generating these particles, may persist in its host for many cell generations as a relatively innocuous passenger, either as a separate intracellular fragment of DNA, known as a *plasmid*, or as a sequence inserted into the cell's regular genome. In their travels, viruses can accidentally pick up fragments of DNA from the genome of one host cell and ferry them into another cell. Such transfers of genetic material are very common in prokaryotes.

Horizontal transfers of genes between eukaryotic cells of different species are very rare, and they do not seem to have played a significant part in eukaryote evolution (although massive transfers from bacterial to eukaryotic genomes have occurred in the evolution of mitochondria and chloroplasts, as we discuss below).

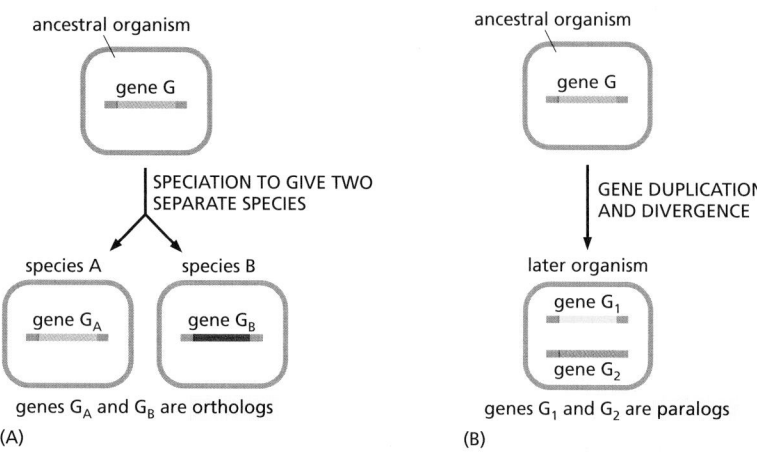

(A)

ancestral organism

gene G

SPECIATION TO GIVE TWO SEPARATE SPECIES

species A — gene G$_A$

species B — gene G$_B$

genes G$_A$ and G$_B$ are orthologs

(B)

ancestral organism

gene G

GENE DUPLICATION AND DIVERGENCE

later organism

gene G$_1$

gene G$_2$

genes G$_1$ and G$_2$ are paralogs

Figure 1–21 Paralogous genes and orthologous genes: two types of gene homology based on different evolutionary pathways. (A) Orthologs. (B) Paralogs.

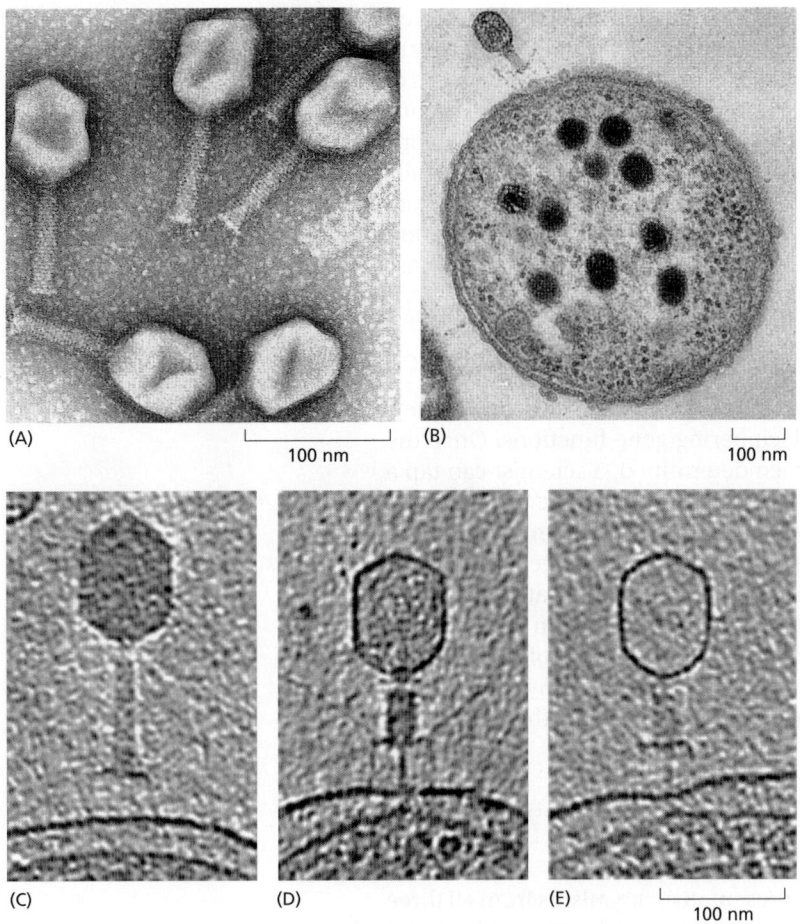

Figure 1–22 The viral transfer of DNA into a cell. (A) An electron micrograph of particles of a bacterial virus, the T4 bacteriophage. The head of this virus contains the viral DNA; the tail contains the apparatus for injecting the DNA into a host bacterium. (B) A cross section of an *E. coli* bacterium with a T4 bacteriophage latched onto its surface. The large dark objects inside the bacterium are the heads of new T4 particles in the course of assembly. When they are mature, the bacterium will burst open to release them. (C–E) The process of DNA injection into the bacterium, as visualized in unstained, frozen samples by cryoelectron microscopy. (C) Attachment begins. (D) Attached state during DNA injection. (E) Virus head has emptied all of its DNA into the bacterium. (A, courtesy of James Paulson; B, courtesy of Jonathan King and Erika Hartwig from G. Karp, Cell and Molecular Biology, 2nd ed. New York: John Wiley & Sons, 1999; C–E, courtesy of Ian Molineux, University of Texas at Austin and Jun Liu.)

In contrast, horizontal gene transfers occur much more frequently between different species of prokaryotes. Many prokaryotes have a remarkable capacity to take up even nonviral DNA molecules from their surroundings and thereby capture the genetic information these molecules carry. By this route, or by virus-mediated transfer, bacteria and archaea in the wild can acquire genes from neighboring cells relatively easily. Genes that confer resistance to an antibiotic or an ability to produce a toxin, for example, can be transferred from species to species and provide the recipient bacterium with a selective advantage. In this way, new and sometimes dangerous strains of bacteria have been observed to evolve in the bacterial ecosystems that inhabit hospitals or the various niches in the human body. For example, horizontal gene transfer is responsible for the spread, over the past 40 years, of penicillin-resistant strains of *Neisseria gonorrhoeae,* the bacterium that causes gonorrhea. On a longer time scale, the results can be even more profound; it has been estimated that at least 18% of all of the genes in the present-day genome of *E. coli* have been acquired by horizontal transfer from another species within the past 100 million years.

Sex Results in Horizontal Exchanges of Genetic Information Within a Species

Horizontal gene transfer among prokaryotes has a parallel in a phenomenon familiar to us all: sex. In addition to the usual vertical transfer of genetic material from parent to offspring, sexual reproduction causes a large-scale horizontal transfer of genetic information between two initially separate cell lineages—those of the father and the mother. A key feature of sex, of course, is that the genetic exchange normally occurs only between individuals of the same species. But no matter whether they occur within a species or between species, horizontal gene transfers leave a characteristic imprint: they result in individuals who are related more closely to one set of relatives with respect to some genes, and more closely to another set of relatives with respect to others. By comparing the DNA sequences of individual human genomes, an intelligent visitor from outer space could deduce that humans reproduce sexually, even if it knew nothing about human behavior.

Sexual reproduction is widespread (although not universal), especially among eukaryotes. Even bacteria indulge from time to time in controlled sexual exchanges of DNA with other members of their own species. Natural selection has clearly favored organisms that can reproduce sexually, although evolutionary theorists dispute precisely what that selective advantage is.

The Function of a Gene Can Often Be Deduced from Its Sequence

Family relationships among genes are important not just for their historical interest, but because they simplify the task of deciphering gene functions. Once the sequence of a newly discovered gene has been determined, a scientist can tap a few keys on a computer to search the entire database of known gene sequences for genes related to it. In many cases, the function of one or more of these homologs will have been already determined experimentally. Since gene sequence determines gene function, one can frequently make a good guess at the function of the new gene: it is likely to be similar to that of the already known homologs.

In this way, it is possible to decipher a great deal of the biology of an organism simply by analyzing the DNA sequence of its genome and using the information we already have about the functions of genes in other organisms that have been more intensively studied.

More Than 200 Gene Families Are Common to All Three Primary Branches of the Tree of Life

Given the complete genome sequences of representative organisms from all three domains—archaea, bacteria, and eukaryotes—we can search systematically for homologies that span this enormous evolutionary divide. In this way we can begin to take stock of the common inheritance of all living things. There are considerable difficulties in this enterprise. For example, individual species have often lost some of the ancestral genes; other genes have almost certainly been acquired by horizontal transfer from another species and therefore are not truly ancestral, even though shared. In fact, genome comparisons strongly suggest that both lineage-specific gene loss and horizontal gene transfer, in some cases between evolutionarily distant species, have been major factors of evolution, at least among prokaryotes. Finally, in the course of 2 or 3 billion years, some genes that were initially shared will have changed beyond recognition through mutation.

Because of all these vagaries of the evolutionary process, it seems that only a small proportion of ancestral gene families has been universally retained in a recognizable form. Thus, out of 4873 protein-coding gene families defined by comparing the genomes of 50 species of bacteria, 13 archaea, and 3 unicellular eukaryotes, only 63 are truly ubiquitous (that is, represented in all the genomes analyzed). The great majority of these universal families include components of the translation and transcription systems. This is not likely to be a realistic approximation of an ancestral gene set. A better—though still crude—idea of the latter can be obtained by tallying the gene families that have representatives in multiple, but not necessarily all, species from all three major domains. Such an analysis reveals 264 ancient conserved families. Each family can be assigned a function (at least in terms of general biochemical activity, but usually with more precision). As shown in Table 1–1, the largest number of shared gene families are involved in translation and in amino acid metabolism and transport. However, this set of highly conserved gene families represents only a very rough sketch of the common inheritance of all modern life. A more precise reconstruction of the gene complement of the last universal common ancestor will hopefully become feasible with further genome sequencing and more sophisticated forms of comparative analysis.

TABLE 1–1 The Number of Gene Families, Classified by Function, Common to All Three Domains of the Living World				
Information processing		**Metabolism**		
Translation	63	Energy production and conversion		19
Transcription	7	Carbohydrate transport and metabolism		16
Replication, recombination, and repair	13	Amino acid transport and metabolism		43
Cellular processes and signaling		Nucleotide transport and metabolism		15
Cell-cycle control, mitosis, and meiosis	2	Coenzyme transport and metabolism		22
Defense mechanisms	3	Lipid transport and metabolism		9
Signal transduction mechanisms	1	Inorganic ion transport and metabolism		8
Cell wall/membrane biogenesis	2	Secondary metabolite biosynthesis, transport, and catabolism		5
Intracellular trafficking and secretion	4	**Poorly characterized**		
Post-translational modification, protein turnover, chaperones	8	General biochemical function predicted; specific biological role unknown		24

For the purpose of this analysis, gene families are defined as "universal" if they are represented in the genomes of at least two diverse archaea (*Archaeoglobus fulgidus* and *Aeropyrum pernix*), two evolutionarily distant bacteria (*Escherichia coli* and *Bacillus subtilis*), and one eukaryote (yeast, *Saccharomyces cerevisiae*). (Data from R.L. Tatusov, E.V. Koonin and D.J. Lipman, *Science* 278:631–637, 1997; R.L. Tatusov et al., *BMC Bioinformatics* 4:41, 2003; and the COGs database at the US National Library of Medicine.)

Mutations Reveal the Functions of Genes

Without additional information, no amount of gazing at genome sequences will reveal the functions of genes. We may recognize that gene B is like gene A, but how do we discover the function of gene A in the first place? And even if we know the function of gene A, how do we test whether the function of gene B is truly the same as the sequence similarity suggests? How do we connect the world of abstract genetic information with the world of real living organisms?

The analysis of gene functions depends on two complementary approaches: genetics and biochemistry. Genetics starts with the study of mutants: we either find or make an organism in which a gene is altered, and then examine the effects on the organism's structure and performance (**Figure 1–23**). Biochemistry more directly examines the functions of molecules: here we extract molecules from an organism and then study their chemical activities. By combining genetics and biochemistry, it is possible to find those molecules whose production depends on a given gene. At the same time, careful studies of the performance of the mutant organism show us what role those molecules have in the operation of the organism as a whole. Thus, genetics and biochemistry used in combination with cell biology provide the best way to relate genes and molecules to the structure and function of an organism.

In recent years, DNA sequence information and the powerful tools of molecular biology have accelerated progress. From sequence comparisons, we can often identify particular subregions within a gene that have been preserved nearly unchanged over the course of evolution. These conserved subregions are likely to be the most important parts of the gene in terms of function. We can test their individual contributions to the activity of the gene product by creating in the laboratory mutations of specific sites within the gene, or by constructing artificial hybrid genes that combine part of one gene with part of another. Organisms can be engineered to make either the RNA or the protein specified by the gene in large quantities to facilitate biochemical analysis. Specialists in molecular structure can determine the three-dimensional conformation of the gene product, revealing the exact position of every atom in it. Biochemists can determine how each of the

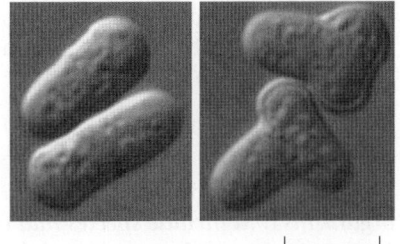

5 μm

Figure 1–23 **A mutant phenotype reflecting the function of a gene.** A normal yeast (of the species *Schizosaccharomyces pombe*) is compared with a mutant in which a change in a single gene has converted the cell from a cigar shape *(left)* to a T shape *(right)*. The mutant gene therefore has a function in the control of cell shape. But how, in molecular terms, does the gene product perform that function? That is a harder question, and it needs biochemical analysis to answer it. (Courtesy of Kenneth Sawin and Paul Nurse.)

parts of the genetically specified molecule contributes to its chemical behavior. Cell biologists can analyze the behavior of cells that are engineered to express a mutant version of the gene.

There is, however, no one simple recipe for discovering a gene's function, and no simple standard universal format for describing it. We may discover, for example, that the product of a given gene catalyzes a certain chemical reaction, and yet have no idea how or why that reaction is important to the organism. The functional characterization of each new family of gene products, unlike the description of the gene sequences, presents a fresh challenge to the biologist's ingenuity. Moreover, we will never fully understand the function of a gene until we learn its role in the life of the organism as a whole. To make ultimate sense of gene functions, therefore, we have to study whole organisms, not just molecules or cells.

Molecular Biology Began with a Spotlight on *E. coli*

Because living organisms are so complex, the more we learn about any particular species, the more attractive it becomes as an object for further study. Each discovery raises new questions and provides new tools with which to tackle general questions in the context of the chosen organism. For this reason, large communities of biologists have become dedicated to studying different aspects of the same **model organism**.

In the early days of molecular biology, the spotlight focused intensely on just one species: the *Escherichia coli*, or *E. coli*, bacterium (see Figures 1–13 and 1–14). This small, rod-shaped bacterial cell normally lives in the gut of humans and other vertebrates, but it can be grown easily in a simple nutrient broth in a culture bottle. It adapts to variable chemical conditions and reproduces rapidly, and it can evolve by mutation and selection at a remarkable speed. As with other bacteria, different strains of *E. coli*, though classified as members of a single species, differ genetically to a much greater degree than do different varieties of a sexually reproducing organism such as a plant or animal. One *E. coli* strain may possess many hundreds of genes that are absent from another, and the two strains could have as little as 50% of their genes in common. The standard laboratory strain *E. coli* K-12 has a genome of approximately 4.6 million nucleotide pairs, contained in a single circular molecule of DNA that codes for about 4300 different kinds of proteins (**Figure 1–24**).

In molecular terms, we know more about *E. coli* than about any other living organism. Most of our understanding of the fundamental mechanisms of life—for example, how cells replicate their DNA, or how they decode the instructions represented in the DNA to direct the synthesis of specific proteins—initially came from studies of *E. coli*. The basic genetic mechanisms have turned out to be highly conserved throughout evolution: these mechanisms are essentially the same in our own cells as in *E. coli*.

Summary

Prokaryotes (cells without a distinct nucleus) are biochemically the most diverse organisms and include species that can obtain all their energy and nutrients from inorganic chemical sources, such as the reactive mixtures of minerals released at hydrothermal vents on the ocean floor—the sort of diet that may have nourished the first living cells 3.5 billion years ago. DNA sequence comparisons reveal the family relationships of living organisms and show that the prokaryotes fall into two groups that diverged early in the course of evolution: the bacteria (or eubacteria) and the archaea. Together with the eukaryotes (cells with a membrane-enclosed nucleus), these constitute the three primary branches of the tree of life.

Most bacteria and archaea are small unicellular organisms with compact genomes comprising 1000–6000 genes. Many of the genes within a single organism show strong family resemblances in their DNA sequences, implying that they originated from the same ancestral gene through gene duplication and divergence. Family resemblances (homologies) are also clear when gene sequences are compared between different species, and more than 200 gene families have been so highly

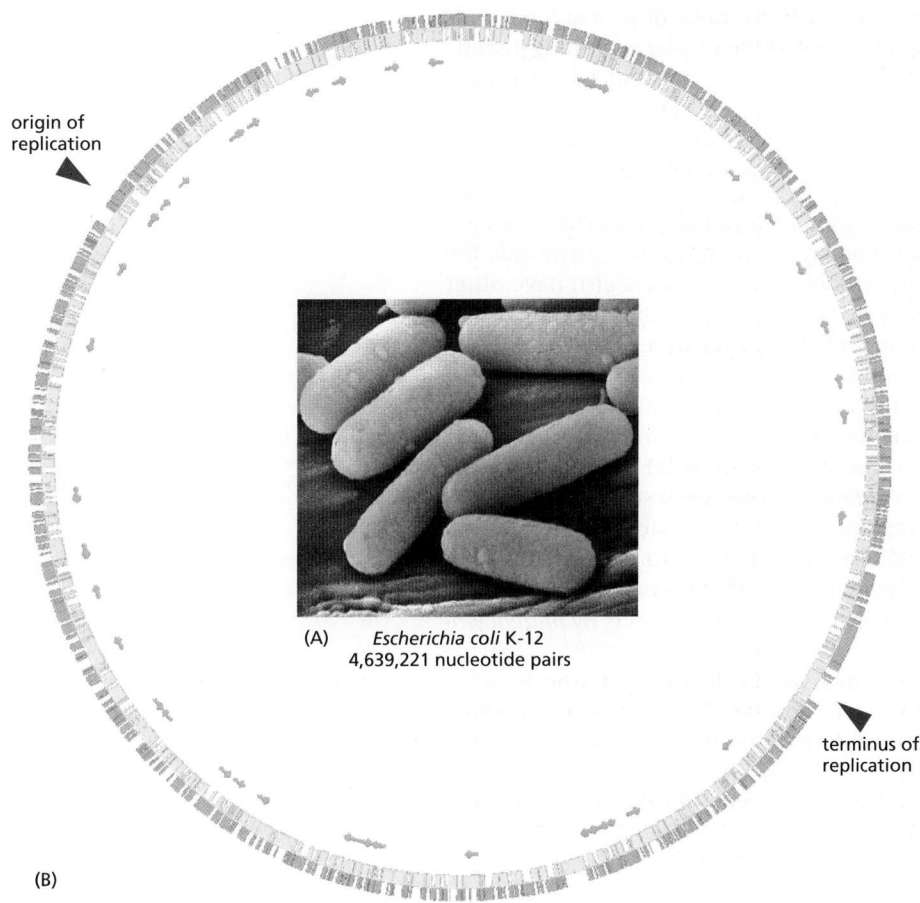

(A) *Escherichia coli* K-12
4,639,221 nucleotide pairs

(B)

Figure 1–24 The genome of *E. coli.* (A) A cluster of *E. coli* cells. (B) A diagram of the genome of *E. coli* strain K-12. The diagram is circular because the DNA of *E. coli*, like that of other prokaryotes, forms a single, closed loop. Protein-coding genes are shown as *yellow* or *orange bars*, depending on the DNA strand from which they are transcribed; genes encoding only RNA molecules are indicated by *green arrows*. Some genes are transcribed from one strand of the DNA double helix (in a clockwise direction in this diagram), others from the other strand (counterclockwise). (A, Dr. Tony Brain & David Parker/Science Photo Library/Science Source; B, adapted from F. R. Blattner et al., *Science* 277:1453–1462, 1997.)

conserved that they can be recognized as common to most species from all three domains of the living world. Thus, given the DNA sequence of a newly discovered gene, it is often possible to deduce the gene's function from the known function of a homologous gene in an intensively studied model organism, such as the bacterium E. coli.

GENETIC INFORMATION IN EUKARYOTES

Eukaryotic cells, in general, are bigger and more elaborate than prokaryotic cells, and their genomes are bigger and more elaborate, too. The greater size is accompanied by radical differences in cell structure and function. Moreover, many classes of eukaryotic cells form multicellular organisms that attain levels of complexity unmatched by any prokaryote.

Because they are so complex, eukaryotes confront molecular biologists with a special set of challenges that will concern us in the rest of this book. Increasingly, biologists attempt to meet these challenges through the analysis and manipulation of the genetic information within cells and organisms. It is therefore important at the outset to know something of the special features of the eukaryotic genome. We begin by briefly discussing how eukaryotic cells are organized, how this reflects their way of life, and how their genomes differ from those of prokaryotes. This leads us to an outline of the strategy by which cell biologists, by exploiting genetic and biochemical information, are attempting to discover how eukaryotic organisms work.

Eukaryotic Cells May Have Originated as Predators

By definition, eukaryotic cells keep their DNA in an internal compartment called the nucleus. The *nuclear envelope*, a double layer of membrane, surrounds the nucleus and separates the DNA from the cytoplasm. Eukaryotes also have other features that set them apart from prokaryotes (**Figure 1–25**). Their cells are, typically, 10 times bigger in linear dimension and 1000 times larger in volume. They have an elaborate *cytoskeleton*—a system of protein filaments crisscrossing the cytoplasm and forming, together with the many proteins that attach to them, a system of girders, ropes, and motors that gives the cell mechanical strength, controls its shape, and drives and guides its movements (**Movie 1.1**). And the nuclear envelope is only one part of a set of *internal membranes*, each structurally similar to the plasma membrane and enclosing different types of spaces inside the cell, many of them involved in digestion and secretion. Lacking the tough cell wall of most bacteria, animal cells and the free-living eukaryotic cells called *protozoa* can change their shape rapidly and engulf other cells and small objects by *phagocytosis* (**Figure 1–26**).

How all of the unique properties of eukaryotic cells evolved, and in what sequence, is still a mystery. One plausible view, however, is that they are all reflections of the way of life of a primordial cell that was a predator, living by capturing other cells and eating them (**Figure 1–27**). Such a way of life requires a large cell with a flexible plasma membrane, as well as an elaborate cytoskeleton to support

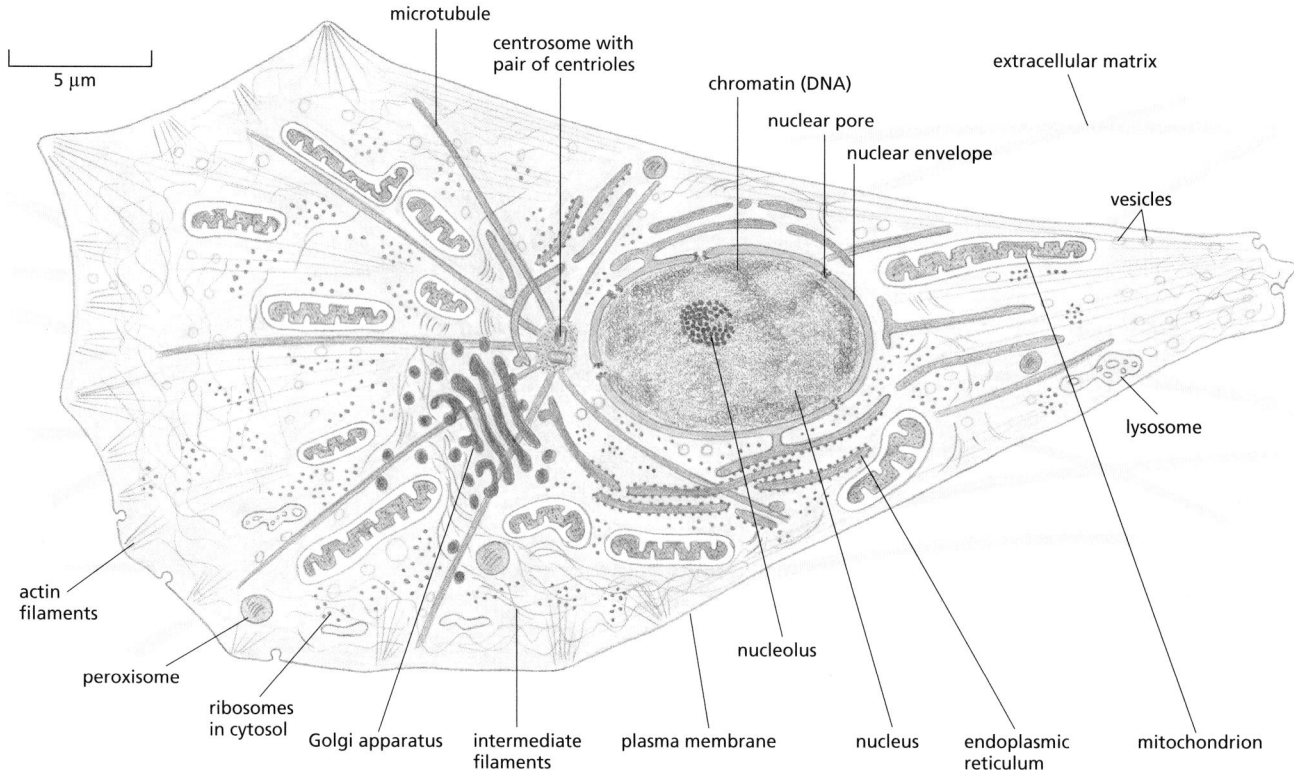

Figure 1–25 **The major features of eukaryotic cells.** The drawing depicts a typical animal cell, but almost all the same components are found in plants and fungi as well as in single-celled eukaryotes such as yeasts and protozoa. Plant cells contain chloroplasts in addition to the components shown here, and their plasma membrane is surrounded by a tough external wall formed of cellulose.

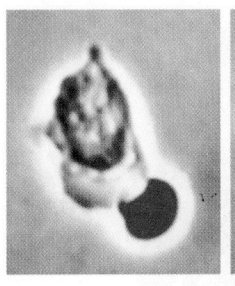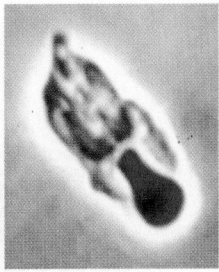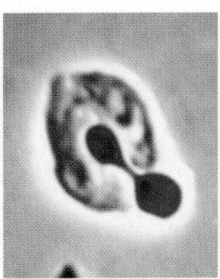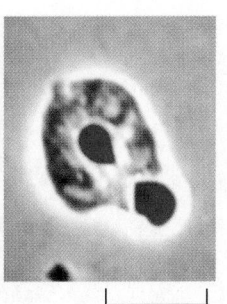

10 μm

Figure 1–26 **Phagocytosis.** This series of stills from a movie shows a human white blood cell (a neutrophil) engulfing a red blood cell (artificially colored *red*) that has been treated with an antibody that marks it for destruction (see Movie 13.5). (Courtesy of Stephen E. Malawista and Anne de Boisfleury Chevance.)

and move this membrane. It may also require that the cell's long, fragile DNA molecules be sequestered in a separate nuclear compartment, to protect the genome from damage by the movements of the cytoskeleton.

Modern Eukaryotic Cells Evolved from a Symbiosis

A predatory way of life helps to explain another feature of eukaryotic cells. All such cells contain (or at one time did contain) *mitochondria* (**Figure 1–28**). These small bodies in the cytoplasm, enclosed by a double layer of membrane, take up oxygen and harness energy from the oxidation of food molecules—such as sugars—to produce most of the ATP that powers the cell's activities. Mitochondria are similar in size to small bacteria, and, like bacteria, they have their own genome in the form of a circular DNA molecule, their own ribosomes that differ from those elsewhere in the eukaryotic cell, and their own transfer RNAs. It is now generally accepted that mitochondria originated from free-living oxygen-metabolizing (*aerobic*) bacteria that were engulfed by an ancestral cell that could otherwise make no such use of oxygen (that is, was *anaerobic*). Escaping digestion, these bacteria evolved in symbiosis with the engulfing cell and its progeny, receiving

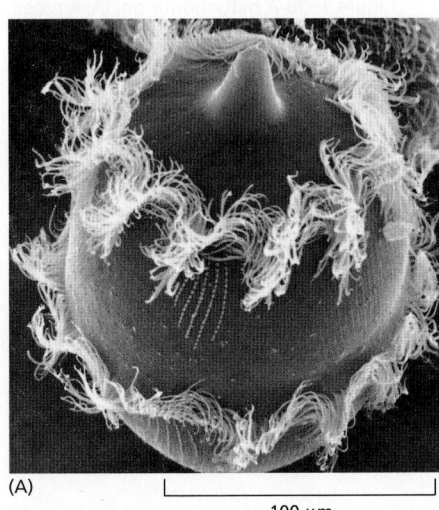

(A)

100 μm

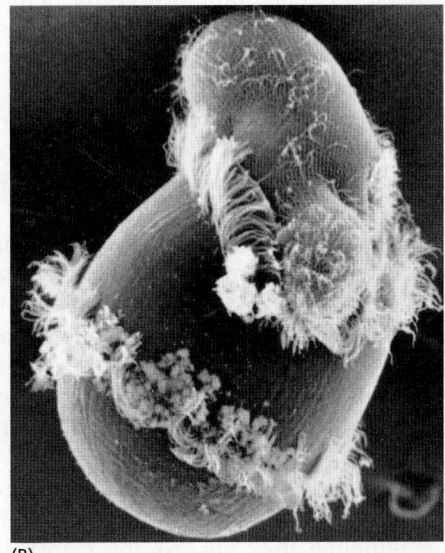

(B)

Figure 1–27 **A single-celled eukaryote that eats other cells.** (A) *Didinium* is a carnivorous protozoan, belonging to the group known as *ciliates*. It has a globular body, about 150 μm in diameter, encircled by two fringes of cilia—sinuous, whiplike appendages that beat continually; its front end is flattened except for a single protrusion, rather like a snout. (B) A *Didinium* engulfing its prey. *Didinium* normally swims around in the water at high speed by means of the synchronous beating of its cilia. When it encounters a suitable prey (*yellow*), usually another type of protozoan, it releases numerous small paralyzing darts from its snout region. Then, the *Didinium* attaches to and devours the other cell by phagocytosis, inverting like a hollow ball to engulf its victim, which can be almost as large as itself. (Courtesy of D. Barlow.)

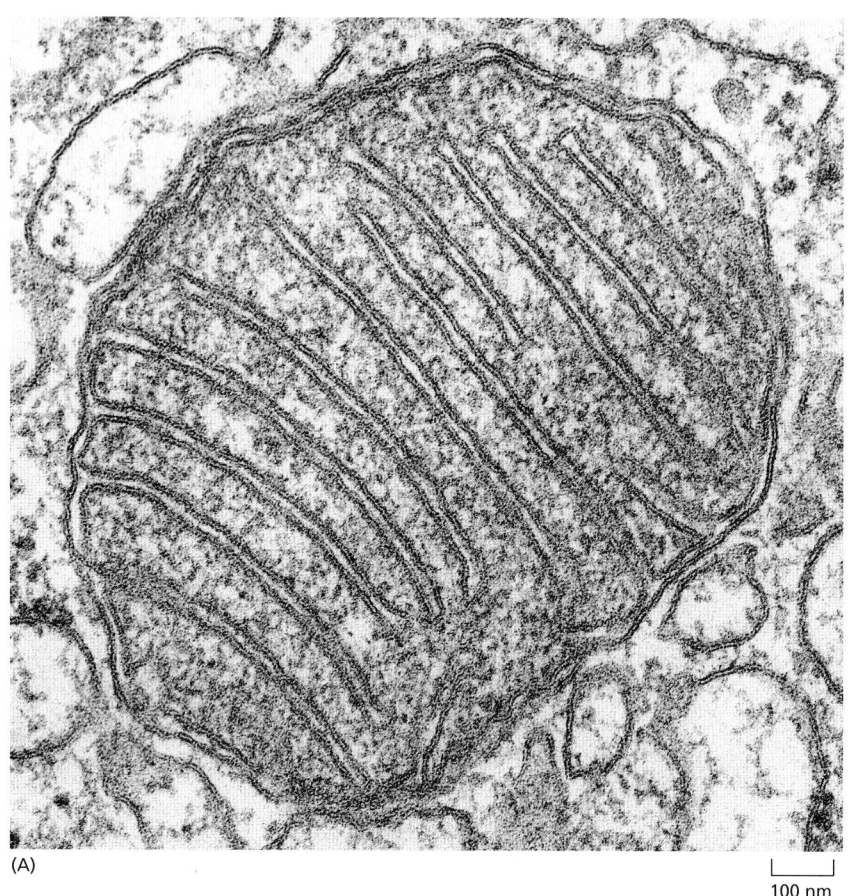

(A)

100 nm

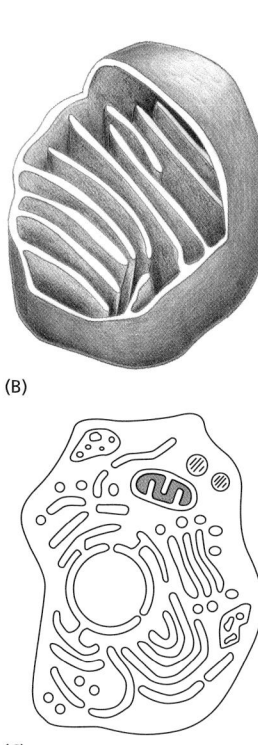

(B)

(C)

shelter and nourishment in return for the power generation they performed for their hosts. This partnership between a primitive anaerobic predator cell and an aerobic bacterial cell is thought to have been established about 1.5 billion years ago, when the Earth's atmosphere first became rich in oxygen.

As indicated in **Figure 1–29**, recent genomic analyses suggest that the first eukaryotic cells formed after an archaeal cell engulfed an aerobic bacterium. This would explain why all eukaryotic cells today, including those that live as strict anaerobes show clear evidence that they once contained mitochondria.

Many eukaryotic cells—specifically, those of plants and algae—also contain another class of small membrane-enclosed organelles somewhat similar to mitochondria—the *chloroplasts* (**Figure 1–30**). Chloroplasts perform photosynthesis, using the energy of sunlight to synthesize carbohydrates from atmospheric carbon dioxide and water, and deliver the products to the host cell as food. Like mitochondria, chloroplasts have their own genome. They almost certainly originated as symbiotic photosynthetic bacteria, acquired by eukaryotic cells that already possessed mitochondria (**Figure 1–31**).

A eukaryotic cell equipped with chloroplasts has no need to chase after other cells as prey; it is nourished by the captive chloroplasts it has inherited from its ancestors. Correspondingly, plant cells, although they possess the cytoskeletal equipment for movement, have lost the ability to change shape rapidly and to engulf other cells by phagocytosis. Instead, they create around themselves a tough, protective cell wall. If the first eukaryotic cells were predators on other organisms, we can view plant cells as cells that have made the transition from hunting to farming.

Fungi represent yet another eukaryotic way of life. Fungal cells, like animal cells, possess mitochondria but not chloroplasts; but in contrast with animal cells and protozoa, they have a tough outer wall that limits their ability to move rapidly

Figure 1–28 A mitochondrion. (A) A cross section, as seen in the electron microscope. (B) A drawing of a mitochondrion with part of it cut away to show the three-dimensional structure (**Movie 1.2**). (C) A schematic eukaryotic cell, with the interior space of a mitochondrion, containing the mitochondrial DNA and ribosomes, colored. Note the smooth outer membrane and the convoluted inner membrane, which houses the proteins that generate ATP from the oxidation of food molecules. (A, courtesy of Daniel S. Friend and by permission of E.L. Bearer.)

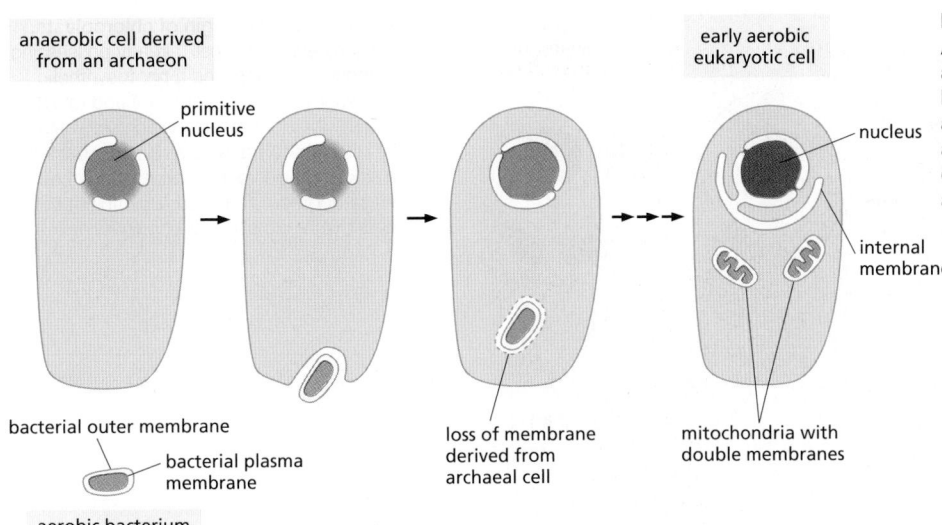

anaerobic cell derived
from an archaeon

primitive
nucleus

early aerobic
eukaryotic cell

nucleus

internal
membranes

bacterial outer membrane

bacterial plasma
membrane

loss of membrane
derived from
archaeal cell

mitochondria with
double membranes

aerobic bacterium

Figure 1–29 The origin of mitochondria.
An ancestral anaerobic predator cell (an archaeon) is thought to have engulfed the bacterial ancestor of mitochondria, initiating a symbiotic relationship. Clear evidence of a dual bacterial and archaeal inheritance can be discerned today in the genomes of all eukaryotes.

or to swallow up other cells. Fungi, it seems, have turned from hunters into scavengers: other cells secrete nutrient molecules or release them upon death, and fungi feed on these leavings—performing whatever digestion is necessary extracellularly, by secreting digestive enzymes to the exterior.

Eukaryotes Have Hybrid Genomes

The genetic information of eukaryotic cells has a hybrid origin—from the ancestral anaerobic archaeal cell, and from the bacteria that it adopted as symbionts. Most of this information is stored in the nucleus, but a small amount remains inside the mitochondria and, for plant and algal cells, in the chloroplasts. When mitochondrial DNA and the chloroplast DNA are separated from the nuclear DNA and individually analyzed and sequenced, the mitochondrial and chloroplast genomes are found to be degenerate, cut-down versions of the corresponding bacterial genomes. In a human cell, for example, the mitochondrial genome consists of only 16,569 nucleotide pairs, and codes for only 13 proteins, 2 ribosomal RNA components, and 22 transfer RNAs.

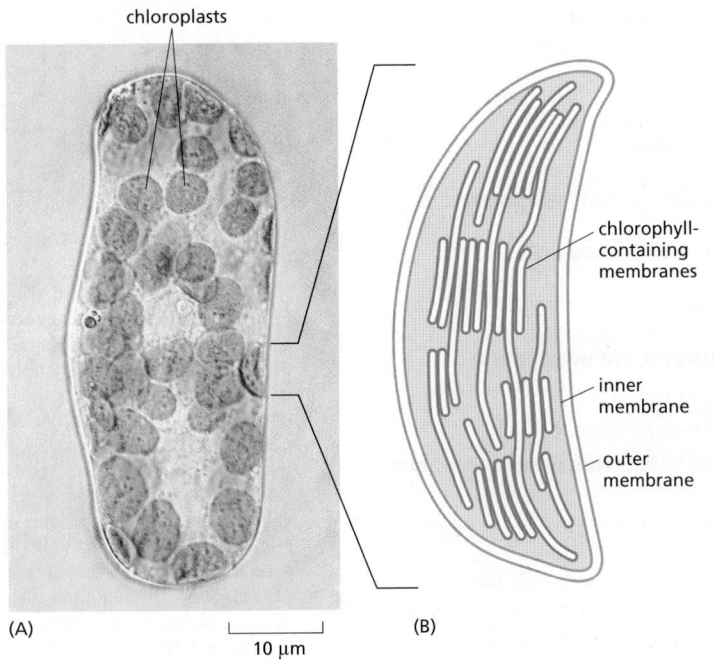

chloroplasts

chlorophyll-
containing
membranes

inner
membrane

outer
membrane

(A)

10 µm

(B)

Figure 1–30 Chloroplasts. These organelles capture the energy of sunlight in plant cells and some single-celled eukaryotes. (A) A single cell isolated from a leaf of a flowering plant, seen in the light microscope, showing the green chloroplasts (**Movie 1.3** and see Movie 14.9). (B) A drawing of one of the chloroplasts, showing the highly folded system of internal membranes containing the chlorophyll molecules by which light is absorbed. (A, courtesy of Preeti Dahiya.)

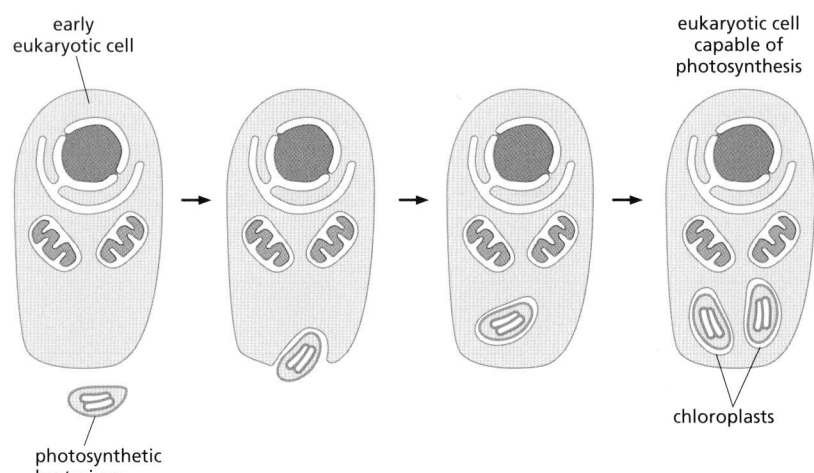

early
eukaryotic cell

eukaryotic cell
capable of
photosynthesis

chloroplasts

photosynthetic
bacterium

Figure 1–31 **The origin of chloroplasts.**
An early eukaryotic cell, already possessing
mitochondria, engulfed a photosynthetic
bacterium (a cyanobacterium) and retained
it in symbiosis. Present-day chloroplasts
are thought to trace their ancestry back to
a single species of cyanobacterium that
was adopted as an internal symbiont (an
endosymbiont) over a billion years ago.

Many of the genes that are missing from the mitochondria and chloroplasts have not been lost; instead, they have moved from the symbiont genome into the DNA of the host cell nucleus. The nuclear DNA of humans contains many genes coding for proteins that serve essential functions inside the mitochondria; in plants, the nuclear DNA also contains many genes specifying proteins required in chloroplasts. In both cases, the DNA sequences of these nuclear genes show clear evidence of their origin from the bacterial ancestor of the respective organelle.

Eukaryotic Genomes Are Big

Natural selection has evidently favored mitochondria with small genomes. By contrast, the nuclear genomes of most eukaryotes seem to have been free to enlarge. Perhaps the eukaryotic way of life has made large size an advantage: predators typically need to be bigger than their prey, and cell size generally increases in proportion to genome size. Whatever the reason, aided by a massive accumulation of DNA segments derived from parasitic transposable elements (discussed in Chapter 5), the genomes of most eukaryotes have become orders of magnitude larger than those of bacteria and archaea (**Figure 1–32**).

The freedom to be extravagant with DNA has had profound implications. Eukaryotes not only have more genes than prokaryotes; they also have vastly more DNA that does not code for protein. The human genome contains 1000 times as many nucleotide pairs as the genome of a typical bacterium, perhaps 10 times as

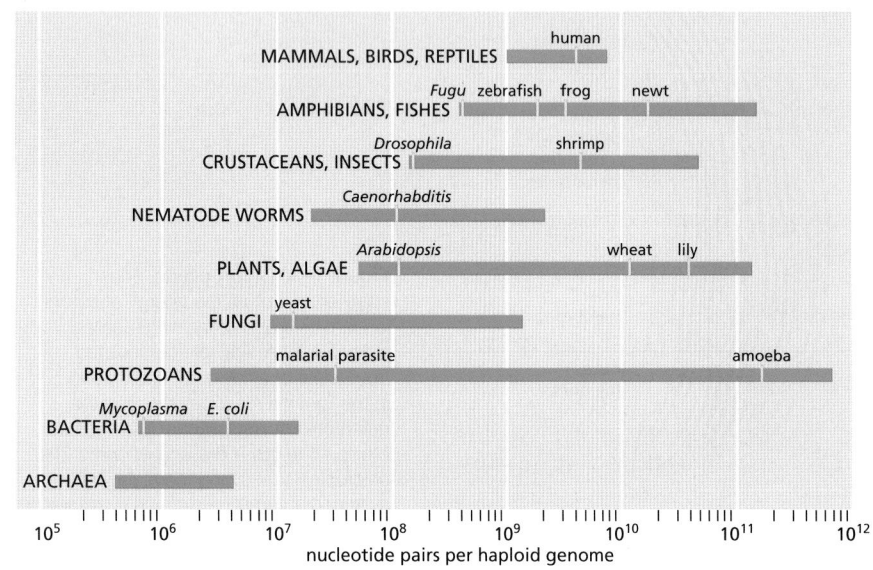

Figure 1–32 **Genome sizes compared.**
Genome size is measured in nucleotide
pairs of DNA per haploid genome, that is,
per single copy of the genome. (The cells
of sexually reproducing organisms such as
ourselves are generally diploid: they contain
two copies of the genome, one inherited
from the mother, the other from the father.)
Closely related organisms can vary widely
in the quantity of DNA in their genomes,
even though they contain similar numbers
of functionally distinct genes. (Data from
W.H. Li, Molecular Evolution, pp. 380–383.
Sunderland, MA: Sinauer, 1997.)

TABLE 1–2 Some Model Organisms and Their Genomes

Organism	Genome size* (nucleotide pairs)	Approximate number of genes
Escherichia coli (bacterium)	4.6×10^6	4300
Saccharomyces cerevisiae (yeast)	13×10^6	6600
Caenorhabditis elegans (roundworm)	130×10^6	21,000
Arabidopsis thaliana (plant)	220×10^6	29,000
Drosophila melanogaster (fruit fly)	200×10^6	15,000
Danio rerio (zebrafish)	1400×10^6	32,000
Mus musculus (mouse)	2800×10^6	30,000
Homo sapiens (human)	3200×10^6	30,000

*Genome size includes an estimate for the amount of highly repeated DNA sequence not in genome databases.

many genes, and a great deal more noncoding DNA (~98.5% of the genome for a human does not code for proteins, as opposed to 11% of the genome for the bacterium *E. coli*). The estimated genome sizes and gene numbers for some eukaryotes are compiled for easy comparison with *E. coli* in **Table 1–2**; we shall discuss how each of these eukaryotes serves as a model organism shortly.

Eukaryotic Genomes Are Rich in Regulatory DNA

Much of our noncoding DNA is almost certainly dispensable junk, retained like a mass of old papers because, when there is little pressure to keep an archive small, it is easier to retain everything than to sort out the valuable information and discard the rest. Certain exceptional eukaryotic species, such as the puffer fish, bear witness to the profligacy of their relatives; they have somehow managed to rid themselves of large quantities of noncoding DNA. Yet they appear similar in structure, behavior, and fitness to related species that have vastly more such DNA (see Figure 4–71).

Even in compact eukaryotic genomes such as that of puffer fish, there is more noncoding DNA than coding DNA, and at least some of the noncoding DNA certainly has important functions. In particular, it regulates the expression of adjacent genes. With this *regulatory DNA*, eukaryotes have evolved distinctive ways of controlling when and where a gene is brought into play. This sophisticated gene regulation is crucial for the formation of complex multicellular organisms.

The Genome Defines the Program of Multicellular Development

The cells in an individual animal or plant are extraordinarily varied. Fat cells, skin cells, bone cells, nerve cells—they seem as dissimilar as any cells could be (**Figure 1–33**). Yet all these cell types are the descendants of a single fertilized egg cell, and all (with minor exceptions) contain identical copies of the genome of the species.

The differences result from the way in which the cells make selective use of their genetic instructions according to the cues they get from their surroundings in the developing embryo. The DNA is not just a shopping list specifying the molecules that every cell must have, and the cell is not an assembly of all the items on the list. Rather, the cell behaves as a multipurpose machine, with sensors to receive environmental signals and with highly developed abilities to call different sets of genes into action according to the sequences of signals to which the cell has been exposed. The genome in each cell is big enough to accommodate the

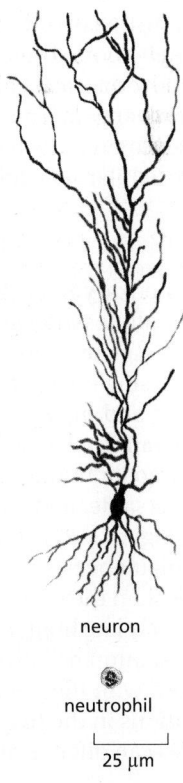

neuron

neutrophil

25 µm

Figure 1–33 Cell types can vary enormously in size and shape. An animal nerve cell is compared here with a neutrophil, a type of white blood cell. Both are drawn to scale.

Figure 1–34 **Genetic control of the program of multicellular development.** The role of a regulatory gene is demonstrated in the snapdragon *Antirrhinum*. In this example, a mutation in a single gene coding for a regulatory protein causes leafy shoots to develop in place of flowers: because a regulatory protein has been changed, the cells adopt characters that would be appropriate to a different location in the normal plant. The mutant is on the left, the normal plant on the right. (Courtesy of Enrico Coen and Rosemary Carpenter.)

information that specifies an entire multicellular organism, but in any individual cell only part of that information is used.

A large number of genes in the eukaryotic genome code for proteins that regulate the activities of other genes. Most of these *transcription regulators* act by binding, directly or indirectly, to the regulatory DNA adjacent to the genes that are to be controlled, or by interfering with the abilities of other proteins to do so. The expanded genome of eukaryotes therefore not only specifies the hardware of the cell, but also stores the software that controls how that hardware is used (**Figure 1–34**).

Cells do not just passively receive signals; rather, they actively exchange signals with their neighbors. Thus, in a developing multicellular organism, the same control system governs each cell, but with different consequences depending on the messages exchanged. The outcome, astonishingly, is a precisely patterned array of cells in different states, each displaying a character appropriate to its position in the multicellular structure.

Many Eukaryotes Live as Solitary Cells

Many species of eukaryotic cells lead a solitary life—some as hunters (the *protozoa*), some as photosynthesizers (the unicellular *algae*), some as scavengers (the unicellular fungi, or *yeasts*). **Figure 1–35** conveys something of the astonishing variety of the single-celled eukaryotes. The anatomy of protozoa, especially, is often elaborate and includes such structures as sensory bristles, photoreceptors, sinuously beating cilia, leglike appendages, mouth parts, stinging darts, and musclelike contractile bundles. Although they are single cells, protozoa can be as intricate, as versatile, and as complex in their behavior as many multicellular organisms (see Figure 1–27, **Movie 1.4**, and **Movie 1.5**).

In terms of their ancestry and DNA sequences, the unicellular eukaryotes are far more diverse than the multicellular animals, plants, and fungi, which arose as three comparatively late branches of the eukaryotic pedigree (see Figure 1–17). As with prokaryotes, humans have tended to neglect them because they are microscopic. Only now, with the help of genome analysis, are we beginning to understand their positions in the tree of life, and to put into context the glimpses these strange creatures can offer us of our distant evolutionary past.

A Yeast Serves as a Minimal Model Eukaryote

The molecular and genetic complexity of eukaryotes is daunting. Even more than for prokaryotes, biologists need to concentrate their limited resources on a few selected model organisms to unravel this complexity.

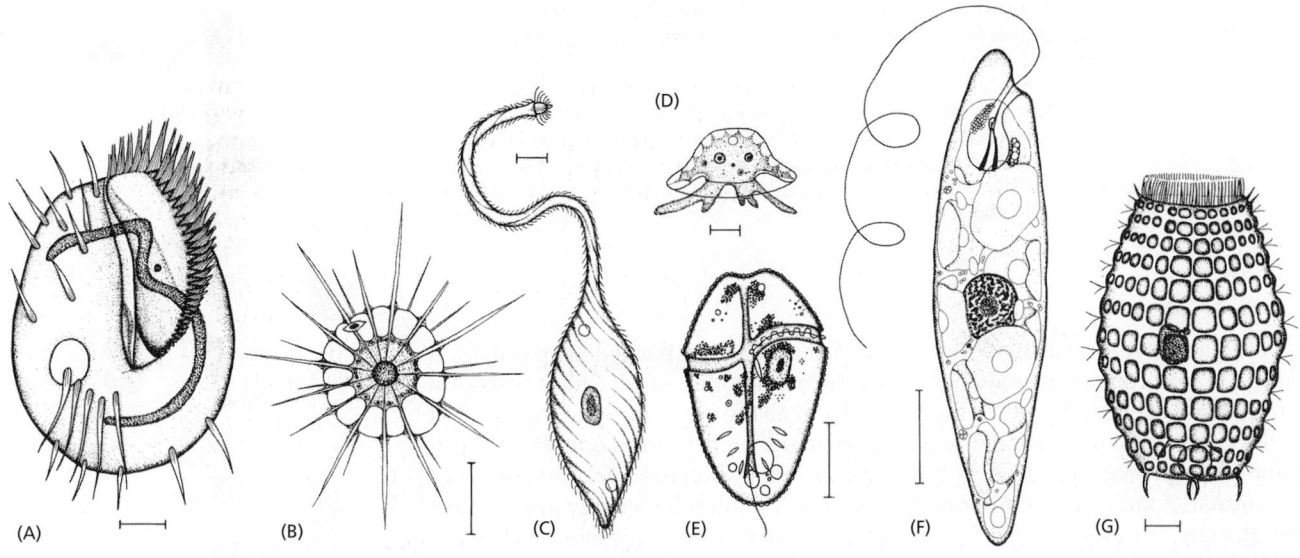

To analyze the internal workings of the eukaryotic cell without the additional problems of multicellular development, it makes sense to use a species that is unicellular and as simple as possible. The popular choice for this role of minimal model eukaryote has been the yeast *Saccharomyces cerevisiae* (**Figure 1–36**)—the same species that is used by brewers of beer and bakers of bread.

S. cerevisiae is a small, single-celled member of the kingdom of fungi and thus, according to modern views, is at least as closely related to animals as it is to plants. It is robust and easy to grow in a simple nutrient medium. Like other fungi, it has a tough cell wall, is relatively immobile, and possesses mitochondria but not chloroplasts. When nutrients are plentiful, it grows and divides almost as rapidly as a bacterium. It can reproduce either vegetatively (that is, by simple cell division), or sexually: two yeast cells that are *haploid* (possessing a single copy of the genome) can fuse to create a cell that is *diploid* (containing a double genome); and the diploid cell can undergo *meiosis* (a reduction division) to produce cells that are once again haploid (**Figure 1–37**). In contrast with higher plants and animals, the yeast can divide indefinitely in either the haploid or the diploid state, and the process leading from one state to the other can be induced at will by changing the growth conditions.

In addition to these features, the yeast has a further property that makes it a convenient organism for genetic studies: its genome, by eukaryotic standards, is exceptionally small. Nevertheless, it suffices for all the basic tasks that every eukaryotic cell must perform. Mutants are available for essentially every gene,

Figure 1–35 **An assortment of protozoa: a small sample of an extremely diverse class of organisms.** The drawings are done to different scales, but in each case the scale bar represents 10 μm. The organisms in (A), (C), and (G) are ciliates; (B) is a heliozoan; (D) is an amoeba; (E) is a dinoflagellate; and (F) is a euglenoid. (Courtesy of Michael Sleigh, from M. A. Sleigh, *The Biology of Protozoa*. Edinburgh: Edward Arnold, 1973.)

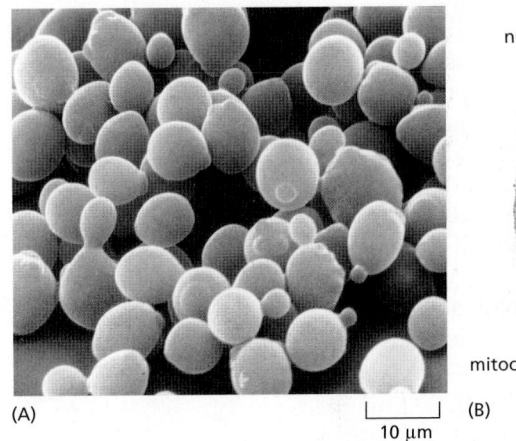

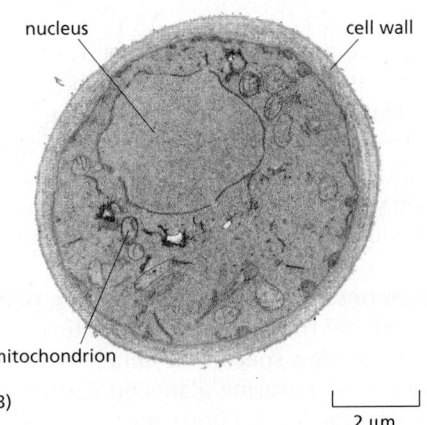

Figure 1–36 **The yeast *Saccharomyces cerevisiae.*** (A) A scanning electron micrograph of a cluster of the cells. This species is also known as budding yeast; it proliferates by forming a protrusion or bud that enlarges and then separates from the rest of the original cell. Many cells with buds are visible in this micrograph. (B) A transmission electron micrograph of a cross section of a yeast cell, showing its nucleus, mitochondrion, and thick cell wall. (A, courtesy of Ira Herskowitz and Eric Schabtach.)

Figure 1–37 The reproductive cycles of the yeast *S. cerevisiae*. Depending on environmental conditions and on details of the genotype, cells of this species can exist in either a diploid (*2n*) state, with a double chromosome set, or a haploid (*n*) state, with a single chromosome set. The diploid form can either proliferate by ordinary cell-division cycles or undergo meiosis to produce haploid cells. The haploid form can either proliferate by ordinary cell-division cycles or undergo sexual fusion with another haploid cell to become diploid. Meiosis is triggered by starvation and gives rise to spores—haploid cells in a dormant state, resistant to harsh environmental conditions.

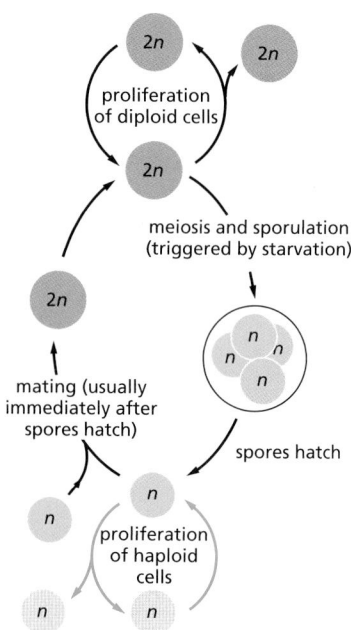

BUDDING YEAST LIFE CYCLE

and studies on yeasts (using both *S. cerevisiae* and other species) have provided a key to many crucial processes, including the eukaryotic cell-division cycle—the critical chain of events by which the nucleus and all the other components of a cell are duplicated and parceled out to create two daughter cells from one. The control system that governs this process has been so well conserved over the course of evolution that many of its components can function interchangeably in yeast and human cells: if a mutant yeast lacking an essential yeast cell-division-cycle gene is supplied with a copy of the homologous cell-division-cycle gene from a human, the yeast is cured of its defect and becomes able to divide normally.

The Expression Levels of All the Genes of An Organism Can Be Monitored Simultaneously

The complete genome sequence of *S. cerevisiae*, determined in 1997, consists of approximately 13,117,000 nucleotide pairs, including the small contribution (78,520 nucleotide pairs) of the mitochondrial DNA. This total is only about 2.5 times as much DNA as there is in *E. coli*, and it codes for only 1.5 times as many distinct proteins (about 6600 in all). The way of life of *S. cerevisiae* is similar in many ways to that of a bacterium, and it seems that this yeast has likewise been subject to selection pressures that have kept its genome compact.

Knowledge of the complete genome sequence of any organism—be it a yeast or a human—opens up new perspectives on the workings of the cell: things that once seemed impossibly complex now seem within our grasp. Using techniques described in Chapter 8, it is now possible, for example, to monitor, simultaneously, the amount of mRNA transcript that is produced from every gene in the yeast genome under any chosen conditions, and to see how this whole pattern of gene activity changes when conditions change. The analysis can be repeated with mRNA prepared from mutant cells lacking a chosen gene—any gene that we care to test. In principle, this approach provides a way to reveal the entire system of control relationships that govern gene expression—not only in yeast cells, but in any organism whose genome sequence is known.

Arabidopsis Has Been Chosen Out of 300,000 Species As a Model Plant

The large multicellular organisms that we see around us—the flowers and trees and animals—seem fantastically varied, but they are much closer to one another in their evolutionary origins, and more similar in their basic cell biology, than the great host of microscopic single-celled organisms. Thus, while bacteria and archaea are separated by perhaps 3.5 billion years of evolution, vertebrates and insects are separated by about 700 million years, fish and mammals by about 450 million years, and the different species of flowering plants by only about 150 million years.

Because of the close evolutionary relationship between all flowering plants, we can, once again, get insight into the cell and molecular biology of this whole class of organisms by focusing on just one or a few species for detailed analysis. Out of the several hundred thousand species of flowering plants on Earth today, molecular biologists have chosen to concentrate their efforts on a small weed, the common Thale cress *Arabidopsis thaliana* (**Figure 1–38**), which can be grown indoors in large numbers and produces thousands of offspring per plant after 8–10 weeks. *Arabidopsis* has a total genome size of approximately 220 million nucleotide pairs, about 17 times the size of yeast's (see Table 1–2).

The World of Animal Cells Is Represented By a Worm, a Fly, a Fish, a Mouse, and a Human

Multicellular animals account for the majority of all named species of living organisms, and for the largest part of the biological research effort. Five species have emerged as the foremost model organisms for molecular genetic studies. In order of increasing size, they are the nematode worm *Caenorhabditis elegans*, the fly *Drosophila melanogaster*, the zebrafish *Danio rerio*, the mouse *Mus musculus*, and the human, *Homo sapiens*. Each has had its genome sequenced.

Caenorhabditis elegans (**Figure 1–39**) is a small, harmless relative of the eel-worm that attacks crops. With a life cycle of only a few days, an ability to survive in a freezer indefinitely in a state of suspended animation, a simple body plan, and an unusual life cycle that is well suited for genetic studies (described in Chapter 21), it is an ideal model organism. *C. elegans* develops with clockwork precision from a fertilized egg cell into an adult worm with exactly 959 body cells (plus a variable number of egg and sperm cells)—an unusual degree of regularity for an animal. We now have a minutely detailed description of the sequence of events by which this occurs, as the cells divide, move, and change their character according to strict and predictable rules. The genome of 130 million nucleotide pairs codes for about 21,000 proteins, and many mutants and other tools are available for the testing of gene functions. Although the worm has a body plan very different from our own, the conservation of biological mechanisms has been sufficient for the worm to be a model for many of the developmental and cell-biological processes that occur in the human body. Thus, for example, studies of the worm have been critical for helping us to understand the programs of cell division and cell death that determine the number of cells in the body—a topic of great importance for both developmental biology and cancer research.

Studies in *Drosophila* Provide a Key to Vertebrate Development

The fruit fly *Drosophila melanogaster* (**Figure 1–40**) has been used as a model genetic organism for longer than any other; in fact, the foundations of classical genetics were built to a large extent on studies of this insect. Over 80 years ago, it provided, for example, definitive proof that genes—the abstract units of hereditary information—are carried on chromosomes, concrete physical objects whose behavior had been closely followed in the eukaryotic cell with the light microscope, but whose function was at first unknown. The proof depended on one of the many features that make *Drosophila* peculiarly convenient for genetics—the giant chromosomes, with characteristic banded appearance, that are visible in

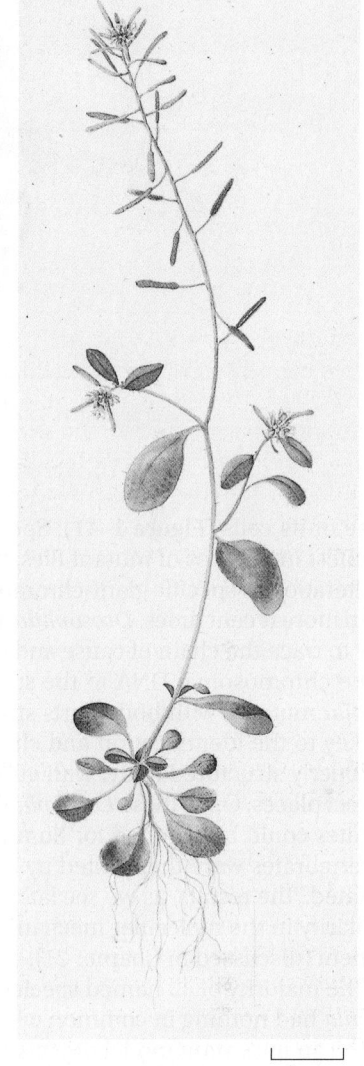

1 cm

Figure 1–38 *Arabidopsis thaliana*, the plant chosen as the primary model for studying plant molecular genetics. (Courtesy of Toni Hayden FLS and the John Innes Foundation.)

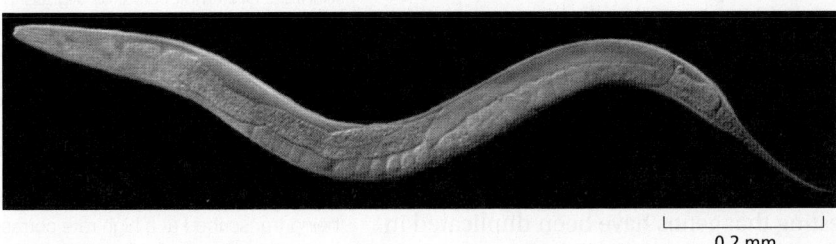

0.2 mm

Figure 1–39 *Caenorhabditis elegans*, the first multicellular organism to have its complete genome sequence determined. This small nematode, about 1 mm long, lives in the soil. Most individuals are hermaphrodites, producing both eggs and sperm. (Courtesy of Maria Gallegos, University of Wisconsin, Madison.)

1 mm

Figure 1–40 *Drosophila melanogaster.* Molecular genetic studies on this fly have provided the main key to understanding how all animals develop from a fertilized egg into an adult. (Courtesy of the Archives, California Institute of Technology.)

some of its cells (**Figure 1–41**). Specific changes in the hereditary information, manifest in families of mutant flies, were found to correlate exactly with the loss or alteration of specific giant-chromosome bands.

In more recent times, *Drosophila*, more than any other organism, has shown us how to trace the chain of cause and effect from the genetic instructions encoded in the chromosomal DNA to the structure of the adult multicellular body. *Drosophila* mutants with body parts strangely misplaced or mispatterned provided the key to the identification and characterization of the genes required to make a properly structured body, with gut, limbs, eyes, and all the other parts in their correct places. Once these *Drosophila* genes were sequenced, the genomes of vertebrates could be scanned for homologs. These were found, and their functions in vertebrates were then tested by analyzing mice in which the genes had been mutated. The results, as we see later in the book, reveal an astonishing degree of similarity in the molecular mechanisms that govern insect and vertebrate development (discussed in Chapter 21).

The majority of all named species of living organisms are insects. Even if *Drosophila* had nothing in common with vertebrates, but only with insects, it would still be an important model organism. But if understanding the molecular genetics of vertebrates is the goal, why not simply tackle the problem head-on? Why sidle up to it obliquely, through studies in *Drosophila*?

Drosophila requires only 9 days to progress from a fertilized egg to an adult; it is vastly easier and cheaper to breed than any vertebrate, and its genome is much smaller—about 200 million nucleotide pairs, compared with 3200 million for a human. This genome codes for about 15,000 proteins, and mutants can now be obtained for essentially any gene. But there is also another, deeper reason why genetic mechanisms that are hard to discover in a vertebrate are often readily revealed in the fly. This relates, as we now explain, to the frequency of gene duplication, which is substantially greater in vertebrate genomes than in the fly genome and has probably been crucial in making vertebrates the complex and subtle creatures that they are.

The Vertebrate Genome Is a Product of Repeated Duplications

Almost every gene in the vertebrate genome has paralogs—other genes in the same genome that are unmistakably related and must have arisen by gene duplication. In many cases, a whole cluster of genes is closely related to similar clusters present elsewhere in the genome, suggesting that genes have been duplicated in linked groups rather than as isolated individuals. According to one hypothesis, at an early stage in the evolution of the vertebrates, the entire genome underwent duplication twice in succession, giving rise to four copies of every gene.

The precise course of vertebrate genome evolution remains uncertain, because many further evolutionary changes have occurred since these ancient events.

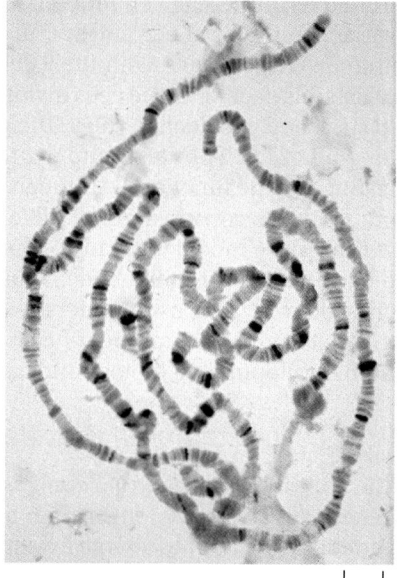

20 μm

Figure 1–41 **Giant chromosomes from salivary gland cells of *Drosophila*.** Because many rounds of DNA replication have occurred without an intervening cell division, each of the chromosomes in these unusual cells contains over 1000 identical DNA molecules, all aligned in register. This makes them easy to see in the light microscope, where they display a characteristic and reproducible banding pattern. Specific bands can be identified as the locations of specific genes: a mutant fly with a region of the banding pattern missing shows a phenotype reflecting loss of the genes in that region. Genes that are being transcribed at a high rate correspond to bands with a "puffed" appearance. The bands stained *dark brown* in the micrograph are sites where a particular regulatory protein is bound to the DNA. (From R. Paro, *Trends Genet.* 6:416–421, 1990. With permission from Elsevier.)

Genes that were once identical have diverged; many of the gene copies have been lost through disruptive mutations; some have undergone further rounds of local duplication; and the genome, in each branch of the vertebrate family tree, has suffered repeated rearrangements, breaking up most of the original gene orderings. Comparison of the gene order in two related organisms, such as the human and the mouse, reveals that—on the time scale of vertebrate evolution—chromosomes frequently fuse and fragment to move large blocks of DNA sequence around. Indeed, it is possible, as discussed in Chapter 4, that the present state of affairs is the result of many separate duplications of fragments of the genome, rather than duplications of the genome as a whole.

There is, however, no doubt that such whole-genome duplications do occur from time to time in evolution, for we can see recent instances in which duplicated chromosome sets are still clearly identifiable as such. The frog genus *Xenopus*, for example, comprises a set of closely similar species related to one another by repeated duplications or triplications of the whole genome. Among these frogs are *X. tropicalis*, with an ordinary diploid genome; the common laboratory species *X. laevis*, with a duplicated genome and twice as much DNA per cell; and *X. ruwenzoriensis*, with a sixfold reduplication of the original genome and six times as much DNA per cell (108 chromosomes, compared with 36 in *X. laevis*, for example). These species are estimated to have diverged from one another within the past 120 million years (**Figure 1–42**).

The Frog and the Zebrafish Provide Accessible Models for Vertebrate Development

Frogs have long been used to study the early steps of embryonic development in vertebrates, because their eggs are big, easy to manipulate, and fertilized outside of the animal, so that the subsequent development of the early embryo is easily followed (**Figure 1–43**). *Xenopus laevis,* in particular, continues to be an important model organism, even though it is poorly suited for genetic analysis (**Movie 1.6** and see Movie 21.1).

The zebrafish *Danio rerio* has similar advantages, but without this drawback. Its genome is compact—only half as big as that of a mouse or a human—and it has a generation time of only about three months. Many mutants are known, and genetic engineering is relatively easy. The zebrafish has the added virtue that it is transparent for the first two weeks of its life, so that one can watch the behavior of individual cells in the living organism (see Movie 21.2). All this has made it an increasingly important model vertebrate (**Figure 1–44**).

The Mouse Is the Predominant Mammalian Model Organism

Mammals have typically two times as many genes as *Drosophila*, a genome that is 16 times larger, and millions or billions of times as many cells in their adult bodies. In terms of genome size and function, cell biology, and molecular mechanisms, mammals are nevertheless a highly uniform group of organisms. Even anatomically, the differences among mammals are chiefly a matter of size and proportions; it is hard to think of a human body part that does not have a counterpart in elephants and mice, and vice versa. Evolution plays freely with quantitative features, but it does not readily change the logic of the structure.

Figure 1–42 **Two species of the frog genus *Xenopus.*** *X. tropicalis*, above, has an ordinary diploid genome; *X. laevis*, below, has twice as much DNA per cell. From the banding patterns of their chromosomes and the arrangement of genes along them, as well as from comparisons of gene sequences, it is clear that the large-genome species have evolved through duplications of the whole genome. These duplications are thought to have occurred in the aftermath of matings between frogs of slightly divergent *Xenopus* species. (From E. Amaya, M. Offield, and R. Grainger, *Trends Genet.* 14:253–255, 1998. With permission from Elsevier.)

hours   0   6   16   34   67   96   284

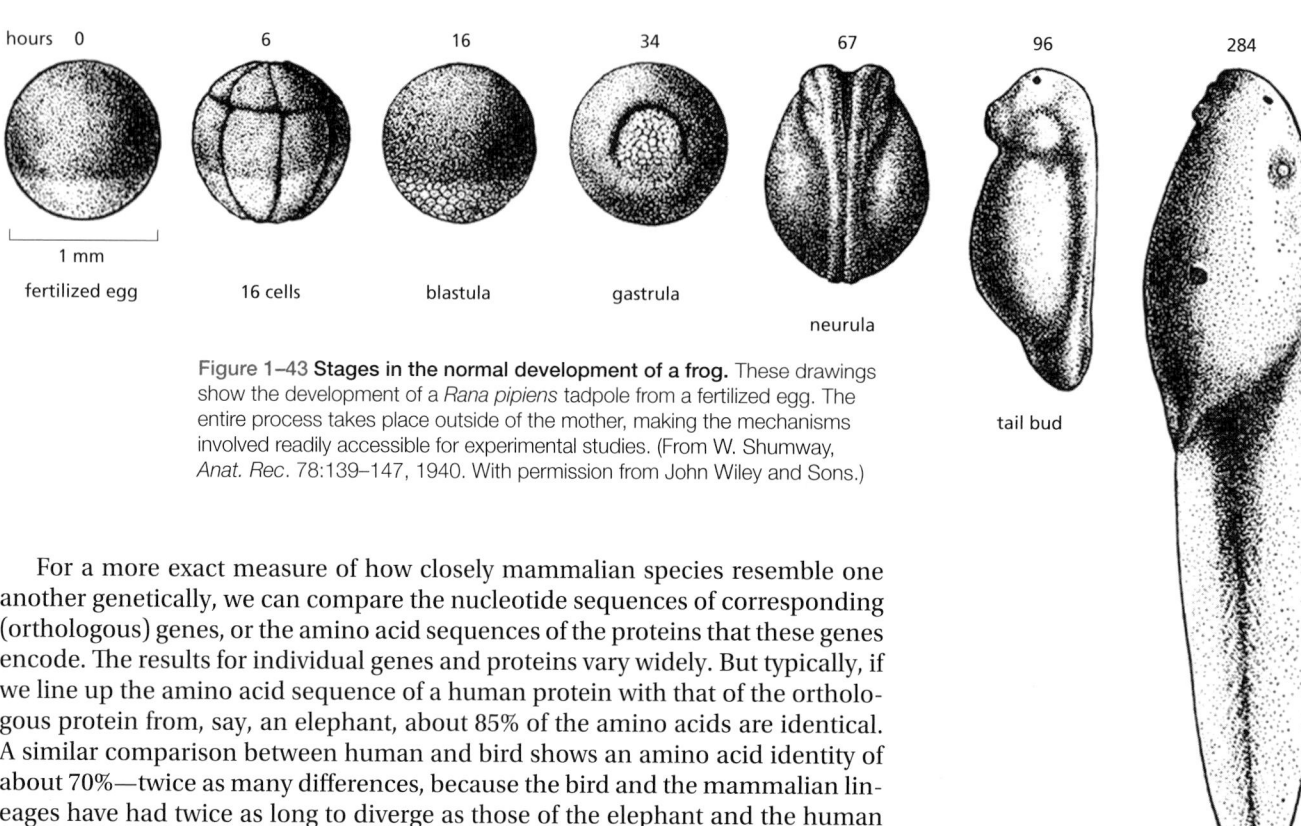

1 mm

fertilized egg   16 cells   blastula   gastrula

neurula

tail bud

Figure 1–43 Stages in the normal development of a frog. These drawings show the development of a *Rana pipiens* tadpole from a fertilized egg. The entire process takes place outside of the mother, making the mechanisms involved readily accessible for experimental studies. (From W. Shumway, *Anat. Rec.* 78:139–147, 1940. With permission from John Wiley and Sons.)

tadpole

For a more exact measure of how closely mammalian species resemble one another genetically, we can compare the nucleotide sequences of corresponding (orthologous) genes, or the amino acid sequences of the proteins that these genes encode. The results for individual genes and proteins vary widely. But typically, if we line up the amino acid sequence of a human protein with that of the orthologous protein from, say, an elephant, about 85% of the amino acids are identical. A similar comparison between human and bird shows an amino acid identity of about 70%—twice as many differences, because the bird and the mammalian lineages have had twice as long to diverge as those of the elephant and the human (**Figure 1–45**).

The mouse, being small, hardy, and a rapid breeder, has become the foremost model organism for experimental studies of vertebrate molecular genetics. Many naturally occurring mutations are known, often mimicking the effects of corresponding mutations in humans (**Figure 1–46**). Methods have been developed, moreover, to test the function of any chosen mouse gene, or of any noncoding portion of the mouse genome, by artificially creating mutations in it, as we explain later in the book.

Just one made-to-order mutant mouse can provide a wealth of information for the cell biologist. It reveals the effects of the chosen mutation in a host of different contexts, simultaneously testing the action of the gene in all the different kinds of cells in the body that could in principle be affected.

Humans Report on Their Own Peculiarities

As humans, we have a special interest in the human genome. We want to know the full set of parts from which we are made, and to discover how they work. But even

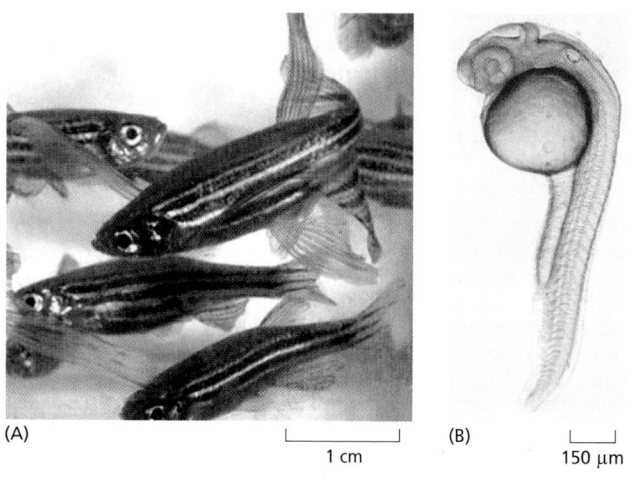

(A)

1 cm

(B)

150 μm

Figure 1–44 Zebrafish as a model for studies of vertebrate development. These small, hardy tropical fish are convenient for genetic studies. Additionally, they have transparent embryos that develop outside of the mother, so that one can clearly observe cells moving and changing their character in the living organism throughout its development. (A) Adult fish. (B) An embryo 24 hours after fertilization. (A, with permission from Steve Baskauf, © 2002 Steven J. Baskauf CC-BY; B, from M. Rhinn et al., *Neur. Dev.* 4:12, 2009. With permission from the authors.)

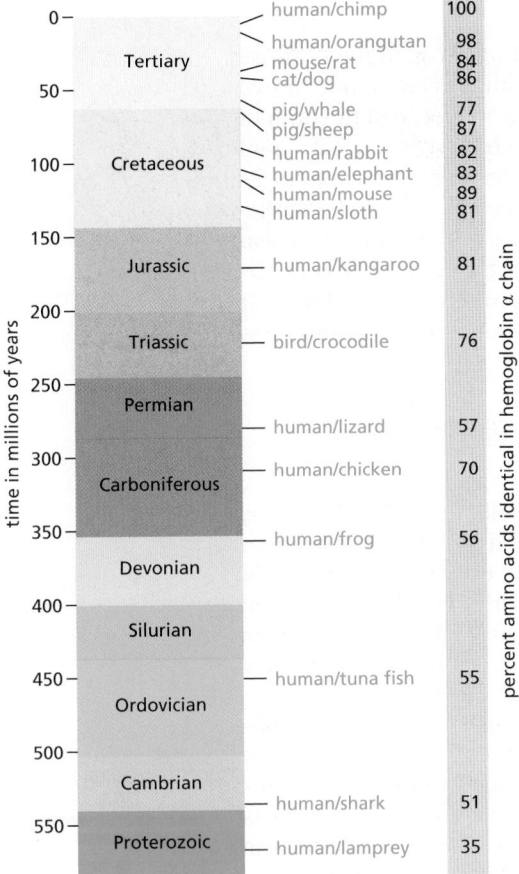

Figure 1–45 **Times of divergence of different vertebrates.** The scale on the left shows the estimated date and geological era of the last common ancestor of each specified pair of animals. Each time estimate is based on comparisons of the amino acid sequences of orthologous proteins; the longer the animals of a pair have had to evolve independently, the smaller the percentage of amino acids that remain identical. The time scale has been calibrated to match the fossil evidence showing that the last common ancestor of mammals and birds lived 310 million years ago.

The figures on the right give data on sequence divergence for one particular protein—the α chain of hemoglobin. Note that although there is a clear general trend of increasing divergence with increasing time for this protein, there are irregularities that are thought to reflect the action of natural selection driving especially rapid changes of hemoglobin sequence when the organisms experienced special physiological demands. Some proteins, subject to stricter functional constraints, evolve much more slowly than hemoglobin, others as much as five times faster. All this gives rise to substantial uncertainties in estimates of divergence times, and some experts believe that the major groups of mammals diverged from one another as much as 60 million years more recently than shown here. (Adapted from S. Kumar and S.B. Hedges, *Nature* 392:917–920, 1998.)

if you were a mouse, preoccupied with the molecular biology of mice, humans would be attractive as model genetic organisms, because of one special property: through medical examinations and self-reporting, we catalog our own genetic (and other) disorders. The human population is enormous, consisting today of some 7 billion individuals, and this self-documenting property means that a huge database of information exists on human mutations. The human genome sequence of more than 3 billion nucleotide pairs has been determined for thousands of different people, making it easier than ever before to identify at a molecular level the precise genetic change responsible for any given human mutant phenotype.

By drawing together the insights from humans, mice, fish, flies, worms, yeasts, plants, and bacteria—using gene sequence similarities to map out the correspondences between one model organism and another—we are enriching our understanding of them all.

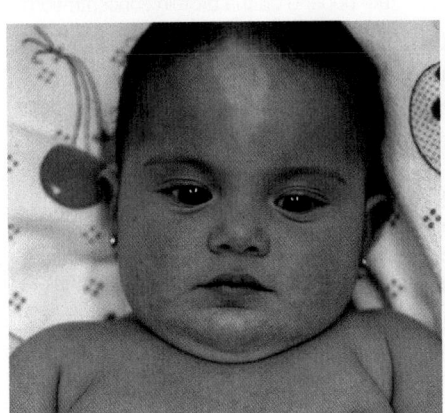

Figure 1–46 **Human and mouse: similar genes and similar development.** The human baby and the mouse shown here have similar white patches on their foreheads because both have mutations in the same gene (called *Kit*), required for the development and maintenance of pigment cells. (Courtesy of R. A. Fleischman, from R. A. Fleischman et al., *Proc. Natl. Acad. Sci. USA* 88:10885–10889, 1991.)

We Are All Different in Detail

What precisely do we mean when we speak of *the* human genome? Whose genome? On average, any two people taken at random differ in about one or two in every 1000 nucleotide pairs in their DNA sequence. The genome of the human species is, properly speaking, a very complex thing, embracing the entire pool of variant genes found in the human population. Knowledge of this variation is helping us to understand, for example, why some people are prone to one disease, others to another; why some respond well to a drug, others badly. It is also providing clues to our history—the population movements and minglings of our ancestors, the infections they suffered, the diets they ate. All these things have left traces in the variant forms of genes that survive today in the human communities that populate the globe.

To Understand Cells and Organisms Will Require Mathematics, Computers, and Quantitative Information

Empowered by knowledge of complete genome sequences, we can list the genes, proteins, and RNA molecules in a cell, and we have methods that allow us to begin to depict the complex web of interactions between them. But how are we to turn all this information into an understanding of how cells work? Even for a single cell type belonging to a single species of organism, the current deluge of data seems overwhelming. The sort of informal reasoning on which biologists usually rely seems totally inadequate in the face of such complexity.

In fact, the difficulty is more than just a matter of information overload. Biological systems are, for example, full of feedback loops, and the behavior of even the simplest of systems with feedback is remarkably difficult to predict by intuition alone (**Figure 1–47**); small changes in parameters can cause radical changes in outcome. To go from a circuit diagram to a prediction of the behavior of the system, we need detailed quantitative information, and to draw deductions from that information we need mathematics and computers.

Such tools for quantitative reasoning are essential, but they are not all-powerful. You might think that, knowing how each protein influences each other protein, and how the expression of each gene is regulated by the products of others, we should soon be able to calculate how the cell as a whole will behave, just as an astronomer can calculate the orbits of the planets, or a chemical engineer can calculate the flows through a chemical plant. But any attempt to perform this feat for anything close to an entire living cell rapidly reveals the limits of our present knowledge. The information we have, plentiful as it is, is full of gaps and uncertainties. Moreover, it is largely qualitative rather than quantitative. Most often, cell biologists studying the cell's control systems sum up their knowledge in simple schematic diagrams—this book is full of them—rather than in numbers, graphs, and differential equations.

To progress from qualitative descriptions and intuitive reasoning to quantitative descriptions and mathematical deduction is one of the biggest challenges for contemporary cell biology. So far, the challenge has been met only for a few very simple fragments of the machinery of living cells—subsystems involving a handful of different proteins, or two or three cross-regulatory genes, where theory and experiment go closely hand in hand. We discuss some of these examples later in the book and devote the entire final section of Chapter 8 to the role of quantitation in cell biology.

Knowledge and understanding bring the power to intervene—with humans, to avoid or prevent disease; with plants, to create better crops; with bacteria, to turn them to our own uses. All these biological enterprises are linked, because the genetic information of all living organisms is written in the same language. The new-found ability of molecular biologists to read and decipher this language has already begun to transform our relationship to the living world. The account of cell biology in the subsequent chapters will, we hope, equip the reader to understand, and possibly to contribute to, the great scientific adventure of the twenty-first century.

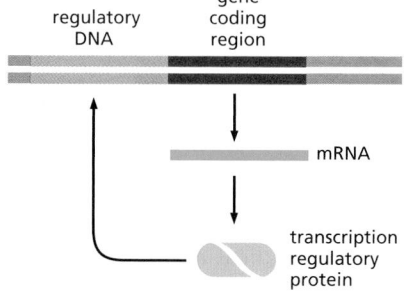

Figure 1–47 A very simple regulatory circuit—a single gene regulating its own expression by the binding of its protein product to its own regulatory DNA. Simple schematic diagrams such as this are found throughout this book. They are often used to summarize what we know, but they leave many questions unanswered. When the protein binds, does it inhibit or stimulate transcription from the gene? How steeply does the transcription rate depend on the protein concentration? How long, on average, does a molecule of the protein remain bound to the DNA? How long does it take to make each molecule of mRNA or protein, and how quickly does each type of molecule get degraded? As explained in Chapter 8, mathematical modeling shows that we need quantitative answers to all these and other questions before we can predict the behavior of even this single-gene system. For different parameter values, the system may settle to a unique steady state; or it may behave as a switch, capable of existing in one or another of a set of alternative states; or it may oscillate; or it may show large random fluctuations.

Summary

Eukaryotic cells, by definition, keep their DNA in a separate membrane-enclosed compartment, the nucleus. They have, in addition, a cytoskeleton for support and movement, elaborate intracellular compartments for digestion and secretion, the capacity (in many species) to engulf other cells, and a metabolism that depends on the oxidation of organic molecules by mitochondria. These properties suggest that eukaryotes may have originated as predators on other cells. Mitochondria—and, in plants, chloroplasts—contain their own genetic material, and they evidently evolved from bacteria that were taken up into the cytoplasm of ancient cells and survived as symbionts.

Eukaryotic cells typically have 3–30 times as many genes as prokaryotes, and often thousands of times more noncoding DNA. The noncoding DNA allows for great complexity in the regulation of gene expression, as required for the construction of complex multicellular organisms. Many eukaryotes are, however, unicellular—among them the yeast Saccharomyces cerevisiae, *which serves as a simple model organism for eukaryotic cell biology, revealing the molecular basis of many fundamental processes that have been strikingly conserved during a billion years of evolution. A small number of other organisms have also been chosen for intensive study: a worm, a fly, a fish, and the mouse serve as "model organisms" for multicellular animals; and a small milkweed serves as a model for plants.*

Powerful new technologies such as genome sequencing are producing striking advances in our knowledge of human beings, and they are helping to advance our understanding of human health and disease. But living systems are incredibly complex, and mammalian genomes contain multiple closely related homologs of most genes. This genetic redundancy has allowed diversification and specialization of genes for new purposes, but it also makes biological mechanisms harder to decipher. For this reason, simpler model organisms have played a key part in revealing universal genetic mechanisms of animal development, and research using these systems remains critical for driving scientific and medical advances.

WHAT WE DON'T KNOW

• What new approaches might provide a clearer view of the anaerobic archaeon that is thought to have formed the nucleus of the first eukaryotic cell? How did its symbiosis with an aerobic bacterium lead to the mitochondrion? Somewhere on Earth, are there cells not yet identified that can fill in the details of how eukaryotic cells originated?

• DNA sequencing has revealed a rich and previously undiscovered world of microbial cells, the vast majority of which fail to grow in a laboratory. How might these cells be made more accessible for detailed study?

• What new model cells or organisms should be developed for scientists to study? Why might a concerted focus on these models speed progress toward understanding a critical aspect of cell function that is poorly understood?

• How did the first cell membranes arise?

PROBLEMS

Which statements are true? Explain why or why not.

1–1 Each member of the human hemoglobin gene family, which consists of seven genes arranged in two clusters on different chromosomes, is an ortholog to all of the other members.

1–2 Horizontal gene transfer is more prevalent in single-celled organisms than in multicellular organisms.

1–3 Most of the DNA sequences in a bacterial genome code for proteins, whereas most of the DNA sequences in the human genome do not.

Discuss the following problems.

1–4 Since it was deciphered four decades ago, some have claimed that the genetic code must be a frozen accident, while others have argued that it was shaped by natural selection. A striking feature of the genetic code is its inherent resistance to the effects of mutation. For example, a change in the third position of a codon often specifies the same amino acid or one with similar chemical properties. The natural code resists mutation more effectively (is less susceptible to error) than most other possible versions, as illustrated in **Figure Q1–1**. Only one in a million computer-generated "random" codes is more error-resistant than the natural genetic code. Does the extraordinary mutation resistance of the genetic code argue in favor of its origin as a frozen accident or as a result of natural selection? Explain your reasoning.

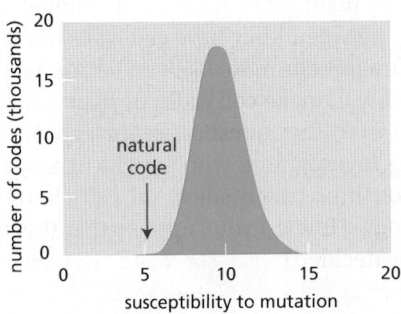

Figure Q1–1 Susceptibility to mutation of the natural code shown relative to that of millions of computer-generated alternative genetic codes (Problem 1–4). Susceptibility measures the average change in amino acid properties caused by random mutations in a genetic code. A small value indicates that mutations tend to cause minor changes. (Data courtesy of Steve Freeland.)

1–5 You have begun to characterize a sample obtained from the depths of the oceans on Europa, one of Jupiter's moons. Much to your surprise, the sample contains a life-form that grows well in a rich broth. Your preliminary analysis shows that it is cellular and contains DNA, RNA, and protein. When you show your results to a colleague, she suggests that your sample was contaminated with an organism from Earth. What approaches might you try to distinguish between contamination and a novel cellular life-form based on DNA, RNA, and protein?

1–6 It is not so difficult to imagine what it means to feed on the organic molecules that living things produce. That is, after all, what we do. But what does it mean to "feed" on sunlight, as phototrophs do? Or, even stranger, to "feed" on rocks, as lithotrophs do? Where is the "food," for example, in the mixture of chemicals (H_2S, H_2, CO, Mn^+, Fe^{2+}, Ni^{2+}, CH_4, and NH_4^+) that spews from a hydrothermal vent?

1–7 How many possible different trees (branching patterns) can in theory be drawn to display the evolution of bacteria, archaea, and eukaryotes, assuming that they all arose from a common ancestor?

1–8 The genes for ribosomal RNA are highly conserved (relatively few sequence changes) in all organisms on Earth; thus, they have evolved very slowly over time. Were ribosomal RNA genes "born" perfect?

1–9 Genes participating in informational processes such as replication, transcription, and translation are transferred between species much less often than are genes involved in metabolism. The basis for this inequality is unclear at present, but one suggestion is that it relates to the underlying complexity of the two types of processes. Informational processes tend to involve large aggregates of different gene products, whereas metabolic reactions are usually catalyzed by enzymes composed of a single protein. Why would the complexity of the underlying process—informational or metabolic—have any effect on the rate of horizontal gene transfer?

1–10 Animal cells have neither cell walls nor chloroplasts, whereas plant cells have both. Fungal cells are somewhere in between; they have cell walls but lack chloroplasts. Are fungal cells more likely to be animal cells that gained the ability to make cell walls, or plant cells that lost their chloroplasts? This question represented a difficult issue for early investigators who sought to assign evolutionary relationships based solely on cell characteristics and morphology. How do you suppose that this question was eventually decided?

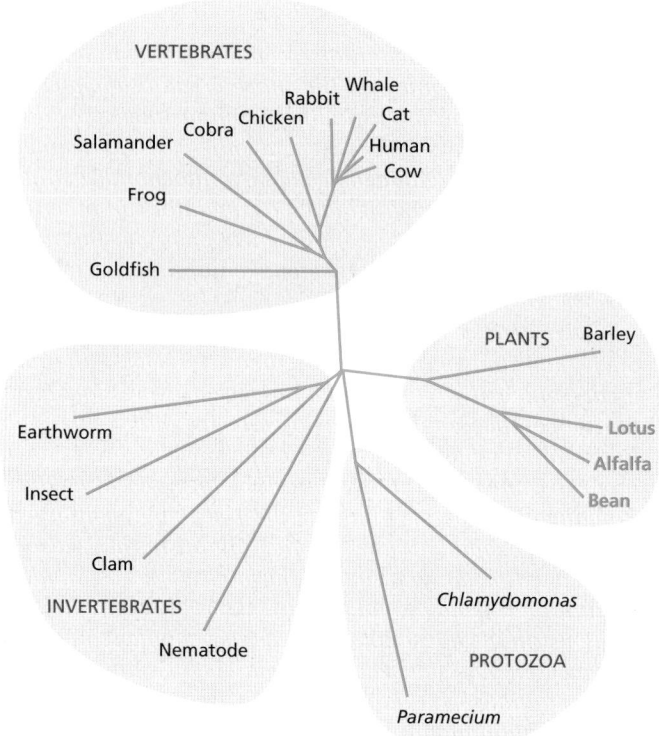

Figure Q1–2 Phylogenetic tree for hemoglobin genes from a variety of species (Problem 1–11). The legumes are highlighted in *green*. The lengths of lines that connect the present-day species represent the evolutionary distances that separate them.

1–11 When plant hemoglobin genes were first discovered in legumes, it was so surprising to find a gene typical of animal blood that it was hypothesized that the plant gene arose by horizontal transfer from an animal. Many more hemoglobin genes have now been sequenced, and a phylogenetic tree based on some of these sequences is shown in **Figure Q1–2**.

A. Does this tree support or refute the hypothesis that the plant hemoglobins arose by horizontal gene transfer?

B. Supposing that the plant hemoglobin genes were originally derived from a parasitic nematode, for example, what would you expect the phylogenetic tree to look like?

1–12 Rates of evolution appear to vary in different lineages. For example, the rate of evolution in the rat lineage is significantly higher than in the human lineage. These rate differences are apparent whether one looks at changes in nucleotide sequences that encode proteins and are subject to selective pressure or at changes in noncoding nucleotide sequences, which are not under obvious selection pressure. Can you offer one or more possible explanations for the slower rate of evolutionary change in the human lineage versus the rat lineage?

REFERENCES

General

Alberts B, Bray D, Hopkin K et al. (2014) Essential Cell Biology, 4th ed. New York: Garland Science.

Barton NH, Briggs DEG, Eisen JA et al. (2007) Evolution. Cold Spring Harbor, NY: Cold Spring Harbor Laboratory Press.

Darwin C (1859) On the Origin of Species. London: Murray.

Graur D & Li W-H (1999) Fundamentals of Molecular Evolution, 2nd ed. Sunderland, MA: Sinauer Associates.

Madigan MT, Martinko JM, Stahl D et al. (2010) Brock Biology of Microorganisms, 13th ed. Menlo Park, CA: Benjamin-Cummings.

Margulis L & Chapman MJ (2009) Kingdoms and Domains: An Illustrated Guide to the Phyla of Life on Earth, 1st ed. San Diego: Academic Press.

Moore JA (1993) Science As a Way of Knowing. Cambridge, MA: Harvard University Press.

Moore JA (1972) Heredity and Development, 2nd ed. New York: Oxford University Press. (Free download at www.nap.edu)

Yang Z (2014) Molecular Evolution: A Statistical Approach. Oxford: Oxford University Press.

The Universal Features of Cells On Earth

Andersson SGE (2006) The bacterial world gets smaller. Science 314, 259–260.

Brenner S, Jacob F & Meselson M (1961) An unstable intermediate carrying information from genes to ribosomes for protein synthesis. Nature 190, 576–581.

Deamer D & Szostak JW eds. (2010) The Origins of Life (Cold Spring Harbor Perspectives in Biology). NY: Cold Spring Harbor Laboratory Press.

Gibson DG, Benders GA, Andrews-Pfannkoch C et al. (2008) Complete chemical synthesis, assembly, and cloning of a Mycoplasma genitalium genome. Science 319, 1215–1220.

Glass JI, Assad-Garcia N, Alperovich N et al. (2006) Essential genes of a minimal bacterium. Proc. Natl Acad. Sci. USA 103, 425–430.

Harris JK, Kelley ST, Spiegelman GB et al. (2003) The genetic core of the universal ancestor. Genome Res. 13, 407–413.

Koonin EV (2005) Orthologs, paralogs, and evolutionary genomics. Annu. Rev. Genet. 39, 309–338.

Noller H (2005) RNA structure: reading the ribosome. Science 309, 1508–1514.

Rinke C, Schwientek P, Sczyrba A et al. (2013) Insights into the phylogeny and coding potential of microbial dark matter. Nature 499, 431–437.

Watson JD & Crick FHC (1953) Molecular structure of nucleic acids. A structure for deoxyribose nucleic acid. Nature 171, 737–738.

The Diversity of Genomes and the Tree of Life

Blattner FR, Plunkett G, Bloch CA et al. (1997) The complete genome sequence of Escherichia coli K-12. Science 277, 1453–1474.

Boucher Y, Douady CJ, Papke RT et al. (2003) Lateral gene transfer and the origins of prokaryotic groups. Annu. Rev. Genet. 37, 283–328.

Cavicchioli R (2010) Archaea–timeline of the third domain. Nat. Rev. Microbiol. 9, 51–61.

Choudhuri S (2014) Bioinformatics for Beginners: Genes, Genomes, Molecular Evolution, Databases and Analytical Tools, 1st ed. San Diego: Academic Press.

Dixon B (1997) Power Unseen: How Microbes Rule the World. Oxford:Oxford University Press.

Handelsman J (2004) Metagenomics: applications of genomics to uncultured microorganisms. Microbiol. Mol. Biol. Rev. 68, 669–685.

Kerr RA (1997) Life goes to extremes in the deep earth—and elsewhere? Science 276, 703–704.

Lee TI, Rinaldi NJ, Robert F et al. (2002) Transcriptional regulatory networks in Saccharomyces cerevisiae. Science 298, 799–804.

Olsen GJ & Woese CR (1997) Archaeal genomics: an overview. Cell 89:991–994.

Williams TA, Foster PG, Cox CJ & Embley TM (2013) An archaeal origin of eukaryotes supports only two primary domains of life. Nature 504, 231–235.

Woese C (1998) The universal ancestor. Proc. Natl Acad. Sci. USA 95, 6854–6859.

Genetic Information in Eukaryotes

Adams MD, Celniker SE, Holt RA et al. (2000) The genome sequence of Drosophila melanogaster. Science 287, 2185–2195.

Amborella Genome Project (2013) The Amborella genome and the evolution of flowering plants. Science 342, 1241089.

Andersson SG, Zomorodipour A, Andersson JO et al. (1998) The genome sequence of Rickettsia prowazekii and the origin of mitochondria. Nature 396, 133–140.

The Arabidopsis Initiative (2000) Analysis of the genome sequence of the flowering plant Arabidopsis thaliana. Nature 408, 796–815.

Carroll SB, Grenier JK & Weatherbee SD (2005) From DNA to Diversity: Molecular Genetics and the Evolution of Animal Design, 2nd ed. Maldon, MA: Blackwell Science.

de Duve C (2007) The origin of eukaryotes: a reappraisal. Nat. Rev. Genet. 8, 395–403.

Delsuc F, Brinkmann H & Philippe H (2005) Phylogenomics and the reconstruction of the tree of life. Nat. Rev. Genet. 6, 361–375.

DeRisi JL, Iyer VR & Brown PO (1997) Exploring the metabolic and genetic control of gene expression on a genomic scale. Science 278, 680–686.

Gabriel SB, Schaffner SF, Nguyen H et al. (2002) The structure of haplotype blocks in the human genome. Science 296, 2225–2229.

Goffeau A, Barrell BG, Bussey H et al. (1996) Life with 6000 genes. Science 274, 546–567.

International Human Genome Sequencing Consortium (2001) Initial sequencing and analysis of the human genome. Nature 409, 860–921.

Keeling PJ & Koonin EV eds. (2014) The Origin and Evolution of Eukaryotes (Cold Spring Harbor Perspectives in Biology). NY: Cold Spring Harbor Laboratory Press.

Lander ES (2011) Initial impact of the sequencing of the human genome. Nature 470, 187–197.

Lynch M & Conery JS (2000) The evolutionary fate and consequences of duplicate genes. Science 290, 1151–1155.

National Center for Biotechnology Information. http://www.ncbi.nlm.nih.gov/

Owens K & King MC (1999) Genomic views of human history. Science 286, 451–453.

Palmer JD & Delwiche CF (1996) Second-hand chloroplasts and the case of the disappearing nucleus. Proc. Natl Acad. Sci. USA 93, 7432–7435.

Reed FA & Tishkoff SA (2006) African human diversity, origins and migrations. Curr. Opin. Genet. Dev. 16, 597–605.

Rine J (2014) A future of the model organism model. Mol. Biol. Cell 25, 549–553.

Rubin GM, Yandell MD, Wortman JR et al. (2000) Comparative genomics of the eukaryotes. Science 287, 2204–2215.

Shen Y, Yue F, McCleary D et al. (2012) A map of the cis-regulatory sequences in the mouse genome. Nature 488, 116–120.

The C. elegans Sequencing Consortium (1998) Genome sequence of the nematode C. elegans: a platform for investigating biology. Science 282, 2012–2018.

Tinsley RC & Kobel HR eds. (1996) The Biology of Xenopus. Oxford: Clarendon Press.

Tyson JJ, Chen KC & Novak B (2003) Sniffers, buzzers, toggles and blinkers: dynamics of regulatory and signaling pathways in the cell. Curr. Opin. Cell Biol. 15, 221–231.

Venter JC, Adams MD, Myers EW et al (2001) The sequence of the human genome. Science 291, 1304–1351.

Cell Chemistry and Bioenergetics

It is at first sight difficult to accept the idea that living creatures are merely chemical systems. Their incredible diversity of form, their seemingly purposeful behavior, and their ability to grow and reproduce all seem to set them apart from the world of solids, liquids, and gases that chemistry normally describes. Indeed, until the nineteenth century animals were believed to contain a Vital Force—an "animus"—that was responsible for their distinctive properties.

We now know that there is nothing in living organisms that disobeys chemical or physical laws. However, the chemistry of life is indeed special. First, it is based overwhelmingly on carbon compounds, the study of which is known as *organic chemistry*. Second, cells are 70% water, and life depends largely on chemical reactions that take place in aqueous solution. Third, and most important, cell chemistry is enormously complex: even the simplest cell is vastly more complicated in its chemistry than any other chemical system known. In particular, although cells contain a variety of small carbon-containing molecules, most of the carbon atoms present are incorporated into enormous polymeric molecules—chains of chemical subunits linked end-to-end. It is the unique properties of these *macromolecules* that enable cells and organisms to grow and reproduce—as well as to do all the other things that are characteristic of life.

THE CHEMICAL COMPONENTS OF A CELL

Living organisms are made of only a small selection of the 92 naturally occurring elements, four of which—carbon (C), hydrogen (H), nitrogen (N), and oxygen (O)—make up 96.5% of an organism's weight (**Figure 2–1**). The atoms of these elements are linked together by **covalent bonds** to form *molecules* (see **Panel 2–1**, pp. 90–91). Because covalent bonds are typically 100 times stronger than the thermal energies within a cell, they resist being pulled apart by thermal motions, and they are normally broken only during specific chemical reactions with other atoms and molecules. Two different molecules can be held together by *noncovalent bonds*,

IN THIS CHAPTER

THE CHEMICAL COMPONENTS OF A CELL

CATALYSIS AND THE USE OF ENERGY BY CELLS

HOW CELLS OBTAIN ENERGY FROM FOOD

Figure 2–1 The main elements in cells, highlighted in the periodic table. When ordered by their atomic number and arranged in this manner, elements fall into vertical columns that show similar properties. Atoms in the same vertical column must gain (or lose) the same number of electrons to attain a filled outer shell, and they thus behave similarly in bond or ion formation. Thus, for example, Mg and Ca tend to give away the two electrons in their outer shells. C, N, and O occur in the same horizontal row, and tend to complete their second shells by sharing electrons.

The four elements highlighted in *red* constitute 99% of the total number of atoms present in the human body. An additional seven elements, highlighted in *blue*, together represent about 0.9% of the total. The elements shown in *green* are required in trace amounts by humans. It remains unclear whether those elements shown in *yellow* are essential in humans. The chemistry of life, it seems, is therefore predominantly the chemistry of lighter elements. The atomic weights shown here are those of the most common isotope of each element.

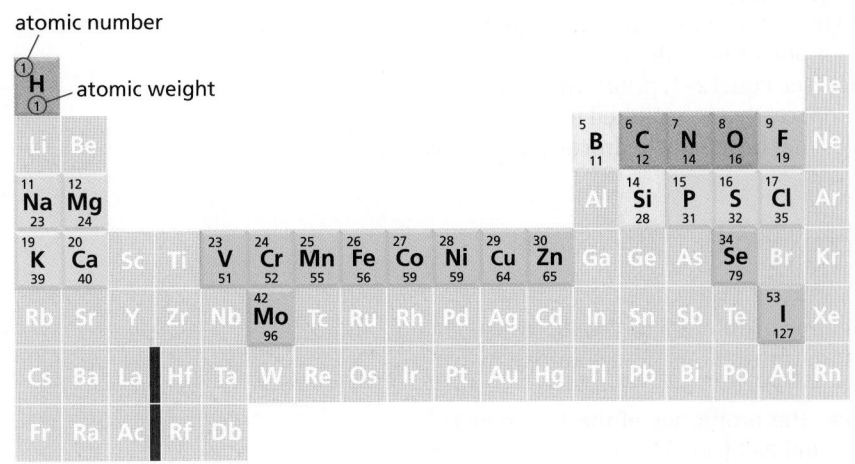

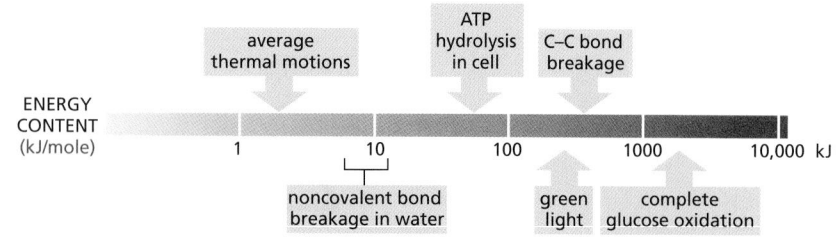

Figure 2–2 **Some energies important for cells.** A crucial property of any bond—covalent or noncovalent—is its strength. *Bond strength* is measured by the amount of energy that must be supplied to break it, expressed in units of either kilojoules per mole (kJ/mole) or kilocalories per mole (kcal/mole). Thus if 100 kJ of energy must be supplied to break 6×10^{23} bonds of a specific type (that is, 1 mole of these bonds), then the strength of that bond is 100 kJ/mole. Note that, in this diagram, energies are compared on a logarithmic scale. Typical strengths and lengths of the main classes of chemical bonds are given in Table 2–1.

One joule (J) is the amount of energy required to move an object a distance of one meter against a force of one Newton. This measure of energy is derived from the SI units (Système Internationale d'Unités) universally employed by physical scientists. A second unit of energy, often used by cell biologists, is the kilocalorie (kcal); one calorie is the amount of energy needed to raise the temperature of 1 gram of water by 1°C. One kJ is equal to 0.239 kcal (1 kcal = 4.18 kJ).

which are much weaker (**Figure 2–2**). We shall see later that noncovalent bonds are important in the many situations where molecules have to associate and dissociate readily to carry out their biological functions.

Water Is Held Together by Hydrogen Bonds

The reactions inside a cell occur in an aqueous environment. Life on Earth began in the ocean, and the conditions in that primeval environment put a permanent stamp on the chemistry of living things. Life therefore hinges on the chemical properties of water, which are reviewed in **Panel 2–2**, pp. 92–93.

In each water molecule (H_2O) the two H atoms are linked to the O atom by covalent bonds. The two bonds are highly polar because the O is strongly attractive for electrons, whereas the H is only weakly attractive. Consequently, there is an unequal distribution of electrons in a water molecule, with a preponderance of positive charge on the two H atoms and of negative charge on the O. When a positively charged region of one water molecule (that is, one of its H atoms) approaches a negatively charged region (that is, the O) of a second water molecule, the electrical attraction between them can result in a *hydrogen bond*. These bonds are much weaker than covalent bonds and are easily broken by the random thermal motions that reflect the heat energy of the molecules. Thus, each bond lasts only a short time. But the combined effect of many weak bonds can be profound. For example, each water molecule can form hydrogen bonds through its two H atoms to two other water molecules, producing a network in which hydrogen bonds are being continually broken and formed. It is only because of the hydrogen bonds that link water molecules together that water is a liquid at room temperature—with a high boiling point and high surface tension—rather than a gas.

Molecules, such as alcohols, that contain polar bonds and that can form hydrogen bonds with water dissolve readily in water. Molecules carrying charges (ions) likewise interact favorably with water. Such molecules are termed **hydrophilic**, meaning that they are water-loving. Many of the molecules in the aqueous environment of a cell necessarily fall into this category, including sugars, DNA, RNA, and most proteins. **Hydrophobic** (water-hating) molecules, by contrast, are uncharged and form few or no hydrogen bonds, and so do not dissolve in water. Hydrocarbons are an important example. In these molecules all of the H atoms are covalently linked to C atoms by a largely nonpolar bond; thus they cannot form effective hydrogen bonds to other molecules (see Panel 2–1, p. 90). This makes the hydrocarbon as a whole hydrophobic—a property that is exploited in cells, whose membranes are constructed from molecules that have long hydrocarbon tails, as we see in Chapter 10.

Four Types of Noncovalent Attractions Help Bring Molecules Together in Cells

Much of biology depends on the specific binding of different molecules caused by three types of noncovalent bonds: **electrostatic attractions** (ionic bonds), **hydrogen bonds**, and **van der Waals attractions**; and on a fourth factor that can push molecules together: the **hydrophobic force**. The properties of the four types of noncovalent attractions are presented in **Panel 2–3** (pp. 94–95). Although each

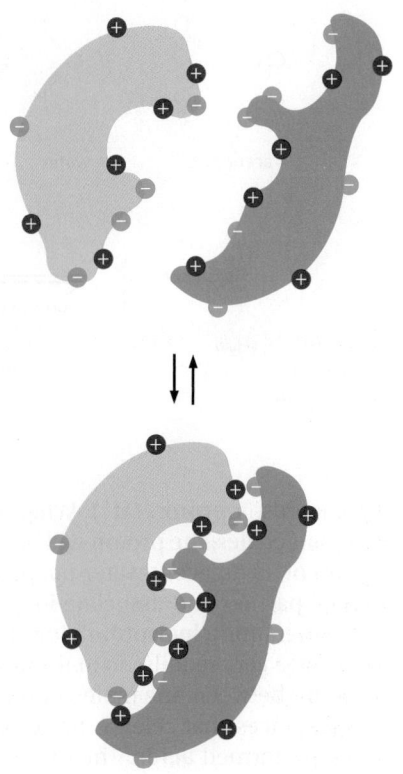

Figure 2–3 **Schematic indicating how two macromolecules with complementary surfaces can bind tightly to one another through noncovalent interactions.** Noncovalent chemical bonds have less than 1/20 the strength of a covalent bond. They are able to produce tight binding only when many of them are formed simultaneously. Although only electrostatic attractions are illustrated here, in reality all four noncovalent forces often contribute to holding two macromolecules together (**Movie 2.1**).

individual noncovalent attraction would be much too weak to be effective in the face of thermal motions, their energies can sum to create a strong force between two separate molecules. Thus sets of noncovalent attractions often allow the complementary surfaces of two macromolecules to hold those two macromolecules together (**Figure 2–3**).

Table 2–1 compares noncovalent bond strengths to that of a typical covalent bond, both in the presence and in the absence of water. Note that, by forming competing interactions with the involved molecules, water greatly reduces the strength of both electrostatic attractions and hydrogen bonds.

The structure of a typical hydrogen bond is illustrated in **Figure 2–4**. This bond represents a special form of polar interaction in which an electropositive hydrogen atom is shared by two electronegative atoms. Its hydrogen can be viewed as a proton that has partially dissociated from a donor atom, allowing it to be shared by a second acceptor atom. Unlike a typical electrostatic interaction, this bond is highly directional—being strongest when a straight line can be drawn between all three of the involved atoms.

The fourth effect that often brings molecules together in water is not, strictly speaking, a bond at all. However, a very important hydrophobic force is caused by a pushing of nonpolar surfaces out of the hydrogen-bonded water network, where they would otherwise physically interfere with the highly favorable interactions between water molecules. Bringing any two nonpolar surfaces together reduces their contact with water; in this sense, the force is nonspecific. Nevertheless, we shall see in Chapter 3 that hydrophobic forces are central to the proper folding of protein molecules.

Some Polar Molecules Form Acids and Bases in Water

One of the simplest kinds of chemical reaction, and one that has profound significance in cells, takes place when a molecule containing a highly polar covalent bond between a hydrogen and another atom dissolves in water. The hydrogen atom in such a molecule has given up its electron almost entirely to the companion atom, and so exists as an almost naked positively charged hydrogen nucleus—in

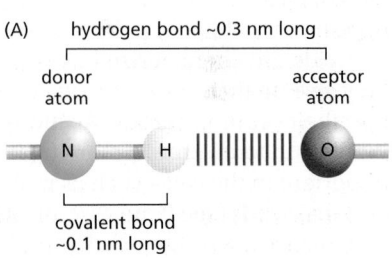

(A)

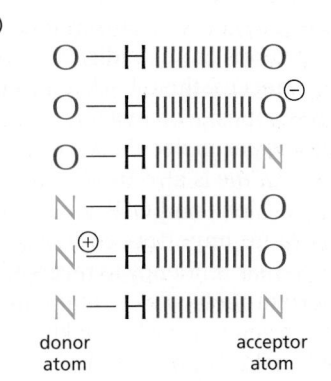

(B)

Figure 2–4 **Hydrogen bonds.** (A) Ball-and-stick model of a typical hydrogen bond. The distance between the hydrogen and the oxygen atom here is less than the sum of their van der Waals radii, indicating a partial sharing of electrons. (B) The most common hydrogen bonds in cells.

TABLE 2–1 **Covalent and Noncovalent Chemical Bonds**				
			Strength kJ/mole**	
Bond type		Length (nm)	in vacuum	in water
Covalent		0.15	377 (90)	377 (90)
Noncovalent	ionic*	0.25	335 (80)	12.6 (3)
	hydrogen	0.30	16.7 (4)	4.2 (1)
	van der Waals attraction (per atom)	0.35	0.4 (0.1)	0.4 (0.1)

*An ionic bond is an electrostatic attraction between two fully charged atoms. **Values in parentheses are kcal/mole. 1 kJ = 0.239 kcal and 1 kcal = 4.18 kJ.

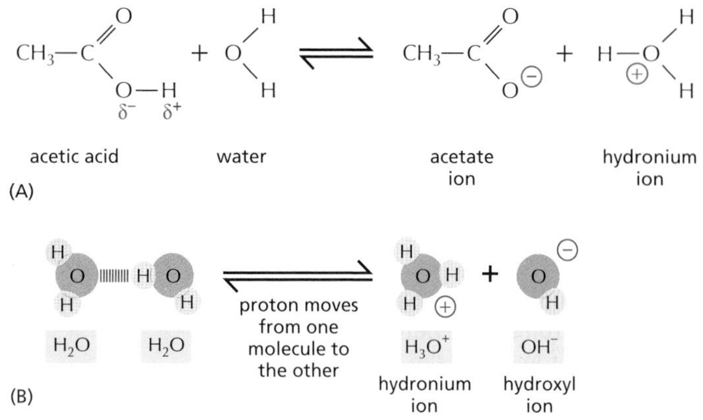

Figure 2–5 **Protons readily move in aqueous solutions.** (A) The reaction that takes place when a molecule of acetic acid dissolves in water. At pH 7, nearly all of the acetic acid is present as acetate ion. (B) Water molecules are continuously exchanging protons with each other to form hydronium and hydroxyl ions. These ions in turn rapidly recombine to form water molecules.

other words, a **proton** (H^+). When the polar molecule becomes surrounded by water molecules, the proton will be attracted to the partial negative charge on the O atom of an adjacent water molecule. This proton can easily dissociate from its original partner and associate instead with the oxygen atom of the water molecule, generating a **hydronium ion** (H_3O^+) (**Figure 2–5**A). The reverse reaction also takes place very readily, so in the aqueous solution protons are constantly flitting to and fro between one molecule and another.

Substances that release protons when they dissolve in water, thus forming H_3O^+, are termed **acids**. The higher the concentration of H_3O^+, the more acidic the solution. H_3O^+ is present even in pure water, at a concentration of 10^{-7} M, as a result of the movement of protons from one water molecule to another (Figure 2-5B). By convention, the H_3O^+ concentration is usually referred to as the H^+ concentration, even though most protons in an aqueous solution are present as H_3O^+. To avoid the use of unwieldy numbers, the concentration of H_3O^+ is expressed using a logarithmic scale called the **pH scale**. Pure water has a pH of 7.0 and is said to be neutral—that is, neither acidic (pH <7) nor basic (pH >7).

Acids are characterized as being strong or weak, depending on how readily they give up their protons to water. Strong acids, such as hydrochloric acid (HCl), lose their protons quickly. Acetic acid, on the other hand, is a weak acid because it holds on to its proton more tightly when dissolved in water. Many of the acids important in the cell—such as molecules containing a carboxyl (COOH) group—are weak acids (see Panel 2-2, pp. 92–93).

Because the proton of a hydronium ion can be passed readily to many types of molecules in cells, altering their character, the concentration of H_3O^+ inside a cell (the acidity) must be closely regulated. Acids—especially weak acids—will give up their protons more readily if the concentration of H_3O^+ in solution is low and will tend to receive them back if the concentration in solution is high.

The opposite of an acid is a **base**. Any molecule capable of accepting a proton from a water molecule is called a base. Sodium hydroxide (NaOH) is basic (the term *alkaline* is also used) because it dissociates readily in aqueous solution to form Na^+ ions and OH^- ions. Because of this property, NaOH is called a strong base. More important in living cells, however, are the weak bases—those that have a weak tendency to reversibly accept a proton from water. Many biologically important molecules contain an amino (NH_2) group. This group is a weak base that can generate OH^- by taking a proton from water: $-NH_2 + H_2O \rightarrow -NH_3^+ + OH^-$ (see Panel 2-2, pp. 92-93).

Because an OH^- ion combines with a H_3O^+ ion to form two water molecules, an increase in the OH^- concentration forces a decrease in the concentration of H_3O^+, and vice versa. A pure solution of water contains an equal concentration (10^{-7} M) of both ions, rendering it neutral. The interior of a cell is also kept close to neutrality by the presence of **buffers**: weak acids and bases that can release or take up protons near pH 7, keeping the environment of the cell relatively constant under a variety of conditions.

A Cell Is Formed from Carbon Compounds

Having reviewed the ways atoms combine into molecules and how these molecules behave in an aqueous environment, we now examine the main classes of small molecules found in cells. We shall see that a few categories of molecules, formed from a handful of different elements, give rise to all the extraordinary richness of form and behavior shown by living things.

If we disregard water and inorganic ions such as potassium, nearly all the molecules in a cell are based on carbon. Carbon is outstanding among all the elements in its ability to form large molecules; silicon is a poor second. Because carbon is small and has four electrons and four vacancies in its outermost shell, a carbon atom can form four covalent bonds with other atoms. Most important, one carbon atom can join to other carbon atoms through highly stable covalent C–C bonds to form chains and rings and hence generate large and complex molecules with no obvious upper limit to their size. The carbon compounds made by cells are called *organic molecules*. In contrast, all other molecules, including water, are said to be *inorganic*.

Certain combinations of atoms, such as the methyl ($-CH_3$), hydroxyl ($-OH$), carboxyl ($-COOH$), carbonyl ($-C=O$), phosphate ($-PO_3^{2-}$), sulfhydryl ($-SH$), and amino ($-NH_2$) groups, occur repeatedly in the molecules made by cells. Each such **chemical group** has distinct chemical and physical properties that influence the behavior of the molecule in which the group occurs. The most common chemical groups and some of their properties are summarized in Panel 2–1, pp. 90–91.

Cells Contain Four Major Families of Small Organic Molecules

The small organic molecules of the cell are carbon-based compounds that have molecular weights in the range of 100–1000 and contain up to 30 or so carbon atoms. They are usually found free in solution and have many different fates. Some are used as *monomer* subunits to construct giant polymeric *macromolecules*—proteins, nucleic acids, and large polysaccharides. Others act as energy sources and are broken down and transformed into other small molecules in a maze of intracellular metabolic pathways. Many small molecules have more than one role in the cell—for example, acting both as a potential subunit for a macromolecule and as an energy source. Small organic molecules are much less abundant than the organic macromolecules, accounting for only about one-tenth of the total mass of organic matter in a cell. As a rough guess, there may be a thousand different kinds of these small molecules in a typical cell.

All organic molecules are synthesized from and are broken down into the same set of simple compounds. As a consequence, the compounds in a cell are chemically related and most can be classified into a few distinct families. Broadly speaking, cells contain four major families of small organic molecules: the *sugars*, the *fatty acids*, the *nucleotides*, and the *amino acids* (**Figure 2–6**). Although many compounds present in cells do not fit into these categories, these four families of small organic molecules, together with the macromolecules made by linking them into long chains, account for a large fraction of the cell mass.

Amino acids and the proteins that they form will be the subject of Chapter 3. A summary of the structures and properties of the remaining three families—sugars, fatty acids, and nucleotides—is presented in **Panels 2–4, 2–5, and 2–6**, respectively (see pages 96–101).

The Chemistry of Cells Is Dominated by Macromolecules with Remarkable Properties

By weight, **macromolecules** are the most abundant carbon-containing molecules in a living cell (**Figure 2–7**). They are the principal building blocks from which a cell is constructed and also the components that confer the most distinctive properties of living things. The macromolecules in cells are polymers that are constructed by covalently linking small organic molecules (called *monomers*) into

A SUGAR

AN AMINO ACID

A FATTY ACID

A NUCLEOTIDE

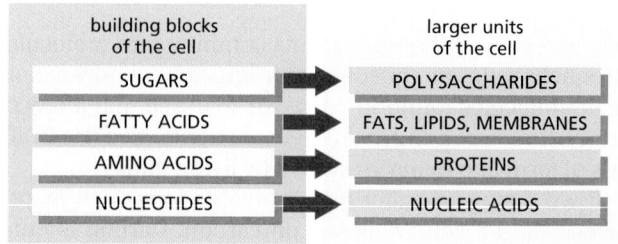

Figure 2–6 The four main families of small organic molecules in cells. These small molecules form the monomeric building blocks, or subunits, for most of the macromolecules and other assemblies of the cell. Some, such as the sugars and the fatty acids, are also energy sources. Their structures are outlined here and shown in more detail in the Panels at the end of this chapter and in Chapter 3.

long chains (**Figure 2–8**). They have remarkable properties that could not have been predicted from their simple constituents.

Proteins are abundant and spectacularly versatile, performing thousands of distinct functions in cells. Many proteins serve as *enzymes*, the catalysts that facilitate the many covalent bond-making and bond-breaking reactions that the cell needs. Enzymes catalyze all of the reactions whereby cells extract energy from food molecules, for example, and an enzyme called ribulose bisphosphate carboxylase helps to convert CO_2 to sugars in photosynthetic organisms, producing most of the organic matter needed for life on Earth. Other proteins are used to build structural components, such as tubulin, a protein that self-assembles to make the cell's long microtubules, or histones, proteins that compact the DNA in chromosomes. Yet other proteins act as molecular motors to produce force and

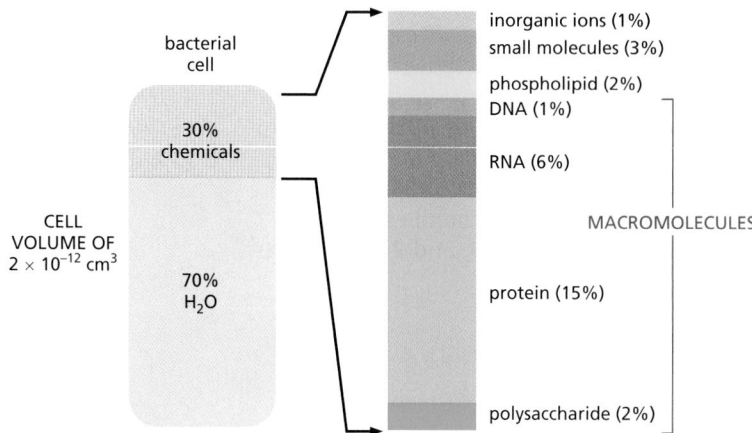

Figure 2–7 The distribution of molecules in cells. The approximate composition of a bacterial cell is shown by weight. The composition of an animal cell is similar, even though its volume is roughly 1000 times greater. Note that macromolecules dominate. The major inorganic ions include Na^+, K^+, Mg^{2+}, Ca^{2+}, and Cl^-.

movement, as for myosin in muscle. Proteins perform many other functions, and we shall examine the molecular basis for many of them later in this book.

Although the chemical reactions for adding subunits to each polymer are different in detail for proteins, nucleic acids, and polysaccharides, they share important features. Each polymer grows by the addition of a monomer onto the end of a growing chain in a *condensation reaction*, in which one molecule of water is lost with each subunit added (**Figure 2–9**). The stepwise polymerization of monomers into a long chain is a simple way to manufacture a large, complex molecule, since the subunits are added by the same reaction performed over and over again by the same set of enzymes. Apart from some of the polysaccharides, most macromolecules are made from a limited set of monomers that are slightly different from one another—for example, the 20 different amino acids from which proteins are made. It is critical to life that the polymer chain is not assembled at random from these subunits; instead the subunits are added in a precise order, or *sequence*. The elaborate mechanisms that allow enzymes to accomplish this task are described in detail in Chapters 5 and 6.

Noncovalent Bonds Specify Both the Precise Shape of a Macromolecule and Its Binding to Other Molecules

Most of the covalent bonds in a macromolecule allow rotation of the atoms they join, giving the polymer chain great flexibility. In principle, this allows a macromolecule to adopt an almost unlimited number of shapes, or *conformations*, as random thermal energy causes the polymer chain to writhe and rotate. However, the shapes of most biological macromolecules are highly constrained because of the many weak *noncovalent bonds* that form between different parts of the same molecule. If these noncovalent bonds are formed in sufficient numbers, the polymer chain can strongly prefer one particular conformation, determined by the linear sequence of monomers in its chain. Most protein molecules and many of the small RNA molecules found in cells fold tightly into a highly preferred conformation in this way (**Figure 2–10**).

The four types of noncovalent interactions important in biological molecules were presented earlier, and they are discussed further in Panel 2–3 (pp. 94–95). In addition to folding biological macromolecules into unique shapes, they can also add up to create a strong attraction between two different molecules (see Figure 2–3). This form of molecular interaction provides for great specificity, inasmuch as the close multipoint contacts required for strong binding make it possible for a macromolecule to select out—through binding—just one of the many thousands of other types of molecules present inside a cell. Moreover, because the strength of the binding depends on the number of noncovalent bonds that are formed, interactions of almost any affinity are possible—allowing rapid dissociation where appropriate.

As we discuss next, binding of this type underlies all biological catalysis, making it possible for proteins to function as enzymes. In addition, noncovalent interactions allow macromolecules to be used as building blocks for the formation of

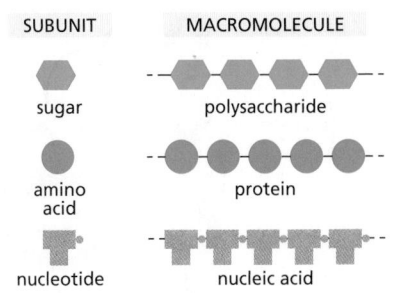

Figure 2–8 **Three families of macromolecules.** Each is a polymer formed from small molecules (called monomers) linked together by covalent bonds.

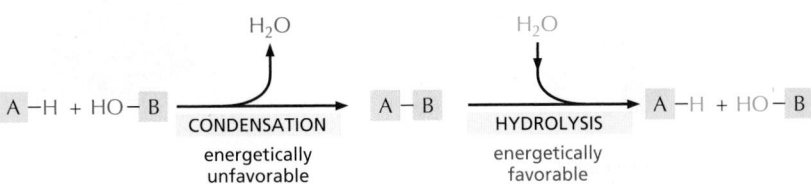

Figure 2–9 **Condensation and hydrolysis as opposite reactions.** The macromolecules of the cell are polymers that are formed from subunits (or monomers) by a condensation reaction, and they are broken down by hydrolysis. The condensation reactions are all energetically unfavorable; thus polymer formation requires an energy input, as will be described in the text.

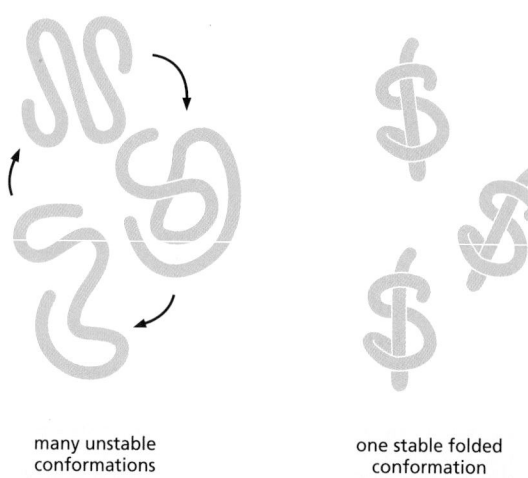

many unstable
conformations

one stable folded
conformation

Figure 2–10 **The folding of proteins and RNA molecules into a particularly stable three-dimensional shape, or conformation.** If the noncovalent bonds maintaining the stable conformation are disrupted, the molecule becomes a flexible chain that loses its biological activity.

larger structures, thereby forming intricate machines with multiple moving parts that perform such complex tasks as DNA replication and protein synthesis (**Figure 2–11**).

Summary

Living organisms are autonomous, self-propagating chemical systems. They are formed from a distinctive and restricted set of small carbon-based molecules that are essentially the same for every living species. Each of these small molecules is composed of a small set of atoms linked to each other in a precise configuration through covalent bonds. The main categories are sugars, fatty acids, amino acids, and nucleotides. Sugars are a primary source of chemical energy for cells and can be incorporated into polysaccharides for energy storage. Fatty acids are also important for energy storage, but their most critical function is in the formation of cell membranes. Long chains of amino acids form the remarkably diverse and versatile macromolecules known as proteins. Nucleotides play a central part in energy transfer, while also serving as the subunits for the informational macromolecules, RNA and DNA.

Most of the dry mass of a cell consists of macromolecules that have been produced as linear polymers of amino acids (proteins) or nucleotides (DNA and RNA), covalently linked to each other in an exact order. Most of the protein molecules and many of the RNAs fold into a unique conformation that is determined by their sequence of subunits. This folding process creates unique surfaces, and it depends on a large set of weak attractions produced by noncovalent forces between atoms.

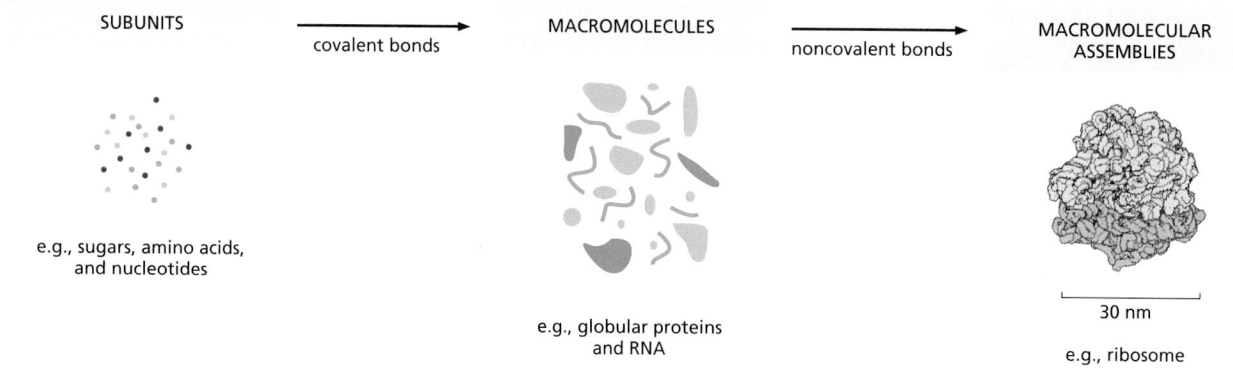

SUBUNITS

covalent bonds

MACROMOLECULES

noncovalent bonds

MACROMOLECULAR
ASSEMBLIES

e.g., sugars, amino acids,
and nucleotides

e.g., globular proteins
and RNA

30 nm

e.g., ribosome

Figure 2–11 **Small molecules become covalently linked to form macromolecules, which in turn assemble through noncovalent interactions to form large complexes.** Small molecules, proteins, and a ribosome are drawn approximately to scale. Ribosomes are a central part of the machinery that the cell uses to make proteins: each ribosome is formed as a complex of about 90 macromolecules (protein and RNA molecules).

These forces are of four types: electrostatic attractions, hydrogen bonds, van der Waals attractions, and an interaction between nonpolar groups caused by their hydrophobic expulsion from water. The same set of weak forces governs the specific binding of other molecules to macromolecules, making possible the myriad associations between biological molecules that produce the structure and the chemistry of a cell.

CATALYSIS AND THE USE OF ENERGY BY CELLS

One property of living things above all makes them seem almost miraculously different from nonliving matter: they create and maintain order, in a universe that is tending always to greater disorder (**Figure 2–12**). To create this order, the cells in a living organism must perform a never-ending stream of chemical reactions. In some of these reactions, small organic molecules—amino acids, sugars, nucleotides, and lipids—are being taken apart or modified to supply the many other small molecules that the cell requires. In other reactions, small molecules are being used to construct an enormously diverse range of proteins, nucleic acids, and other macromolecules that endow living systems with all of their most distinctive properties. Each cell can be viewed as a tiny chemical factory, performing many millions of reactions every second.

Cell Metabolism Is Organized by Enzymes

The chemical reactions that a cell carries out would normally occur only at much higher temperatures than those existing inside cells. For this reason, each reaction requires a specific boost in chemical reactivity. This requirement is crucial, because it allows the cell to control its chemistry. The control is exerted through specialized biological *catalysts*. These are almost always proteins called *enzymes*, although RNA catalysts also exist, called *ribozymes*. Each enzyme accelerates, or *catalyzes*, just one of the many possible kinds of reactions that a particular molecule might undergo. Enzyme-catalyzed reactions are connected in series, so that the product of one reaction becomes the starting material, or *substrate*, for the next (**Figure 2–13**). Long linear reaction pathways are in turn linked to one another, forming a maze of interconnected reactions that enable the cell to survive, grow, and reproduce.

Two opposing streams of chemical reactions occur in cells: (1) the *catabolic* pathways break down foodstuffs into smaller molecules, thereby generating both a useful form of energy for the cell and some of the small molecules that the cell needs as building blocks, and (2) the *anabolic*, or *biosynthetic*, pathways use the

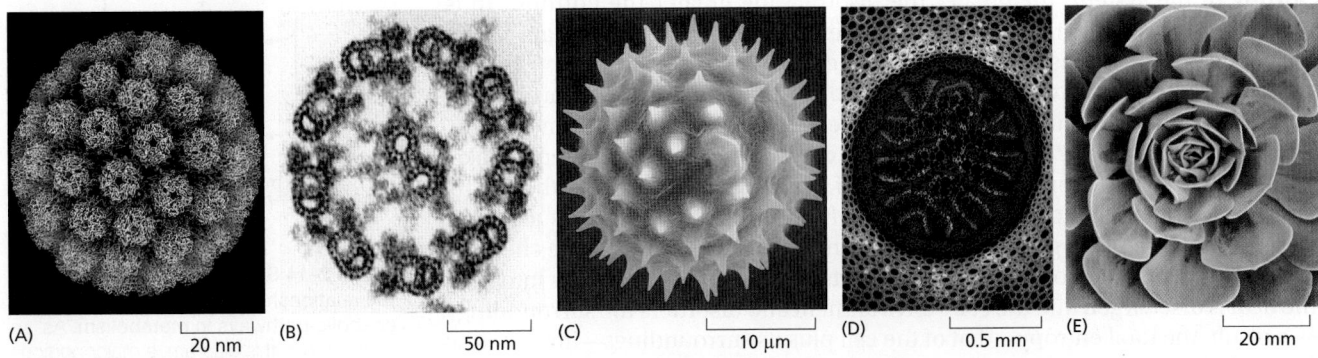

(A) 20 nm (B) 50 nm (C) 10 μm (D) 0.5 mm (E) 20 mm

Figure 2–12 **Biological structures are highly ordered.** Well-defined, ornate, and beautiful spatial patterns can be found at every level of organization in living organisms. In order of increasing size: (A) protein molecules in the coat of a virus (a parasite that, although not technically alive, contains the same types of molecules as those found in living cells); (B) the regular array of microtubules seen in a cross section of a sperm tail; (C) surface contours of a pollen grain (a single cell); (D) cross section of a fern stem, showing the patterned arrangement of cells; and (E) a spiral arrangement of leaves in a succulent plant. (A, courtesy of Robert Grant, Stéphane Crainic, and James M. Hogle; B, courtesy of Lewis Tilney; C, courtesy of Colin MacFarlane and Chris Jeffree; D, courtesy of Jim Haseloff.)

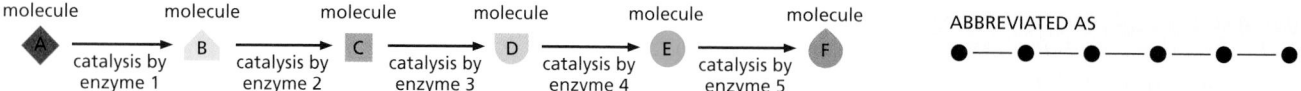

Figure 2–13 How a set of enzyme-catalyzed reactions generates a metabolic pathway. Each enzyme catalyzes a particular chemical reaction, leaving the enzyme unchanged. In this example, a set of enzymes acting in series converts molecule A to molecule F, forming a metabolic pathway. (For a diagram of many of the reactions in a human cell, abbreviated as shown, see Figure 2–63.)

small molecules and the energy harnessed by catabolism to drive the synthesis of the many other molecules that form the cell. Together these two sets of reactions constitute the **metabolism** of the cell (**Figure 2–14**).

The details of cell metabolism form the traditional subject of *biochemistry* and most of them need not concern us here. But the general principles by which cells obtain energy from their environment and use it to create order are central to cell biology. We begin with a discussion of why a constant input of energy is needed to sustain all living things.

Biological Order Is Made Possible by the Release of Heat Energy from Cells

The universal tendency of things to become disordered is a fundamental law of physics—the *second law of thermodynamics*—which states that in the universe, or in any isolated system (a collection of matter that is completely isolated from the rest of the universe), the degree of disorder always increases. This law has such profound implications for life that we will restate it in several ways.

For example, we can present the second law in terms of probability by stating that systems will change spontaneously toward those arrangements that have the greatest probability. If we consider a box of 100 coins all lying heads up, a series of accidents that disturbs the box will tend to move the arrangement toward a mixture of 50 heads and 50 tails. The reason is simple: there is a huge number of possible arrangements of the individual coins in the mixture that can achieve the 50-50 result, but only one possible arrangement that keeps all of the coins oriented heads up. Because the 50-50 mixture is therefore the most probable, we say that it is more "disordered." For the same reason, it is a common experience that one's living space will become increasingly disordered without intentional effort: the movement toward disorder is a *spontaneous process*, requiring a periodic effort to reverse it (**Figure 2–15**).

The amount of disorder in a system can be quantified and expressed as the **entropy** of the system: the greater the disorder, the greater the entropy. Thus, another way to express the second law of thermodynamics is to say that systems will change spontaneously toward arrangements with greater entropy.

Living cells—by surviving, growing, and forming complex organisms—are generating order and thus might appear to defy the second law of thermodynamics. How is this possible? The answer is that a cell is not an isolated system: it takes in energy from its environment in the form of food, or as photons from the sun (or even, as in some chemosynthetic bacteria, from inorganic molecules alone). It then uses this energy to generate order within itself. In the course of the chemical reactions that generate order, the cell converts part of the energy it uses into heat. The heat is discharged into the cell's environment and disorders the surroundings. As a result, the total entropy—that of the cell plus its surroundings—increases, as demanded by the second law of thermodynamics.

To understand the principles governing these energy conversions, think of a cell surrounded by a sea of matter representing the rest of the universe. As the cell lives and grows, it creates internal order. But it constantly releases heat energy as it synthesizes molecules and assembles them into cell structures. Heat is energy in its most disordered form—the random jostling of molecules. When

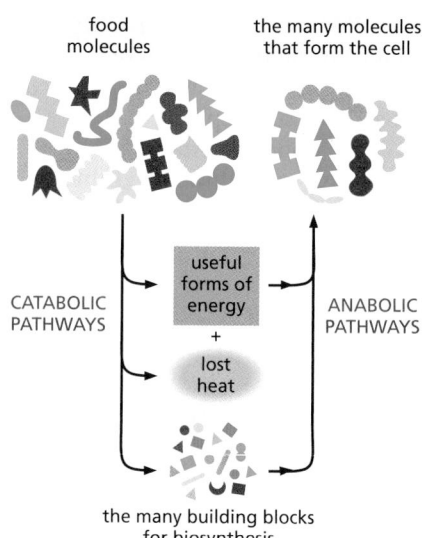

Figure 2–14 Schematic representation of the relationship between catabolic and anabolic pathways in metabolism. As suggested in this diagram, a major portion of the energy stored in the chemical bonds of food molecules is dissipated as heat. In addition, the mass of food required by any organism that derives all of its energy from catabolism is much greater than the mass of the molecules that it can produce by anabolism.

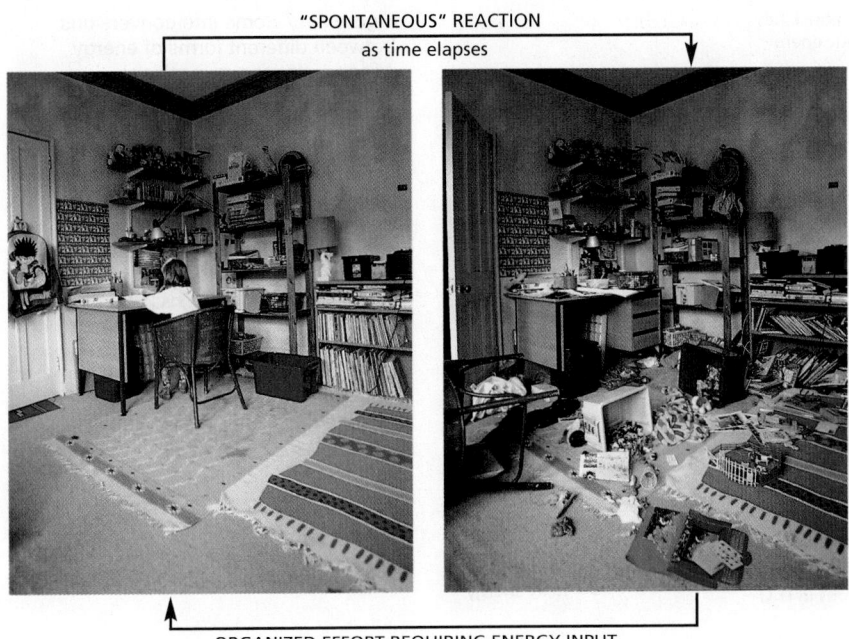

"SPONTANEOUS" REACTION
as time elapses

ORGANIZED EFFORT REQUIRING ENERGY INPUT

Figure 2–15 **An everyday illustration of the spontaneous drive toward disorder.** Reversing this tendency toward disorder requires an intentional effort and an input of energy: it is not spontaneous. In fact, from the second law of thermodynamics, we can be certain that the human intervention required will release enough heat to the environment to more than compensate for the reordering of the items in this room.

the cell releases heat to the sea, it increases the intensity of molecular motions there (thermal motion)—thereby increasing the randomness, or disorder, of the sea. The second law of thermodynamics is satisfied because the increase in the amount of order inside the cell is always more than compensated for by an even greater decrease in order (increase in entropy) in the surrounding sea of matter (**Figure 2–16**).

Where does the heat that the cell releases come from? Here we encounter another important law of thermodynamics. The *first law of thermodynamics* states that energy can be converted from one form to another, but that it cannot be created or destroyed. **Figure 2–17** illustrates some interconversions between different forms of energy. The amount of energy in different forms will change as a result of the chemical reactions inside the cell, but the first law tells us that the total amount of energy must always be the same. For example, an animal cell takes in foodstuffs and converts some of the energy present in the chemical bonds between the atoms of these food molecules (chemical-bond energy) into the random thermal motion of molecules (heat energy).

The cell cannot derive any benefit from the heat energy it releases unless the heat-generating reactions inside the cell are directly linked to the processes that generate molecular order. It is the tight *coupling* of heat production to an increase

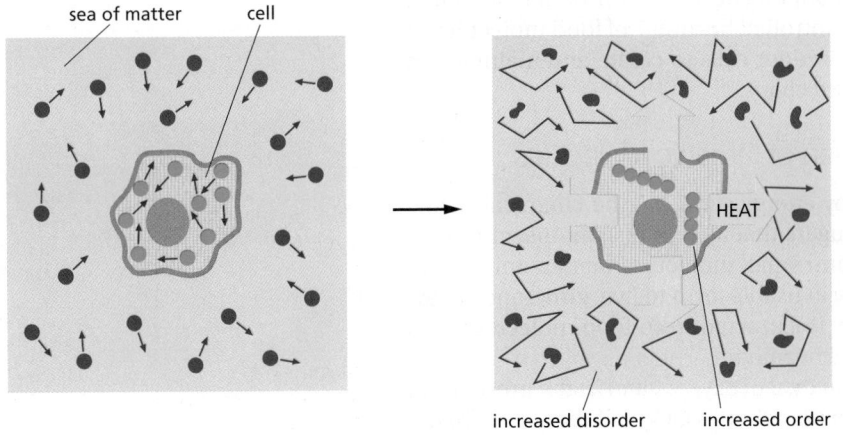

sea of matter   cell

HEAT

increased disorder   increased order

Figure 2–16 **A simple thermodynamic analysis of a living cell.** In the diagram on the left, the molecules of both the cell and the rest of the universe (the sea of matter) are depicted in a relatively disordered state. In the diagram on the right, the cell has taken in energy from food molecules and released heat through reactions that order the molecules the cell contains. The heat released increases the disorder in the environment around the cell (depicted by *jagged arrows* and distorted molecules, indicating increased molecular motions caused by heat). As a result, the second law of thermodynamics—which states that the amount of disorder in the universe must always increase—is satisfied as the cell grows and divides. For a detailed discussion, see Panel 2–7 (pp. 102–103).

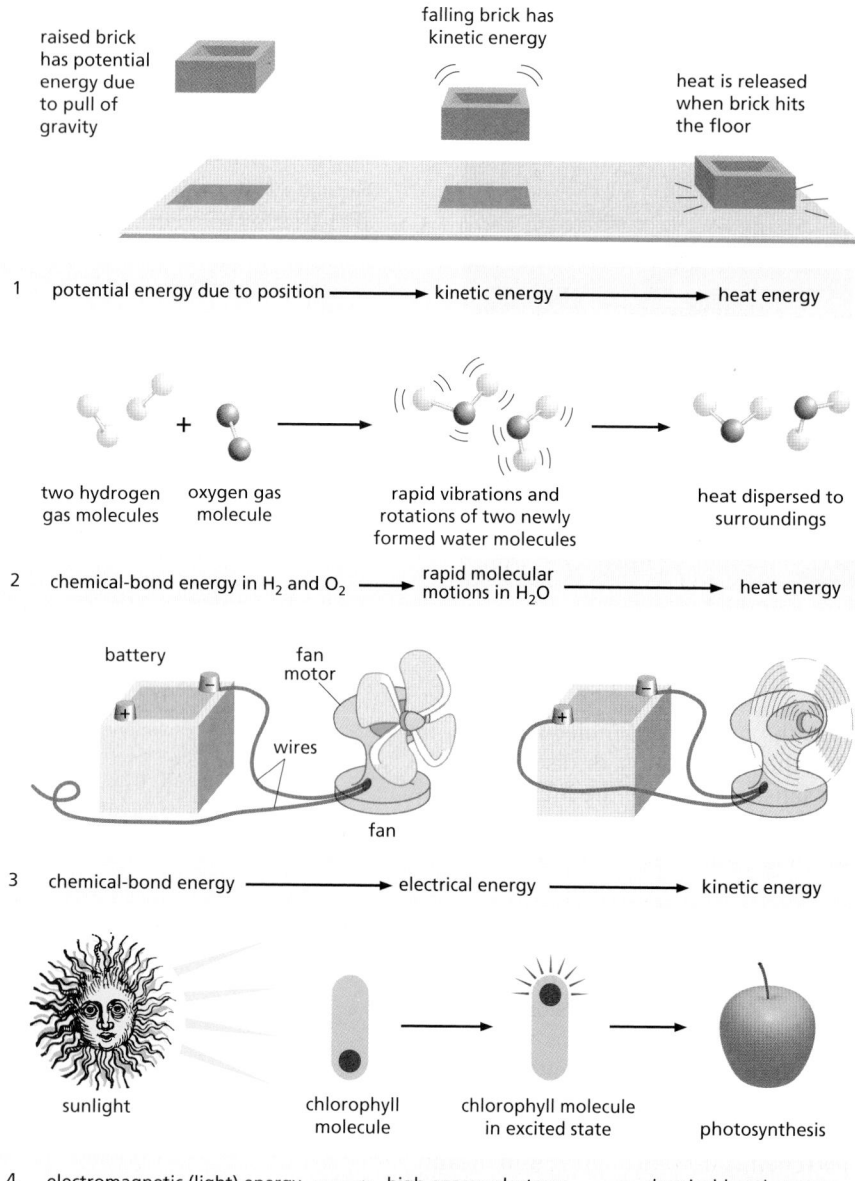

Figure 2–17 Some interconversions between different forms of energy. All energy forms are, in principle, interconvertible. In all these processes the total amount of energy is conserved. Thus, for example, from the height and weight of the brick in (1), we can predict exactly how much heat will be released when it hits the floor. In (2), note that the large amount of chemical-bond energy released when water is formed is initially converted to very rapid thermal motions in the two new water molecules; but collisions with other molecules almost instantaneously spread this kinetic energy evenly throughout the surroundings (heat transfer), making the new molecules indistinguishable from all the rest.

in order that distinguishes the metabolism of a cell from the wasteful burning of fuel in a fire. Later, we illustrate how this coupling occurs. For now, it is sufficient to recognize that a direct linkage of the "controlled burning" of food molecules to the generation of biological order is required for cells to create and maintain an island of order in a universe tending toward chaos.

Cells Obtain Energy by the Oxidation of Organic Molecules

All animal and plant cells are powered by energy stored in the chemical bonds of organic molecules, whether they are sugars that a plant has photosynthesized as food for itself or the mixture of large and small molecules that an animal has eaten. Organisms must extract this energy in usable form to live, grow, and reproduce. In both plants and animals, energy is extracted from food molecules by a process of gradual oxidation, or controlled burning.

The Earth's atmosphere contains a great deal of oxygen, and in the presence of oxygen the most energetically stable form of carbon is CO_2 and that of hydrogen

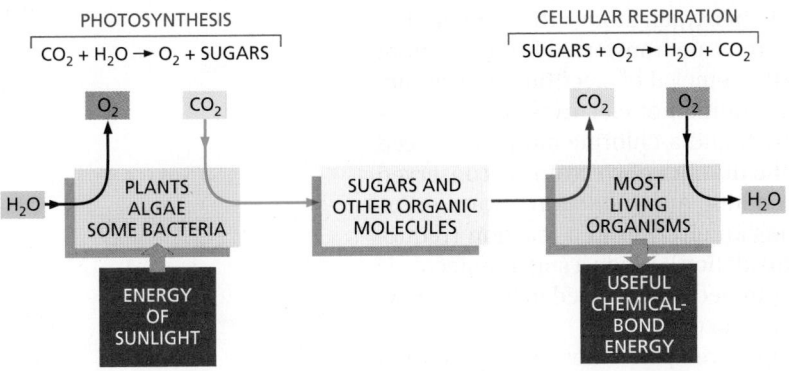

Figure 2–18 **Photosynthesis and respiration as complementary processes in the living world.** Photosynthesis converts the electromagnetic energy in sunlight into chemical-bond energy in sugars and other organic molecules. Plants, algae, and cyanobacteria obtain the carbon atoms that they need for this purpose from atmospheric CO_2 and the hydrogen from water, releasing O_2 gas as a by-product. The organic molecules produced by photosynthesis in turn serve as food for other organisms. Many of these organisms carry out aerobic respiration, a process that uses O_2 to form CO_2 from the same carbon atoms that had been taken up as CO_2 and converted into sugars by photosynthesis. In the process, the organisms that respire obtain the chemical-bond energy that they need to survive.

The first cells on the Earth are thought to have been capable of neither photosynthesis nor respiration (discussed in Chapter 14). However, photosynthesis must have preceded respiration on the Earth, since there is strong evidence that billions of years of photosynthesis were required before O_2 had been released in sufficient quantity to create an atmosphere rich in this gas. (The Earth's atmosphere currently contains 20% O_2.)

is H_2O. A cell is therefore able to obtain energy from sugars or other organic molecules by allowing their carbon and hydrogen atoms to combine with oxygen to produce CO_2 and H_2O, respectively—a process called **aerobic respiration**.

Photosynthesis (discussed in detail in Chapter 14) and respiration are complementary processes (**Figure 2–18**). This means that the transactions between plants and animals are not all one way. Plants, animals, and microorganisms have existed together on this planet for so long that many of them have become an essential part of the others' environments. The oxygen released by photosynthesis is consumed in the combustion of organic molecules during aerobic respiration. And some of the CO_2 molecules that are fixed today into organic molecules by photosynthesis in a green leaf were yesterday released into the atmosphere by the respiration of an animal—or by the respiration of a fungus or bacterium decomposing dead organic matter. We therefore see that carbon utilization forms a huge cycle that involves the *biosphere* (all of the living organisms on Earth) as a whole (**Figure 2–19**). Similarly, atoms of nitrogen, phosphorus, and sulfur move between the living and nonliving worlds in cycles that involve plants, animals, fungi, and bacteria.

Oxidation and Reduction Involve Electron Transfers

The cell does not oxidize organic molecules in one step, as occurs when organic material is burned in a fire. Through the use of enzyme catalysts, metabolism takes these molecules through a large number of reactions that only rarely involve the direct addition of oxygen. Before we consider some of these reactions and their purpose, we discuss what is meant by the process of oxidation.

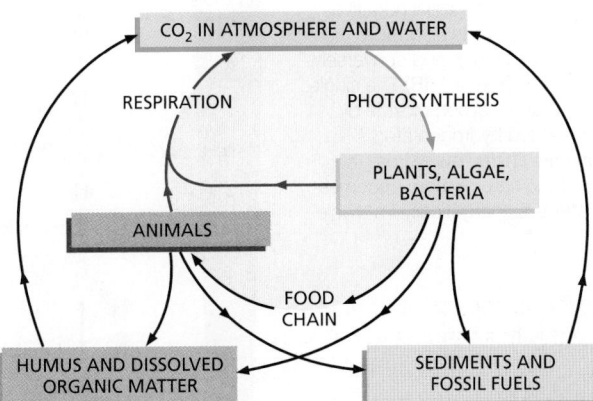

Figure 2–19 **The carbon cycle.** Individual carbon atoms are incorporated into organic molecules of the living world by the photosynthetic activity of bacteria, algae, and plants. They pass to animals, microorganisms, and organic material in soil and oceans in cyclic paths. CO_2 is restored to the atmosphere when organic molecules are oxidized by cells or burned by humans as fuels.

Oxidation refers to more than the addition of oxygen atoms; the term applies more generally to any reaction in which electrons are transferred from one atom to another. Oxidation in this sense refers to the removal of electrons, and **reduction**—the converse of oxidation—means the addition of electrons. Thus, Fe^{2+} is oxidized if it loses an electron to become Fe^{3+}, and a chlorine atom is reduced if it gains an electron to become Cl^-. Since the number of electrons is conserved (no loss or gain) in a chemical reaction, oxidation and reduction always occur simultaneously: that is, if one molecule gains an electron in a reaction (reduction), a second molecule loses the electron (oxidation). When a sugar molecule is oxidized to CO_2 and H_2O, for example, the O_2 molecules involved in forming H_2O gain electrons and thus are said to have been reduced.

The terms "oxidation" and "reduction" apply even when there is only a partial shift of electrons between atoms linked by a covalent bond (**Figure 2–20**). When a carbon atom becomes covalently bonded to an atom with a strong affinity for electrons, such as oxygen, chlorine, or sulfur, for example, it gives up more than its equal share of electrons and forms a *polar* covalent bond. Because the positive charge of the carbon nucleus is now somewhat greater than the negative charge of its electrons, the atom acquires a partial positive charge and is said to be oxidized. Conversely, a carbon atom in a C-H linkage has slightly more than its share of electrons, and so it is said to be reduced.

When a molecule in a cell picks up an electron (e^-), it often picks up a proton (H^+) at the same time (protons being freely available in water). The net effect in this case is to add a hydrogen atom to the molecule.

$$A + e^- + H^+ \rightarrow AH$$

Even though a proton plus an electron is involved (instead of just an electron), such *hydrogenation* reactions are reductions, and the reverse, *dehydrogenation* reactions are oxidations. It is especially easy to tell whether an organic molecule is being oxidized or reduced: reduction is occurring if its number of C-H bonds increases, whereas oxidation is occurring if its number of C-H bonds decreases (see Figure 2–20B).

Cells use enzymes to catalyze the oxidation of organic molecules in small steps, through a sequence of reactions that allows useful energy to be harvested. We now need to explain how enzymes work and some of the constraints under which they operate.

Figure 2–20 **Oxidation and reduction.** (A) When two atoms form a *polar* covalent bond, the atom ending up with a greater share of electrons is said to be reduced, while the other atom acquires a lesser share of electrons and is said to be oxidized. The reduced atom has acquired a partial negative charge (δ^-) as the positive charge on the atomic nucleus is now more than equaled by the total charge of the electrons surrounding it, and conversely, the oxidized atom has acquired a partial positive charge (δ^+). (B) The single carbon atom of methane can be converted to that of carbon dioxide by the successive replacement of its covalently bonded hydrogen atoms with oxygen atoms. With each step, electrons are shifted away from the carbon (as indicated by the *blue* shading), and the carbon atom becomes progressively more oxidized. Each of these steps is energetically favorable under the conditions present inside a cell.

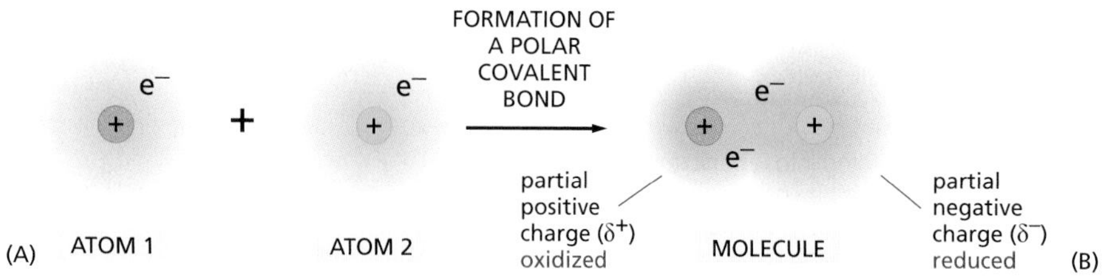

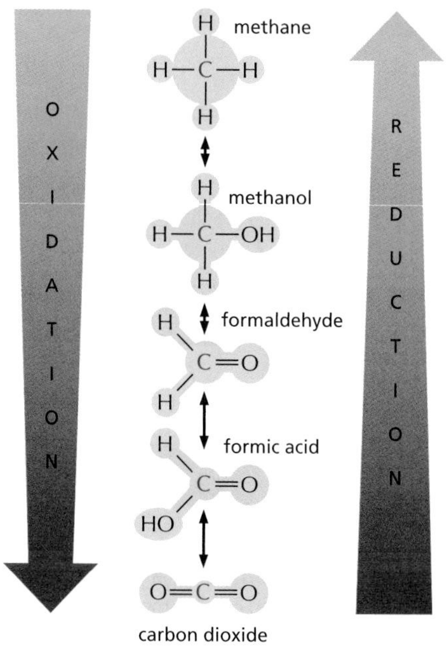

Enzymes Lower the Activation-Energy Barriers That Block Chemical Reactions

Consider the reaction

$$paper + O_2 \rightarrow smoke + ashes + heat + CO_2 + H_2O$$

Once ignited, the paper burns readily, releasing to the atmosphere both energy as heat and water and carbon dioxide as gases. The reaction is irreversible, since the smoke and ashes never spontaneously retrieve these entities from the heated atmosphere and reconstitute themselves into paper. When the paper burns, its chemical energy is dissipated as heat—not lost from the universe, since energy can never be created or destroyed, but irretrievably dispersed in the chaotic random thermal motions of molecules. At the same time, the atoms and molecules of the paper become dispersed and disordered. In the language of thermodynamics, there has been a loss of *free energy*; that is, of energy that can be harnessed to do work or drive chemical reactions. This loss reflects a reduction of orderliness in the way the energy and molecules were stored in the paper.

We shall discuss free energy in more detail shortly, but the general principle is clear enough intuitively: chemical reactions proceed spontaneously only in the direction that leads to a loss of free energy. In other words, the spontaneous direction for any reaction is the direction that goes "downhill," where a "downhill" reaction is one that is *energetically favorable*.

Although the most energetically favorable form of carbon under ordinary conditions is CO_2, and that of hydrogen is H_2O, a living organism does not disappear in a puff of smoke, and the paper book in your hands does not burst into flames. This is because the molecules both in the living organism and in the book are in a relatively stable state, and they cannot be changed to a state of lower energy without an input of energy: in other words, a molecule requires **activation energy**—a kick over an energy barrier—before it can undergo a chemical reaction that leaves it in a more stable state (**Figure 2–21**). In the case of a burning book, the activation energy can be provided by the heat of a lighted match. For the molecules in the watery solution inside a cell, the kick is delivered by an unusually energetic random collision with surrounding molecules—collisions that become more violent as the temperature is raised.

The chemistry in a living cell is tightly controlled, because the kick over energy barriers is greatly aided by a specialized class of proteins—the **enzymes**. Each enzyme binds tightly to one or more molecules, called **substrates**, and holds them in a way that greatly reduces the activation energy of a particular chemical reaction that the bound substrates can undergo. A substance that can lower the activation energy of a reaction is termed a **catalyst**; catalysts increase the rate of chemical reactions because they allow a much larger proportion of the random collisions with surrounding molecules to kick the substrates over the energy barrier, as illustrated in **Figure 2–22**. Enzymes are among the most effective catalysts

Figure 2–21 The important principle of activation energy. (A) Compound Y (a reactant) is in a relatively stable state, and energy is required to convert it to compound X (a product), even though X is at a lower overall energy level than Y. This conversion will not take place, therefore, unless compound Y can acquire enough activation energy (*energy a minus energy b*) from its surroundings to undergo the reaction that converts it into compound X. This energy may be provided by means of an unusually energetic collision with other molecules. For the reverse reaction, X → Y, the activation energy will be much larger (*energy a minus energy c*); this reaction will therefore occur much more rarely. Activation energies are always positive; note, however, that the total energy change for the energetically favorable reaction Y → X is *energy c minus energy b*, a negative number. (B) Energy barriers for specific reactions can be lowered by catalysts, as indicated by the line marked *d*. Enzymes are particularly effective catalysts because they greatly reduce the activation energy for the reactions they perform.

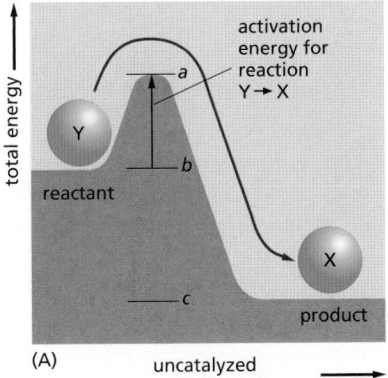

(A) uncatalyzed reaction pathway

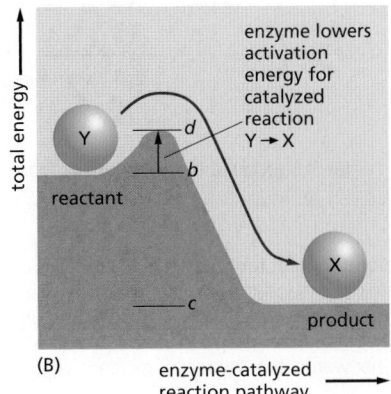

(B) enzyme-catalyzed reaction pathway

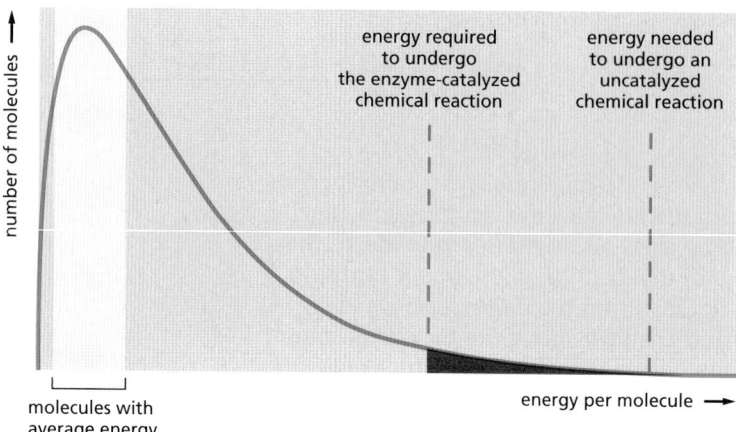

number of molecules

Figure 2–22 Lowering the activation energy greatly increases the probability of a reaction. At any given instant, a population of identical substrate molecules will have a range of energies, distributed as shown on the graph. The varying energies come from collisions with surrounding molecules, which make the substrate molecules jiggle, vibrate, and spin. For a molecule to undergo a chemical reaction, the energy of the molecule must exceed the activation-energy barrier for that reaction *(dashed lines)*. For most biological reactions, this almost never happens without enzyme catalysis. Even with enzyme catalysis, the substrate molecules must experience a particularly energetic collision to react *(red shaded area)*. Raising the temperature will also increase the number of molecules with sufficient energy to overcome the activation energy needed for a reaction; but in marked contrast to enzyme catalysis, this effect is nonselective, speeding up all reactions (Movie 2.2).

known: some are capable of speeding up reactions by factors of 10^{14} or more. Enzymes thereby allow reactions that would not otherwise occur to proceed rapidly at normal temperatures.

Enzymes Can Drive Substrate Molecules Along Specific Reaction Pathways

An enzyme cannot change the equilibrium point for a reaction. The reason is simple: when an enzyme (or any catalyst) lowers the activation energy for the reaction $Y \rightarrow X$, of necessity it also lowers the activation energy for the reaction $X \rightarrow Y$ by exactly the same amount (see Figure 2–21). The forward and backward reactions will therefore be accelerated by the same factor by an enzyme, and the equilibrium point for the reaction will be unchanged (**Figure 2–23**). Thus no matter how much an enzyme speeds up a reaction, it cannot change its direction.

Despite the above limitation, enzymes steer all of the reactions in cells through specific reaction paths. This is because enzymes are both highly selective and very precise, usually catalyzing only one particular reaction. In other words, each enzyme selectively lowers the activation energy of only one of the several possible chemical reactions that its bound substrate molecules could undergo. In this way, sets of enzymes can direct each of the many different molecules in a cell along a particular reaction pathway (**Figure 2–24**).

The success of living organisms is attributable to a cell's ability to make enzymes of many types, each with precisely specified properties. Each enzyme

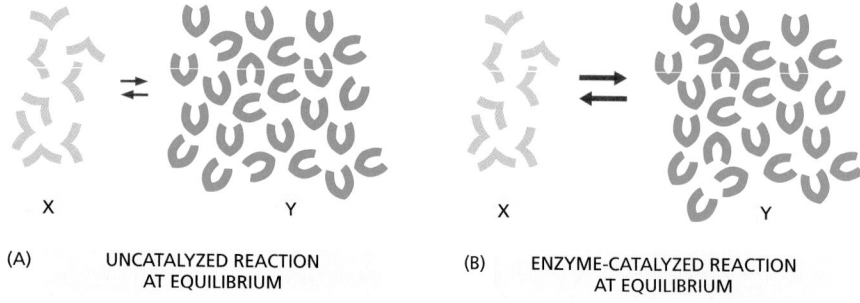

Figure 2–23 Enzymes cannot change the equilibrium point for reactions. Enzymes, like all catalysts, speed up the forward and backward rates of a reaction by the same factor. Therefore, for both the catalyzed and the uncatalyzed reactions shown here, the number of molecules undergoing the transition $X \rightarrow Y$ is equal to the number of molecules undergoing the transition $Y \rightarrow X$ when the ratio of Y molecules to X molecules is 3 to 1. In other words, the two reactions reach equilibrium at exactly the same point.

Figure 2–24 **Directing substrate molecules through a specific reaction pathway by enzyme catalysis.** A substrate molecule in a cell *(green ball)* is converted into a different molecule *(blue ball)* by means of a series of enzyme-catalyzed reactions. As indicated *(yellow box)*, several reactions are energetically favorable at each step, but only one is catalyzed by each enzyme. Sets of enzymes thereby determine the exact reaction pathway that is followed by each molecule inside the cell.

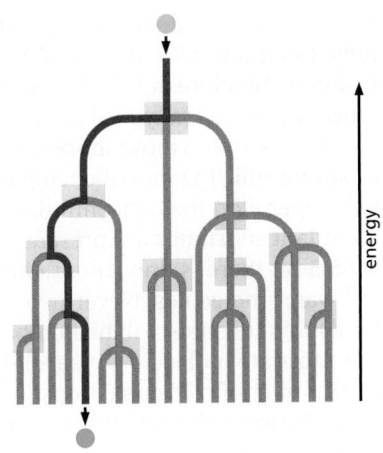

has a unique shape containing an *active site*, a pocket or groove in the enzyme into which only particular substrates will fit (**Figure 2–25**). Like all other catalysts, enzyme molecules themselves remain unchanged after participating in a reaction and therefore can function over and over again. In Chapter 3, we discuss further how enzymes work.

How Enzymes Find Their Substrates: The Enormous Rapidity of Molecular Motions

An enzyme will often catalyze the reaction of thousands of substrate molecules every second. This means that it must be able to bind a new substrate molecule in a fraction of a millisecond. But both enzymes and their substrates are present in relatively small numbers in a cell. How do they find each other so fast? Rapid binding is possible because the motions caused by heat energy are enormously fast at the molecular level. These molecular motions can be classified broadly into three kinds: (1) the movement of a molecule from one place to another (*translational motion*), (2) the rapid back-and-forth movement of covalently linked atoms with respect to one another (vibrations), and (3) rotations. All of these motions help to bring the surfaces of interacting molecules together.

The rates of molecular motions can be measured by a variety of spectroscopic techniques. A large globular protein is constantly tumbling, rotating about its axis about a million times per second. Molecules are also in constant translational motion, which causes them to explore the space inside the cell very efficiently by wandering through it—a process called **diffusion**. In this way, every molecule in a cell collides with a huge number of other molecules each second. As the molecules in a liquid collide and bounce off one another, an individual molecule moves first one way and then another, its path constituting a *random walk* (**Figure 2–26**). In such a walk, the average net distance that each molecule travels (as the "crow flies") from its starting point is proportional to the square root of the time involved: that is, if it takes a molecule 1 second on average to travel 1 µm, it takes 4 seconds to travel 2 µm, 100 seconds to travel 10 µm, and so on.

The inside of a cell is very crowded (**Figure 2–27**). Nevertheless, experiments in which fluorescent dyes and other labeled molecules are injected into cells show that small organic molecules diffuse through the watery gel of the cytosol nearly

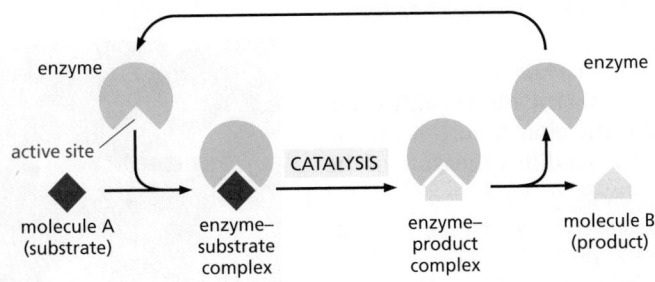

Figure 2–25 **How enzymes work.** Each enzyme has an active site to which one or more *substrate* molecules bind, forming an enzyme–substrate complex. A reaction occurs at the active site, producing an enzyme–product complex. The *product* is then released, allowing the enzyme to bind further substrate molecules.

as rapidly as they do through water. A small organic molecule, for example, takes only about one-fifth of a second on average to diffuse a distance of 10 μm. Diffusion is therefore an efficient way for small molecules to move the limited distances in the cell (a typical animal cell is 15 μm in diameter).

Since enzymes move more slowly than substrates in cells, we can think of them as sitting still. The rate of encounter of each enzyme molecule with its substrate will depend on the concentration of the substrate molecule. For example, some abundant substrates are present at a concentration of 0.5 mM. Since pure water is 55.5 M, there is only about one such substrate molecule in the cell for every 10^5 water molecules. Nevertheless, the active site on an enzyme molecule that binds this substrate will be bombarded by about 500,000 random collisions with the substrate molecule per second. (For a substrate concentration tenfold lower, the number of collisions drops to 50,000 per second, and so on.) A random collision between the active site of an enzyme and the matching surface of its substrate molecule often leads immediately to the formation of an enzyme–substrate complex. A reaction in which a covalent bond is broken or formed can now occur extremely rapidly. When one appreciates how quickly molecules move and react, the observed rates of enzymatic catalysis do not seem so amazing.

Two molecules that are held together by noncovalent bonds can also dissociate. The multiple weak noncovalent bonds that they form with each other will persist until random thermal motion causes the two molecules to separate. In general, the stronger the binding of the enzyme and substrate, the slower their rate of dissociation. In contrast, whenever two colliding molecules have poorly matching surfaces, they form few noncovalent bonds and the total energy of association will be negligible compared with that of thermal motion. In this case, the two molecules dissociate as rapidly as they come together, preventing incorrect and unwanted associations between mismatched molecules, such as between an enzyme and the wrong substrate.

The Free-Energy Change for a Reaction, ΔG, Determines Whether It Can Occur Spontaneously

Although enzymes speed up reactions, they cannot by themselves force energetically unfavorable reactions to occur. In terms of a water analogy, enzymes by themselves cannot make water run uphill. Cells, however, must do just that in order to grow and divide: they must build highly ordered and energy-rich molecules from small and simple ones. We shall see that this is done through enzymes that directly *couple* energetically favorable reactions, which release energy and produce heat, to energetically unfavorable reactions, which produce biological order.

What do cell biologists mean by the term "energetically favorable," and how can this be quantified? According to the second law of thermodynamics the universe tends toward maximum disorder (largest *entropy* or greatest probability). Thus, a chemical reaction can proceed spontaneously only if it results in a net increase in the disorder of the universe (see Figure 2–16). This disorder of the universe can be expressed most conveniently in terms of the *free energy* of a system, a concept we touched on earlier.

Free energy, **G**, is an expression of the *energy available to do work*—for example, the work of driving chemical reactions. The value of G is of interest only when a system undergoes a *change*, denoted Δ**G** (delta G). The change in G is critical because, as explained in **Panel 2–7** (pp. 102–103), ΔG is a direct measure of the

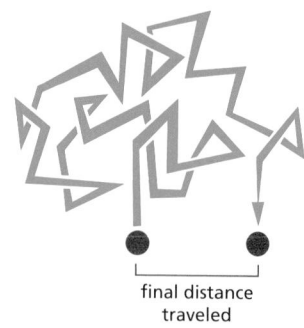

Figure 2–26 A random walk. Molecules in solution move in a random fashion as a result of the continual buffeting they receive in collisions with other molecules. This movement allows small molecules to diffuse rapidly from one part of the cell to another, as described in the text (Movie 2.3).

final distance traveled

Figure 2–27 **The structure of the cytoplasm.** The drawing is approximately to scale and emphasizes the crowding in the cytoplasm. Only the macromolecules are shown: RNAs are shown in *blue*, ribosomes in *green*, and proteins in *red*. Enzymes and other macromolecules diffuse relatively slowly in the cytoplasm, in part because they interact with many other macromolecules; small molecules, by contrast, diffuse nearly as rapidly as they do in water (**Movie 2.4**). (Adapted from D.S. Goodsell, *Trends Biochem. Sci.* 16:203–206, 1991. With permission from Elsevier.)

100 nm

amount of disorder created in the universe when a reaction takes place. *Energetically favorable reactions*, by definition, are those that decrease free energy; in other words, they have a *negative* ΔG and disorder the universe (**Figure 2–28**).

An example of an energetically favorable reaction on a macroscopic scale is the "reaction" by which a compressed spring relaxes to an expanded state, releasing its stored elastic energy as heat to its surroundings; an example on a microscopic scale is salt dissolving in water. Conversely, *energetically unfavorable reactions* with a *positive* ΔG—such as the joining of two amino acids to form a peptide bond—by themselves create order in the universe. Therefore, these reactions can take place only if they are coupled to a second reaction with a *negative* ΔG so large that the ΔG of the overall process is negative (**Figure 2–29**).

The Concentration of Reactants Influences the Free-Energy Change and a Reaction's Direction

As we have just described, a reaction $Y \leftrightarrow X$ will go in the direction $Y \rightarrow X$ when the associated free-energy change, ΔG, is negative, just as a tensed spring left to itself will relax and lose its stored energy to its surroundings as heat. For a chemical reaction, however, ΔG depends not only on the energy stored in each individual molecule, but also on the concentrations of the molecules in the reaction mixture. Remember that ΔG reflects the degree to which a reaction creates a more disordered—in other words, a more probable—state of the universe. Recalling our coin analogy, it is very likely that a coin will flip from a head to a tail orientation if a jiggling box contains 90 heads and 10 tails, but this is a less probable event if the box has 10 heads and 90 tails.

The same is true for a chemical reaction. For a reversible reaction $Y \leftrightarrow X$, a large excess of Y over X will tend to drive the reaction in the direction $Y \rightarrow X$. Therefore, as the ratio of Y to X increases, the ΔG becomes more negative for the transition $Y \rightarrow X$ (and more positive for the transition $X \rightarrow Y$).

The amount of concentration difference that is needed to compensate for a given decrease in chemical-bond energy (and accompanying heat release) is not intuitively obvious. In the late nineteenth century, the relationship was determined through a thermodynamic analysis that makes it possible to separate the concentration-dependent and the concentration-independent parts of the free-energy change, as we describe next.

The Standard Free-Energy Change, $\Delta G°$, Makes It Possible to Compare the Energetics of Different Reactions

Because ΔG depends on the concentrations of the molecules in the reaction mixture at any given time, it is not a particularly useful value for comparing the relative energies of different types of reactions. To place reactions on a comparable basis, we need to turn to the **standard free-energy change** of a reaction, $\Delta G°$. The $\Delta G°$ is the change in free energy under a standard condition, defined as that where the concentrations of all the reactants are set to the same fixed value of 1 mole/liter. Defined in this way, $\Delta G°$ depends only on the intrinsic characters of the reacting molecules.

For the simple reaction $Y \rightarrow X$ at 37°C, $\Delta G°$ is related to ΔG as follows:

$$\Delta G = \Delta G° + RT \ln \frac{[X]}{[Y]}$$

where ΔG is in kilojoules per mole, [Y] and [X] denote the concentrations of Y and X in moles/liter, ln is the *natural logarithm,* and RT is the product of the gas constant, R, and the absolute temperature, T. At 37°C, $RT = 2.58$ J mole^{-1}. (A mole is 6×10^{23} molecules of a substance.)

A large body of thermodynamic data has been collected that has made it possible to determine the standard free-energy change, $\Delta G°$, for the important metabolic reactions of a cell. Given these $\Delta G°$ values, combined with additional information about metabolite concentrations and reaction pathways, it is possible to quantitatively predict the course of most biological reactions.

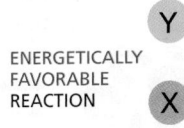

ENERGETICALLY FAVORABLE REACTION

The free energy of Y is greater than the free energy of X. Therefore $\Delta G < 0$, and the disorder of the universe increases during the reaction $Y \rightarrow X$.

this reaction can occur spontaneously

ENERGETICALLY UNFAVORABLE REACTION

If the reaction $X \rightarrow Y$ occurred, ΔG would be > 0, and the universe would become more ordered.

this reaction can occur only if it is coupled to a second, energetically favorable reaction

Figure 2–28 The distinction between energetically favorable and energetically unfavorable reactions.

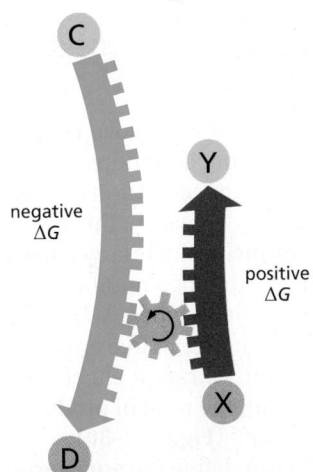

negative ΔG

positive ΔG

the energetically unfavorable reaction $X \rightarrow Y$ is driven by the energetically favorable reaction $C \rightarrow D$, because the net free-energy change for the pair of coupled reactions is less than zero

Figure 2–29 How reaction coupling is used to drive energetically unfavorable reactions.

FOR THE ENERGETICALLY FAVORABLE REACTION Y → X,

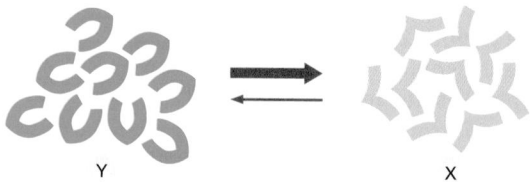

Y   X

when X and Y are at equal concentrations, [Y] = [X], the formation of X
is energetically favored. In other words, the ΔG of Y → X is negative and
the ΔG of X → Y is positive. But because of thermal bombardments,
there will always be some X converting to Y.

THUS, FOR EACH INDIVIDUAL MOLECULE,

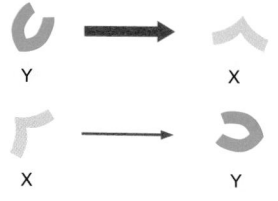

Y   X

conversion of
Y to X will
occur often.

X   Y

Conversion of X to Y
will occur less often
than the transition
Y → X, because it
requires a more
energetic collision.

Therefore the ratio of X to Y
molecules will increase with time

EVENTUALLY,   there will be a large enough excess of X over Y to just
compensate for the slow rate of X → Y, such that the number of Y molecules
being converted to X molecules each second is exactly equal to the number
of X molecules being converted to Y molecules each second. At this point,
the reaction will be at equilibrium.

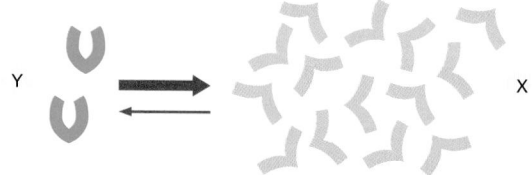

Y   X

AT EQUILIBRIUM,   there is no net change in the ratio of Y to X, and the
ΔG for both forward and backward reactions is zero.

Figure 2–30 **Chemical equilibrium.** When
a reaction reaches equilibrium, the forward
and backward fluxes of reacting molecules
are equal and opposite.

The Equilibrium Constant and ΔG° Are Readily Derived from Each Other

Inspection of the above equation reveals that the ΔG equals the value of ΔG°
when the concentrations of Y and X are equal. But as any favorable reaction proceeds, the concentrations of the products will increase as the concentration of the
substrates decreases. This change in relative concentrations will cause [X]/[Y] to
become increasingly large, making the initially favorable ΔG less and less negative
(the logarithm of a number x is positive for $x > 1$, negative for $x < 1$, and zero for x
$=1$). Eventually, when $ΔG = 0$, a chemical **equilibrium** will be attained; here there
is no net change in free energy to drive the reaction in either direction, inasmuch
as the concentration effect just balances the push given to the reaction by ΔG°.
As a result, the ratio of product to substrate reaches a constant value at chemical
equilibrium (**Figure 2–30**).

We can define the **equilibrium constant, K,** for the reaction Y → X as

$$K = \frac{[X]}{[Y]}$$

where [X] is the concentration of the product and [Y] is the concentration of the
reactant at equilibrium. Remembering that $ΔG = ΔG° + RT \ln [X]/[Y]$, and that
$ΔG = 0$ at equilibrium, we see that

$$ΔG° = -RT \ln \frac{[X]}{[Y]} = -RT \ln K$$

At 37°C, where $RT = 2.58$, the equilibrium equation is therefore:

$$ΔG° = -2.58 \ln K$$

Converting this equation from the natural logarithm (ln) to the more commonly used base 10 logarithm (log), we get

$$\Delta G° = -5.94 \log K$$

The above equation reveals how the equilibrium ratio of X to Y (expressed as the equilibrium constant, K) depends on the intrinsic character of the molecules, (as expressed in the value of $\Delta G°$ in kilojoules per mole). Note that for every 5.94 kJ/mole difference in free energy at 37°C, the equilibrium constant changes by a factor of 10 (Table 2–2). Thus, the more energetically favorable a reaction, the more product will accumulate if the reaction proceeds to equilibrium.

More generally, for a reaction that has multiple reactants and products, such as A + B → C + D,

$$K = \frac{[C][D]}{[A][B]}$$

The concentrations of the two reactants and the two products are multiplied because the rate of the forward reaction depends on the collision of A and B and the rate of the backward reaction depends on the collision of C and D. Thus, at 37°C,

$$\Delta G° = -5.94 \log \frac{[C][D]}{[A][B]}$$

where $\Delta G°$ is in kilojoules per mole, and [A], [B], [C], and [D] denote the concentrations of the reactants and products in moles/liter.

The Free-Energy Changes of Coupled Reactions Are Additive

We have pointed out that unfavorable reactions can be coupled to favorable ones to drive the unfavorable ones forward (see Figure 2–29). In thermodynamic terms, this is possible because the overall free-energy change for a set of coupled reactions is the sum of the free-energy changes in each of its component steps. Consider, as a simple example, two sequential reactions

$$X \rightarrow Y \quad \text{and} \quad Y \rightarrow Z$$

whose $\Delta G°$ values are +5 and –13 kJ/mole, respectively. If these two reactions occur sequentially, the $\Delta G°$ for the coupled reaction will be –8 kJ/mole. This means that, with appropriate conditions, the unfavorable reaction X → Y can be driven by the favorable reaction Y → Z, provided that this second reaction follows the first. For example, several of the reactions in the long pathway that converts sugars into CO_2 and H_2O have positive $\Delta G°$ values. But the pathway nevertheless proceeds because the total $\Delta G°$ for the series of sequential reactions has a large negative value.

Forming a sequential pathway is not adequate for many purposes. Often the desired pathway is simply X → Y, without further conversion of Y to some other product. Fortunately, there are other more general ways of using enzymes to couple reactions together. These often involve the activated carrier molecules that we discuss next.

Activated Carrier Molecules Are Essential for Biosynthesis

The energy released by the oxidation of food molecules must be stored temporarily before it can be channeled into the construction of the many other molecules needed by the cell. In most cases, the energy is stored as chemical-bond energy in a small set of activated "carrier molecules," which contain one or more energy-rich covalent bonds. These molecules diffuse rapidly throughout the cell and thereby carry their bond energy from sites of energy generation to the sites where the energy will be used for biosynthesis and other cell activities (Figure 2–31).

The **activated carriers** store energy in an easily exchangeable form, either as a readily transferable chemical group or as electrons held at a high energy level, and they can serve a dual role as a source of both energy and chemical groups in biosynthetic reactions. For historical reasons, these molecules are also sometimes referred to as *coenzymes*. The most important of the activated carrier molecules

TABLE 2–2 Relationship Between the Standard Free-Energy Change, $\Delta G°$, and the Equilibrium Constant	
Equilibrium constant $\frac{[X]}{[Y]} = K$	Free energy of X minus free energy of Y [kJ/mole (kcal/mole)]
10^5	–29.7 (–7.1)
10^4	–23.8 (–5.7)
10^3	–17.8 (–4.3)
10^2	–11.9 (–2.8)
10^1	–5.9 (–1.4)
1	0 (0)
10^{-1}	5.9 (1.4)
10^{-2}	11.9 (2.8)
10^{-3}	17.8 (4.3)
10^{-4}	23.8 (5.7)
10^{-5}	29.7 (7.1)

Values of the equilibrium constant were calculated for the simple chemical reaction Y ↔ X using the equation given in the text. The $\Delta G°$ given here is in kilojoules per mole at 37°C, with kilocalories per mole in parentheses. One kilojoule (kJ) is equal to 0.239 kilocalories (kcal) (1 kcal = 4.18 kJ). As explained in the text, $\Delta G°$ represents the free-energy difference under standard conditions (where all components are present at a concentration of 1.0 mole/liter). From this table, we see that if there is a favorable standard free-energy change ($\Delta G°$) of –17.8 kJ/mole (–4.3 kcal/mole) for the transition Y → X, there will be 1000 times more molecules in state X than in state Y at equilibrium ($K = 1000$).

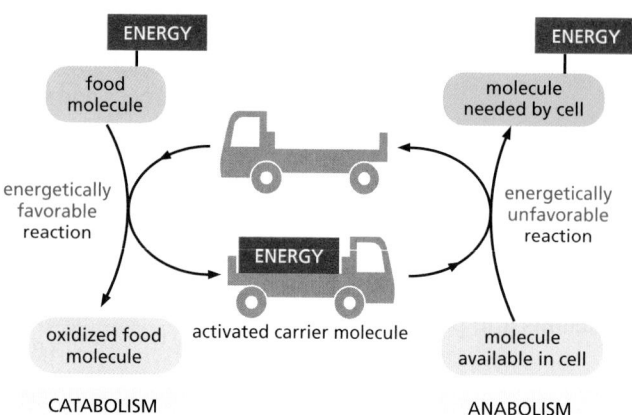

CATABOLISM   ANABOLISM

Figure 2–31 Energy transfer and the role of activated carriers in metabolism. By serving as energy shuttles, activated carrier molecules perform their function as go-betweens that link the breakdown of food molecules and the release of energy (*catabolism*) to the energy-requiring biosynthesis of small and large organic molecules (*anabolism*).

are ATP and two molecules that are closely related to each other, NADH and NADPH. Cells use such activated carrier molecules like money to pay for reactions that otherwise could not take place.

The Formation of an Activated Carrier Is Coupled to an Energetically Favorable Reaction

Coupling mechanisms require enzymes and are fundamental to all the energy transactions of the cell. The nature of a **coupled reaction** is illustrated by a mechanical analogy in **Figure 2–32**, in which an energetically favorable chemical reaction is represented by rocks falling from a cliff. The energy of falling rocks would normally be entirely wasted in the form of heat generated by friction when the rocks hit the ground (see the falling-brick diagram in Figure 2–17). By careful design, however, part of this energy could be used instead to drive a paddle wheel that lifts a bucket of water (Figure 2–32B). Because the rocks can now reach the ground only after moving the paddle wheel, we say that the energetically favorable reaction of rock falling has been directly *coupled* to the energetically unfavorable reaction of lifting the bucket of water. Note that because part of the energy is used to do work in Figure 2–32B, the rocks hit the ground with less velocity than in Figure 2–32A, and correspondingly less energy is dissipated as heat.

Similar processes occur in cells, where enzymes play the role of the paddle wheel. By mechanisms that we discuss later in this chapter, enzymes couple an

Figure 2–32 A mechanical model illustrating the principle of coupled chemical reactions. The spontaneous reaction shown in (A) could serve as an analogy for the direct oxidation of glucose to CO_2 and H_2O, which produces heat only. In (B), the same reaction is coupled to a second reaction; this second reaction is analogous to the synthesis of activated carrier molecules. The energy produced in (B) is in a more useful form than in (A) and can be used to drive a variety of otherwise energetically unfavorable reactions (C).

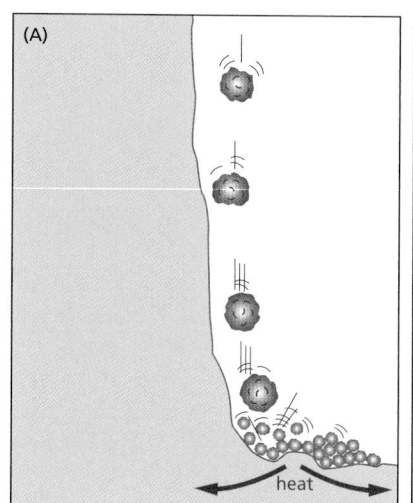

kinetic energy of falling rocks is transformed into heat energy only

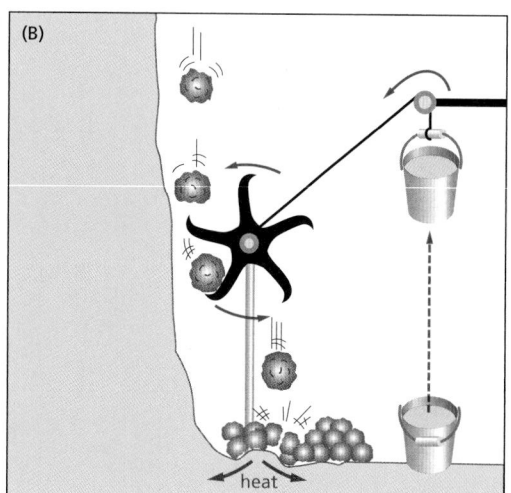

part of the kinetic energy is used to lift a bucket of water, and a correspondingly smaller amount is transformed into heat

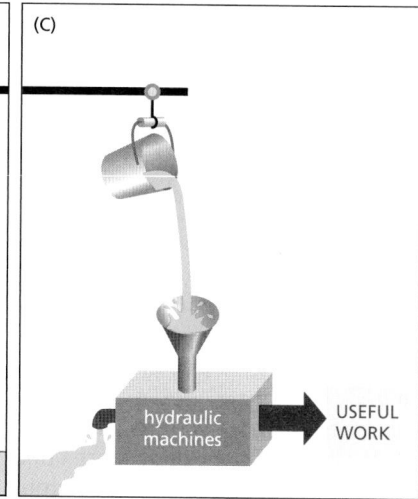

the potential kinetic energy stored in the raised bucket of water can be used to drive hydraulic machines that carry out a variety of useful tasks

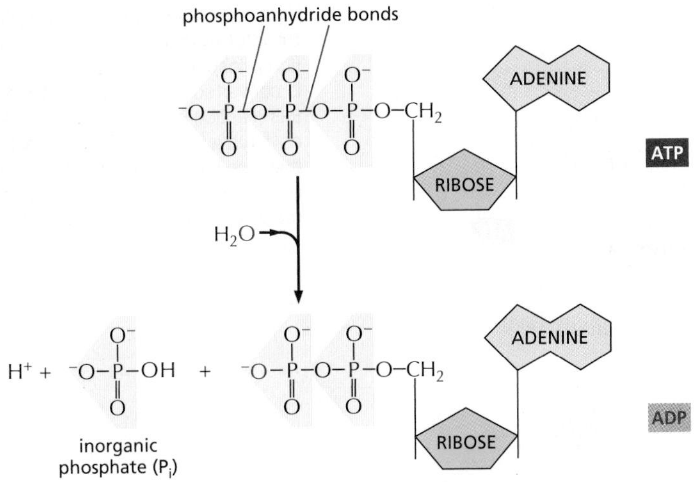

phosphoanhydride bonds

Figure 2–33 **The hydrolysis of ATP to ADP and inorganic phosphate.** The two outermost phosphates in ATP are held to the rest of the molecule by high-energy phosphoanhydride bonds and are readily transferred. As indicated, water can be added to ATP to form ADP and inorganic phosphate (P_i). Hydrolysis of the terminal phosphate of ATP yields between 46 and 54 kJ/mole of usable energy, depending on the intracellular conditions. The large negative ΔG of this reaction arises from several factors: release of the terminal phosphate group removes an unfavorable repulsion between adjacent negative charges, and the inorganic phosphate ion (P_i) released is stabilized by resonance and by favorable hydrogen-bond formation with water.

energetically favorable reaction, such as the oxidation of foodstuffs, to an energetically unfavorable reaction, such as the generation of an activated carrier molecule. In this example, the amount of heat released by the oxidation reaction is reduced by exactly the amount of energy stored in the energy-rich covalent bonds of the activated carrier molecule. And the activated carrier molecule picks up a packet of energy of a size sufficient to power a chemical reaction elsewhere in the cell.

ATP Is the Most Widely Used Activated Carrier Molecule

The most important and versatile of the activated carriers in cells is **ATP** (adenosine triphosphate). Just as the energy stored in the raised bucket of water in Figure 2–32B can drive a wide variety of hydraulic machines, ATP is a convenient and versatile store, or currency, of energy used to drive a variety of chemical reactions in cells. ATP is synthesized in an energetically unfavorable phosphorylation reaction in which a phosphate group is added to **ADP** (adenosine diphosphate). When required, ATP gives up its energy packet through its energetically favorable hydrolysis to ADP and inorganic phosphate (**Figure 2–33**). The regenerated ADP is then available to be used for another round of the phosphorylation reaction that forms ATP.

The energetically favorable reaction of ATP hydrolysis is coupled to many otherwise unfavorable reactions through which other molecules are synthesized. Many of these coupled reactions involve the transfer of the terminal phosphate in ATP to another molecule, as illustrated by the phosphorylation reaction in **Figure 2–34**.

As the most abundant activated carrier in cells, ATP is the principal energy currency. To give just two examples, it supplies energy for many of the pumps that transport substances into and out of the cell (discussed in Chapter 11), and it powers the molecular motors that enable muscle cells to contract and nerve cells to transport materials from one end of their long axons to another (discussed in Chapter 16).

Energy Stored in ATP Is Often Harnessed to Join Two Molecules Together

We have previously discussed one way in which an energetically favorable reaction can be coupled to an energetically unfavorable reaction, X → Y, so as to enable it to occur. In that scheme, a second enzyme catalyzes the energetically favorable reaction Y → Z, pulling all of the X to Y in the process. But when the required product is Y and not Z, this mechanism is not useful.

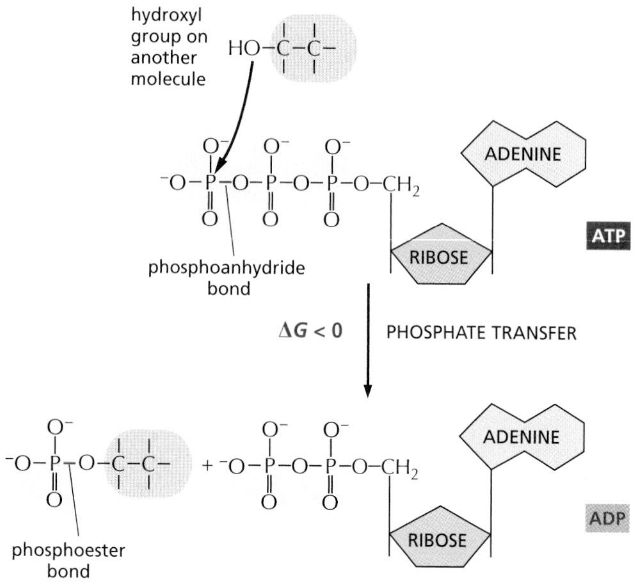

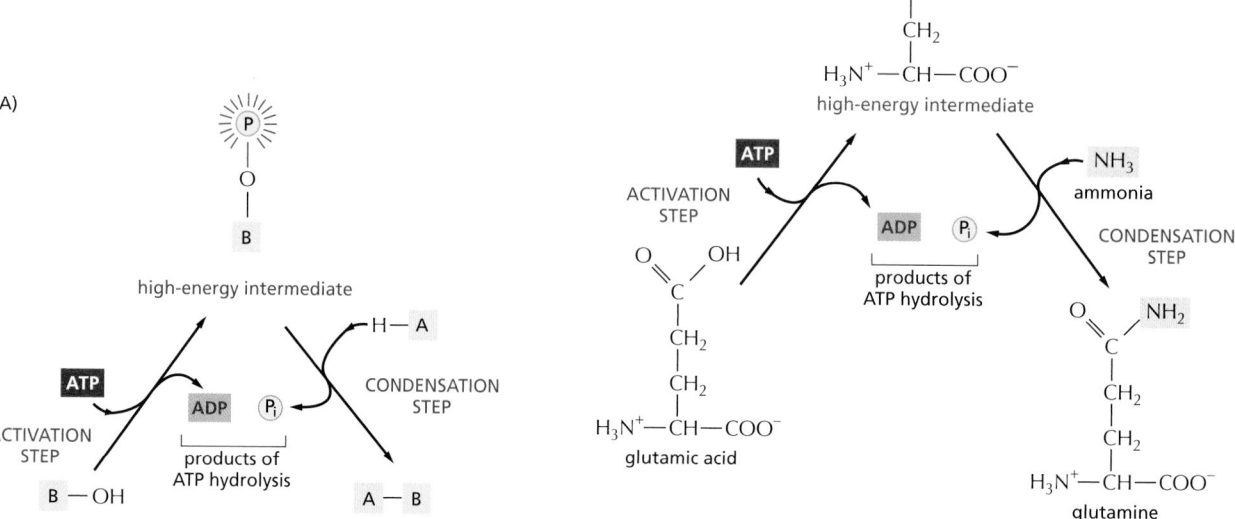

Figure 2–34 An example of a phosphate transfer reaction. Because an energy-rich phosphoanhydride bond in ATP is converted to a phosphoester bond, this reaction is energetically favorable, having a large negative ΔG. Reactions of this type are involved in the synthesis of phospholipids and in the initial steps of reactions that catabolize sugars.

A typical biosynthetic reaction is one in which two molecules, A and B, are joined together to produce A–B in the energetically unfavorable *condensation* reaction

$$A\text{-}H + B\text{-}OH \rightarrow A\text{-}B + H_2O$$

There is an indirect pathway that allows A–H and B–OH to form A–B, in which a coupling to ATP hydrolysis makes the reaction go. Here, energy from ATP hydrolysis is first used to convert B–OH to a higher-energy intermediate compound, which then reacts directly with A–H to give A–B. The simplest possible mechanism involves the transfer of a phosphate from ATP to B–OH to make $B\text{-}O\text{-}PO_3$, in which case the reaction pathway contains only two steps:

1. $B\text{-}OH + ATP \rightarrow B\text{-}O\text{-}PO_3 + ADP$
2. $A\text{-}H + B\text{-}O\text{-}PO_3 \rightarrow A\text{-}B + P_i$
   Net result: $B\text{-}OH + ATP + A\text{-}H \rightarrow A\text{-}B + ADP + P_i$

The condensation reaction, which by itself is energetically unfavorable, is forced to occur by being directly coupled to ATP hydrolysis in an enzyme-catalyzed reaction pathway (**Figure 2–35**A).

A biosynthetic reaction of exactly this type synthesizes the amino acid glutamine (Figure 2–35B). We will see shortly that similar (but more complex) mechanisms are also used to produce nearly all of the large molecules of the cell.

Figure 2–35 An example of an energetically unfavorable biosynthetic reaction driven by ATP hydrolysis. (A) Schematic illustration of the formation of A–B in the condensation reaction described in the text. (B) The biosynthesis of the common amino acid glutamine from glutamic acid and ammonia. Glutamic acid is first converted to a high-energy phosphorylated intermediate (corresponding to the compound $B\text{-}O\text{-}PO_3$ described in the text), which then reacts with ammonia (corresponding to A–H) to form glutamine. In this example, both steps occur on the surface of the same enzyme, *glutamine synthetase*. The high-energy bonds are shaded *red*; here, as elsewhere throughout the book, the symbol $P_i = HPO_4^{2-}$, and a yellow "circled P" = PO_3^{2-}.

NADH and NADPH Are Important Electron Carriers

Other important activated carrier molecules participate in oxidation–reduction reactions and are commonly part of coupled reactions in cells. These activated carriers are specialized to carry electrons held at a high energy level (sometimes called "high-energy" electrons) and hydrogen atoms. The most important of these electron carriers are NAD^+ (nicotinamide adenine dinucleotide) and the closely related molecule $NADP^+$ (nicotinamide adenine dinucleotide phosphate). Each picks up a "packet of energy" corresponding to two electrons plus a proton (H^+), and they are thereby converted to **NADH** (*reduced* nicotinamide adenine dinucleotide) and **NADPH** (*reduced* nicotinamide adenine dinucleotide phosphate), respectively (**Figure 2–36**). These molecules can therefore be regarded as carriers of hydride ions (the H^+ plus two electrons, or H^-).

Like ATP, NADPH is an activated carrier that participates in many important biosynthetic reactions that would otherwise be energetically unfavorable. The NADPH is produced according to the general scheme shown in Figure 2–36A. During a special set of energy-yielding catabolic reactions, two hydrogen atoms are removed from a substrate molecule. Both electrons but just one proton (that is, a hydride ion, H^-) are added to the nicotinamide ring of $NADP^+$ to form NADPH; the second proton (H^+) is released into solution. This is a typical oxidation–reduction reaction, in which the substrate is oxidized and $NADP^+$ is reduced.

NADPH readily gives up the hydride ion it carries in a subsequent oxidation–reduction reaction, because the nicotinamide ring can achieve a more stable arrangement of electrons without it. In this subsequent reaction, which

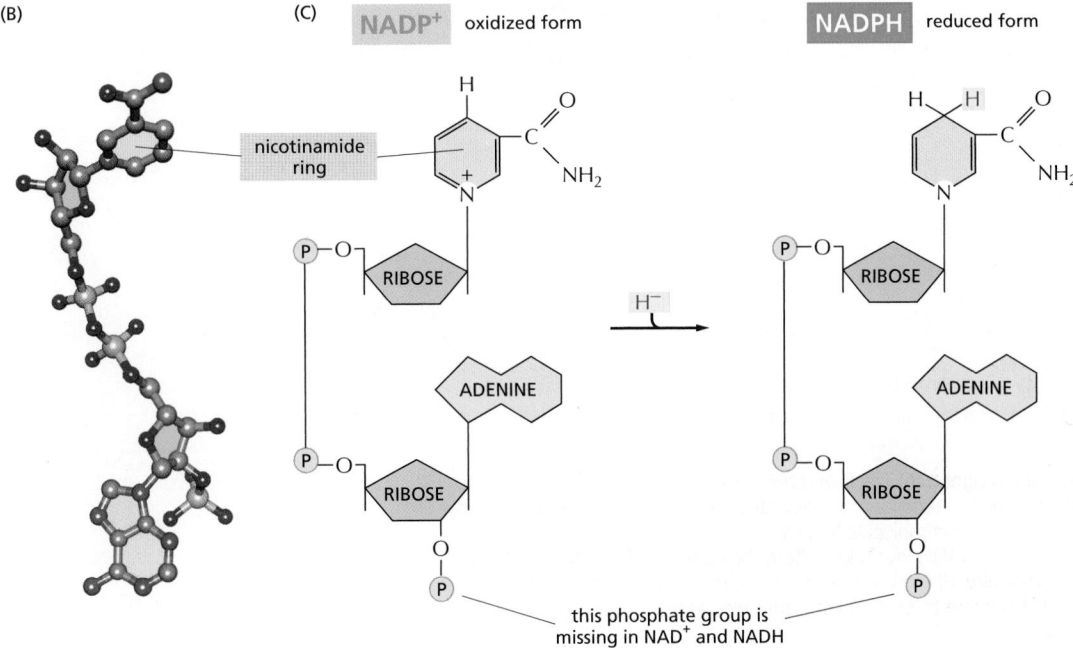

(A)

oxidation of
molecule 1

reduction of
molecule 2

Figure 2–36 NADPH, an important carrier of electrons.
(A) NADPH is produced in reactions of the general type shown on the left, in which two hydrogen atoms are removed from a substrate. The oxidized form of the carrier molecule, $NADP^+$, receives one hydrogen atom plus an electron (a hydride ion); the proton (H^+) from the other H atom is released into solution. Because NADPH holds its hydride ion in a high-energy linkage, the hydride ion can easily be transferred to other molecules, as shown on the right. (B) and (C) The structures of $NADP^+$ and NADPH. The part of the $NADP^+$ molecule known as the nicotinamide ring accepts the hydride ion, H^-, forming NADPH. The molecules NAD^+ and NADH are identical in structure to $NADP^+$ and NADPH, respectively, except that they lack the indicated phosphate group.

(B)

(C) **NADP⁺** oxidized form   **NADPH** reduced form

nicotinamide
ring

RIBOSE   RIBOSE

ADENINE   ADENINE

RIBOSE   RIBOSE

this phosphate group is
missing in NAD⁺ and NADH

regenerates NADP⁺, it is the NADPH that is oxidized and the substrate that is reduced. The NADPH is an effective donor of its hydride ion to other molecules for the same reason that ATP readily transfers a phosphate: in both cases the transfer is accompanied by a large negative free-energy change. One example of the use of NADPH in biosynthesis is shown in **Figure 2–37**.

The extra phosphate group on NADPH has no effect on the electron-transfer properties of NADPH compared with NADH, being far away from the region involved in electron transfer (see Figure 2–36C). It does, however, give a molecule of NADPH a slightly different shape from that of NADH, making it possible for NADPH and NADH to bind as substrates to completely different sets of enzymes. Thus, the two types of carriers are used to transfer electrons (or hydride ions) between two different sets of molecules.

Why should there be this division of labor? The answer lies in the need to regulate two sets of electron-transfer reactions independently. NADPH operates chiefly with enzymes that catalyze anabolic reactions, supplying the high-energy electrons needed to synthesize energy-rich biological molecules. NADH, by contrast, has a special role as an intermediate in the catabolic system of reactions that generate ATP through the oxidation of food molecules, as we will discuss shortly. The genesis of NADH from NAD⁺, and of NADPH from NADP⁺, occur by different pathways and are independently regulated, so that the cell can adjust the supply of electrons for these two contrasting purposes. Inside the cell the ratio of NAD⁺ to NADH is kept high, whereas the ratio of NADP⁺ to NADPH is kept low. This provides plenty of NAD⁺ to act as an oxidizing agent and plenty of NADPH to act as a reducing agent (Figure 2–37B)—as required for their special roles in catabolism and anabolism, respectively.

There Are Many Other Activated Carrier Molecules in Cells

Other activated carriers also pick up and carry a chemical group in an easily transferred, high-energy linkage. For example, coenzyme A carries a readily transferable

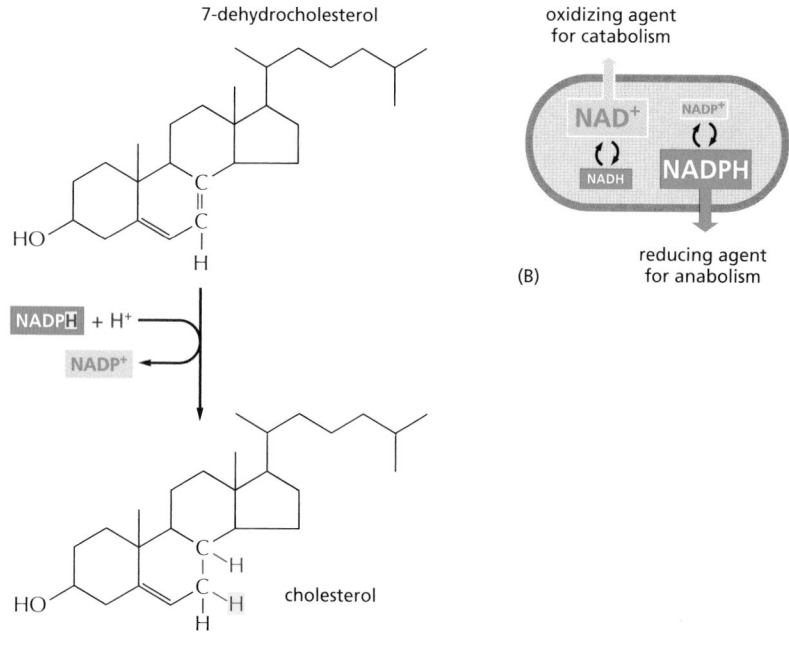

(B)

(A)

Figure 2–37 NADPH as a reducing agent. (A) The final stage in the biosynthetic route leading to cholesterol. As in many other biosynthetic reactions, the reduction of the C=C bond is achieved by the transfer of a hydride ion from the carrier molecule NADPH, plus a proton (H⁺) from the solution. (B) Keeping NADPH levels high and NADH levels low alters their affinities for electrons (see Panel 14–1, p. 765). This causes NADPH to be a much stronger electron donor (reducing agent) than NADH, and NAD⁺ therefore to be a much better electron acceptor (oxidizing agent) than NADP⁺, as indicated.

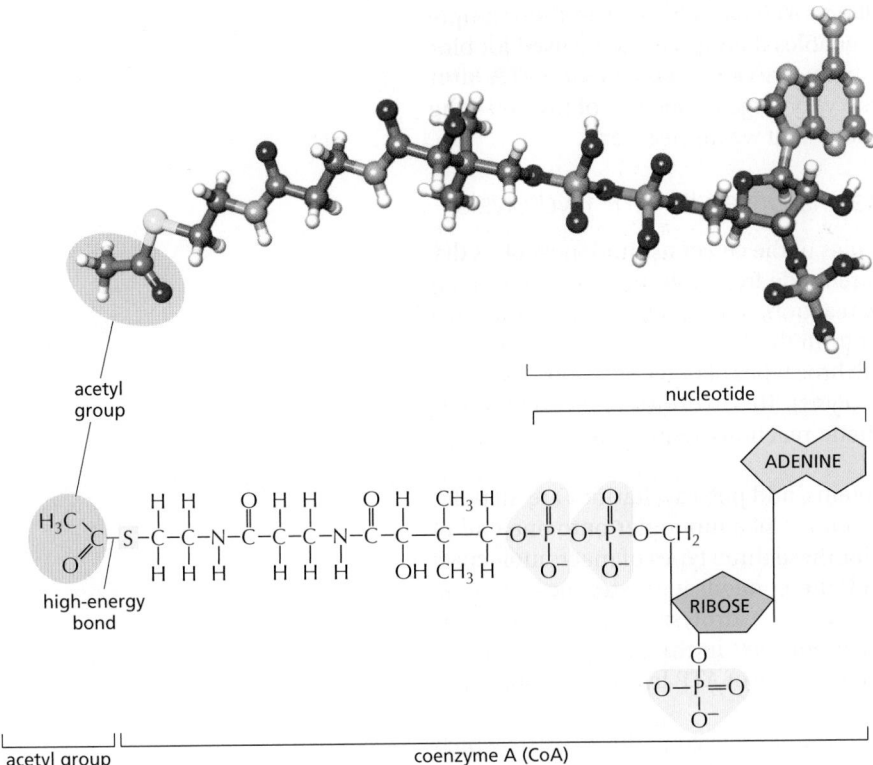

acetyl
group

nucleotide

ADENINE

H₃C

high-energy
bond

RIBOSE

acetyl group

coenzyme A (CoA)

Figure 2–38 **The structure of the important activated carrier molecule acetyl CoA.** A ball-and-stick model is shown above the structure. The sulfur atom *(yellow)* forms a thioester bond to acetate. Because this is a high-energy linkage, releasing a large amount of free energy when it is hydrolyzed, the acetate molecule can be readily transferred to other molecules.

acetyl group in a thioester linkage, and in this activated form is known as **acetyl CoA** (acetyl coenzyme A). Acetyl CoA (**Figure 2–38**) is used to add two carbon units in the biosynthesis of larger molecules.

In acetyl CoA, as in other carrier molecules, the transferable group makes up only a small part of the molecule. The rest consists of a large organic portion that serves as a convenient "handle," facilitating the recognition of the carrier molecule by specific enzymes. As with acetyl CoA, this handle portion very often contains a nucleotide (usually adenosine diphosphate), a curious fact that may be a relic from an early stage of evolution. It is currently thought that the main catalysts for early life-forms—before DNA or proteins—were RNA molecules (or their close relatives), as described in Chapter 6. It is tempting to speculate that many of the carrier molecules that we find today originated in this earlier RNA world, where their nucleotide portions could have been useful for binding them to RNA enzymes (ribozymes).

Thus, ATP transfers phosphate, NADPH transfers electrons and hydrogen, and acetyl CoA transfers two-carbon acetyl groups. **FADH₂** (reduced flavin adenine dinucleotide) is used like NADH in electron and proton transfers (**Figure 2–39**). The reactions of other activated carrier molecules involve the transfer of a methyl, carboxyl, or glucose group for biosyntheses (**Table 2–3**). These activated carriers

(A) FADH₂

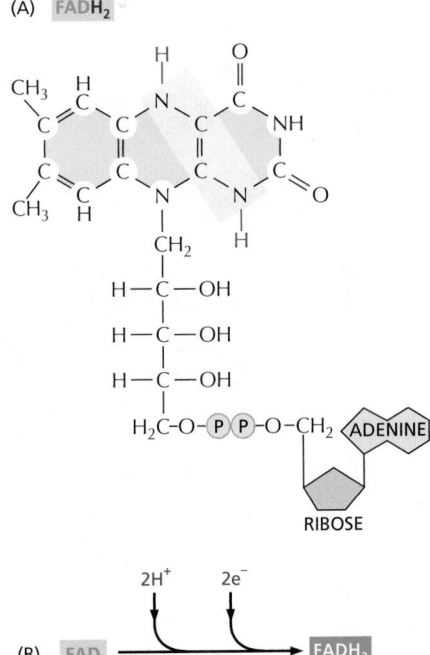

Figure 2–39 **FADH₂ is a carrier of hydrogens and high-energy electrons, like NADH and NADPH.** (A) Structure of FADH₂, with its hydrogen-carrying atoms highlighted in *yellow.* (B) The formation of FADH₂ from FAD.

TABLE 2–3 Some Activated Carrier Molecules Widely Used in Metabolism	
Activated carrier	Group carried in high-energy linkage
ATP	Phosphate
NADH, NADPH, FADH₂	Electrons and hydrogens
Acetyl CoA	Acetyl group
Carboxylated biotin	Carboxyl group
S-Adenosylmethionine	Methyl group
Uridine diphosphate glucose	Glucose

are generated in reactions that are coupled to ATP hydrolysis, as in the example in **Figure 2–40**. Therefore, the energy that enables their groups to be used for biosynthesis ultimately comes from the catabolic reactions that generate ATP. Similar processes occur in the synthesis of the very large molecules of the cell—the nucleic acids, proteins, and polysaccharides—that we discuss next.

The Synthesis of Biological Polymers Is Driven by ATP Hydrolysis

As discussed previously, the macromolecules of the cell constitute most of its dry mass (see Figure 2–7). These molecules are made from subunits (or monomers) that are linked together in a *condensation* reaction, in which the constituents of a water molecule (OH plus H) are removed from the two reactants. Consequently, the reverse reaction—the breakdown of all three types of polymers—occurs by the enzyme-catalyzed addition of water (*hydrolysis*). This hydrolysis reaction is energetically favorable, whereas the biosynthetic reactions require an energy input (see Figure 2–9).

The nucleic acids (DNA and RNA), proteins, and polysaccharides are all polymers that are produced by the repeated addition of a monomer onto one end of a growing chain. The synthesis reactions for these three types of macromolecules are outlined in **Figure 2–41**. As indicated, the condensation step in each case depends on energy from nucleoside triphosphate hydrolysis. And yet, except for the nucleic acids, there are no phosphate groups left in the final product molecules. How are the reactions that release the energy of ATP hydrolysis coupled to polymer synthesis?

For each type of macromolecule, an enzyme-catalyzed pathway exists which resembles that discussed previously for the synthesis of the amino acid glutamine (see Figure 2–35). The principle is exactly the same, in that the –OH group that will

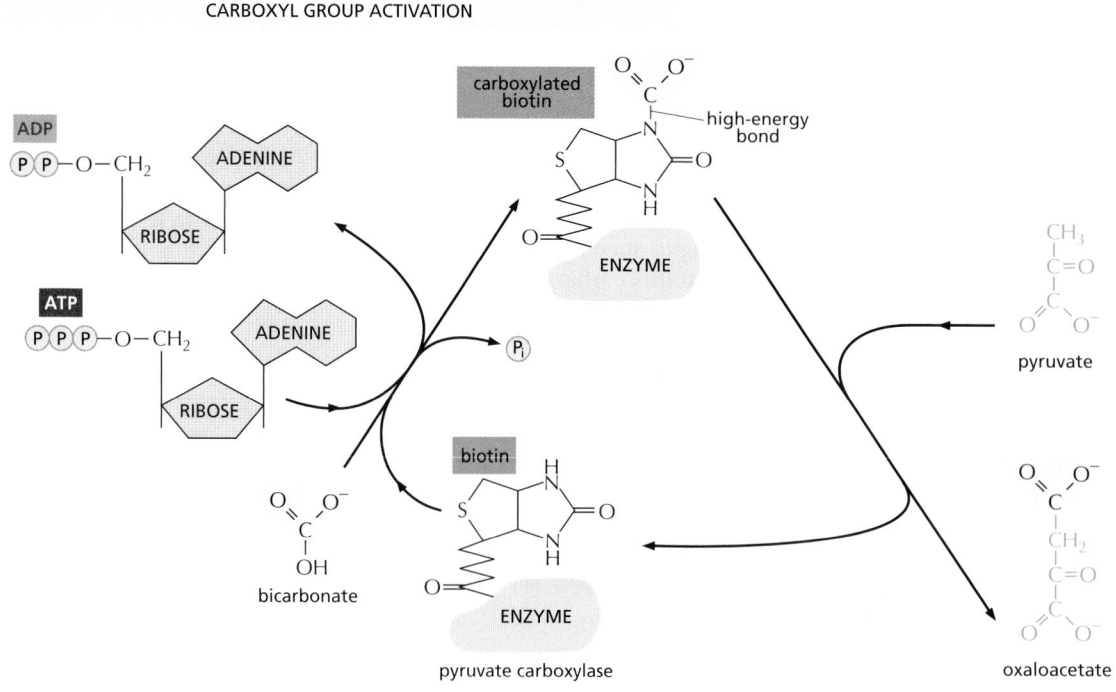

Figure 2–40 A carboxyl group-transfer reaction using an activated carrier molecule. Carboxylated biotin is used by the enzyme *pyruvate carboxylase* to transfer a carboxyl group in the production of oxaloacetate, a molecule needed for the citric acid cycle. The acceptor molecule for this group-transfer reaction is pyruvate. Other enzymes use biotin, a B-complex vitamin, to transfer carboxyl groups to other acceptor molecules. Note that synthesis of carboxylated biotin requires energy that is derived from ATP—a general feature of many activated carriers.

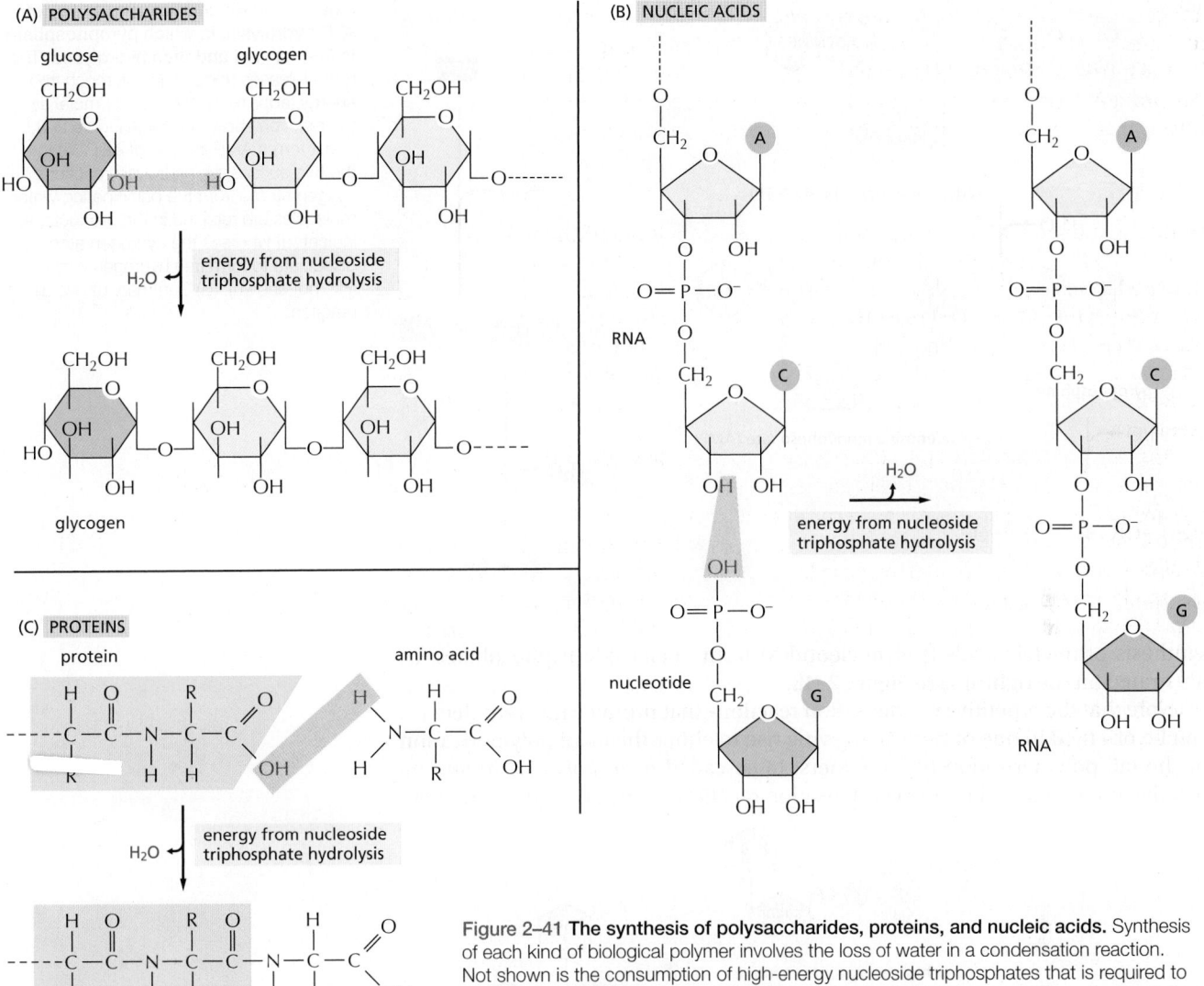

Figure 2–41 **The synthesis of polysaccharides, proteins, and nucleic acids.** Synthesis of each kind of biological polymer involves the loss of water in a condensation reaction. Not shown is the consumption of high-energy nucleoside triphosphates that is required to activate each monomer before its addition. In contrast, the reverse reaction—the breakdown of all three types of polymers—occurs by the simple addition of water (hydrolysis).

be removed in the condensation reaction is first activated by becoming involved in a high-energy linkage to a second molecule. However, the actual mechanisms used to link ATP hydrolysis to the synthesis of proteins and polysaccharides are more complex than that used for glutamine synthesis, since a series of high-energy intermediates is required to generate the final high-energy bond that is broken during the condensation step (discussed in Chapter 6 for protein synthesis).

Each activated carrier has limits in its ability to drive a biosynthetic reaction. The ΔG for the hydrolysis of ATP to ADP and inorganic phosphate (P_i) depends on the concentrations of all of the reactants, but under the usual conditions in a cell it is between –46 and –54 kJ/mole. In principle, this hydrolysis reaction could drive an unfavorable reaction with a ΔG of, perhaps, +40 kJ/mole, provided that a suitable reaction path is available. For some biosynthetic reactions, however, even –50 kJ/mole does not provide enough of a driving force. In these cases, the path of ATP hydrolysis can be altered so that it initially produces AMP and *pyrophosphate* (*PP_i*), which is itself then hydrolyzed in a subsequent step (**Figure 2–42**). The whole process makes available a total free-energy change of about –100 kJ/mole. An important type of biosynthetic reaction that is driven in this way is the

(A)

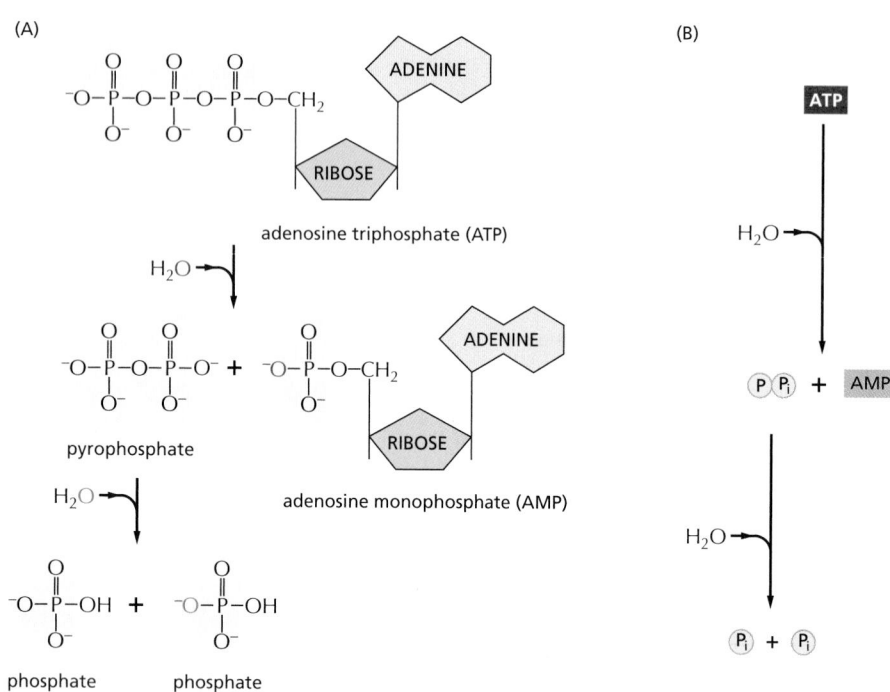

(B)

Figure 2–42 **An alternative pathway of ATP hydrolysis, in which pyrophosphate is first formed and then hydrolyzed.** This route releases about twice as much free energy (approximately –100 kJ/mole) as the reaction shown earlier in Figure 2–33, and it forms AMP instead of ADP. (A) In the two successive hydrolysis reactions, oxygen atoms from the participating water molecules are retained in the products, as indicated, whereas the hydrogen atoms dissociate to form free hydrogen ions (H⁺, not shown). (B) Summary of overall reaction.

synthesis of nucleic acids (polynucleotides) from nucleoside triphosphates, as illustrated on the right side of **Figure 2–43**.

Note that the repetitive condensation reactions that produce macromolecules can be oriented in one of two ways, giving rise to either the head polymerization or the tail polymerization of monomers. In so-called *head polymerization*, the reactive bond required for the condensation reaction is carried on the end of the

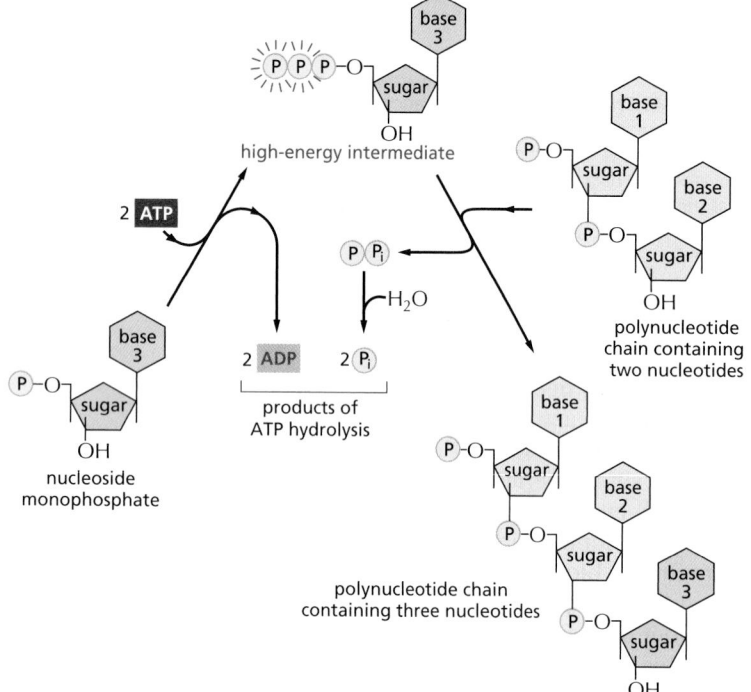

Figure 2–43 **Synthesis of a polynucleotide, RNA or DNA, is a multistep process driven by ATP hydrolysis.** In the first step, a nucleoside monophosphate is activated by the sequential transfer of the terminal phosphate groups from two ATP molecules. The high-energy intermediate formed—a nucleoside triphosphate—exists free in solution until it reacts with the growing end of an RNA or a DNA chain with release of pyrophosphate. Hydrolysis of the latter to inorganic phosphate is highly favorable and helps to drive the overall reaction in the direction of polynucleotide synthesis. For details, see Chapter 5.

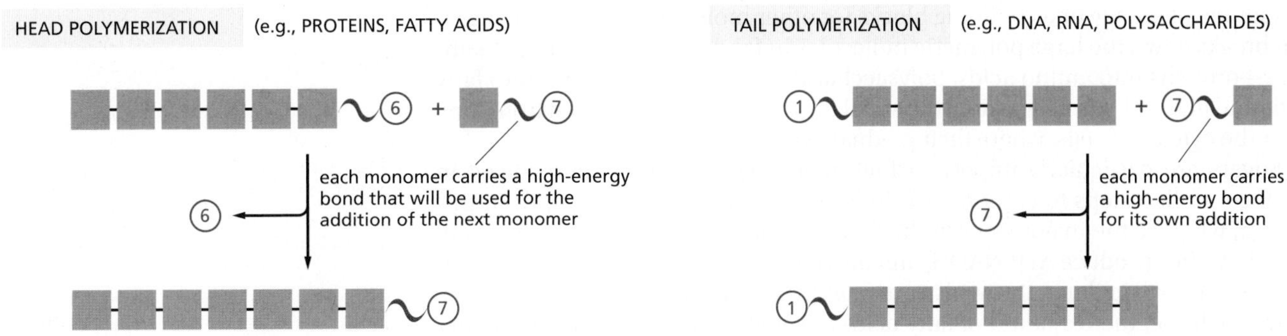

HEAD POLYMERIZATION (e.g., PROTEINS, FATTY ACIDS)   TAIL POLYMERIZATION (e.g., DNA, RNA, POLYSACCHARIDES)

each monomer carries a high-energy bond that will be used for the addition of the next monomer

each monomer carries a high-energy bond for its own addition

Figure 2–44 **The orientation of the active intermediates in the repetitive condensation reactions that form biological polymers.** The head growth of polymers is compared with its alternative, tail growth. As indicated, these two mechanisms are used to produce different types of biological macromolecules.

growing polymer, and it must therefore be regenerated each time that a monomer is added. In this case, each monomer brings with it the reactive bond that will be used in adding the *next* monomer in the series. In *tail polymerization*, the reactive bond carried by each monomer is instead used immediately for its own addition (**Figure 2–44**).

We shall see in later chapters that both of these types of polymerization are used. The synthesis of polynucleotides and some simple polysaccharides occurs by tail polymerization, for example, whereas the synthesis of proteins occurs by a head polymerization process.

Summary

Living cells need to create and maintain order within themselves to survive and grow. This is thermodynamically possible only because of a continual input of energy, part of which must be released from the cells to their environment as heat that disorders the surroundings. The only chemical reactions possible are those that increase the total amount of disorder in the universe. The free-energy change for a reaction, ΔG, measures this disorder, and it must be less than zero for a reaction to proceed spontaneously. This ΔG depends both on the intrinsic properties of the reacting molecules and their concentrations, and it can be calculated from these concentrations if either the equilibrium constant (K) for the reaction or its standard free-energy change, $\Delta G°$, is known.

The energy needed for life comes ultimately from the electromagnetic radiation of the sun, which drives the formation of organic molecules in photosynthetic organisms such as green plants. Animals obtain their energy by eating organic molecules and oxidizing them in a series of enzyme-catalyzed reactions that are coupled to the formation of ATP—a common currency of energy in all cells.

To make possible the continual generation of order in cells, energetically favorable reactions, such as the hydrolysis of ATP, are coupled to energetically unfavorable reactions. In the biosynthesis of macromolecules, ATP is used to form reactive phosphorylated intermediates. Because the energetically unfavorable reaction of biosynthesis now becomes energetically favorable, ATP hydrolysis is said to drive the reaction. Polymeric molecules such as proteins, nucleic acids, and polysaccharides are assembled from small activated precursor molecules by repetitive condensation reactions that are driven in this way. Other reactive molecules, called either activated carriers or coenzymes, transfer other chemical groups in the course of biosynthesis: NADPH transfers hydrogen as a proton plus two electrons (a hydride ion), for example, whereas acetyl CoA transfers an acetyl group.

HOW CELLS OBTAIN ENERGY FROM FOOD

The constant supply of energy that cells need to generate and maintain the biological order that keeps them alive comes from the chemical-bond energy in food molecules.

The proteins, lipids, and polysaccharides that make up most of the food we eat must be broken down into smaller molecules before our cells can use them—either as a source of energy or as building blocks for other molecules. Enzymatic digestion breaks down the large polymeric molecules in food into their monomer subunits—proteins into amino acids, polysaccharides into sugars, and fats into fatty acids and glycerol. After digestion, the small organic molecules derived from food enter the cytosol of cells, where their gradual oxidation begins.

Sugars are particularly important fuel molecules, and they are oxidized in small controlled steps to carbon dioxide (CO_2) and water (**Figure 2–45**). In this section, we trace the major steps in the breakdown, or catabolism, of sugars and show how they produce ATP, NADH, and other activated carrier molecules in animal cells. A very similar pathway also operates in plants, fungi, and many bacteria. As we shall see, the oxidation of fatty acids is equally important for cells. Other molecules, such as proteins, can also serve as energy sources when they are funneled through appropriate enzymatic pathways.

Glycolysis Is a Central ATP-Producing Pathway

The major process for oxidizing sugars is the sequence of reactions known as **glycolysis**—from the Greek *glukus*, "sweet," and *lusis*, "rupture." Glycolysis produces ATP without the involvement of molecular oxygen (O_2 gas). It occurs in the cytosol of most cells, including many anaerobic microorganisms. Glycolysis probably evolved early in the history of life, before photosynthetic organisms introduced oxygen into the atmosphere. During glycolysis, a glucose molecule with six carbon atoms is converted into two molecules of *pyruvate*, each of which contains three carbon atoms. For each glucose molecule, two molecules of ATP are hydrolyzed to provide energy to drive the early steps, but four molecules of ATP are produced in the later steps. At the end of glycolysis, there is consequently a net gain of two molecules of ATP for each glucose molecule broken down. Two molecules of the activated carrier NADH are also produced.

The glycolytic pathway is outlined in **Figure 2–46** and shown in more detail in **Panel 2–8** (pp. 104–105) and **Movie 2.5**. Glycolysis involves a sequence of 10 separate reactions, each producing a different sugar intermediate and each catalyzed by a different enzyme. Like most enzymes, these have names ending in *ase*—such as isomer*ase* and dehydrogen*ase*—to indicate the type of reaction they catalyze.

Although no molecular oxygen is used in glycolysis, oxidation occurs, in that electrons are removed by NAD+ (producing NADH) from some of the carbons derived from the glucose molecule. The stepwise nature of the process releases the energy of oxidation in small packets, so that much of it can be stored in activated carrier molecules rather than all of it being released as heat (see Figure 2–45). Thus, some of the energy released by oxidation drives the direct synthesis of ATP molecules from ADP and P_i, and some remains with the electrons in the electron carrier NADH.

Figure 2–45 Schematic representation of the controlled stepwise oxidation of sugar in a cell, compared with ordinary burning. (A) If the sugar were oxidized to CO_2 and H_2O in a single step, it would release an amount of energy much larger than could be captured for useful purposes. (B) In the cell, enzymes catalyze oxidation via a series of small steps in which free energy is transferred in conveniently sized packets to carrier molecules—most often ATP and NADH. At each step, an enzyme controls the reaction by reducing the activation-energy barrier that has to be surmounted before the specific reaction can occur. The total free energy released is exactly the same in (A) and (B).

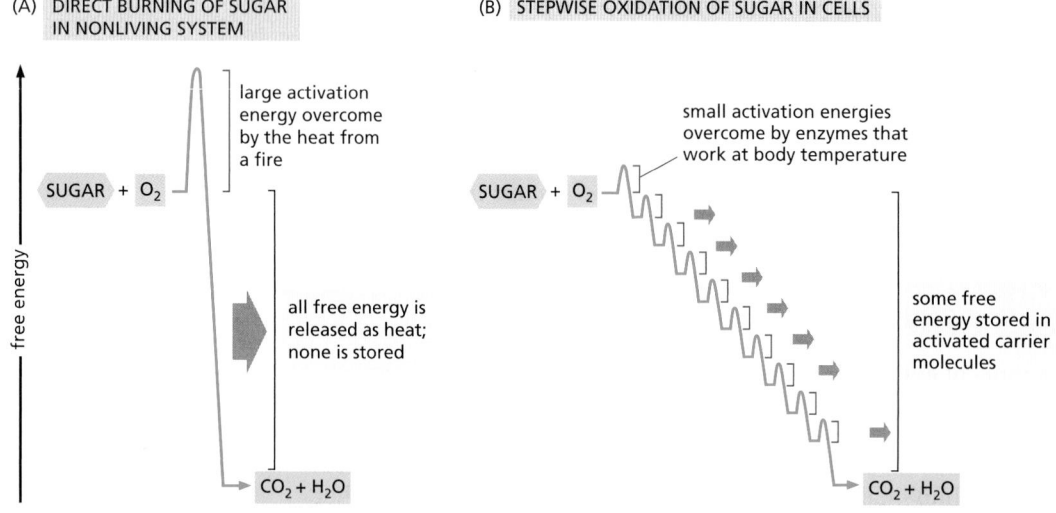

Two molecules of NADH are formed per molecule of glucose in the course of glycolysis. In aerobic organisms, these NADH molecules donate their electrons to the electron-transport chain described in Chapter 14, and the NAD^+ formed from the NADH is used again for glycolysis (see step 6 in Panel 2–8, pp. 104–105).

Fermentations Produce ATP in the Absence of Oxygen

For most animal and plant cells, glycolysis is only a prelude to the final stage of the breakdown of food molecules. In these cells, the pyruvate formed by glycolysis is rapidly transported into the mitochondria, where it is converted into CO_2 plus acetyl CoA, whose acetyl group is then completely oxidized to CO_2 and H_2O.

In contrast, for many anaerobic organisms—which do not utilize molecular oxygen and can grow and divide without it—glycolysis is the principal source of the cell's ATP. Certain animal tissues, such as skeletal muscle, can also continue to function when molecular oxygen is limited. In these anaerobic conditions, the pyruvate and the NADH electrons stay in the cytosol. The pyruvate is converted into products excreted from the cell—for example, into ethanol and CO_2 in the yeasts used in brewing and breadmaking, or into lactate in muscle. In this process, the NADH gives up its electrons and is converted back into NAD^+. This regeneration of NAD^+ is required to maintain the reactions of glycolysis (**Figure 2–47**).

Energy-yielding pathways like these, in which organic molecules both donate and accept electrons (and which are often, as in these cases, anaerobic), are called

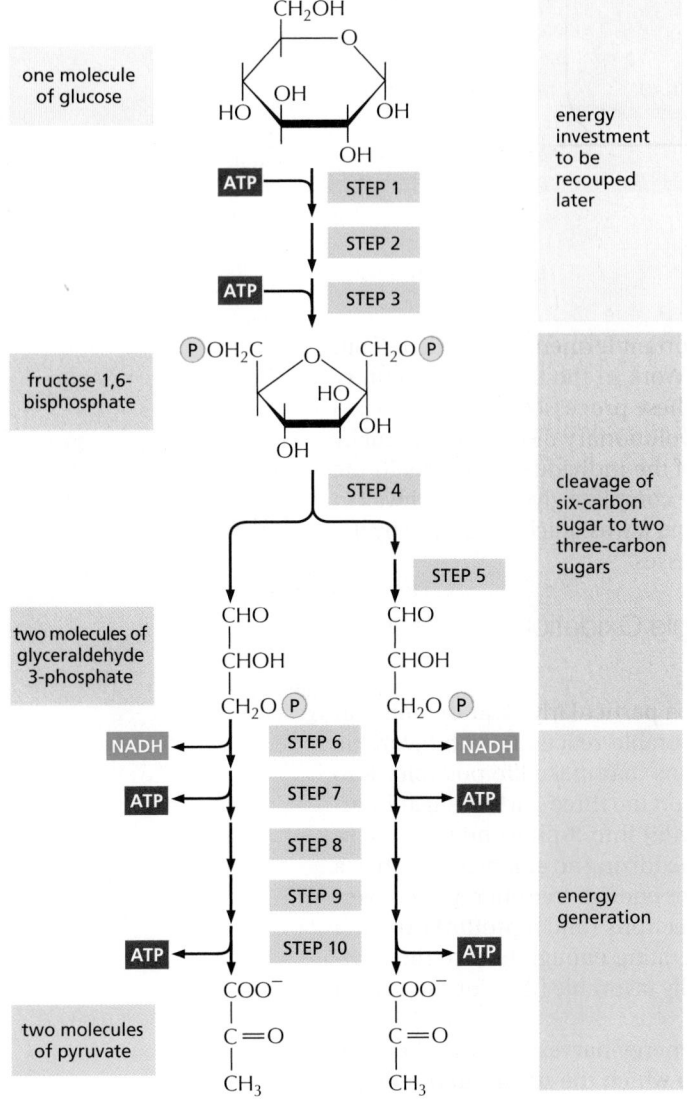

Figure 2–46 An outline of glycolysis. Each of the 10 steps shown is catalyzed by a different enzyme. Note that step 4 cleaves a six-carbon sugar into two three-carbon sugars, so that the number of molecules at every stage after this doubles. As indicated, step 6 begins the energy-generation phase of glycolysis. Because two molecules of ATP are hydrolyzed in the early, energy-investment phase, glycolysis results in the net synthesis of 2 ATP and 2 NADH molecules per molecule of glucose (see also Panel 2–8).

(A)   FERMENTATION LEADING TO EXCRETION OF LACTATE

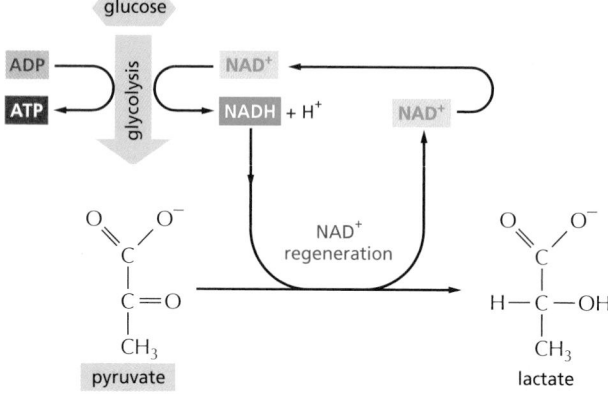

(B)   FERMENTATION LEADING TO EXCRETION OF ETHANOL AND CO_2

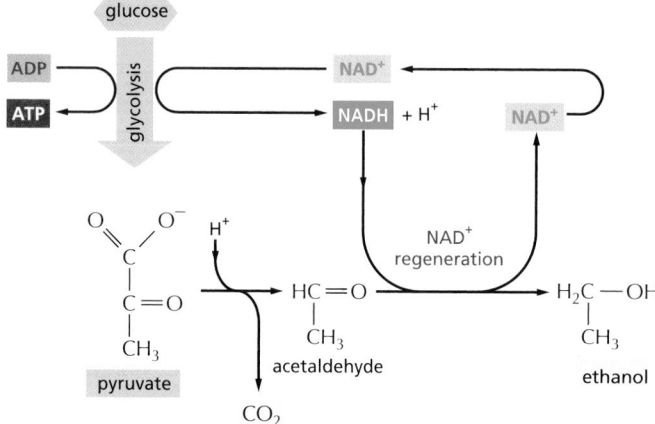

Figure 2–47 Two pathways for the anaerobic breakdown of pyruvate. (A) When there is inadequate oxygen, for example, in a muscle cell undergoing vigorous contraction, the pyruvate produced by glycolysis is converted to lactate as shown. This reaction regenerates the NAD^+ consumed in step 6 of glycolysis, but the whole pathway yields much less energy overall than complete oxidation. (B) In some organisms that can grow anaerobically, such as yeasts, pyruvate is converted via acetaldehyde into carbon dioxide and ethanol. Again, this pathway regenerates NAD^+ from NADH, as required to enable glycolysis to continue. Both (A) and (B) are examples of *fermentations*.

fermentations. Studies of the commercially important fermentations carried out by yeasts inspired much of early biochemistry. Work in the nineteenth century led in 1896 to the then startling recognition that these processes could be studied outside living organisms, in cell extracts. This revolutionary discovery eventually made it possible to dissect out and study each of the individual reactions in the fermentation process. The piecing together of the complete glycolytic pathway in the 1930s was a major triumph of biochemistry, and it was quickly followed by the recognition of the central role of ATP in cell processes.

Glycolysis Illustrates How Enzymes Couple Oxidation to Energy Storage

The formation of ATP during glycolysis provides a particularly clear demonstration of how enzymes couple energetically unfavorable reactions with favorable ones, thereby driving the many chemical reactions that make life possible. Two central reactions in glycolysis (steps 6 and 7) convert the three-carbon sugar intermediate glyceraldehyde 3-phosphate (an aldehyde) into 3-phosphoglycerate (a carboxylic acid; see Panel 2–8, pp. 104–105), thus oxidizing an aldehyde group to a carboxylic acid group. The overall reaction releases enough free energy to convert a molecule of ADP to ATP and to transfer two electrons (and a proton) from the aldehyde to NAD^+ to form NADH, while still liberating enough heat to the environment to make the overall reaction energetically favorable ($\Delta G°$ for the overall reaction is –12.5 kJ/mole).

Figure 2–48 outlines this remarkable feat of energy harvesting. The chemical reactions are precisely guided by two enzymes to which the sugar intermediates

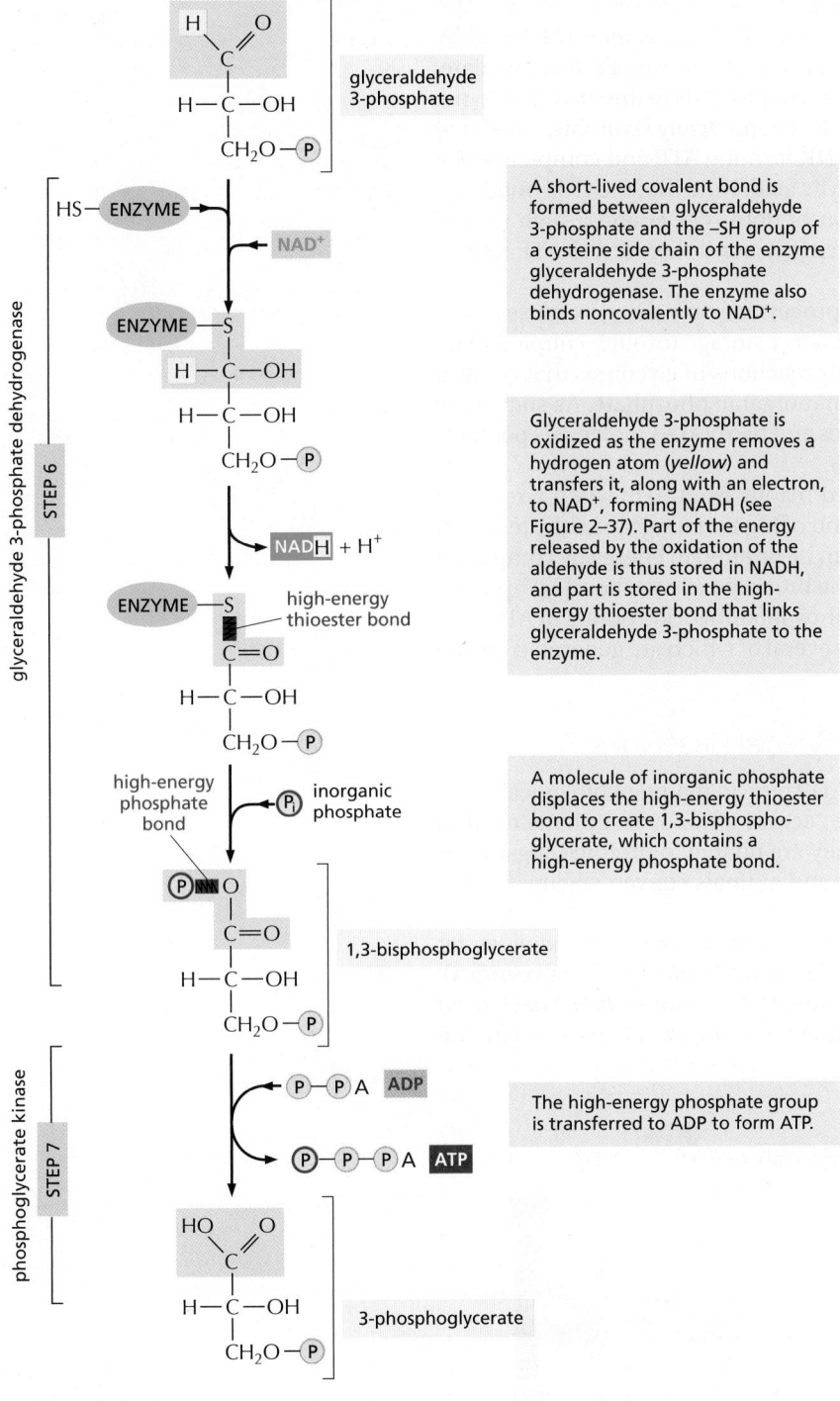

(A) STEPS 6 AND 7 OF GYCOLYSIS

glyceraldehyde 3-phosphate dehydrogenase

STEP 6

A short-lived covalent bond is formed between glyceraldehyde 3-phosphate and the –SH group of a cysteine side chain of the enzyme glyceraldehyde 3-phosphate dehydrogenase. The enzyme also binds noncovalently to NAD⁺.

Glyceraldehyde 3-phosphate is oxidized as the enzyme removes a hydrogen atom (*yellow*) and transfers it, along with an electron, to NAD⁺, forming NADH (see Figure 2–37). Part of the energy released by the oxidation of the aldehyde is thus stored in NADH, and part is stored in the high-energy thioester bond that links glyceraldehyde 3-phosphate to the enzyme.

A molecule of inorganic phosphate displaces the high-energy thioester bond to create 1,3-bisphospho-glycerate, which contains a high-energy phosphate bond.

phosphoglycerate kinase

STEP 7

The high-energy phosphate group is transferred to ADP to form ATP.

(B) SUMMARY OF STEPS 6 AND 7

The oxidation of an aldehyde to a carboxylic acid releases energy, much of which is captured in the activated carriers ATP and NADH.

Figure 2–48 Energy storage in steps 6 and 7 of glycolysis. (A) In step 6, the enzyme glyceraldehyde 3-phosphate dehydrogenase couples the energetically favorable oxidation of an aldehyde to the energetically unfavorable formation of a high-energy phosphate bond. At the same time, it enables energy to be stored in NADH. The formation of the high-energy phosphate bond is driven by the oxidation reaction, and the enzyme thereby acts like the "paddle wheel" coupler in Figure 2–32B. In step 7, the newly formed high-energy phosphate bond in 1,3-bisphosphoglycerate is transferred to ADP, forming a molecule of ATP and leaving a free carboxylic acid group on the oxidized sugar. The part of the molecule that undergoes a change is shaded in *blue*; the rest of the molecule remains unchanged throughout all these reactions. (B) Summary of the overall chemical change produced by reactions 6 and 7.

are tightly bound. As detailed in Figure 2–48, the first enzyme (glyceraldehyde 3-phosphate dehydrogenase) forms a short-lived covalent bond to the aldehyde through a reactive –SH group on the enzyme, and catalyzes its oxidation by NAD⁺ in this attached state. The reactive enzyme–substrate bond is then displaced by an inorganic phosphate ion to produce a high-energy phosphate intermediate, which is released from the enzyme. This intermediate binds to the second enzyme (phosphoglycerate kinase), which catalyzes the energetically favorable transfer of the high-energy phosphate just created to ADP, forming ATP and completing the process of oxidizing an aldehyde to a carboxylic acid. Note that the C–H bond oxidation energy in step 6 drives the formation of both NADH and a high-energy phosphate bond. The breakage of the high-energy bond then drives ATP formation.

We have shown this particular oxidation process in some detail because it provides a clear example of enzyme-mediated energy storage through coupled reactions (**Figure 2–49**). Steps 6 and 7 are the only reactions in glycolysis that create a high-energy phosphate linkage directly from inorganic phosphate. As such, they account for the net yield of two ATP molecules and two NADH molecules per molecule of glucose (see Panel 2–8, pp. 104–105).

As we have just seen, ATP can be formed readily from ADP when a reaction intermediate is formed with a phosphate bond of higher energy than the terminal phosphate bond in ATP. Phosphate bonds can be ordered in energy by comparing the standard free-energy change $(\Delta G°)$ for the breakage of each bond by hydrolysis. **Figure 2–50** compares the high-energy phosphoanhydride bonds in ATP with the energy of some other phosphate bonds, several of which are generated during glycolysis.

Organisms Store Food Molecules in Special Reservoirs

All organisms need to maintain a high ATP/ADP ratio to maintain biological order in their cells. Yet animals have only periodic access to food, and plants need to survive overnight without sunlight, when they are unable to produce sugar from photosynthesis. For this reason, both plants and animals convert sugars and fats to special forms for storage (**Figure 2–51**).

To compensate for long periods of fasting, animals store fatty acids as fat droplets composed of water-insoluble *triacylglycerols* (also called triglycerides). The triacylglycerols in animals are mostly stored in the cytoplasm of specialized fat cells called adipocytes. For shorter-term storage, sugar is stored as glucose

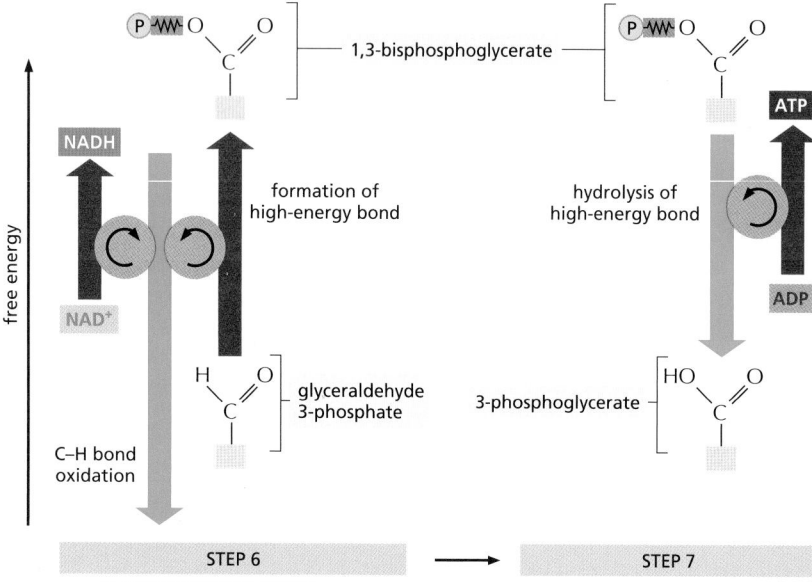

TOTAL ENERGY CHANGE for step 6 followed by step 7 is a favorable –12.5 kJ/mole

Figure 2–49 Schematic view of the coupled reactions that form NADH and ATP in steps 6 and 7 of glycolysis. The C–H bond oxidation energy drives the formation of both NADH and a high-energy phosphate bond. The breakage of the high-energy bond then drives ATP formation.

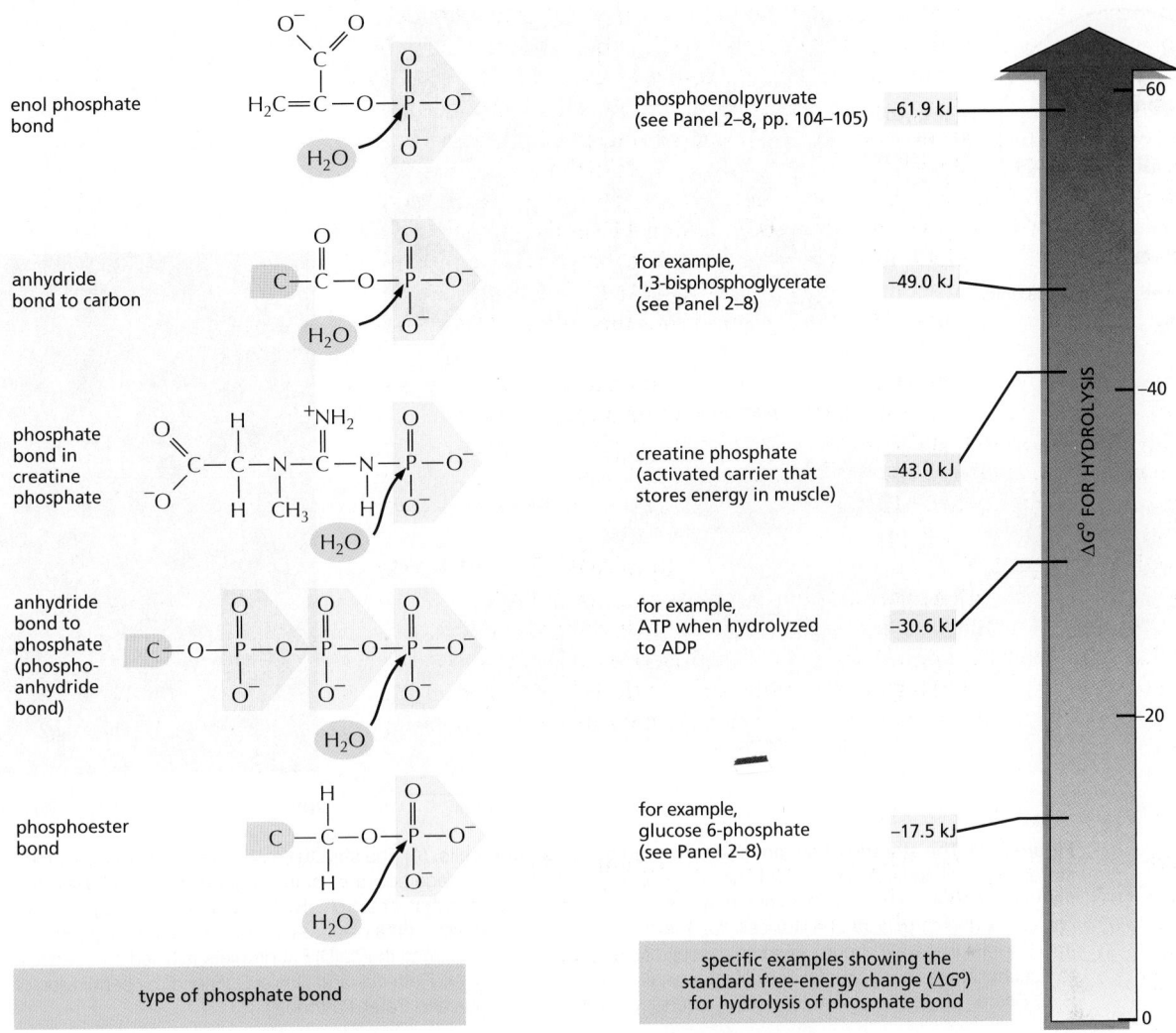

Figure 2–50 Phosphate bonds have different energies. Examples of different types of phosphate bonds with their sites of hydrolysis are shown in the molecules depicted on the left. Those starting with a *gray* carbon atom show only part of a molecule. Examples of molecules containing such bonds are given on the right, with the standard free-energy change for hydrolysis in kilojoules. The transfer of a phosphate group from one molecule to another is energetically favorable if the free-energy change (ΔG) for hydrolysis of the phosphate bond of the first molecule is more negative than that for hydrolysis of the phosphate bond in the second. Thus, under standard conditions, a phosphate group is readily transferred from 1,3-bisphosphoglycerate to ADP to form ATP. (Standard conditions often do *not* pertain to living cells, where the relative concentrations of reactants and products will influence the actual change in free energy.) The hydrolysis reaction can be viewed as the transfer of the phosphate group to water.

subunits in the large branched polysaccharide **glycogen**, which is present as small granules in the cytoplasm of many cells, including liver and muscle. The synthesis and degradation of glycogen are rapidly regulated according to need. When cells need more ATP than they can generate from the food molecules taken in from the bloodstream, they break down glycogen in a reaction that produces glucose 1-phosphate, which is rapidly converted to glucose 6-phosphate for glycolysis (**Figure 2–52**).

Quantitatively, **fat** is far more important than glycogen as an energy store for animals, presumably because it provides for more efficient storage. The oxidation of a gram of fat releases about twice as much energy as the oxidation of a gram of glycogen. Moreover, glycogen differs from fat in binding a great deal of water, producing a sixfold difference in the actual mass of glycogen required to store the same amount of energy as fat. An average adult human stores enough glycogen

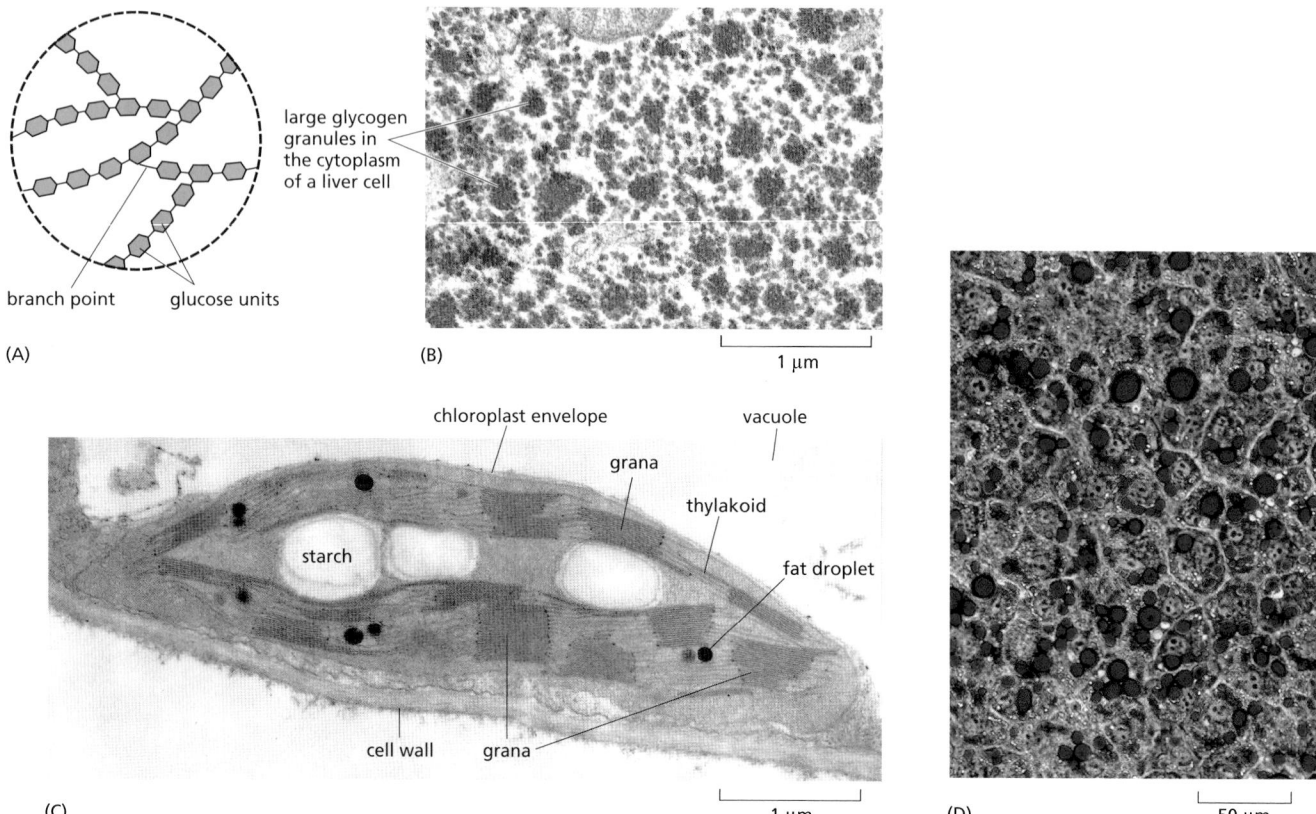

large glycogen
granules in
the cytoplasm
of a liver cell

branch point   glucose units

(A)

(B)

1 μm

chloroplast envelope   vacuole

grana

thylakoid

starch

fat droplet

cell wall   grana

(C)

1 μm

(D)

50 μm

Figure 2–51 The storage of sugars and fats in animal and plant cells. (A) The structures of starch and glycogen, the storage form of sugars in plants and animals, respectively. Both are storage polymers of the sugar glucose and differ only in the frequency of branch points. There are many more branches in glycogen than in starch. (B) An electron micrograph of glycogen granules in the cytoplasm of a liver cell. (C) A thin section of a chloroplast from a plant cell, showing the starch granules and lipid (fat droplets) that have accumulated as a result of the biosyntheses occurring there. (D) Fat droplets (stained *red*) beginning to accumulate in developing fat cells of an animal. (B, courtesy of Robert Fletterick and Daniel S. Friend, by permission of E. L. Bearer; C, courtesy of K. Plaskitt; D, courtesy of Ronald M. Evans and Peter Tontonoz.)

for only about a day of normal activities, but enough fat to last for nearly a month. If our main fuel reservoir had to be carried as glycogen instead of fat, body weight would increase by an average of about 60 pounds.

The sugar and ATP needed by plant cells are largely produced in separate organelles: sugars in chloroplasts (the organelles specialized for photosynthesis),

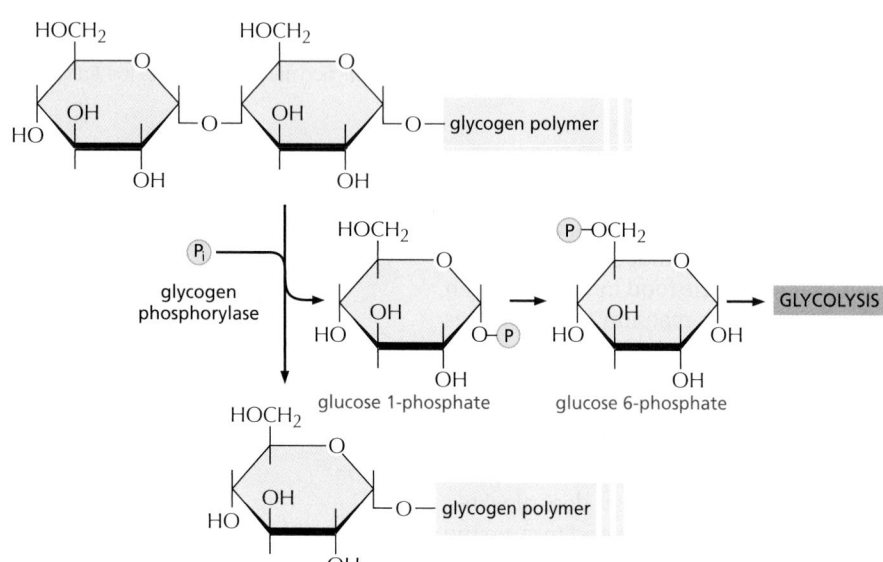

glycogen
phosphorylase

glucose 1-phosphate

glucose 6-phosphate

GLYCOLYSIS

glycogen polymer

Figure 2–52 How sugars are produced from glycogen. Glucose subunits are released from glycogen by the enzyme glycogen phosphorylase. This produces glucose 1-phosphate, which is rapidly converted to glucose 6-phosphate for glycolysis.

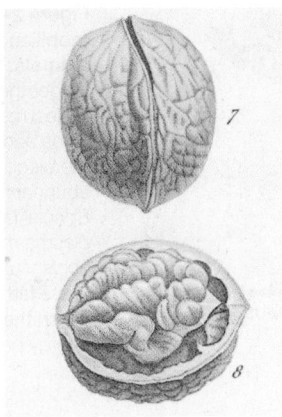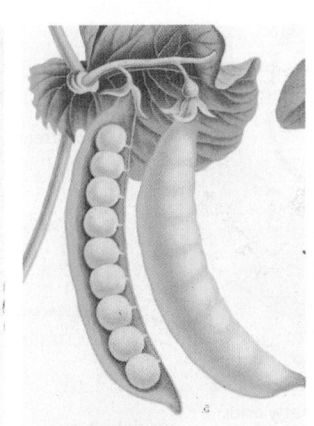

Figure 2–53 **Some plant seeds that serve as important foods for humans.** Corn, nuts, and peas all contain rich stores of starch and fat that provide the young plant embryo in the seed with energy and building blocks for biosynthesis. (Courtesy of the John Innes Foundation.)

and ATP in mitochondria. Although plants produce abundant amounts of both ATP and NADPH in their chloroplasts, this organelle is isolated from the rest of its plant cell by a membrane that is impermeable to both types of activated carrier molecules. Moreover, the plant contains many cells—such as those in the roots—that lack chloroplasts and therefore cannot produce their own sugars. Thus, sugars are exported from chloroplasts to the mitochondria present in all cells of the plant. Most of the ATP needed for general plant cell metabolism is synthesized in these mitochondria, using exactly the same pathways for the oxidative breakdown of sugars as in nonphotosynthetic organisms; this ATP is then passed to the rest of the cell (see Figure 14–42).

During periods of excess photosynthetic capacity during the day, chloroplasts convert some of the sugars that they make into fats and into **starch**, a polymer of glucose analogous to the glycogen of animals. The fats in plants are triacylglycerols (triglycerides), just like the fats in animals, and differ only in the types of fatty acids that predominate. Fat and starch are both stored inside the chloroplast until needed for energy-yielding oxidation during periods of darkness (see Figure 2–51C).

The embryos inside plant seeds must live on stored sources of energy for a prolonged period, until they germinate and produce leaves that can harvest the energy in sunlight. For this reason plant seeds often contain especially large amounts of fats and starch—which makes them a major food source for animals, including ourselves (**Figure 2–53**).

Most Animal Cells Derive Their Energy from Fatty Acids Between Meals

After a meal, most of the energy that an animal needs is derived from sugars obtained from food. Excess sugars, if any, are used to replenish depleted glycogen stores, or to synthesize fats as a food store. But soon the fat stored in adipose tissue is called into play, and by the morning after an overnight fast, fatty acid oxidation generates most of the ATP we need.

Low glucose levels in the blood trigger the breakdown of fats for energy production. As illustrated in **Figure 2–54**, the triacylglycerols stored in fat droplets in adipocytes are hydrolyzed to produce fatty acids and glycerol, and the fatty acids released are transferred to cells in the body through the bloodstream. While animals readily convert sugars to fats, they cannot convert fatty acids to sugars. Instead, the fatty acids are oxidized directly.

Sugars and Fats Are Both Degraded to Acetyl CoA in Mitochondria

In aerobic metabolism, the pyruvate that was produced by glycolysis from sugars in the cytosol is transported into the *mitochondria* of eukaryotic cells. There, it is

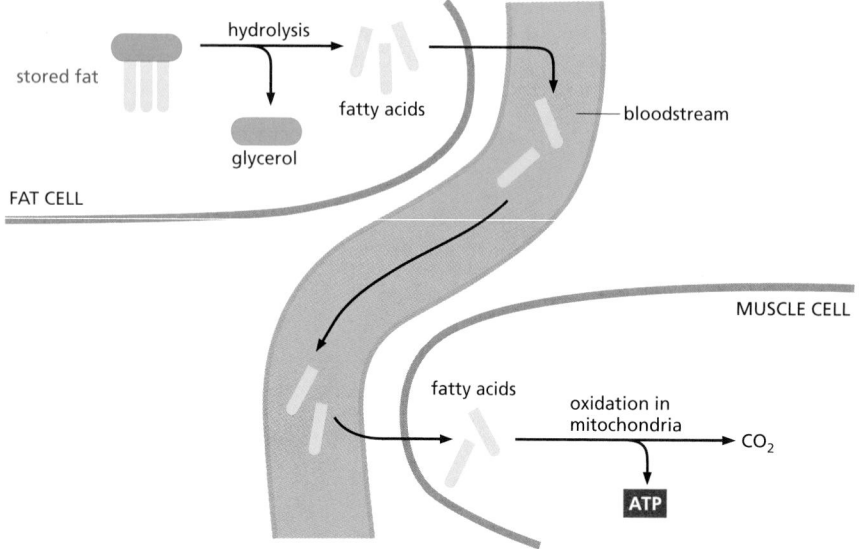

Figure 2–54 How stored fats are mobilized for energy production in animals. Low glucose levels in the blood trigger the hydrolysis of the triacylglycerol molecules in fat droplets to free fatty acids and glycerol. These fatty acids enter the bloodstream, where they bind to the abundant blood protein, serum albumin. Special fatty acid transporters in the plasma membrane of cells that oxidize fatty acids, such as muscle cells, then pass these fatty acids into the cytosol, from which they are moved into mitochondria for energy production.

rapidly decarboxylated by a giant complex of three enzymes, called the *pyruvate dehydrogenase complex*. The products of pyruvate decarboxylation are a molecule of CO_2 (a waste product), a molecule of NADH, and acetyl CoA (see Panel 2–9).

The fatty acids imported from the bloodstream are moved into mitochondria, where all of their oxidation takes place (**Figure 2–55**). Each molecule of fatty acid (as the activated molecule *fatty acyl CoA*) is broken down completely by a cycle of reactions that trims two carbons at a time from its carboxyl end, generating one molecule of acetyl CoA for each turn of the cycle. A molecule of NADH and a molecule of $FADH_2$ are also produced in this process (**Figure 2–56**).

Sugars and fats are the major energy sources for most nonphotosynthetic organisms, including humans. However, most of the useful energy that can be extracted from the oxidation of both types of foodstuffs remains stored in the acetyl CoA molecules that are produced by the two types of reactions just described. The citric acid cycle of reactions, in which the acetyl group ($-COCH_3$) in acetyl CoA is oxidized to CO_2 and H_2O, is therefore central to the energy metabolism of aerobic organisms. In eukaryotes, these reactions all take place in mitochondria. We should therefore not be surprised to discover that the mitochondrion is the place where most of the ATP is produced in animal cells. In contrast, aerobic bacteria carry out all of their reactions, including the citric acid cycle, in a single compartment, the cytosol.

The Citric Acid Cycle Generates NADH by Oxidizing Acetyl Groups to CO_2

In the nineteenth century, biologists noticed that in the absence of air cells produce lactic acid (for example, in muscle) or ethanol (for example, in yeast), while in its presence they consume O_2 and produce CO_2 and H_2O. Efforts to define the pathways of aerobic metabolism eventually focused on the oxidation of pyruvate and led in 1937 to the discovery of the **citric acid cycle**, also known as the

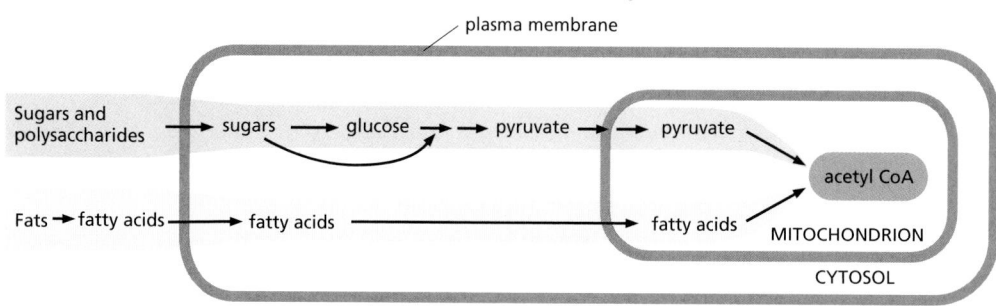

Figure 2–55 Pathways for the production of acetyl CoA from sugars and fats. The mitochondrion in eukaryotic cells is where acetyl CoA is produced from both types of major food molecules. It is therefore the place where most of the cell's oxidation reactions occur and where most of its ATP is made. Amino acids (not shown) can also enter the mitochondria, to be converted there into acetyl CoA or another intermediate of the citric acid cycle. The structure and function of mitochondria are discussed in detail in Chapter 14.

tricarboxylic acid cycle or the *Krebs cycle*. The citric acid cycle accounts for about two-thirds of the total oxidation of carbon compounds in most cells, and its major end products are CO_2 and high-energy electrons in the form of NADH. The CO_2 is released as a waste product, while the high-energy electrons from NADH are passed to a membrane-bound electron-transport chain (discussed in Chapter 14), eventually combining with O_2 to produce H_2O. The citric acid cycle itself does not use gaseous O_2 (it uses oxygen atoms from H_2O). But the cycle does require O_2 in subsequent reactions to keep it going. This is because there is no other efficient way for the NADH to get rid of its electrons and thus regenerate the NAD^+ that is needed.

The citric acid cycle takes place inside mitochondria in eukaryotic cells. It results in the complete oxidation of the carbon atoms of the acetyl groups in acetyl CoA, converting them into CO_2. But the acetyl group is not oxidized directly. Instead, this group is transferred from acetyl CoA to a larger, four-carbon molecule, *oxaloacetate*, to form the six-carbon tricarboxylic acid, *citric acid*, for which the subsequent cycle of reactions is named. The citric acid molecule is then gradually oxidized, allowing the energy of this oxidation to be harnessed to produce energy-rich activated carrier molecules. The chain of eight reactions forms a cycle because at the end the oxaloacetate is regenerated and enters a new turn of the cycle, as shown in outline in **Figure 2–57**.

We have thus far discussed only one of the three types of activated carrier molecules that are produced by the citric acid cycle; NADH, the reduced form of the NAD^+/NADH electron carrier system (see Figure 2–36). In addition to three molecules of NADH, each turn of the cycle also produces one molecule of $FADH_2$ (reduced flavin adenine dinucleotide) from FAD (see Figure 2–39), and one molecule of the ribonucleoside triphosphate **GTP** from GDP. The structure of GTP is illustrated in **Figure 2–58**. GTP is a close relative of ATP, and the transfer of its terminal phosphate group to ADP produces one ATP molecule in each cycle. As we discuss shortly, the energy that is stored in the readily transferred electrons of NADH and $FADH_2$ will be utilized subsequently for ATP production through the

Figure 2–56 The oxidation of fatty acids to acetyl CoA. (A) Electron micrograph of a lipid droplet in the cytoplasm. (B) The structure of fats. Fats are *triacylglycerols*. The glycerol portion, to which three fatty acids are linked through ester bonds, is shown in *blue*. Fats are insoluble in water and form large lipid droplets in the specialized fat cells (adipocytes) in which they are stored. (C) The fatty acid oxidation cycle. The cycle is catalyzed by a series of four enzymes in mitochondria. Each turn of the cycle shortens the fatty acid chain by two carbons (shown in *red*) and generates one molecule of acetyl CoA and one molecule each of NADH and $FADH_2$. (A, courtesy of Daniel S. Friend.)

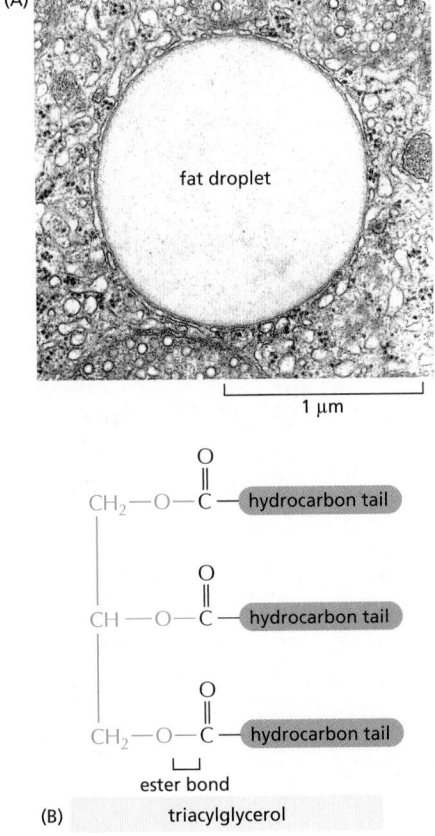

(A)

fat droplet

1 μm

(B) triacylglycerol

(C)

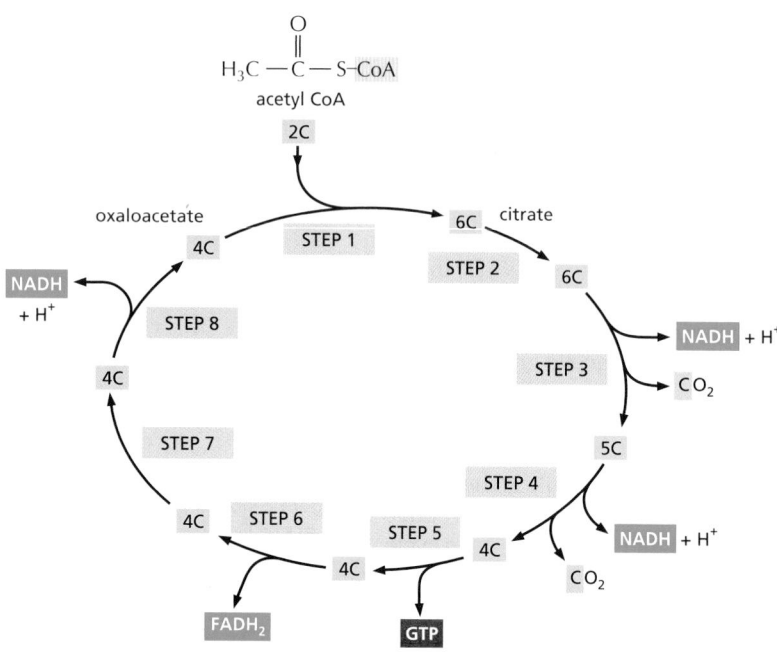

NET RESULT: ONE TURN OF THE CYCLE PRODUCES THREE NADH, ONE GTP, AND ONE FADH$_2$ MOLECULE, AND RELEASES TWO MOLECULES OF CO$_2$

Figure 2–57 Simple overview of the citric acid cycle. The reaction of acetyl CoA with oxaloacetate starts the cycle by producing citrate (citric acid). In each turn of the cycle, two molecules of CO$_2$ are produced as waste products, plus three molecules of NADH, one molecule of GTP, and one molecule of FADH$_2$. The number of carbon atoms in each intermediate is shown in a *yellow box.* For details, see Panel 2–9 (pp. 106–107).

process of *oxidative phosphorylation,* the only step in the oxidative catabolism of foodstuffs that directly requires gaseous oxygen (O$_2$) from the atmosphere.

Panel 2–9 (pp. 106–107) and Movie 2.6 present the complete citric acid cycle. Water, rather than molecular oxygen, supplies the extra oxygen atoms required to make CO$_2$ from the acetyl groups entering the citric acid cycle. As illustrated in the panel, three molecules of water are split in each cycle, and the oxygen atoms of some of them are ultimately used to make CO$_2$.

In addition to pyruvate and fatty acids, some amino acids pass from the cytosol into mitochondria, where they are also converted into acetyl CoA or one of the other intermediates of the citric acid cycle. Thus, in the eukaryotic cell, the mitochondrion is the center toward which all energy-yielding processes lead, whether they begin with sugars, fats, or proteins.

Both the citric acid cycle and glycolysis also function as starting points for important biosynthetic reactions by producing vital carbon-containing intermediates, such as *oxaloacetate* and *α-ketoglutarate.* Some of these substances produced by catabolism are transferred back from the mitochondria to the cytosol, where they serve in anabolic reactions as precursors for the synthesis of many essential molecules, such as amino acids (**Figure 2–59**).

Electron Transport Drives the Synthesis of the Majority of the ATP in Most Cells

Most chemical energy is released in the last stage in the degradation of a food molecule. In this final process, NADH and FADH$_2$ transfer the electrons that they gained when oxidizing food-derived organic molecules to the **electron-transport chain**, which is embedded in the inner membrane of the mitochondrion (see Figure 14–10). As the electrons pass along this long chain of specialized electron acceptor and donor molecules, they fall to successively lower energy states. The energy that the electrons release in this process pumps H$^+$ ions (protons) across the membrane—from the innermost mitochondrial compartment (the matrix) to the intermembrane space (and then to the cytosol)—generating a gradient of H$^+$ ions (**Figure 2–60**). This gradient serves as a major source of energy for cells, being tapped like a battery to drive a variety of energy-requiring reactions. The most prominent of these reactions is the generation of ATP by the phosphorylation of ADP.

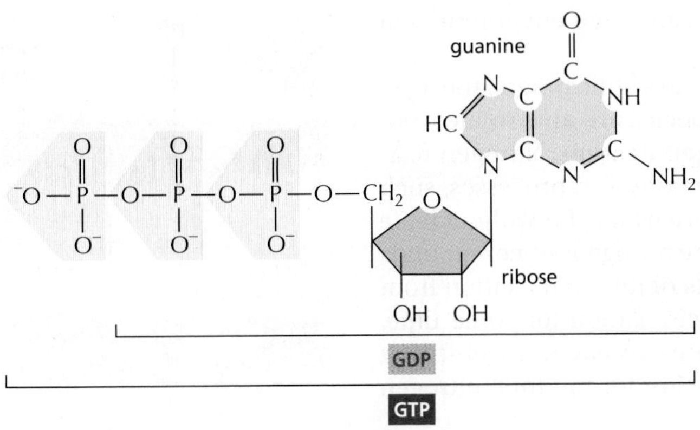

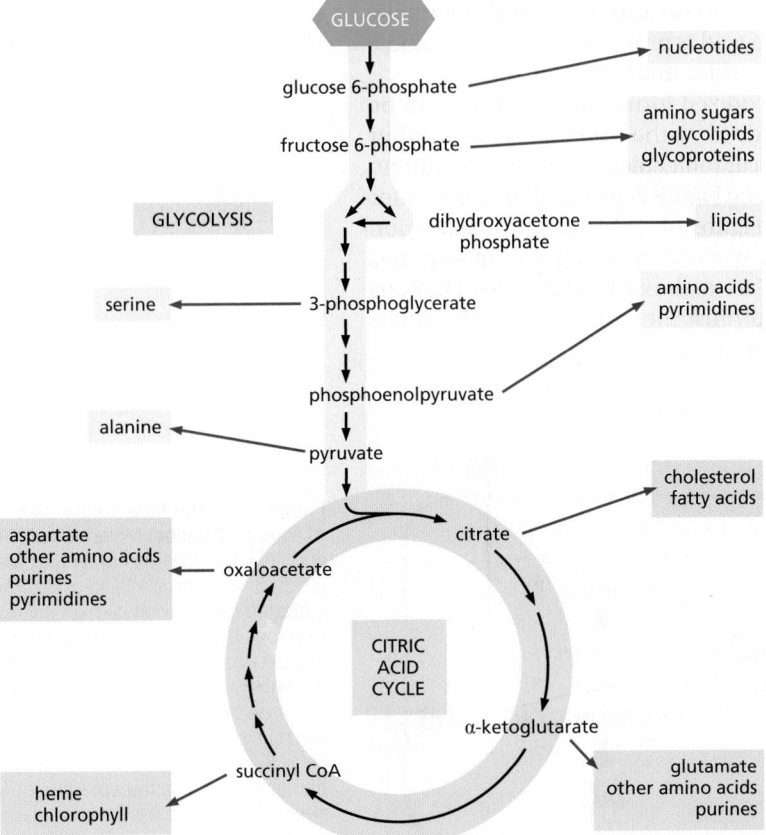

Figure 2–58 **The structure of GTP.** GTP and GDP are close relatives of ATP and ADP, respectively.

At the end of this series of electron transfers, the electrons are passed to molecules of oxygen gas (O_2) that have diffused into the mitochondrion, which simultaneously combine with protons (H^+) from the surrounding solution to produce water. The electrons have now reached a low energy level, and all the available energy has been extracted from the oxidized food molecule. This process, termed **oxidative phosphorylation** (**Figure 2–61**), also occurs in the plasma membrane of bacteria. As one of the most remarkable achievements of cell evolution, it is a central topic of Chapter 14.

In total, the complete oxidation of a molecule of glucose to H_2O and CO_2 is used by the cell to produce about 30 molecules of ATP. In contrast, only 2 molecules of ATP are produced per molecule of glucose by glycolysis alone.

Amino Acids and Nucleotides Are Part of the Nitrogen Cycle

So far we have concentrated mainly on carbohydrate metabolism and have not yet considered the metabolism of nitrogen or sulfur. These two elements are important constituents of biological macromolecules. Nitrogen and sulfur atoms pass

Figure 2–59 **Glycolysis and the citric acid cycle provide the precursors needed to synthesize many important biological molecules.** The amino acids, nucleotides, lipids, sugars, and other molecules—shown here as products—in turn serve as the precursors for the many macromolecules of the cell. Each *black arrow* in this diagram denotes a single enzyme-catalyzed reaction; the *red arrows* generally represent pathways with many steps that are required to produce the indicated products.

from compound to compound and between organisms and their environment in a series of reversible cycles.

Although molecular nitrogen is abundant in the Earth's atmosphere, nitrogen is chemically unreactive as a gas. Only a few living species are able to incorporate it into organic molecules, a process called **nitrogen fixation**. Nitrogen fixation occurs in certain microorganisms and by some geophysical processes, such as lightning discharge. It is essential to the biosphere as a whole, for without it life could not exist on this planet. Only a small fraction of the nitrogenous compounds in today's organisms, however, is due to fresh products of nitrogen fixation from the atmosphere. Most organic nitrogen has been in circulation for some time, passing from one living organism to another. Thus, present-day nitrogen-fixing reactions can be said to perform a "topping-up" function for the total nitrogen supply.

Vertebrates receive virtually all of their nitrogen from their dietary intake of proteins and nucleic acids. In the body, these macromolecules are broken down to amino acids and the components of nucleotides, and the nitrogen they contain is used to produce new proteins and nucleic acids—or other molecules. About half of the 20 amino acids found in proteins are essential amino acids for vertebrates (**Figure 2–62**), which means that they cannot be synthesized from other ingredients of the diet. The other amino acids can be so synthesized, using a variety of raw materials, including intermediates of the citric acid cycle. The essential amino acids are made by plants and other organisms, usually by long and energetically expensive pathways that have been lost in the course of vertebrate evolution.

The nucleotides needed to make RNA and DNA can be synthesized using specialized biosynthetic pathways. All of the nitrogens in the purine and pyrimidine bases (as well as some of the carbons) are derived from the plentiful amino acids glutamine, aspartic acid, and glycine, whereas the ribose and deoxyribose sugars are derived from glucose. There are no "essential nucleotides" that must be provided in the diet.

Amino acids not used in biosynthesis can be oxidized to generate metabolic energy. Most of their carbon and hydrogen atoms eventually form CO_2 or H_2O, whereas their nitrogen atoms are shuttled through various forms and eventually appear as urea, which is excreted. Each amino acid is processed differently, and a whole constellation of enzymatic reactions exists for their catabolism.

Sulfur is abundant on Earth in its most oxidized form, sulfate (SO_4^{2-}). To be useful for life, sulfate must be reduced to sulfide (S^{2-}), the oxidation state of sulfur required for the synthesis of essential biological molecules, including the amino acids methionine and cysteine, coenzyme A (see Figure 2–39), and the iron-sulfur centers essential for electron transport (see Figure 14–16). The sulfur-reduction process begins in bacteria, fungi, and plants, where a special group of enzymes use ATP and reducing power to create a sulfate assimilation pathway. Humans and other animals cannot reduce sulfate and must therefore acquire the sulfur they need for their metabolism in the food that they eat.

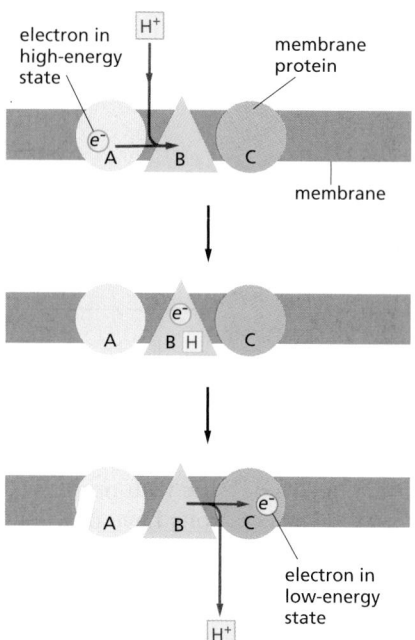

Figure 2–60 The generation of an H⁺ gradient across a membrane by electron-transport reactions. An electron held in a high-energy state (derived, for example, from the oxidation of a metabolite) is passed sequentially by carriers A, B, and C to a lower energy state. In this diagram, carrier B is arranged in the membrane in such a way that it takes up H⁺ from one side and releases it to the other as the electron passes. The result is an H⁺ gradient. As discussed in Chapter 14, this gradient is an important form of energy that is harnessed by other membrane proteins to drive the formation of ATP (for an actual example, see Figure 14–21).

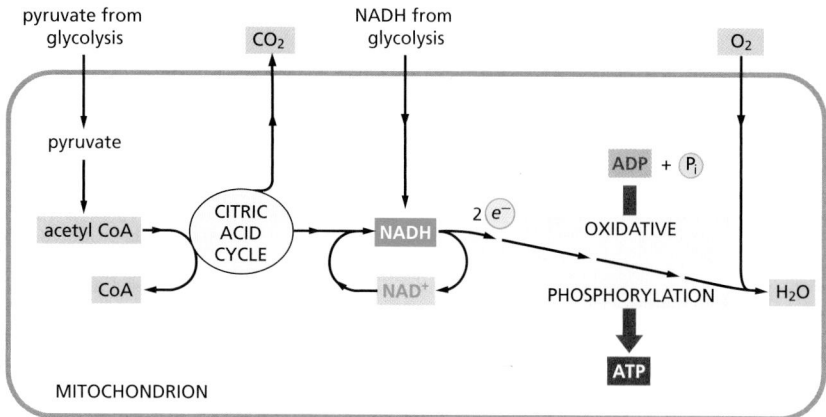

Figure 2–61 The final stages of oxidation of food molecules. Molecules of NADH and FADH₂ (FADH₂ is not shown) are produced by the citric acid cycle. These activated carriers donate high-energy electrons that are eventually used to reduce oxygen gas to water. A major portion of the energy released during the transfer of these electrons along an electron-transfer chain in the mitochondrial inner membrane (or in the plasma membrane of bacteria) is harnessed to drive the synthesis of ATP—hence the name oxidative phosphorylation (discussed in Chapter 14).

Metabolism Is Highly Organized and Regulated

One gets a sense of the intricacy of a cell as a chemical machine from the relation of glycolysis and the citric acid cycle to the other metabolic pathways sketched out in **Figure 2–63**. This chart represents only some of the enzymatic pathways in a human cell. It is obvious that our discussion of cell metabolism has dealt with only a tiny fraction of the broad field of cell chemistry.

All these reactions occur in a cell that is less than 0.1 mm in diameter, and each requires a different enzyme. As is clear from Figure 2–63, the same molecule can often be part of many different pathways. Pyruvate, for example, is a substrate for half a dozen or more different enzymes, each of which modifies it chemically in a different way. One enzyme converts pyruvate to acetyl CoA, another to oxalo-acetate, a third enzyme changes pyruvate to the amino acid alanine, a fourth to lactate, and so on. All of these different pathways compete for the same pyruvate molecule, and similar competitions for thousands of other small molecules go on at the same time.

The situation is further complicated in a multicellular organism. Different cell types will in general require somewhat different sets of enzymes. And different tissues make distinct contributions to the chemistry of the organism as a whole. In addition to differences in specialized products such as hormones or antibodies, there are significant differences in the "common" metabolic pathways among various types of cells in the same organism.

Although virtually all cells contain the enzymes of glycolysis, the citric acid cycle, lipid synthesis and breakdown, and amino acid metabolism, the levels of these processes required in different tissues are not the same. For example, nerve cells, which are probably the most fastidious cells in the body, maintain almost no reserves of glycogen or fatty acids and rely almost entirely on a constant

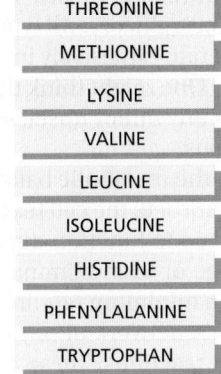

THE ESSENTIAL AMINO ACIDS

THREONINE
METHIONINE
LYSINE
VALINE
LEUCINE
ISOLEUCINE
HISTIDINE
PHENYLALANINE
TRYPTOPHAN

Figure 2–62 **The nine essential amino acids.** These cannot be synthesized by human cells and so must be supplied in the diet.

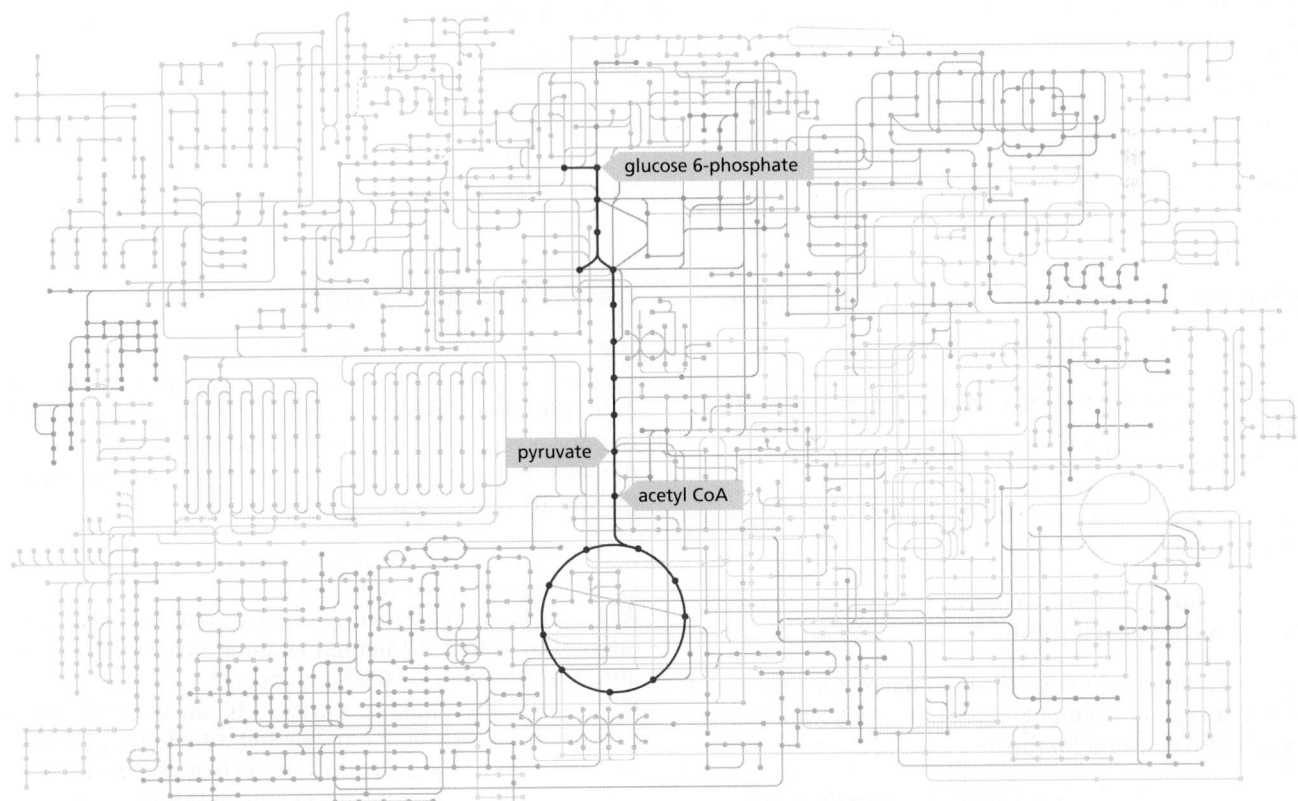

Figure 2–63 **Glycolysis and the citric acid cycle are at the center of an elaborate set of metabolic pathways in human cells.** Some 2000 metabolic reactions are shown schematically with the reactions of glycolysis and the citric acid cycle in *red*. Many other reactions either lead into these two central pathways—delivering small molecules to be catabolized with production of energy—or they lead outward and thereby supply carbon compounds for the purpose of biosynthesis. (Adapted with permission from Kanehisa Laboratories.)

supply of glucose from the bloodstream. In contrast, liver cells supply glucose to actively contracting muscle cells and recycle the lactic acid produced by muscle cells back into glucose. All types of cells have their distinctive metabolic traits, and they cooperate extensively in the normal state, as well as in response to stress and starvation. One might think that the whole system would need to be so finely balanced that any minor upset, such as a temporary change in dietary intake, would be disastrous.

In fact, the metabolic balance of a cell is amazingly stable. Whenever the balance is perturbed, the cell reacts so as to restore the initial state. The cell can adapt and continue to function during starvation or disease. Mutations of many kinds can damage or even eliminate particular reaction pathways, and yet—provided that certain minimum requirements are met—the cell survives. It does so because an elaborate network of *control mechanisms* regulates and coordinates the rates of all of its reactions. These controls rest, ultimately, on the remarkable abilities of proteins to change their shape and their chemistry in response to changes in their immediate environment. The principles that underlie how large molecules such as proteins are built and the chemistry behind their regulation will be our next concern.

Summary

Glucose and other food molecules are broken down by controlled stepwise oxidation to provide chemical energy in the form of ATP and NADH. There are three main sets of reactions that act in series, the products of each being the starting material for the next: glycolysis (which occurs in the cytosol), the citric acid cycle (in the mitochondrial matrix), and oxidative phosphorylation (on the inner mitochondrial membrane). The intermediate products of glycolysis and the citric acid cycle are used both as sources of metabolic energy and to produce many of the small molecules used as the raw materials for biosynthesis. Cells store sugar molecules as glycogen in animals and starch in plants; both plants and animals also use fats extensively as a food store. These storage materials in turn serve as a major source of food for humans, along with the proteins that comprise the majority of the dry mass of most of the cells in the foods we eat.

PROBLEMS

Which statements are true? Explain why or why not.

2–1   A 10^{-8} M solution of HCl has a pH of 8.

2–2   Most of the interactions between macromolecules could be mediated just as well by covalent bonds as by noncovalent bonds.

2–3   Animals and plants use oxidation to extract energy from food molecules.

2–4   If an oxidation occurs in a reaction, it must be accompanied by a reduction.

2–5   Linking the energetically unfavorable reaction A → B to a second, favorable reaction B → C will shift the equilibrium constant for the first reaction.

2–6   The criterion for whether a reaction proceeds spontaneously is ΔG not $\Delta G°$, because ΔG takes into account the concentrations of the substrates and products.

2–7   The oxygen consumed during the oxidation of glucose in animal cells is returned as CO_2 to the atmosphere.

WHAT WE DON'T KNOW

• Did chemiosmosis precede fermentation as the source of biological energy, or did some form of fermentation come first, as had been assumed for many years?

• What is the minimum number of components required to make a living cell from scratch? How might we find out?

• Are other life chemistries possible besides the single one known on Earth (and described in this chapter)? When screening for life on other planets, what type of chemical signatures should we search for?

• Is the shared chemistry inside all living cells a clue for deciphering the environment on Earth where the first cells originated? For example, what might we conclude from the universally shared high K^+/Na^+ ratio, neutral pH, and central role of phosphates?

Discuss the following problems.

2–8   The organic chemistry of living cells is said to be special for two reasons: it occurs in an aqueous environment and it accomplishes some very complex reactions. But do you suppose it is really all that much different from the organic chemistry carried out in the top laboratories in the world? Why or why not?

2–9   The molecular weight of ethanol (CH_3CH_2OH) is 46 and its density is 0.789 g/cm^3.
A.   What is the molarity of ethanol in beer that is 5% ethanol by volume? [Alcohol content of beer varies from about 4% (lite beer) to 8% (stout beer).]
B.   The legal limit for a driver's blood alcohol content varies, but 80 mg of ethanol per 100 mL of blood (usually referred to as a blood alcohol level of 0.08) is typical. What is the molarity of ethanol in a person at this legal limit?
C.   How many 12-oz (355-mL) bottles of 5% beer could a 70-kg person drink and remain under the legal limit? A 70-kg person contains about 40 liters of water. Ignore the metabolism of ethanol, and assume that the water content of the person remains constant.

D. Ethanol is metabolized at a constant rate of about 120 mg per hour per kg body weight, regardless of its concentration. If a 70-kg person were at twice the legal limit (160 mg/100 mL), how long would it take for their blood alcohol level to fall below the legal limit?

2–10 A histidine side chain is known to play an important role in the catalytic mechanism of an enzyme; however, it is not clear whether histidine is required in its protonated (charged) or unprotonated (uncharged) state. To answer this question you measure enzyme activity over a range of pH, with the results shown in **Figure Q2–1**. Which form of histidine is required for enzyme activity?

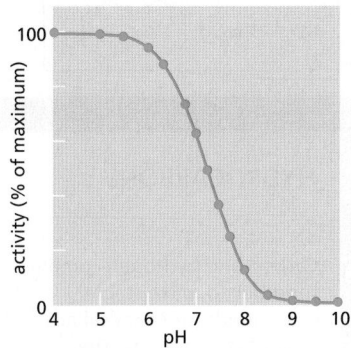

Figure Q2–1 Enzyme activity as a function of pH (Problem 2–10).

2–11 The three molecules in **Figure Q2–2** contain the seven most common reactive groups in biology. Most molecules in the cell are built from these functional groups. Indicate and name the functional groups in these molecules.

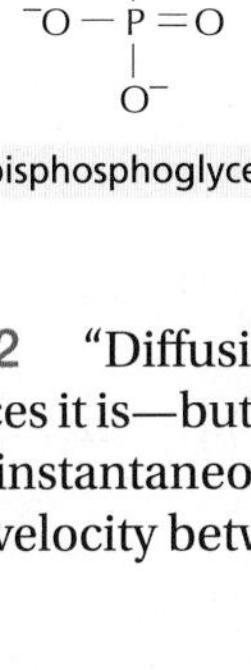

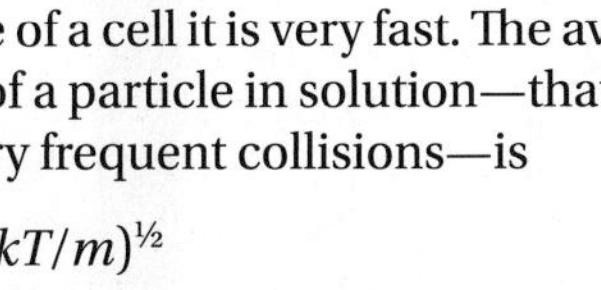

Figure Q2–2 Three molecules that illustrate the seven most common functional groups in biology (Problem 2–11). 1,3-Bisphosphoglycerate and pyruvate are intermediates in glycolysis and cysteine is an amino acid.

1,3-bisphosphoglycerate   pyruvate   cysteine

2–12 "Diffusion" sounds slow—and over everyday distances it is—but on the scale of a cell it is very fast. The average instantaneous velocity of a particle in solution—that is, the velocity between the very frequent collisions—is

$$v = (kT/m)^{1/2}$$

where $k = 1.38 \times 10^{-16}$ g cm^2/K sec^2, T = temperature in K (37°C is 310 K), and m = mass in g/molecule.

Calculate the instantaneous velocity of a water molecule (molecular mass = 18 daltons), a glucose molecule (molecular mass = 180 daltons), and a myoglobin molecule (molecular mass = 15,000 daltons) at 37°C. Just for fun, convert these numbers into kilometers/hour.

Before you do any calculations, try to guess whether the molecules are moving at a slow crawl (<1 km/hr), an easy walk (5 km/hr), or a record-setting sprint (40 km/hr).

2–13 Polymerization of tubulin subunits into microtubules occurs with an increase in the orderliness of the subunits. Yet tubulin polymerization occurs with an increase in entropy (decrease in order). How can that be?

2–14 A 70-kg adult human (154 lb) could meet his or her entire energy needs for one day by eating 3 moles of glucose (540 g). (We do not recommend this.) Each molecule of glucose generates 30 molecules of ATP when it is oxidized to CO_2. The concentration of ATP is maintained in cells at about 2 mM, and a 70-kg adult has about 25 liters of intracellular fluid. Given that the ATP concentration remains constant in cells, calculate how many times per day, on average, each ATP molecule in the body is hydrolyzed and resynthesized.

2–15 Assuming that there are 5×10^{13} cells in the human body and that ATP is turning over at a rate of 10^9 ATP molecules per minute in each cell, how many watts is the human body consuming? (A watt is a joule per second.) Assume that hydrolysis of ATP yields 50 kJ/mole.

2–16 Does a Snickers™ candy bar (65 g, 1360 kJ) provide enough energy to climb from Zermatt (elevation 1660 m) to the top of the Matterhorn (4478 m, **Figure Q2–3**), or might you need to stop at Hörnli Hut (3260 m) to eat another one? Imagine that you and your gear have a mass of 75 kg, and that all of your work is done against gravity (that is, you are just climbing straight up). Remember from your introductory physics course that

work (J) = mass (kg) × g (m/sec^2) × height gained (m)

where g is acceleration due to gravity (9.8 m/sec^2). One joule is 1 kg m^2/sec^2.

What assumptions made here will greatly underestimate how much candy you need?

Figure Q2–3 The Matterhorn (Problem 2–16). (Earth Trotter Photos/Shutterstock.)

2–17 In the absence of oxygen, cells consume glucose at a high, steady rate. When oxygen is added, glucose consumption drops precipitously and is then maintained at the lower rate. Why is glucose consumed at a high rate in the absence of oxygen and at a low rate in its presence?

CARBON SKELETONS

Carbon has a unique role in the cell because of its ability to form strong covalent bonds with other carbon atoms. Thus carbon atoms can join to form:

chains   branched trees   rings

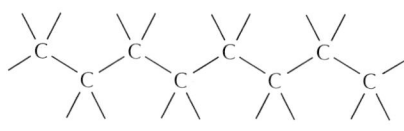

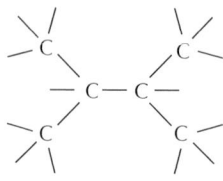

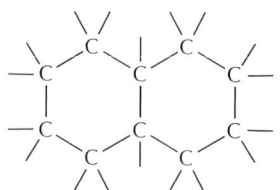

also written as /\/\/\/   also written as ⟩—⟨   also written as ⬡⬡

COVALENT BONDS

A covalent bond forms when two atoms come very close together and share one or more of their electrons. In a single bond, one electron from each of the two atoms is shared; in a double bond, a total of four electrons are shared.

Each atom forms a fixed number of covalent bonds in a defined spatial arrangement. For example, carbon forms four single bonds arranged tetrahedrally, whereas nitrogen forms three single bonds and oxygen forms two single bonds arranged as shown below.

   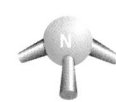   

Double bonds exist and have a different spatial arrangement.

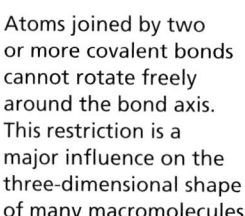

Atoms joined by two or more covalent bonds cannot rotate freely around the bond axis. This restriction is a major influence on the three-dimensional shape of many macromolecules.

HYDROCARBONS

Carbon and hydrogen combine together to make stable compounds (or chemical groups) called hydrocarbons. These are nonpolar, do not form hydrogen bonds, and are generally insoluble in water.

$$
\begin{array}{ccc}
& H & & H \\
| & | & & | \\
H-C-H & & H-C- \\
| & & & | \\
& H & & H
\end{array}
$$

methane   methyl **group**

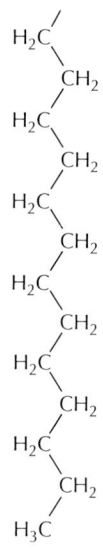

part of the hydrocarbon "tail" of a fatty acid molecule

ALTERNATING DOUBLE BONDS

The carbon chain can include double bonds. If these are on alternate carbon atoms, the bonding electrons move within the molecule, stabilizing the structure by a phenomenon called resonance.

Alternating double bonds in a ring can generate a very stable structure.

the truth is somewhere between these two structures

benzene

often written as ⬡

C–O CHEMICAL GROUPS

Many biological compounds contain a carbon bonded to an oxygen. For example,

alcohol

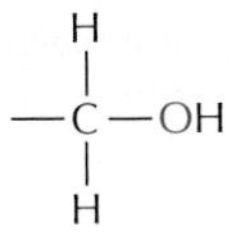

The –OH is called a hydroxyl group.

aldehyde

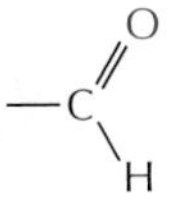

ketone

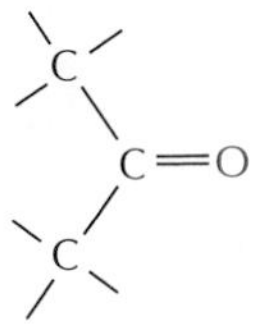

The C=O is called a carbonyl group.

carboxylic acid

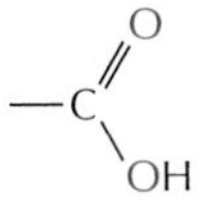

The –COOH is called a carboxyl group. In water this loses an H⁺ ion to become –COO⁻.

esters

Esters are formed by a condensation reaction between an acid and an alcohol.

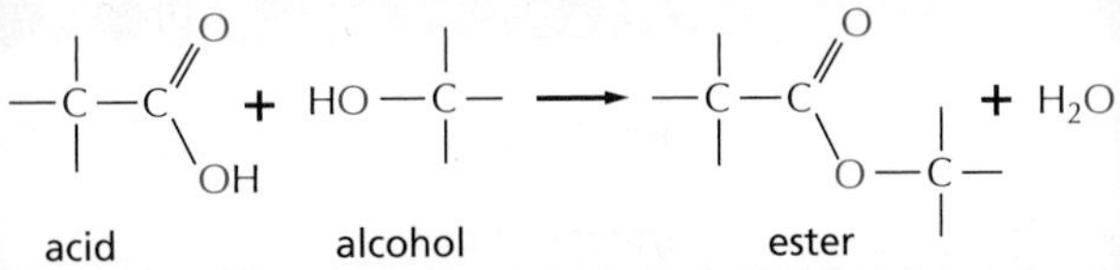

acid   alcohol   ester

C–N CHEMICAL GROUPS

Amines and amides are two important examples of compounds containing a carbon linked to a nitrogen.

Amines in water combine with an H⁺ ion to become positively charged.

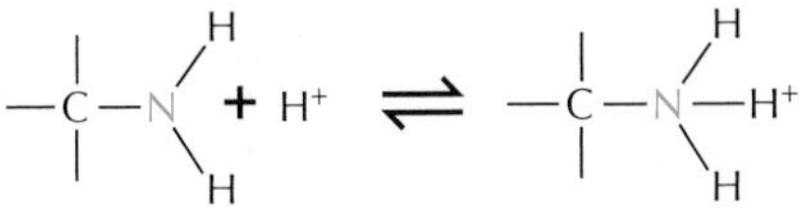

Amides are formed by combining an acid and an amine. Unlike amines, amides are uncharged in water. An example is the peptide bond that joins amino acids in a protein.

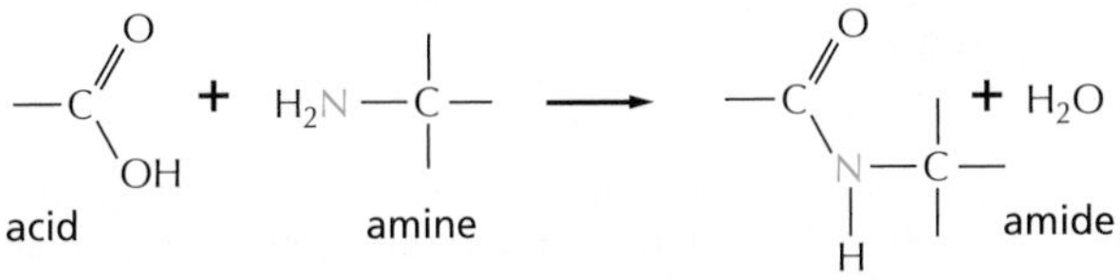

acid   amine   amide

Nitrogen also occurs in several ring compounds, including important constituents of nucleic acids: purines and pyrimidines.

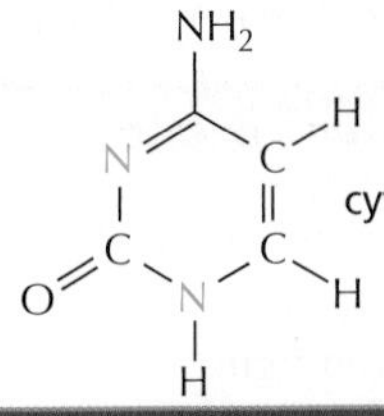

cytosine (a pyrimidine)

SULFHYDRYL GROUP

The $-\overset{|}{\underset{|}{C}}-SH$ is called a sulfhydryl group. In the amino acid cysteine, the sulfhydryl group may exist in the reduced form, $-\overset{|}{\underset{|}{C}}-SH$ or more rarely in an oxidized, cross-bridging form, $-\overset{|}{\underset{|}{C}}-S-S-\overset{|}{\underset{|}{C}}-$

PHOSPHATES

Inorganic phosphate is a stable ion formed from phosphoric acid, H_3PO_4. It is also written as Ⓟ.

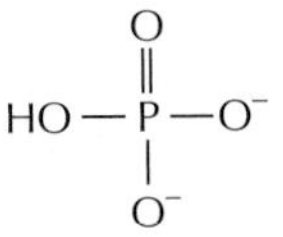

Phosphate esters can form between a phosphate and a free hydroxyl group. Phosphate groups are often attached to proteins in this way.

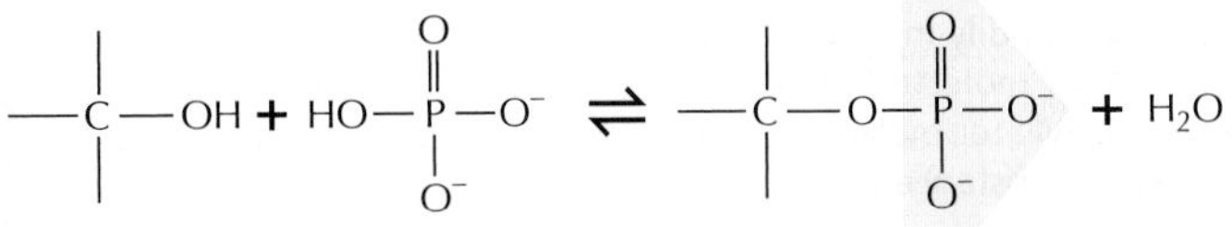

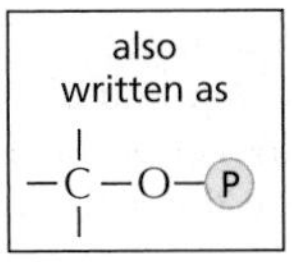

also written as

$-\overset{|}{\underset{|}{C}}-O-$Ⓟ

The combination of a phosphate and a carboxyl group, or two or more phosphate groups, gives an acid anhydride. Because compounds of this kind are easily hydrolysed in the cell, they are sometimes said to contain a "high-energy" bond.

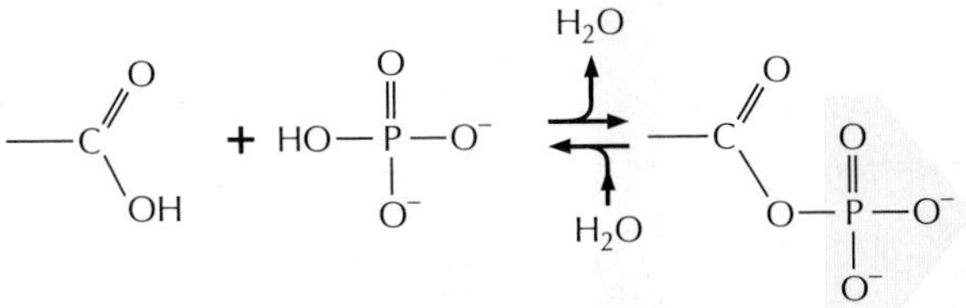

high-energy acyl phosphate bond (carboxylic–phosphoric acid anhydride) found in some metabolites

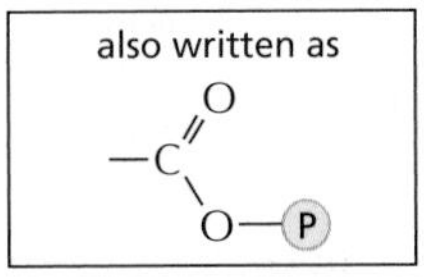

also written as

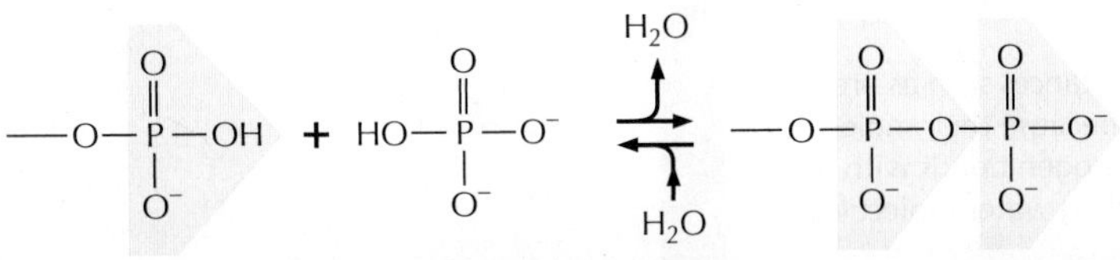

phosphoanhydride—a high-energy bond found in molecules such as ATP

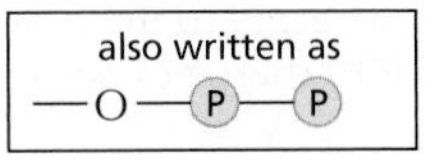

also written as

$-O-$Ⓟ$-$Ⓟ

WATER

Two atoms, connected by a covalent bond, may exert different attractions for the electrons of the bond. In such cases the bond is polar, with one end slightly negatively charged (δ^-) and the other slightly positively charged (δ^+).

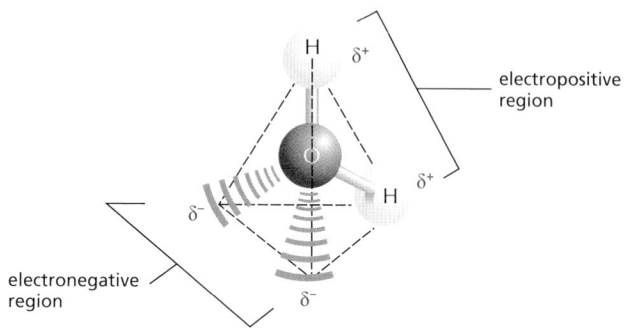

Although a water molecule has an overall neutral charge (having the same number of electrons and protons), the electrons are asymmetrically distributed, which makes the molecule polar. The oxygen nucleus draws electrons away from the hydrogen nuclei, leaving these nuclei with a small net positive charge. The excess of electron density on the oxygen atom creates weakly negative regions at the other two corners of an imaginary tetrahedron.

WATER STRUCTURE

Molecules of water join together transiently in a hydrogen-bonded lattice. Even at 37°C, 15% of the water molecules are joined to four others in a short-lived assembly known as a "flickering cluster."

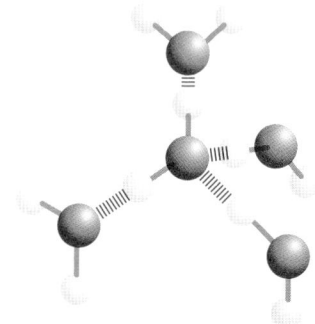

The cohesive nature of water is responsible for many of its unusual properties, such as high surface tension, specific heat, and heat of vaporization.

HYDROGEN BONDS

Because they are polarized, two adjacent H_2O molecules can form a linkage known as a hydrogen bond. Hydrogen bonds have only about 1/20 the strength of a covalent bond.

Hydrogen bonds are strongest when the three atoms lie in a straight line.

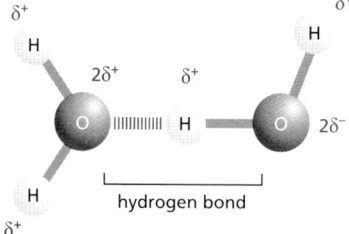

bond lengths

hydrogen bond
0.17 nm

H—O—

0.10 nm
covalent bond

HYDROPHILIC MOLECULES

Substances that dissolve readily in water are termed hydrophilic. They are composed of ions or polar molecules that attract water molecules through electrical charge effects. Water molecules surround each ion or polar molecule on the surface of a solid substance and carry it into solution.

Ionic substances such as sodium chloride dissolve because water molecules are attracted to the positive (Na^+) or negative (Cl^-) charge of each ion.

Polar substances such as urea dissolve because their molecules form hydrogen bonds with the surrounding water molecules.

HYDROPHOBIC MOLECULES

Molecules that contain a preponderance of nonpolar bonds are usually insoluble in water and are termed hydrophobic. This is true, especially, of hydrocarbons, which contain many C–H bonds. Water molecules are not attracted to such molecules and so have little tendency to surround them and carry them into solution.

WATER AS A SOLVENT

Many substances, such as household sugar, dissolve in water. That is, their molecules separate from each other, each becoming surrounded by water molecules.

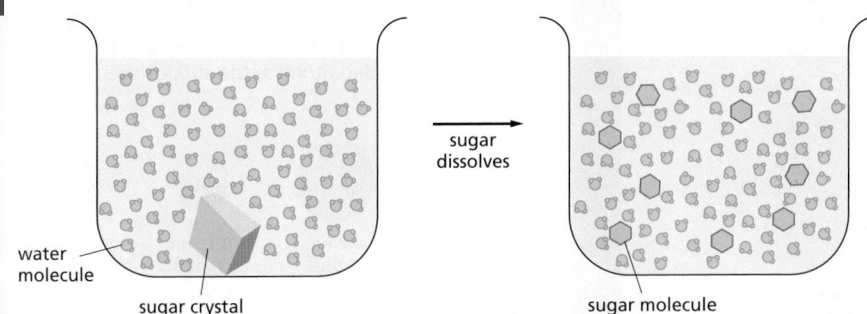

water molecule

sugar crystal

sugar dissolves

sugar molecule

When a substance dissolves in a liquid, the mixture is termed a solution. The dissolved substance (in this case sugar) is the solute, and the liquid that does the dissolving (in this case water) is the solvent. Water is an excellent solvent for many substances because of its polar bonds.

ACIDS

Substances that release hydrogen ions into solution are called acids.

$$HCl \longrightarrow H^+ + Cl^-$$

hydrochloric acid   hydrogen ion   chloride ion
(strong acid)

Many of the acids important in the cell are only partially dissociated, and they are therefore weak acids—for example, the carboxyl group (–COOH), which dissociates to give a hydrogen ion in solution.

$$-C\overset{O}{\underset{OH}{\big\langle}} \rightleftharpoons H^+ + -C\overset{O}{\underset{O^-}{\big\langle}}$$

(weak acid)

Note that this is a reversible reaction.

HYDROGEN ION EXCHANGE

Positively charged hydrogen ions (H^+) can spontaneously move from one water molecule to another, thereby creating two ionic species.

hydronium ion   hydroxyl ion
(water acting as   (water acting as
a weak base)   a weak acid)

often written as: $H_2O \rightleftharpoons H^+ + OH^-$

hydrogen ion   hydroxyl ion

Since the process is rapidly reversible, hydrogen ions are continually shuttling between water molecules. Pure water contains a steady-state concentration of hydrogen ions and hydroxyl ions (both 10^{-7} M).

pH

The acidity of a solution is defined by the concentration of H^+ ions it possesses. For convenience we use the pH scale, where

$$pH = -\log_{10}[H^+]$$

For pure water

$$[H^+] = 10^{-7} \text{ moles/liter}$$

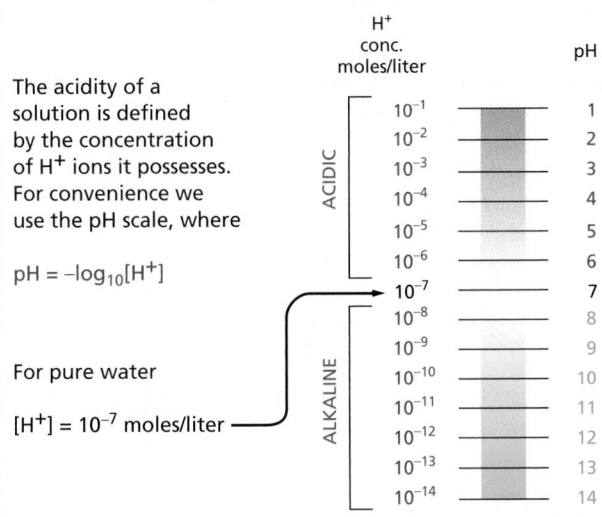

H^+ conc. moles/liter		pH
ACIDIC	10^{-1}	1
	10^{-2}	2
	10^{-3}	3
	10^{-4}	4
	10^{-5}	5
	10^{-6}	6
	10^{-7}	7
ALKALINE	10^{-8}	8
	10^{-9}	9
	10^{-10}	10
	10^{-11}	11
	10^{-12}	12
	10^{-13}	13
	10^{-14}	14

BASES

Substances that reduce the number of hydrogen ions in solution are called bases. Some bases, such as ammonia, combine directly with hydrogen ions.

$$NH_3 + H^+ \longrightarrow NH_4^+$$

ammonia   hydrogen ion   ammonium ion

Other bases, such as sodium hydroxide, reduce the number of H^+ ions indirectly, by making OH^- ions that then combine directly with H^+ ions to make H_2O.

$$NaOH \longrightarrow Na^+ + OH^-$$

sodium hydroxide   sodium ion   hydroxyl ion
(strong base)

Many bases found in cells are partially associated with H^+ ions and are termed weak bases. This is true of compounds that contain an amino group (–NH_2), which has a weak tendency to reversibly accept an H^+ ion from water, increasing the quantity of free OH^- ions.

$$-NH_2 + H^+ \rightleftharpoons -NH_3^+$$

WEAK NONCOVALENT CHEMICAL BONDS

Organic molecules can interact with other molecules through three types of short-range attractive forces known as *noncovalent bonds*: van der Waals attractions, electrostatic attractions, and hydrogen bonds. The repulsion of hydrophobic groups from water is also important for the folding of biological macromolecules.

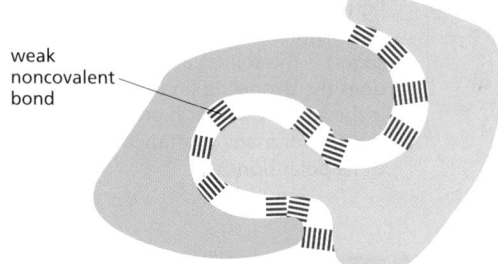

weak noncovalent bond

Weak noncovalent chemical bonds have less than 1/20 the strength of a strong covalent bond. They are strong enough to provide tight binding only when many of them are formed simultaneously.

HYDROGEN BONDS

As already described for water (see Panel 2–2), hydrogen bonds form when a hydrogen atom is "sandwiched" between two electron-attracting atoms (usually oxygen or nitrogen).

Hydrogen bonds are strongest when the three atoms are in a straight line:

O—H ||||||||||O   N—H ||||||||||O

Examples in macromolecules:

Amino acids in a polypeptide chain can be hydrogen-bonded together. These stabilize the structure of folded proteins.

C=O |||||||||| H—N

R—C—H   R—C—H   H—C—R

C=O |||||||||| H—N

Two bases, G and C, are hydrogen-bonded in a DNA double helix.

VAN DER WAALS ATTRACTIONS

If two atoms are too close together they repel each other very strongly. For this reason, an atom can often be treated as a sphere with a fixed radius. The characteristic "size" for each atom is specified by a unique van der Waals radius. The contact distance between any two noncovalently bonded atoms is the sum of their van der Waals radii.

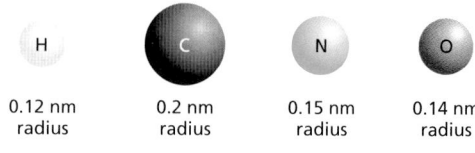

H	C	N	O
0.12 nm radius	0.2 nm radius	0.15 nm radius	0.14 nm radius

At very short distances any two atoms show a weak bonding interaction due to their fluctuating electrical charges. The two atoms will be attracted to each other in this way until the distance between their nuclei is approximately equal to the sum of their van der Waals radii. Although they are individually very weak, van der Waals attractions can become important when two macromolecular surfaces fit very close together, because many atoms are involved.

Note that when two atoms form a covalent bond, the centers of the two atoms (the two atomic nuclei) are much closer together than the sum of the two van der Waals radii. Thus,

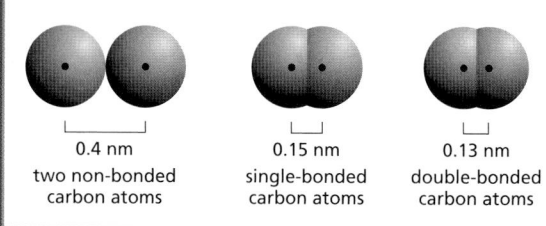

0.4 nm
two non-bonded carbon atoms

0.15 nm
single-bonded carbon atoms

0.13 nm
double-bonded carbon atoms

HYDROGEN BONDS IN WATER

Any molecules that can form hydrogen bonds to each other can alternatively form hydrogen bonds to water molecules. Because of this competition with water molecules, the hydrogen bonds formed between two molecules dissolved in water are relatively weak.

peptide bond

$2H_2O$

$2H_2O$

HYDROPHOBIC FORCES

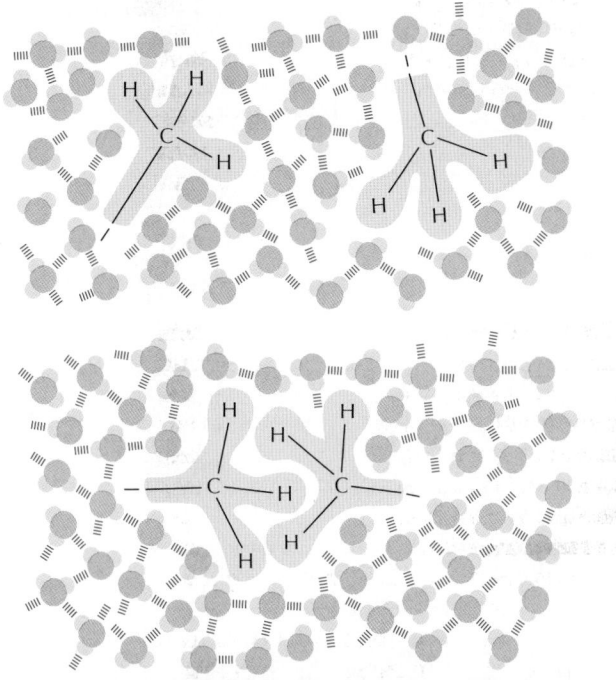

Water forces hydrophobic groups together, because doing so minimizes their disruptive effects on the hydrogen-bonded water network. Hydrophobic groups held together in this way are sometimes said to be held together by "hydrophobic bonds," even though the apparent attraction is actually caused by a repulsion from the water.

ELECTROSTATIC ATTRACTIONS IN AQUEOUS SOLUTIONS

Charged groups are shielded by their interactions with water molecules. Electrostatic attractions are therefore quite weak in water.

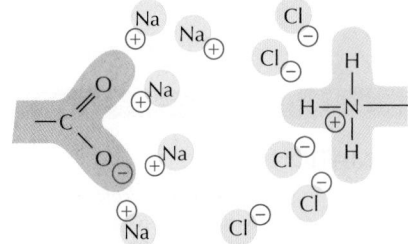

Similarly, ions in solution can cluster around charged groups and further weaken these attractions.

ELECTROSTATIC ATTRACTIONS

Attractive forces occur both between fully charged groups (ionic bond) and between the partially charged groups on polar molecules.

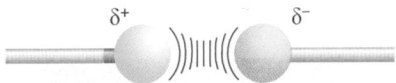

The force of attraction between the two charges, δ^+ and δ^-, falls off rapidly as the distance between the charges increases.

In the absence of water, electrostatic forces are very strong. They are responsible for the strength of such minerals as marble and agate, and for crystal formation in common table salt, NaCl.

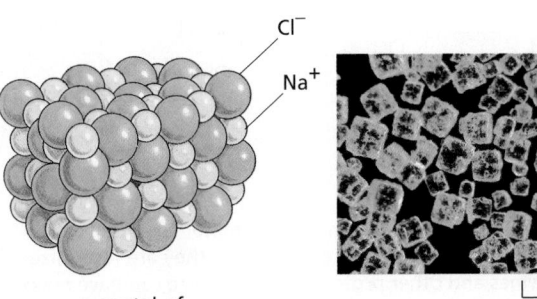

a crystal of salt, NaCl

1 mm

Despite being weakened by water and salt, electrostatic attractions are very important in biological systems. For example, an enzyme that binds a positively charged substrate will often have a negatively charged amino acid side chain at the appropriate place.

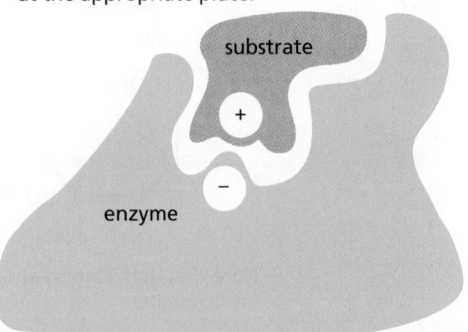

substrate

enzyme

MONOSACCHARIDES

Monosaccharides usually have the general formula $(CH_2O)_n$, where n can be 3, 4, 5, 6, 7, or 8, and have two or more hydroxyl groups. They either contain an aldehyde group ($-C\underset{H}{\overset{O}{\lessgtr}}$) and are called aldoses, or a ketone group ($\overset{}{\underset{}{>}}C=O$) and are called ketoses.

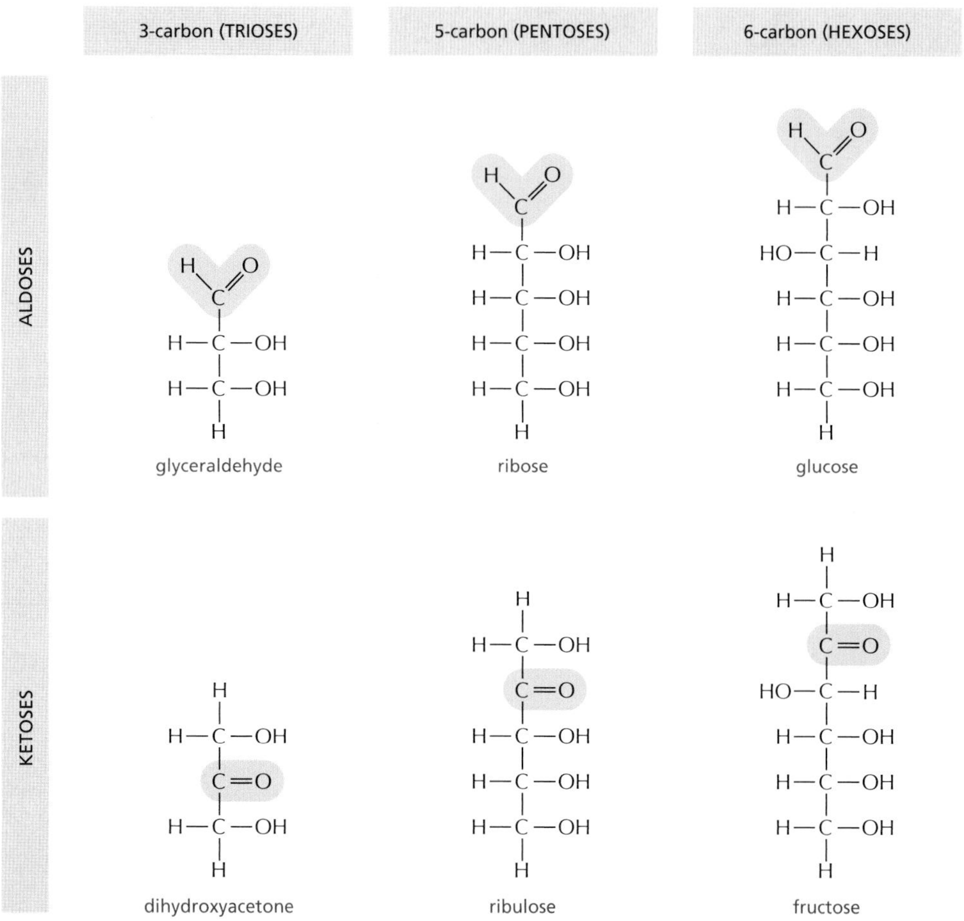

	3-carbon (TRIOSES)	5-carbon (PENTOSES)	6-carbon (HEXOSES)
ALDOSES	glyceraldehyde	ribose	glucose
KETOSES	dihydroxyacetone	ribulose	fructose

RING FORMATION

In aqueous solution, the aldehyde or ketone group of a sugar molecule tends to react with a hydroxyl group of the same molecule, thereby closing the molecule into a ring.

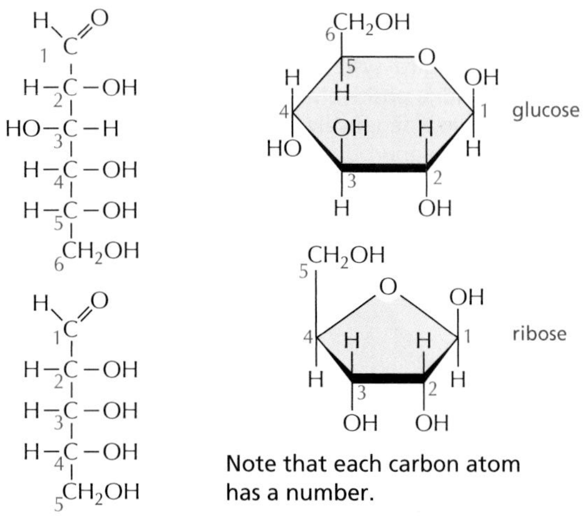

glucose

ribose

Note that each carbon atom has a number.

ISOMERS

Many monosaccharides differ only in the spatial arrangement of atoms—that is, they are isomers. For example, glucose, galactose, and mannose have the same formula ($C_6H_{12}O_6$) but differ in the arrangement of groups around one or two carbon atoms.

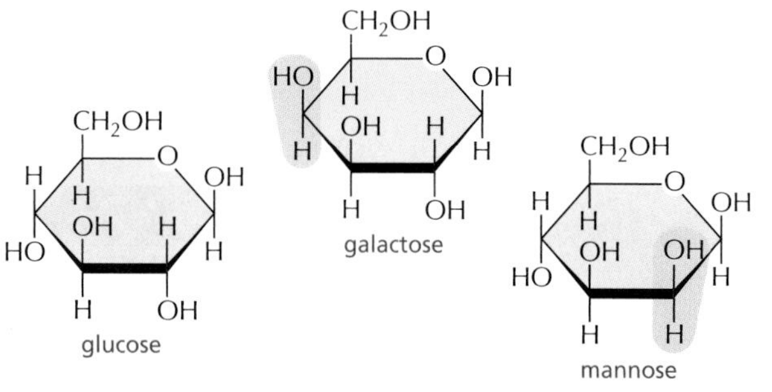

galactose

glucose

mannose

These small differences make only minor changes in the chemical properties of the sugars. But they are recognized by enzymes and other proteins and therefore can have major biological effects.

α AND β LINKS

The hydroxyl group on the carbon that carries the aldehyde or ketone can rapidly change from one position to the other. These two positions are called α and β.

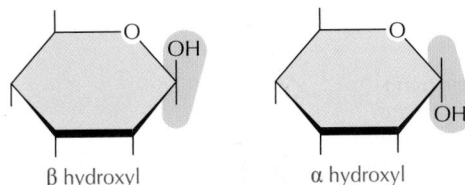

β hydroxyl   α hydroxyl

As soon as one sugar is linked to another, the α or β form is frozen.

SUGAR DERIVATIVES

The hydroxyl groups of a simple monosaccharide such as glucose can be replaced by other groups. For example,

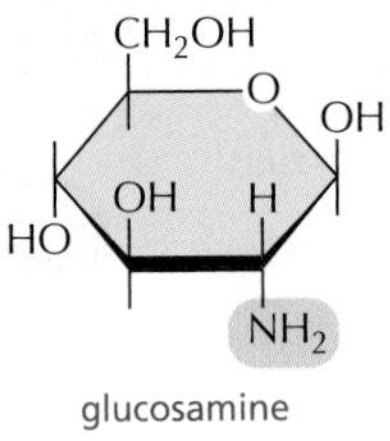

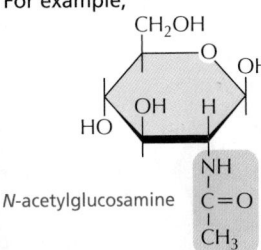

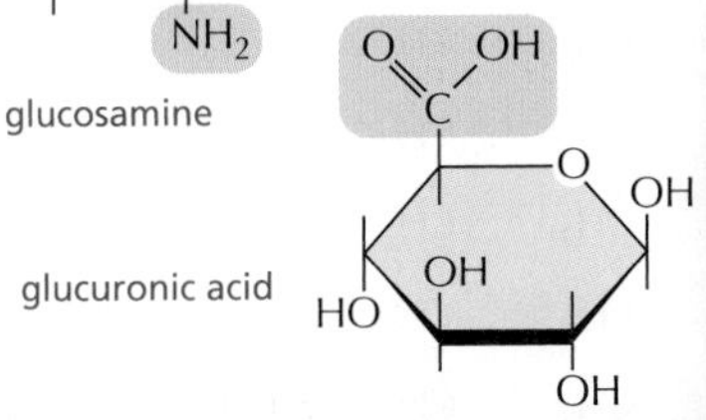

glucosamine

N-acetylglucosamine

glucuronic acid

DISACCHARIDES

The carbon that carries the aldehyde or the ketone can react with any hydroxyl group on a second sugar molecule to form a disaccharide. The linkage is called a glycosidic bond.

Three common disaccharides are

   maltose (glucose + glucose)
   lactose (galactose + glucose)
   sucrose (glucose + fructose)

The reaction forming sucrose is shown here.

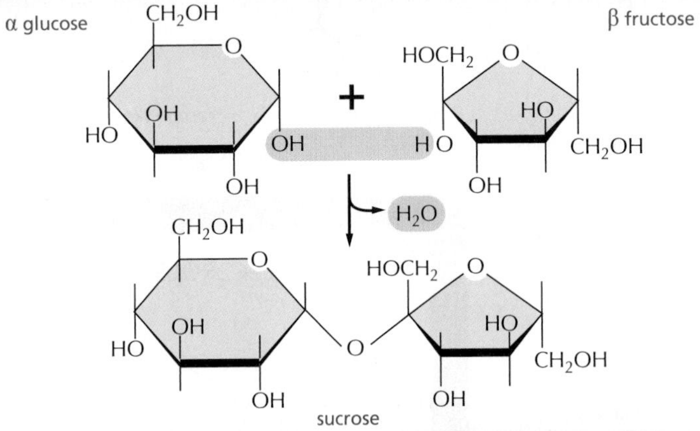

α glucose   β fructose

+

H_2O

sucrose

OLIGOSACCHARIDES AND POLYSACCHARIDES

Large linear and branched molecules can be made from simple repeating sugar subunits. Short chains are called oligosaccharides, while long chains are called polysaccharides. Glycogen, for example, is a polysaccharide made entirely of glucose units joined together.

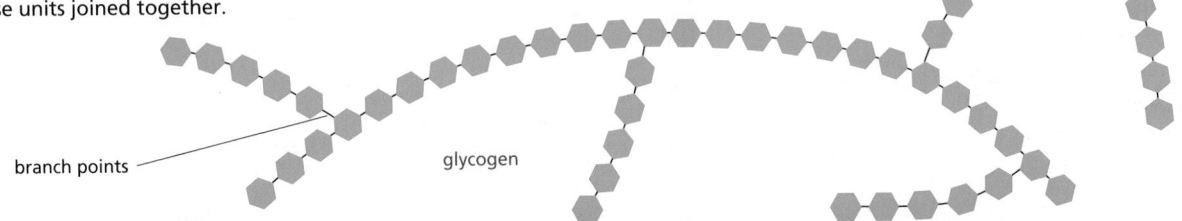

branch points   glycogen

COMPLEX OLIGOSACCHARIDES

In many cases a sugar sequence is nonrepetitive. Many different molecules are possible. Such complex oligosaccharides are usually linked to proteins or to lipids, as is this oligosaccharide, which is part of a cell-surface molecule that defines a particular blood group.

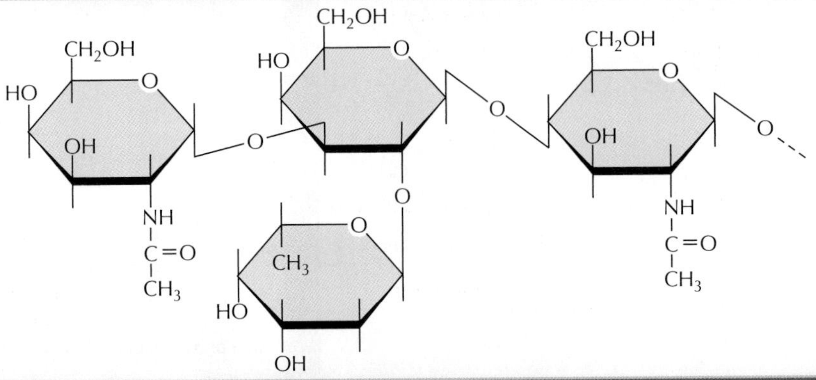

COMMON FATTY ACIDS

These are carboxylic acids with long hydrocarbon tails.

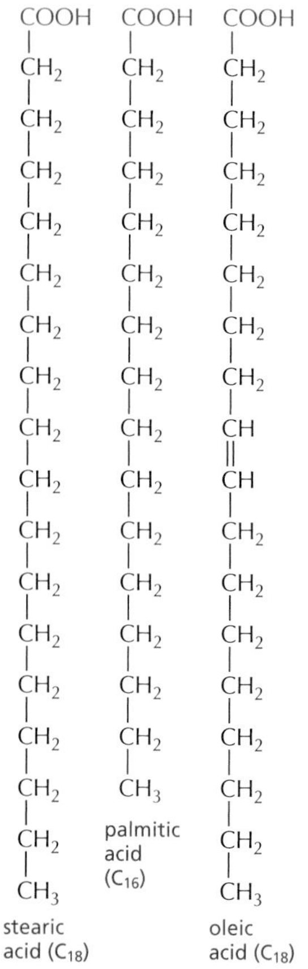

COOH COOH COOH
| | |
CH₂   CH₂   CH₂
| | |
CH₂   CH₂   CH₂
| | |
CH₂   CH₂   CH₂
| | |
CH₂   CH₂   CH₂
| | |
CH₂   CH₂   CH₂
| | |
CH₂   CH₂   CH₂
| | |
CH₂   CH₂   CH₂
| | |
CH₂   CH₂   CH
| | ‖
CH₂   CH₂   CH
| | |
CH₂   CH₂   CH₂
| | |
CH₂   CH₂   CH₂
| | |
CH₂   CH₂   CH₂
| | |
CH₂   CH₂   CH₂
| | |
CH₂   CH₃   CH₂
| palmitic |
CH₂   acid   CH₂
| (C₁₆) |
CH₃   CH₃
stearic   oleic
acid (C₁₈) acid (C₁₈)

Hundreds of different kinds of fatty acids exist. Some have one or more double bonds in their hydrocarbon tail and are said to be unsaturated. Fatty acids with no double bonds are saturated.

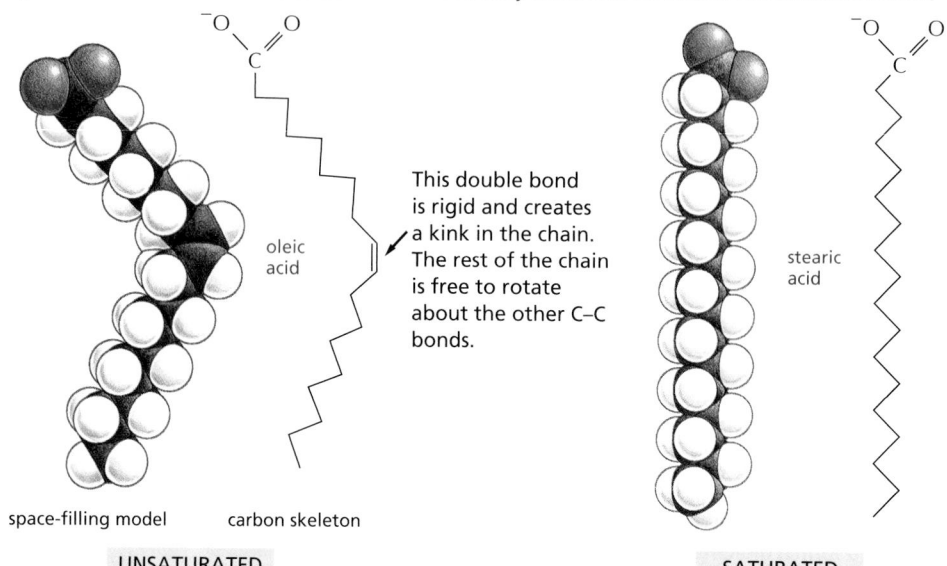

oleic acid

This double bond is rigid and creates a kink in the chain. The rest of the chain is free to rotate about the other C–C bonds.

space-filling model   carbon skeleton

UNSATURATED

stearic acid

SATURATED

TRIACYLGLYCEROLS

Fatty acids are stored as an energy reserve (fats and oils) through an ester linkage to glycerol to form triacylglycerols, also known as triglycerides.

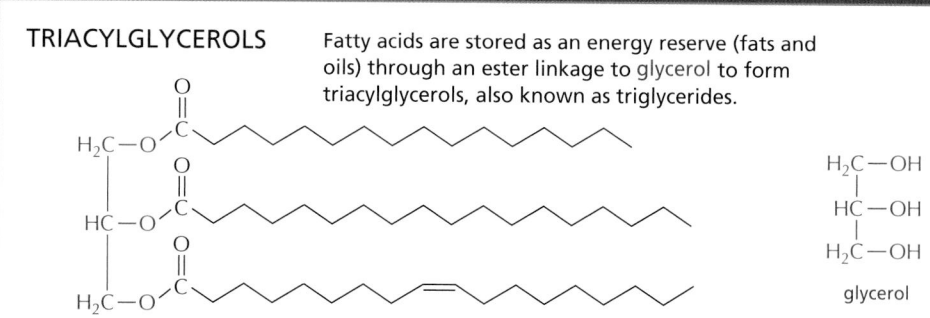

H₂C—OH
|
HC—OH
|
H₂C—OH

glycerol

CARBOXYL GROUP

If free, the carboxyl group of a fatty acid will be ionized.

But more usually it is linked to other groups to form either esters

or amides.

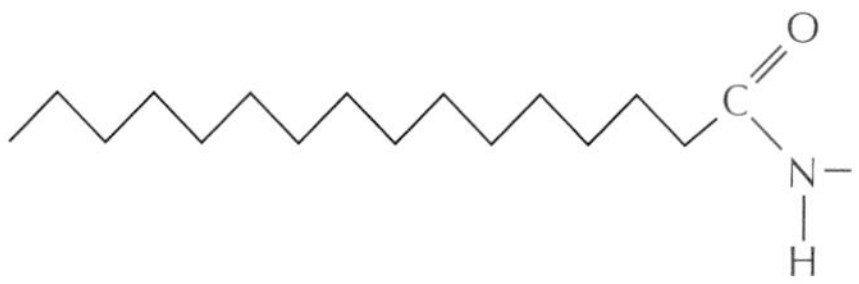

PHOSPHOLIPIDS

Phospholipids are the major constituents of cell membranes.

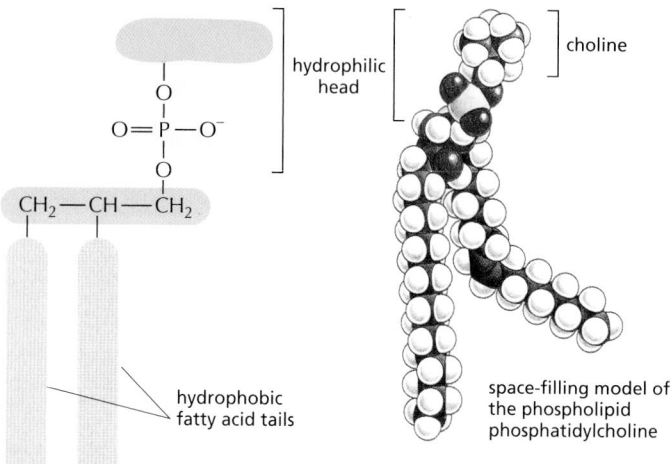

choline

hydrophilic head

O
‖
O=P—O⁻
|
O
|
CH₂—CH—CH₂

hydrophobic fatty acid tails

space-filling model of the phospholipid phosphatidylcholine

general structure of a phospholipid

In phospholipids, two of the –OH groups in glycerol are linked to fatty acids, while the third –OH group is linked to phosphoric acid. The phosphate is further linked to one of a variety of small polar groups, such as choline.

LIPID AGGREGATES

Fatty acids have a hydrophilic head
and a hydrophobic tail.

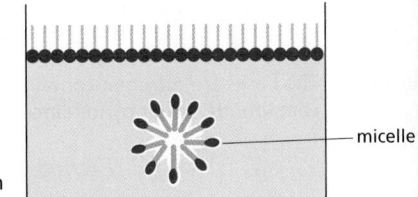

In water they can form a surface film
or form small micelles.

— micelle

Their derivatives can form larger aggregates held together by hydrophobic forces:

Triacylglycerols (triglycerides) can form
large spherical fat droplets in the cell
cytoplasm.

200 nm
or more

Phospholipids and glycolipids form self-sealing lipid
bilayers that are the basis for all cell membranes.

◄— 4 nm —►

OTHER LIPIDS

Lipids are defined as the water-insoluble
molecules in cells that are soluble in organic
solvents. Two other common types of lipids
are steroids and polyisoprenoids. Both are
made from isoprene units.

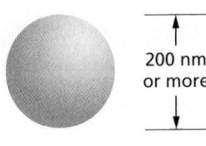

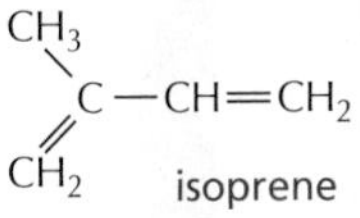

isoprene

STEROIDS

Steroids have a common multiple-ring structure.

HO

cholesterol—found in many membranes

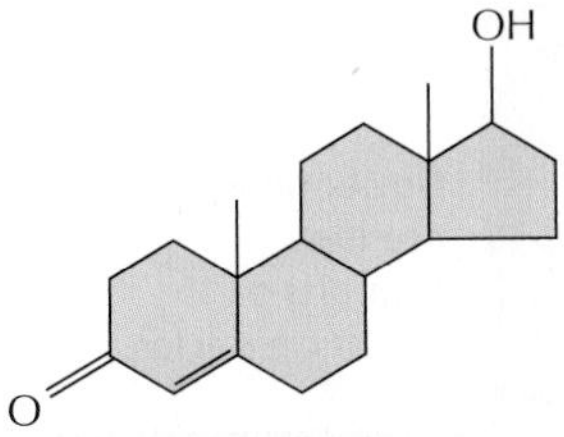

OH

O

testosterone—male steroid hormone

GLYCOLIPIDS

Like phospholipids, these compounds are composed of a hydrophobic
region, containing two long hydrocarbon tails and a polar region,
which contains one or more sugars and, unlike phospholipids,
no phosphate.

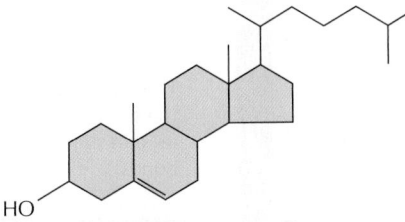

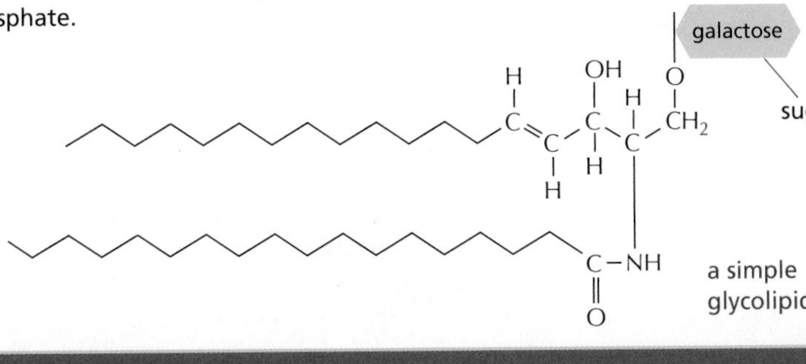

galactose

sugar

a simple
glycolipid

POLYISOPRENOIDS

long-chain polymers of isoprene

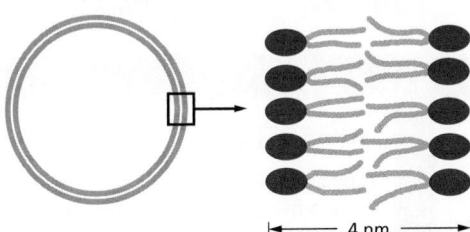

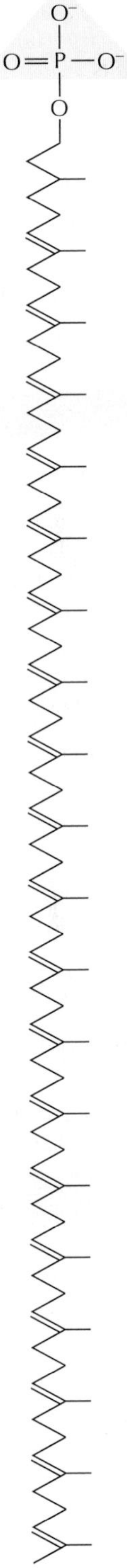

dolichol phosphate—used
to carry activated sugars
in the membrane-associated
synthesis of glycoproteins
and some polysaccharides

BASES

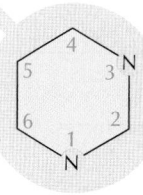

cytosine

uracil

thymine

The bases are nitrogen-containing ring compounds, either pyrimidines or purines.

PYRIMIDINE

PURINE

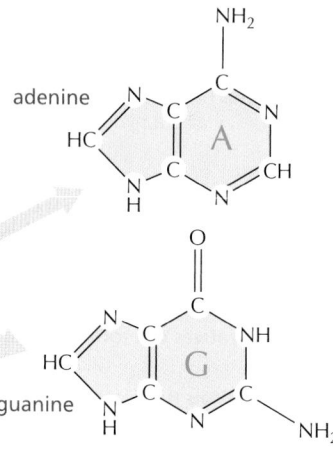

adenine

guanine

PHOSPHATES

The phosphates are normally joined to the C5 hydroxyl of the ribose or deoxyribose sugar (designated 5'). Mono-, di-, and triphosphates are common.

as in AMP

as in ADP

as in ATP

The phosphate makes a nucleotide negatively charged.

NUCLEOTIDES

A nucleotide consists of a nitrogen-containing base, a five-carbon sugar, and one or more phosphate groups.

BASE

PHOSPHATE

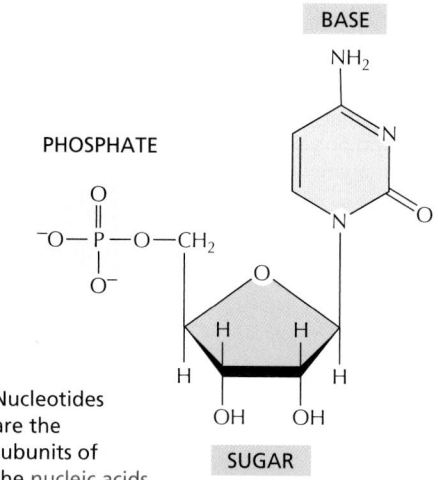

SUGAR

Nucleotides are the subunits of the nucleic acids.

BASIC SUGAR LINKAGE

N-glycosidic bond

BASE

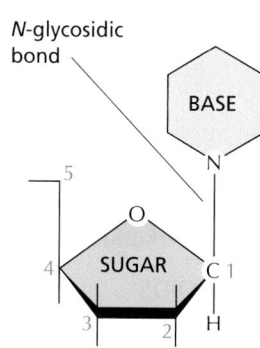

SUGAR

The base is linked to the same carbon (C1) used in sugar–sugar bonds.

SUGARS

PENTOSE

a five-carbon sugar

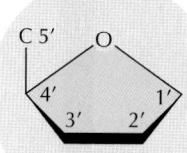

two kinds are used

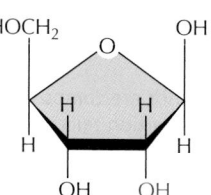

β-D-ribose
used in ribonucleic acid

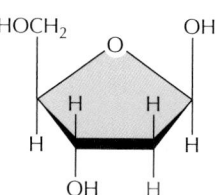

β-D-2-deoxyribose
used in deoxyribonucleic acid

Each numbered carbon on the sugar of a nucleotide is followed by a prime mark; therefore, one speaks of the "5-prime carbon," etc.

NOMENCLATURE

A nucleoside or nucleotide is named according to its nitrogenous base.

Single-letter abbreviations are used variously as shorthand for (1) the base alone, (2) the nucleoside, or (3) the whole nucleotide— the context will usually make clear which of the three entities is meant. When the context is not sufficient, we will add the terms "base", "nucleoside", "nucleotide", or—as in the examples below—use the full 3-letter nucleotide code.

BASE	NUCLEOSIDE	ABBR.
adenine	adenosine	A
guanine	guanosine	G
cytosine	cytidine	C
uracil	uridine	U
thymine	thymidine	T

AMP = adenosine monophosphate
dAMP = deoxyadenosine monophosphate
UDP = uridine diphosphate
ATP = adenosine triphosphate

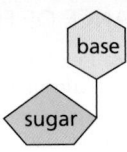

BASE + SUGAR = NUCLEOSIDE

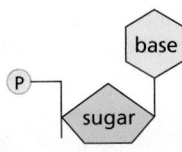

BASE + SUGAR + PHOSPHATE = NUCLEOTIDE

NUCLEIC ACIDS

Nucleotides are joined together by a phosphodiester linkage between 5' and 3' carbon atoms to form nucleic acids. The linear sequence of nucleotides in a nucleic acid chain is commonly abbreviated by a one-letter code, such as A—G—C—T—T—A—C—A, with the 5' end of the chain at the left.

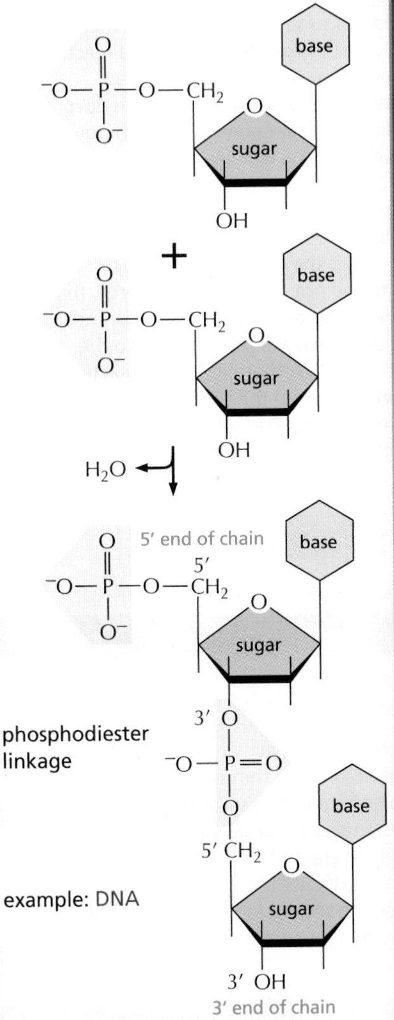

phosphodiester linkage

example: DNA

NUCLEOTIDES HAVE MANY OTHER FUNCTIONS

1 They carry chemical energy in their easily hydrolyzed phosphoanhydride bonds.

phosphoanhydride bonds

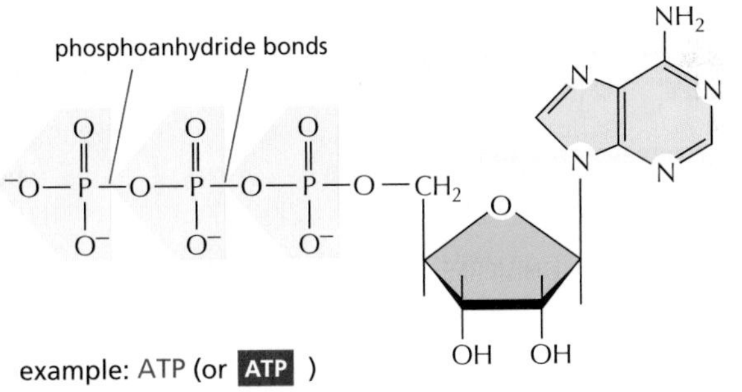

example: ATP (or ATP)

2 They combine with other groups to form coenzymes.

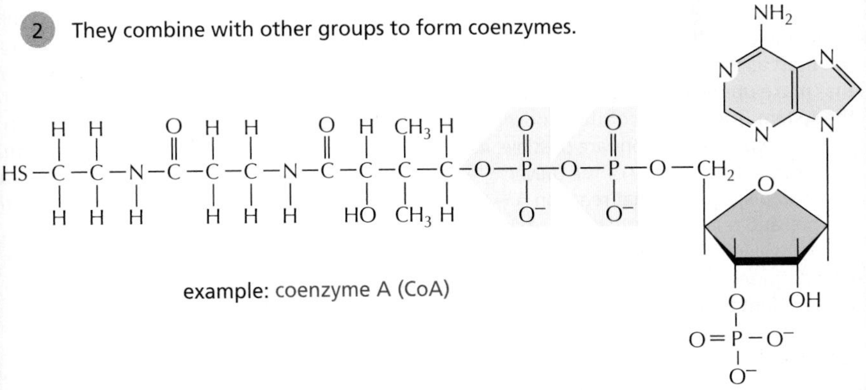

example: coenzyme A (CoA)

3 They are used as specific signaling molecules in the cell.

example: cyclic AMP (cAMP)

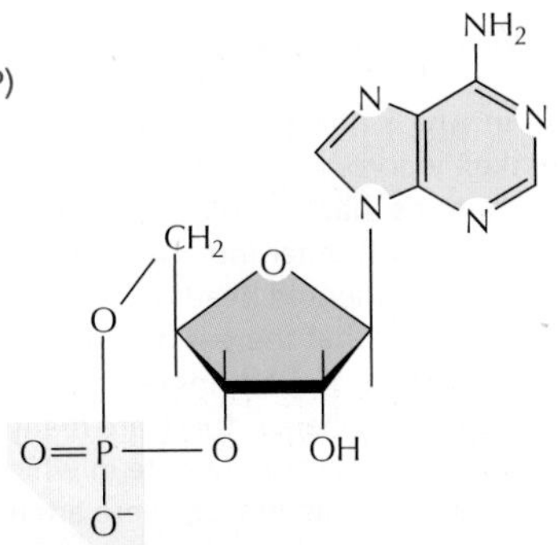

THE IMPORTANCE OF FREE ENERGY FOR CELLS

Life is possible because of the complex network of interacting chemical reactions occurring in every cell. In viewing the metabolic pathways that comprise this network, one might suspect that the cell has had the ability to evolve an enzyme to carry out any reaction that it needs. But this is not so. Although enzymes are powerful catalysts, they can speed up only those reactions that are thermodynamically possible; other reactions proceed in cells only because they are *coupled* to very favorable reactions that drive them. The question of whether a reaction can occur spontaneously, or instead needs to be coupled to another reaction, is central to cell biology. The answer is obtained by reference to a quantity called the *free energy*: the total change in free energy during a set of reactions determines whether or not the entire reaction sequence can occur. In this panel, we shall explain some of the fundamental ideas—derived from a special branch of chemistry and physics called *thermodynamics*—that are required for understanding what free energy is and why it is so important to cells.

ENERGY RELEASED BY CHANGES IN CHEMICAL BONDING IS CONVERTED INTO HEAT

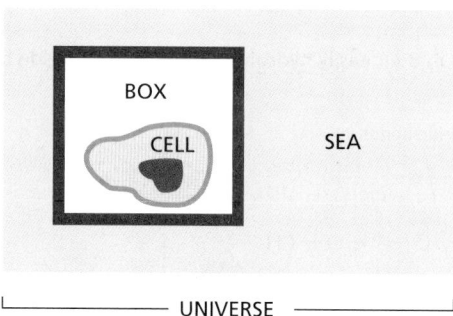

UNIVERSE

An *enclosed system* is defined as a collection of molecules that does not exchange matter with the rest of the universe (for example, the "cell in a box" shown above). Any such system will contain molecules with a total energy E. This energy will be distributed in a variety of ways: some as the translational energy of the molecules, some as their vibrational and rotational energies, but most as the bonding energies between the individual atoms that make up the molecules. Suppose that a reaction occurs in the system. The first law of thermodynamics places a constraint on what types of reactions are possible: it states that "in any process, the total energy of the universe remains constant." For example, suppose that reaction A→B occurs somewhere in the box and releases a great deal of chemical-bond energy. This energy will initially increase the intensity of molecular motions (translational, vibrational, and rotational) in the system, which is equivalent to raising its temperature. However, these increased motions will soon be transferred out of the system by a series of molecular collisions that heat up first the walls of the box and then the outside world (represented by the sea in our example). In the end, the system returns to its initial temperature, by which time all the chemical-bond energy released in the box has been converted into heat energy and transferred out of the box to the surroundings. According to the first law, the change in the energy in the box (ΔE_{box}, which we shall denote as ΔE) must be equal and opposite to the amount of heat energy transferred, which we shall designate as h: that is, $\Delta E = -h$. Thus, the energy in the box (E) decreases when heat leaves the system.

E also can change during a reaction as a result of work being done on the outside world. For example, suppose that there is a small increase in the volume (ΔV) of the box during a reaction. Since the walls of the box must push against the constant pressure (P) in the surroundings in order to expand, this does work on the outside world and requires energy. The energy used is $P(\Delta V)$, which according to the first law must decrease the energy in the box (E) by the same amount. In most reactions, chemical-bond energy is converted into both work and heat. *Enthalpy* (H) is a composite function that includes both of these ($H = E + PV$). To be rigorous, it is the change in enthalpy (ΔH) in an enclosed system, and not the change in energy, that is equal to the heat transferred to the outside world during a reaction. Reactions in which H decreases release heat to the surroundings and are said to be "exothermic," while reactions in which H increases absorb heat from the surroundings and are said to be "endothermic." Thus, $-h = \Delta H$. However, the volume change is negligible in most biological reactions, so to a good approximation

$$-h = \Delta H \cong \Delta E$$

THE SECOND LAW OF THERMODYNAMICS

Consider a container in which 1000 coins are all lying heads up. If the container is shaken vigorously, subjecting the coins to the types of random motions that all molecules experience due to their frequent collisions with other molecules, one will end up with about half the coins oriented heads down. The reason for this reorientation is that there is only a single way in which the original orderly state of the coins can be reinstated (every coin must lie heads up), whereas there are many different ways (about 10^{298}) to achieve a disorderly state in which there is an equal mixture of heads and tails; in fact, there are more ways to achieve a 50-50 state than to achieve any other state. Each state has a probability of occurrence that is proportional to the number of ways it can be realized. The second law of thermodynamics states that "systems will change spontaneously from states of lower probability to states of higher probability." Since states of lower probability are more "ordered" than states of high probability, the second law can be restated: "the universe constantly changes so as to become more disordered."

THE ENTROPY, S

The second law (but not the first law) allows one to predict the *direction* of a particular reaction. But to make it useful for this purpose, one needs a convenient measure of the probability or, equivalently, the degree of disorder of a state. The entropy (S) is such a measure. It is a logarithmic function of the probability such that the *change in entropy* (ΔS) that occurs when the reaction A → B converts one mole of A into one mole of B is

$$\Delta S = R \ln p_B / p_A$$

where p_A and p_B are the probabilities of the two states A and B, R is the gas constant (8.31 J K^{-1} mole^{-1}), and ΔS is measured in entropy units (eu). In our initial example of 1000 coins, the relative probability of all heads (state A) versus half heads and half tails (state B) is equal to the ratio of the number of different ways that the two results can be obtained. One can calculate that $p_A = 1$ and $p_B = 1000!(500! \times 500!) = 10^{299}$. Therefore, the entropy change for the reorientation of the coins when their container is vigorously shaken and an equal mixture of heads and tails is obtained is $R \ln (10^{298})$, or about 1370 eu per mole of such containers (6 x 10^{23} containers). We see that, because ΔS defined above is positive for the transition from state A to state B ($p_B / p_A > 1$), reactions with a large *increase* in S (that is, for which ΔS > 0) are favored and will occur spontaneously.

As discussed in Chapter 2, heat energy causes the random commotion of molecules. Because the transfer of heat from an enclosed system to its surroundings increases the number of different arrangements that the molecules in the outside world can have, it increases their entropy. It can be shown that the release of a fixed quantity of heat energy has a greater disordering effect at low temperature than at high temperature, and that the value of ΔS for the surroundings, as defined above (ΔS$_{sea}$), is precisely equal to *h*, the amount of heat transferred to the surroundings from the system, divided by the absolute temperature (T):

$$\Delta S_{sea} = h/T$$

THE GIBBS FREE ENERGY, G

When dealing with an enclosed biological system, one would like to have a simple way of predicting whether a given reaction will or will not occur spontaneously in the system. We have seen that the crucial question is whether the entropy change for the universe is positive or negative when that reaction occurs. In our idealized system, the cell in a box, there are two separate components to the entropy change of the universe—the entropy change for the system enclosed in the box and the entropy change for the surrounding "sea"—and both must be added together before any prediction can be made. For example, it is possible for a reaction to absorb heat and thereby decrease the entropy of the sea (ΔS$_{sea}$ < 0) and at the same time to cause such a large degree of disordering inside the box (ΔS$_{box}$ > 0) that the total ΔS$_{universe}$ = ΔS$_{sea}$ + ΔS$_{box}$ is greater than 0. In this case, the reaction will occur spontaneously, even though the sea gives up heat to the box during the reaction. An example of such a reaction is the dissolving of sodium chloride in a beaker containing water (the "box"), which is a spontaneous process even though the temperature of the water drops as the salt goes into solution.

Chemists have found it useful to define a number of new "composite functions" that describe *combinations* of physical properties of a system. The properties that can be combined include the temperature (T), pressure (P), volume (V), energy (E), and entropy (S). The enthalpy (H) is one such composite function. But by far the most useful composite function for biologists is the *Gibbs free energy*, G. It serves as an accounting device that allows one to deduce the entropy change of the universe resulting from a chemical reaction in the box, while avoiding any separate consideration of the entropy change in the sea. The definition of G is

$$G = H - TS$$

where, for a box of volume V, H is the enthalpy described above (E + PV), T is the absolute temperature, and S is the entropy. Each of these quantities applies to the inside of the box only. The change in free energy during a reaction in the box (the G of the products minus the G of the starting materials) is denoted as ΔG and, as we shall now demonstrate, it is a direct measure of the amount of disorder that is created in the universe when the reaction occurs.

At constant temperature the change in free energy (ΔG) during a reaction equals ΔH − TΔS. Remembering that ΔH = −h, the heat absorbed from the sea, we have

$$-\Delta G = -\Delta H + T\Delta S$$
$$-\Delta G = h + T\Delta S, \text{ so } -\Delta G/T = h/T + \Delta S$$

But h/T is equal to the entropy change of the sea (ΔS$_{sea}$), and the ΔS in the above equation is ΔS$_{box}$. Therefore

$$-\Delta G/T = \Delta S_{sea} + \Delta S_{box} = \Delta S_{universe}$$

We conclude that the free-energy change is a direct measure of the entropy change of the universe. A reaction will proceed in the direction that causes the change in the free energy (ΔG) to be less than zero, because in this case there will be a positive entropy change in the universe when the reaction occurs.

For a complex set of coupled reactions involving many different molecules, the total free-energy change can be computed simply by adding up the free energies of all the different molecular species after the reaction and comparing this value with the sum of free energies before the reaction; for common substances the required free-energy values can be found from published tables. In this way, one can predict the direction of a reaction and thereby readily check the feasibility of any proposed mechanism. Thus, for example, from the observed values for the magnitude of the electrochemical proton gradient across the inner mitochondrial membrane and the ΔG for ATP hydrolysis inside the mitochondrion, one can be certain that ATP synthase requires the passage of more than one proton for each molecule of ATP that it synthesizes.

The value of ΔG for a reaction is a direct measure of how far the reaction is from equilibrium. The large negative value for ATP hydrolysis in a cell merely reflects the fact that cells keep the ATP hydrolysis reaction as much as 10 orders of magnitude away from equilibrium. If a reaction reaches equilibrium, ΔG = 0, the reaction then proceeds at precisely equal rates in the forward and backward direction. For ATP hydrolysis, equilibrium is reached when the vast majority of the ATP has been hydrolyzed, as occurs in a dead cell.

For each step, the part of the molecule that undergoes a change is shadowed in blue, and the name of the enzyme that catalyzes the reaction is in a yellow box.

Step 1 Glucose is phosphorylated by ATP to form a sugar phosphate. The negative charge of the phosphate prevents passage of the sugar phosphate through the plasma membrane, trapping glucose inside the cell.

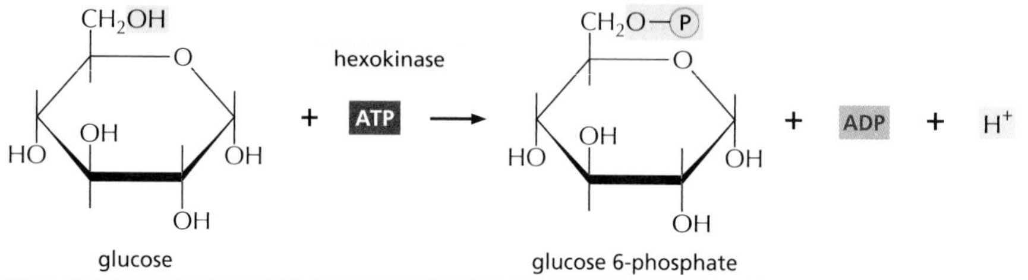

CH_2OH ... glucose + **ATP** $\longrightarrow$ ... glucose 6-phosphate + **ADP** + H^+

hexokinase

glucose
glucose 6-phosphate

Step 2 A readily reversible rearrangement of the chemical structure (isomerization) moves the carbonyl oxygen from carbon 1 to carbon 2, forming a ketose from an aldose sugar. (See Panel 2–4, p. 96.)

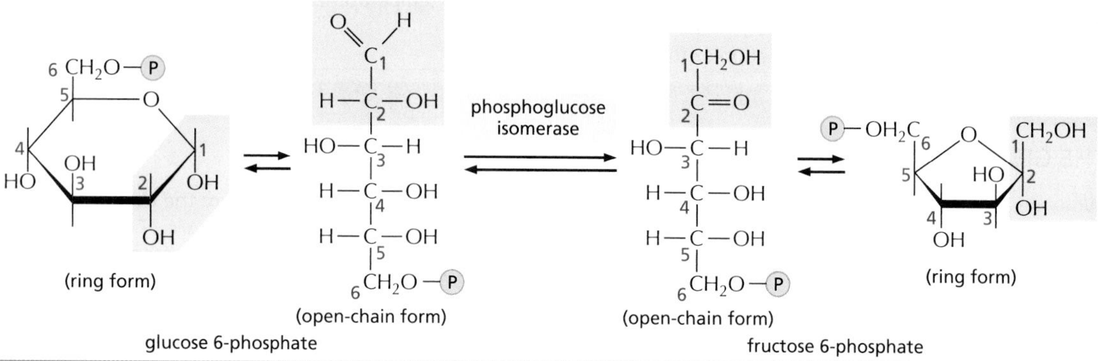

phosphoglucose isomerase

(ring form)
glucose 6-phosphate
(open-chain form)

$6\ CH_2O-P$
O
C_1
$H-C_2-OH$
$HO-C_3-H$
$H-C_4-OH$
$H-C_5-OH$
$6\ CH_2O-P$

$1\ CH_2OH$
$2\ C=O$
$HO-C_3-H$
$H-C_4-OH$
$H-C_5-OH$
$6\ CH_2O-P$
(open-chain form)

(ring form)
fructose 6-phosphate

Step 3 The new hydroxyl group on carbon 1 is phosphorylated by ATP, in preparation for the formation of two three-carbon sugar phosphates. The entry of sugars into glycolysis is controlled at this step, through regulation of the enzyme *phosphofructokinase*.

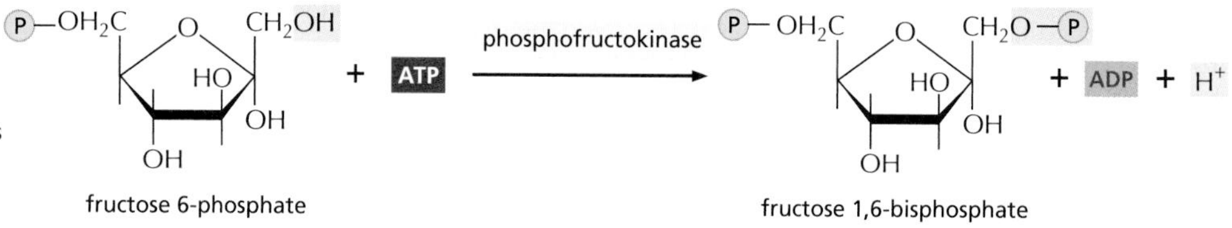

$P-OH_2C$ ... CH_2OH + **ATP** $\longrightarrow$ $P-OH_2C$ ... CH_2O-P + **ADP** + H^+

phosphofructokinase

fructose 6-phosphate
fructose 1,6-bisphosphate

Step 4 The six-carbon sugar is cleaved to produce two three-carbon molecules. Only the glyceraldehyde 3-phosphate can proceed immediately through glycolysis.

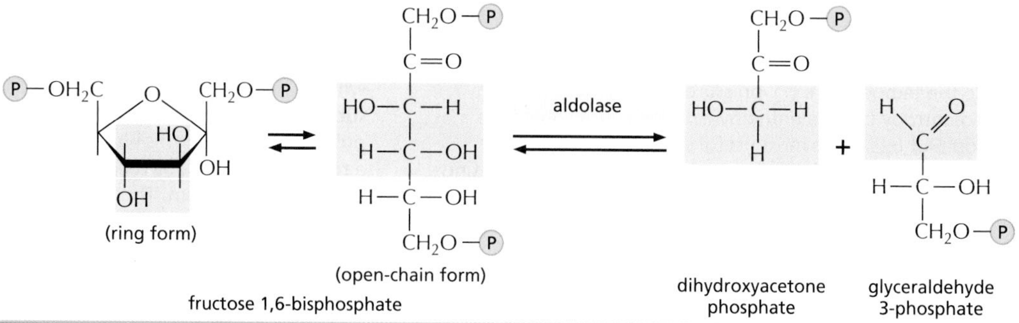

aldolase

$P-OH_2C$ ... CH_2O-P
(ring form)

CH_2O-P
$C=O$
$HO-C-H$
$H-C-OH$
$H-C-OH$
CH_2O-P
(open-chain form)
fructose 1,6-bisphosphate

CH_2O-P
$C=O$
$HO-C-H$
H
dihydroxyacetone phosphate

+

H
$C=O$
$H-C-OH$
CH_2O-P
glyceraldehyde 3-phosphate

Step 5 The other product of step 4, dihydroxyacetone phosphate, is isomerized to form glyceraldehyde 3-phosphate.

triose phosphate isomerase

CH_2OH
$C=O$
CH_2O-P
dihydroxyacetone phosphate

H
$C=O$
$H-C-OH$
CH_2O-P
glyceraldehyde 3-phosphate

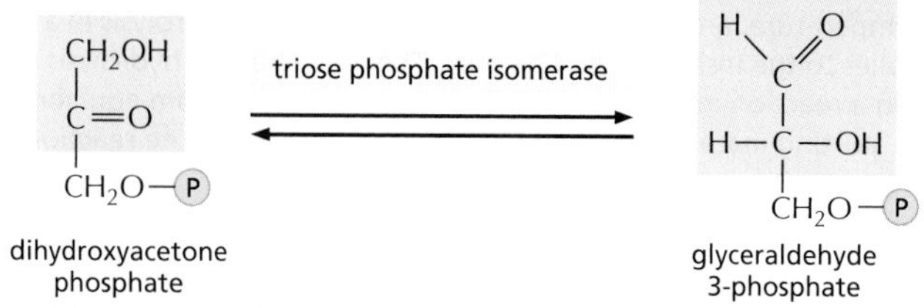

Step 6 The two molecules of glyceraldehyde 3-phosphate are oxidized. The energy-generation phase of glycolysis begins, as NADH and a new high-energy anhydride linkage to phosphate are formed (see Figure 2–46).

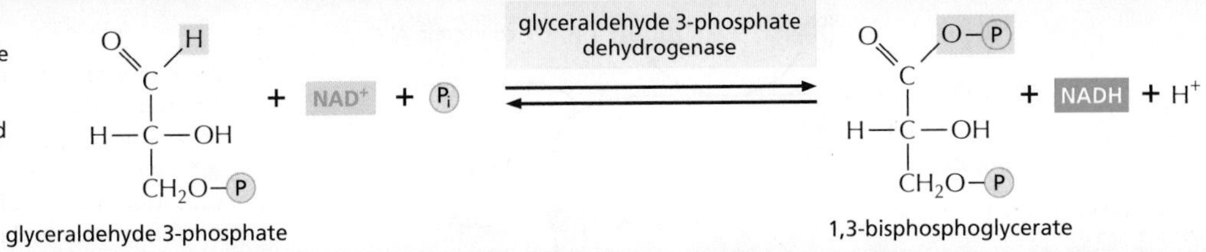

glyceraldehyde 3-phosphate + NAD⁺ + Pᵢ ⇌ [glyceraldehyde 3-phosphate dehydrogenase] ⇌ 1,3-bisphosphoglycerate + NADH + H⁺

glyceraldehyde 3-phosphate

1,3-bisphosphoglycerate

Step 7 The transfer to ADP of the high-energy phosphate group that was generated in step 6 forms ATP.

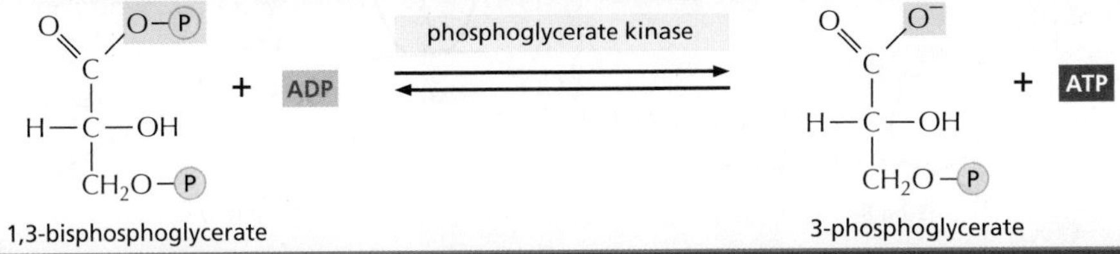

1,3-bisphosphoglycerate + ADP ⇌ [phosphoglycerate kinase] ⇌ 3-phosphoglycerate + ATP

1,3-bisphosphoglycerate

3-phosphoglycerate

Step 8 The remaining phosphate ester linkage in 3-phosphoglycerate, which has a relatively low free energy of hydrolysis, is moved from carbon 3 to carbon 2 to form 2-phosphoglycerate.

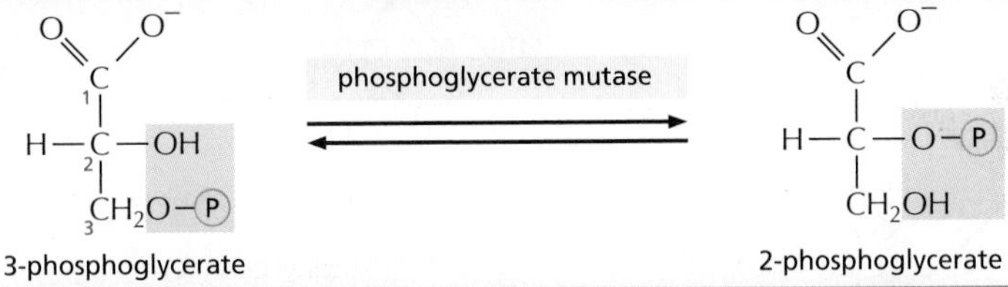

3-phosphoglycerate ⇌ [phosphoglycerate mutase] ⇌ 2-phosphoglycerate

3-phosphoglycerate

2-phosphoglycerate

Step 9 The removal of water from 2-phosphoglycerate creates a high-energy enol phosphate linkage.

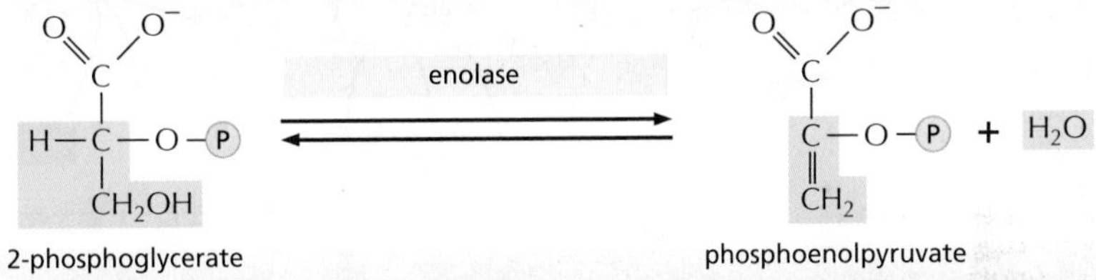

2-phosphoglycerate ⇌ [enolase] ⇌ phosphoenolpyruvate + H₂O

2-phosphoglycerate

phosphoenolpyruvate

Step 10 The transfer to ADP of the high-energy phosphate group that was generated in step 9 forms ATP, completing glycolysis.

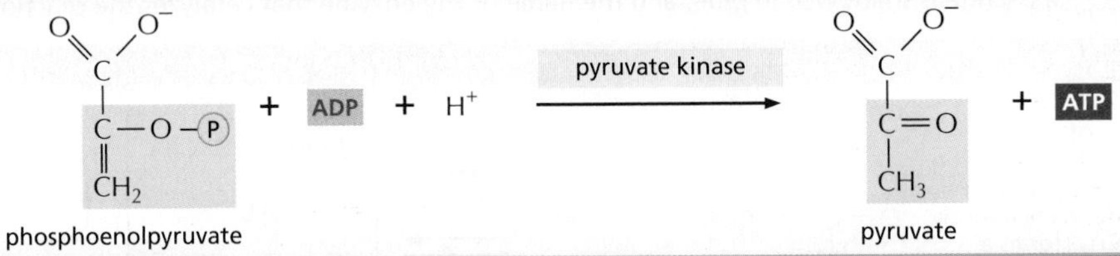

phosphoenolpyruvate + ADP + H⁺ → [pyruvate kinase] → pyruvate + ATP

phosphoenolpyruvate

pyruvate

NET RESULT OF GLYCOLYSIS

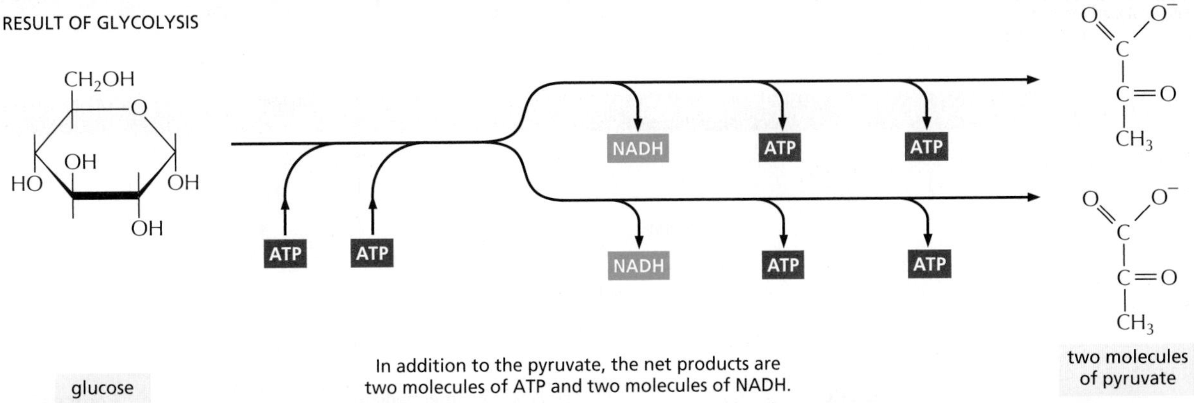

glucose → → → two molecules of pyruvate

ATP   ATP   NADH   ATP   ATP
ATP   ATP   NADH   ATP   ATP

In addition to the pyruvate, the net products are two molecules of ATP and two molecules of NADH.

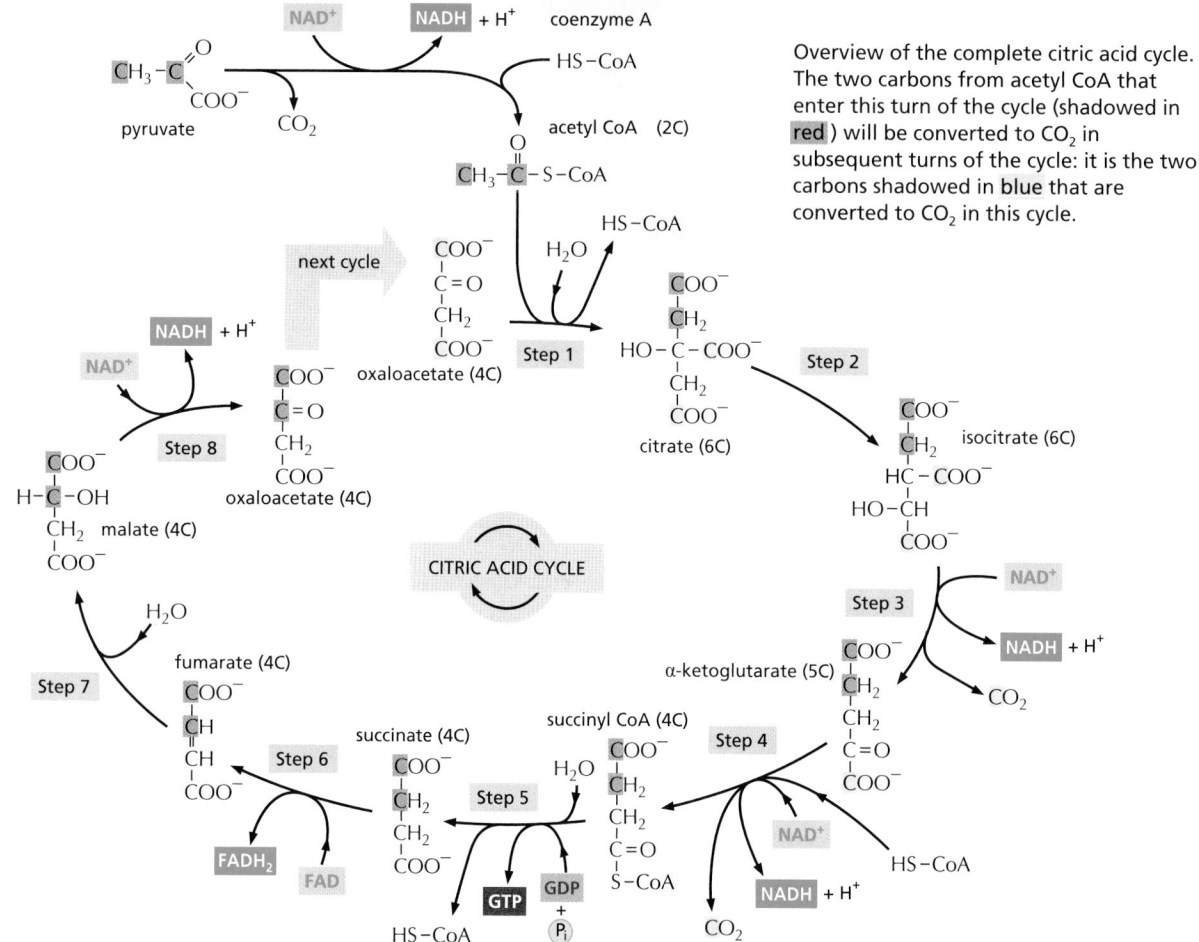

Overview of the complete citric acid cycle. The two carbons from acetyl CoA that enter this turn of the cycle (shadowed in red) will be converted to CO_2 in subsequent turns of the cycle: it is the two carbons shadowed in blue that are converted to CO_2 in this cycle.

Details of these eight steps are shown below. In this part of the panel, for each step, the part of the molecule that undergoes a change is shadowed in blue, and the name of the enzyme that catalyzes the reaction is in a yellow box.

Step 1   After the enzyme removes a proton from the CH_3 group on acetyl CoA, the negatively charged CH_2^- forms a bond to a carbonyl carbon of oxaloacetate. The subsequent loss by hydrolysis of the coenzyme A (HS–CoA) drives the reaction strongly forward.

acetyl CoA + oxaloacetate   →(citrate synthase)   [S-citryl-CoA intermediate] →(H_2O) citrate + HS–CoA + H^+

Step 2   An isomerization reaction, in which water is first removed and then added back, moves the hydroxyl group from one carbon atom to its neighbor.

citrate   →(aconitase, H_2O)   cis-aconitate intermediate   →(H_2O)   isocitrate

Step 3 In the first of four oxidation steps in the cycle, the carbon carrying the hydroxyl group is converted to a carbonyl group. The immediate product is unstable, losing CO_2 while still bound to the enzyme.

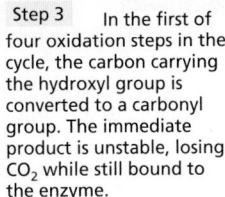

isocitrate → (isocitrate dehydrogenase; NAD^+ → $NADH + H^+$) → [oxalosuccinate intermediate] → (H^+, CO_2) → α-ketoglutarate

Step 4 The *α-ketoglutarate dehydrogenase complex* closely resembles the large enzyme complex that converts pyruvate to acetyl CoA, the *pyruvate dehydrogenase* complex in Figure 3–54D,E. It likewise catalyzes an oxidation that produces NADH, CO_2, and a high-energy thioester bond to coenzyme A (CoA).

α-ketoglutarate + HS–CoA → (α-ketoglutarate dehydrogenase complex; NAD^+ → $NADH + H^+$, CO_2) → succinyl-CoA

Step 5 A phosphate molecule from solution displaces the CoA, forming a high-energy phosphate linkage to succinate. This phosphate is then passed to GDP to form GTP. (In bacteria and plants, ATP is formed instead.)

succinyl-CoA → (succinyl-CoA synthetase; H_2O, P_i, GDP → GTP) → succinate + HS–CoA

Step 6 In the third oxidation step in the cycle, FAD accepts two hydrogen atoms from succinate.

succinate → (succinate dehydrogenase; FAD → $FADH_2$) → fumarate

Step 7 The addition of water to fumarate places a hydroxyl group next to a carbonyl carbon.

fumarate → (fumarase; H_2O) → malate

Step 8 In the last of four oxidation steps in the cycle, the carbon carrying the hydroxyl group is converted to a carbonyl group, regenerating the oxaloacetate needed for step 1.

malate → (malate dehydrogenase; NAD^+ → $NADH + H^+$) → oxaloacetate

REFERENCES

General

Berg JM, Tymoczko JL & Stryer L (2011) Biochemistry, 7th ed. New York: WH Freeman.

Garrett RH & Grisham CM (2012) Biochemistry, 5th ed. Philadelphia: Thomson Brooks/Cole.

Moran LA, Horton HR, Scrimgeour G & Perry M (2011) Principles of Biochemistry, 5th ed. Upper Saddle River, NJ: Prentice Hall.

Nelson DL & Cox MM (2012) Lehninger Principles of Biochemistry, 6th ed. New York: Worth.

van Holde KE, Johnson WC & Ho PS (2005) Principles of Physical Biochemistry, 2nd ed. Upper Saddle River, NJ: Prentice Hall.

Van Vranken D & Weiss G (2013) Introduction to Bioorganic Chemistry and Chemical Biology. New York: Garland Science.

Voet D, Voet JG & Pratt CM (2012) Fundamentals of Biochemistry, 4th ed. New York: Wiley.

The Chemical Components of a Cell

Atkins PW (2003) Molecules, 2nd ed. New York: WH Freeman.

Baldwin RL (2014) Dynamic hydration shell restores Kauzmann's 1959 explanation of how the hydrophobic factor drives protein folding. Proc. Natl Acad. Sci. USA 111, 13052–13056.

Bloomfield VA, Crothers DM & Tinoco I (2000) Nucleic Acids: Structures, Properties, and Functions. Sausalito, CA: University Science Books.

Branden C & Tooze J (1999) Introduction to Protein Structure, 2nd ed. New York: Garland Science.

de Duve C (2005) Singularities: Landmarks on the Pathways of Life. Cambridge: Cambridge University Press.

Dowhan W (1997) Molecular basis for membrane phospholipid diversity: why are there so many lipids? Annu. Rev. Biochem. 66, 199–232.

Eisenberg D & Kauzmann W (1969) The Structure and Properties of Water. Oxford: Oxford University Press.

Franks F (1993) Water. Cambridge: Royal Society of Chemistry.

Henderson LJ (1927) The Fitness of the Environment, 1958 ed. Boston, MA: Beacon.

Neidhardt FC, Ingraham JL & Schaechter M (1990) Physiology of the Bacterial Cell: A Molecular Approach. Sunderland, MA: Sinauer.

Phillips R & Milo R (2009) A feeling for the numbers in biology. Proc. Natl Acad. Sci. USA 106, 21465–21471.

Skinner JL (2010) Following the motions of water molecules in aqueous solutions. Science 328, 985–986.

Taylor ME & Drickamer K (2011) Introduction to Glycobiology, 3rd ed. New York: Oxford University Press.

Vance DE & Vance JE (2008) Biochemistry of Lipids, Lipoproteins and Membranes, 5th ed. Amsterdam: Elsevier.

Catalysis and the Use of Energy by Cells

Atkins PW (1994) The Second Law: Energy, Chaos and Form. New York: Scientific American Books.

Atkins PW & De Paula JD (2011) Physical Chemistry for the Life Sciences, 2nd ed. Oxford: Oxford University Press.

Baldwin JE & Krebs H (1981) The evolution of metabolic cycles. Nature 291, 381–382.

Berg HC (1983) Random Walks in Biology. Princeton, NJ: Princeton University Press.

Dill KA & Bromberg S (2010) Molecular Driving Forces: Statistical Thermodynamics in Biology, Chemistry, Physics, and Nanoscience, 2nd ed. New York: Garland Science.

Einstein A (1956) Investigations on the Theory of the Brownian Movement. New York: Dover.

Fruton JS (1999) Proteins, Enzymes, Genes: The Interplay of Chemistry and Biology. New Haven, CT: Yale University Press.

Hohmann-Marriott MF & Blankenship RE (2011) Evolution of photosynthesis. Annu. Rev. Plant Biol. 62, 515–548.

Karplus M & Petsko GA (1990) Molecular dynamics simulations in biology. Nature 347, 631–639.

Kauzmann W (1967) Thermodynamics and Statistics: with Applications to Gases. In Thermal Properties of Matter, Vol 2. New York: WA Benjamin, Inc.

Kornberg A (1989) For the Love of Enzymes. Cambridge, MA: Harvard University Press.

Lipmann F (1941) Metabolic generation and utilization of phosphate bond energy. Adv. Enzymol. 1, 99–162.

Lipmann F (1971) Wanderings of a Biochemist. New York: Wiley.

Nisbet EG & Sleep NH (2001) The habitat and nature of early life. Nature 409, 1083–1091.

Racker E (1980) From Pasteur to Mitchell: a hundred years of bioenergetics. Fed. Proc. 39, 210–215.

Schrödinger E (1944 & 1958) What is Life? The Physical Aspect of the Living Cell and Mind and Matter, 1992 combined ed. Cambridge: Cambridge University Press.

van Meer G, Voelker DR & Feigenson GW (2008) Membrane lipids: where they are and how they behave. Nat. Rev. Mol. Cell Biol. 9, 112–124.

Walsh C (2001) Enabling the chemistry of life. Nature 409, 226–231.

Westheimer FH (1987) Why nature chose phosphates. Science 235, 1173–1178.

How Cells Obtain Energy from Food

Caetano-Anollés D, Kim KM, Mittenthal JE & Caetano-Anollés G (2011) Proteome evolution and the metabolic origins of translation and cellular life. J. Mol. Evol. 72, 14–33.

Cramer WA & Knaff DB (1990) Energy Transduction in Biological Membranes. New York: Springer-Verlag.

Dismukes GC, Klimov VV, Baranov SV et al. (2001) The origin of atmospheric oxygen on Earth: the innovation of oxygenic photosynthesis. Proc. Natl Acad. Sci. USA 98, 2170–2175.

Fell D (1997) Understanding the Control of Metabolism. London: Portland Press.

Fothergill-Gilmore LA (1986) The evolution of the glycolytic pathway. Trends Biochem. Sci. 11, 47–51.

Friedmann HC (2004) From Butyribacterium to E. coli: an essay on unity in biochemistry. Perspect. Biol. Med. 47, 47–66.

Heinrich R, Meléndez-Hevia E, Montero F et al. (1999) The structural design of glycolysis: an evolutionary approach. Biochem. Soc. Trans. 27, 294–298.

Huynen MA, Dandekar T & Bork P (1999) Variation and evolution of the citric-acid cycle: a genomic perspective. Trends Microbiol. 7, 281–291.

Koonin EV (2014) The origins of cellular life. Antonie van Leeuwenhoek 106, 27–41.

Kornberg HL (2000) Krebs and his trinity of cycles. Nat. Rev. Mol. Cell Biol. 1, 225–228.

Krebs HA (1970) The history of the tricarboxylic acid cycle. Perspect. Biol. Med. 14, 154–170.

Krebs HA & Martin A (1981) Reminiscences and Reflections. Oxford/New York: Clarendon Press/Oxford University Press.

Lane N & Martin WF (2012) The origin of membrane bioenergetics. Cell 151, 1406–1416.

Morowitz HJ (1993) Beginnings of Cellular Life: Metabolism Recapitulates Biogenesis. New Haven, CT: Yale University Press.

Martijn J & Ettema TJG (2013) From archaeon to eukaryote: the evolutionary dark ages of the eukaryotic cell. Biochem. Soc. Trans. 41, 451–457.

Mulkidjanian AY, Bychkov AY, Dibrova DV et al. (2012) Open questions on the origin of life at anoxic geothermal fields. Orig. Life Evol. Biosph. 42, 507–516.

Zimmer C (2009) On the origin of eukaryotes. Science 325, 665–668.

Proteins

When we look at a cell through a microscope or analyze its electrical or bio-chemical activity, we are, in essence, observing proteins. Proteins constitute most of a cell's dry mass. They are not only the cell's building blocks; they also execute the majority of the cell's functions. Thus, proteins that are enzymes provide the intricate molecular surfaces inside a cell that catalyze its many chemical reactions. Proteins embedded in the plasma membrane form channels and pumps that control the passage of small molecules into and out of the cell. Other proteins carry messages from one cell to another, or act as signal integrators that relay sets of signals inward from the plasma membrane to the cell nucleus. Yet others serve as tiny molecular machines with moving parts: *kinesin*, for example, propels organelles through the cytoplasm; *topoisomerase* can untangle knotted DNA molecules. Other specialized proteins act as antibodies, toxins, hormones, antifreeze molecules, elastic fibers, ropes, or sources of luminescence. Before we can hope to understand how genes work, how muscles contract, how nerves conduct electricity, how embryos develop, or how our bodies function, we must attain a deep understanding of proteins.

THE SHAPE AND STRUCTURE OF PROTEINS

From a chemical point of view, proteins are by far the most structurally complex and functionally sophisticated molecules known. This is perhaps not surprising, once we realize that the structure and chemistry of each protein has been developed and fine-tuned over billions of years of evolutionary history. The theoretical calculations of population geneticists reveal that, over evolutionary time periods, a surprisingly small selective advantage is enough to cause a randomly altered protein sequence to spread through a population of organisms. Yet, even to experts, the remarkable versatility of proteins can seem truly amazing.

In this section, we consider how the location of each amino acid in the long string of amino acids that forms a protein determines its three-dimensional shape. Later in the chapter, we use this understanding of protein structure at the atomic level to describe how the precise shape of each protein molecule determines its function in a cell.

The Shape of a Protein Is Specified by Its Amino Acid Sequence

There are 20 different types of amino acids in proteins that are coded for directly in an organism's DNA, each with different chemical properties. A **protein** molecule consists of a long unbranched chain of these amino acids, each linked to its neighbor through a covalent peptide bond. Proteins are therefore also known as *polypeptides*. Each type of protein has a unique sequence of amino acids, and there are many thousands of different proteins in a cell.

The repeating sequence of atoms along the core of the polypeptide chain is referred to as the **polypeptide backbone**. Attached to this repetitive chain are those portions of the amino acids that are not involved in making a peptide bond and that give each amino acid its unique properties: the 20 different amino acid **side chains** (**Figure 3–1**). Some of these side chains are nonpolar and hydrophobic

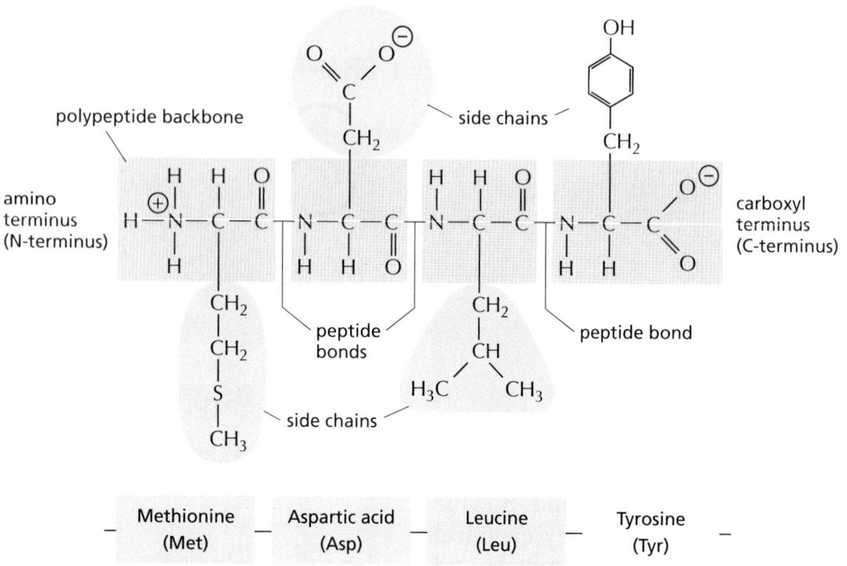

Figure 3–1 **The components of a protein.** A protein consists of a polypeptide backbone with attached side chains. Each type of protein differs in its sequence and number of amino acids; therefore, it is the sequence of the chemically different side chains that makes each protein distinct. The two ends of a polypeptide chain are chemically different: the end carrying the free amino group (NH_3^+, also written NH_2) is the amino terminus, or N-terminus, and that carrying the free carboxyl group (COO^-, also written COOH) is the carboxyl terminus or C-terminus. The amino acid sequence of a protein is always presented in the N-to-C direction, reading from left to right.

("water-fearing"), others are negatively or positively charged, some readily form covalent bonds, and so on. **Panel 3–1** (pp. 112–113) shows their atomic structures and **Figure 3–2** lists their abbreviations.

As discussed in Chapter 2, atoms behave almost as if they were hard spheres with a definite radius (their *van der Waals radius*). The requirement that no two atoms overlap plus other constraints limit the possible bond angles in a polypeptide chain (**Figure 3–3**), severely restricting the possible three-dimensional arrangements (or *conformations*) of atoms. Nevertheless, a long flexible chain such as a protein can still fold in an enormous number of ways.

The folding of a protein chain is also determined by many different sets of weak *noncovalent bonds* that form between one part of the chain and another. These involve atoms in the polypeptide backbone, as well as atoms in the amino acid side chains. There are three types of these weak bonds: *hydrogen bonds, electrostatic attractions*, and *van der Waals attractions*, as explained in Chapter 2 (see p. 44). Individual noncovalent bonds are 30–300 times weaker than the typical covalent bonds that create biological molecules. But many weak bonds acting in parallel can hold two regions of a polypeptide chain tightly together. In this way, the combined strength of large numbers of such noncovalent bonds determines the stability of each folded shape (**Figure 3–4**).

AMINO ACID			SIDE CHAIN	AMINO ACID			SIDE CHAIN
Aspartic acid	Asp	D	negative	Alanine	Ala	A	nonpolar
Glutamic acid	Glu	E	negative	Glycine	Gly	G	nonpolar
Arginine	Arg	R	positive	Valine	Val	V	nonpolar
Lysine	Lys	K	positive	Leucine	Leu	L	nonpolar
Histidine	His	H	positive	Isoleucine	Ile	I	nonpolar
Asparagine	Asn	N	uncharged polar	Proline	Pro	P	nonpolar
Glutamine	Gln	Q	uncharged polar	Phenylalanine	Phe	F	nonpolar
Serine	Ser	S	uncharged polar	Methionine	Met	M	nonpolar
Threonine	Thr	T	uncharged polar	Tryptophan	Trp	W	nonpolar
Tyrosine	Tyr	Y	uncharged polar	Cysteine	Cys	C	nonpolar
POLAR AMINO ACIDS				NONPOLAR AMINO ACIDS			

Figure 3–2 **The 20 amino acids commonly found in proteins.** Each amino acid has a three-letter and a one-letter abbreviation. There are equal numbers of polar and nonpolar side chains; however, some side chains listed here as polar are large enough to have some nonpolar properties (for example, Tyr, Thr, Arg, Lys). For atomic structures, see Panel 3–1 (pp. 112–113).

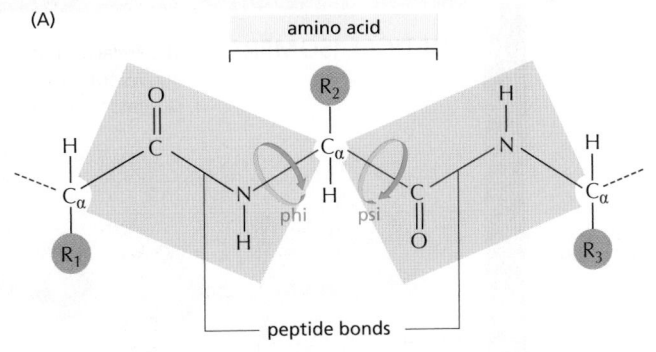

Figure 3–3 **Steric limitations on the bond angles in a polypeptide chain.** (A) Each amino acid contributes three bonds (*red*) to the backbone of the chain. The peptide bond is planar (*gray shading*) and does not permit rotation. By contrast, rotation can occur about the C_α–C bond, whose angle of rotation is called psi (ψ), and about the N–C_α bond, whose angle of rotation is called phi (ϕ). By convention, an R group is often used to denote an amino acid side chain (*purple circles*). (B) The conformation of the main-chain atoms in a protein is determined by one pair of ϕ and ψ angles for each amino acid; because of steric collisions between atoms within each amino acid, most of the possible pairs of ϕ and ψ angles do not occur. In this so-called Ramachandran plot, each dot represents an observed pair of angles in a protein. The three differently shaded clusters of dots reflect three different "secondary structures" repeatedly found in proteins, as will be described in the text. (B, from J. Richardson, *Adv. Prot. Chem.* 34:174–175, 1981. With permission from Elsevier.)

A fourth weak force—a hydrophobic clustering force—also has a central role in determining the shape of a protein. As described in Chapter 2, hydrophobic molecules, including the nonpolar side chains of particular amino acids, tend to be forced together in an aqueous environment in order to minimize their disruptive effect on the hydrogen-bonded network of water molecules (see Panel 2–2, pp. 92–93). Therefore, an important factor governing the folding of any protein is

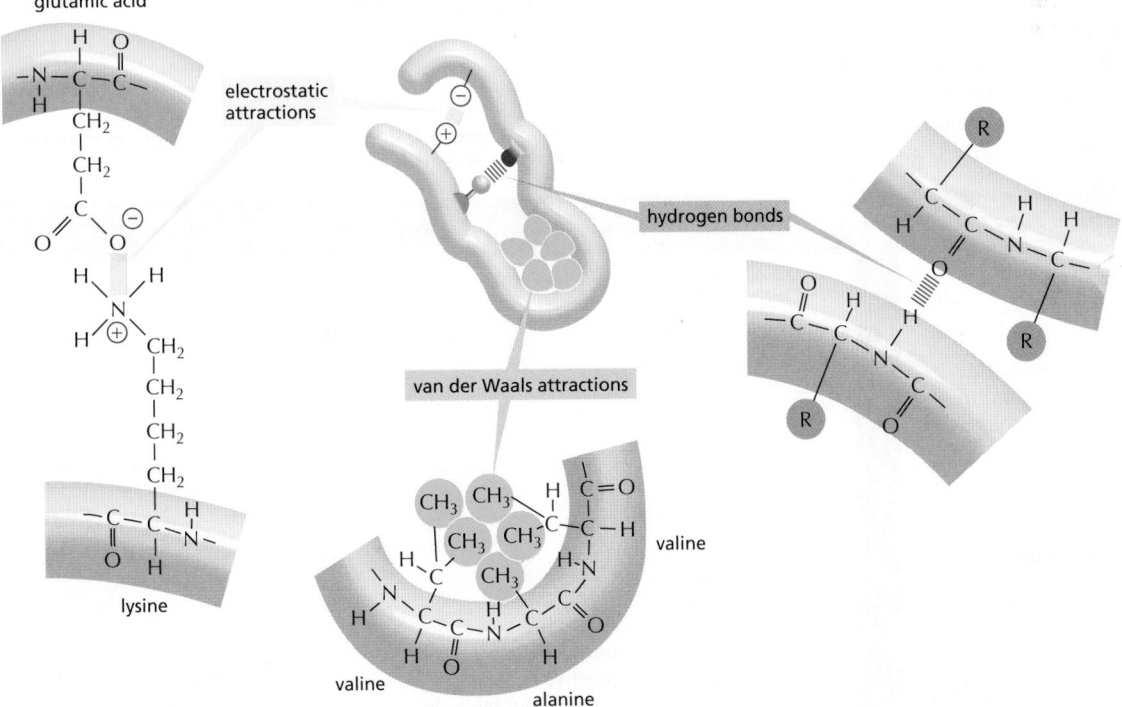

Figure 3–4 **Three types of noncovalent bonds help proteins fold.** Although a single one of these bonds is quite weak, many of them act together to create a strong bonding arrangement, as in the example shown. As in the previous figure, R is used as a general designation for an amino acid side chain.

THE AMINO ACID

The general formula of an amino acid is

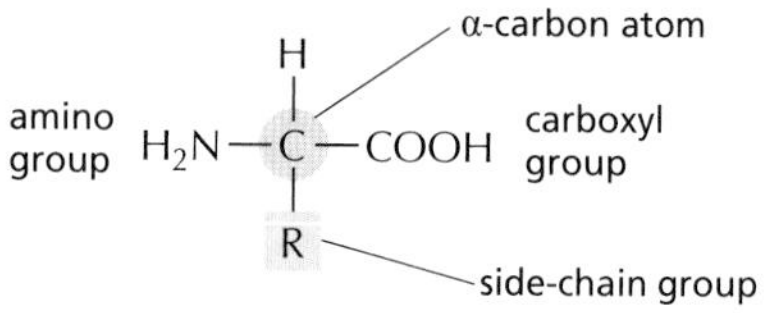

amino group — H₂N — C — COOH — carboxyl group

with H above (α-carbon atom) and R below (side-chain group).

R is commonly one of 20 different side chains. At pH 7 both the amino and carboxyl groups are ionized.

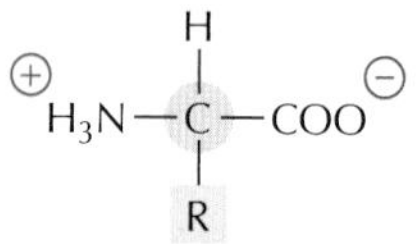

$$\overset{\oplus}{H_3N} - \overset{H}{\underset{R}{C}} - \overset{\ominus}{COO}$$

OPTICAL ISOMERS

The α-carbon atom is asymmetric, which allows for two mirror images (or stereo-) isomers, L and D.

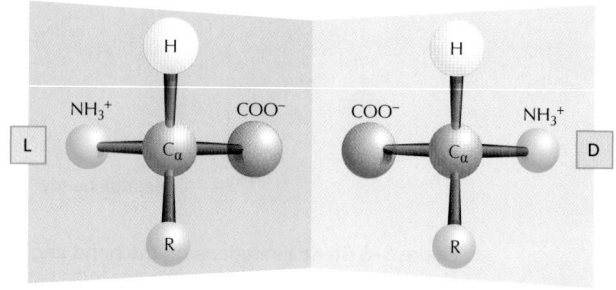

Proteins consist exclusively of L-amino acids.

PEPTIDE BONDS

Amino acids are commonly joined together by an amide linkage, called a peptide bond.

Peptide bond: The four atoms in each *gray box* form a rigid planar unit. There is no rotation around the C–N bond.

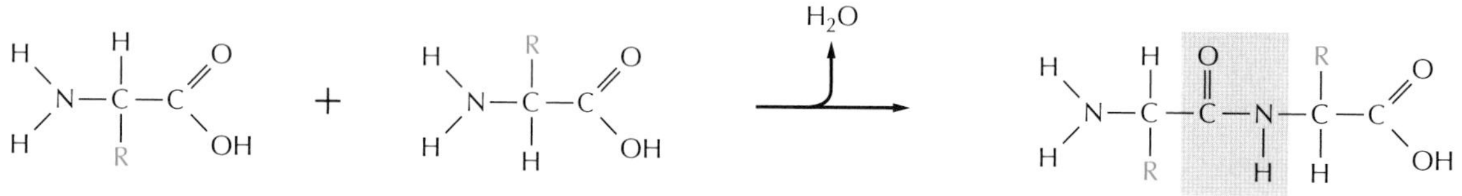

Proteins are long polymers of amino acids linked by peptide bonds, and they are always written with the N-terminus toward the left. The sequence of this tripeptide is histidine-cysteine-valine.

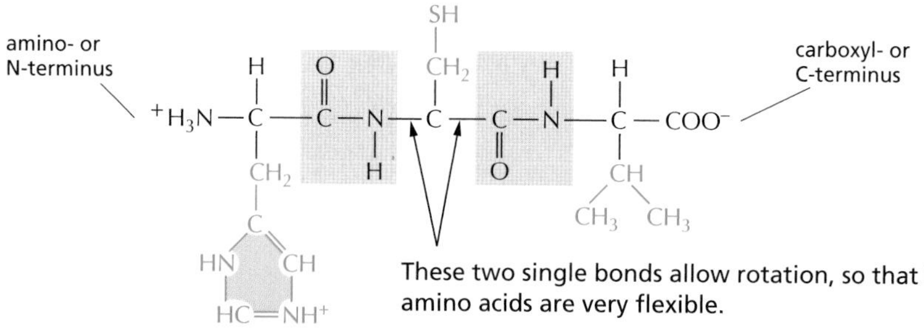

amino- or N-terminus

carboxyl- or C-terminus

These two single bonds allow rotation, so that long chains of amino acids are very flexible.

FAMILIES OF AMINO ACIDS

The common amino acids are grouped according to whether their side chains are

acidic
basic
uncharged polar
nonpolar

These 20 amino acids are given both three-letter and one-letter abbreviations.

Thus: alanine = Ala = A

BASIC SIDE CHAINS

lysine
(Lys, or K)

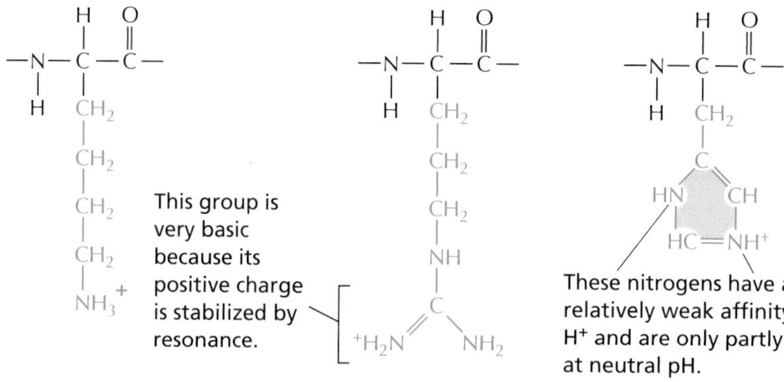

This group is very basic because its positive charge is stabilized by resonance.

arginine
(Arg, or R)

histidine
(His, or H)

These nitrogens have a relatively weak affinity for an H⁺ and are only partly positive at neutral pH.

ACIDIC SIDE CHAINS

aspartic acid	glutamic acid
(Asp, or D)	(Glu, or E)

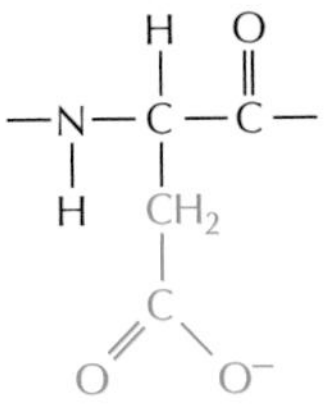

NONPOLAR SIDE CHAINS

alanine	valine
(Ala, or A)	(Val, or V)

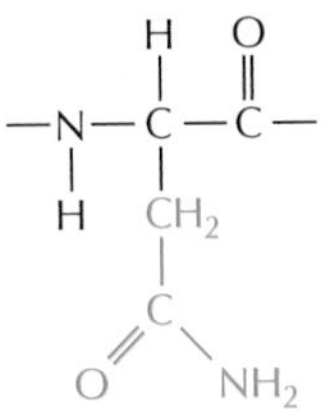

leucine	isoleucine
(Leu, or L)	(Ile, or I)

proline	phenylalanine
(Pro, or P)	(Phe, or F)

(actually an imino acid)

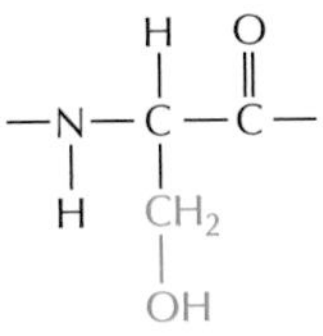

methionine	tryptophan
(Met, or M)	(Trp, or W)

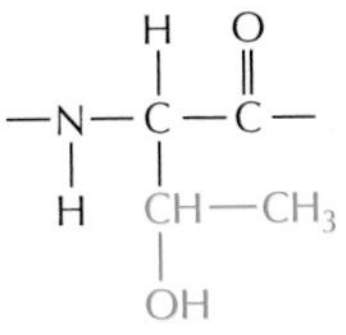

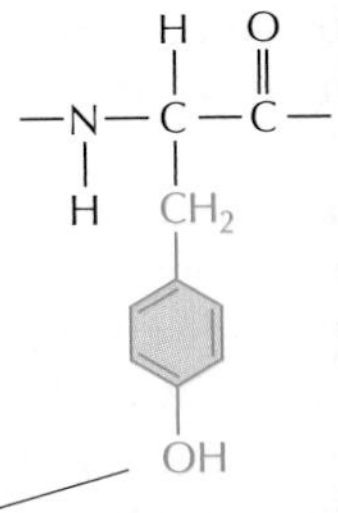

glycine	cysteine
(Gly, or G)	(Cys, or C)

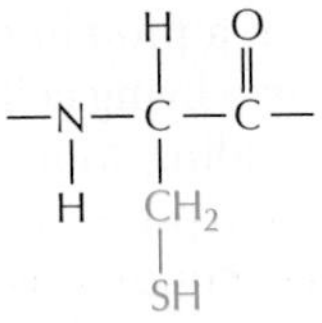

Disulfide bonds can form between two cysteine side chains in proteins.

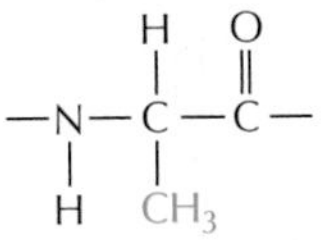

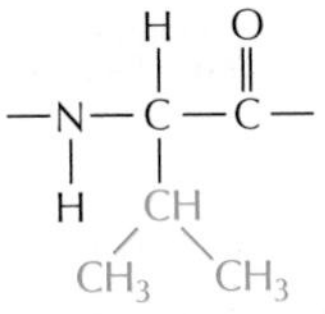

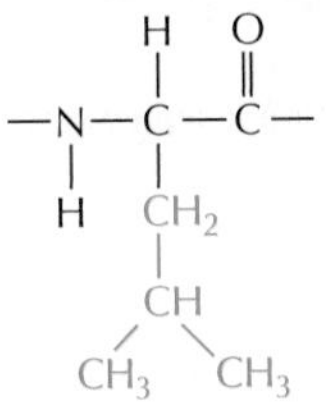

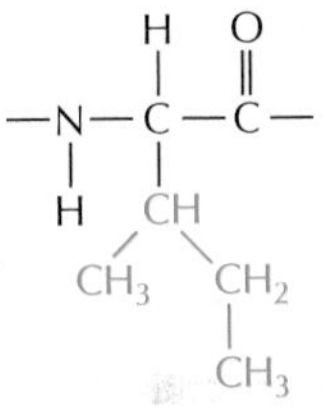

$$--CH_2-S-S-CH_2--$$

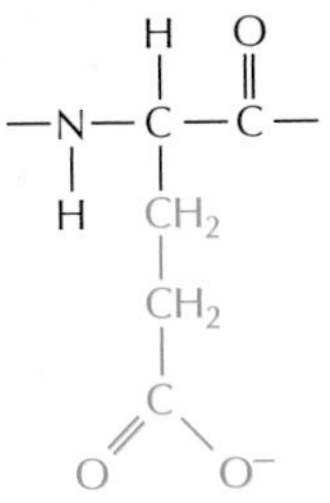

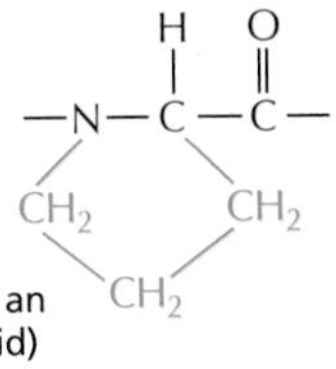

UNCHARGED POLAR SIDE CHAINS

asparagine	glutamine
(Asn, or N)	(Gln, or Q)

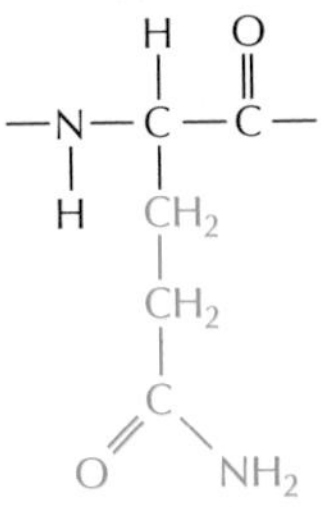

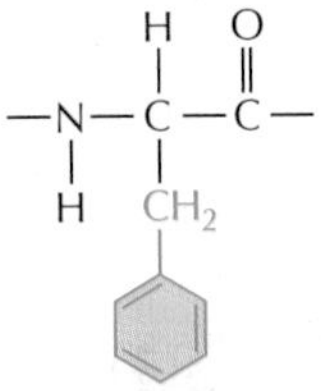

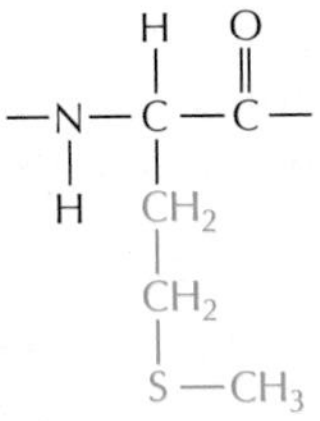

Although the amide N is not charged at neutral pH, it is polar.

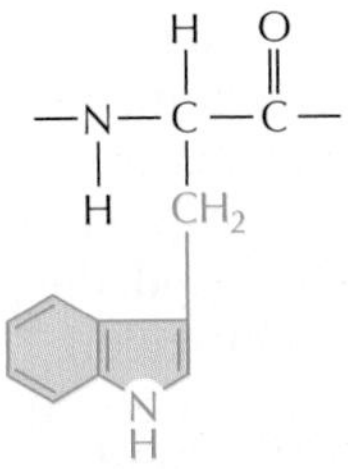

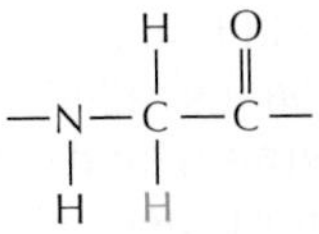

serine	threonine	tyrosine
(Ser, or S)	(Thr, or T)	(Tyr, or Y)

The –OH group is polar.

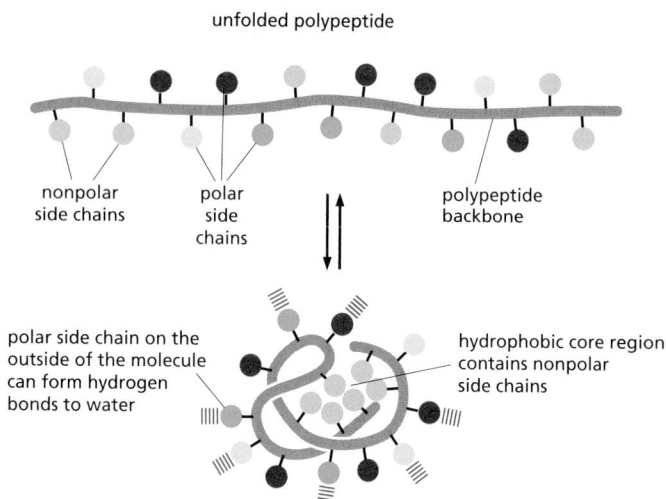

unfolded polypeptide

nonpolar side chains

polar side chains

polypeptide backbone

polar side chain on the outside of the molecule can form hydrogen bonds to water

hydrophobic core region contains nonpolar side chains

folded conformation in aqueous environment

Figure 3–5 How a protein folds into a compact conformation. The polar amino acid side chains tend to lie on the outside of the protein, where they can interact with water; the nonpolar amino acid side chains are buried on the inside forming a tightly packed hydrophobic core of atoms that are hidden from water. In this schematic drawing, the protein contains only about 17 amino acids.

the distribution of its polar and nonpolar amino acids. The nonpolar (hydrophobic) side chains in a protein—belonging to such amino acids as phenylalanine, leucine, valine, and tryptophan—tend to cluster in the interior of the molecule (just as hydrophobic oil droplets coalesce in water to form one large droplet). This enables them to avoid contact with the water that surrounds them inside a cell. In contrast, polar groups—such as those belonging to arginine, glutamine, and histidine—tend to arrange themselves near the outside of the molecule, where they can form hydrogen bonds with water and with other polar molecules (**Figure 3–5**). Polar amino acids buried within the protein are usually hydrogen-bonded to other polar amino acids or to the polypeptide backbone.

Proteins Fold into a Conformation of Lowest Energy

As a result of all of these interactions, most proteins have a particular three-dimensional structure, which is determined by the order of the amino acids in its chain. The final folded structure, or **conformation**, of any polypeptide chain is generally the one that minimizes its free energy. Biologists have studied protein folding in a test tube using highly purified proteins. Treatment with certain solvents, which disrupt the noncovalent interactions holding the folded chain together, unfolds, or *denatures*, a protein. This treatment converts the protein into a flexible polypeptide chain that has lost its natural shape. When the denaturing solvent is removed, the protein often refolds spontaneously, or *renatures*, into its original conformation. This indicates that the amino acid sequence contains all of the information needed for specifying the three-dimensional shape of a protein, a critical point for understanding cell biology.

Most proteins fold up into a single stable conformation. However, this conformation changes slightly when the protein interacts with other molecules in the cell. This change in shape is often crucial to the function of the protein, as we see later.

Although a protein chain can fold into its correct conformation without outside help, in a living cell special proteins called *molecular chaperones* often assist in protein folding. Molecular chaperones bind to partly folded polypeptide chains and help them progress along the most energetically favorable folding pathway. In the crowded conditions of the cytoplasm, chaperones are required to prevent the temporarily exposed hydrophobic regions in newly synthesized protein chains from associating with each other to form protein aggregates (see p. 355). However, the final three-dimensional shape of the protein is still specified by its amino acid sequence: chaperones simply make reaching the folded state more reliable.

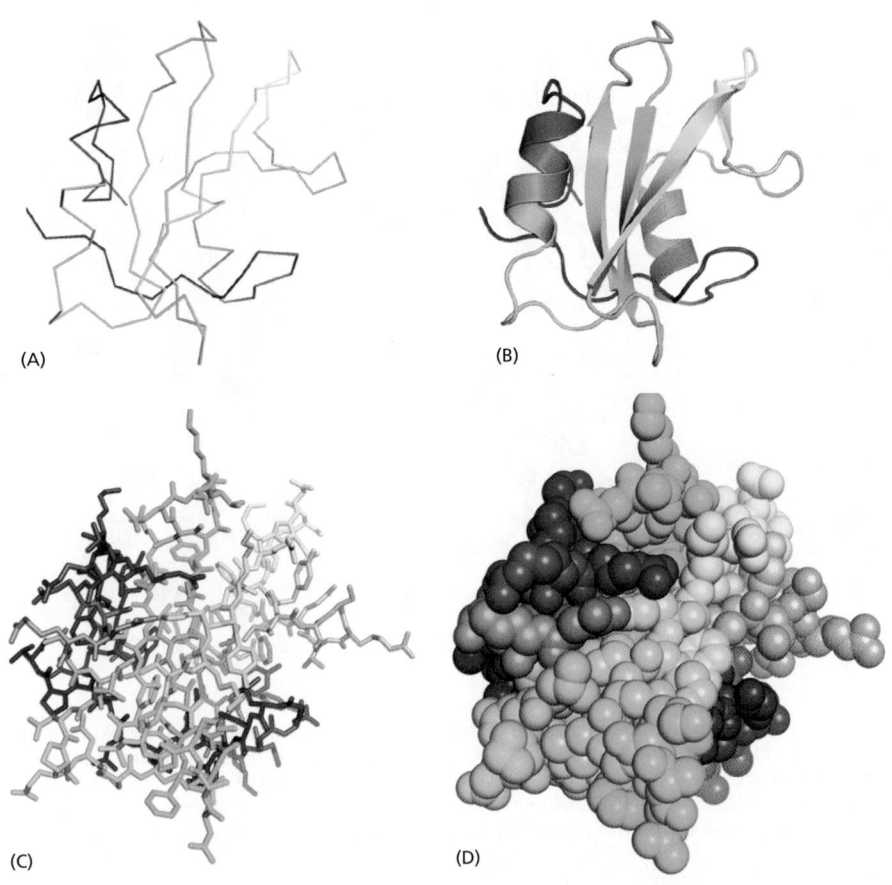

(A)

(B)

(C)

(D)

Figure 3–6 **Four representations describing the structure of a small protein domain.** Constructed from a string of 100 amino acids, the SH2 domain is part of many different proteins (see, for example, Figure 3–63). Here, the structure of the SH2 domain is displayed as (A) a polypeptide backbone model, (B) a ribbon model, (C) a wire model that includes the amino acid side chains, and (D) a space-filling model. These images are colored in a way that allows the polypeptide chain to be followed from its N-terminus *(purple)* to its C-terminus *(red)* (PDB code: 1SHA).

Proteins come in a wide variety of shapes, and most are between 50 and 2000 amino acids long. Large proteins usually consist of several distinct *protein domains*—structural units that fold more or less independently of each other, as we discuss below. The structure of even a small domain is complex, and for clarity, several different representations are conventionally used, each of which emphasizes distinct features. As an example, **Figure 3–6** presents four representations of a protein domain called SH2, a structure present in many different proteins in eukaryotic cells and involved in cell signaling (see Figure 15–46).

Descriptions of protein structures are aided by the fact that proteins are built up from combinations of several common structural motifs, as we discuss next.

The α Helix and the β Sheet Are Common Folding Patterns

When we compare the three-dimensional structures of many different protein molecules, it becomes clear that, although the overall conformation of each protein is unique, two regular folding patterns are often found within them. Both patterns were discovered more than 60 years ago from studies of hair and silk. The first folding pattern to be discovered, called the **α helix**, was found in the protein *α-keratin*, which is abundant in skin and its derivatives—such as hair, nails, and horns. Within a year of the discovery of the α helix, a second folded structure, called a **β sheet**, was found in the protein *fibroin*, the major constituent of silk. These two patterns are particularly common because they result from hydrogen-bonding between the N–H and C=O groups in the polypeptide backbone, without involving the side chains of the amino acids. Thus, although incompatible with some amino acid side chains, many different amino acid sequences can form them. In each case, the protein chain adopts a regular, repeating conformation. **Figure 3–7** illustrates the detailed structures of these two important conformations, which in ribbon models of proteins are represented by a helical ribbon and by a set of aligned arrows, respectively.

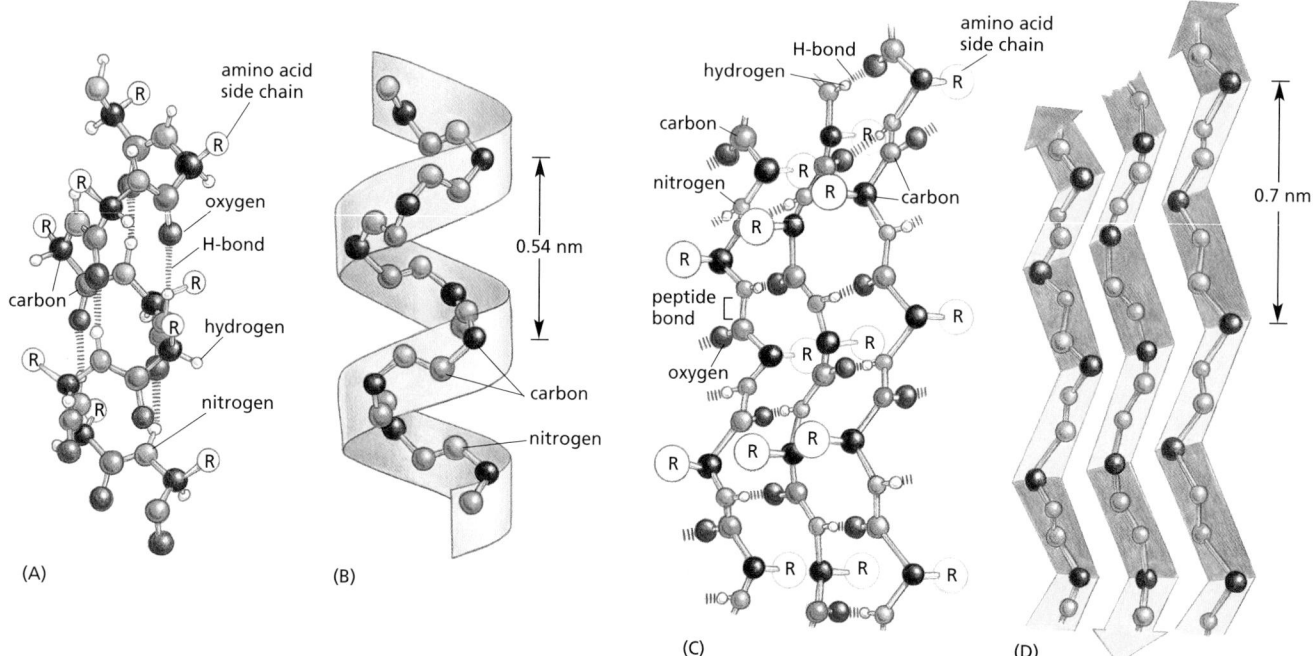

(A)   (B)   (C)   (D)

Figure 3–7 **The regular conformation of the polypeptide backbone in the α helix and the β sheet.** The α helix is shown in (A) and (B). The N–H of every peptide bond is hydrogen-bonded to the C=O of a neighboring peptide bond located four peptide bonds away in the same chain. Note that all of the N–H groups point up in this diagram and that all of the C=O groups point down (toward the C-terminus); this gives a polarity to the helix, with the C-terminus having a partial negative and the N-terminus a partial positive charge (Movie 3.2). The β sheet is shown in (C) and (D). In this example, adjacent peptide chains run in opposite (antiparallel) directions. Hydrogen-bonding between peptide bonds in different strands holds the individual polypeptide chains (strands) together in a β sheet, and the amino acid side chains in each strand alternately project above and below the plane of the sheet (Movie 3.3). (A) and (C) show all the atoms in the polypeptide backbone, but the amino acid side chains are truncated and denoted by R. In contrast, (B) and (D) show only the carbon and nitrogen backbone atoms.

The cores of many proteins contain extensive regions of β sheet. As shown in **Figure 3–8**, these β sheets can form either from neighboring segments of the polypeptide backbone that run in the same orientation (parallel chains) or from a polypeptide backbone that folds back and forth upon itself, with each section of the chain running in the direction opposite to that of its immediate neighbors (antiparallel chains). Both types of β sheet produce a very rigid structure, held together by hydrogen bonds that connect the peptide bonds in neighboring chains (see Figure 3–7C).

An α helix is generated when a single polypeptide chain twists around on itself to form a rigid cylinder. A hydrogen bond forms between every fourth peptide bond, linking the C=O of one peptide bond to the N–H of another (see Figure 3–7A). This gives rise to a regular helix with a complete turn every 3.6 amino acids. The SH2 protein domain illustrated in Figure 3–6 contains two α helices, as well as a three-stranded antiparallel β sheet.

Regions of α helix are abundant in proteins located in cell membranes, such as transport proteins and receptors. As we discuss in Chapter 10, those portions of a transmembrane protein that cross the lipid bilayer usually cross as α helices composed largely of amino acids with nonpolar side chains. The polypeptide backbone, which is hydrophilic, is hydrogen-bonded to itself in the α helix and shielded from the hydrophobic lipid environment of the membrane by its protruding nonpolar side chains.

In other proteins, α helices wrap around each other to form a particularly stable structure, known as a **coiled-coil**. This structure can form when the two (or in some cases, three or four) α helices have most of their nonpolar (hydrophobic) side chains on one side, so that they can twist around each other with these side chains facing inward (**Figure 3–9**). Long rodlike coiled-coils provide the structural framework for many elongated proteins. Examples are α-keratin, which forms the intracellular fibers that reinforce the outer layer of the skin and its appendages, and the myosin molecules responsible for muscle contraction.

Protein Domains Are Modular Units from Which Larger Proteins Are Built

Even a small protein molecule is built from thousands of atoms linked together by precisely oriented covalent and noncovalent bonds. Biologists are aided in visualizing these extremely complicated structures by various graphic and computer-based three-dimensional displays. The student resource site that accompanies this book contains computer-generated images of selected proteins, displayed and rotated on the screen in a variety of formats.

Scientists distinguish four levels of organization in the structure of a protein. The amino acid sequence is known as the **primary structure**. Stretches of polypeptide chain that form α helices and β sheets constitute the protein's **secondary structure**. The full three-dimensional organization of a polypeptide chain is sometimes referred to as the **tertiary structure**, and if a particular protein molecule is formed as a complex of more than one polypeptide chain, the complete structure is designated as the **quaternary structure**.

Studies of the conformation, function, and evolution of proteins have also revealed the central importance of a unit of organization distinct from these four. This is the **protein domain**, a substructure produced by any contiguous part of a polypeptide chain that can fold independently of the rest of the protein into a compact, stable structure. A domain usually contains between 40 and 350 amino acids, and it is the modular unit from which many larger proteins are constructed.

The different domains of a protein are often associated with different functions. **Figure 3–10** shows an example—the Src protein kinase, which functions in signaling pathways inside vertebrate cells (Src is pronounced "sarc"). This protein

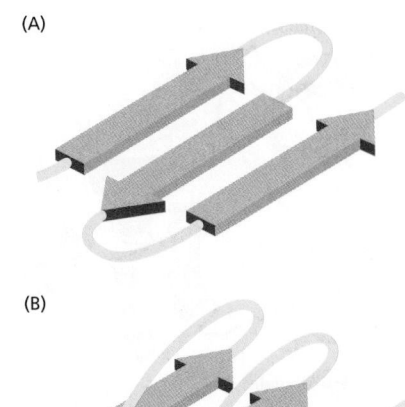

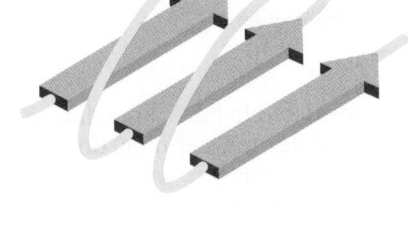

Figure 3–8 Two types of β sheet structures. *(A)* An antiparallel β sheet (see Figure 3–7C). *(B)* A parallel β sheet. Both of these structures are common in proteins.

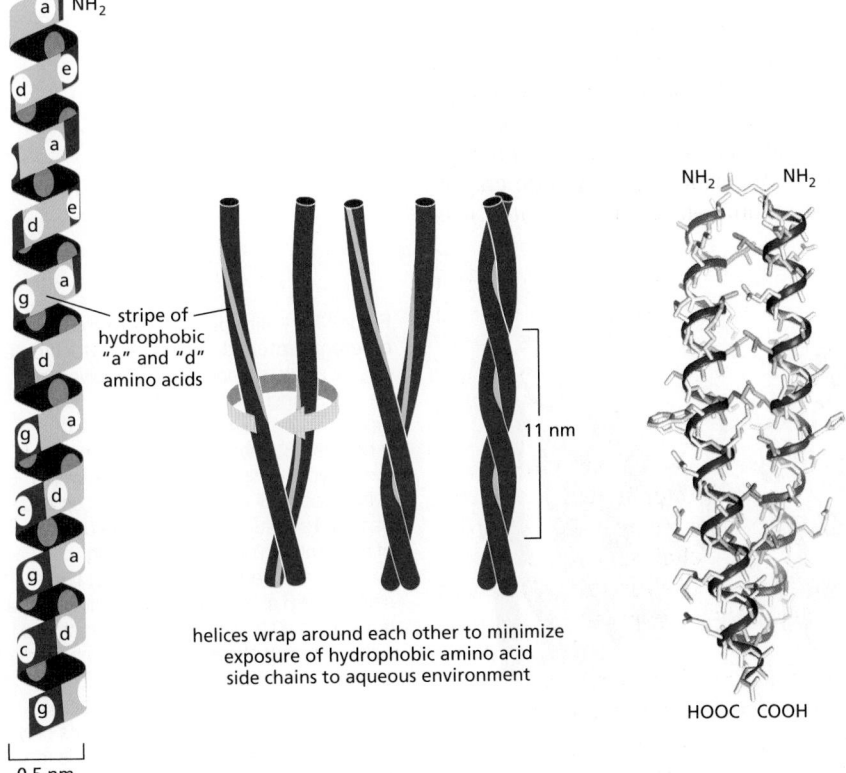

stripe of hydrophobic "a" and "d" amino acids

11 nm

helices wrap around each other to minimize exposure of hydrophobic amino acid side chains to aqueous environment

0.5 nm

(A)   (B)   (C)

Figure 3–9 A coiled-coil. (A) A single α helix, with successive amino acid side chains labeled in a sevenfold sequence, "abcdefg" *(from bottom to top)*. Amino acids "a" and "d" in such a sequence lie close together on the cylinder surface, forming a "stripe" *(green)* that winds slowly around the α helix. Proteins that form coiled-coils typically have nonpolar amino acids at positions "a" and "d." Consequently, as shown in (B), the two α helices can wrap around each other with the nonpolar side chains of one α helix interacting with the nonpolar side chains of the other. (C) The atomic structure of a coiled-coil determined by x-ray crystallography. The alpha helical backbone is shown in *red* and the nonpolar side chains in *green*, while the more hydrophilic amino acid side chains, shown in *gray*, are left exposed to the aqueous environment (**Movie 3.4**). (PDB code: 3NMD.)

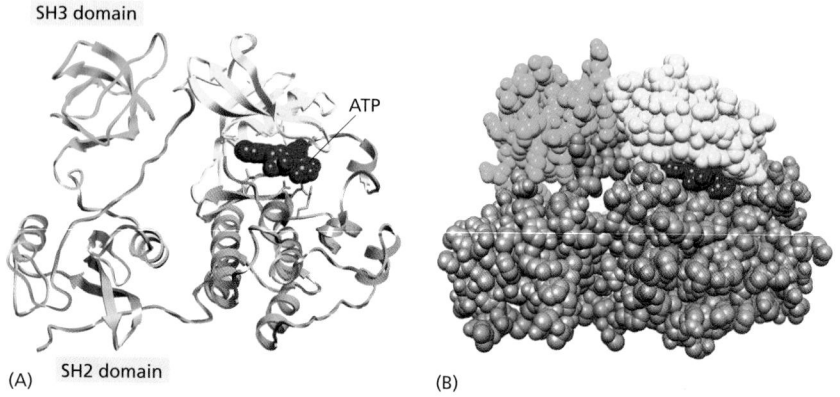

SH3 domain

ATP

(A)   SH2 domain   (B)

Figure 3–10 **A protein formed from multiple domains.** In the Src protein shown, a C-terminal domain with two lobes *(yellow* and *orange)* forms a protein kinase enzyme, while the SH2 and SH3 domains perform regulatory functions. (A) A ribbon model, with ATP substrate in *red*. (B) A space-filling model, with ATP substrate in *red*. Note that the site that binds ATP is positioned at the interface of the two lobes that form the kinase. The structure of the SH2 domain was illustrated in Figure 3–6. (PDB code: 2SRC.)

is considered to have three domains: the SH2 and SH3 domains have regulatory roles, while the C-terminal domain is responsible for the kinase catalytic activity. Later in the chapter, we shall return to this protein, in order to explain how proteins can form molecular switches that transmit information throughout cells.

Figure 3–11 presents ribbon models of three differently organized protein domains. As these examples illustrate, the central core of a domain can be constructed from α helices, from β sheets, or from various combinations of these two fundamental folding elements.

The smallest protein molecules contain only a single domain, whereas larger proteins can contain several dozen domains, often connected to each other by short, relatively unstructured lengths of polypeptide chain that can act as flexible hinges between domains.

Few of the Many Possible Polypeptide Chains Will Be Useful to Cells

Since each of the 20 amino acids is chemically distinct and each can, in principle, occur at any position in a protein chain, there are $20 \times 20 \times 20 \times 20 = 160,000$ different possible polypeptide chains four amino acids long, or 20^n different possible polypeptide chains n amino acids long. For a typical protein length of about 300 amino acids, a cell could theoretically make more than 10^{390} (20^{300}) different polypeptide chains. This is such an enormous number that to produce just one molecule of each kind would require many more atoms than exist in the universe.

Only a very small fraction of this vast set of conceivable polypeptide chains would adopt a stable three-dimensional conformation—by some estimates, less

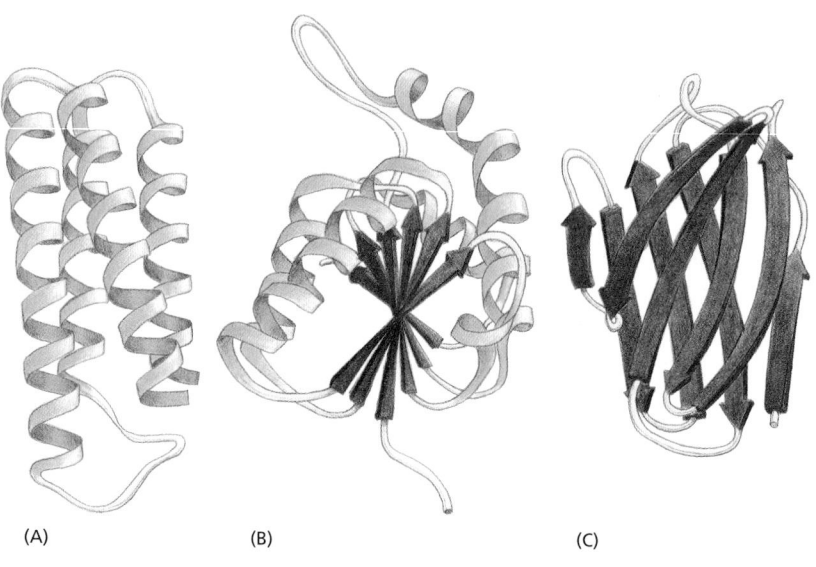

(A)   (B)   (C)

Figure 3–11 **Ribbon models of three different protein domains.** (A) Cytochrome b_{562}, a single-domain protein involved in electron transport in mitochondria. This protein is composed almost entirely of α helices. (B) The NAD-binding domain of the enzyme lactic dehydrogenase, which is composed of a mixture of α helices and parallel β sheets. (C) The variable domain of an immunoglobulin (antibody) light chain, composed of a sandwich of two antiparallel β sheets. In these examples, the α helices are shown in *green*, while strands organized as β sheets are denoted by *red arrows*. Note how the polypeptide chain generally traverses back and forth across the entire domain, making sharp turns only at the protein surface (**Movie 3.5**). It is the protruding loop regions *(yellow)* that often form the binding sites for other molecules.

than one in a billion. And yet the majority of proteins present in cells do adopt unique and stable conformations. How is this possible? The answer lies in natural selection. A protein with an unpredictably variable structure and biochemical activity is unlikely to help the survival of a cell that contains it. Such proteins would therefore have been eliminated by natural selection through the enormously long trial-and-error process that underlies biological evolution.

Because evolution has selected for protein function in living organisms, the amino acid sequence of most present-day proteins is such that a single conformation is stable. In addition, this conformation has its chemical properties finely tuned to enable the protein to perform a particular catalytic or structural function in the cell. Proteins are so precisely built that the change of even a few atoms in one amino acid can sometimes disrupt the structure of the whole molecule so severely that all function is lost. And, as discussed later in this chapter, when certain rare protein misfolding accidents occur, the results can be disastrous for the organisms that contain them.

Proteins Can Be Classified into Many Families

Once a protein had evolved that folded up into a stable conformation with useful properties, its structure could be modified during evolution to enable it to perform new functions. This process has been greatly accelerated by genetic mechanisms that occasionally duplicate genes, allowing one gene copy to evolve independently to perform a new function (discussed in Chapter 4). This type of event has occurred very often in the past; as a result, many present-day proteins can be grouped into protein families, each family member having an amino acid sequence and a three-dimensional conformation that resemble those of the other family members.

Consider, for example, the *serine proteases*, a large family of protein-cleaving (proteolytic) enzymes that includes the digestive enzymes chymotrypsin, trypsin, and elastase, and several proteases involved in blood clotting. When the protease portions of any two of these enzymes are compared, parts of their amino acid sequences are found to match. The similarity of their three-dimensional conformations is even more striking: most of the detailed twists and turns in their polypeptide chains, which are several hundred amino acids long, are virtually identical (**Figure 3–12**). The many different serine proteases nevertheless have distinct enzymatic activities, each cleaving different proteins or the peptide bonds between different types of amino acids. Each therefore performs a distinct function in an organism.

The story we have told for the serine proteases could be repeated for hundreds of other protein families. In general, the structure of the different members of a

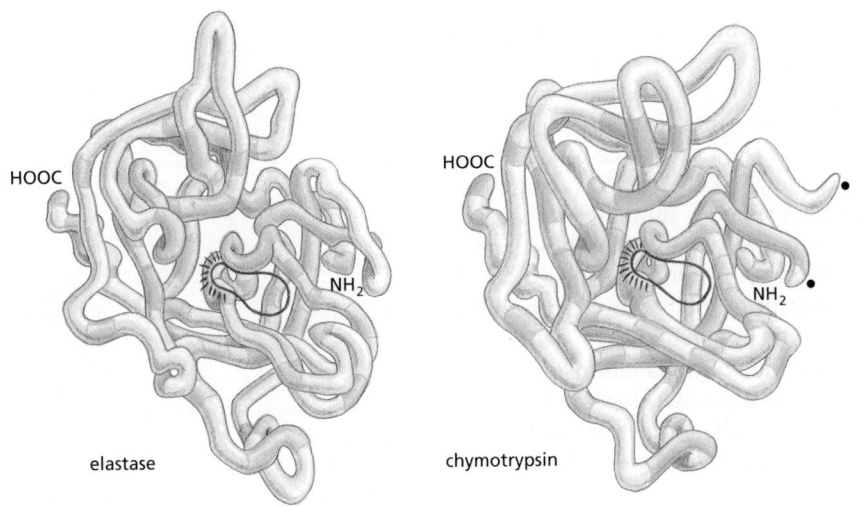

elastase   chymotrypsin

Figure 3–12 **A comparison of the conformations of two serine proteases.** The backbone conformations of elastase and chymotrypsin. Although only those amino acids in the polypeptide chain shaded in *green* are the same in the two proteins, the two conformations are very similar nearly everywhere. The active site of each enzyme is circled in *red*; this is where the peptide bonds of the proteins that serve as substrates are bound and cleaved by hydrolysis. The serine proteases derive their name from the amino acid serine, whose side chain is part of the active site of each enzyme and directly participates in the cleavage reaction. The two dots on the right side of the chymotrypsin molecule mark the new ends created when this enzyme cuts its own backbone.

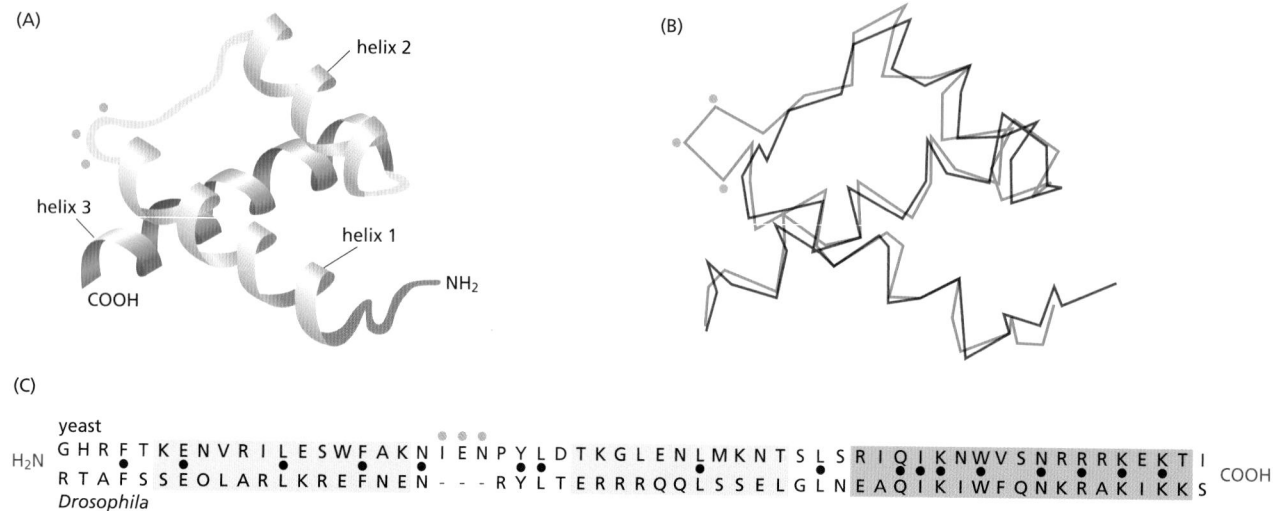

Figure 3–13 **A comparison of a class of DNA-binding domains, called homeodomains, in a pair of proteins from two organisms separated by more than a billion years of evolution.** (A) A ribbon model of the structure common to both proteins. (B) A trace of the α-carbon positions. The three-dimensional structures shown were determined by x-ray crystallography for the yeast α2 protein *(green)* and the *Drosophila* engrailed protein *(red)*. (C) A comparison of amino acid sequences for the region of the proteins shown in (A) and (B). *Black dots* mark sites with identical amino acids. *Orange dots* indicate the position of a three-amino-acid insert in the α2 protein. (Adapted from C. Wolberger et al., *Cell* 67:517–528, 1991.)

protein family has been more highly conserved than has the amino acid sequence. In many cases, the amino acid sequences have diverged so far that we cannot be certain of a family relationship between two proteins without determining their three-dimensional structures. The yeast α2 protein and the *Drosophila* engrailed protein, for example, are both gene regulatory proteins in the homeodomain family (discussed in Chapter 7). Because they are identical in only 17 of their 60 amino acid residues of their homeodomain, their relationship became certain only by comparing their three-dimensional structures (**Figure 3–13**). Many similar examples show that two proteins with more than 25% identity in their amino acid sequences usually share the same overall structure.

The various members of a large protein family often have distinct functions. Some of the amino acid changes that make family members different were no doubt selected in the course of evolution because they resulted in useful changes in biological activity, giving the individual family members the different functional properties they have today. But many other amino acid changes are effectively "neutral," having neither a beneficial nor a damaging effect on the basic structure and function of the protein. In addition, since mutation is a random process, there must also have been many deleterious changes that altered the three-dimensional structure of these proteins sufficiently to harm them. Such faulty proteins would have been lost whenever the individual organisms making them were at enough of a disadvantage to be eliminated by natural selection.

Protein families are readily recognized when the genome of any organism is sequenced; for example, the determination of the DNA sequence for the entire human genome has revealed that we contain about 21,000 protein-coding genes. (Note, however, that as a result of alternative RNA splicing, human cells can produce much more than 21,000 different proteins, as will be explained in Chapter 6.) Through sequence comparisons, we can assign the products of at least 40% of our protein-coding genes to known protein structures, belonging to more than 500 different protein families. Most of the proteins in each family have evolved to perform somewhat different functions, as for the enzymes elastase and chymotrypsin illustrated previously in Figure 3–12. As explained in Chapter 1 (see Figure 1–21), these are sometimes called *paralogs* to distinguish them from the many corresponding proteins in different organisms (*orthologs*, such as mouse and human elastase).

As described in Chapter 8, because of the powerful techniques of x-ray crystallography and nuclear magnetic resonance (NMR), we now know the three-dimensional shapes, or conformations, of more than 100,000 proteins. By carefully comparing the conformations of these proteins, structural biologists (that is, experts on the structure of biological molecules) have concluded that there are a limited number of ways in which protein domains fold up in nature—maybe as few as 2000, if we consider all organisms. For most of these so-called *protein folds*, representative structures have been determined.

The present database of known protein sequences contains more than twenty million entries, and it is growing very rapidly as more and more genomes are sequenced—revealing huge numbers of new genes that encode proteins. The encoded polypeptides range widely in size, from 6 amino acids to a gigantic protein of 33,000 amino acids. Protein comparisons are important because related structures often imply related functions. Many years of experimentation can be saved by discovering that a new protein has an amino acid sequence similarity with a protein of known function. Such sequence relationships, for example, first indicated that certain genes that cause mammalian cells to become cancerous encode protein kinases (discussed in Chapter 20).

Some Protein Domains Are Found in Many Different Proteins

As previously stated, most proteins are composed of a series of protein domains, in which different regions of the polypeptide chain fold independently to form compact structures. Such multidomain proteins are believed to have originated from the accidental joining of the DNA sequences that encode each domain, creating a new gene. In an evolutionary process called *domain shuffling*, many large proteins have evolved through the joining of preexisting domains in new combinations (**Figure 3–14**). Novel binding surfaces have often been created at the juxtaposition of domains, and many of the functional sites where proteins bind to small molecules are found to be located there.

A subset of protein domains has been especially mobile during evolution; these seem to have particularly versatile structures and are sometimes referred to as *protein modules*. The structure of one, the SH2 domain, was illustrated in Figure 3–6. Three other abundant protein domains are illustrated in **Figure 3–15**.

Each of the domains shown has a stable core structure formed from strands of β sheets, from which less-ordered loops of polypeptide chain protrude. The loops are ideally situated to form binding sites for other molecules, as most clearly demonstrated for the immunoglobulin fold, which forms the basis for antibody molecules. Such β-sheet-based domains may have achieved their evolutionary success because they provide a convenient framework for the generation of new binding sites for ligands, requiring only small changes to their protruding loops (see Figure 3–42).

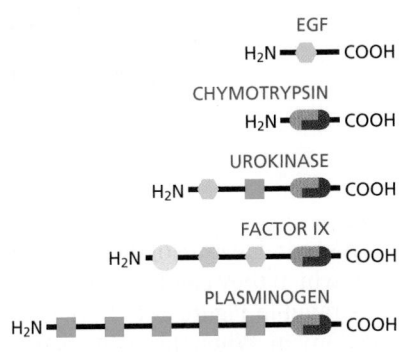

Figure 3–14 Domain shuffling. An extensive shuffling of blocks of protein sequence (protein domains) has occurred during protein evolution. Those portions of a protein denoted by the same shape and color in this diagram are evolutionarily related. Serine proteases like chymotrypsin are formed from two domains *(brown)*. In the three other proteases shown, which are highly regulated and more specialized, these two protease domains are connected to one or more domains that are similar to domains found in epidermal growth factor (EGF; *green*), to a calcium-binding protein *(yellow)*, or to a "kringle" domain *(blue)*. Chymotrypsin is illustrated in Figure 3–12.

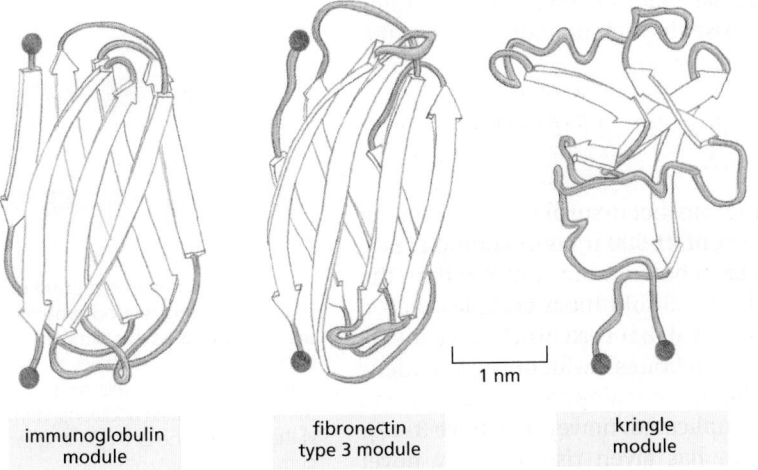

immunoglobulin
module

fibronectin
type 3 module

1 nm

kringle
module

Figure 3–15 The three-dimensional structures of three commonly used protein domains. In these ribbon diagrams, β-sheet strands are shown as *arrows*, and the N- and C-termini are indicated by *red spheres*. Many more such "modules" exist in nature. (Adapted from D. J. Leahy et al., *Science* 258:987–991, 1992. With permission from AAAS.)

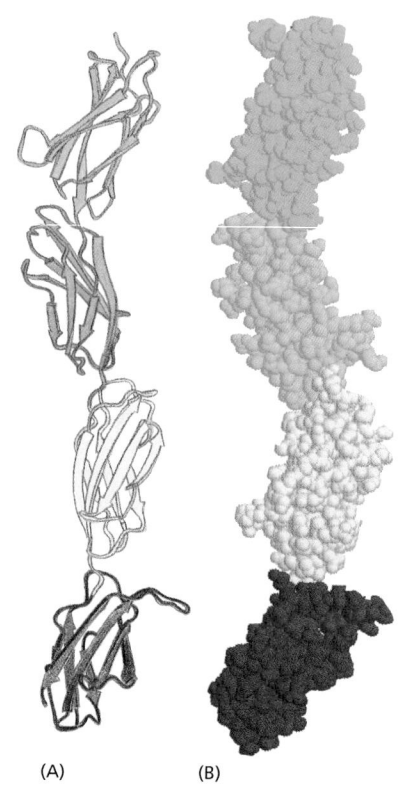

Figure 3–16 **An extended structure formed from a series of protein domains.** Four fibronectin type 3 domains (see Figure 3–15) from the extracellular matrix molecule fibronectin are illustrated in (A) ribbon and (B) space-filling models. (Adapted from D.J. Leahy, I. Aukhil and H.P. Erickson, *Cell* 84:155–164, 1996.)

(A)   (B)

A second feature of these protein domains that explains their utility is the ease with which they can be integrated into other proteins. Two of the three domains illustrated in Figure 3–15 have their N- and C-terminal ends at opposite poles of the domain. When the DNA encoding such a domain undergoes tandem duplication, which is not unusual in the evolution of genomes (discussed in Chapter 4), the duplicated domains with this "in-line" arrangement can be readily linked in series to form extended structures—either with themselves or with other in-line domains (**Figure 3–16**). Stiff extended structures composed of a series of domains are especially common in extracellular matrix molecules and in the extracellular portions of cell-surface receptor proteins. Other frequently used domains, including the kringle domain illustrated in Figure 3–15 and the SH2 domain, are of a "plug-in" type, with their N- and C-termini close together. After genomic rearrangements, such domains are usually accommodated as an insertion into a loop region of a second protein.

A comparison of the relative frequency of domain utilization in different eukaryotes reveals that, for many common domains, such as protein kinases, this frequency is similar in organisms as diverse as yeast, plants, worms, flies, and humans. But there are some notable exceptions, such as the Major Histocompatibility Complex (MHC) antigen-recognition domain (see Figure 24–36) that is present in 57 copies in humans, but absent in the other four organisms just mentioned. Domains such as these have specialized functions that are not shared with the other eukaryotes; they are assumed to have been strongly selected for during recent evolution to produce the multiple copies observed. Similarly, the SH2 domain shows an unusual increase in its numbers in higher eukaryotes; such domains might be assumed to be especially useful for multicellularity.

Certain Pairs of Domains Are Found Together in Many Proteins

We can construct a large table displaying domain usage for each organism whose genome sequence is known. For example, the human genome contains the DNA sequences for about 1000 immunoglobulin domains, 500 protein kinase domains, 250 DNA-binding homeodomains, 300 SH3 domains, and 120 SH2 domains. In addition, we find that more than two-thirds of all proteins consist of two or more domains, and that the same pairs of domains occur repeatedly in the same relative arrangement in a protein. Although half of all domain families are common to archaea, bacteria, and eukaryotes, only about 5% of the two-domain combinations are similarly shared. This pattern suggests that most proteins containing especially useful two-domain combinations arose through domain shuffling relatively late in evolution.

The Human Genome Encodes a Complex Set of Proteins, Revealing That Much Remains Unknown

The result of sequencing the human genome has been surprising, because it reveals that our chromosomes contain only about 21,000 protein-coding genes. Based on this number alone, we would appear to be no more complex than the tiny mustard weed, *Arabidopsis*, and only about 1.3-fold more complex than a nematode worm. The genome sequences also reveal that vertebrates have inherited nearly all of their protein domains from invertebrates—with only 7% of identified human domains being vertebrate-specific.

Each of our proteins is on average more complicated, however (**Figure 3–17**). Domain shuffling during vertebrate evolution has given rise to many novel

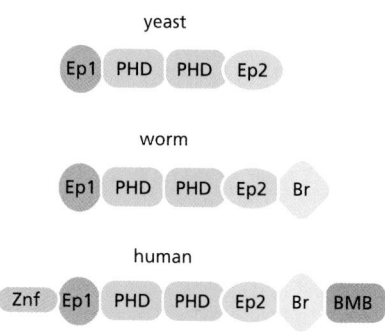

yeast

Ep1 PHD PHD Ep2

worm

Ep1 PHD PHD Ep2 Br

human

Znf Ep1 PHD PHD Ep2 Br BMB

Figure 3–17 **Domain structure of a group of evolutionarily related proteins that are thought to have a similar function.** In general, there is a tendency for the proteins in more complex organisms, such as humans, to contain additional domains—as is the case for the DNA-binding protein compared here.

Figure 3–18 **Two identical protein subunits binding together to form a symmetric protein dimer.** The Cro repressor protein from bacteriophage lambda binds to DNA to turn off a specific subset of viral genes. Its two identical subunits bind head-to-head, held together by a combination of hydrophobic forces (*blue*) and a set of hydrogen bonds (*yellow* region). (Adapted from D.H. Ohlendorf, D.E. Tronrud and B.W. Matthews, *J. Mol. Biol.* 280:129–136, 1998. With permission from Academic Press.)

combinations of protein domains, with the result that there are nearly twice as many combinations of domains found in human proteins as in a worm or a fly. Thus, for example, the trypsinlike serine protease domain is linked to at least 18 other types of protein domains in human proteins, whereas it is found covalently joined to only 5 different domains in the worm. This extra variety in our proteins greatly increases the range of protein–protein interactions possible, but how it contributes to making us human is not known.

The complexity of living organisms is staggering, and it is quite sobering to note that we currently lack even the tiniest hint of what the function might be for more than 10,000 of the proteins that have thus far been identified through examining the human genome. There are certainly enormous challenges ahead for the next generation of cell biologists, with no shortage of fascinating mysteries to solve.

Larger Protein Molecules Often Contain More Than One Polypeptide Chain

The same weak noncovalent bonds that enable a protein chain to fold into a specific conformation also allow proteins to bind to each other to produce larger structures in the cell. Any region of a protein's surface that can interact with another molecule through sets of noncovalent bonds is called a **binding site**. A protein can contain binding sites for various large and small molecules. If a binding site recognizes the surface of a second protein, the tight binding of two folded polypeptide chains at this site creates a larger protein molecule with a precisely defined geometry. Each polypeptide chain in such a protein is called a **protein subunit**.

In the simplest case, two identical folded polypeptide chains bind to each other in a "head-to-head" arrangement, forming a symmetric complex of two protein subunits (a *dimer*) held together by interactions between two identical binding sites. The *Cro repressor protein*—a viral gene regulatory protein that binds to DNA to turn specific viral genes off in an infected bacterial cell—provides an example (**Figure 3–18**). Cells contain many other types of symmetric protein complexes, formed from multiple copies of a single polypeptide chain (for example, see Figure 3–20 below).

Many of the proteins in cells contain two or more types of polypeptide chains. *Hemoglobin*, the protein that carries oxygen in red blood cells, contains two identical α-globin subunits and two identical β-globin subunits, symmetrically arranged (**Figure 3–19**). Such multisubunit proteins are very common in cells, and they can be very large (**Movie 3.6**).

Some Globular Proteins Form Long Helical Filaments

Most of the proteins that we have discussed so far are *globular proteins,* in which the polypeptide chain folds up into a compact shape like a ball with an irregular surface. Some of these protein molecules can nevertheless assemble to form filaments that may span the entire length of a cell. Most simply, a long chain of identical protein molecules can be constructed if each molecule has a binding

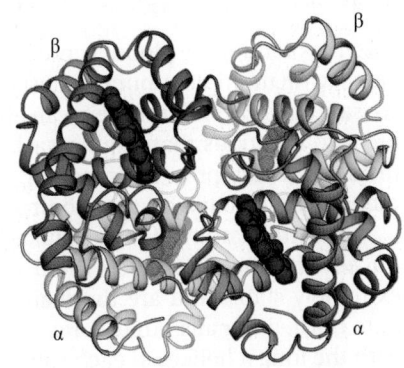

Figure 3–19 **A protein formed as a symmetric assembly using two each of two different subunits.** Hemoglobin is an abundant protein in red blood cells that contains two copies of α-globin (*green*) and two copies of β-globin (*blue*). Each of these four polypeptide chains contains a heme molecule (*red*), which is the site that binds oxygen (O_2). Thus, each molecule of hemoglobin in the blood carries four molecules of oxygen. (PDB code: 2DHB.)

Figure 3–20 **Protein assemblies.** (A) A protein with just one binding site can form a dimer with another identical protein. (B) Identical proteins with two different binding sites often form a long helical filament. (C) If the two binding sites are disposed appropriately in relation to each other, the protein subunits may form a closed ring instead of a helix. (For an example of A, see Figure 3–18; for an example of B, see Figure 3–21; for examples of C, see Figures 5–14 and 14–31.)

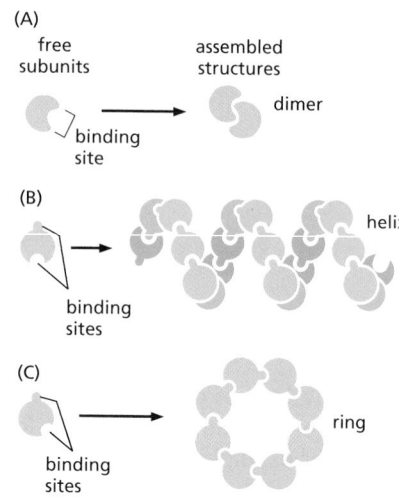

site complementary to another region of the surface of the same molecule (**Figure 3–20**). An actin filament, for example, is a long helical structure produced from many molecules of the protein *actin* (**Figure 3–21**). Actin is a globular protein that is very abundant in eukaryotic cells, where it forms one of the major filament systems of the cytoskeleton (discussed in Chapter 16).

We will encounter many helical structures in this book. Why is a helix such a common structure in biology? As we have seen, biological structures are often formed by linking similar subunits into long, repetitive chains. If all the subunits are identical, the neighboring subunits in the chain can often fit together in only one way, adjusting their relative positions to minimize the free energy of the contact between them. As a result, each subunit is positioned in exactly the same way in relation to the next, so that subunit 3 fits onto subunit 2 in the same way that subunit 2 fits onto subunit 1, and so on. Because it is very rare for subunits to join up in a straight line, this arrangement generally results in a helix—a regular structure that resembles a spiral staircase, as illustrated in **Figure 3–22**. Depending on the twist of the staircase, a helix is said to be either right-handed or left-handed (see Figure 3–22E). Handedness is not affected by turning the helix upside down, but it is reversed if the helix is reflected in the mirror.

The observation that helices occur commonly in biological structures holds true whether the subunits are small molecules linked together by covalent bonds (for example, the amino acids in an α helix) or large protein molecules that are linked by noncovalent forces (for example, the actin molecules in actin filaments). This is not surprising. A helix is an unexceptional structure, and it is generated simply by placing many similar subunits next to each other, each in the same strictly repeated relationship to the one before—that is, with a fixed rotation followed by a fixed translation along the helix axis, as in a spiral staircase.

Many Protein Molecules Have Elongated, Fibrous Shapes

Enzymes tend to be globular proteins: even though many are large and complicated, with multiple subunits, most have an overall rounded shape. In Figure 3–21, we saw that a globular protein can also associate to form long filaments. But there are also functions that require each individual protein molecule to span a large distance. These proteins generally have a relatively simple, elongated three-dimensional structure and are commonly referred to as *fibrous proteins*.

One large family of intracellular fibrous proteins consists of α-keratin, introduced when we presented the α helix, and its relatives. Keratin filaments are extremely stable and are the main component in long-lived structures such as hair, horn, and nails. An α-keratin molecule is a dimer of two identical subunits, with the long α helices of each subunit forming a coiled-coil (see Figure 3–9). The coiled-coil regions are capped at each end by globular domains containing binding sites. This enables this class of protein to assemble into ropelike *intermediate filaments*—an important component of the cytoskeleton that creates the cell's internal structural framework (see Figure 16–67).

Fibrous proteins are especially abundant outside the cell, where they are a main component of the gel-like *extracellular matrix* that helps to bind collections of cells together to form tissues. Cells secrete extracellular matrix proteins into their surroundings, where they often assemble into sheets or long fibrils. *Collagen* is the most abundant of these proteins in animal tissues. A collagen molecule consists of three long polypeptide chains, each containing the nonpolar amino

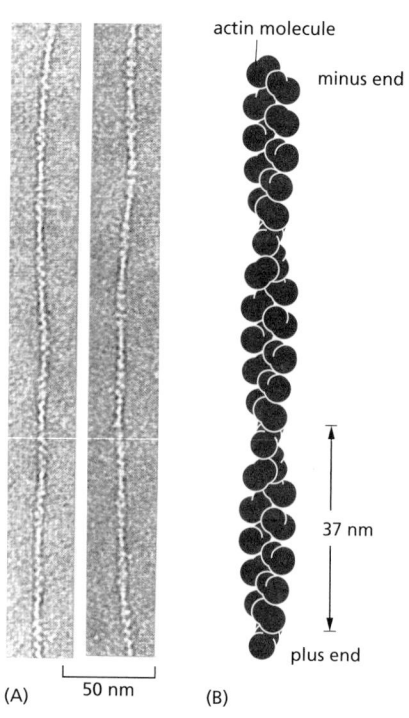

Figure 3–21 **Actin filaments.**
(A) Transmission electron micrographs of negatively stained actin filaments. (B) The helical arrangement of actin molecules in an actin filament. (A, courtesy of Roger Craig.)

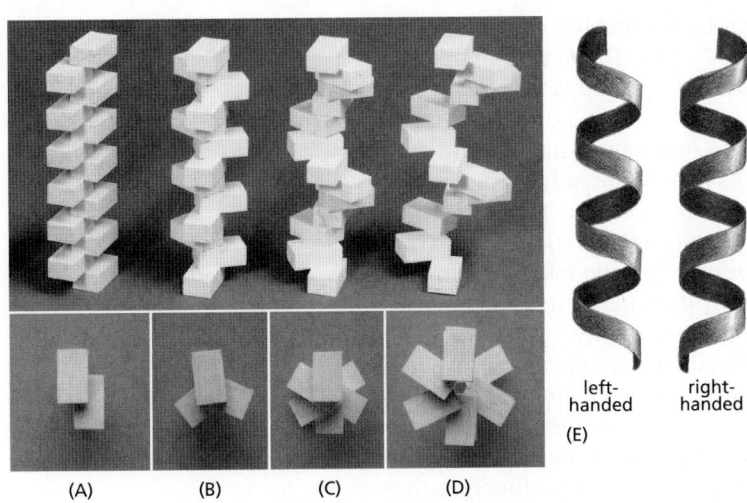

Figure 3–22 **Some properties of a helix.** (A–D) A helix forms when a series of subunits bind to each other in a regular way. At the bottom, each of these helices is viewed from directly above the helix and seen to have two (A), three (B), and six (C and D) subunits per helical turn. Note that the helix in (D) has a wider path than that in (C), but the same number of subunits per turn. (E) As discussed in the text, a helix can be either right-handed or left-handed. As a reference, it is useful to remember that standard metal screws, which insert when turned clockwise, are right-handed. Note that a helix retains the same handedness when it is turned upside down. (PDB code: 2DHB.)

left-handed

right-handed

(E)

acid glycine at every third position. This regular structure allows the chains to wind around one another to generate a long regular triple helix (**Figure 3–23A**). Many collagen molecules then bind to one another side-by-side and end-to-end to create long overlapping arrays—thereby generating the extremely tough collagen fibrils that give connective tissues their tensile strength, as described in Chapter 19.

Proteins Contain a Surprisingly Large Amount of Intrinsically Disordered Polypeptide Chain

It has been well known for a long time that, in complete contrast to collagen, another abundant protein in the extracellular matrix, *elastin*, is formed as a highly disordered polypeptide. This disorder is essential for elastin's function. Its relatively loose and unstructured polypeptide chains are covalently cross-linked to

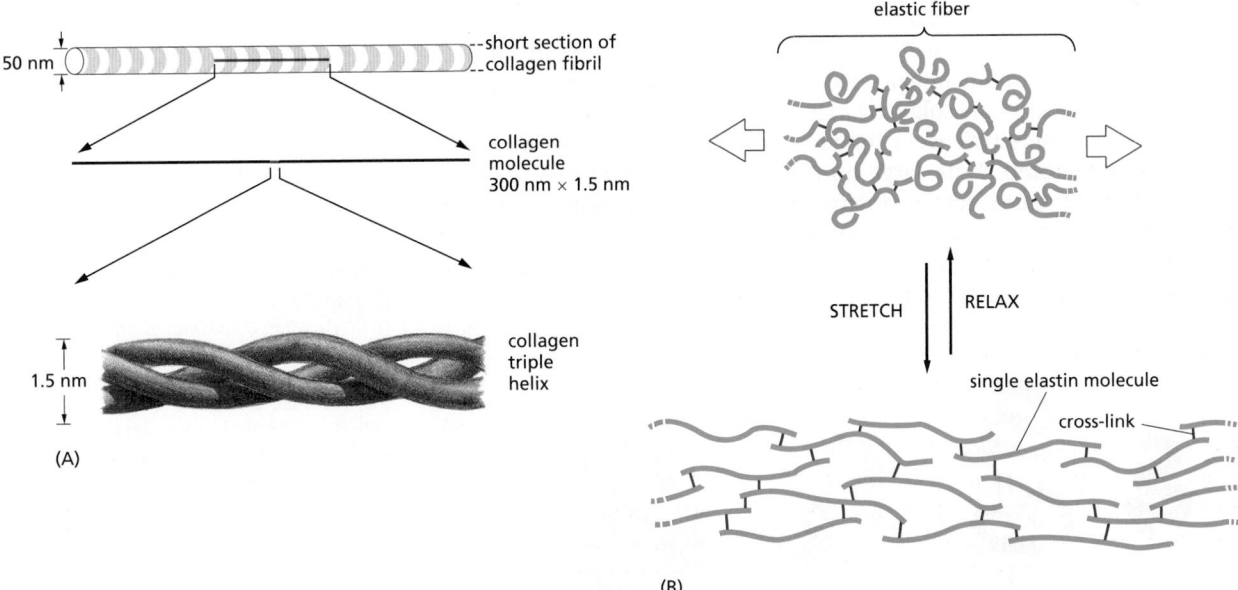

Figure 3–23 **Collagen and elastin.** (A) Collagen is a triple helix formed by three extended protein chains that wrap around one another *(bottom)*. Many rodlike collagen molecules are cross-linked together in the extracellular space to form unextendable collagen fibrils *(top)* that have the tensile strength of steel. The striping on the collagen fibril is caused by the regular repeating arrangement of the collagen molecules within the fibril. (B) Elastin polypeptide chains are cross-linked together in the extracellular space to form rubberlike, elastic fibers. Each elastin molecule uncoils into a more extended conformation when the fiber is stretched and recoils spontaneously as soon as the stretching force is relaxed. The cross-linking in the extracellular space mentioned creates covalent linkages between lysine side chains, but the chemistry is different for collagen and elastin.

produce a rubberlike, elastic meshwork that can be reversibly pulled from one conformation to another, as illustrated in Figure 3–23B. The elastic fibers that result enable skin and other tissues, such as arteries and lungs, to stretch and recoil without tearing.

Intrinsically disordered regions of proteins are frequent in nature, and they have important functions in the interior of cells. As we have already seen, proteins often have loops of polypeptide chain that protrude from the core region of a protein domain to bind other molecules. Some of these loops remain largely unstructured until they bind to a target molecule, adopting a specific folded conformation only when this other molecule is bound. Many proteins were also known to have intrinsically disordered tails at one or the other end of a structured domain (see, for example, the histones in Figure 4–24). But the extent of such disordered structure only became clear when genomes were sequenced. This allowed bioinformatic methods to be used to analyze the amino acid sequences that genes encode, searching for disordered regions based on their unusually low hydrophobicity and relatively high net charge. Combining these results with other data, it is now thought that perhaps a quarter of all eukaryotic proteins can adopt structures that are mostly disordered, fluctuating rapidly between many different conformations. Many such intrinsically disordered regions contain repeated sequences of amino acids. What do these disordered regions do?

Some known functions are illustrated in **Figure 3–24**. One predominant function is to form specific binding sites for other protein molecules that are of high specificity, but readily altered by protein phosphorylation, protein dephosphorylation, or any of the other covalent modifications that are triggered by cell signaling events (Figure 3–24A and B). We shall see, for example, that the eukaryotic RNA polymerase enzyme that produces mRNAs contains a long, unstructured C-terminal tail that is covalently modified as its RNA synthesis proceeds, thereby attracting specific other proteins to the transcription complex at different times (see Figure 6–22). And this unstructured tail interacts with a different type of low complexity domain when the RNA polymerase is recruited to the specific sites on the DNA where it begins synthesis.

As illustrated in Figure 3–24C, an unstructured region can also serve as a "tether" to hold two protein domains in close proximity to facilitate their interaction. For example, it is this tethering function that allows substrates to move between active sites in large multienzyme complexes (see Figure 3–54). A similar tethering function allows large *scaffold proteins* with multiple protein-binding sites to concentrate sets of interacting proteins, both increasing reaction rates and confining their reaction to a particular site in a cell (see Figure 3–78).

Like elastin, other proteins have a function that directly requires that they remain largely unstructured. Thus, large numbers of disordered protein chains in close proximity can create micro-regions of gel-like consistency inside the cell that restrict diffusion. For example, the abundant nucleoporins that coat the inner surface of the nuclear pore complex form a random coil meshwork (Figure 3–24) that is critical for selective nuclear transport (see Figure 12–8).

Figure 3–24 Some important functions for intrinsically disordered protein sequences. (A) Unstructured regions of polypeptide chain often form binding sites for other proteins. Although these binding events are of high specificity, they are often of low affinity due to the free-energy cost of folding the normally unfolded partner (and they are thus readily reversible). (B) Unstructured regions can be easily modified covalently to change their binding preferences, and they are therefore frequently involved in cell signaling processes. In this schematic, multiple sites of protein phosphorylation are indicated. (C) Unstructured regions frequently create "tethers" that hold interacting protein domains in close proximity. (D) A dense network of unstructured proteins can form a diffusion barrier, as the nucleoporins do for the nuclear pore.

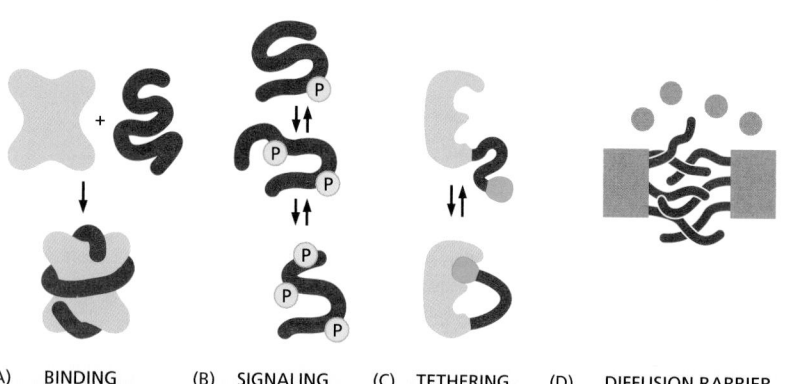

(A)   BINDING   (B)   SIGNALING   (C)   TETHERING   (D)   DIFFUSION BARRIER

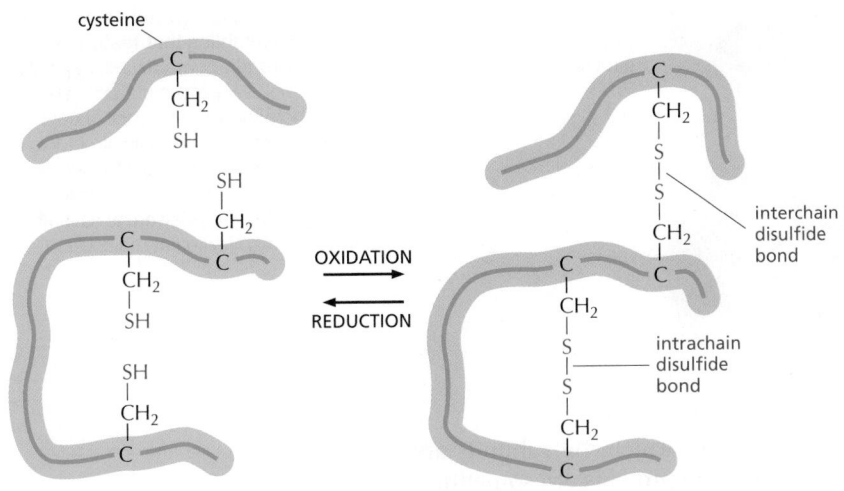

Figure 3–25 Disulfide bonds. Covalent disulfide bonds form between adjacent cysteine side chains. These cross-linkages can join either two parts of the same polypeptide chain or two different polypeptide chains. Since the energy required to break one covalent bond is much larger than the energy required to break even a whole set of noncovalent bonds (see Table 2–1, p. 45), a disulfide bond can have a major stabilizing effect on a protein (Movie 3.7).

Covalent Cross-Linkages Stabilize Extracellular Proteins

Many protein molecules are either attached to the outside of a cell's plasma membrane or secreted to form part of the extracellular matrix. All such proteins are directly exposed to extracellular conditions. To help maintain their structures, the polypeptide chains in such proteins are often stabilized by covalent cross-linkages. These linkages can either tie together two amino acids in the same protein, or connect different polypeptide chains in a multisubunit protein. Although many other types exist, the most common cross-linkages in proteins are covalent sulfur–sulfur bonds. These *disulfide bonds* (also called *S–S bonds*) form as cells prepare newly synthesized proteins for export. As described in Chapter 12, their formation is catalyzed in the endoplasmic reticulum by an enzyme that links together two pairs of –SH groups of cysteine side chains that are adjacent in the folded protein (**Figure 3–25**). Disulfide bonds do not change the conformation of a protein but instead act as atomic staples to reinforce its most favored conformation. For example, lysozyme—an enzyme in tears that dissolves bacterial cell walls—retains its antibacterial activity for a long time because it is stabilized by such cross-linkages.

Disulfide bonds generally fail to form in the cytosol, where a high concentration of reducing agents converts S–S bonds back to cysteine –SH groups. Apparently, proteins do not require this type of reinforcement in the relatively mild environment inside the cell.

Protein Molecules Often Serve as Subunits for the Assembly of Large Structures

The same principles that enable a protein molecule to associate with itself to form rings or a long filament also operate to generate much larger structures formed from a set of different macromolecules, such as enzyme complexes, ribosomes, viruses, and membranes. These large objects are not made as single, giant, covalently linked molecules. Instead they are formed by the noncovalent assembly of many separately manufactured molecules, which serve as the subunits of the final structure.

The use of smaller subunits to build larger structures has several advantages:

1. A large structure built from one or a few repeating smaller subunits requires only a small amount of genetic information.

2. Both assembly and disassembly can be readily controlled reversible processes, because the subunits associate through multiple bonds of relatively low energy.

3. Errors in the synthesis of the structure can be more easily avoided, since correction mechanisms can operate during the course of assembly to exclude malformed subunits.

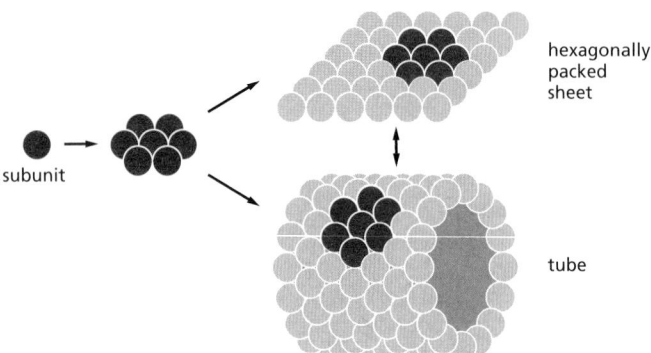

Figure 3–26 **Single protein subunits form protein assemblies that feature multiple protein–protein contacts.** Hexagonally packed globular protein subunits are shown here forming either flat sheets or tubes. Generally, such large structures are not considered to be single "molecules." Instead, like the actin filament described previously, they are viewed as assemblies formed of many different molecules.

Some protein subunits assemble into flat sheets in which the subunits are arranged in hexagonal patterns. Specialized membrane proteins are sometimes arranged this way in lipid bilayers. With a slight change in the geometry of the individual subunits, a hexagonal sheet can be converted into a tube (**Figure 3–26**) or, with more changes, into a hollow sphere. Protein tubes and spheres that bind specific RNA and DNA molecules in their interior form the coats of viruses.

The formation of closed structures, such as rings, tubes, or spheres, provides additional stability because it increases the number of bonds between the protein subunits. Moreover, because such a structure is created by mutually dependent, cooperative interactions between subunits, a relatively small change that affects each subunit individually can cause the structure to assemble or disassemble. These principles are dramatically illustrated in the protein coat or *capsid* of many simple viruses, which takes the form of a hollow sphere based on an icosahedron (**Figure 3–27**). Capsids are often made of hundreds of identical protein subunits that enclose and protect the viral nucleic acid (**Figure 3–28**). The protein in such a capsid must have a particularly adaptable structure: not only must it make several different kinds of contacts to create the sphere, it must also change this arrangement to let the nucleic acid out to initiate viral replication once the virus has entered a cell.

Many Structures in Cells Are Capable of Self-Assembly

The information for forming many of the complex assemblies of macromolecules in cells must be contained in the subunits themselves, because purified subunits can spontaneously assemble into the final structure under the appropriate conditions. The first large macromolecular aggregate shown to be capable of self-assembly from its component parts was *tobacco mosaic virus* (*TMV*). This virus is a long rod in which a cylinder of protein is arranged around a helical RNA core (**Figure 3–29**). If the dissociated RNA and protein subunits are mixed together in solution, they recombine to form fully active viral particles. The assembly process is unexpectedly complex and includes the formation of double rings of protein, which serve as intermediates that add to the growing viral coat.

Another complex macromolecular aggregate that can reassemble from its component parts is the bacterial ribosome. This structure is composed of about 55 different protein molecules and 3 different rRNA molecules. Incubating a mixture of the individual components under appropriate conditions in a test tube causes them to spontaneously re-form the original structure. Most importantly, such reconstituted ribosomes are able to catalyze protein synthesis. As might be expected, the reassembly of ribosomes follows a specific pathway: after certain proteins have bound to the RNA, this complex is then recognized by other proteins, and so on, until the structure is complete.

It is still not clear how some of the more elaborate self-assembly processes are regulated. Many structures in the cell, for example, seem to have a precisely defined length that is many times greater than that of their component macromolecules. How such length determination is achieved is in many cases a mystery. In

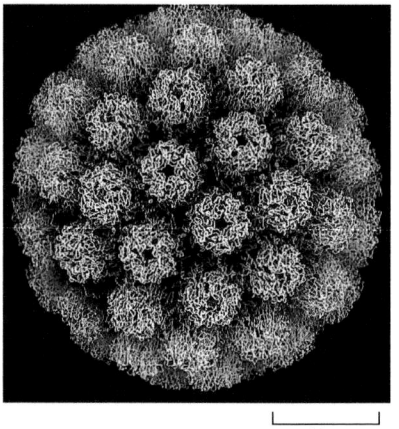

20 nm

Figure 3–27 **The protein capsid of a virus.** The structure of the simian virus SV40 capsid has been determined by x-ray crystallography and, as for the capsids of many other viruses, it is known in atomic detail. (Courtesy of Robert Grant, Stephan Crainic, and James M. Hogle.)

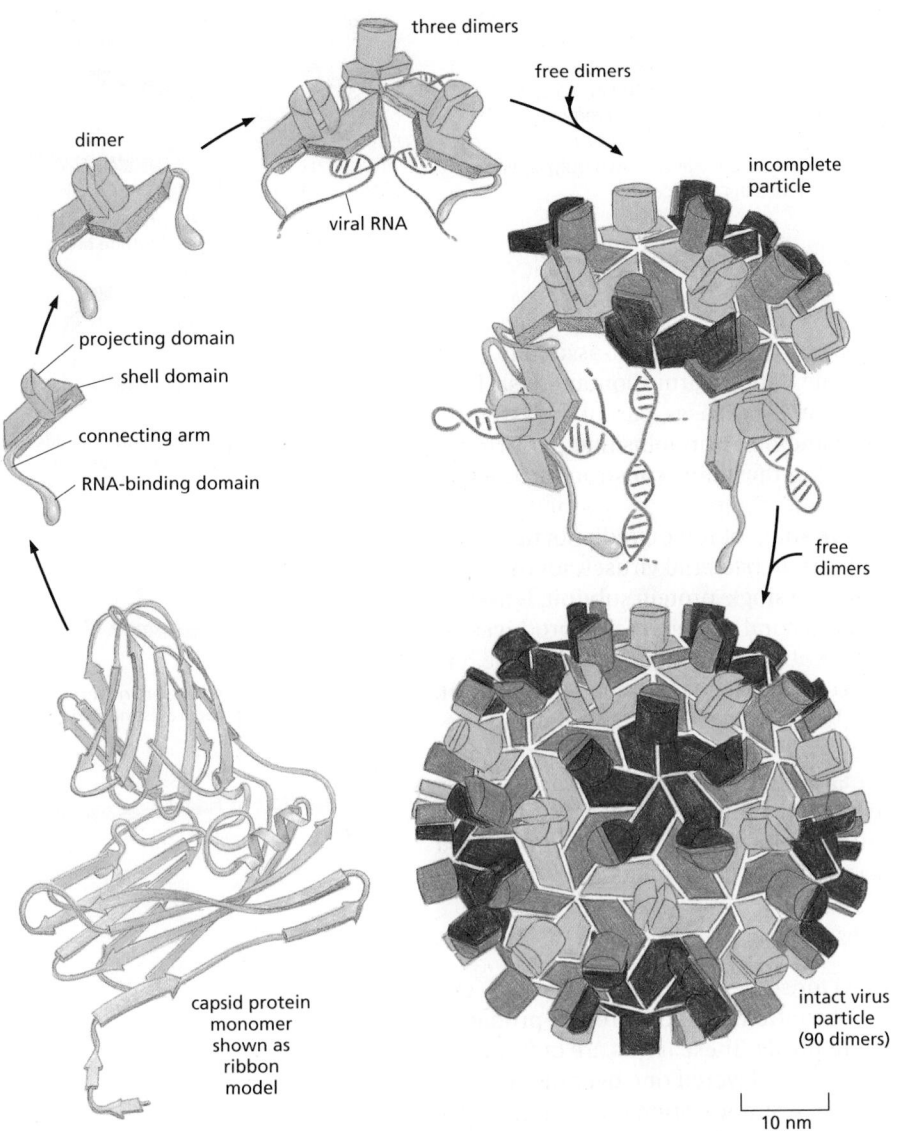

Figure 3–28 The structure of a spherical virus. In viruses, many copies of a single protein subunit often pack together to create a spherical shell (a capsid). This capsid encloses the viral genome, composed of either RNA or DNA (see also Figure 3–27). For geometric reasons, no more than 60 identical subunits can pack together in a precisely symmetric way. If slight irregularities are allowed, however, more subunits can be used to produce a larger capsid that retains icosahedral symmetry. The tomato bushy stunt virus (TBSV) shown here, for example, is a spherical virus about 33 nm in diameter formed from 180 identical copies of a 386-amino-acid capsid protein plus an RNA genome of 4500 nucleotides. To construct such a large capsid, the protein must be able to fit into three somewhat different environments. This requires three slightly different conformations, each of which is differently colored in the virus particle shown here. The postulated pathway of assembly is shown; the precise three-dimensional structure has been determined by x-ray diffraction. (Courtesy of Steve Harrison.)

the simplest case, a long core protein or other macromolecule provides a scaffold that determines the extent of the final assembly. This is the mechanism that determines the length of the TMV particle, where the RNA chain provides the core. Similarly, a core protein interacting with actin is thought to determine the length of the thin filaments in muscle.

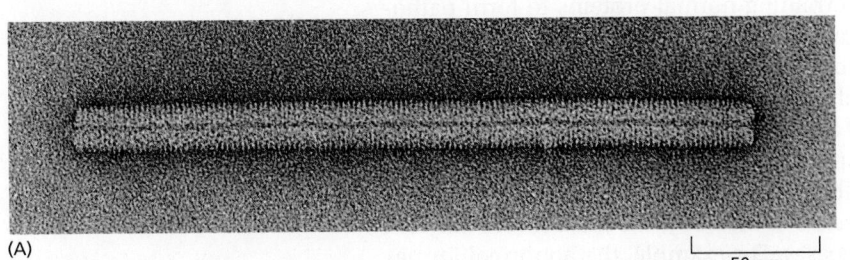
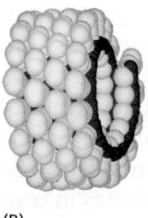

Figure 3–29 The structure of tobacco mosaic virus (TMV). (A) An electron micrograph of the viral particle, which consists of a single long RNA molecule enclosed in a cylindrical protein coat composed of identical protein subunits. (B) A model showing part of the structure of TMV. A single-stranded RNA molecule of 6395 nucleotides is packaged in a helical coat constructed from 2130 copies of a coat protein 158 amino acids long. Fully infective viral particles can self-assemble in a test tube from purified RNA and protein molecules. (A, courtesy of Robley Williams; B, courtesy of Richard J. Feldmann.)

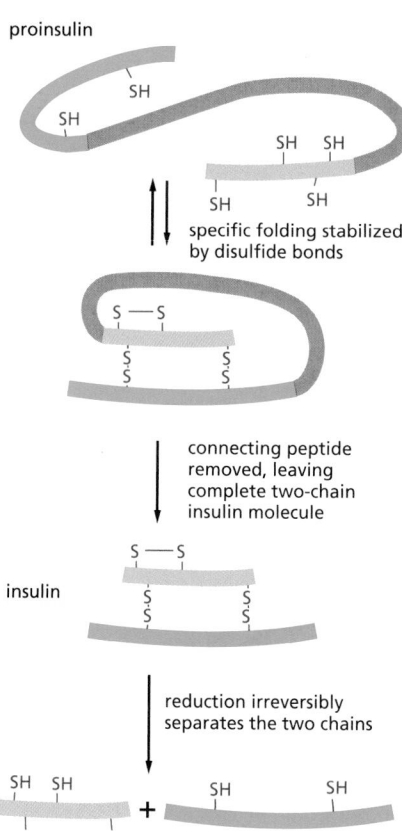

Figure 3–30 Proteolytic cleavage in insulin assembly. The polypeptide hormone insulin cannot spontaneously re-form efficiently if its disulfide bonds are disrupted. It is synthesized as a larger protein (*proinsulin*) that is cleaved by a proteolytic enzyme after the protein chain has folded into a specific shape. Excision of part of the proinsulin polypeptide chain removes some of the information needed for the protein to fold spontaneously into its normal conformation. Once insulin has been denatured and its two polypeptide chains have separated, its ability to reassemble is lost.

Assembly Factors Often Aid the Formation of Complex Biological Structures

Not all cellular structures held together by noncovalent bonds self-assemble. A cilium, or a myofibril of a muscle cell, for example, cannot form spontaneously from a solution of its component macromolecules. In these cases, part of the assembly information is provided by special enzymes and other proteins that perform the function of templates, serving as *assembly factors* that guide construction but take no part in the final assembled structure.

Even relatively simple structures may lack some of the ingredients necessary for their own assembly. In the formation of certain bacterial viruses, for example, the head, which is composed of many copies of a single protein subunit, is assembled on a temporary scaffold composed of a second protein that is produced by the virus. Because the second protein is absent from the final viral particle, the head structure cannot spontaneously reassemble once it has been taken apart. Other examples are known in which proteolytic cleavage is an essential and irreversible step in the normal assembly process. This is even the case for some small protein assemblies, including the structural protein collagen and the hormone insulin (**Figure 3–30**). From these relatively simple examples, it seems certain that the assembly of a structure as complex as a cilium will involve a temporal and spatial ordering that is imparted by numerous other components.

Amyloid Fibrils Can Form from Many Proteins

A special class of protein structures, utilized for some normal cell functions, can also contribute to human diseases when not controlled. These are self-propagating, stable β-sheet aggregates called **amyloid fibrils**. These fibrils are built from a series of identical polypeptide chains that become layered one over the other to create a continuous stack of β sheets, with the β strands oriented perpendicular to the fibril axis to form a *cross-beta filament* (**Figure 3–31**). Typically, hundreds of monomers will aggregate to form an unbranched fibrous structure that is several micrometers long and 5 to 15 nm in width. A surprisingly large fraction of proteins have the potential to form such structures, because the short segment of the polypeptide chain that forms the spine of the fibril can have a variety of different sequences and follow one of several different paths (**Figure 3–32**). However, very few proteins will actually form this structure inside cells.

In normal humans, the quality control mechanisms governing proteins gradually decline with age, occasionally permitting normal proteins to form pathological aggregates. The protein aggregates may be released from dead cells and accumulate as amyloid in the extracellular matrix. In extreme cases, the accumulation of such amyloid fibrils in the cell interior can kill the cells and damage tissues. Because the brain is composed of a highly organized collection of nerve cells that cannot regenerate, the brain is especially vulnerable to this sort of cumulative damage. Thus, although amyloid fibrils may form in different tissues, and are known to cause pathologies in several sites in the body, the most severe amyloid pathologies are neurodegenerative diseases. For example, the abnormal formation of highly stable amyloid fibrils is thought to play a central causative role in both Alzheimer's and Parkinson's diseases.

Prion diseases are a special type of these pathologies. They have attained special notoriety because, unlike Parkinson's or Alzheimer's, prion diseases can spread from one organism to another, providing that the second organism

Figure 3–31 Detailed structure of the core of an amyloid fibril. Illustrated here is the cross-beta spine of the amyloid fibril that is formed by a peptide of seven amino acids from the protein Sup35, an extensively studied yeast prion. Consisting of the sequence glycine-asparagine-asparagine-glutamine-glutamine-asparagine-tyrosine (GNNQQNY), its structure was determined by X-ray crystallography. Although the cross-beta spines of other amyloids are similar, being composed of two long β sheets held together by a "steric zipper," different detailed structures are observed depending on the short peptide sequence involved. (A) One half of the spine is illustrated. Here, a standard parallel β-sheet structure (see p. 116) is held together by a set of hydrogen bonds between two side chains plus hydrogen bonds between two backbone atoms, as illustrated (oxygen atoms *red* and nitrogen atoms *blue*). Note that in this example, the adjacent peptides are exactly in register. Although only five layers are shown (each layer depicted as an *arrow*), the actual structure would extend for many tens of thousands of layers in the plane of the paper. (B) The complete cross-beta spine. A second, identical β-sheet is paired with the first one to form a two-sheet motif that runs the entire length of the fibril. (C) View of the complete spine in (B) from the top. The closely interdigitated side chains form a tight, water-free junction known as a *steric zipper*. (Based on R. Nelson et al., *Nature* 435:773–778, published 2005 by Nature Publishing Group. Reproduced with permission of SNCSC.)

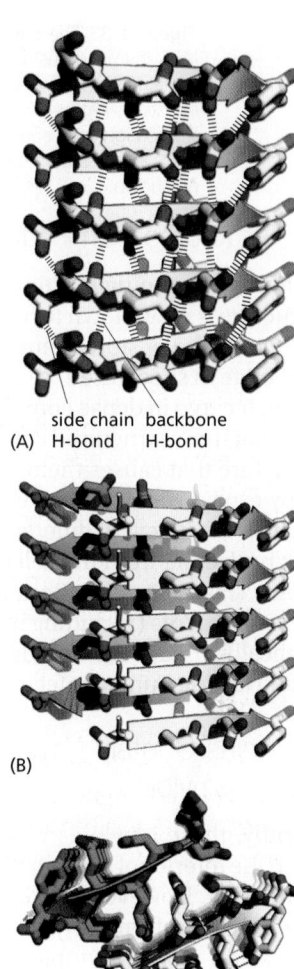

(A) side chain backbone
H-bond   H-bond

(B)

(C)

eats a tissue containing the protein aggregate. A set of closely related diseases—scrapie in sheep, Creutzfeldt–Jakob disease (CJD) in humans, Kuru in humans, and bovine spongiform encephalopathy (BSE) in cattle—are caused by a misfolded, aggregated form of a particular protein called PrP (for prion protein). PrP is normally located on the outer surface of the plasma membrane, most prominently in neurons, and it has the unfortunate property of forming amyloid fibrils that are "infectious" because they convert normally folded molecules of PrP to the same pathological form (**Figure 3–33**). This property creates a positive feedback loop that propagates the abnormal form of PrP, called PrP*, and allows the pathological conformation to spread rapidly from cell to cell in the brain, eventually causing death. It can be dangerous to eat the tissues of animals that contain PrP*, as witnessed by the spread of BSE (commonly referred to as "mad cow disease") from cattle to humans. Fortunately, in the absence of PrP*, PrP is extraordinarily difficult to convert to its abnormal form.

A closely related "protein-only inheritance" has been observed in yeast cells. The ability to study infectious proteins in yeast has clarified another remarkable feature of prions. These protein molecules can form several distinctively different types of amyloid fibrils from the same polypeptide chain. Moreover, each type of aggregate can be infectious, forcing normal protein molecules to adopt the same type of abnormal structure. Thus, several different "strains" of infectious particles can arise from the same polypeptide chain.

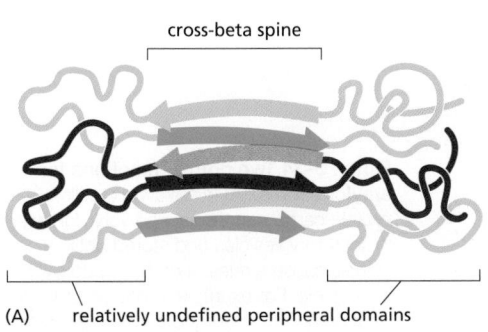

cross-beta spine

(A) relatively undefined peripheral domains

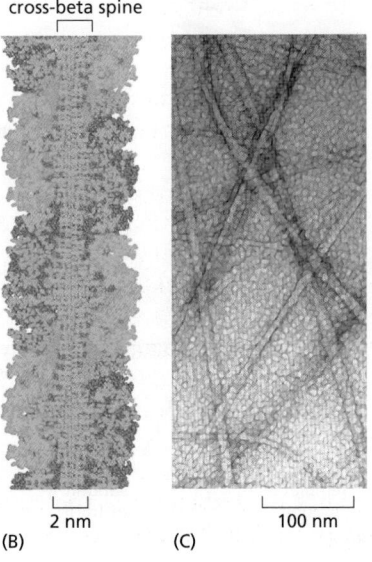

cross-beta spine

2 nm   100 nm

(B)   (C)

Figure 3–32 The structure of an amyloid fibril. (A) Schematic diagram of the structure of a amyloid fibril that is formed by the aggregation of a protein. Only the cross-beta spine of an amyloid fibril resembles the structure shown in Figure 3–31. (B) A cut-away view of a structure proposed for the amyloid fibril that can be formed in a test tube by the enzyme ribonuclease A, illustrating how the core of the fibril—formed by a short segment—relates to the rest of the structure. (C) Electron micrograph of amyloid fibrils. (A, adapted from L. Esposito, C. Pedone and L. Vitagliano, *Proc. Natl Acad. Sci. USA* 103:11533–11538, 2006; B, from S. Sambashivan et al., *Nature* 437:266–269, published 2005 by Nature Publishing Group. Reproduced with permission of SNCSC; C, courtesy of David Eisenberg.)

Figure 3–33 **The special protein aggregates that cause prion diseases.**
(A) Schematic illustration of the type of conformational change in the PrP
protein (prion protein) that produces material for an amyloid fibril. (B) The self-
infectious nature of the protein aggregation that is central to prion diseases.
PrP is highly unusual because the misfolded version of the protein, called
PrP*, induces the normal PrP protein it contacts to change its conformation,
as shown.

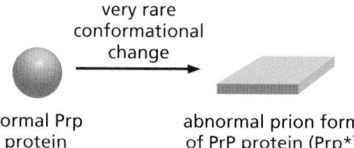

(A) prion protein can adopt an abnormal,
misfolded form

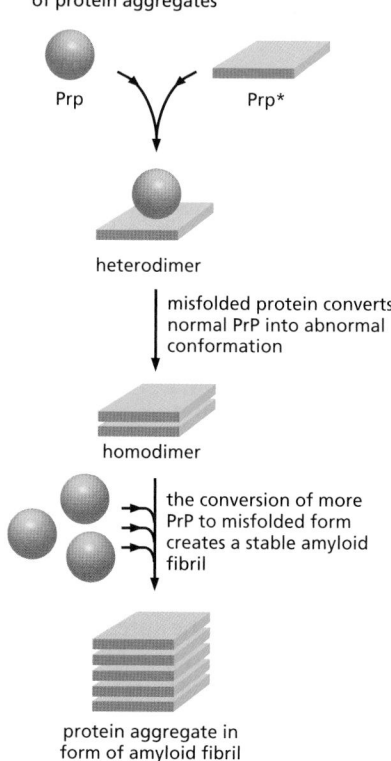

(B) misfolded protein can induce formation
of protein aggregates

Amyloid Structures Can Perform Useful Functions in Cells

Amyloid fibrils were initially studied because they cause disease. But the same
type of structure is now known to be exploited by cells for useful purposes. Eukaryotic cells, for example, store many different peptide and protein hormones that
they will secrete in specialized "secretory vesicles," which package a high concentration of their cargo in dense cores with a regular structure (see Figure 13–65). We
now know that these structured cores consist of amyloid fibrils, which in this case
have a structure that causes them to dissolve to release soluble cargo after being
secreted by exocytosis to the cell exterior (**Figure 3–34**A). Many bacteria use the
amyloid structure in a very different way, secreting proteins that form long amyloid fibrils projecting from the cell exterior that help to bind bacterial neighbors
into biofilms (Figure 3–34B). Because these biofilms help bacteria to survive in
adverse environments (including in humans treated with antibiotics), new drugs
that specifically disrupt the fibrous networks formed by bacterial amyloids have
promise for treating human infections.

Many Proteins Contain Low-complexity Domains that Can Form "Reversible Amyloids"

Until recently, those amyloids with useful functions were thought to be either
confined to the interior of specialized vesicles or expressed on the exterior of cells,
as in Figure 3–34. However, new experiments reveal that a large set of *low complexity domains* can form amyloid fibers that have functional roles in both the
cell nucleus and the cell cytoplasm. These domains are normally unstructured
and consist of stretches of amino acid sequence that can span hundreds of amino
acids, while containing only a small subset of the 20 different amino acids. In contrast to the disease-associated amyloid in Figure 3–33, these newly discovered
structures are held together by weaker noncovalent bonds and readily dissociate
in response to signals—hence their name *reversible amyloids*.

Many proteins with such domains also contain a different set of domains that
bind to specific other protein or RNA molecules. Thus, their controlled aggregation

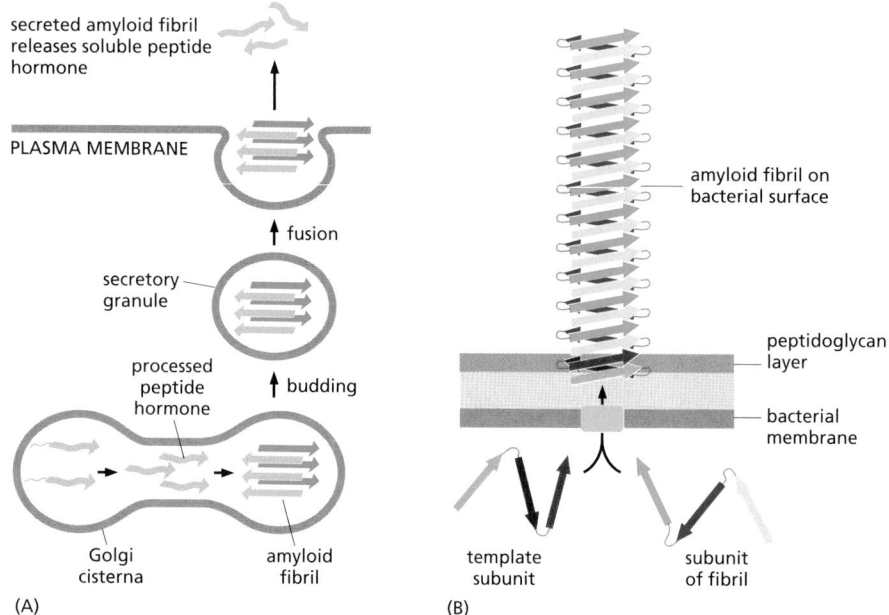

(A)

(B)

Figure 3–34 **Two normal functions for
amyloid fibrils.** (A) In eukaryotic cells,
protein cargo can be packed very densely
in secretory vesicles and stored until
signals cause a release of this cargo by
exocytosis. For example, proteins and
peptide hormones of the endocrine system,
such as glucagon and calcitonin, are
efficiently stored as short amyloid fibrils,
which dissociate when they reach the cell
exterior. (B) Bacteria produce amyloid fibrils
on their surface by secreting the precursor
proteins; these fibrils then create biofilms
that link together, and help to protect, large
numbers of individual bacteria.

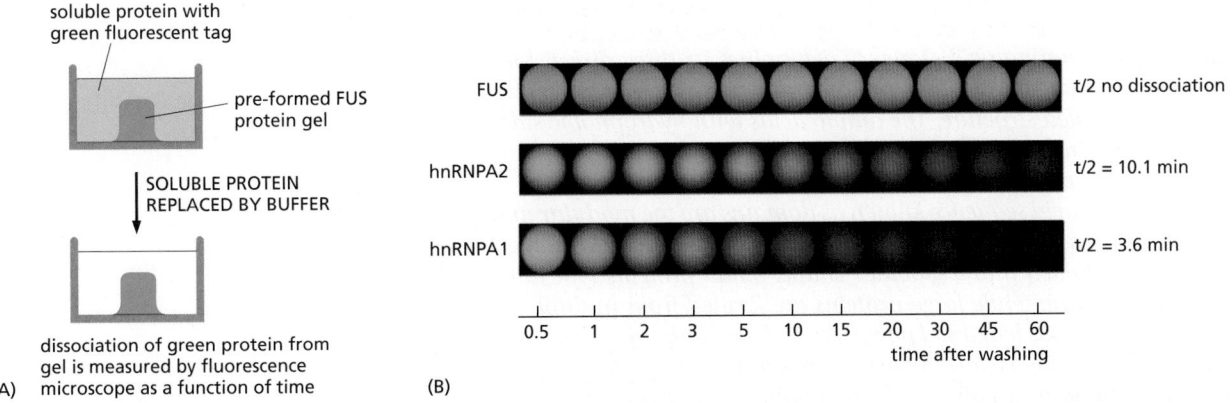

Figure 3–35 **Measuring the association between "reversible amyloids."** (A) Experimental setup. The fiber-forming domains from proteins that contain a low-complexity domain were produced in large quantities by cloning the DNA sequence that encodes them into an *E. coli* plasmid so as to allow overproduction of that domain (see p. 483). After these protein domains were purified by affinity chromatography, a tiny droplet of concentrated solution of one of the domains (here the FUS low-complexity domain) was deposited onto a microscope dish and allowed to gel. The gel was then soaked in a dilute solution of a fluorescently labeled low-complexity domain from the same or a different protein, making the gel fluorescent. After replacing the dilute protein solution with buffer, the relative strength of binding of the various domains to each other could then be measured by the decay of fluorescence, as indicated. (B) Results. The low-complexity domain from the FUS protein binds more tightly to itself than it does to the low-complexity domains from the proteins hnRNPA1 or hnRNPA2. A separate experiment reveals that these three different RNA binding proteins associate by forming mixed amyloid fibrils. (Adapted from M.Kato et al., *Cell* 149: 753-767, 2012).

in the cell can form a hydrogel that pulls these and other molecules into punctate structures called *intracellular bodies,* or *granules.* Specific mRNAs can be sequestered in such granules, where they are stored until made available by a controlled disassembly of the core amyloid structure that holds them together.

Consider the FUS protein, an essential nuclear protein with roles in the transcription, processing, and transport of specific mRNA molecules. Over 80 percent of its C-terminal domain of two hundred amino acids is composed of only four amino acids: glycine, serine, glutamine, and tyrosine. This low complexity domain is attached to several other domains that bind to RNA molecules. At high enough concentrations in a test tube, it forms a hydrogel that will associate with either itself or with the low complexity domains from other proteins. As illustrated by the experiment in **Figure 3–35**, although different low complexity domains bind to each other, homotypic interactions appear to be of greatest affinity (thus, the FUS low complexity domain binds most tightly to itself). Further experiments reveal that both the homotypic and the heterotypic bindings are mediated through a β-sheet core structure forming amyloid fibrils, and that these structures bind to other types of repeat sequences in the manner indicated in **Figure 3–36**. Many of these interactions appear to be controlled by the phosphorylation of serine side chains in the one or both of the interacting partners. However, a great deal remains to be learned concerning these newly discovered structures and the varied roles that they play in the cell biology of eukaryotic cells.

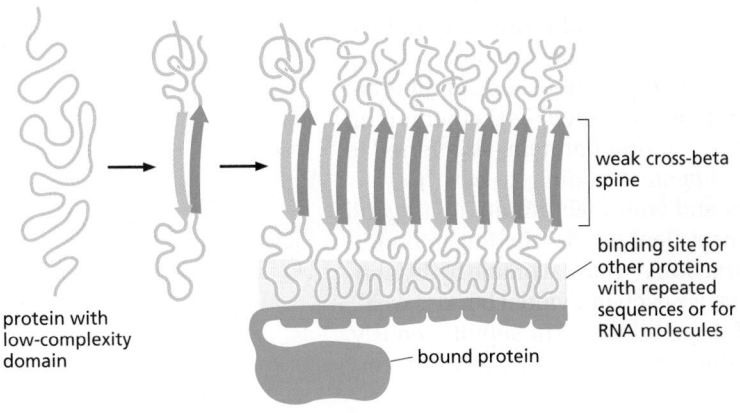

Figure 3–36 **One type of complex that is formed by reversible amyloids.** The structure shown is based on the observed interaction of RNA polymerase with a low-complexity domain of a protein that regulates DNA transcription. (Adapted from I. Kwon et al., *Cell* 155:1049–1060, 2013.)

Summary

A protein molecule's amino acid sequence determines its three-dimensional conformation. Noncovalent interactions between different parts of the polypeptide chain stabilize its folded structure. The amino acids with hydrophobic side chains tend to cluster in the interior of the molecule, and local hydrogen-bond interactions between neighboring peptide bonds give rise to α helices and β sheets.

Regions of amino acid sequence known as domains are the modular units from which many proteins are constructed. Such domains generally contain 40–350 amino acids, often folded into a globular shape. Small proteins typically consist of only a single domain, while large proteins are formed from multiple domains linked together by various lengths of polypeptide chain, some of which can be relatively disordered. As proteins have evolved, domains have been modified and combined with other domains to construct large numbers of new proteins.

Proteins are brought together into larger structures by the same noncovalent forces that determine protein folding. Proteins with binding sites for their own surface can assemble into dimers, closed rings, spherical shells, or helical polymers. The amyloid fibril is a long unbranched structure assembled through a repeating aggregate of β sheets. Although some mixtures of proteins and nucleic acids can assemble spontaneously into complex structures in a test tube, not all structures in the cell are capable of spontaneous reassembly after they have been dissociated into their component parts, because many biological assembly processes involve assembly factors that are not present in the final structure.

PROTEIN FUNCTION

We have seen that each type of protein consists of a precise sequence of amino acids that allows it to fold up into a particular three-dimensional shape, or conformation. But proteins are not rigid lumps of material. They often have precisely engineered moving parts whose mechanical actions are coupled to chemical events. It is this coupling of chemistry and movement that gives proteins the extraordinary capabilities that underlie the dynamic processes in living cells.

In this section, we explain how proteins bind to other selected molecules and how a protein's activity depends on such binding. We show that the ability to bind to other molecules enables proteins to act as catalysts, signal receptors, switches, motors, or tiny pumps. The examples we discuss in this chapter by no means exhaust the vast functional repertoire of proteins. You will encounter the specialized functions of many other proteins elsewhere in this book, based on similar principles.

All Proteins Bind to Other Molecules

A protein molecule's physical interaction with other molecules determines its biological properties. Thus, antibodies attach to viruses or bacteria to mark them for destruction, the enzyme hexokinase binds glucose and ATP so as to catalyze a reaction between them, actin molecules bind to each other to assemble into actin filaments, and so on. Indeed, all proteins stick, or *bind*, to other molecules. In some cases, this binding is very tight; in others it is weak and short-lived. But the binding always shows great *specificity*, in the sense that each protein molecule can usually bind just one or a few molecules out of the many thousands of different types it encounters. The substance that is bound by the protein—whether it is an ion, a small molecule, or a macromolecule such as another protein—is referred to as a **ligand** for that protein (from the Latin word *ligare*, meaning "to bind").

The ability of a protein to bind selectively and with high affinity to a ligand depends on the formation of a set of weak noncovalent bonds—hydrogen bonds, electrostatic attractions, and van der Waals attractions—plus favorable hydrophobic interactions (see Panel 2–3, pp. 94–95). Because each individual bond is weak, effective binding occurs only when many of these bonds form simultaneously. Such binding is possible only if the surface contours of the ligand molecule fit very closely to the protein, matching it like a hand in a glove (**Figure 3–37**).

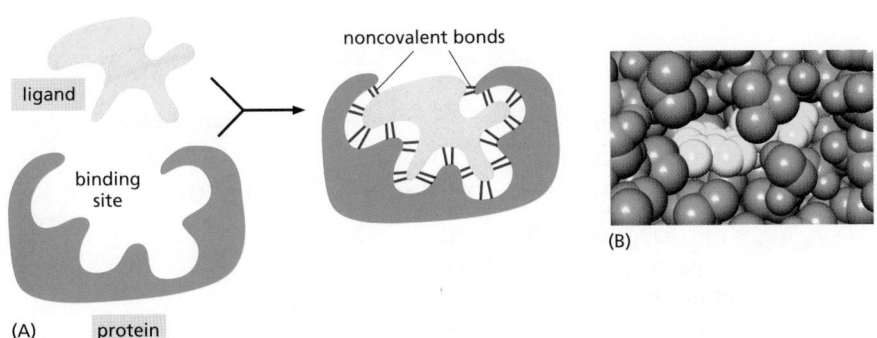

Figure 3–37 The selective binding of a protein to another molecule. Many weak bonds are needed to enable a protein to bind tightly to a second molecule, or *ligand*. A ligand must therefore fit precisely into a protein's binding site, like a hand into a glove, so that a large number of noncovalent bonds form between the protein and the ligand. (A) Schematic; (B) space-filling model. (PDB code: 1G6N.)

The region of a protein that associates with a ligand, known as the ligand's *binding site*, usually consists of a cavity in the protein surface formed by a particular arrangement of amino acids. These amino acids can belong to different portions of the polypeptide chain that are brought together when the protein folds (**Figure 3–38**). Separate regions of the protein surface generally provide binding sites for different ligands, allowing the protein's activity to be regulated, as we shall see later. And other parts of the protein act as a handle to position the protein in the cell—an example is the SH2 domain discussed previously, which often moves a protein containing it to particular intracellular sites in response to signals.

Although the atoms buried in the interior of the protein have no direct contact with the ligand, they form the framework that gives the surface its contours and its chemical and mechanical properties. Even small changes to the amino acids in the interior of a protein molecule can change its three-dimensional shape enough to destroy a binding site on the surface.

The Surface Conformation of a Protein Determines Its Chemistry

The impressive chemical capabilities of proteins often require that the chemical groups on their surface interact in ways that enhance the chemical reactivity of one or more amino acid side chains. These interactions fall into two main categories.

First, the interaction of neighboring parts of the polypeptide chain may restrict the access of water molecules to that protein's ligand-binding sites. Because water molecules readily form hydrogen bonds that can compete with ligands for sites

Figure 3–38 The binding site of a protein. (A) The folding of the polypeptide chain typically creates a crevice or cavity on the protein surface. This crevice contains a set of amino acid side chains disposed in such a way that they can form noncovalent bonds only with certain ligands. (B) A close-up of an actual binding site, showing the hydrogen bonds and electrostatic interactions formed between a protein and its ligand. In this example, a molecule of cyclic AMP is the bound ligand.

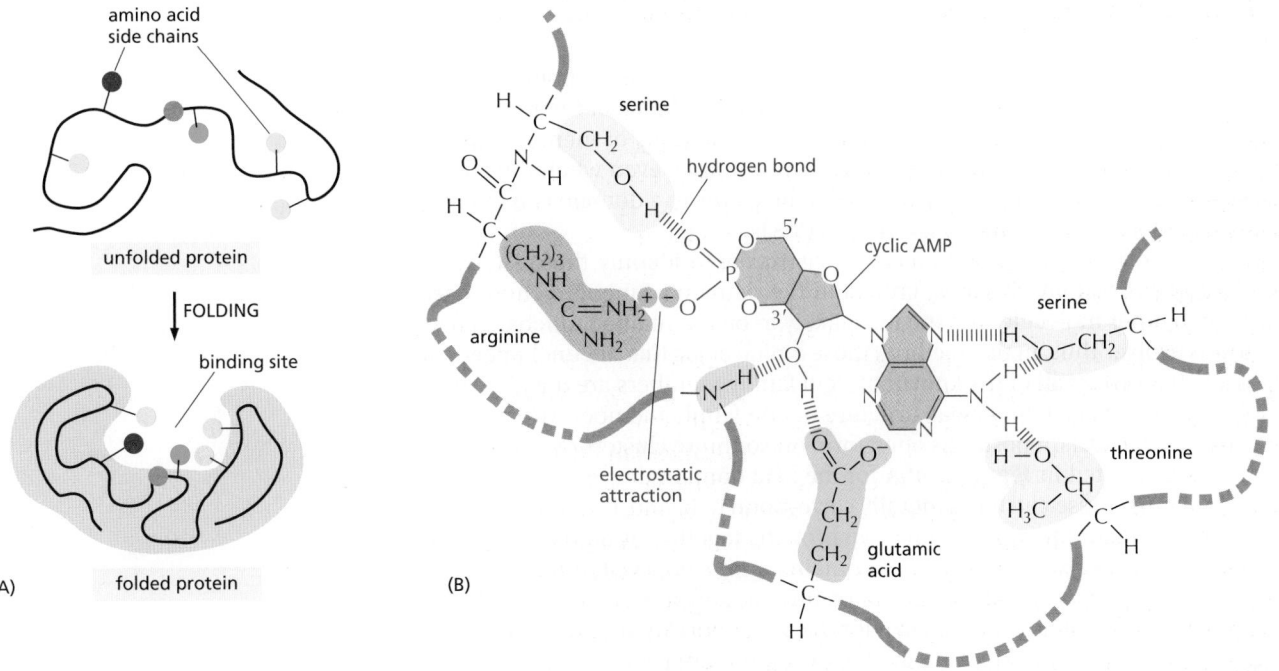

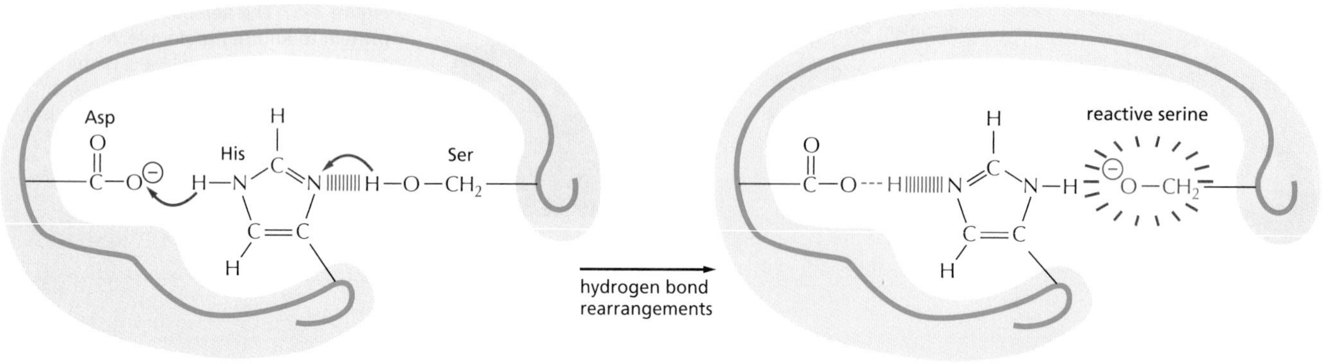

Figure 3–39 **An unusually reactive amino acid at the active site of an enzyme.** This example is the "catalytic triad" Asp-His-Ser found in chymotrypsin, elastase, and other serine proteases (see Figure 3–12). The aspartic acid side chain (Asp) induces the histidine (His) to remove the proton from a particular serine (Ser). This activates the serine and enables it to form a covalent bond with an enzyme substrate, hydrolyzing a peptide bond. The many convolutions of the polypeptide chain are omitted here.

on the protein surface, a ligand will form tighter hydrogen bonds (and electrostatic interactions) with a protein if water molecules are kept away. It might be hard to imagine a mechanism that would exclude a molecule as small as water from a protein surface without affecting the access of the ligand itself. However, because of the strong tendency of water molecules to form water–water hydrogen bonds, water molecules exist in a large hydrogen-bonded network (see Panel 2–2, pp. 92–93). In effect, a protein can keep a ligand-binding site dry, increasing that site's reactivity, because it is energetically unfavorable for individual water molecules to break away from this network—as they must do to reach into a crevice on a protein's surface.

Second, the clustering of neighboring polar amino acid side chains can alter their reactivity. If protein folding forces together a number of negatively charged side chains against their mutual repulsion, for example, the affinity of the site for a positively charged ion is greatly increased. In addition, when amino acid side chains interact with one another through hydrogen bonds, normally unreactive groups (such as the –CH$_2$OH on the serine shown in **Figure 3–39**) can become reactive, enabling them to be used to make or break selected covalent bonds.

The surface of each protein molecule therefore has a unique chemical reactivity that depends not only on which amino acid side chains are exposed, but also on their exact orientation relative to one another. For this reason, two slightly different conformations of the same protein molecule can differ greatly in their chemistry.

Sequence Comparisons Between Protein Family Members Highlight Crucial Ligand-Binding Sites

As we have described previously, genome sequences allow us to group many of the domains in proteins into families that show clear evidence of their evolution from a common ancestor. The three-dimensional structures of members of the same domain family are remarkably similar. For example, even when the amino acid sequence identity falls to 25%, the backbone atoms in a domain can follow a common protein fold within 0.2 nanometers (2 Å).

We can use a method called *evolutionary tracing* to identify those sites in a protein domain that are the most crucial to the domain's function. Those sites that bind to other molecules are the most likely to be maintained, unchanged as organisms evolve. Thus, in this method, those amino acids that are unchanged, or nearly unchanged, in all of the known protein family members are mapped onto a model of the three-dimensional structure of one family member. When this is done, the most invariant positions often form one or more clusters on the protein surface, as illustrated in **Figure 3–40**A for the SH2 domain described previously (see Figure 3–6). These clusters generally correspond to ligand-binding sites.

The SH2 domain functions to link two proteins together. It binds the protein containing it to a second protein that contains a phosphorylated tyrosine side chain in a specific amino acid sequence context, as shown in Figure 3–40B. The amino acids located at the binding site for the phosphorylated polypeptide have been the slowest to change during the long evolutionary process that produced

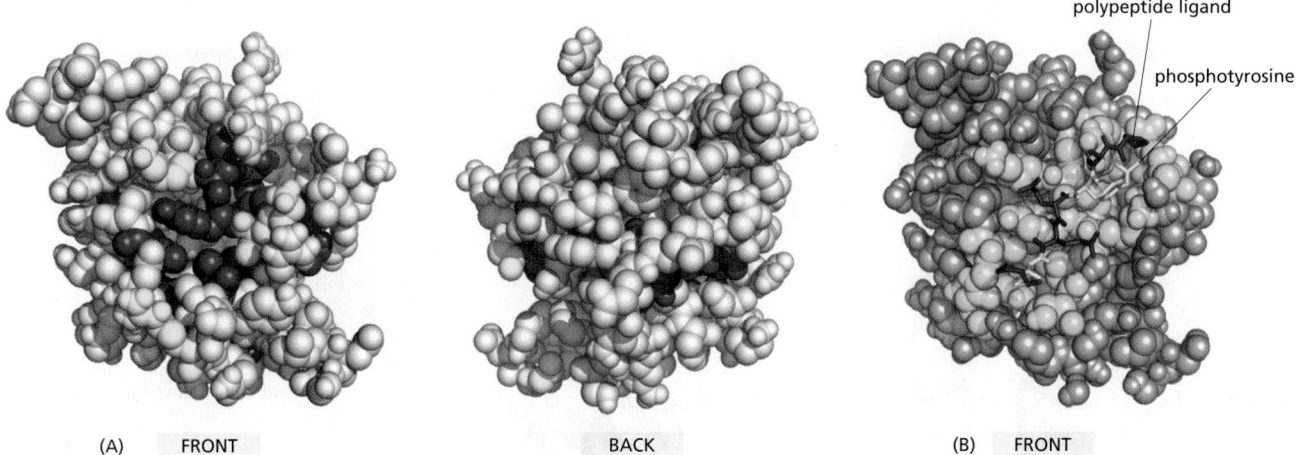

(A) FRONT

BACK

polypeptide ligand

phosphotyrosine

(B) FRONT

the large SH2 family of peptide recognition domains. Mutation is a random process; survival is not. Thus, natural selection (random mutation followed by nonrandom survival) produces the sequence conservation by preferentially eliminating organisms whose SH2 domains become altered in a way that inactivates the SH2 binding site, destroying SH2 function.

Genome sequencing has revealed huge numbers of proteins whose functions are unknown. Once a three-dimensional structure has been determined for one member of a protein family, evolutionary tracing allows biologists to determine binding sites for the members of that family, providing a useful start in deciphering protein function.

Proteins Bind to Other Proteins Through Several Types of Interfaces

Proteins can bind to other proteins in multiple ways. In many cases, a portion of the surface of one protein contacts an extended loop of polypeptide chain (a "string") on a second protein (**Figure 3–41**A). Such a surface–string interaction, for example, allows the SH2 domain to recognize a phosphorylated polypeptide loop on a second protein, as just described, and it also enables a protein kinase to recognize the proteins that it will phosphorylate (see below).

A second type of protein–protein interface forms when two α helices, one from each protein, pair together to form a coiled-coil (Figure 3–41B). This type of protein interface is found in several families of gene regulatory proteins, as discussed in Chapter 7.

The most common way for proteins to interact, however, is by the precise matching of one rigid surface with that of another (Figure 3–41C). Such interactions can be very tight, since a large number of weak bonds can form between two surfaces that match well. For the same reason, such surface–surface interactions can be extremely specific, enabling a protein to select just one partner from the many thousands of different proteins found in a cell.

Figure 3–40 The evolutionary trace method applied to the SH2 domain. (A) Front and back views of a space-filling model of the SH2 domain, with evolutionarily conserved amino acids on the protein surface colored *yellow*, and those more toward the protein interior colored *red*. (B) The structure of one specific SH2 domain with its bound polypeptide. Here, those amino acids located within 0.4 nm of the bound ligand are colored *blue*. The two key amino acids of the ligand are *yellow*, and the others are *purple*. Note the high degree of correspondence between (A) and (B). (Adapted from O. Lichtarge, H.R. Bourne and F.E. Cohen, *J. Mol. Biol.* 257:342–358, 1996. PDB codes: 1SPR, 1SPS.)

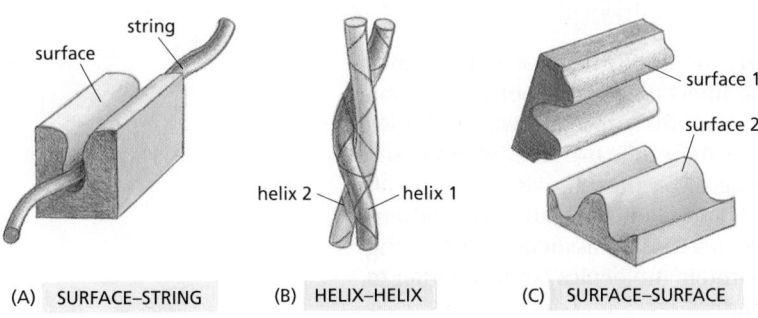

(A) SURFACE–STRING

surface

string

(B) HELIX–HELIX

helix 2   helix 1

(C) SURFACE–SURFACE

surface 1

surface 2

Figure 3–41 Three ways in which two proteins can bind to each other. Only the interacting parts of the two proteins are shown. (A) A rigid surface on one protein can bind to an extended loop of polypeptide chain (a "string") on a second protein. (B) Two α helices can bind together to form a coiled-coil. (C) Two complementary rigid surfaces often link two proteins together. Binding interactions can also involve the pairing of β strands (see, for example, Figure 3–18).

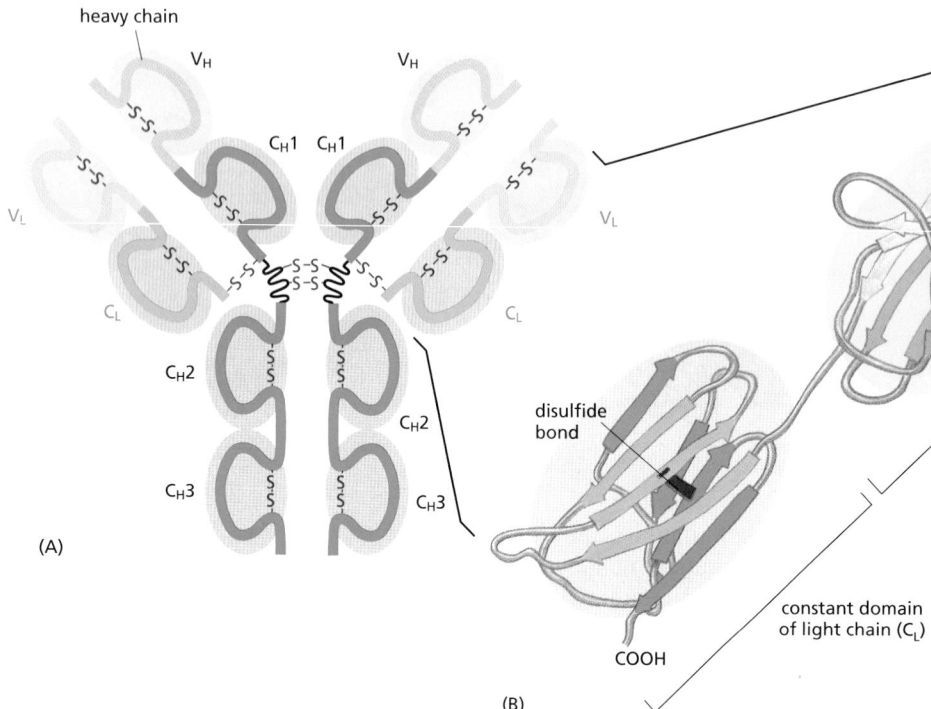

Figure 3–42 An antibody molecule. A typical antibody molecule is Y-shaped and has two identical binding sites for its antigen, one on each arm of the Y. As explained in Chapter 24, the protein is composed of four polypeptide chains (two identical heavy chains and two identical and smaller light chains) held together by disulfide bonds. Each chain is made up of several different immunoglobulin domains, here shaded either *blue* or *gray*. The antigen-binding site is formed where a heavy-chain variable domain (V_H) and a light-chain variable domain (V_L) come close together. These are the domains that differ most in their sequence and structure in different antibodies. At the end of each of the two arms of the antibody molecule, these two domains form loops that bind to the antigen (see Movie 24.5).

Antibody Binding Sites Are Especially Versatile

All proteins must bind to particular ligands to carry out their various functions. The antibody family is notable for its capacity for tight, highly selective binding (discussed in detail in Chapter 24).

Antibodies, or immunoglobulins, are proteins produced by the immune system in response to foreign molecules, such as those on the surface of an invading microorganism. Each antibody binds tightly to a particular target molecule, thereby either inactivating the target molecule directly or marking it for destruction. An antibody recognizes its target (called an **antigen**) with remarkable specificity. Because there are potentially billions of different antigens that humans might encounter, we have to be able to produce billions of different antibodies.

Antibodies are Y-shaped molecules with two identical binding sites that are complementary to a small portion of the surface of the antigen molecule. A detailed examination of the antigen-binding sites of antibodies reveals that they are formed from several loops of polypeptide chain that protrude from the ends of a pair of closely juxtaposed protein domains (**Figure 3–42**). Different antibodies generate an enormous diversity of antigen-binding sites by changing only the length and amino acid sequence of these loops, without altering the basic protein structure.

Loops of this kind are ideal for grasping other molecules. They allow a large number of chemical groups to surround a ligand so that the protein can link to it with many weak bonds. For this reason, loops often form the ligand-binding sites in proteins.

The Equilibrium Constant Measures Binding Strength

Molecules in the cell encounter each other very frequently because of their continual random thermal movements. Colliding molecules with poorly matching surfaces form few noncovalent bonds with one another, and the two molecules dissociate as rapidly as they come together. At the other extreme, when many noncovalent bonds form between two colliding molecules, the association can persist for a very long time (**Figure 3–43**). Strong interactions occur in cells whenever a biological function requires that molecules remain associated for a long time—for example, when a group of RNA and protein molecules come together to make a subcellular structure such as a ribosome.

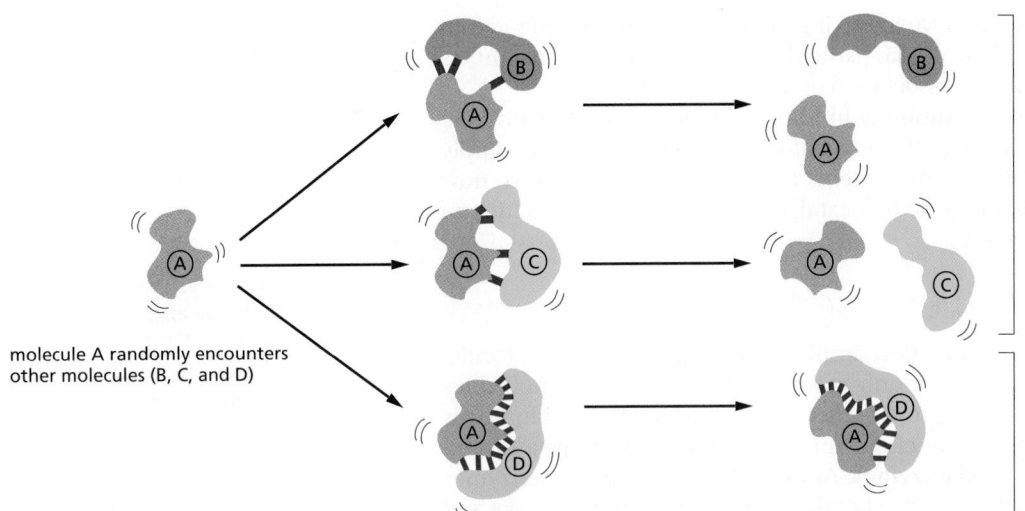

the surfaces of molecules A and B, and A and C, are a poor match and are capable of forming only a few weak bonds; thermal motion rapidly breaks them apart

molecule A randomly encounters other molecules (B, C, and D)

the surfaces of molecules A and D match well and therefore can form enough weak bonds to withstand thermal jolting; they therefore stay bound to each other

We can measure the strength with which any two molecules bind to each other. As an example, consider a population of identical antibody molecules that suddenly encounters a population of ligands diffusing in the fluid surrounding them. At frequent intervals, one of the ligand molecules will bump into the binding site of an antibody and form an antibody–ligand complex. The population of antibody–ligand complexes will therefore increase, but not without limit: over time, a second process, in which individual complexes break apart because of thermally induced motion, will become increasingly important. Eventually, any population of antibody molecules and ligands will reach a steady state, or equilibrium, in which the number of binding (association) events per second is precisely equal to the number of "unbinding" (dissociation) events (see page 62).

From the concentrations of the ligand, antibody, and antibody–ligand complex at equilibrium, we can calculate a convenient measure of the strength of binding—the **equilibrium constant** (*K*)—(**Figure 3–44**A). This constant was described in detail in Chapter 2, where its connection to free energy differences was derived (see p. 62). The equilibrium constant for a reaction in which two molecules (A and B) bind to each other to form a complex (AB) has units of liters/mole, and half of the binding sites will be occupied by ligand when that ligand's concentration (in moles/liter) reaches a value that is equal to $1/K$. This equilibrium constant is larger the greater the binding strength, and it is a direct measure of the free-energy difference between the bound and free states (Figure 3–44B). Even a change

Figure 3–43 How noncovalent bonds mediate interactions between macromolecules (see Movie 2.1).

Figure 3–44 Relating standard free-energy difference (ΔG°) to the equilibrium constant (K). (A) The equilibrium between molecules A and B and the complex AB is maintained by a balance between the two opposing reactions shown in panels 1 and 2. Molecules A and B must collide if they are to react, and the association rate is therefore proportional to the product of their individual concentrations [A] × [B]. (Square brackets indicate concentration.) As shown in panel 3, the ratio of the rate constants for the association and the dissociation reactions is equal to the equilibrium constant (K) for the reaction (see also p. 63). (B) The equilibrium constant in panel 3 is that for the reaction A + B ↔ AB, and the larger its value, the stronger the binding between A and B. Note that for every 5.91 kJ/mole decrease in standard free energy, the equilibrium constant increases by a factor of 10 at 37°C.

The equilibrium constant here has units of liters/mole; for simple binding interactions it is also called the *affinity constant* or *association constant*, denoted K_a. The reciprocal of K_a is called the *dissociation constant*, K_d (in units of moles/liter).

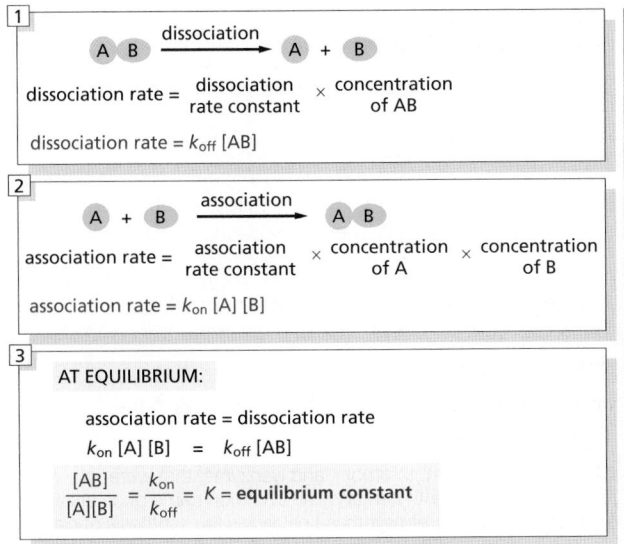

1

A B →(dissociation) A + B

dissociation rate = dissociation rate constant × concentration of AB

dissociation rate = k_{off} [AB]

2

A + B →(association) A B

association rate = association rate constant × concentration of A × concentration of B

association rate = k_{on} [A] [B]

3

AT EQUILIBRIUM:

association rate = dissociation rate

k_{on} [A] [B] = k_{off} [AB]

$\dfrac{[AB]}{[A][B]} = \dfrac{k_{on}}{k_{off}} = K$ = equilibrium constant

(A)

The relationship between standard free-energy differences (ΔG°) and equilibrium constants (37°C)	
equilibrium constant	standard free-energy difference
$\dfrac{[AB]}{[A][B]} = K$ (liters/mole)	of AB minus free energy of A + B (kJ/mole)
1	0
10	−5.9
10^2	−11.9
10^3	−17.8
10^4	−23.7
10^5	−29.7
10^6	−35.6
10^7	−41.5
10^8	−47.4
10^9	−53.4
10^{10}	−59.4

(B)

of a few noncovalent bonds can have a striking effect on a binding interaction, as shown by the example in **Figure 3–45**. (Note that the equilibrium constant, as defined here, is also known as the association or affinity constant, K_a.)

We have used the case of an antibody binding to its ligand to illustrate the effect of binding strength on the equilibrium state, but the same principles apply to any molecule and its ligand. Many proteins are enzymes, which, as we now discuss, first bind to their ligands and then catalyze the breakage or formation of covalent bonds in these molecules.

Enzymes Are Powerful and Highly Specific Catalysts

Many proteins can perform their function simply by binding to another molecule. An actin molecule, for example, need only associate with other actin molecules to form a filament. There are other proteins, however, for which ligand binding is only a necessary first step in their function. This is the case for the large and very important class of proteins called **enzymes**. As described in Chapter 2, enzymes are remarkable molecules that cause the chemical transformations that make and break covalent bonds in cells. They bind to one or more ligands, called **substrates**, and convert them into one or more chemically modified *products*, doing this over and over again with amazing rapidity. Enzymes speed up reactions, often by a factor of a million or more, without themselves being changed—that is, they act as **catalysts** that permit cells to make or break covalent bonds in a controlled way. It is the catalysis of organized sets of chemical reactions by enzymes that creates and maintains the cell, making life possible.

We can group enzymes into functional classes that perform similar chemical reactions (**Table 3–1**). Each type of enzyme within such a class is highly specific,

Consider 1000 molecules of A and 1000 molecules of B in a eukaryotic cell. The concentration of both will be about 10^{-9} M.

If the equilibrium constant (K) for A + B $\rightleftharpoons$ AB is 10^{10}, then one can calculate that at equilibrium there will be

270	270	730
A molecules	B molecules	AB molecules

If the equilibrium constant is a little weaker at 10^8, which represents a loss of 11.9 kilojoule/mole of binding energy from the example above, or 2–3 fewer hydrogen bonds, then there will be

915	915	85
A molecules	B molecules	AB molecules

Figure 3–45 **Small changes in the number of weak bonds can have drastic effects on a binding interaction.** This example illustrates the dramatic effect of the presence or absence of a few weak noncovalent bonds in a biological context.

TABLE 3–1 **Some Common Types of Enzymes**	
Enzyme	**Reaction catalyzed**
Hydrolases	General term for enzymes that catalyze a hydrolytic cleavage reaction; *nucleases* and *proteases* are more specific names for subclasses of these enzymes
Nucleases	Break down nucleic acids by hydrolyzing bonds between nucleotides. *Endo–* and *exonucleases* cleave nucleic acids *within* and *from the ends of* the polynucleotide chains, respectively
Proteases	Break down proteins by hydrolyzing bonds between amino acids
Synthases	Synthesize molecules in anabolic reactions by condensing two smaller molecules together
Ligases	Join together (ligate) two molecules in an energy-dependent process. DNA ligase, for example, joins two DNA molecules together end-to-end through phosphodiester bonds
Isomerases	Catalyze the rearrangement of bonds within a single molecule
Polymerases	Catalyze polymerization reactions such as the synthesis of DNA and RNA
Kinases	Catalyze the addition of phosphate groups to molecules. Protein kinases are an important group of kinases that attach phosphate groups to proteins
Phosphatases	Catalyze the hydrolytic removal of a phosphate group from a molecule
Oxido-Reductases	General name for enzymes that catalyze reactions in which one molecule is oxidized while the other is reduced. Enzymes of this type are often more specifically named *oxidases*, *reductases*, or *dehydrogenases*
ATPases	Hydrolyze ATP. Many proteins with a wide range of roles have an energy-harnessing ATPase activity as part of their function; for example, motor proteins such as *myosin* and membrane transport proteins such as the *sodium–potassium pump*
GTPases	Hydrolyze GTP. A large family of GTP-binding proteins are GTPases with central roles in the regulation of cell processes
Enzyme names typically end in "-ase," with the exception of some enzymes, such as pepsin, trypsin, thrombin, and lysozyme, that were discovered and named before the convention became generally accepted at the end of the nineteenth century. The common name of an enzyme usually indicates the substrate or product and the nature of the reaction catalyzed. For example, citrate synthase catalyzes the synthesis of citrate by a reaction between acetyl CoA and oxaloacetate.	

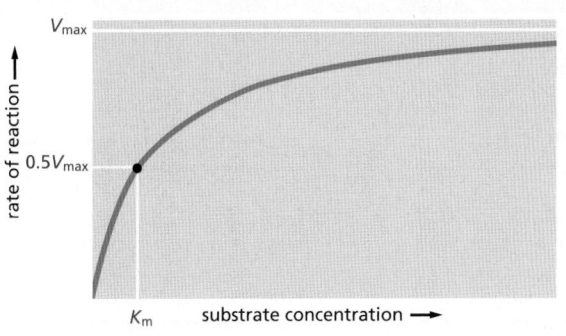

Figure 3–46 **Enzyme kinetics.** The rate of an enzyme reaction (V) increases as the substrate concentration increases until a maximum value (V_{max}) is reached. At this point all substrate-binding sites on the enzyme molecules are fully occupied, and the rate of reaction is limited by the rate of the catalytic process on the enzyme surface. For most enzymes, the concentration of substrate at which the reaction rate is half-maximal (K_m) is a measure of how tightly the substrate is bound, with a large value of K_m corresponding to weak binding.

catalyzing only a single type of reaction. Thus, *hexokinase* adds a phosphate group to D-glucose but ignores its optical isomer L-glucose; the blood-clotting enzyme *thrombin* cuts one type of blood protein between a particular arginine and its adjacent glycine and nowhere else, and so on. As discussed in detail in Chapter 2, enzymes work in teams, with the product of one enzyme becoming the substrate for the next. The result is an elaborate network of metabolic pathways that provides the cell with energy and generates the many large and small molecules that the cell needs (see Figure 2–63).

Substrate Binding Is the First Step in Enzyme Catalysis

For a protein that catalyzes a chemical reaction (an enzyme), the binding of each substrate molecule to the protein is an essential prelude. In the simplest case, if we denote the enzyme by E, the substrate by S, and the product by P, the basic reaction path is E + S → ES → EP → E + P. There is a limit to the amount of substrate that a single enzyme molecule can process in a given time. Although an increase in the concentration of substrate increases the rate at which product is formed, this rate eventually reaches a maximum value (**Figure 3–46**). At that point the enzyme molecule is saturated with substrate, and the rate of reaction (V_{max}) depends only on how rapidly the enzyme can process the substrate molecule. This maximum rate divided by the enzyme concentration is called the *turnover number*. Turnover numbers are often about 1000 substrate molecules processed per second per enzyme molecule, although turnover numbers between 1 and 10,000 are known.

The other kinetic parameter frequently used to characterize an enzyme is its K_m, the concentration of substrate that allows the reaction to proceed at one-half its maximum rate (0.5 V_{max}) (see Figure 3–46). A *low* K_m value means that the enzyme reaches its maximum catalytic rate at a *low concentration* of substrate and generally indicates that the enzyme binds to its substrate very tightly, whereas a *high* K_m value corresponds to weak binding. The methods used to characterize enzymes in this way are explained in **Panel 3–2** (pp. 142–143).

Enzymes Speed Reactions by Selectively Stabilizing Transition States

Enzymes achieve extremely high rates of chemical reaction—rates that are far higher than for any synthetic catalysts. There are several reasons for this efficiency. First, when two molecules need to react, the enzyme greatly increases the local concentration of both of these substrate molecules at the catalytic site, holding them in the correct orientation for the reaction that is to follow. More importantly, however, some of the binding energy contributes directly to the catalysis. Substrate molecules must pass through a series of intermediate states of altered geometry and electron distribution before they form the ultimate products of the reaction. The free energy required to attain the most unstable intermediate state, called the **transition state**, is known as the *activation energy* for the reaction, and it is the major determinant of the reaction rate. Enzymes have a much higher affinity for the transition state of the substrate than they have for the stable form.

WHY ANALYZE THE KINETICS OF ENZYMES?

Enzymes are the most selective and powerful catalysts known. An understanding of their detailed mechanisms provides a critical tool for the discovery of new drugs, for the large-scale industrial synthesis of useful chemicals, and for appreciating the chemistry of cells and organisms. A detailed study of the rates of the chemical reactions that are catalyzed by a purified enzyme—more specifically how these rates change with changes in conditions such as the concentrations of substrates, products, inhibitors, and regulatory ligands—allows

biochemists to figure out exactly how each enzyme works. For example, this is the way that the ATP-producing reactions of glycolysis, shown previously in Figure 2–48, were deciphered—allowing us to appreciate the rationale for this critical enzymatic pathway.

In this Panel, we introduce the important field of enzyme kinetics, which has been indispensable for deriving much of the detailed knowledge that we now have about cell chemistry.

STEADY-STATE ENZYME KINETICS

Many enzymes have only one substrate, which they bind and then process to produce products according to the scheme outlined in Figure 3–50A. In this case, the reaction is written as

$$E + S \underset{k_{-1}}{\overset{k_1}{\rightleftharpoons}} ES \xrightarrow{k_{cat}} E + P$$

Here we have assumed that the reverse reaction, in which E + P recombine to form EP and then ES, occurs so rarely that we can ignore it. In this case, EP need not be represented, and we can express the rate of the reaction—known as its velocity, V, as

$$V = k_{cat} [ES]$$

where [ES] is the concentration of the enzyme–substrate complex, and k_{cat} is the turnover number, a rate constant that has a value equal to the number of substrate molecules processed per enzyme molecule each second.

But how does the value of [ES] relate to the concentrations that we know directly, which are the total concentration of the enzyme, [E$_o$], and the concentration of the substrate, [S]? When enzyme and substrate are first mixed, the concentration [ES] will rise rapidly from zero to a so-called steady-state level, as illustrated below.

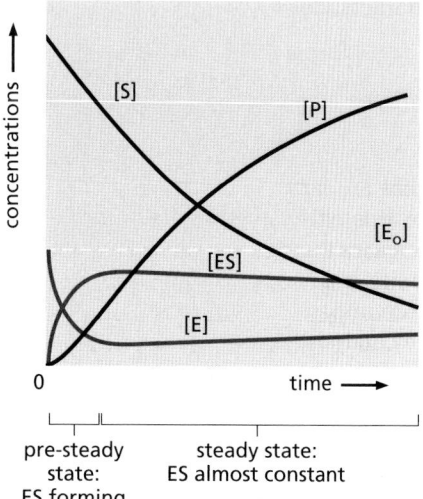

pre-steady state: ES forming

steady state: ES almost constant

At this steady state, [ES] is nearly constant, so that

rate of ES breakdown		rate of ES formation
$k_{-1} [ES] + k_{cat} [ES]$	$=$	$k_1 [E][S]$

or, since the concentration of the free enzyme, [E], is equal to [E$_o$] – [ES],

$$[ES] = \left(\frac{k_1}{k_{-1} + k_{cat}}\right)[E][S] = \left(\frac{k_1}{k_{-1} + k_{cat}}\right)\left([E_o] - [ES]\right)[S]$$

Rearranging, and defining the constant K_m as

$$\frac{k_{-1} + k_{cat}}{k_1}$$

we get

$$[ES] = \frac{[E_o][S]}{K_m + [S]}$$

or, remembering that $V = k_{cat}$ [ES], we obtain the famous Michaelis–Menten equation

$$V = \frac{k_{cat} [E_o][S]}{K_m + [S]}$$

As [S] is increased to higher and higher levels, essentially all of the enzyme will be bound to substrate at steady state; at this point, a maximum rate of reaction, V_{max}, will be reached where $V = V_{max} = k_{cat} [E_o]$. Thus, it is convenient to rewrite the Michaelis–Menten equation as

$$V = \frac{V_{max} [S]}{K_m + [S]}$$

THE DOUBLE-RECIPROCAL PLOT

A typical plot of V versus $[S]$ for an enzyme that follows Michaelis–Menten kinetics is shown below. From this plot, neither the value of V_{max} nor of K_m is immediately clear.

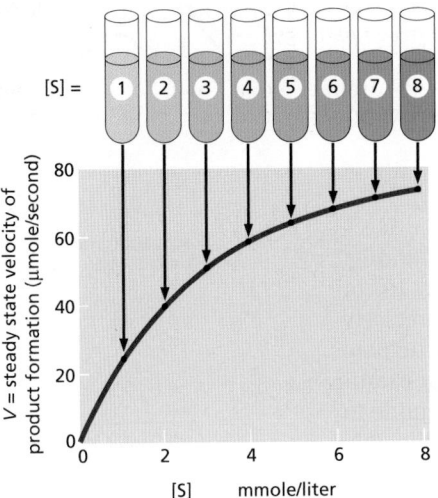

To obtain V_{max} and K_m from such data, a double-reciprocal plot is often used, in which the Michaelis–Menten equation has merely been rearranged, so that $1/V$ can be plotted versus $1/[S]$.

$$1/V = \left(\frac{K_m}{V_{max}}\right)\left(\frac{1}{[S]}\right) + 1/V_{max}$$

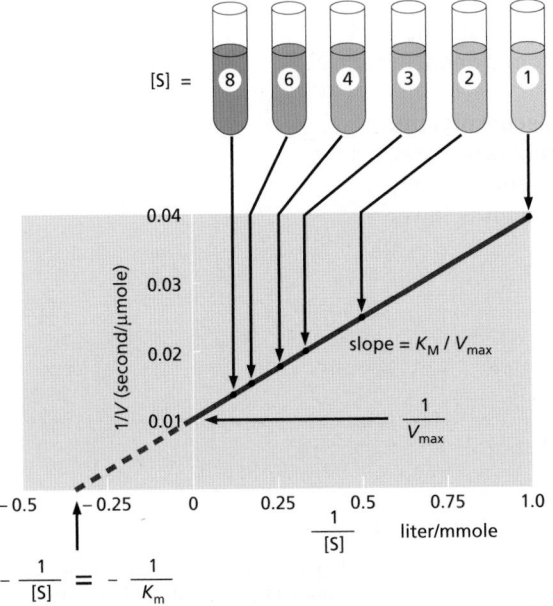

THE SIGNIFICANCE OF K_m, k_{cat}, and k_{cat}/K_m

As described in the text, K_m is an approximate measure of substrate affinity for the enzyme: it is numerically equal to the concentration of $[S]$ at $V = 0.5\ V_{max}$. In general, a lower value of K_m means tighter substrate binding. In fact, for those cases where k_{cat} is much smaller than k_{-1}, the K_m will be equal to K_d, the dissociation constant for substrate binding to the enzyme ($K_d = 1/K_a$; see Figure 3–44).

We have seen that k_{cat} is the turnover number for the enzyme. At very low substrate concentrations, where $[S] \ll K_m$, most of the enzyme is free. Thus we can think of $[E] = [E_o]$, so that the Michaelis–Menten equation becomes $V = k_{cat}/K_m\ [E][S]$. Thus, the ratio k_{cat}/K_m is equivalent to the rate constant for the reaction between free enzyme and free substrate.

A comparison of k_{cat}/K_m for the same enzyme with different substrates, or for two enzymes with their different substrates, is widely used as a measure of enzyme effectiveness.

For simplicity, in this Panel we have discussed enzymes that have only one substrate, such as the lysozyme enzyme described in the text (see p. 144). Most enzymes have two substrates, one of which is often an active carrier molecule—such as NADH or ATP.

A similar, but more complex, analysis is used to determine the kinetics of such enzymes—allowing the order of substrate binding and the presence of covalent intermediates along the pathway to be revealed.

SOME ENZYMES ARE DIFFUSION LIMITED

The values of k_{cat}, K_m, and k_{cat}/K_m for some selected enzymes are given below:

enzyme	substrate	k_{cat} (sec^{-1})	K_m (M)	k_{cat}/K_m (sec^{-1}M^{-1})
acetylcholinesterase	acetylcholine	1.4×10^4	9×10^{-5}	1.6×10^8
catalase	H_2O_2	4×10^7	1	4×10^7
fumarase	fumarate	8×10^2	5×10^{-6}	1.6×10^8

Because an enzyme and its substrate must collide before they can react, k_{cat}/K_m has a maximum possible value that is limited by collision rates. If every collision forms an enzyme–substrate complex, one can calculate from diffusion theory that k_{cat}/K_m will be between 10^8 and 10^9 sec^{-1}M^{-1}, in the case where all subsequent steps proceed immediately. Thus, it is claimed that enzymes like acetylcholinesterase and fumarase are "perfect enzymes," each enzyme having evolved to the point where nearly every collision with its substrate converts the substrate to a product.

Figure 3–47 **Enzymatic acceleration of chemical reactions by decreasing the activation energy.** There is a single transition state in this example. However, often both the uncatalyzed reaction (A) and the enzyme-catalyzed reaction (B) go through a series of transition states. In that case, it is the transition state with the highest energy (S^T and ES^T) that determines the activation energy and limits the rate of the reaction. (S = substrate; P = product of the reaction; ES = enzyme–substrate complex; EP = enzyme–product complex.)

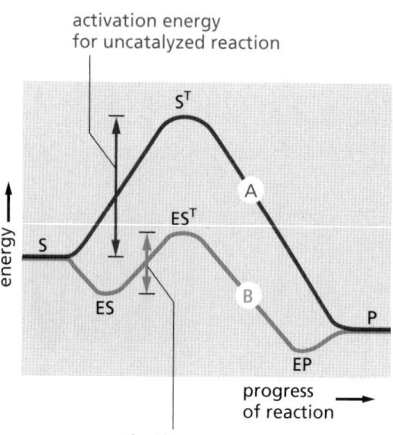

Because this tight binding greatly lowers the energy of the transition state, the enzyme greatly accelerates a particular reaction by lowering the activation energy that is required (**Figure 3–47**).

Enzymes Can Use Simultaneous Acid and Base Catalysis

Figure 3–48 compares the spontaneous reaction rates and the corresponding enzyme-catalyzed rates for five enzymes. Rate accelerations range from 10^9 to 10^{23}. Enzymes not only bind tightly to a transition state, they also contain precisely positioned atoms that alter the electron distributions in the atoms that participate directly in the making and breaking of covalent bonds. Peptide bonds, for example, can be hydrolyzed in the absence of an enzyme by exposing a polypeptide to either a strong acid or a strong base. Enzymes are unique, however, in being able to use acid and base catalysis simultaneously, because the rigid framework of the protein constrains the acidic and basic residues and prevents them from combining with each other, as they would do in solution (**Figure 3–49**).

The fit between an enzyme and its substrate needs to be precise. A small change introduced by genetic engineering in the active site of an enzyme can therefore have a profound effect. Replacing a glutamic acid with an aspartic acid in one enzyme, for example, shifts the position of the catalytic carboxylate ion by only 1 Å (about the radius of a hydrogen atom); yet this is enough to decrease the activity of the enzyme a thousandfold.

Lysozyme Illustrates How an Enzyme Works

To demonstrate how enzymes catalyze chemical reactions, we examine an enzyme that acts as a natural antibiotic in egg white, saliva, tears, and other secretions. **Lysozyme** catalyzes the cutting of polysaccharide chains in the cell walls of bacteria. The bacterial cell is under pressure from osmotic forces, and cutting even a small number of these chains causes the cell wall to rupture and the cell to burst. A relatively small and stable protein that can be easily isolated in large quantities, lysozyme was the first enzyme to have its structure worked out in atomic detail by x-ray crystallography (in the mid-1960s).

The reaction that lysozyme catalyzes is a hydrolysis: it adds a molecule of water to a single bond between two adjacent sugar groups in the polysaccharide chain, thereby causing the bond to break (see Figure 2–9). The reaction is energetically favorable because the free energy of the severed polysaccharide chain is lower

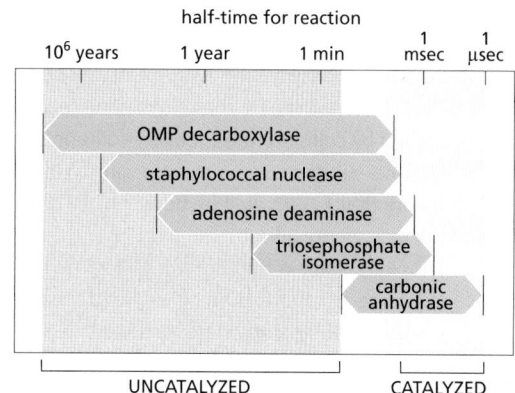

Figure 3–48 **The rate accelerations caused by five different enzymes.** (Adapted from A. Radzicka and R. Wolfenden, *Science* 267:90–93, 1995.)

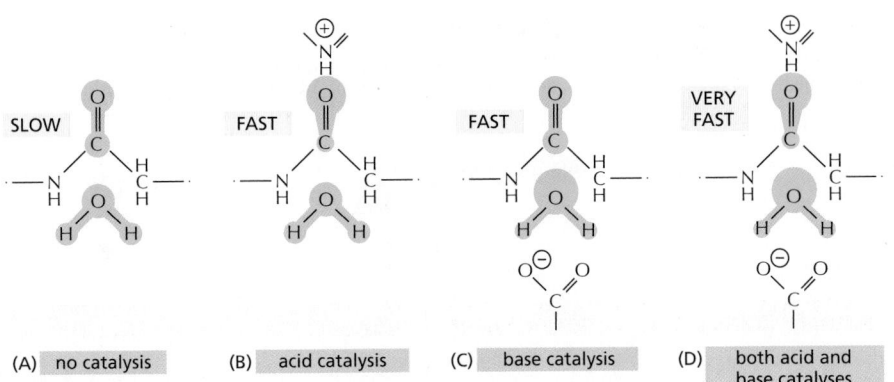

Figure 3–49 **Acid catalysis and base catalysis.** (A) The start of the uncatalyzed reaction that hydrolyzes a peptide bond, with *blue* shading used to indicate electron distribution in the water and carbonyl bonds. (B) An acid likes to donate a proton (H+) to other atoms. By pairing with the carbonyl oxygen, an acid causes electrons to move away from the carbonyl carbon, making this atom much more attractive to the electronegative oxygen of an attacking water molecule. (C) A base likes to take up H+. By pairing with a hydrogen of the attacking water molecule, a base causes electrons to move toward the water oxygen, making it a better attacking group for the carbonyl carbon. (D) By having appropriately positioned atoms on its surface, an enzyme can perform both acid catalysis and base catalysis at the same time.

than the free energy of the intact chain. However, there is an energy barrier to the reaction, and a colliding water molecule can break a bond linking two sugars only if the polysaccharide molecule is distorted into a particular shape—the transition state—in which the atoms around the bond have an altered geometry and electron distribution. Because of this requirement, random collisions must supply a very large activation energy for the reaction to take place. In an aqueous solution at room temperature, the energy of collisions almost never exceeds the activation energy. The pure polysaccharide can therefore remain for years in water without being hydrolyzed to any detectable degree.

This situation changes drastically when the polysaccharide binds to lysozyme. The active site of lysozyme, because its substrate is a polymer, is a long groove that holds six linked sugars at the same time. As soon as the polysaccharide binds to form an enzyme–substrate complex, the enzyme cuts the polysaccharide by adding a water molecule across one of its sugar–sugar bonds. The product chains are then quickly released, freeing the enzyme for further cycles of reaction (**Figure 3–50**).

An impressive increase in hydrolysis rate is possible because conditions are created in the microenvironment of the lysozyme active site that greatly reduce the activation energy necessary for the hydrolysis to take place. In particular, lysozyme distorts one of the two sugars connected by the bond to be broken from its normal, most stable conformation. The bond to be broken is also held close to two amino acids with acidic side chains (a glutamic acid and an aspartic acid) that participate directly in the reaction. **Figure 3–51** shows the three central steps in this enzymatically catalyzed reaction, which occurs millions of times faster than uncatalyzed hydrolysis.

Other enzymes use similar mechanisms to lower activation energies and speed up the reactions they catalyze. In reactions involving two or more reactants, the active site also acts like a template, or mold, that brings the substrates together in the proper orientation for a reaction to occur between them (**Figure 3–52A**). As we saw for lysozyme, the active site of an enzyme contains precisely positioned

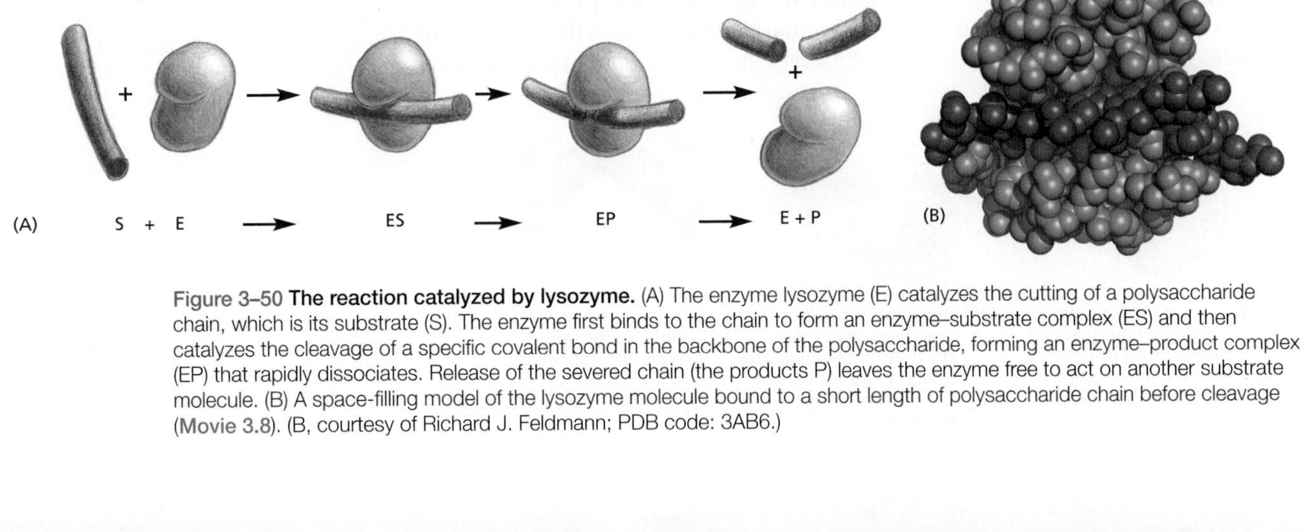

Figure 3–50 **The reaction catalyzed by lysozyme.** (A) The enzyme lysozyme (E) catalyzes the cutting of a polysaccharide chain, which is its substrate (S). The enzyme first binds to the chain to form an enzyme–substrate complex (ES) and then catalyzes the cleavage of a specific covalent bond in the backbone of the polysaccharide, forming an enzyme–product complex (EP) that rapidly dissociates. Release of the severed chain (the products P) leaves the enzyme free to act on another substrate molecule. (B) A space-filling model of the lysozyme molecule bound to a short length of polysaccharide chain before cleavage (Movie 3.8). (B, courtesy of Richard J. Feldmann; PDB code: 3AB6.)

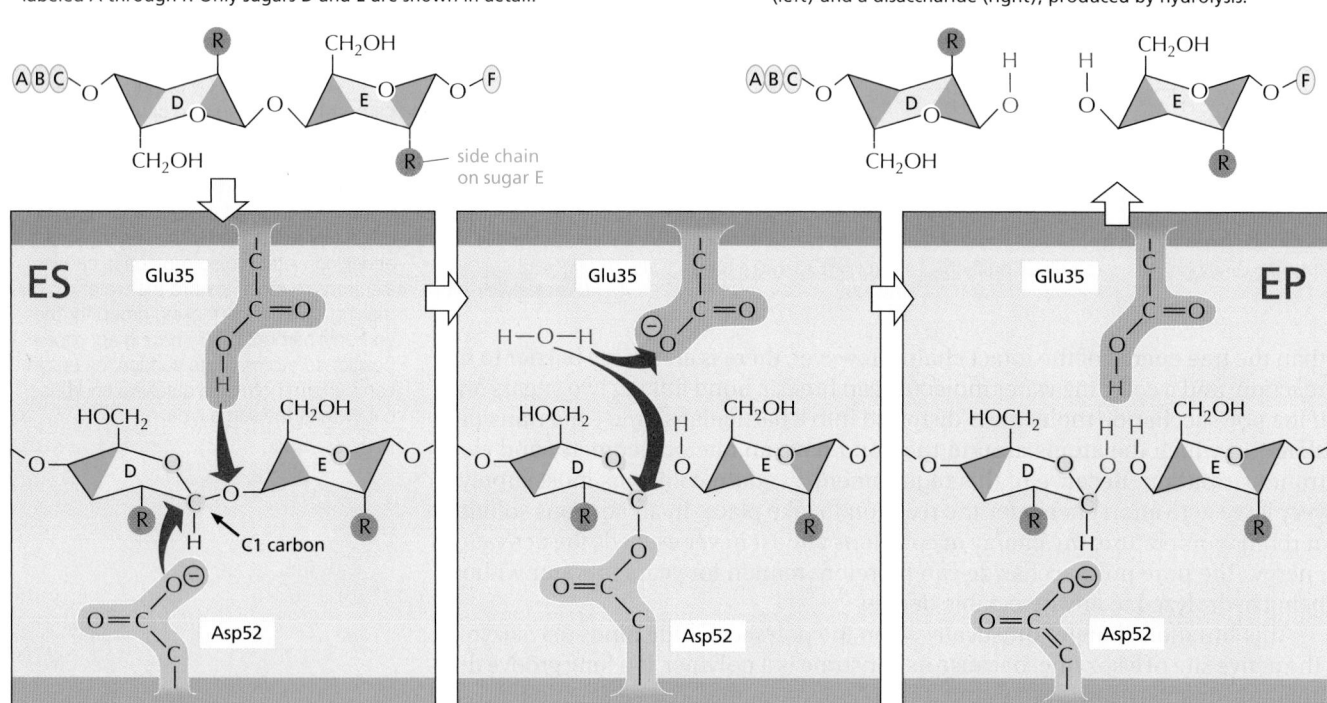

SUBSTRATE

This substrate is an oligosaccharide of six sugars, labeled A through F. Only sugars D and E are shown in detail.

PRODUCTS

The final products are an oligosaccharide of four sugars (left) and a disaccharide (right), produced by hydrolysis.

In the enzyme–substrate complex (ES), the enzyme forces sugar D into a strained conformation. The Glu35 in the enzyme is positioned to serve as an acid that attacks the adjacent sugar–sugar bond by donating a proton (H⁺) to sugar E; Asp52 is poised to attack the C1 carbon atom.

The Asp52 has formed a covalent bond between the enzyme and the C1 carbon atom of sugar D. The Glu35 then polarizes a water molecule (red), so that its oxygen can readily attack the C1 carbon atom and displace Asp52.

The reaction of the water molecule (red) completes the hydrolysis and returns the enzyme to its initial state, forming the final enzyme–product complex (EP).

atoms that speed up a reaction by using charged groups to alter the distribution of electrons in the substrates (Figure 3–52B). And as we have also seen, when a substrate binds to an enzyme, bonds in the substrate are often distorted, changing the substrate shape. These changes, along with mechanical forces, drive a substrate toward a particular transition state (Figure 3–52C). Finally, like lysozyme, many enzymes participate intimately in the reaction by transiently forming a covalent bond between the substrate and a side chain of the enzyme. Subsequent steps in the reaction restore the side chain to its original state, so that the enzyme remains unchanged after the reaction (see also Figure 2–48).

Tightly Bound Small Molecules Add Extra Functions to Proteins

Although we have emphasized the versatility of enzymes—and proteins in general—as chains of amino acids that perform remarkable functions, there are many instances in which the amino acids by themselves are not enough. Just as humans

Figure 3–51 Events at the active site of lysozyme. The top left and top right drawings show the free substrate and the free products, respectively, whereas the other three drawings show the sequential events at the enzyme active site. Note the change in the conformation of sugar D in the enzyme–substrate complex; this shape change stabilizes the oxocarbenium ion-like transition states required for formation and hydrolysis of the covalent intermediate shown in the middle panel. It is also possible that a carbonium ion intermediate forms in step 2, but the covalent intermediate shown in the middle panel has been detected with a synthetic substrate (**Movie 3.9**). (See D.J. Vocadlo et al., *Nature* 412:835–838, 2001.)

(A) enzyme binds to two substrate molecules and orients them precisely to encourage a reaction to occur between them

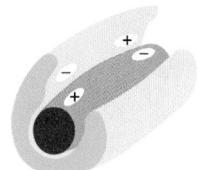

(B) binding of substrate to enzyme rearranges electrons in the substrate, creating partial negative and positive charges that favor a reaction

(C) enzyme strains the bound substrate molecule, forcing it toward a transition state to favor a reaction

Figure 3–52 Some general strategies of enzyme catalysis. (A) Holding substrates together in a precise alignment. (B) Charge stabilization of reaction intermediates. (C) Applying forces that distort bonds in the substrate to increase the rate of a particular reaction.

TABLE 3–2 Many Vitamin Derivatives Are Critical Coenzymes for Human Cells

Vitamin	Coenzyme	Enzyme-catalyzed reactions requiring these coenzymes
Thiamine (vitamin B_1)	Thiamine pyrophosphate	Activation and transfer of aldehydes
Riboflavin (vitamin B_2)	FADH	Oxidation–reduction
Niacin	NADH, NADPH	Oxidation–reduction
Pantothenic acid	Coenzyme A	Acyl group activation and transfer
Pyridoxine	Pyridoxal phosphate	Amino acid activation; also glycogen phosphorylase
Biotin	Biotin	CO_2 activation and transfer
Lipoic acid	Lipoamide	Acyl group activation; oxidation–reduction
Folic acid	Tetrahydrofolate	Activation and transfer of single carbon groups
Vitamin B_{12}	Cobalamin coenzymes	Isomerization and methyl group transfers

employ tools to enhance and extend the capabilities of their hands, enzymes and other proteins often use small nonprotein molecules to perform functions that would be difficult or impossible to do with amino acids alone. Thus, enzymes frequently have a small molecule or metal atom tightly associated with their active site that assists with their catalytic function. *Carboxypeptidase*, for example, an enzyme that cuts polypeptide chains, carries a tightly bound zinc ion in its active site. During the cleavage of a peptide bond by carboxypeptidase, the zinc ion forms a transient bond with one of the substrate atoms, thereby assisting the hydrolysis reaction. In other enzymes, a small organic molecule serves a similar purpose. Such organic molecules are often referred to as **coenzymes**. An example is *biotin*, which is found in enzymes that transfer a carboxylate group (–COO⁻) from one molecule to another (see Figure 2–40). Biotin participates in these reactions by forming a transient covalent bond to the –COO⁻ group to be transferred, being better suited to this function than any of the amino acids used to make proteins. Because it cannot be synthesized by humans, and must therefore be supplied in small quantities in our diet, biotin is a *vitamin*. Many other coenzymes are either vitamins or derivatives of vitamins (Table 3–2).

Other proteins also frequently require specific small-molecule adjuncts to function properly. Thus, the signal receptor protein *rhodopsin*, which is made by the photoreceptor cells in the retina, detects light by means of a small molecule, *retinal*, embedded in the protein (Figure 3–53A). Retinal, which is derived from vitamin A, changes its shape when it absorbs a photon of light, and this change causes the protein to trigger a cascade of enzymatic reactions that eventually lead to an electrical signal being carried to the brain.

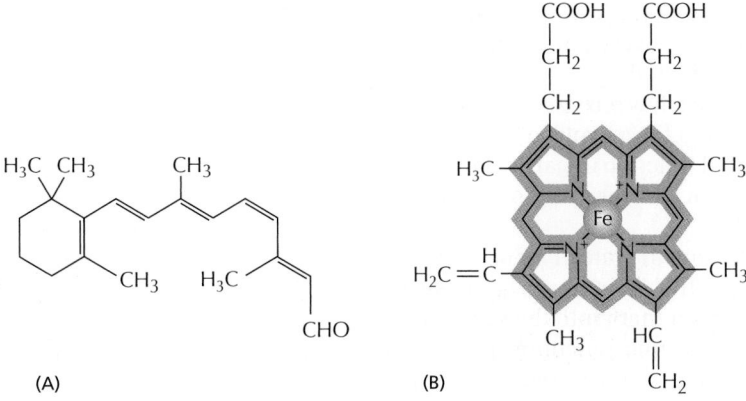

(A) (B)

Figure 3–53 Retinal and heme. (A) The structure of retinal, the light-sensitive molecule attached to rhodopsin in the eye. The structure shown isomerizes when it absorbs light. (B) The structure of a heme group. The carbon-containing heme ring is *red* and the iron atom at its center is *orange*. A heme group is tightly bound to each of the four polypeptide chains in hemoglobin, the oxygen-carrying protein whose structure is shown in Figure 3–19.

Another example of a protein with a nonprotein portion is hemoglobin (see Figure 3–19). Each molecule of hemoglobin carries four *heme* groups, ring-shaped molecules each with a single central iron atom (Figure 3–53B). Heme gives hemoglobin (and blood) its red color. By binding reversibly to oxygen gas through its iron atom, heme enables hemoglobin to pick up oxygen in the lungs and release it in the tissues.

Sometimes these small molecules are attached covalently and permanently to their protein, thereby becoming an integral part of the protein molecule itself. We shall see in Chapter 10 that proteins are often anchored to cell membranes through covalently attached lipid molecules. And membrane proteins exposed on the surface of the cell, as well as proteins secreted outside the cell, are often modified by the covalent addition of sugars and oligosaccharides.

Multienzyme Complexes Help to Increase the Rate of Cell Metabolism

The efficiency of enzymes in accelerating chemical reactions is crucial to the maintenance of life. Cells, in effect, must race against the unavoidable processes of decay, which—if left unattended—cause macromolecules to run downhill toward greater and greater disorder. If the rates of desirable reactions were not greater than the rates of competing side reactions, a cell would soon die. We can get some idea of the rate at which cell metabolism proceeds by measuring the rate of ATP utilization. A typical mammalian cell "turns over" (i.e., hydrolyzes and restores by phosphorylation) its entire ATP pool once every 1 or 2 minutes. For each cell, this turnover represents the utilization of more than 10^7 molecules of ATP per second (or, for the human body, about 30 grams of ATP every minute).

The rates of reactions in cells are rapid because enzyme catalysis is so effective. Some enzymes have become so efficient that there is no possibility of further useful improvement. The factor that limits the reaction rate is no longer the enzyme's intrinsic speed of action; rather, it is the frequency with which the enzyme collides with its substrate. Such a reaction is said to be *diffusion-limited* (see Panel 3–2, pp. 142–143).

The amount of product produced by an enzyme will depend on the concentration of both the enzyme and its substrate. If a sequence of reactions is to occur extremely rapidly, each metabolic intermediate and enzyme involved must be present in high concentration. However, given the enormous number of different reactions performed by a cell, there are limits to the concentrations that can be achieved. In fact, most metabolites are present in micromolar (10^{-6} M) concentrations, and most enzyme concentrations are much lower. How is it possible, therefore, to maintain very fast metabolic rates?

The answer lies in the spatial organization of cell components. The cell can increase reaction rates without raising substrate concentrations by bringing the various enzymes involved in a reaction sequence together to form a large protein assembly known as a *multienzyme complex* (**Figure 3–54**). Because this assembly is organized in a way that allows the product of enzyme A to be passed directly to enzyme B, and so on, diffusion rates need not be limiting, even when the concentrations of the substrates in the cell as a whole are very low. It is perhaps not surprising, therefore, that such enzyme complexes are very common, and they are involved in nearly all aspects of metabolism—including the central genetic processes of DNA, RNA, and protein synthesis. In fact, few enzymes in eukaryotic cells diffuse freely in solution; instead, most seem to have evolved binding sites that concentrate them with other proteins of related function in particular regions of the cell, thereby increasing the rate and efficiency of the reactions that they catalyze.

Eukaryotic cells have yet another way of increasing the rate of metabolic reactions: using their intracellular membrane systems. These membranes can segregate particular substrates and the enzymes that act on them into the same membrane-enclosed compartment, such as the endoplasmic reticulum or the cell nucleus. If, for example, a compartment occupies a total of 10% of the volume of

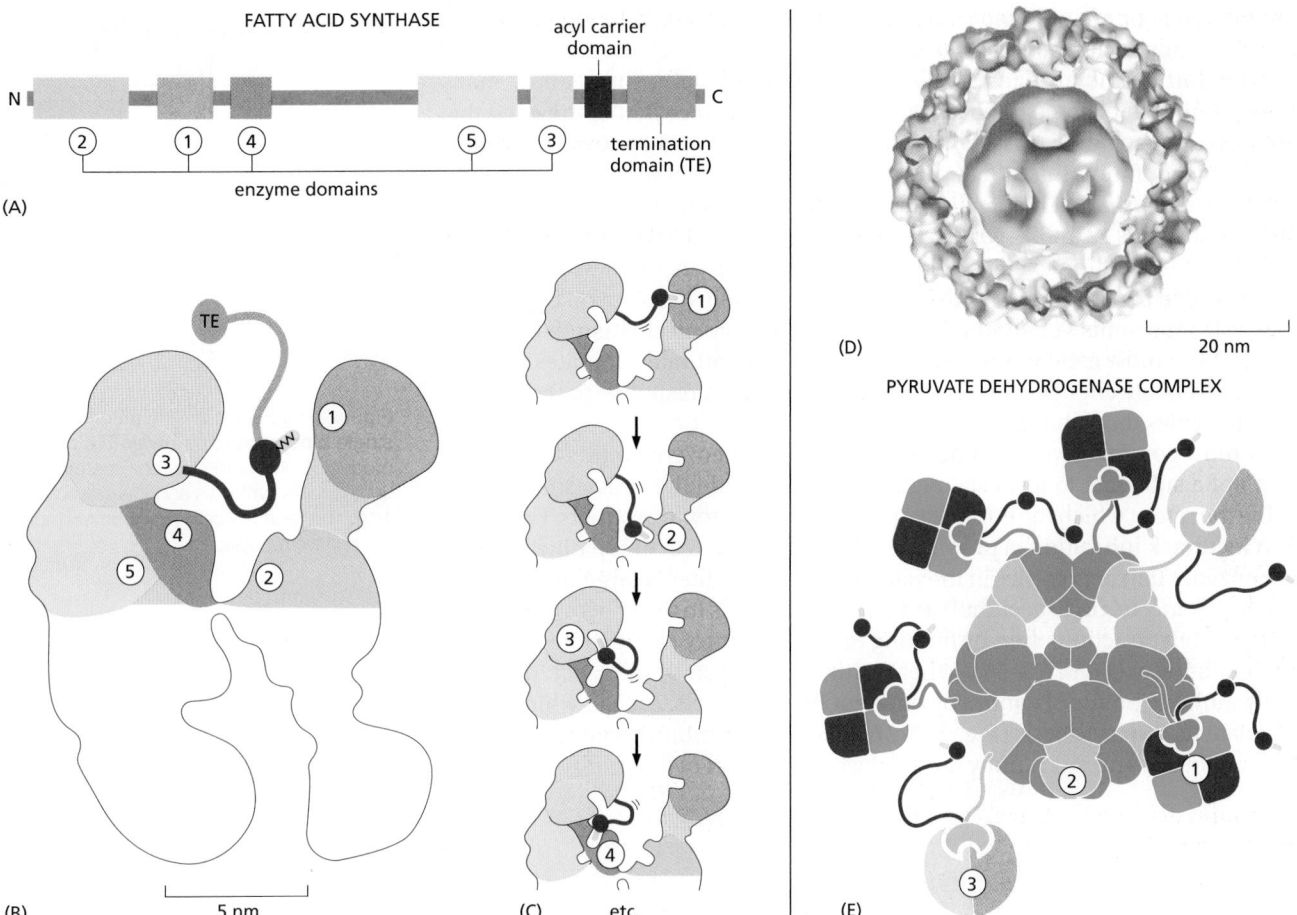

Figure 3–54 **How unstructured regions of polypeptide chain serving as tethers allow reaction intermediates to be passed from one active site to another in large multienzyme complexes.** (A–C) The fatty acid synthase in mammals. (A) The location of seven protein domains with different activities in this 270 kilodalton protein. The numbers refer to the order in which each enzyme domain must function to complete each two-carbon addition step. After multiple cycles of two-carbon addition, the termination domain releases the final product once the desired length of fatty acid has been synthesized. (B) The structure of the dimeric enzyme, with the location of the five active sites in one monomer indicated. (C) How a flexible tether allows the substrate that remains linked to the acyl carrier domain (red) to be passed from one active site to another in each monomer, sequentially elongating and modifying the bound fatty acid intermediate (yellow). The five steps are repeated until the final length of fatty acid chain has been synthesized. (Only steps 1 through 4 are illustrated here.)

(D) Multiple tethered subunits in the giant pyruvate dehydrogenase complex (9500 kilodaltons, larger than a ribosome) that catalyzes the conversion of pyruvate to acetyl CoA. (E) As in (C), a covalently bound substrate held on a flexible tether (red balls with yellow substrate) is serially passed through active sites on subunits (here labeled 1 through 3) to produce the final products. Here, subunit 1 catalyzes the decarboxylation of pyruvate accompanied by the reductive acetylation of a lipoyl group linked to one of the red balls. Subunit 2 transfers this acetyl group to CoA, forming acetyl CoA, and subunit 3 reoxidizes the lipoyl group to prepare it for the next cycle. Only one-tenth of the subunits labeled 1 and 3, attached to the core formed by subunit 2, are illustrated here. This important reaction takes place in the mammalian mitochondrion, as part of the pathway that oxidizes sugars to CO_2 and H_2O (see page 82). (A–C, adapted from T. Maier et al., *Quart. Rev. Biophys.* 43:373–422, 2010; D, from J.L.S. Milne et al., *J. Biol. Chem.* 281:4364–4370, 2006.)

the cell, the concentration of reactants in that compartment may be increased by 10 times compared with a cell with the same number of enzyme and substrate molecules, but no compartmentalization. Reactions limited by the speed of diffusion can thereby be speeded up by a factor of 10.

The Cell Regulates the Catalytic Activities of Its Enzymes

A living cell contains thousands of enzymes, many of which operate at the same time and in the same small volume of the cytosol. By their catalytic action, these enzymes generate a complex web of metabolic pathways, each composed of chains of chemical reactions in which the product of one enzyme becomes the substrate of the next. In this maze of pathways, there are many branch points (nodes) where different enzymes compete for the same substrate. The system is complex (see Figure 2–63), and elaborate controls are required to regulate when and how rapidly each reaction occurs.

Regulation occurs at many levels. At one level, the cell controls how many molecules of each enzyme it makes by regulating the expression of the gene that encodes that enzyme (discussed in Chapter 7). The cell also controls enzymatic activities by confining sets of enzymes to particular subcellular compartments, whether by enclosing them in a distinct membrane-bounded compartment (discussed in Chapters 12 and 14) or by concentrating them on a protein scaffold (see Figure 3–78). As will be explained later in this chapter, enzymes are also covalently modified to control their activity. The rate of protein destruction by targeted proteolysis represents yet another important regulatory mechanism (see Figure 6–86). But the most general process that adjusts reaction rates operates through a direct, reversible change in the activity of an enzyme in response to the specific small molecules that it binds.

The most common type of control occurs when an enzyme binds a molecule that is not a substrate to a special regulatory site outside the active site, thereby altering the rate at which the enzyme converts its substrates to products. For example, in **feedback inhibition**, a product produced late in a reaction pathway inhibits an enzyme that acts earlier in the pathway. Thus, whenever large quantities of the final product begin to accumulate, this product binds to the enzyme and slows down its catalytic action, thereby limiting the further entry of substrates into that reaction pathway (**Figure 3–55**). Where pathways branch or intersect, there are usually multiple points of control by different final products, each of which works to regulate its own synthesis (**Figure 3–56**). Feedback inhibition can work almost instantaneously, and it is rapidly reversed when the level of the product falls.

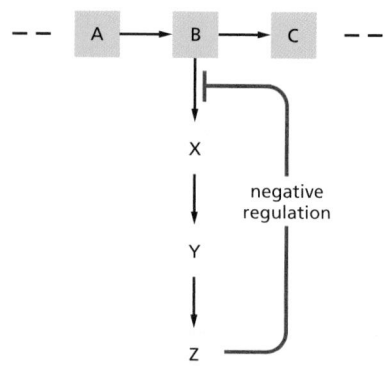

Figure 3–55 Feedback inhibition of a single biosynthetic pathway. The end product Z inhibits the first enzyme that is unique to its synthesis and thereby controls its own level in the cell. This is an example of negative regulation.

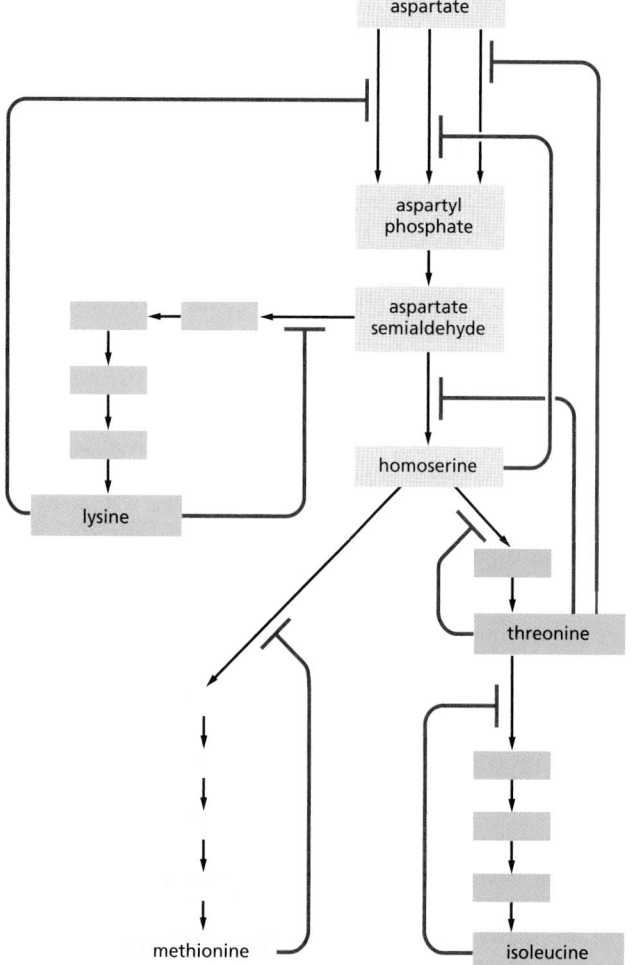

Figure 3–56 Multiple feedback inhibition. In this example, which shows the biosynthetic pathways for four different amino acids in bacteria, the *red lines* indicate positions at which products feed back to inhibit enzymes. Each amino acid controls the first enzyme specific to its own synthesis, thereby controlling its own levels and avoiding a wasteful, or even dangerous, buildup of intermediates. The products can also separately inhibit the initial set of reactions common to all the syntheses; in this case, three different enzymes catalyze the initial reaction, each inhibited by a different product.

Feedback inhibition is *negative regulation*: it prevents an enzyme from acting. Enzymes can also be subject to *positive regulation*, in which a regulatory molecule stimulates the enzyme's activity rather than shutting the enzyme down. Positive regulation occurs when a product in one branch of the metabolic network stimulates the activity of an enzyme in another pathway. As one example, the accumulation of ADP activates several enzymes involved in the oxidation of sugar molecules, thereby stimulating the cell to convert more ADP to ATP.

Allosteric Enzymes Have Two or More Binding Sites That Interact

A striking feature of both positive and negative feedback regulation is that the regulatory molecule often has a shape totally different from the shape of the substrate of the enzyme. This is why the effect on a protein is termed **allostery** (from the Greek words *allos*, meaning "other," and *stereos*, meaning "solid" or "three-dimensional"). As biologists learned more about feedback regulation, they recognized that the enzymes involved must have at least two different binding sites on their surface—an **active site** that recognizes the substrates, and a **regulatory site** that recognizes a regulatory molecule. These two sites must somehow communicate so that the catalytic events at the active site can be influenced by the binding of the regulatory molecule at its separate site on the protein's surface.

The interaction between separated sites on a protein molecule is now known to depend on a *conformational change* in the protein: binding at one of the sites causes a shift from one folded shape to a slightly different folded shape. During feedback inhibition, for example, the binding of an inhibitor at one site on the protein causes the protein to shift to a conformation that incapacitates its active site located elsewhere in the protein.

It is thought that most protein molecules are allosteric. They can adopt two or more slightly different conformations, and a shift from one to another caused by the binding of a ligand can alter their activity. This is true not only for enzymes but also for many other proteins, including receptors, structural proteins, and motor proteins. In all instances of allosteric regulation, each conformation of the protein has somewhat different surface contours, and the protein's binding sites for ligands are altered when the protein changes shape. Moreover, as we discuss next, each ligand will stabilize the conformation that it binds to most strongly, and thus—at high enough concentrations—will tend to "switch" the protein toward the conformation that the ligand prefers.

Two Ligands Whose Binding Sites Are Coupled Must Reciprocally Affect Each Other's Binding

The effects of ligand binding on a protein follow from a fundamental chemical principle known as **linkage**. Suppose, for example, that a protein that binds glucose also binds another molecule, X, at a distant site on the protein's surface. If the binding site for X changes shape as part of the conformational change in the protein induced by glucose binding, the binding sites for X and for glucose are said to be *coupled*. Whenever two ligands prefer to bind to the *same* conformation of an allosteric protein, it follows from basic thermodynamic principles that each ligand must increase the affinity of the protein for the other. For example, if the shift of a protein to a conformation that binds glucose best also causes the binding site for X to fit X better, then the protein will bind glucose more tightly when X is present than when X is absent. In other words, X will positively regulate the protein's binding of glucose (**Figure 3–57**).

Conversely, linkage operates in a negative way if two ligands prefer to bind to *different* conformations of the same protein. In this case, the binding of the first ligand discourages the binding of the second ligand. Thus, if a shape change caused by glucose binding decreases the affinity of a protein for molecule X, the binding of X must also decrease the protein's affinity for glucose (**Figure 3–58**). The linkage relationship is quantitatively reciprocal, so that, for example, if glucose has a very large effect on the binding of X, X has a very large effect on the binding of glucose.

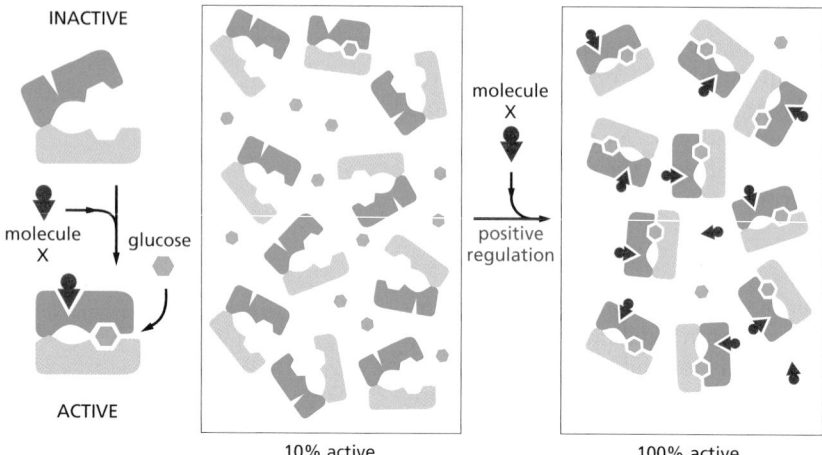

Figure 3–57 Positive regulation caused by conformational coupling between two separate binding sites. In this example, both glucose and molecule X bind best to the *closed* conformation of a protein with two domains. Because both glucose and molecule X drive the protein toward its closed conformation, each ligand helps the other to bind. Glucose and molecule X are therefore said to bind *cooperatively* to the protein.

The relationships shown in Figures 3–57 and 3–58 apply to all proteins, and they underlie all of cell biology. The principle seems so obvious in retrospect that we now take it for granted. But the discovery of linkage in studies of a few enzymes in the 1950s, followed by an extensive analysis of allosteric mechanisms in proteins in the early 1960s, had a revolutionary effect on our understanding of biology. Since molecule X in these examples binds at a site on the enzyme that is distinct from the site where catalysis occurs, it need not have any chemical relationship to the substrate that binds at the active site. Moreover, as we have just seen, for enzymes that are regulated in this way, molecule X can either turn the enzyme on (positive regulation) or turn it off (negative regulation). By such a mechanism, **allosteric proteins** serve as general switches that, in principle, can allow one molecule in a cell to affect the fate of any other.

Symmetric Protein Assemblies Produce Cooperative Allosteric Transitions

A single-subunit enzyme that is regulated by negative feedback can at most decrease from 90% to about 10% activity in response to a 100-fold increase in the concentration of an inhibitory ligand that it binds (**Figure 3–59**, *red line*). Responses of this type are apparently not sharp enough for optimal cell regulation, and most enzymes that are turned on or off by ligand binding consist of symmetric assemblies of identical subunits. With this arrangement, the binding of a molecule of ligand to a single site on one subunit can promote an allosteric change in the entire assembly that helps the neighboring subunits bind the same ligand. As a result, a *cooperative allosteric transition* occurs (Figure 3–59, *blue line*), allowing

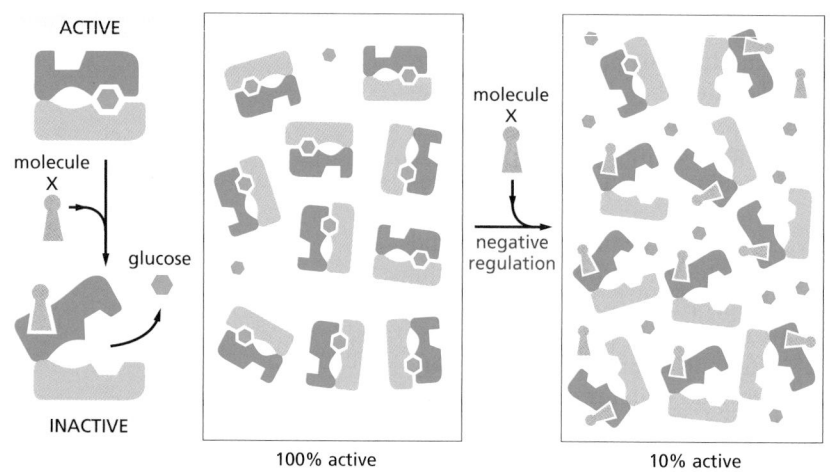

Figure 3–58 Negative regulation caused by conformational coupling between two separate binding sites. The scheme here resembles that in the previous figure, but here molecule X prefers the *open* conformation, while glucose prefers the *closed* conformation. Because glucose and molecule X drive the protein toward opposite conformations (closed and open, respectively), the presence of either ligand interferes with the binding of the other.

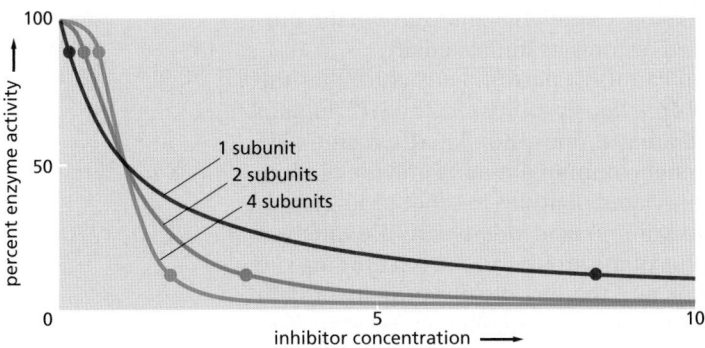

Figure 3–59 **Enzyme activity versus the concentration of inhibitory ligand for single-subunit and multisubunit allosteric enzymes.** For an enzyme with a single subunit *(red line)*, a drop from 90% enzyme activity to 10% activity (indicated by the two dots on the curve) requires a 100-fold increase in the concentration of inhibitor. The enzyme activity is calculated from the simple equilibrium relationship $K = [IP]/[I][P]$, where P is active protein, I is inhibitor, and IP is the inactive protein bound to inhibitor. An identical curve applies to any simple binding interaction between two molecules, A and B. In contrast, a multisubunit allosteric enzyme can respond in a switchlike manner to a change in ligand concentration: the steep response is caused by a cooperative binding of the ligand molecules, as explained in Figure 3–60. Here, the *green line* represents the idealized result expected for the cooperative binding of two inhibitory ligand molecules to an allosteric enzyme with two subunits, and the *blue line* shows the idealized response of an enzyme with four subunits. As indicated by the two dots on each of these curves, the more complex enzymes drop from 90% to 10% activity over a much narrower range of inhibitor concentration than does the enzyme composed of a single subunit.

a relatively small change in ligand concentration in the cell to switch the whole assembly from an almost fully active to an almost fully inactive conformation (or vice versa).

The principles involved in a cooperative "all-or-none" transition are the same for all proteins, whether or not they are enzymes. Thus, for example, they are critical for the efficient uptake and release of O_2 by hemoglobin in our blood. But they are perhaps easiest to visualize for an enzyme that forms a symmetric dimer. In the example shown in **Figure 3–60**, the first molecule of an inhibitory ligand binds with great difficulty since its binding disrupts an energetically favorable interaction between the two identical monomers in the dimer. A second molecule of inhibitory ligand now binds more easily, however, because its binding restores the energetically favorable monomer–monomer contacts of a symmetric dimer (this also completely inactivates the enzyme).

As an alternative to this *induced fit* model for a cooperative allosteric transition, we can view such a symmetric enzyme as having only two possible conformations, corresponding to the "enzyme on" and "enzyme off" structures in Figure 3–60. In this view, ligand binding perturbs an all-or-none equilibrium between these two states, thereby changing the proportion of active molecules. Both models represent true and useful concepts.

Many Changes in Proteins Are Driven by Protein Phosphorylation

Proteins are regulated by more than the reversible binding of other molecules. A second method that eukaryotic cells use extensively to regulate a protein's function is the covalent addition of a smaller molecule to one or more of its amino acid side chains. The most common such regulatory modification in higher eukaryotes is the addition of a phosphate group. We shall therefore use protein phosphorylation to illustrate some of the general principles involved in the control of protein function through the modification of amino acid side chains.

A phosphorylation event can affect the protein that is modified in three important ways. First, because each phosphate group carries two negative charges, the enzyme-catalyzed addition of a phosphate group to a protein can cause a major conformational change in the protein by, for example, attracting a cluster of positively charged amino acid side chains. This can, in turn, affect the binding of ligands elsewhere on the protein surface, dramatically changing the

Figure 3–60 **A cooperative allosteric transition in an enzyme composed of two identical subunits.** This diagram illustrates how the conformation of one subunit can influence that of its neighbor. The binding of a single molecule of an inhibitory ligand *(yellow)* to one subunit of the enzyme occurs with difficulty because it changes the conformation of this subunit and thereby disrupts the symmetry of the enzyme. Once this conformational change has occurred, however, the energy gained by restoring the symmetric pairing interaction between the two subunits makes it especially easy for the second subunit to bind the inhibitory ligand and undergo the same conformational change. Because the binding of the first molecule of ligand increases the affinity with which the other subunit binds the same ligand, the response of the enzyme to changes in the concentration of the ligand is much steeper than the response of an enzyme with only one subunit (see Figure 3–59 and Movie 3.10).

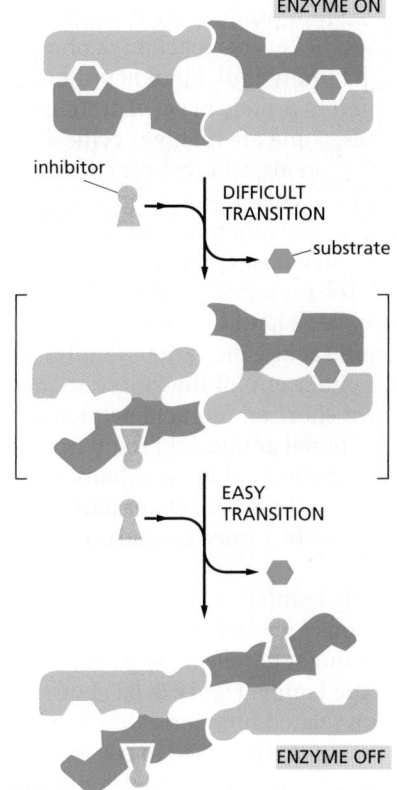

protein's activity. When a second enzyme removes the phosphate group, the protein returns to its original conformation and restores its initial activity.

Second, an attached phosphate group can form part of a structure that the binding sites of other proteins recognize. As previously discussed, the SH2 domain binds to a short peptide sequence containing a phosphorylated tyrosine side chain (see Figure 3–40B). More than ten other common domains provide binding sites for attaching their protein to phosphorylated peptides in other protein molecules, each recognizing a phosphorylated amino acid side chain in a different protein context. Third, the addition of a phosphate group can mask a binding site that otherwise holds two proteins together, and thereby disrupt protein–protein interactions. As a result, protein phosphorylation and dephosphorylation very often drive the regulated assembly and disassembly of protein complexes (see, for example, Figure 15–11).

Reversible protein phosphorylation controls the activity, structure, and cellular localization of enzymes and many other types of proteins in eukaryotic cells. In fact, this regulation is so extensive that more than one-third of the 10,000 or so proteins in a typical mammalian cell are thought to be phosphorylated at any given time—many with more than one phosphate. As might be expected, the addition and removal of phosphate groups from specific proteins often occur in response to signals that specify some change in a cell's state. For example, the complicated series of events that takes place as a eukaryotic cell divides is largely timed in this way (discussed in Chapter 17), and many of the signals mediating cell–cell interactions are relayed from the plasma membrane to the nucleus by a cascade of protein phosphorylation events (discussed in Chapter 15).

A Eukaryotic Cell Contains a Large Collection of Protein Kinases and Protein Phosphatases

Protein phosphorylation involves the enzyme-catalyzed transfer of the terminal phosphate group of an ATP molecule to the hydroxyl group on a serine, threonine, or tyrosine side chain of the protein (**Figure 3–61**). A **protein kinase** catalyzes this reaction, and the reaction is essentially unidirectional because of the large amount of free energy released when the phosphate–phosphate bond in ATP is broken to produce ADP (discussed in Chapter 2). A **protein phosphatase** catalyzes the reverse reaction of phosphate removal, or *dephosphorylation*. Cells contain hundreds of different protein kinases, each responsible for phosphorylating a different protein or set of proteins. There are also many different protein phosphatases; some are highly specific and remove phosphate groups from only one or a few proteins, whereas others act on a broad range of proteins and are targeted to specific substrates by regulatory subunits. The state of phosphorylation of a protein at any moment, and thus its activity, depends on the relative activities of the protein kinases and phosphatases that modify it.

The protein kinases that phosphorylate proteins in eukaryotic cells belong to a very large family of enzymes that share a catalytic (kinase) sequence of about 290 amino acids. The various family members contain different amino acid sequences on either end of the kinase sequence (for example, see Figure 3–10), and often have short amino acid sequences inserted into loops within it. Some of these additional amino acid sequences enable each kinase to recognize the specific set of proteins it phosphorylates, or to bind to structures that localize it in specific regions of the cell. Other parts of the protein regulate the activity of each kinase, so it can be turned on and off in response to different specific signals, as described below.

By comparing the number of amino acid sequence differences between the various members of a protein family, we can construct an "evolutionary tree" that is thought to reflect the pattern of gene duplication and divergence that gave rise to the family. **Figure 3–62** shows an evolutionary tree of protein kinases. Kinases with related functions are often located on nearby branches of the tree: the protein kinases involved in cell signaling that phosphorylate tyrosine side chains, for example, are all clustered in the top left corner of the tree. The other kinases shown

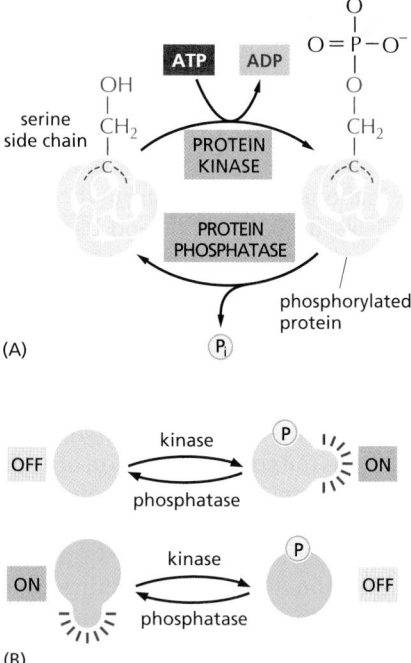

Figure 3–61 Protein phosphorylation. Many thousands of proteins in a typical eukaryotic cell are modified by the covalent addition of a phosphate group. (A) The general reaction transfers a phosphate group from ATP to an amino acid side chain of the target protein by a protein kinase. Removal of the phosphate group is catalyzed by a second enzyme, a protein phosphatase. In this example, the phosphate is added to a serine side chain; in other cases, the phosphate is instead linked to the –OH group of a threonine or a tyrosine in the protein. (B) The phosphorylation of a protein by a protein kinase can either increase or decrease the protein's activity, depending on the site of phosphorylation and the structure of the protein.

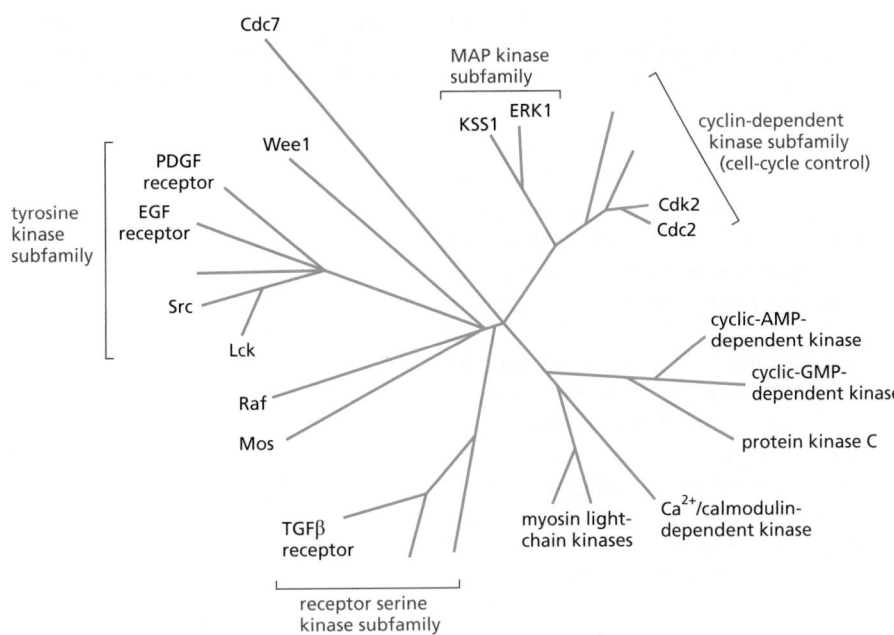

Figure 3–62 **An evolutionary tree of selected protein kinases.** A higher eukaryotic cell contains hundreds of such enzymes, and the human genome codes for more than 500. Note that only some of these, those discussed in this book, are shown.

phosphorylate either a serine or a threonine side chain, and many are organized into clusters that seem to reflect their function—in transmembrane signal transduction, intracellular signal amplification, cell-cycle control, and so on.

As a result of the combined activities of protein kinases and protein phosphatases, the phosphate groups on proteins are continually turning over—being added and then rapidly removed. Such phosphorylation cycles may seem wasteful, but they are important in allowing the phosphorylated proteins to switch rapidly from one state to another: the more rapid the cycle, the faster a population of protein molecules can change its state of phosphorylation in response to a sudden change in its phosphorylation rate (see Figure 15–14). The energy required to drive this phosphorylation cycle is derived from the free energy of ATP hydrolysis, one molecule of which is consumed for each phosphorylation event.

The Regulation of the Src Protein Kinase Reveals How a Protein Can Function as a Microprocessor

The hundreds of different protein kinases in a eukaryotic cell are organized into complex networks of signaling pathways that help to coordinate the cell's activities, drive the cell cycle, and relay signals into the cell from the cell's environment. Many of the extracellular signals involved need to be both integrated and amplified by the cell. Individual protein kinases (and other signaling proteins) serve as input–output devices, or "microprocessors," in the integration process. An important part of the input to these signal-processing proteins comes from the control that is exerted by phosphates added and removed from them by protein kinases and protein phosphatases, respectively.

The Src family of protein kinases (see Figure 3–10) exhibits such behavior. The *Src protein* (pronounced "sarc" and named for the type of tumor, a sarcoma, that its deregulation can cause) was the first tyrosine kinase to be discovered. It is now known to be part of a subfamily of nine very similar protein kinases, which are found only in multicellular animals. As indicated by the evolutionary tree in Figure 3–62, sequence comparisons suggest that tyrosine kinases as a group were a relatively late innovation that branched off from the serine/threonine kinases, with the Src subfamily being only one subgroup of the tyrosine kinases created in this way.

The Src protein and its relatives contain a short N-terminal region that becomes covalently linked to a strongly hydrophobic fatty acid, which anchors the kinase at the cytoplasmic face of the plasma membrane. Next along the linear sequence of

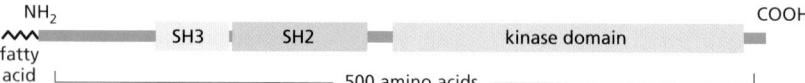

Figure 3–63 The domain structure of the Src family of protein kinases, mapped along the amino acid sequence. For the three-dimensional structure of Src, see Figure 3–10.

amino acids come two peptide-binding domains, a Src homology 3 (SH3) domain and an SH2 domain, followed by the kinase catalytic domain (**Figure 3–63**). These kinases normally exist in an inactive conformation, in which a phosphorylated tyrosine near the C-terminus is bound to the SH2 domain, and the SH3 domain is bound to an internal peptide in a way that distorts the active site of the enzyme and helps to render it inactive.

As shown in **Figure 3–64**, turning the kinase on involves at least two specific inputs: removal of the C-terminal phosphate and the binding of the SH3 domain by a specific activating protein. In this way, the activation of the Src kinase signals the completion of a particular set of separate upstream events (**Figure 3–65**). Thus, the Src family of protein kinases serves as specific signal integrators, contributing to the web of information-processing events that enable the cell to compute useful responses to a complex set of different conditions.

Proteins That Bind and Hydrolyze GTP Are Ubiquitous Cell Regulators

We have described how the addition or removal of phosphate groups on a protein can be used by a cell to control the protein's activity. In the example just discussed, a kinase transfers a phosphate from an ATP molecule to an amino acid side chain of a target protein. Eukaryotic cells also have another way to control protein activity by phosphate addition and removal. In this case, the phosphate is not attached directly to the protein; instead, it is a part of the guanine nucleotide GTP, which binds very tightly to a class of proteins known as *GTP-binding proteins*. In general, proteins regulated in this way are in their active conformations with GTP bound. The loss of a phosphate group occurs when the bound GTP is hydrolyzed to GDP in a reaction catalyzed by the protein itself, and in its GDP-bound state the protein is inactive. In this way, GTP-binding proteins act as on–off switches whose activity is determined by the presence or absence of an additional phosphate on a bound GDP molecule (**Figure 3–66**).

GTP-binding proteins (also called **GTPases** because of the GTP hydrolysis they catalyze) comprise a large family of proteins that all contain variations on the same GTP-binding globular domain. When a tightly bound GTP is hydrolyzed by the GTP-binding protein to GDP, this domain undergoes a conformational

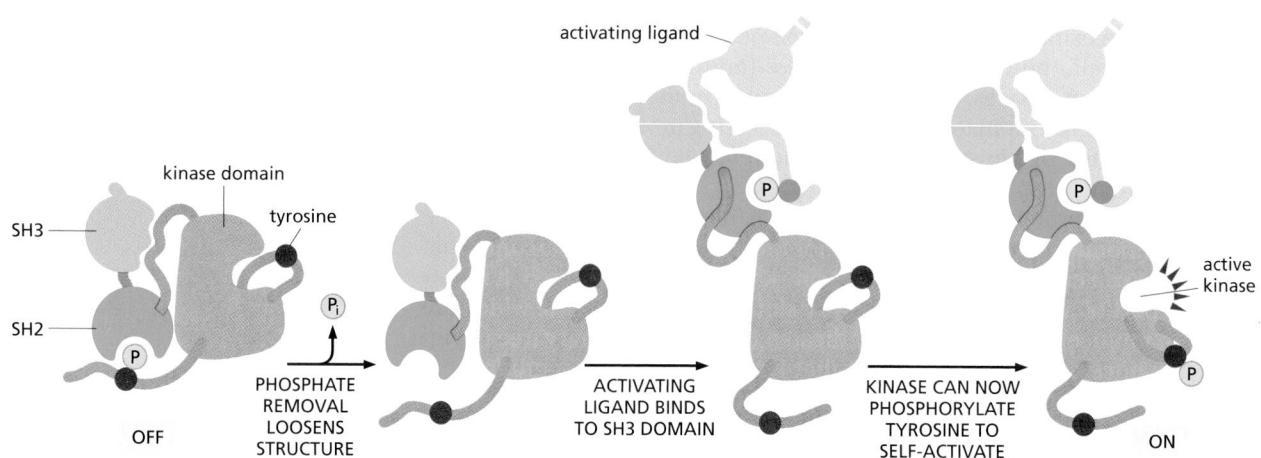

Figure 3–64 The activation of a Src-type protein kinase by two sequential events. As described in the text, the requirement for multiple upstream events to trigger these processes allows the kinase to serve as a signal integrator (**Movie 3.11**). (Adapted from S.C. Harrison et al., *Cell* 112:737–740, 2003.)

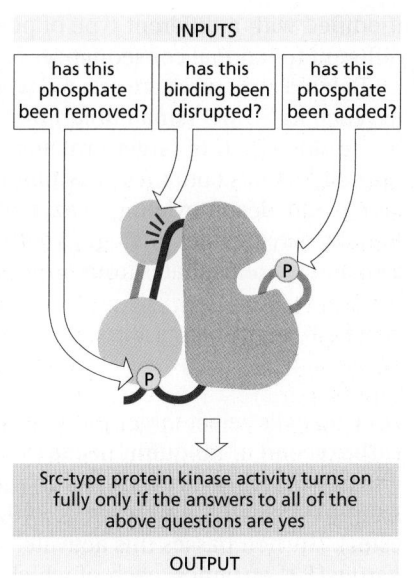

Figure 3–65 **How a Src-type protein kinase acts as a signal-integrating device.** A disruption of the inhibitory interaction illustrated for the SH3 domain *(green)* occurs when its binding to the indicated *orange* linker region is replaced with its higher-affinity binding to an activating ligand.

change that inactivates the protein. The three-dimensional structure of a prototypical member of this family, the monomeric GTPase called Ras, is shown in **Figure 3–67**.

The *Ras protein* has an important role in cell signaling (discussed in Chapter 15). In its GTP-bound form, it is active and stimulates a cascade of protein phosphorylations in the cell. Most of the time, however, the protein is in its inactive, GDP-bound form. It becomes active when it exchanges its GDP for a GTP molecule in response to extracellular signals, such as growth factors, that bind to receptors in the plasma membrane (see Figure 15–47).

Regulatory Proteins GAP and GEF Control the Activity of GTP-Binding Proteins by Determining Whether GTP or GDP Is Bound

GTP-binding proteins are controlled by regulatory proteins that determine whether GTP or GDP is bound, just as phosphorylated proteins are turned on and off by protein kinases and protein phosphatases. Thus, Ras is inactivated by a *GTPase-activating protein* (*GAP*), which binds to the Ras protein and induces Ras to hydrolyze its bound GTP molecule to GDP—which remains tightly bound—and inorganic phosphate (P_i), which is rapidly released. The Ras protein stays in its inactive, GDP-bound conformation until it encounters a *guanine nucleotide exchange factor* (*GEF*), which binds to GDP-Ras and causes Ras to release its GDP. Because the empty nucleotide-binding site is immediately filled by a GTP molecule (GTP is present in large excess over GDP in cells), the GEF activates Ras by *indirectly* adding back the phosphate removed by GTP hydrolysis. Thus, in a sense, the roles of GAP and GEF are analogous to those of a protein phosphatase and a protein kinase, respectively (**Figure 3–68**).

Proteins Can Be Regulated by the Covalent Addition of Other Proteins

Cells contain a special family of small proteins whose members are covalently attached to many other proteins to determine the activity or fate of the second protein. In each case, the carboxyl end of the small protein becomes linked to the amino group of a lysine side chain of a "target" protein through an isopeptide bond. The first such protein discovered, and the most abundantly used, is **ubiquitin** (**Figure 3–69A**). Ubiquitin can be covalently attached to target proteins in a variety of ways, each of which has a different meaning for cells. The major form of ubiquitin addition produces *polyubiquitin* chains in which—once the first ubiquitin molecule is attached to the target—each subsequent ubiquitin molecule links to Lys48 of the previous ubiquitin, creating a chain of Lys48-linked ubiquitins that are attached to a single lysine side chain of the target protein. This form of polyubiquitin directs the target protein to the interior of a proteasome, where it is digested to small peptides (see Figure 6–84). In other circumstances, only single molecules of ubiquitin are added to proteins. In addition, some target proteins are

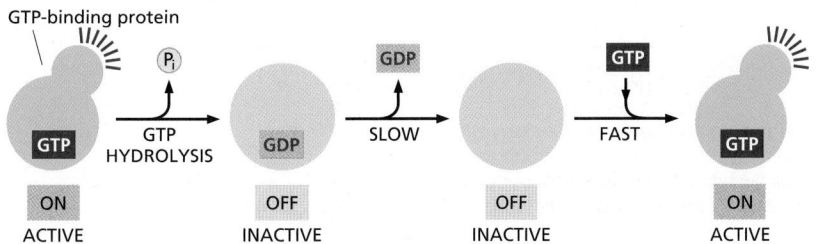

Figure 3–66 **GTP-binding proteins as molecular switches.** The activity of a GTP-binding protein (also called a GTPase) generally requires the presence of a tightly bound GTP molecule (switch "on"). Hydrolysis of this GTP molecule by the GTP-binding protein produces GDP and inorganic phosphate (P_i), and it causes the protein to convert to a different, usually inactive, conformation (switch "off"). Resetting the switch requires that the tightly bound GDP dissociates. This is a slow step that is greatly accelerated by specific signals; once the GDP has dissociated, a molecule of GTP is quickly rebound.

modified with a different type of polyubiquitin chain. These modifications have different functional consequences for the protein that is targeted (Figure 3–69B).

Related structures are created when a different member of the ubiquitin family, such as SUMO (**s**mall **u**biquitin-related **mo**difier), is covalently attached to a lysine side chain of target proteins. Not surprisingly, all such modifications are reversible. Cells contain sets of ubiquitylating and deubiquitylating (and sumoylating and desumoylating) enzymes that manipulate these covalent adducts, thereby playing roles analogous to the protein kinases and phosphatases that add and remove phosphates from protein side chains.

An Elaborate Ubiquitin-Conjugating System Is Used to Mark Proteins

How do cells select target proteins for ubiquitin addition? As an initial step, the carboxyl end of ubiquitin needs to be activated. This activation is accomplished when a protein called a *ubiquitin-activating enzyme* (E1) uses ATP hydrolysis energy to attach ubiquitin to itself through a high-energy covalent bond (a thioester). E1 then passes this activated ubiquitin to one of a set of *ubiquitin-conjugating* (E2) enzymes, each of which acts in conjunction with a set of accessory (E3) proteins called **ubiquitin ligases**. There are roughly 30 structurally similar but distinct E2 enzymes in mammals, and hundreds of different E3 proteins that form complexes with specific E2 enzymes.

Figure 3–70 illustrates the process used to mark proteins for proteasomal degradation. [Similar mechanisms are used to attach ubiquitin (and SUMO) to other types of target proteins.] Here, the ubiquitin ligase binds to specific degradation signals, called *degrons*, in protein substrates, thereby helping E2 to form a *polyubiquitin* chain linked to a lysine of the substrate protein. This polyubiquitin chain on a target protein will then be recognized by a specific receptor in the proteasome, causing the target protein to be destroyed. Distinct ubiquitin ligases recognize different degradation signals, thereby targeting distinct subsets of intracellular proteins for destruction, often in response to specific signals (see Figure 6–86).

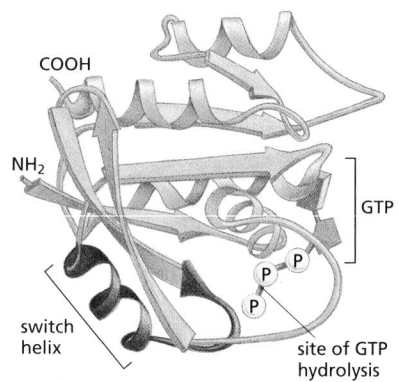

Figure 3–67 The structure of the Ras protein in its GTP-bound form. This monomeric GTPase illustrates the structure of a GTP-binding domain, which is present in a large family of GTP-binding proteins. The *red* regions change their conformation when the GTP molecule is hydrolyzed to GDP and inorganic phosphate by the protein; the GDP remains bound to the protein, while the inorganic phosphate is released. The special role of the "switch helix" in proteins related to Ras is explained in the text (see Figure 3–72 and Movie 15.7).

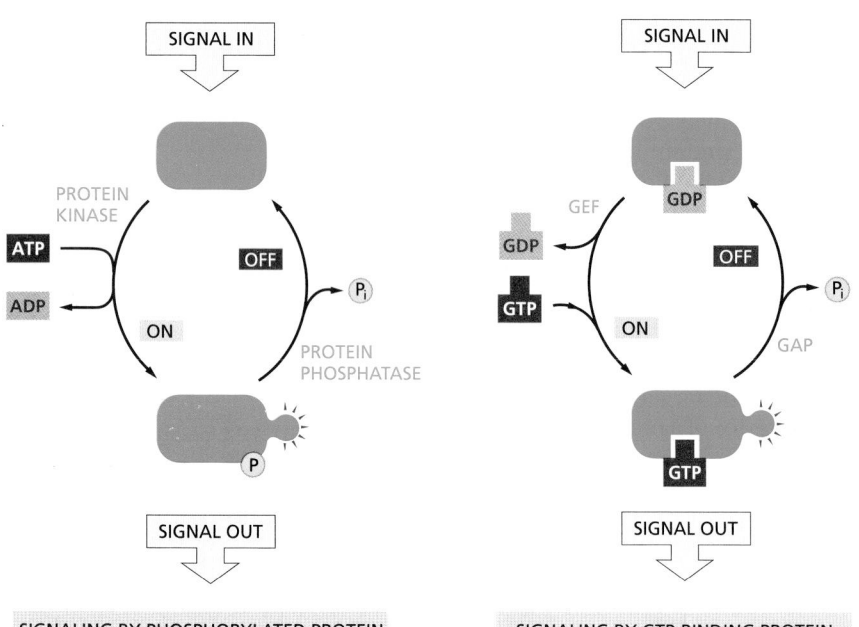

Figure 3–68 A comparison of two major intracellular signaling mechanisms in eukaryotic cells. In both cases, a signaling protein is activated by the addition of a phosphate group and inactivated by the removal of this phosphate. Note that the addition of a phosphate to a protein can also be inhibitory. (Adapted from E.R. Kantrowitz and W.N. Lipscomb, *Trends Biochem. Sci.* 15:53–59, 1990.)

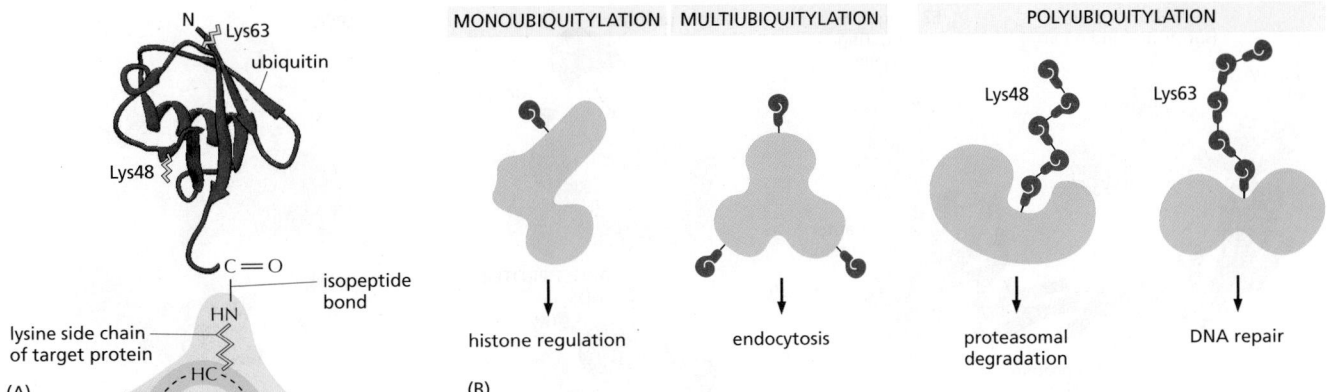

Figure 3–69 **The marking of proteins by ubiquitin.** (A) The three-dimensional structure of ubiquitin, a small protein of 76 amino acids. A family of special enzymes couples its carboxyl end to the amino group of a lysine side chain in a target protein molecule, forming an isopeptide bond. (B) Some modification patterns that have specific meanings to the cell. Note that the two types of polyubiquitylation differ in the way the ubiquitin molecules are linked together. Linkage through Lys48 signifies degradation by the proteasome (see Figure 6–84), whereas that through Lys63 has other meanings. Ubiquitin markings are "read" by proteins that specifically recognize each type of modification.

Protein Complexes with Interchangeable Parts Make Efficient Use of Genetic Information

The *SCF ubiquitin ligase* is a protein complex that binds different "target proteins" at different times in the cell cycle, covalently adding polyubiquitin chains to these targets. Its C-shaped structure is formed from five protein subunits, the largest of which serves as a scaffold on which the rest of the complex is built. The structure underlies a remarkable mechanism (**Figure 3–71**). At one end of the C is an E2 ubiquitin-conjugating enzyme. At the other end is a substrate-binding arm, a subunit known as an *F-box protein*. These two subunits are separated by a gap of about 5 nm. When this protein complex is activated, the F-box protein binds to a specific site on a target protein, positioning the protein in the gap so that some of its lysine side chains contact the ubiquitin-conjugating enzyme. The enzyme can then catalyze repeated additions of ubiquitin to these lysines (see Figure 3–71C), producing polyubiquitin chains that mark the target proteins for rapid destruction in a proteasome.

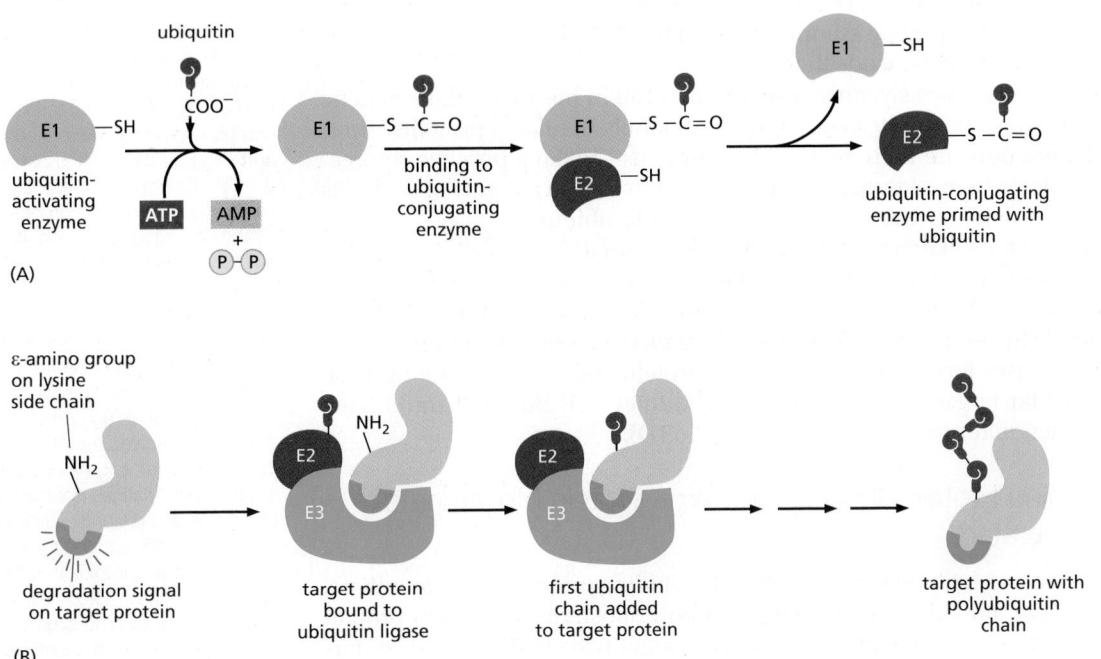

Figure 3–70 **The marking of proteins with ubiquitin.** (A) The C-terminus of ubiquitin is initially activated by being linked via a high-energy thioester bond to a cysteine side chain on the E1 protein. This reaction requires ATP, and it proceeds via a covalent AMP-ubiquitin intermediate. The activated ubiquitin on E1, also known as the ubiquitin-activating enzyme, is then transferred to the cysteine on an E2 molecule. (B) The addition of a polyubiquitin chain to a target protein. In a mammalian cell, there are several hundred distinct E2–E3 complexes. The E2s are called ubiquitin-conjugating enzymes. The E3s are referred to as ubiquitin ligases. (Adapted from D.R. Knighton et al., *Science* 253:407–414, 1991.)

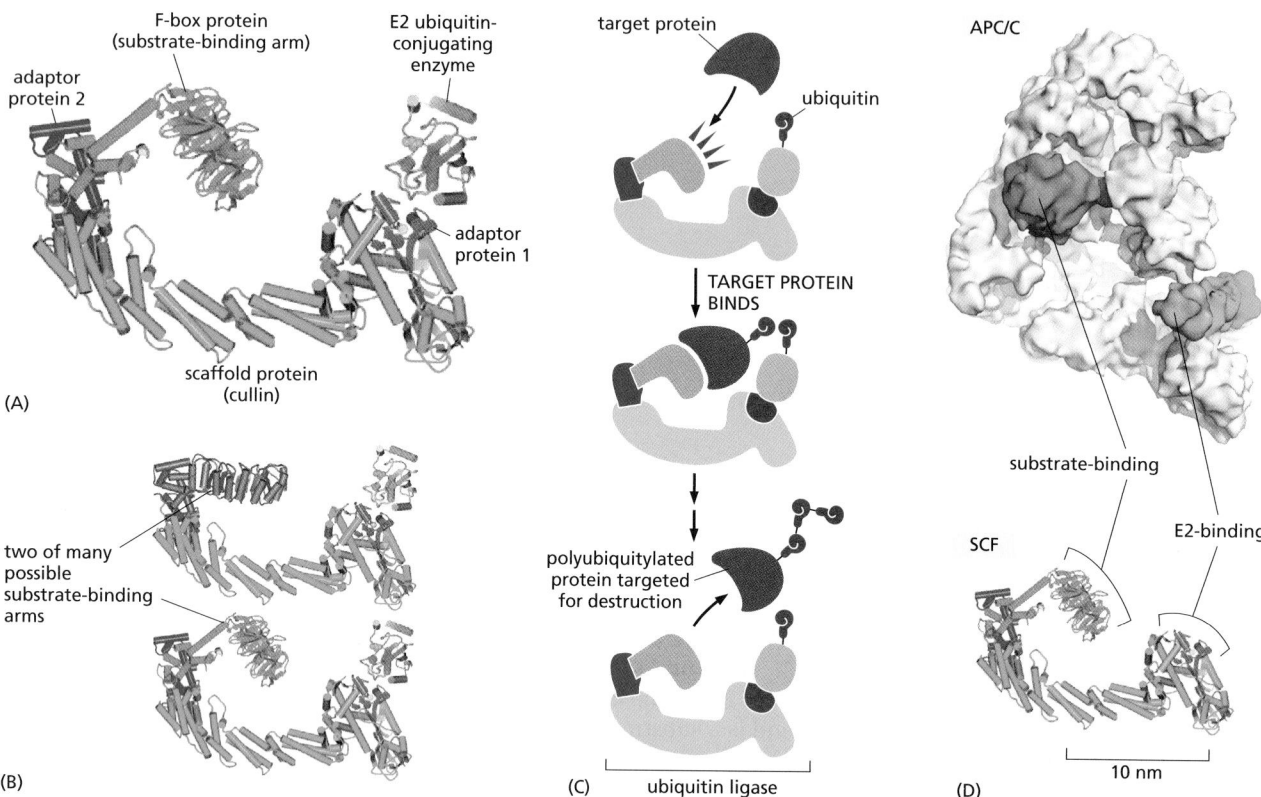

(A)

(B)

two of many possible substrate-binding arms

adaptor protein 2

F-box protein (substrate-binding arm)

E2 ubiquitin-conjugating enzyme

adaptor protein 1

scaffold protein (cullin)

(C)   target protein

ubiquitin

TARGET PROTEIN BINDS

polyubiquitylated protein targeted for destruction

ubiquitin ligase

(D)   APC/C

substrate-binding

SCF

E2-binding

10 nm

In this manner, specific proteins are targeted for rapid destruction in response to specific signals, thereby helping to drive the cell cycle (discussed in Chapter 17). The timing of the destruction often involves creating a specific pattern of phosphorylation on the target protein that is required for its recognition by the F-box subunit. It also requires the activation of an SCF ubiquitin ligase that carries the appropriate substrate-binding arm. Many of these arms (the F-box subunits) are interchangeable in the protein complex (see Figure 3–71B), and there are more than 70 human genes that encode them.

As emphasized previously, once a successful protein has evolved, its genetic information tends to be duplicated to produce a family of related proteins. Thus, for example, not only are there many F-box proteins—making possible the recognition of different sets of target proteins—but there is also a family of scaffolds (known as cullins) that give rise to a family of SCF-like ubiquitin ligases.

A protein machine like the SCF ubiquitin ligase, with its interchangeable parts, makes economical use of the genetic information in cells. It also creates opportunities for "rapid" evolution, inasmuch as new functions can evolve for the entire complex simply by producing an alternative version of one of its subunits.

Ubiquitin ligases form a diverse family of protein complexes. Some of these complexes are far larger and more complicated than SCF, but their underlying enzymatic function remains the same (Figure 3–71D).

A GTP-Binding Protein Shows How Large Protein Movements Can Be Generated

Detailed structures obtained for one of the GTP-binding protein family members, the *EF-Tu protein*, provide a good example of how allosteric changes in protein conformations can produce large movements by amplifying a small, local conformational change. As will be discussed in Chapter 6, EF-Tu is an abundant molecule that serves as an elongation factor (hence the EF) in protein synthesis, loading each aminoacyl-tRNA molecule onto the ribosome. EF-Tu contains a Ras-like domain (see Figure 3–67), and the tRNA molecule forms a tight complex with its GTP-bound form. This tRNA molecule can transfer its amino acid to the growing

Figure 3–71 **The structure and mode of action of an SCF ubiquitin ligase.** (A) The structure of the five-protein ubiquitin ligase complex that includes an E2 ubiquitin-conjugating enzyme. Four proteins form the E3 portion. The protein denoted here as adaptor protein 1 is the Rbx1/Hrt1 protein, adaptor protein 2 is the Skp1 protein, and the cullin is the Cul1 protein. One of the many different F-box proteins completes the complex. (B) Comparison of the same complex with two different substrate-binding arms, the F-box proteins Skp2 *(top)* and β-trCP1 *(bottom)*, respectively. (C) The binding and ubiquitylation of a target protein by the SCF ubiquitin ligase. If, as indicated, a chain of ubiquitin molecules is added to the same lysine of the target protein, that protein is marked for rapid destruction by the proteasome. (D) Comparison of SCF *(bottom)* with a low-resolution electron microscopy structure of a ubiquitin ligase called the anaphase-promoting complex (APC/C; *top*) at the same scale. The APC/C is a large, 15-protein complex. As discussed in Chapter 17, its ubiquitylations control the late stages of mitosis. It is distantly related to SCF and contains a cullin subunit *(green)* that lies along the side of the complex at right, only partly visible in this view. E2 proteins are not shown here, but their binding sites are indicated in *orange,* along with substrate-binding sites in *purple.* (A and B, adapted from G. Wu et al., *Mol. Cell* 11:1445–1456, 2003. D, adapted from P. da Fonseca et al., *Nature* 470:274–278, 2011.)

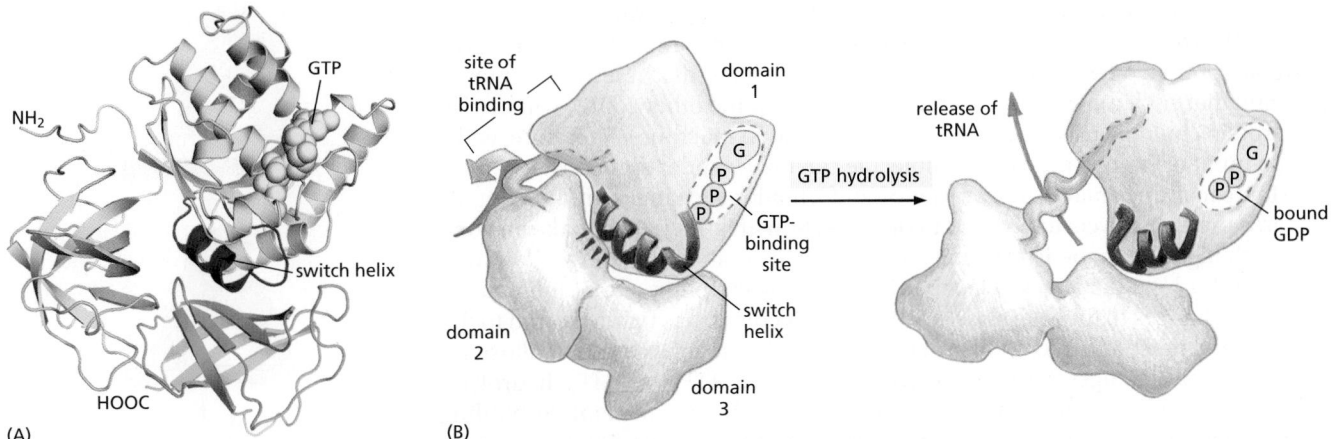

Figure 3–72 **The large conformational change in EF-Tu caused by GTP hydrolysis.** (A and B) The three-dimensional structure of EF-Tu with GTP bound. The domain at the top has a structure similar to the Ras protein, and its red α helix is the switch helix, which moves after GTP hydrolysis. (C) The change in the conformation of the switch helix in domain 1 allows domains 2 and 3 to rotate as a single unit by about 90 degrees toward the viewer, which releases the tRNA that was bound to this structure (see also Figure 3–73). (A, adapted from H. Berchtold et al., *Nature* 365:126–132, 1993. B, courtesy of Mathias Sprinzl and Rolf Hilgenfeld. PDB code: 1EFT.)

polypeptide chain only after the GTP bound to EF-Tu is hydrolyzed, dissociating the EF-Tu. Since this GTP hydrolysis is triggered by a proper fit of the tRNA to the mRNA molecule on the ribosome, the EF-Tu serves as a factor that discriminates between correct and incorrect mRNA–tRNA pairings (see Figure 6–65).

By comparing the three-dimensional structure of EF-Tu in its GTP-bound and GDP-bound forms, we can see how the repositioning of the tRNA occurs. The dissociation of the inorganic phosphate group (P_i), which follows the reaction GTP → GDP + P_i, causes a shift of a few tenths of a nanometer at the GTP-binding site, just as it does in the Ras protein. This tiny movement, equivalent to a few times the diameter of a hydrogen atom, causes a conformational change to propagate along a crucial piece of α helix, called the *switch helix*, in the Ras-like domain of the protein. The switch helix seems to serve as a latch that adheres to a specific site in another domain of the molecule, holding the protein in a "shut" conformation. The conformational change triggered by GTP hydrolysis causes the switch helix to detach, allowing separate domains of the protein to swing apart, through a distance of about 4 nm (**Figure 3–72**). This releases the bound tRNA molecule, allowing its attached amino acid to be used (**Figure 3–73**).

Notice in this example how cells have exploited a simple chemical change that occurs on the surface of a small protein domain to create a movement 50 times larger. Dramatic shape changes of this type also cause the very large movements that occur in motor proteins, as we discuss next.

Motor Proteins Produce Large Movements in Cells

We have seen that conformational changes in proteins have a central role in enzyme regulation and cell signaling. We now discuss proteins whose major function is to move other molecules. These **motor proteins** generate the forces responsible for muscle contraction and the crawling and swimming of cells. Motor proteins also power smaller-scale intracellular movements: they help to move chromosomes to opposite ends of the cell during mitosis (discussed in Chapter 17), to move organelles along molecular tracks within the cell (discussed in Chapter 16), and to move enzymes along a DNA strand during the synthesis of a new DNA molecule (discussed in Chapter 5). All these fundamental processes depend on proteins with moving parts that operate as force-generating machines.

How do these machines work? In other words, how do cells use shape changes in proteins to generate directed movements? If, for example, a protein is required to walk along a narrow thread such as a DNA molecule, it can do this by undergoing a series of conformational changes, such as those shown in **Figure 3–74**. But with nothing to drive these changes in an orderly sequence, they are perfectly

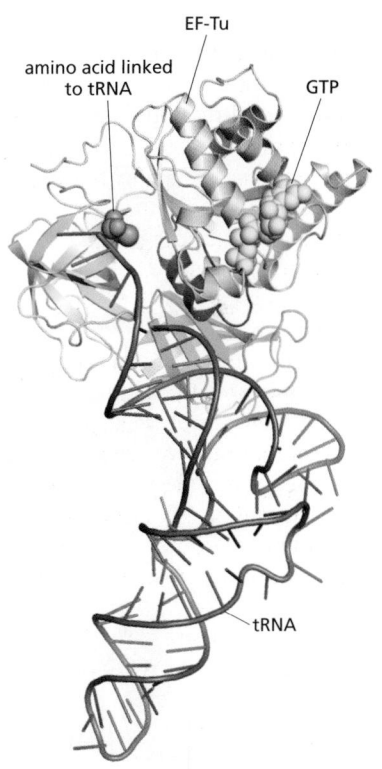

Figure 3–73 **An aminoacyl tRNA molecule bound to EF-Tu.** Note how the bound protein blocks the use of the tRNA-linked amino acid *(green)* for protein synthesis until GTP hydrolysis triggers the conformational changes shown in Figure 3–72C, dissociating the protein-tRNA complex. EF-Tu is a bacterial protein; however, a very similar protein exists in eukaryotes, where it is called EF-1 (**Movie 3.12**). (Coordinates determined by P. Nissen et al., *Science* 270:1464–1472, 1995. PDB code: 1B23.)

reversible, and the protein can only wander randomly back and forth along the thread. We can look at this situation in another way. Since the directional movement of a protein does work, the laws of thermodynamics (discussed in Chapter 2) demand that such movement use free energy from some other source (otherwise the protein could be used to make a perpetual motion machine). Therefore, without an input of energy, the protein molecule can only wander aimlessly.

How can the cell make such a series of conformational changes unidirectional? To force the entire cycle to proceed in one direction, it is enough to make any one of the changes in shape irreversible. Most proteins that are able to walk in one direction for long distances achieve this motion by coupling one of the conformational changes to the hydrolysis of an ATP molecule that is tightly bound to the protein. The mechanism is similar to the one just discussed that drives allosteric protein shape changes by GTP hydrolysis. Because ATP (or GTP) hydrolysis releases a great deal of free energy, it is very unlikely that the nucleotide-binding protein will undergo the reverse shape change needed for moving backward—since this would require that it also reverse the ATP hydrolysis by adding a phosphate molecule to ADP to form ATP.

In the model shown in **Figure 3–75**A, ATP binding shifts a motor protein from conformation 1 to conformation 2. The bound ATP is then hydrolyzed to produce ADP and inorganic phosphate (Pi), causing a change from conformation 2 to conformation 3. Finally, the release of the bound ADP and Pi drives the protein back to conformation 1. Because the energy provided by ATP hydrolysis drives the transition 2 → 3, this series of conformational changes is effectively irreversible. Thus, the entire cycle goes in only one direction, causing the protein molecule to walk continuously to the right in this example.

Many motor proteins generate directional movement through the use of a similar unidirectional ratchet, including the muscle motor protein *myosin*,

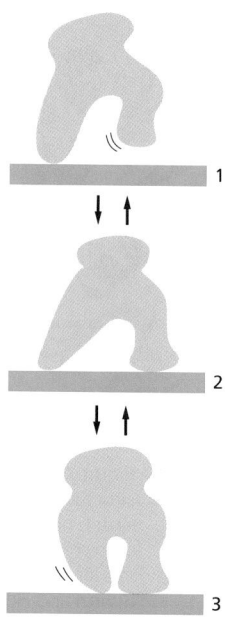

Figure 3–74 An allosteric "walking" protein. Although its three different conformations allow it to wander randomly back and forth while bound to a thread or a filament, the protein cannot move uniformly in a single direction.

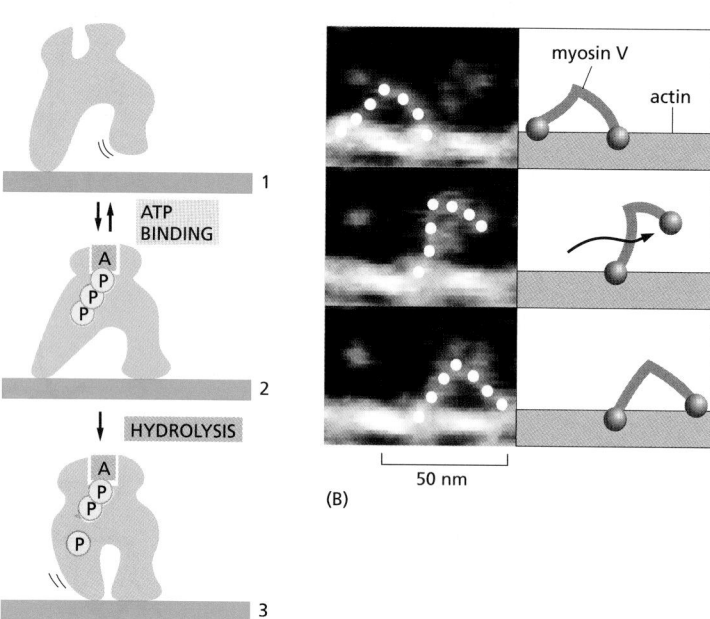

(B)

50 nm

(A)   direction of movement

Figure 3–75 How a protein can walk in one direction. (A) An allosteric motor protein driven by ATP hydrolysis. The transition between three different conformations includes a step driven by the hydrolysis of a bound ATP molecule, creating a "unidirectional ratchet" that makes the entire cycle essentially irreversible. By repeated cycles, the protein therefore moves continuously to the right along the thread. (B) Direct visualization of a walking myosin motor protein by high-speed atomic force microscopy; the elapsed time between steps was less than 0.5 sec (see Movie 16.3). (B, adapted from N. Kodera et al., *Nature* 468:72–76, published 2010 by Macmillan Publishers Ltd. Reproduced with permission of SNCSC.)

which walks along actin filaments (Figure 3–75B), and the *kinesin* proteins that walk along microtubules (both discussed in Chapter 16). These movements can be rapid: some of the motor proteins involved in DNA replication (the DNA helicases) propel themselves along a DNA strand at rates as high as 1000 nucleotides per second.

Membrane-Bound Transporters Harness Energy to Pump Molecules Through Membranes

We have thus far seen how proteins that undergo allosteric shape changes can act as microprocessors (Src family kinases), as assembly factors (EF-Tu), and as generators of mechanical force and motion (motor proteins). Allosteric proteins can also harness energy derived from ATP hydrolysis, ion gradients, or electron-transport processes to pump specific ions or small molecules across a membrane. We consider one example here that will be discussed in more detail in Chapter 11.

The ABC transporters (**ATP-b**inding **c**assette transporters) constitute an important class of membrane-bound pump proteins. In humans, at least 48 different genes encode them. These transporters mostly function to export hydrophobic molecules from the cytoplasm, serving to remove toxic molecules at the mucosal surface of the intestinal tract, for example, or at the blood–brain barrier. The study of ABC transporters is of intense interest in clinical medicine, because the overproduction of proteins in this class contributes to the resistance of tumor cells to chemotherapeutic drugs. In bacteria, the same types of proteins primarily function to import essential nutrients into the cell.

A typical ABC transporter contains a pair of membrane-spanning domains linked to a pair of ATP-binding domains located just below the plasma membrane. As in other examples we have discussed, the hydrolysis of the bound ATP molecules drives conformational changes in the protein, transmitting forces that cause the transporter to move its bound molecule across the lipid bilayer (**Figure 3–76**).

Humans have invented many different types of mechanical pumps, and it should not be surprising that cells also contain membrane-bound pumps that

Figure 3–76 The ABC (ATP-binding cassette) transporter, a protein machine that pumps molecules through a membrane. (A) How this large family of transporters pumps molecules into the cell in bacteria. As indicated, the binding of two molecules of ATP causes the two ATP-binding domains to clamp together tightly, opening a channel to the cell exterior. The binding of a substrate molecule to the extracellular face of the protein complex then triggers ATP hydrolysis followed by ADP release, which opens the cytoplasmic gate; the pump is then reset for another cycle. (B) As discussed in Chapter 11, in eukaryotes an opposite process occurs, causing selected substrate molecules to be pumped out of the cell. (C) The structure of a bacterial ABC transporter (see Movie 11.5). (C, from R.J. Dawson and K.P. Locher, *Nature* 443:180–185, published 2006 by Nature Publishing Group. Reproduced with permission of SNCSC; PDB code: 2HYD).

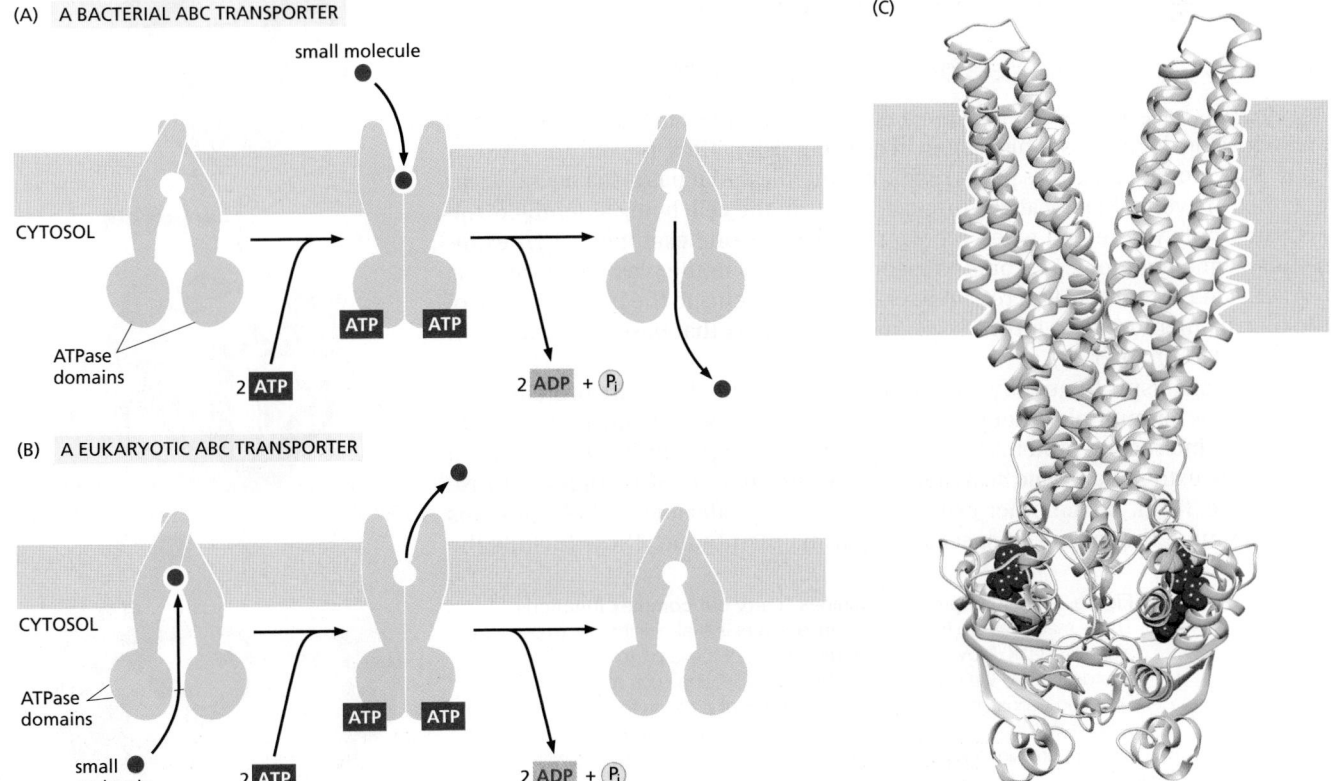

(A)   A BACTERIAL ABC TRANSPORTER

small molecule

CYTOSOL

ATPase domains

2 ATP   2 ADP + Pᵢ

(B)   A EUKARYOTIC ABC TRANSPORTER

CYTOSOL

ATPase domains

small molecule

2 ATP   2 ADP + Pᵢ

(C)

function in other ways. Among the most notable are the rotary pumps that couple the hydrolysis of ATP to the transport of H$^+$ ions (protons). These pumps resemble miniature turbines, and they are used to acidify the interior of lysosomes and other eukaryotic organelles. Like other ion pumps that create ion gradients, they can function in reverse to catalyze the reaction ADP + P$_i$ → ATP, if the gradient across their membrane of the ion that they transport is steep enough.

One such pump, the ATP synthase, harnesses a gradient of proton concentration produced by electron-transport processes to produce most of the ATP used in the living world. This ubiquitous pump has a central role in energy conversion, and we shall discuss its three-dimensional structure and mechanism in Chapter 14.

Proteins Often Form Large Complexes That Function as Protein Machines

Large proteins formed from many domains are able to perform more elaborate functions than small, single-domain proteins. But large protein assemblies formed from many protein molecules linked together by noncovalent bonds perform the most impressive tasks. Now that it is possible to reconstruct most biological processes in cell-free systems in the laboratory, it is clear that each of the central processes in a cell—such as DNA replication, protein synthesis, vesicle budding, or transmembrane signaling—is catalyzed by a highly coordinated, linked set of 10 or more proteins. In most such *protein machines*, an energetically favorable reaction such as the hydrolysis of bound nucleoside triphosphates (ATP or GTP) drives an ordered series of conformational changes in one or more of the individual protein subunits, enabling the ensemble of proteins to move coordinately. In this way, each enzyme can be moved directly into position, as the machine catalyzes successive reactions in a series (**Figure 3–77**). This is what occurs, for example, in protein synthesis on a ribosome (discussed in Chapter 6)—or in DNA replication, where a large multiprotein complex moves rapidly along the DNA (discussed in Chapter 5).

Cells have evolved protein machines for the same reason that humans have invented mechanical and electronic machines. For accomplishing almost any task, manipulations that are spatially and temporally coordinated through linked processes are much more efficient than the use of many separate tools.

Scaffolds Concentrate Sets of Interacting Proteins

As scientists have learned more of the details of cell biology, they have recognized an increasing degree of sophistication in cell chemistry. Thus, not only do we now know that protein machines play a predominant role, but it has also become clear that they are very often localized to specific sites in the cell, being assembled and activated only where and when they are needed. As one example, when extracellular signaling molecules bind to receptor proteins in the plasma membrane, the activated receptors often recruit a set of other proteins to the inside surface of the plasma membrane to form a large protein complex that passes the signal on (discussed in Chapter 15).

The mechanisms frequently involve **scaffold proteins**. These are proteins with binding sites for multiple other proteins, and they serve both to link together specific sets of interacting proteins and to position them at specific locations inside a cell. At one extreme are rigid scaffolds, such as the cullin in SCF ubiquitin ligase (see Figure 3–71). At the other extreme are the large, flexible scaffold proteins that often underlie regions of specialized plasma membrane. These include the

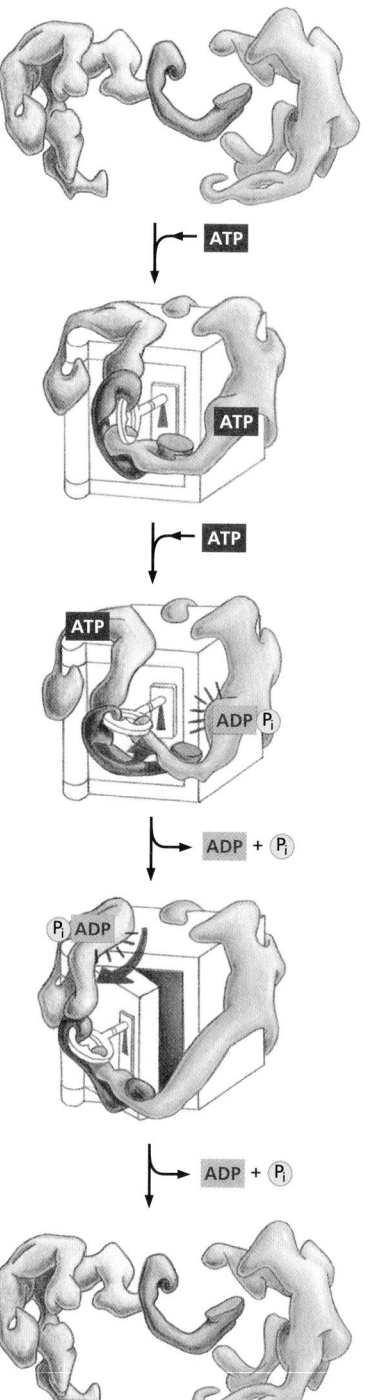

Figure 3–77 How "protein machines" carry out complex functions. These machines are made of individual proteins that collaborate to perform a specific task (**Movie 3.13**). The movement of these proteins is often coordinated by the hydrolysis of a bound nucleotide such as ATP or GTP. Directional allosteric conformational changes of proteins that are driven in this way often occur in a large protein assembly in which the activities of several different protein molecules are coordinated by such movements within the complex.

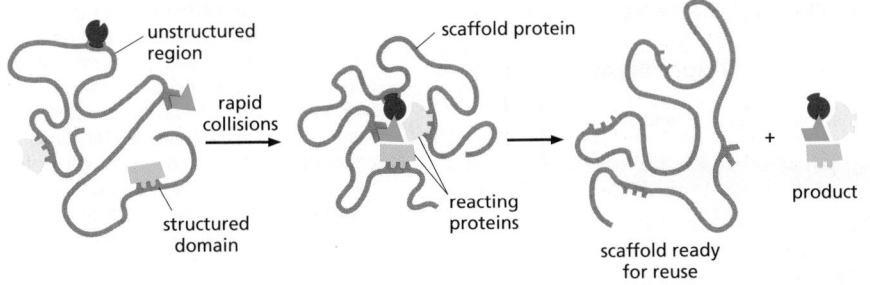

Figure 3–78 **How the proximity created by scaffold proteins can greatly speed reactions in a cell.** In this example, long unstructured regions of polypeptide chain in a large scaffold protein connect a series of structured domains that bind a set of reacting proteins. The unstructured regions serve as flexible "tethers" that greatly speed reaction rates by causing a rapid, random collision of all of the proteins that are bound to the scaffold. (For specific examples of protein tethering, see Figure 3–54 and Figure 16–18; for scaffold RNA molecules, see Figure 7–49B.)

Discs-large protein (Dlg), a protein of about 900 amino acids that is concentrated in special regions beneath the plasma membrane in epithelial cells and at synapses. Dlg contains binding sites for at least seven other proteins, interspersed with regions of more flexible polypeptide chain. An ancient protein, conserved in organisms as diverse as sponges, worms, flies, and humans, Dlg derives its name from the mutant phenotype of the organism in which it was first discovered; the cells in the imaginal discs of a *Drosophila* embryo with a mutation in the *Dlg* gene fail to stop proliferating when they should, and they produce unusually large discs whose epithelial cells can form tumors.

Although incompletely studied, Dlg and a large number of similar scaffold proteins are thought to function like the protein that is schematically illustrated in **Figure 3–78**. By binding a specific set of interacting proteins, these scaffolds can enhance the rate of critical reactions, while also confining them to the particular region of the cell that contains the scaffold. For similar reasons, cells also make extensive use of *scaffold RNA molecules*, as discussed in Chapter 7.

Many Proteins Are Controlled by Covalent Modifications That Direct Them to Specific Sites Inside the Cell

We have thus far described only a few ways in which proteins are post-translationally modified. A large number of other such modifications also occur, more than 200 distinct types being known. To give a sense of the variety, **Table 3–3** presents

TABLE 3–3 **Some Molecules Covalently Attached to Proteins Regulate Protein Function**

Modifying group	Some prominent functions
Phosphate on Ser, Thr, or Tyr	Drives the assembly of a protein into larger complexes (see Figure 15–11)
Methyl on Lys	Helps to create distinct regions in chromatin through forming either mono-, di-, or trimethyl lysine in histones (see Figure 4–36)
Acetyl on Lys	Helps to activate genes in chromatin by modifying histones (see Figure 4–33)
Palmityl group on Cys	This fatty acid addition drives protein association with membranes (see Figure 10–18)
N-acetylglucosamine on Ser or Thr	Controls enzyme activity and gene expression in glucose homeostasis
Ubiquitin on Lys	Monoubiquitin addition regulates the transport of membrane proteins in vesicles (see Figure 13–50)
	A polyubiquitin chain targets a protein for degradation (see Figure 3–70)

Ubiquitin is a 76-amino-acid polypeptide; there are at least 10 other ubiquitin-related proteins in mammalian cells.

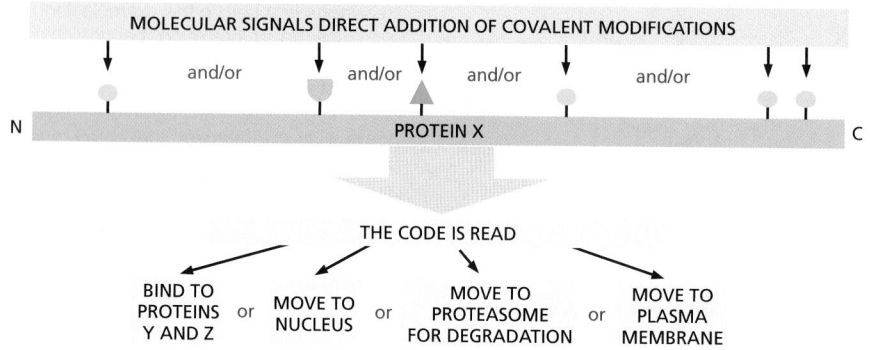

(A) A SPECTRUM OF COVALENT MODIFICATIONS PRODUCES A REGULATORY PROTEIN CODE

MOLECULAR SIGNALS DIRECT ADDITION OF COVALENT MODIFICATIONS

and/or   and/or   and/or   and/or

N   PROTEIN X   C

THE CODE IS READ

BIND TO PROTEINS Y AND Z   or   MOVE TO NUCLEUS   or   MOVE TO PROTEASOME FOR DEGRADATION   or   MOVE TO PLASMA MEMBRANE

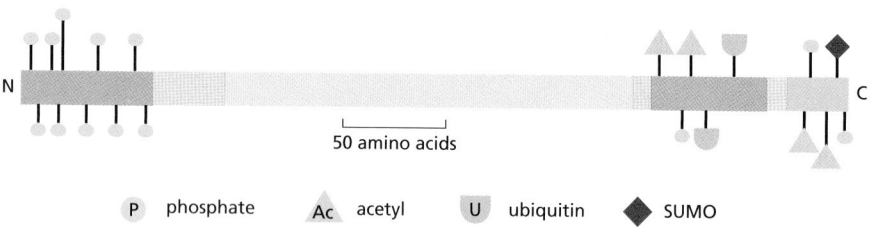

(B)   SOME KNOWN MODIFICATIONS OF PROTEIN p53

N   C

50 amino acids

P phosphate   Ac acetyl   U ubiquitin   ◆ SUMO

Figure 3–79 Multisite protein modification and its effects. (A) A protein that carries a post-translational addition to more than one of its amino acid side chains can be considered to carry a combinatorial regulatory code. Multisite modifications are added to (and removed from) a protein through signaling networks, and the resulting combinatorial regulatory code on the protein is read to alter its behavior in the cell. (B) The pattern of some covalent modifications to the protein p53.

a few of the modifying groups with known regulatory roles. As in phosphate and ubiquitin additions described previously, these groups are added and then removed from proteins according to the needs of the cell.

A large number of proteins are now known to be modified on more than one amino acid side chain, with different regulatory events producing a different pattern of such modifications. A striking example is the protein p53, which plays a central part in controlling a cell's response to adverse circumstances (see Figure 17–62). Through one of four different types of molecular additions, this protein can be modified at 20 different sites. Because an enormous number of different combinations of these 20 modifications are possible, the protein's behavior can in principle be altered in a huge number of ways. Such modifications will often create a site on the modified protein that binds it to a scaffold protein in a specific region of the cell, thereby connecting it—via the scaffold—to the other proteins required for a reaction at that site.

One can view each protein's set of covalent modifications as a *combinatorial regulatory code*. Specific modifying groups are added to or removed from a protein in response to signals, and the code then alters protein behavior—changing the activity or stability of the protein, its binding partners, and/or its specific location within the cell (**Figure 3–79**). As a result, the cell is able to respond rapidly and with great versatility to changes in its condition or environment.

A Complex Network of Protein Interactions Underlies Cell Function

There are many challenges facing cell biologists in this information-rich era when a large number of complete genome sequences are known. One is the need to dissect and reconstruct each one of the thousands of protein machines that exist in an organism such as ourselves. To understand these remarkable protein complexes, each will need to be reconstituted from its purified protein parts, so that we can study its detailed mode of operation under controlled conditions in a test tube, free from all other cell components. This alone is a massive task. But we now know that each of these subcomponents of a cell also interacts with other sets of macromolecules, creating a large network of protein–protein and protein–nucleic acid interactions throughout the cell. To understand the cell, therefore, we will need to analyze most of these other interactions as well.

We can gain some idea of the complexity of intracellular protein networks from a particularly well-studied example described in Chapter 16: the many dozens of proteins that interact with the actin cytoskeleton to control actin filament behavior (see Panel 16–3, p. 905).

The extent of such protein–protein interactions can also be estimated more generally. An enormous amount of valuable information is now freely available in protein databases on the Internet: tens of thousands of three-dimensional protein structures plus tens of millions of protein sequences derived from the nucleotide sequences of genes. Scientists have been developing new methods for mining this great resource to increase our understanding of cells. In particular, computer-based bioinformatics tools are being combined with robotics and other technologies to allow thousands of proteins to be investigated in a single set of experiments. **Proteomics** is a term that is often used to describe such research focused on the analysis of large sets of proteins, analogous to the term *genomics* describing the large-scale analysis of DNA sequences and genes.

A biochemical method based on affinity tagging and mass spectroscopy has proven especially powerful for determining the direct binding interactions between the many different proteins in a cell (discussed in Chapter 8). The results are being tabulated and organized in Internet databases. This allows a cell biologist studying a small set of proteins to readily discover which other proteins in the same cell are likely to bind to, and thus interact with, that set of proteins. When displayed graphically as a *protein interaction map*, each protein is represented by a box or dot in a two-dimensional network, with a straight line connecting those proteins that have been found to bind to each other.

When hundreds or thousands of proteins are displayed on the same map, the network diagram becomes bewilderingly complicated, serving to illustrate the enormous challenges that face scientists attempting to understand the cell (**Figure 3–80**). Much more useful are small subsections of these maps, centered on a few proteins of interest.

We have previously described the structure and mode of action of the SCF ubiquitin ligase, using it to illustrate how protein complexes are constructed from interchangeable parts (see Figure 3–71). **Figure 3–81** shows a network of protein–protein interactions for the five proteins that form this protein complex in a yeast cell. Four of the subunits of this ligase are located at the bottom right of this figure. The remaining subunit, the F-box protein that serves as its substrate-binding arm, appears as a set of 15 different gene products that bind to adaptor protein 2 (the Skp1 protein). Along the top and left of the figure are sets of additional protein interactions marked with *yellow* and *green* shading: as indicated, these protein sets function at the origin of DNA replication, in cell cycle regulation, in methionine synthesis, in the kinetochore, and in vacuolar H⁺-ATPase assembly. We shall use this figure to explain how such protein interaction maps are used, and what they do and do not mean.

1. Protein interaction maps are useful for identifying the likely function of previously uncharacterized proteins. Examples are the products of the genes that have thus far only been inferred to exist from the yeast genome sequence, which are the three proteins in the figure that lack a simple three-letter abbreviation (*white letters* beginning with Y). The three in this diagram are F-box proteins that bind to Skp1; these are therefore likely to function as part of the ubiquitin ligase, serving as substrate-binding arms that recognize different target proteins. However, as we discuss next, neither assignment can be considered certain without additional data.

2. Protein interaction networks need to be interpreted with caution because, as a result of evolution making efficient use of each organism's genetic information, the same protein can be used as part of different protein complexes that have different types of functions. Thus, although protein A binds to protein B and protein B binds to protein C, proteins A and C need not function in the same process. For example, we know from detailed biochemical studies that the functions of Skp1 in the kinetochore and in

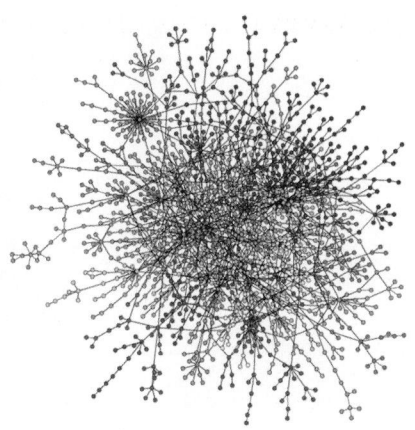

Figure 3–80 A network of protein-binding interactions in a yeast cell. Each line connecting a pair of dots (proteins) indicates a protein–protein interaction. (From A. Guimerá and M. Sales–Pardo, *Mol. Syst. Biol.* 2:42, 2006. With permission from John Wiley and Sons.)

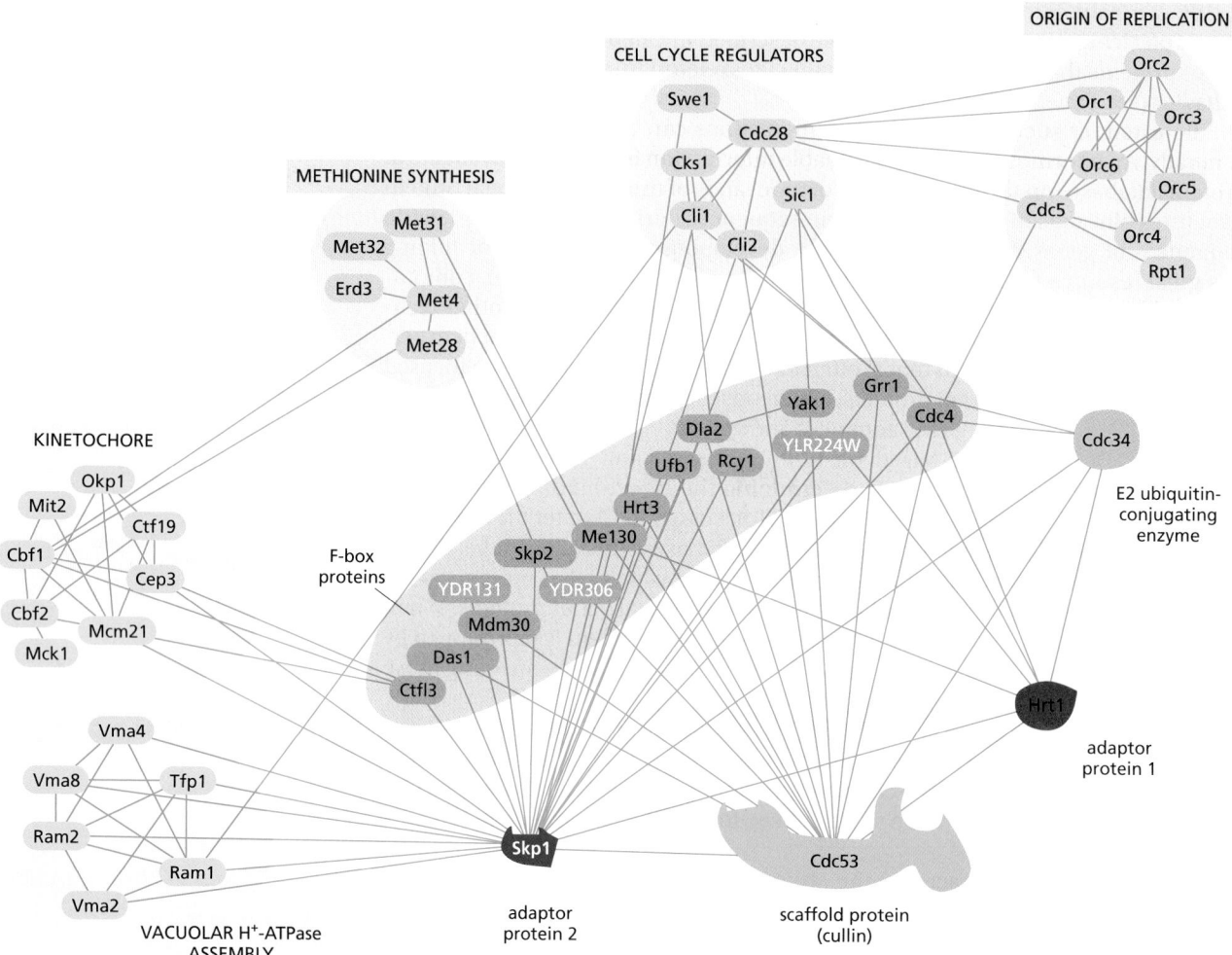

Figure 3–81 A map of some protein–protein interactions of the SCF ubiquitin ligase and other proteins in the yeast *S. cerevisiae.* The symbols and/or colors used for the five proteins of the ligase are those in Figure 3–71. Note that 15 different F-box proteins are shown *(purple)*; those with *white* lettering (beginning with Y) are known from the genome sequence as open reading frames. For additional details, see text. (Courtesy of Peter Bowers and David Eisenberg, UCLA-DOE Institute for Genomics and Proteomics, UCLA.)

vacuolar H⁺-ATPase assembly (*yellow shading*) are separate from its function in the SCF ubiquitin ligase. In fact, only the remaining three functions of Skp1 illustrated in the diagram—methionine synthesis, cell cycle regulation, and origin of replication (*green shading*)—involve ubiquitylation.

3. In cross-species comparisons, those proteins displaying similar patterns of interactions in the two protein interaction maps are likely to have the same function in the cell. Thus, as scientists generate more and more highly detailed maps for multiple organisms, the results will become increasingly useful for inferring protein function. These map comparisons will be a particularly powerful tool for deciphering the functions of human proteins, because a vast amount of direct information about protein function can be obtained from genetic engineering, mutational, and genetic analyses in experimental organisms—such as yeast, worms, and flies—that are not feasible in humans.

What does the future hold? There are likely to be on the order of 10,000 different proteins in a typical human cell, each of which interacts with 5 to 10 different partners. Despite the enormous progress made in recent years, we cannot yet claim to understand even the simplest known cells, such as the small *Mycoplasma* bacterium formed from only about 500 gene products (see Figure 1–10). How then can we hope to understand a human? Clearly, a great deal of new biochemistry will be essential, in which each protein in a particular interacting set is purified so that its chemistry and interactions can be dissected in a test tube. But in addition, more powerful ways of analyzing networks will be needed based on mathematical and computational tools not yet invented, as we shall emphasize in Chapter 8. Clearly, there are many wonderful challenges that remain for future generations of cell biologists.

Summary

Proteins can form enormously sophisticated chemical devices, whose functions largely depend on the detailed chemical properties of their surfaces. Binding sites for ligands are formed as surface cavities in which precisely positioned amino acid side chains are brought together by protein folding. In this way, normally unreactive amino acid side chains can be activated to make and break covalent bonds. Enzymes are catalytic proteins that greatly speed up reaction rates by binding the high-energy transition states for a specific reaction path; they also can perform acid catalysis and base catalysis simultaneously. The rates of enzyme reactions are often so fast that they are limited only by diffusion. Rates can be further increased only if enzymes that act sequentially on a substrate are joined into a single multienzyme complex, or if the enzymes and their substrates are attached to protein scaffolds, or otherwise confined to the same part of the cell.

Proteins reversibly change their shape when ligands bind to their surface. The allosteric changes in protein conformation produced by one ligand affect the binding of a second ligand, and this linkage between two ligand-binding sites provides a crucial mechanism for regulating cell processes. Metabolic pathways, for example, are controlled by feedback regulation: some small molecules inhibit and other small molecules activate enzymes early in a pathway. Enzymes controlled in this way generally form symmetric assemblies, allowing cooperative conformational changes to create a steep response to changes in the concentrations of the ligands that regulate them.

The expenditure of chemical energy can drive unidirectional changes in protein shape. By coupling allosteric shape changes to ATP hydrolysis, for example, proteins can do useful work, such as generating a mechanical force or moving for long distances in a single direction. The three-dimensional structures of proteins have revealed how a small local change caused by nucleoside triphosphate hydrolysis is amplified to create major changes elsewhere in the protein. By such means, these proteins can serve as input–output devices that transmit information, as assembly factors, as motors, or as membrane-bound pumps. Highly efficient protein machines are formed by incorporating many different protein molecules into larger assemblies that coordinate the allosteric movements of the individual components. Such machines perform most of the important reactions in cells.

Proteins are subjected to many reversible, post-translational modifications, such as the covalent addition of a phosphate or an acetyl group to a specific amino acid side chain. The addition of these modifying groups is used to regulate the activity of a protein, changing its conformation, its binding to other proteins, and its location inside the cell. A typical protein in a cell will interact with more than five different partners. Through proteomics, biologists can analyze thousands of proteins in one set of experiments. One important result is the production of detailed protein interaction maps, which aim at describing all of the binding interactions between the thousands of distinct proteins in a cell. However, understanding these networks will require new biochemistry, through which small sets of interacting proteins can be purified and their chemistry dissected in detail. In addition, new computational techniques will be required to deal with the enormous complexity.

WHAT WE DON'T KNOW

• What are the functions of the surprisingly large amount of unfolded polypeptide chain found in proteins?

• How many types of protein functions remain to be discovered? What are the most promising approaches for discovering them?

• When will scientists be able to take any amino acid sequence and accurately predict both that protein's three-dimensional conformations and its chemical properties? What breakthroughs will be needed to accomplish this important goal?

• Are there ways to reveal the detailed workings of a protein machine that do not require the purification of each of its component parts in large amounts, so that the machine's functions can be reconstituted and dissected using chemical techniques in a test tube?

• What are the roles of the dozens of different types of covalent modifications of proteins that have been found in addition to those listed in Table 3–3? Which ones are critical for cell function and why?

• Why is amyloid toxic to cells and how does it contribute to neurodegenerative diseases such as Alzheimer's disease?

PROBLEMS

Which statements are true? Explain why or why not.

3–1   Each strand in a β sheet is a helix with two amino acids per turn.

3–2   Intrinsically disordered regions of proteins can be identified using bioinformatic methods to search genes for encoded amino acid sequences that possess high hydrophobicity and low net charge.

3–3   Loops of polypeptide that protrude from the surface of a protein often form the binding sites for other molecules.

3–4   An enzyme reaches a maximum rate at high substrate concentration because it has a fixed number of active sites where substrate binds.

3–5   Higher concentrations of enzyme give rise to a higher turnover number.

3–6   Enzymes that undergo cooperative allosteric transitions invariably consist of symmetric assemblies of multiple subunits.

3–7   Continual addition and removal of phosphates by protein kinases and protein phosphatases is wasteful of energy—since their combined action consumes ATP—but it is a necessary consequence of effective regulation by phosphorylation.

Discuss the following problems.

3–8   Consider the following statement. "To produce one molecule of each possible kind of polypeptide chain, 300 amino acids in length, would require more atoms than exist in the universe." Given the size of the universe, do you suppose this statement could possibly be correct? Since counting atoms is a tricky business, consider the problem from the standpoint of mass. The mass of the observable universe is estimated to be about 10^{80} grams, give or take an order of magnitude or so. Assuming that the average mass of an amino acid is 110 daltons, what would be the mass of one molecule of each possible kind of polypeptide chain 300 amino acids in length? Is this greater than the mass of the universe?

3–9   A common strategy for identifying distantly related homologous proteins is to search the database using a short signature sequence indicative of the particular protein function. Why is it better to search with a short sequence than with a long sequence? Do you not have more chances for a "hit" in the database with a long sequence?

3–10   The so-called kelch motif consists of a four-stranded β sheet shaped like the blade of a propeller. It is usually found to be repeated four to seven times, forming a β propeller, or kelch repeat domain, in a multidomain protein. One such kelch repeat domain is shown in **Figure Q3–1**. Would you classify this domain as an "in-line" or "plug-in" type domain?

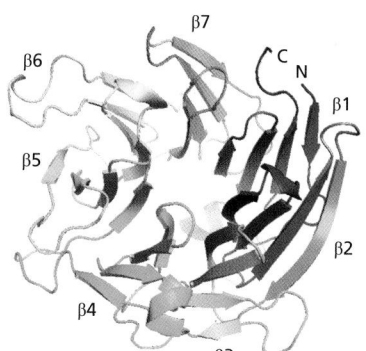

Figure Q3–1 The kelch repeat domain of galactose oxidase from *D. dendroides* (Problem 3–10). The seven individual blades of the β propeller are *color coded* and labeled. The N- and C-termini are indicated by N and C.

3–11   Titin, which has a molecular weight of about 3×10^6, is the largest polypeptide yet described. Titin molecules extend from muscle thick filaments to the Z disc; they are thought to act as springs to keep the thick filaments centered in the sarcomere. Titin is composed of a large number of repeated immunoglobulin (Ig) sequences of 89 amino acids, each of which is folded into a domain about 4 nm in length (**Figure Q3–2A**).

You suspect that the springlike behavior of titin is caused by the sequential unfolding (and refolding) of individual Ig domains. You test this hypothesis using the atomic force microscope, which allows you to pick up one end of a protein molecule and pull with an accurately measured force. For a fragment of titin containing seven repeats of the Ig domain, this experiment gives the sawtooth force-versus-extension curve shown in Figure Q3–2B. If the experiment is repeated in a solution of 8 M urea (a protein denaturant), the peaks disappear and the measured extension becomes much longer for a given force. If the experiment is repeated after the protein has been cross-linked by treatment with glutaraldehyde, once again the peaks disappear but the extension becomes much smaller for a given force.

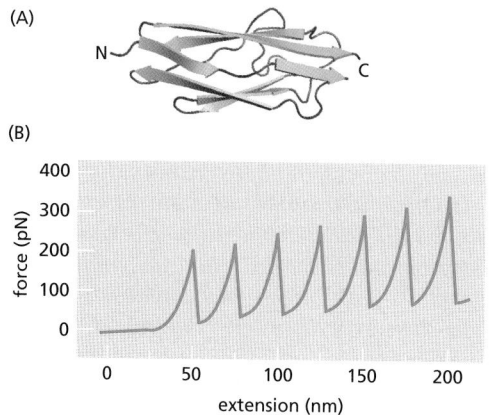

Figure Q3–2 Springlike behavior of titin (Problem 3–11). (A) The structure of an individual Ig domain. (B) Force in piconewtons versus extension in nanometers obtained by atomic force microscopy.

A.   Are the data consistent with your hypothesis that titin's springlike behavior is due to the sequential unfolding of individual Ig domains? Explain your reasoning.

B.   Is the extension for each putative domain-unfolding event the magnitude you would expect? (In an extended polypeptide chain, amino acids are spaced at intervals of 0.34 nm.)

C.   Why is each successive peak in Figure Q3–2B a little higher than the one before?

D.   Why does the force collapse so abruptly after each peak?

3–12   Rous sarcoma virus (RSV) carries an oncogene called *Src*, which encodes a continuously active protein tyrosine kinase that leads to unchecked cell proliferation. Normally, Src carries an attached fatty acid (myristoylate) group that allows it to bind to the cytoplasmic side of the plasma membrane. A mutant version of Src that does not allow attachment of myristoylate does not bind to the membrane. Infection of cells with RSV encoding either the normal or the mutant form of Src leads to the same high level of protein tyrosine kinase activity, but the mutant Src does not cause cell proliferation.

A.   Assuming that the normal Src is all bound to the plasma membrane and that the mutant Src is distributed throughout the cytoplasm, calculate their relative concentrations in the neighborhood of the plasma membrane. For the purposes of this calculation, assume that the cell is a sphere with a radius (r) of 10 μm and that the mutant Src is distributed throughout the cell, whereas the normal Src is confined to a 4-nm-thick layer immediately beneath the membrane. [For this problem, assume that the membrane has no thickness. The volume of a sphere is $(4/3)\pi r^3$.]

B.   The target (X) for phosphorylation by Src resides in the membrane. Explain why the mutant Src does not cause cell proliferation.

3–13   An antibody binds to another protein with an equilibrium constant, K, of 5×10^9 M^{-1}. When it binds to a second, related protein, it forms three fewer hydrogen bonds, reducing its binding affinity by 11.9 kJ/mole. What is the K for its binding to the second protein? (Free-energy change is related to the equilibrium constant by the equation $\Delta G° = -2.3\ RT \log K$, where R is 8.3×10^{-3} kJ/(mole K) and T is 310 K.)

3–14   The protein SmpB binds to a special species of tRNA, tmRNA, to eliminate the incomplete proteins made from truncated mRNAs in bacteria. If the binding of SmpB to tmRNA is plotted as fraction tmRNA bound versus SmpB concentration, one obtains a symmetrical S-shaped curve as shown in **Figure Q3–3**. This curve is a visual display of a very useful relationship between K_d and concentration, which has broad applicability. The general expression for fraction of ligand bound is derived from the equation for K_d ($K_d = [Pr][L]/[Pr–L]$) by substituting $([L]_{TOT} – [L])$ for $[Pr–L]$ and rearranging. Because the total concentration of ligand ($[L]_{TOT}$) is equal to the free ligand ($[L]$) plus bound ligand ($[Pr–L]$),

$$\text{fraction bound} = [Pr–L]/[L]_{TOT} = [Pr]/([Pr] + K_d)$$

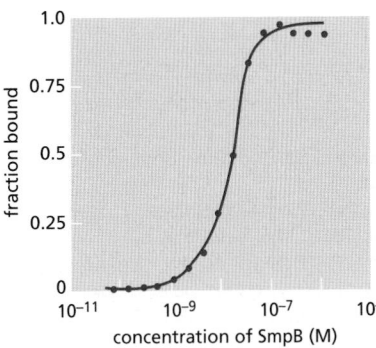

Figure Q3–3 Fraction of tmRNA bound versus SmpB concentration (Problem 3–14). (From A. W. Karzai, M. M. Susskind and R. T. Sauer, *Embo J.* 18:3793-3799, 1999. With permission from John Wiley and Sons.)

For SmpB and tmRNA, the fraction bound = [SmpB–tmRNA]/[tmRNA]$_{TOT}$ = [SmpB]/([SmpB] + K_d). Using this relationship, calculate the fraction of tmRNA bound for SmpB concentrations equal to $10^4 K_d$, $10^3 K_d$, $10^2 K_d$, $10^1 K_d$, K_d, $10^{-1} K_d$, $10^{-2} K_d$, $10^{-3} K_d$, and $10^{-4} K_d$.

3–15   Many enzymes obey simple Michaelis–Menten kinetics, which are summarized by the equation

$$\text{rate} = V_{max} [S]/([S] + K_m)$$

where V_{max} = maximum velocity, $[S]$ = concentration of substrate, and K_m = the Michaelis constant.

It is instructive to plug a few values of $[S]$ into the equation to see how rate is affected. What are the rates for $[S]$ equal to zero, equal to K_m, and equal to infinite concentration?

3–16   The enzyme hexokinase adds a phosphate to D-glucose but ignores its mirror image, L-glucose. Suppose that you were able to synthesize hexokinase entirely from D-amino acids, which are the mirror image of the normal L-amino acids.

A.   Assuming that the "D" enzyme would fold to a stable conformation, what relationship would you expect it to bear to the normal "L" enzyme?

B.   Do you suppose the "D" enzyme would add a phosphate to L-glucose, and ignore D-glucose?

3–17   How do you suppose that a molecule of hemoglobin is able to bind oxygen efficiently in the lungs, and yet release it efficiently in the tissues?

3–18   Synthesis of the purine nucleotides AMP and GMP proceeds by a branched pathway starting with ribose 5-phosphate (R5P), as shown schematically in **Figure Q3–4**. Using the principles of feedback inhibition, propose a regulatory strategy for this pathway that ensures an adequate supply of both AMP and GMP and minimizes the buildup of the intermediates (*A–I*) when supplies of AMP and GMP are adequate.

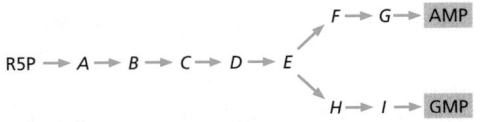

Figure Q3–4 Schematic diagram of the metabolic pathway for synthesis of AMP and GMP from R5P (Problem 3–18).

REFERENCES

General

Berg JM, Tymoczko JL & Stryer L (2011) Biochemistry, 7th ed. New York: WH Freeman.

Branden C & Tooze J (1999) Introduction to Protein Structure, 2nd ed. New York: Garland Science.

Dickerson RE (2005) Present at the Flood: How Structural Molecular Biology Came About. Sunderland, MA: Sinauer.

Kuriyan J, Konforti B & Wemmer D (2013) The Molecules of Life: Physical and Chemical Principles. New York: Garland Science.

Perutz M (1992) Protein Structure: New Approaches to Disease and Therapy. New York: WH Freeman.

Petsko GA & Ringe D (2004) Protein Structure and Function. London: New Science Press.

Williamson M (2011) How Proteins Work. New York: Garland Science.

The Shape and Structure of Proteins

Anfinsen CB (1973) Principles that govern the folding of protein chains. Science 181, 223–230.

Caspar DLD & Klug A (1962) Physical principles in the construction of regular viruses. Cold Spring Harb. Symp. Quant. Biol. 27, 1–24.

Dill KA & MacCallum JL (2012) The protein-folding problem, 50 years on. Science 338, 1042–1046.

Eisenberg D (2003) The discovery of the alpha-helix and beta-sheet, the principal structural features of proteins. Proc. Natl Acad. Sci. USA 100, 11207–11210.

Fraenkel-Conrat H & Williams RC (1955) Reconstitution of active tobacco mosaic virus from its inactive protein and nucleic acid components. Proc. Natl Acad. Sci. USA 41, 690–698.

Goodsell DS & Olson AJ (2000) Structural symmetry and protein function. Annu. Rev. Biophys. Biomol. Struct. 29, 105–153.

Greenwald J & Riek R (2010) Biology of amyloid: structure, function, and regulation. Structure 18, 1244–1260.

Harrison SC (1992) Viruses. Curr. Opin. Struct. Biol. 2, 293–299.

Hudder A, Nathanson L & Deutscher MP (2003) Organization of mammalian cytoplasm. Mol. Cell. Biol. 23, 9318–9326.

Kato M, Han TW, Xie S et al. (2012) Cell-free formation of RNA granules: low complexity sequence domains form dynamic fibers within hydrogels. Cell 149, 753–767.

Koga N, Tatsumi-Koga R, Liu G et al. (2012) Principles for designing ideal protein structures. Nature 491, 222–227.

Li P, Banjade S, Cheng H-C et al. (2012) Phase transitions in the assembly of multivalent signalling proteins. Nature 483, 336–340.

Lindquist SL & Kelly JW (2011) Chemical and biological approaches for adapting proteostasis to ameliorate protein misfolding and aggregation diseases—progress and prognosis. Cold Spring Harb. Perspect. Biol. 3, a004507.

Maji SK, Perrin MH, Sawaya MR et al. (2009) Functional amyloids as natural storage of peptide hormones in pituitary secretory granules. Science 325, 328–332.

Nelson R, Sawaya MR, Balbirnie M et al. (2005) Structure of the cross-β spine of amyloid-like fibrils. Nature 435, 773–778.

Nomura M (1973) Assembly of bacterial ribosomes. Science 179, 864–873.

Oldfield CJ & Dunker AK (2014) Intrinsically disordered proteins and intrinsically disordered protein regions. Annu. Rev. Biochem. 83, 553–584.

Orengo CA & Thornton JM (2005) Protein families and their evolution—a structural perspective. Annu. Rev. Biochem. 74, 867–900.

Pauling L & Corey RB (1951) Configurations of polypeptide chains with favored orientations around single bonds: two new pleated sheets. Proc. Natl Acad. Sci. USA 37, 729–740.

Pauling L, Corey RB & Branson HR (1951) The structure of proteins: two hydrogen-bonded helical configurations of the polypeptide chain. Proc. Natl Acad. Sci. USA 37, 205–211.

Prusiner SB (1998) Prions. Proc. Natl Acad. Sci. USA 95, 13363–13383.

Toyama BH & Weissman JS (2011) Amyloid structure: conformational diversity and consequences. Annu. Rev. Biochem. 80, 557–585.

Zhang C & Kim SH (2003) Overview of structural genomics: from structure to function. Curr. Opin. Chem. Biol. 7, 28–32.

Protein Function

Alberts B (1998) The cell as a collection of protein machines: preparing the next generation of molecular biologists. Cell 92, 291–294.

Benkovic SJ (1992) Catalytic antibodies. Annu. Rev. Biochem. 61, 29–54.

Berg OG & von Hippel PH (1985) Diffusion-controlled macromolecular interactions. Annu. Rev. Biophys. Biophys. Chem. 14, 131–160.

Bourne HR (1995) GTPases: a family of molecular switches and clocks. Philos. Trans. R. Soc. Lond. B Biol. Sci. 349, 283–289.

Costanzo M, Baryshnikova A, Bellay J et al. (2010) The genetic landscape of a cell. Science 327, 425–431.

Deshaies RJ & Joazeiro CAP (2009) RING domain E3 ubiquitin ligases. Annu. Rev. Biochem. 78, 399–434.

Dickerson RE & Geis I (1983) Hemoglobin: Structure, Function, Evolution, and Pathology. Menlo Park, CA: Benjamin Cummings.

Fersht AR (1999) Structure and Mechanism in Protein Science: A Guide to Enzyme Catalysis and Protein Folding. New York: WH Freeman.

Haucke V, Neher E & Sigrist SJ (2011) Protein scaffolds in the coupling of synaptic exocytosis and endocytosis. Nat. Rev. Neurosci. 12, 127–138.

Hua Z & Vierstra RD (2011) The cullin-RING ubiquitin-protein ligases. Annu. Rev. Plant Biol. 62, 299–334.

Hunter T (2012) Why nature chose phosphate to modify proteins. Philos. Trans. R. Soc. Lond. B Biol. Sci. 367, 2513–2516.

Johnson LN & Lewis RJ (2001) Structural basis for control by phosphorylation. Chem. Rev. 101, 2209–2242.

Kantrowitz ER & Lipscomb WN (1988) Escherichia coli aspartate transcarbamylase: the relation between structure and function. Science 241, 669–674.

Kerscher O, Felberbaum R & Hochstrasser M (2006) Modification of proteins by ubiquitin and ubiquitin-like proteins. Annu. Rev. Cell Dev. Biol. 22, 159–180.

Kim E & Sheng M (2004) PDZ domain proteins of synapses. Nat. Rev. Neurosci. 5, 771–781.

Koshland DE Jr (1984) Control of enzyme activity and metabolic pathways. Trends Biochem. Sci. 9, 155–159.

Krogan NJ, Cagney G, Yu H et al. (2006) Global landscape of protein complexes in the yeast Saccharomyces cerevisiae. Nature 440, 637–643.

Lichtarge O, Bourne HR & Cohen FE (1996) An evolutionary trace method defines binding surfaces common to protein families. J. Mol. Biol. 257, 342–358.

Maier T, Leibundgut M & Ban N (2008) The crystal structure of a mammalian fatty acid synthase. Science 321, 1315–1322.

Monod J, Changeux JP & Jacob F (1963) Allosteric proteins and cellular control systems. J. Mol. Biol. 6, 306–329.

Perutz M (1990) Mechanisms of Cooperativity and Allosteric Regulation in Proteins. Cambridge: Cambridge University Press.

Radzicka A & Wolfenden R (1995) A proficient enzyme. Science 267, 90–93.

Schramm VL (2011) Enzymatic transition states, transition-state analogs, dynamics, thermodynamics, and lifetimes. Annu. Rev. Biochem. 80, 703–732.

Scott JD & Pawson T (2009) Cell signaling in space and time: where proteins come together and when they're apart. Science 326, 1220–1224.

Taylor SS, Keshwani MM, Steichen JM & Kornev AP (2012) Evolution of the eukaryotic protein kinases as dynamic molecular switches. Philos. Trans. R. Soc. Lond. B Biol. Sci. 367, 2517–2528.

Vale RD & Milligan RA (2000) The way things move: looking under the hood of molecular motor proteins. Science 288, 88–95.

Wilson MZ & Gitai Z (2013) Beyond the cytoskeleton: mesoscale assemblies and their function in spatial organization. Curr. Opin. Microbiol. 16, 177–183.

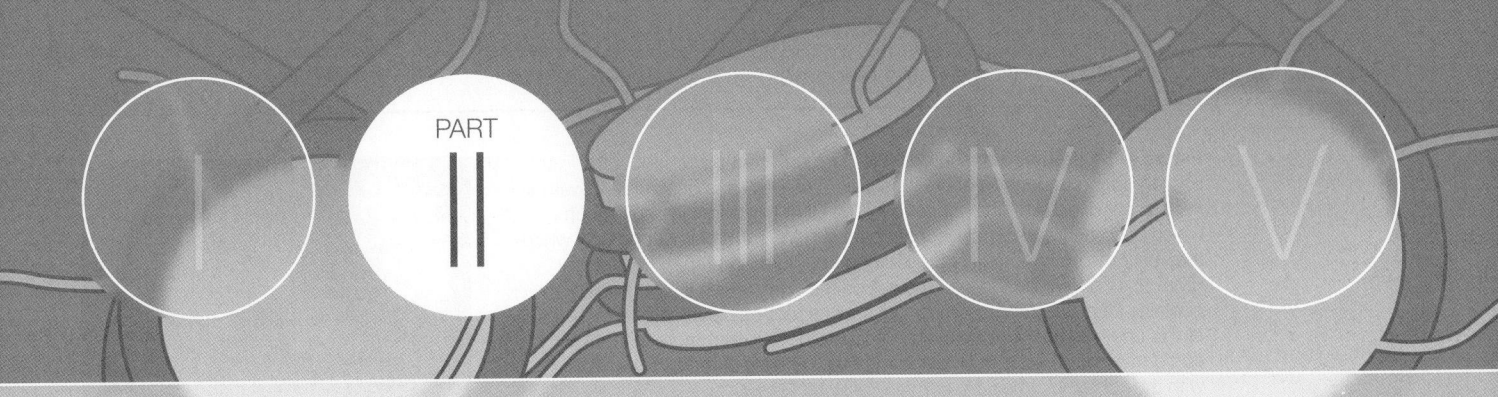

DNA, Chromosomes, and Genomes

Life depends on the ability of cells to store, retrieve, and translate the genetic instructions required to make and maintain a living organism. This *hereditary information* is passed on from a cell to its daughter cells at cell division, and from one generation of an organism to the next through the organism's reproductive cells. The instructions are stored within every living cell as its **genes**, the information-containing elements that determine the characteristics of a species as a whole and of the individuals within it.

As soon as genetics emerged as a science at the beginning of the twentieth century, scientists became intrigued by the chemical structure of genes. The information in genes is copied and transmitted from cell to daughter cell millions of times during the life of a multicellular organism, and it survives the process essentially unchanged. What form of molecule could be capable of such accurate and almost unlimited replication and also be able to exert precise control, directing multicellular development as well as the daily life of every cell? What kind of instructions does the genetic information contain? And how can the enormous amount of information required for the development and maintenance of an organism fit within the tiny space of a cell?

The answers to several of these questions began to emerge in the 1940s. At this time researchers discovered, from studies in simple fungi, that genetic information consists largely of instructions for making proteins. Proteins are phenomenally versatile macromolecules that perform most cell functions. As we saw in Chapter 3, they serve as building blocks for cell structures and form the enzymes that catalyze most of the cell's chemical reactions. They also regulate gene expression (Chapter 7), and they enable cells to communicate with each other (Chapter 15) and to move (Chapter 16). The properties and functions of cells and organisms are determined to a great extent by the proteins that they are able to make.

Painstaking observations of cells and embryos in the late nineteenth century had led to the recognition that the hereditary information is carried on *chromosomes*—threadlike structures in the nucleus of a eukaryotic cell that become visible by light microscopy as the cell begins to divide (**Figure 4–1**). Later, when biochemical analysis became possible, chromosomes were found to consist of deoxyribonucleic acid (DNA) and protein, with both being present in roughly the same amounts. For many decades, the DNA was thought to be merely a structural

IN THIS CHAPTER

THE STRUCTURE AND FUNCTION OF DNA

CHROMOSOMAL DNA AND ITS PACKAGING IN THE CHROMATIN FIBER

CHROMATIN STRUCTURE AND FUNCTION

THE GLOBAL STRUCTURE OF CHROMOSOMES

HOW GENOMES EVOLVE

Figure 4–1 **Chromosomes in cells.** (A) Two adjacent plant cells photographed through a light microscope. The DNA has been stained with a fluorescent dye (DAPI) that binds to it. The DNA is present in chromosomes, which become visible as distinct structures in the light microscope only when they become compact, sausage-shaped structures in preparation for cell division, as shown on the left. The cell on the right, which is not dividing, contains identical chromosomes, but they cannot be clearly distinguished at this phase in the cell's life cycle, because they are in a more extended conformation. (B) Schematic diagram of the outlines of the two cells along with their chromosomes. (A, courtesy of Peter Shaw.)

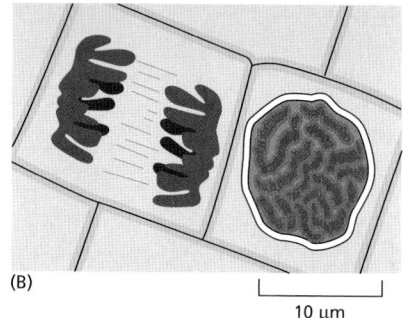

(A)   dividing cell   nondividing cell

(B)

10 μm

element. However, the other crucial advance made in the 1940s was the identification of DNA as the likely carrier of genetic information. This breakthrough in our understanding of cells came from studies of inheritance in bacteria (**Figure 4–2**). But still, as the 1950s began, both how proteins could be specified by instructions in the DNA and how this information might be copied for transmission from cell to cell seemed completely mysterious. The puzzle was suddenly solved in 1953, when James Watson and Francis Crick derived the mechanism from their model of DNA structure. As outlined in Chapter 1, the determination of the double-helical structure of DNA immediately solved the problem of how the information in this molecule might be copied, or *replicated*. It also provided the first clues as to how a molecule of DNA might use the sequence of its subunits to encode the instructions for making proteins. Today, the fact that DNA is the genetic material is so fundamental to biological thought that it is difficult to appreciate the enormous intellectual gap that was filled by this breakthrough discovery.

We begin this chapter by describing the structure of DNA. We see how, despite its chemical simplicity, the structure and chemical properties of DNA make it ideally suited as the raw material of genes. We then consider how the many proteins in chromosomes arrange and package this DNA. The packing has to be done in an orderly fashion so that the chromosomes can be replicated and apportioned correctly between the two daughter cells at each cell division. And it must also allow access to chromosomal DNA, both for the enzymes that repair DNA damage and for the specialized proteins that direct the expression of its many genes.

In the past two decades, there has been a revolution in our ability to determine the exact order of subunits in DNA molecules. As a result, we now know the

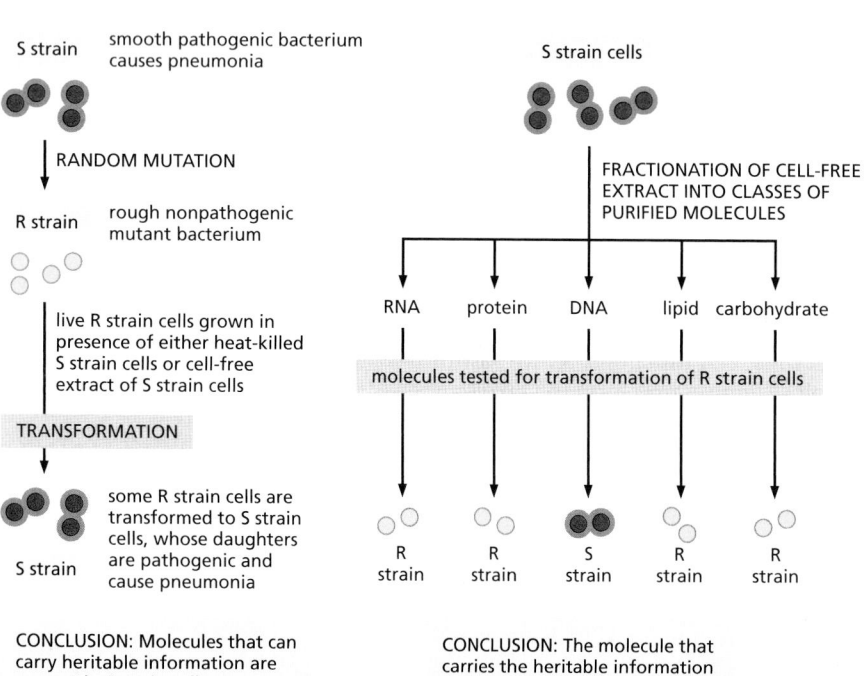

Figure 4–2 **The first experimental demonstration that DNA is the genetic material.** These experiments, carried out in the 1920s (A) and 1940s (B), showed that adding purified DNA to a bacterium changed the bacterium's properties and that this change was faithfully passed on to subsequent generations. Two closely related strains of the bacterium *Streptococcus pneumoniae* differ from each other in both their appearance under the microscope and their pathogenicity. One strain appears smooth (S) and causes death when injected into mice, and the other appears rough (R) and is nonlethal. (A) An initial experiment shows that some substance present in the S strain can change (or transform) the R strain into the S strain and that this change is inherited by subsequent generations of bacteria. (B) This experiment, in which the R strain has been incubated with various classes of biological molecules purified from the S strain, identifies the active substance as DNA.

sequence of the 3.2 billion nucleotide pairs that provide the information for producing a human adult from a fertilized egg, as well as having the DNA sequences for thousands of other organisms. Detailed analyses of these sequences are providing exciting insights into the process of evolution, and it is with this subject that the chapter ends.

This is the first of four chapters that deal with basic genetic mechanisms—the ways in which the cell maintains, replicates, and expresses the genetic information carried in its DNA. In the next chapter (Chapter 5), we shall discuss the mechanisms by which the cell accurately replicates and repairs DNA; we also describe how DNA sequences can be rearranged through the process of genetic recombination. Gene expression—the process through which the information encoded in DNA is interpreted by the cell to guide the synthesis of proteins—is the main topic of Chapter 6. In Chapter 7, we describe how this gene expression is controlled by the cell to ensure that each of the many thousands of proteins and RNA molecules encrypted in its DNA is manufactured only at the proper time and place in the life of a cell.

THE STRUCTURE AND FUNCTION OF DNA

Biologists in the 1940s had difficulty in conceiving how DNA could be the genetic material. The molecule seemed too simple: a long polymer composed of only four types of nucleotide subunits, which resemble one another chemically. Early in the 1950s, DNA was examined by x-ray diffraction analysis, a technique for determining the three-dimensional atomic structure of a molecule (discussed in Chapter 8). The early x-ray diffraction results indicated that DNA was composed of two strands of the polymer wound into a helix. The observation that DNA was double-stranded provided one of the major clues that led to the Watson–Crick model for DNA structure that, as soon as it was proposed in 1953, made DNA's potential for replication and information storage apparent.

A DNA Molecule Consists of Two Complementary Chains of Nucleotides

A **deoxyribonucleic acid** (**DNA**) molecule consists of two long polynucleotide chains composed of four types of nucleotide subunits. Each of these chains is known as a *DNA chain*, or a *DNA strand*. The chains run antiparallel to each other, and *hydrogen bonds* between the base portions of the nucleotides hold the two chains together (**Figure 4–3**). As we saw in Chapter 2 (Panel 2–6, pp. 100–101), nucleotides are composed of a five-carbon sugar to which are attached one or more phosphate groups and a nitrogen-containing base. In the case of the nucleotides in DNA, the sugar is deoxyribose attached to a single phosphate group (hence the name deoxyribonucleic acid), and the base may be either *adenine* (*A*), *cytosine* (*C*), *guanine* (*G*), or *thymine* (*T*). The nucleotides are covalently linked together in a chain through the sugars and phosphates, which thus form a "backbone" of alternating sugar–phosphate–sugar–phosphate. Because only the base differs in each of the four types of nucleotide subunit, each polynucleotide chain in DNA is analogous to a sugar-phosphate necklace (the backbone), from which hang the four types of beads (the bases A, C, G, and T). These same symbols (A, C, G, and T) are commonly used to denote either the four bases or the four entire nucleotides—that is, the bases with their attached sugar and phosphate groups.

The way in which the nucleotides are linked together gives a DNA strand a chemical polarity. If we think of each sugar as a block with a protruding knob (the 5′ phosphate) on one side and a hole (the 3′ hydroxyl) on the other (see Figure 4–3), each completed chain, formed by interlocking knobs with holes, will have all of its subunits lined up in the same orientation. Moreover, the two ends of the chain will be easily distinguishable, as one has a hole (the 3′ hydroxyl) and the other a knob (the 5′ phosphate) at its terminus. This polarity in a DNA chain is indicated by referring to one end as the *3′ end* and the other as the *5′ end*, names derived from the orientation of the deoxyribose sugar. With respect to DNA's

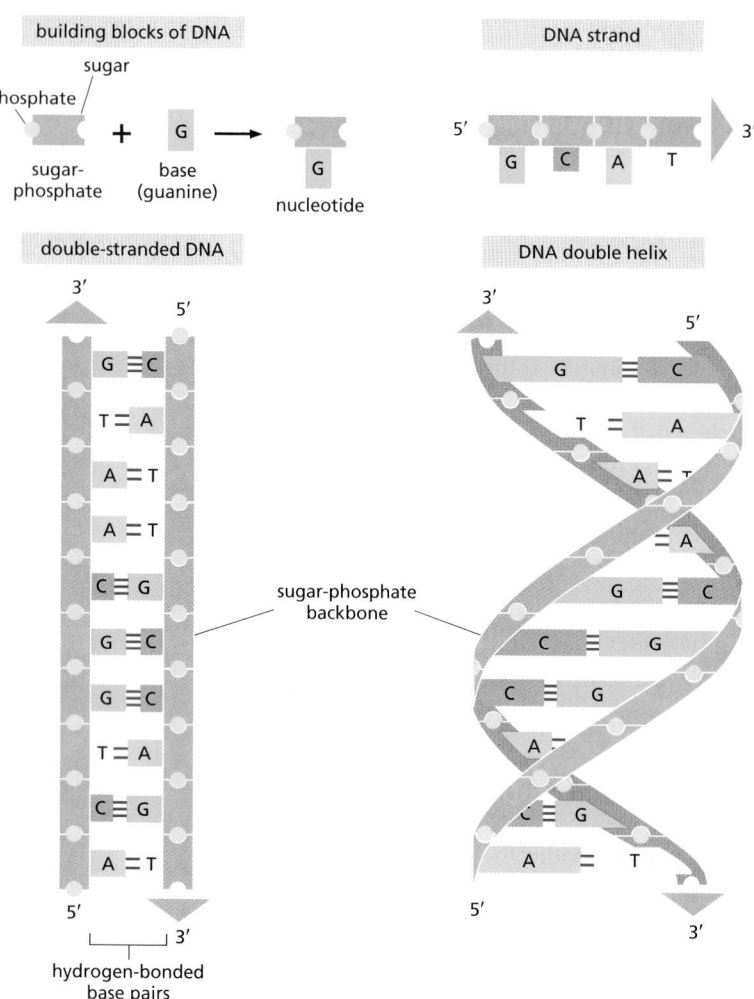

Figure 4–3 DNA and its building blocks. DNA is made of four types of nucleotides, which are linked covalently into a polynucleotide chain (a DNA strand) with a sugar-phosphate backbone from which the bases (A, C, G, and T) extend. A DNA molecule is composed of two antiparallel DNA strands held together by hydrogen bonds between the paired bases. The *arrowheads* at the ends of the DNA strands indicate the polarities of the two strands. In the diagram at the bottom left of the figure, the DNA molecule is shown straightened out; in reality, it is twisted into a double helix, as shown on the right. For details, see Figure 4–5 and Movie 4.1.

information-carrying capacity, the chain of nucleotides in a DNA strand, being both directional and linear, can be read in much the same way as the letters on this page.

The three-dimensional structure of DNA—the DNA **double helix**—arises from the chemical and structural features of its two polynucleotide chains. Because these two chains are held together by hydrogen-bonding between the bases on the different strands, all the bases are on the inside of the double helix, and the sugar-phosphate backbones are on the outside (see Figure 4–3). In each case, a bulkier two-ring base (a purine; see Panel 2–6, pp. 100–101) is paired with a single-ring base (a pyrimidine): A always pairs with T, and G with C (**Figure 4–4**). This *complementary base-pairing* enables the **base pairs** to be packed in the energetically most favorable arrangement in the interior of the double helix. In this arrangement, each base pair is of similar width, thus holding the sugar-phosphate backbones a constant distance apart along the DNA molecule. To maximize the efficiency of base-pair packing, the two sugar-phosphate backbones wind around each other to form a right-handed double helix, with one complete turn every ten base pairs (**Figure 4–5**).

The members of each base pair can fit together within the double helix only if the two strands of the helix are **antiparallel**—that is, only if the polarity of one strand is oriented opposite to that of the other strand (see Figures 4–3 and 4–4). A consequence of DNA's structure and base-pairing requirements is that each strand of a DNA molecule contains a sequence of nucleotides that is exactly **complementary** to the nucleotide sequence of its partner strand.

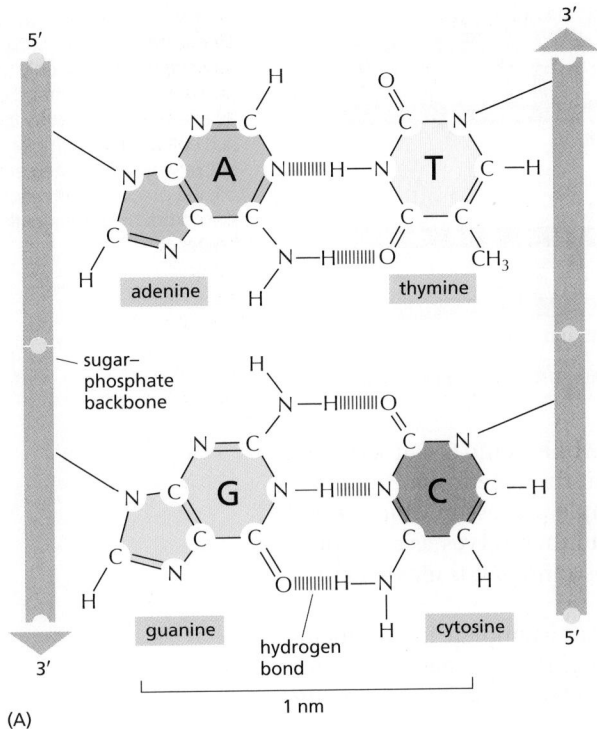

(A)

Figure 4–4 Complementary base pairs in the DNA double helix. The shapes and chemical structures of the bases allow hydrogen bonds to form efficiently only between A and T and between G and C, because atoms that are able to form hydrogen bonds (see Panel 2–3, pp. 94–95) can then be brought close together without distorting the double helix. As indicated, two hydrogen bonds form between A and T, while three form between G and C. The bases can pair in this way only if the two polynucleotide chains that contain them are antiparallel to each other.

The Structure of DNA Provides a Mechanism for Heredity

The discovery of the structure of DNA immediately suggested answers to the two most fundamental questions about heredity. First, how could the information to specify an organism be carried in a chemical form? And second, how could this information be duplicated and copied from generation to generation?

The answer to the first question came from the realization that DNA is a linear polymer of four different kinds of monomer, strung out in a defined sequence like the letters of a document written in an alphabetic script.

The answer to the second question came from the double-stranded nature of the structure: because each strand of DNA contains a sequence of nucleotides that is exactly complementary to the nucleotide sequence of its partner strand, each strand can act as a **template**, or mold, for the synthesis of a new complementary strand. In other words, if we designate the two DNA strands as S and S', strand

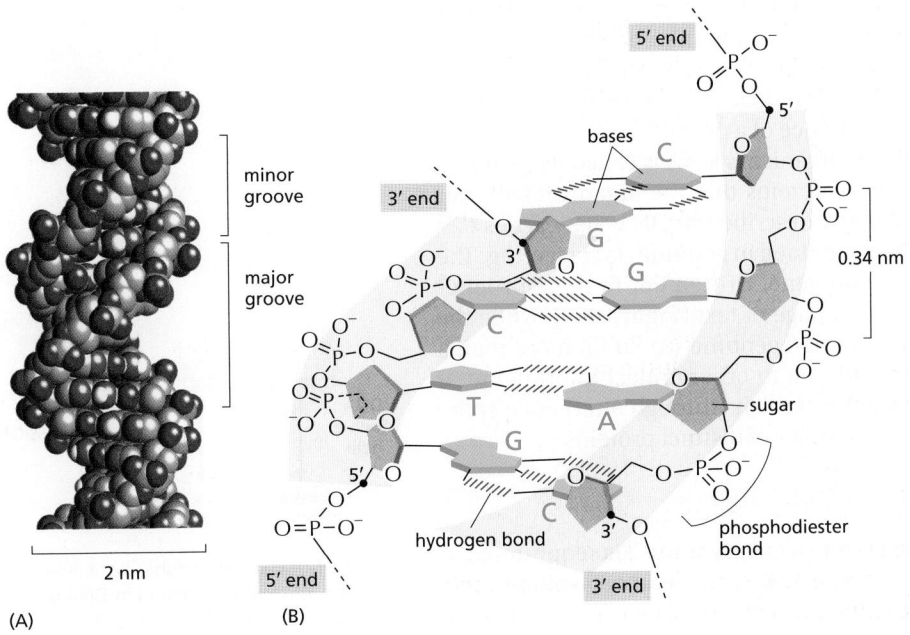

(A)

(B)

Figure 4–5 The DNA double helix.
(A) A space-filling model of 1.5 turns of the DNA double helix. Each turn of DNA is made up of 10.4 nucleotide pairs, and the center-to-center distance between adjacent nucleotide pairs is 0.34 nm. The coiling of the two strands around each other creates two grooves in the double helix: the wider groove is called the major groove, and the smaller the minor groove, as indicated. (B) A short section of the double helix viewed from its side, showing four base pairs. The nucleotides are linked together covalently by phosphodiester bonds that join the 3'-hydroxyl (–OH) group of one sugar to the 5'-hydroxyl group of the next sugar. Thus, each polynucleotide strand has a chemical polarity; that is, its two ends are chemically different. The 5' end of the DNA polymer is by convention often illustrated carrying a phosphate group, while the 3' end is shown with a hydroxyl.

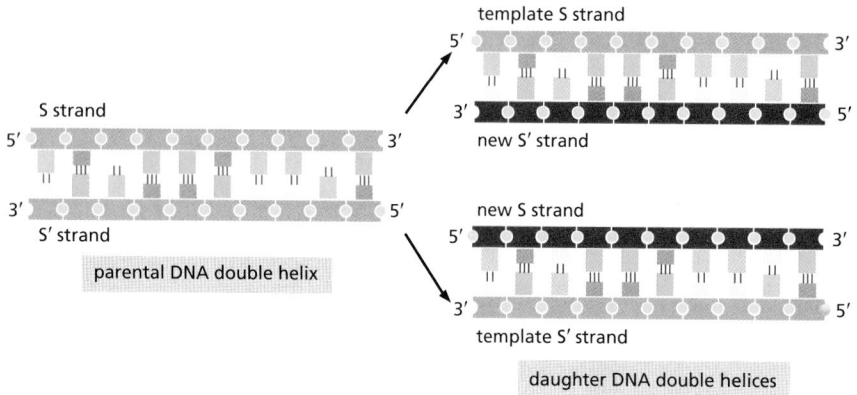

template S strand

new S' strand

new S strand

template S' strand

parental DNA double helix

daughter DNA double helices

Figure 4–6 DNA as a template for its own duplication. Because the nucleotide A successfully pairs only with T, and G pairs with C, each strand of DNA can act as a template to specify the sequence of nucleotides in its complementary strand. In this way, double-helical DNA can be copied precisely, with each parental DNA helix producing two identical daughter DNA helices.

S can serve as a template for making a new strand S′, while strand S′ can serve as a template for making a new strand S (**Figure 4–6**). Thus, the genetic information in DNA can be accurately copied by the beautifully simple process in which strand S separates from strand S′, and each separated strand then serves as a template for the production of a new complementary partner strand that is identical to its former partner.

The ability of each strand of a DNA molecule to act as a template for producing a complementary strand enables a cell to copy, or *replicate*, its genome before passing it on to its descendants. We shall describe the elegant machinery that the cell uses to perform this task in Chapter 5.

Organisms differ from one another because their respective DNA molecules have different nucleotide sequences and, consequently, carry different biological messages. But how is the nucleotide alphabet used to make messages, and what do they spell out?

As discussed above, it was known well before the structure of DNA was determined that genes contain the instructions for producing proteins. If genes are made of DNA, the DNA must therefore somehow encode proteins (**Figure 4–7**). As discussed in Chapter 3, the properties of a protein, which are responsible for its biological function, are determined by its three-dimensional structure. This structure is determined in turn by the linear sequence of the amino acids of which it is composed. The linear sequence of nucleotides in a gene must therefore somehow spell out the linear sequence of amino acids in a protein. The exact correspondence between the four-letter nucleotide alphabet of DNA and the twenty-letter amino acid alphabet of proteins—the *genetic code*—is not at all obvious from the DNA structure, and it took over a decade after the discovery of the double helix before it was worked out. In Chapter 6, we will describe this code in detail in the course of elaborating the process of *gene expression*, through which a cell converts the nucleotide sequence of a gene first into the nucleotide sequence of an RNA molecule, and then into the amino acid sequence of a protein.

The complete store of information in an organism's DNA is called its **genome**, and it specifies all the RNA molecules and proteins that the organism will ever synthesize. (The term genome is also used to describe the DNA that carries this information.) The amount of information contained in genomes is staggering. The nucleotide sequence of a very small human gene, written out in the four-letter nucleotide alphabet, occupies a quarter of a page of text (**Figure 4–8**), while the complete sequence of nucleotides in the human genome would fill more than a thousand books the size of this one. In addition to other critical information, it includes roughly 21,000 protein-coding genes, which (through alternative splicing; see p. 415) give rise to a much greater number of distinct proteins.

In Eukaryotes, DNA Is Enclosed in a Cell Nucleus

As described in Chapter 1, nearly all the DNA in a eukaryotic cell is sequestered in a nucleus, which in many cells occupies about 10% of the total cell volume. This compartment is delimited by a *nuclear envelope* formed by two concentric lipid

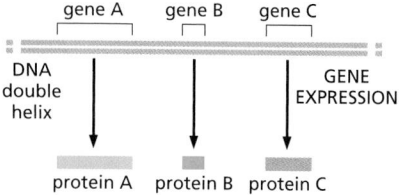

gene A   gene B   gene C

DNA double helix

GENE EXPRESSION

protein A   protein B   protein C

Figure 4–7 The relationship between genetic information carried in DNA and proteins. (Discussed in Chapter 1.)

Figure 4–8 The nucleotide sequence of the human β-globin gene. By convention, a nucleotide sequence is written from its 5′ end to its 3′ end, and it should be read from left to right in successive lines down the page as though it were normal English text. This gene carries the information for the amino acid sequence of one of the two types of subunits of the hemoglobin molecule; a different gene, the α-globin gene, carries the information for the other. (Hemoglobin, the protein that carries oxygen in the blood, has four subunits, two of each type.) Only one of the two strands of the DNA double helix containing the β-globin gene is shown; the other strand has the exact complementary sequence. The DNA sequences highlighted in *yellow* show the three regions of the gene that specify the amino acid sequence for the β-globin protein. We shall see in Chapter 6 how the cell splices these three sequences together at the level of messenger RNA in order to synthesize a full-length β-globin protein.

```
CCCTGTGGAGCCACACCCTAGGGTTGGCCA
ATCTACTCCCAGGAGCAGGGAGGGCAGGAG
CCAGGGCTGGGCATAAAAGTCAGGGCAGAG
CCATCTATTGCTTACATTTGCTTCTGACAC
AACTGTGTTCACTAGCAACTCAAACAGACA
CCATGGTGCACCTGACTCCTGAGGAGAAGT
CTGCCGTTACTGCCCTGTGGGGCAAGGTGA
ACGTGGATGAAGTTGGTGGTGAGGCCCTGG
GCAGGTTGGTATCAAGGTTACAAGACAGGT
TTAAGGAGACCAATAGAAACTGGGCATGTG
GAGACAGAGAAGACTCTTGGGTTTCTGATA
GGCACTGACTCTCTCTGCCTATTGGTCTAT
TTTCCCACCCTTAGGCTGCTGGTGGTCTAC
CCTTGGACCCAGAGGTTCTTTGAGTCCTTT
GGGGATCTGTCCACTCCTGATGCTGTTATG
GGCAACCCTAAGGTGAAGGCTCATGGCAAG
AAAGTGCTCGGTGCCTTTAGTGATGGCCTG
GCTCACCTGGACAACCTCAAGGGCACCTTT
GCCACACTGAGTGAGCTGCACTGTGACAAG
CTGCACGTGGATCCTGAGAACTTCAGGGTG
AGTCTATGGGACCCTTGATGTTTTCTTTCC
CCTTCTTTTCTATGGTTAAGTTCATGTCAT
AGGAAGGGGAGAAGTAACAGGGTACAGTTT
AGAATGGGAAACAGACGAATGATTGCATCA
GTGTGGAAGTCTCAGGATCGTTTTAGTTTC
TTTTATTTGCTGTTCATAACAATTGTTTTC
TTTTGTTTAATTCTTGCTTTCTTTTTTTTT
CTTCTCCGCAATTTTTACTATTATACTTAA
TGCCTTAACATTGTGTATAACAAAAGGAAA
TATCTCTGAGATACATTAAGTAACTTAAAA
AAAAACTTTACACAGTCTGCCTAGTACATT
ACTATTTGGAATATATGTGTGCTTATTTGC
ATATTCATAATCTCCCTACTTTATTTTCTT
TTATTTTTAATTGATACATAATCATTATAC
ATATTTATGGGTTAAAGTGTAATGTTTTAA
TATGTGTACACATATTGACCAAATCAGGGT
AATTTTGCATTTGTAATTTTTAAAAAATGCT
TTCTTCTTTTAATATACTTTTTTGTTTTATC
TTATTTCTAATACTTTCCCTAATCTCTTTC
TTTCAGGGCAATAATGATACAATGTATCAT
GCCTCTTTGCACCATTCTAAAGAATAACAG
TGATAATTTCTGGGTTAAGGCAATAGCAAT
ATTTCTGCATATAAATATTTCTGCATATAA
ATTGTAACTGATGTAAGAGGTTTCATATTG
CTAATAGCAGCTACAATCCAGCTACCATTC
TGCTTTTATTTTATGGTTGGGATAAGGCTG
GATTATTCTGAGTCCAAGCTAGGCCCTTTT
GCTAATCATGTTCATACCTCTTATCTTCCT
CCCACAGCTCCTGGGCAACGTGCTGGTCTG
TGTGCTGGCCCATCACTTTGGCAAAGAATT
CACCCCACCAGTGCAGGCTGCCTATCAGAA
AGTGGTGGCTGGTGTGGCTAATGCCCTGGC
CCACAAGTATCACTAAGCTCGCTTTCTTGC
TGTCCAATTTCTATTAAAGGTTCCTTTGTT
CCCTAAGTCCAACTACTAAACTGGGGGGATA
TTATGAAGGGCCTTGAGCATCTGGATTCTG
CCTAATAAAAAACATTTATTTTCATTGCAA
TGATGTATTTAAATTATTTCTGAATATTTT
ACTAAAAAGGGAATGTGGGAGGTCAGTGCA
TTTAAAACATAAAGAAATGATGAGCTGTTC
AAACCTTGGGAAAATACACTATATCTTAAA
CTCCATGAAAGAAGGTGAGGCTGCAACCAG
CTAATGCACATTGGCAACAGCCCCTGATGC
CTATGCCTTATTCATCCCTCAGAAAAGGAT
TCTTGTAGAGGCTTGATTTGCAGGTTAAAG
TTTTGCTATGCTGTATTTTACATTACTTAT
TGTTTTAGCTGTCCTCATGAATGTCTTTTC
```

bilayer membranes (**Figure 4–9**). These membranes are punctured at intervals by large nuclear pores, through which molecules move between the nucleus and the cytosol. The nuclear envelope is directly connected to the extensive system of intracellular membranes called the *endoplasmic reticulum*, which extend out from it into the cytoplasm. And it is mechanically supported by a network of intermediate filaments called the *nuclear lamina*—a thin feltlike mesh just beneath the inner nuclear membrane (see Figure 4–9B).

The nuclear envelope allows the many proteins that act on DNA to be concentrated where they are needed in the cell, and, as we see in subsequent chapters, it also keeps nuclear and cytosolic enzymes separate, a feature that is crucial for the proper functioning of eukaryotic cells.

Summary

Genetic information is carried in the linear sequence of nucleotides in DNA. Each molecule of DNA is a double helix formed from two complementary antiparallel strands of nucleotides held together by hydrogen bonds between G-C and A-T base pairs. Duplication of the genetic information occurs by the use of one DNA strand as a template for the formation of a complementary strand. The genetic information stored in an organism's DNA contains the instructions for all the RNA molecules and proteins that the organism will ever synthesize and is said to comprise its genome. In eukaryotes, DNA is contained in the cell nucleus, a large membrane-bound compartment.

CHROMOSOMAL DNA AND ITS PACKAGING IN THE CHROMATIN FIBER

The most important function of DNA is to carry genes, the information that specifies all the RNA molecules and proteins that make up an organism—including information about when, in what types of cells, and in what quantity each RNA molecule and protein is to be made. The nuclear DNA of eukaryotes is divided up into chromosomes, and in this section we see how genes are typically arranged on each chromosome. In addition, we describe the specialized DNA sequences that are required for a chromosome to be accurately duplicated as a separate entity and passed on from one generation to the next.

We also confront the serious challenge of DNA packaging. If the double helices comprising all 46 chromosomes in a human cell could be laid end to end, they would reach approximately 2 meters; yet the nucleus, which contains the DNA, is only about 6 μm in diameter. This is geometrically equivalent to packing 40 km (24 miles) of extremely fine thread into a tennis ball. The complex task of packaging DNA is accomplished by specialized proteins that bind to the DNA and fold it, generating a series of organized coils and loops that provide increasingly higher levels of organization, and prevent the DNA from becoming an unmanageable tangle. Amazingly, although the DNA is very tightly compacted, it nevertheless remains accessible to the many enzymes in the cell that replicate it, repair it, and use its genes to produce RNA molecules and proteins.

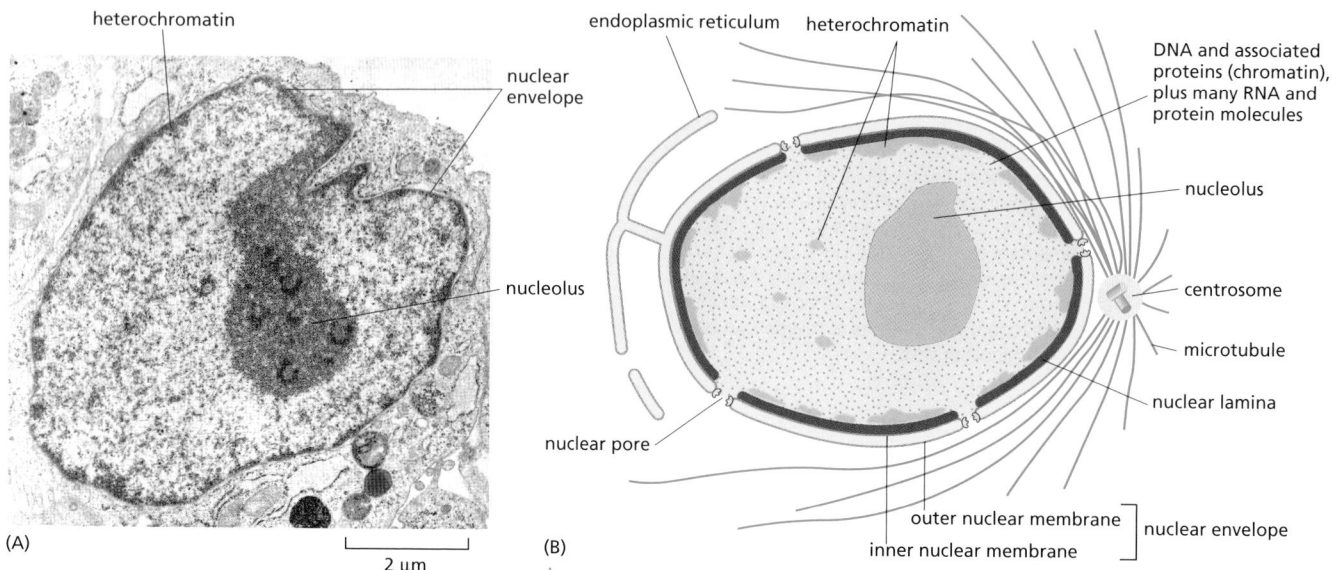

Figure 4–9 A cross-sectional view of a typical cell nucleus. (A) Electron micrograph of a thin section through the nucleus of a human fibroblast. (B) Schematic drawing, showing that the nuclear envelope consists of two membranes, the outer one being continuous with the endoplasmic reticulum (ER) membrane (see also Figure 12–7). The space inside the endoplasmic reticulum (the ER lumen) is colored *yellow*; it is continuous with the space between the two nuclear membranes. The lipid bilayers of the inner and outer nuclear membranes are connected at each nuclear pore. A sheetlike network of intermediate filaments *(brown)* inside the nucleus forms the nuclear lamina *(brown)*, providing mechanical support for the nuclear envelope (for details, see Chapter 12). The dark-staining heterochromatin contains specially condensed regions of DNA that will be discussed later. (A, courtesy of E.G. Jordan and J. McGovern.)

Eukaryotic DNA Is Packaged into a Set of Chromosomes

Each **chromosome** in a eukaryotic cell consists of a single, enormously long linear DNA molecule along with the proteins that fold and pack the fine DNA thread into a more compact structure. In addition to the proteins involved in packaging, chromosomes are also associated with many other proteins (as well as numerous RNA molecules). These are required for the processes of gene expression, DNA replication, and DNA repair. The complex of DNA and tightly bound protein is called *chromatin* (from the Greek *chroma*, "color," because of its staining properties).

Bacteria lack a special nuclear compartment, and they generally carry their genes on a single DNA molecule, which is often circular (see Figure 1–24). This DNA is also associated with proteins that package and condense it, but they are different from the proteins that perform these functions in eukaryotes. Although the bacterial DNA with its attendant proteins is often called the bacterial "chromosome," it does not have the same structure as eukaryotic chromosomes, and less is known about how the bacterial DNA is packaged. Therefore, our discussion of chromosome structure will focus almost entirely on eukaryotic chromosomes.

With the exception of the gametes (eggs and sperm) and a few highly specialized cell types that cannot multiply and either lack DNA altogether (for example, red blood cells) or have replicated their DNA without completing cell division (for example, megakaryocytes), each human cell nucleus contains two copies of each chromosome, one inherited from the mother and one from the father. The maternal and paternal chromosomes of a pair are called **homologous chromosomes** (**homologs**). The only nonhomologous chromosome pairs are the sex chromosomes in males, where a *Y chromosome* is inherited from the father and an *X chromosome* from the mother. Thus, each human cell contains a total of 46 chromosomes—22 pairs common to both males and females, plus two so-called sex chromosomes (X and Y in males, two Xs in females). These human chromosomes can be readily distinguished by "painting" each one a different color using a technique based on *DNA hybridization* (**Figure 4–10**). In this method (described in detail in Chapter 8), a short strand of nucleic acid tagged with a fluorescent dye serves as a "probe" that picks out its complementary DNA sequence, lighting up the target chromosome at any site where it binds. Chromosome painting is most

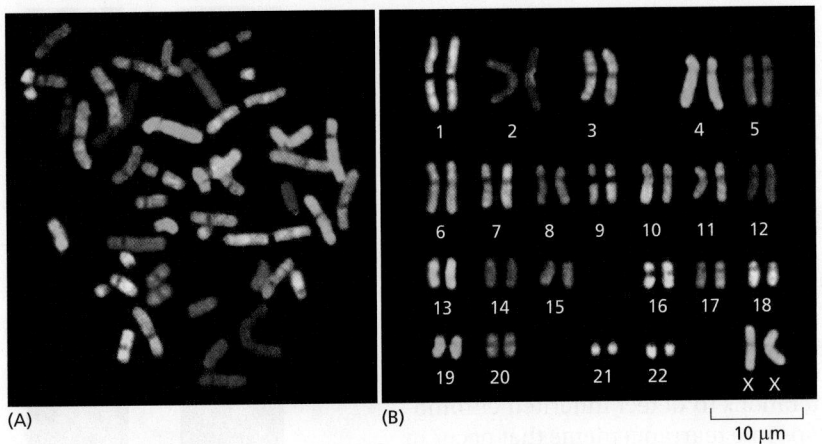

(A)   (B)   10 μm

Figure 4–10 **The complete set of human chromosomes.** These chromosomes, from a female, were isolated from a cell undergoing nuclear division (mitosis) and are therefore highly compacted. Each chromosome has been "painted" a different color to permit its unambiguous identification under the fluorescence microscope, using a technique called "spectral karyotyping." Chromosome painting can be performed by exposing the chromosomes to a large collection of DNA molecules whose sequence matches known DNA sequences from the human genome. The set of sequences matching each chromosome is coupled to a different combination of fluorescent dyes. DNA molecules derived from chromosome 1 are labeled with one specific dye combination, those from chromosome 2 with another, and so on. Because the labeled DNA can form base pairs, or hybridize, only to the chromosome from which it was derived, each chromosome becomes labeled with a different combination of dyes. For such experiments, the chromosomes are subjected to treatments that separate the two strands of double-helical DNA in a way that permits base-pairing with the single-stranded labeled DNA, but keeps the overall chromosome structure relatively intact. (A) The chromosomes visualized as they originally spilled from the lysed cell. (B) The same chromosomes artificially lined up in their numerical order. This arrangement of the full chromosome set is called a karyotype. (Adapted from N. McNeil and T. Ried, *Expert Rev. Mol. Med.* 2:1–14, 2000. With permission from Cambridge University Press.)

frequently done at the stage in the cell cycle called mitosis, when chromosomes are especially compacted and easy to visualize (see below).

Another more traditional way to distinguish one chromosome from another is to stain them with dyes that reveal a striking and reproducible pattern of bands along each mitotic chromosome (**Figure 4–11**). These banding patterns presumably reflect variations in chromatin structure, but their basis is not well understood. Nevertheless, the pattern of bands on each type of chromosome is unique, and it provided the initial means to identify and number each human chromosome reliably.

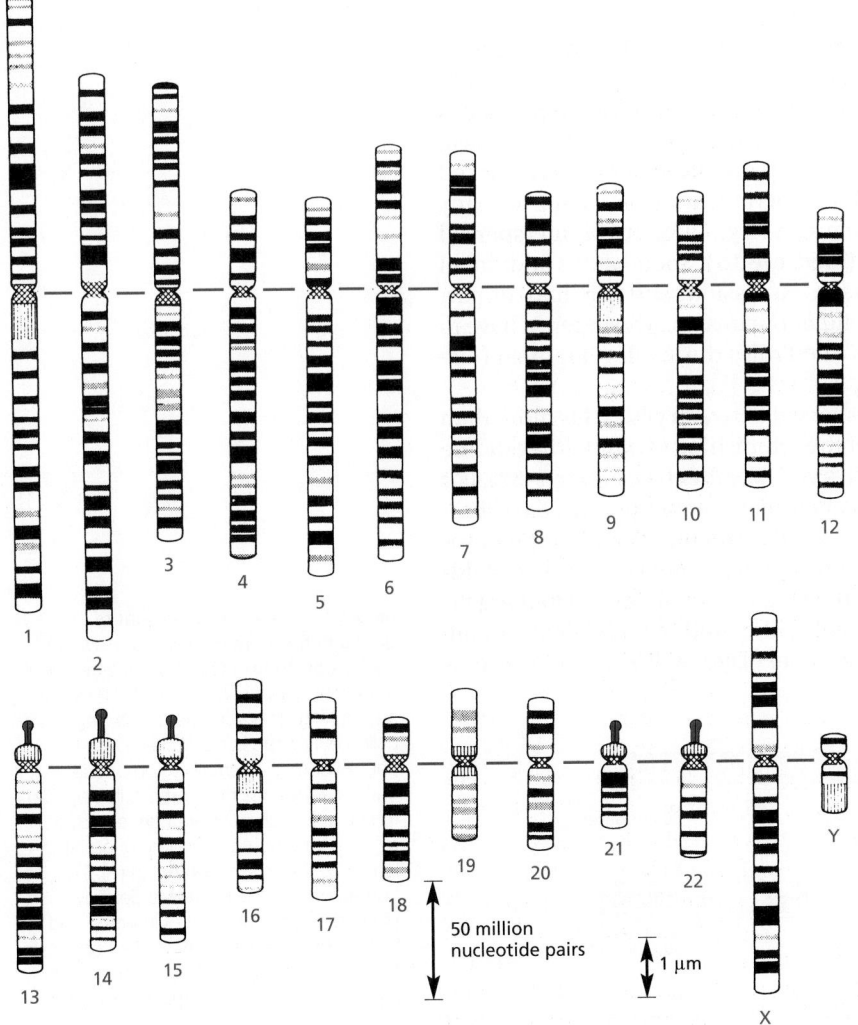

50 million nucleotide pairs

1 μm

Figure 4–11 **The banding patterns of human chromosomes.** Chromosomes 1–22 are numbered in approximate order of size. A typical human cell contains two of each of these chromosomes, plus two sex chromosomes—two X chromosomes in a female, one X and one Y chromosome in a male. The chromosomes used to make these maps were stained at an early stage in mitosis, when the chromosomes are incompletely compacted. The *horizontal red line* represents the position of the centromere (see Figure 4–19), which appears as a constriction on mitotic chromosomes. The red knobs on chromosomes 13, 14, 15, 21, and 22 indicate the positions of genes that code for the large ribosomal RNAs (discussed in Chapter 6). These banding patterns are obtained by staining chromosomes with Giemsa stain, and they can be observed under the light microscope. (Adapted from U. Francke, *Cytogenet. Cell Genet.* 31:24–32, 1981.)

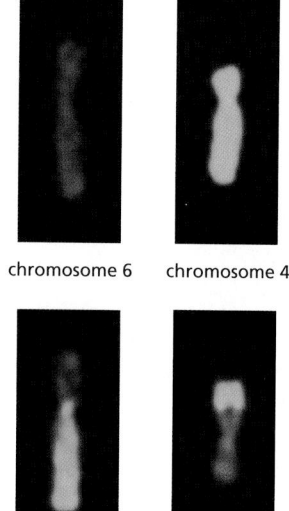

Figure 4–12 Aberrant human chromosomes. (A) Two normal human chromosomes, 4 and 6. (B) In an individual carrying a balanced *chromosomal translocation*, the DNA double helix in one chromosome has crossed over with the DNA double helix in the other chromosome due to an abnormal recombination event. The chromosome painting technique used on the chromosomes in each of the sets allows the identification of even short pieces of chromosomes that have become translocated, a frequent event in cancer cells. (Courtesy of Zhenya Tang and the NIGMS Human Genetic Cell Repository at the Coriell Institute for Medical Research: GM21880.)

(A)   chromosome 6   chromosome 4

(B)   reciprocal chromosomal translocation

The display of the 46 human chromosomes at mitosis is called the human **karyotype.** If parts of chromosomes are lost or are switched between chromosomes, these changes can be detected either by changes in the banding patterns or—with greater sensitivity—by changes in the pattern of chromosome painting (**Figure 4–12**). Cytogeneticists use these alterations to detect inherited chromosome abnormalities and to reveal the chromosome rearrangements that occur in cancer cells as they progress to malignancy (discussed in Chapter 20).

Chromosomes Contain Long Strings of Genes

Chromosomes carry genes—the functional units of heredity. A gene is often defined as a segment of DNA that contains the instructions for making a particular protein (or a set of closely related proteins), but this definition is too narrow. Genes that code for protein are indeed the majority, and most of the genes with clear-cut mutant phenotypes fall under this heading. In addition, however, there are many "RNA genes"—segments of DNA that generate a functionally significant RNA molecule, instead of a protein, as their final product. We shall say more about the RNA genes and their products later.

As might be expected, some correlation exists between the complexity of an organism and the number of genes in its genome (see Table 1–2, p. 29). For example, some simple bacteria have only 500 genes, compared to about 30,000 for humans. Bacteria, archaea, and some single-celled eukaryotes, such as yeast, have concise genomes, consisting of little more than strings of closely packed genes. However, the genomes of multicellular plants and animals, as well as many other eukaryotes, contain, in addition to genes, a large quantity of interspersed DNA whose function is poorly understood (**Figure 4–13**). Some of this additional DNA is crucial for the proper control of gene expression, and this may in part explain why there is so much of it in multicellular organisms, whose genes have to be switched on and off according to complicated rules during development (discussed in Chapters 7 and 21).

Differences in the amount of DNA interspersed between genes, far more than differences in numbers of genes, account for the astonishing variations in genome size that we see when we compare one species with another (see Figure 1–32). For example, the human genome is 200 times larger than that of the yeast *Saccharomyces cerevisiae*, but 30 times smaller than that of some plants and amphibians and 200 times smaller than that of a species of amoeba. Moreover, because of differences in the amount of noncoding DNA, the genomes of closely related organisms (bony fish, for example) can vary several hundredfold in their DNA content, even though they contain roughly the same number of genes. Whatever the excess

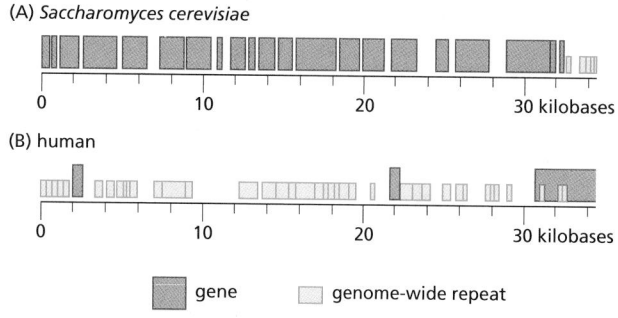

(A) *Saccharomyces cerevisiae*

0   10   20   30 kilobases

(B) human

0   10   20   30 kilobases

■ gene   ▫ genome-wide repeat

Figure 4–13 The arrangement of genes in the genome of *S. cerevisiae* compared to humans. (A) *S. cerevisiae* is a budding yeast widely used for brewing and baking. The genome of this single-celled eukaryote is distributed over 16 chromosomes. A small region of one chromosome has been arbitrarily selected to show its high density of genes. (B) A region of the human genome of equal length to the yeast segment in (A). The human genes are much less densely packed and the amount of interspersed DNA sequence is far greater. Not shown in this sample of human DNA is the fact that most human genes are much longer than yeast genes (see Figure 4–15).

Chinese muntjac

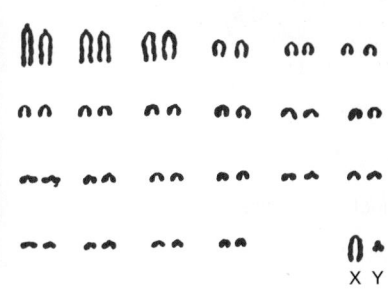
X Y

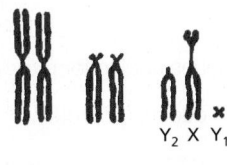
Y₂ X Y₁

Indian muntjac

DNA may do, it seems clear that it is not a great handicap for a eukaryotic cell to carry a large amount of it.

How the genome is divided into chromosomes also differs from one eukaryotic species to the next. For example, while the cells of humans have 46 chromosomes, those of some small deer have only 6, while those of the common carp contain over 100. Even closely related species with similar genome sizes can have very different numbers and sizes of chromosomes (**Figure 4–14**). Thus, there is no simple relationship between chromosome number, complexity of the organism, and total genome size. Rather, the genomes and chromosomes of modern-day species have each been shaped by a unique history of seemingly random genetic events, acted on by poorly understood selection pressures over long evolutionary times.

The Nucleotide Sequence of the Human Genome Shows How Our Genes Are Arranged

With the publication of the full DNA sequence of the human genome in 2004, it became possible to see in detail how the genes are arranged along each of our chromosomes (**Figure 4–15**). It will be many decades before the information contained in the human genome sequence is fully analyzed, but it has already stimulated new experiments that have had major effects on the content of every chapter in this book.

Figure 4–14 Two closely related species of deer with very different chromosome numbers. In the evolution of the Indian muntjac, initially separate chromosomes fused, without having a major effect on the animal. These two species contain a similar number of genes. (Chinese muntjac photo courtesy of Deborah Carreno, Natural Wonders Photography; Indian muntjac photo courtesy of Beatrice Bourgery.)

Figure 4–15 The organization of genes on a human chromosome. (A) Chromosome 22, one of the smallest human chromosomes, contains 48×10^6 nucleotide pairs and makes up approximately 1.5% of the human genome. Most of the left arm of chromosome 22 consists of short repeated sequences of DNA that are packaged in a particularly compact form of chromatin (heterochromatin) discussed later in this chapter. (B) A tenfold expansion of a portion of chromosome 22, with about 40 genes indicated. Those in *dark brown* are known genes and those in *red* are predicted genes. (C) An expanded portion of (B) showing four genes. (D) The intron–exon arrangement of a typical gene is shown after a further tenfold expansion. Each exon (*red*) codes for a portion of the protein, while the DNA sequence of the introns (*gray*) is relatively unimportant, as discussed in detail in Chapter 6.

The human genome (3.2×10^9 nucleotide pairs) is the totality of genetic information belonging to our species. Almost all of this genome is distributed over the 22 different autosomes and 2 sex chromosomes (see Figures 4–10 and 4–11) found within the nucleus. A minute fraction of the human genome (16,569 nucleotide pairs—in multiple copies per cell) is found in the mitochondria (introduced in Chapter 1, and discussed in detail in Chapter 14). The term *human genome sequence* refers to the complete nucleotide sequence of DNA in the 24 nuclear chromosomes and the mitochondria. Being diploid, a human somatic cell nucleus contains roughly twice the haploid amount of DNA, or 6.4×10^9 nucleotide pairs, when not duplicating its chromosomes in preparation for division. (Adapted from International Human Genome Sequencing Consortium, *Nature* 409:860–921, 2001.)

(A) human chromosome 22 in its mitotic conformation, composed of two double-stranded DNA molecules, each 48×10^6 nucleotide pairs long

heterochromatin

×10

(B) 10% of chromosome arm ~40 genes

×10

(C) 1% of chromosome arm containing 4 genes

×10

(D) one gene of 3.4×10^4 nucleotide pairs

regulatory DNA sequences   exon   intron   gene expression

RNA

protein

folded protein

TABLE 4–1 Some Vital Statistics for the Human Genome

Human genome	
DNA length	3.2×10^9 nucleotide pairs*
Number of genes coding for proteins	Approximately 21,000
Largest gene coding for protein	2.4×10^6 nucleotide pairs
Mean size for protein-coding genes	27,000 nucleotide pairs
Smallest number of exons per gene	1
Largest number of exons per gene	178
Mean number of exons per gene	10.4
Largest exon size	17,106 nucleotide pairs
Mean exon size	145 nucleotide pairs
Number of noncoding RNA genes	Approximately 9000**
Number of pseudogenes***	More than 20,000
Percentage of DNA sequence in exons (protein-coding sequences)	1.5%
Percentage of DNA in other highly conserved sequences****	3.5%
Percentage of DNA in high-copy-number repetitive elements	Approximately 50%

* The sequence of 2.85 billion nucleotides is known precisely (error rate of only about 1 in 100,000 nucleotides). The remaining DNA primarily consists of short sequences that are tandemly repeated many times over, with repeat numbers differing from one individual to the next. These highly repetitive blocks are hard to sequence accurately.
** This number is only a very rough estimate.
*** A pseudogene is a DNA sequence closely resembling that of a functional gene, but containing numerous mutations that prevent its proper expression or function. Most pseudogenes arise from the duplication of a functional gene followed by the accumulation of damaging mutations in one copy.
**** These conserved functional regions include DNA encoding 5′ and 3′ UTRs (untranslated regions of mRNA), DNA specifying structural and functional RNAs, and DNA with conserved protein-binding sites.

The first striking feature of the human genome is how little of it (only about 1.5 percent) codes for proteins (Table 4–1 and Figure 4–16). It is also notable that nearly half of the chromosomal DNA is made up of mobile pieces of DNA that have gradually inserted themselves in the chromosomes over evolutionary time, multiplying like parasites in the genome (see Figure 4–62). We discuss these *transposable elements* in detail in later chapters.

A second notable feature of the human genome is the large average gene size—about 27,000 nucleotide pairs. As discussed above, a typical gene carries in its linear sequence of nucleotides the information for the linear sequence of the amino acids of a protein. Only about 1300 nucleotide pairs are required to encode a protein of average size (about 430 amino acids in humans). Most of the remaining sequence in a gene consists of long stretches of noncoding DNA that interrupt the relatively short segments of DNA that code for protein. As will be discussed in detail in Chapter 6, the coding sequences are called **exons**; the intervening (noncoding) sequences in genes are called **introns** (see Figure 4–15 and Table 4–1). The majority of human genes thus consist of a long string of alternating exons and introns, with most of the gene consisting of introns. In contrast, the majority of genes from organisms with concise genomes lack introns. This accounts for the much smaller size of their genes (about one-twentieth that of human genes), as well as for the much higher fraction of coding DNA in their chromosomes.

(A)

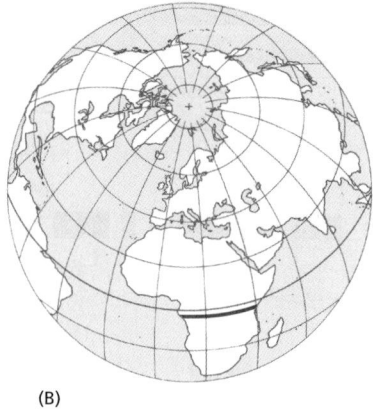

(B)

Figure 4–16 Scale of the human genome. If drawn with a 1 mm space between each nucleotide pair, as in (A), the human genome would extend 3200 km (approximately 2000 miles), far enough to stretch across the center of Africa, the site of our human origins (*red line* in B). At this scale, there would be, on average, a protein-coding gene every 150 m. An average gene would extend for 30 m, but the coding sequences in this gene would add up to only just over a meter.

In addition to introns and exons, each gene is associated with *regulatory DNA sequences*, which are responsible for ensuring that the gene is turned on or off at the proper time, expressed at the appropriate level, and only in the proper type of cell. In humans, the regulatory sequences for a typical gene are spread out over tens of thousands of nucleotide pairs. As would be expected, these regulatory sequences are much more compressed in organisms with concise genomes. We discuss how regulatory DNA sequences work in Chapter 7.

Research in the last decade has surprised biologists with the discovery that, in addition to 21,000 protein-coding genes, the human genome contains many thousands of genes that encode RNA molecules that do not produce proteins, but instead have a variety of other important functions. What is thus far known about these molecules will be presented in Chapters 6 and 7. Last, but not least, the nucleotide sequence of the human genome has revealed that the archive of information needed to produce a human seems to be in an alarming state of chaos. As one commentator described our genome, "In some ways it may resemble your garage/bedroom/refrigerator/life: highly individualistic, but unkempt; little evidence of organization; much accumulated clutter (referred to by the uninitiated as 'junk'); virtually nothing ever discarded; and the few patently valuable items indiscriminately, apparently carelessly, scattered throughout." We shall discuss how this is thought to have come about in the final sections of this chapter entitled "How Genomes Evolve."

Each DNA Molecule That Forms a Linear Chromosome Must Contain a Centromere, Two Telomeres, and Replication Origins

To form a functional chromosome, a DNA molecule must be able to do more than simply carry genes: it must be able to replicate, and the replicated copies must be separated and reliably partitioned into daughter cells at each cell division. This process occurs through an ordered series of stages, collectively known as the **cell cycle**, which provides for a temporal separation between the duplication of chromosomes and their segregation into two daughter cells. The cell cycle is briefly summarized in **Figure 4–17**, and it is discussed in detail in Chapter 17. Briefly, during a long *interphase*, genes are expressed and chromosomes are replicated, with the two replicas remaining together as a pair of *sister chromatids*. Throughout this time, the chromosomes are extended and much of their chromatin exists as long threads in the nucleus so that individual chromosomes cannot be easily distinguished. It is only during a much briefer period of mitosis that each chromosome condenses so that its two sister chromatids can be separated and distributed to the two daughter nuclei. The highly condensed chromosomes in a dividing cell are known as *mitotic chromosomes* (**Figure 4–18**). This is the form in which chromosomes are most easily visualized; in fact, the images of chromosomes shown so far in the chapter are of chromosomes in mitosis.

Each chromosome operates as a distinct structural unit: for a copy to be passed on to each daughter cell at division, each chromosome must be able to replicate, and the newly replicated copies must subsequently be separated and partitioned

Figure 4–17 A simplified view of the eukaryotic cell cycle. During interphase, the cell is actively expressing its genes and is therefore synthesizing proteins. Also, during interphase and before cell division, the DNA is replicated and each chromosome is duplicated to produce two closely paired sister DNA molecules (called sister chromatids). A cell with only one type of chromosome, present in maternal and paternal copies, is illustrated here. Once DNA replication is complete, the cell can enter *M phase*, when mitosis occurs and the nucleus is divided into two daughter nuclei. During this stage, the chromosomes condense, the nuclear envelope breaks down, and the mitotic spindle forms from microtubules and other proteins. The condensed mitotic chromosomes are captured by the mitotic spindle, and one complete set of chromosomes is then pulled to each end of the cell by separating the members of each sister-chromatid pair. A nuclear envelope re-forms around each chromosome set, and in the final step of M phase, the cell divides to produce two daughter cells. Most of the time in the cell cycle is spent in interphase; M phase is brief in comparison, occupying only about an hour in many mammalian cells.

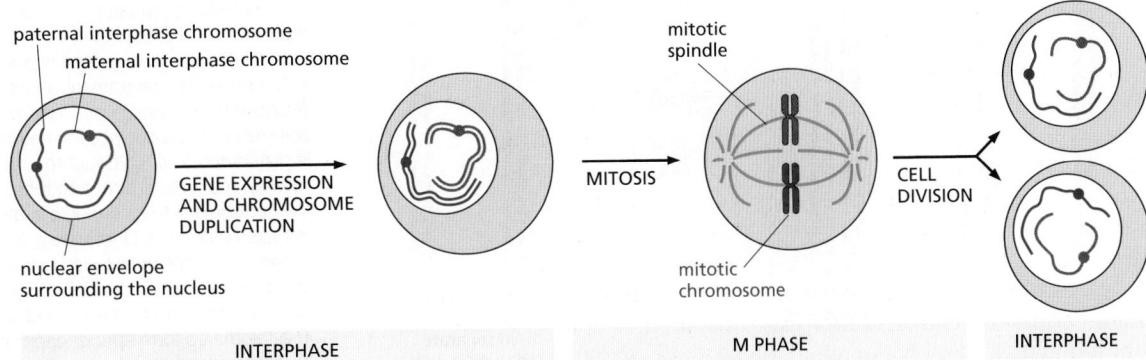

correctly into the two daughter cells. These basic functions are controlled by three types of specialized nucleotide sequences in the DNA, each of which binds specific proteins that guide the machinery that replicates and segregates chromosomes (Figure 4–19).

Experiments in yeasts, whose chromosomes are relatively small and easy to manipulate, have identified the minimal DNA sequence elements responsible for each of these functions. One type of nucleotide sequence acts as a DNA **replication origin**, the location at which duplication of the DNA begins. Eukaryotic chromosomes contain many origins of replication to ensure that the entire chromosome can be replicated rapidly, as discussed in detail in Chapter 5.

After DNA replication, the two sister chromatids that form each chromosome remain attached to one another and, as the cell cycle proceeds, are condensed further to produce mitotic chromosomes. The presence of a second specialized DNA sequence, called a **centromere**, allows one copy of each duplicated and condensed chromosome to be pulled into each daughter cell when a cell divides. A protein complex called a *kinetochore* forms at the centromere and attaches the duplicated chromosomes to the mitotic spindle, allowing them to be pulled apart (discussed in Chapter 17).

The third specialized DNA sequence forms **telomeres**, the ends of a chromosome. Telomeres contain repeated nucleotide sequences that enable the ends of chromosomes to be efficiently replicated. Telomeres also perform another function: the repeated telomere DNA sequences, together with the regions adjoining them, form structures that protect the end of the chromosome from being mistaken by the cell for a broken DNA molecule in need of repair. We discuss both this type of repair and the structure and function of telomeres in Chapter 5.

In yeast cells, the three types of sequences required to propagate a chromosome are relatively short (typically less than 1000 base pairs each) and therefore use only a tiny fraction of the information-carrying capacity of a chromosome. Although telomere sequences are fairly simple and short in all eukaryotes, the DNA sequences that form centromeres and replication origins in more complex organisms are much longer than their yeast counterparts. For example, experiments suggest that a human centromere can contain up to a million nucleotide pairs and that it may not require a stretch of DNA with a defined nucleotide sequence. Instead, as we shall discuss later in this chapter, a human centromere is thought to consist of a large, regularly repeating protein–nucleic acid structure that can be inherited when a chromosome replicates.

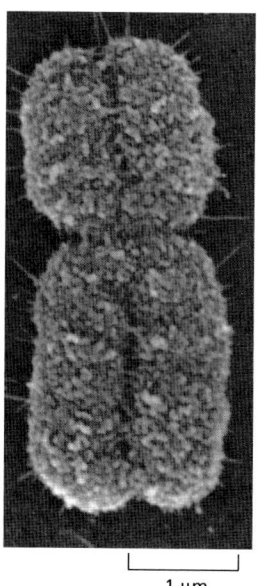

|—————|
1 μm

Figure 4–18 A mitotic chromosome. A mitotic chromosome is a condensed duplicated chromosome in which the two new chromosomes, called sister chromatids, are still linked together (see Figure 4–17). The constricted region indicates the position of the centromere. (Courtesy of Terry D. Allen.)

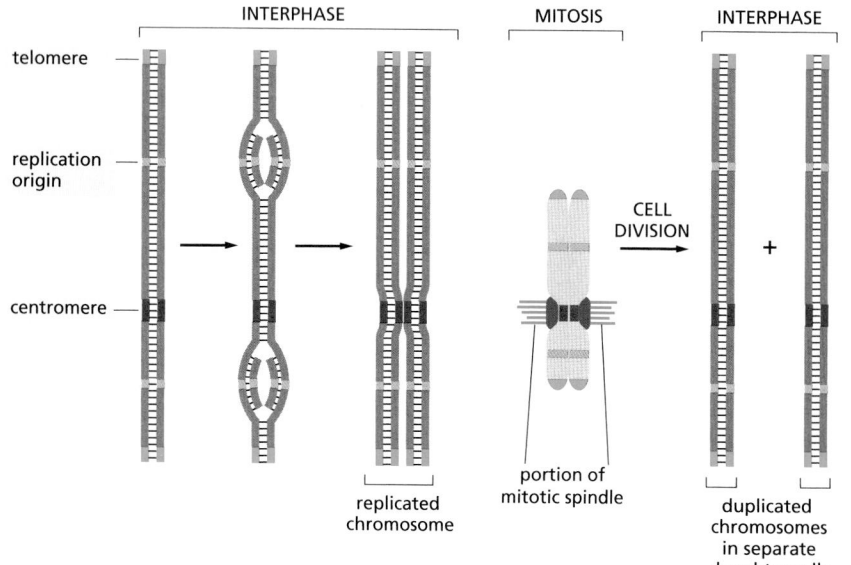

replicated chromosome

portion of mitotic spindle

duplicated chromosomes in separate daughter cells

Figure 4–19 The three DNA sequences required to produce a eukaryotic chromosome that can be replicated and then segregated accurately at mitosis. Each chromosome has multiple origins of replication, one centromere, and two telomeres. Shown here is the sequence of events that a typical chromosome follows during the cell cycle. The DNA replicates in interphase, beginning at the origins of replication and proceeding bidirectionally from the origins across the chromosome. In M phase, the centromere attaches the duplicated chromosomes to the mitotic spindle so that a copy of the entire genome is distributed to each daughter cell during mitosis; the special structure that attaches the centromere to the spindle is a protein complex called the kinetochore *(dark green)*. The centromere also helps to hold the duplicated chromosomes together until they are ready to be moved apart. The telomeres form special caps at each chromosome end.

DNA Molecules Are Highly Condensed in Chromosomes

All eukaryotic organisms have special ways of packaging DNA into chromosomes. For example, if the 48 million nucleotide pairs of DNA in human chromosome 22 could be laid out as one long perfect double helix, the molecule would extend for about 1.5 cm if stretched out end to end. But chromosome 22 measures only about 2 μm in length in mitosis (see Figures 4–10 and 4–11), representing an end-to-end compaction ratio of over 7000-fold. This remarkable feat of compression is performed by proteins that successively coil and fold the DNA into higher and higher levels of organization. Although much less condensed than mitotic chromosomes, the DNA of human interphase chromosomes is still tightly packed.

In reading these sections it is important to keep in mind that chromosome structure is dynamic. We have seen that each chromosome condenses to an extreme degree in the M phase of the cell cycle. Much less visible, but of enormous interest and importance, specific regions of interphase chromosomes decondense to allow access to specific DNA sequences for gene expression, DNA repair, and replication—and then recondense when these processes are completed. The packaging of chromosomes is therefore accomplished in a way that allows rapid localized, on-demand access to the DNA. In the next sections, we discuss the specialized proteins that make this type of packaging possible.

Nucleosomes Are a Basic Unit of Eukaryotic Chromosome Structure

The proteins that bind to the DNA to form eukaryotic chromosomes are traditionally divided into two classes: the **histones** and the *non-histone chromosomal proteins*, each contributing about the same mass to a chromosome as the DNA. The complex of both classes of protein with the nuclear DNA of eukaryotic cells is known as **chromatin** (**Figure 4–20**).

Histones are responsible for the first and most basic level of chromosome packing, the **nucleosome**, a protein–DNA complex discovered in 1974. When interphase nuclei are broken open very gently and their contents examined under the electron microscope, most of the chromatin appears to be in the form of a fiber with a diameter of about 30 nm (**Figure 4–21**A). If this chromatin is subjected to treatments that cause it to unfold partially, it can be seen under the electron microscope as a series of "beads on a string" (Figure 4–21B). The string is DNA, and each bead is a "nucleosome core particle" that consists of DNA wound around a histone core (**Movie 4.2**).

The structural organization of nucleosomes was determined after first isolating them from unfolded chromatin by digestion with particular enzymes (called nucleases) that break down DNA by cutting between the nucleosomes. After digestion for a short period, the exposed DNA between the nucleosome core particles, the *linker DNA*, is degraded. Each individual nucleosome core particle consists of a complex of eight histone proteins—two molecules each of histones H2A,

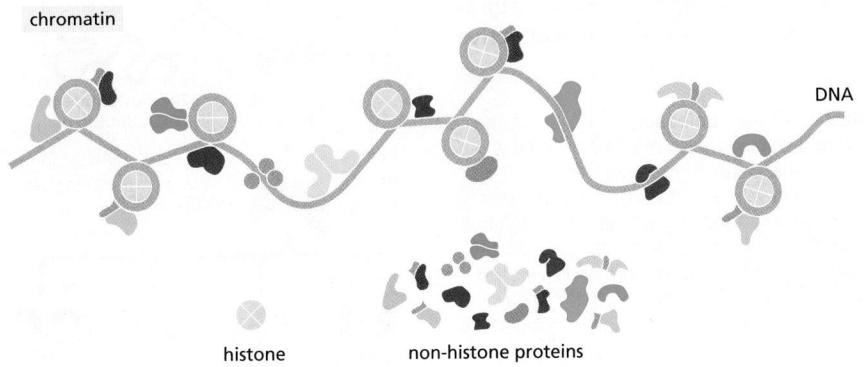

chromatin

DNA

histone   non-histone proteins

Figure 4–20 **Chromatin.** As illustrated, chromatin consists of DNA bound to both histone and non-histone proteins. The mass of histone protein present is about equal to the total mass of non-histone protein, but—as schematically indicated here—the latter class is composed of an enormous number of different species. In total, a chromosome is about one-third DNA and two-thirds protein by mass.

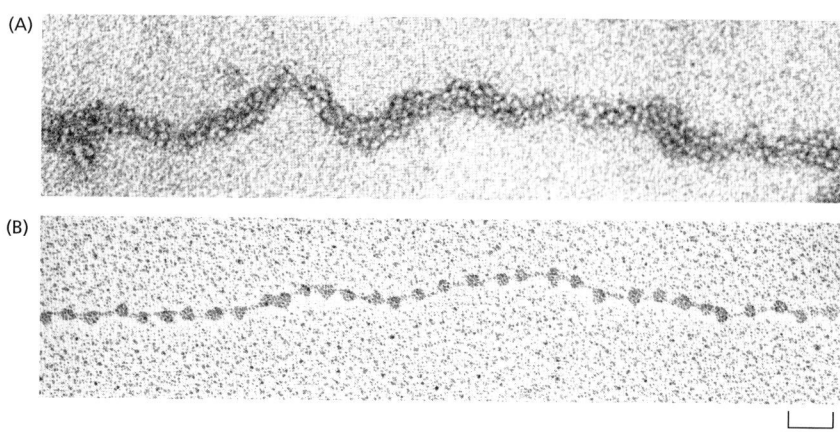

(A)

(B)

50 nm

Figure 4–21 Nucleosomes as seen in the electron microscope. (A) Chromatin isolated directly from an interphase nucleus appears in the electron microscope as a thread about 30 nm thick. (B) This electron micrograph shows a length of chromatin that has been experimentally unpacked, or decondensed, after isolation to show the nucleosomes. (A, courtesy of Barbara Hamkalo; B, courtesy of Victoria Foe.)

H2B, H3, and H4—and double-stranded DNA that is 147 nucleotide pairs long. The *histone octamer* forms a protein core around which the double-stranded DNA is wound (**Figure 4–22**).

The region of linker DNA that separates each nucleosome core particle from the next can vary in length from a few nucleotide pairs up to about 80. (The term *nucleosome* technically refers to a nucleosome core particle plus one of its adjacent DNA linkers, but it is often used synonymously with nucleosome core particle.) On average, therefore, nucleosomes repeat at intervals of about 200 nucleotide pairs. For example, a diploid human cell with 6.4×10^9 nucleotide pairs contains approximately 30 million nucleosomes. The formation of nucleosomes converts a DNA molecule into a chromatin thread about one-third of its initial length.

The Structure of the Nucleosome Core Particle Reveals How DNA Is Packaged

The high-resolution structure of a nucleosome core particle, solved in 1997, revealed a disc-shaped histone core around which the DNA was tightly wrapped in a left-handed coil of 1.7 turns (**Figure 4–23**). All four of the histones that make up the core of the nucleosome are relatively small proteins (102–135 amino acids), and they share a structural motif, known as the *histone fold*, formed from three α helices connected by two loops (**Figure 4–24**). In assembling a nucleosome, the histone folds first bind to each other to form H3–H4 and H2A–H2B dimers, and the H3–H4 dimers combine to form tetramers. An H3–H4 tetramer then further combines with two H2A–H2B dimers to form the compact octamer core, around which the DNA is wound.

The interface between DNA and histone is extensive: 142 hydrogen bonds are formed between DNA and the histone core in each nucleosome. Nearly half of these bonds form between the amino acid backbone of the histones and the sugar-phosphate backbone of the DNA. Numerous hydrophobic interactions and salt linkages also hold DNA and protein together in the nucleosome. More than one-fifth of the amino acids in each of the core histones are either lysine or arginine (two amino acids with basic side chains), and their positive charges can effectively

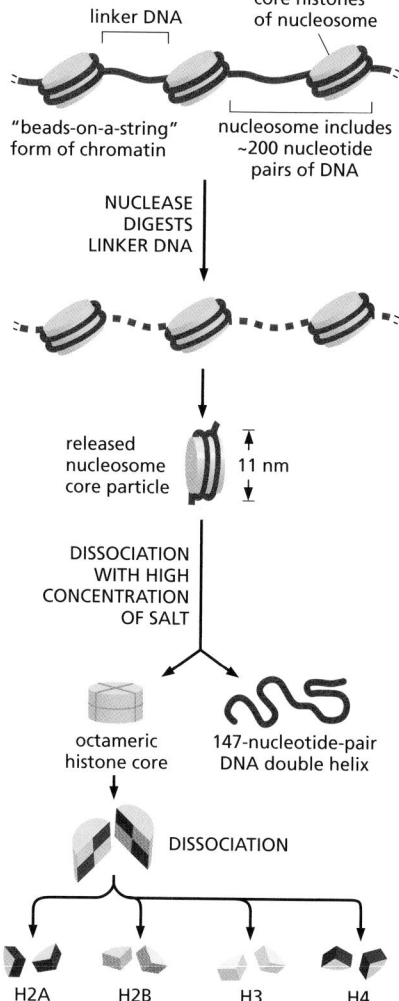

linker DNA

core histones of nucleosome

"beads-on-a-string" form of chromatin

nucleosome includes ~200 nucleotide pairs of DNA

NUCLEASE DIGESTS LINKER DNA

released nucleosome core particle

11 nm

DISSOCIATION WITH HIGH CONCENTRATION OF SALT

octameric histone core

147-nucleotide-pair DNA double helix

DISSOCIATION

H2A   H2B   H3   H4

Figure 4–22 Structural organization of the nucleosome. A nucleosome contains a protein core made of eight histone molecules. In biochemical experiments, the nucleosome core particle can be released from isolated chromatin by digestion of the linker DNA with a nuclease, an enzyme that breaks down DNA. (The nuclease can degrade the exposed linker DNA but cannot attack the DNA wound tightly around the nucleosome core.) After dissociation of the isolated nucleosome into its protein core and DNA, the length of the DNA that was wound around the core can be determined. This length of 147 nucleotide pairs is sufficient to wrap 1.7 times around the histone core.

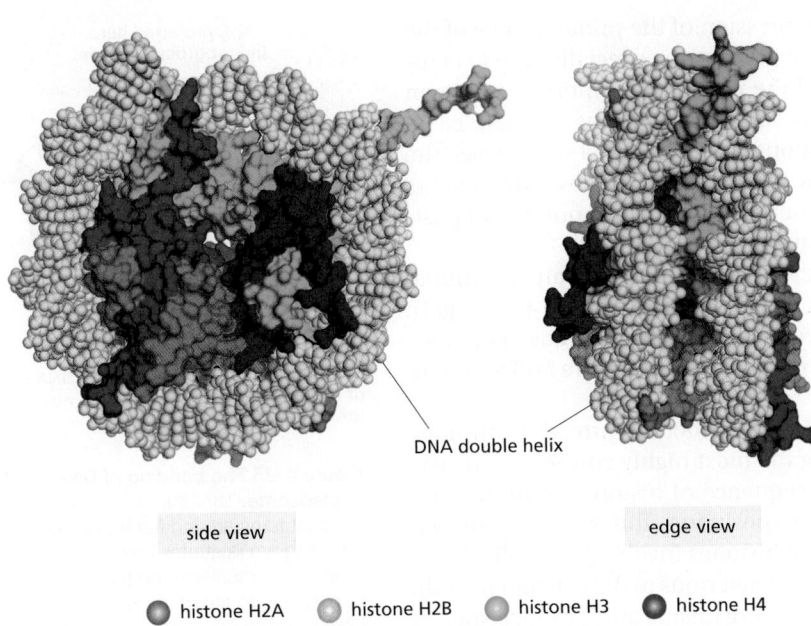

Figure 4–23 **The structure of a nucleosome core particle, as determined by x-ray diffraction analyses of crystals.** Each histone is colored according to the scheme in Figure 4–22, with the DNA double helix in *light gray*. (Adapted from K. Luger et al., *Nature* 389:251–260, 1997.)

DNA double helix

side view

edge view

○ histone H2A   ○ histone H2B   ○ histone H3   ● histone H4

neutralize the negatively charged DNA backbone. These numerous interactions explain in part why DNA of virtually any sequence can be bound on a histone octamer core. The path of the DNA around the histone core is not smooth; rather, several kinks are seen in the DNA, as expected from the nonuniform surface of the

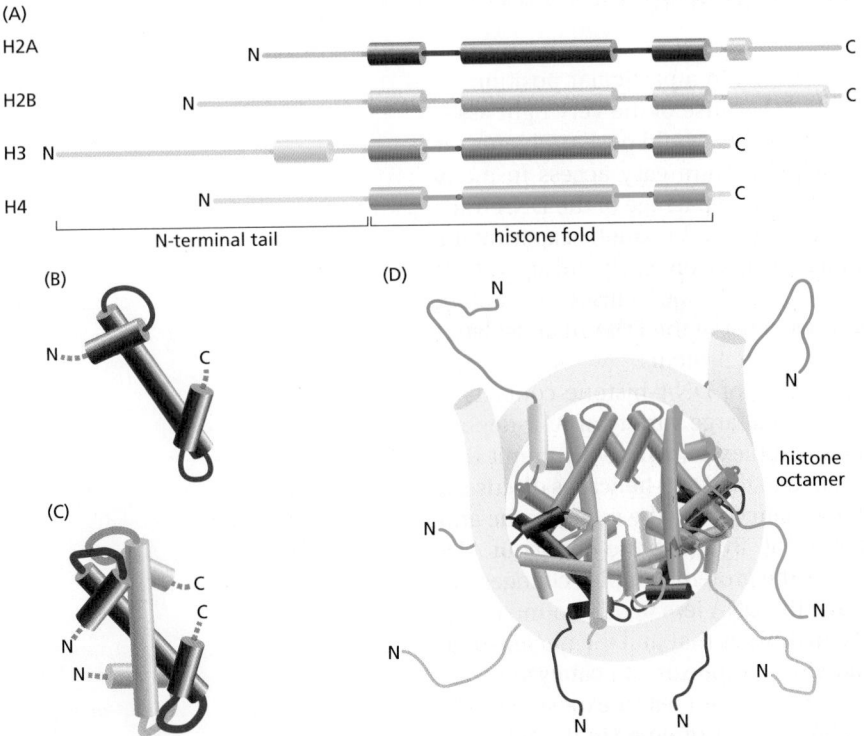

Figure 4–24 **The overall structural organization of the core histones.** (A) Each of the core histones contains an N-terminal tail, which is subject to several forms of covalent modification, and a histone fold region, as indicated. (B) The structure of the histone fold, which is formed by all four of the core histones. (C) Histones 2A and 2B form a dimer through an interaction known as the "handshake." Histones H3 and H4 form a dimer through the same type of interaction. (D) The final histone octamer on DNA. Note that all eight N-terminal tails of the histones protrude from the disc-shaped core structure. Their conformations are highly flexible, and they serve as binding sites for sets of other proteins.

core. The bending requires a substantial compression of the minor groove of the DNA helix. Certain dinucleotides in the minor groove are especially easy to compress, and some nucleotide sequences bind the nucleosome more tightly than others (**Figure 4–25**). This probably explains some striking, but unusual, cases of very precise positioning of nucleosomes along a stretch of DNA. However, the sequence preference of nucleosomes must be weak enough to allow other factors to dominate, inasmuch as nucleosomes can occupy any one of a number of positions relative to the DNA sequence in most chromosomal regions.

In addition to its histone fold, each of the core histones has an N-terminal amino acid "tail," which extends out from the DNA–histone core (see Figure 4–24D). These histone tails are subject to several different types of covalent modifications that in turn control critical aspects of chromatin structure and function, as we shall discuss shortly.

As a reflection of their fundamental role in DNA function through controlling chromatin structure, the histones are among the most highly conserved eukaryotic proteins. For example, the amino acid sequence of histone H4 from a pea differs from that of a cow at only 2 of the 102 positions. This strong evolutionary conservation suggests that the functions of histones involve nearly all of their amino acids, so that a change in any position is deleterious to the cell. But in addition to this remarkable conservation, eukaryotic organisms also produce smaller amounts of specialized variant core histones that differ in amino acid sequence from the main ones. As discussed later, these variants, combined with the surprisingly large number of covalent modifications that can be added to the histones in nucleosomes, give rise to a variety of chromatin structures in cells.

Nucleosomes Have a Dynamic Structure, and Are Frequently Subjected to Changes Catalyzed by ATP-Dependent Chromatin Remodeling Complexes

For many years biologists thought that, once formed in a particular position on DNA, a nucleosome would remain fixed in place because of the very tight association between its core histones and DNA. If true, this would pose problems for genetic readout mechanisms, which in principle require easy access to many specific DNA sequences. It would also hinder the rapid passage of the DNA transcription and replication machinery through chromatin. But kinetic experiments show that the DNA in an isolated nucleosome unwraps from each end at a rate of about four times per second, remaining exposed for 10 to 50 milliseconds before the partially unwrapped structure recloses. Thus, most of the DNA in an isolated nucleosome is in principle available for binding other proteins.

For the chromatin in a cell, a further loosening of DNA–histone contacts is clearly required, because eukaryotic cells contain a large variety of ATP-dependent *chromatin remodeling complexes*. These complexes include a subunit that hydrolyzes ATP (an ATPase evolutionarily related to the DNA helicases discussed in Chapter 5). This subunit binds both to the protein core of the nucleosome and to the double-stranded DNA that winds around it. By using the energy of ATP hydrolysis to move this DNA relative to the core, the protein complex changes the structure of a nucleosome temporarily, making the DNA less tightly bound to the histone core. Through repeated cycles of ATP hydrolysis that pull the nucleosome core along the DNA double helix, the remodeling complexes can catalyze *nucleosome sliding*. In this way, they can reposition nucleosomes to expose specific regions of DNA, thereby making them available to other proteins in the cell (**Figure 4–26**). In addition, by cooperating with a variety of other proteins that bind to histones and serve as *histone chaperones*, some remodeling complexes are able to remove either all or part of the nucleosome core from a nucleosome—catalyzing either an exchange of its H2A–H2B histones, or the complete removal of the octameric core from the DNA (**Figure 4–27**). As a result of such processes, measurements reveal that a typical nucleosome is replaced on the DNA every one or two hours inside the cell.

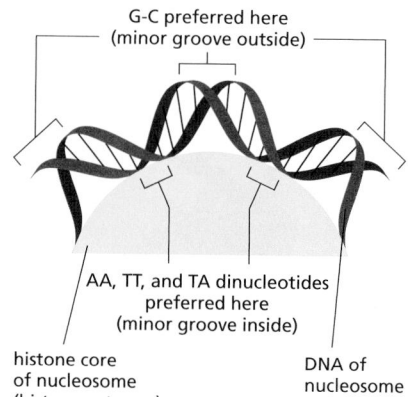

Figure 4–25 The bending of DNA in a nucleosome. The DNA helix makes 1.7 tight turns around the histone octamer. This diagram illustrates how the minor groove is compressed on the inside of the turn. Owing to structural features of the DNA molecule, the indicated dinucleotides are preferentially accommodated in such a narrow minor groove, which helps to explain why certain DNA sequences will bind more tightly than others to the nucleosome core.

Cells contain dozens of different ATP-dependent chromatin remodeling complexes that are specialized for different roles. Most are large protein complexes that can contain 10 or more subunits, some of which bind to specific modifications on histones (see Figure 4–26C). The activity of these complexes is carefully controlled by the cell. As genes are turned on and off, chromatin remodeling complexes are brought to specific regions of DNA where they act locally to influence chromatin structure (discussed in Chapter 7; see also Figure 4–40, below).

Although some DNA sequences bind more tightly than others to the nucleosome core (see Figure 4–25), the most important influence on nucleosome positioning appears to be the presence of other tightly bound proteins on the DNA. Some bound proteins favor the formation of a nucleosome adjacent to them. Others create obstacles that force the nucleosomes to move elsewhere. The exact positions of nucleosomes along a stretch of DNA therefore depend mainly on the presence and nature of other proteins bound to the DNA. And due to the presence of ATP-dependent chromatin remodeling complexes, the arrangement of nucleosomes on DNA can be highly dynamic, changing rapidly according to the needs of the cell.

Nucleosomes Are Usually Packed Together into a Compact Chromatin Fiber

Although enormously long strings of nucleosomes form on the chromosomal DNA, chromatin in a living cell probably rarely adopts the extended "beads-on-a-string" form. Instead, the nucleosomes are packed on top of one another, generating arrays in which the DNA is even more highly condensed. Thus, when nuclei are very gently lysed onto an electron microscope grid, much of the chromatin is seen to be in the form of a fiber with a diameter of about 30 nm, which is considerably wider than chromatin in the "beads-on-a-string" form (see Figure 4–21).

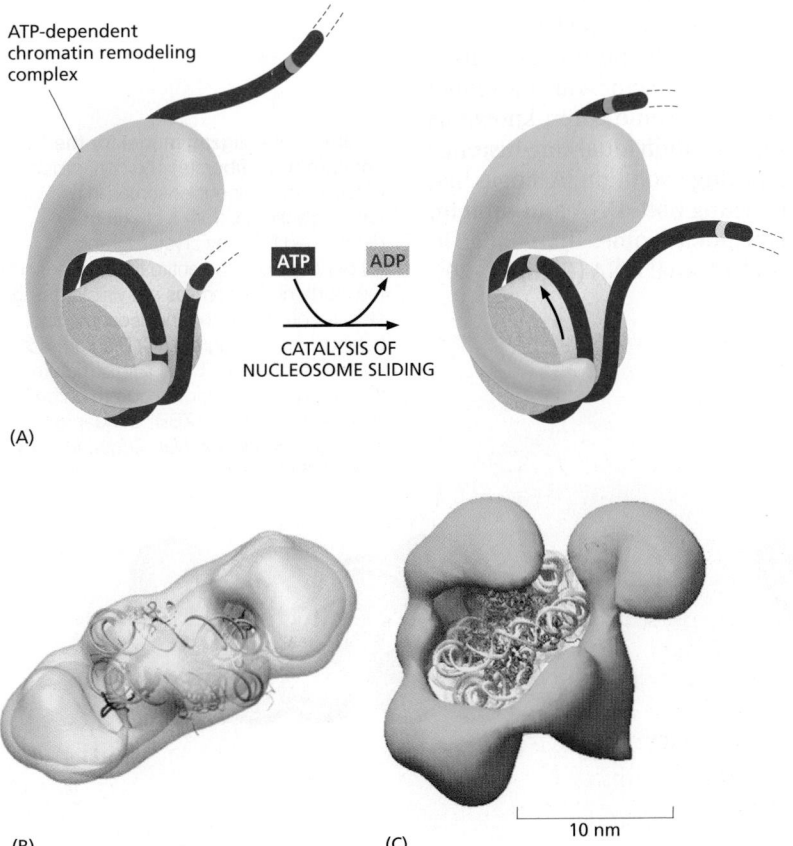

ATP-dependent chromatin remodeling complex

ATP → ADP

CATALYSIS OF NUCLEOSOME SLIDING

(A)

(B)   (C)

10 nm

Figure 4–26 **The nucleosome sliding catalyzed by ATP-dependent chromatin remodeling complexes.** (A) Using the energy of ATP hydrolysis, the remodeling complex is thought to push on the DNA of its bound nucleosome and loosen its attachment to the nucleosome core. Each cycle of ATP binding, ATP hydrolysis, and release of the ADP and P_i products thereby moves the DNA with respect to the histone octamer in the direction of the arrow in this diagram. It requires many such cycles to produce the nucleosome sliding shown. (B) The structure of a nucleosome-bound dimer of the two identical ATPase subunits *(green)* that slide nucleosomes back and forth in the ISW1 family of chromatin remodeling complexes. (C) The structure of a large chromatin remodeling complex, showing how it is thought to wrap around a nucleosome. Modeled in *green* is the yeast RSC complex, which contains 15 subunits— including an ATPase and at least four subunits with domains that recognize specific covalently modified histones. (B, from L. R. Racki et al., *Nature* 462:1016–1021, published 2009 by Nature Publishing Group. Reproduced with permission of SNCSC; C, adapted from A. E. Leschziner et al., *Proc. Natl Acad. Sci. USA* 104:4913–4918, 2007.)

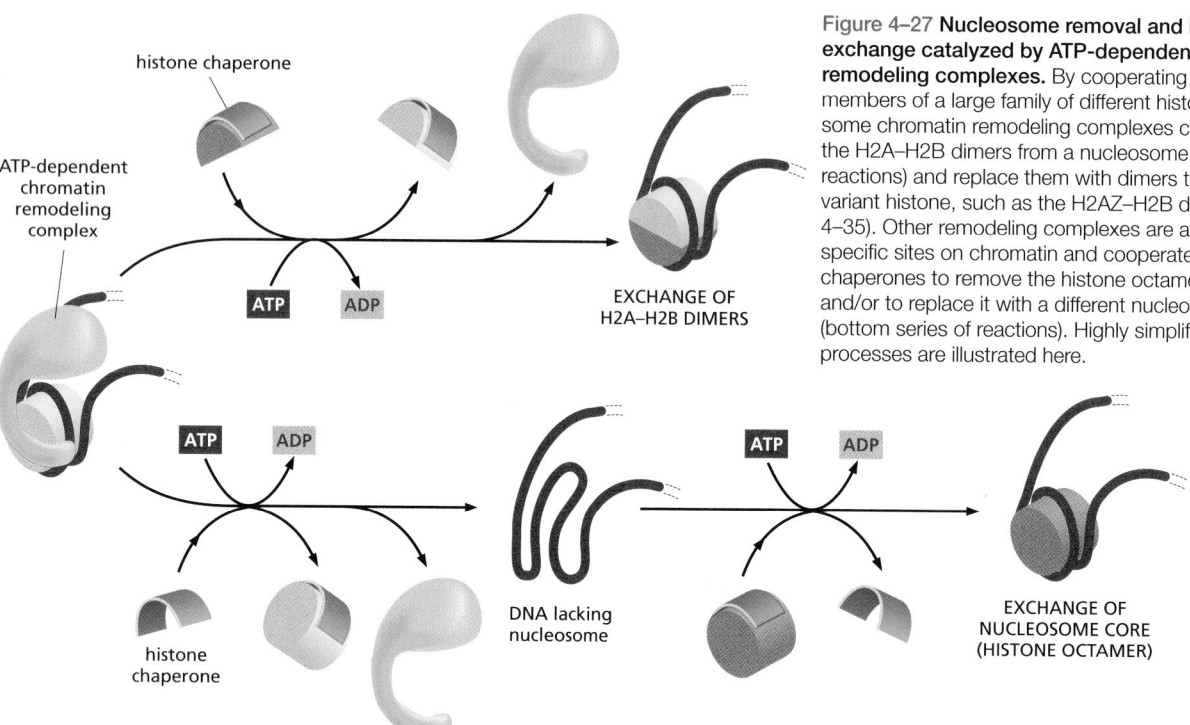

Figure 4–27 **Nucleosome removal and histone exchange catalyzed by ATP-dependent chromatin remodeling complexes.** By cooperating with specific members of a large family of different histone chaperones, some chromatin remodeling complexes can remove the H2A–H2B dimers from a nucleosome (top series of reactions) and replace them with dimers that contain a variant histone, such as the H2AZ–H2B dimer (see Figure 4–35). Other remodeling complexes are attracted to specific sites on chromatin and cooperate with histone chaperones to remove the histone octamer completely and/or to replace it with a different nucleosome core (bottom series of reactions). Highly simplified views of the processes are illustrated here.

How nucleosomes are organized into condensed arrays is unclear. The structure of a tetranucleosome (a complex of four nucleosomes) obtained by x-ray crystallography and high-resolution electron microscopy of reconstituted chromatin have been used to support a zigzag model for the stacking of nucleosomes in a 30-nm fiber (**Figure 4–28**). But cryoelectron microscopy of carefully prepared nuclei suggests that most regions of chromatin are less regularly structured.

What causes nucleosomes to stack so tightly on each other? Nucleosome-to-nucleosome linkages that involve histone tails, most notably the H4 tail, constitute one important factor (**Figure 4–29**). Another important factor is an additional histone that is often present in a 1-to-1 ratio with nucleosome cores, known as **histone H1**. This so-called *linker histone* is larger than the individual core histones and it has been considerably less well conserved during evolution. A single histone H1 molecule binds to each nucleosome, contacting both DNA and protein, and changing the path of the DNA as it exits from the nucleosome. This change in the exit path of DNA is thought to help compact nucleosomal DNA (**Figure 4–30**).

Figure 4–28 **A zigzag model for the 30-nm chromatin fiber.** (A) The conformation of two of the four nucleosomes in a tetranucleosome, from a structure determined by x-ray crystallography. (B) Schematic of the entire tetranucleosome; the fourth nucleosome is not visible, being stacked on the bottom nucleosome and behind it in this diagram. (C) Diagrammatic illustration of a possible zigzag structure that could account for the 30-nm chromatin fiber. (A, PDB code: 1ZBB; C, adapted from C.L. Woodcock, *Nat. Struct. Mol. Biol.* 12:639–640, 2005.)

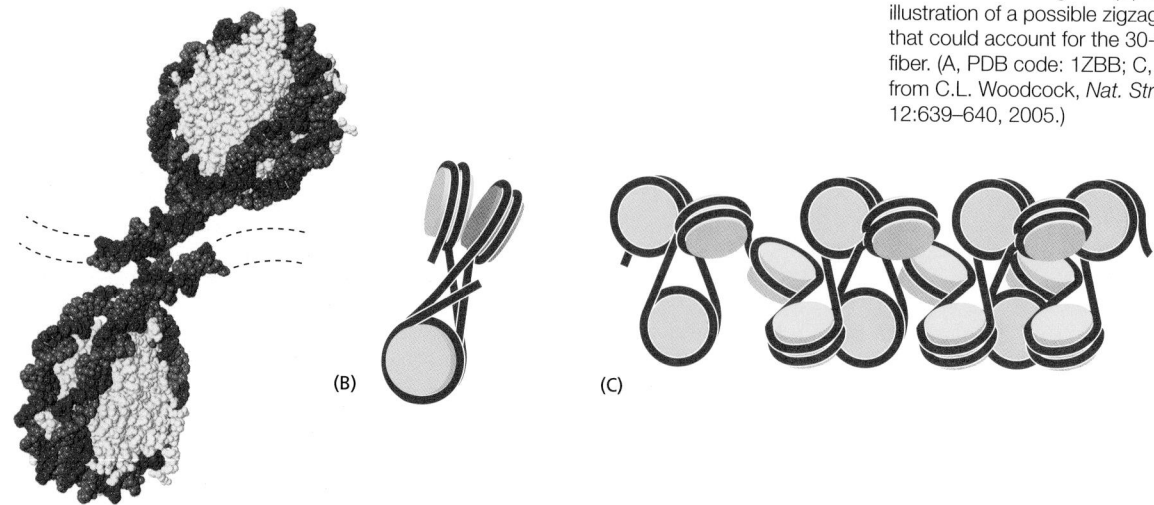

(A)   (B)   (C)

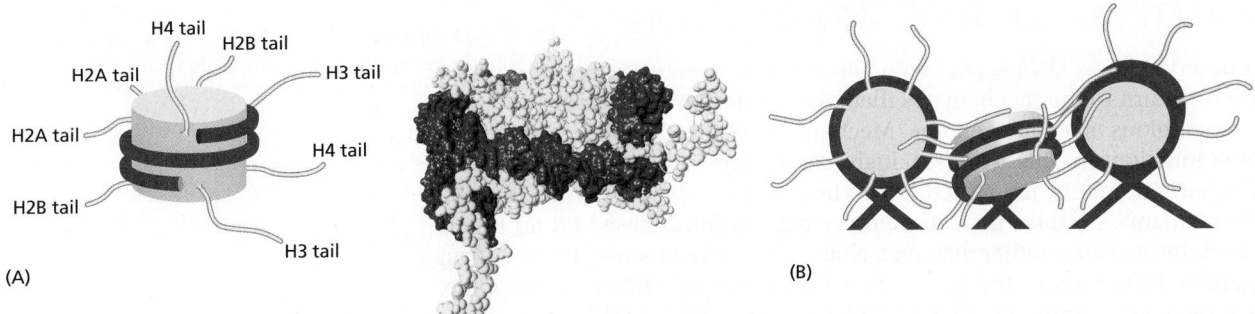

(A)   (B)

Most eukaryotic organisms make several histone H1 proteins of related but quite distinct amino acid sequences. The presence of many other DNA-binding proteins, as well as proteins that bind directly to histones, is certain to add important additional features to any array of nucleosomes.

Summary

A gene is a nucleotide sequence in a DNA molecule that acts as a functional unit for the production of a protein, a structural RNA, or a catalytic or regulatory RNA molecule. In eukaryotes, protein-coding genes are usually composed of a string of alternating introns and exons associated with regulatory regions of DNA. A chromosome is formed from a single, enormously long DNA molecule that contains a linear array of many genes, bound to a large set of proteins. The human genome contains 3.2×10^9 DNA nucleotide pairs, divided between 22 different autosomes (present in two copies each) and 2 sex chromosomes. Only a small percentage of this DNA codes for proteins or functional RNA molecules. A chromosomal DNA molecule also contains three other types of important nucleotide sequences: replication origins and telomeres allow the DNA molecule to be efficiently replicated, while a centromere attaches the sister DNA molecules to the mitotic spindle, ensuring their accurate segregation to daughter cells during the M phase of the cell cycle.

The DNA in eukaryotes is tightly bound to an equal mass of histones, which form repeated arrays of DNA–protein particles called nucleosomes. The nucleosome is composed of an octameric core of histone proteins around which the DNA double helix is wrapped. Nucleosomes are spaced at intervals of about 200 nucleotide pairs, and they are usually packed together (with the aid of histone H1 molecules) into quasi-regular arrays to form a 30-nm chromatin fiber. Even though compact, the structure of chromatin must be highly dynamic to allow access to the DNA. There is some spontaneous DNA unwrapping and rewrapping in the nucleosome itself; however, the general strategy for reversibly changing local chromatin structure features ATP-driven chromatin remodeling complexes. Cells contain a large set of such complexes, which are targeted to specific regions of chromatin at appropriate times. The remodeling complexes collaborate with histone chaperones to allow nucleosome cores to be repositioned, reconstituted with different histones, or completely removed to expose the underlying DNA.

Figure 4–29 A model for the role played by histone tails in the compaction of chromatin. (A) A schematic diagram shows the approximate exit points of the eight histone tails, one from each histone protein, that extend from each nucleosome. The actual structure is shown to its right. In the high-resolution structure of the nucleosome, the tails are largely unstructured, suggesting that they are highly flexible. (B) As indicated, the histone tails are thought to be involved in interactions between nucleosomes that help to pack them together. (A, PDB code: 1KX5.)

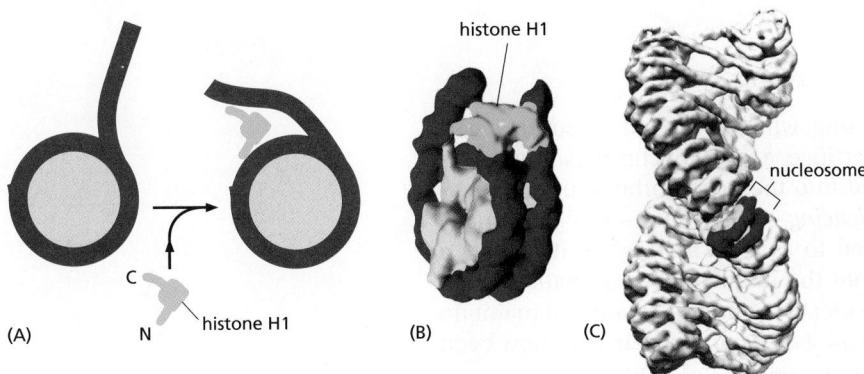

(A)   (B)   (C)

Figure 4–30 How the linker histone binds to the nucleosome. The position and structure of histone H1 is shown. The H1 core region constrains an additional 20 nucleotide pairs of DNA where it exits from the nucleosome core and is important for compacting chromatin. (A) Schematic, and (B) structure inferred for a single nucleosome from a structure determined by high-resolution electron microscopy of a reconstituted chromatin fiber (C). (B and C, adapted from F. Song et al., *Science* 344:376–380, 2014.)

CHROMATIN STRUCTURE AND FUNCTION

Having described how DNA is packaged into nucleosomes to create a chromatin fiber, we now turn to the mechanisms that create different chromatin structures in different regions of a cell's genome. Mechanisms of this type have a variety of important functions in cells. Most strikingly, certain types of chromatin structure can be inherited; that is, the structure can be directly passed down from a cell to its descendants. Because the cell memory that results is based on an inherited chromatin structure rather than on a change in DNA sequence, this is a form of **epigenetic inheritance**. The prefix *epi* is Greek for "on"; this is appropriate, because epigenetics represents a form of inheritance that is superimposed on the genetic inheritance based on DNA.

In Chapter 7, we shall introduce the many different ways in which the expression of genes is regulated. There we discuss epigenetic inheritance in detail and present several different mechanisms that can produce it. Here, we are concerned with only one, that based on chromatin structure. We begin this section by reviewing the observations that first demonstrated that chromatin structures can be inherited. We then describe some of the chemistry that makes this possible—the covalent modification of histones in nucleosomes. These modifications have many functions, inasmuch as they serve as recognition sites for protein domains that link specific protein complexes to different regions of chromatin. Histones thereby have effects on gene expression, as well as on many other DNA-linked processes. Through such mechanisms, chromatin structure plays an important role in the development, growth, and maintenance of all eukaryotic organisms, including ourselves.

Heterochromatin Is Highly Organized and Restricts Gene Expression

Light-microscope studies in the 1930s distinguished two types of chromatin in the interphase nuclei of many higher eukaryotic cells: a highly condensed form, called **heterochromatin**, and all the rest, which is less condensed, called **euchromatin**. Heterochromatin represents an especially compact form of chromatin (see Figure 4–9), and we are finally beginning to understand its molecular properties. It is highly concentrated in certain specialized regions, most notably at the centromeres and telomeres introduced previously (see Figure 4–19), but it is also present at many other locations along chromosomes—locations that can vary according to the physiological state of the cell. In a typical mammalian cell, more than 10% of the genome is packaged in this way.

The DNA in heterochromatin typically contains few genes, and when euchromatic regions are converted to a heterochromatic state, their genes are generally switched off as a result. However, we know now that the term *heterochromatin* encompasses several distinct modes of chromatin compaction that have different implications for gene expression. Thus, heterochromatin should not be thought of as simply encapsulating "dead" DNA, but rather as a descriptor for compact chromatin domains that share the common feature of being unusually resistant to gene expression.

The Heterochromatic State Is Self-Propagating

Through chromosome breakage and rejoining, whether brought about by a natural genetic accident or by experimental artifice, a piece of chromosome that is normally euchromatic can be translocated into the neighborhood of heterochromatin. Remarkably, this often causes *silencing*—inactivation—of the normally active genes. This phenomenon is referred to as a *position effect*. It reflects a spreading of the heterochromatic state into the originally euchromatic region, and it has provided important clues to the mechanisms that create and maintain heterochromatin. First recognized in *Drosophila*, position effects have now been observed in many eukaryotes, including yeasts, plants, and humans.

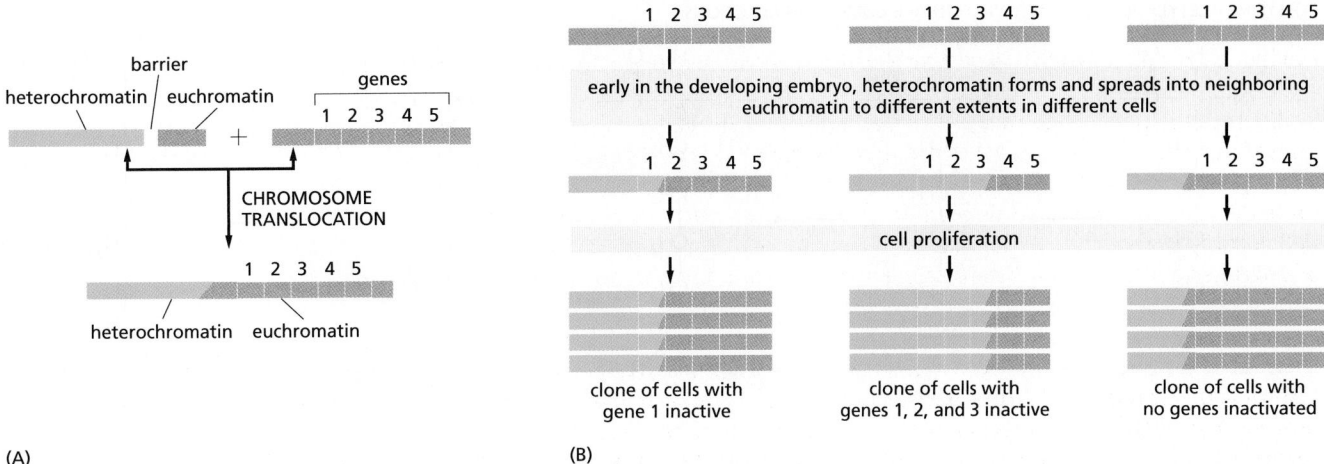

(A)

(B)

Figure 4–31 **The cause of position effect variegation in *Drosophila*.** (A) Heterochromatin *(green)* is normally prevented from spreading into adjacent regions of euchromatin *(red)* by *barrier* DNA sequences, which we shall discuss shortly. In flies that inherit certain chromosomal rearrangements, however, this barrier is no longer present. (B) During the early development of such flies, heterochromatin can spread into neighboring chromosomal DNA, proceeding for different distances in different cells. This spreading soon stops, but the established pattern of heterochromatin is subsequently inherited, so that large clones of progeny cells are produced that have the same neighboring genes condensed into heterochromatin and thereby inactivated (hence the "variegated" appearance of some of these flies; see Figure 4–32). Although "spreading" is used to describe the formation of new heterochromatin close to previously existing heterochromatin, the term may not be wholly accurate. There is evidence that during expansion, the condensation of DNA into heterochromatin can "skip over" some regions of chromatin, sparing the genes that lie within them from repressive effects.

In chromosome breakage-and-rejoining events of the sort just described, the zone of silencing, where euchromatin is converted to a heterochromatic state, spreads for different distances in different early cells in the fly embryo. Remarkably, these differences then are perpetuated for the rest of the animal's life: in each cell, once the heterochromatic condition is established on a piece of chromatin, it tends to be stably inherited by all of that cell's progeny (**Figure 4–31**). This remarkable phenomenon, called **position effect variegation**, was first recognized through a detailed genetic analysis of the mottled loss of red pigment in the fly eye (**Figure 4–32**). It shares features with the extensive spread of heterochromatin that inactivates one of the two X chromosomes in female mammals. There too, a random process acts in each cell of the early embryo to dictate which X chromosome will be inactivated, and that same X chromosome then remains inactive in all the cell's progeny, creating a mosaic of different clones of cells in the adult body (see Figure 7–50).

These observations, taken together, point to a fundamental strategy of heterochromatin formation: heterochromatin begets more heterochromatin. This positive feedback can operate both in space, causing the heterochromatic state to spread along the chromosome, and in time, across cell generations, propagating the heterochromatic state of the parent cell to its daughters. The challenge is to explain the molecular mechanisms that underlie this remarkable behavior.

Figure 4–32 **The discovery of position effects on gene expression.** The *White* gene in the fruit fly *Drosophila* controls eye pigment production and is named after the mutation that first identified it. Wild-type flies with a normal *White* gene (*White*[+]) have normal pigment production, which gives them red eyes, but if the *White* gene is mutated and inactivated, the mutant flies (*White*[−]) make no pigment and have white eyes. In flies in which a normal *White* gene has been moved near a region of heterochromatin, the eyes are mottled, with both red and white patches. The white patches represent cell lineages in which the *White* gene has been silenced by the effects of the heterochromatin. In contrast, the red patches represent cell lineages in which the *White* gene is expressed. Early in development, when the heterochromatin is first formed, it spreads into neighboring euchromatin to different extents in different embryonic cells (see Figure 4–31). The presence of large patches of red and white cells reveals that the state of transcriptional activity, as determined by the packaging of this gene into chromatin in those ancestor cells, is inherited by all daughter cells.

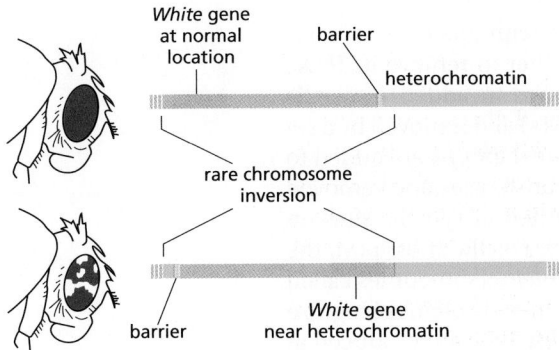

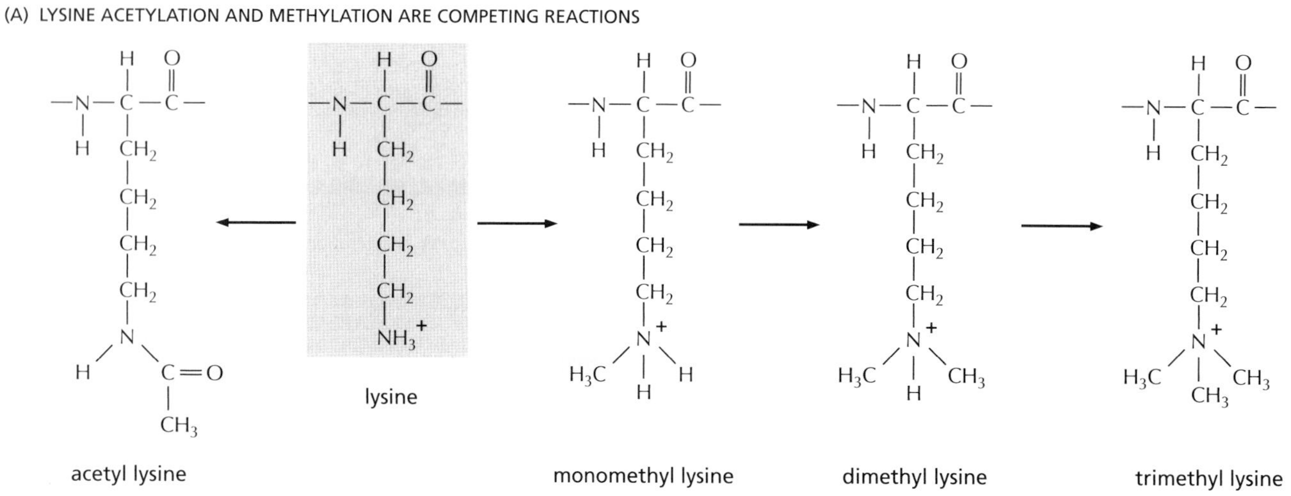

(A) LYSINE ACETYLATION AND METHYLATION ARE COMPETING REACTIONS

acetyl lysine   lysine   monomethyl lysine   dimethyl lysine   trimethyl lysine

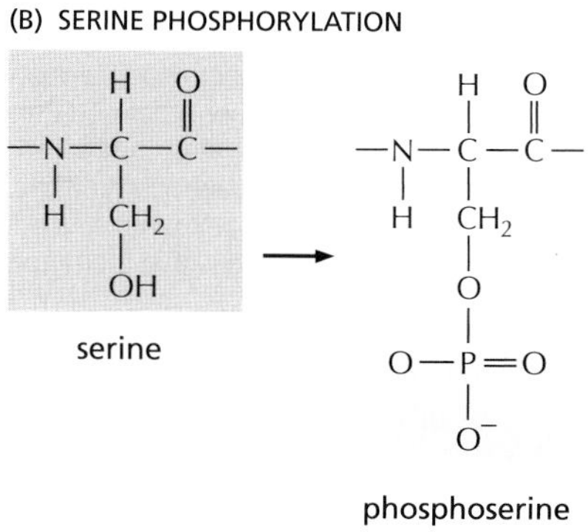

Figure 4–33 Some prominent types of covalent amino acid side-chain modifications found on nucleosomal histones. (A) Three different levels of lysine methylation are shown; each can be recognized by a different binding protein and thus each can have a different significance for the cell. Note that acetylation removes the plus charge on lysine, and that, most importantly, an acetylated lysine cannot be methylated, and vice versa. (B) Serine phosphorylation adds a negative charge to a histone. Modifications of histones not shown here include the mono- or dimethylation of an arginine, the phosphorylation of a threonine, the addition of ADP-ribose to a glutamic acid, and the addition of a ubiquityl, sumoyl, or biotin group to a lysine.

(B) SERINE PHOSPHORYLATION

serine

phosphoserine

As a first step, one can carry out a search for the molecules that are involved. This has been done by means of *genetic screens*, in which large numbers of mutants are generated, after which one picks out those that show an abnormality of the process in question. Extensive genetic screens in *Drosophila*, fungi, and mice have identified more than 100 genes whose products either enhance or suppress the spread of heterochromatin and its stable inheritance—in other words, genes that serve as either enhancers or suppressors of position effect variegation. Many of these genes turn out to code for non-histone chromosomal proteins that interact with histones and are involved in modifying or maintaining chromatin structure. We shall discuss how they work in the sections that follow.

The Core Histones Are Covalently Modified at Many Different Sites

The amino acid side chains of the four histones in the nucleosome core are subjected to a remarkable variety of covalent modifications, including the acetylation of lysines, the mono-, di-, and trimethylation of lysines, and the phosphorylation of serines (**Figure 4–33**). A large number of these side-chain modifications occur on the eight relatively unstructured N-terminal "histone tails" that protrude from the nucleosome (**Figure 4–34**). However, there are also more than 20 specific side-chain modifications on the nucleosome's globular core.

All of the above types of modifications are reversible, with one enzyme serving to create a particular type of modification, and another to remove it. These enzymes are highly specific. Thus, for example, acetyl groups are added to specific lysines by a set of different *histone acetyl transferases* (HATs) and removed by a set of *histone deacetylase complexes* (HDACs). Likewise, methyl groups are added to lysine side chains by a set of different histone methyl transferases and removed by a set of histone demethylases. Each enzyme is recruited to specific sites on the chromatin at defined times in each cell's life history. For the most part, the initial recruitment depends on *transcription regulator proteins* (sometimes called "transcription factors"). As we shall explain in Chapter 7, these proteins recognize and bind to specific DNA sequences in the chromosomes. They are produced at

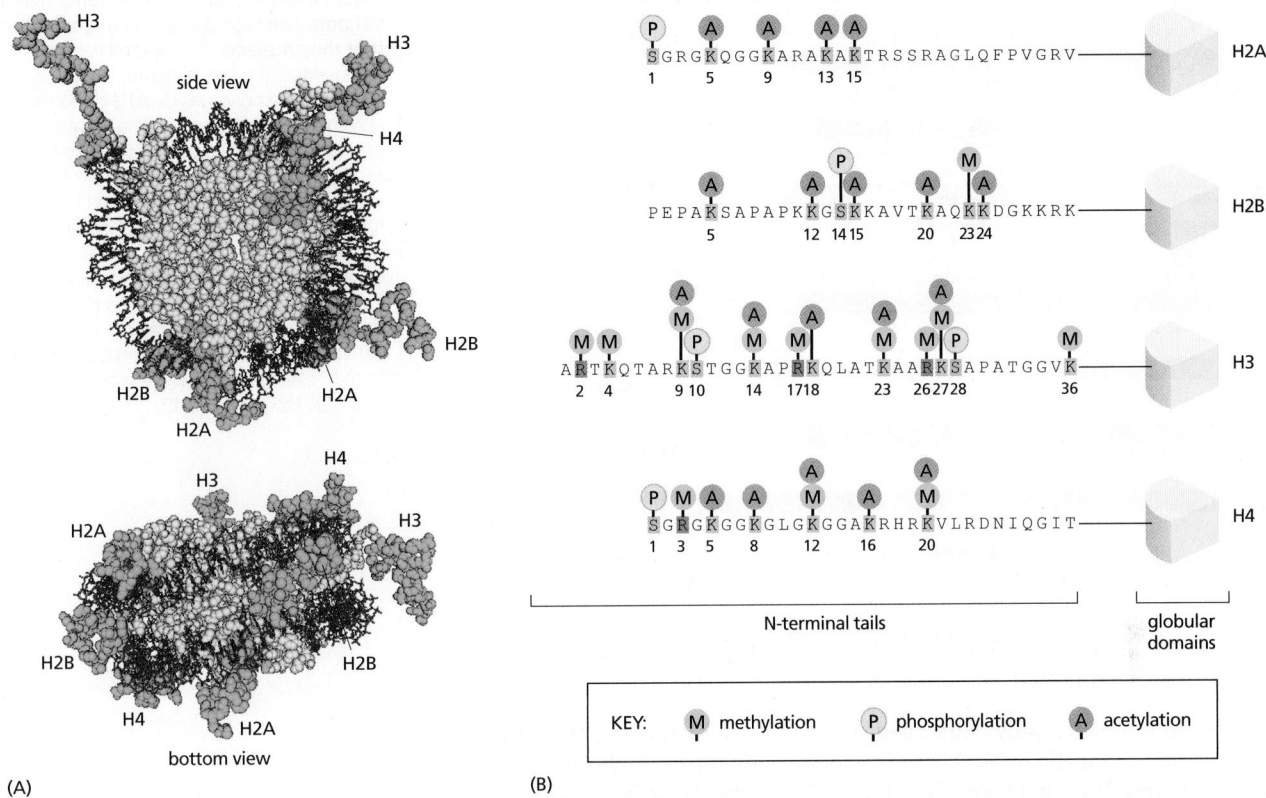

Figure 4–34 The covalent modification of core histone tails. (A) The structure of the nucleosome highlighting the location of the first 30 or so amino acids in each of its eight N-terminal histone tails *(green)*. These tails are unstructured and highly mobile, and thus will change their conformation depending on other bound proteins. (B) Well-documented modifications of the four histone core proteins are indicated. Although only a single symbol is used here for methylation (M), each lysine (K) or arginine (R) can be methylated in several different ways. Note also that some positions (e.g., lysine 9 of H3) can be modified either by methylation or by acetylation, but not both. Most of the modifications shown add a relatively small molecule onto the histone tails; the exception is ubiquitin, a 76-amino-acid protein also used for other cell processes (see Figure 3–69). Not shown are more than 20 possible modifications located in the globular core of the histones. (A, PDB: 1KX5; B, adapted from H. Santos-Rosa and C. Caldas, *Eur. J. Cancer* 41:2381–2402, 2005. With permission from Elsevier.)

different times and places in the life of an organism, thereby determining where and when the chromatin-modifying enzymes will act. In this way, the DNA sequence ultimately determines how histones are modified. But in at least some cases, the covalent modifications on nucleosomes can persist long after the transcription regulator proteins that first induced them have disappeared, thereby providing the cell with a memory of its developmental history. Most remarkably, as in the related phenomenon of position effect variegation discussed above, this memory can be transmitted from one cell generation to the next.

Very different patterns of covalent modification are found on different groups of nucleosomes, depending both on their exact position in the genome and on the history of the cell. The modifications of the histones are carefully controlled, and they have important consequences. The acetylation of lysines on the N-terminal tails loosens chromatin structure, in part because adding an acetyl group to lysine removes its positive charge, thereby reducing the affinity of the tails for adjacent nucleosomes. However, the most profound effects of the histone modifications lie in their ability to recruit specific other proteins to the modified stretch of chromatin. Trimethylation of one specific lysine on the histone H3 tail, for instance, attracts the heterochromatin-specific protein HP1 and contributes to the establishment and spread of heterochromatin. More generally, the recruited proteins act with the modified histones to determine how and when genes will be expressed, as well as other chromosome functions. In this way, the precise structure of each domain of chromatin governs the readout of the genetic information that it contains, and thereby the structure and function of the eukaryotic cell.

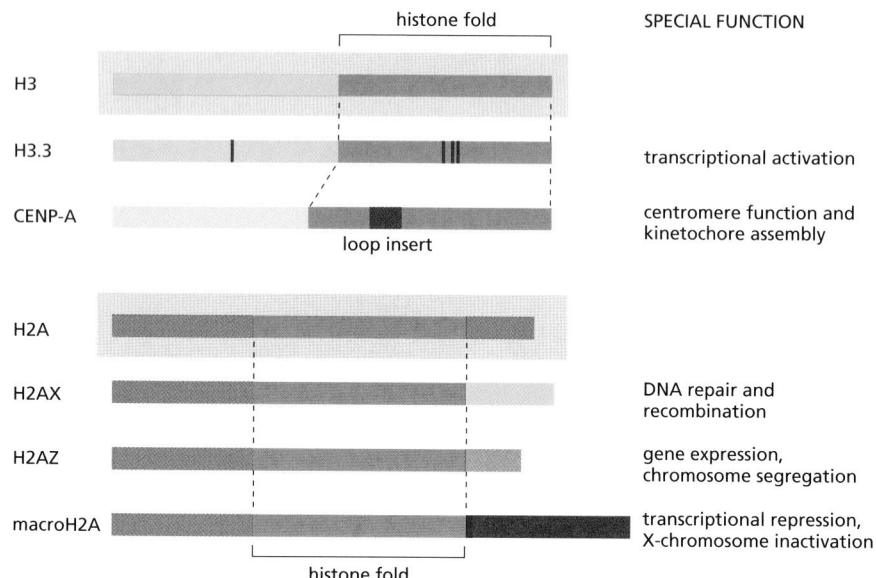

Figure 4–35 The structure of some histone variants compared with the major histone that they replace. The histone variants are inserted into nucleosomes at specific sites on chromosomes by ATP-dependent chromatin remodeling enzymes that act in concert with histone chaperones (see Figure 4–27). The CENP-A (Centromere Protein-A) variant of histone H3 is discussed later in this chapter (see Figure 4–42); other variants are discussed in Chapter 7. The sequences in each variant that are colored differently (compared to the major histone above it) denote regions with an amino acid sequence different from this major histone. (Adapted from K. Sarma and D. Reinberg, *Nat. Rev. Mol. Cell Biol.* 6:139–149, 2005.)

Chromatin Acquires Additional Variety Through the Site-Specific Insertion of a Small Set of Histone Variants

In addition to the four highly conserved standard core histones, eukaryotes also contain a few variant histones that can also assemble into nucleosomes. These histones are present in much smaller amounts than the major histones, and they have been less well conserved over long evolutionary times. Variants are known for each of the core histones with the exception of H4; some examples are shown in **Figure 4–35**.

The major histones are synthesized primarily during the S phase of the cell cycle and assembled into nucleosomes on the daughter DNA helices just behind the replication fork (see Figure 5–32). In contrast, most histone variants are synthesized throughout interphase. They are often inserted into already-formed chromatin, which requires a histone-exchange process catalyzed by the ATP-dependent chromatin remodeling complexes discussed previously. These remodeling complexes contain subunits that cause them to bind both to specific sites on chromatin and to histone chaperones that carry a particular variant. As a result, each histone variant is inserted into chromatin in a highly selective manner (see Figure 4–27).

Covalent Modifications and Histone Variants Act in Concert to Control Chromosome Functions

The number of possible distinct markings on an individual nucleosome is in principle enormous, and this potential for diversity is still greater when we allow for nucleosomes that contain histone variants. However, the histone modifications are known to occur in coordinated sets. More than 15 such sets can be identified in mammalian cells. However, it is not yet clear how many different types of chromatin are functionally important in cells.

Some combinations are known to have a specific meaning for the cell in the sense that they determine how and when the DNA packaged in the nucleosomes is to be accessed or manipulated—a fact that led to the idea of a *"histone code."* For example, one type of marking signals that a stretch of chromatin has been newly replicated, another signals that the DNA in that chromatin has been damaged and needs repair, while others signal when and how gene expression should take place. Various regulatory proteins contain small domains that bind to specific marks, recognizing, for example, a trimethylated lysine 4 on histone H3 (**Figure 4–36**). These domains are often linked together as modules in a single large

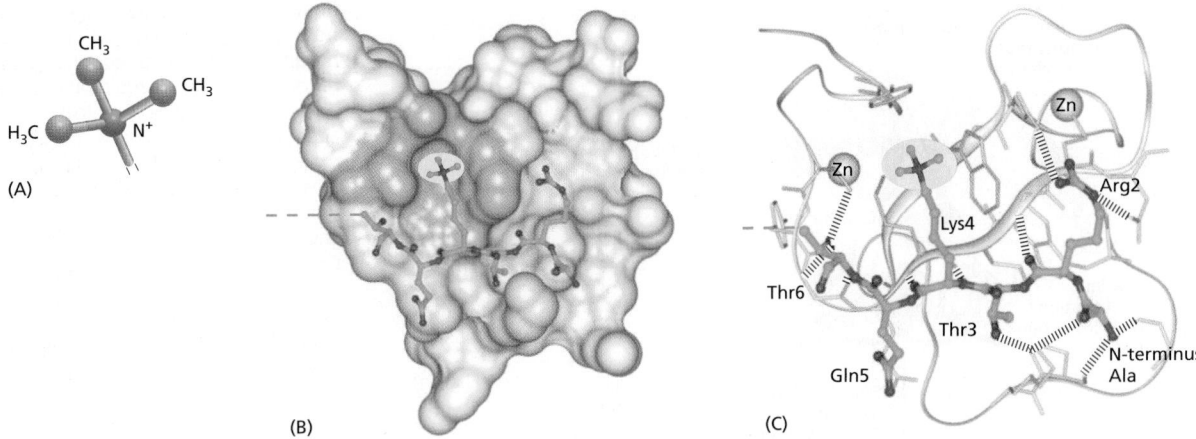

Figure 4–36 How a mark on a nucleosome is read. The figure shows the structure of a protein module (called an ING PHD domain) that specifically recognizes histone H3 trimethylated on lysine 4. (A) A trimethyl group. (B) Space-filling model of an ING PHD domain bound to a histone tail (*green*, with the trimethyl group highlighted in *yellow*). (C) A ribbon model showing how the N-terminal six amino acids in the H3 tail are recognized. The *red lines* represent hydrogen bonds. This is one of a family of PHD domains that recognize methylated lysines on histones; different members of the family bind tightly to lysines located at different positions, and they can discriminate between a mono-, di-, and trimethylated lysine. In a similar way, other small protein modules recognize specific histone side chains that have been marked with acetyl groups, phosphate groups, and so on. (Adapted from P.V. Peña et al., *Nature* 442(7098):100–103, published 2006 by Nature Publishing Group. Reproduced with permission of SNCSC.)

protein or protein complex, which thereby recognizes a specific combination of histone modifications (**Figure 4–37**). The result is a *reader complex* that allows particular combinations of markings on chromatin to attract additional proteins, so as to execute an appropriate biological function at the right time (**Figure 4–38**).

The marks on nucleosomes due to covalent additions to histones are dynamic, being constantly removed and added at rates that depend on their chromosomal locations. Because the histone tails extend outward from the nucleosome core and are likely to be accessible even when chromatin is condensed, they would seem to provide an especially suitable format for creating marks that can be readily altered as a cell's needs change. Although much remains to be learned about the meaning of the different histone modifications, a few well-studied examples of the information that can be encoded in the histone H3 tail are listed in **Figure 4–39**.

A Complex of Reader and Writer Proteins Can Spread Specific Chromatin Modifications Along a Chromosome

The phenomenon of position effect variegation described previously requires that some modified forms of chromatin have the ability to spread for substantial distances along a chromosomal DNA molecule (see Figure 4–31). How is this possible?

The enzymes that add or remove modifications to histones in nucleosomes are part of multisubunit complexes. They can initially be brought to a particular region of chromatin by one of the sequence-specific DNA-binding proteins (transcription regulators) discussed in Chapters 6 and 7 (for a specific example,

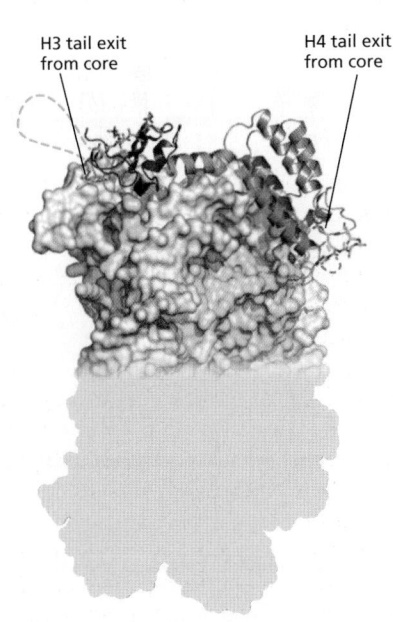

Figure 4–37 Recognition of a specific combination of marks on a nucleosome. In the example shown, two adjacent domains that are part of the NURF (Nucleosome Remodeling Factor) chromatin remodeling complex bind to the nucleosome, with the PHD domain *(red)* recognizing a methylated H3 lysine 4 and another domain (a bromodomain, *blue*) recognizing an acetylated H4 lysine 16. These two histone marks constitute a unique histone modification pattern that occurs in subsets of nucleosomes in human cells. Here the two histone tails are indicated by *green dotted* lines, and only half of one nucleosome is shown. (Adapted from A.J. Ruthenburg et al., *Cell* 145:692–706, 2011.)

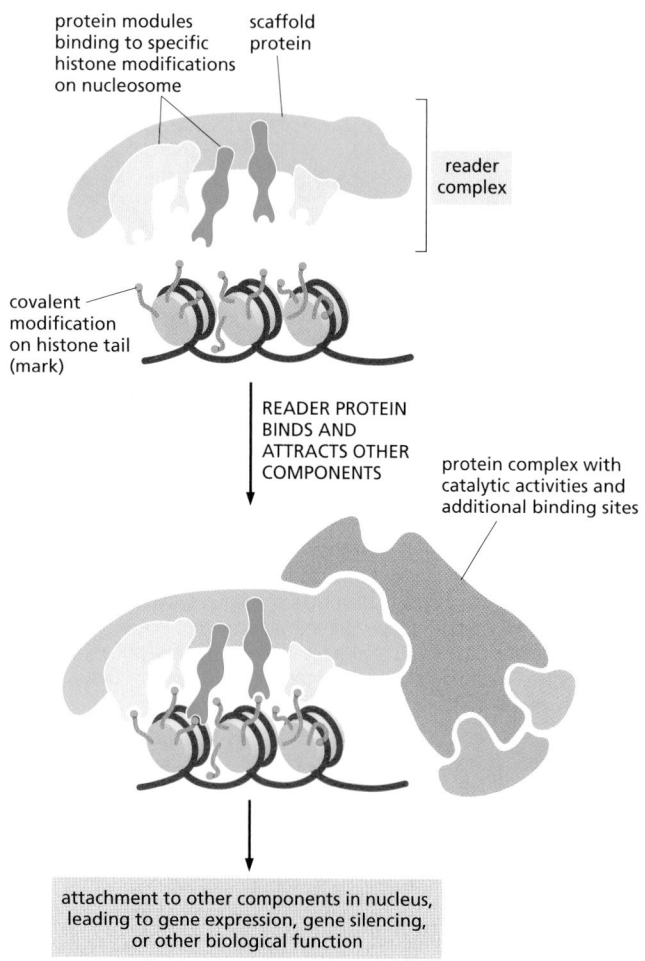

Figure 4–38 Schematic diagram showing how a particular combination of histone modifications can be recognized by a reader complex. A large protein complex that contains a series of protein modules, each of which recognizes a specific histone mark, is schematically illustrated *(green)*. This "reader complex" will bind tightly only to a region of chromatin that contains several of the different histone marks that it recognizes. Therefore, only a specific combination of marks will cause the complex to bind to chromatin and attract the additional protein complexes *(purple)* needed to catalyze a biological function.

see Figure 7–20). But after a modifying enzyme "writes" its mark on one or a few neighboring nucleosomes, events that resemble a chain reaction can ensue. In such a case, the "writer enzyme" works in concert with a "reader protein" located in the same protein complex. The reader protein contains a module that recognizes the mark and binds tightly to the newly modified nucleosome (see Figure

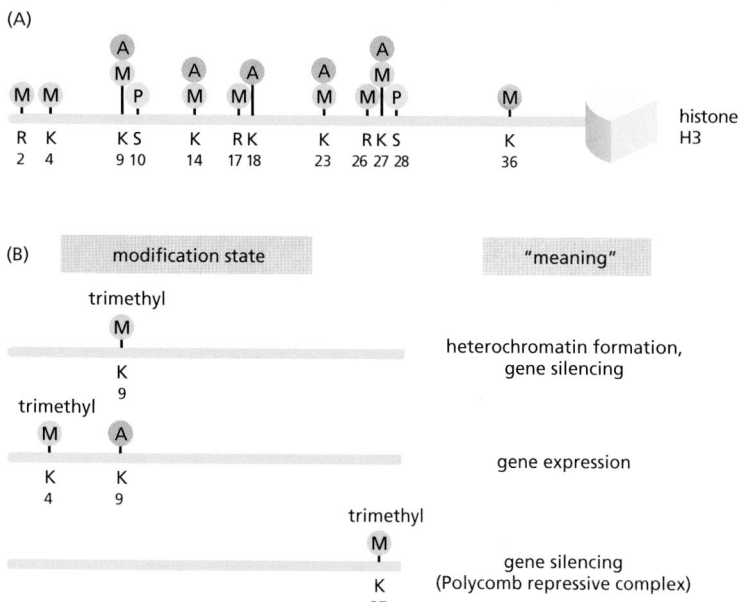

Figure 4–39 Some specific meanings of histone modifications. (A) The modifications on the histone H3 N-terminal tail are shown, repeated from Figure 4–34. (B) The H3 tail can be marked by different sets of modifications that act in combination to convey a specific meaning. Only a small number of the meanings are known, including the three examples shown. Not illustrated is the fact that, as just implied (see Figure 4–38), reading a histone mark generally involves the joint recognition of marks at other sites on the nucleosome along with the indicated H3 tail recognition. In addition, specific levels of methylation (mono-, di-, or trimethyl groups) are generally required. Thus, for example, the trimethylation of lysine 9 attracts the heterochromatin-specific protein HP1, which induces a spreading wave of further lysine 9 trimethylation followed by further HP1 binding, according to the general scheme that will be illustrated shortly (see Figure 4–40). Also important in this process, however, is a synergistic trimethylation of the histone H4 N-terminal tail on lysine 20.

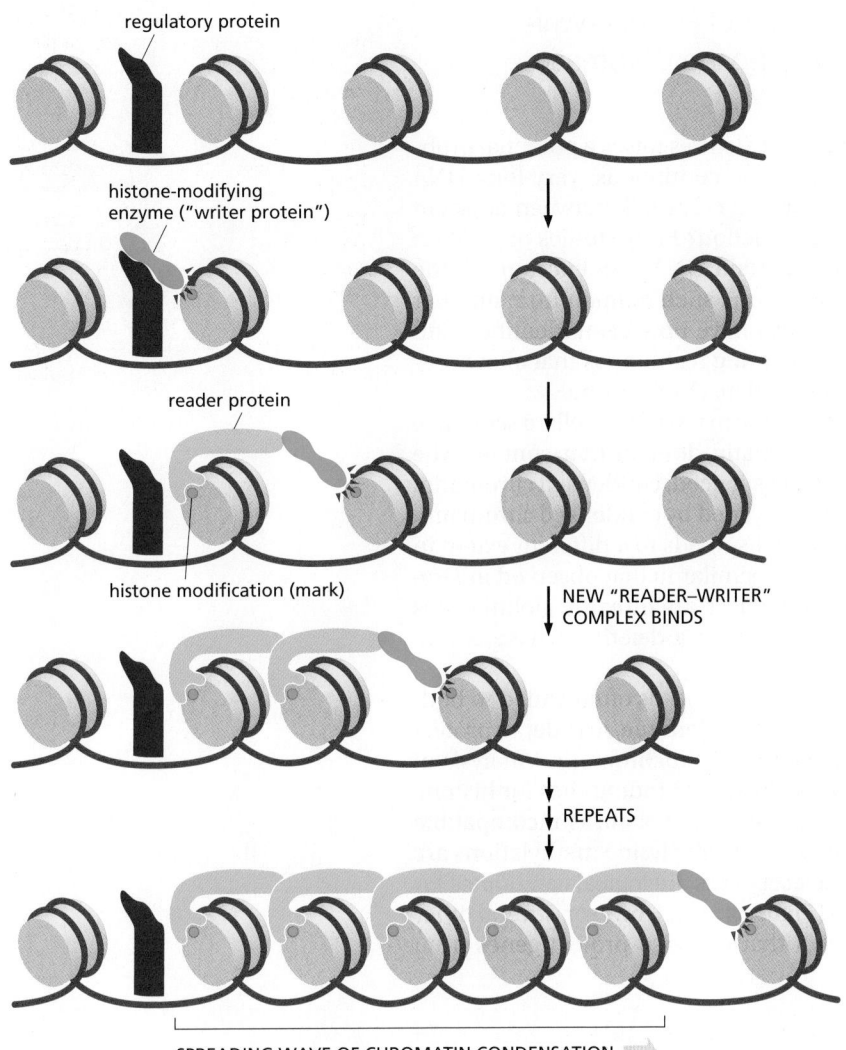

regulatory protein

histone-modifying enzyme ("writer protein")

reader protein

histone modification (mark)

NEW "READER–WRITER" COMPLEX BINDS

REPEATS

SPREADING WAVE OF CHROMATIN CONDENSATION

Figure 4–40 How the recruitment of a reader–writer complex can spread chromatin changes along a chromosome. The writer is an enzyme that creates a specific modification on one or more of the four nucleosomal histones. After its recruitment to a specific site on a chromosome by a transcription regulatory protein, the writer collaborates with a reader protein to spread its mark from nucleosome to nucleosome by means of the indicated reader–writer complex. For this mechanism to work, the reader must recognize the same histone modification mark that the writer produces; its binding to that mark can be shown to activate the writer. In this schematic example, a spreading wave of chromatin condensation is thereby induced. Not shown are the additional proteins involved, including an ATP-dependent chromatin remodeling complex required to reposition the modified nucleosomes.

4–36), activating an attached writer enzyme and positioning it near an adjacent nucleosome. Through many such read–write cycles, the reader protein can carry the writer enzyme along the DNA—spreading the mark in a hand-over-hand manner along the chromosome (**Figure 4–40**).

In reality, the process is more complicated than the scheme just described. Both readers and writers are part of a protein complex that is likely to contain multiple readers and writers, and to require multiple marks on the nucleosome to spread. Moreover, many of these reader–writer complexes also contain an ATP-dependent chromatin remodeling protein (see Figure 4–26C), and the reader, writer, and remodeling proteins can work in concert to either decondense or condense long stretches of chromatin as the reader moves progressively along the nucleosome-packaged DNA.

A similar process is used to remove histone modifications from specific regions of the DNA; in this case, an "eraser enzyme," such as a histone demethylase or histone deacetylase, is recruited to the complex. As for the writer complex in Figure 4–40, sequence-specific DNA-binding proteins (transcription regulators) direct where such modifications occur (discussed in Chapter 7).

Some idea of the complexity of the above processes can be derived from the results of genetic screens for genes that either enhance or suppress the spreading and stability of heterochromatin, as manifest in effects on position effect variegation in *Drosophila* (see Figure 4–32). As pointed out previously, more than 100 such genes are known, and most of them are likely to code for subunits in one or more reader–writer–remodeling protein complexes.

Barrier DNA Sequences Block the Spread of Reader–Writer Complexes and thereby Separate Neighboring Chromatin Domains

The above mechanism for spreading chromatin structures raises a potential problem. Inasmuch as each chromosome contains one continuous, very long DNA molecule, what prevents a cacophony of confusing cross-talk between adjacent chromatin domains of different structure and function? Early studies of position effect variegation had suggested an answer: certain DNA sequences mark the boundaries of chromatin domains and separate one such domain from another (see Figure 4–31). Several such *barrier sequences* have now been identified and characterized through the use of genetic engineering techniques that allow specific DNA segments to be deleted from, or inserted in, chromosomes.

For example, in cells that are destined to give rise to red blood cells, a sequence called HS4 normally separates the active chromatin domain that contains the human β-globin locus from an adjacent region of silenced, condensed chromatin. If this sequence is deleted, the β-globin locus is invaded by condensed chromatin. This chromatin silences the genes it covers, and it spreads to a different extent in different cells, causing position effect variegation similar to that observed in *Drosophila*. As described in Chapter 7, the consequences are dire: the globin genes are poorly expressed, and individuals who carry such a deletion have a severe form of anemia.

In genetic engineering experiments, the HS4 sequence is often added to both ends of a gene that is to be inserted into a mammalian genome, in order to protect that gene from the silencing caused by spreading heterochromatin. Analysis of this barrier sequence reveals that it contains a cluster of binding sites for histone acetylase enzymes. Since the acetylation of a lysine side chain is incompatible with the methylation of the same side chain, and specific lysine methylations are required to spread heterochromatin, histone acetylases are logical candidates for the formation of DNA barriers to spreading (**Figure 4–41**). However, several other types of chromatin modifications are known that can also protect genes from silencing.

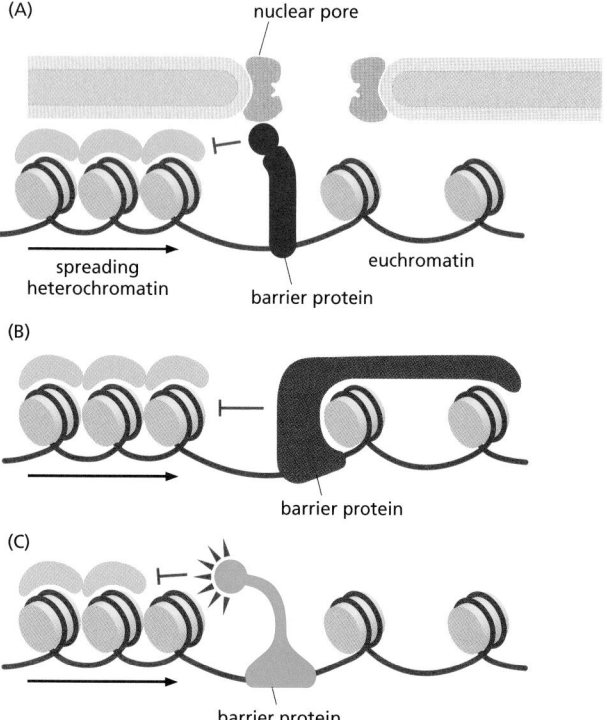

Figure 4–41 **Some mechanisms of barrier action.** These models are derived from experimental analyses of barrier action, and a combination of several of them may function at any one site. (A) The tethering of a region of chromatin to a large fixed site, such as the nuclear pore complex illustrated here, can form a barrier that stops the spread of heterochromatin. (B) The tight binding of barrier proteins to a group of nucleosomes can make this chromatin resistant to heterochromatin spreading. (C) By recruiting a group of highly active histone-modifying enzymes, barriers can erase the histone marks that are required for heterochromatin to spread. For example, a potent acetylation of lysine 9 on histone H3 will compete with lysine 9 methylation, thereby preventing the binding of the HP1 protein needed to form a major form of heterochromatin. (Based on A.G. West and P. Fraser, *Hum. Mol. Genet.* 14:R101–R111, 2005.)

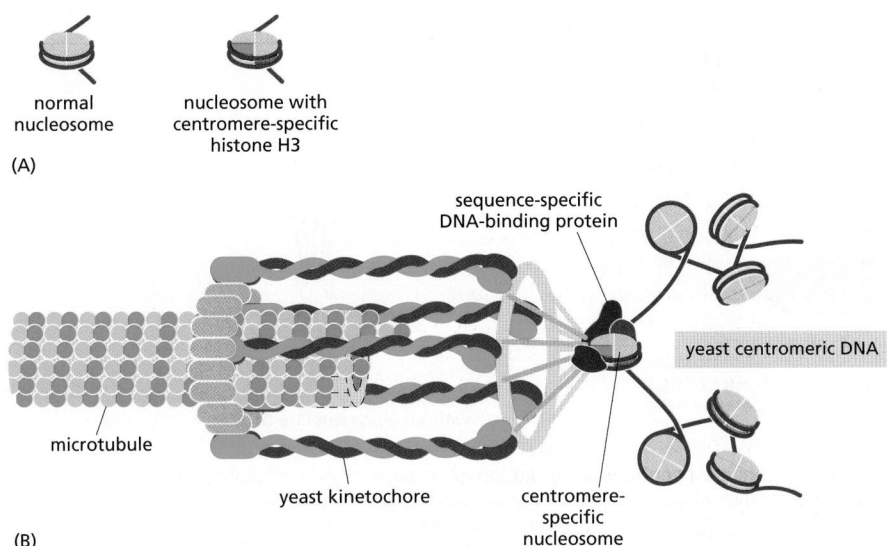

Figure 4–42 **A model for the structure of a simple centromere.** (A) In the yeast *Saccharomyces cerevisiae*, a special centromeric DNA sequence assembles a single nucleosome in which two copies of an H3 variant histone (called CENP-A in most organisms) replace the normal H3. (B) How peptide sequences unique to this variant histone (see Figure 4–35) help to assemble additional proteins, some of which form a kinetochore. The yeast kinetochore is unusual in capturing only a single microtubule; humans have much larger centromeres and form kinetochores that can capture 20 or more microtubules (see Figure 4–43). The kinetochore is discussed in detail in Chapter 17. (Adapted from A. Joglekar et al., *Nat. Cell Biol.* 8:581–585, 2006.)

The Chromatin in Centromeres Reveals How Histone Variants Can Create Special Structures

Nucleosomes carrying histone variants have a distinctive character and are thought to be able to produce marks in chromatin that are unusually long-lasting. An important example is seen in the formation and inheritance of the specialized chromatin structure at the centromere, the region of each chromosome required for attachment to the mitotic spindle and orderly segregation of the duplicated copies of the genome into daughter cells each time a cell divides. In many complex organisms, including humans, each centromere is embedded in a stretch of special *centromeric chromatin* that persists throughout interphase, even though the centromere-mediated attachment to the spindle and movement of DNA occur only during mitosis. This chromatin contains a centromere-specific variant H3 histone, known as CENP-A (Centromere Protein-A; see Figure 4–35), plus additional proteins that pack the nucleosomes into particularly dense arrangements and form the kinetochore, the special structure required for attachment of the mitotic spindle (see Figure 4–19).

A specific DNA sequence of approximately 125 nucleotide pairs is sufficient to serve as a centromere in the yeast *S. cerevisiae*. Despite its small size, more than a dozen different proteins assemble on this DNA sequence; the proteins include the CENP-A histone H3 variant, which, along with the three other core histones, forms a centromere-specific nucleosome. The additional proteins at the yeast centromere attach this nucleosome to a single microtubule from the yeast mitotic spindle (**Figure 4–42**).

The centromeres in more complex organisms are considerably larger than those in budding yeasts. For example, fly and human centromeres extend over hundreds of thousands of nucleotide pairs and, while they contain CENP-A, they do not seem to contain a centromere-specific DNA sequence. These centromeres largely consist of short, repeated DNA sequences, known as *alpha satellite DNA* in humans. But the same repeat sequences are also found at other (non-centromeric) positions on chromosomes, indicating that they are not sufficient to direct centromere formation. Most strikingly, in some unusual cases, new human centromeres (called neocentromeres) have been observed to form spontaneously on fragmented chromosomes. Some of these new positions were originally euchromatic and lack alpha satellite DNA altogether (**Figure 4–43**). It seems that centromeres in complex organisms are defined by an assembly of proteins, rather than by a specific DNA sequence.

Inactivation of some centromeres and genesis of others *de novo* seem to have played an essential part in evolution. Different species, even when quite closely

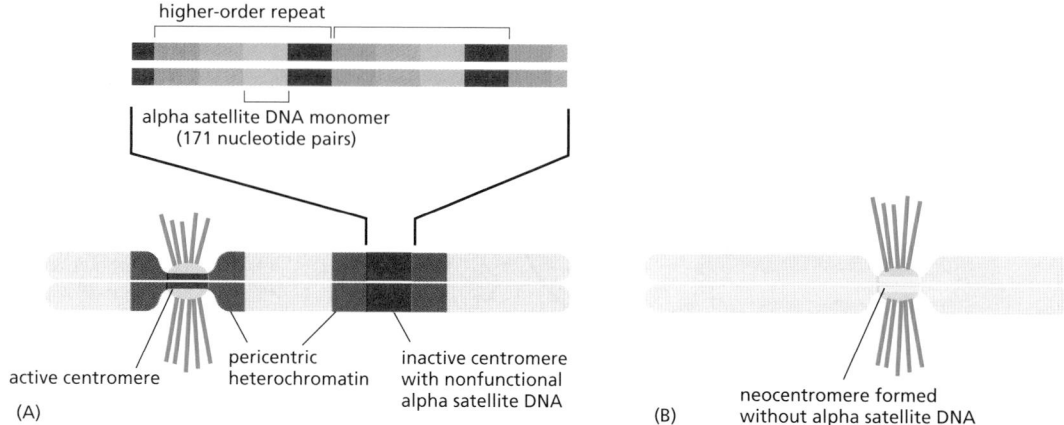

Figure 4–43 Evidence for the plasticity of human centromere formation. (A) A series of A-T-rich alpha satellite DNA sequences is repeated many thousands of times at each human centromere *(red)*, and is surrounded by *pericentric heterochromatin (brown)*. However, due to an ancient chromosome breakage-and-rejoining event, some human chromosomes contain two blocks of alpha satellite DNA, each of which presumably functioned as a centromere in its original chromosome. Usually, chromosomes with two functional centromeres are not stably propagated because they attach improperly to the spindle and are broken apart during mitosis. In chromosomes that do survive, however, one of the centromeres has somehow become inactivated, even though it contains all the necessary DNA sequences. This allows the chromosome to be stably propagated. (B) In a small fraction (1/2000) of human births, extra chromosomes are observed in cells of the offspring. Some of these extra chromosomes, which have formed from a breakage event, lack alpha satellite DNA altogether, yet new centromeres (neocentromeres) have arisen from what was originally euchromatic DNA.

The complexity of centromeric chromatin is not illustrated in these diagrams. The alpha satellite DNA that forms centromeric chromatin in humans is packaged into alternating blocks of chromatin. One block is formed from a long string of nucleosomes containing the CENP-A H3 variant histone; the other block contains nucleosomes that are specially marked with dimethyl lysine 4 on the normal H3 histone. Each block is more than a thousand nucleosomes long. This centromeric chromatin is flanked by pericentric heterochromatin, as shown. The pericentric chromatin contains methylated lysine 9 on its H3 histones, along with HP1 protein, and it is an example of "classical" heterochromatin (see Figure 4–39).

related, often have different numbers of chromosomes; see Figure 4–14 for an extreme example. As we shall discuss below, detailed genome comparisons show that in many cases the changes in chromosome numbers have arisen through chromosome breakage-and-rejoining events, creating novel chromosomes, some of which must initially have contained abnormal numbers of centromeres—either more than one, or none at all. Yet stable inheritance requires that each chromosome should contain one centromere, and one only. It seems that surplus centromeres must have been inactivated, and/or new centromeres created, so as to allow the rearranged chromosome sets to be stably maintained.

Some Chromatin Structures Can Be Directly Inherited

The changes in centromere activity just discussed, once established, need to be perpetuated through subsequent cell generations. What could be the mechanism of this type of epigenetic inheritance?

It has been proposed that *de novo* centromere formation requires an initial seeding event, involving the formation of a specialized DNA–protein structure that contains nucleosomes formed with the CENP-A variant of histone H3. In humans, this seeding event happens more readily on arrays of alpha satellite DNA than on other DNA sequences. The H3–H4 tetramers from each nucleosome on the parental DNA helix are directly inherited by the sister DNA helices at a replication fork (see Figure 5–32). Therefore, once a set of CENP-A-containing nucleosomes has been assembled on a stretch of DNA, it is easy to understand how a new centromere could be generated in the same place on both daughter chromosomes following each round of cell division. One need only assume that the presence of the CENP-A histone in an inherited nucleosome selectively recruits more CENP-A histone to its newly formed neighbors.

There are some striking similarities between the formation and maintenance of centromeres and the formation and maintenance of some other regions of

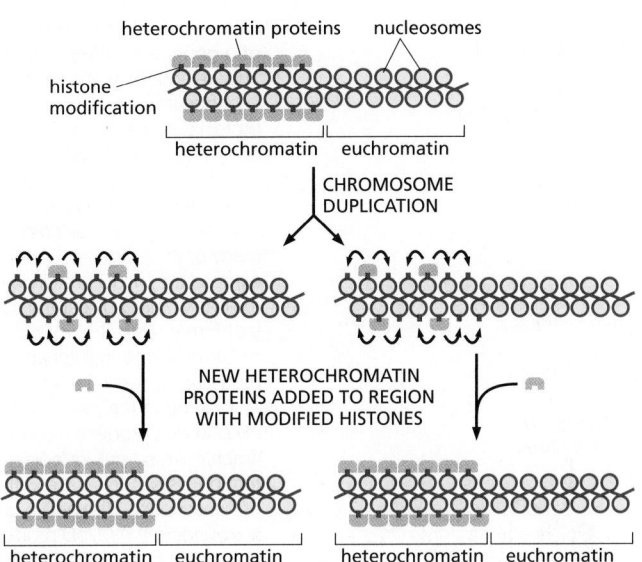

Figure 4–44 How the packaging of DNA in chromatin can be inherited following chromosome replication. In this model, some of the specialized chromatin components are distributed to each sister chromosome after DNA duplication, along with the specially marked nucleosomes that they bind. After DNA replication, the inherited nucleosomes that are specially modified, acting in concert with the inherited chromatin components, change the pattern of histone modification on the newly formed nucleosomes nearby. This creates new binding sites for the same chromatin components, which then assemble to complete the structure. The latter process is likely to involve reader–writer–remodeling complexes operating in a manner similar to that previously illustrated in Figure 4–40.

heterochromatin. In particular, the entire centromere forms as an all-or-none entity, suggesting that the creation of centromeric chromatin is a highly cooperative process, spreading out from an initial seed in a manner reminiscent of the phenomenon of position effect variegation that we discussed earlier. In both cases, a particular chromatin structure, once formed, seems to be directly inherited on the DNA following each round of chromosome replication. A cooperative recruitment of proteins, along with the action of reader–writer complexes, can thus not only account for the spreading of specific forms of chromatin in space along the chromosome, but also for its propagation across cell generations—from parent cell to daughter cell (**Figure 4–44**).

Experiments with Frog Embryos Suggest that both Activating and Repressive Chromatin Structures Can Be Inherited Epigenetically

Epigenetic inheritance plays a central part in the creation of multicellular organisms. Their differentiated cell types become established during development, and persist thereafter even through repeated cell-division cycles. The daughters of a liver cell persist as liver cells, those of an epidermal cell as epidermal cells, and so on, even though they all contain the same genome; and this is because distinctive patterns of gene expression are passed on faithfully from parent cell to daughter cell. Chromatin structure has a role in this epigenetic transmission of information from one cell generation to the next.

One type of evidence comes from studies in which the nucleus of a cell from a frog or tadpole is transplanted into a frog egg whose own nucleus has been removed (an enucleated egg). In a classic set of experiments performed in 1968, it was shown that a nucleus taken from a differentiated donor cell can be reprogrammed in this way to support development of a whole new tadpole (see Figure 7–2). But this reprogramming occurs only with difficulty, and it becomes less and less efficient as nuclei from older animals are used. Thus, for example, less than 2% of the enucleated eggs injected with a nucleus from a tadpole epithelial cell developed to the swimming tadpole stage, compared with 35% when the donor nuclei were taken instead from an early (gastrula-stage) embryo. With new experimental tools, the cause of this resistance to reprogramming can now be traced. It arises, at least in part, because specific chromatin structures in the original differentiated nucleus tend to persist and be transmitted through the many cell-division cycles required for embryonic development. In experiments with *Xenopus* embryos, specific forms of either repressive or active chromatin structures could be demonstrated to persist through as many as 24 cell divisions, causing the misplaced expression of genes. **Figure 4–45** briefly describes one such experiment,

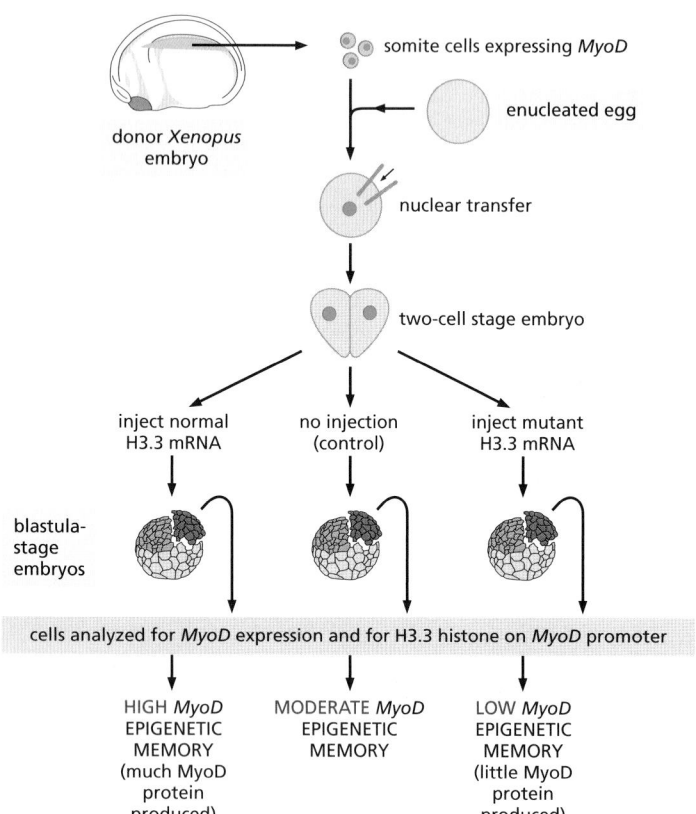

somite cells expressing *MyoD*

enucleated egg

donor *Xenopus* embryo

nuclear transfer

two-cell stage embryo

inject normal H3.3 mRNA   no injection (control)   inject mutant H3.3 mRNA

blastula-stage embryos

cells analyzed for *MyoD* expression and for H3.3 histone on *MyoD* promoter

HIGH *MyoD* EPIGENETIC MEMORY (much MyoD protein produced)   MODERATE *MyoD* EPIGENETIC MEMORY   LOW *MyoD* EPIGENETIC MEMORY (little MyoD protein produced)

Figure 4–45 Evidence for the inheritance of a gene-activating chromatin state. The well-characterized *MyoD* gene encodes a master transcription regulatory protein for muscle, MyoD (see p. 399). This gene is normally turned on in the indicated region of the young embryo where somites form. When a nucleus from this region is injected into an enucleated egg as shown, many of the progeny cell nuclei abnormally express the MyoD protein in non-muscle regions of the "nuclear transplant embryo" that forms. This abnormal expression can be attributed to maintenance of the *MyoD* promoter region in its active chromatin state through the many cycles of cell division that produce the blastula-stage embryo—a so-called "epigenetic memory" that persists in this case in the absence of transcription. The active chromatin surrounding the *MyoD* promoter contains the variant histone H3.3 (see Figure 4–35) in a Lys4 methylated form. As indicated, an overproduction of this histone caused by injecting excess mRNA encoding the normal H3.3 protein increases both H3.3 occupancy on the *MyoD* promoter and the epigenetic MyoD production, whereas injection of an mRNA producing a mutant form of H3.3 that cannot be methylated at Lys4 reduces the epigenetic MyoD production. Such experiments provide evidence that an inherited chromatin state underlies the epigenetic memory observed. (Adapted from R.K. Ng and J.B. Gurdon, *Nat. Cell Biol.* 10:102–109, 2008.)

focused on chromatin containing the histone variant, H3.3. We shall return to these phenomena in the final section of Chapter 22, where we discuss stem cells and the ways in which one cell type can be converted into another.

Chromatin Structures Are Important for Eukaryotic Chromosome Function

Although a great deal remains to be learned about the functions of different chromatin structures, the packaging of DNA into nucleosomes was probably crucial for the evolution of eukaryotes like ourselves. To form a complex multicellular organism, the cells in different lineages must specialize by changing the accessibility and activity of many hundreds of genes. As described in Chapter 21, this process depends on cell memory: each cell holds a record of its past developmental history in the regulatory circuits that control its many genes. That record, it seems, is partly stored in the structure of the chromatin.

Although bacteria also have cell memory mechanisms, the complexity of the memory circuits in higher eukaryotes is unparalleled. Strategies based on local variations in chromatin structure, unique to eukaryotes, can enable individual genes, once they are switched on or switched off, to stay in that state until some new factor acts to reverse it. At one extreme are structures like centromeric chromatin that, once established, are stably inherited from one cell generation to the next. Likewise, the major "classical" type of heterochromatin, which contains long arrays of the HP1 protein (see Figure 4–39), can persist stably throughout life. In contrast, a form of condensed chromatin that is created by the Polycomb group of proteins serves to silence genes that must be kept inactive in some conditions, but are active in others. The latter mechanism governs the expression of a large number of genes that encode transcription regulators important in early embryonic development, as discussed in Chapter 21. There are many other variant forms of chromatin, some with much shorter lifetimes, often less than the division time of the cell. We shall say more about the variety of chromatin types in the next section.

Summary

In the chromosomes of eukaryotes, DNA is uniformly assembled into nucleosomes, but a variety of different chromatin structures is possible. This variety is based on a large set of reversible covalent modifications of the four histones in the nucleosome core. These modifications include the mono-, di-, and trimethylation of many different lysine side chains, an important reaction that is incompatible with the acetylation that can occur on the same lysines. Specific combinations of the modifications mark many nucleosomes, governing their interactions with other proteins. These marks are read when protein modules that are part of a larger protein complex bind to the modified nucleosomes in a region of chromatin. These reader proteins then attract additional proteins that perform various functions.

Some reader protein complexes contain a histone-modifying enzyme, such as a histone lysine methylase, that "writes" the same mark that the reader recognizes. A reader–writer–remodeling complex of this type can spread a specific form of chromatin along a chromosome. In particular, large regions of condensed heterochromatin are thought to be formed in this way. Heterochromatin is commonly found around centromeres and near telomeres, but it is also present at many other positions in chromosomes. The tight packaging of DNA into heterochromatin usually silences the genes within it.

The phenomenon of position effect variegation provides strong evidence for the inheritance of condensed states of chromatin from one cell generation to the next. A similar mechanism appears to be responsible for maintaining the specialized chromatin at centromeres. More generally, the ability to propagate specific chromatin structures across cell generations makes possible an epigenetic cell memory process that plays a role in maintaining the set of different cell states required by complex multicellular organisms.

THE GLOBAL STRUCTURE OF CHROMOSOMES

Having discussed the DNA and protein molecules from which the chromatin fiber is made, we now turn to the organization of the chromosome on a more global scale and the way in which its various domains are arranged in space. As a 30-nm fiber, a typical human chromosome would still be 0.1 cm in length and able to span the nucleus more than 100 times. Clearly, there must be a still higher level of folding, even in interphase chromosomes. Although the molecular details are still largely a mystery, this higher-order packaging almost certainly involves the folding of the chromatin into a series of loops and coils. This chromatin packing is fluid, frequently changing in response to the needs of the cell.

We begin this section by describing some unusual interphase chromosomes that can be easily visualized. Exceptional though they are, these special cases reveal features that are thought to be representative of all interphase chromosomes. Moreover, they provide ways to investigate some fundamental aspects of chromatin structure that we have touched on in the previous section. Next, we describe how a typical interphase chromosome is arranged in the mammalian cell nucleus. Finally, we shall discuss the additional tenfold compaction that chromosomes undergo in the passage from interphase to mitosis.

Chromosomes Are Folded into Large Loops of Chromatin

Insight into the structure of the chromosomes in interphase cells has come from studies of the stiff and enormously extended chromosomes in growing amphibian oocytes (immature eggs). These very unusual **lampbrush chromosomes** (the largest chromosomes known), paired in preparation for meiosis, are clearly visible even in the light microscope, where they are seen to be organized into a series of large chromatin loops emanating from a linear chromosomal axis (**Figure 4–46** and **Figure 4–47**).

In these chromosomes, a given loop always contains the same DNA sequence that remains extended in the same manner as the oocyte grows. These chromosomes are producing large amounts of RNA for the oocyte, and most of the genes

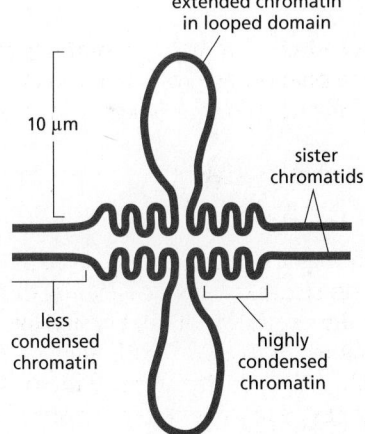

Figure 4–46 A model for the chromatin domains in a lampbrush chromosome. Shown is a small portion of one pair of sister chromatids. Here, two identical DNA double helices are aligned side by side, packaged into different types of chromatin. The set of lampbrush chromosomes in many amphibians contains a total of about 10,000 loops resembling those shown here. The rest of the DNA in each chromosome (the great majority) remains highly condensed. Four copies of each loop are present in the cell, since each lampbrush chromosome consists of two aligned sets of paired chromatids. This four-stranded structure is characteristic of this stage of development of the oocyte, which has arrested at the diplotene stage of meiosis; see Figure 17–56.

present in the DNA loops are being actively expressed. The majority of the DNA, however, is not in loops but remains highly condensed on the chromosome axis, where genes are generally not expressed.

It is thought that the interphase chromosomes of all eukaryotes are similarly arranged in loops. Although these loops are normally too small and fragile to be easily observed in a light microscope, other methods can be used to infer their presence. For example, modern DNA technologies have made it possible to assess the frequency with which any two loci along an interphase chromosome are held together, thus revealing likely candidates for the sites on chromatin that form the bases of loop structures (**Figure 4–48**). These experiments and others suggest that the DNA in human chromosomes is likely to be organized into loops of various lengths. A typical loop might contain between 50,000 and 200,000 nucleotide pairs of DNA, although loops of a million nucleotide pairs have also been suggested (**Figure 4–49**).

Polytene Chromosomes Are Uniquely Useful for Visualizing Chromatin Structures

Further insight has come from another unusual class of cells—the *polytene cells* of flies, such as the fruit fly *Drosophila*. Some types of cells, in many organisms, grow abnormally large through multiple cycles of DNA synthesis without cell division. Such cells, containing increased numbers of standard chromosomes, are said to be *polyploid*. In the salivary glands of fly larvae, this process is taken to an extreme degree, creating huge cells that contain hundreds or thousands of copies of the

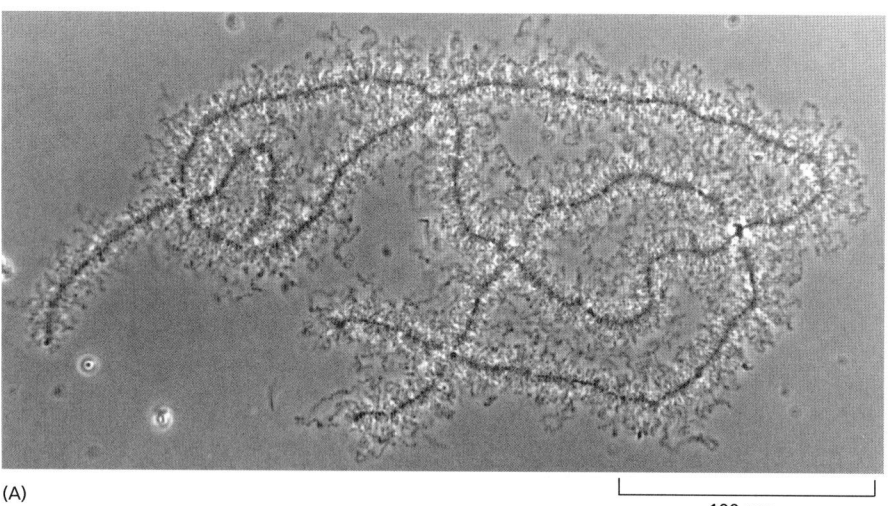

(A)

100 μm

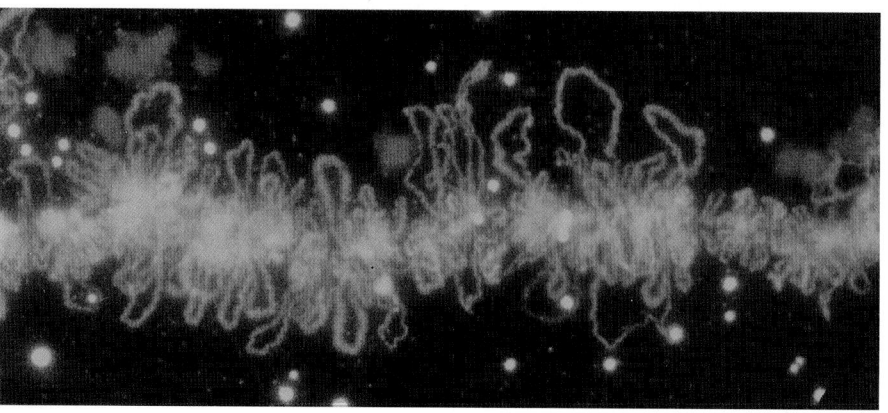

(B)

20 μm

Figure 4–47 Lampbrush chromosomes. (A) A light micrograph of lampbrush chromosomes in an amphibian oocyte. Early in oocyte differentiation, each chromosome replicates to begin meiosis, and the homologous replicated chromosomes pair to form this highly extended structure containing a total of four replicated DNA double helices, or chromatids. The lampbrush chromosome stage persists for months or years, while the oocyte builds up a supply of materials required for its ultimate development into a new individual. (B) An enlarged region of a similar chromosome, stained with a fluorescent reagent that makes the loops active in RNA synthesis clearly visible. (Courtesy of Joseph G. Gall.)

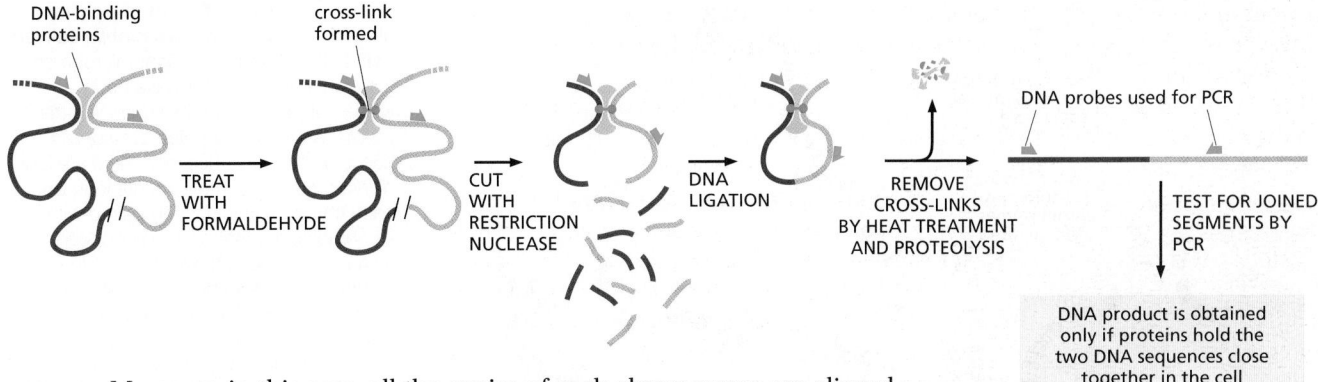

genome. Moreover, in this case, all the copies of each chromosome are aligned side by side in exact register, like drinking straws in a box, to create giant **polytene chromosomes**. These allow features to be detected that are thought to be shared with ordinary interphase chromosomes, but are normally hard to see.

When polytene chromosomes from a fly's salivary glands are viewed in the light microscope, distinct alternating dark *bands* and light *interbands* are visible (**Figure 4–50**), each formed from a thousand identical DNA sequences arranged side by side in register. About 95% of the DNA in polytene chromosomes is in bands, and 5% is in interbands. A very thin band can contain 3000 nucleotide pairs, while a thick band may contain 200,000 nucleotide pairs in each of its chromatin strands. The chromatin in each band appears dark because the DNA is more condensed than the DNA in interbands; it may also contain a higher concentration of proteins (**Figure 4–51**). This banding pattern seems to reflect the same sort of organization detected in the amphibian lampbrush chromosomes described earlier.

There are approximately 3700 bands and 3700 interbands in the complete set of *Drosophila* polytene chromosomes. The bands can be recognized by their different thicknesses and spacings, and each one has been given a number to generate a chromosome "map" that has been indexed to the finished genome sequence of this fly.

The *Drosophila* polytene chromosomes provide a good starting point for examining how chromatin is organized on a large scale. In the previous section, we saw that there are many forms of chromatin, each of which contains nucleosomes with a different combination of modified histones. Specific sets of non-histone proteins assemble on these nucleosomes to affect biological function in different ways. Recruitment of some of these non-histone proteins can spread for long distances along the DNA, imparting a similar chromatin structure to broad tracts

Figure 4–48 A method for determining the position of loops in interphase chromosomes. In this technique, known as the chromosome conformation capture (3C) method, cells are treated with formaldehyde to create the indicated covalent DNA–protein and DNA–DNA cross-links. The DNA is then treated with an enzyme (a restriction nuclease) that chops the DNA into many pieces, cutting at strictly defined nucleotide sequences and forming sets of identical "cohesive ends" (see Figure 8–28). The cohesive ends can be made to join through their complementary base-pairing. Importantly, prior to the ligation step shown, the DNA is diluted so that the fragments that have been kept in close proximity to each other (through cross-linking) are the ones most likely to join. Finally, the cross-links are reversed and the newly ligated fragments of DNA are identified and quantified by PCR (the polymerase chain reaction, described in Chapter 8). From the results, combined with DNA sequence information, one can derive models for the interphase conformation of chromosomes.

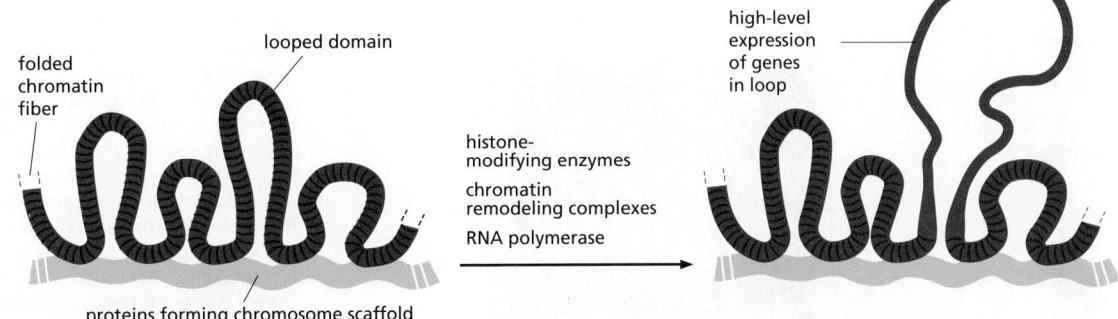

Figure 4–49 A model for the organization of an interphase chromosome. A section of an interphase chromosome is shown folded into a series of looped domains, each containing perhaps 50,000–200,000 or more nucleotide pairs of double-helical DNA condensed into a chromatin fiber. The chromatin in each individual loop is further condensed through poorly understood folding processes that are reversed when the cell requires direct access to the DNA packaged in the loop. Neither the composition of the postulated chromosomal axis nor how the folded chromatin fiber is anchored to it is clear. However, in mitotic chromosomes, the bases of the chromosomal loops are enriched both in condensins (discussed below) and in DNA topoisomerase II enzymes (discussed in Chapter 5), two proteins that may form much of the axis at metaphase.

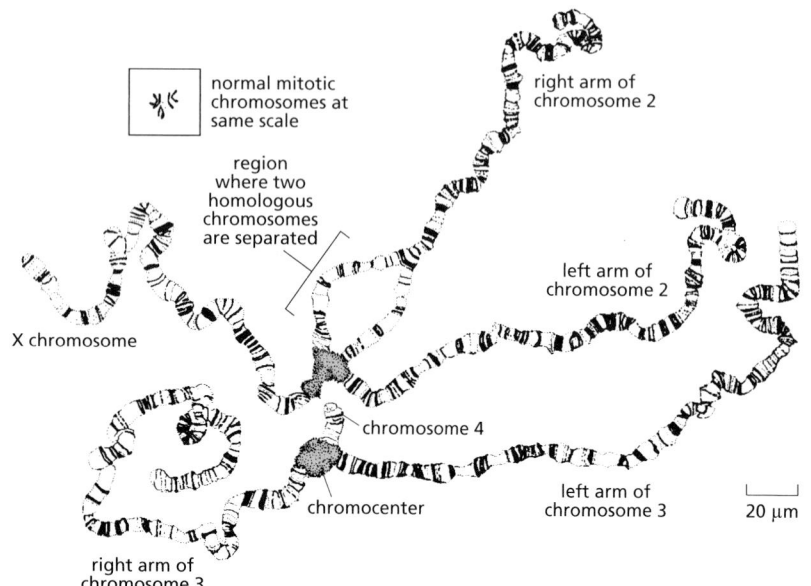

normal mitotic chromosomes at same scale

region where two homologous chromosomes are separated

right arm of chromosome 2

left arm of chromosome 2

X chromosome

chromosome 4

chromocenter

left arm of chromosome 3

20 μm

right arm of chromosome 3

Figure 4–50 The entire set of polytene chromosomes in one *Drosophila* salivary cell. In this drawing of a light micrograph, the giant chromosomes have been spread out for viewing by squashing them against a microscope slide. *Drosophila* has four chromosomes, and there are four different chromosome pairs present. But each chromosome is tightly paired with its homolog (so that each pair appears as a single structure), which is not true in most nuclei (except in meiosis). Each chromosome has undergone multiple rounds of replication, and the homologs and all their duplicates have remained in exact register with each other, resulting in huge chromatin cables many DNA strands thick.

The four polytene chromosomes are normally linked together by heterochromatic regions near their centromeres that aggregate to create a single large chromocenter *(pink region)*. In this preparation, however, the chromocenter has been split into two halves by the squashing procedure used. (Adapted from T.S. Painter, *J. Hered.* 25:465–476, 1934. With permission from Oxford University Press.)

of the genome (see Figure 4–40). Such regions, where all of the chromatin has a similar structure, are separated from neighboring domains by barrier proteins (see Figure 4–41). At low resolution, the interphase chromosome can therefore be considered as a mosaic of chromatin structures, each containing particular nucleosome modifications associated with a particular set of non-histone proteins. Polytene chromosomes allow us to see details of this mosaic of domains in the light microscope, as well as to observe some of the changes associated with gene expression.

There Are Multiple Forms of Chromatin

By staining *Drosophila* polytene chromosomes with antibodies, or by using a more recent technique called ChIP (chromatin immunoprecipitation) analysis (see Chapter 8), the locations of the histone and non-histone proteins in chromatin can be mapped across the entire DNA sequence of an organism's genome. Such an analysis in *Drosophila* has thus far localized more than 50 different chromatin proteins and histone modifications. The results suggest that three major types of repressive chromatin predominate in this organism, along with two major types of chromatin on actively transcribed genes, and that each type is associated with a different complex of non-histone proteins. Thus, classical heterochromatin contains more than six such proteins, including heterochromatin protein 1 (HP1),

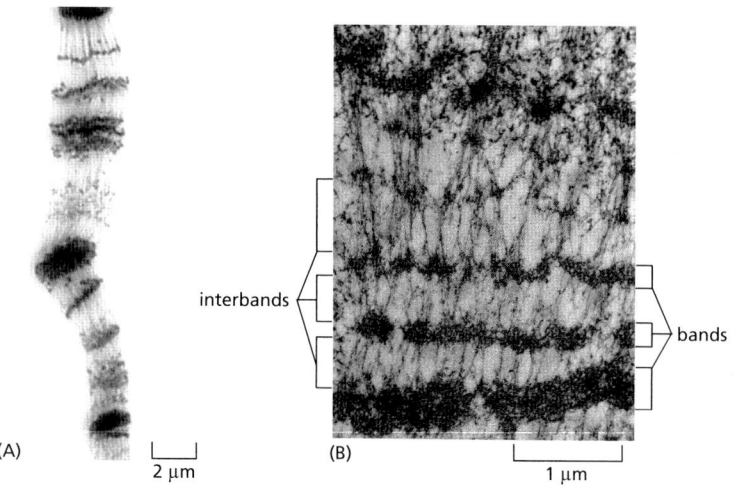

interbands

bands

(A)   2 μm

(B)   1 μm

Figure 4–51 Micrographs of polytene chromosomes from *Drosophila* salivary glands. (A) Light micrograph of a portion of a chromosome. The DNA has been stained with a fluorescent dye, but a reverse image is presented here that renders the DNA *black* rather than *white*; the bands are clearly seen to be regions of increased DNA concentration. This chromosome has been processed by a high-pressure treatment so as to show its distinct pattern of bands and interbands more clearly. (B) An electron micrograph of a small section of a *Drosophila* polytene chromosome seen in thin section. Bands of very different thickness can be readily distinguished, separated by interbands, which contain less condensed chromatin. (A, adapted from D. V. Novikov, I. Kireev, and A. S. Belmont, *Nat. Methods* 4:483–485, published 2007 by Nature Publishing Group. Reproduced with permission of SNCSC; B, courtesy of Veikko Sorsa.)

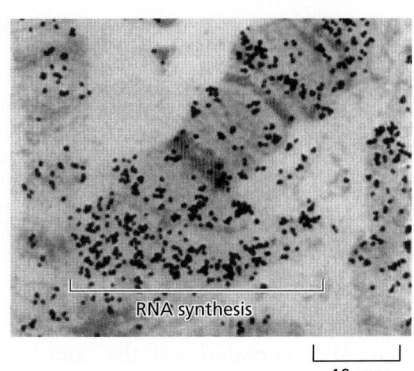

Figure 4–52 RNA synthesis in polytene chromosome puffs. An autoradiograph of a single puff in a polytene chromosome from the salivary glands of the freshwater midge *Chironomus tentans*. As outlined in Chapter 1 and described in detail in Chapter 6, the first step in gene expression is the synthesis of an RNA molecule using the DNA as a template. The decondensed portion of the chromosome is undergoing RNA synthesis and has become labeled with ^{3}H-uridine, an RNA precursor molecule that is incorporated into growing RNA chains. (From J. J. Bonner and M. L. Pardue, *Cell* 12:227–234, 1977. With permission from Elsevier.)

whereas the so-called Polycomb form of heterochromatin contains a similar number of proteins of a different set (PcG proteins). In addition to the five major chromatin types, other more minor forms of chromatin appear to be present, each of which may be differently regulated and have distinct roles in the cell.

The set of proteins bound as part of the chromatin at a given locus varies depending on the cell type and its stage of development. These variations make the accessibility of specific genes different in different tissues, helping to generate the cell diversification that accompanies embryonic development (described in Chapter 21).

Chromatin Loops Decondense When the Genes Within Them Are Expressed

When an insect progresses from one developmental stage to another, distinctive *chromosome puffs* arise and old puffs recede in its polytene chromosomes as new genes become expressed and old ones are turned off (**Figure 4–52**). From inspection of each puff when it is relatively small and the banding pattern is still discernible, it seems that most puffs arise from the decondensation of a single chromosome band.

The individual chromatin fibers that make up a puff can be visualized with an electron microscope. In favorable cases, loops are seen, much like those observed in amphibian lampbrush chromosomes. When genes in the loop are not expressed, the loop assumes a thickened structure, possibly that of a folded 30-nm fiber, but when gene expression is occurring, the loop becomes more extended. In electron micrographs, the chromatin located on either side of the decondensed loop appears considerably more compact, suggesting that a loop constitutes a distinct functional domain of chromatin structure.

Observations in human cells also suggest that highly folded loops of chromatin expand to occupy an increased volume when a gene within them is expressed. For example, quiescent chromosome regions from 0.4 to 2 million nucleotide pairs in length appear as compact dots in an interphase nucleus when visualized by fluorescence microscopy. However, the same DNA is seen to occupy a larger territory when its genes are expressed, with elongated, punctate structures replacing the original dot.

New ways of visualizing individual chromosomes have shown that each of the 46 interphase chromosomes in a human cell tends to occupy its own discrete territory within the nucleus: that is, the chromosomes are not extensively entangled with one another (**Figure 4–53**). However, pictures such as these present only an average view of the DNA in each chromosome. Experiments that specifically localize the heterochromatic regions of a chromosome reveal that they are often

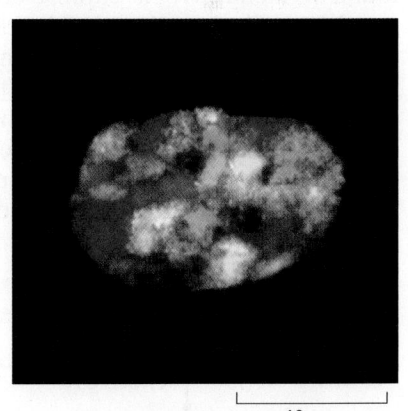

Figure 4–53 Simultaneous visualization of the chromosome territories for all of the human chromosomes in a single interphase nucleus. Here, a mixture of DNA probes for each chromosome has been labeled so as to fluoresce with a different spectra; this allows DNA–DNA hybridization to be used to detect each chromosome, as in Figure 4–10. Three-dimensional reconstructions were then produced. Below the micrograph, each chromosome is identified in a schematic of the actual image. Note that homologous chromosomes (e.g., the two copies of chromosome 9) are not in general co-located. (From M.R. Speicher and N.P. Carter, *Nat. Rev. Genet.* 6:782–793, published 2005 by Nature Publishing Group. Reproduced with permission of SNCSC.)

Figure 4–54 **The distribution of gene-rich regions of the human genome in an interphase nucleus.** Gene-rich regions have been visualized with a fluorescent probe that hybridizes to the *Alu* interspersed repeat, which is present in more than a million copies in the human genome (see page 292). For unknown reasons, these sequences cluster in chromosomal regions rich in genes. In this representation, regions enriched for the *Alu* sequence are *green*, regions depleted for these sequences are *red*, while the average regions are *yellow*. The gene-rich regions are seen to be largely absent in the DNA near the nuclear envelope. (From A. Bolzer et al., *PLoS Biol.* 3:826–842, 2005. With permission from the authors.)

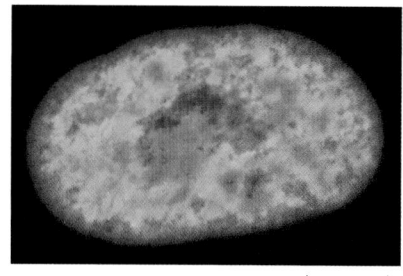

5 μm

closely associated with the nuclear lamina, regardless of the chromosome examined. And DNA probes that preferentially stain gene-rich regions of human chromosomes produce a striking picture of the interphase nucleus that presumably reflects different average positions for active and inactive genes (**Figure 4–54**).

How is most of the chromatin in each interphase chromosome condensed when its genes are not being expressed? A powerful extension of the chromosome conformation capture method described previously (see Figure 4–48), which exploits a high-throughput DNA sequencing technology called massive parallel sequencing (see Panel 8–1, pp. 478–481), allows the connections between all of the different one-megabase (1 Mb) segments of the human genome to be mapped in human interphase chromosomes. The results reveal that most regions of our chromosomes are folded into a conformation referred to as a *fractal globule*: a knot-free arrangement that facilitates maximally dense packing while, at the same time, preserving the ability of the chromatin fiber to unfold and fold (**Figure 4–55**).

Chromatin Can Move to Specific Sites Within the Nucleus to Alter Gene Expression

A variety of different types of experiments has led to the conclusion that the position of a gene in the interior of the nucleus changes when it becomes highly expressed. Thus, a region that becomes very actively transcribed is sometimes found to extend out of its chromosome territory, as if in an extended loop (**Figure 4–56**). We will see in Chapter 6 that the initiation of transcription—the first step in gene expression—requires the assembly of over 100 proteins, and it makes sense that this would be facilitated in regions of the nucleus enriched in these proteins.

More generally, it is clear that the nucleus is very heterogeneous, with functionally different regions to which portions of chromosomes can move as they are subjected to different biochemical processes—such as when their gene expression changes. It is this issue that we discuss next.

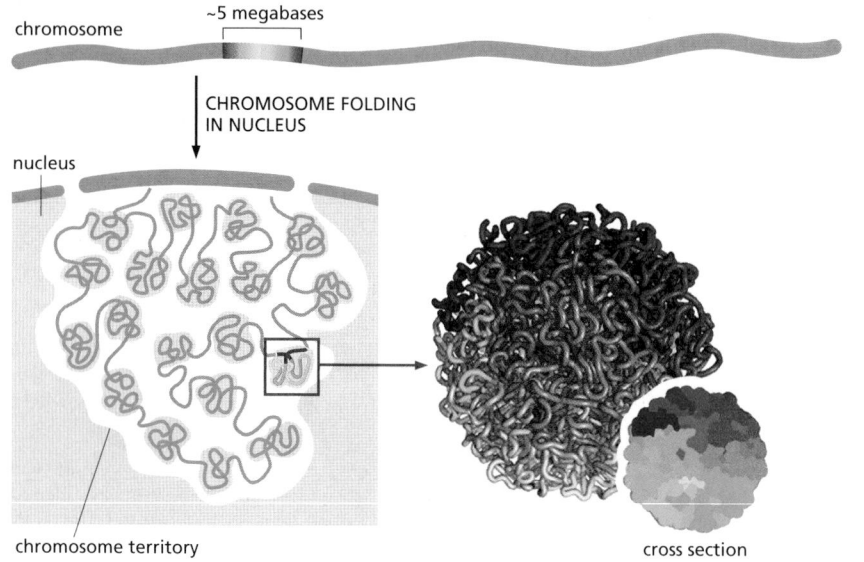

Figure 4–55 **A fractal globule model for interphase chromatin.** An extension of the 3C method in Figure 4–48, called Hi-C, was used to measure the extent to which each of the three thousand 1 Mb segments in the human genome was located adjacent to any other of these segments. The results support the type of model shown. In the enlarged fractal globule illustrated, a region of 5 million base pairs is seen to fold in a way that keeps regions that are neighbors along the one-dimensional DNA helix as neighbors in three dimensions; this gives rise to monochromatic blocks in this representation that are obvious both on the surface and in cross section. The fractal globule is a knot-free conformation of the DNA that permits dense packing, yet retains an ability to easily fold and unfold any genomic locus. (Adapted from E. Lieberman-Aiden et al., *Science* 326:289–293, 2009.)

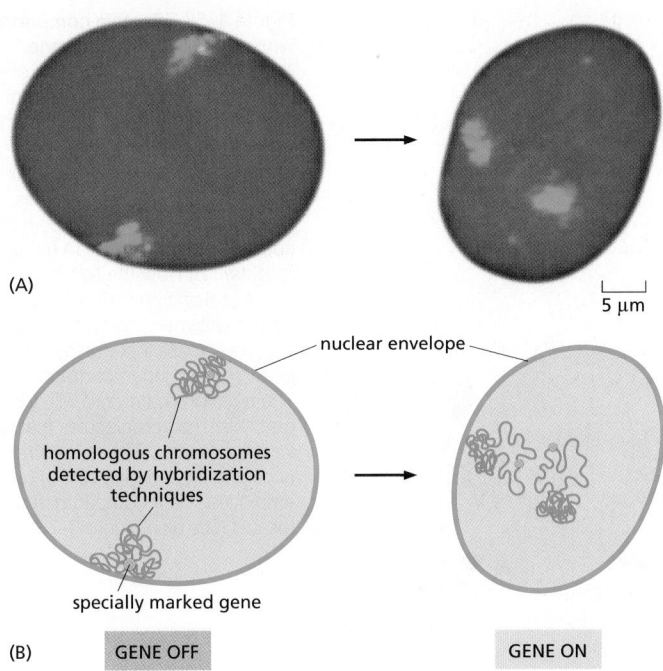

(A)

5 µm

nuclear envelope

homologous chromosomes
detected by hybridization
techniques

specially marked gene

(B)   GENE OFF   GENE ON

Figure 4–56 An effect of high levels of gene expression on the intranuclear location of chromatin. (A) Fluorescence micrographs of human nuclei showing how the position of a gene changes when it becomes highly transcribed. The region of the chromosome adjacent to the gene (red) is seen to leave its chromosomal territory (green) only when it is highly active. (B) Schematic representation of a large loop of chromatin that expands when the gene is on, and contracts when the gene is off. Other genes that are less actively expressed can be shown by the same methods to remain inside their chromosomal territory when transcribed. (From J.R. Chubb and W.A. Bickmore, Cell 112:403–406, 2003. With permission from Elsevier.)

Networks of Macromolecules Form a Set of Distinct Biochemical Environments inside the Nucleus

In Chapter 6, we shall describe the function of a variety of subcompartments that are present within the nucleus. The largest and most obvious of these is the **nucleolus**, a structure well known to microscopists even in the nineteenth century (see Figure 4–9). The nucleolus is the cell's site of ribosome subunit formation, as well as the place where many other specialized reactions occur (see Figure 6–42): it consists of a network of RNAs and proteins concentrated around ribosomal RNA genes that are being actively transcribed. In eukaryotes, the genome contains multiple copies of the ribosomal RNA genes, and although they are typically clustered together in a single nucleolus, they are often located on several separate chromosomes.

A variety of less obvious organelles are also present inside the nucleus. For example, spherical structures called Cajal bodies and interchromatin granule clusters are present in most plant and animal cells (**Figure 4–57**). Like the nucleolus, these organelles are composed of selected protein and RNA molecules that bind together to create networks that are highly permeable to other protein and RNA molecules in the surrounding nucleoplasm.

Structures such as these can create distinct biochemical environments by immobilizing select groups of macromolecules, as can other networks of proteins and RNA molecules associated with nuclear pores and with the nuclear envelope. In principle, this allows other molecules that enter these spaces to be processed with great efficiency through complex reaction pathways. Highly permeable, fibrous networks of this sort can thereby impart many of the kinetic advantages of compartmentalization (see p. 164) to reactions that take place in subregions of the nucleus (**Figure 4–58A**). However, unlike the membrane-bound compartments in the cytoplasm (discussed in Chapter 12), these nuclear subcompartments—lacking a lipid bilayer membrane—can neither concentrate nor exclude specific small molecules.

The cell has a remarkable ability to construct distinct environments to perform complex biochemical tasks efficiently. Those that we have mentioned in the nucleus facilitate various aspects of gene expression, and will be further discussed in Chapter 6. These subcompartments, including the nucleolus, appear to form

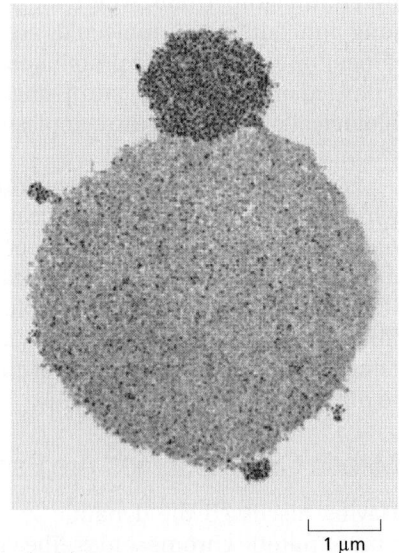

1 µm

Figure 4–57 Electron micrograph showing two very common fibrous nuclear subcompartments. The large sphere here is a Cajal body. The smaller darker sphere is an interchromatin granule cluster, also known as a speckle (see also Figure 6–46). These "subnuclear organelles" are from the nucleus of a Xenopus oocyte. (From K.E. Handwerger and J.G. Gall, Trends Cell Biol. 16:19–26, 2006. With permission from Elsevier.)

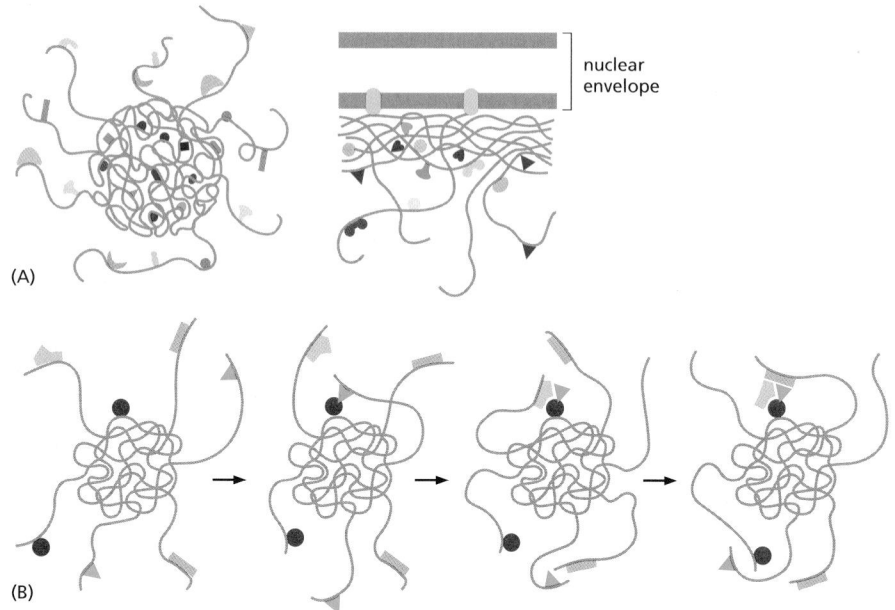

(A)

(B)

Figure 4–58 Effective compartmentalization without a bilayer membrane. (A) Schematic illustration of the organization of a spherical subnuclear organelle *(left)* and of a postulated similarly organized subcompartment just beneath the nuclear envelope *(right)*. In both cases, RNAs and/or proteins *(gray)* associate to form highly porous, gel-like structures that contain binding sites for other specific proteins and RNA molecules *(colored objects)*. (B) How the tethering of a selected set of proteins and RNA molecules to long flexible polymer chains, as in (A), can create "staging areas" that greatly speed the rates of reactions in subcompartments of the nucleus. The reactions catalyzed will depend on the particular macromolecules that are localized by the tethering. The same strategy for accelerating complex sets of reactions is also employed in subcompartments elsewhere in the cell (see also Figure 3–78).

nuclear envelope

only as needed, and they create a high local concentration of the many different enzymes and RNA molecules needed for a particular process. In an analogous way, when DNA is damaged by irradiation, the set of enzymes needed to carry out DNA repair are observed to congregate in discrete foci inside the nucleus, creating "repair factories" (see Figure 5–52). And nuclei often contain hundreds of discrete foci representing factories for DNA or RNA synthesis (see Figure 6–47).

It seems likely that all of these entities make use of the type of tethering illustrated in Figure 4–58B, where long flexible lengths of polypeptide chain and/or long noncoding RNA molecules are interspersed with specific binding sites that concentrate the multiple proteins and other molecules that are needed to catalyze a particular process. Not surprisingly, tethers are similarly used to help to speed biological processes in the cytoplasm, increasing specific reaction rates there (for example, see Figure 16–18).

Is there also an intranuclear framework, analogous to the cytoskeleton, on which chromosomes and other components of the nucleus are organized? The *nuclear matrix*, or *scaffold*, has been defined as the insoluble material left in the nucleus after a series of biochemical extraction steps. Many of the proteins and RNA molecules that form this insoluble material are likely to be derived from the fibrous subcompartments of the nucleus just discussed, while others may be proteins that help to form the base of chromosomal loops or to attach chromosomes to other structures in the nucleus.

Mitotic Chromosomes Are Especially Highly Condensed

Having discussed the dynamic structure of interphase chromosomes, we now turn to mitotic chromosomes. The chromosomes from nearly all eukaryotic cells become readily visible by light microscopy during mitosis, when they coil up to form highly condensed structures. This condensation reduces the length of a typical interphase chromosome only about tenfold, but it produces a dramatic change in chromosome appearance.

Figure 4–59 depicts a typical **mitotic chromosome** at the metaphase stage of mitosis (for the stages of mitosis, see Figure 17–3). The two DNA molecules produced by DNA replication during interphase of the cell-division cycle are separately folded to produce two sister chromosomes, or *sister chromatids*, held together at their centromeres, as mentioned earlier. These chromosomes are normally covered with a variety of molecules, including large amounts of RNA–protein

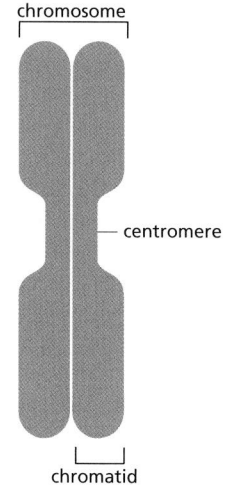

chromosome

centromere

chromatid

Figure 4–59 A typical mitotic chromosome at metaphase. Each sister chromatid contains one of two identical sister DNA molecules generated earlier in the cell cycle by DNA replication (see also Figure 17–21).

Figure 4–60 **A scanning electron micrograph of a region near one end of a typical mitotic chromosome.** Each knoblike projection is believed to represent the tip of a separate looped domain. Note that the two identical paired chromatids (drawn in Figure 4–59) can be clearly distinguished. (From M.P. Marsden and U.K. Laemmli, *Cell* 17:849–858, 1979. With permission from Elsevier.)

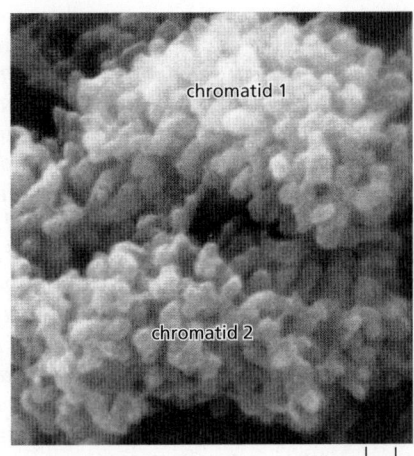

0.1 μm

complexes. Once this covering has been stripped away, each chromatid can be seen in electron micrographs to be organized into loops of chromatin emanating from a central scaffolding (**Figure 4–60**). Experiments using DNA hybridization to detect specific DNA sequences demonstrate that the order of visible features along a mitotic chromosome at least roughly reflects the order of genes along the DNA molecule. Mitotic chromosome condensation can thus be thought of as the final level in the hierarchy of chromosome packaging (**Figure 4–61**).

The compaction of chromosomes during mitosis is a highly organized and dynamic process that serves at least two important purposes. First, when condensation is complete (in metaphase), sister chromatids have been disentangled from each other and lie side by side. Thus, the sister chromatids can easily separate when the mitotic apparatus begins pulling them apart. Second, the compaction of chromosomes protects the relatively fragile DNA molecules from being broken as they are pulled to separate daughter cells.

The condensation of interphase chromosomes into mitotic chromosomes begins in early M phase, and it is intimately connected with the progression of the cell cycle. During M phase, gene expression shuts down, and specific modifications are made to histones that help to reorganize the chromatin as it compacts. Two classes of ring-shaped proteins, called *cohesins* and *condensins*, aid this compaction. How they help to produce the two separately folded chromatids of a mitotic chromosome will be discussed in Chapter 17, along with the details of the cell cycle.

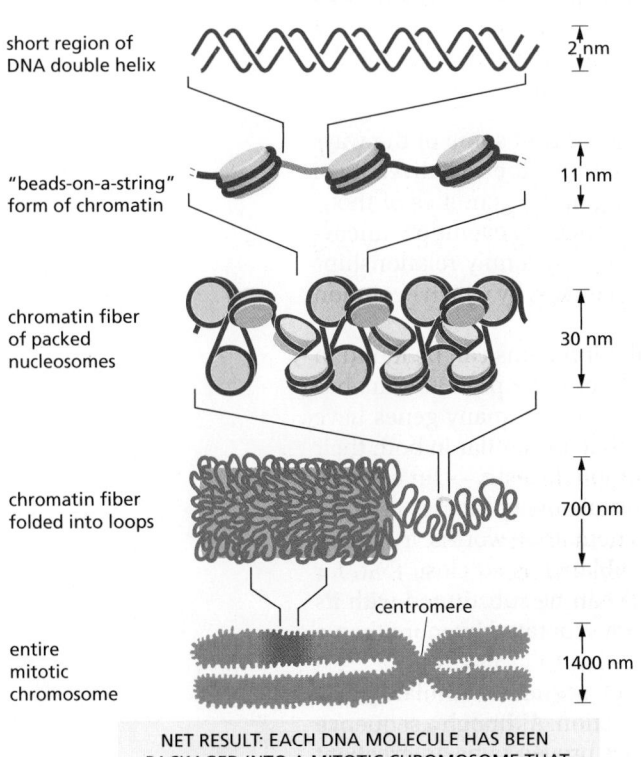

short region of DNA double helix — 2 nm

"beads-on-a-string" form of chromatin — 11 nm

chromatin fiber of packed nucleosomes — 30 nm

chromatin fiber folded into loops — 700 nm

centromere

entire mitotic chromosome — 1400 nm

NET RESULT: EACH DNA MOLECULE HAS BEEN PACKAGED INTO A MITOTIC CHROMOSOME THAT IS 10,000-FOLD SHORTER THAN ITS FULLY EXTENDED LENGTH

Figure 4–61 **Chromatin packing.** This model shows some of the many levels of chromatin packing postulated to give rise to the highly condensed mitotic chromosome.

Summary

Chromosomes are generally decondensed during interphase, so that the details of their structure are difficult to visualize. Notable exceptions are the specialized lampbrush chromosomes of vertebrate oocytes and the polytene chromosomes in the giant secretory cells of insects. Studies of these two types of interphase chromosomes suggest that each long DNA molecule in a chromosome is divided into a large number of discrete domains organized as loops of chromatin that are compacted by further folding. When genes contained in a loop are expressed, the loop unfolds and allows the cell's machinery access to the DNA.

Interphase chromosomes occupy discrete territories in the cell nucleus; that is, they are not extensively intertwined. Euchromatin makes up most of interphase chromosomes and, when not being transcribed, it probably exists as tightly folded fibers of compacted nucleosomes. However, euchromatin is interrupted by stretches of heterochromatin, in which the nucleosomes are subjected to additional packing that usually renders the DNA resistant to gene expression. Heterochromatin exists in several forms, some of which are found in large blocks in and around centromeres and near telomeres. But heterochromatin is also present at many other positions on chromosomes, where it can serve to help regulate developmentally important genes.

The interior of the nucleus is highly dynamic, with heterochromatin often positioned near the nuclear envelope and loops of chromatin moving away from their chromosome territory when genes are very highly expressed. This reflects the existence of nuclear subcompartments, where different sets of biochemical reactions are facilitated by an increased concentration of selected proteins and RNAs. The components involved in forming a subcompartment can self-assemble into discrete organelles such as nucleoli or Cajal bodies; they can also be tethered to fixed structures such as the nuclear envelope.

During mitosis, gene expression shuts down and all chromosomes adopt a highly condensed conformation in a process that begins early in M phase to package the two DNA molecules of each replicated chromosome as two separately folded chromatids. The condensation is accompanied by histone modifications that facilitate chromatin packing, but satisfactory completion of this orderly process, which reduces the end-to-end distance of each DNA molecule from its interphase length by an additional factor of ten, requires additional proteins.

HOW GENOMES EVOLVE

In this final section of the chapter, we provide an overview of some of the ways that genes and genomes have evolved over time to produce the vast diversity of modern-day life-forms on our planet. The sequencing of the genomes of thousands of organisms is revolutionizing our view of the process of evolution, uncovering an astonishing wealth of information about not only family relationships among organisms, but also about the molecular mechanisms by which evolution has proceeded.

It is perhaps not surprising that genes with similar functions can be found in a diverse range of living things. But the great revelation of the past 30 years has been the extent to which the actual nucleotide sequences of many genes have been conserved. **Homologous** genes—that is, genes that are similar in both their nucleotide sequence and function because of a common ancestry—can often be recognized across vast phylogenetic distances. Unmistakable homologs of many human genes are present in organisms as diverse as nematode worms, fruit flies, yeasts, and even bacteria. In many cases, the resemblance is so close that, for example, the protein-coding portion of a yeast gene can be substituted with its human homolog—even though humans and yeast are separated by more than a billion years of evolutionary history.

As emphasized in Chapter 3, the recognition of sequence similarity has become a major tool for inferring gene and protein function. Although a sequence match does not guarantee similarity in function, it has proved to be an excellent clue. Thus, it is often possible to predict the function of genes in humans for which no biochemical or genetic information is available simply by comparing their nucleotide sequences with the sequences of genes that have been characterized in other more readily studied organisms.

In general, the sequences of individual genes are much more tightly conserved than is overall genome structure. Features of genome organization such as genome size, number of chromosomes, order of genes along chromosomes, abundance and size of introns, and amount of repetitive DNA are found to differ greatly when comparing distant organisms, as does the number of genes that each organism contains.

Genome Comparisons Reveal Functional DNA Sequences by their Conservation Throughout Evolution

A first obstacle in interpreting the sequence of the 3.2 billion nucleotide pairs in the human genome is the fact that much of it is probably functionally unimportant. The regions of the genome that code for the amino acid sequences of proteins (the exons) are typically found in short segments (average size about 145 nucleotide pairs), small islands in a sea of DNA whose exact nucleotide sequence is thought to be mostly of little consequence. This arrangement makes it difficult to identify all the exons in a stretch of DNA, and it is often hard too to determine exactly where a gene begins and ends.

One very important approach to deciphering our genome is to search for DNA sequences that are closely similar between different species, on the principle that DNA sequences that have a function are much more likely to be conserved than those without a function. For example, humans and mice are thought to have diverged from a common mammalian ancestor about 80×10^6 years ago, which is long enough for the majority of nucleotides in their genomes to have been changed by random mutational events. Consequently, the only regions that will have remained closely similar in the two genomes are those in which mutations would have impaired function and put the animals carrying them at a disadvantage, resulting in their elimination from the population by natural selection. Such closely similar pieces of DNA sequence are known as *conserved regions*. In addition to revealing those DNA sequences that encode functionally important exons and RNA molecules, these conserved regions will include regulatory DNA sequences as well as DNA sequences with functions that are not yet known. In contrast, most *nonconserved regions* will reflect DNA whose sequence is much less likely to be critical for function.

The power of this method can be increased by including in such comparisons the genomes of large numbers of species whose genomes have been sequenced, such as rat, chicken, fish, dog, and chimpanzee, as well as mouse and human. By revealing in this way the results of a very long natural "experiment," lasting for hundreds of millions of years, such comparative DNA sequencing studies have highlighted the most interesting regions in our genome. The comparisons reveal that roughly 5% of the human genome consists of "multispecies conserved sequences." To our great surprise, only about one-third of these sequences code for proteins (see Table 4–1, p. 184). Many of the remaining conserved sequences consist of DNA containing clusters of protein-binding sites that are involved in gene regulation, while others produce RNA molecules that are not translated into protein but are important for other known purposes. But, even in the most intensively studied species, the function of the majority of these highly conserved sequences remains unknown. This remarkable discovery has led scientists to conclude that we understand much less about the cell biology of vertebrates than we had thought. Certainly, there are enormous opportunities for new discoveries, and we should expect many more surprises ahead.

Genome Alterations Are Caused by Failures of the Normal Mechanisms for Copying and Maintaining DNA, as well as by Transposable DNA Elements

Evolution depends on accidents and mistakes followed by nonrandom survival. Most of the genetic changes that occur result simply from failures in the normal mechanisms by which genomes are copied or repaired when damaged, although the movement of transposable DNA elements (discussed below) also plays an important part. As we will explain in Chapter 5, the mechanisms that maintain DNA sequences are remarkably precise—but they are not perfect. DNA sequences are inherited with such extraordinary fidelity that typically, along a given line of descent, only about one nucleotide pair in a thousand is randomly changed in the germ line every million years. Even so, in a population of 10,000 diploid individuals, every possible nucleotide substitution will have been "tried out" on about 20 occasions in the course of a million years—a short span of time in relation to the evolution of species.

Errors in DNA replication, DNA recombination, or DNA repair can lead either to simple local changes in DNA sequence—so-called *point mutations* such as the substitution of one base pair for another—or to large-scale genome rearrangements such as deletions, duplications, inversions, and translocations of DNA from one chromosome to another. In addition to these failures of the genetic machinery, genomes contain mobile DNA elements that are an important source of genomic change (see Table 5–3, p. 267). These transposable DNA elements (*transposons*) are parasitic DNA sequences that can spread within the genomes they colonize. In the process, they often disrupt the function or alter the regulation of existing genes. On occasion, they have created altogether novel genes through fusions between transposon sequences and segments of existing genes. Over long periods of evolutionary time, DNA transposition events have profoundly affected genomes, so much so that nearly half of the DNA in the human genome consists of recognizable relics of past transposition events (**Figure 4–62**). Even more of our genome is thought to have been derived from transpositions that occurred so long ago ($>10^8$ years) that the sequences can no longer be traced to transposons.

The Genome Sequences of Two Species Differ in Proportion to the Length of Time Since They Have Separately Evolved

The differences between the genomes of species alive today have accumulated over more than 3 billion years. Although we lack a direct record of changes over time, scientists can reconstruct the process of genome evolution from detailed comparisons of the genomes of contemporary organisms.

The basic organizing framework for comparative genomics is the phylogenetic tree. A simple example is the tree describing the divergence of humans from the great apes (**Figure 4–63**). The primary support for this tree comes from comparisons of gene or protein sequences. For example, comparisons between the sequences of human genes or proteins and those of the great apes typically reveal the fewest differences between human and chimpanzee and the most between human and orangutan.

For closely related organisms such as humans and chimpanzees, it is relatively easy to reconstruct the gene sequences of the extinct, last common ancestor of the two species (**Figure 4–64**). The close similarity between human and chimpanzee genes is mainly due to the short time that has been available for the accumulation of mutations in the two diverging lineages, rather than to functional constraints

Figure 4–62 A representation of the nucleotide sequence content of the sequenced human genome. The LINEs (long interspersed nuclear elements), SINEs (short interspersed nuclear elements), retroviral-like elements, and DNA-only transposons are mobile genetic elements that have multiplied in our genome by replicating themselves and inserting the new copies in different positions. These mobile genetic elements are discussed in Chapter 5 (see Table 5–3, p. 267). Simple sequence repeats are short nucleotide sequences (less than 14 nucleotide pairs) that are repeated again and again for long stretches. Segmental duplications are large blocks of DNA sequence (1000–200,000 nucleotide pairs) that are present at two or more locations in the genome. The most highly repeated blocks of DNA in heterochromatin have not yet been completely sequenced; therefore about 10% of human DNA sequences are not represented in this diagram. (Data courtesy of E. Margulies.)

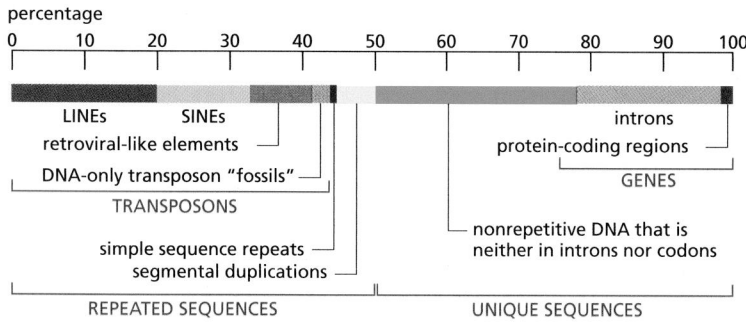

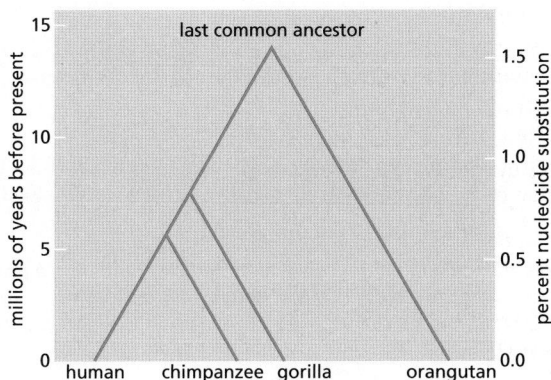

Figure 4–63 **A phylogenetic tree showing the relationship between humans and the great apes based on nucleotide sequence data.** As indicated, the sequences of the genomes of all four species are estimated to differ from the sequence of the genome of their last common ancestor by a little over 1.5%. Because changes occur independently on both diverging lineages, pairwise comparisons reveal twice the sequence divergence from the last common ancestor. For example, human–orangutan comparisons typically show sequence divergences of a little over 3%, while human–chimpanzee comparisons show divergences of approximately 1.2%. (Modified from F.C. Chen and W.H. Li, *Am. J. Hum. Genet.* 68:444–456, 2001.)

that have kept the sequences the same. Evidence for this view comes from the observation that the human and chimpanzee genomes are nearly identical even where there is no functional constraint on the nucleotide sequence—such as in the third position of "synonymous" codons (codons specifying the same amino acid but differing in their third nucleotide).

For much less closely related organisms, such as humans and chickens (which have evolved separately for about 300 million years), the sequence conservation found in genes is almost entirely due to **purifying selection** (that is, selection that eliminates individuals carrying mutations that interfere with important genetic functions), rather than to an inadequate time for mutations to occur.

Phylogenetic Trees Constructed from a Comparison of DNA Sequences Trace the Relationships of All Organisms

Phylogenetic trees based on molecular sequence data can be compared with the fossil record, and we get our best view of evolution by integrating the two approaches. The fossil record remains essential as a source of absolute dates,

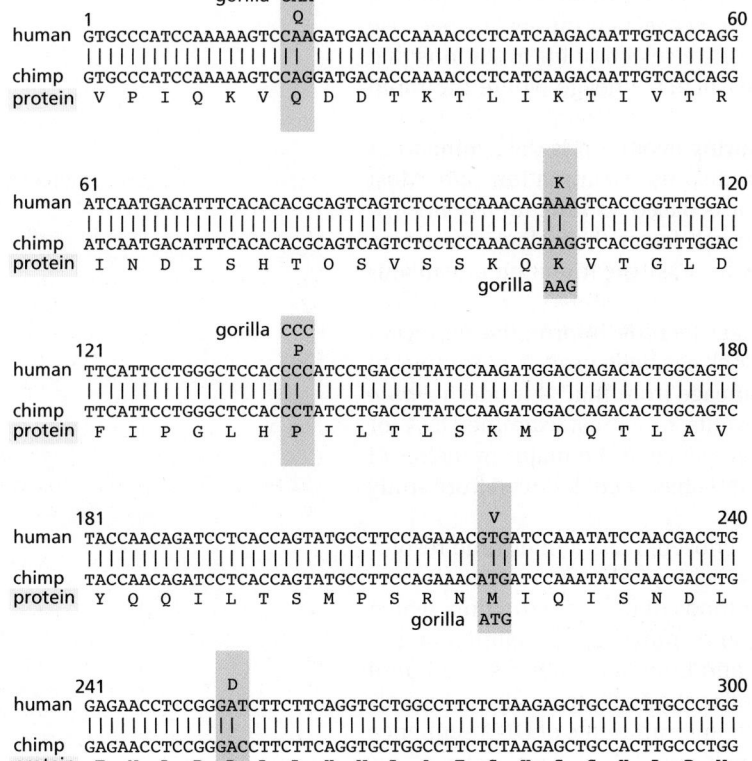

Figure 4–64 **Tracing the ancestral sequence from a sequence comparison of the coding regions of human and chimpanzee leptin genes.** Reading left to right and top to bottom, a continuous 300-nucleotide segment of a leptin-coding gene is illustrated. Leptin is a hormone that regulates food intake and energy utilization in response to the adequacy of fat reserves. As indicated by the codons boxed in green, only 5 nucleotides (of 441 total) differ between the two species. Moreover, in only one of the five positions does the difference in nucleotide lead to a difference in the encoded amino acid. For each of the five variant nucleotide positions, the corresponding sequence in the gorilla is also indicated. In two cases, the gorilla sequence agrees with the human sequence, while in three cases it agrees with the chimpanzee sequence.

What was the sequence of the leptin gene in the last common ancestor? The most economical assumption is that evolution has followed a pathway requiring the minimum number of mutations consistent with the data. Thus, it seems likely that the leptin sequence of the last common ancestor was the same as the human and chimpanzee sequences when they agree; when they disagree, the gorilla sequence would be used as a tiebreaker. For convenience, only the first 300 nucleotides of the leptin-coding sequences are given. The remaining 141 are identical between humans and chimpanzees.

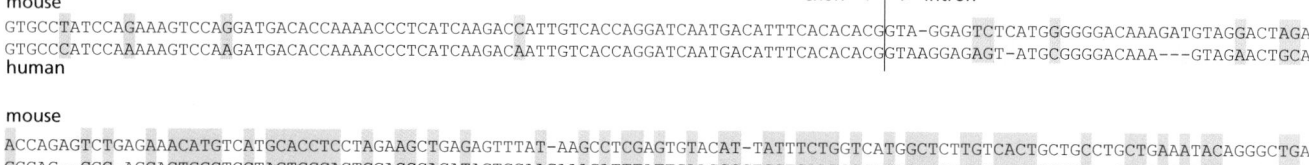

Figure 4–65 The very different rates of evolution of exons and introns, as illustrated by comparing a portion of the mouse and human leptin genes. Positions where the sequences differ by a single nucleotide substitution are boxed in *green*, and positions that differ by the addition or deletion of nucleotides are boxed in *yellow*. Note that, thanks to purifying selection, the coding sequence of the exon is much more conserved than is the adjacent intron sequence.

based on radioisotope decay in the rock formations in which fossils are found. Because the fossil record has many gaps, however, precise divergence times between species are difficult to establish, even for species that leave good fossils with distinctive morphology.

Phylogenetic trees whose timing has been calibrated according to the fossil record suggest that changes in the sequences of particular genes or proteins tend to occur at a nearly constant rate, although rates that differ from the norm by as much as twofold are observed in particular lineages. This provides us with a *molecular clock* for evolution—or rather a set of molecular clocks corresponding to different categories of DNA sequence. As in the example in **Figure 4–65**, the clock runs most rapidly and regularly in sequences that are not subject to purifying selection. These include portions of introns that lack splicing or regulatory signals, the third position in synonymous codons, and genes that have been irreversibly inactivated by mutation (the so-called pseudogenes). The clock runs most slowly for sequences that are subject to strong functional constraints—for example, the amino acid sequences of proteins that engage in specific interactions with large numbers of other proteins and whose structure is therefore highly constrained, or the nucleotide sequences that encode the RNA subunits of the ribosome, on which all protein synthesis depends.

Occasionally, rapid change is seen in a previously highly conserved sequence. As discussed later in this chapter, such episodes are especially interesting because they are thought to reflect periods of strong positive selection for mutations that have conferred a selective advantage in the particular lineage where the rapid change occurred.

The pace at which molecular clocks run during evolution is determined not only by the degree of purifying selection, but also by the mutation rate. Most notably, in animals, although not in plants, clocks based on functionally unconstrained mitochondrial DNA sequences run much faster than clocks based on functionally unconstrained nuclear sequences, because the mutation rate in animal mitochondria is exceptionally high.

Categories of DNA for which the clock runs fast are most informative for recent evolutionary events; the mitochondrial DNA clock has been used, for example, to chronicle the divergence of the Neanderthal lineage from that of modern *Homo sapiens*. To study more ancient evolutionary events, one must examine DNA for which the clock runs more slowly; thus the divergence of the major branches of the tree of life—bacteria, archaea, and eukaryotes—has been deduced from study of the sequences specifying ribosomal RNA.

In general, molecular clocks, appropriately chosen, have a finer time resolution than the fossil record, and they are a more reliable guide to the detailed structure of phylogenetic trees than are classical methods of tree construction, which are based on family resemblances in anatomy and embryonic development. For example, the precise family tree of great apes and humans was not settled until sufficient molecular sequence data accumulated in the 1980s to produce the pedigree shown previously in Figure 4–63. And with huge amounts of DNA sequence now determined from a wide variety of mammals, much better estimates of our relationship to them are being obtained (**Figure 4–66**).

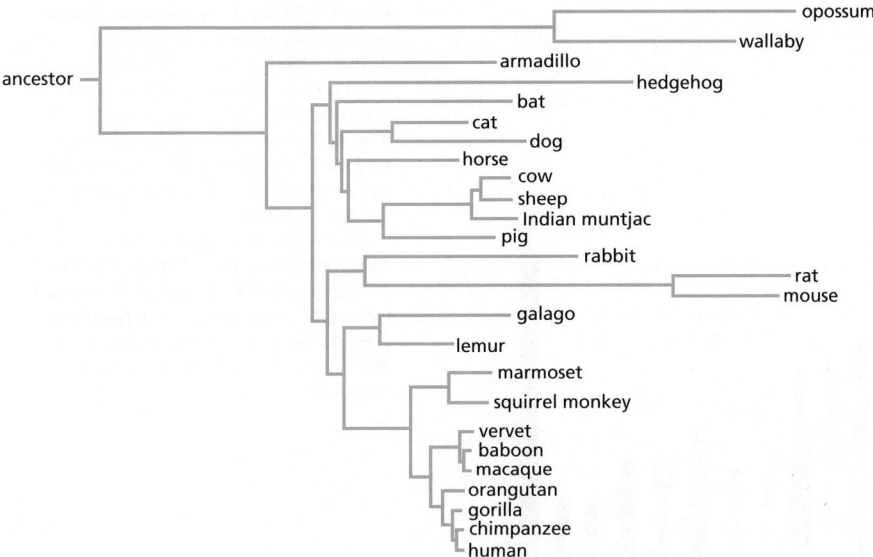

Figure 4–66 A phylogenetic tree showing the evolutionary relationships of some present-day mammals. The length of each line is proportional to the number of "neutral substitutions"—that is, nucleotide changes at sites where there is assumed to be no purifying selection. (Adapted from G.M. Cooper et al., *Genome Res.* 15:901–913, 2005. With permission from Cold Spring Harbor Laboratory Press.)

A Comparison of Human and Mouse Chromosomes Shows How the Structures of Genomes Diverge

As would be expected, the human and chimpanzee genomes are much more alike than are the human and mouse genomes, even though all three genomes are roughly the same size and contain nearly identical sets of genes. Mouse and human lineages have had approximately 80 million years to diverge through accumulated mutations, versus 6 million years for humans and chimpanzees. In addition, as indicated in Figure 4–66, rodent lineages (represented by the rat and the mouse) have unusually fast molecular clocks, and have diverged from the human lineage more rapidly than otherwise expected.

While the way that the genome is organized into chromosomes is almost identical between humans and chimpanzees, this organization has diverged greatly between humans and mice. According to rough estimates, a total of about 180 breakage-and-rejoining events have occurred in the human and mouse lineages since these two species last shared a common ancestor. In the process, although the number of chromosomes is similar in the two species (23 per haploid genome in the human versus 20 in the mouse), their overall structures differ greatly. Nonetheless, even after the extensive genomic shuffling, there are many large blocks of DNA in which the gene order is the same in the human and the mouse. These stretches of conserved gene order in chromosomes are referred to as regions of *synteny*. **Figure 4–67** illustrates how segments of the different mouse chromosomes map onto the human chromosome set. For much more distantly related vertebrates, such as chicken and human, the number of breakage-and-rejoining events has been much greater and the regions of synteny are much shorter; in addition, they are often hard to discern because of the divergence of the DNA sequences that they contain.

An unexpected conclusion from a detailed comparison of the complete mouse and human genome sequences, confirmed by subsequent comparisons between the genomes of other vertebrates, is that small blocks of DNA sequence are being deleted from and added to genomes at a surprisingly rapid rate. Thus, if we assume that our common ancestor had a genome of human size (about 3.2 billion nucleotide pairs), mice would have lost a total of about 45% of that genome from accumulated deletions during the past 80 million years, while humans would have lost about 25%. However, substantial sequence gains from many small chromosome duplications and from the multiplication of transposons have compensated for these deletions. As a result, our genome size is thought to be practically unchanged from that of the last common ancestor of humans and mice, while the mouse genome is smaller by only about 0.3 billion nucleotides.

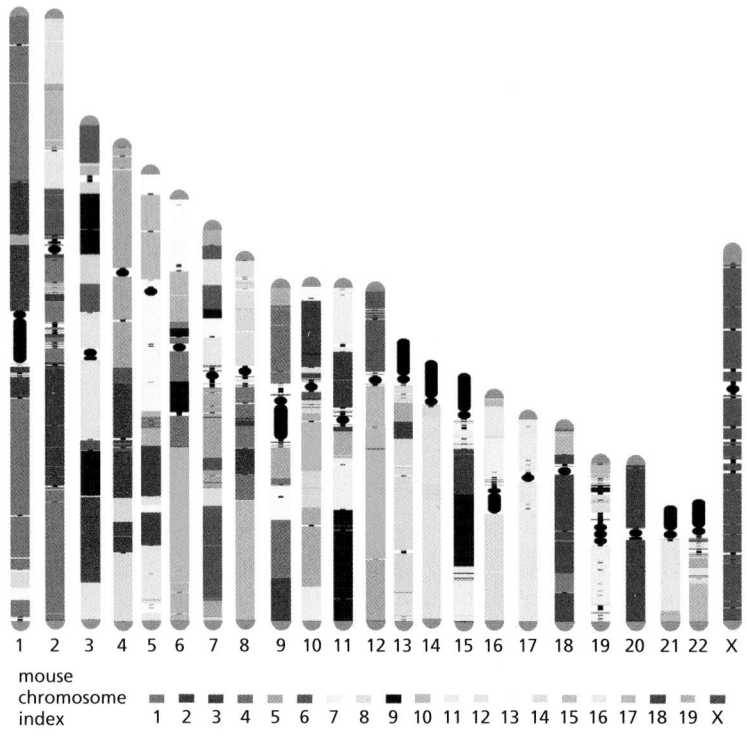

1 2 3 4 5 6 7 8 9 10 11 12 13 14 15 16 17 18 19 20 21 22 X

mouse
chromosome
index

1 2 3 4 5 6 7 8 9 10 11 12 13 14 15 16 17 18 19 X

Figure 4–67 Synteny between human and mouse chromosomes. In this diagram, the human chromosome set is shown, with each part of each chromosome colored according to the mouse chromosome with which it is syntenic. The color coding used for each mouse chromosome is shown at the bottom of the figure.

Heterochromatic highly repetitive regions (such as centromeres) that are difficult to sequence cannot be mapped in this way; these are colored *black*. (Adapted from E.E. Eichler and D. Sankoff, *Science* 301:793–797, 2003. With permission from AAAS.)

Good evidence for the loss of DNA sequences in small blocks during evolution can be obtained from a detailed comparison of regions of synteny in the human and mouse genomes. The comparative shrinkage of the mouse genome can be clearly seen from such comparisons, with the net loss of sequences scattered throughout the long stretches of DNA that are otherwise homologous (**Figure 4–68**).

DNA is added to genomes both by the spontaneous duplication of chromosomal segments that are typically tens of thousands of nucleotide pairs long (as will be discussed shortly) and by insertion of new copies of active transposons. Most transposition events are duplicative, because the original copy of the transposon stays where it was when a copy inserts at the new site; see, for example, Figure 5-63. Comparison of the DNA sequences derived from transposons in the human and the mouse readily reveals some of the sequence additions (**Figure 4–69**).

It remains a mystery why all mammals have maintained genome sizes of roughly 3 billion nucleotide pairs that contain nearly identical sets of genes, even though only approximately 150 million nucleotide pairs appear to be under sequence-specific functional constraints.

The Size of a Vertebrate Genome Reflects the Relative Rates of DNA Addition and DNA Loss in a Lineage

In more distantly related vertebrates, genome size can vary considerably, apparently without a drastic effect on the organism or its number of genes. Thus, the chicken genome, at one billion nucleotide pairs, is only about one-third the size

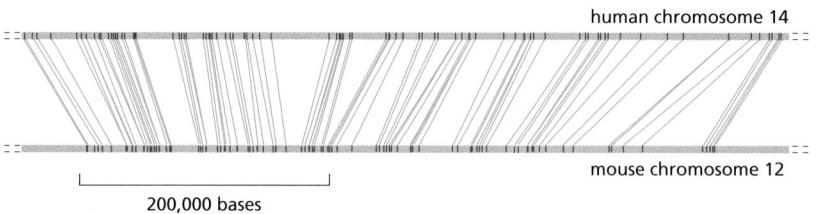

human chromosome 14

mouse chromosome 12

200,000 bases

Figure 4–68 Comparison of a syntenic portion of mouse and human genomes. About 90% of the two genomes can be aligned in this way. Note that while there is an identical order of the matched index sequences *(red marks)*, there has been a net loss of DNA in the mouse lineage that is interspersed throughout the entire region. This type of net loss is typical for all such regions, and it accounts for the fact that the mouse genome contains 14% less DNA than does the human genome. (Adapted from Mouse Genome Sequencing Consortium, *Nature* 420:520–562, published 2002 by Nature Publishing Group. Reproduced with permission of SNCSC.)

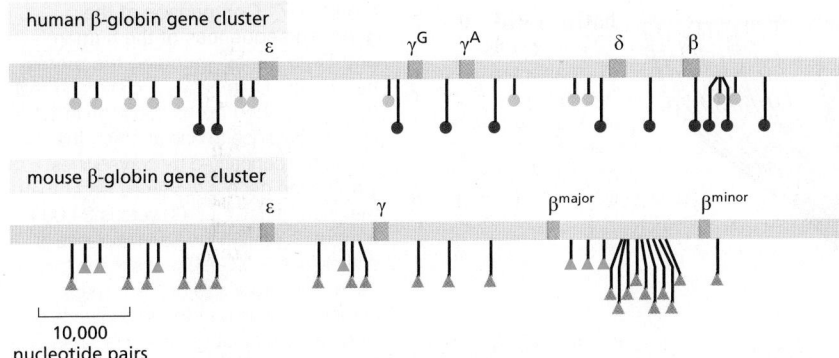

human β-globin gene cluster

ε   γ^G   γ^A   δ   β

mouse β-globin gene cluster

ε   γ   β^major   β^minor

├─────┤
10,000
nucleotide pairs

Figure 4–69 A comparison of the β-globin gene cluster in the human and mouse genomes, showing the locations of transposable elements. This stretch of the human genome contains five functional β-globin-like genes *(orange)*; the comparable region from the mouse genome has only four. The positions of the human *Alu* sequences are indicated by *green circles*, and the human *L1* sequences by *red circles*. The mouse genome contains different but related transposable elements: the positions of B1 elements (which are related to the human *Alu* sequences) are indicated by *blue triangles*, and the positions of the mouse *L1* elements (which are related to the human *L1* sequences) are indicated by *orange triangles*. The absence of transposable elements from the globin structural genes can be attributed to purifying selection, which would have eliminated any insertion that compromised gene function. (Courtesy of Ross Hardison and Webb Miller.)

of the mammalian genome. An extreme example is the puffer fish, *Fugu rubripes* (**Figure 4–70**), which has a tiny genome for a vertebrate (0.4 billion nucleotide pairs compared to 1 billion or more for many other fish). The small size of the *Fugu* genome is largely due to the small size of its introns and intergenic regions. Specifically, *Fugu* introns, as well as other noncoding segments of the *Fugu* genome, lack the repetitive DNA that makes up a large portion of the genomes of most well-studied vertebrates. Nevertheless, the positions of the *Fugu* introns between the exons of each gene are almost the same as in mammalian genomes (**Figure 4–71**).

While initially a mystery, we now have a simple explanation for such large differences in genome size between similar organisms: because all vertebrates experience a continuous process of DNA loss and DNA addition, the size of a genome merely depends on the balance between these opposing processes acting over millions of years. Suppose, for example, that in the lineage leading to *Fugu*, the rate of DNA addition happened to slow greatly. Over long periods of time, this would result in a major "cleansing" from this fish genome of those DNA sequences whose loss could be tolerated. The result is an unusually compact genome, relatively free of junk and clutter, but retaining through purifying selection the vertebrate DNA sequences that are functionally important. This makes *Fugu*, with its 400 million nucleotide pairs of DNA, a valuable resource for genome research aimed at understanding humans.

We Can Infer the Sequence of Some Ancient Genomes

The genomes of ancestral organisms can be inferred, but most can never be directly observed. DNA is very stable compared with most organic molecules, but it is not perfectly stable, and its progressive degradation, even under the best circumstances, means that it is virtually impossible to extract sequence information from fossils that are more than a million years old. Although a modern organism such as the horseshoe crab looks remarkably similar to fossil ancestors that lived 200 million years ago, there is every reason to believe that the horseshoe-crab genome has been changing during all that time in much the same way as in other evolutionary lineages, and at a similar rate. Selection must have maintained key functional properties of the horseshoe-crab genome to account for the morphological stability of the lineage. However, comparisons between different present-day organisms show that the fraction of the genome subject to purifying selection is small; hence, it is fair to assume that the genome of the modern horseshoe crab, while preserving features critical for function, must differ greatly from that of its extinct ancestors, known to us only through the fossil record.

It is possible to get direct sequence information by examining DNA samples from ancient materials if these are not too old. In recent years, technical advances have allowed DNA sequencing from exceptionally well-preserved bone fragments that date from more than 100,000 years ago. Although any DNA this old will be imperfectly preserved, a sequence of the Neanderthal genome has been reconstructed from many millions of short DNA sequences, revealing—among other things—that our human ancestors interbred with Neanderthals in Europe and

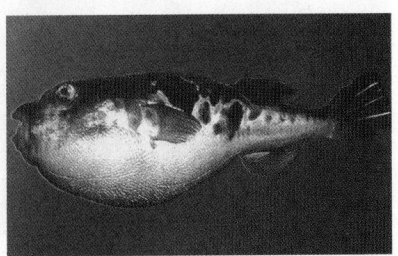

Figure 4–70 The puffer fish, *Fugu rubripes*. (Courtesy of Byrappa Venkatesh.)

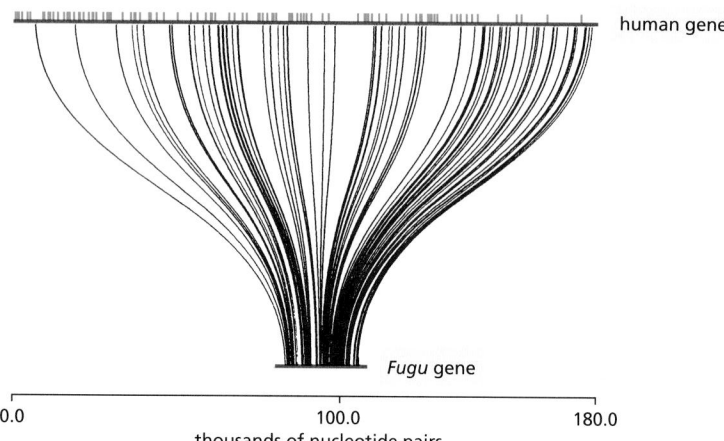

human gene

Fugu gene

| 0.0 | 100.0 | 180.0 |

thousands of nucleotide pairs

Figure 4–71 Comparison of the genomic sequences of the human and *Fugu* genes encoding the protein huntingtin. Both genes (indicated in *red*) contain 67 short exons that align in 1:1 correspondence to one another; these exons are connected by curved lines. The human gene is 7.5 times larger than the *Fugu* gene (180,000 versus 24,000 nucleotide pairs). The size difference is entirely due to larger introns in the human gene. The larger size of the human introns is due in part to the presence of retrotransposons (discussed in Chapter 5), whose positions are represented by green vertical lines; the *Fugu* introns lack retrotransposons. In humans, mutation of the huntingtin gene causes Huntington's disease, an inherited neurodegenerative disorder. (Adapted from S. Baxendale et al., *Nat. Genet.* 10:67–76, published 1995 by Nature Publishing Group. Reproduced with permission of SNCSC.)

that modern humans have inherited specific genes from them (**Figure 4–72**). The average difference in DNA sequence between humans and Neanderthals shows that our two lineages diverged somewhere between 270,000 and 440,000 years ago, well before the time that humans are believed to have migrated out of Africa.

But what about deciphering the genomes of much older ancestors, those for which no useful DNA samples can be isolated? For organisms that are as closely related as human and chimpanzee, we saw that this may not be difficult: reference to the gorilla sequence can be used to sort out which of the few sequence differences between human and chimpanzee are inherited from our common ancestor some 6 million years ago (see Figure 4–64). And for an ancestor that has produced a large number of different organisms alive today, the DNA sequences of many species can be compared simultaneously to unscramble much of the ancestral sequence, allowing scientists to derive DNA sequences much farther back in time. For example, from the genome sequences currently being obtained for dozens of modern placental mammals, it should be possible to infer much of the genome sequence of their 100 million-year-old common ancestor—the precursor of species as diverse as dog, mouse, rabbit, armadillo, and human (see Figure 4–66).

Multispecies Sequence Comparisons Identify Conserved DNA Sequences of Unknown Function

The mass of DNA sequence now in databases (hundreds of billions of nucleotide pairs) provides a rich resource that scientists can mine for many purposes. This information can be used not only to unscramble the evolutionary pathways that have led to modern organisms, but also to provide insights into how cells and organisms function. Perhaps the most remarkable discovery in this realm comes from the observation that a striking amount of DNA sequence that does not code for protein has been conserved during mammalian evolution (see Table 4–1, p. 184). This is most clearly revealed when we align and compare DNA synteny

Figure 4–72 The Neanderthals. (A) Map of Europe showing the location of the cave in Croatia where most of the bones used to isolate the DNA used to derive the Neanderthal genome sequence were discovered. (B) Photograph of the Vindija cave. (C) Photograph of the 38,000-year-old bones from Vindija. More recent studies have succeeded in extracting DNA sequence information from hominid remains that are considerably older (see Movie 8.3). (B, courtesy of Johannes Krause; C, from R.E. Green et al., *Science* 328: 710–722, 2010. Reprinted with permission from AAAS.)

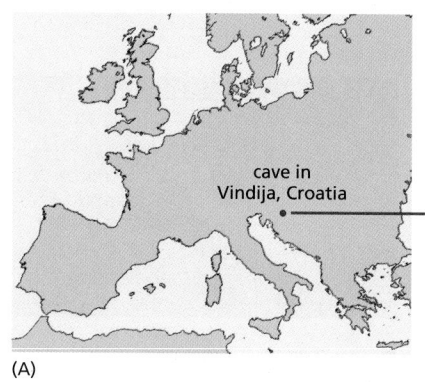

cave in
Vindija, Croatia

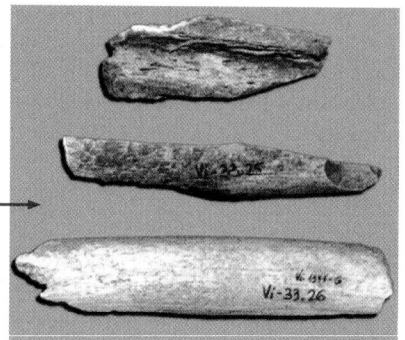

| 5 cm |

(A) (B) (C)

blocks from many different species, thereby identifying large numbers of so-called *multispecies conserved sequences*: some of these code for protein, but most of them do not (**Figure 4–73**).

Most of the noncoding conserved sequences discovered in this way turn out to be relatively short, containing between 50 and 200 nucleotide pairs. Among the most mysterious are the so-called "ultraconserved" noncoding sequences, exemplified by more than 5000 DNA segments over 100 nucleotides long that are exactly the same in human, mouse, and rat. Most have undergone little or no change since mammalian and bird ancestors diverged about 300 million years ago. The strict conservation implies that even though the sequences do not encode proteins, each nevertheless has an important function maintained by purifying selection. The puzzle is to unravel what those functions are.

Many of the conserved sequences that do not code for protein are now known to produce untranslated RNA molecules, such as the thousands of *long noncoding RNAs* (*lncRNAs*) that are thought to have important functions in regulating gene transcription. As we shall also see in Chapter 7, others are short regions of DNA scattered throughout the genome that directly bind proteins involved in gene regulation. But it is uncertain how much of the conserved noncoding DNA can be accounted for in these ways, and the function of most of it remains a mystery. This enigma highlights how much more we need to learn about the fundamental biological mechanisms that operate in animals and other complex organisms, and its solution is certain to have profound consequences for medicine.

How can cell biologists tackle the mystery of noncoding conserved DNA? Traditionally, attempts to determine the function of a puzzling DNA sequence begin by looking at the consequences of its experimental disruption. But many DNA sequences that are crucial for an organism in the wild can be expected to have no noticeable effect on its phenotype under laboratory conditions: what is required for a mouse to survive in a laboratory cage is very much less than what is required

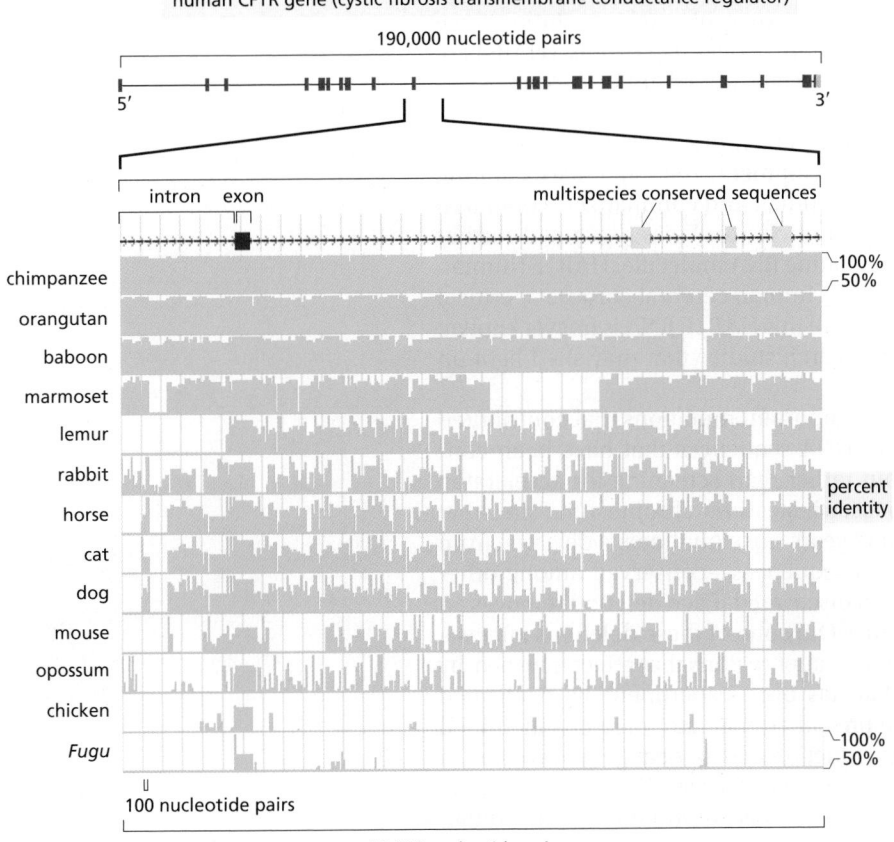

Figure 4–73 The detection of multispecies conserved sequences. In this example, genome sequences for each of the organisms shown have been compared with the indicated region of the human CFTR (cystic fibrosis transmembrane conductance regulator) gene; this region contains one exon plus a large amount of intronic DNA. For each organism, the percent identity with human for each 25-nucleotide block is plotted in *green*. In addition, a computational algorithm has been used to detect the sequences within this region that are most highly conserved when the sequences from all of the organisms are taken into account. Besides the exon (*dark blue* on the line at the top of the figure), the positions of three other blocks of multispecies conserved sequences are indicated (*pale blue*). The function of most such sequences in the human genome is not known. (Courtesy of Eric D. Green.)

for it to succeed in nature. Moreover, calculations based on population genetics reveal that just a tiny selective advantage—less than a 0.1% difference in survival—can be enough to strongly favor retaining a particular DNA sequence over evolutionary time spans. One should therefore not be surprised to find that many DNA sequences that are ultraconserved can be deleted from the mouse genome without any noticeable effect on that mouse in a laboratory.

A second important approach for discovering the function of a mysterious noncoding DNA sequence uses biochemical techniques to identify proteins or RNA molecules that bind to it—and/or to any RNA molecules that it produces. Most of this task still lies before us, but a start has been made (see p. 435).

Changes in Previously Conserved Sequences Can Help Decipher Critical Steps in Evolution

Given genome sequence information, we can tackle another intriguing question: What alterations in our DNA have made humans so different from other animals—or for that matter, what makes any individual species so different from its relatives? For example, as soon as both the human and the chimpanzee genome sequences became available, scientists began searching for DNA sequence changes that might account for the striking differences between us and chimpanzees. With 3.2 billion nucleotide pairs to compare in the two species, this might seem an impossible task. But the job was made much easier by confining the search to 35,000 clearly defined multispecies conserved sequences (a total of about 5 million nucleotide pairs), representing parts of the genome that are most likely to be functionally important. Though these sequences are conserved strongly, they are not conserved perfectly, and when the version in one species is compared with that in another they are generally found to have drifted apart by a small amount corresponding simply to the time elapsed since the last common ancestor. In a small proportion of cases, however, one sees signs of a sudden evolutionary spurt. For example, some DNA sequences that have been highly conserved in other mammalian species are found to have accumulated nucleotide changes exceptionally rapidly during the 6 million years of human evolution since we diverged from the chimpanzees. These *human accelerated regions* (HARs) are thought to reflect functions that have been especially important in making us different in some useful way.

About 50 such sites were identified in one study, one-fourth of which were located near genes associated with neural development. The sequence exhibiting the most rapid change (18 changes between human and chimpanzee, compared to only two changes between chimpanzee and chicken) was examined further and found to encode a 118-nucleotide noncoding RNA molecule, HAR1F (human accelerated region 1F), that is produced in the human cerebral cortex at a critical time during brain development. The function of this HAR1F RNA is not yet known, but findings of this type are stimulating research studies that may shed light on crucial features of the human brain.

A related approach in the search for the important mutations that contributed to human evolution likewise begins with DNA sequences that have been conserved during mammalian evolution, but rather than screening for accelerated changes in individual nucleotides, it focuses instead on chromosome sites that have experienced deletions in the 6 million years since our lineage diverged from that of chimpanzees. More than 500 such sequences—conserved among other species but deleted in humans—have been discovered. Each deletion removes an average of 95 nucleotides of DNA sequence. Only one of these deletions affects a protein-coding region: the rest are thought to alter regions that affect how nearby genes are expressed, an expectation that has been experimentally confirmed in a few cases. A large proportion of the presumed regulatory regions identified in this way lie near genes that affect neural function and/or near genes involved in steroid signaling, suggesting that changes in the nervous system and in immune or reproductive functions have played an especially important role in human evolution.

Mutations in the DNA Sequences That Control Gene Expression Have Driven Many of the Evolutionary Changes in Vertebrates

The vast hoard of genomic sequence data now being accumulated can be explored in many other ways to reveal events that happened even hundreds of millions of years ago. For example, one can attempt to trace the origins of the regulatory elements in DNA that have played critical parts in vertebrate evolution. One such study began with the identification of nearly 3 million noncoding sequences, averaging 28 base pairs in length, that have been conserved in recent vertebrate evolution while being absent in more ancient ancestors. Each of these special non-coding sequences is likely to represent a functional innovation peculiar to a particular branch of the vertebrate family tree, and most of them are thought to consist of regulatory DNA that governs the expression of a neighboring gene. Given full genome sequences, one can identify the genes that lie closest and thus appear most likely to have fallen under the sway of these novel regulatory elements. By comparing many different species, with known divergence times, one can also estimate when each such regulatory element came into existence as a conserved feature. The findings suggest remarkable evolutionary differences between the various functional classes of genes (Figure 4–74). Conserved regulatory elements that originated early in vertebrate evolution—that is, more than about 300 million years ago, which is when the mammalian lineage split from the lineage leading to birds and reptiles—seem to be mostly associated with genes that code for transcription regulator proteins and for proteins with roles in organizing embryonic development. Then came an era when the regulatory DNA innovations arose next to genes coding for receptors for extracellular signals. Finally, over the course of the past 100 million years, the regulatory innovations seem to have been concentrated in the neighborhood of genes coding for proteins (such as protein kinases) that function to modify other proteins post-translationally.

Many questions remain to be answered about these phenomena and what they mean. One possible interpretation is that the logic—the circuit diagram—of the gene regulatory network in vertebrates was established early, and that more recent evolutionary change has mainly occurred through the tuning of quantitative parameters. This could help to explain why, among the mammals, for example, the basic body plan—the topology of the tissues and organs—has been largely conserved.

Gene Duplication Also Provides an Important Source of Genetic Novelty During Evolution

Evolution depends on the creation of new genes, as well as on the modification of those that already exist. How does this occur? When we compare organisms that seem very different—a primate with a rodent, for example, or a mouse with a fish—we rarely encounter genes in the one species that have no homolog in the

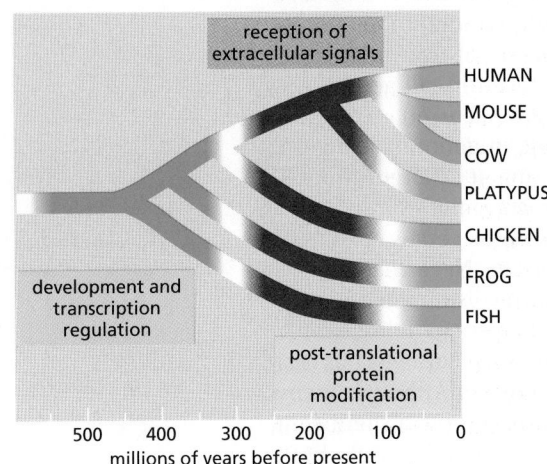

Figure 4–74 **The types of changes in gene regulation inferred to have predominated during the evolution of our vertebrate ancestors.** To produce the information summarized in this plot, wherever possible the type of gene regulated by each conserved noncoding sequence was inferred from the identity of its closest protein-coding gene. The fixation time for each conserved sequence was then used to derive the conclusions shown. (Based on C.B. Lowe et al., *Science* 333:1019–1024, 2011.)

other. Genes without homologous counterparts are relatively scarce even when we compare such divergent organisms as a mammal and a worm. On the other hand, we frequently find gene families that have different numbers of members in different species. To create such families, genes have been repeatedly duplicated, and the copies have then diverged to take on new functions that often vary from one species to another.

Gene duplication occurs at high rates in all evolutionary lineages, contributing to the vigorous process of DNA addition discussed previously. In a detailed study of spontaneous duplications in yeast, duplications of 50,000 to 250,000 nucleotide pairs were commonly observed, most of which were tandemly repeated. These appeared to result from DNA replication errors that led to the inexact repair of double-strand chromosome breaks. A comparison of the human and chimpanzee genomes reveals that, since the time that these two organisms diverged, such *segmental duplications* have added about 5 million nucleotide pairs to each genome every million years, with an average duplication size being about 50,000 nucleotide pairs (although there are some duplications five times larger). In fact, if one counts nucleotides, duplication events have created more differences between our two species than have single-nucleotide substitutions.

Duplicated Genes Diverge

What is the fate of newly duplicated genes? In most cases, there is presumed to be little or no selection—at least initially—to maintain the duplicated state since either copy can provide an equivalent function. Hence, many duplication events are likely to be followed by loss-of-function mutations in one or the other gene. This cycle would functionally restore the one-gene state that preceded the duplication. Indeed, there are many examples in contemporary genomes where one copy of a duplicated gene can be seen to have become irreversibly inactivated by multiple mutations. Over time, the sequence similarity between such a **pseudogene** and the functional gene whose duplication produced it would be expected to be eroded by the accumulation of many mutations in the pseudogene—the homologous relationship eventually becoming undetectable.

An alternative fate for gene duplications is for both copies to remain functional, while diverging in their sequence and pattern of expression, thus taking on different roles. This process of "duplication and divergence" almost certainly explains the presence of large families of genes with related functions in biologically complex organisms, and it is thought to play a critical role in the evolution of increased biological complexity. An examination of many different eukaryotic genomes suggests that the probability that any particular gene will undergo a duplication event that spreads to most or all individuals in a species is approximately 1 percent every million years.

Whole-genome duplications offer particularly dramatic examples of the duplication–divergence cycle. A whole-genome duplication can occur quite simply: all that is required is one round of genome replication in a germ-line cell lineage without a corresponding cell division. Initially, the chromosome number simply doubles. Such abrupt increases in the ploidy of an organism are common, particularly in fungi and plants. After a whole-genome duplication, all genes exist as duplicate copies. However, unless the duplication event occurred so recently that there has been little time for subsequent alterations in genome structure, the results of a series of segmental duplications—occurring at different times— are hard to distinguish from the end product of a whole-genome duplication. In mammals, for example, the role of whole-genome duplications versus a series of piecemeal duplications of DNA segments is quite uncertain. Nevertheless, it is clear that a great deal of gene duplication has occurred in the distant past.

Analysis of the genome of the zebrafish, in which at least one whole-genome duplication is thought to have occurred hundreds of millions of years ago, has cast some light on the process of gene duplication and divergence. Although many duplicates of zebrafish genes appear to have been lost by mutation, a significant fraction—perhaps as many as 30–50%—have diverged functionally while both

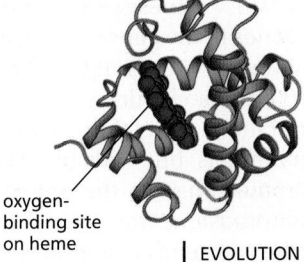

single-chain globin binds
one oxygen molecule

Figure 4–75 **A comparison of the structure of one-chain and four-chain globins.** The four-chain globin shown is hemoglobin, which is a complex of two α-globin and two β-globin chains. The one-chain globin present in some primitive vertebrates represents an intermediate in the evolution of the four-chain globin. With oxygen bound it exists as a monomer; without oxygen it dimerizes.

oxygen-
binding site
on heme

EVOLUTION OF A
SECOND GLOBIN
CHAIN BY
GENE DUPLICATION
FOLLOWED BY
MUTATION

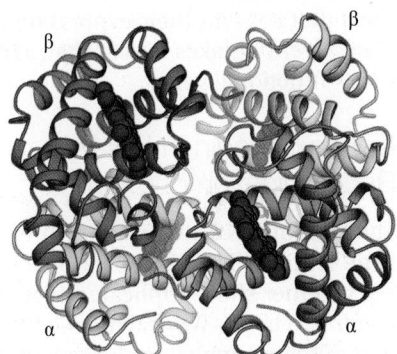

four-chain globin binds four
oxygen molecules in a
cooperative manner

copies have remained active. In many cases, the most obvious functional difference between the duplicated genes is that they are expressed in different tissues or at different stages of development. One attractive theory to explain such an end result imagines that different, mildly deleterious mutations occur quickly in both copies of a duplicated gene set. For example, one copy might lose expression in a particular tissue as a result of a regulatory mutation, while the other copy loses expression in a second tissue. Following such an occurrence, both gene copies would be required to provide the full range of functions that were once supplied by a single gene; hence, both copies would now be protected from loss through inactivating mutations. Over a longer period, each copy could then undergo further changes through which it could acquire new, specialized features.

The Evolution of the Globin Gene Family Shows How DNA Duplications Contribute to the Evolution of Organisms

The globin gene family provides an especially good example of how DNA duplication generates new proteins, because its evolutionary history has been worked out particularly well. The unmistakable similarities in amino acid sequence and structure among the present-day globins indicate that they all must derive from a common ancestral gene, even though some are now encoded by widely separated genes in the mammalian genome.

We can reconstruct some of the past events that produced the various types of oxygen-carrying hemoglobin molecules by considering the different forms of the protein in organisms at different positions on the tree of life. A molecule like hemoglobin was necessary to allow multicellular animals to grow to a large size, since large animals cannot simply rely on the diffusion of oxygen through the body surface to oxygenate their tissues adequately. But oxygen plays a vital part in the life of nearly all living organisms, and oxygen-binding proteins homologous to hemoglobin can be recognized even in plants, fungi, and bacteria. In animals, the most primitive oxygen-carrying molecule is a globin polypeptide chain of about 150 amino acids that is found in many marine worms, insects, and primitive fish. The hemoglobin molecule in more complex vertebrates, however, is composed of two kinds of globin chains. It appears that about 500 million years ago, during the continuing evolution of fish, a series of gene mutations and duplications occurred. These events established two slightly different globin genes in the genome of each individual, coding for α- and β-globin chains that associate to form a hemoglobin molecule consisting of two α chains and two β chains (**Figure 4–75**). The four oxygen-binding sites in the $\alpha_2\beta_2$ molecule interact, allowing a cooperative allosteric change in the molecule as it binds and releases oxygen, which enables hemoglobin to take up and release oxygen more efficiently than the single-chain version.

Still later, during the evolution of mammals, the β-chain gene apparently underwent duplication and mutation to give rise to a second β-like chain that is synthesized specifically in the fetus. The resulting hemoglobin molecule has a higher affinity for oxygen than adult hemoglobin and thus helps in the transfer of oxygen from the mother to the fetus. The gene for the new β-like chain subsequently duplicated and mutated again to produce two new genes, ε and γ, the ε chain being produced earlier in development (to form $\alpha_2\varepsilon_2$) than the fetal γ chain, which forms $\alpha_2\gamma_2$. A duplication of the adult β-chain gene occurred still later, during primate evolution, to give rise to a δ-globin gene and thus to a minor form of hemoglobin ($\alpha_2\delta_2$) that is found only in adult primates (**Figure 4–76**).

Each of these duplicated genes has been modified by point mutations that affect the properties of the final hemoglobin molecule, as well as by changes in regulatory regions that determine the timing and level of expression of the gene.

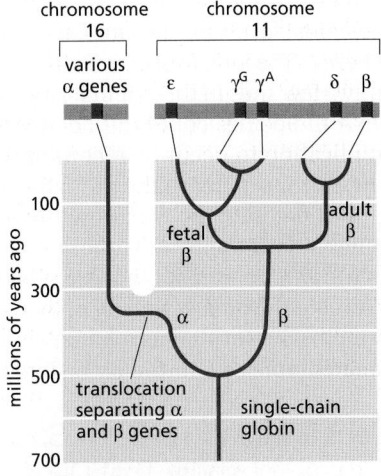

Figure 4–76 **An evolutionary scheme for the globin chains that carry oxygen in the blood of animals.** The scheme emphasizes the β-like globin gene family. A relatively recent gene duplication of the γ-chain gene produced γ^G and γ^A, which are fetal β-like chains of identical function. The location of the globin genes in the human genome is shown at the top of the figure.

As a result, each globin is made in different amounts at different times of human development.

The history of these gene duplications is reflected in the arrangement of hemoglobin genes in the genome. In the human genome, the genes that arose from the original β gene are arranged as a series of homologous DNA sequences located within 50,000 nucleotide pairs of one another on a single chromosome. A similar cluster of human α-globin genes is located on a separate chromosome. Not only other mammals, but birds too have their α- and β-globin gene clusters on separate chromosomes. In the frog *Xenopus*, however, they are together, suggesting that a chromosome translocation event in the lineage of birds and mammals separated the two gene clusters about 300 million years ago, soon after our ancestors diverged from amphibians (see Figure 4–76).

There are several duplicated globin DNA sequences in the α- and β-globin gene clusters that are not functional genes but pseudogenes. These have a close sequence similarity to the functional genes but have been disabled by mutations that prevent their expression as functional proteins. The existence of such pseudogenes makes it clear that, as expected, not every DNA duplication leads to a new functional gene.

Genes Encoding New Proteins Can Be Created by the Recombination of Exons

The role of DNA duplication in evolution is not confined to the expansion of gene families. It can also act on a smaller scale to create single genes by stringing together short duplicated segments of DNA. The proteins encoded by genes generated in this way can be recognized by the presence of repeating similar protein domains, which are covalently linked to one another in series. The immunoglobulins (**Figure 4–77**), for example, as well as most fibrous proteins (such as collagens) are encoded by genes that have evolved by repeated duplications of a primordial DNA sequence.

In genes that have evolved in this way, as well as in many other genes, each separate exon often encodes an individual protein folding unit, or domain. It is believed that the organization of DNA coding sequences as a series of such exons separated by long introns has greatly facilitated the evolution of new proteins. The duplications necessary to form a single gene coding for a protein with repeating domains, for example, can easily occur by breaking and rejoining the DNA anywhere in the long introns on either side of an exon; without introns there would be only a few sites in the original gene at which a recombinational exchange between DNA molecules could duplicate the domain and not disrupt it. By enabling the duplication to occur by recombination at many potential sites rather than just a few, introns increase the probability of a favorable duplication event.

More generally, we know from genome sequences that the various parts of genes—both their individual exons and their regulatory elements—have served as modular elements that have been duplicated and moved about the genome to create the great diversity of living things. Thus, for example, many present-day proteins are formed as a patchwork of domains from different origins, reflecting their complex evolutionary history (see Figure 3–17).

Neutral Mutations Often Spread to Become Fixed in a Population, with a Probability That Depends on Population Size

In comparisons between two species that have diverged from one another by millions of years, it makes little difference which individuals from each species are

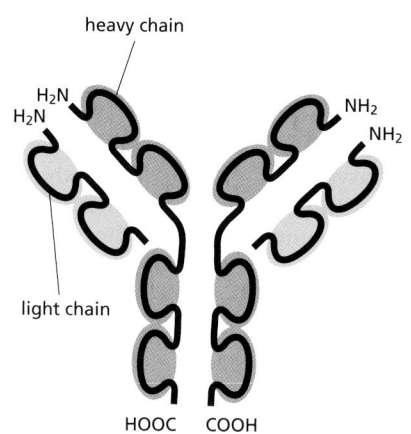

Figure 4–77 Schematic view of an antibody (immunoglobulin) molecule. This molecule is a complex of two identical heavy chains and two identical light chains. Each heavy chain contains four similar, covalently linked domains, while each light chain contains two such domains. Each domain is encoded by a separate exon, and all of the exons are thought to have evolved by the serial duplication of a single ancestral exon.

compared. For example, typical human and chimpanzee DNA sequences differ from one another by about 1%. In contrast, when the same region of the genome is sampled from two randomly chosen humans, the differences are typically about 0.1%. For more distantly related organisms, the interspecies differences outshine intraspecies variation even more dramatically. However, each "fixed difference" between the human and the chimpanzee (in other words, each difference that is now characteristic of all or nearly all individuals of each species) started out as a new mutation in a single individual. If the size of the interbreeding population in which the mutation occurred is N, the initial allele frequency for a new mutation would be $1/(2N)$ for a diploid organism. How does such a rare mutation become fixed in the population, and hence become a characteristic of the species rather than of a few scattered individuals?

The answer to this question depends on the functional consequences of the mutation. If the mutation has a significantly deleterious effect, it will simply be eliminated by purifying selection and will not become fixed. (In the most extreme case, the individual carrying the mutation will die without producing progeny.) Conversely, the rare mutations that confer a major reproductive advantage on individuals who inherit them can spread rapidly in the population. Because humans reproduce sexually and genetic recombination occurs each time a gamete is formed (discussed in Chapter 5), the genome of each individual who has inherited the mutation will be a unique recombinational mosaic of segments inherited from a large number of ancestors. The selected mutation along with a modest amount of neighboring sequence—ultimately inherited from the individual in which the mutation occurred—will simply be one piece of this huge mosaic.

The great majority of mutations that are not harmful are not beneficial either. These selectively neutral mutations can also spread and become fixed in a population, and they make a large contribution to evolutionary change in genomes. For example, as we saw earlier, they account for most of the DNA sequence differences between apes and humans. The spread of neutral mutations is not as rapid as the spread of the rare strongly advantageous mutations. It depends on a random variation in the number of mutation-bearing progeny produced by each mutation-bearing individual, causing changes in the relative frequency of the mutant allele in the population. Through a sort of "random walk" process, the mutant allele may eventually become extinct, or it may become commonplace. This can be modeled mathematically for an idealized interbreeding population, on the assumption of constant population size and random mating, as well as selective neutrality for the mutations. While neither of the first two assumptions is a good description of human population history, study of this idealized case reveals the general principles in a clear and simple way.

When a new neutral mutation occurs in a population of constant size N that is undergoing random mating, the probability that it will ultimately become fixed is approximately $1/(2N)$. This is because there are $2N$ copies of the gene in the diploid population, and each of them has an equal chance of becoming the predominant version in the long run. For those mutations that do become fixed, the mathematics shows that the average time to fixation is approximately $4N$ generations. Detailed analyses of data on human genetic variation have suggested an ancestral population size of approximately 10,000 at the time when the current pattern of genetic variation was largely established. With a population that has reached this size, the probability that a new, selectively neutral mutation would become fixed is small (1/20,000), while the average time to fixation would be on the order of 800,000 years (assuming a 20-year generation time). Thus, while we know that the human population has grown enormously since the development of agriculture approximately 15,000 years ago, most of the present-day set of common human genetic variants reflects the mixture of variants that was already present long before this time, when the human population was still small.

Similar arguments explain another phenomenon with important practical implications for genetic counseling. In an isolated community descended from a small group of founders, such as the people of Iceland or the Jews of Eastern

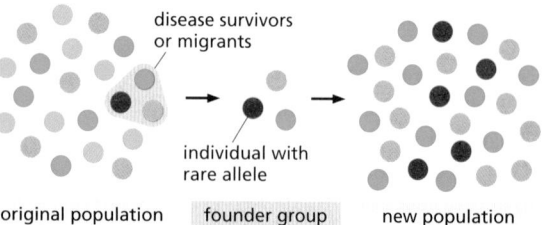

Figure 4–78 **How founder effects determine the set of genetic variants in a population of individuals belonging to the same species.** This example illustrates how a rare allele *(red)* can become established in an isolated population, even though the mutation that produced it has no selective advantage—or is mildly deleterious.

Europe, genetic variants that are rare in the human population as a whole can often be present at a high frequency, even if those variants are mildly deleterious (**Figure 4–78**).

A Great Deal Can Be Learned from Analyses of the Variation Among Humans

Even though the common variant gene alleles among modern humans originate from variants present in a comparatively tiny group of ancestors, the total number of variants now encountered, including those that are individually rare, is very large. New neutral mutations are constantly occurring and accumulating, even though no single one of them has had enough time to become fixed in the vast modern human population.

From detailed comparisons of the DNA sequences of a large number of modern humans located around the globe, scientists can estimate how many generations have elapsed since the origin of a particular neutral mutation. From such data, it has been possible to map the routes of ancient human migrations. For example, by combining this type of genetic analysis with archaeological findings, scientists have been able to deduce the most probable routes that our ancestors took when they left Africa 60,000 to 80,000 years ago (**Figure 4–79**).

We have been focusing on mutations that affect a single gene, but these are not the only source of variation. Another source, perhaps even more important but missed for many years, lies in the many duplications and deletions of large blocks of human DNA. When one compares any individual human with the standard reference genome in the database, one will generally find roughly 100 differences involving gain or loss of long sequence blocks, totaling perhaps 3 million nucleotide pairs. Some of these **copy number variations** (**CNVs**) will be very common, presumably reflecting relatively ancient origins, while others will be present in only a small minority of people (**Figure 4–80**). On average, nearly half of the CNVs contain known genes. CNVs have been implicated in many human traits, including color blindness, infertility, hypertension, and a wide variety of disease susceptibilities. In retrospect, this type of variation is not surprising, given the prominent role of DNA addition and DNA loss in vertebrate evolution.

The intraspecies variations that have been most extensively characterized, however, are **single-nucleotide polymorphisms** (**SNPs**). These are simply points in the genome sequence where one large fraction of the human population has one nucleotide, while another substantial fraction has another. To qualify as

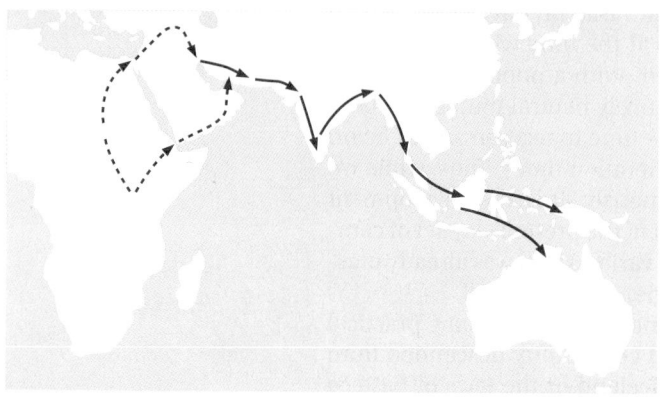

Figure 4–79 **Tracing the course of human history by analyses of genome sequences.** The map shows the routes of the earliest successful human migrations. Dotted lines indicate two alternative routes that our ancestors are thought to have taken out of Africa. DNA sequence comparisons suggest that modern Europeans descended from a small ancestral population. In agreement, archaeological findings suggest that the ancestors of modern native Australians *(solid red arrows)*—and of modern European and Middle Eastern populations—reached their destinations about 45,000 years ago. Even more recent studies, comparing the genome sequences of living humans with those of Neanderthals and another extinct population from southern Siberia (the Denisovans), suggest that our exit from Africa was a bit more convoluted, while also revealing that a number of our ancestors interbred with these hominid neighbors as they made their way across the globe. (Modified from P. Forster and S. Matsumura, *Science* 308:965–966, 2005.)

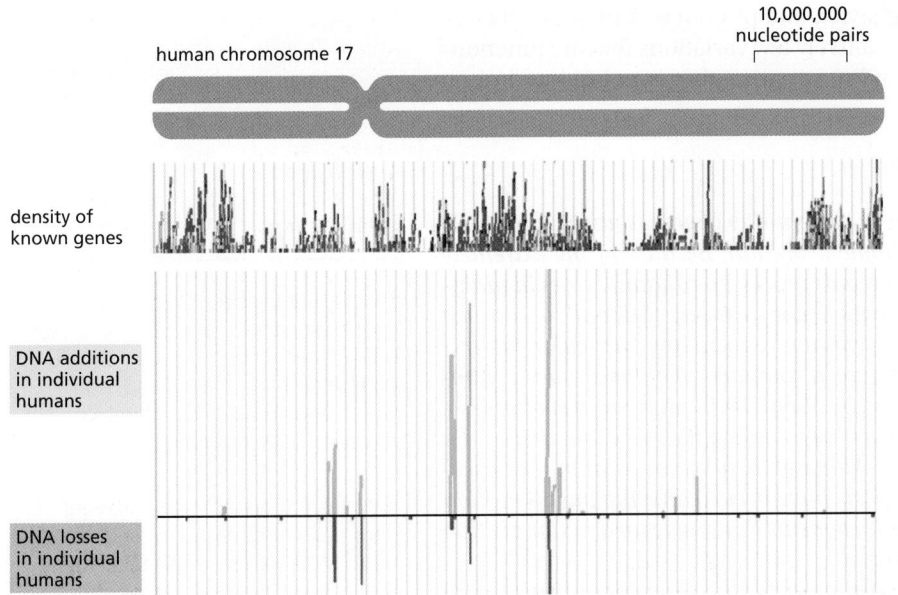

human chromosome 17

10,000,000
nucleotide pairs

density of
known genes

DNA additions
in individual
humans

DNA losses
in individual
humans

Figure 4–80 Detection of copy number variations on human chromosome 17. When 100 individuals were tested by a DNA microarray analysis that detects the copy number of DNA sequences throughout the entire length of this chromosome, the indicated distributions of DNA additions (green bars) and DNA losses (red bars) were observed compared with an arbitrary human reference sequence. The shortest red and green bars represent a single occurrence among the 200 chromosomes examined, whereas the longer bars indicate that the addition or loss was correspondingly more frequent. The results show preferred regions where the variations occur, and these tend to be in or near regions that already contain blocks of segmental duplications. Many of the changes include known genes. (Adapted from J.L. Freeman et al., *Genome Res.* 16:949–961, 2006.)

a polymorphism, the variants must be common enough to give a reasonably high probability that the genomes of two randomly chosen individuals will differ at the given site; a probability of 1% is commonly chosen as the cutoff. Two human genomes sampled from the modern world population at random will differ at approximately 2.5×10^6 such sites (1 per 1300 nucleotide pairs). As will be described in the overview of genetics in Chapter 8, SNPs in the human genome can be extremely useful for genetic mapping analyses, in which one attempts to associate specific traits (phenotypes) with specific DNA sequences for medical or scientific purposes (see p. 493). But while useful as genetic markers, there is good evidence that most of these SNPs have little or no effect on human fitness. This is as expected, since deleterious variants will have been selected against during human evolution and, unlike SNPs, should therefore be rare.

Against the background of ordinary SNPs inherited from our prehistoric ancestors, certain sequences with exceptionally high mutation rates stand out. A dramatic example is provided by CA repeats, which are ubiquitous in the human genome and in the genomes of other eukaryotes. Sequences with the motif $(CA)_n$ are replicated with relatively low fidelity because of a slippage that occurs between the template and the newly synthesized strands during DNA replication; hence, the precise value of n can vary over a considerable range from one genome to the next. These repeats make ideal DNA-based genetic markers, since most humans are heterozygous, having inherited one repeat length (n) from their mother and a different repeat length from their father. While the value of n changes sufficiently rarely that most parent–child transmissions propagate CA repeats faithfully, the changes are sufficiently frequent to maintain high levels of heterozygosity in the human population. These and some other simple repeats that display exceptionally high variability therefore provide the basis for identifying individuals by DNA analysis in crime investigations, paternity suits, and other forensic applications (see Figure 8–39).

While most of the SNPs and CNVs in the human genome sequence are thought to have little or no effect on phenotype, a subset of the genome sequence variations must be responsible for the heritable aspects of human individuality. We know that even a single nucleotide change that alters one amino acid in a protein can cause a serious disease, as for example in sickle-cell anemia, which is caused by such a mutation in hemoglobin (Movie 4.3). We also know that gene dosage—a doubling or halving of the copy number of some genes—can have a profound effect on development by altering the level of gene product, as can changes in regulatory DNA sequences. There is therefore every reason to suppose that some of the many differences between any two human beings will have substantial effects on human health, physiology, behavior, and physique. A major challenge in human genetics is to recognize those relatively few variations that are functionally important against a large background of variation that is neutral and of no consequence.

Summary

Comparisons of the nucleotide sequences of present-day genomes have revolutionized our understanding of gene and genome evolution. Because of the extremely high fidelity of DNA replication and DNA repair processes, random errors in maintaining the nucleotide sequences in genomes occur so rarely that only about one nucleotide in a thousand is altered in every million years in any particular eukaryotic line of descent. Not surprisingly, therefore, a comparison of human and chimpanzee chromosomes—which are separated by about 6 million years of evolution—reveals very few changes. Not only are our genes essentially the same, but their order on each chromosome is almost identical. Although a substantial number of segmental duplications and segmental deletions have occurred in the past 6 million years, even the positions of the transposable elements that make up a major portion of our noncoding DNA are mostly unchanged.

When one compares the genomes of two more distantly related organisms—such as a human and a mouse, separated by about 80 million years—one finds many more changes. Now the effects of natural selection can be clearly seen: through purifying selection, essential nucleotide sequences—both in regulatory regions and in coding sequences (exons)—have been highly conserved. In contrast, nonessential sequences (for example, much of the DNA in introns) have been altered to such an extent that one can no longer see any family resemblance.

Because of purifying selection, the comparison of the genome sequences of multiple related species is an especially powerful way to find DNA sequences with important functions. Although about 5% of the human genome has been conserved as a result of purifying selection, the function of the majority of this DNA (tens of thousands of multispecies conserved sequences) remains mysterious. Future experiments characterizing its functions should teach us many new lessons about vertebrate biology.

Other sequence comparisons show that a great deal of the genetic complexity of present-day organisms is due to the expansion of ancient gene families. DNA duplication followed by sequence divergence has clearly been a major source of genetic novelty during evolution. On a more recent time scale, the genomes of any two humans will differ from each other both because of nucleotide substitutions (single-nucleotide polymorphisms, or SNPs) and because of inherited DNA gains and DNA losses that cause copy number variations (CNVs). Understanding the effects of these differences will improve both medicine and our understanding of human biology.

WHAT WE DON'T KNOW

• How many different types of chromatin structure are important for cells? How is each of these structures established and maintained, and which ones tend to be inherited following DNA replication?

• Why are there so many different chromatin remodeling complexes in cells? What are their essential roles, and how do they get loaded onto chromatin at specific places and at specific times?

• How do chromosomal loops form during interphase, and what happens to these loops in condensed mitotic chromosomes?

• What genetic changes made us uniquely human? What further aspects of our recent evolutionary development can be reconstructed by sequencing DNA from remains of ancient hominids?

• How much of the enormous complexity that we find in cell biology is unnecessary, having evolved by random drift?

PROBLEMS

Which statements are true? Explain why or why not.

4–1   Human females have 23 different chromosomes, whereas human males have 24.

4–2   The four core histones are relatively small proteins with a very high proportion of positively charged amino acids; the positive charge helps the histones bind tightly to DNA, regardless of its nucleotide sequence.

4–3   Nucleosomes bind DNA so tightly that they cannot move from the positions where they are first assembled.

4–4   In a comparison between the DNAs of related organisms such as humans and mice, identifying the con-served DNA sequences facilitates the search for functionally important regions.

4–5   Gene duplication and divergence is thought to have played a critical role in the evolution of increased biological complexity.

Discuss the following problems.

4–6   DNA isolated from the bacterial virus M13 contains 25% A, 33% T, 22% C, and 20% G. Do these results strike you as peculiar? Why or why not? How might you explain these values?

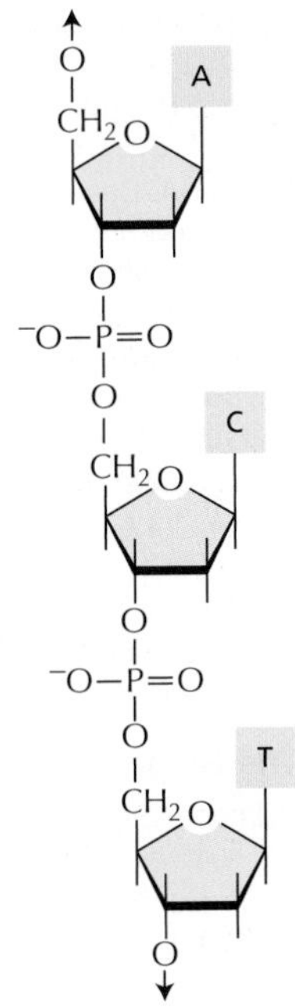

Figure Q4–1 Three nucleotides from the interior of a single strand of DNA (Problem 4–7). *Arrows at the ends of the DNA strand indicate that the structure continues in both directions.*

4–7 A segment of DNA from the interior of a single strand is shown in **Figure Q4–1**. What is the polarity of this DNA from top to bottom?

4–8 Human DNA contains 20% C on a molar basis. What are the mole percents of A, G, and T?

4–9 Chromosome 3 in orangutans differs from chromosome 3 in humans by two inversion events that occurred in the human lineage (**Figure Q4–2**). Draw the intermediate chromosome that resulted from the first inversion and explicitly indicate the segments included in each inversion.

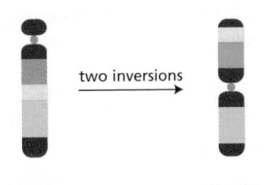

two inversions

orangutan   human

Figure Q4–2 Chromosome 3 in orangutans and humans (Problem 4–9). Differently *colored blocks* indicate segments of the chromosomes that are homologous in DNA sequence.

4–10 Assuming that the 30-nm chromatin fiber contains about 20 nucleosomes (200 bp/nucleosome) per 50 nm of length, calculate the degree of compaction of DNA associated with this level of chromatin structure. What fraction of the 10,000-fold condensation that occurs at mitosis does this level of DNA packing represent?

4–11 In contrast to histone acetylation, which always correlates with gene activation, histone methylation can lead to either transcriptional activation or repression. How do you suppose that the same modification—methylation—can mediate different biological outcomes?

4–12 Why is a chromosome with two centromeres (a dicentric chromosome) unstable? Would a backup centromere not be a good thing for a chromosome, giving it two chances to form a kinetochore and attach to microtubules during mitosis? Would that not help to ensure that the chromosome did not get left behind at mitosis?

4–13 Look at the two yeast colonies in **Figure Q4–3**. Each of these colonies contains about 100,000 cells descended from a single yeast cell, originally somewhere in the middle of the clump. A white colony arises when the *Ade2* gene is expressed from its normal chromosomal location. When the *Ade2* gene is moved to a location near a telomere, it is packed into heterochromatin and inactivated in most cells, giving rise to colonies that are mostly red. In these largely red colonies, white sectors fan out from the middle of the colony. In both the red and white sectors, the *Ade2*

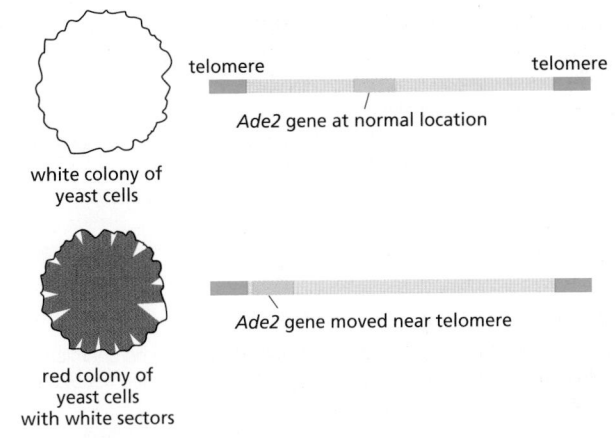

white colony of yeast cells

Ade2 gene at normal location

red colony of yeast cells with white sectors

Ade2 gene moved near telomere

Figure Q4–3 Position effect on expression of the yeast *Ade2* gene (Problem 4–13). The *Ade2* gene codes for one of the enzymes of adenosine biosynthesis, and the absence of the *Ade2* gene product leads to the accumulation of a red pigment. Therefore a colony of cells that express *Ade2* is *white*, and one composed of cells in which the *Ade2* gene is not expressed is *red*.

gene is still located near telomeres. Explain why white sectors have formed near the rim of the red colony. Based on the patterns observed, what can you conclude about the propagation of the transcriptional state of the *Ade2* gene from mother to daughter cells in this experiment?

4–14 Mobile pieces of DNA—transposable elements—that insert themselves into chromosomes and accumulate during evolution make up more than 40% of the human genome. Transposable elements of four types—long interspersed nuclear elements (LINEs), short interspersed nuclear elements (SINEs), long terminal repeat (LTR) retrotransposons, and DNA transposons—are inserted more-or-less randomly throughout the human genome. These elements are conspicuously rare at the four homeobox gene clusters, *HoxA*, *HoxB*, *HoxC*, and *HoxD*, as illustrated for *HoxD* in **Figure Q4–4**, along with an equivalent region of chromosome 22, which lacks a *Hox* cluster. Each *Hox* cluster is about 100 kb in length and contains 9 to 11 genes, whose differential expression along the anteroposterior axis of the developing embryo establishes the basic body plan for humans (and for other animals). Why do you suppose that transposable elements are so rare in the *Hox* clusters?

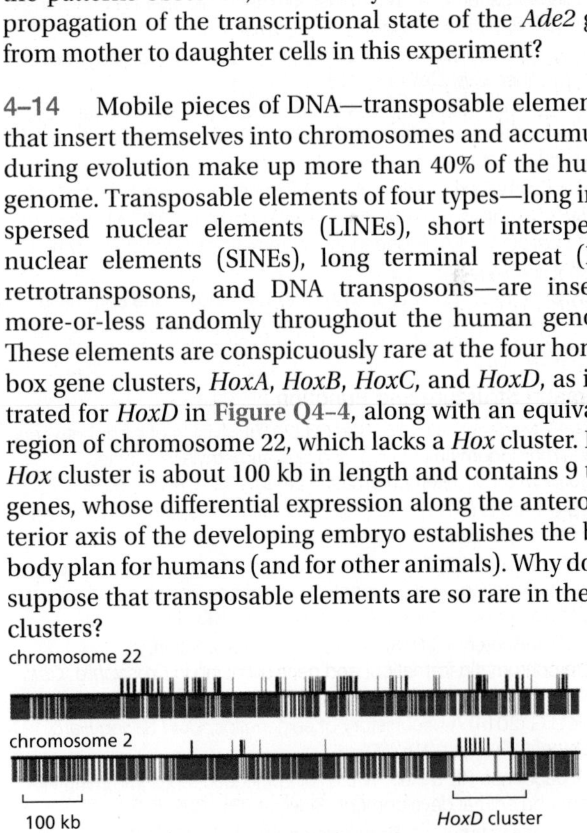

chromosome 22

chromosome 2

100 kb   *HoxD* cluster

Figure Q4–4 Transposable elements and genes in 1-Mb regions of chromosomes 2 and 22 (Problem 4–14). *Blue lines* that project *upward* indicate exons of known genes. *Red lines* that project *downward* indicate transposable elements; they are so numerous (constituting more than 40% of the human genome) that they merge into nearly a solid block outside the *Hox* clusters. (From International Human Genome Sequencing Consortium, *Nature* 409:860–921, published 2001 by Macmillan Magazines Ltd. Reproduced with permission of SNCSC.)

REFERENCES

General

Armstrong L (2014) Epigenetics. New York: Garland Science.

Hartwell L, Hood L, Goldberg ML et al. (2010) Genetics: From Genes to Genomes, 4th ed. Boston, MA: McGraw Hill.

Jobling M, Hollox E, Hurles M et al. (2014) Human Evolutionary Genetics, 2nd ed. New York: Garland Science.

Strachan T & Read AP (2010) Human Molecular Genetics, 4th ed. New York: Garland Science.

The Structure and Function of DNA

Avery OT, MacLeod CM & McCarty M (1944) Studies on the chemical nature of the substance inducing transformation of pneumococcal types. *J. Exp. Med.* 79, 137–158.

Meselson M & Stahl FW (1958) The replication of DNA in *Escherichia coli*. *Proc. Natl Acad. Sci. USA* 44, 671–682.

Watson JD & Crick FHC (1953) Molecular structure of nucleic acids. A structure for deoxyribose nucleic acid. *Nature* 171, 737–738.

Chromosomal DNA and Its Packaging in the Chromatin Fiber

Andrews AJ & Luger K (2011) Nucleosome structure(s) and stability: variations on a theme. *Annu. Rev. Biophys.* 40, 99–117.

Avvakumov N, Nourani A & Côté J (2011) Histone chaperones: modulators of chromatin marks. *Mol. Cell* 41, 502–514.

Deal RB, Henikoff JG & Henikoff S (2010) Genome-wide kinetics of nucleosome turnover determined by metabolic labeling of histones. *Science* 328, 1161–1164.

Grigoryev SA & Woodcock CL (2012) Chromatin organization—the 30 nm fiber. *Exp. Cell Res.* 318, 1448–1455.

Li G, Levitus M, Bustamante C & Widom J (2005) Rapid spontaneous accessibility of nucleosomal DNA. *Nat. Struct. Mol. Biol.* 12, 46–53.

Luger K, Mäder AW, Richmond RK et al. (1997) Crystal structure of the nucleosome core particle at 2.8 Å resolution. *Nature* 389, 251–260.

Narlikar GJ, Sundaramoorthy R & Owen-Hughes T (2013) Mechanisms and functions of ATP-dependent chromatin-remodeling enzymes. *Cell* 154, 490–503.

Song F, Chen P, Sun D et al. (2014) Cryo-EM study of the chromatin fiber reveals a double helix twisted by tetranucleosomal units. *Science* 344, 376–380.

Chromatin Structure and Function

Al-Sady B, Madhani HD & Narlikar GJ (2013) Division of labor between the chromodomains of HP1 and Suv39 methylase enables coordination of heterochromatin spread. *Mol. Cell* 51, 80–91.

Beisel C & Paro R (2011) Silencing chromatin: comparing modes and mechanisms. *Nat. Rev. Genet.* 12, 123–135.

Black BE, Jansen LET, Foltz DR & Cleveland DW (2011) Centromere identity, function, and epigenetic propagation across cell divisions. *Cold Spring Harb. Symp. Quant. Biol.* 75, 403–418.

Elgin SCR & Reuter G (2013) Position-effect variegation, heterochromatin formation, and gene silencing in *Drosophila*. *Cold Spring Harb. Perspect. Biol.* 5, a017780.

Felsenfeld G (2014) A brief history of epigenetics. *Cold Spring Harb. Perspect. Biol.* 6, a018200.

Feng S, Jacobsen SE & Reik W (2010) Epigenetic reprogramming in plant and animal development. *Science* 330, 622–627.

Filion GJ, van Bemmel JG, Braunschweig U et al. (2010) Systematic protein location mapping reveals five principal chromatin types in *Drosophila* cells. *Cell* 143, 212–224.

Fodor BD, Shukeir N, Reuter G & Jenuwein T (2010) Mammalian Su(var) genes in chromatin control. *Annu. Rev. Cell Dev. Biol.* 26, 471–501.

Giles KE, Gowher H, Ghirlando R et al. (2010) Chromatin boundaries, insulators, and long-range interactions in the nucleus. *Cold Spring Harb. Symp. Quant. Biol.* 75, 79–85.

Gohl D, Aoki T, Blanton J et al. (2011) Mechanism of chromosomal boundary action: roadblock, sink, or loop? *Genetics* 187, 731–748.

Mellone B, Erhardt S & Karpen GH (2006) The ABCs of centromeres. *Nat. Cell Biol.* 8, 427–429.

Morris SA, Baek S, Sung M-H et al. (2014) Overlapping chromatin-remodeling systems collaborate genome wide at dynamic chromatin transitions. *Nat. Struct. Mol. Biol.* 21, 73–81.

Politz JCR, Scalzo D & Groudine M (2013) Something silent this way forms: the functional organization of the repressive nuclear compartment. *Annu. Rev. Cell Dev. Biol.* 29, 241–270.

Rothbart SB & Strahl BD (2014) Interpreting the language of histone and DNA modifications. *Biochim. Biophys. Acta* 1839, 627–643.

Weber CM & Henikoff S (2014) Histone variants: dynamic punctuation in transcription. *Genes Dev.* 28, 672–682.

Xu M, Long C, Chen X et al. (2010) Partitioning of histone H3-H4 tetramers during DNA replication-dependent chromatin assembly. *Science* 328, 94–98.

The Global Structure of Chromosomes

Belmont AS (2014) Large-scale chromatin organization: the good, the surprising, and the still perplexing. *Curr. Opin. Cell Biol.* 26, 69–78.

Bickmore W (2013) The spatial organization of the human genome. *Annu. Rev. Genomics Hum. Genet.* 14, 67–84.

Callan HG (1982) Lampbrush chromosomes. *Proc. R. Soc. Lond. B Biol. Sci.* 214, 417–448.

Cheutin T, Bantignies F, Leblanc B & Cavalli G (2010) Chromatin folding: from linear chromosomes to the 4D nucleus. *Cold Spring Harb. Symp. Quant. Biol.* 75, 461–473.

Cremer T & Cremer M (2010) Chromosome territories. *Cold Spring Harb. Perspect. Biol.* 2, a003889.

Lieberman-Aiden E, van Berkum NL, Williams L et al. (2009) Comprehensive mapping of long-range interactions reveals folding principles of the human genome. *Science* 326, 289–293.

Maeshima K & Laemmli UK (2003) A two-step scaffolding model for mitotic chromosome assembly. *Dev. Cell* 4, 467–480.

Moser SC & Swedlow JR (2011) How to be a mitotic chromosome. *Chromosome Res.* 19, 307–319.

Nizami ZF, Deryusheva S & Gall JG (2010) Cajal bodies and histone locus bodies in *Drosophila* and *Xenopus*. *Cold Spring Harb. Symp. Quant. Biol.* 75, 313–320.

Zhimulev IF (1997) Polytene chromosomes, heterochromatin, and position effect variegation. *Adv. Genet.* 37, 1–566.

How Genomes Evolve

Batzer MA & Deininger PL (2002) Alu repeats and human genomic diversity. *Nat. Rev. Genet.* 3, 370–379.

Feuk L, Carson AR & Scherer S (2006) Structural variation in the human genome. *Nat. Rev. Genet.* 7, 85–97.

Green RE, Krause J, Briggs AW et al. (2010) A draft sequence of the Neandertal genome. *Science* 328, 710–722.

International Human Genome Sequencing Consortium (2001) Initial sequencing and analysis of the human genome. *Nature* 409, 860–921.

International Human Genome Sequencing Consortium (2004) Finishing the euchromatic sequence of the human genome. *Nature* 431, 931–945.

Kellis M, Wold B, Snyder MP et al. (2014) Defining functional DNA elements in the human genome. *Proc. Natl Acad. Sci. USA* 111, 6131–6138.

Lander ES (2011) Initial impact of the sequencing of the human genome. *Nature* 470, 187–197.

Lee C & Scherer SW (2010) The clinical context of copy number variation in the human genome. *Expert Rev. Mol. Med.* 12, e8.

Mouse Genome Sequencing Consortium (2002) Initial sequencing and comparative analysis of the mouse genome. *Nature* 420, 520–562.

Pollard KS, Salama SR, Lambert N et al. (2006) An RNA gene expressed during cortical development evolved rapidly in humans. *Nature* 443, 167–172.

DNA Replication, Repair, and Recombination

The ability of cells to maintain a high degree of order in a chaotic universe depends upon the accurate duplication of vast quantities of genetic information carried in chemical form as DNA. This process, called *DNA replication*, must occur before a cell can produce two genetically identical daughter cells. Maintaining order also requires the continued surveillance and repair of this genetic information, because DNA inside cells is repeatedly damaged by chemicals and radiation from the environment, as well as by thermal accidents and reactive molecules generated inside the cell. In this chapter, we describe the protein machines that replicate and repair the cell's DNA. These machines catalyze some of the most rapid and accurate processes that take place within cells, and their mechanisms illustrate the elegance and efficiency of cell chemistry.

While the short-term survival of a cell can depend on preventing changes in its DNA, the long-term survival of a species requires that DNA sequences be changeable over many generations to permit evolutionary adaptation to changing circumstances. We shall see that despite the great efforts that cells make to protect their DNA, occasional changes in DNA sequences do occur. Over time, these changes provide the genetic variation upon which selection pressures act during the evolution of organisms.

We begin this chapter with a brief discussion of the changes that occur in DNA as it is passed down from generation to generation. Next, we discuss the cell mechanisms—DNA replication and DNA repair—that are responsible for minimizing these changes. Finally, we consider some of the most intriguing pathways that alter DNA sequences—in particular, those of DNA recombination including the movement within chromosomes of special DNA sequences called transposable elements.

THE MAINTENANCE OF DNA SEQUENCES

Although, as just pointed out, occasional genetic changes enhance the long-term survival of a species through evolution, the survival of the individual demands a high degree of genetic stability. Only rarely do the cell's DNA-maintenance processes fail, resulting in permanent change in the DNA. Such a change is called a **mutation**, and it can destroy an organism if it occurs in a vital position in the DNA sequence.

Mutation Rates Are Extremely Low

The **mutation rate**, the rate at which changes occur in DNA sequences, can be determined directly from experiments carried out with a bacterium such as *Escherichia coli*—a resident of our intestinal tract and a commonly used laboratory organism (see Figure 1–24). Under laboratory conditions, *E. coli* divides about once every 30 minutes, and a single cell can generate a very large population—several billion—in less than a day. In such a population, it is possible to detect the small fraction of bacteria that have suffered a damaging mutation in a particular gene, if that gene is not required for the bacterium's survival. For example, the mutation rate of a gene specifically required for cells to use the sugar lactose as an energy source can be determined by growing the cells in the presence of a different sugar, such as glucose, and testing them subsequently to see how many have lost the ability to survive on a lactose diet. The fraction of damaged genes underestimates the actual mutation rate because many mutations are *silent* (for example, those that change a codon but not the amino acid it specifies, or those that change an amino acid without affecting the activity of the protein coded for by the gene). After correcting for these silent mutations, one finds that a single gene that encodes an average-sized protein ($\sim 10^3$ coding nucleotide pairs) accumulates a mutation (not necessarily one that would inactivate the protein) approximately once in about 10^6 bacterial cell generations. Stated differently, bacteria display a mutation rate of about three nucleotide changes per 10^{10} nucleotides per cell generation.

Recently, it has become possible to measure the germ-line mutation rate directly in more complex, sexually reproducing organisms such as humans. In this case, the complete genomes from a family—parents and offspring—were directly sequenced, and a careful comparison revealed that approximately 70 new single-nucleotide mutations arose in the germ lines of each offspring. Normalized to the size of the human genome, the mutation rate is one nucleotide change per 10^8 nucleotides per human generation. This is a slight underestimate because some mutations will be lethal and will therefore be absent from progeny; however, because relatively little of the human genome carries critical information, this consideration has only a small effect on the true mutation rate. It is estimated that approximately 100 cell divisions occur in the germ line from the time of conception to the time of production of the eggs and sperm that go on to make the next generation. Thus, the human mutation rate, expressed in terms of cell divisions (instead of human generations), is approximately 1 mutation/10^{10} nucleotides/cell division.

Although *E. coli* and humans differ greatly in their modes of reproduction and in their generation times, when the mutation rates of each are normalized to a single round of DNA replication, they are both extremely low and within a factor of three of each other. We shall see later in the chapter that the basic mechanisms that ensure these low rates of mutation have been conserved since the very early history of cells on Earth.

Low Mutation Rates Are Necessary for Life as We Know It

Since many mutations are deleterious, no species can afford to allow them to accumulate at a high rate in its germ cells. Although the observed mutation frequency is low, it is nevertheless thought to limit the number of essential proteins that any organism can depend upon to perhaps 30,000. More than this, and the probability that at least one critical component will suffer a damaging mutation becomes catastrophically high. By an extension of the same argument, a mutation frequency tenfold higher would limit an organism to about 3000 essential genes. In this case, evolution would have been limited to organisms considerably less complex than a fruit fly.

The cells of a sexually reproducing animal or plant are of two types: **germ cells** and **somatic cells**. The germ cells transmit genetic information from parent to offspring; the somatic cells form the body of the organism (**Figure 5–1**). We have seen that germ cells must be protected against high rates of mutation to maintain the species. However, the somatic cells of multicellular organisms must also be protected from genetic change to properly maintain the organized structure of the body. Nucleotide changes in somatic cells can give rise to variant cells, some of which, through "local" natural selection, proliferate rapidly at the expense of the rest of the organism. In an extreme case, the result is the uncontrolled cell proliferation that we know as cancer, a disease that causes (in Europe and North America) more than 20% of human deaths each year. These deaths are due largely to an accumulation of changes in the DNA sequences of somatic cells, as discussed in Chapter 20. A significant increase in the mutation frequency would presumably cause a disastrous increase in the incidence of cancer by accelerating the rate at which somatic-cell variants arise. Thus, both for the perpetuation of a species

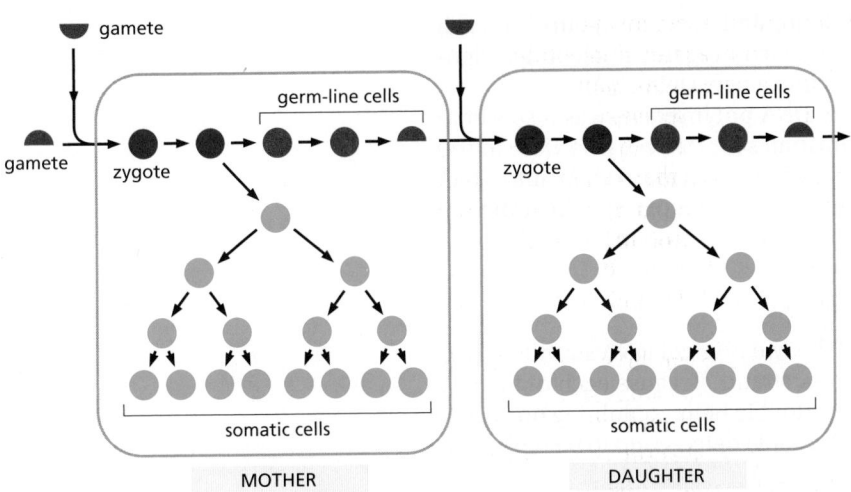

Figure 5–1 **Germ-line cells and somatic cells carry out fundamentally different functions.** In sexually reproducing organisms, the germ-line cells *(red)* propagate genetic information into the next generation. Somatic cells *(blue)*, which form the body of the organism, are necessary for the survival of germ-line cells but do not themselves leave any progeny.

with a large number of genes (germ-cell stability) and for the prevention of cancer resulting from mutations in somatic cells (somatic-cell stability), multicellular organisms like ourselves depend on the remarkably high fidelity with which their DNA sequences are replicated and maintained.

Summary

In all cells, DNA sequences are maintained and replicated with high fidelity. The mutation rate, approximately one nucleotide change per 10^{10} nucleotides each time the DNA is replicated, is roughly the same for organisms as different as bacteria and humans. Because of this remarkable accuracy, the sequence of the human genome (approximately 3.2×10^9 nucleotide pairs) is unchanged or changed by only a few nucleotides each time a typical human cell divides. This allows most humans to pass accurate genetic instructions from one generation to the next, and also to avoid the changes in somatic cells that lead to cancer.

DNA REPLICATION MECHANISMS

All organisms duplicate their DNA with extraordinary accuracy before each cell division. In this section, we explore how an elaborate "replication machine" achieves this accuracy, while duplicating DNA at rates as high as 1000 nucleotides per second.

Base-Pairing Underlies DNA Replication and DNA Repair

As introduced in Chapter 1, *DNA templating* is the mechanism the cell uses to copy the nucleotide sequence of one DNA strand into a complementary DNA sequence (**Figure 5–2**). This process requires the separation of the DNA helix into two template strands, and entails the recognition of each nucleotide in the DNA *template strands* by a free (unpolymerized) complementary nucleotide. The separation of

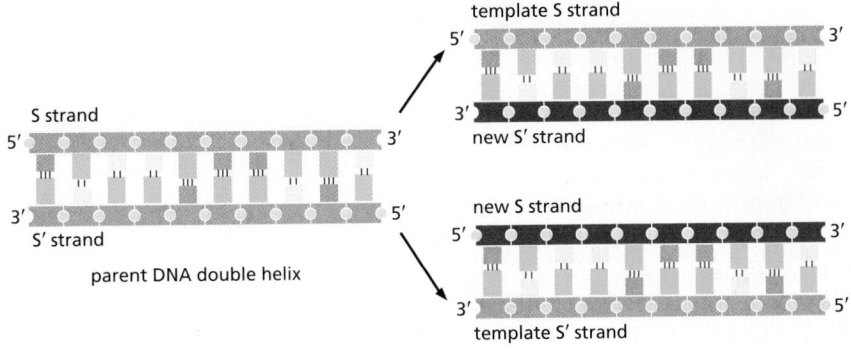

Figure 5–2 **The DNA double helix acts as a template for its own duplication.** Because the nucleotide A will pair successfully only with T, and G only with C, each strand of DNA can serve as a template to specify the sequence of nucleotides in its complementary strand by DNA base-pairing. In this way, a double-helical DNA molecule can be copied precisely.

the DNA helix exposes the hydrogen-bond donor and acceptor groups on each DNA base for base-pairing with the appropriate incoming free nucleotide, aligning it for its enzyme-catalyzed polymerization into a new DNA chain.

The first nucleotide-polymerizing enzyme, **DNA polymerase**, was discovered in 1957. The free nucleotides that serve as substrates for this enzyme were found to be deoxyribonucleoside triphosphates, and their polymerization into DNA required a single-strand DNA template. **Figure 5–3** and **Figure 5–4** illustrate the stepwise mechanism of this reaction.

The DNA Replication Fork Is Asymmetrical

During DNA replication inside a cell, each of the two original DNA strands serves as a template for the formation of an entire new strand. Because each of the two daughters of a dividing cell inherits a new DNA double helix containing one original and one new strand (**Figure 5–5**), the DNA double helix is said to be replicated "semiconservatively." How is this feat accomplished?

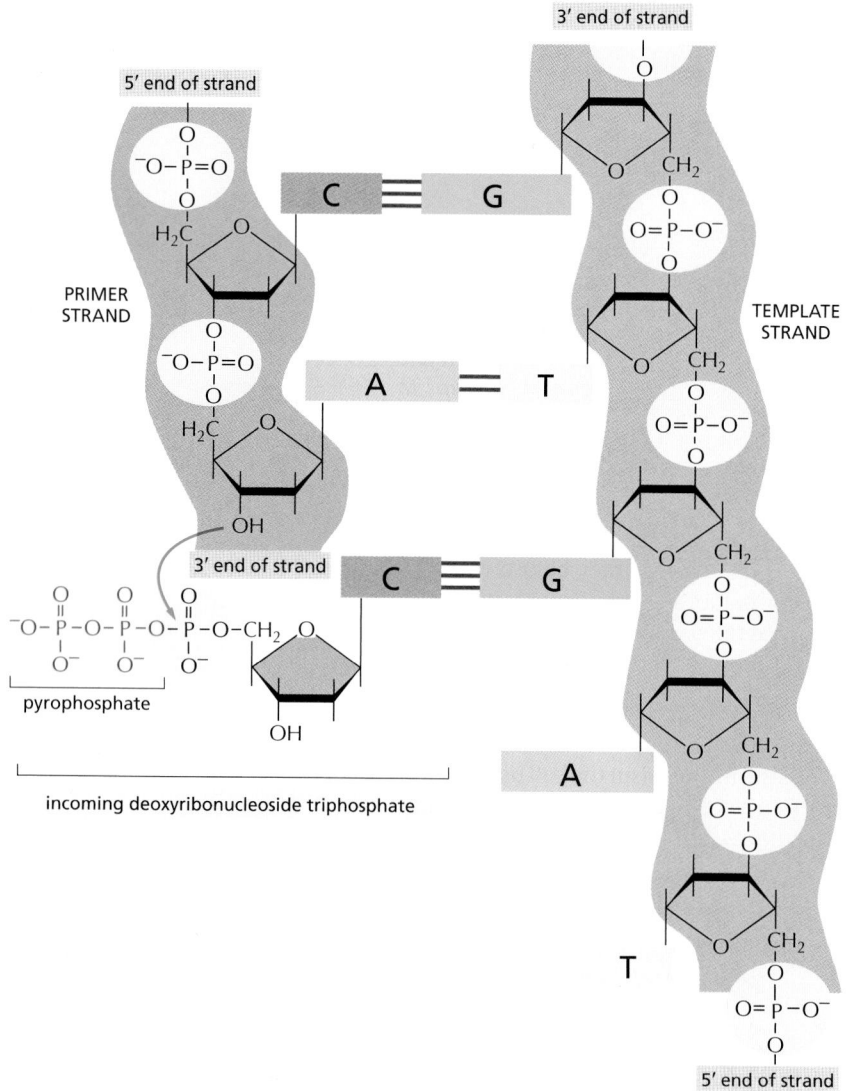

Figure 5–3 The chemistry of DNA synthesis. The addition of a deoxyribonucleotide to the 3′ end of a polynucleotide chain (the *primer strand*) is the fundamental reaction by which DNA is synthesized. As shown, base-pairing between an incoming deoxyribonucleoside triphosphate and an existing strand of DNA (the *template strand*) guides the formation of the new strand of DNA and causes it to have a complementary nucleotide sequence. The way in which complementary nucleotides base-pair is shown in Figure 4–4.

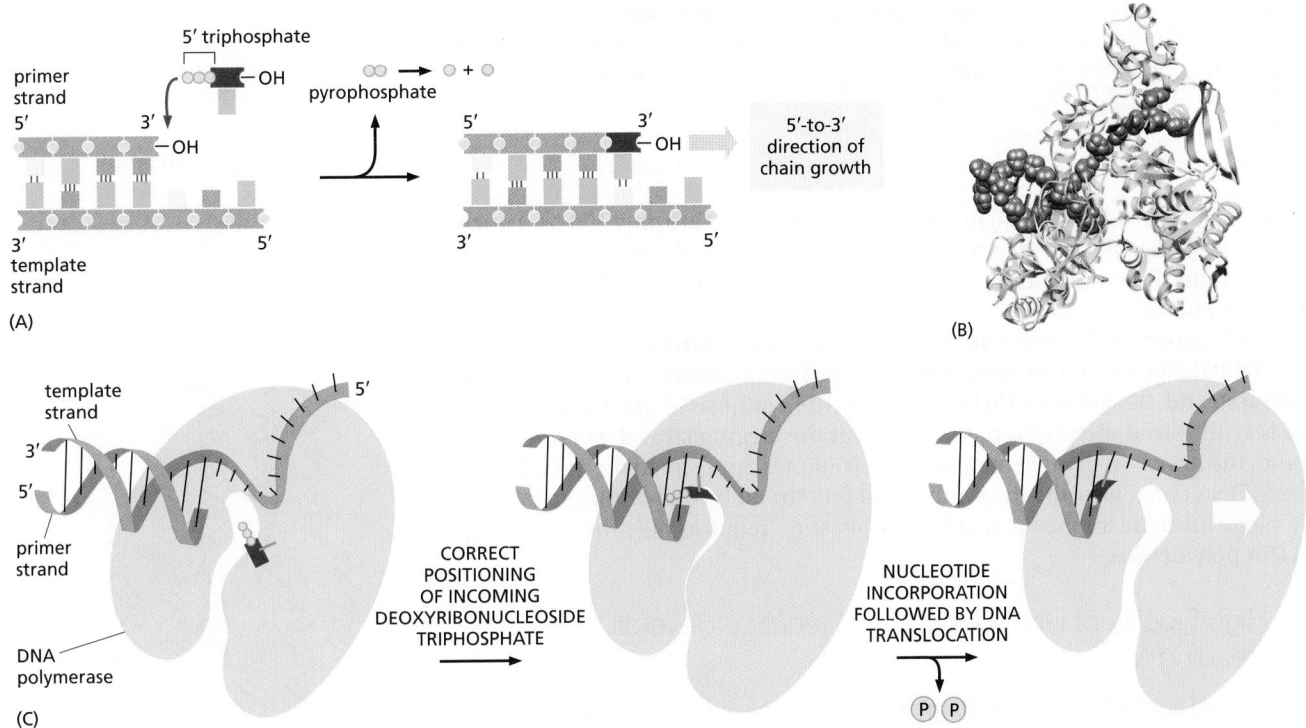

Figure 5–4 DNA synthesis catalyzed by DNA polymerase. (A) DNA polymerase catalyzes the stepwise addition of a deoxyribonucleotide to the 3′-OH end of a polynucleotide chain, the growing *primer strand* that is paired to an existing *template strand*. The newly synthesized DNA strand therefore polymerizes in the 5′-to-3′ direction as shown also in the previous figure. Because each incoming deoxyribonucleoside triphosphate must pair with the template strand to be recognized by the DNA polymerase, this strand determines which of the four possible deoxyribonucleotides (A, C, G, or T) will be added. The reaction is driven by a large, favorable free-energy change, caused by the release of pyrophosphate and its subsequent hydrolysis to two molecules of inorganic phosphate. (B) Structure of DNA polymerase complexed with DNA *(orange)*, as determined by x-ray crystallography (Movie 5.1). The template DNA strand is the longer strand and the newly synthesized DNA is the shorter. (C) Schematic diagram of DNA polymerase, based on the structure in (B). The proper base-pair geometry of a correct incoming deoxyribonucleoside triphosphate causes the polymerase to tighten around the base pair, thereby initiating the nucleotide addition reaction *(middle diagram* (C)). Dissociation of pyrophosphate relaxes the polymerase, allowing translocation of the DNA by one nucleotide so the active site of the polymerase is ready to receive the next deoxyribonucleoside triphosphate.

Analyses carried out in the early 1960s on whole replicating chromosomes revealed a localized region of replication that moves progressively along the parental DNA double helix. Because of its Y-shaped structure, this active region is called a **replication fork** (**Figure 5–6**). At the replication fork, a multienzyme complex that contains the DNA polymerase synthesizes the DNA of both new daughter strands.

Initially, the simplest mechanism of DNA replication seemed to be the continuous growth of both new strands, nucleotide by nucleotide, at the replication fork as it moves from one end of a DNA molecule to the other. But because of the antiparallel orientation of the two DNA strands in the DNA double helix (see Figure 5–2), this mechanism would require one daughter strand to polymerize in the 5′-to-3′ direction and the other in the 3′-to-5′ direction. Such a replication fork would require two distinct types of DNA polymerase enzymes. However, as attractive as this model might be, the DNA polymerases at replication forks can synthesize only in the 5′-to-3′ direction.

How, then, can a DNA strand grow in the 3′-to-5′ direction? The answer was first suggested by the results of an experiment performed in the late 1960s. Researchers added highly radioactive ³H-thymidine to dividing bacteria for a few seconds, so that only the most recently replicated DNA—that just behind the replication fork—became radiolabeled. This experiment revealed the transient existence of pieces of DNA that were 1000–2000 nucleotides long, now commonly known as *Okazaki fragments,* at the growing replication fork. (Similar replication

Figure 5–5 **The semiconservative nature of DNA replication.** In a round of replication, each of the two strands of DNA is used as a template for the formation of a complementary DNA strand. The original strands therefore remain intact through many cell generations.

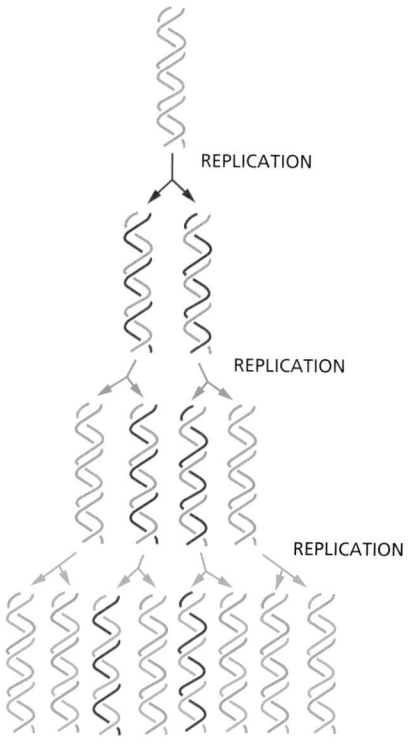

REPLICATION

REPLICATION

REPLICATION

intermediates were later found in eukaryotes, where they are only 100–200 nucleotides long.) The Okazaki fragments were shown to be polymerized only in the 5′-to-3′ chain direction and to be joined together after their synthesis to create long DNA chains.

A replication fork therefore has an asymmetric structure (**Figure 5–7**). The DNA daughter strand that is synthesized continuously is known as the **leading strand**. Its synthesis slightly precedes the synthesis of the daughter strand that is synthesized discontinuously, known as the **lagging strand**. For the lagging strand, the direction of nucleotide polymerization is opposite to the overall direction of DNA chain growth. The synthesis of this strand by a discontinuous "back-stitching" mechanism means that DNA replication requires only the 5′-to-3′ type of DNA polymerase.

The High Fidelity of DNA Replication Requires Several Proofreading Mechanisms

As discussed above, the fidelity of copying DNA during replication is such that only about one mistake occurs for every 10^{10} nucleotides copied. This fidelity is much higher than one would expect from the accuracy of complementary base-pairing. The standard complementary base pairs (see Figure 4–4) are not the only ones possible. For example, with small changes in helix geometry, two hydrogen bonds can form between G and T in DNA. In addition, rare tautomeric forms of the four DNA bases occur transiently in ratios of 1 part to 10^4 or 10^5. These forms mispair without a change in helix geometry: the rare tautomeric form of C pairs with A instead of G, for example.

If the DNA polymerase did nothing special when a mispairing occurred between an incoming deoxyribonucleoside triphosphate and the DNA template, the wrong nucleotide would often be incorporated into the new DNA chain, producing frequent mutations. The high fidelity of DNA replication, however, depends not only on the initial base-pairing but also on several "proofreading" mechanisms that act sequentially to correct any initial mispairings that might have occurred.

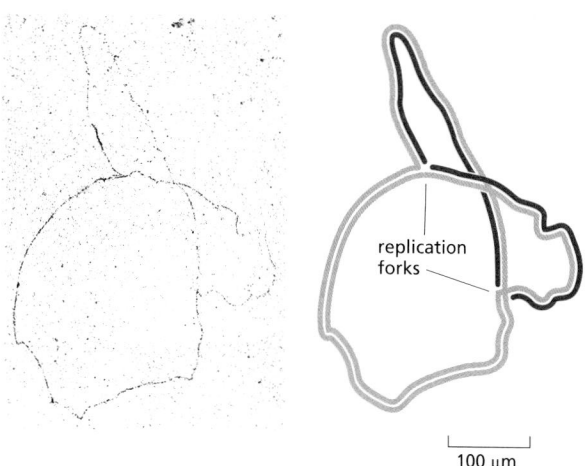

replication forks

100 μm

Figure 5-6 **Two replication forks moving in opposite directions on a circular DNA molecule.** Each replication fork has a Y-shaped structure and moves progressively along the DNA, spinning out newly replicated DNA behind it. The stem of the Y is the parental DNA double helix, and the two arms of the Y contain the newly synthesized DNA. The image was obtained by feeding *E. coli* radioactive thymine for several hours, gently isolating the DNA on filter paper, and placing a piece of photographic film next to the DNA. Because radioactivity exposes photographic film, an image of the DNA was captured when the film was developed and viewed under a light microscope. The diagram is an interpretation of the result, with parental DNA strands in *orange* and newly synthesized DNA strands in *red*. During its isolation, the DNA folded on itself, accounting for the crossing of the double helix. (From J. Cairns, *Cold Spring Harb. Symp. Quant. Biol.* 28:43–46, 1963. With permission from Cold Spring Harbor Laboratory Press.)

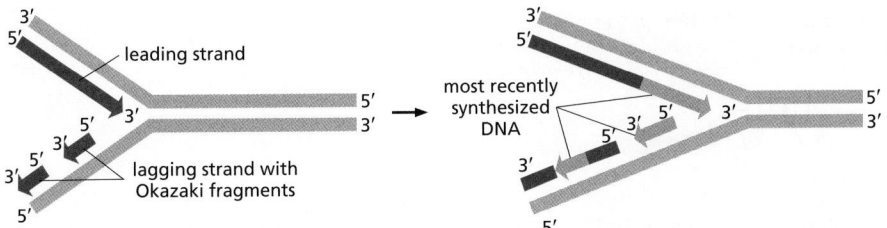

Figure 5–7 The structure of a DNA replication fork. Left, replication fork with newly synthesized DNA in red and arrows indicating the 5′-to-3′ direction of DNA synthesis. Because both daughter DNA strands are polymerized in the 5′-to-3′ direction, the DNA synthesized on the lagging strand must be made initially as a series of short DNA molecules, called *Okazaki fragments*, named after the scientist who discovered them. Right, the same fork a short time later. On the lagging strand, the Okazaki fragments are synthesized sequentially, with those nearest the fork being the most recently made.

DNA polymerase performs the first proofreading step just before a new nucleotide is covalently added to the growing chain. Our knowledge of this mechanism comes from studies of several different DNA polymerases, including one produced by a bacterial virus, T7, that replicates inside *E. coli*. The correct nucleotide has a higher affinity for the moving polymerase than does the incorrect nucleotide, because the correct pairing is more energetically favorable. Moreover, after nucleotide binding, but before the nucleotide is covalently added to the growing chain, the enzyme must undergo a conformational change in which its "grip" tightens around the active site (see Figure 5–4). Because this change occurs more readily with correct than incorrect base-pairing, it allows the polymerase to "double-check" the exact base-pair geometry before it catalyzes the addition of the nucleotide. Incorrectly paired nucleotides are harder to add and therefore more likely to diffuse away before the polymerase can mistakenly add them.

The next error-correcting reaction, known as *exonucleolytic proofreading*, takes place immediately after those rare instances in which an incorrect nucleotide is covalently added to the growing chain. DNA polymerase enzymes are highly discriminating in the types of DNA chains they will elongate: they require a previously formed, base-paired 3′-OH end of a *primer strand* (see Figure 5–4). Those DNA molecules with a mismatched (improperly base-paired) nucleotide at the 3′-OH end of the primer strand are not effective as templates because the polymerase has difficulty extending such a strand. DNA polymerase molecules correct such a mismatched primer strand by means of a separate catalytic site (either in a separate subunit or in a separate domain of the polymerase molecule, depending on the polymerase). This *3′-to-5′ proofreading exonuclease* clips off any unpaired or mispaired residues at the primer terminus, continuing until enough nucleotides have been removed to regenerate a correctly base-paired 3′-OH terminus that can prime DNA synthesis. In this way, DNA polymerase functions as a "self-correcting" enzyme that removes its own polymerization errors as it moves along the DNA (**Figure 5–8** and **Figure 5–9**).

The self-correcting properties of the DNA polymerase depend on its requirement for a perfectly base-paired primer terminus, and it is apparently not possible for such an enzyme to start synthesis *de novo*, without an existing primer. By contrast, the RNA polymerase enzymes involved in gene transcription do not need such an efficient exonucleolytic proofreading mechanism: errors in making RNA are not passed on to the next generation, and the occasional defective RNA molecule that is produced has no long-term significance. RNA polymerases are thus able to start new polynucleotide chains without a primer.

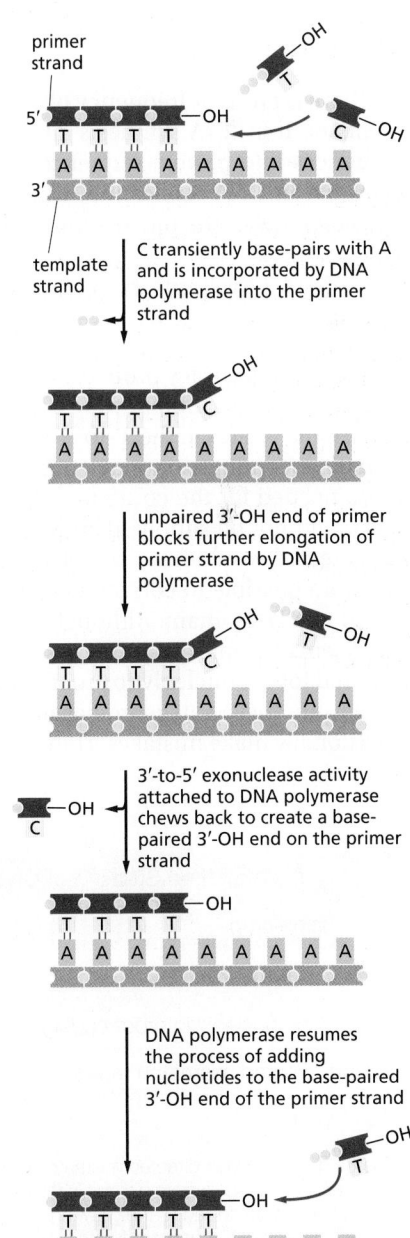

primer strand

template strand

C transiently base-pairs with A and is incorporated by DNA polymerase into the primer strand

unpaired 3′-OH end of primer blocks further elongation of primer strand by DNA polymerase

3′-to-5′ exonuclease activity attached to DNA polymerase chews back to create a base-paired 3′-OH end on the primer strand

DNA polymerase resumes the process of adding nucleotides to the base-paired 3′-OH end of the primer strand

Figure 5–8 Exonucleolytic proofreading by DNA polymerase during DNA replication. In this example, a C is accidentally incorporated at the growing 3′-OH end of a DNA chain. The part of DNA polymerase that removes the misincorporated nucleotide is a specialized member of a large class of enzymes, known as *exonucleases*, that cleave nucleotides one at a time from the ends of polynucleotides.

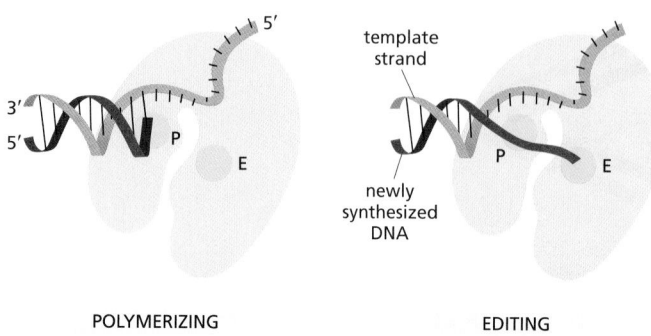

POLYMERIZING   EDITING

Figure 5–9 Editing by DNA polymerase. DNA polymerase complexed with the DNA template in the polymerizing mode *(left)* and the editing mode *(right)*. The catalytic sites for the exonucleolytic (E) and the polymerization (P) reactions are indicated. In the editing mode, the newly synthesized DNA transiently unpairs from the template and enters the editing site where the most recently added nucleotide is catalytically removed.

There is an error frequency of about one mistake for every 10^4 polymerization events both in RNA synthesis and in the separate process of translating mRNA sequences into protein sequences. This error rate is over 100,000 times greater than that in DNA replication, where, as we have seen, a series of proofreading processes makes the process unusually accurate (Table 5–1).

Only DNA Replication in the 5′-to-3′ Direction Allows Efficient Error Correction

The need for accuracy probably explains why DNA replication occurs only in the 5′-to-3′ direction. If there were a DNA polymerase that added deoxyribonucleoside triphosphates in the 3′-to-5′ direction, the growing 5′ end of the chain, rather than the incoming mononucleotide, would have to provide the activating triphosphate needed for the covalent linkage. In this case, the mistakes in polymerization could not be simply hydrolyzed away, because the bare 5′ end of the chain thus created would immediately terminate DNA synthesis (see Figure 5–3). It is therefore possible to correct a mismatched base only if it has been added to the 3′ end of a DNA chain. Although the backstitching mechanism for DNA replication seems complex, it preserves the 5′-to-3′ direction of polymerization that is required for exonucleolytic proofreading.

Despite these safeguards against DNA replication errors, DNA polymerases occasionally make mistakes. However, as we shall see later, cells have yet another

TABLE 5–1 The Three Steps That Give Rise to High-Fidelity DNA Synthesis

Replication step	Errors per nucleotide added
5′ → 3′ polymerization	1 in 10^5
3′ → 5′ exonucleolytic proofreading	1 in 10^2
Strand-directed mismatch repair	1 in 10^3
Combined	1 in 10^{10}

The third step, strand-directed mismatch repair, is described later in this chapter. For the polymerization step, "errors per nucleotide added" describes the probability that an incorrect nucleotide will be added to the growing chain. For the other two steps, "errors per nucleotide added" describes the probability that an error will not be corrected. Each step therefore reduces the chance of a final error by the factor shown.

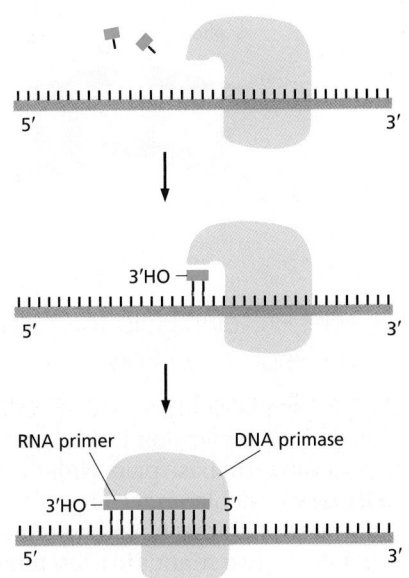

Figure 5–10 **RNA primer synthesis.** A schematic view of the reaction catalyzed by *DNA primase*, the enzyme that synthesizes the short RNA primers made on the lagging strand using DNA as a template. Unlike DNA polymerase, this enzyme can start a new polynucleotide chain by joining two nucleoside triphosphates together. The primase synthesizes a short polynucleotide in the 5′-to-3′ direction and then stops, making the 3′ end of this primer available for the DNA polymerase.

chance to correct these errors by a process called *strand-directed mismatch repair*. Before discussing this mechanism, however, we describe the other types of proteins that function at the replication fork.

A Special Nucleotide-Polymerizing Enzyme Synthesizes Short RNA Primer Molecules on the Lagging Strand

For the leading strand, a primer is needed only at the start of replication: once a replication fork is established, the DNA polymerase is continuously presented with a base-paired chain end on which to add new nucleotides. On the lagging side of the fork, however, each time the DNA polymerase completes a short DNA Okazaki fragment (which takes a few seconds), it must start synthesizing a completely new fragment at a site further along the template strand (see Figure 5–7). A special mechanism produces the base-paired primer strand required by the DNA polymerase molecules. The mechanism depends on an enzyme called **DNA primase**, which uses ribonucleoside triphosphates to synthesize short **RNA primers** on the lagging strand (**Figure 5–10**). In eukaryotes, these primers are about 10 nucleotides long and are made at intervals of 100–200 nucleotides on the lagging strand.

The chemical structure of RNA was introduced in Chapter 1 and is described in detail in Chapter 6. Here, we note only that RNA is very similar in structure to DNA. A strand of RNA can form base pairs with a strand of DNA, generating a DNA–RNA hybrid double helix if the two nucleotide sequences are complementary. Thus, the same templating principle used for DNA synthesis guides the synthesis of RNA primers. Because an RNA primer contains a properly base-paired nucleotide with a 3′-OH group at one end, it can be elongated by the DNA polymerase at this end to begin an Okazaki fragment. The synthesis of each Okazaki fragment ends when this DNA polymerase runs into the RNA primer attached to the 5′ end of the previous fragment. To produce a continuous DNA chain from the many DNA fragments made on the lagging strand, a special DNA repair system acts quickly to erase the old RNA primer and replace it with DNA. An enzyme called **DNA ligase** then joins the 3′ end of the new DNA fragment to the 5′ end of the previous one to complete the process (**Figure 5–11** and **Figure 5–12**).

Why might an erasable RNA primer be preferred to a DNA primer that would not need to be erased? The argument that a self-correcting polymerase cannot start chains *de novo* also implies the converse: an enzyme that starts chains anew cannot be efficient at self-correction. Thus, any enzyme that primes the synthesis of Okazaki fragments will of necessity make a relatively inaccurate copy (at least one error in 10^5). Even if the copies retained in the final product constituted as little as 5% of the total genome (for example, 10 nucleotides per 200-nucleotide DNA fragment), the resulting increase in the overall mutation rate would be enormous. It therefore seems likely that the use of RNA rather than DNA for priming brings a powerful advantage to the cell: the ribonucleotides in the primer automatically mark these sequences as "suspect copy" to be efficiently removed and replaced.

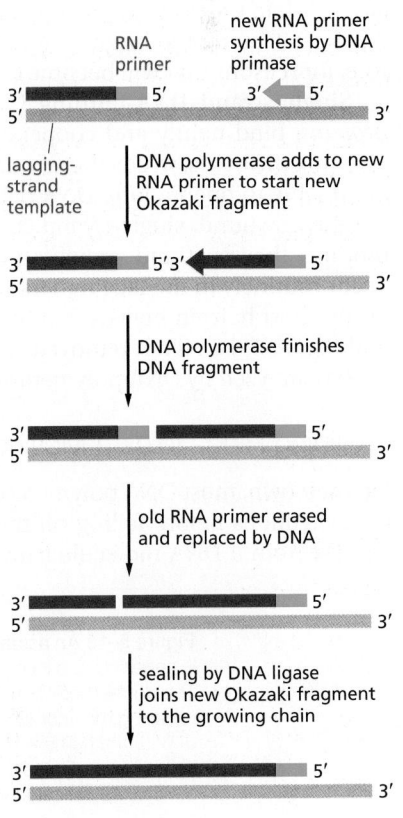

Figure 5–11 **The synthesis of one of many DNA fragments on the lagging strand.** In eukaryotes, RNA primers are made at intervals spaced by about 200 nucleotides on the lagging strand, and each RNA primer is approximately 10 nucleotides long. This primer is erased by a special DNA repair enzyme (an RNAse H) that recognizes an RNA strand in an RNA/DNA helix and fragments it; this leaves gaps that are filled in by DNA polymerase and DNA ligase.

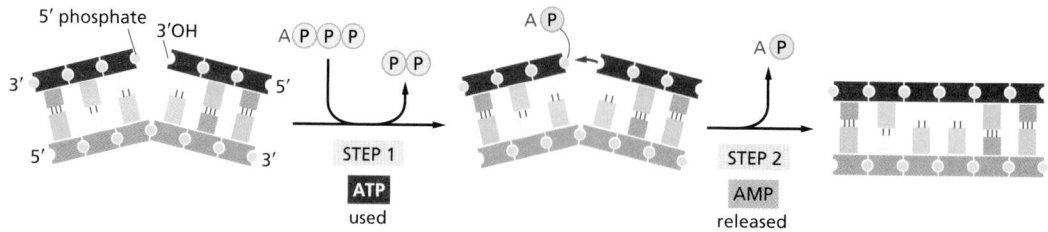

Special Proteins Help to Open Up the DNA Double Helix in Front of the Replication Fork

For DNA synthesis to proceed, the DNA double helix must be opened up ("melted") ahead of the replication fork so that the incoming deoxyribonucleoside triphosphates can form base pairs with the template strands. However, the DNA double helix is very stable under physiological conditions; the base pairs are locked in place so strongly that it requires temperatures approaching that of boiling water to separate the two strands in a test tube. For this reason, two additional types of replication proteins—DNA helicases and single-strand DNA-binding proteins—are needed to open the double helix and provide the appropriate single-strand DNA templates for the DNA polymerase to copy.

DNA helicases were first isolated as proteins that hydrolyze ATP when they are bound to single strands of DNA. As described in Chapter 3, the hydrolysis of ATP can change the shape of a protein molecule in a cyclical manner that allows the protein to perform mechanical work. DNA helicases use this principle to propel themselves rapidly along a DNA single strand. When they encounter a region of double helix, they continue to move along their strand, thereby prying apart the helix at rates of up to 1000 nucleotide pairs per second (**Figure 5–13** and **Figure 5–14**).

The two strands of DNA have opposite polarities, and, in principle, a helicase could unwind the DNA double helix by moving in the 5′-to-3′ direction along one strand or in the 3′-to-5′ direction along the other. In fact, both types of DNA helicase exist. In the best-understood replication systems in bacteria, a helicase moving 5′ to 3′ along the lagging-strand template appears to have the predominant role, for reasons that will become clear shortly.

Single-strand DNA-binding (SSB) proteins, also called *helix-destabilizing proteins*, bind tightly and cooperatively to exposed single-strand DNA without covering the bases, which therefore remain available as templates. These proteins are unable to open a long DNA helix directly, but they aid helicases by stabilizing the unwound, single-strand conformation. In addition, through cooperative binding, they coat and straighten out the regions of single-strand DNA, which occur routinely in the lagging-strand template, thereby preventing the formation of the short hairpin helices that readily form in single-strand DNA (**Figure 5–15** and **Figure 5–16**). If not removed, these hairpin helices can impede the DNA synthesis catalyzed by DNA polymerase.

A Sliding Ring Holds a Moving DNA Polymerase Onto the DNA

On their own, most DNA polymerase molecules will synthesize only a short string of nucleotides before falling off the DNA template. The tendency to dissociate quickly from a DNA molecule allows a DNA polymerase molecule that has just

Figure 5–12 The reaction catalyzed by DNA ligase. This enzyme seals a broken phosphodiester bond. As shown, DNA ligase uses a molecule of ATP to activate the 5′ end at the nick (step 1) before forming the new bond (step 2). In this way, the energetically unfavorable nick-sealing reaction is driven by being coupled to the energetically favorable process of ATP hydrolysis.

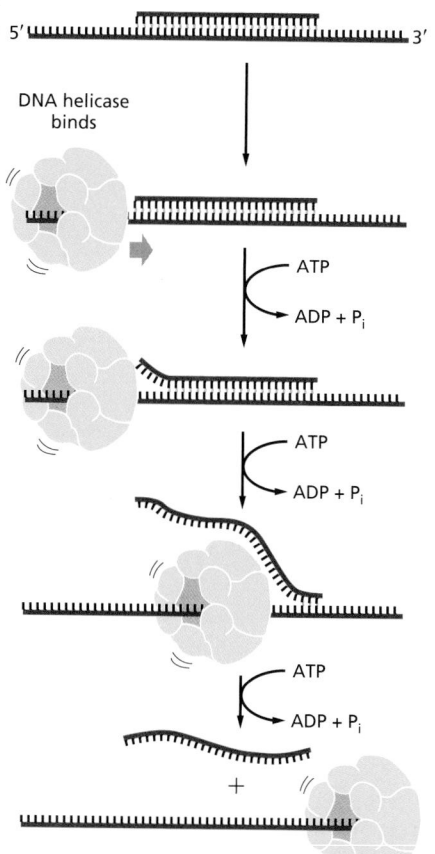

Figure 5–13 An assay for DNA helicase enzymes. A short DNA fragment is annealed to a long DNA single strand to form a region of DNA double helix. The double helix is melted as the helicase runs along the DNA single strand, releasing the short DNA fragment in a reaction that requires the presence of both the helicase protein and ATP. The rapid stepwise movement of the helicase is powered by its ATP hydrolysis (shown schematically in Figure 3–75A). As indicated, many DNA helicases are composed of six subunits.

Figure 5–14 **The structure of a DNA helicase.** (A) Diagram of the protein as a hexameric ring drawn to scale with a replication fork. (B) Detailed structure of the bacteriophage T7 replicative helicase, as determined by x-ray diffraction. Six identical subunits bind and hydrolyze ATP in an ordered fashion to propel this molecule, like a rotary engine, along a DNA single strand that passes through the central hole. *Red* indicates bound ATP molecules in the structure (**Movie 5.2**). (PDB code: 1E0J.)

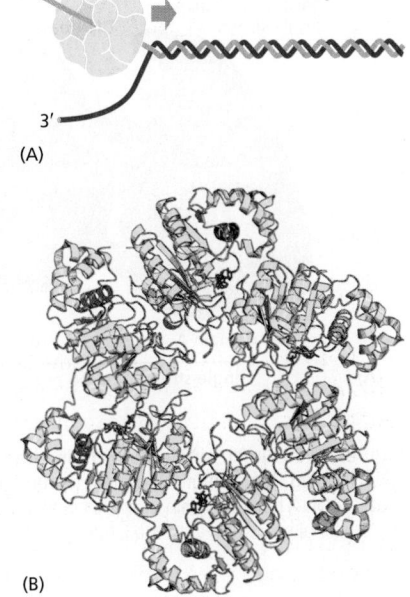

(A)

(B)

finished synthesizing one Okazaki fragment on the lagging strand to be recycled quickly, so as to begin the synthesis of the next Okazaki fragment on the same strand. This rapid dissociation, however, would make it difficult for the polymerase to synthesize the long DNA strands produced at a replication fork were it not for an accessory protein (called PCNA in eukaryotes) that functions as a regulated **sliding clamp**. This clamp keeps the polymerase firmly on the DNA when it is moving, but releases it as soon as the polymerase runs into a double-strand region of DNA.

How can a sliding clamp prevent the polymerase from dissociating without at the same time impeding the polymerase's rapid movement along the DNA molecule? The three-dimensional structure of the clamp protein, determined by x-ray diffraction, revealed it to be a large ring around the DNA double helix. One face of the ring binds to the back of the DNA polymerase, and the whole ring slides freely along the DNA as the polymerase moves. The assembly of the clamp around the DNA requires ATP hydrolysis by a special protein complex, the **clamp loader**, which hydrolyzes ATP as it loads the clamp on to a primer–template junction (**Figure 5–17**).

On the leading-strand template, the moving DNA polymerase is tightly bound to the clamp, and the two remain associated for a very long time. The DNA polymerase on the lagging-strand template also makes use of the clamp, but each time the polymerase reaches the 5′ end of the preceding Okazaki fragment, the polymerase releases itself from the clamp and dissociates from the template. This polymerase molecule then associates with a new clamp that is assembled on the RNA primer of the next Okazaki fragment.

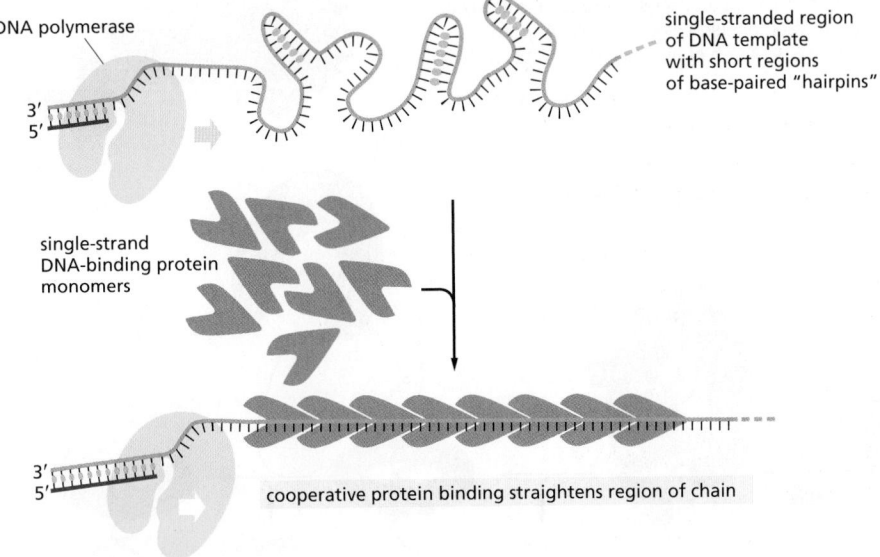

DNA polymerase

3′
5′

single-stranded region of DNA template with short regions of base-paired "hairpins"

single-strand DNA-binding protein monomers

3′
5′

cooperative protein binding straightens region of chain

Figure 5–15 **The effect of single-strand DNA-binding proteins (SSB proteins) on the structure of single-strand DNA.** Because each protein molecule prefers to bind next to a previously bound molecule, long rows of this protein form on a DNA single strand. This *cooperative binding* straightens out the DNA template and facilitates the DNA polymerization process. The "hairpin helices" shown in the bare, single-strand DNA result from a chance matching of short regions of complementary nucleotide sequence; they are similar to the short helices that typically form in RNA molecules (see Figure 1–6).

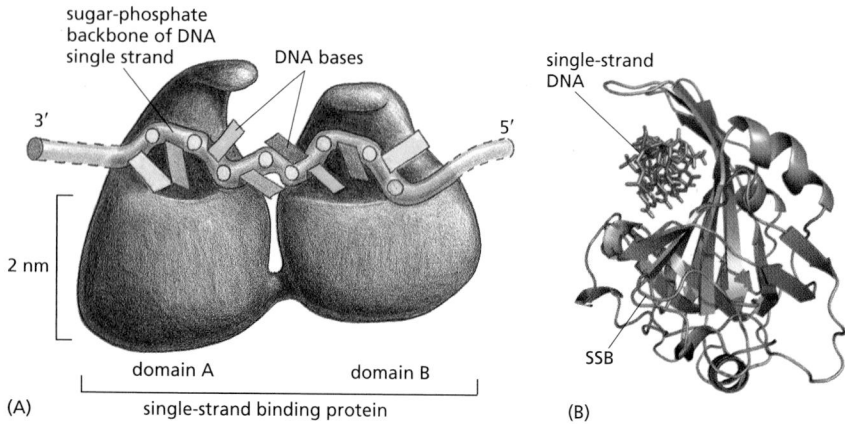

sugar-phosphate backbone of DNA single strand

DNA bases

3′

2 nm

domain A   domain B

(A)   single-strand binding protein

single-strand DNA

SSB

(B)

Figure 5–16 Human single-strand binding protein bound to DNA. (A) Front view of the two DNA-binding domains of the protein (called RPA) which cover a total of eight nucleotides. Note that the DNA bases remain exposed in this protein–DNA complex. (B) Diagram showing the three-dimensional structure, with the DNA strand *(orange)* viewed end-on. (PDB code: 1JMC.)

Figure 5–17 The regulated sliding clamp that holds DNA polymerase on the DNA. (A) The structure of the clamp protein from *E. coli*, as determined by x-ray crystallography, with a DNA helix added to indicate how the protein fits around DNA (Movie 5.3). (B) Schematic illustration showing how the clamp (with *red* and *yellow* subunits) is loaded onto DNA to serve as a tether for a moving DNA polymerase molecule. The structure of the clamp loader *(dark green)* resembles a screw nut, with its threads matching the grooves of double-stranded DNA. The loader binds to a free clamp molecule, forcing a gap in its ring of subunits so that this ring is able to slip around DNA. The clamp loader, thanks to its screw-nut structure, recognises the region of DNA that is double-stranded and latches onto it, tightening around the complex of a template strand with a freshly synthesized elongating (primer) strand. It carries the clamp along this double-stranded region until it encounters the 3′ end of the primer, at which point the loader hydrolyzes ATP and releases the clamp, allowing it to close around the DNA and bind to DNA polymerase. In the simplified reaction shown here, the clamp loader dissociates into solution once the clamp has been assembled. At a true replication fork, the clamp loader remains close to the polymerase so that, on the lagging strand, it is ready to assemble a new clamp at the start of each new Okazaki fragment (see Figure 5–18). (A, from X.P. Kong et al., *Cell* 69:425–437, 1992; B, adapted from B.A. Kelch et al., *Science* 334:1675–1680, 2011. PDB code: 3BEP.)

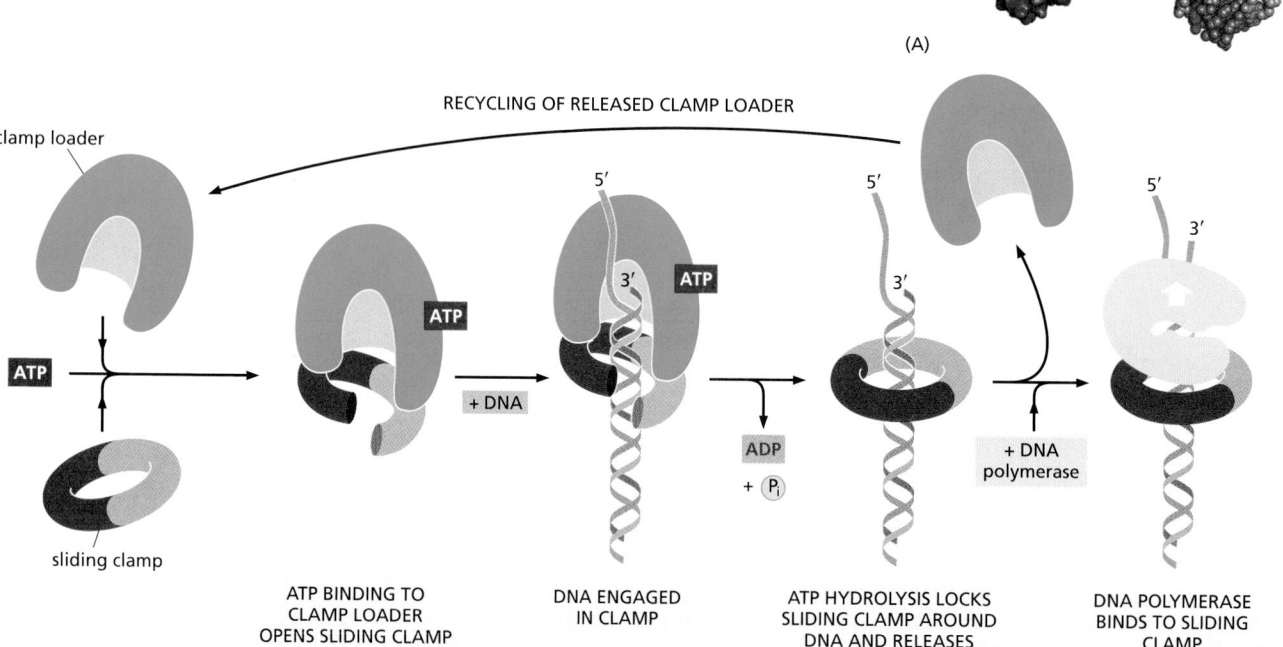

(A)

RECYCLING OF RELEASED CLAMP LOADER

clamp loader

ATP

ATP

+ DNA

ATP

5′

3′

ADP + P$_i$

5′

3′

+ DNA polymerase

5′

3′

sliding clamp

ATP BINDING TO CLAMP LOADER OPENS SLIDING CLAMP

DNA ENGAGED IN CLAMP

ATP HYDROLYSIS LOCKS SLIDING CLAMP AROUND DNA AND RELEASES CLAMP LOADER

DNA POLYMERASE BINDS TO SLIDING CLAMP

(B)

The Proteins at a Replication Fork Cooperate to Form a Replication Machine

Although we have discussed DNA replication as though it were performed by a mix of proteins all acting independently, in reality most of the proteins are held together in a large and orderly multienzyme complex that rapidly synthesizes DNA. This complex can be likened to a tiny sewing machine composed of protein parts and powered by nucleoside triphosphate hydrolysis. Like a sewing machine, the replication complex probably remains stationary with respect to its immediate surroundings; the DNA can be thought of as a long strip of cloth being rapidly threaded through it. Although the replication complex has been most intensively studied in *E. coli* and several of its viruses, a very similar complex also operates in eukaryotes, as we see below.

Figure 5–18 summarizes the functions of the subunits of the replication machine. At the front of the replication fork, DNA helicase opens the DNA helix. Two DNA polymerase molecules work at the fork, one on the leading strand and one on the lagging strand. Whereas the DNA polymerase molecule on the leading strand can operate in a continuous fashion, the DNA polymerase molecule on the lagging strand must restart at short intervals, using a short RNA primer made by a DNA primase molecule. The close association of all these protein components increases the efficiency of replication and is made possible by a folding back of the lagging strand as shown in Figure 5–18A. This arrangement also facilitates the loading of the polymerase clamp each time that an Okazaki fragment is synthesized: the clamp loader and the lagging-strand DNA polymerase molecule are kept in place as a part of the protein machine even when they detach from their DNA template. The replication proteins are thus linked together into a single large

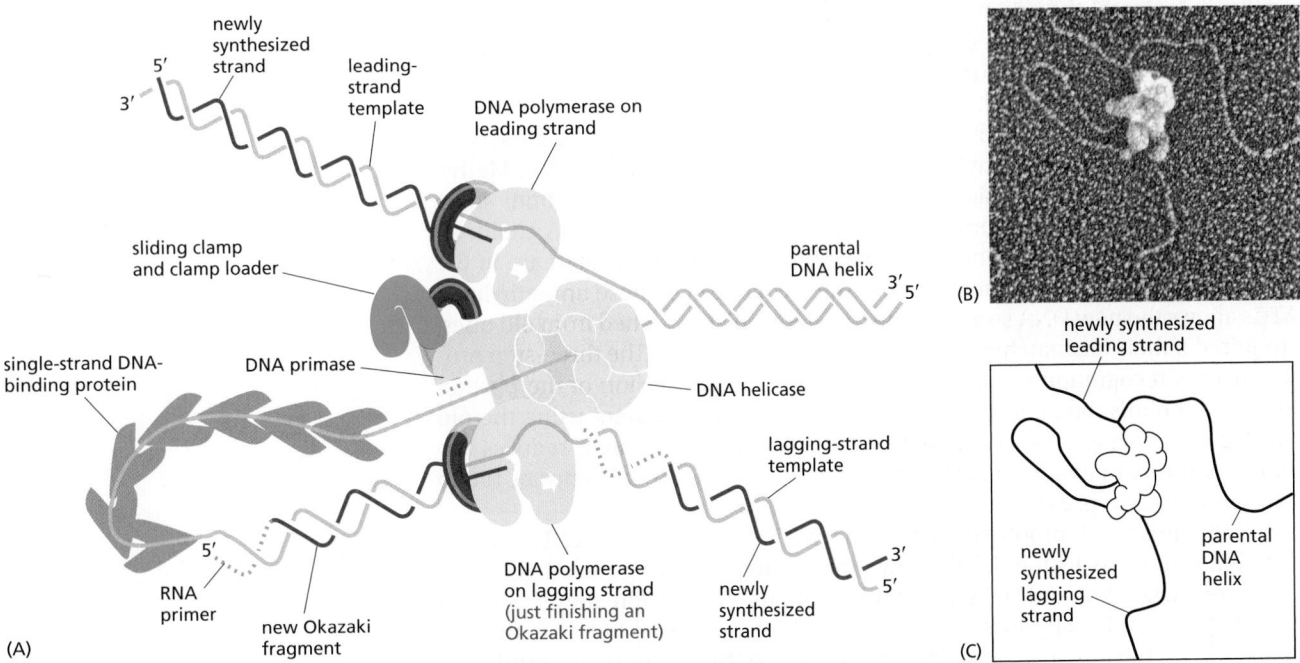

Figure 5–18 **A bacterial replication fork.** (A) This schematic diagram shows a current view of the arrangement of replication proteins at a replication fork when DNA is being synthesized. The lagging-strand DNA is folded to bring the lagging-strand DNA polymerase molecule into a complex with the leading-strand DNA polymerase molecule. This folding also brings the 3′ end of each completed Okazaki fragment close to the start site for the next Okazaki fragment. Because the lagging-strand DNA polymerase molecule remains bound to the rest of the replication proteins, it can be reused to synthesize successive Okazaki fragments. In this diagram, it is about to let go of its completed DNA fragment and move to the RNA primer that is just being synthesized. Additional proteins (not shown) help to hold the different protein components of the fork together, enabling them to function as a well-coordinated protein machine (**Movie 5.4** and **Movie 5.5**). (B) An electron micrograph showing the replication machine from the bacteriophage T4 as it moves along a template synthesizing DNA behind it. (C) An interpretation of the micrograph is given in the sketch: note especially the DNA loop on the lagging strand. Apparently, the replication proteins became partly detached from the very front of the replication fork during the preparation of this sample for electron microscopy. (B, from P.D. Chastain et al., *J. Biol. Chem.* 278:21276–21825, 2003. With permission from American Society for Biochemistry and Molecular Biology.)

unit (total molecular mass $>10^6$ daltons), enabling DNA to be synthesized on both sides of the replication fork in a coordinated and efficient manner.

On the lagging strand, the DNA replication machine leaves behind a series of unsealed Okazaki fragments, which still contain the RNA that primed their synthesis at their 5′ ends. As discussed earlier, this RNA is removed and the resulting gap is filled in by DNA repair enzymes that operate behind the replication fork (see Figure 5–11).

A Strand-Directed Mismatch Repair System Removes Replication Errors That Escape from the Replication Machine

As stated previously, bacteria such as *E. coli* are capable of dividing once every 30 minutes, making it relatively easy to screen large populations to find a rare mutant cell that is altered in a specific process. One interesting class of mutants consists of those with alterations in so-called *mutator genes*, which greatly increase the rate of spontaneous mutation. Not surprisingly, one such mutant makes a defective form of the 3′-to-5′ proofreading exonuclease that is a part of the DNA polymerase enzyme (see Figures 5–8 and 5–9). The mutant DNA polymerase no longer proofreads effectively, and many replication errors that would otherwise have been removed accumulate in the DNA.

The study of other *E. coli* mutants exhibiting abnormally high mutation rates has uncovered a proofreading system that removes replication errors made by the polymerase that have been missed by the proofreading exonuclease. This **strand-directed mismatch repair** system detects the potential for distortion in the DNA helix from the misfit between noncomplementary base pairs.

If the proofreading system simply recognized a mismatch in newly replicated DNA and randomly corrected one of the two mismatched nucleotides, it would mistakenly "correct" the original template strand to match the error exactly half the time, thereby failing to lower the overall error rate. To be effective, such a proofreading system must be able to distinguish and remove the mismatched nucleotide only on the newly synthesized strand, where the replication error occurred.

The strand-distinction mechanism used by the mismatch proofreading system in *E. coli* depends on the methylation of selected A residues in the DNA. Methyl groups are added to all A residues in the sequence GATC, but not until some time after the A has been incorporated into a newly synthesized DNA chain. As a result, the only GATC sequences that have not yet been methylated are in the new strands just behind a replication fork. The recognition of these unmethylated GATCs allows the new DNA strands to be transiently distinguished from old ones, as required if their mismatches are to be selectively removed. The three-step process involves recognition of a newly synthesized strand, excision of the portion containing the mismatch, and resynthesis of the excised segment using the old strand as a template. This strand-directed mismatch repair system reduces the number of errors made during DNA replication by an additional factor of 100 to 1000 (see Table 5–1, p. 244).

A similar mismatch proofreading system functions in eukaryotic cells but uses a different strategy to distinguish the new strand from the old (Figure 5–19). Newly synthesized lagging-strand DNA transiently contains *nicks* (before they are sealed by DNA ligase) and such nicks (also called *single-strand breaks*) provide the signal that directs the mismatch proofreading system to the appropriate strand. This strategy also requires that the newly synthesized DNA on the leading strand be transiently nicked; how this occurs is uncertain.

The importance of mismatch proofreading in humans is seen in individuals who inherit one defective copy of a mismatch repair gene (along with a functional gene on the other copy of the chromosome). These people have a marked predisposition for certain types of cancers. For example, in a type of colon cancer called *hereditary nonpolyposis colon cancer* (*HNPCC*), spontaneous mutation of the one functional gene produces a clone of somatic cells that, because they are deficient in mismatch proofreading, accumulate mutations unusually rapidly. Most cancers arise in cells that have accumulated multiple mutations (see pp. 1096–1097),

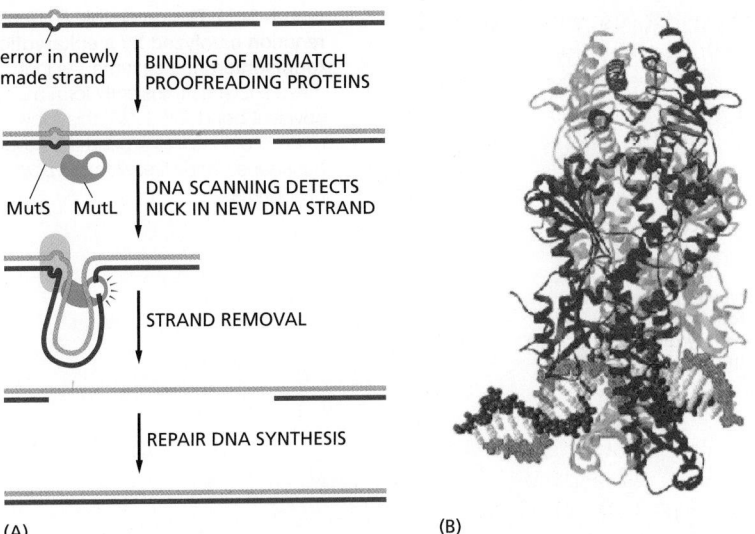

(A)

(B)

Figure 5–19 **Strand-directed mismatch repair.** (A) The two proteins shown are present in both bacteria and eukaryotic cells: MutS binds specifically to a mismatched base pair, while MutL scans the nearby DNA for a nick. Once MutL finds a nick, it triggers the degradation of the nicked strand all the way back through the mismatch. Because nicks are largely confined to newly replicated strands in eukaryotes, replication errors are selectively removed. In bacteria, an additional protein in the complex (MutH) nicks unmethylated (and therefore newly replicated) GATC sequences, thereby beginning the process illustrated here. In eukaryotes, MutL contains a DNA nicking activity that aids in the removal of the damaged strand. (B) The structure of the MutS protein bound to a DNA mismatch. This protein is a dimer, which grips the DNA double helix as shown, kinking the DNA at the mismatched base pair. It seems that the MutS protein scans the DNA for mismatches by testing for sites that can be readily kinked, which are those with an abnormal base pair. (PDB code: 1EWQ.)

In the figure (A), the steps are labeled:
- error in newly made strand — BINDING OF MISMATCH PROOFREADING PROTEINS
- MutS   MutL — DNA SCANNING DETECTS NICK IN NEW DNA STRAND
- STRAND REMOVAL
- REPAIR DNA SYNTHESIS

and cells deficient in mismatch proofreading therefore have a greatly enhanced chance of becoming cancerous. Fortunately, most of us inherit two good copies of each gene that encodes a mismatch proofreading protein; this protects us, because it is highly unlikely for both copies to become mutated in the same cell.

DNA Topoisomerases Prevent DNA Tangling During Replication

As a replication fork moves along double-strand DNA, it creates what has been called the "winding problem." The two parental strands, which are wound around each other, must be unwound and separated for replication to occur. For every 10 nucleotide pairs replicated at the fork, one complete turn of the parental double helix must be unwound. In principle, this unwinding can be achieved by rapidly rotating the entire chromosome ahead of a moving fork; however, this is energetically highly unfavorable (particularly for long chromosomes) and, instead, the DNA in front of a replication fork becomes overwound (**Figure 5–20**). The overwinding, in turn, is continually relieved by proteins known as **DNA topoisomerases**.

A DNA topoisomerase can be viewed as a reversible nuclease that adds itself covalently to a DNA backbone phosphate, thereby breaking a phosphodiester bond in a DNA strand. This reaction is reversible, and the phosphodiester bond re-forms as the protein leaves.

One type of topoisomerase, called *topoisomerase I*, produces a transient single-strand break; this break in the phosphodiester backbone allows the two sections of DNA helix on either side of the nick to rotate freely relative to each other, using the phosphodiester bond in the strand opposite the nick as a swivel point (**Figure 5–21**). Any tension in the DNA helix will drive this rotation in the direction that relieves the tension. As a result, DNA replication can occur with the rotation of only a short length of helix—the part just ahead of the fork. Because the covalent linkage that joins the DNA topoisomerase protein to a DNA phosphate retains

Figure 5–20 **The "winding problem" that arises during DNA replication.** (A) For a bacterial replication fork moving at 500 nucleotides per second, the parental DNA helix ahead of the fork must rotate at 50 revolutions per second. (B) If the ends of the DNA double helix remain fixed (or difficult to rotate), tension builds up in front of the replication fork as it becomes overwound. Some of this tension can be taken up by supercoiling, whereby the DNA double helix twists around itself (see Figure 6–19). However, if the tension continues to build up, the replication fork will eventually stop because further unwinding requires more energy than the helicase can provide. Note that in (A), the dotted line represents about 20 turns of DNA.

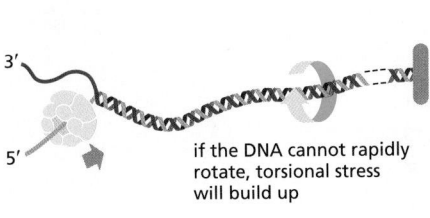

3′

5′

if the DNA cannot rapidly rotate, torsional stress will build up

(A)

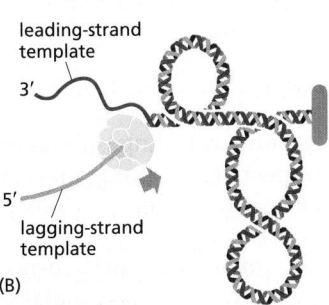

leading-strand template

3′

5′

lagging-strand template

(B)

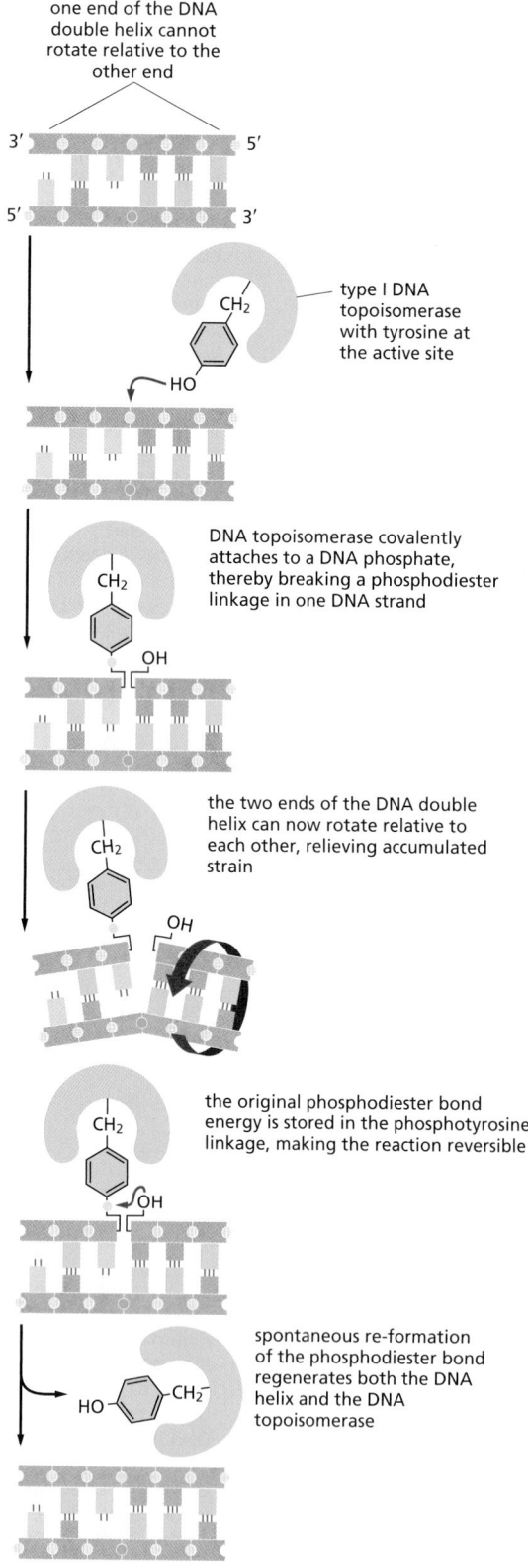

one end of the DNA double helix cannot rotate relative to the other end

3′   5′

5′   3′

type I DNA topoisomerase with tyrosine at the active site

DNA topoisomerase covalently attaches to a DNA phosphate, thereby breaking a phosphodiester linkage in one DNA strand

the two ends of the DNA double helix can now rotate relative to each other, relieving accumulated strain

the original phosphodiester bond energy is stored in the phosphotyrosine linkage, making the reaction reversible

spontaneous re-formation of the phosphodiester bond regenerates both the DNA helix and the DNA topoisomerase

Figure 5–21 The reversible DNA nicking reaction catalyzed by a eukaryotic DNA topoisomerase I enzyme. As indicated, these enzymes transiently form a single covalent bond with DNA; this allows free rotation of the DNA around the covalent backbone bonds linked to the *blue* phosphate.

the energy of the cleaved phosphodiester bond, resealing is rapid and does not require additional energy input. In this respect, the rejoining mechanism differs from that catalyzed by the enzyme DNA ligase, discussed previously (see Figure 5–12).

A second type of DNA topoisomerase, *topoisomerase II*, forms a covalent linkage to both strands of the DNA helix at the same time, making a transient

Figure 5–22 The DNA-helix-passing reaction catalyzed by DNA topoisomerase II. Unlike type I topoisomerases, type II enzymes hydrolyze ATP *(red)*, which is needed to release and reset the enzyme after each cycle. Type II topoisomerases are largely confined to proliferating cells in eukaryotes; partly for that reason, they have been effective targets for anticancer drugs. Some of these drugs inhibit topoisomerase II at the third step in the figure and thereby produce high levels of double-strand breaks that kill rapidly dividing cells. The small *yellow circles* represent the phosphates in the DNA backbone that become covalently bonded to the topoisomerase (see Figure 5–21).

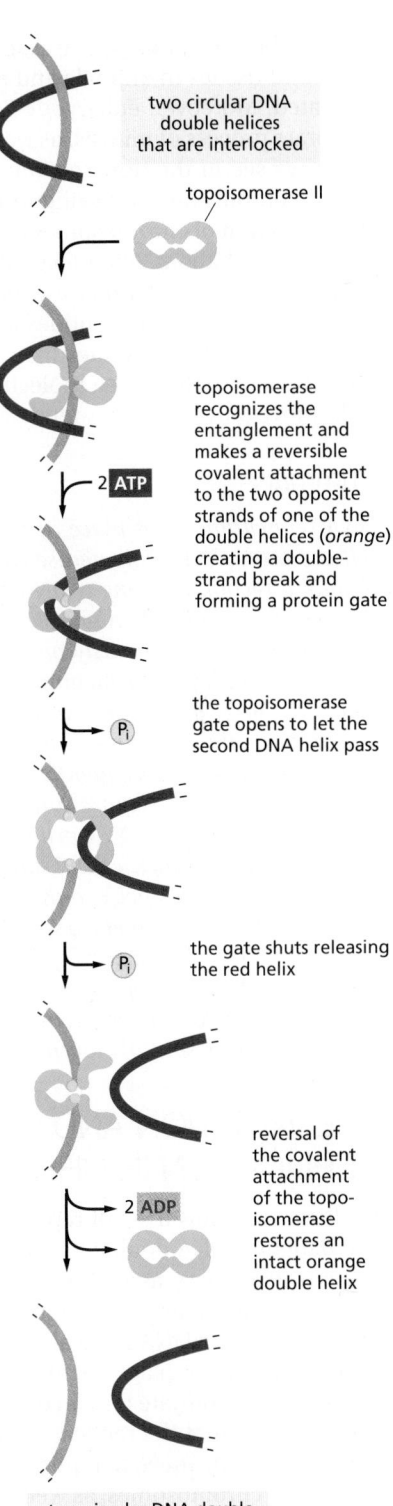

two circular DNA double helices that are interlocked

topoisomerase II

topoisomerase recognizes the entanglement and makes a reversible covalent attachment to the two opposite strands of one of the double helices *(orange)* creating a double-strand break and forming a protein gate

the topoisomerase gate opens to let the second DNA helix pass

the gate shuts releasing the red helix

reversal of the covalent attachment of the topo-isomerase restores an intact orange double helix

two circular DNA double helices that are separated

double-strand break in the helix. These enzymes are activated by sites on chromosomes where two double helices cross over each other such as those generated by supercoiling in front of a replication fork (see Figure 5–20). Once a topoisomerase II molecule binds to such a crossing site, the protein uses ATP hydrolysis to perform the following set of reactions efficiently: (1) it breaks one double helix reversibly to create a DNA "gate"; (2) it causes the second, nearby double helix to pass through this opening; and (3) it then reseals the break and dissociates from the DNA. At crossover points generated by supercoiling, passage of the double helix through the gate occurs in the direction that will reduce supercoiling. In this way, type II topoisomerases can relieve the overwinding tension generated in front of a replication fork. Their reaction mechanism also allows type II DNA topoisomerases to efficiently separate two interlocked DNA circles (**Figure 5–22**).

Topoisomerase II also prevents the severe DNA tangling problems that would otherwise arise during DNA replication. This role is nicely illustrated by mutant yeast cells that produce, in place of the normal topoisomerase II, a version that is inactive above 37°C. When the mutant cells are warmed to this temperature, their daughter chromosomes remain intertwined after DNA replication and are unable to separate. The enormous usefulness of topoisomerase II for untangling chromosomes can readily be appreciated by anyone who has struggled to remove a tangle from a fishing line without the aid of scissors.

DNA Replication Is Fundamentally Similar in Eukaryotes and Bacteria

Much of what we know about DNA replication was first derived from studies of purified bacterial and bacteriophage multienzyme systems capable of DNA replication *in vitro*. The development of these systems in the 1970s was greatly facilitated by the prior isolation of mutants in a variety of replication genes; these mutants were exploited to identify and purify the corresponding replication proteins. The first mammalian replication system that accurately replicated DNA *in vitro* was described in the mid-1980s, and mutations in genes encoding nearly all of the replication components have now been isolated and analyzed in the yeast *Saccharomyces cerevisiae*. As a result, much is known about the detailed enzymology of DNA replication in eukaryotes, and it is clear that the fundamental features of DNA replication—including replication-fork geometry and the use of a multiprotein replication machine—have been conserved during the long evolutionary process that separated bacteria from eukaryotes.

There are more protein components in eukaryotic replication machines than there are in the bacterial analogs, even though the basic functions are the same. Thus, for example, the eukaryotic single-strand binding (SSB) protein is formed from three subunits, whereas only a single subunit is found in bacteria. Similarly, the eukaryotic DNA primase is incorporated into a multisubunit enzyme that also contains a polymerase called DNA polymerase α-primase. This protein complex begins each Okazaki fragment on the lagging strand with RNA and then extends the RNA primer with a short length of DNA. At this point, the two main eukaryotic replicative DNA polymerases, Polδ and Polε, come into play: Polδ completes each Okazaki fragment on the lagging strand and Polε extends the leading strand. The increased complexity of eukaryotic replication machinery probably reflects more elaborate controls. For example, the orderly maintenance of different cell types and tissues in animals and plants requires that DNA replication be tightly regulated. Moreover, eukaryotic DNA replication must be coordinated with the elaborate process of mitosis, as we discuss in Chapter 17.

As we see in the next section, the eukaryotic replication machinery has the added complication of having to replicate through nucleosomes, the repeating structural unit of chromosomes discussed in Chapter 4. Nucleosomes are spaced at intervals of about 200 nucleotide pairs along the DNA, which, as we will see, explains why new Okazaki fragments are synthesized on the lagging strand at intervals of 100–200 nucleotides in eukaryotes, instead of 1000–2000 nucleotides as in bacteria. Nucleosomes may also act as barriers that slow down the movement of DNA polymerase molecules, which may be why eukaryotic replication forks move only about one-tenth as fast as bacterial replication forks.

Summary

DNA replication takes place at a Y-shaped structure called a replication fork. A self-correcting DNA polymerase enzyme catalyzes nucleotide polymerization in a 5′-to-3′ direction, copying a DNA template strand with remarkable fidelity. Since the two strands of a DNA double helix are antiparallel, this 5′-to-3′ DNA synthesis can take place continuously on only one of the strands at a replication fork (the leading strand). On the lagging strand, short DNA fragments must be made by a "backstitching" process. Because the self-correcting DNA polymerase cannot start a new chain, these lagging-strand DNA fragments are primed by short RNA primer molecules that are subsequently erased and replaced with DNA.

DNA replication requires the cooperation of many proteins. These include (1) DNA polymerase and DNA primase to catalyze nucleoside triphosphate polymerization; (2) DNA helicases and single-strand DNA-binding (SSB) proteins to help in opening up the DNA helix so that it can be copied; (3) DNA ligase and an enzyme that degrades RNA primers to seal together the discontinuously synthesized lagging-strand DNA fragments; and (4) DNA topoisomerases to help to relieve helical winding and DNA tangling problems. Many of these proteins associate with each other at a replication fork to form a highly efficient "replication machine," through which the activities and spatial movements of the individual components are coordinated.

THE INITIATION AND COMPLETION OF DNA REPLICATION IN CHROMOSOMES

We have seen how a set of replication proteins rapidly and accurately generates two daughter DNA double helices behind a replication fork. But how is this replication machinery assembled in the first place, and how are replication forks created on an intact, double-strand DNA molecule? In this section, we discuss how cells initiate DNA replication and how they carefully regulate this process to ensure that it takes place not only at the proper positions on the chromosome but also at the appropriate time in the life of the cell. We also discuss a few of the special problems that the replication machinery in eukaryotic cells must overcome. These include the need to replicate the enormously long DNA molecules found in eukaryotic chromosomes, as well as the difficulty of copying DNA molecules that are tightly complexed with histones in nucleosomes.

DNA Synthesis Begins at Replication Origins

As discussed previously, the DNA double helix is normally very stable: the two DNA strands are locked together firmly by many hydrogen bonds formed between the bases on each strand. To begin DNA replication, the double helix must first be opened up and the two strands separated to expose unpaired bases. As we shall see, the process of DNA replication is begun by special *initiator proteins* that bind to double-strand DNA and pry the two strands apart, breaking the hydrogen bonds between the bases.

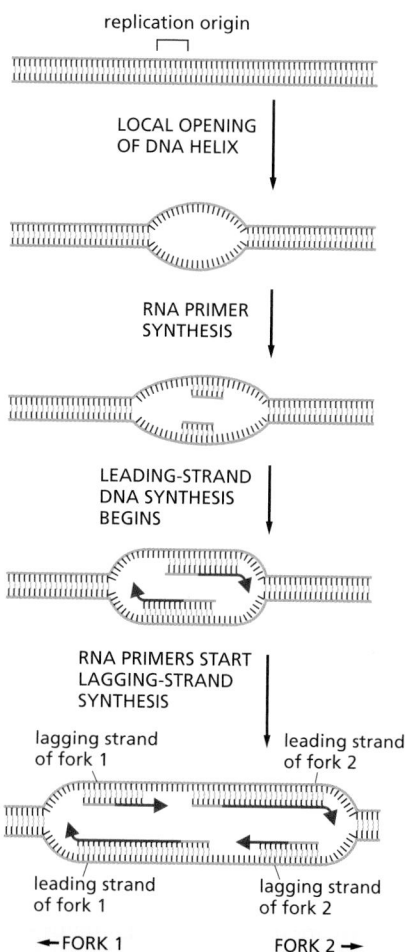

Figure 5–23 A replication bubble formed by replication-fork initiation. This diagram outlines the major steps in the initiation of replication forks at replication origins. The structure formed at the last step, in which both strands of the parental DNA helix have been separated from each other and serve as templates for DNA synthesis, is called a *replication bubble.*

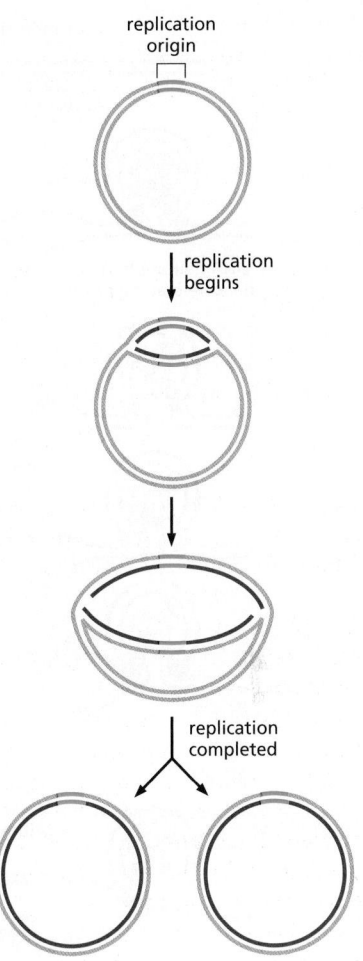

Figure 5–24 **DNA replication of a bacterial genome.** It takes *E. coli* about 30 minutes to duplicate its genome of 4.6×10^6 nucleotide pairs. For simplicity, no Okazaki fragments are shown on the lagging strand. What happens as the two replication forks approach each other and collide at the end of the replication cycle is not well understood, although the replication machines are disassembled as part of the process.

The positions at which the DNA helix is first opened are called **replication origins** (**Figure 5–23**). In simple cells like those of bacteria or yeast, origins are specified by DNA sequences several hundred nucleotide pairs in length. This DNA contains both short sequences that attract initiator proteins and stretches of DNA that are especially easy to open. We saw in Figure 4–4 that an A-T base pair is held together by fewer hydrogen bonds than a G-C base pair. Therefore, DNA rich in A-T base pairs is relatively easy to pull apart, and regions of DNA enriched in A-T base pairs are typically found at replication origins.

Although the basic process of replication-fork initiation depicted in Figure 5–23 is fundamentally the same for bacteria and eukaryotes, the detailed way in which this process is performed and regulated differs between these two groups of organisms. We first consider the simpler and better-understood case in bacteria and then turn to the more complex situation found in yeasts, mammals, and other eukaryotes.

Bacterial Chromosomes Typically Have a Single Origin of DNA Replication

The genome of *E. coli* is contained in a single circular DNA molecule of 4.6×10^6 nucleotide pairs. DNA replication begins at a single origin of replication, and the two replication forks assembled there proceed (at approximately 1000 nucleotides per second) in opposite directions until they meet up roughly halfway around the chromosome (**Figure 5–24**). The only point at which *E. coli* can control DNA replication is initiation: once the forks have been assembled at the origin, they synthesize DNA at relatively constant speed until replication is finished. Therefore, it is not surprising that the initiation of DNA replication is highly regulated. The process begins when initiator proteins (in their ATP-bound state) bind in multiple copies to specific DNA sites located at the replication origin, wrapping the DNA around the proteins to form a large protein–DNA complex that destabilizes the adjacent double helix. This complex then attracts two DNA helicases, each bound to a helicase loader, and these are placed around adjacent DNA single strands whose bases have been exposed by the assembly of the initiator protein–DNA complex. The helicase loader is analogous to the clamp loader we encountered above; it has the additional job of keeping the helicase in an inactive form until it is properly loaded onto a nascent replication fork. Once the helicases are loaded, the loaders dissociate and the helicases begin to unwind DNA, exposing enough single-strand DNA for DNA primase to synthesize the first RNA primers (**Figure 5–25**). This quickly leads to the assembly of remaining proteins to create two replication forks, with replication machines that move, with respect to the replication origin, in opposite directions. They continue to synthesize DNA until all of the DNA template downstream of each fork has been replicated.

In *E. coli*, the interaction of the initiator protein with the replication origin is carefully regulated, with initiation occurring only when sufficient nutrients are available for the bacterium to complete an entire round of replication. Initiation is also controlled to ensure that only one round of DNA replication occurs for each cell division. After replication is initiated, the initiator protein is inactivated by hydrolysis of its bound ATP molecule, and the origin of replication experiences a "refractory period." The refractory period is caused by a delay in the methylation of newly incorporated A nucleotides in the origin (**Figure 5–26**). Initiation cannot occur again until the A's are methylated and the initiator protein is restored to its ATP-bound state.

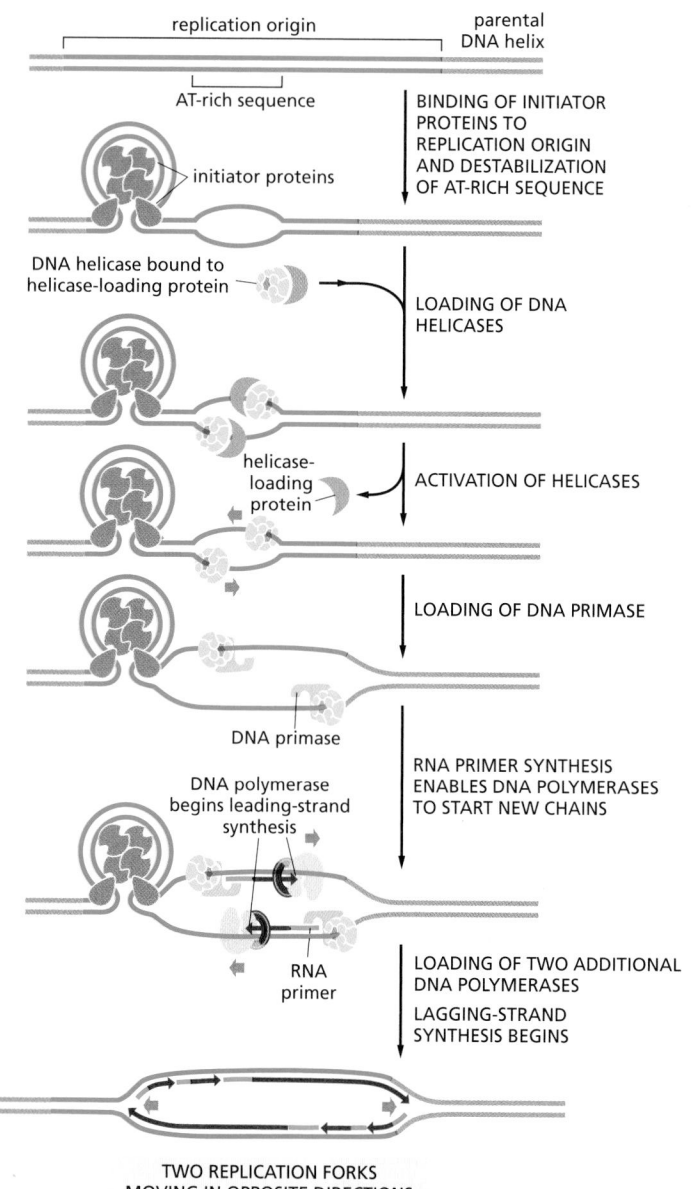

replication origin

parental DNA helix

AT-rich sequence

initiator proteins

BINDING OF INITIATOR PROTEINS TO REPLICATION ORIGIN AND DESTABILIZATION OF AT-RICH SEQUENCE

DNA helicase bound to helicase-loading protein

LOADING OF DNA HELICASES

helicase-loading protein

ACTIVATION OF HELICASES

LOADING OF DNA PRIMASE

DNA primase

DNA polymerase begins leading-strand synthesis

RNA PRIMER SYNTHESIS ENABLES DNA POLYMERASES TO START NEW CHAINS

RNA primer

LOADING OF TWO ADDITIONAL DNA POLYMERASES

LAGGING-STRAND SYNTHESIS BEGINS

TWO REPLICATION FORKS MOVING IN OPPOSITE DIRECTIONS

Figure 5–25 The proteins that initiate DNA replication in bacteria. The mechanism shown was established by studies *in vitro* with mixtures of highly purified proteins. For *E. coli* DNA replication, the major initiator protein, the helicase, and the primase are the dnaA, dnaB, and dnaG proteins, respectively. In the first step, several molecules of the initiator protein bind to specific DNA sequences at the replication origin and destabilize the double helix by forming a compact structure in which the DNA is tightly wrapped around the protein. Next, two helicases are brought in by helicase-loading proteins (the dnaC proteins), which inhibit the helicases until they are properly loaded at the replication origin. Helicase-loading proteins prevent the replicative DNA helices from inappropriately entering other single-strand stretches of DNA in the bacterial genome. Aided by single-strand binding protein (not shown), the loaded helicases open up the DNA, thereby enabling primases to enter and synthesize initial primers. In subsequent steps, two complete replication forks are assembled at the origin and move off in opposite directions. The initiator proteins are displaced as the left-hand fork moves through them (not shown).

Eukaryotic Chromosomes Contain Multiple Origins of Replication

We have seen how two replication forks begin at a single replication origin in bacteria and proceed in opposite directions, moving away from the origin until all of the DNA in the single circular chromosome is replicated. The bacterial genome is sufficiently small for these two replication forks to duplicate the genome in about 30 minutes. Because of the much greater size of most eukaryotic chromosomes, a different strategy is required to allow their replication in a timely manner.

A method for determining the general pattern of eukaryotic chromosome replication was developed in the early 1960s. Human cells growing in culture are labeled for a short time with ³H-thymidine so that the DNA synthesized during this period becomes highly radioactive. The cells are then gently lysed, and the DNA is stretched on the surface of a glass slide coated with a photographic emulsion. Development of the emulsion reveals the pattern of labeled DNA through a technique known as *autoradiography*. The time allotted for radioactive labeling is chosen to allow each replication fork to move several micrometers along the DNA, so that the replicated DNA can be detected in the light microscope as lines of silver grains, even though the DNA molecule itself is too thin to be visible.

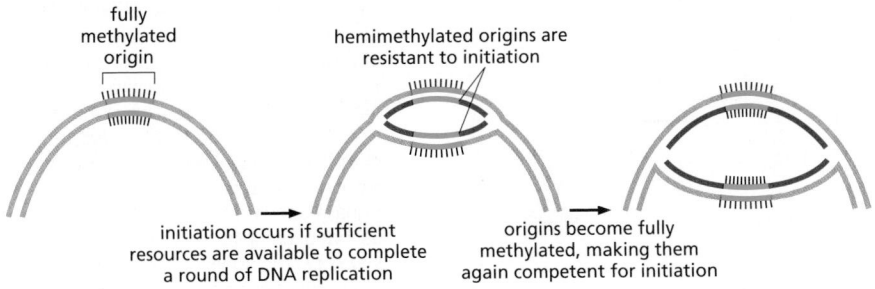

fully methylated origin

hemimethylated origins are resistant to initiation

initiation occurs if sufficient resources are available to complete a round of DNA replication

origins become fully methylated, making them again competent for initiation

In this way, both the rate and the direction of replication-fork movement can be determined (**Figure 5–27**). From the rate at which tracks of replicated DNA increase in length with increasing labeling time, the eukaryotic replication forks are estimated to travel at about 50 nucleotides per second. This is approximately twentyfold slower than the rate at which bacterial replication forks move, possibly reflecting the increased difficulty of replicating DNA that is packaged tightly in chromatin.

An average-size human chromosome contains a single linear DNA molecule of about 150 million nucleotide pairs. It would take 0.02 seconds/nucleotide × 150 × 10^6 nucleotides = 3.0 × 10^6 seconds (about 35 days) to replicate such a DNA molecule from end to end with a single replication fork moving at a rate of 50 nucleotides per second. As expected, therefore, the autoradiographic experiments just described reveal that many forks, belonging to separate replication bubbles, are moving simultaneously on each eukaryotic chromosome.

Much faster and more sophisticated methods now exist for monitoring DNA replication initiation and tracking the movement of DNA replication forks across whole genomes. One approach uses DNA microarrays—grids the size of a postage stamp studded with hundreds of thousands of fragments of known DNA sequence. As we will see in detail in Chapter 8, each different DNA fragment is placed at a unique position on the microarray, and whole genomes can thereby be represented in an orderly manner. If a DNA sample from a group of replicating cells is broken up and hybridized to a microarray representing that organism's genome, the amount of each DNA sequence can be determined. Because a segment of a genome that has been replicated will contain twice as much DNA as an unreplicated segment, replication-fork initiation and fork movement can be accurately monitored across an entire genome (**Figure 5–28**).

Experiments of this type have shown the following: (1) Approximately 30,000–50,000 origins of replication are used each time a human cell divides. (2) The human genome has many more (perhaps tenfold more) potential origins than this, and different cell types use different sets of origins. This may allow a cell to coordinate its active origins with other features of its chromosomes such as which

Figure 5–26 Methylation of the *E. coli* replication origin creates a refractory period for DNA initiation. DNA methylation occurs at GATC sequences, 11 of which are found in the origin of replication (spanning approximately 250 nucleotide pairs). In its hemimethylated state, the origin of replication is bound by an inhibitor protein (Seq A, not shown), which blocks the ability of the initiator proteins to unwind the origin DNA. Eventually (about 15 minutes after replication is initiated), the hemimethylated origins become fully methylated by a DNA methylase enzyme; Seq A then dissociates.

A single enzyme, the *Dam* methylase, is responsible for methylating all *E. coli* GATC sequences. A lag in methylation after the replication of GATC sequences is also used by the *E. coli* mismatch proofreading system to distinguish the newly synthesized DNA strand from the parental DNA strand; in that case, the relevant GATC sequences are scattered throughout the chromosome, and they are not bound by Seq A.

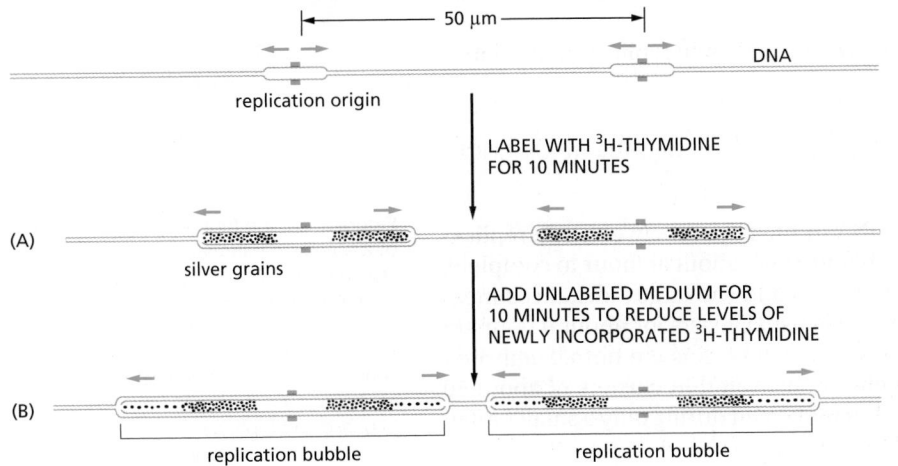

50 μm

DNA

replication origin

(A)

LABEL WITH ^{3}H-THYMIDINE FOR 10 MINUTES

silver grains

ADD UNLABELED MEDIUM FOR 10 MINUTES TO REDUCE LEVELS OF NEWLY INCORPORATED ^{3}H-THYMIDINE

(B)

replication bubble

replication bubble

Figure 5–27 The experiments that demonstrated the pattern in which replication forks are formed and move on eukaryotic chromosomes. The new DNA made in human cells in culture was labeled briefly with a pulse of highly radioactive thymidine (^{3}H-thymidine). (A) In this experiment, the cells were lysed, and the DNA was stretched out on a glass slide that was subsequently covered with a photographic emulsion. After several months, the emulsion was developed, revealing a line of silver grains over the radioactive DNA. The *brown* DNA in this figure is shown only to help with the interpretation of the autoradiograph; the unlabeled DNA is invisible in such experiments. (B) This experiment was the same except that a further incubation in unlabeled medium allowed additional DNA, with a lower level of radioactivity, to be replicated. The pairs of dark tracks in (B) were found to have silver grains tapering off in opposite directions, demonstrating bidirectional fork movement from a central replication origin where a replication bubble forms (see Figure 5–23). A replication fork is thought to stop only when it encounters a replication fork moving in the opposite direction or when it reaches the end of the chromosome; in this way, all the DNA is eventually replicated.

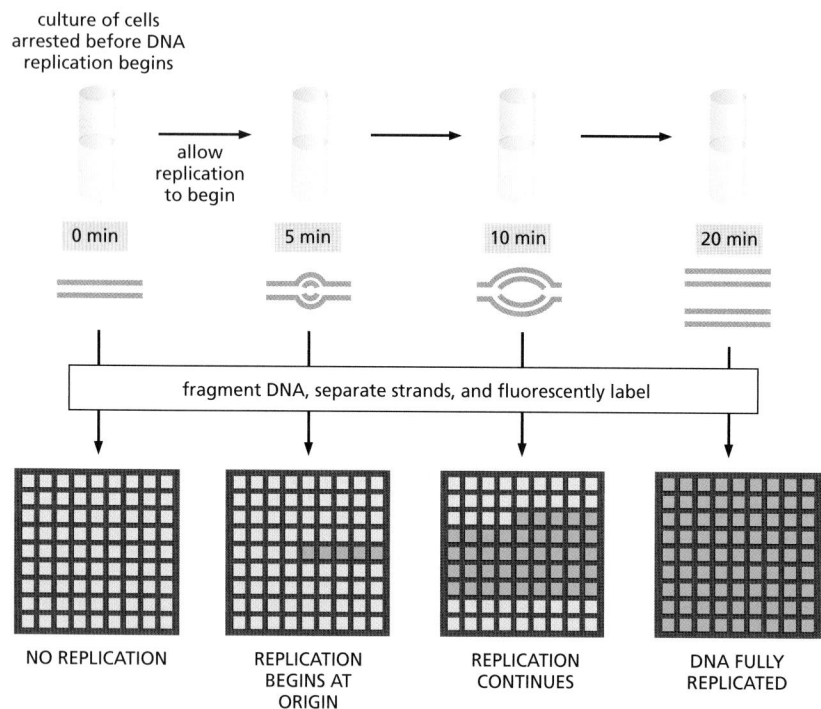

culture of cells
arrested before DNA
replication begins

allow
replication
to begin

0 min   5 min   10 min   20 min

fragment DNA, separate strands, and fluorescently label

NO REPLICATION | REPLICATION BEGINS AT ORIGIN | REPLICATION CONTINUES | DNA FULLY REPLICATED

Figure 5–28 Use of DNA microarrays to monitor the formation and progress of replication forks. For this experiment, a population of cells is synchronized so that they all begin replication at the same time. DNA is collected and hybridized to the microarray; DNA that has been replicated once gives a hybridization signal *(dark green squares)* twice as high as that of unreplicated DNA *(light green squares)*. The spots on these microarrays represent consecutive sequences along a segment of a chromosome arranged left to right, top to bottom. Only 81 spots are shown here, but the actual arrays contain hundreds of thousands of sequences that span an entire genome. As can be seen, replication begins at an origin and proceeds bidirectionally. For simplicity, only one origin is shown here. In human cells, replication begins at 30,000–50,000 origins located throughout the genome. Using this approach it is possible to observe the formation and progress of every replication fork across a genome.

genes are being expressed. The excess origins also provide "backups" in case a primary origin fails. (3) As in bacteria, replication forks are formed in pairs and create a replication bubble as they move in opposite directions away from a common point of origin, stopping only when they collide head-on with a replication fork moving in the opposite direction or when they reach a chromosome end. In this way, many replication forks operate independently on each chromosome and yet form two complete daughter DNA helices.

In Eukaryotes, DNA Replication Takes Place During Only One Part of the Cell Cycle

When growing rapidly, bacteria replicate their DNA nearly continuously. In contrast, DNA replication in most eukaryotic cells occurs only during a specific part of the cell-division cycle, called the *DNA synthesis phase* or **S phase** (Figure 5–29). In a mammalian cell, the S phase typically lasts for about 8 hours; in simpler eukaryotic cells such as yeasts, the S phase can be as short as 40 minutes. By its end, each chromosome has been replicated to produce two complete copies, which remain joined together at their centromeres until the *M phase* (M for *mitosis),* which soon follows. In Chapter 17, we describe the control system that runs the cell cycle, and we explain why entry into each phase of the cycle requires the cell to have successfully completed the previous phase.

In the following sections, we explore how chromosome replication is coordinated within the S phase of the cell cycle.

Different Regions on the Same Chromosome Replicate at Distinct Times in S Phase

In mammalian cells, the replication of DNA in the region between one replication origin and the next should normally require only about an hour to complete, given the rate at which a replication fork moves and the largest distances measured between replication origins. Yet S phase usually lasts for about 8 hours in a mammalian cell. This implies that the replication origins are not all activated simultaneously; indeed, replication origins are activated in clusters of about 50 adjacent replication origins, each of which is replicated during only a small part of the total S-phase interval.

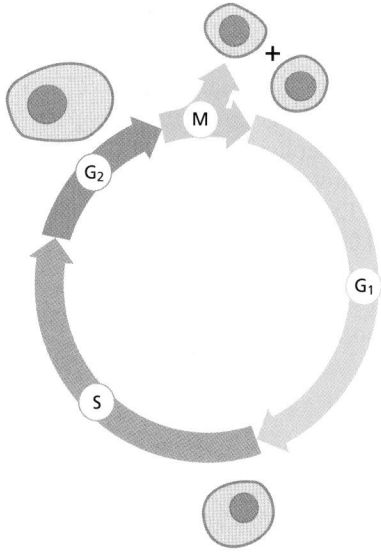

Figure 5–29 The four successive phases of a standard eukaryotic cell cycle. During the G_1, S, and G_2 phases, the cell grows continuously. During M phase growth stops, the nucleus divides, and the cell divides in two. DNA replication is confined to the part of the cell cycle known as S phase. G_1 is the gap between M phase and S phase; G_2 is the gap between S phase and M phase.

It seems that the order in which replication origins are activated depends, in part, on the chromatin structure in which the origins reside. We saw in Chapter 4 that heterochromatin is a particularly condensed state of chromatin, while euchromatin, where most transcription occurs, has a less condensed conformation. Heterochromatin tends to be replicated very late in S phase, suggesting that the timing of replication is related to the packing of the DNA in chromatin.

Once initiated, however, replication forks seem to move at comparable rates throughout S phase, so the extent of chromosome condensation seems to influence the time at which replication forks are initiated, rather than their speed once formed.

A Large Multisubunit Complex Binds to Eukaryotic Origins of Replication

Having seen that a eukaryotic chromosome is replicated using many origins of replication, each of which "fires" at a characteristic time in S phase of the cell cycle, we turn to the nature of these origins of replication. We saw earlier in this chapter that replication origins have been precisely defined in bacteria as specific DNA sequences that attract initiator proteins, which then assemble the DNA replication machinery. We shall see that this is the case for the single-cell budding yeast S. cerevisiae, but it appears not to be strictly true for most other eukaryotes.

For budding yeast, the location of every origin of replication on each chromosome has been determined. The particular chromosome shown in **Figure 5–30**—chromosome III from *S. cerevisiae*—is one of the smallest chromosomes known, with a length less than 1/100 that of a typical human chromosome. Its major origins are spaced an average of 30,000 nucleotide pairs apart, but only a subset of these origins is used by a given cell. Nonetheless, this chromosome can be replicated in about 15 minutes.

The minimal DNA sequence required for directing DNA replication initiation in *S. cerevisiae* has been determined by taking a segment of DNA that spans an origin of replication and testing smaller and smaller DNA fragments for their ability to function as origins. Most DNA sequences that can serve as an origin of replication are found to contain (1) a binding site for a large, multisubunit initiator protein called **ORC**, for **origin recognition complex**; (2) a stretch of DNA that is rich in As and Ts and therefore easy to melt; and (3) at least one binding site for proteins that facilitate ORC binding, probably by adjusting chromatin structure.

In bacteria, once the initiator protein is properly bound to the single origin of replication, the assembly of the replication forks seems to follow more or less automatically. In eukaryotes, the situation is significantly different because of a profound problem eukaryotes have in replicating chromosomes: with so many places to begin replication, how is the process regulated to ensure that all the DNA is copied once and only once?

The answer lies in the sequential manner in which the replicative helicase is first loaded onto origins and is then activated to initiate DNA replication. This matter is discussed in detail in Chapter 17, where we consider the machinery that underlies the cell-division cycle. In brief, during G_1 phase, the replicative helicases are loaded onto DNA next to ORC to create a *prereplicative complex*. Then, upon passage from G_1 phase to S phase, specialized protein kinases come into play to activate the helicases. The resulting opening of the double helix allows the loading of the remaining replication proteins, including the DNA polymerases.

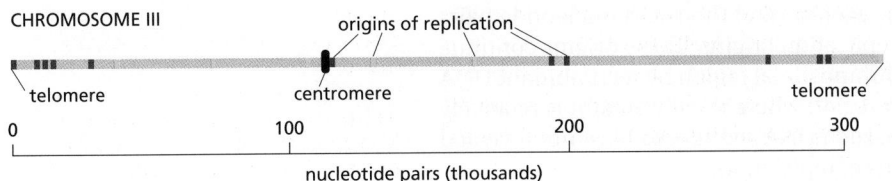

CHROMOSOME III

origins of replication

telomere   centromere   telomere

0   100   200   300

nucleotide pairs (thousands)

Figure 5–30 **The origins of DNA replication on chromosome III of the yeast *S. cerevisiae*.** This chromosome, one of the smallest eukaryotic chromosomes known, carries a total of 180 genes. As indicated, it contains 18 replication origins, although they are used with different frequencies. Those in *red* are typically used in less than 10% of cell divisions, while those in *green* are used about 90% of the time.

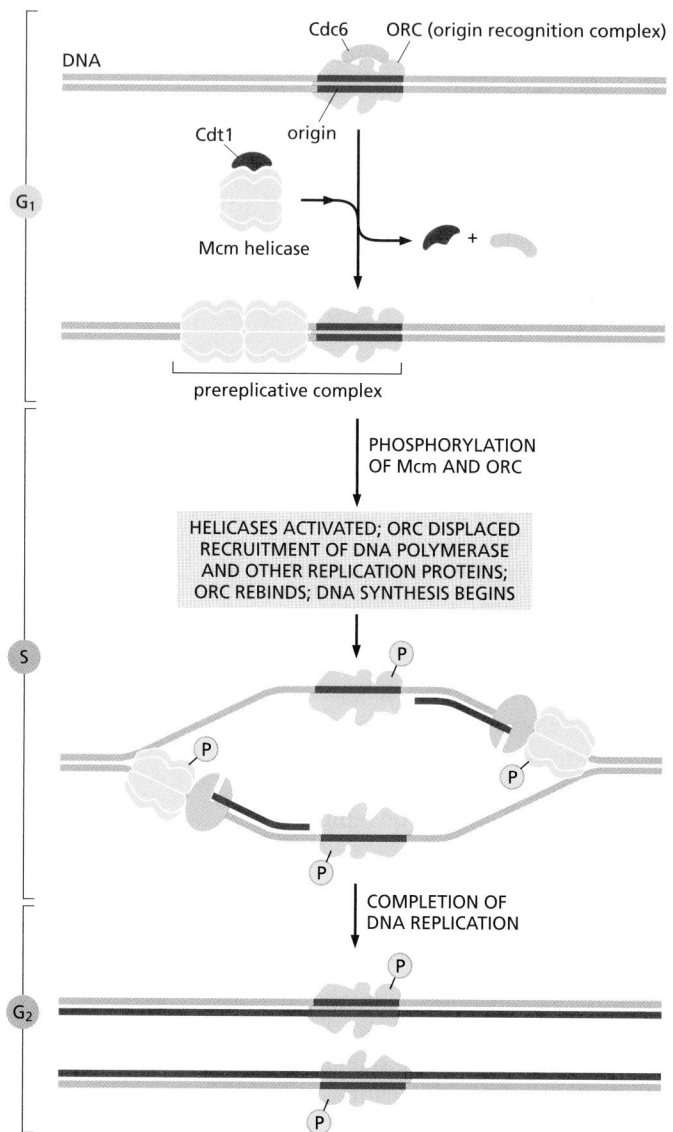

Figure 5–31 **DNA replication initiation in eukaryotes.** This mechanism ensures that each origin of replication is activated only once per cell cycle. An origin of replication can be used only if a prereplicative complex forms there in G_1 phase. At the beginning of S phase, specialized kinases phosphorylate Mcm and ORC, activating the former and inactivating the latter. A new prereplicative complex cannot form at the origin until the cell progresses to the next G_1 phase, when the bound ORC has been dephosphorylated. Note that the eukaryotic Mcm helicase moves along the leading-strand template, whereas the bacterial helicase moves along the lagging-strand template (see Figure 5–25). As the forks begin to move, ORC is displaced, and new ORCs rapidly bind to the newly replicated origins.

The protein kinases that trigger DNA replication simultaneously prevent assembly of new prereplicative complexes until the next M phase resets the entire cycle (for details, see pp. 974–975). They do this, in part, by phosphorylating ORC, rendering it unable to accept new helicases. This strategy provides a single window of opportunity for prereplicative complexes to form (G_1 phase, when kinase activity is low) and a second window for them to be activated and subsequently disassembled (S phase, when kinase activity is high). Because these two phases of the cell cycle are mutually exclusive and occur in a prescribed order, each origin of replication can fire once and only once during each cell cycle (**Figure 5–31**).

Features of the Human Genome That Specify Origins of Replication Remain to Be Discovered

Compared with the situation in budding yeast, the determinants of replication origins in other eukaryotes have been difficult to discover. It has been possible to identify specific human DNA sequences, each several thousand nucleotide pairs in length, that are sufficient to serve as replication origins. These origins continue to function when moved to a different chromosomal region by recombinant DNA methods, as long as they are placed in a region where the chromatin is relatively uncondensed. However, comparisons of such DNA sequences have not revealed specific DNA sequences that mark origins of replication.

Despite this, a human ORC that is very similar to the yeast ORC binds to origins of replication and initiates DNA replication in humans. Many of the other proteins that function in the initiation process in yeast likewise have central roles in humans. It therefore seems likely that the yeast and human initiation mechanisms are similar in outline, but chromatin structure, transcriptional activity, or some property of the genome other than a specific DNA sequence has the central role in attracting ORC and specifying mammalian origins of replication. These ideas could also help to explain how a given mammalian cell chooses which of the many possible origins to use when it replicates its genome and how this choice could differ from cell to cell. Clearly, we have a great deal to discover about the fundamental process of DNA replication initiation.

New Nucleosomes Are Assembled Behind the Replication Fork

Several additional aspects of DNA replication are specific to eukaryotes. As discussed in Chapter 4, eukaryotic chromosomes are composed of roughly equal mixtures of DNA and protein. Chromosome duplication therefore requires not only the replication of DNA, but also the synthesis and assembly of new chromosomal proteins onto the DNA behind each replication fork. Although we are far from understanding this process in detail, we are beginning to learn how the fundamental unit of chromatin packaging, the nucleosome, is duplicated. The cell requires a large amount of new histone protein, approximately equal in mass to the newly synthesized DNA, to make the new nucleosomes in each cell cycle. For this reason, most eukaryotic organisms possess multiple copies of the gene for each histone. Vertebrate cells, for example, have about 20 repeated gene sets, most containing the genes that encode all five histones (H1, H2A, H2B, H3, and H4).

Unlike most proteins, which are made continuously, histones are synthesized mainly in S phase, when the level of histone mRNA increases about fiftyfold as a result of both increased transcription and decreased mRNA degradation. The major histone mRNAs are degraded within minutes when DNA synthesis stops at the end of S phase. The mechanism depends on special properties of the 3′ ends of these mRNAs, as discussed in Chapter 7. In contrast, the histone proteins themselves are remarkably stable and may survive for the entire life of a cell. The tight linkage between DNA synthesis and histone synthesis appears to reflect a feedback mechanism that monitors the level of free histone to ensure that the amount of histone made exactly matches the amount of new DNA synthesized.

As a replication fork advances, it must pass through the parental nucleosomes. In the cell, efficient replication requires chromatin remodeling complexes (discussed in Chapter 4) to destabilize the DNA–histone interfaces. Aided by such complexes, replication forks can transit even highly condensed chromatin efficiently.

As a replication fork passes through chromatin, the histones are transiently displaced leaving about 600 nucleotide pairs of non-nucleosomal DNA in its wake. The reestablishment of nucleosomes behind a moving fork occurs in an intriguing way. When a nucleosome is traversed by a replication fork, the histone octamer appears to be broken into an H3-H4 tetramer and two H2A-H2B dimers (discussed in Chapter 4). The H3-H4 tetramer remains loosely associated with DNA and is distributed at random to one or the other daughter duplex, but the H2A-H2B dimers are released completely from DNA. Freshly made H3-H4 tetramers are added to the newly synthesized DNA to fill in the "spaces," and H2A-H2B dimers—half of which are old and half new—are then added at random to complete the nucleosomes (**Figure 5–32**). The formation of new nucleosomes behind a replication fork has an important consequence for the process of DNA replication itself. As DNA polymerase δ discontinuously synthesizes the lagging strand (see pp. 253–254), the length of each Okazaki fragment is determined by the point at which DNA polymerase δ is blocked by a newly formed nucleosome. This tight coupling between nucleosome duplication and DNA replication explains why the length of Okazaki fragments in eukaryotes (~200 nucleotides) is approximately the same as the nucleosome repeat length.

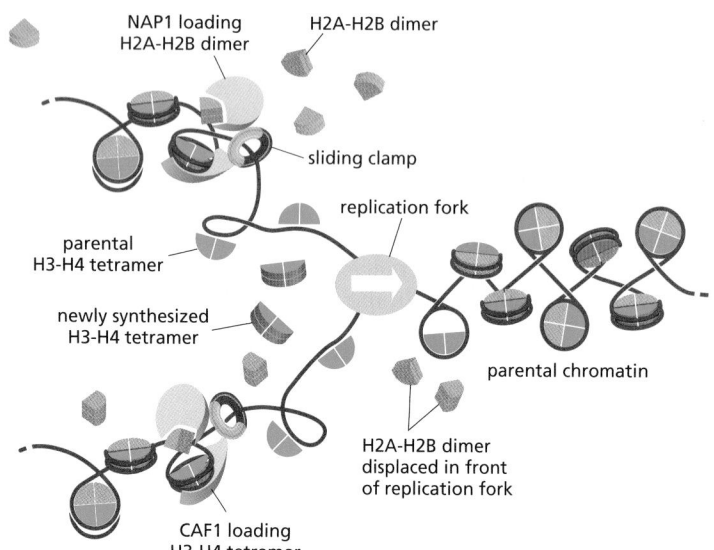

Figure 5–32 Formation of nucleosomes behind a replication fork. Parental H3-H4 tetramers are distributed at random to the daughter DNA molecules, with roughly equal numbers inherited by each daughter. In contrast, H2A-H2B dimers are released from the DNA as the replication fork passes. This release begins just in front of the replication fork and is facilitated by chromatin remodeling complexes that move with the fork. Histone chaperones (NAP1 and CAF1) restore the full complement of histones to daughter molecules using both parental and newly synthesized histones. Although some daughter nucleosomes contain only parental histones or only newly synthesized histones, most are hybrids of old and new. For simplicity, the DNA double helix shown as a single *red* line. (Adapted from J.D. Watson et al., Molecular Biology of the Gene, 5th ed. Cold Spring Harbor: Cold Spring Harbor Laboratory Press, 2004.)

The orderly and rapid addition of new H3-H4 tetramers and H2A-H2B dimers behind a replication fork requires **histone chaperones** (also called *chromatin assembly factors*). These multisubunit complexes bind the highly basic histones and release them for assembly only in the appropriate context. The histone chaperones, along with their cargoes, are directed to newly replicated DNA through a specific interaction with the eukaryotic sliding clamp called *PCNA* (see Figure 5–32). These clamps are left behind moving replication forks and remain on the DNA long enough for the histone chaperones to complete their tasks.

Telomerase Replicates the Ends of Chromosomes

We saw earlier that synthesis of the lagging strand at a replication fork must occur discontinuously through a backstitching mechanism that produces short DNA fragments. This mechanism encounters a special problem when the replication fork reaches an end of a linear chromosome. The final RNA primer synthesized on the lagging-strand template cannot be replaced by DNA because there is no 3′-OH end available for the repair polymerase. Without a mechanism to deal with this problem, DNA would be lost from the ends of all chromosomes each time a cell divides.

Bacteria solve this "end-replication" problem by having circular DNA molecules as chromosomes (see Figure 5–24). Eukaryotes solve it in a different way: they have specialized nucleotide sequences at the ends of their chromosomes that are incorporated into structures called *telomeres* (discussed in Chapter 4). Telomeres contain many tandem repeats of a short sequence that is similar in organisms as diverse as protozoa, fungi, plants, and mammals. In humans, the sequence of the repeat unit is GGGTTA, and it is repeated roughly a thousand times at each telomere.

Telomere DNA sequences are recognized by sequence-specific DNA-binding proteins that attract an enzyme, called **telomerase**, that replenishes these sequences each time a cell divides. Telomerase recognizes the tip of an existing telomere DNA repeat sequence and elongates it in the 5′-to-3′ direction, using an RNA template that is a component of the enzyme itself to synthesize new copies of the repeat (**Figure 5–33**). The enzymatic portion of telomerase resembles other *reverse transcriptases*, proteins that synthesize DNA using an RNA template, although, in this case, the telomerase RNA also contributes functional groups to make the catalysis more efficient. After extension of the parental DNA strand by telomerase, replication of the lagging strand at the chromosome end can be completed by the conventional DNA polymerases, using these extensions as a template to synthesize the complementary strand (**Figure 5–34**).

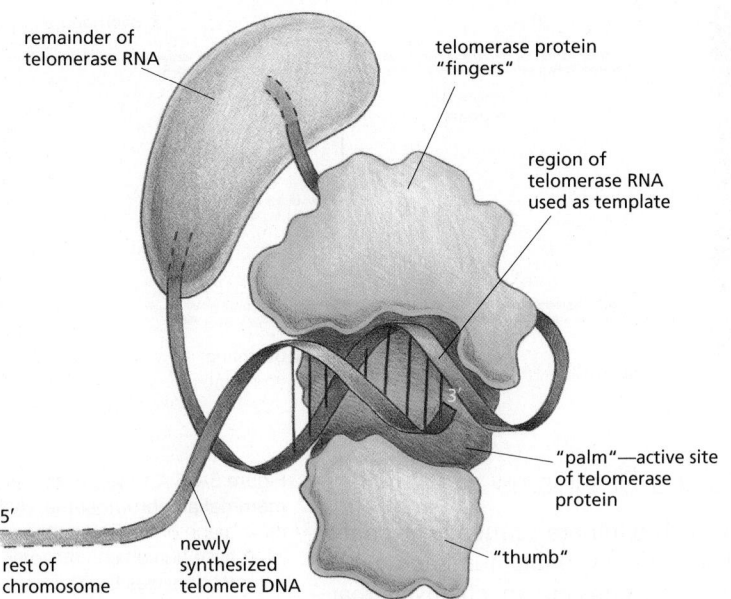

Figure 5–33 Structure of a portion of telomerase. Telomerase is a large protein–RNA complex. The RNA *(blue)* contains a templating sequence for synthesizing new DNA telomere repeats. The synthesis reaction itself is carried out by the reverse transcriptase domain of the protein, shown in *green*. A reverse transcriptase is a special form of polymerase enzyme that uses an RNA template to make a DNA strand; telomerase is unique in carrying its own RNA template with it. Telomerase also has several additional protein domains (not shown) that are needed to assemble the enzyme at the ends of chromosomes. (Modified from J. Lingner and T.R. Cech, *Curr. Opin. Genet. Dev.* 8:226–232, 1998.)

Telomeres Are Packaged Into Specialized Structures That Protect the Ends of Chromosomes

The ends of chromosomes present cells with an additional problem. As we will see in the next part of this chapter, when a chromosome is accidently broken, the break is rapidly repaired (see Figure 5–45). Telomeres must clearly be distinguished from these accidental breaks; otherwise the cell will attempt to "repair" telomeres, causing chromosome fusions and other genetic abnormalities. Telomeres have several features to prevent this from happening.

A specialized nuclease chews back the 5′ end of a telomere leaving a protruding single-strand end. This protruding end—in combination with the GGGTTA repeats in telomeres—attracts a group of proteins that form a protective chromosome cap known as *shelterin*. In particular, shelterin "hides" telomeres from the cell's damage detectors that continually monitor DNA. When human telomeres are artificially cross-linked and viewed by electron microscopy, structures known as "t-loops" are observed in which the protruding end of the telomere loops back and tucks itself into the duplex DNA of the telomere repeat sequence (**Figure 5–35**). It is believed that t-loops are regulated by shelterin and provide additional protection for the ends of chromosomes.

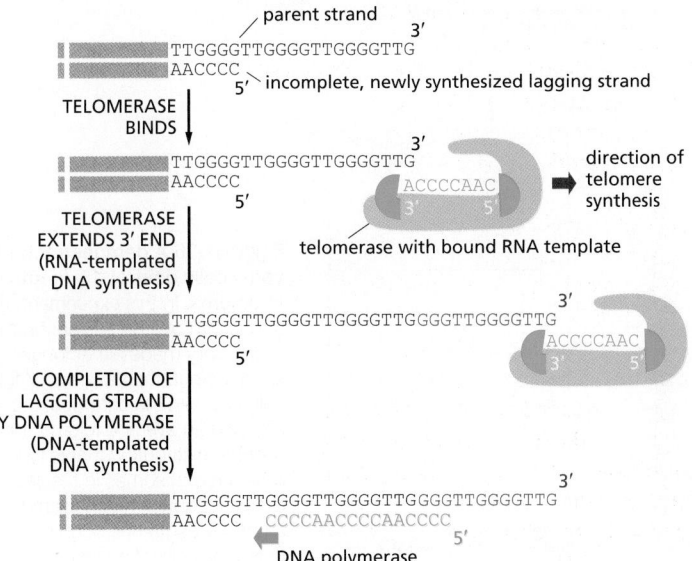

Figure 5–34 Telomere replication. Shown here are the reactions that synthesize the repeating sequences that form the ends of the chromosomes (telomeres) of diverse eukaryotic organisms. The 3′ end of the parental DNA strand is extended by RNA-templated DNA synthesis; this allows the incomplete daughter DNA strand that is paired with it to be extended in its 5′ direction. This incomplete, lagging strand is presumed to be completed by DNA polymerase α, which carries a DNA primase as one of its subunits (**Movie 5.6**). The telomere sequence illustrated is that of the ciliate *Tetrahymena,* in which these reactions were first discovered.

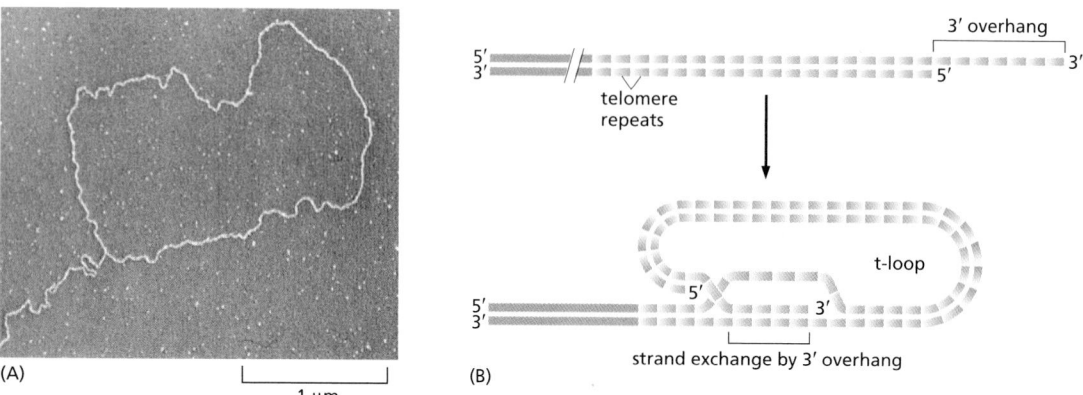

(A)

1 μm

(B)

Telomere Length Is Regulated by Cells and Organisms

Because the processes that grow and shrink each telomere sequence are only approximately balanced, a chromosome end contains a variable number of telomeric repeats. Not surprisingly, many cells have homeostatic mechanisms that maintain the number of these repeats within a limited range (**Figure 5–36**).

In most of the dividing somatic cells of humans, however, telomeres gradually shorten, and it has been proposed that this provides a counting mechanism that helps prevent the unlimited proliferation of wayward cells in adult tissues. In its simplest form, this idea holds that our somatic cells start off in the embryo with a full complement of telomeric repeats. These are then eroded to different extents in different cell types. Some stem cells, notably those in tissues that must be replenished at a high rate throughout life—bone marrow or gut lining, for example—retain full telomerase activity. However, in many other types of cells, the level of telomerase is turned down so that the enzyme cannot quite keep up with chromosome duplication. Such cells lose 100–200 nucleotides from each telomere every time they divide. After many cell generations, the descendant cells will inherit chromosomes that lack telomere function, and, as a result of this defect, activate a DNA-damage response causing them to withdraw permanently from the cell cycle and cease dividing—a process called *replicative cell senescence* (discussed in Chapter 17). In theory, such a mechanism could provide a safeguard against the uncontrolled cell proliferation of abnormal cells in somatic tissues, thereby helping to protect us from cancer.

Figure 5–35 A t-loop at the end of a mammalian chromosome. (A) Electron micrograph of the DNA at the end of an interphase human chromosome. The chromosome was fixed, deproteinated, and artificially thickened before viewing. The loop seen here is approximately 15,000 nucleotide pairs in length. (B) Structure of a t-loop. The insertion of the single-strand 3′ end into the duplex repeats is carried out, and the structure maintained, by specialized proteins. (From J.D. Griffith et al., *Cell* 97:503–514, 1999. With permission from Elsevier.)

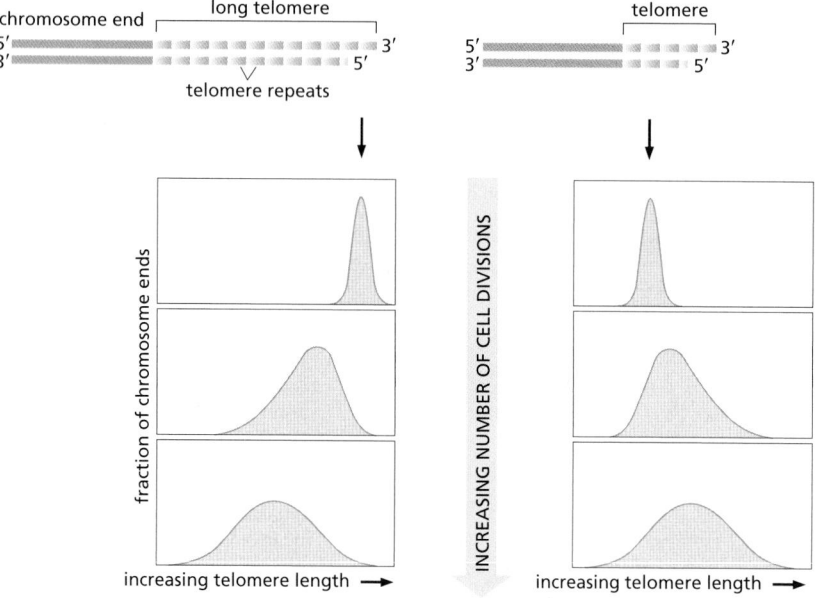

Figure 5–36 A demonstration that yeast cells control the length of their telomeres. In this experiment, the telomere at one end of a particular chromosome is artificially made either longer *(left)* or shorter *(right)* than average. After many cell divisions, the chromosome recovers, showing an average telomere length and a length distribution that is typical of the other chromosomes in the yeast cell. A similar feedback mechanism for controlling telomere length has been proposed for the germ-line cells of animals.

The idea that telomere length acts as a "measuring stick" to count cell divisions and thereby regulate the lifetime of the cell lineage has been tested in several ways. For certain types of human cells grown in tissue culture, the experimental results support such a theory. Human fibroblasts normally proliferate for about 60 cell divisions in culture before undergoing replicative cell senescence. Like most other somatic cells in humans, fibroblasts produce only low levels of telomerase, and their telomeres gradually shorten each time they divide. When telomerase is provided to the fibroblasts by inserting an active telomerase gene, telomere length is maintained and many of the cells now continue to proliferate indefinitely.

It has been proposed that this type of control on cell proliferation may contribute to the aging of animals like ourselves. These ideas have been tested by producing transgenic mice that lack telomerase entirely. The telomeres in mouse chromosomes are about five times longer than human telomeres, and the mice must therefore be bred through three or more generations before their telomeres have shrunk to the normal human length. It is therefore perhaps not surprising that the first generations of mice develop normally. However, the mice in later generations develop progressively more defects in some of their highly proliferative tissues. In addition, these mice show signs of premature aging and have a pronounced tendency to develop tumors. In these and other respects these mice resemble humans with the genetic disease *dyskeratosis congenita*. Individuals afflicted with this disease carry one functional and one nonfunctional copy of the telomerase RNA gene; they have prematurely shortened telomeres and typically die of progressive bone marrow failure. They also develop lung scarring and liver cirrhosis and show abnormalities in various epidermal structures including skin, hair follicles, and nails.

The above observations demonstrate that controlling cell proliferation by telomere shortening poses a risk to an organism, because not all of the cells that begin losing the ends of their chromosomes will stop dividing. Some apparently become genetically unstable, but continue to divide, giving rise to variant cells that can lead to cancer. Clearly, the use of telomere shortening as a regulating mechanism is not foolproof and, like many mechanisms in the cell, seems to strike a balance between benefit and risk.

Summary

The proteins that initiate DNA replication bind to DNA sequences at a replication origin to catalyze the formation of a replication bubble with two outward-moving replication forks. The process begins when an initiator protein–DNA complex is formed that subsequently loads a DNA helicase onto the DNA template. Other proteins are then added to form the multienzyme "replication machine" that catalyzes DNA synthesis at each replication fork.

In bacteria and some simple eukaryotes, replication origins are specified by specific DNA sequences that are only several hundred nucleotide pairs long. In other eukaryotes, such as humans, the sequences needed to specify an origin of DNA replication seem to be less well defined, and the origin can span several thousand nucleotide pairs.

Bacteria typically have a single origin of replication in a circular chromosome. With fork speeds of up to 1000 nucleotides per second, they can replicate their genome in less than an hour. Eukaryotic DNA replication takes place in only one part of the cell cycle, the S phase. The replication fork in eukaryotes moves about 10 times more slowly than the bacterial replication fork, and the much longer eukaryotic chromosomes each require many replication origins to complete their replication in an S phase, which typically lasts for 8 hours in human cells. The different replication origins in these eukaryotic chromosomes are activated in a sequence, determined in part by the structure of the chromatin, with the most condensed regions of chromatin typically beginning their replication last. After the replication fork has passed, chromatin structure is re-formed by the addition of new histones to the old histones that are directly inherited by each daughter DNA molecule.

Eukaryotes solve the problem of replicating the ends of their linear chromosomes with a specialized end structure, the telomere, maintained by a special nucleotide polymerizing enzyme called telomerase. Telomerase extends one of the DNA strands at the end of a chromosome by using an RNA template that is an integral part of the enzyme itself, producing a highly repeated DNA sequence that typically extends for thousands of nucleotide pairs at each chromosome end. Telomeres have specialized structures that distinguish them from broken ends of chromosomes, ensuring that they are not mistakenly repaired.

DNA REPAIR

Maintaining the genetic stability that an organism needs for its survival requires not only an extremely accurate mechanism for replicating DNA, but also mechanisms for repairing the many accidental lesions that DNA continually suffers. Most such spontaneous changes in DNA are temporary because they are immediately corrected by a set of processes that are collectively called **DNA repair**. Of the tens of thousands of random changes created every day in the DNA of a human cell by heat, metabolic accidents, radiation of various sorts, and exposure to substances in the environment, only a few (less than 0.02%) accumulate as permanent mutations in the DNA sequence. The rest are eliminated with remarkable efficiency by DNA repair.

The importance of DNA repair is evident from the large investment that cells make in the enzymes that carry it out: several percent of the coding capacity of most genomes is devoted solely to DNA repair functions. The importance of DNA repair is also demonstrated by the increased rate of mutation that follows the inactivation of a DNA repair gene. Many DNA repair proteins and the genes that encode them—which we now know operate in a wide range of organisms, including humans—were originally identified in bacteria by the isolation and characterization of mutants that displayed an increased mutation rate or an increased sensitivity to DNA-damaging agents.

Recent studies of the consequences of a diminished capacity for DNA repair in humans have linked many human diseases with decreased repair (Table 5–2). Thus, we saw previously that defects in a human gene whose product normally functions to repair the mismatched base pairs resulting from DNA replication errors can lead to an inherited predisposition to cancers of the colon and some other organs, reflecting an increased mutation rate. In another human disease,

TABLE 5–2 Some Inherited Human Syndromes with Defects in DNA Repair

Name	Phenotype	Enzyme or process affected
MSH2, 3, 6, MLH1, PMS2	Colon cancer	Mismatch repair
Xeroderma pigmentosum (XP) groups A–G	Skin cancer, UV sensitivity, neurological abnormalities	Nucleotide excision repair
Cockayne syndrome	UV sensitivity; developmental abnormalities	Coupling of nucleotide excision repair to transcription
XP variant	UV sensitivity, skin cancer	Translesion synthesis by DNA polymerase ν
Ataxia telangiectasia (AT)	Leukemia, lymphoma, γ-ray sensitivity, genome instability	ATM protein, a protein kinase activated by double-strand breaks
BRCA1	Breast and ovarian cancer	Repair by homologous recombination
BRCA2	Breast, ovarian, and prostate cancer	Repair by homologous recombination
Werner syndrome	Premature aging, cancer at several sites, genome instability	Accessory 3′-exonuclease and DNA helicase used in repair
Bloom syndrome	Cancer at several sites, stunted growth, genome instability	DNA helicase needed for recombination
Fanconi anemia groups A–G	Congenital abnormalities, leukemia, genome instability	DNA interstrand cross-link repair
46 BR patient	Hypersensitivity to DNA-damaging agents, genome instability	DNA ligase I

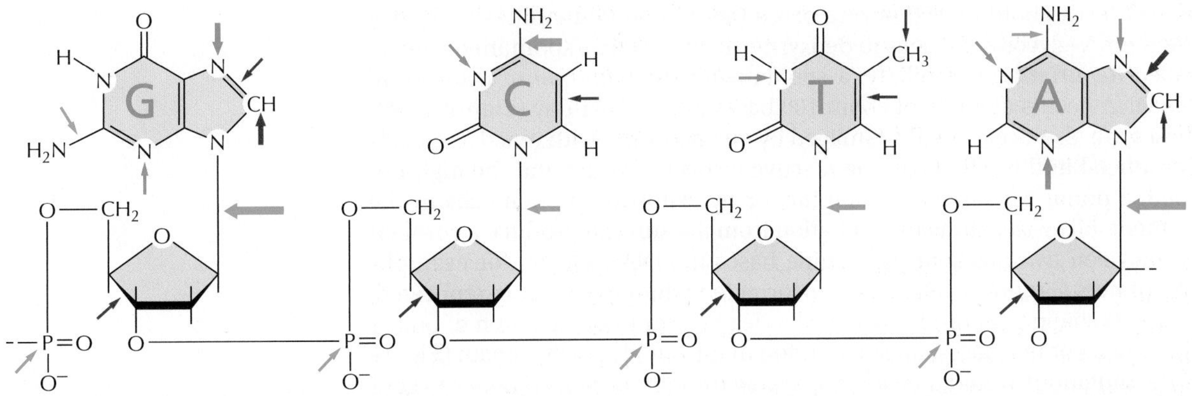

Figure 5–37 **A summary of spontaneous alterations that require DNA repair.** The sites on each nucleotide modified by spontaneous oxidative damage *(red arrows)*, hydrolytic attack *(blue arrows)*, and methylation *(green arrows)* are shown, with the width of each arrow indicating the relative frequency of each event (see Table 5–3). (After T. Lindahl, *Nature* 362:709–715, 1993.)

xeroderma pigmentosum (XP), the afflicted individuals have an extreme sensitivity to ultraviolet radiation because they are unable to repair certain DNA photoproducts. This repair defect results in an increased mutation rate that leads to serious skin lesions and an increased susceptibility to skin cancers. Finally, mutations in the *Brca1* and *Brca2* genes compromise a type of DNA repair known as *homologous recombination* and are a cause of hereditary breast and ovarian cancer.

Without DNA Repair, Spontaneous DNA Damage Would Rapidly Change DNA Sequences

Although DNA is a highly stable material—as required for the storage of genetic information—it is a complex organic molecule that is susceptible, even under normal cell conditions, to spontaneous changes that would lead to mutations if left unrepaired (**Figure 5–37** and see **Table 5–3**). For example, the DNA of each

TABLE 5–3 **Endogenous DNA Lesions Arising and Repaired in a Diploid Mammalian Cell in 24 Hours**

DNA lesion	Number repaired in 24 h
Hydrolysis	
Depurination	18,000
Depyrimidination	600
Cytosine deamination	100
5-Methylcytosine deamination	10
Oxidation	
8-oxo G	1500
Ring-saturated pyrimidines (thymine glycol, cytosine hydrates)	2000
Lipid peroxidation products (M1G, etheno-A, etheno-C)	1000
Nonenzymatic methylation by *S*-adenosylmethionine	
7-Methylguanine	6000
3-Methyladenine	1200
Nonenzymatic methylation by nitrosated polyamines and peptides	
O^6-Methylguanine	20–100

The DNA lesions listed in the table are the result of the normal chemical reactions that take place in cells. Cells that are exposed to external chemicals and radiation suffer greater and more diverse forms of DNA damage. (From T. Lindahl and D.E. Barnes, *Cold Spring Harb. Symp. Quant. Biol.* 65:127–133, 2000.)

human cell loses about 18,000 purine bases (adenine and guanine) every day because their *N*-glycosyl linkages to deoxyribose hydrolyze, a spontaneous reaction called *depurination.* Similarly, a spontaneous *deamination* of cytosine to uracil in DNA occurs at a rate of about 100 bases per cell per day (**Figure 5–38**). DNA bases are also occasionally damaged by an encounter with reactive metabolites produced in the cell, including reactive forms of oxygen and the high-energy methyl donor *S*-adenosylmethionine, or by exposure to chemicals in the environment. Likewise, ultraviolet radiation from the sun can produce a covalent linkage between two adjacent pyrimidine bases in DNA to form, for example, thymine dimers (**Figure 5–39**). If left uncorrected when the DNA is replicated, most of these changes would be expected to lead either to the deletion of one or more base pairs or to a base-pair substitution in the daughter DNA chain (**Figure 5–40**). The mutations would then be propagated throughout subsequent cell generations. Such a high rate of random changes in the DNA sequence would have disastrous consequences.

The DNA Double Helix Is Readily Repaired

The double-helical structure of DNA is ideally suited for repair because it carries two separate copies of all the genetic information—one in each of its two strands. Thus, when one strand is damaged, the complementary strand retains an intact copy of the same information, and this copy is generally used to restore the correct nucleotide sequences to the damaged strand.

An indication of the importance of a double-strand helix to the safe storage of genetic information is that all cells use it; only a few small viruses use single-strand DNA or RNA as their genetic material. The types of repair processes described in this section cannot operate on such nucleic acids, and once damaged, the chance of a permanent nucleotide change occurring in these single-strand genomes of viruses is thus very high. It seems that only organisms with tiny genomes (and therefore tiny targets for DNA damage) can afford to encode their genetic information in any molecule other than a DNA double helix.

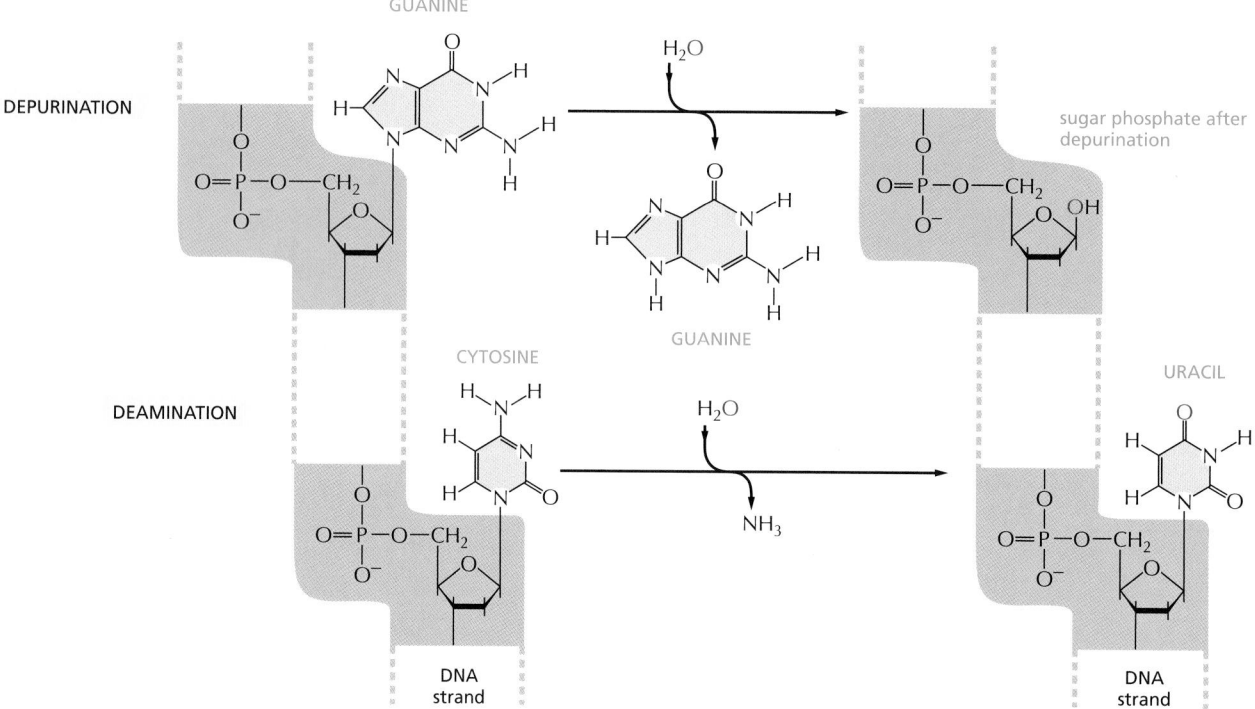

Figure 5–38 Depurination and deamination. These reactions are two of the most frequent spontaneous chemical reactions that create serious DNA damage in cells. Depurination can release guanine (shown here), as well as adenine, from DNA. The major type of deamination reaction converts cytosine to an altered DNA base, uracil (shown here), but deamination occurs on other bases as well. These reactions normally take place in double-helical DNA; for convenience, only one strand is shown.

Figure 5–39 **The most common type of thymine dimer.** This type of damage occurs in the DNA of cells exposed to ultraviolet irradiation (as in sunlight). A similar dimer will form between any two neighboring pyrimidine bases (C or T residues) in DNA.

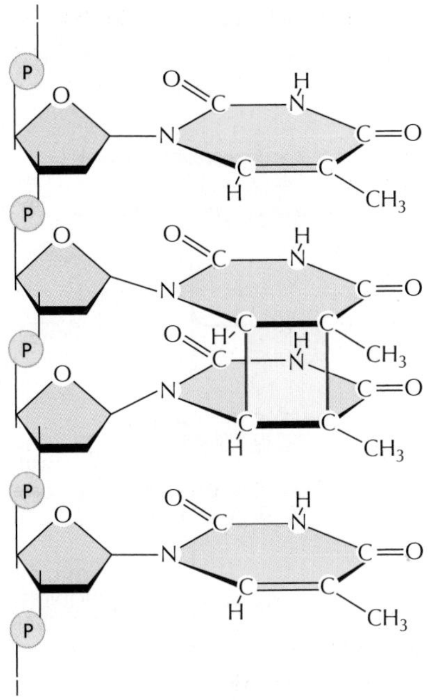

DNA Damage Can Be Removed by More Than One Pathway

Cells have multiple pathways to repair their DNA using different enzymes that act upon different kinds of lesions. **Figure 5–41** shows two of the most common pathways. In both, the damage is excised, the original DNA sequence is restored by a DNA polymerase that uses the undamaged strand as its template, and a remaining break in the double helix is sealed by DNA ligase (see Figure 5–12).

The two pathways differ in the way in which they remove the damage from DNA. The first pathway, called **base excision repair**, involves a battery of enzymes called *DNA glycosylases*, each of which can recognize a specific type of altered base in DNA and catalyze its hydrolytic removal. There are at least six types of these enzymes, including those that remove deaminated Cs, deaminated As, different types of alkylated or oxidized bases, bases with opened rings, and bases in which a carbon–carbon double bond has been accidentally converted to a carbon–carbon single bond. How is an altered base detected within the context of the double helix? A key step is an enzyme-mediated "flipping-out" of the altered nucleotide from the helix, which allows the DNA glycosylase to probe all faces of the base for damage (**Figure 5–42**). It is thought that these enzymes travel along DNA using base-flipping to evaluate the status of each base. Once an enzyme finds the damaged base that it recognizes, it removes that base from its sugar.

The "missing tooth" created by DNA glycosylase action is recognized by an enzyme called *AP endonuclease* (AP for *apurinic* or *apyrimidinic*, *endo* to signify that the nuclease cleaves within the polynucleotide chain), which cuts the phosphodiester backbone, after which the resulting gap is repaired (see Figure 5–41A). Depurination, which is by far the most frequent type of damage suffered by DNA, also leaves a deoxyribose sugar with a missing base. Depurinations are directly repaired beginning with AP endonuclease, following the bottom half of the pathway in Figure 5–41A.

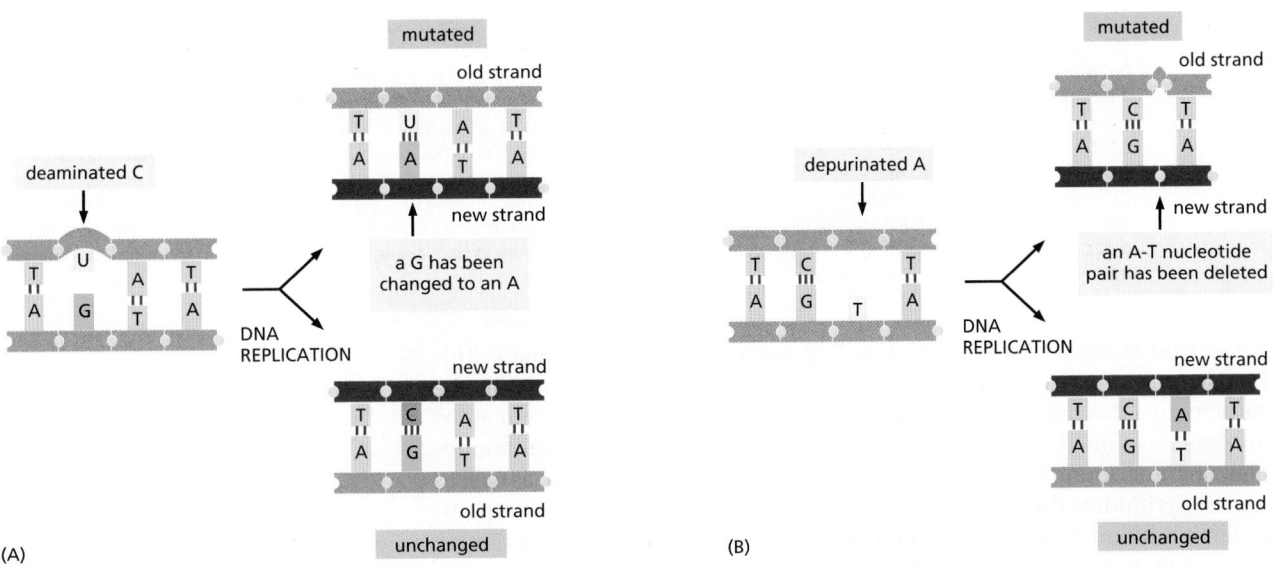

(A) (B)

Figure 5–40 **How chemical modifications of nucleotides produce mutations.** (A) Deamination of cytosine, if uncorrected, results in the substitution of one base for another when the DNA is replicated. As shown in Figure 5–38, deamination of cytosine produces uracil. Uracil differs from cytosine in its base-pairing properties and preferentially base-pairs with adenine. The DNA replication machinery therefore adds an adenine when it encounters a uracil on the template strand. (B) Depurination can lead to the loss of a nucleotide pair. When the replication machinery encounters a missing purine on the template strand, it may skip to the next complete nucleotide as illustrated here, thus producing a nucleotide deletion in the newly synthesized strand. Many other types of DNA damage (see Figure 5–37), if left uncorrected, also produce mutations when the DNA is replicated.

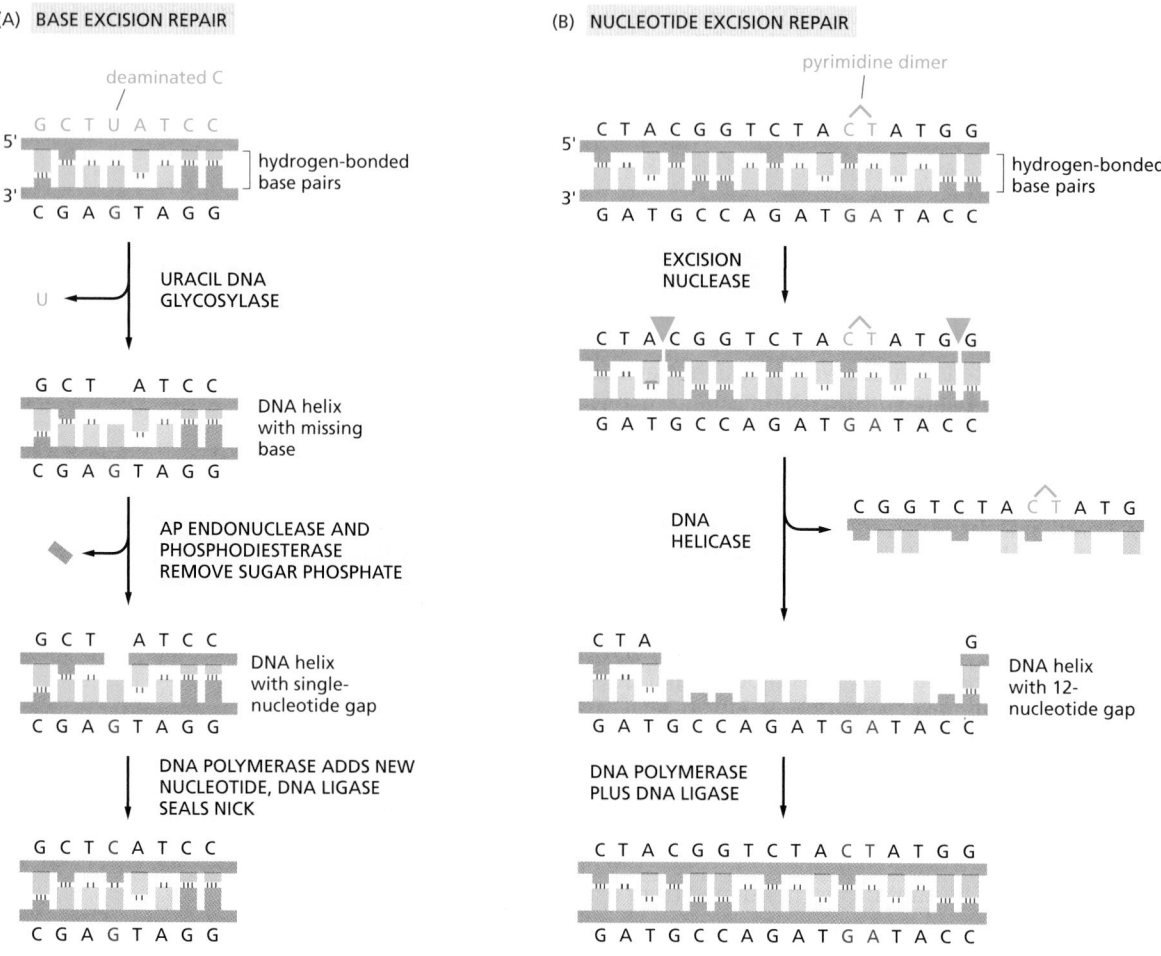

Figure 5–41 A comparison of two major DNA repair pathways. (A) *Base excision repair.* This pathway starts with a DNA glycosylase. Here, the enzyme uracil DNA glycosylase removes an accidentally deaminated cytosine in DNA. After the action of this glycosylase (or another DNA glycosylase that recognizes a different kind of damage), the sugar phosphate with the missing base is cut out by the sequential action of AP endonuclease and a phosphodiesterase. (These same enzymes begin the repair of depurinated sites directly.) The gap of a single nucleotide is then filled by DNA polymerase and DNA ligase. The net result is that the U that was created by accidental deamination is restored to a C. AP endonuclease is so-named because it recognizes any site in the DNA helix that contains a deoxyribose sugar with a missing base; such sites can arise either by the loss of a purine *(ap*urinic sites) or by the loss of a pyrimidine *(ap*yrimidinic sites). (B) *Nucleotide excision repair.* In bacteria, after a multienzyme complex has recognized a lesion such as a pyrimidine dimer (see Figure 5–39), one cut is made on each side of the lesion, and an associated DNA helicase then removes the entire portion of the damaged strand. The excision repair machinery in bacteria leaves the gap of 12 nucleotides shown. In humans, once the damaged DNA is recognized, a helicase is recruited to unwind the DNA duplex locally. Next, the excision nuclease enters and cleaves on either side of the damage, leaving a gap of about 30 nucleotides. The nucleotide excision repair machinery in both bacteria and humans can recognize and repair many different types of DNA damage.

The second major repair pathway is called **nucleotide excision repair**. This mechanism can repair the damage caused by almost any large change in the structure of the DNA double helix. Such "bulky lesions" include those created by the covalent reaction of DNA bases with large hydrocarbons (such as the carcinogen benzopyrene, found in tobacco smoke, coal tar, and diesel exhaust), as well as the various pyrimidine dimers (T-T, T-C, and C-C) caused by sunlight. In this pathway, a large multienzyme complex scans the DNA for a distortion in the double helix, rather than for a specific base change. Once it finds a lesion, it cleaves the phosphodiester backbone of the abnormal strand on both sides of the distortion, and a DNA helicase peels away the single-strand oligonucleotide containing the lesion. The large gap produced in the DNA helix is then repaired by DNA polymerase and DNA ligase (see Figure 5–41B).

An alternative to base and nucleotide excision repair processes is direct chemical reversal of DNA damage, and this strategy is selectively employed for the rapid

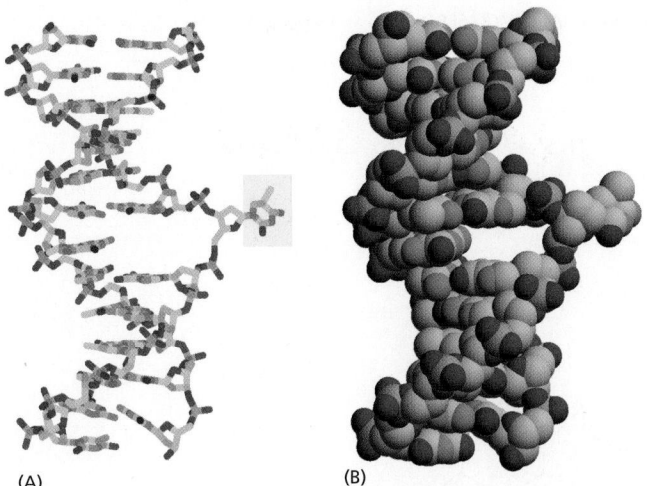

(A)   (B)

Figure 5–42 **The recognition of an unusual nucleotide in DNA by base-flipping.** The DNA glycosylase family of enzymes recognizes specific inappropriate bases in the conformation shown. Each of these enzymes cleaves the glycosyl bond that connects a particular recognized base *(yellow)* to the backbone sugar, removing it from the DNA. (A) Stick model; (B) space-filling model.

removal of certain highly mutagenic or cytotoxic lesions. For example, the alkylation lesion O^6-methylguanine has its methyl group removed by direct transfer to a cysteine residue in the repair protein itself, which is destroyed in the reaction. In another example, methyl groups in the alkylation lesions 1-methyladenine and 3-methylcytosine are "burnt off" by an iron-dependent demethylase, with release of formaldehyde from the methylated DNA and regeneration of the native base.

Coupling Nucleotide Excision Repair to Transcription Ensures That the Cell's Most Important DNA Is Efficiently Repaired

All of a cell's DNA is under constant surveillance for damage, and the repair mechanisms we have described act on all parts of the genome. However, cells have a way of directing DNA repair to the DNA sequences that are most urgently needed. They do this by linking RNA polymerase, the enzyme that transcribes DNA into RNA as the first step in gene expression, to the nucleotide excision repair pathway. As discussed above, this repair system can correct many different types of DNA damage. RNA polymerase stalls at DNA lesions and, through the use of coupling proteins, directs the excision repair machinery to these sites. In bacteria, where genes are relatively short, the stalled RNA polymerase can be dissociated from the DNA; the DNA is repaired, and the gene is transcribed again from the beginning. In eukaryotes, where genes can be enormously long, a more complex reaction is used to "back up" the RNA polymerase, repair the damage, and then restart the polymerase.

The importance of transcription-coupled excision repair is seen in people with Cockayne syndrome, which is caused by a defect in this coupling. These individuals suffer from growth retardation, skeletal abnormalities, progressive neural retardation, and severe sensitivity to sunlight. Most of these problems are thought to arise from RNA polymerase molecules that become permanently stalled at sites of DNA damage that lie in important genes.

The Chemistry of the DNA Bases Facilitates Damage Detection

The DNA double helix seems optimal for repair. As noted above, it contains a backup copy of all genetic information. Equally importantly, the nature of the four bases in DNA makes the distinction between undamaged and damaged bases very clear. For example, every possible deamination event in DNA yields an "unnatural" base, which can be directly recognized and removed by a specific DNA glycosylase. Hypoxanthine, for example, is the simplest purine base capable of pairing specifically with C, but hypoxanthine is the direct deamination product of A (**Figure 5–43A**). The addition of a second amino group to hypoxanthine

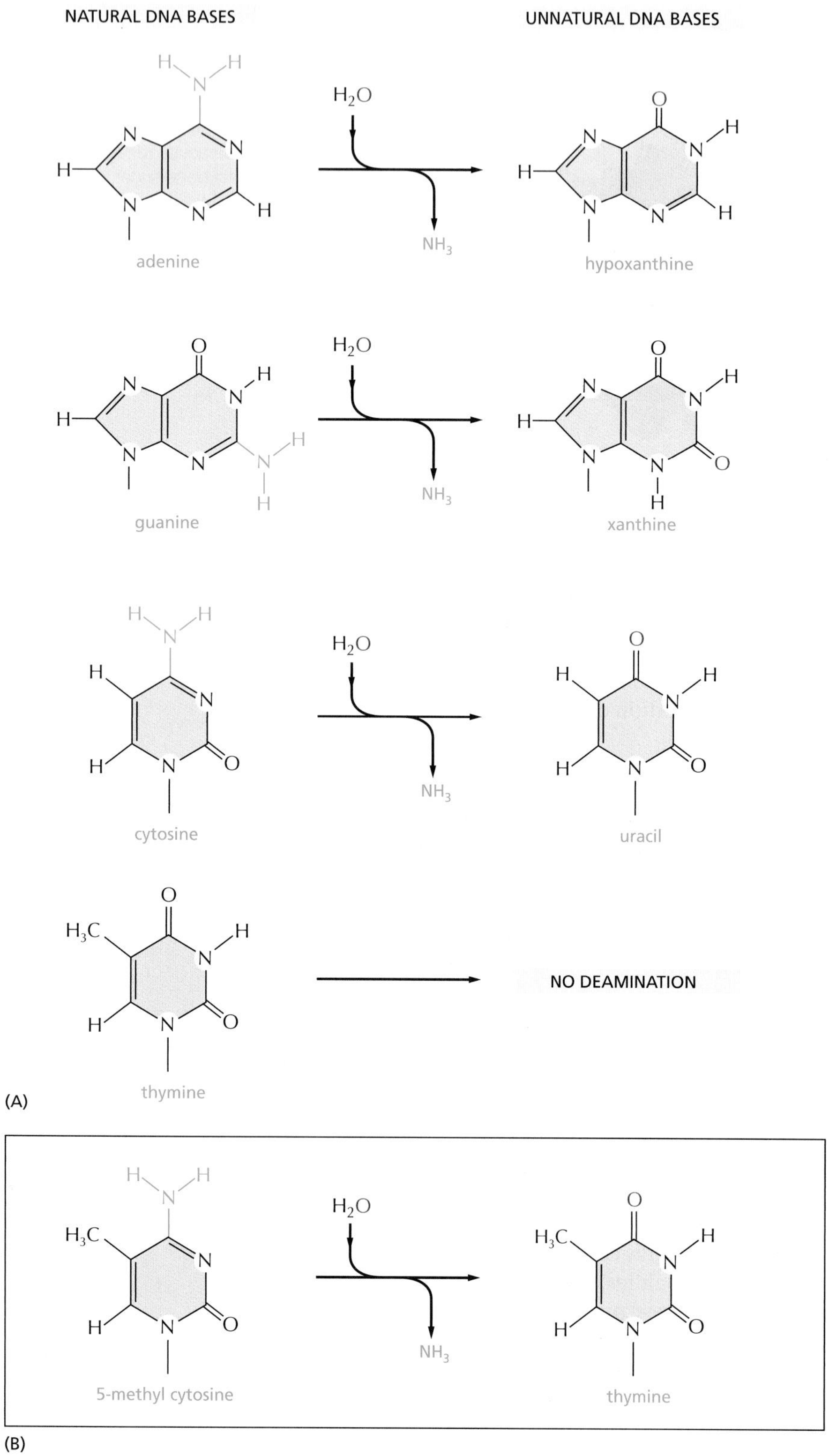

NATURAL DNA BASES

UNNATURAL DNA BASES

(A)

(B)

Figure 5–43 The deamination of DNA nucleotides. In each case, the oxygen atom that is added in this reaction with water is colored *red*. (A) The spontaneous deamination products of A and G are recognizable as unnatural when they occur in DNA and thus are readily found and repaired. The deamination of C to U was also illustrated in Figure 5–38; T has no amino group to remove. (B) About 3% of the C nucleotides in vertebrate DNAs are methylated to help in controlling gene expression (discussed in Chapter 7). When these 5-methyl C nucleotides are accidentally deaminated, they form the natural nucleotide T. However, this T will be paired with a G on the opposite strand, forming a mismatched base pair.

produces G, which cannot be formed from A by spontaneous deamination, and whose deamination product (xanthine) is likewise unique.

As discussed in Chapter 6, RNA is thought, on an evolutionary time scale, to have served as the genetic material before DNA, and it seems likely that the genetic code was initially carried in the four nucleotides A, C, G, and U. This raises the question of why the U in RNA was replaced in DNA by T (which is 5-methyl U). We have seen that the spontaneous deamination of C converts it to U, but that this event is rendered relatively harmless by uracil DNA glycosylase. However, if DNA contained U as a natural base, the repair system would not be able to distinguish a deaminated C from a naturally occurring U.

A special situation occurs in vertebrate DNA, in which selected C nucleotides are methylated at specific CG sequences that are associated with inactive genes (discussed in Chapter 7). The accidental deamination of these methylated C nucleotides produces the natural nucleotide T (Figure 5–43B) in a mismatched base pair with a G on the opposite DNA strand. To help in repairing deaminated methylated C nucleotides, a special DNA glycosylase recognizes a mismatched base pair involving T in the sequence T-G and removes the T. This DNA repair mechanism must be relatively ineffective, however, because methylated C nucleotides are exceptionally common sites for mutations in vertebrate DNA. It is striking that, even though only about 3% of the C nucleotides in human DNA are methylated, mutations in these methylated nucleotides account for about one-third of the single-base mutations that have been observed in inherited human diseases.

Special Translesion DNA Polymerases Are Used in Emergencies

If a cell's DNA suffers heavy damage, the repair mechanisms that we have discussed are often insufficient to cope with it. In these cases, a different strategy is called into play, one that entails some risk to the cell. The highly accurate replicative DNA polymerases stall when they encounter damaged DNA, and in emergencies cells employ versatile, but less accurate, backup polymerases, known as *translesion polymerases*, to replicate through the DNA damage.

Human cells have seven translesion polymerases, some of which can recognize a specific type of DNA damage and correctly add the nucleotide required to restore the initial sequence. Others make only "good guesses," especially when the template base has been extensively damaged. These enzymes are not as accurate as the normal replicative polymerases when they copy a normal DNA sequence. For one thing, the translesion polymerases lack exonucleolytic proofreading activity; in addition, many are much less discriminating than the replicative polymerase in choosing which nucleotide to incorporate initially. Presumably for this reason, each such translesion polymerase is given a chance to add only one or a few nucleotides before the highly accurate replicative polymerase resumes DNA synthesis.

Despite their usefulness in allowing heavily damaged DNA to be replicated, these translesion polymerases do, as noted above, pose risks to the cell. They are probably responsible for most of the base-substitution and single-nucleotide deletion mutations that accumulate in genomes; although they generally produce mutations when copying damaged DNA (see Figure 5–40), they probably also create mutations—at a low level—on undamaged DNA. Clearly, it is important for the cell to tightly regulate these polymerases, releasing them only at sites of DNA damage. Exactly how this happens for each translesion polymerase remains to be discovered, but a conceptual model is given in **Figure 5–44**. The principle of this model applies to many of the DNA repair processes discussed in this chapter: because the enzymes that carry out these reactions are potentially dangerous to the genome, they must be brought into play only at sites of damage.

Double-Strand Breaks Are Efficiently Repaired

An especially dangerous type of DNA damage occurs when both strands of the double helix are broken, leaving no intact template strand to enable accurate

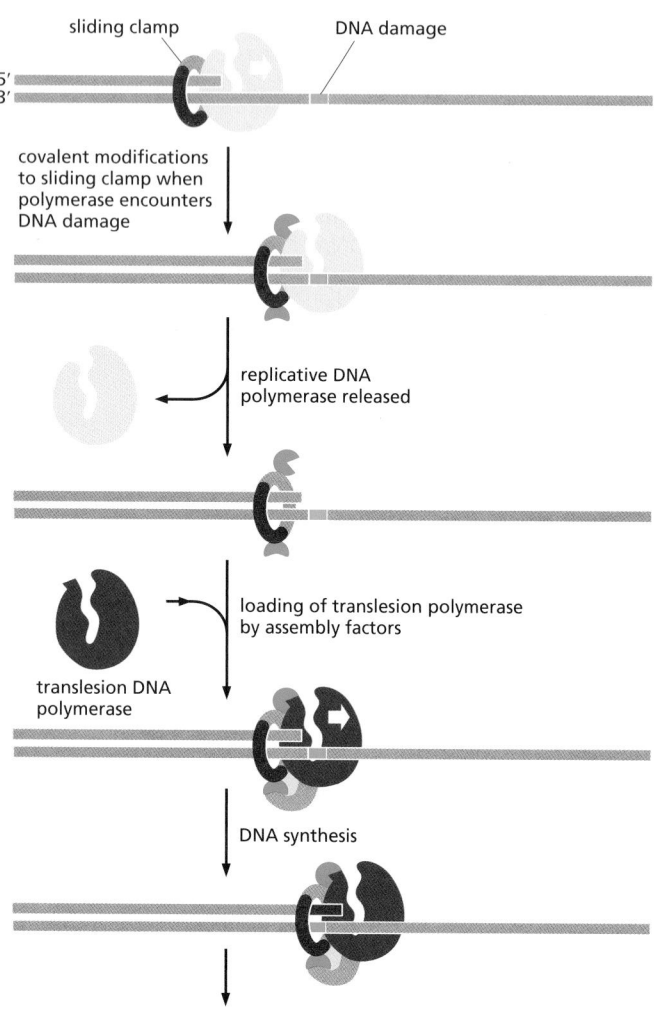

Figure 5–44 Translesion DNA polymerases can use damaged templates. According to this model, a replicative polymerase stalled at a site of DNA damage is recognized by the cell as needing rescue. Specialized enzymes covalently modify the sliding clamp (typically, it is ubiquitylated—see Figure 3–69) which releases the replicative DNA polymerase and, together with damaged DNA, attracts a translesion polymerase specific to that type of damage. Once the damaged DNA is bypassed, the covalent modification of the clamp is removed, the translesion polymerase dissociates, and the replicative polymerase is brought back into play.

repair. Ionizing radiation, replication errors, oxidizing agents, and other metabolites produced in the cell cause breaks of this type. If these lesions were left unrepaired, they would quickly lead to the breakdown of chromosomes into smaller fragments and to loss of genes when the cell divides. However, two distinct mechanisms have evolved to deal with this type of damage (**Figure 5–45**). The simplest to understand is **nonhomologous end joining**, in which the broken ends are simply brought together and rejoined by DNA ligation, generally with the loss of nucleotides at the site of joining (**Figure 5–46**). This end-joining mechanism, which can be seen as a "quick and dirty" solution to the repair of double-strand breaks, is common in mammalian somatic cells. Although a change in the DNA sequence (a mutation) results at the site of breakage, so little of the mammalian genome is essential for life that this mechanism is apparently an acceptable solution to the problem of rejoining broken chromosomes. By the time a human reaches the age of 70, the typical somatic cell contains over 2000 such "scars," distributed throughout its genome, representing places where DNA has been inaccurately repaired by nonhomologous end joining. But nonhomologous end joining presents another danger: because there seems to be no mechanism to ensure that two ends being joined were originally next to each other in the genome, nonhomologous end joining can occasionally generate rearrangements in which one broken chromosome becomes covalently attached to another. This can result in chromosomes with two centromeres and chromosomes lacking centromeres altogether; both

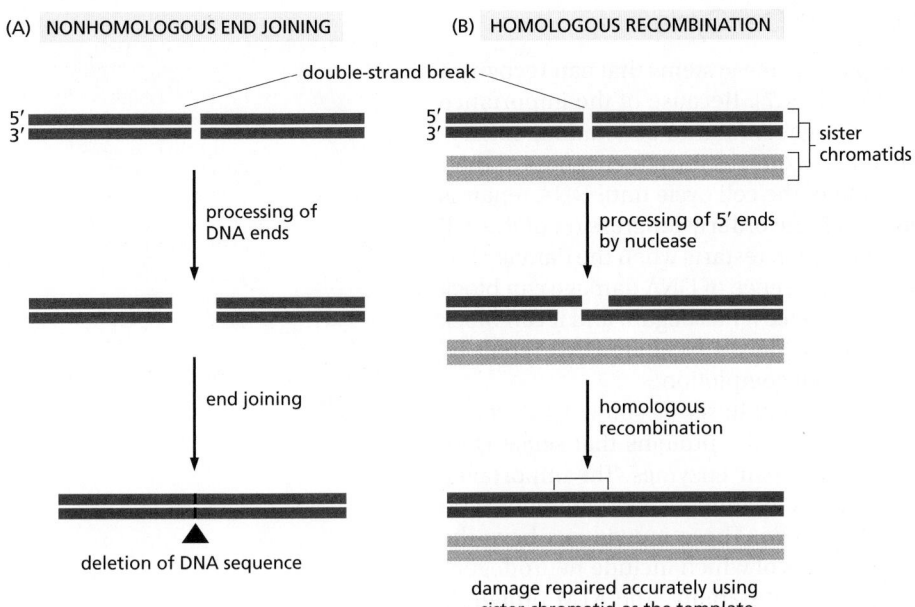

(A) NONHOMOLOGOUS END JOINING

(B) HOMOLOGOUS RECOMBINATION

double-strand break

processing of DNA ends

end joining

deletion of DNA sequence

sister chromatids

processing of 5′ ends by nuclease

homologous recombination

damage repaired accurately using sister chromatid as the template

Figure 5–45 **Two ways to repair double-strand breaks.** (A) Nonhomologous end joining alters the original DNA sequence when repairing a broken chromosome. The initial degradation of the broken DNA ends is important because the nucleotides at the site of the initial break are often damaged and cannot be ligated. Nonhomologous end joining usually takes place when cells have not yet duplicated their DNA. (B) Repairing double-strand breaks by homologous recombination is more difficult to accomplish but restores the original DNA sequence. It typically takes place after the DNA has been duplicated (when a duplex template is available) but before the cell has divided. Details of the homologous recombination pathway are presented in the following section (see Figure 5–48).

types of aberrant chromosomes are missegregated during cell division. As previously discussed, the specialized structure of telomeres prevents the natural ends of chromosomes from being mistaken for broken DNA and "repaired" in this way.

A much more accurate type of double-strand break repair occurs in newly replicated DNA (Figure 5–45B). Here, the DNA is repaired using the sister chromatid as a template. This reaction is an example of *homologous recombination*, and we consider its mechanism later in this chapter. Most organisms employ both nonhomologous end joining and homologous recombination to repair double-strand breaks in DNA. Nonhomologous end joining predominates in humans; homologous recombination is used only during and shortly after DNA replication (in S and G_2 phases), when sister chromatids are available to serve as templates.

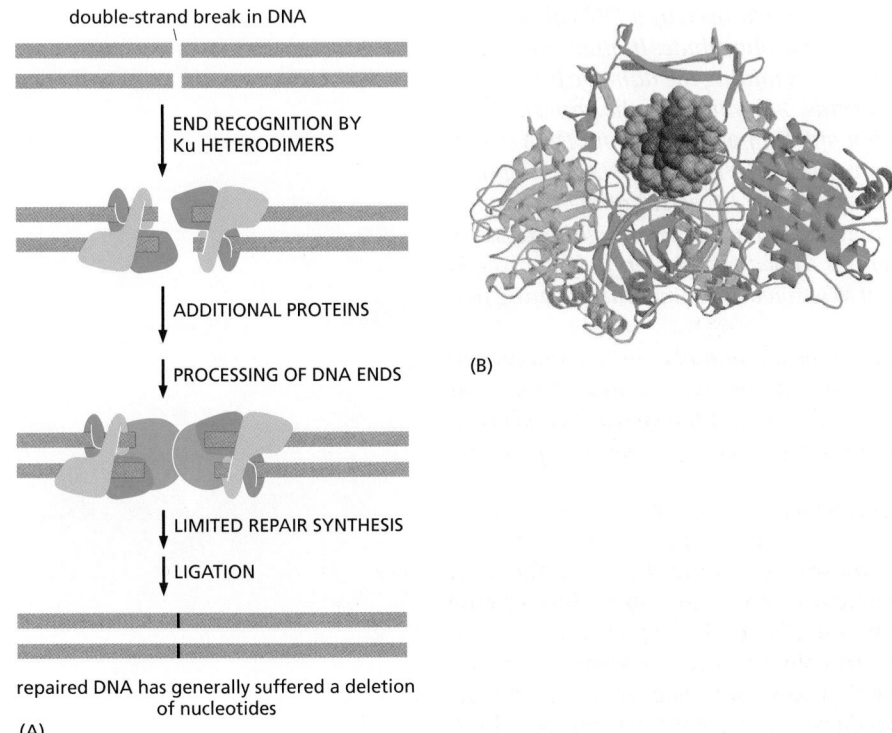

double-strand break in DNA

END RECOGNITION BY Ku HETERODIMERS

ADDITIONAL PROTEINS

PROCESSING OF DNA ENDS

LIMITED REPAIR SYNTHESIS

LIGATION

repaired DNA has generally suffered a deletion of nucleotides

(A)

(B)

Figure 5–46 **Nonhomologous end joining.** (A) A central role is played by the Ku protein, a heterodimer that grasps the broken chromosome ends. The additional proteins shown are needed to hold the broken ends together while they are processed and eventually joined covalently. (B) Three-dimensional structure of a Ku heterodimer bound to the end of a duplex DNA fragment. The Ku protein is also essential for V(D)J joining, a specific recombination process through which antibody and T cell receptor diversity is generated in developing B and T cells (discussed in Chapter 24). V(D)J joining and nonhomologous end joining show many similarities in mechanism but the former relies on specific double-strand breaks produced deliberately by the cell. (B, from J.R. Walker, R.A. Corpina, and J. Goldberg, *Nature* 412:607–614, published 2001 by Nature Publishing Group. Reproduced with permission of SNCSC.)

DNA Damage Delays Progression of the Cell Cycle

We have just seen that cells contain multiple enzyme systems that can recognize and repair many types of DNA damage (Movie 5.7). Because of the importance of maintaining intact, undamaged DNA from generation to generation, eukaryotic cells have an additional mechanism that maximizes the effectiveness of their DNA repair enzymes: they delay progression of the cell cycle until DNA repair is complete. As discussed in detail in Chapter 17, the orderly progression of the cell cycle is stopped if damaged DNA is detected, and it restarts when the damage has been repaired. Thus, in mammalian cells, the presence of DNA damage can block entry from G_1 into S phase, it can slow S phase once it has begun, and it can block the transition from G_2 phase to M phase. These delays facilitate DNA repair by providing the time needed for the repair to reach completion.

DNA damage also results in an increased synthesis of some DNA repair enzymes. This response depends on special signaling proteins that sense DNA damage and up-regulate the appropriate DNA repair enzymes. The importance of this mechanism is revealed by the phenotype of humans who are born with defects in the gene that encodes the *ATM protein*. These individuals have the disease *ataxia telangiectasia* (*AT*), the symptoms of which include neurodegeneration, a predisposition to cancer, and genome instability. The ATM protein is a large kinase needed to generate the intracellular signals that sound the alarm in response to many types of spontaneous DNA damage (see Figure 17–62), and individuals with defects in this protein therefore suffer from the effects of unrepaired DNA lesions.

Summary

Genetic information can be stored stably in DNA sequences only because a large set of DNA repair enzymes continuously scan the DNA and replace any damaged nucleotides. Most types of DNA repair depend on the presence of a separate copy of the genetic information in each of the two strands of the DNA double helix. An accidental lesion on one strand can therefore be cut out by a repair enzyme and a corrected strand resynthesized by reference to the information in the undamaged strand.

Most of the damage to DNA bases is excised by one of two major DNA repair pathways. In base excision repair, the altered base is removed by a DNA glycosylase enzyme, followed by excision of the resulting sugar phosphate. In nucleotide excision repair, a small section of the DNA strand surrounding the damage is removed from the DNA double helix as an oligonucleotide. In both cases, the gap left in the DNA helix is filled in by the sequential action of DNA polymerase and DNA ligase, using the undamaged DNA strand as the template. Some types of DNA damage can be repaired by a different strategy—the direct chemical reversal of the damage—which is carried out by specialized repair proteins. When DNA damage is excessive, a special class of inaccurate DNA polymerases, called translesion polymerases, is used to bypass the damage, allowing the cell to survive but sometimes creating permanent mutations at the sites of damage.

Other critical repair systems—based on either nonhomologous end joining or homologous recombination—reseal the accidental double-strand breaks that occur in the DNA helix. In most cells, an elevated level of DNA damage causes a delay in the cell cycle, which ensures that DNA damage is repaired before a cell divides.

HOMOLOGOUS RECOMBINATION

In the two preceding sections, we discussed the mechanisms that allow the DNA sequences in cells to be maintained from generation to generation with very little change. In this section, we further explore one of the DNA repair mechanisms, a diverse set of reactions known collectively as *homologous recombination*. The key feature of **homologous recombination** (also known as *general recombination*) is an exchange of DNA strands between a pair of homologous duplex DNA sequences, that is, segments of double helix that are very similar or identical in nucleotide sequence. This exchange allows one stretch of duplex DNA to act as a template to restore lost or damaged information on a second stretch of duplex DNA. Because the template for repair is not limited to the strand complementary to that containing the damage, homologous recombination can repair many types of DNA damage. It is, for example, the main way to accurately repair double-strand breaks, as introduced in the previous section (see Figure 5–45B). Double-strand breaks can result from radiation and reactive chemicals, but most of the time they arise from DNA replication forks that become stalled or broken independently of any such external cause. Homologous recombination accurately corrects these accidents and, because they occur during nearly every round of DNA replication, this repair mechanism is essential for every proliferating cell. Homologous recombination is perhaps the most versatile DNA repair mechanism available to the cell; the "all-purpose" nature of recombinational repair probably explains why its mechanism and the proteins that carry it out have been conserved in virtually all cells on Earth.

Additionally, we shall see that homologous recombination plays a special role in sexually reproducing organisms. During meiosis, a key step in gamete (sperm and egg) production, it catalyzes the orderly exchange of bits of genetic information between corresponding (homologous) maternal and paternal chromosomes to create new combinations of DNA sequences in the chromosomes passed to the offspring.

Homologous Recombination Has Common Features in All Cells

The current view of homologous recombination as a critical DNA repair mechanism in all cells evolved slowly from its original discovery as a key component in the specialized process of meiosis in plants and animals. The subsequent recognition that homologous recombination also occurs in unicellular organisms made it much more amenable to molecular analyses. Thus, most of what we know about the biochemistry of genetic recombination was originally derived from studies of bacteria, especially of *E. coli* and its viruses, as well as from experiments with simple eukaryotes such as yeasts. For these organisms with short generation times and relatively small genomes, it was possible to isolate a large set of mutants with defects in their recombination processes. The protein altered in each mutant was then identified and, ultimately, studied biochemically. Close relatives of these proteins have been found in more complex eukaryotes including flies, mice, and humans, and more recently, it has been possible to directly analyze homologous recombination in these species as well. These studies reveal that the fundamental processes that catalyze homologous recombination are common to all cells.

DNA Base-Pairing Guides Homologous Recombination

The hallmark of homologous recombination is that it takes place only between DNA duplexes that have extensive regions of sequence similarity (homology). Not surprisingly, base-pairing underlies this requirement, and two DNA duplexes that are undergoing homologous recombination "sample" each other's DNA sequence by engaging in extensive base-pairing between a single strand from one DNA duplex and the complementary single strand from the other. The match need not be perfect, but it must be very close for homologous recombination to succeed.

In its simplest form, this type of base-pairing interaction can be mimicked in a test tube by allowing a DNA double helix to re-form from its separated single strands. This process, called *DNA renaturation* or **hybridization**, occurs when a rare random collision juxtaposes complementary nucleotide sequences on two matching DNA single strands, allowing the formation of a short stretch of double helix between them. This relatively slow helix-nucleation step is followed by a very rapid "zippering" step, as the region of double helix is extended to maximize the number of base-pairing interactions (**Figure 5–47**).

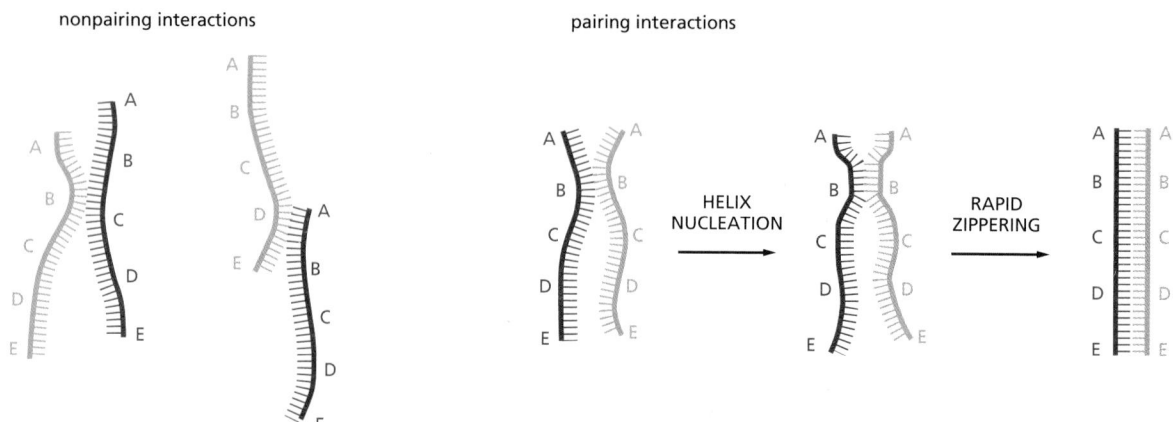

nonpairing interactions   pairing interactions

Figure 5–47 **DNA hybridization.** DNA double helices can re-form from their separated strands in a reaction that depends on the random collision of two complementary DNA strands. The vast majority of such collisions are not productive, as shown on the left, but a few result in a short region where complementary base pairs have formed (helix nucleation). A rapid zippering then leads to the formation of a complete double helix. Through this trial-and-error process, a DNA strand will find its complementary partner even in the midst of millions of nonmatching DNA strands.

DNA hybridization can create a region of DNA double helix consisting of strands that originate from two different duplex DNA molecules as long as they are complementary, or nearly so. As we will see shortly, the formation of such a hybrid molecule, known as a heteroduplex, is an essential feature of homologous recombination. DNA hybridization and heteroduplex formation is also the basis for many of the methods used to study cells, and we will discuss these uses in Chapter 8.

The DNA in a living cell is almost all in the stable double-helical form, so the reaction depicted in Figure 5–47 rarely occurs *in vivo*. Instead, as we shall see, homologous recombination is brought about through a carefully controlled set of reactions that allow two DNA duplexes to sample each other's sequences without fully dissociating into single strands.

Homologous Recombination Can Flawlessly Repair Double-Strand Breaks in DNA

We saw in the previous section that nonhomologous end-joining occurs without a template and usually leaves a mutation at the site at which a double-strand break is repaired. In contrast, homologous recombination can repair double-strand breaks accurately, without any loss or alteration of nucleotides at the site of repair. For homologous recombination to do this repair job, the broken DNA has to be brought into proximity with homologous but unbroken DNA, which can serve as a template for repair. For this reason, homologous recombination often occurs just after DNA replication, when the two daughter DNA molecules lie close together and one can serve as a template for repair of the other. As we shall see, the process of DNA replication itself creates a special risk of accidents requiring this sort of repair.

The simplest pathway through which homologous recombination can repair double-strand breaks is shown in **Figure 5–48**. In essence, the broken DNA duplex and the template duplex carry out a "strand dance" so that one of the damaged strands can use the complementary strand of the intact DNA duplex as a template for repair. First, the ends of the broken DNA are chewed back, or "resected," by specialized nucleases to produce overhanging, single-strand 3′ ends. The next step is **strand exchange** (also called *strand invasion*), during which one of the single-strand 3′ ends from the damaged DNA molecule worms its way into the template duplex and searches it for homologous sequences through base-pairing. We describe this remarkable reaction in detail in the next section. Once stable base-pairing is established (which completes the strand exchange step), an accurate DNA polymerase extends the invading strand by using the information provided by the undamaged template molecule, thus restoring the damaged DNA. The last steps—strand displacement, further repair synthesis, and ligation—restore the two original DNA double helices and complete the repair process. Homologous recombination resembles other DNA repair reactions in that a

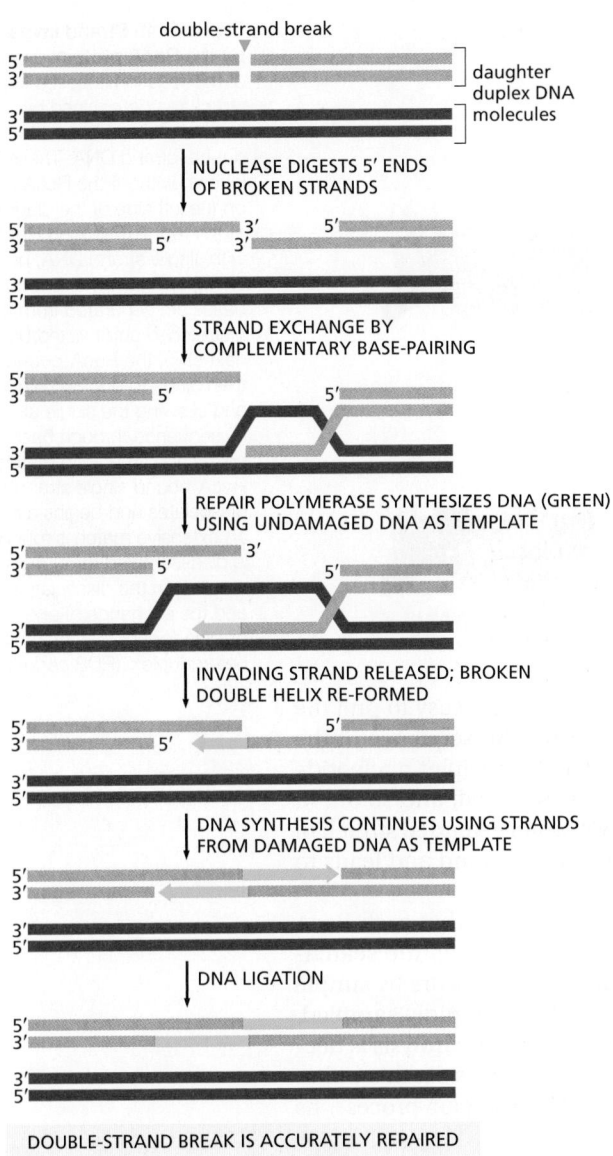

double-strand break

5′
3′ } daughter duplex DNA molecules
3′
5′

NUCLEASE DIGESTS 5′ ENDS OF BROKEN STRANDS

5′ 3′ 5′
3′ 5′ 3′

3′
5′

STRAND EXCHANGE BY COMPLEMENTARY BASE-PAIRING

5′
3′ 5′ 5′

3′
5′

REPAIR POLYMERASE SYNTHESIZES DNA (GREEN) USING UNDAMAGED DNA AS TEMPLATE

5′ 3′
3′ 5′ 5′

3′
5′

INVADING STRAND RELEASED; BROKEN DOUBLE HELIX RE-FORMED

5′ 5′
3′ 5′

3′
5′

DNA SYNTHESIS CONTINUES USING STRANDS FROM DAMAGED DNA AS TEMPLATE

5′
3′

3′
5′

DNA LIGATION

5′
3′

3′
5′

DOUBLE-STRAND BREAK IS ACCURATELY REPAIRED

Figure 5–48 Mechanism of double-strand break repair by homologous recombination. This is the preferred method for repairing DNA double-strand breaks that arise shortly after the DNA has been replicated, while the daughter DNA molecules are still held close together. In general, homologous recombination can be regarded as a flexible series of reactions, with the exact pathway differing from one case to the next. For example, the length of the repair "patch" can vary considerably depending on the extent of 5′ processing and new DNA synthesis, indicated in *green*.

DNA polymerase utilizes a pristine template to restore damaged DNA. However, instead of using the partner complementary strand as a template, as occurs in most DNA repair pathways, homologous recombination exploits a complementary strand from a separate DNA duplex.

Strand Exchange Is Carried Out by the RecA/Rad51 Protein

Of all the steps of homologous recombination, strand exchange is the most difficult to imagine. How does the invading single strand rapidly sample a DNA duplex for homology? Once the homology is found, how does the exchange occur? How is the inherent stability of the template double helix overcome?

The answers to these questions came from biochemical and structural studies of the protein that carries out these feats, called **RecA** in *E. coli* and **Rad51** in virtually all eukaryotic organisms. To catalyze strand exchange, RecA first binds cooperatively to the invading single strand, forming a protein–DNA filament that forces the DNA into an unusual conformation: groups of three consecutive nucleotides are held as though they were in a conventional DNA double helix but, between adjacent triplets, the DNA backbone is untwisted and stretched out (**Figure 5–49**). This unusual protein–DNA filament then binds to duplex DNA

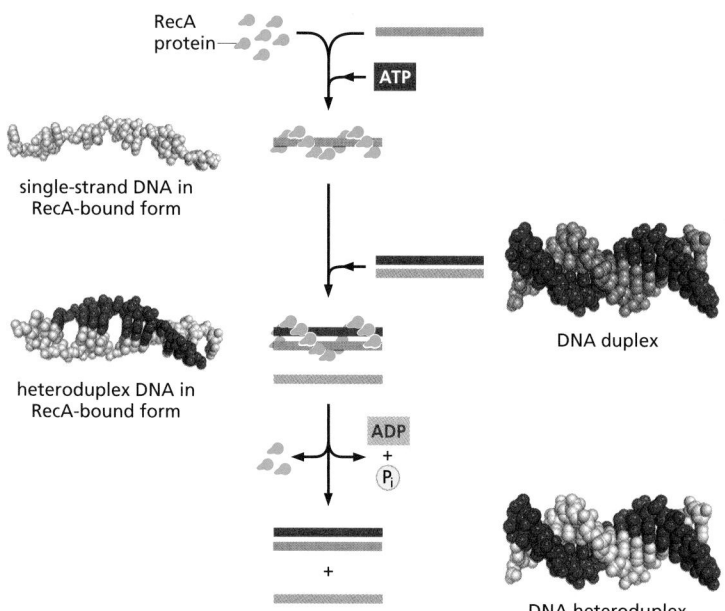

single-strand DNA in
RecA-bound form

heteroduplex DNA in
RecA-bound form

DNA duplex

DNA heteroduplex

Figure 5–49 **Strand invasion catalyzed by the RecA protein.** Our understanding of this reaction is based in part on structures determined by x-ray diffraction studies of RecA bound to single- and double-strand DNA. These DNA structures (shown without the RecA protein) are on the left side of the diagram. Starting at the top, ATP-bound RecA associates with single-strand DNA, holding it in an elongated form where groups of three bases are separated from each other by a stretched and twisted backbone. In the next step, the RecA-bound single strand then binds to duplex DNA, destabilizing it and allowing the single strand to sample its sequence through base-pairing, three bases at a time. If no match is found, the RecA-bound single strand of DNA rapidly dissociates and begins a new search. If an extensive match is found, the structure is disassembled through ATP hydrolysis, resulting in the dissociation of RecA and the exchange of one single strand of DNA for another, thereby forming a heteroduplex. (PDB code: 3CMX.)

in a way that stretches the duplex, destabilizing it and making it easy to pull the strands apart. The invading single strand then can sample the sequence of the duplex by conventional base-pairing. This sampling occurs in triplet nucleotide blocks: if a triplet match is found, the adjacent triplet is sampled, and so on. In this way, mismatches quickly lead to dissociation and only an extended stretch of base-pairing (at least 15 nucleotides) stabilizes the invading strand and leads to strand exchange.

RecA hydrolyzes ATP, and the steps described above require that each RecA monomer along the filament be in the ATP-bound state. However, the searching itself does not require ATP hydrolysis; instead, the process occurs by simple molecular collision, allowing many potential sequences to be rapidly sampled. Once the strand-exchange reaction is completed, however, ATP hydrolysis is necessary to disassemble RecA from the complex of DNA molecules. At this point, repair DNA polymerases and DNA ligase can complete the repair process, as shown in Figure 5–48.

Homologous Recombination Can Rescue Broken DNA Replication Forks

Although accurately repairing double-strand breaks, which can arise from radiation or chemical reactions, is a crucial function of homologous recombination, perhaps its most important role is in rescuing stalled or broken DNA replication forks. Many types of events can cause a replication fork to break, and here we consider just one example: a single-strand nick or gap in the parental DNA helix just ahead of a replication fork. When the fork reaches this lesion, it falls apart—resulting in one broken and one intact daughter chromosome. The broken fork can be flawlessly repaired (**Figure 5–50**) using the same basic homologous recombination reactions we discussed above for the repair of double-strand breaks. With slight modifications, the set of reactions depicted in Figures 5–48 and 5–50—known collectively as homologous recombination—can accurately repair many different types of DNA damage.

Cells Carefully Regulate the Use of Homologous Recombination in DNA Repair

Although homologous recombination neatly solves the problem of accurately repairing double-strand breaks and other types of DNA damage, it does present

Figure 5–50 Repair of a broken replication fork by homologous recombination. When a moving replication fork encounters a single-strand break, it will collapse, but can be repaired by homologous recombination. The process uses many of the same reactions shown in Figure 5–48 and proceeds through the same basic steps. *Green* strands represent the new DNA synthesis that takes place after the replication fork has broken. This pathway allows the fork to move past the site that was nicked on the original template by using the undamaged duplex as a template to synthesize DNA. (Adapted from M.M. Cox, *Proc. Natl Acad. Sci. USA* 98:8173–8180, 2001. Copyright 2001 National Academy of Sciences, USA. With permission from National Academy of Sciences.)

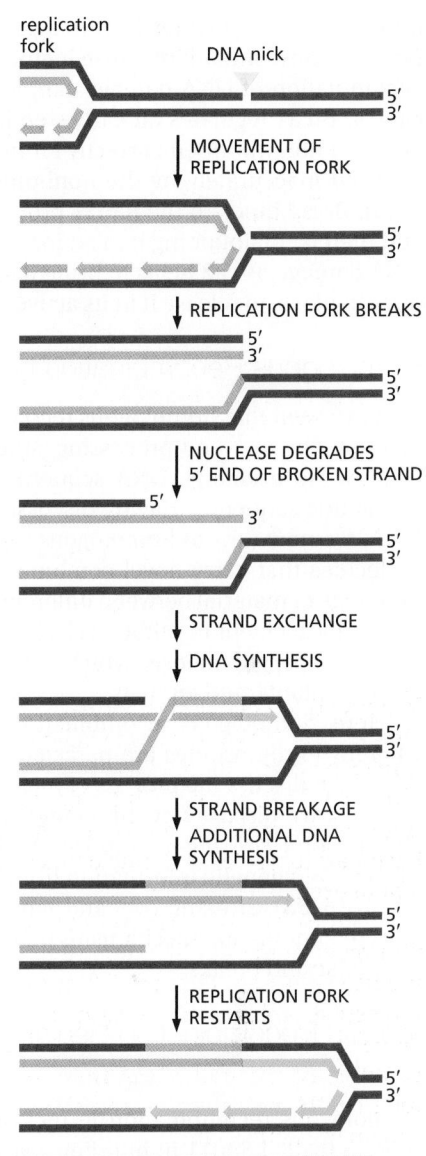

some dangers to the cell as it sometimes "repairs" damage using the wrong bit of the genome as the template. For example, sometimes a broken human chromosome is "repaired" using the homolog from the other parent instead of the sister chromatid as the template. Because maternal and paternal chromosomes differ in DNA sequence at many positions along their lengths, this type of repair can convert the sequence of the repaired DNA from the maternal to the paternal sequence or vice versa. The result of this type of errant recombination is known as **loss of heterozygosity**. It can have severe consequences if the homolog used for repair contains a deleterious mutation, because the recombination event destroys the "good" copy. Loss of heterozygosity, although rare, is a critical step in the formation of many cancers (discussed in Chapter 20).

Cells go to great lengths to minimize the risk of mishaps of these types; indeed, nearly every step of homologous recombination is carefully regulated. For example, the first step, processing of the broken ends, is coordinated with the cell cycle: the nuclease enzymes that carry out this process are activated (in part, by phosphorylation) only in the S and G_2 phases of the cell cycle, when a daughter duplex (either as a partially replicated chromosome or a fully replicated sister chromatid) can serve as a template for repair (see Figure 5–50). The close proximity of the two daughter chromosomes disfavors the use of other genome sequences in the repair process.

The loading of RecA or Rad52 onto the processed DNA ends and the subsequent strand-exchange reaction are also tightly controlled. Although these proteins alone can carry out these steps *in vitro*, a series of accessory proteins, including Rad52, is needed in eukaryotic cells to ensure that homologous recombination is efficient and accurate (**Figure 5–51**). There are many such accessory proteins, and exactly how they coordinate and control homologous recombination remains a mystery. We do know that the enzymes that catalyze recombinational repair are made at relatively high levels in eukaryotes and are dispersed throughout the nucleus in an inactive form. In response to DNA damage, they rapidly converge on the sites of DNA damage, become activated, and form "repair factories" where many lesions are apparently brought together and repaired (**Figure 5–52**).

In Chapter 20, we shall see that both too much and too little homologous recombination can lead to cancer in humans, the former through repair using the "wrong" template (as described above) and the latter through an increased mutation rate caused by inefficient DNA repair. Clearly, a delicate balance has evolved that keeps this process in check on undamaged DNA, while still allowing it to act efficiently and rapidly on DNA lesions as soon as they arise.

Not surprisingly, mutations in the components that carry out and regulate homologous recombination are responsible for several inherited forms of cancer. Two of these, the Brca1 and Brca2 proteins, were first discovered because

Figure 5–51 Structure of a portion of the Rad52 protein. This doughnut-shaped structure is composed of 11 subunits. Single-strand DNA has been modeled into the deep groove running along the protein surface. Rad52 helps load Rad51 onto single-strand DNA to form the nucleoprotein filament that carries out strand exchange. Rad52 also acts later to re-form the double helix and complete the homologous recombination reaction. (From M.R. Singleton et al., *Proc. Natl Acad. Sci. USA* 99:13492–13497, 2002. Copyright 2002 National Academy of Sciences, USA. With permission from National Academy of Sciences.)

mutations in their genes lead to a greatly increased frequency of breast cancer. Because these mutations cause inefficient repair by homologous recombination, accumulation of DNA damage can, in a small proportion of cells, give rise to a cancer. Brca1 regulates an early step in broken-end processing; without it, such ends are not processed correctly for homologous recombination and instead are repaired inaccurately by the nonhomologous end-joining pathway (see Figure 5–45). Brca2 binds to the Rad51 protein, preventing its polymerization on DNA, and thereby maintaining it in an inactive form until it is needed. Normally, upon DNA damage, Brca2 helps to bring Rad51 protein rapidly to sites of damage and, once in place, to release it in its active form onto single-strand DNA.

Homologous Recombination Is Crucial for Meiosis

We have seen that homologous recombination comprises a group of reactions—including broken-end processing, strand exchange, limited DNA synthesis, and ligation—to exchange DNA sequences between two double helices of similar nucleotide sequence. Having discussed its role in accurately repairing damaged DNA, we now turn to homologous recombination as a means to generate DNA molecules that carry novel combinations of genes as a result of the deliberate exchange of material between different chromosomes. Although this occasionally occurs by accident in mitotic cells (and is often detrimental), it is a frequent and necessary part of meiosis, which occurs in sexually reproducing organisms such as fungi, plants, and animals.

Here, homologous recombination occurs as an integral part of the process whereby chromosomes are parceled out to germ cells (sperm and eggs in animals). We discuss the process of meiosis in detail in Chapter 17; in the following sections, we discuss how homologous recombination during meiosis produces chromosome *crossing-over* and *gene conversion,* resulting in hybrid chromosomes that contain genetic information from both the maternal and paternal homologs (**Figure 5–53**). Crossing-over and gene conversion are both generated by homologous recombination mechanisms that, at their core, resemble those used to repair double-strand breaks.

Meiotic Recombination Begins with a Programmed Double-Strand Break

Homologous recombination in meiosis starts with a bold stroke: a specialized protein (called Spo11 in budding yeast) breaks both strands of the DNA double helix in one of the recombining chromosomes (**Figure 5–54**). Like a topoisomerase, Spo11, after catalyzing this reaction, remains covalently bound to the broken

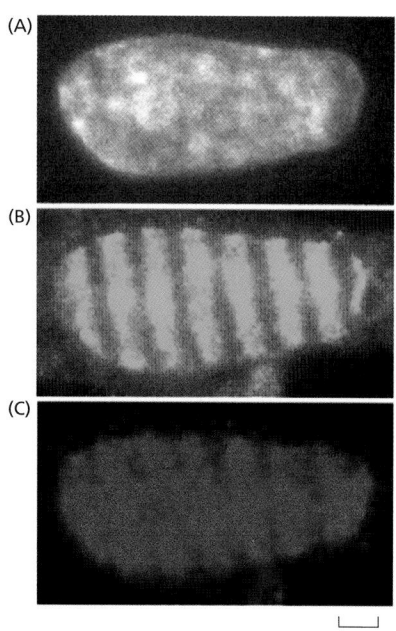

(A)

(B)

(C)

1 µm

Figure 5–52 Experiment demonstrating the rapid localization of repair proteins to DNA double-strand breaks. Human fibroblasts were x-irradiated to produce DNA double-strand breaks. Before the x-rays struck the cells, they were passed through a microscopic grid with x-ray-absorbing "bars" spaced 1 µm apart. This produced a striped pattern of DNA damage, allowing a comparison of damaged and undamaged DNA in the same nucleus. (A) Total DNA in a fibroblast nucleus stained with the dye DAPI. (B) Sites of new DNA synthesis due to repair of DNA damage, indicated by incorporation of BudR (a thymidine analog) and subsequent staining with fluorescently labeled antibodies to BudR *(green)*. (C) Localization of the Mre11 complex to damaged DNA as visualized by antibodies against the Mre11 subunit *(red)*. Mre11 is a nuclease that processes damaged DNA in preparation for homologous recombination (see Figure 5–48). (A), (B), and (C) were processed 30 minutes after x-irradiation. (From B.E. Nelms et al., *Science* 280:590–592, 1998. With permission from AAAS.)

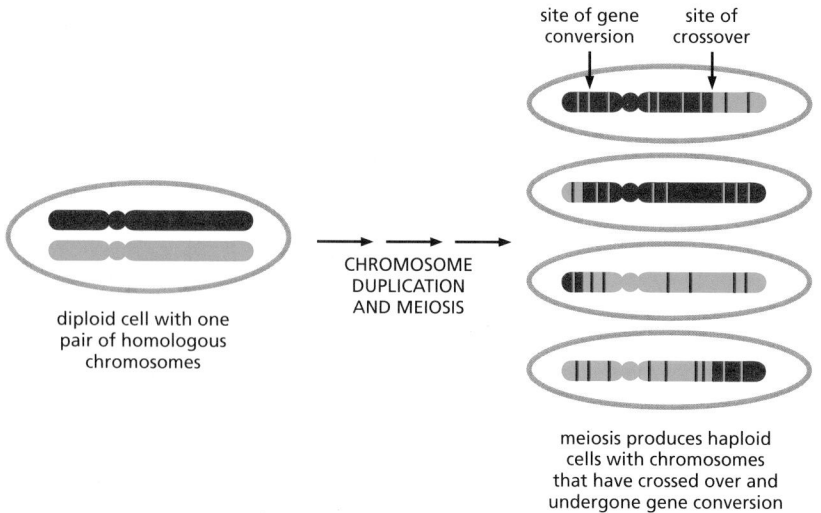

site of gene conversion   site of crossover

diploid cell with one pair of homologous chromosomes

CHROMOSOME DUPLICATION AND MEIOSIS

meiosis produces haploid cells with chromosomes that have crossed over and undergone gene conversion

Figure 5–53 Chromosome crossing-over occurs in meiosis. Meiosis is the process by which a diploid cell gives rise to four haploid germ cells, as described in detail in Chapter 17. Meiosis produces germ cells in which the paternal and maternal genetic information (*red* and *blue*) has been reassorted through chromosome crossovers. In addition, many short regions of gene conversion occur, as indicated.

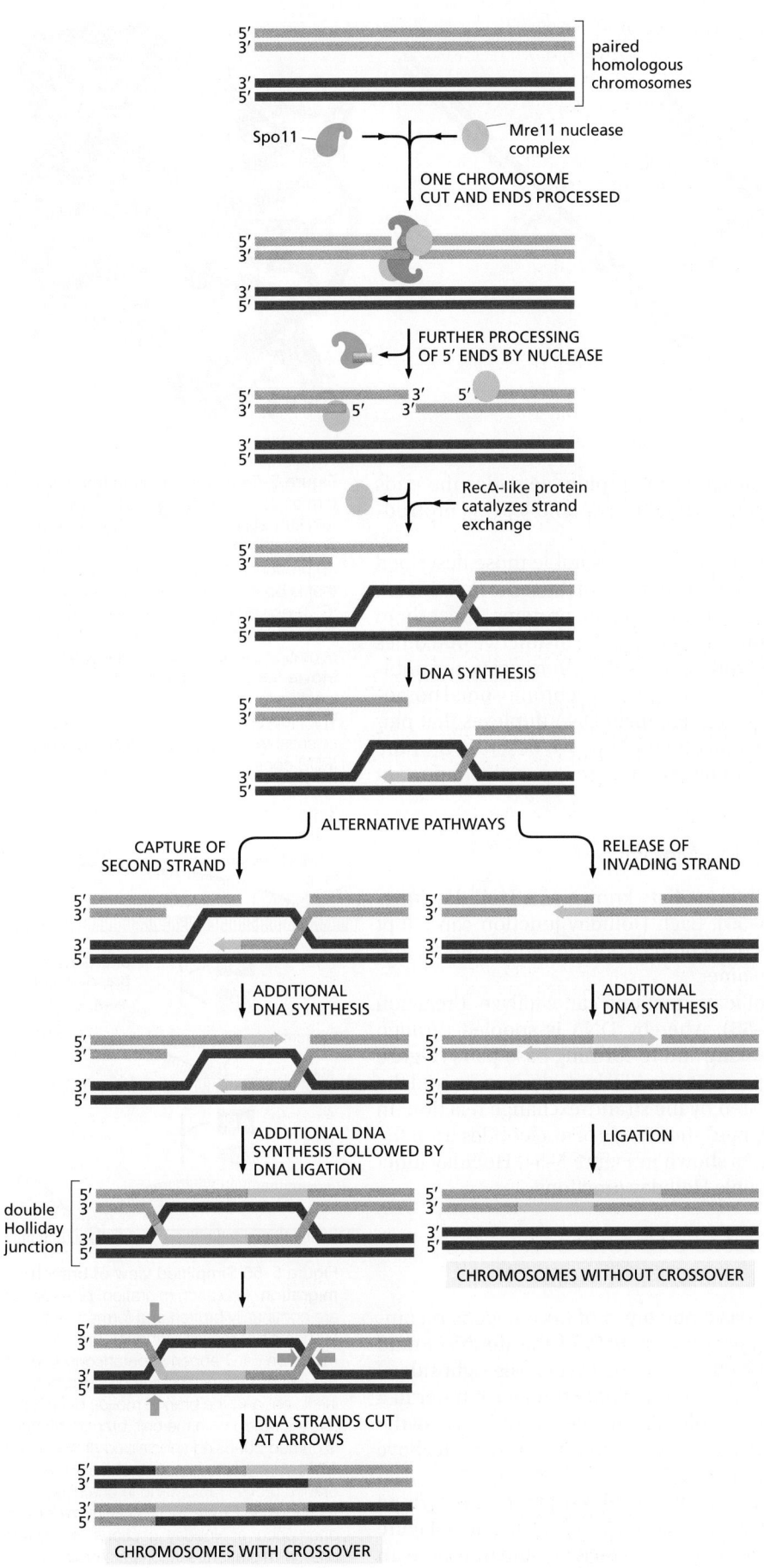

Figure 5–54 **Homologous recombination during meiosis can generate chromosome crossovers.** Once the meiosis-specific protein Spo11 and the Mre11 complex break the duplex DNA and process the ends, homologous recombination can proceed along alternative pathways. One (right side of figure) closely resembles the double-strand break repair reaction shown in Figure 5–48 and results in chromosomes that have been "repaired" but have not crossed over. The other (left side with strand breaks as shown by the *blue arrows*) proceeds through a double Holliday junction and produces two chromosomes that have crossed over. During meiosis, homologous recombination takes place between maternal and paternal chromosome homologs when they are held tightly together (see Figure 17–54).

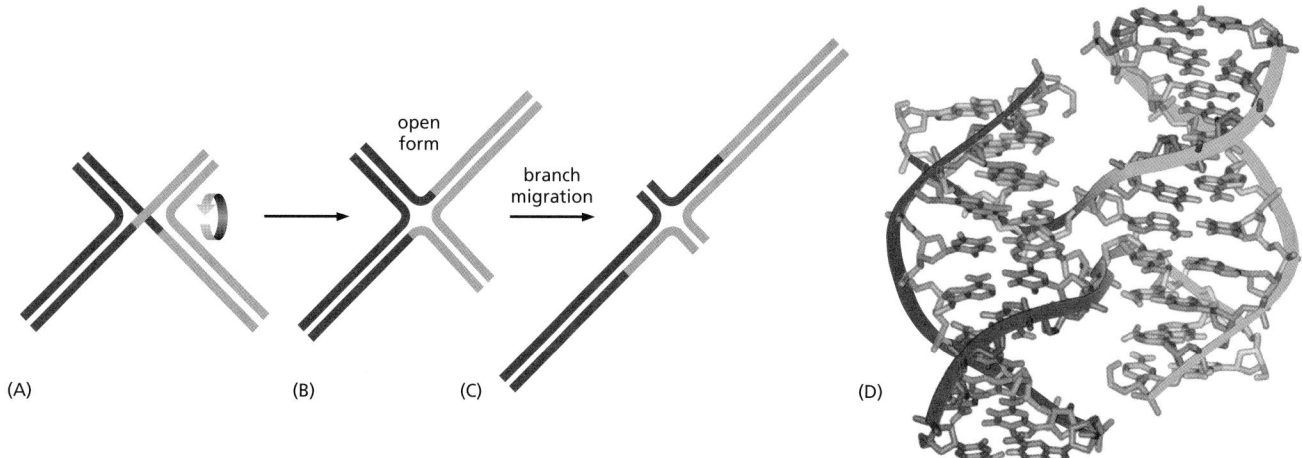

(A)   (B)   (C)   (D)

DNA (see Figure 5–21). A specialized nuclease then rapidly degrades the ends bound by Spo11, removing the protein along with the DNA and leaving protruding 3′ single-strand ends.

At this point, many of the recombination reactions resemble those described above for the repair of double-strand breaks; indeed, some of the same proteins are used for both processes. However, several meiosis-specific proteins direct them to perform their tasks somewhat differently, resulting in the distinctive outcomes observed for meiosis. Another important difference is that, in meiosis, recombination occurs preferentially between maternal and paternal chromosomal homologs rather than between the newly replicated, identical DNA duplexes that pair in double-strand break repair. In the sections that follow, we describe in more detail those aspects of homologous recombination that are especially important for meiosis.

Holliday Junctions Are Formed During Meiosis

Of special importance in meiosis is an intermediate known as a **Holliday junction** or *cross-strand exchange* (**Figure 5–55**). Each Holliday junction can adopt multiple conformations and a special set of recombination proteins binds to, and thereby stabilizes, the open, symmetric isomer.

Specialized proteins that bind to Holliday junctions can catalyze a reaction known as *branch migration* (**Figure 5–56**), whereby DNA is spooled through the Holliday junction by continually breaking and re-forming base pairs (**Figure 5–57**). In this way, the Holliday junction proteins use ATP hydrolysis to expand the region of heteroduplex DNA initially created by the strand-exchange reaction. In meiosis, heteroduplex regions often "migrate" thousands of nucleotides from the original site of the double-strand break. As shown in Figure 5–54, Holliday junctions usually occur in pairs, known as double Holliday junctions.

Homologous Recombination Produces Both Crossovers and Non-Crossovers During Meiosis

As shown in Figure 5–54, there are two basic outcomes of homologous recombination during meiosis. In humans, approximately 90% of the double-strand breaks produced during meiosis are resolved as non-crossovers (see right side of Figure 5–54). Here, the two original DNA duplexes separate from each other in a form unaltered except for a region of heteroduplex that formed near the site of the original double-strand break. This set of reactions resembles that described above for the repair of double-strand breaks (see Figure 5–48).

The other outcome is more profound: a double Holliday junction is formed and is cleaved by specialized enzymes to create a *crossover* (see left side of Figure 5–54). The two original portions of each chromosome upstream and downstream

Figure 5–55 A Holliday junction. The initially formed structure (A) is usually drawn with two strands crossing, as in Figure 5–54. An isomerization of the Holliday junction (B) produces an open, symmetrical structure that is bound by specialized proteins. (C) These proteins "move" the Holliday junctions by a coordinated set of branch-migration reactions (see Figure 5–57 and **Movie 5.8**). (D) Structure of the Holliday junction in the open form depicted in (B). The Holliday junction is named for the scientist who first proposed its formation. (PDB code: 1DCW.)

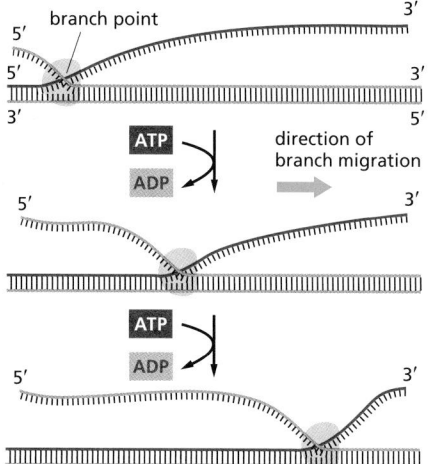

Figure 5–56 Simplified view of branch migration. In branch migration, base pairs are continually broken and formed as the branch point moves. Although branch migration can happen spontaneously on naked DNA molecules, the process is inefficient and the branch moves back and forth at random. In the cell, branch migration is carried out using specialized proteins and ATP hydrolysis to ensure that, as shown, the branch moves rapidly and in one direction. As shown in Figure 5–57, branch migrations often occur at Holliday junctions, where two branch-migration reactions are coupled.

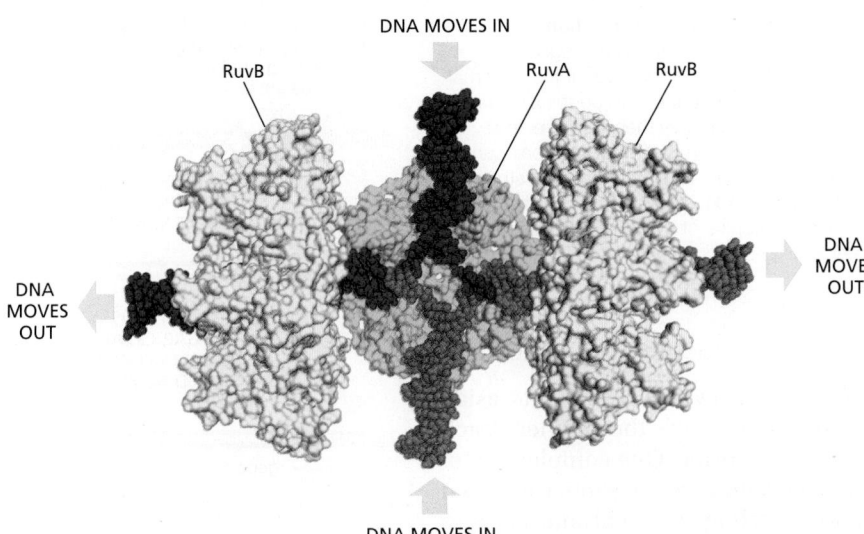

DNA MOVES IN

RuvB   RuvA   RuvB

DNA MOVES OUT

DNA MOVES OUT

DNA MOVES OUT

DNA MOVES IN

Figure 5–57 Enzyme-catalyzed branch movement at a Holliday junction by branch migration. In *E. coli*, a tetramer of the RuvA protein *(green)* and two hexamers of the RuvB protein *(yellow)* bind to the open form of the junction. The RuvB protein, which resembles the hexameric helicases used in DNA replication (Figure 5–14), uses the energy of ATP hydrolysis to spool DNA rapidly through the Holliday junction, extending the heteroduplex region as shown. The RuvA protein coordinates this movement, threading the DNA strands to avoid tangling. (PDB codes: 1IXR, 1C7Y.)

from the two Holliday junctions are thereby swapped, creating two chromosomes that have crossed over.

How does the cell decide which Spo11-induced double-strand breaks to resolve as crossovers? The answer is not yet known, but we know the decision is an important one. The relatively few crossovers that do form are distributed along chromosomes in such a way that a crossover in one position inhibits crossing-over in neighboring regions. Termed *crossover control*, this fascinating but poorly understood regulatory mechanism ensures the roughly even distribution of crossover points along chromosomes. It also ensures that each chromosome—no matter how small—undergoes at least one crossover every meiosis. For many organisms, roughly two crossovers per chromosome occur during each meiosis, one on each arm. As discussed in detail in Chapter 17, these crossovers play an important mechanical role in the proper segregation of chromosomes during meiosis.

Whether a meiotic recombination event is resolved as a crossover or a non-crossover, the recombination machinery leaves behind a *heteroduplex region* where a strand with the DNA sequence of the paternal homolog is base-paired with a strand from the maternal homolog (**Figure 5–58**). These heteroduplex regions can tolerate a small percentage of mismatched base pairs, and because of branch migration, they often extend for thousands of nucleotide pairs. The many non-crossover events that occur in meiosis thereby produce scattered sites in the germ cells where short DNA sequences from one homolog have been pasted into the other homolog. Heteroduplex regions mark sites of potential *gene conversion*—where the four haploid chromosomes produced by meiosis contain three copies of a DNA sequence from one homolog and only one copy of this sequence from the other homolog (see Figure 5–53), as explained next.

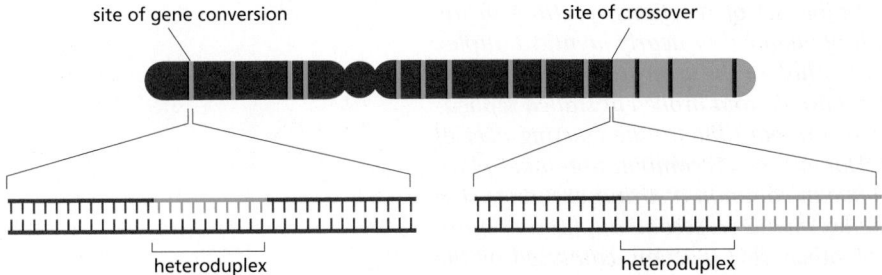

site of gene conversion

site of crossover

heteroduplex

heteroduplex

Figure 5–58 Heteroduplexes formed during meiosis. Heteroduplex DNA is present at sites of recombination that are resolved either as crossovers or non-crossovers. Because the DNA sequences of maternal and paternal chromosomes differ at many positions along their lengths, heteroduplexes often contain a small number of base-pair mismatches.

Figure 5–59 Gene conversion caused by mismatch correction. In this process, heteroduplex DNA is formed at the sites of homologous recombination between maternal and paternal chromosomes. If the maternal and paternal DNA sequences are slightly different, the heteroduplex region will include some mismatched base pairs, which may then be corrected by the DNA mismatch repair machinery (see Figure 5–19). Such repair can "erase" nucleotide sequences on either the paternal or the maternal strand. The consequence of this mismatch repair is gene conversion, detected as a deviation from the segregation of equal copies of maternal and paternal alleles that normally occurs in meiosis.

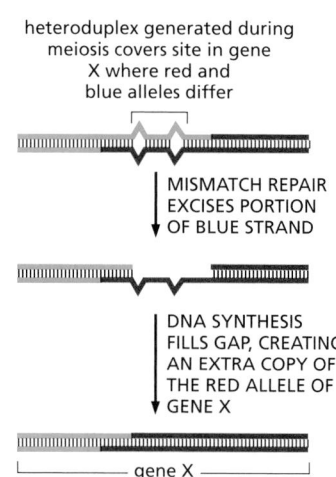

heteroduplex generated during meiosis covers site in gene X where red and blue alleles differ

MISMATCH REPAIR EXCISES PORTION OF BLUE STRAND

DNA SYNTHESIS FILLS GAP, CREATING AN EXTRA COPY OF THE RED ALLELE OF GENE X

gene X

Homologous Recombination Often Results in Gene Conversion

In sexually reproducing organisms, it is a fundamental law of genetics that—aside from mitochondrial DNA, which is inherited only through the mother—each parent makes an equal genetic contribution to an offspring. One complete set of nuclear genes is inherited from the father and one complete set is inherited from the mother. Underlying this law is the accurate parceling out of chromosomes to the germ cells (eggs and sperm) that takes place during meiosis. Thus, when a diploid cell in a parent undergoes meiosis to produce four haploid germ cells, exactly half of the genes distributed among these four cells should be maternal (genes inherited from the mother of this parent) and the other half paternal (genes inherited from the father of this parent). In some organisms (fungi, for example), it is possible to recover and analyze all four of the haploid gametes produced from a single cell by meiosis. Studies in such organisms have revealed rare cases in which the parceling out of genes violates the standard genetic rules. Occasionally, for example, meiosis yields three copies of the maternal version of a gene and only one copy of the paternal allele. Alternative versions of the same gene are called **alleles**, and it is the divergence from their expected distribution during meiosis that is known as **gene conversion**. Genetic studies show that only small sections of DNA typically undergo gene conversion, and in many cases only a part of a gene is changed.

Several pathways in the cell can lead to gene conversion, but one of the most important arises from a particular consequence of recombination during meiosis. We have seen that both crossovers and non-crossovers produce heteroduplex regions of DNA. If the two strands that make up a heteroduplex region do not have identical nucleotide sequences, mismatched base pairs are formed, and these are often repaired by the cell's mismatch repair system (see Figure 5–19). However, the mismatch repair system cannot distinguish between the paternal and maternal strands and will randomly choose the strand to be used as a template. As a consequence, one allele will be lost and the other duplicated (**Figure 5–59**), resulting in net "conversion" of one allele to the other. Thus, gene conversion, originally regarded as a mysterious deviation from the rules of genetics, can be seen as a straightforward consequence of the mechanisms of homologous recombination.

Summary

Homologous recombination describes a flexible set of reactions resulting in the exchange of DNA sequences between a pair of identical or nearly identical duplex DNA molecules. In all cells, this process is essential for the error-free repair of chromosome damage, particularly double-strand breaks and broken or stalled replication forks. Homologous recombination is also responsible for the crossing-over of chromosomes that occurs during meiosis. Homologous recombination takes place through a variety of pathways, but they have in common a strand-exchange step whereby a single strand from one DNA duplex invades a second duplex and base-pairs with one strand while displacing the other. This reaction, catalyzed by the RecA/Rad51 family of proteins, can only occur if the invading strand can form a short stretch of consecutive nucleotide pairs with one of the strands of the duplex. This requirement ensures that homologous recombination occurs only between identical or very similar DNA sequences.

When used as a repair mechanism, homologous recombination occurs between a damaged DNA molecule and its recently duplicated sister molecule, with the undamaged duplex acting as a template to repair the damaged copy flawlessly.

In meiosis, homologous recombination is initiated by deliberate, carefully regulated double-strand breaks and occurs preferentially between the homologous chromosomes rather than the newly replicated sister chromatids. The outcome can be either two chromosomes that have crossed over (that is, chromosomes in which the DNA on either side of the site of DNA pairing originates from two different homologs) or two non-crossover chromosomes. In the latter case, the two chromosomes that result are identical to the original two homologs, except for relatively minor DNA sequence changes at the site of recombination.

TRANSPOSITION AND CONSERVATIVE SITE-SPECIFIC RECOMBINATION

We have seen that homologous recombination can result in the exchange of DNA sequences between chromosomes. However, the order of genes on the interacting chromosomes typically remains the same following homologous recombination, inasmuch as the recombining sequences must be very similar for the process to occur. In this section, we describe two very different types of recombination—**transposition** (also called *transpositional recombination*) and **conservative site-specific recombination**—that do not require substantial regions of DNA homology. These two types of recombination reactions can alter gene order along a chromosome and can cause unusual types of mutations that introduce whole blocks of DNA sequence into the genome.

Transposition and conservative site-specific recombination are largely dedicated to moving a wide variety of specialized segments of DNA—collectively termed *mobile genetic elements*—from one position in a genome to another. We will see that mobile genetic elements can range in size from a few hundred to tens of thousands of nucleotide pairs, and each typically carries a unique set of genes. Often, one of these genes encodes a specialized enzyme that catalyzes the movement of only that element, thereby making this type of recombination possible.

Virtually all cells contain mobile genetic elements (known informally as "jumping genes"). As explained in Chapter 4, over evolutionary time scales, they have had a profound effect on the shaping of modern genomes. For example, nearly half of the human genome can be traced to these elements (see Figure 4–62). Over time, random mutation has altered their nucleotide sequences, and, as a result, only a few of the many copies of these elements in our DNA are still active and capable of movement. The remainder are molecular fossils whose existence provides striking clues to our evolutionary history.

Mobile genetic elements are often considered to be molecular parasites (they are also termed "selfish DNA") that persist because cells cannot get rid of them; they certainly have come close to overrunning our own genome. However, mobile DNA elements can provide benefits to the cell. For example, the genes they carry are sometimes advantageous, as in the case of antibiotic resistance in bacterial cells, discussed below. The movement of mobile genetic elements also produces many of the genetic variants upon which evolution depends, because, in addition to moving themselves, mobile genetic elements occasionally rearrange neighboring sequences of the host genome. Thus, spontaneous mutations observed in *Drosophila*, humans, and other organisms are often due to the movement of mobile genetic elements. While many of these mutations will be deleterious to the organism, some will be advantageous and may spread throughout the population. It is almost certain that much of the variety of life we see around us originally arose from the movement of mobile genetic elements.

In this section, we introduce mobile genetic elements and describe the mechanisms that enable them to move around a genome. We shall see that some of these elements move through transposition mechanisms and others through conservative site-specific recombination. We begin with transposition, as there are many more known examples of this type of movement.

Through Transposition, Mobile Genetic Elements Can Insert Into Any DNA Sequence

Mobile elements that move by way of transposition are called **transposons**, or **transposable elements**. In transposition, a specific enzyme, usually encoded by the transposon itself and typically called a *transposase*, acts on specific DNA sequences at each end of the transposon, causing it to insert into a new target DNA site. Most transposons are only modestly selective in choosing their target site, and they can therefore insert themselves into many different locations in a genome. In particular, there is no general requirement for sequence similarity between the ends of the element and the target sequence. Most transposons move only rarely. In bacteria, where it is possible to measure the frequency accurately, transposons typically move once every 10^5 cell divisions. More frequent movement would probably destroy the host cell's genome.

On the basis of their structure and transposition mechanism, transposons can be grouped into three large classes: *DNA-only transposons, retroviral-like retrotransposons*, and *nonretroviral retrotransposons*. The differences among them are briefly outlined in Table 5–4, and each class will be discussed in turn.

DNA-Only Transposons Can Move by a Cut-and-Paste Mechanism

DNA-only transposons, so named because they exist only as DNA during their movement, predominate in bacteria, and they are largely responsible for the spread of antibiotic resistance in bacterial strains. When antibiotics like penicillin and streptomycin first became widely available in the 1950s, most bacteria that caused human disease were susceptible to them. Now, the situation is different—antibiotics such as penicillin (and its modern derivatives) are no longer effective against many modern bacterial strains, including those causing gonorrhea and bacterial pneumonia. The spread of antibiotic resistance is due largely to genes

TABLE 5–4 Three Major Classes of Transposable Elements

Class description and structure	Specialized enzymes required for movement	Mode of movement	Examples
DNA-only transposons			
Short inverted repeats at each end	Transposase	Moves as DNA, either by cut-and-paste or replicative pathways	P element *(Drosophila)*, Ac-Ds (maize), Tn3 and Tn10 *(E. coli)*, Tam3 (snapdragon)
Retroviral-like retrotransposons			
Directly repeated long terminal repeats (LTRs) at each end	Reverse transcriptase and integrase	Moves via an RNA intermediate whose production is driven by a promoter in the LTR	Copia *(Drosophila)*, Ty1 (yeast), THE1 (human), Bs1 (maize)
Nonretroviral retrotransposons			
Poly A at 3′ end of RNA transcript; 5′ end is often truncated	Reverse transcriptase and endonuclease	Moves via an RNA intermediate that is often synthesized from a neighboring promoter	F element *(Drosophila)*, L1 (human), Cin4 (maize)

These elements range in length from 1000 to about 12,000 nucleotide pairs. Each family contains many members, only a few of which are listed here. Some viruses can also move in and out of host-cell chromosomes by transpositional mechanisms. These viruses are related to the first two classes of transposons.

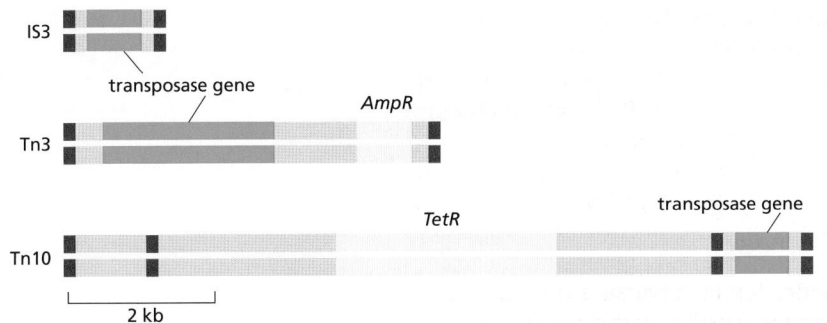

Figure 5–60 Three of the many DNA-only transposons found in bacteria. Each of these mobile DNA elements contains a gene that encodes a *transposase*, an enzyme that carries out the DNA breakage and joining reactions needed for the element to move. Each transposon also carries short DNA sequences (indicated in *red*) that are recognized only by the transposase encoded by that element and are necessary for movement of the element. In addition, two of the three mobile elements shown carry genes that encode enzymes that inactivate the antibiotics ampicillin *(AmpR)*—a penicillin derivative—and tetracycline *(TetR)*. The transposable element Tn10, shown in the bottom diagram, is thought to have evolved from the chance landing of two much shorter mobile elements on either side of a tetracycline-resistance gene.

that encode antibiotic-inactivating enzymes that are carried on transposons (**Figure 5–60**). Although these mobile elements can transpose only within cells that already carry them, they can be moved from one cell to another through other mechanisms known collectively as horizontal gene transfer (see Figure 1–19). Once introduced into a new cell, a transposon can insert itself into the genome and be faithfully passed on to all progeny cells through the normal processes of DNA replication and cell division.

DNA-only transposons can relocate from a donor site to a target site by *cut-and-paste transposition* (**Figure 5–61**). Here, the transposon is literally excised from one spot on a genome and inserted into another. This reaction produces a short duplication of the target DNA sequence at the insertion site; these direct repeat sequences that flank the transposon serve as convenient records of prior transposition events. Such "signatures" often provide valuable clues in identifying transposons in genome sequences.

When a cut-and-paste DNA-only transposon is excised from its original location, it leaves behind a "hole" in the chromosome. This lesion can be perfectly healed by recombinational double-strand break repair (see Figure 5–48), provided that the chromosome has just been replicated and an identical copy of the damaged host sequence is available. Alternatively, a nonhomologous end-joining reaction can reseal the break; in this case, the DNA sequence that originally flanked the transposon is altered, producing a mutation at the chromosomal site from which the transposon was excised (see Figure 5–45).

Remarkably, the same mechanism used to excise cut-and-paste transposons from DNA has been found to operate in developing immune systems of

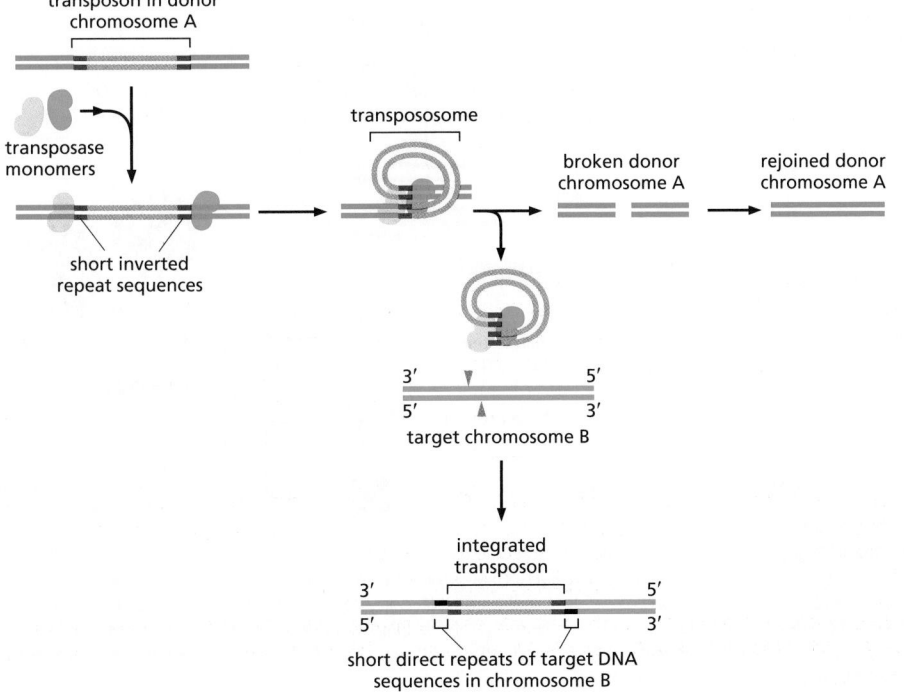

Figure 5–61 Cut-and-paste transposition. DNA-only transposons can be recognized in chromosomes by the "inverted repeat DNA sequences" *(red)* present at their ends. These sequences, which can be as short as 20 nucleotides, are all that is necessary for the DNA between them to be transposed by the particular transposase enzyme associated with the element. The cut-and-paste movement of a DNA-only transposable element from one chromosomal site to another begins when the transposase brings the two inverted DNA sequences together, forming a DNA loop. Insertion into the target chromosome, also catalyzed by the transposase, occurs at a random site through the creation of staggered breaks in the target chromosome *(purple arrowheads)*. Following the transposition reaction, the single-strand gaps created by the staggered breaks are repaired by DNA polymerase and ligase *(black)*. As a result, the insertion site is marked by a short direct repeat of the target DNA sequence, as shown. Although the break in the donor chromosome *(green)* is repaired, this process often alters the DNA sequence, causing a mutation at the original site of the excised transposable element (not shown).

vertebrates, catalyzing the DNA rearrangements that produce antibody and T cell receptor diversity. Known as *V(D)J recombination*, this process will be discussed in Chapter 24. Found only in vertebrates, V(D)J recombination is a relatively recent evolutionary novelty, but it is believed to be derived from the much more ancient cut-and-paste transposons.

Some Viruses Use a Transposition Mechanism to Move Themselves Into Host-Cell Chromosomes

Certain viruses are considered mobile genetic elements because they use transposition mechanisms to integrate their genomes into that of their host cell. However, unlike transposons, these viruses encode proteins that package their genetic information into virus particles that can infect other cells. Many of the viruses that insert themselves into a host chromosome do so by employing one of the first two mechanisms listed in Table 5–4; namely, by behaving like DNA-only transposons or like retroviral-like retrotransposons. Indeed, much of our knowledge of these mechanisms has come from studies of particular viruses that employ them.

Transposition has a key role in the life cycle of many viruses. Most notable are the **retroviruses**, which include the human AIDS virus, HIV. Outside the cell, a retrovirus exists as a single-strand RNA genome packed into a protein shell or *capsid* along with a virus-encoded **reverse transcriptase** enzyme. During the infection process, the viral RNA enters a cell and is converted to a double-strand DNA molecule by the action of this crucial enzyme, which is able to polymerize DNA on either an RNA or a DNA template (**Figure 5–62**). The term *retrovirus* refers to the virus's ability to reverse the usual flow of genetic information, which normally is from DNA to RNA (see Figure 1–4).

Once the reverse transcriptase has produced a double-strand DNA molecule, specific sequences near its two ends are recognized by a virus-encoded

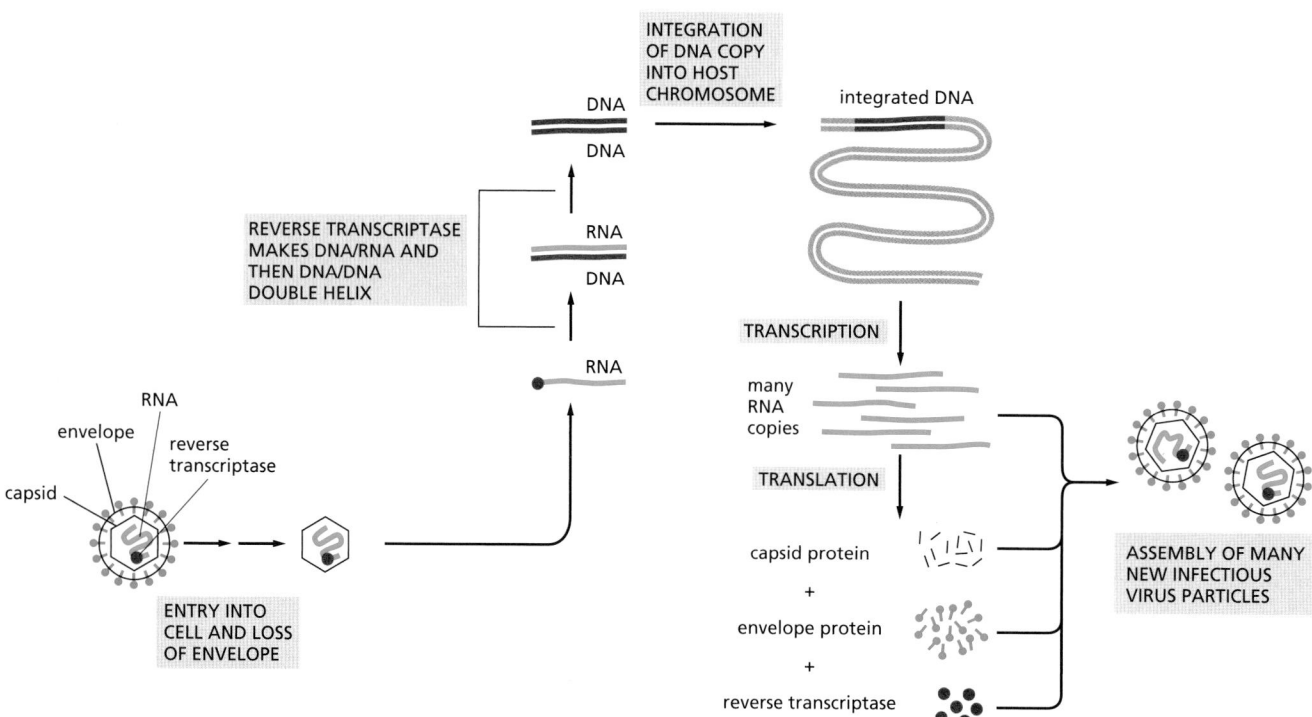

Figure 5–62 The life cycle of a retrovirus. The retrovirus genome consists of an RNA molecule *(blue)* that is typically between 7000 and 12,000 nucleotides in length. It is packaged inside a protein capsid, which is surrounded by a lipid-based envelope that contains virus-encoded envelope proteins *(green)*. Inside an infected cell, the enzyme reverse transcriptase *(red circle)* first makes a DNA copy of the viral RNA molecule and then a second DNA strand, generating a double-strand DNA copy of the RNA genome. The integration of this DNA double helix into the host chromosome is then catalyzed by a virus-encoded integrase enzyme. This integration is required for the synthesis of new viral RNA molecules by the host-cell RNA polymerase, the enzyme that transcribes DNA into RNA (discussed in Chapter 6).

transposase called *integrase*. Integrase then inserts the viral DNA into the chromosome by a mechanism similar to that used by the cut-and-paste DNA-only transposons (see Figure 5–61).

Retroviral-like Retrotransposons Resemble Retroviruses, but Lack a Protein Coat

A large family of transposons called **retroviral-like retrotransposons** (see Table 5–4) move themselves in and out of chromosomes by a mechanism that is similar to that used by retroviruses. These elements are present in organisms as diverse as yeasts, flies, and mammals; unlike viruses, they have no intrinsic ability to leave their resident cell but are passed along to all descendants of that cell through the normal processes of DNA replication and cell division. The first step in their transposition is the transcription of the entire transposon, producing an RNA copy of the element that is typically several thousand nucleotides long. This transcript, which is translated as a messenger RNA by the host cell, encodes a reverse transcriptase enzyme. This enzyme makes a double-strand DNA copy of the RNA molecule via an RNA–DNA hybrid intermediate, precisely mirroring the early stages of infection by a retrovirus (see Figure 5–62). Like a retrovirus, the linear, double-strand DNA molecule then integrates into a site on the chromosome using an integrase enzyme that is also encoded by the element. The structure and mechanisms of these integrases closely resemble those of the transposases of DNA-only transposons.

A Large Fraction of the Human Genome Is Composed of Nonretroviral Retrotransposons

A significant fraction of many vertebrate chromosomes is made up of repeated DNA sequences. In human chromosomes, these repeats are mostly mutated and truncated versions of **nonretroviral retrotransposons**, the third major type of transposon (see Table 5–4). Although most of these transposons in the human genome are immobile, a few retain the ability to move. Relatively recent movements of the *L1 element* (sometimes referred to as a LINE or long interspersed nuclear element) have been identified, some of which result in human disease; for example, a particular type of hemophilia results from an *L1* insertion into the gene encoding the blood-clotting protein Factor VIII (see Figure 6–24).

Nonretroviral retrotransposons are found in many organisms and move via a distinct mechanism that requires a complex of an endonuclease and a reverse transcriptase. As illustrated in **Figure 5–63**, the RNA and reverse transcriptase have a much more direct role in the recombination event than they do in the retroviral-like retrotransposons described above.

Inspection of the human genome sequence reveals that the bulk of nonretroviral retrotransposons—for example, the many copies of the *Alu* element, a member of the SINE (short interspersed nuclear element) family—do not carry their own endonuclease or reverse transcriptase genes. Nonetheless, they have successfully amplified themselves to become major constituents of our genome, presumably by pirating enzymes encoded by other transposons. Together the LINEs and SINEs make up over 30% of the human genome (see Figure 4–62); there are 500,000 copies of the former and over a million of the latter.

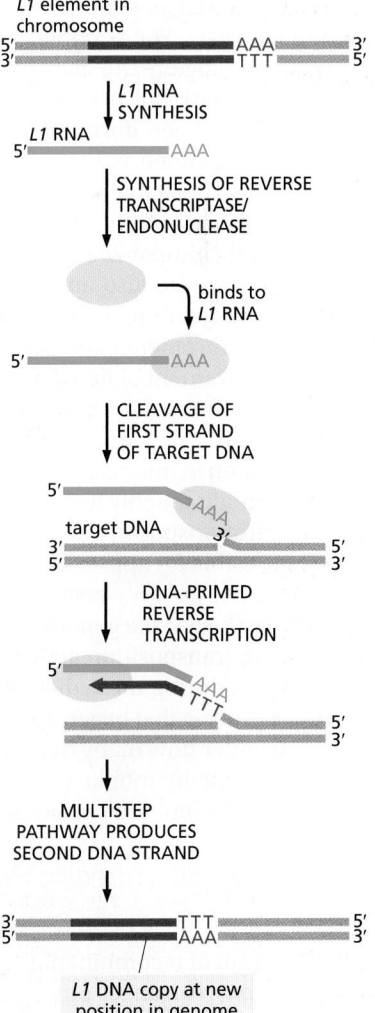

Figure 5–63 Transposition by a nonretroviral retrotransposon. Transposition of the *L1* element *(red)* begins when an endonuclease attached to the *L1* reverse transcriptase *(green)* and the *L1* RNA *(blue)* nick the target DNA at the point at which insertion will occur. This cleavage releases a 3′-OH DNA end in the target DNA, which is then used as a primer for the reverse transcription step shown. This generates a single-strand DNA copy of the element that is directly linked to the target DNA. In subsequent reactions, further processing of the single-strand DNA copy results in the generation of a new double-strand DNA copy of the *L1* element that is inserted at the site of the initial nick.

Different Transposable Elements Predominate in Different Organisms

We have described several types of transposable elements: (1) DNA-only transposons, the movement of which is based on DNA breaking and joining reactions; (2) retroviral-like retrotransposons, which also move via DNA breakage and joining, but where RNA has a key role as a template to generate the DNA recombination substrate; and (3) nonretroviral retrotransposons, in which an RNA copy of the element is central to the incorporation of the element into the target DNA, acting as a direct template for a DNA target-primed reverse transcription event.

Intriguingly, different types of transposons predominate in different organisms. For example, the vast majority of bacterial transposons are DNA-only types, with a few related to the nonretroviral retrotransposons also present. In yeasts, the main mobile elements are retroviral-like retrotransposons. In *Drosophila*, DNA-based, retroviral, and nonretroviral transposons are all found. Finally, the human genome contains all three types of transposon, but as discussed below, their evolutionary histories are strikingly different.

Genome Sequences Reveal the Approximate Times at Which Transposable Elements Have Moved

The nucleotide sequence of the human genome provides a rich fossil record of the activity of transposons over evolutionary time spans. By carefully comparing the nucleotide sequences of the approximately 3 million transposable element remnants in the human genome, it has been possible to broadly reconstruct the movements of transposons in our ancestors' genomes over the past several hundred million years. For example, the DNA-only transposons appear to have been very active well before the divergence of humans and Old World monkeys (25–35 million years ago), but because they gradually accumulated inactivating mutations, they have been dormant in the human lineage since that time. Likewise, although our genome is littered with relics of retroviral-like retrotransposons, none appear to be active today. Only a single family of retroviral-like retrotransposons is believed to have transposed in the human genome since the divergence of human and chimpanzee approximately 6 million years ago. The nonretroviral retrotransposons are also ancient, but in contrast to other types, some are still moving in our genome, as mentioned previously. For example, it is estimated that *de novo* movement of an *Alu* element is seen once in every 100–200 human births. The movement of nonretroviral retrotransposons is responsible for a small but significant fraction of new human mutations—perhaps two mutations out of every thousand.

The situation in mice is significantly different. Although the mouse and human genomes contain roughly the same density of the three types of transposons, both types of retrotransposons are still actively transposing in the mouse genome, being responsible for approximately 10% of new mutations.

Although we are only beginning to understand how the movements of transposons have shaped the genomes of present-day mammals, it has been proposed that bursts in transposition activity could have been responsible for critical speciation events during the radiation of the mammalian lineages from a common ancestor, a process that began approximately 170 million years ago. At present, we can only wonder how many of our uniquely human qualities arose from the past activity of the many mobile genetic elements whose remnants are found today scattered throughout our chromosomes.

Conservative Site-Specific Recombination Can Reversibly Rearrange DNA

A different kind of recombination mechanism, known as *conservative site-specific recombination*, rearranges other types of mobile DNA elements. In this pathway, breakage and joining occur at two special sites, one on each participating DNA

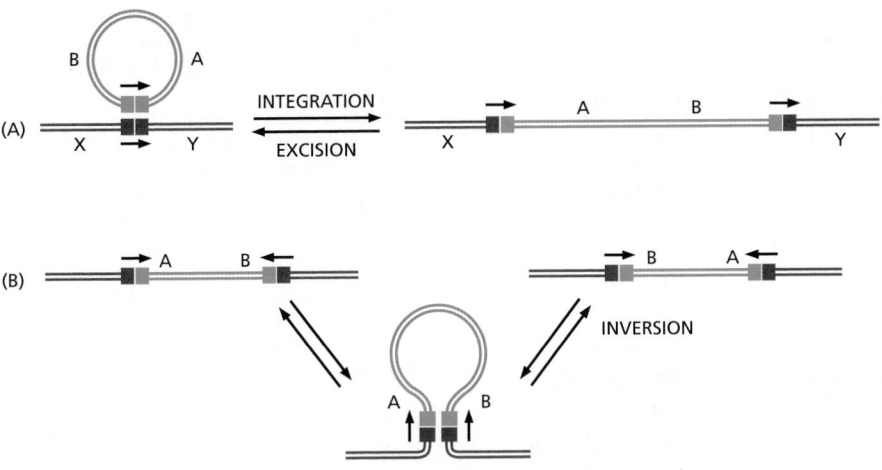

Figure 5–64 **Two types of DNA rearrangement produced by conservative site-specific recombination.** The only difference between the reactions in (A) and (B) is the relative orientation of the two short DNA sites (indicated by *arrows*) at which a site-specific recombination event occurs. (A) Through an integration reaction, a circular DNA molecule can become incorporated into a second DNA molecule; by the reverse reaction (excision), it can exit to re-form the original DNA circle. Many bacterial viruses move in and out of their host chromosomes in this way. (B) Conservative site-specific recombination can also invert a specific segment of DNA in a chromosome. A well-studied example of DNA inversion through site-specific recombination occurs in the bacterium *Salmonella typhimurium*, an organism that is a major cause of food poisoning in humans; as described in the following section, the inversion of a DNA segment changes the type of flagellum that is produced by the bacterium.

molecule. Depending on the positions and relative orientations of the two recombination sites, DNA integration, DNA excision, or DNA inversion can occur (**Figure 5–64**). Conservative site-specific recombination is carried out by specialized enzymes that break and rejoin two DNA double helices at specific sequences on each DNA molecule. The same enzyme system that joins two DNA molecules can often take them apart again, precisely restoring the sequence of the two original DNA molecules (see Figure 5–64A).

Conservative site-specific recombination is often used by DNA viruses to move their genomes in and out of the genomes of their host cells. When integrated into its host genome, the viral DNA is replicated along with the host DNA and is faithfully passed on to all descendent cells. If the host cell suffers damage (for example, by UV irradiation), the virus can reverse the site-specific recombination reaction, excise its genome, and package it into a virus particle. In this way, many viruses can replicate themselves passively as a component of the host genome, but can also "leave the sinking ship" by excising their genomes and packaging them in a protective coat until a new, healthy host cell is encountered.

Several features distinguish conservative site-specific recombination from transposition. First, conservative site-specific recombination requires specialized DNA sequences on both the donor and recipient DNA (hence the term site-specific). These sequences contain recognition sites for the particular recombinase that will catalyze the rearrangement. In contrast, transposition requires only that the transposon have a specialized sequence; for most transposons, the recipient DNA can be of any sequence. Second, the reaction mechanisms are fundamentally different. The recombinases that catalyze conservative site-specific recombination resemble topoisomerases in the sense that they form transient high-energy covalent bonds with the DNA and use this energy to complete the DNA rearrangements (see Figure 5–21). Thus, all the phosphate bonds that are broken during a recombination event are restored upon its completion (hence the term conservative). Transposition, in contrast, does not proceed through a covalently joined protein–DNA intermediate, and this process leaves gaps in the DNA that must be repaired by DNA polymerases.

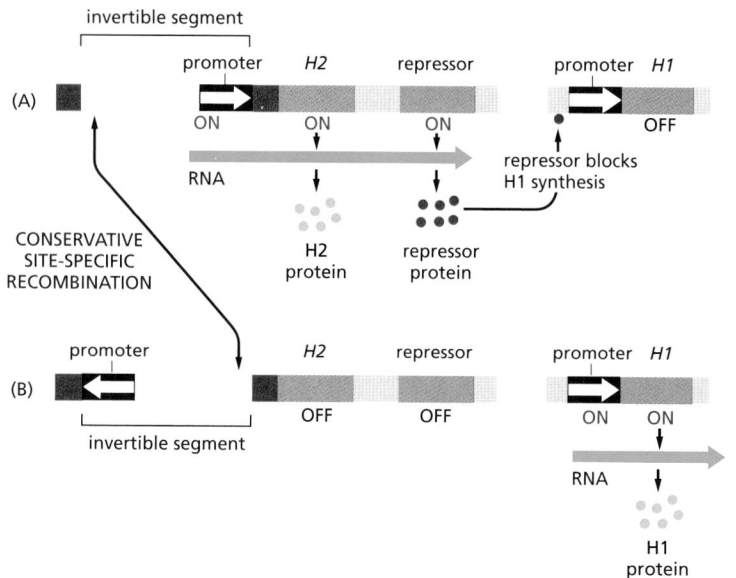

Figure 5–65 **Switching gene expression by DNA inversion in bacteria.**
Alternating transcription of two flagellin genes in a *Salmonella* bacterium is caused by a conservative site-specific recombination event that inverts a small DNA segment containing a promoter. (A) In one orientation, the promoter activates transcription of the *H2* flagellin gene as well as that of a repressor protein that blocks the expression of the *H1* flagellin gene. Promoters and repressors are described in detail in Chapter 7; here we note simply that a promoter is needed to express a gene into protein and that a repressor blocks this from happening. (B) When the promoter is inverted, it no longer turns on *H2* or the repressor, and the *H1* gene, which is thereby released from repression, is expressed instead. The inversion reaction requires specific DNA sequences *(red)* and a recombinase enzyme that is encoded in the invertible DNA segment. This site-specific recombination mechanism is activated only rarely (about once in every 10^5 cell divisions). Therefore, the production of one or the other flagellin tends to be faithfully inherited in each clone of cells.

Conservative Site-Specific Recombination Can Be Used to Turn Genes On or Off

Many bacteria use conservative site-specific recombination to control the expression of particular genes. A well-studied example occurs in *Salmonella* bacteria and is known as **phase variation**. The switch in gene expression results from the occasional inversion of a specific 1000-nucleotide-pair piece of DNA, brought about by a conservative site-specific recombinase encoded in the *Salmonella* genome. This change alters the expression of the cell-surface protein flagellin, for which the bacterium has two different genes (**Figure 5–65**). The DNA inversion changes the orientation of a promoter (a DNA sequence that directs transcription of a gene) that is located within the inverted DNA segment. With the promoter in one orientation, the bacteria synthesize one type of flagellin; with the promoter in the other orientation, they synthesize the other type. The recombination reaction is reversible, allowing bacterial populations to switch back and forth between the two types of flagellin. Inversions occur only rarely, and because such changes in the genome will be copied faithfully during all subsequent replication cycles, entire clones of bacteria will have one type of flagellin or the other.

Phase variation helps protect the bacterial population against the immune response of its vertebrate host. If the host makes antibodies against one type of flagellin, a few bacteria whose flagellin has been altered by gene inversion will still be able to survive and multiply.

Bacterial Conservative Site-Specific Recombinases Have Become Powerful Tools for Cell and Developmental Biologists

Like many of the mechanisms used by cells and viruses, site-specific recombination has been put to work by scientists to study a wide variety of problems. To decipher the roles of specific genes and proteins in complex multicellular organisms, genetic engineering techniques are used to produce worms, flies, and mice carrying a gene encoding a site-specific recombination enzyme plus a carefully designed target DNA with the DNA sites that this enzyme recognizes. At an appropriate time, the gene encoding the enzyme can be activated to rearrange the target DNA sequence. Such a rearrangement is widely used to delete a specific gene in a particular tissue of a multicellular organism (**Figure 5–66**). It is particularly useful when the gene of interest plays a key role in the early development of many tissues, and a complete deletion of the gene from the germ line would cause death

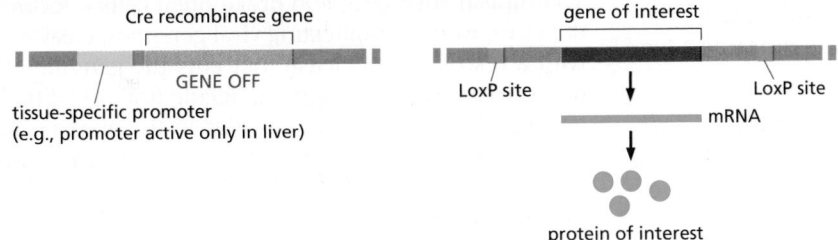

IN SPECIFIC TISSUE (e.g., LIVER)

Cre recombinase gene

GENE ON

mRNA

Cre recombinase made
only in liver cells

gene of interest

LoxP site   LoxP site

+

gene of interest deleted from chromosome
and lost as liver cells divide

IN OTHER TISSUES, THE GENE OF INTEREST IS EXPRESSED NORMALLY

Cre recombinase gene

GENE OFF

tissue-specific promoter
(e.g., promoter active only in liver)

gene of interest

LoxP site   LoxP site

mRNA

protein of interest

Figure 5–66 How a conservative site-specific recombination enzyme from bacteria is used to delete specific genes from particular mouse tissues. This approach requires the insertion of two specially engineered DNA molecules into the animal's germ line. The first contains the gene for a recombinase (in this case, the Cre recombinase from the bacteriophage P1) under the control of a tissue-specific promoter, which ensures that the recombinase is expressed only in that tissue. The second DNA molecule contains the gene of interest flanked by recognition sites (in this case, LoxP sites) for the recombinase. The mouse is engineered so that this is the only copy of this gene. Therefore, if the recombinase is expressed only in the liver, the gene of interest will be deleted there, and only there. The reaction that excises the gene is the same as that shown in Figure 5–64A. As described in Chapter 7, many tissue-specific promoters are known; moreover, many of these promoters are active only at specific times in development. Thus, it is possible to study the effect of deleting specific genes at different times during the development of each tissue.

very early in development. The same strategy can also be used to inappropriately express any specific gene in a tissue of interest; here, the triggered deletion joins a strong transcriptional promoter to the gene of interest. With this tool one can in principle determine the influence of any protein in any desired tissue of an intact animal.

Summary

The genomes of nearly all organisms contain mobile genetic elements that can move from one position in the genome to another by either transpositional or conservative site-specific recombination processes. In most cases, this movement is random and happens at a very low frequency. Mobile genetic elements include transposons, which move within a single cell (and its descendants), plus those viruses whose genomes can integrate into the genomes of their host cells.

There are three classes of transposons: the DNA-only transposons, the retroviral-like retrotransposons, and the nonretroviral retrotransposons. All but the last have close relatives among the viruses. Although viruses and transposable elements can be viewed as parasites, many of the new arrangements of DNA sequences that their site-specific recombination events produce have played an important part in creating the genetic variation crucial for the evolution of cells and organisms.

WHAT WE DON'T KNOW

• How does DNA replication contend with all the other processes that occur simultaneously on chromosomes, including DNA repair and gene transcription?

• What is the basis for the low frequency of errors in DNA replication observed in all cells? Is this the best that cells can do given the speed of replication and the limits of molecular diffusion? Was this mutation rate selected in evolution to provide genetic variation?

• Cells have only one fundamental way of replicating DNA but many different ways of repairing it. Are there still other, undiscovered ways that cells have for repairing DNA?

• Do the many "dead" transposons in the human genome provide any benefits to humans?

PROBLEMS

Which statements are true? Explain why or why not.

5–1   The different cells in your body rarely have genomes with the identical nucleotide sequence.

5–2   In *E. coli,* where the replication fork travels at 500 nucleotide pairs per second, the DNA ahead of the fork—in the absence of topoisomerase—would have to rotate at nearly 3000 revolutions per minute.

5–3   In a replication bubble, the same parental DNA strand serves as the template strand for leading-strand synthesis in one replication fork and as the template for lagging-strand synthesis in the other fork.

5–4   When bidirectional replication forks from adjacent origins meet, a leading strand always runs into a lagging strand.

5–5   DNA repair mechanisms all depend on the existence of two copies of the genetic information, one in each of the two homologous chromosomes.

Discuss the following problems.

5–6   To determine the reproducibility of mutation frequency measurements, you do the following experiment. You inoculate each of 10 cultures with a single *E. coli* bacterium, allow the cultures to grow until each contains 10^6 cells, and then measure the number of cells in each culture that carry a mutation in your gene of interest. You were so surprised by the initial results that you repeated the experiment to confirm them. Both sets of results display the same extreme variability, as shown in Table Q5–1. Assuming that the rate of mutation is constant, why do you suppose there is so much variation in the frequencies of mutant cells in different cultures?

TABLE Q5–1 Frequencies of mutant cells in multiple cultures (Problem 5–6)

Experiment	Culture (mutant cells/10^6 cells)									
	1	2	3	4	5	6	7	8	9	10
1	4	0	257	1	2	32	0	0	2	1
2	128	0	1	4	0	0	66	5	0	2

5–7   DNA repair enzymes preferentially repair mismatched bases on the newly synthesized DNA strand, using the old DNA strand as a template. If mismatches were instead repaired without regard for which strand served as template, would mismatch repair reduce replication errors? Would such a mismatch repair system result in fewer mutations, more mutations, or the same number of mutations as there would have been without any repair at all? Explain your answers.

5–8   Discuss the following statement: "Primase is a sloppy enzyme that makes many mistakes. Eventually, the RNA primers it makes are replaced with DNA made by a polymerase with higher fidelity. This is wasteful. It would be more energy-efficient if a DNA polymerase made an accurate copy in the first place."

5–9   If DNA polymerase requires a perfectly paired primer in order to add the next nucleotide, how is it that any mismatched nucleotides "escape" this requirement and become substrates for mismatch repair enzymes?

5–10   The laboratory you joined is studying the life cycle of an animal virus that uses circular, double-strand DNA as its genome. Your project is to define the location of the origin(s) of replication and to determine whether replication proceeds in one or both directions away from an origin (unidirectional or bidirectional). To accomplish your goal, you broke open cells infected with the virus, isolated replicating viral genomes, cleaved them with a restriction nuclease that cuts the genome at only one site to produce a linear molecule from the circle, and examined the resulting molecules in the electron microscope. Some of the molecules you observed are illustrated schematically in Figure Q5–1. (Note that it is impossible to distinguish the orientation of one DNA molecule relative to another in the electron microscope.)

You must present your conclusions to the rest of the lab tomorrow. How will you answer the two questions your advisor posed for you? Is there a single, unique origin of replication or several origins? Is replication unidirectional or bidirectional?

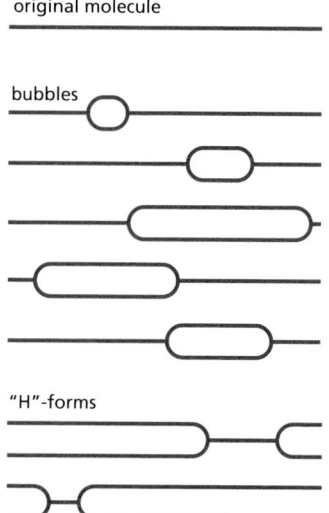

Figure Q5–1 Parental and replicating forms of an animal virus (Problem 5–10).

5–11   You are investigating DNA synthesis in tissue-culture cells, using ^{3}H-thymidine to radioactively label the replication forks. By breaking open the cells in a way that allows some of the DNA strands to be stretched out, very long DNA strands can be isolated intact and examined. You overlay the DNA with a photographic emulsion, and expose it for 3 to 6 months, a procedure known as autoradiography. Because the emulsion is sensitive to radioactive emissions, the ^{3}H-labeled DNA shows up as tracks of silver grains. Because the stretching collapses replication

(A)

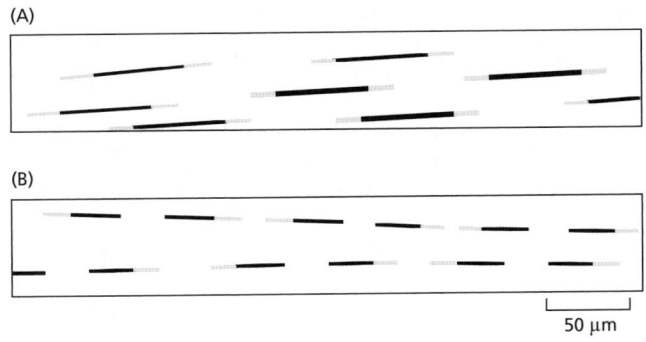

(B)

50 μm

Figure Q5–2 Autoradiographic investigation of DNA replication in cultured cells (Problem 5–11). (A) Addition of ³H-labeled thymidine immediately after release from the synchronizing block. (B) Addition of ³H-labeled thymidine 30 minutes after release from the synchronizing block.

bubbles, the daughter duplexes lie side by side and cannot be distinguished from each other.

You pretreat the cells to synchronize them at the beginning of S phase. In the first experiment, you release the synchronizing block and add ³H-thymidine immediately. After 30 minutes, you wash the cells and change the medium so that the total concentration of thymidine is the same as it was, but only one-third of it is radioactive. After an additional 15 minutes, you prepare DNA for autoradiography. The results of this experiment are shown in Figure Q5–2A. In the second experiment, you release the synchronizing block and then wait 30 minutes before adding ³H-thymidine. After 30 minutes in the presence of ³H-thymidine, you once again change the medium to reduce the concentration of radioactive thymidine and incubate the cells for an additional 15 minutes. The results of the second experiment are shown in Figure Q5–2B.

A. Explain why, in both experiments, some regions of the tracks are dense with silver grains (dark), whereas others are less dense (light).

B. In the first experiment, each track has a central dark section with light sections at each end. In the second experiment, the dark section of each track has a light section at only one end. Explain the reason for this difference.

C. Estimate the rate of fork movement (μm/min) in these experiments. Do the estimates from the two experiments agree? Can you use this information to gauge how long it would take to replicate the entire genome?

5–12 If you compare the frequency of the sixteen possible dinucleotide sequences in the *E. coli* and human genomes, there are no striking differences except for one dinucleotide, 5′-CG-3′. The frequency of CG dinucleotides in the human genome is significantly lower than in *E. coli* and significantly lower than expected by chance. Why do you suppose that CG dinucleotides are underrepresented in the human genome?

5–13 With age, somatic cells are thought to accumulate genomic "scars" as a result of the inaccurate repair of double-strand breaks by nonhomologous end joining (NHEJ).

Estimates based on the frequency of breaks in primary human fibroblasts suggest that by age 70, each human somatic cell may carry some 2000 NHEJ-induced mutations due to inaccurate repair. If these mutations were distributed randomly around the genome, how many protein-coding genes would you expect to be affected? Would you expect cell function to be compromised? Why or why not? (Assume that 2% of the genome—1.5% protein-coding and 0.5% regulatory—is crucial information.)

5–14 Draw the structure of the double Holliday junction that would result from strand invasion by both ends of the broken duplex into the intact homologous duplex shown in Figure Q5–3. Label the left end of each strand in the Holliday junction 5′ or 3′ so that the relationship to the parental and recombinant duplexes is clear. Indicate how DNA synthesis would be used to fill in any single-strand gaps in your double Holliday junction.

5′　　　　　　　　　　3′

5′　　　　　　　　　　3′

Figure Q5–3 A broken duplex with single-strand tails ready to invade an intact homologous duplex (Problem 5–14).

5–15 In addition to correcting DNA mismatches, the mismatch repair system functions to prevent homologous recombination from taking place between similar but not identical sequences. Why would recombination between similar, but nonidentical sequences pose a problem for human cells?

5–16 Cre recombinase is a site-specific enzyme that catalyzes recombination between two LoxP DNA sites. Cre recombinase pairs two LoxP sites in the same orientation, breaks both duplexes at the same point in each LoxP site, and joins the ends with new partners so that each LoxP site is regenerated, as shown schematically in Figure Q5–4A. Based on this mechanism, predict the arrangement of sequences that will be generated by Cre-mediated site-specific recombination for each of the two DNAs shown in Figure Q5–4B.

(A)

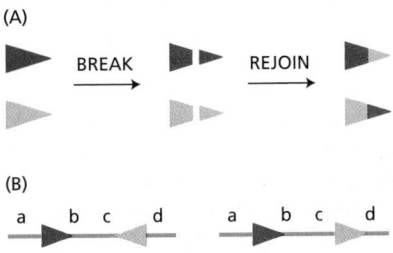

BREAK　　　　REJOIN

(B)

a　　b　c　　d　　　a　　b　c　　d

Figure Q5–4 Cre recombinase-mediated site-specific recombination (Problem 5–16). (A) Schematic representation of Cre/LoxP site-specific recombination. The LoxP sequences in the DNA are represented by *triangles* that are colored so that the site-specific recombination event can be followed more readily. In reality their DNA sequences are identical. (B) DNA substrates containing two arrangements of LoxP sites.

REFERENCES

General

Brown TA (2007) Genomes 3. New York: Garland Science.

Friedberg EC, Walker GC, Siede W et al. (2005) DNA Repair and Mutagenesis. Washington, DC: ASM Press.

Haber JE (2013) Genome Stability: DNA Repair and Recombination. New York: Garland Science.

Hartwell L, Hood L, Goldberg ML et al. (2010) Genetics: from Genes to Genomes. Boston: McGraw Hill.

Stent GS (1971) Molecular Genetics: An Introductory Narrative. San Francisco: WH Freeman.

Watson J, Baker T, Bell S et al. (2013) Molecular Biology of the Gene, 7th ed. Menlo Park, CA: Benjamin Cummings.

The Maintenance of DNA Sequences

Conrad DF, Keebler J, DePristo M et al. (2011) Variation in genome-wide mutation rates within and between human families. *Nat. Genet.* 43, 712–714.

Catarina D & Eichler EE (2013) Properties and rates of germline mutations in humans. *Trends Genet.* 29, 575–584.

Cooper GM, Brudno M, Stone ES et al. (2004) Characterization of evolutionary rates and constraints in three mammalian genomes. *Genome Res.* 14, 539–548.

Hedges SB (2002) The origin and evolution of model organisms. *Nat. Rev. Genet.* 3, 838–849.

King MC & Wilson AC (1965) Evolution at two levels in humans and chimpanzees. *Science* 188, 107–116.

DNA Replication Mechanisms

Alberts B (1998) The cell as a collection of protein machines: preparing the next generation of molecular biologists. *Cell* 92, 291–294.

Kelch BA, Makino DL, O'Donnell M et al. (2011) How a DNA polymerase clamp loader opens a sliding clamp. *Science* 334, 1675–1680.

Kornberg A (1960) Biological synthesis of DNA. *Science* 131, 1503–1508.

Li JJ & Kelly TJ (1984) SV40 DNA replication *in vitro. Proc. Natl. Acad. Sci. USA* 81, 6973–6977.

Meselson M & Stahl FW (1958) The replication of DNA in *E. coli. Proc. Natl. Acad. Sci. USA* 44, 671–682.

Modrich P & Lahue R (1996) Mismatch repair in replication fidelity, genetic recombination, and cancer biology. *Annu. Rev. Biochem.* 65, 101–133.

O'Donnell M, Langston L & Stillman B (2013) Principals and concepts of DNA replication in Bacteria, Archaea, and Eukarya. *Cold Spring Harb. Lab. Perspect. Biol.* 195, 1231–1240.

Okazaki R, Okazaki T, Sakabe K et al. (1968) Mechanism of DNA chain growth. I. Possible discontinuity and unusual secondary structure of newly synthesized chains. *Proc. Natl. Acad. Sci. USA* 59, 598–605.

Raghuraman MK, Winzeler EA, Collingwood D et al. (2001) Replication dynamics of the yeast genome. *Science* 294, 115–121.

Rao PN & Johnson RT (1970) Mammalian cell fusion: studies on the regulation of DNA synthesis and mitosis. *Nature* 225, 159.

Vos SM, Tretter EM, Schmidt BH et al. (2011) All tangled up: how cells direct, manage and exploit topoisomerase function. *Nat. Rev. Mol. Cell Biol.* 12, 827–841.

The Initiation and Completion of DNA Replication in Chromosomes

Chan SR & Blackburn EH (2004) Telomeres and telomerase. *Philos. Trans. R. Soc. Lond. B Bio. Sci.* 359, 109–121.

Gilbert DM (2010) Evaluating genome-scale approaches to eukaryotic DNA replication. *Nat. Rev. Genet.* 11, 673–684.

deLang T (2009) How telomeres solve the end-protection problem. *Science* 326, 948–952.

Mechali M (2010) Eukaryotic DNA replication origins: many choices for appropriate answers. *Nat. Rev. Mol. Cell Biol.* 11, 728–738.

Nandakumar J & Cech T (2013) Finding the end: recruitment of telomerase to telomeres. *Nat. Rev. Mol. Cell Biol.* 14, 69–82.

DNA Repair

Goodman MF & Woodgate, R (2013) Translesion DNA polymerases. *Cold Spring Harb. Perspect. Biol.* 5, a010363.

Hanawalt PC & Spivak G (2008) Transcription-coupled DNA repair: two decades of progress and surprises. *Nat. Rev. Mol. Cell Biol.* 9, 958–970.

Lindahl T (1993) Instability and decay of the primary structure of DNA. *Nature* 362, 709–715.

Malkova A & Haber JE (2012) Mutations arising during repair of chromosome breaks. *Annu. Rev. Genet.* 46, 455–473.

Prakash S, Johnson RE & Prakash L (2005) Eukaryotic translesion synthesis DNA polymerases: specificity of structure and function. *Annu. Rev. Biochem.* 74, 317–353.

Reardon JT & Sancar A (2005) Nucleotide excision repair. *Prog. Nucleic Acid Res. Mol. Biol.* 79, 183–235.

Homologous Recombination

Chen Z, Yang H & Pavletich NP (2008) Mechanism of homologous recombination from the RecA-ssDNA/dsDNA structures. *Nature* 453, 489–494.

Cox MM (2001) Historical overview: searching for replication help in all of the rec places. *Proc. Natl. Acad. Sci. USA* 98, 8173–8180.

Heyer WD, Ehmsen KT & Liu J (2010) Regulation of homologous recombination in eukaryotes. *Annu. Rev. Genet.* 44, 113–139.

Holliday R (1990) The history of the DNA heteroduplex. *BioEssays* 12, 133–142.

Hunter N (2006) Meiotic recombination. In *Topics in Current Genetics, Molecular Genetics of Recombination*, Aguilera A & Rothstein R (eds), pp. 381–422. Springer-Verlag: Heidelberg.

de Massy B (2013) Initiation of meiotic recombination: how and where? Conservation and specificities among eukaryotes. *Annu. Rev. Genet.* 47, 563–599.

Michel B, Gromponee G, Florès MJ & Bidnenko V (2004) Multiple pathways process stalled replication forks. *Proc. Natl. Acad. Sci. USA* 101, 12783–12788.

Moynahan ME & Jasin M (2010) Mitotic homologous recombination maintains genomic stability and suppresses tumorigenesis. *Nat. Rev. Mol. Cell Biol.* 11, 196–207.

Szostak JW, Orr-Weaver TK, Rothstein RJ et al. (1983) The double-strand break repair model for recombination. *Cell* 33, 25–35.

West SC (2003) Molecular views of recombination proteins and their control. *Nat. Rev. Mol. Cell Biol.* 4(6), 435–445.

Yeeles JY, Poli J, Marians KJ et al. (2013) Rescuing stalled or damaged replication forks. *Cold Spring Harb. Perspect. Biol.* 5, a012815.

Zickler D & Kleckner N (1999) Meiotic chromosomes: integrating structure and function. *Annu. Rev. Genet.* 33, 603–754.

Transposition and Conservative Site-specific Recombination

Comfort NC (2001) From controlling elements to transposons: Barbara McClintock and the Nobel Prize. *Trends Biochem. Sci.* 26, 454–457.

Grindley ND, Whiteson KL & Rice PA (2006) Mechanisms of site-specific recombination. *Annu. Rev. Biochem.* 75, 567–605.

Huang, CR, Burns KH & Boeke JD (2012) Active transposition in genomes. *Annu. Rev. Genet.* 46, 651–675.

Varmus H (1988) Retroviruses. *Science* 240, 1427–1435.

How Cells Read the Genome: From DNA to Protein

Since the structure of DNA was discovered in the early 1950s, progress in cell and molecular biology has been astounding. We now know the complete genome sequences for thousands of different organisms, revealing fascinating details of their biochemistry as well as important clues as to how these organisms evolved. Complete genome sequences have also been obtained for thousands of individual humans, as well as for a few of our now-extinct relatives, such as the Neanderthals. Knowing the maximum amount of information that is required to produce a complex organism like ourselves puts constraints on the biochemical and structural features of cells and makes it clear that biology is not infinitely complex.

As discussed in Chapter 1, the DNA in genomes does not direct protein synthesis itself, but instead uses RNA as an intermediary. When the cell needs a particular protein, the nucleotide sequence of the appropriate portion of the immensely long DNA molecule in a chromosome is first copied into RNA (a process called *transcription*). It is these RNA copies of segments of the DNA that are used directly as templates to direct the synthesis of the protein (a process called *translation*). The flow of genetic information in cells is therefore from DNA to RNA to protein (**Figure 6–1**). All cells, from bacteria to humans, express their genetic information in this way—a principle so fundamental that it is termed the *central dogma* of molecular biology. Despite the universality of the central dogma of molecular biology, there are important variations between organisms in the way in which information flows from DNA to protein. Principal among these is that RNA transcripts in eukaryotic cells are subject to a series of processing steps in the nucleus, including *RNA splicing*, before they are permitted to exit from the nucleus and be translated into protein. As we discuss in this chapter, these processing steps can critically change the "meaning" of an RNA molecule and are therefore crucial for understanding how eukaryotic cells read their genome.

Although we focus on the production of the proteins encoded by the genome in this chapter, we see that for many genes, RNA is the final product. Like proteins, some of these RNAs fold into precise three-dimensional structures that have structural and catalytic roles in the cell. Other RNAs, as we discuss in the next chapter, act primarily as regulators of gene expression. But the roles of many noncoding RNAs are not yet known.

One might have predicted that the information present in genomes would be arranged in an orderly fashion, resembling a dictionary or a telephone directory. But it turns out that the genomes of most multicellular organisms are surprisingly disorderly, reflecting their chaotic evolutionary histories. The genes in these organisms largely consist of a long string of alternating short exons and long introns, as discussed in Chapter 4 (see Figure 4–15D). Moreover, small bits of DNA sequence that code for protein are interspersed with large blocks of seemingly meaningless DNA. Some sections of the genome contain many genes and others lack genes altogether. Proteins that work closely with one another in the cell often have their genes located on different chromosomes, and adjacent genes typically encode proteins that have little to do with each other in the cell. Decoding genomes is therefore no simple matter. Even with the aid of powerful computers, it is difficult for researchers to locate definitively the beginning and end of genes, much less to decipher when and where each gene is expressed in the life of the

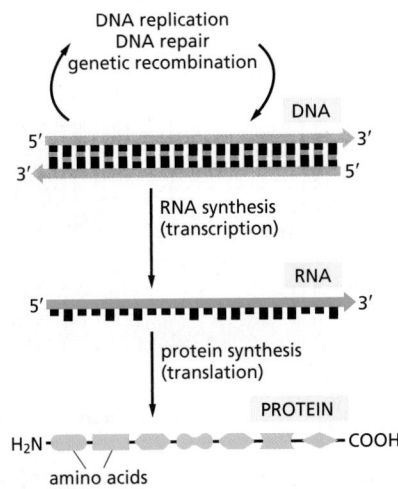

Figure 6–1 The pathway from DNA to protein. The flow of genetic information from DNA to RNA (transcription) and from RNA to protein (translation) occurs in all living cells.

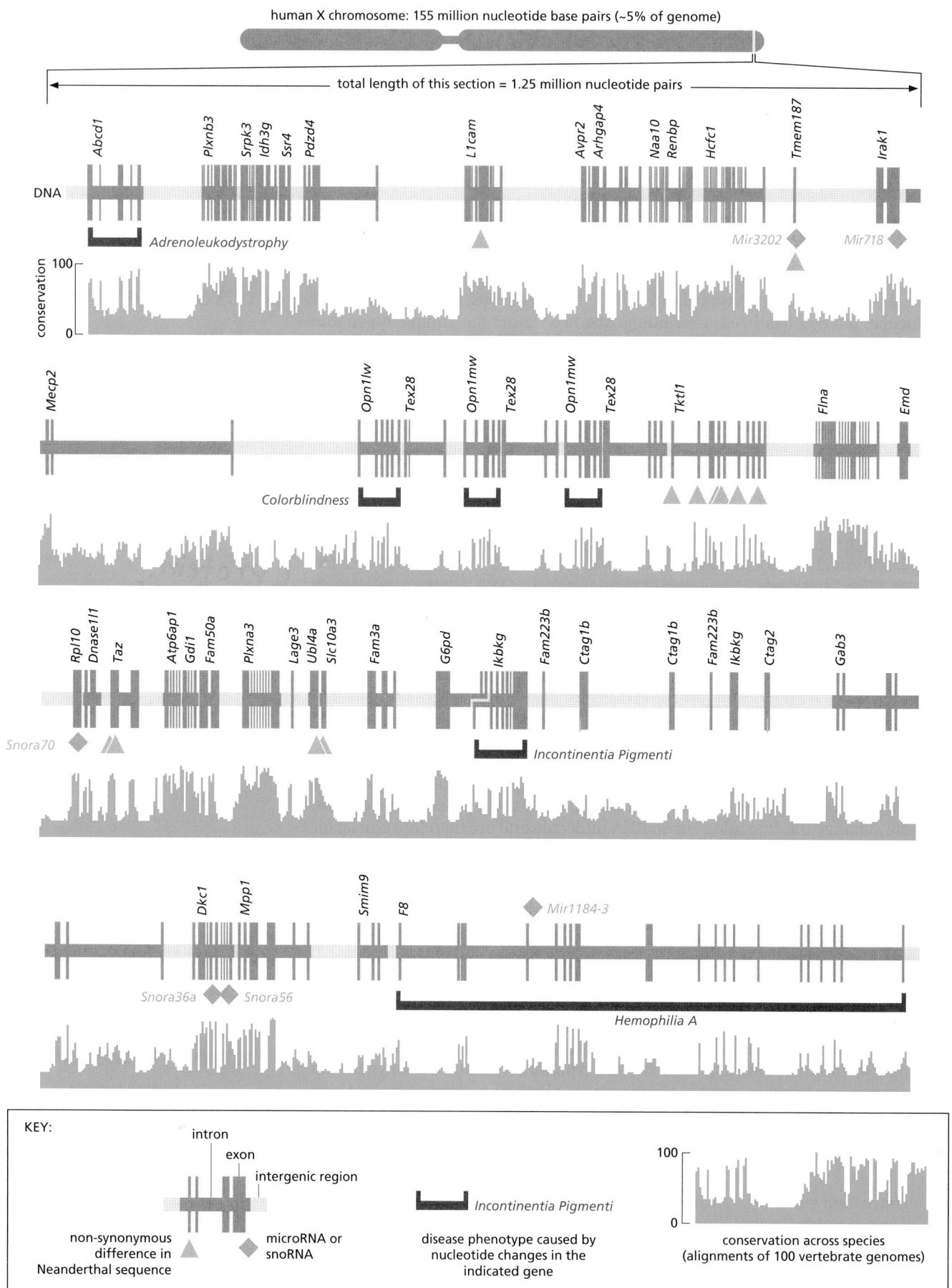

human X chromosome: 155 million nucleotide base pairs (~5% of genome)

total length of this section = 1.25 million nucleotide pairs

KEY:

intron

exon

intergenic region

non-synonymous difference in Neanderthal sequence

microRNA or snoRNA

disease phenotype caused by nucleotide changes in the indicated gene

Incontinentia Pigmenti

conservation across species (alignments of 100 vertebrate genomes)

Figure 6–2 **Schematic depiction of a small portion of the human X chromosome.** As summarized in the key, the known protein-coding genes (starting with *Abcd1* and ending with *F8*) are shown in *dark gray*, with coding regions (exons) indicated by bars that extend above and below the central line. Noncoding RNAs with known functions are indicated by *purple diamonds*. *Yellow triangles* indicate positions within protein-coding regions where the Neanderthal genome sequences codes for a different amino acid than the human genome. The stretch of *yellow triangles* in the *Txtl1* gene appear to have been positively selected for since the divergence of *Homo sapiens* from Neanderthals some 200,000 years ago. Note that most of the proteins are identical between us and our extinct relative. The *blue* histogram indicates the extent to which portions of the human genome are conserved with other vertebrate species. It is likely that additional genes, currently unrecognized, also lie within this portion of the human genome.

Genes whose mutation causes an inherited human condition are indicated by *red brackets*. The *Abcd1* gene codes for a protein that imports fatty acids into the peroxisome; mutations in the gene cause demylination of nerves which can result in cognition and movement disorders. *Incontinentia pigmenti* is a disease of the skin, hair, nails, teeth, and eyes. *Hemophilia A* is a bleeding disorder caused by mutations in the Factor VIII gene, which codes for a blood-clotting protein. Because males have only a single copy of the X chromosome, most of the conditions shown here affect only males; females that inherit one of these defective genes are often asymptomatic because a functional protein is made from their other X chromosome. (Courtesy of Alex Williams, data obtained from the University of California, Genome Browser, http://genome.ucsc.edu)

organism. Yet the cells in our body do this automatically, thousands of times a second.

The problems that cells face in decoding genomes can be appreciated by considering a tiny portion of the human genome (**Figure 6–2**). The region illustrated represents less than 1/2000th of our genome and includes at least 48 genes that encode proteins and 6 genes for noncoding RNAs. When we consider the entire human genome, we can only marvel at the capacity of our cells to rapidly and accurately handle such large amounts of information.

In this chapter, we explain how cells decode and use the information in their genomes. Much has been learned about how the genetic instructions written in an alphabet of just four "letters"—the four different nucleotides in DNA—direct the formation of a bacterium, a fruit fly, or a human. Nevertheless, we still have a great deal to discover about how the information stored in an organism's genome produces even the simplest unicellular bacterium with 500 genes, let alone how it directs the development of a human with approximately 30,000 genes. An enormous amount of ignorance remains; many fascinating challenges therefore await the next generation of cell biologists.

FROM DNA TO RNA

Transcription and translation are the means by which cells read out, or express, the genetic instructions in their genes. Because many identical RNA copies can be made from the same gene, and each RNA molecule can direct the synthesis of many identical protein molecules, cells can synthesize a large amount of protein from a gene when necessary. But genes can be transcribed and translated with different efficiencies, allowing the cell to make vast quantities of some proteins and tiny amounts of others (**Figure 6–3**). Moreover, as we see in the next chapter,

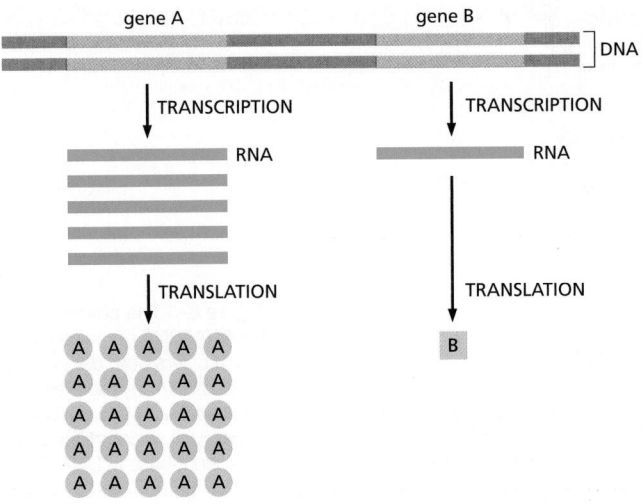

Figure 6–3 **Genes can be expressed with different efficiencies.** In this example, gene A is transcribed much more efficiently than gene B and each RNA molecule that it produces is also translated more frequently. This causes the amount of protein A in the cell to be much greater than that of protein B.

a cell can change (or regulate) the expression of each of its genes according to its needs—most commonly by controlling the production of its RNA.

RNA Molecules Are Single-Stranded

The first step a cell takes in reading out a needed part of its genetic instructions is to copy a particular portion of its DNA nucleotide sequence—a gene—into an RNA nucleotide sequence (Figure 6–4). The information in RNA, although copied into another chemical form, is still written in essentially the same language as it is in DNA—the language of a nucleotide sequence. Hence the name given to producing RNA molecules on DNA is *transcription*.

Like DNA, RNA is a linear polymer made of four different types of nucleotide subunits linked together by phosphodiester bonds (see Figure 6–4). It differs from DNA chemically in two respects: (1) the nucleotides in RNA are *ribonucleotides*—that is, they contain the sugar ribose (hence the name *ribo*nucleic acid) rather than deoxyribose; (2) although, like DNA, RNA contains the bases adenine (A), guanine (G), and cytosine (C), it contains the base uracil (U) instead of the thymine (T) in DNA (Figure 6–5). Since U, like T, can base-pair by hydrogen-bonding with A (Figure 6–6), the complementary base-pairing properties described for DNA in Chapters 4 and 5 apply also to RNA (in RNA, G pairs with C, and A pairs with U). We also find other types of base pairs in RNA: for example, G occasionally pairs with U.

Although these chemical differences are slight, DNA and RNA differ quite dramatically in overall structure. Whereas DNA always occurs in cells as a double-stranded helix, RNA is single-stranded. An RNA chain can therefore fold up into a particular shape, just as a polypeptide chain folds up to form the final shape of a protein (Figure 6–7). As we see later in this chapter, the ability to fold into complex three-dimensional shapes allows some RNA molecules to have precise structural and catalytic functions.

Transcription Produces RNA Complementary to One Strand of DNA

The RNA in a cell is made by **DNA transcription,** a process that has certain similarities to the process of DNA replication discussed in Chapter 5. Transcription begins with the opening and unwinding of a small portion of the DNA double helix to expose the bases on each DNA strand. One of the two strands of the DNA double helix then acts as a template for the synthesis of an RNA molecule. As in DNA replication, the nucleotide sequence of the RNA chain is determined by the complementary base-pairing between incoming nucleotides and the DNA

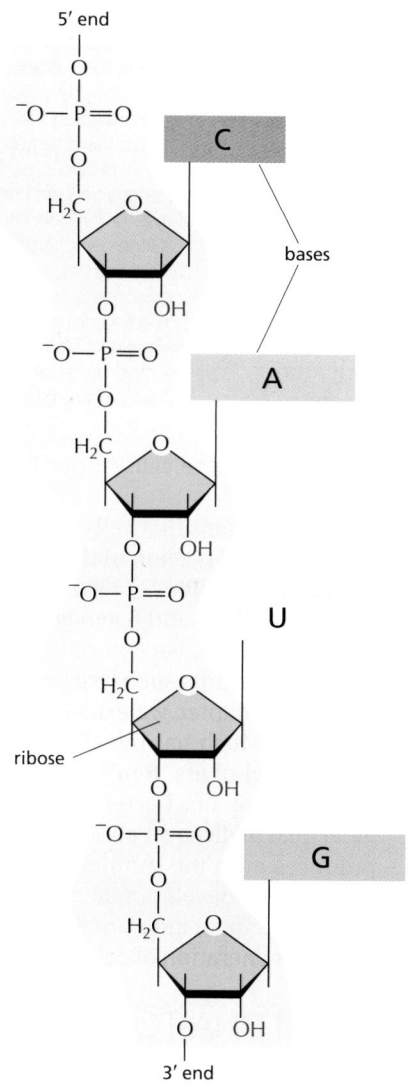

Figure 6–4 A short length of RNA. The phosphodiester chemical linkage between nucleotides in RNA is the same as that in DNA.

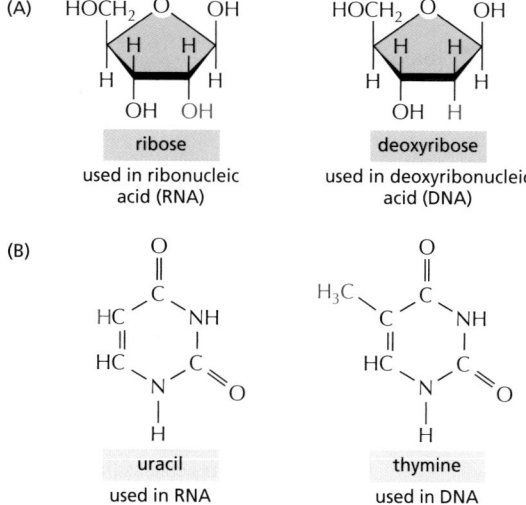

Figure 6–5 The chemical structure of RNA. (A) RNA contains the sugar ribose, which differs from deoxyribose, the sugar used in DNA, by the presence of an additional –OH group. (B) RNA contains the base uracil, which differs from thymine, the equivalent base in DNA, by the absence of a –CH₃ group.

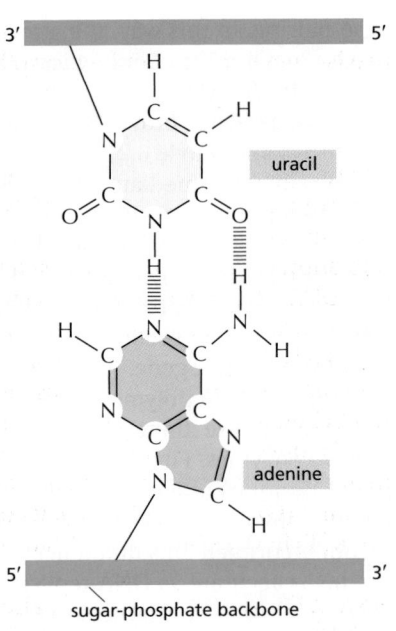

Figure 6–6 **Uracil forms base pairs with adenine.** The absence of a methyl group in U has no effect on base-pairing; thus, U-A base pairs closely resemble T-A base pairs (see Figure 4–4).

sugar-phosphate backbone

template. When a good match is made (A with T, U with A, G with C, and C with G), the incoming ribonucleotide is covalently linked to the growing RNA chain in an enzymatically catalyzed reaction. The RNA chain produced by transcription—the *transcript*—is therefore elongated one nucleotide at a time, and it has a nucleotide sequence that is exactly complementary to the strand of DNA used as the template (Figure 6–8).

Transcription, however, differs from DNA replication in several crucial ways. Unlike a newly formed DNA strand, the RNA strand does not remain hydrogen-bonded to the DNA template strand. Instead, just behind the region where the ribonucleotides are being added, the RNA chain is displaced and the DNA helix re-forms. Thus, the RNA molecules produced by transcription are released from the DNA template as single strands. In addition, because they are copied from only a limited region of the DNA, RNA molecules are much shorter than DNA molecules. A DNA molecule in a human chromosome can be up to 250 million nucleotide-pairs long; in contrast, most RNAs are no more than a few thousand nucleotides long, and many are considerably shorter.

RNA Polymerases Carry Out Transcription

The enzymes that perform transcription are called **RNA polymerases**. Like the DNA polymerase that catalyzes DNA replication (discussed in Chapter 5), RNA polymerases catalyze the formation of the phosphodiester bonds that link the nucleotides together to form a linear chain. The RNA polymerase moves stepwise along the DNA, unwinding the DNA helix just ahead of the active site for polymerization to expose a new region of the template strand for complementary

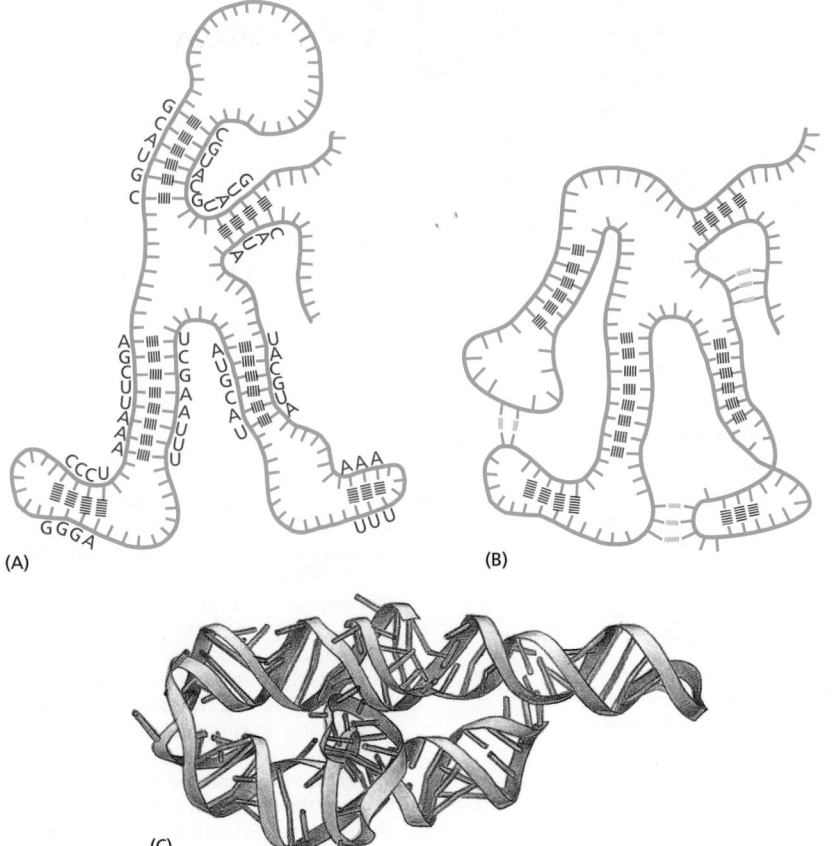

(A)

(B)

(C)

Figure 6–7 **RNA can fold into specific structures.** RNA is largely single-stranded, but it often contains short stretches of nucleotides that can form conventional base pairs with complementary sequences found elsewhere on the same molecule. These interactions, along with additional "nonconventional" base-pair interactions, allow an RNA molecule to fold into a three-dimensional structure that is determined by its sequence of nucleotides. (A) Diagram of a folded RNA structure showing only conventional base-pair interactions. (B) Structure with both conventional (*red*) and nonconventional (*green*) base-pair interactions. (C) Structure of an actual RNA, one that catalyzes its own splicing (see p. 324). Each conventional base-pair interaction is indicated by a "rung" in the double helix. Bases in other configurations are indicated by broken rungs.

base-pairing. In this way, the growing RNA chain is extended by one nucleotide at a time in the 5'-to-3' direction (Figure 6–9). The substrates are ribonucleoside triphosphates (ATP, CTP, UTP, and GTP); as in DNA replication, the hydrolysis of high-energy bonds provides the energy needed to drive the reaction forward (see Figure 5–4 and Movie 6.2).

The almost immediate release of the RNA strand from the DNA as it is synthesized means that many RNA copies can be made from the same gene in a relatively short time, with the synthesis of additional RNA molecules being started before the previous RNA molecules are completed (Figure 6–10). When RNA polymerase molecules follow hard on each other's heels in this way, each moving at about 50 nucleotides per second, over a thousand transcripts can be synthesized in an hour from a single gene.

Although RNA polymerase catalyzes essentially the same chemical reaction as DNA polymerase, there are some important differences between the activities of the two enzymes. First, and most obviously, RNA polymerase catalyzes the linkage of ribonucleotides, not deoxyribonucleotides. Second, unlike the DNA polymerases involved in DNA replication, RNA polymerases can start an RNA chain without a primer. This difference is thought possible because transcription need not be as accurate as DNA replication (see Table 5–1, p. 244). RNA polymerases make about one mistake for every 10^4 nucleotides copied into RNA (compared with an error rate for direct copying by DNA polymerase of about one in 10^7 nucleotides); and the consequences of an error in RNA transcription are much less significant as RNA does not permanently store genetic information in cells. Finally, unlike DNA polymerases, which make their products in segments that are later stitched together, RNA polymerases are absolutely processive; that is, the same RNA polymerase that begins an RNA molecule must finish it without dissociating from the DNA template.

Although not nearly as accurate as the DNA polymerases that replicate DNA, RNA polymerases nonetheless have a modest proofreading mechanism. If an incorrect ribonucleotide is added to the growing RNA chain, the polymerase can back up, and the active site of the enzyme can perform an excision reaction that resembles the reverse of the polymerization reaction, except that a water molecule replaces the pyrophosphate and a nucleoside monophosphate is released.

Given that DNA and RNA polymerases both carry out template-dependent nucleotide polymerization, it might be expected that the two types of enzymes would be structurally related. However, x-ray crystallographic studies reveal that, other than containing a critical Mg^{2+} ion at the catalytic site, the two enzymes are quite different. Template-dependent nucleotide-polymerizing enzymes seem to have arisen at least twice during the early evolution of cells. One lineage led to the

RNA Polymerase Structure & Function

Correction of RNA chain

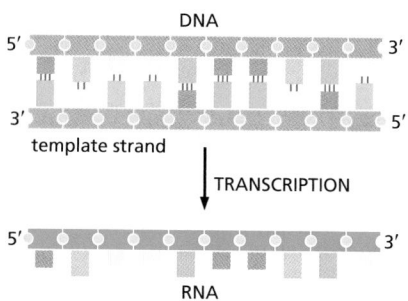

Figure 6–8 DNA transcription produces a single-stranded RNA molecule that is complementary to one strand of the DNA double helix. Note that the sequence of bases in the RNA molecule produced is the same as the sequence of bases in the non-template DNA strand, except that a U replaces every T base in the DNA.

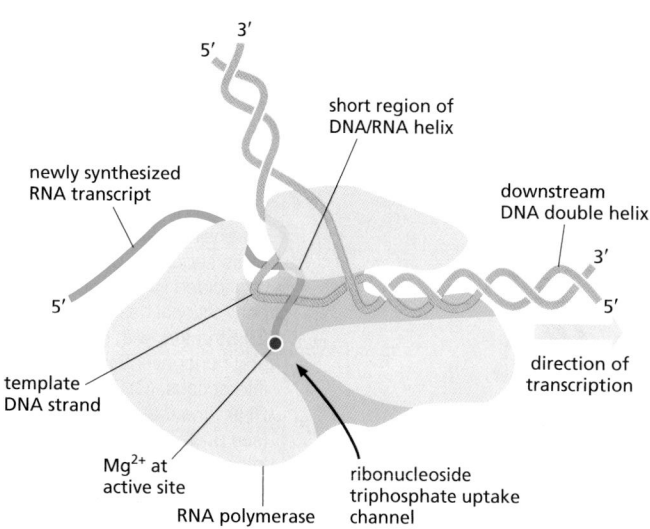

Figure 6–9 DNA is transcribed by the enzyme RNA polymerase. The RNA polymerase (pale blue) moves stepwise along the DNA, unwinding the DNA helix at its active site indicated by the Mg^{2+} (red), which is required for catalysis. As it progresses, the polymerase adds nucleotides one by one to the RNA chain at the polymerization site, using an exposed DNA strand as a template. The RNA transcript is thus a complementary copy of one of the two DNA strands. A short region of DNA/RNA helix (approximately nine nucleotide pairs in length) is formed only transiently, and a "window" of DNA/RNA helix therefore moves along the DNA with the polymerase as the DNA double helix reforms behind it. The incoming nucleotides are in the form of ribonucleoside triphosphates (ATP, UTP, CTP, and GTP), and the energy stored in their phosphate–phosphate bonds provides the driving force for the polymerization reaction (see Figure 5–4). The figure, based on an x-ray crystallographic structure, shows a cutaway view of the polymerase: the part facing the viewer has been sliced away to reveal the interior (Movie 6.3). (Adapted from P. Cramer et al., *Science* 288:640–649, 2000; PDB code: 1HQM.)

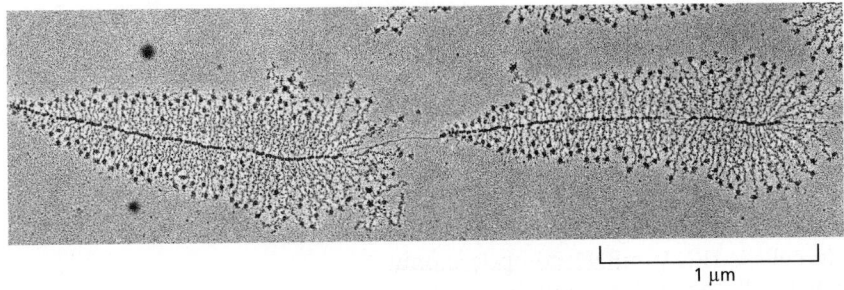

Figure 6–10 Transcription of two genes as observed under the electron microscope. The micrograph shows many molecules of RNA polymerase simultaneously transcribing each of two adjacent genes. Molecules of RNA polymerase are visible as a series of dots along the DNA with the newly synthesized transcripts (fine threads) attached to them. The RNA molecules (ribosomal RNAs) shown in this example are not translated into protein but are instead used directly as components of ribosomes, the machines on which translation takes place. The particles at the 5′ end (the free end) of each rRNA transcript are believed to reflect the beginnings of ribosome assembly. From the relative lengths of the newly synthesized transcripts, it can be deduced that the RNA polymerase molecules are transcribing from left to right. (Courtesy of Ulrich Scheer.)

1 µm

modern DNA polymerases and reverse transcriptases discussed in Chapter 5, as well as to a few RNA polymerases from viruses. The other lineage formed all of the modern RNA polymerases that we discuss in this chapter.

Cells Produce Different Categories of RNA Molecules

The majority of genes carried in a cell's DNA specify the amino acid sequence of proteins; the RNA molecules that are copied from these genes (which ultimately direct the synthesis of proteins) are called **messenger RNA (mRNA)** molecules. The final product of other genes, however, is the RNA molecule itself. These RNAs are known as **noncoding RNAs** because they do not code for protein. In a well-studied, single-celled eukaryote, the yeast *Saccharomyces cerevisiae*, over 1200 genes (more than 15% of the total) produce RNA as their final product. Humans may produce on the order of ten thousand noncoding RNAs. These RNAs, like proteins, serve as enzymatic, structural, and regulatory components for a wide variety of processes in the cell. In Chapter 5, we encountered one of them as the template carried by the enzyme telomerase. Although many of the noncoding RNAs are still mysterious, we shall see in this chapter that *small nuclear RNA (snRNA)* molecules direct the splicing of pre-mRNA to form mRNA, that *ribosomal RNA (rRNA)* molecules form the core of ribosomes, and that *transfer RNA (tRNA)* molecules form the adaptors that select amino acids and hold them in place on a ribosome for incorporation into protein. In Chapter 7, we shall see that *microRNA (miRNA)* molecules and *small interfering RNA (siRNA)* molecules serve as key regulators of eukaryotic gene expression, and that *piwi-interacting RNAs (piRNAs)* protect animal germ lines from transposons; we also discuss the *long noncoding RNAs (lncRNAs)*, a diverse set of RNAs whose functions are just being discovered (Table 6–1).

> How each RNA type Function with each other

> Important

TABLE 6–1 Principal Types of RNAs Produced in Cells	
Type of RNA	Function
mRNAs	Messenger RNAs, code for proteins
rRNAs	Ribosomal RNAs, form the basic structure of the ribosome and catalyze protein synthesis
tRNAs	Transfer RNAs, central to protein synthesis as adaptors between mRNA and amino acids
snRNAs	Small nuclear RNAs, function in a variety of nuclear processes, including the splicing of pre-mRNA
snoRNAs	Small nucleolar RNAs, help to process and chemically modify rRNAs
miRNAs	MicroRNAs, regulate gene expression by blocking translation of specific mRNAs and cause their degradation
siRNAs	Small interfering RNAs, turn off gene expression by directing the degradation of selective mRNAs and the establishment of compact chromatin structures
piRNAs	Piwi-interacting RNAs, bind to piwi proteins and protect the germ line from transposable elements
lncRNAs	Long noncoding RNAs, many of which serve as scaffolds; they regulate diverse cell processes, including X-chromosome inactivation

Each transcribed segment of DNA is called a *transcription unit.* In eukaryotes, a transcription unit typically carries the information of just one gene, and therefore codes for either a single RNA molecule or a single protein (or group of related proteins if the initial RNA transcript is spliced in more than one way to produce different mRNAs). In bacteria, a set of adjacent genes is often transcribed as a unit; the resulting mRNA molecule therefore carries the information for several distinct proteins.

Overall, RNA makes up a few percent of a cell's dry weight, whereas proteins comprise about 50%. Most of the RNA in cells is rRNA; mRNA comprises only 3–5% of the total RNA in a typical mammalian cell. The mRNA population is made up of tens of thousands of different species, and there are on average only 10–15 molecules of each species of mRNA present in each cell.

Signals Encoded in DNA Tell RNA Polymerase Where to Start and Stop

To transcribe a gene accurately, RNA polymerase must recognize where on the genome to start and where to finish. The way in which RNA polymerases perform these tasks differs somewhat between bacteria and eukaryotes. Because the processes in bacteria are simpler, we discuss them first.

The initiation of transcription is an especially important step in gene expression because it is the main point at which the cell regulates which proteins are to be produced and at what rate. The bacterial RNA polymerase core enzyme is a multisubunit complex that synthesizes RNA using the DNA template as a guide. An additional subunit called *sigma (σ) factor* associates with the core enzyme and assists it in reading the signals in the DNA that tell it where to begin transcribing (**Figure 6–11**). Together, σ factor and core enzyme are known as the *RNA polymerase holoenzyme*; this complex adheres only weakly to bacterial DNA when the two collide, and a holoenzyme typically slides rapidly along the long DNA molecule and then dissociates. However, when the polymerase holoenzyme slides into a special sequence of nucleotides indicating the starting point for RNA synthesis called a **promoter**, the polymerase binds tightly, because its σ factor makes specific contacts with the edges of bases exposed on the outside of the DNA double helix (step 1 in Figure 6–11A).

The tightly bound RNA polymerase holoenzyme at a promoter opens up the double helix to expose a short stretch of nucleotides on each strand (step 2 in Figure 6–11A). The region of unpaired DNA (about 10 nucleotides) is called the *transcription bubble* and it is stabilized by the binding of σ factor to the unpaired bases on one of the exposed strands. The other exposed DNA strand then acts as a template for complementary base-pairing with incoming ribonucleotides, two of which are joined together by the polymerase to begin an RNA chain (step 3 in Figure 6–11A). The first ten or so nucleotides of RNA are synthesized using a "scrunching" mechanism, in which RNA polymerase remains bound to the promoter and pulls the upstream DNA into its active site, thereby expanding the transcription bubble. This process creates considerable stress and the short RNAs are often released, thereby relieving the stress and forcing the polymerase, which remains in place, to begin synthesis over again. Eventually this process of *abortive initiation* is overcome and the stress generated by scrunching helps the core enzyme to break free of its interactions with the promoter DNA (step 4 in Figure 6–11A) and discard the σ factor (step 5 in Figure 6–11A). At this point, the polymerase begins to move down the DNA, synthesizing RNA, in a stepwise fashion: the polymerase moves forward one base pair for every nucleotide added. During this process, the transcription bubble continually expands at the front of the polymerase and contracts at its rear. Chain elongation continues (at a speed of approximately 50 nucleotides/sec for bacterial RNA polymerases) until the enzyme encounters a second signal, the **terminator** (step 6 in Figure 6–11A), where the polymerase halts and releases both the newly made RNA molecule and the DNA template (step 7 in Figure 6–11A). The free polymerase core enzyme then reassociates with a free σ factor to form a holoenzyme that can begin the process of transcription again (step 8 in Figure 6–11A).

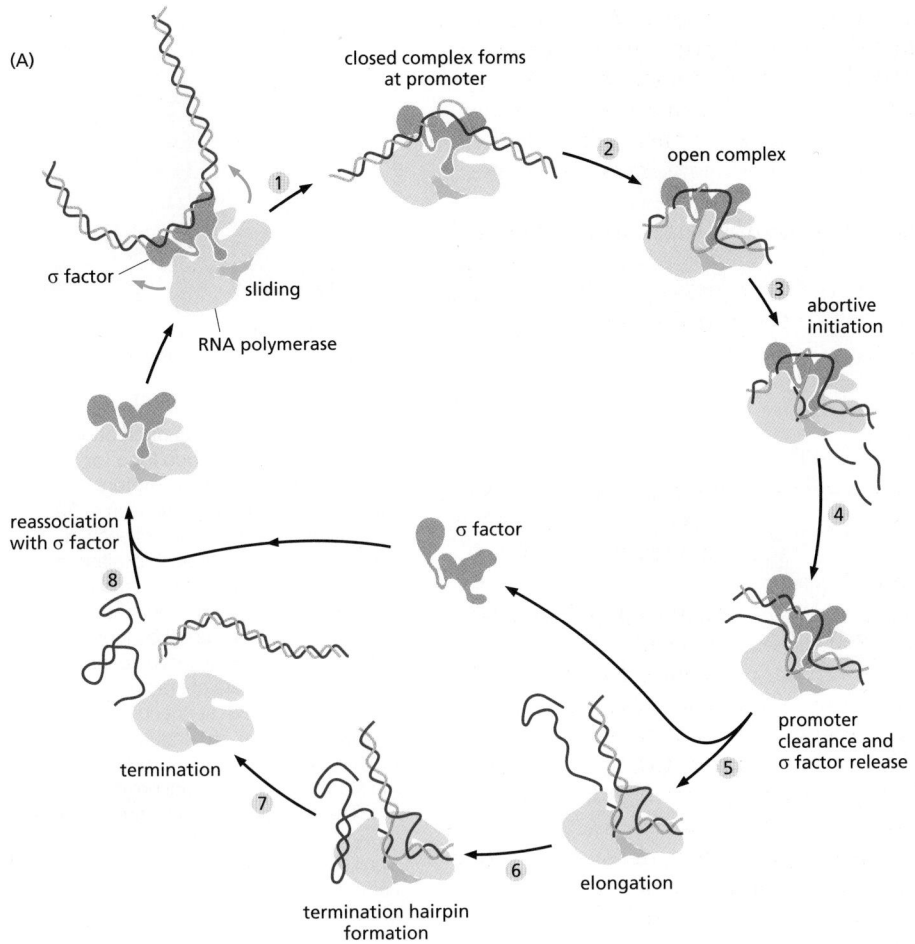

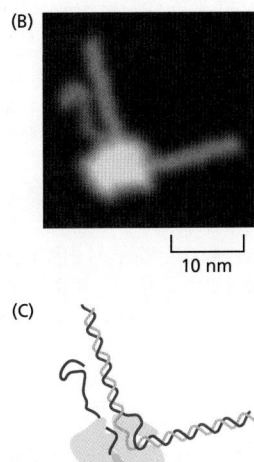

Figure 6–11 **The transcription cycle of bacterial RNA polymerase.** (A) In step 1, the RNA polymerase holoenzyme (polymerase core enzyme plus σ factor) assembles and then locates a promoter DNA sequence (see Figure 6–12). The polymerase opens (unwinds) the DNA at the position at which transcription is to begin (step 2) and begins transcribing (step 3). This initial RNA synthesis (abortive initiation) is relatively inefficient as short, unproductive transcripts are often released. However, once RNA polymerase has managed to synthesize about 10 nucleotides of RNA, it breaks its interactions with the promoter DNA (step 4) and eventually releases σ factor—as the polymerase tightens around the DNA and shifts to the elongation mode of RNA synthesis, moving along the DNA (step 5). During the elongation mode, transcription is highly processive, with the polymerase leaving the DNA template and releasing the newly transcribed RNA only when it encounters a termination signal (steps 6 and 7). Termination signals are typically encoded in DNA, and many function by forming an RNA hairpin-like structure that destabilizes the polymerase's hold on the RNA.

In bacteria, all RNA molecules are synthesized by a single type of RNA polymerase, and the cycle depicted in the figure therefore applies to the production of mRNAs as well as structural and catalytic RNAs. (B) Two-dimensional image of an elongating bacterial RNA polymerase, as determined by atomic force microscopy (see Figure 9–33). (C) Interpretation of the image in (B). (B, from K.M. Herbert et al., *Annu. Rev. Biochem.* 77:149–176, 2008. With permission from Annual Reviews.)

The process of transcription initiation is complicated and requires that the RNA polymerase holoenzyme and the DNA undergo a series of conformational changes. We can view these changes as opening up and positioning the DNA in the active site followed by a successive tightening of the enzyme around the DNA and RNA to ensure that it does not dissociate before it has finished transcribing a gene. If an RNA polymerase does dissociate prematurely, it must start over again at the promoter.

How do the termination signals in the DNA stop the elongating polymerase? For most bacterial genes, a termination signal consists of a string of A-T nucleotide pairs preceded by a twofold symmetric DNA sequence, which, when transcribed into RNA, folds into a "hairpin" structure through Watson–Crick base-pairing (see Figure 6–11A). As the polymerase transcribes across a terminator, the formation of the hairpin helps to disengage the RNA transcript from the active site (step 7 in Figure 6–11A). The process of termination provides an example of a common theme in this chapter: the folding of RNA into specific structures affects many steps in decoding the genome.

Transcription Start and Stop Signals Are Heterogeneous in Nucleotide Sequence

As we have just seen, the processes of transcription initiation and termination involve a complicated series of structural transitions in protein, DNA, and RNA molecules. The signals encoded in DNA that specify these transitions are often difficult for researchers to recognize. Indeed, a comparison of many different bacterial promoters reveals a surprising degree of variation. Nevertheless, they all contain related sequences, reflecting aspects of the DNA that are recognized directly

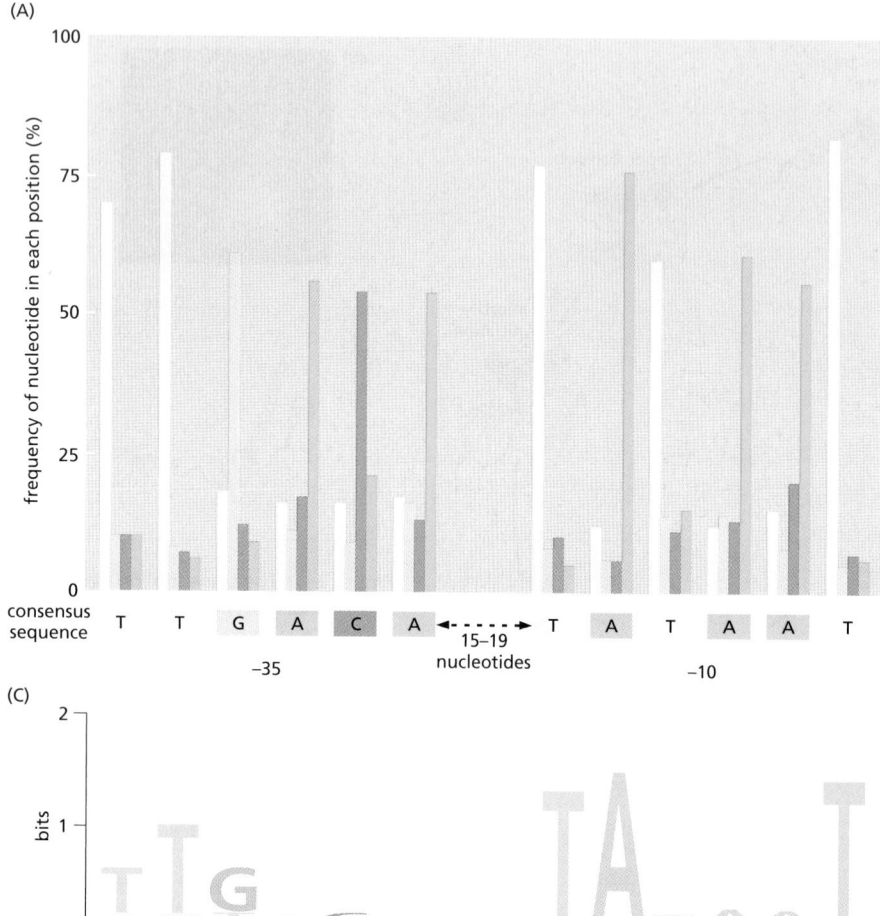

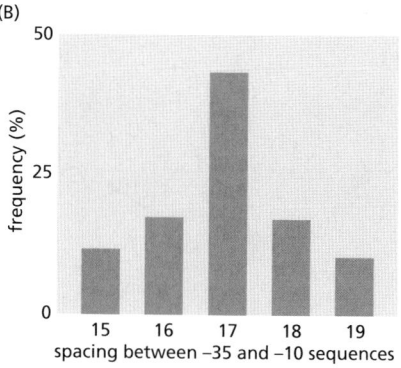

Figure 6–12 Consensus nucleotide sequence and sequence logo for the major class of *E. coli* promoters.
(A) On the basis of a comparison of 300 promoters, the frequencies of each of the four nucleotides at each position in the promoter are given. The consensus sequence, shown *below* the graph, reflects the most common nucleotide found at each position in the collection of promoters. These promoters are characterized by two hexameric DNA sequences—the –35 sequence and the –10 sequence, named for their approximate location relative to the start point of transcription (designated +1). The sequence of nucleotides between the –35 and –10 hexamers shows no significant similarities among promoters. For convenience, the nucleotide sequence of a single strand of DNA is shown; in reality, promoters are double-stranded DNA. The nucleotides shown in the figure are recognized by σ factor, a subunit of the RNA polymerase holoenzyme.
(B) The distribution of spacing between the –35 and –10 hexamers found in *E. coli* promoters. (C) A *sequence logo* displaying the same information as in panel (A). Here, the height of each letter is proportional to the frequency at which that base occurs at that position across a wide variety of promoter sequences. The total height of all the letters at each position is proportional to the information content (expressed in bits) at that position. For example, the total information content of a position that can tolerate several different bases is small (see the last three bases of the –35 sequences), but statistically greater than random.

by the σ factor. These common features are often summarized in the form of a *consensus sequence* (**Figure 6–12**). A **consensus nucleotide sequence** is derived by comparing many sequences with the same basic function and tallying up the most common nucleotides found at each position. It therefore serves as a summary or "average" of a large number of individual nucleotide sequences. A more accurate way of displaying the range of DNA sequences recognized by a protein is through the use of a *sequence logo*, which reveals the relative frequencies of each nucleotide at each position (Figure 6–12C).

The DNA sequences of individual bacterial promoters differ in ways that determine their strength (the number of initiation events per unit time of the promoter). Evolutionary processes have fine-tuned each to initiate as often as necessary and have thereby created a wide spectrum of promoter strengths. Promoters for genes that code for abundant proteins are much stronger than those associated with genes that encode rare proteins, and the nucleotide sequences of their promoters are responsible for these differences.

Like bacterial promoters, transcription terminators also have a wide range of sequences, with the potential to form a simple hairpin RNA structure being the most important common feature. Since an almost unlimited number of nucleotide sequences have this potential, terminator sequences are even more heterogeneous than promoter sequences.

We have discussed bacterial promoters and terminators in some detail to illustrate an important point regarding the analysis of genome sequences. Although we know a great deal about bacterial promoters and terminators and can construct consensus sequences that summarize their most salient features, their variation in nucleotide sequence makes it difficult to definitively locate them simply

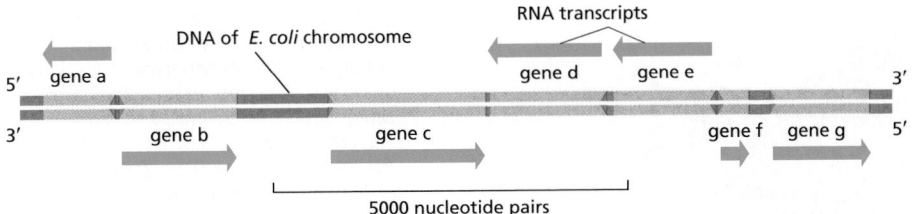

Figure 6–13 **Directions of transcription along a short portion of a bacterial chromosome.** Some genes are transcribed using one DNA strand as a template, while others are transcribed using the other DNA strand. The direction of transcription is determined by the promoter at the beginning of each gene (*green arrowheads*). This diagram shows approximately 0.2% (9000 base pairs) of the *E. coli* chromosome. The genes transcribed from *left* to *right* use the bottom DNA strand as the template; those transcribed from *right* to *left* use the top strand as the template.

by analysis of the nucleotide sequence of a genome. It is even more difficult to locate analogous sequences in eukaryotic genomes, due in part to the excess DNA carried in these genomes. Often we need additional information, some of it from direct experimentation, to locate and accurately interpret the short DNA signals in genomes.

As shown in Figure 6–11, promoter sequences are asymmetric, ensuring that RNA polymerase can bind in only one orientation. Because the polymerase can synthesize RNA only in the 5′-to-3′ direction, the promoter orientation specifies the strand to be used as a template. Genome sequences reveal that the DNA strand that is used as the template for RNA synthesis varies from gene to gene, depending on the orientation of the promoter (Figure 6–13).

Having considered transcription in bacteria, we now turn to the situation in eukaryotes, where the synthesis of RNA molecules is a much more elaborate affair.

Transcription Initiation in Eukaryotes Requires Many Proteins

In contrast to bacteria, which contain a single type of RNA polymerase, eukaryotic nuclei have three: *RNA polymerase I*, *RNA polymerase II*, and *RNA polymerase III*. The three polymerases are structurally similar to one another and share some common subunits, but they transcribe different categories of genes (Table 6–2). RNA polymerases I and III transcribe the genes encoding transfer RNA, ribosomal RNA, and various small RNAs. RNA polymerase II transcribes most genes, including all those that encode proteins, and our subsequent discussion therefore focuses on this enzyme.

Eukaryotic RNA polymerase II has many structural similarities to bacterial RNA polymerase (Figure 6–14). But there are several important differences in the way in which the bacterial and eukaryotic enzymes function, two of which concern us immediately.

1. While bacterial RNA polymerase requires only a single transcription- initiation factor (σ) to begin transcription, eukaryotic RNA polymerases require many such factors, collectively called the *general transcription factors.*

2. Eukaryotic transcription initiation must take place on DNA that is packaged into nucleosomes and higher-order forms of chromatin structure (described in Chapter 4), features that are absent from bacterial chromosomes.

TABLE 6–2 The Three RNA Polymerases in Eukaryotic Cells	
Type of polymerase	Genes transcribed
RNA polymerase I	5.8S, 18S, and 28S rRNA genes
RNA polymerase II	All protein-coding genes, plus snoRNA genes, miRNA genes, siRNA genes, lncRNA genes, and most snRNA genes
RNA polymerase III	tRNA genes, 5S rRNA genes, some snRNA genes, and genes for other small RNAs

The rRNAs were named according to their "S" values, which refer to their rate of sedimentation in an ultracentrifuge. The larger the S value, the larger the rRNA.

Figure 6–14 **Structural similarity between a bacterial RNA polymerase and a eukaryotic RNA polymerase II.** Regions of the two RNA polymerases that have similar structures are indicated in *green*. The eukaryotic polymerase is larger than the bacterial enzyme (12 subunits instead of 5), and some of the additional regions are shown in *gray*. The *blue* spheres represent Zn atoms that serve as structural components of the polymerases, and the *red* sphere represents the Mg atom present at the active site, where polymerization takes place. The RNA polymerases in all modern-day cells (bacteria, archaea, and eukaryotes) are closely related, indicating that the basic features of the enzyme were in place before the divergence of the three major branches of life. (Courtesy of P. Cramer and R. Kornberg.)

RNA Polymerase II Requires a Set of General Transcription Factors

The **general transcription factors** help to position eukaryotic RNA polymerase correctly at the promoter, aid in pulling apart the two strands of DNA to allow transcription to begin, and release RNA polymerase from the promoter to start its elongation mode. The proteins are "general" because they are needed at nearly all promoters used by RNA polymerase II. They consist of a set of interacting proteins denoted arbitrarily as TFIIA, TFIIB, TFIIC, TFIID, and so on (TFII standing for "transcription factor for polymerase II"). In a broad sense, the eukaryotic general transcription factors carry out functions equivalent to those of the σ factor in bacteria; indeed, portions of TFIIF have the same three-dimensional structure as the equivalent portions of σ.

Figure 6–15 illustrates how the general transcription factors assemble at promoters used by RNA polymerase II, and **Table 6–3** summarizes their activities. The assembly process begins when TFIID binds to a short double-helical DNA sequence primarily composed of T and A nucleotides. For this reason, this sequence is known as the TATA sequence, or **TATA box**, and the subunit of TFIID that recognizes it is called TBP (for TATA-binding protein). The TATA box is typically located 25 nucleotides upstream from the transcription start site. It is not the only DNA sequence that signals the start of transcription (**Figure 6–16**), but for most polymerase II promoters it is the most important. The binding of TFIID

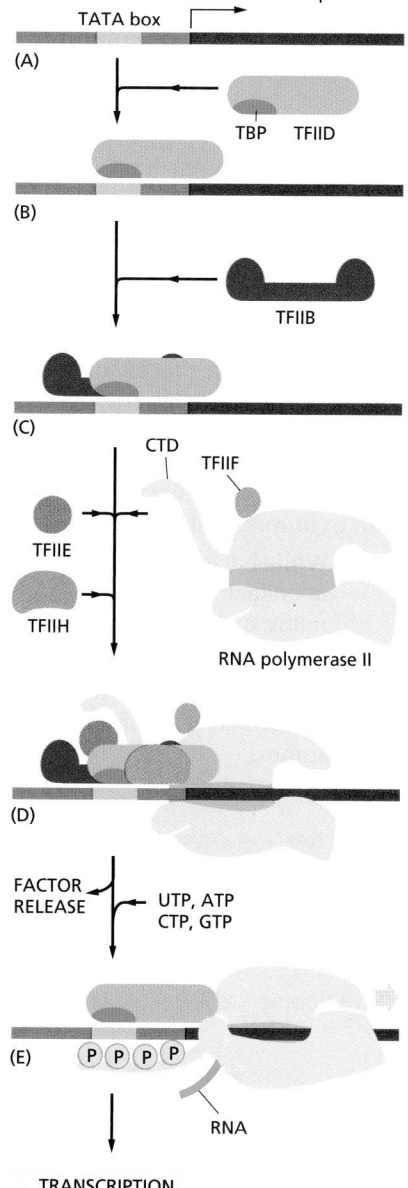

Figure 6–15 **Initiation of transcription of a eukaryotic gene by RNA polymerase II.** To begin transcription, RNA polymerase requires several general transcription factors. (A) The promoter contains a DNA sequence called the TATA box, which is located 25 nucleotides away from the site at which transcription is initiated. (B) Through its subunit TBP, TFIID recognizes and binds the TATA box, which then enables the adjacent binding of TFIIB (C). For simplicity the DNA distortion produced by the binding of TFIID (see Figure 6–17) is not shown. (D) The rest of the general transcription factors, as well as the RNA polymerase itself, assemble at the promoter. (E) TFIIH then uses energy from ATP hydrolysis to pry apart the DNA double helix at the transcription start point, locally exposing the template strand. TFIIH also phosphorylates RNA polymerase II, changing its conformation so that the polymerase is released from the general factors and can begin the elongation phase of transcription. As shown, the site of phosphorylation is a long C-terminal polypeptide tail, also called the C-terminal domain (CTD), that extends from the polymerase molecule. The assembly scheme shown in the figure was deduced from experiments performed *in vitro*, and the exact order in which the general transcription factors assemble on promoters probably varies from gene to gene *in vivo*. The general transcription factors are highly conserved; some of those from human cells can be replaced in biochemical experiments by the corresponding factors from simple yeasts.

TABLE 6–3 The General Transcription Factors Needed for Transcription Initiation by Eukaryotic RNA Polymerase II

Name	Number of subunits	Roles in transition initiation

TFIID
  TBP subunit
  TAF subunits
	
	1
	~11
		Recognizes TATA box
		Recognizes other DNA sequences near the transcription start point; regulates DNA-binding by TBP

TFIIB	1	Recognizes BRE element in promoters; accurately positions RNA polymerase at the start site of transcription
TFIIF	3	Stabilizes RNA polymerase interaction with TBP and TFIIB; helps attract TFIIE and TFIIH
TFIIE	2	Attracts and regulates TFIIH
TFIIH	9	Unwinds DNA at the transcription start point, phosphorylates Ser5 of the RNA polymerase CTD; releases RNA polymerase from the promoter

TFIID is composed of TBP and ~11 additional subunits called TAFs (TBP-associated factors); CTD, C-terminal domain.

causes a large distortion in the DNA of the TATA box (**Figure 6–17**). This distortion is thought to serve as a physical landmark for the location of an active promoter in the midst of a very large genome, and it brings DNA sequences on both sides of the distortion closer together to allow for subsequent protein assembly steps. Other factors then assemble, along with RNA polymerase II, to form a complete *transcription initiation complex* (see Figure 6–15). The most complicated of the general transcription factors is TFIIH. Consisting of nine subunits, it is nearly as large as RNA polymerase II itself and, as we shall see shortly, performs several enzymatic steps needed for the initiation of transcription.

After forming a transcription initiation complex on the promoter DNA, RNA polymerase II must gain access to the template strand at the transcription start point. TFIIH, which contains a DNA helicase as one of its subunits, makes this step possible by hydrolyzing ATP and unwinding the DNA, thereby exposing the template strand. Next, RNA polymerase II, like the bacterial polymerase, remains at the promoter synthesizing short lengths of RNA until it undergoes a series of conformational changes that allow it to move away from the promoter and enter the elongation phase of transcription. A key step in this transition is the addition of phosphate groups to the "tail" of the RNA polymerase (known as the CTD or C-terminal domain). In humans, the CTD consists of 52 tandem repeats of a

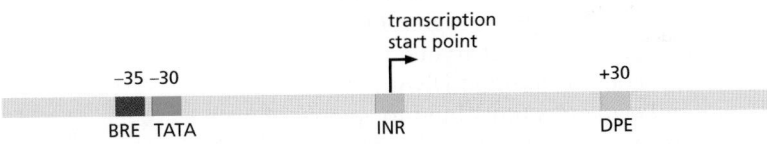

element	consensus sequence	general transcription factor
BRE	G/C G/C G/A C G C C	TFIIB
TATA	T A T A A/T A A/T	TBP subunit of TFIID
INR	C/T C/T A N T/A C/T C/T	TFIID
DPE	A/G G A/T C G T G	TFIID

Figure 6–16 Consensus sequences found in the vicinity of eukaryotic RNA polymerase II start points. The name given to each consensus sequence *(first column)* and the general transcription factor that recognizes it *(last column)* are indicated. N indicates any nucleotide, and two nucleotides separated by a slash indicate an equal probability of either nucleotide at the indicated position. In reality, each consensus sequence is a shorthand representation of a histogram similar to that of Figure 6–12.

For most RNA polymerase II transcription start points, only two or three of the four sequences are present. For example, many polymerase II promoters have a TATA box sequence, but those that do not typically have a "strong" INR sequence. Although most of the DNA sequences that influence transcription initiation are located upstream of the transcription start point, a few, such as the DPE shown in the figure, are located in the transcribed region.

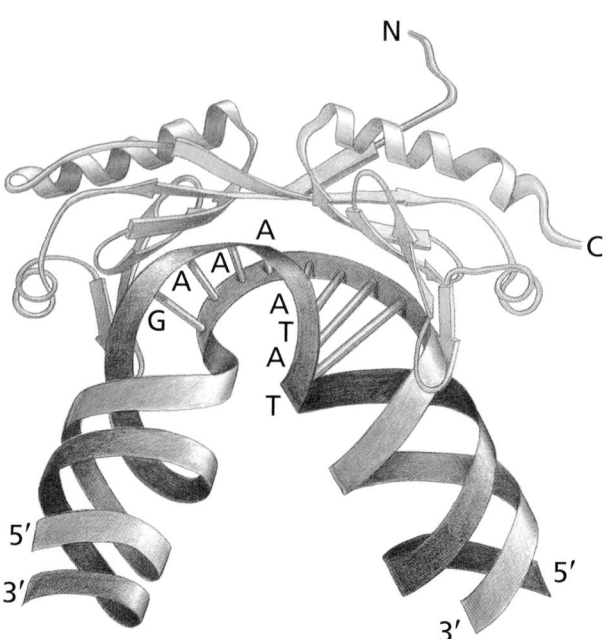

Figure 6–17 **Three-dimensional structure of TBP (TATA-binding protein) bound to DNA.** The TBP is the subunit of the general transcription factor TFIID that is responsible for recognizing and binding to the TATA box sequence in the DNA *(red)*. The unique DNA bending caused by TBP—kinks in the double helix separated by partly unwound DNA—is thought to serve as a landmark that helps to attract the other general transcription factors (Movie 6.4). TBP is a single polypeptide chain that is folded into two very similar domains *(blue* and *green)*. (Adapted from J.L. Kim et al., *Nature* 365:520–527, 1993.)

seven-amino-acid sequence, which extend from the RNA polymerase core structure. During transcription initiation, the serine located at the fifth position in the repeat sequence (Ser5) is phosphorylated by TFIIH, which contains a protein kinase in one of its subunits (see Figure 6–15D and E). The polymerase can then disengage from the cluster of general transcription factors. During this process, it undergoes a series of conformational changes that tighten its interaction with DNA, and it acquires new proteins that allow it to transcribe for long distances, in some cases for many hours, without dissociating from DNA.

Once the polymerase II has begun elongating the RNA transcript, most of the general transcription factors are released from the DNA so that they are available to initiate another round of transcription with a new RNA polymerase molecule. As we see shortly, the phosphorylation of the tail of RNA polymerase II has an additional function: it causes components of the RNA-processing machinery to load onto the polymerase and thus be positioned to modify the newly transcribed RNA as it emerges from the polymerase.

Polymerase II Also Requires Activator, Mediator, and Chromatin-Modifying Proteins

Studies of RNA polymerase II and its general transcription factors acting on DNA templates in purified *in vitro* systems established the model for transcription initiation just described. However, as discussed in Chapter 4, DNA in eukaryotic cells is packaged into nucleosomes, which are further arranged in higher-order chromatin structures. As a result, transcription initiation in a eukaryotic cell is more complex and requires more proteins than it does on purified DNA. First, gene regulatory proteins known as *transcriptional activators* must bind to specific sequences in DNA (called *enhancers*) and help to attract RNA polymerase II to the start point of transcription (**Figure 6–18**). We discuss the role of these activators in Chapter 7, because they are one of the main ways in which cells regulate expression of their genes. Here we simply note that their presence on DNA is required for transcription initiation in a eukaryotic cell. Second, eukaryotic transcription initiation *in vivo* requires the presence of a large protein complex known as *Mediator*, which allows the activator proteins to communicate properly with the polymerase II and with the general transcription factors. Finally, transcription initiation in a eukaryotic cell typically requires the recruitment of chromatin-modifying enzymes, including chromatin remodeling complexes and

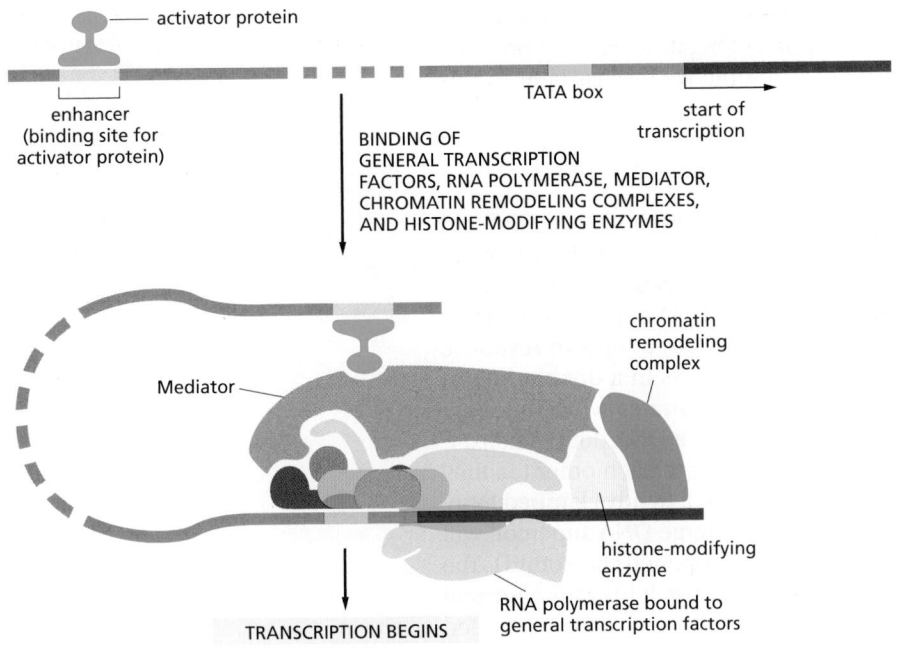

activator protein

enhancer
(binding site for
activator protein)

TATA box

start of
transcription

BINDING OF
GENERAL TRANSCRIPTION
FACTORS, RNA POLYMERASE, MEDIATOR,
CHROMATIN REMODELING COMPLEXES,
AND HISTONE-MODIFYING ENZYMES

chromatin
remodeling
complex

Mediator

histone-modifying
enzyme

TRANSCRIPTION BEGINS

RNA polymerase bound to
general transcription factors

Figure 6–18 Transcription initiation by RNA polymerase II in a eukaryotic cell. Transcription initiation *in vivo* requires the presence of transcription activator proteins. As described in Chapter 7, these proteins bind to specific short sequences in DNA. Although only one is shown here, a typical eukaryotic gene utilizes many transcription activator proteins, which in combination determine its rate and pattern of transcription. Sometimes acting from a distance of several thousand nucleotide pairs (indicated by the dashed DNA molecule), these proteins help RNA polymerase, the general transcription factors, and Mediator all to assemble at the promoter. In addition, activators attract ATP-dependent chromatin remodeling complexes and histone-modifying enzymes. One of the main roles of Mediator is to coordinate the assembly of all these proteins at the promoter so that transcription can begin. As discussed in Chapter 4, the "default" state of chromatin is a condensed fiber (see Figure 4–28), and this is likely to be the form of DNA upon which most transcription is initiated. For simplicity, the chromatin is not shown in this figure.

histone-modifying enzymes. As discussed in Chapter 4, both types of enzymes can increase access to the DNA in chromatin, and by doing so they facilitate the assembly of the transcription initiation machinery onto DNA.

As illustrated in Figure 6–18, many proteins (well over 100 individual subunits) must assemble at the start point of transcription to initiate transcription in a eukaryotic cell. The order of assembly of these proteins does not seem to follow a prescribed pathway; rather, the order differs from gene to gene. Indeed, some of these different protein complexes may be brought to DNA as preformed subassemblies.

To begin transcribing, RNA polymerase II must be released from this large complex of proteins. In addition to the steps described in Figure 6–14, this release often requires the *in situ* proteolysis of the activator protein. We shall return to some of these issues, including the role of chromatin remodeling complexes and histone-modifying enzymes, in Chapter 7, where we discuss how eukaryotic cells regulate the process of transcription initiation.

Transcription Elongation in Eukaryotes Requires Accessory Proteins

Once RNA polymerase has initiated transcription, it moves jerkily, pausing at some DNA sequences and rapidly transcribing through others. Elongating RNA polymerases, both bacterial and eukaryotic, are associated with a series of *elongation factors*, proteins that decrease the likelihood that RNA polymerase will dissociate before it reaches the end of a gene. These factors typically associate with RNA polymerase shortly after initiation and help the polymerase move through the wide variety of different DNA sequences that are found in genes. Eukaryotic RNA polymerases must also contend with chromatin structure as they move along a DNA template, and they are typically aided by ATP-dependent chromatin remodeling complexes that either move with the polymerase or may simply seek out and rescue the occasional stalled polymerase. In addition, histone chaperones help by partially disassembling nucleosomes in front of a moving RNA polymerase and assembling them behind.

As RNA polymerase moves along a gene, some of the enzymes bound to it modify the histones, leaving behind a record of where the polymerase has been. Although it is not clear exactly how the cell uses this information, it may aid in transcribing a gene over and over again once it has become active for the first time. It may also be used to coordinate transcription elongation with the processing of RNA as it emerges from RNA polymerase, a topic we discuss later in this chapter.

Transcription Creates Superhelical Tension

There is yet another barrier to elongating RNA polymerases, both bacterial and eukaryotic, one that also applies to DNA polymerases, as discussed in Chapter 5 (see Figure 5–20). To describe this issue in more detail, we need first to consider a subtle property inherent in the DNA double helix called **DNA supercoiling**. DNA supercoiling is the name given to a conformation that DNA adopts in response to superhelical tension; alternatively, creating loops or coils in a double-helical DNA molecule can create such tension. **Figure 6–19** illustrates why. There are approximately 10 nucleotide pairs for every helical turn in a DNA double helix. If we imagine a helix whose two ends are fixed with respect to each other (as they are in a DNA circle, such as a bacterial chromosome, or in a tightly clamped loop, as is thought to exist in eukaryotic chromosomes), one large DNA supercoil will form to compensate for each 10 nucleotide pairs that are opened (unwound). The formation of this supercoil is energetically favorable because it restores a normal helical twist to the base-paired regions that remain, which would otherwise need to be overwound because of the fixed ends.

RNA polymerase creates superhelical tension as it moves along a stretch of DNA that is anchored at its ends (see Figure 6–19C). As long as the polymerase is not free to rotate rapidly (and such rotation is unlikely given the size of RNA polymerases and their attached transcripts), a moving polymerase generates positive superhelical tension in the DNA in front of it and negative helical tension behind it. For eukaryotes, this situation is thought to provide a bonus: although the positive superhelical tension ahead of the polymerase makes the DNA helix

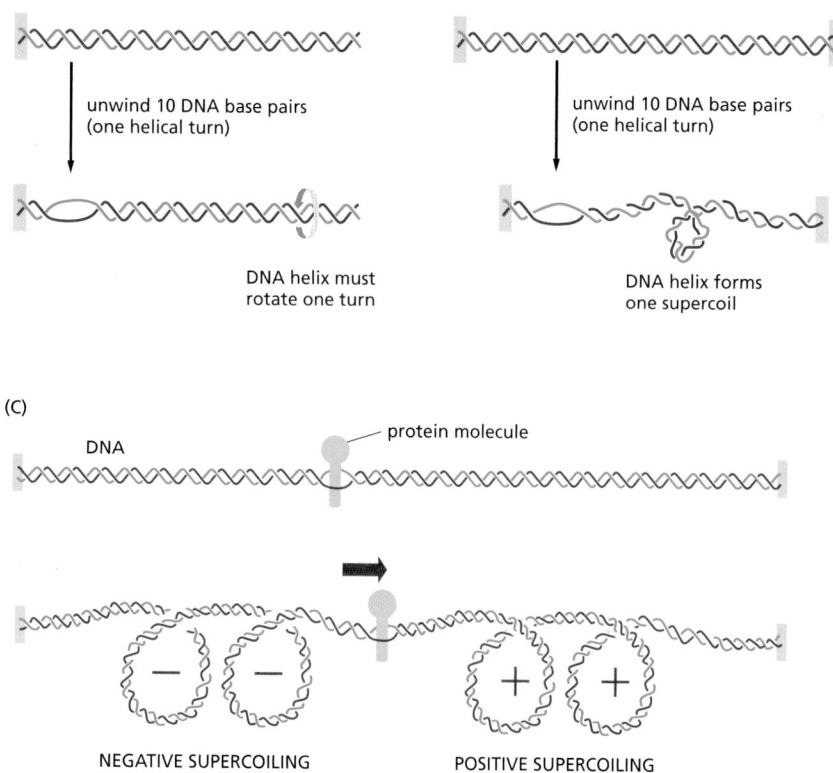

(A)
DNA with free end

unwind 10 DNA base pairs
(one helical turn)

DNA helix must
rotate one turn

(B)
DNA with fixed ends

unwind 10 DNA base pairs
(one helical turn)

DNA helix forms
one supercoil

(C)

DNA   protein molecule

NEGATIVE SUPERCOILING
helix opening facilitated

POSITIVE SUPERCOILING
helix opening hindered

Figure 6–19 Superhelical tension in DNA causes DNA supercoiling. (A) For a DNA molecule with one free end (or a nick in one strand that serves as a swivel), the DNA double helix rotates by one turn for every 10 nucleotide pairs opened. (B) If rotation is prevented, superhelical tension is introduced into the DNA by helix opening. In the example shown, the DNA helix contains 10 helical turns, one of which is opened. One way of accommodating the tension created would be to increase the helical twist from 10 to 11 nucleotide pairs per turn in the double helix that remains. The DNA helix, however, resists such a deformation in a springlike fashion, preferring to relieve the superhelical tension by bending into supercoiled loops. As a result, one DNA supercoil forms in the DNA double helix for every 10 nucleotide pairs opened. The supercoil formed in this case is a positive supercoil. (C) Supercoiling of DNA is induced by a protein tracking through the DNA double helix. The two ends of the DNA shown here are unable to rotate freely relative to each other, and the protein molecule is assumed also to be prevented from rotating freely as it moves. Under these conditions, the movement of the protein causes an excess of helical turns to accumulate in the DNA helix ahead of the protein and a deficit of helical turns to arise in the DNA behind the protein, as shown.

(A) EUKARYOTES

(B) PROKARYOTES

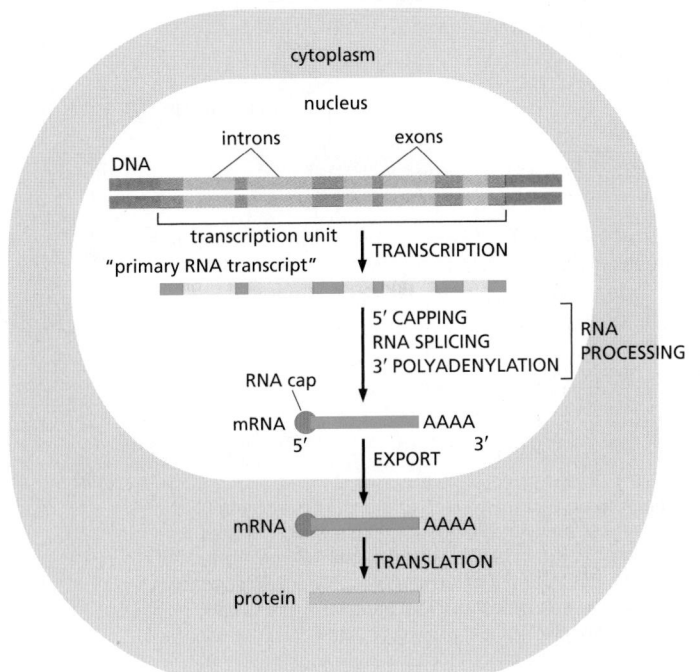

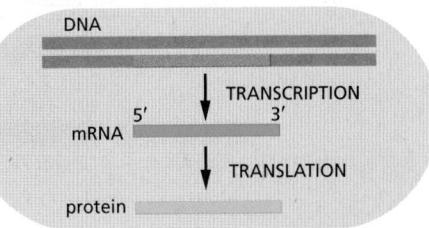

Figure 6–20 **Comparison of the steps leading from gene to protein in eukaryotes and bacteria.** The final level of a protein in the cell depends on the efficiency of each step and on the rates of degradation of the RNA and protein molecules. (A) In eukaryotic cells, the mRNA molecule resulting from transcription contains both coding (exon) and noncoding (intron) sequences. Before it can be translated into protein, the two ends of the RNA are modified, the introns are removed by an enzymatically catalyzed RNA splicing reaction, and the resulting mRNA is transported from the nucleus to the cytoplasm. For convenience, the steps in this figure are depicted as occurring one at a time; in reality, many occur concurrently. For example, the RNA cap is added and splicing begins before transcription has been completed. Because of the coupling between transcription and RNA processing, intact primary transcripts—the full-length RNAs that would, in theory, be produced if no processing had occurred—are found only rarely. (B) In prokaryotes, the production of mRNA is much simpler. The 5′ end of an mRNA molecule is produced by the initiation of transcription, and the 3′ end is produced by the termination of transcription. Since prokaryotic cells lack a nucleus, transcription and translation take place in a common compartment, and the translation of a bacterial mRNA often begins before its synthesis has been completed.

more difficult to open, the tension should facilitate the partial unwrapping of the DNA in nucleosomes, inasmuch as the release of DNA from the histone core helps to relax this tension.

Any protein that propels itself alone along a DNA strand of a double helix, such as a DNA helicase or an RNA polymerase, tends to generate superhelical tension. In eukaryotes, DNA topoisomerase enzymes rapidly remove this superhelical tension (see pp. 251–253). But in bacteria a specialized topoisomerase called *DNA gyrase* uses the energy of ATP hydrolysis to pump supercoils continuously into the DNA, thereby maintaining the DNA under constant tension. These are *negative supercoils*, having the opposite handedness from the *positive supercoils* that form when a region of DNA helix opens (see Figure 6–19B). Whenever a region of helix opens, it removes these negative supercoils from bacterial DNA, reducing the superhelical tension. DNA gyrase therefore makes the opening of the DNA helix in bacteria energetically favorable compared with helix opening in DNA that is not supercoiled. For this reason, it facilitates those genetic processes in bacteria, such as the initiation of transcription by bacterial RNA polymerase, that require helix opening (see Figure 6–11).

Transcription Elongation in Eukaryotes Is Tightly Coupled to RNA Processing

We have seen that bacterial mRNAs are synthesized by the RNA polymerase starting and stopping at specific spots on the genome. The situation in eukaryotes is substantially different. In particular, transcription is only the first of several steps needed to produce a mature mRNA molecule. Other critical steps are the covalent modification of the ends of the RNA and the removal of *intron sequences* that are discarded from the middle of the RNA transcript by the process of *RNA splicing* (**Figure 6–20**).

Both ends of eukaryotic mRNAs are modified: by *capping* on the 5′ end and by *polyadenylation* of the 3′ end (**Figure 6–21**). These special ends allow the cell to assess whether both ends of an mRNA molecule are present (and if the message is therefore intact) before it exports the RNA from the nucleus and translates it

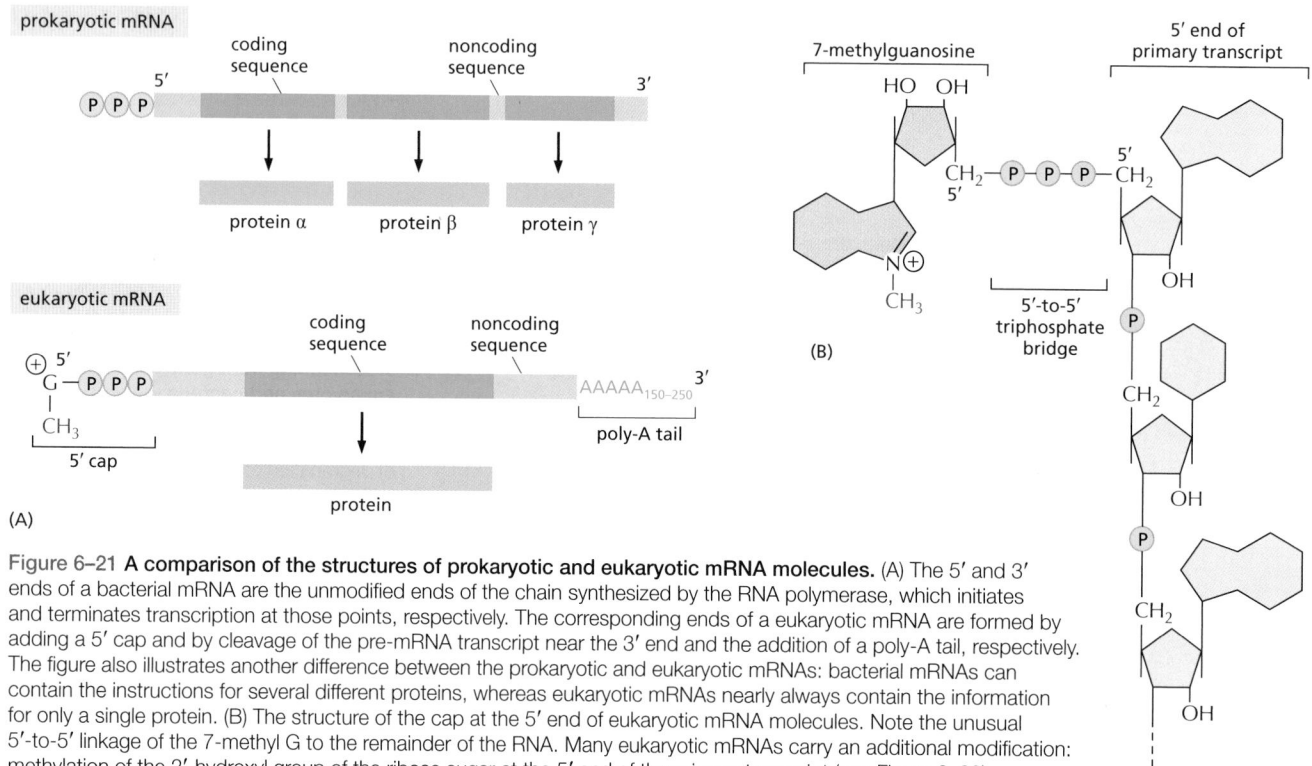

Figure 6–21 A comparison of the structures of prokaryotic and eukaryotic mRNA molecules. (A) The 5′ and 3′ ends of a bacterial mRNA are the unmodified ends of the chain synthesized by the RNA polymerase, which initiates and terminates transcription at those points, respectively. The corresponding ends of a eukaryotic mRNA are formed by adding a 5′ cap and by cleavage of the pre-mRNA transcript near the 3′ end and the addition of a poly-A tail, respectively. The figure also illustrates another difference between the prokaryotic and eukaryotic mRNAs: bacterial mRNAs can contain the instructions for several different proteins, whereas eukaryotic mRNAs nearly always contain the information for only a single protein. (B) The structure of the cap at the 5′ end of eukaryotic mRNA molecules. Note the unusual 5′-to-5′ linkage of the 7-methyl G to the remainder of the RNA. Many eukaryotic mRNAs carry an additional modification: methylation of the 2′-hydroxyl group of the ribose sugar at the 5′ end of the primary transcript (see Figure 6–23).

into protein. RNA splicing joins together the different portions of a protein-coding sequence, and it provides eukaryotes with the ability to synthesize several different proteins from the same gene.

A simple strategy has evolved to couple all of the above RNA processing steps to transcription elongation. As discussed previously, a key step in transcription initiation by RNA polymerase II is the phosphorylation of the RNA polymerase II tail, also called the CTD (C-terminal domain). This phosphorylation, which proceeds gradually as the RNA polymerase initiates transcription and moves along the DNA, not only helps dissociate the RNA polymerase II from other proteins present at the start point of transcription, but also allows a new set of proteins to associate with the RNA polymerase tail that function in transcription elongation and RNA processing. As discussed next, some of these processing proteins are thought to "hop" from the polymerase tail onto the nascent RNA molecule to begin processing it as it emerges from the RNA polymerase. Thus, we can view RNA polymerase II in its elongation mode as an RNA factory that not only moves along the DNA synthesizing an RNA molecule, but also processes the RNA that it produces (**Figure 6–22**). Fully extended, the CTD is nearly 10 times longer than the remainder of RNA polymerase. As a flexible protein domain, it serves as a scaffold or tether, holding a variety of proteins close by so that they can rapidly act when needed. This strategy, which greatly speeds up the overall rate of a series of consecutive reactions, is one that is commonly utilized in the cell (see Figures 4–58 and 16–18).

RNA Capping Is the First Modification of Eukaryotic Pre-mRNAs

As soon as RNA polymerase II has produced about 25 nucleotides of RNA, the 5′ end of the new RNA molecule is modified by addition of a cap that consists of a modified guanine nucleotide (see Figure 6–21B). Three enzymes, acting in succession, perform the capping reaction: one (a phosphatase) removes a phosphate from the 5′ end of the nascent RNA, another (a guanyl transferase) adds a GMP in

Figure 6–22 **Eukaryotic RNA polymerase II as an "RNA factory."** As the polymerase transcribes DNA into RNA, it carries RNA-processing proteins on its tail that are transferred to the nascent RNA at the appropriate time. The tail contains 52 tandem repeats of a seven-amino-acid sequence, and there are two serines in each repeat. The capping proteins first bind to the RNA polymerase tail when it is phosphorylated on Ser5 of the heptad repeat late in the process of transcription initiation (see Figure 6–15). This strategy ensures that the RNA molecule is efficiently capped as soon as its 5′ end emerges from the RNA polymerase. As the polymerase continues transcribing, its tail is extensively phosphorylated on the Ser2 positions by a kinase associated with the elongating polymerase and is eventually dephosphorylated at Ser5 positions. These further modifications attract splicing and 3′-end processing proteins to the moving polymerase, positioning them to act on the newly synthesized RNA as it emerges from the RNA polymerase. There are many RNA-processing enzymes, and not all travel with the polymerase. For RNA splicing, for example, the tail carries only a few critical components; once transferred to an RNA molecule, they serve as a nucleation site for the remaining components.

When RNA polymerase II finishes transcribing a gene, it is released from DNA, soluble phosphatases remove the phosphates on its tail, and it can reinitiate transcription. Only the fully dephosphorylated form of RNA polymerase II is competent to begin RNA synthesis at a promoter.

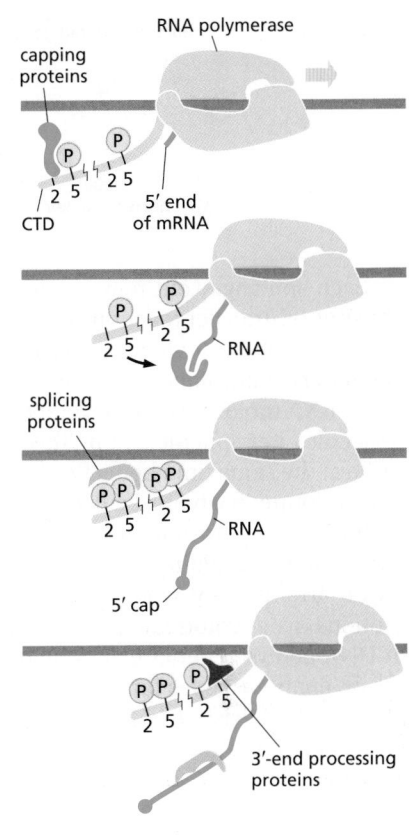

a reverse linkage (5′ to 5′ instead of 5′ to 3′), and a third (a methyl transferase) adds a methyl group to the guanosine (**Figure 6–23**). Because all three enzymes bind to the RNA polymerase tail phosphorylated at the Ser5 position—the modification added by TFIIH during transcription initiation— they are poised to modify the 5′ end of the nascent transcript as soon as it emerges from the polymerase.

The 5′-methyl cap signifies the 5′ end of eukaryotic mRNAs, and this landmark helps the cell to distinguish mRNAs from the other types of RNA molecules present in the cell. For example, RNA polymerases I and III produce uncapped RNAs during transcription, in part because these polymerases lack a CTD. In the nucleus, the cap binds a protein complex called CBC (cap-binding complex), which, as we discuss in subsequent sections, helps a future mRNA be further processed and exported. The 5′-methyl cap also has an important role in the translation of mRNAs in the cytosol, as we discuss later in the chapter.

RNA Splicing Removes Intron Sequences from Newly Transcribed Pre-mRNAs

As discussed in Chapter 4, the protein-coding sequences of eukaryotic genes are typically interrupted by noncoding intervening sequences (introns). Discovered in 1977, this feature of eukaryotic genes came as a surprise to scientists, who had been, until that time, familiar only with bacterial genes, which typically consist of a continuous stretch of coding DNA that is directly transcribed into mRNA. In marked contrast, eukaryotic genes were found to be broken up into small pieces of coding sequence (*expressed sequences* or **exons**) interspersed with much longer *intervening sequences* or **introns**; thus, the coding portion of a eukaryotic gene is often only a small fraction of the length of the gene (**Figure 6–24**).

Both intron and exon sequences are transcribed into RNA. The intron sequences are removed from the newly synthesized RNA through the process of **RNA splicing**. The vast majority of RNA splicing that takes place in cells functions in the production of mRNA, and our discussion of splicing focuses on this so-called precursor-mRNA (or pre-mRNA) splicing. Only after 5′- and 3′-end processing and splicing have taken place is such RNA termed mRNA.

Figure 6–23 **The reactions that cap the 5′ end of each RNA molecule synthesized by RNA polymerase II.** The final cap contains a novel 5′-to-5′ linkage between the positively charged 7-methyl G residue and the 5′ end of the RNA transcript (see Figure 6–21B). The letter N represents any one of the four ribonucleotides, although the nucleotide that starts an RNA chain is usually a purine (an A or a G). (After A.J. Shatkin, *BioEssays* 7:275–277, 1987. With permission from John Wiley and Sons.)

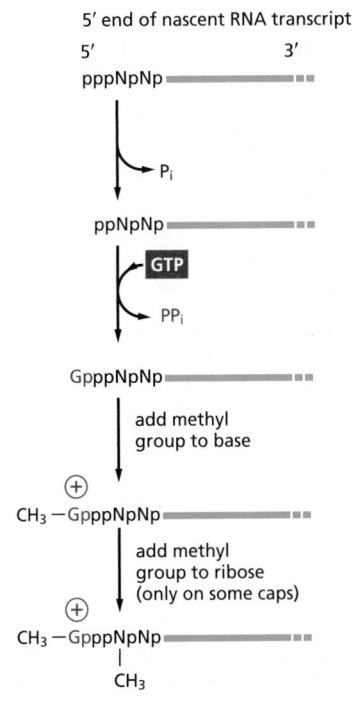

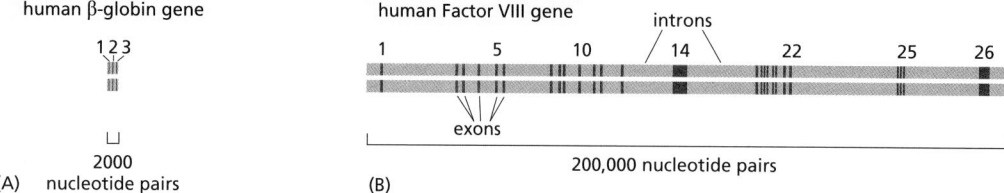

Each splicing event removes one intron, proceeding through two sequential phosphoryl-transfer reactions known as transesterifications; these join two exons together while removing the intron between them as a "lariat" (**Figure 6–25**). The machinery that catalyzes pre-mRNA splicing is complex, consisting of five additional RNA molecules and several hundred proteins, and it hydrolyzes many ATP molecules per splicing event. This complexity ensures that splicing is accurate, while at the same time being flexible enough to deal with the enormous variety of introns found in a typical eukaryotic cell.

It may seem wasteful to remove large numbers of introns by RNA splicing. In attempting to explain why it occurs, scientists have pointed out that the exon–intron arrangement would seem to facilitate the emergence of new and useful proteins over evolutionary time scales. Thus, the presence of numerous introns in DNA allows genetic recombination to readily combine the exons of different genes, enabling genes for new proteins to evolve more easily by the combination of parts of preexisting genes. The observation, described in Chapter 3, that many proteins in present-day cells resemble patchworks composed from a common set of protein *domains*, supports this idea (see pp. 121–122).

RNA splicing also has a present-day advantage. The transcripts of many eukaryotic genes (estimated at 95% of genes in humans) are spliced in more than one way, thereby allowing the same gene to produce a corresponding set of different proteins (**Figure 6–26**). Rather than being the wasteful process it may have seemed at first sight, RNA splicing enables eukaryotes to increase the coding potential of their genomes. We shall return to this idea again in this chapter and the next, but we first need to describe the cellular machinery that performs this remarkable task.

Figure 6–24 Structure of two human genes showing the arrangement of exons and introns. (A) The relatively small β-globin gene, which encodes a subunit of the oxygen-carrying protein hemoglobin, contains 3 exons (see also Figure 4–7). (B) The much larger Factor VIII gene contains 26 exons; it codes for a protein (Factor VIII) that functions in the blood-clotting pathway. The most prevalent form of hemophilia results from mutations in this gene.

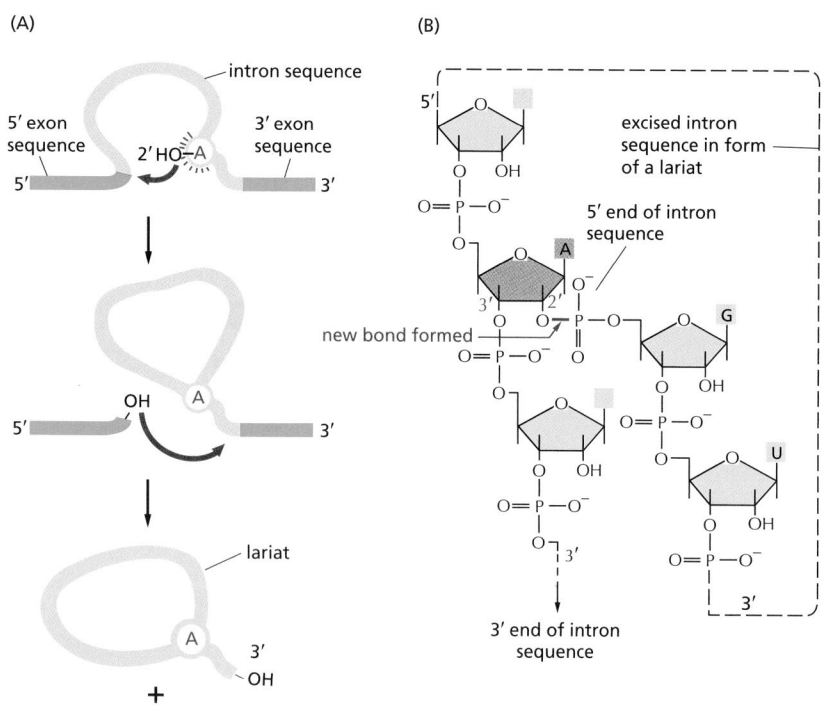

Figure 6–25 The pre-mRNA splicing reaction. (A) In the first step, a specific adenine nucleotide in the intron sequence (indicated in *red*) attacks the 5′ splice site and cuts the sugar-phosphate backbone of the RNA at this point. The cut 5′ end of the intron becomes covalently linked to the adenine nucleotide, as shown in detail in (B), thereby creating a loop in the RNA molecule. The released free 3′-OH end of the exon sequence then reacts with the start of the next exon sequence, joining the two exons together and releasing the intron sequence in the shape of a *lariat*. The two exon sequences thereby become joined into a continuous coding sequence. The released intron sequence is eventually broken down into single nucleotides, which are recycled.

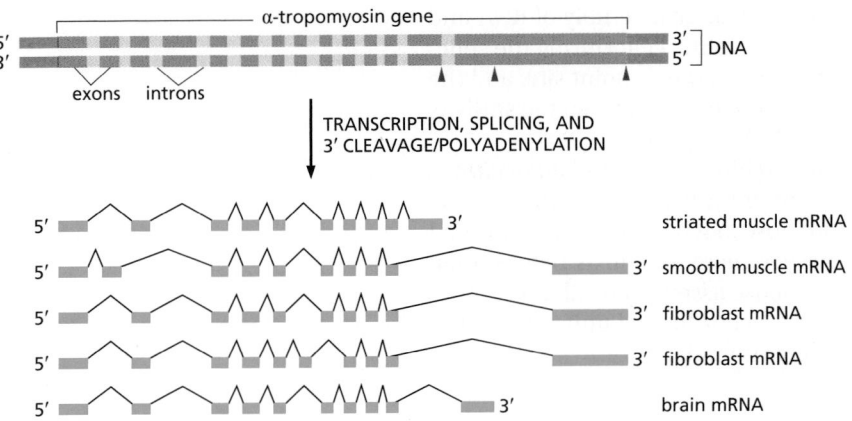

Figure 6–26 **Alternative splicing of the α-tropomyosin gene from rat.** α-Tropomyosin is a coiled-coil protein (see Figure 3–9) that carries out several tasks, most notably the regulation of contraction in muscle cells. The primary transcript can be spliced in different ways, as indicated in the figure, to produce distinct mRNAs, which then give rise to variant proteins. Some of the splicing patterns are specific for certain types of cells. For example, the α-tropomyosin made in striated muscle is different from that made from the same gene in smooth muscle. The arrowheads in the top part of the figure mark the sites where cleavage and poly-A addition form the 3′ ends of the mature mRNAs.

Nucleotide Sequences Signal Where Splicing Occurs

The mechanism of pre-mRNA splicing shown in Figure 6–25 requires that the splicing machinery recognize three portions of the precursor RNA molecule: the 5′ splice site, the 3′ splice site, and the branch point in the intron sequence that forms the base of the excised lariat. Not surprisingly, each site has a consensus nucleotide sequence that is similar from intron to intron and provides the cell with cues for where splicing is to take place (**Figure 6–27**). However, these consensus sequences are relatively short and can accommodate extensive sequence variability; as we shall see shortly, the cell incorporates additional types of information to ultimately choose exactly where, on each RNA molecule, splicing is to take place.

The high variability of the splicing consensus sequences presents a special challenge for scientists attempting to decipher genome sequences. Introns range in size from about 10 nucleotides to over 100,000 nucleotides, and choosing the precise borders of each intron is a difficult task even with the aid of powerful computers. The possibility of alternative splicing compounds the problem of predicting protein sequences solely from a genome sequence. This difficulty is one of the main barriers to identifying all of the genes in a complete genome sequence, and it is one of the primary reasons why we know only the approximate number of different proteins produced by the human genome.

RNA Splicing Is Performed by the Spliceosome

Unlike the other steps of mRNA production we have discussed, key steps in RNA splicing are performed by RNA molecules rather than proteins. Specialized RNA molecules recognize the nucleotide sequences that specify where splicing is to occur and also catalyze the chemistry of splicing. These RNA molecules are relatively short (less than 200 nucleotides each), and there are five of them, U1, U2, U4, U5, and U6. Known as **snRNAs** (**small nuclear RNAs**), each is complexed with at least seven protein subunits to form an snRNP (small nuclear ribonucleoprotein).

Figure 6–27 **The consensus nucleotide sequences in an RNA molecule that signal the beginning and the end of most introns in humans.** The three blocks of nucleotide sequences shown are required to remove an intron sequence. Here A, G, U, and C are the standard RNA nucleotides; R stands for purines (A or G); and Y stands for pyrimidines (C or U). The A highlighted in *red* forms the branch point of the lariat produced by splicing (see Figure 6–25). Only the GU at the start of the intron and the AG at its end are invariant nucleotides in the splicing consensus sequences. Several different nucleotides can occupy the remaining positions, although the indicated nucleotides are preferred. The distances along the RNA between the three splicing consensus sequences are highly variable; however, the distance between the branch point and 3′ splice junction is typically much shorter than that between the 5′ splice junction and the branch point.

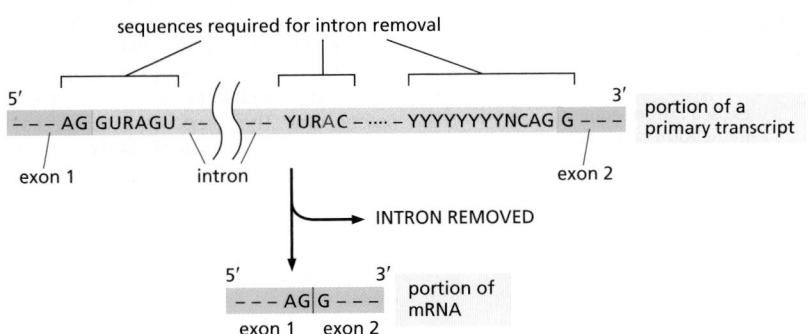

These snRNPs form the core of the **spliceosome**, the large assembly of RNA and protein molecules that performs pre-mRNA splicing in the cell. During the splicing reaction, recognition of the 5′ splice junction, the branch-point site, and the 3′ splice junction is performed largely through base-pairing between the snRNAs and the consensus RNA sequences in the pre-mRNA substrate.

The spliceosome is a complex and dynamic machine. When studied *in vitro*, a few components of the spliceosome assemble on pre-mRNA and, as the splicing reaction proceeds, new components enter and those that have already performed their tasks are jettisoned (**Figure 6–28**). However, many scientists believe that, inside the cell, the spliceosome is a preexisting, loose assembly of all the components—capturing, splicing, and releasing RNA as a coordinated unit, and undergoing extensive rearrangements each time a splice is made.

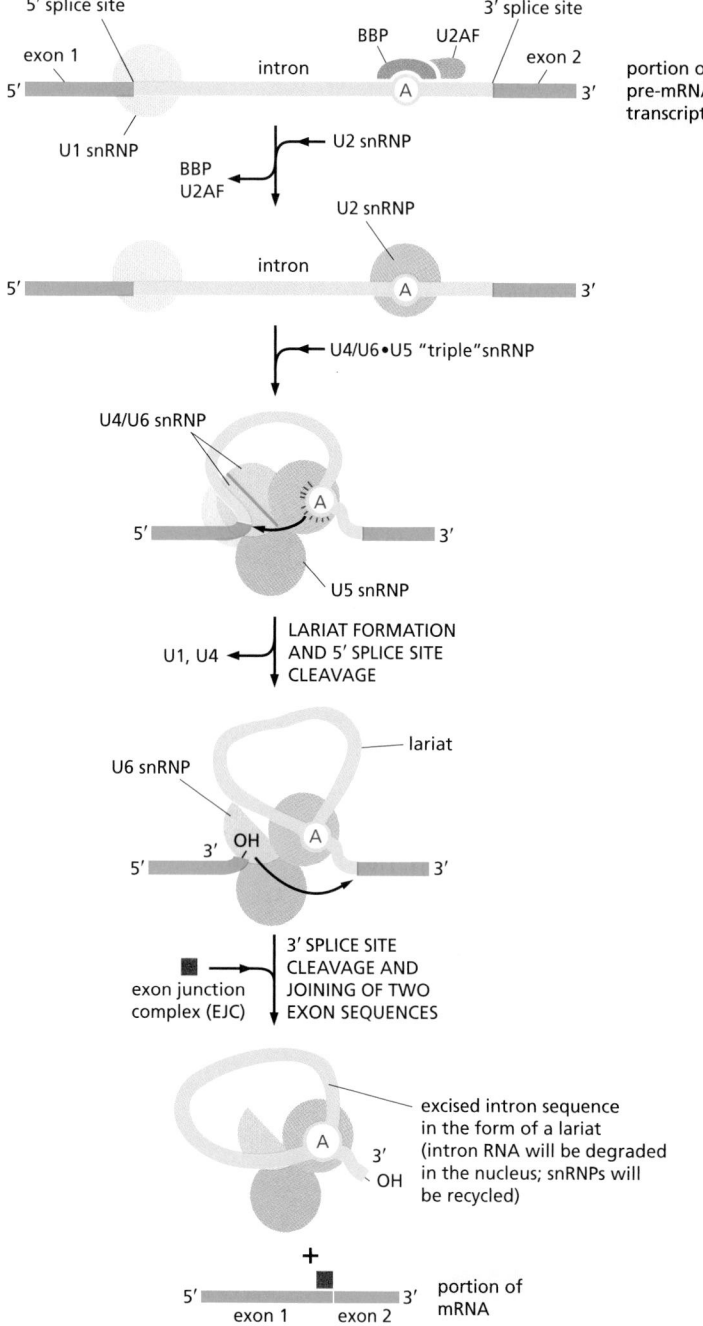

The U1 snRNP forms base pairs with the 5′ splice junction (see Figure 6–29) and the BBP (branch-point binding protein) and U2AF (U2 auxilliary factor) recognize the branch-point site.

The U2 snRNP displaces BBP and U2AF and forms base pairs with the branch-point site consensus sequence.

The U4/U6•U5 "triple" snRNP enters the reaction. In this triple snRNP, the U4 and U6 snRNAs are held firmly together by base-pair interactions. Subsequent rearrangements break apart the U4/U6 base pairs, allowing U6 to displace U1 at the 5′ splice junction (see Figure 6–29). This creates the active site that catalyzes the first phosphoryl-transferase reaction.

Additional RNA–RNA rearrangements create the active site for the second phosphoryl-transferase reaction, which then completes the splice (see Figure 6–25A).

Figure 6–28 The pre-mRNA splicing mechanism. RNA splicing is catalyzed by an assembly of snRNPs (shown as *colored circles*) plus other proteins (most of which are not shown), which together constitute the spliceosome. The spliceosome recognizes the splicing signals on a pre-mRNA molecule, brings the two ends of the intron together, and provides the enzymatic activity for the two reaction steps required (see Figure 6–25A and Movie 6.5). As indicated, a set of proteins called the exon junction complex (EJC) remains on the spliced mRNA molecule; its subsequent role will be discussed shortly.

The Spliceosome Uses ATP Hydrolysis to Produce a Complex Series of RNA–RNA Rearrangements

ATP hydrolysis is not required for the chemistry of RNA splicing *per se* since the two transesterification reactions preserve the high-energy phosphate bonds. However, extensive ATP hydrolysis is required for the assembly and rearrangements of the spliceosome. Some of the additional proteins that make up the spliceosome use the energy of ATP hydrolysis to break existing RNA–RNA interactions to allow the formation of new ones. Each successful splice requires approximately 200 proteins, if we include those that form the snRNPs.

What is the purpose of these rearrangements? First, they allow the splicing signals on the pre-RNA to be examined by snRNPs several times during the course of splicing. For example, the U1 snRNP initially recognizes the 5′ splice site through conventional base-pairing; as splicing proceeds, these base pairs are broken (using the energy of ATP hydrolysis) and U1 is replaced by U6 (**Figure 6–29**). This type of RNA–RNA rearrangement (in which the formation of one RNA–RNA interaction requires the disruption of another) occurs several times during splicing and allows the spliceosomes to check and recheck the splicing signals, thereby increasing the overall accuracy of splicing. Second, the rearrangements that take place in the spliceosome create the active sites for the two transesterification reactions. These two active sites are created, one after the other, and only after the splicing signals on the pre-mRNA have been checked several times. This orderly progression ensures that splicing accidents occur only rarely.

One of the most surprising features of the spliceosome is the nature of the catalytic sites: they are formed by both protein and RNA molecules, although the RNA molecules catalyze the actual chemistry of splicing. In the last section of this chapter, we discuss in general terms the structural and chemical properties of RNA molecules that allow them to act as catalysts.

Once the splicing chemistry is completed, the snRNPs remain bound to the lariat. The disassembly of these snRNPs from the lariat (and from each other) requires another series of RNA–RNA rearrangements that require ATP hydrolysis, thereby returning the snRNAs to their original configuration so that they can be used again in a new reaction. At the completion of a splice, the spliceosome directs a set of proteins to bind to the mRNA near the position formerly occupied by the intron. Called the *exon junction complex* (*EJC*), these proteins mark the site of a successful splicing event and, as we shall see later in this chapter, influence the subsequent fate of the mRNA.

Other Properties of Pre-mRNA and Its Synthesis Help to Explain the Choice of Proper Splice Sites

As we have seen, intron sequences vary enormously in size, with some being in excess of 100,000 nucleotides. If splice-site selection were determined solely by the snRNPs acting on a preformed, protein-free RNA molecule, we would expect frequent splicing mistakes—such as exon skipping and the use of "cryptic" splice sites (**Figure 6–30**). The fidelity mechanisms built into the spliceosome to suppress errors, however, are supplemented by two additional strategies that further increase the accuracy of splicing. The first is a simple consequence of splicing being coupled to transcription. As transcription proceeds, the phosphorylated tail of RNA polymerase carries several components of the spliceosome (see Figure

Figure 6–29 **One of the many rearrangements that take place in the spliceosome during pre-mRNA splicing.** This example comes from the yeast *Saccharomyces cerevisiae*, in which the nucleotide sequences involved are slightly different from those in human cells. The exchange of U1 snRNP for U6 snRNP occurs just before the first phosphoryl-transfer reaction (see Figure 6–28). This exchange requires the 5′ splice site to be read by two different snRNPs, thereby increasing the accuracy of 5′ splice-site selection by the spliceosome.

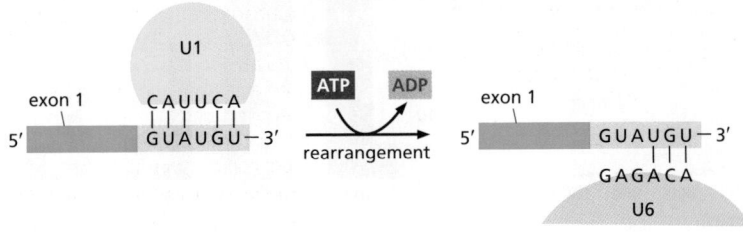

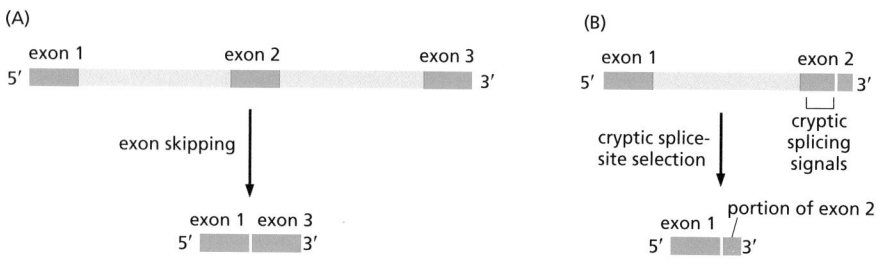

Figure 6–30 **Two types of splicing errors.** (A) Exon skipping. (B) Cryptic splice-site selection. Cryptic splicing signals are nucleotide sequences of RNA that closely resemble true splicing signals and are sometimes mistakenly used by the spliceosome.

6–22), and these components are transferred directly from the polymerase to the RNA as the RNA emerges from the polymerase. This strategy helps the cell keep track of introns and exons: for example, the snRNPs that assemble at a 5′ splice site are initially presented only with the single 3′ splice site that emerges next from the polymerase; the potential sites further downstream have not yet been synthesized. The coordination of transcription with splicing is especially important in preventing inappropriate exon skipping.

A strategy called "exon definition" also helps cells choose the appropriate splice sites. Exon size tends to be much more uniform than intron size, averaging about 150 nucleotide pairs across a wide variety of eukaryotic organisms (**Figure 6–31**). Through exon definition, the splicing machinery can seek out the relatively homogeneously sized exon sequences. As RNA synthesis proceeds, a group of additional components (most notably SR proteins, so-named because they contain a domain rich in serines and arginines) assemble on exon sequences and help to mark off each 3′ and 5′ splice site, starting at the 5′ end of the RNA (**Figure 6–32**). These proteins, in turn, recruit U1 snRNA, which marks the downstream exon boundary, and U2 snRNA, which specifies the upstream one. By specifically marking the exons in this way and thereby taking advantage of the relatively uniform size of exons, the cell increases the accuracy with which it deposits the initial splicing components on the nascent RNA and thereby avoids "near miss" splice sites. How the SR proteins discriminate exon sequences from intron sequences is not understood in detail; however, it is known that some of the SR proteins bind preferentially to specific RNA sequences in exons, termed *splicing enhancers*. In principle, since any one of several different codons can be used to code for a given amino acid, there is freedom to evolve the exon nucleotide sequence so as to form a binding site for an SR protein, without necessarily affecting the amino acid sequence that the exon specifies.

Both the marking of exon and intron boundaries and the assembly of the spliceosome begin on an RNA molecule while it is still being elongated by RNA polymerase at its 3′ end. However, the actual chemistry of splicing can take place later. This delay means that intron sequences are not necessarily removed from a pre-mRNA molecule in the order in which they occur along the RNA chain.

Figure 6–31 **Variation in intron and exon lengths in the human, worm, and fly genomes.** (A) Size distribution of exons. (B) Size distribution of introns. Note that exon length is much more uniform than intron length. (Adapted from International Human Genome Sequencing Consortium, *Nature* 409:860–921, published 2001 by Macmillan Magazines Ltd. Reproduced with permission of SNCSC.)

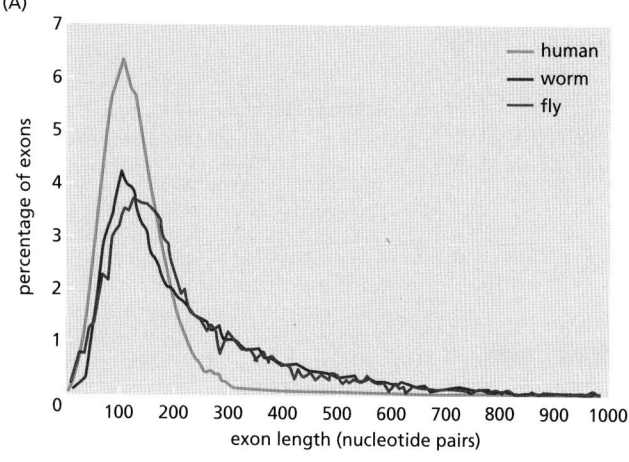

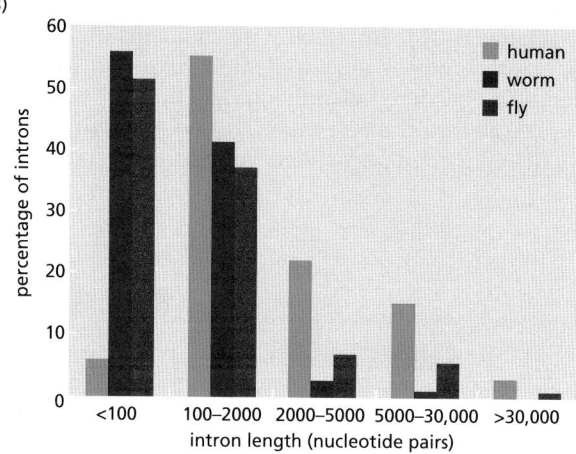

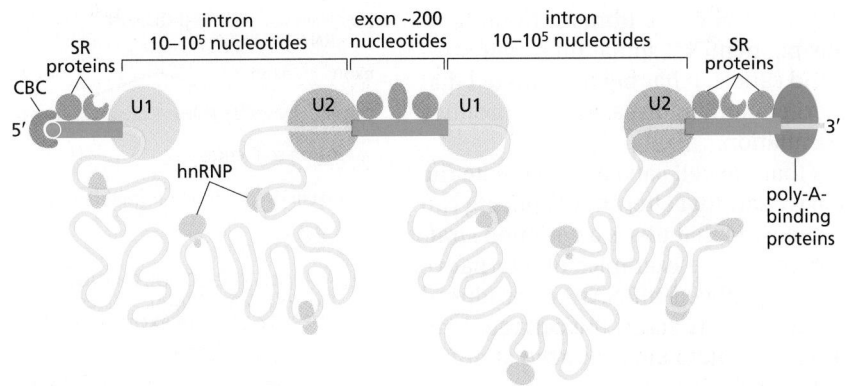

Figure 6–32 **The exon definition hypothesis.** According to this idea, SR proteins bind to each exon sequence in the pre-mRNA and thereby help to guide the snRNPs to the proper intron/exon boundaries. This demarcation of exons by the SR proteins occurs co-transcriptionally, beginning at the CBC (cap-binding complex) at the 5′ end. It has been proposed that a group of proteins known as the heterogeneous nuclear ribonucleoproteins (hnRNPs) may preferentially associate with intron sequences, further helping the spliceosome distinguish introns from exons. (Adapted from R. Reed, *Curr. Opin. Cell Biol.* 12:340–345, 2000. With permission from Elsevier.)

Chromatin Structure Affects RNA Splicing

Although it may seem at first counterintuitive, the way a gene is packaged into chromatin can affect how the RNA transcript of that gene is ultimately spliced. Nucleosomes tend to be positioned over exons (which are, on average, close to the length of DNA in a nucleosome), and it has been proposed that these act as "speed bumps," allowing the proteins responsible for exon definition to assemble on the RNA as it emerges from the polymerase. In addition, changes in chromatin structure are used to alter splicing patterns. There are two ways this can happen. First, because splicing and transcription are coupled, the rate at which RNA polymerase moves along DNA can affect RNA splicing. For example, if polymerase is moving slowly, exon skipping (see Figure 6–30A) is minimized: assembly of the initial spliceosome may be complete before an alternative choice of splice site even emerges from the RNA polymerase. The nucleosomes in condensed chromatin can cause polymerase to pause; the pattern of pauses in turn affects the extent of RNA exposed at any given time to the splicing machinery.

There is a second and more direct way that chromatin structure can affect RNA splicing. Although the details are not yet understood, specific histone modifications attract components of the spliceosome, and, because the chromatin being transcribed is in close association with the nascent RNA, these splicing components can easily be transferred to the emerging RNA. In this way, certain types of histone modifications can affect the final pattern of splicing.

RNA Splicing Shows Remarkable Plasticity

We have seen that the choice of splice sites depends on such features of the pre-mRNA transcript as the strength of the three signals on the RNA (the 5′ and 3′ splice junctions and the branch point) for the splicing machinery, the co-transcriptional assembly of the spliceosome, chromatin structure, and the "bookkeeping" that underlies exon definition. We do not know exactly how accurate splicing normally is because, as we see later, there are several quality control systems that rapidly destroy mRNAs whose splicing goes awry. However, we do know that, compared with other steps in gene expression, splicing is unusually flexible.

Thus, for example, a mutation in a nucleotide sequence critical for splicing of a particular intron does not necessarily prevent splicing of that intron altogether. Instead, the mutation typically creates a new pattern of splicing (**Figure 6–33**). Most commonly, an exon is simply skipped (Figure 6–33B). In other cases, the mutation causes a cryptic splice junction to be efficiently used (Figure 6–33C). Apparently, the splicing machinery has evolved to pick out the best possible pattern of splice junctions, and if the optimal one is damaged by mutation, it will seek out the next best pattern, and so on. This inherent plasticity in the process of RNA splicing suggests that changes in splicing patterns caused by random mutations have been important in the evolution of genes and organisms. It also means that mutations that affect splicing can be severely detrimental to the organism: in addition to the β thalassemia, example presented in Figure 6–33, aberrant splicing plays important roles in the development of cystic fibrosis, frontotemporal dementia, Parkinson's disease, retinitis pigmentosa, spinal muscular atrophy, myotonic dystrophy, premature aging, and cancer. It has been estimated that of the many point mutations that cause inherited human diseases, 10% produce aberrant splicing of the gene containing the mutation.

The plasticity of RNA splicing also means that the cell can easily regulate the pattern of RNA splicing. Earlier in this section we saw that alternative splicing can give rise to different proteins from the same gene and that this is a common strategy to enhance the coding potential of genomes. Some examples of alternative splicing are constitutive; that is, the alternatively spliced mRNAs are produced continuously by cells of an organism. However, in many cases, the cell regulates the splicing patterns so that different forms of the protein are produced at different times and in different tissues (see Figure 6–26). In Chapter 7, we return to this issue to discuss some specific examples of regulated RNA splicing.

Spliceosome-Catalyzed RNA Splicing Probably Evolved from Self-splicing Mechanisms

When the spliceosome was first discovered, it puzzled molecular biologists. Why do RNA molecules instead of proteins perform important roles in splice-site recognition and in the chemistry of splicing? Why is a lariat intermediate used rather than the apparently simpler alternative of bringing the 5′ and 3′ splice sites together in a single step, followed by their direct cleavage and rejoining? The answers to these questions reflect the way in which the spliceosome has evolved.

As discussed briefly in Chapter 1 (and in more detail in the final section of this chapter), it is likely that early cells used RNA molecules rather than proteins as their major catalysts and that they stored their genetic information in RNA rather than in DNA sequences. RNA-catalyzed splicing reactions presumably had critical roles in these early cells. As evidence, some *self-splicing RNA* introns (that is, intron sequences in RNA whose splicing out can occur in the absence of proteins or any other RNA molecules) remain today—for example, in the nuclear rRNA genes of the ciliate *Tetrahymena*, in a few bacteriophage T4 genes, and in some mitochondrial and chloroplast genes. In these cases, the RNA molecule folds into a specific three-dimensional structure that brings the intron/exon junctions together and catalyzes the two transesterification reactions. A self-splicing intron sequence can be identified in a test tube by incubating a pure RNA molecule that contains the intron sequence and observing the splicing reaction. Because the basic chemistry of some self-splicing reactions is so similar to pre-mRNA splicing, it has been proposed that the much more involved process of pre-mRNA splicing evolved from a simpler, ancestral form of RNA self-splicing.

RNA-Processing Enzymes Generate the 3′ End of Eukaryotic mRNAs

We have seen that the 5′ end of the pre-mRNA produced by RNA polymerase II is capped almost as soon as it emerges from the RNA polymerase. Then, as the polymerase continues its movement along a gene, the spliceosome assembles on the RNA and delineates the intron and exon boundaries. The long C-terminal tail of the RNA polymerase coordinates these processes by transferring capping and splicing components directly to the RNA as it emerges from the enzyme. In this section, we shall see that, as RNA polymerase II reaches the end of a gene, a similar mechanism ensures that the 3′ end of the pre-mRNA is appropriately processed.

The position of the 3′ end of each mRNA molecule is specified by signals encoded in the genome (**Figure 6–34**). These signals are transcribed into RNA as the RNA polymerase II moves through them, and they are then recognized (as RNA) by a series of RNA-binding proteins and RNA-processing enzymes (**Figure 6–35**). Two multisubunit proteins, called CstF (cleavage stimulation factor) and CPSF (cleavage and polyadenylation specificity factor), are of special importance.

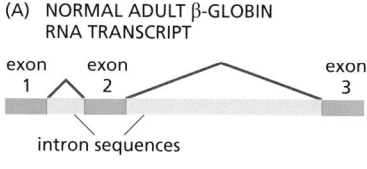

(A) NORMAL ADULT β-GLOBIN
      RNA TRANSCRIPT

intron sequences

normal mRNA is formed from three exons

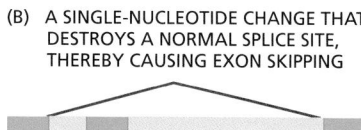

(B) A SINGLE-NUCLEOTIDE CHANGE THAT
      DESTROYS A NORMAL SPLICE SITE,
      THEREBY CAUSING EXON SKIPPING

mRNA with exon 2 missing

(C) A SINGLE-NUCLEOTIDE CHANGE THAT
      DESTROYS A NORMAL SPLICE SITE, THEREBY
      ACTIVATING A CRYPTIC SPLICE SITE

mRNA with extended exon 3

(D) A SINGLE-NUCLEOTIDE CHANGE THAT
      CREATES A NEW SPLICE SITE THEREBY CAUSING
      A NEW EXON TO BE INCORPORATED

mRNA with extra exon inserted
between exon 2 and exon 3

Figure 6–33 Abnormal processing of the β-globin primary RNA transcript in humans with the disease β thalassemia. In the examples shown, the disease (a severe anemia due to aberrant hemoglobin synthesis) is caused by splice-site mutations found in the genomes of affected patients. The *dark blue boxes* represent the three normal exon sequences; the *red lines* connect the 5′ and 3′ splice sites that are used. In (B), (C), and (D), the *light blue boxes* depict new nucleotide sequences included in the final mRNA molecule as a result of the mutation denoted by the *black arrowhead*. Note that when a mutation leaves a normal splice site without a partner, an exon is skipped (B) or one or more abnormal cryptic splice sites nearby is used as the partner site (C).

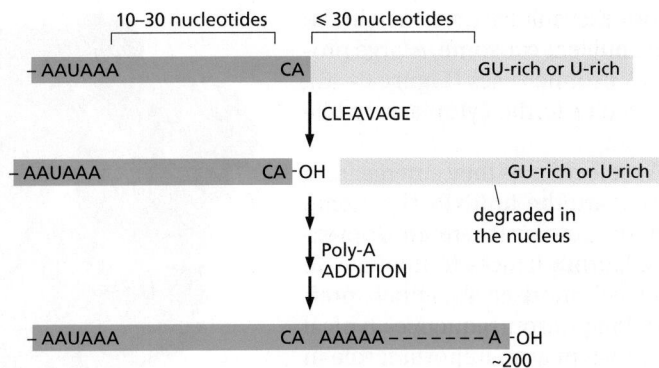

Figure 6–34 **Consensus nucleotide sequences that direct cleavage and polyadenylation to form the 3′ end of a eukaryotic mRNA.** These sequences are encoded in the genome, and specific proteins recognize them—as RNA—after they are transcribed. As shown in Figure 6–35, the hexamer AAUAAA is bound by CPSF and the GU-rich element beyond the cleavage site is bound by CstF; the CA sequence is bound by a third protein factor required for the cleavage step. Like other consensus nucleotide sequences discussed in this chapter (see Figure 6–12), the sequences shown in the figure represent a variety of individual cleavage and polyadenylation signals.

Both of these proteins travel with the RNA polymerase tail and are transferred to the 3′-end processing sequence on an RNA molecule as it emerges from the RNA polymerase.

Once CstF and CPSF bind to their recognition sequences on the emerging RNA molecule, additional proteins assemble with them to create the 3′ end of the mRNA. First, the RNA is cleaved from the polymerase (see Figure 6–35). Next an enzyme called poly-A polymerase (PAP) adds, one at a time, approximately 200 A nucleotides to the 3′ end produced by the cleavage. The nucleotide precursor for these additions is ATP, and the same type of 5′-to-3′ bonds are formed as in conventional RNA synthesis. But unlike other RNA polymerases, poly-A polymerase does not require a template; hence the poly-A tail of eukaryotic mRNAs is not directly encoded in the genome. As the poly-A tail is synthesized, proteins called poly-A-binding proteins assemble onto it and, by a poorly understood mechanism, help determine the final length of the tail.

After the 3′-end of a eukaryotic pre-mRNA molecule has been cleaved, the RNA polymerase II continues to transcribe, in some cases for hundreds of nucleotides. Once 3′-end cleavage has occurred, the newly synthesized RNA that emerges from the polymerases lacks a 5′ cap; this unprotected RNA is rapidly degraded by a 5′ → 3′ exonuclease carried along on the polymerase tail. Apparently, it is this continued RNA degradation that eventually causes the RNA polymerase to release its grip on the template and terminate transcription.

Mature Eukaryotic mRNAs Are Selectively Exported from the Nucleus

Eukaryotic pre-mRNA synthesis and processing take place in an orderly fashion within the cell nucleus. But of the pre-mRNA that is synthesized, only a small fraction—the mature mRNA—is of further use to the cell. Most of the rest—excised introns, broken RNAs, and aberrantly processed pre-mRNAs—is not only useless but potentially dangerous. How does the cell distinguish between the relatively rare mature mRNA molecules it wishes to keep and the overwhelming amount of debris created by RNA processing?

The answer is that, as an RNA molecule is processed, it loses certain proteins and acquires others. For example, we have seen that acquisition of cap-binding complexes, exon junction complexes, and poly-A-binding proteins mark the completion of capping, splicing, and poly-A addition, respectively. A properly completed mRNA molecule is also distinguished by the proteins it lacks. For example, the presence of an snRNP protein would signify incomplete or aberrant splicing. Only when the proteins present on an mRNA molecule collectively signify that processing was successfully completed is the mRNA exported from the nucleus into the cytosol, where it can be translated into protein. Improperly processed mRNAs

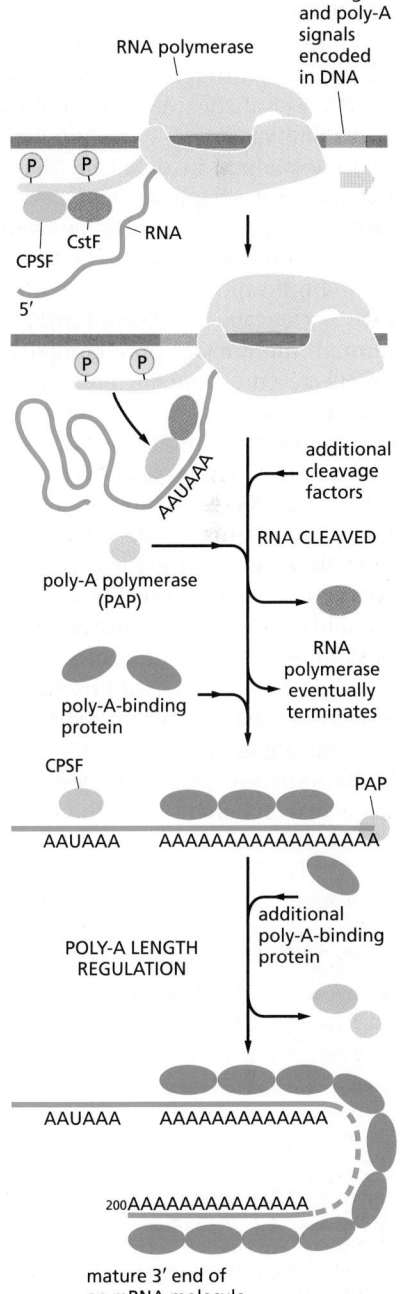

Figure 6–35 **Some of the major steps in generating the 3′ end of a eukaryotic mRNA.** This process is much more complicated than the analogous process in bacteria, where the RNA polymerase simply stops at a termination signal and releases both the 3′ end of its transcript and the DNA template (see Figure 6–11).

and other RNA debris (excised intron sequences, for example) are retained in the nucleus, where they are eventually degraded by the nuclear **exosome**, a large protein complex whose interior is rich in 3′-to-5′ RNA exonucleases (**Figure 6–36**). Eukaryotic cells thus export only useful RNA molecules to the cytoplasm, while debris is disposed of in the nucleus.

Of all the proteins that assemble on pre-mRNA molecules as they emerge from transcribing RNA polymerases, the most abundant are the hnRNPs (heterogeneous nuclear ribonuclear proteins). Some of these proteins (there are approximately 30 different ones in humans) unwind the hairpin helices in the RNA so that splicing and other signals on the RNA can be read more easily. Others preferentially package the RNA contained in the very long intron sequences typical in complex organisms (see Figure 6–31) and these may play an important role in distinguishing mature mRNA from the debris left over from RNA processing.

Successfully processed mRNAs are guided through the **nuclear pore complexes** (NPCs)—aqueous channels in the nuclear membrane that directly connect the nucleoplasm and cytosol (**Figure 6–37**). Small molecules (less than 60,000 daltons) can diffuse freely through these channels. However, most of the macromolecules in cells, including mRNAs complexed with proteins, are far too large to pass through the channels without a special process. The cell uses energy to actively transport such macromolecules in both directions through the nuclear pore complexes.

As explained in detail in Chapter 12, macromolecules are moved through nuclear pore complexes by *nuclear transport receptors,* which, depending on the identity of the macromolecule, escort it from the nucleus to the cytoplasm or vice versa. For mRNA export to occur, a specific nuclear transport receptor must be loaded onto the mRNA, a step that, in many organisms, takes place in concert with 3′ cleavage and polyadenylation. Once it helps to move an RNA molecule through the nuclear pore complex, the transport receptor dissociates from the mRNA, re-enters the nucleus, and is then used to export a new mRNA molecule.

The export of mRNA–protein complexes from the nucleus can be readily observed with the electron microscope for the unusually abundant mRNA of the insect *Balbiani Ring genes.* As these genes are transcribed, the newly formed RNA is seen to be packaged by proteins, including hnRNPs, SR proteins, and components of the spliceosome. This protein–RNA complex undergoes a series of structural transitions, probably reflecting RNA processing events, culminating in a curved fiber (see Figure 6–37). This curved fiber moves through the nucleoplasm and enters the nuclear pore complex (with its 5′ cap proceeding first), and it then undergoes another series of structural transitions as it moves through the pore. These and other observations reveal that the pre-mRNA–protein and mRNA–protein complexes are dynamic structures that gain and lose numerous specific proteins during RNA synthesis, processing, and export (**Figure 6–38**).

The analysis just described has been complemented by new methods that allow researchers to track the fate of more typical mRNA molecules, which can

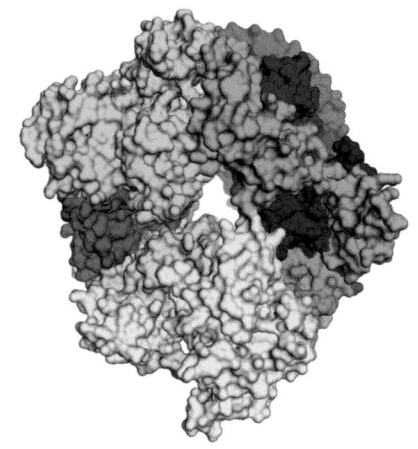

Figure 6–36 Structure of the core of human RNA exosome. RNA is fed into one end of the central pore and is degraded by RNAses that associate with the other end. Nine different protein subunits (each represented by a different color) make up this large ring structure. Eukaryotic cells have both a nuclear exosome and a cytoplasmic exosome; both forms include the core exosome shown here and additional subunits (including specialized RNAses) that differentiate the two forms. The nuclear exosome degrades aberrant RNAs before they are exported to the cytosol. It also processes certain types of RNA (for example, the ribosomal RNAs) to produce their final form. The cytoplasmic form of the exosome is responsible for degrading mRNAs in the cytosol, and is thus crucial in determining the lifetime of each mRNA molecule. (PDB code: 2NN6.)

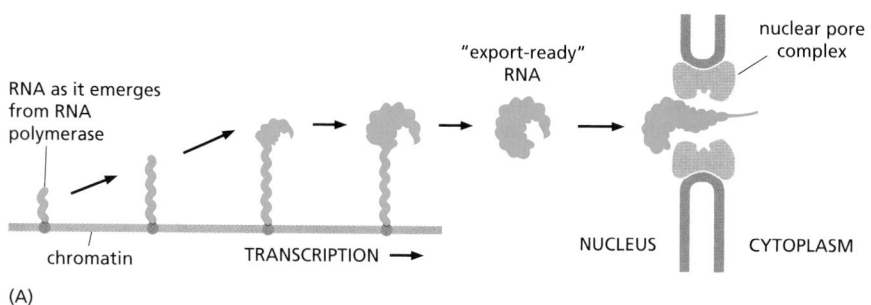

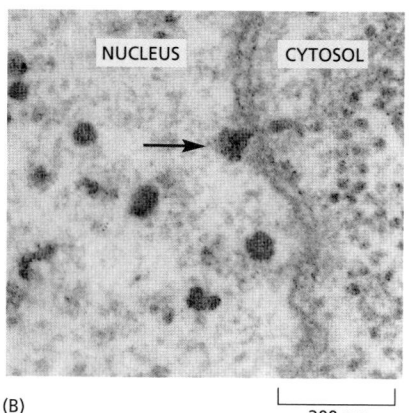

Figure 6–37 Transport of a large mRNA molecule through the nuclear pore complex. (A) The maturation of an mRNA molecule as it is synthesized by RNA polymerase and packaged by a variety of nuclear proteins. This drawing of an unusually large and abundant insect RNA, called the Balbiani Ring mRNA, is based on electron microscope micrographs such as that shown in (B). (A, adapted from B. Daneholt, *Cell* 88:585–588, 1997; B, © 1966 B.J. Stevens and H. Swift. Originally published in *J. Cell Biol.* https:doi.org/10.1083/jcb.31.1.55. With permission from Rockefeller University Press.)

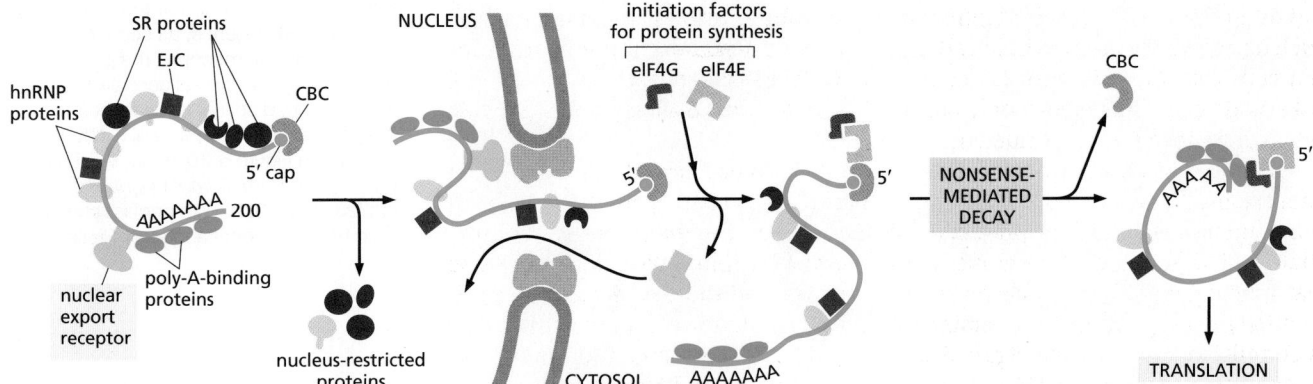

Figure 6–38 Schematic illustration of an export-ready mRNA molecule and its transport through the nuclear pore. As indicated, some proteins travel with the mRNA as it moves through the pore, whereas others remain in the nucleus. The nuclear export receptor for mRNAs is a complex of proteins that binds to an mRNA molecule once it has been correctly spliced and polyadenylated. After the mRNA has been exported to the cytosol, this export receptor dissociates from the mRNA and is re-imported into the nucleus, where it can be used again. The final check indicated here, called *nonsense-mediated decay*, will be described later in the chapter.

be fluorescently labeled and observed individually. A typical RNA molecule is released from its site of transcription and spends several minutes diffusing to a nuclear pore complex. During this time it is likely that RNA processing events continue and that the RNA sheds previously bound proteins and acquires new ones. Once it arrives at the entrance to the pore, the "export-ready" mRNA hovers for several seconds, during which time the completion of processing may occur, and then is transported through the pore very rapidly, in tens of milliseconds. Some mRNA–protein complexes are very large, and how they move through the nuclear pore complexes so rapidly remains a mystery.

Some of the proteins deposited on the mRNA while it is still in the nucleus can affect the fate of the RNA after it is transported to the cytosol. Thus, the stability of an mRNA in the cytosol, the efficiency with which it is translated into protein, and its ultimate destination in the cytosol can all be determined by proteins acquired in the nucleus that remain bound to the RNA after it leaves the nucleus.

But before discussing what happens to mRNAs in the cytosol, we briefly consider how the synthesis and processing of some noncoding RNA molecules occurs. There are many types of noncoding RNAs produced by cells (see Table 6–1, p. 305), but here we focus on the rRNAs, which are critically important for the translation of mRNAs into protein.

Noncoding RNAs Are Also Synthesized and Processed in the Nucleus

Only a few percent of the dry weight of a mammalian cell is RNA; of that, only about 3–5% is mRNA. The bulk of the RNA in cells performs structural and catalytic functions (see Table 6–1). The most abundant RNAs in cells are the ribosomal RNAs (rRNAs), constituting approximately 80% of the RNA in rapidly dividing cells. As discussed later in this chapter, these RNAs form the core of the ribosome. Unlike bacteria—in which a single RNA polymerase synthesizes all RNAs in the cell—eukaryotes have a separate, specialized polymerase, RNA polymerase I, that is dedicated to producing rRNAs. RNA polymerase I is similar structurally to the RNA polymerase II discussed previously; however, the absence of a C-terminal tail in polymerase I helps to explain why its transcripts are neither capped nor polyadenylated.

Because multiple rounds of translation of each mRNA molecule can provide an enormous amplification in the production of protein molecules, many of the proteins that are very abundant in a cell can be synthesized from genes that are present in a single copy per haploid genome (see Figure 6–3). In contrast, the RNA components of the ribosome are final gene products, and a growing mammalian cell must synthesize approximately 10 million copies of each type of ribosomal RNA in each cell generation to construct its 10 million ribosomes. The cell can produce adequate quantities of ribosomal RNAs only because it contains multiple copies of the **rRNA genes** that code for **ribosomal RNAs (rRNAs)**. Even *E. coli* needs seven copies of its rRNA genes to meet the cell's need for ribosomes. Human cells contain about 200 rRNA gene copies per haploid genome, spread

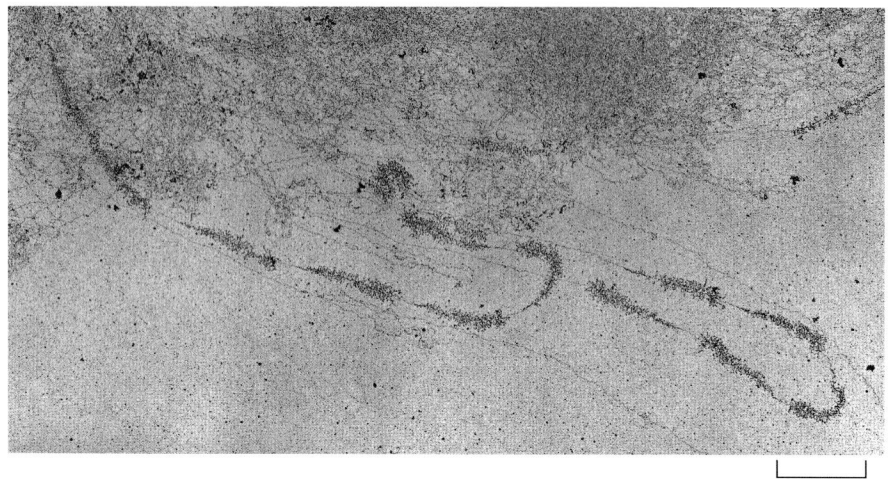

2 μm

Figure 6–39 Transcription from tandemly arranged rRNA genes, as seen in the electron microscope. The pattern of alternating transcribed gene and nontranscribed spacer is readily seen. A higher-magnification view of rRNA genes is shown in Figure 6–10. (From V.E. Foe, *Cold Spring Harb. Symp. Quant. Biol.* 42:723–740, 1978. With permission from Cold Spring Harbor Laboratory Press.)

out in small clusters on five different chromosomes (see Figure 4–11), while cells of the frog *Xenopus* contain about 600 rRNA gene copies per haploid genome in a single cluster on one chromosome (**Figure 6–39**).

There are four types of eukaryotic rRNAs, each present in one copy per ribosome. Three of the four rRNAs (18S, 5.8S, and 28S) are made by chemically modifying and cleaving a single large precursor rRNA (**Figure 6–40**); the fourth (5S RNA) is synthesized from a separate cluster of genes by a different polymerase, RNA polymerase III, and does not require chemical modification.

Extensive chemical modifications occur in the 13,000-nucleotide-long precursor rRNA before the rRNAs are cleaved out of it and assembled into ribosomes. These include about 100 methylations of the 2′-OH positions on nucleotide sugars and 100 isomerizations of uridine nucleotides to pseudouridine (**Figure 6–41**A). The functions of these modifications are not understood in detail, but they probably aid in the folding and assembly of the final rRNAs, or subtly alter the function of ribosomes. Each modification is made at a specific position in the precursor rRNA, specified by "guide RNAs," which position themselves on the precursor rRNA through base-pairing and thereby bring an RNA-modifying enzyme to the appropriate position (Figure 6–41B). Other guide RNAs promote cleavage of the precursor rRNAs into the mature rRNAs, probably by causing conformational changes in the precursor rRNA that expose these sites to nucleases. All of these guide RNAs are members of a large class of RNAs called **small nucleolar RNAs (or snoRNAs)**, so named because these RNAs perform their functions in a subcompartment of the nucleus called the nucleolus. Many snoRNAs are encoded in

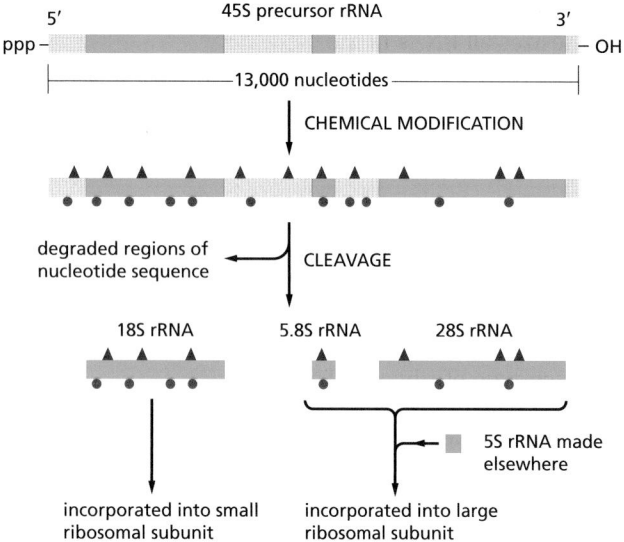

Figure 6–40 The chemical modification and nucleolytic processing of a eukaryotic 45S precursor rRNA molecule into three separate ribosomal RNAs. Two types of chemical modifications (color-coded as indicated in Figure 6–41) are made to the precursor rRNA before it is cleaved. Nearly half of the nucleotide sequences in this precursor rRNA are discarded and degraded in the nucleus by the exosome. The rRNAs are named according to their "S" values, which refer to their rate of sedimentation in an ultracentrifuge. The larger the S value, the larger the rRNA.

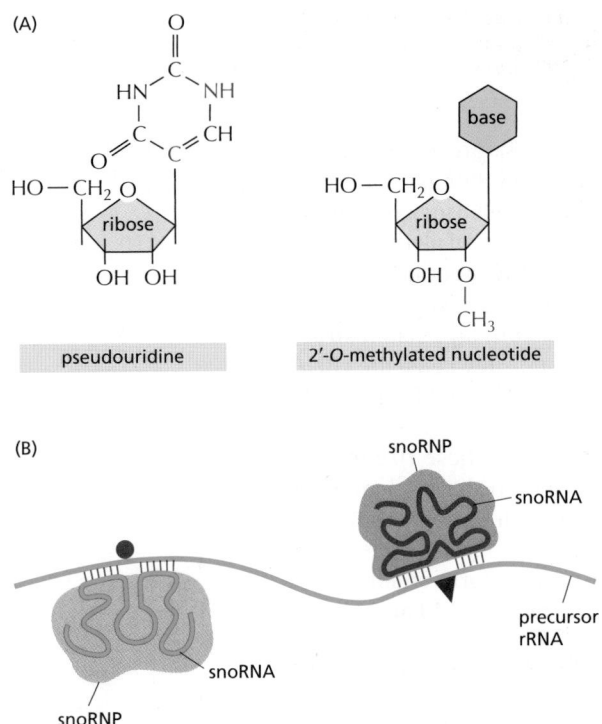

Figure 6–41 **Modifications of the precursor rRNA by guide RNAs.** (A) Two prominent covalent modifications made to rRNA; the differences from the initially incorporated nucleotide are indicated by *red* atoms. Pseudouridine is an isomer of uridine; the base has been "rotated," and is attached to the *red* C rather than to the *red* N of the sugar (compare to Figure 6–5B). (B) As indicated, snoRNAs determine the sites of modification by base-pairing to complementary sequences on the precursor rRNA. The snoRNAs are bound to proteins, and the complexes are called snoRNPs (small nucleolar ribonucleoproteins). snoRNPs contain both the guide sequences and the enzymes that modify the rRNA.

the introns of other genes, especially those encoding ribosomal proteins. They are synthesized by RNA polymerase II and processed from excised intron sequences.

The Nucleolus Is a Ribosome-Producing Factory

The nucleolus is the most obvious structure seen in the nucleus of a eukaryotic cell when viewed in the light microscope. It was so closely scrutinized by early cytologists that an 1898 review could list some 700 references. We now know that the nucleolus is the site for the processing of rRNAs and their assembly into ribosome subunits. Unlike many of the major organelles in the cell, the nucleolus is not bound by a membrane (**Figure 6–42**); instead, it is a huge aggregate

Figure 6–42 **Electron micrograph of a thin section of a nucleolus in a human fibroblast, showing its three distinct zones.** (A) View of entire nucleus. (B) Higher-power view of the nucleolus. It is believed that transcription of the rRNA genes takes place between the fibrillar center and the dense fibrillar component and that processing of the rRNAs and their assembly into the two subunits of the ribosome proceeds outward from the dense fibrillar component to the surrounding granular components. (Courtesy of E.G. Jordan and J. McGovern.)

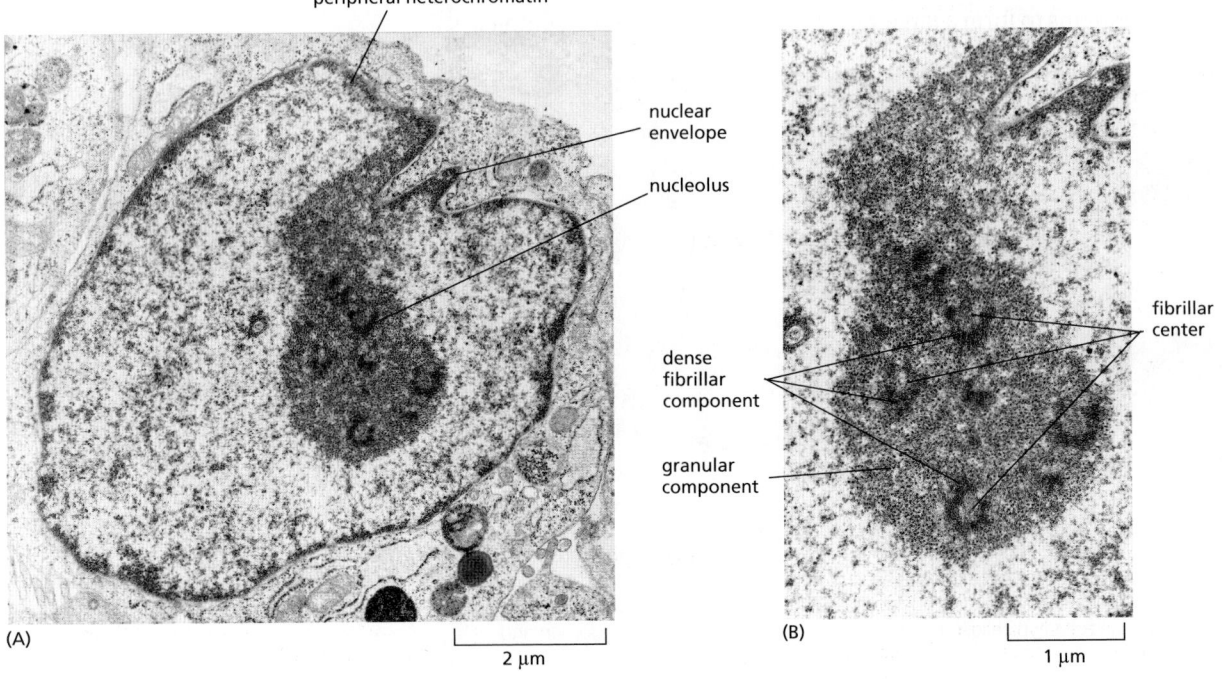

Figure 6–43 Changes in the appearance of the nucleolus in a human cell during the cell cycle. Only the cell nucleus is represented in this diagram. In most eukaryotic cells, the nuclear envelope breaks down during mitosis, as indicated by the *dashed circles.*

of macromolecules, including the rRNA genes themselves, precursor rRNAs, mature rRNAs, rRNA-processing enzymes, snoRNPs, a large set of assembly factors (including ATPases, GTPases, protein kinases, and RNA helicases), ribosomal proteins, and partly assembled ribosomes. The close association of all these components allows the assembly of ribosomes to occur rapidly and smoothly.

Various types of RNA molecules play a central part in the chemistry and structure of the nucleolus, suggesting that it may have evolved from an ancient structure present in cells dominated by RNA catalysis. In present-day cells, the rRNA genes have an important role in forming the nucleolus. In a diploid human cell, the rRNA genes are distributed into 10 clusters, located near the tips of five different chromosome pairs (see Figure 4–11). During interphase, these 10 chromosomes contribute DNA loops (containing the rRNA genes) to the nucleolus; in M phase, when the chromosomes condense, the nucleolus fragments and then disappears. Then, in the telophase part of mitosis, as chromosomes return to their semi-dispersed state, the tips of the 10 chromosomes reform small nucleoli, which progressively coalesce into a single nucleolus (**Figure 6–43** and **Figure 6–44**). As might be expected, the size of the nucleolus reflects the number of ribosomes that the cell is producing. Its size therefore varies greatly in different cells and can change in a single cell, occupying 25% of the total nuclear volume in cells that are making unusually large amounts of protein.

Ribosome assembly is a complex process, the most important features of which are outlined in **Figure 6–45**. In addition to its central role in ribosome biogenesis, the nucleolus is the site where other noncoding RNAs are produced and other RNA–protein complexes are assembled. For example, the U6 snRNP, which functions in pre-mRNA splicing (see Figure 6–28), is composed of one RNA molecule and at least seven proteins. The U6 snRNA is chemically modified by snoRNAs in the nucleolus before its final assembly there into the U6 snRNP. Other important RNA–protein complexes, including telomerase (encountered in Chapter 5) and the signal-recognition particle (which we discuss in Chapter 12), are assembled at the nucleolus. Finally, the tRNAs (transfer RNAs) that carry the amino acids for protein synthesis are processed there as well; like the rRNA genes, the genes encoding tRNAs are clustered in the nucleolus. Thus, the nucleolus can be thought of as a large factory at which different noncoding RNAs are transcribed, processed, and assembled with proteins to form a large variety of ribonucleoprotein complexes.

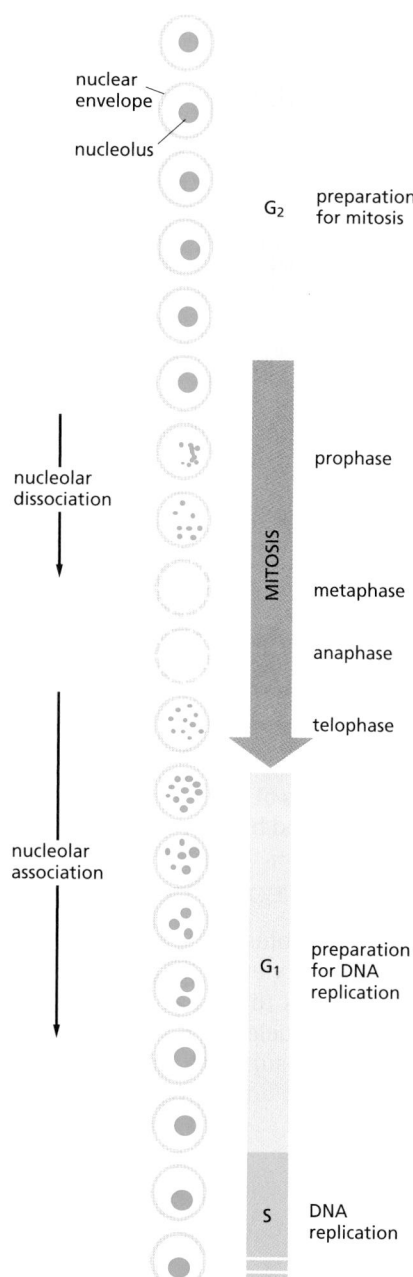

nuclear envelope

nucleolus

G₂ preparation for mitosis

MITOSIS

prophase

nucleolar dissociation

metaphase

anaphase

telophase

nucleolar association

G₁ preparation for DNA replication

S DNA replication

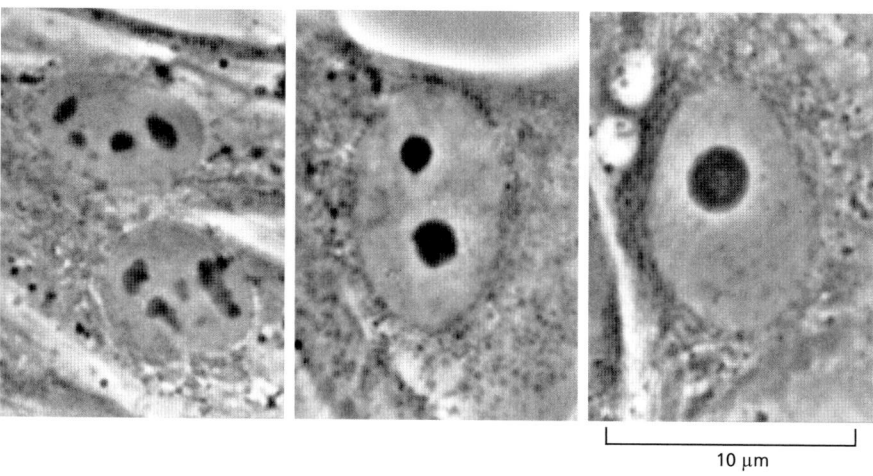

10 μm

Figure 6–44 Nucleolar fusion. These light micrographs of human fibroblasts grown in culture show various stages of nucleolar fusion. After mitosis, each of the 10 human chromosomes that carry a cluster of rRNA genes begins to form a tiny nucleolus, but these rapidly coalesce as they grow to form the single large nucleolus typical of many interphase cells. (Courtesy of E.G. Jordan and J. McGovern.)

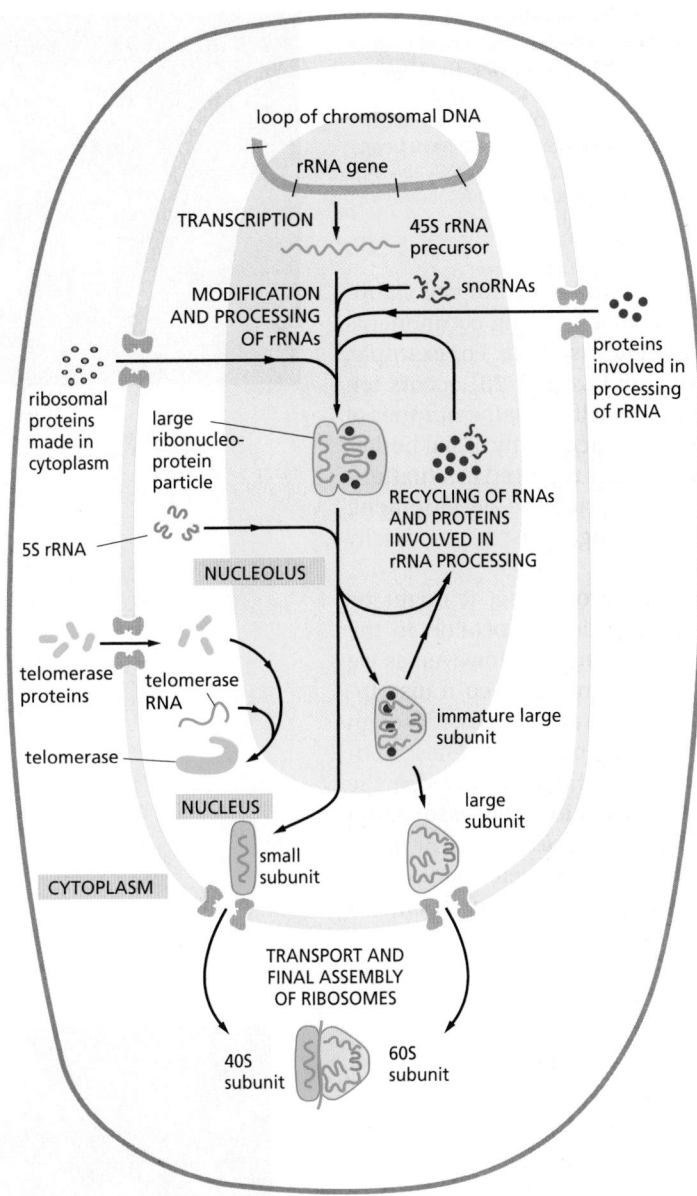

loop of chromosomal DNA

rRNA gene

TRANSCRIPTION

45S rRNA precursor

MODIFICATION AND PROCESSING OF rRNAs

snoRNAs

proteins involved in processing of rRNA

ribosomal proteins made in cytoplasm

large ribonucleo-protein particle

RECYCLING OF RNAs AND PROTEINS INVOLVED IN rRNA PROCESSING

5S rRNA

NUCLEOLUS

telomerase proteins

telomerase RNA

telomerase

immature large subunit

NUCLEUS

large subunit

small subunit

CYTOPLASM

TRANSPORT AND FINAL ASSEMBLY OF RIBOSOMES

40S subunit

60S subunit

Figure 6–45 The function of the nucleolus in ribosome and other ribonucleoprotein synthesis. The 45S precursor rRNA is packaged in a large ribonucleoprotein particle containing many ribosomal proteins imported from the cytoplasm. While this particle remains at the nucleolus, selected components are added and others discarded as it is processed into immature large and small ribosomal subunits. The two ribosomal subunits attain their final functional form only after each is individually transported through the nuclear pores into the cytoplasm. Other ribonucleoprotein complexes, including telomerase shown here, are also assembled in the nucleolus.

The Nucleus Contains a Variety of Subnuclear Aggregates

Although the nucleolus is the most prominent structure in the nucleus, several other nuclear bodies have been observed and studied (**Figure 6–46**). These include Cajal bodies (named for the scientist who first described them in 1906) and interchromatin granule clusters (also called "speckles"). Like the nucleolus, these other nuclear structures lack membranes and are highly dynamic depending on the needs of the cell. Their assembly is likely mediated by the association of low complexity protein domains, as described in Chapter 3 (see Figure 3–36). Their appearance is the result of the tight association of protein and RNA components involved in the synthesis, assembly, and storage of macromolecules involved in gene expression. Cajal bodies are sites where the snRNPs and snoRNPs undergo their final maturation steps, and where the snRNPs are recycled and their RNAs are "reset" after the rearrangements that occur during splicing (see p. 321). In contrast, the interchromatin granule clusters have been proposed to be stockpiles of fully mature snRNPs and other RNA processing components that are ready to be used in the production of mRNA.

Scientists have had difficulties in working out the function of these small subnuclear structures, in part because their appearances can change dramatically as cells traverse the cell cycle or respond to changes in their environment. Moreover,

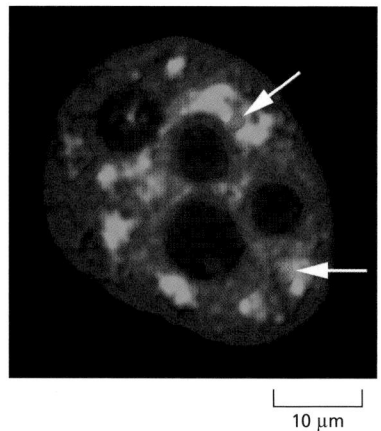

Figure 6–46 **Visualization of some prominent nuclear bodies.** The protein fibrillarin *(red)*, a component of several snoRNPs, is present at both nucleoli and Cajal bodies; the latter are indicated by the arrows. The Cajal bodies (but not the nucleoli) are also highlighted by staining one of their main components, the protein coilin; the superposition of the snoRNP and coilin stains appears *pink*. Interchromatin granule clusters *(green)* have been revealed by using antibodies against a protein involved in pre-mRNA splicing. DNA is stained blue by the dye DAPI. (From J.R. Swedlow and A.I. Lamond, *Gen. Biol.* 2:1–7, 2001. Micrograph courtesy of Judith Sleeman.)

10 µm

disrupting a particular type of nuclear body often has little effect on cell viability. It seems that the main function of these aggregates is to bring components together at high concentration in order to speed up their assembly. For example, it is estimated that assembly of the U4/U6 snRNP (see Figure 6–28) occurs ten times more rapidly in Cajal bodies than would be the case if the same number of components were dispersed throughout the nucleus. Consequently, Cajal bodies appear dispensable in many types of cells but are absolutely required in situations where cells must proliferate rapidly, such as in early vertebrate development. Here, protein synthesis (which depends on RNA splicing) must be especially rapid, and delays can be lethal.

Given the prominence of nuclear bodies in RNA processing, it might be expected that pre-mRNA splicing would occur in a particular location in the nucleus, as it requires numerous RNA and protein components. However, as we have seen, the assembly of splicing components on pre-mRNA is co-transcriptional; thus, splicing must occur at many locations along chromosomes. Although a typical mammalian cell may be expressing on the order of 15,000 genes, transcription and RNA splicing takes place in only several thousand sites in the nucleus. These sites are highly dynamic and probably result from the association of transcription and splicing components to create small *factories*, the name given to specific aggregates containing a high local concentration of selected components that create biochemical assembly lines (**Figure 6–47**). Interchromatin

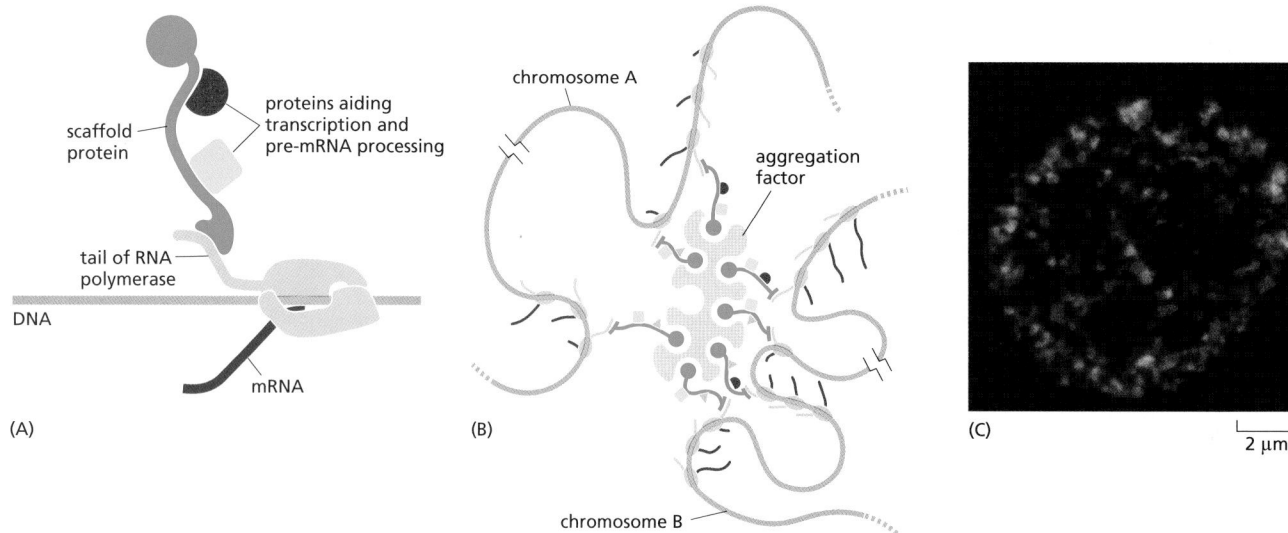

(A)   (B)   (C)   2 µm

Figure 6–47 **A model for an mRNA production factory.** mRNA production is made more efficient in the nucleus by an aggregation of the many components needed for transcription and pre-mRNA processing, thereby producing a specialized biochemical factory. In (A), a postulated scaffold protein holds various components in the proximity of a transcribing RNA polymerase. Other key components are bound directly to the RNA polymerase tail, which likewise serves as a scaffold (see Figure 6–22), but for simplicity these are not shown here. In (B), a large number of such scaffolds have been brought together to form an aggregate that is highly enriched in the many components needed for the synthesis and processing of pre-mRNAs. Such a scaffold model can account for the several thousand sites of active RNA transcription and processing typically observed in the nucleus of a mammalian cell, each of which has a diameter of roughly 100nm and is estimated to contain, on average, about 10 RNA polymerase II molecules in addition to many other proteins. (C) Here, mRNA production factories and DNA replication factories have been visualized in the same mammalian cell by briefly incorporating differently modified nucleotides into each nucleic acid and detecting the RNA and DNA produced using antibodies, one *(green)* detecting the newly synthesized DNA and the other *(red)* detecting the newly synthesized RNA. (C, from D.G. Wansink et al., *J. Cell Sci.* 107:1449–1456, 1994. With permission from The Company of Biologists.)

granule clusters—which contain stockpiles of RNA processing components—are often observed next to these sites of transcription, as though poised to replenish supplies. We can thus view the nucleus as organized into subdomains, with snRNPs, snoRNPs, and other nuclear components moving among them in an orderly fashion according to the needs of the cell.

Summary

Before the synthesis of a particular protein can begin, the corresponding mRNA molecule must be produced by transcription. Bacteria contain a single type of RNA polymerase (the enzyme that carries out the transcription of DNA into RNA). An mRNA molecule is produced after this enzyme initiates transcription at a promoter, synthesizes the RNA by chain elongation, stops transcription at a terminator, and releases both the DNA template and the completed mRNA molecule. In eukaryotic cells, the process of transcription is much more complex, and there are three RNA polymerases—polymerase I, II, and III—that are related evolutionarily to one another and to the bacterial polymerase.

RNA polymerase II synthesizes eukaryotic mRNA. This enzyme requires a set of additional proteins, both the general transcription factors, and specific transcriptional activator proteins, to initiate transcription on a DNA template. It requires still more proteins (including chromatin remodeling complexes and histone-modifying enzymes) to initiate transcription on its chromatin templates inside the cell.

During the elongation phase of transcription, the nascent RNA undergoes three types of processing events: a special nucleotide is added to its 5′ end (capping), intron sequences are removed from the middle of the RNA molecule (splicing), and the 3′ end of the RNA is generated (cleavage and polyadenylation). Each of these processes is initiated by proteins that travel along with RNA polymerase II by binding to sites on its long, extended C-terminal tail. Splicing is unusual in that many of its key steps are carried out by specialized RNA molecules rather than proteins. Only properly processed mRNAs are passed through nuclear pore complexes into the cytosol, where they are translated into protein.

For many genes, RNA, rather than protein, is the final product. In eukaryotes, these genes are usually transcribed by either RNA polymerase I or RNA polymerase III. RNA polymerase I makes the ribosomal RNAs. After their synthesis as a large precursor, the rRNAs are chemically modified, cleaved, and assembled into the two ribosomal subunits in the nucleolus—a distinct subnuclear structure that also helps to process some smaller RNA–protein complexes in the cell. Additional subnuclear structures (including Cajal bodies and interchromatin granule clusters) are sites where components involved in RNA processing are assembled, stored, and recycled. The high concentration of components in such "factories" ensures that the processes being catalyzed are rapid and efficient.

FROM RNA TO PROTEIN

In the preceding section, we have seen that the final product of some genes is an RNA molecule itself, such as the RNAs present in the snRNPs and in ribosomes. However, most genes in a cell produce mRNA molecules that serve as intermediaries on the pathway to proteins. In this section, we examine how the cell converts the information carried in an mRNA molecule into a protein molecule. This feat of translation was a strong focus of attention for biologists in the late 1950s, when it was posed as the "coding problem": how is the information in a linear sequence of nucleotides in RNA translated into the linear sequence of a chemically quite different set of units—the amino acids in proteins? This fascinating question stimulated great excitement. Here was a cryptogram set up by nature that, after more than 3 billion years of evolution, could finally be solved by one of the products of evolution—human beings. And indeed, not only was the code cracked step by step, but in the year 2000 the structure of the elaborate machinery by which cells read this code—the ribosome—was finally revealed in atomic detail.

	Arg									Leu					Ser				Val	
	AGA									UUA					AGC					
	AGG									UUG					AGU					
GCA	CGA					GGA				CUA				CCA	UCA	ACA			GUA	UAA
GCC	CGC					GGC		AUA	CUC				CCC	UCC	ACC			GUC	UAG
GCG	CGG	GAC	AAC	UGC	GAA	CAA	GGG	CAC	AUC	CUG	AAA		UUC	CCG	UCG	ACG		UAC	GUG	UGA
GCU	CGU	GAU	AAU	UGU	GAG	CAG	GGU	CAU	AUU	CUU	AAG	AUG	UUU	CCU	UCU	ACU	UGG	UAU	GUU	

Ala	Arg	Asp	Asn	Cys	Glu	Gln	Gly	His	Ile	Leu	Lys	Met	Phe	Pro	Ser	Thr	Trp	Tyr	Val	stop
A	R	D	N	C	E	Q	G	H	I	L	K	M	F	P	S	T	W	Y	V	

An mRNA Sequence Is Decoded in Sets of Three Nucleotides

Once an mRNA has been produced by transcription and processing, the information present in its nucleotide sequence is used to synthesize a protein. Transcription is simple to understand as a means of information transfer: since DNA and RNA are chemically and structurally similar, the DNA can act as a direct template for the synthesis of RNA by complementary base-pairing. As the term *transcription* signifies, it is as if a message written out by hand is being converted, say, into a typewritten text. The language itself and the form of the message do not change, and the symbols used are closely related.

In contrast, the conversion of the information in RNA into protein represents a **translation** of the information into another language that uses quite different symbols. Moreover, since there are only 4 different nucleotides in mRNA and 20 different types of amino acids in a protein, this translation cannot be accounted for by a direct one-to-one correspondence between a nucleotide in RNA and an amino acid in protein. The nucleotide sequence of a gene, through the intermediary of mRNA, is instead translated into the amino acid sequence of a protein by rules that are known as the **genetic code**. This code was deciphered in the early 1960s.

The sequence of nucleotides in the mRNA molecule is read in consecutive groups of three. RNA is a linear polymer of four different nucleotides, so there are $4 \times 4 \times 4 = 64$ possible combinations of three nucleotides: the triplets AAA, AUA, AUG, and so on. However, only 20 different amino acids are commonly found in proteins. Either some nucleotide triplets are never used, or the code is redundant and some amino acids are specified by more than one triplet. The second possibility is, in fact, the correct one, as shown by the completely deciphered genetic code in **Figure 6–48**. Each group of three consecutive nucleotides in RNA is called a **codon**, and each codon specifies either one amino acid or a stop to the translation process.

This genetic code is used universally in all present-day organisms. Although a few slight differences in the code have been found, these are chiefly in the DNA of mitochondria. Mitochondria have their own transcription and protein-synthesis systems that operate quite independently from those of the rest of the cell, and it is understandable that their tiny genomes have been able to accommodate minor changes to the code (discussed in Chapter 14).

In principle, an RNA sequence can be translated in any one of three different **reading frames**, depending on where the decoding process begins (**Figure 6–49**). However, only one of the three possible reading frames in an mRNA encodes the required protein. We see later how a special punctuation signal at the beginning of each RNA message sets the correct reading frame at the start of protein synthesis.

tRNA Molecules Match Amino Acids to Codons in mRNA

The codons in an mRNA molecule do not directly recognize the amino acids they specify: the group of three nucleotides does not, for example, bind directly to the amino acid. Rather, the translation of mRNA into protein depends on *adaptor* molecules that can recognize and bind both to the codon and, at another site on their surface, to the amino acid. These adaptors consist of a set of small RNA molecules known as **transfer RNAs** (**tRNAs**), each about 80 nucleotides in length.

Figure 6–48 The genetic code. The standard one-letter abbreviation for each amino acid is presented below its three-letter abbreviation (see Panel 3–1, pp. 112–113, for the full name of each amino acid and its structure). By convention, codons are always written with the 5′-terminal nucleotide to the left. Note that most amino acids are represented by more than one codon, and that there are some regularities in the set of codons that specifies each amino acid: codons for the same amino acid tend to contain the same nucleotides at the first and second positions, and vary at the third position. Three codons do not specify any amino acid but act as termination sites (stop codons), signaling the end of the protein-coding sequence. One codon—AUG—acts both as an initiation codon, signaling the start of a protein-coding message, and also as the codon that specifies methionine.

Figure 6–49 The three possible reading frames in protein synthesis. In the process of translating a nucleotide sequence *(blue)* into an amino acid sequence *(red)*, the sequence of nucleotides in an mRNA molecule is read from the 5′ end to the 3′ end in consecutive sets of three nucleotides. In principle, therefore, the same RNA sequence can specify three completely different amino acid sequences, depending on the reading frame. In reality, however, only one of these reading frames contains the actual message.

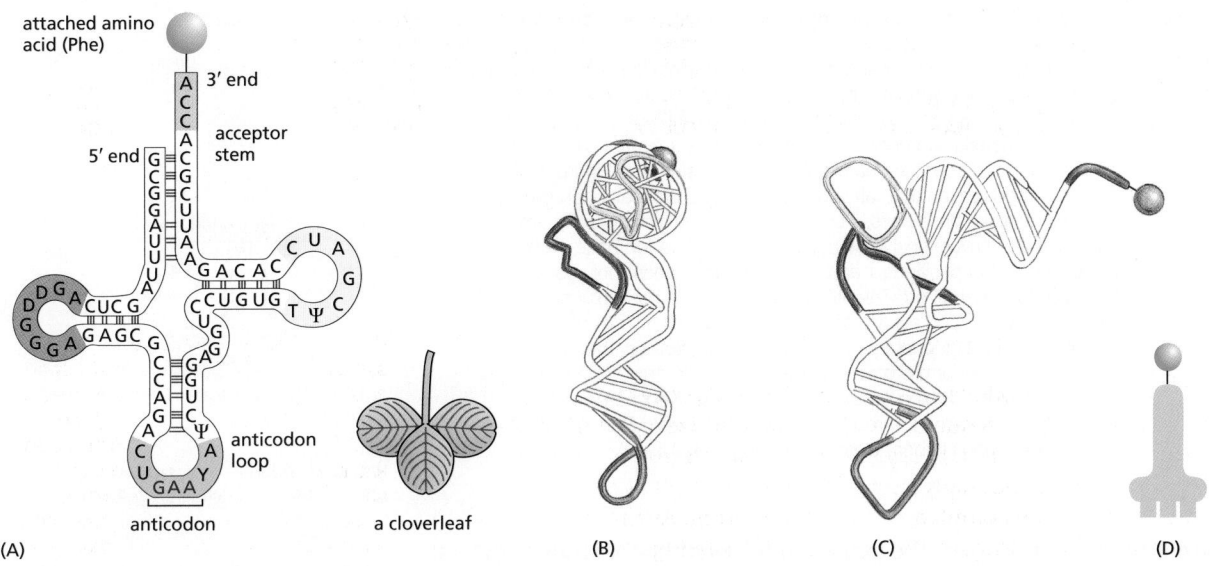

5′ GCGGAUUUAGCUCAGDDGGGAGAGCGCCAGACUGAAYAΨCUGGAGGUCCUGUGTΨCGAUCCACAGAAUUCGCACCA 3′
(E)   anticodon

Figure 6–50 **A tRNA molecule.** A tRNA specific for the amino acid phenylalanine (Phe) is depicted in various ways. (A) The cloverleaf structure showing the complementary base-pairing *(red lines)* that creates the double-helical regions of the molecule. The anticodon is the sequence of three nucleotides that base-pairs with a codon in mRNA. The amino acid matching the codon/anticodon pair is attached at the 3′ end of the tRNA. tRNAs contain some unusual bases, which are produced by chemical modification after the tRNA has been synthesized. For example, the bases denoted ψ (pseudouridine—see Figure 6–41) and D (dihydrouridine—see Figure 6–53) are derived from uracil. (B and C) Views of the L-shaped molecule, based on x-ray diffraction analysis. Although this diagram shows the tRNA for the amino acid phenylalanine, all other tRNAs have similar structures. (D) The tRNA icon we use in this book. (E) The linear nucleotide sequence of the molecule, color-coded to match (A), (B), and (C).

We saw earlier in this chapter that RNA molecules can fold into precise three-dimensional structures, and the tRNA molecules provide a striking example. Four short segments of the folded tRNA are double-helical, producing a molecule that looks like a cloverleaf when drawn schematically (**Figure 6–50**). For example, a 5′-GCUC-3′ sequence in one part of a polynucleotide chain can form a relatively strong association with a 5′-GAGC-3′ sequence in another region of the same molecule. The cloverleaf undergoes further folding to form a compact L-shaped structure that is held together by additional hydrogen bonds between different regions of the molecule (see Figure 6–50B and C).

Two regions of unpaired nucleotides situated at either end of the L-shaped molecule are crucial to the function of tRNA in protein synthesis. One of these regions forms the **anticodon**, a set of three consecutive nucleotides that pairs with the complementary codon in an mRNA molecule. The other is a short single-stranded region at the 3′ end of the molecule; this is the site where the amino acid that matches the codon is attached to the tRNA.

We saw above that the genetic code is redundant; that is, several different codons can specify a single amino acid. This redundancy implies either that there is more than one tRNA for many of the amino acids or that some tRNA molecules can base-pair with more than one codon. In fact, both situations occur. Some amino acids have more than one tRNA and some tRNAs are constructed so that they require accurate base-pairing only at the first two positions of the codon and can tolerate a mismatch (or *wobble*) at the third position (**Figure 6–51**). This wobble base-pairing explains why so many of the alternative codons for an amino acid differ only in their third nucleotide (see Figure 6–48). In bacteria, wobble base-pairings make it possible to fit the 20 amino acids to their 61 codons with as

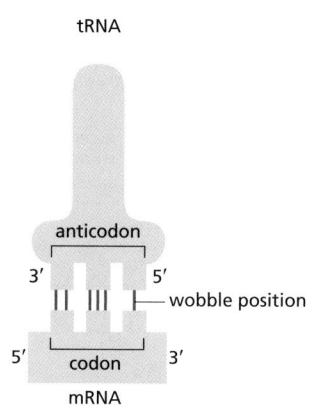

Figure 6–51 Wobble base-pairing between codons and anticodons. If the nucleotide listed in the first column is present at the third, or wobble, position of the codon, it can base-pair with any of the nucleotides listed in the second column. Thus, for example, when inosine (I) is present in the wobble position of the tRNA anticodon, the tRNA can recognize any one of three different codons in bacteria and either of two codons in eukaryotes. The inosine in tRNAs is formed from the deamination of adenosine (see Figure 6–53), a chemical modification that takes place after the tRNA has been synthesized. The nonstandard base pairs, including those made with inosine, are generally weaker than conventional base pairs. Codon–anticodon base-pairing is more stringent at positions 1 and 2 of the codon, where only conventional base pairs are permitted. The differences in wobble base-pairing interactions between bacteria and eukaryotes presumably result from subtle structural differences between bacterial and eukaryotic ribosomes, the molecular machines that perform protein synthesis. (Adapted from C. Guthrie and J. Abelson, in The Molecular Biology of the Yeast *Saccharomyces*: Metabolism and Gene Expression, pp. 487–528. Cold Spring Harbor, New York: Cold Spring Harbor Laboratory Press, 1982.)

bacteria

wobble codon base	possible anticodon bases
U	A, G, or I
C	G or I
A	U or I
G	C or U

eukaryotes

wobble codon base	possible anticodon bases
U	A, G, or I
C	G or I
A	U
G	C

few as 31 kinds of tRNA molecules. The exact number of different kinds of tRNAs, however, differs from one species to the next. For example, humans have nearly 500 tRNA genes, and among them 48 different anticodons are represented.

tRNAs Are Covalently Modified Before They Exit from the Nucleus

Like most other eukaryotic RNAs, tRNAs are covalently modified before they are allowed to exit from the nucleus. Eukaryotic tRNAs are synthesized by RNA polymerase III. Both bacterial and eukaryotic tRNAs are typically synthesized as larger precursor tRNAs, which are then trimmed to produce the mature tRNA. In addition, some tRNA precursors (from both bacteria and eukaryotes) contain introns that must be spliced out. This splicing reaction differs chemically from pre-mRNA splicing; rather than generating a lariat intermediate, tRNA splicing uses a cut-and-paste mechanism that is catalyzed by proteins (**Figure 6–52**). Trimming and splicing both require the precursor tRNA to be correctly folded in its cloverleaf configuration. Because misfolded tRNA precursors will not be processed properly, the trimming and splicing reactions serve as quality-control steps in the generation of tRNAs.

All tRNAs are modified chemically—nearly 1 in 10 nucleotides in each mature tRNA molecule is an altered version of a standard G, U, C, or A ribonucleotide. Over 50 different types of tRNA modifications are known; a few are shown in **Figure 6–53**. Some of the modified nucleotides—most notably inosine, produced by the deamination of adenosine—affect the conformation and base-pairing of the anticodon and thereby facilitate the recognition of the appropriate mRNA codon by the tRNA molecule (see Figure 6–51). Others affect the accuracy with which the tRNA is attached to the correct amino acid.

Specific Enzymes Couple Each Amino Acid to Its Appropriate tRNA Molecule

We have seen that, to read the genetic code in DNA, cells make a series of different tRNAs. We now consider how each tRNA molecule becomes linked to the one amino acid in 20 that is its appropriate partner. Recognition and attachment of the correct amino acid depends on enzymes called **aminoacyl-tRNA synthetases**, which covalently couple each amino acid to its appropriate set of tRNA molecules (**Figure 6–54** and **Figure 6–55**). Most cells have a different synthetase enzyme for each amino acid (that is, 20 synthetases in all); one attaches glycine to all tRNAs that recognize codons for glycine, another attaches alanine to all tRNAs that recognize codons for alanine, and so on. Many bacteria, however, have fewer than 20 synthetases, and the same synthetase enzyme is responsible for coupling more than one amino acid to the appropriate tRNAs. In these cases, a single synthetase places the identical amino acid on two different types of tRNAs, only one of which

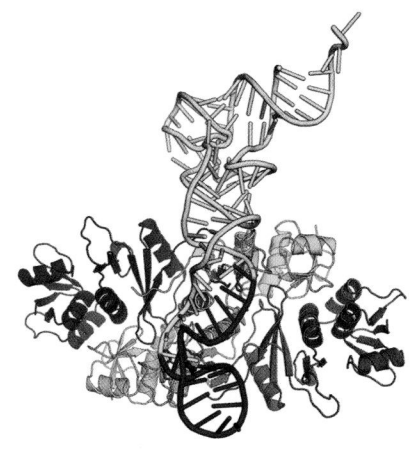

Figure 6–52 Structure of a tRNA-splicing endonuclease docked to a precursor tRNA. The endonuclease (a four-subunit enzyme) removes the tRNA intron *(dark blue, bottom)*. A second enzyme, a multifunctional tRNA ligase (not shown), then joins the two tRNA halves together. (Courtesy of Hong Li, Christopher Trotta, and John Abelson; PDB code: 2A9L.)

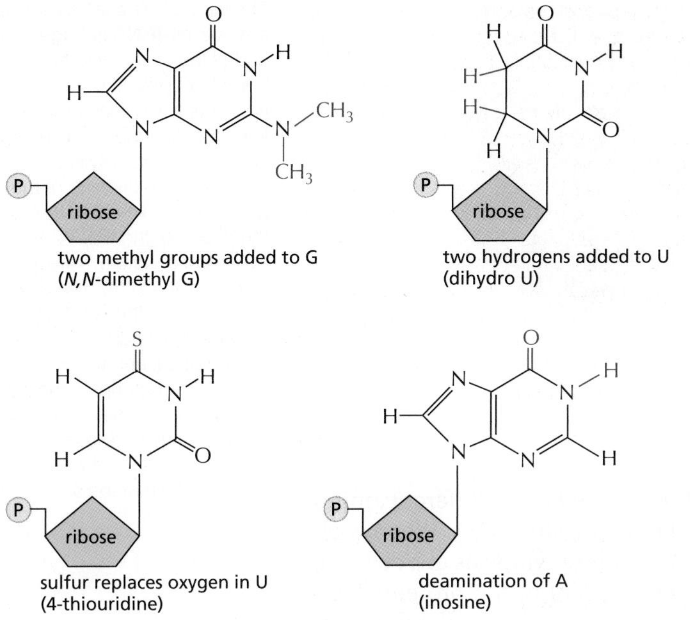

two methyl groups added to G
(*N,N*-dimethyl G)

two hydrogens added to U
(dihydro U)

sulfur replaces oxygen in U
(4-thiouridine)

deamination of A
(inosine)

Figure 6–53 **A few of the unusual nucleotides found in tRNA molecules.** These nucleotides are produced by covalent modification of a normal nucleotide after it has been incorporated into an RNA chain. Two other types of modified nucleotides are shown in Figure 6–41. In most tRNA molecules, about 10% of the nucleotides are modified (see Figure 6–50). As shown in Figure 6–51, inosine is sometimes present at the wobble position in the tRNA anticodon.

has an anticodon that matches the amino acid. A second enzyme then chemically modifies each "incorrectly" attached amino acid so that it now corresponds to the anticodon displayed by its covalently linked tRNA.

The synthetase-catalyzed reaction that attaches the amino acid to the 3′ end of the tRNA is one of many reactions coupled to the energy-releasing hydrolysis of ATP (see pp. 64–65), and it produces a high-energy bond between the tRNA and the amino acid. The energy of this bond is used at a later stage in protein synthesis to link the amino acid covalently to the growing polypeptide chain.

The aminoacyl-tRNA synthetase enzymes and the tRNAs are equally important in the decoding process (**Figure 6–56**). This was established by an experiment in

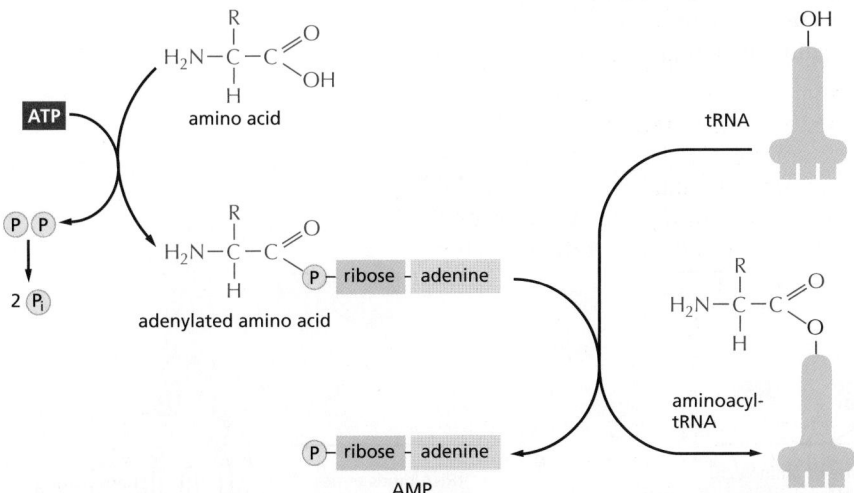

amino acid

adenylated amino acid

aminoacyl-tRNA

AMP

Figure 6–54 **Amino acid activation by synthetase enzymes.** An amino acid is activated for protein synthesis by an aminoacyl-tRNA synthetase enzyme in two steps. As indicated, the energy of ATP hydrolysis is used to attach each amino acid to its tRNA molecule in a high-energy linkage. The amino acid is first activated through the linkage of its carboxyl group directly to AMP, forming an *adenylated amino acid*; the linkage of the AMP, normally an unfavorable reaction, is driven by the hydrolysis of the ATP molecule that donates the AMP. Without leaving the synthetase enzyme, the AMP-linked carboxyl group on the amino acid is then transferred to a hydroxyl group on the sugar at the 3′ end of the tRNA molecule. This transfer joins the amino acid by an activated ester linkage to the tRNA and forms the final aminoacyl-tRNA molecule. The synthetase enzyme is not shown in this diagram.

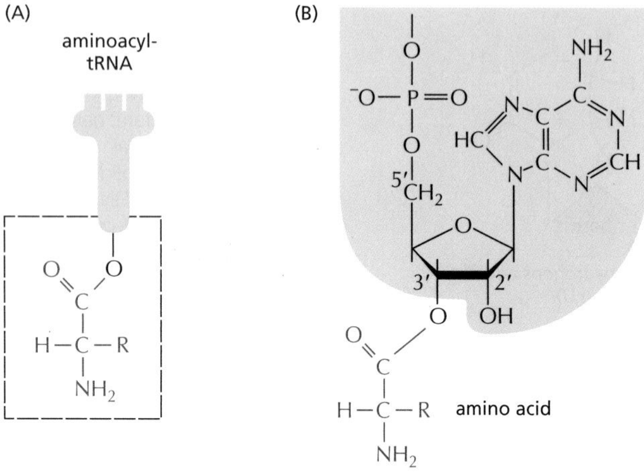

(A)

aminoacyl-
tRNA

(B)

Figure 6–55 The structure of the aminoacyl-tRNA linkage. The carboxyl end of the amino acid forms an ester bond to ribose. Because the hydrolysis of this ester bond is associated with a large favorable change in free energy, an amino acid held in this way is said to be activated. (A) Schematic drawing of the structure. The amino acid is linked to the nucleotide at the 3′ end of the tRNA (see Figure 6–50). (B) Actual structure corresponding to the boxed region in (A). There are two major classes of synthetase enzymes: one links the amino acid directly to the 3′-OH group of the ribose, and the other links it initially to the 2′-OH group. In the latter case, a subsequent transesterification reaction shifts the amino acid to the 3′ position. As in Figure 6–54, the "R group" indicates the side chain of the amino acid.

which one amino acid (cysteine) was chemically converted into a different amino acid (alanine) after it already had been attached to its specific tRNA. When such "hybrid" aminoacyl-tRNA molecules were used for protein synthesis in a cell-free system, the wrong amino acid was inserted at every point in the protein chain where that tRNA was used. Although, as we shall see, cells have several quality control mechanisms to avoid this type of mishap, the experiment did establish that the genetic code is translated by two sets of adaptors that act sequentially. Each matches one molecular surface to another with great specificity, and it is their combined action that associates each sequence of three nucleotides in the mRNA molecule—that is, each codon—with its particular amino acid.

Editing by tRNA Synthetases Ensures Accuracy

Several mechanisms working together ensure that an aminoacyl-tRNA synthetase links the correct amino acid to each tRNA. Most synthetase enzymes select the correct amino acid by a two-step mechanism. The correct amino acid has the highest affinity for the active-site pocket of its synthetase and is therefore favored over the other 19; in particular, amino acids larger than the correct one are excluded from the active site. However, accurate discrimination between two similar amino acids, such as isoleucine and valine (which differ by only a methyl

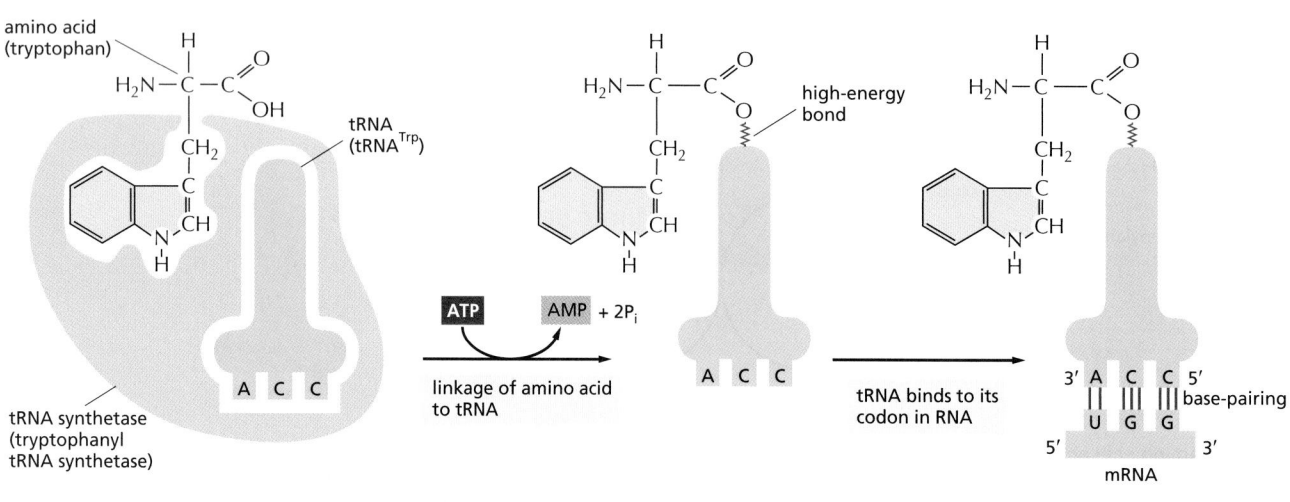

Figure 6–56 The genetic code is translated by means of two adaptors that act one after another. The first adaptor is the aminoacyl-tRNA synthetase, which couples a particular amino acid to its corresponding tRNA; the second adaptor is the tRNA molecule itself, whose *anticodon* forms base pairs with the appropriate *codon* on the mRNA. An error in either step would cause the wrong amino acid to be incorporated into a protein chain (Movie 6.6). In the sequence of events shown, the amino acid tryptophan (Trp) is selected by the codon UGG on the mRNA.

(A)

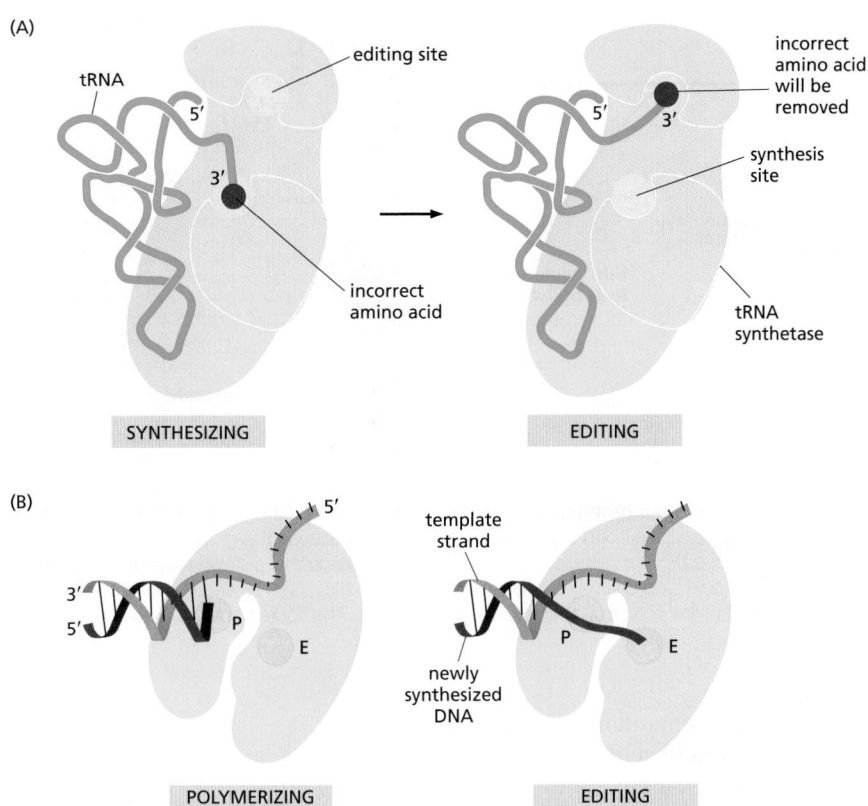

Figure 6–57 **Hydrolytic editing.**
(A) Aminoacyl tRNA synthetases correct their own coupling errors through hydrolytic editing of incorrectly attached amino acids. As described in the text, the correct amino acid is rejected by the editing site.
(B) The error-correction process performed by DNA polymerase has similarities; however, it differs because the removal process depends strongly on a mispairing with the template (see Figure 5–8). (P, polymerization site; E, editing site.)

group), is very difficult to achieve in a single step. A second discrimination step occurs after the amino acid has been covalently linked to AMP (see Figure 6–54): when tRNA binds, the synthetase tries to force the adenylated amino acid into a second editing pocket in the enzyme. The precise dimensions of this pocket exclude the correct amino acid, while allowing access by closely related amino acids. In the editing pocket, an amino acid is removed from the AMP (or from the tRNA itself if the aminoacyl-tRNA bond has already formed) by hydrolysis. This hydrolytic editing, which is analogous to the exonucleolytic proofreading by DNA polymerases, increases the overall accuracy of tRNA charging to approximately one mistake in 40,000 couplings (**Figure 6–57**).

The tRNA synthetase must also recognize the correct set of tRNAs, and extensive structural and chemical complementarity between the synthetase and the tRNA allows the synthetase to probe various features of the tRNA (**Figure 6–58**). Most tRNA synthetases directly recognize the matching tRNA anticodon; these synthetases contain three adjacent nucleotide-binding pockets, each of which is complementary in shape and charge to a nucleotide in the anticodon. For other synthetases, the nucleotide sequence of the amino acid-accepting arm (acceptor stem) is the key recognition determinant. In most cases, however, the synthetase "reads" the nucleotides at several different positions on the tRNA.

Amino Acids Are Added to the C-terminal End of a Growing Polypeptide Chain

Having seen that each amino acid is first coupled to specific tRNA molecules, we now turn to the mechanism that joins these amino acids together to form proteins. The fundamental reaction of protein synthesis is the formation of a peptide bond between the carboxyl group at the end of a growing polypeptide chain and a free amino group on an incoming amino acid. Consequently, a protein is synthesized stepwise from its N-terminal end to its C-terminal end. Throughout the entire process, the growing carboxyl end of the polypeptide chain remains activated by its covalent attachment to a tRNA molecule (forming a *peptidyl-tRNA*). Each

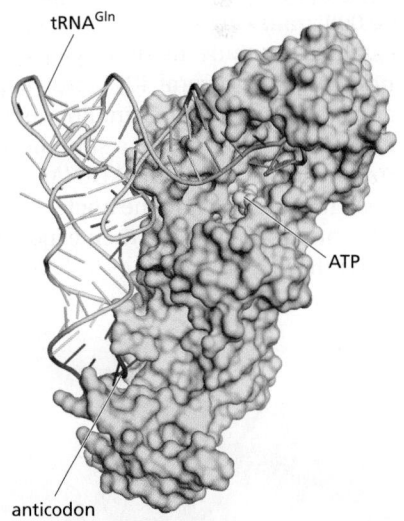

Figure 6–58 **The recognition of a tRNA molecule by its aminoacyl-tRNA synthetase.** For this tRNA (tRNA^Gln), specific nucleotides in both the anticodon *(dark blue)* and the amino acid-accepting arm *(green)* allow the correct tRNA to be recognized by the synthetase enzyme *(yellow-green).* A bound ATP molecule is *yellow.* (Courtesy of Tom Steitz; PDB code: 1QRS.)

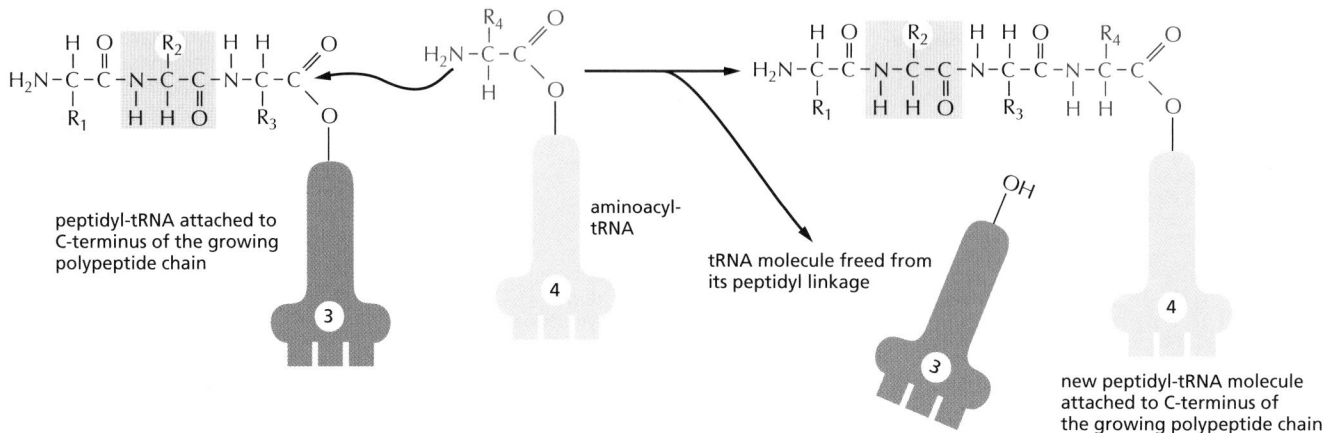

Figure 6–59 **The incorporation of an amino acid into a protein.** A polypeptide chain grows by the stepwise addition of amino acids to its C-terminal end. The formation of each peptide bond is energetically favorable because the growing C-terminus has been activated by the covalent attachment of a tRNA molecule. The peptidyl-tRNA linkage that activates the growing end is regenerated during each addition. The amino acid side chains have been abbreviated as R_1, R_2, R_3, and R_4; as a reference point, all of the atoms in the second amino acid in the polypeptide chain are shaded *gray*. The figure shows the addition of the fourth amino acid *(red)* to the growing chain.

addition disrupts this high-energy covalent linkage, but immediately replaces it with an identical linkage on the most recently added amino acid (**Figure 6–59**). In this way, each amino acid added carries with it the activation energy for the addition of the next amino acid rather than the energy for its own addition—an example of the "head growth" type of polymerization described in Figure 2–44.

The RNA Message Is Decoded in Ribosomes

The synthesis of proteins is guided by information carried by mRNA molecules. To maintain the correct reading frame and to ensure accuracy (about 1 mistake every 10,000 amino acids), protein synthesis is performed in the **ribosome**, a complex catalytic machine made from more than 50 different proteins (the *ribosomal proteins*) and several RNA molecules, the ribosomal RNAs (rRNAs). A typical eukaryotic cell contains millions of ribosomes in its cytoplasm (**Figure 6–60**). The large and small ribosome subunits are assembled at the nucleolus, where newly transcribed and modified rRNAs associate with the ribosomal proteins that have been transported into the nucleus after their synthesis in the cytoplasm. These two ribosomal subunits are then exported to the cytoplasm, where they join together to synthesize proteins.

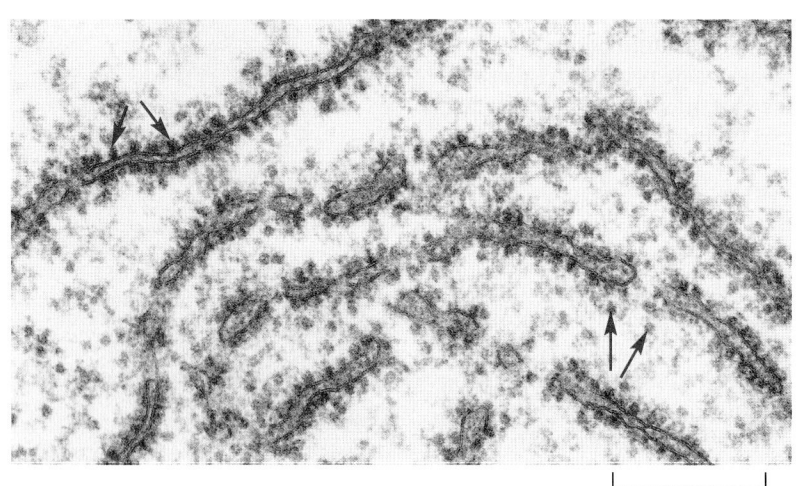

400 nm

Figure 6–60 **Ribosomes in the cytoplasm of a eukaryotic cell.** This electron micrograph shows a thin section of a small region of cytoplasm. The ribosomes appear as black dots *(red arrows)*. Some are free in the cytosol; others are attached to membranes of the endoplasmic reticulum. (Courtesy of Daniel S. Friend, by permission of E.L. Bearer.)

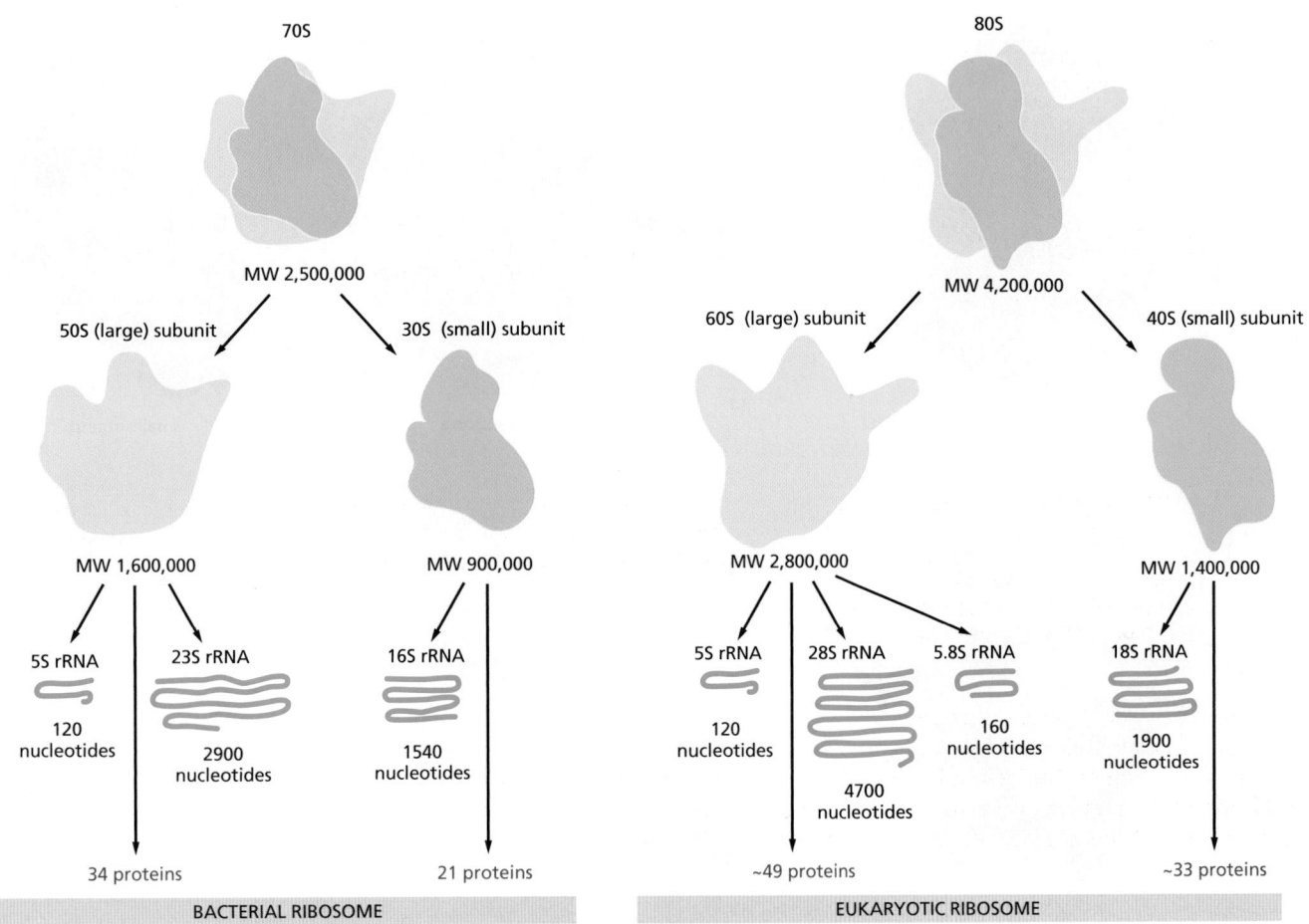

Figure 6–61 **A comparison of bacterial and eukaryotic ribosomes.** Despite differences in the number and size of their rRNA and protein components, both bacterial and eukaryotic ribosomes have nearly the same structure and they function similarly. Although the 18S and 28S rRNAs of the eukaryotic ribosome contain many nucleotides not present in their bacterial counterparts, these nucleotides are present as multiple insertions that form extra domains and leave the basic structure of the rRNA largely unchanged.

Eukaryotic and bacterial ribosomes have similar structures and functions, being composed of one large and one small subunit that fit together to form a complete ribosome with a mass of several million daltons (**Figure 6–61**). The small subunit provides the framework on which the tRNAs are accurately matched to the codons of the mRNA, while the large subunit catalyzes the formation of the peptide bonds that link the amino acids together into a polypeptide chain (see Figure 6–58).

When not actively synthesizing proteins, the two subunits of the ribosome are separate. They join together on an mRNA molecule, usually near its 5′ end, to initiate the synthesis of a protein. The mRNA is then pulled through the ribosome, three nucleotides at a time. As its codons enter the core of the ribosome, the mRNA nucleotide sequence is translated into an amino acid sequence using the tRNAs as adaptors to add each amino acid in the correct sequence to the growing end of the polypeptide chain. When a stop codon is encountered, the ribosome releases the finished protein, and its two subunits separate again. These subunits can then be used to start the synthesis of another protein on another mRNA molecule. Ribosomes operate with remarkable efficiency: in one second, a eukaryotic ribosome adds 2 amino acids to a polypeptide chain; the ribosomes of bacterial cells operate even faster, at a rate of about 20 amino acids per second.

To choreograph the many coordinated movements required for efficient translation, a ribosome contains four binding sites for RNA molecules: one is for the mRNA and three (called the A site, the P site, and the E site) are for tRNAs (**Figure 6–62**). A tRNA molecule is held tightly at the A and P sites only if its anticodon

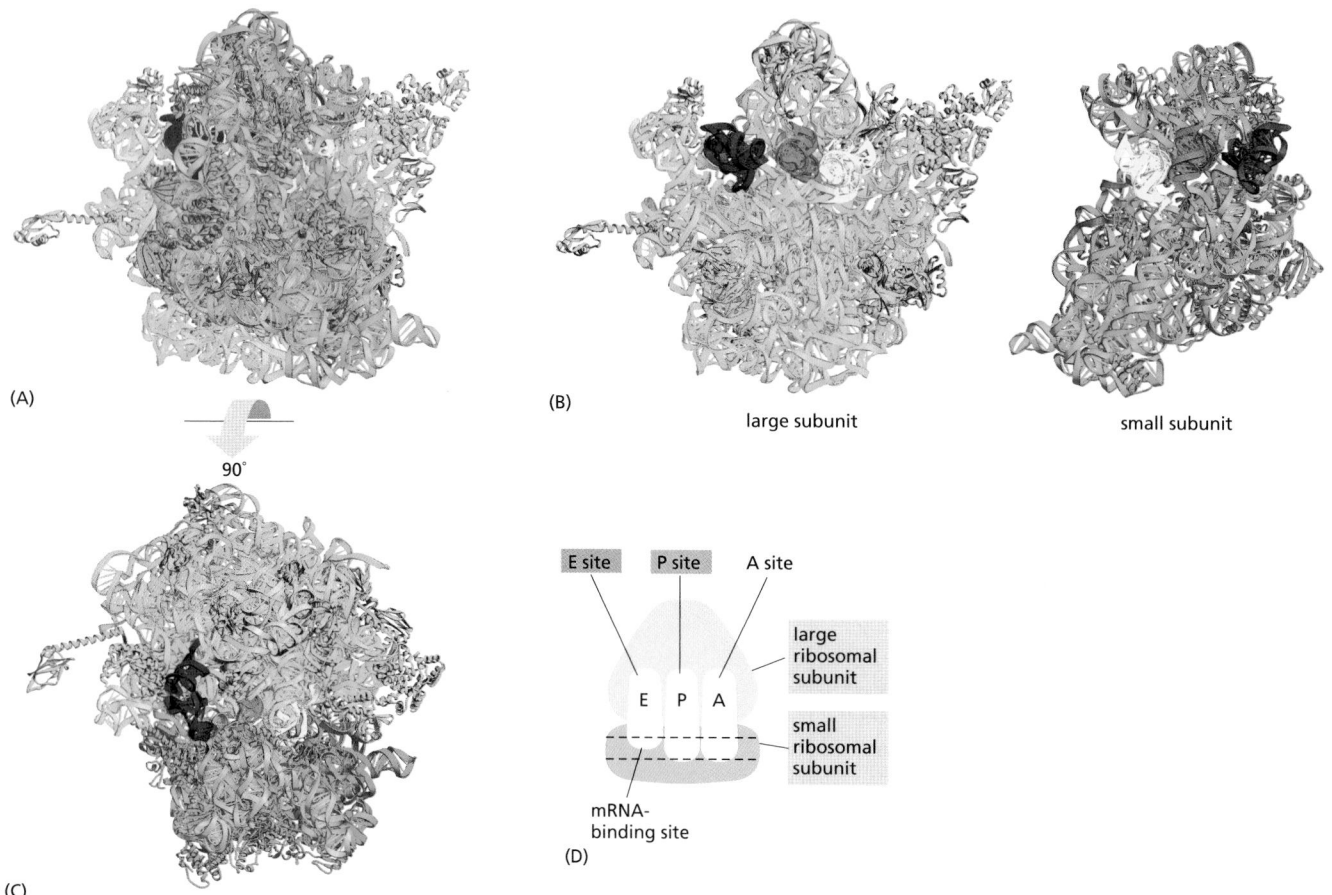

Figure 6–62 **The RNA-binding sites in the ribosome.** Each ribosome has one binding site for mRNA and three binding sites for tRNA: the A, P, and E sites (short for aminoacyl-tRNA, peptidyl-tRNA, and exit, respectively). (A) A bacterial ribosome viewed with the small subunit in the front *(dark green)* and the large subunit in the back *(light green)*. Both the rRNAs and the ribosomal proteins are illustrated. tRNAs are shown bound in the E site *(red)*, the P site *(orange)*, and the A site *(yellow)*. Although all three tRNA sites are shown occupied here, during the process of protein synthesis not more than two of these sites are thought to contain tRNA molecules at any one time (see Figure 6–64). (B) Large and small ribosomal subunits arranged as though the ribosome in (A) were opened like a book. (C) The ribosome in (A) rotated through 90° and viewed with the large subunit on top and small subunit on the bottom. (D) Schematic representation of a ribosome [in the same orientation as (C)], which will be used in subsequent figures. (A, B, and C, adapted from M.M. Yusupov et al., *Science* 292:883–896, 2001. With permission from AAAS.)

forms base pairs with a complementary codon (allowing for wobble) on the mRNA molecule that is threaded through the ribosome (**Figure 6–63**). The A and P sites are close enough together for their two tRNA molecules to be forced to form base pairs with adjacent codons on the mRNA molecule. This feature of the ribosome maintains the correct reading frame on the mRNA.

Once protein synthesis has been initiated, each new amino acid is added to the elongating chain in a cycle of reactions containing four major steps: tRNA binding (step 1), peptide bond formation (step 2), large subunit translocation (step 3), and small subunit translocation (step 4). As a result of the two translocation steps, the entire ribosome moves three nucleotides along the mRNA and is positioned to start the next cycle. **Figure 6–64** illustrates this four-step process, beginning at a point at which three amino acids have already been linked together and there is a tRNA molecule in the P site on the ribosome, covalently joined to the C-terminal end of the short polypeptide. In step 1, a tRNA carrying the next amino acid in the chain binds to the ribosomal A site by forming base pairs with the mRNA codon positioned there, so that the P site and the A site contain adjacent bound tRNAs. In step 2, the carboxyl end of the polypeptide chain is released from the tRNA at the P site (by breakage of the high-energy bond between the tRNA and its amino acid)

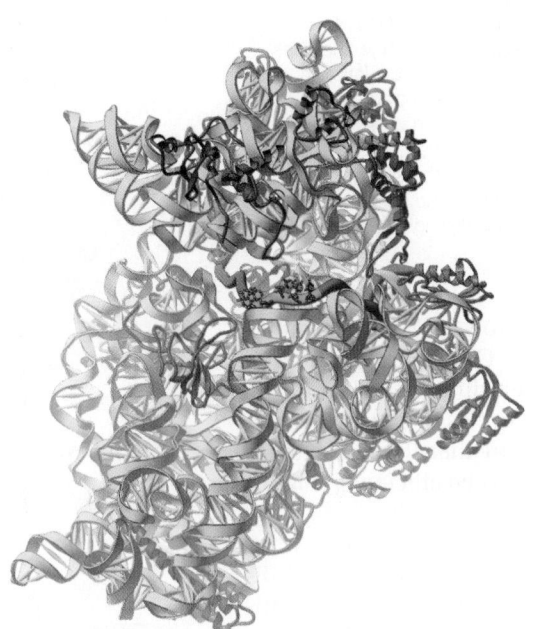

Figure 6–63 The path of mRNA *(blue)* through the small ribosomal subunit. The orientation is the same as that in the right-hand panel of Figure 6–62B. (Courtesy of Harry F. Noller, based on data in G.Z. Yusupova et al., *Cell* 106:233–241, 2001.)

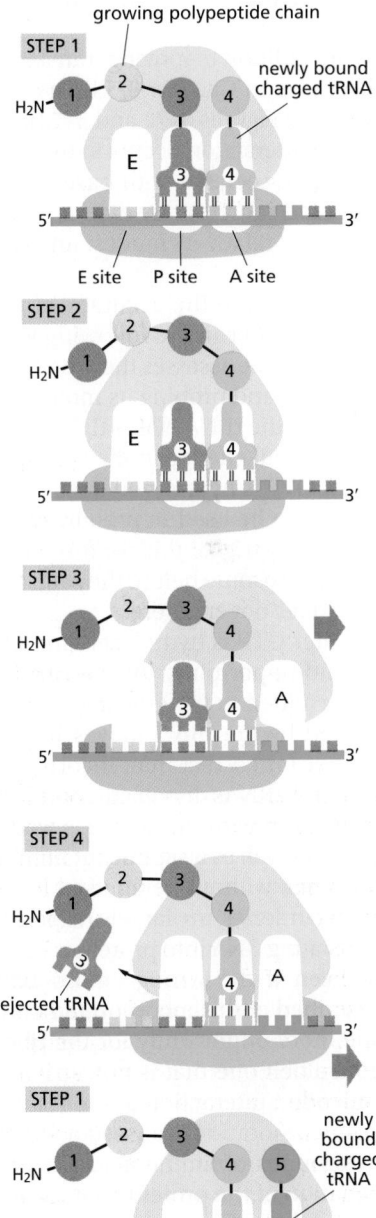

and joined to the free amino group of the amino acid linked to the tRNA at the A site, forming a new peptide bond. This central reaction of protein synthesis is catalyzed by a *peptidyl transferase* contained in the large ribosomal subunit. In step 3, the large subunit moves relative to the mRNA held by the small subunit, thereby shifting the acceptor stems of the two tRNAs to the E and P sites of the large subunit. In step 4, another series of conformational changes moves the small subunit and its bound mRNA exactly three nucleotides, ejecting the spent tRNA from the E site and resetting the ribosome so it is ready to receive the next aminoacyl-tRNA. Step 1 is then repeated with a new incoming aminoacyl-tRNA, and so on.

This four-step cycle is repeated each time an amino acid is added to the polypeptide chain, as the chain grows from its amino to its carboxyl end.

Elongation Factors Drive Translation Forward and Improve Its Accuracy

The basic cycle of polypeptide elongation shown in outline in Figure 6–64 has an additional feature that makes translation especially efficient and accurate. Two *elongation factors* enter and leave the ribosome during each cycle, each hydrolyzing GTP to GDP and undergoing conformational changes in the process. These factors are called EF-Tu and EF-G in bacteria, and EF1 and EF2 in eukaryotes. Under some conditions *in vitro*, ribosomes can be forced to synthesize proteins

Figure 6–64 Translating an mRNA molecule. Each amino acid added to the growing end of a polypeptide chain is selected by complementary base-pairing between the anticodon on its attached tRNA molecule and the next codon on the mRNA chain. Because only one of the many types of tRNA molecules in a cell can base-pair with each codon, the codon determines the specific amino acid to be added to the growing polypeptide chain. The four-step cycle shown is repeated over and over during the synthesis of a protein. In step 1, an aminoacyl-tRNA molecule binds to a vacant A site on the ribosome. In step 2, a new peptide bond is formed. In step 3, the large subunit translocates relative to the small subunit, leaving the two tRNAs in hybrid sites: P on the large subunit and A on the small, for one; E on the large subunit and P on the small, for the other. In step 4, the small subunit translocates carrying its mRNA a distance of three nucleotides through the ribosome. This "resets" the ribosome with a fully empty A site, ready for the next aminoacyl-tRNA molecule to bind. As indicated, the mRNA is translated in the 5′-to-3′ direction, and the N-terminal end of a protein is made first, with each cycle adding one amino acid to the C-terminus of the polypeptide chain (Movie 6.7 and Movie 6.8).

Figure 6–65 Detailed view of the translation cycle. The outline of translation presented in Figure 6–64 has been expanded to show the roles of the two elongation factors EF-Tu and EF-G, which drive translation in the forward direction. As explained in the text, EF-Tu provides opportunities for proofreading of the codon–anticodon match. In this way, incorrectly paired tRNAs are selectively rejected, and the accuracy of translation is improved. The binding of a molecule of EF-G to the ribosome and the subsequent hydrolysis of GTP lead to a rearrangement of the ribosome structure, moving the mRNA being decoded exactly three nucleotides through it (Movie 6.9).

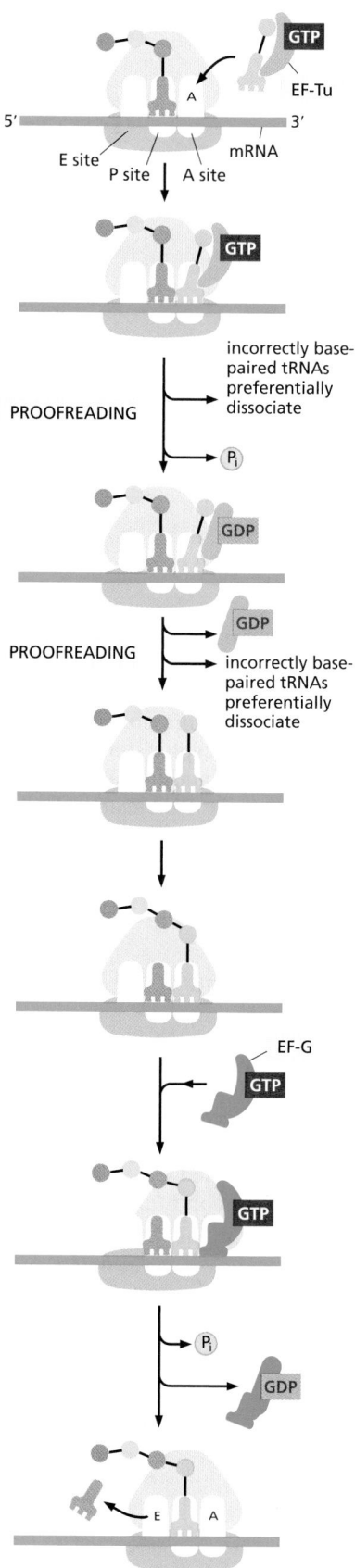

without the aid of these elongation factors and GTP hydrolysis, but this synthesis is very slow, inefficient, and inaccurate. Coupling the GTP hydrolysis-driven changes in the elongation factors to transitions between different states of the ribosome speeds up protein synthesis enormously. The cycles of elongation factor association, GTP hydrolysis, and dissociation also ensure that all such changes occur in the "forward" direction, helping translation to proceed efficiently (Figure 6–65).

In addition to moving translation forward, EF-Tu increases its accuracy. As we discussed in Chapter 3, EF-Tu can simultaneously bind GTP and aminoacyl-tRNAs (see Figures 3–72 and 3–73), and it is in this form that the initial codon–anticodon interaction occurs in the A site of the ribosome. Because of the free-energy change associated with base-pair formation, a correct codon–anticodon match will bind more tightly than an incorrect interaction. However, this difference in affinity is relatively modest and cannot by itself account for the high accuracy of translation.

To increase the accuracy of this binding reaction, the ribosome and EF-Tu work together in the following ways. First, the 16s rRNA in the small subunit of the ribosome assesses the "correctness" of the codon–anticodon match by folding around it and probing its molecular details (Figure 6–66). When a correct match is found, the rRNA closes tightly around the codon–anticodon pair, causing a conformational change in the ribosome that triggers GTP hydrolysis by EF-Tu. Only when GTP is hydrolyzed does EF-Tu release its grip on the aminoacyl-tRNA and allow it to be used in protein synthesis. Incorrect codon–anticodon matches do not readily trigger this conformational change, and these errant tRNAs mostly fall off the ribosome before they can be used in protein synthesis. Proofreading, however, does not end here.

After GTP is hydrolyzed and EF-Tu dissociates from the ribosome, there is a second opportunity for the ribosome to prevent an incorrect amino acid from being added to the growing chain. There is a short time delay as the amino acid carried by the tRNA moves into position on the ribosome. This time delay is shorter for correct than incorrect codon–anticodon pairs. Moreover, incorrectly matched tRNAs dissociate more rapidly than those correctly bound because their interaction with the codon is weaker. Thus, most incorrectly bound tRNA molecules (as well as a significant number of correctly bound molecules) will leave the ribosome without being used for protein synthesis. The two proofreading steps, acting in series, are largely responsible for the 99.99% accuracy of the ribosome in translating RNA into protein.

Even if the wrong amino acid slips through the proofreading steps just described and is incorporated onto the growing polypeptide chain, there is still one more opportunity for the ribosome to detect the error and provide a solution, albeit one that is not, strictly speaking, proofreading. An incorrect codon–anticodon interaction in the P site of the ribosome (which would occur *after* the misincorporation) causes an increased rate of misreading in the A site. Successive rounds of amino acid misincorporation eventually lead to premature termination of the protein by *release factors*, which are described below. Normally, these release factors act when translation of a protein is complete; here, they act early. Although this mechanism does not correct the original error, it releases the flawed protein for degradation, ensuring that no additional peptide synthesis is wasted on it.

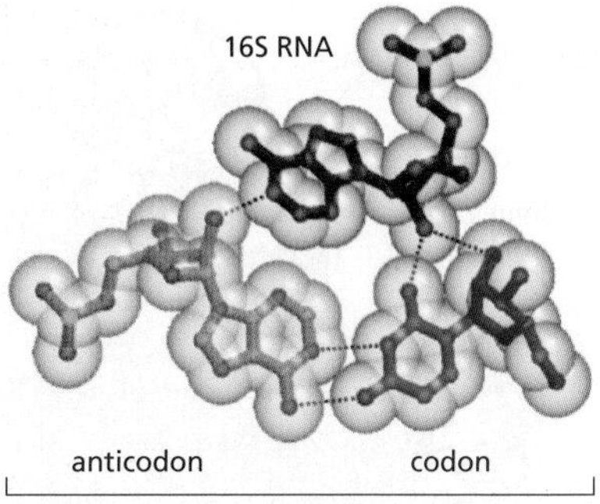

Figure 6–66 **Recognition of correct codon–anticodon matches by the small-subunit rRNA of the ribosome.** Shown here is the interaction between a nucleotide of the small-subunit rRNA and the first nucleotide pair of a correctly paired codon–anticodon. Similar interactions form between other nucleotides of the rRNA and the second and third positions of codon–anticodon pair. The small-subunit rRNA can form this network of hydrogen bonds only when an anticodon is correctly matched to a codon. As explained in the text, this codon–anticodon monitoring by the small-subunit rRNA increases the accuracy of protein synthesis. (From J.M. Ogle et al., *Science* 292:897–902, 2001. With permission from AAAS.)

Many Biological Processes Overcome the Inherent Limitations of Complementary Base-Pairing

We have seen in this and the previous chapter that DNA replication, repair, transcription, and translation all rely on complementary base-pairing—G with C, and A with T (or U). However, if only the difference in hydrogen bonding is considered, a correct versus incorrect match should differ in affinity only by a factor of 10- to 100-fold. These processes have an accuracy much higher than can be accounted for by this difference. Although the mechanisms used to "squeeze out" additional specificity from complementary base-pairing differ from one process to the next, two principles exemplified by the ribosome appear to be general.

The first is **induced fit**. We have seen that, before an amino acid is added to a growing polypeptide chain, the ribosome folds around the codon–anticodon interaction, and only when the match is correct is this folding completed and the reaction allowed to proceed. Thus, the codon–anticodon interaction is thereby checked twice—once by the initial complementary base-pairing and a second time by the folding of the ribosome, which depends on the correctness of the match. This same principle of induced fit is seen in transcription by RNA polymerase; here, an incoming nucleoside triphosphate initially forms a base pair with the template; at this point the enzyme folds around the base pair (thereby assessing its correctness) and, in doing so, creates the active site of the enzyme. The enzyme then covalently adds the nucleotide to the growing chain. Because their geometry is "wrong," incorrect base pairs block this induced fit, and they are therefore likely to dissociate before being incorporated into the growing chain.

A second principle used to increase the specificity of complementary base-pairing is called **kinetic proofreading**. We have seen that after the initial codon-anticodon pairing and conformational change of the ribosome, GTP is hydrolyzed. This creates an irreversible step and starts the clock on a time delay during which the aminoacyl-tRNA moves into the proper position for catalysis. During this delay, those incorrect codon–anticodon pairs that have somehow slipped through the induced-fit scrutiny have a higher likelihood of dissociating than correct pairs. There are two reasons for this: (1) the interaction of the wrong tRNA with the codon is weaker, and (2) the delay is longer for incorrect than correct matches.

In its most general form, kinetic proofreading refers to a time delay that begins with an irreversible step such as ATP or GTP hydrolysis, during which an incorrect substrate is more likely to dissociate than a correct one. In this case, kinetic proofreading thus increases the specificity of complementary base-pairing above what is possible from simple thermodynamic associations alone. The increase in specificity produced by kinetic proofreading comes at an energetic cost in the form of ATP or GTP hydrolysis. Kinetic proofreading is believed to operate in many biological processes, but its role is understood particularly well for translation.

Accuracy in Translation Requires an Expenditure of Free Energy

Translation by the ribosome is a compromise between the opposing constraints of accuracy and speed. We have seen, for example, that the accuracy of translation (1 mistake per 10^4 amino acids joined) requires time delays each time a new amino acid is added to a growing polypeptide chain, producing an overall speed

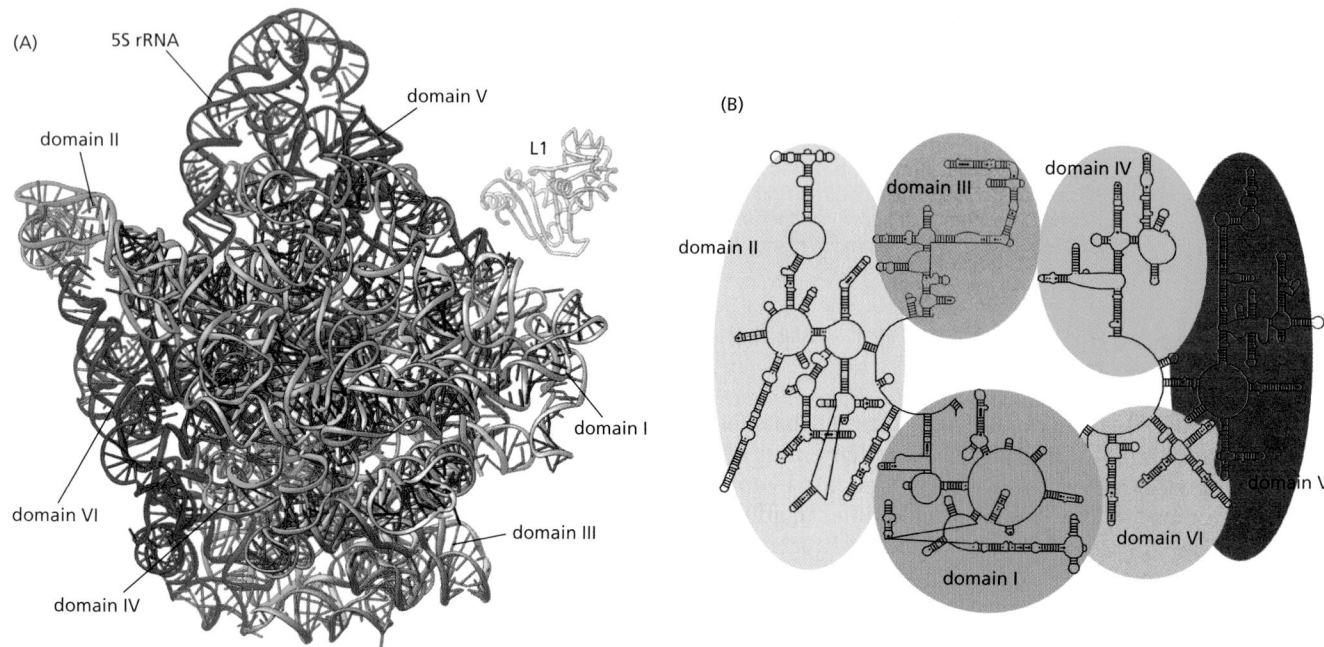

Figure 6–67 **Structure of the rRNAs in the large subunit of a bacterial ribosome, as determined by x-ray crystallography.** (A) Three-dimensional conformations of the large-subunit rRNAs (5S and 23S) as they appear in the ribosome. One of the protein subunits of the ribosome (L1) is also shown as a reference point, since it forms a characteristic protrusion on the ribosome. (B) Schematic diagram of the secondary structure of the 23S rRNA, showing the extensive network of base-pairing. The structure has been divided into six "domains" whose colors correspond to those in (A). The secondary-structure diagram is highly schematized to represent as much of the structure as possible in two dimensions. To do this, several discontinuities in the RNA chain have been introduced, although in reality the 23S rRNA is a single RNA molecule. For example, the base of Domain III is continuous with the base of Domain IV even though a gap appears in the diagram. (Adapted from N. Ban et al., *Science* 289:905–920, 2000. With permission from AAAS.)

of translation of 20 amino acids incorporated per second in bacteria. Mutant bacteria with a specific alteration in the small ribosomal subunit have longer delays and translate mRNA into protein with an accuracy considerably higher than this; however, protein synthesis is so slow in these mutants that the bacteria are barely able to survive.

We have also seen that attaining the observed accuracy of protein synthesis requires the expenditure of a great deal of free energy; this is expected, since, as discussed in Chapter 2, there is a price to be paid for any increase in order in the cell. In most cells, protein synthesis consumes more energy than any other biosynthetic process. At least four high-energy phosphate bonds are split to make each new peptide bond: two are consumed in charging a tRNA molecule with an amino acid (see Figure 6–54), and two more drive steps in the cycle of reactions occurring on the ribosome during protein synthesis itself (see Figure 6–65). In addition, extra energy is consumed each time that an incorrect amino acid linkage is hydrolyzed by a tRNA synthetase (see Figure 6–57) and each time that an incorrect tRNA enters the ribosome, triggers GTP hydrolysis, and is rejected (see Figure 6–65). To be effective, any proofreading mechanism must also allow an appreciable fraction of correct interactions to be removed; for this reason, proofreading is even more costly in energy than it might at first seem.

The Ribosome Is a Ribozyme

The ribosome is a large complex composed of two-thirds RNA and one-third protein. The determination, in 2000, of the entire three-dimensional conformation of its large and small subunits is a major triumph of modern structural biology. The findings confirm earlier evidence that rRNAs—and not proteins—are responsible for the ribosome's overall structure, its ability to position tRNAs on the mRNA, and its catalytic activity in forming covalent peptide bonds. The ribosomal RNAs are folded into highly compact, precise three-dimensional structures that form the compact core of the ribosome and determine its overall shape (**Figure 6–67**).

In marked contrast to the central positions of the rRNAs, the ribosomal proteins are generally located on the surface and fill in the gaps and crevices of the folded RNA (**Figure 6–68**). Some of these proteins send out extended regions of polypeptide chain that penetrate short distances into holes in the RNA core (**Figure 6–69**). The main role of the ribosomal proteins seems to be to stabilize the

RNA core, while permitting the changes in rRNA conformation that are necessary for this RNA to catalyze efficient protein synthesis. The proteins also aid in the initial assembly of the rRNAs that make up the core of the ribosome.

Not only are the A, P, and E binding sites for tRNAs formed primarily by ribosomal RNAs, but the catalytic site for peptide bond formation is also formed by RNA, as the nearest amino acid is located more than 1.8 nm away. This discovery came as a surprise to biologists because, unlike proteins, RNA does not contain easily ionizable functional groups that can be used to catalyze sophisticated reactions like peptide bond formation. Moreover, metal ions, which are often used by RNA molecules to catalyze chemical reactions (as discussed later in the chapter), were not observed at the active site of the ribosome. Instead, it is believed that the 23S rRNA forms a highly structured pocket that, through a network of hydrogen bonds, precisely orients the two reactants (the growing peptide chain and an aminoacyl-tRNA) and thereby greatly accelerates their covalent joining. An additional surprise came from the discovery that the tRNA in the P site contributes an important OH group to the active site and participates directly in the catalysis. This mechanism may ensure that catalysis occurs only when the P site tRNA is properly positioned in the ribosome.

RNA molecules that possess catalytic activity are known as **ribozymes**. We saw earlier in this chapter that some ribozymes function in self-splicing reactions. In the final section of this chapter, we consider what the ability of RNA molecules to function as catalysts might mean for the early evolution of living cells. For now, we merely note that there is good reason to suspect that RNA rather than protein molecules served as the first catalysts for living cells. If so, the ribosome, with its RNA core, may be a relic of an earlier time in life's history—when protein synthesis evolved in cells that were run almost entirely by ribozymes.

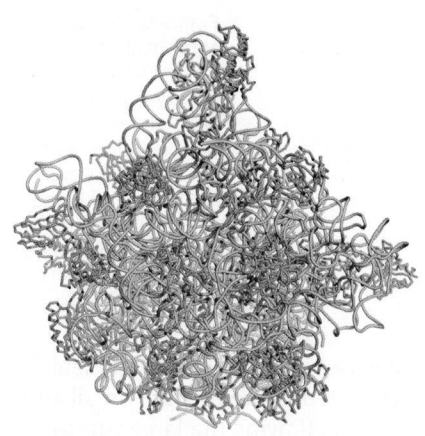

Figure 6–68 **Location of the protein components of the bacterial large ribosomal subunit.** The rRNAs (5S and 23S) are shown in *blue* and the proteins of the large subunit in *green*. This view is toward the outside of the ribosome; the interface with the small subunit is on the opposite face. (PDB code: 1FFK.)

Nucleotide Sequences in mRNA Signal Where to Start Protein Synthesis

The initiation and termination of translation share features of the translation elongation cycle described above. The site at which protein synthesis begins on the mRNA is especially crucial, since it sets the reading frame for the whole length of the message. An error of one nucleotide either way at this stage would cause every subsequent codon in the message to be misread, resulting in a nonfunctional protein with a garbled sequence of amino acids. The initiation step is also important because for most genes it is the last point at which the cell can decide whether the mRNA is to be translated to produce a protein. The rate of this step is thus one determinant of the rate at which any particular protein will be synthesized. We shall see in Chapter 7 how regulation of this step occurs.

The translation of an mRNA begins with the codon AUG, and a special tRNA is required to start translation. This **initiator tRNA** always carries the amino acid methionine (in bacteria, a modified form of methionine—formylmethionine—is used), with the result that all newly made proteins have methionine as the first amino acid at their N-terminus, the end of a protein that is synthesized first. (This methionine is usually removed later by a specific protease.) The initiator tRNA is specially recognized by initiation factors because it has a nucleotide sequence distinct from that of the tRNA that normally carries methionine.

In eukaryotes, the initiator tRNA–methionine complex (Met–tRNAi) is first loaded into the small ribosomal subunit along with additional proteins called **eukaryotic initiation factors**, or **eIFs**. Of all the aminoacyl-tRNAs in the cell, only the methionine-charged initiator tRNA is capable of tightly binding the small ribosome subunit without the complete ribosome being present, and unlike other tRNAs it binds directly to the P site (**Figure 6–70**). Next, the small ribosomal subunit binds to the 5' end of an mRNA molecule, which is recognized by virtue of its 5' cap that has previously bound two initiation factors, eIF4E and eIF4G (see Figure 6–38). The small ribosomal subunit then moves forward (5' to 3') along the mRNA, searching for the first AUG; additional initiation factors that act as ATP-powered

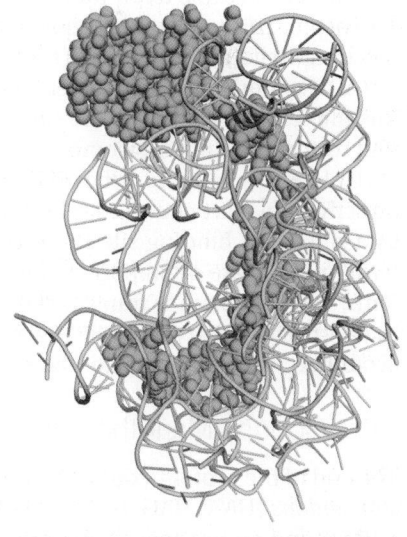

Figure 6–69 **Structure of the L15 protein in the large subunit of the bacterial ribosome.** The globular domain of the protein lies on the surface of the ribosome and an extended region penetrates deeply into the RNA core of the ribosome. The L15 protein is shown in *green* and a portion of the ribosomal RNA core is shown in *blue*. (From D. Klein, P.B. Moore and T.A. Steitz, *J. Mol. Biol.* 340:141–177, 2004. PDB code: 1S72.)

Figure 6–70 The initiation of protein synthesis in eukaryotes. Only three of the many translation initiation factors required for this process are shown. Efficient translation initiation also requires the poly-A tail of the mRNA bound by poly-A-binding proteins, which, in turn, interact with eIF4G (see Figure 6–38). In this way, the translation apparatus ascertains that both ends of the mRNA are intact before initiating protein synthesis. Although only one GTP-hydrolysis event is shown in the figure, a second is known to occur just before the large and small ribosomal subunits join. In the last two steps shown in the figure, the ribosome has begun the standard elongation cycle, depicted in Figure 6–64.

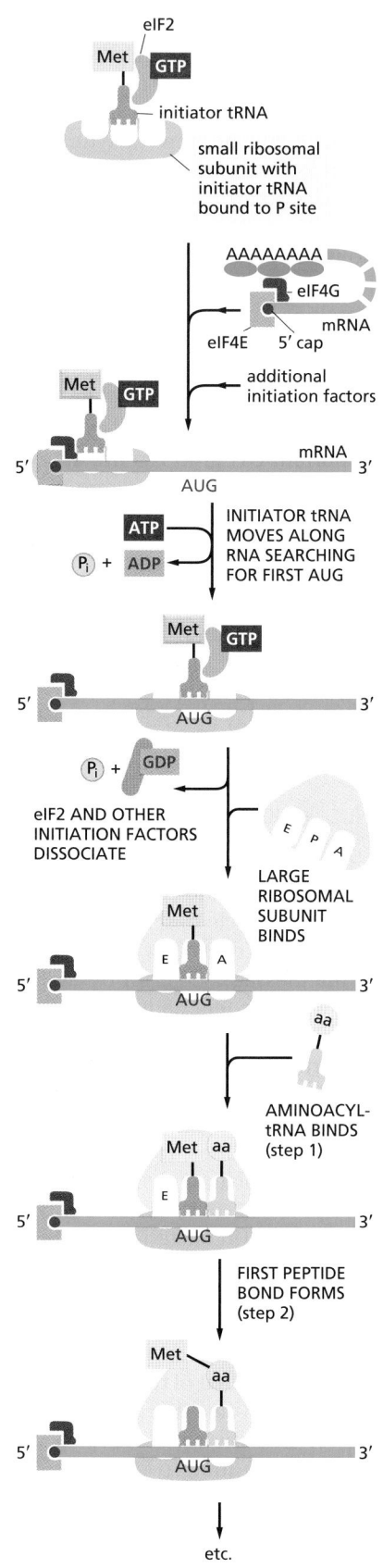

helicases facilitate this movement. In 90% of mRNAs, translation begins at the first AUG encountered by the small subunit. At this point, the initiation factors dissociate, allowing the large ribosomal subunit to assemble with the complex and complete the ribosome. The initiator tRNA remains at the P site, leaving the A site vacant. Protein synthesis is therefore ready to begin (see Figure 6–70).

The nucleotides immediately surrounding the start site in eukaryotic mRNAs influence the efficiency of AUG recognition during the above scanning process. If this recognition site differs substantially from the consensus recognition sequence (5′-ACCAUGG-3′), scanning ribosomal subunits will sometimes ignore the first AUG codon in the mRNA and skip to the second or third AUG codon instead. Cells frequently use this phenomenon, known as "leaky scanning," to produce two or more proteins, differing in their N-termini, from the same mRNA molecule. This mechanism allows some genes to produce the same protein with and without a signal sequence attached at its N-terminus, for example, so that the protein is directed to two different compartments in the cell.

The mechanism for selecting a start codon in bacteria is different. Bacterial mRNAs have no 5′ caps to signal the ribosome where to begin searching for the start of translation. Instead, each bacterial mRNA contains a specific ribosome-binding site (called the Shine–Dalgarno sequence, named after its discoverers) that is located a few nucleotides upstream of the AUG at which translation is to begin. This nucleotide sequence, with the consensus 5′-AGGAGGU-3′, forms base pairs with the 16S rRNA of the small ribosomal subunit to position the initiating AUG codon in the ribosome. A set of translation initiation factors orchestrates this interaction, as well as the subsequent assembly of the large ribosomal subunit to complete the ribosome.

Unlike a eukaryotic ribosome, a bacterial ribosome can readily assemble directly on a start codon that lies in the interior of an mRNA molecule, so long as a ribosome-binding site precedes it by several nucleotides. As a result, bacterial mRNAs are often *polycistronic*—that is, they encode several different proteins, each of which is translated from the same mRNA molecule (**Figure 6–71**). In contrast, a eukaryotic mRNA generally encodes only a single protein, or more accurately, a single set of closely related proteins.

Stop Codons Mark the End of Translation

The end of the protein-coding message is signaled by the presence of one of three *stop codons* (UAA, UAG, or UGA) (see Figure 6–48). These are not recognized by a tRNA and do not specify an amino acid, but instead signal to the ribosome to stop translation. Proteins known as *release factors* bind to any ribosome with a stop codon positioned in the A site, forcing the peptidyl transferase in the ribosome to catalyze the addition of a water molecule instead of an amino acid to the peptidyl-tRNA (**Figure 6–72**). This reaction frees the carboxyl end of the growing polypeptide chain from its attachment to a tRNA molecule, and since only this attachment normally holds the growing polypeptide to the ribosome, the completed protein chain is immediately released into the cytoplasm. The ribosome then releases its bound mRNA molecule and separates into the large and small subunits. These subunits can then assemble on this or another mRNA molecule to begin a new round of protein synthesis.

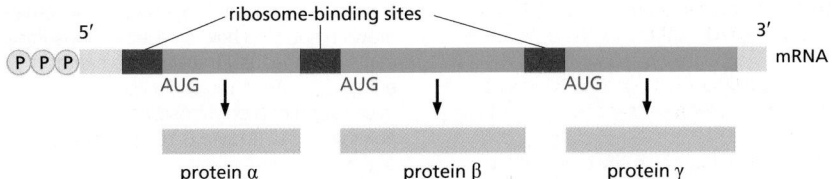

Figure 6–71 Structure of a typical bacterial mRNA molecule. Unlike eukaryotic ribosomes, which typically require a capped 5′ end on the mRNA, prokaryotic ribosomes initiate translation at ribosome-binding sites (Shine–Dalgarno sequences), which can be located anywhere along an mRNA molecule. This property of their ribosomes permits bacteria to synthesize more than one type of protein from a single mRNA molecule.

During translation, the nascent polypeptide moves through a large, water-filled tunnel (approximately 10 nm × 1.5 nm) in the large subunit of the ribosome. The walls of this tunnel, made primarily of 23S rRNA, are a patchwork of tiny hydrophobic surfaces embedded in a more extensive hydrophilic surface. This structure is not complementary to any peptide, and thus provides a "Teflon" coating through which a polypeptide chain can easily slide. The dimensions of the tunnel suggest that nascent proteins are largely unstructured as they pass through the ribosome, although some α-helical regions of the protein can form before leaving the ribosome tunnel. As it leaves the ribosome, a newly synthesized protein must fold into its proper three-dimensional conformation to be useful to the cell. Later in this chapter we discuss how this folding occurs. First, however, we describe several additional aspects of the translation process itself.

Proteins Are Made on Polyribosomes

The synthesis of most protein molecules takes between 20 seconds and several minutes. During this very short period, however, it is usual for multiple initiations to take place on each mRNA molecule being translated. As soon as the preceding ribosome has translated enough of the nucleotide sequence to move out of the way, the 5′ end of the mRNA is threaded into a new ribosome. The mRNA molecules being translated are therefore usually found in the form of *polyribosomes* (or *polysomes*): large cytoplasmic assemblies made up of several ribosomes spaced as close as 80 nucleotides apart along a single mRNA molecule (**Figure 6–73**). These multiple initiations allow the cell to make many more protein molecules in a given time than would be possible if each protein had to be completed before the next could start.

Both bacteria and eukaryotes use polysomes, and both employ additional strategies to speed up the overall rate of protein synthesis. Because bacterial mRNA does not need to be processed and is accessible to ribosomes while it is being made, ribosomes can attach to the free end of a bacterial mRNA molecule and start translating it even before the transcription of that RNA is complete, following closely behind the RNA polymerase as it moves along DNA. In eukaryotes, as we have seen, the 5′ and 3′ ends of the mRNA interact (see Figure 6–73A); therefore, as soon as a ribosome dissociates, its two subunits are in an optimal position to reinitiate translation on the same mRNA molecule.

There Are Minor Variations in the Standard Genetic Code

As discussed in Chapter 1, the genetic code (shown in Figure 6–48) applies to all three major branches of life, providing important evidence for the common ancestry of all life on Earth. Although rare, there are exceptions to this code. For example, *Candida albicans,* the most prevalent human fungal pathogen, translates the codon CUG as serine, whereas nearly all other organisms translate it as leucine. Mitochondria (which have their own genomes and encode much of their translational apparatus) often deviate from the standard code. For example, in mammalian mitochondria AUA is translated as methionine, whereas in the

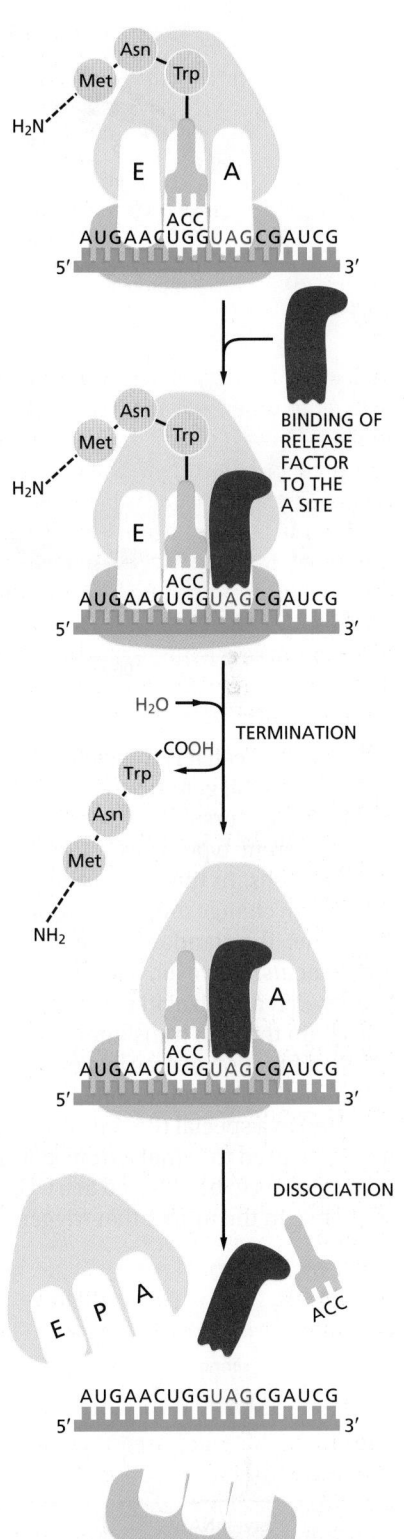

Figure 6–72 The final phase of protein synthesis. The binding of a release factor to an A site bearing a stop codon terminates translation. The completed polypeptide is released and, in a series of reactions that requires additional proteins and GTP hydrolysis (not shown), the ribosome dissociates into its two separate subunits.

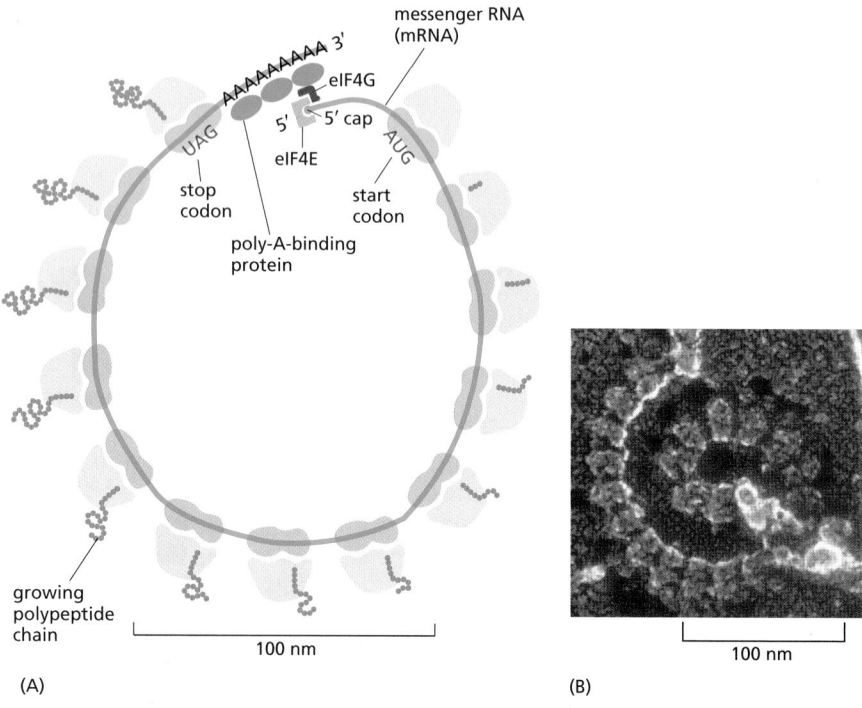

(A)

Figure 6–73 **A polyribosome.** (A) Schematic drawing showing how a series of ribosomes can simultaneously translate the same eukaryotic mRNA molecule. (B) Electron micrograph of a polyribosome from a eukaryotic cell (Movie 6.10). (B, courtesy of John Heuser.)

cytosol of the cell it is translated as isoleucine (see Table 14–3, p. 805). This type of deviation in the genetic code is "hardwired" into the organisms or the organelles in which it occurs.

A different type of variation, sometimes called *translation recoding*, occurs in many cells. In this case, other nucleotide sequence information present in an mRNA can change the meaning of the genetic code at a particular site in the mRNA molecule. The standard code allows cells to manufacture proteins using only 20 amino acids. However, bacteria, archaea, and eukaryotes have available to them a twenty-first amino acid that can be incorporated directly into a growing polypeptide chain through translation recoding. Selenocysteine, which is essential for the efficient function of a variety of enzymes, contains a selenium atom in place of the sulfur atom of cysteine. Selenocysteine is enzymatically produced from a serine attached to a special tRNA molecule that base-pairs with the UGA codon, a codon normally used to signal a translation stop. The mRNAs for proteins in which selenocysteine is to be inserted at a UGA codon carry an additional nearby nucleotide sequence in the mRNA that triggers this recoding event (**Figure 6–74**).

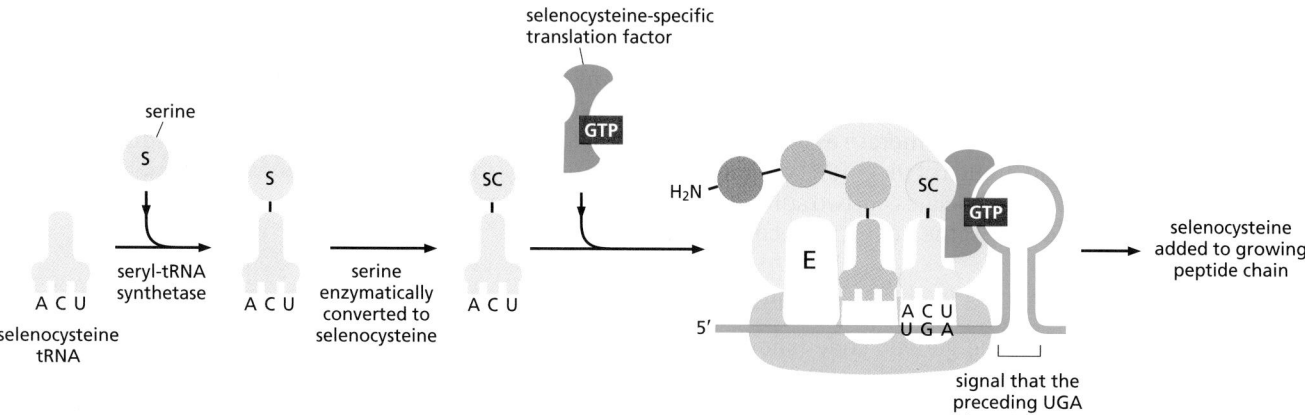

Figure 6–74 **Incorporation of selenocysteine into a growing polypeptide chain.** A specialized tRNA is charged with serine by the normal seryl-tRNA synthetase, and the serine is subsequently converted enzymatically to selenocysteine. A specific RNA structure in the mRNA (a stem and loop structure with a particular nucleotide sequence) signals that selenocysteine is to be inserted at the neighboring UGA codon. As indicated, this event requires the participation of a selenocysteine-specific translation factor. After the addition of selenocysteine, translation continues until a conventional stop codon is encountered.

Inhibitors of Prokaryotic Protein Synthesis Are Useful as Antibiotics

Many of the most effective antibiotics used in modern medicine are compounds made by fungi that inhibit bacterial protein synthesis. Fungi and bacteria compete for many of the same environmental niches, and millions of years of coevolution have resulted in fungi producing potent bacterial inhibitors. Some of these drugs exploit the structural and functional differences between bacterial and eukaryotic ribosomes so as to interfere preferentially with the function of bacterial ribosomes. Thus, humans can take high dosages of some of these compounds without undue toxicity. Many antibiotics lodge in pockets in the ribosomal RNAs and simply interfere with the smooth operation of the ribosome; others block specific parts of the ribosome such as the exit channel (**Figure 6–75**). **Table 6–4** lists some common antibiotics of this kind along with several other inhibitors of protein synthesis, some of which act on eukaryotic cells and therefore cannot be used as antibiotics.

Because they block specific steps in the processes that lead from DNA to protein, many of the compounds listed in Table 6–4 are useful for cell biological studies. Among the most commonly used drugs in such investigations are *chloramphenicol*, *cycloheximide*, and *puromycin*, all of which specifically inhibit protein synthesis. In a eukaryotic cell, for example, chloramphenicol inhibits protein synthesis on ribosomes only in mitochondria (and in chloroplasts in plants), presumably reflecting the prokaryotic origins of these organelles (discussed in Chapter 14). Cycloheximide, in contrast, affects only ribosomes in the cytosol. Puromycin is especially interesting because it is a structural analog of a tRNA molecule linked to an amino acid and is therefore another example of molecular mimicry; the ribosome mistakes it for an authentic amino acid and covalently incorporates it at the C-terminus of the growing peptide chain, thereby causing the premature termination and release of the polypeptide. As might be expected, puromycin inhibits protein synthesis in both prokaryotes and eukaryotes.

Quality Control Mechanisms Act to Prevent Translation of Damaged mRNAs

In eukaryotes, mRNA production involves both transcription and a series of elaborate RNA processing steps; as we have seen, these take place in the nucleus, segregated from ribosomes, and only when the processing is complete are the mRNAs transported to the cytosol to be translated (see Figure 6–38). However, this scheme is not foolproof, and some incorrectly processed mRNAs are inadvertently sent to the cytosol. In addition, mRNAs that were flawless when they left the nucleus can become broken or otherwise damaged in the cytosol. The danger of

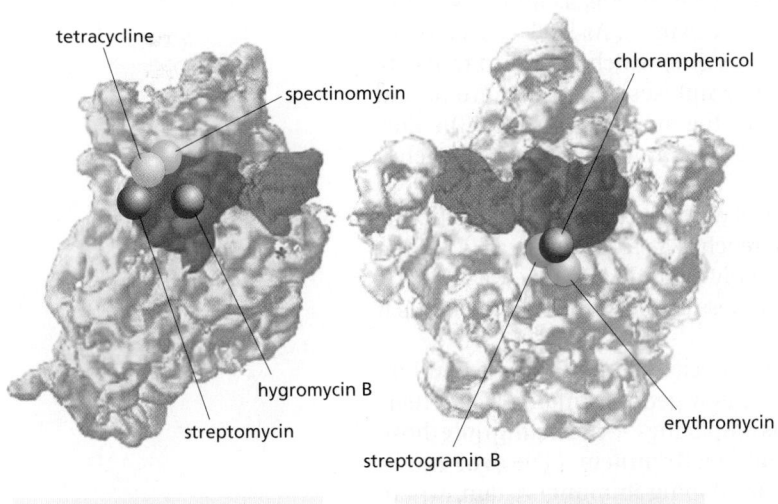

small ribosomal subunit   large ribosomal subunit

Figure 6–75 Binding sites for antibiotics on the bacterial ribosome. The small *(left)* and large *(right)* subunits of the ribosome are arranged as though the ribosome has been opened like a book. Antibiotic binding sites are marked with colored spheres, and the bound tRNA molecules are shown in *purple* (see Figure 6–62). Most of the antibiotics shown bind directly to pockets formed by the ribosomal RNA molecules. Hygromycin B induces errors in translation, spectinomycin blocks the translocation of the peptidyl-tRNA from the A site to the P site, and streptogramin B prevents elongation of nascent peptides. Table 6–4 lists the inhibitory mechanisms of the other antibiotics shown in the figure. (Adapted from J. Poehlsgaard and S. Douthwaite, *Nat. Rev. Microbiol.* 3:870–881, 2005.)

TABLE 6–4 Inhibitors of Protein or RNA Synthesis

Inhibitor	Specific effect
Acting only on bacteria	
Tetracycline	Blocks binding of aminoacyl-tRNA to the A site of ribosome
Streptomycin	Prevents the transition from translation initiation to chain elongation and also causes miscoding
Chloramphenicol	Blocks the peptidyl transferase reaction on ribosomes (step 2 in Figure 6–64)
Erythromycin	Binds in the exit channel of the ribosome and thereby inhibits elongation of the peptide chain
Rifamycin	Blocks initiation of RNA chains by binding to RNA polymerase (prevents RNA synthesis)
Acting on bacteria and eukaryotes	
Puromycin	Causes the premature release of nascent polypeptide chains by its addition to the growing chain end
Actinomycin D	Binds to DNA and blocks the movement of RNA polymerase (prevents RNA synthesis)
Acting on eukaryotes but not bacteria	
Cycloheximide	Blocks the translocation reaction on ribosomes (step 3 in Figure 6–64)
Anisomycin	Blocks the peptidyl transferase reaction on ribosomes (step 2 in Figure 6–64)
α-Amanitin	Blocks mRNA synthesis by binding preferentially to RNA polymerase II

The ribosomes of eukaryotic mitochondria (and chloroplasts) often resemble those of bacteria in their sensitivity to inhibitors. Therefore, some of these antibiotics can have a deleterious effect on human mitochondria.

translating damaged or incompletely processed mRNAs (which would produce truncated or otherwise aberrant proteins) is apparently so great that the cell has several backup measures to prevent this from happening. To avoid translating broken mRNAs, for example, the 5′ cap and the poly-A tail are both recognized by the translation-initiation machinery before translation begins (see Figure 6–70).

The most powerful mRNA surveillance system, called **nonsense-mediated mRNA decay**, eliminates defective mRNAs before they move away from the nucleus. This mechanism is brought into play when the cell determines that an mRNA molecule has a nonsense (stop) codon (UAA, UAG, or UGA) in the "wrong" place. This situation is likely to arise in an mRNA molecule that has been improperly spliced, because aberrant splicing will usually result in the random introduction of a nonsense codon into the reading frame of the mRNA—especially in organisms, such as humans, that have a large average intron size (see Figure 6–31B).

The nonsense-mediated mRNA decay mechanism begins as an mRNA molecule is being transported from the nucleus to the cytosol. As its 5′ end emerges from a nuclear pore, the mRNA is met by a ribosome, which begins to translate it. As translation proceeds, the exon junction complexes (EJCs) that are bound to the mRNA at each splice site are displaced by the moving ribosome. The normal stop codon will lie within the last exon, so by the time the ribosome reaches it and stalls, no more EJCs will be bound to the mRNA. In this case, the mRNA "passes inspection" and is released to the cytosol where it can be translated in earnest (**Figure 6–76**). However, if the ribosome reaches a stop codon earlier, when EJCs remain bound, the mRNA molecule is rapidly degraded. In this way, the first round of translation allows the cell to test the fitness of each mRNA molecule as it exits the nucleus.

Nonsense-mediated decay may have been especially important in evolution, allowing eukaryotic cells to more easily explore new genes formed by DNA rearrangements, mutations, or alternative patterns of splicing—by selecting only those mRNAs for translation that can produce a full-length protein. Nonsense-mediated decay is also important in cells of the developing immune system, where the extensive DNA rearrangements that occur (see Figure 24–28) often generate

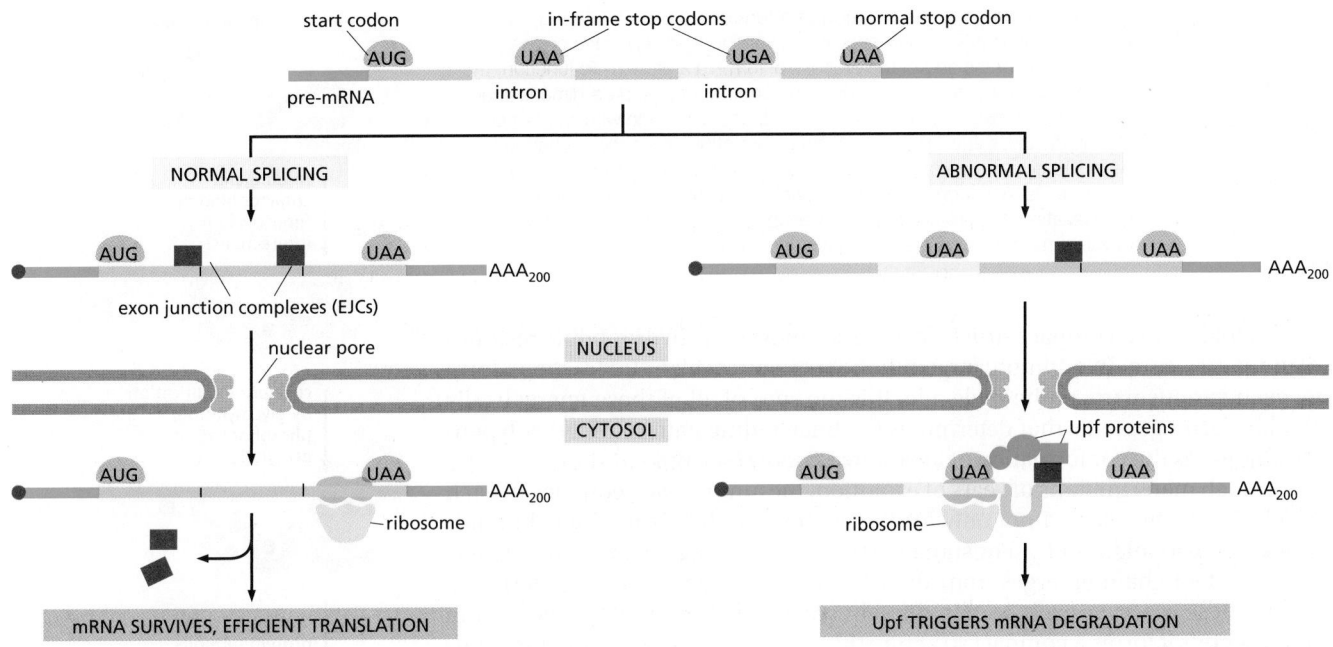

Figure 6–76 **Nonsense-mediated mRNA decay.** As shown on the right, the failure to correctly splice a pre-mRNA often introduces a premature stop codon into the reading frame for the protein. These abnormal mRNAs are destroyed by the nonsense-mediated decay mechanism. To activate this mechanism, an mRNA molecule, bearing exon junction complexes (EJCs) to mark successfully completed splices, is first met by a ribosome that performs a "test" round of translation. As the mRNA passes through the tight channel of the ribosome, the EJCs are stripped off, and successful mRNAs are released to undergo multiple rounds of translation *(left side)*. However, if an in-frame stop codon is encountered before the final EJC is reached *(right side)*, the mRNA undergoes nonsense-mediated decay, which is triggered by the Upf proteins *(green)* that bind to each EJC. Note that this mechanism ensures that nonsense-mediated decay is triggered only when the premature stop codon is in the same reading frame as that of the normal protein. (Adapted from J. Lykke-Andersen et al., *Cell* 103:1121–1131, 2000.)

premature termination codons. The surveillance system degrades the mRNAs produced from such rearranged genes, thereby avoiding the potential toxic effects of truncated proteins.

The nonsense-mediated surveillance pathway also plays an important role in mitigating the symptoms of many inherited human diseases. As we have seen, inherited diseases are usually caused by mutations that spoil the function of a key protein, such as hemoglobin or one of the blood-clotting factors. Approximately one-third of all genetic disorders in humans result from nonsense mutations or mutations (such as frameshift mutations or splice-site mutations) that place nonsense mutations into the gene's reading frame. In individuals that carry one mutant and one functional gene, nonsense-mediated decay eliminates the aberrant mRNA and thereby prevents a potentially toxic protein from being made. Without this safeguard, individuals with one functional and one mutant "disease gene" would likely suffer much more severe symptoms.

Some Proteins Begin to Fold While Still Being Synthesized

The process of gene expression is not over when the genetic code has been used to create the sequence of amino acids that constitutes a protein. To be useful to the cell, this new polypeptide chain must fold up into its unique three-dimensional conformation, bind any small-molecule cofactors required for its activity, be appropriately modified by protein kinases or other protein-modifying enzymes, and assemble correctly with the other protein subunits with which it functions (**Figure 6–77**).

The information needed for all of the steps listed above is ultimately contained in the sequence of amino acids that the ribosome produces when it translates an mRNA molecule into a polypeptide chain. As discussed in Chapter 3, when a

Figure 6–77 **Steps in the creation of a functional protein.** As indicated, translation of an mRNA sequence into an amino acid sequence on the ribosome is not the end of the process of forming a protein. To function, the completed polypeptide chain must fold correctly into its three-dimensional conformation, bind any cofactors required, and assemble with its partner protein chains, if any. Noncovalent bond formation drives these changes. As indicated, many proteins also require covalent modifications of selected amino acids. Although the most frequent modifications are protein glycosylation and protein phosphorylation, over 200 different types of covalent modifications are known (see pp. 165–166).

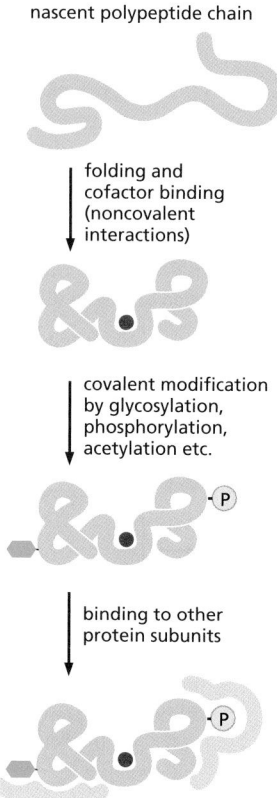

nascent polypeptide chain

folding and cofactor binding (noncovalent interactions)

covalent modification by glycosylation, phosphorylation, acetylation etc.

binding to other protein subunits

mature functional protein

protein folds into a compact structure, it buries most of its hydrophobic residues in an interior core. In addition, large numbers of noncovalent interactions form between various parts of the molecule. It is the sum of all of these energetically favorable arrangements that determines the final folding pattern of the polypeptide chain—as the conformation of lowest free energy (see pp. 114–115).

Through many millions of years of evolution, the amino acid sequence of each protein has been selected not only for the conformation that it adopts but also for an ability to fold rapidly. For some proteins, this folding begins immediately, as the protein chain emerges from the ribosome, starting from the N-terminal end. In these cases, as each protein domain emerges from the ribosome, within a few seconds it forms a compact structure that contains most of the final secondary features (α helices and β sheets) aligned in roughly the right conformation (**Figure 6–78**). For some protein domains, this unusually dynamic and flexible state, called a *molten globule*, is the starting point for a relatively slow process in which many side-chain adjustments occur that eventually form the correct tertiary structure. It takes several minutes to synthesize a protein of average size, and for some proteins much of the folding process is complete by the time the ribosome releases the C-terminal end of a protein (**Figure 6–79**).

Molecular Chaperones Help Guide the Folding of Most Proteins

Most proteins probably do not fold correctly during their synthesis and require a special class of proteins called **molecular chaperones** to do so. Molecular chaperones are useful for cells because there are many different folding paths available to an unfolded or partially folded protein. Without chaperones, some of these pathways would not lead to the correctly folded (and most stable) form because the protein would become "kinetically trapped" in structures that are off-pathway. Some of these off-pathway conformations would aggregate and be left as irreversible dead ends of nonfunctional (and potentially dangerous) structures.

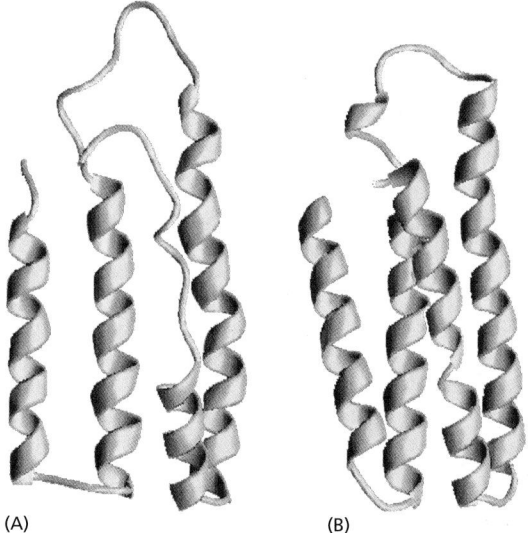

(A)   (B)

Figure 6–78 **The structure of a molten globule.** (A) A molten globule form of cytochrome b_{562} is more open and less highly ordered than the final folded form of the protein, shown in (B). Note that the molten globule contains most of the secondary structure of the final form, although the ends of the α helices are unraveled and one of the helices is only partly formed. (From Y. Feng et al., *Nat. Struct. Mol. Biol.* 1:30–35, published 1994 by Macmillan Magazines Ltd. Reproduced with permission of SNCSC.)

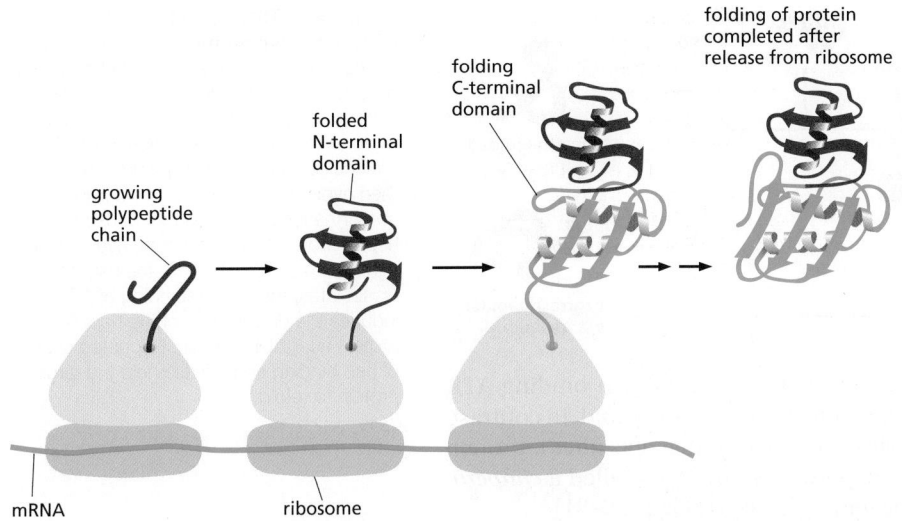

folding of protein completed after release from ribosome

folding C-terminal domain

folded N-terminal domain

growing polypeptide chain

mRNA

ribosome

Figure 6–79 Co-translational protein folding. A growing polypeptide chain is shown acquiring its secondary and tertiary structure as it emerges from a ribosome. The N-terminal domain folds first, while the C-terminal domain is still being synthesized. This protein has not achieved its final conformation at the time it is released from the ribosome. (Modified from A.N. Fedorov and T.O. Baldwin, *J. Biol. Chem.* 272:32715–32718, 1997.)

Molecular chaperones specifically recognize incorrect, off-pathway configurations by their exposure of hydrophobic surfaces, which in correctly folded proteins are typically buried in the interior. The binding of these exposed hydrophobic surfaces to each other is what causes off-pathway conformations to irreversibly aggregate. We saw in Chapter 3 that in some cases of inherited human diseases, aggregates do form and can cause severe symptoms and even death. Chaperones prevent this from happening in normal proteins by binding to the exposed hydrophobic surfaces using hydrophobic surfaces of their own. As we shall see shortly, there are several types of chaperones; once bound to an incorrectly folded protein, they ultimately release it in a way that gives the protein another chance to fold correctly.

Cells Utilize Several Types of Chaperones

Many molecular chaperones are called *heat-shock proteins* (designated *hsp*), because they are synthesized in dramatically increased amounts after a brief exposure of cells to an elevated temperature (for example, 42°C for cells that normally live at 37°C). This reflects the operation of a feedback system that responds to an increase in misfolded proteins (such as those produced by elevated temperatures) by boosting the synthesis of the chaperones that help these proteins refold.

There are several major families of molecular chaperones, including the hsp60 and hsp70 proteins. Different members of these families function in different organelles. Thus, as discussed in Chapter 12, mitochondria contain their own hsp60 and hsp70 molecules that are distinct from those that function in the cytosol; and a special hsp70 (called *BIP*) helps to fold proteins in the endoplasmic reticulum.

The hsp60 and hsp70 proteins each work with their own small set of associated proteins when they help other proteins to fold. These hsps share an affinity for the exposed hydrophobic patches on incompletely folded proteins, and they hydrolyze ATP, often binding and releasing their protein substrate with each cycle of ATP hydrolysis. In other respects, the two types of hsp proteins function differently. The hsp70 machinery acts early in the life of many proteins (often before the protein leaves the ribosome), with each monomer of hsp70 binding to a string

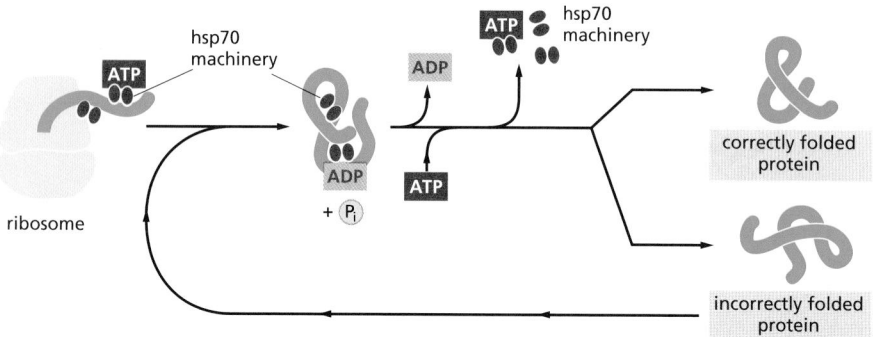

Figure 6–80 **The hsp70 family of molecular chaperones.** These proteins act early, recognizing a small stretch of hydrophobic amino acids on a protein's surface. Aided by a set of smaller hsp40 proteins (not shown), ATP-bound hsp70 molecules grasp their target protein and then hydrolyze ATP to ADP, undergoing conformational changes that cause the hsp70 molecules to associate even more tightly with the target. After the hsp40 dissociates, the rapid rebinding of ATP induces the dissociation of the hsp70 protein after ADP release. Repeated cycles of hsp binding and release help the target protein to refold.

of about four or five hydrophobic amino acids (**Figure 6–80**). On binding ATP, hsp70 releases the protein into solution allowing it a chance to re-fold. In contrast, hsp60-like proteins form a large barrel-shaped structure that acts after a protein has been fully synthesized. This type of chaperone, sometimes called a *chaperonin*, forms an "isolation chamber" for the folding process (**Figure 6–81**).

To enter a chamber, a substrate protein is first captured via the hydrophobic entrance to the chamber. The protein is then released into the interior of the chamber, which is lined with hydrophilic surfaces, and the chamber is sealed with a lid, a step requiring ATP. Here, the substrate is allowed to fold into its final conformation in isolation, where there are no other proteins with which to aggregate. When ATP is hydrolyzed, the lid pops off, and the substrate protein, whether folded or not, is released from the chamber.

The chaperones shown in Figures 6–80 and 6–81 often need many cycles of ATP hydrolysis to fold a single polypeptide chain correctly. This energy is used to perform mechanical movements of the hsp60 and hsp70 "machines," converting them from binding forms to releasing forms. Just as we saw for transcription, splicing, and translation, the expenditure of free energy can be used by cells to improve the accuracy of a biological process. In the case of protein folding, ATP hydrolysis allows chaperones to recognize a wide variety of misfolded structures, to halt any further misfolding, and to recommence the folding of a protein in an orderly way.

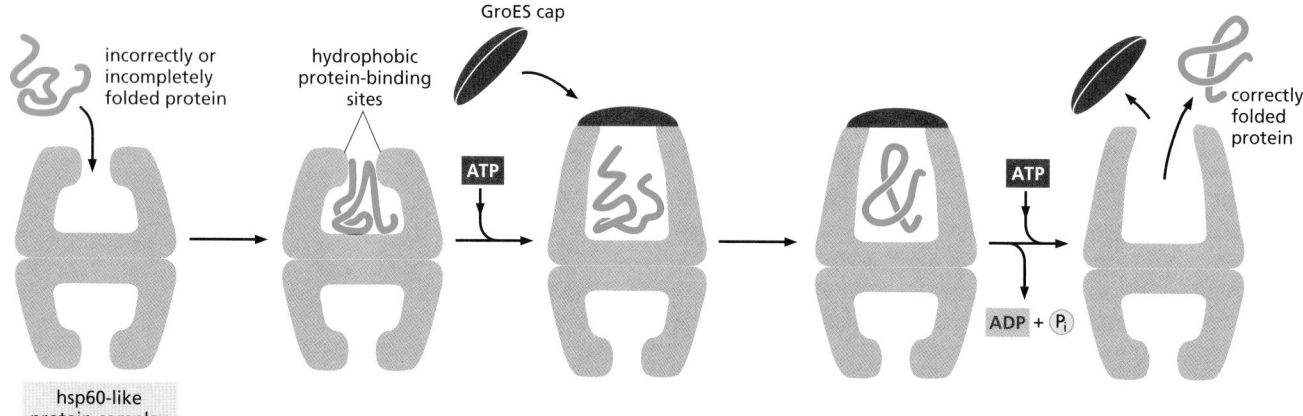

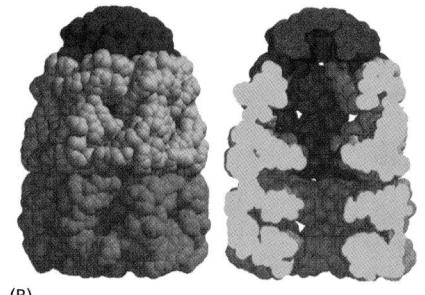

Figure 6–81 **The structure and function of the hsp60 family of molecular chaperones.** (A) A misfolded protein is initially captured by hydrophobic interactions with the exposed surface of the opening. The initial binding often helps to unfold a misfolded protein. The subsequent binding of ATP and a cap releases the substrate protein into an enclosed space, where it has a new opportunity to fold. After about 10 seconds, ATP hydrolysis occurs, weakening the binding of the cap. Subsequent binding of additional ATP molecules ejects the cap, and the protein is released. As indicated, only half of the symmetric barrel operates on a client protein at any one time. This type of molecular chaperone is also known as a chaperonin; it is designated as hsp60 in mitochondria, TCP1 in the cytosol of vertebrate cells, and GroEL in bacteria. (B) The structure of GroEL bound to its GroES cap, as determined by x-ray crystallography. On the *left* is shown the outside of the barrel-like structure, and on the *right* a cross section through its center. (B, adapted from B. Bukau and A.L. Horwich, *Cell* 92:351–366, 1998. With permission from Elsevier.)

Although our discussion focuses on only two types of chaperones, the cell has a variety of others. The enormous diversity of proteins in cells presumably requires a wide range of chaperones with versatile surveillance and correction capabilities.

Exposed Hydrophobic Regions Provide Critical Signals for Protein Quality Control

If radioactive amino acids are added to cells for a brief period, the newly synthesized proteins can be followed as they mature into their final functional forms. This type of experiment demonstrates that the hsp70 proteins act first, beginning when a protein is still being synthesized on a ribosome, and the hsp60-like proteins act only later to help fold completed proteins. We have seen that the cell distinguishes misfolded proteins, which require additional rounds of ATP-catalyzed refolding, from those with correct structures through the recognition of hydrophobic surfaces.

Usually, if a protein has a sizable exposed patch of hydrophobic amino acids on its surface, it is abnormal: it has either failed to fold correctly after leaving the ribosome, suffered an accident that partly unfolded it at a later time, or failed to find its normal partner subunit in a larger protein complex. Such a protein is not merely useless to the cell, it can be dangerous.

Proteins that rapidly fold correctly on their own do not display such patterns and generally bypass the chaperones. For the others, the chaperones can carry out "protein repair" by giving them additional chances to fold while, at the same time, preventing their aggregation.

Figure 6–82 outlines all of the quality-control choices that a cell makes for a difficult-to-fold, newly synthesized protein. As indicated, when attempts to refold a protein fail, an additional mechanism is called into play that completely destroys the protein by proteolysis. This proteolytic pathway begins with the recognition of an abnormal hydrophobic patch on a protein's surface, and it ends with the delivery of the entire protein to a protein-destruction machine, a complex protease known as the *proteasome*. As described next, this process depends on an elaborate protein-marking system that also carries out other central functions in the cell by destroying selected normal proteins.

The Proteasome Is a Compartmentalized Protease with Sequestered Active Sites

The proteolytic machinery and the chaperones compete with one another to recognize a misfolded protein. If a newly synthesized protein folds rapidly, at most only a small fraction of it is degraded. In contrast, a slowly folding protein is vulnerable to the proteolytic machinery for a longer time, and many more of its molecules may be destroyed before the remainder attain the proper folded state. Due to mutations or to errors in transcription, RNA splicing, and translation, some proteins never fold properly, and it is particularly important that the cell destroy these potentially harmful proteins.

The apparatus that deliberately destroys aberrant proteins is the **proteasome**, an abundant ATP-dependent protease that constitutes nearly 1% of cell protein.

Figure 6–82 **The processes that monitor protein quality following protein synthesis.** A newly synthesized protein sometimes folds correctly and assembles on its own with its partner proteins, in which case the quality control mechanisms leave it alone. Incompletely folded proteins are helped to properly fold by molecular chaperones: first by a family of hsp70 proteins, and then, in some cases, by hsp60-like proteins. For both types of chaperones, the substrate proteins are recognized by an abnormally exposed patch of hydrophobic amino acids on their surface. These "protein-rescue" processes compete with another mechanism that, upon recognizing an abnormally exposed hydrophobic patch, marks the protein for destruction by the proteasome. The combined activity of all of these processes is needed to prevent massive protein aggregation in a cell, which can occur when many hydrophobic regions on proteins clump together nonspecifically.

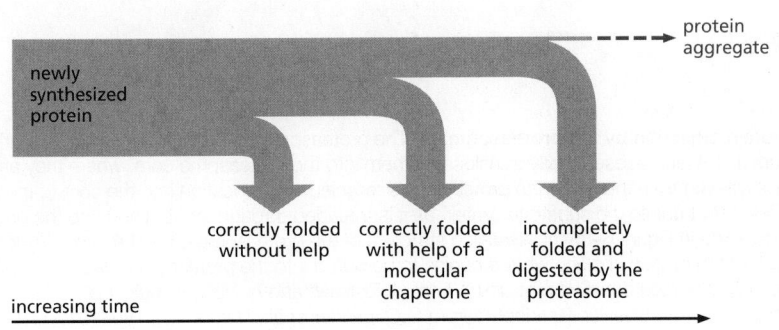

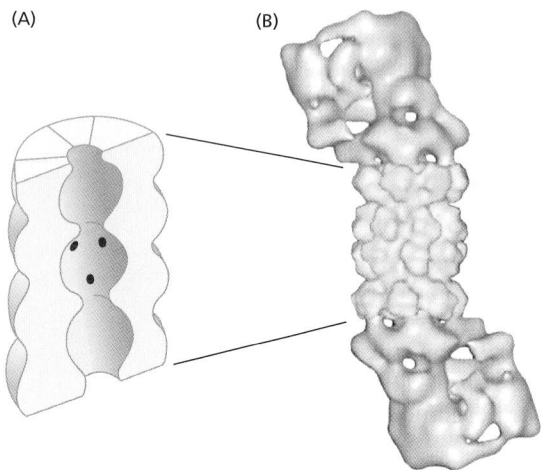

Figure 6–83 **The proteasome.** (A) A cutaway view of the structure of the central 20S cylinder, as determined by x-ray crystallography, with the active sites of the proteases indicated by *red dots.* (B) The entire proteasome, in which the central cylinder *(yellow)* is supplemented by a 19S cap *(blue)* at each end. The complex cap (also called the regulatory particle) selectively binds proteins that have been marked by ubiquitin for destruction; it then uses ATP hydrolysis to unfold their polypeptide chains and feed them through a narrow channel (see Figure 6–85) into the inner chamber of the 20S cylinder for digestion to short peptides. (B, from W. Baumeister et al., *Cell* 92:367–380, 1998. With permission from Elsevier.)

Present in many copies dispersed throughout the cytosol and the nucleus, the proteasome also destroys aberrant proteins that have entered the endoplasmic reticulum (ER). In the latter case, an ER-based surveillance system detects proteins that have failed either to fold or to be assembled properly after they enter the ER, and *retrotranslocates* them back to the cytosol for degradation by the proteasome (discussed in Chapter 12).

Each proteasome consists of a central hollow cylinder (the 20S core proteasome) formed from multiple protein subunits that assemble as a stack of four heptameric rings (**Figure 6–83**). Some of the subunits are proteases whose active sites face the cylinder's inner chamber, thus preventing them from running rampant through the cell. Each end of the cylinder is normally associated with a large protein complex (the 19S cap) that contains a six-subunit protein ring through which target proteins are threaded into the proteasome core, where they are degraded (**Figure 6–84**). The threading reaction, driven by ATP hydrolysis, unfolds the target proteins as they move through the cap, exposing them to the proteases lining the proteasome core (**Figure 6–85**). The proteins that make up the ring structure in the proteasome cap belong to a large class of protein "unfoldases" known as *AAA proteins*. Many of them function as hexamers, and they share mechanistic features with the ATP-dependent DNA helicases that unwind DNA (see Figure 5–14).

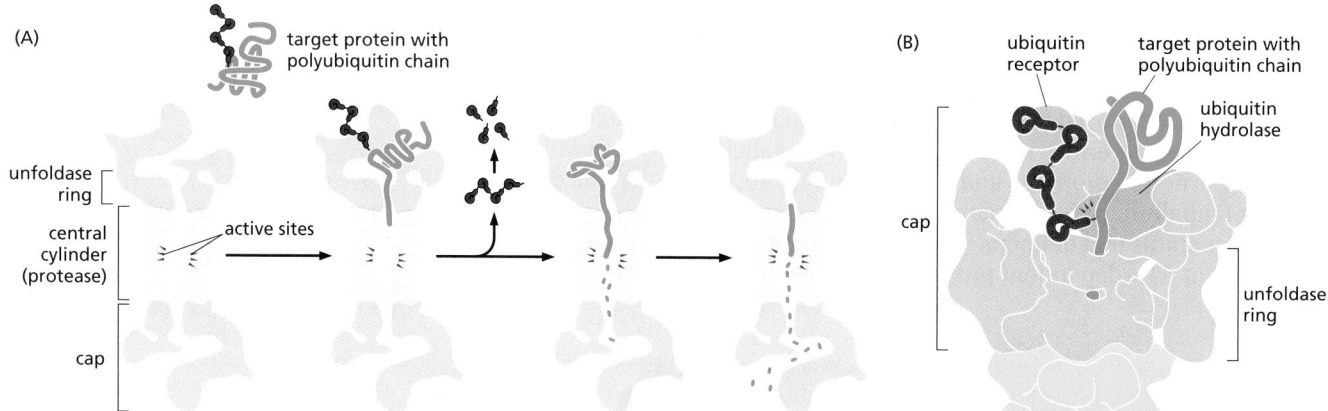

Figure 6–84 **Processive protein digestion by the proteasome.** (A) The proteasome cap recognizes proteins marked by a polyubiquitin chain (see Figure 3–70), and subsequently translocates them into the proteasome core, where they are digested. At an early stage, the ubiquitin is cleaved from the substrate protein and is recycled. Translocation into the core of the proteasome is mediated by a ring of ATPases that unfold the substrate protein as it is threaded through the ring and into the proteasome core. This unfoldase ring is depicted in Figure 6–85. (B) Detailed structure of the proteasome cap. The cap includes a ubiquitin receptor, which holds a ubiquitylated protein in place while it begins to be pulled into the proteasome core, and a ubiquitin hydrolase, which cleaves ubiquitin from the doomed protein. (A, from S. Prakash and A. Matouschek, *Trends Biochem. Sci.* 29:593–600, 2004. With permission from Elsevier. B, adapted from G.C. Lander et al., *Nature* 482:186–191, 2012.)

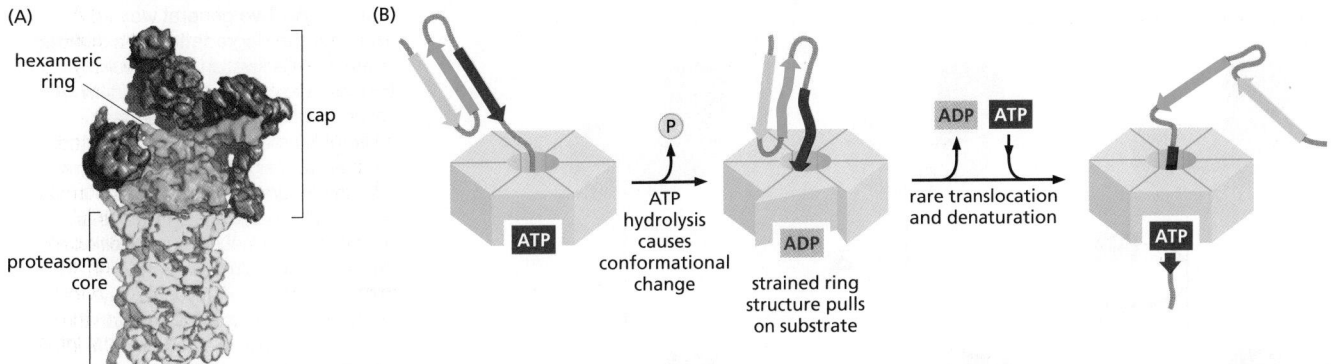

Figure 6–85 **A hexameric protein unfoldase.** (A) The proteasome cap includes proteins *(orange)* that recognize and hydrolyze ubiquitin and a hexameric ring *(blue)* through which ubiquitylated proteins are threaded. The hexameric ring is formed from six subunits, each belonging to the AAA family of proteins. (B) Model for the ATP-dependent unfoldase activity of AAA proteins. The ATP-bound form of a hexameric ring of AAA proteins binds a folded substrate protein that is held in place by its ubiquitin tag. A conformational change, driven by ATP hydrolysis, pulls the substrate into the central core and strains the ring structure. At this point, the substrate protein, which is being tugged upon, can partially unfold and enter further into the pore or it can maintain its structure and partially withdraw. Very stable protein substrates may require hundreds of cycles of ATP hydrolysis and dissociation before they are successfully pulled through the AAA protein ring. Once unfolded (and de-ubiquitylated), the substrate protein moves relatively quickly through the pore by successive rounds of ATP hydrolysis. (A, adapted from G.C. Lander et al., *Nature* 482:186–191, published 2012 by Macmillan Publishers Ltd. Reproduced with permission of SNCSC; B, adapted from R.T. Sauer et al., *Cell* 119:9–18, 2004.)

A crucial property of the proteasome, and one reason for the complexity of its design, is the *processivity* of its mechanism: in contrast to a "simple" protease that cleaves a substrate's polypeptide chain just once before dissociating, the proteasome keeps the entire substrate bound until all of it is converted into short peptides.

One would expect that a machine as efficient as the proteasome would be tightly regulated; in particular, it must be able to distinguish abnormal proteins from those that are properly folded. The 19S cap of the proteasome acts as a gate at the entrance to the inner proteolytic core, and only those proteins marked for destruction are threaded through the cap. The destruction "mark" is the covalent attachment of the small protein ubiquitin. As we saw in Chapter 3, ubiquitin modification of proteins is used for many purposes in the cell. The particular type of ubiquitin linkage that concerns us here is a chain of ubiquitin molecules linked together at lysine 48 (see Figure 3–69); this is the distinguishing feature of the ubiquitin tag that marks a protein for destruction in the proteasome.

A special set of E3 molecules (see Figure 3–70B) is responsible for the ubiquitylation of denatured or otherwise misfolded proteins, as well as proteins containing oxidized or other abnormal amino acids. Abnormal proteins tend to display on their surface hydrophobic amino acid sequences or conformational motifs that are recognized as degradation signals by these E3 molecules; these sequences are buried and therefore inaccessible in the normal, properly folded version. However, a proteolytic pathway that recognizes and destroys abnormal proteins must be able to distinguish *completed* proteins that have "wrong" conformations from the many growing polypeptides on ribosomes (as well as polypeptides just released from ribosomes) that have not yet achieved their normal folded conformation. This is not a trivial problem; in the course of carrying out its main job, the ubiquitin–proteasome system probably destroys many nascent and newly formed protein molecules, not because these proteins are abnormal as such, but because they have transiently exposed degradation signals that are buried in their mature (folded) state.

Many Proteins Are Controlled by Regulated Destruction

One function of intracellular proteolytic mechanisms is to recognize and eliminate misfolded or otherwise abnormal proteins, as just described. Indeed, every protein in the cell eventually accumulates damage and is probably degraded by the proteasome. Yet another function of these proteolytic pathways is to confer short lifetimes on specific normal proteins whose concentrations must change promptly with alterations in the state of a cell. Some of these short-lived proteins are degraded rapidly at all times, while many others are *conditionally* short-lived; that is, they are metabolically stable under some conditions, but become unstable upon a change in the cell's state. For example, mitotic cyclins are long-lived throughout the cell cycle until their sudden degradation at the end of mitosis, as explained in Chapter 17.

(A) ACTIVATION OF A UBIQUITIN LIGASE

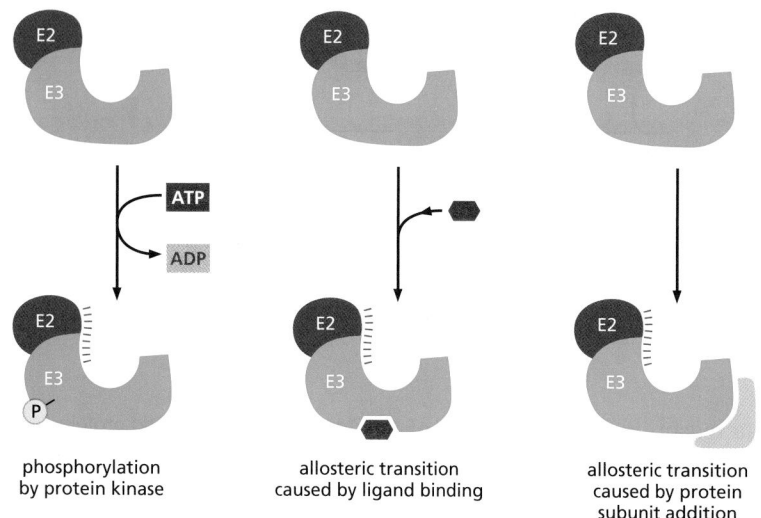

phosphorylation
by protein kinase

allosteric transition
caused by ligand binding

allosteric transition
caused by protein
subunit addition

(B) ACTIVATION OF A DEGRADATION SIGNAL

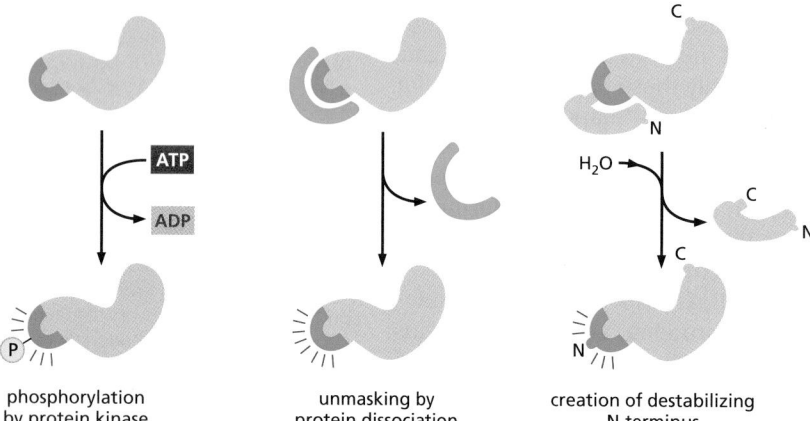

phosphorylation
by protein kinase

unmasking by
protein dissociation

creation of destabilizing
N-terminus

Figure 6–86 Two general ways of inducing the degradation of a specific protein. (A) Activation of a specific E3 molecule creates a new ubiquitin ligase. Eukaryotic cells have many different E3 molecules, each activated by a different signal. (B) Creation of an exposed degradation signal in the protein to be degraded. This signal binds a ubiquitin ligase, causing the addition of a polyubiquitin chain to a nearby lysine on the target protein. All six pathways shown are known to be used by cells to induce the movement of selected proteins into the proteasome.

How is such a regulated destruction of a protein controlled? Several general mechanisms are illustrated in **Figure 6–86**. Specific examples of each mechanism are discussed in later chapters. In one general class of mechanism (Figure 6–86A), the activity of a ubiquitin ligase is turned on either by E3 phosphorylation or by an allosteric transition in an E3 protein caused by its binding to a specific small or large molecule. For example, the anaphase-promoting complex (APC) is a multi-subunit ubiquitin ligase that is activated by a cell-cycle-timed subunit addition at mitosis. The activated APC then causes the degradation of mitotic cyclins and several other regulators of the metaphase–anaphase transition (see Figure 17–15A).

Alternatively, in response either to intracellular signals or to signals from the environment, a degradation signal can be created in a protein, causing its rapid ubiquitylation and destruction by the proteasome (Figure 6–86B). One common way to create such a signal is to phosphorylate a specific site on a protein that unmasks a normally hidden degradation signal. Another way to unmask such a signal is by the regulated dissociation of a protein subunit. Finally, powerful degradation signals can be created by cleaving a single peptide bond, provided that this cleavage creates a new N-terminus that is recognized by a specific E3 protein as a "destabilizing" N-terminal residue. This E3 protein recognizes only certain amino acids at the N-terminus of a protein; thus not all protein-cleavage events will lead to degradation of the C-terminal fragment produced.

In humans, nearly 80% of proteins are acetylated on their N-terminal residue, and we now know that this modification is recognized by a specific E3 enzyme, which directs the ubiquitylation of the protein and sends it to the proteasome for degradation. Thus, the majority of human proteins carry their own signals for destruction. It has been proposed that when a protein is properly folded (and, before that, when it is in contact with a chaperone), this acetylated N-terminus is buried and therefore inaccessible to the E3 enzyme. According to this idea, as a protein ages and becomes damaged (or if it fails to fold correctly from the start), this destruction signal becomes exposed, and the protein is destroyed.

There Are Many Steps From DNA to Protein

We have seen so far in this chapter that many different types of chemical reactions are required to produce a properly folded protein from the information contained in a gene (Figure 6–87). The final level of a properly folded protein in a cell therefore depends upon the efficiency with which each of the many steps is performed. We also now know that the cell devotes enormous resources to selectively degrading proteins, particularly those that fail to fold properly or accumulate damage as they age. It is the balance between the rates of synthesis and degradation that determines the final amount of every protein in the cell.

In the following chapter, we shall see that cells have the ability to change the levels of their proteins according to their needs. In principle, any or all of the steps in Figure 6–87 could be regulated for each individual protein. As we shall see in Chapter 7, there are examples of regulation at each step from gene to protein.

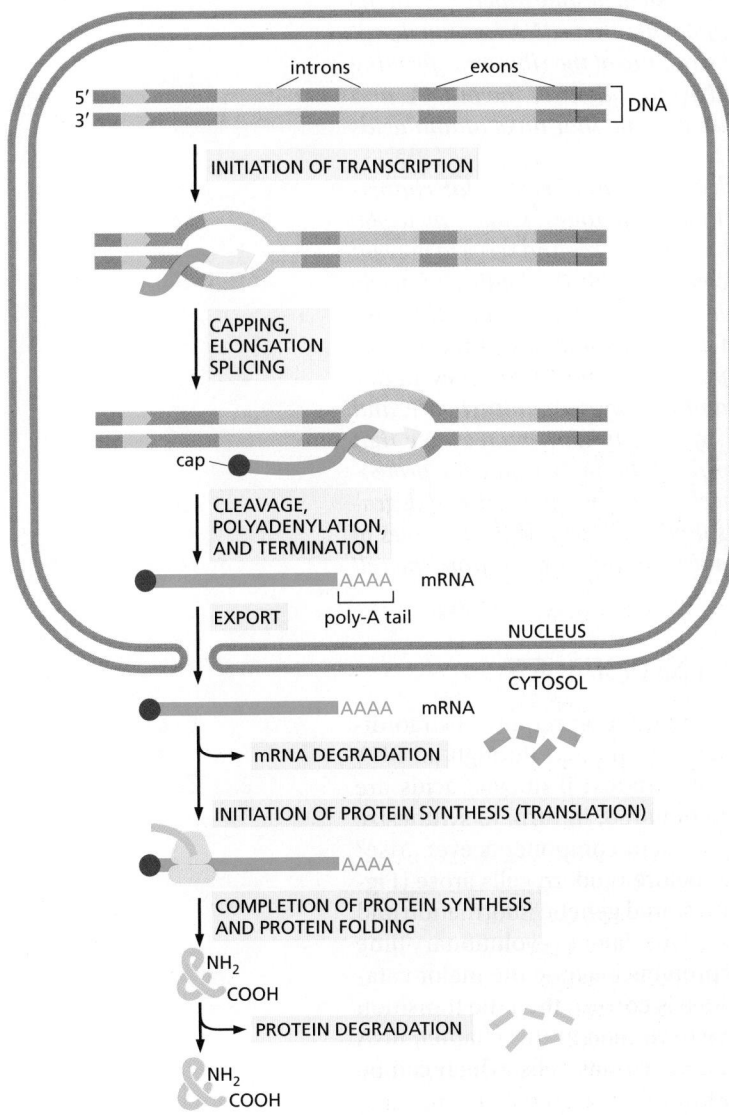

Figure 6–87 **The production of a protein by a eukaryotic cell.** The final level of each protein in a eukaryotic cell depends upon the efficiency of each step depicted.

Summary

The translation of the nucleotide sequence of an mRNA molecule into protein takes place in the cytosol on a large ribonucleoprotein assembly called a ribosome. Each amino acid used for protein synthesis is first attached to a tRNA molecule that recognizes, by complementary base-pair interactions, a particular set of three nucleotides (codons) in the mRNA. As an mRNA is threaded through a ribosome, its sequence of nucleotides is then read from one end to the other in sets of three according to the genetic code.

To initiate translation, a small ribosomal subunit binds to the mRNA molecule at a start codon (AUG) that is recognized by a unique initiator tRNA molecule. A large ribosomal subunit then binds to complete the ribosome and begin protein synthesis. During this phase, aminoacyl-tRNAs—each bearing a specific amino acid—bind sequentially to the appropriate codons in mRNA through complementary base-pairing between tRNA anticodons and mRNA codons. Each amino acid is added to the C-terminal end of the growing polypeptide in four sequential steps: aminoacyl-tRNA binding, followed by peptide bond formation, followed by two ribosome translocation steps. Elongation factors use GTP hydrolysis both to drive these reactions forward and to improve the accuracy of amino acid selection. The mRNA molecule progresses codon by codon through the ribosome in the 5'-to-3' direction until it reaches one of three stop codons. A release factor then binds to the ribosome, terminating translation and releasing the completed polypeptide.

Eukaryotic and bacterial ribosomes are closely related, despite differences in the number and size of their rRNA and protein components. The rRNA has the dominant role in translation, determining the overall structure of the ribosome, forming the binding sites for the tRNAs, matching the tRNAs to codons in the mRNA, and creating the active site of the peptidyl transferase enzyme that links amino acids together during translation.

In the final steps of protein synthesis, two distinct types of molecular chaperones guide the folding of polypeptide chains. These chaperones, known as hsp60 and hsp70, recognize exposed hydrophobic patches on proteins and serve to prevent the protein aggregation that would otherwise compete with the folding of newly synthesized proteins into their correct three-dimensional conformations. This protein-folding process must also compete with an elaborate quality control mechanism that destroys proteins with abnormally exposed hydrophobic patches. In this case, ubiquitin is covalently added to a misfolded protein by a ubiquitin ligase, and the resulting polyubiquitin chain is recognized by the cap on a proteasome that unfolds the protein and threads it into the interior of the proteasome for proteolytic degradation. A closely related proteolytic mechanism, based on special degradation signals recognized by ubiquitin ligases, is used to determine the lifetimes of many normally folded proteins as well as to remove selected proteins from the cell in response to specific signals.

THE RNA WORLD AND THE ORIGINS OF LIFE

We have seen that the expression of hereditary information requires extraordinarily complex machinery and proceeds from DNA to protein through an RNA intermediate. This machinery presents a central paradox: if nucleic acids are required to synthesize proteins and proteins are required, in turn, to synthesize nucleic acids, how did such a system of interdependent components ever arise? One view is that an **RNA world** existed on Earth before modern cells arose (**Figure 6–88**). According to this hypothesis, RNA both stored genetic information and catalyzed the chemical reactions in primitive cells. Only later in evolutionary time did DNA take over as the genetic material and proteins become the major catalysts and structural components of cells. If this idea is correct, then the transition out of the RNA world was never complete; as we have seen in this chapter, RNA still catalyzes several fundamental reactions in modern-day cells, which can be viewed as molecular fossils from an earlier world.

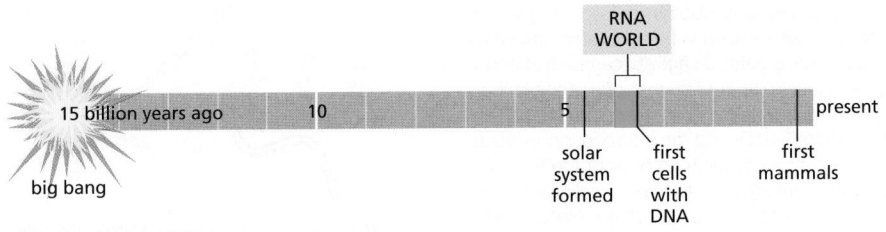

Figure 6–88 Time line for the universe, suggesting the early existence of an RNA world of living systems.

The RNA world hypothesis relies on the fact that, among present-day biological molecules, RNA is unique in being able to act as both a carrier of genetic information and as a ribozyme to catalyze chemical reactions. In this section, we discuss these properties of RNA and how they may have been especially important in early cells.

Single-Stranded RNA Molecules Can Fold into Highly Elaborate Structures

We have seen in this chapter that RNA can carry genetic information in mRNAs, and we saw in Chapter 5 that the genomes of some viruses are composed solely of RNA. We have also seen that complementary base-pairing and other types of hydrogen bonds can occur between nucleotides in the same chain of RNA, causing an RNA molecule to fold up in a unique way determined by its nucleotide sequence (see, for example, Figures 6–50 and 6–67). Comparisons of many RNA structures have revealed conserved motifs, short structural elements that are used over and over again as parts of larger structures (Figure 6–89).

Protein catalysts require a surface with unique contours and chemical properties on which a given set of substrates can react (discussed in Chapter 3). In exactly the same way, an RNA molecule with an appropriately folded shape can serve as a catalyst (Figure 6–90). Like some proteins, many of these ribozymes work by positioning metal ions at their active sites. This feature gives them a wider range of catalytic activities than provided by the limited chemical groups of a polynucleotide chain.

Much of our inference about the RNA world has come from experiments in which large pools of RNA molecules of random nucleotide sequences are generated in the laboratory. Those rare RNA molecules with a property specified by the experimenter are then selected out and studied (Figure 6–91). Such experiments have created RNAs that can catalyze a wide variety of biochemical reactions (Table 6–5), with reaction rate enhancements only a few orders of magnitude lower than

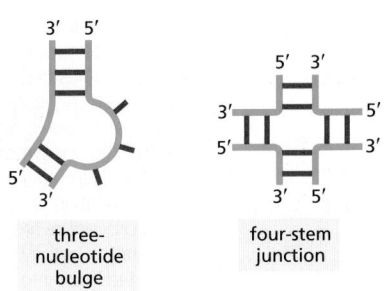

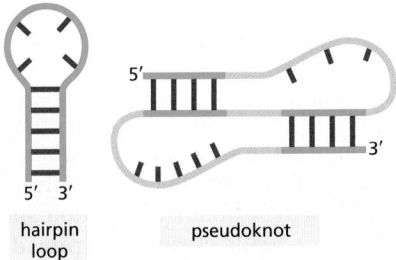

Figure 6–89 Some common elements of RNA structure. Conventional, complementary base-pairing interactions are indicated by *red* "rungs" in double-helical portions of the RNA.

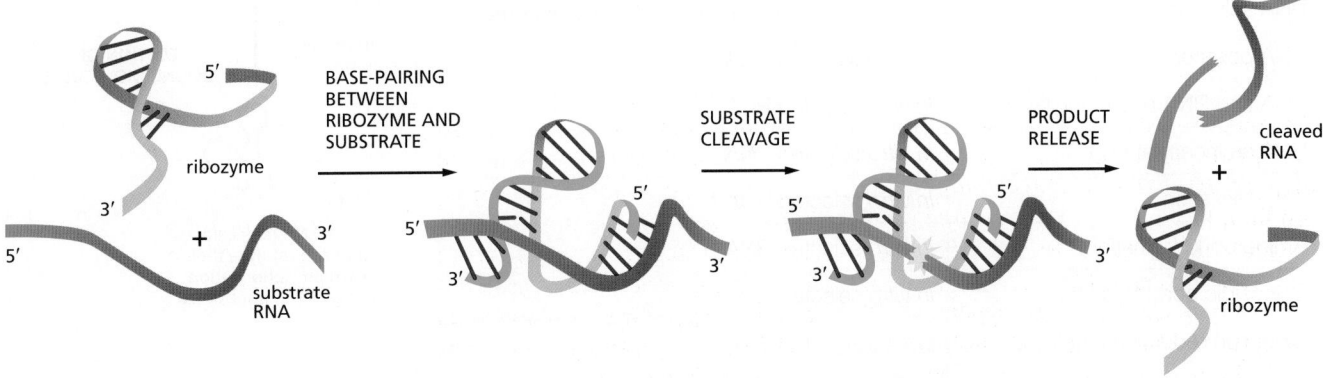

Figure 6–90 A ribozyme. This simple RNA molecule catalyzes the cleavage of a second RNA at a specific site. This ribozyme is found embedded in larger RNA genomes—called viroids—which infect plants. The cleavage, which occurs in nature at a distant location on the same RNA molecule that contains the ribozyme, is a step in the replication of the viroid genome. Although not shown in the figure, the reaction requires a magnesium ion positioned at the active site. (Adapted from T.R. Cech and O.C. Uhlenbeck, *Nature* 372:39–40, 1994.)

Figure 6–91 *In vitro* **selection of a synthetic ribozyme.** Beginning with a large pool of nucleic acid molecules synthesized in the laboratory, those rare RNA molecules that possess a specified catalytic activity can be isolated and studied. Although a specific example (that of an autophosphorylating ribozyme) is shown, variations of this procedure have been used to generate many of the ribozymes listed in Table 6–5. During the autophosphorylation step, the RNA molecules are kept sufficiently dilute to prevent the "cross"-phosphorylation of additional RNA molecules. In reality, several repetitions of this procedure are necessary to select the very rare RNA molecules with this catalytic activity. Thus, the material initially eluted from the column is converted back into DNA, amplified many fold (using reverse transcriptase and PCR, as explained in Chapter 8), transcribed back into RNA, and subjected to repeated rounds of selection. (Adapted from J.R. Lorsch and J.W. Szostak, *Nature* 371:31–36, 1994.)

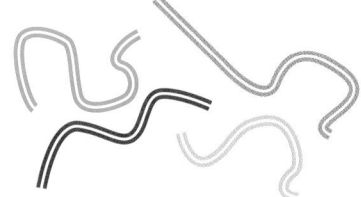

large pool of double-stranded DNA molecules, each with a different, randomly generated nucleotide sequence

TRANSCRIPTION BY RNA POLYMERASE AND FOLDING OF RNA MOLECULES

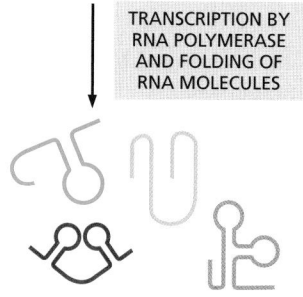

large pool of single-stranded RNA molecules, each with a different, randomly generated nucleotide sequence

ADDITION OF ATP DERIVATIVE CONTAINING A SULFUR IN PLACE OF AN OXYGEN

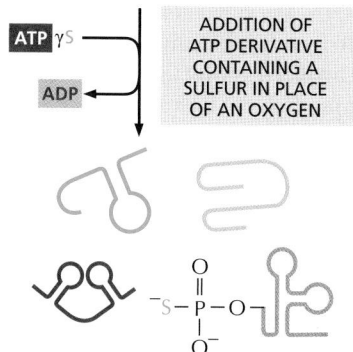

only the rare RNA molecules able to phosphorylate themselves incorporate sulfur

CAPTURE OF PHOSPHORYLATED MATERIAL ON COLUMN MATERIAL THAT BINDS TIGHTLY TO THE SULFUR GROUP

discard RNA molecules that fail to bind to the column

ELUTION OF BOUND MOLECULES

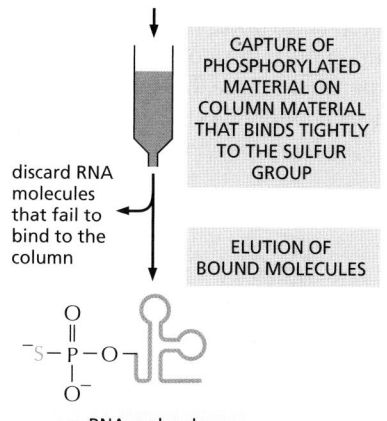

rare RNA molecules that can catalyze their own phosphorylation using ATP as a substrate

those of the "fastest" protein enzymes. Given these findings, it is not clear why protein catalysts greatly outnumber ribozymes in modern cells. Experiments have shown, however, that RNA molecules may have more difficulty than proteins in binding to flexible, hydrophobic substrates. In any case, the availability of 20 types of amino acids presumably provides proteins with a greater number of catalytic strategies.

RNA Can Both Store Information and Catalyze Chemical Reactions

RNA molecules have one property that contrasts with those of polypeptides: they can directly guide the formation of copies of their own sequence. This capacity depends on complementary base-pairing of their nucleotide subunits, which enables one RNA to act as a template for the formation of another. As we have seen in this and the preceding chapter, these complementary templating mechanisms lie at the heart of DNA replication and transcription in modern-day cells.

TABLE 6–5 Some Biochemical Reactions That Can Be Catalyzed by Ribozymes

Activity	Ribozymes
Peptide bond formation in protein synthesis	Ribosomal RNA
RNA cleavage, RNA ligation	Self-splicing RNAs; RNAse P; also *in vitro* selected RNA
DNA cleavage	Self-splicing RNAs
RNA splicing	Self-splicing RNAs, RNAs of the spliceosome
RNA polymerizaton	*In vitro* selected RNA
RNA and DNA phosphorylation	*In vitro* selected RNA
RNA aminoacylation	*In vitro* selected RNA
RNA alkylation	*In vitro* selected RNA
Amide bond formation	*In vitro* selected RNA
Glycosidic bond formation	*In vitro* selected RNA
Oxidation/reduction reactions	*In vitro* selected RNA
Carbon–carbon bond formation	*In vitro* selected RNA
Phosphoamide bond formation	*In vitro* selected RNA
Disulfide exchange	*In vitro* selected RNA

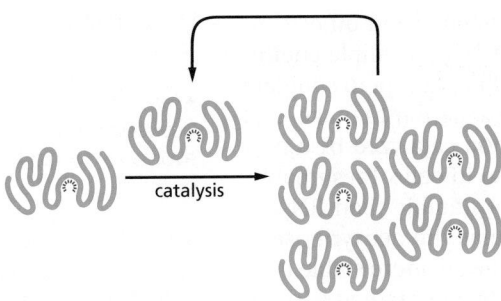

Figure 6–92 **An RNA molecule that can catalyze its own synthesis.** This hypothetical process would require catalysis both of the production of a second RNA strand of complementary nucleotide sequence (not shown) and the use of this second RNA molecule as a template to form many molecules of RNA with the original sequence. The *red* rays represent the active site of this hypothetical RNA enzyme.

But the efficient synthesis of RNA by such complementary templating mechanisms requires catalysts to promote the polymerization reaction: without catalysts, polymer formation is slow, error-prone, and inefficient.

Because RNA has all the properties required of a molecule that could catalyze a variety of chemical reactions, including those that lead to its own synthesis (**Figure 6–92**), it has been proposed that RNAs served long ago as the catalysts for template-dependent RNA synthesis. Although self-replicating systems of RNA molecules have not been found in nature, scientists have made significant progress toward constructing them in the laboratory. While such demonstrations would not prove that self-replicating RNA molecules were central to the origin of life on Earth, they would establish that such a scenario is plausible.

How Did Protein Synthesis Evolve?

The molecular processes underlying protein synthesis in present-day cells seem inextricably complex. Although we understand most of them, they do not make conceptual sense in the way that DNA transcription, DNA repair, and DNA replication do. It is especially difficult to imagine how protein synthesis evolved because it is now performed by a complex interlocking system of protein and RNA molecules; obviously the proteins could not have existed until an early version of the translation apparatus was already in place. As attractive as the RNA world idea is for envisioning early life, it does not explain how the modern-day system of protein synthesis arose. Although we can only speculate on the origins of the genetic code, several experimental observations have provided plausible scenarios.

In modern cells, some short peptides (such as antibiotics) are synthesized without the ribosome; peptide synthetase enzymes assemble these peptides, with their proper sequence of amino acids, without mRNAs to guide their synthesis. It is plausible that this noncoded, primitive version of protein synthesis first developed in the RNA world, where it would have been catalyzed by RNA molecules. This idea presents no conceptual difficulties because, as we have seen, rRNA catalyzes peptide bond formation in present-day cells. However, it leaves unexplained how the genetic code—which lies at the core of protein synthesis in today's cells—might have arisen. We know that ribozymes created in the laboratory can perform specific aminoacylation reactions; that is, they can match specific amino acids to specific tRNAs. It is therefore possible that tRNA-like adaptors, each matched to a specific amino acid, could have arisen in the RNA world, marking the beginnings of a genetic code.

Once coded protein synthesis evolved, the transition to a protein-dominated world could proceed, with proteins eventually taking over the majority of catalytic and structural tasks because of their greater versatility, with 20 rather than 4 different subunits. Although these ideas are highly speculative, they are consistent with the known properties of RNA and protein molecules.

All Present-Day Cells Use DNA as Their Hereditary Material

If the evolutionary speculations embodied in the RNA world hypothesis are correct, early cells would have differed fundamentally from the cells we know today in having their hereditary information stored in RNA rather than in DNA (**Figure 6–93**). Evidence that RNA arose before DNA in evolution can be found

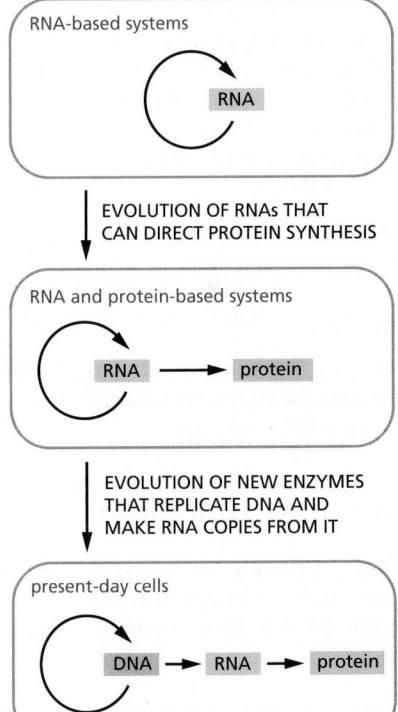

Figure 6–93 **The hypothesis that RNA preceded DNA and proteins in evolution.** In the earliest cells, RNA molecules (or their close analogs) would have had combined genetic, structural, and catalytic functions. In present-day cells, DNA is the repository of genetic information, and proteins perform the vast majority of catalytic functions in cells. RNA primarily functions today as a go-between in protein synthesis, although it remains a catalyst for a small number of crucial reactions.

in the chemical differences between them. Ribose, like glucose and other simple carbohydrates, can be formed from formaldehyde (HCHO), a simple chemical which is readily produced in laboratory experiments that attempt to simulate conditions on the primitive Earth. The sugar deoxyribose is harder to make, and in present-day cells it is produced from ribose in a reaction catalyzed by a protein enzyme, suggesting that ribose pre-dates deoxyribose in cells. Presumably, DNA appeared on the scene later, but then proved more suitable than RNA as a permanent repository of genetic information. In particular, the deoxyribose in its sugar-phosphate backbone makes chains of DNA chemically more stable than chains of RNA, so that much greater lengths of DNA can be maintained without breakage.

The other differences between RNA and DNA—the double-helical structure of DNA and the use of thymine rather than uracil—further enhance DNA stability by making the many unavoidable accidents that occur to the molecule much easier to repair, as discussed in detail in Chapter 5 (pp. 271–273).

Summary

From our knowledge of present-day organisms and the molecules they contain, it seems likely that the development of the distinctive autocatalytic mechanisms fundamental to living systems began with the evolution of families of RNA molecules that could catalyze their own replication. DNA is likely to have been a late addition: as the accumulation of protein catalysts allowed more efficient and complex cells to evolve, the DNA double helix replaced RNA as a more stable molecule for storing the increased amounts of genetic information required by such cells.

WHAT WE DON'T KNOW

• How did the present relationships between nucleic acids and proteins evolve? How did the genetic code originate?

• The information carried in genomes specifies the sequences of all proteins and RNA molecules in the cell, and it determines when and where these molecules are synthesized. Do genomes carry other types of information that we have not yet discovered?

• Cells go to great length to correct mistakes in the processes of DNA replication, transcription, splicing, and translation. Are there analogous strategies to correct mistakes in the selection of which genes are to be expressed in a given cell type? Could the great complexity of transcription initiation in animals and plants reflect such a strategy?

PROBLEMS

Which statements are true? Explain why or why not.

6–1   The consequences of errors in transcription are less severe than those of errors in DNA replication.

6–2   Since introns are largely genetic "junk," they do not have to be removed precisely from the primary transcript during RNA splicing.

6–3   Wobble pairing occurs between the first position in the codon and the third position in the anticodon.

6–4   During protein synthesis, the thermodynamics of base-pairing between tRNAs and mRNAs sets the upper limit for the accuracy with which protein molecules are made.

6–5   Protein enzymes are thought to greatly outnumber ribozymes in modern cells because they can catalyze a much greater variety of reactions and all of them have faster rates than any ribozyme.

Discuss the following problems.

6–6   In which direction along the template must the RNA polymerase in **Figure Q6–1** be moving to have generated the supercoiled structures that are shown? Would you expect supercoils to be generated if the RNA polymerase were free to rotate about the axis of the DNA as it progressed along the template?

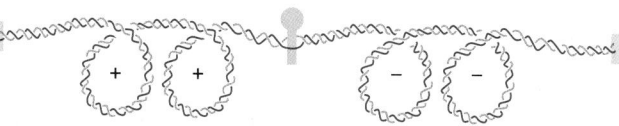

Figure Q6–1 Supercoils around a moving RNA polymerase (Problem 6–6).

6–7   You have attached an RNA polymerase molecule to a glass slide and have allowed it to initiate transcription on a template DNA that is tethered to a magnetic bead as shown in **Figure Q6–2**. If the DNA with its attached magnetic bead moves relative to the RNA polymerase as indicated in the figure, in which direction will the bead rotate?

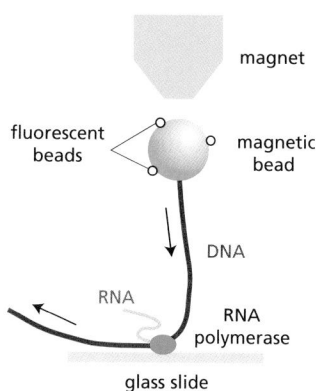

Figure Q6–2 System for measuring the rotation of DNA caused by RNA polymerase (Problem 6–7). The magnet holds the bead upright (but doesn't interfere with its rotation), and the attached tiny fluorescent beads allow the direction of motion to be visualized under the microscope. RNA polymerase is held in place by attachment to the glass slide.

(A) HUMAN α-TROPOMYOSIN GENE

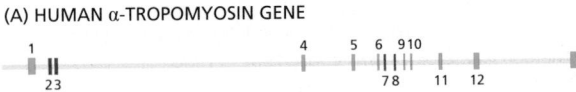

(B) FOUR DIFFERENT SPLICE VARIANTS

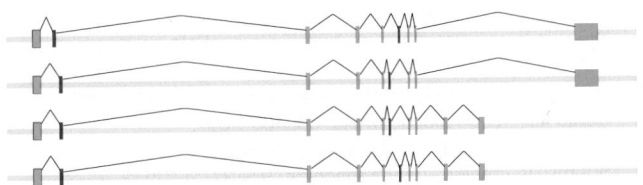

Figure Q6–3 Alternatively spliced mRNAs from the human α-tropomyosin gene (Problem 6–8). (A) Exons in the human α-tropomyosin gene. The locations and relative sizes of exons are shown by the *blue* and *red rectangles*, with alternative exons in *red*. (B) Splicing patterns for four α-tropomyosin mRNAs. Splicing is indicated by *lines* connecting the exons that are included in the mRNA.

6–8 The human α-tropomyosin gene is alternatively spliced to produce different forms of α-tropomyosin mRNA in different cell types (**Figure Q6–3**). For all forms of the mRNA, the protein sequences encoded by exon 1 are the same, as are the protein sequences encoded by exon 10. Exons 2 and 3 are alternative exons used in different mRNAs, as are exons 7 and 8. Which of the following statements about exons 2 and 3 is the most accurate? Is that statement also the most accurate one for exons 7 and 8? Explain your answers.
A.   Exons 2 and 3 must have the same number of nucleotides.
B.   Exons 2 and 3 must each contain an integral number of codons (that is, the number of nucleotides divided by 3 must be an integer).
C.   Exons 2 and 3 must each contain a number of nucleotides that when divided by 3 leaves the same remainder (that is, 0, 1, or 2).

6–9 After treating cells with a chemical mutagen, you isolate two mutants. One carries alanine and the other carries methionine at a site in the protein that normally contains valine (**Figure Q6–4**). After treating these two mutants again with the mutagen, you isolate mutants from each that now carry threonine at the site of the original valine (Figure Q6–4). Assuming that all mutations involve single-nucleotide changes, deduce the codons that are used for valine, methionine, threonine, and alanine at the affected site. Would you expect to be able to isolate valine-to-threonine mutants in one step?

6–10 Which of the following mutational changes would you predict to be the most deleterious to gene function? Explain your answers.

1.   Insertion of a single nucleotide near the end of the coding sequence.
2.   Removal of a single nucleotide near the beginning of the coding sequence.
3.   Deletion of three consecutive nucleotides in the middle of the coding sequence.
4.   Substitution of one nucleotide for another in the middle of the coding sequence.

6–11 Prokaryotes and eukaryotes both protect against the dangers of translating broken mRNAs. What dangers do partial mRNAs pose for the cell?

6–12 Both hsp60-like and hsp70 molecular chaperones share an affinity for exposed hydrophobic patches on proteins, using them as indicators of incomplete folding. Why do you suppose hydrophobic patches serve as critical signals for the folding status of a protein?

6–13 Most proteins require molecular chaperones to assist in their correct folding. How do you suppose the chaperones themselves manage to fold correctly?

6–14 What is so special about RNA that it is hypothesized to be an evolutionary precursor to DNA and protein? What is it about DNA that makes it a better material than RNA for storage of genetic information?

6–15 If an RNA molecule could form a hairpin with a symmetric internal loop, as shown in **Figure Q6–5**, could the complement of this RNA form a similar structure? If so, would there be any regions of the two structures that are identical? Which ones?

```
            C–U
5'–G–C–A        C–C–G
   ‖  ‖  ‖       ‖  ‖  ‖ ∖U
3'–C–G–U        G–G–C
            A–C
```

Figure Q6–5 An RNA hairpin with a symmetric internal loop (Problem 6–15).

6–16 Imagine a warm pond on the primordial Earth. Chance processes have just assembled a single copy of an RNA molecule with a catalytic site that can carry out RNA replication. This RNA molecule folds into a structure that is capable of linking nucleotides according to instructions in an RNA template. Given an adequate supply of nucleotides, will this single RNA molecule be able to use itself as a template to catalyze its own replication? Why or why not?

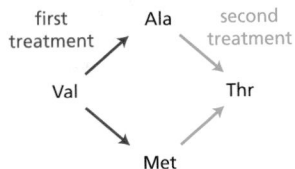

Figure Q6–4 Two rounds of mutagenesis and the altered amino acids at a single position in a protein (Problem 6–9).

REFERENCES

General

Atkins JF, Gesteland RF & Cech TR (eds) (2011) The RNA Worlds: From Life's Origins to Diversity in Gene Regulation. Cold Spring Harbor, NY: Cold Spring Harbor Laboratory Press.

Berg JM, Tymoczko JL & Stryer L (2012) Biochemistry, 7th ed. New York: WH Freeman.

Brown TA (2007) Genomes 3. New York: Garland Science.

Darnell J (2011) RNA: Life's Indispensable Molecule. Cold Spring Harbor, NY: Cold Spring Harbor Laboratory Press.

Hartwell L, Hood L, Goldberg ML et al. (2011) Genetics: from Genes to Genomes, 4th ed. Boston: McGraw Hill.

Judson HF (1996) The Eighth Day of Creation, 25th anniversary ed. Cold Spring Harbor, NY: Cold Spring Harbor Laboratory Press.

Lodish H, Berk A, Kaiser C et al. (2012) Molecular Cell Biology, 7th ed. New York: WH Freeman.

Stent GS (1971) Molecular Genetics: An Introductory Narrative. San Francisco: WH Freeman.

The Genetic Code (1966) Cold Spring Harb. Symp. Quant. Biol. 31.

The Ribosome (2001) Cold Spring Harb. Symp. Quant. Biol. 66.

Watson JD, Baker TA, Bell SP et al. (2013) Molecular Biology of the Gene, 7th ed. Menlo Park, CA: Benjamin Cummings.

From DNA to RNA

Berget SM, Moore C & Sharp PA (1977) Spliced segments at the 5′ terminus of adenovirus 2 late mRNA. Proc. Natl. Acad. Sci. USA 74, 3171–3175.

Brenner S, Jacob F & Meselson M (1961) An unstable intermediate carrying information from genes to ribosomes for protein synthesis. Nature 190, 576–581.

Chow LT, Gelinas RE, Broker TR et al. (1977) An amazing sequence arrangement at the 5′ ends of adenovirus 2 messenger RNA. Cell 12, 1–8.

Conaway CC & Conaway JW (2011) Function and regulation of the Mediator complex. Curr. Opin. Genet. Dev. 21, 225–230.

Cooper TA, Wan L & Dreyfuss G (2009) RNA and disease. Cell 136, 777–793.

Cramer P, Armache KJ, Baumli S et al. (2008) Structure of eukaryotic RNA polymerases. Annu. Rev. Biophys. 37, 337–352.

Fica SM, Tuttle N, Novak T et al. (2013) RNA catalyses nuclear pre-mRNA splicing. Nature 503, 229–234.

Grunberg S & Hahn S (2013) Structural insights into transcription initiation by RNA polymerases II. Trends Biochem. Sci. 38, 603–611.

Grunwald D, Singer RH & Rout M (2011) Nuclear export dynamics of TNA-protein complexes. Nature 475, 333–341.

Kornblihtt AR, Schor IE, Allo M et al. (2013) Alternative splicing: a pivotal step between eukaryotic transcription and translation. Nature 14, 153–165.

Liu X, Bushnell DA & Kornberg RD (2012) RNA polymerase II transcription: Structure and mechanism. Biochim. Biophys. Acta 1829, 2–8.

Makino DL, Halibach F & Conti E (2013) The RNA exosome and proteasome: common principles of degradation control. Nature 14, 654–660.

Malik S & Roeder RC (2010) The metazoan mediator co-activator complex as an integrative hub for transcriptional regulation. Nat. Rev. Genet. 11, 761–772.

Mao YS, Zhang B & Spector DL (2011) Biogenesis and function of nuclear bodies. Trends Genet. 27, 295–306.

Matera AG & Wang Z (2014) A day in the life of the spliceosome. Nature 15, 108–121.

Matsui T, Segall J, Weil PA & Roeder RG (1980) Multiple factors required for accurate initiation of transcription by purified RNA polymerase II. J. Biol. Chem. 255, 11992–11996.

Opalka N, Brown J, Lane WJ et al. (2010) Complete structural model of Escherichia coli RNA polymerase from a hybrid approach. PLoS Biol. 9, 1–16.

Ruskin B, Krainer AR, Maniatis T et al. (1984) Excision of an intact intron as a novel lariat structure during pre-mRNA splicing in vitro. Cell 38, 317–331.

Schneider C & Tollervey D (2013) Threading the barrel of the RNA exosome. Trends Biochem. Sci. 38, 485–493.

Semlow DR & Staley JP (2012) Staying on message: ensuring fidelity in pre-mRNA splicing. Trends Biochem. Sci. 37, 263–273.

From RNA to Protein

Anfinsen CB (1973) Principles that govern the folding of protein chains. Science 181, 223–230.

Crick FHC (1966) The genetic code: III. Sci. Am. 215, 55–62.

Forster F, Unverdorben P, Sledz P et al. (2013) Unveiling the long-held secrets of the 26S proteasome. Structure 21, 1551–1562.

Hershko A, Ciechanover A & Varshavsky A (2000) The ubiquitin system. Nat. Med. 6, 1073–1081.

Horwich AL, Fenton WA, Chapman E et al. (2007) Two families of chaperonin: physiology and mechanism. Annu. Rev. Cell Dev. Biol. 23, 115–145.

Ling J, Reynolds N & Ibba M (2009) Aminoacyl-tRNA synthesis and translational quality control. Annu Rev. Microbiol. 63, 61–78.

Moore PB (2012) How should we think about the ribosome? Annu. Rev. Biophys. 41, 1–19.

Noller HF (2005) RNA structure: reading the ribosome. Science 309, 1508–1514.

Popp MW & Maquat LE (2013) Organizing principles of mammalian nonsense-mediated mRNA decay. Annu. Rev. Genet. 47, 139–165.

Saibil H (2013) Chaperone machines for protein folding, unfolding and disaggregation. Nature 14, 630–642.

Schmidt M & Finley D (2013) Regulation of proteasome activity in health and disease. Biochim. Biophys. Acta 1843, 13–25.

Steitz TA (2008) A structural understanding of the dynamic ribosome machine. Nature 9, 242–253.

Varshavsky A (2012) The ubiquitin system, an immense realm. Annu. Rev. Biochem. 81, 167–176.

Voorhees RM & Ramakrishnan V (2013) Structural basis of the translational elongation cycle. Annu. Rev. Biochem. 82, 203–236.

Wilson DN (2014) Ribosome-targeting antibiotics and mechanisms of bacterial resistance. Nat. Rev. Microbiol. 12, 35–48.

Zaher HS & Green R (2009) Fidelity at the molecular level: Lessons from protein synthesis. Cell 136, 746–762.

The RNA World and the Origins of Life

Blain JC & Szostak JW (2014) Progress Towards Synthetic Cells. Annu. Rev. Biochem. 83, 615–640.

Cech TR (2009) Crawling out of the RNA world. Cell 136, 599–602.

Kruger K, Grabowski P, Zaug P et al. (1982) Self-splicing RNA: Autoexcision and autocyclization of the ribosomal RNA intervening seuence of Tetrahymena. Cell 31, 147–157.

Orgel L (2000) Origin of life. A simpler nucleic acid. Science 290, 1306–1307.

Robertson MP& Joyce GF (2012) The origins of the RNA world. Cold Spring Harb. Perspect. Biol. 4, a003608.

Control of Gene Expression

An organism's DNA encodes all of the RNA and protein molecules required to construct its cells. Yet a complete description of the DNA sequence of an organism—be it the few million nucleotides of a bacterium or the few billion nucleotides of a human—no more enables us to reconstruct the organism than a list of English words enables us to reconstruct a play by Shakespeare. In both cases, the problem is to know how the elements in the DNA sequence or the words on the list are used. Under what conditions is each gene product made, and, once made, what does it do?

In this chapter, we focus on the first half of this problem—the rules and mechanisms that enable a subset of genes to be selectively expressed in each cell. These mechanisms operate at many levels, and we shall discuss each level in turn. But first we present some of the basic principles involved.

AN OVERVIEW OF GENE CONTROL

The different cell types in a multicellular organism differ dramatically in both structure and function. If we compare a mammalian neuron with a liver cell, for example, the differences are so extreme that it is difficult to imagine that the two cells contain the same genome (Figure 7–1). For this reason, and because cell differentiation often seemed irreversible, biologists originally suspected that genes might be selectively lost when a cell differentiates. We now know, however, that cell differentiation generally occurs without changes in the nucleotide sequence of a cell's genome.

The Different Cell Types of a Multicellular Organism Contain the Same DNA

The cell types in a multicellular organism become different from one another because they synthesize and accumulate different sets of RNA and protein molecules. The initial evidence that they do this without altering the sequence of their DNA came from a classic set of experiments in frogs. When the nucleus of a fully differentiated frog cell is injected into a frog egg whose nucleus has been removed, the injected donor nucleus is capable of directing the recipient egg to

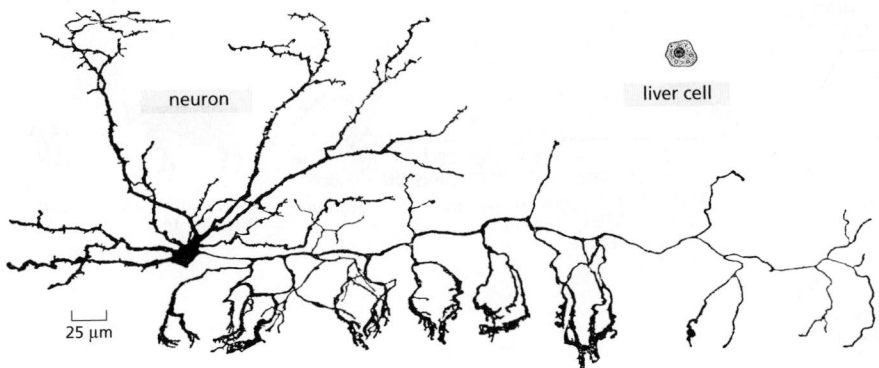

Figure 7–1 **A neuron and a liver cell share the same genome.** The long branches of this neuron from the retina enable it to receive electrical signals from many other neurons and convey them to neighboring neurons. The liver cell, which is drawn to the same scale, is involved in many metabolic processes, including digestion and the detoxification of alcohol and other drugs. Both of these mammalian cells contain the same genome, but they express different sets of RNAs and proteins. (Neuron adapted from S. Ramón y Cajal, Histologie du Systeme Nerveux de l'Homme et de Vertebres, 1909–1911. Paris: Maloine; reprinted, Madrid: C.S.I.C, 1972.)

neuron

liver cell

25 μm

produce a normal tadpole (**Figure 7–2**A). The tadpole contains a full range of differentiated cells that derived their DNA sequences from the nucleus of the original donor cell. Thus, the differentiated donor cell cannot have lost any important DNA sequences. A similar conclusion came from experiments performed with plants. When differentiated pieces of plant tissue are placed in culture and then dissociated into single cells, often one of these individual cells can regenerate an entire adult plant (Figure 7–2B). And the same principle has been more recently demonstrated in mammals that include sheep, cattle, pigs, goats, dogs, and mice (Figure 7–2C).

Most recently, detailed DNA sequencing has confirmed the conclusion that the changes in gene expression that underlie the development of multicellular organisms do not generally involve changes in the DNA sequence of the genome.

Different Cell Types Synthesize Different Sets of RNAs and Proteins

As a first step in understanding cell differentiation, we would like to know how many differences there are between any one cell type and another. Although we

Figure 7–2 **Differentiated cells contain all the genetic instructions necessary to direct the formation of a complete organism.** (A) The nucleus of a skin cell from an adult frog transplanted into an enucleated egg can give rise to an entire tadpole. The *broken arrow* indicates that, to give the transplanted genome time to adjust to an embryonic environment, a further transfer step is required in which one of the nuclei is taken from an early embryo that begins to develop and is put back into a second enucleated egg. (B) In many types of plants, differentiated cells retain the ability to "de-differentiate," so that a single cell can form a clone of progeny cells that later give rise to an entire plant. (C) A nucleus removed from a differentiated cell from an adult cow and introduced into an enucleated egg from a different cow can give rise to a calf. Different calves produced from the same differentiated cell donor are all clones of the donor and are therefore genetically identical. (A, modified from J.B. Gurdon, *Sci. Am.* 219:24–35, 1968.)

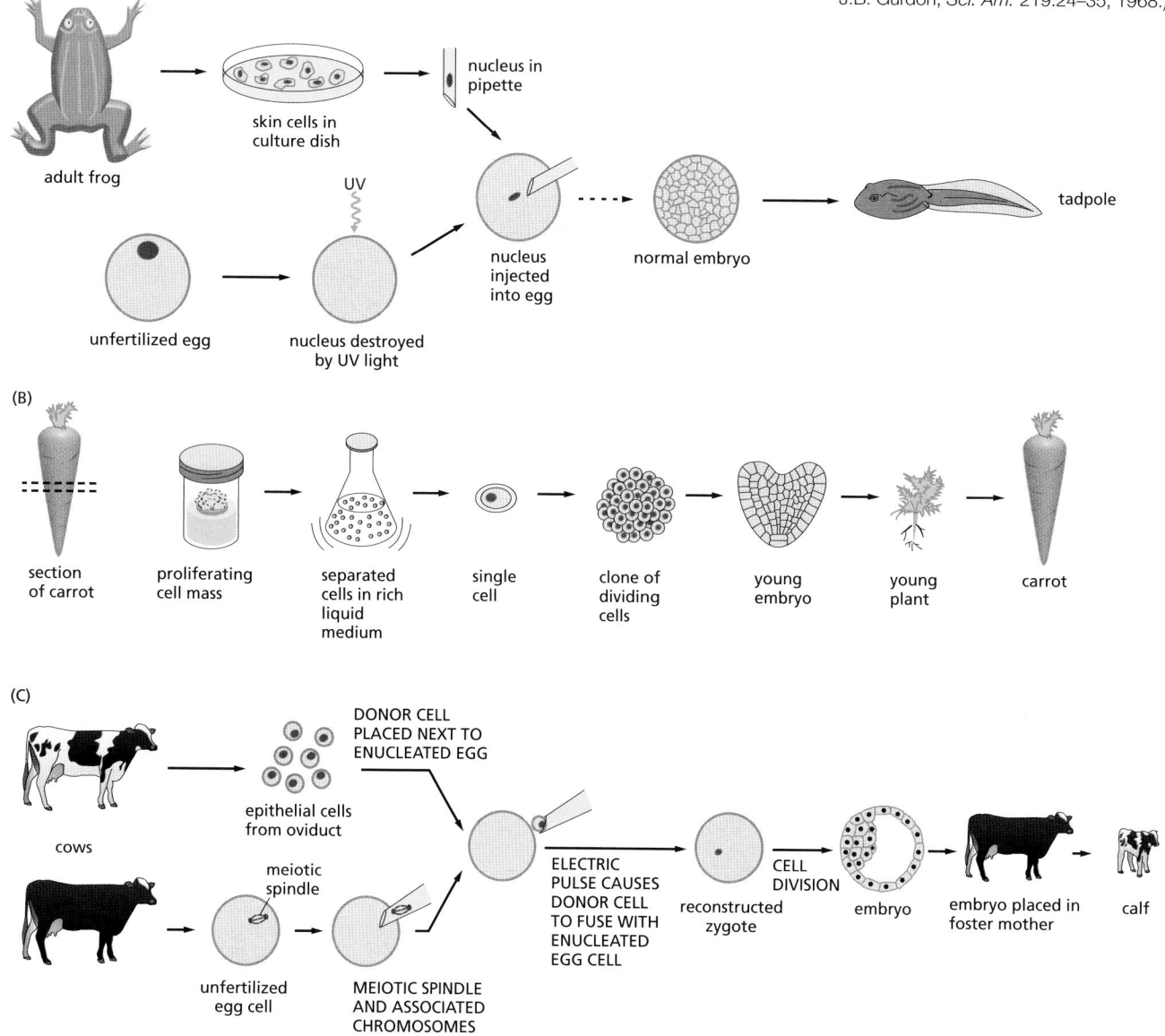

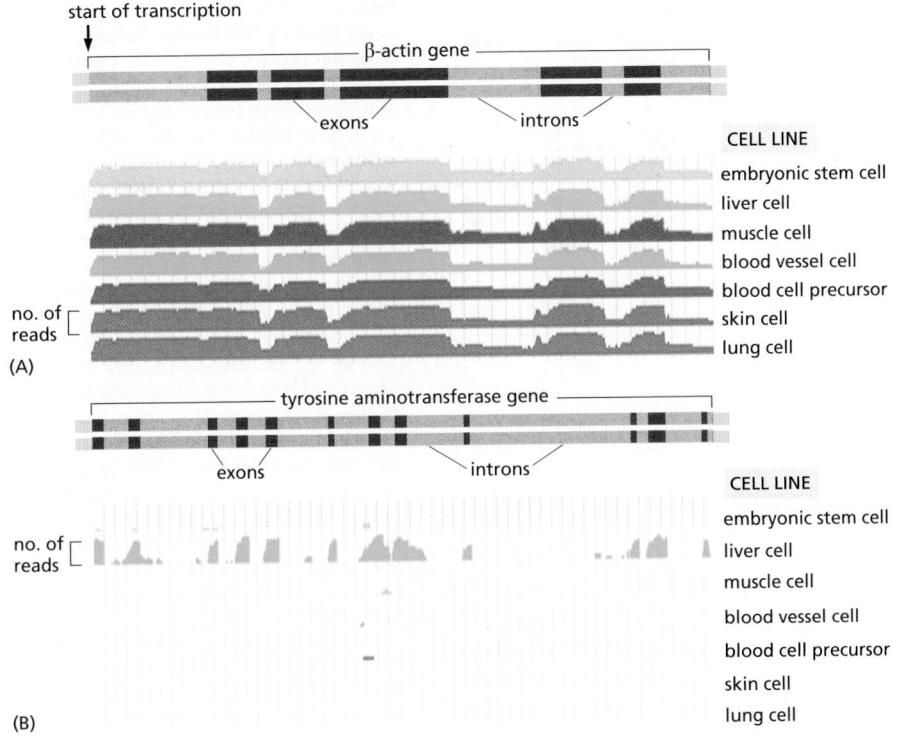

(A)

(B)

Figure 7–3 Differences in RNA levels for two human genes in seven different tissues. To obtain the RNA data by the technique known as *RNA-seq* (see p. 447), RNA was collected from human cell lines grown in culture, derived from each of the seven indicated tissues. Millions of "sequence reads" were obtained and mapped across the human genome by matching RNA sequences to the DNA sequence of the genome. At each position along the genome, the height of the colored trace is proportional to the number of sequence reads that match the genome sequence at that point. As seen in the figure, the exon sequences in transcribed genes are present at high levels, reflecting their presence in mature mRNAs. Intron sequences are present at much lower levels and reflect pre-mRNA molecules that have not yet been spliced plus intron sequences that have been spliced out but not yet degraded. (A) The gene coding for "all-purpose" actin, a major component of the cytoskeleton. Note that the left-hand end of the mature β-actin mRNA is not translated into protein. As explained later in this chapter, many mRNAs have 5′ untranslated regions that regulate their translation into protein. (B) The same type of data displayed for the enzyme tyrosine aminotransferase, which is highly expressed in liver cells but not in the other cell types tested. (Information for both panels from the University of California Santa Cruz, Genome Browser (http://genome.ucsc.edu), which provides this type of information for every human gene. See also S. Djebali et al., *Nature* 489:101–108, 2012.)

still do not have an exact answer to this fundamental question, we can make several general statements.

1. Many processes are common to all cells, and any two cells in a single organism therefore have many gene products in common. These include the structural proteins of chromosomes, RNA and DNA polymerases, DNA repair enzymes, ribosomal proteins and RNAs, the enzymes that catalyze the central reactions of metabolism, and many of the proteins that form the cytoskeleton such as actin (**Figure 7–3A**).

2. Some RNAs and proteins are abundant in the specialized cells in which they function and cannot be detected elsewhere, even by sensitive tests. Hemoglobin, for example, is expressed specifically in red blood cells, where it carries oxygen, and the enzyme tyrosine aminotransferase (which breaks down tyrosine in food) is expressed in liver but not in most other tissues (Figure 7–3B).

3. Studies of the number of different RNAs suggest that, at any one time, a typical human cell expresses 30–60% of its approximately 30,000 genes at some level. There are about 21,000 protein-coding genes and a roughly estimated 9000 noncoding RNA genes in humans. When the patterns of RNA expression in different human cell lines are compared, the level of expression of almost every gene is found to vary from one cell type to another. A few of these differences are striking, like those of hemoglobin and tyrosine aminotransferase noted above, but most are much more subtle. But even those genes that are expressed in all cell types usually vary in their *level* of expression from one cell type to the next.

4. Although there are striking differences in coding RNAs (mRNAs) in specialized cell types, they underestimate the full range of differences in the final pattern of protein production. As we discuss in this chapter, there are many steps after RNA production at which gene expression can be regulated. And, as we saw in Chapter 3, proteins are often covalently modified after they are synthesized. The radical differences in gene expression between cell types are therefore most fully revealed through methods that directly display the levels of proteins along with their post-translational modifications (**Figure 7–4**).

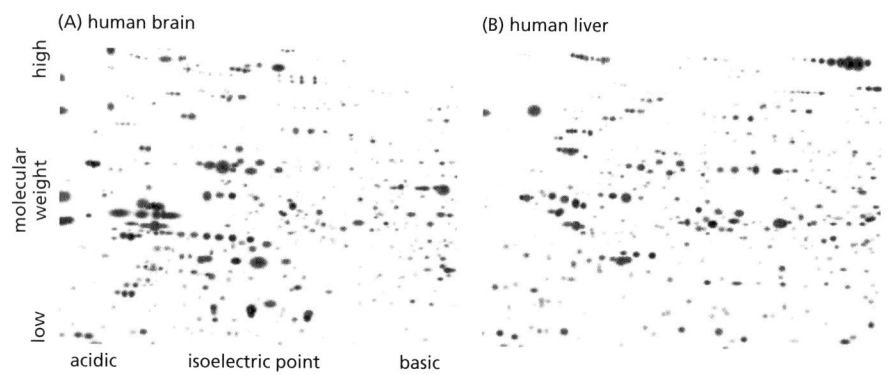

Figure 7–4 **Differences in the proteins expressed by two human tissues, (A) brain and (B) liver.** In each panel, the proteins are displayed using two-dimensional polyacrylamide-gel electrophoresis (see pp. 452–454). The proteins have been separated by molecular weight (*top* to *bottom*) and isoelectric point, the pH at which the protein has no net charge (*right* to *left*). The protein spots artificially colored *red* are common to both samples; those in *blue* are specific to that tissue. The differences between the two tissue samples vastly outweigh their similarities: even for proteins that are shared between the two tissues, their relative abundances are usually different. Note that this technique separates proteins by both size and charge; therefore a protein that has several different phosphorylation states will appear as a series of *horizontal spots* (see *upper right-hand* portion of *right* panel). Only a small portion of the complete protein spectrum is shown for each sample.

Methods based on mass spectrometry (see pp. 455–457) provide much more detailed information, including the identity of each protein, the position of each modification, and the nature of the modification. (Courtesy of Tim Myers and Leigh Anderson, Large Scale Biology Corporation.)

External Signals Can Cause a Cell to Change the Expression of Its Genes

Although the specialized cells in a multicellular organism have characteristic patterns of gene expression, each cell is capable of altering its pattern of gene expression in response to extracellular cues. If a liver cell is exposed to a glucocorticoid hormone, for example, the production of a set of proteins is dramatically increased. Released in the body during periods of starvation or intense exercise, glucocorticoids signal the liver to increase the production of energy from amino acids and other small molecules; the set of proteins whose production is induced includes the enzyme tyrosine aminotransferase, mentioned above. When the hormone is no longer present, the production of these proteins drops to its normal, unstimulated level in liver cells.

Other cell types respond to glucocorticoids differently. Fat cells, for example, reduce the production of tyrosine aminotransferase, while some other cell types do not respond to glucocorticoids at all. These examples illustrate a general feature of cell specialization: different cell types often respond very differently to the same extracellular signal. Other features of the gene expression pattern do not change and give each cell type its permanently distinctive character.

Gene Expression Can Be Regulated at Many of the Steps in the Pathway from DNA to RNA to Protein

If differences among the various cell types of an organism depend on the particular genes that the cells express, at what level is the control of gene expression exercised? As we saw in the previous chapter, there are many steps in the pathway leading from DNA to protein. We now know that all of them can in principle be regulated. Thus a cell can control the proteins it makes by (1) controlling when and how often a given gene is transcribed (**transcriptional control**), (2) controlling the splicing and processing of RNA transcripts (**RNA processing control**), (3) selecting which completed mRNAs are exported from the nucleus to the cytosol and determining where in the cytosol they are localized (**RNA transport and localization control**), (4) selecting which mRNAs in the cytoplasm are translated by ribosomes (**translational control**), (5) selectively destabilizing certain mRNA molecules in the cytoplasm (**mRNA degradation control**), or (6) selectively activating, inactivating, degrading, or localizing specific protein molecules after they have been made (**protein activity control**) (**Figure 7–5**).

For most genes, transcriptional controls are paramount. This makes sense because, of all the possible control points illustrated in Figure 7–5, only transcriptional control ensures that the cell will not synthesize superfluous intermediates. In the following sections, we discuss the DNA and protein components that perform this function by regulating the initiation of gene transcription. We shall then return to the additional ways of regulating gene expression.

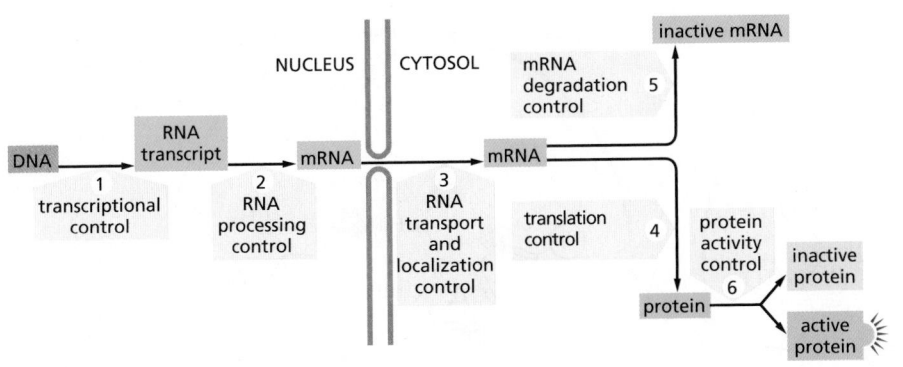

Figure 7–5 Six steps at which eukaryotic gene expression can be controlled. Controls that operate at steps 1 through 5 are discussed in this chapter. Step 6, the regulation of protein activity, occurs largely through covalent post-translational modifications including phosphorylation, acetylation, and ubiquitylation (see Table 3–3, p. 165). Step 6 was introduced in Chapter 3 and is subsequently discussed in many chapters throughout the book.

Summary

The genome of a cell contains in its DNA sequence the information to make many thousands of different protein and RNA molecules. A cell typically expresses only a fraction of its genes, and the different types of cells in multicellular organisms arise because different sets of genes are expressed. Moreover, cells can change the pattern of genes they express in response to changes in their environment, such as signals from other cells. Although all of the steps involved in expressing a gene can in principle be regulated, for most genes the initiation of RNA transcription provides the most important point of control.

CONTROL OF TRANSCRIPTION BY SEQUENCE-SPECIFIC DNA-BINDING PROTEINS

How does a cell determine which of its thousands of genes to transcribe? Perhaps the most important concept, one that applies to all species on Earth, is based on a group of proteins known as **transcription regulators**. These proteins recognize specific sequences of DNA (typically 5–10 nucleotide pairs in length) that are often called *cis-regulatory sequences*, because they must be on the same chromosome (that is, *in cis*) to the genes they control. Transcription regulators bind to these sequences, which are dispersed throughout genomes, and this binding puts into motion a series of reactions that ultimately specify which genes are to be transcribed and at what rate. Approximately 10% of the protein-coding genes of most organisms are devoted to transcription regulators, making them one of the largest classes of proteins in the cell. In most cases, a given transcription regulator recognizes its own *cis*-regulatory sequence, which is different from those recognized by all the other regulators in the cell.

Transcription of each gene is, in turn, controlled by its own collection of *cis*-regulatory sequences. These typically lie near the gene, often in the intergenic region directly upstream from the transcription start point of the gene. Although a few genes are controlled by a single *cis*-regulatory sequence that is recognized by a single transcription regulator, the majority have complex arrangements of *cis*-regulatory sequences, each of which is recognized by a different transcription regulator. It is therefore the positions, identity, and arrangement of *cis*-regulatory sequences—which are an important part of the information embedded in the genome—that ultimately determine the time and place that each gene is transcribed.

We begin our discussion by describing how transcription regulators recognize *cis*-regulatory sequences.

The Sequence of Nucleotides in the DNA Double Helix Can Be Read by Proteins

As discussed in Chapter 4, the DNA in a chromosome consists of a very long double helix that has both a major and a minor groove (**Figure 7–6**). Transcription regulators must recognize short, specific *cis*-regulatory sequences within

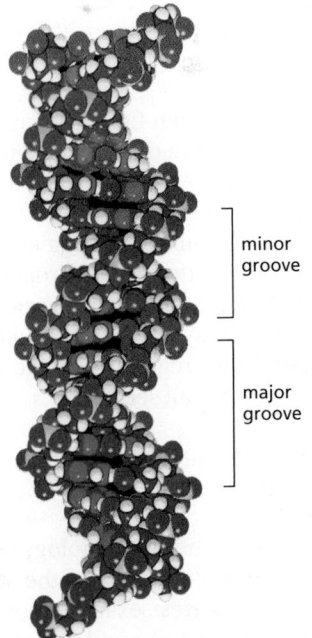

Figure 7–6 Double-helical structure of DNA. A space-filling model of DNA showing the major and minor grooves on the outside of the double helix (see Movie 4.1). The atoms are colored as follows: carbon, *dark blue*; nitrogen, *light blue*; hydrogen, *white*; oxygen, *red*; phosphorus, *yellow*.

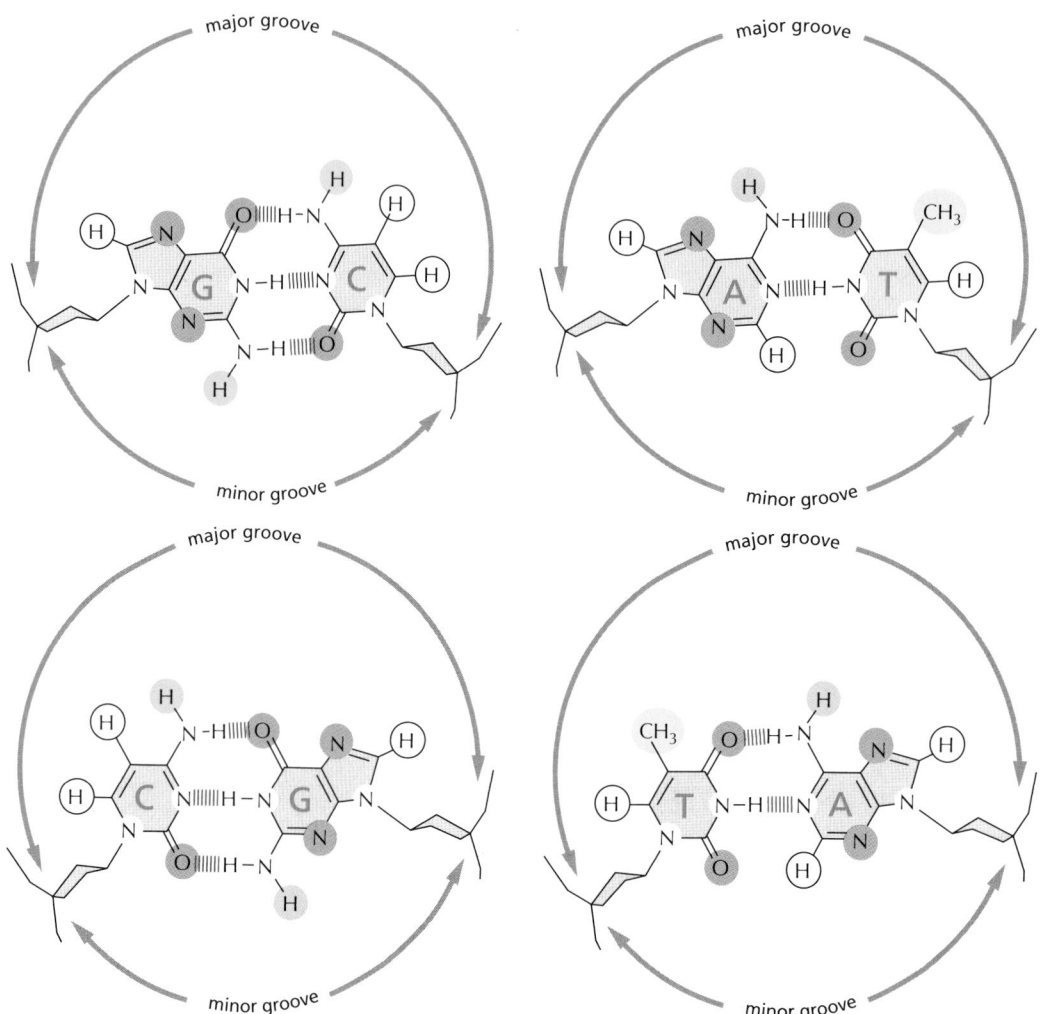

this structure. When first discovered in the 1960s, it was thought that these proteins might require direct access to the interior of the double helix to distinguish between one DNA sequence and another. It is now clear, however, that the outside of the double helix is studded with DNA sequence information that transcription regulators recognize: the edge of each base pair presents a distinctive pattern of hydrogen-bond donors, hydrogen-bond acceptors, and hydrophobic patches in both the major and minor grooves (**Figure 7–7**). Because the major groove is wider and displays more molecular features than does the minor groove, nearly all transcription regulators make the majority of their contacts with the major groove—as we shall see.

Transcription Regulators Contain Structural Motifs That Can Read DNA Sequences

Molecular recognition in biology generally relies on an exact fit between the surfaces of two molecules, and the study of transcription regulators has provided some of the clearest examples of this principle. A transcription regulator recognizes a specific *cis*-regulatory sequence because the surface of the protein is extensively complementary to the special surface features of the double helix that displays that sequence. Each transcription regulator makes a series of contacts with the DNA, involving hydrogen bonds, ionic bonds, and hydrophobic interactions. Although each individual contact is weak, the 20 or so contacts that are typically formed at the protein–DNA interface add together to ensure that the interaction is both highly specific and very strong (**Figure 7–8**). In fact, DNA–protein

Figure 7–7 How the different base pairs in DNA can be recognized from their edges without the need to open the double helix. The four possible configurations of base pairs are shown, with potential hydrogen-bond donors indicated in *blue*, potential hydrogen-bond acceptors in *red*, and hydrogen bonds of the base pairs themselves as a series of short, parallel *red* lines. Methyl groups, which form hydrophobic protuberances, are shown in *yellow*, and hydrogen atoms that are attached to carbons, and are therefore unavailable for hydrogen-bonding, are *white*. From the major groove, each of the four base-pair configurations projects a unique pattern of features. (From C. Branden and J. Tooze, Introduction to Protein Structure, 2nd ed. New York: Garland Publishing, 1999. With permission from Taylor & Francis Group LLC Books.)

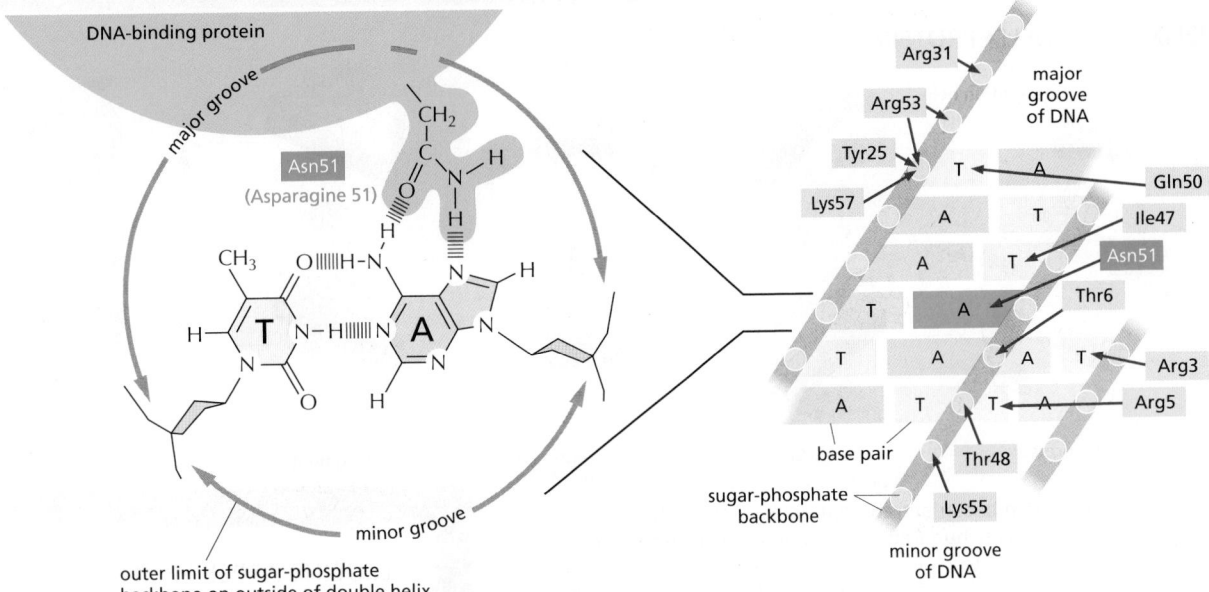

Figure 7–8 **The binding of a transcription regulator to a specific DNA sequence.** On the *left*, a single contact is shown between a transcription regulator and DNA; such contacts allow the protein to "read" the DNA sequence. On the *right*, the complete set of contacts between a transcription regulator (a member of the homeodomain family—see Panel 7–1) and its *cis*-regulatory sequence is shown. The DNA-binding portion of the protein is 60 amino acids long. Although the interactions in the major groove are the most important, the protein is also seen to contact both the minor groove and phosphates in the sugar–phosphate DNA backbone. (See C. Wolberger et al., *Cell* 67:517–528, 1991.)

interactions include some of the tightest and most specific molecular interactions known in biology.

Although each example of protein–DNA recognition is unique in detail, x-ray crystallographic and nuclear magnetic resonance (NMR) spectroscopic studies of hundreds of transcription regulators have revealed that many of them contain one or another of a small set of DNA-binding structural motifs (**Panel 7–1**). These motifs generally use either α helices or β sheets to bind to the major groove of DNA. The amino acid side chains that extend from these protein motifs make the specific contacts with the DNA. Thus, a given structural motif can be used to recognize many different *cis*-regulatory sequences depending on the specific side chains present.

Dimerization of Transcription Regulators Increases Their Affinity and Specificity for DNA

A monomer of a typical transcription regulator recognizes about 6–8 nucleotide pairs of DNA. However, sequence-specific DNA-binding proteins do not bind tightly to a single DNA sequence and reject all others; rather, they recognize a range of closely related sequences, with the affinity of the protein for the DNA varying according to how closely the DNA matches the optimal sequence. Hence, *cis*-regulatory sequences are often depicted as "logos" which display the range of sequences recognized by a particular transcription regulator (**Figure 7–9**A and B). In Chapter 6, we saw this same representation at work for the binding of RNA polymerase to promoters (see Figure 6–12).

The DNA sequence recognized by a monomer does not contain sufficient information to be picked out from the background of such sequences that would occur at random all over the genome. For example, an exact six-nucleotide DNA sequence would be expected to occur by chance approximately once every 4096 nucleotides (4^6), and the range of six-nucleotide sequences described by a typical logo would be expected to occur by chance much more often, perhaps every 1000 nucleotides. Clearly, for a bacterial genome of 4.6×10^6 nucleotide pairs, not to mention a mammalian genome of 3×10^9 nucleotide pairs, this is insufficient information to accurately control the transcription of individual genes. Additional contributions to DNA-binding specificity must therefore be present. Many transcription regulators form dimers, with both monomers making nearly identical contacts with DNA (Figure 7–9C). This arrangement doubles the length of the *cis*-regulatory sequence recognized and greatly increases both the affinity and the specificity of transcription regulator binding. Because the DNA sequence

HELIX–TURN–HELIX PROTEINS

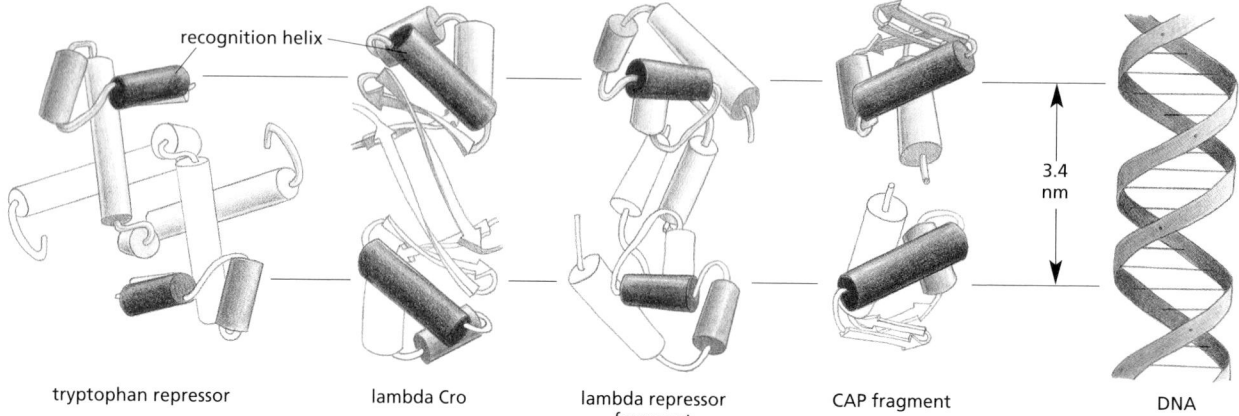

recognition helix

3.4 nm

tryptophan repressor   lambda Cro   lambda repressor fragment   CAP fragment   DNA

Originally identified in bacterial transcription regulators, this motif has since been found in many hundreds of DNA-binding proteins from both eukaryotes and prokaryotes. It is constructed from two α helices (*blue* and *red*) connected by a short extended chain of amino acids, which constitutes the "turn." The two helices are held at a fixed angle, primarily through interactions between the two helices. The more C-terminal helix (in *red*) is called the *recognition helix* because it fits into the major groove of DNA; its amino acid side chains, which differ from protein to protein, play an important part in recognizing the specific DNA sequence to which the protein binds. All of the proteins shown here bind DNA as dimers in which the two copies of the recognition helix (in *red*) are separated by exactly one turn of the DNA helix (3.4 nm); thus both recognition helices of the dimer can fit into the major groove of DNA.

LEUCINE ZIPPER PROTEINS

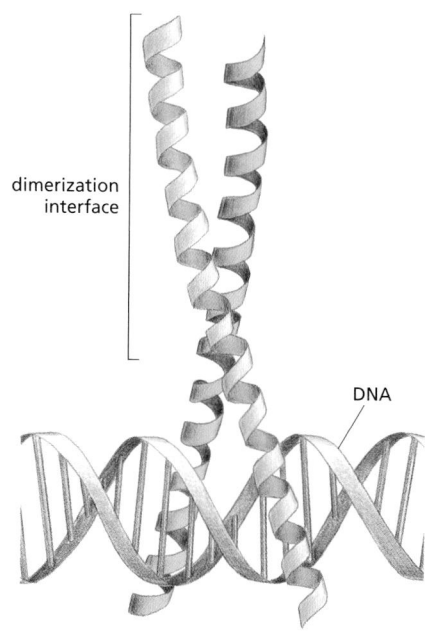

dimerization interface

DNA

The *leucine zipper* motif is named because of the way the two α helices, one from each monomer, are joined together to form a short coiled-coil. These proteins bind DNA as dimers where the two long α helices are held together by interactions between hydrophobic amino acid side chains (often on leucines) that extend from one side of each helix. Just beyond the dimerization interface, the two α helices separate from each other to form a Y-shaped structure, which allows their side chains to contact the major groove of DNA. The dimer thus grips the double helix like a clothespin on a clothesline (Movie 7.2).

HOMEODOMAIN PROTEINS

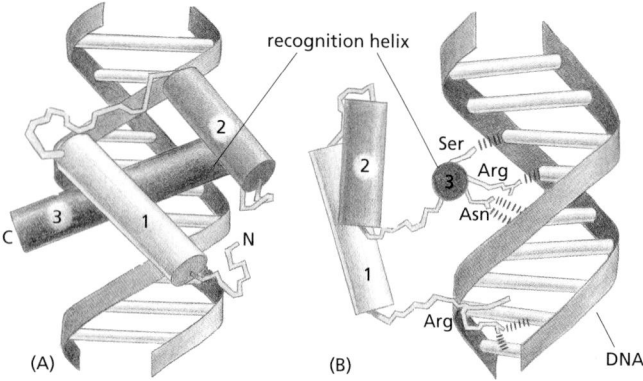

recognition helix

Ser
Arg
Asn
Arg

(A)   (B)   DNA

Not long after the first transcription regulators were discovered in bacteria, genetic analyses of the fruit fly *Drosophila* led to the characterization of an important class of genes, the *homeotic selector genes*, that play a critical part in orchestrating fly development (discussed in Chapter 21). It was later shown that these genes coded for transcription regulators that bound DNA through a structural motif named the homeodomain. Two different views of the same structure are shown. (A) The homeodomain is folded into three α helices, which are packed tightly together by hydrophobic interactions. The part containing helices 2 and 3 closely resembles the helix–turn–helix motif. (B) The recognition helix (helix 3, *red*) forms important contacts with the major groove of DNA. The asparagine (Asn) of helix 3, for example, contacts an adenine, as shown in Figure 7–8. A flexible arm attached to helix 1 forms contacts with nucleotide pairs in the minor groove (Movie 7.1).

β SHEET DNA RECOGNITION PROTEINS

In the other DNA-binding motifs displayed in this panel, α helices are the primary mechanism used to recognize specific DNA sequences. In one large group of transcription regulators, however, a two-stranded β sheet, with amino acid side chains extending from the sheet toward the DNA, reads the information on the surface of the major groove. As in the case of a recognition α helix, this β-sheet motif can be used to recognize many different DNA sequences; the exact DNA sequence recognized depends on the sequence of amino acids that make up the β sheet. Shown is a transcription regulator that binds two molecules of *S*-adenosyl methionine (*red*). On the left is a dimer of the protein; on the right is a simplified diagram showing just the two-stranded β sheet bound to the major groove of DNA.

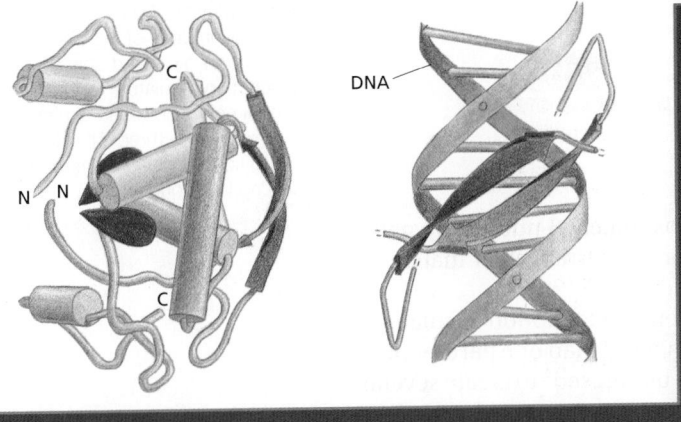

ZINC FINGER PROTEINS

This group of DNA-binding motifs includes one or more zinc atoms as structural components. All such zinc-coordinated DNA-binding motifs are called zinc fingers, referring to their appearance in early schematic drawings (*left*). They fall into several distinct structural groups, only one of which we consider here. It has a simple structure, in which the zinc atom holds an α helix and a β sheet together (*middle*). This type of zinc finger is often found in clusters with the α helix of each finger contacting the major groove of the DNA, forming a nearly continuous stretch of α helices along that groove (Movie 7.3). In this way, a strong and specific DNA–protein interaction is built up through a repeating basic structural unit. Three such fingers are shown on the *right*.

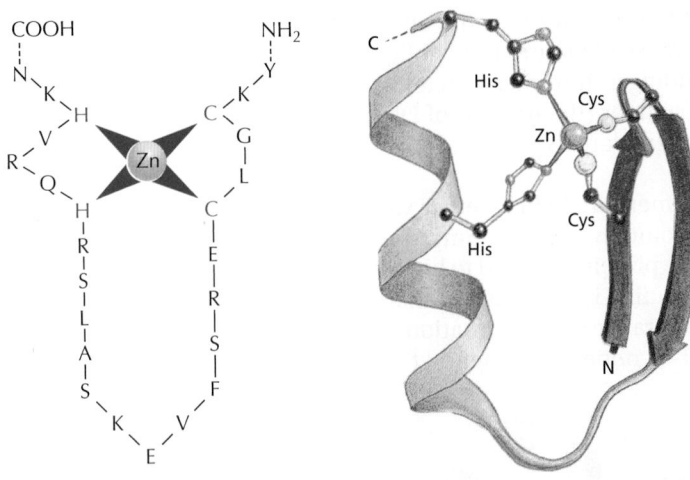

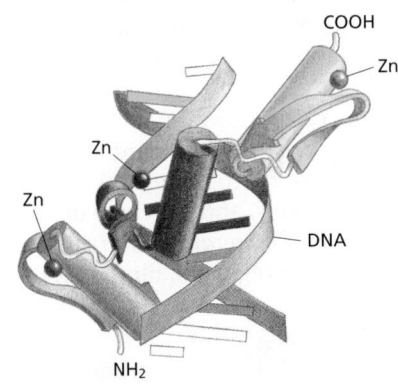

HELIX–LOOP–HELIX PROTEINS

Related to the leucine zipper, the helix–loop–helix motif consists of a short α helix connected by a loop (*red*) to a second, longer α helix. The flexibility of the loop allows one helix to fold back and park against the other thereby forming the dimerization surface. As shown, this two-helix structure binds both to DNA and to the two-helix structure of a second protein to create either a homodimer or a heterodimer. Two α helices that extend from the dimerization interface make specific contacts with the major groove of DNA.

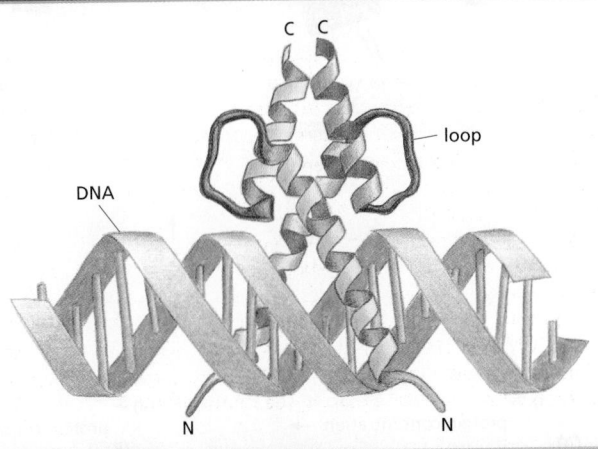

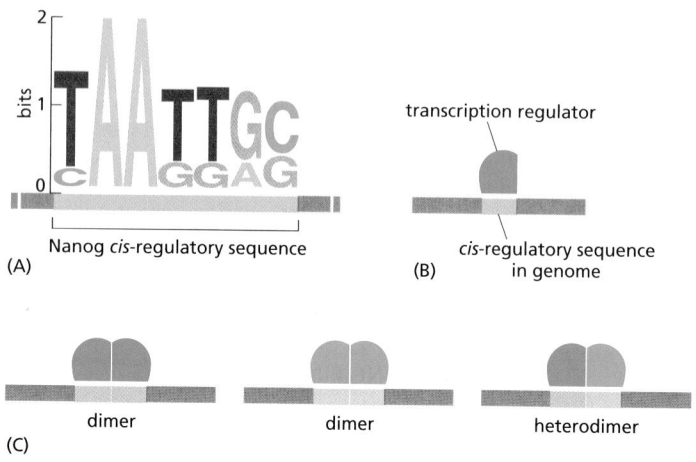

(A) Nanog *cis*-regulatory sequence

(B) *cis*-regulatory sequence in genome

transcription regulator

(C)   dimer   dimer   heterodimer

Figure 7–9 **Transcription regulators and *cis*-regulatory sequences.** (A) Depiction of the *cis*-regulatory sequence for Nanog, a homeodomain family member that is a key regulator in embryonic stem cells. This "logo" form (see Figure 6–12) shows that the protein can recognize a collection of closely related DNA sequences and gives the preferred nucleotide pair at each position. *Cis*-regulatory sequences are "read" as double-stranded DNA, but only one strand typically is shown in a logo. (B) Representation of the *cis*-regulatory sequence as a colored box. (C) Many transcription regulators form dimers (homodimers) and heterodimers. In the example shown, three different DNA-binding specificities are formed from two transcription regulators.

recognized by the protein has increased from approximately 6 nucleotide pairs to 12 nucleotide pairs, there are many fewer random occurrences of matching sequences.

Heterodimers are often formed from two different transcription regulators. Transcription regulators may form heterodimers with more than one partner protein; in this way, the same transcription regulator can be "reused" to create several distinct DNA-binding specificities (see Figure 7–9C).

Transcription Regulators Bind Cooperatively to DNA

In the simplest case, the collection of noncovalent bonds that holds the above dimers or heterodimers together is so extensive that these structures form obligatorily, and never fall apart. In this case, the unit of binding is the dimer or heterodimer, and the binding curve for the transcription regulator (the fraction of DNA bound as a function of protein concentration) has a standard exponential shape (**Figure 7–10A**).

In many cases, however, the dimers and heterodimers are held together very weakly; they exist predominantly as monomers in solution, and yet dimers are observed on the appropriate DNA sequence. Here, the proteins are said to bind to DNA cooperatively, and the curve describing their binding is sigmoidal in shape (Figure 7–10B). *Cooperative binding* means that, over a range of concentrations of the transcription regulator, binding is more of an all-or-none phenomenon than

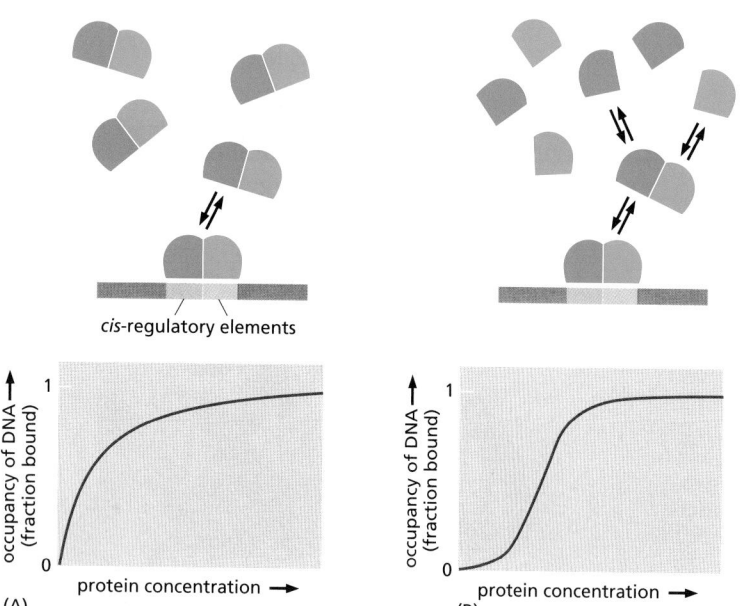

cis-regulatory elements

(A) occupancy of DNA (fraction bound) vs protein concentration →

(B) occupancy of DNA (fraction bound) vs protein concentration →

Figure 7–10 **Occupancy of a *cis*-regulatory sequence by a transcription regulator.** (A) Noncooperative binding by a stable heterodimer. (B) Cooperative binding by components of a heterodimer that are predominantly monomers in solution. The shape of the curve differs from that of (A) because the fraction of protein in a form competent to bind DNA (the heterodimer) increases with increasing protein concentration.

for noncooperative binding; that is, at most protein concentrations, the *cis*-regulatory sequence is either nearly empty or nearly fully occupied and rarely is somewhere in between. A discussion of the mathematics behind cooperative binding is given in Chapter 8 (see Figure 8–79A).

Nucleosome Structure Promotes Cooperative Binding of Transcription Regulators

As we have just seen, cooperative binding of transcription regulators to DNA often occurs because the monomers have only a weak affinity for each other. However, there is a second, indirect mechanism for cooperative binding, one that arises from the nucleosome structure of eukaryotic chromosomes.

In general, transcription regulators bind to DNA in nucleosomes with lower affinity than they do to naked DNA. There are two reasons for this difference. First, the surface of the *cis*-regulatory sequence recognized by the transcription regulator may be facing inward on the nucleosome, toward the histone core, and therefore not be readily available to the regulatory protein. Second, even if the face of the *cis*-regulatory sequence is exposed on the outside of the nucleosome, many transcription regulators subtly alter the conformation of the DNA when they bind, and these changes are generally opposed by the tight wrapping of the DNA around the histone core. For example, many transcription regulators induce a bend or kink in the DNA when they bind.

We saw in Chapter 4 that nucleosome remodeling can alter the structure of the nucleosome, allowing transcription regulators access to the DNA. Even without remodeling, however, transcription regulators can still gain limited access to DNA in a nucleosome. The DNA at the end of a nucleosome "breathes," transiently exposing the DNA and allowing regulators to bind. This breathing happens at a much lower rate in the middle of the nucleosome; therefore, the positions where the DNA exits the nucleosome are much easier to occupy (**Figure 7–11**).

These properties of the nucleosome promote cooperative DNA binding by transcription regulators. If a regulatory protein enters the DNA of a nucleosome and prevents the DNA from tightly rewrapping around the nucleosome core, it will increase the affinity of a second transcription regulator for a nearby *cis*-regulatory sequence. If the two transcription regulators also interact with each other (as described above), the cooperative effect is even greater. In some cases, the combined action of the regulatory proteins can eventually displace the histone core of the nucleosome altogether.

Figure 7–11 **How nucleosomes effect the binding of transcription regulators.**

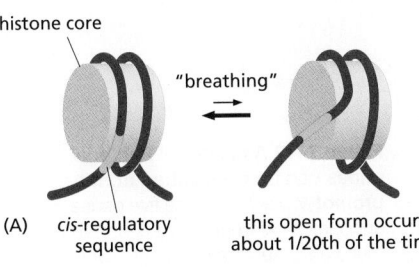

(A) *cis*-regulatory sequence   this open form occurs about 1/20th of the time

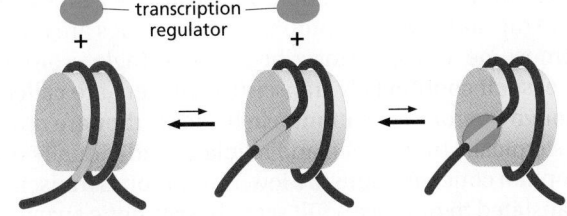

(B) compared to its affinity for naked DNA, a typical transcription regulator will bind with 20 times lower affinity if its *cis*-regulatory sequence is located near the end of a nucleosome

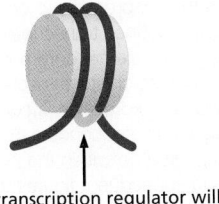

(C) a typical transcription regulator will bind with roughly 200-fold less affinity if its *cis*-regulatory sequence is located in the middle of a nucleosome

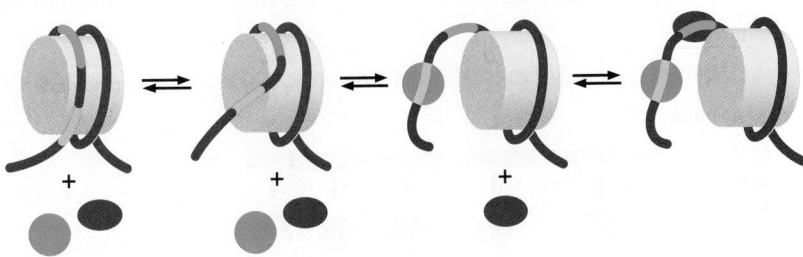

(D) one transcription regulator can destabilize the nucleosome, facilitating binding of another

The cooperation among transcription regulators can become much greater when nucleosome remodeling complexes are involved. If one transcription regulator binds its *cis*-regulatory sequence and attracts a chromatin remodeling complex, the localized action of the remodeling complex can allow a second transcription regulator to efficiently bind nearby. Moreover, we have discussed how transcription regulators can work together in pairs; in reality, larger numbers often cooperate by repeated use of the same principles. A highly cooperative binding of transcription regulators to DNA probably explains why many sites in eukaryotic genomes that are bound by transcription regulators are "nucleosome free."

Summary

Transcription regulators recognize short stretches of double-helical DNA of defined sequence called cis-*regulatory sequences, and thereby determine which of the thousands of genes in a cell will be transcribed. Approximately 10% of the protein-coding genes in most organisms produce transcription regulators, and they control many features of cells. Although each of these transcription regulators has unique features, most bind to DNA as homodimers or heterodimers and recognize DNA through one of a small number of structural motifs. Transcription regulators typically work in groups and bind DNA cooperatively, a feature that has several underlying mechanisms, some of which exploit the packaging of DNA in nucleosomes.*

TRANSCRIPTION REGULATORS SWITCH GENES ON AND OFF

Having seen how transcription regulators bind to *cis*-regulatory sequences embedded in the genome, we can now discuss how, once bound, these proteins influence the transcription of genes. The situation in bacteria is simpler than in eukaryotes (for one thing, chromatin structure is not an issue), and we therefore discuss it first. Following this, we turn to the more complex situation in eukaryotes.

The Tryptophan Repressor Switches Genes Off

The genome of the bacterium *E. coli* consists of a single, circular DNA molecule of about 4.6×10^6 nucleotide pairs. This DNA encodes approximately 4300 proteins, although only a fraction of these are made at any one time. Bacteria regulate the expression of many of their genes according to the food sources that are available in the environment. For example, in *E. coli*, five genes code for enzymes that manufacture the amino acid tryptophan. These genes are arranged in a cluster on the chromosome and are transcribed from a single promoter as one long mRNA molecule; such coordinately transcribed clusters are called *operons* (**Figure 7–12**). Although operons are common in bacteria, they are rare in eukaryotes, where genes are typically transcribed and regulated individually (see Figure 7–3).

When tryptophan concentrations are low, the operon is transcribed; the resulting mRNA is translated to produce a full set of biosynthetic enzymes, which work in tandem to synthesize tryptophan from much simpler molecules. When tryptophan is abundant, however—for example, when the bacterium is in the gut of a mammal that has just eaten a protein-rich meal—the amino acid is imported into the cell and shuts down production of the enzymes, which are no longer needed.

Figure 7–12 A cluster of bacterial genes can be transcribed from a single promoter. Each of these five genes encodes a different enzyme, and all of these enzymes are needed to synthesize the amino acid tryptophan from simpler molecules. The genes are transcribed as a single mRNA molecule, a feature that allows their expression to be coordinated. Clusters of genes transcribed as a single mRNA molecule are common in bacteria. Each of these clusters is called an *operon* because its expression is controlled by a *cis*-regulatory sequence called the *operator* (*green*), situated within the promoter. (In this and subsequent figures, the *yellow* blocks in the promoter represent DNA sequences that bind RNA polymerase; see Figure 6–12).

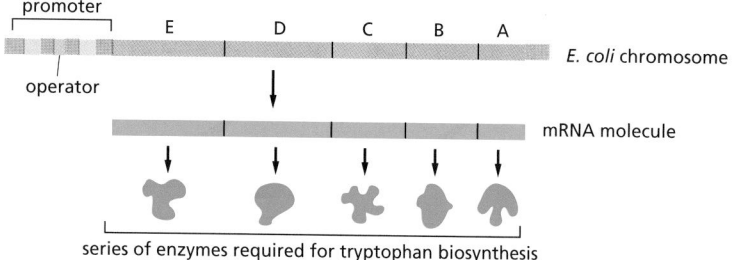

series of enzymes required for tryptophan biosynthesis

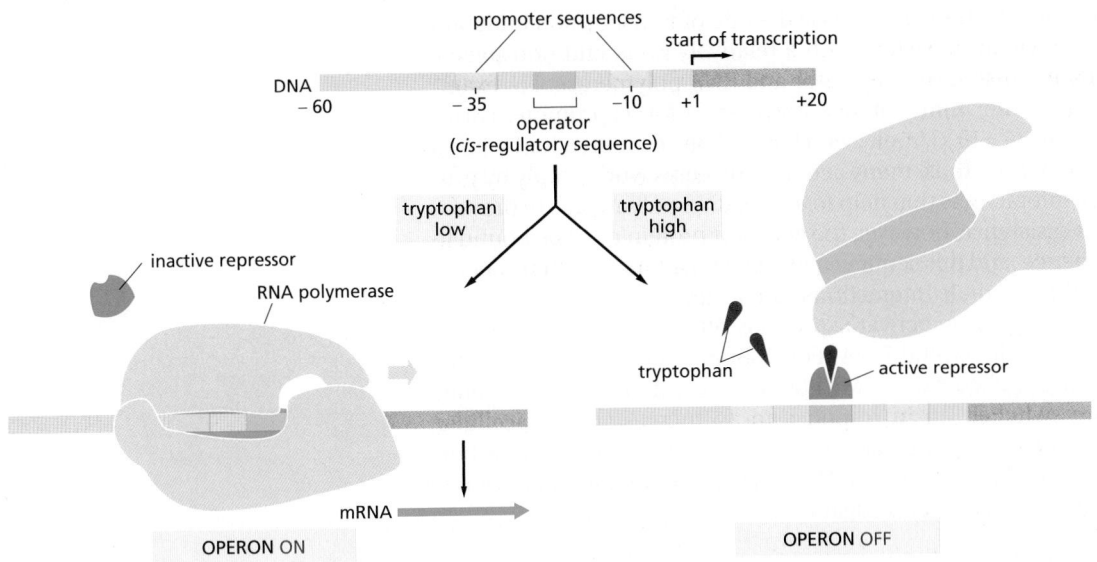

We now understand exactly how this repression of the tryptophan operon comes about. Within the operon's promoter is a *cis*-regulatory sequence that is recognized by a transcription regulator. When this regulator binds to this sequence, it blocks access of RNA polymerase to the promoter, thereby preventing transcription of the operon (and thus production of the tryptophan-producing enzymes). The transcription regulator is known as the *tryptophan repressor* and its *cis*-regulatory sequence is called the *tryptophan operator*. These components are controlled in a simple way: the repressor can bind to DNA only if it has also bound several molecules of tryptophan (**Figure 7–13**).

The tryptophan repressor is an allosteric protein, and the binding of tryptophan causes a subtle change in its three-dimensional structure so that the protein can bind to the operator sequence. Whenever the concentration of free tryptophan in the bacterium drops, tryptophan dissociates from the repressor, the repressor no longer binds to DNA, and the tryptophan operon is transcribed. The repressor is thus a simple device that switches production of a set of biosynthetic enzymes on and off according to the availability of the end product of the pathway that the enzymes catalyze.

The tryptophan repressor protein itself is always present in the cell. The gene that encodes it is continuously transcribed at a low level, so that a small amount of the repressor protein is always being made. Thus the bacterium can respond very rapidly to a rise or fall in tryptophan concentration.

Repressors Turn Genes Off and Activators Turn Them On

The tryptophan repressor, as its name suggests, is a *transcriptional repressor* protein: in its active form, it switches genes off, or *represses* them. Some bacterial transcription regulators do the opposite: they switch genes on, or *activate* them. These *transcriptional activator* proteins work on promoters that—in contrast to the promoter for the tryptophan operon—are only marginally able to bind and position RNA polymerase on their own. However, these poorly functioning promoters can be made fully functional by activator proteins that bind to nearby *cis*-regulatory sequences and contact the RNA polymerase to help it initiate transcription (**Figure 7–14**).

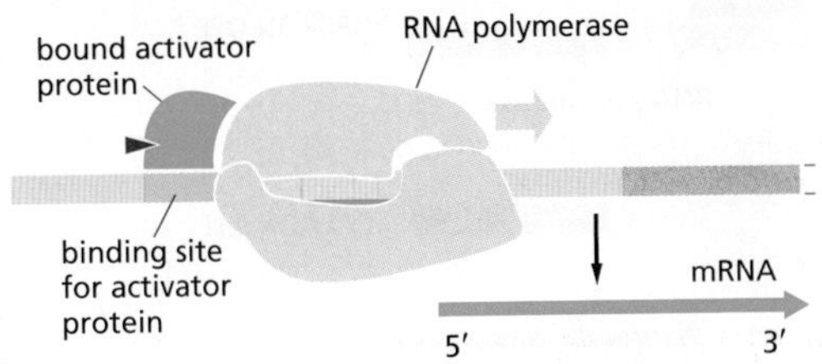

Figure 7–13 Genes can be switched off by repressor proteins. If the concentration of tryptophan inside a bacterium is low *(left)*, RNA polymerase *(blue)* binds to the promoter and transcribes the five genes of the tryptophan operon. However, if the concentration of tryptophan is high *(right)*, the repressor protein *(dark green)* becomes active and binds to the operator *(light green)*, where it blocks the binding of RNA polymerase to the promoter. Whenever the concentration of intracellular tryptophan drops, the repressor falls off the DNA, allowing the polymerase to again transcribe the operon. Although not shown in the figure, the repressor is a stable dimer.

Figure 7–14 Genes can be switched on by activator proteins. An activator protein binds to its *cis*-regulatory sequence on the DNA and interacts with the RNA polymerase to help it initiate transcription. Without the activator, the promoter fails to initiate transcription efficiently. In bacteria, the binding of the activator to DNA is often controlled by the interaction of a metabolite or other small molecule *(red triangle)* with the activator protein. The *Lac* operon works in this manner, as we discuss shortly.

DNA-bound activator proteins can increase the rate of transcription initiation as much as 1000-fold, a value consistent with a relatively weak and nonspecific interaction between the transcription regulator and RNA polymerase. For example, a 1000-fold change in the affinity of RNA polymerase for its promoter corresponds to a change in ΔG of ≈ 18 kJ/mole, which could be accounted for by just a few weak, noncovalent bonds. Thus, many activator proteins work simply by providing a few favorable interactions that help to attract RNA polymerase to the promoter. To provide this assistance, however, the activator protein must be bound to its *cis*-regulatory sequence, and this sequence must be positioned, with respect to the promoter, so that the favorable interactions can occur.

Like the tryptophan repressor, activator proteins often have to interact with a second molecule to be able to bind DNA. For example, the bacterial activator protein *CAP* has to bind cyclic AMP (cAMP) before it can bind to DNA. Genes activated by CAP are switched on in response to an increase in intracellular cAMP concentration, which rises when glucose, the bacterium's preferred carbon source, is no longer available; as a result, CAP drives the production of enzymes that allow the bacterium to digest other sugars.

An Activator and a Repressor Control the *Lac* Operon

In many instances, the activity of a single promoter is controlled by several different transcription regulators. The *Lac* operon in *E. coli*, for example, is controlled by both the *Lac* repressor and the CAP activator that we just discussed. The *Lac* operon encodes proteins required to import and digest the disaccharide lactose. In the absence of glucose, the bacterium makes cAMP, which activates CAP to switch on genes that allow the cell to utilize alternative sources of carbon—including lactose. It would be wasteful, however, for CAP to induce expression of the *Lac* operon if lactose itself were not present. Thus the Lac repressor shuts off the operon in the absence of lactose. This arrangement enables the control region of the *Lac* operon to integrate two different signals, so that the operon is highly expressed only when two conditions are met: glucose must be absent and lactose must be present (**Figure 7–15**). This genetic circuit thus behaves much like

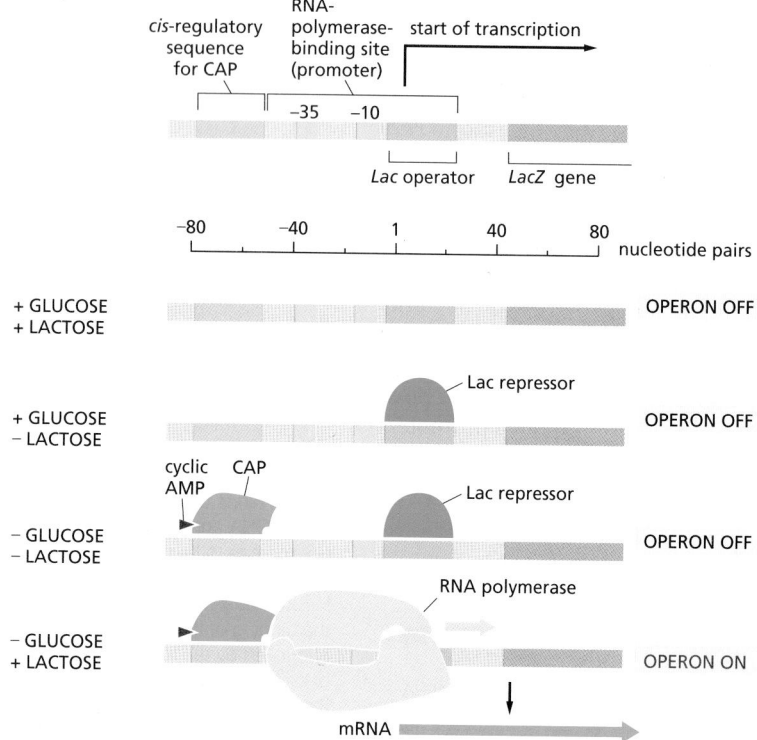

Figure 7–15 **The *Lac* operon is controlled by two transcription regulators, the Lac repressor and CAP.** *LacZ*, the first gene of the operon, encodes the enzyme β-galactosidase, which breaks down lactose to galactose and glucose. When lactose is absent, the Lac repressor binds to a *cis*-regulatory sequence, called the *Lac* operator, and shuts off expression of the operon (Movie 7.4). Addition of lactose increases the intracellular concentration of a related compound, allolactose; allolactose binds to the Lac repressor, causing it to undergo a conformational change that releases its grip on the operator DNA (not shown). When glucose is absent, cyclic AMP *(red triangle)* is produced by the cell, and CAP binds to DNA.

a switch that carries out a logic operation in a computer. When lactose is present AND glucose is absent, the cell executes the appropriate program—in this case, transcription of the genes that permit the uptake and utilization of lactose.

All transcription regulators, whether they are repressors or activators, must be bound to DNA to exert their effects. In this way, each regulatory protein acts selectively, controlling only those genes that bear a *cis*-regulatory sequence recognized by it. The logic of the *Lac* operon first attracted the attention of biologists more than 50 years ago. The way it works was uncovered by a combination of genetics and biochemistry, providing some of the first insights into how transcription is controlled in any organism.

DNA Looping Can Occur During Bacterial Gene Regulation

We have seen that transcription activators help RNA polymerase initiate transcription and repressors hinder it. However, the two types of proteins are very similar to one another. For example, to occupy their *cis*-regulatory sequences, both the tryptophan repressor and the CAP activator protein must bind a small molecule; moreover, they both recognize their *cis*-regulatory sequences using the same structural motif (the helix–turn–helix shown in Panel 7–1). Indeed, some proteins (for example, the CAP protein) can act as both a repressor and an activator, depending on the exact placement of their *cis*-regulatory sequence relative to the promoter: for some genes, the CAP *cis*-regulatory sequence overlaps the promoter, and CAP binding thereby prevents the assembly of RNA polymerase at the promoter.

Most bacteria have small, compact genomes, and the *cis*-regulatory sequences that control the transcription of a gene are typically located very near to the start point of transcription. But there are some exceptions to this generalization— *cis*-regulatory sequences can be located hundreds and even thousands of nucleotide pairs from the bacterial genes they control (**Figure 7–16**). In these cases, the intervening DNA is looped out, allowing a protein bound at a distant site along the DNA to contact RNA polymerase. Here, the DNA acts as a tether, enormously increasing the probability that the proteins will collide, compared with the situation where one protein is bound to DNA and the other is free in solution. We will see shortly that, although it is the exception in bacteria, DNA looping occurs in the regulation of nearly every eukaryotic gene.

A possible explanation for this difference is based on evolutionary considerations. It has been proposed that the compact, simple genetic switches found in bacteria evolved in response to large population sizes where competition for growth put selective pressure on bacteria to maintain small genome sizes. In contrast, there appears to have been little selective pressure to "streamline" the genomes of multicellular organisms.

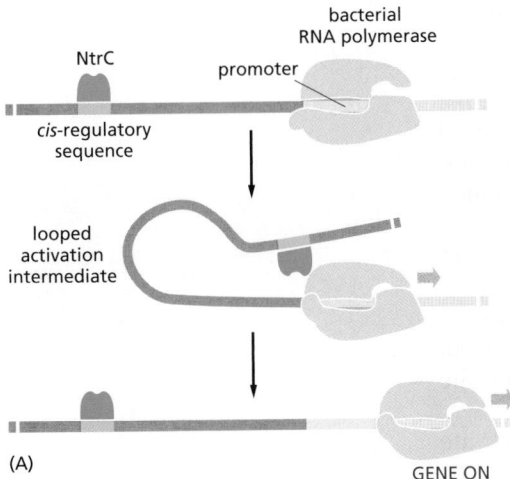

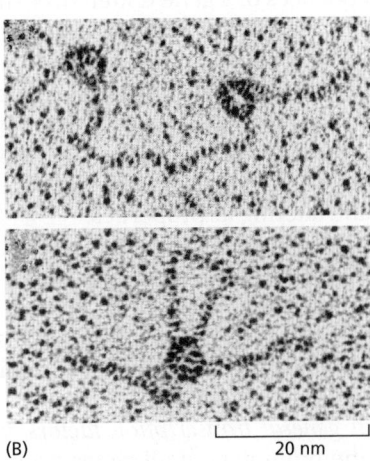

(A)

GENE ON

(B)

20 nm

Figure 7–16 **Transcriptional activation at a distance.** (A) The NtrC protein is a bacterial transcription regulator that activates transcription by directly contacting RNA polymerase. (B) The interaction of NtrC and RNA polymerase, with the intervening DNA looped out, can be seen in the electron microscope. (B, from W. Su et al., *PNAS* 87(14):5504–5508, 1990. Courtesy of Harrison Echols and Sydney Kustu.)

Complex Switches Control Gene Transcription in Eukaryotes

When compared to the situation in bacteria, transcription regulation in eukaryotes involves many more proteins and much longer stretches of DNA. It often seems bewilderingly complex. Yet many of the same principles apply. As in bacteria, the time and place that each gene is to be transcribed is specified by its *cis*-regulatory sequences, which are "read" by the transcription regulators that bind to them. Once bound to DNA, positive transcription regulators (activators) help RNA polymerase begin transcribing genes, and negative regulators (repressors) block this from happening. In bacteria, as we have seen, most of the interactions between DNA-bound transcription regulators and RNA polymerases (whether they activate or repress transcription) are direct. In contrast, these interactions are almost always indirect in eukaryotes: many intermediate proteins, including the histones, act between the DNA-bound transcription regulator and RNA polymerase. Moreover, in multicellular organisms, it is common for dozens of transcription regulators to control a single gene, with *cis*-regulatory sequences spread over tens of thousands of nucleotide pairs. DNA looping allows the DNA-bound regulatory proteins to interact with each other and ultimately with RNA polymerase at the promoter. Finally, because nearly all of the DNA in eukaryotic organisms is compacted by nucleosomes and higher-order structures, transcription initiation in eukaryotes must overcome this inherent block.

In the next sections, we discuss these features of transcription initiation in eukaryotes, emphasizing how they provide extra levels of control not found in bacteria.

A Eukaryotic Gene Control Region Consists of a Promoter Plus Many *cis*-Regulatory Sequences

In eukaryotes, RNA polymerase II transcribes all the protein-coding genes and many noncoding RNA genes, as we saw in Chapter 6. This polymerase requires five general transcription factors (27 subunits *in toto*; see Table 6–3, p. 311), in contrast to bacterial RNA polymerase, which needs only a single general transcription factor (the σ subunit). As we have seen, the stepwise assembly of the general transcription factors at a eukaryotic promoter provides, in principle, multiple steps at which the cell can speed up or slow down the rate of transcription initiation in response to transcription regulators.

Because the many *cis*-regulatory sequences that control the expression of a typical gene are often spread over long stretches of DNA, we use the term **gene control region** to describe the whole expanse of DNA involved in regulating and initiating transcription of a eukaryotic gene. This includes the **promoter**, where the general transcription factors and the polymerase assemble, plus all of the ***cis*-regulatory sequences** to which transcription regulators bind to control the rate of the assembly processes at the promoter (**Figure 7–17**). In animals and plants, it is not unusual to find the regulatory sequences of a gene dotted over stretches of DNA as large as 100,000 nucleotide pairs. Some of this DNA is transcribed (but not translated), and we discuss these long noncoding RNAs (lncRNAs) later in this chapter. For now, we can regard much of this DNA as "spacer" sequences that transcription regulators do not directly recognize. It is important to keep in mind that, like other regions of eukaryotic chromosomes, most of the DNA in gene control regions is packaged into nucleosomes and higher-order forms of chromatin, thereby compacting its overall length and altering its properties.

In this chapter, we shall loosely use the term **gene** to refer to a segment of DNA that is transcribed into a functional RNA molecule, one that either codes for a protein or has a different role in the cell (see Table 6–1, p. 305). However, the classical view of a gene includes the gene control region as well, since mutations in it can produce an altered phenotype. Alternative RNA splicing further complicates the definition of a gene—a point we shall return to later.

In contrast to the small number of *general transcription factors*, which are abundant proteins that assemble on the promoters of all genes transcribed by

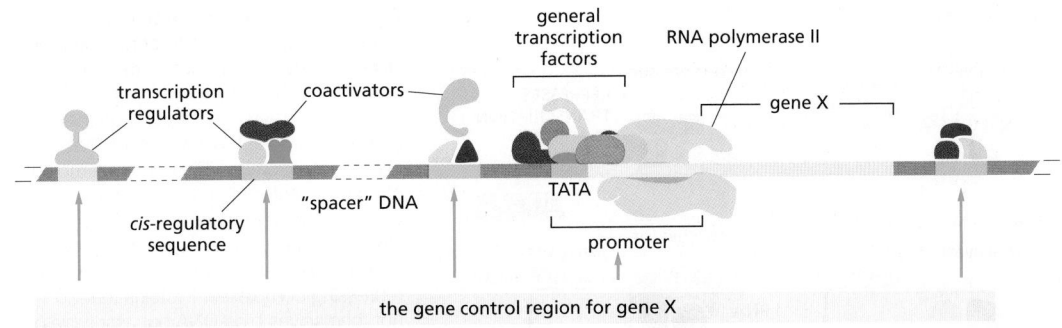

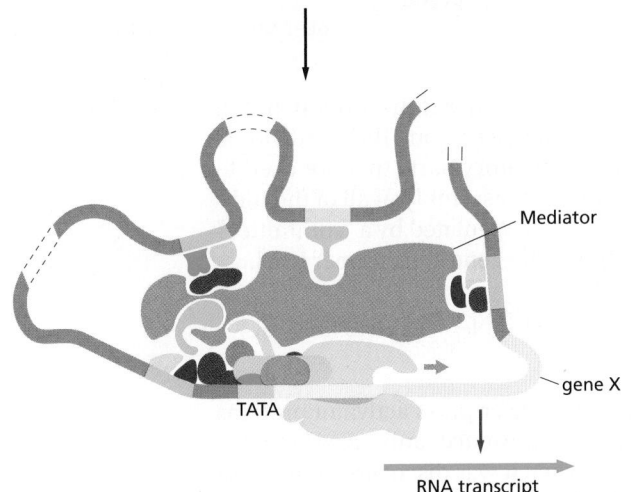

Figure 7–17 The gene control region for a typical eukaryotic gene. The *promoter* is the DNA sequence where the general transcription factors and the polymerase assemble (see Figure 6–15). The *cis-regulatory sequences* are binding sites for transcription regulators, whose presence on the DNA affects the rate of transcription initiation. These sequences can be located adjacent to the promoter, far upstream of it, or even within introns or entirely downstream of the gene. The broken stretches of DNA signify that the length of DNA between the *cis*-regulatory sequences and the start of transcription varies, sometimes reaching tens of thousands of nucleotide pairs in length. The *TATA box* is a DNA recognition sequence for the general transcription factor TFIID. As shown in the lower panel, DNA looping allows transcription regulators bound at any of these positions to interact with the proteins that assemble at the promoter. Many transcription regulators act through Mediator (described in Chapter 6), while some interact with the general transcription factors and RNA polymerase directly. Transcription regulators also act by recruiting proteins that alter the chromatin structure of the promoter (not shown, but discussed below).

Whereas Mediator and the general transcription factors are the same for all RNA polymerase II-transcribed genes, the transcription regulators and the locations of their binding sites relative to the promoter differ for each gene.

RNA polymerase II, there are thousands of different *transcription regulators* devoted to turning individual genes on and off. In eukaryotes, operons—sets of genes transcribed as a unit—are rare, and, instead, each gene is regulated individually. Not surprisingly, the regulation of each gene is different in detail from that of every other gene, and it is difficult to formulate simple rules for gene regulation that apply in every case. We can, however, make some generalizations about how transcription regulators, once bound to gene control regions on DNA, set in motion the series of events that lead to gene activation or repression.

Eukaryotic Transcription Regulators Work in Groups

In bacteria, we saw that proteins such as the tryptophan repressor, the *Lac* repressor, and the CAP protein bind to DNA on their own and directly affect RNA polymerase at the promoter. Eukaryotic transcription regulators, in contrast, usually assemble in groups at their *cis*-regulatory sequences. Often two or more regulators bind cooperatively, as discussed earlier in the chapter. In addition, a broad class of multisubunit proteins termed *coactivators* and *co-repressors* assemble on DNA with them. Typically, these coactivators and co-repressors do not recognize specific DNA sequences themselves; they are brought to those sequences by the transcription regulators. Often the protein–protein interactions between transcription regulators and between regulators and coactivators are too weak for them to assemble in solution; however, the appropriate combination of *cis*-regulatory sequences can "crystallize" the assembly of these complexes on DNA (**Figure 7–18**).

As their names imply, coactivators are typically involved in activating transcription and co-repressors in repressing it. In the following sections, we will see that coactivators and co-repressors can act in a variety of different ways to influence transcription after they have been localized on the genome by transcription regulators.

As shown in Figure 7–18, an individual transcription regulator can often participate in more than one type of regulatory complex. A protein might function,

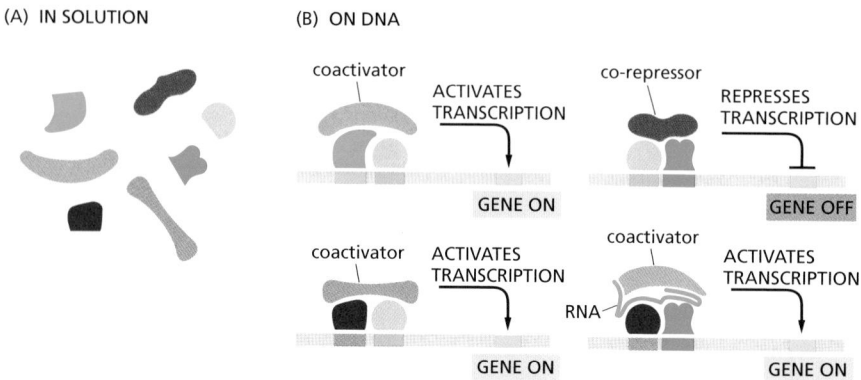

(A) IN SOLUTION

(B) ON DNA

coactivator — ACTIVATES TRANSCRIPTION → GENE ON

co-repressor — REPRESSES TRANSCRIPTION ⊣ GENE OFF

coactivator — ACTIVATES TRANSCRIPTION → GENE ON

coactivator — ACTIVATES TRANSCRIPTION → GENE ON

RNA

Figure 7–18 Eukaryotic transcription regulators assemble into complexes on DNA. (A) Seven transcription regulators are shown. The nature and function of the complex they form depend on the specific *cis*-regulatory sequences that seed their assembly. (B) Some assembled complexes activate gene transcription, while another represses transcription. Note that the *light green* and *dark green* proteins are shared by both activating and repressing complexes. Proteins that do not themselves bind DNA but assemble on other DNA-bound transcription regulators are termed coactivators or co-repressors. In some cases *(lower right)*, RNA molecules are found in these assemblies. As described later in this chapter, these RNAs often act as scaffolds to hold a group of proteins together.

for example, in one case as part of a complex that activates transcription and in another case as part of a complex that represses transcription. Thus, individual eukaryotic transcription regulators function as regulatory parts that are used to build complexes whose function depends on the final assembly of all of the individual components. Each eukaryotic gene is therefore regulated by a "committee" of proteins, all of which must be present to express the gene at its proper level.

Activator Proteins Promote the Assembly of RNA Polymerase at the Start Point of Transcription

The *cis*-regulatory sequences to which eukaryotic transcription activator proteins bind were originally called *enhancers* because their presence "enhanced" the rate of transcription initiation. It came as a surprise when it was discovered that these sequences could be found tens of thousands of nucleotide pairs away from the promoter; as we have seen, DNA looping, which was not widely appreciated at the time, can now explain this initially puzzling observation.

Once bound to DNA, how do assemblies of activator proteins increase the rate of transcription initiation? At most genes, mechanisms work in concert. Their function is both to attract and position RNA polymerase II at the promoter and to release it so that transcription can begin.

Some activator proteins bind directly to one or more of the general transcription factors, accelerating their assembly on a promoter that has been brought in proximity—through DNA looping—to that activator. Most transcription activators, however, attract coactivators that then perform the biochemical tasks needed to initiate transcription. One of the most prevalent coactivators is the large *Mediator* protein complex, composed of more than 30 subunits. About the same size as RNA polymerase itself, Mediator serves as a bridge between DNA-bound transcription activators, RNA polymerase, and the general transcription factors, facilitating their assembly at the promoter (see Figure 7–17).

Eukaryotic Transcription Activators Direct the Modification of Local Chromatin Structure

The eukaryotic general transcription factors and RNA polymerase are unable, on their own, to assemble on a promoter that is packaged in nucleosomes. Thus, in addition to directing the assembly of the transcription machinery at the promoter, eukaryotic transcription activators promote transcription by triggering changes to the chromatin structure of the promoters, making the underlying DNA more accessible.

The most important ways of locally altering chromatin are through covalent histone modifications, nucleosome remodeling, nucleosome removal, and histone replacement (discussed in Chapter 4). Eukaryotic transcription activators use all four of these mechanisms: thus they attract coactivators that include histone modification enzymes, ATP-dependent chromatin remodeling complexes, and histone chaperones, each of which can alter the chromatin structure of

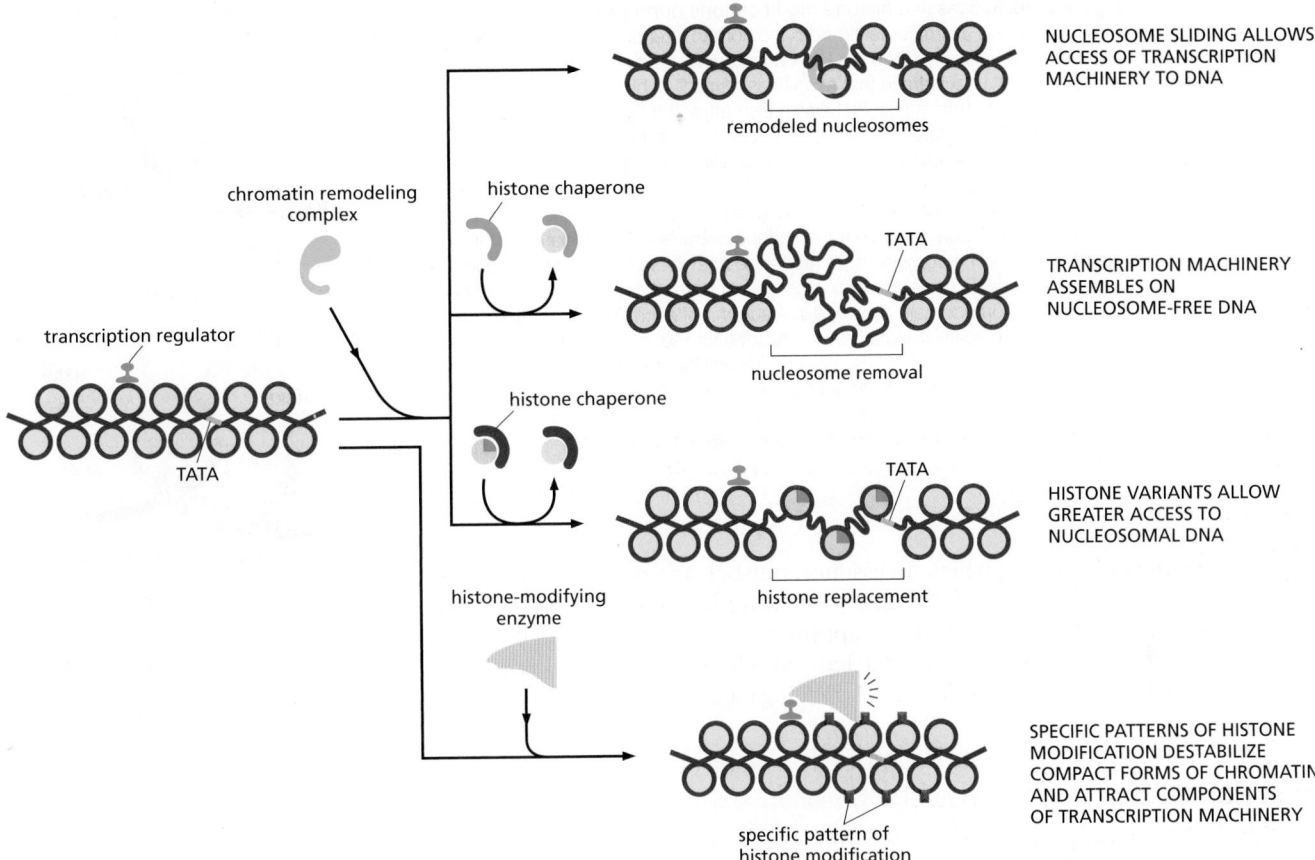

NUCLEOSOME SLIDING ALLOWS ACCESS OF TRANSCRIPTION MACHINERY TO DNA

remodeled nucleosomes

chromatin remodeling complex

histone chaperone

TATA

TRANSCRIPTION MACHINERY ASSEMBLES ON NUCLEOSOME-FREE DNA

nucleosome removal

transcription regulator

TATA

histone chaperone

TATA

HISTONE VARIANTS ALLOW GREATER ACCESS TO NUCLEOSOMAL DNA

histone replacement

histone-modifying enzyme

SPECIFIC PATTERNS OF HISTONE MODIFICATION DESTABILIZE COMPACT FORMS OF CHROMATIN AND ATTRACT COMPONENTS OF TRANSCRIPTION MACHINERY

specific pattern of histone modification

promoters (**Figure 7–19**). These local alterations in chromatin structure provide greater access to DNA, thereby facilitating the assembly of the general transcription factors at the promoter. In addition, some histone modifications specifically attract these proteins to the promoter. These mechanisms often work together during transcription initiation (**Figure 7–20**). Finally, as discussed earlier in this chapter, the local chromatin changes directed by one transcriptional regulator can allow the binding of additional regulators. By repeated use of this principle, large assemblies of proteins can form on control regions of genes to regulate their transcription.

The alterations of chromatin structure that occur during transcription initiation can persist for different lengths of time. In some cases, as soon as the transcription regulator dissociates from DNA, the chromatin modifications are rapidly reversed, restoring the gene to its pre-activated state. This rapid reversal is especially important for genes that the cell must quickly switch on and off in response to external signals. In other cases, the altered chromatin structure persists, even after the transcription regulator that directed its establishment has dissociated from DNA. In principle, this memory can extend into the next cell generation because, as discussed in Chapter 4, chromatin structure can be self-renewing (see Figure 4–44). The fact that different histone modifications persist for different times provides the cell with a mechanism that makes possible both longer- and shorter-term memory of gene expression patterns.

A special type of chromatin modification occurs as RNA polymerase II transcribes through a gene. The histones just ahead of the polymerase can be acetylated by enzymes carried by the polymerase, removed by histone chaperones, and deposited behind the moving polymerase. These histones are then rapidly deacetylated and methylated, also by complexes that are carried by the polymerase, leaving behind nucleosomes that are especially resistant to transcription. This remarkable process seems to prevent spurious transcription reinitiation

Figure 7–19 **Eukaryotic transcription activator proteins direct local alterations in chromatin structure.** Nucleosome remodeling, nucleosome removal, histone replacement, and certain types of histone modifications favor transcription initiation (see Figure 4–39). These alterations increase the accessibility of DNA and facilitate the binding of RNA polymerase and the general transcription factors.

Figure 7–20 **Successive histone modifications during transcription initiation.** In this example, taken from the human interferon gene promoter, a transcription activator binds to DNA packaged into chromatin and attracts a histone acetyl transferase that acetylates lysine 9 of histone H3 and lysine 8 of histone H4. Then a histone kinase, also attracted by the transcription activator, phosphorylates serine 10 of histone H3 but it can only do so after lysine 9 has been acetylated. This serine modification signals the histone acetyl transferase to acetylate position K14 of histone H3. Next, the general transcription factor TFIID and a chromatin remodeling complex bind to the chromatin to promote the subsequent steps of transcription initiation. TFIID and the remodeling complex both recognize acetylated histone tails through a *bromodomain*, a protein domain specialized to read this particular mark on histones; a bromodomain is carried in a subunit of each protein complex.

The histone acetyl transferase, the histone kinase, and the chromatin remodeling complex are all coactivators. The order of events shown applies to a specific promoter; at other genes, the steps may occur in a different order or individual steps may be omitted altogether. (Adapted from T. Agalioti, G. Chen and D. Thanos, *Cell* 111:381–392, 2002.)

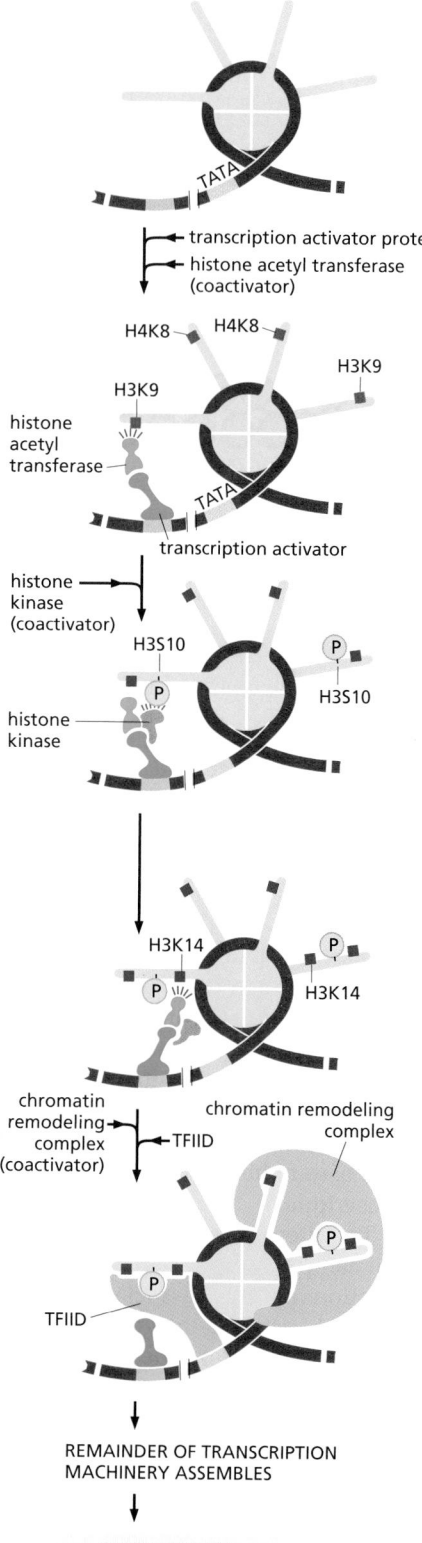

behind a moving polymerase, which, in essence, must clear a path through chromatin as it transcribes. Later in this chapter, when we discuss *RNA interference*, the potential dangers to the cell of such inappropriate transcription will become especially obvious. The modification of nucleosomes behind a moving RNA polymerase also plays an important role in RNA splicing (see p. 323).

Transcription Activators Can Promote Transcription by Releasing RNA Polymerase from Promoters

In some cases, transcription initiation requires that a DNA-bound transcription activator releases RNA polymerase from the promoter so as to allow it to begin transcribing the gene. In other cases, the RNA polymerase halts after transcribing about 50 nucleotides of RNA, and further elongation requires a transcription activator bound behind it (**Figure 7–21**). These paused polymerases are common in humans, where a significant fraction of genes that are not being transcribed have a paused polymerase located just downstream from the promoter.

The release of RNA polymerase can occur in several ways. In some cases, the activator brings in a chromatin remodeling complex that removes a nucleosome block to the elongating RNA polymerase. In other cases, the activator communicates with RNA polymerase (typically through a coactivator), signaling it to move ahead. Finally, as we saw in Chapter 6, RNA polymerase requires *elongation factors* to effectively transcribe through chromatin. In some cases, the key step in gene activation is the loading of these factors onto RNA polymerase, which can be directed by DNA-bound transcription activators. Once loaded, these factors allow the polymerase to move through blocks imposed by chromatin structure and begin transcribing the gene in earnest. Having RNA polymerase already poised on a promoter in the beginning stages of transcription bypasses the step of assembling many components at the promoter, which is often slow. This mechanism can therefore allow cells to begin transcribing a gene as a rapid response to an extracellular signal.

Transcription Activators Work Synergistically

We have seen that complexes of transcription activators and coactivators assemble cooperatively on DNA. We have also seen that these assemblies can promote different steps in transcription initiation. In general, where several factors work together to enhance a reaction rate, the joint effect is not merely the sum of the enhancements that each factor alone contributes, but the product. If, for example, factor A lowers the free-energy barrier for a reaction by a certain amount and thereby speeds up the reaction 100-fold, and factor B, by acting on another aspect of the reaction, does likewise, then A and B acting in parallel will lower the barrier

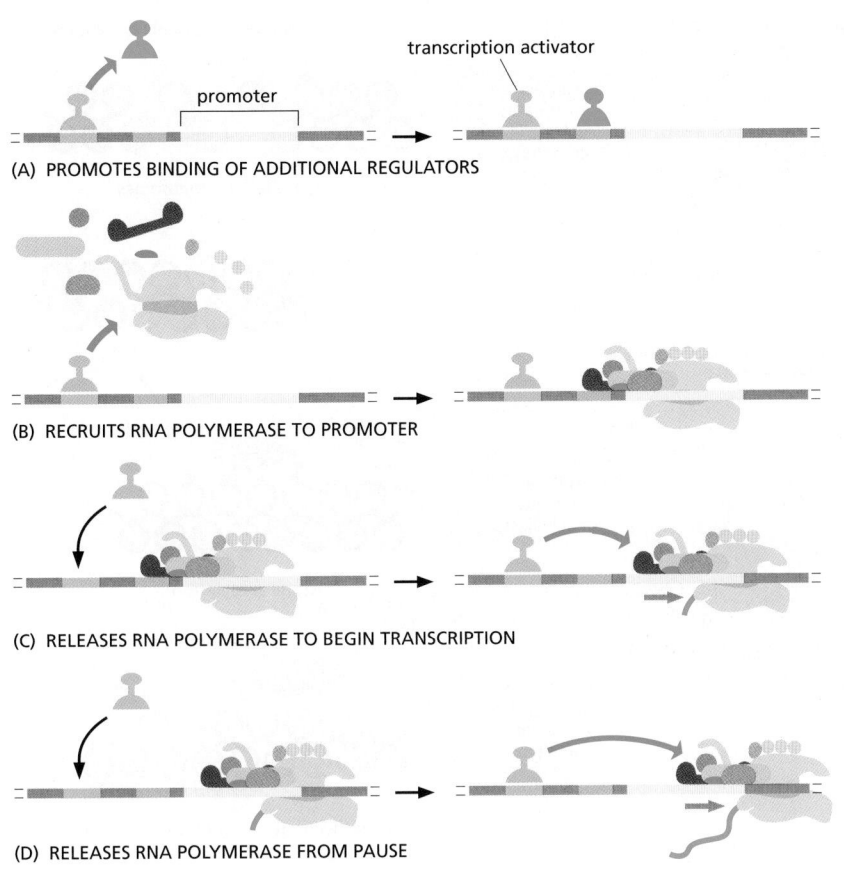

(A) PROMOTES BINDING OF ADDITIONAL REGULATORS

transcription activator

promoter

(B) RECRUITS RNA POLYMERASE TO PROMOTER

(C) RELEASES RNA POLYMERASE TO BEGIN TRANSCRIPTION

(D) RELEASES RNA POLYMERASE FROM PAUSE

Figure 7–21 **Transcription activators can act at different steps.** In addition to (A) promoting binding of additional transcription regulators and (B) assembling RNA polymerase at promoters, transcription activators are often needed (C) to release already assembled RNA polymerases from promoters or (D) to release RNA polymerase molecules that become stalled after transcribing about 50 nucleotides of RNA. The activities shown in Figure 7–19 can affect each of these four steps.

by a double amount and speed up the reaction 10,000-fold. Even if A and B work simply by attracting the same protein, the affinity of that protein for the reaction site increases multiplicatively. Thus, transcription activators often exhibit *transcriptional synergy*, where several DNA-bound activator proteins working together produce a transcription rate that is much higher than the sum of their transcription rates working alone (**Figure 7–22**).

An important point is that a transcription activator protein must be bound to DNA to influence transcription of its target gene. And the rate of transcription of a gene ultimately depends upon the spectrum of regulatory proteins bound upstream and downstream of its transcription start site, along with the coactivator proteins they bring to DNA.

Eukaryotic Transcription Repressors Can Inhibit Transcription in Several Ways

Although the "default" state of eukaryotic DNA packaged into nucleosomes is resistant to transcription, eukaryotes nonetheless use transcription regulators to

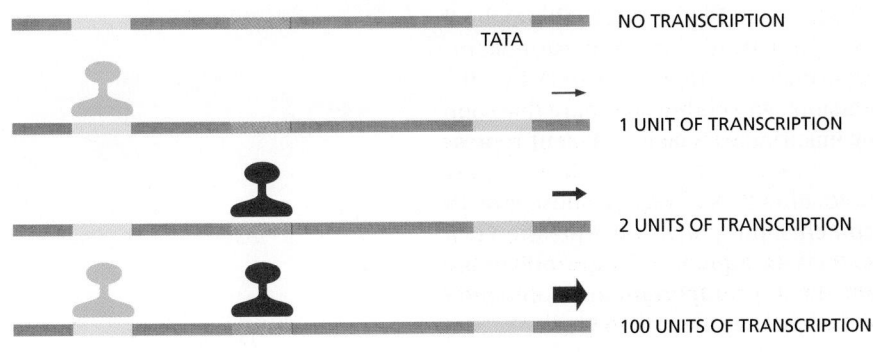

TATA

NO TRANSCRIPTION

1 UNIT OF TRANSCRIPTION

2 UNITS OF TRANSCRIPTION

100 UNITS OF TRANSCRIPTION

Figure 7–22 **Transcriptional synergy.** This experiment compares the rate of transcription produced by three experimentally constructed regulatory regions in a eukaryotic cell and reveals transcriptional synergy, a greater than additive effect of multiple activators working together. For simplicity, coactivators have been omitted from the diagram.

Such transcriptional synergy is not only observed between different transcription activators from the same organism; it is also seen between activator proteins from different eukaryotic species when they are experimentally introduced into the same cell. This last observation reflects the high degree of conservation of the machinery responsible for eukaryotic transcription initiation.

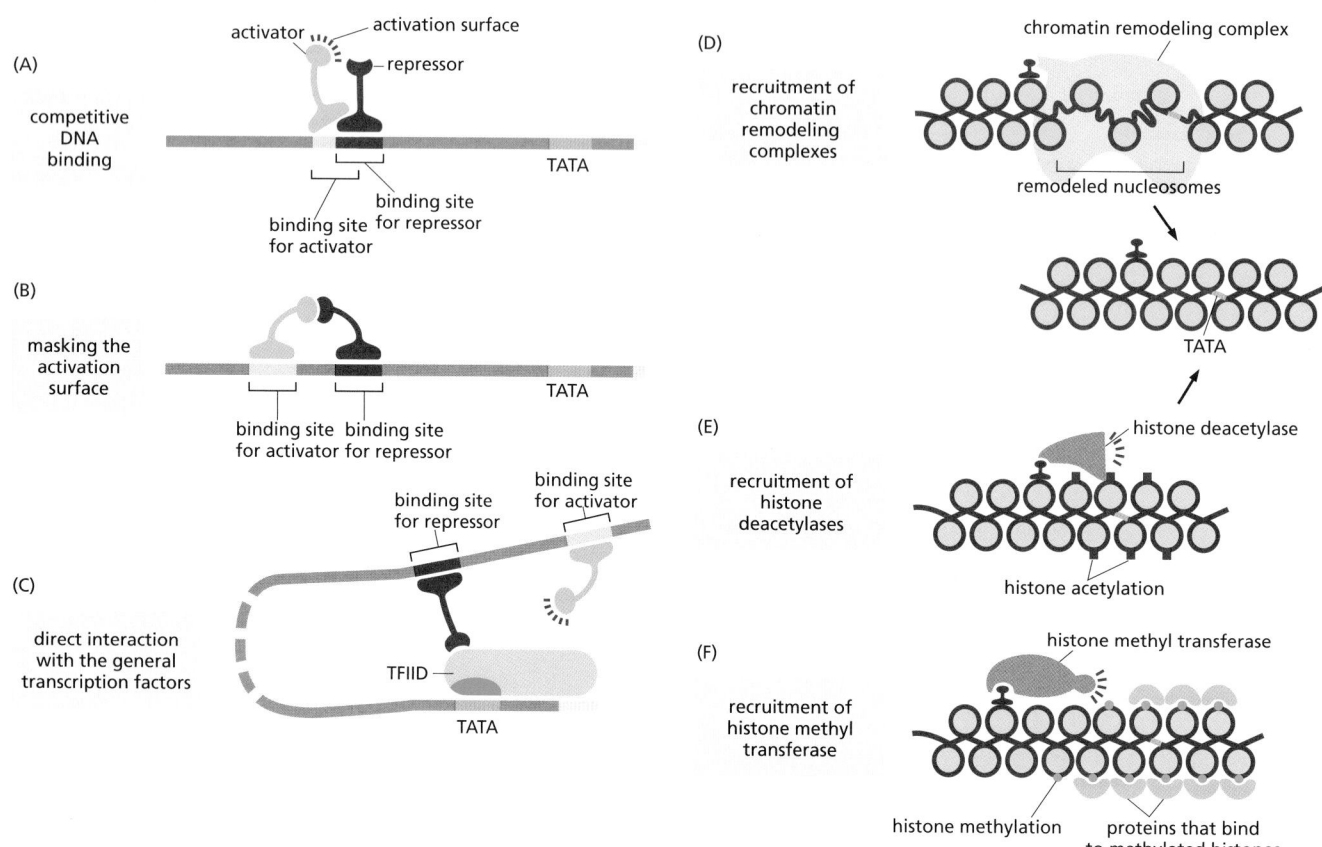

Figure 7–23 Six ways in which eukaryotic repressor proteins can operate. (A) Activator proteins and repressor proteins compete for binding to the same regulatory DNA sequence. (B) Both proteins bind DNA, but the repressor prevents the activator from carrying out its functions. (C) The repressor blocks assembly of the general transcription factors. (D) The repressor recruits a chromatin remodeling complex, which returns the nucleosomal state of the promoter region to its pre-transcriptional form. (E) The repressor attracts a histone deacetylase to the promoter. As we have seen, histone acetylation can stimulate transcription initiation (see Figure 7–20), and the repressor simply reverses this modification. (F) The repressor attracts a histone methyl transferase, which modifies certain positions on histones by attaching methyl groups; the methylated histones, in turn, are bound by proteins that maintain the chromatin in a transcriptionally silent form.

repress the transcription of genes. These transcription repressors can both depress the rate of transcription below the default value and rapidly shut off genes that were previously activated. We saw in Chapter 4 that large regions of the genome can be shut down by the packaging of DNA into especially resistant forms of chromatin. However, eukaryotic genes are rarely organized along the genome according to function, and this strategy is not generally applicable for shutting off a set of genes that work together. Instead, most eukaryotic repressors work on a gene-by-gene basis. Unlike bacterial repressors, eukaryotic repressors do not directly compete with the RNA polymerase for access to the DNA; rather, they use a variety of other mechanisms, some of which are illustrated in **Figure 7–23**. Although all of these mechanisms ultimately block transcription by RNA polymerase, eukaryotic transcription repressors typically act by bringing co-repressors to DNA. Like transcription activation, transcription repression can act through more than one mechanism at a given target gene, thereby ensuring especially efficient repression.

Gene repression is especially important to animals and plants whose growth depends on elaborate and complex developmental programs. Misexpression of a single gene at a critical time can have disastrous consequences for the individual. For this reason, many of the genes encoding the most important developmental regulatory proteins are kept tightly repressed when they are not needed.

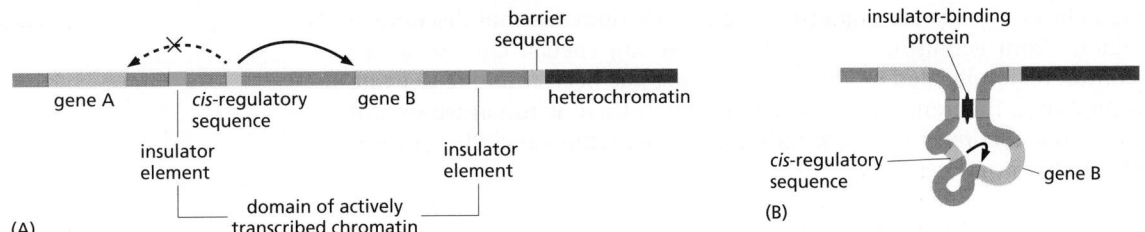

Figure 7–24 Schematic diagram summarizing the properties of insulators and barrier sequences. (A) Insulators directionally block the action of *cis*-regulatory sequences, whereas barrier sequences prevent the spread of heterochromatin. How barrier sequences likely function is depicted in Figure 4–41. (B) Insulator-binding proteins *(purple)* hold chromatin in loops, thereby favoring "correct" *cis*-regulatory sequence–gene associations. Thus, gene B is properly regulated, and gene B's *cis*-regulatory sequences are prevented from influencing the transcription of gene A.

Insulator DNA Sequences Prevent Eukaryotic Transcription Regulators from Influencing Distant Genes

We have seen that all genes have control regions, which dictate at which times, under what conditions, and in what tissues the gene will be expressed. We have also seen that eukaryotic transcription regulators can act across very long stretches of DNA, with the intervening DNA looped out. How, then, are control regions of different genes kept from interfering with one another? For example, what keeps a transcription regulator bound on the control region of one gene from looping in the wrong direction and inappropriately influencing the transcription of an adjacent gene?

To avoid such cross-talk, several types of DNA elements compartmentalize the genome into discrete regulatory domains. In Chapter 4, we discussed *barrier sequences* that prevent the spread of heterochromatin into genes that need to be expressed. A second type of DNA element, called an *insulator*, prevents *cis*-regulatory sequences from running amok and activating inappropriate genes (**Figure 7–24**). Insulators function by forming loops of chromatin, an effect mediated by specialized proteins that bind them (see Figures 4–48 and 7–24B). The loops hold a gene and its control region in rough proximity and help to prevent the control region from "spilling over" to adjacent genes. Importantly, these loops can be in different in different cell types, depending on the particular proteins and chromatin structures that are present.

The distribution of insulators and barrier sequences in a genome is thought to divide it into independent domains of gene regulation and chromatin structure (see pp. 207–208). Aspects of this organization can be visualized by staining whole chromosomes for the specialized proteins that bind these DNA elements (**Figure 7–25**).

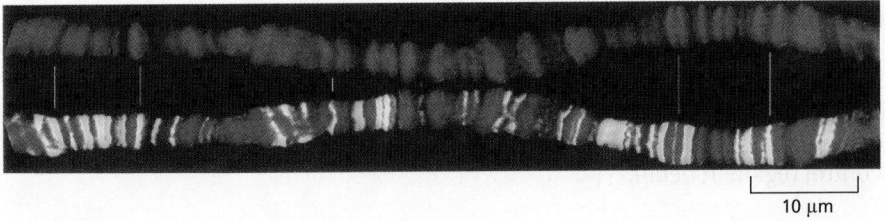

10 µm

Figure 7–25 Localization of a *Drosophila* insulator-binding protein on polytene chromosomes. A polytene chromosome (discussed in Chapter 4) was stained with propidium iodide *(red)* to show its banding patterns, with bands appearing *bright red* and interbands as dark gaps in the pattern (top). The positions on this polytene chromosome that are bound by a particular insulator protein are stained *bright green* using antibodies directed against the protein (bottom). This protein is preferentially localized to interband regions, reflecting its role in organizing chromosomes into structural, as well as functional, domains. For convenience, these two micrographs of the same polytene chromosome are arranged as mirror images. (Courtesy of Uli Laemmli, from K. Zhao et al., *Cell* 81:879–889, 1995. With permission from Elsevier.)

Although chromosomes are organized into orderly domains that discourage control regions from acting indiscriminately, there are special circumstances where a control region located on one chromosome has been found to activate a gene located on a different chromosome. Although there is much we do not understand about this mechanism, it indicates the extreme versatility of transcriptional regulation strategies.

Summary

Transcription regulators switch the transcription of individual genes on and off in cells. In prokaryotes, these proteins typically bind to specific DNA sequences close to the RNA polymerase start site and, depending on the nature of the regulatory protein and the precise location of its binding site relative to the start site, either activate or repress transcription of the gene. The flexibility of the DNA helix, however, also allows proteins bound at distant sites to affect the RNA polymerase at the promoter by the looping out of the intervening DNA. The regulation of higher eukaryotic genes is much more complex, commensurate with a larger genome size and the large variety of cell types that are formed. A single eukaryotic gene is typically controlled by many transcription regulators bound to sequences that can be tens or even hundreds of thousands of nucleotide pairs from the promoter that directs transcription of the gene. Eukaryotic activators and repressors act by a wide variety of mechanisms—generally altering chromatin structure and controlling the assembly of the general transcription factors and RNA polymerase at the promoter. They do this by attracting coactivators and co-repressors, protein complexes that perform the necessary biochemical reactions. The time and place that each gene is transcribed, as well as its rates of transcription under different conditions, are determined by the particular spectrum of transcription regulators that bind to the regulatory region of the gene.

MOLECULAR GENETIC MECHANISMS THAT CREATE AND MAINTAIN SPECIALIZED CELL TYPES

Although all cells must be able to switch genes on and off in response to changes in their environments, the cells of multicellular organisms have evolved this capacity to an extreme degree. In particular, once a cell in a multicellular organism becomes committed to differentiate into a specific cell type, the cell maintains this choice through many subsequent cell generations, which means that it remembers the changes in gene expression involved in the choice. This phenomenon of *cell memory* is a prerequisite for the creation of organized tissues and for the maintenance of stably differentiated cell types. In contrast, other changes in gene expression in eukaryotes, as well as most such changes in bacteria, are only transient. The tryptophan repressor, for example, switches off the tryptophan genes in bacteria only in the presence of tryptophan; as soon as tryptophan is removed from the medium, the genes are switched back on, and the descendants of the cell will have no memory that their ancestors had been exposed to tryptophan.

In this section, we shall examine not only cell memory mechanisms, but also how gene regulatory devices can be combined to create the "logic circuits" through which cells integrate signals and remember events in their past. We begin by considering one such complex gene control region in detail.

Complex Genetic Switches That Regulate *Drosophila* Development Are Built Up from Smaller Molecules

We have seen that transcription regulators can be positioned at multiple sites along long stretches of DNA and that these proteins can bring into play coactivators and co-repressors. Here, we discuss how the numerous transcription regulators that are bound to the control region of a gene can cause the gene to be transcribed at the proper place and time.

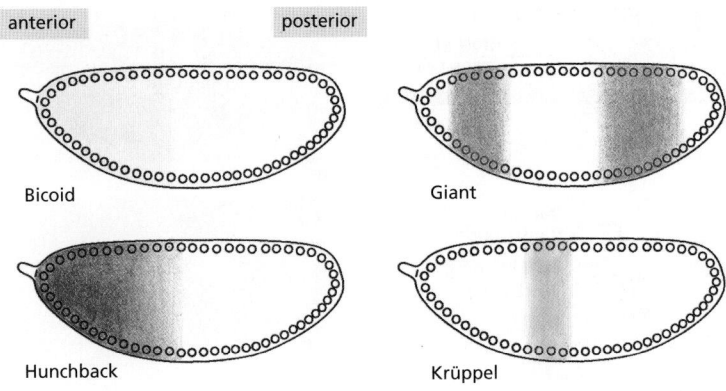

anterior   posterior

Bicoid

Giant

Hunchback

Krüppel

Figure 7–26 **The nonuniform distribution of transcription regulators in an early *Drosophila* embryo.** At this stage, the embryo is a syncytium; that is, multiple nuclei are contained in a common cytoplasm. Although not shown in these drawings, all of these proteins are concentrated in the nuclei. How such differences are established is discussed in Chapter 21.

Consider the *Drosophila Even-skipped* (*Eve*) gene, whose expression plays an important part in the development of the *Drosophila* embryo. If this gene is inactivated by mutation, many parts of the embryo fail to form, and the embryo dies early in development. As discussed in Chapter 21, at the stage of development when *Eve* begins to be expressed, the embryo is a single giant cell containing multiple nuclei in a common cytoplasm. This cytoplasm contains a mixture of transcription regulators that are distributed unevenly along the length of the embryo, thus providing *positional information* that distinguishes one part of the embryo from another (**Figure 7–26**). Although the nuclei are initially identical, they rapidly begin to express different genes because they are exposed to different transcription regulators. For example, the nuclei near the anterior end of the developing embryo are exposed to a set of transcription regulators that is distinct from the set that influences nuclei at the middle or at the posterior end of the embryo.

The regulatory DNA sequences that control the *Eve* gene have evolved to "read" the concentrations of transcription regulators at each position along the length of the embryo, and they cause the *Eve* gene to be expressed in seven precisely positioned stripes, each initially five to six nuclei wide (**Figure 7–27**). How is this remarkable feat of information processing carried out? Although there is still much to learn, several general principles have emerged from studies of *Eve* and other genes that are similarly regulated.

The regulatory region of the *Eve* gene is very large (approximately 20,000 nucleotide pairs). It is formed from a series of relatively simple regulatory modules, each of which contains multiple *cis*-regulatory sequences and is responsible for specifying a particular stripe of *Eve* expression along the embryo. This modular organization of the *Eve* gene control region was revealed by experiments in which a particular regulatory module (say, that specifying stripe 2) is removed from its normal setting upstream of the *Eve* gene, placed in front of a reporter gene, and reintroduced into the *Drosophila* genome. When developing embryos derived from flies carrying this genetic construct are examined, the reporter gene is found to be expressed in precisely the position of stripe 2 (**Figure 7–28**). Similar experiments reveal the existence of other regulatory modules, each of which specifies other stripes.

Figure 7–27 **The seven stripes of the protein encoded by the *Even-skipped* (*Eve*) gene in a developing *Drosophila* embryo.** Two and one-half hours after fertilization, the egg was fixed and stained with antibodies that recognize the Eve protein *(green)* and antibodies that recognize the Giant protein *(red)*. Where Eve and Giant proteins are both present, the staining appears *yellow*. At this stage in development, the egg contains approximately 4000 nuclei. The Eve and Giant proteins are both located in the nuclei, and the Eve stripes are about four nuclei wide. The pattern for the Giant protein is also shown in Figure 7–26. (Courtesy of Michael Levine.)

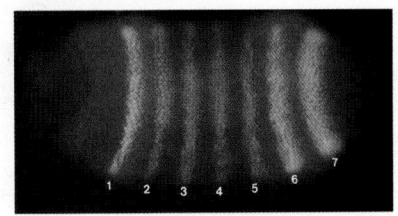

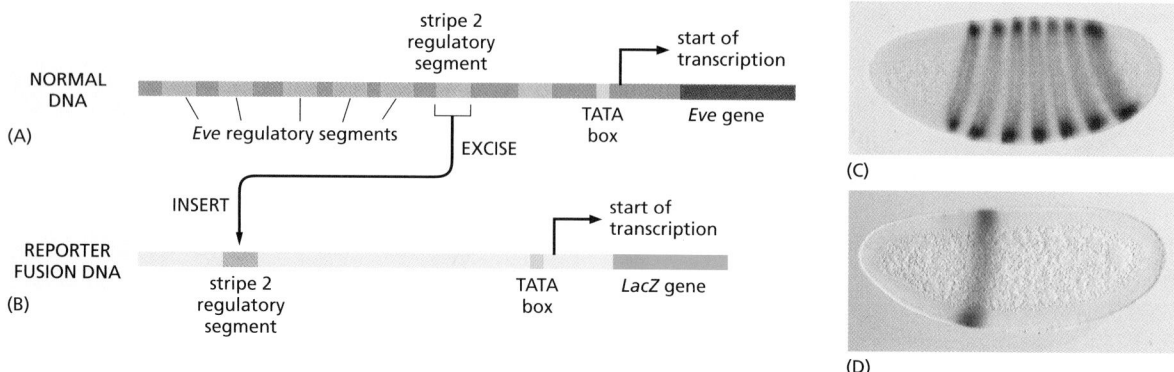

Figure 7–28 Experiment demonstrating the modular construction of the *Eve* gene regulatory region. (A) A 480-nucleotide-pair section of the *Eve* regulatory region was removed and (B) inserted upstream of a test promoter that directs the synthesis of the enzyme β-galactosidase (the product of the *E. coli LacZ* gene—see Figure 7–15). (C, D) When this artificial construct was reintroduced into the genome of *Drosophila* embryos, the embryos (D) expressed β-galactosidase (detectable by histochemical staining) precisely in the position of the second of the seven *Eve* stripes (C). β-Galactosidase is simple to detect and thus provides a convenient way to monitor the expression specified by a gene control region. As used here, β-galactosidase is said to serve as a reporter, since it "reports" the activity of a gene control region. (C and D, courtesy of Stephen Small and Michael Levine.)

The *Drosophila Eve* Gene Is Regulated by Combinatorial Controls

A detailed study of the stripe 2 regulatory module has provided insights into how it reads and interprets positional information. The module contains recognition sequences for two transcription regulators (Bicoid and Hunchback) that activate *Eve* transcription and for two transcription regulators (Krüppel and Giant) that repress it (**Figure 7–29**). The relative concentrations of these four proteins determine whether the protein complexes that form at the stripe 2 module activate transcription of the *Eve* gene. **Figure 7–30** shows the distributions of the four transcription regulators across the region of a *Drosophila* embryo where stripe 2 forms. It is thought that either of the two repressor proteins, when bound to the DNA, will turn off the stripe 2 module, whereas both Bicoid and Hunchback must bind for this module's maximal activation. This simple regulatory scheme suffices to turn on the stripe 2 module (and therefore the expression of the *Eve* gene) only in those nuclei located where the levels of both Bicoid and Hunchback are high and both Krüppel and Giant are absent—a combination that occurs in only one region of the early embryo. It is not known exactly how these four transcription regulators interact with coactivators and co-repressors to specify the final level of transcription across the stripe, but the outcome very likely relies on competition between activators and repressors that act by the mechanisms outlined in Figures 7–17, 7–19, and 7–23.

The stripe 2 element is autonomous, inasmuch as it specifies stripe 2 when isolated from its normal context (see Figure 7–28). The other stripe regulatory modules are thought to be constructed similarly, reading positional information provided by other combinations of transcription regulators. The entire *Eve* gene control region binds more than 20 different transcription regulators. Seven combinations of regulators—one combination for each stripe—specify *Eve* expression, while many other combinations (all those found in the interstripe regions of

Figure 7–29 The *Eve* stripe 2 unit. The segment of the *Eve* gene control region identified in Figure 7–28 contains *cis*-regulatory sequences for four transcription regulators. It is known from genetic experiments that these four regulatory proteins are responsible for the proper expression of *Eve* in stripe 2. Flies that are deficient in the two gene activators Bicoid and Hunchback, for example, fail to efficiently express *Eve* in stripe 2. In flies deficient in either of the two gene repressors, Giant and Krüppel, stripe 2 expands and covers an abnormally broad region of the embryo. As indicated, in some cases the binding sites for the transcription regulators overlap, and the proteins can compete for binding to the DNA. For example, binding of Krüppel and binding of Bicoid to the site at the far right is mutually exclusive.

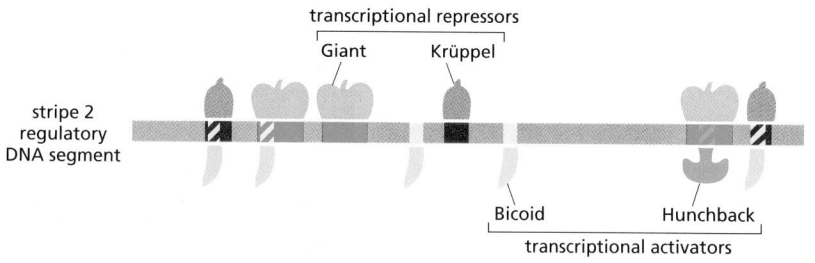

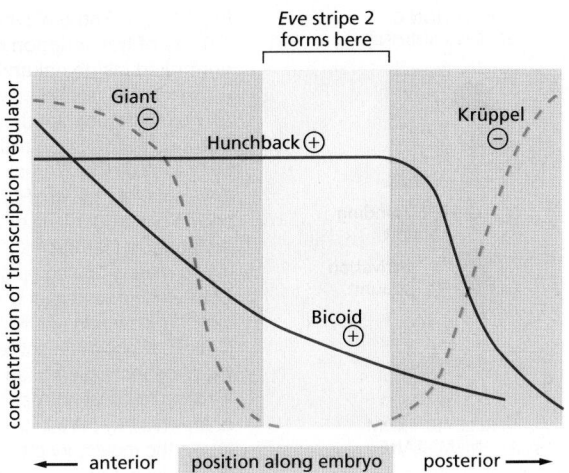

Figure 7–30 **Distribution of the transcription regulators responsible for ensuring that *Eve* is expressed in stripe 2.** The distributions of these proteins were visualized by staining a developing *Drosophila* embryo with antibodies directed against each of the four proteins. The expression of *Eve* in stripe 2 occurs only at the position where the two activators (Bicoid and Hunchback) are present and the two repressors (Giant and Krüppel) are absent. In fly embryos that lack Krüppel, for example, stripe 2 expands posteriorly. Likewise, stripe 2 expands posteriorly if the DNA-binding sites for Krüppel in the stripe 2 module are inactivated by mutation (see also Figures 7–26 and 7–27).

the embryo) keep the stripe elements silent. A large and complex control region is thereby built from a series of smaller modules, each of which consists of a unique arrangement of short *cis*-regulatory sequences recognized by specific transcription regulators.

The *Eve* gene itself encodes a transcription regulator, which, after its pattern of expression is set up in seven stripes, controls the expression of other *Drosophila* genes. As development proceeds, the embryo is thus subdivided into finer and finer regions that eventually give rise to the different body parts of the adult fly, as discussed in Chapter 21.

Eve exemplifies the complex control regions found in plants and animals. As this example shows, control regions can respond to many different inputs, integrate this information, and produce a complex spatial and temporal output as development proceeds. However, exactly how all these mechanisms work together to produce the final output is understood only in broad outline (**Figure 7–31**).

Transcription Regulators Are Brought Into Play by Extracellular Signals

The above example from *Drosophila* clearly illustrates the power of combinatorial control, but this case is unusual in that the nuclei are exposed directly to positional cues in the form of concentrations of transcription regulators. In embryos of most other organisms and in all adults, individual nuclei are in separate cells, and extracellular information (including positional cues) must be passed across the plasma membrane so as to generate signals in the cytosol that cause different transcription regulators to become active in different cell types. Some of the different mechanisms that are known to be used to activate transcription regulators are diagrammed in **Figure 7–32**, and in Chapter 15, we discuss how extracellular signals trigger these changes.

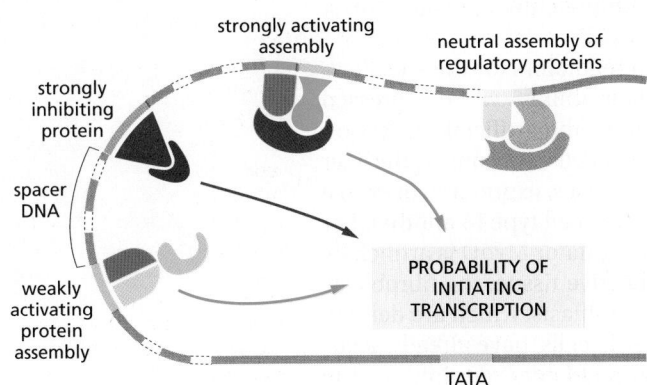

Figure 7–31 **The integration of multiple inputs at a promoter.** Multiple sets of transcription regulators, coactivators, and co-repressors can work together to influence transcription initiation at a promoter, as they do in the *Eve* stripe 2 module illustrated in Figure 7–29. It is not yet understood in detail how the cell achieves integration of multiple inputs, but it is likely that the final transcriptional activity of the gene results from a competition between activators and repressors that act by the mechanisms summarized in Figures 7–17, 7–19, and 7–23.

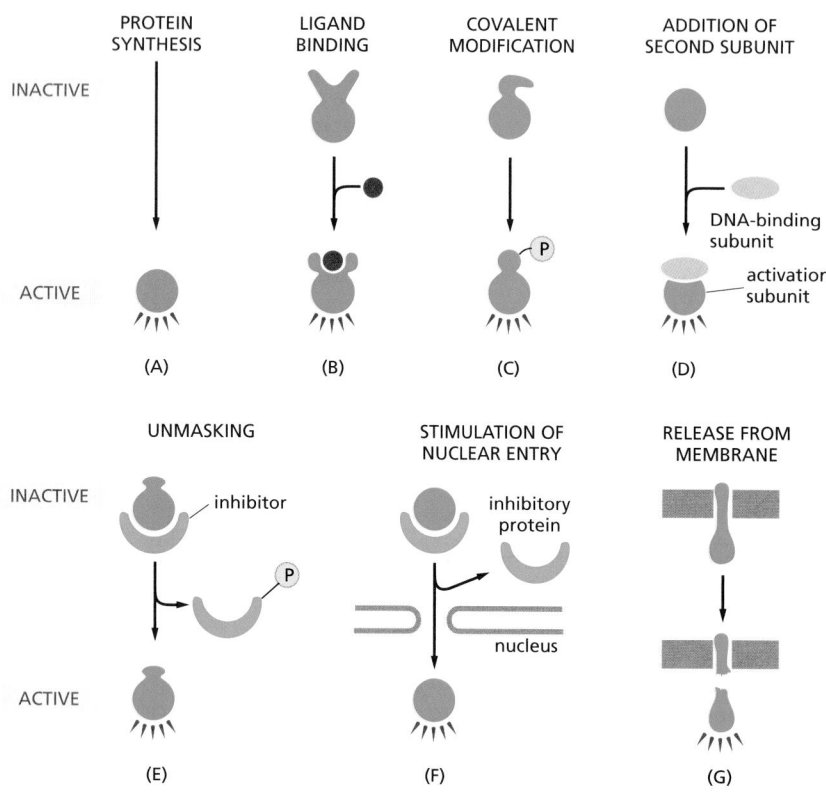

Figure 7–32 **Some ways in which the activity of transcription regulators is controlled inside eukaryotic cells.** (A) The protein is synthesized only when needed and is rapidly degraded by proteolysis so that it does not accumulate. (B) Activation by ligand binding. (C) Activation by covalent modification. Phosphorylation is shown here, but many other modifications are possible (see Table 3–3, p. 165). (D) Formation of a complex between a DNA-binding protein and a separate protein with a transcription-activating domain. (E) Unmasking of an activation domain by the phosphorylation of an inhibitor protein. (F) Stimulation of nuclear entry by removal of an inhibitory protein that otherwise keeps the regulatory protein from entering the nucleus. (G) Release of a transcription regulator from a membrane bilayer by regulated proteolysis.

Combinatorial Gene Control Creates Many Different Cell Types

We have seen that transcription regulators can act in combination to control the expression of an individual gene. It is also generally true that each transcription regulator in an organism contributes to the control of many genes. This point is illustrated schematically in **Figure 7–33**, which shows how combinatorial gene control makes it possible to generate a great deal of biological complexity even with relatively few transcription regulators.

Due to combinatorial control, a given transcription regulator does not necessarily have a single, simply definable function as commander of a particular battery of genes or specifier of a particular cell type. Rather, transcription regulators can be likened to the words of a language: they are used with different meanings in a variety of contexts and rarely alone; it is the well-chosen combination that conveys the information that specifies a gene regulatory event.

Combinatorial gene control causes the effect of adding a new transcription regulator to a cell to depend on that cell's past history, since it is this history that determines which transcription regulators are already present. Thus, during development, a cell can accumulate a series of transcription regulators that need not initially alter gene expression. The addition of the final members of the requisite combination of transcription regulators will complete the regulatory message, and can lead to large changes in gene expression.

The importance of combinations of transcription regulators for the specification of cell types is most easily demonstrated by their ability—when expressed artificially—to convert one type of cell to another. Thus, the artificial expression of three neuron-specific transcription regulators in liver cells can convert the liver cells into functional nerve cells (**Figure 7–34**). In some cases, expression of even a single transcription regulator is sufficient to convert one cell type to another. For example, when the gene encoding the transcription regulator MyoD is artificially introduced into fibroblasts cultured from skin connective tissue, the fibroblasts form muscle-like cells. As discussed in Chapter 22, fibroblasts, which are derived from the same broad class of embryonic cells as muscle cells, have already accumulated many of the other necessary transcription regulators required for the

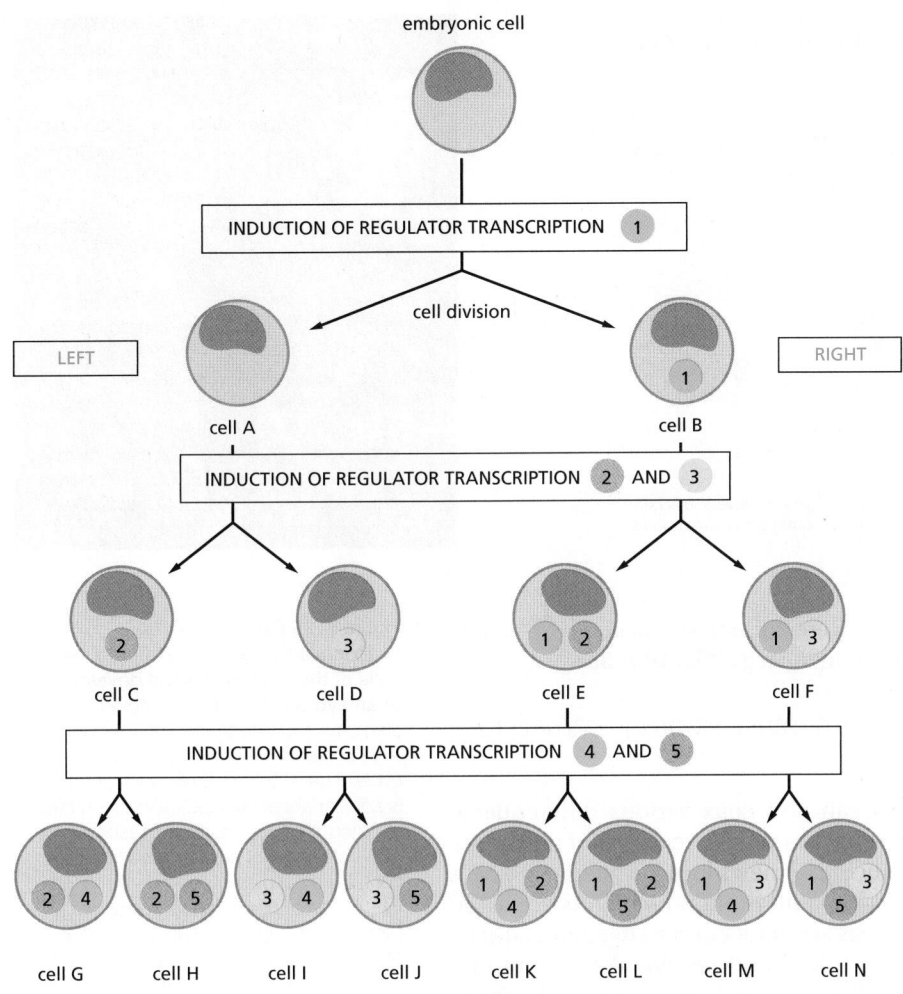

Figure 7–33 The importance of combinatorial gene control for development. Combinations of a few transcription regulators can generate many cell types during development. In this simple, idealized scheme, a "decision" to make one of a pair of different transcription regulators (shown as numbered circles) is made after each cell division. Sensing its relative position in the embryo, the daughter cell toward the *left side* of the embryo is always induced to synthesize the even-numbered protein of each pair, while the daughter cell toward the *right side* of the embryo is induced to synthesize the odd-numbered protein. The production of each transcription regulator is assumed to be self-perpetuating once it has become initiated (see Figure 7–39). In this way, through cell memory, the final combinatorial specification is built up step by step. In this purely hypothetical example, five different transcription regulators have created eight final cell types (G–N).

combinatorial control of the muscle-specific genes, and the addition of MyoD completes the unique combination required to direct the cells to become muscle. An even more striking example is seen by artificially expressing, early in development, a single *Drosophila* transcription regulator (Eyeless) in groups of cells

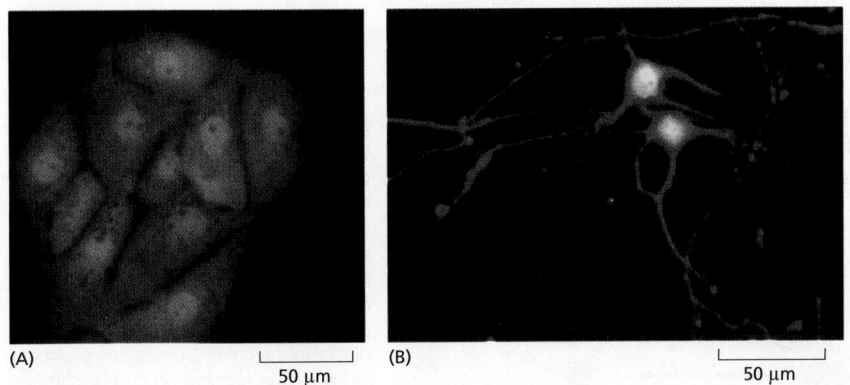

(A)   (B)
⎣_____⎦   ⎣_____⎦
    50 µm   50 µm

Figure 7–34 A small set of transcription regulators can convert one differentiated cell type into another. In this experiment, (A) liver cells grown in culture were converted into (B) neuronal cells via the artificial expression of three nerve-specific transcription regulators. Both types of cells express an artificial *red* fluorescent protein, which is used to visualize them. This conversion involves the activation of many nerve-specific genes as well as the repression of many liver-specific genes. (From S. Marro et al., *Cell Stem Cell* 9:374–382, 2011. With permission from Elsevier.)

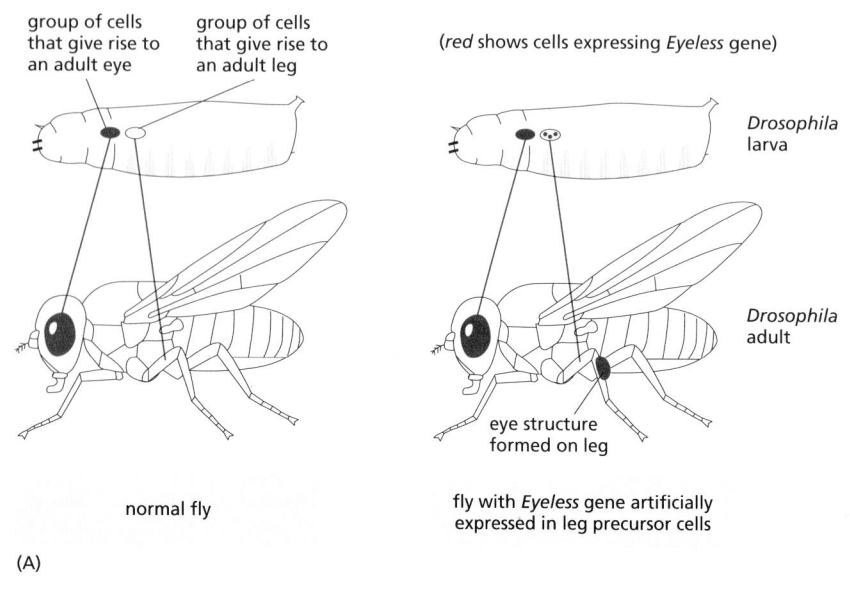

group of cells that give rise to an adult eye

group of cells that give rise to an adult leg

(*red* shows cells expressing *Eyeless* gene)

Drosophila larva

Drosophila adult

eye structure formed on leg

normal fly

fly with *Eyeless* gene artificially expressed in leg precursor cells

(A)

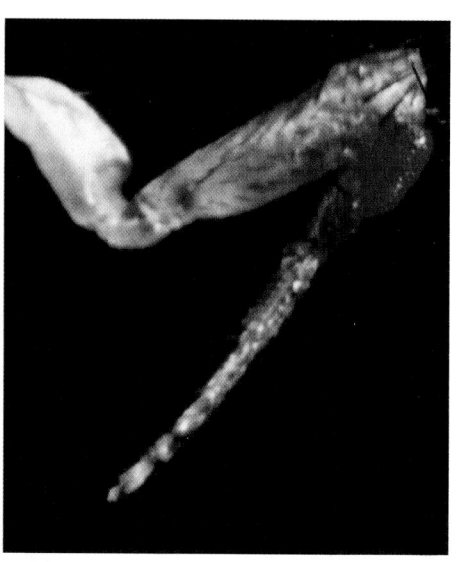

(B)

that would normally go on to form leg parts. Here, this abnormal gene expression change causes eye-like structures to develop in the legs (**Figure 7–35**).

Specialized Cell Types Can Be Experimentally Reprogrammed to Become Pluripotent Stem Cells

Manipulation of transcription regulators can also coax various differentiated cells to *de-differentiate* into pluripotent stem cells that are capable of giving rise to the different cell types in the body, much like the embryonic stem (ES) cells discussed in Chapter 22. When three specific transcription regulators are artificially expressed in cultured mouse fibroblasts, a number of cells become **induced pluripotent stem cells** (**iPS cells**)—cells that look and behave like the pluripotent ES cells that are derived from embryos (**Figure 7–36**). This approach has been adapted to produce iPS cells from a variety of specialized cell types, including cells taken from humans. Such human iPS cells can then be directed to generate a population of differentiated cells for use in the study or treatment of disease, as we discuss in Chapter 22.

Although it was once thought that cell differentiation was irreversible, it is now clear that by manipulating combinations of **master transcription regulators**, cell types and differentiation pathways can be readily altered.

Combinations of Master Transcription Regulators Specify Cell Types by Controlling the Expression of Many Genes

As we saw in the introduction of this chapter, different cell types of multicellular organisms differ enormously in the proteins and RNAs they express. For example, only muscle cells express special types of actin and myosin that form the contractile

Figure 7–35 Expression of the *Drosophila* Eyeless gene in precursor cells of the leg triggers the development of an eye on the leg. (A) Simplified diagrams showing the result when a fruit fly larva contains either the normally expressed *Eyeless* gene *(left)* or an *Eyeless* gene that is additionally expressed artificially in cells that normally give rise to leg tissue *(right)*. (B) Photograph of an abnormal leg that contains a misplaced eye (see also Figure 21–2). The transcription regulator was named Eyeless because its inactivation in otherwise normal flies causes the loss of eyes. (B, courtesy of Universität Basel.)

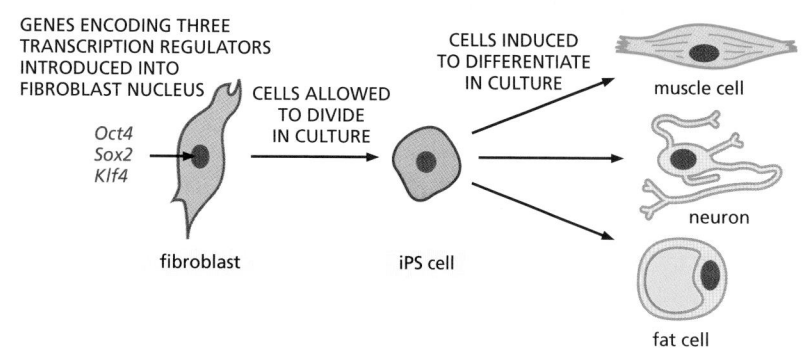

GENES ENCODING THREE TRANSCRIPTION REGULATORS INTRODUCED INTO FIBROBLAST NUCLEUS

CELLS ALLOWED TO DIVIDE IN CULTURE

CELLS INDUCED TO DIFFERENTIATE IN CULTURE

muscle cell

neuron

fat cell

Oct4
Sox2
Klf4

fibroblast

iPS cell

Figure 7–36 A combination of transcription regulators can induce a differentiated cell to de-differentiate into a pluripotent cell. The artificial expression of a set of three genes, each of which encodes a transcription regulator, can reprogram a fibroblast into a pluripotent cell with embryonic stem (ES)-cell-like properties. Like ES cells, such induced pluripotent stem (iPS) cells can proliferate indefinitely in culture and can be stimulated by appropriate extracellular signal molecules to differentiate into almost any cell type found in the body. Transcription regulators such as Oct4, Sox2, and Klf4 are often called *master transcription regulators* because their expression is sufficient to trigger a change in cell identity.

(A)

(B)

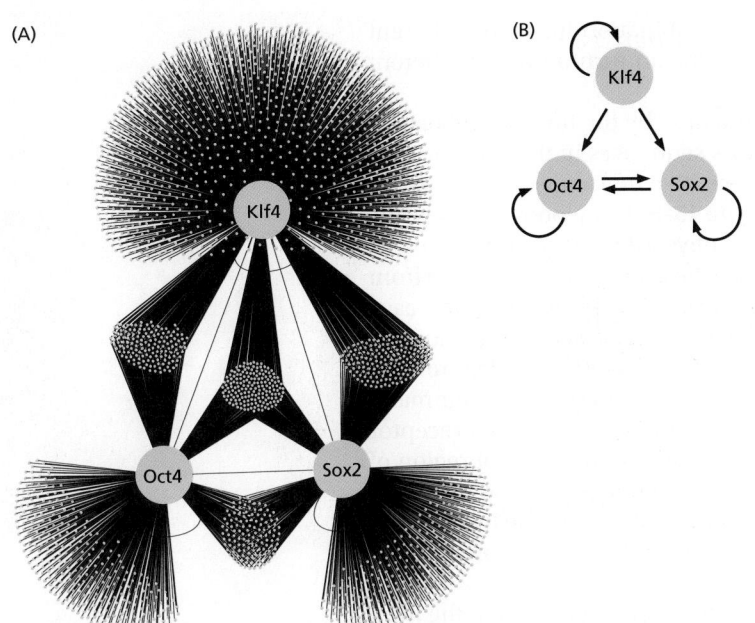

Figure 7–37 **A portion of the transcription network specifying embryonic stem cells.** (A) The three master transcription regulators in Figure 7–36 are shown as large circles. Genes whose *cis*-regulatory sequences are bound by each regulator in embryonic stem cells are indicated by a small dot (representing the gene) connected by a thin line (representing the binding reaction). Note that many of the target genes are bound by more than one of the regulators. (B) The master regulators control their own expression. As shown here, the three transcriptional regulators bind to their own control regions (indicated by feedback loops), as well as those of the other master regulators (indicated by straight arrows). (Courtesy of Trevor Sorrells, based on data from J. Kim et al., *Cell* 132:1049–1061, 2008.)

apparatus, while nerve cells must make and assemble all the proteins needed to form dendrites and synapses. We have seen that these patterns of cell-type-specific expression are orchestrated by a combination of master transcription regulators. In many cases, these proteins bind directly to *cis*-regulatory sequences of the genes particular to that cell type. Thus, MyoD binds directly to *cis*-regulatory sequences located in the control regions of the muscle-specific genes. In other cases, the master regulators control the expression of "downstream" transcription regulators which, in turn, bind to the control regions of other cell-type-specific genes and control their synthesis.

The specification of a particular cell type typically involves changes in the expression of several thousand genes. Genes whose protein products are required in the cell type are expressed at high levels, while those not needed are typically down-regulated. As might be imagined, the pattern of binding between the master regulators and all of the regulated genes can be extremely elaborate (**Figure 7–37**). When we consider that many of these regulated genes have control regions that span tens of thousands of nucleotide pairs, commensurate with the *Eve* example discussed above, we can begin to appreciate the enormous complexity of cell-type specification.

An outstanding question in biology is how the information in a genome is used to specify a multicellular organism. Although we have the general outline of the answer, we are far from understanding how a single cell type is completely specified, let alone a whole organism.

Specialized Cells Must Rapidly Turn Sets of Genes On and Off

Although they generally maintain their identities, specialized cells must constantly respond to changes in their environment. Among the most important changes are signals from other cells that coordinate the behavior of the whole organism. Many of these signals induce transient changes in gene transcription, and we discuss the nature of these signals in detail in Chapter 15. Here, we consider how specialized cell types rapidly and decisively switch groups of genes on and off in response to their environment. Even though control of gene expression is combinatorial, the effect of a single transcription regulator can still be decisive in switching any particular gene on or off, simply by completing the combination needed to maximally activate or repress that gene. This situation is analogous to dialing in the final number of a combination lock: the lock will spring open with only this simple addition if all of the other numbers have been previously entered.

Moreover, the same number can complete the combination for many different locks. Likewise, the addition of a particular protein can turn on many different genes.

An example is the rapid control of gene expression by the human glucocorticoid receptor protein. To bind to its *cis*-regulatory sequences in the genome, this transcription regulator must first form a complex with a molecule of a glucocorticoid steroid hormone, such as cortisol (see Figure 15–64). The body releases this hormone during times of starvation and intense physical activity, and among its other activities, it stimulates liver cells to increase the production of glucose from amino acids and other small molecules. To respond in this way, liver cells increase the expression of many different genes that code for metabolic enzymes, such as tyrosine aminotransferase, as we discussed earlier in this chapter (see Figure 7–3). Although these genes all have different and complex control regions, their maximal expression depends on the binding of the hormone–glucocorticoid receptor complex to its *cis*-regulatory sequence, which is present in the control region of each gene. When the body has recovered and the hormone is no longer present, the expression of each of these genes drops to its normal level in the liver. In this way, a single transcription regulator can rapidly control the expression of many different genes (Figure 7–38).

The effects of the glucocorticoid receptor are not confined to cells of the liver. In other cell types, activation of this transcription regulator by hormone also causes changes in the expression levels of many genes; the genes affected, however, are usually different from those affected in liver cells. As we have seen, each cell type has an individualized set of transcription regulators, and because of combinatorial control, these critically influence the action of the glucocorticoid receptor. Because the receptor is able to assemble with many different sets of cell-type-specific transcription regulators, switching it on with hormone produces a different spectrum of effects in each cell type.

Differentiated Cells Maintain Their Identity

Once a cell has become differentiated into a particular cell type, it will generally remain differentiated, and all its progeny cells will remain that same cell type. Some highly specialized cells, including skeletal muscle cells and neurons, never divide again once they have differentiated—that is, they are *terminally differentiated* (as discussed in Chapter 17). But many other differentiated cells—such as

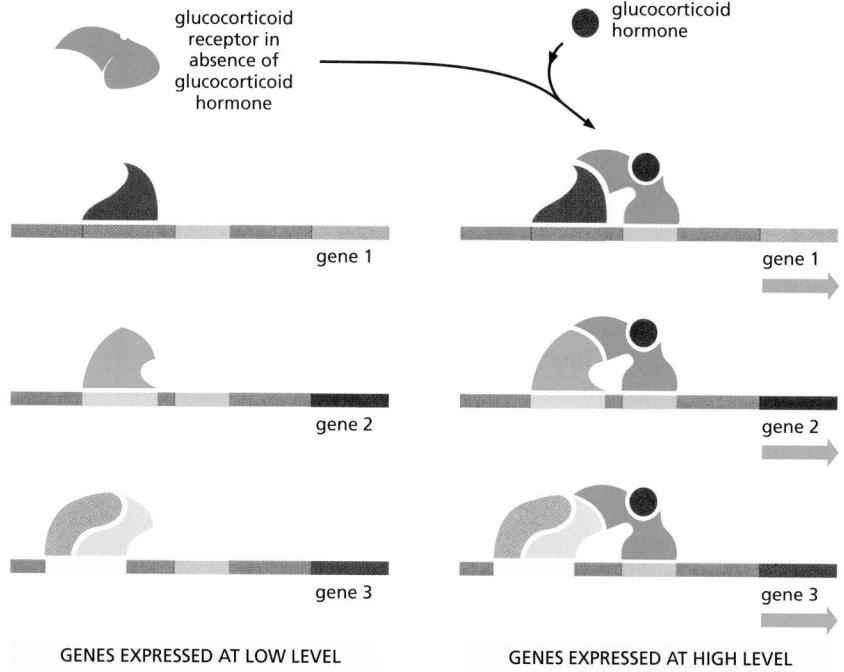

GENES EXPRESSED AT LOW LEVEL

GENES EXPRESSED AT HIGH LEVEL

Figure 7–38 A single transcription regulator can coordinate the expression of many different genes. The action of the glucocorticoid receptor is illustrated schematically. On the *left* is a series of genes, each of which has various transcription regulators bound to its regulatory region. However, these bound proteins are not sufficient on their own to fully activate transcription. On the *right* is shown the effect of adding an additional transcription regulator—the glucocorticoid receptor in a complex with glucocorticoid hormone—that has a *cis*-regulatory sequence in the control region of each gene. The glucocorticoid receptor completes the combination of transcription regulators required for maximal initiation of transcription, and the genes are now switched on as a set. When the hormone is no longer present, the glucocorticoid receptor dissociates from DNA and the genes return to their pre-stimulated levels.

fibroblasts, smooth muscle cells, and liver cells—will divide many times in the life of an individual. When they do, these specialized cell types give rise only to cells like themselves: smooth muscle cells do not give rise to liver cells, nor liver cells to fibroblasts.

For a proliferating cell to maintain its identity—a property called **cell memory**—the patterns of gene expression responsible for that identity must be remembered and passed on to its daughter cells through subsequent cell divisions. Thus, in the model we discussed in Figure 7–33, the production of each transcription regulator, once begun, has to be continued in the daughter cells of each cell division. How is such perpetuation accomplished?

Cells have several ways of ensuring that their daughters "remember" what kind of cells they are. One of the simplest and most important is through a positive feedback loop, where a master cell-type transcription regulator activates transcription of its own gene, in addition to that of other cell-type-specific genes. Each time a cell divides, the regulator is distributed to both daughter cells, where it continues to stimulate the positive feedback loop, making more of itself each division. Positive feedback is crucial for establishing "self-sustaining" circuits of gene expression that allow a cell to commit to a particular fate—and then to transmit that information to its progeny (**Figure 7–39**).

As was previously shown in Figure 7–37B, the master regulators needed to maintain the pluripotency of iPS cells bind to *cis*-regulatory sequences in their own control regions, providing examples of the type of positive feedback loop. In addition, most of these pluripotent cell regulators also activate transcription of other master regulators, resulting in a complex series of indirect feedback loops. For example, if A activates B, and B activates A, this forms a positive feedback loop where A activates its own expression, albeit indirectly. The series of direct and indirect feedback loops observed in the iPS circuit is typical of other specialized cell circuits. Such a network structure strengthens cell memory, increasing the probability that a particular pattern of gene expression is transmitted through successive generations. For example, if the level of A drops below the critical threshold to stimulate its own synthesis, regulator B can rescue it. By successive application of this mechanism, a complex series of positive feedback loops among multiple transcription regulators can stably maintain a differentiated state through many cell divisions.

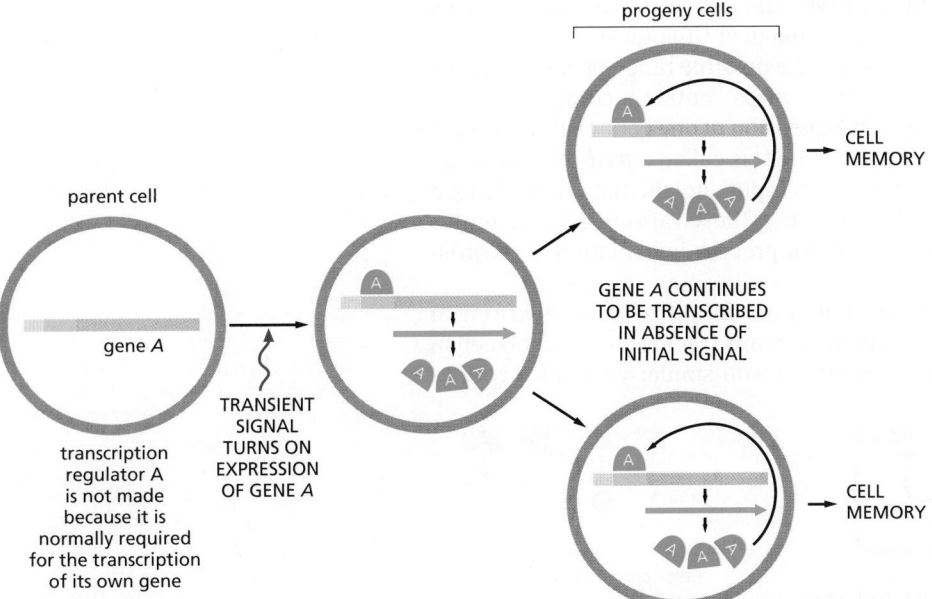

Figure 7–39 **A positive feedback loop can create cell memory.** Protein A is a master transcription regulator that activates the transcription of its own gene—as well as other cell-type-specific genes (not shown). All of the descendants of the original cell will therefore "remember" that the progenitor cell had experienced a transient signal that initiated the production of protein A.

Positive feedback loops formed by transcription regulators are probably the most prevalent way of ensuring that daughter cells remember what kind of cells they are meant to be, and they are found in all species on Earth. For example, many bacteria and single-cell eukaryotes form different types of cells, and positive feedback loops lie at the heart of mechanisms that maintain their cell types through many rounds of cell division. Plants and animals also make extensive use of transcription feedback loops; as we shall discuss later in the chapter, they have additional, more specialized mechanisms for making cell memory even stronger. But first, we briefly consider how combinations of transcription regulators and *cis*-regulatory sequences can be combined to create useful logic devices for the cell.

Transcription Circuits Allow the Cell to Carry Out Logic Operations

Simple gene regulatory switches can be combined to create all sorts of control devices, just as simple electronic switching elements in a computer can be linked to perform different types of operations. An analysis of gene regulatory circuits reveals that certain simple types of arrangements (called *network motifs*) are found over and over again in cells from widely different species. For example, positive and negative feedback loops are common in all cells (Figure 7–40). Whereas the former provides a simple memory device, the latter is often used to keep the expression of a gene close to a standard level despite the variations in biochemical conditions inside a cell. Suppose, for example, that a transcription repressor protein binds to the regulatory region of its own gene and exerts a strong negative feedback, such that transcription falls to a very low rate when the concentration of the repressor protein is above some critical value (determined by its affinity for its DNA binding site). The concentration of the protein can then be held close to the critical value, since any circumstance that causes a fall below that value can lead to a steep increase in synthesis, and any that causes a rise above that value will lead to synthesis being switched off. Such adjustments will, however, take time, so that an abrupt change of conditions will cause a disturbance of gene expression that is strong but transient. If there is a delay in the feedback loop, the result may be spontaneous oscillations in the expression of the gene (see Figure 15–18). The different types of behavior produced by a feedback loop will depend on the details of the system; for example, how tightly the transcription regulator binds to its *cis*-regulatory sequence, its rate of synthesis, and its rate of decay. We discuss these issues in quantitative terms and in more detail in Chapter 8.

With two or more transcription regulators, the possible range of circuit behaviors becomes more complex. Some bacterial viruses contain a common type of two-gene circuit that can flip-flop between expression of one gene and expression of the other. Another common circuit arrangement is called a *feed-forward* loop; such a loop can serve as a filter, responding to input signals that are prolonged but disregarding those that are brief (Figure 7–41). These various network motifs resemble miniature logic devices, and they can process information in surprisingly sophisticated ways.

The simple types of devices just illustrated are found to be interwoven in eukaryotic cells, creating exceedingly complex circuits (Figure 7–42). Each cell in a developing multicellular organism is equipped with similarly complex control

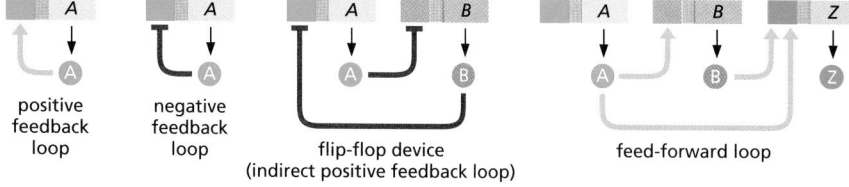

positive feedback loop

negative feedback loop

flip-flop device (indirect positive feedback loop)

feed-forward loop

Figure 7–40 Common types of network motifs in transcription circuits. A and B represent transcription regulators, arrows indicate positive transcription control, while lines with bars depict negative transcription control. In the feed-forward loop, A and B represent transcription regulators that both activate the transcription of target gene Z (see also Figure 8–86).

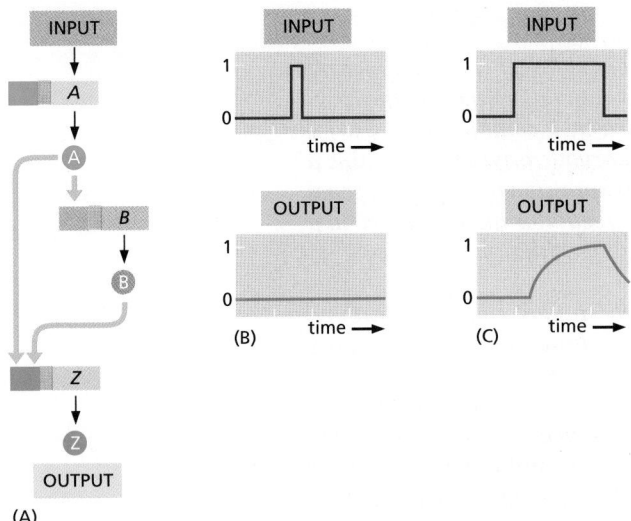

Figure 7–41 **How a feed-forward loop can measure the duration of a signal.**
(A) In this theoretical example, transcription regulators A and B are both required for transcription of *Z*, and A becomes active only when an input signal is present.
(B) If the input signal to A is brief, A does not stay active long enough for B to accumulate, and the *Z* gene is not transcribed. (C) If the signal to A persists, B accumulates, A remains active, and *Z* is transcribed. This arrangement allows the cell to ignore rapid fluctuations of the input signal and respond only to persistent levels. This strategy could be used, for example, to distinguish between random noise and a true signal.

The behavior shown here was computed for one particular set of parameter values describing the quantitative properties of A, B, and the product of Z, along with their syntheses. With different values of these parameters, feed-forward loops can in principle perform other types of "calculations." Many feed-forward loops have been discovered in cells, and theoretical analysis helps researchers to discern—and subsequently test—the different ways in which they may function (see Figure 8–86). (Adapted from S.S. Shen-Orr et al., *Nat. Genet.* 31:64–68, 2002.)

machinery, and it must, in effect, use its intricate system of interlocking transcription switches to compute how it should behave at each time point in response to the many different past and present inputs received. We are only beginning to understand how to study such complex intracellular control networks. Indeed, without new approaches, coupled with quantitative information that is far more precise and complete than we now possess, it will be impossible to predict the behavior of a system such as that shown in Figure 7–42. As explained in Chapter 8, a circuit diagram by itself is not enough.

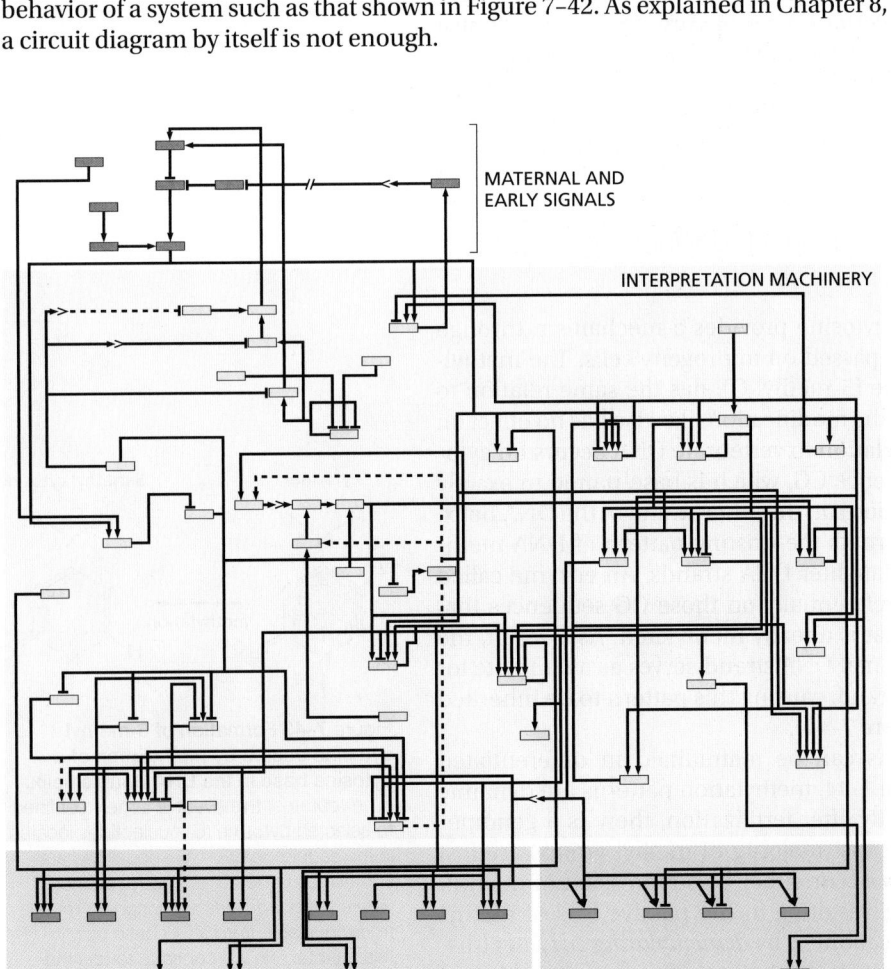

Figure 7–42 **The exceedingly complex gene circuit that specifies a portion of the developing sea urchin embryo.** Each *colored small box* represents a different gene. Those in *yellow* code for transcription regulators and those in *green* and *blue* code for proteins that give cells of the mesoderm and endoderm, respectively, their specialized characteristics. Genes depicted in *gray* are largely active in the mother and provide the egg with cues needed for proper development. As in Figure 7–40, arrows depict instances in which a transcription regulator activates the transcription of another gene. Lines ending in bars indicate examples of gene repression. (I.S. Peter and E.H. Davidson, *Nature* 474:635–639, published 2011 by Macmillan Publishers Ltd. Reproduced with permission of SNCSC.)

Summary

The many types of cells in animals and plants are created largely through mechanisms that cause different sets of genes to be transcribed in different cells. The transcription of any particular gene is generally controlled by a combination of transcription regulators. Each type of cell in a higher eukaryotic organism contains a specific set of transcription regulators that ensures the expression of only those genes appropriate to that type of cell. A given transcription regulator may be active in a variety of circumstances and is typically involved in the regulation of many different genes.

Since specialized animal cells can maintain their unique character through many cell-division cycles, and even when grown in culture, the gene regulatory mechanisms involved in creating them must be stable once established and heritable when the cell divides. These features reflect the cell's memory of its developmental history. Direct or indirect positive feedback loops, which enable transcription regulators to perpetuate their own synthesis, provide the simplest mechanism for cell memory. Transcription circuits also provide the cell with the means to carry out other types of logic operations. Simple transcription circuits combined into large regulatory networks drive highly sophisticated programs of embryonic development that will require new approaches to decipher.

MECHANISMS THAT REINFORCE CELL MEMORY IN PLANTS AND ANIMALS

Thus far in this chapter, we have emphasized the regulation of gene transcription by proteins that associate either directly or indirectly with DNA. However, DNA itself can be covalently modified, and certain types of chromatin states appear to be inherited. In this section, we shall see how these phenomena also provide opportunities for the regulation of gene expression. At the end of this section, we discuss how, in mice and humans, an entire chromosome can be transcriptionally inactivated using such mechanisms, and how this state can be maintained through many cell divisions.

Patterns of DNA Methylation Can Be Inherited When Vertebrate Cells Divide

In vertebrate cells, the methylation of cytosine provides a mechanism through which gene expression patterns can be passed on to progeny cells. The methylated form of cytosine, 5-methyl cytosine (5-methyl C), has the same relation to cytosine that thymine has to uracil, and the modification likewise has no effect on base-pairing (**Figure 7–43**). **DNA methylation** in vertebrate DNA occurs on cytosine (C) nucleotides largely in the sequence CG, which is base-paired to exactly the same sequence (in opposite orientation) on the other strand of the DNA helix. Consequently, a simple mechanism permits the existing pattern of DNA methylation to be inherited directly by the daughter DNA strands. An enzyme called *maintenance methyl transferase* acts preferentially on those CG sequences that are base-paired with a CG sequence that is already methylated. As a result, the pattern of DNA methylation on the parental DNA strand serves as a template for the methylation of the daughter DNA strand, causing this pattern to be inherited directly following DNA replication (**Figure 7–44**).

Although DNA methylation patterns can be maintained in differentiated cells by the mechanism shown in Figure 7–44, methylation patterns are dynamic during mammalian development. Shortly after fertilization, there is a genome-wide wave of demethylation, when the vast majority of methyl groups are lost from the DNA. This demethylation may occur either by suppression of maintenance DNA methyl transferase activity, resulting in the passive loss of methyl groups during each round of DNA replication, or by *demethylating enzymes* (discussed below). Later in development, new methylation patterns are established by several *de novo DNA methyl transferases* that are directed to DNA by sequence

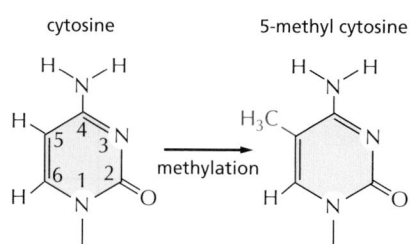

Figure 7–43 Formation of 5-methyl cytosine occurs by methylation of a cytosine base in the DNA double helix. In vertebrates, this event is largely confined to selected cytosine (C) nucleotides located in the sequence CG. CG sequences are sometimes denoted as CpG sequences, where the p indicates a phosphate linkage to distinguish it from a CG base pair. In this chapter, we will continue to use the simpler nomenclature CG to indicate this dinucleotide.

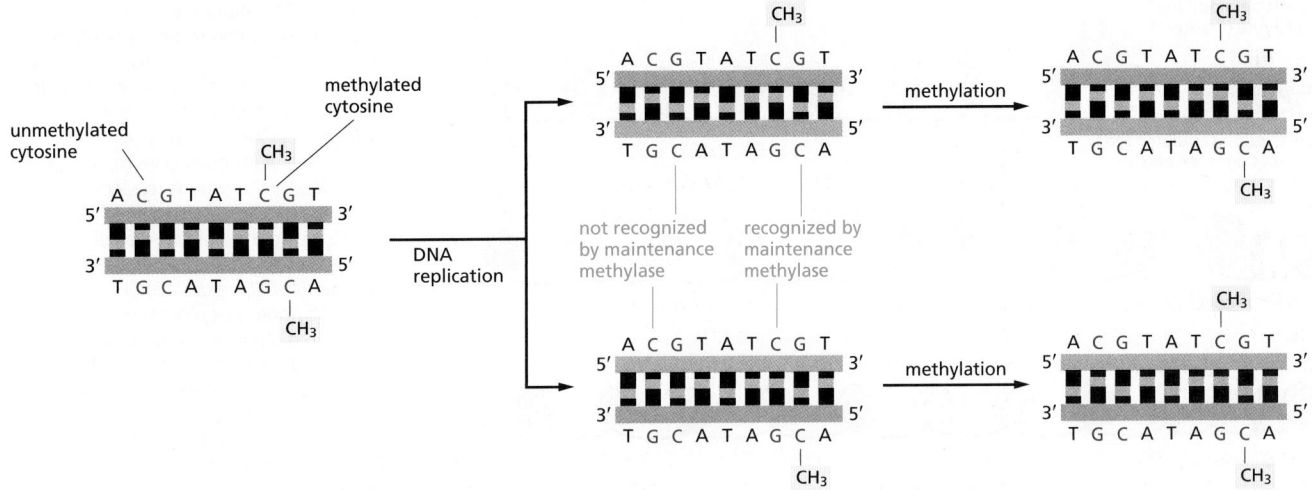

Figure 7–44 **How DNA methylation patterns are faithfully inherited.** In vertebrate DNA, a large fraction of the cytosine nucleotides in the sequence CG is methylated (see Figure 7–43). Because of the existence of a methyl-directed methylating enzyme (the maintenance methyl transferase), once a pattern of DNA methylation is established, that pattern of methylation is inherited in the progeny DNA, as shown.

specific DNA-binding proteins. Once the new patterns of methylation are established, they can be propagated through rounds of DNA replication by the maintenance methyl transferases.

DNA methylation has several uses in the vertebrate cell. A very important role is to work in conjunction with other gene expression control mechanisms to establish a particularly efficient form of gene repression. This combination of mechanisms ensures that unneeded eukaryotic genes can be repressed to very high degrees. For example, the rate at which a vertebrate gene is transcribed can vary 10^6-fold between one tissue and another. The unexpressed vertebrate genes are much less "leaky" in terms of transcription than bacterial genes, in which the largest known differences in transcription rates between expressed and unexpressed gene states are about 1000-fold.

DNA methylation helps to repress transcription in several ways. The methyl groups on methylated cytosines lie in the major groove of DNA and interfere directly with the binding of proteins (transcription regulators as well as the general transcription factors) required for transcription initiation. In addition, the cell contains a repertoire of proteins that bind specifically to methylated DNA. The best characterized of these associate with histone modifying enzymes, leading to a repressive chromatin state where chromatin structure and DNA methylation act synergistically (**Figure 7–45**). One reflection of the importance of DNA methylation to humans is the widespread involvement of "incorrect" DNA methylation patterns in cancer progression (discussed in Chapter 20).

CG-Rich Islands Are Associated with Many Genes in Mammals

Because of the way in which DNA repair enzymes work, methylated C nucleotides in the vertebrate genome tend to be eliminated in the course of evolution. Accidental deamination of an unmethylated C gives rise to U (see Figure 5–38), which is not normally present in DNA and thus is recognized easily by the DNA repair enzyme uracil DNA glycosylase, excised, and then replaced with a C (as discussed in Chapter 5). But accidental deamination of a 5-methyl C cannot be repaired in this way, for the deamination product is a T and so is indistinguishable from the other, nonmutant T nucleotides in the DNA. Although a special repair system exists to remove these mutant T nucleotides, many of the deaminations escape detection, so that those C nucleotides in the genome that are methylated tend to mutate to T over evolutionary time.

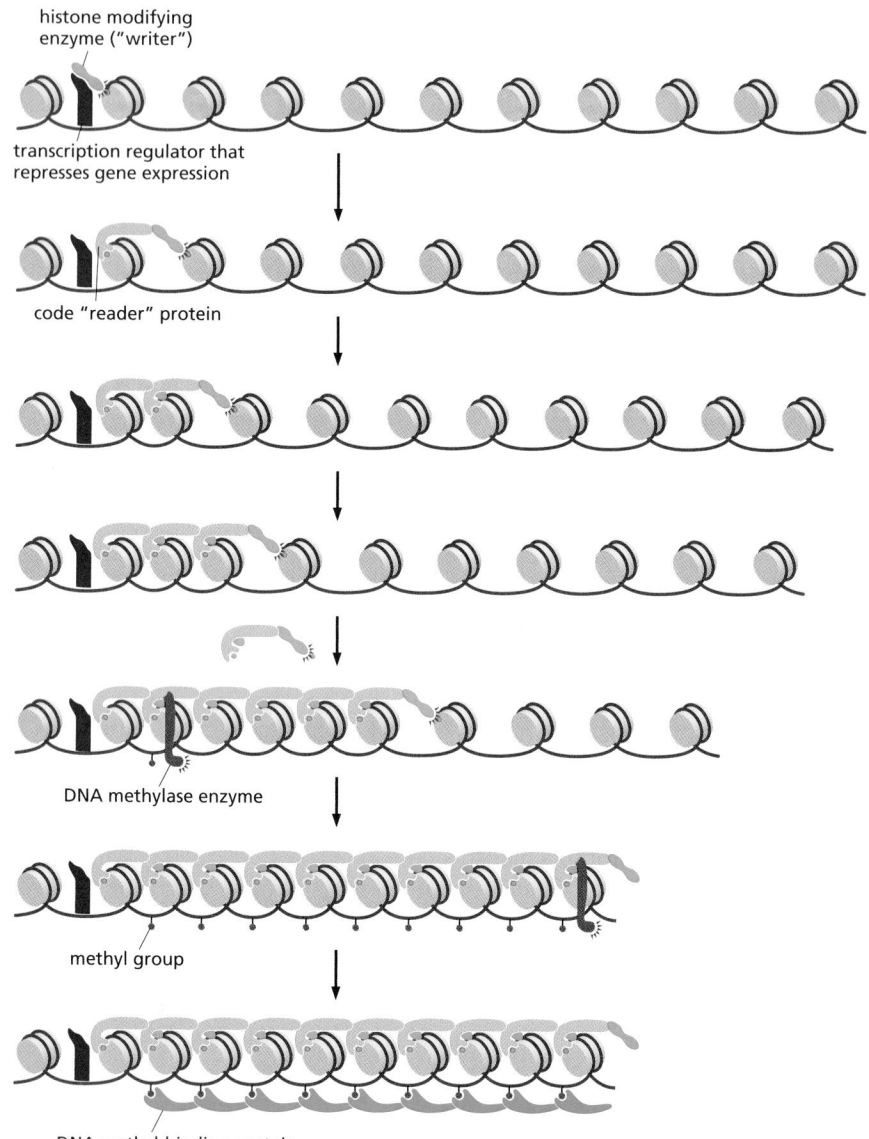

histone modifying
enzyme ("writer")

transcription regulator that
represses gene expression

code "reader" protein

DNA methylase enzyme

methyl group

DNA methyl-binding protein

Figure 7–45 Multiple mechanisms contribute to stable gene repression. In this schematic example, histone reader and writer proteins (discussed in Chapter 4), under the direction of transcription regulators, establish a repressive form of chromatin. A *de novo* DNA methylase is attracted by the histone reader and methylates nearby cytosines in DNA, which are, in turn, bound by DNA methyl-binding proteins. During DNA replication, some of the modified *(blue dot)* histones will be inherited by one daughter chromosome, some by the other, and in each daughter they can induce reconstruction of the same pattern of chromatin modifications (discussed in Chapter 4). At the same time, the mechanism shown in Figure 7–44 will cause both daughter chromosomes to inherit the same methylation pattern. In these cases where DNA methylation stimulates the activity of the histone writer, the two inheritance mechanisms will be mutually reinforcing. This scheme can account for the inheritance by daughter cells of both the histone and the DNA modifications. It can also explain the tendency of some chromatin modifications to spread along a chromosome (see Figure 4–44).

During the course of evolution, more than three out of every four CGs have been lost in this way, leaving vertebrates with a remarkable deficiency of this dinucleotide. The CG sequences that remain are very unevenly distributed in the genome; they are present at 10 times their average density in selected regions, called **CG islands**, which average 1000 nucleotide pairs in length. The human genome contains roughly 20,000 CG islands and they usually include promoters of genes. For example, 60% of human protein-coding genes have promoters embedded in CG islands and these include virtually all the promoters of the so-called *housekeeping genes*—those genes that code for the many proteins that are essential for cell viability and are therefore expressed in nearly all cells (**Figure 7–46**). Over evolutionary timescales, the CG islands were spared the accelerated mutation rate of bulk CG sequences because they remained unmethylated in the germ line (**Figure 7–47**).

CG islands also remain unmethylated in most somatic tissues whether or not the associated gene is expressed. The unmethylated state is maintained by sequence-specific DNA-binding proteins, many of whose *cis*-regulatory sequences contain a CG. By binding to these sequences, which are spread across CG islands, they protect the DNA from methyl transferases. These proteins also recruit *DNA demethylases*, which convert 5-methyl C to hydroxy-methyl C, which

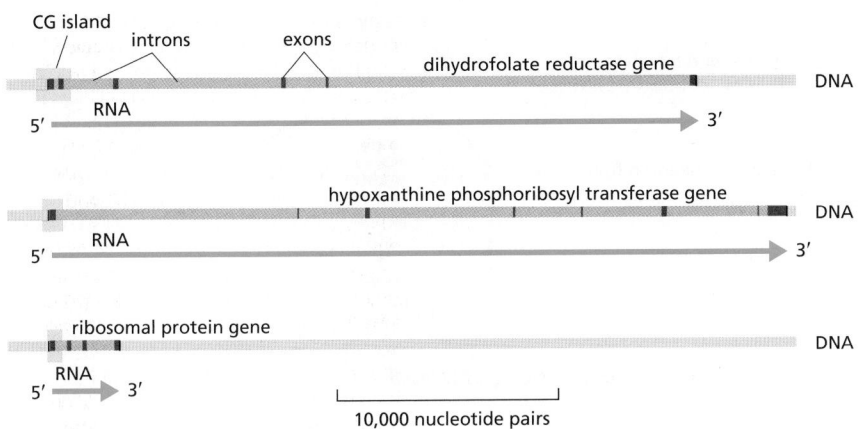

Figure 7–46 **The CG islands surrounding the promoter in three mammalian housekeeping genes.** The *yellow boxes* show the extent of each island. As for most genes in mammals, the exons *(dark red)* are very short relative to the introns *(light red)*. (Adapted from A.P. Bird, *Trends Genet.* 3:342–347, 1987.)

is later replaced by C either through DNA repair (see Figure 5–41A) or, passively, through multiple rounds of DNA replication. Unmethylated CG islands have several properties that make them particularly suitable for promoters. For example, some of the same proteins that bind to CG islands and protect them from methylation recruit histone modifying enzymes that make the islands particularly "promoter friendly." As a result, RNA polymerase is often found bound to promoters within CG islands, even when the associated gene is not being actively transcribed. At unmethylated CG islands, the balance between polymerase and nucleosome assembly is thus tipped toward the former. Additional steps are needed to "push" the bound polymerase into transcribing the adjacent gene, and these are directed by transcription regulators that bind to *cis*-regulatory sequences of DNA (often well upstream from the CG islands). These regulators serve to release the polymerase with the appropriate elongation factors (see Figure 7–21C and D).

Genomic Imprinting Is Based on DNA Methylation

Mammalian cells are diploid, containing one set of genes inherited from the father and one set from the mother. The expression of a small minority of genes depends on whether they have been inherited from the mother or the father: when the paternally inherited gene copy is active, the maternally inherited gene copy is silent, or vice versa. This phenomenon is called **genomic imprinting**.

Roughly 300 genes are imprinted in humans. Because only one copy of an imprinted gene is expressed, imprinting can "unmask" mutations that would normally be covered by the other, functional copy. For example, Angelman syndrome, a disorder of the nervous system in humans that causes reduced mental ability and severe speech impairment, results from a gene deletion on one chromosomal homolog and the silencing, by imprinting, of the intact gene on the other homolog.

The *insulin-like growth factor-2* (*Igf2*) gene in the mouse provides a well-studied example of imprinting. Mice that do not express *Igf2* at all are born half the size of normal mice. However, only the paternal copy of *Igf2* is transcribed, and only this gene copy matters for the phenotype. As a result, mice with a mutated paternally derived *Igf2* gene are stunted, while mice with a mutated maternally derived *Igf2* gene are normal.

Figure 7–47 **A mechanism to explain both the marked overall deficiency of CG sequences and their clustering into CG islands in vertebrate genomes.** *White lines* mark the location of CG dinucleotides in the DNA sequences, while *red circles* indicate the presence of a methyl group on the CG dinucleotide. CG sequences that lie in regulatory sequences of genes that are transcribed in germ cells are unmethylated and therefore tend to be retained in evolution. Methylated CG sequences, on the other hand, tend to be lost through deamination of 5-methyl C to T, unless the CG sequence is critical for survival.

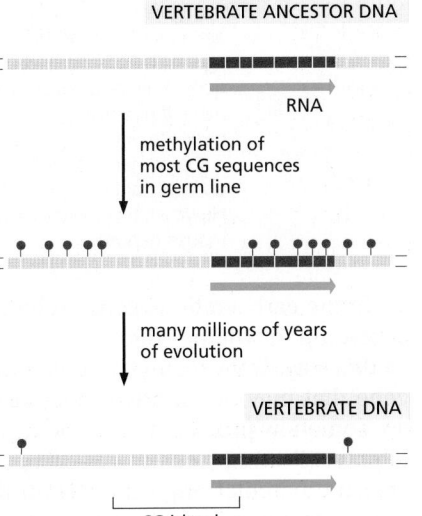

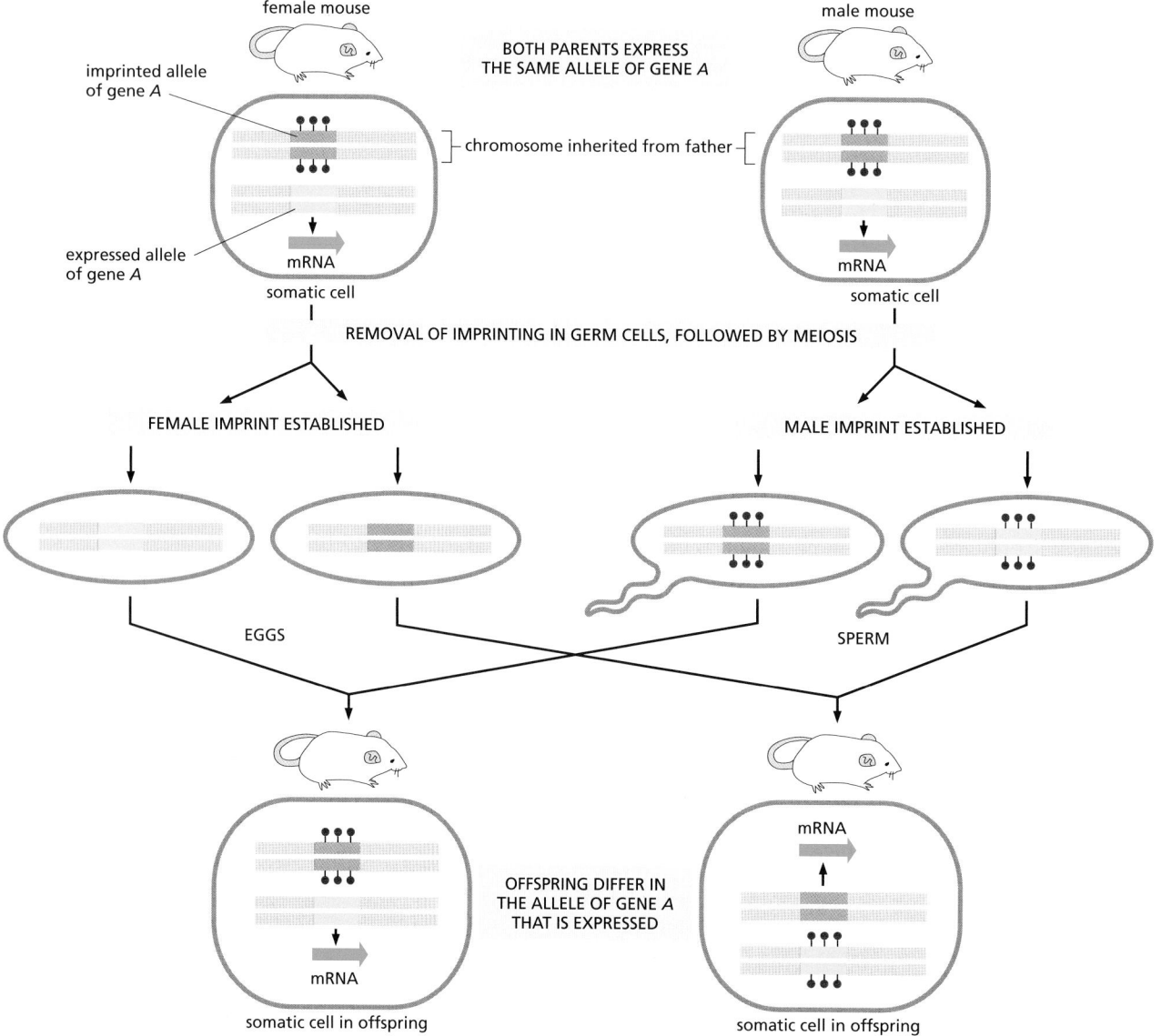

Figure 7–48 **Imprinting in the mouse.** The *top* portion of the figure shows a pair of homologous chromosomes in the somatic cells of two adult mice, one male and one female. In this example, both mice have inherited the top homolog from their father and the bottom homolog from their mother, and the paternal copy of a gene subject to imprinting (indicated in *orange*) is methylated, preventing its expression. The maternally derived copy of the same gene *(yellow)* is expressed. The remainder of the figure shows the outcome of a cross between these two mice. During germ-cell formation, but before meiosis, the imprints are erased and then, much later in germ-cell development, they are reimposed in a sex-specific pattern (*middle* portion of figure). In eggs produced from the female, neither allele of the *A* gene is methylated. In sperm from the male, both alleles of gene *A* are methylated. Shown at the *bottom* of the figure are two of the possible imprinting patterns inherited by the progeny mice; the mouse on the *left* has the same imprinting pattern as each of the parents, whereas the mouse on the *right* has the opposite pattern. If the two alleles of gene *A* are distinct, these different imprinting patterns can cause phenotypic differences in the progeny mice, even though they carry exactly the same DNA sequences of the two *A* gene alleles. Imprinting provides an important exception to classical genetic behavior, and several hundred mouse genes are thought to be affected in this way. However, the majority of mouse genes are not imprinted, and therefore the rules of Mendelian inheritance apply to most of the mouse genome.

In the early embryo, genes subject to imprinting are marked by methylation according to whether they were derived from a sperm or an egg chromosome. In this way, DNA methylation is used as a mark to distinguish two copies of a gene that may be otherwise identical (**Figure 7–48**). Because imprinted genes are somehow protected from the wave of demethylation that takes place shortly after fertilization (see pp. 404–405), this mark enables somatic cells to "remember" the parental origin of each of the two copies of the gene and to regulate their expression accordingly. In most cases, the methyl imprint silences nearby

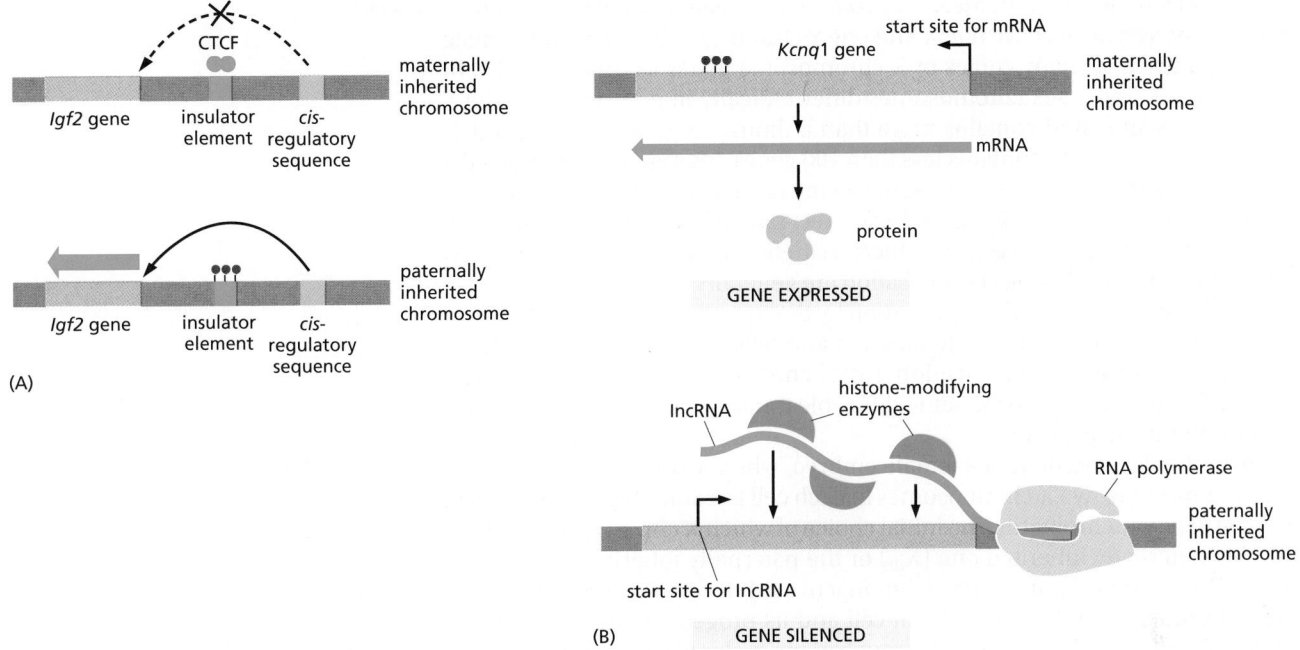

gene expression. In some cases, however, it can activate expression of a gene. In the case of *Igf2*, for example, methylation of an insulator element on the paternally derived chromosome blocks its function and allows distant *cis*-regulatory sequences to activate transcription of the *Igf2* gene. On the maternally derived chromosome, the insulator is not methylated and the *Igf2* gene is therefore not transcribed (**Figure 7–49**A).

Other cases of imprinting involve *long noncoding RNAs*, which are defined as RNA molecules over 200 nucleotides in length that do not code for proteins. We discuss lncRNAs broadly at the end of this chapter; here, we focus on the role of a specific lncRNA in imprinting. In the case of the *Kcnq1* gene, which codes for a voltage-gated calcium channel needed for proper heart function, the lncRNA is made from the paternal allele (which is unmethylated) but it is not released by the RNA polymerase, remaining instead at its site of synthesis on the DNA template. This RNA in turn recruits histone-modifying and DNA-methylating enzymes that direct the formation of repressive chromatin, which silences the protein-coding gene associated on the paternally derived chromosome (Figure 7–49B). The maternally derived gene, on the other hand, is immune to these effects because the specific methylation present from imprinting blocks the synthesis of the lncRNA but allows transcription of the protein-coding gene. Like *Igf2*, the specificity of *Kcnq1* imprinting arises from the inherited methylation patterns; the difference lies in the way these patterns bring about differential expression of the imprinted gene.

Why imprinting should exist at all is a mystery. In vertebrates, it is restricted to placental mammals, and many of the imprinted genes are involved in fetal development. One idea is that imprinting reflects a middle ground in the evolutionary struggle between males to produce larger offspring and females to limit offspring size. Whatever its purpose might be, imprinting provides startling evidence that features of DNA other than its sequence of nucleotides can be inherited.

Figure 7–49 Mechanisms of imprinting. (A) On chromosomes inherited from the female, a protein called CTCF binds to an insulator (see Figure 7–24), blocking communication between *cis*-regulatory sequences *(green)* and the *Igf2* gene *(orange)*. *Igf2* is therefore not expressed from the maternally inherited chromosome. Because of imprinting, the insulator on the male-derived chromosome is methylated *(red circles)*; this inactivates the insulator by blocking the binding of the CTCF protein, and allows the *cis*-regulatory sequences to activate transcription of the *Igf2* gene. In other examples of imprinting, methylation simply blocks gene expression by interfering with the binding of proteins required for a gene's transcription. (B) Imprinting of the mouse *Kcnq1* gene. On the maternally derived chromosome, synthesis of the lncRNA is blocked by methylation of the DNA *(red circles)*, and the *Kcnq1* gene is expressed. On the paternally derived chromosome, the lncRNA is synthesized, remains in place, and by directing alterations in chromatin structure blocks expression of the *Kcnq1* gene. Although shown as directly binding to lncRNA, the histone-modifying enzymes are likely to be recruited indirectly, through additional proteins.

Chromosome-Wide Alterations in Chromatin Structure Can Be Inherited

We have seen that DNA methylation and certain types of chromatin structure can be heritable, preserving patterns of gene expression across cell generations. Perhaps the most striking example of this effect occurs in mammals, in which an alteration in the chromatin structure of an entire chromosome can modulate the levels of expression of most genes on that chromosome.

Males and females differ in their *sex chromosomes*. Females have two X chromosomes, whereas males have one X and one Y chromosome. As a result, female cells contain twice as many copies of X-chromosome genes as do male cells. In mammals, the X and Y sex chromosomes differ radically in gene content: the X chromosome is large and contains more than a thousand genes, whereas the Y chromosome is small and contains less than 100 genes. Mammals have evolved a *dosage compensation* mechanism to equalize the dosage of X-chromosome gene products between males and females. The correct ratio of X chromosome to *autosome* (non-sex chromosome) gene products is carefully controlled, and mutations that interfere with this dosage compensation are generally lethal.

Mammals achieve dosage compensation by the transcriptional inactivation of one of the two X chromosomes in female somatic cells, a process known as **X-inactivation**. As a result of X-inactivation, two X chromosomes can coexist within the same nucleus, exposed to the same diffusible transcription regulators, yet differ entirely in their expression.

Early in the development of a female embryo, when it consists of a few hundred cells, one of the two X chromosomes in each cell becomes highly condensed into a type of heterochromatin. The initial choice of which X chromosome to inactivate, the maternally inherited one (X_m) or the paternally inherited one (X_p), is random. Once either X_p or X_m has been inactivated, it remains silent throughout all subsequent cell divisions of that cell and its progeny, indicating that the inactive state is faithfully maintained through many cycles of DNA replication and mitosis. Because X-inactivation is random and takes place after several hundred cells have already formed in the embryo, every female is a mosaic of clonal groups of cells in which either X_p or X_m is silenced (**Figure 7–50**). These clonal groups are

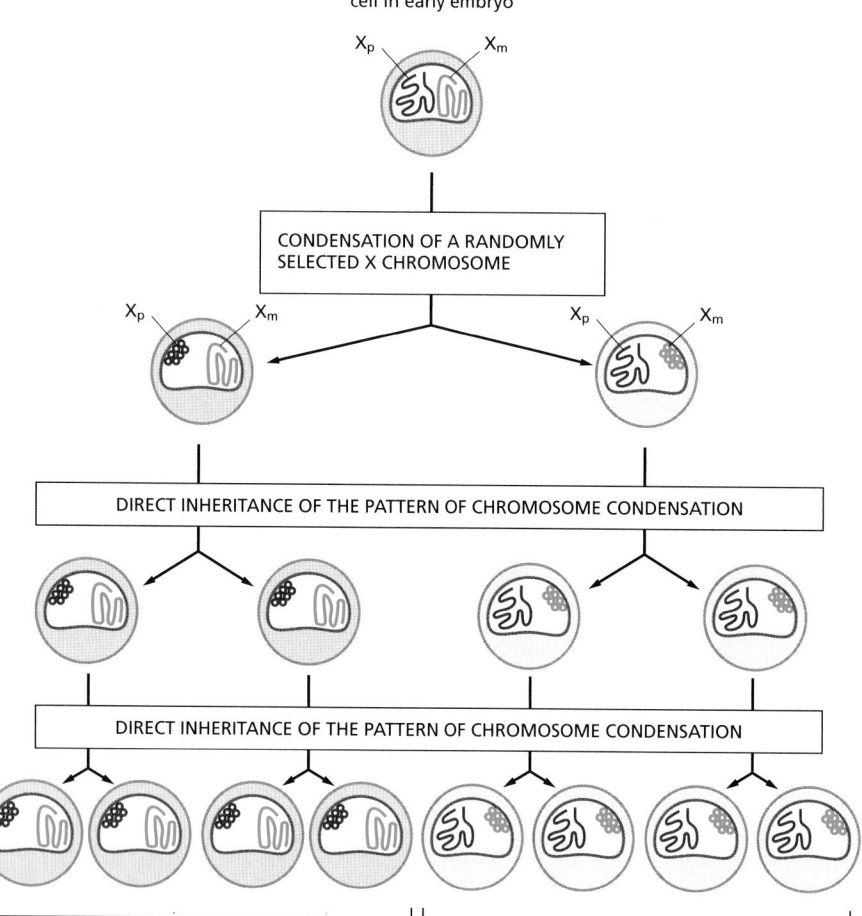

Figure 7–50 **X-inactivation.** The clonal inheritance in female mammals of a condensed, inactive X chromosome.

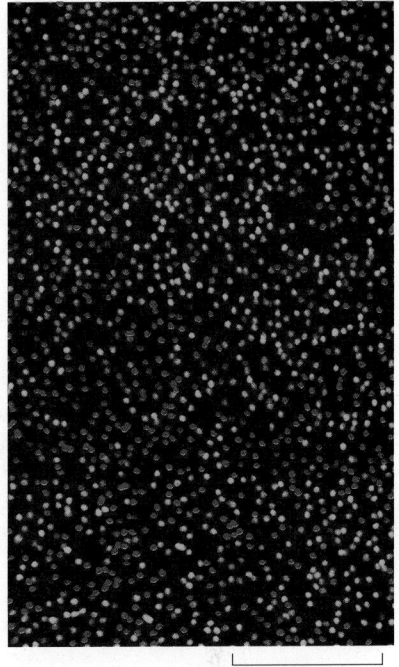

Figure 7–51 **Photoreceptor cells in the retina of a female mouse showing patterns of X-chromosome inactivation.** Using genetic engineering techniques (described in Chapter 8), the germ line of a mouse was modified so that one copy of the X chromosome (if active) makes a green fluorescent protein and the other a red fluorescent protein. Both proteins concentrate in the nucleus and, in the field of cells shown here, it is clear that only one of the two X chromosomes is active in each cell. (From H. Wu et al., *Neuron* 81:103–119, 2014. With permission from Elsevier.)

50 μm

distributed in small clusters in the adult animal because sister cells tend to remain close together during later stages of development (**Figure 7–51**). For example, X-chromosome inactivation causes the orange and black "tortoiseshell" coat coloration of some female cats. In these cats, one X chromosome carries a gene that produces orange hair color, and the other X chromosome carries an allele of the same gene that results in black hair color; it is the random X-inactivation that produces patches of cells of two distinctive colors. In contrast, male cats of this genetic stock are either solid orange or solid black, depending on which X chromosome they inherit from their mothers. Although X-chromosome inactivation is maintained over thousands of cell divisions, it is reversed during germ-cell formation, so that all haploid oocytes contain an active X chromosome and can express X-linked gene products.

How is an entire chromosome transcriptionally inactivated? X-chromosome inactivation is initiated and spreads from a single site near the middle of the X chromosome, the **X-inactivation center** (**XIC**). Within the XIC is a transcribed 20,000-nucleotide lncRNA (called Xist), which is expressed solely from the inactive X chromosome. Xist RNA spreads from the XIC over the entire chromosome and directs gene silencing. Although we do not know exactly how this is accomplished, it likely involves recruitment of histone-modifying enzymes and other proteins to form a repressive form of chromatin analogous to that of Figure 7–45. Curiously, about 10% of the genes on the X chromosome (including *Xist* itself) escape this silencing and remain active.

The spread of Xist RNA along the X chromosome does not proceed linearly along the DNA. Rather, starting at its site of synthesis, it is first handed off across the base of the DNA loops that make up the chromosome; these shortcuts explain how Xist can spread rapidly, by a "hand-over-hand" mechanism, along the X chromosome once the inactivation process begins (**Figure 7–52**). It also helps to explain why the inactivation does not spread to the other, active X chromosome.

Imprinting and X-chromosome inactivation are examples of **monoallelic gene expression**, where in a diploid genome, only one of the two copies of a gene is expressed. In addition to the approximately 1000 genes on the X chromosome and the 300 or so genes that are imprinted, there are another 1000–2000 human genes that exhibit monoallelic expression. Like X-chromosome inactivation (but unlike imprinting), the choice of which copy of the gene is to be expressed and which is to be silenced often appears random. Yet once the choice is made, it can persist for many cell divisions. Because the choice is often made relatively late in development, cells of the same tissue in the same individual can express different copies of a given gene. In other words, somatic tissues are often mosaics, where different clones of cells have subtly different patterns of gene expression. The mechanisms responsible for this type of monoallelic expression are not known in detail, and its general purpose—if any—is poorly understood. Several different mechanisms may contribute to such epigenetic inheritance, as we explain next.

Epigenetic Mechanisms Ensure That Stable Patterns of Gene Expression Can Be Transmitted to Daughter Cells

As we have seen, once a cell in an organism differentiates into a particular cell type, it generally remains specialized in that way; if it divides, its daughters inherit the same specialized character. Perhaps the simplest way for a cell to remember

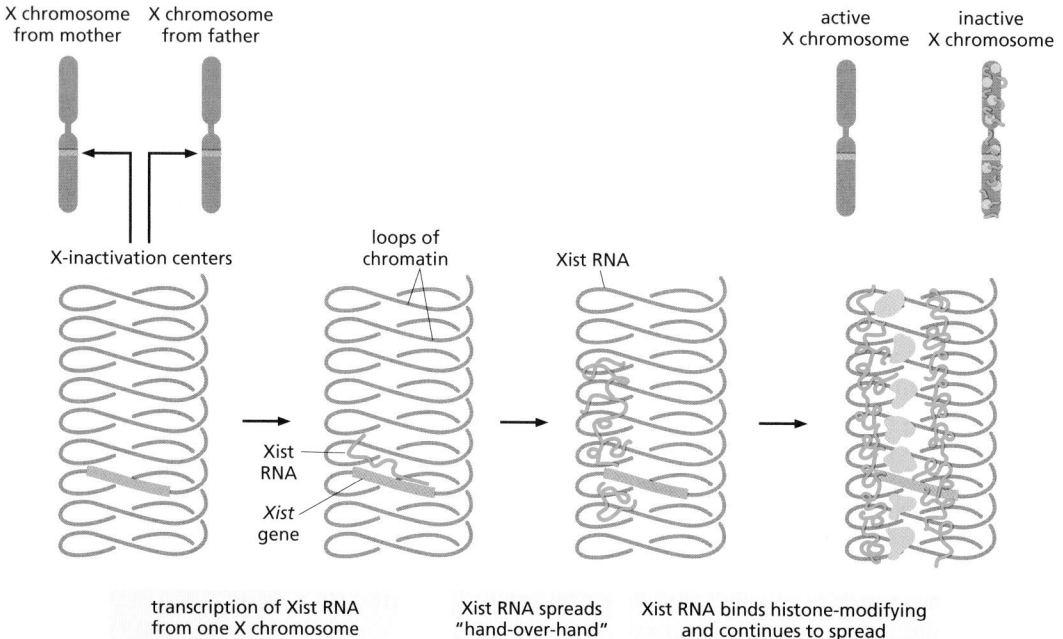

transcription of Xist RNA
from one X chromosome

Xist RNA spreads
"hand-over-hand"

Xist RNA binds histone-modifying
and continues to spread

its identity is through a positive feedback loop in which a key transcription regulator activates, either directly or indirectly, the transcription of its own gene (see Figure 7–39). Interlocking positive feedback loops of the type shown in Figure 7–37 provide greater stability by buffering the circuit against fluctuations in the level of any one transcription regulator. Because transcription regulators are synthesized in the cytosol and diffuse throughout the nucleus, feedback loops based on this mechanism will affect both copies of a gene in a diploid cell. However, as discussed in this section, the expression pattern of a gene on one chromosome can differ from the copy of the same gene on the other chromosome (as in X-chromosome inactivation or in imprinting), and such differences can also be inherited through many cell divisions.

The ability of a daughter cell to retain a memory of the gene expression patterns that were present in the parent cell is an example of **epigenetic inheritance**: a heritable alteration in a cell or organism's phenotype that does not result from changes in the nucleotide sequence of DNA (discussed in Chapter 4). (Unfortunately, the term epigenetic is sometimes also used to refer to all covalent modifications to histones and DNA, whether or not they are self-propagating; many of these modifications are erased each time a cell divides and do not generate cell memory.)

In **Figure 7–53**, we contrast two self-propagating epigenetic mechanisms that work in *cis*, affecting only one chromosomal copy with two self-propagating mechanisms that work in *trans*, affecting both chromosomal copies of a gene. Cells can combine these mechanisms to ensure that patterns of gene expression are maintained and inherited accurately and reliably—over a period of up to a hundred years or more, in our own case.

We can get some idea of the prevalence of epigenetic changes by comparing identical twins. Their genomes have the same sequence of nucleotides, and, obviously, many features of identical twins—such as their appearance—are strongly determined by the genome sequences they inherit. When their gene expression, histone modification, and DNA methylation patterns are compared, however, many differences are observed. Because these differences are roughly correlated not only with age but also with the time that the twins have spent apart from each other, it has been proposed that some of these differences are heritable from cell to cell and are the result of environmental factors. Although these studies are in early stages, the idea that environmental events can be permanently registered as epigenetic changes in our cells is a fascinating one that presents an important challenge to the next generation of biological scientists.

Figure 7–52 Mammalian X-chromosome inactivation. X-chromosome inactivation begins with the synthesis of Xist (X-inactivation specific transcript) RNA from the XIC (X-inactivation center) locus and moves outward to the chromosome ends. According to the model depicted here, the long (≈20,000 nucleotides) Xist RNA has many low-affinity binding sites for the structural components of chromosomes and spreads by releasing its hold on one portion of the chromosome while grasping another. The continued synthesis of Xist from the center of the chromosome drives it to the ends. As shown, Xist RNA does not move linearly along the chromosomal DNA, but, instead, moves first across the base of chromosome loops. It has been proposed that the portions of chromosomal DNA at the tips of long loops contain the 10% of genes that escape X-chromosome inactivation.

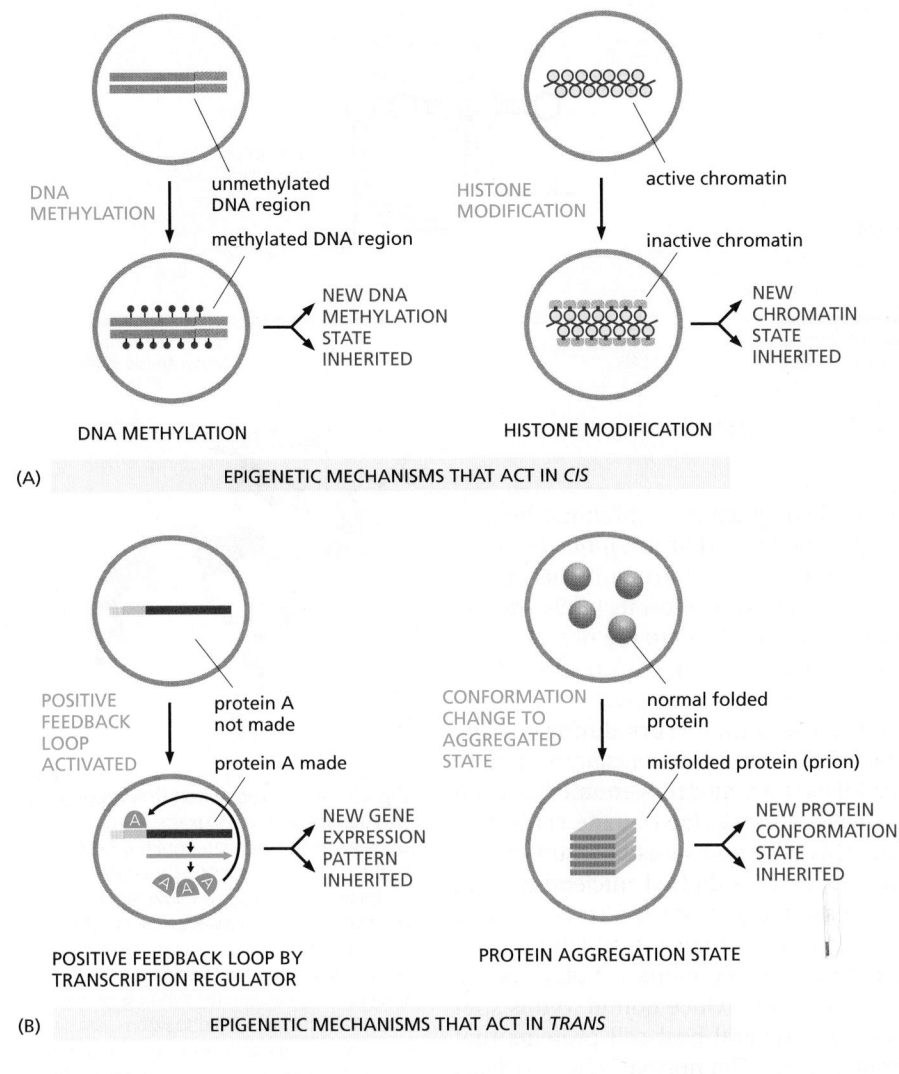

(A) EPIGENETIC MECHANISMS THAT ACT IN *CIS*

DNA METHYLATION

- unmethylated DNA region
- methylated DNA region
- NEW DNA METHYLATION STATE INHERITED

HISTONE MODIFICATION

- active chromatin
- inactive chromatin
- NEW CHROMATIN STATE INHERITED

(B) EPIGENETIC MECHANISMS THAT ACT IN *TRANS*

POSITIVE FEEDBACK LOOP BY TRANSCRIPTION REGULATOR

- protein A not made
- protein A made
- NEW GENE EXPRESSION PATTERN INHERITED

PROTEIN AGGREGATION STATE

- normal folded protein
- misfolded protein (prion)
- NEW PROTEIN CONFORMATION STATE INHERITED

Figure 7–53 Four distinct mechanisms that can produce an epigenetic form of inheritance in an organism. (A) Epigenetic mechanisms that act in *cis*. As discussed in this chapter, a maintenance methylase can propagate specific patterns of cytosine methylation (see Figure 7–44). As discussed in Chapter 4, a histone modifying enzyme that replicates the same modification that attracts it to chromatin can result in the modification being self-propagating (see Figure 4–44). (B) Epigenetic mechanisms that act in *trans*. Positive feedback loops, formed by transcriptional regulators are found in all species and are probably the most common form of cell memory. As discussed in Chapter 3, some proteins can form self-propagating prions (Figure 3–33). If these proteins are involved in gene expression, they can transmit patterns of gene expression to daughter cells.

Summary

Eukaryotic cells can use inherited forms of DNA methylation and inherited states of chromatin condensation as additional mechanisms for generating cell memory of gene expression patterns. An especially dramatic case that involves chromatin condensation is the inactivation of an entire X chromosome in female mammals. DNA methylation underlies the phenomenon in mammals of genomic imprinting, in which the expression of a gene depends on whether it was inherited from the mother or the father.

POST-TRANSCRIPTIONAL CONTROLS

In principle, every step required for the process of gene expression can be controlled. Indeed, one can find examples of each type of regulation, and many genes are regulated by multiple mechanisms. As we have seen, controls on the initiation of gene transcription are a critical form of regulation for all genes. But other controls can act later in the pathway from DNA to protein to modulate the amount of gene product that is made—and in some cases, to determine the exact amino acid sequence of the protein product. These **post-transcriptional controls**, which operate after RNA polymerase has bound to the gene's promoter and has begun RNA synthesis, are crucial for the regulation of many genes.

In the following sections, we consider the varieties of post-transcriptional regulation in temporal order, according to the sequence of events that an RNA molecule might experience after its transcription has begun (**Figure 7–54**).

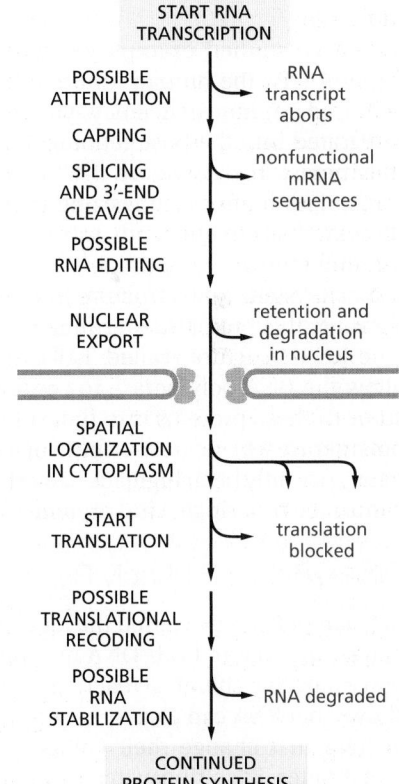

START RNA TRANSCRIPTION

- POSSIBLE ATTENUATION → RNA transcript aborts
- CAPPING
- SPLICING AND 3'-END CLEAVAGE → nonfunctional mRNA sequences
- POSSIBLE RNA EDITING
- NUCLEAR EXPORT → retention and degradation in nucleus
- SPATIAL LOCALIZATION IN CYTOPLASM
- START TRANSLATION → translation blocked
- POSSIBLE TRANSLATIONAL RECODING
- POSSIBLE RNA STABILIZATION → RNA degraded

CONTINUED PROTEIN SYNTHESIS

Figure 7–54 Post-transcriptional controls of gene expression. The final synthesis rate of a protein can, in principle, be controlled at any of the steps listed in capital letters. In addition, RNA splicing, RNA editing, and translation recoding can also alter the sequence of amino acids in a protein, making it possible for the cell to produce more than one protein variant from the same gene. Only a few of the steps depicted here are likely to be critical for the regulation of any one particular protein.

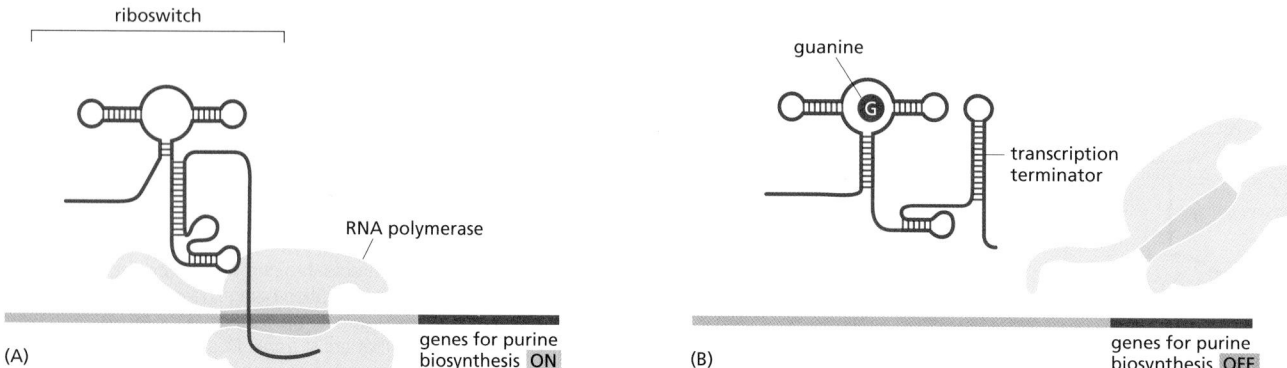

(A)   genes for purine biosynthesis ON

(B)   guanine — transcription terminator — genes for purine biosynthesis OFF

Transcription Attenuation Causes the Premature Termination of Some RNA Molecules

It has long been known that the expression of some genes is inhibited by premature termination of transcription, a phenomenon called *transcription attenuation*. In some of these cases, the nascent RNA chain adopts a structure that causes it to interact with the RNA polymerase in such a way as to abort its transcription. When the gene product is required, regulatory proteins bind to the nascent RNA chain and remove the attenuation, allowing the transcription of a complete RNA molecule.

A well-studied example of transcription attenuation occurs during the life cycle of HIV, the human immunodeficiency virus that is the causative agent of acquired immune deficiency syndrome, or AIDS. Once the HIV genome has been integrated into the host genome, the viral DNA is transcribed by the cell's RNA polymerase II (see Figure 5–62). However, this polymerase usually terminates transcription after synthesizing transcripts of several hundred nucleotides and therefore fails to efficiently transcribe the entire viral genome. When conditions for viral growth are optimal, a virus-encoded protein called Tat, which binds to a specific stem-loop structure in the nascent RNA that contains a "bulged base," prevents this premature termination (see Figure 6–89). Once bound to this specific RNA structure (called TAR), Tat assembles several host-cell proteins that allow the RNA polymerase to continue transcribing. The normal role of at least some of these proteins is to prevent pausing and premature termination by RNA polymerase when it transcribes normal cell genes. Thus, a normal cell mechanism has apparently been highjacked by HIV to permit transcription of its genome to be controlled by a single viral protein.

Riboswitches Probably Represent Ancient Forms of Gene Control

In Chapter 6, we discussed the idea that, before modern cells arose on Earth, RNA played the role of both DNA and proteins, both storing hereditary information and catalyzing chemical reactions (see pp. 362–366). The discovery of *riboswitches* shows that RNA can also form control devices. Riboswitches are short sequences of RNA that change their conformation on binding small molecules, such as metabolites. Each riboswitch recognizes a specific small molecule and the resulting conformational change is used to regulate gene expression. Riboswitches are often located near the 5′ end of mRNAs, and they fold while the mRNA is being synthesized, blocking or permitting progress of the RNA polymerase according to whether the regulatory small molecule is bound (**Figure 7–55**).

Riboswitches are particularly common in bacteria, in which they sense key small metabolites in the cell and adjust gene expression accordingly. Perhaps their most remarkable feature is the high specificity and affinity with which each recognizes only the appropriate small molecule; in many cases, every chemical feature of the small molecule is read by the RNA (Figure 7–55C). Moreover, the binding affinities observed are as tight as those typically observed between small molecules and proteins.

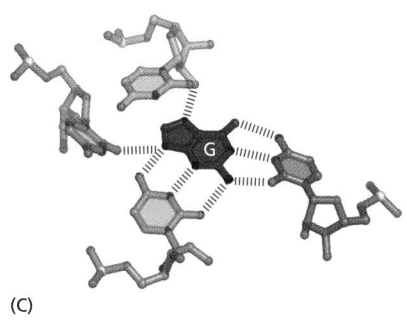

(C)

Figure 7–55 A riboswitch that responds to guanine. (A) In this example from bacteria, the riboswitch controls expression of the purine biosynthetic genes. When guanine levels in cells are low, an elongating RNA polymerase transcribes the purine biosynthetic genes, and the enzymes needed for guanine synthesis are therefore expressed. (B) When guanine is abundant, it binds the riboswitch, causing it to undergo a conformational change that forces the RNA polymerase to terminate transcription (see Figure 6–11). (C) Guanine *(red)* bound to the riboswitch. Only those nucleotides that form the guanine-binding pocket are shown. Many other riboswitches exist, including those that recognize *S*-adenosylmethionine, coenzyme B_{12}, flavin mononucleotide, adenine, lysine, and glycine. (Adapted from M. Mandal and R.R. Breaker, *Nat. Rev. Mol. Cell Biol.* 5:451–463, 2004; and C.K. Vanderpool and S. Gottesman, *Mol. Microbiol.* 54:1076–1089, 2004.)

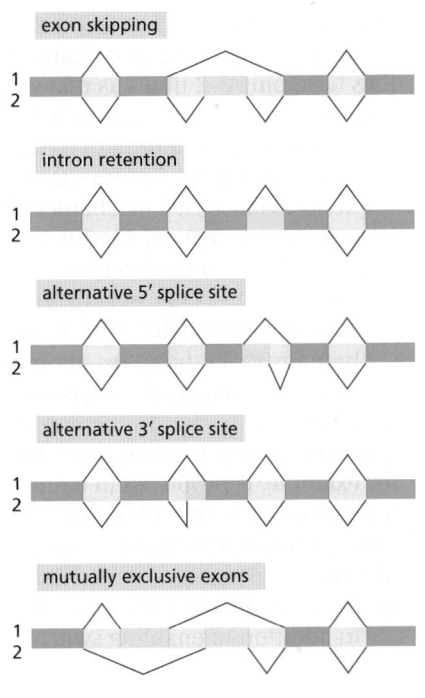

Figure 7–56 **Five patterns of alternative RNA splicing.** In each case, a single type of RNA transcript is spliced in two alternative ways to produce two distinct mRNAs (1 and 2). The *dark blue boxes* mark exon sequences that are retained in both mRNAs. The *light blue boxes* mark possible exon sequences that are included in only one of the mRNAs. The boxes are joined by *red lines* to indicate where intron sequences *(yellow)* are removed. (Adapted from H. Keren et al. *Nat. Rev. Genet.* 11:345–355, 2010.)

Riboswitches are perhaps the most economical examples of gene control devices, inasmuch as they bypass the need for regulatory proteins altogether. In the example shown in Figure 7–55, the riboswitch controls transcription elongation, but they can also regulate other steps in gene expression, as we shall see later in this chapter. Clearly, highly sophisticated gene control devices can be made from short sequences of RNA, a fact that supports the hypothesis of an early "RNA world."

Alternative RNA Splicing Can Produce Different Forms of a Protein from the Same Gene

As discussed in Chapter 6 (see Figure 6–26), RNA splicing shortens the transcripts of many eukaryotic genes by removing the intron sequences from the mRNA precursor. We also saw that a cell can splice an RNA transcript differently and thereby make different polypeptide chains from the same gene—a process called **alternative RNA splicing** (**Figure 7–56**). A substantial proportion of animal genes (estimated at 90% in humans) produce multiple proteins in this way.

When different splicing possibilities exist at several positions in the transcript, a single gene can produce dozens of different proteins. In one extreme case, a *Drosophila* gene may produce as many as 38,000 different proteins from a single gene through alternative splicing (**Figure 7–57**), although only a fraction of these forms have thus far been experimentally observed. Considering that the *Drosophila* genome has approximately 14,000 identified genes, it is clear that the protein complexity of an organism can greatly exceed the number of its genes. This example also illustrates the perils in equating gene number with an organism's complexity. For example, alternative splicing is rare in single-celled budding yeasts

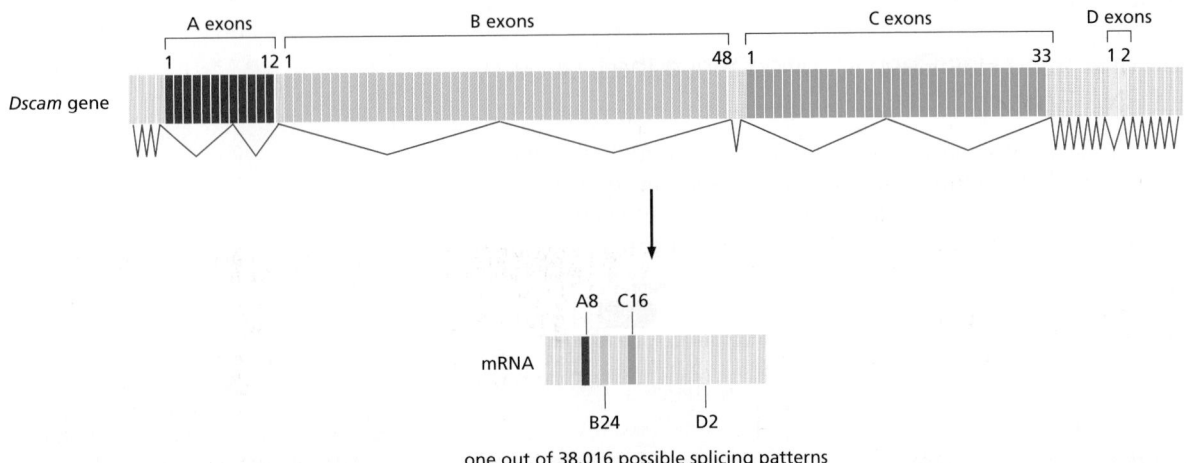

one out of 38,016 possible splicing patterns

Figure 7–57 **Alternative splicing of RNA transcripts of the *Drosophila Dscam* gene.** DSCAM proteins have several different functions. In cells of the fly immune system, they mediate the phagocytosis of bacterial pathogens. In cells of the nervous system, DSCAM proteins are needed for proper wiring of neurons. The final mRNA contains 24 exons, four of which (denoted A, B, C, and D) are present in the *Dscam* gene as arrays of alternative exons. Each RNA contains 1 of 12 alternatives for exon A *(red)*, 1 of 48 alternatives for exon B *(green)*, 1 of 33 alternatives for exon C *(blue)*, and 1 of 2 alternatives for exon D *(yellow)*. This figure shows only one of the many possible splicing patterns (indicated by the *red line* and by the mature mRNA below it). Each variant DSCAM protein would fold into roughly the same structure (predominantly a series of extracellular immunoglobulin-like domains linked to a membrane-spanning region; see Figure 24–48), but the amino acid sequence of the domains vary according to the splicing pattern. The diversity of DSCAM variants contributes to the plasticity of the immune system as well as the formation of complex neural circuits; we take up the specific role of the DSCAM variants in more detail when we describe the development of the nervous system in Chapter 21. (Adapted from D.L. Black, *Cell* 103:367–370, 2000. With permission from Elsevier.)

but very common in flies. Budding yeast has ≈6200 genes, only about 300 of which are subject to splicing, and nearly all of these have only a single intron. To say that flies have only 2–3 times as many genes as yeasts greatly underestimates the difference in complexity of these two genomes.

In some cases, alternative RNA splicing occurs because there is an *intron sequence ambiguity:* the standard spliceosome mechanism for removing intron sequences (discussed in Chapter 6) is unable to distinguish clearly between two or more alternative pairings of 5′ and 3′ splice sites, so that different choices are made by chance on different individual transcripts. Where such constitutive alternative splicing occurs, several versions of the protein encoded by the gene are made in all cells in which the gene is expressed.

In many cases, however, alternative RNA splicing is regulated. In the simplest examples, regulated splicing is used to switch from the production of a nonfunctional protein to the production of a functional one (or the other way around). The transposase that catalyzes the transposition of the *Drosophila* P element, for example, is produced in a functional form in germ cells and a nonfunctional form in somatic cells of the fly, allowing the P element to spread throughout the genome of the fly without causing damage in somatic cells (see Figure 5–61). The difference in transposon activity has been traced to the presence of an intron sequence in the transposase RNA that is removed only in germ cells.

In addition to enabling switching from the production of a functional protein to the production of a nonfunctional one (or vice versa), the regulation of RNA splicing can generate different versions of a protein in different cell types, according to the needs of the cell. Tropomyosin, for example, is produced in specialized forms in different types of cells (see Figure 6–26). Cell-type-specific forms of many other proteins are produced in the same way.

RNA splicing can be regulated either negatively, by a regulatory molecule that prevents the splicing machinery from gaining access to a particular splice site on the RNA, or positively, by a regulatory molecule that helps direct the splicing machinery to an otherwise overlooked splice site (**Figure 7–58**).

Because of the plasticity of RNA splicing, the blocking of a "strong" splicing site will often expose a "weak" site and result in a different pattern of splicing. Thus, the splicing of a pre-mRNA molecule can be thought of as a delicate balance between competing splice sites—a balance that can easily be tipped by effects on splicing of regulatory proteins.

The Definition of a Gene Has Been Modified Since the Discovery of Alternative RNA Splicing

The discovery that eukaryotic genes usually contain introns and that their coding sequences can be assembled in more than one way raised new questions about the definition of a gene. A gene was first clearly defined in molecular terms in the early 1940s from work on the biochemical genetics of the fungus *Neurospora*.

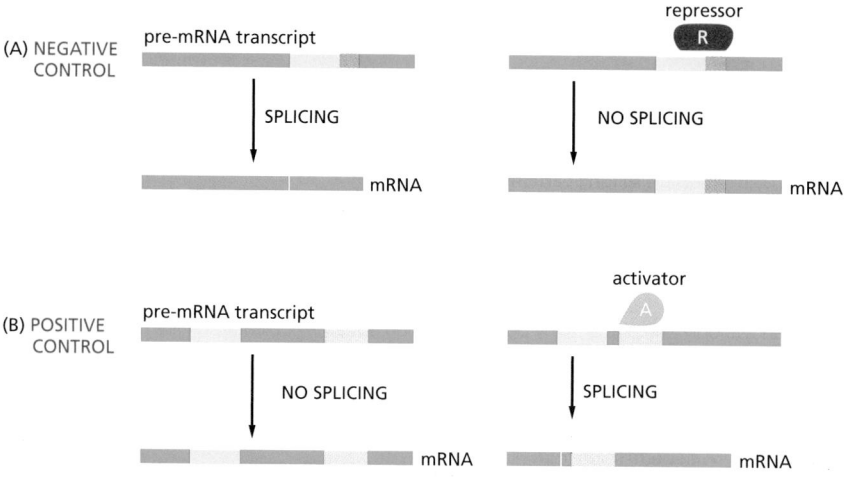

Figure 7–58 Negative and positive control of alternative RNA splicing. (A) In negative control, a repressor protein binds to a specific sequence in the pre-mRNA transcript and blocks access of the splicing machinery to a splice junction. This often results in the use of a secondary splice site, thereby producing an altered pattern of splicing (see Figure 7–56). (B) In positive control, the splicing machinery is unable to remove a particular intron sequence efficiently without assistance from an activator protein. Because RNA is flexible, the nucleotide sequences that bind these activators can be located many nucleotide pairs from the splice junctions they control, and they are often called *splicing enhancers*, by analogy with the transcriptional enhancers mentioned earlier in this chapter.

Until then, a gene had been defined operationally as a region of the genome that segregates as a single unit during meiosis and gives rise to a definable phenotypic trait, such as a red or a white eye in *Drosophila* or a round or wrinkled seed in peas. The work on *Neurospora* showed that most genes correspond to a region of the genome that directs the synthesis of a single enzyme. This led to the hypothesis that one gene encodes one polypeptide chain. The hypothesis proved fruitful for subsequent research; as more was learned about the mechanism of gene expression in the 1960s, a gene became identified as that stretch of DNA that was transcribed into the RNA coding for a single polypeptide chain (or a single structural RNA such as a tRNA or an rRNA molecule). The discovery of split genes and introns in the late 1970s could be readily accommodated by the original definition of a gene, provided that a single polypeptide chain was specified by the RNA transcribed from any one DNA sequence. But it is now clear that many DNA sequences in higher eukaryotic cells can produce a set of distinct (but related) proteins by means of alternative RNA splicing. How, then, is a gene to be defined?

In those relatively rare cases in which a single transcription unit produces two very different eukaryotic proteins, the two proteins are considered to be produced by distinct genes that overlap on the chromosome. It seems unnecessarily complex, however, to consider most of the protein variants produced by alternative RNA splicing as being derived from overlapping genes. A more sensible alternative is to modify the original definition to count a DNA sequence that is transcribed as a single unit and encodes one set of closely related polypeptide chains (protein isoforms) as a single protein-coding gene. This definition also accommodates those DNA sequences that encode protein variants produced by post-transcriptional processes other than RNA splicing, such as transcript cleavage and RNA editing (discussed below).

A Change in the Site of RNA Transcript Cleavage and Poly-A Addition Can Change the C-terminus of a Protein

We saw in Chapter 6 that the 3′ end of a eukaryotic mRNA molecule is not formed by the termination of RNA synthesis by the RNA polymerase, as it is in bacteria. Instead, it results from an RNA cleavage reaction that is catalyzed by additional proteins while the transcript is elongating (see Figure 6–34). A cell can control the site of this cleavage so as to change the C-terminus of the resultant protein. In the simplest cases, one protein variant is simply a truncated version of the other; in many other cases, however, the alternative cleavage and polyadenylation sites lie within intron sequences and the pattern of splicing is thereby altered. This process can produce two closely related proteins differing only in the amino acid sequences at their C-terminal ends. Close analysis of RNAs produced from the human genome in a variety of cell types (see Figure 7–3) indicate that as many as 50% of human protein-coding genes produce mRNA species that differ at their site of polyadenylation.

A well-studied example of regulated polyadenylation is the switch from the synthesis of membrane-bound to secreted antibody molecules that occurs during the development of B lymphocytes (see Figure 24–22). Early in the life history of a B lymphocyte, the antibody it produces is anchored in the plasma membrane, where it serves as a receptor for antigen. Antigen stimulation causes B lymphocytes to multiply and to begin secreting their antibody. The secreted form of the antibody is identical to the membrane-bound form except at the extreme C-terminus. In this part of the protein, the membrane-bound form has a long string of hydrophobic amino acids that traverses the lipid bilayer of the membrane, whereas the secreted form has a much shorter string of hydrophilic amino acids. The switch from membrane-bound to secreted antibody is generated through a change in the site of RNA cleavage and polyadenylation, as shown in **Figure 7–59**.

The change is caused by an increase in the concentration of a subunit of a protein (CstF) that promotes RNA cleavage (see Figure 6–34). The first cleavage/poly-A addition site that a transcribing RNA polymerase encounters is suboptimal and is usually skipped in unstimulated B lymphocytes, leading to production

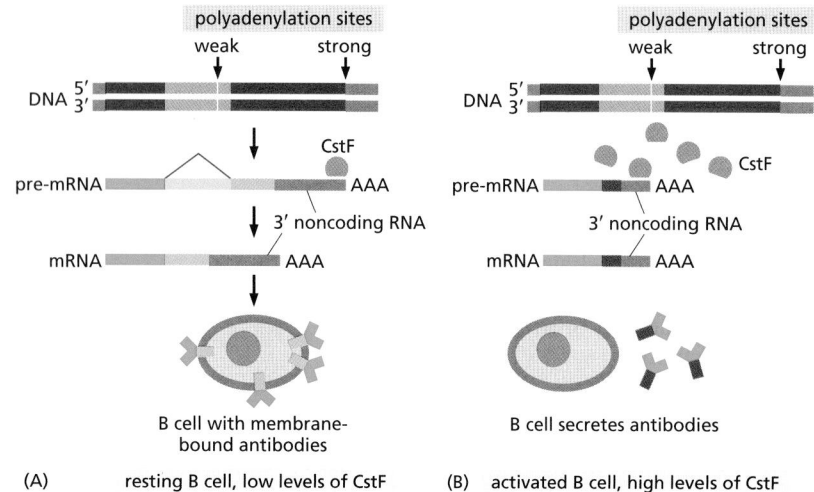

(A)   resting B cell, low levels of CstF   (B)   activated B cell, high levels of CstF

Figure 7–59 **Regulation of the site of RNA cleavage and poly-A addition determines whether an antibody molecule is secreted or remains membrane-bound.** In unstimulated B lymphocytes *(left)*, a long RNA transcript is produced, and the intron sequence *(yellow)* near its 3′ end is removed by RNA splicing to provide an mRNA molecule that codes for a membrane-bound antibody molecule. Only a portion of the antibody gene is shown in the figure; the actual gene and its mRNA would extend further to the left of the diagram. After antigen stimulation *(right)*, the RNA transcript is cleaved and polyadenylated upstream from the intron's 3′ splice site. As a result, some of the intron sequence remains as a coding sequence in the short transcript and specifies the hydrophilic C-terminal portion of the secreted antibody molecule *(brown)*. (Adapted from D. Di Giammartino et al., *Mol. Cell* 43:853–866, 2011.)

of the longer RNA transcript. When activated to produce antibodies, the B lymphocyte increases its CstF concentration; as a result, cleavage now occurs at the suboptimal site, and the shorter transcript is produced. In this way, a change in concentration of a general RNA-processing factor has a dramatic effect on the expression of a particular gene.

RNA Editing Can Change the Meaning of the RNA Message

The molecular mechanisms used by cells are a continual source of surprises. An example is the process of **RNA editing**, which alters the nucleotide sequences of RNA transcripts once they are synthesized and thereby changes the coded message they carry. We saw in Chapter 6 that tRNA and rRNA molecules are chemically modified after they are synthesized: here we focus on changes to mRNAs.

In animals, two principal types of mRNA editing occur: the deamination of adenine to produce inosine (A-to-I editing) and, less frequently, the deamination of cytosine to produce uracil (C-to-U editing), as shown in Figure 5–43. Because these chemical modifications alter the pairing properties of the bases (I pairs with C, and U pairs with A), they can have profound effects on the meaning of the RNA. If the edit occurs in a coding region, it can change the amino acid sequence of the protein or produce a truncated protein by creating a premature stop codon. Edits that occur outside coding sequences can affect the pattern of pre-mRNA splicing, the transport of mRNA from the nucleus to the cytosol, the efficiency with which the RNA is translated, or the base-pairing between microRNAs (miRNAs) and their mRNA targets, a form of regulation that will be discussed later in the chapter.

The process of A-to-I editing is particularly prevalent in humans, where it occurs in approximately 1000 genes. Enzymes called *ADARs* (adenosine deaminases acting on RNA) perform this type of editing; these enzymes recognize a double-stranded RNA structure that is formed through base-pairing between the site to be edited and a complementary sequence located elsewhere on the same RNA molecule, typically in an intron (**Figure 7–60**). The structure of the double-stranded RNA specifies whether the mRNA is to be edited, and if so, where the edit should be made. An especially important example of A-to-I editing takes place in the mRNA that codes for a transmitter-gated ion channel in the brain. A single edit changes a glutamine to an arginine; the affected amino acid lies on the inner wall of the channel, and the editing change alters the Ca^{2+} permeability of the channel. Mutant mice that cannot make this edit are prone to epileptic seizures and die during or shortly after weaning, showing that editing of the ion channel RNA is normally crucial for proper brain development.

C-to-U editing, which is carried out by a different set of enzymes, is also crucial in mammals. For example, in certain cells of the gut, the mRNA for apolipoprotein B undergoes a C-to-U edit that creates a premature stop codon and therefore

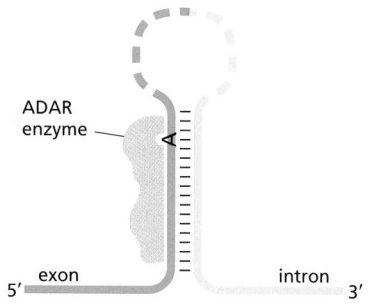

Figure 7–60 **Mechanism of A-to-I RNA editing in mammals.** Typically, a sequence complementary to the position of the edit is present in an intron, and the resulting double-stranded RNA structure attracts an A-to-I editing enzyme (ADAR). In the case illustrated, the edit is made in an exon; in most cases, however, this occurs in noncoding portions of the mRNA. Editing by ADAR takes place in the nucleus, before the pre-mRNA has been fully processed. Mice and humans have two ADAR genes: *ADR1* is expressed in many tissues and is required in the liver for proper red blood cell development; *ADR2* is expressed only in the brain, where it is required for proper brain development.

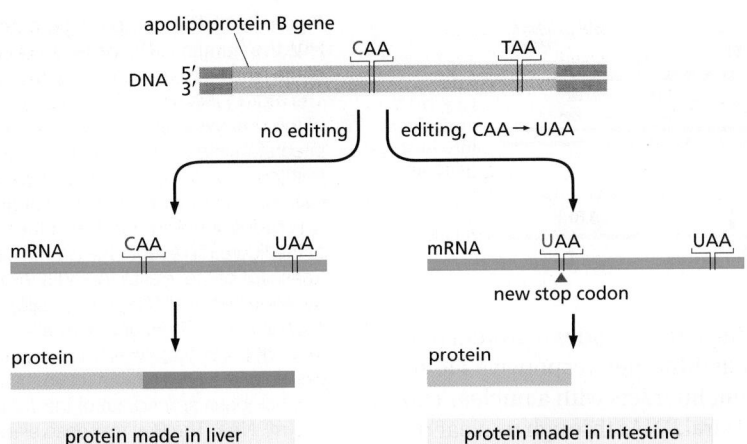

Figure 7–61 **C-to-U RNA editing produces a truncated form of apolipoprotein B.**

produces a shorter form of the protein. In cells of the liver, the editing enzyme is not expressed, and the full-length apolipoprotein B is produced. The two protein isoforms have different properties, and each plays a role in lipid metabolism that is specific to the organ that produces it (**Figure 7–61**).

Why RNA editing exists at all is a mystery. One idea is that it arose in evolution to correct "mistakes" in the genome. Another is that it arose as a somewhat slapdash way for the cell to produce subtly different proteins from the same gene. A third possibility is that RNA editing originally evolved as a defense mechanism against retroviruses and retrotransposons and was later adapted by the cell to change the meanings of certain mRNAs. Indeed, RNA editing still plays important roles in cell defense. Some retroviruses, including HIV, are extensively edited after they infect cells. This hyperediting creates many harmful mutations in the viral RNA genome and also causes viral mRNAs to be retained in the nucleus, where they are eventually degraded. Although some modern retroviruses protect themselves against this defense mechanism, RNA editing presumably helps to hold many viruses in check.

RNA Transport from the Nucleus Can Be Regulated

It has been estimated that in mammals only about one-twentieth of the total mass of RNA synthesized ever leaves the nucleus. We saw in Chapter 6 that most mammalian RNA molecules undergo extensive processing and that the "leftover" RNA fragments (excised introns and RNA sequences 3′ to the cleavage/poly-A site) are degraded in the nucleus. Incompletely processed and otherwise damaged RNAs are also eventually degraded as part of the quality control system for RNA production.

As described in Chapter 6, the export of RNA molecules from the nucleus is delayed until processing has been completed. However, mechanisms that deliberately override this control point can be used to regulate gene expression. This strategy forms the basis for one of the best-understood examples of **regulated nuclear transport** of mRNA, which occurs in the human AIDS virus, HIV.

As we saw in Chapter 5, HIV, once inside the cell, directs the formation of a double-stranded DNA copy of its genome, which is then inserted into the genome of the host (see Figure 5–62). Once inserted, the viral DNA can be transcribed as one long RNA molecule by the host cell's RNA polymerase II. This transcript is then spliced in many different ways to produce over 30 different species of mRNA, which in turn are translated into a variety of different proteins (**Figure 7–62**). In order to make progeny virus, entire, unspliced viral transcripts must be exported from the nucleus to the cytosol, where they are packaged into viral capsids and serve as the viral genome. This large transcript, as well as alternatively spliced HIV mRNAs that the virus needs to move to the cytoplasm for protein synthesis, still carries complete introns. The host cell's normal block to the nuclear export of unspliced RNAs therefore presents a special problem for HIV.

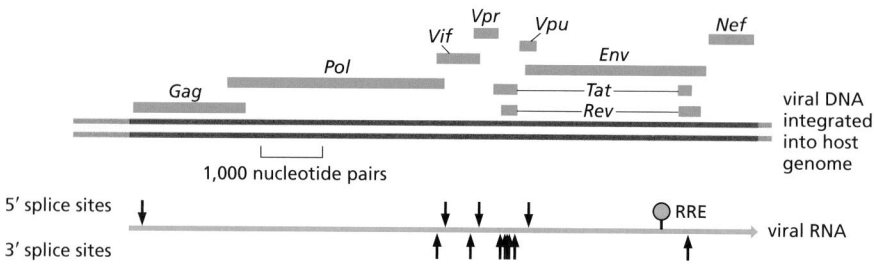

Figure 7–62 **The compact genome of HIV, the human AIDS virus.** The positions of the nine HIV genes are shown in *green*. The *red double line* indicates a DNA copy of the viral genome that has become integrated into the host DNA *(gray)*. Note that the coding regions of many genes overlap, and that those of *Tat* and *Rev* are split by introns. The *blue line* at the bottom of the figure represents the pre-mRNA transcript of the viral DNA and shows the locations of all the possible splice sites *(arrows)*. There are many alternative ways of splicing the viral transcript; for example, the *Env* mRNAs retain the intron that has been spliced out of the *Tat* and *Rev* mRNAs. The Rev response element (RRE) is indicated by a *blue ball and stick*. It is a 234-nucleotide-long stretch of RNA that folds into a defined structure; Rev recognizes a particular hairpin within this larger structure.

The block is overcome in an ingenious way. The virus encodes a protein (called Rev) that binds to a specific RNA sequence (called the Rev responsive element, RRE) located within a viral intron. The Rev protein interacts with a nuclear export receptor (Crm1), which directs the movement of viral RNAs through nuclear pores into the cytosol despite the presence of intron sequences. We discuss in detail the way in which export receptors function in Chapter 12.

The regulation of nuclear export by Rev has several important consequences for HIV growth and pathogenesis. In addition to ensuring the nuclear export of specific unspliced RNAs, it divides the viral infection into an early phase (in which Rev is translated from a fully spliced RNA and all of the intron-containing viral RNAs are retained in the nucleus and degraded) and a late phase (in which unspliced RNAs are exported due to Rev function). This timing helps the virus replicate by providing the gene products in roughly the order in which they are needed (**Figure 7–63**). Regulation by Rev and by Tat, the HIV protein that counteracts premature transcription termination (see p. 414), allows the virus to achieve latency, a condition in which the HIV genome has become integrated into the host-cell genome but the production of viral proteins has temporarily ceased.

The *Gag* gene codes for a protein that is cleaved into several smaller proteins that form the viral capsid. The *Pol* gene codes for a protein that is cleaved to produce reverse transcriptase (which transcribes RNA into DNA), as well as the integrase involved in integrating the viral genome (as double-stranded DNA) into the host genome. The *Env* gene codes for the envelope proteins (see Figure 5–62). Tat, Rev, Vif, Vpr, Vpu, and Nef are small proteins with a variety of functions. For example, Rev regulates nuclear export (see Figure 7–63) and Tat regulates the elongation of transcription across the integrated viral genome (see p. 414).

(A) early HIV synthesis

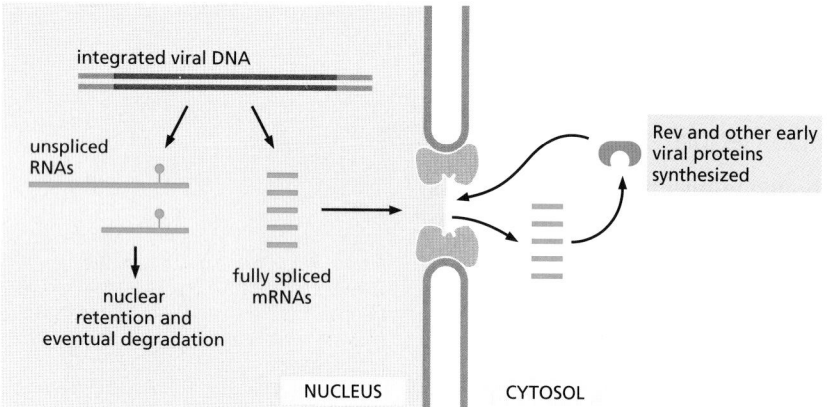

(B) late HIV synthesis

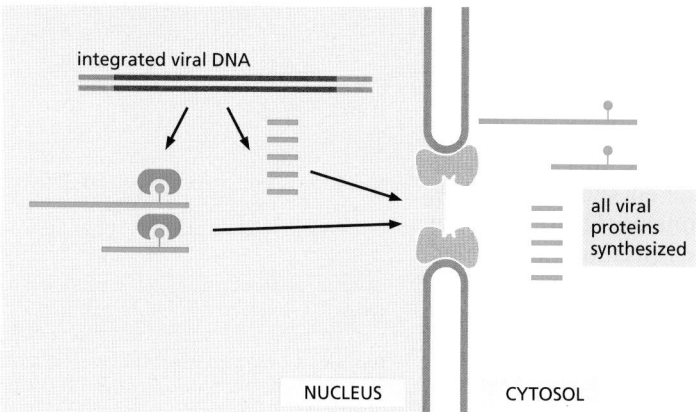

Figure 7–63 **Regulation of nuclear export by the HIV Rev protein.** (A) Early in HIV infection, only the fully spliced RNAs (which contain the coding sequences for Rev, Tat, and Nef) are exported from the nucleus and translated. (B) Once sufficient Rev protein has accumulated and been transported into the nucleus, unspliced viral RNAs can be exported from the nucleus. Many of these RNAs are translated into protein, and the full-length transcripts are packaged into new viral particles.

If, after its initial entry into a host cell, conditions become unfavorable for viral transcription and replication, Rev and Tat are made at levels too low to promote transcription and export of unspliced RNA. This situation stalls the viral growth cycle until conditions improve, whereupon Rev and Tat levels increase, and the virus enters the replication cycle.

Some mRNAs Are Localized to Specific Regions of the Cytosol

Once a newly made eukaryotic mRNA molecule has passed through a nuclear pore and entered the cytosol, it is typically met by ribosomes, which translate it into a polypeptide chain (see Figure 6–8). Once the first round of translation "passes" the nonsense-mediated decay test (see Figure 6–76), the mRNA is usually translated in earnest. If the mRNA encodes a protein that is destined to be secreted or expressed on the cell surface, a signal sequence at the protein's N-terminus will direct it to the endoplasmic reticulum (ER). In this case, as discussed in Chapter 12, components of the cell's protein-sorting apparatus recognize the signal sequence as soon as it emerges from the ribosome and direct the entire complex of ribosome, mRNA, and nascent protein to the membrane of the ER, where the remainder of the polypeptide chain is synthesized. In other cases, free ribosomes in the cytosol synthesize the entire protein, and signals in the completed polypeptide chain may then direct the protein to other sites in the cell.

Many mRNAs are themselves directed to specific intracellular locations before their efficient translation begins, allowing the cell to position its mRNAs close to the sites where the encoded protein is needed. RNA localization has been observed in many organisms, including unicellular fungi, plants, and animals, and it is likely to be a common mechanism that cells use to concentrate high-level production of proteins at specific sites. This strategy also provides the cell with other advantages. For example, it allows the establishment of asymmetries in the cytosol of the cell, a key step in many stages of development. Localized mRNA, coupled with translational control, also allows the cell to regulate gene expression independently in different regions. This feature is particularly important in large, highly polarized cells such as neurons, where it plays a central role in synaptic function.

Several mechanisms for mRNA localization have been discovered (**Figure 7–64**), all of which require specific signals in the mRNA itself. These signals are usually concentrated in the 3′ *untranslated region* (*UTR*), the region of RNA that

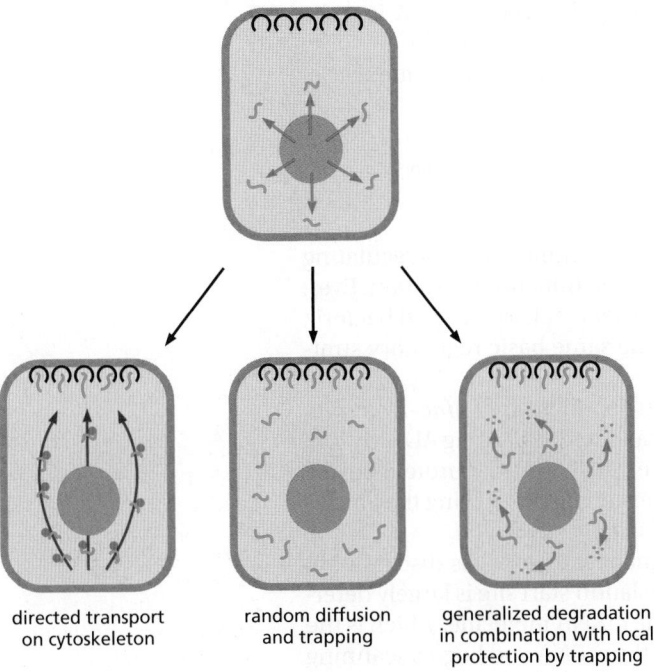

directed transport on cytoskeleton

random diffusion and trapping

generalized degradation in combination with local protection by trapping

Figure 7–64 Mechanisms for the localization of mRNAs. The mRNA to be localized leaves the nucleus through nuclear pores *(top)*. Some localized mRNAs *(left diagram)* travel to their destination by associating with cytoskeletal motors, which use the energy of ATP hydrolysis to move the mRNAs unidirectionally along filaments in the cytoskeleton *(red)* (see Chapter 16). At their destination, the mRNAs are held in place by anchor proteins *(black)*. Other mRNAs randomly diffuse through the cytosol and are simply trapped by anchor proteins and at their sites of localization *(center diagram)*. Some mRNAs *(right diagram)* are degraded in the cytosol unless they have bound, through random diffusion, a localized protein complex that anchors and protects the mRNA from degradation *(black)*. Each mechanism requires specific signals on the mRNA, which are typically located in the 3′ UTR. Additional components can block the translation of the mRNA until it is properly localized. (Adapted from H.D. Lipshitz and C.A. Smibert, *Curr. Opin. Genet. Dev.* 10:476–488, 2000.)

Figure 7–65 **An experiment demonstrating the importance of the 3′ UTR in localizing mRNAs to specific regions of the cytoplasm.** For this experiment, two different fluorescently labeled RNAs were prepared by transcribing DNA *in vitro* in the presence of fluorescently labeled derivatives of UTP. One RNA (labeled with a *red* fluorochrome) contains the coding region for the *Drosophila* Hairy protein and includes the adjacent 3′ UTR (see Figure 6–21). The other RNA (labeled *green*) contains the Hairy coding region with the 3′ UTR deleted. The two RNAs were mixed and injected into a *Drosophila* embryo at a stage of development when multiple nuclei reside in a common cytoplasm (see Figure 7–26). When the fluorescent RNAs were visualized 10 minutes later, the full-length *hairy* RNA *(red)* was localized to the apical side of nuclei *(blue)* but the transcript missing the 3′ UTR *(green)* failed to localize. Hairy is one of many transcriptional regulators that specify positional information in the developing *Drosophila* embryo (discussed in Chapter 21), and the localization of its mRNA (shown in this experiment to depend on its 3′ UTR) is critical for proper fly development. (Courtesy of Simon Bullock and David Ish-Horowicz.)

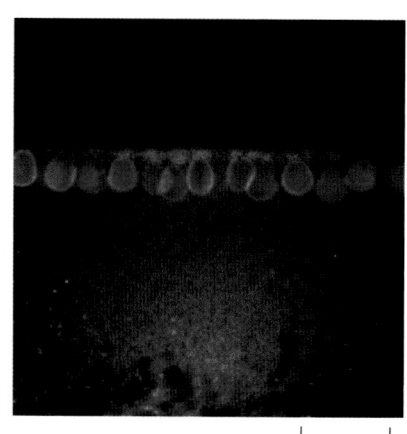

20 μm

extends from the stop codon that terminates protein synthesis to the start of the poly-A tail (**Figure 7–65**). The mRNA localization is usually coupled with translational controls to ensure that the mRNA remains quiescent until it has been moved into place.

The *Drosophila* egg exhibits an especially striking example of mRNA localization. The mRNA encoding the Bicoid transcription regulator is localized by attachment to the cytoskeleton at the anterior tip of the developing egg. When fertilization triggers the translation of this mRNA, it generates a gradient of the Bicoid protein that plays a crucial part in directing the development of the anterior part of the embryo (see Figure 7–26). Many mRNAs in somatic cells are also localized in a similar way. The mRNA that encodes actin, for example, is localized to the actin-filament-rich cell cortex in mammalian fibroblasts by means of a 3′ UTR signal.

We saw in Chapter 6 that mRNA molecules exit from the nucleus bearing numerous markings in the form of RNA modifications (the 5′ cap and the 3′ poly-A tail) and bound proteins (exon-junction complexes, for example) that signify the successful completion of the different pre-mRNA processing steps. As just described, the 3′ UTR of an mRNA can be thought of as a "zip code" that directs mRNAs to different places in the cell. Below, we will also see that mRNAs carry information specifying their average lifetime in the cytosol and the efficiency with which they are translated into protein. In a broad sense, the untranslated regions of eukaryotic mRNAs resemble the transcriptional control regions of genes: their nucleotide sequences contain information specifying the way the RNA is to be used, and proteins interpret this information by binding specifically to these sequences. Thus, over and above the specification of the amino acid sequences of proteins, mRNA molecules are rich with information.

The 5′ and 3′ Untranslated Regions of mRNAs Control Their Translation

Once an mRNA has been synthesized, one of the most common ways of regulating the levels of its protein product is to control the step that initiates translation. Even though the details of translation initiation differ between eukaryotes and bacteria (as we saw in Chapter 6), they each use some of the same basic regulatory strategies.

In bacterial mRNAs, a conserved stretch of nucleotides, the *Shine–Dalgarno sequence*, is always found a few nucleotides upstream of the initiating AUG codon. In bacteria, translational control mechanisms are carried out by proteins or by RNA molecules, and they generally involve either exposing or blocking the Shine–Dalgarno sequence (**Figure 7–66**).

Eukaryotic mRNAs do not contain such a sequence. Instead, as discussed in Chapter 6, the selection of an AUG codon as a translation start site is largely determined by its proximity to the cap at the 5′ end of the mRNA molecule, which is the site at which the small ribosomal subunit binds to the mRNA and begins scanning for an initiating AUG codon. In eukaryotes, translational repressors can bind to the 5′ end of the mRNA and thereby inhibit translation initiation. Other repressors recognize nucleotide sequences in the 3′ UTR of specific mRNAs and decrease translation initiation by interfering with the communication between the 5′ cap and 3′ poly-A tail, a step required for efficient translation (see Figure 6–70). A particularly important type of translational control in eukaryotes relies on small RNAs (termed *microRNAs* or *miRNAs*) that bind to mRNAs and reduce protein output, as described later in this chapter.

The Phosphorylation of an Initiation Factor Regulates Protein Synthesis Globally

Eukaryotic cells decrease their overall rate of protein synthesis in response to a variety of situations, including deprivation of growth factors or nutrients, infection by viruses, and sudden increases in temperature. Much of this decrease is caused by the phosphorylation of the translation initiation factor eIF2 by specific protein kinases that respond to the changes in conditions.

The normal function of eIF2 was outlined in Chapter 6. It forms a complex with GTP and mediates the binding of the methionyl initiator tRNA to the small ribosomal subunit, which then binds to the 5′ end of the mRNA and begins scanning along the mRNA. When an AUG codon is recognized, the eIF2 protein hydrolyzes the bound GTP to GDP, causing a conformational change in the protein and releasing it from the small ribosomal subunit. The large ribosomal subunit then joins the small one to form a complete ribosome that begins protein synthesis.

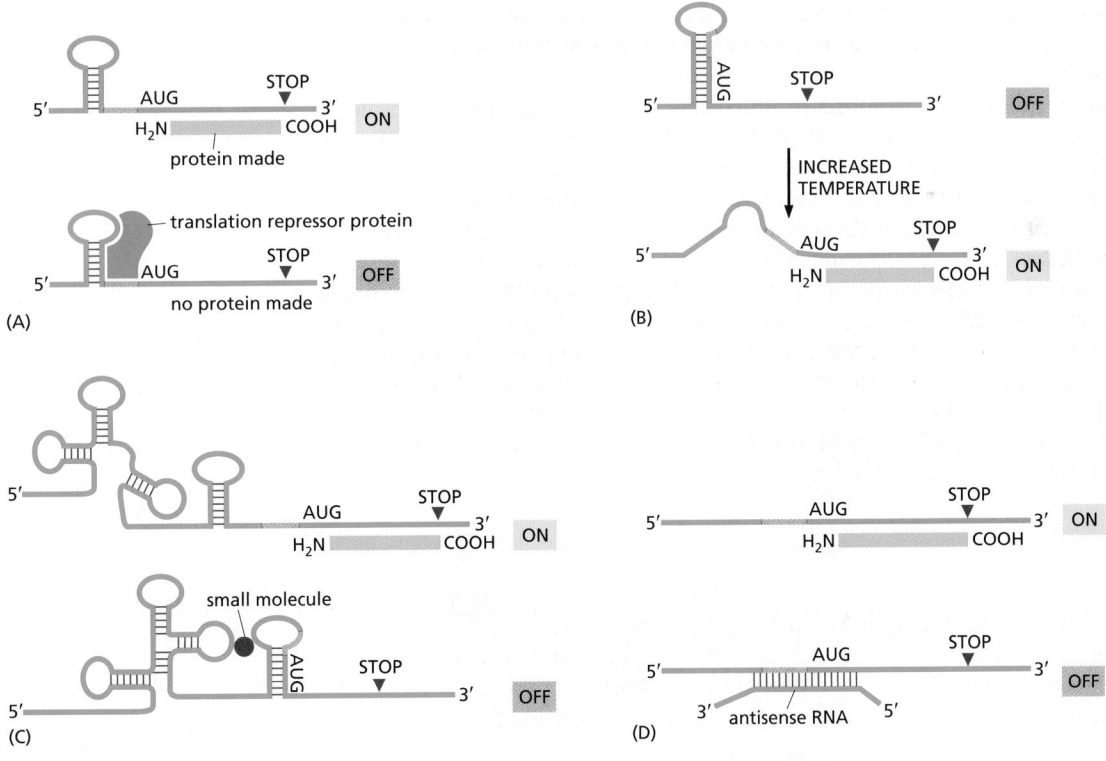

Figure 7–66 Mechanisms of translational control. Although these examples are from bacteria, many of the same principles operate in eukaryotes. (A) Sequence-specific RNA-binding proteins repress translation of specific mRNAs by blocking access of the ribosome to the Shine–Dalgarno sequence *(orange)*. For example, some ribosomal proteins repress translation of their own RNA. This mechanism allows the cell to maintain correctly balanced quantities of the various components needed to form ribosomes. (B) An RNA "thermosensor" permits efficient translation initiation only at elevated temperatures at which the stem-loop structure has been melted. An example occurs in the human pathogen *Listeria monocytogenes*, in which the translation of its virulence genes increases at 37°C, the temperature of the host. (C) Binding of a small molecule to a riboswitch causes a major rearrangement of the RNA forming a different set of stem-loop structures. In the bound structure, the Shine–Dalgarno sequence *(orange)* is sequestered and translation initiation is thereby blocked. In many bacteria, *S*-adenosylmethionine acts in this manner to block production of the enzymes that synthesize it. (D) An "antisense" RNA produced elsewhere from the genome base-pairs with a specific mRNA and blocks its translation. Many bacteria regulate expression of iron-storage proteins in this way.

Because eIF2 binds very tightly to GDP, a guanine nucleotide exchange factor (see p. 157), designated eIF2B, is required to cause GDP release so that a new GTP molecule can bind and eIF2 can be reused (Figure 7–67A). The reuse of eIF2 is inhibited when it is phosphorylated—the phosphorylated eIF2 binds to eIF2B unusually tightly, inactivating eIF2B. There is more eIF2 than eIF2B in cells, and even a fraction of phosphorylated eIF2 can trap nearly all of the eIF2B. This prevents the reuse of the nonphosphorylated eIF2 and greatly slows protein synthesis (Figure 7–67B).

Regulation of the level of active eIF2 is especially important in mammalian cells; eIF2 is part of the mechanism that allows cells to enter a nonproliferating, resting state (called G_0) in which the rate of total protein synthesis is reduced to about one-fifth the rate in proliferating cells.

Initiation at AUG Codons Upstream of the Translation Start Can Regulate Eukaryotic Translation Initiation

We saw in Chapter 6 that eukaryotic translation typically begins at the first AUG downstream of the 5′ end of the mRNA, which is the first AUG encountered by a scanning small ribosomal subunit. But the nucleotides immediately surrounding the AUG also influence the efficiency of translation initiation. If the recognition site is poor enough, scanning ribosomal subunits will sometimes ignore the first AUG codon in the mRNA and skip to the second or third AUG codon instead. This phenomenon, known as "leaky scanning," is a strategy frequently used to produce two or more closely related proteins, differing only in their N-termini, from the same mRNA. A particularly important use of this mechanism is the production of the same protein with and without a signal sequence attached at its N-terminus. This allows the protein to be directed to two different locations in the cell (for example, to both mitochondria and the cytosol). Cells can regulate the relative abundance of the protein isoforms produced by leaky scanning; for example, a cell-type-specific increase in the abundance of the initiation factor eIF4F favors the use of the AUG closest to the 5′ end of the mRNA.

Another type of control found in eukaryotes uses one or more short *open reading frames*—short stretches of DNA that begin with a start codon (ATG) and end with a stop codon, with no stop codons in between—that lie between the 5′ end of the mRNA and the beginning of the gene. Often, the amino acid sequences coded by these upstream open reading frames (uORFs) are not important; rather, the uORFs serve a purely regulatory function. An uORF present on an mRNA molecule will generally decrease translation of the downstream gene by trapping a

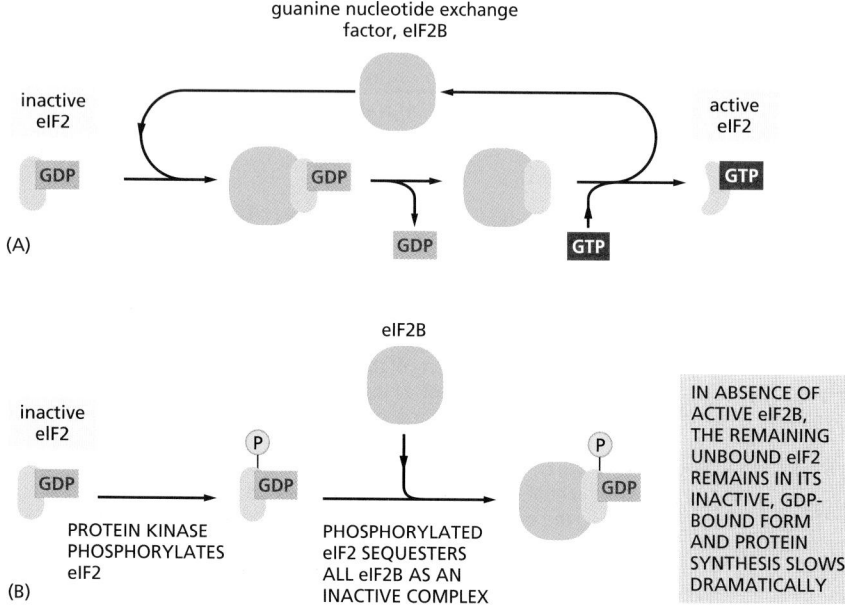

Figure 7–67 **The eIF2 cycle.** (A) The recycling of used eIF2 by a guanine nucleotide exchange factor (eIF2B). (B) eIF2 phosphorylation controls protein synthesis rates by tying up eIF2B.

scanning ribosome initiation complex and causing the ribosome to translate the uORF and dissociate from the mRNA before it reaches the bona fide protein-coding sequence.

When the activity of a general translation factor (such as the eIF2 discussed above) is reduced, one might expect that the translation of all mRNAs would be reduced equally. Contrary to this expectation, however, the phosphorylation of eIF2 can have selective effects, even enhancing the translation of specific mRNAs that contain uORFs. This can enable cells, for example, to adapt to starvation for specific nutrients by shutting down the synthesis of all proteins except those that are required for synthesis of the missing nutrients. The details of this mechanism have been worked out for a specific yeast mRNA that encodes a protein called Gcn4, a transcription regulator that activates many genes that encode proteins that are important for amino acid synthesis.

The *Gcn4* mRNA contains several short uORFs, and when amino acids are abundant, ribosomes translate the uORFs and generally dissociate before they reach the *Gcn4* coding region. A global decrease in eIF2 activity brought about by amino acid starvation makes it more likely that a scanning small ribosomal subunit will move across the uORFs (without translating them) before it acquires a molecule of eIF2 (see Figure 6–70). Such a ribosomal subunit is then free to initiate translation on the actual *Gcn4* sequences. The increased level of this transcription regulator increases production of the amino acid biosynthetic enzymes.

Internal Ribosome Entry Sites Provide Opportunities for Translational Control

Although approximately 90% of eukaryotic mRNAs are translated beginning with the first AUG downstream from the 5′ cap, certain AUGs, as we saw in the previous section, can be skipped over during the scanning process. In this section, we discuss yet another way that cells can initiate translation at positions distant from the 5′ end of the mRNA, using a specialized type of RNA sequence called an **internal ribosome entry site** (**IRES**). In some cases, two distinct protein-coding sequences are carried in tandem on the same eukaryotic mRNA; translation of the first occurs by the usual scanning mechanism, and translation of the second occurs through an IRES. IRESs are typically several hundred nucleotides in length and fold into specific structures that bind many, but not all, of the same proteins that are used to initiate normal 5′ cap-dependent translation (**Figure 7–68**). In fact, different IRESs require different subsets of initiation factors. However, all of them bypass the need for a 5′ cap structure and the translation initiation factor that recognizes it, eIF4E.

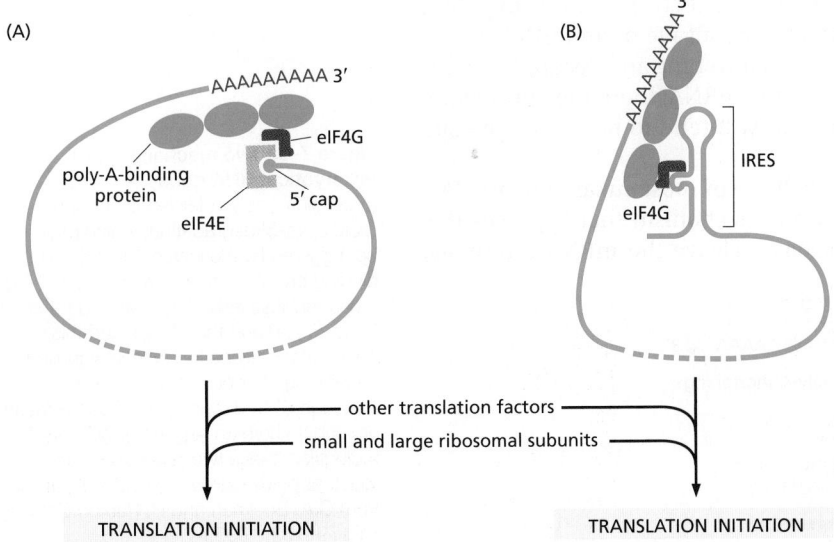

Figure 7–68 **Two mechanisms of translation initiation.** (A) The normal, cap-dependent mechanism requires a set of initiation factors whose assembly on the mRNA is stimulated by the presence of a 5′ cap and a poly-A tail (see also Figure 6–70). (B) The IRES-dependent mechanism, seen mainly in viruses, requires only a subset of the normal translation initiating factors, and these assemble directly on the folded IRES. (Adapted from A. Sachs, *Cell* 101:243–245, 2000.)

Some viruses use IRESs as part of a strategy to get their own mRNA molecules translated while blocking normal 5′ cap-dependent translation of host mRNAs. On infection, these viruses produce a protease (encoded in the viral genome) that cleaves the host-cell translation factor eIF4G, rendering it unable to bind to eIF4E, the cap-binding complex. This shuts down most of the host cell's translation and effectively diverts the translation machinery to the IRES sequences present on the viral mRNAs. (The truncated eIF4G remains competent to initiate translation at these internal sites.)

The many ways in which viruses manipulate their host's protein-synthesis machinery for their own advantage continue to surprise cell biologists. Studying this "arms race" between humans and pathogens has led to many fundamental insights into the workings of the cell, and we revisit this topic in more detail in Chapter 23.

Changes in mRNA Stability Can Regulate Gene Expression

Most mRNAs in a bacterial cell are very unstable, having half-lives of less than a couple of minutes. Exonucleases, which degrade in the 3′-to-5′ direction, are usually responsible for the rapid destruction of these mRNAs. Because its mRNAs are both rapidly synthesized and rapidly degraded, a bacterium can adapt quickly to environmental changes.

As a general rule, the mRNAs in eukaryotic cells are more stable. Some, such as that encoding β globin, have half-lives of more than 10 hours, but most have considerably shorter half-lives, typically less than 30 minutes. The mRNAs that code for proteins such as growth factors and transcription regulators, whose production rates need to change rapidly in cells, have especially short half-lives.

We saw in Chapter 6 that the cell has several mechanisms that rapidly destroy incorrectly processed RNAs; here, we consider the fate of the typical "normal" eukaryotic mRNA. Two general mechanisms exist for eventually destroying each mRNA that is made by the cell. Both begin with the gradual shortening of the poly-A tail by an exonuclease, a process that starts as soon as the mRNA reaches the cytosol. In a broad sense, this poly-A shortening acts as a timer that counts down the lifetime of each mRNA. Once the poly-A tail is reduced to a critical length (about 25 nucleotides in humans), the two pathways diverge. In one, the 5′ cap is removed (a process called decapping) and the "exposed" mRNA is rapidly degraded from its 5′ end. In the other, the mRNA continues to be degraded from the 3′ end, through the poly-A tail, into the coding sequences (**Figure 7–69**).

Nearly all mRNAs are subject to both types of decay, and the specific sequences of each mRNA determine how fast each step occurs and therefore how long each mRNA will persist in the cell and be able to produce protein. The 3′ UTR sequences are especially important in controlling mRNA lifetimes, and they often carry binding sites for specific proteins that increase or decrease the rate of poly-A shortening, decapping, or 3′-to-5′ degradation. The half-life of an mRNA is also affected by how efficiently it is translated. Poly-A shortening and decapping compete directly with the machinery that translates the mRNA; therefore, any factors that affect the translation efficiency of an mRNA will tend to have the opposite effect on its degradation (**Figure 7–70**).

Although poly-A shortening controls the half-life of most eukaryotic mRNAs, some mRNAs can be degraded by a specialized mechanism that bypasses this step altogether. In these cases, specific nucleases cleave the mRNA internally,

Figure 7–69 Two mechanisms of eukaryotic mRNA decay. A critical threshold of poly-A tail length induces rapid 3′-to-5′ degradation, which may be triggered by the loss of the poly-A-binding proteins. As shown in Figure 7–70, a deadenylase associates with both the 3′ poly-A tail and the 5′ cap, and this connection may be involved in signaling decapping after poly-A shortening. Although 5′-to-3′ and 3′-to-5′ degradation are shown here on separate RNA molecules, these two processes can occur together on the same molecule. (Adapted from C.A. Beelman and R. Parker, *Cell* 81:179–183, 1995.)

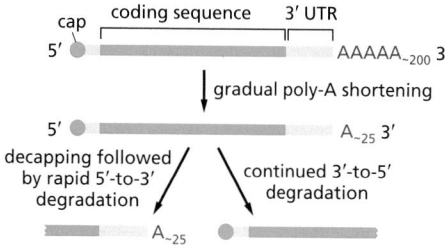

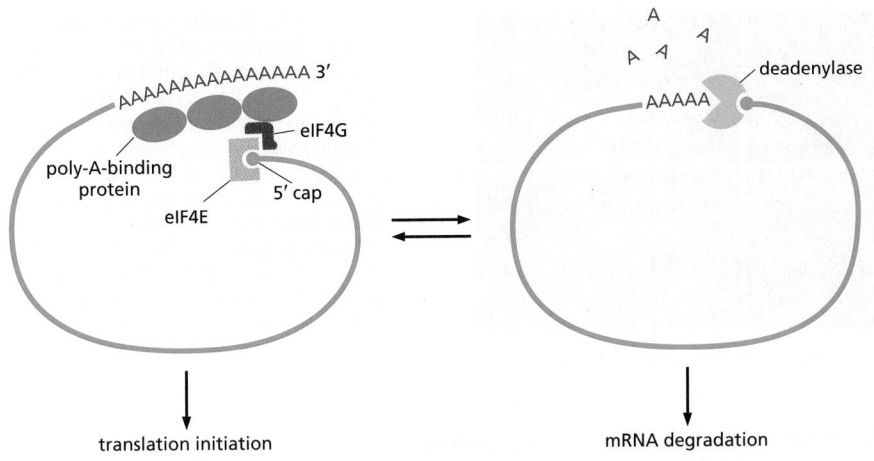

Figure 7–70 The competition between mRNA translation and mRNA decay. The same two features of an mRNA molecule—its 5′ cap and the 3′ poly-A tail—are used in both translation initiation and deadenylation-dependent mRNA decay (see Figure 7–69). The deadenylase that shortens the poly-A tail in the 3′-to-5′ direction associates with the 5′ cap. As described in Chapter 6 (see Figure 6–70), the translation initiation machinery also associates with both the 5′ cap and the poly-A tail. (Adapted from M. Gao et al., *Mol. Cell* 5:479–488, 2000.)

effectively decapping one end and removing the poly-A tail from the other so that both halves are rapidly degraded. The mRNAs that are destroyed in this way carry specific nucleotide sequences, often in the 3′ UTRs, that serve as recognition sequences for these endonucleases. This strategy makes it especially simple to tightly regulate the stability of these mRNAs by blocking or exposing the endonuclease site in response to extracellular signals. For example, the addition of iron to cells decreases the stability of the mRNA that encodes the receptor protein that binds the iron-transporting protein transferrin, causing less of this receptor to be made. This effect is mediated by the iron-sensitive RNA-binding protein aconitase. Aconitase can bind to the 3′ UTR of the transferrin receptor mRNA and increase receptor production by blocking endonucleolytic cleavage of the mRNA. On the addition of iron, aconitase is released from the mRNA, exposing the cleavage site and thereby decreasing the stability of the mRNA (**Figure 7–71**).

Regulation of mRNA Stability Involves P-bodies and Stress Granules

We saw in Chapters 3 and 6 that large aggregates of proteins and nucleic acids that work together are often held in proximity by loose, low-affinity connections (see Figure 3–36). In this way, they function as "organelles" even though they are not surrounded by membranes. Many of the events discussed in the previous

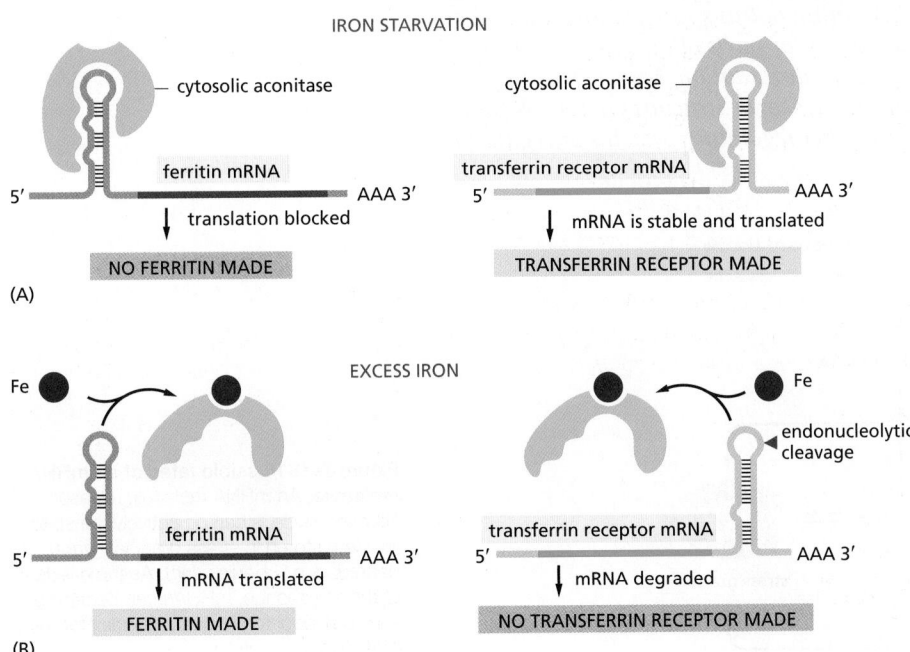

Figure 7–71 Two post-translational controls mediated by iron. (A) During iron starvation, the binding of aconitase to the 5′ UTR of the ferritin mRNA blocks translation initiation; its binding to the 3′ UTR of the transferrin receptor mRNA blocks an endonuclease cleavage site and thereby stabilizes the mRNA. (B) In response to an increase in iron concentration in the cytosol, a cell increases its synthesis of ferritin in order to bind the extra iron and decreases its synthesis of transferrin receptors in order to import less iron across the plasma membrane. Both responses are mediated by the same iron-responsive regulatory protein, aconitase, which recognizes common features in a stem-loop structure in the mRNAs encoding ferritin and the transferrin receptor. Aconitase dissociates from the mRNA when it binds iron. But because the transferrin receptor and ferritin are regulated by different types of mechanisms, their levels respond oppositely to iron concentrations even though they are regulated by the same iron-responsive regulatory protein. (Adapted from M.W. Hentze et al., *Science* 238:1570–1573, 1987 and J.L. Casey et al., *Science* 240:924–928, 1988.)

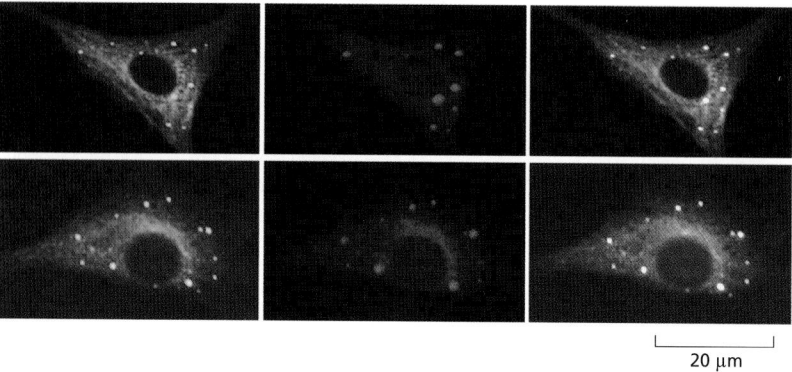

20 μm

Figure 7–72 **Visualization of P-bodies.** Human cells were stained with antibodies to a component of the mRNA decapping enzyme Dcp1a *(left panels)* and to the Argonaute protein *(middle panels).* As described later in this chapter, Argonaute is a key component of RNA interference pathways. The merged image *(right panels)* shows that the two proteins co-localize to P-bodies in the cytoplasm. (Adapted from J. Liu et al., *Nat. Cell Biol.* 7:719–723, published 2005 by Nature Publishing Group. Reproduced with permission of SNCSC.)

section—including decapping and RNA degradation—take place in aggregates known as *Processing-* or *P-bodies,* which are present in the cytosol (**Figure 7–72**).

Although many mRNAs are eventually degraded in P-bodies, some remain intact and are later returned to the pool of translating mRNAs. To be "rescued" in this way, mRNAs move from P-bodies to another type of aggregate known as a *stress granule,* which contains translation initiation factors, poly-A-binding protein, and small ribosomal subunits. Translation itself does not occur in stress granules, but mRNAs can become "translation-ready" as the proteins bound to them in P-bodies are replaced with those in stress granules. The movement of mRNAs between active translation, P-bodies, and stress granules can be seen as an mRNA cycle (**Figure 7–73**) where the competition between translation and mRNA degradation is carefully controlled. Thus, when translation initiation is blocked (by starvation, drugs, or genetic manipulation), stress granules enlarge as more and more nontranslated mRNAs are moved directly into them for storage. Clearly, once a cell has made the large investment in producing a properly processed mRNA molecule, it carefully controls its subsequent fate.

Summary

Many steps in the pathway from RNA to protein are regulated by cells in order to control gene expression. Most genes are regulated at multiple levels, in addition to being controlled at the initiation stage of transcription. The regulatory mechanisms include (1) attenuation of the RNA transcript by its premature termination, (2) alternative RNA splice-site selection, (3) control of 3'-end formation by cleavage and poly-A addition, (4) RNA editing, (5) control of transport from the nucleus to the cytosol, (6) localization of mRNAs to particular parts of the cell, (7) control of translation initiation, and (8) regulated mRNA degradation. Most of these control processes require the recognition of specific sequences or structures in the RNA molecule being regulated, a task performed by either regulatory proteins or regulatory RNA molecules.

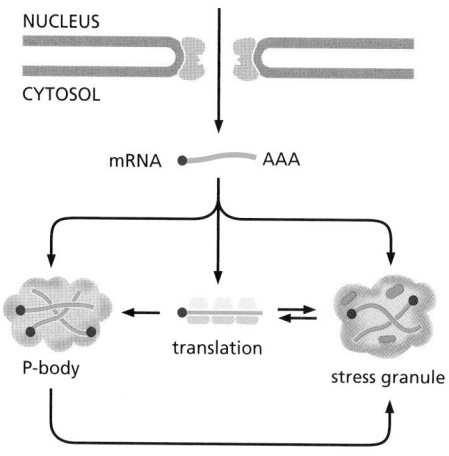

Figure 7–73 **Possible fates of an mRNA molecule.** An mRNA molecule released from the nucleus can be actively translated *(center),* stored in stress granules *(right),* or degraded in P-bodies *(left).* As the needs of the cell change, mRNAs can be shuffled from one pool to the next, as indicated by the arrows.

REGULATION OF GENE EXPRESSION BY NONCODING RNAs

In the previous chapter, we introduced the central dogma, according to which the flow of genetic information proceeds from DNA through RNA to protein (Figure 6–1). But we have seen throughout this book that RNA molecules perform many critical tasks in the cell besides serving as intermediate carriers of genetic information. Among these noncoding RNAs are the rRNA and tRNA molecules, which are responsible for reading the genetic code and synthesizing proteins. The RNA molecule in telomerase serves as a template for the replication of chromosome ends, snoRNAs modify ribosomal RNA, and snRNAs carry out the major events of RNA splicing. And we saw in the previous section that Xist RNA has an important role in inactivating one copy of the X chromosome in females.

A series of recent discoveries has revealed that noncoding RNAs are even more prevalent than previously imagined. We now know that such RNAs play widespread roles in regulating gene expression and in protecting the genome from viruses and transposable elements. These newly discovered RNAs are the subject of this section.

Small Noncoding RNA Transcripts Regulate Many Animal and Plant Genes Through RNA Interference

We begin our discussion with a group of short RNAs that carry out **RNA interference** or **RNAi**. Here, short single-stranded RNAs (20–30 nucleotides) serve as guide RNAs that selectively reorganize and bind—through base-pairing—other RNAs in the cell. When the target is a mature mRNA, the small noncoding RNAs can inhibit its translation or even catalyze its destruction. If the target RNA molecule is in the process of being transcribed, the small noncoding RNA can bind to it and direct the formation of certain types of repressive chromatin on its attached DNA template (**Figure 7–74**). Three classes of small noncoding RNAs work in this way—*microRNAs* (*miRNAs*), *small interfering RNAs* (*siRNAs*), and *piwi-interacting RNAs* (*piRNAs*)—and we discuss them in turn in the next sections. Although they differ in the way the short pieces of single-stranded RNA are generated, all three types of short RNAs locate their targets through RNA–RNA base-pairing, and they generally cause reductions in gene expression.

miRNAs Regulate mRNA Translation and Stability

Over 1000 different **microRNAs (miRNAs)** are produced from the human genome, and these appear to regulate at least one-third of all human protein-coding genes. Once made, miRNAs base-pair with specific mRNAs and fine-tune their translation and stability. The miRNA precursors are synthesized by RNA polymerase II and are capped and polyadenylated. They then undergo a special type of processing, after which the miRNA (typically 23 nucleotides in length) is assembled with a set of proteins to form an *RNA-induced silencing complex* or *RISC*. Once formed, the RISC seeks out its target mRNAs by searching for complementary nucleotide

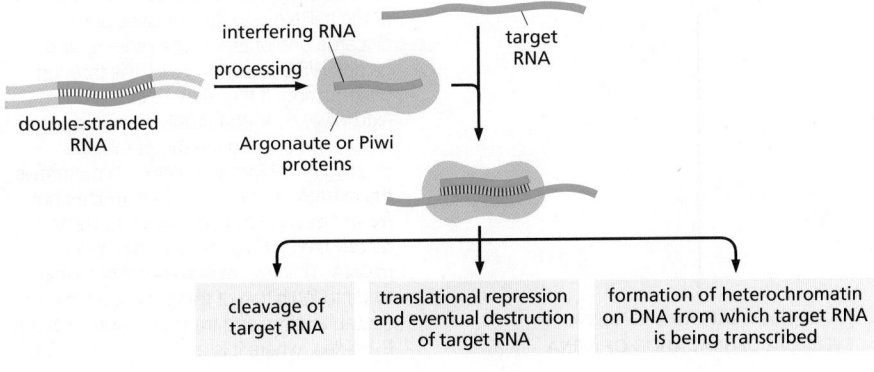

Figure 7–74 **RNA interference in eukaryotes.** Single-stranded interfering RNAs are generated from double-stranded RNA. They locate target RNAs through base-pairing and, at this point, several fates are possible, as shown. As described in the text, there are several types of RNA interference; the way the double-stranded RNA is produced and processed and the ultimate fate of the target RNA depends on the particular system.

sequences (**Figure 7-75**). This search is greatly facilitated by the Argonaute protein, a component of RISC, which holds the 5′ region of the miRNA so that it is optimally positioned for base-pairing to another RNA molecule (**Figure 7-76**). In animals, the extent of base-pairing is typically at least seven nucleotide pairs, and this pairing most often occurs in the 3′ UTR of the target mRNA.

Once an mRNA has been bound by an miRNA, several outcomes are possible. If the base-pairing is extensive (which is unusual in humans but common in many plants), the mRNA is cleaved (*sliced*) by the Argonaute protein, effectively removing the mRNA's poly-A tail and exposing it to exonucleases (see Figure 7-69). Following cleavage of the mRNA, the RISC with its associated miRNA is released, and it can seek out additional mRNAs (see Figure 7-75). Thus, a single miRNA can act catalytically to destroy many complementary mRNAs. These miRNAs can be thus thought of as guide sequences that bring destructive nucleases into contact with specific mRNAs.

If the base-pairing between the miRNA and the mRNA is less extensive (as observed for most human miRNAs), Argonaute does not slice the mRNA; rather, translation of the mRNA is repressed and the mRNA is shuttled to P-bodies (see Figure 7-73) where, sequestered from ribosomes, it eventually undergoes poly-A tail shortening, decapping, and degradation.

Several features make miRNAs especially useful regulators of gene expression. First, a single miRNA can regulate a whole set of different mRNAs, so long as the mRNAs carry a common short sequence in their UTRs. This situation is common in humans, where a single miRNA can control hundreds of different mRNAs. Second, regulation by miRNAs can be combinatorial. When the base-pairing between the miRNA and mRNA fails to trigger cleavage, additional miRNAs binding to the same mRNA lead to further reductions in its translation. As discussed earlier for transcription regulators, combinatorial control greatly expands the possibilities available to the cell by linking gene expression to a combination of different regulators rather than a single regulator. Third, an miRNA occupies relatively little

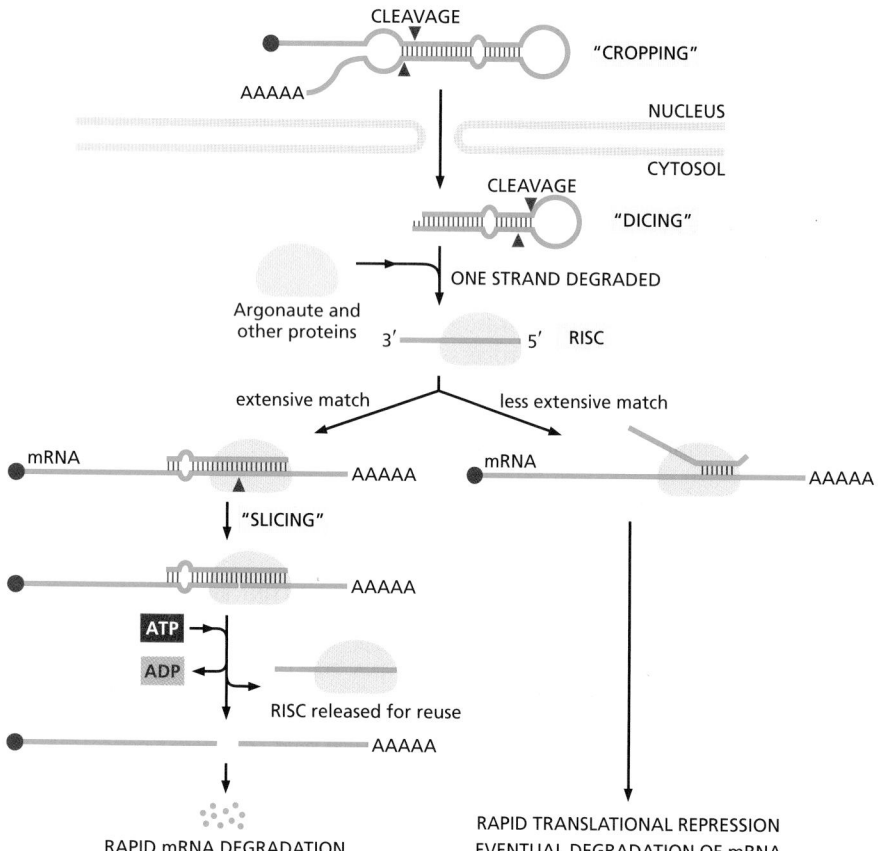

Figure 7-75 miRNA processing and mechanism of action. The precursor miRNA, through complementarity between one part of its sequence and another, forms a double-stranded structure. This RNA is cropped while still in the nucleus and then exported to the cytosol, where it is further cleaved by the Dicer enzyme to form the miRNA proper. Argonaute, in conjunction with other components of RISC, initially associates with both strands of the miRNA and then cleaves and discards one of them. The other strand guides RISC to specific mRNAs through base-pairing. If the RNA–RNA match is extensive, as is commonly seen in plants, Argonaute cleaves the target mRNA, causing its rapid degradation. In mammals, the miRNA–mRNA match often does not extend beyond a short seven-nucleotide "seed" region near the 5′ end of the miRNA. This less extensive base-pairing leads to inhibition of translation, mRNA destabilization, and transfer of the mRNA to P-bodies, where it is eventually degraded.

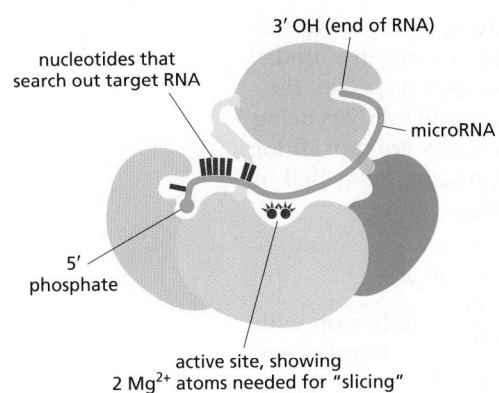

nucleotides that search out target RNA

3′ OH (end of RNA)

microRNA

5′ phosphate

active site, showing 2 Mg²⁺ atoms needed for "slicing"

Figure 7–76 **Human Argonaute protein carrying an miRNA.** The protein is folded into four structural domains, each indicated by a different color. The miRNA is held in an extended form that is optimal for forming RNA–RNA base pairs. The active site of Argonaute that "slices" a target RNA, when it is extensively base-paired with the miRNA, is indicated in *red*. Many Argonaute proteins (three out of the four human proteins, for example) lack the catalytic site and therefore bind target RNAs without slicing them. (Adapted from C.D. Kuhn and L. Joshua-Tor, *Trends Biochem. Sci.* 38:263–271, 2013.)

space in the genome when compared with a protein. Indeed, their small size is one reason that miRNAs were discovered only recently. Although we are only beginning to appreciate the full impact of miRNAs, it is clear that they represent an important part of the cell's equipment for regulating the expression of genes. We discuss specific examples of miRNAs that have key roles in development in Chapter 21.

RNA Interference Is Also Used as a Cell Defense Mechanism

Many of the proteins that participate in the miRNA regulatory mechanisms just described also serve a second function as a defense mechanism: they orchestrate the degradation of foreign RNA molecules, specifically those that occur in double-stranded form. Many transposable elements and viruses produce double-stranded RNA, at least transiently, in their life cycles, and RNA interference helps to keep these potentially dangerous invaders in check. As we shall see, this form of RNAi also provides scientists with a powerful experimental technique to turn off the expression of individual genes.

The presence of double-stranded RNA in the cell triggers RNAi by attracting a protein complex containing *Dicer*, the same nuclease that processes miRNAs (see Figure 7–75). This protein cleaves the double-stranded RNA into small fragments (approximately 23 nucleotide pairs) called **small interfering RNAs (siRNAs)**. These double-stranded siRNAs are then bound by Argonaute and other components of RISC. As we saw above for miRNAs, one strand of the duplex RNA is then cleaved by Argonaute and discarded. The single-stranded siRNA molecule that remains directs RISC back to complementary RNA molecules produced by the virus or transposable element. Because the match is usually exact, Argonaute cleaves these molecules, leading to their rapid destruction.

Each time RISC cleaves a new RNA molecule, the RISC is released; thus, as we saw for miRNAs, a single RNA molecule can act catalytically to destroy many complementary RNAs. Some organisms employ an additional mechanism that amplifies the RNAi response even further. In these organisms, RNA-dependent RNA polymerases use siRNAs as primers to produce additional copies of double-strand RNAs which are then cleaved into siRNAs. This amplification ensures that, once initiated, RNA interference can continue even after all the initiating double-stranded RNA has been degraded or diluted out. For example, it permits progeny cells to continue carrying out the specific RNA interference that was provoked in the parent cells.

In some organisms, the RNA interference activity can be spread by the transfer of RNA fragments from cell to cell. This is particularly important in plants (whose cells are linked by fine connecting channels, as discussed in Chapter 19), because it allows an entire plant to become resistant to an RNA virus after only a few of its cells have been infected. In a broad sense, the RNAi response resembles certain aspects of the animal immune system; in both, an invading organism elicits a customized response, and—through amplification of the "attack" molecules—the host becomes systemically protected.

We have seen that although miRNAs and siRNAs are generated in slightly different ways, they rely on the same proteins and seek out their targets in a fundamentally similar manner. Because siRNAs are found in widespread species, they are believed to be the most ancient form of RNA interference, with miRNAs being a later refinement. These siRNA-mediated defense mechanisms are crucial for plants, worms, and insects. In mammals, a protein-based system (described in Chapter 24) has largely taken over the task of fighting off viruses.

RNA Interference Can Direct Heterochromatin Formation

The siRNA interference pathway just described does not necessarily stop with the destruction of target RNA molecules. In some cases, the RNA interference machinery can also selectively shut off *synthesis* of the target RNAs. For this to occur, the short siRNAs produced by the Dicer protein are assembled with a group of proteins (including Argonaute) to form the RITS (RNA-induced transcriptional silencing) complex. Using single-stranded siRNA as a guide sequence, this complex binds complementary RNA transcripts as they emerge from a transcribing RNA polymerase II (**Figure 7–77**). Positioned on the genome in this way, the RITS complex attracts proteins that covalently modify nearby histones and eventually direct the formation of heterochromatin to prevent further transcription initiation. In some cases, an RNA-dependent RNA polymerase and a Dicer enzyme are also recruited by the RITS complex to continually generate additional siRNAs *in situ*. This positive feedback loop ensures continued repression of the target gene even after the initiating siRNA molecules have disappeared.

RNAi-directed heterochromatin formation is an important cell defense mechanism that limits the spread of transposable elements in genomes by maintaining their DNA sequences in a transcriptionally silent form. However, this same mechanism is also used in some normal processes in the cell. For example, in many organisms, the RNA interference machinery maintains the heterochromatin formed around centromeres. Centromeric DNA sequences are transcribed in both directions, producing complementary RNA transcripts that can basepair to form double-stranded RNA. This double-stranded RNA triggers the RNA

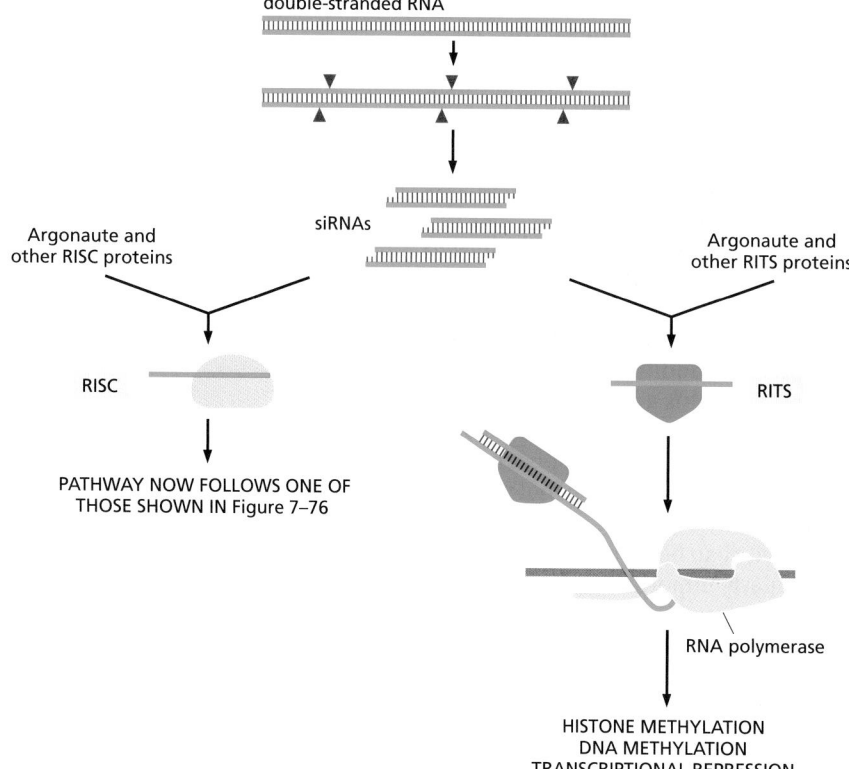

double-stranded RNA

Argonaute and
other RISC proteins

siRNAs

Argonaute and
other RITS proteins

RISC

RITS

PATHWAY NOW FOLLOWS ONE OF
THOSE SHOWN IN Figure 7–76

RNA polymerase

HISTONE METHYLATION
DNA METHYLATION
TRANSCRIPTIONAL REPRESSION

Figure 7–77 RNA interference directed by siRNAs. In many organisms, double-stranded RNA can trigger both the destruction of complementary mRNAs *(left)* and transcriptional silencing *(right)*. The change in chromatin structure induced by the bound RITS (RNA-induced transcriptional silencing) complex resembles that in Figure 7–45.

interference pathway and stimulates formation of the heterochromatin that surrounds centromeres, which is necessary for the centromeres to segregate chromosomes accurately during mitosis.

piRNAs Protect the Germ Line from Transposable Elements

A third system of RNA interference relies on **piRNAs** (**piwi-interacting RNAs**, named for Piwi, a class of proteins related to Argonaute). piRNAs are made specifically in the germ line, where they block the movement of transposable elements. Found in many organisms, including humans, genes coding for piRNAs consist largely of sequence fragments of transposable elements. These clusters of fragments are transcribed and broken up into short, single-stranded piRNAs. The processing differs from that for miRNAs and siRNAs (for one thing, the Dicer enzyme is not involved), and the resulting piRNAs are slightly longer than miRNAs and siRNAs; moreover, they are complexed with Piwi rather than Argonaute proteins. Once formed, the piRNAs seek out RNA targets by base-pairing and, much like siRNAs, transcriptionally silence intact transposon genes and destroy any RNA (including mRNAs) produced by them.

Many mysteries surround piRNAs. Over a million piRNA species are coded in the genomes of many mammals and expressed in the testes, yet only a small fraction seem to be directed against the transposons present in those genomes. Are the piRNAs remnants of past invaders? Do they cover so much "sequence space" that they are broadly protective for any foreign DNA? Another curious feature of piRNAs is that many of them (particularly if base-pairing does not have to be perfect) should, in principle, attack the normal mRNAs made by the organism, yet they do not. It has been proposed that these large numbers of piRNAs may form a system to distinguish "self" RNAs from "foreign" RNAs and attack only the latter. If this is the case, there must be a special way for the cell to spare its own RNAs. One idea is that RNAs produced in the previous generation of an organism are somehow registered and set aside from piRNA attack in subsequent generations. Whether or not this mechanism truly exists, and, if so, how it might work, are questions that demonstrate our incomplete understanding of the full implications of RNA interference.

RNA Interference Has Become a Powerful Experimental Tool

Although it likely arose as a defense mechanism against viruses and transposable elements, RNA interference, as we have seen, has become thoroughly integrated into many aspects of normal cell biology, ranging from the control of gene expression to the structure of chromosomes. It has also been developed by scientists into a powerful experimental tool that allows almost any gene to be inactivated by evoking an RNAi response to it. This technique, which can be readily carried out in cultured cells and, in many cases, whole animals and plants, has made possible new genetic approaches in cell and molecular biology. We shall discuss it in detail in the following chapter where we cover modern genetic methods used to study cells (see pp. 499–501). RNAi also has potential in treating human disease. Since many human disorders result from the misexpression of genes, the ability to turn these genes off by experimentally introducing complementary siRNA molecules holds great medical promise. Although the mechanism of RNA interference was discovered a few decades ago, we are still being surprised by its mechanistic details and by its broad biological implications.

Bacteria Use Small Noncoding RNAs to Protect Themselves from Viruses

Bacteria make up the vast majority of the Earth's biomass and, not surprisingly, viruses that infect bacteria greatly outnumber plant and animal viruses. These viruses generally have DNA genomes. A recent discovery revealed that many species of bacteria (and almost all species of archaebacteria) use a repository of small noncoding RNA molecules to seek out and destroy the DNA of the invading viruses. Many features of this defense mechanism, known as the **CRISPR** system, resemble those we saw above for miRNAs and siRNAs, but there are two important differences. First, when bacteria and archaea are first infected by a virus, they have a mechanism that causes short fragments of that viral DNA to become integrated into their genomes. These serve as "vaccinations," in the sense that they become the templates for producing small noncoding RNAs known as **crRNAs** (CRISPR RNAs) that will thereafter destroy the virus should it reinfect the descendants of the original cell. This aspect of the CRISPR system is similar in principle to adaptive immunity in mammals, in that the cell carries a record of past exposures that is used to protect against future exposures. The second distinguishing feature of the CRISPR system is that these crRNAs then become associated with special proteins that allow them to seek out and destroy double-stranded DNA molecules, rather than single-stranded RNA molecules.

Although many details of CRISPR-mediated immunity remain to be discovered, we can outline the general process in three steps (**Figure 7–78**). In the first, viral DNA sequences are integrated into special regions of the bacterial genome known as CRISPR (clustered regularly interspersed short palindromic repeat) loci, named for the peculiar structure that first drew the attention of scientists. In its simplest form, a CRISPR locus consists of several hundred repeats of a host DNA sequence interspersed with a large collection of sequences (typically 25–70 nucleotide pairs each) that has been derived from prior exposures to viruses and other foreign DNA. The newest viral sequence is always integrated at the 5′ end of the CRISPR locus, the end that is transcribed first. Each locus, therefore, carries a temporal record of prior infections. Many bacterial and archaeal species carry several large CRISPR loci in their genomes and are thus immune to a wide range of viruses.

In the second step, the CRISPR locus is transcribed to produce a long RNA molecule, which is then processed into the much shorter (approximately 30 nucleotides) crRNAs. In the third step, crRNAs complexed with *Cas* (*CRISPR-associated*) proteins seek out complementary viral DNA sequences and direct their destruction by nucleases. Although structurally dissimilar, Cas proteins are analogous to the Argonaute and Piwi proteins discussed above: they hold small single-stranded RNAs in an extended configuration that is optimized, in this case, for seeking and forming complementary base pairs with DNA.

We still have much to learn about CRISPR-based immunity in bacteria and archaebacteria. The mechanism through which viral sequences are first identified and integrated into the host genome is poorly understood, as is the way that the crRNAs find their complementary sequences in double-stranded DNA. Moreover, in different species of bacteria and archaebacteria, crRNAs are processed in different ways, and in some cases, the crRNAs can attack viral RNAs as well as DNAs.

We shall see in the following chapter that bacterial CRISPR systems have already been artificially "moved" into plants and animals, where they have become very powerful experimental tools for manipulating genomes.

Figure 7–78 CRISPR-mediated immunity in bacteria and archaebacteria. After infection by a virus *(left panel)*, a small bit of DNA from the viral genome is inserted into the CRISPR locus. For this to happen, a small fraction of infected cells must survive the initial viral infection. The surviving cells, or more generally their descendants, transcribe the CRISPR locus and process the transcript into crRNAs *(middle panel)*. Upon reinfection with a virus that the population has already been "vaccinated" against, the incoming viral DNA is destroyed by a complementary crRNA *(right panel)*. For a CRISPR system to be effective, the crRNAs must not destroy the CRISPR locus itself, even though the crRNAs are complementary in sequence to it. In many species, in order for crRNAs to attack an invading DNA molecule, there must be additional short nucleotide sequences that are carried by the target molecule. Because these sequences, known as PAMs (protospacer adjacent motifs), lie outside the crRNA sequences, the host CRISPR locus is spared (see Figure 8–55).

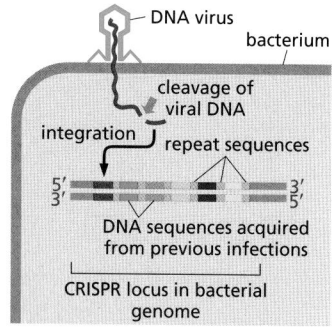

STEP 1: short viral DNA sequence is integrated into CRISPR locus

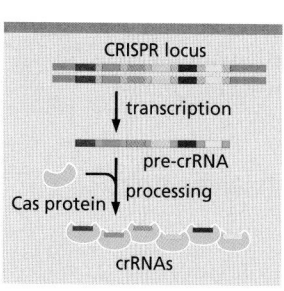

STEP 2: RNA is transcribed from CRISPR locus, processed, and bound to Cas protein

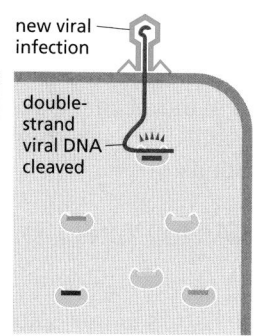

STEP 3: small crRNA in complex with Cas seeks out and destroys viral sequences

Long Noncoding RNAs Have Diverse Functions in the Cell

In this and the preceding chapters, we have seen that noncoding RNA molecules have many functions in the cell. Yet, as is the case with proteins, there remain many noncoding RNAs whose function is still unknown. Many RNAs of unknown function belong to a group known as **long noncoding RNA (lncRNA)**. These are arbitrarily defined as RNAs longer than 200 nucleotides that do not code for protein. As methods have improved for determining the nucleotide sequences of all the RNA molecules produced by a cell line or tissue, the sheer number of lncRNAs (an estimated 8000 for the human genome, for example) came as a surprise to scientists. Most lncRNAs are transcribed by RNA polymerase II and have 5′ caps and poly-A tails, and, in many cases, they are spliced. It has been difficult to annotate lncRNAs because low levels of RNA are now known to be made from 75% of the human genome. Most of these RNAs are thought to result from the background "noise" of transcription and RNA processing. According to this idea, such nonfunctional RNAs provide no fitness advantage or disadvantage to the organism and are a tolerated by-product of the complex patterns of gene expression that need to be produced in multicellular organisms. For these reasons, it is difficult to estimate the number of lncRNAs that are likely to have a function in the cell and to distinguish them from the background transcription.

We have already encountered a few lncRNAs, including the RNA in telomerase (see Figure 5–33), Xist RNA (see Figure 7–52), and an RNA involved in imprinting (see Figure 7–49). Other lncRNAs have been implicated in controlling the enzymatic activity of proteins, inactivating transcription regulators, affecting splicing patterns, and blocking translation of certain mRNAs.

In terms of biological function, lncRNA should be considered a catch-all phrase encompassing a great diversity of functions. Nevertheless, there are two unifying features of lncRNAs that can account for their many roles in the cell. The first is that lncRNAs can function as *scaffold RNA molecules*, holding together groups of proteins to coordinate their functions (**Figure 7–79A**). We have already seen an example in telomerase, where the RNA molecule holds together and organizes protein components. These RNA-based scaffolds are analogous to protein scaffolds we discussed in Chapter 3 (see Figure 3–78) and Chapter 6 (see Figure 6–47). RNA molecules are well suited to act as scaffolds: small bits of RNA sequence, often those portions that form stem-loop structures, can serve as binding sites for proteins, and these can be strung together with random sequences of RNA in between. This property may be one reason that lncRNAs show relatively little primary-sequence conservation across species.

The second key feature of lncRNAs is their ability to serve as guide sequences, binding to specific RNA or DNA target molecules through base-pairing. By doing so, they bring proteins that are bound to them into close proximity with the DNA

Figure 7–79 Roles of long noncoding RNA (lncRNA). (A) lncRNAs can serve as scaffolds, bringing together proteins that function in the same process. As described in Chapter 6, RNAs can fold into specific three-dimensional structures that are often recognized by proteins. (B) In addition to serving as scaffolds, lncRNAs can, through formation of complementary base pairs, localize proteins to specific sequences on RNA or DNA molecules. (C) In some cases, lncRNAs act only *in cis*, for example, when the RNA is held in place by RNA polymerase *(top)*. Other lncRNAs, however, diffuse from their sites of synthesis and therefore act *in trans*.

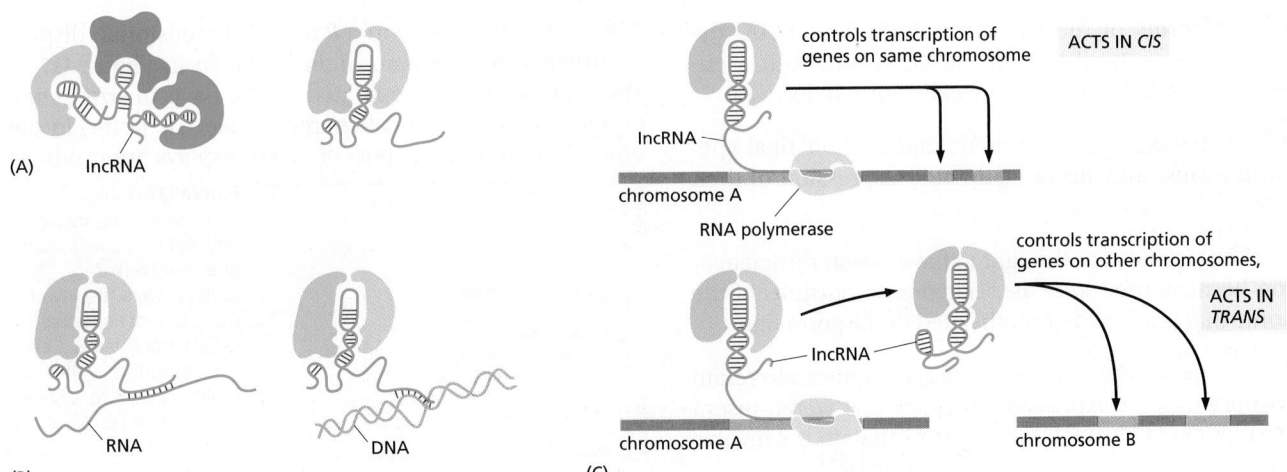

and RNA sequences (Figure 7–79B). This behavior is similar to that of snoRNAs (see Figure 6–41), crRNAs (see Figure 7–78), and miRNAs (see Figure 7–75), all of which act in this way to guide protein enzymes to specific nucleic acid sequences.

In some cases, lncRNAs work simply by base-pairing, without bringing in enzymes or other proteins. For example, a number of lncRNA genes are embedded in protein-coding genes, but they are transcribed in the "wrong direction." These *antisense RNAs* can form complementary base pairs with the mRNA (transcribed in the "correct" direction) and block its translation into protein (see Figure 7–66D). Other antisense lncRNAs base-pair with pre-mRNAs as they are synthesized and change the pattern of RNA splicing by masking splice-site sequences. Still others act as "sponges," base-pairing with miRNAs and thereby reducing their effects.

Finally, we note that some lncRNAs can act only *in cis*; that is, they affect only the chromosome from which they are transcribed. This readily occurs when the transcribed RNA has not yet been released from RNA polymerases (Figure 7–79C). Many lncRNAs, however, diffuse from their site of synthesis and act *in trans*. Although the best understood lncRNAs work in the nucleus, many are found in the cytosol. The functions—if any—of the great majority of these cytosolic lncRNAs remain undiscovered.

Summary

RNA molecules have many uses in the cell besides carrying the information needed to specify the order of amino acids during protein synthesis. Although we have encountered noncoding RNAs in other chapters (tRNAs, rRNAs, snoRNAs, for example), the sheer number of noncoding RNAs produced by cells has surprised scientists. One well understood use of noncoding RNAs occurs in RNA interference, where guide RNAs (miRNAs, siRNAs, piRNAs) base-pair with mRNAs. RNA interference can cause mRNAs to be either destroyed or translationally repressed. It can also cause specific genes to be packaged into heterochromatin suppressing their transcription. In bacteria and archaebacteria, RNA interference is used as an adaptive immune response to destroy viruses that infect them. A large family of large noncoding RNAs (lncRNAs) has recently been discovered. Although the function of most of these RNAs is unknown, some serve as RNA scaffolds to bring specific proteins and RNA molecules together to speed up needed reactions.

WHAT WE DON'T KNOW

• How is the final rate of transcription of a gene specified by the hundreds of proteins that assemble on its control regions? Will we ever be able to predict this rate from inspection of the DNA sequences of control regions?

• How does the collection of *cis*-regulatory sequences embedded in a genome orchestrate the developmental program of a multicellular organism?

• How much of the human genome sequence is functional, and why is the remainder retained?

• Which of the thousands of unstudied noncoding RNAs have functions in the cell, and what are these functions?

• Were introns present in early cells (and subsequently lost in some organisms), or did they arise at later times?

PROBLEMS

Which statements are true? Explain why or why not.

7–1   In terms of the way it interacts with DNA, the helix–loop–helix motif is more closely related to the leucine zipper motif than it is to the helix–turn–helix motif.

7–2   Once cells have differentiated to their final specialized forms, they never again alter expression of their genes.

7–3   CG islands are thought to have arisen during evolution because they were associated with portions of the genome that remained unmethylated in the germ line.

7–4   In most differentiated tissues, daughter cells retain a memory of gene expression patterns that were present in the parent cell through mechanisms that do not involve changes in the sequence of their genomic DNA.

Discuss the following problems.

7–5   A small portion of a two-dimensional display of proteins from human brain is shown in **Figure Q7–1**. These proteins were separated on the basis of size in one dimension and electrical charge (isoelectric point) in the other. Not all protein spots on such displays are products

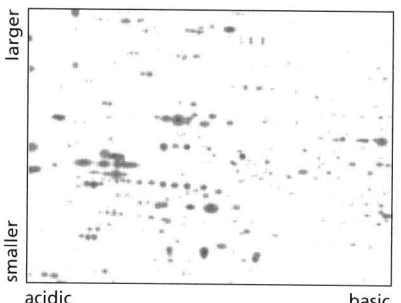

Figure Q7–1 Two-dimensional separation of proteins from the human brain (Problem 7–5). The proteins were displayed using two-dimensional gel electrophoresis. Only a small portion of the protein spectrum is shown. (Courtesy of Tim Myers and Leigh Anderson, Large Scale Biology Corporation.)

of different genes; some represent modified forms of a protein that migrate to different positions. Pick out a couple of sets of spots that could represent proteins that differ by the number of phosphates they carry. Explain the basis for your selection.

7–6 Comparisons of the patterns of mRNA levels across different human cell types show that the level of expression of almost every active gene is different. The patterns of mRNA abundance are so characteristic of cell type that they can be used to determine the tissue of origin of cancer cells, even though the cells may have metastasized to different parts of the body. By definition, however, cancer cells are different from their noncancerous precursor cells. How do you suppose then that patterns of mRNA expression might be used to determine the tissue source of a human cancer?

7–7 What are the two fundamental components of a genetic switch?

7–8 The nucleus of a eukaryotic cell is much larger than a bacterium, and it contains much more DNA. As a consequence, a transcription regulator in a eukaryotic cell must be able to select its specific binding site from among many more unrelated sequences than does a transcription regulator in a bacterium. Does this present any special problems for eukaryotic gene regulation?

Consider the following situation. Assume that the eukaryotic nucleus and the bacterial cell each have a single copy of the same DNA binding site. In addition, assume that the nucleus is 500 times the volume of the bacterium, and has 500 times as much DNA. If the concentration of the transcription regulator that binds the site were the same in the nucleus and in the bacterium, would the regulator occupy its binding site equally as well in the eukaryotic nucleus as it does in the bacterium? Explain your answer.

7–9 Some transcription regulators bind to DNA and cause the double helix to bend at a sharp angle. Such "bending proteins" can affect the initiation of transcription without directly contacting any other protein. Can you devise a plausible explanation for how such proteins might work to modulate transcription? Draw a diagram that illustrates your explanation.

7–10 How is it that protein–protein interactions that are too weak to cause proteins to assemble in solution can nevertheless allow the same proteins to assemble into complexes on DNA?

7–11 Imagine the two situations shown in **Figure Q7–2**. In cell 1, a transient signal induces the synthesis of protein A, which is a transcription activator that turns on many genes including its own. In cell 2, a transient signal induces the synthesis of protein R, which is a transcription repressor that turns off many genes including its own. In which, if either, of these situations will the descendants of the original cell "remember" that the progenitor cell had experienced the transient signal? Explain your reasoning.

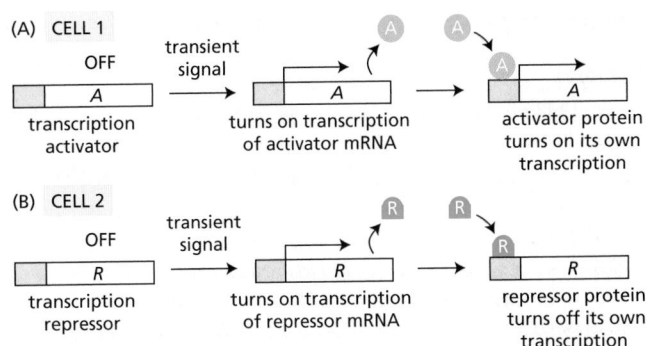

Figure Q7–2 Gene regulatory circuits and cell memory (Problem 7–11). (A) Induction of synthesis of transcription activator A by a transient signal. (B) Induction of synthesis of transcription repressor R by a transient signal.

7–12 Examine the two pedigrees shown in **Figure Q7–3**. One results from deletion of a maternally imprinted autosomal gene. The other pedigree results from deletion of a paternally imprinted autosomal gene. In both pedigrees, affected individuals (*red* symbols) are heterozygous for the deletion. These individuals are affected because one copy of the chromosome carries an imprinted, inactive gene, while the other carries a deletion of the gene. Dotted *yellow* symbols indicate individuals that carry the deleted locus, but do not display the mutant phenotype. Which pedigree is based on paternal imprinting and which on maternal imprinting? Explain your answer.

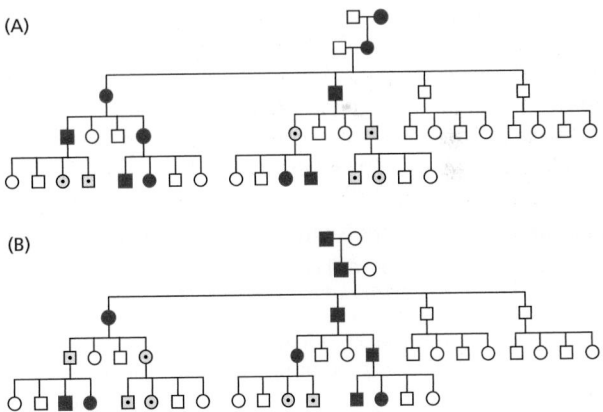

Figure Q7–3 Pedigrees reflecting maternal and paternal imprinting (Problem 7–12). In one pedigree, the gene is paternally imprinted; in the other, it is maternally imprinted. In generations 3 and 4, only one of the two parents in the indicated matings is shown; the other parent is a normal individual from outside this pedigree. Affected individuals are represented by *red circles* for females and *red squares* for males. *Dotted yellow symbols* indicate individuals that carry the deletion but do not display the phenotype.

7–13 If you insert a *β-galactosidase* gene lacking its own transcription control region into a cluster of piRNA genes in *Drosophila*, you find that β-galactosidase expression from a normal copy elsewhere in the genome is strongly inhibited in the fly's germ cells. If the inactive *β-galactosidase* gene is inserted outside the piRNA gene cluster, the normal gene is properly expressed. What do you suppose is the basis for this observation? How would you test your hypothesis?

REFERENCES

General

Brown TA (2007) Genomes 3. New York: Garland Science.

Epigenetics (2004) *Cold Spring Harb. Symp. Quant. Biol.* 69.

Gilbert SF (2013) Developmental Biology, 10th ed. Sunderland, MA: Sinauer Associates, Inc.

Hartwell L, Hood L, Goldberg ML et al. (2010) Genetics: from Genes to Genomes, 4th ed. Boston: McGraw Hill.

McKnight SL & Yamamoto KR (eds) (1993) Transcriptional Regulation. Cold Spring Harbor, NY: Cold Spring Harbor Laboratory Press.

Mechanisms of Transcription (1998) *Cold Spring Harb. Symp. Quant. Biol.* 63.

Ptashne M & Gann A (2002) Genes and Signals. Cold Spring Harbor, NY: Cold Spring Harbor Laboratory Press.

Watson J, Baker T, Bell S et al. (2013) Molecular Biology of the Gene, 7th ed. Menlo Park, CA: Benjamin Cummings.

An Overview of Gene Control

Davidson EH (2006) The Regulatory Genome: Gene Regulatory Networks in Development and Evolution. Burlington, MA: Elsevier.

Gurdon JB (1992) The generation of diversity and pattern in animal development. *Cell* 68, 185–199.

Kellis M, Wold B, Synder MP et al. (2014) Defining functional DNA elements in the human genome. *Proc. Natl. Acad. Sci. USA* 111, 6131–6138.

Control Of Transcription By Sequence-Specific DNA-Binding Proteins

McKnight SL (1991) Molecular zippers in gene regulation. *Sci. Am.* 264, 54–64.

Pabo CO & Sauer RT (1992) Transcription factors: structural families and principles of DNA recognition. *Annu. Rev. Biochem.* 61, 1053–1095.

Seeman NC, Rosenberg JM & Rich A (1976) Sequence-specific recognition of double helical nucleic acids by proteins. *Proc. Natl. Acad. Sci. USA* 73, 804–808.

Weirauch MT & Hughes TR (2011) A catalogue of eukaryotic transcription factor types, their evolutionary origin, and species distribution. In A Handbook of Transcription Factors. New York, NY: Springer Publishing Company.

Transcription Regulators Switch Genes On and Off

Beckwith J (1987) The operon: an historical account. In *Escherichia coli* and *Salmonella typhimurium*: Cellular and Molecular Biology (Neidhart FC, Ingraham JL, Low KB et al. eds), vol. 2, pp. 1439–1443. Washington, DC: ASM Press.

Gilbert W & Müller-Hill B (1967) The lac operator is DNA. *Proc. Natl. Acad. Sci. USA* 58, 2415.

Jacob F & Monod J (1961) Genetic regulatory mechanisms in the synthesis of proteins. *J. Mol. Biol.* 3, 318–356.

Levine M, Cattoglio C & Tjian R (2014) Looping back to leap forward: transcription enters a new era. *Cell* 157, 13–25.

Narlikar GJ, Sundaramoorthy R & Owen-Hughes T (2013) Mechanisms and Functions of ATP-dependent chromatin-remodeling enzymes. *Cell* 154, 490–503.

Ptashne M (2004) A Genetic Switch: Phage and Lambda Revisited, 3rd ed. Cold Spring Harbor, NY: Cold Spring Harbor Laboratory Press.

Ptashne M (1967) Specific binding of the lambda phage repressor to lambda DNA. *Nature* 214, 232–234.

St Johnston D & Nusslein-Volhard C (1992) The origin of pattern and polarity in the *Drosophila* embryo. *Cell* 68, 201–219.

Turner BM (2014) Nucleosome signaling: an evolving concept. *Biochim. Biophys. Acta* 1839, 623–626.

Molecular Genetic Mechanisms that Create and Maintain Specialized Cell Types

Alon U (2007) Network motifs: theory and experimental approaches. *Nature* 8, 450–461.

Buganim Y, Faddah DA & Jaenisch R (2013) Mechanisms and models of somatic cell reprogramming. *Nat. Rev. Genet.* 14, 427–439.

Hobert O (2011) Regulation of Terminal differentiation programs in the nervous system. *Annu. Rev. Cell Dev. Biol.* 27, 681–696.

Lawrence PA (1992) The Making of a Fly: The Genetics of Animal Design. New York: Blackwell Scientific Publications.

Mechanisms That Reinforce Cell Memory in Plants and Animals

Bird A (2011) Putting the DNA back into DNA methylation. *Nat. Genet.* 43, 1050–1051.

Gehring M (2013) Genomic imprinting: insights from plants. *Annu. Rev. Genet.* 47, 187–208.

Lawson HA, Cheverud JM & Wolf JB (2013) Genomic imprinting and parent-of-origin effects on complex traits. *Genetics* 14, 609–617.

Lee JT & Bartolomei MS (2013) X-Inactivation, imprinting, and long noncoding RNAs in Health and disease. *Cell* 152, 1308–1323.

Li E & Zhang Y (2014) DNA methylation in mammals. *Cold Spring Harb. Perspect. Biol.* 6, a019133.

Post-Transcriptional Controls

DiGiammartino DC, Nishida K & Manley JL (2011) Mechanisms and consequences of alternative polyadenylation. *Mol Cell* 43, 853–866.

Gottesman S & Storz G (2011) Bacterial small RNA regulators: versatile roles and rapidly evolving variations. *Cold Spring Harb. Perspect. Biol.* 3, a003798.

Hershey JWB, Sonenberg N & Mathews MB (2012) Principles of translational control: an overview. *Cold Spring Harb. Perspect. Biol.* 4, a011528.

Hundley HA & Bass BL (2010) ADAR editing in double-stranded UTRs and other noncoding RNA sequences. *Trends Biochem. Sci.* 35, 377–383.

Kalsotra A & Cooper TA (2011) Functional consequences of developmentally regulated alternative splicing. *Nat. Rev. Genet.* 12, 715–729.

Kortmann J & Narberhaus F (2012) Bacterial RNA thermometers: molecular zippers and switches. *Nat. Rev. Microbiol.* 10, 255–265.

Parker R (2012) RNA degradation in *Saccharomyces cerevisae*. *Genetics* 191, 671–702.

Popp MW & Maquat LE (2013) Organizing principles of mammalian nonsense-mediated mRNA decay. *Annu. Rev. Genet.* 47, 139–165.

Serganov A & Nudler E (2013) A decade of riboswitches. *Cell* 152, 17–24.

Thompson SR (2012) Tricks an IRES uses to enslave ribosomes. *Trends Microbiol.* 20, 558–566.

Regulation of Gene Expression By Noncoding RNAs

Bhaya D, Davison M & Barrangou R (2011) CRISPR-Cas systems in bacteria and archaea: Versatile small RNAs for adaptive defense and regulation. *Annu. Rev. Genet.* 45, 273–297.

Cech TR & Steitz JA (2014) The noncoding RNA revolution–trashing old rules to forge new ones. *Cell* 157, 77–94.

Fire A, Xu S, Montgomery MK et al (1998) Potent and specific genetic interference by double-stranded RNA in *Caenorhabditis elegans*. *Nature* 391, 806–811.

Guttman M & Rinn JL (2012) Modular regulatory principles of large non-coding RNAs. *Nature* 482, 339–346.

Lee HC, Gu W, Shirayama M et al. (2012) *C. elegans* piRNAs mediate the genome-wide surveillance of germline transcripts. *Cell* 150, 78–87.

Meister G (2013) Argonaute proteins: functional insights and emerging roles. *Nat. Rev. Genet.* 14, 447–459.

Rinn JL & Chang HY (2012) Genome regulation by long noncoding RNAs. *Annu. Rev. Biochem.* 81, 145–166.

tenOever BR (2013) RNA viruses and the host microRNA machinery. *Nat. Rev. Microbiol.* 11, 169–180.

Ulitsky I & Bartel DP (2013) lincRNAs: genomics, evolution, and mechanisms. *Cell* 154, 26–46.

Wiedenheft B, Sternberg SH & Doudna JA (2012) RNA-guided genetic silencing systems in bacteria and archaea. *Nature* 482, 331–338.

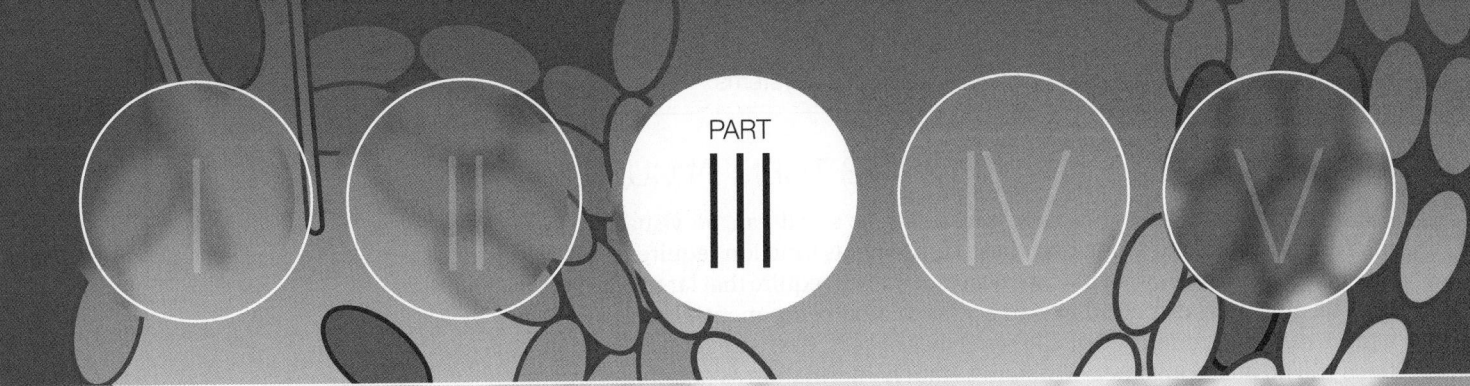

WAYS OF WORKING WITH CELLS

Analyzing Cells, Molecules, and Systems

Progress in science is often driven by advances in technology. The entire field of cell biology, for example, came into being when optical craftsmen learned to grind small lenses of sufficiently high quality to observe cells and their substructures. Innovations in lens grinding, rather than any conceptual or philosophical advance, allowed Hooke and van Leeuwenhoek to discover a previously unseen cellular world, where tiny creatures tumble and twirl in a small droplet of water (**Figure 8–1**).

The twenty-first century is a particularly exciting time for biology. New methods for analyzing cells, proteins, DNA, and RNA are fueling an information explosion and allowing scientists to study cells and their macromolecules in previously unimagined ways. We now have access to the sequences of many billions of nucleotides, providing the complete molecular blueprints for hundreds of organisms—from microbes and mustard weeds to worms, flies, mice, dogs, chimpanzees, and humans. And powerful new techniques are helping us to decipher that information, allowing us not only to compile huge, detailed catalogs of genes and proteins but also to begin to unravel how these components work together to form functional cells and organisms. The long-range goal is nothing short of obtaining a complete understanding of what takes place inside a cell as it responds to its environment and interacts with its neighbors.

In this chapter, we present some of the principal methods used to study cells and their molecular components. We consider how to separate cells of different types from tissues, how to grow cells outside the body, and how to disrupt cells and isolate their organelles and constituent macromolecules in pure form. We also present the techniques used to determine protein structure, function, and interactions, and we discuss the breakthroughs in DNA technology that continue to revolutionize our understanding of cell function. We end the chapter with an overview of some of the mathematical approaches that are helping us deal with the enormous complexity of cells. By considering cells as dynamic systems with many moving parts, mathematical approaches can reveal hidden insights into how the many components of cells work together to produce the special qualities of life.

IN THIS CHAPTER

ISOLATING CELLS AND GROWING THEM IN CULTURE

PURIFYING PROTEINS

ANALYZING PROTEINS

ANALYZING AND MANIPULATING DNA

STUDYING GENE EXPRESSION AND FUNCTION

MATHEMATICAL ANALYSIS OF CELL FUNCTIONS

ISOLATING CELLS AND GROWING THEM IN CULTURE

Although the organelles and large molecules in a cell can be visualized with microscopes, understanding how these components function requires a detailed biochemical analysis. Most biochemical procedures require that large numbers of cells be physically disrupted to gain access to their components. If the sample is a piece of tissue, composed of different types of cells, heterogeneous cell populations will be mixed together. To obtain as much information as possible about the cells in a tissue, biologists have developed ways of dissociating cells from tissues and separating them according to type. These manipulations result in a relatively homogeneous population of cells that can then be analyzed—either directly or after their number has been greatly increased by allowing the cells to proliferate in culture.

Cells Can Be Isolated from Tissues

Intact tissues provide the most realistic source of material, as they represent the actual cells found within the body. The first step in isolating individual cells is to disrupt the extracellular matrix and cell–cell junctions that hold the cells together. For this purpose, a tissue sample is typically treated with proteolytic enzymes (such as trypsin and collagenase) to digest proteins in the extracellular matrix and with agents (such as ethylenediaminetetraacetic acid, or EDTA) that bind, or chelate, the Ca^{2+} on which cell–cell adhesion depends. The tissue can then be teased apart into single cells by gentle agitation.

For some biochemical preparations, the protein of interest can be obtained in sufficient quantity without having to separate the tissue or organ into cell types. Examples include the preparation of histones from calf thymus, actin from rabbit muscle, or tubulin from cow brain. In other cases, obtaining the desired protein requires enrichment for a specific cell type of interest. Several approaches are used to separate the different cell types from a mixed cell suspension. One of the most sophisticated cell-separation techniques uses an antibody coupled to a fluorescent dye to label specific cells. An antibody is chosen that specifically binds to the surface of only one cell type in the tissue. The labeled cells can then be separated from the unlabeled ones in a *fluorescence-activated cell sorter.* In this remarkable machine, individual cells traveling single file in a fine stream pass through a laser beam, and the fluorescence of each cell is rapidly measured. A vibrating nozzle generates tiny droplets, most containing either one cell or no cells. The droplets containing a single cell are automatically given a positive or a negative charge at the moment of formation, depending on whether the cell they contain is fluorescent; they are then deflected by a strong electric field into an appropriate container. Occasional clumps of cells, detected by their increased light scattering, are left uncharged and are discarded into a waste container. Such machines can accurately select 1 fluorescent cell from a pool of 1000 unlabeled cells and sort several thousand cells each second (**Figure 8–2**).

Cells Can Be Grown in Culture

Although molecules can be extracted from whole tissues, this is often not the most convenient or useful source of material. The complexity of intact tissues and organs is an inherent disadvantage when trying to purify particular molecules. Cells grown in culture provide a more homogeneous population of cells from which to extract material, and they are also much more convenient to work with in the laboratory. Given appropriate surroundings, most plant and animal cells can live, multiply, and even express differentiated properties in a culture dish. The cells can be watched continuously under the microscope or analyzed biochemically, and the effects of adding or removing specific molecules, such as hormones or growth factors, can be systematically explored.

Experiments performed on cultured cells are sometimes said to be carried out *in vitro* (literally, "in glass") to contrast them with experiments using intact organisms, which are said to be carried out *in vivo* (literally, "in the living organism").

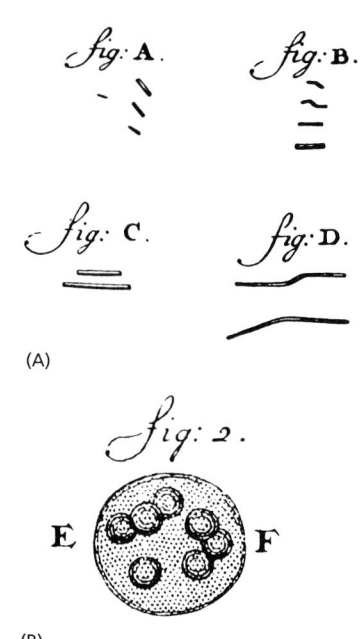

Figure 8–1 Microscopic life. A sample of "diverse animalcules" seen by van Leeuwenhoek using his simple microscope. (A) Bacteria seen in material he excavated from between his teeth. Those in fig. B he described as "swimming first forward and then backwards" (1692). (B) The eukaryotic green alga *Volvox* (1700). (Courtesy of the John Innes Foundation.)

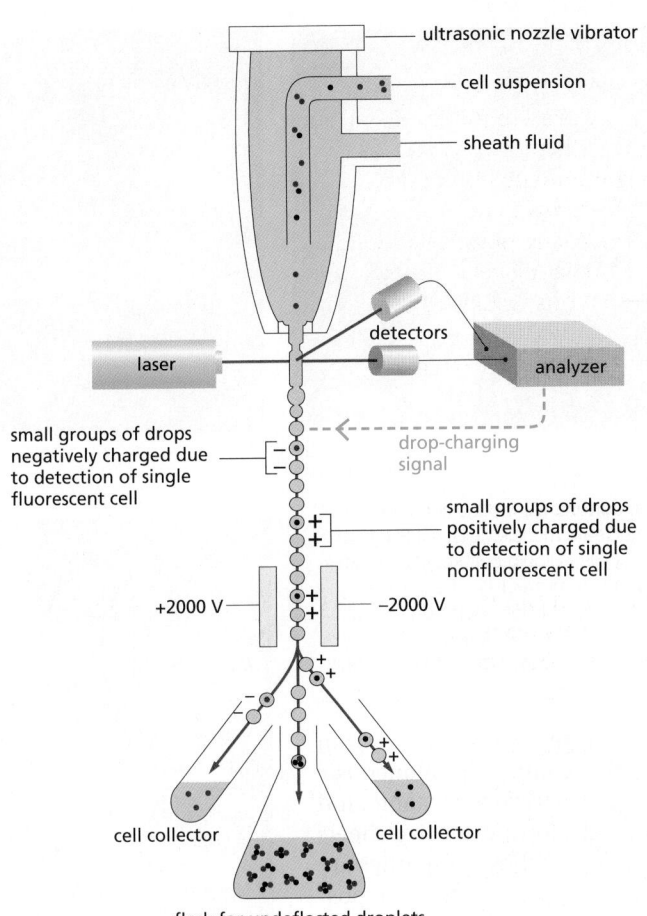

- ultrasonic nozzle vibrator
- cell suspension
- sheath fluid
- detectors
- laser
- analyzer
- small groups of drops negatively charged due to detection of single fluorescent cell
- drop-charging signal
- small groups of drops positively charged due to detection of single nonfluorescent cell
- +2000 V
- −2000 V
- cell collector
- cell collector
- flask for undeflected droplets

Figure 8–2 A fluorescence-activated cell sorter. A cell passing through the laser beam is monitored for fluorescence. Droplets containing single cells are given a negative or positive charge, depending on whether the cell is fluorescent or not. The droplets are then deflected by an electric field into collection tubes according to their charge. Note that the cell concentration must be adjusted so that most droplets contain no cells and flow to a waste container together with any cell clumps.

These terms can be confusing, however, because they are often used in a very different sense by biochemists. In the biochemistry lab, *in vitro* refers to reactions carried out in a test tube in the absence of living cells, whereas *in vivo* refers to any reaction taking place inside a living cell, even if that cell is growing in culture.

Tissue culture began in 1907 with an experiment designed to settle a controversy in neurobiology. The hypothesis under examination was known as the neuronal doctrine, which states that each nerve fiber is the outgrowth of a single nerve cell and not the product of the fusion of many cells. To test this contention, small pieces of spinal cord were placed on clotted tissue fluid in a warm, moist chamber and observed at regular intervals under the microscope. After a day or so, individual nerve cells could be seen extending long, thin filaments (axons) into the clot. Thus, the neuronal doctrine received strong support, and the foundation was laid for the cell-culture revolution.

These original experiments on nerve fibers used cultures of small tissue fragments called explants. Today, cultures are more commonly made from suspensions of cells dissociated from tissues. Unlike bacteria, most tissue cells are not adapted to living suspended in fluid and require a solid surface on which to grow and divide. For cell cultures, this support is usually provided by the surface of a plastic culture dish. Cells vary in their requirements, however, and many do not proliferate or differentiate unless the culture dish is coated with materials that cells like to adhere to, such as polylysine or extracellular matrix components.

Cultures prepared directly from the tissues of an organism are called *primary cultures*. These can be made with or without an initial fractionation step to separate different cell types. In most cases, cells in primary cultures can be removed from the culture dish and recultured repeatedly in so-called secondary cultures; in this way, they can be repeatedly subcultured (*passaged*) for weeks or months. Such cells often display many of the differentiated properties appropriate to their

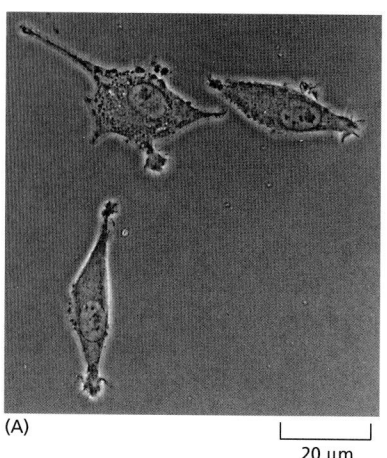

(A)

20 µm

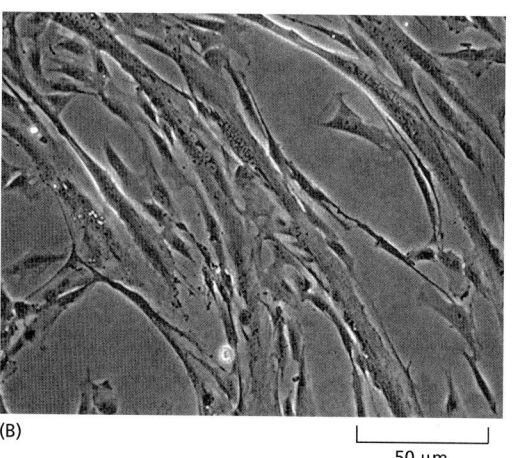

(B)

50 µm

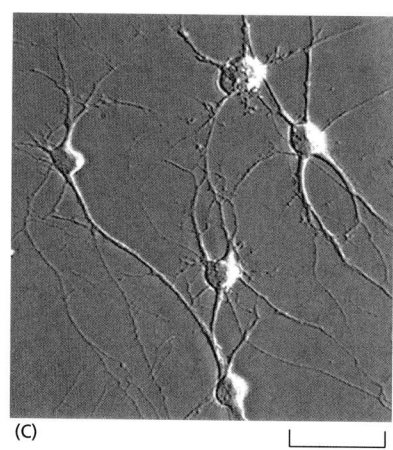

(C)

50 µm

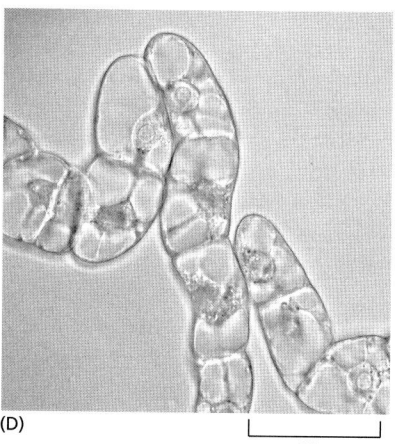

(D)

50 µm

Figure 8–3 Light micrographs of cells in culture. (A) Mouse fibroblasts. (B) Mouse myoblasts fusing to form multinucleate muscle cells. (C) Purified rat retinal ganglion nerve cells. (D) Tobacco cells in liquid culture. (A, courtesy of Daniel Zicha; B, from Cerletti M, Molloy MJ, Tomczak KK, et al., *J. Cell Sci.* 2006;119(Pt 15):3117–3127. doi:10.1242/jcs.03056. Reproduced with permission of The Company of Biologists Ltd.; C, from A. Meyer-Franke et al., *Neuron* 15:805–819, 1995. With permission from Elsevier; D, courtesy of Gethin Roberts.)

origin (Figure 8–3): fibroblasts continue to secrete collagen; cells derived from embryonic skeletal muscle fuse to form muscle fibers that contract spontaneously in the culture dish; nerve cells extend axons that are electrically excitable and make synapses with other nerve cells; and epithelial cells form extensive sheets with many of the properties of an intact epithelium. Because these properties are maintained in culture, they are accessible to study in ways that are often not possible in intact tissues.

Cell culture is not limited to animal cells. When a piece of plant tissue is cultured in a sterile medium containing nutrients and appropriate growth regulators, many of the cells are stimulated to proliferate indefinitely in a disorganized manner, producing a mass of relatively undifferentiated cells called a *callus*. If the nutrients and growth regulators are carefully manipulated, one can induce the formation of a shoot and then root apical meristems within the callus, and, in many species, regenerate a whole new plant. Similar to animal cells, callus cultures can be mechanically dissociated into single cells, which will grow and divide as a suspension culture (see Figure 8–3D).

Eukaryotic Cell Lines Are a Widely Used Source of Homogeneous Cells

The cell cultures obtained by disrupting tissues tend to suffer from a problem—eventually the cells die. Most vertebrate cells stop dividing after a finite number of cell divisions in culture, a process called *replicative cell senescence* (discussed in Chapter 17). Normal human fibroblasts, for example, typically divide only 25–40 times in culture before they stop. In these cells, the limited proliferation capacity reflects a progressive shortening and uncapping of the cell's telomeres, the repetitive DNA sequences and associated proteins that cap the ends of each chromosome (discussed in Chapter 5). Human somatic cells in the body have turned off production of the enzyme, called *telomerase*, that normally maintains the telomeres, which is why their telomeres shorten with each cell division. Human fibroblasts can often be coaxed to proliferate indefinitely by providing them with the gene that encodes the catalytic subunit of telomerase; in this case, they can be propagated as an "immortalized" cell line.

Some human cells, however, cannot be immortalized by this trick. Although their telomeres remain long, they still stop dividing after a limited number of divisions because culture conditions cause excessive mitogenic stimulation, which activates a poorly understood protective mechanism (discussed in Chapter 17) that stops cell division—a process sometimes called "culture shock." To immortalize these cells, one has to do more than introduce telomerase. One must also inactivate the protective mechanisms, which can be done by introducing certain cancer-promoting oncogenes (discussed in Chapter 20). Unlike human cells, most rodent cells do not turn off production of telomerase and therefore their telomeres do not shorten with each cell division. Therefore, if culture shock can be avoided, some rodent cell types will divide indefinitely in culture. In addition, rodent cells often undergo spontaneous genetic changes in culture that inactivate their protective mechanisms, thereby producing immortalized cell lines.

Cell lines can often be most easily generated from cancer cells, but these cultures—referred to as *transformed cell lines*—differ from those prepared from normal cells in several ways. Transformed cell lines often grow without attaching to a surface, for example, and they can proliferate to a much higher density in a culture dish. Similar properties can be induced experimentally in normal cells by transforming them with a tumor-inducing virus or chemical. The resulting transformed cell lines can usually cause tumors if injected into a susceptible animal.

Transformed and nontransformed cell lines are extremely useful in cell research as sources of very large numbers of cells of a uniform type, especially since they can be stored in liquid nitrogen at –196°C for an indefinite period and retain their viability when thawed. It is important to keep in mind, however, that cell lines nearly always differ in important ways from their normal progenitors in the tissues from which they were derived.

Some widely used cell lines are listed in Table 8–1. Different lines have different advantages; for example, the PtK epithelial cell lines derived from the rat

TABLE 8–1 **Some Commonly Used Cell Lines**	
Cell line*	Cell type and origin
3T3	Fibroblast (mouse)
BHK21	Fibroblast (Syrian hamster)
MDCK	Epithelial cell (dog)
HeLa	Epithelial cell (human)
PtK1	Epithelial cell (rat kangaroo)
L6	Myoblast (rat)
PC12	Chromaffin cell (rat)
SP2	Plasma cell (mouse)
COS	Kidney (monkey)
293	Kidney (human); transformed with adenovirus
CHO	Ovary (Chinese hamster)
DT40	Lymphoma cell for efficient targeted recombination (chick)
R1	Embryonic stem cell (mouse)
E14.1	Embryonic stem cell (mouse)
H1, H9	Embryonic stem cell (human)
S2	Macrophage-like cell (*Drosophila*)
BY2	Undifferentiated meristematic cell (tobacco)

*Many of these cell lines were derived from tumors. All of them are capable of indefinite replication in culture and express at least some of the special characteristics of their cells of origin.

kangaroo, unlike many other cell types, remain flat during mitosis, allowing the mitotic apparatus to be readily observed in action.

Hybridoma Cell Lines Are Factories That Produce Monoclonal Antibodies

As we see throughout this book, antibodies are particularly useful tools for cell biology. Their great specificity allows precise visualization of selected proteins among the many thousands that each cell typically produces. Antibodies are often produced by inoculating animals with the protein of interest and subsequently isolating the antibodies specific to that protein from the serum of the animal. However, only limited quantities of antibodies can be obtained from a single inoculated animal, and the antibodies produced will be a heterogeneous mixture of antibodies that recognize a variety of different antigenic sites on a macromolecule that differs from animal to animal. Moreover, antibodies specific for the antigen will constitute only a fraction of the antibodies found in the serum. An alternative technology, which allows the production of an unlimited quantity of identical antibodies and greatly increases the specificity and convenience of antibody-based methods, is the production of monoclonal antibodies by hybridoma cell lines.

This technology, developed in 1975, revolutionized the production of antibodies for use as tools in cell biology, as well as for the diagnosis and treatment of certain diseases, including rheumatoid arthritis and cancer. The procedure requires hybrid cell technology (**Figure 8–4**), and it involves propagating a clone of cells from a single antibody-secreting B lymphocyte to obtain a homogeneous preparation of antibodies in large quantities. B lymphocytes normally have a limited life-span in culture, but individual antibody-producing B lymphocytes from an immunized mouse, when fused with cells derived from a transformed B lymphocyte cell line, can give rise to hybrids that have both the ability to make a particular antibody and the ability to multiply indefinitely in culture. These **hybridomas** are propagated as individual clones, each of which provides a permanent and stable source of a single type of **monoclonal antibody**. Each type of monoclonal antibody recognizes a single type of antigenic site—for example, a particular cluster of five or six amino acid side chains on the surface of a protein. Their uniform specificity makes monoclonal antibodies much more useful than conventional antisera for many purposes.

An important advantage of the hybridoma technique is that monoclonal antibodies can be made against molecules that constitute only a minor component of a complex mixture. In an ordinary antiserum made against such a mixture, the proportion of antibody molecules that recognize the minor component would be too small to be useful. But if the B lymphocytes that produce the various components of this antiserum are made into hybridomas, it becomes possible to screen individual hybridoma clones from the large mixture to select one that produces the desired type of monoclonal antibody and to propagate the selected hybridoma

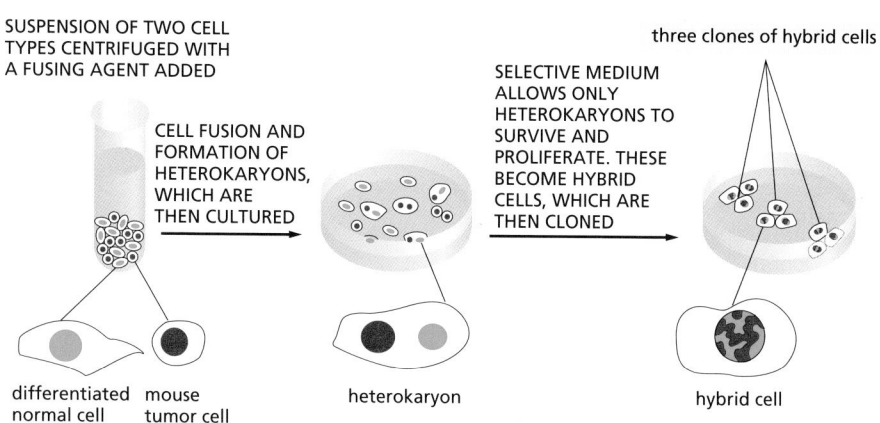

SUSPENSION OF TWO CELL TYPES CENTRIFUGED WITH A FUSING AGENT ADDED

CELL FUSION AND FORMATION OF HETEROKARYONS, WHICH ARE THEN CULTURED

SELECTIVE MEDIUM ALLOWS ONLY HETEROKARYONS TO SURVIVE AND PROLIFERATE. THESE BECOME HYBRID CELLS, WHICH ARE THEN CLONED

three clones of hybrid cells

differentiated normal cell   mouse tumor cell

heterokaryon

hybrid cell

Figure 8–4 **The production of hybrid cells.** It is possible to fuse one cell with another to form a *heterokaryon*, a combined cell with two separate nuclei. Typically, a suspension of cells is treated with certain inactivated viruses or with polyethylene glycol, each of which alters the plasma membranes of cells in a way that induces them to fuse. Eventually, a heterokaryon proceeds to mitosis and produces a hybrid cell in which the two separate nuclear envelopes have been disassembled, allowing all the chromosomes to be brought together in a single large nucleus. Such hybrid cells can give rise to immortal hybrid cell lines. If one of the parent cells was from a tumor cell line, the hybrid cell is called a hybridoma.

indefinitely so as to produce that antibody in unlimited quantities. In principle, therefore, a monoclonal antibody can be made against any protein in a biological sample. Once an antibody has been made, it can be used to localize the protein in cells and tissues, to follow its movement, and to purify the protein to study its structure and function.

Summary

Tissues can be dissociated into their component cells, from which individual cell types can be purified and used for biochemical analysis or for the establishment of cell cultures. Many animal and plant cells survive and proliferate in a culture dish if they are provided with a suitable culture medium containing nutrients and appropriate signal molecules. Although many animal cells stop dividing after a finite number of cell divisions, cells that have been immortalized through spontaneous mutations or genetic manipulation can be maintained indefinitely as cell lines. Hybridoma cells are widely employed to produce unlimited quantities of uniform monoclonal antibodies, which are used to detect and purify cell proteins, as well as to diagnose and treat diseases.

PURIFYING PROTEINS

The challenge of isolating a single type of protein from the thousands of other proteins present in a cell is a formidable one, but must be overcome in order to study protein function *in vitro*. As we shall see later in this chapter, *recombinant DNA technology* can enormously simplify this task by "tricking" cells into producing large quantities of a given protein, thereby making its purification much easier. Whether the source of the protein is an engineered cell or a natural tissue, a purification procedure usually starts with subcellular fractionation to reduce the complexity of the material, and is then followed by purification steps of increasing specificity.

Cells Can Be Separated into Their Component Fractions

To purify a protein, it must first be extracted from inside the cell. Cells can be broken up in various ways: they can be subjected to osmotic shock or ultrasonic vibration, forced through a small orifice, or ground up in a blender. These procedures break many of the membranes of the cell (including the plasma membrane and endoplasmic reticulum) into fragments that immediately reseal to form small closed vesicles. If carefully carried out, however, the disruption procedures leave organelles such as nuclei, mitochondria, the Golgi apparatus, lysosomes, and peroxisomes largely intact. The suspension of cells is thereby reduced to a thick slurry (called a *homogenate* or *extract*) that contains a variety of membrane-enclosed organelles, each with a distinctive size, charge, and density. Provided that the homogenization medium has been carefully chosen (by trial and error for each organelle), the various components—including the vesicles derived from the endoplasmic reticulum, called microsomes—retain most of their original biochemical properties.

The different components of the homogenate must then be separated. Such cell fractionations became possible only after the commercial development in the early 1940s of an instrument known as the *preparative ultracentrifuge*, which rotates extracts of broken cells at high speeds (Figure 8–5). This treatment separates cell components by size and density: in general, the largest objects experience the largest centrifugal force and move the most rapidly. At relatively low speed, large components such as nuclei sediment to form a pellet at the bottom of the centrifuge tube; at slightly higher speed, a pellet of mitochondria is deposited; and at even higher speeds and with longer periods of centrifugation, first the small closed vesicles and then the ribosomes can be collected (Figure 8–6). All of these fractions are impure, but many of the contaminants can be removed by resuspending the pellet and repeating the centrifugation procedure several times.

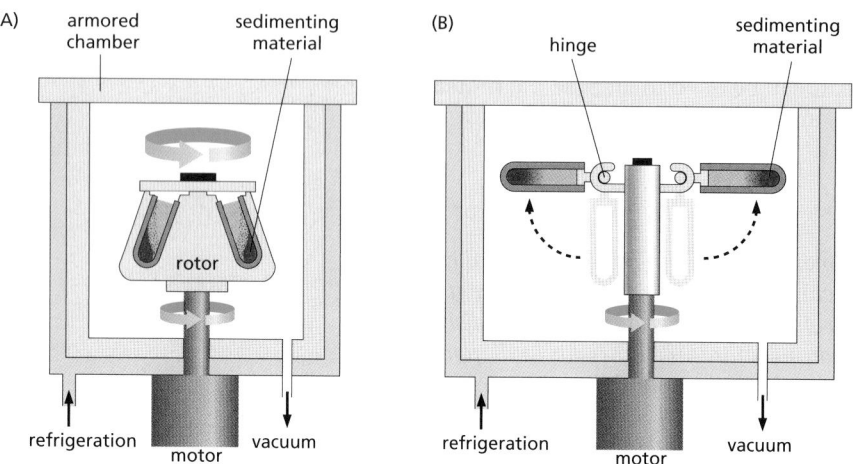

Figure 8–5 The preparative ultracentrifuge. (A) The sample is contained in tubes that are inserted into a ring of angled cylindrical holes in a metal *rotor*. Rapid rotation of the rotor generates enormous centrifugal forces, which cause particles in the sample to sediment against the bottom sides of the sample tubes, as shown here. The vacuum reduces friction, preventing heating of the rotor and allowing the refrigeration system to maintain the sample at 4°C. (B) Some fractionation methods require a different type of rotor called a *swinging-bucket rotor*. In this case, the sample tubes are placed in metal tubes on hinges that allow the tubes to swing outward when the rotor spins. Sample tubes are therefore horizontal during spinning, and samples are sedimented toward the bottom, not the sides, of the tube, providing better separation of differently sized components (see Figures 8–6 and 8–7).

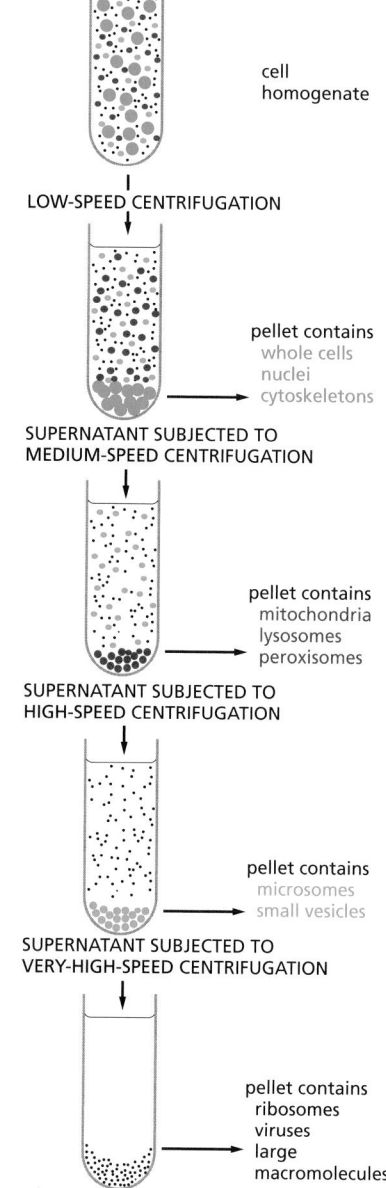

Centrifugation is the first step in most fractionations, but it separates only components that differ greatly in size. A finer degree of separation can be achieved by layering the homogenate in a thin band on top of a salt solution that fills a centrifuge tube. When centrifuged, the various components in the mixture move as a series of distinct bands through the solution, each at a different rate, in a process called *velocity sedimentation* (**Figure 8–7**A). For the procedure to work effectively, the bands must be protected from convective mixing, which would normally occur whenever a denser solution (for example, one containing organelles) finds itself on top of a lighter one (the salt solution). This is achieved by augmenting the solution in the tube with a shallow gradient of sucrose prepared by a special mixing device. The resulting density gradient—with the dense end at the bottom of the tube—keeps each region of the solution denser than any solution above it, and it thereby prevents convective mixing from distorting the separation.

When sedimented through sucrose gradients, different cell components separate into distinct bands that can be collected individually. The relative rate at which each component sediments depends primarily on its size and shape—normally being described in terms of its sedimentation coefficient, or S value. Present-day ultracentrifuges rotate at speeds of up to 80,000 rpm and produce forces as high as 500,000 times gravity. These enormous forces drive even small macromolecules, such as tRNA molecules and simple enzymes, to sediment at an appreciable rate and allow them to be separated from one another by size.

The ultracentrifuge is also used to separate cell components on the basis of their buoyant density, independently of their size and shape. In this case, the

Figure 8–6 Cell fractionation by centrifugation. Repeated centrifugation at progressively higher speeds will fractionate homogenates of cells into their components. In general, the smaller the subcellular component, the greater the centrifugal force required to sediment it. Typical values for the various centrifugation steps referred to in the figure are:
low speed: 1000 times gravity for 10 minutes
medium speed: 20,000 times gravity for 20 minutes
high speed: 80,000 times gravity for 1 hour
very high speed: 150,000 times gravity for 3 hours

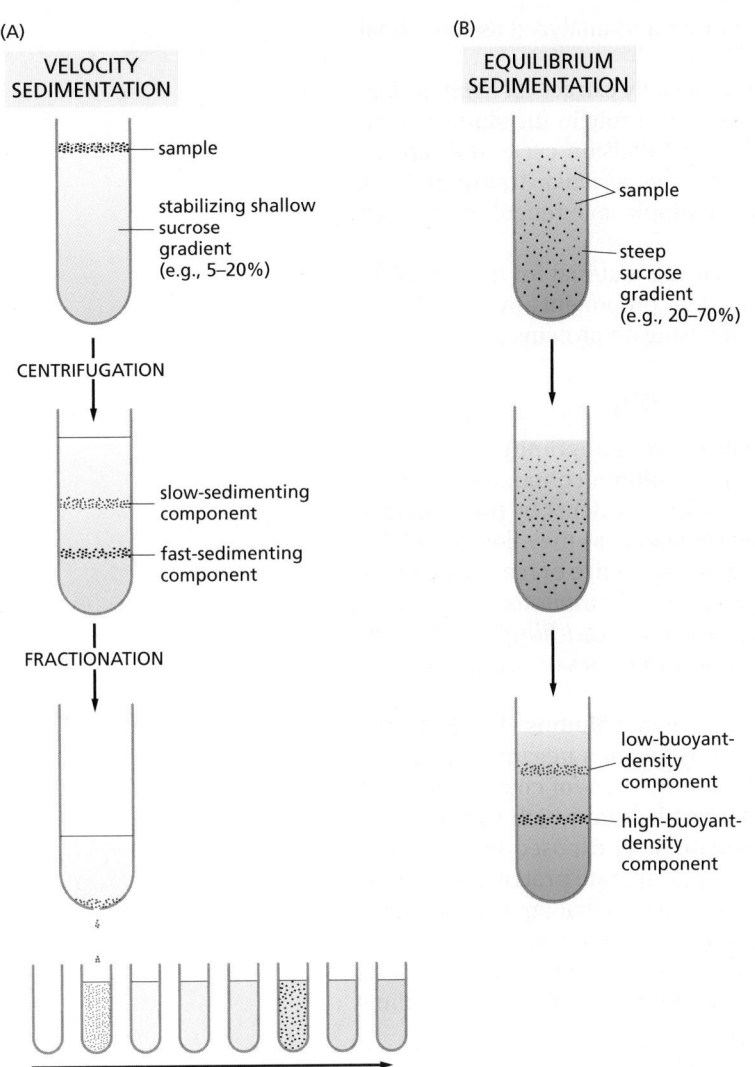

(A) VELOCITY SEDIMENTATION

sample

stabilizing shallow sucrose gradient (e.g., 5–20%)

CENTRIFUGATION

slow-sedimenting component

fast-sedimenting component

FRACTIONATION

(B) EQUILIBRIUM SEDIMENTATION

sample

steep sucrose gradient (e.g., 20–70%)

low-buoyant-density component

high-buoyant-density component

Figure 8–7 Comparison of velocity sedimentation and equilibrium sedimentation. (A) In velocity sedimentation, subcellular components sediment at different speeds according to their size and shape when layered over a solution containing sucrose. To stabilize the sedimenting bands against convective mixing caused by small differences in temperature or solute concentration, the tube contains a continuous shallow gradient of sucrose, which increases in concentration toward the bottom of the tube (typically from 5% to 20% sucrose). After centrifugation, the different components can be collected individually, most simply by puncturing the plastic centrifuge tube with a needle and collecting drops from the bottom, as illustrated here. (B) In equilibrium sedimentation, subcellular components move up or down when centrifuged in a gradient until they reach a position where their density matches their surroundings. Although a sucrose gradient is shown here, denser gradients, which are especially useful for protein and nucleic acid separation, can be formed from cesium chloride. The final bands, at equilibrium, can be collected as in (A).

sample is sedimented through a steep density gradient that contains a very high concentration of sucrose or cesium chloride. Each cell component begins to move down the gradient as in Figure 8–7A, but it eventually reaches a position where the density of the solution is equal to its own density. At this point, the component floats and can move no farther. A series of distinct bands is thereby produced in the centrifuge tube, with the bands closest to the bottom of the tube containing the components of highest buoyant density (Figure 8–7B). This method, called *equilibrium sedimentation*, is so sensitive that it can separate macromolecules that have incorporated heavy isotopes, such as ^{13}C or ^{15}N, from the same macromolecules that contain the lighter, common isotopes (^{12}C or ^{14}N). In fact, the cesium-chloride method was developed in 1957 to separate the labeled from the unlabeled DNA produced after exposure of a growing population of bacteria to nucleotide precursors containing ^{15}N; this classic experiment provided direct evidence for the semiconservative replication of DNA (see Figure 5–5).

Cell Extracts Provide Accessible Systems to Study Cell Functions

Studies of organelles and other large subcellular components isolated in the ultracentrifuge have contributed enormously to our understanding of the functions of different cell components. Experiments on mitochondria and chloroplasts purified by centrifugation, for example, demonstrated the central function of these organelles in converting energy into forms that the cell can use. Similarly, resealed vesicles formed from fragments of rough and smooth endoplasmic reticulum (microsomes) have been separated from each other and analyzed as functional models of these compartments of the intact cell.

Similarly, highly concentrated cell extracts, especially extracts of *Xenopus laevis* (African clawed frog) oocytes, have played a critical role in the study of such complex and highly organized processes as the cell-division cycle, the separation of chromosomes on the mitotic spindle, and the vesicular-transport steps involved in the movement of proteins from the endoplasmic reticulum through the Golgi apparatus to the plasma membrane.

Cell extracts also provide, in principle, the starting material for the complete separation of all of the individual macromolecular components of the cell. We now consider how this separation is achieved, focusing on proteins.

Proteins Can Be Separated by Chromatography

Proteins are most often fractionated by **column chromatography**, in which a mixture of proteins in solution is passed through a column containing a porous solid matrix. Different proteins are retarded to different extents by their interaction with the matrix, and they can be collected separately as they flow out of the bottom of the column (**Figure 8–8**). Depending on the choice of matrix, proteins can be separated according to their charge (*ion-exchange chromatography*), their hydrophobicity (*hydrophobic chromatography*), their size (*gel-filtration chromatography*), or their ability to bind to particular small molecules or to other macromolecules (*affinity chromatography*).

Many types of matrices are available. Ion-exchange columns (**Figure 8–9**A) are packed with small beads that carry either a positive or a negative charge, so that proteins are fractionated according to the arrangement of charges on their surface. Hydrophobic columns are packed with beads from which hydrophobic side chains protrude, selectively retarding proteins with exposed hydrophobic regions. Gel-filtration columns (Figure 8–9B), which separate proteins according to their size, are packed with tiny porous beads: molecules that are small enough to enter the pores linger inside successive beads as they pass down the column, while larger molecules remain in the solution flowing between the beads and therefore move more rapidly, emerging from the column first. Besides providing

COLUMN CHROMATOGRAPHY

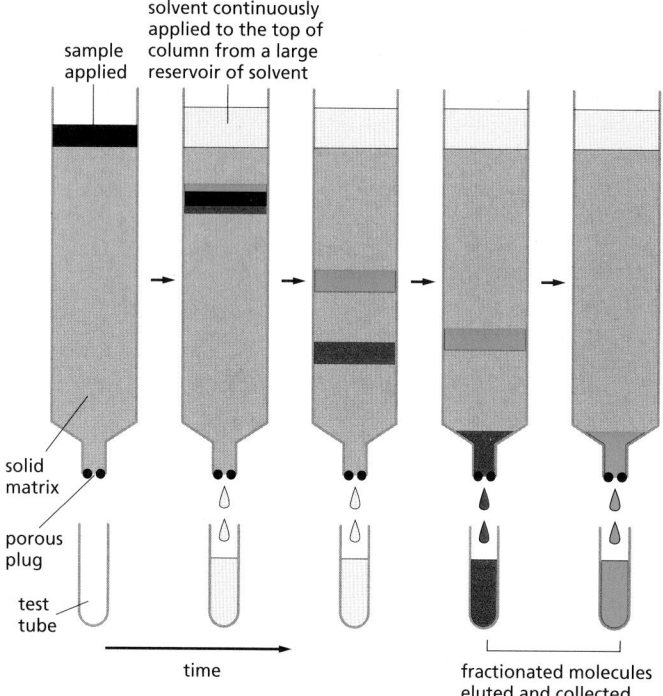

Figure 8–8 **The separation of molecules by column chromatography.** The sample, a solution containing a mixture of different molecules, is applied to the top of a cylindrical glass or plastic column filled with a permeable solid matrix, such as cellulose. A large amount of solvent is then passed slowly through the column and collected in separate tubes as it emerges from the bottom. Because various components of the sample travel at different rates through the column, they are fractionated into different tubes.

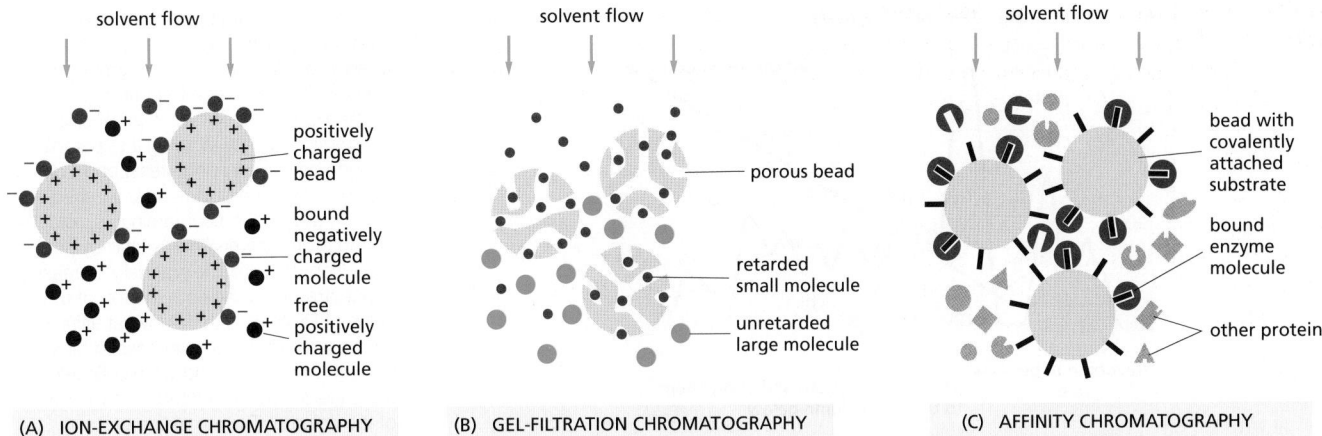

solvent flow

(A) ION-EXCHANGE CHROMATOGRAPHY

positively charged bead

bound negatively charged molecule

free positively charged molecule

solvent flow

(B) GEL-FILTRATION CHROMATOGRAPHY

porous bead

retarded small molecule

unretarded large molecule

solvent flow

(C) AFFINITY CHROMATOGRAPHY

bead with covalently attached substrate

bound enzyme molecule

other proteins

a means of separating molecules, gel-filtration chromatography is a convenient way to estimate their size.

Affinity chromatography (Figure 8–9C) takes advantage of the biologically important binding interactions that occur on protein surfaces. If a substrate molecule is covalently coupled to an inert matrix such as a polysaccharide bead, the enzyme that operates on that substrate will often be specifically retained by the matrix and can then be eluted (washed out) in nearly pure form. Likewise, short DNA oligonucleotides of a specifically designed sequence can be immobilized in this way and used to purify DNA-binding proteins that normally recognize this sequence of nucleotides in chromosomes. Alternatively, specific antibodies can be coupled to a matrix to purify protein molecules recognized by the antibodies. Because of the great specificity of all such affinity columns, 1000- to 10,000-fold purifications can sometimes be achieved in a single pass.

If one starts with a complex mixture of proteins, a single passage through an ion-exchange or a gel-filtration column does not produce very highly purified fractions, since these methods individually increase the proportion of a given protein in the mixture no more than twentyfold. Because most individual proteins represent less than 1/1000 of the total cell protein, it is usually necessary to use several different types of columns in succession to attain sufficient purity, with affinity chromatography being the most efficient (**Figure 8–10**).

Inhomogeneities in the matrices (such as cellulose), which cause an uneven flow of solvent through the column, limit the resolution of conventional column chromatography. Special chromatography resins (usually silica-based) composed of tiny spheres (3–10 μm in diameter) can be packed with a special apparatus to form a uniform column bed. Such **high-performance liquid chromatography (HPLC)** columns attain a high degree of resolution. In HPLC, the solutes equilibrate very rapidly with the interior of the tiny spheres, and so solutes with different affinities for the matrix are efficiently separated from one another even at very fast flow rates. HPLC is therefore the method of choice for separating many proteins and small molecules.

Immunoprecipitation Is a Rapid Affinity Purification Method

Immunoprecipitation is a useful variation on the theme of affinity chromatography. Specific antibodies that recognize the protein to be purified are attached to small agarose beads. Rather than being packed into a column, as in affinity chromatography, a small quantity of the antibody-coated beads is simply added to a protein extract in a test tube and mixed in suspension for a short period of time—thereby allowing the antibodies to bind the desired protein. The beads are then collected by low-speed centrifugation, and the unbound proteins in the supernatant are discarded. This method is commonly used to purify small amounts of enzymes from cell extracts for analysis of enzymatic activity or for studies of associated proteins.

Figure 8–9 Three types of matrices used for chromatography. (A) In ion-exchange chromatography, the insoluble matrix carries ionic charges that retard the movement of molecules of opposite charge. Matrices used for separating proteins include diethylaminoethylcellulose (DEAE-cellulose), which is positively charged, and carboxymethylcellulose (CM-cellulose) and phosphocellulose, which are negatively charged. Analogous matrices based on agarose or other polymers are also frequently used. The strength of the association between the dissolved molecules and the ion-exchange matrix depends on both the ionic strength and the pH of the solution that is passing down the column, which may therefore be varied systematically (as in Figure 8–10) to achieve an effective separation. **(B)** In gel-filtration chromatography, the small beads that form the matrix are inert but porous. Molecules that are small enough to penetrate into the matrix beads are thereby delayed and travel more slowly through the column than larger molecules that cannot penetrate. Beads of cross-linked polysaccharide (dextran, agarose, or acrylamide) are available commercially in a wide range of pore sizes, making them suitable for the fractionation of molecules of various molecular mass, from less than 500 daltons to more than 5×10^6 daltons. **(C)** Affinity chromatography uses an insoluble matrix that is covalently linked to a specific ligand, such as an antibody molecule or an enzyme substrate, that will bind a specific protein. Enzyme molecules that bind to immobilized substrates on such columns can be eluted with a concentrated solution of the free form of the substrate molecule, while molecules that bind to immobilized antibodies can be eluted by dissociating the antibody–antigen complex with concentrated salt solutions or solutions of high or low pH. High degrees of purification can be achieved in a single pass through an affinity column.

(A) ION-EXCHANGE CHROMATOGRAPHY

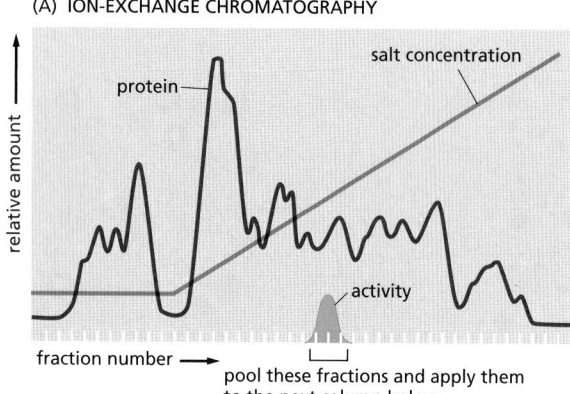

(B) GEL-FILTRATION CHROMATOGRAPHY

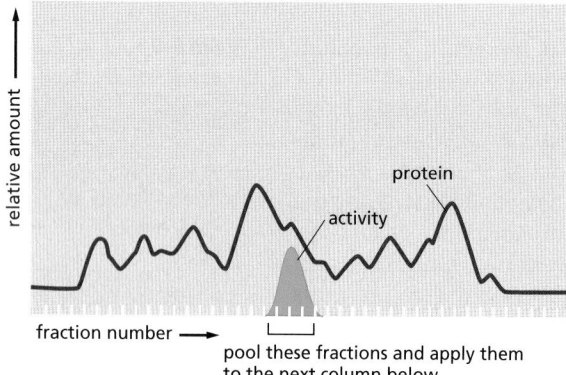

(C) AFFINITY CHROMATOGRAPHY

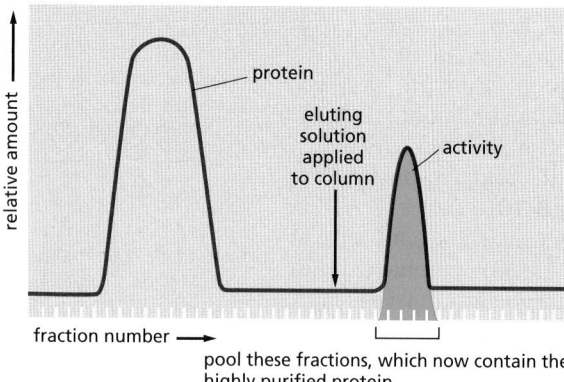

Figure 8–10 Protein purification by chromatography. Typical results obtained when three different chromatographic steps are used in succession to purify a protein. In this example, a homogenate of cells was first fractionated by allowing it to percolate through an ion-exchange resin packed into a column (A). The column was washed to remove all unbound contaminants, and the bound proteins were then eluted by pouring a solution containing a gradually increasing concentration of salt onto the top of the column. Proteins with the lowest affinity for the ion-exchange resin passed directly through the column and were collected in the earliest fractions eluted from the bottom of the column. The remaining proteins were eluted in sequence according to their affinity for the resin—those proteins binding most tightly to the resin requiring the highest concentration of salt to remove them. The protein of interest was eluted in several fractions and was detected by its enzymatic activity. The fractions with activity were pooled and then applied to a gel-filtration column (B). The elution position of the still-impure protein was again determined by its enzymatic activity, and the active fractions were pooled and purified to homogeneity on an affinity column (C) that contained an immobilized substrate of the enzyme.

Genetically Engineered Tags Provide an Easy Way to Purify Proteins

Using the recombinant DNA methods discussed in subsequent sections, any gene can be modified to produce its protein with a special recognition tag attached to it, so as to make subsequent purification of the protein simple and rapid. Often the recognition tag is itself an antigenic determinant, or *epitope*, which can be recognized by a highly specific antibody. The antibody can then be used to purify the protein by affinity chromatography or immunoprecipitation (**Figure 8–11**). Other types of tags are specifically designed for protein purification. For example, a repeated sequence of the amino acid histidine binds to certain metal ions, including nickel and copper. If genetic engineering techniques are used to attach a short string of histidines to one end of a protein, the slightly modified protein can be retained selectively on an affinity column containing immobilized nickel ions. Metal affinity chromatography can thereby be used to purify the modified protein from a complex molecular mixture.

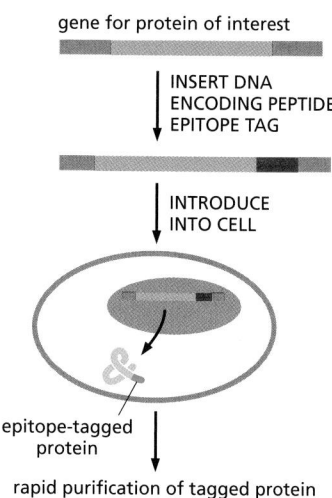

Figure 8–11 Epitope tagging for the purification of proteins. Using standard genetic engineering techniques, a short peptide tag can be added to a protein of interest. If the tag is itself an antigenic determinant, or *epitope*, it can be targeted by an appropriate antibody, which can be used to purify the protein by immunoprecipitation or affinity chromatography.

In other cases, an entire protein is used as the recognition tag. When cells are engineered to synthesize the small enzyme glutathione *S*-transferase (GST) attached to a protein of interest, the resulting **fusion protein** can be purified from the other contents of the cell with an affinity column containing glutathione, a substrate molecule that binds specifically and tightly to GST.

As a further refinement of purification methods using recognition tags, an amino acid sequence that forms a cleavage site for a highly specific proteolytic enzyme can be engineered between the protein of choice and the recognition tag. Because the amino acid sequences at the cleavage site are very rarely found by chance in proteins, the tag can later be cleaved off without destroying the purified protein.

This type of specific cleavage is used in an especially powerful purification methodology known as *tandem affinity purification tagging* (*TAP-tagging*). Here, one end of a protein is engineered to contain two recognition tags that are separated by a protease cleavage site. The tag on the very end of the construct is chosen to bind irreversibly to an affinity column, allowing the column to be washed extensively to remove all contaminating proteins. Protease cleavage then releases the protein, which is then further purified using the second tag. Because this two-step strategy provides an especially high degree of protein purification with relatively little effort, it is used extensively in cell biology. Thus, for example, a set of approximately 6000 yeast strains, each with a different gene fused to DNA that encodes a TAP-tag, has been constructed to allow any yeast protein to be rapidly purified.

Purified Cell-free Systems Are Required for the Precise Dissection of Molecular Functions

Purified cell-free systems provide a means of studying biological processes free from all of the complex side reactions that occur in a living cell. To make this possible, cell homogenates are fractionated with the aim of purifying each of the individual macromolecules that are needed to catalyze a biological process of interest. For example, the experiments to decipher the mechanisms of protein synthesis began with a cell homogenate that could translate RNA molecules to produce proteins. Fractionation of this homogenate, step by step, produced in turn the ribosomes, tRNAs, and various enzymes that together constitute the protein-synthetic machinery. Once individual pure components were available, each could be added or withheld separately to define its exact role in the overall process.

A major goal for cell biologists is the reconstitution of every biological process in a purified cell-free system. Only in this way can we define all of the components needed for the process and control their concentrations, which is required to work out their precise mechanism of action. Although much remains to be done, a great deal of what we know today about the molecular biology of the cell has been discovered by studies in such cell-free systems. They have been used, for example, to decipher the molecular details of DNA replication and DNA transcription, RNA splicing, protein translation, muscle contraction, and particle transport along microtubules, and many other processes that occur in cells.

Summary

Populations of cells can be analyzed biochemically by disrupting them and fractionating their contents, allowing functional cell-free systems to be developed. Highly purified cell-free systems are needed for determining the molecular details of complex cell processes, and the development of such systems requires extensive purification of all the proteins and other components involved. The proteins in soluble cell extracts can be purified by column chromatography; depending on the type of column matrix, biologically active proteins can be separated on the basis of their molecular weight, hydrophobicity, charge characteristics, or affinity for other molecules. In a typical purification, the sample is passed through several different columns in turn, with the enriched fractions obtained from one column being applied to the next. Recombinant DNA techniques (described later) allow special recognition tags to be attached to proteins, thereby greatly simplifying their purification.

ANALYZING PROTEINS

Proteins perform most cellular processes: they catalyze metabolic reactions, use nucleotide hydrolysis to do mechanical work, and serve as the major structural elements of the cell. The great variety of protein structures and functions has stimulated the development of a multitude of techniques to study them.

Proteins Can Be Separated by SDS Polyacrylamide-Gel Electrophoresis

Proteins usually possess a net positive or negative charge, depending on the mixture of charged amino acids they contain. An electric field applied to a solution containing a protein molecule causes the protein to migrate at a rate that depends on its net charge and on its size and shape. The most popular application of this property is **SDS polyacrylamide-gel electrophoresis** (**SDS-PAGE**). It uses a highly cross-linked gel of polyacrylamide as the inert matrix through which the proteins migrate. The gel is prepared by polymerization of monomers; the pore size of the gel can be adjusted so that it is small enough to retard the migration of the protein molecules of interest. The proteins are dissolved in a solution that includes a powerful negatively charged detergent, sodium dodecyl sulfate, or SDS (**Figure 8–12**). Because this detergent binds to hydrophobic regions of the protein molecules, causing them to unfold into extended polypeptide chains, the individual protein molecules are released from their associations with other proteins or lipid molecules and rendered freely soluble in the detergent solution. In addition, a reducing agent such as β-mercaptoethanol (see Figure 8–12) is usually added to break any S–S linkages in the proteins, so that all of the constituent polypeptides in multisubunit proteins can be analyzed separately.

What happens when a mixture of SDS-solubilized proteins is run through a slab of polyacrylamide gel? Each protein molecule binds large numbers of the negatively charged detergent molecules, which mask the protein's intrinsic charge and cause it to migrate toward the positive electrode when a voltage is applied. Proteins of the same size tend to move through the gel with similar speeds because (1) their native structure is completely unfolded by the SDS, so that their shapes are the same, and (2) they bind the same amount of SDS and therefore have the same amount of negative charge. Larger proteins, with more charge, are subjected to larger electrical forces but also to a larger drag. In free solution, the two effects would cancel out, but, in the mesh of the polyacrylamide gel, which acts as a molecular sieve, large proteins are retarded much more than small ones. As a result, a complex mixture of proteins is fractionated into a series of discrete protein bands arranged in order of molecular weight (**Figure 8–13**). The major proteins are readily detected by staining the proteins in the gel with a dye such as Coomassie blue. Even minor proteins are seen in gels treated with a silver stain, so that as little as 10 ng of protein can be detected in a band. For some purposes, specific proteins can also be labeled with a radioactive isotope tag; exposure of the gel to film results in an *autoradiograph* on which the labeled proteins are visible (see Figure 8–16).

SDS-PAGE is widely used because it can separate all types of proteins, including those that are normally insoluble in water—such as the many proteins in membranes. And because the method separates polypeptides by size, it provides information about the molecular weight and the subunit composition of proteins. **Figure 8–14** presents a photograph of a gel that has been used to analyze each of the successive stages in the purification of a protein.

Two-Dimensional Gel Electrophoresis Provides Greater Protein Separation

Because different proteins can have similar sizes, shapes, masses, and overall charges, most separation techniques such as SDS polyacrylamide-gel electrophoresis or ion-exchange chromatography cannot typically separate all the proteins

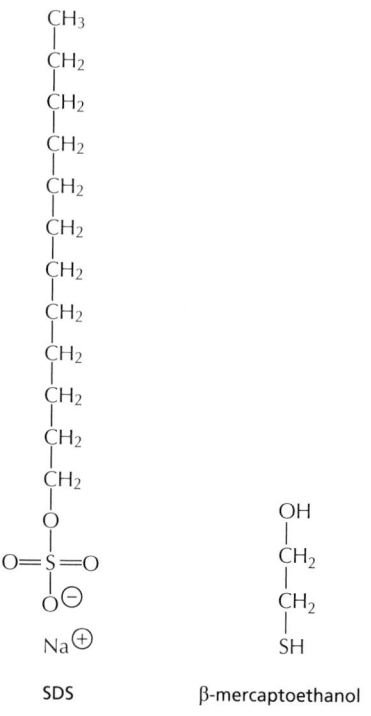

Figure 8–12 The detergent sodium dodecyl sulfate (SDS) and the reducing agent β-mercaptoethanol. These two chemicals are used to solubilize proteins for SDS polyacrylamide-gel electrophoresis. The SDS is shown here in its ionized form.

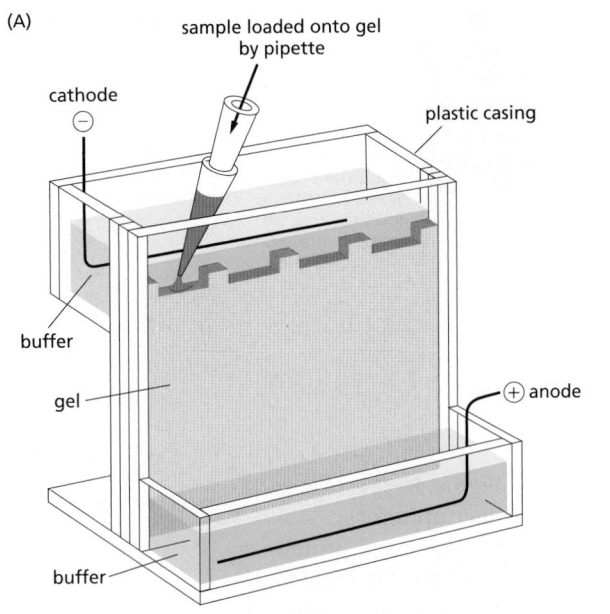

(A)

cathode ⊖

sample loaded onto gel by pipette

plastic casing

buffer

gel

⊕ anode

buffer

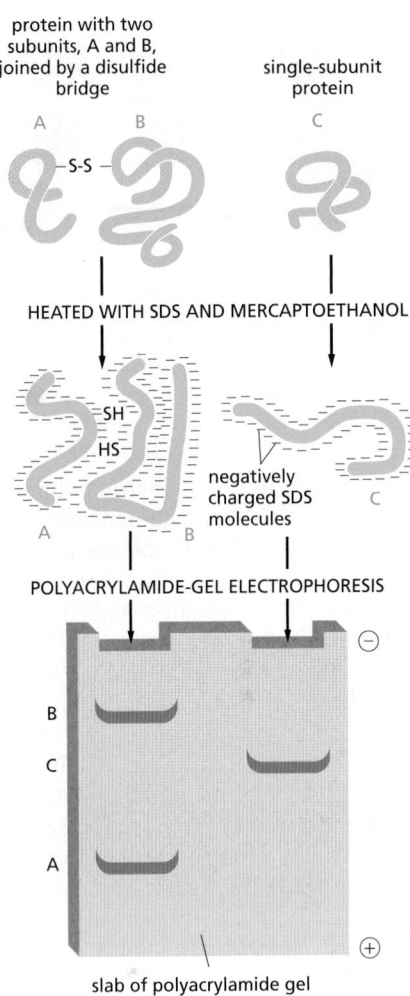

(B)

protein with two subunits, A and B, joined by a disulfide bridge

single-subunit protein

A   B   C

—S-S—

HEATED WITH SDS AND MERCAPTOETHANOL

SH
HS

negatively charged SDS molecules

A   B   C

POLYACRYLAMIDE-GEL ELECTROPHORESIS

⊖

B

C

A

⊕

slab of polyacrylamide gel

Figure 8–13 SDS polyacrylamide-gel electrophoresis (SDS-PAGE). (A) An electrophoresis apparatus. (B) Individual polypeptide chains form a complex with negatively charged molecules of sodium dodecyl sulfate (SDS) and therefore migrate as a negatively charged SDS–protein complex through a porous gel of polyacrylamide. Because the speed of migration under these conditions is greater the smaller the polypeptide, this technique can be used to determine the approximate molecular weight of a polypeptide chain as well as the subunit composition of a protein. If the protein contains a large amount of carbohydrate, however, it will move anomalously on the gel and its apparent molecular weight estimated by SDS-PAGE will be misleading. Other modifications, such as phosphorylation, can also cause small changes in a protein's migration in the gel.

in a cell or even in an organelle. In contrast, **two-dimensional gel electrophoresis**, which combines two different separation procedures, can resolve up to 2000 proteins in the form of a two-dimensional protein map.

In the first step, the proteins are separated by their intrinsic charges. The sample is dissolved in a small volume of a solution containing a nonionic (uncharged) detergent, together with β-mercaptoethanol and the denaturing reagent urea. This solution solubilizes, denatures, and dissociates all the polypeptide chains but leaves their intrinsic charge unchanged. The polypeptide chains are then separated in a pH gradient by a procedure called *isoelectric focusing*, which takes advantage of the variation in the net charge on a protein molecule with the pH of its surrounding solution. Every protein has a characteristic isoelectric point, the pH at which the protein has no net charge and therefore does not migrate in an electric field. In isoelectric focusing, proteins are separated electrophoretically in a narrow tube of polyacrylamide gel in which a gradient of pH is established by a mixture of special buffers. Each protein moves to a position in the gradient that

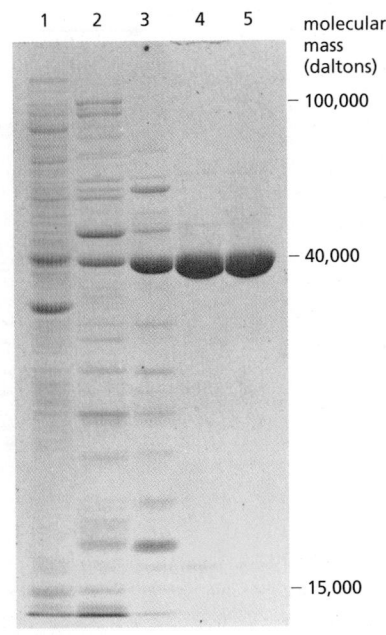

1   2   3   4   5   molecular mass (daltons)

— 100,000

— 40,000

— 15,000

Figure 8–14 Analysis of protein samples by SDS polyacrylamide-gel electrophoresis. The photograph shows a Coomassie-stained gel that has been used to detect the proteins present at successive stages in the purification of an enzyme. The leftmost lane (lane 1) contains the complex mixture of proteins in the starting cell extract, and each succeeding lane analyzes the proteins obtained after a chromatographic fractionation of the protein sample analyzed in the previous lane (see Figure 8–10). The same total amount of protein (10 μg) was loaded onto the gel at the top of each lane. Individual proteins normally appear as sharp, dye-stained bands; a band broadens, however, when it contains a large amount of protein. (From T. Formosa and B.M. Alberts, *J. Biol. Chem.* 261:6107–6118, 1986.)

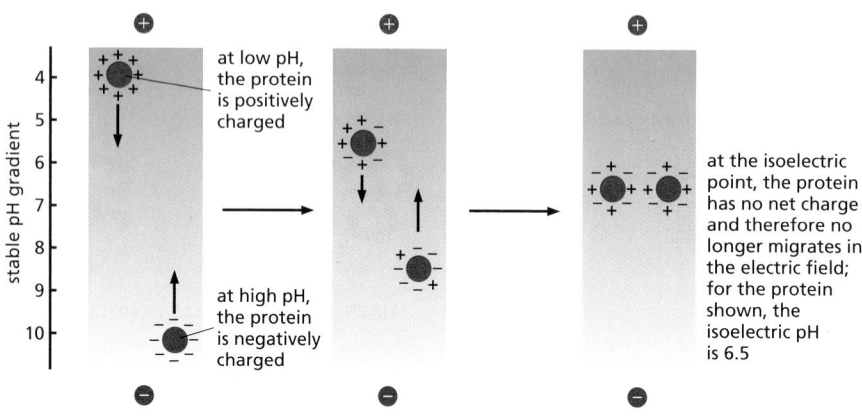

Figure 8–15 Separation of protein molecules by isoelectric focusing. At low pH (high H$^+$ concentration), the carboxylic acid groups of proteins tend to be uncharged (–COOH) and their nitrogen-containing basic groups fully charged (for example, –NH$_3^+$), giving most proteins a net positive charge. At high pH, the carboxylic acid groups are negatively charged (–COO$^-$) and the basic groups tend to be uncharged (for example, –NH$_2$), giving most proteins a net negative charge. At its *isoelectric pH*, a protein has no net charge since the positive and negative charges balance. Thus, when a tube containing a fixed pH gradient is subjected to a strong electric field in the appropriate direction, each protein species migrates until it forms a sharp band at its isoelectric pH, as shown.

corresponds to its isoelectric point and remains there (**Figure 8–15**). This is the first dimension of two-dimensional polyacrylamide-gel electrophoresis.

In the second step, the narrow tube gel containing the separated proteins is again subjected to electrophoresis but in a direction that is at a right angle to the direction used in the first step. This time SDS is added, and the proteins separate according to their size, as in one-dimensional SDS-PAGE: the original tube gel is soaked in SDS and then placed along the top edge of an SDS polyacrylamide-gel slab, through which each polypeptide chain migrates to form a discrete spot. This is the second dimension of two-dimensional polyacrylamide-gel electrophoresis. The only proteins left unresolved are those that have both identical sizes and identical isoelectric points, a relatively rare situation. Even trace amounts of each polypeptide chain can be detected on the gel by various staining procedures—or by autoradiography if the protein sample was initially labeled with a radioisotope (**Figure 8–16**). The technique has such great resolving power that it can distinguish between two proteins that differ in only a single charged amino acid, or a single negatively charged phosphorylation site.

Specific Proteins Can Be Detected by Blotting with Antibodies

A specific protein can be identified after its fractionation on a polyacrylamide gel by exposing all the proteins present on the gel to a specific antibody that has been labeled with a radioactive isotope or a fluorescent dye. This procedure is normally carried out after transferring all of the separated proteins present in the

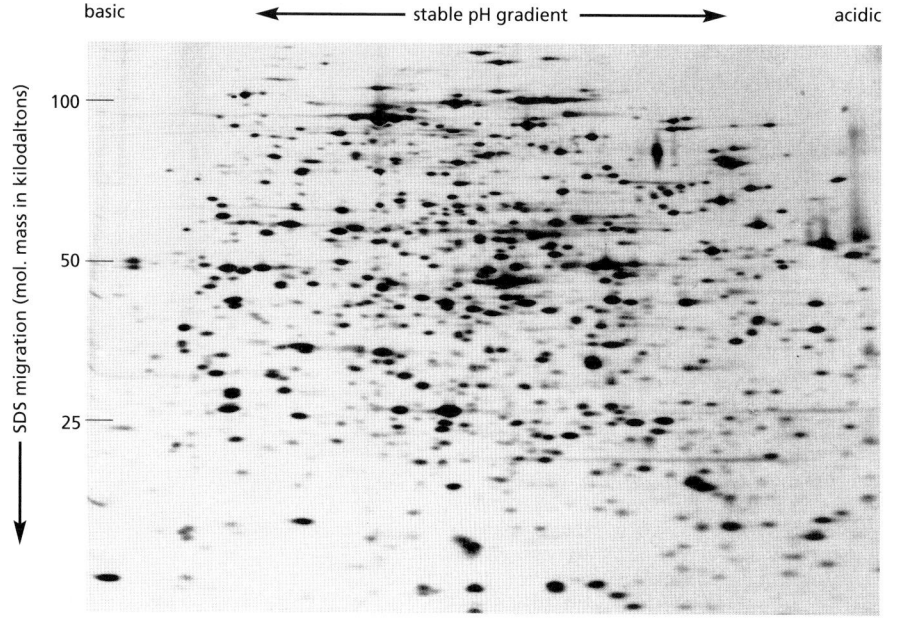

Figure 8–16 Two-dimensional polyacrylamide-gel electrophoresis. All the proteins in an *E. coli* bacterial cell are separated in this gel, in which each spot corresponds to a different polypeptide chain. The proteins were first separated on the basis of their isoelectric points by isoelectric focusing in the horizontal dimension. They were then further fractionated according to their molecular mass by electrophoresis from top to bottom in the presence of SDS. Note that different proteins are present in very different amounts. The bacteria were fed with a mixture of radioisotope-labeled amino acids so that all of their proteins were radioactive and could be detected by autoradiography. (Courtesy of Patrick O'Farrell.)

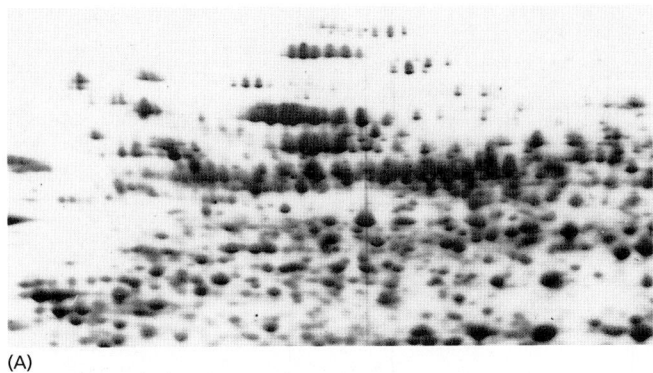

(A)

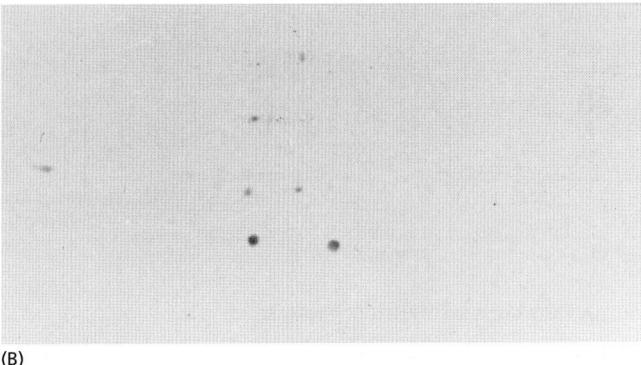

(B)

gel onto a sheet of nitrocellulose paper or nylon membrane. Placing the membrane over the gel and driving the proteins out of the gel with a strong electric current transfers the protein onto the membrane. The membrane is then soaked in a solution of labeled antibody to reveal the protein of interest. This method of detecting proteins is called **Western blotting**, or **immunoblotting** (**Figure 8–17**). Sensitive Western-blotting methods can detect very small amounts of a specific protein (1 nanogram or less) in a total cell extract or some other heterogeneous protein mixture. The method can be very useful when assessing the amounts of a specific protein in the cell or when measuring changes in those amounts under various conditions.

Hydrodynamic Measurements Reveal the Size and Shape of a Protein Complex

Most proteins in a cell act as part of larger complexes, and knowledge of the size and shape of these complexes often leads to insights regarding their function. This information can be obtained in several important ways. Sometimes, a complex can be directly visualized using electron microscopy, as described in Chapter 9. A complementary approach relies on the hydrodynamic properties of a complex; that is, its behavior as it moves through a liquid medium. Usually, two separate measurements are made. One measure is the velocity of a complex as it moves under the influence of a centrifugal field produced by an ultracentrifuge (see Figure 8–7A). The sedimentation coefficient (or S value) obtained depends on both the size and the shape of the complex and does not, by itself, convey especially useful information. However, once a second hydrodynamic measurement is performed—by charting the migration of a complex through a gel-filtration chromatography column (see Figure 8–9B)—both the approximate shape of a complex and its molecular weight can be calculated.

Molecular weight can also be determined more directly by using an analytical ultracentrifuge, a complex device that allows protein absorbance measurements to be made on a sample while it is subjected to centrifugal forces. In this approach, the sample is centrifuged until it reaches equilibrium, where the centrifugal force on a protein complex exactly balances its tendency to diffuse away. Because this balancing point is dependent on a complex's molecular weight but not on its particular shape, the molecular weight can be directly calculated.

Mass Spectrometry Provides a Highly Sensitive Method for Identifying Unknown Proteins

A frequent problem in cell biology and biochemistry is the identification of a protein or collection of proteins that has been obtained by one of the purification procedures discussed in the preceding pages. Because the genome sequences of most experimental organisms are now known, catalogs of all the proteins produced in those organisms are available. The task of identifying an unknown protein (or collection of unknown proteins) thus reduces to matching some of the amino acid

Figure 8–17 Western blotting. All the proteins from dividing tobacco cells in culture were first separated by two-dimensional polyacrylamide-gel electrophoresis. In (A), the positions of the proteins are revealed by a sensitive protein stain. In (B), the separated proteins on an identical gel were then transferred to a sheet of nitrocellulose and exposed to an antibody that recognizes only those proteins that are phosphorylated on threonine residues during mitosis. The positions of the few proteins that are recognized by this antibody are revealed by an enzyme-linked second antibody. (From J.A. Traas et al., *Plant J.* 2:723–732, 1992. With permission from John Wiley and Sons.)

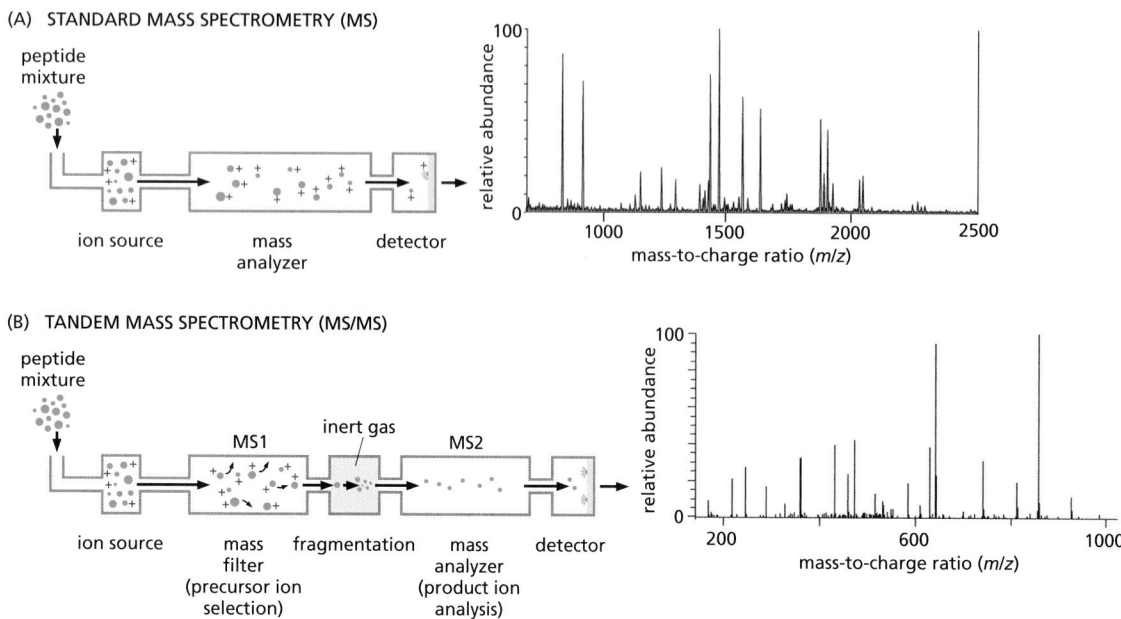

Figure 8–18 The mass spectrometer. (A) Mass spectrometers used in biology contain an ion source that generates gaseous peptides or other molecules under conditions that render most molecules positively charged. The two major types of ion source are MALDI and electrospray, as described in the text. Ions are accelerated into a mass analyzer, which separates the ions on the basis of their mass and charge by one of three major methods: 1. *Time-of-flight* (*TOF*) analyzers determine the mass-to-charge ratio of each ion in the mixture from the rate at which it travels from the ion source to the detector. 2. *Quadropole* mass filters contain a long chamber lined by four electrodes that produce oscillating electric fields that govern the trajectory of ions; by varying the properties of the electric field over a wide range, a spectrum of ions with specific mass-to-charge ratios is allowed to pass through the chamber to the detector, while other ions are discarded. 3. *Ion traps* contain doughnut-shaped electrodes producing a three-dimensional electric field that traps all ions in a circular chamber; the properties of the electric field can be varied over a wide range to eject a spectrum of specific ions to a detector. (B) Tandem mass spectrometry typically involves two mass analyzers separated by a collision chamber containing an inert, high-energy gas. The electric field of the first mass analyzer is adjusted to select a specific peptide ion, called a *precursor ion*, which is then directed to the collision chamber. Collision of the peptide with gas molecules causes random peptide fragmentation, primarily at peptide bonds, resulting in a highly complex mixture of fragments containing one or more amino acids from throughout the original peptide. The second mass analyzer is then used to measure the masses of the fragments (called *product* or *daughter ions*). With computer assistance, the pattern of fragments can be used to deduce the amino acid sequence of the original peptide.

sequences present in the unknown sample with known cataloged genes. This task is now performed almost exclusively by using mass spectrometry in conjunction with computer searches of databases.

Charged particles have very precise dynamics when subjected to electrical and magnetic fields in a vacuum. Mass spectrometry exploits this principle to separate ions according to their mass-to-charge (m/z) ratio. It is an enormously sensitive technique. It requires very little material and is capable of determining the precise mass of intact proteins and of peptides derived from them by enzymatic or chemical cleavage. Masses can be obtained with great accuracy, often with an error of less than one part in a million.

Mass spectrometry is performed using complex instruments with three major components (**Figure 8–18**A). The first is the *ion source*, which transforms tiny amounts of a peptide sample into a gas containing individual charged peptide molecules. These ions are accelerated by an electric field into the second component, the *mass analyzer*, where electric or magnetic fields are used to separate the ions on the basis of their mass-to-charge ratios. Finally, the separated ions collide with a *detector*, which generates a mass spectrum containing a series of peaks representing the masses of the molecules in the sample.

There are many different types of mass spectrometer, varying mainly in the nature of their ion sources and mass analyzers. One of the most common ion sources depends on a technique called *matrix-assisted laser desorption ionization* (*MALDI*). In this approach, the proteins in the sample are first cleaved into short peptides by a protease such as trypsin. These peptides are mixed with an organic acid and then dried onto a metal or ceramic slide. A brief laser burst is directed toward the sample, producing a gaseous puff of ionized peptides, each carrying one or more positive charges. In many cases, the MALDI ion source is coupled to a mass analyzer called a *time-of-flight* (*TOF*) analyzer, which is a long chamber through which the ionized peptides are accelerated by an electric field toward a detector. Their mass and charge determine the time it takes them to reach the detector: large peptides move more slowly, and more highly charged molecules move more quickly. By analyzing those ionized peptides that bear a single charge, the precise masses of peptides present in the original sample can be determined. This information is then used to search genomic databases, in which the masses of all proteins and of all their predicted peptide fragments have been tabulated from the genomic sequences of the organism. An unambiguous match to a particular open reading frame can often be made by knowing the mass of only a few peptides derived from a given protein.

By employing two mass analyzers in tandem (an arrangement known as MS/MS; Figure 8–18B), it is possible to directly determine the amino acid sequences of individual peptides in a complex mixture. The MALDI-TOF instrument described above is not ideal for this method. Instead, MS/MS typically involves an *electrospray* ion source, which produces a continuous thin stream of peptides that are ionized and accelerated into the first mass analyzer. The mass analyzer is typically either a *quadropole* or *ion trap*, which employs large electrodes to produce oscillating electric fields inside the chamber containing the ions. These instruments act as *mass filters*: the electric field is adjusted over a broad range to select a single peptide ion and discard all the others in the peptide mixture. In tandem mass spectrometry, this single ion is then exposed to an inert, high-energy gas, which collides with the peptide, resulting in fragmentation, primarily at peptide bonds. The second mass analyzer then determines the masses of the peptide fragments, which can be used by computational methods to determine the amino acid sequence of the original peptide and thereby identify the protein from which it came.

Tandem mass spectrometry is also useful for detecting and precisely mapping post-translational modifications of proteins, such as phosphorylations or acetylations. Because these modifications impart a characteristic mass increase to an amino acid, they are easily detected during the analysis of peptide fragments in the second mass analyzer, and the precise site of the modification can often be deduced from the spectrum of peptide fragments.

A powerful, "two-dimensional" mass spectrometry technique can be used to determine all of the proteins present in an organelle or another complex mixture of proteins. First, the mixture of proteins present is digested with trypsin to produce short peptides. Next, these peptides are separated by automated high-performance liquid chromatography (LC). Every peptide fraction from the chromatographic column is injected directly into an electrospray ion source on a tandem mass spectrometer (MS/MS), providing the amino acid sequence and post-translational modifications for every peptide in the mixture. This method, often called LC-MS/MS, is used to identify hundreds or thousands of proteins in complex protein mixtures from specific organelles or from whole cells. It can also be used to map all of the phosphorylation sites in the cell, or all of the proteins targeted by other post-translational modifications such as acetylation or ubiquitylation.

Sets of Interacting Proteins Can Be Identified by Biochemical Methods

Because most proteins in the cell function as part of complexes with other proteins, an important way to begin to characterize the biological role of an unknown protein is to identify all of the other proteins to which it specifically binds.

A key method for identifying proteins that bind to one another tightly is *co-immunoprecipitation*. A specific target protein is immunoprecipitated from a cell lysate using specific antibodies coupled to beads, as described earlier. If the target protein is associated tightly enough with another protein when it is captured by the antibody, the partner precipitates as well and can be identified by mass spectrometry. This method is useful for identifying proteins that are part of a complex inside cells, including those that interact only transiently—for example, when extracellular signal molecules stimulate cells (discussed in Chapter 15).

In addition to capturing protein complexes on columns or in test tubes, researchers are developing high-density protein arrays to investigate protein interactions. These arrays, which contain thousands of different proteins or antibodies spotted onto glass slides or immobilized in tiny wells, allow one to examine the biochemical activities and binding profiles of a large number of proteins at once. For example, if one incubates a fluorescently labeled protein with arrays containing thousands of immobilized proteins, the spots that remain fluorescent after extensive washing each contain a protein that specifically binds the labeled protein.

Optical Methods Can Monitor Protein Interactions

Once two proteins—or a protein and a small molecule—are known to associate, it becomes important to characterize their interaction in more detail. Proteins can associate with each other more or less permanently (like the subunits of RNA polymerase or the proteasome), or engage in transient encounters that may last only a few milliseconds (like a protein kinase and its substrate). To understand how a protein functions inside a cell, we need to determine how tightly it binds to other proteins, how rapidly it dissociates from them, and how covalent modifications, small molecules, or other proteins influence these interactions.

As we discussed in Chapter 3 (see Figure 3–44), the extent to which two proteins interact is determined by the rates at which they associate and dissociate. These rates depend, respectively, on the association rate constant (k_{on}) and dissociation rate constant (k_{off}). The kinetic rate constant k_{off} is a particularly useful number because it provides valuable information about how long two proteins remain bound to one another. The ratio of the two kinetic constants (k_{on}/k_{off}) yields another very useful number called the equilibrium constant (K, also known as K_{eq} or K_a), the inverse of which is the more commonly used dissociation constant K_d. The equilibrium constant is useful as a general indicator of the affinity of the interaction, and it can be used to estimate the amount of bound complex at different concentrations of the two protein partners—thereby providing insights into the importance of the interaction at the protein concentrations found inside the cell.

A wide range of methods can be used to determine binding constants for a two-protein complex. In a simple *equilibrium binding experiment*, two proteins are mixed at a range of concentrations, allowed to reach equilibrium, and the amount of bound complex is measured; half of the protein complex will be bound at a concentration that is equal to K_d. Equilibrium experiments often involve the use of radioactive or fluorescent tags on one of the protein partners, coupled with biochemical or optical methods for measuring the amount of bound protein. In a more complex *kinetic binding experiment*, the kinetic rate constants are determined using rapid methods that allow real-time measurement of the formation of a bound complex over time (to determine k_{on}) or the dissociation of a bound complex over time (to determine k_{off}).

Optical techniques provide particularly rapid, convenient, and accurate binding measurements, and in some cases the proteins do not even need to be labeled. Certain amino acids (tryptophan, for example) exhibit weak fluorescence that can be detected with sensitive fluorimeters. In many cases, the fluorescence intensity, or the emission spectrum of fluorescent amino acids located in a protein–protein interface, will change when two proteins associate. When this change can be detected by fluorimetry, it provides a simple and sensitive measure of protein binding that is useful in both equilibrium and kinetic binding experiments. A related but more widely useful optical binding technique is based on *fluorescence anisotropy*, a change in the polarized light that is emitted by a fluorescently tagged protein in the bound and free states (**Figure 8–19**).

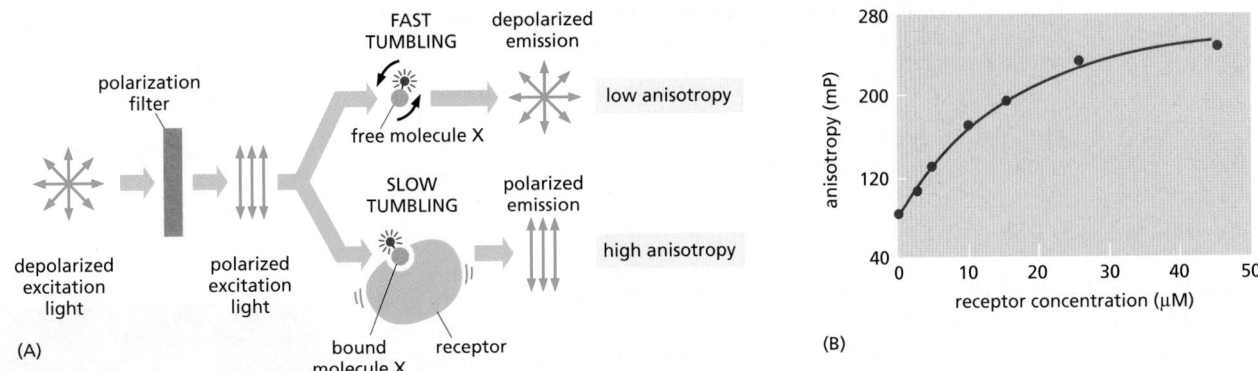

Figure 8–19 Measurement of binding with fluorescence anisotropy. This method depends on a fluorescently tagged protein that is illuminated with polarized light at the appropriate wavelength for excitation; a fluorimeter is used to measure the intensity and polarization of the emitted light. If the fluorescent protein is fixed in position and therefore does not rotate during the brief period between excitation and emission, then the emitted light will be polarized at the same angle as the excitation light. This directional effect is called *fluorescence anisotropy*. Protein molecules in solution rotate or tumble rapidly, however, so that there is a decrease in the amount of anisotropic fluorescence. Larger molecules tumble at a slower rate and therefore have higher fluorescence anisotropy. (A) To measure the binding between a small molecule and a large receptor protein, the smaller molecule is labeled with a fluorophore. In the absence of its binding partner, the molecule tumbles rapidly, resulting in low fluorescence anisotropy (*top*). When the small molecule binds to its larger partner, however, it tumbles less rapidly, resulting in an increase in fluorescence anisotropy (*bottom*). (B) In the equilibrium binding experiment shown here, a small, fluorescent peptide ligand was present at a low concentration, and the amount of fluorescence anisotropy (in millipolarization units, mP) was measured after incubation with various concentrations of a larger protein receptor for the ligand. From the hyperbolic curve that fits the data, it can be seen that 50% binding occurred at about 10 µM, which is equal to the dissociation constant K_d for the binding interaction.

Another optical method for probing protein interactions uses *green fluorescent protein* (discussed in detail below) and its derivatives of different colors. In this application, two proteins of interest are each labeled with a different fluorescent protein, such that the emission spectrum of one fluorescent protein overlaps the absorption spectrum of the second. If the two proteins come very close to each other (within about 1–5 nm), the energy of the absorbed light is transferred from one fluorescent protein to the other. The energy transfer, called *fluorescence resonance energy transfer* (*FRET*), is determined by illuminating the first fluorescent protein and measuring emission from the second (see Figure 9–26). When combined with fluorescence microscopy, this method can be used to characterize protein–protein interactions at specific locations inside living cells (discussed in Chapter 9).

Protein Function Can Be Selectively Disrupted With Small Molecules

Small chemical inhibitors of specific proteins have contributed a great deal to the development of cell biology. For example, the microtubule inhibitor colchicine is routinely used to test whether microtubules are required for a given biological process; it also led to the first purification of tubulin several decades ago. In the past, these small molecules were usually natural products; that is, they were synthesized by living creatures. Although natural products have been extraordinarily useful in science and medicine (see, for example, Table 6–4, p. 352), they act on a limited number of biological processes. However, the recent development of methods to synthesize hundreds of thousands of small molecules and to carry out large-scale automated screens holds the promise of identifying chemical inhibitors for virtually any biological process. In such approaches, large collections of small chemical compounds are simultaneously tested, either on living cells or in cell-free assays. Once an inhibitor is identified, it can be used as a probe to identify, through affinity chromatography or other means, the protein to which the inhibitor binds. This general strategy, sometimes called **chemical biology**, has successfully identified inhibitors of many proteins that carry out key processes in cell biology. An inhibitor of a kinesin protein that functions in mitosis, for

Figure 8–20 Small-molecule inhibitors for manipulating living cells. (A) Chemical structure of monastrol, a kinesin inhibitor identified in a large-scale screen for small molecules that disrupt mitosis. (B) Normal mitotic spindle seen in an untreated cell. The microtubules are stained *green* and chromosomes *blue*. (C) Monopolar spindle that forms in cells treated with monastrol, which inhibits a kinesin protein required for separation of the spindle poles in early mitosis. (B and C, from T.U. Mayer et al., *Science* 286:971–974, 1999. With permission from AAAS.)

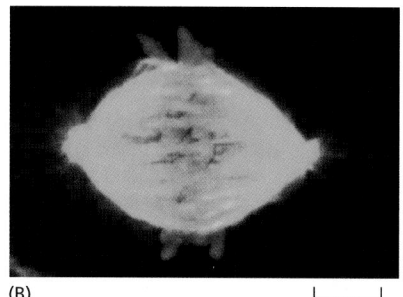

(A)   monastrol

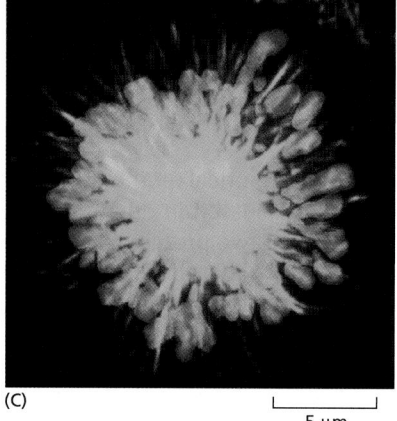

(B)   5 μm

(C)   5 μm

example, was identified by this method (Figure 8–20). Chemical inhibitors give the cell biologist great control over the timing of inhibition, as drugs can be rapidly added to or removed from cells, allowing protein function to be switched on or off quickly.

Protein Structure Can Be Determined Using X-Ray Diffraction

The main technique that has been used to discover the three-dimensional structure of molecules, including proteins, at atomic resolution is **x-ray crystallography**. X-rays, like light, are a form of electromagnetic radiation, but they have a much shorter wavelength, typically around 0.1 nm (the diameter of a hydrogen atom). If a narrow beam of parallel x-rays is directed at a sample of a pure protein, most of the x-rays pass straight through it. A small fraction, however, are scattered by the atoms in the sample. If the sample is a well-ordered crystal, the scattered waves reinforce one another at certain points and appear as diffraction spots when recorded by a suitable detector (Figure 8–21).

The position and intensity of each spot in the x-ray diffraction pattern contain information about the locations of the atoms in the crystal that gave rise to it. Deducing the three-dimensional structure of a large molecule from the diffraction pattern of its crystal is a complex task and was not achieved for a protein molecule until 1960. But in recent years x-ray diffraction analysis has become increasingly automated, and now the slowest step is likely to be the generation of suitable protein crystals. This step requires large amounts of very pure protein and often involves years of trial and error to discover the proper crystallization conditions; the pace has greatly accelerated with the use of recombinant DNA techniques to produce pure proteins and robotic techniques to test large numbers of crystallization conditions.

Analysis of the resulting diffraction pattern produces a complex three-dimensional electron-density map. Interpreting this map—translating its contours into a three-dimensional structure—is a complicated procedure that requires knowledge of the amino acid sequence of the protein. Largely by trial and error, the sequence and the electron-density map are correlated by computer to give the best possible fit. The reliability of the final atomic model depends on the resolution of the original crystallographic data: 0.5 nm resolution might produce a low-resolution map of the polypeptide backbone, whereas a resolution of 0.15 nm allows all of the non-hydrogen atoms in the molecule to be reliably positioned.

A complete atomic model is often too complex to appreciate directly, but simplified versions that show a protein's essential structural features can be readily derived from it (see Panel 3–2, pp. 142–143). The three-dimensional structures of tens of thousands of different proteins have been determined by x-ray crystallography or by NMR spectroscopy (see page 461)—enough to allow the grouping of common structures into families (Movie 8.1). These structures or protein folds often seem to be more conserved in evolution than are the amino acid sequences that form them (see Figure 3–13).

X-ray crystallographic techniques can also be applied to the study of macromolecular complexes. The method was used, for example, to determine the structure of the ribosome, a large and complex machine made of several RNAs and more than 50 proteins (see Figure 6–62). The determination required the use of a synchrotron, a radiation source that generates x-rays with the intensity needed to analyze the crystals of such large macromolecular complexes.

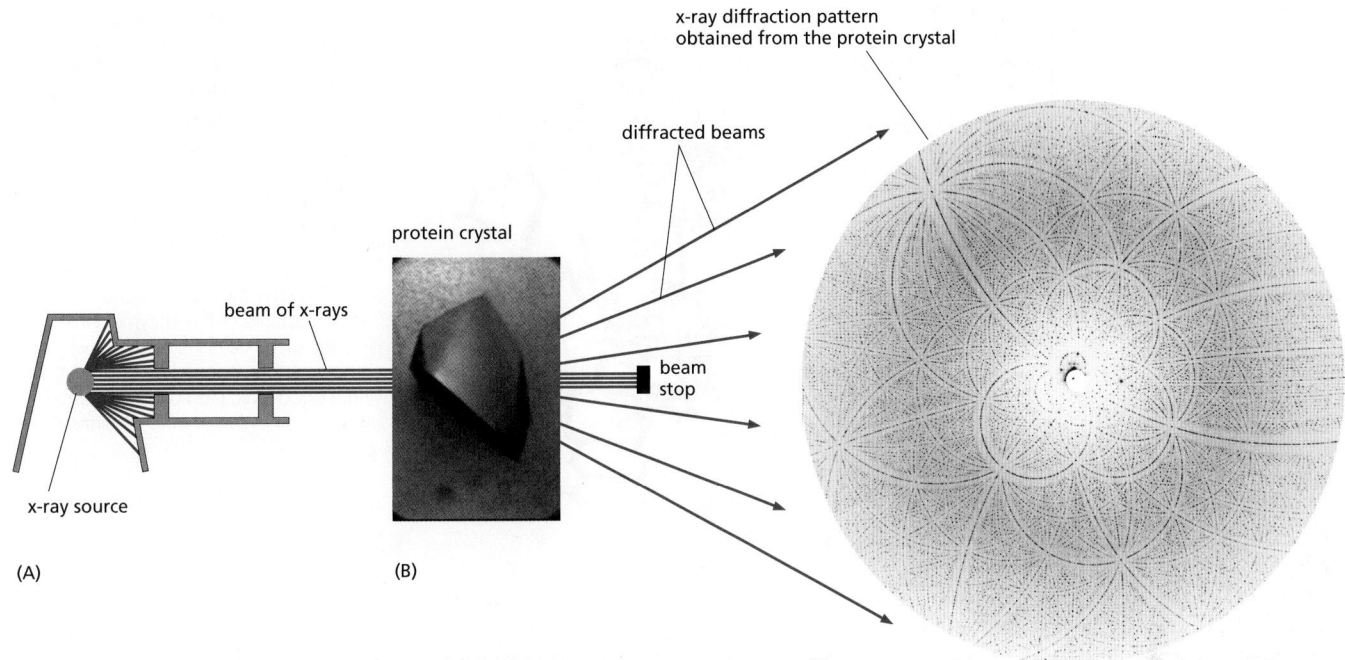

x-ray diffraction pattern
obtained from the protein crystal

diffracted beams

protein crystal

beam of x-rays

beam
stop

x-ray source

(A)

(B)

(C)

Figure 8–21 **X-ray crystallography.** (A) A narrow beam of x-rays is directed at a well-ordered crystal (B). Shown here is a protein crystal of ribulose bisphosphate carboxylase, an enzyme with a central role in CO_2 fixation during photosynthesis. The atoms in the crystal scatter some of the beam, and the scattered waves reinforce one another at certain points and appear as a pattern of diffraction spots (C). This diffraction pattern, together with the amino acid sequence of the protein, can be used to produce an atomic model (D). The complete atomic model is hard to interpret, but this simplified version, derived from the x-ray diffraction data, shows the protein's structural features clearly (α helices, *green*; β strands, *red*). The components pictured in A to D are not shown to scale. (B, courtesy of C. Branden; C, courtesy of J. Hajdu and I. Andersson; D, adapted from original provided by B. Furugren.)

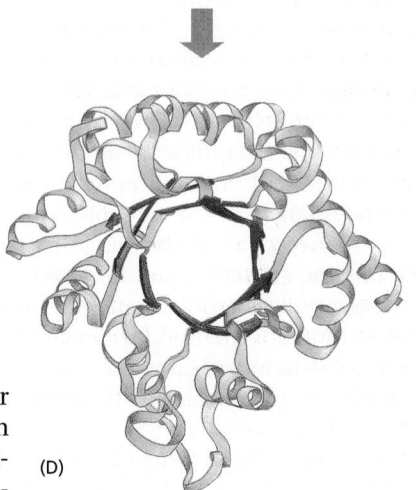

(D)

NMR Can Be Used to Determine Protein Structure in Solution

Nuclear magnetic resonance (NMR) spectroscopy has been widely used for many years to analyze the structure of small molecules, small proteins, or protein domains. Unlike x-ray crystallography, NMR does not depend on having a crystalline sample. It simply requires a small volume of concentrated protein solution that is placed in a strong magnetic field; indeed, it is the main technique that yields detailed evidence about the three-dimensional structure of molecules in solution.

Certain atomic nuclei, particularly hydrogen nuclei, have a magnetic moment or spin: that is, they have an intrinsic magnetization, like a bar magnet. The spin aligns along the strong magnetic field, but it can be changed to a misaligned, excited state in response to applied radiofrequency (RF) pulses of electromagnetic radiation. When the excited hydrogen nuclei return to their aligned state, they emit RF radiation, which can be measured and displayed as a spectrum. The nature of the emitted radiation depends on the environment of each hydrogen nucleus, and if one nucleus is excited, it influences the absorption and emission of radiation by other nuclei that lie close to it. It is consequently possible, by an ingenious elaboration of the basic NMR technique known as two-dimensional NMR, to distinguish the signals from hydrogen nuclei in different amino acid residues, and to identify and measure the small shifts in these signals that occur when these hydrogen nuclei lie close enough together to interact. Because the size of such a shift reveals the distance between the interacting pair of hydrogen atoms, NMR can provide information about the distances between the parts of the protein molecule. By combining this information with a knowledge of the amino acid

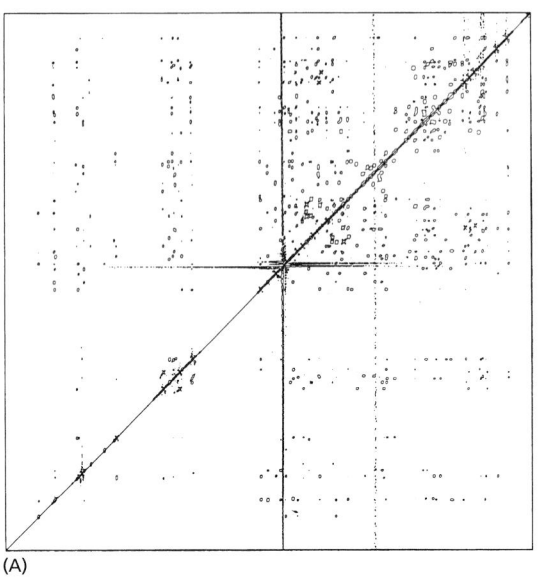

(A)

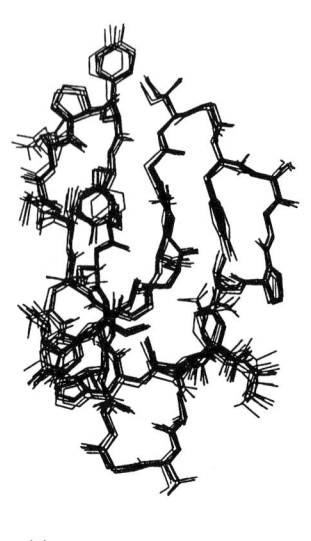

(B)

Figure 8–22 NMR spectroscopy.
(A) An example of the data from an NMR machine. This two-dimensional NMR spectrum is derived from the C-terminal domain of the enzyme cellulase. The spots represent interactions between hydrogen atoms that are near neighbors in the protein and hence reflect the distance that separates them. Complex computing methods, in conjunction with the known amino acid sequence, enable possible compatible structures to be derived.
(B) Ten structures of the enzyme, which all satisfy the distance constraints equally well, are shown superimposed on one another, giving a good indication of the probable three-dimensional structure. (Courtesy of P. Kraulis.)

sequence, it is possible in principle to compute the three-dimensional structure of the protein (**Figure 8–22**).

For technical reasons, the structure of small proteins of about 20,000 daltons or less can be most readily determined by NMR spectroscopy. Resolution decreases as the size of a macromolecule increases. But recent technical advances have now pushed the limit to about 100,000 daltons, thereby making the majority of proteins accessible for structural analysis by NMR.

Because NMR studies are performed in solution, this method also offers a convenient means of monitoring changes in protein structure, for example during protein folding or when the protein binds to another molecule. NMR is also used widely to investigate molecules other than proteins and is valuable, for example, as a method to determine the three-dimensional structures of RNA molecules and the complex carbohydrate side chains of glycoproteins.

A third major method for the determination of protein structure, and particularly the structure of large protein complexes, is single-particle analysis by electron microscopy. We discuss this approach in Chapter 9.

Protein Sequence and Structure Provide Clues About Protein Function

Having discussed methods for purifying and analyzing proteins, we now turn to a common situation in cell and molecular biology: an investigator has identified a gene important for a biological process but has no direct knowledge of the biochemical properties of its protein product.

Thanks to the proliferation of protein and nucleic acid sequences that are cataloged in genome databases, the function of a gene—and its encoded protein—can often be predicted by simply comparing its sequence with those of previously characterized genes. Because amino acid sequence determines protein structure, and structure dictates biochemical function, proteins that share a similar amino acid sequence usually have the same structure and usually perform similar biochemical functions, even when they are found in distantly related organisms. In modern cell biology, the study of a newly discovered protein usually begins with a search for previously characterized proteins that are similar in their amino acid sequences.

Searching a collection of known sequences for similar genes or proteins is typically done over the Internet, and it simply involves selecting a database and entering the desired sequence. A sequence-alignment program—the most popular is BLAST—scans the database for similar sequences by sliding the submitted

```
Score =  399 bits (1025), Expect = e-111
Identities = 198/290 (68%), Positives = 241/290 (82%),  Gaps = 1/290

Query: 57   MENFQKVEKIGEGTYGVVYKARNKLTGEVVALKKIRLDTETEGVPSTAIREISLLKELNH 116
            ME ++KVEKIGEGTYGVVYKA +K T E +ALKKIRL+ E EGVPSTAIREISLLKE+NH
Sbjct: 1    MEQYEKVEKIGEGTYGVVYKALDKATNETIALKKIRLEQEDEGVPSTAIREISLLKEMNH 60

Query: 117  PNIVKLLDVIHTENKLYLVFEFLHQDLKKFMDASALTGIPLPLIKSYLFQLLQGLAFCHS 176
            NIV+L DV+H+E ++YLVFE+L  DLKKFMD+       LIKSYL+Q+L G+A+CHS
Sbjct: 61·  GNIVRLHDVVHSEKRLYLVFEYLDLDLKKFMDSCPEFAKNPTLIKSYLYQILHGVAYCHS 120

Query: 177  HRVLHRDLKPQNLLINTE-GAIKLADFGLARAFGVPVRTYTHEVVTLWYRAPEILLGCKY 235
            HRVLHRDLKPQNLLI+    A+KLADFGLARAFG+PVRT+THEVVTLWYRAPEILLG +
Sbjct: 121  HRVLHRDLKPQNLLIDRRTNALKLADFGLARAFGIPVRTFTHEVVTLWYRAPEILLGARQ 180

Query: 236  YSTAVDIWSLGCIFAEMVTRRALFPGDSEIDQLFRIFRTLGTPDEVVWPGVTSMPDYKPS 295
            YST VD+WS+GCIFAEMV ++ LFPGDSEID+LF+IFR LGTP+E  WPGV+ +PD+K +
Sbjct: 181  YSTPVDVWSVGCIFAEMVNQKPLFPGDSEIDELFKIFRILGTPNEQSWPGVSCLPDFKTA 240

Query: 296  FPKWARQDFSKVVPPLDEDGRSLLSQMLHYDPNKRISAKAALAHPFFQDV 345
            FP+W  QD + VVP LD  G  LLS+ML Y+P+KRI+A+A+ AL H +F+D+
Sbjct: 241  FPRWQAQDLATVVPNLDPAGLDLLSKMLRYEPSKRITARQALEHEYFKDL 290
```

sequence along the archived sequences until a cluster of residues falls into full or partial alignment (**Figure 8–23**). Such comparisons can predict the functions of individual proteins, families of proteins, or even most of the protein complement of a newly sequenced organism.

As was explained in Chapter 3, many proteins that adopt the same conformation and have related functions are too distantly related to be identified from a comparison of their amino acid sequences alone (see Figure 3–13). Thus, an ability to reliably predict the three-dimensional structure of a protein from its amino acid sequence would improve our ability to infer protein function from the sequence information in genomic databases. In recent years, major progress has been made in predicting the precise structure of a protein. These predictions are based, in part, on our knowledge of the thousands of protein structures that have already been determined by x-ray crystallography and NMR spectroscopy and, in part, on computations using our knowledge of the physical forces acting on the atoms. However, it remains a substantial and important challenge to predict the structures of proteins that are large or have multiple domains, or to predict structures at the very high levels of resolution needed to assist in computer-based drug discovery.

While finding related sequences and structures for a new protein will provide many clues about its function, it is usually necessary to test these insights through direct experimentation. However, the clues generated from sequence comparisons typically point the investigator in the correct experimental direction, and their use has therefore become one of the most important strategies in modern cell biology.

Summary

Many methods exist for identifying proteins and analyzing their biochemical properties, structures, and interactions with other proteins. Small-molecule inhibitors allow the functions of proteins they act upon to be studied in living cells. Because proteins with similar structures often have similar functions, the biochemical activity of a protein can often be predicted by searching databases for previously characterized proteins that are similar in their amino acid sequences.

ANALYZING AND MANIPULATING DNA

Until the early 1970s, DNA was the most difficult biological molecule for the biochemist to analyze. Enormously long and chemically monotonous, the string of nucleotides that forms the genetic material of an organism could be examined only indirectly, by protein sequencing or by genetic analysis. Today, the situation has changed entirely. From being the most difficult macromolecule of the cell to

Figure 8–23 Results of a BLAST search. Sequence databases can be searched to find similar amino acid or nucleic acid sequences. Here, a search for proteins similar to the human cell-cycle regulatory protein Cdc2 *(Query)* locates maize Cdc2 *(Sbjct)*, which is 68% identical to human Cdc2 in its amino acid sequence. The alignment begins at residue 57 of the Query protein, suggesting that the human protein has an N-terminal region that is absent from the maize protein. The *green* blocks indicate differences in sequence, and the *yellow* bar summarizes the similarities: when the two amino acid sequences are identical, the residue is shown; similar amino acid substitutions are indicated by a plus sign (+). Only one small gap has been introduced—indicated by the *red arrow* at position 194 in the Query sequence—to align the two sequences maximally. The alignment score *(Score)*, which is expressed in two different types of units, takes into account penalties for substitutions and gaps; the higher the alignment score, the better the match. The significance of the alignment is reflected in the *Expectation* (E) value, which specifies how often a match this good would be expected to occur by chance. The lower the E value, the more significant the match; the extremely low value here (e^{-111}) indicates certain significance. E values much higher than 0.1 are unlikely to reflect true relatedness. For example, an E value of 0.1 means there is a 1 in 10 likelihood that such a match would arise solely by chance.

analyze, DNA has become the easiest. It is now possible to determine the entire nucleotide sequence of a bacterial or fungal genome in a matter of hours, and the sequence of an individual human genome in less than a day. Once the nucleotide sequence of a genome is known, any individual gene can be easily isolated, and large quantities of the gene product (be it RNA or protein) can be made either by introducing the gene into bacteria or animal cells and coaxing these cells to over-express the foreign gene or by synthesizing the gene product *in vitro*. In this way, proteins and RNA molecules that might be present in only tiny amounts in living cells can be produced in large quantities for biochemical and structural analyses. And this approach can also be used to produce large quantities of human proteins (such as insulin, or interferon, or blood-clotting proteins) for use as human phar-maceuticals. As we will see later in this chapter, it is also possible for scientists to alter an isolated gene and transfer it back into the germ line of an animal or plant, so as to become a functional and heritable part of the organism's genome. In this way, the biological roles of any gene can be assessed by observing—in the whole organism—the results of modifying it.

The ability to manipulate DNA with precision in a test tube or an organism, known as **recombinant DNA technology** has had a dramatic impact on all aspects of cell and molecular biology, allowing us to routinely study cells and their mac-romolecules in ways that were unimaginable even twenty years ago. Central to the technology are the following manipulations:

1. Cleavage of DNA at specific sites by restriction nucleases, which greatly facilitates the isolation and manipulation of individual pieces of a genome.

2. DNA ligation, which makes it possible to seamlessly join together DNA molecules from widely different sources.

3. DNA cloning (through the use of either cloning vectors or the polymerase chain reaction) in which a portion of a genome (often an individual gene) is "purified" away from the remainder of the genome by repeatedly copying it to generate many billions of identical molecules.

4. Nucleic acid hybridization, which makes it possible to identify any specific sequence of DNA or RNA with great accuracy and sensitivity based on its ability to selectively bind a complementary nucleic acid sequence.

5. DNA synthesis, which makes it possible to chemically synthesize DNA molecules with any sequence of nucleotides, whether or not the sequence occurs in nature.

6. Rapid determination of the sequence of nucleotides of any DNA or RNA molecule.

In the following sections, we describe each of these basic techniques which, together, have revolutionized the study of cell and molecular biology.

Restriction Nucleases Cut Large DNA Molecules into Specific Fragments

Unlike a protein, a gene does not exist as a discrete entity in cells, but rather as a small region of a much longer DNA molecule. Although the DNA molecules in a cell can be randomly broken into small pieces by mechanical force, a fragment containing a single gene in a mammalian genome would still be only one among a hundred thousand or more DNA fragments, indistinguishable in their average size. How could such a gene be separated from all the others? Because all DNA molecules consist of an approximately equal mixture of the same four nucleo-tides, they cannot be readily separated, as proteins can, on the basis of their dif-ferent charges and biochemical properties. The solution to this problem began to emerge with the discovery of **restriction nucleases**. These enzymes, which are purified from bacteria, cut the DNA double helix at specific sites defined by the local nucleotide sequence, thereby cleaving a long, double-stranded DNA mole-cule into fragments of strictly defined sizes.

Like many of the tools of recombinant DNA technology, restriction nucleases were discovered by researchers trying to understand an intriguing biological

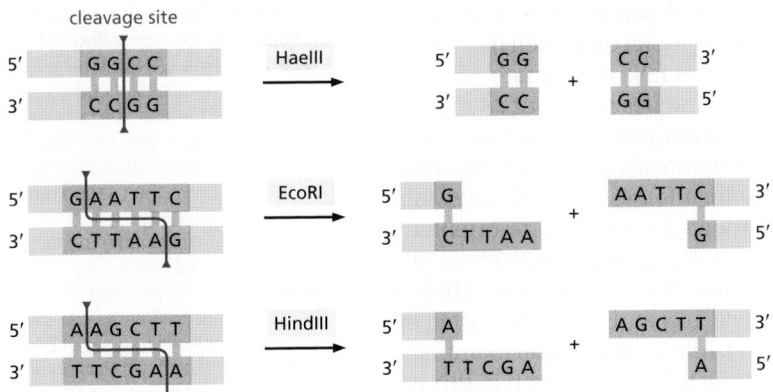

cleavage site

Figure 8–24 Restriction nucleases cleave DNA at specific nucleotide sequences.
Like the sequence-specific DNA-binding proteins we encountered in Chapter 7 (see Figure 7–8), restriction enzymes often work as dimers, and the DNA sequence they recognize and cleave is often symmetrical around a central point. Here, both strands of the DNA double helix are cut at specific points within the target sequence *(orange)*. Some enzymes, such as HaeIII, cut straight across the double helix and leave two blunt-ended DNA molecules; with others, such as EcoRI and HindIII, the cuts on each strand are staggered. These staggered cuts generate "sticky ends"—short, single-stranded overhangs that help the cut DNA molecules join back together through complementary base-pairing. This rejoining of DNA molecules becomes important for DNA cloning, as we discuss below. Restriction nucleases are usually obtained from bacteria, and their names reflect their origins: for example, the enzyme EcoRI comes from *Escherichia coli*. There are currently hundreds of different restriction enzymes available; they can be ordered from companies that commercially produce them.

phenomenon. It had been observed that certain bacteria always degraded "foreign" DNA that was introduced into them experimentally. A search for the mechanism responsible revealed a then unanticipated class of bacterial nucleases that cleave DNA at specific nucleotide sequences. The bacterium's own DNA is protected from cleavage by methylation of these same sequences, thereby protecting a bacterium's own genome from being overrun by foreign DNA. Because these enzymes restrict the transfer of DNA into bacteria, they were called restriction nucleases. The pursuit of this seemingly arcane biological puzzle set off the development of technologies that have forever changed the way cell and molecular biologists study living things.

Different bacterial species produce different restriction nucleases, each cutting at a different, specific nucleotide sequence (**Figure 8–24**). Because these target sequences are short—generally four to eight nucleotide pairs—many sites of cleavage will occur, purely by chance, in any long DNA molecule. The reason restriction nucleases are so useful in the laboratory is that each enzyme will always cut a particular DNA molecule at the same sites. Thus for a given sample of DNA (which contains many identical molecules), a particular restriction nuclease will reliably generate the same set of DNA fragments.

The size of the resulting fragments depends on the length of the target sequences of the restriction nucleases. As shown in Figure 8–24, the enzyme HaeIII cuts at a sequence of four nucleotide pairs; a sequence this long would be expected to occur purely by chance approximately once every 256 nucleotide pairs (1 in 4^4). In comparison, a restriction nuclease with a target sequence that is eight nucleotides long would be expected to cleave DNA on average once every 65,536 nucleotide pairs (1 in 4^8). This difference in sequence selectivity makes it possible to cleave a long DNA molecule into the fragment sizes that are most suitable for a given application.

Gel Electrophoresis Separates DNA Molecules of Different Sizes

The same types of gel-electrophoresis methods that have proved so useful in the analysis of proteins (see Figure 8–13) can be applied to DNA molecules. The procedure is actually simpler than for proteins: because each nucleotide in a nucleic acid molecule already carries a single negative charge (on the phosphate group), there is no need to add the negatively charged detergent SDS that is required to make protein molecules move uniformly toward the positive electrode. Larger DNA fragments will migrate more slowly because their progress is impeded to a greater extent by the gel matrix. Over several hours, the DNA fragments become spread out across the gel according to size, forming a ladder of discrete bands, each composed of a collection of DNA molecules of identical length (**Figure 8–25A and B**). To separate DNA molecules longer than 500 nucleotide pairs, the gel is made of a diluted solution of agarose (a polysaccharide isolated from seaweed). For DNA fragments less than 500 nucleotides long, specially designed polyacrylamide gels allow the separation of molecules that differ in length by as little as a single nucleotide (see Figure 8–25C).

A variation of agarose-gel electrophoresis, called *pulsed-field gel electrophoresis*, makes it possible to separate extremely long DNA molecules, even those found in whole chromosomes. Ordinary gel electrophoresis fails to separate very large DNA molecules because the steady electric field stretches them out so that they travel end-first through the gel in snakelike configurations at a rate that is independent of their length. In pulsed-field gel electrophoresis, by contrast, the direction of the electric field changes periodically, which forces the molecules to re-orient before continuing to move snakelike through the gel. This re-orientation takes much more time for larger molecules, so that longer molecules move more slowly than shorter ones. As a consequence, entire bacterial or yeast chromosomes separate into discrete bands in pulsed-field gels and so can be sorted and identified on the basis of their size (Figure 8–25D). Although a typical mammalian chromosome of 10^8 nucleotide pairs is still too long to be sorted even in this way, large segments of these chromosomes are readily separated and identified if the chromosomal DNA is first cut with a restriction nuclease selected to recognize sequences that occur only rarely.

The DNA bands on agarose or polyacrylamide gels are invisible unless the DNA is labeled or stained in some way. A particularly sensitive method of staining DNA is to soak the gel in the dye *ethidium bromide,* which fluoresces under ultraviolet light when it is bound to DNA (see Figure 8–25B and D). Even more sensitive detection methods incorporate a radioisotope or chemical marker into the DNA molecules before electrophoresis, as we next describe.

Figure 8–25 DNA molecules can be separated by size using gel electrophoresis. (A) Schematic illustration comparing the results of cutting the same DNA molecule (in this case, the genome of a virus that infects wasps) with two different restriction nucleases, EcoRI (*middle*) and HindIII (*right*). The fragments are then separated by gel electrophoresis using a gel matrix of agarose. Because larger fragments migrate more slowly than smaller ones, the lowermost bands on the gel contain the smallest DNA fragments. The sizes of the fragments can be estimated by comparing them to a set of DNA fragments of known sizes (*left*). (B) Photograph of an actual agarose gel showing DNA "bands" that have been stained with ethidium bromide. (C) A polyacrylamide gel with small pores was used to separate short DNA molecules that differ by only a single nucleotide. Shown here are the results of a dideoxy sequencing reaction, explained later in this chapter. From left to right, the bands in the four lanes were produced by adding G, A, T, and C chain-terminating nucleotides (see Panel 8–1). The DNA molecules were labeled with ^{32}P, and the image shown was produced by laying a piece of photographic film over the gel and allowing the ^{32}P to expose the film, producing the dark bands observed when the film was developed. (D) The technique of pulsed-field agarose-gel electrophoresis was used to separate the 16 different chromosomes of the yeast species *Saccharomyces cerevisiae*, which range in size from 220,000 to 2.5 million nucleotide pairs. The DNA was stained as in (B). DNA molecules as large as 10^7 nucleotide pairs can be separated in this way. (B, from U. Albrecht et al., *J. Gen. Virol.* 75:3353–3363, 1994. With permission from Microbiology Society; C, courtesy of Leander Lauffer and Peter Walter; D, from D. Vollrath and R.W. Davis, *Nucleic Acids Res.* 15:7865–7876, 1987. With permission from Oxford University Press.)

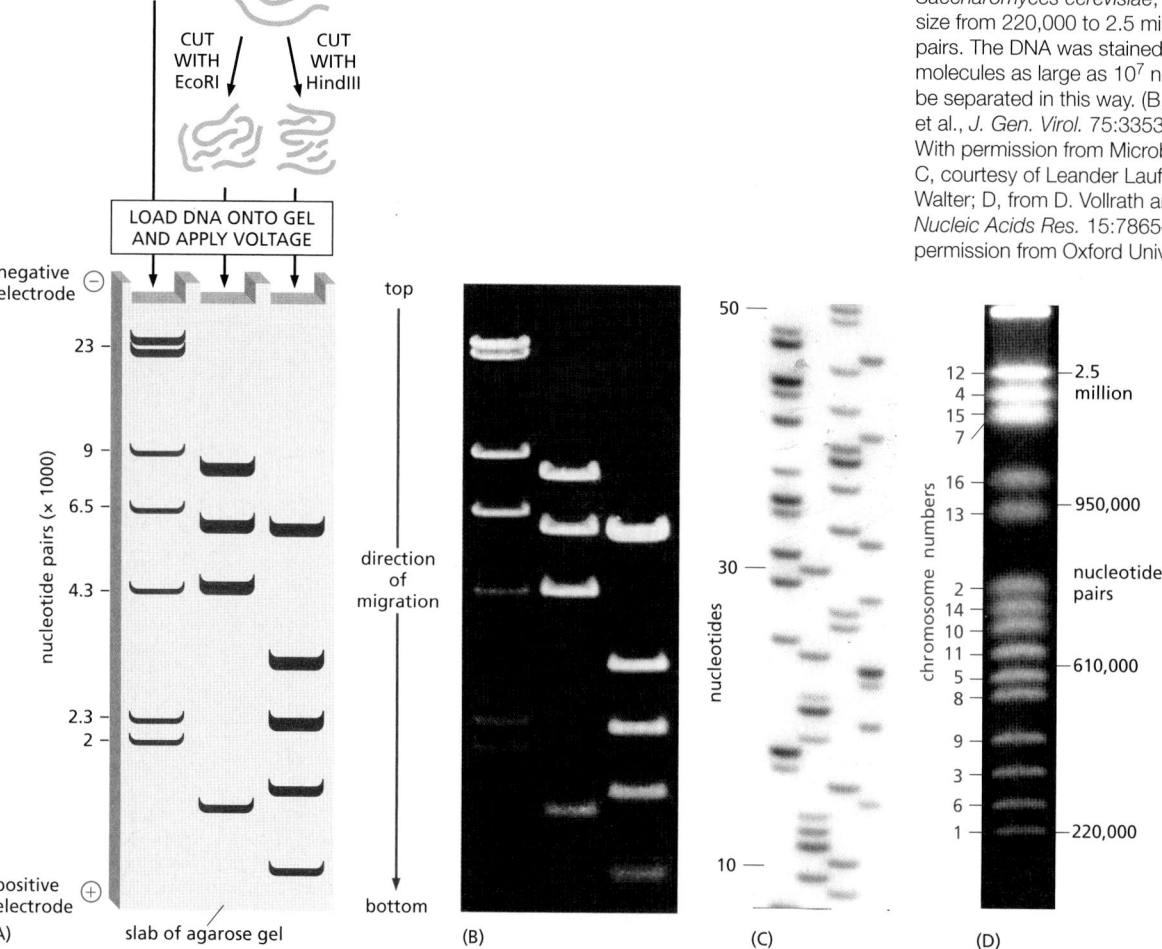

Purified DNA Molecules Can Be Specifically Labeled with Radioisotopes or Chemical Markers *in vitro*

The DNA polymerases that synthesize and repair DNA (discussed in Chapter 5) have become important tools in experimentally manipulating DNA. Because they synthesize sequences complementary to an existing DNA molecule, they are often used in the test tube to create exact copies of existing DNA molecules. The copies can include specially modified nucleotides (**Figure 8–26**). To synthesize DNA in this way, the DNA polymerase is presented with a template and a pool of nucleotide precursors that contain the modification. As long as the polymerase can use these precursors, it automatically makes new, modified molecules that match the sequence of the template. Modified DNA molecules have many uses. DNA labeled with the radioisotope ^{32}P can be detected following gel electrophoresis by placing the gel next to a piece of photographic film (see Figure 8–25C). The ^{32}P atoms emit β particles which expose the film, producing a visible record of every band on the gel. Alternatively, the gel can be scanned by a detector that measures the β emissions directly. Other types of modified DNA, such as that labeled by digoxigenin (see Figure 8–26B), are useful for visualizing DNA molecules in whole cells, a topic we discuss later in this chapter.

Genes Can Be Cloned Using Bacteria

Any DNA fragment can be cloned. In molecular biology, the term **DNA cloning** is used in two senses. It literally refers to the act of making many identical copies (typically billions) of a DNA molecule—the amplification of a particular DNA sequence. However, the term also describes the isolation of a particular stretch of DNA (often a particular gene) from the rest of the cell's genome; the same term

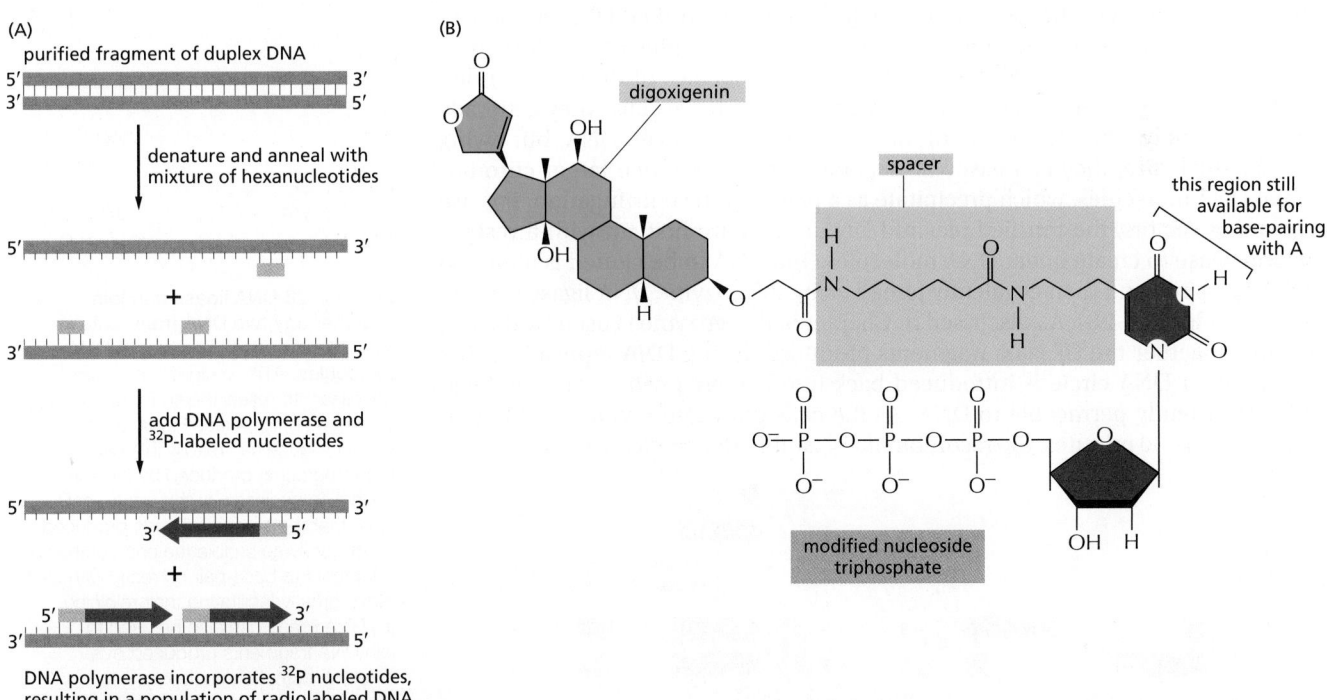

Figure 8–26 Methods for labeling DNA molecules *in vitro*. (A) A purified DNA polymerase enzyme can incorporate radiolabeled nucleotides as it synthesizes new DNA molecules. In this way, radiolabeled versions of any DNA sequence can be prepared in the laboratory. (B) The method in (A) is also used to produce nonradioactive DNA molecules that carry a specific chemical marker that can be detected with an appropriate antibody. The base on the nucleoside triphosphate shown is an analog of thymine, in which the methyl group on T has been replaced by a spacer arm linked to the plant steroid digoxigenin. An anti-digoxigenin antibody coupled to a visible marker such as a fluorescent dye is then used to visualize the DNA. Other chemical labels, such as biotin, can be attached to nucleotides and used in the same way. The only requirements are that the modified nucleotides properly base-pair and appear "normal" to the DNA polymerase.

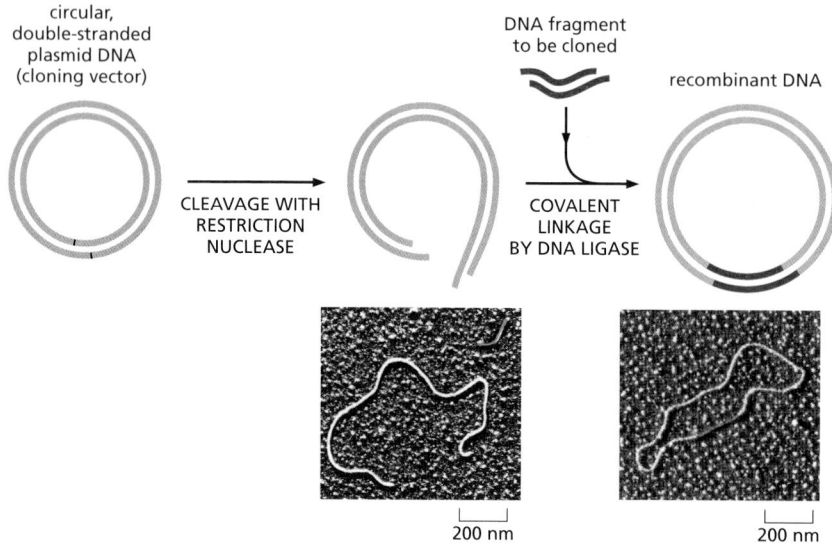

Figure 8–27 **The insertion of a DNA fragment into a bacterial plasmid with the enzyme DNA ligase.** The plasmid is cut open with a restriction nuclease (in this case, one that produces staggered ends) and is mixed with the DNA fragment to be cloned (which has been prepared with the same restriction nuclease). DNA ligase and ATP are added. The staggered ends base-pair, and DNA ligase seals the nicks in the DNA backbone, producing a complete recombinant DNA molecule. In the accompanying micrographs, the inserted DNA is colored *red*. (Micrographs courtesy of Huntington Potter and David Dressler.)

is used because this isolation is usually accomplished by making many identical copies of only the DNA of interest. We note that elsewhere in the book, cloning, particularly when used in the context of developmental biology, can also refer to the generation of many genetically identical cells starting from a single cell or even to the generation of genetically identical organisms (see, for example, Figure 7–2). In all cases, cloning refers to the act of making many identical copies, and in this section, we use the term to refer to methods designed to generate many identical copies of a defined segment of nucleic acid.

DNA cloning can be accomplished in several ways. One of the simplest involves inserting a particular fragment of DNA into the purified DNA genome of a self-replicating genetic element—usually a plasmid. The **plasmid vectors** most widely used for gene cloning are small, circular molecules of double-stranded DNA derived from plasmids that occur naturally in bacterial cells. They generally account for only a minor fraction of the total host bacterial cell DNA, but owing to their small size, they can easily be separated from the much larger chromosomal DNA molecules, which precipitate as a pellet upon centrifugation. For use as cloning vectors, the purified plasmid DNA circles are first cut with a restriction nuclease to create linear DNA molecules. The DNA to be cloned is added to the cut plasmid and then covalently joined using the enzyme DNA ligase (**Figure 8–27** and **Figure 8–28**). As discussed in Chapter 5, this enzyme is used by the cell to stitch together the Okazaki fragments produced during DNA replication. The recombinant DNA circle is introduced back into bacterial cells that have been made transiently permeable to DNA. As the cells grow and divide, doubling in number every 30 minutes, the recombinant plasmids also replicate to produce an

Figure 8–28 **DNA ligase can join together any two DNA fragments *in vitro* to produce recombinant DNA molecules.** ATP provides the energy necessary to reseal the sugar-phosphate backbone of DNA (see Figure 5–12). (A) DNA ligase can readily join two DNA fragments produced by the same restriction nuclease, in this case EcoRI. Note that the staggered ends produced by this enzyme enable the ends of the two fragments to base-pair correctly with each other, greatly facilitating their rejoining. (B) DNA ligase can also be used to join DNA fragments produced by different restriction nucleases—for example, EcoRI and HaeIII. In this case, before the fragments undergo ligation, DNA polymerase plus a mixture of deoxyribonucleoside triphosphates (dNTPs) are used to fill in the staggered cut produced by EcoRI. Each DNA fragment shown in the figure is oriented so that its 5′ ends are at the left end of the upper strand and the right end of the lower strand, as indicated.

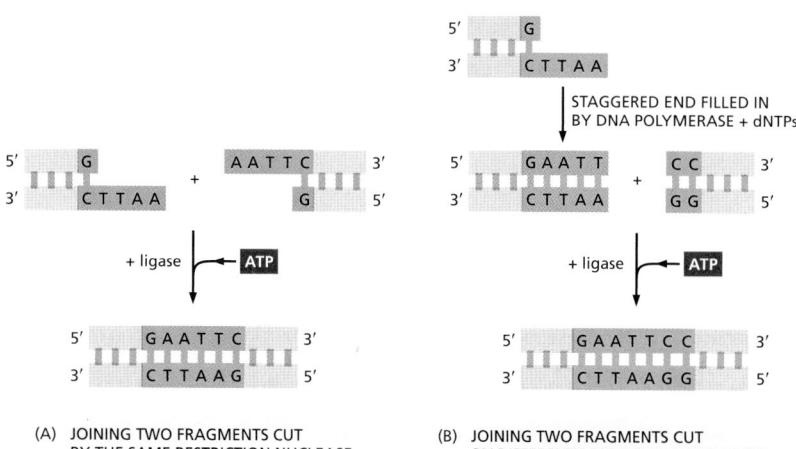

(A) JOINING TWO FRAGMENTS CUT BY THE SAME RESTRICTION NUCLEASE

(B) JOINING TWO FRAGMENTS CUT BY DIFFERENT RESTRICTION NUCLEASES

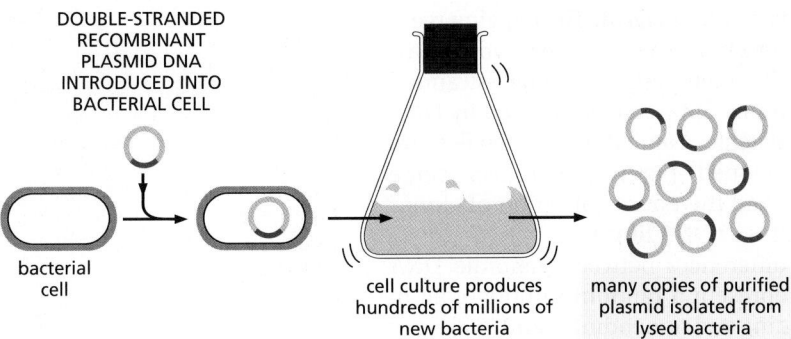

DOUBLE-STRANDED
RECOMBINANT
PLASMID DNA
INTRODUCED INTO
BACTERIAL CELL

bacterial
cell

cell culture produces
hundreds of millions of
new bacteria

many copies of purified
plasmid isolated from
lysed bacteria

Figure 8–29 **A DNA fragment can be replicated inside a bacterial cell.** To clone a particular fragment of DNA, it is first inserted into a plasmid vector, as shown in Figure 8–27. The resulting recombinant plasmid DNA is then introduced into a bacterium, where it is replicated many millions of times as the bacterium multiplies. For simplicity, the genome of the bacterial cell is not shown.

enormous number of copies of DNA circles containing the foreign DNA (**Figure 8–29**). Once the cells are lysed and the plasmid DNA isolated, the cloned DNA fragment can be readily recovered by cutting it out of the plasmid DNA with the same restriction nuclease that was used to insert it, and then separating it from the plasmid DNA by gel electrophoresis. Together, these steps allow the amplification and purification of any segment of DNA from the genome of any organism.

A particularly useful plasmid vector is based on the naturally occurring F plasmid of *E. coli*. Unlike smaller bacterial plasmids, the F plasmid—and its engineered derivative, the **bacterial artificial chromosome (BAC)**—is present in only one or two copies per *E. coli* cell. The fact that BACs are kept in such low numbers means that they can stably maintain very long DNA sequences, up to 1 million nucleotide pairs in length. With only a few BACs present per bacterium, it is less likely that the cloned DNA fragments will become scrambled by recombination with sequences carried on other copies of the plasmid. Because of their stability, ability to accept large DNA inserts, and ease of handling, BACs are now the preferred vector for handling large fragments of foreign DNA. As we will see below, BACs were instrumental in determining the complete nucleotide sequence of the human genome.

An Entire Genome Can Be Represented in a DNA Library

Often it is useful to break up a genome into much smaller fragments and clone every fragment, separately, using a plasmid vector. This approach is useful because it allows scientists to work with easily managed, discrete pieces of a genome instead of whole, unwieldy chromosomes.

This strategy involves cleaving genomic DNA into small pieces using a restriction nuclease (or, in some cases, by mechanically shearing the DNA) and ligating the entire collection of DNA fragments into plasmid vectors, using conditions that favor the insertion of a single DNA fragment into each plasmid molecule. These recombinant plasmids are then introduced into *E. coli* at a concentration that ensures that no more than one plasmid molecule is taken up by each bacterium. The collection of cloned plasmid molecules is known as a **DNA library**. Because the DNA fragments were derived directly from the chromosomal DNA of the organism of interest, the resulting collection—called a **genomic library**—will represent the entire genome of that organism (**Figure 8–30**), spread out over tens of thousands of individual bacterial colonies.

An alternative strategy, one that enriches for protein-coding genes, is to begin the cloning process by selecting only those DNA sequences that are transcribed into mRNA and thus correspond to protein-encoding genes. This is done by extracting the mRNA from cells and then making a DNA copy of each mRNA

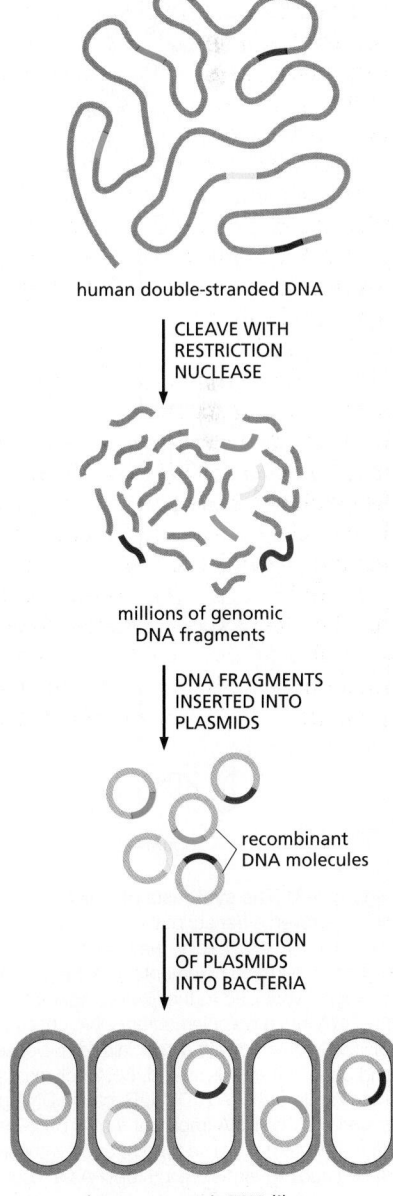

human double-stranded DNA

CLEAVE WITH
RESTRICTION
NUCLEASE

millions of genomic
DNA fragments

DNA FRAGMENTS
INSERTED INTO
PLASMIDS

recombinant
DNA molecules

INTRODUCTION
OF PLASMIDS
INTO BACTERIA

human genomic DNA library

Figure 8–30 **Human genomic libraries containing DNA fragments that represent the whole human genome can be constructed using restriction nucleases and DNA ligase.** Such a genomic library consists of a set of bacteria, each carrying a different fragment of human DNA. For simplicity, only the *colored* DNA fragments are shown in the library; in reality, all of the different *gray* fragments will also be represented.

molecule present—a so-called *complementary DNA,* or *cDNA.* The copying reaction is catalyzed by the reverse transcriptase enzyme of retroviruses, which synthesizes a complementary DNA chain on an RNA template. The single-stranded cDNA molecules synthesized by the reverse transcriptase are converted by DNA polymerase into double-stranded cDNA molecules, and these molecules are inserted into a plasmid or virus vector and cloned (**Figure 8–31**). Each clone obtained in this way is called a **cDNA clone**, and the entire collection of clones derived from one mRNA preparation constitutes a **cDNA library**.

Figure 8–32 illustrates some important differences between genomic DNA clones and cDNA clones. Genomic clones represent a random sample of all of the DNA sequences in an organism—both coding and noncoding—and, with very rare exceptions, are the same regardless of the cell type used to prepare them. By contrast, cDNA clones contain only those regions of the genome that have been transcribed into mRNA. Because the cells of different tissues produce distinct sets of mRNA molecules, a distinct cDNA library is obtained for each type of cell used to prepare the library.

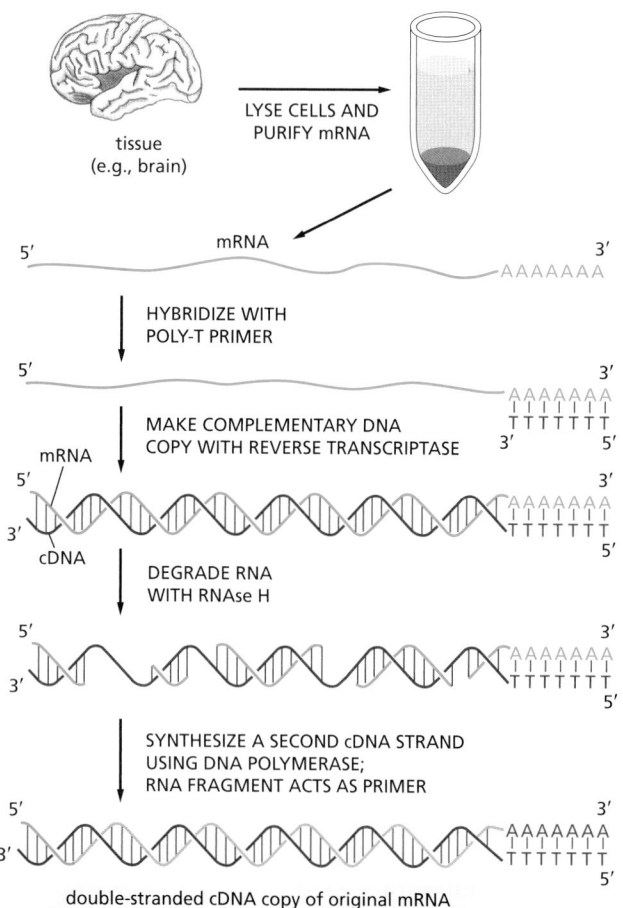

Figure 8–31 The synthesis of cDNA. Total mRNA is extracted from a particular tissue, and the enzyme reverse transcriptase (see Figure 5–62) is used to produce DNA copies (cDNA) of the mRNA molecules. For simplicity, the copying of just one of these mRNAs into cDNA is illustrated. A short oligonucleotide complementary to the poly-A tail at the 3′ end of the mRNA (discussed in Chapter 6) is first hybridized to the RNA to act as a primer for the reverse transcriptase, which then copies the RNA into a complementary DNA chain, thereby forming a DNA–RNA hybrid helix. Treating the DNA–RNA hybrid with a specialized nuclease (RNAse H) that attacks only the RNA produces nicks and gaps in the RNA strand. DNA polymerase then copies the remaining single-stranded cDNA into double-stranded cDNA. Because DNA polymerase can synthesize through the bound RNA molecules, the RNA fragment that is base-paired to the 3′ end of the first DNA strand usually acts as the primer for the second strand synthesis, as shown. Any remaining RNA is eventually degraded during subsequent cloning steps. As a result, the nucleotide sequences at the extreme 5′ ends of the original mRNA molecules are often absent from cDNA libraries.

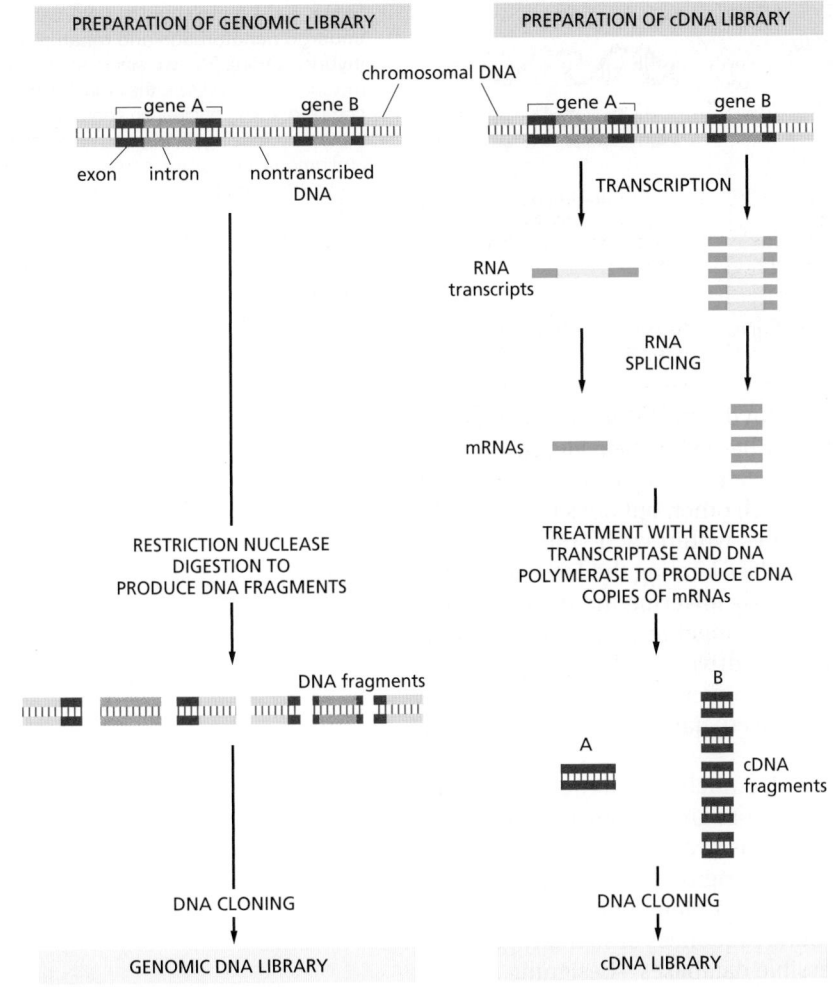

PREPARATION OF GENOMIC LIBRARY

PREPARATION OF cDNA LIBRARY

chromosomal DNA

gene A   gene B

exon   intron   nontranscribed
DNA

gene A   gene B

TRANSCRIPTION

RNA
transcripts

RNA
SPLICING

mRNAs

RESTRICTION NUCLEASE
DIGESTION TO
PRODUCE DNA FRAGMENTS

TREATMENT WITH REVERSE
TRANSCRIPTASE AND DNA
POLYMERASE TO PRODUCE cDNA
COPIES OF mRNAs

DNA fragments

B

A

cDNA
fragments

DNA CLONING

GENOMIC DNA LIBRARY

DNA CLONING

cDNA LIBRARY

Figure 8–32 The differences between cDNA clones and genomic DNA clones derived from the same region of DNA. In this example, gene A is infrequently transcribed, whereas gene B is frequently transcribed, and both genes contain introns *(orange)*. In the genomic DNA library, both the introns and the nontranscribed DNA *(gray)* are included in the clones, and most clones contain, at most, only part of the coding sequence of a gene *(red)*. In the cDNA clones, the intron sequences *(yellow)* have been removed by RNA splicing during the formation of the mRNA *(blue)*, and a continuous coding sequence is therefore present in each clone. Because gene B is transcribed more frequently than gene A in the cells from which the cDNA library was made, it is represented much more frequently than A in the cDNA library. In contrast, A and B are represented equally in the genomic DNA library.

Genomic and cDNA Libraries Have Different Advantages and Drawbacks

Genomic libraries are especially useful in determining the nucleotide sequences of a whole genome. For example, to determine the nucleotide sequence of the human genome, it was broken up into roughly 100,000-nucleotide-pair pieces, each of which was inserted into a BAC plasmid and amplified in *E. coli*. The resulting genomic library consisted of tens of thousands of bacterial colonies, each containing a different human DNA insert. The nucleotide sequence of each insert was determined separately and the sequence of the entire genome was stitched together from the pieces.

The most important advantage of cDNA clones, over genomic clones, is that they contain the uninterrupted coding sequence of a gene. When the aim of the cloning, for example, is to produce the protein in large quantities by expressing the cloned gene in a bacterial or yeast cell, it is more preferable to start with cDNA.

Genomic and cDNA libraries are inexhaustible resources, which are widely shared among investigators. Today, many such libraries are also available from commercial sources. Because the identity of each insert in a library is often known (through sequencing the insert), it is often possible to order a particular region of a chromosome (or, in the case of cDNA, a complete, intron-less protein-coding gene) and have it delivered by mail.

Cloning DNA by using bacteria revolutionized the study of genomes and is still in wide use today. However, there is an even simpler way to clone DNA, one that can be carried out entirely *in vitro*. We discuss this approach, called the *polymerase chain reaction*, below. However, first we need to review a fundamental, far-reaching property of DNA and RNA called *hybridization*.

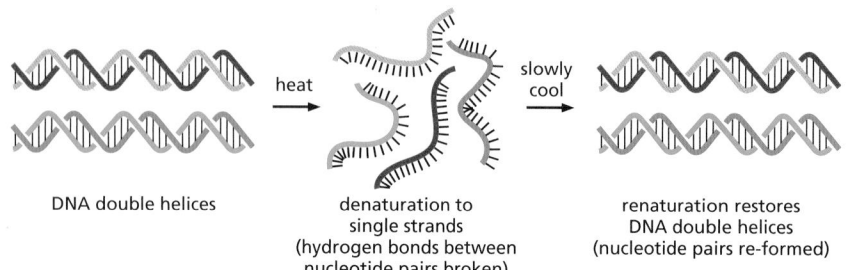

DNA double helices

denaturation to
single strands
(hydrogen bonds between
nucleotide pairs broken)

renaturation restores
DNA double helices
(nucleotide pairs re-formed)

Figure 8–33 **A molecule of DNA can
undergo denaturation and renaturation
(hybridization).** For two single-stranded
molecules to hybridize, they must have
complementary nucleotide sequences that
allow base-pairing. In this example, the *red*
and *orange* strands are complementary to
each other, and the *blue* and *green* strands
are complementary to each other. Although
denaturation by heating is shown, DNA can
also be renatured after being denatured by
alkali treatment.

Hybridization Provides a Powerful, But Simple Way to Detect Specific Nucleotide Sequences

Under normal conditions, the two strands of a DNA double helix are held together by hydrogen bonds between the complementary base pairs (see Figure 4–3). But these relatively weak, noncovalent bonds can be fairly easily broken. Such *DNA denaturation* will release the two strands from each other, but does not break the covalent bonds that link together the nucleotides within each strand. Perhaps the simplest way to achieve this separation involves heating the DNA to around 90°C. When the conditions are reversed—by slowly lowering the temperature—the complementary strands will readily come back together to re-form a double helix. This **hybridization**, or *DNA renaturation*, is driven by the re-formation of the hydrogen bonds between complementary base pairs (**Figure 8–33**). We saw in Chapter 5 that DNA hybridization underlies the crucial process of homologous recombination (see Figure 5–47).

This fundamental capacity of a single-stranded nucleic acid molecule, either DNA or RNA, to form a double helix with a single-stranded molecule of a complementary sequence provides a powerful and sensitive technique for detecting specific nucleotide sequences. Today, one simply designs a short, single-stranded DNA molecule (often called a *DNA probe*) that is complementary to the nucleotide sequence of interest. Because the nucleotide sequences of so many genomes are known—and are stored in publicly accessible databases—designing a probe to hybridize anywhere in a genome is straightforward. Probes are single-stranded, typically 30 nucleotides in length, and are usually synthesized chemically by a commercial service for pennies per nucleotide. A DNA sequence of 30 nucleotides will occur by chance only once every 1×10^{18} nucleotides (4^{30}); so, even in the human genome of 3×10^9 nucleotide pairs, a DNA probe designed to match a particular 30-nucleotide sequence will be highly unlikely to hybridize—by chance—anywhere else on the genome. This, of course, presumes that the sequence complementary to the probe does not occur multiple times in the genome, a condition that can be checked beforehand by scanning the genomic sequence *in silico* (using a computer) and designing probes that match only one spot. The hybridization conditions can be set so that even a single mismatch will prevent hybridization to "near-miss" sequences. The exquisite specificity of nucleic acid hybridization can be easily appreciated by the *in situ* (Latin for "in place") *hybridization* experiment shown in **Figure 8–34**. As we will see throughout this chapter, nucleic acid

Figure 8–34 *In situ* **hybridization can be used to locate genes on
isolated chromosomes.** Here, six different DNA probes have been used
to mark the locations of their complementary nucleotide sequences on
human Chromosome 5, isolated from a mitotic cell in metaphase (see
Figure 4–59 and Panel 17–1, pp. 980–981). The DNA probes have been
labeled with different chemical groups (see Figure 8–26B) and are detected
using fluorescent antibodies specific for those groups. The chromosomal
DNA has been partially denatured to allow the probes to base-pair with
their complementary sequences. Both the maternal and paternal copies of
Chromosome 5 are shown, aligned side by side. Each probe produces two
dots on each chromosome because chromosomes undergoing mitosis have
already replicated their DNA; therefore, each chromosome contains two
identical DNA helices. The technique employed here is nicknamed FISH, for
fluorescence in situ hybridization. (Courtesy of David C. Ward.)

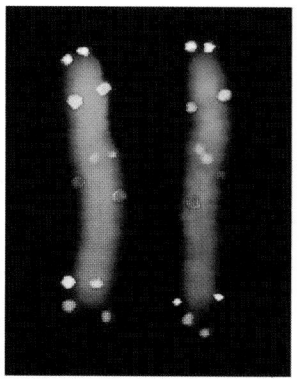

hybridization has many uses in modern cell and molecular biology; one of the most powerful is in the cloning of DNA by the polymerase chain reaction, as we next discuss.

Genes Can Be Cloned *in vitro* Using PCR

Genomic and cDNA libraries were once the only route to cloning genes and they are still used for cloning very large genes and for sequencing whole genomes. However, a powerful and versatile method for amplifying DNA, known as the **polymerase chain reaction** (**PCR**), provides a more rapid and straightforward approach to DNA cloning, particularly in organisms whose complete genome sequence is known. Today, since genome sequences are abundant, most cloning is carried out by PCR.

Invented in the 1980s, PCR revolutionized the way that DNA and RNA are analyzed. The technique can amplify any nucleotide sequence selectively and is performed entirely in a test tube. Eliminating the need for bacteria makes PCR convenient and rapid—billions of copies of a nucleotide can be generated in a matter of hours. Starting with an entire genome, PCR allows DNA from a specified region—selected by the experimenter—to be greatly amplified, effectively "purifying" this DNA away from the remainder of the genome, which remains unamplified. Because of its power to greatly amplify nucleic acids, PCR is remarkably sensitive: the method can be used to detect the trace amounts of DNA in a drop of blood left at a crime scene or in a few copies of a viral genome in a patient's blood sample.

The success of PCR depends both on the selectivity of DNA hybridization and on the ability of DNA polymerase to copy a DNA template faithfully through repeated rounds of replication *in vitro*. As discussed in Chapter 5, this enzyme adds nucleotides to the 3′ end of a growing strand of DNA (see Figure 5–4). To copy DNA, the polymerase requires a primer—a short nucleotide sequence that provides a 3′ end from which synthesis can begin. For PCR, the primers are designed by the experimenter, synthesized chemically, and, by hybridizing to genomic DNA, "tell" the polymerase which part of the genome to copy. As discussed in the previous section, *DNA primers* (in essence, the same type of molecules as DNA probes but without a radioactive or fluorescent label) can be designed to uniquely locate any position on a genome.

PCR is an iterative process in which the cycle of amplification is repeated dozens of times. At the start of each cycle, the two strands of the double-stranded DNA template are separated and a different primer is annealed to each. These primers mark the right and left boundaries of the DNA to be amplified. DNA polymerase is then allowed to replicate each strand independently (**Figure 8–35**). In

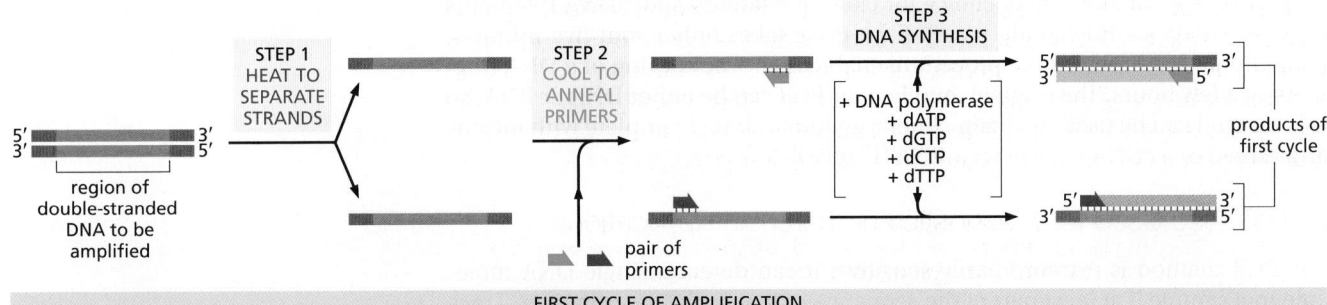

Figure 8–35 A pair of primers directs the synthesis of a desired segment of DNA in a test tube. Each cycle of PCR includes three steps: (1) The double-stranded DNA is heated briefly to separate the two strands. (2) The DNA is exposed to a large excess of a pair of specific primers—designed to bracket the region of DNA to be amplified—and the sample is cooled to allow the primers to hybridize to complementary sequences in the two DNA strands. (3) This mixture is incubated with DNA polymerase and the four deoxyribonucleoside triphosphates so that DNA can be synthesized, starting from the two primers. To amplify the DNA, the cycle is repeated many times by reheating the sample to separate the newly synthesized DNA strands (see Figure 8–36).
The technique depends on the use of a special DNA polymerase isolated from a thermophilic bacterium; this polymerase is stable at much higher temperatures than eukaryotic DNA polymerases, so it is not denatured by the heat treatment shown in step 1. The enzyme therefore does not have to be added again after each cycle.

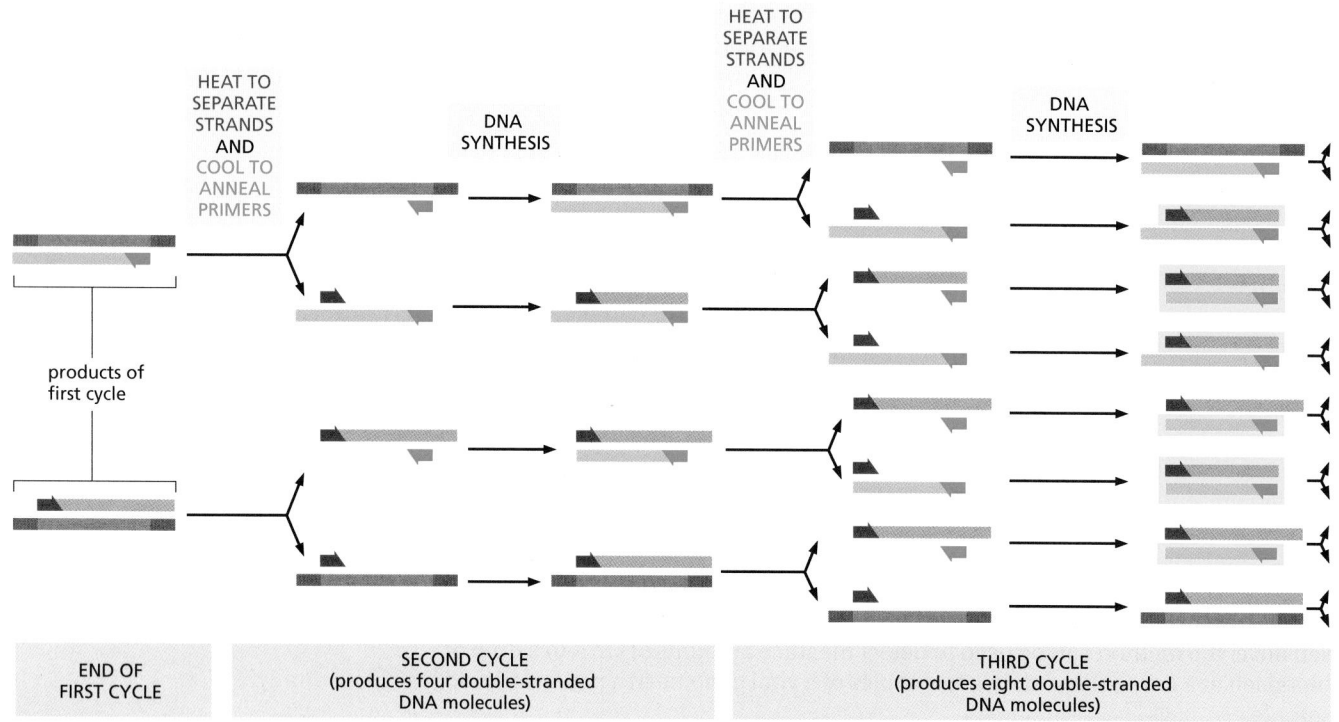

Figure 8–36 **PCR uses repeated rounds of strand separation, hybridization, and synthesis to amplify DNA.** As the procedure outlined in Figure 8–35 is repeated, all the newly synthesized fragments serve as templates in their turn. Because the polymerase and the primers remain in the sample after the first cycle, PCR involves simply heating and then cooling the same sample, in the same test tube, again and again. Each cycle doubles the amount of DNA synthesized in the previous cycle, so that within a few cycles, the predominant DNA is identical to the sequence bracketed by and including the two primers in the original template. In the example illustrated here, three cycles of reaction produce 16 DNA chains, 8 of which (boxed in *yellow*) correspond exactly to one or the other strand of the original bracketed sequence. After four more cycles, 240 of the 256 DNA chains will correspond exactly to the original sequence, and after several more cycles, essentially all of the DNA strands will be this length. Typically, 20–30 cycles are carried out to effectively clone a region of DNA starting from genomic DNA; the rest of the genome remains unamplified, and its concentration is therefore negligible compared with that of the amplified region (Movie 8.2).

subsequent cycles, all the newly synthesized DNA molecules produced by the polymerase serve as templates for the next round of replication (**Figure 8–36**). Through this iterative amplification process, many copies of the original sequence can be made—billions after about 20 to 30 cycles.

PCR is now the method of choice for cloning relatively short DNA fragments (say, under 10,000 nucleotide pairs). Each cycle takes only about five minutes, and automation of the whole procedure enables cell-free cloning of a DNA fragment in a few hours. The original template for PCR can be either DNA or RNA, so this method can be used to obtain either a genomic clone (complete with introns and exons) or a cDNA copy of an mRNA (**Figure 8–37**).

PCR Is Also Used for Diagnostic and Forensic Applications

The PCR method is extraordinarily sensitive; it can detect a single DNA molecule in a sample if at least part of the sequence of that molecule is known. Trace amounts of RNA can be analyzed in the same way by first transcribing them into DNA with reverse transcriptase. For these reasons, PCR is frequently employed for uses that go beyond simple cloning. For example, it can be used to detect invading pathogens at very early stages of infection. In this case, short sequences complementary to a segment of the infectious agent's genome are used as primers and following many cycles of amplification, even a few copies of an invading bacterial or viral genome in a patient's sample can be detected (**Figure 8–38**). For many infections, PCR has replaced the use of antibodies against microbial molecules to

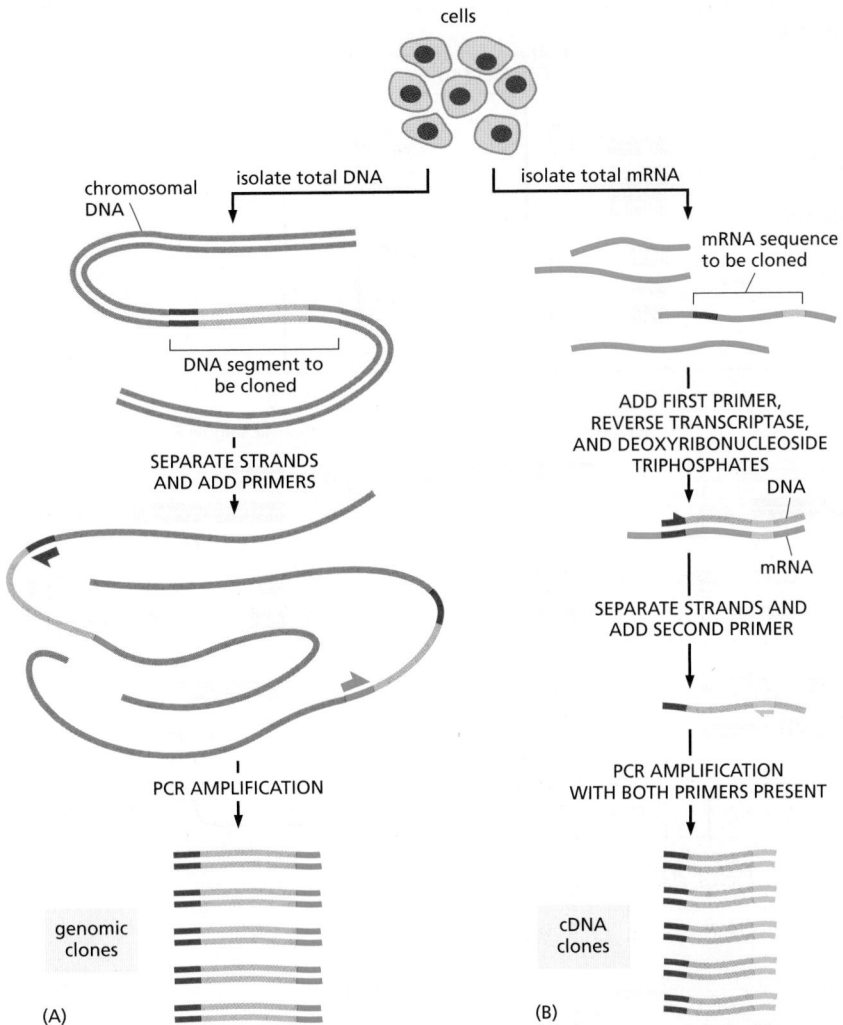

cells

chromosomal DNA

isolate total DNA

DNA segment to be cloned

SEPARATE STRANDS AND ADD PRIMERS

PCR AMPLIFICATION

genomic clones

(A)

isolate total mRNA

mRNA sequence to be cloned

ADD FIRST PRIMER, REVERSE TRANSCRIPTASE, AND DEOXYRIBONUCLEOSIDE TRIPHOSPHATES

DNA

mRNA

SEPARATE STRANDS AND ADD SECOND PRIMER

PCR AMPLIFICATION WITH BOTH PRIMERS PRESENT

cDNA clones

(B)

Figure 8–37 PCR can be used to obtain either genomic or cDNA clones. (A) To use PCR to clone a segment of chromosomal DNA, total genomic DNA is first purified from cells. PCR primers that flank the stretch of DNA to be cloned are added, and many cycles of PCR are completed (see Figure 8–36). Because only the DNA between (and including) the primers is amplified, PCR provides a way to obtain selectively any short stretch of chromosomal DNA in an effectively pure form. (B) To use PCR to obtain a cDNA clone of a gene, total mRNA is first purified from cells. The first primer is added to the population of mRNAs, and reverse transcriptase is used to make a DNA strand complementary to the specific RNA sequence of interest. The second primer is then added, and the DNA molecule is amplified through many cycles of PCR.

detect the presence of the invader. It is also used to verify the authenticity of a food source—for example, whether a sample of beef actually came from a cow.

Finally, PCR is now widely used in forensics. The method's extreme sensitivity allows forensic investigators to isolate DNA from minute traces of human blood or other tissue to obtain a *DNA fingerprint* of the person who left the sample behind.

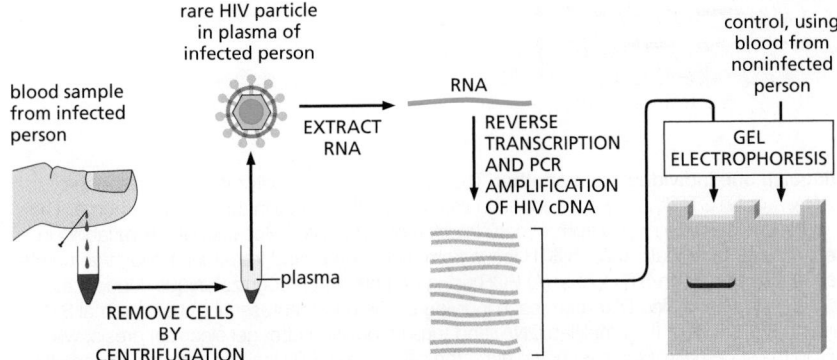

rare HIV particle in plasma of infected person

blood sample from infected person

control, using blood from noninfected person

RNA

EXTRACT RNA

REVERSE TRANSCRIPTION AND PCR AMPLIFICATION OF HIV cDNA

GEL ELECTROPHORESIS

plasma

REMOVE CELLS BY CENTRIFUGATION

Figure 8–38 PCR can be used to detect the presence of a viral genome in a sample of blood. Because of its ability to amplify enormously the signal from a single molecule of nucleic acid, PCR is an extraordinarily sensitive method for detecting trace amounts of virus in a sample of blood or tissue, without the need to purify the virus. For HIV, the virus that causes AIDS, the genome is a single-stranded molecule of RNA, as illustrated here. In addition to HIV, many other viruses that infect humans are now detected in this way.

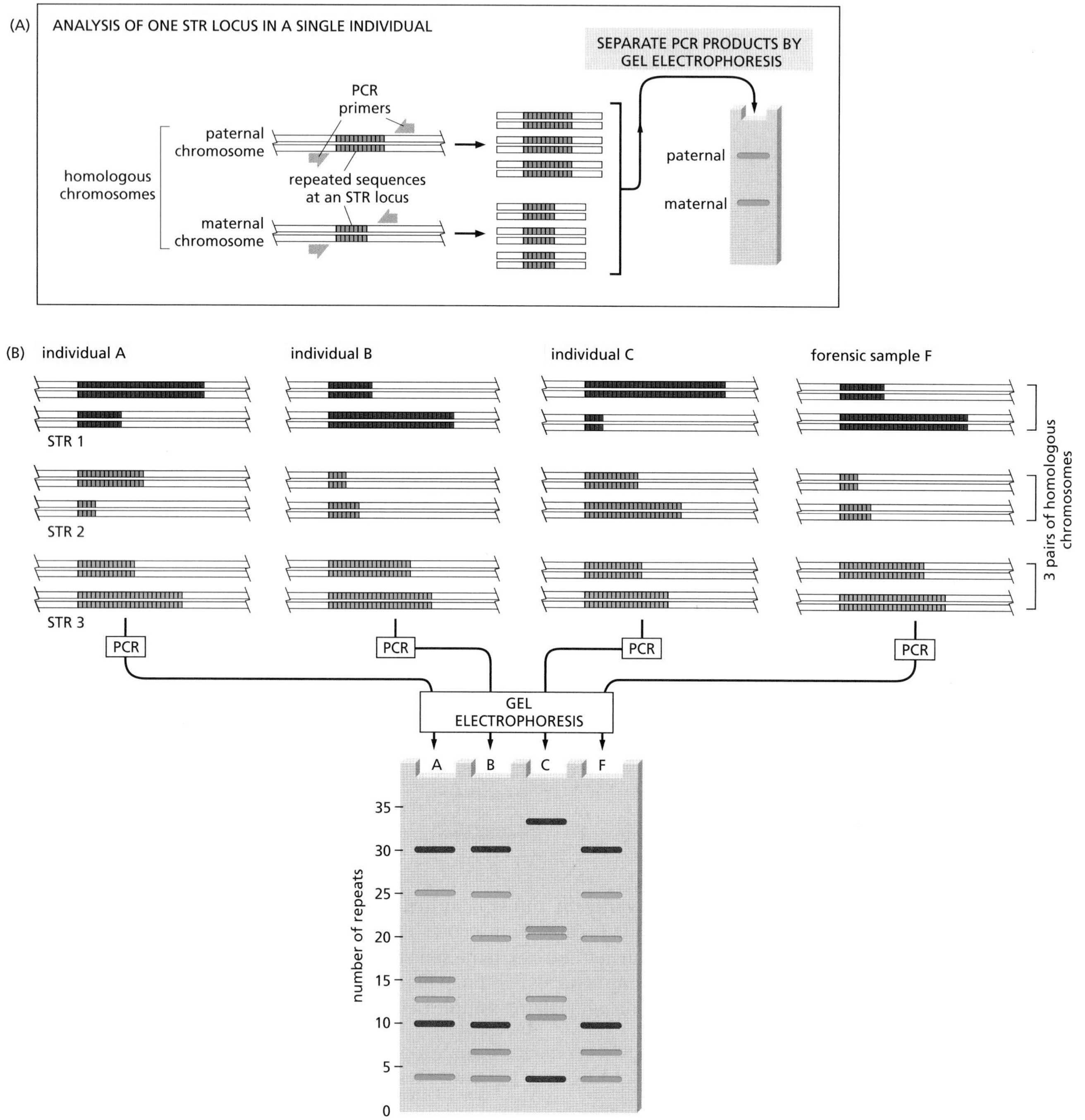

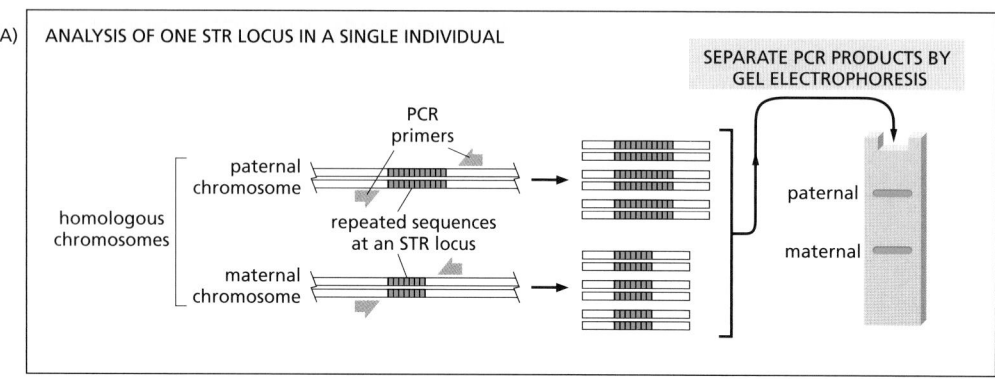

(A) ANALYSIS OF ONE STR LOCUS IN A SINGLE INDIVIDUAL

(B)   individual A   individual B   individual C   forensic sample F

STR 1

STR 2

STR 3

3 pairs of homologous chromosomes

Figure 8–39 PCR is used in forensic science to distinguish one individual from another. The DNA sequences analyzed are short tandem repeats (STRs) composed of sequences such as CACACA… or GTGTGT… STRs are found in various positions (loci) in the human genome. The number of repeats in each STR locus is highly variable in the population, ranging from 4 to 40 in different individuals. Because of the variability in these sequences, individuals will usually inherit a different number of repeats at each STR locus from their mother and from their father; two unrelated individuals, therefore, rarely contain the same pair of sequences at a given STR locus. (A) PCR using primers that recognize unique sequences on either side of one particular STR locus produces a pair of bands of amplified DNA from each individual, one band representing the maternal STR variant and the other representing the paternal STR variant. The length of the amplified DNA, and thus its position after gel electrophoresis, will depend on the exact number of repeats at the locus. (B) In the schematic example shown here, the same three STR loci are analyzed in samples from three suspects (individuals A, B, and C), producing six bands for each individual. Although different people can have several bands in common, the overall pattern is quite distinctive for each person. The band pattern can therefore serve as a DNA fingerprint to identify an individual nearly uniquely. The fourth lane (F) contains the products of the same PCR amplifications carried out on a hypothetical forensic DNA sample, which could have been obtained from a single hair or a tiny spot of blood left at a crime scene.

The more loci that are examined, the more confident one can be about the results. When examining the variability at 5–10 different STR loci, the odds that two random individuals would share the same fingerprint by chance are approximately one in 10 billion. In the case shown here, individuals A and C can be eliminated from inquiries, while B is a clear suspect. A similar approach is used routinely in paternity testing.

With the possible exception of identical twins, the genome of each human differs in DNA sequence from that of every other person on Earth. Using primer pairs targeted at genome sequences that are known to be highly variable in the human population, PCR makes it possible to generate a distinctive DNA fingerprint for any individual (**Figure 8–39**). Such forensic analyses can be used not only to help identify those who have done wrong, but also—equally important—to exonerate those who have been wrongfully accused.

Both DNA and RNA Can Be Rapidly Sequenced

Most current methods of manipulating DNA, RNA, and proteins rely on prior knowledge of the nucleotide sequence of the genome of interest. But how were these sequences determined in the first place? And how are new DNA and RNA molecules sequenced today? In the late 1970s, researchers developed several strategies for determining, simply and quickly, the nucleotide sequence of any purified DNA fragment. The one that became the most widely used is called **dideoxy sequencing** or **Sanger sequencing** (**Panel 8–1**). This method was used to determine the nucleotide sequence of many genomes, including those of *E. coli*, fruit flies, nematode worms, mice, and humans. Today, cheaper and faster methods are routinely used to sequence DNA, and even more efficient strategies are being developed (see Panel 8–1). The original "reference" sequence of the human genome, completed in 2003, cost over $1 billion and required many scientists from around the world working together for 13 years. The enormous progress made in the past decade makes it possible for a single person to complete the sequence of an individual human genome in less than a day.

The methods summarized in Panel 8–1 for rapidly sequencing DNA can also be applied to RNA. Although methods are being developed to sequence RNA directly, it is most commonly carried out by converting the RNA to complementary DNA (using reverse transcriptase) and using one of the methods described for DNA sequencing. It is important to keep in mind that although genomes remain the same from cell to cell and from tissue to tissue, the RNA produced from the genome can vary enormously. We will see later in this chapter that sequencing the entire repertoire of RNA from a cell or tissue (known as **deep RNA sequencing**, or **RNA-seq**) is a powerful way to understand how the information present in the genome is used by different cells under different circumstances. In the next section, we shall see how RNA-seq has also become a valuable tool for annotating genomes.

To Be Useful, Genome Sequences Must Be Annotated

Long strings of nucleotides, at first glance, reveal nothing about how this genetic information directs the development of a living organism—or even what types of DNA, protein, and RNA molecules are produced by a genome. The process of **genome annotation** attempts to mark out all the genes (both protein-coding and noncoding) in a genome and ascribe a role to each. It also seeks to understand more subtle types of genome information, such as the *cis*-regulatory sequences that specify the time and place that a given gene is expressed and whether its mRNA undergoes alternative splicing to produce different protein isotypes. Clearly, this is a daunting task, and we are far short of completing it for any form of life, even the simplest bacterium. For many organisms, we know the approximate number of genes, and, for very simple organisms, we understand the functions of about half their genes.

In this section, we discuss broadly how genes are identified in genome sequences and what clues we can discern about their roles from simply inspecting their sequences. Later in the chapter, we turn to the more difficult problem of experimentally determining gene function.

How does one begin to make sense of a genome sequence? The first step is usually to translate *in silico* the entire genome into protein. There are six different reading frames for any piece of double-stranded DNA (three on each strand). We saw in Chapter 6 that a random sequence of nucleotides, read in frame, will

DNA SEQUENCING

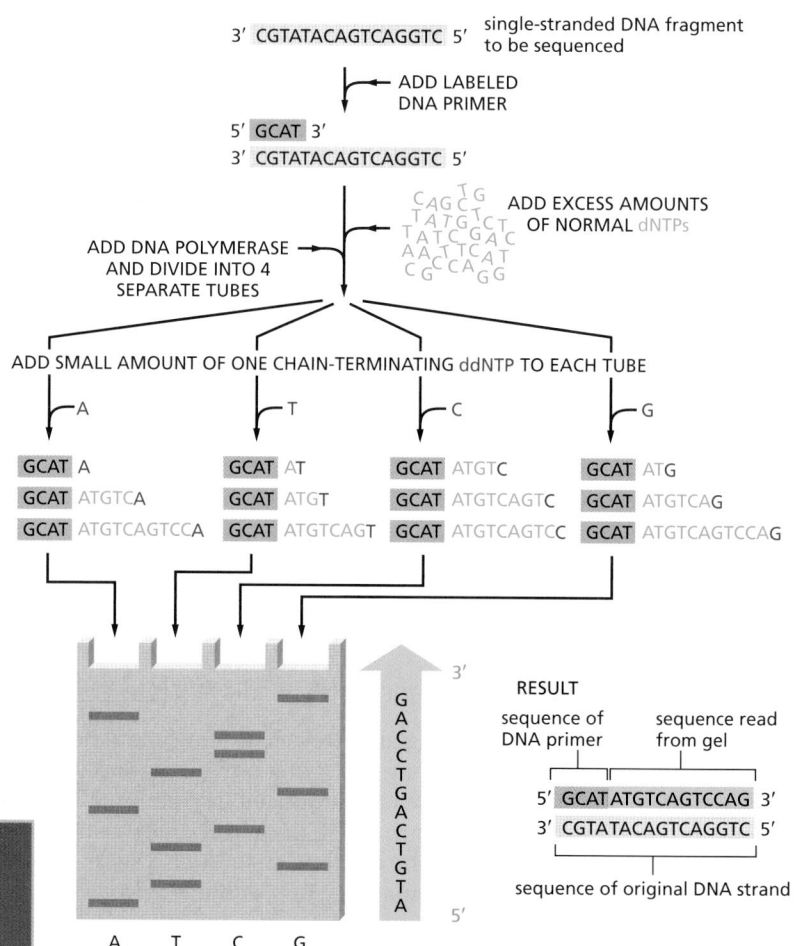

normal deoxyribonucleoside
triphosphate (dNTP)

3′ OH allows strand extension at 3′ end

chain-terminating dideoxyribonucleoside
triphosphate (ddNTP)

3′ H prevents strand extension at 3′ end

Dideoxy sequencing, or **Sanger sequencing** (named after the scientist who invented it), uses DNA polymerase, along with special chain-terminating nucleotides called dideoxyribonucleoside triphosphates (*left*), to make partial copies of the DNA fragment to be sequenced. These ddNTPs are derivatives of the normal deoxyribonucleoside triphosphates that lack the 3′ hydroxyl group. When incorporated into a growing DNA strand, they block further elongation of that strand.

MANUAL DIDEOXY SEQUENCING

To determine the complete sequence of a single-stranded fragment of DNA *(gray)*, the DNA is first hybridized with a short DNA primer *(orange)* that is labeled with a fluorescent dye or radioisotope. DNA polymerase and an excess of all four normal deoxyribonucleoside triphosphates (*blue* A, C, G, or T) are added to the primed DNA, which is then divided into four reaction tubes. Each of these tubes receives a small amount of a single chain-terminating dideoxyribonucleoside triphosphate (*red* A, C, G, or T). Because these will be incorporated only occasionally, each reaction produces a set of DNA copies that terminate at different points in the sequence. The products of these four reactions are separated by electrophoresis in four parallel lanes of a polyacrylamide gel (labeled here A, T, C, and G). In each lane, the bands represent fragments that have terminated at a given nucleotide but at different positions in the DNA. By reading off the bands in order, starting at the bottom of the gel and reading across all lanes, the DNA sequence of the newly synthesized strand can be determined (see Figure 8–25C). The sequence, which is given in the green arrow to the right of the gel, is complementary to the sequence of the original *gray* single-stranded DNA.

3′ CGTATACAGTCAGGTC 5′   single-stranded DNA fragment to be sequenced

ADD LABELED DNA PRIMER

5′ GCAT 3′
3′ CGTATACAGTCAGGTC 5′

ADD EXCESS AMOUNTS OF NORMAL dNTPs

ADD DNA POLYMERASE AND DIVIDE INTO 4 SEPARATE TUBES

ADD SMALL AMOUNT OF ONE CHAIN-TERMINATING ddNTP TO EACH TUBE

A

GCAT A
GCAT ATGTCA
GCAT ATGTCAGTCCA

T

GCAT AT
GCAT ATGT
GCAT ATGTCAGT

C

GCAT ATGTC
GCAT ATGTCAGTC
GCAT ATGTCAGTCC

G

GCAT ATG
GCAT ATGTCAG
GCAT ATGTCAGTCCAG

A   T   C   G

3′
G A C C T G A C T G T A
5′

RESULT

sequence of DNA primer   sequence read from gel

5′ GCAT ATGTCAGTCCAG 3′
3′ CGTATACAGTCAGGTC 5′

sequence of original DNA strand

AUTOMATED DIDEOXY SEQUENCING

mixture of DNA products, each containing a chain-terminating ddNTP labeled with a different fluorescent marker

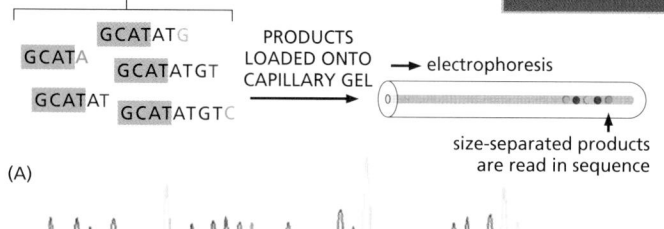

GCATA
GCATAT
GCATATG
GCATATGT
GCATATGTC

PRODUCTS LOADED ONTO CAPILLARY GEL → electrophoresis

size-separated products are read in sequence

(A)

(B) TTCTATAGTGTCACCTAAATAGCTTGGCGTAATCATGGT

Fully automated machines can run dideoxy sequencing reactions. (A) The automated method uses an excess amount of normal dNTPs plus a mixture of four different chain-terminating ddNTPs, each of which is labeled with a fluorescent tag of a different color. The reaction products are loaded onto a long, thin capillary gel and separated by electrophoresis. A camera (not shown) reads the color of each band as it moves through the gel and feeds the data to a computer that assembles the sequence. (B) A tiny part of the data from such an automated sequencing run. Each colored peak represents a nucleotide in the DNA sequence.

SEQUENCING WHOLE GENOMES

Shotgun sequencing: To determine the nucleotide sequence of a whole genome, the genomic DNA is first fragmented into small pieces and a genomic library is constructed, typically using plasmids and bacteria (see Figure 8–30). In *shotgun sequencing*, the nucleotide sequence of tens of thousands of individual clones is determined; the full genome sequence is then reconstructed by stitching together (*in silico*) the nucleotide sequence of each clone, using the overlaps between clones as a guide. The shotgun method works well for small genomes (such as those of viruses and bacteria) that lack repetitive DNA.

BAC clones: Most plant and animal genomes are large (often over 10^9 nucleotide pairs) and contain extensive amounts of repetitive DNA spread throughout the genome. Because a nucleotide sequence of a fragment of repetitive DNA will "overlap" every instance of the repeated DNA, it is difficult, if not impossible, to assemble the fragments into a unique order solely by the shotgun method.

To circumvent this problem, the human genome was first broken down into very large DNA fragments (each approximately 100,000 nucleotide pairs) and cloned into BACs (see p. 469). The order of the BACs along a chromosome was determined by comparing the pattern of restriction enzyme cleavage sites in a given BAC clone with that of the whole genome. In this way, a given BAC clone can be mapped, say, to the left arm of human Chromosome 3. Once a collection of BAC clones was obtained that spanned the entire genome, each individual BAC was sequenced by the shotgun method. At the end, the sequences of all the BAC inserts were stitched together using the knowledge of the position of each BAC insert in the human genome. In all, approximately 30,000 BAC clones were sequenced to complete the human genome.

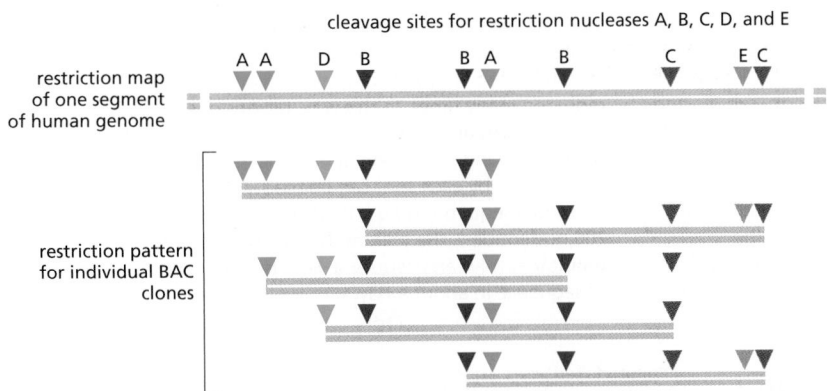

Thousands of genomes from individual humans have now been sequenced and it is not necessary to painstakingly reconstruct the order of DNA sequence "reads" each time; they are simply assembled using the order determined from the original human genome sequencing project. For this reason, *resequencing*, the term applied when the genome of a species is sequenced again (even though it may be from a different individual), is far easier than the original sequencing.

SECOND-GENERATION SEQUENCING TECHNOLOGIES

The dideoxy method made it possible to sequence the genomes of humans and most of the other organisms discussed in this book. But newer methods, developed since 2005, have made genome sequencing even more rapid—and very much cheaper. With these so-called second-generation sequencing methods, the cost of sequencing DNA has decreased dramatically. Not surprisingly, the number of genomes that have been sequenced has increased enormously. These rapid methods allow multiple genomes to be sequenced in parallel in a matter of weeks, enabling investigators to examine thousands of individual human genomes, catalog the variation in nucleotide sequences from people around the world, and uncover the mutations that increase the risk of various diseases, from cancer to autism. These methods have also made it possible to determine the genome sequence of extinct species, including Neanderthal man and the wooly mammoth (Movie 8.3). By sequencing genomes from many closely related species, they have also made it possible to understand the molecular basis of key evolutionary events in the tree of life, such as the "inventions" of multicellularity, vision, and language. The ability to rapidly sequence DNA has had major impacts on all branches of biology and medicine; it is almost impossible to imagine where we would be without it.

ILLUMINA® SEQUENCING

Several second-generation sequencing methods are now in wide use, and we will discuss two of the most common. Both rely on the construction of libraries of DNA fragments that represent—*in toto*—the DNA of the genome. Instead of using bacterial cells to generate these libraries, as we saw in Figure 8–30), they are made using PCR amplification of billions of DNA fragments, each attached to a solid support. The amplification is carried out so that the PCR-generated copies, instead of floating away in solution, remain bound in proximity to the original DNA fragment. This process generates clusters of DNA fragments, where each cluster contains about 1000 identical copies of a small bit of the genome. These clusters—a billion of which can fit in a single slide or plate—are then sequenced at the same time; that is, in parallel.

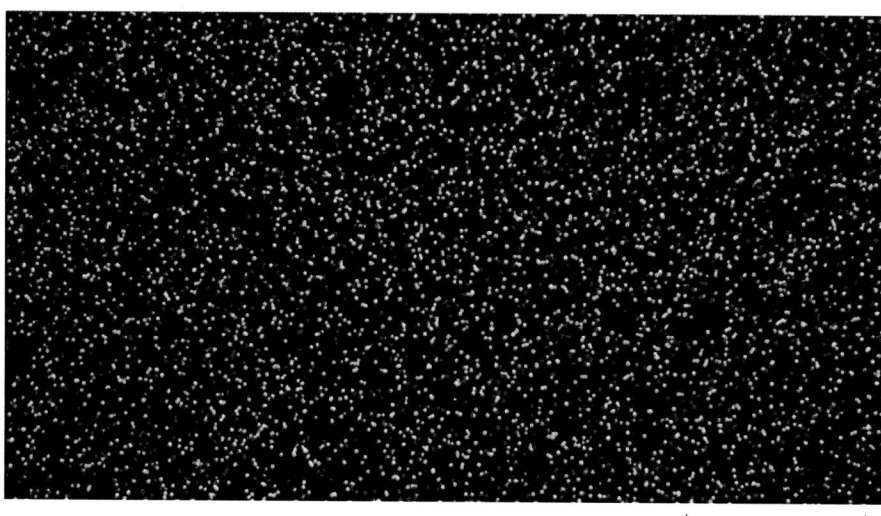

A slide showing individual clusters of PCR-generated DNA molecules. Each cluster carries about 1000 identical DNA molecules; the four colors are produced by incorporation of C, G, A, or T, each of which has a different color fluorophore. The image has been taken just after a fluorescent nucleotide has been incorporated into each growing DNA chain. (Courtesy of Illumina Inc.)

100 µm

One method, known as *Illumina sequencing*, is based on the dideoxy method described above, but it incorporates several innovations. Here, each nucleotide is attached to a removable fluorescent molecule (a different color for each of the four bases) as well as a special chain-terminating chemical adduct: instead of a 3′-OH group, as in conventional dideoxy sequencing, the nucleotides carry a chemical group that blocks elongation by DNA polymerase but which can be removed chemically. Sequencing is then carried out as follows: the four fluorescently labeled nucleotides along with DNA polymerase are added to billions of DNA clusters immobilized on a slide. Only the appropriate nucleotide (that is complementary to the next nucleotide in the template) is covalently incorporated at each cluster; the unincorporated nucleotides are washed away, and a high-resolution digital camera takes an image that registers which of the four nucleotides was added to the chain at each cluster. The fluorescent label and the 3′-OH blocking group are then removed enzymatically, washed away, and the process is repeated many times. In this way, billions of sequencing reactions are carried out simultaneously. By keeping track of the color changes occuring at each cluster, the DNA sequence represented by each cluster can be read. Although each individual sequence read is relatively short (approximately 200 nucleotides), the billions that are carried out simultaneously can produce several human genomes worth of sequence in about a day.

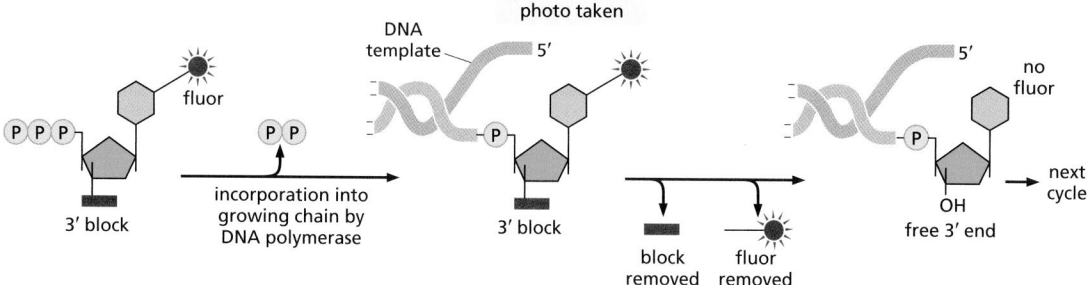

Principle behind Illumina sequencing. This reaction is carried out stepwise, on billions of DNA clusters at once. The method relies on a color digital camera that rapidly scans all the DNA clusters after each round of modified nucleotide incorporation. The DNA sequence of each cluster is then determined by the sequence of color changes it undergoes as the elongation reaction proceeds stepwise. Each round of modified nucleotide incorporation, image acquisition, and removal of the 3′ block and the fluorescent group takes less than an hour. Each cluster on the slide contains many copies of different, random bits of a genome; in preparing the clusters, a DNA sequence (specified by the experimenter) is joined to each copy in every cluster, and a primer complementary to this sequence is used to begin the elongation reaction by DNA polymerase.

ION TORRENT™ SEQUENCING

Another widely used strategy for rapid DNA sequencing is called the *ion torrent* method. Here, a genome is fragmented, and the individual fragments are attached to microscopic beads. Using PCR, each DNA fragment is then amplified so that copies of it eventually coat the bead to which it was initially attached. This process produces a library of billions of individual beads, each covered with identical copies of a particular DNA fragment.

Like eggs in a carton, the beads are placed into individual wells on an array that can hold a billion beads in a square inch. Beginning with a primer, DNA synthesis is then initiated on each bead. A hydrogen ion (H$^+$) is released (along with pyrophosphate) each time a nucleotide is incorporated into a growing DNA chain (see Figure 5–3), and the ion torrent method is based on this simple fact. Each of the four nucleotides is washed in, one at a time, over the array of beads; when a nucleotide is incorporated in the DNA of a given bead, the release of an H$^+$ ion changes the pH, which is registered by a semiconductor chip placed beneath the array of wells. In this way, the DNA sequence on a given bead can be read from the pattern of pH changes observed as nucleotides are washed over them. Like a high-resolution sensor in a digital camera, the ion torrent semiconductor chip can register enormous amounts of information and can thus keep track of billions of parallel sequencing reactions. Using this technology it is currently possible, using a single chip, to determine the nucleotide sequences of several human genomes in just a few hours.

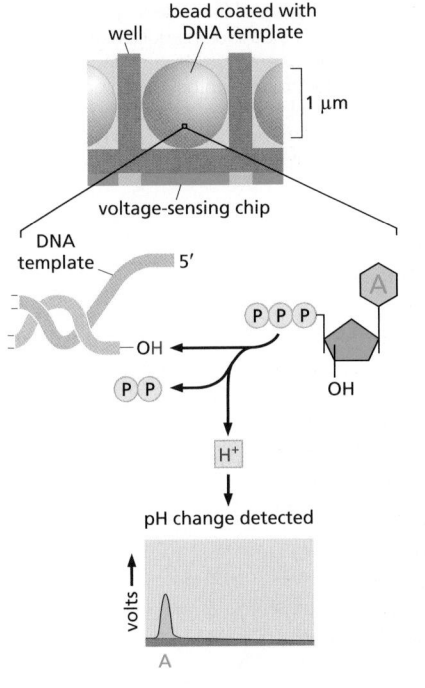

DNA sequencing by the ion torrent method. Beads, each coated with a DNA molecule that has been amplified many times, are placed in wells along with primers and DNA polymerase. As nucleotides are sequentially washed over the beads, those incorporated by the polymerase cause a pH change. In the example shown, an A is incorporated; thus, the template must have a T in this position. As the four nucleotides are sequentially washed over the beads, the sequence of the DNA on each bead can be "read" by the pattern of pH fluctuations. Billions of beads are monitored at once by a voltage-sensitive semiconductor chip placed below the array of beads.

THE FUTURE OF DNA SEQUENCING

Even newer, potentially faster, methods of sequencing DNA are being developed. Some of these "third-generation" technologies bypass the DNA amplification steps altogether and determine the sequence of single molecules of DNA. In one technique, a DNA molecule is pushed through a tiny channel, like a thread through the eye of a needle. As the DNA molecule moves through the pore, it generates electrical currents that depend on its sequence of nucleotides; the pattern of currents can then be used to deduce the nucleotide sequence. Other methods visualize single DNA molecules using electron or atomic force microscopy; the nucleotide sequence is read from the small differences in the "appearance" of the DNA as it is scanned. Finally, another method is based on immobilizing a single DNA polymerase molecule (with a template) and measuring the "dwell" time of each of the four nucleotides, which are labeled with different removable fluorescent dyes. Nucleotides that reside longer on the polymerase (before their dye is removed) are those incorporated by the polymerase. Although the two methods we have described in detail (Illumina and ion torrent) are now used extensively, it is likely that faster and cheaper methods will continue to be developed.

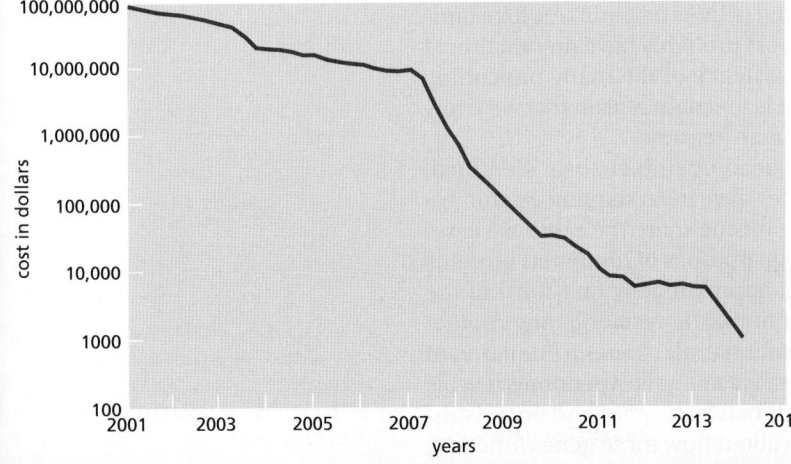

Shown here are the costs of sequencing a human genome, which was $100 million in 2001 and about a thousand dollars by the end of 2014. (Data from the National Human Genome Research Initiative.)

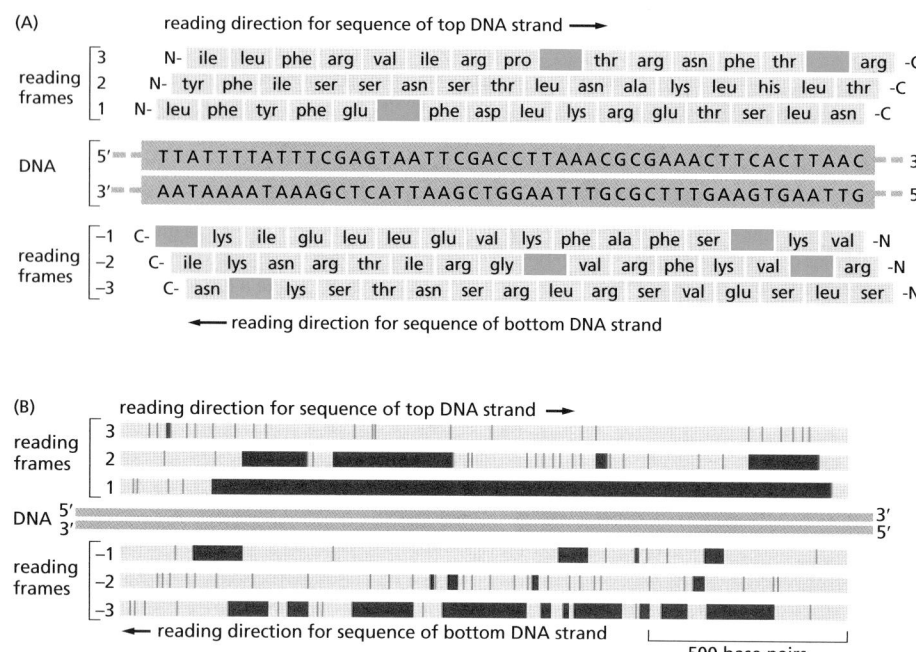

Figure 8–40 Finding the regions in a DNA sequence that encode a protein. (A) Any region of the DNA sequence can, in principle, code for six different amino acid sequences, because any one of three different reading frames can be used to interpret the nucleotide sequence on each strand. Note that a nucleotide sequence is always read in the 5′-to-3′ direction and encodes a polypeptide from the N-terminus to the C-terminus. For a random nucleotide sequence read in a particular frame, a stop signal for protein synthesis is encountered, on average, about once every 20 amino acids. In this sample sequence of 48 base pairs, each such signal (*stop codon*) is colored *blue*, and only reading frame 2 lacks a stop signal. (B) Search of a 1700-base-pair DNA sequence for a possible protein-encoding sequence. The information is displayed as in (A), with each stop signal for protein synthesis denoted by a *blue line*. In addition, all of the regions between possible start and stop signals for protein synthesis (see pp. 347–349) are displayed as *red bars*. Only reading frame 1 actually encodes a protein, which is 475 amino acid residues long.

contain a stop codon about every 20 amino acids; protein-coding regions will, in contrast, usually contain much longer stretches without stop codons (**Figure 8–40**). Known as **open reading frames** (**ORFs**), these usually signify bona fide protein-coding genes. This assignment is often "double-checked" by comparing the ORF amino acid sequence to the many databases of documented proteins from other species. If a match is found, even as imperfect one, it is very likely that the ORF will code for a functional protein (see Figure 8–23).

This strategy works very well for compact genomes, where intron sequences are rare and ORFs often extend for many hundreds of amino acids. However, in many animals and plants, the average exon size is 150–200 nucleotide pairs (see Figure 6–31) and additional information is usually required to unambiguously locate all the exons of a gene. Although it is possible to search genomes for splicing signals and other features that help to identify exons (codon bias, for example), one of the most powerful methods is simply to sequence the total RNA produced from the genome in living cells. As can be seen in Figure 7–3, this RNA-seq information, when mapped onto the genome sequence, can be used to accurately locate all the introns and exons of even complex genes. By sequencing total RNA from different cell types, it is also possible to identify cases of alternative splicing (see Figure 6–26).

RNA-seq also identifies noncoding RNAs produced by a genome. Although the function of some of them can be readily recognized (tRNAs or snoRNAs, for example), many have unknown functions and still others probably have no function at all (discussed in Chapter 7, pp. 429–436). The existence of the many noncoding RNAs and our relative ignorance of their function is the main reason that we know only the approximate number of genes in the human genome.

But even for protein-coding genes that have been unambiguously identified, we still have much to learn. Thousands of genomes have been sequenced, and we know from *comparative genomics* that many organisms share the same basic set of proteins. However, the functions of a very large number of identified proteins remain unknown. Depending on the organism, approximately one-third of the proteins encoded by a sequenced genome do not clearly resemble any protein that has been studied biochemically. This observation underscores a limitation of the emerging field of genomics: although comparative analysis of genomes reveals a great deal of information about the relationships between genes and organisms, it often does not provide immediate information about how these genes function, or what roles they have in the physiology of an organism. Comparison of the full gene complement of several thermophilic bacteria, for example, does not reveal why these bacteria thrive at temperatures exceeding 70°C. And examination of the genome of the incredibly radioresistant bacterium *Deinococcus radiodurans* does not explain how this organism can survive a blast of radiation that can shatter glass. Further biochemical and genetic studies, like those described in the other sections of this chapter, are required to determine how genes, and the proteins they produce, function in the context of living organisms.

DNA Cloning Allows Any Protein to be Produced in Large Amounts

In the last section, we saw how protein-coding genes can be identified in genome sequences. Using the genetic code (and provided the intron and exon boundaries are known), the amino acid sequence of any protein coded in a genome can be deduced. As was discussed earlier, this sequence can often provide an important clue to the protein's function if found to be similar to the amino acid sequence of a protein that has already been studied (see Figure 8–23). Although this strategy is often successful, it typically provides only the likely biochemical function of the protein; for example, whether the protein resembles a kinase or a protease. It usually remains for the experimenter to verify (or refute) this assignment and, most importantly, to discover the protein's biological function in the whole organism; that is, to what attributes of the organism does the kinase or the protease contribute and in what molecular pathways does it function? Nowadays, most new proteins are "discovered" through genome sequencing, and it often remains a great challenge to ascertain their functions.

An important approach in determining gene function is to alter the gene (or in some cases, its expression pattern), to put the altered copy back into the germ line of the organism, and to deduce the function of the normal gene by the changes caused by its alteration. Various techniques to implement this strategy are discussed in the next section of this chapter. But it is equally important to study the biochemical and structural properties of a gene product, as outlined in the first part of this chapter. One of the most important contributions of DNA cloning to cell and molecular biology is the ability to produce any protein, even the rare ones, in nearly unlimited amounts—as long as the gene coding for it is known. Such high-level production is usually carried out in living cells using expression vectors (**Figure 8–41**). These are generally plasmids that have been designed to produce a large amount of stable mRNA that can be efficiently translated into protein when the plasmid is introduced into bacterial, yeast, insect, or mammalian cells. To prevent the high level of the foreign protein from interfering with the cell's growth, the expression vector is often designed to delay the synthesis of the foreign mRNA and protein until shortly before the cells are harvested and lysed (**Figure 8–42**).

Because the desired protein made from an expression vector is produced inside a cell, it must be purified away from the host-cell proteins by chromatography following cell lysis; but because it is such a plentiful species in the cell (often 1–10% of the total cell protein), the purification is usually easy to accomplish in only a few steps. As we saw in the first part of this chapter, many expression

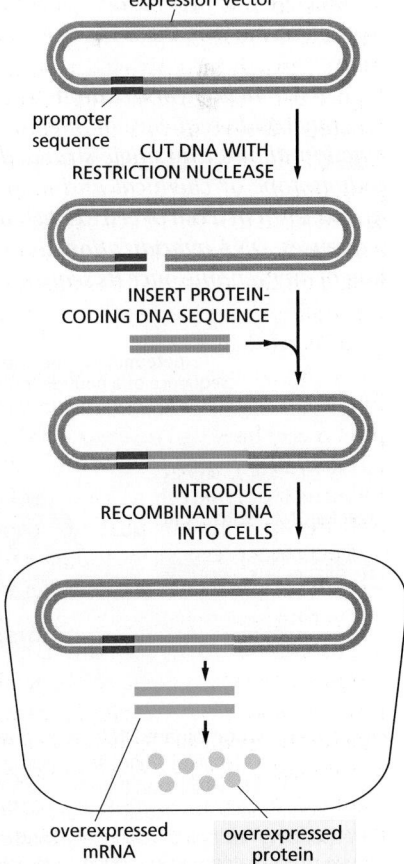

Figure 8–41 Production of large amounts of a protein from a protein-coding DNA sequence cloned into an expression vector and introduced into cells. A plasmid vector has been engineered to contain a highly active promoter, which causes unusually large amounts of mRNA to be produced from an adjacent protein-coding gene inserted into the plasmid vector. Depending on the characteristics of the cloning vector, the plasmid is introduced into bacterial, yeast, insect, or mammalian cells, where the inserted gene is efficiently transcribed and translated into protein. If the gene to be overexpressed has no introns (typical for genes from bacteria, archaea, and simple eukaryotes), it can simply be cloned from genomic DNA by PCR. For cloned animal and plant genes, it is often more convenient to obtain the gene as cDNA, either from a cDNA library (see Figure 8–32) or cloned directly by PCR from RNA isolated from the organism (see Figure 8–37). Alternatively, the DNA coding for the protein can be made by chemical synthesis (see p. 472).

vectors have been designed to add a molecular tag—a cluster of histidine residues or a small marker protein—to the expressed protein to facilitate easy purification by affinity chromatography (see Figure 8–11). A variety of expression vectors is available, each engineered to function in the type of cell in which the protein is to be made.

This technology is also used to make large amounts of many medically useful proteins, including hormones (such as insulin and growth factors) used as human pharmaceuticals, and viral coat proteins for use in vaccines. Expression vectors also allow scientists to produce many proteins of biological interest in large enough amounts for detailed structural studies. Nearly all three-dimensional protein structures depicted in this book are of proteins produced in this way. Recombinant DNA techniques thus allow scientists to move with ease from protein to gene, and vice versa, so that the functions of both can be explored on multiple fronts (**Figure 8–43**).

Summary

DNA cloning allows a copy of any specific part of a DNA or RNA sequence to be selected from the millions of other sequences in a cell and produced in unlimited amounts in pure form. DNA sequences can be amplified after breaking up chromosomal DNA and inserting the resulting DNA fragments into the chromosome of a self-replicating genetic element such as a plasmid. The resulting "genomic DNA library" is housed in millions of bacterial cells, each carrying a different cloned DNA fragment. Individual cells from this library that are allowed to proliferate produce large amounts of a single cloned DNA fragment. Bypassing cloning vectors and bacterial cells altogether, the polymerase chain reaction (PCR) allows DNA cloning to be performed directly with a DNA polymerase and DNA primers—provided that the DNA sequence of interest is already known.

The procedures used to obtain DNA clones that correspond in sequence to mRNA molecules are the same, except that a DNA copy of the mRNA sequence, called cDNA, is first made. Unlike genomic DNA clones, cDNA clones lack intron sequences, making them the clones of choice for analyzing the protein product of a gene.

Nucleic acid hybridization reactions provide a sensitive means of detecting any nucleotide sequence of interest. The enormous specificity of this hybridization reaction allows any single-stranded sequence of nucleotides to be labeled with a radioisotope or chemical and used as a probe to find a complementary partner strand, even in a cell or cell extract that contains millions of different DNA and RNA sequences. DNA hybridization also makes it possible to use PCR to amplify any section of any genome once its sequence is known.

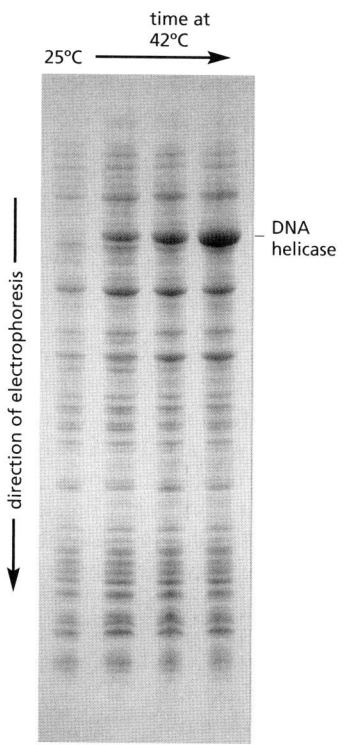

Figure 8–42 Production of large amounts of a protein by using a plasmid expression vector. In this example, an expression vector that overproduces a DNA helicase has been introduced into bacteria. In this expression vector, transcription from this coding sequence is under the control of a viral promoter that becomes active only at a temperature of 37°C or higher. The total cell protein, either from bacteria grown at 25°C (no helicase protein made) or after a shift of the same bacteria to 42°C for up to 2 hours (helicase protein has become the most abundant protein species in the lysate), has been analyzed by SDS polyacrylamide-gel electrophoresis. (Courtesy of Kevin Hacker.)

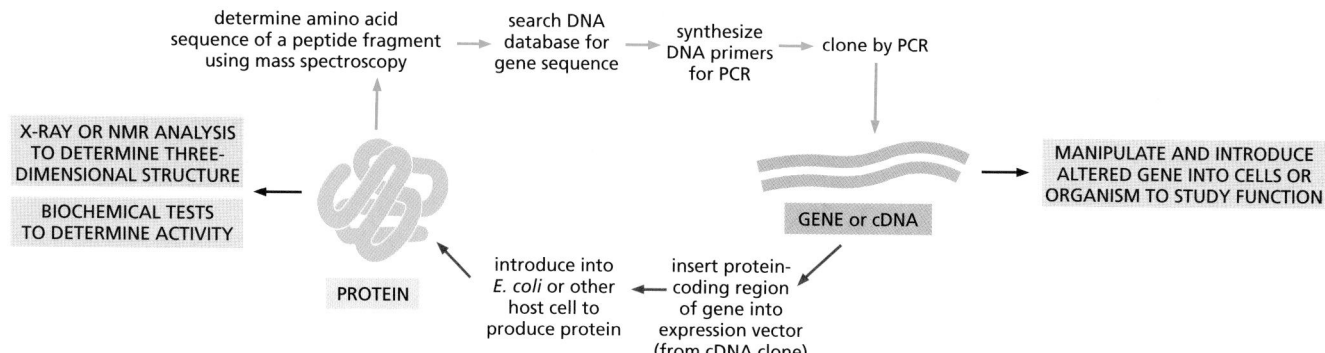

Figure 8–43 Recombinant DNA techniques make it possible to move experimentally from gene to protein and from protein to gene. If a gene has been identified (*right*), its protein-coding sequence can be inserted into an expression vector to produce large quantities of the protein (see Figure 8–41), which can then be studied biochemically or structurally. If a protein has been purified based on its biochemical properties, mass spectrometry (see Figure 8–18) can be used to obtain a partial amino acid sequence, which is used to search a genome sequence for the corresponding nucleotide sequence. The complete gene can then be cloned by PCR from a sequenced genome (see Figure 8–37). The gene can also be manipulated and introduced into cells or organisms to study its function, a topic covered in the next section of this chapter.

The nucleotide sequence of any genome can be determined rapidly and simply by using highly automated techniques based on several different strategies. Comparison of the genome sequences of different organisms allows us to trace the evolutionary relationships among genes and organisms, and it has proved valuable for discovering new genes and predicting their functions.

Taken together, these techniques for analyzing and manipulating DNA have made it possible to sequence, identify, and isolate genes from any organism of interest. Related technologies allow scientists to produce the protein products of these genes in the large quantities needed for detailed analyses of their structure and function, as well as for medical purposes.

STUDYING GENE EXPRESSION AND FUNCTION

Ultimately, one wishes to determine how genes—and the proteins they encode—function in the intact organism. Although it may seem counterintuitive, one of the most direct ways to find out what a gene does is to see what happens to the organism when that gene is missing. Studying mutant organisms that have acquired changes or deletions in their nucleotide sequences is a time-honored practice in biology and forms the basis of the important field of **genetics**. Because mutations can disrupt cell processes, mutants often hold the key to understanding gene function. In the classical genetic approach, one begins by isolating mutants that have an interesting or unusual appearance: fruit flies with white eyes or curly wings, for example. Working backward from the **phenotype**—the appearance or behavior of the individual—one then determines the organism's **genotype**, the form of the gene responsible for that characteristic (**Panel 8–2**).

Today, with numerous genome sequences available, the exploration of gene function often begins with a DNA sequence. Here, the challenge is to translate sequence into function. One approach, discussed earlier in the chapter, is to search databases for well-characterized proteins that have similar amino acid sequences to the protein encoded by a new gene. From there, the protein (or for noncoding genes, the RNA molecule) can be overexpressed and purified and the methods described in the first part of this chapter can be employed to study its three-dimensional structure and its biochemical properties. But to determine directly a gene's function in a cell or organism, the most effective approach involves studying mutants that either lack the gene or express an altered version of it. Determining which cell processes have been disrupted or compromised in such mutants will usually shed light on a gene's biological role.

In this section, we describe several approaches to determining a gene's function, starting either from an individual with an interesting phenotype or from a DNA sequence. We begin with the classical genetic approach, which starts with a *genetic screen* for isolating mutants of interest and then proceeds toward identification of the gene or genes responsible for the observed phenotype. We then describe the set of techniques that are collectively called *reverse genetics*, in which one begins with a gene or gene sequence and attempts to determine its function. This approach often involves some intelligent guesswork—searching for similar sequences in other organisms or determining when and where a gene is expressed—as well as generating mutant organisms and characterizing their phenotype.

Classical Genetics Begins by Disrupting a Cell Process by Random Mutagenesis

Before the advent of gene cloning technology, most genes were identified by the abnormalities produced when the gene was mutated. Indeed, the very concept of the gene was deduced from the heritability of such abnormalities. This classical genetic approach—identifying the genes responsible for mutant phenotypes—is most easily performed in organisms that reproduce rapidly and are amenable to genetic manipulation, such as bacteria, yeasts, nematode worms, and fruit flies. Although spontaneous mutants can sometimes be found by examining extremely

GENES AND PHENOTYPES

Gene: a functional unit of inheritance, usually corresponding to the segment of DNA coding for a single protein.

Genome: all of an organism's DNA sequences.

locus: the site of the gene in the genome

alleles: alternative forms of a gene

Wild-type: the normal, naturally occurring type

Mutant: differing from the wild-type because of a genetic change (a mutation)

GENOTYPE: the specific set of alleles forming the genome of an individual

PHENOTYPE: the visible character of the individual

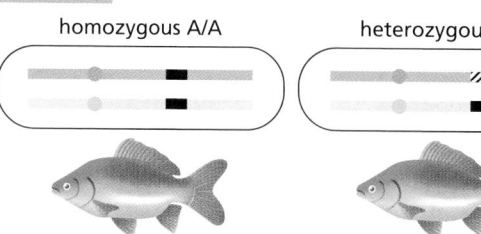

homozygous A/A   heterozygous a/A   homozygous a/a

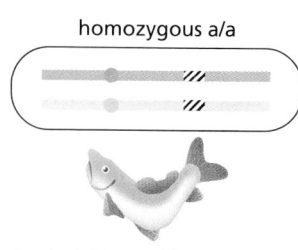

allele A is dominant (relative to a); allele a is recessive (relative to A)

In the example above, the phenotype of the heterozygote is the same as that of one of the homozygotes; in cases where it is different from both, the two alleles are said to be co-dominant.

CHROMOSOMES

a chromosome at the beginning of the cell cycle, in G$_1$ phase; the single long bar represents one long double helix of DNA

centromere

short "p" arm   long "q" arm

a chromosome near the end of the cell cycle, in metaphase; it is duplicated and condensed, consisting of two identical sister chromatids (each containing one DNA double helix) joined at the centromere.

short "p" arm   long "q" arm

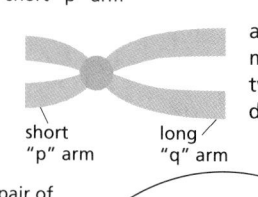

pair of autosomes

maternal 1

paternal 1

paternal 3

maternal 3

paternal 2

maternal 2

Y

X

sex chromosomes

A normal diploid chromosome set, as seen in a metaphase spread, prepared by bursting open a cell at metaphase and staining the scattered chromosomes. In the example shown schematically here, there are three pairs of autosomes (chromosomes inherited symmetrically from both parents, regardless of sex) and two sex chromosomes—an X from the mother and a Y from the father. The numbers and types of sex chromosomes and their role in sex determination are variable from one class of organisms to another, as is the number of pairs of autosomes.

THE HAPLOID–DIPLOID CYCLE OF SEXUAL REPRODUCTION

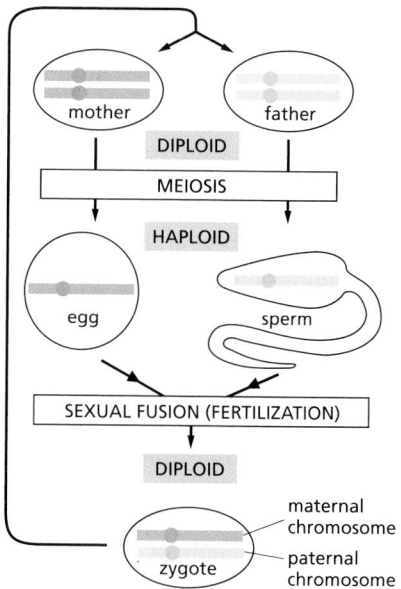

mother   father

DIPLOID

MEIOSIS

HAPLOID

egg   sperm

SEXUAL FUSION (FERTILIZATION)

DIPLOID

maternal chromosome

paternal chromosome

zygote

For simplicity, the cycle is shown for only one chromosome/chromosome pair.

MEIOSIS AND GENETIC RECOMBINATION

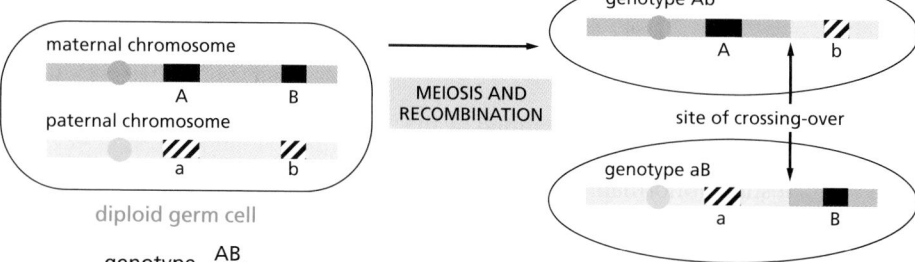

maternal chromosome

A   B

paternal chromosome

a   b

diploid germ cell

genotype $\dfrac{AB}{ab}$

MEIOSIS AND RECOMBINATION

genotype Ab

A   b

site of crossing-over

genotype aB

a   B

haploid gametes (eggs or sperm)

The greater the distance between two loci on a single chromosome, the greater is the chance that they will be separated by crossing over occurring at a site between them. If two genes are thus reassorted in x% of gametes, they are said to be separated on a chromosome by a genetic map distance of x map units (or x centimorgans).

TYPES OF MUTATIONS

POINT MUTATION: maps to a single site in the genome, corresponding to a single nucleotide pair or a very small part of a single gene

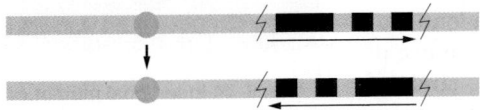

INVERSION: inverts a segment of a chromosome

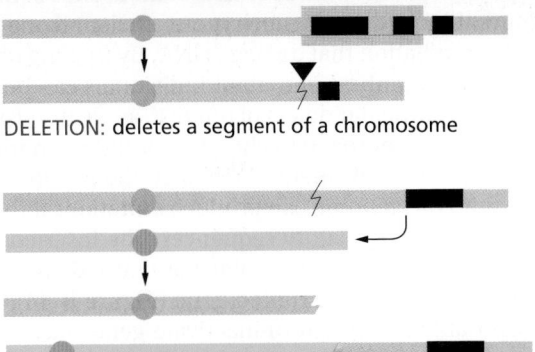

DELETION: deletes a segment of a chromosome

TRANSLOCATION: breaks off a segment from one chromosome and attaches it to another

lethal mutation: causes the developing organism to die prematurely.

conditional mutation: produces its phenotypic effect only under certain conditions, called the *restrictive* conditions. Under other conditions—the *permissive* conditions—the effect is not seen. For a *temperature-sensitive* mutation, the restrictive condition typically is high temperature, while the permissive condition is low temperature.

loss-of-function mutation: either reduces or abolishes the activity of the gene. These are the most common class of mutations. Loss-of-function mutations are usually *recessive*—the organism can usually function normally as long as it retains at least one normal copy of the affected gene.

null mutation: a loss-of-function mutation that completely abolishes the activity of the gene.

gain-of-function mutation: increases the activity of the gene or makes it active in inappropriate circumstances; these mutations are usually *dominant.*

dominant-negative mutation: dominant-acting mutation that blocks gene activity, causing a loss-of-function phenotype even in the presence of a normal copy of the gene. This phenomenon occurs when the mutant gene product interferes with the function of the normal gene product.

suppressor mutation: suppresses the phenotypic effect of another mutation, so that the double mutant seems normal. An *intragenic* suppressor mutation lies within the gene affected by the first mutation; an *extragenic* suppressor mutation lies in a second gene—often one whose product interacts directly with the product of the first.

TWO GENES OR ONE?

Given two mutations that produce the same phenotype, how can we tell whether they are mutations in the same gene? If the mutations are recessive (as they most often are), the answer can be found by a complementation test.

In the simplest type of complementation test, an individual who is homozygous for one mutation is mated with an individual who is homozygous for the other. The phenotype of the offspring gives the answer to the question.

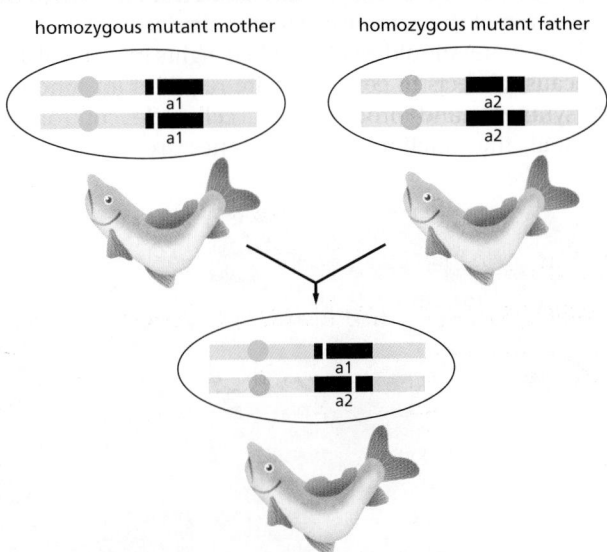

COMPLEMENTATION: MUTATIONS IN TWO DIFFERENT GENES	NONCOMPLEMENTATION: TWO INDEPENDENT MUTATIONS IN THE SAME GENE

homozygous mutant mother   homozygous mutant father

homozygous mutant mother   homozygous mutant father

hybrid offspring shows normal phenotype: one normal copy of each gene is present

hybrid offspring shows mutant phenotype: no normal copies of the mutated gene are present

large populations—thousands or tens of thousands of individual organisms—isolating mutant individuals is much more efficient if one generates mutations with chemicals or radiation that damage DNA. By treating organisms with such mutagens, very large numbers of mutant individuals can be created quickly and then screened for a particular defect of interest, as we discuss shortly.

An alternative approach to chemical or radiation mutagenesis is called *insertional mutagenesis*. This method relies on the fact that exogenous DNA inserted randomly into the genome can produce mutations if the inserted fragment interrupts a gene or its regulatory sequences. The inserted DNA, whose sequence is known, then serves as a molecular tag that aids in the subsequent identification and cloning of the disrupted gene (**Figure 8–44**). In *Drosophila*, the use of the transposable P element to inactivate genes has revolutionized the study of gene function in the fly. Transposable elements (see Table 5–4, p. 288) have also been used to generate mutations in bacteria, yeast, mice, and the flowering plant *Arabidopsis*.

Genetic Screens Identify Mutants with Specific Abnormalities

Once a collection of mutants in a model organism such as yeast or fly has been produced, one generally must examine thousands of individuals to find the altered phenotype of interest. Such a search is called a **genetic screen**, and the larger the genome, the less likely it is that any particular gene will be mutated. Therefore, the larger the genome of an organism, the bigger the screening task becomes. The phenotype being screened for can be simple or complex. Simple phenotypes are easiest to detect: one can screen many organisms rapidly, for example, for mutations that make it impossible for the organism to survive in the absence of a particular amino acid or nutrient.

More complex phenotypes, such as defects in learning or behavior, may require more elaborate screens (**Figure 8–45**). But even genetic screens that are used to dissect complex physiological systems can be simple in design, which permits the simultaneous examination of large numbers of mutants. As an example, one particularly elegant screen was designed to search for genes involved in visual processing in zebrafish. The basis of this screen, which monitors the fishes' response to motion, is a change in behavior. Wild-type fish tend to swim in the direction of a perceived motion, whereas mutants with defects in their visual processing systems swim in random directions—a behavior that is easily detected. One mutant discovered in this screen is called *lakritz*, which is missing 80% of the retinal ganglion cells that help to relay visual signals from the eye to the brain. As the cellular organization of the zebrafish retina is similar to that of all vertebrates, the study of such mutants should also provide insights into visual processing in humans.

Because defects in genes that are required for fundamental cell processes—RNA synthesis and processing or cell-cycle control, for example—are usually lethal, the functions of these genes are often studied in individuals with

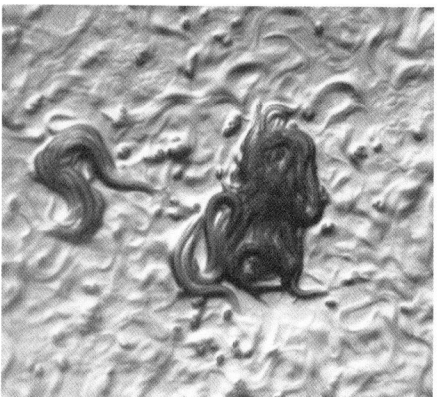

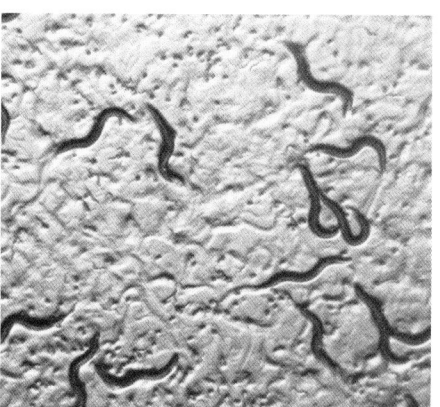

Figure 8–44 **Insertional mutant of the snapdragon,** *Antirrhinum*. A mutation in a single gene coding for a regulatory protein causes leafy shoots (*left*) to develop in place of flowers, which occur in the normal plant (*right*). The mutation causes cells to adopt a character that would be appropriate to a different part of the normal plant, so instead of a flower, the cells produce a leafy shoot. (Courtesy of Enrico Coen and Rosemary Carpenter.)

1 mm

Figure 8–45 **A behavioral phenotype detected in a genetic screen.** (A) Wild-type *C. elegans* engage in social feeding. The worms migrate around until they encounter their neighbors and commence feeding on bacteria. (B) Mutant animals feed by themselves. (Courtesy of Cornelia Bargmann, *Cell* 94: cover, 1998.)

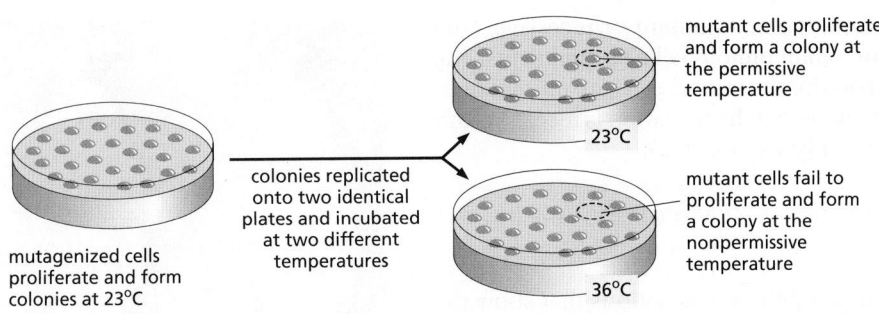

Figure 8–46 **Screening for temperature-sensitive bacterial or yeast mutants.** Mutagenized cells are plated out at the permissive temperature. They divide and form colonies, which are transferred to two identical Petri dishes by replica plating. One of these plates is incubated at the permissive temperature, the other at the nonpermissive temperature. Cells containing a temperature-sensitive mutation in a gene essential for proliferation can divide at the normal, permissive temperature but fail to divide at the elevated, nonpermissive temperature. Temperature-sensitive mutations of this type were especially useful for identifying genes needed for DNA replication, an essential process.

conditional mutations. The mutant individuals function normally as long as "permissive" conditions prevail, but demonstrate abnormal gene function when subjected to "nonpermissive" (restrictive) conditions. In organisms with *temperature-sensitive mutations*, for example, the abnormality can be switched on and off experimentally simply by changing the ambient temperature; thus, a cell containing a temperature-sensitive mutation in a gene essential for survival will die at a nonpermissive temperature but proliferate normally at the permissive temperature (**Figure 8–46**). The temperature-sensitive gene in such a mutant usually contains a point mutation that causes a subtle change in its protein product; for example, the mutant protein may function normally at low temperatures but unfold at higher temperatures.

Temperature-sensitive mutations were crucial to find the bacterial genes that encode the proteins required for DNA replication. The mutants were identified by screening populations of mutagen-treated bacteria for cells that stop making DNA when they are warmed from 30°C to 42°C. These mutants were later used to identify and characterize the corresponding DNA replication proteins (discussed in Chapter 5). Similarly, screens for temperature-sensitive mutations led to the identification of many proteins involved in regulating the cell cycle, as well as many proteins involved in moving proteins through the secretory pathway in yeast. Related screening approaches demonstrated the function of enzymes involved in the principal metabolic pathways of bacteria and yeast (discussed in Chapter 2) and identified many of the gene products responsible for the orderly development of the *Drosophila* embryo (discussed in Chapter 21).

Mutations Can Cause Loss or Gain of Protein Function

Gene mutations are generally classed as "loss of function" or "gain of function." A loss-of-function mutation results in a gene product that either does not work or works too little; thus, it can reveal the normal function of the gene. A gain-of-function mutation results in a gene product that works too much, works at the wrong time or place, or works in a new way (**Figure 8–47**).

An important early step in the genetic analysis of any mutant cell or organism is to determine whether the mutation causes a loss or a gain of function. A standard test is to determine whether the mutation is *dominant* or *recessive*. A dominant mutation is one that still causes the mutant phenotype in the presence of a single copy of the wild-type gene. A recessive mutation is one that is no longer able to cause the mutant phenotype in the presence of a single wild-type copy of the gene. Although cases have been described in which a loss-of-function mutation is dominant or a gain-of-function mutation is recessive, in the vast majority of cases, recessive mutations are loss of function and dominant mutations are gain

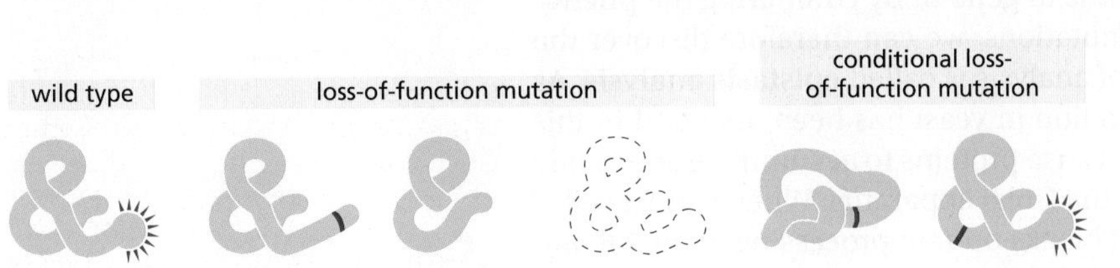

Figure 8–47 **Gene mutations that affect their protein product in different ways.** In this example, the wild-type protein has a specific cell function denoted by the *red rays*. Mutations that eliminate this function or inactivate it at higher temperatures are shown. The conditional mutant protein carries an amino acid substitution *(red)* that prevents its proper folding at 37°C, but allows the protein to fold and function normally at 25°C. Such temperature-sensitive conditional mutations are especially useful for studying essential genes; the organism can be grown under the permissive condition and then be moved to the nonpermissive condition to study the consequences of losing the gene product.

wild type	loss-of-function mutation			conditional loss-of-function mutation	
	point mutation	truncation	deletion	37°C	25°C

of function. It is easy to determine if a mutation is dominant or recessive. One simply mates a mutant with a wild type to obtain diploid cells or organisms. The progeny from the mating will be heterozygous for the mutation. If the mutant phenotype is no longer observed, one can conclude that the mutation is recessive and is very likely to be a loss-of-function mutation (see Panel 8–2).

Complementation Tests Reveal Whether Two Mutations Are in the Same Gene or Different Genes

A large-scale genetic screen can turn up many different mutations that show the same phenotype. These defects might lie in different genes that function in the same process, or they might represent different mutations in the same gene. Alternative forms of the same gene are known as **alleles**. The most common difference between alleles is a substitution of a single nucleotide pair, but different alleles can also bear deletions, substitutions, and duplications. How can we tell, then, whether two mutations that produce the same phenotype occur in the same gene or in different genes? If the mutations are recessive—if, for example, they represent a loss of function of a particular gene—a **complementation test** can be used to ascertain whether the mutations fall in the same gene or in different genes. To test complementation in a diploid organism, an individual that is homozygous for one mutation—that is, it possesses two identical alleles of the mutant gene in question—is mated with an individual that is homozygous for the other mutation. If the two mutations are in the same gene, the offspring show the mutant phenotype, because they still will have no normal copies of the gene in question (**Figure 8–48**). If, in contrast, the mutations fall in different genes, the resulting offspring show a normal phenotype, because they retain one normal copy (and one mutant copy) of each gene; the mutations thereby complement one another and restore a normal phenotype. Complementation testing of mutants identified during genetic screens has revealed, for example, that 5 different genes are required for yeast to digest the sugar galactose, 20 genes are needed for *E. coli* to build a functional flagellum, 48 genes are involved in assembling bacteriophage T4 viral particles, and hundreds of genes are involved in the development of an adult nematode worm from a fertilized egg.

Gene Products Can Be Ordered in Pathways by Epistasis Analysis

Once a set of genes involved in a particular biological process has been identified, the next step is often to determine in which order the genes function. Gene order is perhaps easiest to explain for metabolic pathways, where, for example, enzyme A is necessary to produce the substrate for enzyme B. In this case, we would say that the gene encoding enzyme A acts before (upstream of) the gene encoding enzyme B in the pathway. Similarly, where one protein regulates the activity of another protein, we would say that the former gene acts before the latter. Gene order can, in many cases, be determined purely by genetic analysis without any knowledge of the mechanism of action of the gene products involved.

Suppose we have a biosynthetic process consisting of a sequence of steps, such that performance of step B is conditional on completion of the preceding step A; and suppose gene *A* is required for step A, and gene *B* is required for step B. Then a null mutation (a mutation that abolishes function) in gene *A* will arrest the process at step A, regardless of whether gene *B* is functional or not, whereas a null mutation in gene *B* will cause arrest at step B only if gene *A* is still active. In such a case, gene *A* is said to be *epistatic* to gene *B*. By comparing the phenotypes of the different combinations of mutations, we can therefore discover the order in which the genes act. This type of analysis is called **epistasis analysis**. As an example, the pathway of protein secretion in yeast has been analyzed in this way. Different mutations in this pathway cause proteins to accumulate aberrantly in the endoplasmic reticulum (ER) or in the Golgi apparatus. When a yeast cell is engineered to carry both a mutation that blocks protein processing in the ER *and* a mutation that blocks processing in the Golgi apparatus, proteins accumulate in

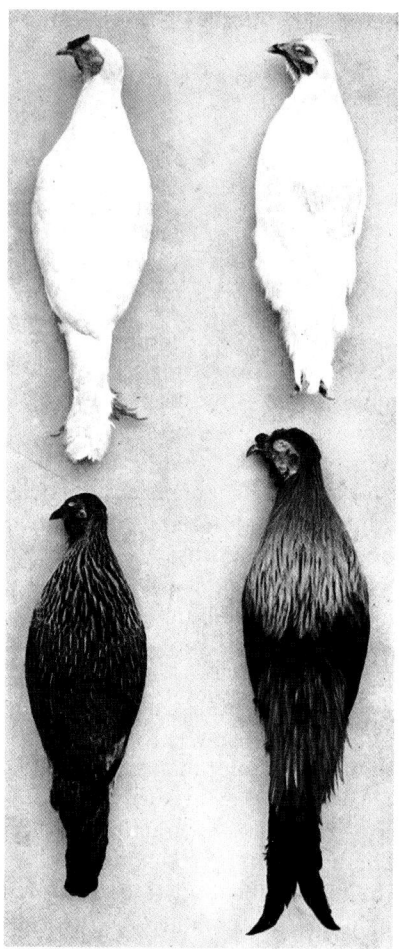

Figure 8–48 A complementation test can reveal that mutations in two different genes are responsible for the same abnormal phenotype. When an albino (white) bird from one strain is bred with an albino from a different strain, the resulting offspring (bottom) have normal coloration. This restoration of the wild-type plumage indicates that the two white breeds lack color because of recessive mutations in different genes. (From W. Bateson, Mendel's Principles of Heredity, 1st ed. Cambridge, UK: Cambridge University Press, 1913.)

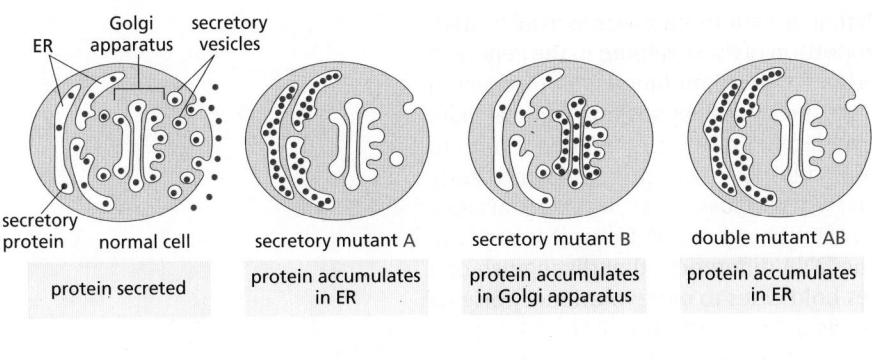

secretory protein

normal cell

protein secreted

secretory mutant A

protein accumulates in ER

secretory mutant B

protein accumulates in Golgi apparatus

double mutant AB

protein accumulates in ER

Figure 8–49 **Using genetics to determine the order of function of genes.** In normal cells, secretory proteins are loaded into vesicles, which fuse with the plasma membrane to secrete their contents into the extracellular medium. Two mutants, A and B, fail to secrete proteins. In mutant A, secretory proteins accumulate in the ER. In mutant B, secretory proteins accumulate in the Golgi. In the double mutant AB, proteins accumulate in the ER; this indicates that the gene defective in mutant A acts before the gene defective in mutant B in the secretory pathway.

the ER. This indicates that proteins must pass through the ER before being sent to the Golgi before secretion (**Figure 8–49**). Strictly speaking, an epistasis analysis can only provide information about gene order in a pathway when both mutations are null alleles. When the mutations retain partial function, their epistasis interactions can be difficult to interpret.

Sometimes, a double mutant will show a new or more severe phenotype than either single mutant alone. This type of genetic interaction is called a *synthetic phenotype*, and if the phenotype is death of the organism, it is called *synthetic lethality*. In most cases, a synthetic phenotype indicates that the two genes act in two different parallel pathways, either of which is capable of mediating the same cell process. Thus, when both pathways are disrupted in the double mutant, the process fails altogether, and the synthetic phenotype is observed.

Mutations Responsible for a Phenotype Can Be Identified Through DNA Analysis

Once a collection of mutant organisms with interesting phenotypes has been obtained, the next task is to identify the gene or genes responsible for the altered phenotype. If the phenotype has been produced by insertional mutagenesis, locating the disrupted gene is fairly simple. DNA fragments containing the insertion (a transposon or a retrovirus, for example) are amplified by PCR, and the nucleotide sequence of the flanking DNA is determined. The gene affected by the insertion can then be identified by a computer-aided search of the complete genome sequence of the organism.

If a DNA-damaging chemical was used to generate the mutations, identifying the inactivated gene is often more laborious, but there are several powerful strategies available. If the genome size of the organism is small (for example, for bacteria or simple eukaryotes), it is possible to simply determine the genome sequence of the mutant organism and identify the affected gene by comparison with the wild-type sequence. Because of the continuous accumulation of neutral mutations, there will probably be differences between the two genome sequences in addition to the mutation responsible for the phenotype. One way of proving that a mutation is causative is to introduce the putative mutation back into a normal organism and determine whether or not it causes the mutant phenotype. We will discuss how this is accomplished later in the chapter.

Rapid and Cheap DNA Sequencing Has Revolutionized Human Genetic Studies

Genetic screens in model experimental organisms have been spectacularly successful in identifying genes and relating them to various phenotypes, including many that are conserved between these organisms and humans. But how can we study humans directly? They do not reproduce rapidly, cannot be treated with mutagens, and, if they have a defect in an essential process such as DNA replication, would die long before birth.

Despite their limitations compared to model organisms, humans are becoming increasingly attractive subjects for genetic studies. Because the human population is so large, spontaneous, nonlethal mutations have arisen in all human genes—many times over. A substantial proportion of these remain in the genomes of present-day humans. The most deleterious of these mutations are discovered when the mutant individuals call attention to themselves by seeking medical help.

With the recent advances that have enabled the sequencing of entire human genomes cheaply and quickly, we can now identify such mutations and study their evolution and inheritance in ways that were impossible even a few years ago. By comparing the sequences of thousands of human genomes from all around the world, we can begin to identify directly the DNA differences that distinguish one individual from another. These differences hold clues to our evolutionary origins and can be used to explore the roots of disease.

Linked Blocks of Polymorphisms Have Been Passed Down from Our Ancestors

When we compare the sequences of multiple human genomes, we find that any two individuals will differ in roughly 1 nucleotide pair in 1000. Most of these variations are common and relatively harmless. When two sequence variants coexist in the population and both are common, the variants are called **polymorphisms**. The majority of polymorphisms are due to the substitution of a single nucleotide, called **single-nucleotide polymorphisms** or **SNPs** (**Figure 8–50**). The rest are due largely to insertions or deletions—called *indels* when the change is small, or *copy number variations* (*CNVs*) when it is large. Although these common variants can be found throughout the genome, they are not scattered randomly—or even independently. Instead, they tend to travel in groups called **haplotype blocks**—combinations of polymorphisms that are inherited as a unit.

To understand why such haplotype blocks exist, we need to consider our evolutionary history. It is thought that modern humans expanded from a relatively small population—perhaps around 10,000 individuals—that existed in Africa about 60,000 years ago. Among that small group of our ancestors, some individuals will have carried one set of genetic variants, others a different set. The chromosomes of a present-day human represent a shuffled combination of chromosome segments from different members of this small ancestral group of people. Because only about two thousand generations separate us from them, large segments of these ancestral chromosomes have passed from parent to child, unbroken by the crossover events that occur during meiosis. As described in Chapter 5, only a few crossovers occur between each set of homologous chromosomes during each meiosis (see Figure 5–53).

As a result, certain sets of DNA sequences—and their associated polymorphisms—have been inherited in linked groups, with little genetic rearrangement across the generations. These are the haplotype blocks. Like genes that exist in different allelic forms, haplotype blocks also come in a limited number of variants that are common in the human population, each representing a combination of DNA polymorphisms passed down from a particular ancestor long ago.

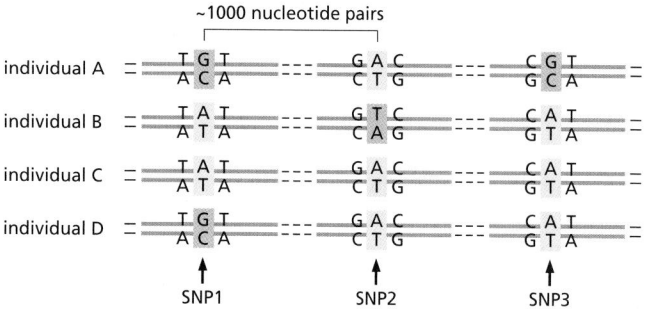

Figure 8–50 Single-nucleotide polymorphisms (SNPs) are sites in the genome where two or more alternative choices of a nucleotide are common in the population. Most such variations in the human genome occur at locations where they do not significantly affect a gene's function.

Polymorphisms Can Aid the Search for Mutations Associated with Disease

Mutations that give rise, in a reproducible way, to rare but clearly defined abnormalities, such as albinism, hemophilia, or congenital deafness, can often be identified by studies of affected families. Such single-gene, or monogenic, disorders are often referred to as *Mendelian* because their pattern of inheritance is easy to track. Moreover, individuals who inherit the causative mutation will exhibit the abnormality irrespective of environmental factors such as diet or exercise. But for many common diseases, the genetic roots are more complex. Instead of a single allele of a single gene, such disorders stem from a combination of contributions from multiple genes. And often, environmental factors have strong influences on the severity of the disorder. For these *multigenic* conditions, such as diabetes or arthritis, population studies are often helpful in tracking down the genes that increase the risk of getting the disease.

In population studies, investigators collect DNA samples from a large number of people who have the disease and compare them to samples from a group of people who do not have the disease. They look for variants—SNPs, for example—that are more common among the people who have the disease. Because DNA sequences that are close together on a chromosome tend to be inherited together, the presence of such SNPs could indicate that an allele that increases the risk of the disease might lie nearby (**Figure 8–51**). Although, in principle, the disease could be caused by the SNP itself, the culprit is much more likely to be a change that is merely linked to the SNP as part of a haplotype block.

Such *genome-wide association studies* have been used to search for genes that predispose individuals to common diseases, including diabetes, coronary artery disease, rheumatoid arthritis, and even depression. For many of these conditions, the DNA polymorphisms identified increase the risk of disease only slightly. Moreover, environmental factors (diet, exercise, for example) play an important role in the onset and severity of the disease. Nonetheless, the identification of genes affected by these polymorphisms is leading to a mechanistic understanding of some of our most common disorders.

Genomics Is Accelerating the Discovery of Rare Mutations That Predispose Us to Serious Disease

The genetic variants that have thus far allowed us to identify some of the genes that increase our risk of disease are common ones. They arose long ago in our evolutionary past and are now present, in one form or another, in a substantial portion (1% or more) of the population. Such polymorphisms are thought to account

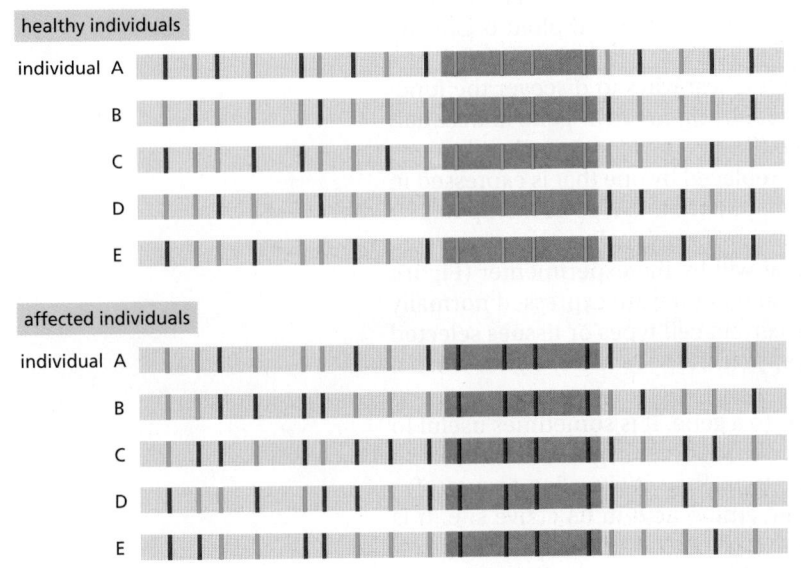

Figure 8–51 Genes that affect the risk of developing a common disease can often be tracked down through linkage to SNPs. Here, the patterns of SNPs are compared between two sets of individuals—a set of healthy controls and a set affected by a particular common disease. A segment of a typical chromosome is shown. For most polymorphic sites in this segment, it is a random matter whether an individual has one SNP variant (*red* vertical bars) or another (*blue* vertical bars); this same randomness is seen both for the control group and for the affected individuals. However, in the part of the chromosome that is shaded in *darker gray*, a bias is seen: most normal individuals have the blue SNP variants, whereas most affected individuals have the red SNP variants. This suggests that this region contains, or is close to, a gene that is genetically linked to these red SNP variants and that predisposes individuals to the disease. Using carefully selected controls and thousands of affected individuals, this approach can help track down disease-related genes, even when they confer only a slight increase in the risk of developing the disease.

healthy individuals

individual A
B
C
D
E

affected individuals

individual A
B
C
D
E

for about 90% of the differences between one person's genome and another's. But when we try to tie these common variants to differences in disease susceptibility or other heritable traits, such as height, we find that they do not have as much predictive power as we had anticipated: thus, for example, most confer relatively small increases—less than twofold—in the risk of developing a common disease.

In contrast to polymorphisms, rare DNA variants—those much less frequent in humans than SNPs—can have large effects on the risk of developing some common diseases. For example, a number of different loss-of-function mutations, each individually rare, have been found to increase greatly the predisposition to autism and schizophrenia. Many of these are *de novo* mutations, which arose spontaneously in the germ-line cells of one or the other parent. The fact that these mutations arise spontaneously with some frequency could help explain why these common disorders—each observed in about 1% of the population—remain with us, even though the affected individuals leave few or no descendants. These rare mutations may arise in any one of hundreds of different genes, which could explain much of the clinical variability of autism and schizophrenia. Because they are kept rare by natural selection, most such variants with a large effect on risk would be missed by genome-wide association studies.

Now that DNA sequencing has become fast and inexpensive, the most efficient and cost-effective way to identify these rare, large-effect mutations is by sequencing the genomes of affected individuals, along with those of their parents and siblings as controls.

Reverse Genetics Begins with a Known Gene and Determines Which Cell Processes Require Its Function

As we have seen, classical genetics starts with a mutant phenotype (or, in the case of humans, a range of characteristics) and identifies the mutations, and consequently the genes, responsible for it. Recombinant DNA technology has made possible a different type of genetic approach, one that is used widely in a variety of genetically tractable species. Instead of beginning with a mutant organism and using it to identify a gene and its protein, an investigator can start with a particular gene and proceed to make mutations in it, creating mutant cells or organisms so as to analyze the gene's function. Because this approach reverses the traditional direction of genetic discovery—proceeding from genes to mutations, rather than vice versa—it is commonly referred to as **reverse genetics**. And because the genome of the organism is deliberately altered in a particular way, this approach is also called *genome engineering* or *genome editing*. We shall see in this chapter that this approach can be scaled upward so that whole collections of organisms can be created, each of which has a different gene altered.

There are several ways a gene of interest can be altered. In the simplest, the gene can simply be deleted from the genome, although in a diploid organism, this requires that both copies—one on each chromosome homolog—be deleted. Although somewhat counterintuitive, one of the best ways to discover the function of a gene is by observing the effects of not having it. Such "gene knockouts" are especially useful if the gene is not essential. Through reverse genetics, the gene in question (even if it is essential) can also be replaced by one that is expressed in the wrong tissue or at the wrong time in development; this type of manipulation often provides important clues to the gene's normal function. For example, a gene of interest can be modified to be expressed at will by the experimenter (**Figure 8–52**). Finally, genes can also be engineered so that they are expressed normally in most cell types and tissues but deleted in certain cell types or tissues selected by the experimenter (see Figure 5–66). This approach is especially useful when a gene has different roles in different tissues.

It is also possible to make subtler changes to a gene. It is sometimes useful to make slight changes in a protein's structure so that one can begin to dissect which portions of a protein are important for its function. The activity of an enzyme, for example, can be studied by changing a single amino acid in its active site. It is also possible, through genome engineering, to create new types of proteins in an

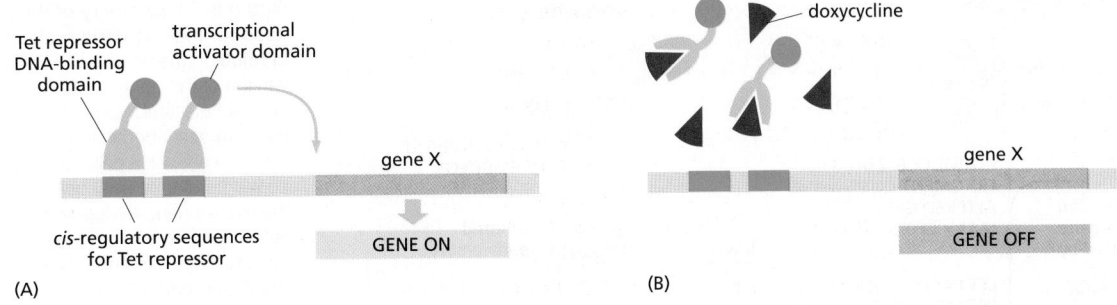

Figure 8–52 Engineered genes can be turned on and off with small molecules. Here, the DNA-binding portion of a bacterial protein (the tetracycline, Tet, repressor) has been fused to a portion of a mammalian transcriptional activator and expressed in cultured mammalian cells. The engineered gene *X*, present in place of the normal gene, has its usual gene control region replaced by *cis*-regulatory sequences recognized by the tetracycline repressor. In the absence of doxycycline (a particularly stable version of tetracycline), the engineered gene is expressed; in the presence of doxycycline, the gene is turned off because the drug causes the tetracycline repressor to dissociate from the DNA. This strategy can also be used in mice by incorporating the engineered genes into the germ line. In many tissues, the gene can be turned on and off simply by adding or removing doxycycline from the animal's water. If the tetracycline repressor construct is placed under the control of a tissue-specific gene control region, the engineered gene will be turned on and off only in that tissue.

animal. For example, a gene can be fused to the gene for a fluorescent protein. When this altered gene is introduced into the genome, the protein can be tracked in the living organism by monitoring its fluorescence.

Altered genes can be created in several ways. Perhaps the simplest is to chemically synthesize the DNA that makes up the gene. In this way, the investigator can specify any type of variant of the normal gene. It is also possible to construct altered genes using recombinant DNA technology, as described earlier in this chapter. Once obtained, altered genes can be introduced into cells in a variety of ways. DNA can be microinjected into mammalian cells with a glass micropipette or introduced by a virus that has been engineered to carry foreign genes. In plant cells, genes are frequently introduced by a technique called particle bombardment: DNA samples are painted onto tiny gold beads and then literally shot through the cell wall with a specially modified gun. *Electroporation* is the method of choice for introducing DNA into bacteria and some other cells. In this technique, a brief electric shock renders the cell membrane temporarily permeable, allowing foreign DNA to enter the cytoplasm.

To be most useful to experimenters, the altered gene, once it is introduced into a cell, must recombine with the cell's genome so that the normal gene is replaced. In simple organisms such as bacteria and yeasts, this process occurs with high frequency using the cell's own homologous recombination machinery, as described in Chapter 5. In more complex organisms that have elaborate developmental programs, the procedure is more complicated because the altered gene must be introduced into the germ line, as we next describe.

Animals and Plants Can Be Genetically Altered

Animals and plants that have been genetically engineered by gene deletion or gene replacement are called **transgenic organisms**, and any foreign or modified genes that are added are called **transgenes**. We discuss transgenic plants later in this chapter and, for now, concentrate our discussion on transgenic mice, as enormous progress has been made in this area. If a DNA molecule carrying a mutated mouse gene is transferred into a mouse cell, it often inserts into the chromosomes at random, but methods have been developed to direct the mutant gene to replace the normal gene by homologous recombination. By exploiting these "gene targeting" events, any specific gene can be altered or inactivated in a mouse cell by a direct gene replacement. In the case in which both copies of the gene of interest are completely inactivated or deleted, the resulting animal is called a *"knockout"* mouse. The technique is summarized in **Figure 8–53**.

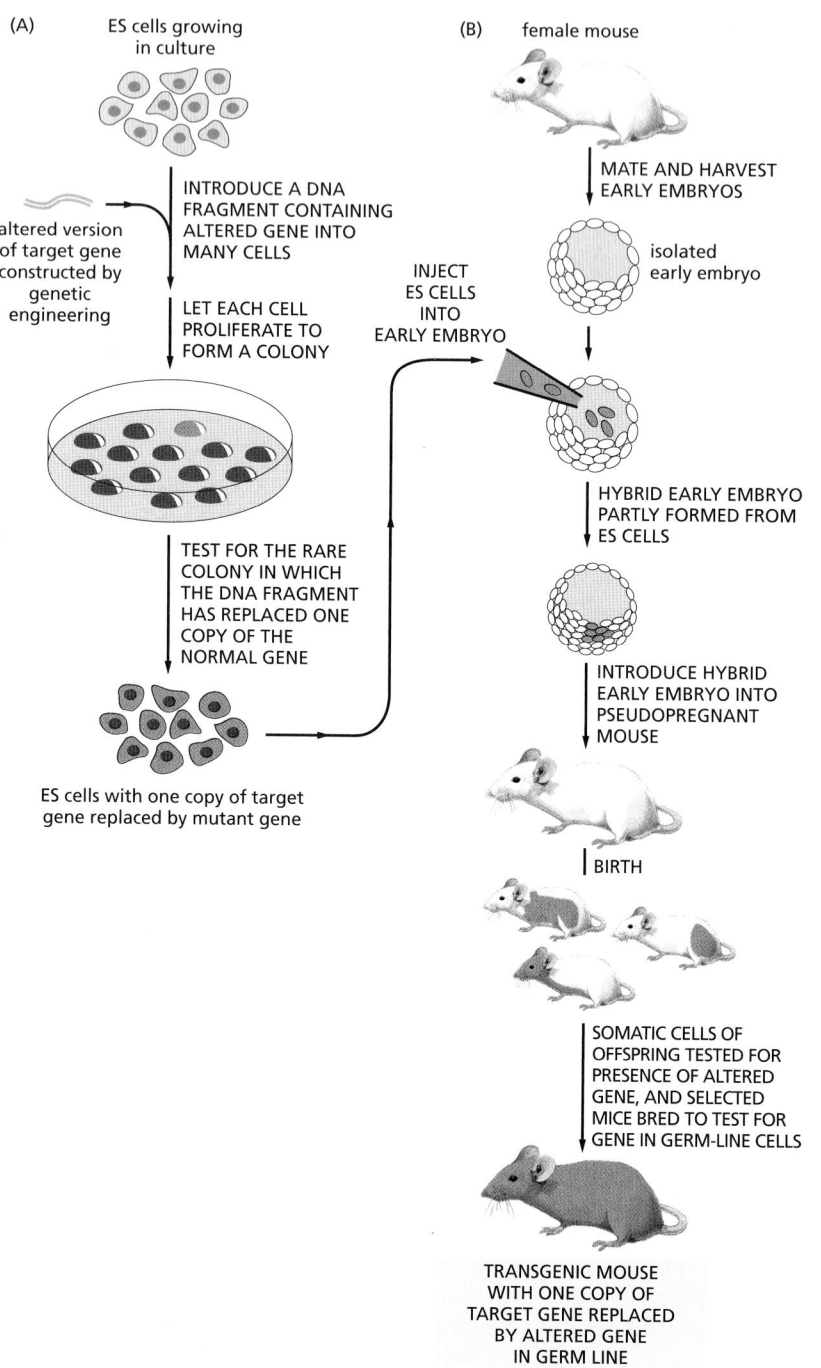

(A) ES cells growing in culture

altered version of target gene constructed by genetic engineering

INTRODUCE A DNA FRAGMENT CONTAINING ALTERED GENE INTO MANY CELLS

LET EACH CELL PROLIFERATE TO FORM A COLONY

TEST FOR THE RARE COLONY IN WHICH THE DNA FRAGMENT HAS REPLACED ONE COPY OF THE NORMAL GENE

ES cells with one copy of target gene replaced by mutant gene

(B) female mouse

MATE AND HARVEST EARLY EMBRYOS

isolated early embryo

INJECT ES CELLS INTO EARLY EMBRYO

HYBRID EARLY EMBRYO PARTLY FORMED FROM ES CELLS

INTRODUCE HYBRID EARLY EMBRYO INTO PSEUDOPREGNANT MOUSE

BIRTH

SOMATIC CELLS OF OFFSPRING TESTED FOR PRESENCE OF ALTERED GENE, AND SELECTED MICE BRED TO TEST FOR GENE IN GERM-LINE CELLS

TRANSGENIC MOUSE WITH ONE COPY OF TARGET GENE REPLACED BY ALTERED GENE IN GERM LINE

Figure 8–53 Summary of the procedures used for making gene replacements in mice. In the first step (A), an altered version of the gene is introduced into cultured ES (embryonic stem) cells. These cells are described in detail in Chapter 22. Only a few ES cells will have their corresponding normal genes replaced by the altered gene through a homologous recombination event. These cells can be identified by PCR and cultured to produce many descendants, each of which carries an altered gene in place of one of its two normal corresponding genes. In the next step of the procedure (B), these altered ES cells are injected into a very early mouse embryo; the cells are incorporated into the growing embryo, and a mouse produced by such an embryo will contain some somatic cells (indicated by *orange*) that carry the altered gene. Some of these mice will also contain germ-line cells that contain the altered gene; when bred with a normal mouse, some of the progeny of these mice will contain one copy of the altered gene in all of their cells.

The mice with the transgene in their germ line are then bred to produce both a male and a female animal, each heterozygous for the gene replacement (that is, they have one normal and one mutant copy of the gene). When these two mice are mated (not shown), one-fourth of their progeny will be homozygous for the altered gene.

The ability to prepare transgenic mice lacking a known normal gene has been a major advance, and the technique has been used to determine the functions of many mouse genes (**Figure 8–54**). If the gene functions in early development, a knockout mouse will usually die before it reaches adulthood. These lethal defects can be carefully analyzed to help determine the function of the missing gene. As described in Chapter 5, an especially useful type of transgenic animal takes advantage of a site-specific recombination system to excise—and thus disable— the target gene in a particular place or at a particular time (see Figure 5–66). In this case, the target gene in embryonic stem (ES) cells is replaced by a fully functional version of the gene that is flanked by a pair of the short DNA sequences, called *lox sites*, that are recognized by the *Cre recombinase* protein. The transgenic mice that result are phenotypically normal. They are then mated with transgenic mice that express the Cre recombinase gene under the control of an inducible promoter. In

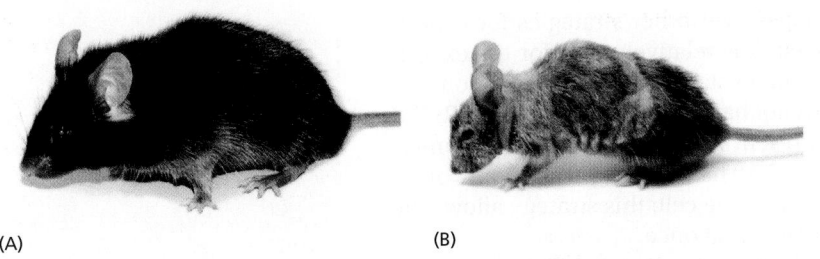

(A)   (B)

Figure 8–54 **Transgenic mice engineered to express a mutant DNA helicase show premature aging.** The helicase, encoded by the *Xpd* gene, is involved in both transcription and DNA repair. Compared with a wild-type mouse of the same age (A), a transgenic mouse that expresses a defective version of *Xpd* (B) exhibits many of the symptoms of premature aging, including osteoporosis, emaciation, early graying, infertility, and reduced life-span. The mutation in *Xpd* used here impairs the activity of the helicase and mimics a mutation that in humans causes trichothiodystrophy, a disorder characterized by brittle hair, skeletal abnormalities, and a very reduced life expectancy. These results indicate that an accumulation of DNA damage can contribute to the aging process in both humans and mice. (From J. de Boer et al., *Science* 296:1276–1279, 2002. With permission from AAAS.)

the specific cells or tissues in which Cre is switched on, it catalyzes recombination between the lox sequences—excising a target gene and eliminating its activity (see Figure 22–5).

The Bacterial CRISPR System Has Been Adapted to Edit Genomes in a Wide Variety of Species

One of the difficulties in making transgenic mice by the procedure just described is that the introduced DNA molecule (bearing the experimentally altered gene) often inserts at random in the genome, and many ES cells must therefore be screened individually to find one that has the "correct" gene replacement.

Creative use of the CRISPR system, discovered in bacteria as a defense against viruses, has largely solved this problem. As described in Chapter 7, the CRISPR system uses a guide RNA sequence to target (through complementary base-pairing) double-stranded DNA, which it then cleaves (see Figure 7–78). The gene coding for the key component of this system, the bacterial Cas9 protein, has been transferred into a variety of organisms, where it greatly simplifies the process of making transgenic organisms (**Figure 8–55**A and B). The basic strategy is as follows: Cas9 protein is expressed in ES cells along with a guide RNA designed by the experimenter to target a particular location on the genome. The Cas9 and guide RNA associate, the complex is brought to the matching sequence on the genome, and the Cas9 protein makes a double-strand break. As we saw in Chapter 5, double-strand breaks are often repaired by homologous recombination; here, the template chosen by the cell to repair the damage is often the altered gene, which is introduced to ES cells by the experimenter. In this way, the normal gene can be selectively damaged by the CRISPR system and replaced at high efficiency by the experimentally altered gene.

The CRISPR system has a variety of other uses. Its particular power lies with its ability to target Cas9 to thousands of different positions across a genome through the simple rules of complementary base-pairing. Thus, if a catalytically inactive Cas9 protein is fused to a transcription activator or repressor, it is possible, in principle, to turn any gene on or off (Figure 8–55C and D).

Figure 8–55 **Use of CRISPR to study gene function in a wide variety of species.** (A) The Cas9 protein (artificially expressed in the species of interest) binds to a guide RNA, designed by the experimenter and also expressed. The portion of RNA in *light blue* is needed for associations with Cas9; that in *dark blue* is specified by the experimenter to match a position on the genome. The only other requirement is that the adjacent genome sequence includes a short PAM (protospacer adjacent motif) that is needed for Cas9 to cleave. As described in Chapter 7, this sequence is how the CRISPR system in bacteria distinguishes its own genome from that of invading viruses. (B) When directed to make double-strand breaks, the CRISPR system greatly improves the ability to replace an endogenous gene with an experimentally altered gene since the altered gene is used to "repair" the double-strand break. (C, D) By using a mutant form of Cas9 that can no longer cleave DNA, Cas9 can be used to activate a normally dormant gene (C) or turn off an actively expressed gene (D). (Adapted from P. Mali et al., *Nat. Methods* 10:957–963, 2013.)

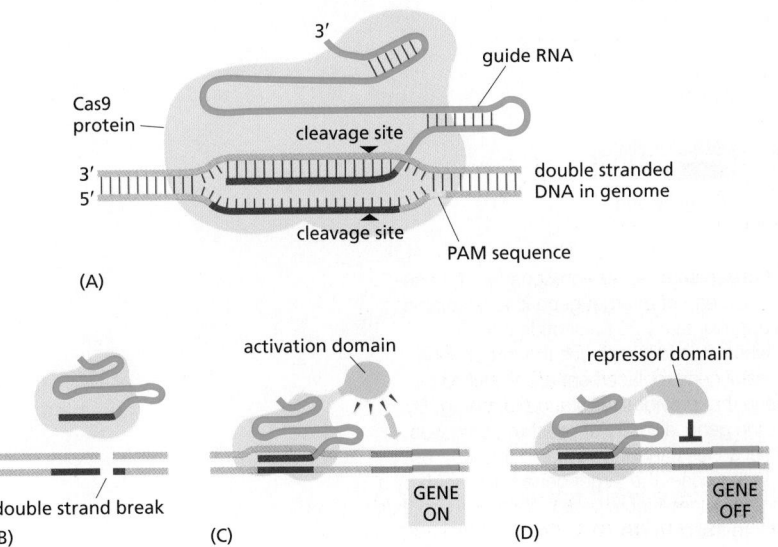

(A)

(B) double strand break

(C) GENE ON

(D) GENE OFF

The CRISPR system has several advantages over other strategies for experimentally manipulating gene expression. First, it is relatively easy for the experimenter to design the guide RNA: it simply follows standard base pairing convention. Second, the gene to be controlled does not have to be modified; the CRISPR strategy exploits DNA sequences already present in the genome. Third, numerous genes can be controlled simultaneously. Cas9 has to be expressed only once, but many guide RNAs can be expressed in the same cell; this strategy allows the experimenter to turn on or off a whole set of genes at once.

The export of the CRISPR system from bacteria to virtually all other experimental organisms (including mice, zebrafish, worms, flies, rice, and wheat) has revolutionized the study of gene function. Like the earlier discovery of restriction enzymes, this breakthrough came from scientists studying a fascinating phenomenon in bacteria without—at first—realizing the enormous impact these discoveries would have on all aspects of biology.

Large Collections of Engineered Mutations Provide a Tool for Examining the Function of Every Gene in an Organism

Extensive collaborative efforts have produced comprehensive libraries of mutations in a variety of model organisms, including *S. cerevisiae, C. elegans, Drosophila, Arabidopsis*, and even the mouse. The ultimate aim in each case is to produce a collection of mutant strains in which every gene in the organism has been systematically deleted or altered in such a way that it can be conditionally disrupted. Collections of this type provide an invaluable resource for investigating gene function on a genomic scale. For example, a large collection of mutant organisms can be screened for a particular phenotype. Like the classic genetic approaches described earlier, this is one of the most powerful ways to identify the genes responsible for a particular phenotype. Unlike the classical genetic approach, however, the set of mutants is "pre-engineered," so that there is no need to rely on chance events such as spontaneous mutations or transposon insertions. In addition, each of the individual mutations within the collection is often engineered to contain a distinct molecular "barcode"—in the form of a unique DNA sequence—designed to make identification of the altered gene rapid and routine (**Figure 8–56**).

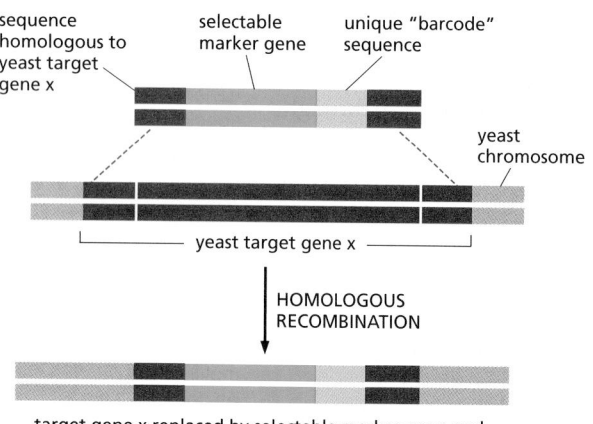

Figure 8–56 **Making barcoded collections of mutant organisms.** A deletion construct for use in yeast contains DNA sequences *(red)* homologous to each end of a target gene x, a selectable marker gene *(blue)*, and a unique "barcode" sequence approximately 20 nucleotide pairs in length *(green)*. This DNA is introduced into yeast cells, where it readily replaces the target gene by homologous recombination. Cells that carry a successful gene replacement are identified by expression of the selectable marker gene, typically a gene that provides resistance to a drug. By using a collection of such constructs, each specific for one gene, a library of yeast mutants was constructed containing a mutant for every gene. Essential genes cannot be studied this way, as their deletion from the genome causes the cells to die. In this case, the target gene is replaced by a version of the gene that can be regulated by the experimenter (see Figure 8–52). The gene can then be turned off and the effect of this can be monitored before the cells die.

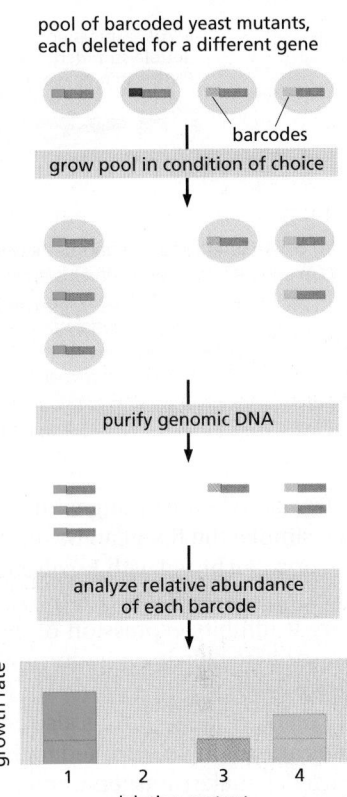

pool of barcoded yeast mutants, each deleted for a different gene

barcodes

grow pool in condition of choice

purify genomic DNA

analyze relative abundance of each barcode

growth rate

deletion mutant

Figure 8–57 Genome-wide screens for fitness using a large pool of barcoded yeast deletion mutants. A large pool of yeast mutants, each with a different gene deleted and present in equal amounts, is grown under conditions selected by the experimenter. Some mutants *(blue)* grow normally, but others show reduced growth *(orange and green)* or no growth at all *(red)*. The fitness of each mutant is experimentally determined in the following way. After the growth phase is completed, genomic DNA (isolated from the mixture of strains) is purified and the relative abundance of each mutant is determined by quantifying the level of the DNA barcode matched to each deletion. This can be done by sequencing the pooled genomic DNA or hybridizing it to microarrays (see Figure 8–64) that contain DNA oligonucleotides complementary to each barcode. In this way, the contribution of every gene to growth under the specified condition can be rapidly ascertained. This type of study has revealed that of the approximately 6000 coding genes in yeast, only about 1000 are essential under standard growth conditions.

In *S. cerevisiae*, the task of generating a complete set of 6000 mutants, each missing only one gene, was accomplished several years ago. Because each mutant strain has an individual barcode sequence embedded in its genome, a large mixture of engineered strains can be grown under various selective test conditions—such as nutritional deprivation, a temperature shift, or the presence of various drugs—and the cells that survive can be rapidly identified by the unique sequence tags present in their genomes. By assessing how well each mutant in the mixture fares, one can begin to discern which genes are essential, useful, or irrelevant for growth under the various conditions (**Figure 8–57**).

The insights generated by examining mutant libraries can be considerable. For example, studies of an extensive collection of mutants in *Mycoplasma genitalium*—the organism with the smallest known genome—have identified the minimum complement of genes essential for cellular life. Growth under laboratory conditions requires about three-quarters of the 480 protein-coding genes in *M. genitalium*. Approximately 100 of these essential genes are of unknown function, which suggests that a surprising number of the basic molecular mechanisms that underlie life have yet to be discovered.

Collections of mutant organisms are also available for many animal and plant species. For example, it is possible to "order," by phone or email from a consortium of investigators, a deletion or insertion mutant for almost all coding genes in *Drosophila*. Likewise, a nearly complete set of mutants exists for the "model" plant *Arabidopsis*. And the adaptation of the CRISPR system for use in mice means that, in the near future, we can expect to be able to turn on or off—at will—each gene in the mouse genome. Although we are still ignorant about the function of most genes in most organisms, these technologies allow an exploration of gene function on a scale that was unimaginable a decade ago.

RNA Interference Is a Simple and Rapid Way to Test Gene Function

Although knocking out (or conditionally expressing) a gene in an organism and studying the consequences is the most powerful approach for understanding the functions of the gene, *RNA interference* (*RNAi*, for short), is an alternative, particularly convenient approach. As discussed in Chapter 7, this method exploits a natural mechanism used in many plants, animals, and fungi to protect themselves against viruses and transposable elements. The technique introduces into a cell or organism a double-stranded RNA molecule whose nucleotide sequence matches that of part of the gene to be inactivated. After the RNA is processed, it hybridizes with the target-gene RNA (either mRNA or noncoding RNA) and reduces its expression by the mechanisms shown in Figure 7–75.

RNAi is frequently used to inactivate genes in *Drosophila* and mammalian cell culture lines. Indeed, a set of 15,000 *Drosophila* RNAi molecules (one for every coding gene) allows researchers, in several months, to test the role of every fly gene in any process that can be monitored using cultured cells. RNAi has also been widely used to study gene function in whole organisms, including the nematode

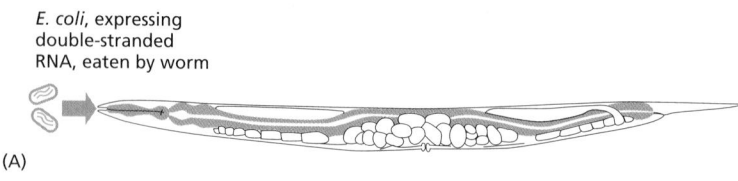

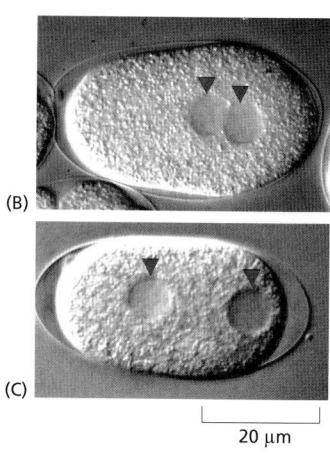

(B)

(C)

20 μm

Figure 8–58 Gene function can be tested by RNA interference. (A) Double-stranded RNA (dsRNA) can be introduced into *C. elegans* by (1) feeding the worms *E. coli* that express the dsRNA or (2) injecting the dsRNA directly into the animal's gut. (B) In a wild-type worm embryo, the egg and sperm pronuclei (*red* arrowheads) come together in the posterior half of the embryo shortly after fertilization. (C) In an embryo in which a particular gene has been inactivated by RNAi, the pronuclei fail to migrate. This experiment revealed an important but previously unknown function of this gene in embryonic development. (B and C, from P. Göczy et al., *Nature* 408:331–336, published 2000 by Macmillan Magazines Ltd. Reproduced with permission of SNCSC.)

C. elegans. When working with worms, introducing the double-stranded RNA is quite simple: the RNA can be injected directly into the intestine of the animal, or the worm can be fed with *E. coli* engineered to produce the RNA (**Figure 8–58**). The RNA is amplified (see p. 431) and distributed throughout the body of the worm, where it inhibits expression of the target gene in different tissue types. RNAi is being used to help in assigning functions to the entire complement of worm genes (**Figure 8–59**).

A related technique has also been applied to mice. In this case, the RNAi molecules are not injected or fed to the mouse; rather, recombinant DNA techniques are used to make transgenic animals that express the RNAi under the control of an inducible promoter. Often this is a specially designed RNA that can fold back on itself and, through base-pairing, produce a double-stranded region that is recognized by the RNAi machinery. In the simplest cases, the process inactivates only the genes that exactly match the RNAi sequence. Depending on the inducible promoter used, the RNAi can be produced only in a specified tissue or only at a particular time in development, allowing the functions of the target genes to be analyzed in elaborate detail.

RNAi has made reverse genetics simple and efficient in many organisms, but it has several potential limitations compared with true genetic knockouts. For

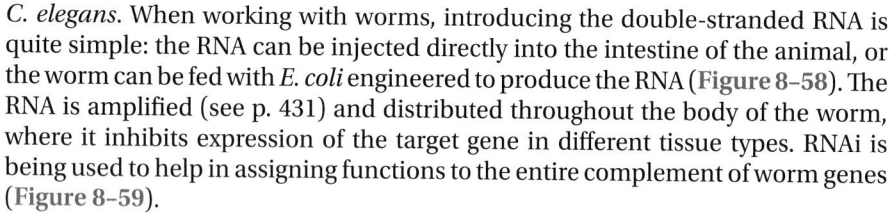

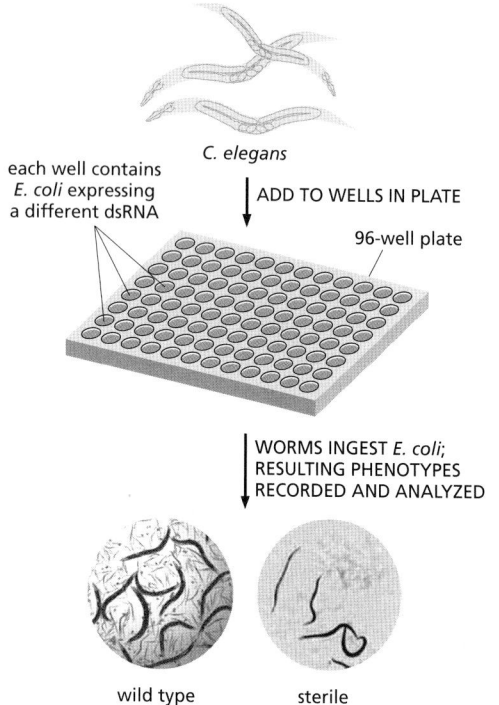

each well contains
E. coli expressing
a different dsRNA

C. elegans

ADD TO WELLS IN PLATE

96-well plate

WORMS INGEST *E. coli*;
RESULTING PHENOTYPES
RECORDED AND ANALYZED

wild type   sterile

Figure 8–59 RNA interference provides a convenient method for conducting genome-wide genetic screens. In this experiment, each well in this 96-well plate is filled with *E. coli* that produce a different double-stranded RNA. Each interfering RNA matches the nucleotide sequence of a single *C. elegans* gene, thereby inactivating it. About 10 worms are added to each well, where they ingest the genetically modified bacteria. The plate is incubated for several days, which gives the RNAs time to inactivate their target genes—and the worms time to grow, mate, and produce offspring. The plate is then examined in a microscope, which can be controlled robotically, to screen for genes that affect the worms' ability to survive, reproduce, develop, and behave. Shown here are normal worms alongside worms that show an impaired ability to reproduce due to inactivation of a particular "fertility" gene. (From B. Lehner et al., *Nat. Genet.* 38:896–903, published 2006 by Nature Publishing Group. Reproduced with permission of SNCSC.)

unknown reasons, RNAi does not efficiently inactivate all genes. Moreover, within whole organisms, certain tissues may be resistant to the action of RNAi (for example, neurons in nematodes). Another problem arises because many organisms contain large gene families, the members of which exhibit sequence similarity. RNAi therefore sometimes produces "off-target" effects, inactivating related genes in addition to the targeted gene. One strategy to avoid such problems is to use multiple small RNA molecules matched to different regions of the same gene. Ultimately, the results of any RNAi experiment must be viewed as a strong clue to, but not necessarily a proof of, normal gene function.

Reporter Genes Reveal When and Where a Gene Is Expressed

In the preceding section, we discussed how genetic approaches can be used to assess a gene's function in cultured cells or, even better, in the intact organism. Although this information is crucial to understanding gene function, it does not generally reveal the molecular mechanisms through which the gene product works in the cell. For example, genetics on its own rarely tells us all the places in the organism where the gene is expressed, or how its expression is controlled. It does not necessarily reveal whether the gene acts in the nucleus, the cytosol, on the cell surface, or in one of the numerous other compartments of the cell. And it does not reveal how a gene product might change its location or its expression pattern when the external environment of the cell changes. Key insights into gene function can be obtained by simply observing when and where a gene is expressed. A variety of approaches, most involving some form of genetic engineering, can easily provide this critical information.

As discussed in detail in Chapter 7, *cis*-regulatory DNA sequences, located upstream or downstream of the coding region, control gene transcription. These regulatory sequences, which determine precisely when and where the gene is expressed, can be easily studied by placing a reporter gene under their control and introducing these recombinant DNA molecules into cells (**Figure 8–60**). In this way, the normal expression pattern of a gene can be determined, as well as

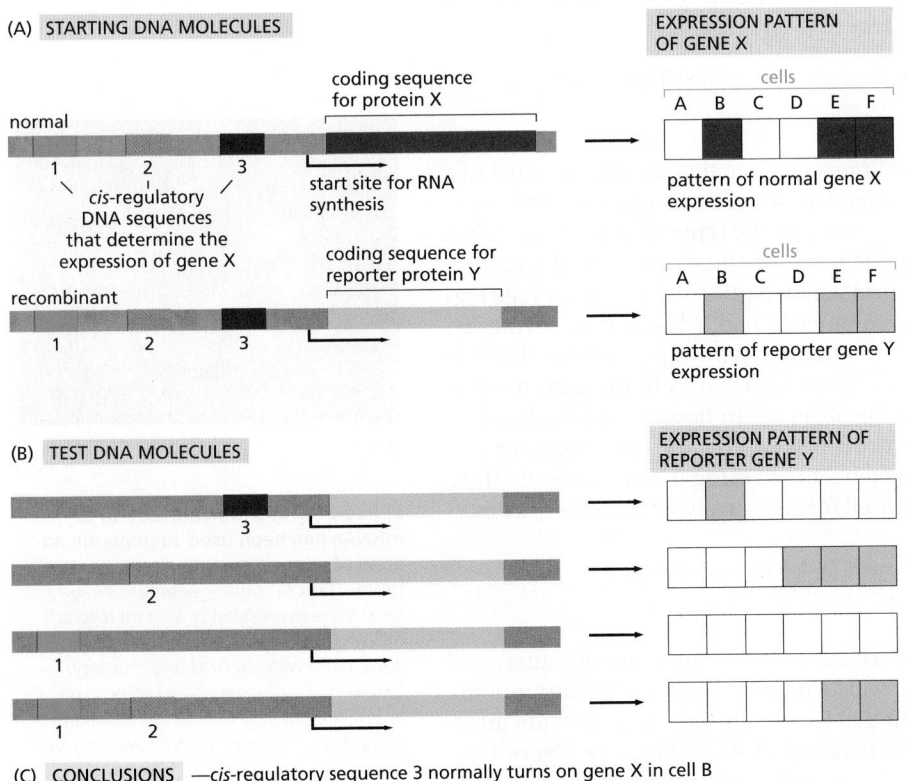

Figure 8–60 Using a reporter protein to determine the pattern of a gene's expression. (A) In this example, the coding sequence for protein X is replaced by the coding sequence for reporter protein Y. The expression patterns for X and Y are the same. (B) Various fragments of DNA containing candidate *cis*-regulatory sequences are added in combinations to produce test DNA molecules encoding reporter gene Y. These recombinant DNA molecules are then tested for expression after introducing them into a variety of different types of mammalian cells. The results are summarized in (C).

For experiments in eukaryotic cells, two commonly used reporter proteins are the enzyme β-galactosidase *(β-gal)* (see Figure 7–28) and green fluorescent protein (GFP) (see Figure 9–22).

(A) STARTING DNA MOLECULES

EXPRESSION PATTERN OF GENE X

coding sequence for protein X

normal

cis-regulatory DNA sequences that determine the expression of gene X

start site for RNA synthesis

coding sequence for reporter protein Y

recombinant

cells
A B C D E F

pattern of normal gene X expression

cells
A B C D E F

pattern of reporter gene Y expression

(B) TEST DNA MOLECULES

EXPRESSION PATTERN OF REPORTER GENE Y

(C) CONCLUSIONS —*cis*-regulatory sequence 3 normally turns on gene X in cell B
—*cis*-regulatory sequence 2 normally turns on gene X in cells D, E, and F
—*cis*-regulatory sequence 1 normally turns off gene X in cell D

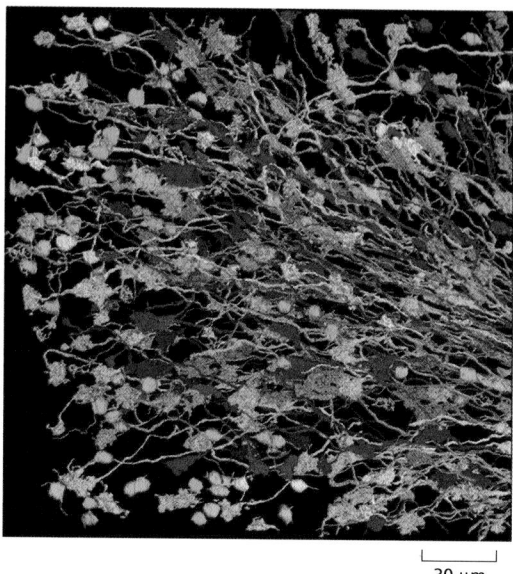

30 μm

Figure 8–61 GFPs that fluoresce at different wavelengths help reveal the connections that individual neurons make within the brain. This image shows differently colored neurons in one region of a mouse brain. The neurons randomly express different combinations of differently colored GFPs (see Figure 9–13), making it possible to distinguish and trace many individual neurons within a population. These images were obtained by genetically engineering the genes for four different fluorescent proteins, each flanked by loxP sites of recombination (see Figure 5–66), and integrating them into the mouse germ line. When crossed to a mouse that produced the Cre recombinase in neuronal cells, the fluorescent protein genes were randomly excised, producing neurons that express many different combinations of the four fluorescent proteins. Over 100 combinations of fluorescent protein can be produced, allowing scientists to distinguish one neuron from the next. The stunning appearance of these labeled neurons has earned these animals the colorful nickname "brainbow mice." (From J. Livet et al., *Nature* 450:56–62, published 2007 by Nature Publishing Group. Reproduced with permission of SNCSC.)

the contribution of individual *cis*-regulatory sequences in establishing this pattern (see also Figure 7–29).

Reporter genes also allow any protein to be tracked over time in living cells. Here, the reporter gene typically encodes a fluorescent protein, often **green fluorescent protein** (**GFP**), the molecule that gives luminescent jellyfish their greenish glow. The GFP is simply attached—in the coding frame—to the protein-coding gene of interest. The resulting *GFP fusion protein* often behaves in the same way the normal protein does and its location can be monitored by fluorescence microscopy, a topic that is discussed in the next chapter (see Figure 9–25). GFP fusion has become a standard strategy for tracking not only the location but also the movement of specific proteins in living cells. In addition, the use of multiple GFP variants that fluoresce at different wavelengths can provide insights into how different cells interact in a living tissue (**Figure 8–61**).

In situ Hybridization Can Reveal the Location of mRNAs and Noncoding RNAs

It is also possible to directly observe the time and place that an RNA product of a gene is expressed using *in situ hybridization*. For protein-coding genes, this strategy often provides the same general information as the reporter gene approaches described above; however, it is crucial for genes whose final product is RNA rather than protein. We encountered *in situ* hybridization earlier in the chapter (see Figure 8–34); it relies on the basic principles of nucleic acid hybridization. Typically, tissues are gently fixed so that their RNA is retained in an exposed form that can hybridize with a labeled complementary DNA or RNA probe. In this way, the patterns of differential gene expression can be observed in tissues, and the location of specific RNAs can be determined (**Figure 8–62**). An advantage of *in situ* hybridization over other approaches is that genetic engineering is not required. Thus, it is often simpler and faster and can be used for genetically intractable species.

Expression of Individual Genes Can Be Measured Using Quantitative RT-PCR

Although reporter genes and *in situ* hybridization accurately reveal patterns of gene expression, they are not the most powerful methods for quantifying amounts of individual RNAs in cells. We have seen that RNA sequencing can provide information about the relative abundance of different RNA molecules (see Figure 7–3). Here, the number of "sequence reads" (short bits of nucleotide sequence) is proportional to the abundance of the RNA species. But this method is limited to RNAs

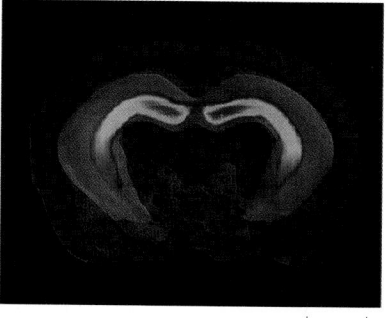

2 mm

Figure 8–62 *In situ* hybridization to mRNAs has been used to generate an atlas of gene expression in the mouse brain. This computer-generated image shows the expression of several different mRNAs specific to an area of the brain associated with learning and memory. Similar maps of expression patterns of all known genes in the mouse brain are compiled in the brain atlas project, which is available online. (From M. Hawrylycz et al., *PLoS Comput. Biol.* 7:e1001065, 2011. With permission from the authors.)

Figure 8–63 **RNA levels can be measured by quantitative RT-PCR.** The fluorescence measured is generated by a dye that fluoresces only when bound to the double-stranded DNA products of the RT-PCR (see Figure 8–36). The red sample has a higher concentration of the mRNA being measured than does the blue sample, since it requires fewer PCR cycles to reach the same half-maximal concentration of double-stranded DNA. Based on this difference, the relative amounts of the mRNA in the two samples can be precisely determined.

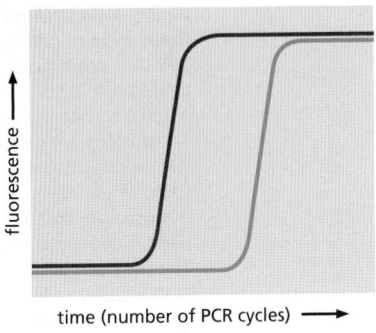

time (number of PCR cycles) ⟶

that are expressed at reasonably high levels, and it is difficult to quantify (or even identify) rare RNAs. A more accurate method is based on the principles of PCR (**Figure 8–63**). Called **quantitative RT-PCR** (reverse transcription–polymerase chain reaction), this method begins with the total population of RNA molecules purified from a tissue or a cell culture. It is important that no DNA be present in the preparation; it must be purified away or enzymatically degraded. Two DNA primers that specifically match the mRNA of interest are added, along with reverse transcriptase, DNA polymerase, and the four deoxyribonucleoside triphosphates needed for DNA synthesis. The first round of synthesis is the reverse transcription of the RNA into DNA using one of the primers. Next, a series of heating and cooling cycles allows the amplification of that DNA strand by PCR (see Figure 8–36). The quantitative part of this method relies on a direct relationship between the rate at which the PCR product is generated and the original concentration of the mRNA species of interest. By adding chemical dyes to the PCR that fluoresce only when bound to double-stranded DNA, a simple fluorescence measurement can be used to track the progress of the reaction and thereby accurately deduce the starting concentration of the mRNA that is amplified. Although it seems complicated, this quantitative RT-PCR technique is relatively fast and simple to perform in the laboratory; it is currently the method of choice for accurately quantifying mRNA levels from any given gene.

Analysis of mRNAs by Microarray or RNA-seq Provides a Snapshot of Gene Expression

As discussed in Chapter 7, a cell expresses only a subset of the many thousands of genes available in its genome; moreover, this subset differs from one cell type to another or, in the same cell, from one environment to the next. One way to determine which genes are being expressed by a population of cells or a tissue is to analyze which mRNAs are being produced.

The first tool that allowed investigators to analyze simultaneously the thousands of different RNAs produced by cells or tissues was the **DNA microarray**. Developed in the 1990s, DNA microarrays are glass microscope slides that contain hundreds of thousands of DNA fragments, each of which serves as a probe for the mRNA produced by a specific gene. Such microarrays allow investigators to monitor the expression of every gene in a genome in a single experiment. To do the analysis, mRNAs are extracted from cells or tissues and converted to cDNAs (see Figure 8–31). The cDNAs are fluorescently labeled and allowed to hybridize to the fragments bound to the microarray. An automated fluorescence microscope then determines which mRNAs were present in the original sample based on the array positions to which the cDNAs are bound (**Figure 8–64**).

Although microarrays are relatively inexpensive and easy to use, they suffer from one obvious drawback: the sequences of the mRNA samples to be analyzed must be known in advance and represented by a corresponding probe on the array. With the development of improved sequencing technologies, investigators increasingly use *RNA-seq*, discussed earlier, as a more direct approach for cataloging the RNAs produced by a cell. For example, this approach can readily detect alternative RNA splicing, RNA editing, and the many noncoding RNAs produced from a complex genome.

DNA microarrays and RNA-seq analysis have been used to examine everything from the changes in gene expression that make strawberries ripen to the gene expression "signatures" of different types of human cancer cells; or from changes

Figure 8–64 DNA microarrays are used to analyze the production of thousands of different mRNAs in a single experiment. In this example, mRNA is collected from two different cell samples—for example, cells treated with a hormone and untreated cells of the same type—to allow for a direct comparison of the specific genes expressed under both conditions. The mRNAs are converted to cDNAs that are labeled with a red fluorescent dye for one sample and a green fluorescent dye for the other. The labeled samples are mixed and then allowed to hybridize to the microarray. Each microscopic spot on the microarray is a 50-nucleotide DNA molecule of defined sequence made by chemical synthesis and spotted on the array. The DNA sequence represented by each spot is different, and the hundreds of thousands of such spots are designed to span the sequence of the genome. The DNA sequence of each spot is kept track of by computer. After incubation, the array is washed and the fluorescence scanned. Only a small proportion of the microarray, representing 676 genes, is shown. *Red* spots indicate that the gene in sample 1 is expressed at a higher level than the corresponding gene in sample 2, and the *green* spots indicate the opposite. *Yellow* spots reveal genes that are expressed at about equal levels in both cell samples. The intensity of the fluorescence provides an estimate of how much RNA is present from a gene. *Dark* spots indicate little or no expression of the gene whose probe is located at that position in the array.

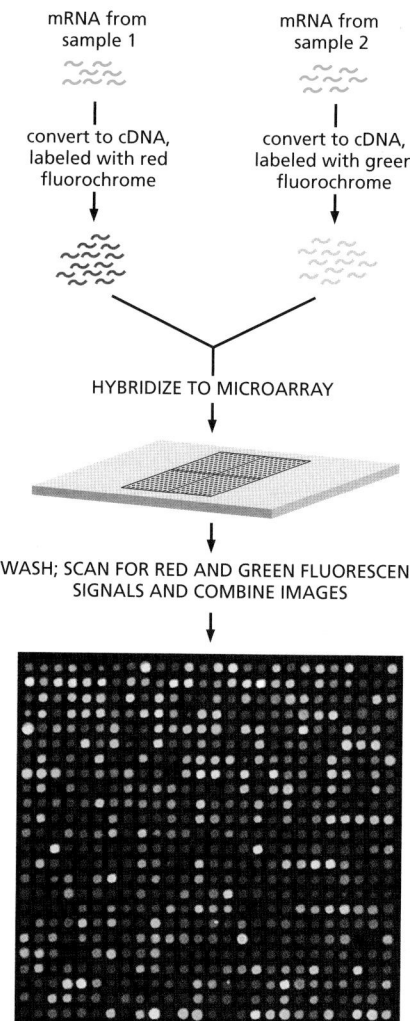

HYBRIDIZE TO MICROARRAY

WASH; SCAN FOR RED AND GREEN FLUORESCENT SIGNALS AND COMBINE IMAGES

small region of microarray representing 676 genes

that occur as cells progress through the cell cycle to those made in response to sudden shifts in temperature. Indeed, because these approaches allow the simultaneous monitoring of large numbers of RNAs, they can detect subtle changes in a cell, changes that might not be manifested in its outward appearance or behavior.

Comprehensive studies of gene expression also provide information that is useful for predicting gene function. Earlier in this chapter, we discussed how identifying a protein's interaction partners can yield clues about that protein's function. A similar principle holds true for genes: information about a gene's function can be deduced by identifying genes that share its expression pattern. Using an approach called *cluster analysis*, one can identify sets of genes that are coordinately regulated. Genes that are turned on or turned off together under different circumstances are likely to work in concert in the cell: they may encode proteins that are part of the same multiprotein machine, or proteins that are involved in a complex coordinated activity, such as DNA replication or RNA splicing. Characterizing a gene whose function is unknown by grouping it with known genes that share its transcriptional behavior is sometimes called "guilt by association." Cluster analyses have been used to analyze the gene expression profiles that underlie many interesting biological processes, including wound healing in humans (**Figure 8–65**).

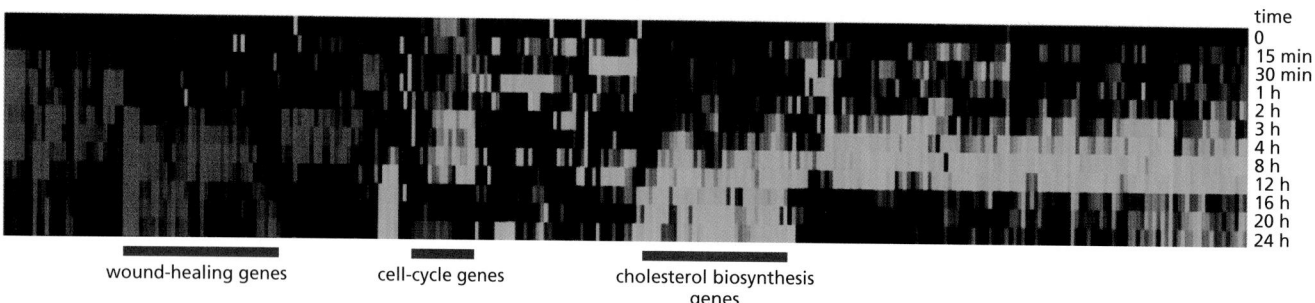

wound-healing genes   cell-cycle genes   cholesterol biosynthesis genes

time
0
15 min
30 min
1 h
2 h
3 h
4 h
8 h
12 h
16 h
20 h
24 h

Figure 8–65 Using cluster analysis to identify sets of genes that are coordinately regulated. Genes that have the same expression pattern are likely to be involved in common pathways or processes. To perform a cluster analysis, RNA-seq or microarray data are obtained from cell samples exposed to a variety of different conditions, and genes that show coordinate changes in their expression pattern are grouped together. In this experiment, human fibroblasts were deprived of serum for 48 hours; serum was then added back to the cultures at time 0 and the cells were harvested for microarray analysis at different time points. Of the 8600 genes depicted here (each represented by a thin, vertical line), just over 300 showed threefold or greater variation in their expression patterns in response to serum reintroduction. Here, *red* indicates an increase in expression; *green* is a decrease in expression. On the basis of the results of many other experiments, the 8600 genes have been grouped in clusters based on similar patterns of expression. The results of this analysis show that genes involved in wound healing are turned on in response to serum, while genes involved in regulating cell-cycle progression and cholesterol biosynthesis are shut down. (From M.B. Eisen et al., *Proc. Natl Acad. Sci. USA* 94:14863–14868, 1998. Copyright 1998 National Academy of Sciences, USA. With permission from National Academy of Sciences.)

Figure 8–66 **Chromatin immunoprecipitation.** This method allows the identification of all the sites in a genome that a transcription regulator occupies *in vivo*. The identities of the precipitated, amplified DNA fragments are determined by DNA sequencing.

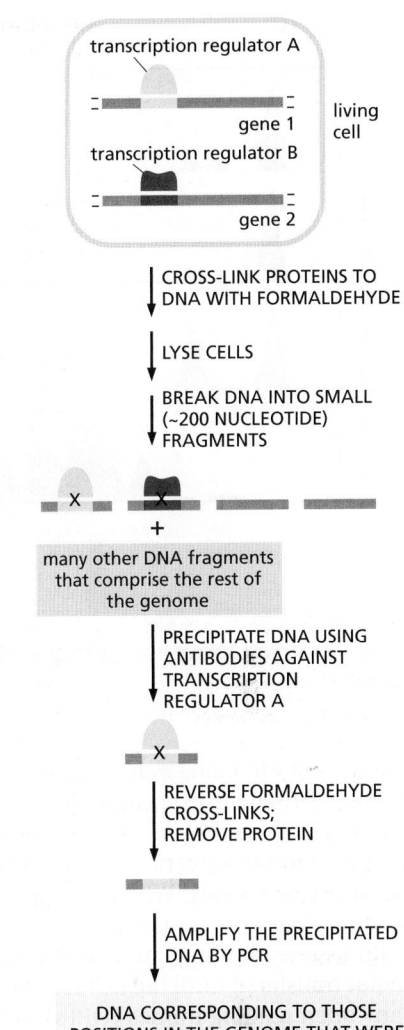

transcription regulator A

gene 1

living cell

transcription regulator B

gene 2

CROSS-LINK PROTEINS TO DNA WITH FORMALDEHYDE

LYSE CELLS

BREAK DNA INTO SMALL (~200 NUCLEOTIDE) FRAGMENTS

X   X

+

many other DNA fragments that comprise the rest of the genome

PRECIPITATE DNA USING ANTIBODIES AGAINST TRANSCRIPTION REGULATOR A

X

REVERSE FORMALDEHYDE CROSS-LINKS; REMOVE PROTEIN

AMPLIFY THE PRECIPITATED DNA BY PCR

DNA CORRESPONDING TO THOSE POSITIONS IN THE GENOME THAT WERE OCCUPIED BY TRANSCRIPTION REGULATOR A IN THE CELLS

Genome-wide Chromatin Immunoprecipitation Identifies Sites on the Genome Occupied by Transcription Regulators

We have discussed several strategies to measure the levels of individual RNAs in a cell and to monitor changes in their levels in response to external signals. But this information does not tell us how such changes are brought about. We saw in Chapter 7 that transcription regulators, by binding to *cis*-regulatory sequences in DNA, are responsible for establishing and changing patterns of transcription. Typically, these proteins do not occupy all of their potential *cis*-regulatory sequences in the genome under all conditions. For example, in some cell types, the regulatory protein may not be expressed, or it may be present but lack an obligatory partner protein, or it may be excluded from the nucleus until an appropriate signal is received from the cell's environment. Even if the protein is present in the nucleus and is competent to bind DNA, other transcription regulators or components of chromatin can occupy overlapping DNA sequences and thereby occlude some of its *cis*-regulatory sequences in the genome.

Chromatin immunoprecipitation provides a way to experimentally determine all the *cis*-regulatory sequences in a genome that are occupied by a given transcription regulator under a particular set of conditions (**Figure 8–66**). In this approach, proteins are covalently cross-linked to DNA in living cells, the cells are broken open, and the DNA is mechanically sheared into small fragments. Antibodies directed against a given transcription regulator are then used to purify the DNA that became covalently cross-linked to that protein in the cell. This DNA is then sequenced using the rapid methods discussed earlier; the precise location of each precipitated DNA fragment along the genome is determined by comparing its DNA sequence to that of the whole genome sequence (**Figure 8–67**). In this way, all of the sites occupied by the transcription regulator in the cell sample can be mapped across the cell's genome (see Figure 7–37). In combination with microarray or RNA-seq information, chromatin immunoprecipitation can identify the key transcriptional regulator responsible for specifying a particular pattern of gene expression.

Chromatin immunoprecipitation can also be used to deduce the *cis*-regulatory sequences recognized by a given transcription regulator. Here, all the DNA sequences precipitated by the regulator are lined up (by computer) and features in common are tabulated to produce the spectrum of *cis*-regulatory sequences recognized by the protein (see Figure 7–9A). Chromatin immunoprecipitation is also used routinely to identify the positions along a genome that are bound by the various types of modified histones discussed in Chapter 4. In this case, antibodies specific to the particular histone modification are employed (see Figure 8–67). A variation of the technique can also be used to map positions of chromosomes that are in physical proximity (see Figure 4–48).

Ribosome Profiling Reveals Which mRNAs Are Being Translated in the Cell

In preceding sections, we discussed several ways that RNA levels in the cell can be monitored. But for mRNAs, this represents only one step in gene expression, and we are often more interested in the final level of the protein produced by the gene. As described in the first part of this chapter, mass-spectroscopy methods can be used to monitor the levels of all proteins in the cell, including modified forms of the proteins. However, if we want to understand *how* synthesis of proteins is controlled by the cell, we need to consider the translation step of gene expression.

An approach called *ribosome profiling* provides an instantaneous map of the position of ribosomes on each mRNA in the cell and thereby identifies those

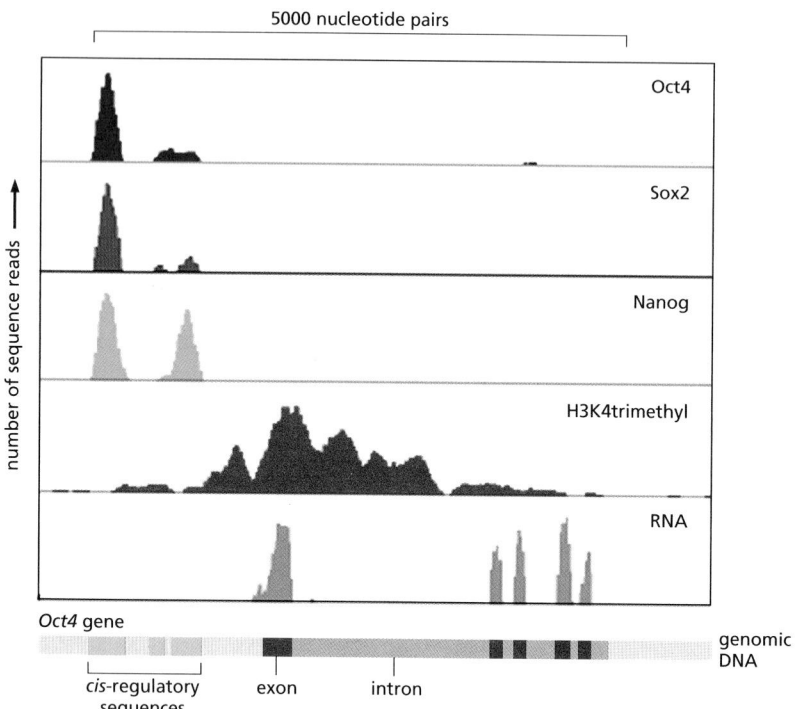

5000 nucleotide pairs

number of sequence reads

Oct4

Sox2

Nanog

H3K4trimethyl

RNA

Oct4 gene

genomic DNA

cis-regulatory sequences   exon   intron

Figure 8–67 Results of several chromatin immunoprecipitations showing proteins bound to the control region that control expression of the *Oct4* gene. In this series of chromatin immunoprecipitation experiments, antibodies directed against a transcription regulator (first three panels) or a particular histone modification (fourth panel) were used to precipitate bound, cross-linked DNA. Precipitated DNA was sequenced, and the positions across the genome were mapped. (Only the small part of the mouse genome containing the *Oct4* gene is shown.) The results show that, in the embryonic stem cells analyzed in these experiments, Oct4 binds upstream of its own gene and that Sox2 and Nanog are bound in close proximity. Oct4, Sox2, and Nanog are key regulators in embryonic stem cells (discussed in Chapter 22) and this experiment reveals the position on the genome through which they exert their effects on *Oct4* expression. In the fourth panel, the positions of a histone modification associated with actively transcribed genes is shown (see Figure 4–39). Finally, the bottom panel shows the RNA produced from the *Oct4* gene under the same conditions used for the chromatin immunoprecipitations. Note that the introns and exons are relatively easy to identify from these RNA-seq data.

mRNAs that are being actively translated. To accomplish this, total RNA from a cell line or tissue is exposed to RNAses under conditions where only those RNA sequences covered by ribosomes are spared. The protected RNAs are released from ribosomes, converted to DNA, and the nucleotide sequence of each is determined (**Figure 8–68**). When these sequences are mapped on the genome, the position of ribosomes across each mRNA species can be ascertained.

Ribosome profiling has revealed many cases where mRNAs are abundant but are not translated until the cell receives an external signal. It has also shown that many open reading frames (ORFs) that were too short to be annotated as genes are actively translated and probably encode functional, albeit very small, proteins (**Figure 8–69**). Finally, ribosome profiling has revealed the ways that cells rapidly and globally change their translation patterns in response to sudden changes in temperature, nutrient availability, or chemical stress.

Recombinant DNA Methods Have Revolutionized Human Health

We have seen that nucleic acid methodologies developed in the past 40 years have completely changed the way that cell and molecular biology is studied. But they have also had a profound effect on our day-to-day lives. Many human pharmaceuticals in routine use (insulin, human growth hormone, blood-clotting factors, and interferon, for example) are based on cloning human genes and expressing the encoded proteins in large amounts. As DNA sequencing continues to drop in cost, more and more individuals will elect to have their genome sequenced; this information can be used to predict susceptibility to diseases (often with the option of minimizing this possibility by appropriate behavior) or to predict the way an individual will respond to a given drug. The genomes of tumor cells from an individual can be sequenced to determine the best type of anticancer treatment. And mutations that cause or greatly increase the risk of disease continue to be identified at an unprecedented pace. Using the recombinant DNA technologies discussed in this chapter, these mutations can then be introduced into animals, such as mice, that can be studied in the laboratory. The resulting transgenic animals, which often mimic some of the phenotypic abnormalities associated with the condition in patients, can be used to explore the cellular and molecular basis of the disease and to screen for drugs that could potentially be used therapeutically in humans.

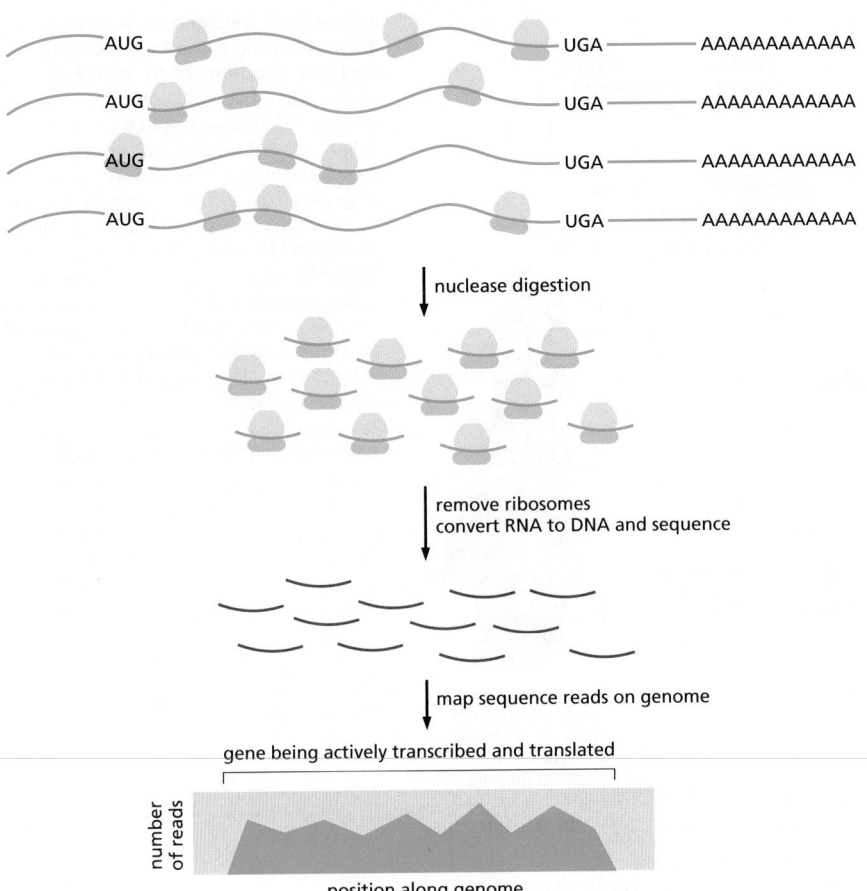

Figure 8–68 **Ribosome profiling.** RNA is purified from cells and digested with an RNAse to leave only those portions of the mRNAs that are protected by a bound ribosome. These short pieces of protected RNA (approximately 20 nucleotides in length) are converted to DNA and sequenced. The resulting information is displayed as the number of sequence reads along each position of the genome. In the diagram here, the data for only one gene, whose mRNA is being efficiently translated, are shown. Ribosome profiling provides this type of information for every mRNA produced by the cell.

Transgenic Plants Are Important for Agriculture

Although we tend to think of recombinant DNA research in terms of animal biology, these techniques have also had a profound impact on the study of plants. In fact, certain features of plants make them especially amenable to recombinant DNA methods.

When a piece of plant tissue is cultured in a sterile medium containing nutrients and appropriate growth regulators, some of the cells are stimulated to proliferate indefinitely in a disorganized manner, producing a mass of relatively undifferentiated cells called a *callus*. If the nutrients and growth regulators are carefully manipulated, one can induce the formation of a shoot within the callus, and in many species a whole new plant can be regenerated from such shoots. In a number of plants—including tobacco, petunia, carrot, potato, and *Arabidopsis*—a single cell from such a callus (known as a *totipotent cell*) can be grown into a small clump of cells from which a whole plant can be regenerated (see Figure 7–2B). Just as mutant mice can be derived by the genetic manipulation of embryonic stem

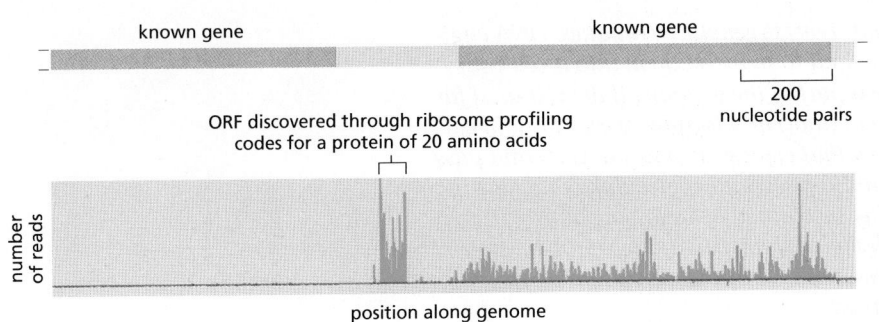

Figure 8–69 **Ribosome profiling can identify new genes.** This experiment shows the discovery of a previously unrecognized gene—one that encodes a protein of only 20 amino acids. At the top is shown a portion of a viral genome with two previously annotated genes. Below are the results of a ribosome profiling experiment, displayed across the same section of the genome, after the virus was infected into human cells. The results show that the left-hand gene is not expressed under these conditions, the right-hand gene is expressed at low levels, and a previously unrecognized gene that lies between them is expressed at high levels.

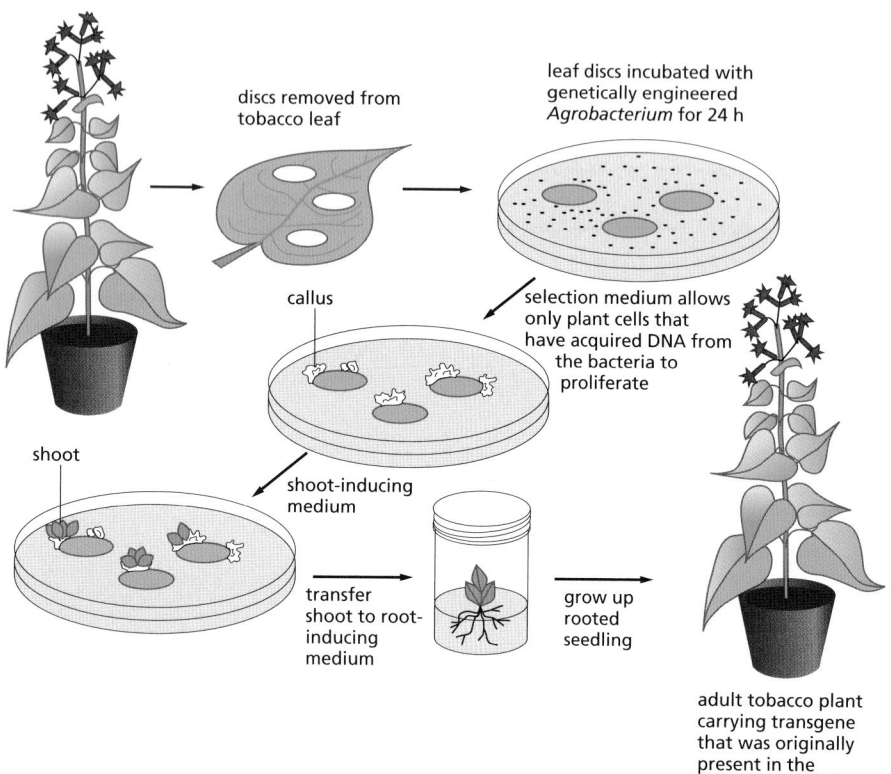

leaf discs incubated with
genetically engineered
Agrobacterium for 24 h

discs removed from
tobacco leaf

callus

selection medium allows
only plant cells that
have acquired DNA from
the bacteria to
proliferate

shoot

shoot-inducing
medium

transfer
shoot to root-
inducing
medium

grow up
rooted
seedling

adult tobacco plant
carrying transgene
that was originally
present in the
bacterial plasmid

Figure 8–70 Transgenic plants can be made using recombinant DNA techniques optimized for plants. A disc is cut out of a leaf and incubated in a culture of *Agrobacterium* that carries a recombinant plasmid with both a selectable marker and a desired genetically engineered gene. The wounded plant cells at the edge of the disc release substances that attract the bacteria, which inject their DNA into the plant cells. Only those plant cells that take up the appropriate DNA and express the selectable marker gene survive and proliferate and form a callus. The manipulation of growth factors supplied to the callus induces it to form shoots, which subsequently root and grow into adult plants carrying the engineered gene.

cells in culture, so transgenic plants can be created from plant cells transfected with DNA in culture (**Figure 8–70**).

The ability to produce transgenic plants has greatly accelerated progress in many areas of plant cell biology. It has played an important part, for example, in isolating receptors for growth regulators and in analyzing the mechanisms of morphogenesis and of gene expression in plants. These techniques have also opened up many new possibilities in agriculture that could benefit both the farmer and the consumer. They have made it possible, for example, to modify the ratio of lipid, starch, and protein in seeds, to impart pest and virus resistance to plants, and to create modified plants that tolerate extreme habitats such as salt marshes or water-stressed soil. One variety of rice has been genetically engineered to produce β-carotene, the precursor of vitamin A. Were it to replace conventional rice, this "golden rice"—so-called because of its faint yellow color—could help to alleviate severe vitamin A deficiency, which causes blindness in hundreds of thousands of children in the developing world each year.

Summary

Genetics and genetic engineering provide powerful tools for understanding the function of individual genes in cells and organisms. In the classical genetic approach, random mutagenesis is coupled with screening to identify mutants that are deficient in a particular biological process. These mutants are then used to locate and study the genes responsible for that process.

Gene function can also be ascertained by reverse genetic techniques. DNA engineering methods can be used to alter genes and to re-insert them into a cell's chromosomes so that they become a permanent part of the genome. If the cell used for this gene transfer is a fertilized egg (for an animal) or a totipotent plant cell in culture, transgenic organisms can be produced that express the mutant gene and pass it on to their progeny. Especially important for cell and molecular biology is the ability to alter cells and organisms in highly specific ways—allowing one to discern the effect on the cell or the organism of a designed change in a single protein or RNA molecule. For example, genomes can be altered so that the expression of any gene can be switched on or off by the experimenter.

Many of these methods are being expanded to investigate gene function on a genome-wide scale. The generation of mutant libraries in which every gene in an organism has been systematically deleted, disrupted, or made controllable by the experimenter provides invaluable tools for exploring the role of each gene in the elaborate molecular collaboration that gives rise to life. Technologies such as RNA-seq and DNA microarrays can monitor the expression of tens of thousands of genes simultaneously, providing detailed, comprehensive snapshots of the dynamic patterns of gene expression that underlie complex cell processes.

MATHEMATICAL ANALYSIS OF CELL FUNCTIONS

Quantitative experiments combined with mathematical theory mark the beginning of modern science. Galileo, Kepler, Newton, and their contemporaries did more than set out some rules of mechanics and offer an explanation of the movements of the planets around the Sun: they showed how a quantitative mathematical approach could provide a depth and precision of understanding, at least for physical systems, that had never before been dreamed to be possible.

What is it that gives mathematics this almost magical power to explain the natural world, and why has mathematics played so much more important a part in physical sciences than in biology? What do biologists need to know about mathematics?

Mathematics can be viewed as a tool for deriving logical consequences from propositions. It differs from ordinary intuitive reasoning in its insistence on rigorous, accurate logic and the precise treatment of quantitative information. If the initial propositions are correct, then the deductions drawn from them by mathematics will be true. The surprising power of mathematics comes from the length of the chains of reasoning that rigorous logic and mathematical arguments make possible, and from the unexpectedness of the conclusions that can be reached, often revealing connections that one would not otherwise have guessed at. Reversing the argument, mathematics provides a way to test experimental hypotheses: if mathematical reasoning from a given hypothesis leads to a prediction that is not true, then the hypothesis is not true.

Clearly, mathematics is not much use unless we can frame our ideas—our initial hypotheses—about the given system in a precise, quantitative form. A mathematical edifice raised on a rickety or—even worse—a vague or overcomplicated set of propositions is likely to lead us astray. For mathematics to be useful, we must focus our analysis on simple subsystems in which we can pick out key quantitative parameters and frame well-defined hypotheses. This approach has been used with great success in physics for centuries, but it has been less common in biology. But times are changing, and more and more it is becoming possible for biologists to exploit the power of quantitative mathematical analysis.

In this final section of our methods chapter, we do not attempt to teach readers every way in which mathematics can be fruitfully applied to biological problems. Rather, we simply aim to give a sense of what mathematics and quantitative approaches can do for us in modern biology. We focus primarily on the important principles that mathematics teaches us about the dynamics of molecular interactions, and how mathematics can unveil surprising and useful features of complex systems containing feedback. We will illustrate these principles using the regulation of gene expression by transcription regulators like those discussed in Chapter 7. The same principles apply to the post-transcriptional regulatory systems that govern cell signaling (Chapter 15), cell-cycle control (Chapter 17), and essentially all cell processes.

Regulatory Networks Depend on Molecular Interactions

Cell function and regulation depend on transient interactions among thousands of different macromolecules in the cell. We often summarize these interactions in this book with schematic cartoons. These diagrams are useful, but a complete picture requires a deeper, more *quantitative* level of understanding. To meaningfully assess the biological impact of any interaction in the cell, we need to know in precise terms how the molecules interact, how they catalyze reactions, and, most importantly, how the behaviors of the molecules change over time. If a cartoon shows that protein A activates protein B, for example, we cannot judge the importance of this relationship without quantitative details about the concentrations, affinities, and kinetic behaviors of proteins A and B.

Let us begin by defining two different types of regulatory interaction in our cartoons: one designating inhibition and the other designating activation. If the protein product of gene X is a transcription repressor that inhibits the expression of gene Z, we depict the relationship as a *red bar-headed line* ($\dashv$) drawn between genes X and Z (**Figure 8–71**). If the protein product of gene Y is a transcription activator that induces the expression of gene Z, then a *green arrow* ($\rightarrow$) is drawn between genes Y and Z.

The regulation of one gene's expression by another is more complicated than a single arrow connecting them, and a complete understanding of this regulation requires that we tease apart the underlying biochemical processes. **Figure 8–72**A sketches some of the biochemical steps in the activation of gene expression by a transcription activator. A gene encoding the activator, designated as gene A, will produce its product, protein A, via an RNA intermediate. This protein A will then bind to p_X, the regulatory *promoter* of gene X, to form the complex $A{:}p_X$. Once the $A{:}p_X$ complex forms, it stimulates the production of an RNA transcript that is subsequently translated to produce protein X.

We will focus here on the binding interaction that lies at the heart of this regulatory system: the interaction between protein A and the promoter p_X. Any molecule of protein A that is bound to p_X can also dissociate from it. The steps represented by the green activation arrow in Figure 8–72A include both the binding of A to p_X and the dissociation of the complex $A{:}p_X$ to re-form A and p_X, as illustrated

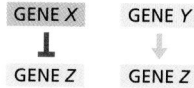

Figure 8–71 Diagrams that summarize biochemical relationships. Here, a simple cartoon indicates that gene X represses gene Z (*left*) whereas gene Y activates gene Z (*right*).

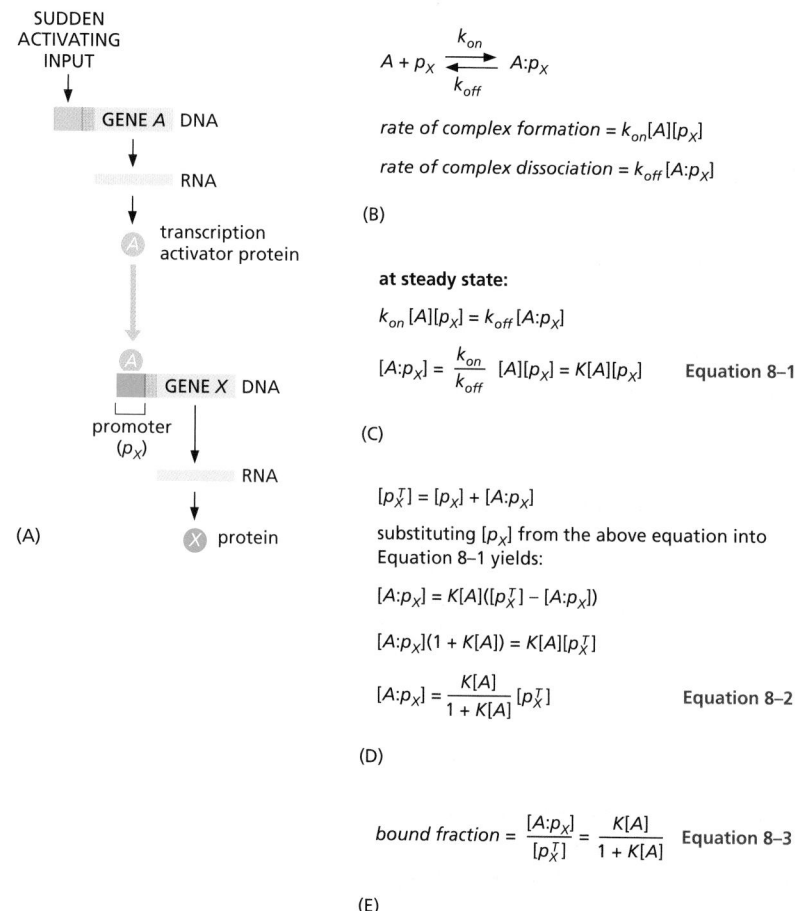

Figure 8–72 A simple transcriptional interaction. (A) Genes A and X each produce a protein, with the product of gene A serving as a transcription activator to stimulate expression of gene X. As indicated by the *green arrow*, stimulation depends in part on the binding of protein A to the promoter region of gene X, designated as p_X. (B) The binding of protein A to the gene promoter is determined by the concentrations of the two binding partners (denoted as $[A]$ and $[p_X]$, in units of mol/liter, or M), the association rate constant k_{on} (in units of $sec^{-1} M^{-1}$), and the dissociation rate constant k_{off} (in units of sec^{-1}). (C) At steady state, the rates of association and dissociation are equal, and the concentration of the bound complex is determined by Equation 8–1, in which the two rate constants are combined in the equilibrium constant K. (D) Equation 8–2 can be derived to calculate the steady-state concentration of bound complex at a known total concentration of the promoter $[p_X^T]$. (E) Rearrangement of Equation 8–2 yields Equation 8–3, which allows calculation of the fraction of promoter p_X that is occupied by protein A.

by the notation in Figure 8–72B. This reaction notation is more informative than the diagrams in our figures, but has its own limitations. Suppose that the concentration of A increases by a factor of ten as a response to an environmental input. If A increases, we intuitively know that $A{:}p_X$ should increase too, but we cannot determine the amount of the increase without additional information. We need to know the affinity of the binding interaction and the concentrations of the components. With this information in hand, we can rigorously derive the answer.

As discussed earlier and in Chapter 3 (see Figure 3–44), we know that the formation of a complex between two binding partners, such as A and p_X, depends on a rate constant k_{on}, which describes how many productive collisions occur per unit time per protein at a given concentration of p_X. The rate of complex formation equals the product of this rate constant k_{on} and the concentrations of A and p_X (see Figure 8–72B). Complex dissociation occurs at a rate k_{off} multiplied by the concentration of the complex. The rate constant k_{off} can differ by orders of magnitude for different DNA sequences because it depends on the strength of the noncovalent bonds formed between A and p_X.

We are primarily interested in understanding the amount of bound promoter complex at equilibrium or *steady state*, where the rate of complex formation equals the rate of complex dissociation. Under these conditions, the concentration of the promoter complex is specified by a simple equation that combines the two rate constants into a single equilibrium constant $K = k_{on}/k_{off}$ (Equation 8–1; Figure 8–72C). K is sometimes called the association constant, K_a. The larger this constant K, the stronger the interaction between A and p_X (see Figure 3–44). The reciprocal of K is the dissociation constant, K_d.

To calculate the steady-state concentration of promoter complex using Equation 8–1, we need to account for another complication: both A and p_X exist in two forms—free in solution and bound to each other. In most cases, we know the total concentration of p_X and not the free or bound concentrations, so we must find a way to use the total concentration in our calculations. To do this, we first specify that the total concentration of p_X ($[p_X^T]$) is the sum of the concentrations of free ($[p_X]$) and bound ($[A{:}p_X]$) forms (Figure 8–72D). This leads to a new equation that allows us to use $[p_X^T]$ to calculate the steady-state concentration of the promoter complex ($[A{:}p_X]$) (Equation 8–2, Figure 8–72D).

Protein A also exists in two forms: free ($[A]$) and bound to p_X ($[A{:}p_X]$). In a cell, there are typically one or two copies of p_X (assuming there is only one gene X per haploid genome) and multiple copies of A. As a result, we can safely assume that from the viewpoint of A, $[A{:}p_X]$ is negligible relative to the total $[A^T]$. This means that $[A] \approx [A^T]$, and we can just plug in the values of total $[A^T]$ in Equation 8–2 without incurring appreciable error in the calculation of $[A{:}p_X]$.

Now, we are ready to determine the effects of increasing the concentration of A. Suppose that $K = 10^8$ M^{-1}, which is a typical value for many such interactions. The starting concentration of A is $[A^T] = 10^{-9}$ M, and $[p_X^T] = 10^{-10}$ M (assuming there is one copy of gene X in a haploid yeast cell, for example, with a volume of around 2×10^{-14} L). Using Equation 8–2, we find that a tenfold increase in the concentration of A causes the amount of promoter complex $[A{:}p_X]$ to increase 5.5-fold, from 0.09×10^{-10} M to 0.5×10^{-10} M at steady state. The effects of a tenfold increase in the concentration of A will vary dramatically depending on its starting concentration relative to the equilibrium constant. Only through this mathematical approach can we achieve a thorough understanding of what these effects will be and what impact they will have on the biological response.

To assess the biological impact of a change in transcription activator levels, it is also important in many cases to determine the fraction of the target gene promoter that is bound by the activator, since this number will be directly proportional to the activity of the gene's promoter. In our case, we can calculate the fraction of the gene X promoter, p_X, that has protein A bound to it by rearranging Equation 8–2 (Equation 8–3, Figure 8–72E). This fraction can be viewed as the probability that promoter p_X is occupied, averaged over time. It is also equal to the average occupancy across a large population of cells at any instant in time. When there is no protein A present, p_X is always free, the bound fraction is zero, and transcription is off. When $[A] = 1/K$, the promoter p_X has a 50% chance of being occupied. When $[A]$ greatly exceeds $1/K$, the bound fraction is almost equal to one, meaning that p_X is fully occupied and transcription is maximal.

Differential Equations Help Us Predict Transient Behavior

The most important and basic insights for which we, as biologists, depend on mathematics concern the behavior of regulatory systems over time. This is the central theme of dynamics, and it was for the solution of problems in dynamics that the techniques of calculus were developed, by Newton and Leibniz, in the seventeenth century. Briefly, the general problem is this: if we are given the rates of change of a set of variables that characterize the system at any instant, how can we compute its future state? The problem becomes especially interesting, and the predictions often remarkable, when the rates of change themselves depend on the values of the state variables, as in systems with feedback.

Let us return to Equation 8–2 (Figure 8–72D), which tells us that when $[A]$ changes, $[A{:}p_X]$ at steady state will also change to a new concentration that we can calculate with precision. However, $[A{:}p_X]$ does not change instantaneously to this value. If we hope to understand the behavior of this system in detail, we must also ask how long it takes $[A{:}p_X]$ to get to its new steady-state value inside the cell. Equation 8–2 cannot answer this question. We need calculus.

The most common strategy for solving this problem is to use ordinary differential equations. The equations that describe biochemical reactions have a simple premise: the rate of change in the concentration of any molecular species X (that is, $d[X]/dt$) is given by the balance of the rate of its appearance with that of its disappearance. For our example, the rate of change in the concentration of the bound promoter complex, $[A{:}p_X]$, is determined by the rates of complex assembly and disassembly. We can incorporate these rates into the differential equation shown in **Figure 8–73**A (Equation 8–4). When $[A]$ changes, Equation 8–4 can be solved to generate the concentration of $[A{:}p_X]$ as a function of time. Notice that when $k_{on}[A][p_X] = k_{off}[A{:}p_X]$, then $d[A{:}p_X]/dt = 0$ and $[A{:}p_X]$ stops changing. At this point, the system has reached the steady state.

Calculation of all $[A{:}p_X]$ values as a function of time, using Equation 8–4, allows us to determine the rate at which $[A{:}p_X]$ reaches its steady-state value. Because this value is attained asymptotically, it is often most useful to compare the times needed to get to 50, 90, or 99 percent of this new steady state. The simplest way to determine these values is to solve Equation 8–4 with a method called numerical integration, which involves plugging in values for all of the parameters (k_{on}, k_{off}, etc.) and then using a computer to determine the values of $[A{:}p_X]$ over time, starting from given initial concentrations of $[A]$ and $[p_X]$. For $k_{on} = 0.5 \times 10^7$ sec^{-1} M^{-1}, $k_{off} = 0.5 \times 10^{-1}$ sec^{-1} ($K = 10^8$ M^{-1} as above), and $[p_X^T] = 10^{-10}$ M, it takes $[A{:}p_X]$ about 5, 20, and 40 seconds to reach 50, 90, and 99 percent of the new steady-state value following a sudden tenfold change in $[A]$ (Figure 8–73B). Thus, a sudden jump in $[A]$ does not have instantaneous effects, as we might have assumed from looking at the cartoon in Figure 8–72A.

Differential equations therefore allow us to understand the transient dynamics of biochemical reactions. This tool is critical for achieving a deep understanding of cell behavior, in part because it allows us to determine the dependence of the dynamics inside cells on parameters that are specific to the particular molecules involved. For example, if we double the values of both k_{on} and k_{off}, then Equation 8–1 (Figure 8–72C) indicates that the steady-state value of $[A{:}p_X]$ does not change. However, the time it takes to reach 50% of this steady state after a ten-fold

Figure 8–73 Using differential equations to study the dynamics and steady-state behavior of a biological system. (A) Equation 8–4 is an ordinary differential equation for calculating the rate of change in the formation of bound promoter complex in response to a change in other components. (B) Formation of $[A{:}p_X]$ after a tenfold increase in $[A]$, as determined by solving Equation 8–4. In *blue* is the solution corresponding to $k_{on} = 0.5 \times 10^7$ sec^{-1} M^{-1} and $k_{off} = 0.5 \times 10^{-1}$ sec^{-1}. In this case, it takes $[A{:}p_X]$ about 5, 20, and 40 seconds to reach 50, 90, and 99 percent of the new steady-state value. For the *red curve*, the k_{on} and k_{off} values are doubled, and the system reaches the same steady state more rapidly.

$$\frac{d[A{:}p_X]}{dt} = \textit{rate of complex formation} - \textit{rate of complex dissociation}$$

$$\frac{d[A{:}p_X]}{dt} = k_{on}[A][p_X] - k_{off}[A{:}p_X] \qquad \textbf{Equation 8–4}$$

(A)

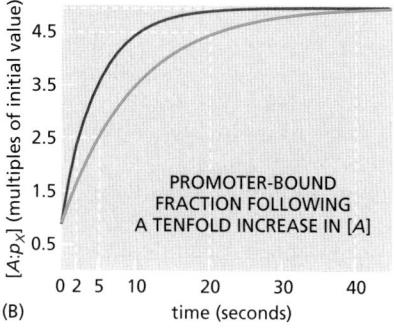

(B)

PROMOTER-BOUND FRACTION FOLLOWING A TENFOLD INCREASE IN $[A]$

$[A{:}p_X]$ (multiples of initial value)

time (seconds)

$$transcription\ rate = \beta\,\frac{K[A]}{1 + K[A]}$$

$$protein\ production\ rate = \beta\cdot m\,\frac{K[A]}{1 + K[A]}$$

$$protein\ degradation\ rate = \frac{[X]}{\tau_X}$$

(A)

$$\frac{d[X]}{dt} = protein\ production\ rate - protein\ degradation\ rate$$

$$\frac{d[X]}{dt} = \beta\cdot m\,\frac{K[A]}{1 + K[A]} - \frac{[X]}{\tau_X} \qquad \textbf{Equation 8–5}$$

(B)

at steady state:

$$[X_{st}] = \beta\cdot m\,\frac{K[A]}{1 + K[A]}\cdot \tau_X \qquad \textbf{Equation 8–6}$$

(C)

$$[X](t) = [X_{st}](1 - e^{-\frac{t}{\tau_X}})$$

(D)

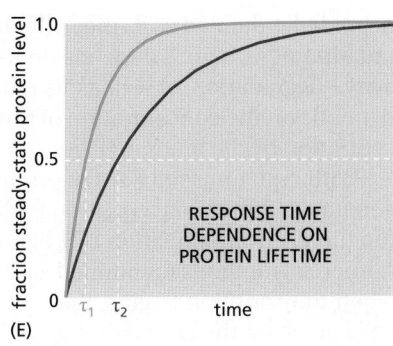

(E)

Figure 8–74 Effect of protein lifetime on the timing of the response.
(A) Equations for calculation of the rates of gene X transcription, protein X production, and protein X degradation, as explained in the text. (B) Equation 8–5 is an ordinary differential equation for calculating the rate of change in protein X in response to changes in other components. (C) When the rate of change in protein X is zero (steady state), its concentration can be calculated with Equation 8–6, revealing a direct relationship with protein lifetime (τ). (D) The solution of Equation 8–5 specifies the concentration of protein X over time as it approaches its steady-state concentration. (E) Response time depends on protein lifetime. As described in the text, the time that it takes a protein to reach a new steady state is greater when the protein is more stable. Here, the *blue line* corresponds to a protein with a lifetime that is 2.5-fold shorter than the lifetime of the protein in *red*.

change in [A] in our example changes from about 5 seconds to 2 seconds (see Figure 8–73B). These insights are not accessible from either cartoons or equilibrium equations. This is an unusually simple example; mathematical descriptions such as differential equations become more indispensable for understanding biological interactions as the number of interactions increases.

Both Promoter Activity and Protein Degradation Affect the Rate of Change of Protein Concentration

To understand our gene regulatory system further, we also need to describe the dynamics of protein X production in response to changes in the amount of transcription activator protein A. Here again, we use an ordinary differential equation for the rate of change of protein X concentration—determined by the balance of the rate of production of protein X through expression of gene X and the protein's rate of degradation.

Let us begin with the rate of protein X production, which is determined primarily by the occupancy of the promoter of gene X by protein A. The binding and dissociation of a transcription regulator at a promoter generally occurs on a much faster time scale than transcription initiation, causing many binding and unbinding events to occur before transcription proceeds. As a result, we can assume that the binding reaction is at equilibrium on the time scale of transcription, and we can calculate promoter occupancy by protein A using the equilibrium equation discussed earlier (Equation 8–3, Figure 8–72E). To determine transcription rate, we simply multiply the occupied promoter fraction by a *transcription rate constant*, β, that represents the binding of RNA polymerase and the subsequent steps that lead to production of mRNA and protein (**Figure 8–74**A). If each mRNA molecule produces, on average, m molecules of protein product, then we can determine protein production rate by multiplying the transcription rate by m (Figure 8–74A).

Now let us consider the factors that influence protein X degradation and its dilution due to cell growth. Degradation generally results in an exponential decline in protein levels, and the average time required for a specific protein to be degraded is defined as its mean lifetime, τ. In our current example, the rate of degradation of protein X depends on its mean lifetime τ_X, which takes into account active degradation as well as its dilution as the cell grows. The degradation rate depends on the concentration of protein X and is calculated by dividing this concentration by the lifetime (Figure 8–74A).

With equations for rates of production and degradation in hand, we can now generate a differential equation to determine the rate of change of protein X as a function of time (Equation 8–5, Figure 8–74B). This equation can be solved by the numerical methods mentioned earlier. According to the solution of this equation, when transcription begins, the concentration of protein X rises to a steady-state level at which the concentration of X is not changing anymore; that is, its rate of change is zero. When this occurs, rearrangement of Equation 8–5 yields an equation that can be used to determine the steady-state value of X, $[X_{st}]$ (Equation 8–6, Figure 8–74C). An important concept emerges from the mathematics: the steady-state concentration of a gene product is directly proportional to its lifetime. If lifetime doubles, protein concentration doubles as well.

The Time Required to Reach Steady State Depends on Protein Lifetime

We can see from Equation 8–6 (see Figure 8–74C) that when the concentration of protein A rises, protein X increases to a new steady-state value, $[X_{st}]$. But this cannot happen instantaneously. Instead, X changes dynamically according to the solution of its differential rate equation (Equation 8–5). The solution of this equation reveals that the concentration of X over time is related to its steady-state concentration according to the equation in Figure 8–74D. Once again, mathematics uncovers a simple but important concept that is not intuitively obvious: following a sudden increase in $[A]$, $[X]$ rises to a new steady state at an exponential rate that is inversely related to its lifetime; the faster X is degraded, the less time it takes it to reach its new steady-state value (Figure 8–74E). The faster response time comes at a higher metabolic cost, however, since proteins with a rapid response time must be produced and degraded at a high rate. For proteins that are not rapidly turned over, the response time is very long, and protein concentration is determined primarily by the dilution that results from cell growth and division.

Quantitative Methods Are Similar for Transcription Repressors and Activators

Positive control is not the only mechanism that cells use to regulate the expression of their genes. As we discussed in Chapter 7, cells also actively shut off genes, often by employing transcription repressor proteins that bind to specific sites on target genes, thereby blocking access to RNA polymerase. We can analyze the function of these repressors by the same quantitative methods described above for transcription activators. If a repressor protein R binds to the regulatory region of gene X and represses its transcription, then the fraction of gene binding sites occupied by the repressor is specified by the same equation we used earlier for the transcription activator (**Figure 8–75**A). In this case, however, it is only when the DNA is free that RNA polymerase can bind to the promoter and transcribe the gene. Thus, the quantity of interest is the unbound fraction, which can be viewed as the probability that the site is free, averaged over multiple binding and unbinding events. When the repressor concentration is zero, the unbound fraction is 1 and the promoter is fully active; when the repressor concentration greatly exceeds $1/K$, the unbound fraction approaches zero. Figures 8–75B and C compare these relationships for a transcription activator and a transcription repressor.

We can create a differential equation that provides the rate of change in protein X when repressor concentrations change (Equation 8–7, Figure 8–75D). As in the case of the transcription activator, the steady-state concentration of protein X increases as its lifetime increases, but it decreases as the concentration of the transcription repressor increases.

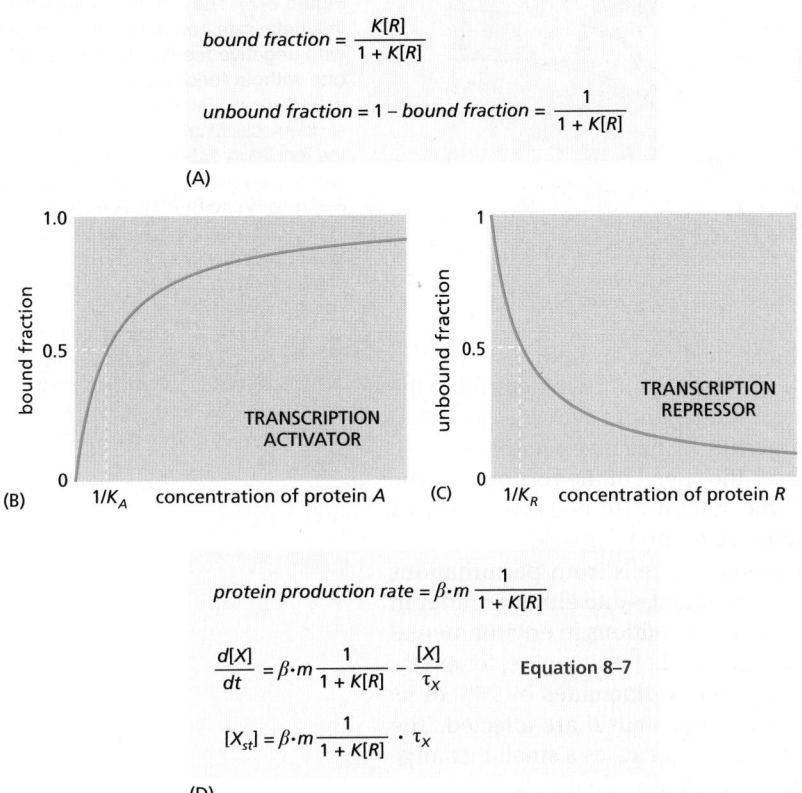

$$\text{bound fraction} = \frac{K[R]}{1 + K[R]}$$

$$\text{unbound fraction} = 1 - \text{bound fraction} = \frac{1}{1 + K[R]}$$

(A)

(B) bound fraction — TRANSCRIPTION ACTIVATOR — $1/K_A$ concentration of protein A

(C) unbound fraction — TRANSCRIPTION REPRESSOR — $1/K_R$ concentration of protein R

$$\text{protein production rate} = \beta \cdot m \frac{1}{1 + K[R]}$$

$$\frac{d[X]}{dt} = \beta \cdot m \frac{1}{1 + K[R]} - \frac{[X]}{\tau_X} \qquad \textbf{Equation 8–7}$$

$$[X_{st}] = \beta \cdot m \frac{1}{1 + K[R]} \cdot \tau_X$$

(D)

Figure 8–75 How promoter occupancy depends on the binding affinity of a transcription regulator protein. (A) The fraction of a binding site that is occupied by a transcription repressor R is determined by an equation that is similar to the one we used for a transcription activator (see Figure 8–72E), except that in the case of a repressor we are interested primarily in the unbound fraction. (B) For a transcription activator A, half of the promoters are occupied when $[A] = 1/K_A$. Gene activity is proportional to this bound fraction. (C) For a transcription repressor R, gene activity is proportional to the unbound fraction of promoters. As indicated, this fraction is reduced to half of its maximal value when $[R]=1/K_R$. (D) As in the case of the transcription activator A (see Figure 8–74), we can derive equations to assess the timing of protein X production as a function of repressor concentrations.

Negative Feedback Is a Powerful Strategy in Cell Regulation

Thus far, we have considered simple regulatory systems of just a few components. In most of the complex regulatory systems that govern cell behaviors, multiple modules are linked to produce larger circuits that we call *network motifs*, which can produce surprisingly complex and biologically useful responses whose properties become apparent only through mathematical analysis. A particularly common and important network motif is the negative feedback loop, which can have dramatically different functions depending on how it is structured.

We take as a first example a network motif consisting of two linked modules (**Figure 8–76**A). Here, an input signal initiates the transcription of gene A, which produces a transcription activator protein A. This activates gene R, which synthesizes a transcription repressor protein R. Protein R in turn binds to the promoter of gene A to inhibit its expression. This cyclical organization creates a negative feedback loop that one can intuitively understand as a mechanism to prevent proteins from accumulating to high levels. But what can we learn about negative feedback loops, and their value in biology, by using mathematics to model them?

The negative feedback loop in Figure 8–76A can be modeled using Equation 8–7 (see Figure 8–75D) for the repression of gene A and Equation 8–5 (see Figure 8–74B) for the activation of gene R. Thus, for proteins A and R, we use the set of differential equations (Equation set 8–8) shown in Figure 8–76B. The two equations in this set are coupled, which means that they must be solved together to describe

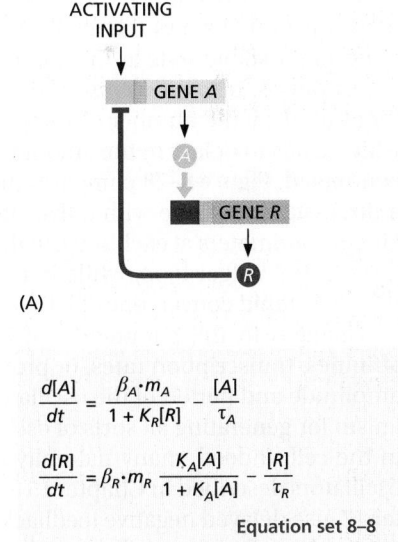

(A) ACTIVATING INPUT → GENE A → A → GENE R → R

$$\frac{d[A]}{dt} = \frac{\beta_A \cdot m_A}{1 + K_R[R]} - \frac{[A]}{\tau_A}$$

$$\frac{d[R]}{dt} = \beta_R \cdot m_R \frac{K_A[A]}{1 + K_A[A]} - \frac{[R]}{\tau_R}$$

Equation set 8–8

(B)

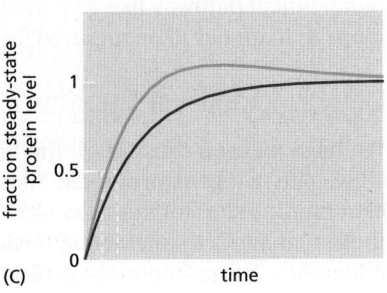

(C)

Figure 8–76 A simple negative feedback motif. (A) Gene A negatively regulates its own expression by activating gene R. The product of gene R is a transcription repressor that inhibits gene A. (B) Equation set 8–8 can be solved to determine the dynamics of system components over time. (C) A system with negative feedback *(blue)* reaches its steady state faster than a system with no feedback *(red)*. The plots indicate the levels of protein A, expressed as a fraction of the steady-state level. The *blue line* reflects the solution of Equation set 8–8, which includes negative feedback of gene A by the repressor R. The *red line* represents the solution when the rate of synthesis of A was set to a constant value that is unaffected by the repressor R.

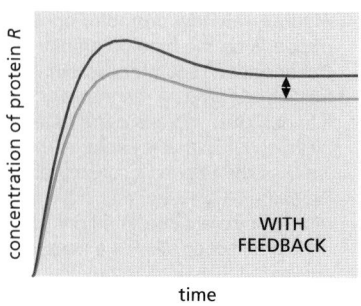

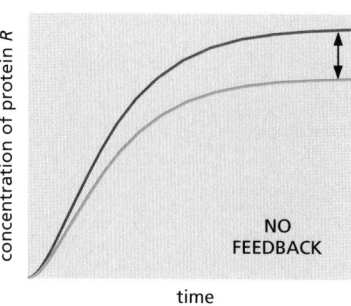

Figure 8–77 **The effect of fluctuations in kinetic rate constants on a system with negative feedback compared to one without feedback.** The plot at *left* represents the levels of protein R after a sudden activating stimulus, according to the regulatory scheme in Figure 8–76A and determined by the solution of Equation set 8–8 (see Figure 8–76B). A perturbation was induced by changing β_A from 4 M/min *(red line)* to 3 M/min *(blue line)*. The plot at *right* shows the results when negative feedback was removed. The system with negative feedback deviates less from its normal operation as β changes than does the system with no feedback. Notice that, as in Figure 8–76C, the system with negative feedback also reaches its steady state more rapidly.

the behavior of A and R over time for any value of the input. As before, we plug in values for the parameters (β_R, τ_R, etc.) and then use a computer to determine the values of [A] and [R] as a function of time after a sudden input activates gene A.

The results reveal several important properties of negative feedback. First, rather surprisingly, negative feedback increases the speed of the response to the activating input. As shown in Figure 8–76C, the system with negative feedback reaches its new steady state faster than the system with no feedback.

Second, negative feedback is useful for protecting cells from perturbations that continuously arise in the cell's internal environment—due either to random variations in the birth and death of molecules or to fluctuations in environmental variables such as temperature and nutritional supplies. Let us imagine, for example, that β_A, the transcription rate constant for gene A, fluctuates by 25% of its value and ask whether and how much the levels of protein R are affected. The results, shown in **Figure 8–77**, reveal that a change in β_A causes a smaller change in the steady-state value of R when the network has negative feedback.

Delayed Negative Feedback Can Induce Oscillations

A beautiful thing happens when a negative feedback loop contains some delay mechanism that slows the feedback signal through the loop: rather than generating a new stable state as in a rapid negative feedback loop, a delayed loop generates pulses, or *oscillations*, in the levels of its components. This can be seen, for example, if the number of components in a negative feedback loop increases, which leads to delays in the amount of time required for the cycle of signals to be completed. **Figure 8–78** compares the behavior of two network motifs—one with a three-stage and one with a five-stage negative feedback loop. Using the same kinetic parameters at each stage in the two loops, one finds that stable oscillations arise in the longer loop, while in the shorter loop the same parameters lead to relatively rapid convergence to a stable steady state.

Changes in the parameters of a delayed negative feedback loop—binding affinities, transcription rates, or protein stabilities, for example—can change the amplitude and period of the oscillations, providing a remarkably versatile mechanism for generating all sorts of oscillators that can be used for various purposes in the cell. Indeed, many naturally occurring oscillators, including the calcium oscillators described in Chapter 15 and the cell-cycle network described in Chapter 17, use delayed negative feedback as the basis for biologically important oscillations. Not all of the oscillations observed in cells are thought to have a function, however. Oscillations become inevitable in a highly complex, multicomponent biochemical pathway like glycolysis, due simply to the large number of feedback loops that appear to be required for its regulation.

DNA Binding By a Repressor or an Activator Can Be Cooperative

We have focused thus far on the binding of a single transcription regulator to a single site in a gene promoter. Many promoters, however, contain multiple adjacent binding sites for the same transcription regulator, and it is not uncommon for these regulators to interact with each other on the DNA to form dimers or larger oligomers. These interactions can result in a *cooperative* form of DNA binding,

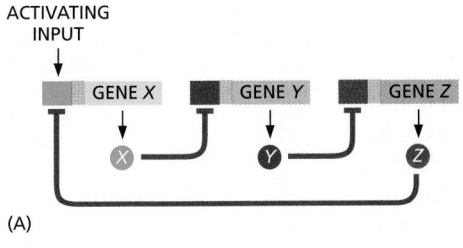

(A)

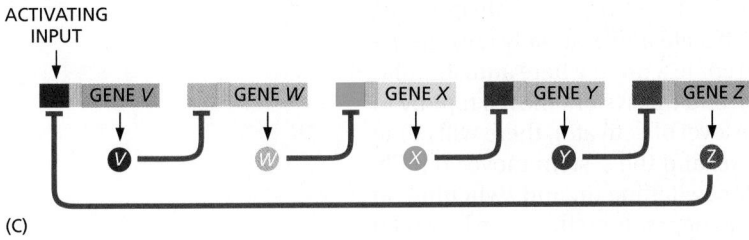

(C)

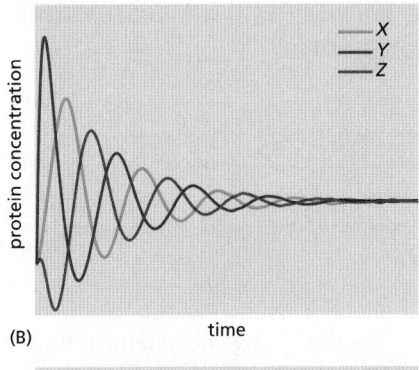

(B)

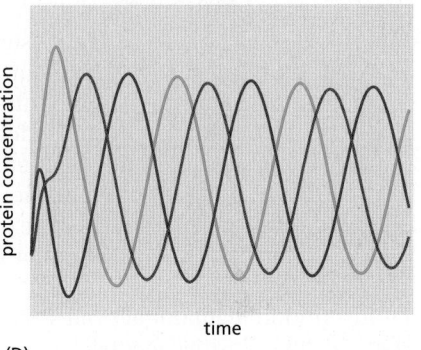

(D)

Figure 8–78 **Oscillations arising from delayed negative feedback.** A transcriptional circuit with three components (A, B) is less likely to oscillate than a transcriptional circuit with five components (C, D). The *X (light blue)*, *Y (dark blue)*, and *Z (brown)* here represent transcription regulatory proteins. For the simulations in (B) and (D), the system was initiated from random initial conditions for *X*, *Y*, and *Z*. Oscillations are produced by a delay induced as the signal propagates through the loop.

such that DNA-binding affinity increases at higher concentrations of the transcription regulator. Cooperativity produces a steeper transcriptional response to increasing regulator concentration than the response that can be generated by the binding of a monomeric protein to a single site. A steep transcriptional response of this sort, when present in conjunction with positive feedback, is an important ingredient for producing systems with the ability to switch between different discrete phenotypic states. To begin to understand how this occurs, we need to modify our equations to include cooperativity.

Cooperative binding events can produce steep S-shaped (or *sigmoidal*) relationships between the concentration of regulatory protein and the amount bound on the DNA (see Figure 15–16). In this case, a number called the *Hill coefficient* (*h*) describes the degree of cooperativity, and we can include this coefficient in our equations for calculating the bound fraction of promoter (**Figure 8–79**A). As the Hill coefficient increases, the dependence of binding on protein concentration becomes steeper (Figure 8–79B). In principle, the Hill coefficient is similar to the number of molecules that must come together to generate a reaction. In practice, however, cooperativity is rarely complete, and the Hill coefficient does not reach this number.

Figure 8–79 **How the cooperative binding of transcription regulatory proteins affects the fraction of promoters bound.** (A) Cooperativity is incorporated into our mathematical models by including a Hill coefficient (*h*) in the equations used previously to determine the fraction of bound promoter (see Figures 8–72E and 8–75A). When *h* is 1, the equations shown here become identical to the equations used previously, and there is no cooperativity. (B) The *left panel* depicts a cooperatively bound transcription activator and the *right panel* depicts a cooperatively bound transcription repressor. Recall from Figure 8–75B that gene activity is proportional to bound activator (*left panel*) or unbound repressor (*right panel*). Note that the plots get steeper as the Hill coefficient increases.

$$bound\ fraction = \frac{(K_A[A])^h}{1 + (K_A[A])^h}\ \text{for activators, or}\ \frac{(K_R[R])^h}{1 + (K_R[R])^h}\ \text{for repressors}$$

(A)

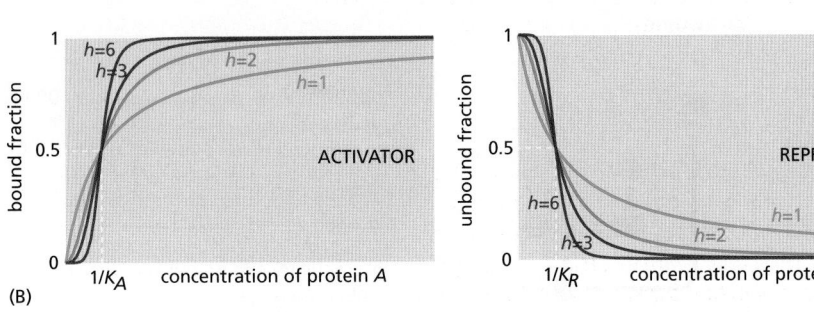

(B)

Positive Feedback Is Important for Switchlike Responses and Bistability

We turn now to positive feedback and its very important consequences. First and foremost, positive feedback can make a system *bistable*, enabling it to persist in either of two (or more) alternative steady states. The idea is simple and can be conveyed by drawing an analogy with a candle, which can exist either in a burning state or in an unlit state. The burning state is maintained by positive feedback: the heat generated by burning keeps the flame alight. The unlit state is maintained by the absence of this feedback signal: so long as sufficient heat has never been applied, the candle will stay unlit.

For the biological system, as for the candle, bistability has an important corollary: it means that the system has a memory, such that its present state depends on its history. If we start with the system in an Off state and gradually rack up the concentration of the activator protein, there will come a point where autostimulation becomes self-sustaining (the candle lights), and the system moves rapidly to an On state. If we now intervene to decrease the level of activator, there will come a point where the same thing happens in reverse, and the system moves rapidly back to an Off state. But the transition points for switching on and switching off are different, and so the current state of the system depends on the route by which it has been taken in the past—a phenomenon called *hysteresis*.

A simple case of positive feedback can be seen in a regulatory system in which a transcription regulator activates (directly or indirectly) its own expression, as in **Figure 8–80**A. Positive feedback can also arise in a circuit with many intervening repressors or activators, so long as the net overall effect of the interactions is activation (Figure 8–80B and C).

To illustrate how positive feedback can generate stable states, let us focus on a simple positive feedback loop containing two repressors, X and Y, each of which inhibits expression of the other (**Figure 8–81**A). As we saw with Equation set 8–8 (Figure 8–76B) earlier, we can create differential equations describing the rate of change of $[X]$ and $[Y]$ (Equation set 8–9, Figure 8–81B). We can further modify these equations to include cooperativity by adding Hill coefficients. As we did earlier, we can then create equations for calculating the concentrations of $[X]$ and $[Y]$ when the system reaches a steady state (that is, when $(d[X]/dt) = 0$ and $(d[Y]/dt) = 0$; Equations 8–10 and 8–11, Figure 8–81C).

Equations 8–10 and 8–11 can be used to carry out an intriguing mathematical procedure called a *nullcline* analysis. These equations define the relationships between the concentration of X at steady state, $[X_{st}]$, and the concentration of Y at steady state, $[Y_{st}]$, which must be simultaneously satisfied. We can plug in different values for $[Y_{st}]$ in Equation 8–10, and calculate the corresponding $[X_{st}]$ for each of these values. We can then graph $[X_{st}]$ as a function of $[Y_{st}]$. Next, we repeat the process by varying $[X_{st}]$ in Equation 8–11 to graph the resulting $[Y_{st}]$. The intersections of these two graphs determine the theoretically possible steady states of the system. For systems in which the Hill coefficients h_X and h_Y are much larger than 1, the lines in the two graphs intersect at three locations (Figure 8–81D). In other systems that have the same arrangement of regulators but different parameters, there might only be one intersection, indicating the presence of only a single

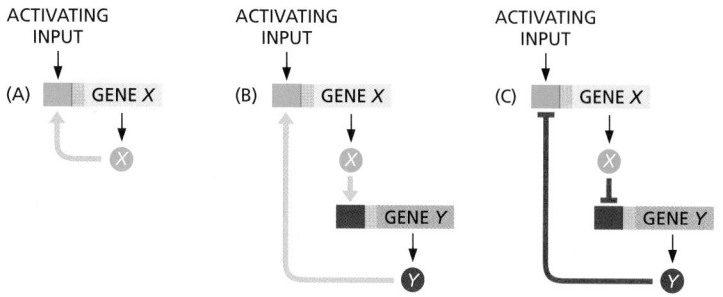

Figure 8–80 Positive feedback of a gene onto itself through serially connected interactions. A sequence of activators and repressors of any length can be connected to produce a positive feedback loop, as long as the overall sign is positive. Because the negative of a negative is positive, not only circuit (A) and (B) but also circuit (C) create positive feedback.

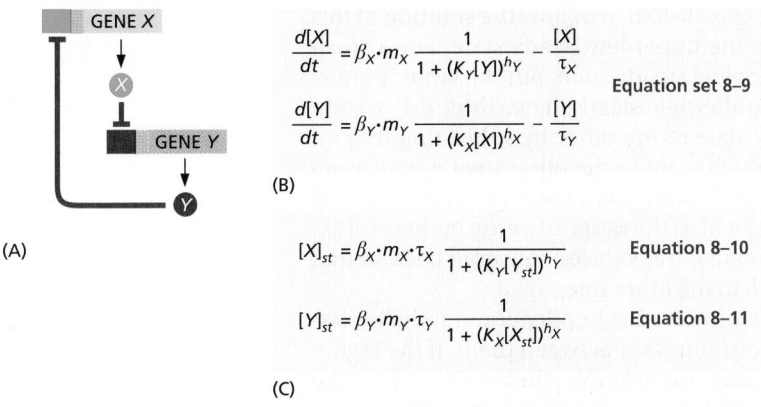

(A)

(B)

$$\frac{d[X]}{dt} = \beta_X \cdot m_X \frac{1}{1 + (K_Y[Y])^{h_Y}} - \frac{[X]}{\tau_X}$$

$$\frac{d[Y]}{dt} = \beta_Y \cdot m_Y \frac{1}{1 + (K_X[X])^{h_X}} - \frac{[Y]}{\tau_Y}$$

Equation set 8–9

$$[X]_{st} = \beta_X \cdot m_X \cdot \tau_X \frac{1}{1 + (K_Y[Y_{st}])^{h_Y}}$$

Equation 8–10

$$[Y]_{st} = \beta_Y \cdot m_Y \cdot \tau_Y \frac{1}{1 + (K_X[X_{st}])^{h_X}}$$

Equation 8–11

(C)

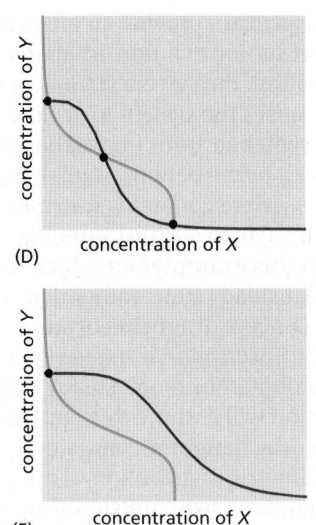

(D)

(E)

Figure 8–81 A graphical nullcline analysis. (A) X inhibits Y and Y inhibits X, resulting in a positive feedback loop. (B) Equation set 8–9 can be used to determine the rate of change in the concentrations of proteins X and Y. (C) Equations 8–10 and 8–11 provide the concentrations of proteins X and Y, respectively, when these concentrations reach a steady state. (D, E) *Blue curves* (called nullclines) are plots of $[X_{st}]$ calculated from Equation 8–10 over a range of concentrations of $[Y_{st}]$. *Red curves* indicate values of $[Y_{st}]$ calculated from Equation 8–11 over a range of concentrations of $[X_{st}]$. At an intersection of the two lines, both $[X]$ and $[Y]$ are at steady state. For plot (D), the binding of both proteins to their target gene promoters was cooperative (h_X and h_Y much larger than 1), resulting in the presence of multiple intersections of the nullclines—suggesting that the system can assume multiple discrete steady states. In plot (E), the binding of protein X to the promoter of gene Y was not cooperative (h_X close to 1), resulting in only one nullcline intersection and thus just one likely steady state.

steady state. For example, when there is a low cooperativity of protein X binding to the promoter of gene Y (that is, a small Hill coefficient, h_X, in Equation 8–11), the plot of $[Y]$ is less curved (Figure 8–81E), and it is less likely that there will be multiple intersections of the two curves.

We emphasized earlier that positive feedback typically generates a bistable system with two stable steady states. Why does the system modeled in Figure 8–81D have three? This conundrum can be explained by solving the reaction rate equations (Equation set 8–9, Figure 8–81B) for various different starting conditions of $[X]$ and $[Y]$, determining all values of $[X]$ and $[Y]$ as a function of time. Starting with each set of initial concentrations of $[X]$ and $[Y]$, these calculations produce a so-called *trajectory* of points, each indicated by a curved green line on **Figure 8–82A**. A fascinating pattern emerges: each trajectory moves across the plot and settles in one of two steady states, but never in the third (middle steady state). We conclude that the middle steady state is *unstable* because it cannot "attract" any trajectories. The system therefore has only two *stable* steady states. Thus, the number of stable steady states in a system need not be equal to the total number of its theoretically possible steady states. In fact, stable steady states are usually separated by unstable ones, as in our example.

Once this system adopts a fate by settling in one of the two steady states, does it have the ability to switch to the other state? The numerical solution of Equation

Figure 8–82 Analysis of the stability of a system's steady states. (A) The dotted lines are the nullclines for the system shown in Figure 8–81. Also shown are dynamic trajectories *(green)* that show the changes over time in $[X]$ and $[Y]$, starting at a variety of different initial concentrations (determined by solution of Equation set 8–9; see Figure 8–81B). By plotting $[X]$ versus $[Y]$ at each time point, we find that, although there are three possible steady states in this system, the dynamic trajectories converge on only two of them. The middle steady state is avoided: it is unstable, being unable to attract any trajectories. (B) Imagine that the system is at the upper-left steady state and experiences a perturbation *(black arrows)*, such as a random fluctuation in the production rates of X and/or Y. If the perturbation is small *(arrow 1)*, the system will return to the same steady state. On the other hand, a perturbation that drives the system beyond the unstable (middle) steady state *(arrow 2)* causes it to switch to the lower-right steady state. The set of perturbations that a system can withstand without switching from one steady state to the other is known as the region of attraction of that steady state.

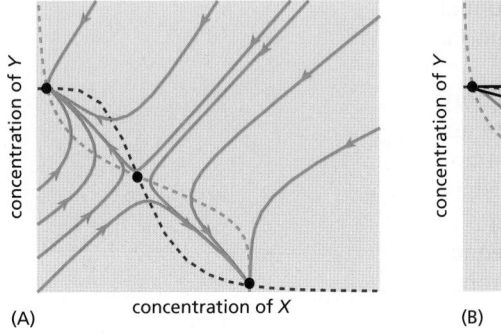

(A)

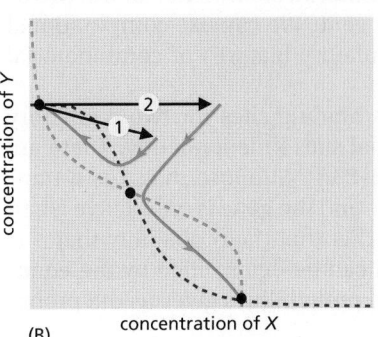

(B)

set 8–9 can again provide an answer. In Figure 8–82B, we show the solution of this equation set for two perturbations from the upper-left steady state. For a small perturbation, the system returns to its original steady state. But the larger perturbation causes the system to switch to the alternate steady state. Thus, this system can be switched from one stable steady state to the other by subjecting it to an input (or a perturbation) that is large enough to make the other steady state more attractive. More generally, every stable steady state has a corresponding *region of attraction*, which can be intuitively thought of as the range of perturbations (of [X] or [Y] in this example) for which the dynamic trajectories converge back to that particular steady state, rather than switch to the other one.

The concept of a region of attraction has interesting implications for the heritability of transcriptional states and the transition rates between them. If the region of attraction around one steady state is large, for example, then most cells in the population will assume this particular state. Furthermore, this state is likely to be inherited by daughter cells, since minor perturbations, like those ensuing from an asymmetric distribution of molecules during cell division, will rarely be sufficient to induce switching to the other steady state. We should expect that the use of positive feedback, coupled to cooperativity, will quite often be associated with systems requiring stable cell memory.

Robustness Is an Important Characteristic of Biological Networks

Biological regulatory systems are exposed to frequent and sometimes extreme variations in external conditions or the concentrations or activities of key components. The ability of these systems to function normally in the face of such perturbations is called **robustness**. If we understand a complex system to the extent that we can reproduce its behavior with a computational model, then the robustness of the system can be assessed by determining how well its normal function persists following changes in various parameters, such as rate constants and component concentrations. We have already seen, for example, how the presence of negative feedback reduces the sensitivity of the steady state to changes in the values of the system's parameters (see Figure 8–77). Considerations of robustness also apply to dynamic behaviors. Thus, for example, when discussing negative feedback, we described how the behavior of a system tends to become more oscillatory as the number of components that constitute the feedback loop increases. If we use different values of the parameters in models derived for systems like those in Figure 8–78, we find that the system with the longer loop tends to exhibit stable oscillations within a much broader range of parameters, indicating that this system provides a more robust oscillator. We can perform similar calculations to determine the ability of different systems to achieve robust bistability arising from positive feedback. Thus, one benefit of computational models is that they allow us to probe the robustness of biological networks in a systematic and rigorous way.

Two Transcription Regulators That Bind to the Same Gene Promoter Can Exert Combinatorial Control

Thus far, we have discussed how one transcription regulator can modulate the expression level of a gene. Most genes, however, are controlled by more than one type of transcription regulator, providing *combinatorial control* that allows two or more inputs to influence the expression of one gene. We can use computational methods to unveil some of the important regulatory features of combinatorial control systems.

Consider a gene whose promoter contains binding sites for two regulatory proteins, A and R, which bind to their individual sites independently. There are four possible binding configurations (**Figure 8–83**A). Suppose that A is a transcription activator, R is a transcription repressor, and the gene is only active when A is bound and R is not bound. We learned earlier that the probability that A is bound and the probability that R is not bound can be determined by the equations in **Figure 8–84**A. The product of these two probabilities gives us the probability of gene activation.

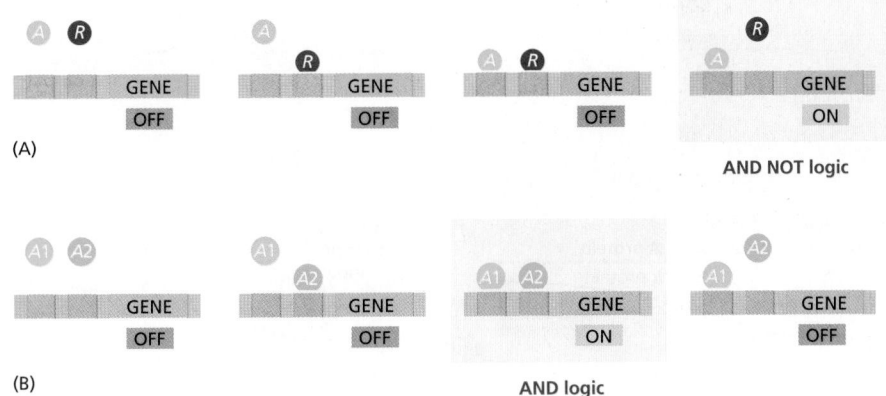

(A)

AND NOT logic

(B)

AND logic

Figure 8–83 Combinatorial control of gene expression. There are many ways in which gene expression can be controlled by two transcription regulators. To define precisely the relationship between the two inputs and the gene expression output, a regulatory circuit is often described as a specific type of *logic gate*, a term borrowed from electronic circuit design. A simple example is the OR logic gate (not shown here), in which a gene is controlled by two transcription activators, and one or the other can activate gene expression. (A) In a system with an activator A and repressor R, if transcription is turned on only when A is bound and R is not, then the result is an AND NOT logic gate. We saw an example of this logic in Chapter 7 (Figure 7–15). (B) An AND gate results when two transcription activators, A1 and A2, are both required to turn on a gene.

This example illustrates an AND NOT logic function (A and not R) (see Figure 8–83A). Maximal activation of this gene is accomplished when $[A]$ is high and $[R]$ is zero. However, intermediate levels of gene activation are also possible depending on the levels of A and R and also on the binding affinities of $[A]$ and $[R]$ for their respective sites (that is, K_A and K_R). When $K_A \gg K_R$, even a small concentration of $[A]$ is capable of overcoming repression by R. Conversely, if $K_A \ll K_R$, then much more $[A]$ is needed to activate the gene (Figure 8–84B and C).

Many other logic functions can govern combinatorial gene regulation. For example, an AND logic gate results when two activators, $A1$ and $A2$, are both required for a gene to be transcribed (Figures 8–83B and 8–84D). In *E. coli* cells, the *AraJ* gene controls some aspects of arabinose sugar metabolism: its expression requires two transcription regulators, one activated by arabinose and the other activated by the small molecule cAMP (Figure 8–84E).

$$\text{fraction of } A \text{ bound} = \frac{K_A[A]}{1 + K_A[A]}$$

$$\text{fraction of } R \text{ not bound} = \frac{1}{1 + K_R[R]}$$

$$P(A,R) = \frac{K_A[A]}{1 + K_A[A]} \cdot \frac{1}{1 + K_R[R]} = \frac{K_A[A]}{1 + K_A[A] + K_R[R] + K_A K_R[A][R]}$$

(A)

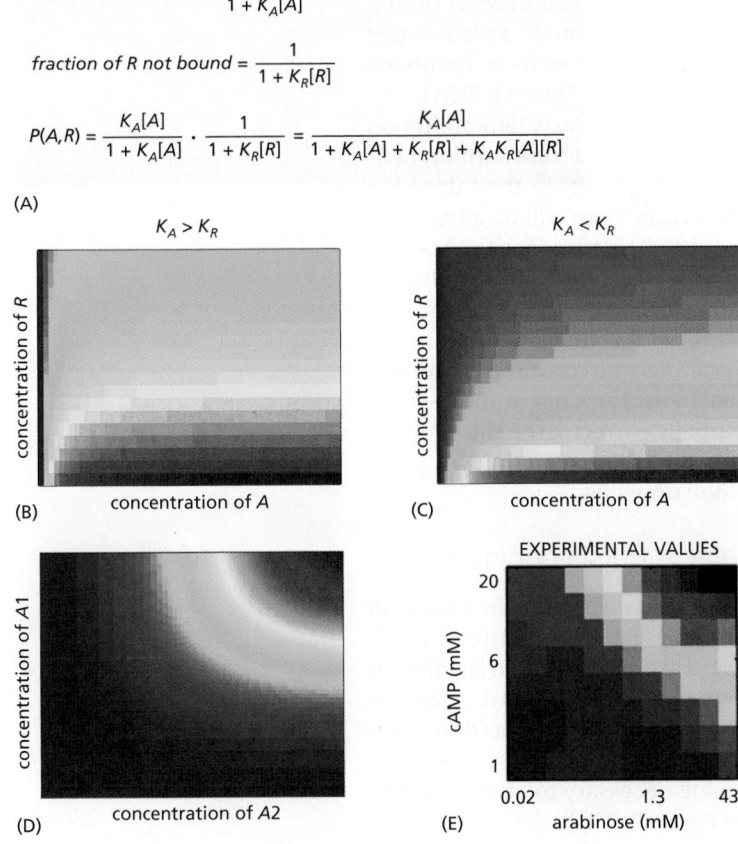

$K_A > K_R$

$K_A < K_R$

concentration of R

concentration of R

(B)   concentration of A

(C)   concentration of A

concentration of A1

EXPERIMENTAL VALUES

20

6

1

cAMP (mM)

0.02   1.3   43

(D)   concentration of A2

(E)   arabinose (mM)

Figure 8–84 How the quantitative output of a gene depends on both its combinatorial logic and the affinities of transcription regulators. (A) In a combinatorial gene regulatory system like that illustrated in Figure 8–83A, the fraction of promoters bound by activator A and not bound by repressor R are each determined as shown here. The product of these probabilities provides the probability, P(A, R), that a gene promoter is active. (B–E) In these four panels, *red* indicates high gene expression and *blue* indicates low gene expression. (B) and (C) depict gene expression from the system described in panel (A). The two panels demonstrate how the system behaves when the relative affinities of the two transcription regulators change as indicated above each panel. (D) Gene expression in a case where the gene turns on only at high levels of both activating inputs (A1 and A2), as shown in Figure 8–83B. (E) Experimental data showing measured expression of a gene in *E. coli* that is combinatorially regulated by two inputs: arabinose and cAMP. Note the close resemblance to panel (D). (E, adapted from S. Kaplan et al., *Mol. Cell* 29:786–792, 2008. With permission from Elsevier.)

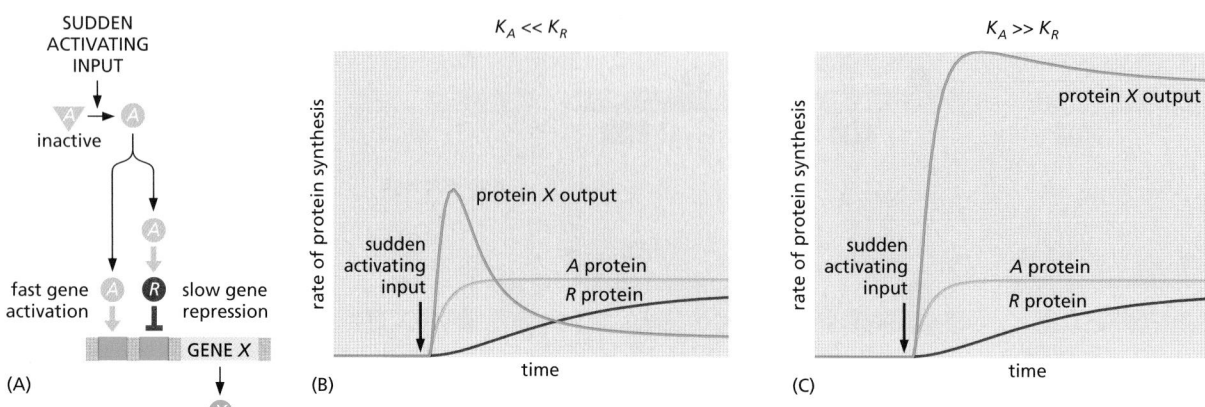

An Incoherent Feed-forward Interaction Generates Pulses

Imagine that a sudden input signal immediately activates a transcription activator A and that the same input signal induces the much slower synthesis of a transcription repressor protein R that acts on the same gene X. If A and R control gene expression by an AND NOT logic function like that described above, our intuition tells us that this system should be able to generate a pulse of transcription: when A is activated (and R is absent), the transcription of gene X will begin and cause an increase in the concentration of protein X, but then transcription will shut off when the concentration of R increases to a sufficiently high value.

Arrangements of this type are common in the cell. In *E. coli*, for example, galactose metabolic genes are positively regulated by the catabolite activator protein (CAP), which is activated at high levels of cAMP. The same genes are repressed by the GalS repressor protein, which is encoded by a gene whose transcription is likewise activated by CAP. Thus, an increase in input (cAMP) activates A (CAP), and transcription of the galactose genes begins. But activation of A also causes a subsequent buildup of R (GalS), which causes the same genes to be repressed after a delay. This results in an *incoherent feed-forward motif* (**Figure 8–85**A).

The response of the incoherent feed-forward motif will vary, depending on the parameters of the system. Suppose, for example, that the transcription activator protein A binds more weakly to the gene regulatory region than does the transcription repressor protein R ($K_A < K_R$). In this case, there will be a transient burst of protein synthesized by the affected gene (gene X) in response to a sudden activating input (Figure 8–85B). In contrast, the output will be more sustained if K_A is much larger than K_R, because the repression will be too weak to overcome the gene activation (Figure 8–85C). Other properties of this network, such as the dependence of the amplitude of the pulse on the various rate constants in the system, can be explored with the same computational tools. Thus, our intuitive guess about how this system would behave was only partially correct; even the simplest of networks depends on precise interaction strengths, demonstrating yet again why mathematics is needed to complement cartoon drawings.

A Coherent Feed-forward Interaction Detects Persistent Inputs

In the bacterium *E. coli*, the sugar arabinose is only consumed when the preferred sugar, glucose, is scarce. The strategy that cells use to assess the presence of arabinose and absence of glucose involves a feed-forward arrangement that is different from the one just described. In this case, depletion of glucose causes an increase of cAMP, which is sensed by the CAP transcription activator protein, as described previously. In this case, however, CAP also induces the synthesis of a second transcription activator, AraC. Both activator proteins are necessary to activate arabinose metabolic genes (the AND logic function in Figure 8–83B).

Figure 8–85 How an incoherent feed-forward motif can generate a brief pulse of gene activation in response to a sustained input. (A) Diagram of an incoherent feed-forward motif in which the transcription activator A and the repressor R control the expression of gene X using the AND NOT logic of Figure 8–83A. (B) When $K_A \ll K_R$, this motif generates a pulse of protein X expression, such that the output goes back down even if the input remains high. (C) When $K_A \gg K_R$, the same motif responds to a sustained input by generating a sustained output.

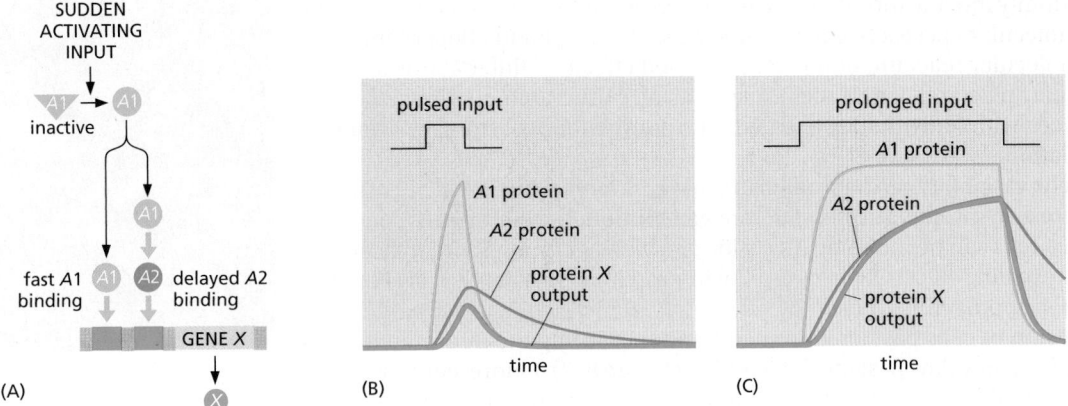

(A)

(B)

(C)

This arrangement, known as a *coherent feed-forward motif*, has the interesting characteristics illustrated in **Figure 8–86**. Imagine that two activators, *A1* and *A2*, are both required to initiate transcription of a gene. The input to the network activates *A1* directly, but only activates *A2* through this *A1* activation. Thus, for a protein to be synthesized from this gene, long-term inputs are required that allow both *A1* and *A2* to be produced in active form. Brief input pulses are either ignored or produce small outputs. The requirement for a long input is important if assurances about a signal are needed before a costly cellular program is triggered. For example, glucose is the sugar on which *E. coli* cells grow best. Before cells trigger arabinose metabolism in the example above, it might be beneficial to be sure that glucose has been depleted (a sustained CAP pulse), rather than inducing the arabinose program during a transient glucose fluctuation.

Figure 8–86 How a coherent feed-forward motif responds to various inputs. (A) Diagram of a coherent feed-forward motif in which the transcription activators *A1* and *A2* together activate expression of gene *X* using the AND logic of Figure 8–83B. (B) The response to a brief input can be either weak (as shown) or nonexistent. This allows the motif to ignore random fluctuations in the concentration of signaling molecules. (C) A prolonged input produces a strong response that can turn off rapidly.

The Same Network Can Behave Differently in Different Cells Due to Stochastic Effects

Up to this point, we have assumed that all cells in a population produce identical behaviors if they contain the same network. It is important, however, to account for the fact that cells often show considerable individuality in their responses. Consider a situation in which a single mother cell divides into two daughter cells of equal volume. If the mother cell has only one molecule of a given protein, then only one daughter will inherit it. The daughters, though genetically identical, are already different. This variability is most pronounced for molecules that are present in small numbers. Nevertheless, even when there are many copies of a particular protein (or RNA), it is very unlikely that both daughter cells will end up with exactly the same number of molecules.

This is just one illustration of a universal feature of cells: their behaviors are often **stochastic**, meaning that they display variability in their protein content and therefore exhibit variations in phenotypes. In addition to the asymmetric partitioning of molecules following cell division, variability can originate from many chemical reactions. Imagine, for example, that our mother cell contains a simple gene regulatory circuit with a positive feedback loop like that shown in Figure 8–80B. Even if both daughter cells receive a copy of this circuit, including one copy of the initial transcription activator protein, there will be variability in the time required for promoter binding—and it will be statistically nearly impossible for the genes in the two daughter cells to become activated at precisely the same time. If the system is bistable and poised near a switching point, then variability in the response might flip the switch in only one daughter cell. Two daughter cells that were born identical can thereby acquire, by chance, a dramatic difference in phenotype.

More generally, isogenic populations of cells grown in the same environment display diversity in size, shape, cell-cycle position, and gene expression. These differences arise because biochemical reactions require probabilistic collisions between randomly moving molecules, with each event resulting in changes in the number of molecular species by integer amounts. The amplified effect of fluctuations in a molecular reactant, or the compounded effects of fluctuations across many molecular reactants, often accumulates as an observable phenotype. This can endow a cell with individuality and generate non-genetic cell-to-cell variability in a population.

Non-genetic variability can be studied in the laboratory by single-cell measurements of fluorescent proteins expressed from genes under the control of a specific promoter. Live cells can be mounted on a slide and viewed through a fluorescence microscope, revealing the striking variability in protein expression levels (**Figure 8–87**). Another approach is to use flow cytometry, which works by streaming a dilute suspension of cells past an illuminator and measuring the fluorescence of individual cells as they flow past the detector (see Figure 8–2). Fluorescence values can be used to build histograms that reveal the variability in a process across a population of cells, with a broad histogram indicating higher variability.

Figure 8–87 Different levels of gene expression in individual cells within a population of *E. coli* bacteria. For this experiment, two different reporter proteins (one fluorescing *green*, the other *red*), controlled by a copy of the same promoter, have been introduced into all of the bacteria. Some cells express only one gene copy, and so appear either *red* or *green*, while others express both gene copies, and so appear *yellow*. This experiment reveals variable levels of fluorescence, indicating variable levels of gene expression within an apparently uniform population of cells. (From M.B. Elowitz et al., *Science* 297:1183–1186, 2002. With permission from AAAS.)

Several Computational Approaches Can Be Used to Model the Reactions in Cells

We have focused primarily on the use of ordinary differential equations to model the dynamics of simple regulatory circuits. These models are called *deterministic*, because they do not incorporate stochastic variability and will always produce the same result from a specific set of parameters. As we have seen, such models can provide useful insights, particularly in the detailed mechanistic analysis of small regulatory circuits. However, other types of computational approaches are also needed to comprehend the great complexity of cell behavior. *Stochastic models*, for example, attempt to account for the very important problem of random variability in molecular networks. These models do not provide deterministic predictions about the behavior of molecules; instead, they incorporate random variation into molecule numbers and interactions, and the purpose of these models is to obtain a better understanding of the probability that a system will exist in a certain state over time.

Numerous other modeling strategies have been or are being developed. *Boolean networks* are used for the qualitative analysis of complex gene regulatory networks containing large numbers of interacting components. In these models, each molecule is a node that can exist in either the active or inactive state, thereby affecting the state of the nodes it is linked to. Models of this sort provide insights into the flow of information through a network, and they were useful in helping us understand the complex gene regulatory network that controls the early development of the sea urchin (see Figure 7–42). Boolean networks therefore reduce complex networks to a highly simplified (and potentially inaccurate) form. At the other extreme are *agent-based simulations*, in which thousands of molecules (or "agents") in a system are modeled individually, and their probable behaviors and interactions with each other over time are calculated on the basis of predicted physical and chemical behaviors, often while taking stochastic variation into account. Agent-based approaches are computationally demanding but have the potential to generate highly lifelike simulations of real biological systems.

Statistical Methods Are Critical For the Analysis of Biological Data

Dynamics, differential equations, and theoretical modeling are not the be-all and end-all of mathematics. Other branches of the subject are no less important for biologists. Statistics—the mathematics of probabilistic processes and noisy datasets—is an inescapable part of every biologist's life.

This is true in two main ways. First, imperfect measurement devices and other errors generate experimental noise in our data. Second, all cell-biological processes depend on the stochastic behavior of individual molecules, as we just discussed, and this results in biological noise in our results. How, in the face of all this noise, do we come to conclusions about the truth of hypotheses? The answer is statistical analysis, which shows how to move from one level of description to another: from a set of erratic individual data points to a simpler description of the key features of the data.

Statistics teaches us that the more times we repeat our measurements, the better and more refined the conclusions we can draw from them. Given many repetitions, it becomes possible to describe our data in terms of variables that summarize the features that matter: the mean value of the measured variable, taken over the set of data points; the magnitude of the noise (the standard deviation of the set of data points); the likely error in our estimate of the mean value (the standard error of the mean); and, for specialists, the details of the probability distribution describing the likelihood that an individual measurement will yield a given value. For all these things, statistics provides recipes and quantitative formulas that biologists must understand if they are to make rigorous conclusions on the basis of variable results.

Summary

Quantitative mathematical analysis can provide a powerful extra dimension in our understanding of cell regulation and function. Cell regulatory systems often depend on macromolecular interactions, and mathematical analysis of the dynamics of these interactions can unveil important insights into the importance of binding affinities and protein stability in the generation of transcriptional or other signals. Regulatory systems often employ network motifs that generate useful behaviors: a rapid negative feedback loop dampens the response to input signals; a delayed negative feedback loop creates a biochemical oscillator; positive feedback yields a system that alternates between two stable states; and feed-forward motifs provide systems that generate transient signal pulses or respond only to sustained inputs. The dynamic behavior of these network motifs can be dissected in detail with deterministic and stochastic mathematical modeling.

WHAT WE DON'T KNOW

• Many of the tools that revolutionized DNA technology were discovered by scientists studying basic biological problems that had no obvious applications. What are the best strategies to ensure that such crucially important technologies will continue to be discovered?

• As the cost of DNA sequencing decreases and the amount of sequence data accumulates, how are we going to keep track of and meaningfully analyze this vast amount of information? What new questions will this information allow us to answer?

• Can we develop tools to analyze each of the post-transcriptional modifications on the proteins in living cells, so as to follow all of their changes in real time?

• Can we develop mathematical models to accurately describe the enormous complexity of cellular networks and to predict undiscovered components and mechanisms?

PROBLEMS

Which statements are true? Explain why or why not.

8–1 Because a monoclonal antibody recognizes a specific antigenic site (epitope), it binds only to the specific protein against which it was made.

8–2 Given the inexorable march of technology, it seems inevitable that the sensitivity of detection of molecules will ultimately be pushed beyond the yoctomole level (10^{-24} mole).

8–3 If each cycle of PCR doubles the amount of DNA synthesized in the previous cycle, then 10 cycles will give a 10^3-fold amplification, 20 cycles will give a 10^6-fold amplification, and 30 cycles will give a 10^9-fold amplification.

8–4 To judge the biological importance of an interaction between protein A and protein B, we need to know quantitative details about their concentrations, affinities, and kinetic behaviors.

8–5 The rate of change in the concentration of any molecular species X is given by the balance between its rate of appearance and its rate of disappearance.

8–6 After a sudden increase in transcription, a protein with a slow rate of degradation will reach a new steady state level more quickly than a protein with a rapid rate of degradation.

Discuss the following problems.

8–7 A common step in the isolation of cells from a sample of animal tissue is to treat the tissue with trypsin, collagenase, and EDTA. Why is such a treatment necessary, and what does each component accomplish? And why does this treatment not kill the cells?

8–8 Tropomyosin, at 93 kd, sediments at 2.6S, whereas the 65-kd protein, hemoglobin, sediments at 4.3S. (The sedimentation coefficient S is a linear measure of the rate of sedimentation.) These two proteins are drawn to scale in **Figure Q8–1**. How is it that the bigger protein sediments more slowly than the smaller one? Can you think of an analogy from everyday experience that might help you with this problem?

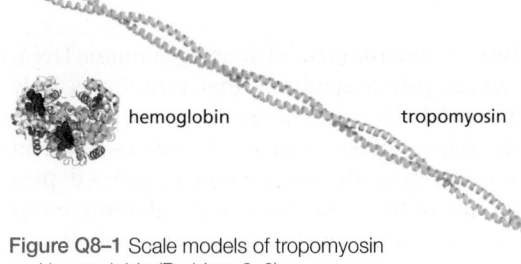

hemoglobin   tropomyosin

Figure Q8–1 Scale models of tropomyosin and hemoglobin (Problem 8–8).

8–9   Hybridoma technology allows one to generate monoclonal antibodies to virtually any protein. Why is it, then, that genetically tagging proteins with epitopes is such a commonly used technique, especially since an epitope tag has the potential to interfere with the function of the protein?

8–10   How many copies of a protein need to be present in a cell in order for it to be visible as a band on an SDS gel? Assume that you can load 100 μg of cell extract onto a gel and that you can detect 10 ng in a single band by silver staining the gel. The concentration of protein in cells is about 200 mg/mL, and a typical mammalian cell has a volume of about 1000 μm³ and a typical bacterium a volume of about 1 μm³. Given these parameters, calculate the number of copies of a 120-kd protein that would need to be present in a mammalian cell and in a bacterium in order to give a detectable band on a gel. You might try an order-of-magnitude guess before you make the calculations.

8–11   You have isolated the proteins from two adjacent spots after two-dimensional polyacrylamide-gel electrophoresis and digested them with trypsin. When the masses of the peptides were measured by MALDI-TOF mass spectrometry, the peptides from the two proteins were found to be identical except for one (**Figure Q8–2**). For this peptide, the mass-to-charge (m/z) values differed by 80, a value that does not correspond to a difference in amino acid sequence. (For example, glutamic acid instead of valine at one position would give an m/z difference of around 30.) Can you suggest a possible difference between the two peptides that might account for the observed m/z difference?

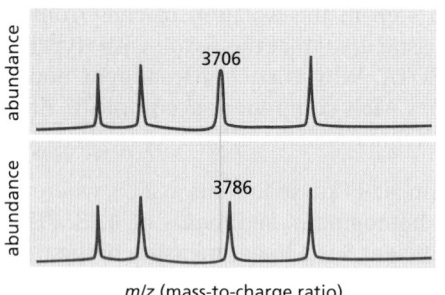

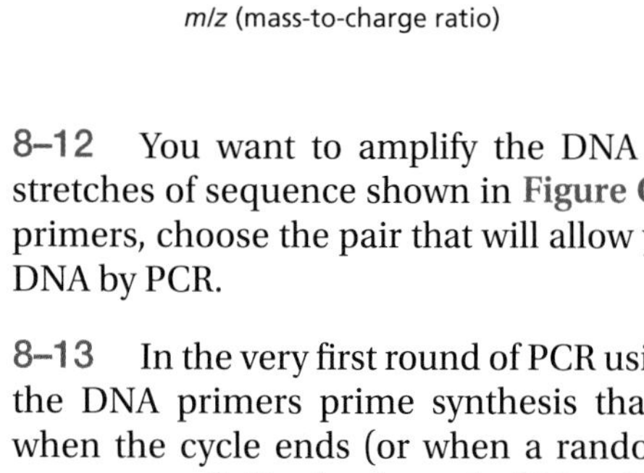

Figure Q8–2 Masses of peptides measured by MALDI-TOF mass spectrometry (Problem 8–11). Only the numbered peaks differ between the two protein samples.

3706

3786

m/z (mass-to-charge ratio)

8–12   You want to amplify the DNA between the two stretches of sequence shown in **Figure Q8–3**. Of the listed primers, choose the pair that will allow you to amplify the DNA by PCR.

8–13   In the very first round of PCR using genomic DNA, the DNA primers prime synthesis that terminates only when the cycle ends (or when a random end of DNA is encountered). Yet, by the end of 20 to 30 cycles—a typical amplification—the only visible product is defined precisely by the ends of the DNA primers. In what cycle is a double-stranded fragment of the correct size first generated?

DNA to be amplified

5′-GACCTGTGGAAGC ———————— CATACGGGATTGA-3′
3′-CTGGACACCTTCG ———————— GTATGCCCTAACT-5′

primers

(1) 5′-GACCTGTGGAAGC-3'   (5) 5′-CATACGGGATTGA-3'
(2) 5′-CTGGACACCTTCG-3'   (6) 5′-GTATGCCCTAACT-3'
(3) 5′-CGAAGGTGTCCAG-3'   (7) 5′-TGTTAGGGCATAC-3'
(4) 5′-GCTTCCACAGGTC-3'   (8) 5′-TCAATCCCGTATG-3'

Figure Q8–3 DNA to be amplified and potential PCR primers (Problem 8–12).

8–14   Explain the difference between a gain-of-function mutation and a dominant-negative mutation. Why are both these types of mutation usually dominant?

8–15   Discuss the following statement: "We would have no idea today of the importance of insulin as a regulatory hormone if its absence were not associated with the human disease diabetes. It is the dramatic consequences of its absence that focused early efforts on the identification of insulin and the study of its normal role in physiology."

8–16   You have received the results from an RNA-seq analysis of mRNAs from liver. You had anticipated counting the number of reads of each mRNA to determine the relative abundance of different mRNAs. But you are puzzled because many of the mRNAs have given you results like those shown in **Figure Q8–4**. How is it that different parts of an mRNA can be represented at different levels?

8–17   Examine the network motifs in **Figure Q8–5**. Decide which ones are negative feedback loops and which are positive. Explain your reasoning.

8–18   Imagine that a random perturbation positions a bistable system precisely at the boundary between two stable states (at the *orange dot* in **Figure Q8–6**). How would the system respond?

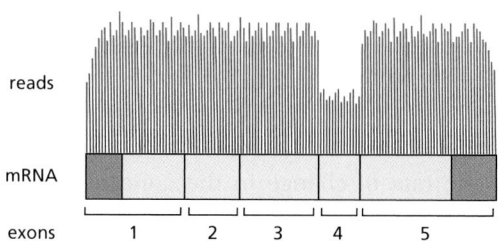

reads

mRNA

exons   1   2   3   4   5

Figure Q8–4 RNA-seq reads for a liver mRNA (Problem 8–16). The exon structure of the mRNA is indicated, with protein-coding segments indicated in *light blue* and untranslated regions in *dark blue*. The numbers of sequencing reads are indicated by the heights of the vertical lines above the mRNA.

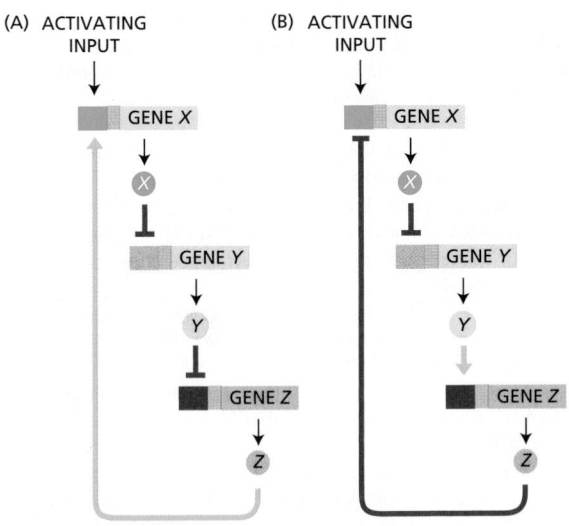

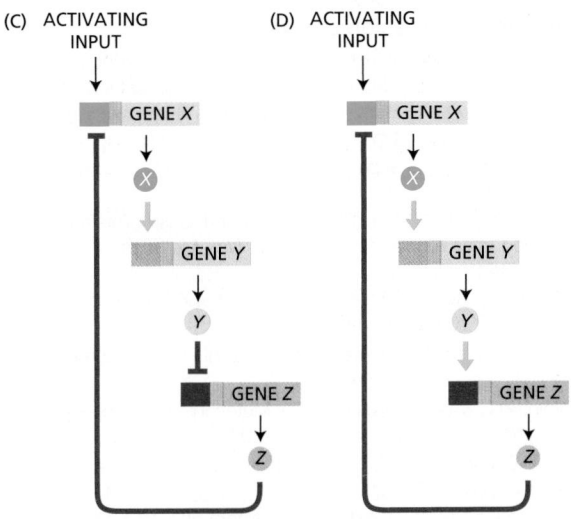

Figure Q8–5 Network motifs composed of transcription activators and repressors (Problem 8–17).

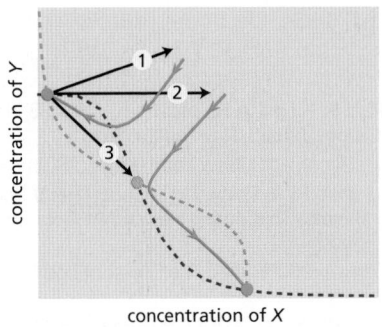

Figure Q8–6 Perturbations of a bistable system (Problem 8–18). As shown by the *green lines*, after perturbation 1 the system returns to its original stable state (*green dot* at left), and after perturbation 2, the system moves to the other stable state (*green dot* at right). Perturbation 3 moves the system to the precise boundary between the two stable states (*orange dot*).

A.   Which single operator site is the most important for repression? How can you tell?

B.   Do combinations of operator sites (Figure Q8–7, constructs 1, 2, 3, and 5) substantially increase repression by the dimeric repressor? Do combinations of operator sites substantially increase repression by the tetrameric repressor? If the two repressors behave differently, offer an explanation for the difference.

C.   The wild-type repressor binds O_3 very weakly when it is by itself on a segment of DNA. However, if O_1 is included on the same segment of DNA, the repressor binds O_3 quite well. How can that be?

8–19   Detailed analysis of the regulatory region of the *Lac* operon has revealed surprising complexity. Instead of a single binding site for the *Lac* repressor, as might be expected, there are three sites termed operators: O_1, O_2, and O_3, arrayed along the DNA as shown in **Figure Q8–7**. To probe the functions of these three sites, you make a series of constructs in which various combinations of operator sites are present. You examine their ability to repress expression of β-galactosidase, using either tetrameric (wild type) or dimeric (mutant) forms of the *Lac* repressor. The dimeric form of the repressor can bind to a single operator (with the same affinity as the tetramer) with each monomer binding to half the site. The tetramer, the form normally expressed in cells, can bind to two sites simultaneously. When you measure repression of β-galactosidase expression, you find the results shown in Figure Q8–7, with higher numbers indicating more effective repression.

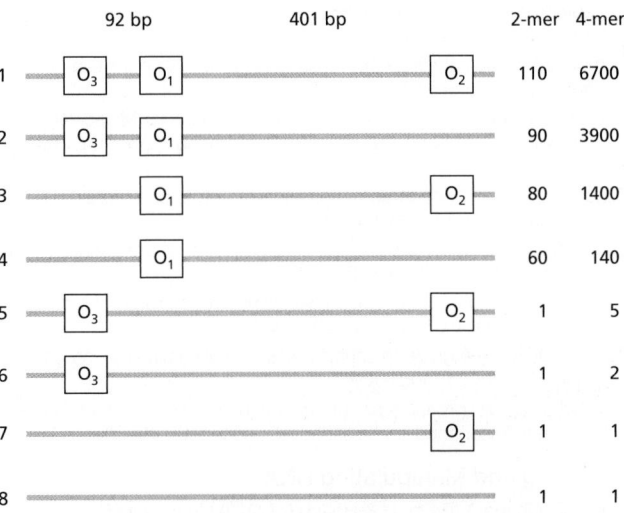

Figure Q8–7 Repression of β-galactosidase by promoter regions that contain different combinations of *Lac* repressor binding sites (Problem 8–19). The base-pair (bp) separation of the three operator sites is shown. Numbers at *right* refer to the level of repression, with higher numbers indicating more effective repression by dimeric (2-mer) or tetrameric (4-mer) repressors. (From S. Oehler et al., *EMBO J.* 9:973–979, 1990. With permission from John Wiley and Sons.)

REFERENCES

General

Ausubel FM, Brent R, Kingston RE et al. (eds) (2002) Short Protocols in Molecular Biology, 5th ed. New York: Wiley.

Brown TA (2007) Genomes 3. New York: Garland Science Publishing.

Spector DL, Goldman RD & Leinwand LA (eds) (1998) Cells: A Laboratory Manual. Cold Spring Harbor, NY: Cold Spring Harbor Laboratory Press.

Watson JD & Berry A (2008) DNA: The Secret of Life. New York: Alfred A Knopf.

Watson JD, Myers RM & Caudy AA (2007) Recombinant DNA: Genes and Genomes – A Short Course, 3rd ed. Cold Spring Harbor, NY: Cold Spring Harbor Laboratory Press.

Isolating Cells and Growing Them in Culture

Ham RG (1965) Clonal growth of mammalian cells in a chemically defined, synthetic medium. *Proc. Natl Acad. Sci. USA* 53, 288–293.

Harlow E & Lane D (1999) Using Antibodies: A Laboratory Manual. Cold Spring Harbor, NY: Cold Spring Harbor Laboratory Press.

Herzenberg LA, Sweet RG & Herzenberg LA (1976) Fluorescence-activated cell sorting. *Sci. Am.* 234, 108–116.

Milstein C (1980) Monoclonal antibodies. *Sci. Am.* 243, 66–74.

Purifying Proteins

de Duve C & Beaufay H (1981) A short history of tissue fractionation. *J. Cell Biol.* 91, 293s–299s.

Laemmli UK (1970) Cleavage of structural proteins during the assembly of the head of bacteriophage T4. *Nature* 227, 680–685.

Scopes RK (1994) Protein Purification: Principles and Practice, 3rd ed. New York: Springer-Verlag.

Simpson RJ, Adams PD & Golemis EA (2008) Basic Methods in Protein Purification and Analysis: A Laboratory Manual. Cold Spring Harbor, NY: Cold Spring Harbor Laboratory Press.

Wood DW (2014) New trends and affinity tag designs for recombinant protein purification. *Curr. Opin. Struct. Biol.* 26, 54–61.

Analyzing Proteins

Choudhary C & Mann M (2010) Decoding signalling networks by mass spectrometry-based proteomics. *Nat. Rev. Mol. Cell Biol.* 11, 427–439.

Domon B & Aebersold R (2006) Mass spectrometry and protein analysis. *Science* 312, 212–217.

Goodrich JA & Kugel JF (2007) Binding and Kinetics for Molecular Biologists. Cold Spring Harbor, NY: Cold Spring Harbor Laboratory Press.

Kendrew JC (1961) The three-dimensional structure of a protein molecule. *Sci. Am.* 205, 96–111.

Knight ZA & Shokat KM (2007) Chemical genetics: where genetics and pharmacology meet. *Cell* 128, 425–430.

O'Farrell PH (1975) High resolution two-dimensional electrophoresis of proteins. *J. Biol. Chem.* 250, 4007–4021.

Pollard TD (2010) A guide to simple and informative binding assays. *Mol. Biol. Cell* 21, 4061–4067.

Wüthrich K (1989) Protein structure determination in solution by nuclear magnetic resonance spectroscopy. *Science* 243, 45–50.

Analyzing and Manipulating DNA

Cohen S, Chang A, Boyer H & Helling R (1973) Construction of biologically functional bacterial plasmids *in vitro*. *Proc. Natl Acad. Sci. USA* 70, 3240–3244.

Green MR & Sambrook J (2012) Molecular Cloning: A Laboratory Manual, 4th ed. Cold Spring Harbor, NY: Cold Spring Harbor Laboratory Press.

International Human Genome Sequencing Consortium (2001) Initial sequencing and analysis of the human genome. *Nature* 409, 860–921.

Jackson D, Symons R & Berg P (1972) Biochemical method for inserting new genetic information into DNA of Simian Virus 40: circular SV40 DNA molecules containing lambda phage genes and the galactose operon of *Escherichia coli*. *Proc. Natl Acad. Sci. USA* 69, 2904–2909.

Kosuri S & Church GM (2014) Large-scale de novo DNA synthesis: technologies and applications. *Nat. Methods* 11, 499–507.

Maniatis T, Hardison RC, Lacy E et al. (1978) The isolation of structural genes from libraries of eucaryotic DNA. *Cell* 15, 687–701.

Mullis KB (1990) The unusual origin of the polymerase chain reaction. *Sci. Am.* 262, 56–61.

Nathans D & Smith HO (1975) Restriction endonucleases in the analysis and restructuring of DNA molecules. *Annu. Rev. Biochem.* 44, 273–293.

Saiki RK, Gelfand DH, Stoffel S et al. (1988) Primer-directed enzymatic amplification of DNA with a thermostable DNA polymerase. *Science* 239, 487–491.

Sanger F, Nicklen S & Coulson AR (1977) DNA sequencing with chain-terminating inhibitors. *Proc. Natl Acad. Sci. USA* 74, 5463–5467.

Shendure J & Lieberman Aiden E (2012) The expanding scope of DNA sequencing. *Nat. Biotechnol.* 30, 1084–1094.

Studying Gene Expression and Function

Botstein D, White RL, Skolnick M & Davis RW (1980) Construction of a genetic linkage map in man using restriction fragment length polymorphisms. *Am. J. Hum. Genet.* 32, 314–331.

DeRisi JL, Iyer VR & Brown PO (1997) Exploring the metabolic and genetic control of gene expression on a genomic scale. *Science* 278, 680–686.

Esvelt KM, Mali P, Braff JL et al. (2013) Orthogonal Cas9 proteins for RNA-guided gene regulation and editing. *Nat. Methods* 10, 1116–1121.

Fellmann C & Lowe SW (2014) Stable RNA interference rules for silencing. *Nat. Cell Biol.* 16, 10–18.

Mello CC & Conte D (2004) Revealing the world of RNA interference. *Nature* 431, 338–342.

Nüsslein-Volhard C & Wieschaus E (1980) Mutations affecting segment number and polarity in *Drosophila*. *Nature* 287, 795–801.

Palmiter RD & Brinster RL (1985) Transgenic mice. *Cell* 41, 343–345.

Weigel D & Glazebrook J (2002) *Arabidopsis*: A Laboratory Manual. Cold Spring Harbor, NY: Cold Spring Harbor Laboratory Press.

Wilson RC & Doudna JA (2013) Molecular mechanisms of RNA interference. *Annu. Rev. Biophys.* 42, 217–239.

Mathematical Analysis of Cell Functions

Alon U (2006) An Introduction to Systems Biology: Design Principles of Biological Circuits. Boca Raton, FL: Chapman & Hall/CRC.

Alon U (2007) Network motifs: theory and experimental approaches. *Nat. Rev. Genet.* 8, 450–461.

Ferrell JE Jr (2002) Self-perpetuating states in signal transduction: positive feedback, double-negative feedback and bistability. *Curr. Opin. Cell Biol.* 14, 140–148.

Ferrell JE Jr, Tsai TY & Yang Q (2011) Modeling the cell cycle: why do certain circuits oscillate? *Cell* 144, 874–885.

Gunawardena J (2014) Models in biology: 'accurate descriptions of our pathetic thinking'. *BMC Biol.* 12, 29.

Lewis J (2008) From signals to patterns: space, time, and mathematics in developmental biology. *Science* 322, 399–403.

Mogilner A, Allard J & Wollman R (2012) Cell polarity: quantitative modeling as a tool in cell biology. *Science* 336, 175–179.

Novak B & Tyson JJ (2008) Design principles of biochemical oscillators. *Nat. Rev. Mol. Cell Biol.* 9, 981–991.

Silva-Rocha R & de Lorenzo V (2008) Mining logic gates in prokaryotic transcriptional regulation networks. *FEBS Lett.* 582, 1237–1244.

Tyson JJ, Chen KC & Novak B (2003) Sniffers, buzzers, toggles and blinkers: dynamics of regulatory and signaling pathways in the cell. *Curr. Opin. Cell Biol.* 15, 221–231.

Visualizing Cells

Understanding the structural organization of cells is essential for learning how they function. In this chapter, we briefly describe some of the principal microscopy methods used to study cells. Optical microscopy will be our starting point because cell biology began with the light microscope, and it is still an indispensible tool. The development of methods for the specific labeling and imaging of individual cellular constituents and the reconstruction of their three-dimensional architecture has meant that, far from falling into disuse, optical microscopy continues to increase in importance. One advantage of optical microscopy is that light is relatively nondestructive. By tagging specific cell components with fluorescent probes, such as intrinsically fluorescent proteins, we can watch their movement, dynamics, and interactions in living cells. Although conventional optical microscopy is limited in resolution by the wavelength of visible light, new methods cleverly bypass this limitation and allow the position of even single molecules to be mapped. By using a beam of electrons instead of visible light, electron microscopy can image the interior of cells, and their macromolecular components, at almost atomic resolution, and in three dimensions.

This chapter is intended as a companion, rather than an introduction, to the chapters that follow; readers may wish to refer back to it as they encounter applications of microscopy to basic biological problems in the later pages of the book.

LOOKING AT CELLS IN THE LIGHT MICROSCOPE

A typical animal cell is 10–20 μm in diameter, which is about one-fifth the size of the smallest object that we can normally see with the naked eye. Only after good light microscopes became available in the early part of the nineteenth century did Schleiden and Schwann propose that all plant and animal tissues were aggregates of individual cells. Their proposal in 1838, known as the **cell doctrine**, marks the formal birth of cell biology.

Animal cells are not only tiny, but they are also colorless and translucent. The discovery of their main internal features, therefore, depended on the development, in the late nineteenth century, of a variety of stains that provided sufficient contrast to make those features visible. Similarly, the far more powerful electron microscope introduced in the early 1940s required the development of new techniques for preserving and staining cells before the full complexities of their internal fine structure could begin to emerge. To this day, microscopy often relies as much on techniques for preparing the specimen as on the performance of the microscope itself. In the following discussions, we therefore consider both instruments and specimen preparation, beginning with the light microscope.

The images in **Figure 9–1** illustrate a stepwise progression from a thumb to a cluster of atoms. Each successive image represents a tenfold increase in magnification. The naked eye can see features in the first two panels, the light microscope allows us to see details corresponding to about the fourth or fifth panel, and the electron microscope takes us to about the seventh or eighth panel. **Figure 9–2** shows the sizes of various cellular and subcellular structures and the ranges of size that different types of microscopes can visualize.

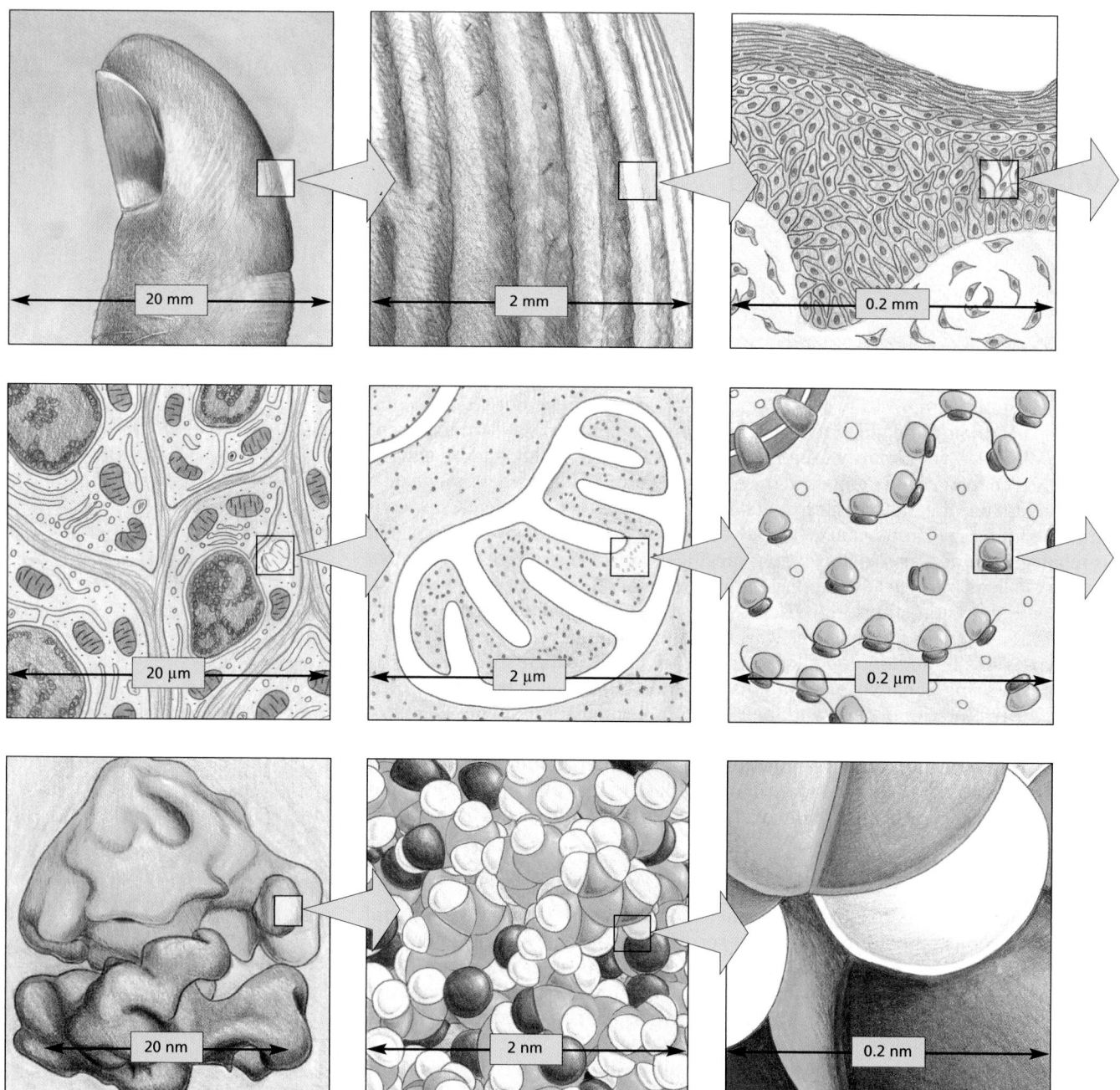

The Light Microscope Can Resolve Details 0.2 μm Apart

For well over 100 years, all microscopes were constrained by a fundamental limitation: that a given type of radiation cannot be used to probe structural details much smaller than its own wavelength. A limit to the resolution of a light microscope was therefore set by the wavelength of visible light, which ranges from about 0.4 μm (for violet) to 0.7 μm (for deep red). In practical terms, bacteria and mitochondria, which are about 500 nm (0.5 μm) wide, are generally the smallest objects whose shape we can clearly discern in the **light microscope**; details smaller than this are obscured by effects resulting from the wavelike nature of light. To understand why this occurs, we must follow the behavior of a beam of light as it passes through the lenses of a microscope (**Figure 9–3**).

Because of its wave nature, light does not follow the idealized straight ray paths that geometrical optics predicts. Instead, light waves travel through an optical system by many slightly different routes, like ripples in water, so that they

Figure 9–1 **A sense of scale between living cells and atoms.** Each diagram shows an image magnified by a factor of ten in an imaginary progression from a thumb, through skin cells, to a ribosome, to a cluster of atoms forming part of one of the many protein molecules in our body. Atomic details of biological macromolecules, as shown in the last two panels, are usually beyond the power of the electron microscope. While color has been used here in all the panels, it is not a feature of objects much smaller than the wavelength of light, so the last five panels should really be in black and white.

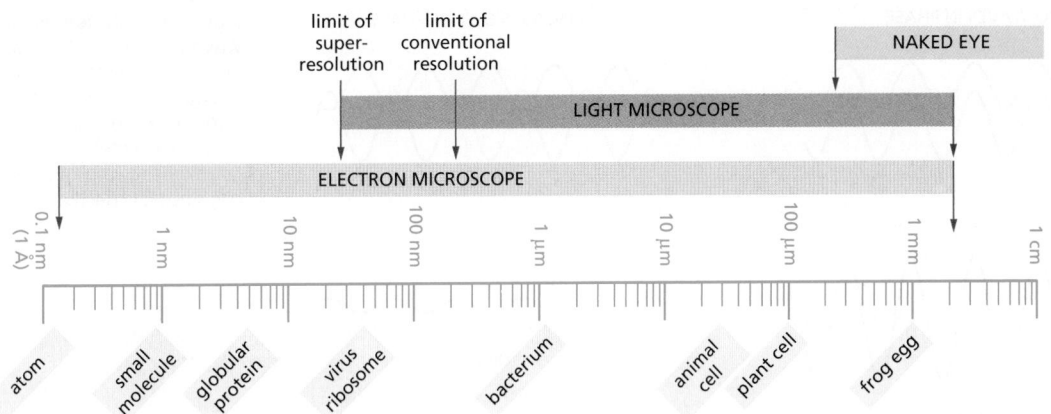

Figure 9–2 Resolving power. Sizes of cells and their components are drawn on a logarithmic scale, indicating the range of objects that can be readily resolved by the naked eye and in the light and electron microscopes. Note that new superresolution microscopy techniques, discussed in detail later, allow an improvement in resolution by an order of magnitude compared with conventional light microscopy.

interfere with one another and cause *optical diffraction* effects. If two trains of waves reaching the same point by different paths are precisely *in phase*, with crest matching crest and trough matching trough, they will reinforce each other so as to increase brightness. In contrast, if the trains of waves are *out of phase*, they will interfere with each other in such a way as to cancel each other partly or entirely (**Figure 9–4**). The interaction of light with an object changes the phase relationships of the light waves in a way that produces complex interference effects. At high magnification, for example, the shadow of an edge that is evenly illuminated with light of uniform wavelength appears as a set of parallel lines (**Figure 9–5**), whereas that of a circular spot appears as a set of concentric rings. For the same reason, a single point seen through a microscope appears as a blurred disc, and two point objects close together give overlapping images and may merge into one.

The following units of length are commonly employed in microscopy:

μm (micrometer) = 10^{-6} m

nm (nanometer) = 10^{-9} m

Å (Ångström unit) = 10^{-10} m

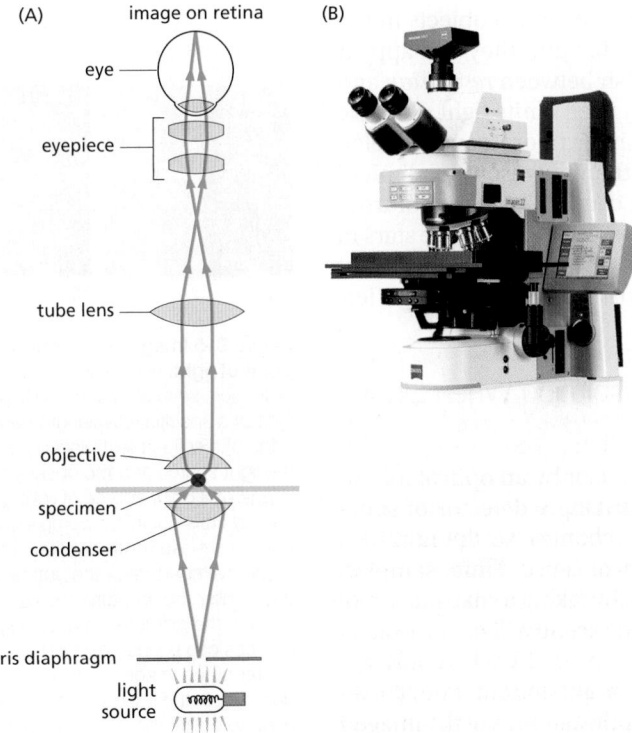

Figure 9–3 A light microscope.
(A) Diagram showing the light path in a compound microscope. Light is focused on the specimen by lenses in the condenser. A combination of objective lenses, tube lenses, and eyepiece lenses is arranged to focus an image of the illuminated specimen in the eye. (B) A modern research light microscope. (B, courtesy of Carl Zeiss Microscopy GmbH.)

TWO WAVES IN PHASE

TWO WAVES OUT OF PHASE

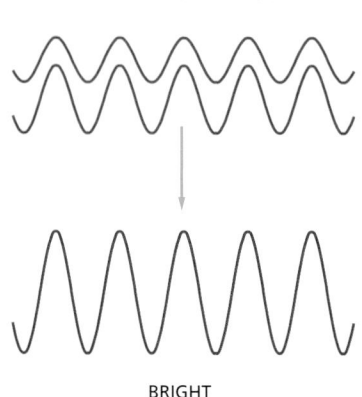

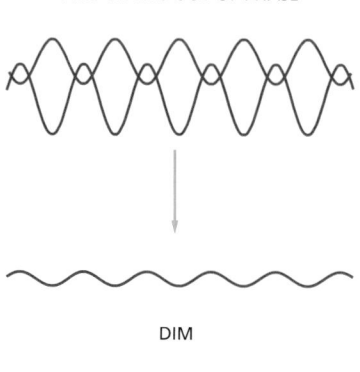

BRIGHT

DIM

Figure 9–4 **Interference between light waves.** When two light waves combine in phase, the amplitude of the resultant wave is larger and the brightness is increased. Two light waves that are out of phase cancel each other partly and produce a wave whose amplitude, and therefore brightness, is decreased.

Although no amount of refinement of the lenses can overcome the diffraction limit imposed by the wavelike nature of light, other ways of cleverly bypassing this limit have emerged, creating so-called superresolution imaging techniques that can even detect the position of single molecules.

The limiting separation at which two objects appear distinct—the so-called **limit of resolution**—depends on both the wavelength of the light and the *numerical aperture* of the lens system used. The numerical aperture affects the light-gathering ability of the lens and is related both to the angle of the cone of light that can enter it and to the refractive index of the medium the lens is operating in; the wider the microscope opens its eye, so to speak, the more sharply it can see (**Figure 9–6**). The *refractive index* is the ratio of the speed of light in a vacuum to the speed of light in a particular transparent medium. For example, for water this is 1.33, meaning that light travels 1.33 times slower in water than in a vacuum. Under the best conditions, with violet light (wavelength = 0.4 μm) and a numerical aperture of 1.4, the basic light microscope can theoretically achieve a limit of resolution of about 0.2 μm, or 200 nm. Some microscope makers at the end of the nineteenth century achieved this resolution, but it is routinely matched in contemporary, factory-produced microscopes. Although it is possible to *enlarge* an image as much as we want—for example, by projecting it onto a screen—it is not possible, in a conventional light microscope, to resolve two objects in the light microscope that are separated by less than about 0.2 μm; they will appear as a single object. It is important, however, to distinguish between *resolution* and *detection*. If a small object, below the resolution limit, itself emits light, then we may still be able to see or detect it. Thus, we can see a single fluorescently labeled microtubule even though it is about ten times thinner than the resolution limit of the light microscope. Diffraction effects, however, will cause it to appear blurred and at least 0.2 μm thick (see Figure 9–16). In a similar way, we can see the stars in the night sky, even though their diameters are far below the angular resolution of our unaided eyes: they all appear as similar, slightly blurred points of light, differing only in their color and brightness.

Photon Noise Creates Additional Limits to Resolution When Light Levels Are Low

Any image, whether produced by an electron microscope or by an optical microscope, is made by particles—electrons or photons—striking a detector of some sort. But these particles are governed by quantum mechanics, so the numbers reaching the detector are predictable only in a statistical sense. Finite samples, collected by imaging for a limited period of time (that is, by taking a snapshot), will show random variation: successive snapshots of the same scene will not be exactly identical. Moreover, every detection method has some level of background signal or noise, adding to the statistical uncertainty. With bright illumination, corresponding to very large numbers of photons or electrons, the features of the imaged

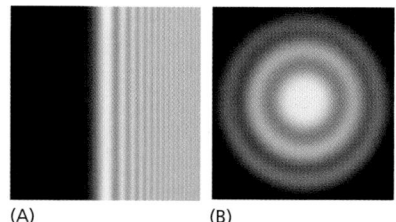

(A)   (B)

Figure 9–5 **Images of an edge and of a point of light.** (A) The interference effects, or fringes, seen at high magnification when light of a specific wavelength passes the edge of a solid object placed between the light source and the observer. (B) The image of a point source of light. Diffraction spreads this out into a complex, circular pattern, whose width depends on the numerical aperture of the optical system: the smaller the aperture, the bigger (more blurred) the diffracted image. Two point sources can be just resolved when the center of the image of one lies on the first dark ring in the image of the other: this is used to define the limit of resolution.

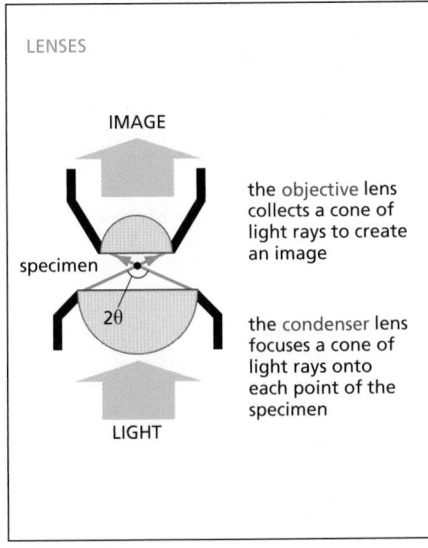

LENSES

IMAGE

the objective lens collects a cone of light rays to create an image

specimen

2θ

the condenser lens focuses a cone of light rays onto each point of the specimen

LIGHT

RESOLUTION: the resolving power of the microscope depends on the width of the cone of illumination and therefore on both the condenser and the objective lens. It is calculated using the formula

$$\text{resolution} = \frac{0.61\,\lambda}{n\,\sin\theta}$$

where:

θ = half the angular width of the cone of rays collected by the objective lens from a typical point in the central region of the specimen (since the maximum width is 180°, $\sin\theta$ has a maximum value of 1)

n = the refractive index of the medium (usually air or oil) separating the specimen from the objective and condenser lenses

λ = the wavelength of light used (for white light a figure of 0.53 μm is commonly assumed)

NUMERICAL APERTURE: $n\sin\theta$ in the equation above is called the numerical aperture of the lens and is a function of its light-collecting ability. For dry lenses this cannot be more than 1, but for oil-immersion lenses it can be as high as 1.4. The higher the numerical aperture, the greater the resolution and the brighter the image (brightness is important in fluorescence microscopy). However, this advantage does necessitate very short working distances and a very small depth of field.

Figure 9–6 Numerical aperture. The path of light rays passing through a transparent specimen in a microscope illustrates the concept of numerical aperture and its relation to the limit of resolution.

specimen are accurately determined based on the distribution of these particles at the detector. However, with smaller numbers of particles, the structural details of the specimen are obscured by the statistical fluctuations in the numbers of particles detected in each region, which give the image a speckled appearance and limit its precision. The term *noise* describes this random variability.

Living Cells Are Seen Clearly in a Phase-Contrast or a Differential-Interference-Contrast Microscope

There are many ways in which contrast in a specimen can be generated (**Figure 9–7**A). While fixing and staining a specimen can generate contrast through color, microscopists have always been challenged by the possibility that some components of the cell may be lost or distorted during specimen preparation. The only certain way to avoid the problem is to examine cells while they are alive, without fixing or freezing. For this purpose, light microscopes with special optical systems are especially useful.

In the normal **bright-field microscope**, light passing through a cell in culture forms the image directly. Another system, **dark-field microscopy**, exploits the fact that light rays can be scattered in all directions by small objects in their path. If oblique lighting from the condenser is arranged, which does not directly enter the objective, focused but unstained objects in a living cell can scatter the rays, some of which then enter the objective to create a bright image against a black background (Figure 9–7B).

When light passes through a living cell, the phase of the light wave is changed according to the cell's refractive index: a relatively thick or dense part of the cell, such as a nucleus, slows the light passing through it. The phase of the light, consequently, is shifted relative to light that has passed through an adjacent thinner region of the cytoplasm (Figure 9–7C). The **phase-contrast microscope** and, in a more complex way, the **differential-interference-contrast microscope** increase these phase differences so that the waves are more nearly out of phase, producing amplitude differences when the sets of waves recombine, thereby creating an image of the cell's structure. Both types of light microscopy are widely used to visualize living cells (see Movie 17.2). **Figure 9–8** compares images of the same cell obtained by four kinds of light microscopy.

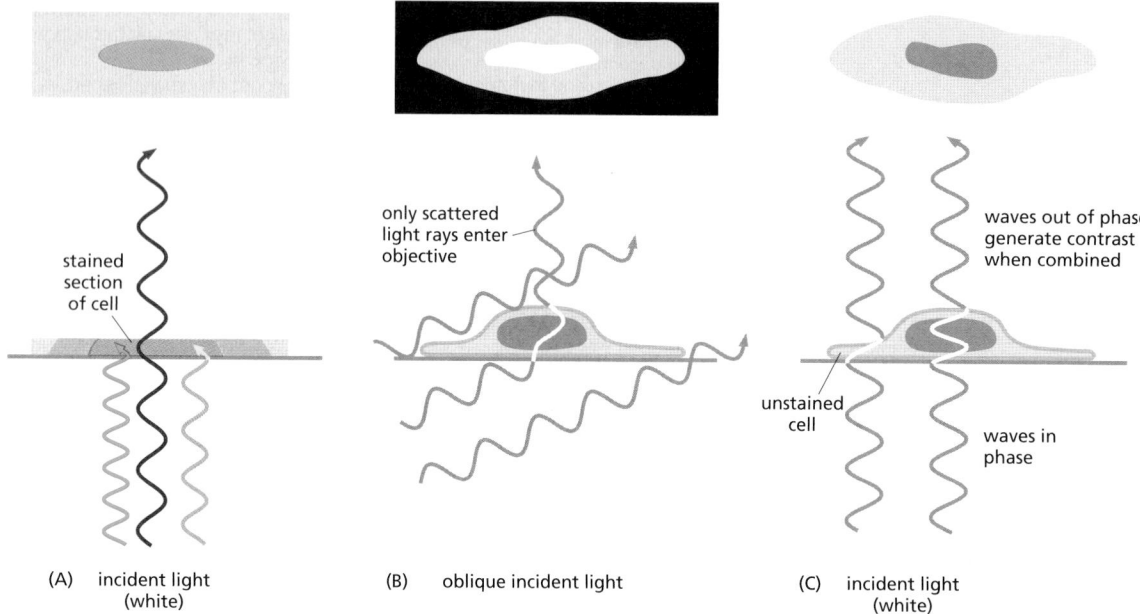

Figure 9–7 Contrast in light microscopy. (A) The stained portion of the cell will absorb light of some wavelengths, which depends on the stain, but will allow other wavelengths to pass through it. A colored image of the cell is thereby obtained that is visible in the normal bright-field light microscope. (B) In the dark-field microscope, oblique rays of light focused on the specimen do not enter the objective lens, but light that is scattered by components in the living cell can be collected to produce a bright image on a dark background. (C) Light passing through the unstained living cell experiences very little change in amplitude, and the structural details cannot be seen even if the image is highly magnified. The phase of the light, however, is altered by its passage through either thicker or denser parts of the cell, and small phase differences can be made visible by exploiting interference effects using a phase-contrast or a differential-interference-contrast microscope.

Phase-contrast, differential-interference-contrast, and dark-field microscopy make it possible to watch the movements involved in such processes as mitosis and cell migration. Since many cellular motions are too slow to be seen in real time, it is often helpful to make time-lapse movies in which the camera records successive frames separated by a short time delay, so that when the resulting picture series is played at normal speed, events appear greatly speeded up.

Images Can Be Enhanced and Analyzed by Digital Techniques

In recent years, electronic, or digital, imaging systems, and the associated technology of **image processing**, have had a major impact on light microscopy. Certain practical limitations of microscopes relating to imperfections in the optical system have been largely overcome. Electronic imaging systems have also circumvented two fundamental limitations of the human eye: the eye cannot see well in extremely dim light, and it cannot perceive small differences in light intensity against a bright background. To increase our ability to observe cells in these difficult conditions, we can attach a sensitive digital camera to a microscope. These cameras detect light by means of charge-coupled devices (CCDs), or high sensitivity complementary metal-oxide semiconductor (CMOS) sensors, similar to those found in digital cameras. Such image sensors are 10 times more sensitive than the human eye and can detect 100 times more intensity levels. It is therefore possible to observe cells for long periods at very low light levels, thereby avoiding the damaging effects of prolonged bright light (and heat). Such low-light cameras are especially important for viewing fluorescent molecules in living cells, as explained below.

Because images produced by digital cameras are in electronic form, they can be processed in various ways to extract latent information. Such image processing makes it possible to compensate for several optical faults in microscopes. Moreover, by digital image processing, contrast can be greatly enhanced to overcome

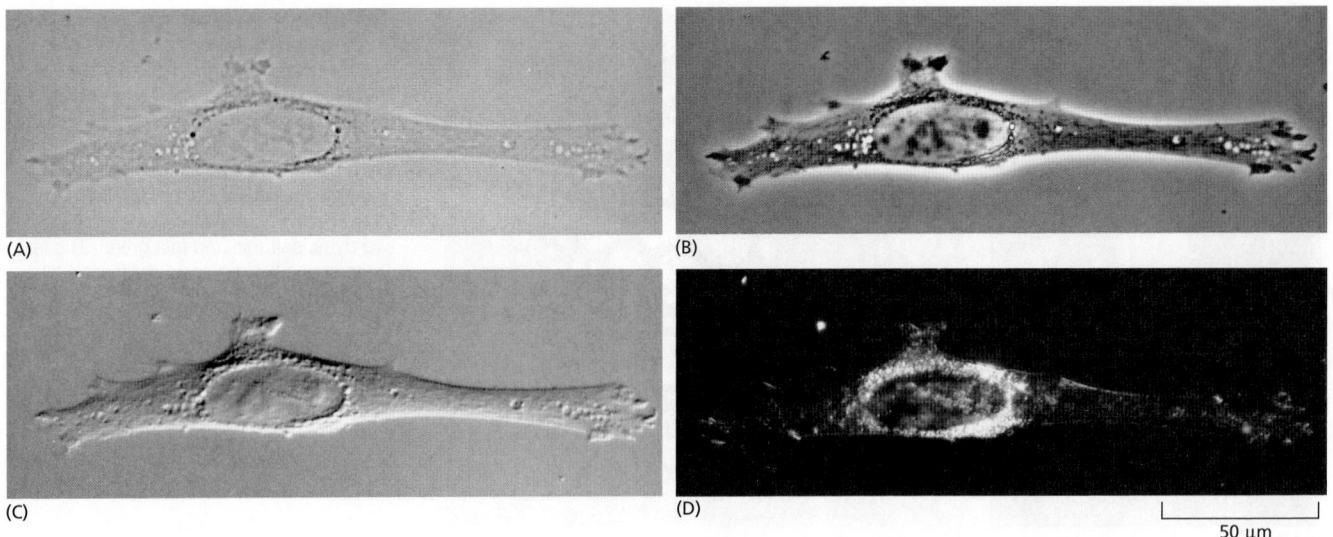

(A)

(B)

(C)

(D)

50 µm

Figure 9–8 Four types of light microscopy. Four images are shown of the same fibroblast cell in culture. All images can be obtained with most modern microscopes by interchanging optical components. (A) Bright-field microscopy, in which light is transmitted straight through the specimen. (B) Phase-contrast microscopy, in which phase alterations of light transmitted through the specimen are translated into brightness changes. (C) Differential-interference-contrast microscopy, which highlights edges where there is a steep change of refractive index. (D) Dark-field microscopy, in which the specimen is lit from the side and only the scattered light is seen.

the eye's limitations in detecting small differences in light intensity, and background irregularities in the optical system can be digitally subtracted. This procedure reveals small transparent objects that were previously impossible to distinguish from the background.

Intact Tissues Are Usually Fixed and Sectioned Before Microscopy

Because most tissue samples are too thick for their individual cells to be examined directly at high resolution, they are often cut into very thin transparent slices, or *sections*. To preserve the cells within the tissue they must be treated with a *fixative*. Common fixatives include glutaraldehyde, which forms covalent bonds with the free amino groups of proteins, cross-linking them so they are stabilized and locked into position.

Because tissues are generally soft and fragile, even after fixation, they need to be either frozen or embedded in a supporting medium before being sectioned. The usual embedding media are waxes or resins. In liquid form, these media both permeate and surround the fixed tissue; they can then be hardened (by cooling or by polymerization) to form a solid block, which is readily sectioned with a microtome. This is a machine with a sharp blade, usually of steel or glass, which operates like a meat-slicer (**Figure 9–9**). The sections (typically 0.5–10 µm thick) are then laid flat on the surface of a glass microscope slide.

There is little in the contents of most cells (which are 70% water by weight) to impede the passage of light rays. Thus, most cells in their natural state, even if fixed and sectioned, are almost invisible in an ordinary light microscope. We have seen that cellular components can be made visible by techniques such as phase-contrast and differential-interference-contrast microscopy, but these methods tell us almost nothing about the underlying chemistry. There are three main approaches to working with thin tissue sections that reveal differences in types of molecules that are present.

First, and traditionally, sections can be stained with organic dyes that have some specific affinity for particular subcellular components. The dye hematoxylin, for example, has an affinity for negatively charged molecules and therefore reveals the distribution of DNA, RNA, and acidic proteins in a cell (**Figure 9–10**). The chemical basis for the specificity of many dyes, however, is not known.

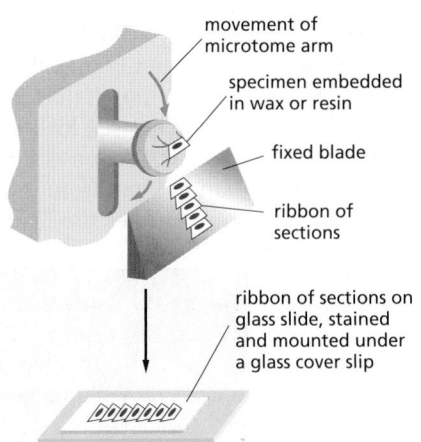

movement of microtome arm

specimen embedded in wax or resin

fixed blade

ribbon of sections

ribbon of sections on glass slide, stained and mounted under a glass cover slip

Figure 9–9 Making tissue sections. This illustration shows how an embedded tissue is sectioned with a microtome in preparation for examination in the light microscope.

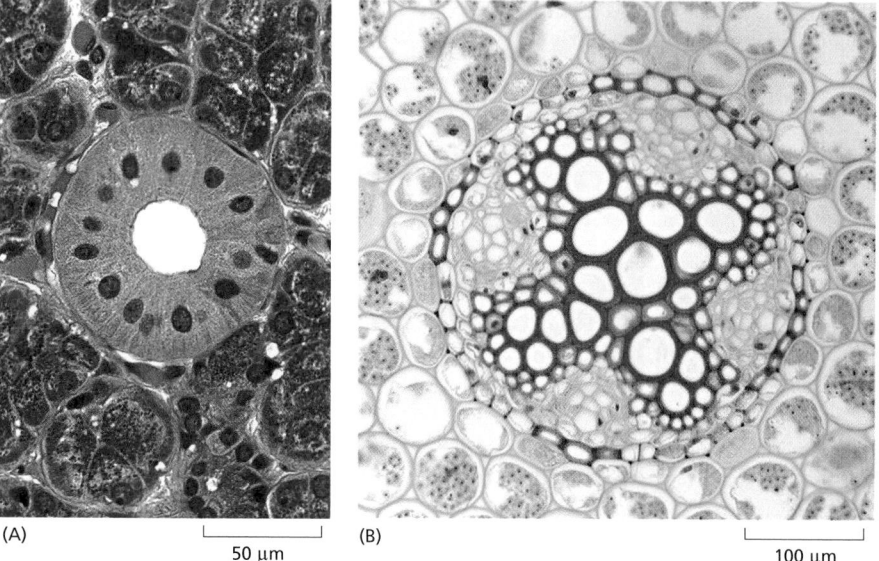

(A)

⌞ 50 μm ⌟

(B)

⌞ 100 μm ⌟

Figure 9–10 **Staining of cell components.** (A) This section of cells in a salivary gland was stained with hematoxylin and eosin, two dyes commonly used in histology. The central duct is made of closely packed cells with nuclei stained purple and cytoplasm stained red. The duct is surrounded by groups of saliva secreting cells. (B) This section of a young plant root is stained with two dyes, safranin and fast green. The fast green stains the cellulosic cell walls while the safranin stains the lignified xylem cell walls bright red. (A, from R.L. Sorenson and T.C. Brelje, *Atlas of Human Histology: A Guide to Microscopic Structure of Cells, Tissues and Organs,* 3rd Ed. (2014). With permission from the authors; B, courtesy of University of Wisconsin Plant Teaching Collection.)

Second, sectioned tissues can be used to visualize specific patterns of differential gene expression. *In situ* hybridization, discussed earlier (see Figure 8–34), reveals the cellular distribution and abundance of specific expressed RNA molecules in sectioned material or in whole mounts of small organisms or organs. This is particularly effective when used in conjunction with fluorescent probes (**Figure 9–11**).

A third and very sensitive approach, generally and widely applicable for localizing proteins of interest, also depends on using fluorescent probes and markers, as we explain next.

Specific Molecules Can Be Located in Cells by Fluorescence Microscopy

Fluorescent molecules absorb light at one wavelength and emit it at another, longer wavelength (**Figure 9–12**A). If we illuminate such a molecule at its absorbing wavelength and then view it through a filter that allows only light of the emitted wavelength to pass, it will glow against a dark background. Because the background is dark, even a minute amount of the glowing fluorescent dye can be detected. In contrast, the same number of molecules of a nonfluorescent stain, viewed conventionally, would be practically indiscernible because the absorption of light by molecules in the stain would result in only the faintest tinge of color in the light transmitted through that part of the specimen.

The fluorescent dyes used for staining cells are visualized with a **fluorescence microscope.** This microscope is similar to an ordinary light microscope except that the illuminating light, from a very powerful source, is passed through two sets of filters—one to filter the light before it reaches the specimen and one to

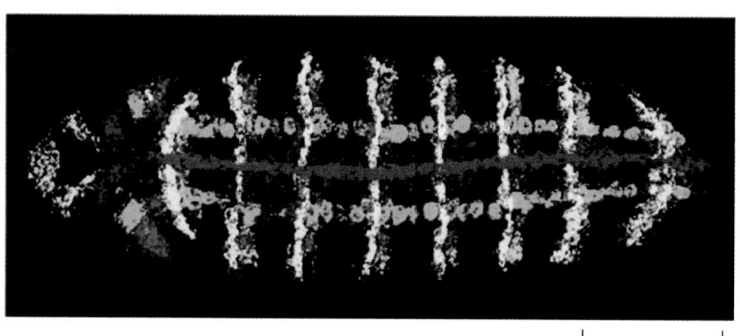

⌞ 100 μm ⌟

Figure 9–11 **RNA *in situ* hybridization.** As described in Chapter 8 (see Figure 8–62), it is possible to visualize the distribution of different RNAs in tissues using *in situ* hybridization. Here, the transcription pattern of five different genes involved in patterning the early fly embryo is revealed in a single embryo. Each RNA probe has been fluorescently labeled in a different way, some directly and some indirectly; the resulting images are displayed each in a different color ("false-colored") and combined to give an image where different color combinations represent different sets of genes expressed. The genes whose expression pattern is revealed here are *wingless (yellow), engrailed (blue), short gastrulation (red), intermediate neuroblasts defective (green),* and *muscle specific homeobox (purple).* (From D. Kosman et al., *Science* 305:846, 2004. With permission from AAAS.)

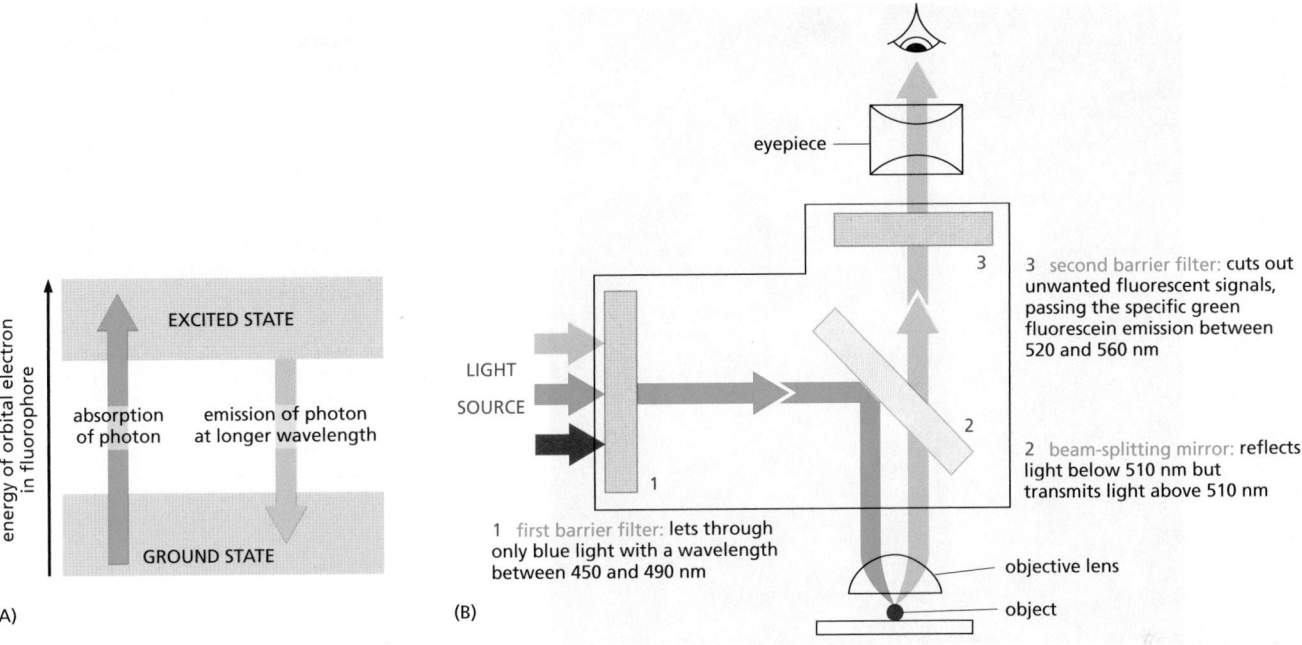

EXCITED STATE

absorption of photon

emission of photon at longer wavelength

energy of orbital electron in fluorophore

GROUND STATE

(A)

LIGHT SOURCE

eyepiece

3 second barrier filter: cuts out unwanted fluorescent signals, passing the specific green fluorescein emission between 520 and 560 nm

2 beam-splitting mirror: reflects light below 510 nm but transmits light above 510 nm

1 first barrier filter: lets through only blue light with a wavelength between 450 and 490 nm

objective lens

object

(B)

Figure 9–12 **Fluorescence and the fluorescence microscope.** (A) An orbital electron of a fluorochrome molecule can be raised to an excited state following the absorption of a photon. **Fluorescence** occurs when the electron returns to its ground state and emits a photon of light at a longer wavelength. Too much exposure to light, or too bright a light, can also destroy the fluorochrome molecule, in a process called *photobleaching*. (B) In the fluorescence microscope, a filter set consists of two barrier filters (1 and 3) and a dichroic (beam-splitting) mirror (2). This example shows the filter set for detection of the fluorescent molecule fluorescein. High-numerical-aperture objective lenses are especially important in this type of microscopy because, for a given magnification, the brightness of the fluorescent image is proportional to the fourth power of the numerical aperture (see also Figure 9–6).

filter the light obtained from the specimen. The first filter passes only the wavelengths that excite the particular fluorescent dye, while the second filter blocks out this light and passes only those wavelengths emitted when the dye fluoresces (Figure 9–12B).

Fluorescence microscopy is most often used to detect specific proteins or other molecules in cells and tissues. A very powerful and widely used technique is to couple fluorescent dyes to antibody molecules, which then serve as highly specific and versatile staining reagents that bind selectively to the particular macromolecules they recognize in cells or in the extracellular matrix. Two fluorescent dyes that have been commonly used for this purpose are *fluorescein*, which emits an intense green fluorescence when excited with blue light, and *rhodamine*, which emits deep red fluorescence when excited with green–yellow light (**Figure 9–13**). By coupling one antibody to fluorescein and another to rhodamine, the distributions of different molecules can be compared in the same cell; the two molecules are visualized separately in the microscope by switching back and forth between two sets of filters, each specific for one dye. As shown in **Figure 9–14**, three fluorescent dyes can be used in the same way to distinguish among three types of molecules in the same cell. Many newer fluorescent dyes, such as Cy3, Cy5, and the Alexa dyes, have been specifically developed for fluorescence microscopy

Figure 9–13 **Fluorescent probes.** The maximum excitation and emission wavelengths of several commonly used fluorescent probes are shown in relation to the corresponding colors of the spectrum. The photon emitted by a fluorescent molecule is necessarily of lower energy (longer wavelength) than the absorbed photon and this accounts for the difference between the excitation and emission peaks. CFP, GFP, YFP, and RFP are cyan, green, yellow, and red fluorescent proteins, respectively. DAPI is widely used as a general fluorescent DNA probe, which absorbs ultraviolet light and fluoresces bright blue. FITC is an abbreviation for fluorescein isothiocyanate, a widely used derivative of fluorescein, which fluoresces bright green. The other probes are all commonly used to fluorescently label antibodies and other proteins. The use of fluorescent proteins will be discussed later in the chapter.

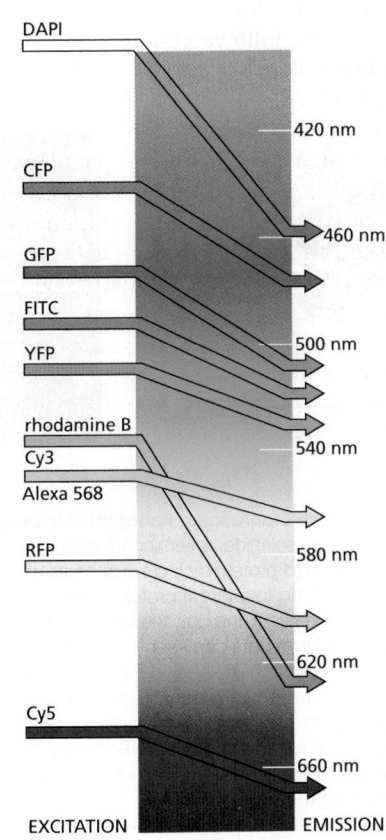

DAPI

420 nm

CFP

460 nm

GFP

FITC

YFP

500 nm

rhodamine B

Cy3

540 nm

Alexa 568

RFP

580 nm

620 nm

Cy5

660 nm

EXCITATION

EMISSION

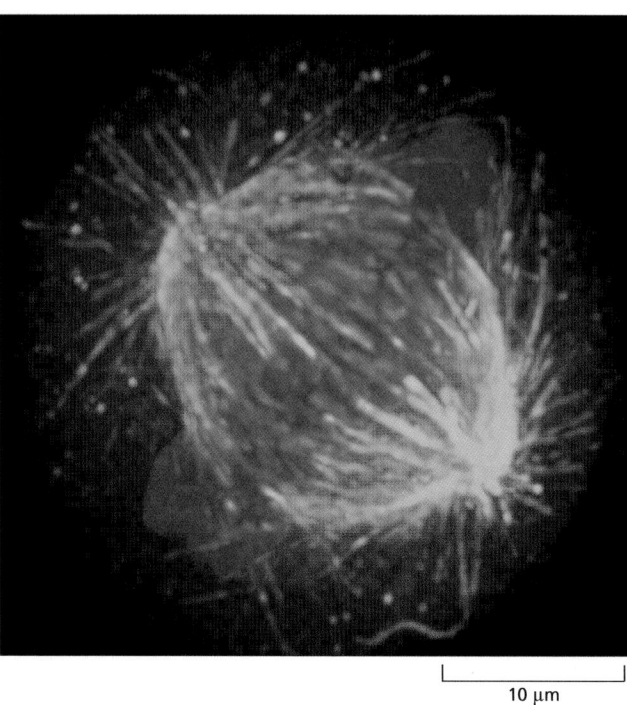

10 µm

Figure 9–14 **Different fluorescent probes can be visualized in the same cell.** In this composite micrograph of a cell in mitosis, three different fluorescent probes have been used to label three different cellular components (Movie 9.1). The spindle microtubules are revealed with a *green* fluorescent antibody, centromeres with a *red* fluorescent antibody, and the DNA of the condensed chromosomes with the *blue* fluorescent dye DAPI. (Courtesy of Kevin F. Sullivan.)

(see Figure 9–13) but, like many organic fluorochromes, they fade fairly rapidly when continuously illuminated. More stable fluorochromes have been developed based on inorganic chemistry. Tiny crystals of semiconductor material, called nanoparticles, or *quantum dots*, can be excited to fluoresce by a broad spectrum of blue light. Their emitted light has a color that depends on the exact size of the nanocrystal, between 2 and 10 nm in diameter, and additionally the fluorescence fades only slowly with time (Figure 9–15). These nanoparticles, when coupled to other probes such as antibodies, are therefore ideal for tracking molecules over time. If introduced into a living cell, in an embryo for example, the progeny of that cell can be followed many days later by their fluorescence, allowing cell lineages to be tracked.

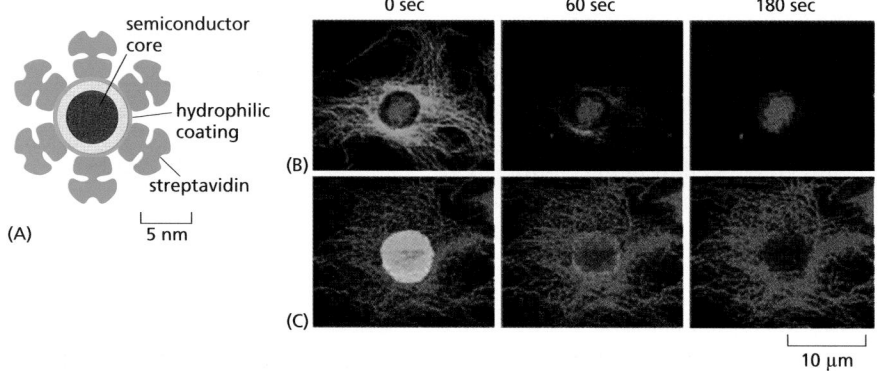

Figure 9–15 **Fluorescent nanoparticles or quantum dots.** (A) Quantum dots are tiny particles of cadmium selenide, a semiconductor, with a coating to make them water-soluble. They can be coupled to protein molecules such as antibodies or streptavidin and, when introduced into a cell, will bind to a target protein of interest. Different-sized quantum dots emit light of different colors—the larger the dot, the longer the wavelength—but they are all excited by the same blue light. Quantum dots can keep shining for weeks, unlike most fluorescent organic dyes. (B) In this cell, microtubules are labeled *(green)* with an organic fluorescent dye (Alexa 488), while a nuclear protein is stained *(red)* with quantum dots bound to streptavidin. On continuous exposure to strong blue light, the fluorescent dye fades quickly while the quantum dots continue to shine. (C) In this cell, the labeling pattern is reversed; a nuclear protein is labeled *(green)* with an organic fluorescent dye (Alexa 488), while microtubules are labeled *(red)* with quantum dots. Again, the quantum dots far outlast the fluorescent dye. (B and C, from L. Medintz et al., *Nat. Mater.* 4:435–446, published 2005 by Nature Publishing Group. Reproduced with permission of SNCSC.)

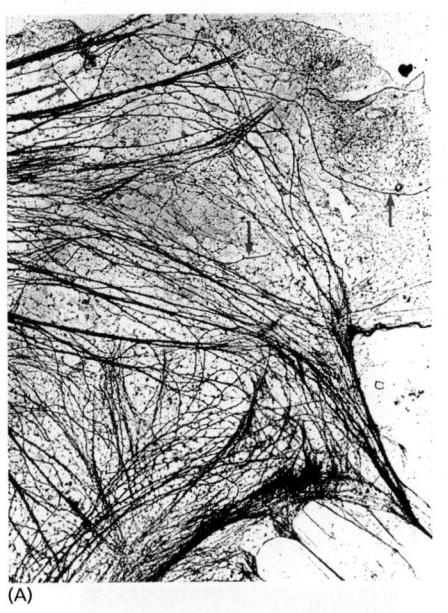

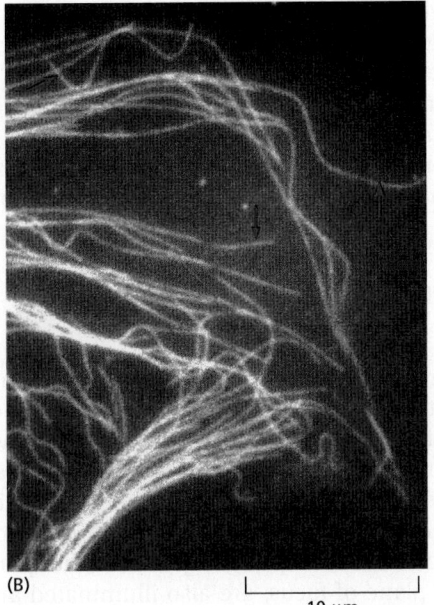

(A)

(B)

10 μm

Figure 9–16 **Immunofluorescence.**
(A) A transmission electron micrograph of the periphery of a cultured epithelial cell showing the distribution of microtubules and other filaments. (B) The same area stained with fluorescent antibodies against tubulin, the protein that assembles to form microtubules, using the technique of indirect immunocytochemistry (see Figure 9–17). *Red arrows* indicate individual microtubules that are readily recognizable in both images. Note that, because of diffraction effects, the microtubules in the light microscope appear 0.2 μm wide rather than their true width of 0.025 μm. (© 1978 M. Osborn et al. Originally published in *J. Cell Biol.* http://doi .org/10.1083/jcb.77.3.R27. With permission from Rockefeller University Press.)

Later in the chapter, additional fluorescence microscopy methods will be discussed that can be used to monitor changes in the concentration and location of specific molecules inside *living* cells.

Antibodies Can Be Used to Detect Specific Molecules

Antibodies are proteins produced by the vertebrate immune system as a defense against infection (discussed in Chapter 24). They are unique among proteins in that they are made in billions of different forms, each with a different binding site that recognizes a specific target molecule (or *antigen*). The precise antigen specificity of antibodies makes them powerful tools for the cell biologist. When labeled with fluorescent dyes, antibodies are invaluable for locating specific molecules in cells by fluorescence microscopy (**Figure 9–16**); labeled with electron-dense particles such as colloidal gold spheres, they are used for similar purposes in the electron microscope (discussed below). The antibodies employed in microscopy are commonly either purified from antiserum so as to remove all nonspecific antibodies, or they are specific monoclonal antibodies that only recognize the target molecule.

When we use antibodies as probes to detect and assay specific molecules in cells, we frequently use chemical methods to amplify the fluorescent signal they produce. For example, although a marker molecule such as a fluorescent dye can be linked directly to an antibody—the *primary antibody*—a stronger signal is achieved by using an unlabeled primary antibody and then detecting it with a group of labeled *secondary antibodies* that bind to it (**Figure 9–17**). This process is called *indirect immunocytochemistry*.

Some amplification methods use an enzyme as a marker molecule attached to the secondary antibody. The enzyme alkaline phosphatase, for example, in the presence of appropriate chemicals, produces inorganic phosphate that in turn

Figure 9–17 **Indirect immunocytochemistry.** This detection method is very sensitive because many molecules of the secondary antibody recognize each primary antibody. The secondary antibody is covalently coupled to a marker molecule that makes it readily detectable. Commonly used marker molecules include fluorescent probes (for fluorescence microscopy), the enzyme horseradish peroxidase (for either conventional light microscopy or electron microscopy), colloidal gold spheres (for electron microscopy), and the enzymes alkaline phosphatase or peroxidase (for biochemical detection).

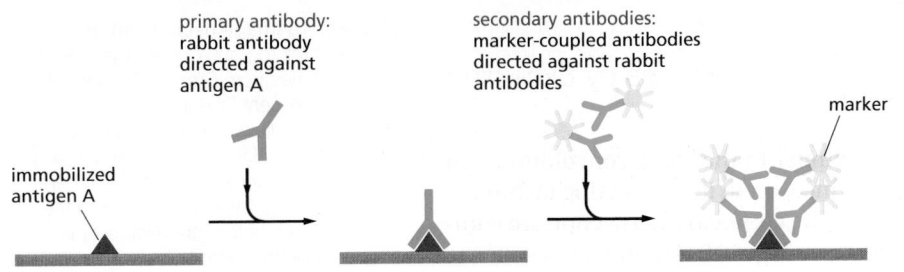

primary antibody:
rabbit antibody
directed against
antigen A

secondary antibodies:
marker-coupled antibodies
directed against rabbit
antibodies

marker

immobilized
antigen A

leads to the local formation of a colored precipitate. This reveals the location of the secondary antibody and hence the location of the antibody–antigen complex. Since each enzyme molecule acts catalytically to generate many thousands of molecules of product, even tiny amounts of antigen can be detected. Although the enzyme amplification makes enzyme-linked methods sensitive, diffusion of the colored precipitate away from the enzyme limits the spatial resolution of this method for microscopy, and fluorescent labels are usually used for the most sensitive and precise optical localization.

Imaging of Complex Three-Dimensional Objects Is Possible with the Optical Microscope

For ordinary light microscopy, as we have seen, a tissue has to be sliced into thin sections to be examined; the thinner the section, the crisper the image. Since information about the third dimension is lost upon sectioning, how, then, can we get a picture of the three-dimensional architecture of a cell or tissue, and how can we view the microscopic structure of a specimen that, for one reason or another, cannot first be sliced into sections? Although an optical microscope is focused on a particular focal plane within a three-dimensional specimen, all the other parts of the specimen, above and below the plane of focus, are also illuminated and the light originating from these regions contributes to the image as "out-of-focus" blur. This can make it very hard to interpret the image in detail and can lead to fine image structure being obscured by the out-of-focus light.

Two distinct but complementary approaches solve this problem: one is computational, the other optical. These three-dimensional microscopic imaging methods make it possible to focus on a chosen plane in a thick specimen while rejecting the light that comes from out-of-focus regions above and below that plane. Thus one sees a crisp, thin *optical section*. From a series of such optical sections taken at different depths and stored in a computer, a three-dimensional image can be reconstructed. The methods do for the microscopist what the computed tomography (CT) scanner does (by different means) for the radiologist investigating a human body: both machines give detailed sectional views of the interior of an intact structure.

The computational approach is often called *image deconvolution*. To understand how it works, remember that the wavelike nature of light means that the microscope lens system produces a small blurred disc as the image of a point light source (see Figure 9–5), with increased blurring if the point source lies above or below the focal plane. This blurred image of a point source is called the *point spread function* (see Figure 9–36). An image of a complex object can then be thought of as being built up by replacing each point of the specimen by a corresponding blurred disc, resulting in an image that is blurred overall. For deconvolution, we first obtain a series of (blurred) images, usually with a cooled CCD camera or more recently a CMOS camera, focusing the microscope in turn on a series of focal planes—in effect, a (blurred) three-dimensional image. Digital processing of the stack of digital images then removes as much of the blur as possible. In essence, the computer program uses the measured point spread function of a point source of light from that microscope to determine what the effect of the blurring would have been on the image, and then applies an equivalent "deblurring" (deconvolution), turning the blurred three-dimensional image into a series of clean optical sections, albeit still constrained by the diffraction limit. **Figure 9–18** shows an example.

The Confocal Microscope Produces Optical Sections by Excluding Out-of-Focus Light

The confocal microscope achieves a result similar to that of deconvolution, but does so by manipulating the light before it is measured; it is an analog technique rather than a digital one. The optical details of the **confocal microscope** are complex, but the basic idea is simple, as illustrated in **Figure 9–19**, and the results are

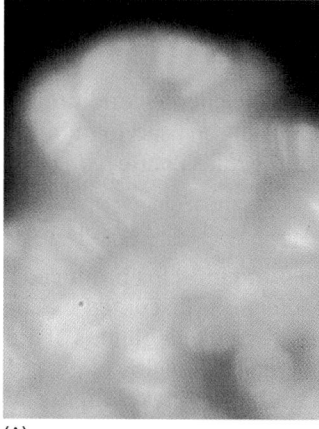

(A)

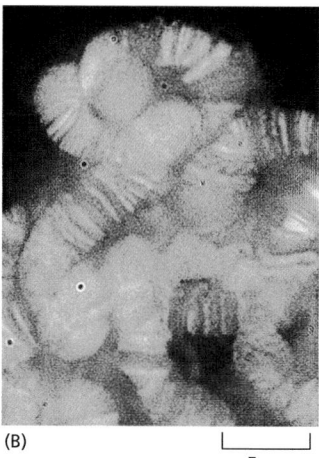

(B)

5 µm

Figure 9–18 Image deconvolution.
(A) A light micrograph of the large polytene chromosomes from *Drosophila*, stained with a fluorescent DNA-binding dye.
(B) The same field of view after image deconvolution clearly reveals the banding pattern on the chromosomes. Each band is about 0.25 µm thick, approaching the diffraction limit of the light microscope. (Courtesy of the John Sedat Laboratory.)

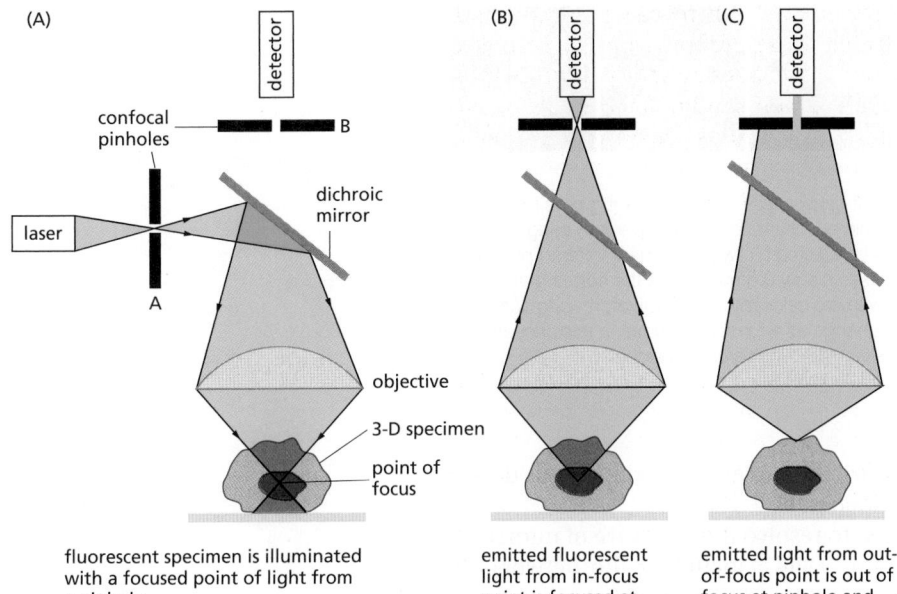

(A) fluorescent specimen is illuminated with a focused point of light from a pinhole

(B) emitted fluorescent light from in-focus point is focused at pinhole and reaches detector

(C) emitted light from out-of-focus point is out of focus at pinhole and is largely excluded from detector

Figure 9–19 The confocal fluorescence microscope. (A) This simplified diagram shows that the basic arrangement of optical components is similar to that of the standard fluorescence microscope shown in Figure 9–12, except that a laser is used to illuminate a small pinhole whose image is focused at a single point in the three-dimensional (3-D) specimen. (B) Emitted fluorescence from this focal point in the specimen is focused at a second (confocal) pinhole. (C) Emitted light from elsewhere in the specimen is not focused at the pinhole and therefore does not contribute to the final image. By scanning the beam of light across the specimen, a very sharp two-dimensional image of the exact plane of focus is built up that is not significantly degraded by light from other regions of the specimen.

far superior to those obtained by conventional light microscopy (Figure 9–20A and B).

The confocal microscope is generally used with fluorescence optics (see Figure 9–12), but instead of illuminating the whole specimen at once, in the usual way, the optical system at any instant focuses a spot of light onto a single point at a specific depth in the specimen. This requires a source of pinpoint illumination that is usually supplied by a laser whose light has been passed through a pinhole. The fluorescence emitted from the illuminated material is collected at a suitable light detector and used to generate an image. A pinhole aperture is placed in front of the detector, at a position that is *confocal* with the illuminating pinhole—that is, precisely where the rays emitted from the illuminated point in the specimen come to a focus. Thus, the light from this point in the specimen converges on this aperture and enters the detector.

By contrast, the light from regions out of the plane of focus of the spotlight is also out of focus at the pinhole aperture and is therefore largely excluded from the detector (see Figure 9–19). To build up a two-dimensional image, data from each point in the plane of focus are collected sequentially by scanning across the field from left to right in a regular pattern of pixels and are displayed on a computer screen. Although not shown in Figure 9–19, the scanning is usually done by deflecting the beam with an oscillating mirror placed between the dichroic mirror and the objective lens in such a way that the illuminating spotlight and the

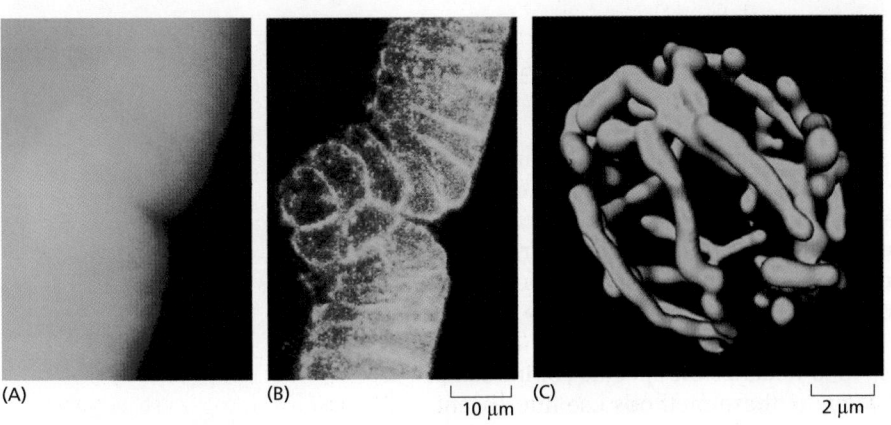

(A) (B) (C)

10 μm   2 μm

Figure 9–20 Confocal fluorescence microscopy produces clear optical sections and three-dimensional data sets. The first two micrographs are of the same intact gastrula-stage *Drosophila* embryo, which has been stained with a fluorescent probe for actin filaments. (A) The conventional, unprocessed image is blurred by the presence of fluorescent structures above and below the plane of focus. (B) In the confocal image, this out-of-focus information is removed, resulting in a crisp optical section of the cells in the embryo. (C) A three-dimensional reconstruction of an object can be assembled from a stack of such optical sections. In this case, the complex branching structure of the mitochondrial compartment in a single live yeast cell is shown. (A and B, courtesy of Richard Warn and Peter Shaw; C, courtesy of Stefan Hell.)

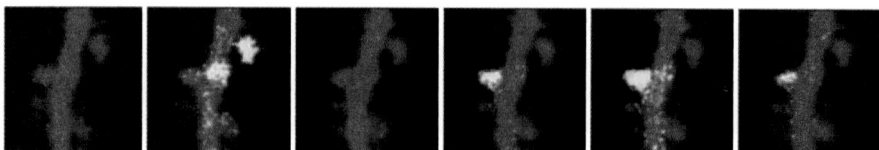

Figure 9–21 Multiphoton imaging. Infrared laser light causes less damage to living cells than visible light and can also penetrate further, allowing microscopists to peer deeper into living tissues. The two-photon effect, in which a fluorochrome can be excited by two coincident infrared photons instead of a single high-energy photon, allows us to see nearly 0.5 mm inside the cortex of a live mouse brain. A dye, whose fluorescence changes with the calcium concentration, reveals active synapses *(yellow)* on the dendritic spines *(red)* that change as a function of time; in this case, there is a day between each image. (Courtesy of Thomas Oertner and Karel Svoboda.)

confocal pinhole at the detector remain strictly in register. Variations in design now allow the rapid collection of data at video rates.

The confocal microscope has been used to resolve the structure of numerous complex three-dimensional objects (Figure 9–20C) including the networks of cytoskeletal fibers in the cytoplasm and the arrangements of chromosomes and genes in the nucleus.

The relative merits of deconvolution methods and confocal microscopy for three-dimensional optical microscopy depend on the specimen being imaged. Confocal microscopes tend to be better for thicker specimens with high levels of out-of-focus light. They are also generally easier to use than deconvolution systems and the final optical sections can be seen quickly. In contrast, the cooled CCD or CMOS cameras used for deconvolution systems are extremely efficient at collecting small amounts of light, and they can be used to make detailed three-dimensional images from specimens that are too weakly stained or too easily damaged by the bright light used for confocal microscopy.

Both methods, however, have another drawback; neither is good at coping with very thick specimens. Deconvolution methods quickly become ineffective any deeper than about 40 μm into a specimen, while confocal microscopes can only obtain images up to a depth of about 150 μm. Special microscopes can now take advantage of the way in which fluorescent molecules are excited, to probe even deeper into a specimen. Fluorescent molecules are usually excited by a single high-energy photon, of shorter wavelength than the emitted light, but they can in addition be excited by the absorption of two (or more) photons of lower energy, as long as they both arrive within a femtosecond or so of each other. The use of this longer-wavelength excitation has some important advantages. In addition to reducing background noise, red or near-infrared light can penetrate deeper into a specimen. Multiphoton microscopes, constructed to take advantage of this *two-photon* effect, can obtain sharp images, sometimes even at a depth of 250 μm within a specimen. This is particularly valuable for studies of living tissues, notably in imaging the dynamic activity of synapses and neurons just below the surface of living brains (**Figure 9–21**).

Individual Proteins Can Be Fluorescently Tagged in Living Cells and Organisms

Even the most stable cell structures must be assembled, disassembled, and reorganized during the cell's life cycle. Other structures, often enormous on the molecular scale, rapidly change, move, and reorganize themselves as the cell conducts its internal affairs and responds to its environment. Complex, highly organized pieces of molecular machinery move components around the cell, controlling traffic into and out of the nucleus, from one organelle to another, and into and out of the cell itself.

Various techniques have been developed to visualize the specific components involved in such dynamic phenomena. Many of these methods use fluorescent proteins, and they require a trade-off between structural preservation and efficient labeling. All of the fluorescent molecules discussed so far are made outside the cell and then artificially introduced into it. But use of genes coding for protein molecules that are themselves inherently fluorescent also enables the creation of organisms and cell lines that make their own visible tags and labels, without the introduction of foreign molecules. These cellular exhibitionists display their inner workings in glowing fluorescent color.

Foremost among the fluorescent proteins used for these purposes by cell biologists is the **green fluorescent protein** (**GFP**), isolated from the jellyfish *Aequorea victoria*. This protein is encoded by a single gene, which can be cloned and introduced into cells of other species. The freshly translated protein is not fluorescent, but within an hour or so (less for some alleles of the gene, more for others) it undergoes a self-catalyzed post-translational modification to generate an efficient fluorochrome, shielded within the interior of a barrel-like protein, which will now fluoresce when illuminated appropriately with blue light (**Figure 9–22**). Extensive site-directed mutagenesis performed on the original gene sequence has resulted in multiple variants that can be used effectively in organisms ranging from animals and plants to fungi and microbes. The fluorescence efficiency has also been improved, and variants have been generated with altered absorption and emission spectra from the blue–green, like blue fluorescent protein or BFP, to the far visible red. Other, related fluorescent proteins have since been discovered (for example, in corals) that also extend the range into the red region of the spectrum, like red fluorescent protein or RFP.

One of the simplest uses of GFP is as a reporter molecule, a fluorescent probe to monitor gene expression. A transgenic organism can be made with the GFP-coding sequence placed under the transcriptional control of the promoter belonging to a gene of interest, giving a directly visible readout of the gene's expression pattern in the living organism (**Figure 9–23**). In another application, a peptide location signal can be added to the GFP to direct it to a particular cell compartment, such as the endoplasmic reticulum or a mitochondrion, lighting up these organelles so they can be observed in the living state (see Figure 12–31).

The GFP DNA coding sequence can also be inserted at the beginning or end of the gene for another protein, yielding a chimeric product consisting of that protein with a GFP domain attached. In many cases, this GFP fusion protein behaves in the same way as the original protein, directly revealing its location and activities by means of its genetically encoded fluorescence (**Figure 9–24**). It is often possible to prove that the GFP fusion protein is functionally equivalent to the untagged protein, for example by using it to rescue a mutant lacking that protein. GFP tagging is the clearest and most unequivocal way of showing the distribution and dynamics of a protein in a living organism (**Figure 9–25** and see Movie 16.8).

Protein Dynamics Can Be Followed in Living Cells

Fluorescent proteins are now exploited not just to see where in a cell a particular protein is located, but also to uncover its kinetic properties and to find out whether it might interact with other molecules. We now describe three techniques in which fluorescent proteins are used in this way.

First, interactions between one protein and another can be monitored by **fluorescence resonance energy transfer**, also called **Förster resonance energy**

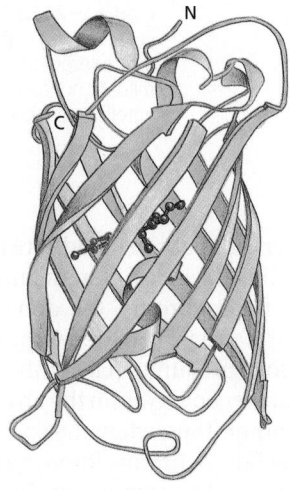

Figure 9–22 **Green fluorescent protein (GFP).** The structure of GFP, shown here schematically, highlights the eleven β strands that form the staves of a barrel. Buried within the barrel is the active chromophore *(dark green)* that is formed post-translationally from the protruding side chains of three amino acid residues. (From M. Ormö et al., *Science* 273:1392–1395, 1996. With permission from AAAS.)

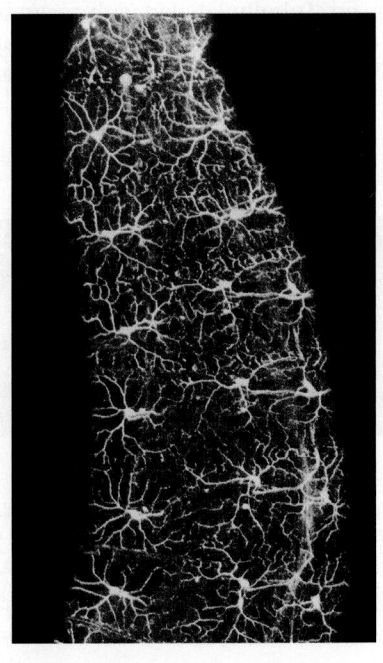

Figure 9–23 **Green fluorescent protein (GFP) as a reporter.** For this experiment, carried out in the fruit fly, the GFP gene was joined (using recombinant DNA techniques) to a fly promoter that is active only in a specialized set of neurons. This image of a live fly embryo was captured by a fluorescence microscope and shows approximately 20 neurons, each with long projections (axons and dendrites) that communicate with other (nonfluorescent) cells. These neurons are located just under the surface of the animal and allow it to sense its immediate environment. (From W.B. Grueber et al., *Curr. Biol.* 13:618–626, 2003. With permission from Elsevier.)

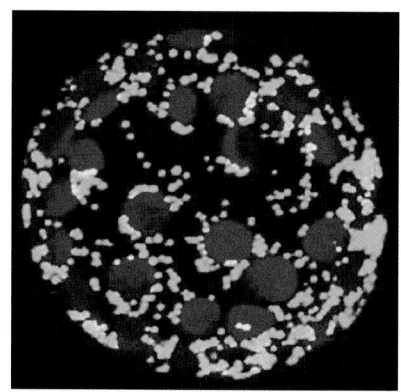

Figure 9–24 **GFP-tagged proteins.** This living cell from a tobacco plant is expressing high levels of green fluorescent protein, fused to a protein that is targeted to mitochondria, which accordingly appear *green*. The mitochondria are seen to cluster around the chloroplasts, whose chlorophyll autofluorescence marks them out in *red*. (Courtesy of Olivier Grandjean.)

transfer, both abbreviated **FRET**. In this technique, two molecules of interest are each labeled with a different fluorochrome, chosen so that the emission spectrum of one fluorochrome, the donor, overlaps with the absorption spectrum of the other, the acceptor. If the two proteins bind so as to bring their fluorochromes into very close proximity (closer than about 5 nm), one fluorochrome, when excited, can transfer energy from the absorbed light directly (by resonance, nonradiatively) to the other. Thus, when the complex is illuminated at the excitation wavelength of the first fluorochrome, fluorescent light is produced at the emission wavelength of the second. This method can be used with two different spectral variants of GFP as fluorochromes to monitor processes such as the interaction of signaling molecules with their receptors, or proteins in macromolecular complexes at specific locations inside living cells (**Figure 9–26**). The FRET can be measured by quantifying the reduction of the donor fluorescence in the presence of the acceptor.

A second example of a fluorescence-tagging technique that allows detailed observations of proteins within cells involves synthesizing an inactive form of the fluorescent molecule of interest, introducing it into the cell, and then activating it suddenly at a chosen site in the cell by focusing a spot of light on it. This process is referred to as **photoactivation**. Many inactive photosensitive precursors of this type, often called *caged molecules*, have been made based on a variety of fluorescent molecules. A microscope can be used to focus a strong pulse of light from a laser on any tiny region of the cell, so that the experimenter can control exactly where and when the fluorescent molecule is photoactivated. The technique allows us to follow complex and rapid intracellular processes, such as the actions of signaling molecules or the movements of cytoskeletal proteins.

When a photoactivatable fluorescent tag is attached to a purified protein, it is important that the modified protein remain biologically active: labeling with a caged fluorescent dye adds a bulky group to the surface of a protein, which can easily change the protein's properties. A satisfactory labeling protocol is usually found by trial and error. Once a biologically active labeled protein has been produced, it needs to be introduced into the living cell where its behavior can be followed. Tubulin labeled with caged fluorescein, for example, can be injected into a dividing cell, where it is incorporated into microtubules of the mitotic spindle. When a small region of the spindle is illuminated with a laser, the labeled tubulin

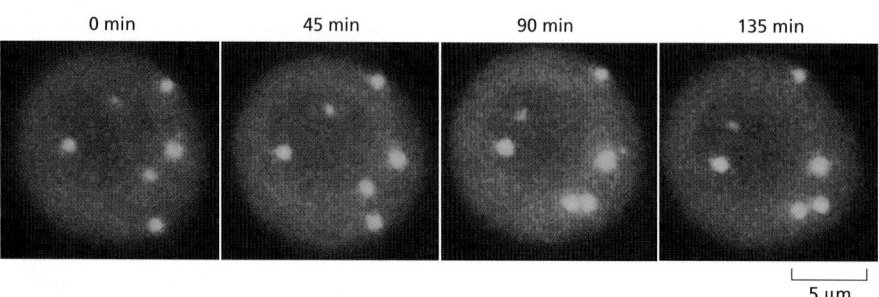

|| 0 min | 45 min | 90 min | 135 min ||

5 μm

Figure 9–25 **Dynamics of GFP tagging.** This sequence of micrographs shows a set of three-dimensional images of a living nucleus taken over the course of 135 minutes. Tobacco cells have been stably transformed with GFP fused to a spliceosomal protein that is concentrated in small nuclear bodies called Cajal bodies (see Figure 6–46). The fluorescent Cajal bodies, easily visible in a living cell with confocal microscopy, are dynamic structures that move around within the nucleus. (Courtesy of Kurt Boudonck, Liam Dolan, and Peter Shaw.)

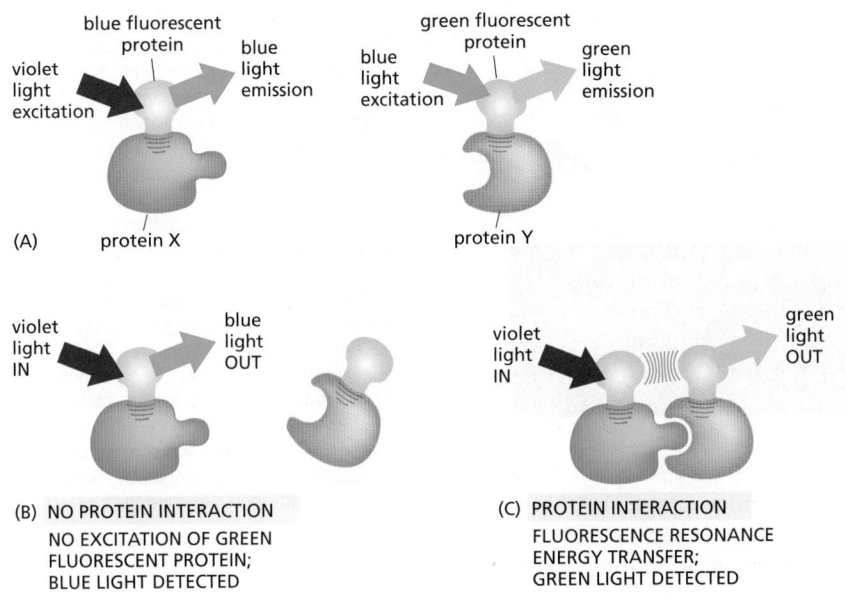

(A)

(B) NO PROTEIN INTERACTION
NO EXCITATION OF GREEN
FLUORESCENT PROTEIN;
BLUE LIGHT DETECTED

(C) PROTEIN INTERACTION
FLUORESCENCE RESONANCE
ENERGY TRANSFER;
GREEN LIGHT DETECTED

Figure 9–26 Fluorescence resonance energy transfer (FRET). To determine whether (and when) two proteins interact inside a cell, the proteins are first produced as fusion proteins attached to different color variants of green fluorescent protein (GFP). (A) In this example, protein X is coupled to a blue fluorescent protein, which is excited by violet light (370–440 nm) and emits blue light (440–480 nm); protein Y is coupled to a green fluorescent protein, which is excited by blue light (440–480 nm) and emits green light (510 nm). (B) If protein X and Y do not interact, illuminating the sample with violet light yields fluorescence from the blue fluorescent protein only. (C) When protein X and protein Y interact, the resonance transfer of energy, FRET, can now occur. Illuminating the sample with violet light excites the blue fluorescent protein, which transfers its energy to the green fluorescent protein, resulting in an emission of green light. The fluorochromes must be quite close together—within about 1–5 nm of one another—for FRET to occur. Because not every molecule of protein X and protein Y is bound at all times, some blue light may still be detected. But as the two proteins begin to interact, emission from the donor blue fluorescent protein falls as the emission from the acceptor GFP rises.

becomes fluorescent, so that its movement along the spindle microtubules can be readily followed (**Figure 9–27**).

A further development in photoactivation is the discovery that the genes encoding GFP and related fluorescent proteins can be engineered to produce protein variants, usually with one or more amino acid changes, that fluoresce only weakly under normal excitation conditions, but can be induced to fluoresce either more strongly or with a color shift (for example, from green to red) by activating them with a strong pulse of light at a different wavelength. In principle, the microscopist can then follow the local *in vivo* behavior of any protein that can be expressed as a fusion with one of these GFP variants. These genetically encoded, photoactivatable fluorescent proteins allow the lifetime and behavior of any protein to be studied independently of other newly synthesized proteins (**Figure 9–28**).

A third way to exploit GFP fused to a protein of interest is known as **fluorescence recovery after photobleaching (FRAP)**. Here, one uses a strong focused beam of light from a laser to extinguish the GFP fluorescence in a specified region of the cell, after which one can analyze the way in which remaining unbleached fluorescent protein molecules move into the bleached area as a function of time.

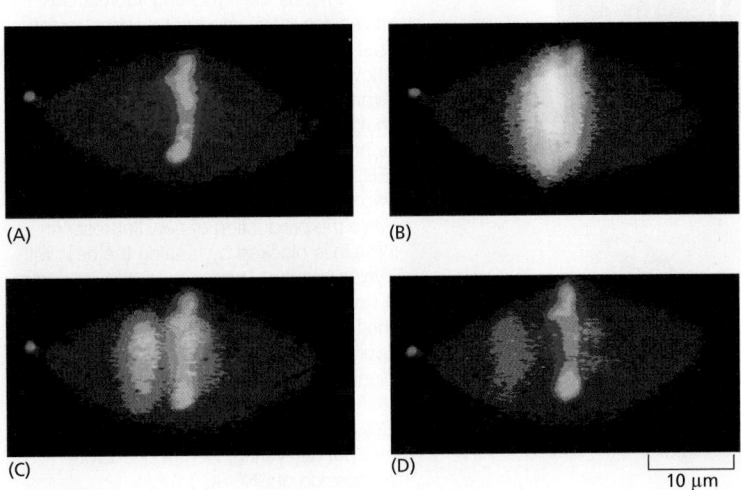

(A)

(B)

(C)

(D)

10 μm

Figure 9–27 Determining microtubule flux in the mitotic spindle with caged fluorescein linked to tubulin.
(A) A metaphase spindle formed *in vitro* from an extract of *Xenopus* eggs has incorporated three fluorescent markers: rhodamine-labeled tubulin *(red)* to mark all the microtubules, a *blue* DNA-binding dye that labels the chromosomes, and caged-fluorescein-labeled tubulin, which is also incorporated into all the microtubules but is invisible because it is nonfluorescent until activated by ultraviolet (UV) light. (B) A beam of UV light activates, or "uncages," the caged-fluorescein-labeled tubulin locally, mainly just to the left side of the metaphase plate. Over the next few minutes—after 1.5 minutes in (C) and after 2.5 minutes in (D)—the uncaged-fluorescein–tubulin signal moves toward the left spindle pole, indicating that tubulin is continuously moving poleward even though the spindle (visualized by the *red* rhodamine-labeled tubulin fluorescence) remains largely unchanged. (© 1991 K.E. Sawin and T.J. Mitchison. Originally published in *J. Cell Biol.* https://doi.org/10.1083/jcb.112.5.941. With permission from Rockefeller University Press.)

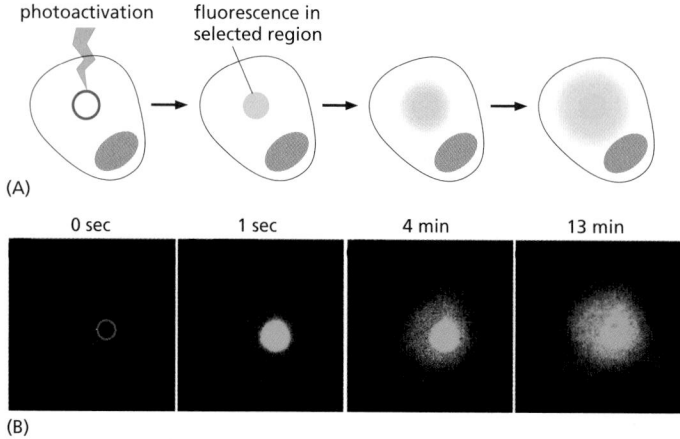

(A)

| 0 sec | 1 sec | 4 min | 13 min |

(B)

Figure 9–28 Photoactivation.
Photoactivation is the light-induced
activation of an inert molecule to an
active state. In this experiment, shown
schematically in (A), a photoactivatable
variant of GFP is expressed in a cultured
animal cell. (B) Before activation (time
0 sec), little or no GFP fluorescence is
detected in the selected region *(red circle)*
when excited by blue light at 488 nm. After
activation of the GFP with an ultraviolet
laser pulse at 413 nm, it rapidly fluoresces
brightly in the selected region *(green)*. The
movement of GFP, as it diffuses out of this
region, can be measured. Since only the
photoactivated proteins are fluorescent
within the cell, the trafficking, turnover, and
degradative pathways of proteins can be
monitored. (B, from J. Lippincott-Schwartz
and G.H. Patterson, *Science* 300:87–91,
2003.)

This technique is usually carried out with a confocal microscope and, like photoactivation, can deliver valuable quantitative data about a protein's kinetic parameters, such as diffusion coefficients, active transport rates, or binding and dissociation rates from other proteins (**Figure 9–29**).

Light-Emitting Indicators Can Measure Rapidly Changing Intracellular Ion Concentrations

One way to study the chemistry of a single living cell is to insert the tip of a fine,
glass, ion-sensitive **microelectrode** directly into the cell interior through the
plasma membrane. This technique is used to measure the intracellular concentrations of common inorganic ions, such as H^+, Na^+, K^+, Cl^-, and Ca^{2+}. However,
ion-sensitive microelectrodes reveal the ion concentration only at one point
in a cell, and for an ion present at a very low concentration, such as Ca^{2+}, their
responses are slow and somewhat erratic. Thus, these microelectrodes are not
ideally suited to record the rapid and transient changes in the concentration of
cytosolic Ca^{2+} that have an important role in allowing cells to respond to extracellular signals. Such changes can be analyzed with **ion-sensitive indicators**, whose
light emission reflects the local concentration of the ion. Some of these indicators are luminescent (emitting light spontaneously), while others are fluorescent
(emitting light on exposure to light).

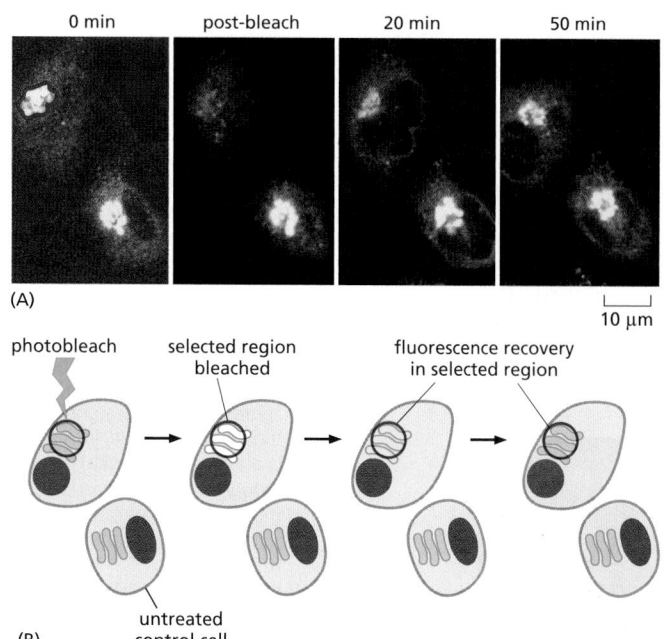

(A)

| 0 min | post-bleach | 20 min | 50 min |

⊢———⊣
10 μm

(B)

photobleach → selected region bleached → fluorescence recovery in selected region →

untreated control cell

**Figure 9–29 Fluorescence recovery
after photobleaching (FRAP).** A
strong focused pulse of laser light will
extinguish, or bleach, the fluorescence
of GFP. By selectively photobleaching
a set of fluorescently tagged protein
molecules within a defined region of a cell,
the microscopist can monitor recovery
over time, as the remaining fluorescent
molecules move into the bleached region
(see Movie 10.6). (A) The experiment
shown uses monkey cells in culture that
express galactosyltransferase, an enzyme
that constantly recycles between the Golgi
apparatus and the endoplasmic reticulum
(ER). The Golgi apparatus in one of the
two cells is selectively photobleached,
while the production of new fluorescent
protein is blocked by treating the cells with
cycloheximide. The recovery, resulting from
fluorescent enzyme molecules moving from
the ER to the Golgi, can then be followed
over a period of time. (B) Schematic
diagram of the experiment shown in (A).
(A, from J. Lippincott-Schwartz, *Histochem.
Cell Biol.* 116(2):97–107, published 2001
by Springer-Verlag. Reproduced with
permission of SNCSC.)

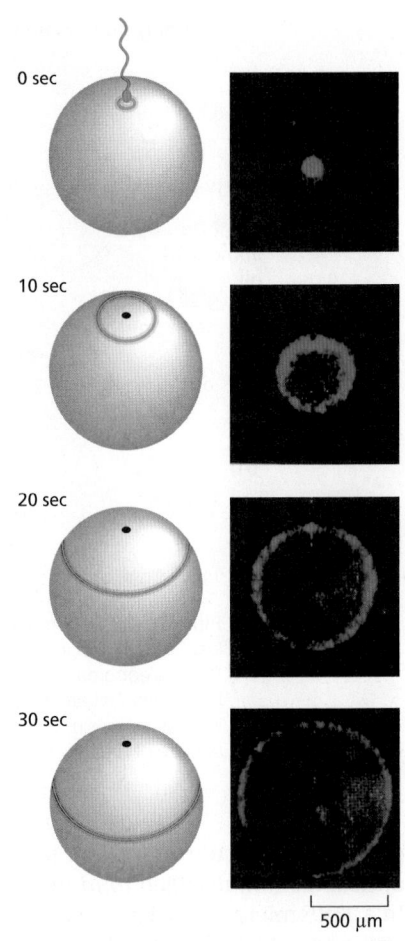

Figure 9–30 **Aequorin, a luminescent protein.** The luminescent protein aequorin emits blue light in the presence of free Ca^{2+}. Here, an egg of the medaka fish has been injected with aequorin, which has diffused throughout the cytosol, and the egg has then been fertilized with a sperm and examined with the help of a very sensitive camera. The four photographs were taken looking down on the site of sperm entry at intervals of 10 seconds and reveal a wave of release of free Ca^{2+} into the cytosol from internal stores just beneath the plasma membrane. This wave sweeps across the egg starting from the site of sperm entry, as indicated in the diagrams on the *left*. (© 1978 J.C. Gilkey et al. Originally published in *J. Cell Biol*. https://doi.org/10.1083 /jcb.76.2.448. With permission from Rockefeller University Press.)

Aequorin is a luminescent protein isolated from the same marine jellyfish that produces GFP. It emits blue light in the presence of Ca^{2+} and responds to changes in Ca^{2+} concentration in the range of 0.5–10 µM. If microinjected into an egg, for example, aequorin emits a flash of light in response to the sudden localized release of free Ca^{2+} into the cytoplasm that occurs when the egg is fertilized (**Figure 9–30**). Aequorin has also been expressed transgenically in plants and other organisms to provide a method of monitoring Ca^{2+} in all their cells without the need for microinjection, which can be a difficult procedure.

Bioluminescent molecules like aequorin emit tiny amounts of light—at best, a few photons per indicator molecule—that are difficult to measure. Fluorescent indicators produce orders of magnitude more photons per molecule; they are therefore easier to measure and can give better spatial resolution. Genetically encoded fluorescent Ca^{2+} indicators have been synthesized that bind Ca^{2+} tightly and are excited by or emit light at slightly different wavelengths when they are free of Ca^{2+} than when they are in their Ca^{2+}-bound form. By measuring the ratio of fluorescence intensity at two excitation or emission wavelengths, we can determine the concentration ratio of the Ca^{2+}-bound indicator to the Ca^{2+}-free indicator, thereby providing an accurate measurement of the free Ca^{2+} concentration (see Movie 15.4). Indicators of this type are widely used for second-by-second monitoring of changes in intracellular Ca^{2+} concentration, or other ion concentrations, in the different parts of a cell viewed in a fluorescence microscope (**Figure 9–31**).

Similar fluorescent indicators measure other ions; some detect H^+, for example, and hence measure intracellular pH. Some of these indicators can enter cells by diffusion and thus need not be microinjected; this makes it possible to monitor large numbers of individual cells simultaneously in a fluorescence microscope. New types of indicators, used in conjunction with modern image-processing methods, make possible similarly rapid and precise methods for analyzing changes in the concentrations of many types of small molecules in cells.

Single Molecules Can Be Visualized by Total Internal Reflection Fluorescence Microscopy

In ordinary microscopes, single fluorescent molecules such as tagged proteins cannot be reliably detected. The limitation has nothing to do with the resolution limit, but instead arises from the strong background due to light emitted or scattered by out-of-focus molecules. This tends to blot out the fluorescence from the

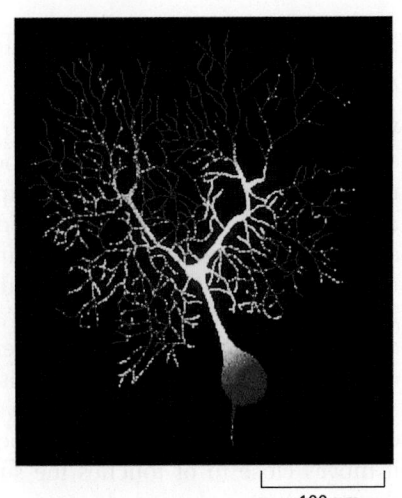

Figure 9–31 **Visualizing intracellular Ca^{2+} concentrations by using a fluorescent indicator.** The branching tree of dendrites of a Purkinje cell in the cerebellum receives more than 100,000 synapses from other neurons. The output from the cell is conveyed along the single axon seen leaving the cell body at the bottom of the picture. This image of the intracellular Ca^{2+} concentration in a single Purkinje cell (from the brain of a guinea pig) was taken with a low-light camera and the Ca^{2+}-sensitive fluorescent indicator fura-2. The concentration of free Ca^{2+} is represented by different colors, *red* being the highest and *blue* the lowest. The highest Ca^{2+} levels are present in the thousands of dendritic branches. (Courtesy of D.W. Tank, J.A. Connor, M. Sugimori, and R.R. Llinas.)

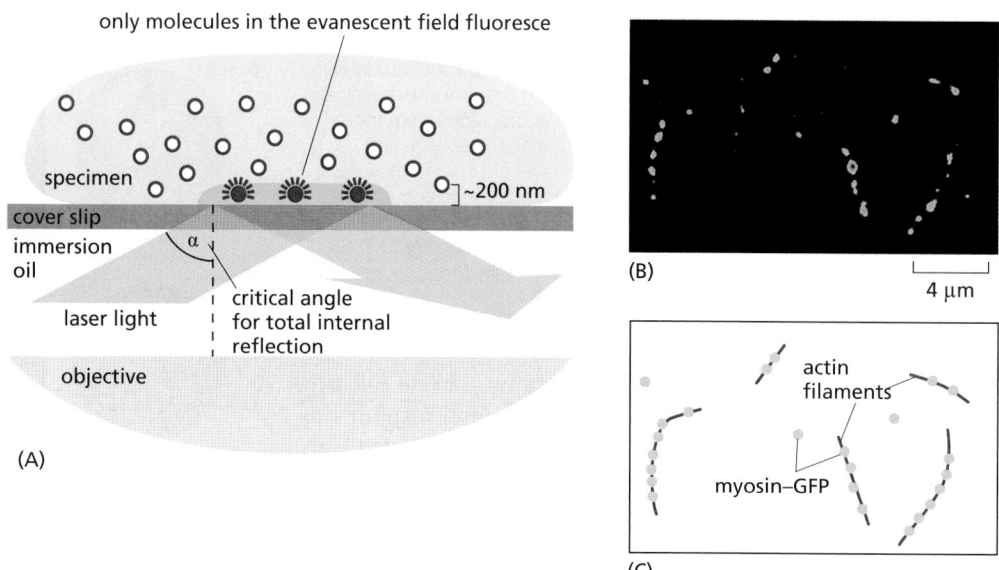

Figure 9–32 TIRF microscopy allows the detection of single fluorescent molecules. (A) TIRF microscopy uses excitatory laser light to illuminate the cover-slip surface at the critical angle at which all the light is reflected by the glass–water interface. Some electromagnetic energy extends a short distance across the interface as an evanescent wave that excites just those molecules that are attached to the cover slip or are very close to its surface. (B) TIRF microscopy is used here to image individual myosin–GFP molecules *(green dots)* attached to nonfluorescent actin filaments (C), which are invisible but stuck to the surface of the cover slip. (Courtesy of Dmitry Cherny and Clive R. Bagshaw.)

particular molecule of interest. This problem can be solved by the use of a special optical technique called *total internal reflection fluorescence* (*TIRF*) microscopy. In a TIRF microscope, laser light shines onto the cover-slip surface at the precise critical angle at which total internal reflection occurs (**Figure 9–32A**). Because of total internal reflection, the light does not enter the sample, and the majority of fluorescent molecules are not, therefore, illuminated. However, electromagnetic energy does extend, as an evanescent field, for a very short distance beyond the surface of the cover slip and into the specimen, allowing just those molecules in the layer closest to the surface to become excited. When these molecules fluoresce, their emitted light is no longer competing with out-of-focus light from the overlying molecules, and can now be detected. TIRF has allowed several dramatic experiments, for instance imaging of single motor proteins moving along microtubules or single actin filaments forming and branching. At present, the technique is restricted to a thin layer within only 100–200 nm of the cell surface (Figure 9–32B and C).

Individual Molecules Can Be Touched, Imaged, and Moved Using Atomic Force Microscopy

While TIRF allows single molecules to be visualized under certain conditions, it is strictly a passive observation method. In order to probe molecular function, it is ultimately useful to be able to manipulate individual molecules themselves, and *atomic force microscopy* (*AFM*) provides a method to do just that. In an AFM device, an extremely small and sharply pointed tip, often of silicon or silicon nitride, is made using nanofabrication methods similar to those used in the semiconductor industry. The tip of the AFM probe is attached to a springy cantilever arm mounted on a highly precise positioning system that allows it to be moved over very small distances. In addition to this precise movement capability, the AFM device is able to collect information about a variety of forces that it encounters—including electrostatic, van der Waals, and mechanical forces—which are felt by its tip as it moves close to or touches the surface (**Figure 9–33A**). When AFM was first developed, it was intended as an imaging technology to measure molecular-scale

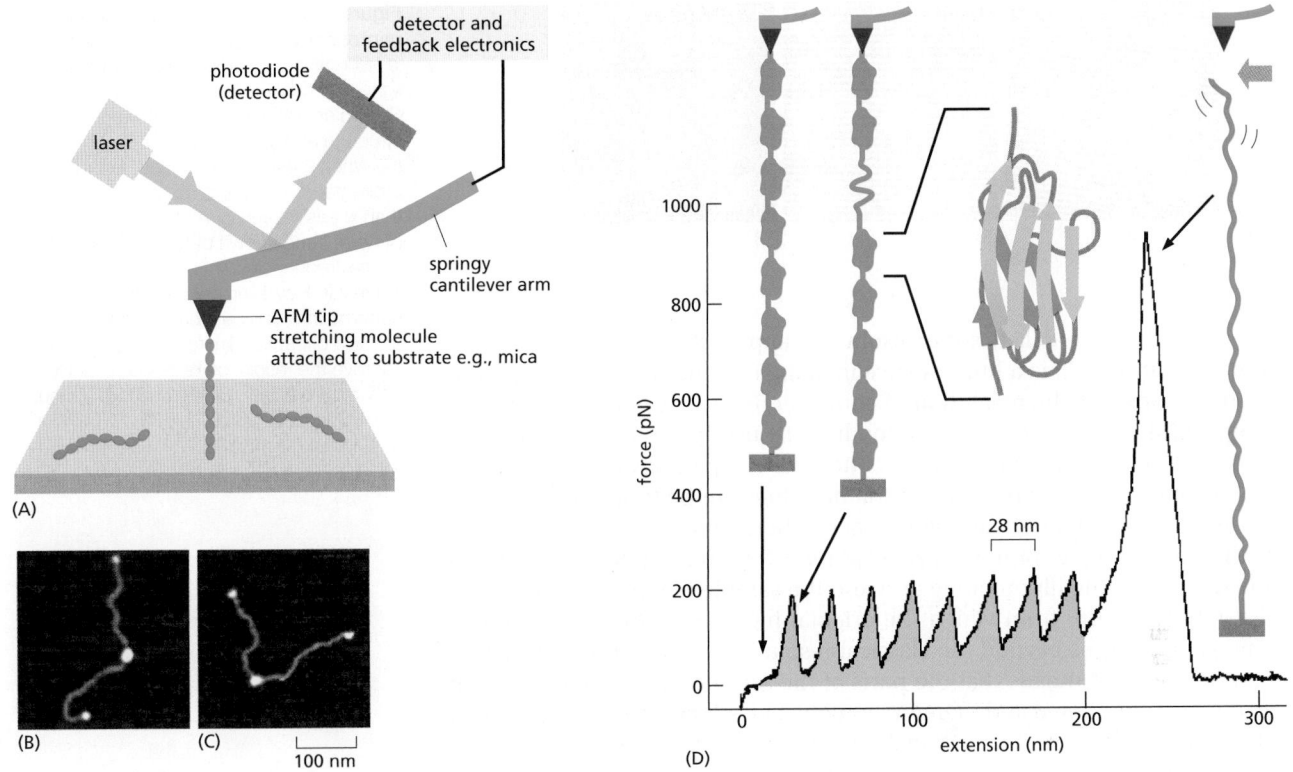

Figure 9–33 Single molecules can be imaged and manipulated by atomic force microscopy. (A) Schematic diagram of the key components of an atomic force microscope (AFM), showing the force-sensing tip attached to one end of a single protein molecule, as in the experiment described in (D). (B) and (C) An AFM in imaging mode created these images of a single heteroduplex DNA molecule with a MutS protein dimer *(larger white regions)* bound near its center, at the point of a mismatched base pair. MutS is the first protein that binds to DNA when the mismatch repair process is initiated (see Figure 5–19). The smaller *white dots* are single streptavidin molecules, used to label the two ends of each DNA molecule. (D) Titin is an enormous protein molecule that provides muscle with its passive elasticity (see Figure 16–34). The extensibility of this protein can be directly tested using a short, artificially produced protein that contains eight repeated immunoglobulin (Ig) domains from one region of the titin protein. In this experiment, the tip of the AFM is used to pick up, and progressively stretch, a single molecule until it eventually ruptures. As force is applied, each Ig domain suddenly begins to unfold, and the force needed in each case (about 200 pN) can be recorded. The region of the force–extension curve shaded *green* records the sequential unfolding event for each of the eight protein domains. (B and C, from Y. Jiang and P.E. Marszalek, *EMBO J.* 30:2881–2893, 2011. Reprinted with permission of John Wiley & Sons; D, adapted from W.A. Linke et al., *J. Struct. Biol.* 137:194–205, 2002.)

features on a surface. When used in this mode, the probe is scanned over the surface, moving up and down as necessary to maintain a constant interaction force with the surface, thus revealing any objects such as proteins or other molecules that might be present on the otherwise flat surface (Figure 9–33B and C). AFM is not limited to simply imaging surfaces, however, and can also be used to pick up and move single molecules that adsorb strongly to the tip. Using this technology, the mechanical properties of individual protein molecules can be measured in detail. For example, AFM has been used to unfold a single protein molecule in order to measure the energetics of domain folding (Figure 9–33 D).

Superresolution Fluorescence Techniques Can Overcome Diffraction-Limited Resolution

The variations on light microscopy we have described so far are all constrained by the classic diffraction limit to resolution described earlier; that is, to about 200 nm (see Figure 9–6). Yet many cellular structures—from nuclear pores to nucleosomes and clathrin-coated pits—are much smaller than this and so are unresolvable by conventional light microscopy. Several approaches, however, are now available that bypass the limit imposed by the diffraction of light, and successfully allow objects as small as 20 nm to be imaged and clearly resolved: a remarkable, order-of-magnitude improvement.

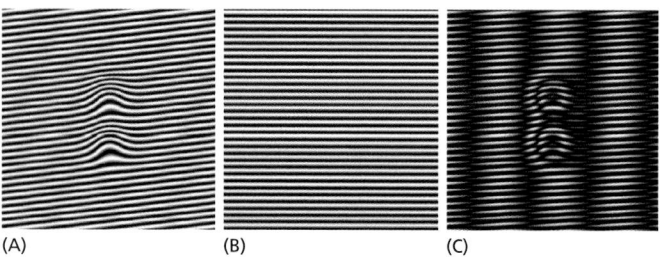

(A)   (B)   (C)

The first of these so-called **superresolution** approaches, *structured illumination microscopy* (*SIM*), is a fluorescence imaging method with a resolution of about 100 nm, or twice the resolution of conventional bright-field and confocal microscopy. SIM overcomes the diffraction limit by using a grated or structured pattern of light to illuminate the sample. The microscope's physical set-up and operation is quite complex, but the general principle can be thought of as similar to creating a moiré pattern, an interference pattern created by overlaying two grids with different angles or mesh sizes (**Figure 9–34**). In a similar way to creating a moiré pattern, the illuminating grid and the sample features combine into an interference pattern, from which the original high-resolution contributions to the image of features beyond the classical resolution limit can be calculated. Illumination by a grid means that the parts of the sample in the dark stripes of the grid are not illuminated and therefore not imaged, so the imaging is repeated several times (usually three) after translating the grid through a fraction of the grid spacing between each image. As the interference effect is strongest for image components close to the direction of the grid bars, the whole process is repeated with the grid pattern rotated through a series of angles to obtain an equivalent enhancement in all directions. Finally, mathematically combining all these separate images by computer creates an enhanced superresolution image. SIM is versatile because it can be used with any fluorescent dye or protein, and combining SIM images captured at consecutive focal planes can create three-dimensional data sets (**Figure 9–35**).

Figure 9–34 Structured illumination microscopy. The principle, illustrated here, is to illuminate a sample with patterned light and measure the moiré pattern. Shown are (A) the pattern from an unknown structure and (B) a known pattern. (C) When these are combined, the resulting moiré pattern contains more information than is easily seen in (A), the original pattern. If the known pattern (B) has higher spatial frequencies, then better resolution will result. However, because the spatial patterns that can be created optically are also diffraction-limited, SIM can only improve the resolution by about a factor of two. (From B.O. Leung and K.C. Chou, *Appl. Spectrosc.* 65:967–980, 2011. With permission from SAGE.)

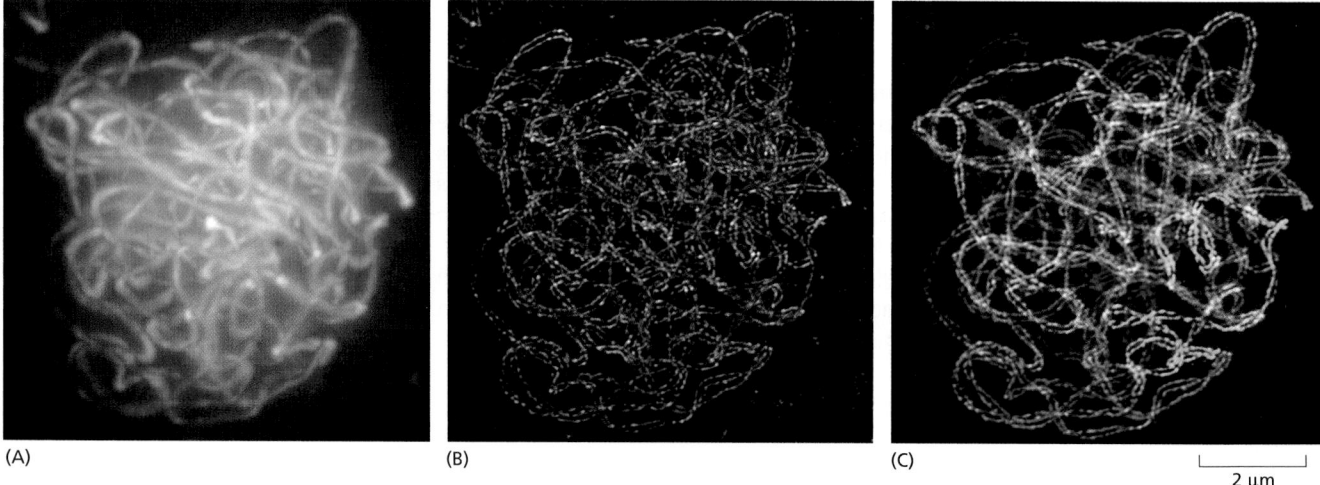

(A)   (B)   (C)

2 μm

Figure 9–35 Structured illumination microscopy can be used to create three-dimensional data. These three-dimensional projections of the meiotic chromosomes at pachytene in a maize cell show the paired lateral elements of the synaptonemal complexes. (A) The chromosome set has been stained with a fluorescent antibody to cohesin and is viewed here by conventional fluorescence microscopy. Because the distance between the two lateral elements is about 200 nm, the diffraction limit, the two lateral elements that make up each complex are not resolved. (B) In the three-dimensional SIM image, the improved resolution enables each lateral element, about 100 nm across, to be clearly resolved, and the two chromosomes can clearly be seen to coil around each other. (C) Because the complete three-dimensional data set for the whole nucleus is available, the path of each separate pair of chromosomes can be traced and artificially assigned a different color. (Courtesy of C.J. Rachel Wang, Peter Carlton and Zacheus Cande.)

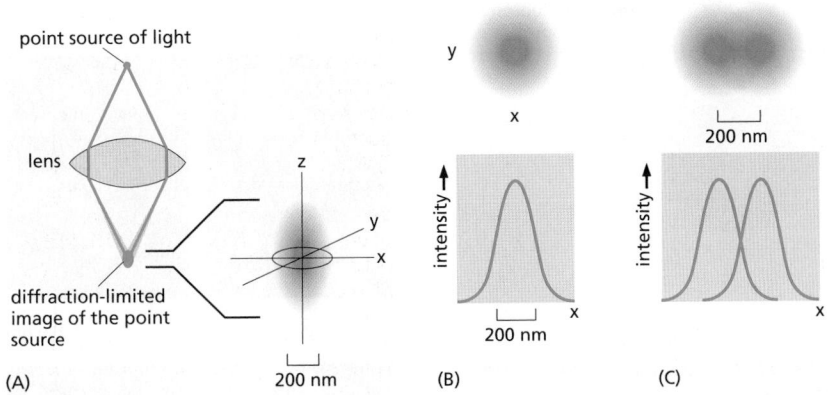

point source of light

lens

diffraction-limited image of the point source

(A)

200 nm

y

x

z

y

x

200 nm

(B)

intensity

x

200 nm

(C)

intensity

x

200 nm

Figure 9–36 The point spread function of a lens determines resolution. (A) When a point source of light is brought to a focus by a lens system, diffraction effects mean that, instead of being imaged as a point, it is blurred in all dimensions. (B) In the plane of the image, the distribution of light approximates a Gaussian distribution, whose width at half-maximum height under ideal conditions is about 200 nm. (C) Two point sources that are about 200 nm apart can still just be distinguished as separate objects in the image, but if they are any nearer than that, their images will overlap and not be resolvable.

To get around the diffraction limit, the other two superresolution techniques exploit aspects of the point spread function, a property of the optical system mentioned earlier. The *point spread function* is the distribution of light intensity within the three-dimensional, blurred image that is formed when a single point source of light is brought to a focus with a lens. Instead of being identical to the point source, the image has an intensity distribution that is approximately described by a Gaussian distribution, which in turn determines the resolution of the lens system (Figure 9–36). Two points that are closer than the width at half-maximum height of this distribution will become hard to resolve because their images overlap too much (see Figure 9–36C).

In fluorescence microscopy, the excitation light is focused to a spot on the specimen by the objective lens, which then captures the photons emitted by any fluorescent molecule that the beam has raised from a ground state to an excited state. Because the excitation spot is blurred according to the point spread function, fluorescent molecules that are closer than about 200 nm will be imaged as a single blurred spot. One approach to increasing the resolution is to switch all the fluorescent molecules at the periphery of the blurry excitation spot back to their ground state, or to a state where they no longer fluoresce in the normal way, leaving only those at the very center to be recorded. This can be done in practice by adding a second, very bright laser beam that wraps around the excitation beam like a torus. The wavelength and intensity of this second beam are adjusted so as to switch the fluorescent molecules off everywhere except at the very center of the point spread function, a region that can be as small as 20 nm across (Figure 9–37). The fluorescent probes used must be in a special class that is photoswitchable: their emission can be reversibly switched on and off with lights of different wavelengths. As the specimen is scanned with this arrangement of lasers, fluorescent molecules are switched on and off, and the small point spread function at each location is recorded. The diffraction limit is breached because the technique ensures that similar but very closely spaced molecules are in one of two different states, either fluorescing or dark. This approach is called *STED (stimulated emission depletion microscopy)* and various microscopes using versions of the general method are now in wide use. Resolutions of 20 nm have been achieved in biological specimens, and even higher resolution attained with nonbiological specimens (see Figure 9–37).

Superresolution Can Also be Achieved Using Single-Molecule Localization Methods

If a single fluorescent molecule is imaged, it appears as a circular blurry disc, but if sufficient photons have contributed to this image, the precise mathematical center of the disclike image can be determined very accurately, often to within a few nanometers. But the problem with a specimen that contains a large number

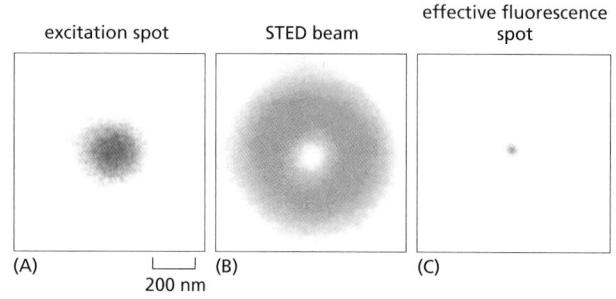

excitation spot   STED beam   effective fluorescence spot

(A)   200 nm   (B)   (C)

(D)   (E)   250 nm

(F)   (G)   2 μm

Figure 9–37 Superresolution microscopy can be achieved by reducing the size of the point spread function. (A) The size of a normal focused beam of excitatory light. (B) An extremely strong superimposed laser beam, at a different wavelength and in the shape of a torus, depletes emitted fluorescence everywhere in the specimen except right in the center of the beam, reducing the effective width of the point spread function (C). As the specimen is scanned, this small point spread function can then build up a crisp image in a process called STED (stimulated emission depletion microscopy). (D) Synaptic vesicles in live cultured neurons, fluorescently labeled and imaged by ordinary confocal microscopy, with a resolution of 260 nm. (E) The same vesicles imaged by STED, with a resolution of 60 nm, which allows single vesicles to be resolved. (F) Fluorescently labeled replication factories in the nucleus of a cultured cell, imaged by ordinary confocal microscopy. (G) The same replication factories imaged by STED. Single, discrete replication sites can be resolved by STED that cannot be seen in the confocal image. (A, B, and C, from G. Donnert et al., *Proc. Natl Acad. Sci. USA* 103:11440–11445, 2006. Copyright 2006 National Academy of Sciences, USA. With permission from National Academy of Sciences; D and E, from V. Westphal et al., *Science* 320:246–249, 2008. With permission from AAAS; F and G, from Z. Cseresnyes, U. Schwarz and C.M. Green, *BMC Cell Biol.* 10:88, 2009. With permission from the authors.)

of adjacent fluorescent molecules, as we saw earlier, is that they each contribute blurry, overlapping point spread functions to the image, making the exact position of any one molecule impossible to resolve. Another way round this limitation is to arrange for only a very few, clearly separated molecules to actively fluoresce at any one moment. The exact position of each of these can then be computed, before subsequent sets of molecules are examined.

In practice, this can be achieved by using lasers to sequentially switch on a sparse subset of fluorescent molecules in a specimen containing photoactivatable or photoswitchable fluorescent labels. Labels are activated, for example, by illumination with near-ultraviolet light, which modifies a small subset of molecules so that they fluoresce when exposed to an excitation beam at another wavelength. These are then imaged before bleaching quenches their fluorescence and a new subset is activated. Each molecule emits a few thousand photons in response to the excitation before switching off, and the switching process can be repeated hundreds or even thousands of times, allowing the exact coordinates of a very large set of single molecules to be determined. The full set can be combined and digitally displayed as an image in which the computed location of each individual molecule is exactly marked (**Figure 9–38**). This class of methods has been variously termed *photoactivated localization microscopy* (*PALM*) or *stochastic optical reconstruction microscopy* (*STORM*).

By switching the fluorophores off and on sequentially in different regions of the specimen as a function of time, all the superresolution imaging methods described above allow the resolution of molecules that are much closer together than the 200 nm diffraction limit. In STED, the locations of the molecules are determined by using optical methods to define exactly where their fluorescence will be on or off. In PALM and STORM, individual fluorescent molecules are switched on and off at random over a period of time, allowing their positions to be accurately determined. PALM and STORM techniques have depended on the

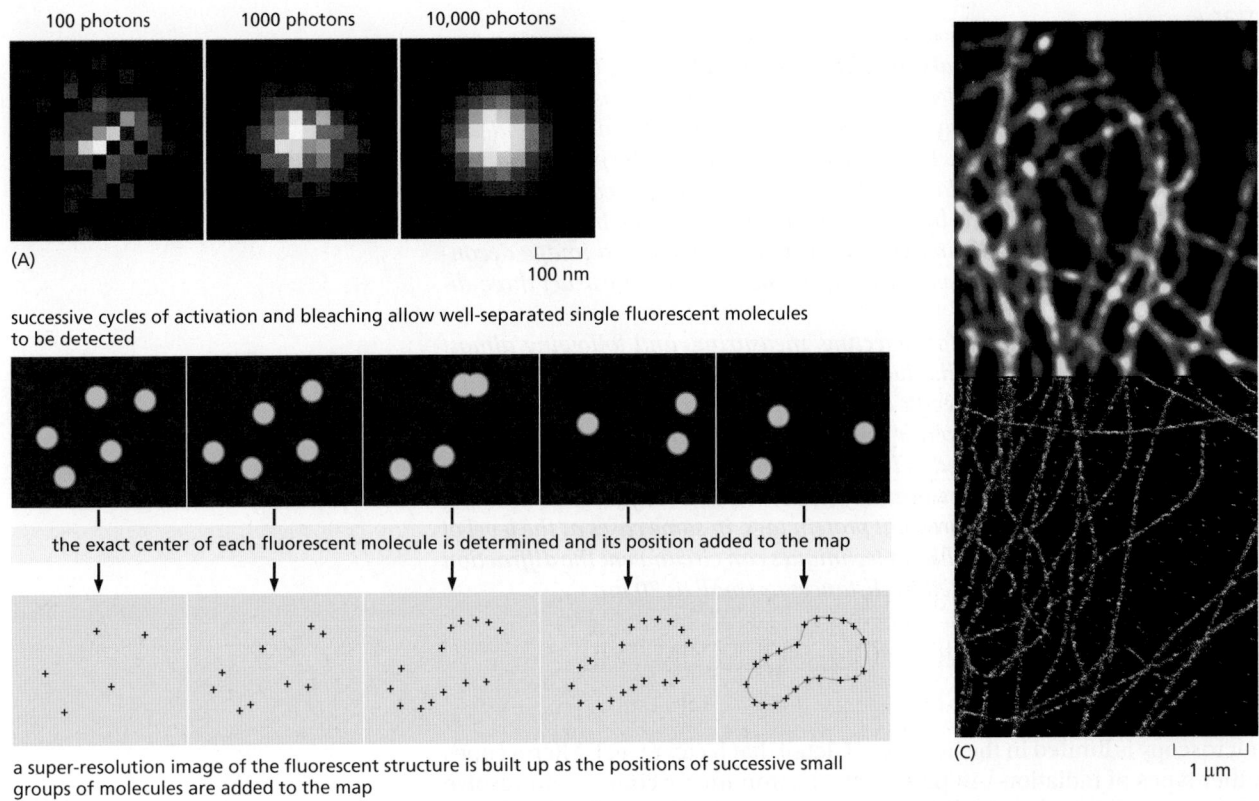

100 photons   1000 photons   10,000 photons

(A)

100 nm

successive cycles of activation and bleaching allow well-separated single fluorescent molecules to be detected

the exact center of each fluorescent molecule is determined and its position added to the map

a super-resolution image of the fluorescent structure is built up as the positions of successive small groups of molecules are added to the map

(B)

(C)

1 µm

Figure 9–38 Single fluorescent molecules can be located with great accuracy. (A) Determining the exact mathematical center of the blurred image of a single fluorescent molecule becomes more accurate the more photons contribute to the final image. The point spread function described in the text dictates that the size of the molecular image is about 200 nm across, but in very bright specimens, the position of its center can be pinpointed to within a nanometer. (B) In this imaginary specimen, sparse subsets of fluorescent molecules are individually switched on briefly and then bleached. The exact positions of all these well-spaced molecules can be gradually built up into an image at superresolution. (C) In this portion of a cell, the microtubules have been fluorescently labeled and imaged at the top in a TIRF microscope (see Figure 9–32) and below, at superresolution, in a PALM microscope. The diameter of the microtubules in the lower panel now resembles their true size, about 25 nm, rather than the 250 nm in the blurred image at the top. (A, from A.L. McEvoy et al., *BMC Biol.* 8:106, 2010. With permission from the authors; C, courtesy of Shinsuke Niwa.)

development of novel fluorescent probes that exhibit the appropriate switching behavior. All these methods are now being extended to incorporate multicolor imaging, three-dimensional imaging (**Figure 9–39**), and live-cell imaging in real time. Ending the long reign of the diffraction limit has certainly reinvigorated light microscopy and its place in cell biology research.

Figure 9–39 Small fluorescent structures can be imaged in three dimensions with superresolution. (A) The image of two touching 180-nm-diameter clathrin-coated pits on the plasma membrane of a cultured cell is diffraction-limited, and the individual pits cannot be distinguished in this conventional fluorescence image. (B) Using STORM superresolution microscopy, however, the pits are clearly resolvable. Not only can such pits be imaged using probes of different colors, but additional three-dimensional information can also be obtained. (C) and (D) Shown are two different orthogonal views of one single coated pit. The clathrin is labeled *red* and transferrin—the cargo within the pit—is labeled *green*. Images of this sort can be acquired in less than one second, making possible dynamic observations on living cells. These techniques depend heavily on the development of new, very fast-switching, and extremely bright fluorescent probes. (A and B, from M. Bates et al., *Science* 317:1749–1753, 2007. With permission from AAAS; C and D, from S.A. Jones et al., *Nat. Methods* 8:499–508, published 2011 by Macmillan Publishers Ltd. Reproduced with permission of SNCSC.)

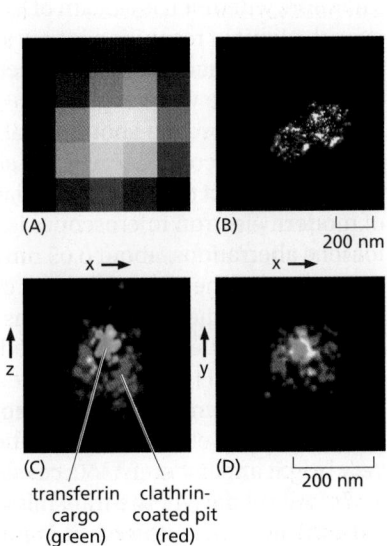

(A)

(B)

200 nm

x ⟶

x ⟶

z

y

(C)

(D)

200 nm

transferrin   clathrin-
cargo   coated pit
(green)   (red)

Summary

Many light-microscope techniques are available for observing cells. Cells that have been fixed and stained can be studied in a conventional light microscope, whereas antibodies coupled to fluorescent dyes can be used to locate specific molecules in cells in a fluorescence microscope. Living cells can be seen with phase-contrast, differential-interference-contrast, dark-field, or bright-field microscopes. All forms of light microscopy are facilitated by digital image-processing techniques, which enhance sensitivity and refine the image. Confocal microscopy and image deconvolution both provide thin optical sections and can be used to reconstruct three-dimensional images.

Techniques are now available for detecting, measuring, and following almost any desired molecule in a living cell. Fluorescent indicator dyes can be introduced to measure the concentrations of specific ions in individual cells or in different parts of a cell. Virtually any protein of interest can be genetically engineered as a fluorescent fusion protein, and then imaged in living cells by fluorescence microscopy. The dynamic behavior and interactions of many molecules can be followed in living cells by variations on the use of fluorescent protein tags, in some cases at the level of single molecules. Various superresolution techniques can circumvent the diffraction limit and resolve molecules separated by distances as small as 20 nm.

LOOKING AT CELLS AND MOLECULES IN THE ELECTRON MICROSCOPE

Light microscopy is limited in the fineness of detail that it can reveal. Microscopes using other types of radiation—in particular, electron microscopes—can resolve much smaller structures than is possible with visible light. This higher resolution comes at a cost: specimen preparation for electron microscopy is complex and it is harder to be sure that what we see in the image corresponds precisely to the original living structure. It is possible, however, to use very rapid freezing to preserve structures faithfully for electron microscopy. Digital image analysis can be used to reconstruct three-dimensional objects by combining information either from many individual particles or from multiple tilted views of a single object. Together, these approaches extend the resolution and scope of electron microscopy to the point at which we can faithfully image the structures of individual macromolecules and the complexes they form.

The Electron Microscope Resolves the Fine Structure of the Cell

The formal relationship between the diffraction limit to resolution and the wavelength of the illuminating radiation (see Figure 9–6) holds true for any form of radiation, whether it is a beam of light or a beam of electrons. With electrons, however, the limit of resolution is very small. The wavelength of an electron decreases as its velocity increases. In an **electron microscope** with an accelerating voltage of 100,000 V, the wavelength of an electron is 0.004 nm. In theory, the resolution of such a microscope should be about 0.002 nm, which is 100,000 times that of the light microscope. Because the aberrations of an electron lens are considerably harder to correct than those of a glass lens, however, the practical resolving power of modern electron microscopes is, even with careful image processing to correct for lens aberrations, about 0.05 nm (0.5 Å) (**Figure 9–40**). This is because only the very center of the electron lenses can be used, and the effective numerical aperture is tiny. Furthermore, problems of specimen preparation, contrast, and radiation damage have generally limited the normal effective resolution for biological objects to 1 nm (10 Å). This is nonetheless about 200 times better than the resolution of the light microscope. Moreover, the performance of electron microscopes is improved by electron illumination sources called field emission guns. These very bright and coherent sources substantially improve the resolution achieved.

In overall design, the transmission electron microscope (TEM) is similar to a light microscope, although it is much larger and "upside down" (**Figure 9–41**).

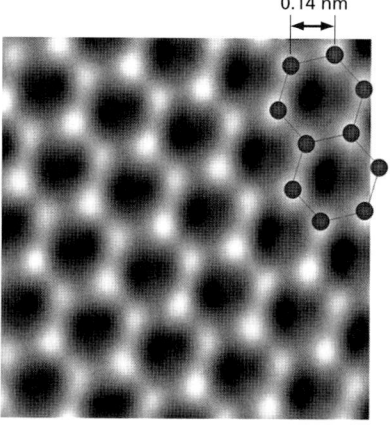

Figure 9–40 The resolution of the electron microscope. This transmission electron micrograph of a monolayer of graphene resolves the individual carbon atoms as bright spots in a hexagonal lattice. Graphene is a single isolated atomic plane of graphite and forms the basis of carbon nanotubes. The distance between adjacent bonded carbon atoms is 0.14 nm (1.4 Å). Such resolution can only be obtained in a specially built transmission electron microscope in which all lens aberrations are carefully corrected, and with optimal specimens; it cannot be achieved with most conventional biological specimens. (From A. Dato et al., *Chem. Commun.* 40:6095–6097, 2009. With permission from The Royal Society of Chemistry.)

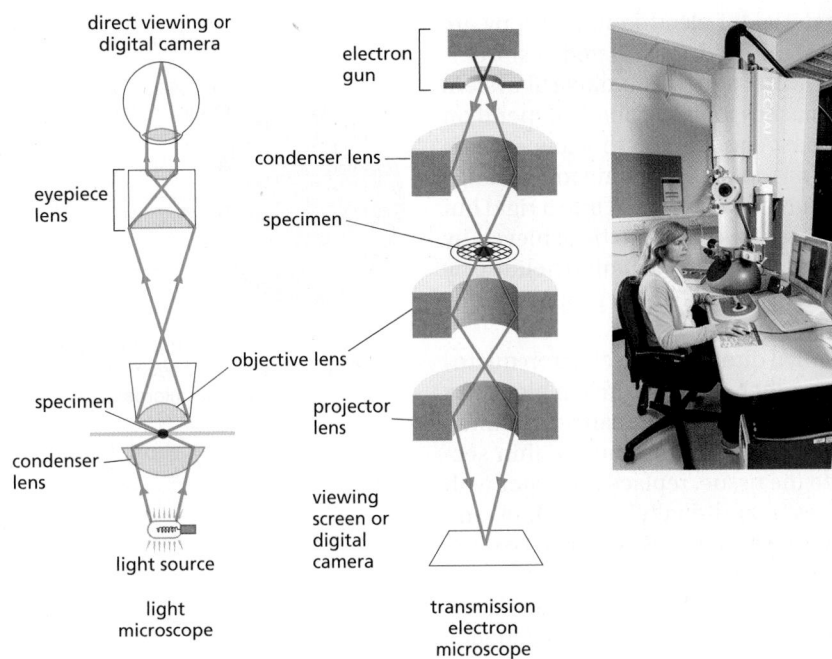

Figure 9–41 The principal features of a light microscope and a transmission electron microscope. These drawings emphasize the similarities of overall design. Whereas the lenses in the light microscope are made of glass, those in the electron microscope are magnetic coils. The electron microscope requires that the specimen be placed in a vacuum. The inset shows a transmission electron microscope in use. (Photograph courtesy of Andrew Davis.)

The source of illumination is a filament or cathode that emits electrons at the top of a cylindrical column about 2 m high. Since electrons are scattered by collisions with air molecules, air must first be pumped out of the column to create a vacuum. The electrons are then accelerated from the filament by a nearby anode and allowed to pass through a tiny hole to form an electron beam that travels down the column. Magnetic coils placed at intervals along the column focus the electron beam, just as glass lenses focus the light in a light microscope. The specimen is put into the vacuum, through an airlock, into the path of the electron beam. As in light microscopy, the specimen is usually stained—in this case, with *electron-dense* material. Some of the electrons passing through the specimen are scattered by structures stained with the electron-dense material; the remainder are focused to form an image, in a manner analogous to the way an image is formed in a light microscope. The image can be observed on a phosphorescent screen or recorded with a high-resolution digital camera. Because the scattered electrons are lost from the beam, the dense regions of the specimen show up in the image as areas of reduced electron flux, which look dark.

Biological Specimens Require Special Preparation for Electron Microscopy

In the early days of its application to biological materials, the electron microscope revealed many previously unimagined structures in cells. But before these discoveries could be made, electron microscopists had to develop new procedures for embedding, cutting, and staining tissues.

Since the specimen is exposed to a very high vacuum in the electron microscope, living tissue is usually killed and preserved by fixation—first with *glutaraldehyde*, which covalently cross-links protein molecules to their neighbors, and then with *osmium tetroxide*, which binds to and stabilizes lipid bilayers as well as proteins (**Figure 9–42**). Because electrons have very limited penetrating power, the fixed tissues normally have to be cut into extremely thin sections (25–100 nm thick, about 1/200 the thickness of a single cell) before they are viewed. This is achieved by dehydrating the specimen, permeating it with a monomeric resin that polymerizes to form a solid block of plastic, then cutting the block with a fine glass or diamond knife on a special microtome. The resulting *thin sections*, free of water and other volatile solvents, are supported on a small metal grid for viewing in the microscope (**Figure 9–43**).

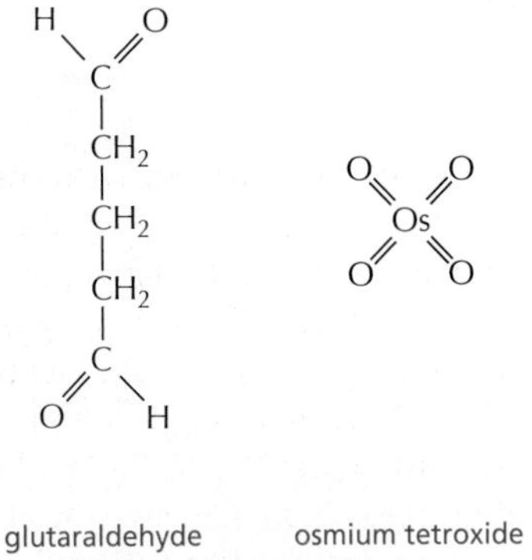

glutaraldehyde   osmium tetroxide

Figure 9–42 Two common chemical fixatives used for electron microscopy. The two reactive aldehyde groups of glutaraldehyde enable it to cross-link various types of molecules, forming covalent bonds between them. Osmium tetroxide forms cross-linked complexes with many organic compounds, and in the process becomes reduced. This reaction is especially useful for fixing cell membranes, since the C=C double bonds present in many fatty acids react with osmium tetroxide.

The steps required to prepare biological material for electron microscopy are challenging. How can we be sure that the image of the fixed, dehydrated, resin-embedded specimen bears any relation to the delicate, aqueous biological system present in the living cell? The best current approaches to this problem depend on rapid freezing. If an aqueous system is cooled fast enough and to a low enough temperature, the water and other components in it do not have time to rearrange themselves or crystallize into ice. Instead, the water is supercooled into a rigid but noncrystalline state—a "glass"—called vitreous ice. This state can be achieved by slamming the specimen onto a polished copper block cooled by liquid helium, by plunging it into or spraying it with a jet of a coolant such as liquid propane, or by cooling it at high pressure.

Some rapidly frozen specimens can be examined directly in the electron microscope using a special cooled specimen holder. In other cases, the frozen block can be fractured to reveal interior cell surfaces, or the surrounding ice can be sublimed away to expose external surfaces. However, we often want to examine thin sections. A compromise is therefore to rapid-freeze the tissue, replace the water with organic solvents, embed the tissue in plastic resin, and finally cut sections and stain. Although technically still difficult, this approach stabilizes and preserves the tissue in a condition very close to its original living state (**Figure 9–44**).

Image clarity in an electron micrograph depends upon having a range of contrasting electron densities within the specimen. Electron density in turn depends on the atomic number of the atoms that are present: the higher the atomic number, the more electrons are scattered and the darker that part of the image. Biological tissues are composed mainly of atoms of very low atomic number (primarily carbon, oxygen, nitrogen, and hydrogen). To make them visible, tissues are usually impregnated (before or after sectioning) with the salts of heavy metals such as uranium, lead, and osmium. The degree of impregnation, or "staining," with these salts will vary for different cell constituents. Lipids, for example, tend to stain darkly after osmium fixation, revealing the location of cell membranes.

Specific Macromolecules Can Be Localized by Immunogold Electron Microscopy

We have seen how antibodies can be used in conjunction with fluorescence microscopy to localize specific macromolecules. An analogous method—**immunogold**

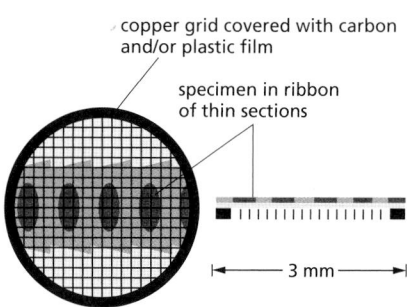

Figure 9–43 **The metal grid that supports the thin sections of a specimen in a transmission electron microscope.**

copper grid covered with carbon and/or plastic film

specimen in ribbon of thin sections

3 mm

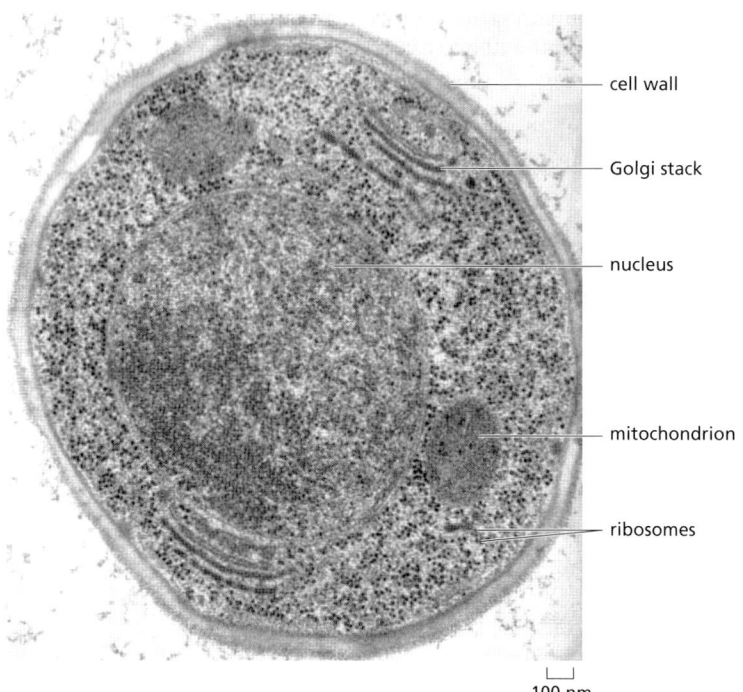

cell wall

Golgi stack

nucleus

mitochondrion

ribosomes

100 nm

Figure 9–44 **Thin section of a cell.** This thin section is of a yeast cell that has been very rapidly frozen and the vitreous ice replaced by organic solvents and then by plastic resin. The nucleus, mitochondria, cell wall, Golgi stacks, and ribosomes can all be readily seen in a state that is presumed to be as lifelike as possible. (Courtesy of Andrew Staehelin.)

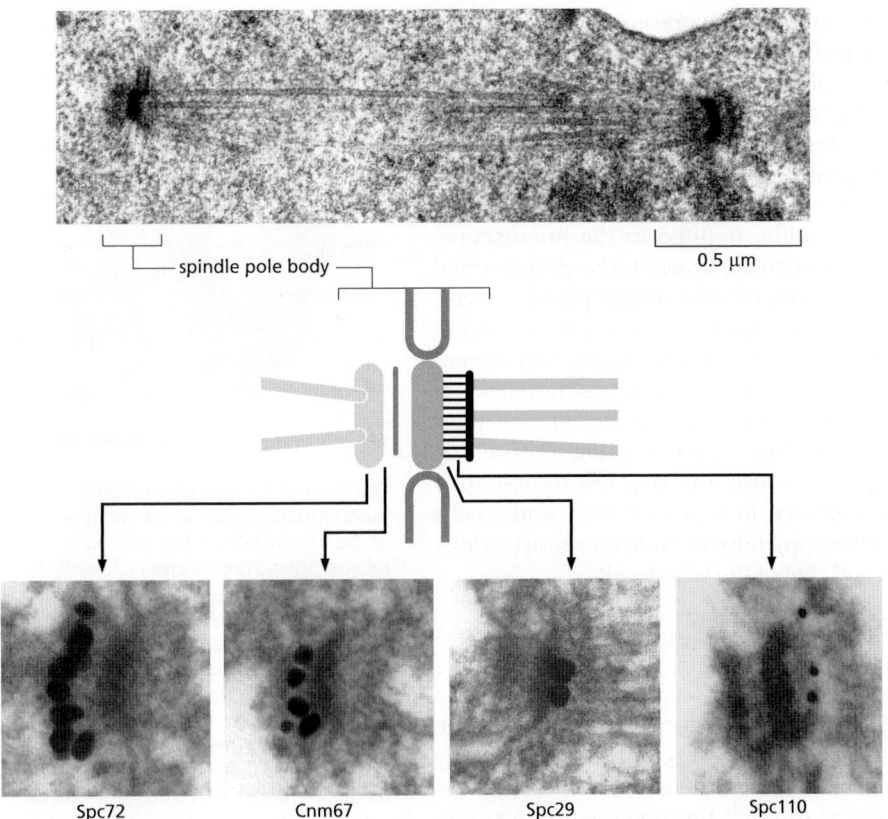

spindle pole body

0.5 µm

Spc72   Cnm67   Spc29   Spc110

Figure 9–45 **Localizing proteins in electron microscopy.** Immunogold electron microscopy is used here to find the specific location of four different protein components within the spindle pole body of yeast. At the top is a thin section of a yeast mitotic spindle showing the spindle microtubules that cross the nucleus and connect at each end to spindle pole bodies embedded in the nuclear envelope. A diagram of the components of a single spindle pole body is shown below. On separate sections, antibodies against four different proteins of the spindle pole body are used, together with colloidal gold particles *(black dots)*, to reveal where within the complex structure each protein is located. (Courtesy of John Kilmartin.)

electron microscopy—can be used in the electron microscope. The usual procedure is to incubate a thin section first with a specific primary antibody, and then with a secondary antibody to which a colloidal gold particle has been attached. The gold particle is electron-dense and can be seen as a black dot in the electron microscope (**Figure 9–45**). Different antibodies can be conjugated to different sized gold particles so multiple proteins can be localized in a single sample.

A complication for immunogold labeling is that the antibodies and colloidal gold particles do not penetrate into the resin used for embedding; therefore, they detect antigens only at the surface of the section. This means that the method's sensitivity is low, since antigen molecules in the deeper parts of the section are not detected. Furthermore, we may get a false impression regarding which structures contain the antigen and which do not. One solution is to label the specimen before embedding it in plastic, when cells and tissues are still fully accessible to labeling reagents. Extremely small gold particles, about 1 nm in diameter, work best for this procedure. Such small gold particles are usually not easily visible in the final sections, so additional silver or gold is nucleated around the tiny 1 nm gold particles in a chemical process very much like photographic development.

Different Views of a Single Object Can Be Combined to Give a Three-Dimensional Reconstruction

Thin sections often fail to convey the three-dimensional arrangement of cellular components viewed in a TEM, and the image can be very misleading: a linear structure such as a microtubule may appear in section as a pointlike object, for example, and a section through protruding parts of a single irregularly shaped solid body may give the appearance of two or more separate objects (**Figure 9–46**). The third dimension can be reconstructed from serial sections, but this is a lengthy and tedious process. Even thin sections, however, have a significant depth compared with the resolution of the electron microscope, so the TEM image can also be misleading in an opposite way, through the superimposition of objects that lie at different depths.

Because of the large depth of field of electron microscopes, all the parts of the three-dimensional specimen are in focus, and the resulting image is a projection (a superimposition of layers) of the structure along the viewing direction. The lost information in the third dimension can be recovered if we have views of the same specimen from many different directions. The computational methods for this technique are widely used in medical CT scans. In a CT scan, the imaging equipment is moved around the patient to generate the different views. In **electron-microscope (EM) tomography**, the specimen holder is tilted in the microscope, which achieves the same result. In this way, we can arrive at a three-dimensional reconstruction, in a chosen standard orientation, by combining different views of a single object. Each individual view will be very noisy but by combining them in three dimensions and taking an average, the noise can be largely eliminated. Starting with thick plastic sections of embedded material, three-dimensional reconstructions, or *tomograms*, are used extensively to describe the detailed anatomy of specific regions of the cell, such as the Golgi apparatus (**Figure 9–47**) or the cytoskeleton. Increasingly, microscopists are also applying EM tomography to unstained frozen, hydrated sections, and even to rapidly frozen whole cells or organelles (**Figure 9–48**). Electron microscopy now provides a robust bridge between the scale of the single molecule and that of the whole cell.

Images of Surfaces Can Be Obtained by Scanning Electron Microscopy

A **scanning electron microscope (SEM)** directly produces an image of the three-dimensional structure of the surface of a specimen. The SEM is usually smaller, simpler, and cheaper than a transmission electron microscope. Whereas the TEM uses the electrons that have passed through the specimen to form an

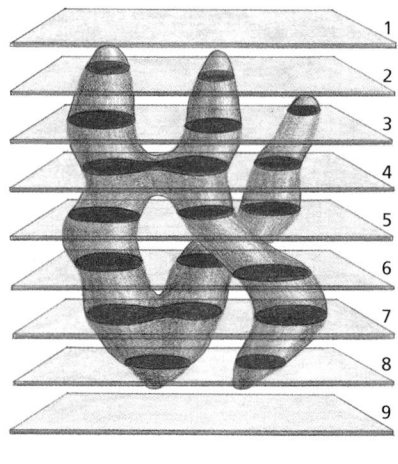

Figure 9–46 A three-dimensional reconstruction from serial sections. Single thin sections in the electron microscope sometimes give misleading impressions. In this example, most sections through a cell containing a branched mitochondrion seem to contain two or three separate mitochondria (compare Figure 9–44). Sections 4 and 7, moreover, might be interpreted as showing a mitochondrion in the process of dividing. The true three-dimensional shape can be reconstructed from a complete set of serial sections.

(A)

(B)

(C)

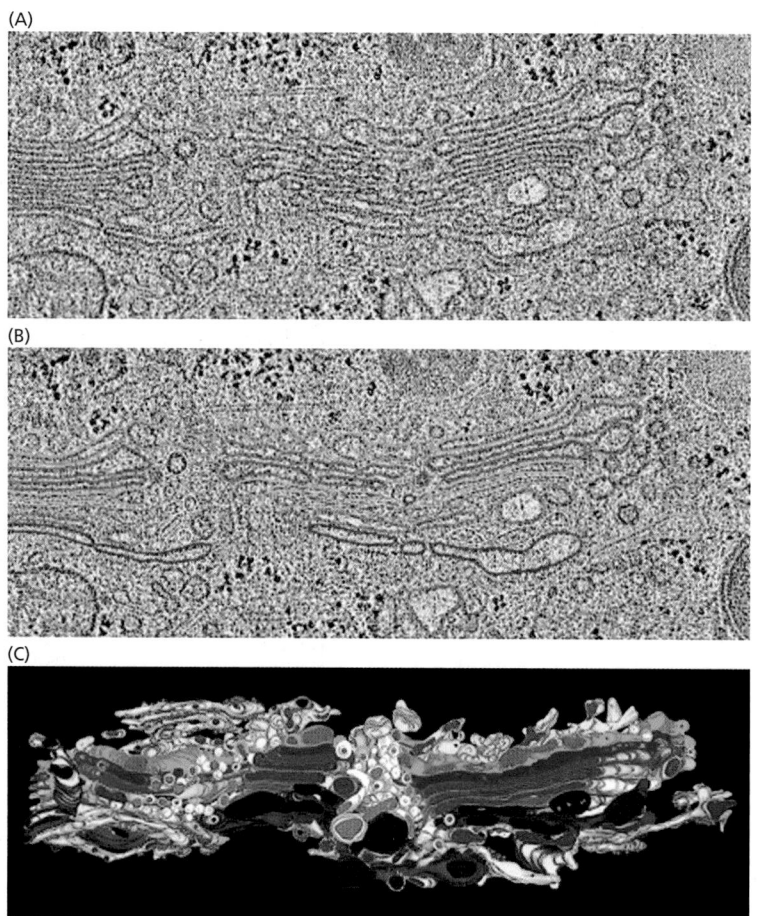

250 nm

Figure 9–47 Electron-microscope (EM) tomography. Samples that have been rapidly frozen, and then freeze-substituted and embedded in plastic, preserve their structure in a condition that is very close to their original living state (Movie 9.2). This example shows the three-dimensional structure of the Golgi apparatus from a rat kidney cell. Several thick sections (250 nm) of the cell were tilted in a high-voltage electron microscope, along two different axes, and about 160 different views recorded. The digital data allow individual thin slices of the complete three-dimensional data set, or tomogram, to be viewed; for example, the serial slices, each only 4 nm thick, are shown in (A) and (B). Very little changes from one slice to the next, but using the full data set, and manually color-coding the membranes (B), one can obtain a full three-dimensional reconstruction, at a resolution of about 7 nm, of the complete Golgi complex and its associated vesicles (C). (© 1999 M.S. Ladinsky et al. Originally published in *J. Cell Biol.* https://doi.org/10.1083/jcb.144.6.1135. With permission from Rockefeller University Press.)

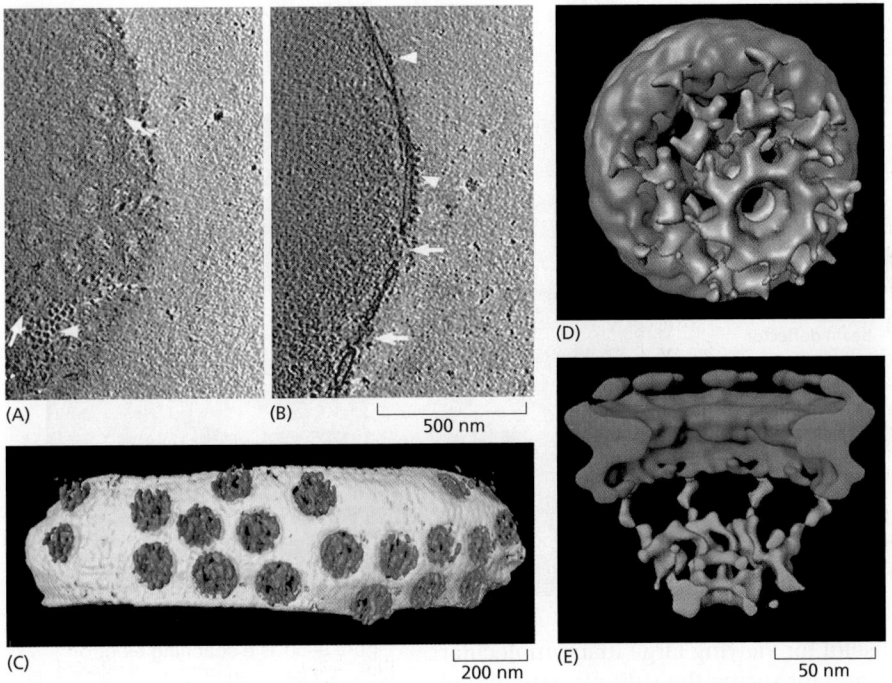

(A)

(B)

500 nm

(C)

200 nm

(D)

(E)

50 nm

Figure 9–48 **Combining cryoelectron-microscope tomography and single-particle reconstruction.** Small, unfixed, rapidly frozen specimens can be examined while still frozen. In this example, the small nuclei of the amoeba *Dictyostelium* were gently isolated and then very rapidly frozen before a series of angled views were recorded with the aid of a tilting microscope stage. These digital views are combined by EM tomographs to produce a three-dimensional tomogram. Two thin digital slices (10 nm) through this tomogram show (A) top views and (B) side views of individual nuclear pores (*white arrows*). (C) In the three-dimensional model, a surface rendering of the pores (*blue*) is seen embedded in the nuclear envelope (*yellow*). From a series of tomograms it was possible to extract data sets for nearly 300 separate nuclear pores, whose structures could then be averaged using the techniques of single-particle reconstruction. The surface-rendered view of one of these reconstructed pores is shown (D) from the nuclear face and (E) in cross section (compare with Figure 12–8). The pore complex is colored *blue* and the nuclear basket *brown*. (From M. Beck et al., *Science* 306:1387–1390, 2004. With permission from AAAS.)

image, the SEM uses electrons that are scattered or emitted from the specimen's surface. The specimen to be examined is fixed, dried, and coated with a thin layer of heavy metal. Alternatively, it can be rapidly frozen, and then transferred to a cooled specimen stage for direct examination in the microscope. Often an entire plant part or small animal can be put into the microscope with very little preparation (**Figure 9–49**). The specimen is scanned with a very narrow beam of electrons. The quantity of electrons scattered or emitted as this primary beam bombards each successive point of the metallic surface is measured and used to control the intensity of a second beam, which moves in synchrony with the primary beam and forms an image on a computer screen. Eventually a highly enlarged image of the surface as a whole is built up (**Figure 9–50**).

The SEM technique provides great depth of field; moreover, since the amount of electron scattering depends on the angle of the surface relative to the beam, the image has highlights and shadows that give it a three-dimensional appearance (see Figure 9–49 and **Figure 9–51**). Only surface features can be examined, however, and in most forms of SEM, the resolution attainable is not very high (about 10 nm, with an effective magnification of up to 20,000 times). As a result, the technique is usually used to study whole cells and tissues rather than subcellular organelles (see Movie 21.3). Very-high-resolution SEMs have, however, been developed with a bright coherent-field emission gun as the electron source. This type of SEM can produce images that rival the resolution possible with a TEM (**Figure 9–52**).

Negative Staining and Cryoelectron Microscopy Both Allow Macromolecules to Be Viewed at High Resolution

If they are shadowed with a heavy metal to provide contrast, isolated macromolecules such as DNA or large proteins can be visualized readily in the electron microscope, but **negative staining** allows finer detail to be seen. In this technique, the molecules are supported on a thin film of carbon and mixed with a solution of a heavy-metal salt such as uranyl acetate. After the sample has dried, a very thin film of metal salt covers the carbon film everywhere except where it has been excluded by the presence of an adsorbed macromolecule. Because the macromolecule allows electrons to pass through it much more readily than does the surrounding heavy-metal stain, a reverse or negative image of the molecule is

1 mm

Figure 9–49 A developing wheat flower, or spike. This delicate flower spike was rapidly frozen, coated with a thin metal film, and examined in the frozen state with an SEM. This low-magnification micrograph demonstrates the large depth of focus of an SEM. (Courtesy of Kim Findlay.)

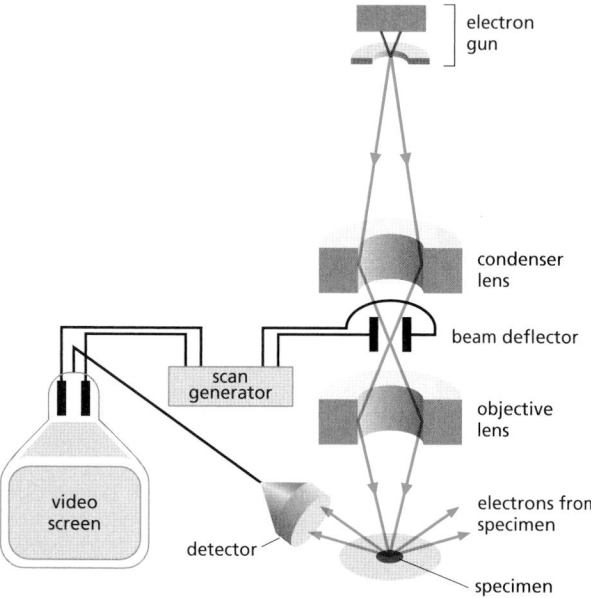

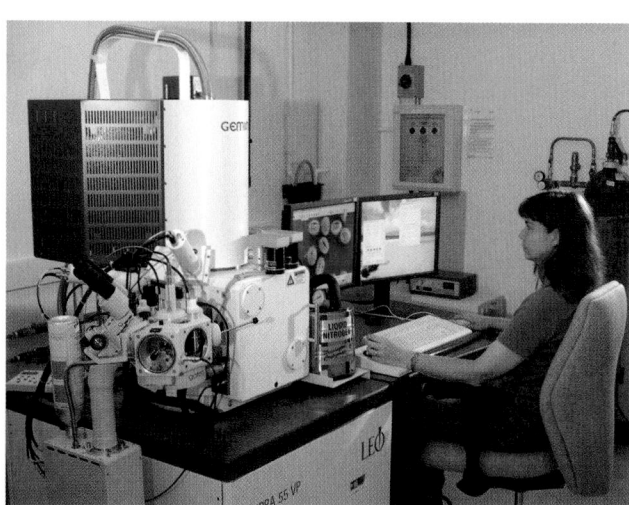

created. Negative staining is especially useful for viewing large macromolecular aggregates such as viruses or ribosomes, and for seeing the subunit structure of protein filaments (**Figure 9–53**).

Shadowing and negative staining can provide high-contrast surface views of small macromolecular assemblies, but the size of the smallest metal particles in the shadow or stain used limits the resolution of both techniques. An alternative that allows us to visualize directly at high resolution even the interior features of three-dimensional structures such as viruses and organelles is **cryoelectron microscopy**, in which rapid freezing to form vitreous ice is again the key. A very thin (about 100 nm) film of an aqueous suspension of virus or purified macromolecular complex is prepared on a microscope grid and is then rapidly frozen by

Figure 9–50 The scanning electron microscope. In an SEM, the specimen is scanned by a beam of electrons brought to a focus on the specimen by the electromagnetic coils that act as lenses. The detector measures the quantity of electrons scattered or emitted as the beam bombards each successive point on the surface of the specimen and controls the intensity of successive points in an image built up on a screen. The SEM creates striking images of three-dimensional objects with great depth of focus and a resolution between 3 nm and 20 nm depending on the instrument. (Photograph courtesy of Andrew Davis.)

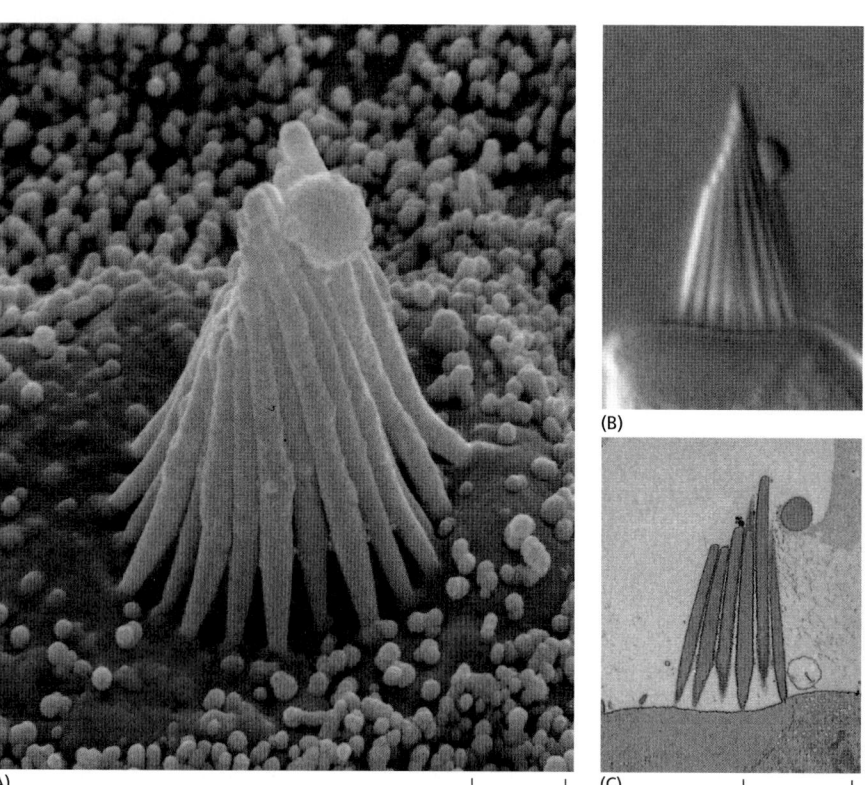

Figure 9–51 Scanning electron microscopy. (A) A scanning electron micrograph of the stereocilia projecting from a hair cell in the inner ear of a bullfrog. For comparison, the same structure is shown by (B) differential-interference-contrast light microscopy (**Movie 9.3**) and (C) thin-section transmission electron microscopy. (Courtesy of Richard Jacobs and James Hudspeth.)

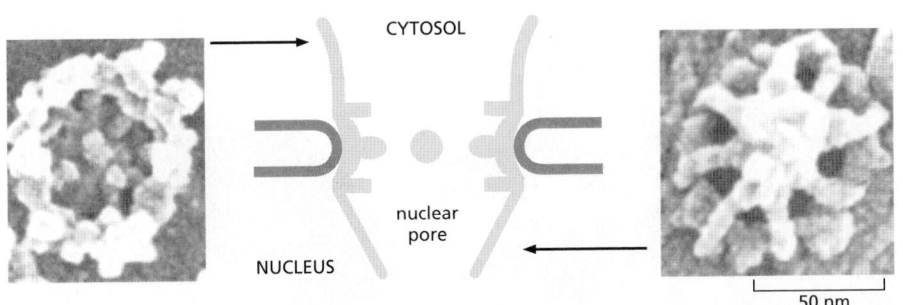

CYTOSOL

nuclear
pore

NUCLEUS

50 nm

Figure 9–52 **The nuclear pore.** Rapidly frozen nuclear envelopes were imaged in a high-resolution SEM, equipped with a field emission gun as the electron source. These views of each side of a nuclear pore represent the limit of resolution of the SEM (compare with Figure 12–8). (Courtesy of Martin Goldberg and Terry Allen.)

being plunged into a coolant. A special sample holder keeps this hydrated specimen at –160°C in the vacuum of the microscope, where it can be viewed directly without fixation, staining, or drying. Unlike negative staining, in which what we see is the envelope of stain exclusion around the particle, hydrated cryoelectron microscopy produces an image from the macromolecular structure itself. However, the contrast in this image is very low, and to extract the maximum amount of structural information, special image-processing techniques must be used, as we describe next.

Multiple Images Can Be Combined to Increase Resolution

As we saw earlier (p. 532), noise is important in light microscopy at low light levels, but it is a particularly severe problem for electron microscopy of unstained macromolecules. A protein molecule can tolerate a dose of only a few tens of electrons per square nanometer without damage, and this dose is orders of magnitude below what is needed to define an image at atomic resolution.

The solution is to obtain images of many identical molecules—perhaps tens of thousands of individual images—and combine them to produce an averaged image, revealing structural details that are hidden by the noise in the original images. This procedure is called **single-particle reconstruction**. Before combining all the individual images, however, they must be aligned with each other. Sometimes it is possible to induce proteins and complexes to form crystalline arrays, in which each molecule is held in the same orientation in a regular lattice. In this case, the alignment problem is easily solved, and several protein structures have been determined at atomic resolution by this type of electron crystallography. In principle, however, crystalline arrays are not absolutely required. With the help of a computer, the digital images of randomly distributed and unaligned molecules can be processed and combined to yield high-resolution reconstructions (see Movie 13.1). Although structures that have some intrinsic symmetry make the task of alignment easier and more accurate, this technique has also been used for objects like ribosomes, with no symmetry. **Figure 9–54** shows the structure of

WHAT WE DON'T KNOW

• We know in detail about many cell processes, such as DNA replication and transcription and RNA translation, but will we ever be able to visualize such rapid molecular processes in action in cells?

• Will we ever be able to image intracellular structures at the resolution of the electron microscope in living cells?

• How can we improve crystallization and single-particle cryoelectron microscopy techniques to obtain high-resolution structures of all important membrane channels and transporters? What new concepts might these structures reveal?

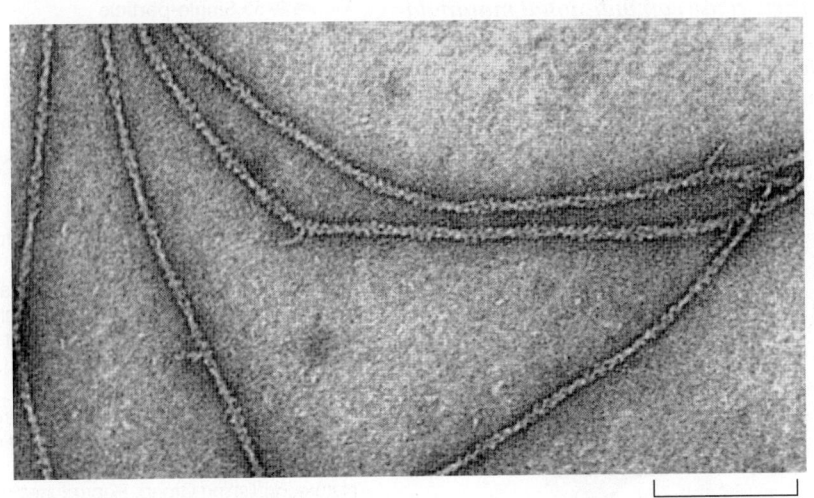

100 nm

Figure 9–53 **Negatively stained actin filaments.** In this transmission electron micrograph, each filament is about 8 nm in diameter and is seen, on close inspection, to be composed of a helical chain of globular actin molecules. (Courtesy of Roger Craig.)

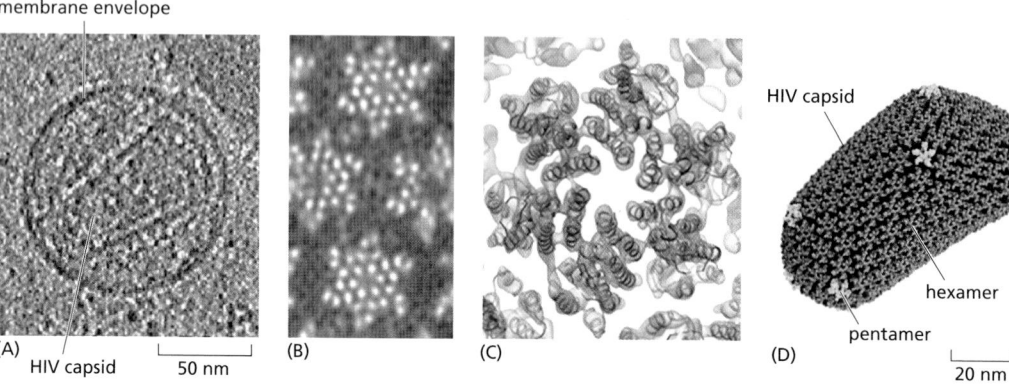

membrane envelope

(A)   HIV capsid   50 nm

(B)

(C)

HIV capsid

hexamer

pentamer

(D)   20 nm

the protein capsid inside a human immunodeficiency virus (HIV) that has been determined at high resolution by the combination of many particles, multiple views, and molecular modeling.

A resolution of 0.3 nm has been achieved by electron microscopy—enough to begin to see the internal atomic arrangements in a protein and to rival x-ray crystallography in resolution. Although electron microscopy is unlikely to supersede x-ray crystallography (discussed in Chapter 8) as a method for macromolecular structure determination, it has some very clear advantages. First, it does not absolutely require crystalline specimens. Second, it can deal with extremely large complexes—structures that may be too large or too variable to crystallize satisfactorily. Third, it allows the rapid analysis of different conformations of protein complexes.

The analysis of large and complex macromolecular structures is helped considerably if the atomic structure of one or more of the subunits is known, for example from x-ray crystallography. Molecular models can then be mathematically "fitted" into the envelope of the structure determined at lower resolution using the electron microscope (see Figures 16–16D and 16–46). **Figure 9–55** shows the structure of a ribosome with the location of a bound release factor displayed in this way (see also Figure 6–72).

Summary

Discovering the detailed structure of membranes and organelles requires the higher resolution attainable in a transmission electron microscope. Specific macromolecules can be localized after being labeled with colloidal gold linked to antibodies. Three-dimensional views of the surfaces of cells and tissues are obtained by scanning electron microscopy. The shapes of isolated molecules can be readily determined by electron microscopy techniques involving fast freezing or negative staining. Electron tomography and single-particle reconstruction use computational manipulations of data obtained from multiple images and multiple viewing angles to produce detailed reconstructions of macromolecules and molecular complexes. The resolution obtained with these methods means that atomic structures of individual

Figure 9–54 Single-particle reconstruction. The structure of a complete human immunodeficiency virus (HIV) capsid has been determined by a combination of cryoelectron microscopy, protein structure determination, and modeling. (A) A single 4 nm slice from an EM tomographic model (see also Figure 9–48) of an intact HIV particle with its membrane outer envelope and its internal, irregularly shaped protein capsid that houses its RNA genome. (B) Electron microscopy of capsid subunits that have self-assembled into a helical tube can be used to derive an electron-density map at a resolution of 8 nm, in which details of the hexamers are clearly visible. (C) Using the known atomic coordinates of a single subunit of the hexamer, the structure has been modeled into the electron-density map from (B). (D) A molecular reconstruction of the entire HIV capsid, based on the detailed structures shown in (A) and (C). This capsid contains 216 hexamers *(blue)* and 12 pentamers *(yellow)*. (Adapted from G. Zhao et al., *Nature* 497:643–646, published 2013 by Macmillan Publishers Ltd. Reproduced with permission of SNCSC. C, PDB code: 3J34.)

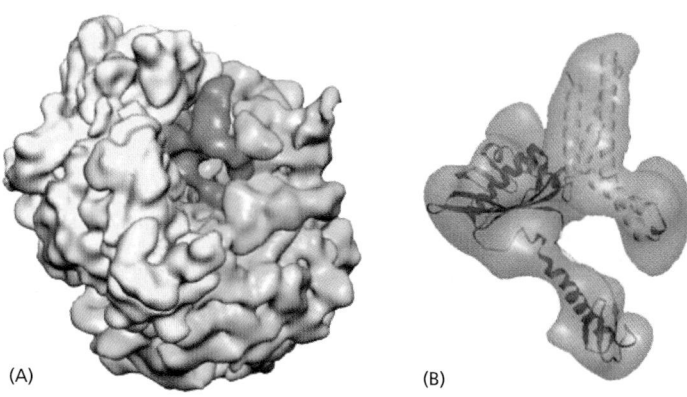

(A)

(B)

Figure 9–55 Single-particle reconstruction and molecular model fitting. Bacterial ribosomes, with and without the release factor required for peptide release from the ribosome, were used to derive high-resolution, three-dimensional cryoelectron microscopy maps at a resolution of better than 1 nm. Images of nearly 20,000 separate ribosomes preserved in ice were used to produce single-particle reconstructions. (A) The 30S ribosomal subunit *(yellow)* and the 50S subunit *(blue)* can be distinguished from the additional electron density that can be attributed to the release factor RF2 *(purple)*. (B) The known molecular structure of RF2 modeled into the electron density from (A). (From U.B.S. Rawat et al., *Nature* 421:87–90, published 2003 by Nature Publishing Group. Reproduced with permission of SNCSC.)

macromolecules can often be "fitted" to the images derived by electron microscopy. In this way, the TEM is increasingly able to bridge the gap between structures discovered by x-ray crystallography and those discovered with the light microscope.

PROBLEMS

Which statements are true? Explain why or why not.

9–1 Because the DNA double helix is only 2 nm wide—well below the limit of resolution of the light microscope—it is impossible to see chromosomes in living cells without special stains.

9–2 A fluorescent molecule, having absorbed a single photon of light at one wavelength, always emits it at a longer wavelength.

Discuss the following problems.

9–3 The diagrams in **Figure Q9-1** show the paths of light rays passing through a specimen with a dry lens and with an oil-immersion lens. Offer an explanation for why oil-immersion lenses should give better resolution. Air, glass, and oil have refractive indices of 1.00, 1.51, and 1.51, respectively.

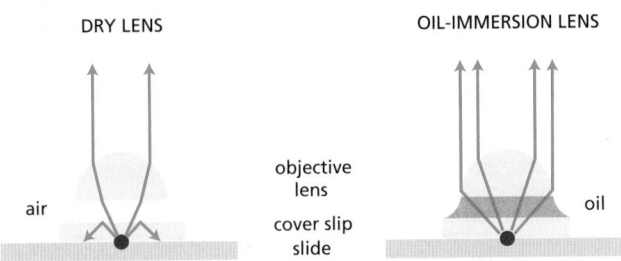

Figure Q9-1 Paths of light rays through dry and oil-immersion lenses (Problem 9–3). The *red circle* at the origin of the light rays is the specimen.

9–4 **Figure Q9-2** shows a diagram of the human eye. The refractive indices of the components in the light path are: cornea 1.38, aqueous humor 1.33, crystalline lens 1.41, and vitreous humor 1.38. Where does the main refraction—the main focusing—occur? What role do you suppose the lens plays?

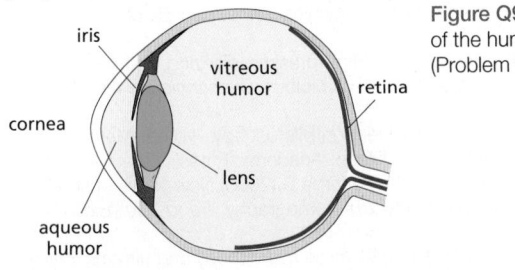

Figure Q9-2 Diagram of the human eye (Problem 9–4).

9–5 Why do humans see so poorly under water? And why do goggles help?

9–6 Explain the difference between resolution and magnification.

9–7 Antibodies that bind to specific proteins are important tools for defining the locations of molecules in cells. The sensitivity of the primary antibody—the antibody that reacts with the target molecule—is often enhanced by using labeled secondary antibodies that bind to it. What are the advantages and disadvantages of using secondary antibodies that carry fluorescent tags versus those that carry bound enzymes?

9–8 **Figure Q9-3** shows a series of modified fluorescent proteins that emit light in a range of colors. How do you suppose the exact same chromophore can fluoresce at so many different wavelengths?

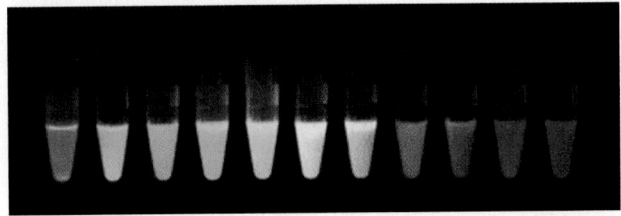

Figure Q9-3 A rainbow of colors produced by modified fluorescent proteins (Problem 9–8). (Courtesy of Nathan Shaner, Paul Steinbach and Roger Tsien.)

9–9 Consider a fluorescent detector designed to report the cellular location of active protein tyrosine kinases. A blue (cyan) fluorescent protein (CFP) and a yellow fluorescent protein (YFP) were fused to either end of a hybrid protein domain. The hybrid protein segment consisted of a substrate peptide recognized by the Abl protein tyrosine kinase and a phosphotyrosine-binding domain (**Figure Q9-4A**). Stimulation of the CFP domain does not cause emission by the YFP domain when the domains are

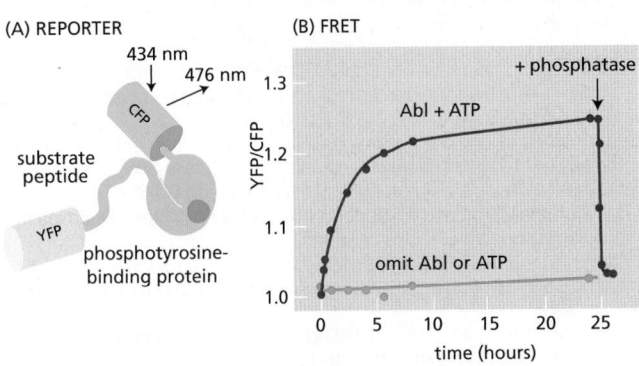

Figure Q9-4 Fluorescent reporter protein designed to detect tyrosine phosphorylation (Problem 9–9). (A) Domain structure of reporter protein. Four domains are indicated: CFP, YFP, tyrosine kinase substrate peptide, and a phosphotyrosine-binding domain. (B) FRET assay. YFP/CFP is normalized to 1.0 at time zero. The reporter was incubated in the presence (or absence) of Abl and ATP for the indicated times. *Arrow* indicates time of addition of a tyrosine phosphatase. (From A.Y. Ting, K.H. Kain, R.L. Klemke and R.Y. Tsien, *Proc. Natl Acad. Sci. USA* 98:15003–15008, 2001. Copyright 2001 National Academy of Sciences, USA. With permission from National Academy of Sciences.)

separated. When the CFP and YFP domains are brought close together, however, fluorescence resonance energy transfer (FRET) allows excitation of CFP to stimulate emission by YFP. FRET shows up experimentally as an increase in the ratio of emission at 526 nm versus 476 nm (YFP/CFP) when CFP is excited by 434 nm light.

Incubation of the reporter protein with Abl protein tyrosine kinase in the presence of ATP gave an increase in YFP/CFP emission (**Figure Q9–4B**). In the absence of ATP or the Abl protein, no FRET occurred. FRET was also eliminated by addition of a tyrosine phosphatase (Figure Q9–4B). Describe as best you can how the reporter protein detects active Abl protein tyrosine kinase.

REFERENCES

General

Celis JE, Carter N, Simons K et al. (eds) (2005) Cell Biology: A Laboratory Handbook, 3rd ed. San Diego: Academic Press. (Volume 3 of this four-volume set covers the practicalities of most of the current light and electron imaging methods that are used in cell biology.)

Pawley BP (ed) (2006) Handbook of Biological Confocal Microscopy, 3rd ed. New York: Springer Science.

Wayne R (2014) Light and Video Microscopy. San Diego: Academic Press.

Looking at Cells in the Light Microscope

Adams MC, Salmon WC, Gupton SL et al. (2003) A high-speed multispectral spinning-disk confocal microscope system for fluorescent speckle microscopy of living cells. *Methods* 29, 29–41.

Agard DA, Hiraoka Y, Shaw P & Sedat JW (1989) Fluorescence microscopy in three dimensions. In Methods in Cell Biology, Vol. 30: Fluorescence Microscopy of Living Cells in Culture, part B (DL Taylor, Y-L Wang eds). San Diego: Academic Press.

Burnette DT, Sengupta P, Dai Y et al. (2011) Bleaching/blinking assisted localization microscopy for superresolution imaging using standard fluorescent molecules. *Proc. Natl Acad. Sci. USA* 108, 21081–21086.

Chalfie M, Tu Y, Euskirchen G et al. (1994) Green fluorescent protein as a marker for gene expression. *Science* 263, 802–805.

Giepmans BN, Adams SR, Ellisman MH & Tsien RY (2006) The fluorescent toolbox for assessing protein location and function. *Science* 312, 217–224.

Harlow E & Lane D (1998) Using Antibodies: A Laboratory Manual. Cold Spring Harbor, NY: Cold Spring Harbor Laboratory Press.

Hell S (2009) Microscopy and its focal switch. *Nat. Methods* 6, 24–32.

Huang B, Babcock H & Zhuang X (2010) Breaking the diffraction barrier: super-resolution imaging of cells. *Cell* 143, 1047–1058.

Huang B, Bates M & Zhuang X (2009) Super-resolution fluorescence microscopy. *Annu. Rev. Biochem.* 78, 993–1016.

Jaiswal JK & Simon SM (2004) Potentials and pitfalls of fluorescent quantum dots for biological imaging. *Trends Cell Biol.* 14, 497–504.

Klar TA, Jakobs S, Dyba M et al. (2000) Fluorescence microscopy with diffraction resolution barrier broken by stimulated emission. *Proc. Natl Acad. Sci. USA* 97, 8206–8210.

Lippincott-Schwartz J & Patterson GH (2003) Development and use of fluorescent protein markers in living cells. *Science* 300, 87–91.

Lippincott-Schwartz J, Altan-Bonnet N & Patterson G (2003) Photobleaching and photoactivation: following protein dynamics in living cells. *Nat. Cell Biol.* 5(Suppl), S7–S14.

McEvoy AL, Greenfield D, Bates M & Liphardt J (2010) Q&A: Single-molecule localization microscopy for biological imaging. *BMC Biol.* 8, 106.

Minsky M (1988) Memoir on inventing the confocal scanning microscope. *Scanning* 10, 128–138.

Miyawaki A, Sawano A & Kogure T (2003) Lighting up cells: labelling proteins with fluorophores. *Nat. Cell Biol.* 5(Suppl), S1–S7.

Parton RM & Read ND (1999) Calcium and pH imaging in living cells. In Light Microscopy in Biology. A Practical Approach, 2nd ed. (Lacey AJ ed.). Oxford: Oxford University Press.

Patterson G, Davidson M, Manley S & Lippincott-Schwartz J (2010) Superresolution imaging using single-molecule localization. *Annu. Rev. Phys. Chem.* 61, 345–367.

Rust MJ, Bates M & Zhuang X (2006) Sub-diffraction-limit imaging by stochastic optical reconstruction microscopy (STORM). *Nat. Methods* 3, 793–795.

Sako Y & Yanagida T (2003) Single-molecule visualization in cell biology. *Nat. Rev. Mol. Cell Biol.* 4(Suppl), SS1–SS5.

Schermelleh L, Heintzmann R & Leonhardt H (2010) A guide to super-resolution fluorescence microscopy. *J. Cell Biol.* 190, 165–175.

Shaner NC, Steinbach PA & Tsien RY (2005) A guide to choosing fluorescent proteins. *Nat. Methods* 2, 905–909.

Sluder G & Wolf DE (2007) Digital Microscopy, 3rd ed: Methods in Cell Biology, Vol 81. San Diego: Academic Press.

Stephens DJ & Allan VJ (2003) Light microscopy techniques for live cell imaging. *Science* 300, 82–86.

Tsien RY (2008) Constructing and exploiting the fluorescent protein paintbox. Nobel Prize Lecture. www.nobelprize.org

White JG, Amos WB & Fordham M (1987) An evaluation of confocal versus conventional imaging of biological structures by fluorescence light microscopy. *J. Cell Biol.* 105, 41–48.

Zernike F (1955) How I discovered phase contrast. *Science* 121, 345–349.

Looking at Cells and Molecules in the Electron Microscope

Allen TD & Goldberg MW (1993) High-resolution SEM in cell biology. *Trends Cell Biol.* 3, 205–208.

Baumeister W (2002) Electron tomography: towards visualizing the molecular organization of the cytoplasm. *Curr. Opin. Struct. Biol.* 12, 679–684.

Böttcher B, Wynne SA & Crowther RA (1997) Determination of the fold of the core protein of hepatitis B virus by electron cryomicroscopy. *Nature* 386, 88–91.

Dubochet J, Adrian M, Chang J-J et al. (1988) Cryo-electron microscopy of vitrified specimens. *Q. Rev. Biophys.* 21, 129–228.

Frank J (2003) Electron microscopy of functional ribosome complexes. *Biopolymers* 68, 223–233.

Frank J (2009) Single-particle reconstruction of biological macromolecules in electron microscopy—30 years. *Quart. Rev. Biophys.* 42, 139–158.

Hayat MA (2000) Principles and Techniques of Electron Microscopy, 4th ed. Cambridge: Cambridge University Press.

Heuser J (1981) Quick-freeze, deep-etch preparation of samples for 3-D electron microscopy. *Trends Biochem. Sci.* 6, 64–68.

Liao M, Cao E, Julius D & Cheng Y (2014) Single particle electron cryo-microscopy of a mammalian ion channel. *Curr. Opin. Struct. Biol.* 27, 1–7.

Lucic V, Förster F & Baumeister W (2005) Structural studies by electron tomography: from cells to molecules. *Annu. Rev. Biochem.* 74, 833–865.

McDonald KL & Auer M (2006) High-pressure freezing, cellular tomography, and structural cell biology. *Biotechniques* 41, 137–139.

McIntosh JR (2007) Cellular Electron Microscopy. 3rd ed: Methods in Cell Biology, Vol 79. San Diego: Academic Press.

McIntosh R, Nicastro D & Mastronarde D (2005) New views of cells in 3D: an introduction to electron tomography. *Trends Cell Biol.* 15, 43–51.

Pease DC & Porter KR (1981) Electron microscopy and ultramicrotomy. *J. Cell Biol.* 91, 287s–292s.

Unwin PNT & Henderson R (1975) Molecular structure determination by electron microscopy of unstained crystalline specimens. *J. Mol. Biol.* 94, 425–440.

Zhou ZH (2008) Towards atomic resolution structural determination buy single particle cryo-electron microscopy. *Curr. Opin. Struct. Biol.* 18, 218–228.

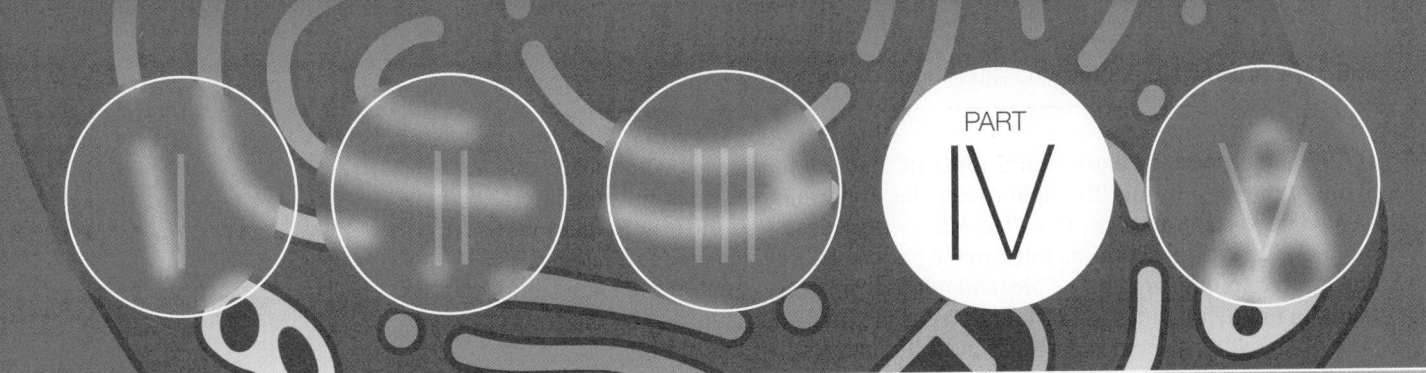
CHAPTER

10

Membrane Structure

Cell membranes are crucial to the life of the cell. The **plasma membrane** encloses the cell, defines its boundaries, and maintains the essential differences between the cytosol and the extracellular environment. Inside eukaryotic cells, the membranes of the nucleus, endoplasmic reticulum, Golgi apparatus, mitochondria, and other membrane-enclosed organelles maintain the characteristic differences between the contents of each organelle and the cytosol. Ion gradients across membranes, established by the activities of specialized membrane proteins, can be used to synthesize ATP, to drive the transport of selected solutes across the membrane, or, as in nerve and muscle cells, to produce and transmit electrical signals. In all cells, the plasma membrane also contains proteins that act as sensors of external signals, allowing the cell to change its behavior in response to environmental cues, including signals from other cells; these protein sensors, or *receptors*, transfer information—rather than molecules—across the membrane.

Despite their differing functions, all biological membranes have a common general structure: each is a very thin film of lipid and protein molecules, held together mainly by noncovalent interactions (**Figure 10-1**). Cell membranes

IN THIS CHAPTER

THE LIPID BILAYER

MEMBRANE PROTEINS

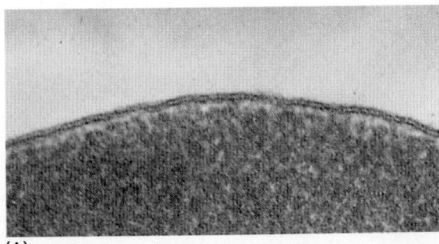

(A)

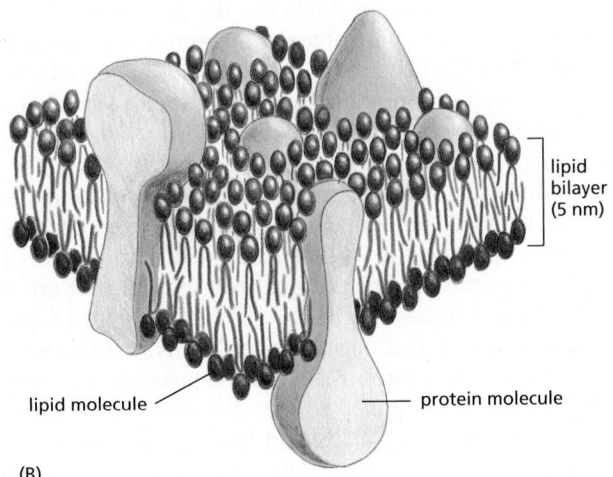

lipid molecule

protein molecule

lipid bilayer (5 nm)

(B)

Figure 10–1 **Two views of a cell membrane.** (A) An electron micrograph of a segment of the plasma membrane of a human red blood cell seen in cross section, showing its bilayer structure. (B) A three-dimensional schematic view of a cell membrane and the general disposition of its lipid and protein constituents. (A, courtesy of Daniel S. Friend, reused by permission of E. L. Bearer.)

are dynamic, fluid structures, and most of their molecules move about in the plane of the membrane. The lipid molecules are arranged as a continuous double layer about 5 nm thick. This *lipid bilayer* provides the basic fluid structure of the membrane and serves as a relatively impermeable barrier to the passage of most water-soluble molecules. Most *membrane proteins* span the lipid bilayer and mediate nearly all of the other functions of the membrane, including the transport of specific molecules across it, and the catalysis of membrane-associated reactions such as ATP synthesis. In the plasma membrane, some transmembrane proteins serve as structural links that connect the cytoskeleton through the lipid bilayer to either the extracellular matrix or an adjacent cell, while others serve as receptors to detect and transduce chemical signals in the cell's environment. It takes many kinds of membrane proteins to enable a cell to function and interact with its environment, and it is estimated that about 30% of the proteins encoded in an animal's genome are membrane proteins.

In this chapter, we consider the structure and organization of the two main constituents of biological membranes—the lipids and the proteins. Although we focus mainly on the plasma membrane, most concepts discussed apply to the various internal membranes of eukaryotic cells as well. The functions of cell membranes are considered in later chapters: their role in energy conversion and ATP synthesis, for example, is discussed in Chapter 14; their role in the transmembrane transport of small molecules in Chapter 11; and their roles in cell signaling and cell adhesion in Chapters 15 and 19, respectively. In Chapters 12 and 13, we discuss the internal membranes of the cell and the protein traffic through and between them.

THE LIPID BILAYER

The **lipid bilayer** provides the basic structure for all cell membranes. It is easily seen by electron microscopy, and its bilayer structure is attributable exclusively to the special properties of the lipid molecules, which assemble spontaneously into bilayers even under simple artificial conditions. In this section, we discuss the different types of lipid molecules found in cell membranes and the general properties of lipid bilayers.

Phosphoglycerides, Sphingolipids, and Sterols Are the Major Lipids in Cell Membranes

Lipid molecules constitute about 50% of the mass of most animal cell membranes, nearly all of the remainder being protein. There are approximately 5×10^6 lipid molecules in a 1 µm × 1 µm area of lipid bilayer, or about 10^9 lipid molecules in the plasma membrane of a small animal cell. All of the lipid molecules in cell membranes are **amphiphilic**—that is, they have a **hydrophilic** ("water-loving") or *polar* end and a **hydrophobic** ("water-fearing") or *nonpolar* end.

The most abundant membrane lipids are the **phospholipids**. These have a polar head group containing a phosphate group and two hydrophobic *hydrocarbon tails*. In animal, plant, and bacterial cells, the tails are usually fatty acids, and they can differ in length (they normally contain between 14 and 24 carbon atoms). One tail typically has one or more *cis*-double bonds (that is, it is *unsaturated*), while the other tail does not (that is, it is *saturated*). As shown in **Figure 10–2**, each *cis*-double bond creates a kink in the tail. Differences in the length and saturation of the fatty acid tails influence how phospholipid molecules pack against one another, thereby affecting the fluidity of the membrane, as we discuss later.

The main phospholipids in most animal cell membranes are the **phosphoglycerides**, which have a three-carbon *glycerol* backbone (see Figure 10–2). Two long-chain fatty acids are linked through ester bonds to adjacent carbon atoms of the glycerol, and the third carbon atom of the glycerol is attached to a phosphate group, which in turn is linked to one of several types of head group. By combining several different fatty acids and head groups, cells make many different phosphoglycerides. *Phosphatidylethanolamine, phosphatidylserine,* and

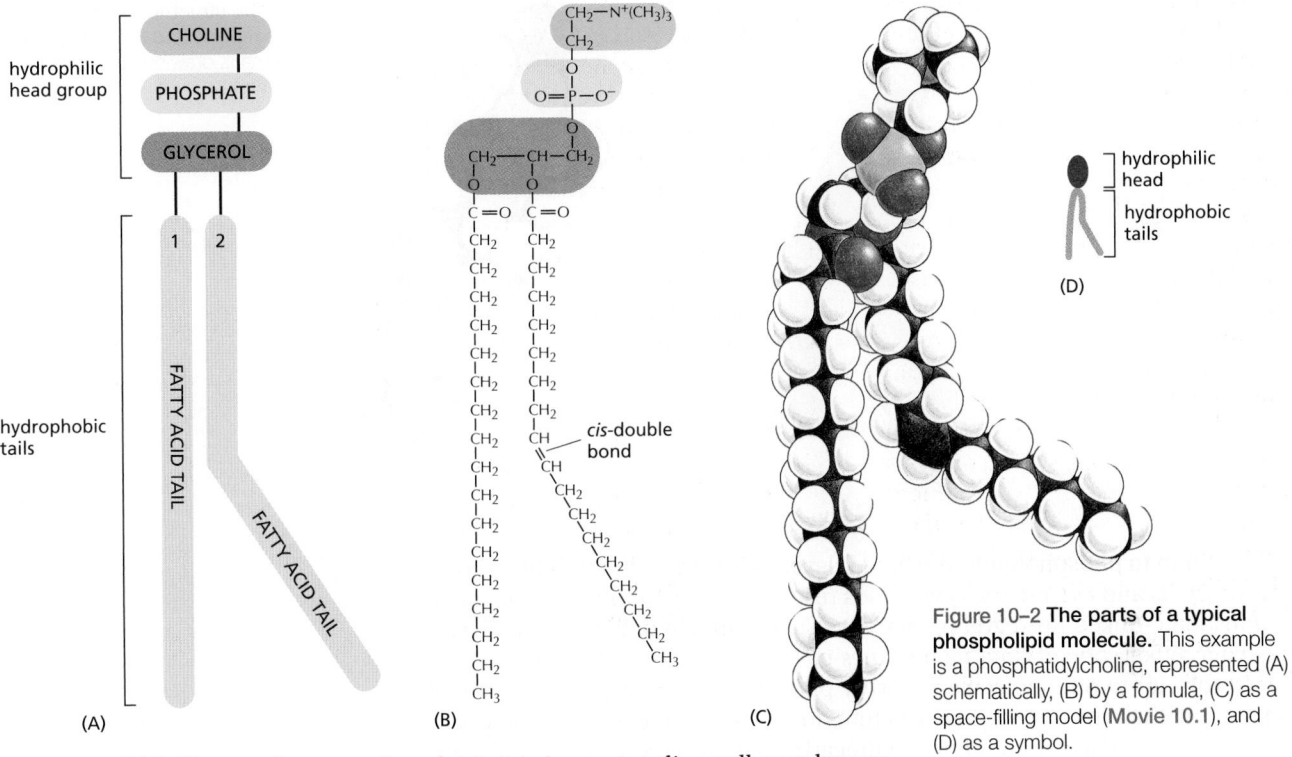

Figure 10–2 **The parts of a typical phospholipid molecule.** This example is a phosphatidylcholine, represented (A) schematically, (B) by a formula, (C) as a space-filling model (**Movie 10.1**), and (D) as a symbol.

phosphatidylcholine are the most abundant ones in mammalian cell membranes (**Figure 10–3A–C**).

Another important class of phospholipids are the *sphingolipids*, which are built from *sphingosine* rather than glycerol (Figure 10–3D–E). Sphingosine is a long hydrocarbon chain with an amino group (NH_2) and two hydroxyl groups (OH) at one end. In sphingomyelin, the most common sphingolipid, a fatty acid tail is attached to the amino group, and a phosphocholine group is attached to the terminal hydroxyl group. Together, the phospholipids phosphatidylcholine, phosphatidylethanolamine, phosphatidylserine, and sphingomyelin constitute more than half the mass of lipid in most mammalian cell membranes (see Table 10–1, p. 571).

Figure 10–3 **Four major phospholipids in mammalian plasma membranes.** Different head groups are represented by different colors in the symbols. The lipid molecules shown in (A–C) are phosphoglycerides, which are derived from glycerol. The molecule in (D) is sphingomyelin, which is derived from sphingosine (E) and is therefore a sphingolipid. Note that only phosphatidylserine carries a net negative charge, the importance of which we discuss later; the other three are electrically neutral at physiological pH, carrying one positive and one negative charge.

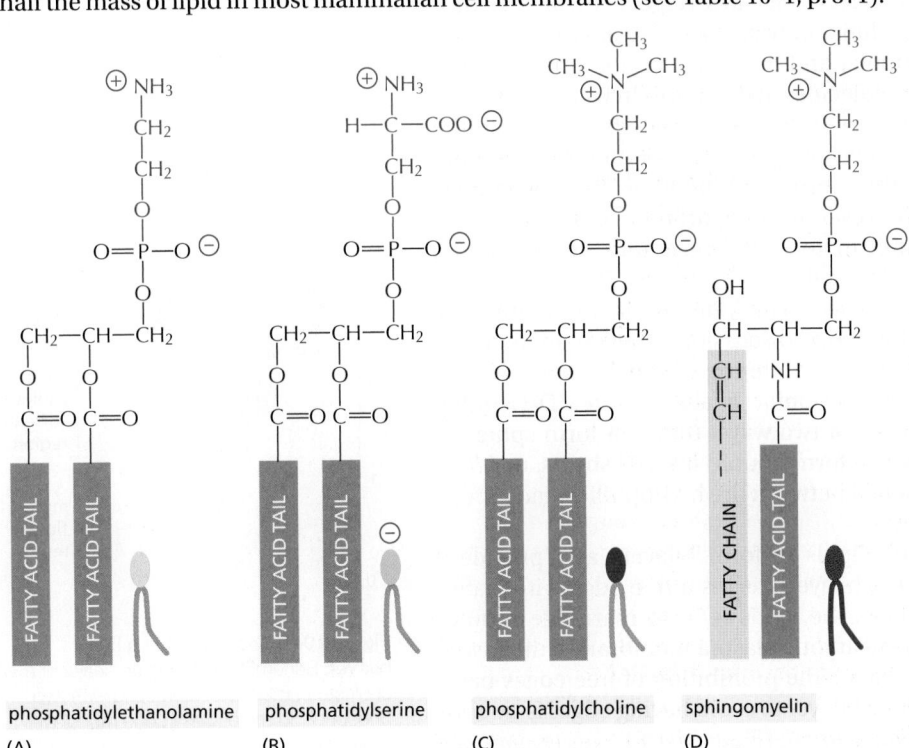

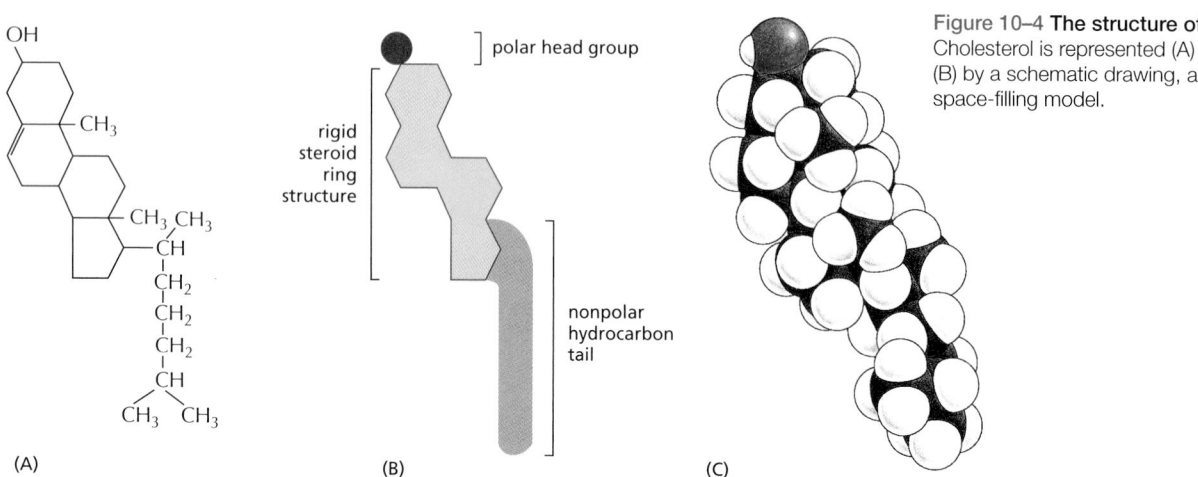

(A)

(B) polar head group

rigid steroid ring structure

nonpolar hydrocarbon tail

(C)

Figure 10–4 **The structure of cholesterol.** Cholesterol is represented (A) by a formula, (B) by a schematic drawing, and (C) as a space-filling model.

In addition to phospholipids, the lipid bilayers in many cell membranes contain *glycolipids* and *cholesterol*. Glycolipids resemble sphingolipids, but, instead of a phosphate-linked head group, they have sugars attached. We discuss glycolipids later. Eukaryotic plasma membranes contain especially large amounts of **cholesterol**—up to one molecule for every phospholipid molecule. Cholesterol is a sterol. It contains a rigid ring structure, to which is attached a single polar hydroxyl group and a short nonpolar hydrocarbon chain (**Figure 10–4**). The cholesterol molecules orient themselves in the bilayer with their hydroxyl group close to the polar head groups of adjacent phospholipid molecules (**Figure 10–5**).

Phospholipids Spontaneously Form Bilayers

The shape and amphiphilic nature of the phospholipid molecules cause them to form bilayers spontaneously in aqueous environments. As discussed in Chapter 2, hydrophilic molecules dissolve readily in water because they contain charged groups or uncharged polar groups that can form either favorable electrostatic interactions or hydrogen bonds with water molecules (**Figure 10–6A**). Hydrophobic molecules, by contrast, are insoluble in water because all, or almost all, of their atoms are uncharged and nonpolar and therefore cannot form energetically favorable interactions with water molecules. If dispersed in water, they force the adjacent water molecules to reorganize into icelike cages that surround the hydrophobic molecule (Figure 10–6B). Because these cage structures are more ordered than the surrounding water, their formation increases the free energy. This free-energy cost is minimized, however, if the hydrophobic molecules (or the hydrophobic portions of amphiphilic molecules) cluster together so that the smallest number of water molecules is affected.

When amphiphilic molecules are exposed to an aqueous environment, they behave as you would expect from the above discussion. They spontaneously aggregate to bury their hydrophobic tails in the interior, where they are shielded from the water, and they expose their hydrophilic heads to water. Depending on their shape, they can do this in either of two ways: they can form spherical *micelles*, with the tails inward, or they can form double-layered sheets, or *bilayers*, with the hydrophobic tails sandwiched between the hydrophilic head groups (**Figure 10–7**).

The same forces that drive phospholipids to form bilayers also provide a self-sealing property. A small tear in the bilayer creates a free edge with water; because this is energetically unfavorable, the lipids tend to rearrange spontaneously to eliminate the free edge. (In eukaryotic plasma membranes, the fusion of intracellular vesicles repairs larger tears.) The prohibition of free edges has a profound consequence: the only way for a bilayer to avoid having edges is by closing in on itself and forming a sealed compartment (**Figure 10–8**). This remarkable

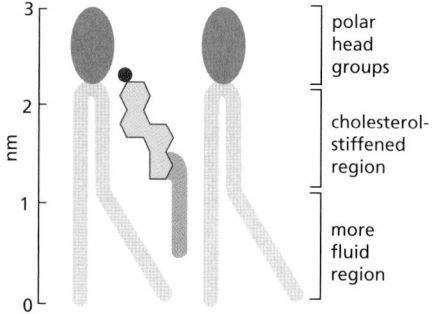

polar head groups

cholesterol-stiffened region

more fluid region

Figure 10–5 **Cholesterol in a lipid bilayer.** Schematic drawing (to scale) of a cholesterol molecule interacting with two phospholipid molecules in one monolayer of a lipid bilayer.

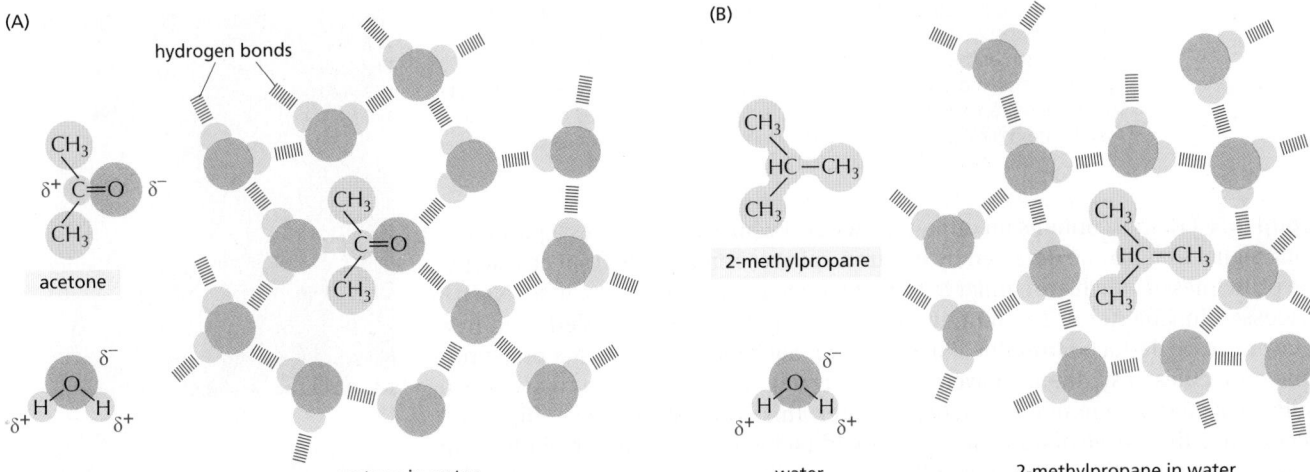

behavior, fundamental to the creation of a living cell, follows directly from the shape and amphiphilic nature of the phospholipid molecule.

A lipid bilayer also has other characteristics that make it an ideal structure for cell membranes. One of the most important of these is its fluidity, which is crucial to many membrane functions (**Movie 10.2**).

The Lipid Bilayer Is a Two-dimensional Fluid

Around 1970, researchers first recognized that individual lipid molecules are able to diffuse freely within the plane of a lipid bilayer. The initial demonstration came from studies of synthetic (artificial) lipid bilayers, which can be made in the form of spherical vesicles, called **liposomes** (**Figure 10–9**); or in the form of planar bilayers formed across a hole in a partition between two aqueous compartments or on a solid support.

Various techniques have been used to measure the motion of individual lipid molecules and their components. One can construct a lipid molecule, for example, with a fluorescent dye or a small gold particle attached to its polar head group and follow the diffusion of even individual molecules in a membrane. Alternatively, one can modify a lipid head group to carry a "spin label," such as a nitroxide

Figure 10–6 How hydrophilic and hydrophobic molecules interact differently with water. (A) Because acetone is polar, it can form hydrogen bonds (red) and favorable electrostatic interactions (yellow) with water molecules, which are also polar. Thus, acetone readily dissolves in water. (B) By contrast, 2-methyl propane is entirely hydrophobic. Because it cannot form favorable interactions with water, it forces adjacent water molecules to reorganize into icelike cage structures, which increases the free energy. This compound is therefore virtually insoluble in water. The symbol δ^- indicates a partial negative charge, and δ^+ indicates a partial positive charge. Polar atoms are shown in color and nonpolar groups are shown in *gray*.

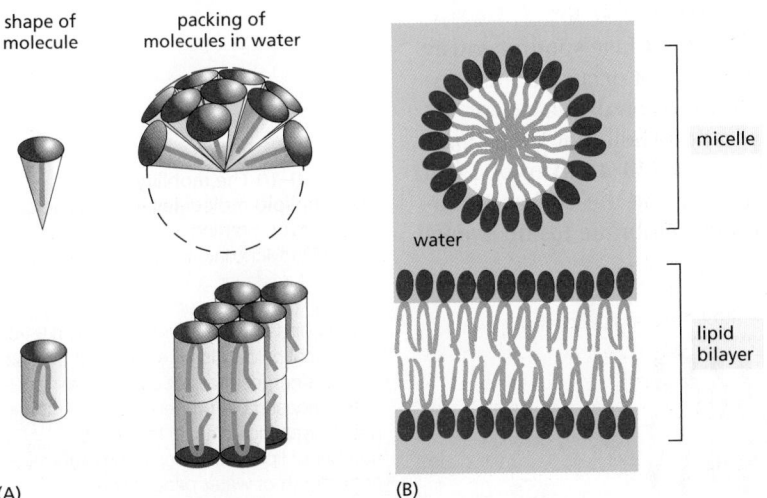

Figure 10–7 Packing arrangements of amphiphilic molecules in an aqueous environment. (A) These molecules spontaneously form micelles or bilayers in water, depending on their shape. Cone-shaped amphiphilic molecules (above) form micelles, whereas cylinder-shaped amphiphilic molecules such as phospholipids (below) form bilayers. (B) A micelle and a lipid bilayer seen in cross section. Note that micelles of amphiphilic molecules are thought to be much more irregular than drawn here (see Figure 10–26C).

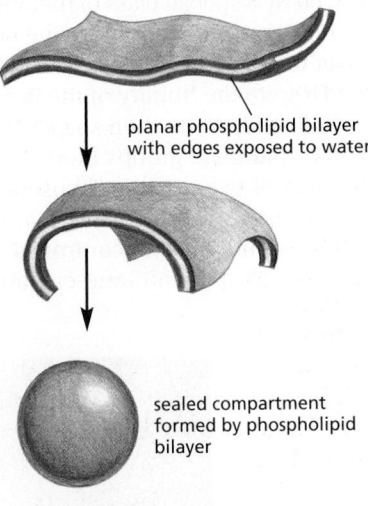

Figure 10–8 The spontaneous closure of a phospholipid bilayer to form a sealed compartment. The closed structure is stable because it avoids the exposure of the hydrophobic hydrocarbon tails to water, which would be energetically unfavorable.

Figure 10–9 Liposomes. (A) An electron micrograph of unfixed, unstained, synthetic phospholipid vesicles—liposomes—in water, which have been rapidly frozen at liquid-nitrogen temperature. (B) A drawing of a small spherical liposome seen in cross section. Liposomes are commonly used as model membranes in experimental studies, especially to study incorporated membrane proteins. (A, from P. Frederik and D. Hubert, *Methods Enzymol.* 391:431–448, 2005. With permission from Elsevier.)

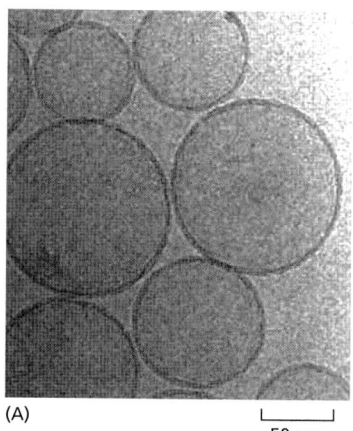

(A)

⌞_____⌟
50 nm

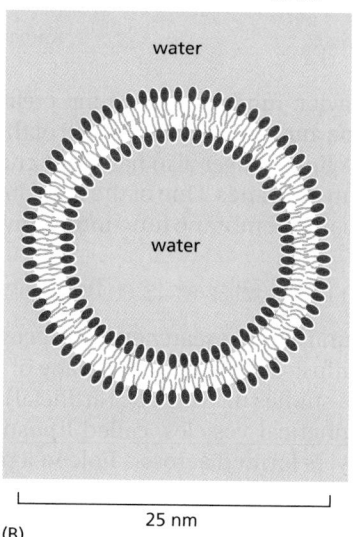

water

water

⌞_____⌟
25 nm

(B)

group (=N–O); this contains an unpaired electron whose spin creates a paramagnetic signal that can be detected by electron spin resonance (ESR) spectroscopy, the principles of which are similar to those of nuclear magnetic resonance (NMR), discussed in Chapter 8. The motion and orientation of a spin-labeled lipid in a bilayer can be deduced from the ESR spectrum. Such studies show that phospholipid molecules in synthetic bilayers very rarely migrate from the monolayer (also called a *leaflet*) on one side to that on the other. This process, known as "flip-flop," occurs on a time scale of hours for any individual molecule, although cholesterol is an exception to this rule and can flip-flop rapidly. In contrast, lipid molecules rapidly exchange places with their neighbors *within* a monolayer (~10^7 times per second). This gives rise to a rapid lateral diffusion, with a diffusion coefficient (D) of about 10^{-8} cm²/sec, which means that an average lipid molecule diffuses the length of a large bacterial cell (~2 μm) in about 1 second. These studies have also shown that individual lipid molecules rotate very rapidly about their long axis and have flexible hydrocarbon chains. Computer simulations show that lipid molecules in synthetic bilayers are very disordered, presenting an irregular surface of variously spaced and oriented head groups to the water phase on either side of the bilayer (**Figure 10–10**).

Similar mobility studies on labeled lipid molecules in isolated biological membranes and in living cells give results similar to those in synthetic bilayers. They demonstrate that the lipid component of a biological membrane is a two-dimensional liquid in which the constituent molecules are free to move laterally. As in synthetic bilayers, individual phospholipid molecules are normally confined to their own monolayer. This confinement creates a problem for their synthesis. Phospholipid molecules are manufactured in only one monolayer of a membrane, mainly in the cytosolic monolayer of the endoplasmic reticulum membrane. If none of these newly made molecules could migrate reasonably promptly to the noncytosolic monolayer, new lipid bilayer could not be made. The problem is solved by a special class of membrane proteins called *phospholipid translocators*, or *flippases*, which catalyze the rapid flip-flop of phospholipids from one monolayer to the other, as discussed in Chapter 12.

Despite the fluidity of the lipid bilayer, liposomes do not fuse spontaneously with one another when suspended in water. Fusion does not occur because the polar lipid head groups bind water molecules that need to be displaced for the bilayers of two different liposomes to fuse. The hydration shell that keeps liposomes apart also insulates the many internal membranes in a eukaryotic cell and prevents their uncontrolled fusion, thereby maintaining the compartmental integrity of membrane-enclosed organelles. All cell membrane fusion events

(A)

(B)

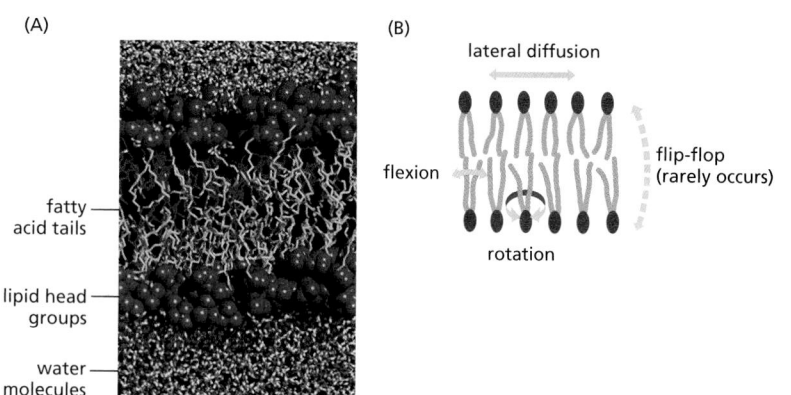

fatty acid tails

lipid head groups

water molecules

lateral diffusion

flexion

flip-flop (rarely occurs)

rotation

Figure 10–10 The mobility of phospholipid molecules in an artificial lipid bilayer. Starting with a model of 100 phosphatidylcholine molecules arranged in a regular bilayer, a computer calculated the position of every atom after 300 picoseconds of simulated time. From these theoretical calculations, a model of the lipid bilayer emerges that accounts for almost all of the measurable properties of a synthetic lipid bilayer, including its thickness, number of lipid molecules per membrane area, depth of water penetration, and unevenness of the two surfaces. Note that the tails in one monolayer can interact with those in the other monolayer, if the tails are long enough. (B) The different motions of a lipid molecule in a bilayer. (A, based on S.W. Chiu et al., *Biophys. J.* 69: 1230–1245, 1995.)

are catalyzed by tightly regulated fusion proteins, which force appropriate membranes into tight proximity, squeezing out the water layer that keeps the bilayers apart, as we discuss in Chapter 13.

The Fluidity of a Lipid Bilayer Depends on Its Composition

The fluidity of cell membranes has to be precisely regulated. Certain membrane transport processes and enzyme activities, for example, cease when the bilayer viscosity is experimentally increased beyond a threshold level.

The fluidity of a lipid bilayer depends on both its composition and its temperature, as is readily demonstrated in studies of synthetic lipid bilayers. A synthetic bilayer made from a single type of phospholipid changes from a liquid state to a two-dimensional rigid crystalline (or gel) state at a characteristic temperature. This change of state is called a *phase transition*, and the temperature at which it occurs is lower (that is, the membrane becomes more difficult to freeze) if the hydrocarbon chains are short or have double bonds. A shorter chain length reduces the tendency of the hydrocarbon tails to interact with one another, in both the same and opposite monolayer, and *cis*-double bonds produce kinks in the chains that make them more difficult to pack together, so that the membrane remains fluid at lower temperatures (**Figure 10–11**). Bacteria, yeasts, and other organisms whose temperature fluctuates with that of their environment adjust the fatty acid composition of their membrane lipids to maintain a relatively constant fluidity. As the temperature falls, for instance, the cells of those organisms synthesize fatty acids with more *cis*-double bonds, thereby avoiding the decrease in bilayer fluidity that would otherwise result from the temperature drop.

Cholesterol modulates the properties of lipid bilayers. When mixed with phospholipids, it enhances the permeability-barrier properties of the lipid bilayer. Cholesterol inserts into the bilayer with its hydroxyl group close to the polar head groups of the phospholipids, so that its rigid, platelike steroid rings interact with—and partly immobilize—those regions of the hydrocarbon chains closest to the polar head groups (see Figure 10–5 and Movie 10.3). By decreasing the mobility of the first few CH_2 groups of the chains of the phospholipid molecules, cholesterol makes the lipid bilayer less deformable in this region and thereby decreases the permeability of the bilayer to small water-soluble molecules. Although cholesterol tightens the packing of the lipids in a bilayer, it does not make membranes any less fluid. At the high concentrations found in most eukaryotic plasma membranes, cholesterol also prevents the hydrocarbon chains from coming together and crystallizing.

Table 10–1 compares the lipid compositions of several biological membranes. Note that bacterial plasma membranes are often composed of one main type of phospholipid and contain no cholesterol. In archaea, lipids usually contain

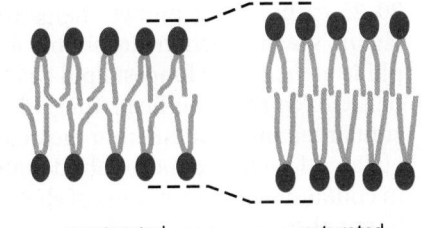

unsaturated hydrocarbon chains with *cis*-double bonds

saturated hydrocarbon chains

Figure 10–11 **The influence of *cis*-double bonds in hydrocarbon chains.** The double bonds make it more difficult to pack the chains together, thereby making the lipid bilayer more difficult to freeze. In addition, because the hydrocarbon chains of unsaturated lipids are more spread apart, lipid bilayers containing them are thinner than bilayers formed exclusively from saturated lipids.

TABLE 10–1 Approximate Lipid Compositions of Different Cell Membranes						
	Percentage of total lipid by weight					
Lipid	Liver cell plasma membrane	Red blood cell plasma membrane	Myelin	Mitochondrion (inner and outer membranes)	Endoplasmic reticulum	*E. coli* bacterium
Cholesterol	17	23	22	3	6	0
Phosphatidylethanolamine	7	18	15	28	17	70
Phosphatidylserine	4	7	9	2	5	trace
Phosphatidylcholine	24	17	10	44	40	0
Sphingomyelin	19	18	8	0	5	0
Glycolipids	7	3	28	trace	trace	0
Others	22	14	8	23	27	30

20–25-carbon-long prenyl chains instead of fatty acids; prenyl and fatty acid chains are similarly hydrophobic and flexible (see Figure 10–20F); in thermophilic archaea, the longest lipid chains span both leaflets, making the membrane particularly stable to heat. Thus, lipid bilayers can be built from molecules with similar features but different molecular designs. The plasma membranes of most eukaryotic cells are more varied than those of prokaryotes and archaea, not only in containing large amounts of cholesterol but also in containing a mixture of different phospholipids.

Analysis of membrane lipids by mass spectrometry has revealed that the lipid composition of a typical eukaryotic cell membrane is much more complex than originally thought. These membranes contain a bewildering variety of perhaps 500–2000 different lipid species with even the simple plasma membrane of a red blood cell containing well over 150. While some of this complexity reflects the combinatorial variation in head groups, hydrocarbon chain lengths, and desaturation of the major phospholipid classes, some membranes also contain many structurally distinct minor lipids, at least some of which have important functions. The *inositol phospholipids*, for example, are present in small quantities in animal cell membranes and have crucial functions in guiding membrane traffic and in cell signaling (discussed in Chapters 13 and 15, respectively). Their local synthesis and destruction are regulated by a large number of enzymes, which create both small intracellular signaling molecules and lipid docking sites on membranes that recruit specific proteins from the cytosol, as we discuss later.

Despite Their Fluidity, Lipid Bilayers Can Form Domains of Different Compositions

Because a lipid bilayer is a two-dimensional fluid, we might expect most types of lipid molecules in it to be well mixed and randomly distributed in their own monolayer. The van der Waals attractive forces between neighboring hydrocarbon tails are not selective enough to hold groups of phospholipid molecules together. With certain lipid mixtures in artificial bilayers, however, one can observe phase segregations in which specific lipids come together in separate domains (**Figure 10–12**).

There has been a long debate among cell biologists about whether the lipid molecules in the plasma membrane of living cells similarly segregate into specialized domains, called **lipid rafts**. Although many lipids and membrane proteins are not distributed uniformly, large-scale lipid phase segregations are rarely seen in living cell membranes. Instead, specific membrane proteins and lipids are seen to concentrate in a more temporary, dynamic fashion facilitated by protein–protein interactions that allow the transient formation of specialized membrane regions (**Figure 10–13**). Such clusters can be tiny nanoclusters on a scale of a few molecules, or larger assemblies that can be seen with electron microscopy, such as the *caveolae* involved in endocytosis (discussed in Chapter 13). The tendency of mixtures of lipids to undergo phase partitioning, as seen in artificial bilayers (see Figure 10–12), may help create rafts in living cell membranes—organizing and concentrating membrane proteins either for transport in membrane vesicles

Figure 10–12 Lateral phase separation in artificial lipid bilayers. (A) Giant liposomes produced from a 1:1 mixture of phosphatidylcholine and sphingomyelin form uniform bilayers. (B) By contrast, liposomes produced from a 1:1:1 mixture of phosphatidylcholine, sphingomyelin, and cholesterol form bilayers with two separate phases. The liposomes are stained with trace concentrations of a fluorescent dye that preferentially partitions into one of the two phases. The average size of the domains formed in these giant artificial liposomes is much larger than that expected in cell membranes, where "lipid rafts" (see text) may be as small as a few nanometers in diameter. (A, from N. Kahya et al., *J. Struct. Biol.* 147:77–89, 2004. With permission from Elsevier; B, courtesy of Schwille Lab, MPG.)

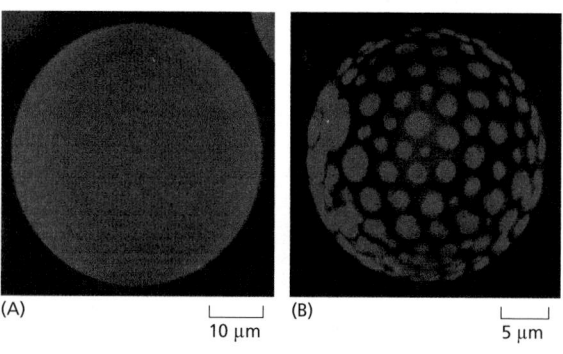

(A)   10 μm   (B)   5 μm

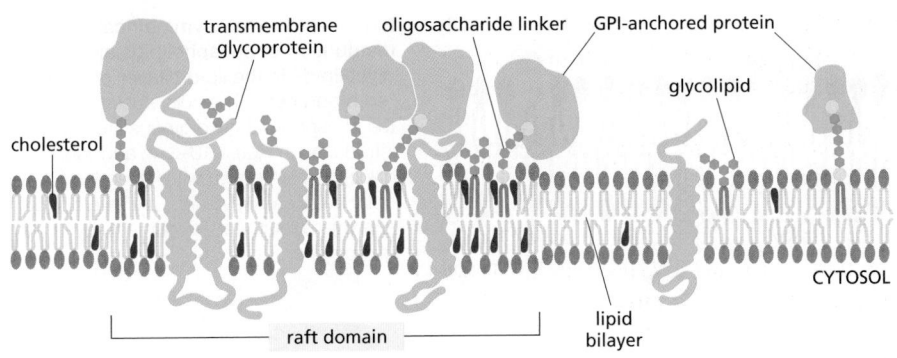

Figure 10–13 **A model of a raft domain.**
Weak protein–protein, protein–lipid, and
lipid–lipid interactions reinforce one another
to partition the interacting components into
raft domains. Cholesterol, sphingolipids,
glycolipids, glycosylphosphatidylinositol
(GPI)-anchored proteins, and some
transmembrane proteins are enriched
in these domains. Note that because of
their composition, raft domains have an
increased membrane thickness. We discuss
glycolipids, GPI-anchored proteins, and
oligosaccharide linkers later. (Adapted
from D. Lingwood and K. Simons, *Science*
327:46–50, 2010.)

(discussed in Chapter 13) or for working together in protein assemblies, such as when they convert extracellular signals into intracellular ones (discussed in Chapter 15).

Lipid Droplets Are Surrounded by a Phospholipid Monolayer

Most cells store an excess of lipids in **lipid droplets**, from where they can be retrieved as building blocks for membrane synthesis or as a food source. Fat cells, or *adipocytes*, are specialized for lipid storage. They contain a giant lipid droplet that fills up most of their cytoplasm. Most other cells have many smaller lipid droplets, the number and size varying with the cell's metabolic state. Fatty acids can be liberated from lipid droplets on demand and exported to other cells through the bloodstream. Lipid droplets store neutral lipids, such as triacylglycerols and cholesterol esters, which are synthesized from fatty acids and cholesterol by enzymes in the endoplasmic reticulum membrane. Because these lipids do not contain hydrophilic head groups, they are exclusively hydrophobic molecules, and therefore aggregate into three-dimensional droplets rather than into bilayers.

Lipid droplets are unique organelles in that they are surrounded by a single monolayer of phospholipids, which contains a large variety of proteins. Some of the proteins are enzymes involved in lipid metabolism, but the functions of most are unknown. Lipid droplets form rapidly when cells are exposed to high concentrations of fatty acids. They are thought to form from discrete regions of the endoplasmic reticulum membrane where many enzymes of lipid metabolism are concentrated. **Figure 10–14** shows one model of how lipid droplets may form and acquire their surrounding monolayer of phospholipids and proteins.

The Asymmetry of the Lipid Bilayer Is Functionally Important

The lipid compositions of the two monolayers of the lipid bilayer in many membranes are strikingly different. In the human red blood cell (erythrocyte) membrane, for example, almost all of the phospholipid molecules that have choline—$(CH_3)_3N^+CH_2CH_2OH$—in their head group (phosphatidylcholine and

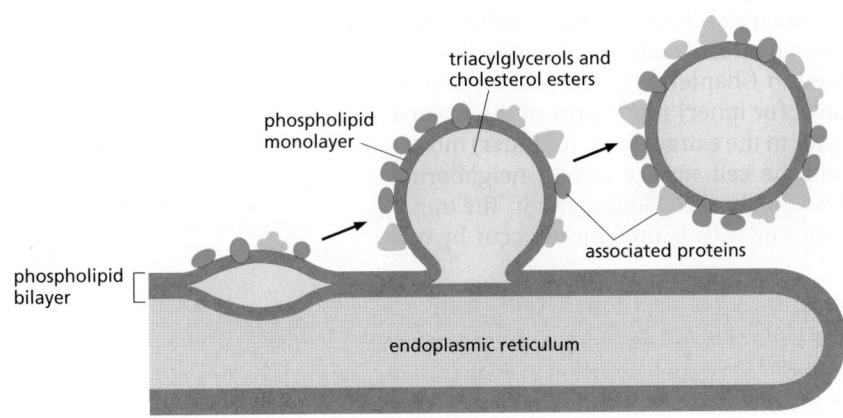

Figure 10–14 **A model for the formation of lipid droplets.** Neutral lipids are deposited between the two monolayers of the endoplasmic reticulum membrane. There, they aggregate into a three-dimensional droplet, which buds and pinches off from the endoplasmic reticulum membrane as a unique organelle, surrounded by a single monolayer of phospholipids and associated proteins. (Adapted from S. Martin and R.G. Parton, *Nat. Rev. Mol. Cell Biol.* 7:373–378, 2006.)

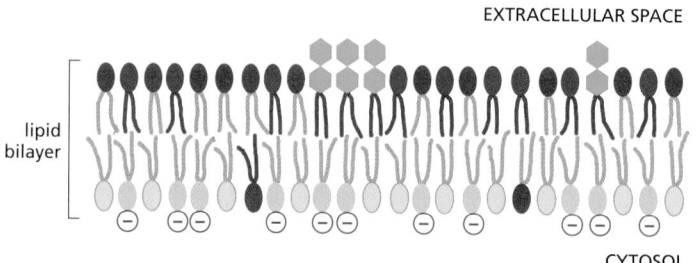

EXTRACELLULAR SPACE

lipid bilayer

CYTOSOL

Figure 10–15 **The asymmetrical distribution of phospholipids and glycolipids in the lipid bilayer of human red blood cells.** The colors used for the phospholipid head groups are those introduced in Figure 10–3. In addition, glycolipids are drawn with hexagonal polar head groups *(blue)*. Cholesterol (not shown) is distributed roughly equally in both monolayers.

sphingomyelin) are in the outer monolayer, whereas almost all that contain a terminal primary amino group (phosphatidylethanolamine and phosphatidylserine) are in the inner monolayer (**Figure 10–15**). Because the negatively charged phosphatidylserine is located in the inner monolayer, there is a significant difference in charge between the two halves of the bilayer. We discuss in Chapter 12 how membrane-bound phospholipid translocators generate and maintain lipid asymmetry.

Lipid asymmetry is functionally important, especially in converting extracellular signals into intracellular ones (discussed in Chapter 15). Many cytosolic proteins bind to specific lipid head groups found in the cytosolic monolayer of the lipid bilayer. The enzyme *protein kinase C* (*PKC*), for example, which is activated in response to various extracellular signals, binds to the cytosolic face of the plasma membrane, where phosphatidylserine is concentrated, and requires this negatively charged phospholipid for its activity.

In other cases, specific lipid head groups must first be modified to create protein-binding sites at a particular time and place. One example is *phosphatidylinositol* (PI), one of the minor phospholipids that are concentrated in the cytosolic monolayer of cell membranes (see Figure 13–10A–C). Various lipid kinases can add phosphate groups at distinct positions on the inositol ring, creating binding sites that recruit specific proteins from the cytosol to the membrane. An important example of such a lipid kinase is *phosphoinositide 3-kinase* (*PI 3-kinase*), which is activated in response to extracellular signals and helps to recruit specific intracellular signaling proteins to the cytosolic face of the plasma membrane (see Figure 15–53). Similar lipid kinases phosphorylate inositol phospholipids in intracellular membranes and thereby help to recruit proteins that guide membrane transport.

Phospholipids in the plasma membrane are used in yet another way to convert extracellular signals into intracellular ones. The plasma membrane contains various *phospholipases* that are activated by extracellular signals to cleave specific phospholipid molecules, generating fragments of these molecules that act as short-lived intracellular mediators. *Phospholipase C*, for example, cleaves an inositol phospholipid in the cytosolic monolayer of the plasma membrane to generate two fragments, one of which remains in the membrane and helps activate protein kinase C, while the other is released into the cytosol and stimulates the release of Ca^{2+} from the endoplasmic reticulum (see Figure 15–28).

Animals exploit the phospholipid asymmetry of their plasma membranes to distinguish between live and dead cells. When animal cells undergo apoptosis (a form of programmed cell death, discussed in Chapter 18), phosphatidylserine, which is normally confined to the cytosolic (or inner) monolayer of the plasma membrane lipid bilayer, rapidly translocates to the extracellular (or outer) monolayer. The phosphatidylserine exposed on the cell surface signals neighboring cells, such as macrophages, to phagocytose the dead cell and digest it. The translocation of the phosphatidylserine in apoptotic cells is thought to occur by two mechanisms:

1. The phospholipid translocator that normally transports this lipid from the outer monolayer to the inner monolayer is inactivated.

2. A "scramblase" that transfers phospholipids nonspecifically in both directions between the two monolayers is activated.

Glycolipids Are Found on the Surface of All Eukaryotic Plasma Membranes

Sugar-containing lipid molecules called **glycolipids** have the most extreme asymmetry in their membrane distribution: these molecules, whether in the plasma membrane or in intracellular membranes, are found exclusively in the monolayer facing away from the cytosol. In animal cells, they are made from sphingosine, just like sphingomyelin (see Figure 10–3). These intriguing molecules tend to self-associate, partly through hydrogen bonds between their sugars and partly through van der Waals forces between their long and straight hydrocarbon chains, which causes them to partition preferentially into lipid raft phases (see Figure 10–13). The asymmetric distribution of glycolipids in the bilayer results from the addition of sugar groups to the lipid molecules in the lumen of the Golgi apparatus. Thus, the compartment in which they are manufactured is topologically equivalent to the exterior of the cell (discussed in Chapter 12). As they are delivered to the plasma membrane, the sugar groups are exposed at the cell surface (see Figure 10–15), where they have important roles in interactions of the cell with its surroundings.

Glycolipids probably occur in all eukaryotic cell plasma membranes, where they generally constitute about 5% of the lipid molecules in the outer monolayer. They are also found in some intracellular membranes. The most complex of the glycolipids, the **gangliosides**, contain oligosaccharides with one or more sialic acid moieties, which give gangliosides a net negative charge (**Figure 10–16**). The most abundant of the more than 40 different gangliosides that have been identified are in the plasma membrane of nerve cells, where gangliosides constitute 5–10% of the total lipid mass; they are also found in much smaller quantities in other cell types.

Hints as to the functions of glycolipids come from their localization. In the plasma membrane of epithelial cells, for example, glycolipids are confined to the exposed apical surface, where they may help to protect the membrane against the harsh conditions frequently found there (such as low pH and high concentrations of degradative enzymes). Charged glycolipids, such as gangliosides, may be important because of their electrical effects: their presence alters the electrical field across the membrane and the concentrations of ions—especially Ca^{2+}—at the membrane surface. Glycolipids also function in cell-recognition processes,

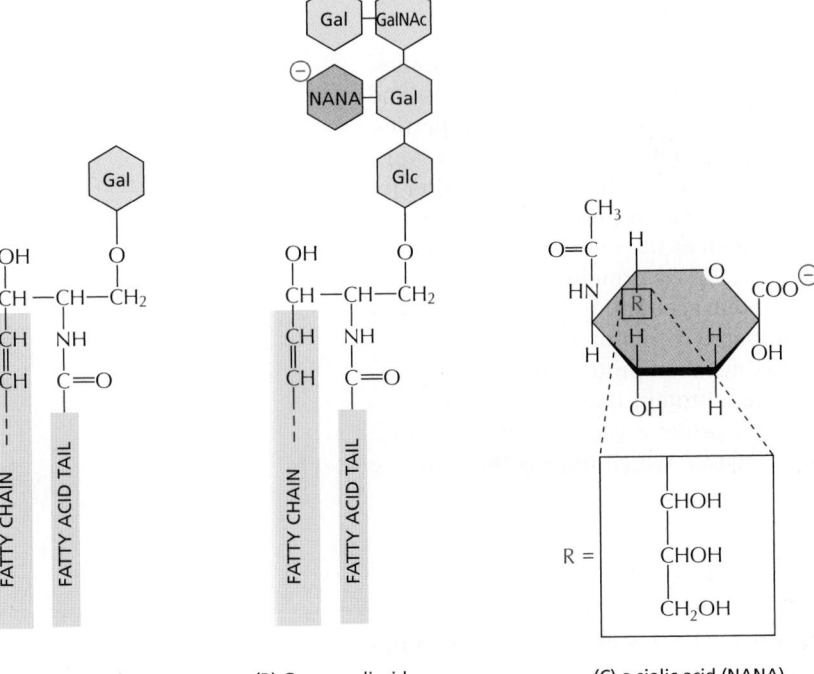

Figure 10–16 Glycolipid molecules.
(A) Galactocerebroside is called a *neutral glycolipid* because the sugar that forms its head group is uncharged. (B) A ganglioside always contains one or more negatively charged sialic acid moiety. There are various types of sialic acid; in human cells, it is mostly *N*-acetylneuraminic acid, or NANA), whose structure is shown in (C). Whereas in bacteria and plants almost all glycolipids are derived from glycerol, as are most phospholipids, in animal cells almost all glycolipids are based on sphingosine, as is the case for sphingomyelin (see Figure 10–3). Gal = galactose; Glc = glucose, GalNAc = *N*-acetylgalactosamine; these three sugars are uncharged.

(A) galactocerebroside   (B) G$_{M1}$ ganglioside   (C) a sialic acid (NANA)

in which membrane-bound carbohydrate-binding proteins (*lectins*) bind to the sugar groups on both glycolipids and glycoproteins in the process of cell–cell adhesion (discussed in Chapter 19). Mutant mice that are deficient in all of their complex gangliosides show abnormalities in the nervous system, including axonal degeneration and reduced myelination.

Some glycolipids provide entry points for certain bacterial toxins and viruses. The ganglioside G_{M1} (see Figure 10–16), for example, acts as a cell-surface receptor for the bacterial toxin that causes the debilitating diarrhea of cholera. Cholera toxin binds to and enters only those cells that have G_{M1} on their surface, including intestinal epithelial cells. Its entry into a cell leads to a prolonged increase in the concentration of intracellular cyclic AMP (discussed in Chapter 15), which in turn causes a large efflux of Cl^-, leading to the secretion of Na^+, K^+, HCO_3^-, and water into the intestine. Polyomaviruses also enter the cell after binding initially to gangliosides.

Summary

Biological membranes consist of a continuous double layer of lipid molecules in which membrane proteins are embedded. This lipid bilayer is fluid, with individual lipid molecules able to diffuse rapidly within their own monolayer. The membrane lipid molecules are amphiphilic. When placed in water, they assemble spontaneously into bilayers, which form sealed compartments.

Although cell membranes can contain hundreds of different lipid species, the plasma membrane in animal cells contains three major classes—phospholipids, cholesterol, and glycolipids. Because of their different backbone structure, phospholipids fall into two subclasses—phosphoglycerides and sphingolipids. The lipid compositions of the inner and outer monolayers are different, reflecting the different functions of the two faces of a cell membrane. Different mixtures of lipids are found in the membranes of cells of different types, as well as in the various membranes of a single eukaryotic cell. Inositol phospholipids are a minor class of phospholipids, which in the cytosolic leaflet of the plasma membrane lipid bilayer play an important part in cell signaling: in response to extracellular signals, specific lipid kinases phosphorylate the head groups of these lipids to form docking sites for cytosolic signaling proteins, whereas specific phospholipases cleave certain inositol phospholipids to generate small intracellular signaling molecules.

MEMBRANE PROTEINS

Although the lipid bilayer provides the basic structure of biological membranes, the membrane proteins perform most of the membrane's specific tasks and therefore give each type of cell membrane its characteristic functional properties. Accordingly, the amounts and types of proteins in a membrane are highly variable. In the myelin membrane, which serves mainly as electrical insulation for nerve-cell axons, less than 25% of the membrane mass is protein. By contrast, in the membranes involved in ATP production (such as the internal membranes of mitochondria and chloroplasts), approximately 75% is protein. A typical plasma membrane is somewhere in between, with protein accounting for about half of its mass. Because lipid molecules are small compared with protein molecules, however, there are always many more lipid molecules than protein molecules in cell membranes—about 50 lipid molecules for each protein molecule in cell membranes that are 50% protein by mass. Membrane proteins vary widely in structure and in the way they associate with the lipid bilayer, which reflects their diverse functions.

Membrane Proteins Can Be Associated with the Lipid Bilayer in Various Ways

Figure 10–17 shows the different ways in which proteins can associate with the membrane. Like their lipid neighbors, **membrane proteins** are amphiphilic,

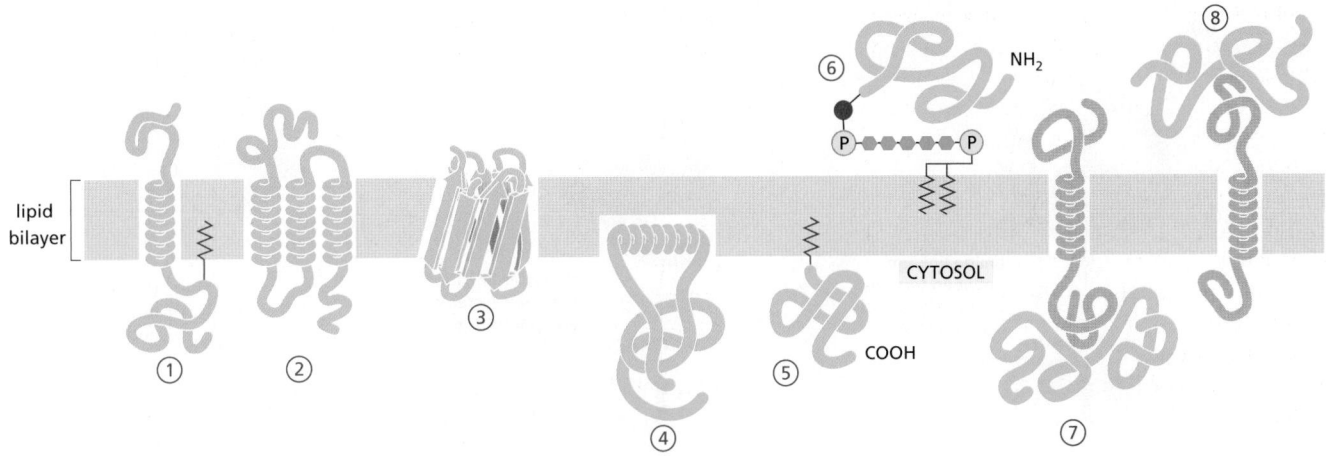

lipid bilayer

NH₂

CYTOSOL

COOH

① ② ③ ④ ⑤ ⑥ ⑦ ⑧

having hydrophobic and hydrophilic regions. Many membrane proteins extend through the lipid bilayer, and hence are called **transmembrane proteins**, with part of their mass on either side (Figure 10–17, examples 1, 2, and 3). Their hydrophobic regions pass through the membrane and interact with the hydrophobic tails of the lipid molecules in the interior of the bilayer, where they are sequestered away from water. Their hydrophilic regions are exposed to water on either side of the membrane. The covalent attachment of a fatty acid chain that inserts into the cytosolic monolayer of the lipid bilayer increases the hydrophobicity of some of these transmembrane proteins (see Figure 10–17, example 1).

Other membrane proteins are located entirely in the cytosol and are attached to the cytosolic monolayer of the lipid bilayer, either by an amphiphilic α helix exposed on the surface of the protein (Figure 10–17, example 4) or by one or more covalently attached lipid chains (Figure 10–17, example 5). Yet other membrane proteins are entirely exposed at the external cell surface, being attached to the lipid bilayer only by a covalent linkage (via a specific oligosaccharide) to a lipid anchor in the outer monolayer of the plasma membrane (Figure 10–17, example 6).

The lipid-linked proteins in example 5 in Figure 10–17 are made as soluble proteins in the cytosol and are subsequently anchored to the membrane by the covalent attachment of the lipid group. The proteins in example 6, however, are made as single-pass membrane proteins in the endoplasmic reticulum (ER). While still in the ER, the transmembrane segment of the protein is cleaved off and a **glycosylphosphatidylinositol (GPI) anchor** is added, leaving the protein bound to the noncytosolic surface of the ER membrane solely by this anchor (discussed in Chapter 12); transport vesicles eventually deliver the protein to the plasma membrane (discussed in Chapter 13).

By contrast to these examples, **membrane-associated proteins** do not extend into the hydrophobic interior of the lipid bilayer at all; they are instead bound to either face of the membrane by noncovalent interactions with other membrane proteins (Figure 10–17, examples 7 and 8). Many of the proteins of this type can be released from the membrane by relatively gentle extraction procedures, such as exposure to solutions of very high or low ionic strength or of extreme pH, which interfere with protein–protein interactions but leave the lipid bilayer intact; these proteins are often referred to as *peripheral membrane proteins*. Transmembrane proteins and many proteins held in the bilayer by lipid groups or hydrophobic polypeptide regions that insert into the hydrophobic core of the lipid bilayer cannot be released in these ways.

Lipid Anchors Control the Membrane Localization of Some Signaling Proteins

How a membrane protein is associated with the lipid bilayer reflects the function of the protein. Only transmembrane proteins can function on both sides of

Figure 10–17 Various ways in which proteins associate with the lipid bilayer. Most membrane proteins are thought to extend across the bilayer as (1) a single α helix, (2) as multiple α helices, or (3) as a rolled-up β sheet (a β barrel). Some of these "single-pass" and "multipass" proteins have a covalently attached fatty acid chain inserted in the cytosolic lipid monolayer (1). Other membrane proteins are exposed at only one side of the membrane. (4) Some of these are anchored to the cytosolic surface by an amphiphilic α helix that partitions into the cytosolic monolayer of the lipid bilayer through the hydrophobic face of the helix. (5) Others are attached to the bilayer solely by a covalently bound lipid chain—either a fatty acid chain or a prenyl group (see Figure 10–18)—in the cytosolic monolayer or, (6) via an oligosaccharide linker, to phosphatidylinositol in the noncytosolic monolayer—called a GPI anchor. (7, 8) Finally, membrane-associated proteins are attached to the membrane only by noncovalent interactions with other membrane proteins. The way in which the structure in (5) is formed is illustrated in Figure 10–18, while the way in which the GPI anchor shown in (6) is formed is illustrated in Figure 12–52. The details of how membrane proteins become associated with the lipid bilayer are discussed in Chapter 12.

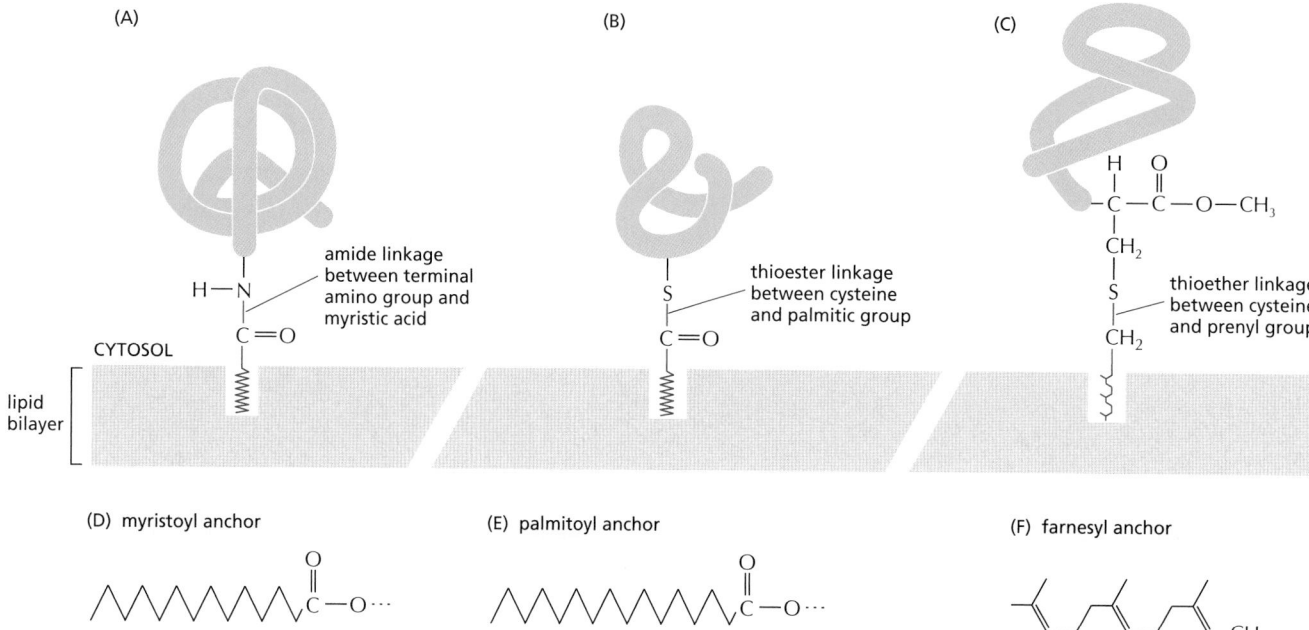

(A)

amide linkage
between terminal
amino group and
myristic acid

H—N

C=O

CYTOSOL

lipid
bilayer

(B)

thioester linkage
between cysteine
and palmitic group

S

C=O

(C)

$$-\overset{H}{\underset{|}{C}} - \overset{O}{\underset{||}{C}} - O - CH_3$$

CH₂

S

thioether linkage
between cysteine
and prenyl group

CH₂

(D) myristoyl anchor

O
‖
C—O···

(E) palmitoyl anchor

O
‖
C—O···

(F) farnesyl anchor

CH₂·····

the bilayer or transport molecules across it. Cell-surface receptors, for example, are usually transmembrane proteins that bind signal molecules in the extracellular space and generate different intracellular signals on the opposite side of the plasma membrane. To transfer small hydrophilic molecules across a membrane, a membrane transport protein must provide a path for the molecules to cross the hydrophobic permeability barrier of the lipid bilayer; the molecular architecture of multipass transmembrane proteins (Figure 10–17, examples 2 and 3) is ideally suited for this task, as we discuss in Chapter 11.

Proteins that function on only one side of the lipid bilayer, by contrast, are often associated exclusively with either the lipid monolayer or a protein domain on that side. Some intracellular signaling proteins, for example, that help relay extracellular signals into the cell interior are bound to the cytosolic half of the plasma membrane by one or more covalently attached lipid groups, which can be fatty acid chains or *prenyl groups* (Figure 10–18). In some cases, myristic acid, a saturated 14-carbon fatty acid, is added to the N-terminal amino group of the protein during its synthesis on a ribosome. All members of the *Src family* of cytoplasmic protein tyrosine kinases (discussed in Chapter 15) are myristoylated in this way. Membrane attachment through a single lipid anchor is not very strong, however, and a second lipid group is often added to anchor proteins more firmly to a membrane. For most Src kinases, the second lipid modification is the attachment of palmitic acid, a saturated 16-carbon fatty acid, to a cysteine side chain of the protein. This modification occurs in response to an extracellular signal and helps recruit the kinases to the plasma membrane. When the signaling pathway is turned off, the palmitic acid is removed, allowing the kinase to return to the cytosol. Other intracellular signaling proteins, such as the Ras family small GTPases (discussed in Chapter 15), use a combination of prenyl group and palmitic acid attachment to recruit the proteins to the plasma membrane.

Many proteins attach to membranes transiently. Some are classical peripheral membrane proteins that associate with membranes by regulated protein–protein interactions. Others undergo a transition from soluble to membrane protein by a conformational change that exposes a hydrophobic peptide or covalently attached lipid anchor. Many of the small GTPases of the Rab protein family that regulate intracellular membrane traffic (discussed in Chapter 13), for example, switch depending on the nucleotide that is bound to the protein. In their GDP-bound state they are soluble and free in the cytosol, whereas in their GTP-bound state their lipid anchor is exposed and tethers them to membranes. They are

Figure 10–18 Membrane protein attachment by a fatty acid chain or a prenyl group. The covalent attachment of either type of lipid can help localize a water-soluble protein to a membrane after its synthesis in the cytosol. (A) A fatty acid chain (myristic acid) is attached via an amide linkage to an N-terminal glycine. (B) A fatty acid chain (palmitic acid) is attached via a thioester linkage to a cysteine. (C) A prenyl chain (either farnesyl or a longer geranylgeranyl chain) is attached via a thioether linkage to a cysteine residue that is initially located four residues from the protein's C-terminus. After prenylation, the terminal three amino acids are cleaved off, and the new C-terminus is methylated before insertion of the anchor into the membrane (not shown). The structures of the lipid anchors are shown below: (D) a myristoyl anchor (derived from a 14-carbon saturated fatty acid chain), (E) a palmitoyl anchor (a 16-carbon saturated fatty acid chain), and (F) a farnesyl anchor (a 15-carbon unsaturated hydrocarbon chain).

Figure 10–19 **A segment of a membrane-spanning polypeptide chain crossing the lipid bilayer as an α helix.** Only the α-carbon backbone of the polypeptide chain is shown, with the hydrophobic amino acids in *green* and *yellow*. The polypeptide segment shown is part of the bacterial photosynthetic reaction center, the structure of which was determined by x-ray diffraction. (Based on data from J. Deisenhofer et al., *Nature* 318:618–624, 1985, and H. Michel et al., *EMBO J.* 5:1149–1158, 1986.)

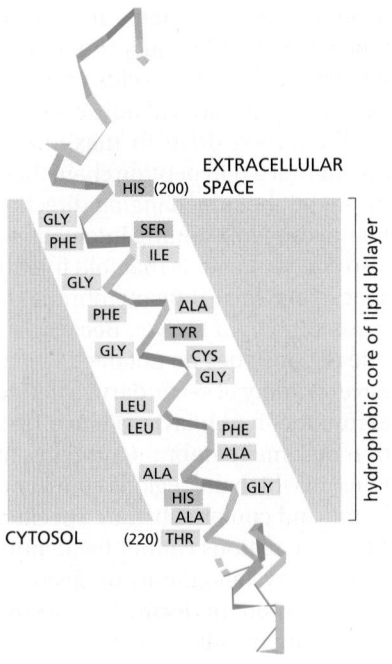

membrane proteins at one moment and soluble proteins at the next. Such highly dynamic interactions greatly expand the repertoire of membrane functions.

In Most Transmembrane Proteins, the Polypeptide Chain Crosses the Lipid Bilayer in an α-Helical Conformation

A transmembrane protein always has a unique orientation in the membrane. This reflects both the asymmetric manner in which it is inserted into the lipid bilayer in the ER during its biosynthesis (discussed in Chapter 12) and the different functions of its cytosolic and noncytosolic domains. These domains are separated by the membrane-spanning segments of the polypeptide chain, which contact the hydrophobic environment of the lipid bilayer and are composed largely of amino acids with nonpolar side chains. Because the peptide bonds themselves are polar and because water is absent, all peptide bonds in the bilayer are driven to form hydrogen bonds with one another. The hydrogen-bonding between peptide bonds is maximized if the polypeptide chain forms a regular α helix as it crosses the bilayer, and this is how most membrane-spanning segments of polypeptide chains traverse the bilayer (**Figure 10–19**).

In **single-pass transmembrane proteins**, the polypeptide chain crosses only once (see Figure 10–17, example 1), whereas in **multipass transmembrane proteins**, the polypeptide chain crosses multiple times (see Figure 10–17, example 2). An alternative way for the peptide bonds in the lipid bilayer to satisfy their hydrogen-bonding requirements is for multiple transmembrane strands of a polypeptide chain to be arranged as a β sheet that is rolled up into a cylinder (a so-called *β barrel*; see Figure 10–17, example 3). This protein architecture is seen in the *porin proteins* that we discuss later.

Progress in the x-ray crystallography of membrane proteins has enabled the determination of the three-dimensional structure of many of them. The structures confirm that it is often possible to predict from the protein's amino acid sequence which parts of the polypeptide chain extend across the lipid bilayer. Segments containing about 20–30 amino acids, with a high degree of hydrophobicity, are long enough to span a lipid bilayer as an α helix, and they can often be identified in *hydropathy plots* (**Figure 10–20**). From such plots, it is estimated that about 30%

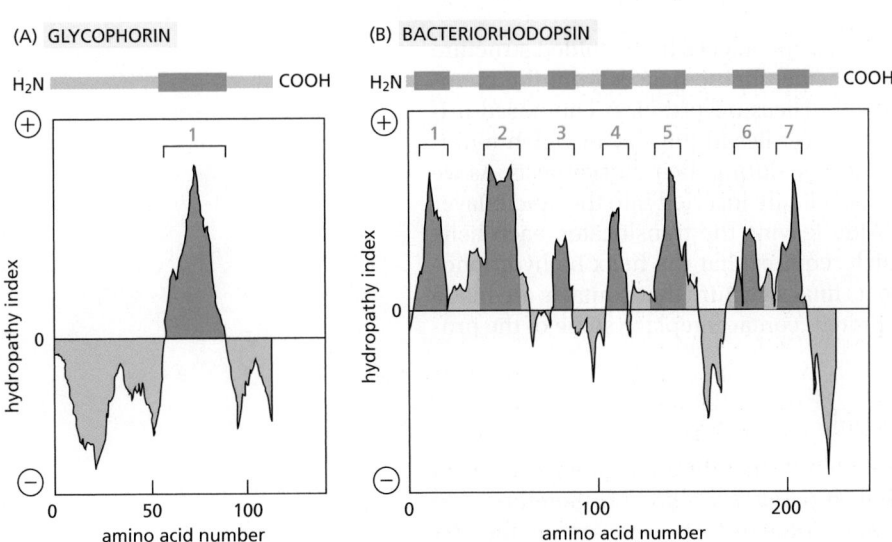

Figure 10–20 **Using hydropathy plots to localize potential α-helical membrane-spanning segments in a polypeptide chain.** The free energy needed to transfer successive segments of a polypeptide chain from a nonpolar solvent to water is calculated from the amino acid composition of each segment using data obtained from model compounds. This calculation is made for segments of a fixed size (usually around 10–20 amino acids), beginning with each successive amino acid in the chain. The "hydropathy index" of the segment is plotted on the Y axis as a function of its location in the chain. A positive value indicates that free energy is required for transfer to water (i.e., the segment is hydrophobic), and the value assigned is an index of the amount of energy needed. Peaks in the hydropathy index appear at the positions of hydrophobic segments in the amino acid sequence. (A and B) Hydropathy plots for two membrane proteins that are discussed later in this chapter. Glycophorin (A) has a single membrane-spanning α helix, and one corresponding peak in the hydropathy plot. Bacteriorhodopsin (B) has seven membrane-spanning α helices and seven corresponding peaks in the hydropathy plot. (A, adapted from D. Eisenberg, *Annu. Rev. Biochem.* 53:595–624, 1984.)

of an organism's proteins are transmembrane proteins, emphasizing their importance. Hydropathy plots cannot identify the membrane-spanning segments of a β barrel, as 10 amino acids or fewer are sufficient to traverse a lipid bilayer as an extended β strand and only every other amino acid side chain is hydrophobic.

The strong drive to maximize hydrogen-bonding in the absence of water means that a polypeptide chain that enters the lipid bilayer is likely to pass entirely through it before changing direction, since chain bending requires a loss of regular hydrogen-bonding interactions. But multipass transmembrane proteins can also contain regions that fold into the membrane from either side, squeezing into spaces between transmembrane α helices without contacting the hydrophobic core of the lipid bilayer. Because such regions interact only with other polypeptide regions, they do not need to maximize hydrogen-bonding; they can therefore have a variety of secondary structures, including helices that extend only part way across the lipid bilayer (**Figure 10–21**). Such regions are important for the function of some membrane proteins, including water channel and ion channel proteins, in which the regions contribute to the walls of the pores traversing the membrane and confer substrate specificity on the channels, as we discuss in Chapter 11. These regions cannot be identified in hydropathy plots and are only revealed by x-ray crystallography or electron crystallography (a technique similar to x-ray diffraction but performed on two-dimensional arrays of proteins) of the protein's three-dimensional structure.

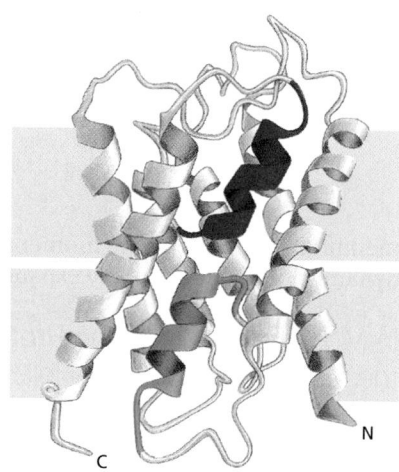

Figure 10–21 **Two short α helices in the aquaporin water channel, each of which spans only halfway through the lipid bilayer.** In the plasma membrane, four monomers, one of which is shown here, form a tetramer. Each monomer has a hydrophilic pore at its center, which allows water molecules to cross the membrane in single file (see Figure 11–20 and Movie 11.6). The two short colored helices are buried at an interface formed by protein–protein interactions. The mechanism by which the channel allows the passage of water molecules is discussed in more detail in Chapter 11.

Transmembrane α Helices Often Interact with One Another

The transmembrane α helices of many single-pass membrane proteins do not contribute to the folding of the protein domains on either side of the membrane. As a consequence, it is often possible to engineer cells to produce just the cytosolic or extracellular domains of these proteins as water-soluble molecules. This approach has been invaluable for studying the structure and function of these domains, especially the domains of transmembrane receptor proteins (discussed in Chapter 15). A transmembrane α helix, even in a single-pass membrane protein, however, often does more than just anchor the protein to the lipid bilayer. Many single-pass membrane proteins form homo- or heterodimers that are held together by noncovalent, but strong and highly specific, interactions between the two transmembrane α helices; the sequence of the hydrophobic amino acids of these helices contains the information that directs the protein–protein interaction.

Similarly, the transmembrane α helices in multipass membrane proteins occupy specific positions in the folded protein structure that are determined by interactions between the neighboring helices. These interactions are crucial for the structure and function of the many channels and transporters that move molecules across cell membranes.

In these proteins, neighboring transmembrane helices in the folded structure of the protein shield many of the other transmembrane helices from the membrane lipids. Why, then, are these shielded helices nevertheless composed primarily of hydrophobic amino acids? The answer lies in the way in which multipass proteins are integrated into the membrane during their biosynthesis. As we discuss in Chapter 12, transmembrane α helices are inserted into the lipid bilayer sequentially by a protein translocator. After leaving the translocator, each helix is transiently surrounded by lipids, which requires that the helix be hydrophobic. It is only as the protein folds up into its final structure that contacts are made between adjacent helices, and protein–protein contacts replace some of the protein–lipid contacts (**Figure 10–22**).

Some β Barrels Form Large Channels

Multipass membrane proteins that have their transmembrane segments arranged as *β barrels* rather than as α helices are comparatively rigid and therefore tend to form crystals readily when isolated. Thus, some of them were among the first

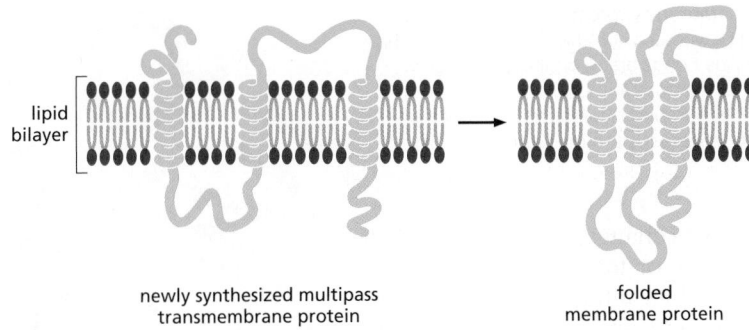

lipid bilayer

newly synthesized multipass transmembrane protein

folded membrane protein

Figure 10–22 **Steps in the folding of a multipass transmembrane protein.** When a newly synthesized transmembrane α helix is released into the lipid bilayer, it is initially surrounded by lipid molecules. As the protein folds, contacts between the helices displace some of the lipid molecules surrounding the helices.

multipass membrane protein structures to be determined by x-ray crystallography. The number of β strands in a β barrel varies widely, from as few as 8 strands to as many as 22 (**Figure 10–23**).

β-barrel proteins are abundant in the outer membranes of bacteria, mitochondria, and chloroplasts. Some are pore-forming proteins, which create water-filled channels that allow selected small hydrophilic molecules to cross the membrane. The porins are well-studied examples (example 3 in Figure 10–23C). Many porin barrels are formed from a 16-strand, antiparallel β sheet rolled up into a cylindrical structure. Polar amino acid side chains line the aqueous channel on the inside, while nonpolar side chains project from the outside of the barrel to interact with the hydrophobic core of the lipid bilayer. Loops of the polypeptide chain often protrude into the **lumen** of the channel, narrowing it so that only certain solutes can pass. Some porins are therefore highly selective: *maltoporin*, for example, preferentially allows maltose and maltose oligomers to cross the outer membrane of *E. coli*.

The *FepA protein* is a more complex example of a β barrel transport protein (Figure 10–23D). It transports iron ions across the bacterial outer membrane. It is constructed from 22 β strands, and a large globular domain completely fills the inside of the barrel. Iron ions bind to this domain, which by an unknown mechanism moves or changes its conformation to transfer the iron across the membrane.

Not all β-barrel proteins are transport proteins. Some form smaller barrels that are completely filled by amino acid side chains that project into the center of the barrel. These proteins function as receptors or enzymes (Figure 10–23A and B); the barrel serves as a rigid anchor, which holds the protein in the membrane and orients the cytosolic loops that form binding sites for specific intracellular molecules.

Most multipass membrane proteins in eukaryotic cells and in the bacterial plasma membrane are constructed from transmembrane α helices. The helices

Figure 10–23 **β barrels formed from different numbers of β strands.** (A) The *E. coli* OmpA protein serves as a receptor for a bacterial virus. (B) The *E. coli* OMPLA protein is an enzyme (a lipase) that hydrolyzes lipid molecules. The amino acids that catalyze the enzymatic reaction (shown in *red*) protrude from the outside surface of the barrel. (C) A porin from the bacterium *Rhodobacter capsulatus* forms a water-filled pore across the outer membrane. The diameter of the channel is restricted by loops (shown in *blue*) that protrude into the channel. (D) The *E. coli* FepA protein transports iron ions. The inside of the barrel is completely filled by a globular protein domain (shown in *blue*) that contains an iron-binding site (not shown).

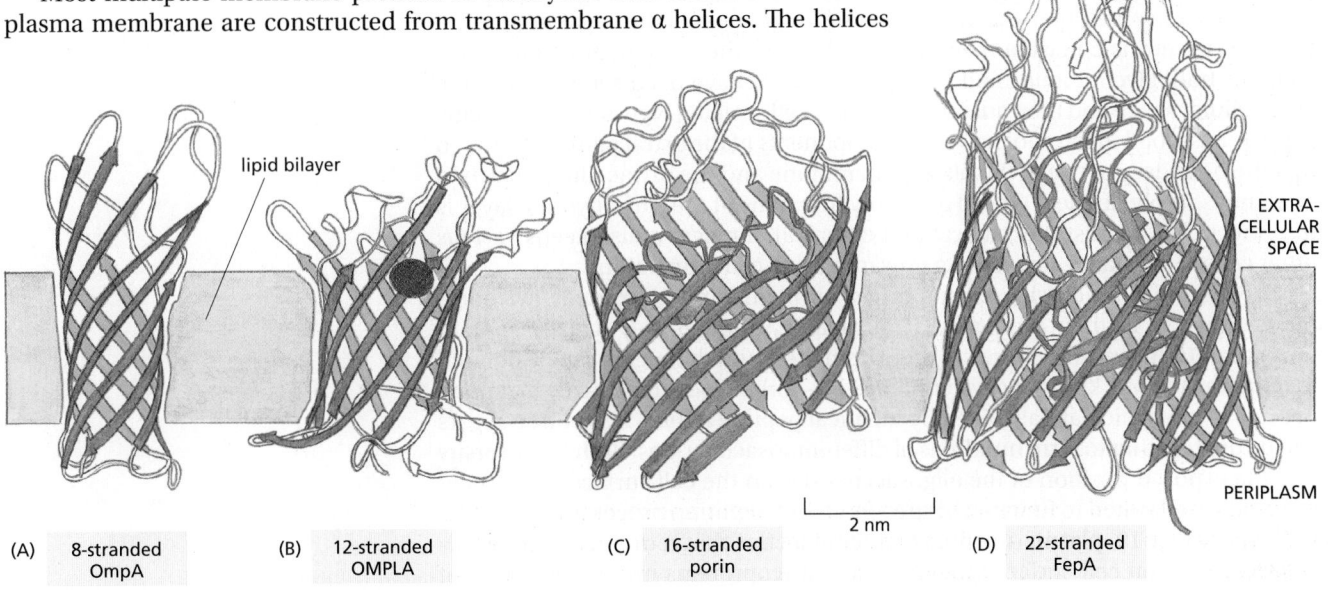

lipid bilayer

EXTRA-CELLULAR SPACE

PERIPLASM

2 nm

(A) 8-stranded OmpA

(B) 12-stranded OMPLA

(C) 16-stranded porin

(D) 22-stranded FepA

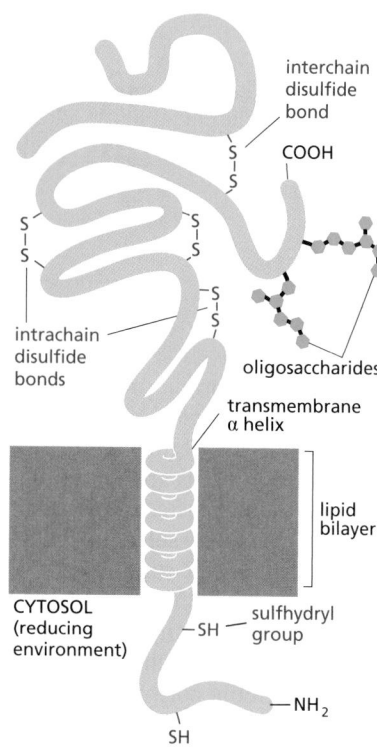

Figure 10–24 A single-pass transmembrane protein. Note that the polypeptide chain traverses the lipid bilayer as a right-handed α helix and that the oligosaccharide chains and disulfide bonds are all on the noncytosolic surface of the membrane. The sulfhydryl groups in the cytosolic domain of the protein do not normally form disulfide bonds because the reducing environment in the cytosol maintains these groups in their reduced (–SH) form.

can slide against each other, allowing conformational changes in the protein that can open and shut ion channels, transport solutes, or transduce extracellular signals into intracellular ones. In β-barrel proteins, by contrast, hydrogen bonds bind each β strand rigidly to its neighbors, making conformational changes within the wall of the barrel unlikely.

Many Membrane Proteins Are Glycosylated

Most transmembrane proteins in animal cells are glycosylated. As in glycolipids, the sugar residues are added in the lumen of the ER and the Golgi apparatus (discussed in Chapters 12 and 13). For this reason, the oligosaccharide chains are always present on the noncytosolic side of the membrane. Another important difference between proteins (or parts of proteins) on the two sides of the membrane results from the reducing environment of the cytosol. This environment decreases the likelihood that intrachain or interchain disulfide (S–S) bonds will form between cysteines on the cytosolic side of membranes. These bonds form on the noncytosolic side, where they can help stabilize either the folded structure of the polypeptide chain or its association with other polypeptide chains (**Figure 10–24**).

Because the extracellular part of most plasma membrane proteins are glycosylated, carbohydrates extensively coat the surface of all eukaryotic cells. These carbohydrates occur as oligosaccharide chains covalently bound to membrane proteins (glycoproteins) and lipids (glycolipids). They also occur as the polysaccharide chains of integral membrane *proteoglycan* molecules. Proteoglycans, which consist of long polysaccharide chains linked covalently to a protein core, are found mainly outside the cell, as part of the extracellular matrix (discussed in Chapter 19). But, for some proteoglycans, the protein core either extends across the lipid bilayer or is attached to the bilayer by a glycosylphosphatidylinositol (GPI) anchor.

The terms cell coat or glycocalyx are sometimes used to describe the carbohydrate-rich zone on the cell surface. This **carbohydrate layer** can be visualized by various stains, such as ruthenium red (**Figure 10–25A**), as well as by its affinity for carbohydrate-binding proteins called **lectins**, which can be labeled with a fluorescent dye or some other visible marker. Although most of the sugar groups are attached to intrinsic plasma membrane molecules, the carbohydrate layer also contains both glycoproteins and proteoglycans that have been secreted into the extracellular space and then adsorbed onto the cell surface (Figure 10–25B). Many of these adsorbed macromolecules are components of the extracellular matrix, so that the boundary between the plasma membrane and the extracellular matrix is often not sharply defined. One of the many functions of the carbohydrate layer is to protect cells against mechanical and chemical damage; it also keeps various other cells at a distance, preventing unwanted cell–cell interactions.

The oligosaccharide side chains of glycoproteins and glycolipids are enormously diverse in their arrangement of sugars. Although they usually contain fewer than 15 sugars, the chains are often branched, and the sugars can be bonded together by various kinds of covalent linkages—unlike the amino acids in a polypeptide chain, which are all linked by identical peptide bonds. Even three sugars can be put together to form hundreds of different trisaccharides. Both the diversity and the exposed position of the oligosaccharides on the cell surface make them especially well suited to function in specific cell-recognition processes. As we discuss in Chapter 19, plasma-membrane-bound lectins that recognize specific oligosaccharides on cell-surface glycolipids and glycoproteins mediate a variety of

(A)

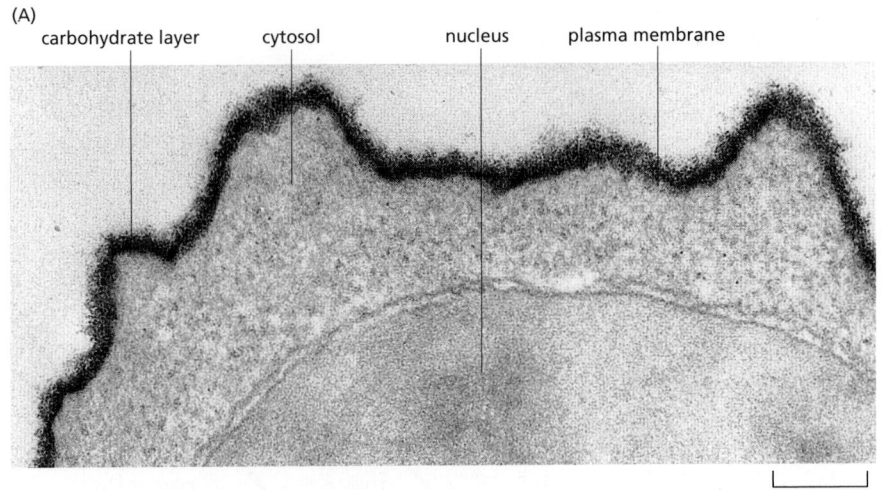

carbohydrate layer   cytosol   nucleus   plasma membrane

200 nm

Figure 10–25 **The carbohydrate layer on the cell surface.** (A) This electron micrograph of the surface of a lymphocyte stained with ruthenium red emphasizes the thick carbohydrate-rich layer surrounding the cell. (B) The carbohydrate layer is made up of the oligosaccharide side chains of membrane glycolipids and membrane glycoproteins and the polysaccharide chains on membrane proteoglycans. In addition, adsorbed glycoproteins, and adsorbed proteoglycans (not shown), contribute to the carbohydrate layer in many cells. Note that all of the carbohydrate is on the extracellular surface of the membrane. (A, courtesy of Audrey M. Glauert and G.M.W. Cook.)

(B)

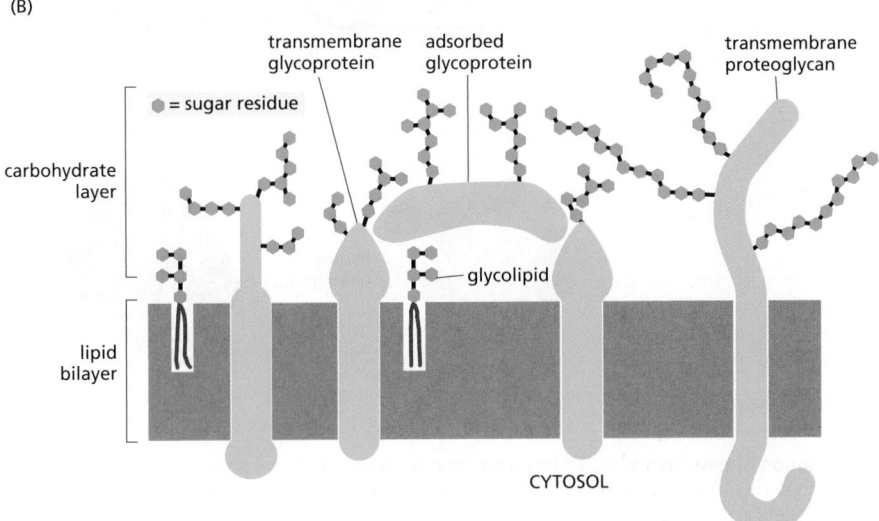

transmembrane glycoprotein   adsorbed glycoprotein   transmembrane proteoglycan

= sugar residue

carbohydrate layer

glycolipid

lipid bilayer

CYTOSOL

transient cell–cell adhesion processes, including those occurring in lymphocyte recirculation and inflammatory responses (see Figure 19–28).

Membrane Proteins Can Be Solubilized and Purified in Detergents

In general, only agents that disrupt hydrophobic associations and destroy the lipid bilayer can solubilize membrane proteins. The most useful of these for the membrane biochemist are **detergents**, which are small amphiphilic molecules of variable structure (**Movie 10.4**). Detergents are much more soluble in water than lipids. Their polar (hydrophilic) ends can be either charged (ionic), as in *sodium dodecyl sulfate* (*SDS*), or uncharged (nonionic), as in *octylglucoside* and Triton (**Figure 10–26**A). At low concentration, detergents are monomeric in solution, but when their concentration is increased above a threshold, called the *critical micelle concentration* (*CMC*), they aggregate to form micelles (Figure 10–26B–D). Above the CMC, detergent molecules rapidly diffuse in and out of micelles, keeping the concentration of monomer in the solution constant, no matter how many micelles are present. Both the CMC and the average number of detergent molecules in a micelle are characteristic properties of each detergent, but they also depend on the temperature, pH, and salt concentration. Detergent solutions are therefore complex systems and are difficult to study.

When mixed with membranes, the hydrophobic ends of detergents bind to the hydrophobic regions of the membrane proteins, where they displace lipid molecules with a collar of detergent molecules. Since the other end of the detergent

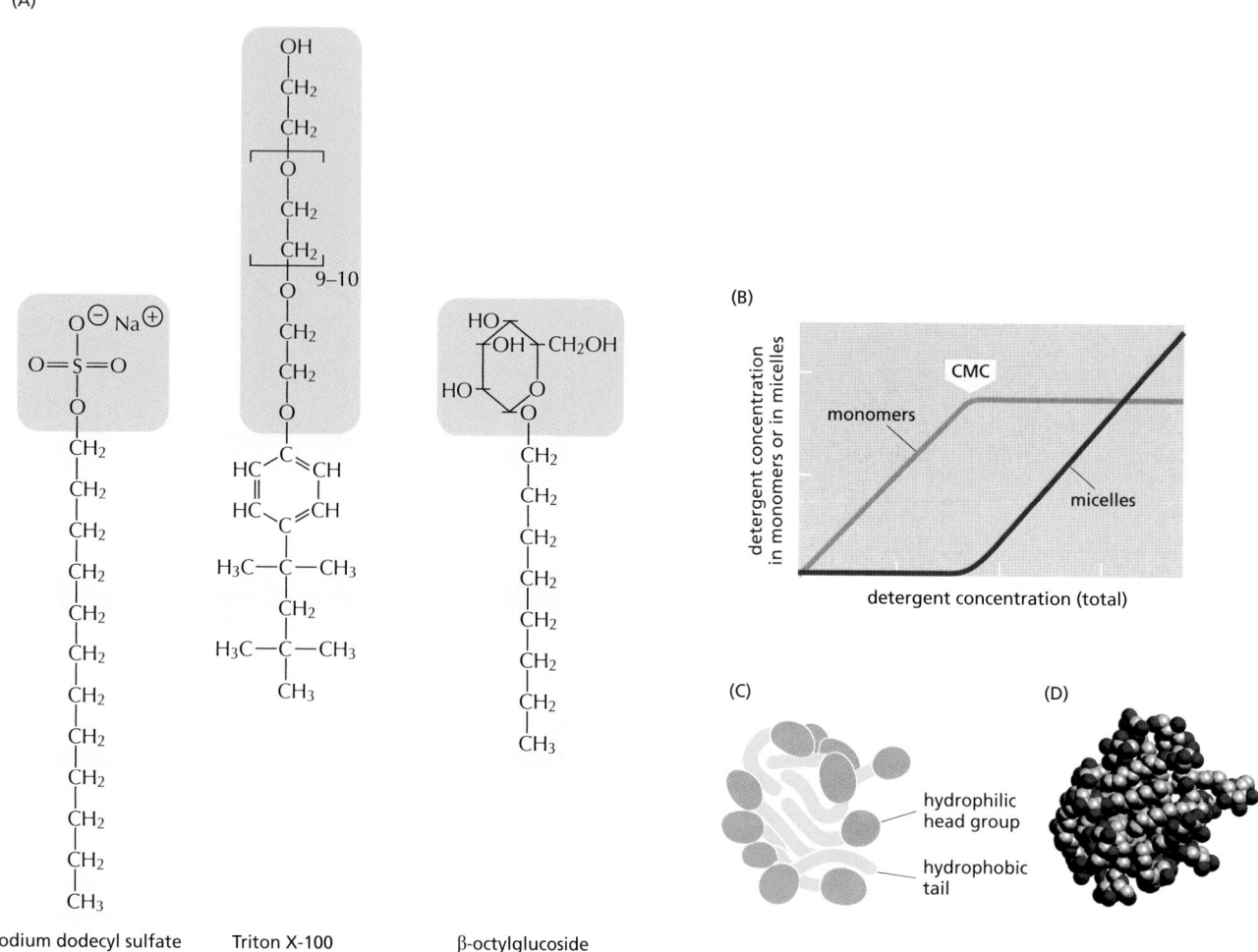

Figure 10–26 The structure and function of detergents. (A) Three commonly used detergents are sodium dodecyl sulfate (SDS), an anionic detergent, and Triton X-100 and β-octylglucoside, two nonionic detergents. Triton X-100 is a mixture of compounds in which the region in brackets is repeated between 9 and 10 times. The hydrophobic portion of each detergent is shown in *yellow*, and the hydrophilic portion is shown in *orange*. (B) At low concentration, detergent molecules are monomeric in solution. As their concentration is increased beyond the critical micelle concentration (CMC), some of the detergent molecules form micelles. Note that the concentration of detergent monomer stays constant above the CMC. (C) Because they have both polar and nonpolar ends, detergent molecules are amphiphilic; and because they are cone-shaped, they form micelles rather than bilayers (see Figure 10–7). Detergent micelles are thought to have irregular shapes, and, due to packing constraints, the hydrophobic tails are partially exposed to water. (D) The space-filling model shows the structure of a micelle composed of 20 β-octylglucoside molecules, predicted by molecular dynamics calculations. The head groups are shown in *red* and the hydrophobic tails in *gray*. (B, adapted with permission from G. Gunnarsson, B. Jönsson and H. Wennerström, *J. Phys. Chem. A* 84:3114–3121, 1980. Copyright 1980 American Chemical Society; C, from S. Bogusz, R. M. Venable and R. W. Pastor, *J. Phys. Chem. B* 104:5462–5470, 2000.)

molecule is polar, this binding tends to bring the membrane proteins into solution as detergent–protein complexes (**Figure 10–27**). Usually, some lipid molecules also remain attached to the protein.

Strong ionic detergents, such as SDS, can solubilize even the most hydrophobic membrane proteins. This allows the proteins to be analyzed by *SDS polyacrylamide-gel electrophoresis* (discussed in Chapter 8), a procedure that has revolutionized the study of proteins. Such strong detergents, however, unfold (denature) proteins by binding to their internal "hydrophobic cores," thereby rendering the proteins inactive and unusable for functional studies. Nonetheless, proteins can be readily separated and purified in their SDS-denatured form. In some cases, removal of the SDS allows the purified protein to renature, with recovery of functional activity.

Many membrane proteins can be solubilized and then purified in an active form by the use of mild detergents. These detergents cover the hydrophobic regions on membrane-spanning segments that become exposed after lipid

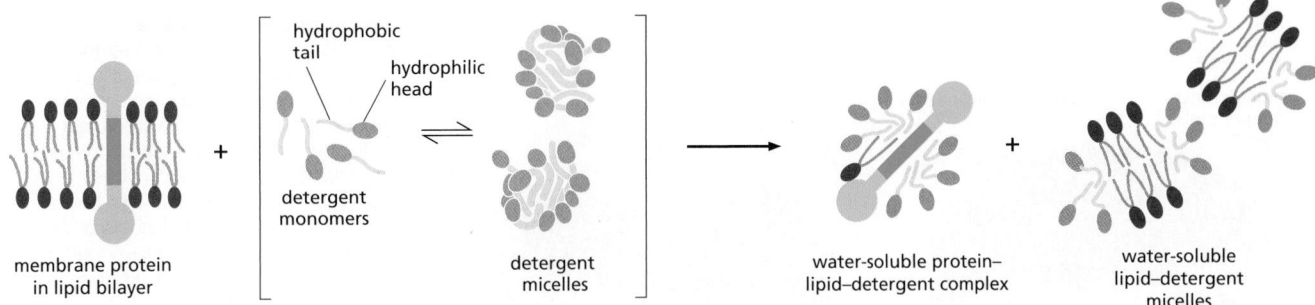

removal but do not unfold the protein. If the detergent concentration of a solution of solubilized membrane proteins is reduced (by dilution, for example), membrane proteins do not remain soluble. In the presence of an excess of phospholipid molecules in such a solution, however, membrane proteins incorporate into small liposomes that form spontaneously. In this way, functionally active membrane protein systems can be reconstituted from purified components, providing a powerful means of analyzing the activities of membrane transporters, ion channels, signaling receptors, and so on (**Figure 10–28**). Such functional reconstitution, for example, provided proof for the hypothesis that the enzymes that make

Figure 10–27 Solubilizing a membrane protein with a mild nonionic detergent. The detergent disrupts the lipid bilayer and brings the protein into solution as protein–lipid–detergent complexes. The phospholipids in the membrane are also solubilized by the detergent, as lipid-detergent micelles.

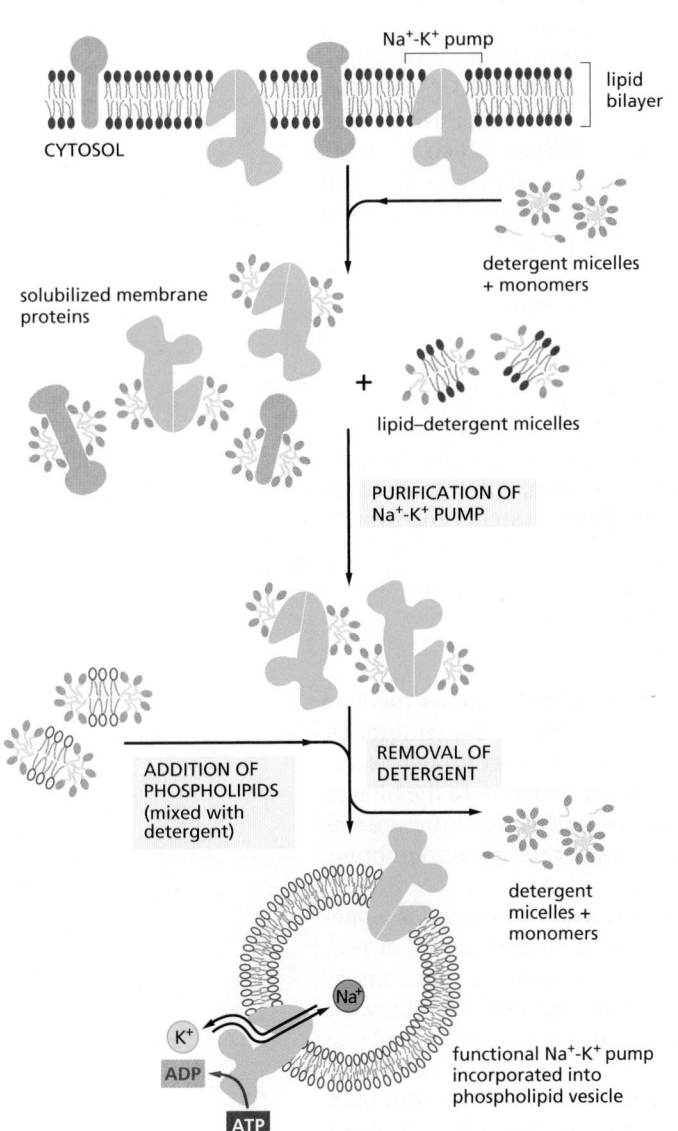

Figure 10–28 The use of mild nonionic detergents for solubilizing, purifying, and reconstituting functional membrane protein systems. In this example, functional Na+-K+ pump molecules are purified and incorporated into phospholipid vesicles. This pump is present in the plasma membrane of most animal cells, where it uses the energy of ATP hydrolysis to pump Na+ out of the cell and K+ in, as discussed in Chapter 11.

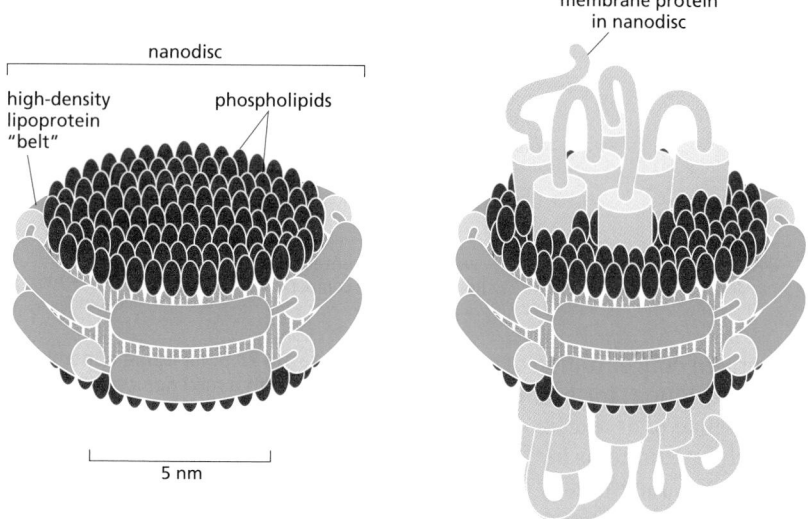

nanodisc

high-density
lipoprotein
"belt"

phospholipids

membrane protein
in nanodisc

5 nm

Figure 10–29 **Model of a membrane protein reconstituted into a nanodisc.** When detergent is removed from a solution containing a multipass membrane protein, lipids, and a protein subunit of the high-density lipoprotein (HDL), the membrane protein becomes embedded in a small patch of lipid bilayer, which is surrounded by a belt of the HDL protein. In such nanodiscs, the hydrophobic edges of the bilayer patch are shielded by the protein belt, which renders the assembly water-soluble.

ATP (ATP synthases) use H^+ gradients in mitochondrial, chloroplast, and bacterial membranes to produce ATP.

Membrane proteins can also be reconstituted from detergent solution into nanodiscs, which are small, uniformly sized patches of membrane that are surrounded by a belt of protein, which covers the exposed edge of the bilayer to keep the patch in solution (Figure 10–29). The belt is derived from high-density lipoproteins (HDL), which keep lipids soluble for transport in the blood. In nanodiscs the membrane protein of interest can be studied in its native lipid environment and is experimentally accessible from both sides of the bilayer, which is useful, for example, for ligand-binding experiments. Proteins contained in nanodiscs can also be analyzed by single particle electron microscopy techniques to determine their structure. By this rapidly improving technique (discussed in Chapter 9), the structure of a membrane protein can be determined to high resolution without a requirement of the protein of interest to crystallize into a regular lattice, which is often hard to achieve for membrane proteins.

Detergents have also played a crucial part in the purification and crystallization of membrane proteins. The development of new detergents and new expression systems that produce large quantities of membrane proteins from cDNA clones has led to a rapid increase in the number of three-dimensional structures of membrane proteins and protein complexes that are known, although they are still few compared to the known structures of water-soluble proteins and protein complexes.

Bacteriorhodopsin Is a Light-driven Proton (H^+) Pump That Traverses the Lipid Bilayer as Seven α Helices

In Chapter 11, we consider how multipass transmembrane proteins mediate the selective transport of small hydrophilic molecules across cell membranes. But a detailed understanding of how such a membrane transport protein works requires precise information about its three-dimensional structure in the bilayer. *Bacteriorhodopsin* was the first membrane transport protein whose structure was determined, and it has remained the prototype of many multipass membrane proteins with a similar structure.

The "purple membrane" of the archaeon *Halobacterium salinarum* is a specialized patch in the plasma membrane that contains a single species of protein molecule, **bacteriorhodopsin** (Figure 10–30A). The protein functions as a light-activated H^+ pump that transfers H^+ out of the archaeal cell. Because the bacteriorhodopsin molecules are tightly packed and arranged as a planar two-dimensional crystal (FIgure 10–30B and C), it was possible to determine their three-dimensional structure by combining electron microscopy and electron diffraction analysis—a procedure called *electron crystallography*, which we

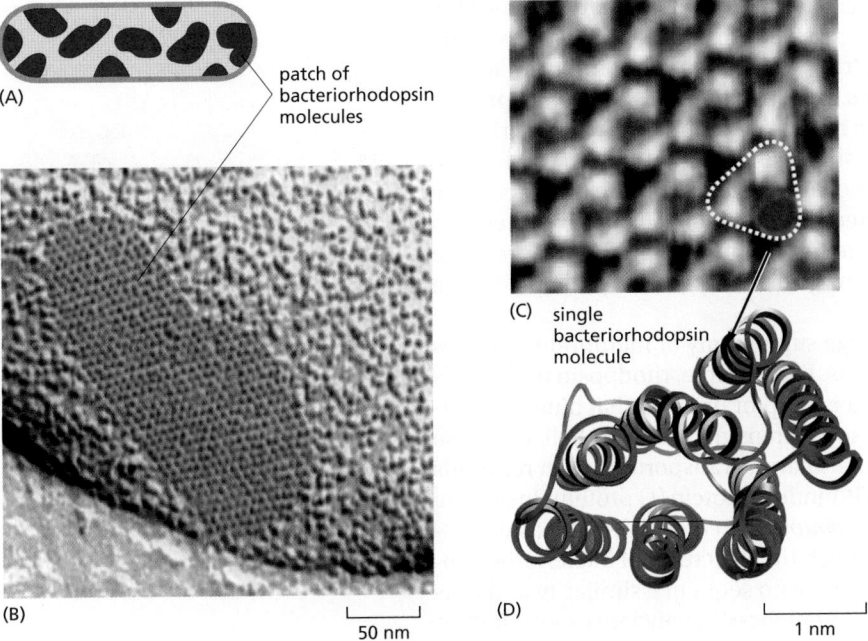

(A)

patch of bacteriorhodopsin molecules

(B)

50 nm

(C) single bacteriorhodopsin molecule

(D)

1 nm

Figure 10–30 **Patches of purple membrane, which contain bacteriorhodopsin in the archaeon *Halobacterium salinarum*.** (A) These archaea live in saltwater pools, where they are exposed to sunlight. They have evolved a variety of light-activated proteins, including bacteriorhodopsin, which is a light-activated H$^+$ pump in the plasma membrane. (B) The bacteriorhodopsin molecules in the purple membrane patches are tightly packed into two-dimensional crystalline arrays. (C) Details of the molecular surface visualized by atomic force microscopy. With this technique, individual bacteriorhodopsin molecules can be seen. (D) Outline of the approximate location of the bacteriorhodopsin monomer and the individual α helices in the image shown in (C). (B–C, courtesy of Dieter Oesterhelt; D, PDB code: 2BRD.)

mentioned earlier. This method has provided the first structural views of many membrane proteins that were found to be difficult to crystallize from detergent solutions. For bacteriorhodopsin, the structure was later confirmed and extended to very high resolution by x-ray crystallography.

Each bacteriorhodopsin molecule is folded into seven closely packed transmembrane α helices and contains a single light-absorbing group, or chromophore (in this case, *retinal*), which gives the protein its purple color. Retinal is vitamin A in its aldehyde form and is identical to the chromophore found in *rhodopsin* of the photoreceptor cells of the vertebrate eye (discussed in Chapter 15). Retinal is covalently linked to a lysine side chain of the bacteriorhodopsin protein. When activated by a single photon of light, the excited chromophore changes its shape and causes a series of small conformational changes in the protein, resulting in the transfer of one H$^+$ from the inside to the outside of the cell (**Figure 10–31A**). In bright light, each bacteriorhodopsin molecule can pump several hundred protons per second. The light-driven proton transfer establishes an H$^+$ gradient across the plasma membrane, which in turn drives the production of ATP by a second protein in the cell's plasma membrane. The energy stored in the H$^+$ gradient also drives other energy-requiring processes in the cell. Thus, bacteriorhodopsin converts solar energy into a H$^+$ gradient, which provides energy to the archaeal cell.

The high-resolution crystal structure of bacteriorhodopsin reveals many lipid molecules bound in specific places on the protein surface (Figure 10–31B).

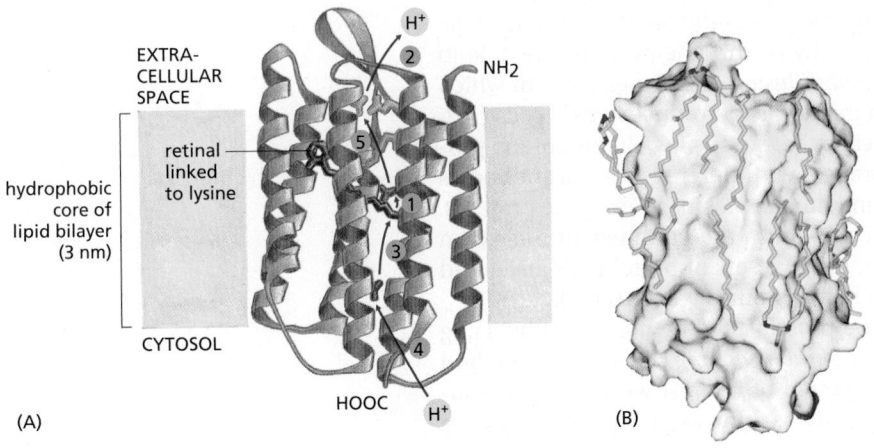

EXTRA-CELLULAR SPACE

NH$_2$

retinal linked to lysine

hydrophobic core of lipid bilayer (3 nm)

CYTOSOL

HOOC

H$^+$

(A)

(B)

Figure 10–31 **The three-dimensional structure of a bacteriorhodopsin molecule.** (Movie 10.5) (A) The polypeptide chain crosses the lipid bilayer seven times as α helices. The location of the retinal chromophore *(purple)* and the probable pathway taken by H$^+$ during the light-activated pumping cycle are shown. The first and key step is the passing of an H$^+$ from the chromophore to the side chain of aspartic acid 85 *(red, located next to the chromophore)* that occurs upon absorption of a photon by the chromophore. Subsequently, other H$^+$ transfers—in the numerical order indicated and utilizing the hydrophilic amino acid side chains that line a path through the membrane—complete the pumping cycle and return the enzyme to its starting state. Color code: glutamic acid *(orange)*, aspartic acid *(red)*, arginine *(blue)*. (B) The high-resolution crystal structure of bacteriorhodopsin shows many lipid molecules *(yellow* with *red* head groups) that are tightly bound to specific places on the surface of the protein. (A, adapted from H. Luecke et al., *Science* 286:255–261, 1999; B, from H. Luecke et al., *J. Mol. Biol.* 291:899–911, 1999. With permission from Academic Press.)

Interactions with specific lipids are thought to help stabilize many membrane proteins, which work best and sometimes crystallize more readily if some of the lipids remain bound during detergent extraction, or if specific lipids are added back to the proteins in detergent solutions. The specificity of these lipid–protein interactions helps explain why eukaryotic membranes contain such a variety of lipids, with head groups that differ in size, shape, and charge. We can think of the membrane lipids as constituting a two-dimensional solvent for the proteins in the membrane, just as water constitutes a three-dimensional solvent for proteins in an aqueous solution: some membrane proteins can function only in the presence of specific lipid head groups, just as many enzymes in aqueous solution require a particular ion for activity.

Bacteriorhodopsin is a member of a large superfamily of membrane proteins with similar structures but different functions. For example, rhodopsin in rod cells of the vertebrate retina and many cell-surface receptor proteins that bind extracellular signal molecules are also built from seven transmembrane α helices. These proteins function as signal transducers rather than as transporters: each responds to an extracellular signal by activating a GTP-binding protein (G protein) inside the cell and they are therefore called *G-protein-coupled receptors* (*GPCRs*), as we discuss in Chapter 15 (see Figure 15–6B). Although the structures of bacteriorhodopsins and GPCRs are strikingly similar, they show no sequence similarity and thus probably belong to two evolutionarily distant branches of an ancient protein family. A related class of membrane proteins, the *channelrhodopsins* that green algae use to detect light, form ion channels when they absorb a photon. When engineered so that they are expressed in animal brains, these proteins have become invaluable tools in neurobiology because they allow specific neurons to be stimulated experimentally by shining light on them, as we discuss in Chapter 11 (Figure 11–32).

Membrane Proteins Often Function as Large Complexes

Many membrane proteins function as part of multicomponent complexes, several of which have been studied by x-ray crystallography. One is a bacterial *photosynthetic reaction center*, which was the first membrane protein complex to be crystallized and analyzed by x-ray diffraction. In Chapter 14, we discuss how such photosynthetic complexes function to capture light energy and use it to pump H^+ across the membrane. Many of the membrane protein complexes involved in photosynthesis, proton pumping, and electron transport are even larger than the photosynthetic reaction center. The enormous photosystem II complex from cyanobacteria, for example, contains 19 protein subunits and well over 60 transmembrane helices (see Figure 14–49). Membrane proteins are often arranged in large complexes, not only for harvesting various forms of energy, but also for transducing extracellular signals into intracellular ones (discussed in Chapter 15).

Many Membrane Proteins Diffuse in the Plane of the Membrane

Like most membrane lipids, membrane proteins do not tumble (*flip-flop*) across the lipid bilayer, but they do rotate about an axis perpendicular to the plane of the bilayer (*rotational diffusion*). In addition, many membrane proteins are able to move laterally within the membrane (*lateral diffusion*). An experiment in which mouse cells were artificially fused with human cells to produce hybrid cells (*heterokaryons*) provided the first direct evidence that some plasma membrane proteins are mobile in the plane of the membrane. Two differently labeled antibodies were used to distinguish selected mouse and human plasma membrane proteins. Although at first the mouse and human proteins were confined to their own halves of the newly formed heterokaryon, the two sets of proteins diffused and mixed over the entire cell surface in about half an hour (**Figure 10–32**).

The lateral diffusion rates of membrane proteins can be measured by using the technique of *fluorescence recovery after photobleaching* (*FRAP*). The method usually involves marking the membrane protein of interest with a specific fluorescent group. This can be done either with a fluorescent ligand such as a

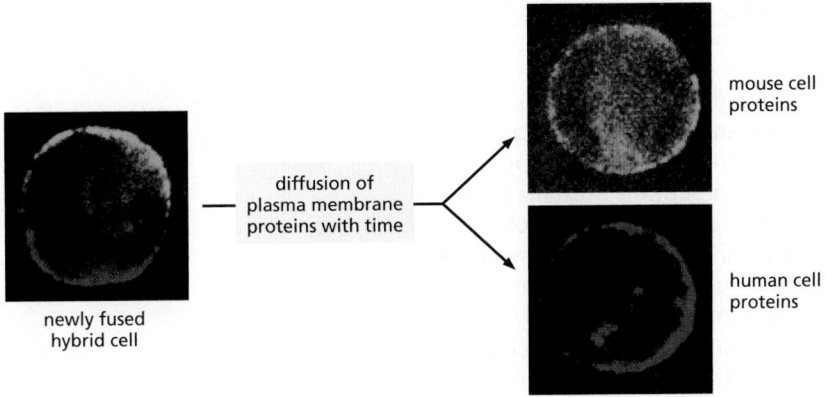

newly fused
hybrid cell

diffusion of
plasma membrane
proteins with time

mouse cell
proteins

human cell
proteins

Figure 10–32 An experiment demonstrating the diffusion of proteins in the plasma membrane of mouse–human hybrid cells. In this experiment, a mouse and a human cell were fused to create a hybrid cell, which was then stained with two fluorescently labeled antibodies. One antibody (labeled with a *green* dye) detects mouse plasma membrane proteins, the other antibody (labeled with a *red* dye) detects human plasma membrane proteins. When cells were stained immediately after fusion, mouse and human plasma membrane proteins are still found in the membrane domains originating from the mouse and human cell, respectively. After a short time, however, the plasma membrane proteins diffuse over the entire cell surface and completely intermix. (From L.D. Frye and M. Edidin, *J. Cell Sci.* 7:319–335, 1970. With permission from The Company of Biologists.)

fluorophore-labeled antibody that binds to the protein or with recombinant DNA technology to express the protein fused to a fluorescent protein such as green fluorescent protein (GFP) (discussed in Chapter 9). The fluorescent group is then bleached in a small area of membrane by a laser beam, and the time taken for adjacent membrane proteins carrying unbleached ligand or GFP to diffuse into the bleached area is measured (**Figure 10–33**). From FRAP measurements, we can estimate the diffusion coefficient for the marked cell-surface protein. The values of the diffusion coefficients for different membrane proteins in different cells are highly variable, because interactions with other proteins impede the diffusion of the proteins to varying degrees. Measurements of proteins that are minimally impeded in this way indicate that cell membranes have a viscosity comparable to that of olive oil.

One drawback to the FRAP technique is that it monitors the movement of large populations of molecules in a relatively large area of membrane; one cannot follow individual protein molecules. If a protein fails to migrate into a bleached area, for example, one cannot tell whether the molecule is truly immobile or just restricted in its movement to a very small region of membrane—perhaps by cytoskeletal proteins. *Single-particle tracking* techniques overcome this problem by labeling individual membrane molecules with antibodies coupled to fluorescent dyes or tiny gold particles and tracking their movement by video microscopy. Using single-particle tracking, one can record the diffusion path of a single membrane protein molecule over time. Results from all of these techniques indicate that plasma membrane proteins differ widely in their diffusion characteristics, as we now discuss.

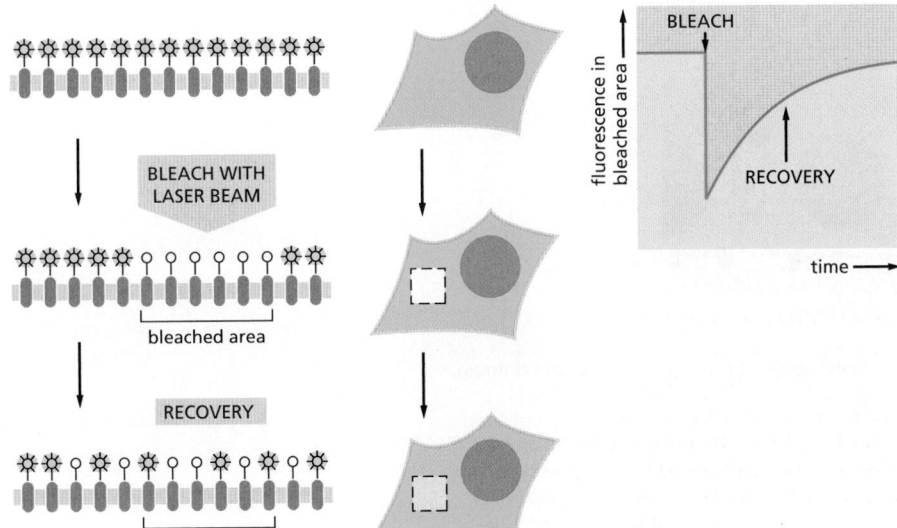

BLEACH WITH
LASER BEAM

bleached area

RECOVERY

BLEACH

fluorescence in
bleached area

RECOVERY

time

Figure 10–33 Measuring the rate of lateral diffusion of a membrane protein by fluorescence recovery after photobleaching. A specific protein of interest can be expressed as a fusion protein with green fluorescent protein (GFP), which is intrinsically fluorescent. The fluorescent molecules are bleached in a small area using a laser beam. The fluorescence intensity recovers as the bleached molecules diffuse away and unbleached molecules diffuse into the irradiated area (shown here in side and top views). The diffusion coefficient is calculated from a graph of the rate of recovery: the greater the diffusion coefficient of the membrane protein, the faster the recovery (**Movie 10.6**).

Cells Can Confine Proteins and Lipids to Specific Domains Within a Membrane

The recognition that biological membranes are two-dimensional fluids was a major advance in understanding membrane structure and function. It has become clear, however, that the picture of a membrane as a lipid sea in which all proteins float freely is greatly oversimplified. Most cells confine membrane proteins to specific regions in a continuous lipid bilayer. We have already discussed how bacteriorhodopsin molecules in the purple membrane of *Halobacterium* assemble into large two-dimensional crystals, in which individual protein molecules are relatively fixed in relationship to one another (see Figure 10–30). ATP synthase complexes in the inner mitochondrial membrane also associate into long double rows, as we discuss in Chapter 14 (see Figure 14–32). Large aggregates of this kind diffuse very slowly.

In epithelial cells, such as those that line the gut or the tubules of the kidney, certain plasma membrane enzymes and transport proteins are confined to the apical surface of the cells, whereas others are confined to the basal and lateral surfaces (**Figure 10–34**). This asymmetric distribution of membrane proteins is often essential for the function of the epithelium, as we discuss in Chapter 11 (see Figure 11–11). The lipid compositions of these two membrane domains are also different, demonstrating that epithelial cells can prevent the diffusion of lipid as well as protein molecules between the domains. The barriers set up by a specific type of intercellular junction (called a *tight junction*, discussed in Chapter 19; see Figure 19–18) maintain the separation of both protein and lipid molecules. Clearly, the membrane proteins that form these intercellular junctions cannot be allowed to diffuse laterally in the interacting membranes.

A cell can also create membrane domains without using intercellular junctions. As we already discussed, regulated protein–protein interactions in membranes are thought to create nanoscale raft domains that function in signaling and membrane trafficking. A more extreme example is seen in the mammalian spermatozoon, a single cell that consists of several structurally and functionally distinct parts covered by a continuous plasma membrane. When a sperm cell is examined by immunofluorescence microscopy with a variety of antibodies, each of which reacts with a specific cell-surface molecule, the plasma membrane is found to consist of at least three distinct domains (**Figure 10–35**). Some of the membrane molecules are able to diffuse freely within the confines of their own domain. The molecular nature of the "fence" that prevents the molecules from

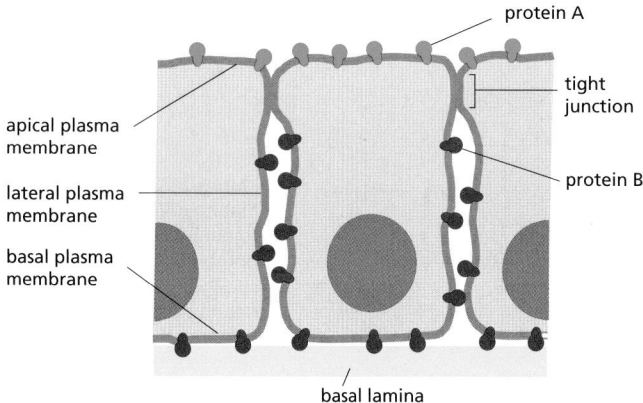

Figure 10–34 How membrane molecules can be restricted to a particular membrane domain. In this drawing of an epithelial cell, protein A (in the apical domain of the plasma membrane) and protein B (in the basal and lateral domains) can diffuse laterally in their own domains but are prevented from entering the other domain, at least partly by the specialized cell–cell junction called a tight junction. Lipid molecules in the outer (extracellular) monolayer of the plasma membrane are likewise unable to diffuse between the two domains; lipids in the inner (cytosolic) monolayer, however, are able to do so (not shown). The basal lamina is a thin mat of extracellular matrix that separates epithelial sheets from other tissues (discussed in Chapter 19).

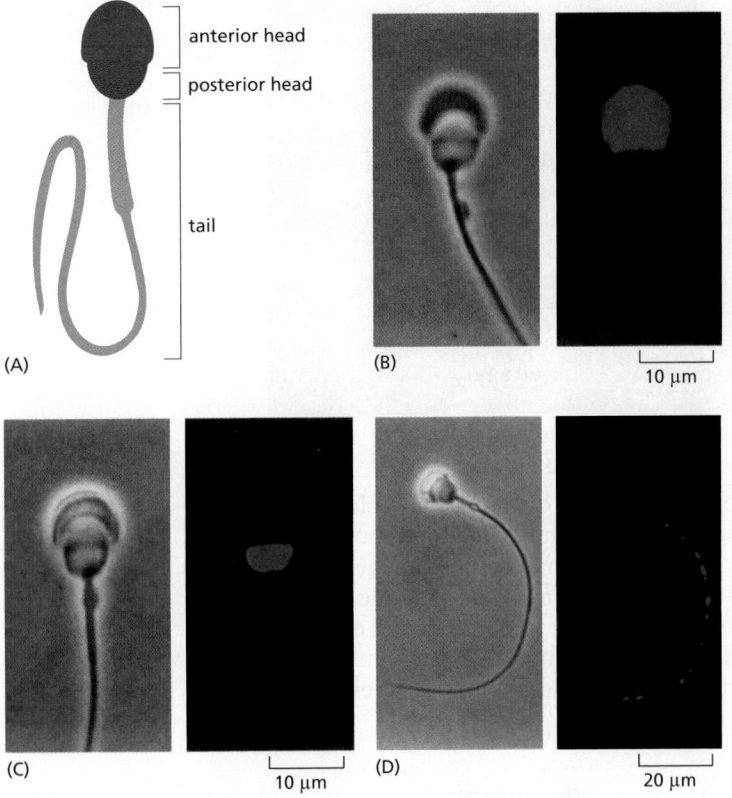

(A)

anterior head

posterior head

tail

(B)

10 µm

(C)

10 µm

(D)

20 µm

Figure 10–35 Three domains in the plasma membrane of a guinea pig sperm. (A) A drawing of a guinea pig sperm. (B–D) In the three pairs of micrographs, phase-contrast micrographs are on the *left*, and the same cell is shown with cell-surface immunofluorescence staining on the *right*. Different monoclonal antibodies selectively label cell-surface molecules on (B) the anterior head, (C) the posterior head, and (D) the tail. (From D.G. Myles, P. Primakoff, and A.R. Bellvé, *Cell* 23:434–439, 1981. With permission from Elsevier.)

leaving their domain is not known. Many other cells have similar membrane fences that confine membrane protein diffusion to certain membrane domains. The plasma membrane of nerve cells, for example, contains a domain enclosing the cell body and dendrites, and another enclosing the axon; it is thought that a belt of actin filaments tightly associated with the plasma membrane at the cell-body–axon junction forms part of the barrier.

Figure 10–36 shows four common ways of immobilizing specific membrane proteins through protein–protein interactions.

The Cortical Cytoskeleton Gives Membranes Mechanical Strength and Restricts Membrane Protein Diffusion

As shown in Figure 10–36B and C, a common way in which a cell restricts the lateral mobility of specific membrane proteins is to tether them to macromolecular assemblies on either side of the membrane. The characteristic biconcave shape of a red blood cell (**Figure 10–37**), for example, results from interactions of its plasma membrane proteins with an underlying *cytoskeleton*, which consists mainly of a meshwork of the filamentous protein **spectrin**. Spectrin is a long, thin, flexible rod about 100 nm in length. As the principal component of the red cell cytoskeleton, it maintains the structural integrity and shape of the plasma membrane, which is the red cell's only membrane, as the cell has no nucleus or other organelles. The spectrin cytoskeleton is riveted to the membrane through various membrane proteins. The final result is a deformable, netlike meshwork that covers the entire cytosolic surface of the red cell membrane (**Figure 10–38**). This spectrin-based cytoskeleton enables the red cell to withstand the stress on its membrane as it is forced through narrow capillaries. Mice and humans with genetic abnormalities in spectrin are anemic and have red cells that are spherical (instead of concave) and fragile; the severity of the anemia increases with the degree of spectrin deficiency.

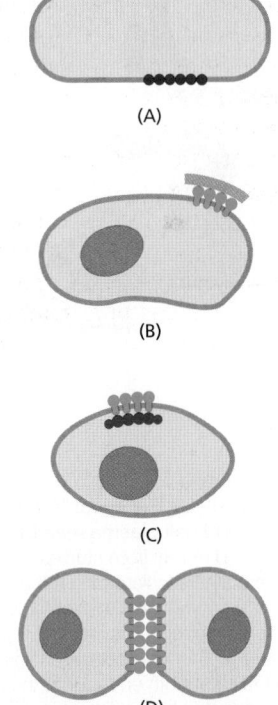

(A)

(B)

(C)

(D)

Figure 10–36 Four ways of restricting the lateral mobility of specific plasma membrane proteins. (A) The proteins can self-assemble into large aggregates (as seen for bacteriorhodopsin in the purple membrane of *Halobacterium salinarum*); they can be tethered by interactions with assemblies of macromolecules (B) outside or (C) inside the cell; or (D) they can interact with proteins on the surface of another cell.

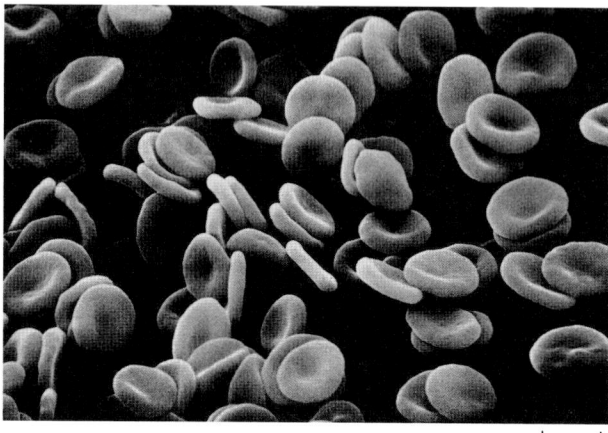

Figure 10–37 **A scanning electron micrograph of human red blood cells.** The cells have a biconcave shape and lack a nucleus and other organelles (Movie 10.7). (Courtesy of Bernadette Chailley.)

5 µm

An analogous but much more elaborate and highly dynamic cytoskeletal network exists beneath the plasma membrane of most other cells in our body. This network, which constitutes the **cortex** of the cell, is rich in actin filaments, which are attached to the plasma membrane in numerous ways. The dynamic remodeling of the cortical actin network provides a driving force for many essential cell functions, including cell movement, endocytosis, and the formation of transient, mobile plasma membrane structures such as filopodia and lamellopodia

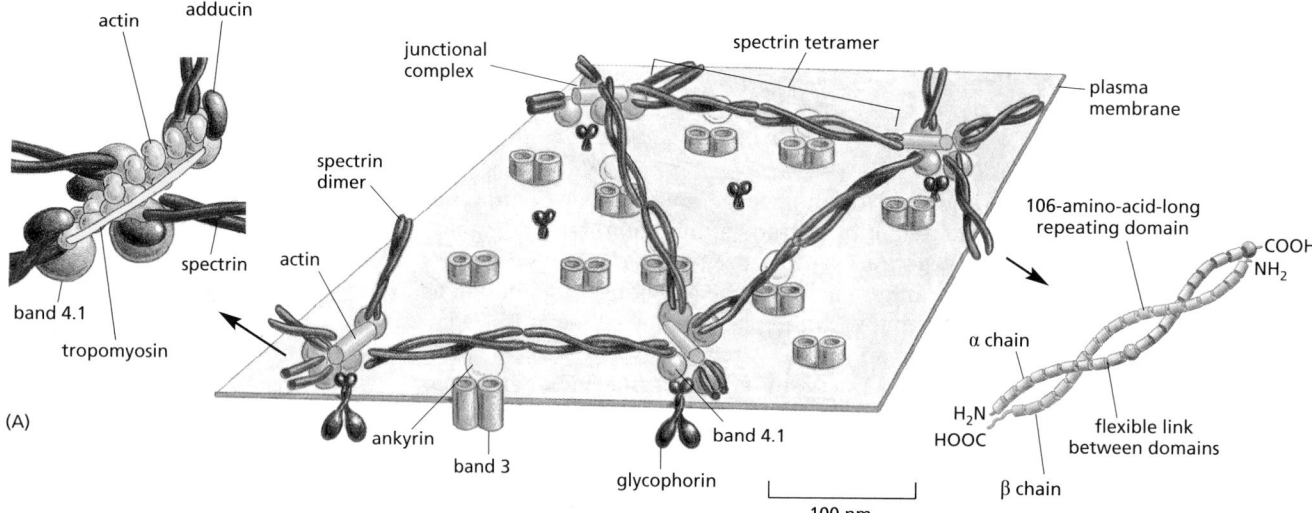

(A)

100 nm

Figure 10–38 **The spectrin-based cytoskeleton on the cytosolic side of the human red blood cell plasma membrane.** (A) The arrangement shown in the drawing has been deduced mainly from studies on the interactions of purified proteins *in vitro*. Spectrin heterodimers (enlarged in the drawing on the *right*) are linked together into a netlike meshwork by "junctional complexes" (enlarged in the drawing on the *left*). Each spectrin heterodimer consists of two antiparallel, loosely intertwined, flexible polypeptide chains called α and β. The two spectrin chains are attached noncovalently to each other at multiple points, including at both ends. Both the α and β chains are composed largely of repeating domains. Two spectrin heterodimers join end-to-end to form tetramers.

The junctional complexes are composed of short actin filaments (containing 13 actin monomers) and these proteins—*band 4.1*, *adducin*, and a *tropomyosin* molecule that probably determines the length of the actin filaments. The cytoskeleton is linked to the membrane through two transmembrane proteins—a multipass protein called *band 3* and a single-pass protein called *glycophorin*. The spectrin tetramers bind to some band 3 proteins via *ankyrin* molecules, and to glycophorin and band 3 (not shown) via band 4.1 proteins.

(B) The electron micrograph shows the cytoskeleton on the cytosolic side of a red blood cell membrane after fixation and negative staining. The spectrin meshwork has been purposely stretched out to allow the details of its structure to be seen. In a normal cell, the meshwork shown would be much more crowded and occupy only about one-tenth of this area. (B, courtesy of T. Byers and D. Branton, *Proc. Natl Acad. Sci. USA* 82:6153–6157, 1985.)

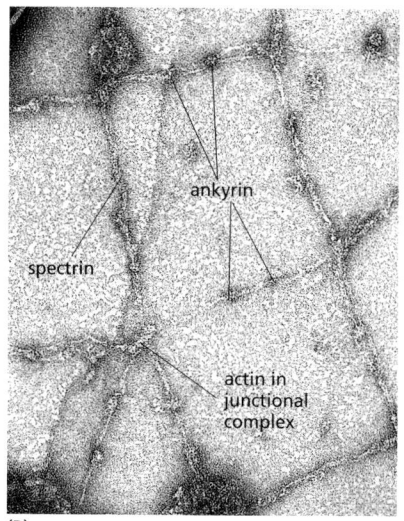

(B)

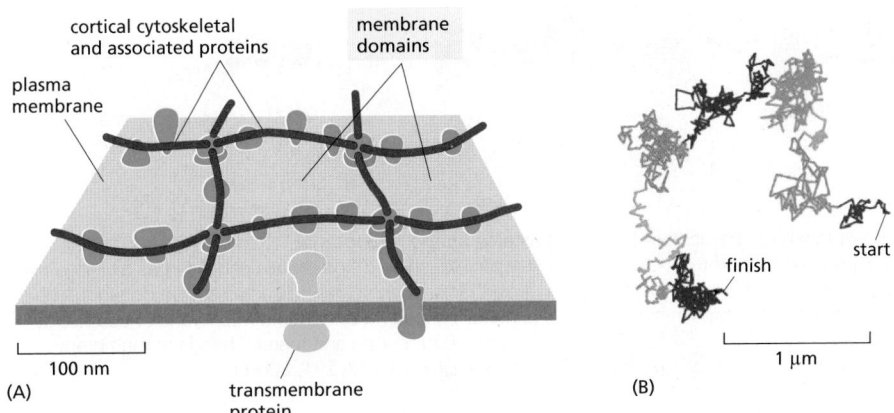

cortical cytoskeletal
and associated proteins

membrane
domains

plasma
membrane

100 nm

(A)

transmembrane
protein

finish

start

1 μm

(B)

Figure 10–39 **Corralling plasma membrane proteins by cortical cytoskeletal filaments.** (A) The filaments are thought to provide diffusion barriers that divide the membrane into small domains, or corrals. (B) High-speed, single-particle tracking was used to follow the path of single fluorescently labeled membrane protein of one type over time. The trace shows that the individual protein molecules (the movement of each shown in a different color) diffuse within a tightly delimited membrane domain and only infrequently escape into a neighboring domain. (Adapted from A. Kusumi et al., *Annu. Rev. Biophys. Biomol. Struct.* 34:351–378, 2005. With permission from Annual Reviews.)

discussed in Chapter 16. The cortex of nucleated cells also contains proteins that are structurally homologous to spectrin and the other components of the red cell cytoskeleton. We discuss the cortical cytoskeleton in nucleated cells and its interactions with the plasma membrane in Chapter 16.

The cortical cytoskeletal network restricts diffusion of not only the plasma membrane proteins that are directly anchored to it. Because the cytoskeletal filaments are often closely apposed to the cytosolic surface of the plasma membrane, they can form mechanical barriers that obstruct the free diffusion of proteins in the membrane. These barriers partition the membrane into small domains, or *corrals* (**Figure 10–39A**), which can be either permanent, as in the sperm (see Figure 10–35), or transient. The barriers can be detected when the diffusion of individual membrane proteins is followed by high-speed, single-particle tracking. The proteins diffuse rapidly but are confined within an individual corral (Figure 10–39B); occasionally, however, thermal motions cause a few cortical filaments to detach transiently from the membrane, allowing the protein to escape into an adjacent corral.

The extent to which a transmembrane protein is confined within a corral depends on its association with other proteins and the size of its cytoplasmic domain; proteins with a large cytosolic domain will have a harder time passing through cytoskeletal barriers. When a cell-surface receptor binds its extracellular signal molecules, for example, large protein complexes build up on the cytosolic domain of the receptor, making it more difficult for the receptor to escape from its corral. It is thought that corralling helps concentrate such signaling complexes, increasing the speed and efficiency of the signaling process (discussed in Chapter 15).

Membrane-bending Proteins Deform Bilayers

Cell membranes assume many different shapes, as illustrated by the elaborate and varied structures of cell-surface protrusions and membrane-enclosed organelles in eukaryotic cells. Flat sheets, narrow tubules, round vesicles, fenestrated sheets, and pitta bread-shaped cisternae are all part of the repertoire: often, a variety of shapes will be present in different regions of the same continuous bilayer. Membrane shape is controlled dynamically, as many essential cell processes—including vesicle budding, cell movement, and cell division—require elaborate transient membrane deformations. In many cases, membrane shape is influenced by dynamic pushing and pulling forces exerted by cytoskeletal or extracellular structures, as we discuss in Chapters 13 and 16). A crucial part in producing these deformations is played by **membrane-bending proteins**, which control local membrane curvature. Often, cytoskeletal dynamics and membrane-bending-protein forces work together. Membrane-bending proteins attach to specific membrane regions as needed and act by one or more of three principal mechanisms:

1. Some insert hydrophobic protein domains or attached lipid anchors into one of the leaflets of a lipid bilayer. Increasing the area of only one leaflet

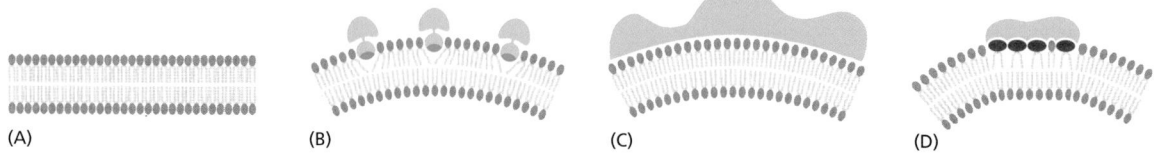

(A)   (B)   (C)   (D)

Figure 10–40 **Three ways in which membrane-bending proteins shape membranes.** Lipid bilayers are blue and proteins are green. (A) Bilayer without protein bound. (B) A hydrophobic region of the protein can insert as a wedge into one monolayer to pry lipid head groups apart. Such regions can either be amphiphilic helices as shown or hydrophobic hairpins. (C) The curved surface of the protein can bind to lipid head groups and deform the membrane or stabilize its curvature. (D) A protein can bind to and cluster lipids that have large head groups and thereby bend the membrane. (Adapted from W.A. Prinz and J.E. Hinshaw, *Crit. Rev. Biochem. Mol. Biol.* 44:278–291, 2009.)

causes the membrane to bend (**Figure 10–40**B). The proteins that shape the convoluted network of narrow ER tubules are thought to work in this way.

2. Some membrane-bending proteins form rigid scaffolds that deform the membrane or stabilize an already bent membrane (Figure 10–40C). The coat proteins that shape the budding vesicles in intracellular transport fall into this class.

3. Some membrane-bending proteins cause particular membrane lipids to cluster together, thereby inducing membrane curvature. The ability of a lipid to induce positive or negative membrane curvature is determined by the relative cross-sectional areas of its head group and its hydrocarbon tails. For example, the large head group of phosphoinositides make these lipid molecules wedge-shaped, and their accumulation in a domain of one leaflet of a bilayer therefore induces positive curvature (Figure 10–40D). By contrast, phospholipases that remove lipid head groups produce inversely shaped lipid molecules that induce negative curvature.

Often, different membrane-bending proteins collaborate to achieve a particular curvature, as in shaping a budding transport vesicle, as we discuss in Chapter 13.

Summary

Whereas the lipid bilayer determines the basic structure of biological membranes, proteins are responsible for most membrane functions, serving as specific receptors, enzymes, transporters, and so on. Transmembrane proteins extend across the lipid bilayer. Some of these membrane proteins are single-pass proteins, in which the polypeptide chain crosses the bilayer as a single α helix. Others are multipass proteins, in which the polypeptide chain crosses the bilayer multiple times—either as a series of α helices or as a β sheet rolled up into the shape of a barrel. All proteins responsible for the transport of ions and other small water-soluble molecules through the membrane are multipass proteins. Some membrane proteins do not span the bilayer but instead are attached to either side of the membrane: some are attached to the cytosolic side by an amphipathic α helix on the protein surface or by the covalent attachment of one or more lipid chains, others are attached to the noncytosolic side by a GPI anchor. Some membrane-associated proteins are bound by noncovalent interactions with transmembrane proteins. In the plasma membrane of all eukaryotic cells, most of the proteins exposed on the cell surface and some of the lipid molecules in the outer lipid monolayer have oligosaccharide chains covalently attached to them. Like the lipid molecules in the bilayer, many membrane proteins are able to diffuse rapidly in the plane of the membrane. However, cells have ways of immobilizing specific membrane proteins, as well as ways of confining both membrane protein and lipid molecules to particular domains in a continuous lipid bilayer. The dynamic association of membrane-bending proteins confers on membranes their characteristic three-dimensional shapes.

WHAT WE DON'T KNOW

• Given the highly complex lipid composition of cell membranes, what are the variations within different organelle membranes in an animal cell? What are the functional consequences of these differences, and what are the roles of the minor lipid species?

• Is the biophysical tendency of lipids to partition into separate phases within a lipid bilayer functionally utilized in cell membranes? If so, how is it regulated and what membrane functions does it control?

• How commonly do specific lipid molecules associate with membrane proteins to regulate their function?

• Given that the structure of only a tiny fraction of all membrane proteins has been determined, what new principles of membrane protein structure remain to be discovered?

PROBLEMS

Which statements are true? Explain why or why not.

10–1   Although lipid molecules are free to diffuse in the plane of the bilayer, they cannot flip-flop across the bilayer unless enzyme catalysts called phospholipid translocators are present in the membrane.

10–2   Whereas all the carbohydrate in the plasma membrane faces outward on the external surface of the cell, all the carbohydrate on internal membranes faces toward the cytosol.

10–3   Although membrane domains with different protein compositions are well known, there are at present no examples of membrane domains that differ in lipid composition.

Discuss the following problems.

10–4   When a lipid bilayer is torn, why does it not seal itself by forming a "hemi-micelle" cap at the edges, as shown in **Figure Q10–1**?

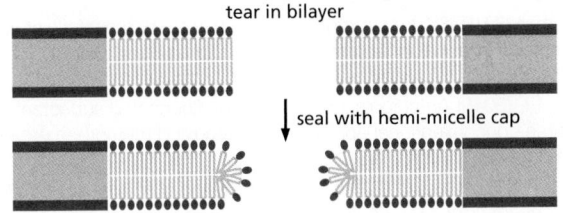

Figure Q10–1 A torn lipid bilayer sealed with a hypothetical "hemi-micelle" cap (Problem 10–4).

10–5   Margarine is made from vegetable oil by a chemical process. Do you suppose this process converts saturated fatty acids to unsaturated ones, or vice versa? Explain your answer.

10–6   If a lipid raft is typically 70 nm in diameter and each lipid molecule has a diameter of 0.5 nm, about how many lipid molecules would there be in a lipid raft composed entirely of lipid? At a ratio of 50 lipid molecules per protein molecule (50% protein by mass), how many proteins would be in a typical raft? (Neglect the loss of lipid from the raft that would be required to accommodate the protein.)

10–7   Monomeric single-pass transmembrane proteins span a membrane with a single α helix that has characteristic chemical properties in the region of the bilayer. Which of the three 20-amino-acid sequences listed below is the most likely candidate for such a transmembrane segment? Explain the reasons for your choice. (See back of book for one-letter amino acid code; FAMILY VW is a convenient mnemonic for hydrophobic amino acids.)

A. I T L I Y F G V M A G V I G T I L L I S
B. I T P I Y F G P M A G V I G T P L L I S
C. I T E I Y F G R M A G V I G T D L L I S

10–8   You are studying the binding of proteins to the cytoplasmic face of cultured neuroblastoma cells and have found a method that gives a good yield of inside-out vesicles from the plasma membrane. Unfortunately, your preparations are contaminated with variable amounts of right-side-out vesicles. Nothing you have tried avoids this problem. A friend suggests that you pass your vesicles over an affinity column made of lectin coupled to solid beads. What is the point of your friend's suggestion?

10–9   Glycophorin, a protein in the plasma membrane of the red blood cell, normally exists as a homodimer that is held together entirely by interactions between its transmembrane domains. Since transmembrane domains are hydrophobic, how is it that they can associate with one another so specifically?

10–10   Three mechanisms by which membrane-binding proteins bend a membrane are illustrated in **Figure Q10–2A, B, and C**. As shown, each of these cytosolic membrane-bending proteins would induce an invagination of the plasma membrane. Could similar kinds of cytosolic proteins induce a protrusion of the plasma membrane (Figure Q10–2D)? Which ones? Explain how they might work.

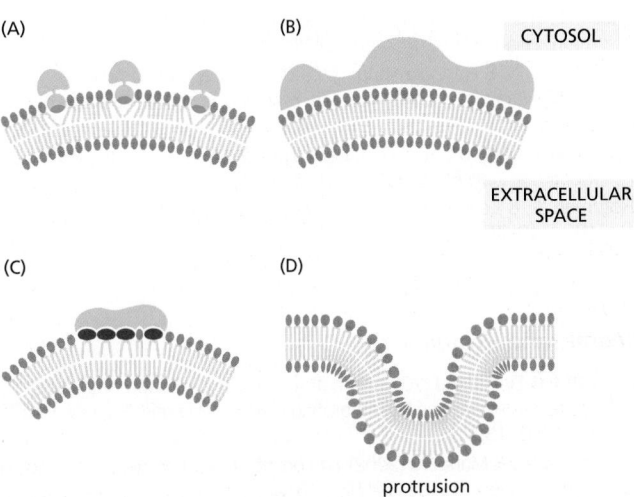

Figure Q10–2 Bending of the plasma membrane by cytosolic proteins (Problem 10–10). (A) Insertion of a protein "finger" into the cytosolic leaflet of the membrane. (B) Binding of lipids to the curved surface of a membrane-binding protein. (C) Binding of membrane proteins to membrane lipids with large head groups. (D) A segment of the plasma membrane showing a protrusion.

REFERENCES

General

Bretscher MS (1973) Membrane structure: some general principles. *Science* 181, 622–629.

Edidin M (2003) Lipids on the frontier: a century of cell-membrane bilayers. *Nat. Rev. Mol. Cell Biol.* 4, 414–418.

Goñi FM (2014) The basic structure and dynamics of cell membranes: an update of the Singer-Nicolson model. *Biochim. Biophys. Acta* 1838, 1467–1476.

Lipowsky R & Sackmann E (eds) (1995) Structure and Dynamics of Membranes. Amsterdam: Elsevier.

Singer SJ & Nicolson GL (1972) The fluid mosaic model of the structure of cell membranes. *Science* 175, 720–731.

Tanford C (1980) The Hydrophobic Effect: Formation of Micelles and Biological Membranes. New York: Wiley.

The Lipid Bilayer

Bevers EM, Comfurius P, Dekkers DW & Zwaal RF (1999) Lipid translocation across the plasma membrane of mammalian cells. *Biochim. Biophys. Acta* 1439, 317–330.

Brügger B (2014) Lipidomics: analysis of the lipid composition of cells and subcellular organelles by electrospray ionization mass spectrometry. *Annu. Rev. Biochem.* 83, 79–98.

Contreras FX, Sánchez-Magraner L, Alonso A & Goñi FM (2010) Transbilayer (flip-flop) lipid motion and lipid scrambling in membranes. *FEBS Lett.* 584, 1779–1786.

Hakomori SI (2002) The glycosynapse. *Proc. Natl Acad. Sci. USA* 99, 225–232.

Ichikawa S & Hirabayashi Y (1998) Glucosylceramide synthase and glycosphingolipid synthesis. *Trends Cell Biol.* 8, 198–202.

Klose C, Surma MA & Simons K (2013) Organellar lipidomics— background and perspectives. *Curr. Opin. Cell Biol.* 25, 406–413.

Kornberg RD & McConnell HM (1971) Lateral diffusion of phospholipids in a vesicle membrane. *Proc. Natl Acad. Sci. USA* 68, 2564–2568.

Lingwood D & Simons K (2010) Lipid rafts as a membrane-organizing principle. *Science* 327, 46–50.

Mansilla MC, Cybulski LE, Albanesi D & de Mendoza D (2004) Control of membrane lipid fluidity by molecular thermosensors. *J. Bacteriol.* 186, 6681–6688.

Maxfield FR & van Meer G (2010) Cholesterol, the central lipid of mammalian cells. *Curr. Opin. Cell Biol.* 22, 422–429.

McConnell HM & Radhakrishnan A (2003) Condensed complexes of cholesterol and phospholipids. *Biochim. Biophys. Acta* 1610, 159–173.

Pomorski T & Menon AK (2006) Lipid flippases and their biological functions. *Cell. Mol. Life Sci.* 63, 2908–2921.

Rothman JE & Lenard J (1977) Membrane asymmetry. *Science* 195, 743–753.

Walther TC & Farese RV Jr (2012) Lipid droplets and cellular lipid metabolism. *Annu. Rev. Biochem.* 81, 687–714.

Membrane Proteins

Bennett V & Baines AJ (2001) Spectrin and ankyrin-based pathways: metazoan inventions for integrating cells into tissues. *Physiol. Rev.* 81, 1353–1392.

Bijlmakers MJ & Marsh M (2003) The on-off story of protein palmitoylation. *Trends Cell Biol.* 13, 32–42.

Branden C & Tooze J (1999) Introduction to Protein Structure, 2nd ed. New York: Garland Science.

Bretscher MS & Raff MC (1975) Mammalian plasma membranes. *Nature* 258, 43–49.

Buchanan SK (1999) Beta-barrel proteins from bacterial outer membranes: structure, function and refolding. *Curr. Opin. Struct. Biol.* 9, 455–461.

Chen Y, Lagerholm BC, Yang B & Jacobson K (2006) Methods to measure the lateral diffusion of membrane lipids and proteins. *Methods* 39, 147–153.

Curran AR & Engelman DM (2003) Sequence motifs, polar interactions and conformational changes in helical membrane proteins. *Curr. Opin. Struct. Biol.* 13, 412–417.

Deisenhofer J & Michel H (1991) Structures of bacterial photosynthetic reaction centers. *Annu. Rev. Cell Biol.* 7, 1–23.

Drickamer K & Taylor ME (1993) Biology of animal lectins. *Annu. Rev. Cell Biol.* 9, 237–264.

Drickamer K & Taylor ME (1998) Evolving views of protein glycosylation. *Trends Biochem. Sci.* 23, 321–324.

Frye LD & Edidin M (1970) The rapid intermixing of cell surface antigens after formation of mouse-human heterokaryons. *J. Cell Sci.* 7, 319–335.

Helenius A & Simons K (1975) Solubilization of membranes by detergents. *Biochim. Biophys. Acta* 415, 29–79.

Henderson R & Unwin PN (1975) Three-dimensional model of purple membrane obtained by electron microscopy. *Nature* 257, 28–32.

Kyte J & Doolittle RF (1982) A simple method for displaying the hydropathic character of a protein. *J. Mol. Biol.* 157, 105–132.

Lee AG (2003) Lipid-protein interactions in biological membranes: a structural perspective. *Biochim. Biophys. Acta* 1612, 1–40.

Marchesi VT, Furthmayr H & Tomita M (1976) The red cell membrane. *Annu. Rev. Biochem.* 45, 667–698.

Nakada C, Ritchie K, Oba Y et al. (2003) Accumulation of anchored proteins forms membrane diffusion barriers during neuronal polarization. *Nat. Cell Biol.* 5, 626–632.

Oesterhelt D (1998) The structure and mechanism of the family of retinal proteins from halophilic archaea. *Curr. Opin. Struct. Biol.* 8, 489–500.

Popot J-L (2010) Amphipols, nanodiscs, and fluorinated surfactants: three nonconventional approaches to studying membrane proteins in aqueous solution. *Annu. Rev. Biochem.* 79, 737–775.

Prinz WA & Hinshaw JE (2009) Membrane-bending proteins. *Crit. Rev. Biochem. Mol. Biol.* 44, 278–291.

Rao M & Mayor S (2014) Active organization of membrane constituents in living cells. *Curr. Opin. Cell Biol.* 29, 126–132.

Reig N & van der Goot FG (2006) About lipids and toxins. *FEBS Lett.* 580, 5572–5579.

Sharon N & Lis H (2004) History of lectins: from hemagglutinins to biological recognition molecules. *Glycobiology* 14, 53R–62R.

Sheetz MP (2001) Cell control by membrane-cytoskeleton adhesion. *Nat. Rev. Mol. Cell Biol.* 2, 392–396.

Shibata Y, Hu J, Kozlov MM & Rapoport TA (2009) Mechanisms shaping the membranes of cellular organelles. *Annu. Rev. Cell Dev. Biol.* 25, 329–354.

Steck TL (1974) The organization of proteins in the human red blood cell membrane. A review. *J. Cell Biol.* 62, 1–19.

Subramaniam S (1999) The structure of bacteriorhodopsin: an emerging consensus. *Curr. Opin. Struct. Biol.* 9, 462–468.

Viel A & Branton D (1996) Spectrin: on the path from structure to function. *Curr. Opin. Cell Biol.* 8, 49–55.

Vinothkumar KR & Henderson R (2010) Structures of membrane proteins. *Q. Rev. Biophys.* 43, 65–158.

von Heijne G (2011) Membrane proteins: from bench to bits. *Biochem. Soc. Trans.* 39, 747–750.

White SH & Wimley WC (1999) Membrane protein folding and stability: physical principles. *Annu. Rev. Biophys. Biomol. Struct.* 28, 319–365.

Membrane Transport of Small Molecules and the Electrical Properties of Membranes

Because of its hydrophobic interior, the lipid bilayer of cell membranes restricts the passage of most polar molecules. This barrier function allows the cell to maintain concentrations of solutes in its cytosol that differ from those in the extracellular fluid and in each of the intracellular membrane-enclosed compartments. To benefit from this barrier, however, cells have had to evolve ways of transferring specific water-soluble molecules and ions across their membranes in order to ingest essential nutrients, excrete metabolic waste products, and regulate intracellular ion concentrations. Cells use specialized *membrane transport proteins* to accomplish this goal. The importance of such small molecule transport is reflected in the large number of genes in all organisms that code for the transmembrane transport proteins involved, which make up 15–30% of the membrane proteins in all cells. Some mammalian cells, such as nerve and kidney cells, devote up to two-thirds of their total metabolic energy consumption to such transport processes.

Cells can also transfer macromolecules and even large particles across their membranes, but the mechanisms involved in most of these cases differ from those used for transferring small molecules, and they are discussed in Chapters 12 and 13.

We begin this chapter by describing some general principles of how small water-soluble molecules traverse cell membranes. We then consider, in turn, the two main classes of membrane proteins that mediate this transmembrane traffic: *transporters*, which undergo sequential conformational changes to transport specific small molecules across membranes, and *channels*, which form narrow pores, allowing passive transmembrane movement, primarily of water and small inorganic ions. Transporters can be coupled to a source of energy to catalyze active transport, which together with selective passive permeability, creates large differences in the composition of the cytosol compared with that of either the extracellular fluid (Table 11–1) or the fluid within membrane-enclosed organelles. By generating inorganic ion-concentration differences across the lipid bilayer, cell membranes can store potential energy in the form of electrochemical gradients, which drive various transport processes, convey electrical signals in electrically excitable cells, and (in mitochondria, chloroplasts, and bacteria) make most of the cell's ATP. We focus our discussion mainly on transport across the plasma membrane, but similar mechanisms operate across the other membranes of the eukaryotic cell, as discussed in later chapters.

In the last part of the chapter, we concentrate mainly on the functions of ion channels in neurons (nerve cells). In these cells, channel proteins perform at their highest level of sophistication, enabling networks of neurons to carry out all the astonishing feats your brain is capable of.

PRINCIPLES OF MEMBRANE TRANSPORT

We begin this section by describing the permeability properties of protein-free, synthetic lipid bilayers. We then introduce some of the terms used to describe the various forms of membrane transport and some strategies for characterizing the proteins and processes involved.

TABLE 11–1 A Comparison of Inorganic Ion Concentrations Inside and Outside a Typical Mammalian Cell*

Component	Cytoplasmic concentration (mM)	Extracellular concentration (mM)
Cations		
Na^+	5–15	145
K^+	140	5
Mg^{2+}	0.5	1–2
Ca^{2+}	10^{-4}	1–2
H^+	7×10^{-5} ($10^{-7.2}$ M or pH 7.2)	4×10^{-5} ($10^{-7.4}$ M or pH 7.4)
Anions		
Cl^-	5–15	110

*The cell must contain equal quantities of positive and negative charges (that is, it must be electrically neutral). Thus, in addition to Cl^-, the cell contains many other anions not listed in this table; in fact, most cell constituents are negatively charged (HCO_3^-, PO_4^{3-}, nucleic acids, metabolites carrying phosphate and carboxyl groups, etc.). The concentrations of Ca^{2+} and Mg^{2+} given are for the free ions: although there is a total of about 20 mM Mg^{2+} and 1–2 mM Ca^{2+} in cells, both ions are mostly bound to other substances (such as proteins, free nucleotides, RNA, etc.) and, for Ca^{2+}, stored within various organelles.

Protein-Free Lipid Bilayers Are Impermeable to Ions

Given enough time, virtually any molecule will diffuse across a protein-free lipid bilayer down its concentration gradient. The rate of diffusion, however, varies enormously, depending partly on the size of the molecule but mostly on its relative hydrophobicity (solubility in oil). In general, the smaller the molecule and the more hydrophobic, or nonpolar, it is, the more easily it will diffuse across a lipid bilayer. Small nonpolar molecules, such as O_2 and CO_2, readily dissolve in lipid bilayers and therefore diffuse rapidly across them. Small uncharged polar molecules, such as water or urea, also diffuse across a bilayer, albeit much more slowly (**Figure 11–1** and see Movie 10.3). By contrast, lipid bilayers are essentially impermeable to charged molecules (ions), no matter how small: the charge and high degree of hydration of such molecules prevents them from entering the hydrocarbon phase of the bilayer (**Figure 11–2**).

There Are Two Main Classes of Membrane Transport Proteins: Transporters and Channels

Like synthetic lipid bilayers, cell membranes allow small nonpolar molecules to permeate by diffusion. Cell membranes, however, also have to allow the passage of various polar molecules, such as ions, sugars, amino acids, nucleotides, water, and many cell metabolites that cross synthetic lipid bilayers only very slowly. Special **membrane transport proteins** transfer such solutes across cell membranes. These proteins occur in many forms and in all types of biological membranes. Each protein often transports only a specific molecular species or sometimes a class of molecules (such as ions, sugars, or amino acids). Studies in the 1950s found that bacteria with a single-gene mutation were unable to transport sugars across their plasma membrane, thereby demonstrating the specificity of membrane transport proteins. We now know that humans with similar mutations suffer from various inherited diseases that hinder the transport of a specific solute or solute class in the kidney, intestine, or other cell type. Individuals with the inherited disease *cystinuria*, for example, cannot transport certain amino acids (including cystine, the disulfide-linked dimer of cysteine) from either the urine or the intestine into the

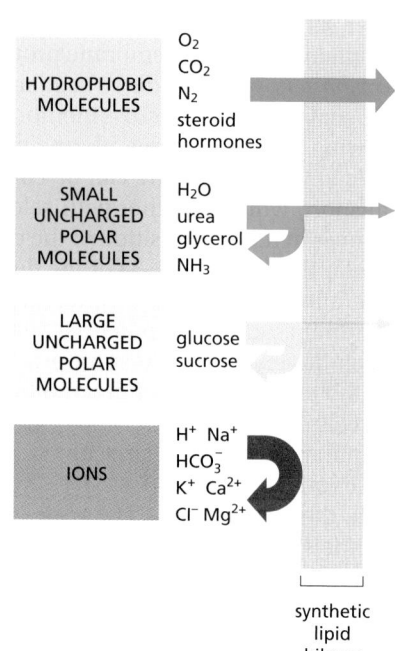

Figure 11–1 The relative permeability of a synthetic lipid bilayer to different classes of molecules. The smaller the molecule and, more importantly, the less strongly it associates with water, the more rapidly the molecule diffuses across the bilayer.

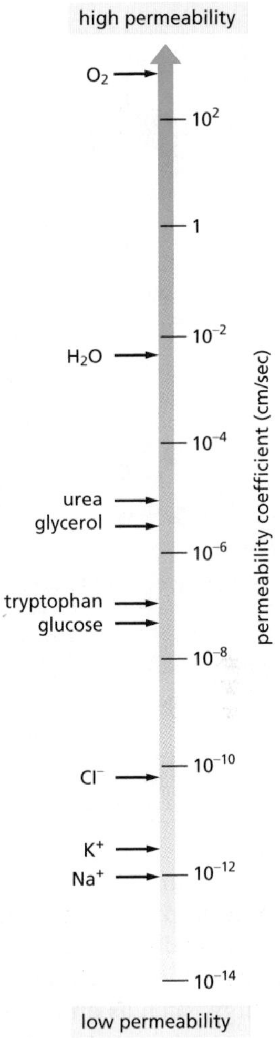

Figure 11–2 Permeability coefficients for the passage of various molecules through synthetic lipid bilayers. The rate of flow of a solute across the bilayer is directly proportional to the difference in its concentration on the two sides of the membrane. Multiplying this concentration difference (in mol/cm^3) by the permeability coefficient (in cm/sec) gives the flow of solute in moles per second per square centimeter of bilayer. A concentration difference of tryptophan of 10^{-4} mol/cm^3 (10^{-4} mol / 10^{-3} L = 0.1 M), for example, would cause a flow of 10^{-4} mol/cm^3 × 10^{-7} cm/sec = 10^{-11} mol/sec through 1 cm^2 of bilayer, or 6 × 10^4 molecules/sec through 1 μm^2 of bilayer.

high permeability

O_2

10^2

1

10^{-2} H_2O

10^{-4}

urea
glycerol

10^{-6}

tryptophan
glucose

10^{-8}

Cl^- 10^{-10}

K^+
Na^+ 10^{-12}

10^{-14}

low permeability

permeability coefficient (cm/sec)

blood; the resulting accumulation of cystine in the urine leads to the formation of cystine stones in the kidneys.

All membrane transport proteins that have been studied in detail are multi-pass transmembrane proteins—that is, their polypeptide chains traverse the lipid bilayer multiple times. By forming a protein-lined pathway across the membrane, these proteins enable specific hydrophilic solutes to cross the membrane without coming into direct contact with the hydrophobic interior of the lipid bilayer.

Transporters and channels are the two major classes of membrane transport proteins (**Figure 11–3**). **Transporters** (also called *carriers*, or *permeases*) bind the specific solute to be transported and undergo a series of conformational changes that alternately expose solute-binding sites on one side of the membrane and then on the other to transfer the solute across it. **Channels**, by contrast, interact with the solute to be transported much more weakly. They form continuous pores that extend across the lipid bilayer. When open, these pores allow specific solutes (such as inorganic ions of appropriate size and charge and in some cases small molecules, including water, glycerol, and ammonia) to pass through them and thereby cross the membrane. Not surprisingly, transport through channels occurs at a much faster rate than transport mediated by transporters. Although water can slowly diffuse across synthetic lipid bilayers, cells use dedicated channel proteins (called *water channels*, or *aquaporins*) that greatly increase the permeability of their membranes to water, as we discuss later.

Active Transport Is Mediated by Transporters Coupled to an Energy Source

All channels and many transporters allow solutes to cross the membrane only passively ("downhill"), a process called **passive transport**. In the case of transport of a single uncharged molecule, the difference in the concentration on the two sides of the membrane—its *concentration gradient*—drives passive transport and determines its direction (**Figure 11–4A**). If the solute carries a net charge, however, both its concentration gradient and the electrical potential difference across the membrane, the *membrane potential*, influence its transport. The concentration gradient and the electrical gradient combine to form a net driving force, the **electrochemical gradient**, for each charged solute (Figure 11–4B). We discuss electrochemical gradients in more detail later and in Chapter 14. In fact, almost all plasma membranes have an electrical potential (i.e., a voltage) across them, with the inside usually negative with respect to the outside. This potential favors the entry of positively charged ions into the cell but opposes the entry of negatively charged ions (see Figure 11–4B); it also opposes the efflux of positively charged ions.

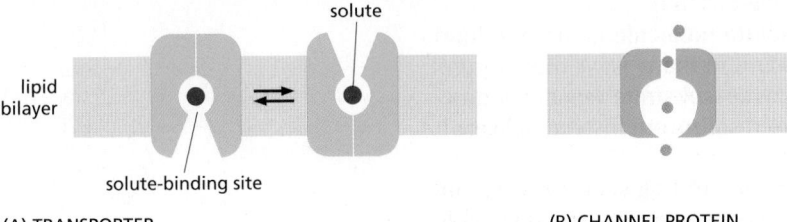

(A) TRANSPORTER

(B) CHANNEL PROTEIN

solute

lipid bilayer

solute-binding site

Figure 11–3 Transporters and channel proteins. (A) A transporter alternates between two conformations, so that the solute-binding site is sequentially accessible on one side of the bilayer and then on the other. (B) In contrast, a channel protein forms a pore across the bilayer through which specific solutes can passively diffuse.

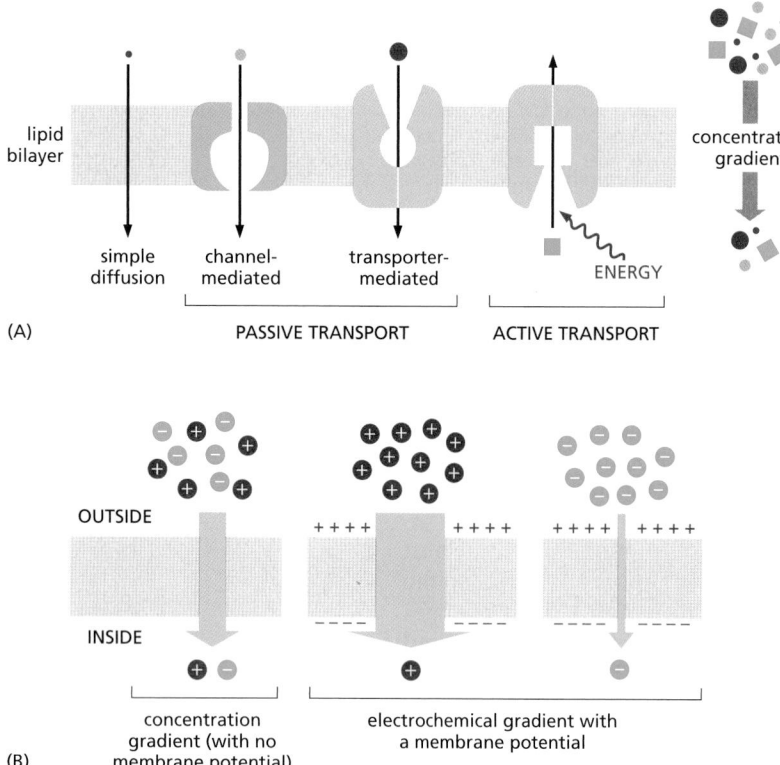

Figure 11–4 **Different forms of membrane transport and the influence of the membrane.** Passive transport down a concentration gradient (or an electrochemical gradient—see B below) occurs spontaneously, by diffusion, either through the lipid bilayer directly or through channels or passive transporters. By contrast, active transport requires an input of metabolic energy and is always mediated by transporters that pump the solute against its concentration or electrochemical gradient. (B) The electrochemical gradient of a charged solute (an ion) affects its transport. This gradient combines the membrane potential and the concentration gradient of the solute. The electrical and chemical gradients can work additively to increase the driving force on an ion across the membrane (*middle*) or can work against each other (*right*).

As shown in Figure 11-4A, in addition to passive transport, cells need to be able to actively pump certain solutes across the membrane "uphill," against their electrochemical gradients. Such **active transport** is mediated by transporters whose pumping activity is directional because it is tightly coupled to a source of metabolic energy, such as an ion gradient or ATP hydrolysis, as discussed later. Transmembrane movement of small molecules mediated by transporters can be either active or passive, whereas that mediated by channels is always passive (see Figure 11-4A).

Summary

Lipid bilayers are virtually impermeable to most polar molecules. To transport small water-soluble molecules into or out of cells or intracellular membrane-enclosed compartments, cell membranes contain various membrane transport proteins, each of which is responsible for transferring a particular solute or class of solutes across the membrane. There are two classes of membrane transport proteins—transporters and channels. Both form protein pathways across the lipid bilayer. Whereas transmembrane movement mediated by transporters can be either active or passive, solute flow through channel proteins is always passive. Both active and passive ion transport is influenced by the ion's concentration gradient and the membrane potential—that is, its electrochemical gradient.

TRANSPORTERS AND ACTIVE MEMBRANE TRANSPORT

The process by which a transporter transfers a solute molecule across the lipid bilayer resembles an enzyme–substrate reaction, and in many ways transporters behave like enzymes. By contrast to ordinary enzyme–substrate reactions, however, the transporter does not modify the transported solute but instead delivers it unchanged to the other side of the membrane.

Each type of transporter has one or more specific binding sites for its solute (substrate). It transfers the solute across the lipid bilayer by undergoing reversible

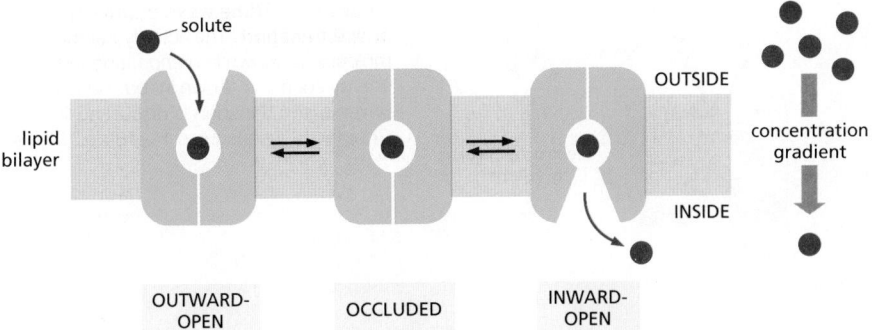

Figure 11–5 A model of how a conformational change in a transporter mediates the passive movement of a solute. The transporter is shown in three conformational states: in the outward-open state, the binding sites for solute are exposed on the outside; in the occluded state, the same sites are not accessible from either side; and in the inward-open state, the sites are exposed on the inside. The transitions between the states occur randomly. They are completely reversible and do not depend on whether the solute-binding site is occupied. Therefore, if the solute concentration is higher on the outside of the bilayer, more solute binds to the transporter in the outward-open conformation than in the inward-open conformation, and there is a net transport of solute down its concentration gradient (or, if the solute is an ion, down its electrochemical gradient).

conformational changes that alternately expose the solute-binding site first on one side of the membrane and then on the other—but never on both sides at the same time. The transition occurs through an intermediate state in which the solute is inaccessible, or occluded, from either side of the membrane (**Figure 11–5**). When the transporter is saturated (that is, when all solute-binding sites are occupied), the rate of transport is maximal. This rate, referred to as V_{max} (V for velocity), is characteristic of the specific carrier. V_{max} measures the rate at which the carrier can flip between its conformational states. In addition, each transporter has a characteristic affinity for its solute, reflected in the K_m of the reaction, which is equal to the concentration of solute when the transport rate is half its maximum value (**Figure 11–6**). As with enzymes, the binding of solute can be blocked by either competitive inhibitors (which compete for the same binding site and may or may not be transported) or noncompetitive inhibitors (which bind elsewhere and alter the structure of the transporter).

As we discuss shortly, it requires only a relatively minor modification of the model shown in Figure 11–5 to link a transporter to a source of energy in order to pump a solute uphill against its electrochemical gradient. Cells carry out such active transport in three main ways (**Figure 11–7**):

1. *Coupled transporters* harness the energy stored in concentration gradients to couple the uphill transport of one solute across the membrane to the downhill transport of another.

2. *ATP-driven pumps* couple uphill transport to the hydrolysis of ATP.

3. *Light- or redox-driven pumps*, which are known in bacteria, archaea, mitochondria, and chloroplasts, couple uphill transport to an input of energy from light, as with bacteriorhodopsin (discussed in Chapter 10), or from a redox reaction, as with cytochrome *c* oxidase (discussed in Chapter 14).

Amino acid sequence and three-dimensional structure comparisons suggest that, in many cases, there are strong similarities in structure between transporters that mediate active transport and those that mediate passive transport. Some bacterial transporters, for example, that use the energy stored in the H^+ gradient across the plasma membrane to drive the active uptake of various sugars are structurally similar to the transporters that mediate passive glucose transport into most animal cells. This suggests an evolutionary relationship between various transporters. Given the importance of small metabolites and sugars as energy sources, it is not surprising that the superfamily of transporters is an ancient one.

We begin our discussion of active membrane transport by considering a class of coupled transporters that are driven by ion concentration gradients. These proteins have a crucial role in the transport of small metabolites across membranes in all cells. We then discuss ATP-driven pumps, including the Na^+-K^+ pump that is found in the plasma membrane of most animal cells. Examples of the third class of active transport—light- or redox-driven pumps—are discussed in Chapter 14.

Active Transport Can Be Driven by Ion-Concentration Gradients

Some transporters simply passively mediate the movement of a single solute from one side of the membrane to the other at a rate determined by their V_{max} and

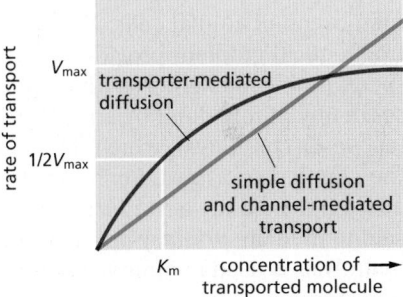

Figure 11–6 The kinetics of simple diffusion compared with transporter-mediated diffusion. Whereas the rate of diffusion and channel-mediated transport is directly proportional to the solute concentration (within the physical limits imposed by total surface area or total channels available), the rate of transporter-mediated diffusion reaches a maximum (V_{max}) when the transporter is saturated. The solute concentration when the transport rate is at half its maximal value approximates the binding constant (K_m) of the transporter for the solute and is analogous to the K_m of an enzyme for its substrate. Note that, while the graph illustrates the shape of the curves, their absolute scale on the Y-axis is very different. A typical channel conducts water or ions at rates of up to 10^8 per second, whereas a typical transporter moves solutes at rates between 10^2–10^4 molecules per second.

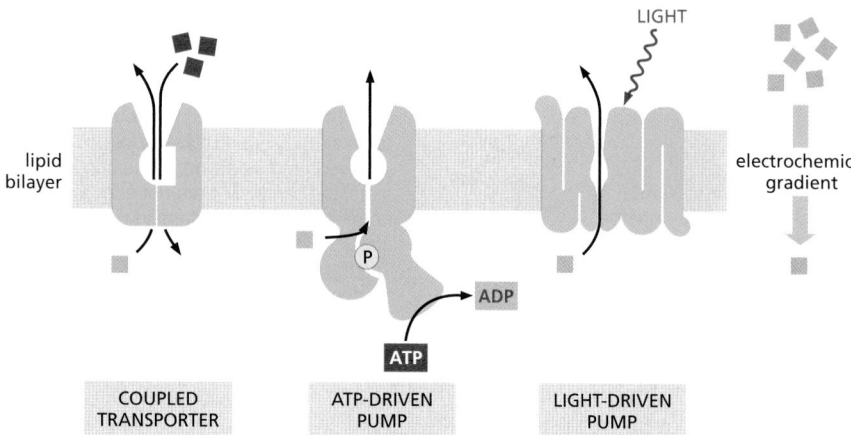

Figure 11–7 Three ways of driving active transport. The actively transported molecule is shown in *orange*, and the energy source is shown in *red*. Redox driven active transport is discussed in Chapter 14 (see Figures 14–18 and 14–19).

K_m; they are called **uniporters**. Others function as *coupled transporters*, in which the transfer of one solute strictly depends on the transport of a second. Coupled transport involves either the simultaneous transfer of a second solute in the same direction, performed by **symporters** (also called *co-transporters*), or the transfer of a second solute in the opposite direction, performed by **antiporters** (also called *exchangers*) (**Figure 11–8**).

The tight coupling between the transfer of two solutes allows the coupled transporters to harvest the energy stored in the electrochemical gradient of one solute, typically an inorganic ion, to transport the other. In this way, the free energy released during the movement of an inorganic ion down an electrochemical gradient is used as the driving force to pump other solutes uphill, against their electrochemical gradient. This strategy can work in either direction; some coupled transporters function as symporters, others as antiporters. In the plasma membrane of animal cells, Na^+ is the usual co-transported ion because its electrochemical gradient provides a large driving force for the active transport of a second molecule. The Na^+ that enters the cell during coupled transport is subsequently pumped out by an ATP-driven Na^+-K^+ pump in the plasma membrane (as we discuss later), which, by maintaining the Na^+ gradient, indirectly drives the coupled transport. Such ion-driven coupled transporters as just described are said to mediate *secondary active transport*. In contrast, ATP-driven pumps are said to mediate *primary active transport* because in these the free energy of ATP hydrolysis is used to directly drive the transport of a solute against its concentration gradient.

Intestinal and kidney epithelial cells contain a variety of symporters that are driven by the Na^+ gradient across the plasma membrane. Each Na^+-driven symporter is specific for importing a small group of related sugars or amino acids into the cell. Because the Na^+ tends to move into the cell down its electrochemical gradient, the sugar or amino acid is, in a sense, "dragged" into the cell with it. The greater the electrochemical gradient for Na^+, the more solute is pumped

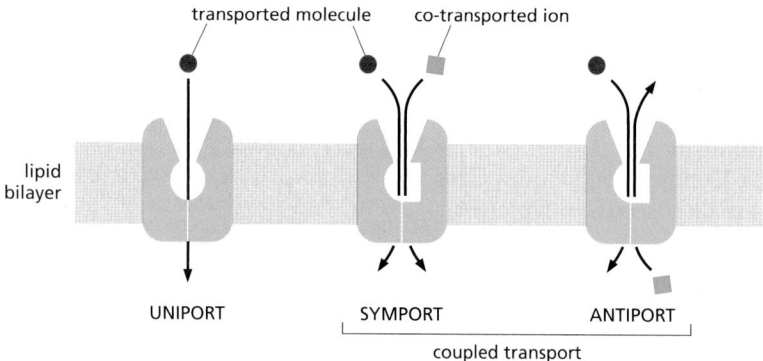

Figure 11–8 This schematic diagram shows transporters functioning as uniporters, symporters, and antiporters (Movie 11.1).

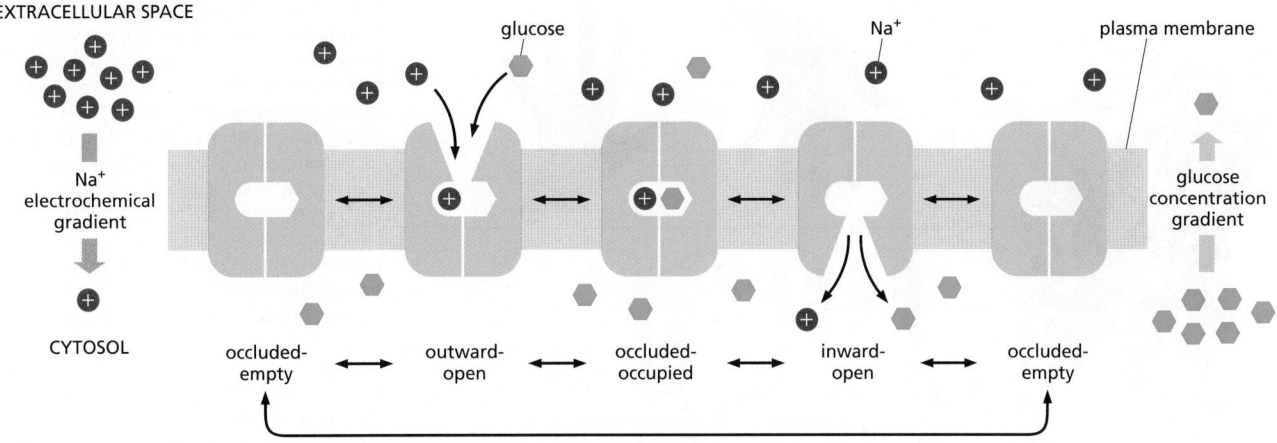

Figure 11–9 Mechanism of glucose transport fueled by a Na⁺ gradient. As in the model shown in Figure 11–5, the transporter alternates between inward-open and outward-open states via an occluded intermediate state. Binding of Na⁺ and glucose is cooperative—that is, the binding of either solute increases the protein's affinity for the other. Since the Na⁺ concentration is much higher in the extracellular space than in the cytosol, glucose is more likely to bind to the transporter in the outward-facing state. The transition to the occluded state occurs only when both Na⁺ and glucose are bound; their precise interactions in the solute-binding sites slightly stabilize the occluded state and thereby make this transition energetically favorable. Stochastic fluctuations caused by thermal energy drive the transporter randomly into the inward-open or outward-open conformation. If it opens outwardly, nothing is achieved, and the process starts all over. However, whenever it opens inwardly, Na⁺ dissociates quickly in the low-Na⁺-concentration environment of the cytosol. Glucose dissociation is likewise enhanced when Na⁺ is lost, because of cooperativity in binding of the two solutes. The overall result is the net transport of both Na⁺ and glucose into the cell. Because the occluded state is not formed when only one of the solutes is bound, the transporter switches conformation only when it is fully occupied or fully empty, thereby assuring strict coupling of the transport of Na⁺ and glucose.

into the cell (**Figure 11–9**). Neurotransmitters (released by nerve cells to signal at synapses—as we discuss later) are taken up again by Na⁺ symporters after their release. These neurotransmitter transporters are important drug targets: stimulants, such as cocaine and antidepressants, inhibit them and thereby prolong signaling by the neurotransmitters, which are not cleared efficiently.

Despite their great variety, transporters share structural features that can explain how they function and how they evolved. Transporters are typically built from bundles of 10 or more α helices that span the membrane. Solute- and ion-binding sites are located midway through the membrane, where some helices are broken or distorted and amino acid side chains and polypeptide backbone atoms form ion- and solute-binding sites. In the inward-open and outward-open conformations, these binding sites are accessible by passageways from one side of the membrane but not the other. In switching between the two conformations, the transporter protein transiently adopts an occluded conformation, in which both passageways are closed; this prevents the driving ion and the transported solute from crossing the membrane unaccompanied, which would deplete the cell's energy store to no purpose. Because only transporters with both types of binding sites appropriately filled change their conformation, tight coupling between ion and solute transport is assured.

Like enzymes, transporters can work in the reverse direction if ion and solute gradients are appropriately adjusted experimentally. This chemical symmetry is mirrored in their physical structure. Crystallographic analyses have revealed that transporters are built from *inverted repeats*: the packing of the transmembrane α helices in one half of the helix bundle is structurally similar to the packing in the other half, but the two halves are inverted in the membrane relative to each other. Transporters are therefore said to be pseudosymmetric, and the passageways that open and close on either side of the membrane have closely similar geometries, allowing alternating access to the ion- and solute-binding sites in the center (**Figure 11–10**). It is thought that the two halves evolved by gene duplication of a smaller ancestor protein.

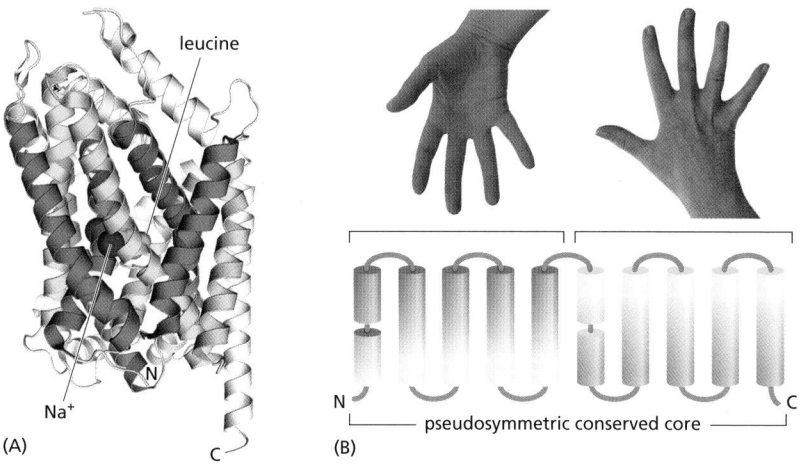

(A)

(B) pseudosymmetric conserved core

Figure 11–10 Transporters are built from inverted repeats. (A) LeuT, a bacterial leucine/Na$^+$ symporter related to human neurotransmitter transporters, such as the serotonin transporter, is shown. The core of the transporter is built from two bundles, each composed of five α helices (*blue* and *yellow*). The helices shown in *gray* differ among members of this transporter family and are thought to play regulatory roles, which are specific to a particular transporter. (B) Both core helix bundles are packed in a similar arrangement (shown as a hand, with the broken helix as the thumb), but the second bundle is inverted with respect to the first. The transporter's structural pseudosymmetry reflects its functional symmetry: the transporter can work in either direction, depending on the direction of the ion gradient. (Adapted from K.R. Vinothkumar and R. Henderson, *Q. Rev. Biophys.* 43:65–158, 2010. With permission from Cambridge University Press. PDB code: 3F3E.)

Some other types of important membrane transport proteins are also built from inverted repeats. Examples even include channel proteins such as the aquaporin water channel (discussed later) and the Sec61 channel through which nascent polypeptides move into the endoplasmic reticulum (discussed in Chapter 12). It is thought that these channels evolved from coupled transporters, in which the gating functions were lost, allowing them to open toward both sides of the membrane simultaneously to provide a continuous path across the membrane.

In bacteria, yeasts, and plants, as well as in many membrane-enclosed organelles of animal cells, most ion-driven active transport systems depend on H$^+$ rather than Na$^+$ gradients, reflecting the predominance of H$^+$ pumps in these membranes. An electrochemical H$^+$ gradient across the bacterial plasma membrane, for example, drives the inward active transport of many sugars and amino acids.

Transporters in the Plasma Membrane Regulate Cytosolic pH

Most proteins operate optimally at a particular pH. Lysosomal enzymes, for example, function best at the low pH (~5) found in lysosomes, whereas cytosolic enzymes function best at the close-to-neutral pH (~7.2) found in the cytosol. It is therefore crucial that cells control the pH of their intracellular compartments.

Most cells have one or more types of Na$^+$-driven antiporters in their plasma membrane that help to maintain the cytosolic pH at about 7.2. These transporters use the energy stored in the Na$^+$ gradient to pump out excess H$^+$, which either leaks in or is produced in the cell by acid-forming reactions. Two mechanisms are used: either H$^+$ is directly transported out of the cell or HCO$_3^-$ is brought into the cell to neutralize H$^+$ in the cytosol (according to the reaction HCO$_3^-$ + H$^+$ → H$_2$O + CO$_2$). One of the antiporters that uses the first mechanism is a *Na$^+$-H$^+$ exchanger*, which couples an influx of Na$^+$ to an efflux of H$^+$. Another, which uses a combination of the two mechanisms, is a *Na$^+$-driven Cl$^-$-HCO$_3^-$ exchanger* that couples an influx of Na$^+$ and HCO$_3^-$ to an efflux of Cl$^-$ and H$^+$ (so that NaHCO$_3$ comes in and HCl goes out). The Na$^+$-driven Cl$^-$-HCO$_3^-$ exchanger is twice as effective as the Na$^+$-H$^+$ exchanger: it pumps out one H$^+$ and neutralizes another for each Na$^+$ that enters the cell. If HCO$_3^-$ is available, as is usually the case, this antiporter is the most important transporter regulating the cytosolic pH. The pH inside the cell regulates both exchangers; when the pH in the cytosol falls, both exchangers increase their activity.

A *Na$^+$-independent Cl$^-$-HCO$_3^-$ exchanger* adjusts the cytosolic pH in the reverse direction. Like the Na$^+$-dependent transporters, pH regulates the Na$^+$-independent Cl$^-$-HCO$_3^-$ exchanger, but the exchanger's activity increases as the cytosol becomes too alkaline. The movement of HCO$_3^-$ in this case is normally out of the cell, down its electrochemical gradient, which decreases the pH of the cytosol. A Na^+-independent Cl^-–HCO_3^- exchanger in the membrane of red blood cells (called band 3 protein—see Figure 10–38) facilitates the quick discharge of CO_2 (as HCO_3^-) as the cells pass through capillaries in the lung.

The intracellular pH is not entirely regulated by transporters in the plasma membrane: ATP-driven H^+ pumps are used to control the pH of many intracellular compartments. As discussed in Chapter 13, H^+ pumps maintain the low pH in lysosomes, as well as in endosomes and secretory vesicles. These H^+ pumps use the energy of ATP hydrolysis to pump H^+ into these organelles from the cytosol.

An Asymmetric Distribution of Transporters in Epithelial Cells Underlies the Transcellular Transport of Solutes

In epithelial cells, such as those that absorb nutrients from the gut, transporters are distributed nonuniformly in the plasma membrane and thereby contribute to the **transcellular transport** of absorbed solutes. By the actions of the transporters in these cells, solutes are moved across the epithelial cell layer into the extracellular fluid from where they pass into the blood. As shown in **Figure 11–11**, Na^+-linked symporters located in the apical (absorptive) domain of the plasma membrane actively transport nutrients into the cell, building up substantial concentration gradients for these solutes across the plasma membrane. Uniporters in the basal and lateral (basolateral) domains allow the nutrients to leave the cell passively down these concentration gradients.

In many of these epithelial cells, the plasma membrane area is greatly increased by the formation of thousands of microvilli, which extend as thin, fingerlike projections from the apical surface of each cell. Such microvilli can increase the total absorptive area of a cell as much as 25-fold, thereby enhancing its transport capabilities.

As we have seen, ion gradients have a crucial role in driving many essential transport processes in cells. Ion pumps that use the energy of ATP hydrolysis establish and maintain these gradients, as we discuss next.

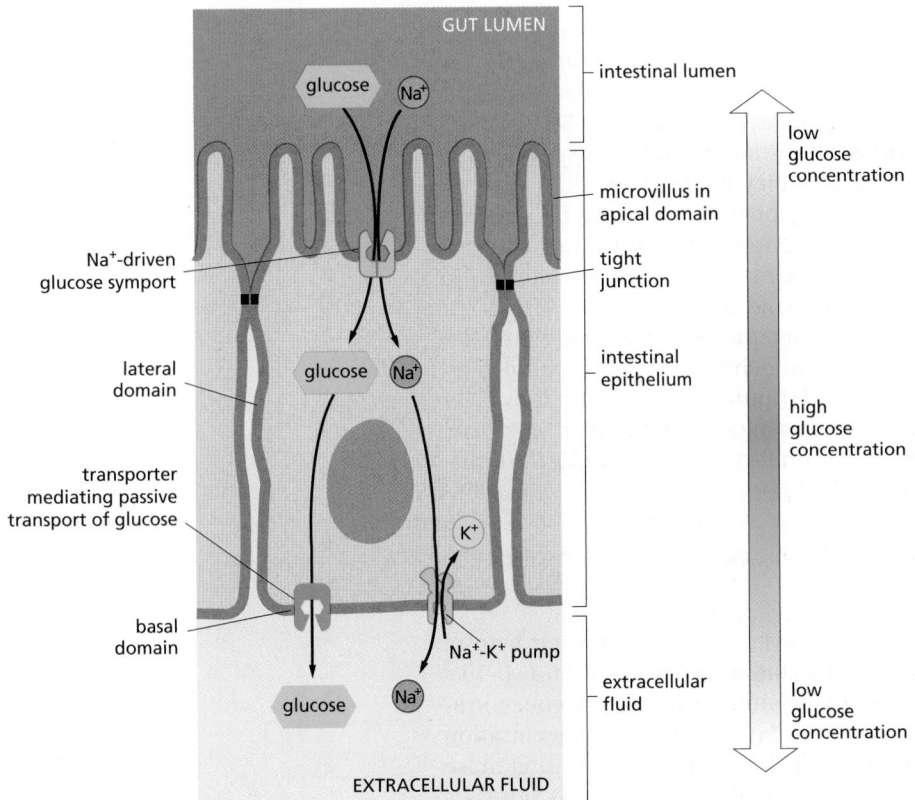

Figure 11–11 **Transcellular transport.** The transcellular transport of glucose across an intestinal epithelial cell depends on the nonuniform distribution of transporters in the cell's plasma membrane. The process shown here results in the transport of glucose from the intestinal lumen to the extracellular fluid (from where it passes into the blood). Glucose is pumped into the cell through the apical domain of the membrane by a Na^+-powered glucose symporter. Glucose passes out of the cell (down its concentration gradient) by passive movement through a glucose uniporter in the basal and lateral membrane domains. The Na^+ gradient driving the glucose symport is maintained by the Na^+-K^+ pump in the basal and lateral plasma membrane domains, which keeps the internal concentration of Na^+ low (Movie 11.2). Adjacent cells are connected by impermeable tight junctions, which have a dual function in the transport process illustrated: they prevent solutes from crossing the epithelium between cells, allowing a concentration gradient of glucose to be maintained across the cell sheet (see Figure 19–18). They also serve as diffusion barriers (fences) within the plasma membrane, which help confine the various transporters to their respective membrane domains (see Figure 10–34).

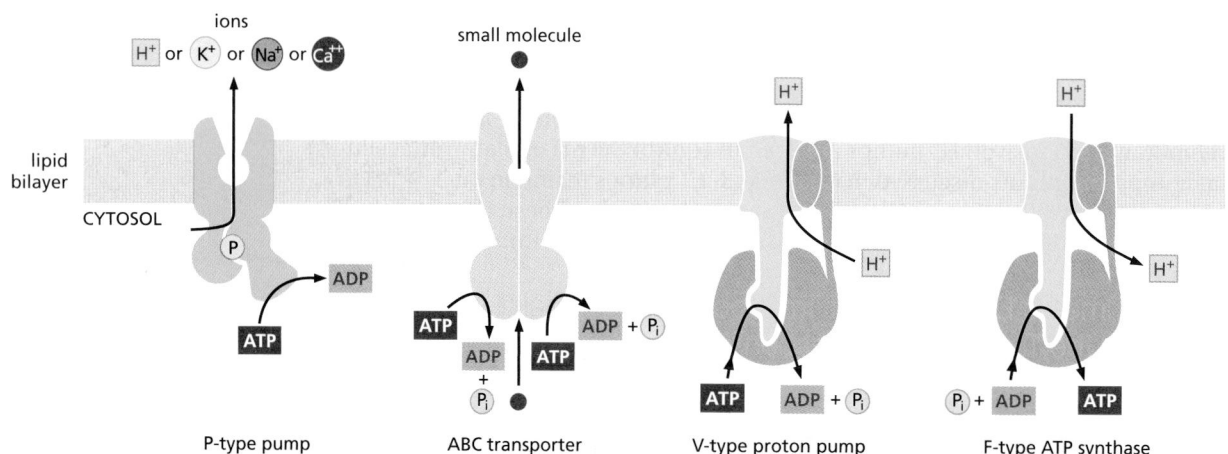

There Are Three Classes of ATP-Driven Pumps

ATP-driven pumps are often called *transport ATPases* because they hydrolyze ATP to ADP and phosphate and use the energy released to pump ions or other solutes across a membrane. There are three principal classes of ATP-driven pumps (**Figure 11–12**), and representatives of each are found in all prokaryotic and eukaryotic cells.

1. **P-type pumps** are structurally and functionally related multipass transmembrane proteins. They are called "P-type" because they phosphorylate themselves during the pumping cycle. This class includes many of the ion pumps that are responsible for setting up and maintaining gradients of Na^+, K^+, H^+, and Ca^{2+} across cell membranes.

2. **ABC transporters** (**ATP-B**inding **C**assette transporters) differ structurally from P-type ATPases and primarily pump small molecules across cell membranes.

3. **V-type pumps** are turbine-like protein machines, constructed from multiple different subunits. The V-type proton pump transfers H^+ into organelles such as lysosomes, synaptic vesicles, and plant or yeast vacuoles (V = vacuolar), to acidify the interior of these organelles (see Figure 13–37).

Structurally related to the V-type pumps is a distinct family of *F-type ATPases*, more commonly called *ATP synthases* because they normally work in reverse: instead of using ATP hydrolysis to drive H^+ transport, they use the H^+ gradient across the membrane to drive the synthesis of ATP from ADP and phosphate (see Figure 14–30). ATP synthases are found in the plasma membrane of bacteria, the inner membrane of mitochondria, and the thylakoid membrane of chloroplasts. The H^+ gradient is generated either during the electron-transport steps of oxidative phosphorylation (in aerobic bacteria and mitochondria), during photosynthesis (in chloroplasts), or by the light-driven H^+ pump (bacteriorhodopsin) in *Halobacterium*. We discuss some of these proteins in detail in Chapter 14.

For the remainder of this section, we focus on P-type pumps and ABC transporters.

A P-type ATPase Pumps Ca^{2+} into the Sarcoplasmic Reticulum in Muscle Cells

Eukaryotic cells maintain very low concentrations of free Ca^{2+} in their cytosol ($\sim10^{-7}$ M) in the face of a very much higher extracellular Ca^{2+} concentration ($\sim10^{-3}$ M). Therefore, even a small influx of Ca^{2+} significantly increases the concentration of free Ca^{2+} in the cytosol, and the flow of Ca^{2+} down its steep concentration gradient in response to extracellular signals is one means of transmitting these signals rapidly across the plasma membrane (discussed in Chapter 15). It is thus

Figure 11–12 Three types of ATP-driven pumps. Like any enzyme, all ATP-driven pumps can work in either direction, depending on the electrochemical gradients of their solutes and the ATP/ADP ratio. When the ATP/ADP ratio is high, they hydrolyze ATP; when the ATP/ADP ratio is low, they can synthesize ATP. The F-type ATPase in mitochondria normally works in this "reverse" mode to make most of the cell's ATP.

important that the cell maintains a steep Ca^{2+} gradient across its plasma membrane. Ca^{2+} transporters that actively pump Ca^{2+} out of the cell help maintain the gradient. One of these is a P-type Ca^{2+} ATPase; the other is an antiporter (called a *Na^+–Ca^{2+} exchanger*) that is driven by the Na^+ electrochemical gradient (discussed in Chapter 15).

The **Ca^{2+} pump**, or **Ca^{2+} ATPase**, in the *sarcoplasmic reticulum* (SR) membrane of skeletal muscle cells is a well-understood P-type transport ATPase. The SR is a specialized type of endoplasmic reticulum that forms a network of tubular sacs in the muscle cell cytoplasm, and it serves as an intracellular store of Ca^{2+}. When an action potential depolarizes the muscle cell plasma membrane, Ca^{2+} is released into the cytosol from the SR through *Ca^{2+}-release channels*, stimulating the muscle to contract (discussed in Chapters 15 and 16). The Ca^{2+} pump, which accounts for about 90% of the membrane protein of the SR, moves Ca^{2+} from the cytosol back into the SR. The endoplasmic reticulum of nonmuscle cells contains a similar Ca^{2+} pump, but in smaller quantities.

Enzymatic studies and analyses of the three-dimensional structures of transport intermediates of the SR Ca^{2+} pump and related pumps have revealed the molecular mechanism of P-type transport ATPases in great detail. They all have similar structures, containing 10 transmembrane α helices connected to three cytosolic domains (**Figure 11–13**). In the Ca^{2+} pump, amino acid side chains protruding from the transmembrane helices form two centrally positioned binding sites for Ca^{2+}. As shown in **Figure 11–14**, in the pump's ATP-bound nonphosphorylated state, these binding sites are accessible only from the cytosolic side of the SR membrane. Ca^{2+} binding triggers a series of conformational changes that close the passageway to the cytosol and activate a phosphotransfer reaction in which the terminal phosphate of the ATP is transferred to an aspartate that is highly conserved among all P-type ATPases. The ADP then dissociates and is replaced with a fresh ATP, causing another conformational change that opens a passageway to the SR lumen through which the two Ca^{2+} ions exit. They are replaced by two H^+ ions and a water molecule that stabilize the empty Ca^{2+}-binding sites and close the passageway to the SR lumen. Hydrolysis of the labile phosphoryl-aspartate bond returns the pump to the initial conformation, and the cycle starts again. The transient self-phosphorylation of the pump during its cycle is an essential characteristic of all P-type pumps.

The Plasma Membrane Na^+-K^+ Pump Establishes Na^+ and K^+ Gradients Across the Plasma Membrane

The concentration of K^+ is typically 10–30 times higher inside cells than outside, whereas the reverse is true of Na^+ (see Table 11-1, p. 598). A **Na^+-K^+ pump**, or **Na^+-K^+ ATPase**, found in the plasma membrane of virtually all animal cells maintains

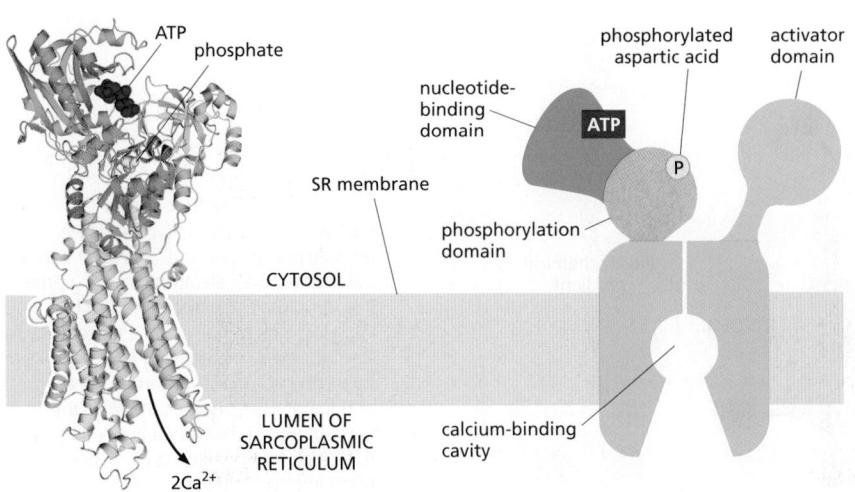

Figure 11–13 **The structure of the sarcoplasmic reticulum Ca^{2+} pump**. The ribbon model (*left*), derived from x-ray crystallographic analyses, shows the pump in its phosphorylated, ATP-bound state. The three globular cytosolic domains of the pump—the nucleotide-binding domain (*dark green*), the activator domain (*blue*), and the phosphorylation domain (*red*), also shown schematically on the *right*—change conformation dramatically during the pumping cycle. These changes in turn alter the arrangement of the transmembrane helices, which allows the Ca^{2+} to be released from its binding cavity into the SR lumen (**Movie 11.3**). (PDB code: 3B9B.)

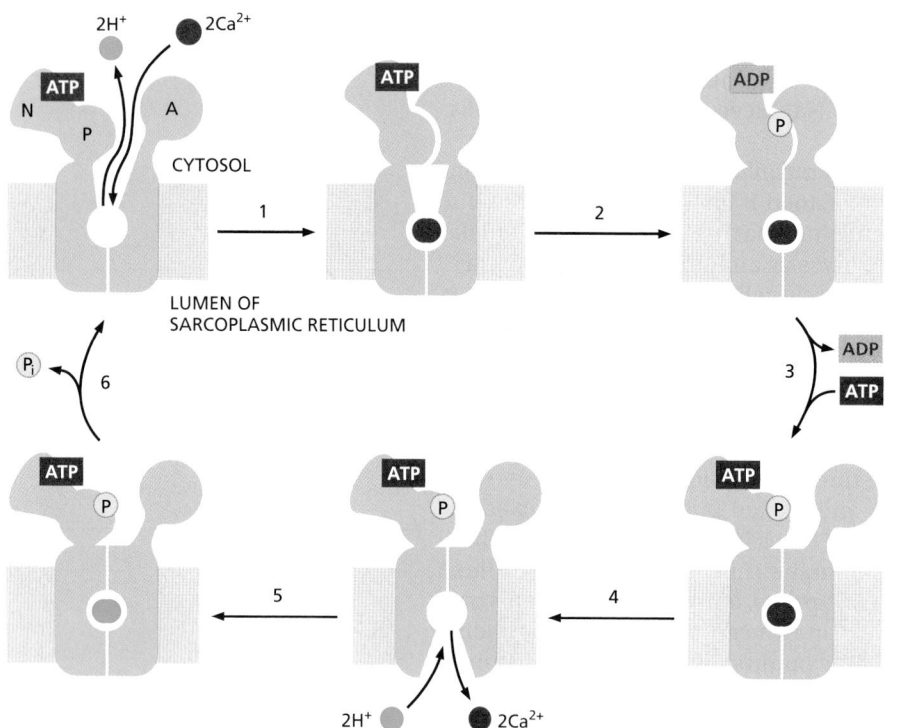

Figure 11–14 The pumping cycle of the sarcoplasmic reticulum Ca²⁺ pump. Ion pumping proceeds by a series of stepwise conformational changes in which movements of the pump's three cytosolic domains [the nucleotide-binding domain (N), the phosphorylation domain (P), and the activator domain (A)] are mechanically coupled to movements of the transmembrane α helices. Helix movement opens and closes passageways through which Ca²⁺ enters from the cytosol and binds to the two centrally located Ca²⁺ binding sites. The two Ca²⁺ then exit into the SR lumen and are replaced by two H⁺, which are transported in the opposite direction. The Ca²⁺-dependent phosphorylation and H⁺-dependent dephosphorylation of aspartic acid are universally conserved steps in the reaction cycle of all P-type pumps: they cause the conformational transitions to occur in an orderly manner, enabling the proteins to do useful work. (Adapted from C. Toyoshima et al., *Nature* 432:361–368, 2004 and J.V. Møller et al., *Q. Rev. Biophys.* 43:501–566, 2010.)

these concentration differences. Like the Ca²⁺ pump, the Na⁺-K⁺ pump belongs to the family of P-type ATPases and operates as an ATP-driven antiporter, actively pumping Na⁺ out of the cell against its steep electrochemical gradient and pumping K⁺ in (**Figure 11–15**).

We mentioned earlier that the Na⁺ gradient produced by the Na⁺-K⁺ pump drives the transport of most nutrients into animal cells and also has a crucial role in regulating cytosolic pH. A typical animal cell devotes almost one-third of its energy to fueling this pump, and the pump consumes even more energy in nerve cells and in cells that are dedicated to transport processes, such as those forming kidney tubules.

Since the Na⁺-K⁺ pump drives three positively charged ions out of the cell for every two it pumps in, it is *electrogenic*: it drives a net electric current across the membrane, tending to create an electrical potential, with the cell's inside being negative relative to the outside. This electrogenic effect of the pump, however, seldom directly contributes more than 10% to the membrane potential. The remaining 90%, as we discuss later, depends only indirectly on the Na⁺-K⁺ pump.

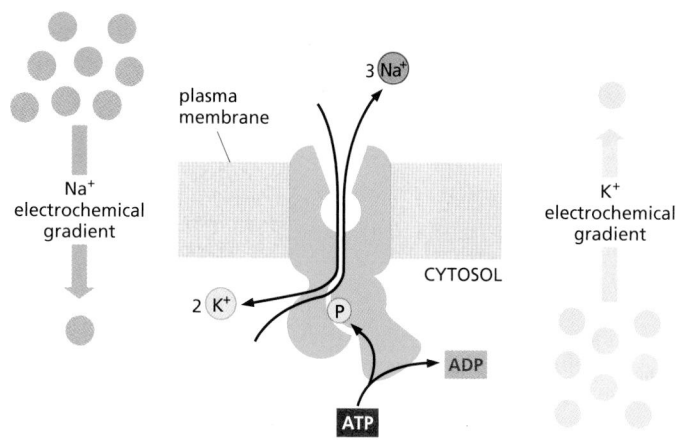

Figure 11–15 The function of the Na⁺-K⁺ pump. This P-type ATPase actively pumps Na⁺ out of and K⁺ into a cell against their electrochemical gradients. It is structurally closely related to the Ca²⁺ ATPase but differs in its selectivity for ions: for every molecule of ATP hydrolyzed by the pump, three Na⁺ are pumped out and two K⁺ are pumped in. As in the Ca²⁺ pump, an aspartate is phosphorylated and dephosphorylated during the pumping cycle (**Movie 11.4**).

ABC Transporters Constitute the Largest Family of Membrane Transport Proteins

The last type of transport ATPase that we discuss is the family of the ABC transporters, so named because each member contains two highly conserved ATPase domains, or **A**TP-**B**inding "**C**assettes," on the cytosolic side of the membrane. ATP binding brings together the two ATPase domains, and ATP hydrolysis leads to their dissociation (**Figure 11–16**). These movements of the cytosolic domains are transmitted to the transmembrane segments, driving cycles of conformational changes that alternately expose solute-binding sites on one side of the membrane and then on the other, as we have seen for other transporters. In this way, ABC transporters harvest the energy released upon ATP binding and hydrolysis to drive transport of solutes across the bilayer. The transport is directional toward inside or toward outside, depending on the particular conformational change in the solute binding site that is linked to ATP hydrolysis (see Figure 11–16).

ABC transporters constitute the largest family of membrane transport proteins and are of great clinical importance. The first of these proteins to be characterized was found in bacteria. We have already mentioned that the plasma membranes of all bacteria contain transporters that use the H⁺ gradient across the membrane to actively transport a variety of nutrients into the cell. In addition, bacteria use ABC transporters to import certain small molecules. In bacteria such as *E. coli* that have double membranes (**Figure 11–17**), the ABC transporters are located in the inner membrane, and an auxiliary mechanism operates to capture the nutrients and deliver them to the transporters (**Figure 11–18**).

In *E. coli,* 78 genes (an amazing 5% of the bacterium's genes) encode ABC transporters, and animal genomes encode an even larger number. Although each transporter is thought to be specific for a particular molecule or class of molecules, the variety of substrates transported by this superfamily is great and includes inorganic ions, amino acids, mono- and polysaccharides, peptides, lipids, drugs, and, in some cases, even proteins that can be larger than the transporter itself.

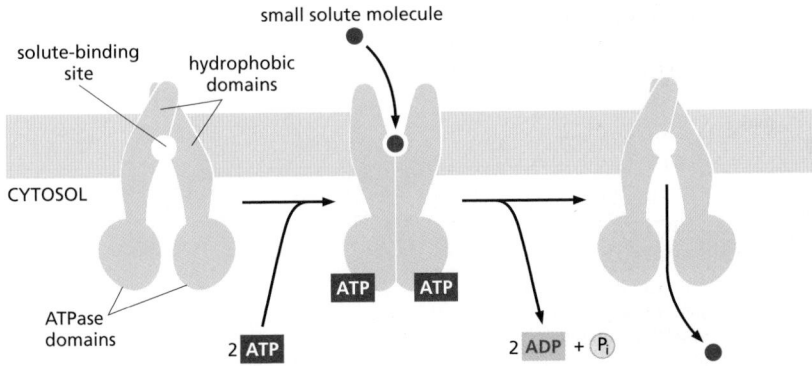

(A) A BACTERIAL ABC TRANSPORTER

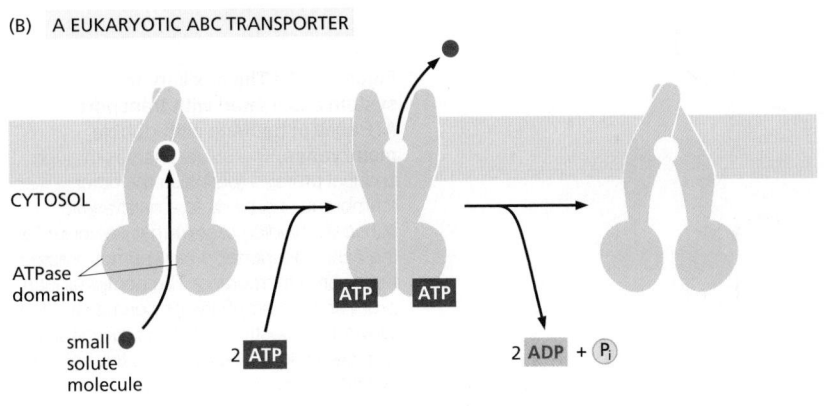

(B) A EUKARYOTIC ABC TRANSPORTER

Figure 11–16 Small-molecule transport by typical ABC transporters. ABC transporters consist of multiple domains. Typically, two hydrophobic domains, each built of six membrane-spanning α helices, together form the translocation pathway and provide substrate specificity. Two ATPase domains protrude into the cytosol. In some cases, the two halves of the transporter are formed by a single polypeptide, whereas in other cases they are formed by two or more separate polypeptides that assemble into a similar structure. Without ATP bound, the transporter exposes a substrate-binding site on one side of the membrane. ATP binding induces a conformational change that exposes the substrate-binding site on the opposite side; ATP hydrolysis followed by ADP dissociation returns the transporter to its original conformation. Most individual ABC transporters are unidirectional. (A) Both importing and exporting ABC transporters are found in bacteria; an ABC importer is shown in this cartoon. The crystal structure of a bacterial ABC transporter is shown in Figure 3–76. (B) In eukaryotes, most ABC transporters export substances—either from the cytosol to the extracellular space or from the cytosol to a membrane-bound intracellular compartment such as the endoplasmic reticulum—or from the mitochondrial matrix to the cytosol.

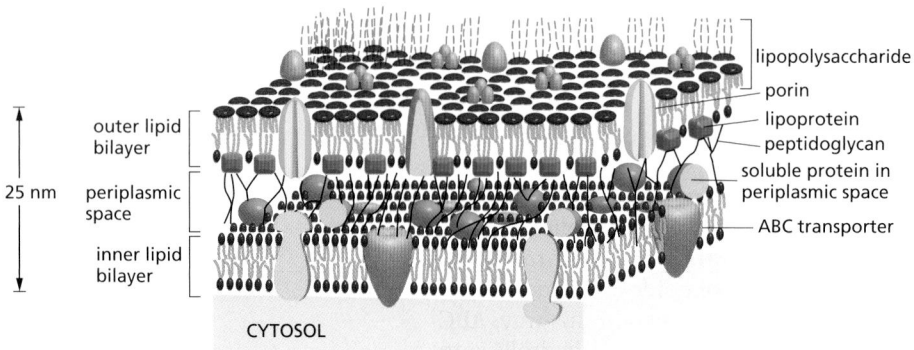

Figure 11–17 **A small section of the double membrane of an *E. coli* bacterium.** The inner membrane is the cell's plasma membrane. Between the inner and outer membranes is a highly porous, rigid peptidoglycan layer, composed of protein and polysaccharide that constitute the bacterial cell wall. It is attached to lipoprotein molecules in the outer membrane and fills the *periplasmic space* (only a little of the peptidoglycan layer is shown). This space also contains a variety of soluble protein molecules. The dashed threads (shown in *green*) at the top represent the polysaccharide chains of the special lipopolysaccharide molecules that form the external monolayer of the outer membrane; for clarity, only a few of these chains are shown. Bacteria with double membranes are called *Gram-negative* because they do not retain the dark blue dye used in Gram staining. Bacteria with single membranes (but thicker peptidoglycan cell walls), such as staphylococci and streptococci, retain the blue dye and are therefore called *Gram-positive*; their single membrane is analogous to the inner (plasma) membrane of Gram-negative bacteria.

The first eukaryotic ABC transporters identified were discovered because of their ability to pump hydrophobic drugs out of the cytosol. One of these transporters is the **multidrug resistance (MDR) protein**, also called P-glycoprotein. It is present at elevated levels in many human cancer cells and makes the cells simultaneously resistant to a variety of chemically unrelated cytotoxic drugs that are widely used in cancer chemotherapy. Treatment with any one of these drugs can result in the selective survival and overgrowth of those cancer cells that express an especially large amount of the MDR transporter. These cells pump drugs out of the cell very efficiently and are therefore relatively resistant to the drugs' toxic effects (Movie 11.5). Selection for cancer cells with resistance to one drug can thereby lead to resistance to a wide variety of anticancer drugs. Some studies indicate that up to 40% of human cancers develop multidrug resistance, making it a major hurdle in the battle against cancer.

A related and equally sinister phenomenon occurs in the protist *Plasmodium falciparum*, which causes malaria. More than 200 million people are infected worldwide with this parasite, which remains a major cause of human death, killing almost a million people every year. The development of resistance to the antimalarial drug *chloroquine* has hampered the control of malaria. The resistant *P. falciparum* have amplified a gene encoding an ABC transporter that pumps out the chloroquine.

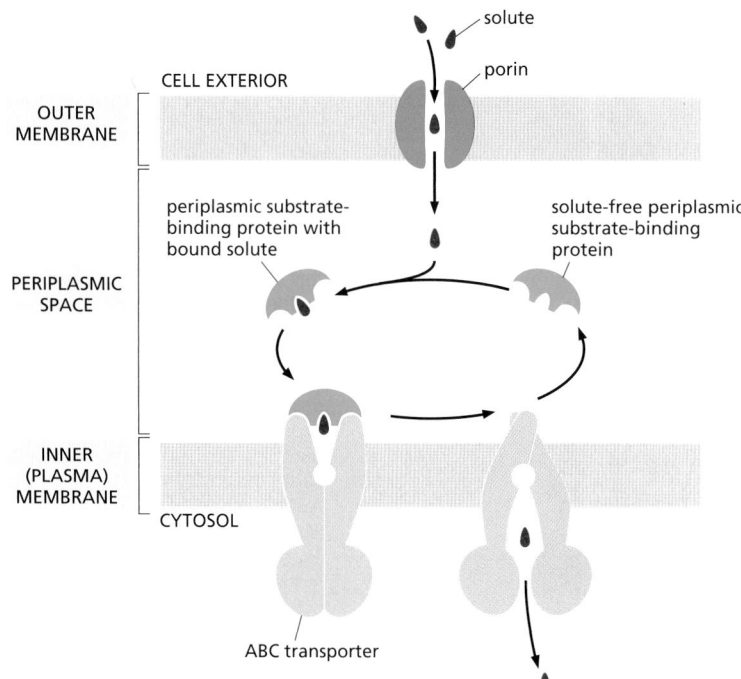

Figure 11–18 **The auxiliary transport system associated with transport ATPases in bacteria with double membranes.** The solute diffuses through channel proteins (porins) in the outer membrane and binds to a *periplasmic substrate-binding protein* that delivers it to the ABC transporter, which pumps it across the plasma membrane. The peptidoglycan is omitted for simplicity; its porous structure allows the substrate-binding proteins and water-soluble solutes to move through it by diffusion.

In most vertebrate cells, an ABC transporter in the endoplasmic reticulum (ER) membrane (named *transporter associated with antigen processing*, or *TAP transporter*) actively pumps a wide variety of peptides from the cytosol into the ER lumen. These peptides are produced by protein degradation in proteasomes (discussed in Chapter 6). They are carried from the ER to the cell surface, where they are displayed for scrutiny by cytotoxic T lymphocytes, which kill the cell if the peptides are derived from a virus or other microorganism lurking in the cytosol of an infected cell (discussed in Chapter 24).

Yet another member of the ABC transporter family is the *cystic fibrosis trans-membrane conductance regulator* protein (CFTR), which was discovered through studies of the common genetic disease *cystic fibrosis*. This disease is caused by a mutation in the gene encoding CFTR, a Cl^- transport protein in the plasma membrane of epithelial cells. CFTR regulates ion concentrations in the extracellular fluid, especially in the lung. One in 27 Caucasians carries a gene encoding a mutant form of this protein; in 1 in 2900, both copies of the gene are mutated, causing the disease. In contrast to other ABC transporters, ATP binding and hydrolysis in the CFTR protein do not drive the transport process. Instead, they control the opening and closing of a continuous channel, which provides a passive conduit for Cl^- to move down its electrochemical gradient. Thus, some ABC proteins can function as transporters and others as gated channels.

Summary

Transporters bind specific solutes and transfer them across the lipid bilayer by undergoing conformational changes that alternately expose the solute-binding site on one of the membrane and then on the other. Some transporters move a single solute "downhill," whereas others can act as pumps to move a solute "uphill" against its electrochemical gradient, using energy provided by ATP hydrolysis, by a downhill flow of another solute (such as Na^+ or H^+), or by light to drive the requisite series of conformational changes in an orderly manner. Transporters belong to a small number of protein families. Each family evolved from a common ancestral protein, and its members all operate by a similar mechanism. The family of P-type transport ATPases, which includes Ca^{2+} and Na^+-K^+ pumps, is an important example; each of these ATPases sequentially phosphorylates and dephosphorylates itself during the pumping cycle. The superfamily of ABC transporters is the largest family of membrane transport proteins and is especially important clinically. It includes proteins that are responsible for cystic fibrosis, for drug resistance in both cancer cells and malaria-causing parasites, and for pumping pathogen-derived peptides into the ER for cytotoxic lymphocytes to reorganize on the surface of infected cells.

CHANNELS AND THE ELECTRICAL PROPERTIES OF MEMBRANES

Unlike transporters, channels form pores across membranes. One class of channel proteins found in virtually all animals forms *gap junctions* between adjacent cells; each plasma membrane contributes equally to the formation of the channel, which connects the cytoplasm of the two cells. These channels are discussed in Chapter 19 and will not be considered further here. Both gap junctions and *porins*, the channels in the outer membranes of bacteria, mitochondria, and chloroplasts (discussed in Chapter 10), have relatively large and permissive pores, and it would be disastrous if they directly connected the inside of a cell to an extracellular space. Indeed, many bacterial toxins do exactly that to kill other cells (discussed in Chapter 24).

In contrast, most channels in the plasma membrane of animal and plant cells that connect the cytosol to the cell exterior necessarily have narrow, highly selective pores that can open and close rapidly. Because these proteins are concerned specifically with inorganic ion transport, they are referred to as **ion channels**. For transport efficiency, ion channels have an advantage over transporters, in that they can pass up to 100 million ions through one open channel each second—a rate 10^5 times greater than the fastest rate of transport mediated by any known transporter. As discussed earlier, however, channels cannot be coupled to an energy source to perform active transport, so the transport they mediate is always passive (downhill). Thus, the function of ion channels is to allow specific inorganic ions—primarily Na^+, K^+, Ca^{2+}, or Cl^-—to diffuse rapidly down their electrochemical gradients across the lipid bilayer. In this section, we will see that the ability to control ion fluxes through these channels is essential for many cell functions. Nerve cells (neurons), in particular, have made a specialty of using ion channels, and we will consider how they use many different ion channels to receive, conduct, and transmit signals. Before we discuss ion channels, however, we briefly consider the aquaporin water channels that we mentioned earlier.

Aquaporins Are Permeable to Water But Impermeable to Ions

Because cells are mostly water (typically ~70% by weight), water movement across cell membranes is fundamentally important for life. Cells also contain a high concentration of solutes, including numerous negatively charged organic molecules that are confined inside the cell (the so-called *fixed anions*) and their accompanying cations that are required for charge balance. This creates an osmotic gradient, which mostly is balanced by an opposite osmotic gradient due to a high concentration of inorganic ions—chiefly Na^+ and Cl^-—in the extracellular fluid. The small remaining osmotic force tends to "pull" water into the cell, causing it to swell until the forces are balanced. Because all biological membranes are moderately permeable to water (see Figure 11–2), cell volume equilibrates in minutes or less in response to an osmotic gradient. For most animal cells, however, osmosis has only a minor role in regulating cell volume. This is because most of the cytoplasm is in a gel-like state and resists large changes in its volume in response to changes in osmolarity.

In addition to the direct diffusion of water across the lipid bilayer, some prokaryotic and eukaryotic cells have **water channels**, or **aquaporins**, embedded in their plasma membrane to allow water to move more rapidly. Aquaporins are particularly abundant in animal cells that must transport water at high rates, such as the epithelial cells of the kidney or exocrine cells that must transport or secrete large volumes of fluids, respectively (**Figure 11–19**).

Aquaporins must solve a problem that is opposite to that facing ion channels. To avoid disrupting ion gradients across membranes, they have to allow the rapid passage of water molecules while completely blocking the passage of ions. The three-dimensional structure of an aquaporin reveals how it achieves this remarkable selectivity. The channels have a narrow pore that allows water molecules to traverse the membrane in single file, following the path of carbonyl oxygens that line one side of the pore (**Figure 11–20**A and B). Hydrophobic amino acids line the other side of the pore. The pore is too narrow for any hydrated ion to enter, and the energy cost of dehydrating an ion would be enormous because the hydrophobic wall of the pore cannot interact with a dehydrated ion to compensate for the loss of water. This design readily explains why the aquaporins cannot conduct K^+,

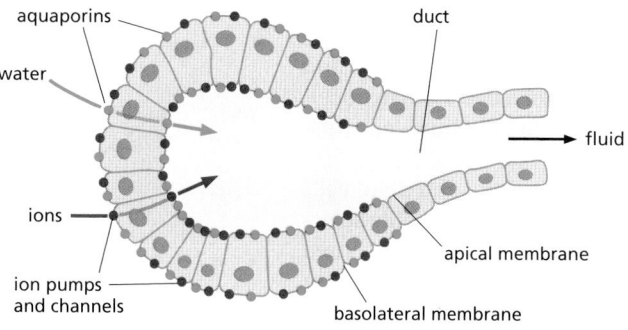

Figure 11–19 The role of aquaporins in fluid secretion. Cells lining the ducts of exocrine glands (as found, for example, in the pancreas and liver, and in mammary, sweat, and salivary glands) secrete large volumes of body fluids. These cells are organized into epithelial sheets in which their apical plasma membrane faces the lumen of the duct. Ion pumps and channels situated in the basolateral and apical plasma membrane move ions (mostly Na^+ and Cl^-) into the ductal lumen, creating an osmotic gradient between the surrounding tissue and the duct. Water molecules rapidly follow the osmotic gradient through aquaporins that are present in high concentrations in both the apical and basolateral membranes.

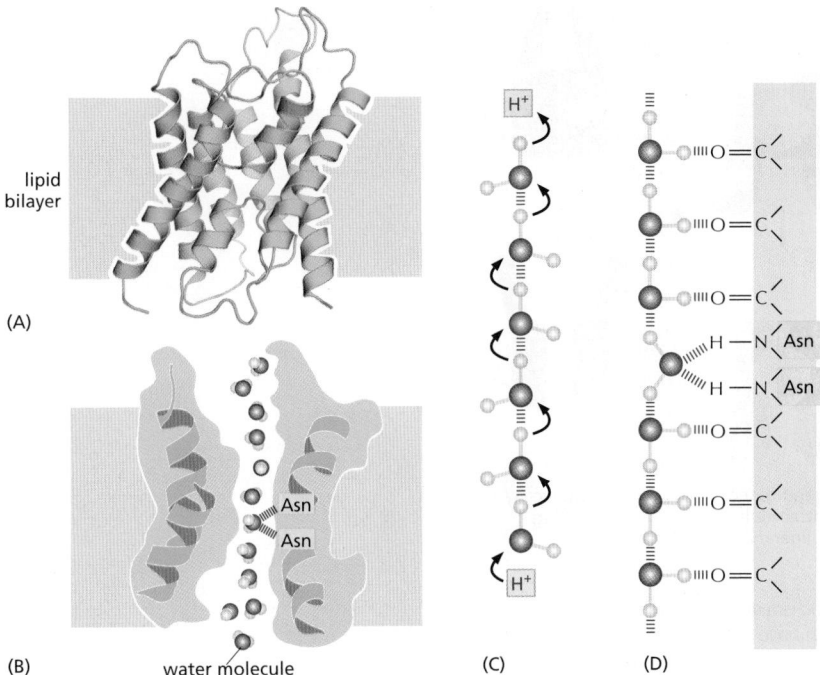

(A)

(B)

water molecule

(C)

(D)

H+

Asn

Asn

H—N Asn

H—N Asn

H+

Figure 11–20 **The structure of aquaporins.** (A) A ribbon diagram of an aquaporin monomer. In the membrane, aquaporins form tetramers, with each monomer containing an aqueous pore in its center (not shown). Each individual aquaporin channel passes about 10^9 water molecules per second. (B) A longitudinal cross section through one aquaporin monomer, in the plane of the central pore. One face of the pore is lined with hydrophilic amino acids, which provide transient hydrogen bonds to water molecules; these bonds help line up the transiting water molecules in a single row and orient them as they traverse the pore. (C and D) A model explaining why aquaporins are impermeable to H+. (C) In water, H+ diffuses extremely rapidly by being relayed from one water molecule to the next. (D) Carbonyl groups (C=O) lining the hydrophilic face of the pore align water molecules, and two strategically placed asparagines in the center help tether a central water molecule such that both valences on its oxygen are occupied. This arrangement bipolarizes the entire line of water molecules, with each water molecule acting as a hydrogen-bond acceptor from its inner neighbor (**Movie 11.6**). (A and B, adapted from R.M. Stroud et al., *Curr. Opin. Struct. Biol.* 13:424–431, 2003.)

Na+, Ca^{2+}, or Cl$^-$ ions. These channels are also impermeable to H+, which is mainly present in cells as H$_3$O+. These hydronium ions diffuse through water extremely rapidly, using a molecular relay mechanism that requires the making and breaking of hydrogen bonds between adjacent water molecules (Figure 11–20C). Aquaporins contain two strategically placed asparagines, which bind to the oxygen atom of the central water molecule in the line of water molecules traversing the pore, imposing a bipolarity on the entire column of water molecules (Figure 11–20C and D). This makes it impossible for the "making and breaking" sequence of hydrogen bonds (shown in Figure 11–20C) to get past the central asparagine-bonded water molecule. Because both valences of this central oxygen are unavailable for hydrogen-bonding, the central water molecule cannot participate in an H+ relay, and the pore is therefore impermeable to H+.

We now turn to ion channels, the subject of the rest of the chapter.

Ion Channels Are Ion-Selective and Fluctuate Between Open and Closed States

Two important properties distinguish ion channels from aqueous pores. First, they show *ion selectivity*, permitting some inorganic ions to pass, but not others. This suggests that their pores must be narrow enough in places to force permeating ions into intimate contact with the walls of the channel so that only ions of appropriate size and charge can pass. The permeating ions have to shed most or all of their associated water molecules to pass, often in single file, through the narrowest part of the channel, which is called the *selectivity filter*; this limits their rate of passage (**Figure 11–21**). Thus, as the ion concentration increases, the flux of the ion through a channel increases proportionally but then levels off (saturates) at a maximum rate.

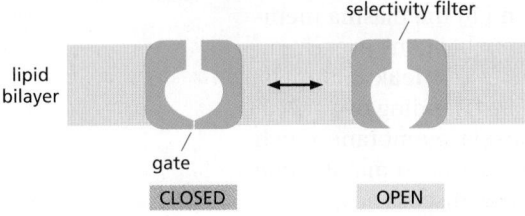

selectivity filter

lipid bilayer

gate

CLOSED

OPEN

Figure 11–21 **A typical ion channel, which fluctuates between closed and open conformations.** The ion channel shown here in cross section forms a pore across the lipid bilayer only in the "open" conformational state. The pore narrows to atomic dimensions in one region (the selectivity filter), where the ion selectivity of the channel is largely determined. Another region of the channel forms the gate.

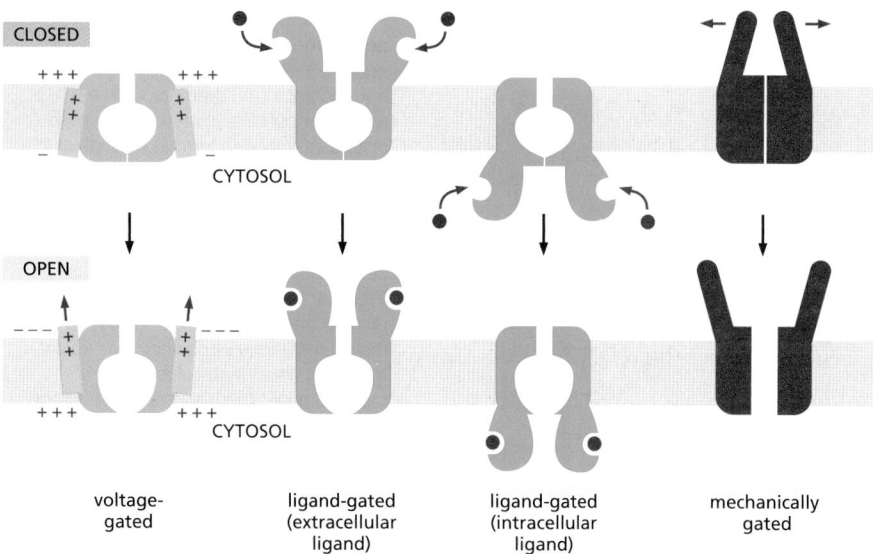

Figure 11–22 The gating of ion channels. This schematic drawing shows several kinds of stimuli that open ion channels. Mechanically gated channels often have cytoplasmic extensions (not shown) that link the channel to the cytoskeleton.

The second important distinction between ion channels and aqueous pores is that ion channels are not continuously open. Instead, they are *gated*, which allows them to open briefly and then close again. Moreover, with prolonged (chemical or electrical) stimulation, most ion channels go into a closed "desensitized," or "inactivated," state, in which they are refractory to further opening until the stimulus has been removed, as we discuss later. In most cases, the gate opens in response to a specific stimulus. As shown in **Figure 11–22**, the main types of stimuli that are known to cause ion channels to open are a change in the voltage across the membrane (*voltage-gated channels*), a mechanical stress (*mechanically gated channels*), or the binding of a ligand (*ligand-gated channels*). The ligand can be either an extracellular mediator—specifically, a neurotransmitter (*transmitter-gated channels*)—or an intracellular mediator such as an ion (*ion-gated channels*) or a nucleotide (*nucleotide-gated channels*). In addition, protein phosphorylation and dephosphorylation regulates the activity of many ion channels; this type of channel regulation is discussed, together with nucleotide-gated ion channels, in Chapter 15.

More than 100 types of ion channels have been identified thus far, and new ones are still being discovered, each characterized by the ions it conducts, the mechanism by which it is gated, and its abundance and localization in the cell and in specific cells. Ion channels are responsible for the electrical excitability of muscle cells, and they mediate most forms of electrical signaling in the nervous system. A single neuron typically contains 10 or more kinds of ion channels, located in different domains of its plasma membrane. But ion channels are not restricted to electrically excitable cells. They are present in all animal cells and are found in plant cells and microorganisms: they propagate the leaf-closing response of the mimosa plant, for example (**Movie 11.7**), and allow the single-celled *Paramecium* to reverse direction after a collision.

Ion channels that are permeable mainly to K^+ are found in the plasma membrane of almost all cells. An important subset of K^+ channels opens even in an unstimulated or "resting" cell, and hence these are called K^+ **leak channels**. Although this term applies to many different K^+ channels, depending on the cell type, they serve a common purpose: by making the plasma membrane much more permeable to K^+ than to other ions, they have a crucial role in maintaining the membrane potential across all plasma membranes, as we discuss next.

The Membrane Potential in Animal Cells Depends Mainly on K⁺ Leak Channels and the K⁺ Gradient Across the Plasma Membrane

A **membrane potential** arises when there is a difference in the electrical charge on the two sides of a membrane, due to a slight excess of positive ions over negative ones on one side and a slight deficit on the other. Such charge differences can result both from active electrogenic pumping (see p. 608) and from passive ion diffusion. As we discuss in Chapter 14, electrogenic H^+ pumps in the mitochondrial inner membrane generate most of the membrane potential across this membrane. Electrogenic pumps also generate most of the electrical potential across the plasma membrane in plants and fungi. In typical animal cells, however, passive ion movements make the largest contribution to the electrical potential across the plasma membrane.

As explained earlier, due to the action of the Na^+-K^+ pump, there is little Na^+ inside the cell, and other intracellular inorganic cations have to be plentiful enough to balance the charge carried by the cell's fixed anions—the negatively charged organic molecules that are confined inside the cell. This balancing role is performed largely by K^+, which is actively pumped into the cell by the Na^+-K^+ pump and can also move freely in or out through the *K⁺ leak channels* in the plasma membrane. Because of the presence of these channels, K^+ comes almost to equilibrium, where an electrical force exerted by an excess of negative charges attracting K^+ into the cell balances the tendency of K^+ to leak out down its concentration gradient. The membrane potential (of the plasma membrane) is the manifestation of this electrical force, and we can calculate its equilibrium value from the steepness of the K^+ concentration gradient. The following argument may help to make this clear.

Suppose that initially there is no voltage gradient across the plasma membrane (the membrane potential is zero) but the concentration of K^+ is high inside the cell and low outside. K^+ will tend to leave the cell through the K^+ leak channels, driven by its concentration gradient. As K^+ begins to move out, each ion leaves behind an unbalanced negative charge, thereby creating an electrical field, or membrane potential, which will tend to oppose the further efflux of K^+. The net efflux of K^+ halts when the membrane potential reaches a value at which this electrical driving force on K^+ exactly balances the effect of its concentration gradient—that is, when the electrochemical gradient for K^+ is zero. Although Cl^- ions also equilibrate across the membrane, the membrane potential keeps most of these ions out of the cell because their charge is negative.

The equilibrium condition, in which there is no net flow of ions across the plasma membrane, defines the **resting membrane potential** for this idealized cell. A simple but very important formula, the **Nernst equation**, quantifies the equilibrium condition and, as explained in Panel 11–1, makes it possible to calculate the theoretical resting membrane potential if we know the ratio of internal and external ion concentrations. As the plasma membrane of a real cell is not exclusively permeable to K^+ and Cl^-, however, the actual resting membrane potential is usually not exactly equal to that predicted by the Nernst equation for K^+ or Cl^-.

The Resting Potential Decays Only Slowly When the Na⁺-K⁺ Pump Is Stopped

Movement of only a minute number of inorganic ions across the plasma membrane through ion channels suffices to set up the membrane potential. Thus, we can think of the membrane potential as arising from movements of charge that leave ion *concentrations* practically unaffected and result in only a very slight discrepancy in the number of positive and negative ions on the two sides of the membrane (Figure 11–23). Moreover, these movements of charge are generally rapid, taking only a few milliseconds or less.

Consider the change in the membrane potential in a real cell after the sudden inactivation of the Na^+-K^+ pump. A slight drop in the membrane potential occurs immediately. This is because the pump is electrogenic and, when active, makes a

THE NERNST EQUATION AND ION FLOW

The flow of any inorganic ion through a membrane channel is driven by the electrochemical gradient for that ion. This gradient represents the combination of two influences: the voltage gradient and the concentration gradient of the ion across the membrane. When these two influences just balance each other, the electrochemical gradient for the ion is zero, and there is no *net* flow of the ion through the channel. The voltage gradient (membrane potential) at which this equilibrium is reached is called the equilibrium potential for the ion. It can be calculated from an equation that will be derived below, called the Nernst equation.

The Nernst equation is $\qquad V = \dfrac{RT}{zF} \ln \dfrac{C_o}{C_i}$

where

    V = the equilibrium potential in volts (internal potential minus external potential)

    C_o and C_i = outside and inside concentrations of the ion, respectively

    R = the gas constant ($8.3 \, \text{J mol}^{-1} \text{K}^{-1}$)

    T = the absolute temperature (K)

    F = Faraday's constant ($9.6 \times 10^4 \, \text{J V}^{-1} \text{mol}^{-1}$)

    z = the valence (charge) of the ion

    $\ln$ = logarithm to the base e

The Nernst equation is derived as follows:

A molecule in solution (a solute) tends to move from a region of high concentration to a region of low concentration simply due to the random movement of molecules, which results in their equilibrium. Consequently, movement down a concentration gradient is accompanied by a favorable free-energy change ($\Delta G < 0$), whereas movement up a concentration gradient is accompanied by an unfavorable free-energy change ($\Delta G > 0$). (Free energy is introduced in Chapter 2 and discussed in the context of redox reactions in Panel 14–1, p. 765.)

The free-energy change per mole of solute moved across the plasma membrane (ΔG_{conc}) is equal to $-RT \ln C_o/C_i$.

If the solute is an ion, moving it into a cell across a membrane whose inside is at a voltage V relative to the outside will cause an additional free-energy change (per mole of solute moved) of $\Delta G_{volt} = zFV$.

At the point where the concentration and voltage gradients just balance,

$$\Delta G_{conc} + \Delta G_{volt} = 0$$

and the ion distribution is at equilibrium across the membrane.

Thus,

$$zFV - RT \ln \frac{C_o}{C_i} = 0$$

and, therefore,

$$V = \frac{RT}{zF} \ln \frac{C_o}{C_i}$$

or, using the constant that converts natural logarithms to base 10,

$$V = 2.3 \frac{RT}{zF} \log_{10} \frac{C_o}{C_i}$$

For a univalent cation,

$$2.3 \frac{RT}{F} = 58 \text{ mV at } 20°C \quad \text{and} \quad 61.5 \text{ mV at } 37°C.$$

Thus, for such an ion at 37°C,

$$V = +61.5 \text{ mV for } C_o / C_i = 10,$$

whereas

$$V = 0 \text{ for } C_o / C_i = 1.$$

The K^+ equilibrium potential (V_K), for example, is
$$61.5 \log_{10}([K^+]_o / [K^+]_i) \text{ millivolts}$$
(-89 mV for a typical cell, where $[K^+]_o = 5$ mM and $[K^+]_i = 140$ mM).

At V_K, there is no net flow of K^+ across the membrane.

Similarly, when the membrane potential has a value of
$$61.5 \log_{10}([Na^+]_o / [Na^+]_i),$$
the Na^+ equilibrium potential (V_{Na}),
there is no net flow of Na^+.

For any particular membrane potential, V_M, the net force tending to drive a particular type of ion out of the cell, is proportional to the difference between V_M and the equilibrium potential for the ion: hence,

for K^+ it is $V_M - V_K$

and for Na^+ it is $V_M - V_{Na}$.

When there is a voltage gradient across the membrane, the ions responsible for it—the positive ions on one side and the negative ions on the other—are concentrated in thin layers on either side of the membrane because of the attraction between positive and negative electric charges. The number of ions that go to form the layer of charge adjacent to the membrane is minute compared with the total number inside the cell. For example, the movement of 6000 Na^+ ions across 1 μm^2 of membrane will carry sufficient charge to shift the membrane potential by about 100 mV.

Because there are about 3×10^7 Na^+ ions in a typical cell (1 μm^3 of bulk cytoplasm), such a movement of charge will generally have a negligible effect on the ion concentration gradients across the membrane.

exact balance of charges on each side of the membrane; membrane potential = 0

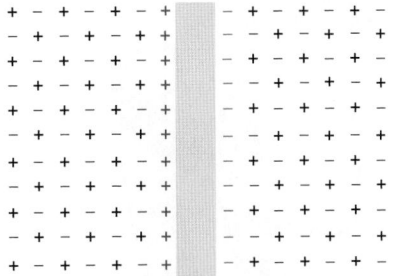

a few of the positive ions *(red)* cross the membrane from right to left, leaving their negative counterions *(red)* behind; this sets up a nonzero membrane potential

Figure 11–23 **The ionic basis of a membrane potential.** A small flow of inorganic ions through an ion channel carries sufficient charge to cause a large change in the membrane potential. The ions that give rise to the membrane potential lie in a thin (< 1 nm) surface layer close to the membrane, held there by their electrical attraction to their oppositely charged counterparts (counterions) on the other side of the membrane. For a typical cell, 1 microcoulomb of charge (6×10^{12} monovalent ions) per square centimeter of membrane, transferred from one side of the membrane to the other, changes the membrane potential by roughly 1 V. This means, for example, that in a spherical cell of diameter 10 μm, the number of K^+ ions that have to flow out to alter the membrane potential by 100 mV is only about 1/100,000 of the total number of K^+ ions in the cytosol. This amount is so minute that the intracellular K^+ concentration remains virtually unchanged.

small direct contribution to the membrane potential by pumping out three Na^+ for every two K^+ that it pumps in (see Figure 11–15). However, switching off the pump does not abolish the major component of the resting potential, which is generated by the K^+ equilibrium mechanism just described. This component of the membrane potential persists as long as the Na^+ concentration inside the cell stays low and the K^+ ion concentration high—typically for many minutes. But the plasma membrane is somewhat permeable to all small ions, including Na^+. Therefore, without the Na^+-K^+ pump, the ion gradients set up by the pump will eventually run down, and the membrane potential established by diffusion through the K^+ leak channels will fall as well. As Na^+ enters, the cell eventually comes to a new resting state where Na^+, K^+, and Cl^- are all at equilibrium across the membrane. The membrane potential in this state is much less than it was in the normal cell with an active Na^+-K^+ pump.

The resting potential of an animal cell varies between –20 mV and –120 mV, depending on the organism and cell type. Although the K^+ gradient always has a major influence on this potential, the gradients of other ions (and the disequilibrating effects of ion pumps) also have a significant effect: the more permeable the membrane for a given ion, the more strongly the membrane potential tends to be driven toward the equilibrium value for that ion. Consequently, changes in a membrane's permeability to ions can cause significant changes in the membrane potential. This is one of the key principles relating the electrical excitability of cells to the activities of ion channels.

To understand how ion channels select their ions and how they open and close, we need to know their atomic structure. The first ion channel to be crystallized and studied by x-ray diffraction was a bacterial K^+ channel. The details of its structure revolutionized our understanding of ion channels.

The Three-Dimensional Structure of a Bacterial K^+ Channel Shows How an Ion Channel Can Work

Scientists were puzzled by the remarkable ability of ion channels to combine exquisite ion selectivity with a high conductance. K^+ leak channels, for example, conduct K^+ 10,000-fold faster than Na^+, yet the two ions are both featureless spheres and have similar diameters (0.133 nm and 0.095 nm, respectively). A single amino acid substitution in the pore of an animal cell K^+ channel can result in a loss of ion selectivity and cell death. We cannot explain the normal K^+ selectivity by pore size, because Na^+ is smaller than K^+. Moreover, the high conductance rate is incompatible with the channel's having selective, high-affinity K^+-binding sites, as the binding of K^+ ions to such sites would greatly slow their passage.

The puzzle was solved when the structure of a *bacterial K^+ channel* was determined by x-ray crystallography. The channel is made from four identical transmembrane subunits, which together form a central pore through the membrane. Each subunit contributes two transmembrane α helices, which are tilted outward in the membrane and together form a cone, with its wide end facing the outside of

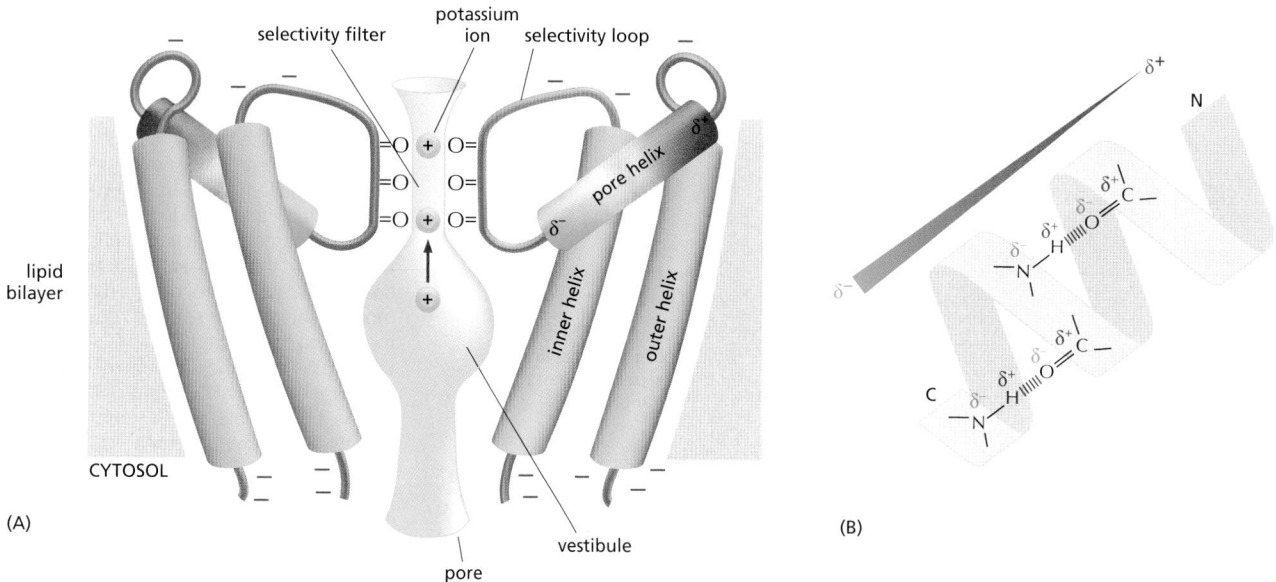

Figure 11–24 The structure of a bacterial K+ channel. (A) The transmembrane α helices from only two of the four identical subunits are shown. From the cytosolic side, the pore (schematically shaded in *blue*) opens up into a vestibule in the middle of the membrane. The pore vestibule facilitates transport by allowing the K+ ions to remain hydrated even though they are more than halfway across the membrane. The narrow selectivity filter of the pore links the vestibule to the outside of the cell. Carbonyl oxygens line the walls of the selectivity filter and form transient binding sites for dehydrated K+ ions. Two K+ ions occupy different sites in the selectivity filter, while a third K+ ion is located in the center of the vestibule, where it is stabilized by electrical interactions with the more negatively charged ends of the pore helices. The ends of the four short "pore helices" (only two of which are shown) point precisely toward the center of the vestibule, thereby guiding K+ ions into the selectivity filter (**Movie 11.8**). (B) Peptide bonds have an electric dipole, with more negative charge accumulated at the oxygen of the C=O bond and at the nitrogen of the N–H bond. In an α helix, hydrogen bonds (*red*) align the dipoles. As a consequence, every α helix has an electric dipole along its axis, resulting from summation of the dipoles of the individual peptide bonds, with a more negatively charged C-terminal end (δ^-) and a more positively charged N-terminal end (δ^+). (A, adapted from D.A. Doyle et al., *Science* 280:69–77, 1998.)

the cell where K+ ions exit from the channel (**Figure 11–24**). The polypeptide chain that connects the two transmembrane helices forms a short α helix (the *pore helix*) and a crucial loop that protrudes into the wide section of the cone to form the **selectivity filter**. The selectivity loops from the four subunits form a short, rigid, narrow pore, which is lined by the carbonyl oxygen atoms of their polypeptide backbones. Because the selectivity loops of all known K+ channels have similar amino acid sequences, it is likely that they form a closely similar structure.

The structure of the selectivity filter explains the ion selectivity of the channel. A K+ ion must lose almost all of its bound water molecules to enter the filter, where it interacts instead with the carbonyl oxygens lining the filter; the oxygens are rigidly spaced at the exact distance to accommodate a K+ ion. A Na+ ion, in contrast, cannot enter the filter because the carbonyl oxygens are too far away from the smaller Na+ ion to compensate for the energy expense associated with the loss of water molecules required for entry (**Figure 11–25**).

Structural studies of K+ channels and other ion channels have also indicated some general principles of how these channels open and close. The gating involves movement of the helices in the membrane so that they either obstruct or open the path for ion movement. Depending on the particular type of channel, helices tilt, rotate, or bend during gating. The structure of a closed K+ channel shows that by tilting the inner helices, the pore constricts like a diaphragm at its cytosolic end (**Figure 11–26**). Bulky hydrophobic amino acid side chains block the small opening that remains, preventing the entry of ions.

Many other ion channels operate on similar principles: the channel's gating helices are allosterically coupled to domains that form the ion-conducting pathway; and a conformational change in the gate—in response, say, to ligand binding or altered membrane potential—brings about conformational change in the conducting pathway, either opening it or blocking it off.

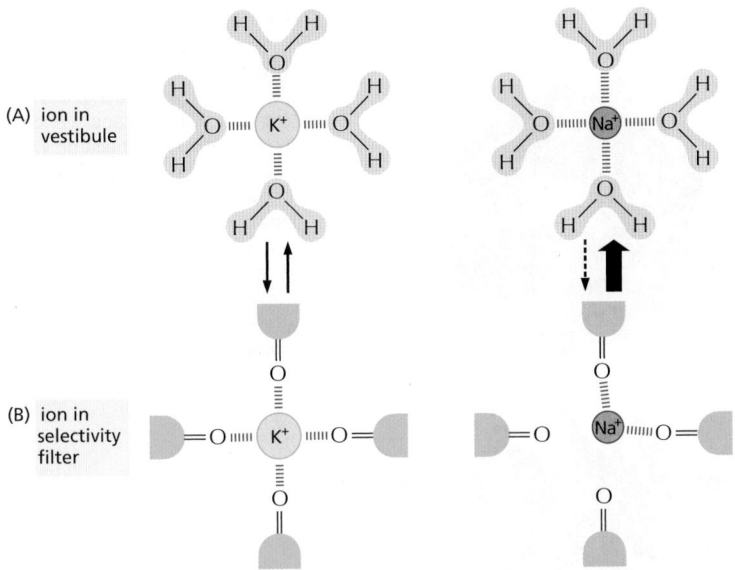

Figure 11–25 **K⁺ specificity of the selectivity filter in a K⁺ channel.** The drawings show K⁺ and Na⁺ ions (A) in the vestibule and (B) in the selectivity filter of the pore, viewed in cross section. In the vestibule, the ions are hydrated. In the selectivity filter, they have lost their water, and the carbonyl oxygens are placed to accommodate a dehydrated K⁺ ion. The dehydration of the K⁺ ion requires energy, which is precisely balanced by the energy regained by the interaction of the ion with all of the carbonyl oxygens that serve as surrogate water molecules. Because the Na⁺ ion is too small to interact with the oxygens, it can enter the selectivity filter only at a great energetic expense. The filter therefore selects K⁺ ions with high specificity. (A, adapted from Y. Zhou et al., *Nature* 414:43–48, 2001.)

Mechanosensitive Channels Protect Bacterial Cells Against Extreme Osmotic Pressures

All organisms, from single-cell bacteria to multicellular animals and plants, must sense and respond to mechanical forces in their external environment (such as sound, touch, pressure, shear forces, and gravity) and in their internal environment (such as osmotic pressure and membrane bending). Numerous proteins are known to be capable of responding to such mechanical forces, and a large subset of those proteins has been identified as possible **mechanosensitive channels**, but very few of the candidate proteins have been shown directly to be mechanically activated ion channels. One reason for this dearth in our knowledge is that most such channels are extremely rare. Auditory hair cells in the human cochlea, for example, contain extraordinarily sensitive mechanically gated ion channels, but each of the approximately 15,000 individual hair cells is thought to have a total of only 50–100 of them (**Movie 11.9**). Additional difficulties arise because the gating mechanisms of many mechanosensitive channel types require the channels to be embedded in complex architectures that require attachment to the extracellular matrix or to the cytoskeleton and are difficult to reconstitute in the test tube. The study of mechanosensitive receptors is a field of active investigation.

A well-studied class of mechanosensitive channels is found in the bacterial plasma membrane. These channels open in response to mechanical stretching of the lipid bilayer in which they are embedded. When a bacterium experiences a low-ionic-strength external environment (hypotonic conditions), such as

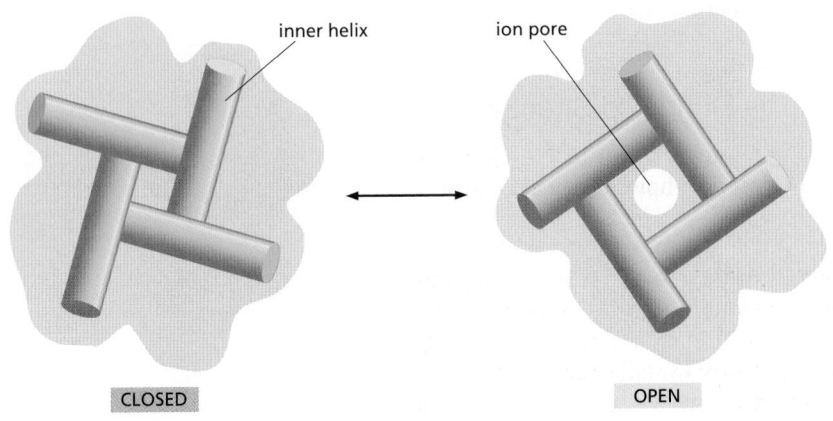

Figure 11–26 **A model for the gating of a bacterial K⁺ channel.** The channel is viewed in cross section. To adopt the closed conformation, the four inner transmembrane helices that line the pore on the cytosolic side of the selectivity filter (see Figure 11–24) rearrange to close the cytosolic entrance to the channel. (Adapted from E. Perozo et al., *Science* 285:73–78, 1999.)

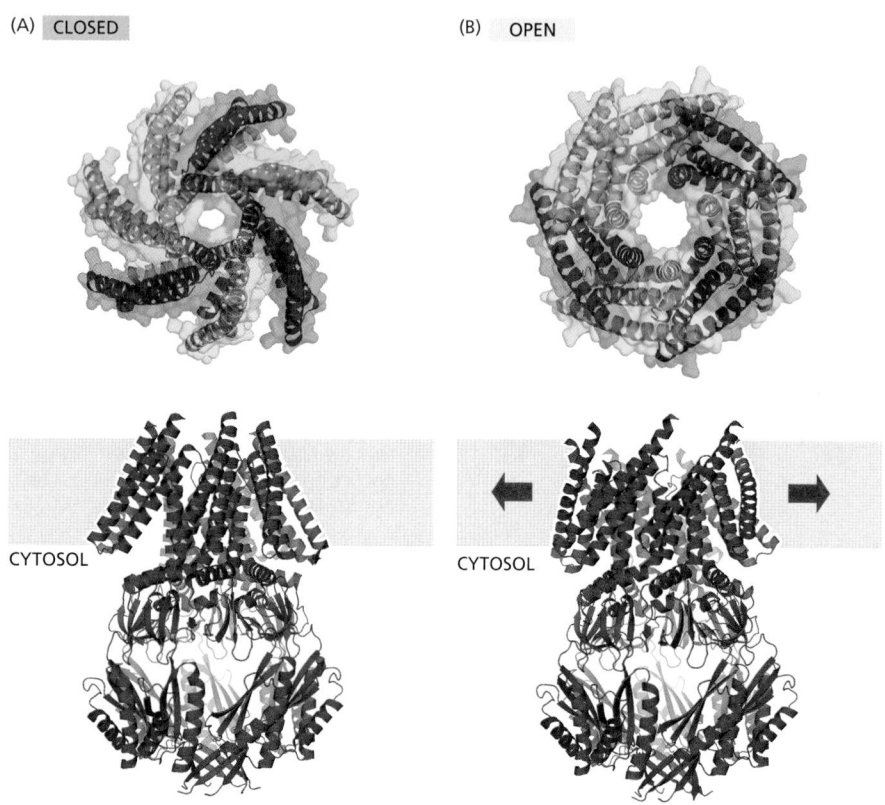

(A) CLOSED   (B) OPEN

CYTOSOL   CYTOSOL

Figure 11–27 **The structure of mechanosensitive channels.** The crystal structures of MscS in its (A) closed and (B) open conformation are shown. The side views (lower panels) show the entire protein, including the large intracellular domain. The face views (upper panels) show the transmembrane domains only. The open structure occupies more area in the lipid bilayer and is energetically favored when a membrane is stretched. This may explain why the MscS channel opens as pressure builds up inside the cell. (PDB codes: 2OAU, 2VV5.)

rainwater, the cell swells as water seeps in due to an increase in the osmotic pressure. If the pressure rises to dangerous levels, the cell opens mechanosensitive channels that allow small molecules to leak out. Bacteria that are experimentally placed in fresh water can rapidly lose more than 95% of their small molecules in this manner, including amino acids, sugars, and potassium ions. However, they keep their macromolecules safely inside and thus can recover quickly after environmental conditions return to normal.

Mechanical gating has been demonstrated using biophysical techniques in which force is exerted on pure lipid bilayers containing the bacterial mechanosensitive channels; for example, by applying suction with a micropipette. Such measurements demonstrate that the cell has several different channels that open at different levels of pressure. The mechanosensitive channel of small conductance, called the MscS channel, opens at low and moderate pressures (**Figure 11–27**). It is composed of seven identical subunits, which in the open state form a pore about 1.3 nm in diameter—just big enough to pass ions and small molecules. Large cytoplasmic domains limit the size of molecules that can reach the pore. The mechanosensitive channel of large conductance, called the MscL channel, opens to over 3 nm in diameter when the pressure gets so high that the cell might burst.

The Function of a Neuron Depends on Its Elongated Structure

The cells that make most sophisticated use of channels are neurons. Before discussing how they do so, we digress briefly to describe how a typical neuron is organized.

The fundamental task of a **neuron**, or **nerve cell**, is to receive, conduct, and transmit signals. To perform these functions, neurons are often extremely elongated. In humans, for example, a single neuron extending from the spinal cord to a muscle in the foot may be as long as 1 meter. Every neuron consists of a cell body (containing the nucleus) with a number of thin processes radiating outward from it. Usually one long **axon** conducts signals away from the cell body toward distant

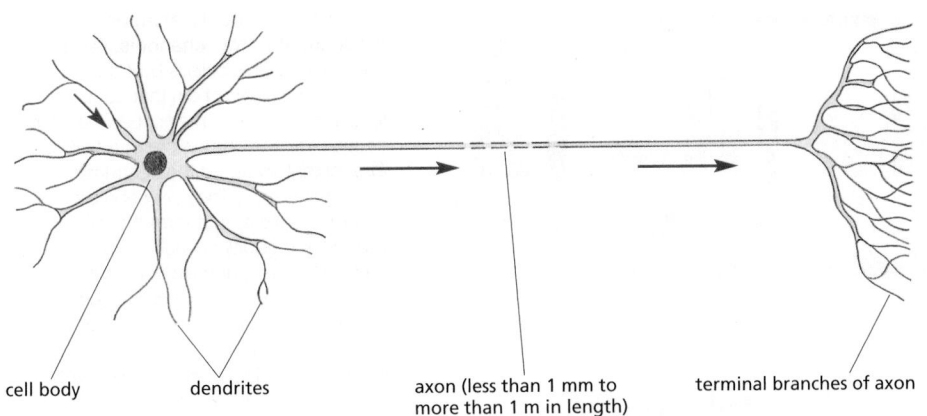

Figure 11–28 **A typical vertebrate neuron.** The arrows indicate the direction in which signals are conveyed. The single axon conducts signals away from the cell body, while the multiple dendrites (and the cell body) receive signals from the axons of other neurons. The axon terminals end on the dendrites or cell body of other neurons or on other cell types, such as muscle or gland cells.

cell body   dendrites   axon (less than 1 mm to more than 1 m in length)   terminal branches of axon

targets, and several shorter, branching **dendrites** extend from the cell body like antennae, providing an enlarged surface area to receive signals from the axons of other neurons (**Figure 11–28**), although the cell body itself also receives such signals. A typical axon divides at its far end into many branches, passing on its message to many target cells simultaneously. Likewise, the extent of branching of the dendrites can be very great—in some cases sufficient to receive as many as 100,000 inputs on a single neuron.

Despite the varied significance of the signals carried by different classes of neurons, the form of the signal is always the same, consisting of changes in the electrical potential across the neuron's plasma membrane. The signal spreads because an electrical disturbance produced in one part of the membrane spreads to other parts, although the disturbance becomes weaker with increasing distance from its source, unless the neuron expends energy to amplify it as it travels. Over short distances, this attenuation is unimportant; in fact, many small neurons conduct their signals passively, without amplification. For long-distance communication, however, such passive spread is inadequate. Thus, larger neurons employ an active signaling mechanism, which is one of their most striking features. An electrical stimulus that exceeds a certain threshold strength triggers an explosion of electrical activity that propagates rapidly along the neuron's plasma membrane and is sustained by automatic amplification all along the way. This traveling wave of electrical excitation, known as an **action potential**, or *nerve impulse*, can carry a message without attenuation from one end of a neuron to the other at speeds of 100 meters per second or more. Action potentials are the direct consequence of the properties of voltage-gated cation channels, as we now discuss.

Voltage-Gated Cation Channels Generate Action Potentials in Electrically Excitable Cells

The plasma membrane of all electrically excitable cells—not only neurons, but also muscle, endocrine, and egg cells—contains **voltage-gated cation channels**, which are responsible for generating the action potentials. An action potential is triggered by a **depolarization** of the plasma membrane—that is, by a shift in the membrane potential to a less negative value inside. (We shall see later how the action of a neurotransmitter causes depolarization.) In nerve and skeletal muscle cells, a stimulus that causes sufficient depolarization promptly opens the **voltage-gated Na$^+$ channels**, allowing a small amount of Na$^+$ to enter the cell down its electrochemical gradient. The influx of positive charge depolarizes the membrane further, thereby opening more Na$^+$ channels, which admit more Na$^+$ ions, causing still further depolarization. This self-amplification process (an example of *positive feedback*, discussed in Chapters 8 and 15) continues until, within a fraction of a millisecond, the electrical potential in the local region of membrane has shifted from its resting value of about –70 mV (in squid giant axon; about –40 mV in human) to almost as far as the Na$^+$ equilibrium potential of about +50 mV (see

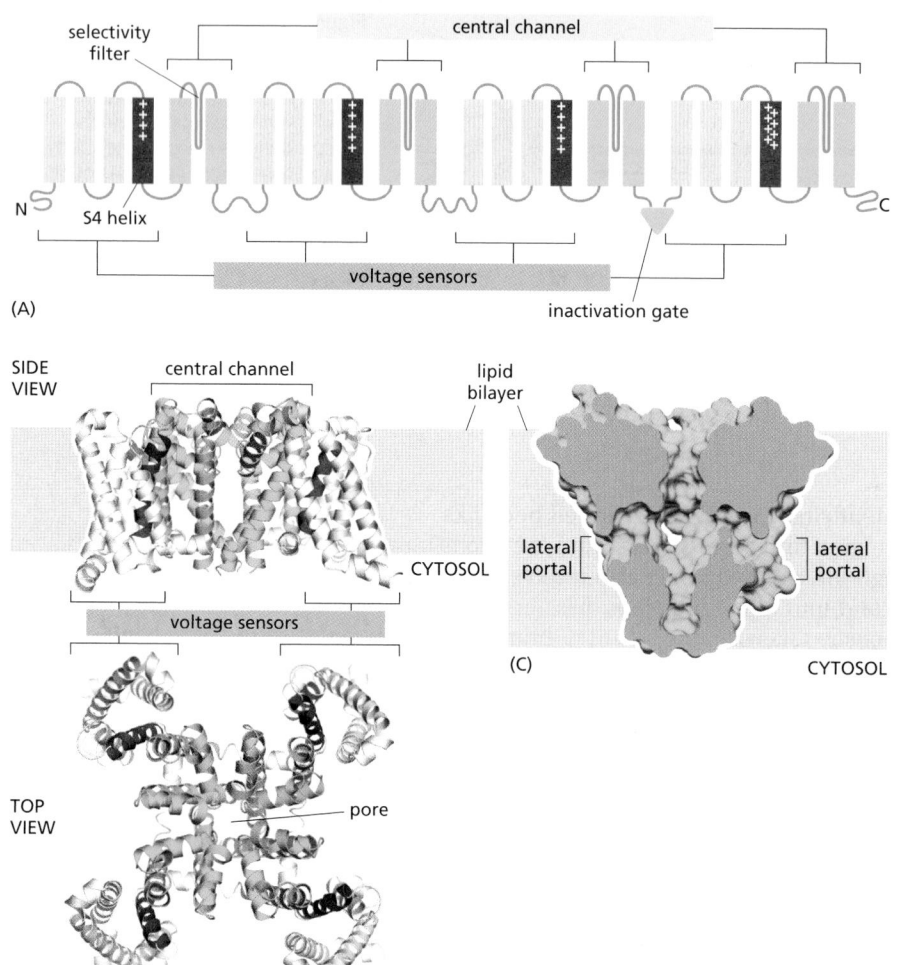

(A)

(B)

(C)

SIDE VIEW

TOP VIEW

Figure 11–29 Structural models of voltage-gated Na⁺ channels. (A) The channel in animal cells is built from a single polypeptide chain that contains four homologous domains. Each domain contains two transmembrane α helices *(green)* that surround the central ion-conducting pore. They are separated by sequences *(blue)* that form the selectivity filter. Four α additional helices *(gray and red)* in each domain constitute the voltage sensor. The S4 helices *(red)* are unique in that they contain an abundance of positively charged arginines. An inactivation gate that is part of a flexible loop connecting the third and fourth domains acts as a plug that obstructs the pore in the channel's inactivated state, as shown in Figure 11–30. (B) Side and top views of a homologous bacterial channel protein showing its arrangement within the membrane. (C) A cross section of the pore domain of the channel shown in (B) shows lateral portals, through which the central cavity is accessible from the hydrophobic core of the lipid bilayer. In the crystals, lipid acyl chains were found to intrude into the pore. These lateral portals are large enough to allow entry of small, hydrophobic, pore-blocking drugs that are commonly used as anesthetics and block ion conductance. (PDB code: 3RVZ.)

Panel 11–1, p. 616). At this point, when the net electrochemical driving force for the flow of Na⁺ is almost zero, the cell would come to a new resting state, with all of its Na⁺ channels permanently open, if the open conformation of the channel were stable. Two mechanisms act in concert to save the cell from such a permanent electrical spasm: the Na⁺ channels automatically inactivate and **voltage-gated K⁺ channels** open to restore the membrane potential to its initial negative value.

The Na⁺ channel is built from a single polypeptide chain that contains four structurally very similar domains. It is thought that these domains evolved by gene duplication followed by fusion into a single large gene (**Figure 11–29A**). In bacteria, in fact, the Na⁺ channel is a tetramer of four identical polypeptide chains, supporting this evolutionary idea.

Each domain contributes to the central channel, which is very similar to the K⁺ channel. Each domain also contains a *voltage sensor* that is characterized by an unusual transmembrane helix, S4, that contains many positively charged amino acids. As the membrane depolarizes, the S4 helices experience an electrostatic pulling force that attracts them to the now negatively charged extracellular side of the plasma membrane. The resulting conformational change opens the channel. The structure of a bacterial voltage-gated Na⁺ channel provides insights how the structural elements are arranged in the membrane (Figure 11–29B and C).

The Na⁺ channels also have an automatic inactivating mechanism, which causes the channels to reclose rapidly even though the membrane is still depolarized (see Figure 11–30). The Na⁺ channels remain in this *inactivated* state, unable to reopen, until after the membrane potential has returned to its initial negative value. The time necessary for a sufficient number of Na⁺ channels to recover from inactivation to support a new action potential, termed the *refractory period*, limits

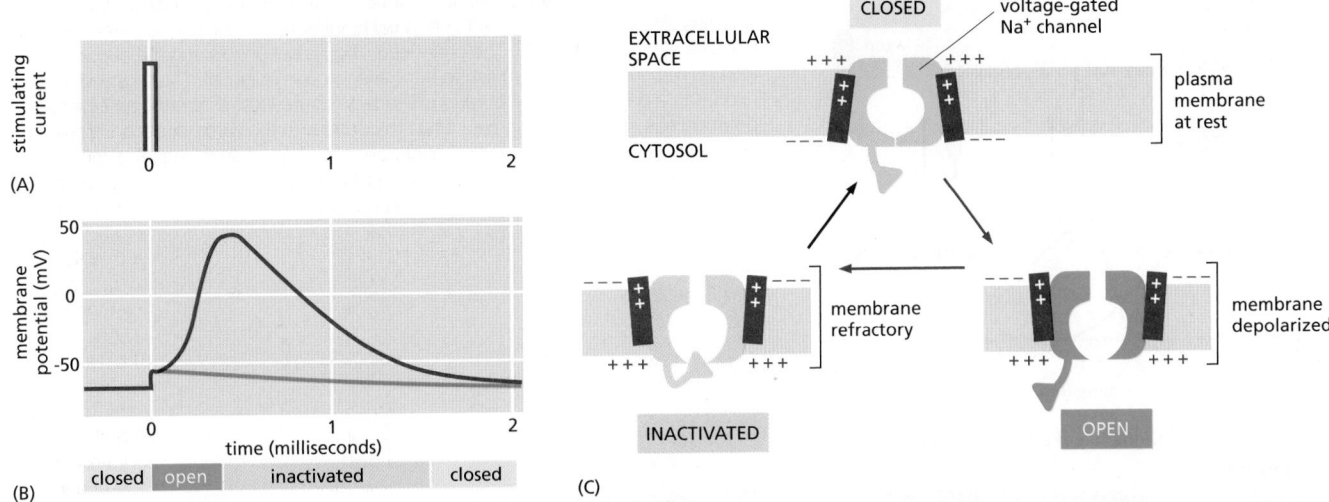

Figure 11–30 **Na+ channels and an action potential.** (A) An action potential is triggered by a brief pulse of current, which (B) partially depolarizes the membrane, as shown in the plot of membrane potential versus time. The *green curve* shows how the membrane potential would have simply relaxed back to the resting value after the initial depolarizing stimulus if there had been no voltage-gated Na+ channels in the membrane. The *red curve* shows the course of the action potential that is caused by the opening and subsequent inactivation of voltage-gated Na+ channels. The states of the Na+ channels are indicated in (B). The membrane cannot fire a second action potential until the Na+ channels have returned from the inactivated to the closed conformation; until then, the membrane is refractory to stimulation. (C) The three states of the Na+ channel. When the membrane is at rest (highly polarized), the closed conformation of the channel has the lowest free energy and is therefore most stable; when the membrane is depolarized, the energy of the *open* conformation is lower, so the channel has a high probability of opening. But the free energy of the *inactivated* conformation is lower still; therefore, after a randomly variable period spent in the open state, the channel becomes inactivated. Thus, the open conformation corresponds to a metastable state that can exist only transiently when the membrane depolarizes (**Movie 11.10**).

the repetitive firing rate of a neuron. The cycle from initial stimulus to the return to the original resting state takes a few milliseconds or less. The Na+ channel can therefore exist in three distinct states—closed, open, and inactivated—which contribute to the rise and fall of the action potential (**Figure 11–30**).

This description of an action potential applies only to a small patch of plasma membrane. The self-amplifying depolarization of the patch, however, is sufficient to depolarize neighboring regions of membrane, which then go through the same cycle. In this way, the action potential sweeps like a wave from the initial site of depolarization over the entire plasma membrane, as shown in **Figure 11–31**.

The Use of Channelrhodopsins Has Revolutionized the Study of Neural Circuits

Channelrhodopsins are photosensitive ion channels that open in response to light. They evolved as sensory receptors in photosynthetic green algae to allow the algae to swim toward light. The structure of channelrhodopsin closely resembles that of bacteriorhodopsin (see Figure 10–31). It contains a covalently bound retinal group that absorbs light and undergoes an isomerization reaction, which triggers a conformational change in the protein, opening an ion channel in the plasma membrane. In contrast to bacteriorhodopsin, which is a light-driven proton pump, channelrhodopsin is a light-driven cation channel.

Using genetic engineering techniques, channelrhodopsin can be expressed in virtually any cell type in vertebrates and invertebrates. Researchers first introduced the gene into cultured neurons and showed that flashing light could now activate the channelrhodopsin and induce the neurons to fire action potentials. Because the frequency of the light flashes determined the frequency of the action potentials, one can control the frequency of neuronal firing with millisecond precision.

Next, neurobiologists used the approach to activate specific neurons in the brain of experimental animals. Using a tiny fiber optic cable implanted near the

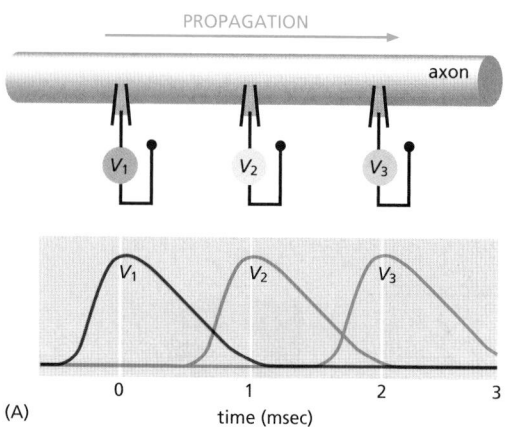

(A)

Figure 11–31 **The propagation of an action potential along an axon.** (A) The voltages that would be recorded from a set of intracellular electrodes placed at intervals along the axon. (B) The changes in the Na$^+$ channels and the current flows (*curved red arrows*) that give rise to a traveling action potential. The region of the axon with a depolarized membrane is shaded in *blue.* Note that once an action potential has started to progress, it has to continue in the same direction, traveling only away from the site of depolarization, because Na$^+$-channel inactivation prevents the depolarization from spreading backward.

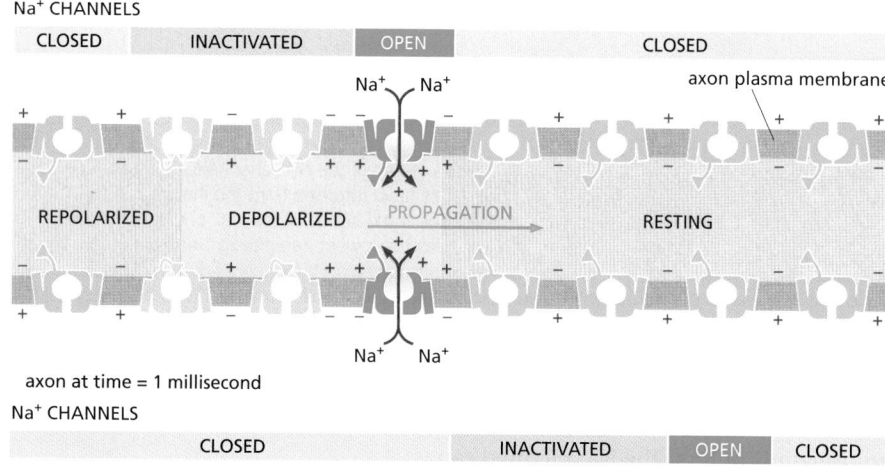

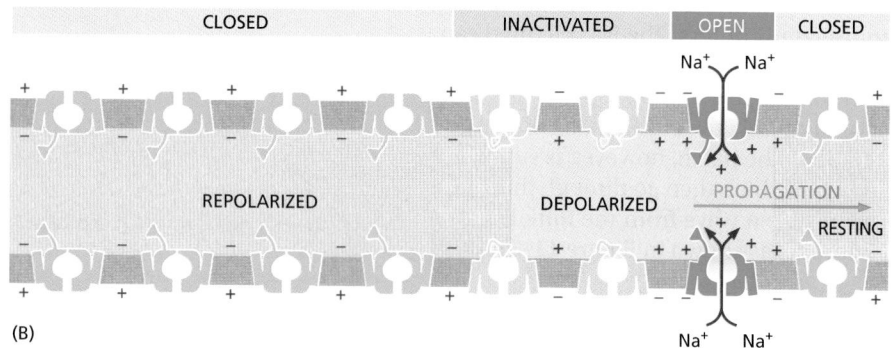

(B)

relevant brain region, they could flash light to specifically activate the channelrhodopsin-containing neurons to fire action potentials. One group of researchers expressed channelrhodopsin in a subset of mouse neurons thought to be involved in aggression: when these cells were activated by light, the mouse immediately attacked anything in its environment—including other mice or even an inflated rubber glove (**Figure 11–32**); when the light was switched off, the neurons fell silent and the mouse's behavior returned to normal.

Since these pioneering studies, researchers have engineered additional light-responsive ion channels and transporters, including some that can rapidly

Figure 11–32 **Optogenetic control of aggression neurons in a living mouse.** A gene encoding channelrhodopsin was introduced into a subpopulation of neurons in the hypothalamus of a mouse. When the neurons were exposed to flashing blue light using a tiny, implanted fiber optic cable, the channelrhodopsin channels opened, depolarizing and activating the cells. When the light was switched on, the mouse immediately became aggressive and attacked the inflated rubber glove; when the light was switched off, its behavior immediately returned to normal (**Movie 11.11**). (Adapted from D. Lin et al., *Nature* 470:221–226, 2011. With permission from the authors.)

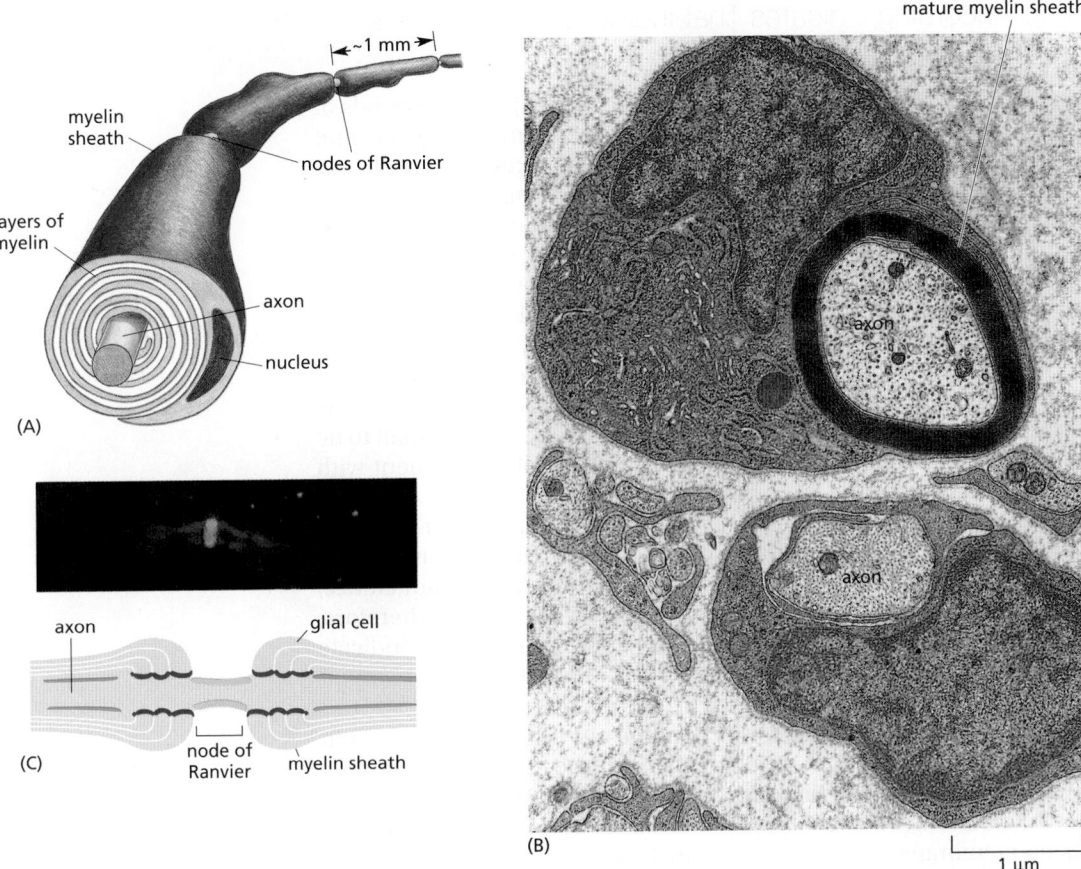

Figure 11–33 **Myelination.**
(A) A myelinated axon from a peripheral nerve. Each Schwann cell wraps its plasma membrane concentrically around the axon to form a segment of myelin sheath about 1 mm long. For clarity, the membrane layers of the myelin are shown less compacted than they are in reality (see part B). (B) An electron micrograph of a nerve in the leg of a young rat. Two Schwann cells can be seen: one near the bottom is just beginning to myelinate its axon; the one above it has formed an almost mature myelin sheath. (C) Fluorescence micrograph and diagram of individual myelinated axons teased apart in a rat optic nerve, showing the confinement of the voltage-gated Na^+ channels (*green*) in the axonal membrane at the node of Ranvier. A protein called Caspr (*red*) marks the junctions where the myelinating glial cell plasma membrane tightly abuts the axon on either side of the node. Voltage-gated K^+ channels (*blue*) localize to regions in the axon plasma membrane well away from the node. (B, from Cedric S. Raine, in *Myelin* [P. Morell, ed.]. New York: Plenum, 1976. With permission from Springer Nature; C, from M.N. Rasband and P. Shrager, *J. Physiol.* 525:63–73, 2004. With permission from John Wiley & Sons.)

inactivate specific neurons. It is therefore now possible to transiently activate or inhibit specific neurons in the brains of awake animals with remarkable spatial and temporal precision. In this way, the rapidly expanding new field of **optogenetics** is revolutionizing neurobiology, allowing neuroscientists to analyze the neurons and circuits underlying even the most complex behaviors in experimental animals, including nonhuman primates.

Myelination Increases the Speed and Efficiency of Action Potential Propagation in Nerve Cells

The axons of many vertebrate neurons are insulated by a **myelin sheath**, which greatly increases the rate at which an axon can conduct an action potential. The importance of myelination is dramatically demonstrated by the demyelinating disease *multiple sclerosis*, in which the immune system destroys myelin sheaths in some regions of the central nervous system; in the affected regions, nerve impulse propagation greatly slows or even fails, often with devastating neurological consequences.

Myelin is formed by specialized non-neuronal supporting cells called **glial cells**. **Schwann cells** are the glial cells that myelinate axons in peripheral nerves, and **oligodendrocytes** do so in the central nervous system. These myelinating glial cells wrap layer upon layer of their own plasma membrane in a tight spiral around the axon (**Figure 11–33**A and B), thereby insulating the axonal membrane so that little current can leak across it. The myelin sheath is interrupted at regularly spaced *nodes of Ranvier*, where almost all the Na^+ channels in the axon are concentrated (Figure 11–33C). This arrangement allows an action potential to propagate along a myelinated axon by jumping from node to node, a process called *saltatory conduction*. This type of conduction has two main advantages: action potentials travel very much faster, and metabolic energy is conserved because the active excitation is confined to the small regions of axonal plasma membrane at nodes of Ranvier.

Patch-Clamp Recording Indicates That Individual Ion Channels Open in an All-or-Nothing Fashion

Neuron and skeletal muscle cell plasma membranes contain many thousands of voltage-gated Na^+ channels, and the current crossing the membrane is the sum of the currents flowing through all of these. An intracellular microelectrode can record this aggregate current, as shown in Figure 11–31A. Remarkably, however, it is also possible to record current flowing through individual channels. **Patch-clamp recording**, developed in the 1970s and 1980s, revolutionized the study of ion channels and made it possible to examine transport through a single channel in a small patch of membrane covering the mouth of a micropipette (**Figure 11–34**). With this simple but powerful technique, one can study the detailed properties of ion channels in all sorts of cell types. This work led to the discovery that even cells that are not electrically excitable usually have a variety of ion channels in their plasma membrane. Many of these cells, such as yeasts, are too small to be investigated by the traditional electrophysiologist's method of impalement with an intracellular microelectrode.

Patch-clamp recording indicates that individual ion channels open in an all-or-nothing fashion. For example, a voltage-gated Na^+ channel opens and closes at random, but when open, the channel always has the same large conductance, allowing more than 1000 ions to pass per millisecond (**Figure 11–35**). Therefore, the aggregate current crossing the membrane of an entire cell does not indicate the *degree* to which a typical individual channel is open but rather the *total number* of channels in its membrane that are open at any one time.

Some simple physical principles allow us to refine our understanding of voltage-gating from the perspective of a single Na^+ channel. The interior of the resting neuron or muscle cell is at an electrical potential about 40–100 mV more negative than the external medium. Although this potential difference seems small, it exists across a plasma membrane only about 5 nm thick, so that the resulting voltage gradient is about 100,000 V/cm. Charged proteins in the membrane such as Na^+ channels are thus subjected to a very large electrical field that can profoundly affect their conformation. Each conformation can "flip" to another conformation if given a sufficient jolt by the random thermal movements of the surroundings, and it is the relative stability of the closed, open, and inactivated conformations against flipping that is altered by changes in the membrane potential (see Figure 11–30C).

Voltage-Gated Cation Channels Are Evolutionarily and Structurally Related

Na^+ channels are not the only kind of voltage-gated cation channel that can generate an action potential. The action potentials in some muscle, egg, and endocrine cells, for example, depend on *voltage-gated Ca^{2+} channels* rather than on Na^+ channels.

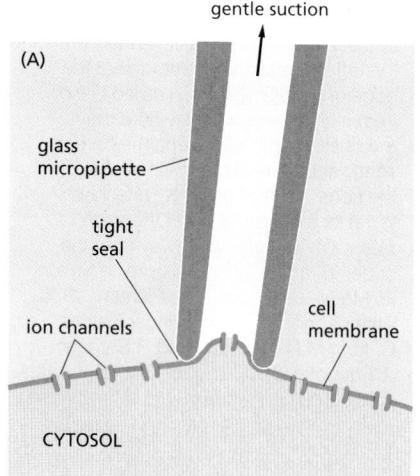

(A)

gentle suction

glass micropipette

tight seal

ion channels

cell membrane

CYTOSOL

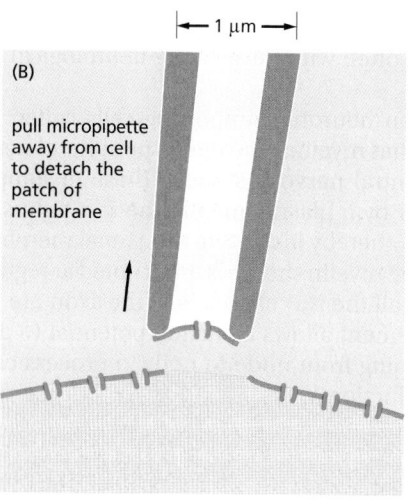

(B)

|← 1 μm →|

pull micropipette away from cell to detach the patch of membrane

Figure 11–34 The technique of patch-clamp recording. Because of the extremely tight seal between the micropipette and the membrane, current can enter or leave the micropipette only by passing through the ion channels in the *patch* of membrane covering its tip. The term *clamp* is used because an electronic device is employed to maintain, or "clamp," the membrane potential at a set value while recording the ionic current through individual channels. The current through these channels can be recorded with the patch still attached to the rest of the cell, as in (A), or detached, as in (B). The advantage of the detached patch is that it is easy to alter the composition of the solution on either side of the membrane to test the effect of various solutes on channel behavior. A detached patch can also be produced with the opposite orientation, so that the cytoplasmic surface of the membrane faces the inside of the pipette.

There is a surprising amount of structural and functional diversity within each of the different classes of voltage-gated cation channels, generated both by multiple genes and by the alternative splicing of RNA transcripts produced from the same gene. Nonetheless, the amino acid sequences of the known voltage-gated Na$^+$, K$^+$, and Ca^{2+} channels show striking similarities, demonstrating that they all belong to a large superfamily of evolutionarily and structurally related proteins and share many of the design principles. Whereas the single-celled yeast *S. cerevisiae* contains a single gene that codes for a voltage-gated K$^+$ channel, the genome of the worm *C. elegans* contains 68 genes that encode different but related K$^+$ channels. This complexity indicates that even a simple nervous system made up of only 302 neurons uses a large number of different ion channels to compute its responses.

Humans who inherit mutant genes encoding ion channels can suffer from a variety of nerve, muscle, brain, or heart diseases, depending in which cells the channel encoded by the mutant gene normally functions. Mutations in genes that encode voltage-gated Na$^+$ channels in skeletal muscle cells, for example, can cause *myotonia*, a condition in which there is a delay in muscle relaxation after voluntary contraction, causing painful muscle spasms. In some cases, this occurs because the abnormal channels fail to inactivate normally; as a result, Na$^+$ entry persists after an action potential finishes and repeatedly reinitiates membrane depolarization and muscle contraction. Similarly, mutations that affect Na$^+$ or K$^+$ channels in the brain can cause *epilepsy*, in which excessive synchronized firing of large groups of neurons causes epileptic seizures (convulsions, or fits).

The particular combination of ion channels conducting Na$^+$, K$^+$, and Ca^{2+} that are expressed in a neuron largely determines how the cell fires repetitive sequences of action potentials. Some nerve cells can repeat action potentials up to 300 times per second; other neurons fire short bursts of action potentials separated by periods of silence; while others rarely fire more than one action potential at a time. There is a remarkable diversity of neurons in the brain.

Different Neuron Types Display Characteristic Stable Firing Properties

It is estimated that the human brain contains about 10^{11} neurons and 10^{14} synaptic connections. To make matters more complex, neural circuitry is continuously sculpted in response to experience, modified as we learn and store memories, and irreversibly altered by the gradual loss of neurons and their connections as we age. How can a system so complex be subject to such change and yet continue to function stably? One emerging theory suggests that individual neurons are self-tuning devices, constantly adjusting the expression of ion channels and neurotransmitter receptors in order to maintain a stable function. How might this work?

Neurons can be categorized into functionally different types, based in part on their propensity to fire action potentials and their pattern of firing. For example, some neurons fire action potentials at high frequencies, while others fire rarely. The firing properties of each neuron type are determined to a large extent by the ion channels that the cell expresses. The number of ion channels in a neuron's membrane is not fixed: as conditions change, a neuron can modify the numbers of depolarizing (Na$^+$ and Ca^{2+}) and hyperpolarizing (K$^+$) channels and keep their proportions adjusted so as to maintain its characteristic firing behavior—a remarkable example of homeostatic control. The molecular mechanisms involved remain an important mystery.

Transmitter-Gated Ion Channels Convert Chemical Signals into Electrical Ones at Chemical Synapses

Neuronal signals are transmitted from cell to cell at specialized sites of contact known as **synapses**. The usual mechanism of transmission is indirect. The cells are

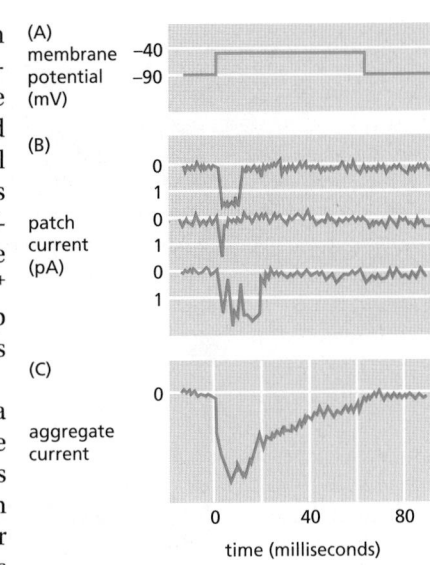

Figure 11–35 Patch-clamp measurements for a single voltage-gated Na$^+$ channel. A tiny patch of plasma membrane was detached from an embryonic rat muscle cell, as in Figure 11–34. (A) The membrane was depolarized by an abrupt shift of potential from –90 to about –40 mV. (B) Three current records from three experiments performed on the same patch of membrane. Each major current step in (B) represents the opening and closing of a single channel. A comparison of the three records shows that, whereas the durations of channel opening and closing vary greatly, the rate at which current flows through an open channel (its conductance) is practically constant. The minor fluctuations in the current records arise largely from electrical noise in the recording apparatus. Current flowing into the cell, measured in picoamperes (pA), is shown as a downward deflection of the curve. By convention, the electrical potential on the outside of the cell is defined as zero. (C) The sum of the currents measured in 144 repetitions of the same experiment. This aggregate current is equivalent to the usual Na$^+$ current that would be observed flowing through a relatively large region of membrane containing 144 channels. A comparison of (B) and (C) reveals that the time course of the aggregate current reflects the probability that any individual channel will be in the open state; this probability decreases with time as the channels in the depolarized membrane adopt their inactivated conformation. (Data from J. Patlak and R. Horn, *J. Gen. Physiol.* 79:333–351, 1982.)

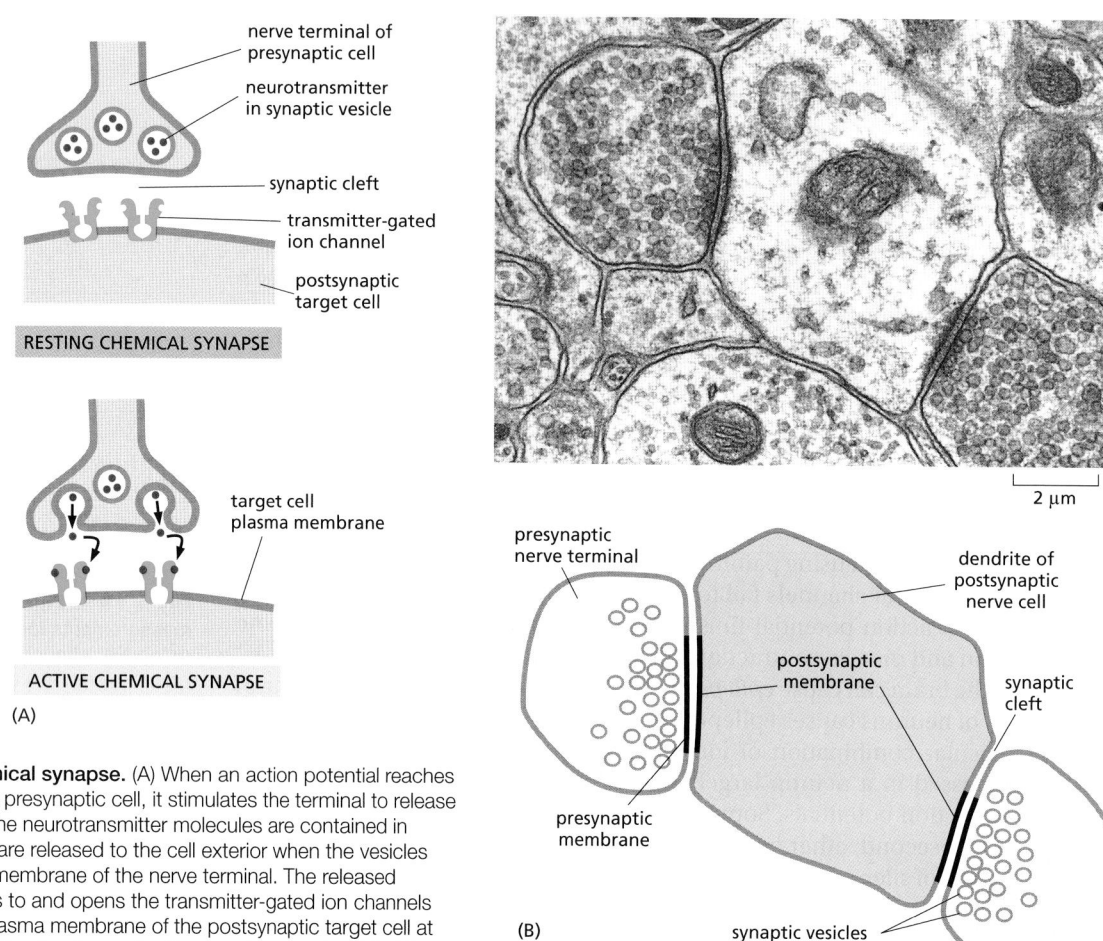

Figure 11–36 A chemical synapse. (A) When an action potential reaches the nerve terminal in a presynaptic cell, it stimulates the terminal to release its neurotransmitter. The neurotransmitter molecules are contained in synaptic vesicles and are released to the cell exterior when the vesicles fuse with the plasma membrane of the nerve terminal. The released neurotransmitter binds to and opens the transmitter-gated ion channels concentrated in the plasma membrane of the postsynaptic target cell at the synapse. The resulting ion flows alter the membrane potential of the postsynaptic membrane, thereby transmitting a signal from the excited nerve (Movie 11.12). (B) A thin-section electron micrograph of two nerve terminal synapses on a dendrite of a postsynaptic cell. (B, courtesy of Cedric Raine.)

electrically isolated from one another, the *presynaptic cell* being separated from the *postsynaptic cell* by a narrow *synaptic cleft*. When an action potential arrives at the presynaptic site, the depolarization of the membrane opens voltage-gated Ca^{2+} channels that are clustered in the presynaptic membrane. Ca^{2+} influx triggers the release into the cleft of small signal molecules known as **neurotransmitters**, which are stored in membrane-enclosed *synaptic vesicles* and released by exocytosis (discussed in Chapter 13). The neurotransmitter diffuses rapidly across the synaptic cleft and provokes an electrical change in the postsynaptic cell by binding to and opening *transmitter-gated ion channels* (**Figure 11–36**). After the neurotransmitter has been secreted, it is rapidly removed: it is either destroyed by specific enzymes in the synaptic cleft or taken up by the presynaptic nerve terminal or by surrounding glial cells. Reuptake is mediated by a variety of Na^+-dependent neurotransmitter symporters (see Figure 11–8); in this way, neurotransmitters are recycled, allowing cells to keep up with high rates of release. Rapid removal ensures both spatial and temporal precision of signaling at a synapse. It decreases the chances that the neurotransmitter will influence neighboring cells, and it clears the synaptic cleft before the next pulse of neurotransmitter is released, so that the timing of repeated, rapid signaling events can be accurately communicated to the postsynaptic cell. As we shall see, signaling via such *chemical synapses* is far more versatile and adaptable than direct electrical coupling via gap junctions at *electrical synapses* (discussed in Chapter 19), which are also used by neurons but to a much smaller extent.

Transmitter-gated ion channels, also called **ionotropic receptors**, are built for rapidly converting extracellular chemical signals into electrical signals at chemical synapses. The channels are concentrated in a specialized region of the postsynaptic plasma membrane at the synapse and open transiently in response to the binding of neurotransmitter molecules, thereby producing a brief permeability change in the membrane (see Figure 11–36A). Unlike the voltage-gated channels responsible for action potentials, transmitter-gated channels are relatively insensitive to the membrane potential and therefore cannot by themselves produce a self-amplifying excitation. Instead, they produce local permeability increases, and hence changes of membrane potential, that are graded according to the amount of neurotransmitter released at the synapse and how long it persists there. Only if the summation of small depolarizations at this site opens sufficient numbers of nearby voltage-gated cation channels can an action potential be triggered. This may require the opening of transmitter-gated ion channels at numerous synapses in close proximity on the target nerve cell.

Chemical Synapses Can Be Excitatory or Inhibitory

Transmitter-gated ion channels differ from one another in several important ways. First, as receptors, they have highly selective binding sites for the neurotransmitter that is released from the presynaptic nerve terminal. Second, as channels, they are selective in the type of ions that they let pass across the plasma membrane; this determines the nature of the postsynaptic response. **Excitatory neurotransmitters** open cation channels, causing an influx of Na^+, and in many cases Ca^{2+}, that depolarizes the postsynaptic membrane toward the threshold potential for firing an action potential. **Inhibitory neurotransmitters**, by contrast, open either Cl^- channels or K^+ channels, and this suppresses firing by making it harder for excitatory neurotransmitters to depolarize the postsynaptic membrane. Many transmitters can be either excitatory or inhibitory, depending on where they are released, what receptors they bind to, and the ionic conditions that they encounter. *Acetylcholine*, for example, can either excite or inhibit, depending on the type of acetylcholine receptors it binds to. Usually, however, acetylcholine, *glutamate*, and *serotonin* are used as excitatory transmitters, and γ-*aminobutyric acid* (*GABA*) and *glycine* are used as inhibitory transmitters. Glutamate, for instance, mediates most of the excitatory signaling in the vertebrate brain.

We have already discussed how the opening of Na^+ or Ca^{2+} channels depolarizes a membrane. The opening of K^+ channels has the opposite effect because the K^+ concentration gradient is in the opposite direction—high concentration inside the cell, low outside. Opening K^+ channels tends to keep the cell close to the equilibrium potential for K^+, which, as we discussed earlier, is normally close to the resting membrane potential because at rest K^+ channels are the main type of channel that is open. When additional K^+ channels open, it becomes harder to drive the cell away from the resting state. We can understand the effect of opening Cl^- channels similarly. The concentration of Cl^- is much higher outside the cell than inside (see Table 11–1, p. 598), but the membrane potential opposes its influx. In fact, for many neurons, the equilibrium potential for Cl^- is close to the resting potential—or even more negative. For this reason, opening Cl^- channels tends to buffer the membrane potential; as the membrane starts to depolarize, more negatively charged Cl^- ions enter the cell and counteract the depolarization. Thus, the opening of Cl^- channels makes it more difficult to depolarize the membrane and hence to excite the cell. Some powerful toxins act by blocking the action of inhibitory neurotransmitters: strychnine, for example, binds to glycine receptors and prevents their inhibitory action, causing muscle spasms, convulsions, and death.

However, not all chemical signaling in the nervous system operates through these ionotropic ligand-gated ion channels. In fact, most neurotransmitter molecules that are secreted by nerve terminals, including a large variety of neuropeptides, bind to **metabotropic receptors**, which regulate ion channels only indirectly through the action of small intracellular signal molecules (discussed in Chapter 15). All neurotransmitter receptors fall into one or other of these two major classes—ionotropic or metabotropic—on the basis of their signaling mechanisms:

1. Ionotropic receptors are ion channels and feature at fast chemical synapses. Acetylcholine, glycine, glutamate, and GABA all act on transmitter-gated ion channels, mediating excitatory or inhibitory signaling that is generally immediate, simple, and brief.

2. Metabotropic receptors are *G-protein-coupled receptors* (discussed in Chapter 15) that bind to all other neurotransmitters (and, confusingly, also acetylcholine, glutamate, and GABA). Signaling mediated by ligand-binding to metabotropic receptors tends to be far slower and more complex than that at ionotropic receptors, and longer-lasting in its consequences.

The Acetylcholine Receptors at the Neuromuscular Junction Are Excitatory Transmitter-Gated Cation Channels

A well-studied example of a transmitter-gated ion channel is the **acetylcholine receptor** of skeletal muscle cells. This channel is opened transiently by acetylcholine released from the nerve terminal at a **neuromuscular junction**—the specialized chemical synapse between a motor neuron and a skeletal muscle cell (**Figure 11–37**). This synapse has been intensively investigated because it is readily accessible to electrophysiological study, unlike most of the synapses in the central nervous system, that is, the brain and spinal cord in vertebrates. Moreover, the acetylcholine receptors are densely packed in the muscle cell plasma membrane at a neuromuscular junction (about 20,000 such receptors per μm^2), with relatively few receptors elsewhere in the same membrane.

The receptors are composed of five transmembrane polypeptides, two of one kind and three others, encoded by four separate genes (**Figure 11–38A**). The four genes are strikingly similar in sequence, implying that they evolved from a single ancestral gene. The two identical polypeptides in the pentamer each contribute one acetylcholine-binding site. When two acetylcholine molecules bind to the pentameric complex, they induce a conformational change that opens the channel. With ligand bound, the channel still flickers between open and closed states, but now it has a 90% probability of being open. This state continues—with acetylcholine binding and unbinding—until hydrolysis of the free acetylcholine by the enzyme *acetylcholinesterase* lowers its concentration at the neuromuscular junction sufficiently. Once freed of its bound neurotransmitter, the acetylcholine receptor reverts to its initial resting state. If the presence of acetylcholine persists for a prolonged time as a result of excessive nerve stimulation, the channel inactivates. Normally, the acetylcholine is rapidly hydrolyzed and the channel closes within about 1 millisecond, well before significant desensitization occurs. Desensitization would occur after about 20 milliseconds in the continued presence of acetylcholine.

The five subunits of the acetylcholine receptor are arranged in a ring, forming a water-filled transmembrane channel that consists of a narrow pore through the lipid bilayer, which widens into vestibules at both ends. Acetylcholine binding opens the channel by causing the helices that line the pore to rotate outward, thus disrupting a ring of hydrophobic amino acids that blocks ion flow in the closed state. Clusters of negatively charged amino acids at either end of the pore help to exclude negative ions and encourage any positive ion of diameter less than 0.65 nm to pass through (Figure 11–38B). The normal through-traffic consists chiefly of Na^+ and K^+, together with some Ca^{2+}. Thus, unlike voltage-gated cation channels, such as the K^+ channel discussed earlier, there is little selectivity among cations, and the relative contributions of the different cations to the current through the channel depend chiefly on their concentrations and on the electrochemical driving forces. When the muscle cell membrane is at its resting potential, the net driving force for K^+ is near zero, since the voltage gradient nearly balances the K^+ concentration gradient across the membrane (see Panel 11–1, p. 616). For Na^+, in contrast, the voltage gradient and the concentration gradient both act in the same direction to drive the ion into the cell. (The same is true for Ca^{2+}, but the

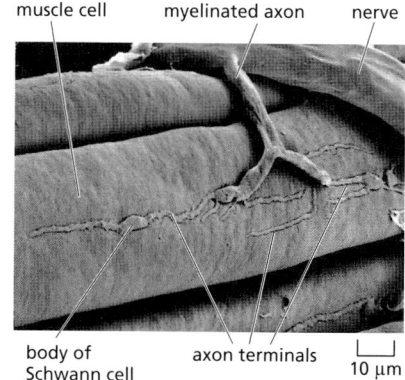

muscle cell   myelinated axon   nerve

body of   axon terminals
Schwann cell   10 µm

Figure 11–37 A low-magnification scanning electron micrograph of a neuromuscular junction in a frog. The termination of a single axon on a skeletal muscle cell is shown. (From J. Desaki and Y. Uehara, *J. Neurocytol.* 10:101–110, 1981. With permission from Springer Nature.)

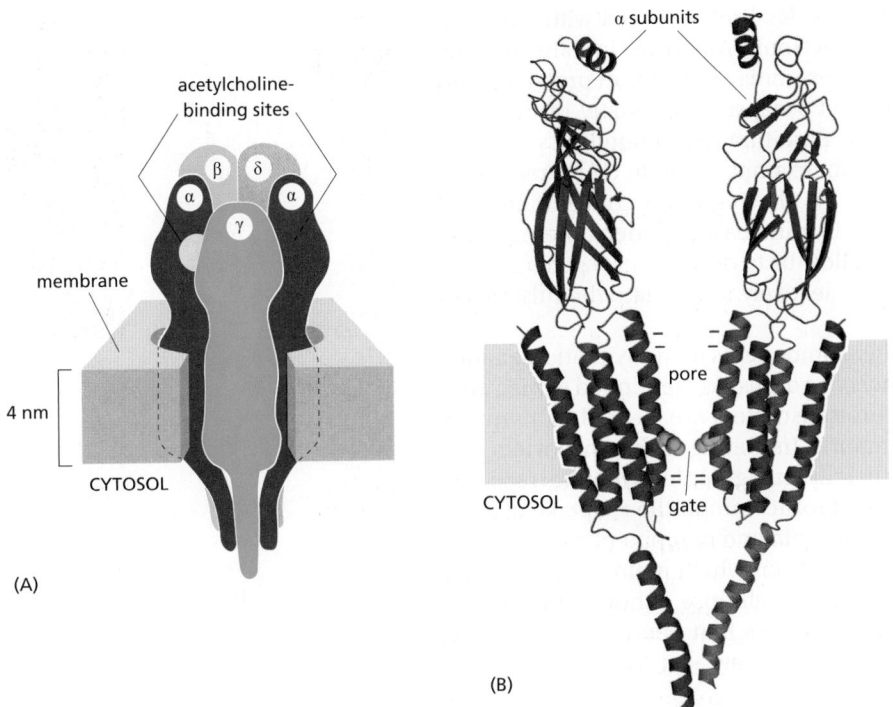

Figure 11–38 **A model for the structure of the skeletal muscle acetylcholine receptor.** (A) Five homologous subunits (α, α, β, γ, δ) combine to form a transmembrane pore. Both of the α subunits contribute an acetylcholine-binding site nestled between adjoining subunits. (B) The pore is lined by a ring of five transmembrane α helices, one contributed by each subunit (just the two α subunits are shown). In its closed conformation, the pore is occluded by the hydrophobic side chains of five leucines (*green*), one from each α helix, which form a gate near the middle of the lipid bilayer. When acetylcholine binds to both α subunits, the channel undergoes a conformational change that opens the gate by an outward rotation of the helices containing the occluding leucines. Negatively charged side chains (indicated by the "–" signs) at either end of the pore ensure that only positively charged ions pass through the channel. (PDB code: 2BG9.)

extracellular concentration of Ca^{2+} is so much lower than that of Na^+ that Ca^{2+} makes only a small contribution to the total inward current.) Therefore, the opening of the acetylcholine-receptor channels leads to a large net influx of Na^+ (a peak rate of about 30,000 ions per channel each millisecond). This influx causes a membrane depolarization that signals the muscle to contract, as discussed below.

Neurons Contain Many Types of Transmitter-Gated Channels

The ion channels that open directly in response to the neurotransmitters acetylcholine, serotonin, GABA, and glycine contain subunits that are structurally similar and probably form transmembrane pores in the same way as the ionotropic acetylcholine receptor, even though they have distinct neurotransmitter-binding specificities and ion selectivities. These channels are all built from homologous polypeptide subunits, which assemble as a pentamer. Glutamate-gated ion channels are an exception, in that they are constructed from a distinct family of subunits and form tetramers resembling the K^+ channels discussed earlier (see Figure 11–24A).

For each class of transmitter-gated ion channel, there are alternative forms of each type of subunit, which may be encoded by distinct genes or else generated by alternative RNA splicing of a single gene product. The subunits assemble in different combinations to form an extremely diverse set of distinct channel subtypes, with different ligand affinities, different channel conductances, different rates of opening and closing, and different sensitivities to drugs and toxins. Some vertebrate neurons, for example, have acetylcholine-gated ion channels that differ from those of muscle cells in that they are formed from two subunits of one type and three of another; but there are at least nine genes coding for different versions of the first type of subunit and at least three coding for different versions of the second. Subsets of such neurons performing different functions in the brain express different combinations of the genes for these subunits. In principle, and already to some extent in practice, it is possible to design drugs targeted against these narrowly defined subsets, thereby specifically influencing particular brain functions.

Many Psychoactive Drugs Act at Synapses

Transmitter-gated ion channels have for a long time been important drug targets. A surgeon, for example, can relax muscles for the duration of an operation by blocking the acetylcholine receptors on skeletal muscle cells with *curare*, a plant-derived drug that was originally used by South American Indians to make poison arrows. Most drugs used to treat insomnia, anxiety, depression, and schizophrenia exert their effects at chemical synapses, and many of these act by binding to transmitter-gated channels. Barbiturates, tranquilizers such as Valium, and sleeping pills such as Ambien, for example, bind to GABA receptors, potentiating the inhibitory action of GABA by allowing lower concentrations of this neurotransmitter to open Cl⁻ channels. Our increasing understanding of the molecular biology of ion channels should allow us to design a new generation of psychoactive drugs that will act still more selectively to alleviate the miseries of mental illness.

In addition to ion channels, many other components of the synaptic signaling machinery are potential targets for psychoactive drugs. As mentioned earlier, after release into the synaptic cleft, many neurotransmitters are cleared by reuptake mechanisms mediated by Na⁺-driven symports. Inhibiting such transporters prolongs the effect of the neurotransmitter, thereby strengthening synaptic transmission. Many antidepressant drugs, including Prozac, inhibit the reuptake of serotonin; others inhibit the reuptake of both serotonin and norepinephrine.

Ion channels are the basic molecular units from which neuronal devices for signaling and computation are built. To provide a glimpse of how sophisticated these devices can be, we consider several examples that demonstrate how the coordinated activities of groups of ion channels allow you to move, feel, and remember.

Neuromuscular Transmission Involves the Sequential Activation of Five Different Sets of Ion Channels

The following process, in which a nerve impulse stimulates a muscle cell to contract, illustrates the importance of ion channels to electrically excitable cells. This apparently simple response requires the sequential activation of at least five different sets of ion channels, all within a few milliseconds (**Figure 11–39**).

1. The process is initiated when a nerve impulse reaches the nerve terminal and depolarizes the plasma membrane of the terminal. The depolarization transiently opens voltage-gated Ca²⁺ channels in this presynaptic membrane. As the Ca²⁺ concentration outside cells is more than 1000 times

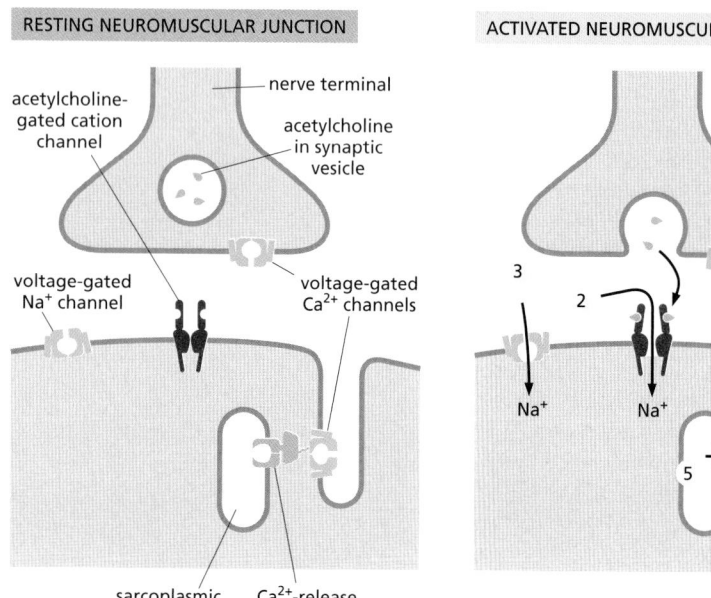

Figure 11–39 **The system of ion channels at a neuromuscular junction.** These gated ion channels are essential for the stimulation of muscle contraction by a nerve impulse. The various channels are numbered in the sequence in which they are activated, as described in the text.

greater than the free Ca^{2+} concentration inside, Ca^{2+} flows into the nerve terminal. The increase in Ca^{2+} concentration in the cytosol of the nerve terminal triggers the local release of acetylcholine by exocytosis into the synaptic cleft.

2. The released acetylcholine binds to acetylcholine receptors in the muscle cell plasma membrane, transiently opening the cation channels associated with them. The resulting influx of Na^+ causes a local membrane depolarization.

3. The local depolarization opens voltage-gated Na^+ channels in this membrane, allowing more Na^+ to enter, which further depolarizes the membrane. This, in turn, opens neighboring voltage-gated Na^+ channels and results in a self-propagating depolarization (an action potential) that spreads to involve the entire plasma membrane (see Figure 11–31).

4. The generalized depolarization of the muscle cell plasma membrane activates voltage-gated Ca^{2+} channels in the transverse tubules (T tubules—discussed in Chapter 16) of this membrane.

5. This in turn causes *Ca^{2+}-release channels* in an adjacent region of the sarcoplasmic reticulum (SR) membrane to open transiently and release Ca^{2+} stored in the SR into the cytosol. The T-tubule and SR membranes are closely apposed with the two types of channel joined together in a specialized structure, in which activation of the voltage-sensitive Ca^{2+} channel in the T-tubule plasma membrane causes a channel conformational change that is mechanically transmitted to the Ca^{2+}-release channel in the SR membrane, opening it and allowing Ca^{2+} to flow from the SR lumen into the cytoplasm (see Figure 16–35). The sudden increase in the cytosolic Ca^{2+} concentration causes the myofibrils in the muscle cell to contract.

Whereas the initiation of muscle contraction by a motor neuron is complex, an even more sophisticated interplay of ion channels is required for a neuron to integrate a large number of input signals at its synapses and compute an appropriate output, as we now discuss.

Single Neurons Are Complex Computation Devices

In the central nervous system, a single neuron can receive inputs from thousands of other neurons, and it can in turn form synapses with many thousands of other cells. Several thousand nerve terminals, for example, make synapses on an average motor neuron in the spinal cord, almost completely covering its cell body and dendrites (**Figure 11–40**). Some of these synapses transmit signals from the brain or spinal cord; others bring sensory information from muscles or from the skin. The motor neuron must combine the information received from all these sources and react, either by firing action potentials along its axon or by remaining quiet.

Of the many synapses on a neuron, some tend to excite it, while others inhibit it. Neurotransmitter released at an excitatory synapse causes a small depolarization in the postsynaptic membrane called an *excitatory postsynaptic potential* (*excitatory PSP*), whereas neurotransmitter released at an inhibitory synapse generally causes a small hyperpolarization called an *inhibitory PSP*. The plasma membrane of the dendrites and cell body of most neurons contains a relatively low density of voltage-gated Na^+ channels, and so an individual excitatory PSP is generally too small to trigger an action potential. Instead, each incoming signal initiates a local PSP, which decreases with distance from the site of the synapse. If signals arrive simultaneously at several synapses in the same region of the dendritic tree, the total PSP in that neighborhood will be roughly the sum of the individual PSPs, with inhibitory PSPs making a negative contribution to the total. The PSPs from each neighborhood spread passively and converge on the cell body. For long-distance transmission, the combined magnitude of the PSP is then translated, or *encoded*, into the *frequency* of firing of action potentials: the greater the stimulation (depolarization), the higher the frequency of action potentials.

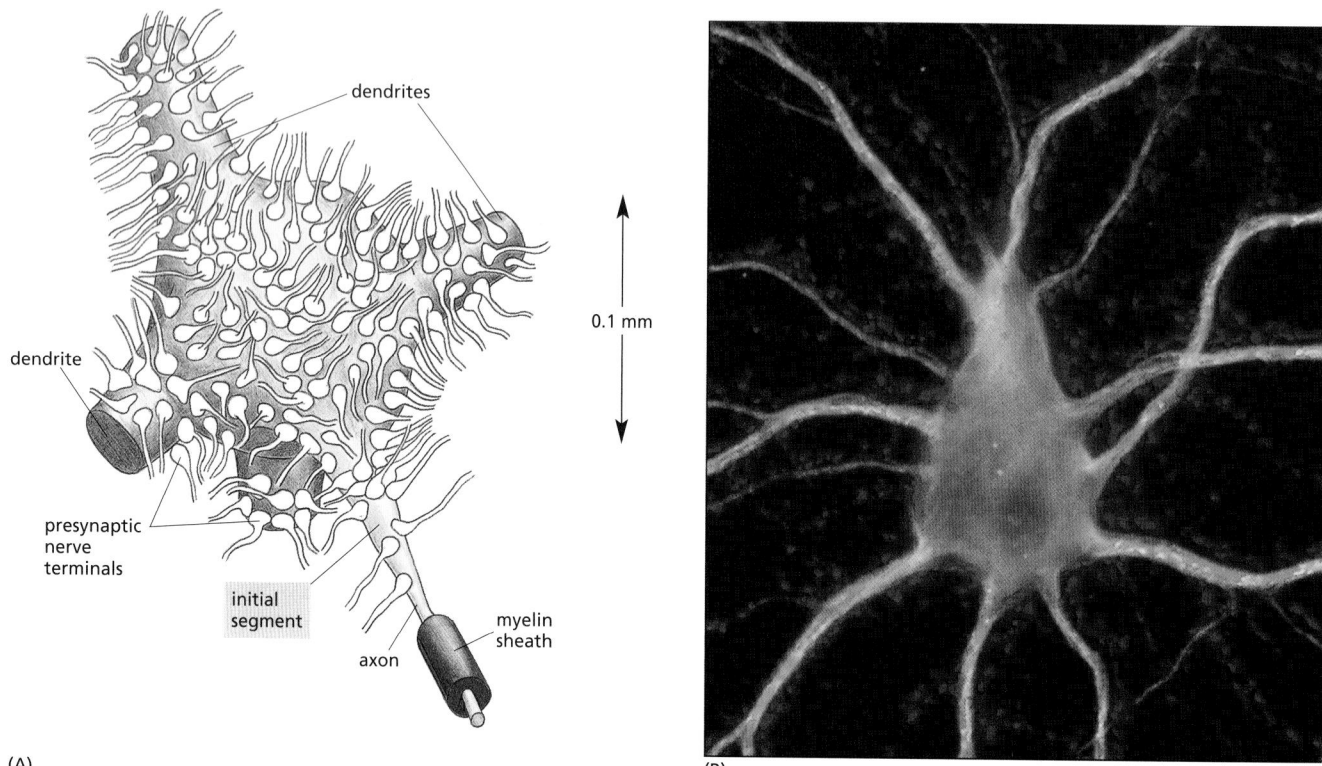

dendrites

0.1 mm

dendrite

presynaptic
nerve
terminals

initial
segment

myelin
sheath

axon

(A)

(B)

Neuronal Computation Requires a Combination of at Least Three Kinds of K⁺ Channels

The intensity of stimulation that a neuron receives is encoded by that neuron into action potential frequency for long-distance transmission. The encoding takes place at a specialized region of the axonal membrane known as the **initial segment**, or *axon hillock*, at the junction of the axon and the cell body (see Figure 11–40). This membrane is rich in voltage-gated Na⁺ channels; but it also contains at least four other classes of ion channels—three selective for K⁺ and one selective for Ca²⁺—all of which contribute to the axon hillock's encoding function. The three varieties of K⁺ channels have different properties; we shall refer to them as *delayed, rapidly inactivating*, and *Ca²⁺-activated K⁺ channels*.

To understand the need for multiple types of channels, consider first what would happen if the only voltage-gated ion channels present in the nerve cell were the Na⁺ channels. Below a certain threshold level of synaptic stimulation, the depolarization of the initial-segment membrane would be insufficient to trigger an action potential. With gradually increasing stimulation, the threshold would be crossed, the Na⁺ channels would open, and an action potential would fire. The action potential would be terminated by inactivation of the Na⁺ channels. Before another action potential could fire, these channels would have to recover from their inactivation. But that would require a return of the membrane voltage to a very negative value, which would not occur as long as the strong depolarizing stimulus (from PSPs) was maintained. An additional channel type is needed, therefore, to repolarize the membrane after each action potential to prepare the cell to fire again.

The **delayed K⁺ channels** perform this task, as discussed previously in relation to the propagation of the action potential (see Figure 11–31). They are voltage-gated, but because of their slower kinetics they open only during the falling phase of the action potential, when the Na⁺ channels are inactive. Their opening permits an efflux of K⁺ that drives the membrane back toward the K⁺ equilibrium potential, which is so negative that the Na⁺ channels rapidly recover from their inactivated state. Repolarization of the membrane also closes the delayed K⁺ channels. The initial segment is now reset so that the depolarizing stimulus from

Figure 11–40 A motor neuron in the spinal cord. (A) Many thousands of nerve terminals synapse on the cell body and dendrites. These deliver signals from other parts of the organism to control the firing of action potentials along the single axon of this large cell. (B) Fluorescence micrograph showing a nerve cell body and its dendrites stained with a fluorescent antibody that recognizes a cytoskeletal protein *(green)* that is not present in axons. Thousands of axon terminals *(red)* from other nerve cells (not visible) make synapses on the cell body and dendrites; the terminals are stained with a fluorescent antibody that recognizes a protein in synaptic vesicles. (B, courtesy of Olaf Mundigl and Pietro de Camilli.)

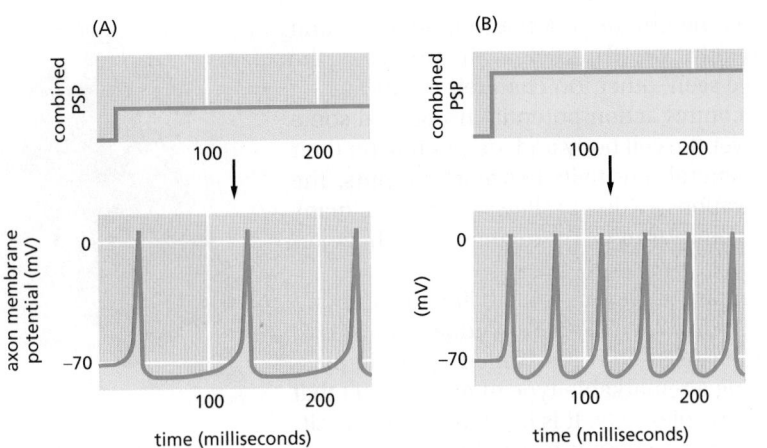

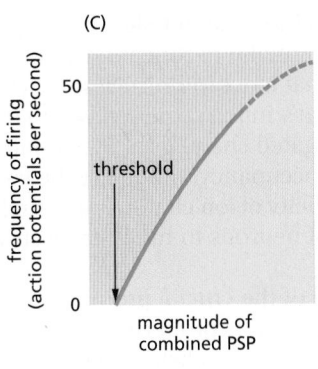

Figure 11–41 **The magnitude of the combined postsynaptic potential (PSP) is reflected in the frequency of firing of action potentials.** The mix of excitatory and inhibitory PSPs produces a *combined PSP* at the initial segment. A comparison of (A) and (B) shows how the firing frequency of an axon increases with an increase in the combined PSP, while (C) summarizes the general relationship.

synaptic inputs can fire another action potential. In this way, sustained stimulation of the dendrites and cell body leads to repetitive firing of the axon.

Repetitive firing in itself, however, is not enough. The frequency of firing has to reflect the intensity of stimulation, and a simple system of Na$^+$ channels and delayed K$^+$ channels is inadequate for this purpose. Below a certain threshold level of steady stimulation, the cell will not fire at all; above that threshold level, it will abruptly begin to fire at a relatively rapid rate. The **rapidly inactivating K$^+$ channels** solve the problem. These, too, are voltage-gated and open when the membrane is depolarized, but their specific voltage sensitivity and kinetics of inactivation are such that they act to reduce the rate of firing at levels of stimulation that are only just above the threshold required for firing. Thus, they remove the discontinuity in the relationship between the firing rate and the intensity of stimulation. The result is a firing rate that is proportional to the strength of the depolarizing stimulus over a very broad range (**Figure 11–41**).

The process of encoding is usually further modulated by the two other types of ion channels in the initial segment that were mentioned earlier—voltage-gated Ca^{2+} channels and Ca^{2+}-activated K$^+$ channels. They act together to decrease the response of the cell to an unchanging, prolonged stimulation—a process called **adaptation**. These Ca^{2+} channels are similar to the Ca^{2+} channels that mediate the release of neurotransmitter from presynaptic axon terminals; they open when an action potential fires, transiently allowing Ca^{2+} into the axon cytosol at the initial segment.

The **Ca^{2+}-activated K$^+$ channel** opens in response to a raised concentration of Ca^{2+} at the channel's cytoplasmic face (**Figure 11–42**). Prolonged, strong depolarizing stimuli will trigger a long train of action potentials, each of which permits a brief influx of Ca^{2+} through the voltage-gated Ca^{2+} channels, so that local cytosolic Ca^{2+} concentration gradually builds up to a level high enough to open the Ca^{2+}-activated K$^+$ channels. Because the resulting increased permeability of the membrane to K$^+$ makes the membrane harder to depolarize, the delay between one action potential and the next is increased. In this way, a neuron that is stimulated continuously for a prolonged period becomes gradually less responsive to the constant stimulus.

Such adaptation, which can also occur by other mechanisms, allows a neuron—indeed, the nervous system generally—to react sensitively to *change*, even against a high background level of steady stimulation. It is one of the computational strategies that help us, for example, to feel a light touch on the shoulder and yet ignore the constant pressure of our clothing. We discuss adaptation as a general feature in cell signaling processes in more detail in Chapter 15.

Other neurons do different computations, reacting to their synaptic inputs in myriad ways, reflecting the different assortments of ion channels in their membrane. There are several hundred genes that code for ion channels in the human genome, with over 150 encoding voltage-gated channels alone. Further complexity is introduced by alternative splicing of RNA transcripts and assembling channel subunits in different combinations. Moreover, ion channels are selectively localized to different sites in the plasma membrane of a neuron. Some K^+ and Ca^{2+} channels are concentrated in the dendrites and participate in processing the input that a neuron receives. As we have seen, other ion channels are located at the axon's initial segment, where they control action potential firing; and some ligand-gated channels are distributed over the cell body and, depending on their ligand occupancy, modulate the cell's general sensitivity to synaptic inputs. The multiplicity of ion channels and their locations evidently allows each of the many types of neurons to tune the electrical behavior to the particular tasks they perform.

One of the crucial properties of the nervous system is its ability to learn and remember. This property depends in part on the ability of individual synapses to strengthen or weaken depending on their use—a process called **synaptic plasticity**. We end this chapter by considering a remarkable type of ion channel that has a special role in some forms of synaptic plasticity. It is located at many excitatory synapses in the central nervous system, where it is gated by both voltage and the excitatory neurotransmitter glutamate. It is also the site of action of the psychoactive drug phencyclidine, or angel dust.

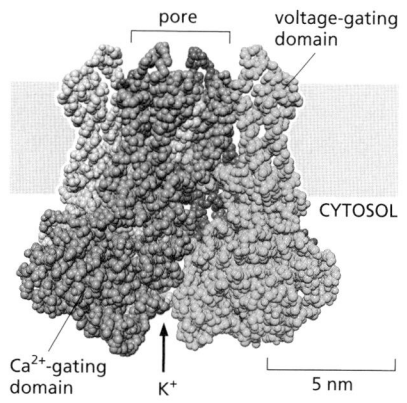

Figure 11–42 **Structure of a Ca^{2+}-activated K^+ channel.** The channel contains four identical subunits (which are shown in different colors for clarity). It is both voltage- and Ca^{2+}-gated. The structure shown is a composite of the cytosolic and membrane portions of the channel that were separately crystallized. (PDB codes: 2R99, 1LNQ.)

Long-Term Potentiation (LTP) in the Mammalian Hippocampus Depends on Ca^{2+} Entry Through NMDA-Receptor Channels

Practically all animals can learn, but mammals seem to learn exceptionally well (or so we like to think). In a mammal's brain, the region called the *hippocampus* has a special role in learning. When it is destroyed on both sides of the brain, the ability to form new memories is largely lost, although previous long-established memories remain. Some synapses in the hippocampus show a striking form of synaptic plasticity with repeated use: whereas occasional single action potentials in the presynaptic cells leave no lasting trace, a short burst of repetitive firing causes **long-term potentiation** (**LTP**), such that subsequent single action potentials in the presynaptic cells evoke a greatly enhanced response in the postsynaptic cells. The effect lasts hours, days, or weeks, according to the number and intensity of the bursts of repetitive firing. Only the synapses that were activated exhibit LTP; synapses that have remained quiet on the same postsynaptic cell are not affected. However, while the cell is receiving a burst of repetitive stimulation via one set of synapses, if a single action potential is delivered at *another* synapse on its surface, that latter synapse also will undergo LTP, even though a single action potential delivered there at another time would leave no such lasting trace.

The underlying rule in such events seems to be that *LTP occurs on any occasion when a presynaptic cell fires (once or more) at a time when the postsynaptic membrane is strongly depolarized* (either through recent repetitive firing of the same presynaptic cell or by other means). This rule reflects the behavior of a particular class of ion channels in the postsynaptic membrane. Glutamate is the main excitatory neurotransmitter in the mammalian central nervous system, and glutamate-gated ion channels are the most common of all transmitter-gated channels in the brain. In the hippocampus, as elsewhere, most of the depolarizing current responsible for excitatory PSPs is carried by glutamate-gated ion channels called **AMPA receptors**, which operate in the standard way (**Figure 11–43**). But the current has, in addition, a second and more intriguing component, which is mediated by a separate subclass of glutamate-gated ion channels known as **NMDA receptors**, so named because they are selectively activated by the artificial glutamate analog *N*-methyl-D-aspartate. The NMDA-receptor channels are doubly gated, opening only when two conditions are satisfied simultaneously: glutamate must be bound to the receptor, and the membrane must be strongly depolarized. The second condition is required for releasing the Mg^{2+} that normally blocks the resting channel. This means that NMDA receptors are normally activated only when AMPA receptors are activated as well and depolarize the membrane. The NMDA receptors are critical for LTP. When they are selectively blocked with a specific inhibitor or inactivated genetically, LTP does not occur, even though ordinary synaptic transmission continues, indicating the importance of NMDA receptors

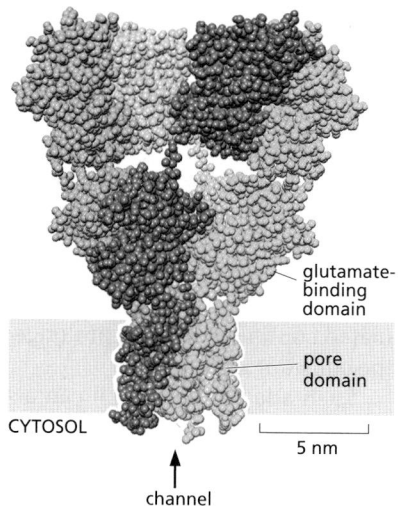

Figure 11–43 **The structure of the AMPA receptor.** This ionotropic glutamate receptor (named after the glutamate analog α-Amino 3-hydroxy 5-Methyl 4-isoxazole Propionic Acid) is the most common mediator of fast, excitatory synaptic transmission in the central nervous system (CNS). (PDB code: 3KG2.)

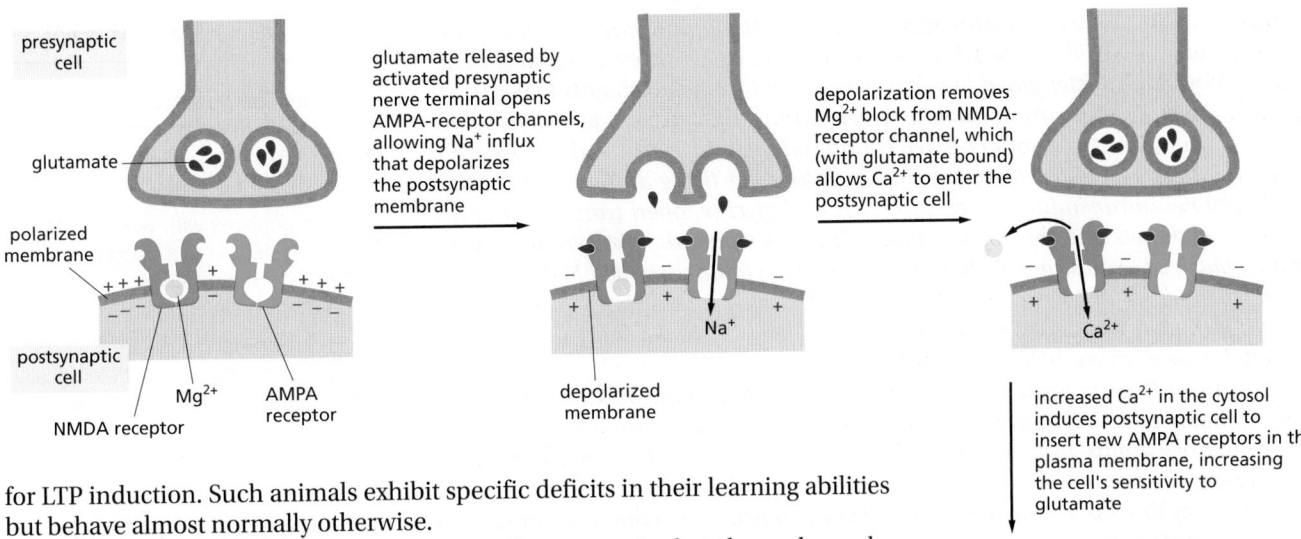

presynaptic cell

glutamate

polarized membrane

postsynaptic cell

NMDA receptor   Mg²⁺   AMPA receptor

glutamate released by activated presynaptic nerve terminal opens AMPA-receptor channels, allowing Na⁺ influx that depolarizes the postsynaptic membrane

depolarized membrane

Na⁺

depolarization removes Mg²⁺ block from NMDA-receptor channel, which (with glutamate bound) allows Ca²⁺ to enter the postsynaptic cell

Ca²⁺

increased Ca²⁺ in the cytosol induces postsynaptic cell to insert new AMPA receptors in the plasma membrane, increasing the cell's sensitivity to glutamate

Figure 11–44 **The signaling events in long-term potentiation.** Although not shown, transmission-enhancing changes can also occur in the presynaptic nerve terminals in LTP, which may be induced by retrograde signals from the postsynaptic cell.

for LTP induction. Such animals exhibit specific deficits in their learning abilities but behave almost normally otherwise.

How do NMDA receptors mediate LTP? The answer is that these channels, when open, are highly permeable to Ca^{2+}, which acts as an intracellular signal in the postsynaptic cell, triggering a cascade of changes that are responsible for LTP. Thus, LTP is prevented when Ca^{2+} levels are held artificially low in the postsynaptic cell by injecting the Ca^{2+} chelator EGTA into it, and LTP can be induced by artificially raising intracellular Ca^{2+} levels in the cell. Among the long-term changes that increase the sensitivity of the postsynaptic cell to glutamate is the insertion of new AMPA receptors into the plasma membrane (**Figure 11–44**). In some forms of LTP, changes occur in the presynaptic cell as well, so that it releases more glutamate than normal when it is activated subsequently.

If synapses were capable only of LTP they would quickly become saturated, and thus be of limited value as an information-storage device. In fact, they also exhibit **long-term depression (LTD)**, with the long-term effect of reducing the number of AMPA receptors in the post-synaptic membrane. This feat is accomplished by degrading AMPA receptors after their selective endocytosis. Surprisingly, LTD also requires NMDA receptor activation and a rise in Ca^{2+}. How does Ca^{2+} trigger opposite effects at the same synapse? It turns out that this bidirectional control of synaptic strength depends on the magnitude of the rise in Ca^{2+}: high Ca^{2+} levels activate protein kinases and LTP, whereas modest Ca^{2+} levels activate protein phosphatases and LTD.

There is evidence that NMDA receptors have an important role in synaptic plasticity and learning in other parts of the brain, as well as in the hippocampus. Moreover, they have a crucial role in adjusting the anatomical pattern of synaptic connections in the light of experience during the development of the nervous system.

Thus, neurotransmitters released at synapses, besides relaying transient electrical signals, can also alter concentrations of intracellular mediators that bring about lasting changes in the efficacy of synaptic transmission. However, it is still uncertain how these changes endure for weeks, months, or a lifetime in the face of the normal turnover of cell constituents.

Summary

Ion channels form aqueous pores across the lipid bilayer and allow inorganic ions of appropriate size and charge to cross the membrane down their electrochemical gradients at rates about 1000 times greater than those achieved by any known transporter. The channels are "gated" and usually open transiently in response to a specific perturbation in the membrane, such as a change in membrane potential (voltage-gated channels), or the binding of a neurotransmitter to the channel (transmitter-gated channels).

K^+-selective leak channels have an important role in determining the resting membrane potential across the plasma membrane in most animal cells.

Voltage-gated cation channels are responsible for the amplification and propagation of action potentials in electrically excitable cells, such as neurons and skeletal muscle cells. Transmitter-gated ion channels convert chemical signals to electrical signals at chemical synapses. Excitatory neurotransmitters, such as acetylcholine and glutamate, open transmitter-gated cation channels and thereby depolarize the postsynaptic membrane toward the threshold level for firing an action potential. Inhibitory neurotransmitters, such as GABA and glycine, open transmitter-gated Cl^- or K^+ channels and thereby suppress firing by keeping the postsynaptic membrane polarized. A subclass of glutamate-gated ion channels, called NMDA-receptor channels, is highly permeable to Ca^{2+}, which can trigger the long-term changes in synapse efficacy (synaptic plasticity) such as LTP and LTD that are thought to be involved in some forms of learning and memory.

Ion channels work together in complex ways to control the behavior of electrically excitable cells. A typical neuron, for example, receives thousands of excitatory and inhibitory inputs, which combine by spatial and temporal summation to produce a combined postsynaptic potential (PSP) at the initial segment of its axon. The magnitude of the PSP is translated into the rate of firing of action potentials by a mixture of cation channels in the initial segment membrane.

WHAT WE DON'T KNOW

• How do individual neurons establish and maintain their characteristic intrinsic firing properties?

• Even organisms with very simple nervous systems have dozens of different K^+ channels. Why is it important to have so many?

• Why do cells that are not electrically active contain voltage-gated ion channels?

• How are memories stored for so many years in the human brain?

PROBLEMS

Which statements are true? Explain why or why not.

11–1   Transport by transporters can be either active or passive, whereas transport by channels is always passive.

11–2   Transporters saturate at high concentrations of the transported molecule when all their binding sites are occupied; channels, on the other hand, do not bind the ions they transport and thus the flux of ions through a channel does not saturate.

11–3   The membrane potential arises from movements of charge that leave ion concentrations practically unaffected, causing only a very slight discrepancy in the number of positive and negative ions on the two sides of the membrane.

Discuss the following problems.

11–4   Order Ca^{2+}, CO_2, ethanol, glucose, RNA, and H_2O according to their ability to diffuse through a lipid bilayer, beginning with the one that crosses the bilayer most readily. Explain your order.

11–5   How is it possible for some molecules to be at equilibrium across a biological membrane and yet not be at the same concentration on both sides?

11–6   Ion transporters are "linked" together—not physically, but as a consequence of their actions. For example, cells can raise their intracellular pH, when it becomes too acidic, by exchanging external Na^+ for internal H^+, using a Na^+-H^+ antiporter. The change in internal Na^+ is then redressed using the Na^+-K^+ pump.
A.   Can these two transporters, operating together, normalize both the H^+ and the Na^+ concentrations inside the cell?

B.   Does the linked action of these two pumps cause imbalances in either the K^+ concentration or the membrane potential? Why or why not?

11–7   Microvilli increase the surface area of intestinal cells, providing more efficient absorption of nutrients. Microvilli are shown in profile and cross section in **Figure Q11–1**. From the dimensions given in the figure, estimate the increase in surface area that microvilli provide (for the portion of the plasma membrane in contact with the lumen of the gut) relative to the corresponding surface of a cell with a "flat" plasma membrane.

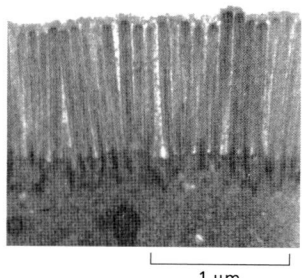
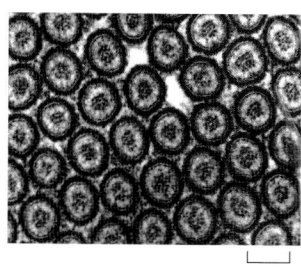

1 µm	0.1 µm

Figure Q11–1 Microvilli of intestinal epithelial cells in profile and cross section (Problem 11–7). (Left panel, from Rippel Electron Microscope Facility, Dartmouth College; right panel, from David Burgess.)

11–8   According to Newton's laws of motion, an ion exposed to an electric field in a vacuum would experience a constant acceleration from the electric driving force, just as a falling body in a vacuum constantly accelerates due to gravity. In water, however, an ion moves at constant velocity in an electric field. Why do you suppose that is?

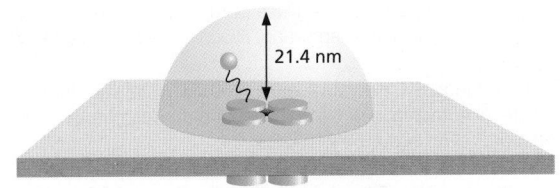

Figure Q11–2 A "ball" tethered by a "chain" to a voltage-gated K$^+$ channel (Problem 11–9).

TABLE Q11–1 **Ionic composition of seawater and of the cytosol in the squid giant axon** (Problem 11–10).

Ion	Cytosol	Seawater
Na$^+$	65 mM	430 mM
K$^+$	344 mM	9 mM

11–9 In a subset of voltage-gated K$^+$ channels, the N-terminus of each subunit acts like a tethered ball that occludes the cytoplasmic end of the pore soon after it opens, thereby inactivating the channel. This "ball-and-chain" model for the rapid inactivation of voltage-gated K$^+$ channels has been elegantly supported for the *shaker* K$^+$ channel from *Drosophila melanogaster*. (The *shaker* K$^+$ channel in *Drosophila* is named after a mutant form that causes excitable behavior—even anesthetized flies keep twitching.) Deletion of the N-terminal amino acids from the normal *shaker* channel gives rise to a channel that opens in response to membrane depolarization, but stays open instead of rapidly closing as the normal channel does. A peptide (MAAVAGLYGLGEDRQHRKKQ) that corresponds to the deleted N-terminus can inactivate the open channel at 100 μM.

Is the concentration of free peptide (100 μM) that is required to inactivate the defective K$^+$ channel anywhere near the local concentration of the tethered ball on a normal channel? Assume that the tethered ball can explore a hemisphere [volume = $(2/3)\pi r^3$] with a radius of 21.4 nm, which is the length of the polypeptide "chain" (**Figure Q11–2**). Calculate the concentration for one ball in this hemisphere. How does that value compare with the concentration of free peptide needed to inactivate the channel?

11–10 The giant axon of the squid (**Figure Q11–3**) occupies a unique position in the history of our understanding of cell membrane potentials and nerve action. When an electrode is stuck into an intact giant axon, the membrane potential registers –70 mV. When the axon, suspended in a bath of seawater, is stimulated to conduct a nerve impulse, the membrane potential changes transiently from –70 mV to +40 mV.

For univalent ions and at 20°C (293 K), the Nernst equation reduces to

$$V = 58 \text{ mV} \times \log(C_o/C_i)$$

where C_o and C_i are the concentrations outside and inside, respectively.

Using this equation, calculate the potential across the resting membrane (1) assuming that it is due solely to K$^+$ and (2) assuming that it is due solely to Na$^+$. (The Na$^+$ and K$^+$ concentrations in the axon cytosol and in seawater are given in **Table Q11–1**.) Which calculation is closer to the measured resting potential? Which calculation is closer to the measured action potential? Explain why these assumptions approximate the measured resting and action potentials.

11–11 Acetylcholine-gated cation channels at the neuromuscular junction open in response to acetylcholine released by the nerve terminal and allow Na$^+$ ions to enter the muscle cell, which causes membrane depolarization and ultimately leads to muscle contraction.
A. Patch-clamp measurements show that young rat muscles have cation channels that respond to acetylcholine (**Figure Q11–4**). How many kinds of channel are there? How can you tell?
B. For each kind of channel, calculate the number of ions that enter in one millisecond. (One ampere is a current of one coulomb per second; one pA equals 10^{-12} ampere. An ion with a single charge such as Na$^+$ carries a charge of 1.6×10^{-19} coulomb.)

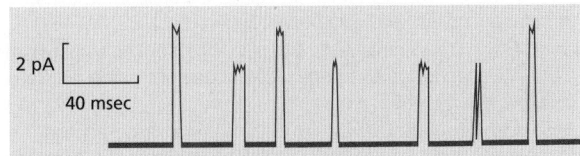

Figure Q11–4 Patch-clamp measurements of acetylcholine-gated cation channels in young rat muscle (Problem 11–11).

Figure Q11–3 The squid *Sepioteuthus lessoniana* (Problem 11–10). This squid can grow up to 30 cm in length. (Courtesy of Tim Hunt.)

REFERENCES

General

Engel A & Gaub HE (2008) Structure and mechanics of membrane proteins. *Annu. Rev. Biochem.* 77, 127–148.

Hille B (2001) Ionic Channels of Excitable Membranes, 3rd ed. Sunderland, MA: Sinauer.

Stein WD (2014) Channels, Carriers, and Pumps: An Introduction to Membrane Transport, 2nd ed. San Diego, CA: Academic Press.

Vinothkumar KR & Henderson R (2010) Structures of membrane proteins. *Q. Rev. Biophys.* 43, 65–158.

Principles of Membrane Transport

Al-Awqati Q (1999) One hundred years of membrane permeability: does Overton still rule? *Nat. Cell Biol.* 1, E201–E202.

Forrest LR & Sansom MS (2000) Membrane simulations: bigger and better? *Curr. Opin. Struct. Biol.* 10, 174–181.

Gouaux E & MacKinnon R (2005) Principles of selective ion transport in channels and pumps. *Science* 310, 1461–1465.

Mitchell P (1977) Vectorial chemiosmotic processes. *Annu. Rev. Biochem.* 46, 996–1005.

Tanford C (1983) Mechanism of free energy coupling in active transport. *Annu. Rev. Biochem.* 52, 379–409.

Transporters and Active Membrane Transport

Almers W & Stirling C (1984) Distribution of transport proteins over animal cell membranes. *J. Membr. Biol.* 77, 169–186.

Baldwin SA & Henderson PJ (1989) Homologies between sugar transporters from eukaryotes and prokaryotes. *Annu. Rev. Physiol.* 51, 459–471.

Doige CA & Ames GF (1993) ATP-dependent transport systems in bacteria and humans: relevance to cystic fibrosis and multidrug resistance. *Annu. Rev. Microbiol.* 47, 291–319.

Forrest LR & Rudnick G (2009) The rocking bundle: a mechanism for ion-coupled solute flux by symmetrical transporters. *Physiology* 24, 377–386.

Gadsby DC (2009) Ion channels versus ion pumps: the principal difference, in principle. *Nat. Rev. Mol. Cell Biol.* 10, 344–352.

Higgins CF (2007) Multiple molecular mechanisms for multidrug resistance transporters. *Nature* 446, 749–757.

Kaback HR, Sahin-Tóth M & Weinglass AB (2001) The kamikaze approach to membrane transport. *Nat. Rev. Mol. Cell Biol.* 2, 610–620.

Kühlbrandt W (2004) Biology, structure and mechanism of P-type ATPases. *Nat. Rev. Mol. Cell Biol.* 5, 282–295.

Lodish HF (1986) Anion-exchange and glucose transport proteins: structure, function, and distribution. *Harvey Lect.* 82, 19–46.

Møller JV, Olesen C, Winther AML & Nissen P (2010) The sarcoplasmic Ca^{2+}-ATPase: design of a perfect chemi-osmotic pump. *Q. Rev. Biophys.* 43, 501–566.

Perez C, Koshy C, Yildiz O & Ziegler C (2012) Alternating-access mechanism in conformationally asymmetric trimers of the betaine transporter BetP. *Nature* 490, 126–130.

Rees D, Johnson E & Lewinson O (2009) ABC transporters: the power to change. *Nat. Rev. Mol. Cell Biol.* 10, 218–227.

Romero MF & Boron WF (1999) Electrogenic Na^+/HCO_3^- cotransporters: cloning and physiology. *Annu. Rev. Physiol.* 61, 699–723.

Rudnick G (2011) Cytoplasmic permeation pathway of neurotransmitter transporters. *Biochemistry* 50, 7462–7475.

Saier MH Jr (2000) Vectorial metabolism and the evolution of transport systems. *J. Bacteriol.* 182, 5029–5035.

Stein WD (2002) Cell volume homeostasis: ionic and nonionic mechanisms. The sodium pump in the emergence of animal cells. *Int. Rev. Cytol.* 215, 231–258.

Toyoshima C (2009) How Ca^{2+}-ATPase pumps ions across the sarcoplasmic reticulum membrane. *Biochim. Biophys. Acta* 1793, 941–946.

Yamashita A, Singh SK, Kawate T et al. (2005) Crystal structure of a bacterial homologue of Na^+/Cl^--dependent neurotransmitter transporters. *Nature* 437, 215–223.

Channels and the Electrical Properties of Membranes

Armstrong C (1998) The vision of the pore. *Science* 280, 56–57.

Arnadóttir J & Chalfie M (2010) Eukaryotic mechanosensitive channels. *Annu. Rev. Biophys.* 39, 111–137.

Bezanilla F (2008) How membrane proteins sense voltage. *Nat. Rev. Mol. Cell Biol.* 9, 323–332.

Catterall WA (2010) Ion channel voltage sensors: structure, function, and pathophysiology. *Neuron* 67, 915–928.

Davis GW (2006) Homeostatic control of neural activity: from phenomenology to molecular design. *Annu. Rev. Neurosci.* 29, 307–323.

Greengard P (2001) The neurobiology of slow synaptic transmission. *Science* 294, 1024–1030.

Hodgkin AL & Huxley AF (1952) A quantitative description of membrane current and its application to conduction and excitation in nerve. *J. Physiol.* 117, 500–544.

Hodgkin AL & Huxley AF (1952) Currents carried by sodium and potassium ions through the membrane of the giant axon of *Loligo*. *J. Physiol.* 116, 449–472.

Jessell TM & Kandel ER (1993) Synaptic transmission: a bidirectional and self-modifiable form of cell–cell communication. *Cell* 72(Suppl), 1–30.

Julius D (2013) TRP channels and pain. *Annu. Rev. Cell Dev. Biol.* 29, 355–384.

Katz B (1966) Nerve, Muscle and Synapse. New York: McGraw-Hill.

King LS, Kozono D & Agre P (2004) From structure to disease: the evolving tale of aquaporin biology. *Nat. Rev. Mol. Cell Biol.* 5, 687–698.

Liao M, Cao E, Julius D & Cheng Y (2014) Single particle electron cryo-microscopy of a mammalian ion channel. *Curr. Opin. Struct. Biol.* 27, 1–7.

MacKinnon R (2003) Potassium channels. *FEBS Lett.* 555, 62–65.

Miesenböck G (2011) Optogenetic control of cells and circuits. *Annu. Rev. Cell Dev. Biol.* 27, 731–758.

Moss SJ & Smart TG (2001) Constructing inhibitory synapses. *Nat. Rev. Neurosci.* 2, 240–250.

Neher E & Sakmann B (1992) The patch clamp technique. *Sci. Am.* 266, 44–51.

Nicholls JG, Fuchs PA, Martin AR & Wallace BG (2000) From Neuron to Brain, 4th ed. Sunderland, MA: Sinauer.

Numa S (1987) A molecular view of neurotransmitter receptors and ionic channels. *Harvey Lect.* 83, 121–165.

Payandeh J, Scheuer T, Zheng N & Catterall WA (2011) The crystal structure of a voltage-gated sodium channel. *Nature* 475, 353–358.

Scannevin RH & Huganir RL (2000) Postsynaptic organization and regulation of excitatory synapses. *Nat. Rev. Neurosci.* 1, 133–141.

Snyder SH (1996) Drugs and the Brain. New York: WH Freeman/Scientific American Books.

Sobolevsky AI, Rosconi MP & Gouaux E (2009) X-ray structure, symmetry and mechanism of an AMPA-subtype glutamate receptor. *Nature* 462, 745–756.

Stevens CF (2004) Presynaptic function. *Curr. Opin. Neurobiol.* 14, 341–345.

Verkman AS (2013) Aquaporins. *Curr. Biol.* 23, R52–R55.

Intracellular Compartments and Protein Sorting

Unlike a bacterium, which generally consists of a single intracellular compartment surrounded by a plasma membrane, a eukaryotic cell is elaborately subdivided into functionally distinct, membrane-enclosed compartments. Each compartment, or **organelle**, contains its own characteristic set of enzymes and other specialized molecules, and complex distribution systems transport specific products from one compartment to another. To understand the eukaryotic cell, it is essential to know how the cell creates and maintains these compartments, what occurs in each of them, and how molecules move between them.

Proteins confer upon each compartment its characteristic structural and functional properties. They catalyze the reactions that occur there and selectively transport small molecules into and out of the compartment. For membrane-enclosed organelles in the cytoplasm, proteins also serve as organelle-specific surface markers that direct new deliveries of proteins and lipids to the appropriate organelle.

An animal cell contains about 10 billion (10^{10}) protein molecules of perhaps 10,000 kinds, and the synthesis of almost all of them begins in the **cytosol**, the space of the cytoplasm outside the membrane-enclosed organelles. Each newly synthesized protein is then delivered specifically to the organelle that requires it. The intracellular transport of proteins is the central theme of both this chapter and the next. By tracing the protein traffic from one compartment to another, one can begin to make sense of the otherwise bewildering maze of intracellular membranes.

THE COMPARTMENTALIZATION OF CELLS

In this brief overview of the compartments of the cell and the relationships between them, we organize the organelles conceptually into a small number of discrete families, discuss how proteins are directed to specific organelles, and explain how proteins cross organelle membranes.

All Eukaryotic Cells Have the Same Basic Set of Membrane-enclosed Organelles

Many vital biochemical processes take place in membranes or on their surfaces. Membrane-bound enzymes, for example, catalyze lipid metabolism; and oxidative phosphorylation and photosynthesis both require a membrane to couple the transport of H^+ to the synthesis of ATP. In addition to providing increased membrane area to host biochemical reactions, intracellular membrane systems form enclosed compartments that are separate from the cytosol, thus creating functionally specialized aqueous spaces within the cell. In these spaces, subsets of molecules (proteins, reactants, ions) are concentrated to optimize the biochemical reactions in which they participate. Because the lipid bilayer of cell membranes is impermeable to most hydrophilic molecules, the membrane of an organelle must contain membrane transport proteins to import and export specific metabolites. Each organelle membrane must also have a mechanism for importing, and incorporating into the organelle, the specific proteins that make the organelle unique.

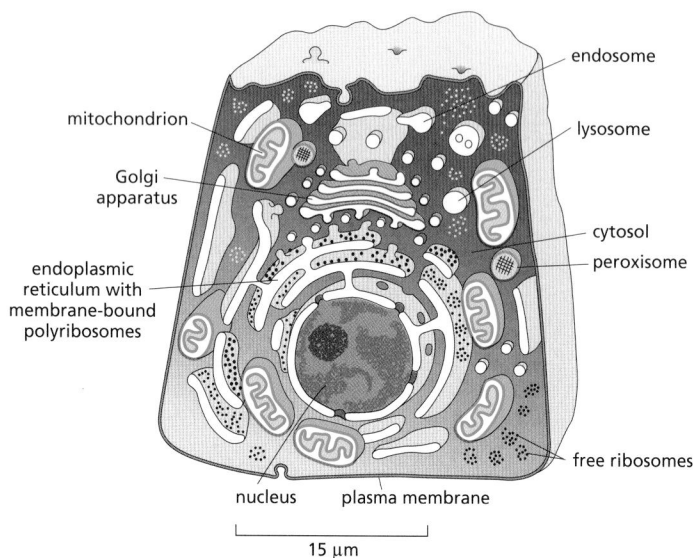

nucleus   plasma membrane

15 μm

Figure 12–1 The major intracellular compartments of an animal cell. The cytosol *(gray)*, endoplasmic reticulum, Golgi apparatus, nucleus, mitochondrion, endosome, lysosome, and peroxisome are distinct compartments isolated from the rest of the cell by at least one selectively permeable membrane (see Movie 9.2).

Figure 12–1 illustrates the major intracellular compartments common to eukaryotic cells. The *nucleus* contains the genome (aside from mitochondrial and chloroplast DNA), and it is the principal site of DNA and RNA synthesis. The surrounding **cytoplasm** consists of the cytosol and the cytoplasmic organelles suspended in it. The cytosol constitutes a little more than half the total volume of the cell, and it is the main site of protein synthesis and degradation. It also performs most of the cell's *intermediary metabolism*—that is, the many reactions that degrade some small molecules and synthesize others to provide the building blocks for macromolecules (discussed in Chapter 2).

About half the total area of membrane in a eukaryotic cell encloses the labyrinthine spaces of the *endoplasmic reticulum* (*ER*). The *rough ER* has many ribosomes bound to its cytosolic surface. Ribosomes are organelles that are not membrane-enclosed; they synthesize both soluble and integral membrane proteins, most of which are destined either for secretion to the cell exterior or for other organelles. We shall see that, whereas proteins are transported into other membrane-enclosed organelles only after their synthesis is complete, they are transported into the ER as they are synthesized. This explains why the ER membrane is unique in having ribosomes tethered to it. The ER also produces most of the lipid for the rest of the cell and functions as a store for Ca^{2+} ions. Regions of the ER that lack bound ribosomes are called *smooth ER*. The ER sends many of its proteins and lipids to the *Golgi apparatus*, which often consists of organized stacks of disc-like compartments called *Golgi cisternae*. The Golgi apparatus receives lipids and proteins from the ER and dispatches them to various destinations, usually covalently modifying them *en route*.

Mitochondria and *chloroplasts* generate most of the ATP that cells use to drive reactions requiring an input of free energy; chloroplasts are a specialized version of *plastids* (present in plants, algae, and some protozoa), which can also have other functions, such as the storage of food or pigment molecules. *Lysosomes* contain digestive enzymes that degrade defunct intracellular organelles, as well as macromolecules and particles taken in from outside the cell by endocytosis. On the way to lysosomes, endocytosed material must first pass through a series of organelles called *endosomes*. Finally, *peroxisomes* are small vesicular compartments that contain enzymes used in various oxidative reactions.

In general, each membrane-enclosed organelle performs the same set of basic functions in all cell types. But to serve the specialized functions of cells, these organelles vary in abundance and can have additional properties that differ from cell type to cell type.

On average, the membrane-enclosed compartments together occupy nearly half the volume of a cell (Table 12–1), and a large amount of intracellular membrane is required to make them. In liver and pancreatic cells, for example, the endoplasmic reticulum has a total membrane surface area that is, respectively, 25 times and 12 times that of the plasma membrane (**Table 12–2**). The membrane-enclosed organelles are packed tightly in the cytoplasm, and, in terms of area and mass, the plasma membrane is only a minor membrane in most eukaryotic cells (**Figure 12–2**).

The abundance and shape of membrane-enclosed organelles are regulated to meet the needs of the cell. This is particularly apparent in cells that are highly specialized and therefore disproportionately rely on specific organelles. Plasma cells, for example, which secrete their own weight every day in antibody molecules into the bloodstream, contain vastly amplified amounts of rough ER, which is found in large, flat sheets. Cells that specialize in lipid synthesis also expand their ER, but in this case the organelle forms a network of convoluted tubules. Moreover, membrane-enclosed organelles are often found in characteristic positions in the cytoplasm. In most cells, for example, the Golgi apparatus is located close to the nucleus, whereas the network of ER tubules extends from the nucleus throughout the entire cytosol. These characteristic distributions depend on interactions of the organelles with the cytoskeleton. The localization of both the ER and the Golgi apparatus, for instance, depends on an intact microtubule array; if the microtubules are experimentally depolymerized with a drug, the Golgi apparatus fragments and disperses throughout the cell, and the ER network collapses toward the cell center (discussed in Chapter 16). The size, shape, composition, and location are all important and regulated features of these organelles that ultimately contribute to the organelle's function.

Evolutionary Origins May Help Explain the Topological Relationships of Organelles

To understand the relationships between the compartments of the cell, it is helpful to consider how they might have evolved. The precursors of the first eukaryotic cells are thought to have been relatively simple cells that—like most bacterial and

TABLE 12–1 Relative Volumes Occupied by the Major Intracellular Compartments in a Liver Cell (Hepatocyte)

Intracellular compartment	Percentage of total cell volume
Cytosol	54
Mitochondria	22
Rough ER cisternae	9
Smooth ER cisternae plus Golgi cisternae	6
Nucleus	6
Peroxisomes	1
Lysosomes	1
Endosomes	1

TABLE 12–2 Relative Amounts of Membrane Types in Two Kinds of Eukaryotic Cells

Membrane Type	Percentage of total cell membrane	
	Liver hepatocyte*	Pancreatic exocrine cell*
Plasma membrane	2	5
Rough ER membrane	35	60
Smooth ER membrane	16	<1
Golgi apparatus membrane	7	10
Mitochondria Outer membrane Inner membrane	7 32	4 17
Nucleus Inner membrane	0.2	0.7
Secretory vesicle membrane	Not determined	3
Lysosome membrane	0.4	Not determined
Peroxisome membrane	0.4	Not determined
Endosome membrane	0.4	Not determined

*These two cells are of very different sizes: the average hepatocyte has a volume of about 5000 μm³ compared with 1000 μm³ for the pancreatic exocrine cell. Total cell membrane areas are estimated at about 110,000 μm² and 13,000 μm², respectively.

rough endoplasmic reticulum   nucleus   lysosomes

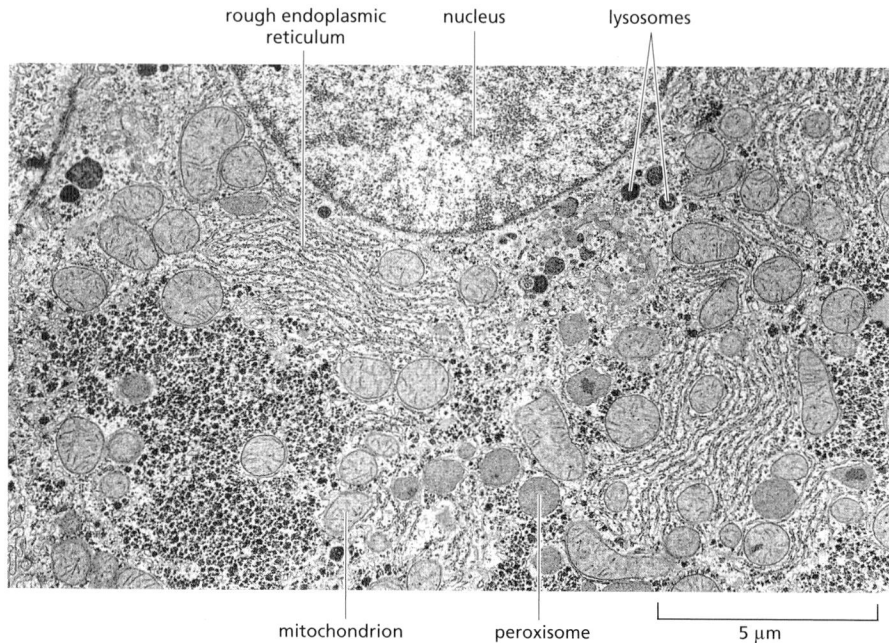

mitochondrion   peroxisome   5 µm

Figure 12–2 An electron micrograph of part of a liver cell seen in cross section. Examples of most of the major intracellular organelles are indicated. (Reused by permission of E.L. Bearer and Daniel S. Friend.)

archaeal cells—have a plasma membrane but no internal membranes. The plasma membrane in such cells provides all membrane-dependent functions, including the pumping of ions, ATP synthesis, protein secretion, and lipid synthesis. Typical present-day eukaryotic cells are 10–30 times larger in linear dimension and 1000–10,000 times greater in volume than a typical bacterium such as *E. coli*. The profusion of internal membranes can be regarded, in part, as an adaptation to this increase in size: the eukaryotic cell has a much smaller ratio of surface area to volume, and its plasma membrane therefore presumably has too small an area to sustain the many vital functions that membranes perform. The extensive internal membrane systems of a eukaryotic cell alleviate this problem.

The evolution of internal membranes evidently went hand-in-hand with the specialization of membrane function. A hypothetical scheme for how the first eukaryotic cells, with a nucleus and ER, might have evolved by the invagination and pinching off of the plasma membrane of an ancestral cell is illustrated in **Figure 12–3**. This process would create membrane-enclosed organelles with an interior or **lumen** that is topologically equivalent to the exterior of the cell. We shall see that this topological relationship holds for all of the organelles involved in the secretory and endocytic pathways, including the ER, Golgi apparatus, endosomes, lysosomes, and peroxisomes. We can therefore think of all of these organelles as members of the same topologically equivalent compartment. As we discuss in detail in the next chapter, their interiors communicate extensively with one another and with the outside of the cell via *transport vesicles*, which bud off from one organelle and fuse with another (**Figure 12–4**).

As described in Chapter 14, mitochondria and plastids differ from the other membrane-enclosed organelles because they contain their own genomes. The nature of these genomes, and the close resemblance of the proteins in these organelles to those in some present-day bacteria, strongly suggest that mitochondria and plastids evolved from bacteria that were engulfed by other cells with which they initially lived in symbiosis (see Figures 1–29 and 1–31): the inner membrane of mitochondria and plastids presumably corresponds to the original plasma membrane of the bacterium, while the lumen of these organelles evolved from the bacterial cytosol. Like the bacteria from which they were derived, both mitochondria and plastids are enclosed by a double membrane and they remain isolated from the extensive vesicular traffic that connects the interiors of most of the other membrane-enclosed organelles to each other and to the outside of the cell.

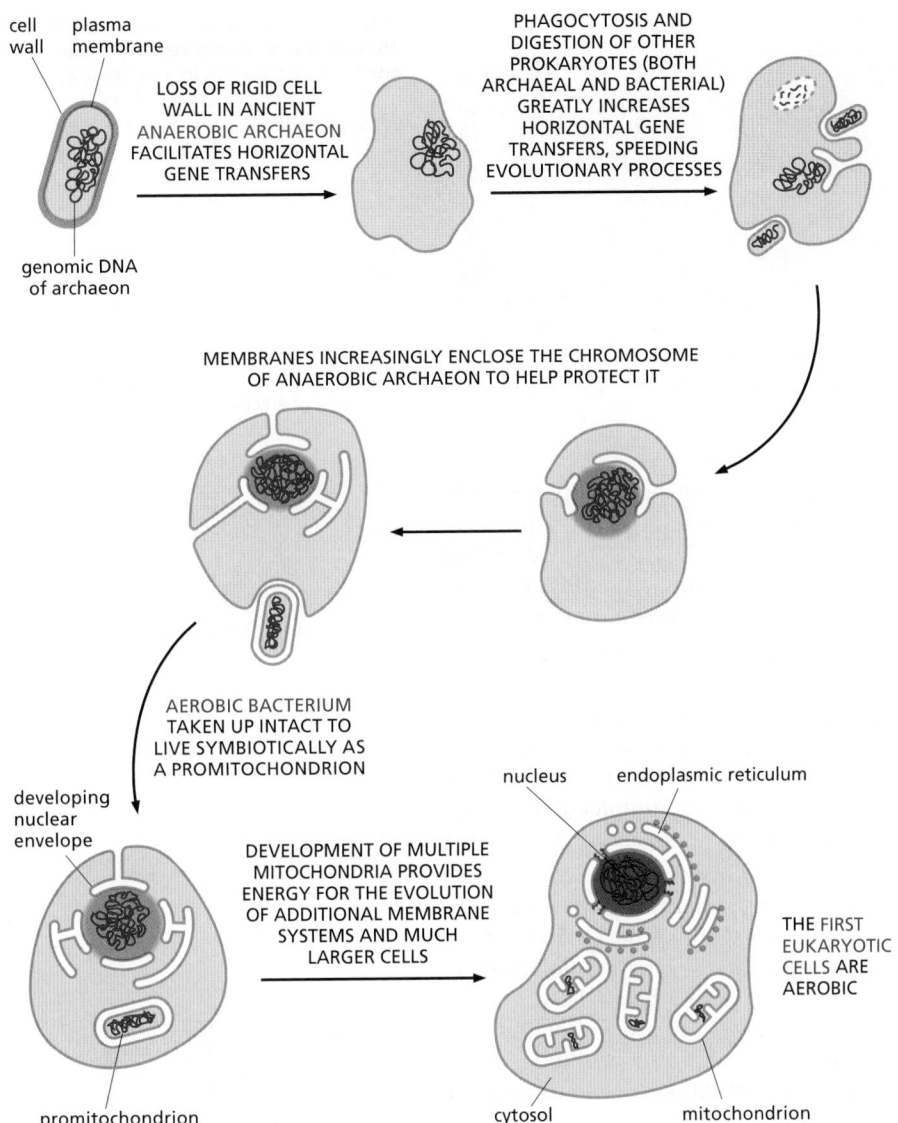

cell wall   plasma membrane

LOSS OF RIGID CELL WALL IN ANCIENT ANAEROBIC ARCHAEON FACILITATES HORIZONTAL GENE TRANSFERS

genomic DNA of archaeon

PHAGOCYTOSIS AND DIGESTION OF OTHER PROKARYOTES (BOTH ARCHAEAL AND BACTERIAL) GREATLY INCREASES HORIZONTAL GENE TRANSFERS, SPEEDING EVOLUTIONARY PROCESSES

MEMBRANES INCREASINGLY ENCLOSE THE CHROMOSOME OF ANAEROBIC ARCHAEON TO HELP PROTECT IT

AEROBIC BACTERIUM TAKEN UP INTACT TO LIVE SYMBIOTICALLY AS A PROMITOCHONDRION

developing nuclear envelope

DEVELOPMENT OF MULTIPLE MITOCHONDRIA PROVIDES ENERGY FOR THE EVOLUTION OF ADDITIONAL MEMBRANE SYSTEMS AND MUCH LARGER CELLS

promitochondrion

nucleus   endoplasmic reticulum

THE FIRST EUKARYOTIC CELLS ARE AEROBIC

cytosol   mitochondrion

Figure 12–3 One suggested pathway for the evolution of the eukaryotic cell and its internal membranes As discussed in Chapter 1, there is evidence that the nuclear genome of a eukaryotic cell evolved from an ancient archaeon. For example, clear homologs of actin, tubulin, histones, and the nuclear DNA replication system are found in archaea, but not in bacteria. Thus, it is now thought that the first eukaryotic cells arose when an ancient anaerobic archaeon joined forces with an aerobic bacterium roughly 1.6 billion years ago. As indicated, the nuclear envelope may have originated from an invagination of the plasma membrane of this ancient archaeon—an invagination that protected its chromosome while still allowing access of the DNA to the cytosol (as required for DNA to direct protein synthesis). This envelope may have later pinched off completely from the plasma membrane, so as to produce a separate nuclear compartment surrounded by a double membrane. Because this double membrane is penetrated by nuclear pore complexes, the nuclear compartment is topologically equivalent to the cytosol. In contrast, the lumen of the ER is continuous with the space between the inner and outer nuclear membranes, and it is topologically equivalent to the extracellular space (see Figure 12–4). (Adapted from J. Martijn and T.J.G. Ettema, *Biochem. Soc. Trans.* 41: 451–457, 2013.)

The evolutionary schemes just described group the intracellular compartments in eukaryotic cells into four distinct families: (1) the nucleus and the cytosol, which communicate with each other through *nuclear pore complexes* and are thus topologically continuous (although functionally distinct); (2) all organelles that function in the secretory and endocytic pathways—including the ER, Golgi apparatus, endosomes, and lysosomes, the numerous classes of transport intermediates such as transport vesicles that move between them, and peroxisomes; (3) the mitochondria; and (4) the plastids (in plants only).

Proteins Can Move Between Compartments in Different Ways

The synthesis of all proteins begins on ribosomes in the cytosol, except for the few that are synthesized on the ribosomes of mitochondria and plastids. Their subsequent fate depends on their amino acid sequence, which can contain **sorting signals** that direct their delivery to locations outside the cytosol or to organelle surfaces. Some proteins do not have a sorting signal and consequently remain in the cytosol as permanent residents. Many others, however, have specific sorting signals that direct their transport from the cytosol into the nucleus, the ER, mitochondria, plastids, or peroxisomes; sorting signals can also direct the transport of proteins from the ER to other destinations in the cell.

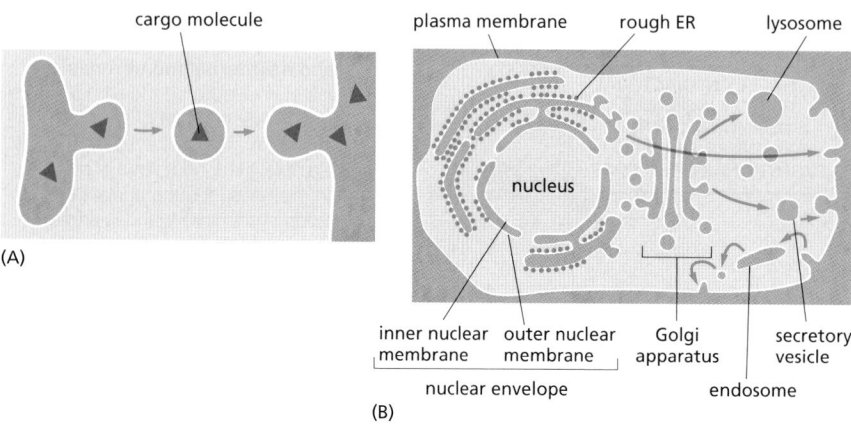

(A)

(B)

Figure 12–4 Topologically equivalent compartments in the secretory and endocytic pathways in a eukaryotic cell. Compartments are said to be *topologically equivalent* if they can communicate with one another, in the sense that molecules can get from one to the other without having to cross a membrane. Topologically equivalent spaces are shown in *red*. (A) Molecules can be carried from one compartment to another topologically equivalent compartment by vesicles that bud from one and fuse with the other. (B) In principle, cycles of membrane budding and fusion permit the lumen of any of the organelles shown to communicate with any other and with the cell exterior by means of transport vesicles. *Blue arrows* indicate the extensive outbound and inbound vesicular traffic (discussed in Chapter 13). Some organelles, most notably mitochondria and (in plant cells) plastids, do not take part in this communication and are isolated from the vesicular traffic between organelles shown here.

To understand the general principles by which sorting signals operate, it is important to distinguish three fundamentally different ways by which proteins move from one compartment to another. These three mechanisms are described below, and the transport steps at which they operate are outlined in **Figure 12–5.** We discuss the first two mechanisms (gated transport and transmembrane transport) in this chapter, and the third (vesicular transport, *green arrows* in Figure 12–5) in Chapter 13.

1. In **gated transport**, proteins and RNA molecules move between the cytosol and the nucleus through nuclear pore complexes in the nuclear envelope. The nuclear pore complexes function as selective gates that support the active transport of specific macromolecules and macromolecular assemblies between the two topologically equivalent spaces, although they also allow free diffusion of smaller molecules.

2. In **protein translocation**, transmembrane *protein translocators* directly transport specific proteins across a membrane from the cytosol into a space that is topologically distinct. The transported protein molecule usually must unfold to snake through the translocator. The initial transport of selected proteins from the cytosol into the ER lumen or mitochondria, for example, occurs in this way. Integral membrane proteins often use the same translocators but translocate only partially across the membrane, so that the protein becomes embedded in the lipid bilayer.

3. In **vesicular transport**, membrane-enclosed transport intermediates—which may be small, spherical transport vesicles or larger, irregularly shaped organelle fragments—ferry proteins from one topologically equivalent compartment to another. The transport vesicles and fragments become loaded with a cargo of molecules derived from the lumen of one compartment as they bud and pinch off from its membrane; they discharge their cargo into a second compartment by fusing with the membrane enclosing that compartment (**Figure 12–6**). The transfer of soluble proteins from the ER to the Golgi apparatus, for example, occurs in this way. Because the

Figure 12–5 A simplified "roadmap" of protein traffic within a eukaryotic cell. Proteins can move from one compartment to another by gated transport *(red)*, protein translocation *(blue)*, or vesicular transport *(green)*. The sorting signals that direct a given protein's movement through the system, and thereby determine its eventual location in the cell, are contained in each protein's amino acid sequence. The journey begins with the synthesis of a protein on a ribosome in the cytosol and, for many proteins, terminates when the protein reaches its final destination. Other proteins shuttle back and forth between the nucleus and cytosol. At each intermediate station *(boxes)*, a decision is made as to whether the protein is to be retained in that compartment or transported further. A sorting signal may direct either retention in or exit from a compartment.

We shall refer to this figure often as a guide in this chapter and the next, highlighting in color the particular pathway being discussed.

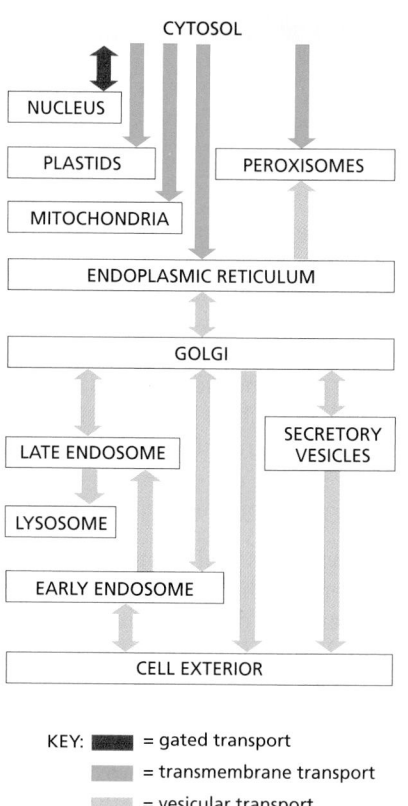

Figure 12–6 Vesicle budding and fusion during vesicular transport. Transport vesicles bud from one compartment (donor) and fuse with another topologically equivalent (target) compartment. In the process, soluble components *(red dots)* are transferred from lumen to lumen. Note that membrane is also transferred and that the original orientation of both proteins and lipids in the donor compartment membrane is preserved in the target compartment membrane. Thus, membrane proteins retain their asymmetric orientation, with the same domains always facing the cytosol.

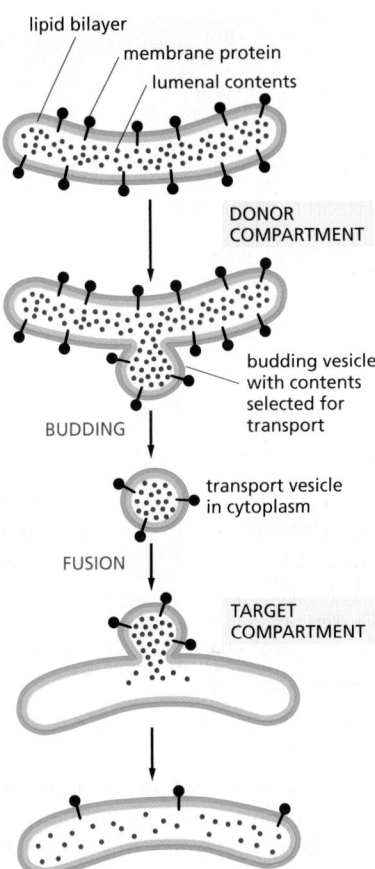

lipid bilayer
membrane protein
lumenal contents

DONOR COMPARTMENT

budding vesicle with contents selected for transport

BUDDING

transport vesicle in cytoplasm

FUSION

TARGET COMPARTMENT

transported proteins do not cross a membrane, vesicular transport can move proteins only between compartments that are topologically equivalent (see Figure 12–4).

Each mode of protein transfer is usually guided by sorting signals in the transported protein, which are recognized by complementary *sorting receptors*. If a large protein is to be imported into the nucleus, for example, it must possess a sorting signal that receptor proteins recognize to guide it through the nuclear pore complex. If a protein is to be transferred directly across a membrane, it must possess a sorting signal that the translocator recognizes. Likewise, if a protein is to be loaded into a certain type of vesicle or retained in certain organelles, a complementary receptor in the appropriate membrane must recognize its sorting signal.

Signal Sequences and Sorting Receptors Direct Proteins to the Correct Cell Address

Most protein sorting signals involved in transmembrane transport reside in a stretch of amino acid sequence, typically 15–60 residues long. Such **signal sequences** are often found at the N-terminus of the polypeptide chain, and in many cases specialized **signal peptidases** remove the signal sequence from the finished protein once the sorting process is complete. Signal sequences can also be internal stretches of amino acids, which remain part of the protein. Such signals are used in gated transport into the nucleus. Sorting signals can also be composed of multiple internal amino acid sequences that form a specific three-dimensional arrangement of atoms on the protein's surface; such **signal patches** are sometimes used for nuclear import and in vesicular transport.

Each signal sequence specifies a particular destination in the cell. Proteins destined for initial transfer to the ER usually have a signal sequence at their N-terminus that characteristically includes a sequence composed of about 5–10 hydrophobic amino acids. Many of these proteins will in turn pass from the ER to the Golgi apparatus, but those with a specific signal sequence of four amino acids at their C-terminus are recognized as ER residents and are returned to the ER. Proteins destined for mitochondria have signal sequences of yet another type, in which positively charged amino acids alternate with hydrophobic ones. Finally, many proteins destined for peroxisomes have a signal sequence of three characteristic amino acids at their C-terminus.

Table 12–3 presents some specific signal sequences. Experiments in which the peptide is transferred from one protein to another by genetic engineering techniques have demonstrated the importance of each of these signal sequences for protein targeting. Placing the N-terminal ER signal sequence at the beginning of a cytosolic protein, for example, redirects the protein to the ER; removing or mutating the signal sequence of an ER protein causes its retention in the cytosol. Signal sequences are therefore both necessary and sufficient for protein targeting. Even though their amino acid sequences can vary greatly, the signal sequences of proteins having the same destination are functionally interchangeable; physical properties, such as hydrophobicity, often seem to be more important in the signal-recognition process than the exact amino acid sequence.

Signal sequences are recognized by complementary sorting receptors that guide proteins to their appropriate destination, where the receptors unload their cargo. The receptors function catalytically: after completing one round of targeting, they return to their point of origin to be reused. Most sorting receptors

TABLE 12–3 Some Typical Signal Sequences	
Function of signal sequence	Example of signal sequence
Import into nucleus	-Pro-Pro-Lys-Lys-Lys-Arg-Lys-Val-
Export from nucleus	-Met-Glu-Glu-Leu-Ser-Gln-Ala-Leu-Ala-Ser-Ser-Phe-
Import into mitochondria	^+H_3N-Met-Leu-Ser-Leu-Arg-Gln-Ser-Ile-Arg-Phe-Phe-Lys-Pro-Ala-Thr-Arg-Thr-Leu-Cys-Ser-Ser-Arg-Tyr-Leu-Leu-
Import into plastid	^+H_3N-Met-Val-Ala-Met-Ala-Met-Ala-Ser-Leu-Gln-Ser-Ser-Met-Ser-Ser-Leu-Ser-Leu-Ser-Ser-Asn-Ser-Phe-Leu-Gly-Gln-Pro-Leu-Ser-Pro-Ile-Thr-Leu-Ser-Pro-Phe-Leu-Gln-Gly-
Import into peroxisomes	-Ser-Lys-Leu-COO$^-$
Import into ER	^+H_3N-Met-Met-Ser-Phe-Val-Ser-Leu-Leu-Leu-Val-Gly-Ile-Leu-Phe-Trp-Ala-Thr-Glu-Ala-Glu-Gln-Leu-Thr-Lys-Cys-Glu-Val-Phe-Gln-
Return to ER	-Lys-Asp-Glu-Leu-COO$^-$

Some characteristic features of the different classes of signal sequences are highlighted in color. Where they are known to be important for the function of the signal sequence, positively charged amino acids are shown in *red* and negatively charged amino acids are shown in *green*. Similarly, important hydrophobic amino acids are shown in *orange* and important hydroxylated amino acids are shown in *blue*. ^+H_3N indicates the N-terminus of a protein; COO$^-$ indicates the C-terminus.

recognize classes of proteins rather than an individual protein species. They can therefore be viewed as public transportation systems, dedicated to delivering numerous different components to their correct location in the cell.

Most Organelles Cannot Be Constructed *De Novo:* They Require Information in the Organelle Itself

When a cell reproduces by division, it has to duplicate its organelles, in addition to its chromosomes. In general, cells do this by incorporating new molecules into the existing organelles, thereby enlarging them; the enlarged organelles then divide and are distributed to the two daughter cells. Thus, each daughter cell inherits a complete set of specialized cell membranes from its mother. This inheritance is essential because a cell could not make such membranes from scratch. If the ER were completely removed from a cell, for example, how could the cell reconstruct it? As we discuss later, the membrane proteins that define the ER and perform many of its functions are themselves products of the ER. A new ER could not be made without an existing ER or, at least, a membrane that specifically contains the protein translocators required to import selected proteins into the ER from the cytosol (including the ER-specific translocators themselves). The same is true for mitochondria and plastids.

Thus, it seems that the information required to construct an organelle does not reside exclusively in the DNA that specifies the organelle's proteins. Information in the form of at least one distinct protein that preexists in the organelle membrane is also required, and this information is passed from parent cell to daughter cells in the form of the organelle itself. Presumably, such information is essential for the propagation of the cell's compartmental organization, just as the information in DNA is essential for the propagation of the cell's nucleotide and amino acid sequences.

As we discuss in more detail in Chapter 13, however, the ER buds off a constant stream of transport vesicles that incorporate only a subset of ER proteins and therefore have a composition different from the ER itself. Similarly, the plasma membrane constantly buds off various types of specialized endocytic vesicles. Thus, some organelles can form from other organelles and do not have to be inherited at cell division.

Summary

Eukaryotic cells contain intracellular membrane-enclosed organelles that make up nearly half the cell's total volume. The main ones present in all eukaryotic cells are the endoplasmic reticulum, Golgi apparatus, nucleus, mitochondria, lysosomes, endosomes, and peroxisomes; plant cells also contain plastids such as chloroplasts. These organelles contain distinct sets of proteins, which mediate each organelle's unique function.

Each newly synthesized organelle protein must find its way from a ribosome in the cytosol, where the protein is made, to the organelle where it functions. It does so by following a specific pathway, guided by sorting signals in its amino acid sequence that function as either signal sequences or signal patches. Sorting signals are recognized by complementary sorting receptors, which deliver the protein to the appropriate target organelle. Proteins that function in the cytosol do not contain sorting signals and therefore remain there after they are synthesized.

During cell division, organelles such as the ER and mitochondria are distributed to each daughter cell. These organelles contain information that is required for their construction, and so they cannot be made de novo.

THE TRANSPORT OF MOLECULES BETWEEN THE NUCLEUS AND THE CYTOSOL

The **nuclear envelope** encloses the DNA and defines the *nuclear compartment*. This envelope consists of two concentric membranes, which are penetrated by nuclear pore complexes (**Figure 12–7**). Although the inner and outer nuclear membranes are continuous, they maintain distinct protein compositions. The **inner nuclear membrane** contains proteins that act as binding sites for chromosomes and for the *nuclear lamina,* a protein meshwork that provides structural support for the nuclear envelope; the lamina also acts as an anchoring site for chromosomes and the cytoplasmic cytoskeleton (via protein complexes that span the nuclear envelope). The inner membrane is surrounded by the **outer nuclear membrane**, which is continuous with the membrane of the ER. Like the ER membrane (discussed later), the outer nuclear membrane is studded with ribosomes engaged in protein synthesis. The proteins made on these ribosomes are transported into the space between the inner and outer nuclear membranes (the *perinuclear space*), which is continuous with the ER lumen (see Figure 12–7).

Bidirectional traffic occurs continuously between the cytosol and the nucleus. The many proteins that function in the nucleus—including histones, DNA polymerases, RNA polymerases, transcriptional regulators, and RNA-processing proteins—are selectively imported into the nuclear compartment from the cytosol, where they are made. At the same time, almost all RNAs—including mRNAs, rRNAs, tRNAs, miRNAs, and snRNAs—are synthesized in the nuclear compartment and then exported to the cytosol. Like the import process, the export process is selective; mRNAs, for example, are exported only after they have been properly modified by RNA-processing reactions in the nucleus. In some cases, the transport process is complex. Ribosomal proteins, for instance, are made in the cytosol and imported into the nucleus, where they assemble with newly made ribosomal RNA into particles. The particles are then exported to the cytosol, where they assemble into ribosomes. Each of these steps requires selective transport across the nuclear envelope.

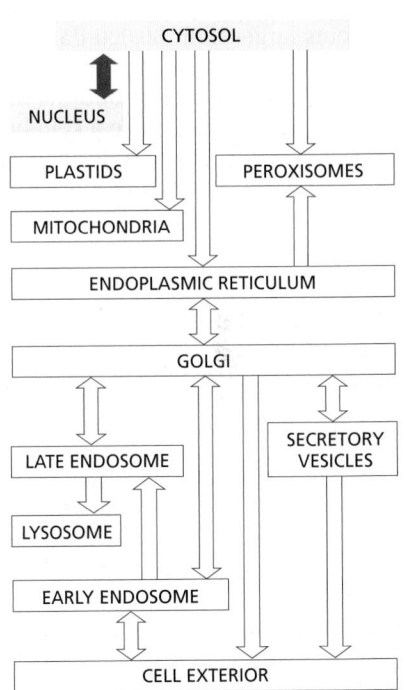

Nuclear Pore Complexes Perforate the Nuclear Envelope

Large and elaborate **nuclear pore complexes (NPCs)** perforate the nuclear envelope in all eukaryotes. Each NPC is composed of a set of approximately 30 different proteins, or **nucleoporins**. Reflecting the high degree of internal symmetry, each nucleoporin is present in multiple copies, resulting in 500–1000 protein molecules in the fully assembled NPC, with an estimated mass of 66 million daltons in yeast and 125 million daltons in vertebrates (**Figure 12–8**). Most nucleoporins are composed of repetitive protein domains of only a few different types, which have evolved through extensive gene duplication. Some of the scaffold nucleoporins (see Figure 12–8) are structurally related to vesicle coat protein complexes, such as clathrin and COPII coatomer (discussed in Chapter 13), which shape transport vesicles; and one protein is used as a common building block in both NPCs and vesicle coats. These similarities suggest a common evolutionary origin for NPCs and vesicle coats: they may derive from an early membrane-bending protein module that helped shape the elaborate membrane systems of eukaryotic cells, and in present-day cells stabilize the sharp membrane bends required to form a nuclear pore.

The nuclear envelope of a typical mammalian cell contains 3000–4000 NPCs, although that number varies widely, from a few hundred in glial cells to almost 20,000 in Purkinje neurons. The total traffic that passes through each NPC is enormous: each NPC can transport up to 1000 macromolecules per second and can transport in both directions at the same time. How it coordinates the bidirectional flow of macromolecules to avoid congestion and head-on collisions is not known.

Each NPC contains aqueous passages, through which small water-soluble molecules can diffuse passively. Researchers have determined the effective size of these passages by injecting labeled water-soluble molecules of different sizes into the cytosol and then measuring their rate of diffusion into the nucleus. Small molecules (5000 daltons or less) diffuse in so fast that we can consider the nuclear envelope freely permeable to them. Large proteins, however, diffuse in much more slowly, and the larger a protein, the more slowly it passes through the NPC. Proteins larger than 60,000 daltons cannot enter by passive diffusion. This size cut-off to free diffusion is thought to result from the NPC structure (see Figure 12–8). The channel nucleoporins with extensive unstructured regions form a disordered tangle (much like a kelp bed in the ocean) that restricts the diffusion of large macromolecules while allowing smaller molecules to pass.

Because many cell proteins are too large to diffuse passively through the NPCs, the nuclear compartment and the cytosol can maintain different protein compositions. Mature cytosolic ribosomes, for example, are about 30 nm in diameter and thus cannot diffuse through the NPC, confining protein synthesis to the cytosol. But how does the nucleus export newly made ribosomal subunits or import large molecules, such as DNA polymerases and RNA polymerases, which have subunit molecular masses of 100,000–200,000 daltons? As we discuss next, these and most other transported protein and RNA molecules bind to specific receptor proteins that actively ferry large molecules through NPCs. Even small proteins like histones frequently use receptor-mediated mechanisms to cross the NPC, thereby increasing transport efficiency.

Nuclear Localization Signals Direct Nuclear Proteins to the Nucleus

When proteins are experimentally extracted from the nucleus and reintroduced into the cytosol, even the very large ones reaccumulate efficiently in the nucleus. Sorting signals called **nuclear localization signals** (**NLSs**) are responsible for the selectivity of this active nuclear import process. The signals have been precisely defined by using recombinant DNA technology for numerous nuclear proteins, as well as for proteins that enter the nucleus only transiently (**Figure 12–9**). In many nuclear proteins, the signals consist of one or two short sequences that are rich in the positively charged amino acids lysine and arginine (see Table 12–3, p. 648), with the precise sequence varying for different proteins. Other nuclear proteins contain different signals, some of which are not yet characterized.

Nuclear localization signals can be located almost anywhere in the amino acid sequence and are thought to form loops or patches on the protein surface. Many function even when linked as short peptides to lysine side chains on the surface of a cytosolic protein, suggesting that the precise location of the signal within the amino acid sequence of a nuclear protein is not important. Moreover, as long as one of the protein subunits of a multicomponent complex displays a nuclear localization signal, the entire complex will be imported into the nucleus.

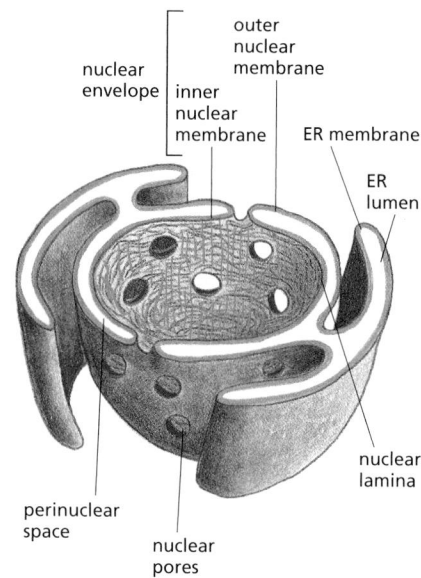

Figure 12–7 The nuclear envelope. The double-membrane envelope is penetrated by pores in which nuclear pore complexes (not shown) are positioned. The outer nuclear membrane is continuous with the endoplasmic reticulum (ER). The ribosomes that are normally bound to the cytosolic surface of the ER membrane and outer nuclear membrane are not shown. The nuclear lamina is a fibrous protein meshwork underlying the inner membrane.

One can visualize the transport of nuclear proteins through NPCs by coating gold particles with a nuclear localization signal, injecting the particles into the cytosol, and then following their fate by electron microscopy (**Figure 12–10**). The particles bind to the tentaclelike fibrils that extend from the scaffold nucleoporins at the rim of the NPC into the cytosol, and then proceed through the center of the NPC. Presumably, the unstructured regions of the nucleoporins that form a diffusion barrier for large molecules (mentioned earlier) are pushed away to allow the coated gold particles to squeeze through.

Macromolecular transport across NPCs differs fundamentally from the transport of proteins across the membranes of other organelles, in that it occurs

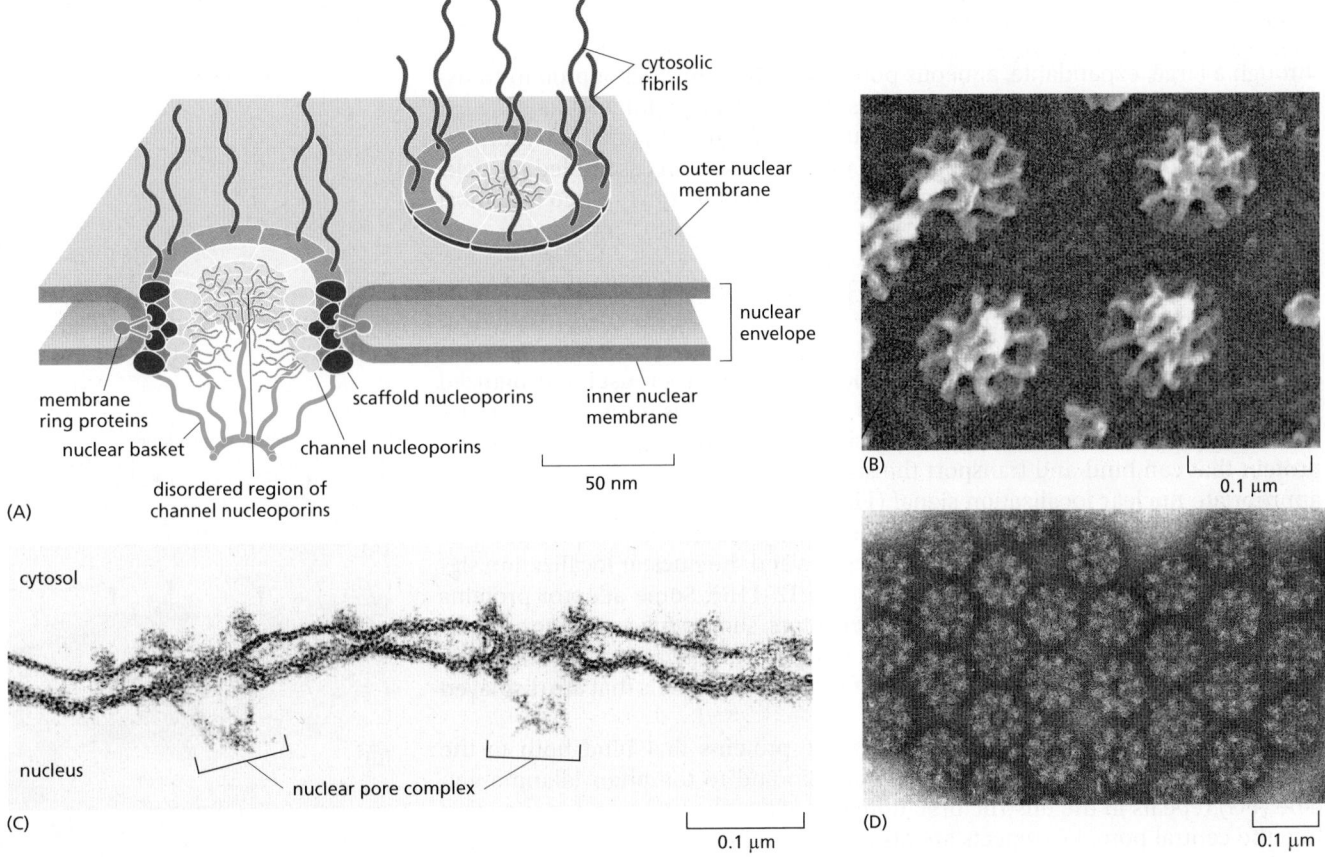

Figure 12–8 The arrangement of NPCs in the nuclear envelope. (A) In a vertebrate NPC, nucleoporins are arranged with striking eightfold rotational symmetry. In addition, immunoelectron microscopic studies show that the proteins that make up the central portion of the NPC are oriented symmetrically across the nuclear envelope, so that the nuclear and cytosolic sides look identical. The eightfold rotational and twofold transverse symmetry explains how such a huge structure can be formed from only about 30 different proteins: many of the nucleoporins are present in 8, 16, or 32 copies. Based on their approximate localization in the central portion of the NPC, nucleoporins can be classified into (1) transmembrane ring proteins that span the nuclear envelope and anchor the NPC to the envelope; (2) scaffold nucleoporins that form layered ring structures. Some scaffold nucleoporins are membrane-bending proteins that stabilize the sharp membrane curvature where the nuclear envelope is penetrated; and (3) channel nucleoporins that line a central pore. In addition to folded domains that anchor the proteins in specific places, many channel nucleoporins contain extensive unstructured regions, where the polypeptide chains are intrinsically disordered. The central pore is filled with a tangled mesh of these disordered domains that blocks the passive diffusion of large macromolecules. The disordered regions contain a large number of phenylalanine-glycine (FG) repeats. Fibrils protrude from both the cytosolic and the nuclear sides of the NPC. By contrast to the twofold transverse symmetry of the NPC core, the fibrils facing the cytosol and nucleus are different: on the nuclear side, the fibrils converge at their distal end to form a basketlike structure. The precise arrangement of individual nucleoporins in the assembled NPC is still a matter of intense debate, because atomic resolution analyses have been hindered by the sheer size and flexible nature of the NPC, and by difficulties in purifying sufficient amounts of homogeneous material. A combination of electron microscopy, computational analyses, and crystal structures of nucleoporin subcomplexes has been used to develop the current models of the NPC architecture. (B) A scanning electron micrograph of the nuclear side of the nuclear envelope of an oocyte (see also Figure 9–52). (C) An electron micrograph showing a side view of two NPCs *(brackets)*; note that the inner and outer nuclear membranes are continuous at the edges of the pore. (D) An electron micrograph showing face-on views of negatively stained NPCs. The membrane has been removed by detergent extraction. Note that some of the NPCs contain material in their center, which is thought to be trapped macromolecules in transit through these NPCs. (A, adapted from A. Hoelz, E.W. Debler and G. Blobel, *Annu. Rev. Biochem.* 80:613–643, 2011; B, © 1992 M.W. Goldberg and T.D. Allen. Originally published in *J. Cell Biol.* https://doi.org/10.1083/jcb.119.6.1429. With permission from Rockefeller University Press; C, courtesy of Werner Franke and Ulrich Scheer; D, courtesy of Ron Milligan.)

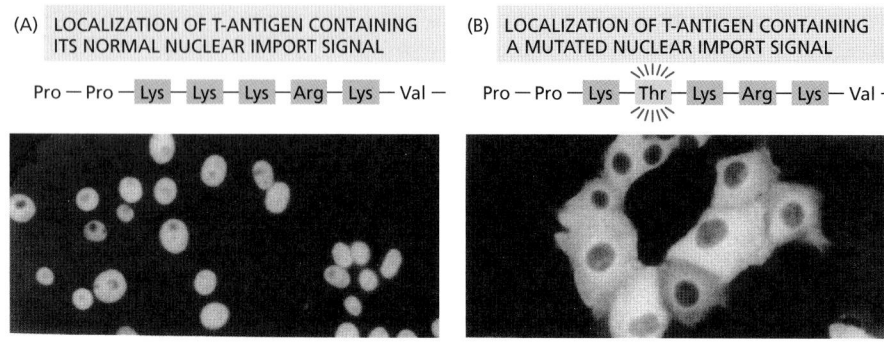

(A) LOCALIZATION OF T-ANTIGEN CONTAINING ITS NORMAL NUCLEAR IMPORT SIGNAL

Pro — Pro — Lys — Lys — Lys — Arg — Lys — Val —

(B) LOCALIZATION OF T-ANTIGEN CONTAINING A MUTATED NUCLEAR IMPORT SIGNAL

Pro — Pro — Lys — Thr — Lys — Arg — Lys — Val —

Figure 12–9 The function of a nuclear localization signal. Immunofluorescence micrographs showing the cell location of SV40 virus T-antigen containing or lacking a short sequence that serves as a nuclear localization signal. (A) The normal T-antigen protein contains the lysine-rich sequence indicated and is imported to its site of action in the nucleus, as indicated by immunofluorescence staining with antibodies against the T-antigen. (B) T-antigen with an altered nuclear localization signal (a threonine replacing a lysine) remains in the cytosol. (From D. Kalderon, B. Roberts, W. Richardson and A. Smith, *Cell* 39:499–509, 1984. With permission from Elsevier.)

through a large, expandable, aqueous pore, rather than through a protein transporter spanning one or more lipid bilayers. For this reason, fully folded nuclear proteins can be transported into the nucleus through an NPC, and newly formed ribosomal subunits are transported out of the nucleus as an assembled particle. By contrast, proteins have to be extensively unfolded to be transported into most other organelles, as we discuss later.

Nuclear Import Receptors Bind to Both Nuclear Localization Signals and NPC Proteins

To initiate nuclear import, most nuclear localization signals must be recognized by **nuclear import receptors**, sometimes called *importins*, most of which are encoded by a family of related genes. Each family member encodes a receptor protein that can bind and transport the subset of cargo proteins containing the appropriate nuclear localization signal (**Figure 12–11**A). Nuclear import receptors do not always bind to nuclear proteins directly. Additional adaptor proteins can form a bridge between the import receptors and the nuclear localization signals on the proteins to be transported (Figure 12–11B). Some adaptor proteins are structurally related to nuclear import receptors, suggesting a common evolutionary origin. By using a variety of import receptors and adaptors, cells are able to recognize the broad repertoire of nuclear localization signals that are displayed on nuclear proteins.

The import receptors are soluble cytosolic proteins that bind both to the nuclear localization signal on the cargo protein and to the phenylalanine-glycine (FG) repeats in the unstructured domains of the channel nucleoporins that line the central pore. FG-repeats are also found in the cytoplasmic and nuclear fibrils. FG-repeats in the unstructured tangle of the pore are thought to do double duty. They interact weakly, which gives the protein tangle gel-like properties that impose a permeability barrier to large macromolecules, and they serve as docking sites for nuclear import receptors. FG-repeats line the path through the NPCs taken by the import receptors and their bound cargo proteins. According to one model of nuclear transport, the receptor–cargo complexes move along the transport path by repeatedly binding, dissociating, and then re-binding to adjacent FG-repeat sequences. In this way, the complexes may hop from one nucleoporin to another to traverse the tangled interior of the NPC in a random walk. As import receptors bind to FG-repeats during this journey, they would disrupt interaction between the repeats and locally dissolve the gel phase of the protein tangle that fills the pore, allowing the passage of the receptor–cargo complex. Once inside the nucleus, the import receptors dissociate from their cargo and return to the cytosol. As we will see, this dissociation only occurs on the nuclear side of the NPC and thereby confers directionality to the import process.

Nuclear Export Works Like Nuclear Import, But in Reverse

The nuclear export of large molecules, such as new ribosomal subunits and RNA molecules, occurs through NPCs and also depends on a selective transport

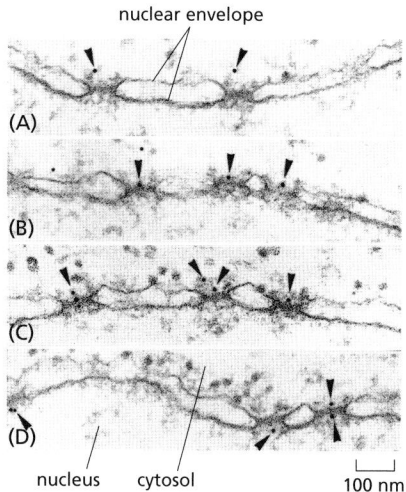

nuclear envelope

(A)

(B)

(C)

(D)

nucleus   cytosol   100 nm

Figure 12–10 Visualizing active import through NPCs. This series of electron micrographs shows colloidal gold spheres *(arrowheads)* coated with peptides containing nuclear localization signals entering the nucleus through NPCs. The gold particles were injected into the cytosol of living cells, which then were fixed and prepared for electron microscopy at various times after injection. (A) Gold particles are first seen in proximity to the cytosolic fibrils of the NPCs. (B, C) They are then seen at the center of the NPCs, exclusively on the cytosolic face. (D) They then appear on the nuclear face. These gold particles have much larger diameters than the diffusion channels in the NPC and are imported by active transport. (From N. Panté and U. Aebi, *Science* 273:1729–1732, 1996. With permission from AAAS.)

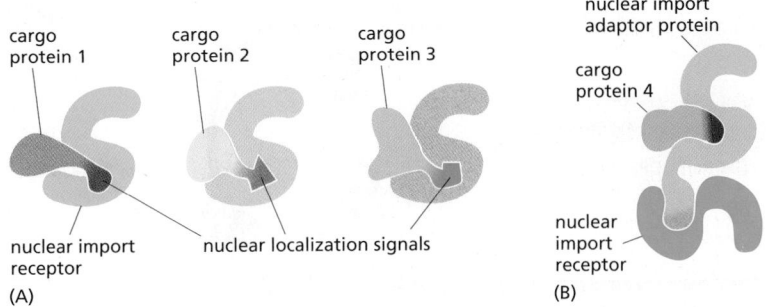

Figure 12–11 Nuclear import receptors (importins). (A) Different nuclear import receptors bind different nuclear localization signals and thereby different cargo proteins. (B) Cargo protein 4 requires an adaptor protein to bind to its nuclear import receptor. The adaptors are structurally related to nuclear import receptors and recognize nuclear localization signals on cargo proteins. They also contain a nuclear localization signal that binds them to an import receptor, but this signal only becomes exposed when they are loaded with a cargo protein.

system. The transport system relies on **nuclear export signals** on the macromolecules to be exported, as well as on complementary **nuclear export receptors**, or *exportins*. These receptors bind to both the export signal and NPC proteins to guide their cargo through the NPC to the cytosol.

Many nuclear export receptors are structurally related to nuclear import receptors, and they are encoded by the same gene family of **nuclear transport receptors**, or *karyopherins*. In yeast, there are 14 genes encoding karyopherins; in animal cells, the number is significantly larger. It is often not possible to tell from their amino acid sequence alone whether a particular family member works as a nuclear import or nuclear export receptor. As might be expected, therefore, the import and export transport systems work in similar ways but in opposite directions: the import receptors bind their cargo molecules in the cytosol, release them in the nucleus, and are then exported to the cytosol for reuse, while the export receptors function in the opposite fashion.

The Ran GTPase Imposes Directionality on Transport Through NPCs

The import of nuclear proteins through NPCs concentrates specific proteins in the nucleus and thereby increases order in the cell. The cell fuels this ordering process by harnessing energy stored in concentration gradients of the GTP-bound form of the monomeric GTPase **Ran**, which is required for both nuclear import and export.

Like other GTPases, Ran is a molecular switch that can exist in two conformational states, depending on whether GDP or GTP is bound (discussed in Chapter 3). Two Ran-specific regulatory proteins trigger the conversion between the two states: a cytosolic *GTPase-activating protein* (*GAP*) triggers GTP hydrolysis and thus converts Ran-GTP to Ran-GDP, and a nuclear *guanine exchange factor* (*GEF*) promotes the exchange of GDP for GTP and thus converts Ran-GDP to Ran-GTP. Because *Ran-GAP* is located in the cytosol and *Ran-GEF* is located in the nucleus where it is anchored to chromatin, the cytosol contains mainly Ran-GDP, and the nucleus contains mainly Ran-GTP (**Figure 12–12**).

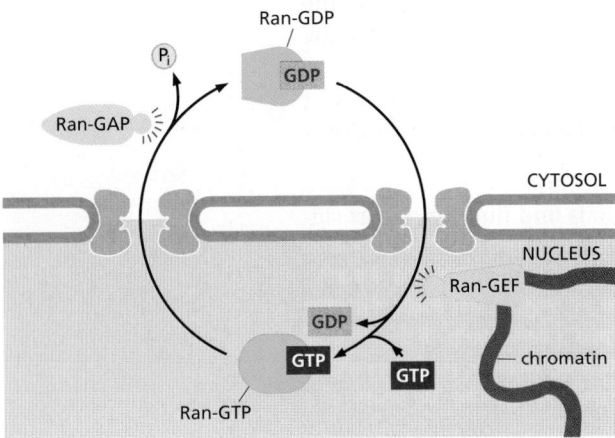

Figure 12–12 The compartmentalization of Ran-GDP and Ran-GTP. Localization of Ran-GDP in the cytosol and Ran-GTP in the nucleus results from the localization of two Ran regulatory proteins: Ran GTPase-activating protein (Ran-GAP) is located in the cytosol, and Ran guanine nucleotide exchange factor (Ran-GEF) binds to chromatin and is therefore located in the nucleus.

Ran-GDP is imported into the nucleus by its own import receptor, which is specific for the GDP-bound conformation of Ran. The Ran-GDP receptor is structurally unrelated to the main family of nuclear transport receptors. However, it also binds to FG-repeats in NPC channel nucleoporins.

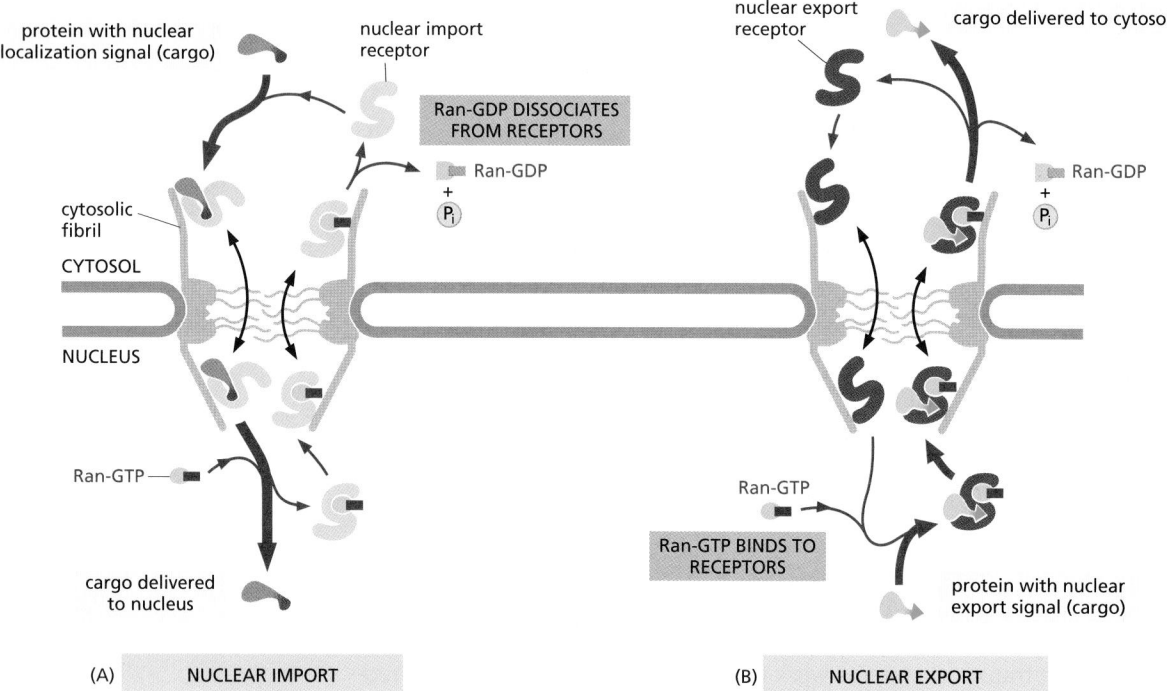

(A)   NUCLEAR IMPORT

(B)   NUCLEAR EXPORT

This gradient of the two conformational forms of Ran drives nuclear transport in the appropriate direction. Docking of nuclear import receptors to FG-repeats on the cytosolic side of the NPC, for example, occurs whether or not these receptors are loaded with appropriate cargo. Import receptors, facilitated by FG-repeat binding, then enter the channel. If they reach the nuclear side of the pore complex, Ran-GTP binds to them, and, if the receptors arrive loaded with cargo molecules, the Ran-GTP binding causes the receptors to release their cargo (**Figure 12–13**A). Because the Ran-GDP in the cytosol does not bind to import (or export) receptors, unloading occurs only on the nuclear side of the NPC. In this way, the nuclear localization of Ran-GTP creates the directionality of the import process.

Having discharged its cargo in the nucleus, the empty import receptor with Ran-GTP bound is transported back through the pore complex to the cytosol. There, Ran-GAP triggers Ran-GTP to hydrolyze its bound GTP, thereby converting it to Ran-GDP, which dissociates from the receptor. The receptor is then ready for another cycle of nuclear import.

Nuclear export occurs by a similar mechanism, except that Ran-GTP in the nucleus promotes cargo binding to the export receptor, rather than promoting cargo dissociation. Once the export receptor moves through the pore to the cytosol, it encounters Ran-GAP, which induces the receptor to hydrolyze its GTP to GDP. As a result, the export receptor releases both its cargo and Ran-GDP in the cytosol. Free export receptors are then returned to the nucleus to complete the cycle (Figure 12–13B).

Transport Through NPCs Can Be Regulated by Controlling Access to the Transport Machinery

Some proteins contain both nuclear localization signals and nuclear export signals. These proteins continually shuttle back and forth between the nucleus and the cytosol. The relative rates of their import and export determine the steady-state localization of such *shuttling proteins*: if the rate of import exceeds the rate of export, a protein will be located mainly in the nucleus; conversely, if the rate of export exceeds the rate of import, a protein will be located mainly in the cytosol. Thus, changing the rate of import, export, or both, can change the location of a protein.

Figure 12–13 How GTP hydrolysis by Ran in the cytosol provides directionality to nuclear transport. Movement through the NPC of loaded nuclear transport receptors occurs along the FG-repeats displayed by certain NPC proteins. The differential localization of Ran-GTP in the nucleus and Ran-GDP in the cytosol provides directionality *(red arrows)* to both nuclear import (A) and nuclear export (B). Ran-GAP stimulates the hydrolysis of GTP to produce Ran-GDP on the cytosolic side of the NPC (see Figure 12–12).

Some shuttling proteins move continuously into and out of the nucleus. In other cases, however, the transport is stringently controlled. As discussed in Chapter 7, cells control the activity of some transcription regulators by keeping them out of the nucleus until they are needed there (**Figure 12–14**). In many cases, cells control transport by regulating nuclear localization and export signals—turning them on or off, often by phosphorylation of amino acids close to the signal sequences (**Figure 12–15**).

Other transcription regulators are bound to inhibitory cytosolic proteins that either anchor them in the cytosol (through interactions with the cytoskeleton or specific organelles) or mask their nuclear localization signals so that they cannot interact with nuclear import receptors. An appropriate stimulus releases the gene regulatory protein from its cytosolic anchor or mask, and it is then transported into the nucleus. One important example is the latent gene regulatory protein that controls the expression of proteins involved in cholesterol metabolism. The protein is made and stored in an inactive form as a transmembrane protein in the ER. When a cell is deprived of cholesterol, the protein is transported from the ER to the Golgi apparatus where it encounters specific proteases that cleave off the cytosolic domain, releasing it into the cytosol. This domain is then imported into the nucleus, where it activates the transcription of genes required for both cholesterol uptake and synthesis (**Figure 12–16**).

As we discuss in detail in Chapter 6, cells control the export of RNAs from the nucleus in a similar way. snRNAs, miRNAs, and tRNAs bind to the same family of nuclear export receptors just discussed, and they use the same Ran-GTP gradient to fuel the transport process. By contrast, the export of mRNAs out of the nucleus uses a different mechanism. mRNAs are exported as large assemblies, which can be as large as 100 million daltons (see Figure 6–37) and can contain hundreds of proteins of a few dozen different types. These mRNA ribonucleoprotein complexes (mRNPs) first dock at the nuclear side of the NPC, where they are extensively remodeled. Although Ran-GTP is indirectly involved in the export (because it imports the proteins that bind to the mRNA molecules), the translocation across the NPC is thought to be driven by ATP hydrolysis. How export directionality is assured is unclear. It is likely that the many accessory proteins tethered to the NPC's nuclear and cytoplasmic fibrils have important roles in remodeling the mRNPs as they pass through the pores, in particular stripping away nuclear proteins as the mRNPs exit on the cytosolic side of the NPC, thereby ensuring that transport is unidirectional. Upon entry into the cytosol, these nuclear mRNP proteins are rapidly returned to the nucleus.

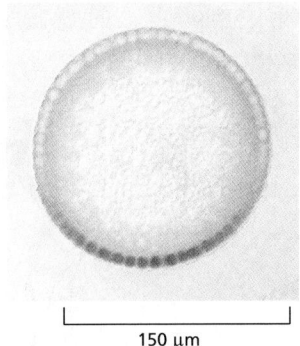

150 μm

Figure 12–14 The control of nuclear transport in the early *Drosophila* embryo. The embryo at this stage is a syncytium, shown here in cross section, with many nuclei in a common cytoplasm, arranged around the periphery, just beneath the plasma membrane. The transcription regulatory protein Dorsal is produced uniformly throughout the peripheral cytoplasm, but it can act only when inside the nuclei. The Dorsal protein has been stained with an enzyme-coupled antibody that yields a brown product, revealing that Dorsal is excluded from the nuclei at the dorsal side *(top)* of the embryo but is concentrated in the nuclei toward the ventral side *(bottom)* of the embryo. The regulated traffic of Dorsal into the nuclei controls the differential development between the back and belly of the animal. (Courtesy of Siegfried Roth.)

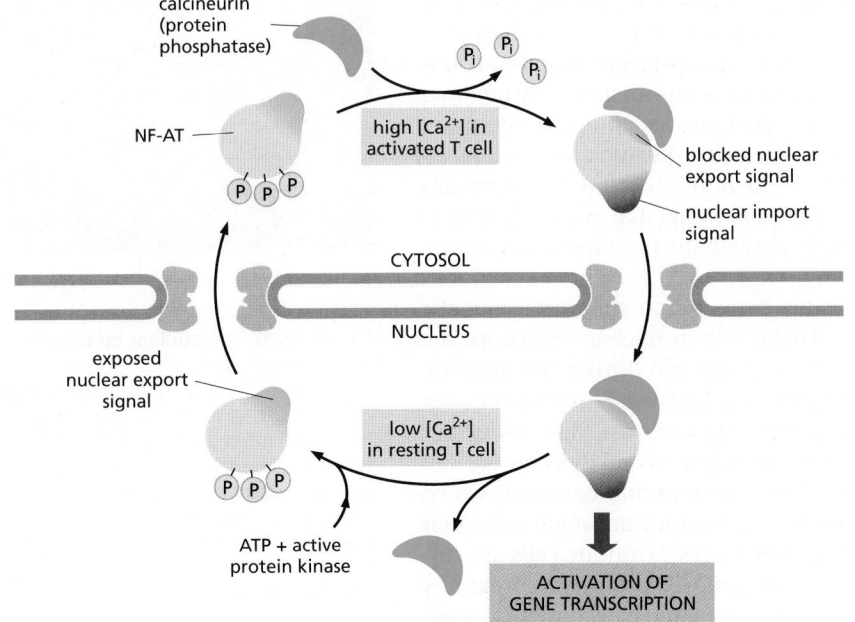

Figure 12–15 The control of nuclear import during T cell activation. The nuclear factor of activated T cells (NF-AT) is a transcription regulatory protein that, in the resting T cell, is found in the cytosol in a phosphorylated state. When T cells are activated by foreign antigen (discussed in Chapter 24), the intracellular Ca^{2+} concentration increases. In high Ca^{2+}, the protein phosphatase calcineurin binds to NF-AT and dephosphorylates it. The dephosphorylation exposes nuclear import signals and blocks a nuclear export signal. The complex of NF-AT and calcineurin is therefore imported into the nucleus, where NF-AT activates the transcription of numerous genes required for T cell activation.

The response shuts off when Ca^{2+} levels decrease, releasing NF-AT from calcineurin. Rephosphorylation of NF-AT inactivates the nuclear import signals and re-exposes the nuclear export signal, causing NF-AT to relocate to the cytosol. Some of the most potent immunosuppressive drugs, including cyclosporin A and FK506, inhibit the ability of calcineurin to dephosphorylate NF-AT and thereby block the nuclear accumulation of NF-AT and T cell activation (**Movie 12.1**).

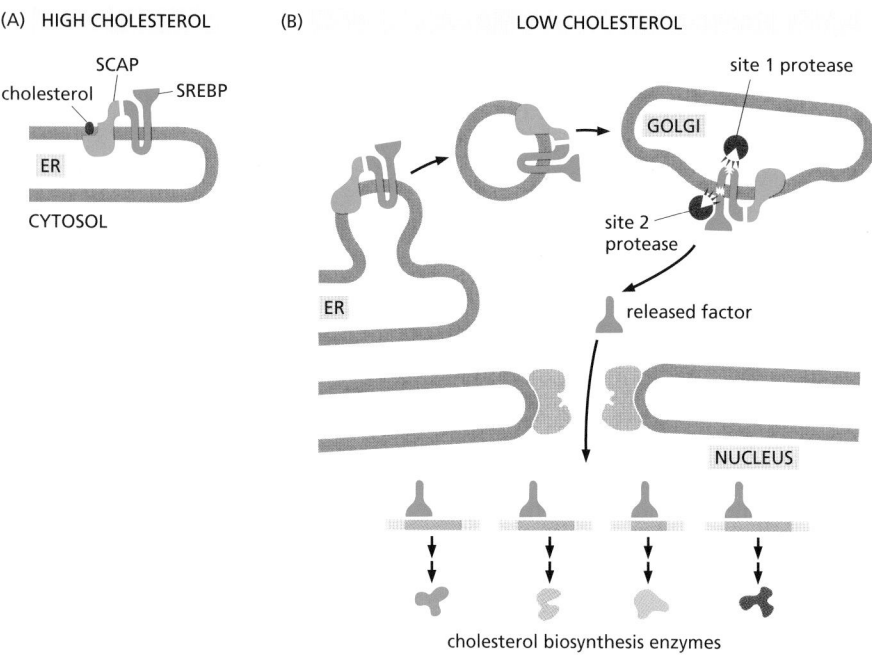

(A) HIGH CHOLESTEROL

(B) LOW CHOLESTEROL

Figure 12–16 Feedback regulation of cholesterol biosynthesis. SREBP (sterol response element binding protein), a latent transcription regulator that controls expression of cholesterol biosynthetic enzymes, is initially synthesized as an ER membrane protein. It is anchored in the ER if there is sufficient cholesterol in the membrane by interaction with another ER membrane protein, called SCAP (SREBP cleavage activation protein), which binds cholesterol. If the cholesterol binding site on SCAP is empty (at low cholesterol concentrations), SCAP changes conformation and is packaged together with SREBP into transport vesicles, which deliver their cargo to the Golgi apparatus, where two Golgi-resident proteases cleave SREBP to free its cytosolic domain from the membrane. The cytosolic domain then moves into the nucleus, where it binds to the promoters of genes that encode proteins involved in cholesterol biosynthesis and activates their transcription. In this way, more cholesterol is made when its concentration falls below a threshold.

During Mitosis the Nuclear Envelope Disassembles

The **nuclear lamina**, located on the nuclear side of the inner nuclear membrane, is a meshwork of interconnected protein subunits called **nuclear lamins**. The lamins are a special class of intermediate filament proteins (discussed in Chapter 16) that polymerize into a two-dimensional lattice (**Figure 12–17**). The nuclear lamina gives shape and stability to the nuclear envelope, to which it is anchored by attachment to both the NPCs and transmembrane proteins of the inner nuclear membrane. The lamina also interacts directly with chromatin, which itself interacts with transmembrane proteins of the inner nuclear membrane. Together with the lamina, these inner membrane proteins provide structural links between the DNA and the nuclear envelope.

When a nucleus is dismantled during mitosis, the NPCs and nuclear lamina disassemble and the nuclear envelope fragments. The dismantling process is at least partly a consequence of direct phosphorylation of nucleoporins and lamins by the cyclin-dependent protein kinase (Cdk) that is activated at the onset of mitosis (discussed in Chapter 17). During this process, some NPC proteins become bound to nuclear import receptors, which play an important part in the reassembly of NPCs at the end of mitosis. Nuclear envelope membrane proteins—no longer tethered to the pore complexes, lamina, or chromatin—disperse throughout the ER membrane. The dynein motor protein, which moves along microtubules (discussed in Chapter 16), actively participates in tearing the nuclear envelope off the chromatin. Together, these processes break down the barriers that normally separate the nucleus and cytosol, and the nuclear proteins that are not bound to membranes or chromosomes intermix completely with the proteins of the cytosol (**Figure 12–18**).

Later in mitosis, the nuclear envelope reassembles on the surface of the daughter chromosomes. In addition to its crucial role in nuclear transport, the Ran GTPase also acts as a positional marker for chromatin during cell division, when the nuclear and cytosolic components intermix. Because Ran-GEF remains bound to chromatin when the nuclear envelope breaks down, Ran molecules close to chromatin are mainly in their GTP-bound conformation. By contrast, Ran molecules further away have a high likelihood of encountering Ran-GAP, which is distributed throughout the cytosol; these Ran molecules are mainly in their GDP-bound conformation. As a result, the chromosomes in mitotic cells are surrounded by a cloud of Ran-GTP. Ran-GTP releases the NPC proteins in proximity to the chromosomes from nuclear import receptors. The free NPC proteins attach

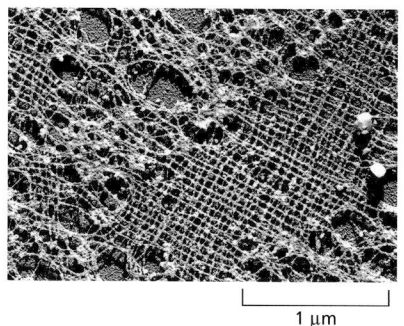

Figure 12–17 The nuclear lamina. An electron micrograph of a portion of the nuclear lamina in a *Xenopus* oocyte prepared by freeze-drying and metal shadowing. The lamina is formed by a regular lattice of specialized intermediate filaments. Lamins are only present in metazoan cells. Other, yet-unknown proteins may serve similar functions in species that lack lamins. (From U. Aebi, *Nature* 323:560–564, published 1986 by Nature Publishing Group. Reproduced with permission of SNCSC.)

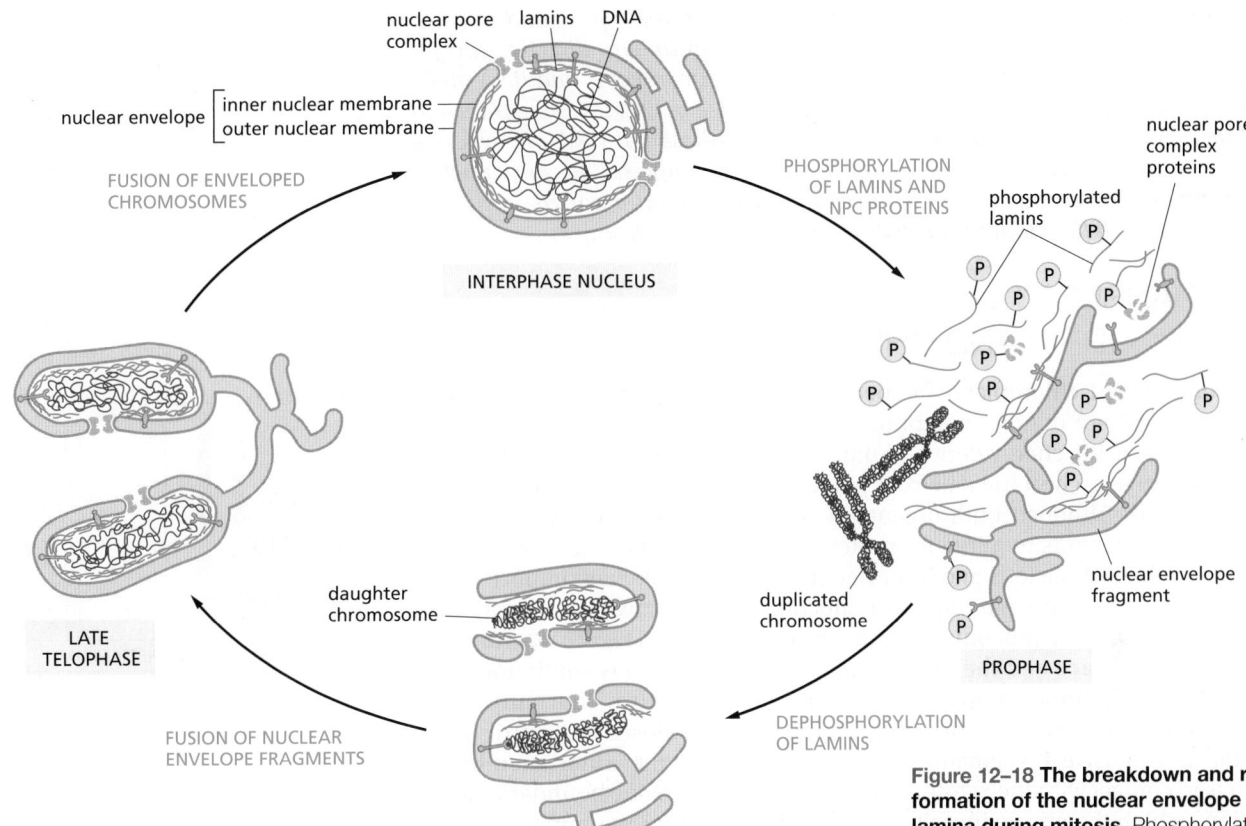

nuclear pore complex   lamins   DNA

nuclear envelope [inner nuclear membrane
outer nuclear membrane]

FUSION OF ENVELOPED
CHROMOSOMES

INTERPHASE NUCLEUS

PHOSPHORYLATION
OF LAMINS AND
NPC PROTEINS

nuclear pore
complex
proteins

phosphorylated
lamins

LATE
TELOPHASE

daughter
chromosome

duplicated
chromosome

nuclear envelope
fragment

PROPHASE

FUSION OF NUCLEAR
ENVELOPE FRAGMENTS

DEPHOSPHORYLATION
OF LAMINS

EARLY TELOPHASE

Figure 12–18 The breakdown and re-formation of the nuclear envelope and lamina during mitosis. Phosphorylation of the lamins triggers the disassembly of the nuclear lamina, which initiates the nuclear envelope to break up. Dephosphorylation of the lamins reverses the process. An analogous phosphorylation and dephosphorylation cycle occurs for some nucleoporins and proteins of the inner nuclear membrane, and some of these dephosphorylations are also involved in the reassembly process. As indicated, the nuclear envelope initially re-forms around individual decondensing daughter chromosomes. Eventually, as decondensation progresses, these structures fuse to form a single complete nucleus.

Mitotic breakdown of the nuclear envelope occurs in all metazoan cells. However, in many other species, such as yeasts, the nuclear envelope remains intact during mitosis, and the nucleus divides by fission.

to the chromosome surface, where they assemble into new NPCs. At the same time, inner nuclear membrane proteins and dephosphorylated lamins bind again to chromatin. ER membranes wrap around groups of chromosomes until they form a sealed nuclear envelope (**Movie 12.2**). During this process, the NPCs start actively re-importing proteins that contain nuclear localization signals. Because the nuclear envelope is initially closely applied to the surface of the chromosomes, the newly formed nucleus excludes all proteins except those initially bound to the mitotic chromosomes and those that are selectively imported through NPCs. In this way, all other large proteins, including ribosomes, are kept out of the newly assembled nucleus.

As we discuss in Chapter 17, the cloud of Ran-GTP surrounding chromatin is also important in assembling the mitotic spindle in a dividing cell.

Summary

The nuclear envelope consists of an inner and an outer nuclear membrane that are continuous with each other and with the ER membrane, and the space between the inner and outer nuclear membrane is continuous with the ER lumen. RNA molecules, which are made in the nucleus, and ribosomal subunits, which are assembled there, are exported to the cytosol; in contrast, all the proteins that function in the nucleus are synthesized in the cytosol and are then imported. The extensive traffic of materials between the nucleus and cytosol occurs through nuclear pore complexes (NPCs), which provide a direct passageway across the nuclear envelope. Small molecules diffuse passively through the NPCs, but large macromolecules have to be actively transported.

Proteins containing nuclear localization signals are actively transported into the nucleus through NPCs, while proteins containing nuclear export signals are transported out of the nucleus to the cytosol. Some proteins, including the nuclear import and export receptors, continually shuttle between the cytosol and nucleus. The monomeric GTPase Ran provides both the free energy and the directionality for nuclear transport. Cells regulate the transport of nuclear proteins and RNA molecules through the NPCs by controlling the access of these molecules to the transport machinery. Newly transcribed messenger RNA and ribosomal RNA are exported from the nucleus as parts of large ribonucleoprotein complexes. Because nuclear localization signals are not removed, nuclear proteins can be imported repeatedly, as is required each time that the nucleus reassembles after mitosis.

THE TRANSPORT OF PROTEINS INTO MITOCHONDRIA AND CHLOROPLASTS

Mitochondria and chloroplasts (a specialized form of plastids in green algae and plant cells) are double-membrane-enclosed organelles. They specialize in ATP synthesis, using energy derived from electron transport and oxidative phosphorylation in mitochondria and from photosynthesis in chloroplasts (discussed in Chapter 14). Although both organelles contain their own DNA, ribosomes, and other components required for protein synthesis, most of their proteins are encoded in the cell nucleus and imported from the cytosol. Each imported protein must reach the particular organelle subcompartment in which it functions.

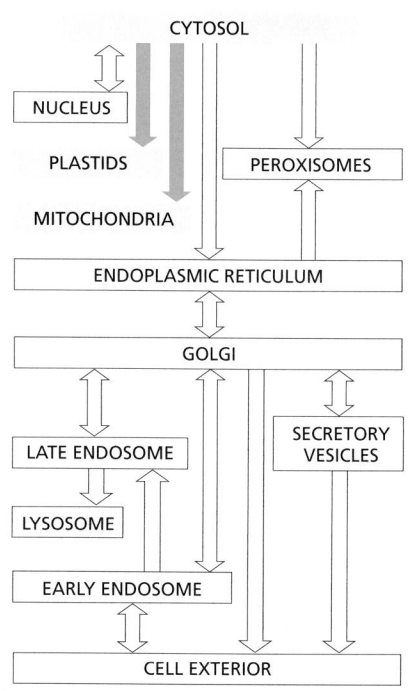

There are different subcompartments in mitochondria (**Figure 12–19**A): the internal **matrix space** and the **intermembrane space**, which is continuous with the cristae space. These compartments are formed by the two concentric mitochondrial membranes: the **inner membrane**, which encloses the matrix space and forms extensive invaginations called *cristae*, and the **outer membrane**, which is in contact with the cytosol. Protein complexes provide boundaries at the junctions where the cristae invaginate and divide the inner membrane into two domains: one inner membrane domain surrounds the cristae space, and the other domain abuts the outer membrane. Chloroplasts also have an outer and inner membrane, which enclose an intermembrane space, and the stroma, which is the chloroplast equivalent of the mitochondrial matrix space (Figure 12–19B). They have an additional subcompartment, the *thylakoid space*, which is surrounded by the *thylakoid membrane*. The thylakoid membrane derives from the inner membrane during plastid development and is pinched off to become discontinuous with it. Each of the subcompartments in mitochondria and chloroplasts contains a distinct set of proteins.

New mitochondria and chloroplasts are produced by the growth of preexisting organelles, followed by fission (discussed in Chapter 14). The growth depends mainly on the import of proteins from the cytosol. The imported proteins must be transported across a number of membranes in succession and end up in the appropriate place. The process of protein movement across membranes is called *protein translocation*. This section explains how it occurs.

(A) MITOCHONDRION

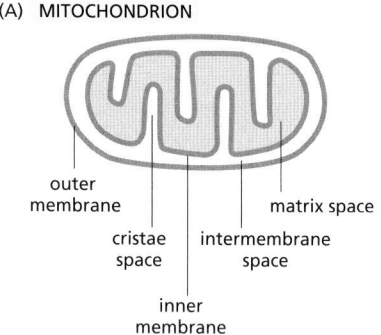

outer membrane
cristae space
intermembrane space
matrix space
inner membrane

(B) CHLOROPLAST

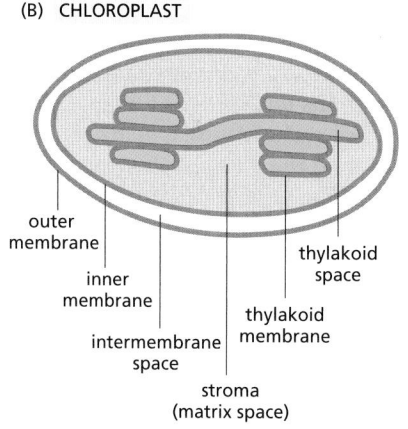

outer membrane
inner membrane
intermembrane space
thylakoid space
thylakoid membrane
stroma (matrix space)

Figure 12–19 The subcompartments of mitochondria and chloroplasts. In contrast to the cristae of mitochondria (A), the thylakoids of chloroplasts (B) are not connected to the inner membrane and therefore form a sealed compartment with a separate internal space.

Translocation into Mitochondria Depends on Signal Sequences and Protein Translocators

Proteins imported into **mitochondria** are usually taken up from the cytosol within seconds or minutes of their release from ribosomes. Thus, in contrast to protein translocation into the ER, which often takes place simultaneously with translation by a ribosome docked on the rough ER membrane (described later), mitochondrial proteins are first fully synthesized as **mitochondrial precursor proteins** in the cytosol and then translocated into mitochondria by a *post-translational* mechanism. One or more signal sequences direct all mitochondrial precursor proteins to their appropriate mitochondrial subcompartment. Many proteins entering the matrix space contain a signal sequence at their N-terminus that a signal peptidase rapidly removes after import. Other imported proteins, including all outer membrane and many inner membrane and intermembrane space proteins, have internal signal sequences that are not removed. The signal sequences are both necessary and sufficient for the import and correct localization of the proteins: when genetic engineering techniques are used to link these signals to a cytosolic protein, the signals direct the protein to the correct mitochondrial subcompartment.

The signal sequences that direct precursor proteins into the mitochondrial matrix space are best understood. They all form an amphiphilic α helix, in which positively charged residues cluster on one side of the helix, while uncharged hydrophobic residues cluster on the opposite side. Specific receptor proteins that initiate protein translocation recognize this configuration rather than the precise amino acid sequence of the signal sequence (**Figure 12–20**).

Multisubunit protein complexes that function as **protein translocators** mediate protein movement across mitochondrial membranes. The **TOM complex** transfers proteins across the outer membrane, and two **TIM complexes** (TIM23 and TIM22) transfer proteins across the inner membrane (**Figure 12–21**). These complexes contain some components that act as receptors for mitochondrial precursor proteins, and other components that form the translocation channels.

The TOM complex is required for the import of all nucleus-encoded mitochondrial proteins. It initially transports their signal sequences into the intermembrane space and helps to insert transmembrane proteins into the outer membrane. β-barrel proteins, which are particularly abundant in the outer membrane, are then passed on to an additional translocator, the **SAM complex**, which helps them to fold properly in the outer membrane. The TIM23 complex transports some soluble proteins into the matrix space and helps to insert transmembrane proteins into the inner membrane. The TIM22 complex mediates the insertion of a subclass of inner membrane proteins, including the transporter that moves ADP, ATP, and phosphate in and out of mitochondria. Yet another protein translocator in the inner mitochondrial membrane, the **OXA complex**, mediates the insertion of

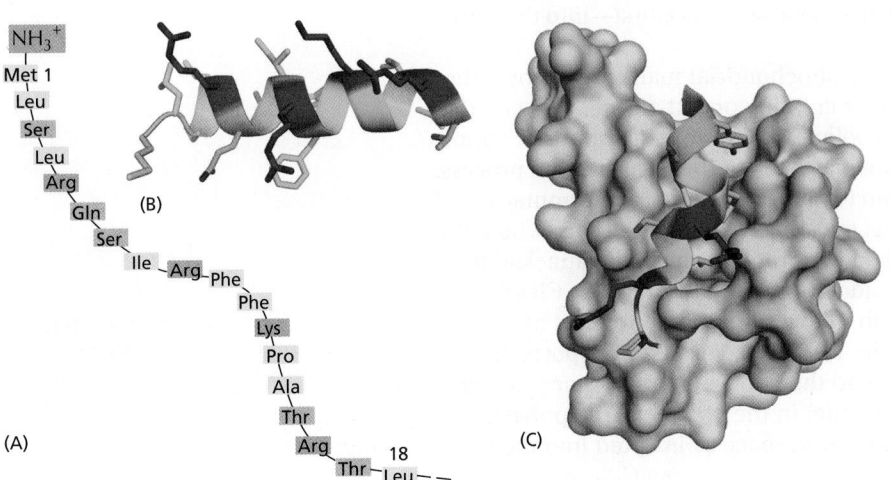

(A) (B) (C)

Figure 12–20 A signal sequence for mitochondrial protein import. Cytochrome oxidase is a large multiprotein complex located in the inner mitochondrial membrane, where it functions as the terminal enzyme in the electron-transport chain (discussed in Chapter 14). (A) The first 18 amino acids of the precursor to subunit IV of this enzyme serve as a signal sequence for import of the subunit into the mitochondrion. (B) When the signal sequence is folded as an α helix, the positively charged amino acids *(red)* are clustered on one face of the helix, while the nonpolar ones *(green)* are clustered primarily on the opposite face. Uncharged polar amino acids are shaded *orange*; nitrogen atoms on the side chains of Arg and Gln are colored *blue*. Signal sequences that direct proteins into the matrix space always have the potential to form such an amphiphilic α helix, which is recognized by specific receptor proteins on the mitochondrial surface. (C) The structure of a signal sequence (of alcohol dehydrogenase, another mitochondrial matrix enzyme), bound to an import receptor *(gray)*, as determined by nuclear magnetic resonance. The amphiphilic α helix binds with its hydrophobic face to a hydrophobic groove in the receptor (PDB code: 1OM2).

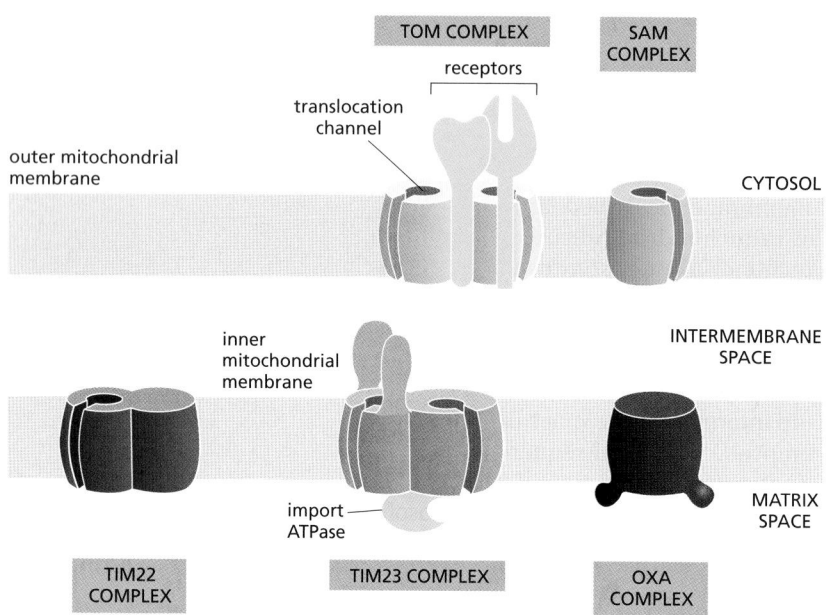

Figure 12–21 The protein translocators in the mitochondrial membranes. The TOM, TIM, SAM, and OXA complexes are multimeric membrane protein assemblies that catalyze protein translocation across mitochondrial membranes. The protein components of the TIM22 and TIM23 complexes that line the import channel are structurally related, suggesting a common evolutionary origin of both TIM complexes. On the matrix side, the TIM23 complex is bound to a multimeric protein complex containing mitochondrial hsp70, which acts as an import ATPase, using ATP hydrolysis to pull proteins through the pore. In animal cells, subtle variations exist in the subunit composition of the translocator complexes to adapt the mitochondrial import machinery to the particular needs of specialized cell types. SAM = Sorting and Assembly Machinery; OXA = cytochrome OXidase Activity; TIM = Translocator of the Inner Mitochondrial membrane; TOM = Translocator of the Outer Membrane.

those inner membrane proteins that are synthesized within mitochondria. It also helps to insert some imported inner membrane proteins that are initially transported into the matrix space by the other complexes.

Mitochondrial Precursor Proteins Are Imported as Unfolded Polypeptide Chains

We have learned almost everything we know about the molecular mechanism of protein import into mitochondria from analyses of cell-free, reconstituted translocation systems, in which purified mitochondria in a test tube import radiolabeled mitochondrial precursor proteins. By changing the conditions in the test tube, it is possible to establish the biochemical requirements for the import.

Mitochondrial precursor proteins do not fold into their native structures after they are synthesized; instead, they remain unfolded in the cytosol through interactions with other proteins. Some of these interacting proteins are general *chaperone proteins* of the *hsp70 family* (discussed in Chapter 6), whereas others are dedicated to mitochondrial precursor proteins and bind directly to their signal sequences. All the interacting proteins help to prevent the precursor proteins from aggregating or folding up spontaneously before they engage with the TOM complex in the outer mitochondrial membrane. As a first step in the import process, the import receptors of the TOM complex bind the signal sequence of the mitochondrial precursor protein. The interacting proteins are then stripped off, and the unfolded polypeptide chain is fed—signal sequence first—into the translocation channel.

In principle, a protein could reach the mitochondrial matrix space by either crossing the two membranes all at once or crossing one at a time. One can distinguish between these possibilities by cooling a cell-free mitochondrial import system to arrest the proteins at an intermediate step in the translocation process. The result is that the arrested proteins no longer contain their N-terminal signal sequence, indicating that the N-terminus must be in the matrix space where the signal peptidase is located, but the bulk of the protein can still be attacked from outside the mitochondria by externally added proteolytic enzymes. Clearly, the precursor proteins can pass through both mitochondrial membranes at once to enter the matrix space (**Figure 12–22**). The TOM complex first transports the signal sequence across the outer membrane to the intermembrane space, where it binds to a TIM complex, opening the channel in the complex. The polypeptide chain is then either translocated into the matrix space or inserted into the inner membrane.

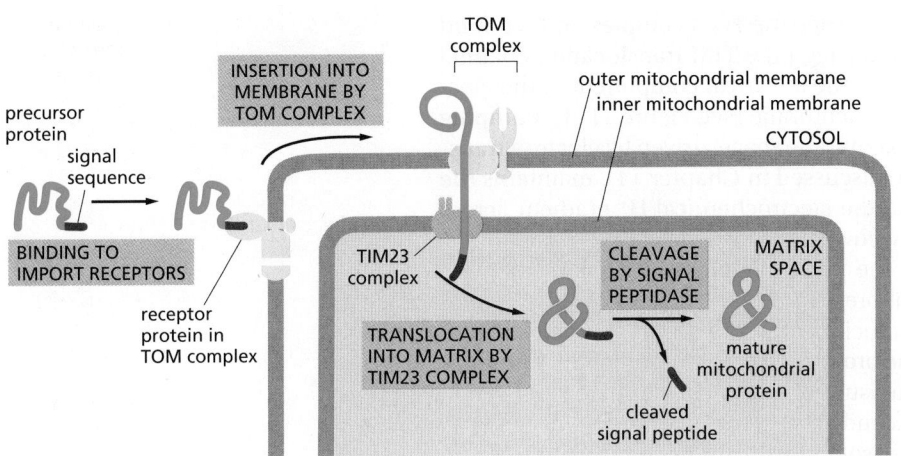

Figure 12–22 Protein import by mitochondria. The N-terminal signal sequence of the mitochondrial precursor protein is recognized by receptors of the TOM complex. The protein is then translocated through the TIM23 complex so that it transiently spans both mitochondrial membranes (**Movie 12.3**). The signal sequence is cleaved off by a signal peptidase in the matrix space to form the mature protein. The free signal sequence is then rapidly degraded (not shown).

Although the TOM and TIM complexes usually work together to translocate precursor proteins across both membranes at the same time, they can work independently. In isolated outer membranes, for example, the TOM complex can translocate the signal sequence of precursor proteins across the membrane. Similarly, if the outer membrane is experimentally disrupted in isolated mitochondria, the exposed TIM23 complex can efficiently import precursor proteins into the matrix space.

ATP Hydrolysis and a Membrane Potential Drive Protein Import Into the Matrix Space

Directional transport requires energy, which in most biological systems is supplied by ATP hydrolysis. ATP hydrolysis fuels mitochondrial protein import at two discrete sites, one outside the mitochondria and one in the matrix space. In addition, protein import requires another energy source, which is the membrane potential across the inner mitochondrial membrane (**Figure 12–23**).

The first requirement for energy occurs at the initial stage of the translocation process, when the unfolded precursor protein, associated with chaperone proteins, interacts with the import receptors of the TOM complex. As discussed in Chapter 6, the binding and release of newly synthesized polypeptides from the chaperone proteins requires ATP hydrolysis.

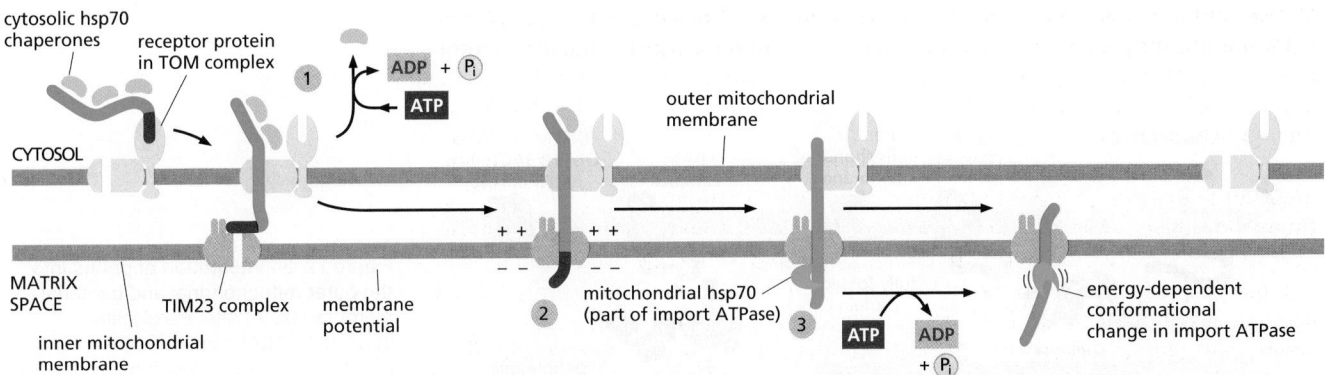

Figure 12–23 The role of energy in protein import into the mitochondrial matrix space. (1) Bound *cytosolic hsp70* chaperone is released from the precursor protein in a step that depends on ATP hydrolysis. After initial insertion of the signal sequence and of adjacent portions of the polypeptide chain into the TOM complex translocation channel, the signal sequence interacts with a TIM complex. (2) The signal sequence is then translocated into the matrix space in a process that requires the energy in the membrane potential across the inner membrane. (3) *Mitochondrial hsp70*, which is part of an import ATPase complex, binds to regions of the polypeptide chain as they become exposed in the matrix space, pulling the protein through the translocation channel, using the energy of ATP hydrolysis.

Once the signal sequence has passed through the TOM complex and is bound to a TIM complex, further translocation through the TIM translocation channel requires the membrane potential, which is the electrical component of the electrochemical H+ gradient across the inner membrane (see Figure 11–4). Pumping of H+ from the matrix space to the intermembrane space, driven by electron transport processes in the inner membrane (discussed in Chapter 14), maintains the electrochemical gradient. The energy in the electrochemical H+ gradient across the inner membrane therefore not only powers most of the cell's ATP synthesis, but it also drives the translocation of the positively charged signal sequences through the TIM complexes by electrophoresis.

Mitochondrial hsp70 also plays a crucial part in the import process. Mitochondria containing mutant forms of the protein fail to import precursor proteins. The mitochondrial hsp70 is part of a multisubunit protein assembly that is bound to the matrix side of the TIM23 complex and acts as a motor to pull the precursor protein into the matrix space. Like its cytosolic cousin, mitochondrial hsp70 has a high affinity for unfolded polypeptide chains, and it binds tightly to an imported protein chain as soon as the chain emerges from the TIM translocator in the matrix space. The hsp70 then undergoes a conformational change and releases the protein chain in an ATP-dependent step, exerting a ratcheting/pulling force on the protein being imported. This energy-driven cycle of binding and subsequent release provides the final driving force needed to complete protein import after a protein has initially inserted into the TIM23 complex (see Figure 12–23).

After the initial interaction with mitochondrial hsp70, many imported matrix proteins are passed on to another chaperone protein, *mitochondrial hsp60*. As discussed in Chapter 6, hsp60 helps the unfolded polypeptide chain to fold by binding and releasing it through cycles of ATP hydrolysis.

Bacteria and Mitochondria Use Similar Mechanisms to Insert Porins into their Outer Membrane

The outer mitochondrial membrane, like the outer membrane of Gram-negative bacteria (see Figure 11–17), contains abundant pore-forming β-barrel proteins called **porins**, and it is thus freely permeable to inorganic ions and metabolites (but not to most proteins). In contrast to other outer membrane proteins, which are anchored in the membrane through transmembrane α-helical regions, the TOM complex cannot integrate porins into the lipid bilayer. Instead, porins are first transported unfolded into the intermembrane space, where they transiently bind specialized chaperone proteins, which keep the porins from aggregating (**Figure 12–24**A). They then bind to the SAM complex in the outer membrane, which both inserts them into the outer membrane and helps them fold properly.

One of the central subunits of the SAM complex is homologous to a bacterial outer membrane protein that helps insert β-barrel proteins into the bacterial outer

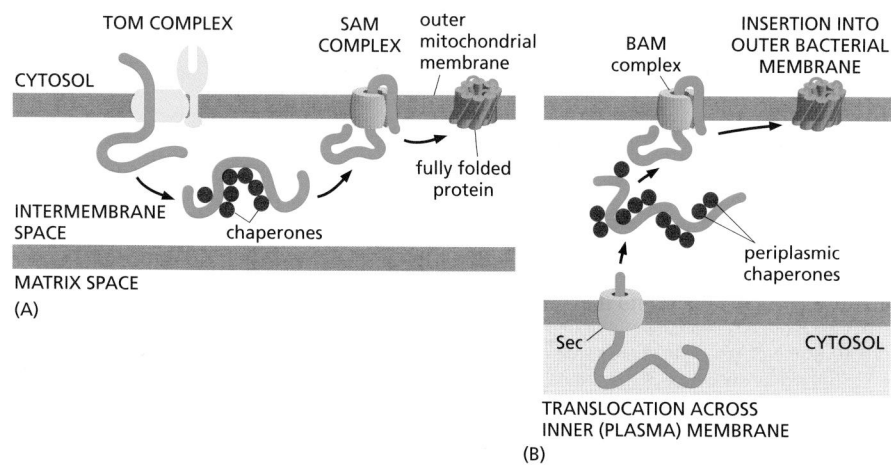

Figure 12–24 Integration of porins into the outer mitochondrial and bacterial membranes. (A) After translocation through the TOM complex in the outer mitochondrial membrane, β-barrel proteins bind to chaperones in the intermembrane space. The SAM complex then inserts the unfolded polypeptide chain into the outer membrane and helps the chain fold. (B) A structurally related BAM complex in the outer membrane of Gram-negative bacteria catalyzes β-barrel protein insertion and folding (see Figure 11–17).

membrane from the periplasmic space (the equivalent of the intermembrane space in mitochondria) (Figure 12–24B). This conserved pathway for inserting β-barrel proteins further underscores the endosymbiotic origin of mitochondria.

Transport Into the Inner Mitochondrial Membrane and Intermembrane Space Occurs Via Several Routes

The same mechanism that transports proteins into the matrix space using the TOM and TIM23 translocators (see Figure 12–22) also mediates the initial translocation of many proteins that are destined for the inner mitochondrial membrane or the intermembrane space. In the most common translocation route, only the N-terminal signal sequence of the transported protein actually enters the matrix space (**Figure 12–25A**). A hydrophobic amino acid sequence, strategically placed after the N-terminal signal sequence, acts as a *stop-transfer sequence*, preventing

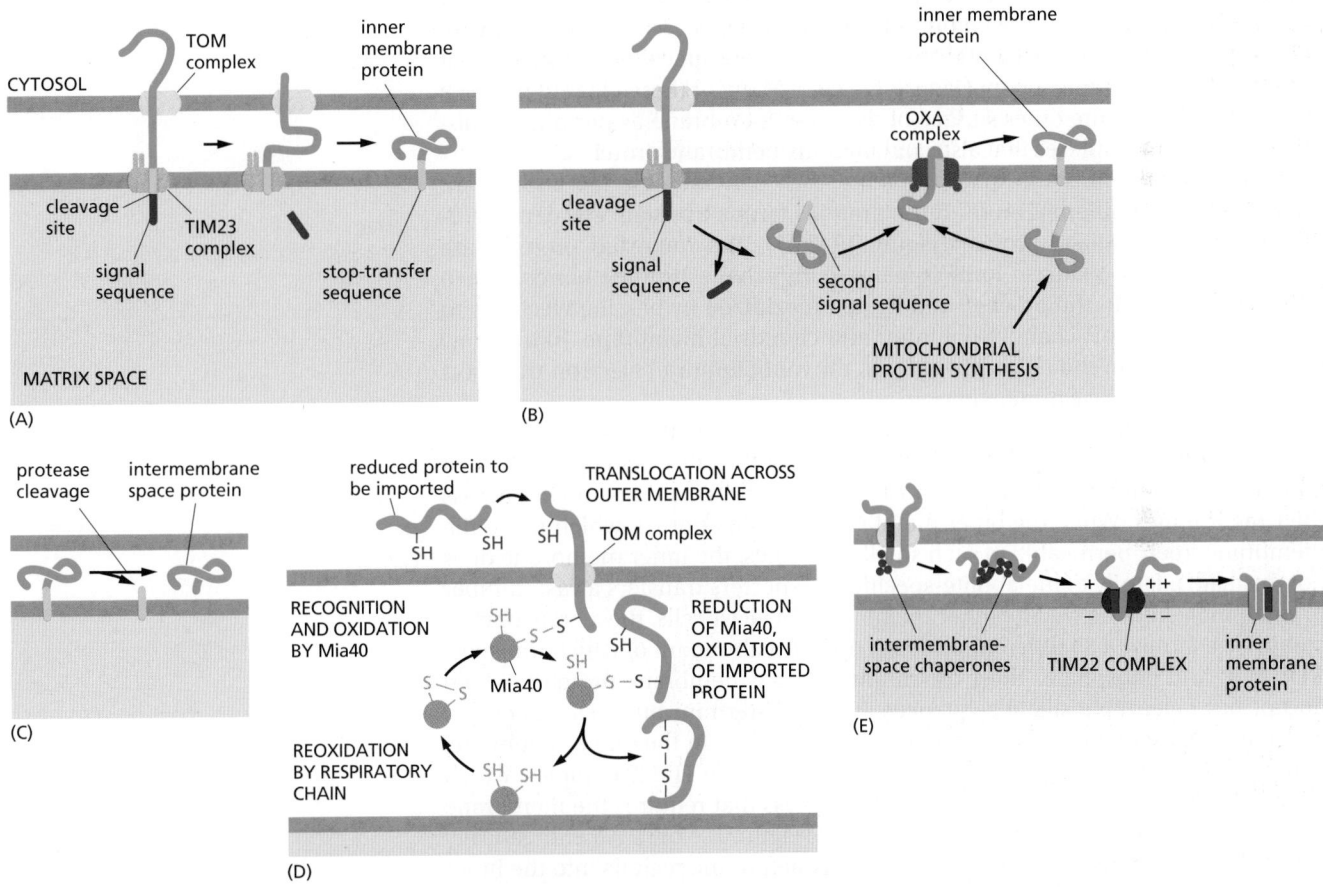

Figure 12–25 Protein import from the cytosol into the inner mitochondrial membrane and intermembrane space. (A) The N-terminal signal sequence *(red)* initiates import into the matrix space (see Figure 12–22). A hydrophobic sequence *(blue)* that follows the matrix-targeting signal sequence binds to the TIM23 translocator *(orange)* in the inner membrane and stops translocation. The remainder of the protein is then pulled into the intermembrane space through the TOM translocator in the outer membrane, and the hydrophobic sequence is released into the inner membrane anchoring the protein there. (B) A second route for protein integration into the inner membrane first delivers the protein completely into the matrix space. Cleavage of the signal sequence *(red)* used for the initial translocation unmasks an adjacent hydrophobic signal sequence *(blue)* at the new N-terminus. This signal then directs the protein into the inner membrane, using the same OXA-dependent pathway that inserts proteins that are encoded by the mitochondrial genome and translated in the matrix space. (C) Some soluble proteins of the intermembrane space also use the pathways shown in (A) and (B) before they are released into the intermembrane space by a second signal peptidase, which has its active site in the intermembrane space and removes the hydrophobic signal sequence. (D) Some soluble intermembrane-space proteins become oxidized by the Mia40 protein (Mia = <u>mi</u>tochondrial <u>i</u>ntermembrane space <u>a</u>ssembly) during import. Mia40 forms a covalent intermediate through an intermolecular disulfide bond, which helps pull the transported protein through the TOM complex. Mia40 becomes reduced in the process, and then is reoxidized by the electron transport chain, so that it can catalyze the next round of import. (E) Multipass inner membrane proteins that function as metabolite transporters contain internal signal sequences and snake through the TOM complex as a loop. They then bind to the chaperones in the intermembrane space, which guide the proteins to the TIM22 complex. The TIM22 complex is specialized for the insertion of multipass inner membrane proteins.

further translocation across the inner membrane. The remainder of the protein then crosses the outer membrane through the TOM complex into the intermembrane space; the signal sequence is cleaved off in the matrix, and the hydrophobic sequence, released from TIM23, remains anchored in the inner membrane.

In another transport route to the inner membrane or intermembrane space, the TIM23 complex initially translocates the entire protein into the matrix space (Figure 12–25B). A matrix signal peptidase then removes the N-terminal signal sequence, exposing a hydrophobic sequence at the new N-terminus. This signal sequence guides the protein to the OXA complex, which inserts the protein into the inner membrane. As mentioned earlier, the OXA complex is primarily used to insert proteins that are encoded and translated in the mitochondrion into the inner membrane, and only a few imported proteins use this pathway. Translocators that are closely related to the OXA complex are found in the plasma membrane of bacteria and in the thylakoid membrane of chloroplasts, where they insert membrane proteins by a similar mechanism.

Many proteins that use these pathways to the inner membrane remain anchored there through their hydrophobic signal sequence (see Figure 12–25A,B). Others, however, are released into the intermembrane space by a protease that removes the membrane anchor (Figure 12–25C). Many of these cleaved proteins remain attached to the outer surface of the inner membrane as peripheral subunits of protein complexes that also contain transmembrane proteins.

Certain intermembrane-space proteins that contain cysteine motifs are imported by a yet different route. These proteins form a transient covalent disulfide bond to the Mia40 protein (Figure 12–25D). The imported proteins are then released in an oxidized form containing intrachain disulfide bonds. Mia40 becomes reduced in the process, and is then reoxidized by passing electrons to the electron transport chain in the inner mitochondrial membrane. In this way, the energy stored in the redox potential in the mitochondrial electron transport chain is tapped to drive protein import.

Mitochondria are the principal sites of ATP synthesis in the cell, but they also contain many metabolic enzymes, such as those of the citric acid cycle. Thus, in addition to proteins, mitochondria must also transport small metabolites across their membranes. While the outer membrane contains porins, which make the membrane freely permeable to such small molecules, the inner membrane does not. Instead, a family of metabolite-specific transporters transfers a vast number of small molecules across the inner membrane. In yeast cells, these transporters comprise a family of 35 different proteins, the most abundant of which transport ATP, ADP, and phosphate. These are multipass transmembrane proteins, which do not have cleavable signal sequences at their N-termini but instead contain internal signal sequences. They cross the TOM complex in the outer membrane, and intermembrane-space chaperones guide them to the TIM22 complex, which inserts them into the inner membrane by a process that requires the membrane potential, but not mitochondrial hsp70 or ATP (Figure 12–25E). An energetically favorable partitioning of the hydrophobic transmembrane regions into the inner membrane is likely to drive this process.

Two Signal Sequences Direct Proteins to the Thylakoid Membrane in Chloroplasts

Protein transport into **chloroplasts** resembles transport into mitochondria. Both processes occur post-translationally, use separate translocation complexes in each membrane, require energy, and use amphiphilic N-terminal signal sequences that are removed after use. With the exception of some of the chaperone molecules, however, the protein components that form the translocation complexes differ. Moreover, whereas mitochondria harness the electrochemical H^+ gradient across their inner membrane to drive transport, chloroplasts, which have an electrochemical H^+ gradient across their thylakoid membrane but not their inner membrane, use GTP and ATP hydrolysis to power import across their

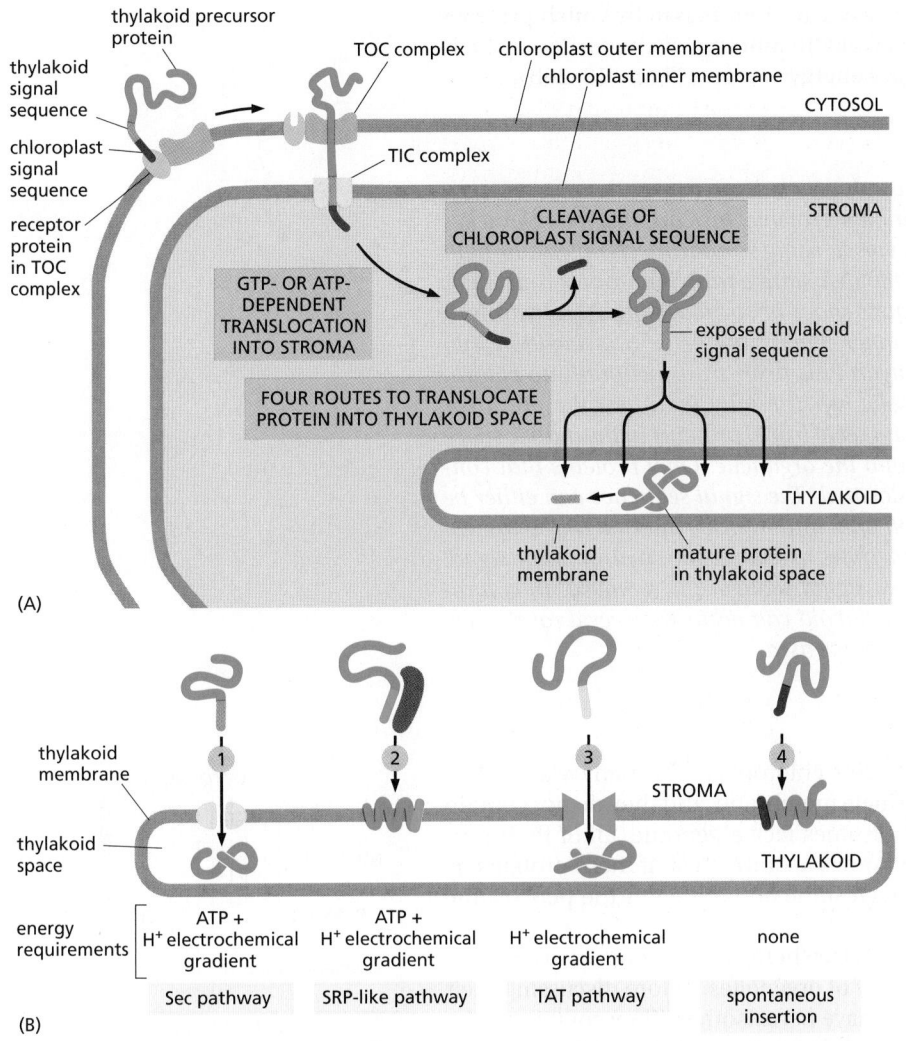

Figure 12–26 Translocation of chloroplast precursor proteins into the thylakoid space. (A) The precursor protein contains an N-terminal chloroplast signal sequence *(red)*, followed immediately by a thylakoid signal sequence *(brown)*. The chloroplast signal sequence initiates translocation into the stroma by a mechanism similar to that used for the translocation of mitochondrial precursor proteins into the matrix space, although the translocator complexes, TOC and TIC, are different. The signal sequence is then cleaved off, unmasking the thylakoid signal sequence, which initiates translocation across the thylakoid membrane. (B) Translocation into the thylakoid space or thylakoid membrane can occur by any one of at least four routes: (1) a *Sec pathway*, so called because it uses components that are homologs of Sec proteins, which mediate protein translocation across the bacterial plasma membrane (discussed later); (2) an *SRP-like pathway*, so called because it uses a chloroplast homolog of the signal-recognition particle, or SRP (discussed later); (3) a *TAT* (twin arginine translocation) *pathway*, so called because two arginines are critical in the signal sequences that direct proteins into this pathway, which depends on the H^+ gradient across the thylakoid membrane; and (4) a *spontaneous insertion pathway* that seems not to require any protein translocator.

double membrane. The functional similarities may thus result from convergent evolution, reflecting the common requirements for translocation across a double membrane.

Although the signal sequences for import into chloroplasts superficially resemble those for import into mitochondria, the same plant cells have both mitochondria and chloroplasts, so proteins must partition appropriately between the two organelles. In plants, for example, a bacterial enzyme can be directed specifically to mitochondria if it is experimentally joined to an N-terminal signal sequence of a mitochondrial protein; the same enzyme joined to an N-terminal signal sequence of a chloroplast protein ends up in chloroplasts. Thus, the import receptors on each organelle distinguish between the different signal sequences.

Chloroplasts have an extra membrane-enclosed compartment, the **thylakoid**. Many chloroplast proteins, including the protein subunits of the photosynthetic system and of the ATP synthase (discussed in Chapter 14), are located in the thylakoid membrane. Like the precursors of some mitochondrial proteins, the precursors of these proteins are translocated from the cytosol to their final destination in two steps. First, they pass across the double membrane into the matrix space (called the **stroma** in chloroplasts), and then they either integrate into the thylakoid membrane or translocate into the thylakoid space (**Figure 12–26**A). The precursors of these proteins have a hydrophobic thylakoid signal sequence following the N-terminal chloroplast signal sequence. After the N-terminal signal sequence has been used to import the protein into the stroma, a stromal signal peptidase removes it, unmasking the thylakoid signal sequence that initiates transport across the thylakoid membrane. There are at least four routes by which proteins cross or become integrated into the thylakoid membrane, distinguished by their need for different stromal chaperones and energy sources (Figure 12–26B).

Summary

Although mitochondria and chloroplasts have their own genetic systems, they produce only a small proportion of their own proteins. Instead, the two organelles import most of their proteins from the cytosol, using similar mechanisms. In both cases, proteins are transported in an unfolded state across both outer and inner membranes simultaneously into the matrix space or stroma. Both ATP hydrolysis and a membrane potential across the inner membrane drive translocation into mitochondria, whereas GTP and ATP hydrolysis drive translocation into chloroplasts. Chaperone proteins of the cytosolic hsp70 family maintain the precursor proteins in an unfolded state, and a second set of hsp70 proteins in the matrix space or stroma pulls the polypeptide chain into the organelle. Only proteins that contain a specific signal sequence are translocated. The signal sequence can either be located at the N-terminus and cleaved off after import or be internal and retained. Transport into the inner membrane sometimes uses a second, hydrophobic signal sequence that is unmasked when the first signal sequence is removed. In chloroplasts, import from the stroma into the thylakoid can occur by several routes, distinguished by the chaperones and energy source used.

PEROXISOMES

Peroxisomes differ from mitochondria and chloroplasts in many ways. Most notably, they are surrounded by only a single membrane, and they do not contain DNA or ribosomes. Thus, because peroxisomes lack a genome, all of their proteins are encoded in the nucleus. Peroxisomes acquire most of these proteins by selective import from the cytosol, although some of them enter the peroxisome membrane via the ER.

Because we do not discuss peroxisomes elsewhere, we shall digress to consider some of the functions of this diverse family of organelles, before discussing their biosynthesis. Virtually all eukaryotic cells have peroxisomes. They contain oxidative enzymes, such as *catalase* and *urate oxidase*, at such high concentrations that, in some cells, the peroxisomes stand out in electron micrographs because of the presence of a crystalloid protein core (**Figure 12–27**).

Like mitochondria, peroxisomes are major sites of oxygen utilization. One hypothesis is that peroxisomes are a vestige of an ancient organelle that performed all the oxygen metabolism in the primitive ancestors of eukaryotic cells. When the oxygen produced by photosynthetic bacteria first accumulated in the atmosphere, it would have been highly toxic to most cells. Peroxisomes might have lowered the intracellular concentration of oxygen, while also exploiting its chemical reactivity to perform useful oxidation reactions. According to this view, the later development of mitochondria rendered peroxisomes largely obsolete because many of the same biochemical reactions—which had formerly been carried out in peroxisomes without producing energy—were now coupled to ATP formation by means of oxidative phosphorylation. The oxidation reactions performed by peroxisomes in present-day cells could therefore partly be those whose functions were not taken over by mitochondria.

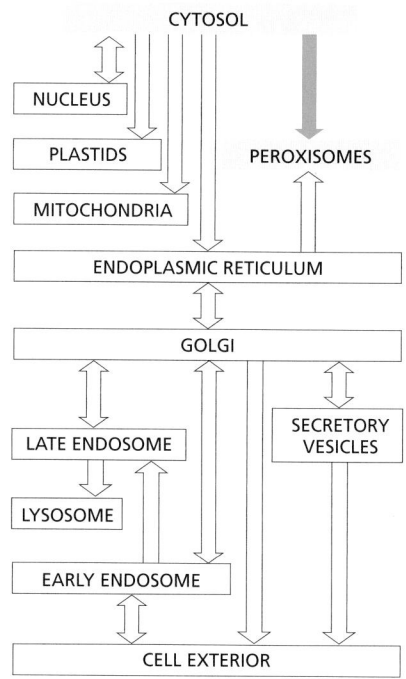

Peroxisomes Use Molecular Oxygen and Hydrogen Peroxide to Perform Oxidation Reactions

Peroxisomes are so named because they usually contain one or more enzymes that use molecular oxygen to remove hydrogen atoms from specific organic substrates (designated here as R) in an oxidation reaction that produces *hydrogen peroxide* (H_2O_2):

$$RH_2 + O_2 \rightarrow R + H_2O_2$$

Catalase uses the H_2O_2 generated by other enzymes in the organelle to oxidize a variety of other substrates—including formic acid, formaldehyde, and alcohol— by the "peroxidation" reaction: $H_2O_2 + R'H_2 \rightarrow R' + 2H_2O$. This type of oxidation reaction is particularly important in liver and kidney cells, where the peroxisomes detoxify various harmful molecules that enter the bloodstream. About 25% of the ethanol we drink is oxidized to acetaldehyde in this way. In addition, when excess H_2O_2 accumulates in the cell, catalase converts it to H_2O through the reaction

$$2H_2O_2 \rightarrow 2H_2O + O_2$$

A major function of the oxidation reactions performed in peroxisomes is the breakdown of fatty acid molecules. The process, called *β oxidation*, shortens the alkyl chains of fatty acids sequentially in blocks of two carbon atoms at a time, thereby converting the fatty acids to acetyl CoA. The peroxisomes then export the acetyl CoA to the cytosol for use in biosynthetic reactions. In mammalian cells, β oxidation occurs in both mitochondria and peroxisomes; in yeast and plant cells, however, this essential reaction occurs exclusively in peroxisomes.

An essential biosynthetic function of animal peroxisomes is to catalyze the first reactions in the formation of *plasmalogens*, which are the most abundant class of phospholipids in myelin (**Figure 12–28**). Plasmalogen deficiencies cause profound abnormalities in the myelination of nerve-cell axons, which is one reason why many peroxisomal disorders lead to neurological disease.

Peroxisomes are unusually diverse organelles, and even in the various cell types of a single organism they may contain different sets of enzymes. They also adapt remarkably to changing conditions. Yeasts grown on sugar, for example, have few small peroxisomes. But when some yeasts are grown on methanol, numerous large peroxisomes are formed that oxidize methanol; and when grown on fatty acids, they develop numerous large peroxisomes that break down fatty acids to acetyl CoA by β oxidation.

Peroxisomes are also important in plants. Two types of plant peroxisomes have been studied extensively. One is present in leaves, where it participates in *photorespiration* (discussed in Chapter 14) (**Figure 12–29**A). The other type of peroxisome is present in germinating seeds, where it converts the fatty acids stored in seed lipids into the sugars needed for the growth of the young plant. Because this conversion of fats to sugars is accomplished by a series of reactions known as the *glyoxylate cycle*, these peroxisomes are also called *glyoxysomes* (Figure 12–29B). In the glyoxylate cycle, two molecules of acetyl CoA produced by fatty acid breakdown in the peroxisome are used to make succinic acid, which then leaves the peroxisome and is converted into glucose in the cytosol. The glyoxylate cycle does not occur in animal cells, and animals are therefore unable to convert the fatty acids in fats into carbohydrates.

A Short Signal Sequence Directs the Import of Proteins into Peroxisomes

A specific sequence of three amino acids (Ser–Lys–Leu) located at the C-terminus of many peroxisomal proteins functions as an import signal (see Table 12–3, p. 648). Other peroxisomal proteins contain a signal sequence near the N-terminus. If either sequence is attached to a cytosolic protein, the protein is imported into peroxisomes. The import signals are first recognized by soluble receptor proteins in the cytosol. Numerous distinct proteins, called **peroxins**, participate in the import process, which is driven by ATP hydrolysis. A complex of at least six different peroxins forms a protein translocator in the peroxisome membrane. Even oligomeric proteins do not have to unfold to be imported. To allow the passage of such compactly folded cargo molecules, the pore formed by the transporter is thought to be dynamic in its dimensions, adapting in size to the particular cargo molecules to be transported. In this respect, the mechanism differs from that used by mitochondria and chloroplasts. One soluble import receptor, the peroxin Pex5 recognizes the C-terminal peroxisomal import signal. It accompanies its cargo all the way into peroxisomes and, after cargo release, cycles back to the cytosol. After

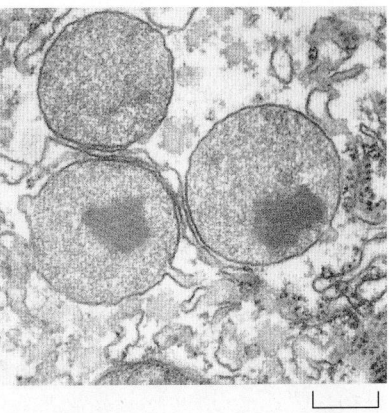

Figure 12–27 An electron micrograph of three peroxisomes in a rat liver cell. The paracrystalline, electron-dense inclusions are composed primarily of the enzyme urate oxidase. (Courtesy of Daniel S. Friend, by permission of E.L. Bearer.)

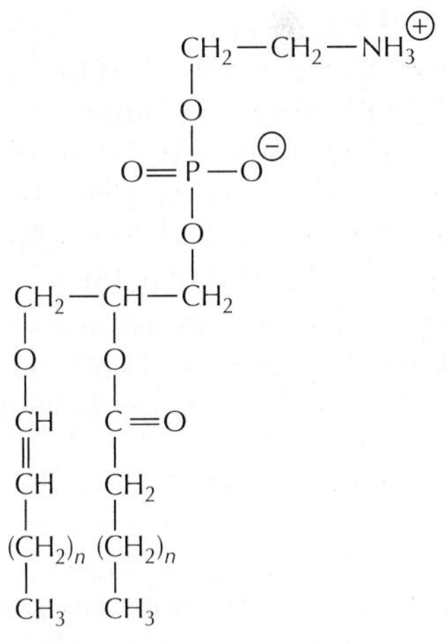

200 nm

Figure 12–28 The structure of a plasmalogen. Plasmalogens are very abundant in the myelin sheaths that insulate the axons of nerve cells. They make up some 80–90% of the myelin membrane phospholipids. In addition to an ethanolamine head group and a long-chain fatty acid attached to the same glycerol phosphate backbone used for phospholipids, plasmalogens contain an unusual fatty alcohol that is attached through an ether linkage *(bottom left)*.

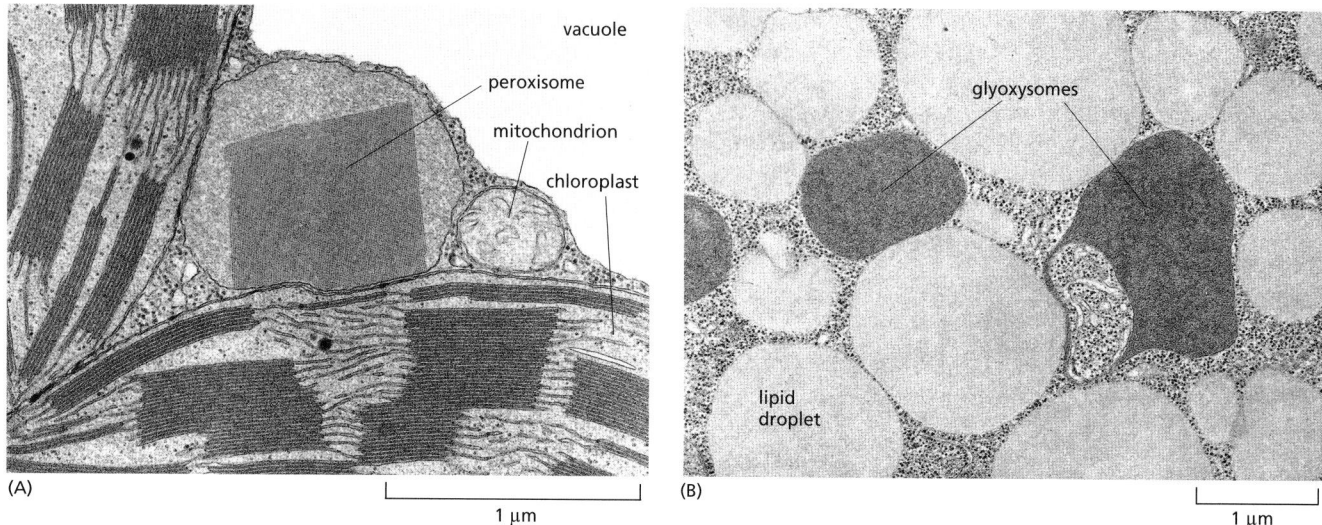

Figure 12–29 Electron micrographs of two types of peroxisomes found in plant cells. (A) A peroxisome with a paracrystalline core in a tobacco leaf mesophyll cell. Its close association with chloroplasts is thought to facilitate the exchange of materials between these organelles during photorespiration. The vacuole in plant cells is equivalent to the lysosome in animal cells. (B) Peroxisomes in a fat-storing cotyledon cell of a tomato seed 4 days after germination. Here the peroxisomes (glyoxysomes) are associated with the lipid droplets that store fat, reflecting their central role in fat mobilization and gluconeogenesis during seed germination. (A, © 1969 S.E. Frederick and E.H. Newcomb. Originally published in *J. Cell Biol.* https://doi.org/10.1083/jcb.43.2.343. With permission from Rockefeller University Press; B, from W.P. Wergin, P.J. Gruber, and E.H. Newcomb, *J. Ultrastruct. Res.* 30:533–557, 1970. With permission from Elsevier.)

delivering its cargo to the peroxisome lumen, Pex5 undergoes ubiquitylation. This modification is required to release Pex5 back into the cytosol, where the ubiquitin is removed. An ATPase composed of Pex1 and Pex6 harnesses the energy of ATP hydrolysis to help release Pex5 from peroxisomes.

The importance of this import process and of peroxisomes is demonstrated by the inherited human disease *Zellweger syndrome*, in which a defect in importing proteins into peroxisomes leads to a profound peroxisomal deficiency. These individuals, whose cells contain "empty" peroxisomes, have severe abnormalities in their brain, liver, and kidneys, and they die soon after birth. A mutation in the gene encoding peroxin Pex5 causes one form of the disease. A defect in Pex7, the receptor for the N-terminal import signal, causes a milder peroxisomal disease.

It has long been debated whether new peroxisomes arise from preexisting ones by organelle growth and fission—as mentioned earlier for mitochondria and plastids—or whether they derive as a specialized compartment from the endoplasmic reticulum (ER). Aspects of both views are true (**Figure 12–30**). Most peroxisomal membrane proteins are made in the cytosol and insert into the membrane of

Figure 12–30 A model that explains how peroxisomes proliferate and how new peroxisomes arise. Peroxisomal precursor vesicles bud from the ER. At least two peroxisomal membrane proteins, Pex3 and Pex15, follow this route. The machinery that drives the budding reaction and that selects only peroxisomal proteins for packaging into these vesicles depends on Pex19 and other cytosolic proteins that are still unknown. Peroxisomal precursor vesicles may then fuse with one another or with preexisting peroxisomes. The peroxisome membrane contains import receptors and protein translocators that are required for the import of peroxisomal proteins made on cytosolic ribosomes, including new copies of the import receptors and translocator components. Presumably, the lipids required for growth are also imported, although some may derive directly from the ER in the membrane of peroxisomal precursor vesicles.

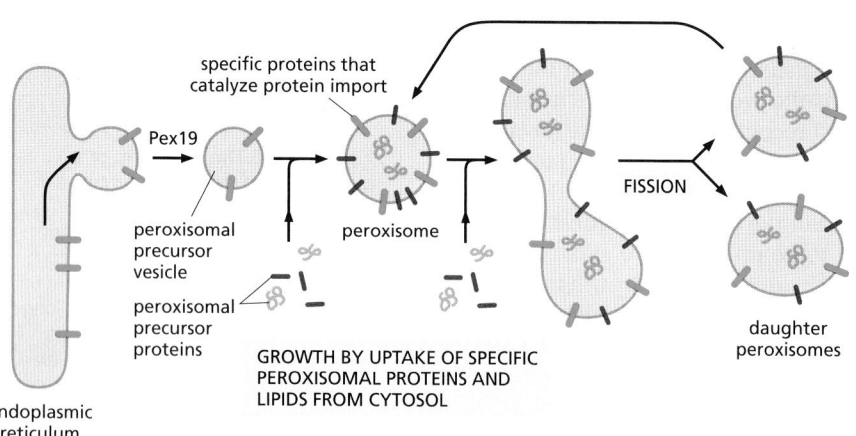

specific proteins that catalyze protein import

Pex19

peroxisomal precursor vesicle

peroxisomal precursor proteins

peroxisome

FISSION

daughter peroxisomes

GROWTH BY UPTAKE OF SPECIFIC PEROXISOMAL PROTEINS AND LIPIDS FROM CYTOSOL

endoplasmic reticulum

preexisting peroxisomes, but others are first integrated into the ER membrane, where they are packaged into specialized peroxisomal precursor vesicles. New precursor vesicles may then fuse with one another and begin importing additional peroxisomal proteins, using their own protein import machinery to grow into mature peroxisomes, which can undergo cycles of growth and fission.

Summary

Peroxisomes are specialized for carrying out oxidation reactions using molecular oxygen. They generate hydrogen peroxide, which they employ for oxidative purposes—and contain catalase to destroy the excess. Like mitochondria and plastids, peroxisomes are self-replicating organelles. Because they do not contain DNA or ribosomes, however, all of their proteins are encoded in the cell nucleus. Some of these proteins are conveyed to peroxisomes via peroxisomal precursor vesicles that bud from the ER, but most are synthesized in the cytosol and directly imported. A specific sequence of three amino acids near the C-terminus of many of the latter proteins functions as a peroxisomal import signal. The mechanism of protein import differs from that of mitochondria and chloroplasts, in that even oligomeric proteins are imported from the cytosol without unfolding.

THE ENDOPLASMIC RETICULUM

All eukaryotic cells have an **endoplasmic reticulum** (**ER**). Its membrane typically constitutes more than half of the total membrane of an average animal cell (see Table 12–2, p. 643). The ER is organized into a netlike labyrinth of branching tubules and flattened sacs that extends throughout the cytosol (**Figure 12–31** and **Movie 12.4**). The tubules and sacs interconnect, and their membrane is continuous with the outer nuclear membrane; the compartment that they enclose therefore is also continuous with the space between the inner and outer nuclear membranes. Thus, the ER and nuclear membranes form a continuous sheet enclosing a single internal space, called the **ER lumen** or the *ER cisternal space*, which often occupies more than 10% of the total cell volume (see Table 12–1, p. 643).

As mentioned at the beginning of this chapter, the ER has a central role in both lipid and protein biosynthesis, and it also serves as an intracellular Ca^{2+} store that is used in many cell signaling responses (discussed in Chapter 15). The ER membrane is the site of production of all the transmembrane proteins and lipids for most of the cell's organelles, including the ER itself, the Golgi apparatus, lysosomes, endosomes, secretory vesicles, and the plasma membrane. The ER membrane is also the site at which most of the lipids for mitochondrial and peroxisomal membranes are made. In addition, almost all of the proteins that will be secreted to the cell exterior—plus those destined for the lumen of the ER, Golgi apparatus, or lysosomes—are initially delivered to the ER lumen.

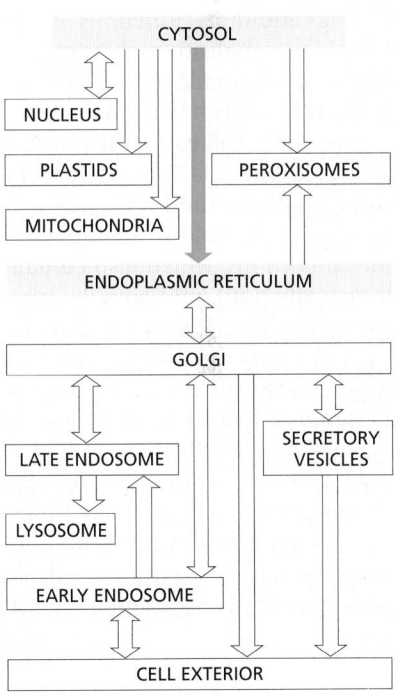

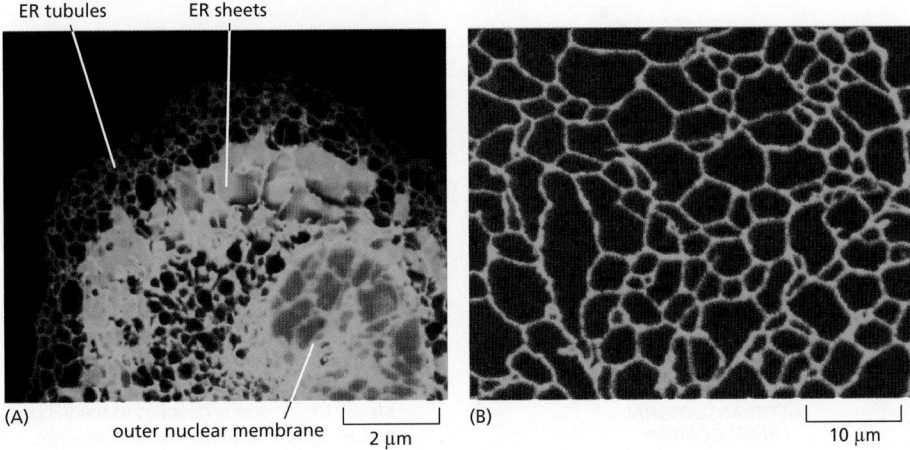

ER tubules   ER sheets

(A)   outer nuclear membrane   2 µm

(B)   10 µm

Figure 12–31 **Fluorescent micrographs of the endoplasmic reticulum.** (A) An animal cell in tissue culture that was genetically engineered to express an ER membrane protein fused to a fluorescent protein. The ER extends as a network of tubules and sheets throughout the entire cytosol, so that all regions of the cytosol are close to some portion of the ER membrane. The outer nuclear membrane, which is continuous with the ER, is also stained. (B) Part of an ER network in a living plant cell that was genetically engineered to express a fluorescent protein in the ER. (A, courtesy of Patrick Chitwood and Gia Voeltz; B, courtesy of Petra Boevink and Chris Hawes.)

The ER Is Structurally and Functionally Diverse

While the various functions of the ER are essential to every cell, their relative importance varies greatly between individual cell types. To meet different functional demands, distinct regions of the ER become highly specialized. We observe such functional specialization as dramatic changes in ER structure, and different cell types can therefore possess characteristically different types of ER membrane. One of the most remarkable ER specializations is the *rough ER*.

Mammalian cells begin to import most proteins into the ER before complete synthesis of the polypeptide chain—that is, import is a **co-translational** process (**Figure 12–32**A). In contrast, the import of proteins into mitochondria, chloroplasts, nuclei, and peroxisomes is a **post-translational** process (Figure 12–32B). In co-translational transport, the ribosome that is synthesizing the protein is attached directly to the ER membrane, enabling one end of the protein to be translocated into the ER while the rest of the polypeptide chain is being synthesized. These membrane-bound ribosomes coat the surface of the ER, creating regions termed **rough endoplasmic reticulum**, or **rough ER**; regions of ER that lack bound ribosomes are called **smooth endoplasmic reticulum**, or **smooth ER** (**Figure 12–33**).

Most cells have scanty regions of smooth ER, and the ER is often partly smooth and partly rough. Areas of smooth ER from which transport vesicles carrying newly synthesized proteins and lipids bud off for transport to the Golgi apparatus are called *transitional ER*. In certain specialized cells, the smooth ER is abundant and has additional functions. It is prominent, for example, in cells that specialize in lipid metabolism, such as cells that synthesize steroid hormones from cholesterol; the expanded smooth ER accommodates the enzymes that make cholesterol and modify it to form the hormones (see Figure 12–33B).

The main cell type in the liver, the *hepatocyte*, also has a substantial amount of smooth ER. It is the principal site of production of *lipoprotein particles*, which carry lipids via the bloodstream to other parts of the body. The enzymes that synthesize the lipid components of the particles are located in the membrane of the smooth ER, which also contains enzymes that catalyze a series of reactions to detoxify both lipid-soluble drugs and various harmful compounds produced by metabolism. The most extensively studied of these *detoxification reactions* are carried out by the *cytochrome P450* family of enzymes, which catalyze a series of reactions in which water-insoluble drugs or metabolites that would otherwise accumulate to toxic levels in cell membranes are rendered sufficiently water-soluble to leave the cell and be excreted in the urine. Because the rough ER alone cannot house enough of these and other necessary enzymes, a substantial portion of the membrane in a hepatocyte normally consists of smooth ER (see Table 12–2).

Another crucially important function of the ER in most eukaryotic cells is to sequester Ca^{2+} from the cytosol. The release of Ca^{2+} into the cytosol from the ER, and its subsequent reuptake, occurs in many rapid responses to extracellular

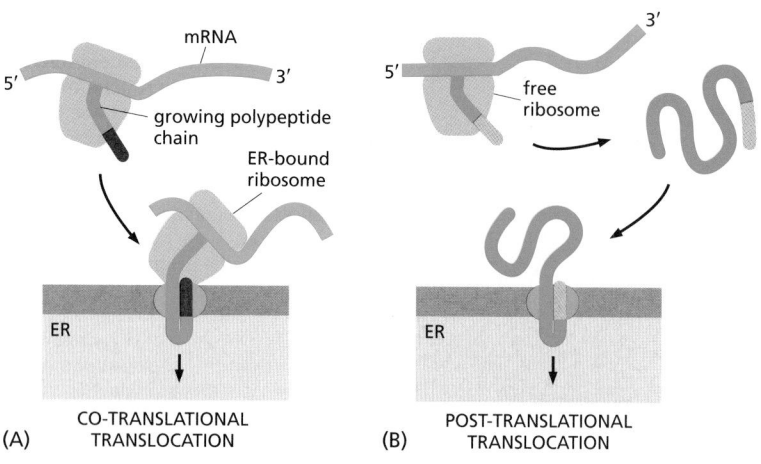

Figure 12–32 Co-translational and post-translational protein translocation.
(A) Ribosomes bind to the ER membrane during co-translational translocation. (B) By contrast, cytosolic ribosomes complete the synthesis of a protein and release it prior to post-translational translocation. In both cases, the protein is directed to the ER by an ER signal sequence (*red* and *orange*).

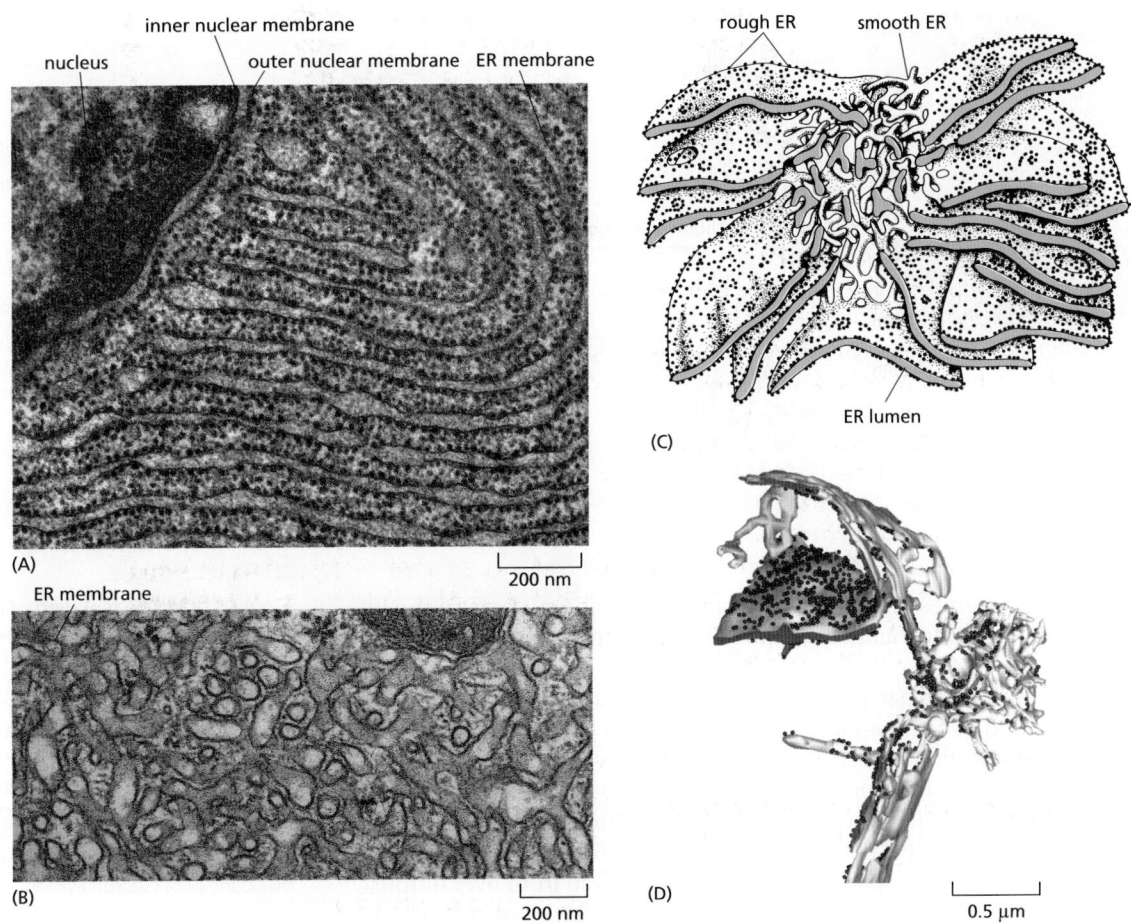

Figure 12–33 The rough and smooth ER. (A) An electron micrograph of the rough ER in a pancreatic exocrine cell that makes and secretes large amounts of digestive enzymes every day. The cytosol is filled with closely packed sheets of ER membrane that is studded with ribosomes. At the top left is a portion of the nucleus and its nuclear envelope; note that the outer nuclear membrane, which is continuous with the ER, is also studded with ribosomes. (B) Abundant smooth ER in a steroid-hormone-secreting cell. This electron micrograph is of a testosterone-secreting Leydig cell in the human testis. (C) A three-dimensional reconstruction of a region of smooth ER and rough ER in a liver cell. The rough ER forms oriented stacks of flattened cisternae, each having a lumenal space 20–30 nm wide. The smooth ER membrane is connected to these cisternae and forms a fine network of tubules 30–60 nm in diameter. The ER lumen is colored *green*. (D) A tomographic reconstruction of a portion of the ER network in a yeast cell. Membrane-bound ribosomes *(tiny dark spheres)* are seen in both flat sheets and tubular regions of irregular diameter, demonstrating that the ribosomes bind to ER membranes of different curvature in these cells. (A, courtesy of Lelio Orci; B, courtesy of Daniel S. Friend, by permission of E.L. Bearer; C, after R.V. Krstić, *Ultrastructure of the Mammalian Cell.* New York: Springer-Verlag, 1979. With permission from Springer Nature; D, © 2011 M. West et al. Originally published in *J. Cell Biol.* https://doi.org/10.1083/jcb.201011039. With permission from Rockefeller University Press.)

signals, as discussed in Chapter 15. A Ca^{2+} pump transports Ca^{2+} from the cytosol into the ER lumen. A high concentration of Ca^{2+}-binding proteins in the ER facilitates Ca^{2+} storage. In some cell types, and perhaps in most, specific regions of the ER are specialized for Ca^{2+} storage. Muscle cells have an abundant, modified smooth ER called the *sarcoplasmic reticulum*. The release and reuptake of Ca^{2+} by the sarcoplasmic reticulum trigger myofibril contraction and relaxation, respectively, during each round of muscle contraction (discussed in Chapter 16).

To study the functions and biochemistry of the ER, it is necessary to isolate it. This may seem to be a hopeless task because the ER is intricately interleaved with other components of the cytoplasm. Fortunately, when tissues or cells are disrupted by homogenization, the ER breaks into fragments, which reseal to form small (~100–200 nm in diameter) closed vesicles called **microsomes**. Microsomes are relatively easy to purify. To the biochemist, microsomes represent small authentic versions of the ER, still capable of protein translocation, protein glycosylation (discussed later), Ca^{2+} uptake and release, and lipid synthesis. Microsomes derived from rough ER are studded with ribosomes and are called *rough*

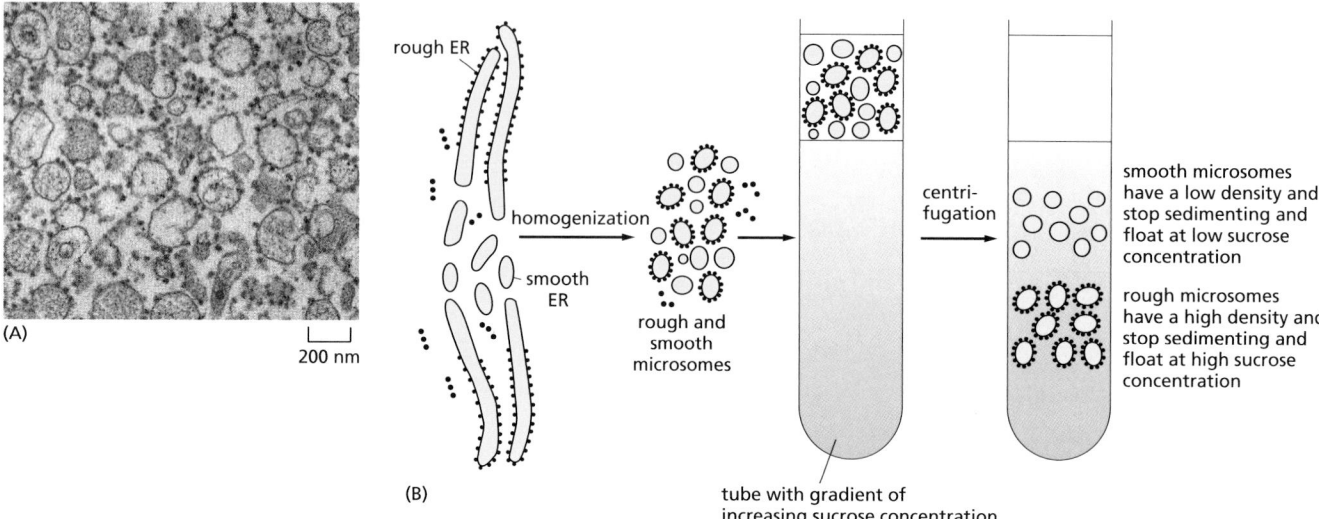

(A)

200 nm

(B)

rough ER

smooth ER

homogenization

rough and smooth microsomes

tube with gradient of increasing sucrose concentration

centri-fugation

smooth microsomes have a low density and stop sedimenting and float at low sucrose concentration

rough microsomes have a high density and stop sedimenting and float at high sucrose concentration

Figure 12–34 The isolation of purified rough and smooth microsomes from the ER. (A) A thin section electron micrograph of the purified rough ER fraction shows an abundance of ribosome-studded vesicles. (B) When sedimented to equilibrium through a gradient of sucrose, the two types of microsomes separate from each other on the basis of their different densities. Note that the smooth fraction will also contain non-ER-derived material. (A, courtesy of George Palade.)

microsomes. The ribosomes are always found on the outside surface, so the interior of the microsome is biochemically equivalent to the lumen of the ER (**Figure 12–34A**).

Many vesicles similar in size to rough microsomes, but lacking attached ribosomes, are also found in cell homogenates. Such *smooth microsomes* are derived in part from smooth portions of the ER and in part from vesiculated fragments of the plasma membrane, Golgi apparatus, endosomes, and mitochondria (the ratio depending on the tissue). Thus, whereas rough microsomes are clearly derived from rough portions of ER, it is not easy to separate smooth microsomes derived from different organelles. The smooth microsomes prepared from liver or muscle cells are an exception. Because of the unusually large quantities of smooth ER or sarcoplasmic reticulum, respectively, most of the smooth microsomes in homogenates of these tissues are derived from the smooth ER or sarcoplasmic reticulum. The ribosomes attached to rough microsomes make them more dense than smooth microsomes. As a result, we can use equilibrium centrifugation to separate the rough and smooth microsomes (Figure 12–34B). Microsomes have been invaluable in elucidating the molecular aspects of ER function, as we discuss next.

Signal Sequences Were First Discovered in Proteins Imported into the Rough ER

The ER captures selected proteins from the cytosol as they are being synthesized. These proteins are of two types: *transmembrane proteins*, which are only partly translocated across the ER membrane and become embedded in it, and *water-soluble proteins*, which are fully translocated across the ER membrane and are released into the ER lumen. Some of the transmembrane proteins function in the ER, but many are destined to reside in the plasma membrane or the membrane of another organelle. The water-soluble proteins are destined either for secretion or for residence in the lumen of the ER or of another organelle. All of these proteins, regardless of their subsequent fate, are directed to the ER membrane by an **ER signal sequence**, which initiates their translocation by a common mechanism.

Signal sequences (and the signal sequence strategy of protein sorting) were first discovered in the early 1970s in secreted proteins that are translocated across the ER membrane as a first step toward their eventual discharge from the cell. In the key experiment, the mRNA encoding a secreted protein was translated by ribosomes *in vitro*. When microsomes were omitted from this cell-free system, the protein synthesized was slightly larger than the normal secreted protein. In the presence of microsomes derived from the rough ER, however, a protein of the correct size was produced. According to the *signal hypothesis*, the size difference reflects the initial presence of a signal sequence that directs the secreted protein

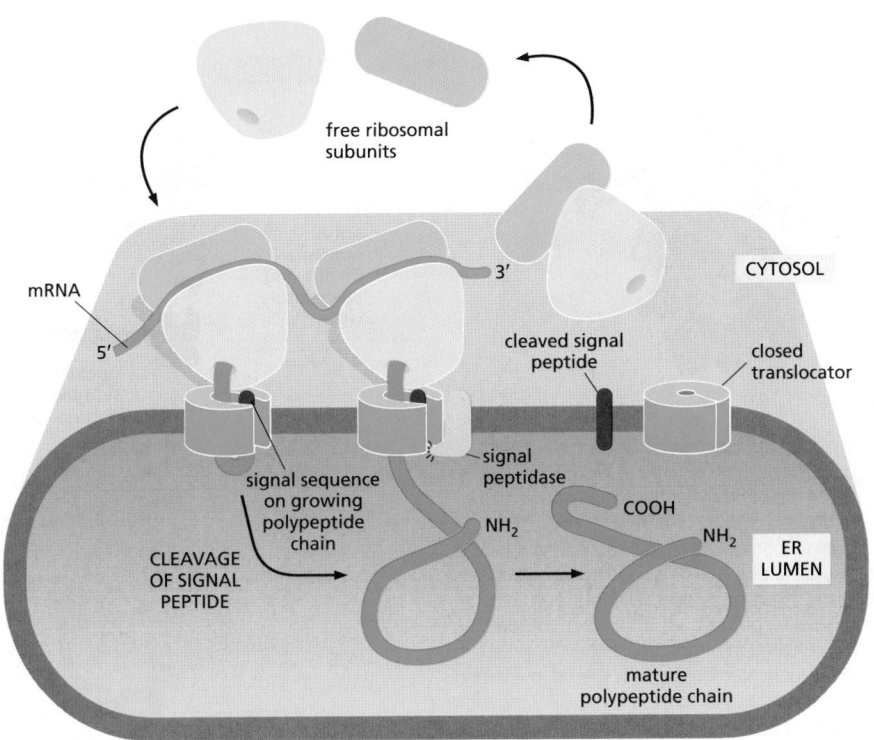

Figure 12–35 The signal hypothesis.
A simplified view of protein translocation across the ER membrane, as originally proposed. When the ER signal sequence emerges from the ribosome, it directs the ribosome to a translocator on the ER membrane that forms a pore in the membrane through which the polypeptide is translocated. A signal peptidase is closely associated with the translocator and clips off the signal sequence during translation, and the mature protein is released into the lumen of the ER immediately after its synthesis is completed. The translocator is closed until the ribosome has bound, so that the permeability barrier of the ER membrane is maintained at all times.

to the ER membrane and is then cleaved off by a *signal peptidase* in the ER membrane before the polypeptide chain has been completed (**Figure 12–35**). Cell-free systems in which proteins are imported into microsomes have provided powerful procedures for identifying, purifying, and studying the various components of the molecular machinery responsible for the ER import process.

A Signal-Recognition Particle (SRP) Directs the ER Signal Sequence to a Specific Receptor in the Rough ER Membrane

The ER signal sequence is guided to the ER membrane by at least two components: a **signal-recognition particle** (**SRP**), which cycles between the ER membrane and the cytosol and binds to the signal sequence, and an **SRP receptor** in the ER membrane. The SRP is a large complex; in animal cells, it consists of six different polypeptide chains bound to a single small RNA molecule. While the SRP and SRP receptor have fewer subunits in bacteria, homologs are present in all cells, indicating that this protein-targeting mechanism arose early in evolution and has been conserved.

ER signal sequences vary greatly in amino acid sequence, but each has eight or more nonpolar amino acids at its center (see Table 12–3, p. 648). How can the SRP bind specifically to so many different sequences? The answer has come from the crystal structure of the SRP protein, which shows that the signal-sequence-binding site is a large hydrophobic pocket lined by methionines. Because methionines have unbranched, flexible side chains, the pocket is sufficiently plastic to accommodate hydrophobic signal sequences of different sequences, sizes, and shapes.

The SRP is a rodlike structure, which wraps around the large ribosomal subunit, with one end binding to the ER signal sequence as it emerges from the ribosome as part of the newly made polypeptide chain; the other end blocks the elongation factor binding site at the interface between the large and small ribosomal subunits (**Figure 12–36**). This block halts protein synthesis as soon as the signal peptide has emerged from the ribosome. The transient pause presumably gives the ribosome enough time to bind to the ER membrane before completion of the polypeptide chain, thereby ensuring that the protein is not released into the cytosol. This safety

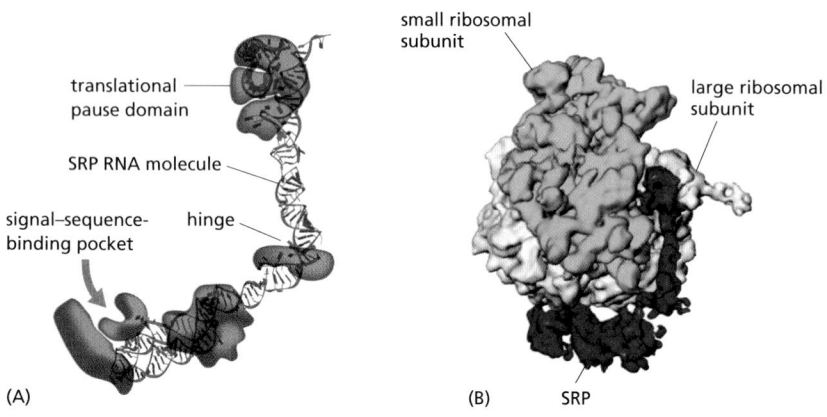

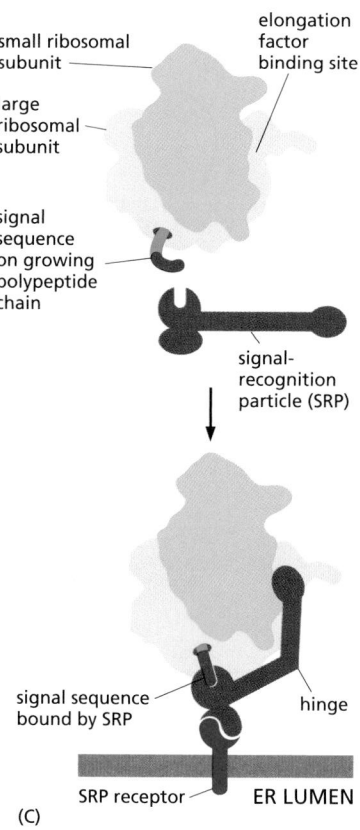

Figure 12–36 The signal-recognition particle (SRP). (A) A mammalian SRP is a rodlike ribonucleoprotein complex containing six protein subunits *(brown)* and one RNA molecule *(blue)*. The SRP RNA forms a backbone that links the protein domain containing the signal-sequence-binding pocket to the domain responsible for pausing translation. Crystal structures of various SRP pieces from different species are assembled here into a composite model to approximate the structure of a complete SRP. (B) The three-dimensional outline of the SRP bound to a ribosome was determined by cryoelectron microscopy. SRP binds to the large ribosomal subunit so that its signal-sequence-binding pocket is positioned near the growing polypeptide chain exit site, and its translational pause domain is positioned at the interface between the ribosomal subunits, where it interferes with elongation factor binding. (C) As a signal sequence emerges from the ribosome and binds to the SRP, a conformational change in the SRP exposes a binding site for the SRP receptor. (B, adapted from M. Halic et al., *Nature* 427:808–814, 2004.)

device may be especially important for secreted and lysosomal hydrolases, which could wreak havoc in the cytosol; cells that secrete large amounts of hydrolases, however, take the added precaution of having high concentrations of hydrolase inhibitors in their cytosol. The pause also ensures that large portions of a protein that could fold into a compact structure are not made before reaching the translocator in the ER membrane. Thus, in contrast to the post-translational import of proteins into mitochondria and chloroplasts, chaperone proteins are not required to keep the protein unfolded.

When a signal sequence binds, SRP exposes a binding site for the SRP receptor (see Figure 12–36B,C), which is a transmembrane protein complex in the rough ER membrane. The binding of the SRP to its receptor brings the SRP–ribosome complex to an unoccupied protein translocator in the same membrane. The SRP and SRP receptor are then released, and the translocator transfers the growing polypeptide chain across the membrane (Figure 12–37).

This co-translational transfer process creates two spatially separate populations of ribosomes in the cytosol. **Membrane-bound ribosomes,** attached to the

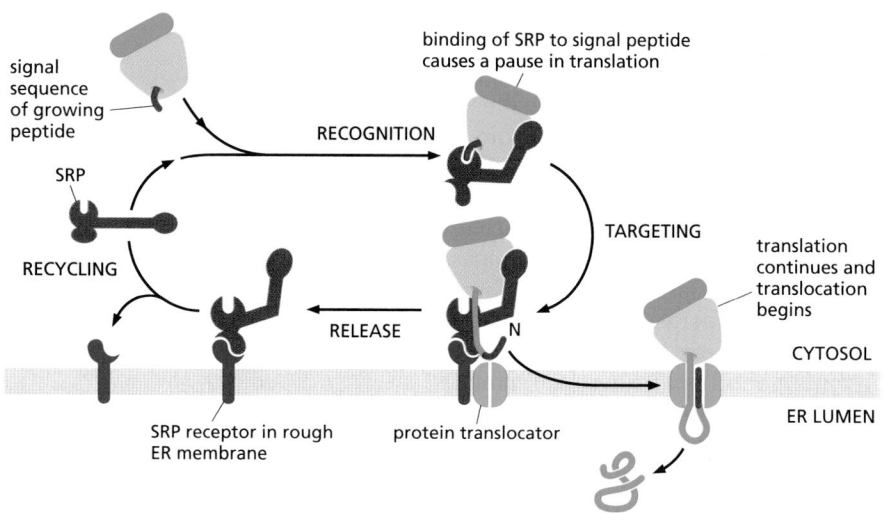

Figure 12–37 How ER signal sequences and SRP direct ribosomes to the ER membrane. The SRP and its receptor act in concert. The SRP binds to both the exposed ER signal sequence and the ribosome, thereby inducing a pause in translation. The SRP receptor in the ER membrane, which in animal cells is composed of two different polypeptide chains, binds the SRP–ribosome complex and directs it to the translocator. In a poorly understood reaction, the SRP and SRP receptor are then released, leaving the ribosome bound to the translocator in the ER membrane. The translocator then inserts the polypeptide chain into the membrane and transfers it across the lipid bilayer. Because one of the SRP proteins and both chains of the SRP receptor contain GTP-binding domains, it is thought that conformational changes that occur during cycles of GTP binding and hydrolysis (discussed in Chapter 15) ensure that SRP release occurs only after the ribosome has become properly engaged with the translocator in the ER membrane. The translocator is closed until the ribosome has bound, so that the permeability barrier of the ER membrane is maintained at all times.

cytosolic side of the ER membrane, are engaged in the synthesis of proteins that are being concurrently translocated into the ER. **Free ribosomes**, unattached to any membrane, synthesize all other proteins encoded by the nuclear genome. Membrane-bound and free ribosomes are structurally and functionally identical. They differ only in the proteins they are making at any given time.

Since many ribosomes can bind to a single mRNA molecule, a **polyribosome** is usually formed. If the mRNA encodes a protein with an ER signal sequence, the polyribosome becomes attached to the ER membrane, directed there by the signal sequences on multiple growing polypeptide chains. The individual ribosomes associated with such an mRNA molecule can return to the cytosol when they finish translation and intermix with the pool of free ribosomes. The mRNA itself, however, remains attached to the ER membrane by a changing population of ribosomes, each transiently held at the membrane by the translocator (**Figure 12–38**).

The Polypeptide Chain Passes Through an Aqueous Channel in the Translocator

It had long been debated whether polypeptide chains are transferred across the ER membrane in direct contact with the lipid bilayer or through a channel in a protein translocator. The debate ended with the identification of the translocator, which was shown to form a water-filled channel in the membrane through

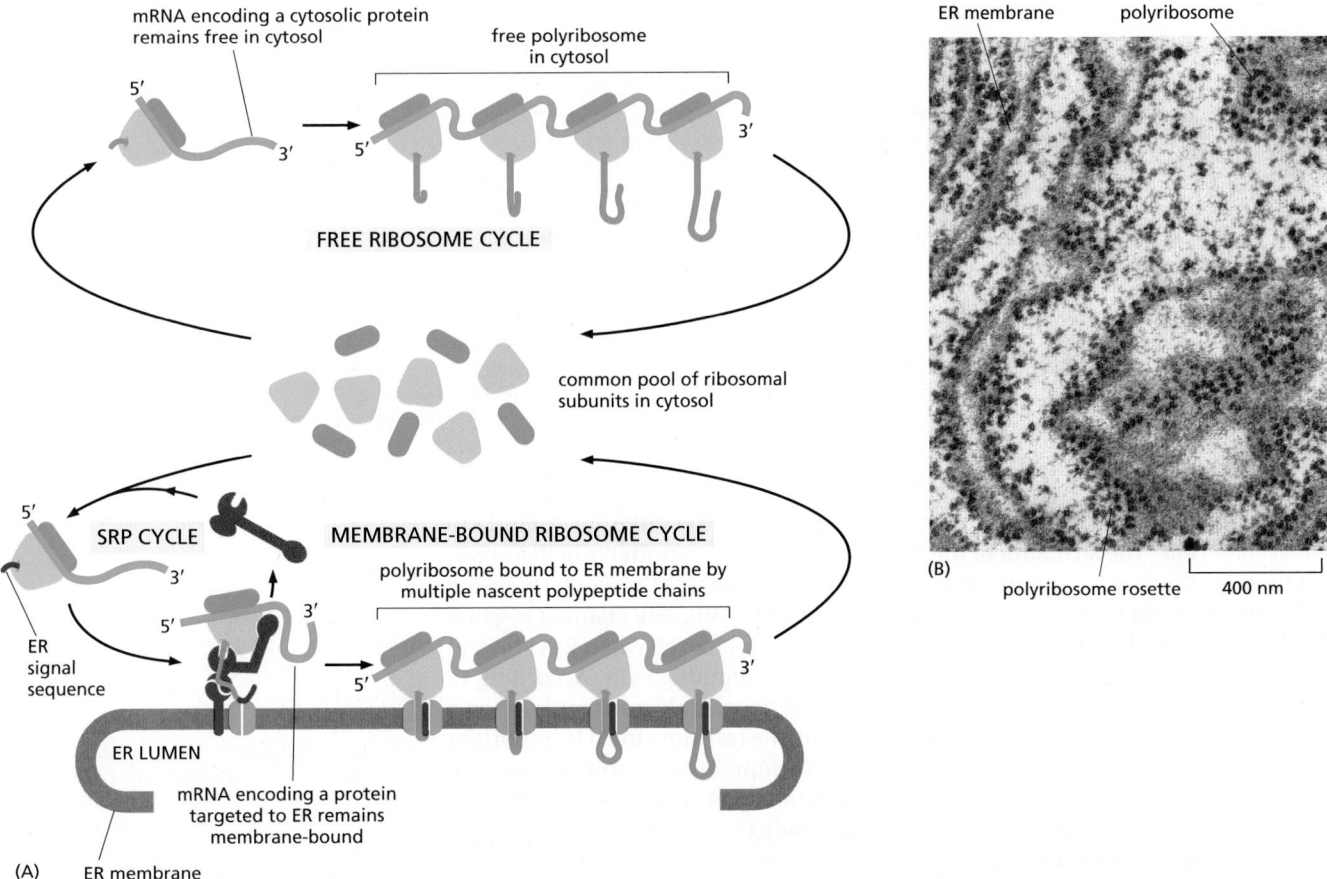

Figure 12–38 Free and membrane-bound polyribosomes. (A) A common pool of ribosomes synthesizes the proteins that stay in the cytosol and those that are transported into the ER. The ER signal sequence on a newly formed polypeptide chain binds to SRP, which directs the translating ribosome to the ER membrane. The mRNA molecule remains permanently bound to the ER as part of a polyribosome, while the ribosomes that move along it are recycled; at the end of each round of protein synthesis, the ribosomal subunits are released and rejoin the common pool in the cytosol. (B) A thin section electron micrograph of polyribosomes attached to the ER membrane. The plane of section in some places cuts through the ER roughly parallel to the membrane, giving a face-on view of the rosettelike pattern of the polyribosomes. (B, courtesy of George Palade.)

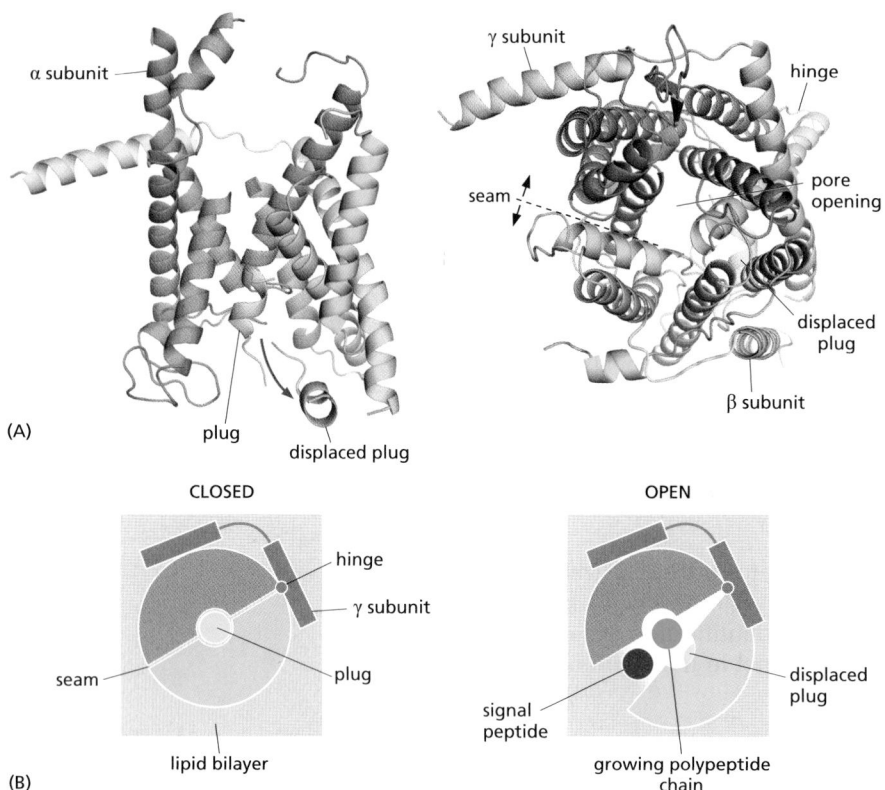

(A)

(B)

Figure 12–39 Structure of the Sec61 complex. (A) A side view (*left*) and a top view (*right*, seen from the cytosol) of the structure of the Sec61 complex of the archaeon *Methanococcus jannaschii*. The Sec61α subunit has an inverted repeat structure (see Figure 11–10) and is shown in *blue* and *beige* to indicate this pseudo-symmetry; the two smaller β and γ subunits are shown in *gray*. In the side view, some helices in front have been omitted to make the inside of the pore visible. The *yellow* short helix is thought to form a plug that seals the pore when the translocator is closed. To open, the complex rearranges itself to move the plug helix out of the way, as indicated by the *red* arrow. A ring of hydrophobic amino acid side chains is thought to form a tight-fitting diaphragm around translocating polypeptide chain to prevent leaks of other molecules across the membrane. The pore of the Sec61 complex can also open sideways at a lateral seam. (B) Models of the closed and open states of the translocator are shown in top view, illustrating how a signal sequence (or a transmembrane segment) could be released into the lipid bilayer after opening of the seam. (PDB codes: 1RH5 and 1RHZ.)

which the polypeptide chain passes. The core of the translocator, called the **Sec61 complex**, is built from three subunits that are highly conserved from bacteria to eukaryotic cells. The structure of the Sec61 complex suggests that α helices contributed by the largest subunit surround a central channel through which the polypeptide chain traverses the membrane (**Figure 12–39**). The channel is gated by a short α helix that is thought to keep the translocator closed when it is idle and to move aside when it is engaged in passing a polypeptide chain. According to this view, the pore is a dynamic gated channel that opens only transiently when a polypeptide chain traverses the membrane. In an idle translocator, it is important to keep the channel closed, so that the membrane remains impermeable to ions, such as Ca^{2+}, which otherwise would leak out of the ER. As a polypeptide chain is translocating, a ring of hydrophobic amino acid side chains is thought to provide a flexible seal to prevent ion leaks.

The structure of the Sec61 complex suggests that the pore can also open along a seam on its side. Indeed, some structures of the translocator show it locked in an open-seam conformation. This opening allows a translocating peptide chain lateral access into the hydrophobic core of the membrane, a process that is important both for the release of a cleaved signal peptide into the membrane (see Figure 12–35) and for the integration of transmembrane proteins into the bilayer, as we discuss later.

Figure 12–40 A ribosome *(green)* bound to the ER protein translocator *(blue).* (A) A side-view reconstruction of the complex from electron microscopic images. (B) A view of the translocator seen from the ER lumen. The translocator contains Sec61, accessory proteins, and detergent used in the preparation. Domains of accessory proteins extend across the membrane and form the lumenal bulge. (C) A schematic drawing of a membrane-bound ribosome attached to the translocator, indicating the location of the tunnel in the large ribosomal subunit through which the growing polypeptide chain exits from the ribosome. The mRNA (not shown) would be located between the small and large ribosomal subunits. (Adapted from J.F. Ménétret et al., *J. Mol. Biol.* 348:445–457, 2005.)

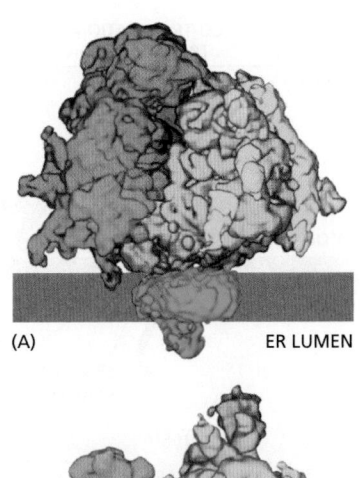

(A)　　　　　　　　　　　ER LUMEN

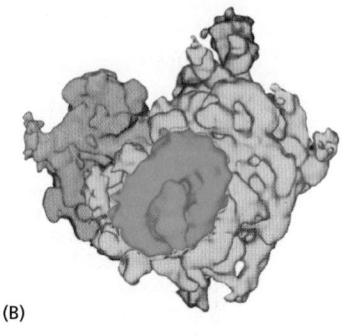

(B)

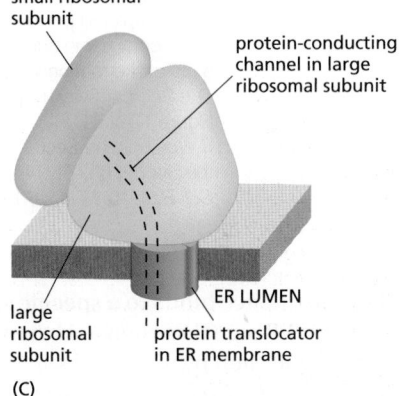

small ribosomal subunit

protein-conducting channel in large ribosomal subunit

ER LUMEN

large ribosomal subunit

protein translocator in ER membrane

(C)

In eukaryotic cells, four Sec61 complexes form a large translocator assembly that can be visualized on ER-bound ribosomes after detergent solubilization of the ER membrane (**Figure 12–40**). It is likely that this assembly includes other membrane complexes that associate with the translocator, such as enzymes that modify the growing polypeptide chain, including oligosaccharide transferase and the signal peptidase. The assembly of a translocator with these accessory components is called the **translocon**.

Translocation Across the ER Membrane Does Not Always Require Ongoing Polypeptide Chain Elongation

As we have seen, translocation of proteins into mitochondria, chloroplasts, and peroxisomes occurs post-translationally, after the protein has been made and released into the cytosol, whereas translocation across the ER membrane usually occurs during translation (co-translationally). This explains why ribosomes are bound to the ER but not to other organelles.

Some completely synthesized proteins, however, are imported into the ER, demonstrating that translocation does not always require ongoing translation. Post-translational protein translocation is especially common across the yeast ER membrane and the bacterial plasma membrane (which is thought to be evolutionarily related to the ER). To function in post-translational translocation, the ER translocator needs accessory proteins that feed the polypeptide chain into the pore and drive translocation (**Figure 12–41**). In bacteria, a translocation motor protein, the *SecA ATPase*, attaches to the cytosolic side of the translocator, where it undergoes cyclic conformational changes driven by ATP hydrolysis. Each time an ATP is hydrolyzed, a portion of the SecA protein inserts into the pore of the translocator, pushing a short segment of the passenger protein with it. As a result of this ratchet mechanism, the SecA ATPase progressively pushes the polypeptide chain of the transported protein across the membrane.

Eukaryotic cells use a different set of accessory proteins that associate with the Sec61 complex. These proteins span the ER membrane and use a small domain on the lumenal side of the ER membrane to deposit an hsp70-like chaperone protein (called *BiP*, for *b*inding *p*rotein) onto the polypeptide chain as it emerges from the pore into the ER lumen. ATP-dependent cycles of BiP binding and release drive unidirectional translocation, as described earlier for the mitochondrial hsp70 proteins that pull proteins across mitochondrial membranes.

Proteins that are transported into the ER by a post-translational mechanism are first released into the cytosol, where they bind to chaperone proteins to prevent folding, as discussed earlier for proteins destined for mitochondria and chloroplasts.

In Single-Pass Transmembrane Proteins, a Single Internal ER Signal Sequence Remains in the Lipid Bilayer as a Membrane-spanning α Helix

The ER signal sequence in the growing polypeptide chain is thought to trigger the opening of the pore in the Sec61 protein translocator: after the signal sequence is released from the SRP and the growing chain has reached a sufficient length, the

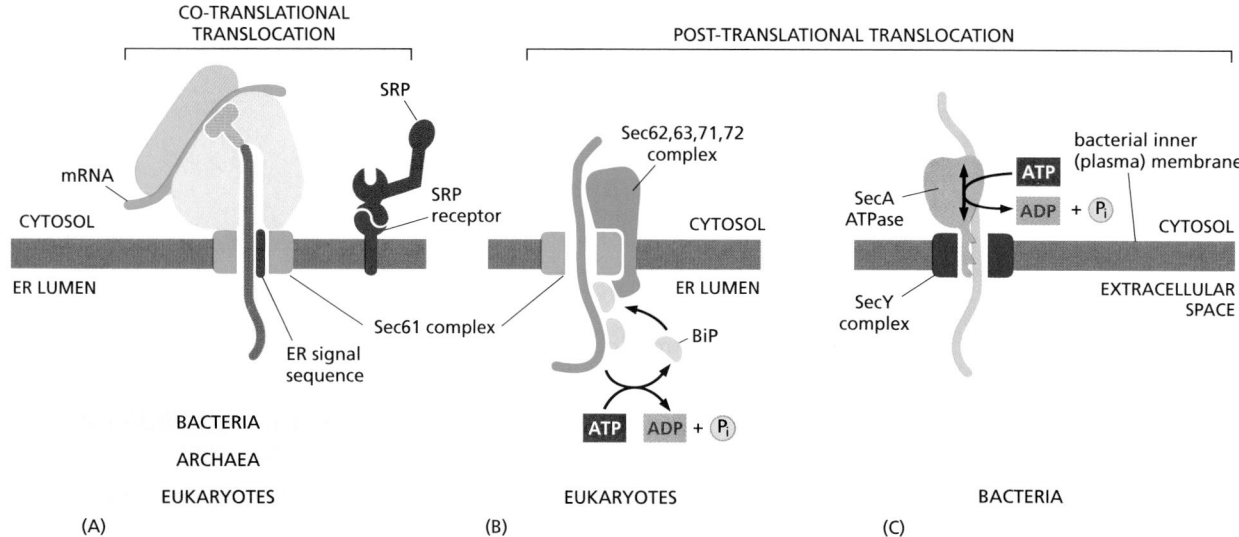

Figure 12–41 Three ways in which protein translocation can be driven through structurally similar translocators.
(A) Co-translational translocation. The ribosome is brought to the membrane by the SRP and SRP receptor and then engages with the Sec61 protein translocator. The growing polypeptide chain is threaded across the membrane as it is made. No additional energy is needed, as the only path available to the growing chain is to cross the membrane. (B) Post-translational translocation in eukaryotic cells requires an additional complex composed of Sec62, Sec63, Sec71, and Sec72 proteins, which is attached to the Sec61 translocator and deposits BiP molecules onto the translocating chain as it emerges from the translocator in the lumen of the ER. ATP-driven cycles of BiP binding and release pull the protein into the lumen, a mechanism that closely resembles the mechanism of mitochondrial import in Figure 12–23. (C) Post-translational translocation in bacteria. The completed polypeptide chain is fed from the cytosolic side into the bacterial homolog of the Sec61 complex (called the SecY complex in bacteria) in the plasma membrane by the SecA ATPase. ATP hydrolysis-driven conformational changes drive a pistonlike motion in SecA, each cycle pushing about 20 amino acids of the protein chain through the pore of the translocator. The Sec pathway used for protein translocation across the thylakoid membrane in chloroplasts uses a similar mechanism (see Figure 12–26B).
 Whereas the Sec61 translocator, SRP, and SRP receptor are found in all organisms, SecA is found exclusively in bacteria, and the Sec62, 63, 71, 72 complex is found exclusively in eukaryotic cells. (Adapted from P. Walter and A.E. Johnson, *Annu. Rev. Cell Biol.* 10:87–119, 1994.)

signal sequence binds to a specific site inside the pore itself, thereby opening the pore. An ER signal sequence is therefore recognized twice: first by an SRP in the cytosol and then by a binding site in the pore of the protein translocator, where it serves as a **start-transfer signal** (or start-transfer peptide) that opens the pore (for example, see Figure 12–35 for how this works for a soluble protein). Dual recognition may help ensure that only appropriate proteins enter the lumen of the ER.

While bound in the translocation pore, a signal sequence is in contact not only with the Sec61 complex, which forms the walls of the pore, but also, along the lateral seam, with the hydrophobic core of the lipid bilayer. This was shown in chemical cross-linking experiments in which the signal sequence and the hydrocarbon chains of lipids were covalently linked together. When the nascent polypeptide chain grew long enough, the ER signal peptidase cleaved off the signal sequence and released it from the pore into the membrane, where it was rapidly degraded to amino acids by other proteases in the ER membrane. To release the signal sequence into the membrane, the translocator opens laterally along the seam (see Figures 12–35 and 12–39). The translocator is therefore gated in two directions: it opens to form a pore across the membrane to let the hydrophilic portions of proteins cross the lipid bilayer, and it opens laterally within the membrane to let hydrophobic portions of proteins partition into the lipid bilayer. Lateral gating of the pore is an essential step during the integration of transmembrane proteins.

The integration of membrane proteins requires that some parts of the polypeptide chain be translocated across the lipid bilayer whereas others are not. Despite this additional complexity, all modes of insertion of membrane proteins are simply variants of the sequence of events just described for transferring a soluble protein into the lumen of the ER. We begin by describing the three ways in

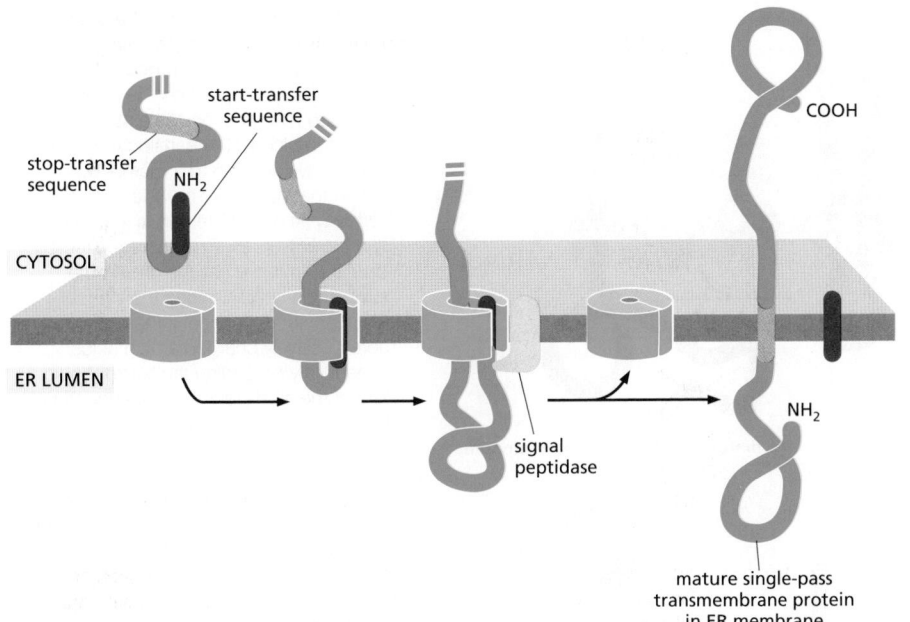

start-transfer sequence

stop-transfer sequence

NH₂

COOH

CYTOSOL

ER LUMEN

signal peptidase

NH₂

mature single-pass transmembrane protein in ER membrane

Figure 12–42 How a single-pass transmembrane protein with a cleaved ER signal sequence is integrated into the ER membrane. In this protein, the co-translational translocation process is initiated by an N-terminal ER signal sequence *(red)* that functions as a start-transfer signal, opening the translocator as in Figure 12–35. In addition to this start-transfer sequence, however, the protein also contains a stop-transfer sequence *(orange)*; when this sequence enters the translocator and interacts with a binding site within the pore, the translocator opens at the seam and discharges the protein laterally into the lipid bilayer, where the stop-transfer sequence remains to anchor the protein in the membrane. (In this figure and the two figures that follow, the ribosomes have been omitted for clarity.)

which **single-pass transmembrane proteins** (see Figure 10–17) become inserted into the ER membrane.

In the simplest case, an N-terminal signal sequence initiates translocation, just as for a soluble protein, but an additional hydrophobic segment in the polypeptide chain stops the transfer process before the entire polypeptide chain is translocated. This **stop-transfer signal** anchors the protein in the membrane after the ER signal sequence (the start-transfer signal) has been cleaved off and released from the translocator (**Figure 12–42**). The lateral gating mechanism transfers the stop-transfer sequence into the bilayer, where it remains as a single α-helical membrane-spanning segment, with the N-terminus of the protein on the lumenal side of the membrane and the C-terminus on the cytosolic side.

In the other two cases, the signal sequence is internal, rather than at the N-terminal end of the protein. As for an N-terminal ER signal sequence, the SRP binds to an internal signal sequence by recognizing its hydrophobic α-helical features. The SRP brings the ribosome making the protein to the ER membrane, and the ER signal sequence then serves as a start-transfer signal that initiates the protein's translocation. After release from the translocator, the internal start-transfer sequence remains in the lipid bilayer as a single membrane-spanning α helix.

Internal start-transfer sequences can bind to the translocation apparatus in either of two orientations; this in turn determines which protein segment (the one preceding or the one following the start-transfer sequence) is moved across the membrane into the ER lumen. In one case, the resulting membrane protein has its C-terminus on the lumenal side (pathway A in **Figure 12–43**), while in the other, it has its N-terminus on the lumenal side (pathway B in Figure 12–43). The orientation of the start-transfer sequence depends on the distribution of nearby charged amino acids, as described in the figure legend.

Combinations of Start-Transfer and Stop-Transfer Signals Determine the Topology of Multipass Transmembrane Proteins

In **multipass transmembrane proteins**, the polypeptide chain passes back and forth repeatedly across the lipid bilayer as hydrophobic α helices (see Figure 10–17). It is thought that an internal signal sequence serves as a start-transfer signal in these proteins to initiate translocation, which continues until the translocator encounters a stop-transfer sequence; in double-pass transmembrane proteins, for example, the polypeptide can then be released into the bilayer

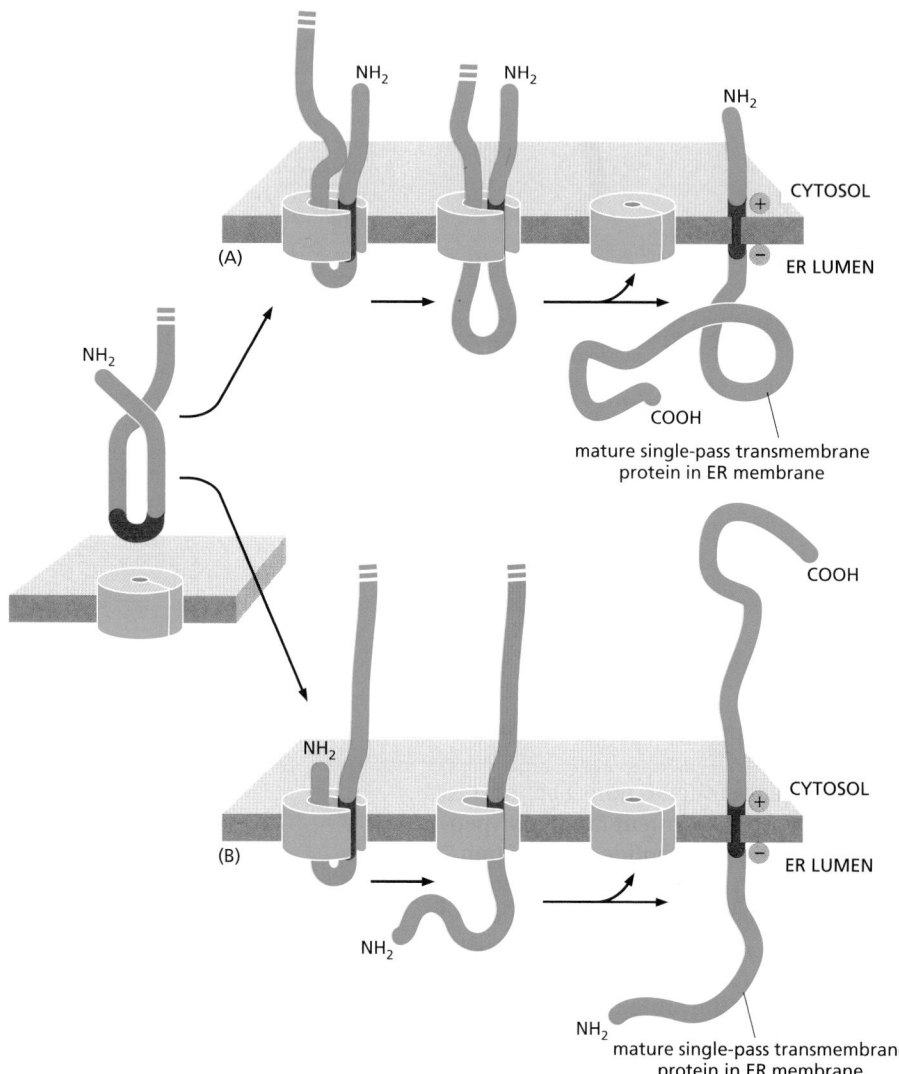

mature single-pass transmembrane
protein in ER membrane

mature single-pass transmembrane
protein in ER membrane

Figure 12–43 Integration of a single-pass transmembrane protein with an internal signal sequence into the ER membrane. An internal ER signal sequence that functions as a start-transfer signal can bind to the translocator in one of two ways, leading to a membrane protein that has either its C-terminus (pathway A) or its N-terminus (pathway B) in the ER lumen. Proteins are directed into either pathway by features in the polypeptide chain flanking the internal start-transfer sequence: if there are more positively charged amino acids immediately *preceding* the hydrophobic core of the start-transfer sequence than there are following it, the membrane protein is inserted into the translocator in the orientation shown in pathway A, whereas if there are more positively charged amino acids immediately *following* the hydrophobic core of the start-transfer sequence than there are preceding it, the membrane protein is inserted into the translocator in the orientation shown in pathway B. Because translocation cannot start before a start-transfer sequence appears outside the ribosome, translocation of the N-terminal portion of the protein shown in (B) can occur only after this portion has been fully synthesized.

Note that there are two ways to insert a single-pass membrane-spanning protein whose N-terminus is located in the ER lumen: that shown in Figure 12–42 and that shown here in (B).

(**Figure 12–44**). In more complex multipass proteins, in which many hydrophobic α helices span the bilayer, a second start-transfer sequence reinitiates translocation further down the polypeptide chain until the next stop-transfer sequence causes polypeptide release, and so on for subsequent start-transfer and stop-transfer sequences (**Figure 12–45** and **Movie 12.5**).

Hydrophobic start-transfer and stop-transfer signal sequences both act to fix the topology of the protein in the membrane by locking themselves into the membrane as membrane-spanning α helices; and they can do this in either orientation. Whether a given hydrophobic signal sequence functions as a start-transfer or stop-transfer sequence must depend on its location in a polypeptide chain, since its function can be switched by changing its location in the protein by using recombinant DNA techniques. Thus, the distinction between start-transfer and stop-transfer sequences results mostly from their relative order in the growing polypeptide chain. It seems that the SRP begins scanning an unfolded polypeptide chain for hydrophobic segments at its N-terminus and proceeds toward the C-terminus, in the direction that the protein is synthesized. By recognizing the first appropriate hydrophobic segment to emerge from the ribosome, the SRP sets the "reading frame" for membrane integration: after the SRP initiates translocation, the translocator recognizes the next appropriate hydrophobic segment in the direction of transfer as a stop-transfer sequence, causing the region of the polypeptide chain in between to be threaded across the membrane. A similar

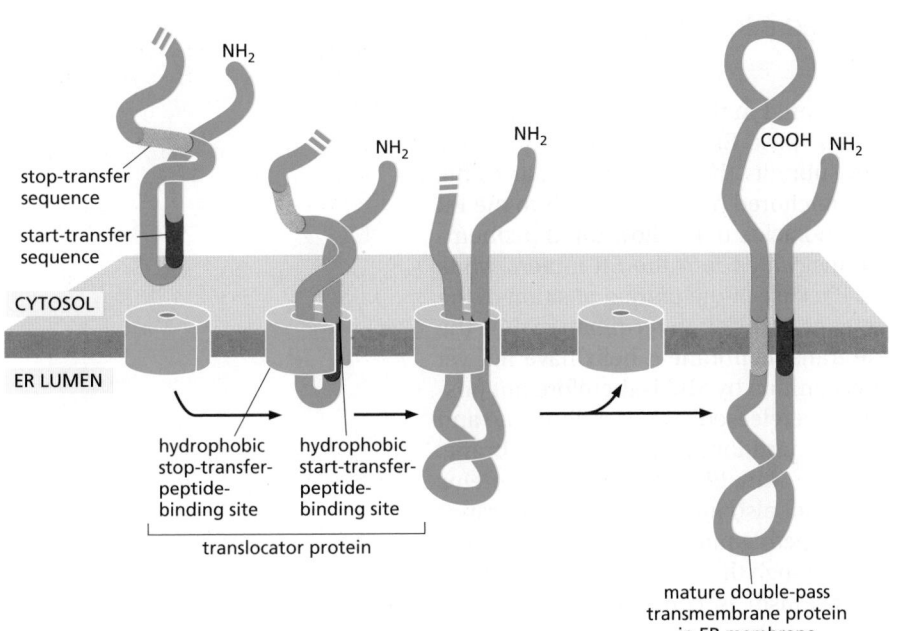

Figure 12–44 Integration of a double-pass transmembrane protein with an internal signal sequence into the ER membrane. In this protein, an internal ER signal sequence acts as a start-transfer signal (as in Figure 12–43) and initiates the transfer of the C-terminal part of the protein. At some point after a stop-transfer sequence has entered the translocator, the translocator discharges the sequence laterally into the membrane.

scanning process continues until all of the hydrophobic regions in the protein have been inserted into the membrane as transmembrane α helices.

Because membrane proteins are always inserted from the cytosolic side of the ER in this programmed manner, all copies of the same polypeptide chain will have the same orientation in the lipid bilayer. This generates an asymmetrical ER membrane in which the protein domains exposed on one side are different from those exposed on the other side. This asymmetry is maintained during the many membrane budding and fusion events that transport the proteins made in the ER to other cell membranes (discussed in Chapter 13). Thus, the way in which a newly synthesized protein is inserted into the ER membrane determines the orientation of the protein in all of the other membranes as well.

When proteins are extracted with detergent from a membrane and then reconstituted into artificial lipid vesicles, a random mixture of right-side-out and inside-out protein orientations usually results. Thus, the protein asymmetry observed in cell membranes seems not to be an inherent property of the proteins, but instead results solely from the process by which proteins are inserted into the ER membrane from the cytosol.

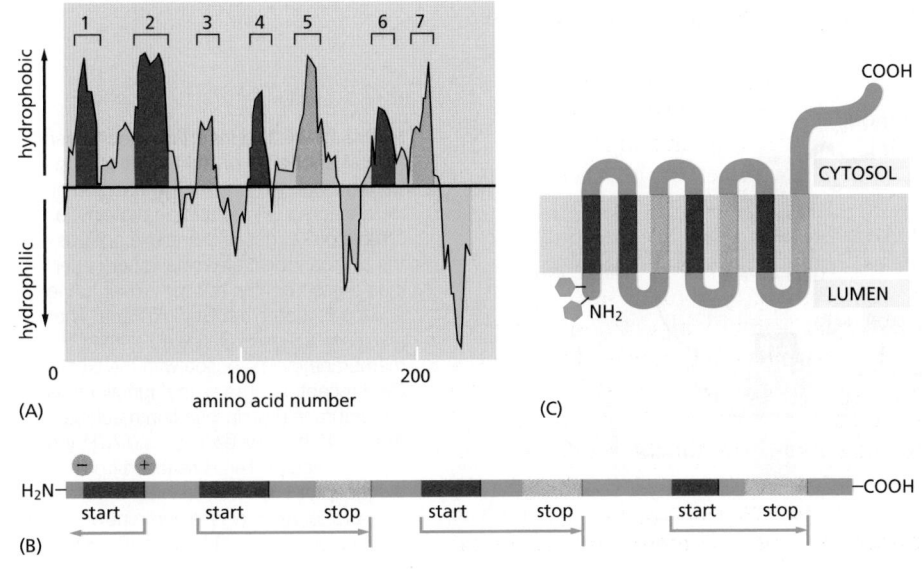

Figure 12–45 The insertion of the multipass membrane protein rhodopsin into the ER membrane. Rhodopsin is the light-sensitive protein in rod photoreceptor cells in the mammalian retina (discussed in Chapter 15). (A) A hydropathy plot (see Figure 10–20) identifies seven short hydrophobic regions in rhodopsin. (B) The hydrophobic region nearest the N-terminus serves as a start-transfer sequence that causes the preceding N-terminal portion of the protein to pass across the ER membrane. Subsequent hydrophobic sequences function in alternation as start-transfer and stop-transfer sequences. The *green arrows* indicate the paired start and stop signals inserted into the translocator. (C) The final integrated rhodopsin has its N-terminus located in the ER lumen and its C-terminus located in the cytosol. The *blue hexagons* represent covalently attached oligosaccharides.

ER Tail-anchored Proteins Are Integrated into the ER Membrane by a Special Mechanism

Many important membrane proteins are anchored in the membrane by a C-terminal transmembrane, hydrophobic α helix. These **ER tail-anchored proteins** include a large number of SNARE protein subunits that guide vesicular traffic (discussed in Chapter 13). When such a tail-anchored protein inserts into the ER membrane from the cytosol, only a few amino acids that follow the transmembrane α helix on its C-terminal side are translocated into the ER lumen, while most of the protein remains in the cytosol. Because of the unique position of the transmembrane α helix in the protein sequence, translation terminates while the C-terminal amino acids that will form the transmembrane α helix have not yet emerged from the ribosome exit tunnel. Recognition by SRP is therefore not possible. It was long thought that these proteins are released from the ribosome and the hydrophobic C-terminal tail spontaneously partitions into the ER membrane. Such a mechanism could not explain, however, why ER tail-anchored proteins insert into the ER membrane selectively and not also into all other membranes in the cell. It is now clear that a specialized targeting machinery is involved that is fueled by ATP hydrolysis (**Figure 12–46**). Although the components and details differ, this post-translational targeting mechanism is conceptually similar to SRP-dependent protein targeting (see Figure 12–37).

Not all tail-anchored proteins are inserted into the ER, however. Some proteins contain a C-terminal membrane anchor that contains additional sorting information that directs the protein to mitochondria or peroxisomes. How these proteins are sorted there remains unknown.

Translocated Polypeptide Chains Fold and Assemble in the Lumen of the Rough ER

Many of the proteins in the lumen of the ER are in transit, *en route* to other destinations; others, however, normally reside there and are present at high concentrations. These **ER resident proteins** contain an **ER retention signal** of four amino acids at their C-terminus that is responsible for retaining the protein in the ER (see Table 12–3. p. 648; discussed in Chapter 13). Some of these proteins function as catalysts that help the many proteins that are translocated into the ER lumen to fold and assemble correctly.

One important ER resident protein is *protein disulfide isomerase (PDI)*, which catalyzes the oxidation of free sulfhydryl (SH) groups on cysteines to form disulfide (S–S) bonds. Almost all cysteines in protein domains exposed to either the extracellular space or the lumen of organelles in the secretory and endocytic pathways are disulfide-bonded. By contrast, disulfide bonds form only very rarely in domains exposed to the cytosol, because of the reducing environment there.

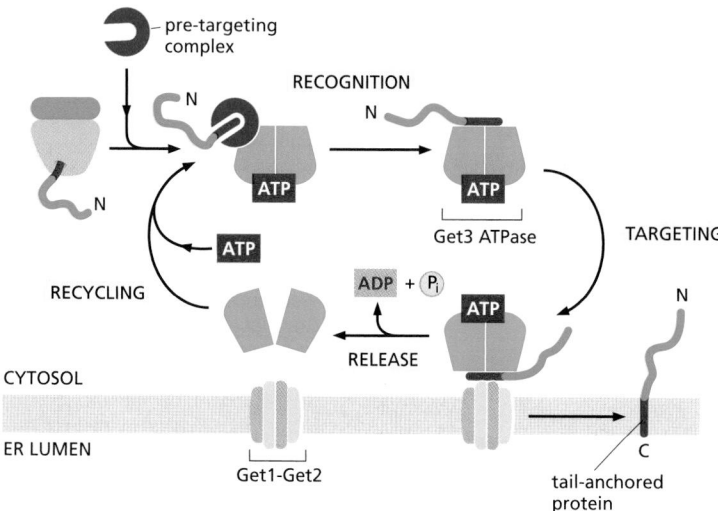

Figure 12–46 The insertion mechanism for tail-anchored proteins. In this post-translational pathway for the insertion of tail-anchored ER membrane proteins, a soluble pre-targeting complex captures the hydrophobic C-terminal α helix after it emerges from the ribosomal exit tunnel and loads it onto the Get3 ATPase. The resulting complex is targeted to the ER membrane by interaction with the Get1–Get2 receptor complex that functions as a membrane protein insertion machine. After Get3 hydrolyzes its bound ATP, the tail-anchored protein is released from the receptor and inserted into the ER membrane. ADP release and renewed ATP binding recycles Get3 back to the cytosol.

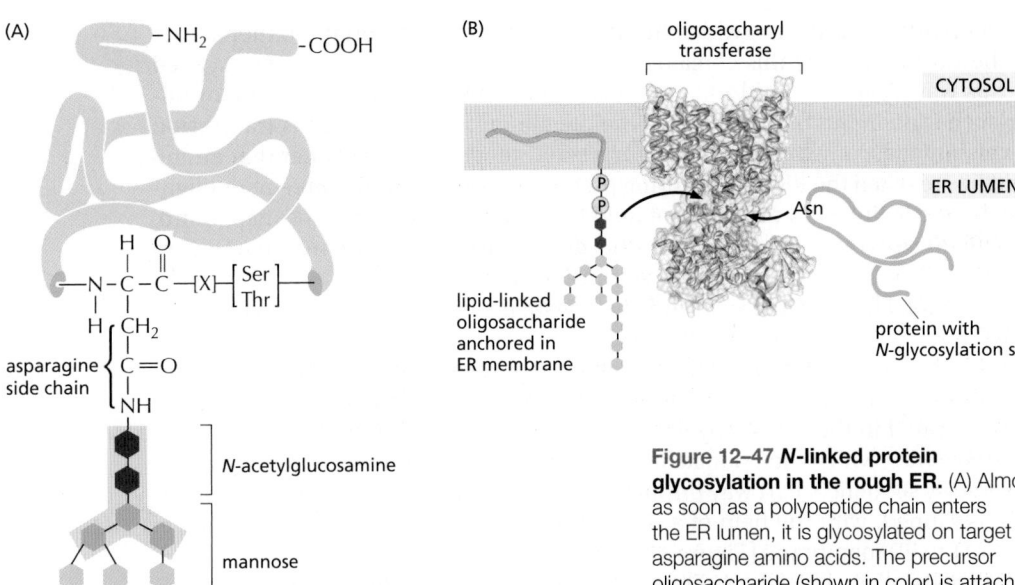

Figure 12–47 N-linked protein glycosylation in the rough ER. (A) Almost as soon as a polypeptide chain enters the ER lumen, it is glycosylated on target asparagine amino acids. The precursor oligosaccharide (shown in color) is attached only to asparagines in the sequences Asn-X-Ser and Asn-X-Thr (where X is any amino acid except proline). These sequences occur much less frequently in glycoproteins than in nonglycosylated cytosolic proteins. Evidently there has been selective pressure against these sequences during protein evolution, presumably because glycosylation at too many sites would interfere with protein folding. The five sugars in the *gray* box form the core region of this oligosaccharide. For many glycoproteins, only the core sugars survive the extensive oligosaccharide trimming that takes place in the Golgi apparatus. (B) The precursor oligosaccharide is transferred from a dolichol lipid anchor to the asparagine as an intact unit in a reaction catalyzed by a transmembrane *oligosaccharyl transferase* enzyme. One copy of this enzyme is associated with each protein translocator in the ER membrane. (The translocator is not shown.) Oligosaccharyl transferase contains 13 transmembrane α helices and a large ER lumenal domain that contains its substrate-binding sites. The asparagine binds a tunnel that penetrates the enzyme interior. There, the amino group of the asparagine is twisted out of the plane that stabilizes the otherwise poorly reactive amide bond, activating it for reaction with the dolichol–oligosaccharide. The structure shown is of a prokaryotic homolog that closely resembles the catalytic subunit of the eukaryotic oligosaccharyl transferase. (PDB code: 3RCE.)

Another ER resident protein is the chaperone protein **BiP**. We have already discussed how BiP pulls proteins post-translationally into the ER through the Sec61 ER translocator. Like other chaperones (discussed in Chapter 13), BiP recognizes incorrectly folded proteins, as well as protein subunits that have not yet assembled into their final oligomeric complexes. It does so by binding to exposed amino acid sequences that would normally be buried in the interior of correctly folded or assembled polypeptide chains. An example of a BiP-binding site is a stretch of alternating hydrophobic and hydrophilic amino acids that would normally be buried in a β sheet with its hydrophobic side oriented towards the hydrophobic core of the folded protein. The bound BiP both prevents the protein from aggregating and helps keep it in the ER (and thus out of the Golgi apparatus and later parts of the secretory pathway). Like some other members of the hsp70 family of chaperone proteins, which bind unfolded proteins and facilitate their import into mitochondria and chloroplasts, BiP hydrolyzes ATP to shuttle between high- and low-affinity binding states, which allow it to hold on to and let go of its substrate proteins in a dynamic cycle.

Most Proteins Synthesized in the Rough ER Are Glycosylated by the Addition of a Common N-Linked Oligosaccharide

The covalent addition of oligosaccharides to proteins is one of the major biosynthetic functions of the ER. About half of the soluble and membrane-bound proteins that are processed in the ER—including those destined for transport to the Golgi apparatus, lysosomes, plasma membrane, or extracellular space—are **glycoproteins** that are modified in this way. Many proteins in the cytosol and nucleus are also glycosylated, but not with oligosaccharides: they carry a much simpler sugar modification, in which a single *N*-acetylglucosamine group is added to a serine or threonine of the protein.

During the most common form of **protein glycosylation** in the ER, a preformed *precursor oligosaccharide* (composed of *N*-acetylglucosamine, mannose, and glucose, and containing a total of 14 sugars) is transferred *en bloc* to proteins. Because this oligosaccharide is transferred to the side-chain NH$_2$ group of an asparagine in the protein, it is said to be *N-linked* or *asparagine-linked* (**Figure 12-47**A). The transfer is catalyzed by a membrane-bound enzyme complex, an *oligosaccharyl transferase*, which has its active site exposed on the lumenal side of the ER membrane; this explains why cytosolic proteins are not glycosylated in this way. A special lipid molecule called **dolichol** anchors the precursor oligosaccharide in the ER membrane. The precursor oligosaccharide is transferred to the target asparagine in a single enzymatic step immediately after that amino acid has reached the ER lumen during protein translocation. The precursor oligosaccharide is linked to the dolichol lipid by a high-energy pyrophosphate bond, which provides the activation energy that drives the glycosylation reaction (Figure 12–47B). One copy of oligosaccharyl transferase is associated with each protein translocator, allowing it to scan and glycosylate the incoming polypeptide chains efficiently.

The precursor oligosaccharide is built up sugar by sugar on the membrane-bound dolichol lipid and is then transferred to a protein. The sugars are first activated in the cytosol by the formation of *nucleotide (UDP or GDP)-sugar intermediates*, which then donate their sugar (directly or indirectly) to the lipid in an orderly sequence. Part way through this process, the lipid-linked oligosaccharide is flipped, with the help of a transporter, from the cytosolic to the lumenal side of the ER membrane (**Figure 12–48**).

All of the diversity of the *N*-linked oligosaccharide structures on mature glycoproteins results from the later modification of the original precursor oligosaccharide. While still in the ER, three glucoses (see Figure 12–47) and one mannose are quickly removed from the oligosaccharides of most glycoproteins. We shall return to the importance of glucose trimming shortly. This oligosaccharide "trimming," or "processing," continues in the Golgi apparatus, as we discuss in Chapter 13.

The *N*-linked oligosaccharides are by far the most common oligosaccharides, being found on 90% of all glycoproteins. Less frequently, oligosaccharides are linked to the hydroxyl group on the side chain of a serine, threonine, or hydroxylysine amino acid. A first sugar of these *O-linked oligosaccharides* is added in the ER and the oligosaccharide is then further extended in the Golgi apparatus (see Figure 13–32).

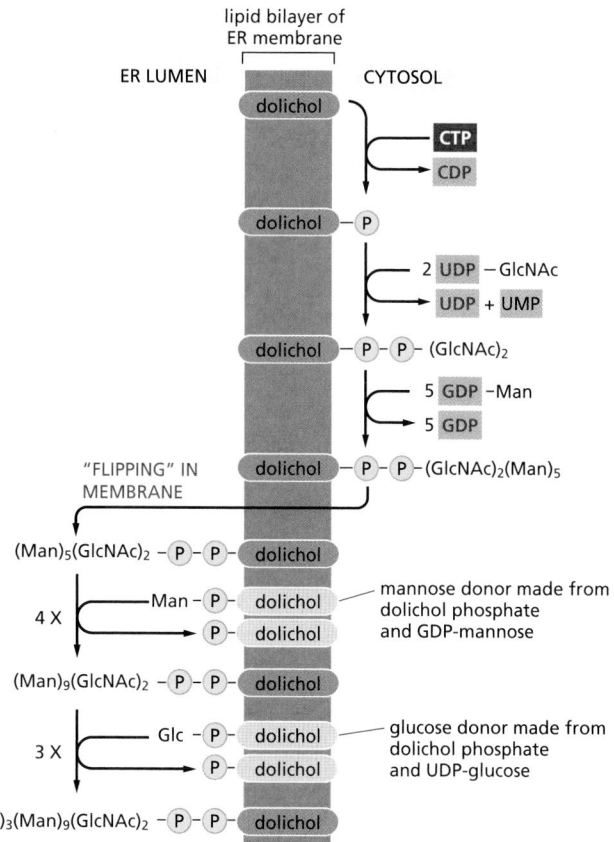

Figure 12–48 Synthesis of the lipid-linked precursor oligosaccharide in the rough ER membrane. The oligosaccharide is assembled sugar by sugar onto the carrier lipid dolichol (a polyisoprenoid; see Panel 2–5, pp. 98–99). Dolichol is long and very hydrophobic: its 22 five-carbon units can span the thickness of a lipid bilayer more than three times, so that the attached oligosaccharide is firmly anchored in the membrane. The first sugar is linked to dolichol by a pyrophosphate bridge. This high-energy bond activates the oligosaccharide for its eventual transfer from the lipid to an asparagine side chain of a growing polypeptide on the lumenal side of the rough ER. As indicated, the synthesis of the oligosaccharide starts on the cytosolic side of the ER membrane and continues on the lumenal face after the $(Man)_5(GlcNAc)_2$ lipid intermediate is flipped across the bilayer by a transporter (which is not shown). All the subsequent glycosyl transfer reactions on the lumenal side of the ER involve transfers from dolichol-P-glucose and dolichol-P-mannose; these activated, lipid-linked monosaccharides are synthesized from dolichol phosphate and UDP-glucose or GDP-mannose (as appropriate) on the cytosolic side of the ER and are then flipped across the ER membrane. GlcNAc = *N*-acetylglucosamine; Man = mannose; Glc = glucose.

Oligosaccharides Are Used as Tags to Mark the State of Protein Folding

It has long been debated why glycosylation is such a common modification of proteins that enter the ER. One particularly puzzling observation has been that some proteins require *N*-linked glycosylation for proper folding in the ER, yet the precise location of the oligosaccharides attached to the protein's surface does not seem to matter. A clue to the role of glycosylation in protein folding came from studies of two ER chaperone proteins, which are called **calnexin** and **calreticulin** because they require Ca²⁺ for their activities. These chaperones are carbohydrate-binding proteins, or *lectins*, which bind to oligosaccharides on incompletely folded proteins and retain them in the ER. Like other chaperones, they prevent incompletely folded proteins from irreversibly aggregating. Both calnexin and calreticulin also promote the association of incompletely folded proteins with another ER chaperone, which binds to cysteines that have not yet formed disulfide bonds.

Calnexin and calreticulin recognize *N*-linked oligosaccharides that contain a single terminal glucose, and they therefore bind proteins only after two of the three glucoses on the precursor oligosaccharide have been removed during glucose trimming by ER glucosidases. When the third glucose has been removed, the glycoprotein dissociates from its chaperone and can leave the ER.

How, then, do calnexin and calreticulin distinguish properly folded from incompletely folded proteins? The answer lies in yet another ER enzyme, a glucosyl transferase that keeps adding a glucose to those oligosaccharides that have lost their last glucose. It adds the glucose, however, only to oligosaccharides that are attached to unfolded proteins. Thus, an unfolded protein undergoes continuous cycles of glucose trimming (by glucosidase) and glucose addition (by glucosyl transferase), maintaining an affinity for calnexin and calreticulin until it has achieved its fully folded state (**Figure 12–49**).

Improperly Folded Proteins Are Exported from the ER and Degraded in the Cytosol

Despite all the help from chaperones, many protein molecules (more than 80% for some proteins) translocated into the ER fail to achieve their properly folded or oligomeric state. Such proteins are exported from the ER back into the cytosol, where they are degraded in proteasomes (discussed in Chapter 6). In many ways, the mechanism of retrotranslocation is similar to other post-translational

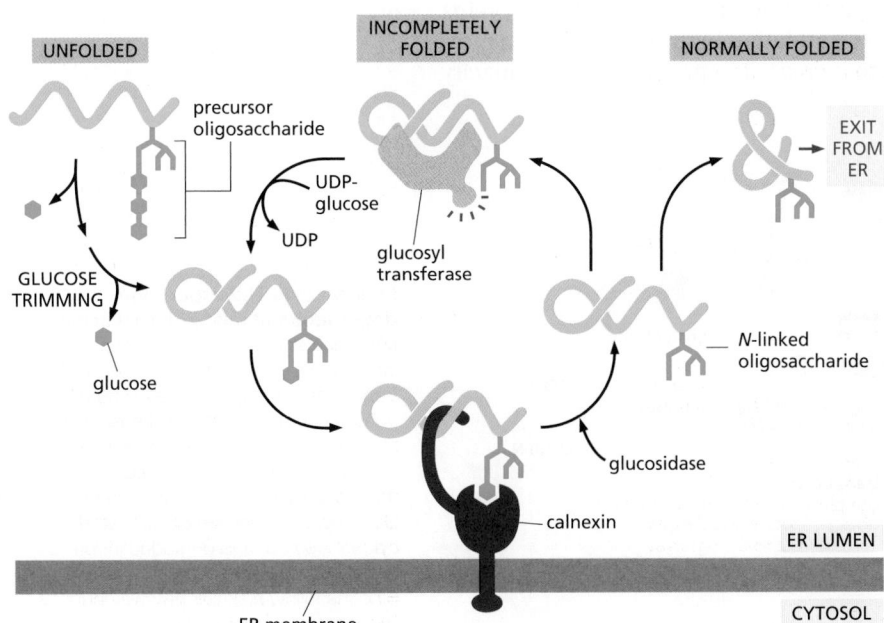

Figure 12–49 The role of *N*-linked glycosylation in ER protein folding. The ER-membrane-bound chaperone protein calnexin binds to incompletely folded proteins containing one terminal glucose on *N*-linked oligosaccharides, trapping the protein in the ER. Removal of the terminal glucose by a glucosidase releases the protein from calnexin. A glucosyl transferase is the crucial enzyme that determines whether the protein is folded properly or not: if the protein is still incompletely folded, the enzyme transfers a new glucose from UDP-glucose to the *N*-linked oligosaccharide, renewing the protein's affinity for calnexin and retaining it in the ER. The cycle repeats until the protein has folded completely. Calreticulin functions similarly, except that it is a soluble ER resident protein. Another ER chaperone, ERp57 (not shown), collaborates with calnexin and calreticulin in retaining an incompletely folded protein in the ER. ERp57 recognizes free sulfhydryl groups, which are a sign of incomplete disulfide bond formation.

modes of translocation. For example, like translocation into mitochondria or chloroplasts, chaperone proteins are necessary to keep the polypeptide chain in an unfolded state prior to and during translocation. Similarly, a source of energy is required to provide directionality to the transport and to pull the protein into the cytosol. Finally, a translocator is necessary.

Selecting proteins from the ER for degradation is a challenging process: misfolded proteins or unassembled protein subunits should be degraded, but folding intermediates of newly made proteins should not. Help in making this distinction comes from the N-linked oligosaccharides, which serve as timers that measure how long a protein has spent in the ER. The slow trimming of a particular mannose on the core oligosaccharide tree by an enzyme (a mannosidase) in the ER creates a new oligosaccharide structure that ER-lumenal lectins of the retrotranslocation apparatus recognize. Proteins that fold and exit from the ER faster than the mannosidase can remove its target mannose therefore escape degradation.

In addition to the lectins in the ER that recognize the oligosaccharides, chaperones and protein disulfide isomerases (enzymes mentioned earlier that catalyze the formation and breakage of S–S bonds) associate with the proteins that must be degraded. The chaperones prevent the unfolded proteins from aggregating, and the disulfide isomerases break disulfide bonds that may have formed incorrectly, so that a linear polypeptide chain can be translocated back into the cytosol.

Multiple translocator complexes move different proteins from the ER membrane or lumen into the cytosol. A common feature is that they each contain an E3 ubiquitin ligase enzyme, which attaches polyubiquitin tags to the unfolded proteins as they emerge into the cytosol, marking them for destruction. Fueled by the energy derived from ATP hydrolysis, a hexomeric ATPase of the family of AAA-ATPases (see Figure 6–85) pulls the unfolded protein through the translocator into the cytosol. An N-glycanase removes its oligosaccharide chains *en bloc*. Guided by its ubiquitin tag, the deglycosylated polypeptide is rapidly fed into proteasomes, where it is degraded (**Figure 12–50**).

Misfolded Proteins in the ER Activate an Unfolded Protein Response

Cells carefully monitor the amount of misfolded protein in various compartments. An accumulation of misfolded proteins in the cytosol, for example, triggers a *heat-shock response* (discussed in Chapter 6), which stimulates the transcription of genes encoding cytosolic chaperones that help to refold the proteins. Similarly, an accumulation of misfolded proteins in the ER triggers an **unfolded protein response**, which includes an increased transcription of genes encoding proteins involved in retrotranslocation and protein degradation in the cytosol, ER chaperones, and many other proteins that help to increase the protein-folding capacity of the ER.

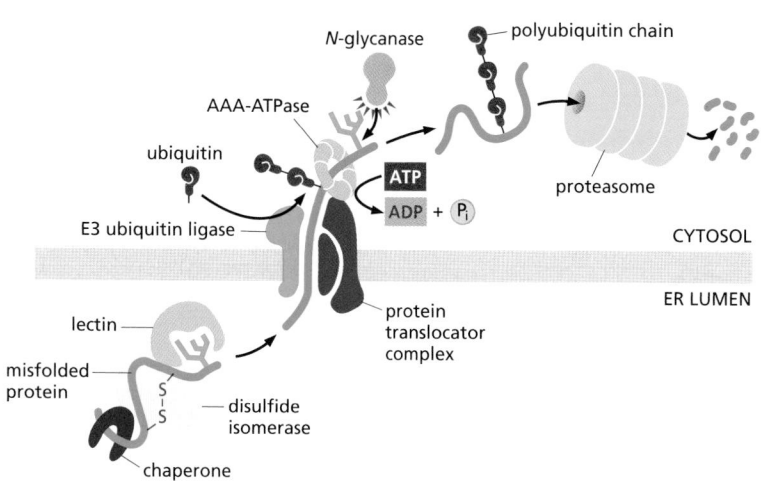

Figure 12–50 The export and degradation of misfolded ER proteins. Misfolded soluble proteins in the ER lumen are recognized and targeted to a translocator complex in the ER membrane. They first interact in the ER lumen with chaperones, disulfide isomerases, and lectins. They are then exported into the cytosol through the translocator. In the cytosol, they are ubiquitylated, deglycosylated, and degraded in proteasomes. Misfolded membrane proteins follow a similar pathway but use a different translocator.

How do misfolded proteins in the ER signal to the nucleus? There are three parallel pathways that execute the unfolded protein response (**Figure 12–51A**). The first pathway, which was initially discovered in yeast cells, is particularly remarkable. Misfolded proteins in the ER activate a transmembrane protein kinase in the ER, called IRE1, which causes the kinase to oligomerize and phosphorylate itself. (Some cell-surface receptor kinases in the plasma membrane are activated in a

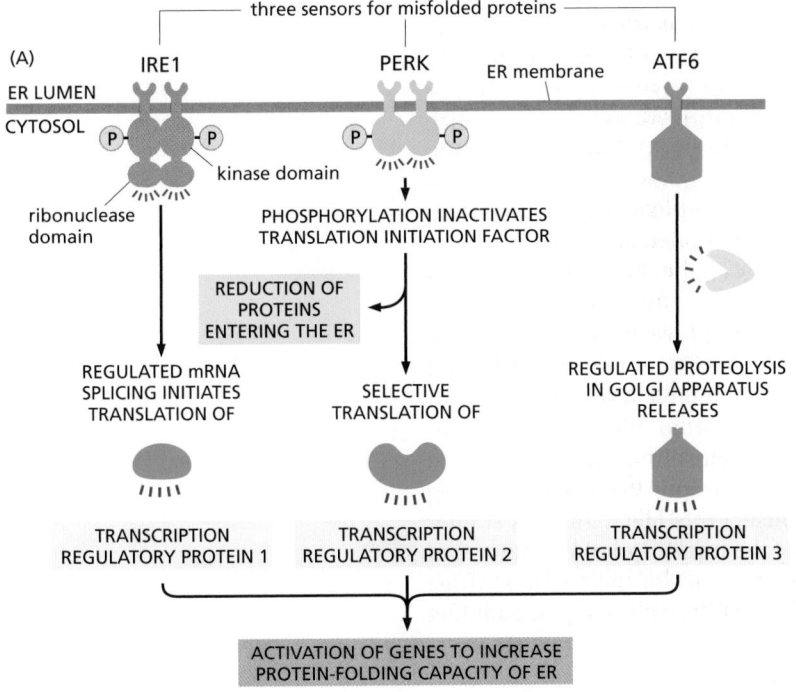

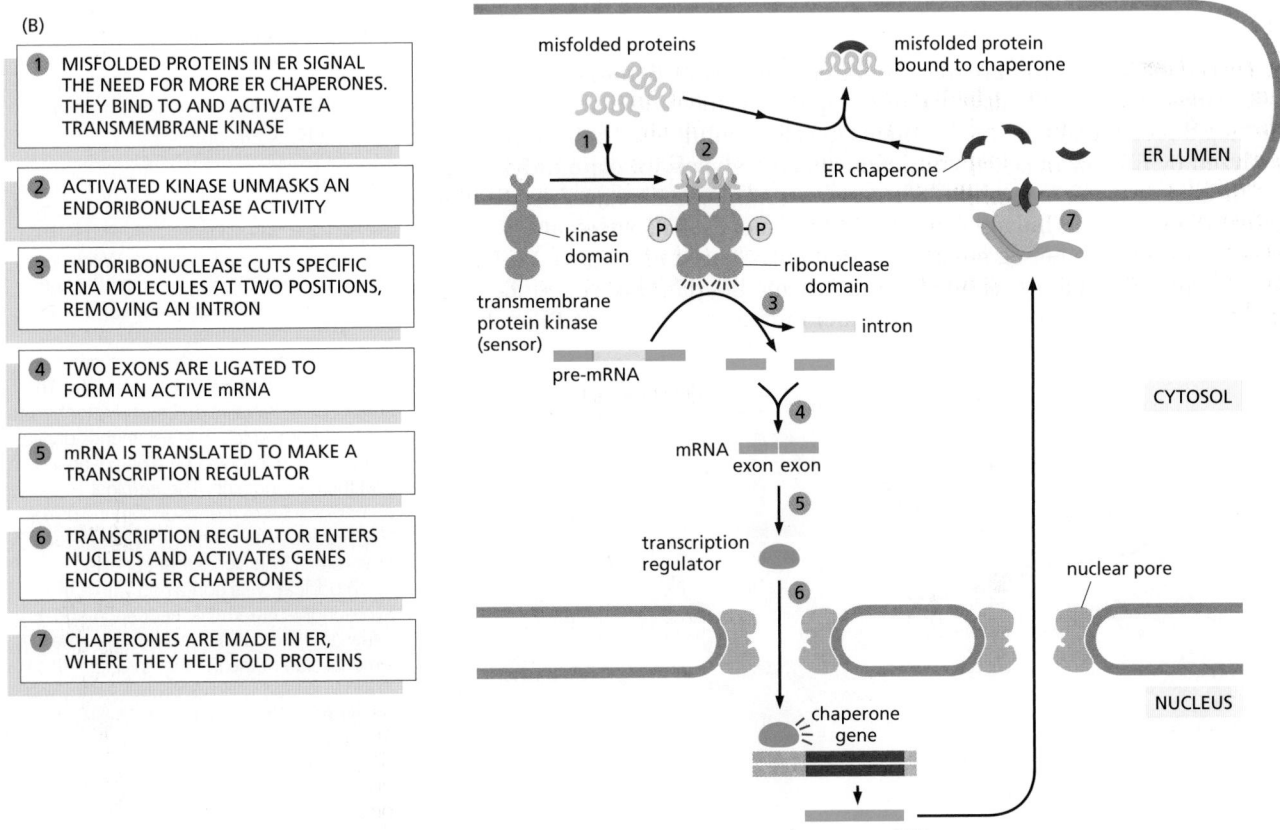

Figure 12–51 The unfolded protein response. (A) By three parallel intracellular signaling pathways, the accumulation of misfolded proteins in the ER lumen signals to the nucleus to activate the transcription of genes that encode proteins that help the cell cope with misfolded proteins in the ER. (B) Regulated RNA splicing is a key regulatory switch in pathway 1 of the unfolded protein response (Movie 12.6).

similar way, as discussed in Chapter 15.) The oligomerization and autophosphorylation of IRE1 activates an endoribonuclease domain in the cytosolic portion of the same molecule, which cleaves a specific cytosolic mRNA molecule at two positions, excising an intron. (This is a unique exception to the rule that introns are spliced out while the RNA is still in the nucleus.) The separated exons are then joined by an RNA ligase, generating a spliced mRNA, which is translated to produce an active transcription regulatory protein. This protein activates the transcription of genes encoding the proteins that help mediate the unfolded protein response (Figure 12–51B).

Misfolded proteins also activate a second transmembrane kinase in the ER, PERK, which inhibits a translation initiation factor by phosphorylating it, thereby reducing the production of new proteins throughout the cell. One consequence of the reduction in protein synthesis is to reduce the flux of proteins into the ER, thereby reducing the load of proteins that need to be folded there. Some proteins, however, are preferentially translated when translation initiation factors are scarce (discussed in Chapter 7, p. 424), and one of these is a transcription regulator that helps activate the transcription of the genes encoding proteins active in the unfolded protein response.

Finally, a third transcription regulator, ATF6, is initially synthesized as a transmembrane ER protein. Because it is embedded in the ER membrane, it cannot activate the transcription of genes in the nucleus. When misfolded proteins accumulate in the ER, however, the ATF6 protein is transported to the Golgi apparatus, where it encounters proteases that cleave off its cytosolic domain, which can now migrate to the nucleus and help activate the transcription of genes encoding proteins involved in the unfolded protein response. (This mechanism is similar to that described in Figure 12–16 for activation of the transcription regulator that controls cholesterol biosynthesis.) The relative importance of each of these three pathways in the unfolded protein response differs in different cell types, enabling each cell type to tailor the unfolded protein response to its particular needs.

Some Membrane Proteins Acquire a Covalently Attached Glycosylphosphatidylinositol (GPI) Anchor

As discussed in Chapter 10, several cytosolic enzymes catalyze the covalent addition of a single fatty acid chain or prenyl group to selected proteins. The attached lipids help direct and attach these proteins to cell membranes. A related process is catalyzed by ER enzymes that covalently attach a **glycosylphosphatidylinositol (GPI) anchor** to the C-terminus of some membrane proteins destined for the plasma membrane. This linkage forms in the lumen of the ER, where, at the same time, the transmembrane segment of the protein is cleaved off (**Figure 12–52**). A large number of plasma membrane proteins are modified in this way. Since they are attached to the exterior of the plasma membrane only by their GPI anchors,

Figure 12–52 The attachment of a GPI anchor to a protein in the ER. GPI-anchored proteins are targeted to the ER membrane by an N-terminal signal sequence (not shown), which is removed (see Figure 12–42). Immediately after the completion of protein synthesis, the precursor protein remains anchored in the ER membrane by a hydrophobic C-terminal sequence of 15–20 amino acids; the rest of the protein is in the ER lumen. Within less than a minute, an enzyme in the ER cuts the protein free from its membrane-bound C-terminus and simultaneously attaches the new C-terminus to an amino group on a preassembled GPI intermediate. The sugar chain contains an inositol attached to the lipid from which the GPI anchor derives its name. It is followed by a glucosamine and three mannoses. The terminal mannose links to a phosphoethanolamine that provides the amino group to attach the protein. The signal that specifies this modification is contained within the hydrophobic C-terminal sequence and a few amino acids adjacent to it on the lumenal side of the ER membrane; if this signal is added to other proteins, they too become modified in this way. Because of the covalently linked lipid anchor, the protein remains membrane-bound, with all of its amino acids exposed initially on the lumenal side of the ER and eventually on the exterior of the plasma membrane.

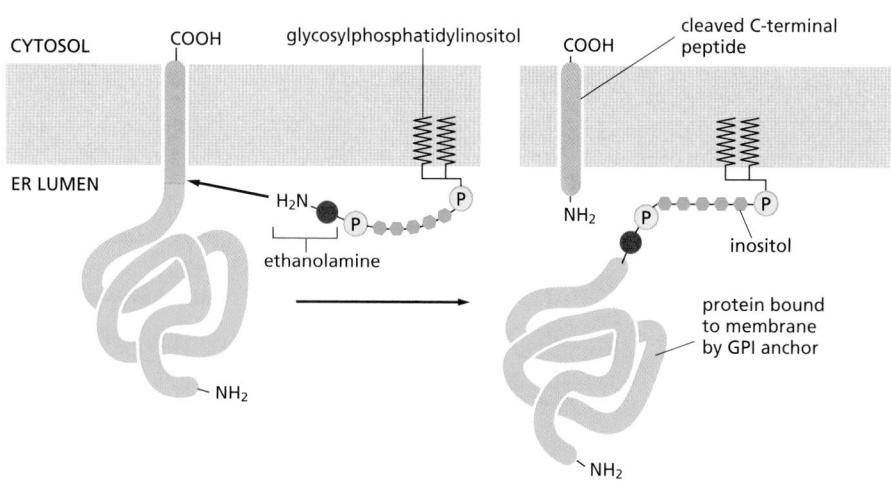

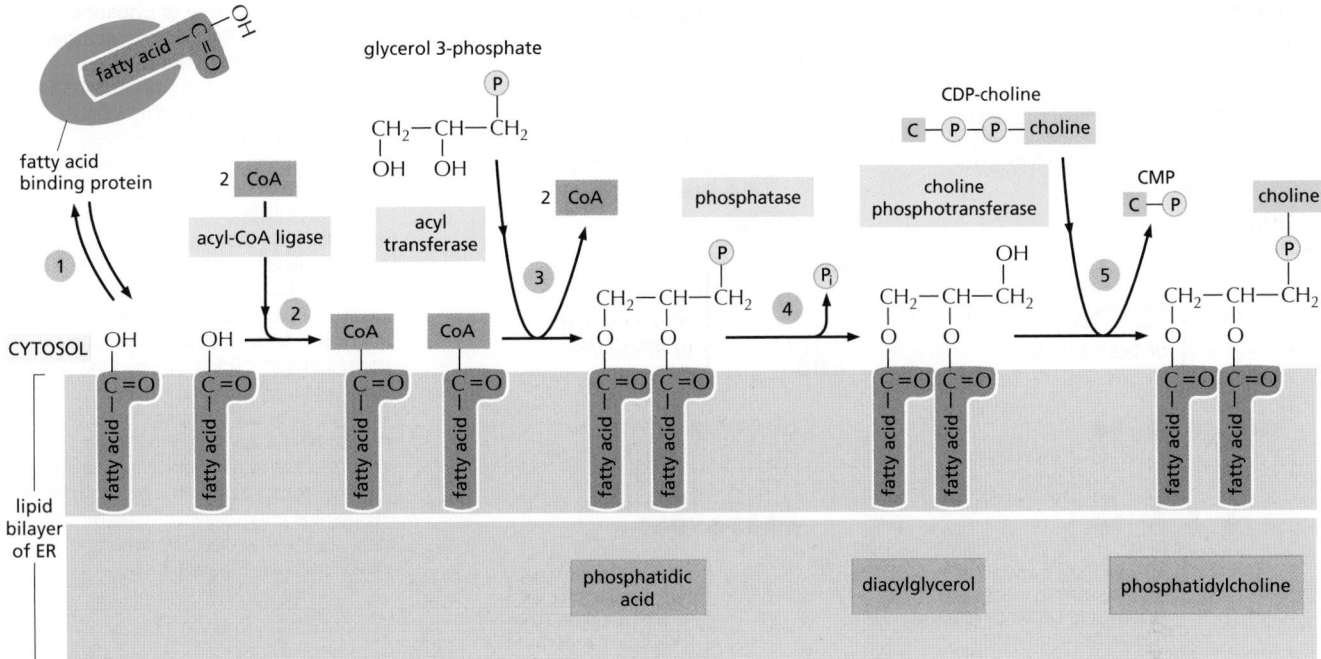

they can in principle be released from cells in soluble form in response to signals that activate a specific phospholipase in the plasma membrane. Trypanosome parasites, for example, use this mechanism to shed their coat of GPI-anchored surface proteins when attacked by the immune system. GPI anchors may also be used to direct plasma membrane proteins into *lipid rafts* and thus segregate the proteins from other membrane proteins (see Figure 10–13).

Figure 12–53 The synthesis of phosphatidylcholine. As illustrated, this phospholipid is synthesized from glycerol 3-phosphate, cytidine-diphosphocholine (CDP-choline), and fatty acids delivered to the ER by a cytosolic fatty acid binding protein.

The ER Assembles Most Lipid Bilayers

The ER membrane is the site of synthesis of nearly all of the cell's major classes of lipids, including both phospholipids and cholesterol, required for the production of new cell membranes. The major phospholipid made is *phosphatidylcholine*, which can be formed in three steps from choline, two fatty acids, and glycerol phosphate (**Figure 12–53**). Each step is catalyzed by enzymes in the ER membrane, which have their active sites facing the cytosol, where all of the required metabolites are found. Thus, phospholipid synthesis occurs exclusively in the cytosolic leaflet of the ER membrane. Because fatty acids are not soluble in water, they are shepherded from their sites of synthesis to the ER by a fatty acid binding protein in the cytosol. After arrival in the ER membrane and activation with CoA, acyl transferases successively add two fatty acids to glycerol phosphate to produce phosphatidic acid. Phosphatidic acid is sufficiently water-insoluble to remain in the lipid bilayer; it cannot be extracted from the bilayer by the fatty acid binding proteins. It is therefore this first step that enlarges the ER lipid bilayer. The later steps determine the head group of a newly formed lipid molecule and therefore the chemical nature of the bilayer, but they do not result in net membrane growth. The two other major membrane phospholipids—phosphatidylethanolamine and phosphatidylserine (see Figure 10–3)—as well as the minor phospholipid phosphatidylinositol (PI), are all synthesized in this way.

Because phospholipid synthesis takes place in the cytosolic leaflet of the ER lipid bilayer, there needs to be a mechanism that transfers some of the newly formed phospholipid molecules to the lumenal leaflet of the bilayer. In synthetic lipid bilayers, lipids do not "flip-flop" in this way (see Figure 10–10). In the ER, however, phospholipids equilibrate across the membrane within minutes, which is almost 100,000 times faster than can be accounted for by spontaneous "flip-flop." This rapid trans-bilayer movement is mediated by a poorly characterized

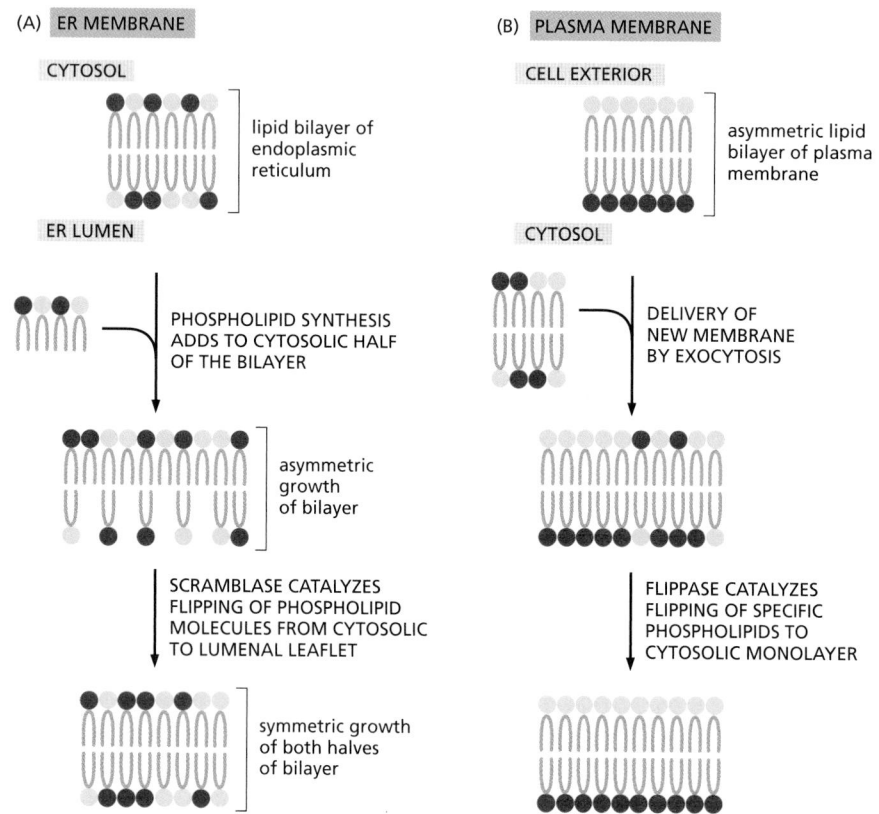

Figure 12–54 The role of phospholipid translocators in lipid bilayer synthesis. (A) Because new lipid molecules are added only to the cytosolic half of the ER membrane bilayer and lipid molecules do not flip spontaneously from one monolayer to the other, a transmembrane phospholipid translocator (called a scramblase) is required to transfer lipid molecules from the cytosolic half to the lumenal half so that the membrane grows as a bilayer. The scramblase is not specific for particular phospholipid head groups and therefore equilibrates the different phospholipids between the two monolayers. (B) Fueled by ATP hydrolysis, a head-group-specific flippase in the plasma membrane actively flips phosphatidylserine and phosphatidylethanolamine directionally from the extracellular to the cytosolic leaflet, creating the characteristically asymmetric lipid bilayer of the plasma membrane of animal cells (see Figure 10–15).

phospholipid translocator called a *scramblase*, which nonselectively equilibrates phospholipids between the two leaflets of the lipid bilayer (**Figure 12–54**). Thus, the different types of phospholipids are thought to be equally distributed between the two leaflets of the ER membrane.

The plasma membrane contains a different type of phospholipid translocator that belongs to the family of P-type pumps (discussed in Chapter 11). These *flippases* specifically recognize those phospholipids that contain free amino groups in their head groups (phosphatidylserine and phosphatidylethanolamine—see Figure 10–3) and transfers them from the extracellular to the cytosolic leaflet, using the energy of ATP hydrolysis. The plasma membrane therefore has a highly asymmetric phospholipid composition, which is actively maintained by the flippases (see Figure 10–15). The plasma membrane also contains a scramblase but, in contrast to the ER scramblase, which is always active, the plasma membrane enzyme is regulated and only activated in some situations, such as in apoptosis and in activated platelets, where it acts to abolish the lipid bilayer asymmetry; the resulting exposure of phosphatidylserine on the surface of apoptotic cells serves as a signal for phagocytic cells to ingest and degrade the dead cell.

The ER also produces cholesterol and ceramide (**Figure 12–55**). *Ceramide* is made by condensing the amino acid serine with a fatty acid to form the amino alcohol *sphingosine* (see Figure 10–3); a second fatty acid is then covalently added to form ceramide. The ceramide is exported to the Golgi apparatus, where it serves as a precursor for the synthesis of two types of lipids: oligosaccharide chains are added to form *glycosphingolipids* (glycolipids; see Figure 10–16), and phosphocholine head groups are transferred from phosphatidylcholine to other ceramide molecules to form *sphingomyelin* (discussed in Chapter 10). Thus, both glycolipids and sphingomyelin are produced relatively late in the process of membrane synthesis. Because they are produced by enzymes that have their active sites exposed to the Golgi lumen, they are found exclusively in the noncytosolic leaflet of the lipid bilayers that contain them.

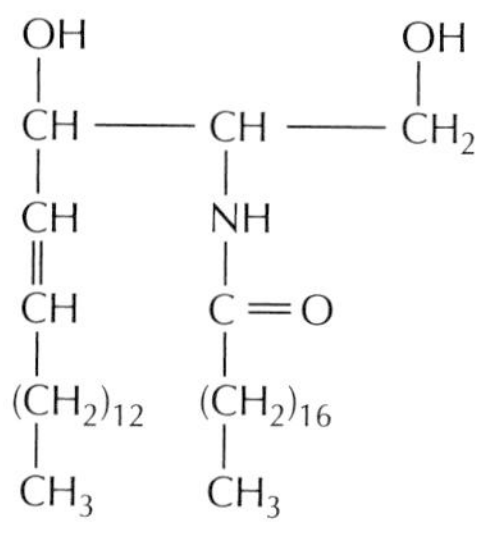

Figure 12–55 The structure of ceramide.

As discussed in Chapter 13, the plasma membrane and the membranes of the Golgi apparatus, lysosomes, and endosomes all form part of a membrane system that communicates with the ER by means of transport vesicles, which transfer both proteins and lipids. Mitochondria and plastids, however, do not belong to this system, and they therefore require different mechanisms to import proteins and lipids for growth. We have already seen that they import most of their proteins from the cytosol. Although mitochondria modify some of the lipids they import, they do not synthesize lipids *de novo*; instead, their lipids have to be imported from the ER, either directly or indirectly by way of other cell membranes. In either case, special mechanisms are required for the transfer.

The details of how lipid distribution between different membranes is catalyzed and regulated are not known. Water-soluble carrier proteins called *phospholipid exchange proteins* (or *phospholipid transfer proteins*) are thought to transfer individual phospholipid molecules between membranes, functioning much like fatty acid binding proteins that shepherd fatty acids through the cytosol (see Figure 12–54). In addition, mitochondria are often seen in close juxtaposition to ER membranes in electron micrographs, and specific junction complexes have been identified that hold the ER and outer mitochondrial membrane in close proximity. It is thought that these junction complexes provide specific contact-dependent lipid transfer mechanisms that operate between these adjacent membranes.

Summary

The extensive ER network serves as a factory for the production of almost all of the cell's lipids. In addition, a major portion of the cell's protein synthesis occurs on the cytosolic surface of the rough ER: virtually all proteins destined for secretion or for the ER itself, the Golgi apparatus, the lysosomes, the endosomes, and the plasma membrane are first imported into the ER from the cytosol. In the ER lumen, the proteins fold and oligomerize, disulfide bonds are formed, and N-linked oligosaccharides are added. The pattern of N-linked glycosylation is used to indicate the extent of protein folding, so that proteins leave the ER only when they are properly folded. Proteins that do not fold or oligomerize correctly are translocated back into the cytosol, where they are deglycosylated, polyubiquitylated, and degraded in proteasomes. If misfolded proteins accumulate in excess in the ER, they trigger an unfolded protein response, which activates appropriate genes in the nucleus to help the ER cope.

Only proteins that carry a special ER signal sequence are imported into the ER. The signal sequence is recognized by a signal-recognition particle (SRP), which binds both the growing polypeptide chain and the ribosome and directs them to a receptor protein on the cytosolic surface of the rough ER membrane. This binding to the ER membrane initiates the translocation process that threads a loop of polypeptide chain across the ER membrane through the hydrophilic pore of a protein translocator.

Soluble proteins—destined for the ER lumen, for secretion, or for transfer to the lumen of other organelles—pass completely into the ER lumen. Transmembrane proteins destined for the ER or for other cell membranes are translocated part way across the ER membrane and remain anchored there by one or more membrane-spanning α-helical segments in their polypeptide chains. These hydrophobic portions of the protein can act either as start-transfer or stop-transfer signals during the translocation process. When a polypeptide contains multiple, alternating start-transfer and stop-transfer signals, it will pass back and forth across the bilayer multiple times as a multipass transmembrane protein.

The asymmetry of protein insertion and glycosylation in the ER establishes the sidedness of the membranes of all the other organelles that the ER supplies with membrane proteins.

WHAT WE DON'T KNOW

• How do nuclear import receptors negotiate the tangled gel-like interior of a nuclear pore complex so efficiently?

• Is the nuclear pore complex a rigid structure or can it expand and contract, depending on the cargo transported?

• Sequence comparisons show that signal sequences for an individual protein such as insulin are quite conserved across species, much more so than would be expected from our current understanding that all that matters for their function are general structural features such as hydrophobicity. What other functions might signal sequences have that could account for their evolutionary sequence conservation?

• How are polyribosomes on the endoplasmic reticulum membrane arranged so that the next initiating ribosome will find an unoccupied translocator?

• Why does the signal-recognition particle have an indispensable RNA subunit?

PROBLEMS

Which statements are true? Explain why or why not.

12–1   Like the lumen of the ER, the interior of the nucleus is topologically equivalent to the outside of the cell.

12–2   ER-bound and free ribosomes, which are structurally and functionally identical, differ only in the proteins they happen to be making at a particular time.

12–3   To avoid the inevitable collisions that would occur if two-way traffic through a single pore were allowed, nuclear pore complexes are specialized so that some mediate import while others mediate export.

12–4   Peroxisomes are found in only a few specialized types of eukaryotic cell.

Discuss the following problems.

12–5   What is the fate of a protein with no sorting signal?

12–6   The rough ER is the site of synthesis of many classes of membrane proteins. Some of these proteins remain in the ER, whereas others are sorted to compartments such as the Golgi apparatus, lysosomes, and the plasma membrane. One measure of the difficulty of the sorting problem is the degree of "purification" that must be achieved during transport from the ER. Are proteins bound for the plasma membrane common or rare among all ER membrane proteins?

A few simple considerations allow one to answer this question. In a typical growing cell that is dividing once every 24 hours, the equivalent of one new plasma membrane must transit the ER every day. If the ER membrane is 20 times the area of a plasma membrane, what is the ratio of plasma membrane proteins to other membrane proteins in the ER? (Assume that all proteins on their way to the plasma membrane remain in the ER for 30 minutes on average before exiting, and that the ratio of proteins to lipids in the ER and plasma membranes is the same.)

12–7   Before nuclear pore complexes were well understood, it was unclear whether nuclear proteins diffused passively into the nucleus and accumulated there by binding to residents of the nucleus such as chromosomes, or whether they were actively imported and accumulated regardless of their affinity for nuclear components.

A classic experiment that addressed this problem used several forms of radioactive nucleoplasmin, which is a large pentameric protein involved in chromatin assembly. In this experiment, either the intact protein or the nucleoplasmin heads, tails, or heads with a single tail were injected into the cytoplasm of a frog oocyte or into the nucleus (**Figure Q12–1**). All forms of nucleoplasmin, except heads, accumulated in the nucleus when injected into the cytoplasm, and all forms were retained in the nucleus when injected there.

A.   What portion of the nucleoplasmin molecule is responsible for localization in the nucleus?

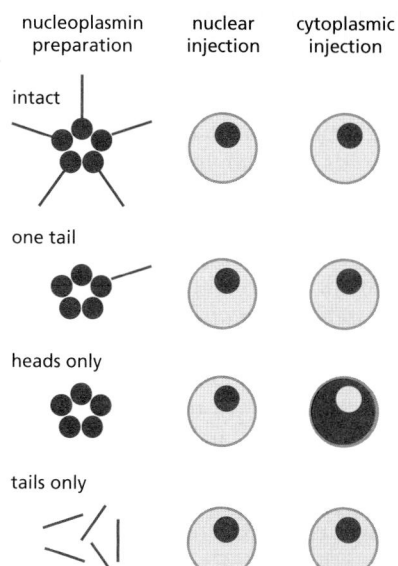

Figure Q12–1 Cellular location of injected nucleoplasmin and nucleoplasmin components (Problem 12–7). Schematic diagrams of autoradiographs show the cytoplasm and nucleus with the location of nucleoplasmin indicated by the *red* areas.

B.   How do these experiments distinguish between active transport, in which a nuclear localization signal triggers transport by the nuclear pore complex, and passive diffusion, in which a binding site for a nuclear component allows accumulation in the nucleus?

12–8   Assuming that 32 million histone octamers are required to package the human genome, how many histone molecules must be transported per second per nuclear pore complex in cells whose nuclei contain 3000 nuclear pores and are dividing once per day?

12–9   The nuclear pore complex (NPC) creates a barrier to the free exchange of molecules between the nucleus and cytosol, but in a way that remains mysterious. In yeast, for example, the central pore of the NPC has a diameter of 35 nm and is 30 nm long, which is somewhat smaller than its vertebrate counterpart. Even so, it is large enough to accommodate virtually all components of the cytosol. Yet the pore allows passive diffusion of molecules only up to about 40 kd; entry of anything larger requires help from a nuclear import receptor. Selective permeability is controlled by protein components of the NPC that have unstructured, polar tails extending into the central pore. These tails are characterized by periodic repeats of the hydrophobic amino acids phenylalanine (F) and glycine (G).

At high enough concentration (~50 mM), the FG-repeat domains of these proteins can form a gel, with a meshwork of interactions between the hydrophobic FG repeats (**Figure Q12–2A**). These gels allow passive diffusion of small molecules, but they prevent entry of larger proteins such as the fluorescent protein mCherry fused to maltose binding protein (MBP) (**Figure Q12–2B**). (The fusion to MBP makes mCherry too large to enter the nucleus by passive diffusion.) However, if the nuclear import receptor, importin, is fused to a similar protein, MBP-GFP, the importin-MBP-GFP fusion readily enters the gel (Figure Q12–2B).

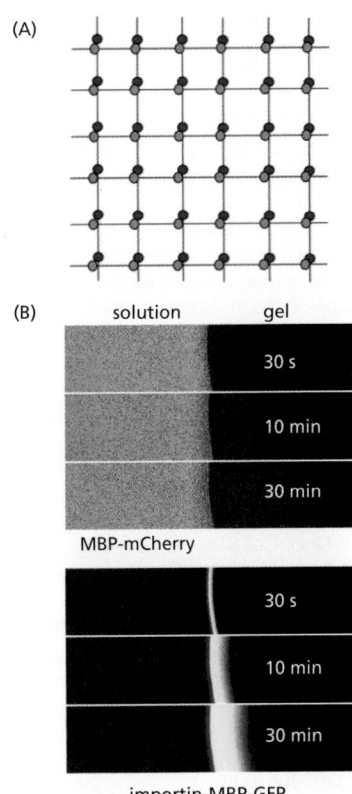

Figure Q12–2 FG-repeat gel and influx of proteins into the nucleus (Problem 12–9). (A) Cartoon of the meshwork (gel) formed by pairwise interactions between hydrophobic FG repeats. For FG-repeats separated by 17 amino acids, as is typical, the mesh formed by extended amino acid side chains would correspond to about 4 nm on a side, which would be large enough to account for the characteristic passive diffusion of proteins through nuclear pores. (B) Diffusion of MBP-mCherry and importin-MBP-GFP into a gel of FG-repeats. In each group, the solution is shown at left and the gel at right. The bright areas indicate regions that contain the fluorescent proteins.

A. FG-repeats only form gels *in vitro* at relatively high concentration (50 mM). Is this concentration reasonable for FG repeats in the NPC core? In yeast, there are about 5000 FG-repeats in each NPC. Given the dimensions of the yeast nuclear pore (35 nm diameter and 30 nm length), calculate the concentration of FG-repeats in the cylindrical volume of the pore. Is this concentration comparable to the one used *in vitro*?

B. A second question is whether the diffusion of importin-MBP-GFP through the FG-repeat gel is fast enough to account for the efficient flow of materials between the nucleus and cytosol. From experiments of the type shown in Figure Q12–2B, the diffusion coefficient (D) of importin-MBP-GFP through the FG-repeat gel was determined to be about 0.1 μm^2/s. The equation for diffusion is $t = x^2/2D$, where t is time and x is distance. Calculate the time it would take importin-MBP-GFP to diffuse through a yeast nuclear pore (30 nm) if the pore consisted of a gel of FG-repeats. Does this time seem fast enough for the needs of a eukaryotic cell?

12–10 Components of the TIM complexes, the multi-subunit protein translocators in the mitochondrial inner membrane, are much less abundant than those of the TOM complex. They were initially identified using a genetic trick. The yeast *Ura3* gene, whose product is an enzyme that is normally located in the cytosol where it is essential for synthesis of uracil, was modified so that the protein carried an import signal for the mitochondrial matrix. A population of cells carrying the modified *Ura3* gene in place of the normal gene was then grown in the absence of uracil. Most cells died, but the rare cells that grew were shown to be defective for mitochondrial import. Explain how this selection identifies cells with defects in components required for import into the mitochondrial matrix. Why don't normal cells with the modified *Ura3* gene grow in the absence of uracil? Why do cells that are defective for mitochondrial import grow in the absence of uracil?

12–11 If the enzyme dihydrofolate reductase (DHFR), which is normally located in the cytosol, is engineered to carry a mitochondrial targeting sequence at its N-terminus, it is efficiently imported into mitochondria. If the modified DHFR is first incubated with methotrexate, which binds tightly to the active site, the enzyme remains in the cytosol. How do you suppose that the binding of methotrexate interferes with mitochondrial import?

12–12 Why do mitochondria need a special translocator to import proteins across the outer membrane, when the membrane already has large pores formed by porins?

12–13 Examine the multipass transmembrane protein shown in **Figure Q12–3**. What would you predict would be the effect of converting the first hydrophobic transmembrane segment to a hydrophilic segment? Sketch the arrangement of the modified protein in the ER membrane.

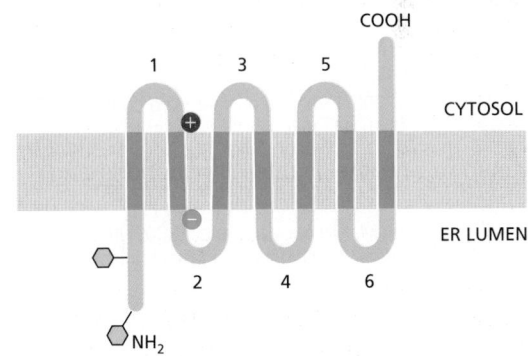

Figure Q12–3 Arrangement of a multipass transmembrane protein in the ER membrane (Problem 12–13). *Blue hexagons* represent covalently attached oligosaccharides. The positions of positively and negatively charged amino acids flanking the second transmembrane segment are shown.

12–14 All new phospholipids are added to the cytosolic leaflet of the ER membrane, yet the ER membrane has a symmetrical distribution of different phospholipids in its two leaflets. By contrast, the plasma membrane, which receives all its membrane components ultimately from the ER, has a very asymmetrical distribution of phospholipids in the two leaflets of its lipid bilayer. How is the symmetry generated in the ER membrane, and how is the asymmetry generated and maintained in the plasma membrane?

REFERENCES

General

Palade G (1975) Intracellular aspects of the process of protein synthesis. *Science* 189, 347–358.

The Compartmentalization of Cells

Blobel G (1980) Intracellular protein topogenesis. *Proc. Natl Acad. Sci. USA* 77, 1496–1500.

Devos DP, Gräf R & Field MC (2014) Evolution of the nucleus. *Curr. Opin. Cell Biol.* 28, 8–15.

Warren G & Wickner W (1996) Organelle inheritance. *Cell* 84, 395–400.

The Transport of Molecules Between the Nucleus and the Cytosol

Adam SA & Gerace L (1991) Cytosolic proteins that specifically bind nuclear location signals are receptors for nuclear import. *Cell* 66, 837–847.

Burke B & Stewart CL (2013) The nuclear lamins: flexibility in function. *Nat. Rev. Mol. Cell Biol.* 14, 13–24.

Cole CN & Scarcelli JJ (2006) Transport of messenger RNA from the nucleus to the cytoplasm. *Curr. Opin. Cell Biol.* 18, 299–306.

Güttinger S, Laurell E & Kutay U (2009) Orchestrating nuclear envelope disassembly and reassembly during mitosis. *Nat. Rev. Mol. Cell Biol.* 10, 178–191.

Hetzer MW & Wente SR (2009) Border control at the nucleus: biogenesis and organization of the nuclear membrane and pore complexes. *Dev. Cell* 17, 606–616.

Hoelz A, Debler EW & Blobel G (2011) The structure of the nuclear pore complex. *Annu. Rev. Biochem.* 80, 613–643.

Hülsmann BB, Labokha AA & Görlich D (2012) The permeability of reconstituted nuclear pores provides direct evidence for the selective phase model. *Cell* 150, 738–751.

Köhler A & Hurt E (2007) Exporting RNA from the nucleus to the cytoplasm. *Nat. Rev. Mol. Cell Biol.* 8, 761–773.

Rothballer A & Kutay U (2013) Poring over pores: nuclear pore complex insertion into the nuclear envelope. *Trends Biochem. Sci.* 38, 292–301.

Strambio-De-Castillia C, Niepel M & Rout MP (2010) The nuclear pore complex: bridging nuclear transport and gene regulation. *Nat. Rev. Mol. Cell Biol.* 11, 490–501.

Tran EJ & Wente SR (2006) Dynamic nuclear pore complexes: life on the edge. *Cell* 125, 1041–1053.

The Transport of Proteins Into Mitochondria and Chloroplasts

Chacinska A, Koehler CM, Milenkovic D et al. (2009) Importing mitochondrial proteins: machineries and mechanisms. *Cell* 138, 628–644.

Jarvis P & Robinson C (2004) Mechanisms of protein import and routing in chloroplasts. *Curr. Biol.* 14, R1064–R1077.

Kessler F & Schnell DJ (2009) Chloroplast biogenesis: diversity and regulation of the protein import apparatus. *Curr. Opin. Cell Biol.* 21, 494–500.

Prakash S & Matouschek A (2004) Protein unfolding in the cell. *Trends Biochem. Sci.* 29, 593–600.

Schleiff E & Becker T (2011) Common ground for protein translocation: access control for mitochondria and chloroplasts. *Nat. Rev. Mol. Cell Biol.* 12, 48–59.

Peroxisomes

Dimitrov L, Lam SK & Schekman R (2013) The role of the endoplasmic reticulum in peroxisome biogenesis. *Cold Spring Harb. Perspect. Biol.* 5, a013243.

Fujiki Y, Yagita Y & Matsuzaki T (2012) Peroxisome biogenesis disorders. *Biochim. Biophys. Acta* 1822, 1337–1342.

Schliebs W, Girzalsky W & Erdmann R (2010) Peroxisomal protein import and ERAD: variations on a common theme. *Nat. Rev. Mol. Cell Biol.* 11, 885–890.

Tabak HF, Braakman I & van der Zand A (2013) Peroxisome formation and maintenance are dependent on the endoplasmic reticulum. *Annu. Rev. Biochem.* 82, 723–744.

The Endoplasmic Reticulum

Akopian D, Shen K, Zhang X & Shan SO (2013) Signal recognition particle: an essential protein-targeting machine. *Annu. Rev. Biochem.* 82, 693–721.

Blobel G & Dobberstein B (1975) Transfer of proteins across membranes. I. Presence of proteolytically processed and unprocessed nascent immunoglobulin light chains on membrane-bound ribosomes of murine myeloma. *J. Cell Biol.* 67, 835–851.

Borgese N, Mok W, Kreibich G & Sabatini DD (1974) Ribosomal-membrane interaction: *in vitro* binding of ribosomes to microsomal membranes. *J. Mol. Biol.* 88, 559–580.

Braakman I & Bulleid NJ (2011) Protein folding and modification in the mammalian endoplasmic reticulum. *Annu. Rev. Biochem.* 80, 71–99.

Brodsky JL & Skach WR (2011) Protein folding and quality control in the endoplasmic reticulum: recent lessons from yeast and mammalian cell systems. *Curr. Opin. Cell Biol.* 23, 464–475.

Chen S, Novick P & Ferro-Novick S (2013) ER structure and function. *Curr. Opin. Cell Biol.* 25, 428–433.

Clark MR (2011) Flippin' lipids. *Nat. Immunol.* 12, 373–375.

Daleke DL (2003) Regulation of transbilayer plasma membrane phospholipid asymmetry. *J. Lipid Res.* 44, 233–242.

Deshaies RJ, Sanders SL, Feldheim DA & Schekman R (1991) Assembly of yeast Sec proteins involved in translocation into the endoplasmic reticulum into a membrane-bound multisubunit complex. *Nature* 349, 806–808.

Gething MJ (1999) Role and regulation of the ER chaperone BiP. *Semin. Cell Dev. Biol.* 10, 465–472.

Görlich D, Prehn S, Hartmann E et al. (1992) A mammalian homolog of SEC61p and SECYp is associated with ribosomes and nascent polypeptides during translocation. *Cell* 71, 489–503.

Hegde RS & Ploegh HL (2010) Quality and quantity control at the endoplasmic reticulum. *Curr. Opin. Cell Biol.* 22, 437–446.

Hegde RS & Keenan RJ (2011) Tail-anchored membrane protein insertion into the endoplasmic reticulum. *Nat. Rev. Mol. Cell Biol.* 12, 787–798.

Levine T & Loewen C (2006) Inter-organelle membrane contact sites: through a glass, darkly. *Curr. Opin. Cell Biol.* 18, 371–378.

López-Marqués RL, Holthuis JCM & Pomorski TG (2011) Pumping lipids with P4-ATPases. *Biol. Chem.* 392, 67–76.

Mamathambika BS & Bardwell JC (2008) Disulfide-linked protein folding pathways. *Annu. Rev. Cell Dev. Biol.* 24, 211–235.

Marciniak SJ & Ron D (2006) Endoplasmic reticulum stress signaling in disease. *Physiol. Rev.* 86, 1133–1149.

Milstein C, Brownlee GG, Harrison TM & Mathews MB (1972) A possible precursor of immunoglobulin light chains. *Nat. New Biol.* 239, 117–120.

Park E & Rapoport TA (2012) Mechanisms of Sec61/SecY-mediated protein translocation across membranes. *Annu. Rev. Biophys.* 41, 21–40.

Römisch K (2005) Endoplasmic reticulum-associated degradation. *Annu. Rev. Cell Dev. Biol.* 21, 435–456.

Rowland AA & Voeltz GK (2012) Endoplasmic reticulum-mitochondria contacts: function of the junction. *Nat. Rev. Mol. Cell Biol.* 13, 607–625.

Trombetta ES & Parodi AJ (2003) Quality control and protein folding in the secretory pathway. *Annu. Rev. Cell Dev. Biol.* 19, 649–676.

Tsai B, Ye Y & Rapoport TA (2002) Retro-translocation of proteins from the endoplasmic reticulum into the cytosol. *Nat. Rev. Mol. Cell Biol.* 3, 246–255.

Walter P & Ron D (2011) The unfolded protein response: from stress pathway to homeostatic regulation. *Science* 334, 1081–1086.

von Heijne G (2011) Introduction to theme "membrane protein folding and insertion". *Annu. Rev. Biochem.* 80, 157–160.

Intracellular Membrane Traffic

Every cell must eat, communicate with the world around it, and quickly respond to changes in its environment. To help accomplish these tasks, cells continually adjust the composition of their plasma membrane and internal compartments in rapid response to need. They use an elaborate internal membrane system to add and remove cell-surface proteins, such as receptors, ion channels, and transporters (**Figure 13–1**). Through the process of *exocytosis*, the secretory pathway delivers newly synthesized proteins, carbohydrates, and lipids either to the plasma membrane or the extracellular space. By the converse process of *endocytosis*, cells remove plasma membrane components and deliver them to internal compartments called *endosomes*, from where they can be recycled to the same or different regions of the plasma membrane or be delivered to lysosomes for degradation. Cells also use endocytosis to capture important nutrients, such as vitamins, cholesterol, and iron; these are taken up together with the macromolecules to which they bind and are then moved on to endosomes and lysosomes, from where they can be transported into the cytosol for use in various biosynthetic processes.

The interior space, or **lumen**, of each membrane-enclosed compartment along the secretory and endocytic pathways is equivalent to the lumen of most other membrane-enclosed compartments and to the cell exterior, in the sense that proteins can travel in this space without having to cross a membrane as they are passed from one compartment to another by means of numerous membrane-enclosed transport containers. These containers are formed from the donor compartment and are either small, spherical *vesicles*, larger irregular vesicles, or tubules. We shall use the term **transport vesicle** to apply to all forms of these containers.

Within a eukaryotic cell, transport vesicles continually bud off from one membrane and fuse with another, carrying membrane components and soluble lumenal molecules, which are referred to as **cargo** (**Figure 13–2**). This vesicular traffic flows along highly organized, directional routes, which allow the cell to secrete, eat, and remodel its plasma membrane and organelles. The *secretory pathway* leads outward from the endoplasmic reticulum (ER) toward the Golgi apparatus and cell surface, with a side route leading to lysosomes, while the *endocytic pathway* leads inward from the plasma membrane. In each case, *retrieval pathways*

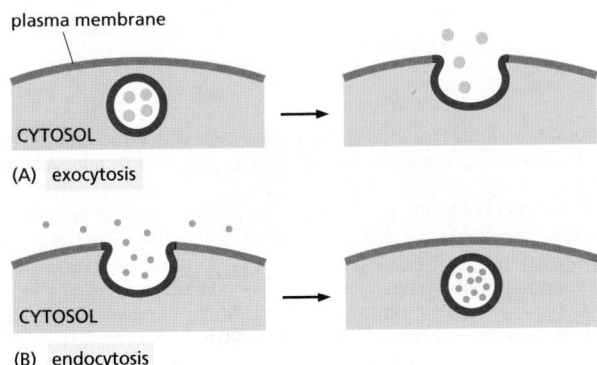

Figure 13–1 **Exocytosis and endocytosis.** (A) In exocytosis, a transport vesicle fuses with the plasma membrane. Its content is released into the extracellular space, while the vesicle membrane *(red)* becomes continuous with the plasma membrane. (B) In endocytosis, a plasma membrane patch *(red)* is internalized, forming a transport vesicle. Its content derives from the extracellular space.

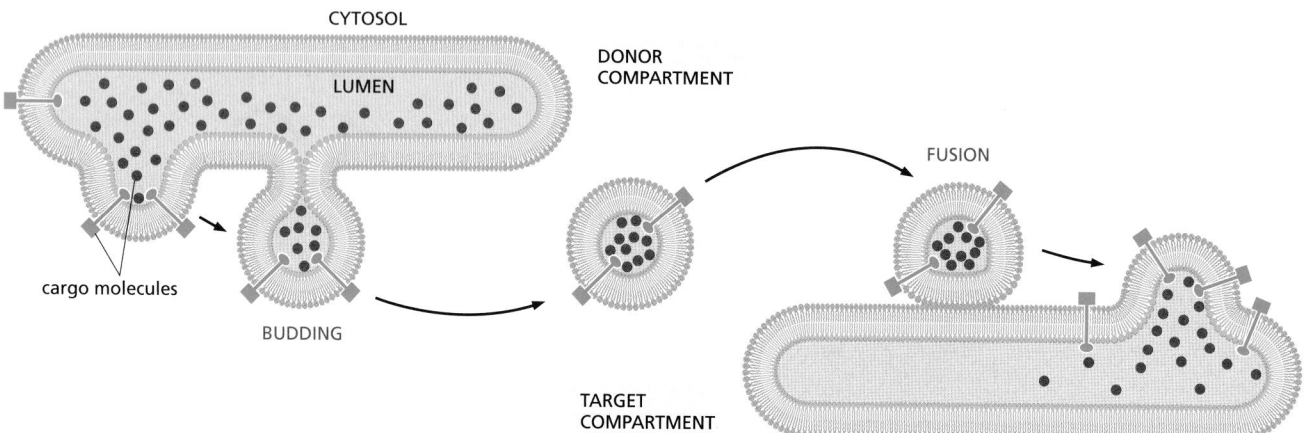

balance the flow of membrane between compartments in the opposite direction, bringing membrane and selected proteins back to the compartment of origin (**Figure 13–3**).

To perform its function, each transport vesicle that buds from a compartment must be selective. It must take up only the appropriate molecules and must fuse only with the appropriate target membrane. A vesicle carrying cargo from the ER to the Golgi apparatus, for example, must exclude most proteins that are to stay in the ER, and it must fuse only with the Golgi apparatus and not with any other organelle.

We begin this chapter by considering the molecular mechanisms of budding and fusion that underlie all vesicle transport. We then discuss the fundamental problem of how, in the face of this transport, the cell maintains the molecular and

Figure 13–2 Vesicle transport. Transport vesicles bud off from one compartment and fuse with another. As they do so, they carry material as cargo from the *lumen* (the space within a membrane-enclosed compartment) and membrane of the donor compartment to the lumen and membrane of the target compartment, as shown.

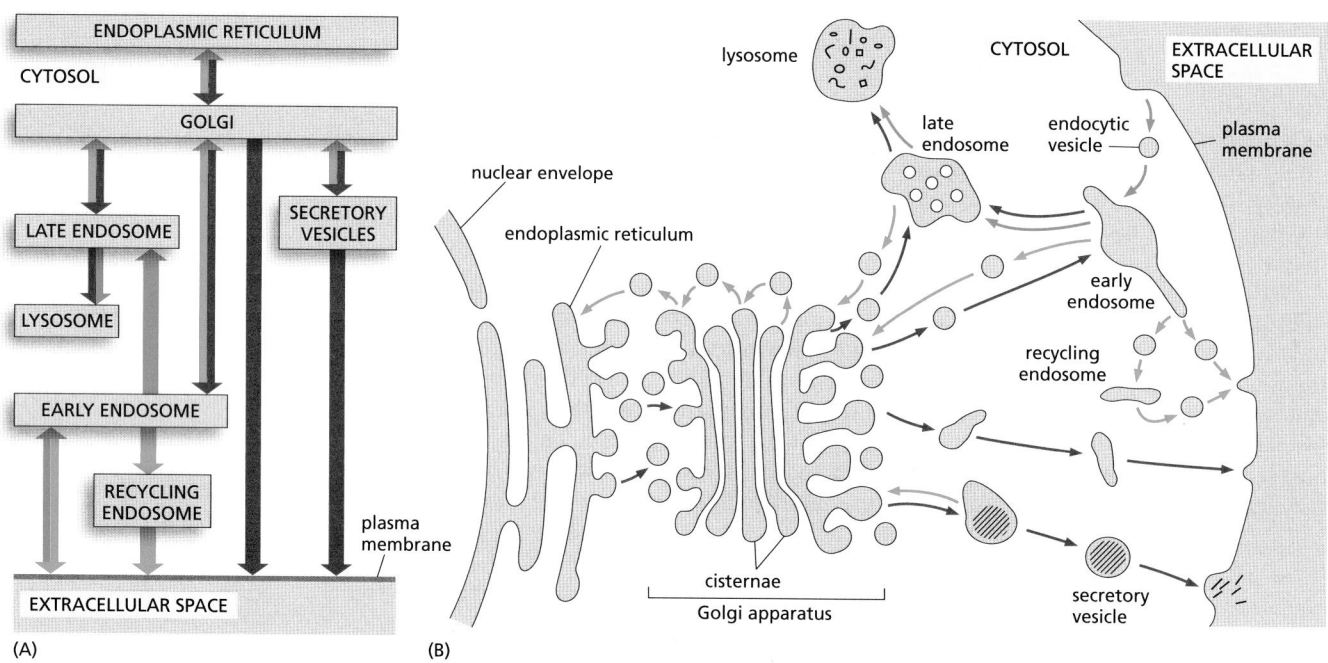

Figure 13–3 **A "road-map" of the secretory and endocytic pathways.** (A) In this schematic roadmap, which was introduced in Chapter 12, the endocytic and secretory pathways are illustrated with *green* and *red arrows*, respectively. In addition, *blue arrows* denote retrieval pathways for the backflow of selected components. (B) The compartments of the eukaryotic cell involved in vesicle transport. The lumen of each membrane-enclosed compartment is topologically equivalent to the outside of the cell. All compartments shown communicate with one another and the outside of the cell by means of transport vesicles. In the secretory pathway *(red arrows)*, protein molecules are transported from the ER to the plasma membrane or (via endosomes) to lysosomes. In the endocytic pathway *(green arrows)*, molecules are ingested in endocytic vesicles derived from the plasma membrane and delivered to early endosomes and then (via late endosomes) to lysosomes. Many endocytosed molecules are retrieved from early endosomes and returned (some via recycling endosomes) to the cell surface for reuse; similarly, some molecules are retrieved from the early and late endosomes and returned to the Golgi apparatus, and some are retrieved from the Golgi apparatus and returned to the ER. All of these retrieval pathways are shown with *blue arrows*, as in part (A).

functional differences between its compartments. Finally, we consider the function of the Golgi apparatus, lysosomes, secretory vesicles, and endosomes, as we trace the pathways that connect these organelles.

THE MOLECULAR MECHANISMS OF MEMBRANE TRANSPORT AND THE MAINTENANCE OF COMPARTMENTAL DIVERSITY

Vesicle transport mediates a continuous exchange of components between the ten or more chemically distinct, membrane-enclosed compartments that collectively comprise the secretory and endocytic pathways. With this massive exchange, how can each compartment maintain its special identity? To answer this question, we must first consider what defines the character of a compartment. Above all, it is the composition of the enclosing membrane: molecular markers displayed on the cytosolic surface of the membrane serve as guidance cues for incoming traffic to ensure that transport vesicles fuse only with the correct compartment. Many of these membrane markers, however, are found on more than one compartment, and it is the specific combination of marker molecules that gives each compartment its molecular address.

How are these membrane markers kept at high concentration on one compartment and at low concentration on another? To answer this question, we need to consider how patches of membrane, enriched or depleted in specific membrane components, bud off from one compartment and transfer to another.

We begin by discussing how cells segregate proteins into separate membrane domains by assembling a special protein coat on the membrane's cytosolic face. We consider how coats form, what they are made of, and how they are used to extract specific cargo components from a membrane and compartment lumen for delivery to another compartment. Finally, we discuss how transport vesicles dock at the appropriate target membrane and then fuse with it to deliver their cargo.

There Are Various Types of Coated Vesicles

Most transport vesicles form from specialized, coated regions of membranes. They bud off as **coated vesicles**, which have a distinctive cage of proteins covering their cytosolic surface. Before the vesicles fuse with a target membrane, they discard their coat, as is required for the two cytosolic membrane surfaces to interact directly and fuse.

The coat performs two main functions that are reflected in a common two-layered structure. First, an inner coat layer concentrates specific membrane proteins in a specialized patch, which then gives rise to the vesicle membrane. In this way, the inner layer selects the appropriate membrane molecules for transport. Second, an outer coat layer assembles into a curved, basketlike lattice that deforms the membrane patch and thereby shapes the vesicle.

There are three well-characterized types of coated vesicles, distinguished by their major coat proteins: *clathrin-coated*, *COPI-coated*, and *COPII-coated* (**Figure 13–4**). Each type is used for different transport steps. Clathrin-coated vesicles, for example, mediate transport from the Golgi apparatus and from the plasma membrane, whereas COPI- and COPII-coated vesicles most commonly mediate transport from the ER and from the Golgi cisternae (**Figure 13–5**). There is, however, much more variety in coated vesicles and their functions than this short list suggests. As we discuss below, there are several types of clathrin-coated vesicles, each specialized for a different transport step, and the COPI- and COPII-coated vesicles may be similarly diverse.

The Assembly of a Clathrin Coat Drives Vesicle Formation

Clathrin-coated vesicles, the first coated vesicles to be identified, transport material from the plasma membrane and between endosomal and Golgi compartments. **COPI-coated vesicles** and **COPII-coated vesicles** transport material early

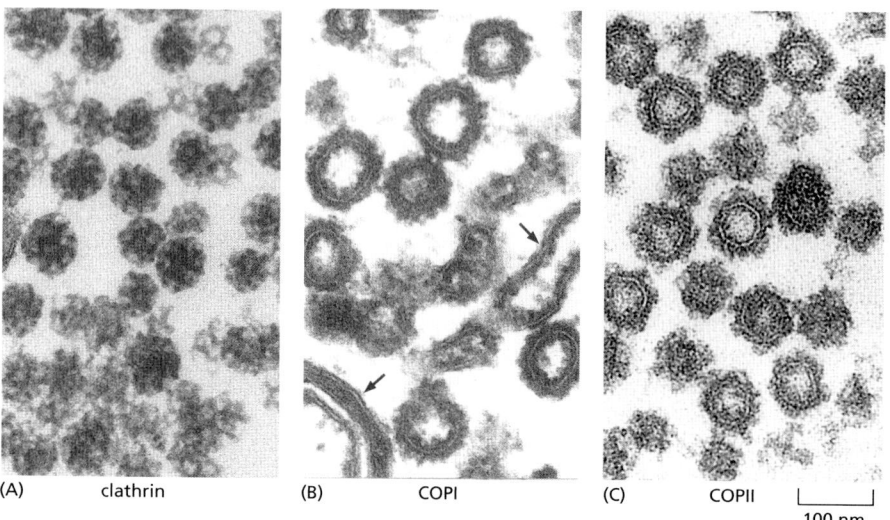

(A)   clathrin   (B)   COPI   (C)   COPII

100 nm

Figure 13–4 **Electron micrographs of clathrin-coated, COPI-coated, and COPII-coated vesicles.** All are shown in electron micrographs at the same scale. (A Clathrin-coated vesicles. (B) COPI-coated vesicles and Golgi cisternae (*red arrows*) from a cell-free system in which COPI-coated vesicles bud in the test tube. (C) COPII-coated vesicles. (A and B, from L. Orci, B. Glick and J. Rothman, *Cell* 46:171–184, 1986. With permission from Elsevier; C, courtesy of Charles Barlowe and Lelio Orci.)

in the secretory pathway: COPI-coated vesicles bud from Golgi compartments, and COPII-coated vesicles bud from the ER (see Figure 13–5). We discuss clathrin-coated vesicles first, as they provide a good example of how vesicles form.

The major protein component of clathrin-coated vesicles is **clathrin** itself, which forms the outer layer of the coat. Each clathrin subunit consists of three large and three small polypeptide chains that together form a three-legged structure called a *triskelion* (**Figure 13–6**A,B). Clathrin triskelions assemble into a basketlike framework of hexagons and pentagons to form coated pits (buds) on the cytosolic surface of membranes (**Figure 13–7**). Under appropriate conditions, isolated triskelions spontaneously self-assemble into typical polyhedral cages in a test tube, even in the absence of the membrane vesicles that these baskets normally enclose (Figure 13–6C,D). Thus, the clathrin triskelions determine the geometry of the clathrin cage (Figure 13–6E).

Adaptor Proteins Select Cargo into Clathrin-Coated Vesicles

Adaptor proteins, another major coat component in clathrin-coated vesicles, form a discrete inner layer of the coat, positioned between the clathrin cage and the membrane. They bind the clathrin coat to the membrane and trap various transmembrane proteins, including transmembrane receptors that capture soluble cargo molecules inside the vesicle—so-called *cargo receptors*. In this way, the adaptor proteins select a specific set of transmembrane proteins, together with the soluble proteins that interact with them, and package them into each newly formed clathrin-coated transport vesicle (**Figure 13–8**).

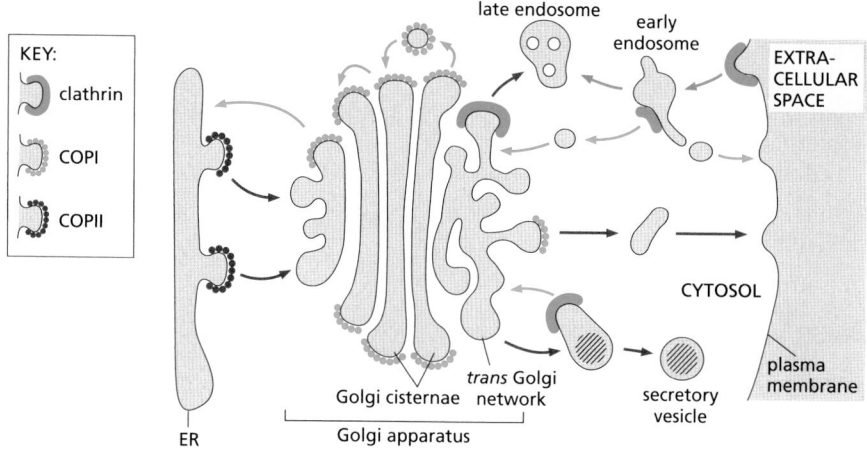

KEY:

clathrin

COPI

COPII

late endosome   early endosome   EXTRACELLULAR SPACE

CYTOSOL

trans Golgi network

Golgi cisternae   secretory vesicle   plasma membrane

Golgi apparatus

ER

Figure 13–5 **Use of different coats for different steps in vesicle traffic.** Different coat proteins select different cargo and shape the transport vesicles that mediate the various steps in the secretory and endocytic pathways. When the same coats function in different places in the cell, they usually incorporate different coat protein subunits that modify their properties (not shown). Many differentiated cells have additional pathways beside those shown here, including a sorting pathway from the *trans* Golgi network to the apical surface of epithelial cells and a specialized recycling pathway for proteins of synaptic vesicles in the nerve terminals of neurons (see Figure 11–36). The arrows are colored as in Figure 13–3.

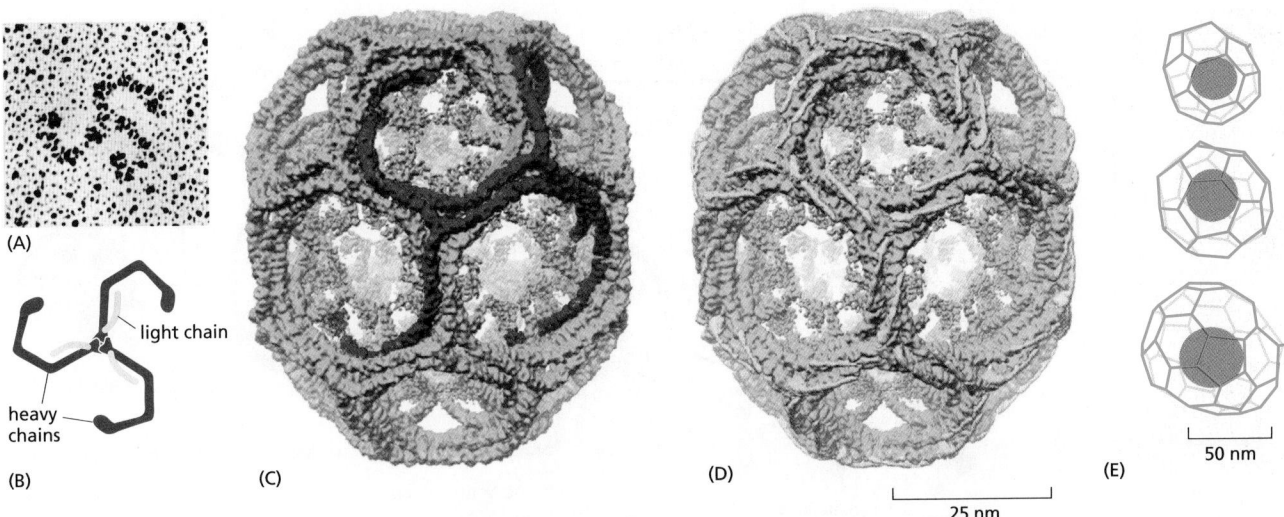

(A)

(B) light chain / heavy chains

(C)

(D)

25 nm

(E)

50 nm

Figure 13–6 The structure of a clathrin coat. (A) Electron micrograph of a clathrin triskelion shadowed with platinum. (B) Each triskelion is composed of three clathrin heavy chains and three clathrin light chains, as shown in the diagram. (C and D) A cryoelectron micrograph taken of a clathrin coat composed of 36 triskelions organized in a network of 12 pentagons and 6 hexagons, with some heavy chains (C) and light chains (D) highlighted (Movie 13.1). The light chains link to the actin cytoskeleton, which helps generate force for membrane budding and vesicle movement, and their phosphorylation regulates clathrin coat assembly. The interwoven legs of the clathrin triskelions form an outer shell from which the N-terminal domains of the triskelions protrude inward. These domains bind to the adaptor proteins shown in Figure 13–8. The coat shown was assembled biochemically from pure clathrin triskelions and is too small to enclose a membrane vesicle. (E) Images of clathrin-coated vesicles isolated from bovine brain. The clathrin coats are constructed in a similar but less regular way, from pentagons, a larger number of hexagons, and sometime heptagons, resembling the architecture of deformed soccer balls. The structures were determined by cryoelectron microscopy and tomographic reconstruction. (A, from E. Ungewickell and D. Branton, *Nature* 289:420–422, published 1981 by Macmillan Journals Ltd.; C and D, from A. Fotin et al., *Nature* 432:573–579, published 2004 by Nature Publishing Group; All reproduced with permission of SNCSC; E, from Y. Cheng et al., *J. Mol. Biol.* 365:892–899, 2007. With permission from Elsevier.)

There are several types of adaptor proteins. The best characterized have four different protein subunits; others are single-chain proteins. Each type of adaptor protein is specific for a different set of cargo receptors. Clathrin-coated vesicles budding from different membranes use different adaptor proteins and thus package different receptors and cargo molecules.

The assembly of adaptor proteins on the membrane is tightly controlled, in part by the cooperative interaction of the adaptor proteins with other components of the coat. The adaptor protein AP2 serves as a well-understood example. When it binds to a specific phosphorylated phosphatidylinositol lipid (a *phosphoinositide*), it alters its conformation, exposing binding sites for cargo receptors in the membrane. The simultaneous binding to the cargo receptors and lipid head groups greatly enhances the binding of AP2 to the membrane (Figure 13–9).

Because several requirements must be met simultaneously to stably bind AP2 proteins to a membrane, the proteins act as *coincidence detectors* that only assemble at the right time and place. Upon binding, they induce membrane curvature, which makes the binding of additional AP2 proteins in its proximity more likely. The cooperative assembly of the AP2 coat layer then is further amplified by clathrin binding, which leads to the formation and budding of a transport vesicle.

Adaptor proteins found in other coats also bind to phosphoinositides, which not only have a major role in directing when and where coats assemble in the cell, but also are used much more widely as molecular markers of compartment identity. This helps to control vesicular traffic, as we now discuss.

Figure 13–7 Clathrin-coated pits and vesicles. This rapid-freeze, deep-etch electron micrograph shows numerous clathrin-coated pits and vesicles on the inner surface of the plasma membrane of cultured fibroblasts. The cells were rapidly frozen in liquid helium, fractured, and deep-etched to expose the cytoplasmic surface of the plasma membrane. (Courtesy of John Heuser.)

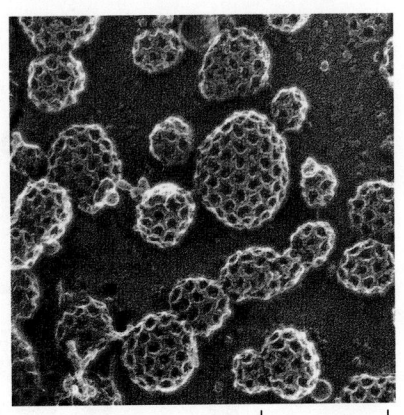

0.2 μm

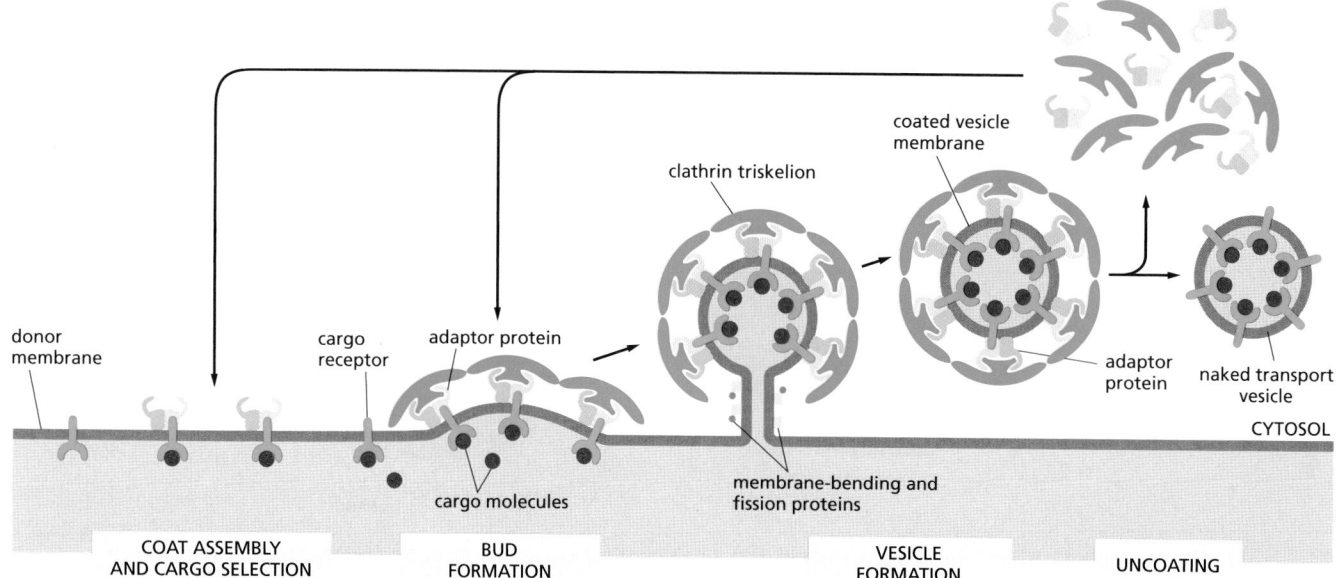

COAT ASSEMBLY
AND CARGO SELECTION

BUD
FORMATION

VESICLE
FORMATION

UNCOATING

Phosphoinositides Mark Organelles and Membrane Domains

Although inositol phospholipids typically comprise less than 10% of the total phospholipids in a membrane, they have important regulatory functions. They can undergo rapid cycles of phosphorylation and dephosphorylation at the 3′, 4′, and 5′ positions of their inositol sugar head groups to produce various types of **phosphoinositides** (**phosphatidylinositol phosphates**, or **PIPs**). The interconversion of phosphatidylinositol (PI) and PIPs is highly compartmentalized: different organelles in the endocytic and secretory pathways have distinct sets of PI and PIP kinases and PIP phosphatases (**Figure 13–10**). The distribution, regulation, and local balance of these enzymes determine the steady-state distribution of each PIP species. As a consequence, the distribution of PIPs varies from organelle to organelle, and often within a continuous membrane from one region to another, thereby defining specialized membrane domains.

Many proteins involved at different steps in vesicle transport contain domains that bind with high specificity to the head groups of particular PIPs, distinguishing one phosphorylated form from another (see Figure 13–10 E and F). Local control of the PI and PIP kinases and PIP phosphatases can therefore be used to rapidly control the binding of proteins to a membrane or membrane domain. The production of a particular type of PIP recruits proteins containing matching PIP-binding domains. The PIP-binding proteins then help regulate vesicle formation and other steps in the control of vesicle traffic (**Figure 13–11**). The same strategy is widely used to recruit specific intracellular signaling proteins to the plasma membrane in response to extracellular signals (discussed in Chapter 15).

Figure 13–8 The assembly and disassembly of a clathrin coat. The assembly of the coat introduces curvature into the membrane, which leads in turn to the formation of a coated bud (called a coated pit if it is in the plasma membrane). The adaptor proteins bind both clathrin triskelions and membrane-bound cargo receptors, thereby mediating the selective recruitment of both membrane and soluble cargo molecules into the vesicle. Other membrane-bending and fission proteins are recruited to the neck of the budding vesicle, where sharp membrane curvature is introduced. The coat is rapidly lost shortly after the vesicle buds off.

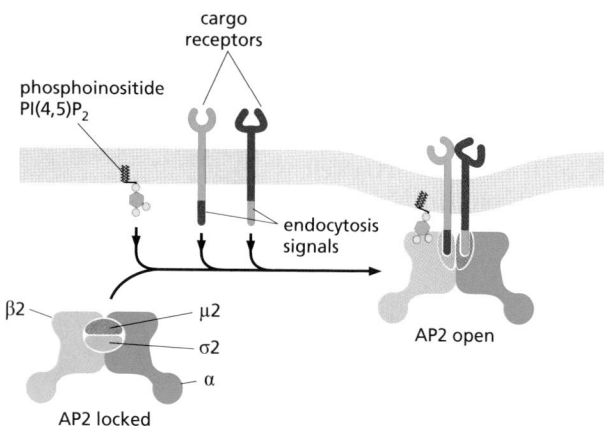

Figure 13–9 Lipid-induced conformation switching of AP2. The AP2 adaptor protein complex has four subunits (α, β2, μ2, and σ2). Upon interaction with the phosphoinositide PI(4,5)P$_2$ (see Figure 13–10) in the cytosolic leaflet of the plasma membrane, AP2 rearranges so that binding sites for cargo receptors become exposed. Each AP2 complex binds four PI(4,5)P$_2$ molecules (for clarity, only one is shown). In the open AP2 complex, the μ2 and σ2 subunits bind the cytosolic tails of cargo receptors that display the appropriate endocytosis signals. These signals consist of short amino acid sequence motifs. When AP2 binds tightly to the membrane, it induces curvature, which favors the binding of additional AP2 complexes in the vicinity.

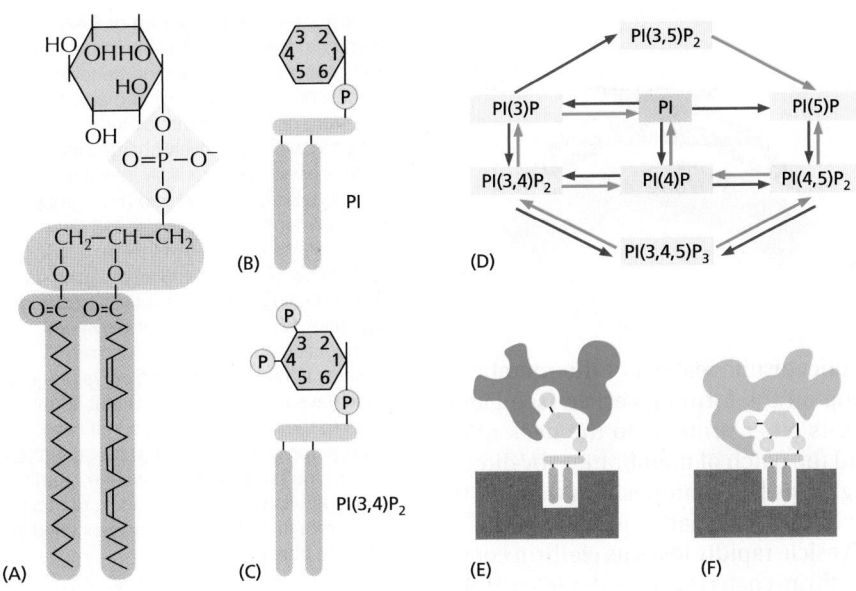

Figure 13–10 Phosphatidylinositol (PI) and phosphoinositides (PIPs). (A, B) The structure of PI shows the free hydroxyl groups in the inositol sugar that can in principle be modified. (C) Phosphorylation of one, two, or three of the hydroxyl groups on PI by PI and PIP kinases produces a variety of PIP species. They are named according to the ring position (in parentheses) and the number of phosphate groups (subscript) added to PI. $PI(3,4)P_2$ is shown. (D) Animal cells have several PI and PIP kinases and a similar number of PIP phosphatases, which are localized to different organelles, where they are regulated to catalyze the production of particular PIPs. The *red* and *green arrows* show the kinase and phosphatase reactions, respectively. (E, F) Phosphoinositide head groups are recognized by protein domains that discriminate between the different forms. In this way, select groups of proteins containing such domains are recruited to regions of membrane in which these phosphoinositides are present. $PI(3)P$ and $PI(4,5)P_2$ are shown. (D, modified from M.A. de Matteis and A. Godi, *Nat. Cell Biol.* 6:487–492, 2004.)

Membrane-Bending Proteins Help Deform the Membrane During Vesicle Formation

The forces generated by clathrin coat assembly alone are not sufficient to shape and pinch off a vesicle from the membrane. Other membrane-bending and force-generating proteins participate at every stage of the process. Membrane-bending proteins that contain crescent-shaped domains, called *BAR domains*, bind to and impose their shape on the underlying membrane via electrostatic interactions with the lipid head groups (**Figure 13–12**; also see Figure 10–40). Such BAR-domain proteins are thought to help AP2 nucleate clathrin-mediated endocytosis by shaping the plasma membrane to allow a clathrin-coated bud to form. Some of these proteins also contain amphiphilic helices that induce membrane curvature after being inserted as wedges into the cytoplasmic leaflet of the membrane. Other BAR-domain proteins are important in shaping the neck of a budding vesicle, where stabilization of sharp membrane bends is essential. Finally, the clathrin machinery nucleates the local assembly of actin filaments that introduce tension to help pinch off and propel the forming vesicle away from the membrane.

Cytoplasmic Proteins Regulate the Pinching-Off and Uncoating of Coated Vesicles

As a clathrin-coated bud grows, soluble cytoplasmic proteins, including **dynamin**, assemble at the neck of each bud (**Figure 13–13**). Dynamin contains a $PI(4,5)$ P_2-binding domain, which tethers the protein to the membrane, and a GTPase domain, which regulates the rate at which vesicles pinch off from the membrane.

Figure 13–11 The intracellular location of phosphoinositides. Different types of PIPs are located in different membranes and membrane domains, where they are often associated with specific vesicle transport events. The membrane of secretory vesicles, for example, contains $PI(4)P$. When the vesicles fuse with the plasma membrane, a PI 5-kinase that is localized there converts the $PI(4)P$ into $PI(4,5)P_2$. The $PI(4,5)P_2$, in turn, helps recruit adaptor proteins, which initiate the formation of a clathrin-coated pit, as the first step in clathrin-mediated endocytosis. Once the clathrin-coated vesicle buds off from the plasma membrane, a PI(5)P phosphatase hydrolyzes $PI(4,5)P_2$, which weakens the binding of the adaptor proteins, promoting vesicle uncoating. We discuss phagocytosis and the distinction between regulated and constitutive exocytosis later in the chapter. (Modified from M.A. de Matteis and A. Godi, *Nat. Cell Biol.* 6:487–492, 2004.)

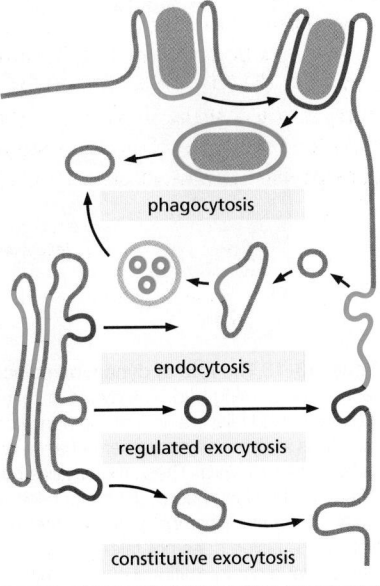

phagocytosis

endocytosis

regulated exocytosis

constitutive exocytosis

KEY: $PI(3)P$ $PI(4)P$ $PI(4,5)P_2$ $PI(3,5)P_2$ $PI(3,4,5)P_3$

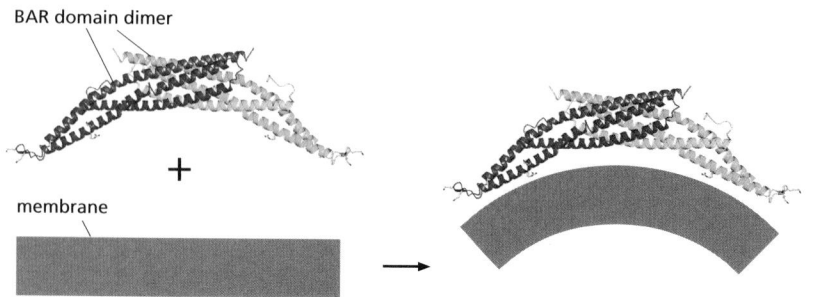

BAR domain dimer

membrane

Figure 13–12 The structure of BAR domains. BAR-domain proteins are diverse and enable many membrane-bending processes in the cell. BAR domains are built from coiled coils that dimerize into modules with a positively charged inner surface, which preferentially interacts with negatively charged lipid head groups to bend membranes. Local membrane deformations caused by BAR-domain proteins facilitate the binding of additional BAR-domain proteins, thereby generating a positive feedback cycle for curvature propagation. Individual BAR-domain proteins contain a distinctive curvature and often have additional features that adapt them to their specific tasks: some have short amphiphilic helices that cause further membrane deformation by wedge insertion; others are flanked by PIP-binding domains that direct them to membranes enriched in cognate phosphoinositides.

The pinching-off process brings the two noncytosolic leaflets of the membrane into close proximity and fuses them, sealing off the forming vesicle (see Figure 13–2). To perform this task, dynamin recruits other proteins to the neck of the bud. Together with dynamin, they help bend the patch of membrane—by directly distorting the bilayer structure, or by changing its lipid composition through the recruitment of lipid-modifying enzymes, or by both mechanisms.

Once released from the membrane, the vesicle rapidly loses its clathrin coat. A PIP phosphatase that is co-packaged into clathrin-coated vesicles depletes PI(4,5)P$_2$ from the membrane, which weakens the binding of the adaptor proteins. In addition, an hsp70 chaperone protein (see Figure 6–80) functions as an uncoating ATPase, using the energy of ATP hydrolysis to peel off the clathrin coat. *Auxilin*, another vesicle protein, is thought to activate the ATPase. The release of the coat, however, must not happen prematurely, so additional control mechanisms must somehow prevent the clathrin from being removed before it has formed a complete vesicle (discussed below).

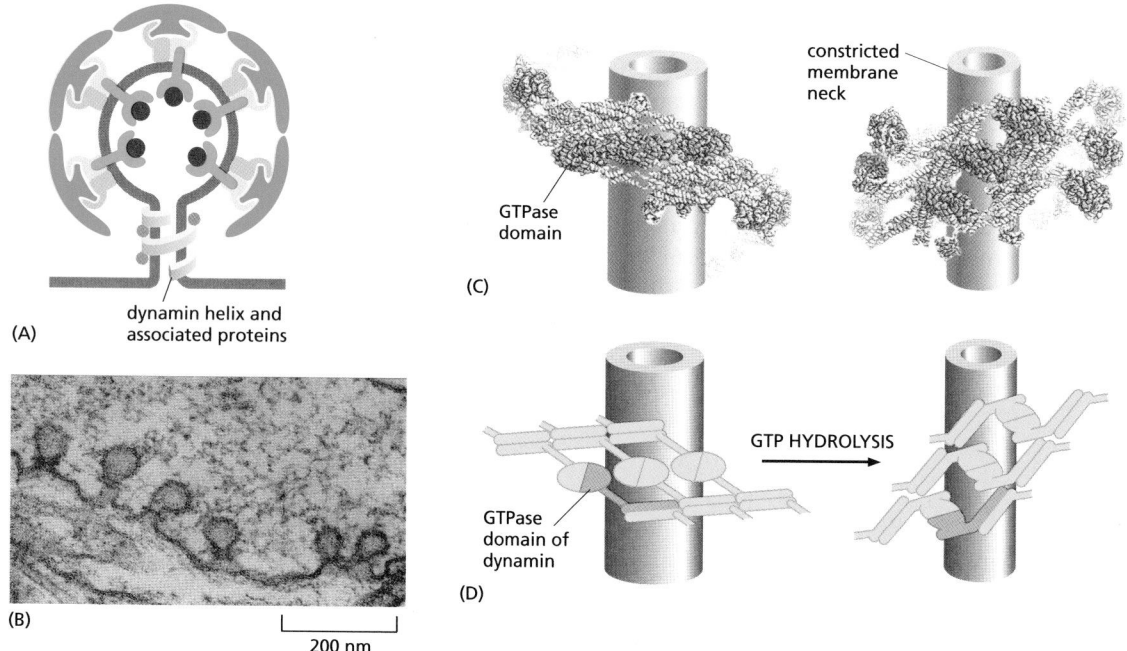

(A) dynamin helix and associated proteins

(B)

200 nm

(C) GTPase domain

constricted membrane neck

GTPase domain of dynamin

(D) GTP HYDROLYSIS

Figure 13–13 The role of dynamin in pinching off clathrin-coated vesicles. (A) Multiple dynamin molecules assemble into a spiral around the neck of the forming bud. The dynamin spiral is thought to recruit other proteins to the bud neck, which, together with dynamin, destabilize the interacting lipid bilayers so that the noncytoplasmic leaflets flow together. The newly formed vesicle then pinches off from the membrane. Specific mutations in dynamin can either enhance or block the pinching-off process. (B) Dynamin was discovered as the protein defective in the *shibire* mutant of *Drosophila*. These mutant flies become paralyzed because clathrin-mediated endocytosis stops, and the synaptic vesicle membrane fails to recycle, blocking neurotransmitter release. Deeply invaginated clathrin-coated pits form in the nerve endings of the fly's nerve cells, with a belt of mutant dynamin assembled around the neck, as shown in this thin-section electron micrograph. The pinching-off process fails because the required membrane fusion does not take place. (C, D) A model of how conformational changes in the GTPase domains of membrane-assembled dynamin may power a conformational change that constricts the neck of the bud. A single dynamin molecule is shown in *orange* in D. (B, from J.H. Koenig and K. Ikeda, *J. Neurosci.* 9:3844–3860, 1989. Copyright 1989 Society for Neuroscience. With permission from the Society for Neuroscience; C and D, adapted from M.G.J. Ford, S. Jenni, and J. Nunnari, *Nature* 477:561–566, 2011.)

Monomeric GTPases Control Coat Assembly

To balance the vesicle traffic to and from a compartment, coat proteins must assemble only when and where they are needed. While local production of PIPs plays a major part in regulating the assembly of clathrin coats on the plasma membrane and Golgi apparatus, cells superimpose additional ways of regulating coat formation. *Coat-recruitment GTPases*, for example, control the assembly of clathrin coats on endosomes and the COPI and COPII coats on Golgi and ER membranes.

Many steps in vesicle transport depend on a variety of GTP-binding proteins that control both the spatial and temporal aspects of vesicle formation and fusion. As discussed in Chapter 3, GTP-binding proteins regulate most processes in eukaryotic cells. They act as molecular switches, which flip between an active state with GTP bound and an inactive state with GDP bound. Two classes of proteins regulate the flipping: *guanine nucleotide exchange factors* (*GEFs*) activate the proteins by catalyzing the exchange of GDP for GTP, and *GTPase-activating proteins* (*GAPs*) inactivate the proteins by triggering the hydrolysis of the bound GTP to GDP (see Figures 3–68 and 15–7). Although both monomeric GTP-binding proteins (monomeric GTPases) and trimeric GTP-binding proteins (G proteins) have important roles in vesicle transport, the roles of the monomeric GTPases are better understood, and we focus on them here.

Coat-recruitment GTPases are members of a family of monomeric GTPases. They include the **ARF proteins**, which are responsible for the assembly of both COPI and clathrin coats assembly at Golgi membranes, and the **Sar1 protein**, which is responsible for the assembly of COPII coats at the ER membrane. Coat-recruitment GTPases are usually found in high concentration in the cytosol in an inactive, GDP-bound state. When a COPII-coated vesicle is to bud from the ER membrane, for example, a specific Sar1-GEF embedded in the ER membrane binds to cytosolic Sar1, causing the Sar1 to release its GDP and bind GTP in its place. (Recall that GTP is present in much higher concentration in the cytosol than GDP and therefore will spontaneously bind after GDP is released.) In its GTP-bound state, the Sar1 protein exposes an amphiphilic helix, which inserts into the cytoplasmic leaflet of the lipid bilayer of the ER membrane. The tightly bound Sar1 now recruits adaptor coat protein subunits to the ER membrane to initiate budding (**Figure 13–14**). Other GEFs and coat-recruitment GTPases operate in a similar way on other membranes.

The coat-recruitment GTPases also have a role in coat disassembly. The hydrolysis of bound GTP to GDP causes the GTPase to change its conformation so that its hydrophobic tail pops out of the membrane, causing the vesicle's coat to disassemble. Although it is not known what triggers the GTP hydrolysis, it has been proposed that the GTPases work like timers, which hydrolyze GTP at slow but predictable rates, to ensure that vesicle formation is synchronized with the requirements of the moment. COPII coats accelerate GTP hydrolysis by Sar1, and a fully formed vesicle will be produced only when bud formation occurs faster than the timed disassembly process; otherwise, disassembly will be triggered before a vesicle pinches off, and the process will have to start again, perhaps at a more appropriate time and place. Once a vesicle pinches off, GTP hydrolysis releases Sar1, but the sealed coat is sufficiently stabilized through many cooperative interactions, including binding to the cargo receptors in the membrane, that it may stay on the vesicle until the vesicle docks at a target membrane. There, a kinase phosphorylates the coat proteins, which completes coat disassembly and readies the vesicle for fusion.

Clathrin- and COPI-coated vesicles, by contrast, shed their coat soon after they pinch off. For COPI vesicles, the curvature of the vesicle membrane serves as a trigger to begin uncoating. An ARF-GAP is recruited to the COPI coat as it assembles. It interacts with the membrane, and senses the lipid packing density. It becomes activated when the curvature of the membrane approaches that of a transport vesicle. It then inactivates ARF, causing the coat to disassemble.

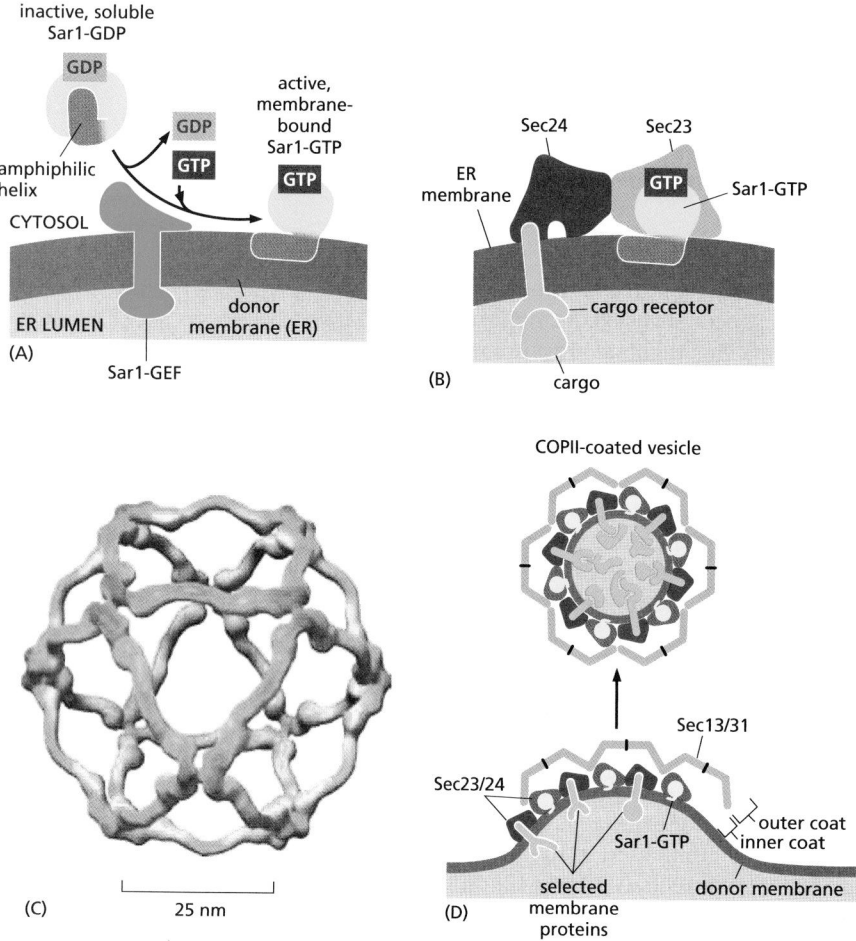

(A)

(B)

COPII-coated vesicle

(C) |—— 25 nm ——|

(D)

Figure 13–14 **Formation of a COPII-coated vesicle.** (A) Inactive, soluble Sar1-GDP binds to a Sar1-GEF in the ER membrane, causing the Sar1 to release its GDP and bind GTP. A GTP-triggered conformational change in Sar1 exposes an amphiphilic helix, which inserts into the cytoplasmic leaflet of the ER membrane, initiating membrane bending (which is not shown). (B) GTP-bound Sar1 binds to a complex of two COPII adaptor coat proteins, called Sec23 and Sec24, which form the inner coat. Sec24 has several different binding sites for the cytosolic tails of cargo receptors. The entire surface of the complex that attaches to the membrane is gently curved, matching the diameter of COPII-coated vesicles. (C) A complex of two additional COPII coat proteins, called Sec13 and Sec 31, forms the outer shell of the coat. Like clathrin, they can assemble on their own into symmetrical cages with appropriate dimensions to enclose a COPII-coated vesicle. (D) Membrane-bound, active Sar1-GTP recruits COPII adaptor proteins to the membrane. They select certain transmembrane proteins and cause the membrane to deform. The adaptor proteins then recruit the outer coat proteins which help form a bud. A subsequent membrane fusion event pinches off the coated vesicle. Other coated vesicles are thought to form in a similar way. (C, adapted from S.M. Stagg et al., *Nature* 439:234–238, published 2006 by Nature Publishing Group. Reproduced with permission of SNCSC.)

Not All Transport Vesicles Are Spherical

Although vesicle-budding is similar at various locations in the cell, each cell membrane poses its own special challenges. The plasma membrane, for example, is comparatively flat and stiff, owing to its cholesterol-rich lipid composition and underlying actin-rich cortex. Thus, the coordinated action of clathrin coats and membrane-bending proteins has to produce sufficient force to introduce curvature, especially at the neck of the bud where sharp bends are required for the pinching-off processes. In contrast, vesicle-budding from many intracellular membranes occurs preferentially at regions where the membranes are already curved, such as the rims of the Golgi cisternae or ends of membrane tubules. In these places, the primary function of the coats is to capture the appropriate cargo proteins rather than to deform the membrane.

Transport vesicles also occur in various sizes and shapes. Diverse COPII vesicles are required for the transport of large cargo molecules. Collagen, for example, is assembled in the ER as 300-nm-long, stiff procollagen rods that then are secreted from the cell where they are cleaved by proteases to collagen, which is embedded into the extracellular matrix (discussed in Chapter 19). Procollagen rods do not fit into the 60–80 nm COPII vesicles normally observed. To circumvent this problem, the procollagen cargo molecules bind to transmembrane *packaging proteins* in the ER, which control the assembly of the COPII coat components (**Figure 13–15**). These events drive the local assembly of much larger COPII vesicles that accommodate the oversized cargo. Human mutations in genes encoding such packaging proteins result in collagen defects with severe consequences, such as skeletal abnormalities and other developmental defects. Similar mechanisms must regulate the sizes of vesicles required to secrete other large macromolecular complexes, including the lipoprotein particles that transport lipids out of cells.

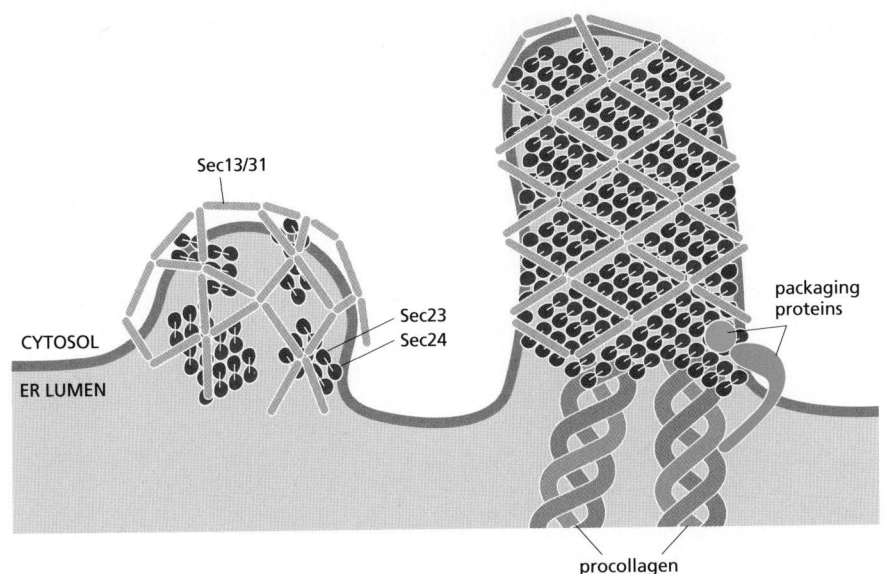

Sec13/31

Sec23
Sec24

CYTOSOL

ER LUMEN

packaging proteins

procollagen

Figure 13–15 Packaging of procollagen into large tubular COPII-coated vesicles. The cartoons show models for two COPII coat assembly modes. The models are based on cryoelectron tomography images of reconstituted COPII vesicles. On a spherical membrane (left), the Sec23/24 inner coat proteins assemble in patches that anchor the Sec13/31 outer coat protein cage. The Sec13/31 rods assemble a cage of triangles, squares, and pentagons. When procollagen needs to be packaged (right), special packaging proteins sense the cargo and modify the coat assembly process. This interaction recruits the COPII inner coat protein Sec24 and locally enhances the rate with which Sar1 cycles on and off the membrane (not shown). In addition, a monoubiquitin is added to the Sec31 protein, changing the assembly properties of the outer cage. Sec23/24 proteins arrange in larger arrays and Sec13/31 arrange in a regular lattice of diamond shapes. As the result, a large tubular vesicle is formed that can accommodate the large cargo molecules. The packaging proteins are not part of the budding vesicle but remain in the ER. (Modified from G. Zanetti et al., *eLife* 2:e00951, 2013. With permission from the authors.)

Many other vesicle budding events likewise involve variations of common mechanisms. When living cells are genetically engineered to express fluorescent membrane components, the endosomes and *trans* Golgi network are seen in a fluorescence microscope to continually send out long tubules. Coat proteins assemble onto the membrane tubules and help recruit specific cargo. The tubules then either regress or pinch off with the help of dynamin-like proteins to form transport vesicles of different sizes and shapes.

Tubules have a higher surface-to-volume ratio than the larger organelles from which they form. They are therefore relatively enriched in membrane proteins compared with soluble cargo proteins. As we discuss later, this property of tubules is an important feature for sorting proteins in endosomes.

Rab Proteins Guide Transport Vesicles to Their Target Membrane

To ensure an orderly flow of vesicle traffic, transport vesicles must be highly accurate in recognizing the correct target membrane with which to fuse. Because of the diversity and crowding of membrane systems in the cytoplasm, a vesicle is likely to encounter many potential target membranes before it finds the correct one. Specificity in targeting is ensured because all transport vesicles display surface markers that identify them according to their origin and type of cargo, and target membranes display complementary receptors that recognize the appropriate markers. This crucial process occurs in two steps. First, *Rab proteins* and *Rab effectors* direct the vesicle to specific spots on the correct target membrane. Second, *SNARE proteins* and *SNARE regulators* mediate the fusion of the lipid bilayers.

Rab proteins play a central part in the specificity of vesicle transport. Like the coat-recruitment GTPases discussed earlier (see Figure 13–14), Rab proteins are also monomeric GTPases. With over 60 known members, the Rab subfamily is the largest of the monomeric GTPase subfamilies. Each Rab protein is associated with one or more membrane-enclosed organelles of the secretory or endocytic pathways, and each of these organelles has at least one Rab protein on its cytosolic surface (**Table 13–1**). Their highly selective distribution on these membrane systems makes Rab proteins ideal molecular markers for identifying each membrane type and guiding vesicle traffic between them. Rab proteins can function on transport vesicles, on target membranes, or both.

Like the coat-recruitment GTPases, Rab proteins cycle between a membrane and the cytosol and regulate the reversible assembly of protein complexes on the membrane. In their GDP-bound state, they are inactive and bound to another protein (*Rab-GDP dissociation inhibitor*, or *GDI*) that keeps them soluble in the

TABLE 13–1 Subcellular Locations of Some Rab Proteins	
Protein	Organelle
Rab1	ER and Golgi complex
Rab2	*cis* Golgi network
Rab3A	Synaptic vesicles, secretory vesicles
Rab4/Rab11	Recycling endosomes
Rab5	Early endosomes, plasma membrane, clathrin-coated vesicles
Rab6	Medial and *trans* Golgi
Rab7	Late endosomes
Rab8	Cilia
Rab9	Late endosomes, *trans* Golgi

cytosol; in their GTP-bound state, they are active and tightly associated with the membrane of an organelle or transport vesicle. Membrane-bound Rab-GEFs activate Rab proteins on both transport vesicle and target membranes; for some membrane fusion events, activated Rab molecules are required on both sides of the reaction. Once in the GTP-bound state and membrane-bound through a now-exposed lipid anchor, Rab proteins bind to other proteins, called **Rab effectors**, which are the downstream mediators of vesicle transport, membrane tethering, and membrane fusion (**Figure 13–16**). The rate of GTP hydrolysis sets the concentration of active Rab and, consequently, the concentration of its effectors on the membrane.

In contrast to the highly conserved structure of Rab proteins, the structures and functions of Rab effectors vary greatly, and the same Rab proteins can often bind to many different effectors. Some Rab effectors are *motor proteins* that propel vesicles along actin filaments or microtubules to their target membrane. Others are *tethering proteins*, some of which have long, threadlike domains that serve as "fishing lines" that can extend to link two membranes more than 200 nm apart; other tethering proteins are large protein complexes that link two membranes that are closer together and interact with a wide variety of other proteins that facilitate the membrane fusion step. The tethering complex that docks COPII-coated

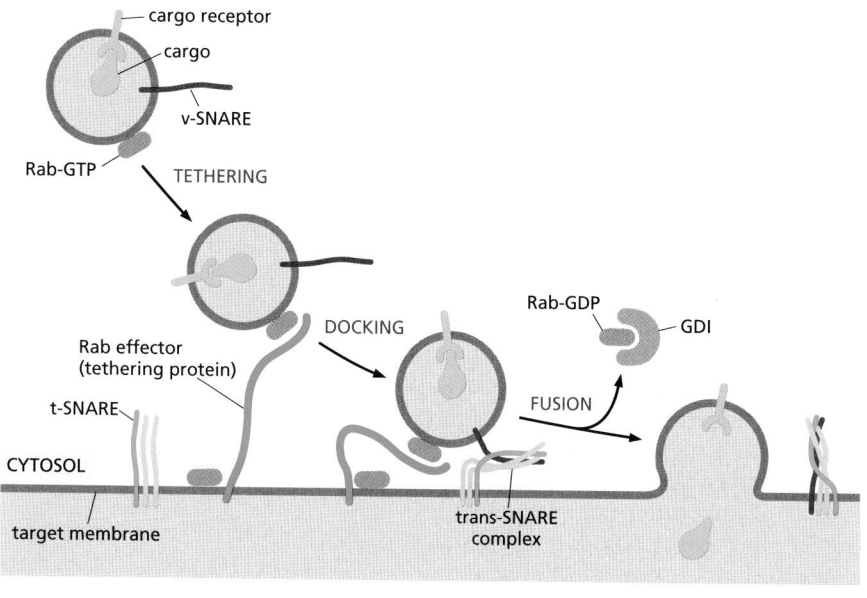

Figure 13–16 Tethering of a transport vesicle to a target membrane. Rab effector proteins interact with active Rab proteins (Rab-GTPs, *yellow*) located on the target membrane, vesicle membrane, or both, to establish the first connection between the two membranes that are going to fuse. In the example shown here, the Rab effector is a filamentous tethering protein *(dark green)*. Next, SNARE proteins on the two membranes *(red* and *blue)* pair, docking the vesicle to the target membrane and catalyzing the fusion of the two apposed lipid bilayers. During docking and fusion, a Rab-GAP (not shown) induces the Rab protein to hydrolyze its bound GTP to GDP, causing the Rab to dissociate from the membrane and return to the cytosol as Rab-GDP, where it is bound by a GDI protein that keeps the Rab soluble and inactive.

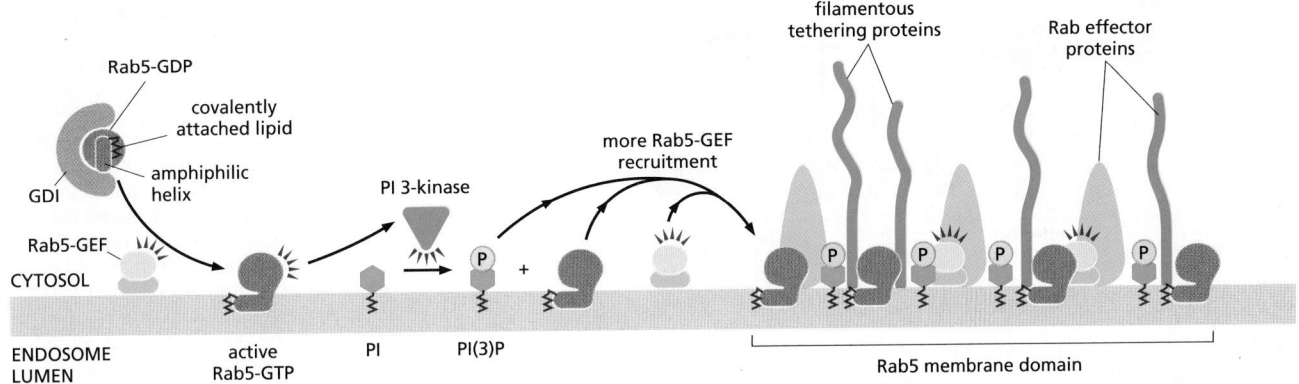

vesicles, for example, contains a protein kinase that phosphorylates the coat proteins to complete the uncoating process. Coupling uncoating to vesicle delivery helps to ensure directionality of the transport process and fusion with the proper membrane. Rab effectors can also interact with SNAREs to couple membrane tethering to fusion (see Figure 13–16).

The assembly of Rab proteins and their effectors on a membrane is cooperative and results in the formation of large, specialized membrane patches. Rab5, for example, assembles on endosomes and mediates the capture of endocytic vesicles arriving from the plasma membrane. The experimental depletion of Rab5 causes disappearance of the entire endosomal and lysosomal membrane system, highlighting the crucial role of Rab proteins in organelle biogenesis and maintenance.

A Rab5 domain concentrates tethering proteins that catch incoming vesicles. Its assembly on endosomal membranes begins when a Rab5-GDP/GDI complex encounters a Rab-GEF. GDI is released and Rab5-GDP is converted to Rab5-GTP. Active Rab5-GTP becomes anchored to the membrane and recruits more Rab5-GEF to the endosome, thereby stimulating the recruitment of more Rab5 to the same site. In addition, active Rab5 activates a PI 3-kinase, which locally converts PI to PI(3)P, which in turn binds some of the Rab effectors including tethering proteins and stabilizes their local membrane attachment (**Figure 13–17**). This type of positive feedback greatly amplifies the assembly process and helps to establish functionally distinct membrane domains within a continuous membrane.

The endosomal membrane provides a striking example of how different Rab proteins and their effectors help to create multiple specialized membrane domains, each fulfilling a particular set of functions. Thus, while the Rab5 membrane domain receives incoming endocytic vesicles from the plasma membrane, distinct Rab11 and Rab4 domains in the same membrane organize the budding of recycling vesicles that return proteins from the endosome to the plasma membrane.

Rab Cascades Can Change the Identity of an Organelle

A Rab domain can be disassembled and replaced by a different Rab domain, changing the identity of an organelle. Such ordered recruitment of sequentially acting Rab proteins is called a **Rab cascade**. Over time, for example, Rab5 domains are replaced by Rab7 domains on endosomal membranes. This converts an early endosome, marked by Rab5, into a late endosome, marked by Rab7. Because the set of Rab effectors recruited by Rab7 is different from that recruited by Rab5, this change reprograms the compartment: as we discuss later, it alters the membrane dynamics, including the incoming and outgoing traffic, and repositions the organelle away from the plasma membrane toward the cell interior. All of the cargo contained in the early endosome that has not been recycled to the plasma membrane is now part of a late endosome. This process is also referred to as *endosome maturation*. The self-amplifying nature of the Rab domains renders the process of endosome maturation unidirectional and irreversible (**Figure 13–18**).

Figure 13–17 The formation of a Rab5 domain on the endosome membrane. A Rab5-GEF on the endosome membrane binds a Rab5 protein and induces it to exchange GDP for GTP. GDI is lost and GTP binding alters the conformation of the Rab protein, exposing an amphiphilic helix and a covalently attached lipid group, which together anchor the Rab5-GTP to the membrane. Active Rab5 activates PI 3-kinase, which converts PI into PI(3)P. PI(3)P and active Rab5 together bind a variety of Rab effector proteins that contain PI(3)P-binding sites, including filamentous tethering proteins that catch incoming clathrin-coated endocytic vesicles from the plasma membrane. With the help of another effector, active Rab5 also recruits more Rab5-GEF, further enhancing the assembly of the Rab5 domain on the membrane.

Controlled cycles of GTP hydrolysis and GDP–GTP exchange dynamically regulate the size and activity of such Rab domains. Unlike SNAREs, which are integral membrane proteins, the GDP/GTP cycle, coupled to the membrane/cytosol translocation cycle, endows the Rab machinery with the ability to undergo assembly and disassembly on the membrane. (Adapted from M. Zerial and H. McBride, *Nat. Rev. Mol. Cell Biol.* 2:107–117, 2001.)

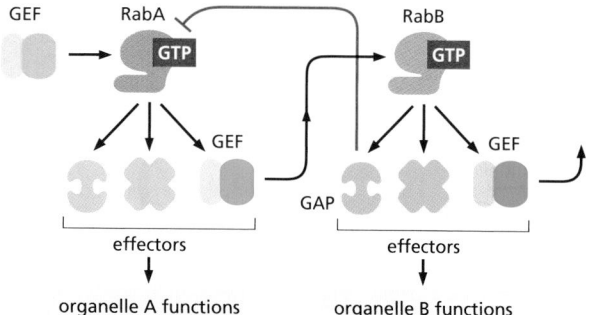

Figure 13–18 **A model for a generic Rab cascade.** The local activation of a RabA-GEF leads to assembly of a RabA domain on the membrane. Active RabA recruits its effector proteins, one of which is a GEF for RabB. The RabB-GEF then recruits RabB to the membrane, which in turn begins to recruit its effectors, among them a GAP for RabA. The RabA-GAP activates RabA GTP hydrolysis leading to the inactivation of the RabA and the disassembly of the RabA domain as the RabB domain grows. In this way, the RabA domain is irreversibly replaced by the RabB domain. In principle, this sequence can be continued by the recruitment of a next GEF by RabB. (Adapted from A.H. Hutagalung and P.J. Novick, *Physiol. Rev.* 91:119–149, 2011.)

SNAREs Mediate Membrane Fusion

Once a transport vesicle has been tethered to its target membrane, it unloads its cargo by membrane fusion. Membrane fusion requires bringing the lipid bilayers of two membranes to within 1.5 nm of each other so that they can merge. When the membranes are in such close apposition, lipids can flow from one bilayer to the other. For this close approach, water must be displaced from the hydrophilic surface of the membrane—a process that is highly energetically unfavorable and requires specialized *fusion proteins* that overcome this energy barrier. We have already discussed the role of dynamin in a related task during the pinching-off of clathrin-coated vesicles (see Figure 13–13).

The **SNARE proteins** (also called **SNAREs**, for short) catalyze the membrane fusion reactions in vesicle transport. There are at least 35 different SNAREs in an animal cell, each associated with a particular organelle in the secretory or endocytic pathways. These transmembrane proteins exist as complementary sets, with **v-SNAREs** usually found on vesicle membranes and **t-SNAREs** usually found on target membranes (see Figure 13–16). A v-SNARE is a single polypeptide chain, whereas a t-SNARE is usually composed of three proteins. The v-SNAREs and t-SNAREs have characteristic helical domains, and when a v-SNARE interacts with a t-SNARE, the helical domains of one wrap around the helical domains of the other to form a very stable four-helix bundle. The resulting *trans-SNARE complex* locks the two membranes together. Biochemical membrane fusion assays with all different SNARE combinations show that v- and t-SNARE pairing is highly specific. The SNAREs thus provide an additional layer of specificity in the transport process by helping to ensure that vesicles fuse only with the correct target membrane.

The trans-SNARE complexes catalyze membrane fusion by using the energy that is freed when the interacting helices wrap around each other to pull the membrane faces together, simultaneously squeezing out water molecules from the interface (**Figure 13–19**). When liposomes containing purified v-SNAREs are mixed with liposomes containing complementary t-SNAREs, their membranes fuse, albeit slowly. In the cell, other proteins recruited to the fusion site, presumably Rab effectors, cooperate with SNAREs to accelerate fusion. Fusion does not always follow immediately after v-SNAREs and t-SNAREs pair. As we discuss later, in the process of regulated exocytosis, fusion is delayed until secretion is triggered by a specific extracellular signal.

Rab proteins, which can regulate the availability of SNARE proteins, exert an additional layer of control. t-SNAREs in target membranes are often associated with inhibitory proteins that must be released before the t-SNARE can function. Rab proteins and their effectors trigger the release of such SNARE inhibitory

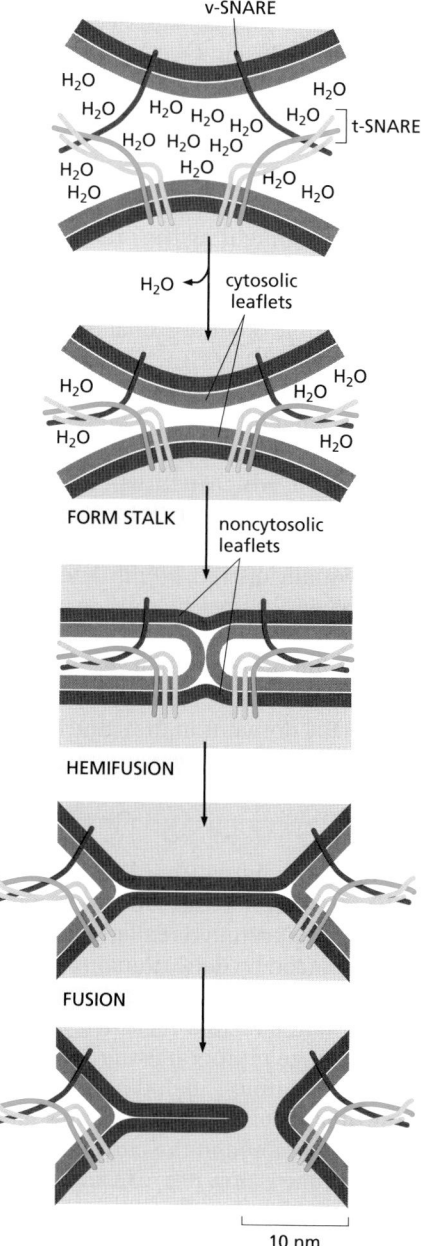

Figure 13–19 **A model for how SNARE proteins may catalyze membrane fusion.** Bilayer fusion occurs in multiple steps. A tight pairing between v- and t-SNAREs forces lipid bilayers into close apposition and expels water molecules from the interface. Lipid molecules in the two interacting (cytosolic) leaflets of the bilayers then flow between the membranes to form a connecting stalk. Lipids of the two noncytosolic leaflets then contact each other, forming a new bilayer, which widens the fusion zone (*hemifusion*, or half-fusion). Rupture of the new bilayer completes the fusion reaction.

proteins. In this way, SNARE proteins are concentrated and activated in the correct location on the membrane, where tethering proteins capture incoming vesicles. Rab proteins thus speed up the process by which appropriate SNARE proteins in two membranes find each other.

For vesicle transport to operate normally, transport vesicles must incorporate the appropriate SNARE and Rab proteins. Not surprisingly, therefore, many transport vesicles will form only if they incorporate the appropriate complement of SNARE and Rab proteins in their membrane. How this crucial control process operates during vesicle budding remains a mystery.

Interacting SNAREs Need to Be Pried Apart Before They Can Function Again

Most SNARE proteins in cells have already participated in multiple rounds of vesicle transport and are sometimes present in a membrane as stable complexes with partner SNAREs. The complexes have to disassemble before the SNAREs can mediate new rounds of transport. A crucial protein called **NSF** cycles between membranes and the cytosol and catalyzes the disassembly process. NSF is a hexameric ATPase of the family of AAA-ATPases (see Figure 6–85) that uses the energy of ATP hydrolysis to unravel the intimate interactions between the helical domains of paired SNARE proteins (**Figure 13–20**). The requirement for NSF-mediated reactivation of SNAREs by SNARE complex disassembly helps prevent membranes from fusing indiscriminately: if the t-SNAREs in a target membrane were always active, any membrane containing an appropriate v-SNARE might fuse whenever the two membranes made contact. It is not known how the activity of NSF is controlled so that the SNARE machinery is activated at the right time and place. It is also not known how v-SNAREs are selectively retrieved and returned to their compartment of origin so that they can be reused in newly formed transport vesicles.

Membrane fusion is important in other processes beside vesicle transport. The plasma membranes of a sperm and an egg fuse during fertilization, and myoblasts fuse with one another during the development of multinucleate muscle fibers (discussed in Chapter 22). Likewise, the ER network and mitochondria fuse and fragment in a dynamic way (discussed in Chapters 12 and 14). All cell membrane fusions require special proteins and are tightly regulated to ensure that only appropriate membranes fuse. The controls are crucial for maintaining both the identity of cells and the individuality of each type of intracellular compartment.

The membrane fusions catalyzed by viral fusion proteins are well understood. These proteins have a crucial role in permitting the entry of enveloped viruses (which have a lipid-bilayer-based membrane coat) into the cells that they infect (discussed in Chapters 5 and 23). For example, viruses such as the human immunodeficiency virus (HIV), which causes AIDS, bind to cell-surface receptors and then fuse with the plasma membrane of the target cell (**Figure 13–21**). This fusion event allows the viral nucleic acid inside the nucleocapsid to enter the cytosol, where it replicates. Other viruses, such as the influenza virus, first enter the cell by receptor-mediated endocytosis (discussed later) and are delivered to endosomes; the low pH in endosomes activates a fusion protein in the viral envelope that catalyzes the fusion of the viral and endosomal membranes, releasing the viral nucleic acid into the cytosol. Viral fusion proteins and SNAREs promote lipid bilayer fusion in similar ways.

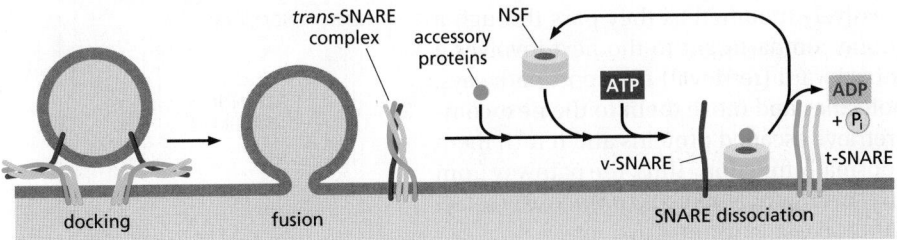

Figure 13–20 **Dissociation of SNARE pairs by NSF after a membrane fusion cycle.** After a v-SNARE and t-SNARE have mediated the fusion of a transport vesicle with a target membrane, NSF binds to the SNARE complex and, with the help of accessory proteins, hydrolyzes ATP to pry the SNAREs apart.

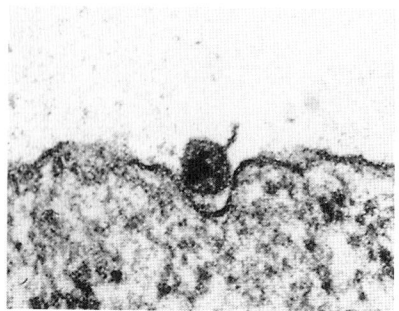 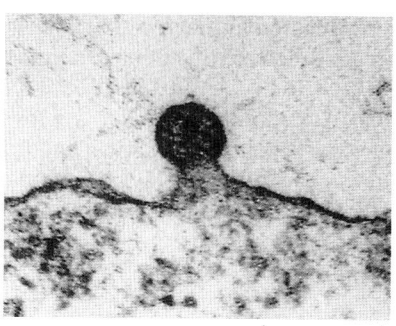

200 nm

Figure 13–21 **The entry of enveloped viruses into cells.** Electron micrographs showing how HIV enters a cell by fusing its membrane with the plasma membrane of the cell. (From B.S. Stein et al., *Cell* 49:659–668, 1987. With permission from Elsevier.)

Summary

Directed and selective transport of particular membrane components from one membrane-enclosed compartment to another in a eukaryotic cell maintains the differences between those compartments. Transport vesicles, which can be spherical, tubular, or irregularly shaped, bud from specialized coated regions of the donor membrane. The assembly of the coat helps to collect specific membrane and soluble cargo molecules for transport and to drive the formation of the vesicle.

There are various types of coated vesicles. The best characterized are clathrin-coated vesicles, which mediate transport from the plasma membrane and the trans Golgi network, and COPI- and COPII-coated vesicles, which mediate transport between Golgi cisternae and between the ER and the Golgi apparatus, respectively. Coats have a common two-layered structure: an inner layer formed of adaptor proteins links the outer layer (or cage) to the vesicle membrane and also traps specific cargo molecules for packaging into the vesicle. The coat is shed before the vesicle fuses with its appropriate target membrane.

Local synthesis of specific phosphoinositides creates binding sites that trigger clathrin coat assembly and vesicle budding. In addition, monomeric GTPases help regulate various steps in vesicle transport, including both vesicle budding and docking. The coat-recruitment GTPases, including Sar1 and the ARF proteins, regulate coat assembly and disassembly. A large family of Rab proteins functions as vesicle-targeting GTPases. Rab proteins are recruited to both, forming transport vesicles and target membranes. The assembly and disassembly of Rab proteins and their effectors in specialized membrane domains are dynamically controlled by GTP binding and hydrolysis. Active Rab proteins recruit Rab effectors, such as motor proteins, which transport vesicles along actin filaments or microtubules, and filamentous tethering proteins, which help ensure that the vesicles deliver their contents only to the appropriate target membrane. Complementary v-SNARE proteins on transport vesicles and t-SNARE proteins on the target membrane form stable trans-SNARE complexes, which force the two membranes into close apposition so that their lipid bilayers can fuse.

TRANSPORT FROM THE ER THROUGH THE GOLGI APPARATUS

As discussed in Chapter 12, newly synthesized proteins cross the ER membrane from the cytosol to enter the secretory pathway. During their subsequent transport, from the ER to the Golgi apparatus and from the Golgi apparatus to the cell surface and elsewhere, these proteins are successively modified as they pass through a series of compartments. Transfer from one compartment to the next involves a delicate balance between forward and backward (retrieval) transport pathways. Some transport vesicles select cargo molecules and move them to the next compartment in the pathway, while others retrieve escaped proteins and return them to a previous compartment where they normally function. Thus, the pathway from the ER to the cell surface consists of many sorting steps, which continuously select membrane and soluble lumenal proteins for packaging and transport.

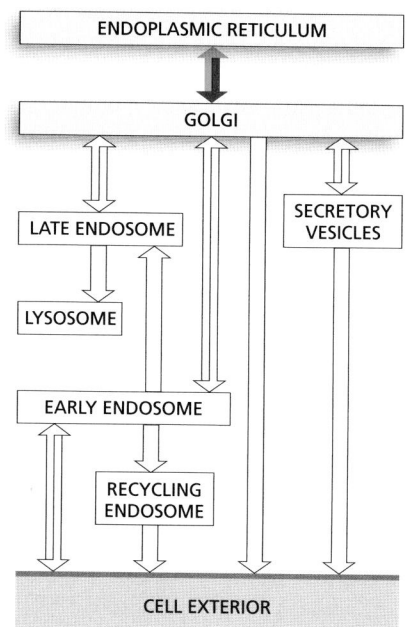

In this section, we focus mainly on the **Golgi apparatus** (also called the **Golgi complex**). It is a major site of carbohydrate synthesis, as well as a sorting and dispatching station for products of the ER. The cell makes many polysaccharides in the Golgi apparatus, including the pectin and hemicellulose of the cell wall in plants and most of the glycosaminoglycans of the extracellular matrix in animals (discussed in Chapter 19). The Golgi apparatus also lies on the exit route from the ER, and a large proportion of the carbohydrates that it makes are attached as oligosaccharide side chains to the many proteins and lipids that the ER sends to it. A subset of these oligosaccharide groups serve as tags to direct specific proteins into vesicles that then transport them to lysosomes. But most proteins and lipids, once they have acquired their appropriate oligosaccharides in the Golgi apparatus, are recognized in other ways for targeting into the transport vesicles going to other destinations.

Proteins Leave the ER in COPII-Coated Transport Vesicles

To initiate their journey along the secretory pathway, proteins that have entered the ER and are destined for the Golgi apparatus or beyond are first packaged into COPII-coated transport vesicles. These vesicles bud from specialized regions of the ER called *ER exit sites*, whose membrane lacks bound ribosomes. Most animal cells have ER exit sites dispersed throughout the ER network.

Entry into vesicles that leave the ER can be a selective process or can happen by default. Many membrane proteins are actively recruited into such vesicles, where they become concentrated. These cargo membrane proteins display exit (transport) signals on their cytosolic surface that adaptor proteins of the inner COPII coat recognize (**Figure 13–22**); some of these components act as cargo receptors and are recycled back to the ER after they have delivered their cargo to the Golgi apparatus. Soluble cargo proteins in the ER lumen, by contrast, have exit signals that attach them to transmembrane cargo receptors. Proteins without exit signals can also enter transport vesicles, including protein molecules that normally function in the ER (so-called *ER resident proteins*), some of which slowly leak out of the ER and are delivered to the Golgi apparatus. Different cargo proteins enter the transport vesicles with substantially different rates and efficiencies, which may result from differences in their folding and oligomerization efficiencies and kinetics, as well as the factors already discussed. The exit step from the ER is a major checkpoint at which quality control is exerted on the proteins that a cell secretes or displays on its surface, as we discussed in Chapter 12.

The exit signals that direct soluble proteins out of the ER for transport to the Golgi apparatus and beyond are not well understood. Some transmembrane proteins that serve as cargo receptors for packaging some secretory proteins into COPII-coated vesicles are lectins that bind to oligosaccharides on the secreted

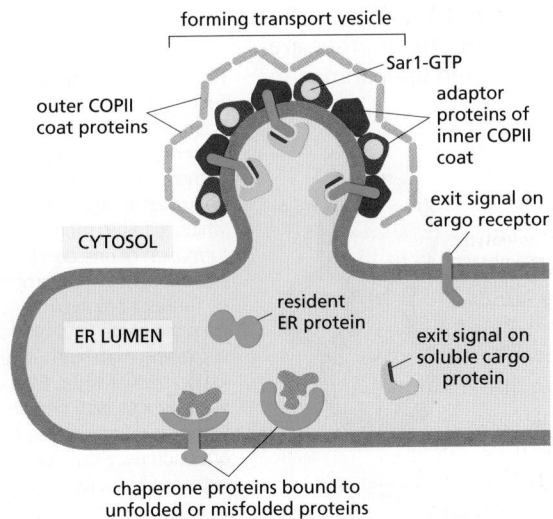

forming transport vesicle

outer COPII coat proteins

Sar1-GTP

adaptor proteins of inner COPII coat

exit signal on cargo receptor

CYTOSOL

ER LUMEN

resident ER protein

exit signal on soluble cargo protein

chaperone proteins bound to unfolded or misfolded proteins

Figure 13–22 The recruitment of membrane and soluble cargo molecules into ER transport vesicles. Membrane proteins are packaged into budding transport vesicles through interactions of exit signals on their cytosolic tails with adaptor proteins of the inner COPII coat. Some of these membrane proteins function as cargo receptors, binding soluble proteins in the ER lumen and helping to package them into vesicles. Other proteins may enter the vesicle by bulk flow. A typical 50 nm transport vesicle contains about 200 membrane proteins, which can be of many different types. As indicated, unfolded or incompletely assembled proteins are bound to chaperones and transiently retained in the ER compartment.

proteins. One such lectin, for example, binds to mannose on two secreted blood-clotting factors (Factor V and Factor VIII), thereby packaging the proteins into transport vesicles in the ER; its role in protein transport was identified because humans who lack it owing to an inherited mutation have lowered serum levels of Factors V and VIII, and they therefore bleed excessively.

Only Proteins That Are Properly Folded and Assembled Can Leave the ER

To exit from the ER, proteins must be properly folded and, if they are subunits of multiprotein complexes, they need to be completely assembled. Those that are misfolded or incompletely assembled transiently remain in the ER, where they are bound to chaperone proteins (discussed in Chapter 6) such as *BiP* or *calnexin*. The chaperones may cover up the exit signals or somehow anchor the proteins in the ER. Such failed proteins are eventually transported back into the cytosol, where they are degraded by proteasomes (discussed in Chapters 6 and 12). This quality-control step prevents the onward transport of misfolded or misassembled proteins that could potentially interfere with the functions of normal proteins. Such failures are surprisingly common. More than 90% of the newly synthesized subunits of the T cell receptor (discussed in Chapter 24) and of the acetylcholine receptor (discussed in Chapter 11), for example, are normally degraded without ever reaching the cell surface where they function. Thus, cells must make a large excess of some protein molecules to produce a select few that fold, assemble, and function properly.

Sometimes, however, there are drawbacks to the stringent quality-control mechanism. The predominant mutations that cause cystic fibrosis, a common inherited disease, result in the production of a slightly misfolded form of a plasma membrane protein important for Cl⁻ transport. Although the mutant protein would function normally if it reached the plasma membrane, it is retained in the ER and then is degraded by cytosolic proteasomes. This devastating disease thus results not because the mutation inactivates the protein but because the active protein is discarded before it reaches the plasma membrane.

Vesicular Tubular Clusters Mediate Transport from the ER to the Golgi Apparatus

After transport vesicles have budded from ER exit sites and have shed their coat, they begin to fuse with one another. The fusion of membranes from the same compartment is called *homotypic fusion*, to distinguish it from *heterotypic fusion*, in which a membrane from one compartment fuses with the membrane of a different compartment. As with heterotypic fusion, homotypic fusion requires a set of matching SNAREs. In this case, however, the interaction is symmetrical, with both membranes contributing v-SNAREs and t-SNAREs (**Figure 13–23**).

The structures formed when ER-derived vesicles fuse with one another are called *vesicular tubular clusters*, because they have a convoluted appearance in

Figure 13–23 Homotypic membrane fusion. In step 1, NSF pries apart identical pairs of v-SNAREs and t-SNAREs in both membranes (see Figure 13–20). In steps 2 and 3, the separated matching SNAREs on adjacent identical membranes interact, which leads to membrane fusion and the formation of one continuous compartment. Subsequently, the compartment grows by further homotypic fusion with vesicles from the same kind of membrane, displaying matching SNAREs. Homotypic fusion occurs when ER-derived transport vesicles fuse with one another, but also when endosomes fuse to generate larger endosomes. Rab proteins help regulate the extent of homotypic fusion and hence the size of a cell's compartments (not shown).

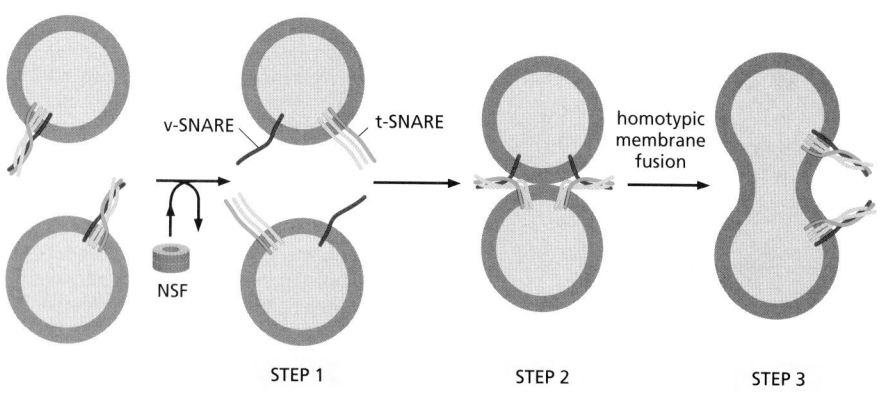

v-SNARE   t-SNARE   homotypic membrane fusion

NSF

STEP 1   STEP 2   STEP 3

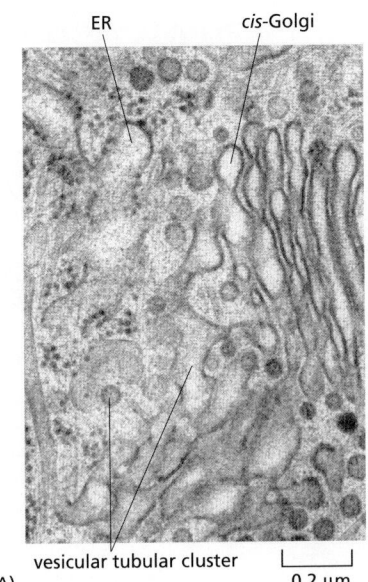

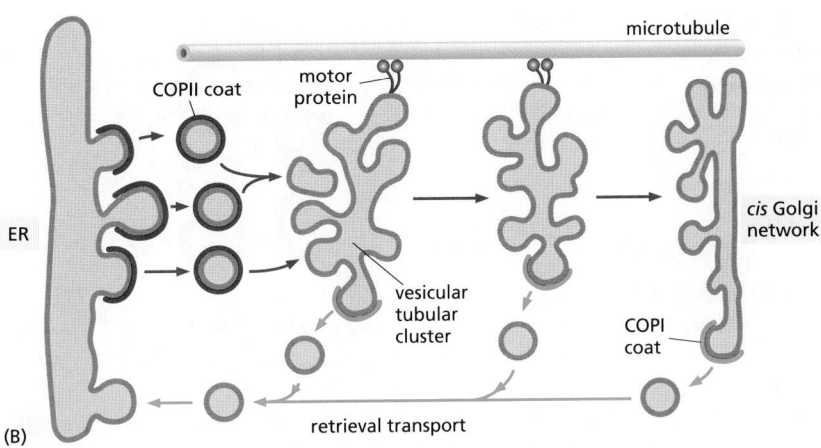

Figure 13–24 Vesicular tubular clusters. (A) An electron micrograph of vesicular tubular clusters forming around an exit site. Many of the vesicle-like structures seen in the micrograph are cross sections of tubules that extend above and below the plane of this thin section and are interconnected. (B) Vesicular tubular clusters move along microtubules to carry proteins from the ER to the Golgi apparatus. COPI-coated vesicles mediate the budding of vesicles that return to the ER from these clusters (and from the Golgi apparatus). (A, from J.A. Martinez-Menárguez et al., *Cell* 98(1):81–90, 1999. With permission from Elsevier.)

the electron microscope (**Figure 13–24A**). These clusters constitute a compartment that is separate from the ER and lacks many of the proteins that function in the ER. They are generated continually and function as transport containers that bring material from the ER to the Golgi apparatus. The clusters move quickly along microtubules to the Golgi apparatus with which they fuse (Figure 13–24B and **Movie 13.2**).

As soon as vesicular tubular clusters form, they begin to bud off transport vesicles of their own. Unlike the COPII-coated vesicles that bud from the ER, these vesicles are COPI-coated (see Figure 13–24A). COPI-coated vesicles are unique in that the components that make up the inner and outer coat layers are recruited as a preassembled complex, called *coatomer*. They function as a *retrieval pathway*, carrying back ER resident proteins that have escaped, as well as proteins such as cargo receptors and SNAREs that participated in the ER budding and vesicle fusion reactions. This retrieval process demonstrates the exquisite control mechanisms that regulate coat assembly reactions. The COPI coat assembly begins only seconds after the COPII coats have been shed, and remains a mystery how this switch in coat assembly is controlled.

The retrieval (or retrograde) transport continues as the vesicular tubular clusters move toward the Golgi apparatus. Thus, the clusters continuously mature, gradually changing their composition as selected proteins are returned to the ER. The retrieval continues from the Golgi apparatus, after the vesicular tubular clusters have delivered their cargo.

The Retrieval Pathway to the ER Uses Sorting Signals

The retrieval pathway for returning escaped proteins back to the ER depends on *ER retrieval signals*. Resident ER membrane proteins, for example, contain signals that bind directly to COPI coats and are thus packaged into COPI-coated transport vesicles for retrograde delivery to the ER. The best-characterized retrieval signal of this type consists of two lysines, followed by any two other amino acids, at the extreme C-terminal end of the ER membrane protein. It is called a *KKXX sequence*, based on the single-letter amino acid code.

Soluble ER resident proteins, such as BiP, also contain a short ER retrieval signal at their C-terminal end, but it is different: it consists of a Lys-Asp-Glu-Leu or a similar sequence. If this signal (called the *KDEL sequence*) is removed from BiP by genetic engineering, the protein is slowly secreted from the cell. If the signal is transferred to a protein that is normally secreted, the protein is now efficiently returned to the ER, where it accumulates.

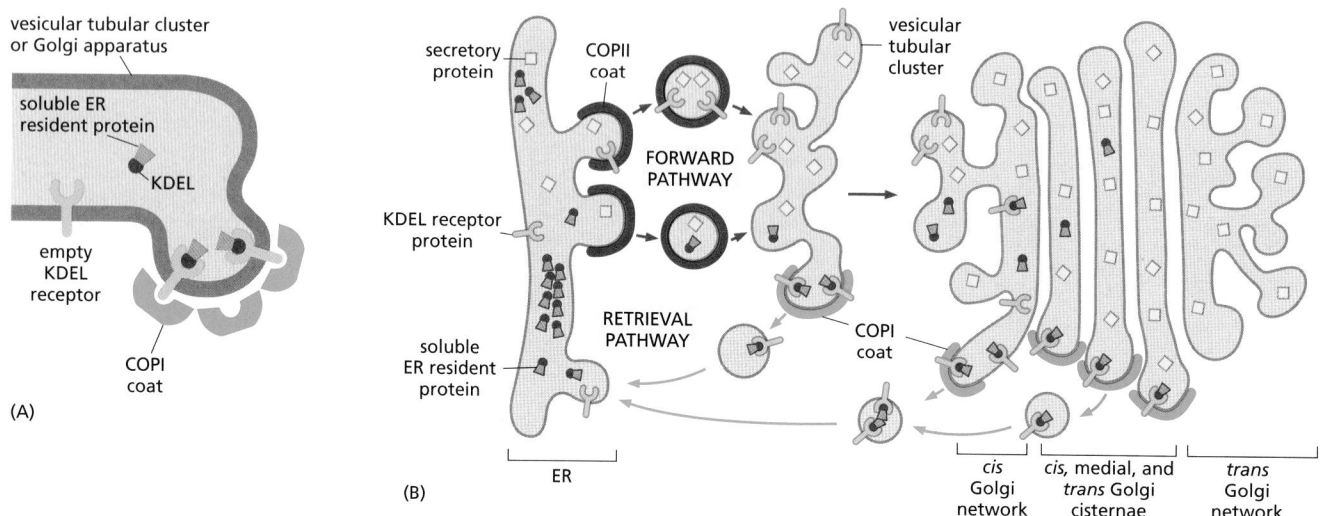

(A)

(B)

ER

cis
Golgi
network

cis, medial, and
trans Golgi
cisternae

trans
Golgi
network

Unlike the retrieval signals on ER membrane proteins, which can interact directly with the COPI coat, soluble ER resident proteins must bind to specialized receptor proteins such as the *KDEL receptor*—a multipass transmembrane protein that binds to the KDEL sequence and packages any protein displaying it into COPI-coated retrograde transport vesicles (**Figure 13–25**). To accomplish this task, the KDEL receptor itself must cycle between the ER and the Golgi apparatus, and its affinity for the KDEL sequence must differ in these two compartments. The receptor must have a high affinity for the KDEL sequence in vesicular tubular clusters and the Golgi apparatus, so as to capture escaped, soluble ER resident proteins that are present there at low concentration. It must have a low affinity for the KDEL sequence in the ER, however, to unload its cargo in spite of the very high concentration of KDEL-containing soluble resident proteins in the ER.

How does the affinity of the KDEL receptor change depending on the compartment in which it resides? The answer is likely related to the lower pH in the Golgi compartments, which is regulated by H^+ pumps. As we discuss later, pH-sensitive protein–protein interactions form the basis for many of the protein sorting steps in the cell.

Most membrane proteins that function at the interface between the ER and Golgi apparatus, including v- and t-SNAREs and some cargo receptors, also enter the retrieval pathway back to the ER.

Many Proteins Are Selectively Retained in the Compartments in Which They Function

The KDEL retrieval pathway only partly explains how ER resident proteins are maintained in the ER. As mentioned, cells that express genetically modified ER resident proteins, from which the KDEL sequence has been experimentally removed, secrete these proteins. But the rate of secretion is much slower than for a normal secretory protein. It seems that a mechanism that is independent of their KDEL signal normally retains ER resident proteins and that only those proteins that escape this retention mechanism are captured and returned via the KDEL receptor. A suggested retention mechanism is that ER resident proteins bind to one another, thus forming complexes that are too big to enter transport vesicles efficiently. Because ER resident proteins are present in the ER at very high concentrations (estimated to be millimolar), relatively low-affinity interactions would suffice to retain most of the proteins in such complexes.

Aggregation of proteins that function in the same compartment is a general mechanism that compartments use to organize and retain their resident proteins. Golgi enzymes that function together, for example, also bind to each other and are thereby restrained from entering transport vesicles leaving the Golgi apparatus.

Figure 13–25 Retrieval of soluble ER resident proteins. ER resident proteins that escape from the ER are returned by vesicle transport. (A) The KDEL receptor present in both vesicular tubular clusters and the Golgi apparatus captures the soluble ER resident proteins and carries them in COPI-coated transport vesicles back to the ER. (Recall that the COPI-coated vesicles shed their coats as soon as they are formed.) Upon binding its ligands in the tubular cluster or Golgi, the KDEL receptor may change conformation, so as to facilitate its recruitment into budding COPI-coated vesicles. (B) The retrieval of ER proteins begins in vesicular tubular clusters and continues from later parts of the Golgi apparatus. In the environment of the ER, the ER resident proteins dissociate from the KDEL receptor, which is then returned to the Golgi apparatus for reuse. We discuss the different compartments of the Golgi apparatus shortly.

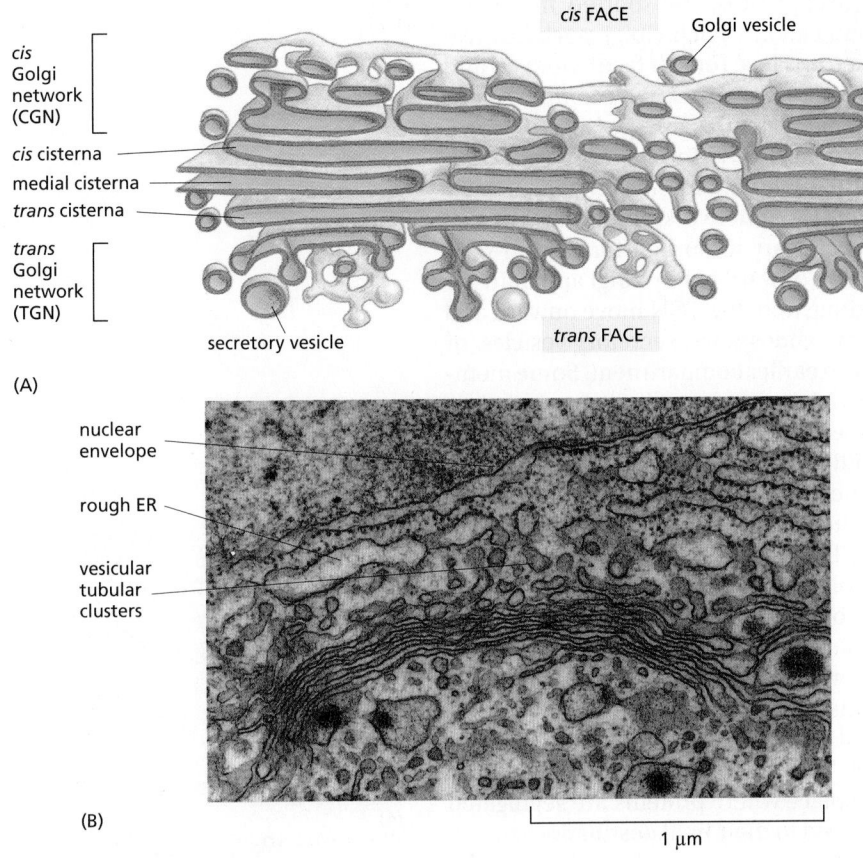

cis FACE
Golgi vesicle

cis Golgi network (CGN)

cis cisterna

medial cisterna

trans cisterna

trans Golgi network (TGN)

secretory vesicle

trans FACE

(A)

nuclear envelope

rough ER

vesicular tubular clusters

(B)

1 µm

Figure 13–26 The Golgi apparatus.
(A) Three-dimensional reconstruction from electron micrographs of the Golgi apparatus in a secretory animal cell. The *cis* face of the Golgi stack is that closest to the ER. (B) A thin-section electron micrograph of an animal cell. In plant cells, the Golgi apparatus is generally more distinct and more clearly separated from other intracellular membranes than in animal cells. (A, redrawn from A. Rambourg and Y. Clermont, *Eur. J. Cell Biol.* 51:189–200, 1990; B, courtesy of Brij L. Gupta.)

The Golgi Apparatus Consists of an Ordered Series of Compartments

Because it could be selectively visualized by silver stains, the Golgi apparatus was one of the first organelles described by early light microscopists. It consists of a collection of flattened, membrane-enclosed compartments called *cisternae*, that somewhat resemble a stack of pita breads. Each Golgi stack typically consists of four to six cisternae (**Figure 13–26**), although some unicellular flagellates can have more than 20. In animal cells, tubular connections between corresponding cisternae link many stacks, thus forming a single complex, which is usually located near the cell nucleus and close to the centrosome (**Figure 13–27**A). This localization depends on microtubules. If microtubules are experimentally depolymerized, the Golgi apparatus reorganizes into individual stacks that are found throughout the cytoplasm, adjacent to ER exit sites. Some cells, including most plant cells, have hundreds of individual Golgi stacks dispersed throughout the cytoplasm where they are typically found adjacent to ER exit sites (Figure 13–27B).

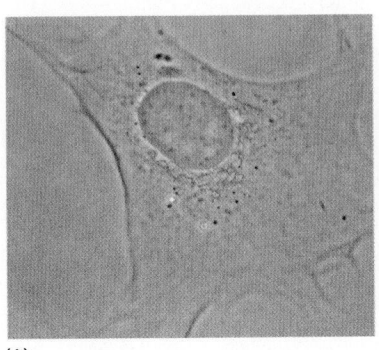

(A)

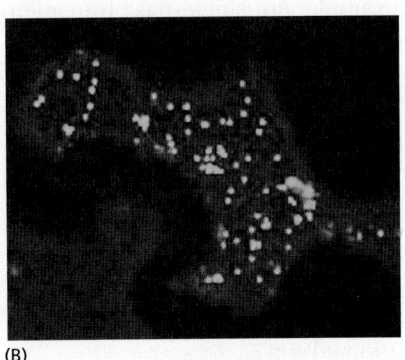

(B)

Figure 13–27 Localization of the Golgi apparatus in animal and plant cells.
(A) The Golgi apparatus in a cultured fibroblast stained with a fluorescent antibody that recognizes a Golgi resident protein *(bright orange)*. The Golgi apparatus is polarized, facing the direction in which the cell was crawling before fixation. (B) The Golgi apparatus in a plant cell that is expressing a fusion protein consisting of a resident Golgi enzyme fused to green fluorescent protein. (A, courtesy of John Henley and Mark McNiven; B, courtesy of Chris Hawes.)

During their passage through the Golgi apparatus, transported molecules undergo an ordered series of covalent modifications. Each Golgi stack has two distinct faces: a *cis* face (or entry face) and a *trans* face (or exit face). Both *cis* and *trans* faces are closely associated with special compartments, each composed of a network of interconnected tubular and cisternal structures: the *cis* **Golgi network** (**CGN**) and the *trans* **Golgi network** (**TGN**), respectively. The CGN is a collection of fused vesicular tubular clusters arriving from the ER. Proteins and lipids enter the *cis* Golgi network and exit from the *trans* Golgi network, bound for the cell surface or another compartment. Both networks are important for protein sorting: proteins entering the CGN can either move onward in the Golgi apparatus or be returned to the ER. Similarly, proteins exiting from the TGN move onward and are sorted according to their next destination: endosomes, secretory vesicles, or the cell surface. They also can be returned to an earlier compartment. Some membrane proteins are retained in the part of the Golgi apparatus where they function.

As described in Chapter 12, a single species of *N-linked oligosaccharide* is attached *en bloc* to many proteins in the ER and then trimmed while the protein is still in the ER. The oligosaccharide intermediates created by the trimming reactions serve to help proteins fold and to help transport misfolded proteins to the cytosol for degradation in proteasomes. Thus, they play an important role in controlling the quality of proteins exiting from the ER. Once these ER functions have been fulfilled, the cell reutilizes the oligosaccharides for new functions. This begins in the Golgi apparatus, which generates the heterogeneous oligosaccharide structures seen in mature proteins. After arrival in the CGN, proteins enter the first of the Golgi processing compartments (the *cis* Golgi cisternae). They then move to the next compartment (the medial cisternae) and finally to the *trans* cisternae, where glycosylation is completed. The lumen of the *trans* cisternae is thought to be continuous with the TGN, the place where proteins are segregated into different transport packages and dispatched to their final destinations.

The oligosaccharide processing steps occur in an organized sequence in the Golgi stack, with each cisterna containing a characteristic mixture of processing enzymes. Proteins are modified in successive stages as they move from cisterna to cisterna across the stack, so that the stack forms a multistage processing unit.

Investigators discovered the functional differences between the *cis*, medial, and *trans* subdivisions of the Golgi apparatus by localizing the enzymes involved in processing *N*-linked oligosaccharides in distinct regions of the organelle, both by physical fractionation of the organelle and by labeling the enzymes in electron microscope sections with antibodies (**Figure 13–28**). The removal of mannose and the addition of *N*-acetylglucosamine, for example, occur in the *cis* and medial cisternae, while the addition of galactose and sialic acid occurs in the *trans* cisterna and *trans* Golgi network. **Figure 13–29** summarizes the functional compartmentalization of the Golgi apparatus.

Oligosaccharide Chains Are Processed in the Golgi Apparatus

Whereas the ER lumen is full of soluble lumenal resident proteins and enzymes, the resident proteins in the Golgi apparatus are all membrane bound, as the enzymatic reactions apparently occur entirely on membrane surfaces. All of the Golgi glycosidases and glycosyl transferases, for example, are single-pass transmembrane proteins, many of which are organized in multienzyme complexes.

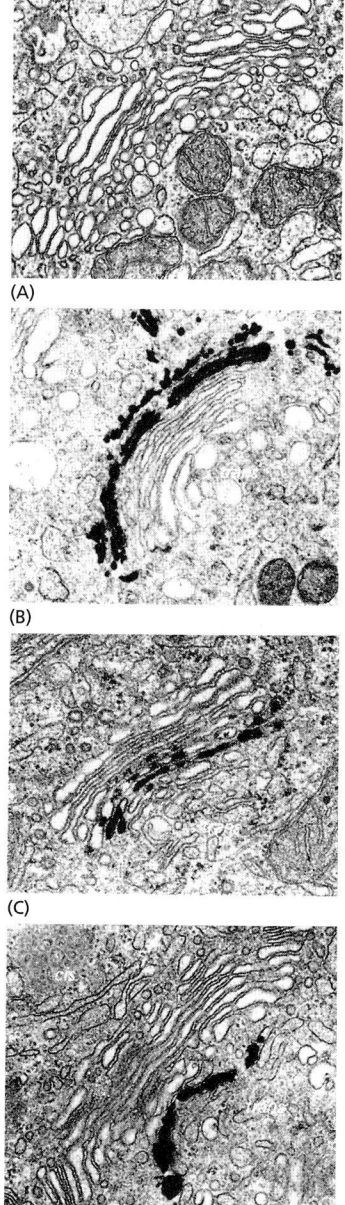

(A)

(B)

(C)

(D)

Figure 13–28 Molecular compartmentalization of the Golgi apparatus. A series of electron micrographs shows the Golgi apparatus (A) unstained, (B) stained with osmium, which preferentially labels the cisternae of the *cis* compartment, and (C and D) stained to reveal the location of specific enzymes. Nucleoside diphosphatase is found in the *trans* Golgi cisternae (C), while acid phosphatase is found in the *trans* Golgi network (D). Note that usually more than one cisterna is stained. The enzymes are therefore thought to be highly enriched rather than precisely localized to a specific cisterna. (Courtesy of Daniel S. Friend and by permission of E.L. Bearer.)

1 μm

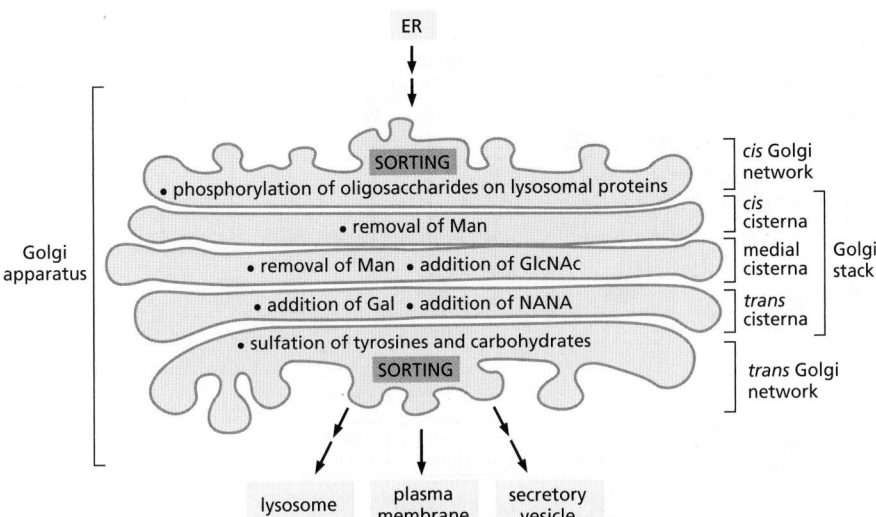

Figure 13–29 Oligosaccharide processing in Golgi compartments. The localization of each processing step shown was determined by a combination of techniques, including biochemical subfractionation of the Golgi apparatus membranes and electron microscopy after staining with antibodies specific for some of the processing enzymes. Processing enzymes are not restricted to a particular cisterna; instead, their distribution is graded across the stack, such that early-acting enzymes are present mostly in the *cis* Golgi cisternae and later-acting enzymes are mostly in the *trans* Golgi cisternae. Man, mannose; GlcNAc, *N*-acetylglucosamine; Gal, galactose; NANA, *N*-acetylneuraminic acid (sialic acid).

Two broad classes of *N*-linked oligosaccharides, the **complex oligosaccharides** and the **high-mannose oligosaccharides**, are attached to mammalian glycoproteins. Sometimes, both types are attached (in different places) to the same polypeptide chain. Complex oligosaccharides are generated when the original *N*-linked oligosaccharide added in the ER is trimmed and further sugars are added; by contrast, high-mannose oligosaccharides are trimmed but have no new sugars added to them in the Golgi apparatus (**Figure 13–30**). The sialic acids in the

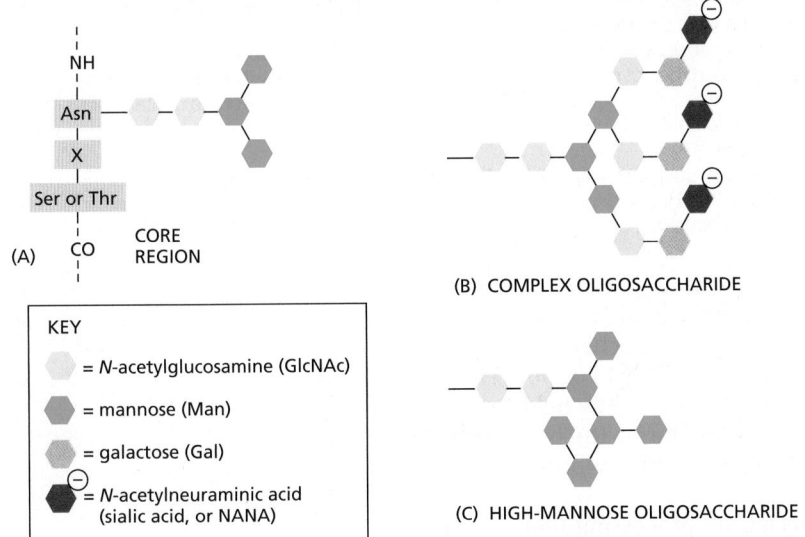

Figure 13–30 The two main classes of asparagine-linked (*N*-linked) oligosaccharides found in mature mammalian glycoproteins. (A) Both complex oligosaccharides and high-mannose oligosaccharides share a common *core region* derived from the original *N*-linked oligosaccharide added in the ER (see Figure 12–50) and typically containing two *N*-acetylglucosamines (GlcNAc) and three mannoses (Man). (B) Each complex oligosaccharide consists of a *core region*, together with a *terminal region* that contains a variable number of copies of a special trisaccharide unit (*N-acetylglucosamine–galactose–sialic acid*) linked to the core mannoses. Frequently, the terminal region is truncated and contains only GlcNAc and galactose (Gal) or just GlcNAc. In addition, a fucose may be added, usually to the core GlcNAc attached to the asparagine (Asn). Thus, although the steps of processing and subsequent sugar addition are rigidly ordered, complex oligosaccharides can be heterogeneous. Moreover, although the complex oligosaccharide shown has three terminal branches, two and four branches are also common, depending on the glycoprotein and the cell in which it is made. (C) High-mannose oligosaccharides are not trimmed back all the way to the core region and contain additional mannoses. Hybrid oligosaccharides with one Man branch and one GlcNAc and Gal branch are also found (not shown).

The three amino acids indicated in (A) constitute the sequence recognized by the oligosaccharyl transferase enzyme that adds the initial oligosaccharide to the protein. Ser, serine; Thr, threonine; X, any amino acid, except proline.

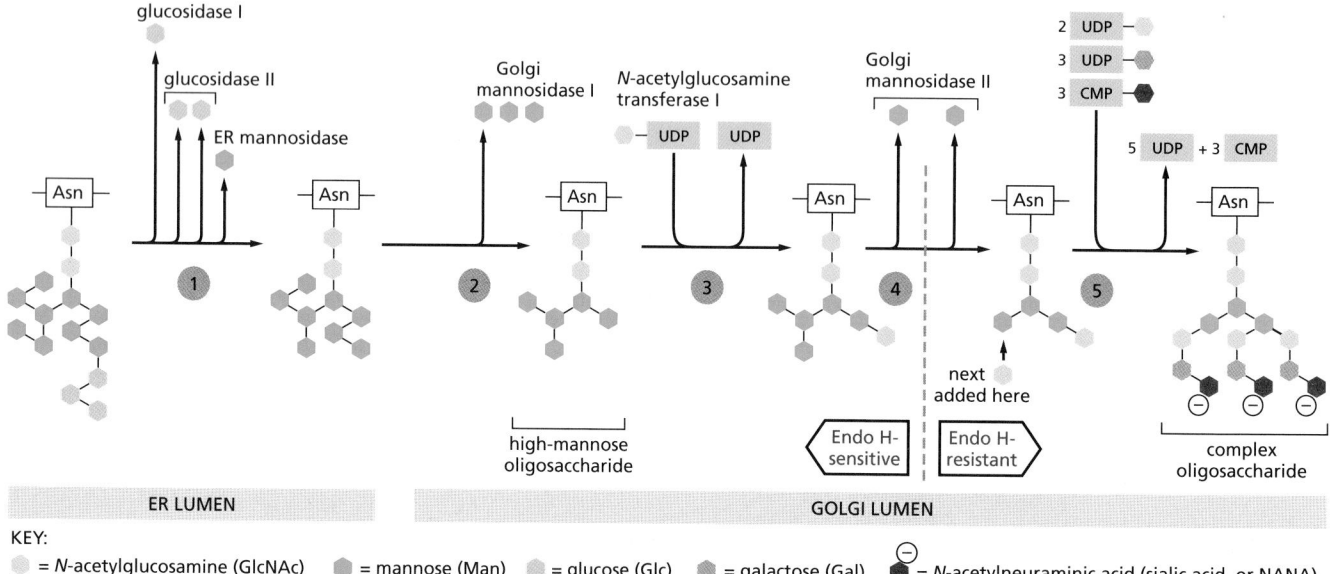

ER LUMEN

GOLGI LUMEN

KEY:

= N-acetylglucosamine (GlcNAc)   = mannose (Man)   = glucose (Glc)   = galactose (Gal)   = N-acetylneuraminic acid (sialic acid, or NANA)

Figure 13–31 Oligosaccharide processing in the ER and the Golgi apparatus. The processing pathway is highly ordered, so that each step shown depends on the previous one. Step 1: Processing begins in the ER with the removal of the glucoses from the oligosaccharide initially transferred to the protein. Then a mannosidase in the ER membrane removes a specific mannose. The remaining steps occur in the Golgi stack. Step 2: Golgi mannosidase I removes three more mannoses. Step 3: N-acetylglucosamine transferase I then adds an N-acetylglucosamine. Step 4: Mannosidase II then removes two additional mannoses. This yields the final core of three mannoses that is present in a complex oligosaccharide. At this stage, the bond between the two N-acetylglucosamines in the core becomes resistant to attack by a highly specific endoglycosidase (*Endo H*). Since all later structures in the pathway are also Endo H-resistant, treatment with this enzyme is widely used to distinguish complex from high-mannose oligosaccharides. Step 5: Finally, as shown in Figure 13–30, additional N-acetylglucosamines, galactoses, and sialic acids are added. These final steps in the synthesis of a complex oligosaccharide occur in the cisternal compartments of the Golgi apparatus: three types of glycosyl transferase enzymes act sequentially, using sugar substrates that have been activated by linkage to the indicated nucleotide; the membranes of the Golgi cisternae contain specific carrier proteins that allow each sugar nucleotide to enter in exchange for the nucleoside phosphates that are released after the sugar is attached to the protein on the lumenal face.

Note that, as a biosynthetic organelle, the Golgi apparatus differs from the ER: all sugars in the Golgi are assembled inside the lumen from sugar nucleotide, whereas in the ER, the N-linked precursor oligosaccharide is assembled partly in the cytosol and partly in the lumen, and most lumenal reactions use dolichol-linked sugars as their substrates (see Figure 12–51).

complex oligosaccharides are of special importance because they bear a negative charge. Whether a given oligosaccharide remains high-mannose or is processed depends largely on its position in the protein. If the oligosaccharide is accessible to the processing enzymes in the Golgi apparatus, it is likely to be converted to a complex form; if it is inaccessible because its sugars are tightly held to the protein's surface, it is likely to remain in a high-mannose form. The processing that generates complex oligosaccharide chains follows the highly ordered pathway shown in **Figure 13–31**.

Beyond these commonalities in oligosaccharide processing that are shared among most cells, the products of the carbohydrate modifications carried out in the Golgi apparatus are highly complex and have given rise to a new field of study called glycobiology. The human genome, for example, encodes hundreds of different Golgi glycosyl transferases and many glycosidases, which are expressed differently from one cell type to another, resulting in a variety of glycosylated forms of a given protein or lipid in different cell types and at varying stages of differentiation, depending on the spectrum of enzymes expressed by the cell. The complexity of modifications is not limited to N-linked oligosaccharides but also occurs on *O-linked sugars*, as we discuss next.

Proteoglycans Are Assembled in the Golgi Apparatus

In addition to the N-linked oligosaccharide alterations made to proteins as they pass through the Golgi cisternae *en route* from the ER to their final destinations, many proteins are also modified in the Golgi apparatus in other ways. Some proteins have sugars added to the hydroxyl groups of selected serines or threonines,

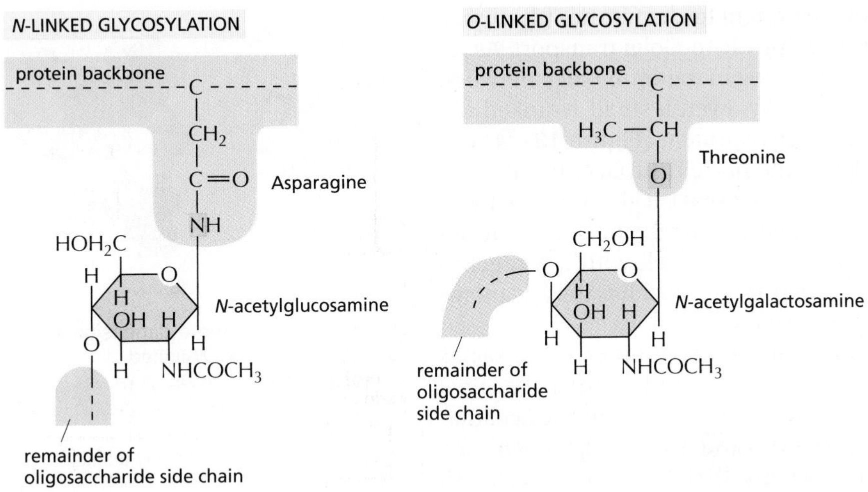

Figure 13–32 *N*- and *O*-linked glycosylation. In each case, only the single sugar group that is directly attached to the protein chain is shown.

or, in some cases—such as collagens—to hydroxylated proline and lysine side chains. This *O*-linked glycosylation (Figure 13–32), like the extension of *N*-linked oligosaccharide chains, is catalyzed by a series of glycosyl transferase enzymes that use the sugar nucleotides in the lumen of the Golgi apparatus to add sugars to a protein one at a time. Usually, *N*-acetylgalactosamine is added first, followed by a variable number of additional sugars, ranging from just a few to 10 or more.

The Golgi apparatus confers the heaviest *O*-linked glycosylation of all on *mucins*, the glycoproteins in mucus secretions, and on *proteoglycan core proteins*, which it modifies to produce **proteoglycans**. As discussed in Chapter 19, this process involves the polymerization of one or more *glycosaminoglycan chains* (long, unbranched polymers composed of repeating disaccharide units; see Figure 19–35) onto serines on a core protein. Many proteoglycans are secreted and become components of the extracellular matrix, while others remain anchored to the extracellular face of the plasma membrane. Still others form a major component of slimy materials, such as the mucus that is secreted to form a protective coating on the surface of many epithelia.

The sugars incorporated into glycosaminoglycans are heavily sulfated in the Golgi apparatus immediately after these polymers are made, thus adding a significant portion of their characteristically large negative charge. Some tyrosines in proteins also become sulfated shortly before they exit from the Golgi apparatus. In both cases, the sulfation depends on the sulfate donor 3′-phosphoadenosine-5′-phosphosulfate (PAPS) (Figure 13–33), which is transported from the cytosol into the lumen of the *trans* Golgi network.

What Is the Purpose of Glycosylation?

There is an important difference between the construction of an oligosaccharide and the synthesis of other macromolecules such as DNA, RNA, and protein. Whereas nucleic acids and proteins are copied from a template in a repeated series of identical steps using the same enzyme or set of enzymes, complex carbohydrates require a different enzyme at each step, each product being recognized as the exclusive substrate for the next enzyme in the series. The vast abundance of glycoproteins and the complicated pathways that have evolved to synthesize them emphasize that the oligosaccharides on glycoproteins and glycosphingolipids have very important functions.

N-linked glycosylation, for example, is prevalent in all eukaryotes, including yeasts. *N*-linked oligosaccharides also occur in a very similar form in archaeal cell wall proteins, suggesting that the whole machinery required for their synthesis is evolutionarily ancient. *N*-linked glycosylation promotes protein folding in two ways. First, it has a direct role in making folding intermediates more soluble, thereby preventing their aggregation. Second, the sequential modifications of the *N*-linked oligosaccharide establish a "glyco-code" that marks the progression of

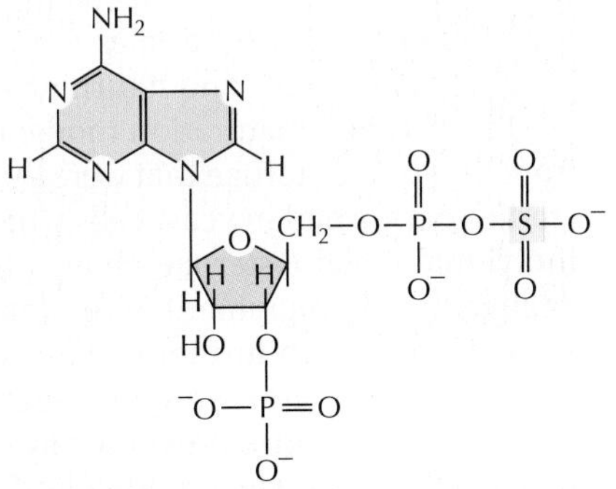

3′-phosphoadenosine-5′-phosphosulfate (PAPS)

Figure 13–33 The structure of PAPS.

protein folding and mediates the binding of the protein to chaperones (discussed in Chapter 12) and lectins—for example, in guiding ER-to-Golgi transport. As we discuss later, lectins also participate in protein sorting in the *trans* Golgi network.

Because chains of sugars have limited flexibility, even a small *N*-linked oligosaccharide protruding from the surface of a glycoprotein (**Figure 13–34**) can limit the approach of other macromolecules to the protein surface. In this way, for example, the presence of oligosaccharides tends to make a glycoprotein more resistant to digestion by proteolytic enzymes. It may be that the oligosaccharides on cell-surface proteins originally provided an ancestral cell with a protective coat; compared to the rigid bacterial cell wall, such a sugar coat has the advantage that it leaves the cell with the freedom to change shape and move.

The sugar chains have since become modified to serve other purposes as well. The mucus coat of lung and intestinal cells, for example, protects against many pathogens. The recognition of sugar chains by *lectins* in the extracellular space is important in many developmental processes and in cell–cell recognition: *selectins*, for example, are transmembrane lectins that function in cell–cell adhesion during blood cell migration, as discussed in Chapter 19. The presence of oligosaccharides may modify a protein's antigenic and functional properties, making glycosylation an important factor in the production of proteins for pharmaceutical purposes.

Glycosylation can also have important regulatory roles. Signaling through the cell-surface signaling receptor Notch, for example, is an important factor in determining the cell's fate in development (discussed in Chapter 21). Notch is a transmembrane protein that is *O*-glycosylated by addition of a single fucose to some serines, threonines, and hydroxylysines. Some cell types express an additional glycosyl transferase that adds an *N*-acetylglucosamine to each of these fucoses in the Golgi apparatus. This addition changes the specificity of Notch for the cell-surface signal proteins that activate it.

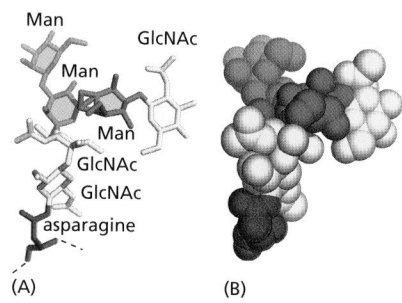

Figure 13–34 The three-dimensional structure of a small *N*-linked oligosaccharide. The structure was determined by x-ray crystallographic analysis of a glycoprotein. This oligosaccharide contains only 6 sugars, whereas there are 14 sugars in the *N*-linked oligosaccharide that is initially transferred to proteins in the ER (see Figure 12–47). (A) A backbone model showing all atoms except hydrogens; only the asparagine of the protein is shown. (B) A space-filling model, with the asparagine and sugars indicated using the same color scheme as in (A). (B, courtesy of Richard Feldmann.)

Transport Through the Golgi Apparatus May Occur by Cisternal Maturation

It is still uncertain how the Golgi apparatus achieves and maintains its polarized structure and how molecules move from one cisterna to another, and it is likely that more than one mechanism is involved in each case. One hypothesis, called the **cisternal maturation model**, views the Golgi cisternae as dynamic structures that mature from early to late by acquiring and then losing specific Golgi-resident proteins. According to this view, new *cis* cisternae continually form as vesicular tubular clusters arrive from the ER and progressively mature to become a medial cisterna and then a *trans* cisterna (**Figure 13–35**A). A cisterna therefore moves through the Golgi stack with cargo in its lumen. Retrograde transport of the Golgi enzymes by budding COPI-coated vesicles explains their characteristic distribution. As we discuss later, when a cisterna finally moves forward to become part of the *trans* Golgi network, various types of coated vesicles bud off it until this network disappears, to be replaced by a maturing cisterna just behind. At the same time, other transport vesicles are continually retrieving membrane from post-Golgi compartments and returning it to the *trans* Golgi network.

The cisternal maturation model is supported by studies using Golgi enzymes from different cisternae that were fluorescently labeled with different colors. Such studies performed in yeast cells where Golgi cisternae are not stacked reveal that individual Golgi cisternae change their color, thereby demonstrating that they change their complement of resident enzymes as they mature, even though they are not stacked. In further support of the model, electron microscopic observations found that large structures such as procollagen rods in fibroblasts and scales in certain algae move progressively through the Golgi stack.

An alternative view holds that Golgi cisternae are long-lived structures that retain their characteristic set of Golgi-resident proteins firmly in place, and cargo proteins are transported from one cisterna to the next by transport vesicles (Figure 13–35B). According to this **vesicle transport model**, retrograde flow of vesicles

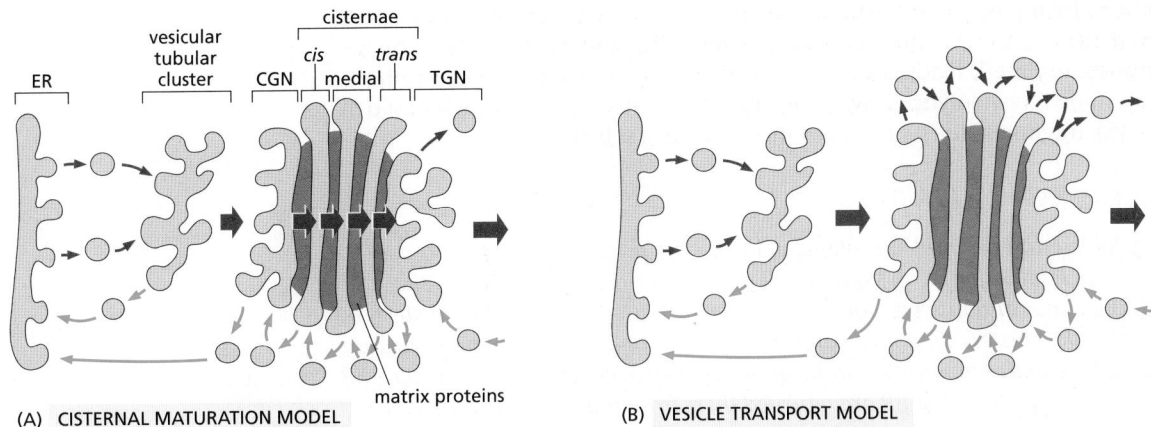

Figure 13–35 Two possible models explaining the organization of the Golgi apparatus and how proteins move through it. It is likely that the transport through the Golgi apparatus in the forward direction *(red arrows)* involves elements of both models. (A) According to the cisternal maturation model, each Golgi cisterna matures as it migrates outward through the stack. At each stage, the Golgi resident proteins that are carried forward in a maturing cisterna are moved backward to an earlier compartment in COPI-coated vesicles. When a newly formed cisterna moves to a medial position, for example, "leftover" *cis* Golgi enzymes would be extracted and transported retrogradely to a new *cis* cisterna behind. Likewise, the medial enzymes would be received by retrograde transport from the cisternae just ahead. In this way, a *cis* cisterna would mature to a medial and then *trans* cisterna as it moves outward. (B) In the vesicle transport model, Golgi cisternae are static compartments, which contain a characteristic complement of resident enzymes. The passing of molecules from *cis* to *trans* through the Golgi is accomplished by forward-moving transport vesicles, which bud from one cisterna and fuse with the next in a *cis*-to-*trans* direction.

retrieves escaped ER and Golgi proteins and returns them to upstream compartments. Directional flow could be achieved because forward-moving cargo molecules are selectively packaged into forward-moving vesicles. Although both forward- and backward-moving vesicles would likely be COPI-coated, the coats may contain different adaptor proteins that confer selectivity on the packaging of cargo molecules. Alternatively, transport vesicles shuttling between Golgi cisternae might not be directional at all, transporting cargo randomly back and forth; directional flow would then occur because of the continual input to the *cis* cisterna and output from the *trans* cisterna.

The vesicle transport model is supported by experiments that show that cargo molecules are present in small COPI-coated vesicles and that these vesicles can deliver them to Golgi cisternae over large distances. In addition, when experimentally aggregated membrane proteins are introduced into Golgi cisternae, they can be observed staying in place, while soluble cargo, even if present as large aggregates, traverses the Golgi at normal rates.

It is likely that aspects of both models are true. A stable core of long-lasting cisternae might exist in the center of each Golgi cisterna, while regions at the rim may undergo continuous maturation, perhaps utilizing Rab cascades that change their identity. As matured pieces of the cisternae are formed, they might break off and fuse with downstream cisternae by homotypic fusion mechanisms, taking large cargo molecules with them. In addition, small COPI-coated vesicles might transport small cargo in the forward direction and retrieve escaped Golgi enzymes and return them to their appropriate upstream cisternae.

Golgi Matrix Proteins Help Organize the Stack

The unique architecture of the Golgi apparatus depends on both the microtubule cytoskeleton, as already mentioned, and cytoplasmic Golgi matrix proteins, which form a scaffold between adjacent cisternae and give the Golgi stack its structural integrity. Some of the matrix proteins, called *golgins*, form long tethers composed of stiff coiled-coil domains with interspersed hinge regions. Golgins form a forest of tentacles that can extend 100–400 nm from the surface of the Golgi stack. They are thought to help retain Golgi transport vesicles close to the organelle through interactions with Rab proteins (**Figure 13–36**). When the cell prepares to divide, mitotic protein kinases phosphorylate the Golgi matrix proteins, causing the Golgi

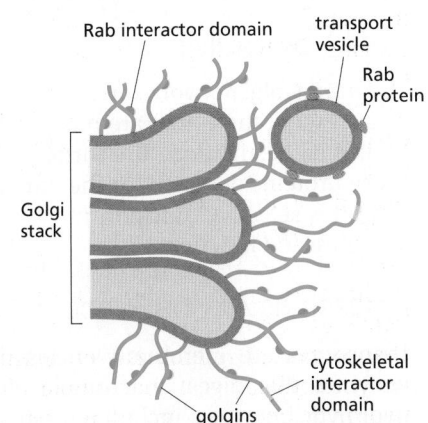

Figure 13–36 A model of golgin function. Filamentous golgins anchored to Golgi membranes capture transport vesicles by binding to Rab proteins on the vesicle surface.

apparatus to fragment and disperse throughout the cytosol. The Golgi fragments are then distributed evenly to the two daughter cells, where the matrix proteins are dephosphorylated, leading to the reassembly of the Golgi stack. Similarly, during apoptosis, proteolytic cleavage of golgins by caspases ensues (discussed in Chapter 18), fragments the Golgi apparatus as the cell self-destructs.

Summary

Correctly folded and assembled proteins in the ER are packaged into COPII-coated transport vesicles that pinch off from the ER membrane. Shortly thereafter, the vesicles shed their coat and fuse with one another to form vesicular tubular clusters. In animal cells, the clusters then move on microtubule tracks to the Golgi apparatus, where they fuse with one another to form the cis *Golgi network. Any resident ER proteins that escape from the ER are returned there from the vesicular tubular clusters and Golgi apparatus by retrograde transport in COPI-coated vesicles.*

The Golgi apparatus, unlike the ER, contains many sugar nucleotides, which glycosyl transferase enzymes use to glycosylate lipid and protein molecules as they pass through the Golgi apparatus. The mannoses on the N-linked *oligosaccharides that are added to proteins in the ER are often initially removed, and further sugars are added. Moreover, the Golgi apparatus is the site where O-linked glycosylation occurs and where glycosaminoglycan chains are added to core proteins to form proteoglycans. Sulfation of the sugars in proteoglycans and of selected tyrosines on proteins also occurs in a late Golgi compartment.*

The Golgi apparatus modifies the many proteins and lipids that it receives from the ER and then distributes them to the plasma membrane, lysosomes, and secretory vesicles. The Golgi apparatus is a polarized organelle, consisting of one or more stacks of disc-shaped cisternae. Each stack is organized as a series of at least three functionally distinct compartments, termed cis, medial, *and* trans *cisternae. The* cis *and* trans *cisternae are each connected to special sorting stations, called the* cis *Golgi network and the* trans *Golgi network, respectively. Proteins and lipids move through the Golgi stack in the* cis-to-trans *direction. This movement may occur by vesicle transport, by progressive maturation of the* cis *cisternae as they migrate continuously through the stack, or, most likely, by a combination of these two mechanisms. Continual retrograde vesicle transport from upstream to more downstream cisternae is thought to keep the enzymes concentrated in the cisternae where they are needed. The finished new proteins end up in the* trans *Golgi network, which packages them in transport vesicles and dispatches them to their specific destinations in the cell.*

TRANSPORT FROM THE *TRANS* GOLGI NETWORK TO LYSOSOMES

The *trans* Golgi network sorts all of the proteins that pass through the Golgi apparatus (except those that are retained there as permanent residents) according to their final destination. The sorting mechanism is especially well understood for those proteins destined for the lumen of lysosomes, and in this section we consider this selective transport process. We begin with a brief account of lysosome structure and function.

Lysosomes Are the Principal Sites of Intracellular Digestion

Lysosomes are membrane-enclosed organelles filled with soluble hydrolytic enzymes that digest macromolecules. Lysosomes contain about 40 types of hydrolytic enzymes, including proteases, nucleases, glycosidases, lipases, phospholipases, phosphatases, and sulfatases. All are **acid hydrolases**; that is, hydrolases that work best at acidic pH. For optimal activity, they need to be activated by proteolytic cleavage, which also requires an acid environment. The lysosome provides this acidity, maintaining an interior pH of about 4.5–5.0. By this arrangement, the contents of the cytosol are doubly protected against attack by the cell's

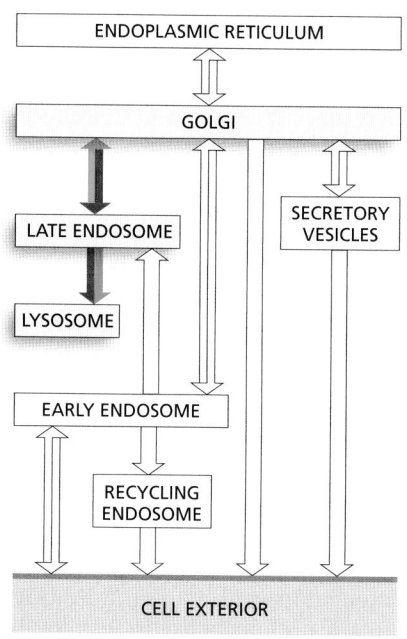

own digestive system: the membrane of the lysosome keeps the digestive enzymes out of the cytosol, but, even if they leak out, they can do little damage at the cytosolic pH of about 7.2.

Like all other membrane-enclosed organelles, the lysosome not only contains a unique collection of enzymes, but also has a unique surrounding membrane. Most of the lysosome membrane proteins, for example, are highly glycosylated, which helps to protect them from the lysosome proteases in the lumen. Transport proteins in the lysosome membrane carry the final products of the digestion of macromolecules—such as amino acids, sugars, and nucleotides—to the cytosol, where the cell can either reuse or excrete them.

A *vacuolar H⁺ ATPase* in the lysosome membrane uses the energy of ATP hydrolysis to pump H⁺ into the lysosome, thereby maintaining the lumen at its acidic pH (**Figure 13–37**). The lysosome H⁺ pump belongs to the family of *V-type ATPases* and has a similar architecture to the mitochondrial and chloroplast ATP synthases (F-type ATPases), which convert the energy stored in H⁺ gradients into ATP (see Figure 11–12). By contrast to these enzymes, however, the vacuolar H⁺ ATPase exclusively works in reverse, pumping H⁺ into the organelle. Similar or identical V-type ATPases acidify all endocytic and exocytic organelles, including lysosomes, endosomes, some compartments of the Golgi apparatus, and many transport and secretory vesicles. In addition to providing a low-pH environment that is suitable for reactions occurring in the organelle lumen, the H⁺ gradient provides a source of energy that drives the transport of small metabolites across the organelle membrane.

Lysosomes Are Heterogeneous

Lysosomes are found in all eukaryotic cells. They were initially discovered by the biochemical fractionation of cell extracts; only later were they seen clearly in the electron microscope. Although extraordinarily diverse in shape and size, staining them with specific antibodies shows they are members of a single family of organelles. They can also be identified by histochemical techniques that reveal which organelles contain acid hydrolase (**Figure 13–38**).

The heterogeneous morphology of lysosomes contrasts with the relatively uniform structures of many other cell organelles. The diversity reflects the wide variety of digestive functions that acid hydrolases mediate, including the breakdown of intra- and extracellular debris, the destruction of phagocytosed microorganisms, and the production of nutrients for the cell. Their morphological diversity, however, also reflects the way lysosomes form. Late endosomes containing material received from both the plasma membrane by endocytosis and newly synthesized lysosomal hydrolases fuse with preexisting lysosomes to form structures that are sometimes referred to as *endolysosomes*, which then fuse with one another (**Figure 13–39**). When the majority of the endocytosed material within an endolysosome has been digested so that only resistant or slowly digestible residues remain, these organelles become "classical" lysosomes. These are relatively dense, round, and small, but they can enter the cycle again by fusing with late endosomes or endolysosomes. Thus, there is no real distinction between endolysosomes and lysosomes: they are the same except that they are in different stages of a maturation cycle. For this reason, lysosomes are sometimes viewed as a heterogeneous collection of distinct organelles, the common feature of which is a high content of

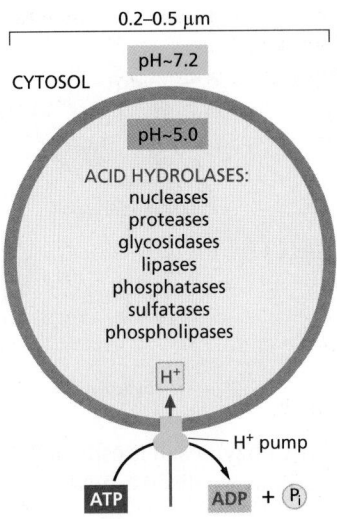

Figure 13–37 Lysosomes. The acid hydrolases are hydrolytic enzymes that are active under acidic conditions. An H⁺ ATPase in the membrane pumps H⁺ into the lysosome, maintaining its lumen at an acidic pH.

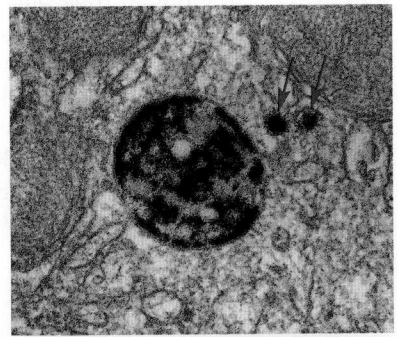
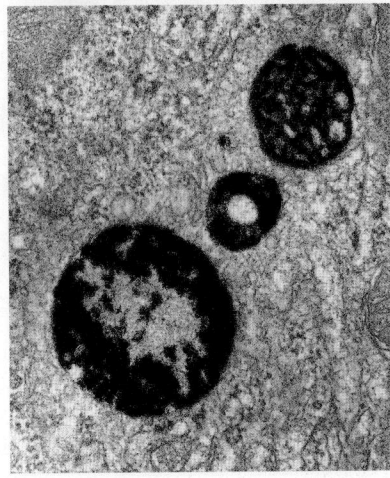

Figure 13–38 Histochemical visualization of lysosomes. These electron micrographs show two sections of a cell stained to reveal the location of acid phosphatase, a marker enzyme for lysosomes. The larger membrane-enclosed organelles, containing dense precipitates of lead phosphate, are lysosomes. Their diverse morphology reflects variations in the amount and nature of the material they are digesting. The precipitates are produced when tissue fixed with glutaraldehyde (to fix the enzyme in place) is incubated with a phosphatase substrate in the presence of lead ions. *Red arrows* in the top panel indicate two small vesicles thought to be carrying acid hydrolases from the Golgi apparatus. (Courtesy of Daniel S. Friend and by permission of E.L. Bearer.)

200 nm

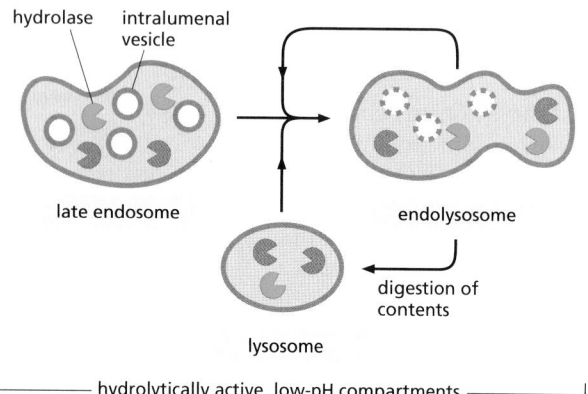

hydrolytic enzymes. It is especially hard to apply a narrower definition than this in plant cells, as we discuss next.

Plant and Fungal Vacuoles Are Remarkably Versatile Lysosomes

Most plant and fungal cells (including yeasts) contain one or several very large, fluid-filled vesicles called **vacuoles**. They typically occupy more than 30% of the cell volume, and as much as 90% in some cell types (**Figure 13–40**). Vacuoles are related to animal cell lysosomes and contain a variety of hydrolytic enzymes, but their functions are remarkably diverse. The plant vacuole can act as a storage organelle for both nutrients and waste products, as a degradative compartment, as an economical way of increasing cell size, and as a controller of *turgor pressure* (the osmotic pressure that pushes outward on the cell wall and keeps the plant from wilting) (**Figure 13–41**). The same cell may have different vacuoles with distinct functions, such as digestion and storage.

The vacuole is important as a homeostatic device, enabling plant cells to withstand wide variations in their environment. When the pH in the environment drops, for example, the flux of H^+ into the cytosol is balanced, at least in part, by an increased transport of H^+ into the vacuole, which tends to keep the pH in the cytosol constant. Similarly, many plant cells maintain an almost constant turgor pressure despite large changes in the tonicity of the fluid in their immediate environment. They do so by changing the osmotic pressure of the cytosol and vacuole—in part by the controlled breakdown and resynthesis of polymers such as polyphosphate in the vacuole, and in part by altering the transport rates of sugars,

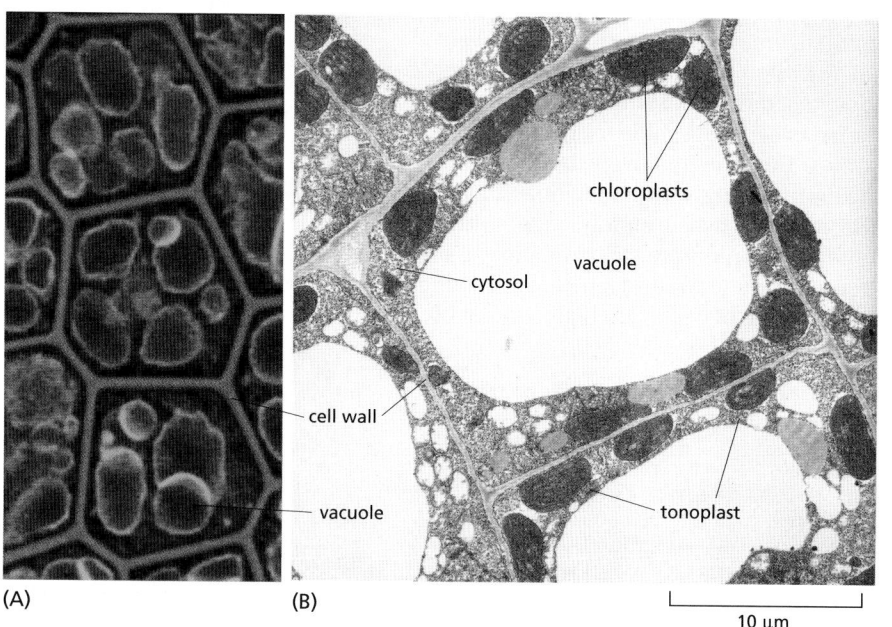

(A) (B)

10 µm

Figure 13–40 **The plant cell vacuole.** (A) A confocal image of cells from an *Arabidopsis* embryo that is expressing an aquaporin—YFP (yellow fluorescent protein) fusion protein in its tonoplast, or vacuole membrane *(green)*; the cell walls have been false-colored *orange*. Each cell contains several large vacuoles. (B) This electron micrograph of cells in a young tobacco leaf shows the cytosol as a thin layer, containing chloroplasts, pressed against the cell wall by the enormous vacuole. (A, courtesy of C. Carroll and L. Frigerio, based on S. Gattolin et al., *Mol. Plant* 4: 180–189, 2011; B, courtesy of J. Burgess.)

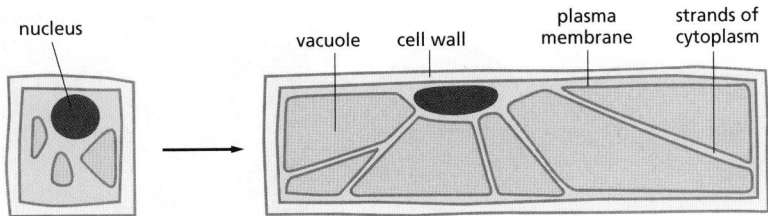

Figure 13–41 The role of the vacuole in controlling the size of plant cells. A plant cell can achieve a large increase in volume without increasing the volume of the cytosol. Localized weakening of the cell wall orients a turgor-driven cell enlargement that accompanies the uptake of water into an expanding vacuole. The cytosol is eventually confined to a thin peripheral layer, which is connected to the nuclear region by strands of cytosol stabilized by bundles of actin filaments (not shown).

amino acids, and other metabolites across the plasma membrane and the vacuolar membrane. The turgor pressure regulates the activities of distinct transporters in each membrane to control these fluxes.

Humans often harvest substances stored in plant vacuoles—from rubber to opium to the flavoring of garlic. Many stored products have a metabolic function. Proteins, for example, can be preserved for years in the vacuoles of the storage cells of many seeds, such as those of peas and beans. When the seeds germinate, these proteins are hydrolyzed, and the resulting amino acids provide a food supply for the developing embryo. Anthocyanin pigments stored in vacuoles color the petals of many flowers so as to attract pollinating insects, while noxious molecules released from vacuoles when a plant is eaten or damaged provide a defense against predators.

Multiple Pathways Deliver Materials to Lysosomes

Lysosomes are meeting places where several streams of intracellular traffic converge. A route that leads outward from the ER via the Golgi apparatus delivers most of the lysosome's digestive enzymes, while at least four paths from different sources feed substances into lysosomes for digestion.

The best studied of these degradation paths is the one followed by macromolecules taken up from extracellular fluid by *endocytosis*. A similar pathway found in phagocytic cells, such as macrophages and neutrophils in vertebrates, is dedicated to the engulfment, or *phagocytosis*, of large particles and microorganisms to form *phagosomes*. A third pathway called *macropinocytosis* specializes in the nonspecific uptake of fluids, membrane, and particles attached to the plasma membrane. We will return to discuss these pathways later in the chapter. A fourth pathway called *autophagy* originates in the cytoplasm of the cell itself and is used to digest cytosol and worn-out organelles, as we discuss next. The four paths to degradation in lysosomes are illustrated in **Figure 13–42**.

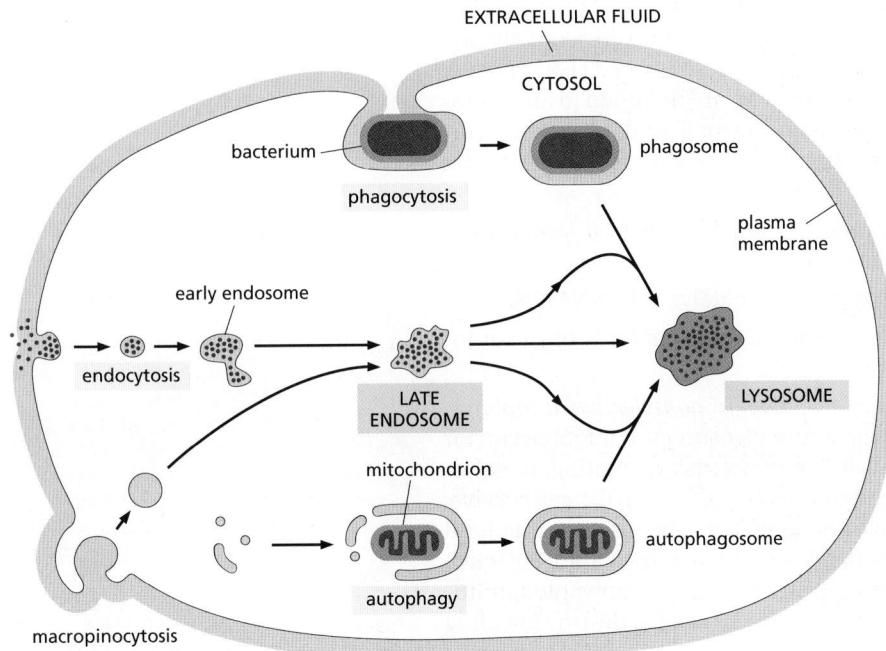

Figure 13–42 Four pathways to degradation in lysosomes. Materials in each pathway are derived from a different source. Note that the autophagosome has a double membrane. In all cases, the final step is the fusion with lysosomes.

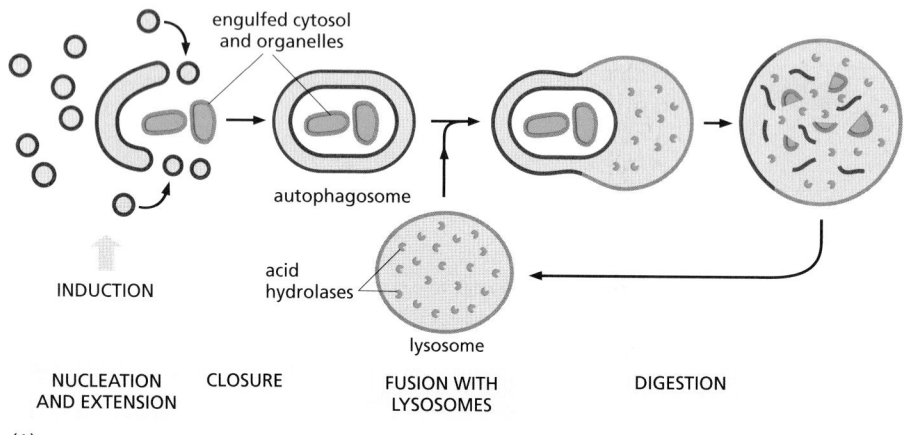

(A)

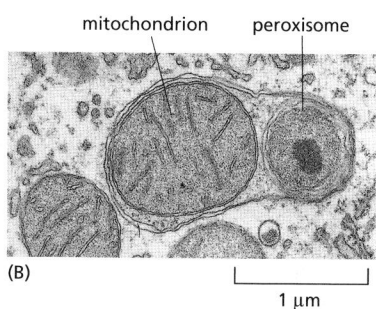

(B)

Figure 13–43 A model of autophagy.
(A) Activation of a signaling pathway initiates a nucleation event in the cytoplasm. A crescent of autophagosomal membrane grows by fusion of vesicles of unknown origin and eventually fuses to form a double-membrane-enclosed autophagosome, which sequesters a portion of the cytoplasm. The autophagosome then fuses with lysosomes containing acid hydrolases that digest its content. During the formation of the autophagosome membrane, a ubiquitin-like protein becomes activated by covalent attachment of a phosphatidylethanolamine lipid anchor. These proteins then mediate vesicle tethering and fusion, leading to the formation of a crescent-shaped membrane structure that assembles around its target (not shown). (B) An electron micrograph of an autophagosome containing a mitochondrion and a peroxisome.
(B, courtesy of Daniel S. Friend and by permission of E.L. Bearer.)

Autophagy Degrades Unwanted Proteins and Organelles

All cell types dispose of obsolete parts by a lysosome-dependent process called **autophagy**, or "self-eating." The degradation process is important during normal cell growth and in development, where it helps restructure differentiating cells, but also in adaptive responses to stresses such as starvation and infection. Autophagy can remove large objects—macromolecules, large protein aggregates, and even whole organelles—that other disposal mechanisms such as proteasomal degradation cannot handle. Defects in autophagy may prevent cells from clearing away invading microbes, unwanted protein aggregates and abnormal proteins, and thereby contribute to diseases ranging from infectious disorders to neurodegeneration and cancer.

In the initial stages of autophagy, cytoplasmic cargo becomes surrounded by a double membrane that assembles by the fusion of small vesicles of unknown origin, forming an **autophagosome** (Figure 13–43). A few tens of different proteins have been identified in yeast and animal cells that participate in the process, which must be tightly regulated: either too little or too much can be deleterious. The whole process occurs in the following sequence of steps:

1. Induction by activation of signaling molecules: Protein kinases (including the mTOR complex 1, discussed in Chapter 15) that relay information about the metabolic status of the cell, become activated and signal to the autophagic machinery.

2. Nucleation and extension of a delimiting membrane into a crescent-shaped cup: Membrane vesicles, characterized by the presence of ATG9, the only transmembrane protein involved in the process, are recruited to an assembly site, where they nucleate autophagosome formation. ATG9 is not incorporated into the autophagosome: a retrieval pathway must remove it from the assembling structure.

3. Closure of the membrane cup around the target to form a sealed double-membrane-enclosed autophagosome.

4. Fusion of the autophagosome with lysosomes, catalyzed by SNAREs.

5. Digestion of the inner membrane and the lumenal contents of the autophagosome.

Autophagy can be either nonselective or selective. In *nonselective autophagy*, a bulk portion of cytoplasm is sequestered in autophagosomes. It might occur, for example, in starvation conditions: when external nutrients are limiting, metabolites derived from the digestion of the captured cytosol might help the cell survive. In *selective autophagy* specific cargo is packaged into autophagosomes that tend to contain little cytosol, and their shape reflects the shape of the cargo. Selective autophagy mediates the degradation of worn out, or otherwise unwanted, mitochondria, peroxisomes, ribosomes, and ER; it can also be used to destroy invading microbes.

The selective autophagy of worn out or damaged mitochondria is called *mitophagy*. As discussed in Chapters 12 and 14, when mitochondria function normally, the inner mitochondrial membrane is energized by an electrochemical H^+ gradient that drives ATP synthesis and the import of mitochondrial precursor proteins and metabolites. Damaged mitochondria cannot maintain the gradient, so protein import is blocked. As a consequence, a protein kinase called Pink1, which is normally imported into mitochondria, is instead retained on the mitochondrial surface where it recruits the ubiquitin ligase Parkin from the cytosol. Parkin ubiquitylates mitochondrial outer membrane proteins, which mark the organelle for selective destruction in autophagosomes. Mutations in Pink1 or Parkin cause a form of early-onset Parkinson's disease, a degenerative disorder of the central nervous system. It is not known why the neurons that die prematurely in this disease are particularly reliant on mitophagy.

A Mannose 6-Phosphate Receptor Sorts Lysosomal Hydrolases in the *Trans* Golgi Network

We now consider the pathway that delivers lysosomal hydrolases from the TGN to lysosomes. The enzymes are first delivered to endosomes in transport vesicles that bud from the TGN, before they move on to endolysosomes and lysosomes (see Figure 13–39). The vesicles that leave the TGN incorporate the lysosomal proteins and exclude the many other proteins being packaged into different transport vesicles for delivery elsewhere.

How are lysosomal hydrolases recognized and selected in the TGN with the required accuracy? In animal cells they carry a unique marker in the form of *mannose 6-phosphate* (*M6P*) groups, which are added exclusively to the *N*-linked oligosaccharides of these soluble lysosomal enzymes as they pass through the lumen of the *cis* Golgi network (**Figure 13–44**). Transmembrane **M6P receptor proteins**, which are present in the TGN, recognize the M6P groups and bind to the lysosomal hydrolases on the lumenal side of the membrane and to adaptor proteins in assembling clathrin coats on the cytosolic side. In this way, the receptors help package the hydrolases into clathrin-coated vesicles that bud from the TGN and deliver their contents to early endosomes.

The M6P receptor protein binds to M6P at pH 6.5–6.7 in the TGN lumen and releases it at pH 6, which is the pH in the lumen of endosomes. Thus, after the receptor is delivered, the lysosomal hydrolases dissociate from the M6P receptors, which are retrieved into transport vesicles that bud from endosomes. These vesicles are coated with *retromer*, a coat protein complex specialized for endosome-to-TGN transport, which returns the receptors to the TGN for reuse (**Figure 13–45**).

Transport in either direction requires signals in the cytoplasmic tail of the M6P receptor that direct this protein to the endosome or back to the TGN. These signals are recognized by the retromer complex that recruits M6P receptors into transport vesicles that bud from endosomes. The recycling of the M6P receptor resembles the recycling of the KDEL receptor discussed earlier, although it differs in the type of coated vesicles that mediate the transport.

Not all the hydrolase molecules that are tagged with M6P get to lysosomes. Some escape the normal packaging process in the *trans* Golgi network and are transported "by default" to the cell surface, where they are secreted into the extracellular fluid. Some M6P receptors, however, also take a detour to the plasma membrane, where they recapture the escaped lysosomal hydrolases and return them by *receptor-mediated endocytosis* (discussed later) to lysosomes via early and late endosomes. As lysosomal hydrolases require an acidic milieu to work, they can do little harm in the extracellular fluid, which usually has a neutral pH of 7.4.

For the sorting system that segregates lysosomal hydrolases and dispatches them to endosomes to work, the M6P groups must be added only to the appropriate glycoproteins in the Golgi apparatus. This requires specific recognition of the hydrolases by the Golgi enzymes responsible for adding M6P. Since all glycoproteins leave the ER with identical *N*-linked oligosaccharide chains, the signal for

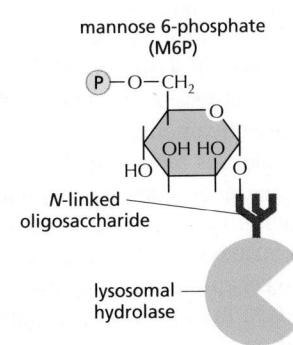

Figure 13–44 The structure of mannose 6-phosphate on a lysosomal hydrolase.

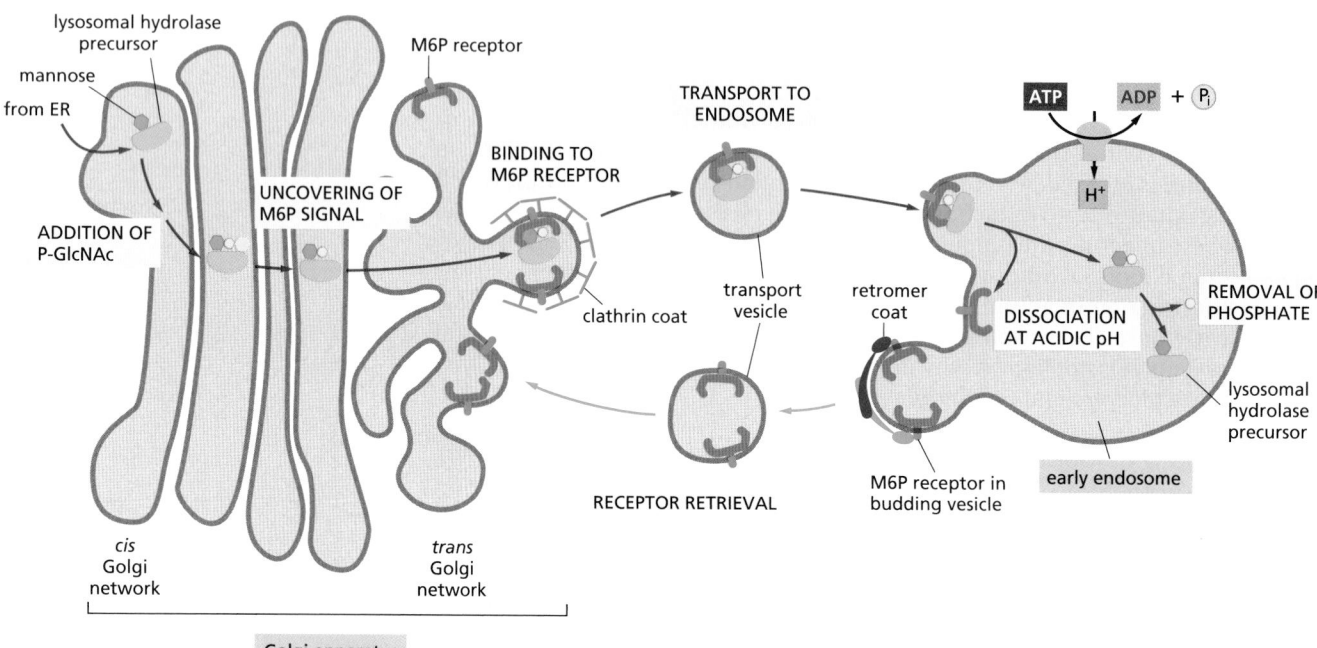

adding the M6P units to oligosaccharides must reside somewhere in the polypeptide chain of each hydrolase. Genetic engineering experiments have revealed that the recognition signal is a cluster of neighboring amino acids on each protein's surface, known as a *signal patch* (**Figure 13–46**). Since most lysosomal hydrolases contain multiple oligosaccharides, they acquire many M6P groups, providing a high-affinity signal for the M6P receptor.

Defects in the GlcNAc Phosphotransferase Cause a Lysosomal Storage Disease in Humans

Genetic defects that affect one or more of the lysosomal hydrolases cause a number of human **lysosomal storage diseases**. The defects result in an accumulation of undigested substrates in lysosomes, with severe pathological consequences, most often in the nervous system. In most cases, there is a mutation in a structural gene that codes for an individual lysosomal hydrolase. This occurs in *Hurler's disease*, for example, in which the enzyme required for the breakdown of certain types of glycosaminoglycan chains is defective or missing. The most severe form of lysosomal storage disease, however, is a very rare inherited metabolic disorder called *inclusion-cell disease* (*I-cell disease*). In this condition, almost all of the hydrolytic enzymes are missing from the lysosomes of many cell types, and their undigested substrates accumulate in these lysosomes, which consequently form large *inclusions* in the cells. The consequent pathology is complex, affecting all organ systems, skeletal integrity, and mental development; individuals rarely live beyond six or seven years.

I-cell disease is due to a single gene defect and, like most genetic enzyme deficiencies, it is recessive—that is, it occurs only in individuals having two copies of the defective gene. In patients with I-cell disease, all the hydrolases missing from lysosomes are found in the blood: because they fail to sort properly in the Golgi apparatus, they are secreted rather than transported to lysosomes. The mis-sorting has been traced to a defective or missing GlcNAc phosphotransferase. Because lysosomal enzymes are not phosphorylated in the *cis* Golgi network, the M6P receptors do not segregate them into the appropriate transport vesicles in the TGN. Instead, the lysosomal hydrolases are carried to the cell surface and secreted.

In I-cell disease, the lysosomes in some cell types, such as hepatocytes, contain a normal complement of lysosomal enzymes, implying that there is another

Figure 13–45 The transport of newly synthesized lysosomal hydrolases to endosomes. The sequential action of two enzymes in the *cis* and *trans* Golgi network adds mannose 6-phosphate (M6P) groups to the precursors of lysosomal enzymes (see Figure 13–46). The M6P-tagged hydrolases then segregate from all other types of proteins in the TGN because adaptor proteins (not shown) in the clathrin coat bind the M6P receptors, which, in turn, bind the M6P-modified lysosomal hydrolases. The clathrin-coated vesicles bud off from the TGN, shed their coat, and fuse with early endosomes. At the lower pH of the endosome, the hydrolases dissociate from the M6P receptors, and the empty receptors are retrieved in retromer-coated vesicles to the TGN for further rounds of transport. In the endosomes, the phosphate is removed from the M6P attached to the hydrolases, which may further ensure that the hydrolases do not return to the TGN with the receptor.

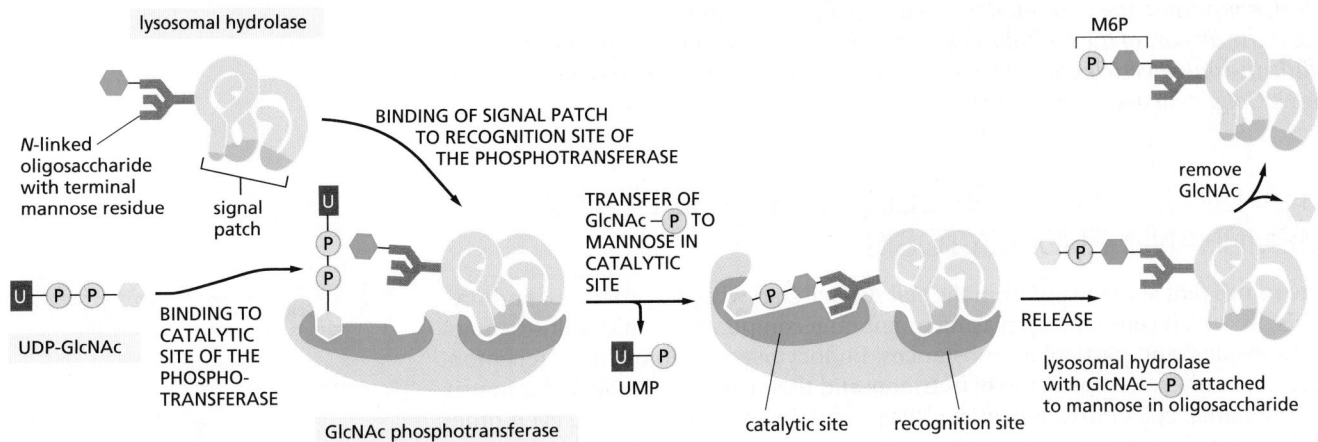

Figure 13–46 **The recognition of
a lysosomal hydrolase.** A GlcNAc
phosphotransferase recognizes lysosomal
hydrolases in the Golgi apparatus. The
enzyme has separate catalytic and
recognition sites. The catalytic site
binds both high-mannose N-linked
oligosaccharides and UDP-GlcNAc. The
recognition site binds to a signal patch that
is present only on the surface of lysosomal
hydrolases. A second enzyme cleaves
off the GlcNAc, leaving the mannose
6-phosphate exposed.

pathway for directing hydrolases to lysosomes that is used by some cell types but not others. Alternative sorting receptors function in these M6P-independent pathways. Similarly, an M6P-independent pathway in all cells sorts the membrane proteins of lysosomes from the TGN for transport to late endosomes, and those proteins are therefore normal in I-cell disease.

Some Lysosomes and Multivesicular Bodies Undergo Exocytosis

Targeting of material to lysosomes is not necessarily the end of the pathway. *Lysosomal secretion* of undigested content enables all cells to eliminate indigestible debris. For most cells, this seems to be a minor pathway, used only when the cells are stressed. Some cell types, however, contain specialized lysosomes that have acquired the necessary machinery for fusion with the plasma membrane. *Melanocytes* in the skin, for example, produce and store pigments in their lysosomes. These pigment-containing *melanosomes* release their pigment into the extracellular space of the epidermis by exocytosis. The pigment is then taken up by keratinocytes, leading to normal skin pigmentation. In some genetic disorders, defects in melanosome exocytosis block this transfer process, leading to forms of hypopigmentation (albinism). Under certain conditions, multivesicular bodies can also fuse with the plasma membrane. If that occurs, their intralumenal vesicles are released from cells. Circulating small vesicles, also called *exosomes*, have been observed in the blood and may be used to transport components between cells, although the importance of such a mechanism of potential communication between distant cells is unknown. Some exosomes may derive from direct vesicle budding events at the plasma membrane, which is a topologically equivalent process (see Figure 13–57).

Summary

Lysosomes are specialized for the intracellular digestion of macromolecules. They contain unique membrane proteins and a wide variety of soluble hydrolytic enzymes that operate best at pH 5, which is the internal pH of lysosomes. An ATP-driven H⁺ pump in the lysosomal membrane maintains this low pH. Newly synthesized lysosomal proteins transported from the lumen of the ER, through the Golgi apparatus; they are then carried from the trans Golgi network to endosomes by means of clathrin-coated transport vesicles, before moving on to lysosomes.

The lysosomal hydrolases contain N-linked oligosaccharides that are covalently modified in a unique way in the cis Golgi so that their mannoses are phosphorylated. These mannose 6-phosphate (M6P) groups are recognized by an M6P receptor protein in the trans Golgi network that segregates the hydrolases and helps package them into budding transport vesicles that deliver their contents to endosomes. The M6P receptors shuttle back and forth between the trans Golgi network and the endosomes. The low pH in endosomes and the removal of the phosphate from the

M6P group cause the lysosomal hydrolases to dissociate from these receptors, making the transport of the hydrolases unidirectional. A separate transport system uses clathrin-coated vesicles to deliver resident lysosomal membrane proteins from the trans Golgi network to endosomes.

TRANSPORT INTO THE CELL FROM THE PLASMA MEMBRANE: ENDOCYTOSIS

The routes that lead inward from the cell surface start with the process of **endocytosis**, by which cells take up plasma membrane components, fluid, solutes, macromolecules, and particulate substances. Endocytosed cargo includes receptor–ligand complexes, a spectrum of nutrients and their carriers, extracellular matrix components, cell debris, bacteria, viruses, and, in specialized cases, even other cells. Through endocytosis, the cell regulates the composition of its plasma membrane in response to changing extracellular conditions.

In endocytosis, the material to be ingested is progressively enclosed by a small portion of the plasma membrane, which first invaginates and then pinches off to form an **endocytic vesicle** containing the ingested substance or particle. Most eukaryotic cells constantly form endocytic vesicles, a process called *pinocytosis* ("cell drinking"); in addition, some specialized cells contain dedicated pathways that take up large particles on demand, a process called *phagocytosis* ("cell eating"). Endocytic vesicles form at the plasma membrane by multiple mechanisms that differ in both the molecular machinery used and how that machinery is regulated.

Once generated at the plasma membrane, most endocytic vesicles fuse with a common receiving compartment, the *early endosome*, where internalized cargo is sorted: some cargo molecules are returned to the plasma membrane, either directly or via a *recycling endosome*, and others are designated for degradation by inclusion in a *late endosome*. Late endosomes form from a bulbous, vacuolar portion of early endosomes by a process called *endosome maturation*. This conversion process changes the protein composition of the endosome membrane, patches of which invaginate and become incorporated within the organelles as *intralumenal vesicles*, while the endosome itself moves from the cell periphery to a location close to the nucleus. As an endosome matures, it ceases to recycle material to the plasma membrane and irreversibly commits its remaining contents to degradation: late endosomes fuse with one another and with lysosomes to form endolysosomes, which degrade their contents, as discussed earlier (**Figure 13–47**).

Each of the stages of endosome maturation—from the early endosome to the endolysosome—is connected through bidirectional vesicle transport pathways to

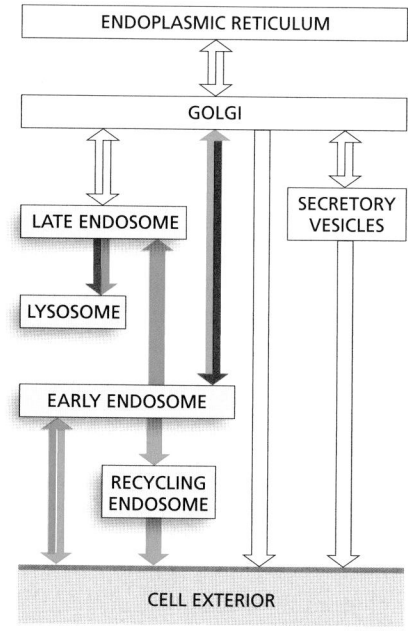

Figure 13–47 Endosome maturation: the endocytic pathway from the plasma membrane to lysosomes. Endocytic vesicles fuse near the cell periphery with an early endosome, which is the primary sorting station. Tubular portions of the early endosome bud off vesicles that recycle endocytosed cargo back to the plasma membrane—either directly, or indirectly via recycling endosomes. Recycling endosomes can store proteins until they are needed. Conversion of early endosomes to late endosomes is accompanied by loss of the tubular projections. Membrane proteins destined for degradation are internalized in intralumenal vesicles. The developing late endosome, or multivesicular body, moves on microtubules to the cell interior. Fully matured late endosomes no longer send vesicles to the plasma membrane, and they fuse with one another and with endolysosomes and lysosomes to degrade their contents. Each stage of endosome maturation is connected via transport vesicles with the TGN, providing a continuous supply of newly synthesized lysosomal proteins.

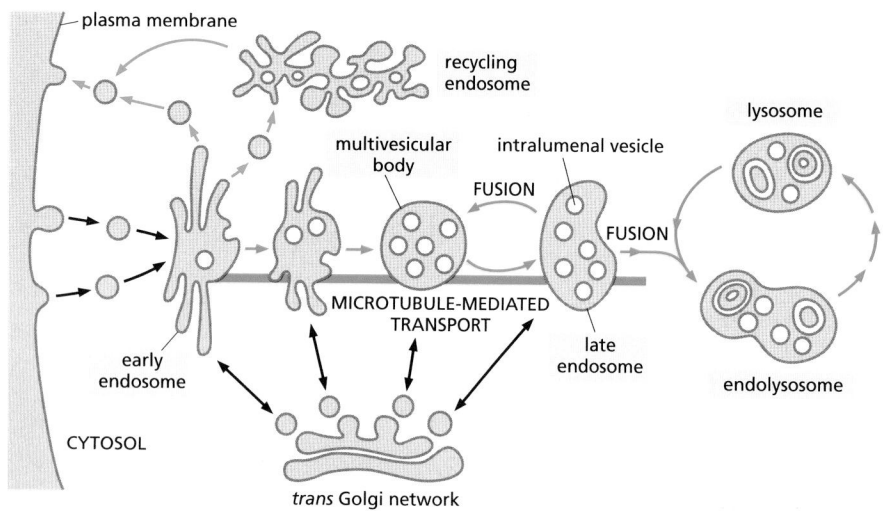

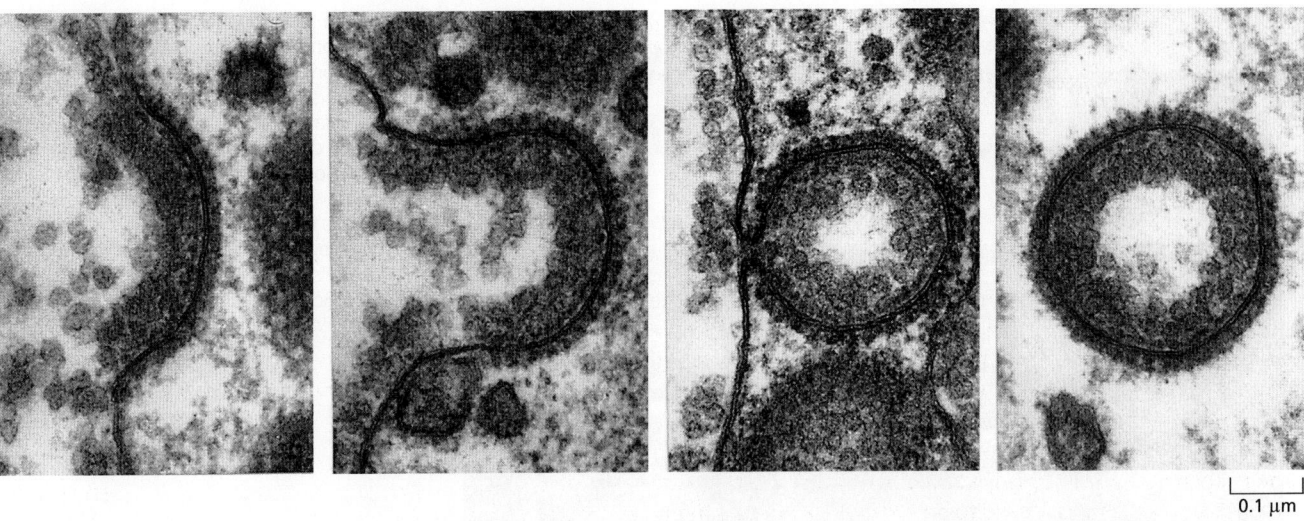

0.1 μm

Figure 13–48 **The formation of clathrin-coated vesicles from the plasma membrane.** These electron micrographs illustrate the probable sequence of events in the formation of a clathrin-coated vesicle from a clathrin-coated pit. The clathrin-coated pits and vesicles shown are larger than those seen in normal-sized cells; they are from a very large hen oocyte and they take up lipoprotein particles to form yolk. The lipoprotein particles bound to their membrane-bound receptors appear as a dense, fuzzy layer on the extracellular surface of the plasma membrane—which is the inside surface of the coated pit and vesicle. (From M.M. Perry and A.B. Gilbert, *J. Cell Sci.* 39:257–272, 1979. With permission from The Company of Biologists.)

the TGN. These pathways allow insertion of newly synthesized materials, such as lysosomal enzymes arriving from the ER, and the retrieval of components, such as the M6P receptor, back into the early parts of the secretory pathway. We next discuss how the cell uses and controls the various features of endocytic trafficking.

Pinocytic Vesicles Form from Coated Pits in the Plasma Membrane

Virtually all eukaryotic cells continually ingest portions of their plasma membrane in the form of small pinocytic (endocytic) vesicles. The rate at which plasma membrane is internalized in this process of **pinocytosis** varies between cell types, but it is usually surprisingly high. A macrophage, for example, ingests 25% of its own volume of fluid each hour. This means that it must ingest 3% of its plasma membrane each minute, or 100% in about half an hour. Fibroblasts endocytose at a somewhat lower rate (1% of their plasma membrane per minute), whereas some amoebae ingest their plasma membrane even more rapidly. Since a cell's surface area and volume remain unchanged during this process, it is clear that the same amount of membrane being removed by endocytosis is being added to the cell surface by the converse process of *exocytosis*. In this sense, endocytosis and exocytosis are linked processes that can be considered to constitute an *endocytic-exocytic cycle*. The coupling between exocytosis and endocytosis is particularly strict in specialized structures characterized by high membrane turnover, such as a nerve terminal.

The endocytic part of the cycle often begins at **clathrin-coated pits**. These specialized regions typically occupy about 2% of the total plasma membrane area. The lifetime of a clathrin-coated pit is short: within a minute or so of being formed, it invaginates into the cell and pinches off to form a clathrin-coated vesicle (**Figure 13–48**). About 2500 clathrin-coated vesicles pinch off from the plasma membrane of a cultured fibroblast every minute. The coated vesicles are even more transient than the coated pits: within seconds of being formed, they shed their coat and fuse with early endosomes.

Not All Pinocytic Vesicles Are Clathrin-Coated

In addition to clathrin-coated pits and vesicles, cells can form other types of pinocytic vesicles, such as **caveolae** (from the Latin for "little cavities"), originally recognized by their ability to transport molecules across endothelial cells that form the inner lining of blood vessels. Caveolae, sometimes seen in the electron microscope as deeply invaginated flasks, are present in the plasma membrane of most vertebrate cell types (**Figure 13–49**). They are thought to form from *lipid rafts* in the plasma membrane (discussed in Chapter 10), which are especially rich

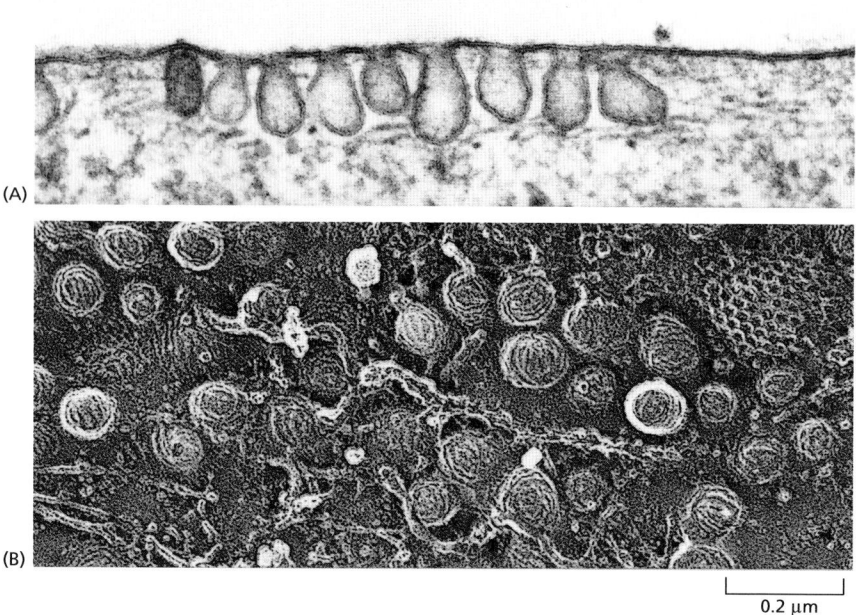

(A)

(B)

_0.2 μm_

Figure 13–49 Caveolae in the plasma membrane of a fibroblast. (A) This electron micrograph shows a plasma membrane with a very high density of caveolae. (B) This rapid-freeze deep-etch image demonstrates the characteristic "cauliflower" texture of the cytosolic face of the caveolae membrane. The characteristic texture is thought to result from aggregates of caveolins and cavins. A clathrin-coated pit is also seen at the upper right. (From K.G. Rothberg et al., _Cell_ 68:673–682, 1992. With permission from Elsevier.)

in cholesterol, glycosphingolipids, and glycosylphosphatidylinositol (GPI)-anchored membrane proteins (see Figure 10–13). The major structural proteins in caveolae are **caveolins**, a family of unusual integral membrane proteins that each insert a hydrophobic loop into the membrane from the cytosolic side but do not extend across the membrane. On their cytosolic side, caveolins are bound to large protein complexes of caving proteins, which are thought to stabilize the membrane curvature.

In contrast to clathrin-coated and COPI- or COPII-coated vesicles, caveolae are usually static structures. Nonetheless, they can be induced to pinch off and serve as endocytic vesicles to transport cargo to early endosomes or to the plasma membrane on the opposite side of a polarized cell (in a process called _transcytosis_, which we discuss later). Some animal viruses such as SV40 and papillomavirus (which causes warts) enter cells in vesicles derived from caveolae. The viruses are first delivered to early endosomes and move from there in transport vesicles to the lumen of the ER. The viral genome exits across the ER membrane into the cytosol, from where it is imported into the nucleus to start the infection cycle. Cholera toxin (discussed in Chapters 15 and 19) also enters the cell through caveolae and is transported to the ER before entering the cytosol.

Macropinocytosis is another clathrin-independent endocytic mechanism that can be activated in practically all animal cells. In most cell types, it does not operate continually but rather is induced for a limited time in response to cell-surface receptor activation by specific cargoes, including growth factors, integrin ligands, apoptotic cell remnants, and some viruses. These ligands activate a complex signaling pathway, resulting in a change in actin dynamics and the formation of cell-surface protrusions, called _ruffles_ (discussed in Chapter 16). When ruffles collapse back onto the cell, large fluid-filled endocytic vesicles form, called _macropinosomes_ (**Figure 13–50**), which transiently increase the bulk fluid uptake of a cell by up to tenfold. Macropinocytosis is a dedicated degradative pathway: macropinosomes acidify and then fuse with late endosomes or endolysosomes, without recycling their cargo back to the plasma membrane.

Cells Use Receptor-Mediated Endocytosis to Import Selected Extracellular Macromolecules

In most animal cells, clathrin-coated pits and vesicles provide an efficient pathway for taking up specific macromolecules from the extracellular fluid. In this process, called **receptor-mediated endocytosis**, the macromolecules bind to

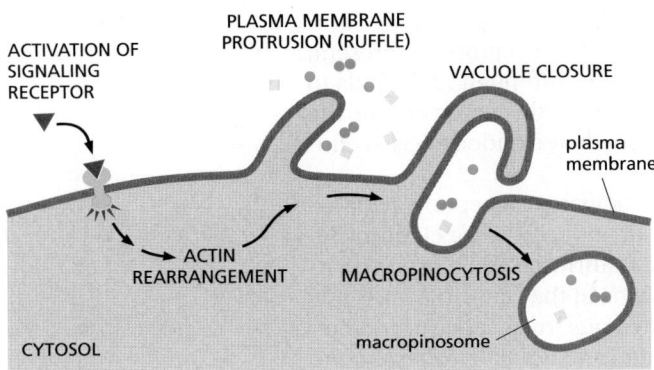

Figure 13–50 **Schematic representation of macropinocytosis.** Cell signaling events lead to a reprogramming of actin dynamics, which in turn triggers the formation of cell-surface ruffles. As the ruffles collapse back onto the cell surface, they nonspecifically trap extracellular fluid and macromolecules and particles contained in it, forming large vacuoles, or macropinosomes, as shown.

complementary transmembrane receptor proteins, which accumulate in coated pits, and then enter the cell as receptor–macromolecule complexes in clathrin-coated vesicles (see Figure 13–48). Because ligands are selectively captured by receptors, receptor-mediated endocytosis provides a selective concentrating mechanism that increases the efficiency of internalization of particular ligands more than a hundredfold. In this way, even minor components of the extracellular fluid can be efficiently taken up in large amounts. A particularly well-understood and physiologically important example is the process that mammalian cells use to import cholesterol.

Many animal cells take up cholesterol through receptor-mediated endocytosis and, in this way, acquire most of the cholesterol they require to make new membrane. If the uptake is blocked, cholesterol accumulates in the blood and can contribute to the formation in blood vessel (artery) walls of *atherosclerotic plaques*, deposits of lipid and fibrous tissue that can cause strokes and heart attacks by blocking arterial blood flow. In fact, it was a study of humans with a strong genetic predisposition for *atherosclerosis* that first revealed the mechanism of receptor-mediated endocytosis.

Most cholesterol is transported in the blood as cholesteryl esters in the form of lipid–protein particles known as **low-density lipoproteins (LDLs)** (Figure 13–51). When a cell needs cholesterol for membrane synthesis, it makes transmembrane receptor proteins for LDL and inserts them into its plasma membrane. Once in the plasma membrane, the *LDL receptors* diffuse until they associate with clathrin-coated pits that are in the process of forming. There, an endocytosis signal in the cytoplasmic tail of the LDL receptors binds the membrane-bound adaptor protein AP2 after its conformation has been locally unlocked by binding to $PI(4,5)P_2$ on the plasma membrane. Coincidence detection, as discussed earlier, thus imparts both efficiency and selectivity to the process (see Figure 13–9). AP2 then recruits clathrin to initiate endocytosis.

Since coated pits constantly pinch off to form coated vesicles, any LDL particles bound to LDL receptors in the coated pits are rapidly internalized in coated vesicles. After shedding their clathrin coats, the vesicles deliver their contents to early endosomes. Once the LDL and LDL receptors encounter the low pH in early endosomes, LDL is released from its receptor and is delivered via late endosomes to lysosomes. There, the cholesteryl esters in the LDL particles are hydrolyzed to free cholesterol, which is now available to the cell for new membrane synthesis (Movie 13.3). If too much free cholesterol accumulates in a cell, the cell shuts off both its own cholesterol synthesis and the synthesis of LDL receptors, so that it ceases both to make or to take up cholesterol.

This regulated pathway for cholesterol uptake is disrupted in individuals who inherit defective genes encoding LDL receptors. The resulting high levels of blood cholesterol predispose these individuals to develop atherosclerosis prematurely, and many would die at an early age of heart attacks resulting from coronary artery disease if they were not treated with drugs such as statins that lower the level of blood cholesterol. In some cases, the receptor is lacking altogether. In others, the receptors are defective—in either the extracellular binding site for LDL or the

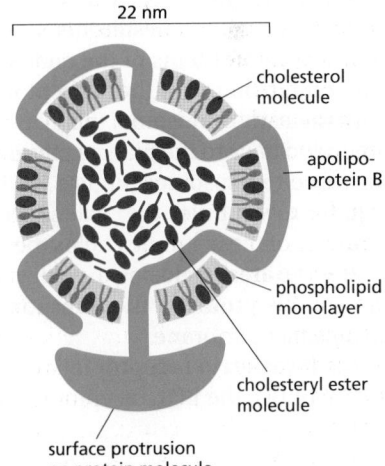

Figure 13–51 **A low-density lipoprotein (LDL) particle.** Each spherical particle has a mass of 3×10^6 daltons. It contains a core of about 1500 cholesterol molecules esterified to long-chain fatty acids. A lipid monolayer composed of about 800 phospholipid and 500 unesterified cholesterol molecules surrounds the core of cholesteryl esters. A single molecule of apolipoprotein B, a 500,000-dalton beltlike protein, organizes the particle and mediates the specific binding of LDL to cell-surface LDL receptors.

intracellular binding site that attaches the receptor AP2 adaptor protein in clathrin-coated pits. In the latter case, normal numbers of LDL receptors are present, but they fail to become localized in clathrin-coated pits. Although LDL binds to the surface of these mutant cells, it is not internalized, directly demonstrating the importance of clathrin-coated pits for the receptor-mediated endocytosis of cholesterol.

More than 25 distinct receptors are known to participate in receptor-mediated endocytosis of different types of molecules. They all apparently use clathrin-dependent internalization routes and are guided into clathrin-coated pits by signals in their cytoplasmic tails that bind to adaptor proteins in the clathrin coat. Many of these receptors, like the LDL receptor, enter coated pits irrespective of whether they have bound their specific ligands. Others enter preferentially when bound to a specific ligand, suggesting that a ligand-induced conformational change is required for them to activate the signal sequence that guides them into the pits. Since most plasma membrane proteins fail to become concentrated in clathrin-coated pits, the pits serve as molecular filters, preferentially collecting certain plasma membrane proteins (receptors) over others.

Electron-microscope studies of cultured cells exposed simultaneously to different labeled ligands demonstrate that many kinds of receptors can cluster in the same coated pit, whereas some other receptors cluster in different clathrin-coated pits. The plasma membrane of one clathrin-coated pit can accommodate more than 100 receptors of assorted varieties.

Specific Proteins Are Retrieved from Early Endosomes and Returned to the Plasma Membrane

Early endosomes are the main sorting station in the endocytic pathway, just as the *cis* and *trans* Golgi networks serve this function in the secretory pathway. In the mildly acidic environment of the early endosome, many internalized receptor proteins change their conformation and release their ligand, as already discussed for the M6P receptors. Those endocytosed ligands that dissociate from their receptors in the early endosome are usually doomed to destruction in lysosomes (although cholesterol is an exception, as just discussed), along with the other soluble contents of the endosome. Some other endocytosed ligands, however, remain bound to their receptors, and thereby share the fate of the receptors.

In the early endosome, the LDL receptor dissociates from its ligand, LDL, and is recycled back to the plasma membrane for reuse, leaving the discharged LDL to be carried to lysosomes (**Figure 13–52**). The recycling transport vesicles bud from long, narrow tubules that extend from the early endosomes. It is likely that the geometry of these tubules helps the sorting process: because tubules have a large membrane area enclosing a small volume, membrane proteins become enriched over soluble proteins. The transport vesicles return the LDL receptor directly to the plasma membrane.

The **transferrin receptor** follows a similar recycling pathway as the LDL receptor, but unlike the LDL receptor it also recycles its ligand. Transferrin is a soluble

Figure 13–52 **The receptor-mediated endocytosis of LDL.** Note that the LDL dissociates from its receptors in the acidic environment of the early endosome. After a number of steps, the LDL ends up in endolysosomes and lysosomes, where it is degraded to release free cholesterol. In contrast, the LDL receptors are returned to the plasma membrane via transport vesicles that bud off from the tubular region of the early endosome, as shown. For simplicity, only one LDL receptor is shown entering the cell and returning to the plasma membrane. Whether it is occupied or not, an LDL receptor typically makes one round trip into the cell and back to the plasma membrane every 10 minutes, making a total of several hundred trips in its 20-hour life-span.

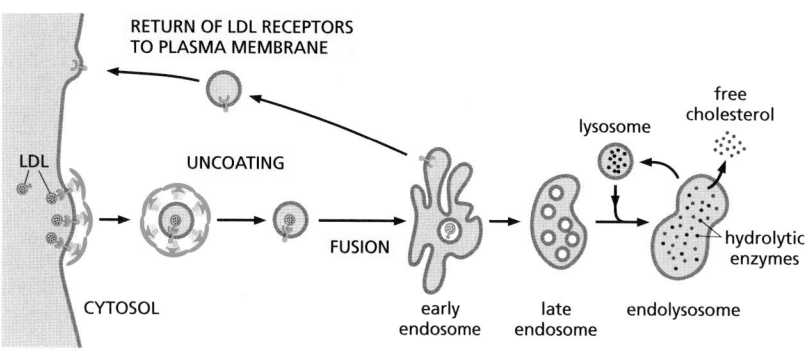

protein that carries iron in the blood. Cell-surface transferrin receptors deliver transferrin with its bound iron to early endosomes by receptor-mediated endocytosis. The low pH in the endosome induces transferrin to release its bound iron, but the iron-free transferrin itself (called apotransferrin) remains bound to its receptor. The receptor–apotransferrin complex enters the tubular extensions of the early endosome and from there is recycled back to the plasma membrane. When the apotransferrin returns to the neutral pH of the extracellular fluid, it dissociates from the receptor and is thereby freed to pick up more iron and begin the cycle again. Thus, transferrin shuttles back and forth between the extracellular fluid and early endosomes, avoiding lysosomes and delivering iron to the cell interior, as needed for cells to grow and proliferate.

Plasma Membrane Signaling Receptors are Down-Regulated by Degradation in Lysosomes

A second pathway that endocytosed receptors can follow from endosomes is taken by many signaling receptors, including opioid receptors and the receptor that binds *epidermal growth factor* (*EGF*). EGF is a small, extracellular signal protein that stimulates epidermal and various other cells to divide. Unlike LDL receptors, EGF receptors accumulate in clathrin-coated pits only after binding their ligand, and most do not recycle but are degraded in lysosomes, along with the ingested EGF. EGF binding therefore first activates intracellular signaling pathways and then leads to a decrease in the concentration of EGF receptors on the cell surface, a process called *receptor downregulation*, that reduces the cell's subsequent sensitivity to EGF (see Figure 15–20).

Receptor downregulation is highly regulated. The activated receptors are first covalently modified on the cytosolic face with the small protein ubiquitin. Unlike *polyubiquitylation*, which adds a chain of ubiquitins that typically targets a protein for degradation in proteasomes (discussed in Chapter 6), ubiquitin tagging for sorting into the clathrin-dependent endocytic pathway adds just one or a few single ubiquitin molecules to the protein—a process called *monoubiquitylation* or *multiubiquitylation*, respectively. Ubiquitin-binding proteins recognize the attached ubiquitin and help direct the modified receptors into clathrin-coated pits. After delivery to the early endosome, other ubiquitin-binding proteins that are part of *ESCRT complexes* (ESCRT = Endosome Sorting Complex Required for Transport) recognize and sort the ubiquitylated receptors into intralumenal vesicles, which are retained in the maturing late endosome (see Figure 13–47). In this way, addition of ubiquitin blocks receptor recycling to the plasma membrane and directs the receptors into the degradation pathway, as we discuss next.

Early Endosomes Mature into Late Endosomes

The endosomal compartments can be made visible in the electron microscope by adding a readily detectable tracer molecule, such as the enzyme peroxidase, to the extracellular medium and allowing varying lengths of time for the cell to endocytose the tracer. The distribution of the molecule after its uptake reveals the sequence of events. Within a minute or so after adding the tracer, it starts to appear in **early endosomes**, just beneath the plasma membrane (**Figure 13–53**). By 5–15 minutes later, it has moved to **late endosomes**, close to the Golgi apparatus and near the nucleus.

How early endosomes arise is not entirely clear, but their membrane and volume are mainly derived from incoming endocytic vesicles that fuse with one another (**Movie 13.4**). Early endosomes are relatively small and patrol the cytoplasm underlying the plasma membrane in jerky back-and-forth movements along microtubules, capturing incoming vesicles. Typically, an early endosome receives incoming vesicles for about 10 minutes, during which time membrane and fluid is rapidly recycled to the plasma membrane. Some of the incoming cargo, however, accumulates over the lifetime of the early endosome, eventually to be included in the late endosome.

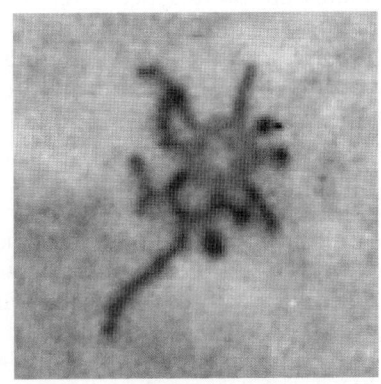

0.5 μm

Figure 13–53 **Electron micrograph of an early endosome.** The endosome is labeled with endocytosed horseradish peroxidase, a widely used enzyme marker, detected in this case by an electron-dense reaction product. Many tubular extensions protrude from the central vacuolar space of the early endosome, which will later mature to give rise to a late endosome. (© 1992 J. Tooze and M. Hollinshead. Originally published in *J. Cell Biol.* https://doi .org/10.1083/jcb.118.4.113. With permission from Rockefeller University Press.)

Early endosomes have tubular and vacuolar domains (see Figure 13–53). Most of the membrane surface is in the tubules and most of the volume is in the vacuolar domain. During **endosome maturation**, the two domains have different fates: the vacuolar portions of the early endosome are retained and transformed into late endosomes; the tubular portions shrink. Maturing endosomes, also called *multivesicular bodies*, migrate along microtubules toward the cell interior, shedding membrane tubules and vesicles that recycle material to the plasma membrane and TGN, and receiving newly synthesized lysosomal proteins. As they concentrate in a perinuclear region of the cell, the multivesicular bodies fuse with each other, and eventually with endolysosomes and lysosomes (see Figure 13–47).

Many changes occur during the maturation process. (1) The endosome changes shape and location, as the tubular domains are lost and the vacuolar domains are thoroughly modified. (2) Rab proteins, phosphoinositide lipids, fusion machinery (SNAREs and tethers), and microtubule motor proteins all participate in a molecular makeover of the cytosolic face of the endosome membrane, changing the functional characteristics of the organelle. (3) A V-type ATPase in the endosome membrane pumps H^+ from the cytosol into the endosome lumen and acidifies the organelle. Crucially, the increasing acidity that accompanies maturation renders lysosomal hydrolases increasingly more active, influencing many receptor–ligand interactions, thereby controlling receptor loading and unloading. (4) Intralumenal vesicles sequester endocytosed signaling receptors inside the endosome, thus halting the receptor signaling activity. (5) Lysosome proteins are delivered from the TGN to the maturing endosome. Most of these events occur gradually but eventually lead to a complete transformation of the endosome into an early endolysosome.

In addition to committing selected cargo for degradation, the maturation process is important for lysosome maintenance. The continual delivery of lysosome components from the TGN to maturing endosomes, ensures a steady supply of new lysosome proteins. The endocytosed materials mix in early endosomes with newly arrived acid hydrolases. Although mild digestion may start here, many hydrolases are synthesized and delivered as proenzymes, called *zymogens*, which contain extra inhibitory domains that keep the hydrolases inactive until these domains are proteolytically removed at later stages of endosome maturation. Moreover, the pH in early endosomes is not low enough to activate lysosomal hydrolases optimally. By these means, cells can retrieve membrane proteins intact from early endosomes and recycle them back to the plasma membrane.

ESCRT Protein Complexes Mediate the Formation of Intralumenal Vesicles in Multivesicular Bodies

As endosomes mature, patches of their membrane invaginate into the endosome lumen and pinch off to form intralumenal vesicles. Because of their appearance in the electron microscope such maturing endosomes are also called **multivesicular bodies** (**Figure 13–54**).

The multivesicular bodies carry endocytosed membrane proteins that are to be degraded. As part of the protein-sorting process, receptors destined for degradation, such as the occupied EGF receptors described previously, selectively partition into the invaginating membrane of the multivesicular bodies. In this way, both the receptors and any signaling proteins strongly bound to them are sequestered away from the cytosol where they might otherwise continue signaling. They also are made fully accessible to the digestive enzymes that eventually will degrade them (**Figure 13–55**). In addition to endocytosed membrane proteins, multivesicular bodies include the soluble content of early endosomes destined for late endosomes and digestion in lysosomes.

As discussed earlier, sorting into intralumenal vesicles requires one or multiple ubiquitin tags, which are added to the cytosolic domains of membrane proteins. These tags initially help guide the proteins into clathrin-coated vesicles in the plasma membrane. Once delivered to the endosomal membrane, the ubiquitin tags are recognized again, this time by a series of cytosolic **ESCRT protein**

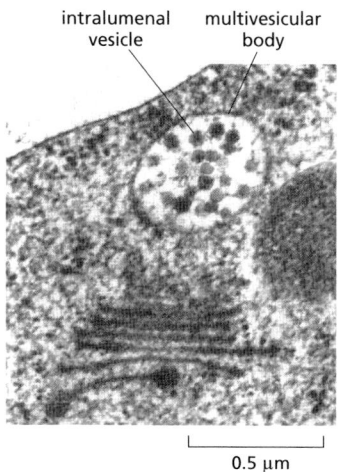

intralumenal   multivesicular
vesicle   body

0.5 μm

Figure 13–54 Electron micrograph of a multivesicular body. The large amount of internal membrane will be delivered to the lysosome, for digestion. (From A. Driouich, A. Jauneau, and L.A. Staehelin; *Plant Physiol.* 113:487–492, 1997. With permission from the American Society of Plant Biologists.)

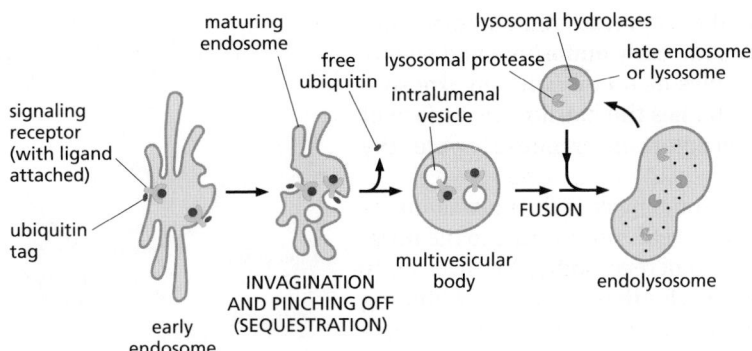

Figure 13–55 **The sequestration of endocytosed proteins into intralumenal vesicles of multivesicular bodies.** Ubiquitylated membrane proteins are sorted into domains on the endosome membrane, which invaginate and pinch off to form intralumenal vesicles. The ubiquitin marker is removed and returned to the cytosol for reuse before the intralumenal vesicle closes. Eventually, proteases and lipases in lysosomes digest all of the internal membranes. The invagination processes are essential for complete digestion of endocytosed membrane proteins: because the outer membrane of the multivesicular body becomes continuous with the lysosomal membrane, which is resistant to lysosomal hydrolases; the hydrolases, for example, could not digest the cytosolic domains of endocytosed transmembrane proteins, such as the EGF receptor shown here, if the protein were not localized in intralumenal vesicles.

complexes, (*ESCRT-0, -I, -II,* and *-III*), which bind sequentially and ultimately mediate the sorting process into the intralumenal vesicles. Membrane invagination into multivesicular bodies also depends on a lipid kinase that phosphorylates phosphatidylinositol to produce PI(3)P, which serves as an additional docking site for the ESCRT complexes; these complexes require both PI(3)P and the presence of ubiquitylated cargo proteins to bind to the endosomal membrane. ESCRT-III forms large multimeric assemblies on the membrane that bend the membrane (**Figure 13–56**).

Mutant cells compromised in ESCRT function display signaling defects. In such cells, activated receptors cannot be down-regulated by endocytosis and packaging into multivesicular bodies. The still-active receptors therefore mediate prolonged signaling, which can lead to uncontrolled cell proliferation and cancer.

Processes that shape membranes often use similar machinery. Because of strong similarities in their protein sequences, researchers think that ESCRT complexes are evolutionarily related to components that mediate cell-membrane deformation in cytokinesis in archaea. Similarly, the ESCRT machinery that drives the internal budding from the endosome membrane to form intralumenal vesicles is also used in animal cell cytokinesis and virus budding, which are topologically equivalent, as both processes involve budding away from the cytosolic surface of the membrane (**Figure 13–57**).

Recycling Endosomes Regulate Plasma Membrane Composition

The fates of endocytosed receptors—and of any ligands remaining bound to them—vary according to the specific type of receptor. As we discussed, most receptors are recycled and returned to the same plasma membrane domain from which they came; some proceed to a different domain of the plasma membrane, thereby mediating **transcytosis**; and some progress to lysosomes, where they are degraded.

Receptors on the surface of polarized epithelial cells can transfer specific macromolecules from one extracellular space to another by transcytosis. A newborn, for example, obtains antibodies from its mother's milk (which help protect it against infection) by transporting them across the epithelium of its gut. The lumen of the gut is acidic, and, at this low pH, the antibodies in the milk bind to specific receptors on the apical (absorptive) surface of the gut epithelial cells. The receptor–antibody complexes are internalized via clathrin-coated pits and

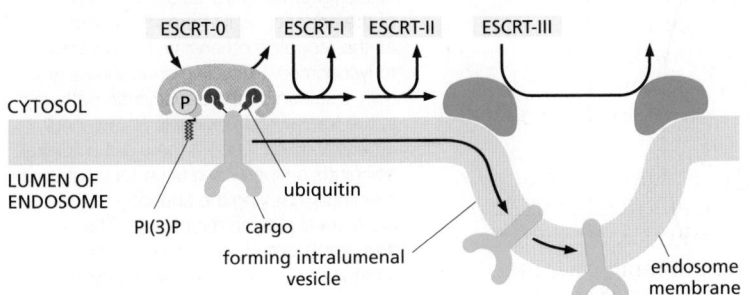

Figure 13–56 **Sorting of endocytosed membrane proteins into the intralumenal vesicles of a multivesicular body.** A series of complex binding events passes the ubiquitylated cargo proteins sequentially from one ESCRT complex (ESCRT-0) to the next, eventually concentrating them in membrane areas that bud into the lumen of the endosome to form intralumenal vesicles. ESCRT-III assembles into expansive multimeric structures and mediates invagination. The mechanisms of how cargo molecules are shepherded into the vesicles and how the vesicles are formed without including the ESCRT complexes themselves remain unknown. ESCRT complexes are soluble in the cytosol, are recruited to the membrane sequentially as needed, and are then released back into the cytosol as the vesicle pinches off.

vesicles and are delivered to early endosomes. The complexes remain intact and are retrieved in transport vesicles that bud from the early endosome and subsequently fuse with the basolateral domain of the plasma membrane. On exposure to the neutral pH of the extracellular fluid that bathes the basolateral surface of the cells, the antibodies dissociate from their receptors and eventually enter the baby's bloodstream.

The transcytotic pathway from the early endosome back to the plasma membrane is not direct. The receptors first move from the early endosome to the **recycling endosome**. The variety of pathways that different receptors follow from early endosomes implies that, in addition to binding sites for their ligands and binding sites for coated pits, many receptors also possess sorting signals that guide them into the appropriate transport pathway (**Figure 13–58**).

Cells can regulate the release of membrane proteins from recycling endosomes, thus adjusting the flux of proteins through the transcytotic pathway according to need. This regulation, the mechanism of which is uncertain, allows recycling endosomes to play an important part in adjusting the concentration of specific plasma membrane proteins. Fat cells and muscle cells, for example, contain large intracellular pools of the glucose transporters that are responsible for the uptake of glucose across the plasma membrane. These membrane transport proteins are stored in specialized recycling endosomes until the hormone *insulin* stimulates the cell to increase its rate of glucose uptake. In response to the insulin signal, transport vesicles rapidly bud from the recycling endosome and deliver large numbers of glucose transporters to the plasma membrane, thereby greatly increasing the rate of glucose uptake into the cell (**Figure 13–59**). Similarly, kidney cells regulate the insertion of aquaporins and V-ATPase into the plasma membrane to increase water and acid excretion, respectively, both in response to hormones.

Specialized Phagocytic Cells Can Ingest Large Particles

Phagocytosis is a special form of endocytosis in which a cell uses large endocytic vesicles called **phagosomes** to ingest large particles such as microorganisms and dead cells. Phagocytosis is distinct, both in purpose and mechanism, from macropinocytosis, which we discussed earlier. In protozoa, phagocytosis is a form of feeding: large particles taken up into phagosomes end up in lysosomes, and the products of the subsequent digestive processes pass into the cytosol to be used as food. However, few cells in multicellular organisms are able to ingest such large particles efficiently. In the gut of animals, for example, extracellular processes break down food particles, and cells import the small products of hydrolysis.

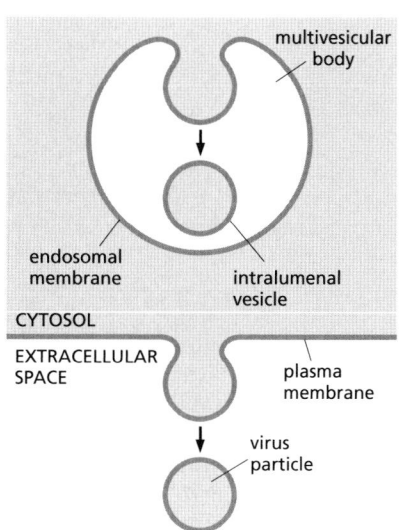

Figure 13–57 Conserved mechanism in multivesicular body formation and virus budding. In the two topologically equivalent processes indicated by the arrows, ESCRT complexes (not shown) shape membranes into buds that bulge away from the cytosol.

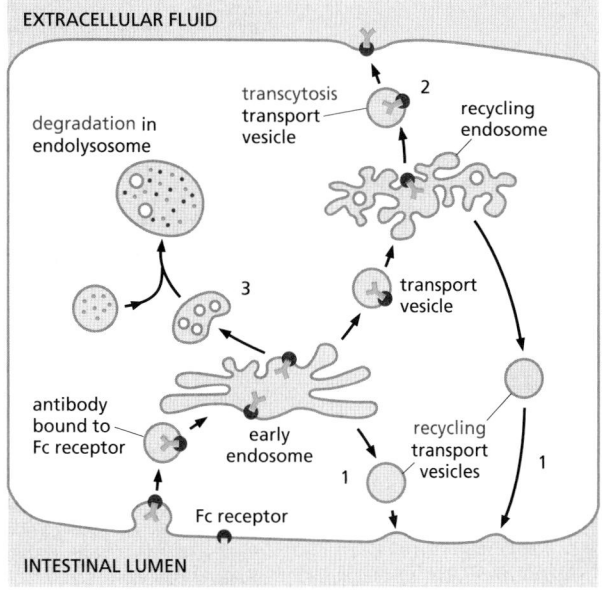

Figure 13–58 Possible fates for transmembrane receptor proteins that have been endocytosed. Three pathways from the early endosomal compartment in an epithelial cell are shown. Retrieved receptors are returned (1) to the same plasma membrane domain from which they came (*recycling*) or (2) via a recycling endosome to a different domain of the plasma membrane (*transcytosis*). (3) Receptors that are not specifically retrieved from early or recycling endosomes follow the pathway from the endosomal compartment to lysosomes, where they are degraded (*degradation*). If the ligand that is endocytosed with its receptor stays bound to the receptor in the acidic environment of the endosome, it shares the same fate as the receptor; otherwise, it is delivered to lysosomes. Recycling endosomes are a way-station on the transcytotic pathway. In the transcytosis example shown here, an antibody Fc receptor on a gut epithelial cell binds antibody and is endocytosed, eventually carrying the antibody to the basolateral plasma membrane. The receptor is called an Fc receptor because it binds the Fc part of the antibody (discussed in Chapter 24).

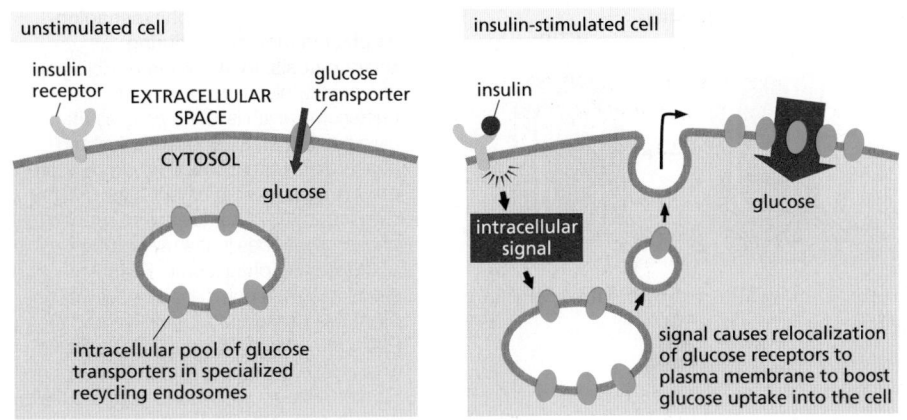

unstimulated cell

insulin receptor

EXTRACELLULAR SPACE

glucose transporter

CYTOSOL

glucose

intracellular pool of glucose transporters in specialized recycling endosomes

insulin-stimulated cell

insulin

intracellular signal

glucose

signal causes relocalization of glucose receptors to plasma membrane to boost glucose uptake into the cell

Figure 13–59 Storage of plasma membrane proteins in recycling endosomes. Recycling endosomes can serve as an intracellular storage site for specialized plasma membrane proteins that can be mobilized when needed. In the example shown, insulin binding to the insulin receptor triggers an intracellular signaling pathway that causes the rapid insertion of glucose transporters into the plasma membrane of a fat or muscle cell, greatly increasing its glucose intake.

Phagocytosis is important in most animals for purposes other than nutrition, and it is carried out mainly by specialized cells—so-called *professional phagocytes.* In mammals, two important classes of white blood cells that act as professional phagocytes are **macrophages** and **neutrophils** (Movie 13.5). These cells develop from hemopoietic stem cells (discussed in Chapter 22), and they ingest invading microorganisms to defend us against infection. Macrophages also have an important role in scavenging senescent cells and cells that have died by apoptosis (discussed in Chapter 18). In quantitative terms, the clearance of senescent and dead cells is by far the most important: our macrophages, for example, phagocytose more than 10^{11} senescent red blood cells in each of us every day.

The diameter of a phagosome is determined by the size of its ingested particles, and those particles can be almost as large as the phagocytic cell itself (Figure 13–60). Phagosomes fuse with lysosomes, and the ingested material is then degraded. Indigestible substances remain in the lysosomes, forming *residual bodies* that can be excreted from cells by exocytosis, as mentioned earlier. Some of the internalized plasma membrane components never reach the lysosome, because they are retrieved from the phagosome in transport vesicles and returned to the plasma membrane.

Some pathogenic bacteria have evolved elaborate mechanisms to prevent phagosome–lysosome fusion. The bacterium *Legionella pneumophila*, for example, which causes Legionnaires' disease (discussed in Chapter 23), injects into its unfortunate host a Rab-modifying enzyme that causes certain Rab proteins to misdirect membrane traffic, thereby preventing phagosome–lysosome fusion. The bacterium, thus spared from lysosomal degradation, remains in the modified phagosome, growing and dividing as an intracellular pathogen, protected from the host's adaptive immune system.

Phagocytosis is a cargo-triggered process. That is, it requires the activation of cell-surface receptors that transmit signals to the cell interior. Thus, to be phagocytosed, particles must first bind to the surface of the phagocyte (although not all particles that bind are ingested). Phagocytes have a variety of cell surface receptors that are functionally linked to the phagocytic machinery of the cell. The best-characterized triggers of phagocytosis are antibodies, which protect us by binding to the surface of infectious microorganisms (pathogens) and initiating a series of events that culminate in the invader being phagocytosed. When antibodies initially attack a pathogen, they coat it with antibody molecules that bind to *Fc receptors* on the surface of macrophages and neutrophils, activating the receptors to induce the phagocytic cell to extend pseudopods, which engulf the particle and fuse at their tips to form a phagosome (Figure 13–61A). Localized actin polymerization, initiated by Rho family GTPases and their activating Rho-GEFs (discussed in Chapters 15 and 16), shapes the pseudopods. The activated Rho GTPases switch on the kinase activity of local PI kinases to produce $PI(4,5)P_2$ in the membrane (see Figure 13–11), which stimulates actin polymerization. To seal off the phagosome and complete the engulfment, actin is depolymerized by a PI 3-kinase that converts the $PI(4,5)P_2$ to $PI(3,4,5)P_3$, which is required for closure

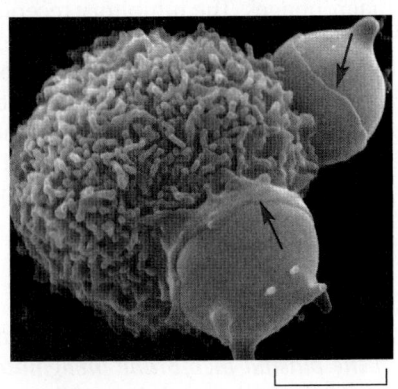

5 μm

Figure 13–60 Phagocytosis by a macrophage. A scanning electron micrograph of a mouse macrophage phagocytosing two chemically altered red blood cells. The *red arrows* point to edges of thin processes (pseudopods) of the macrophage that are extending as collars to engulf the red cells. (Courtesy of Jean Paul Revel.)

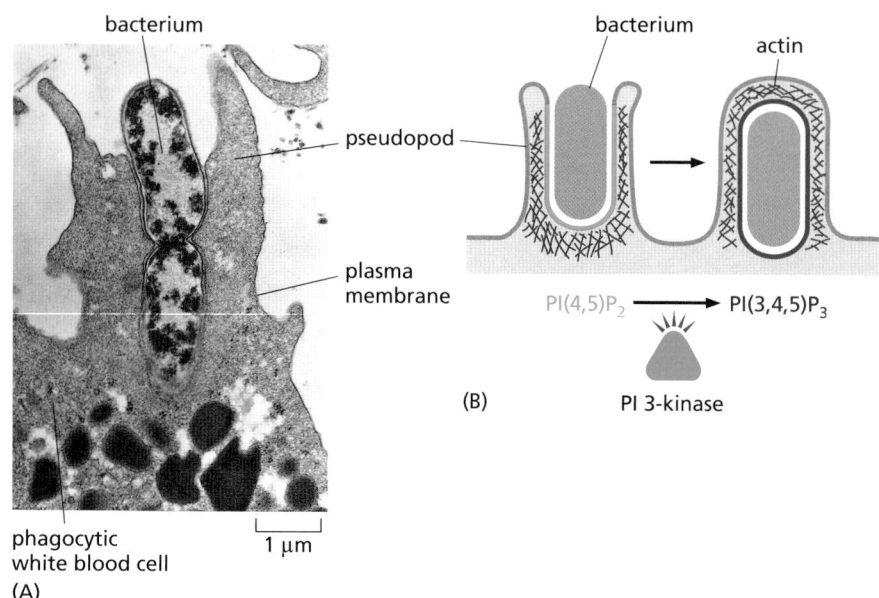

bacterium

pseudopod

plasma membrane

bacterium

actin

PI(4,5)P$_2$ ⟶ PI(3,4,5)P$_3$

PI 3-kinase

(B)

phagocytic white blood cell
(A)

1 μm

Figure 13–61 **A neutrophil reshaping its plasma membrane during phagocytosis.** (A) An electron micrograph of a neutrophil phagocytosing a bacterium, which is in the process of dividing. (B) Pseudopod extension and phagosome formation are driven by actin polymerization and reorganization, which respond to the accumulation of specific phosphoinositides in the membrane of the forming phagosome: PI(4,5)P$_2$ stimulates actin polymerization, which promotes pseudopod formation, and then PI(3,4,5)P$_3$ depolymerizes actin filaments at the base. (A, courtesy of Dorothy F. Bainton, Phagocytic Mechanisms in Health and Disease. New York: Intercontinental Medical Book Corporation, 1971.)

of the phagosome and may also contribute to reshaping the actin network to help drive the invagination of the forming phagosome (Figure 13–61B). In this way, the ordered generation and consumption of specific phosphoinositides guides sequential steps in phagosome formation.

Several other classes of receptors that promote phagocytosis have been characterized. Some recognize *complement* components, which collaborate with antibodies in targeting microbes for destruction (discussed in Chapter 24). Others directly recognize oligosaccharides on the surface of certain pathogens. Still others recognize cells that have died by apoptosis. Apoptotic cells lose the asymmetric distribution of phospholipids in their plasma membrane. As a consequence, negatively charged phosphatidylserine, which is normally confined to the cytosolic leaflet of the lipid bilayer, is now exposed on the outside of the cell, where it helps to trigger the phagocytosis of the dead cell.

Remarkably, macrophages will also phagocytose a variety of inanimate particles—such as glass or latex beads and asbestos fibers—yet they do not phagocytose live cells in their own body. The living cells display "don't-eat-me" signals in the form of cell-surface proteins that bind to inhibiting receptors on the surface of macrophages. The inhibitory receptors recruit tyrosine phosphatases that antagonize the intracellular signaling events required to initiate phagocytosis, thereby locally inhibiting the phagocytic process. Thus phagocytosis, like many other cell processes, depends on a balance between positive signals that activate the process and negative signals that inhibit it. Apoptotic cells are thought both to gain "eat-me" signals (such as extracellularly exposed phosphatidylserine) and to lose their "don't-eat-me" signals, causing them to be very rapidly phagocytosed by macrophages.

Summary

Cells ingest fluid, molecules, and particles by endocytosis, in which localized regions of the plasma membrane invaginate and pinch off to form endocytic vesicles. In most cells, endocytosis internalizes a large fraction of the plasma membrane every hour. The cells remain the same size because most of the plasma membrane components (proteins and lipids) that are endocytosed are continually returned to the cell surface by exocytosis. This large-scale endocytic–exocytic cycle is mediated largely by clathrin-coated pits and vesicles but clathrin-independent endocytic pathways also contribute.

While many of the endocytosed molecules are quickly recycled to the plasma membrane, others eventually end up in lysosomes, where they are degraded. Most of the ligands that are endocytosed with their receptors dissociate from their receptors in the acidic environment of the endosome and eventually end up in lysosomes, while most of the receptors are recycled via transport vesicles back to the cell surface for reuse. Many cell-surface signaling receptors become tagged with ubiquitin when activated by binding their extracellular ligands. Ubiquitylation guides activated receptors into clathrin-coated pits, they and their ligands are efficiently internalized and delivered to early endosomes.

Early endosomes, rapidly mature into late endosomes. During maturation, patches of the endosomal membrane containing ubiquitylated receptors invaginate and pinch off to form intralumenal vesicles. This process is mediated by ESCRT complexes and sequesters the receptors away from the cytosol, which terminates their signaling activity. Late endosomes migrate along microtubules toward the interior of the cell where they fuse with one another and with lysosomes to form endolysosomes, where degradation occurs.

In some cases, both receptor and ligand are transferred to a different plasma membrane domain, causing the ligand to be released at a different surface from where it originated, a process called transcytosis. In some cells, endocytosed plasma membrane proteins and lipids can be stored in recycling endosomes, for as long as necessary until they are needed.

TRANSPORT FROM THE *TRANS* GOLGI NETWORK TO THE CELL EXTERIOR: EXOCYTOSIS

Having considered the cell's endocytic and internal digestive systems and the various types of incoming membrane traffic that converge on lysosomes, we now return to the Golgi apparatus and examine the secretory pathways that lead outward to the cell exterior. Transport vesicles destined for the plasma membrane normally leave the TGN in a steady stream as irregularly shaped tubules. The membrane proteins and the lipids in these vesicles provide new components for the cell's plasma membrane, while the soluble proteins inside the vesicles are secreted to the extracellular space. The fusion of the vesicles with the plasma membrane is called **exocytosis**. This is the route, for example, by which cells secrete most of the proteoglycans and glycoproteins of the *extracellular matrix,* as discussed in Chapter 19.

All cells require this **constitutive secretory pathway**, which operates continuously (**Movie 13.6**). Specialized secretory cells, however, have a second secretory pathway in which soluble proteins and other substances are initially stored in *secretory vesicles* for later release by exocytosis. This is the **regulated secretory pathway**, found mainly in cells specialized for secreting products rapidly on demand—such as hormones, neurotransmitters, or digestive enzymes (**Figure 13–62**). In this section, we consider the role of the Golgi apparatus in both of these pathways and compare the two mechanisms of secretion.

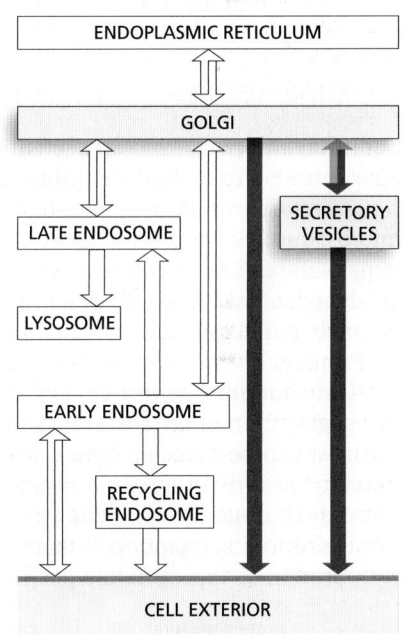

Many Proteins and Lipids Are Carried Automatically from the *Trans* Golgi Network (TGN) to the Cell Surface

A cell capable of regulated secretion must separate at least three classes of proteins before they leave the TGN—those destined for lysosomes (via endosomes), those destined for secretory vesicles, and those destined for immediate delivery to the cell surface (**Figure 13–63**). We have already discussed how proteins destined for lysosomes are tagged with M6P for packaging into specific departing vesicles, and analogous signals are thought to direct *secretory proteins* into secretory vesicles. The nonselective constitutive secretory pathway transports most other proteins directly to the cell surface. Because entry into this pathway does not require a particular signal, it is also called the **default pathway**. Thus, in an unpolarized cell such as a white blood cell or a fibroblast, it seems that any protein in the lumen of the Golgi apparatus is automatically carried by the constitutive pathway to the cell surface unless it is specifically returned to the ER, retained as a resident protein in the Golgi apparatus itself, or selected for the pathways that lead to regulated secretion or to lysosomes. In polarized cells, where different products have to be

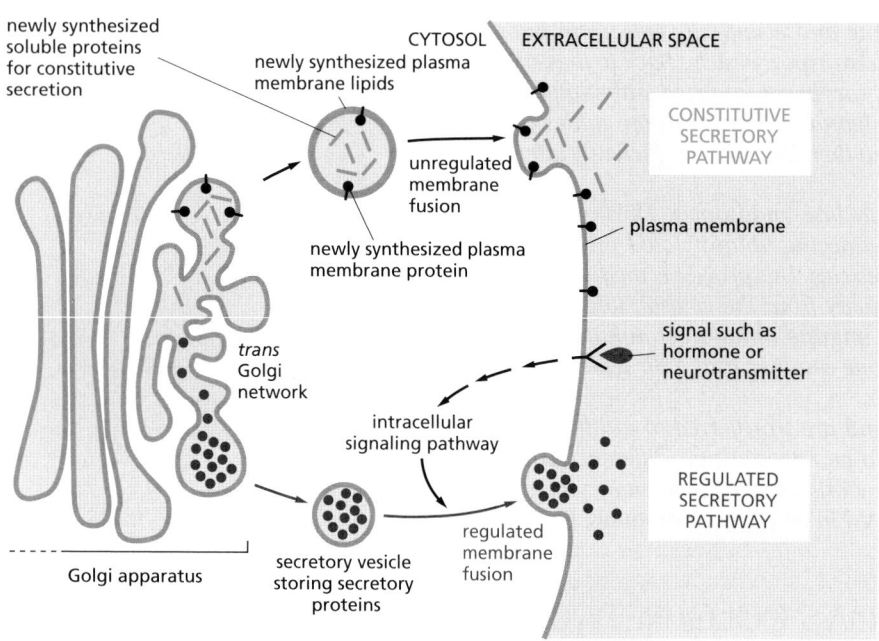

Figure 13–62 **The constitutive and regulated secretory pathways.** The two pathways diverge in the TGN. The constitutive secretory pathway operates in all cells. Many soluble proteins are continually secreted from the cell by this pathway, which also supplies the plasma membrane with newly synthesized membrane lipids and proteins. Specialized secretory cells also have a regulated secretory pathway, by which selected proteins in the TGN are diverted into secretory vesicles, where the proteins are concentrated and stored until an extracellular signal stimulates their secretion. The regulated secretion of small molecules, such as histamine and neurotransmitters, occurs by a similar pathway; these molecules are actively transported from the cytosol into preformed secretory vesicles. There they are often bound to specific macromolecules (proteoglycans, for histamine) so that they can be stored at high concentration without generating an excessively high osmotic pressure.

delivered to different domains of the cell surface, we shall see that the options are more complex.

Secretory Vesicles Bud from the *Trans* Golgi Network

Cells that are specialized for secreting some of their products rapidly on demand concentrate and store these products in **secretory vesicles** (often called *dense-core secretory granules* because they have dense cores when viewed in the electron microscope). Secretory vesicles form from the TGN, and they release their contents to the cell exterior by exocytosis in response to specific signals. The secreted product can be either a small molecule (such as histamine or a neuropeptide) or a protein (such as a hormone or digestive enzyme).

Proteins destined for secretory vesicles (called *secretory proteins*) are packaged into appropriate vesicles in the TGN by a mechanism that involves the selective aggregation of the secretory proteins. Clumps of aggregated, electron-dense material can be detected by electron microscopy in the lumen of the TGN. The signals that direct secretory proteins into such aggregates are not well defined and may be quite diverse. When a gene encoding a secretory protein is artificially expressed in a secretory cell that normally does not make the protein, the foreign protein is appropriately packaged into secretory vesicles. This observation

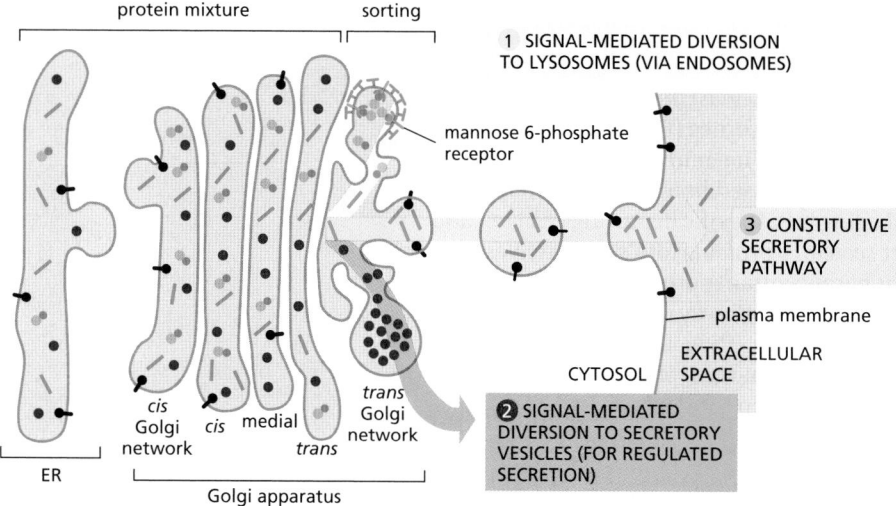

Figure 13–63 **The three best-understood pathways of protein sorting in the** *trans* **Golgi network.** (1) Proteins with the mannose 6-phosphate (M6P) marker (see Figure 13–45) are diverted to lysosomes (via endosomes) in clathrin-coated transport vesicles. (2) Proteins with signals directing them to secretory vesicles are concentrated in such vesicles as part of a regulated secretory pathway that is present only in specialized secretory cells. (3) In unpolarized cells, a constitutive secretory pathway delivers proteins with no special features to the cell surface. In polarized cells, such as epithelial cells, however, secreted and plasma membrane proteins are selectively directed to either the apical or the basolateral plasma membrane domain, so a specific signal must mediate at least one of these two pathways, as we discuss later.

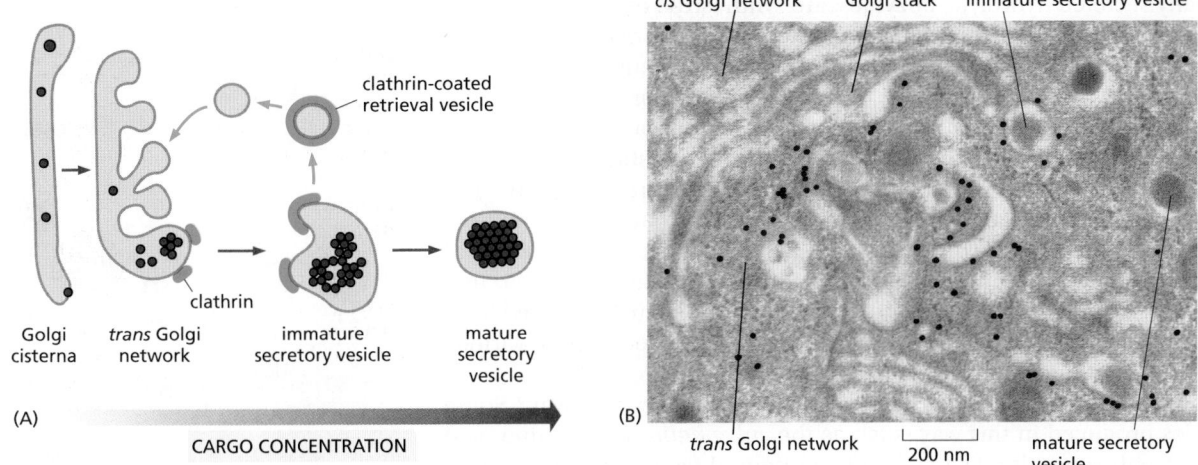

(A)   CARGO CONCENTRATION

(B)   *trans* Golgi network   200 nm   mature secretory vesicle

cis Golgi network   Golgi stack   immature secretory vesicle

shows that, although the proteins that an individual cell expresses and packages in secretory vesicles differ, they contain common sorting signals, which function properly even when the proteins are expressed in cells that do not normally make them.

It is unclear how the aggregates of secretory proteins are segregated into secretory vesicles. Secretory vesicles have unique proteins in their membrane, some of which might serve as receptors for aggregated protein in the TGN. The aggregates are much too big, however, for each molecule of the secreted protein to be bound by its own cargo receptor, as occurs for transport of the lysosomal enzymes. The uptake of the aggregates into secretory vesicles may therefore more closely resemble the uptake of particles by phagocytosis at the cell surface, where the plasma membrane zippers up around large structures.

Initially, most of the membrane of the secretory vesicles that leave the TGN is only loosely wrapped around the clusters of aggregated secretory proteins. Morphologically, these *immature secretory vesicles* resemble dilated *trans* Golgi cisternae that have pinched off from the Golgi stack. As the vesicles mature, they fuse with one another and their contents become concentrated (**Figure 13–64A**), probably as the result of both the continuous retrieval of membrane that is recycled to the TGN, and the progressive acidification of the vesicle lumen that results from the increasing concentration of V-type ATPases in the vesicle membrane that acidify all endocytic and exocytic organelles (see Figure 13–37). The degree of concentration of proteins during the formation and maturation of secretory vesicles is only a small part of the total 200–400-fold concentration of these proteins that occurs after they leave the ER. Secretory and membrane proteins become concentrated as they move from the ER through the Golgi apparatus because of an extensive retrograde retrieval process mediated by COPI-coated transport vesicles that carry soluble ER resident proteins back to the ER, while excluding the secretory and membrane proteins (see Figure 13–25).

Membrane recycling is important for returning Golgi components to the Golgi apparatus, as well as for concentrating the contents of secretory vesicles. The vesicles that mediate this retrieval originate as clathrin-coated buds on the surface of immature secretory vesicles, often being seen even on budding secretory vesicles that have not yet pinched off from the Golgi stack (Figure 13–64B).

Because the final mature secretory vesicles are so densely filled with contents, the secretory cell can disgorge large amounts of material promptly by exocytosis when triggered to do so (**Figure 13–65**).

Figure 13–64 The formation of secretory vesicles. (A) Secretory proteins become segregated and highly concentrated in secretory vesicles by two mechanisms. First, they aggregate in the ionic environment of the TGN; often the aggregates become more condensed as secretory vesicles mature and their lumen becomes more acidic. Second, clathrin-coated vesicles retrieve excess membrane and lumenal content present in immature secretory vesicles as the secretory vesicles mature. (B) This electron micrograph shows secretory vesicles forming from the TGN in an insulin-secreting β cell of the pancreas. Anti-clathrin antibodies conjugated to gold spheres *(black dots)* have been used to locate clathrin molecules. The immature secretory vesicles, which contain insulin precursor protein (proinsulin), contain clathrin patches, which are no longer seen on the mature secretory vesicle. (B, courtesy of Lelio Orci.)

Precursors of Secretory Proteins Are Proteolytically Processed During the Formation of Secretory Vesicles

Concentration is not the only process to which secretory proteins are subjected as the secretory vesicles mature. Many protein hormones and small neuropeptides, as well as many secreted hydrolytic enzymes, are synthesized as inactive precursors. Proteolysis is necessary to liberate the active molecules from these precursor proteins. The cleavages occur in the secretory vesicles and sometimes in the extracellular fluid after secretion. Additionally, many of the precursor proteins have an N-terminal *pro-peptide* that is cleaved off to yield the mature protein. These proteins are synthesized as *pre-pro-proteins*, the *pre-peptide* consisting of the ER signal peptide that is cleaved off earlier in the rough ER (see Figure 12–36). In other cases, peptide signaling molecules are made as *polyproteins* that contain multiple copies of the same amino acid sequence. In still more complex cases, a variety of peptide signaling molecules are synthesized as parts of a single polyprotein that acts as a precursor for multiple end products, which are individually cleaved from the initial polypeptide chain. The same polyprotein may be processed in various ways to produce different peptides in different cell types (Figure 13–66).

Why is proteolytic processing so common in the secretory pathway? Some of the peptides produced in this way, such as the *enkephalins* (five-amino-acid neuropeptides with morphine-like activity), are undoubtedly too short in their mature forms to be co-translationally transported into the ER lumen or to include the necessary signal for packaging into secretory vesicles. In addition, for secreted hydrolytic enzymes—or any other protein whose activity could be harmful inside the cell that makes it—delaying activation of the protein until it reaches a secretory vesicle, or until after it has been secreted, has a clear advantage: the delay prevents the protein from acting prematurely inside the cell in which it is synthesized.

Secretory Vesicles Wait Near the Plasma Membrane Until Signaled to Release Their Contents

Once loaded, a secretory vesicle has to reach the site of secretion, which in some cells is far away from the TGN. Nerve cells are the most extreme example. Secretory proteins, such as peptide neurotransmitters (neuropeptides), which will be released from nerve terminals at the end of the axon, are made and packaged into secretory vesicles in the cell body. They then travel along the axon to the nerve terminals, which can be a meter or more away. As discussed in Chapter 16, motor proteins propel the vesicles along axonal microtubules, whose uniform orientation guides the vesicles in the proper direction. Microtubules also guide transport vesicles to the cell surface for constitutive exocytosis.

Whereas transport vesicles containing materials for constitutive release fuse with the plasma membrane once they arrive there, secretory vesicles in the regulated pathway wait at the membrane until the cell receives a signal to secrete, and they then fuse. The signal can be an electrical nerve impulse (an action potential) or an extracellular signal molecule, such as a hormone: in either case, it leads to a transient increase in the concentration of free Ca^{2+} in the cytosol.

For Rapid Exocytosis, Synaptic Vesicles Are Primed at the Presynaptic Plasma Membrane

Nerve cells (and some endocrine cells) contain two types of secretory vesicles. As for all secretory cells, these cells package proteins and neuropeptides in dense-cored secretory vesicles in the standard way for release by the regulated secretory pathway. In addition, however, they use another specialized class of tiny (≈ 50 nm

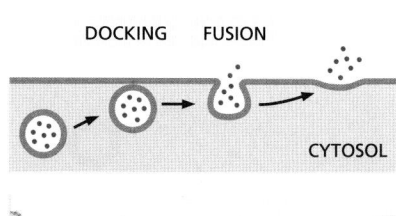

DOCKING   FUSION

CYTOSOL

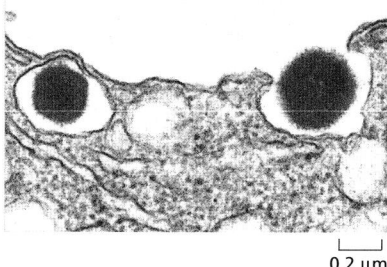

0.2 µm

Figure 13–65 Exocytosis of secretory vesicles. The process is illustrated schematically (top) and in an electron micrograph that shows the release of insulin from a secretory vesicle of a pancreatic β cell. (Courtesy of Lelio Orci, from L. Orci, J.-D. Vassalli and A. Perrelet, *Sci. Am.* 259:85–94, 1988.)

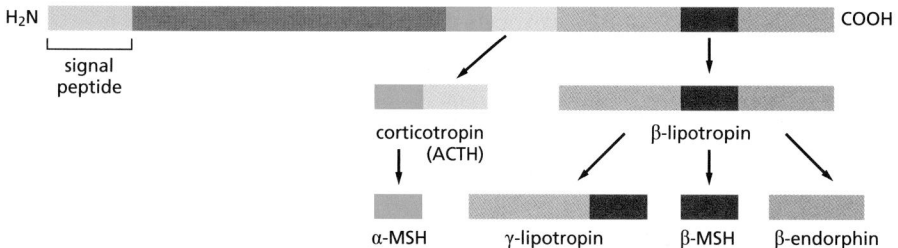

H_2N —— COOH

signal peptide

corticotropin (ACTH)

β-lipotropin

α-MSH   γ-lipotropin   β-MSH   β-endorphin

Figure 13–66 Alternative processing pathways for the prohormone polyprotein proopiomelanocortin. The initial cleavages are made by proteases that cut next to pairs of positively charged amino acids (Lys-Arg, Lys-Lys, Arg-Lys, or Arg-Arg pairs). Trimming reactions then produce the final secreted products. Different cell types produce different concentrations of individual processing enzymes, so that the same prohormone precursor is cleaved to produce different peptide hormones. In the anterior lobe of the pituitary gland, for example, only corticotropin (ACTH) and β-lipotropin are produced from proopiomelanocortin, whereas in the intermediate lobe of the pituitary gland mainly α-melanocyte stimulating hormone (α-MSH), γ-lipotropin, β-MSH, and β-endorphin are produced— α-MSH from ACTH and the other three from β-lipotropin, as shown.

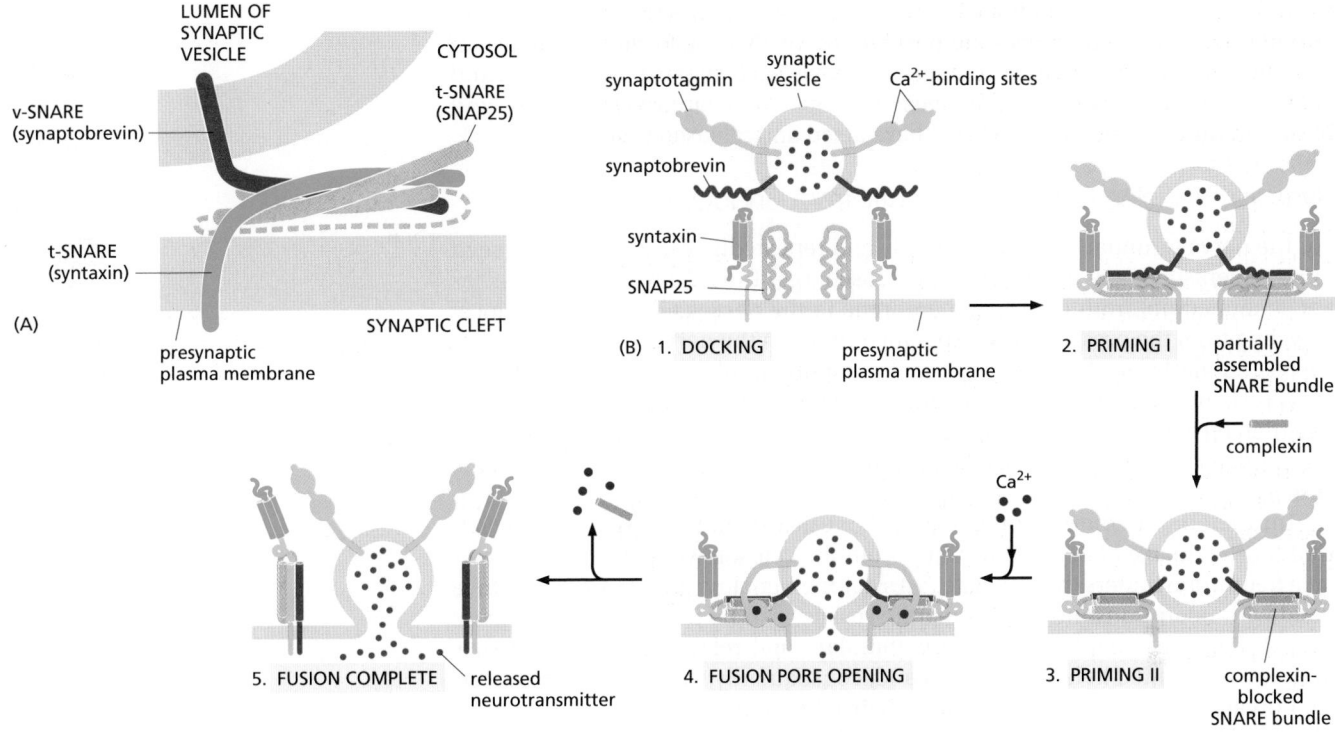

Figure 13–67 Exocytosis of synaptic vesicles. For orientation at a synapse, see Figure 11–36. (A) The trans-SNARE complex responsible for docking synaptic vesicles at the plasma membrane of nerve terminals consists of three proteins. The v-SNARE *synaptobrevin* and the t-SNARE *syntaxin* are both transmembrane proteins, and each contributes one α helix to the complex. By contrast to other SNAREs discussed earlier, the t-SNARE *SNAP25* is a peripheral membrane protein that contributes two α helices to the four-helix bundle; the two helices are connected by a loop *(dashed line)* that lies parallel to the membrane and has fatty acyl chains (not shown) attached to anchor it there. The four α helices are shown as rods for simplicity. (B) At the synapse, the basic SNARE machinery is modulated by the Ca^{2+} sensor *synaptotagmin* and an additional protein called *complexin*. Synaptic vesicles first dock at the membrane (Step 1), and the SNARE bundle partially assembles (Step 2), resulting in a "primed vesicle" that is already drawn close to the membrane. The SNARE bundle assembles further but the additional binding of complexin prevents fusion (Step 3). Upon arrival of an action potential, Ca^{2+} enters the cell and binds to synaptotagmin, which releases the block and opens a fusion pore (Step 4). Further rearrangements complete the fusion reaction (Step 5) and release the fusion machinery, which now can be reused. This elaborate arrangement allows the fusion machinery to respond on the millisecond time scale essential for rapid and repetitive synaptic signaling. (A, adapted from R.B. Sutton et al., *Nature* 395:347–353, 1998; B, adapted from J. Tang et al., *Cell* 126:1175–1187, 2006. With permission from Elsevier.)

diameter) secretory vesicles called **synaptic vesicles**. These vesicles store small *neurotransmitter molecules*, such as acetylcholine, glutamate, glycine, and γ-aminobutyric acid (GABA), which mediate rapid signaling from nerve cell to its target cell at chemical synapses. When an action potential arrives at a nerve terminal, it causes an influx of Ca^{2+} through voltage-gated Ca^{2+} channels, which triggers the synaptic vesicles to fuse with the plasma membrane and release their contents to the extracellular space (see Figure 11–36). Some neurons fire more than 1000 times per second, releasing neurotransmitters each time.

The speed of transmitter release (taking only milliseconds) indicates that the proteins mediating the fusion reaction do not undergo complex, multistep rearrangements. Rather, after vesicles have been docked at the presynaptic plasma membrane, they undergo a priming step, which prepares them for rapid fusion. In the primed state, the SNAREs are partly paired, their helices are not fully wound into the final four-helix bundle required for fusion (**Figure 13–67**). Proteins called *complexins* freeze the SNARE complexes in this metastable state. The brake imposed by the complexins is released by another synaptic vesicle protein, synaptotagmin, which contains Ca^{2+}-binding domains. A rise in cytosolic Ca^{2+} triggers binding of synaptotagmin to phospholipids and to the SNAREs, displacing the complexins. As the SNARE bundle zippers up completely, a fusion pore opens and the neurotransmitters are released. At a typical synapse, only a small number of the docked vesicles are primed and ready for exocytosis. The use of only a small number of vesicles at a time allows each synapse to fire over and over again in quick succession. With each firing, new synaptic vesicles dock and become primed to replace those that have fused and released their contents.

Synaptic Vesicles Can Form Directly from Endocytic Vesicles

For the nerve terminal to respond rapidly and repeatedly, synaptic vesicles need to be replenished very quickly after they discharge. Thus, most synaptic vesicles are generated not from the Golgi membrane in the nerve cell body but by local recycling from the presynaptic plasma membrane in the nerve terminals (**Figure 13–68**). Similarly, newly made membrane components of the synaptic vesicles are initially delivered to the plasma membrane by the constitutive secretory pathway and then retrieved by endocytosis. But instead of fusing with endosomes, most of the endocytic vesicles immediately fill with neurotransmitter to become synaptic vesicles.

The membrane components of a synaptic vesicle include transporters specialized for the uptake of neurotransmitter from the cytosol, where the small-molecule neurotransmitters that mediate fast synaptic signaling are synthesized. Once filled with neurotransmitter, the synaptic vesicles can be used again (see Figure 13–68). Because synaptic vesicles are abundant and relatively uniform in size, they can be purified in large numbers and, consequently, are the best-characterized organelle of the cell, in that all of their membrane components have been identified by quantitative proteomic analyses (**Figure 13–69**). Extending this analysis to a complete presynaptic terminal, allows us to model the crowded environment in which these reactions occur.

Secretory Vesicle Membrane Components Are Quickly Removed from the Plasma Membrane

When a secretory vesicle fuses with the plasma membrane, its contents are discharged from the cell by exocytosis, and its membrane becomes part of the plasma membrane. Although this should greatly increase the surface area of the plasma membrane, it does so only transiently, because membrane components are removed from the surface by endocytosis almost as fast as they are added by exocytosis, a process reminiscent of the endocytic–exocytic cycle discussed earlier. After their removal from the plasma membrane, the proteins of the secretory

Figure 13–68 The formation of synaptic vesicles in a nerve cell. These tiny uniform vesicles are found only in nerve cells and in some endocrine cells, where they store and secrete small-molecule neurotransmitters. The import of neurotransmitter directly into the small endocytic vesicles that form from the plasma membrane is mediated by membrane transporters that function as antiports and are driven by an H⁺ gradient maintained by V-ATPase H⁺ pumps in the vesicle membrane (discussed in Chapter 11).

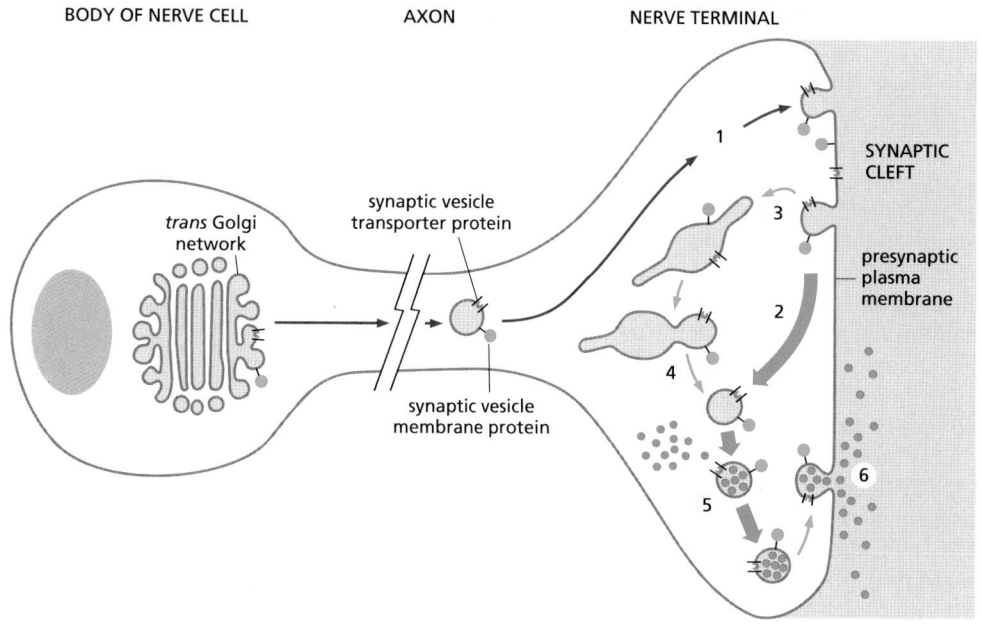

1 DELIVERY OF SYNAPTIC VESICLE MEMBRANE COMPONENTS TO PRESYNAPTIC PLASMA MEMBRANE

2 ENDOCYTOSIS OF SYNAPTIC VESICLE MEMBRANE COMPONENTS TO FORM NEW SYNAPTIC VESICLES DIRECTLY

3 ENDOCYTOSIS OF SYNAPTIC VESICLE MEMBRANE COMPONENTS AND DELIVERY TO ENDOSOME

4 BUDDING OF SYNAPTIC VESICLE FROM ENDOSOME

5 LOADING OF NEUROTRANSMITTER INTO SYNAPTIC VESICLE

6 SECRETION OF NEUROTRANSMITTER BY EXOCYTOSIS IN RESPONSE TO AN ACTION POTENTIAL

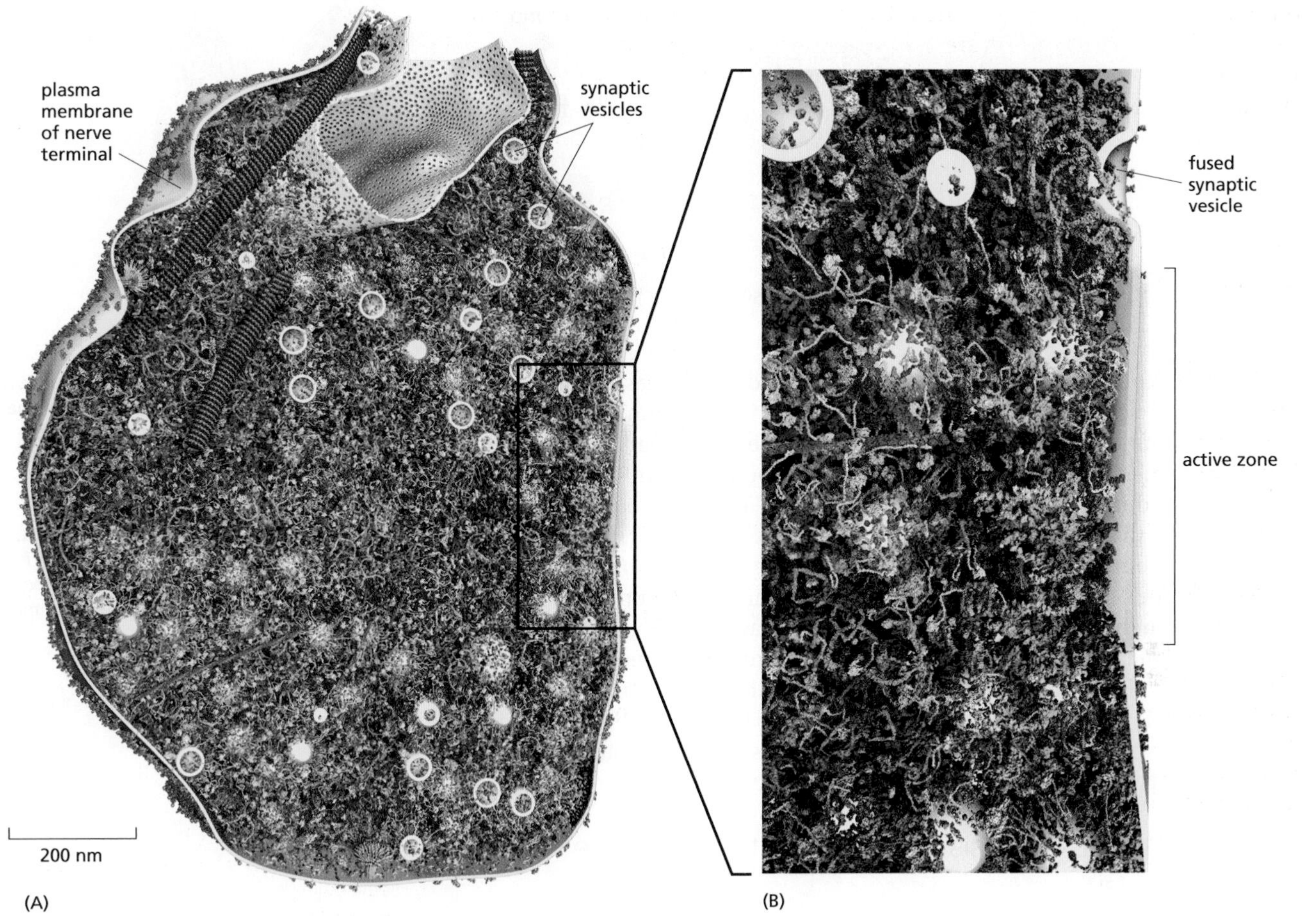

plasma
membrane
of nerve
terminal

synaptic
vesicles

200 nm

(A)

fused
synaptic
vesicle

active zone

(B)

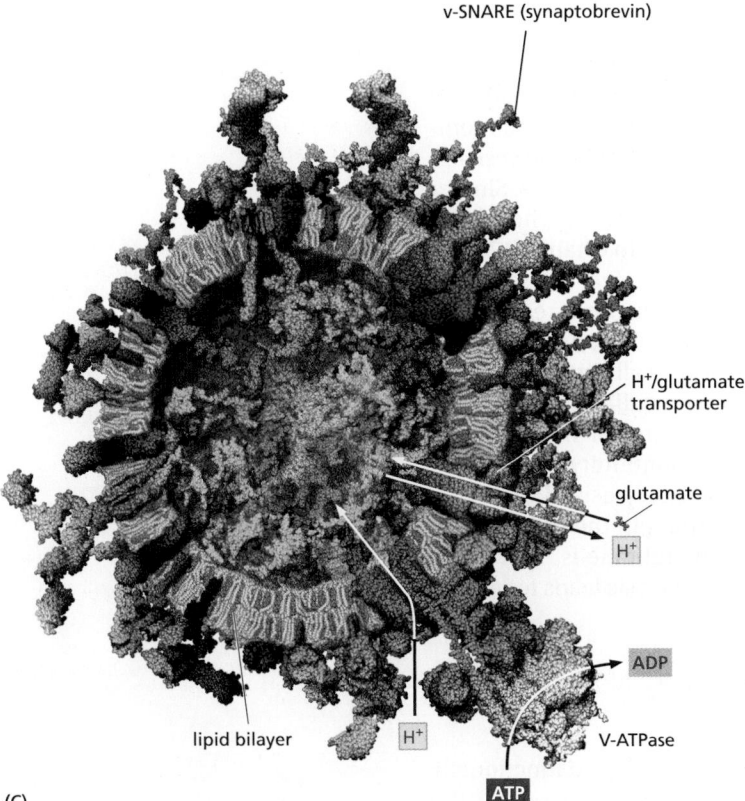

v-SNARE (synaptobrevin)

H⁺/glutamate
transporter

glutamate

H⁺

ADP

lipid bilayer

H⁺

V-ATPase

ATP

(C)

Figure 13–69 Scale models of a brain presynaptic terminal and a synaptic vesicle. The illustrations show sections through a pre-synaptic terminal (A; enlarged in B) and a synaptic vesicle (C) in which proteins and lipids are drawn to scale based on their known stoichiometry and either known or approximated structures. The relative localization of protein molecules in different regions of the presynaptic terminal was inferred from super-resolution imaging and electron microscopy. The model in (A) contains 300,000 proteins of 60 different kinds that vary in abundance from 150 copies to 20,000 copies. In the model in (C), only 70% of the membrane proteins present in the membrane are shown; a complete model would therefore show a membrane that is even more crowded than this picture suggests (Movie 13.7). Each synaptic vesicle membrane contains 7000 phospholipid molecules and 5700 cholesterol molecules. Each also contains close to 50 different integral membrane protein molecules, which vary widely in their relative abundance and together contribute about 600 transmembrane α helices. The transmembrane v-SNARE synaptobrevin is the most abundant protein in the vesicle (~70 copies per vesicle). By contrast, the V-ATPase, which uses ATP hydrolysis to pump H⁺ into the vesicle lumen, is present in 1–2 copies per vesicle. The H⁺ gradient provides the energy for neurotransmitter import by an H⁺/neurotransmitter antiport, which loads each vesicle with 1800 neurotransmitter molecules, such as glutamate, one of which is shown to scale. (A and B, from B.G. Wilhelm et al., *Science* 344:1023–1028, 2014. With permission from AAAS; C, adapted from S. Takamori et al., *Cell* 127:831–846, 2006. With permission from Elsevier.)

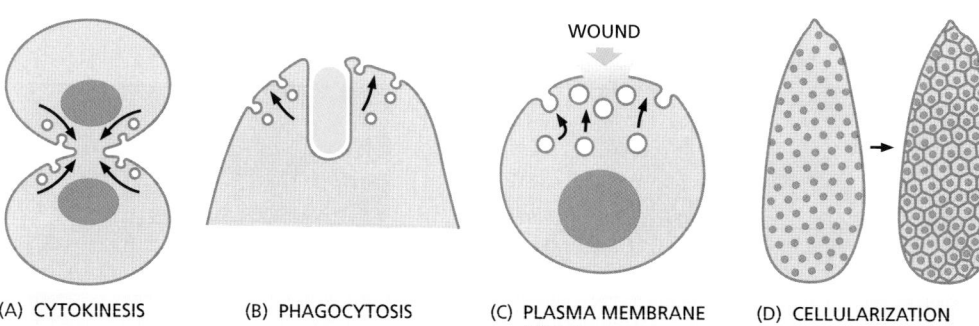

(A) CYTOKINESIS   (B) PHAGOCYTOSIS   (C) PLASMA MEMBRANE REPAIR   (D) CELLULARIZATION

vesicle membrane are either recycled or shuttled to lysosomes for degradation. The amount of secretory vesicle membrane that is temporarily added to the plasma membrane can be enormous: in a pancreatic acinar cell discharging digestive enzymes for delivery to the gut lumen, about 900 μm^2 of vesicle membrane is inserted into the apical plasma membrane (whose area is only 30 μm^2) when the cell is stimulated to secrete.

Control of membrane traffic thus has a major role in maintaining the composition of the various membranes of the cell. To maintain each membrane-enclosed compartment in the secretory and endocytic pathways at a constant size, the balance between the outward and inward flows of membrane needs to be precisely regulated. For cells to grow, however, the forward flow needs to be greater than the retrograde flow, so that the membrane can increase in area. For cells to maintain a constant size, the forward and retrograde flows must be equal. We still know very little about the mechanisms that coordinate these flows.

Some Regulated Exocytosis Events Serve to Enlarge the Plasma Membrane

An important task of regulated exocytosis is to deliver more membrane to enlarge the surface area of a cell's plasma membrane when such a need arises. A spectacular example is the plasma membrane expansion that occurs during the cellularization process in a fly embryo, which initially is a syncytium—a single cell containing about 6000 nuclei surrounded by a single plasma membrane (see Figure 21–15). Within tens of minutes, the embryo is converted into the same number of cells. This process of *cellularization* requires a vast amount of new plasma membrane, which is added by a carefully orchestrated fusion of cytoplasmic vesicles, eventually forming the plasma membranes that enclose the separate cells. Similar vesicle fusion events are required to enlarge the plasma membrane when other animal cells or plant cells divide during *cytokinesis* (discussed in Chapter 17).

Many animal cells, especially those subjected to mechanical stresses, frequently experience small ruptures in their plasma membrane. In a remarkable process thought to involve both homotypic vesicle–vesicle fusion and exocytosis, a temporary cell-surface patch is quickly fashioned from locally available internal-membrane sources, such as lysosomes. In addition to providing an emergency barrier against leaks, the patch reduces membrane tension over the wounded area, allowing the bilayer to flow back together to restore continuity and seal the puncture. This membrane repair process, the fusion and exocytosis of vesicles is triggered by the sudden increase of Ca^{2+}, which is abundant in the extracellular space and rushes into the cell as soon as the plasma membrane is punctured. **Figure 13–70** shows four examples in which regulated exocytosis leads to plasma membrane expansion.

Polarized Cells Direct Proteins from the *Trans* Golgi Network to the Appropriate Domain of the Plasma Membrane

Most cells in tissues are *polarized*, with two or more molecularly and functionally distinct plasma membrane domains. This raises the general problem of how the

Figure 13–70 Four examples of regulated exocytosis leading to plasma membrane enlargement. The vesicles fusing with the plasma membrane during cytokinesis (A) and phagocytosis (B) are thought to be derived from endosomes, whereas those involved in wound repair (C) may be derived from plasma membranes. The vast amount of new plasma membrane inserted during cellularization in a fly embryo occurs by the fusion of cytoplasmic vesicles (D).

delivery of membrane from the Golgi apparatus is organized so as to maintain the differences between one cell-surface domain and another. A typical epithelial cell, for example, has an *apical domain*, which faces either an internal cavity or the outside world and often has specialized features such as cilia or a brush border of microvilli. It also has a *basolateral domain*, which covers the rest of the cell. The two domains are separated by a ring of *tight junctions* (see Figure 19–21), which prevent proteins and lipids (in the outer leaflet of the lipid bilayer) from diffusing between the two domains, so that the differences between the two domains are maintained.

In principle, differences between plasma membrane domains need not depend on the targeted delivery of the appropriate membrane components. Instead, membrane components could be delivered to all regions of the cell surface indiscriminately but then be selectively stabilized in some locations and selectively eliminated in others. Although this strategy of random delivery followed by selective retention or removal seems to be used in certain cases, deliveries are often specifically directed to the appropriate membrane domain. Epithelial cells lining the gut, for example, secrete digestive enzymes and mucus at their apical surface and components of the basal lamina at their basolateral surface. Such cells must have ways of directing vesicles carrying different cargoes to different plasma membrane domains. Proteins from the ER destined for different domains travel together until they reach the TGN, where they are separated and dispatched in secretory or transport vesicles to the appropriate plasma membrane domain (**Figure 13–71**).

The apical plasma membrane of most epithelial cells is greatly enriched in glycosphingolipids, which help protect this exposed surface from damage—for example, from the digestive enzymes and low pH in sites such as the gut or stomach, respectively. Similarly, plasma membrane proteins that are linked to the lipid bilayer by a GPI anchor (see Figure 12–52) are found predominantly in the apical plasma membrane. If recombinant DNA techniques are used to attach a GPI anchor to a protein that would normally be delivered to the basolateral surface, the protein is usually delivered to the apical surface instead. GPI-anchored proteins are thought to be directed to the apical membrane because they associate with glycosphingolipids in lipid rafts that form in the membrane of the TGN. As discussed in Chapter 10, lipid rafts form in the TGN and plasma membrane when glycosphingolipids and cholesterol molecules self-associate (see Figure 10–13).

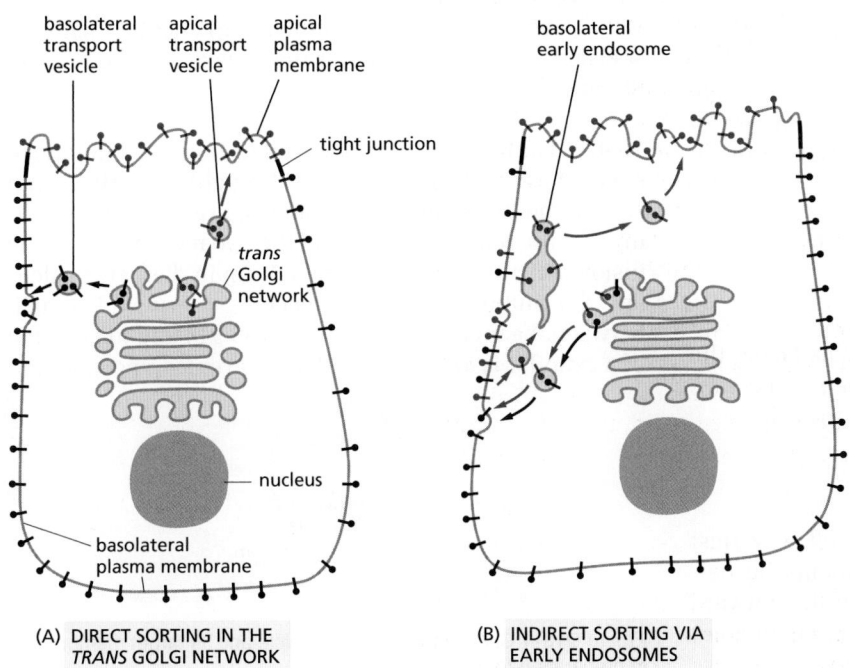

(A) DIRECT SORTING IN THE
TRANS GOLGI NETWORK

(B) INDIRECT SORTING VIA
EARLY ENDOSOMES

Figure 13–71 **Two ways of sorting plasma membrane proteins in a polarized epithelial cell.** (A) In the direct pathway, proteins destined for different plasma membrane domains are sorted and packaged into different transport vesicles. The lipid-raft-dependent delivery system to the apical domain described in the text is an example of the direct pathway. (B) In the indirect pathway, a protein is retrieved from the inappropriate plasma membrane domain by endocytosis and then transported to the correct domain via early endosomes—that is, by transcytosis. The indirect pathway, for example, is used in liver hepatocytes to deliver proteins to the apical domain that lines bile ducts.

Having selected a unique set of cargo molecules, the rafts then bud from the TGN into transport vesicles destined for the apical plasma membrane. This process is similar to the selective partitioning of some membrane proteins into the specialized lipid domains in caveolae at the plasma membrane discussed earlier.

Membrane proteins destined for delivery to the basolateral membrane contain sorting signals in their cytosolic tail. When present in an appropriate structural context, these signals are recognized by coat proteins that package them into appropriate transport vesicles in the TGN. The same basolateral signals that are recognized in the TGN also function in early endosomes to redirect the proteins back to the basolateral plasma membrane after they have been endocytosed.

Summary

Cells can secrete molecules by exocytosis in either a constitutive or a regulated fashion. Whereas the regulated pathways operate only in specialized secretory cells, a constitutive secretory pathway operates in all eukaryotic cells, characterized by continual vesicle transport from the TGN to the plasma membrane. In the regulated pathways, the molecules are stored either in secretory vesicles or in synaptic vesicles, which do not fuse with the plasma membrane to release their contents until they receive an appropriate signal. Secretory vesicles containing proteins for secretion bud from the TGN. The secretory proteins become concentrated during the formation and maturation of the secretory vesicles. Synaptic vesicles, which are confined to nerve cells and some endocrine cells, form from both endocytic vesicles and from endosomes, and they mediate the regulated secretion of small-molecule neurotransmitters at the axon terminals of nerve cells.

Proteins are delivered from the TGN to the plasma membrane by the constitutive pathway unless they are diverted into other pathways or retained in the Golgi apparatus. In polarized cells, the transport pathways from the TGN to the plasma membrane operate selectively to ensure that different sets of membrane proteins, secreted proteins, and lipids are delivered to the different domains of the plasma membrane.

WHAT WE DON'T KNOW

• How are targeting and fusion proteins such as SNAREs regulated, so that they can be returned to their respective donor compartments in an inactive state?

• How does a cell balance exocytic and endocytic events to keep its plasma membrane a constant size?

• Can newly formed daughter cells generate a Golgi apparatus *de novo*, or do they have to inherit it?

• How do lysosomes avoid digesting their own membranes?

• How does a cell maintain the right amount of every component (organelles, molecules), and how does it change these amounts as needed (for example, to greatly expand the endoplasmic reticulum when the cell needs to produce large amounts of secreted proteins)?

PROBLEMS

Which statements are true? Explain why or why not.

13–1 In all events involving fusion of a vesicle to a target membrane, the cytosolic leaflets of the vesicle and target bilayers always fuse together, as do the leaflets that are not in contact with the cytosol.

13–2 There is one strict requirement for the exit of a protein from the ER: it must be correctly folded.

13–3 All the glycoproteins and glycolipids in intracellular membranes have oligosaccharide chains facing the lumenal side, and all those in the plasma membrane have oligosaccharide chains facing the outside of the cell.

Discuss the following problems.

13–4 In a nondividing cell such as a liver cell, why must the flow of membrane between compartments be balanced, so that the retrieval pathways match the outward flow? Would you expect the same balanced flow in a gut epithelial cell, which is actively dividing?

13–5 Enveloped viruses, which have a membrane coat, gain access to the cytosol by fusing with a cell membrane. Why do you suppose that these viruses encode their own special fusion protein, rather than making use of a cell's SNAREs?

13–6 For fusion of a vesicle with its target membrane to occur, the membranes have to be brought to within 1.5 nm so that the two bilayers can join (**Figure Q13–1**). Assuming that the relevant portions of the two membranes at the fusion site are circular regions 1.5 nm in diameter, calculate the number of water molecules that would remain between the membranes. (Water is 55.5 M and the volume of a cylinder is $\pi r^2 h$.) Given that an average phospholipid

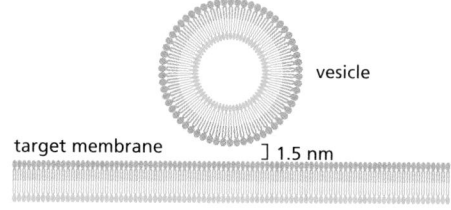

Figure Q13–1 Close approach of a vesicle and its target membrane in preparation for fusion (Problem 13–6).

occupies a membrane surface area of 0.2 nm², how many phospholipids would be present in each of the opposing monolayers at the fusion site? Are there sufficient water molecules to bind to the hydrophilic head groups of this number of phospholipids? (It is estimated that 10–12 water molecules are normally associated with each phospholipid head group at the exposed surface of a membrane.)

13–7 SNAREs exist as complementary partners that carry out membrane fusions between appropriate vesicles and their target membranes. In this way, a vesicle with a particular variety of v-SNARE will fuse only with a membrane that carries the complementary t-SNARE. In some instances, however, fusions of identical membranes (homotypic fusions) are known to occur. For example, when a yeast cell forms a bud, vesicles derived from the mother cell's vacuole move into the bud where they fuse with one another to form a new vacuole. These vesicles carry both v-SNAREs and t-SNAREs. Are both types of SNAREs essential for this homotypic fusion event?

To test this point, you have developed an ingenious assay for fusion of vacuolar vesicles. You prepare vesicles from two different mutant strains of yeast: strain B has a defective gene for vacuolar alkaline phosphatase (Pase); strain A is defective for the protease that converts the precursor of alkaline phosphatase (pro-Pase) into its active form (Pase) (**Figure Q13–2A**). Neither strain has active alkaline phosphatase, but when extracts of the strains are mixed, vesicle fusion generates active alkaline phosphatase, which can be easily measured (Figure Q13–2).

Now you delete the genes for the vacuolar v-SNARE, t-SNARE, or both in each of the two yeast strains. You prepare vacuolar vesicles from each and test them for their ability to fuse, as measured by the alkaline phosphatase assay (Figure Q13–2B).

What do these data say about the requirements for v-SNAREs and t-SNAREs in the fusion of vacuolar vesicles? Does it matter which kind of SNARE is on which vesicle?

13–8 If you were to remove the ER retrieval signal from protein disulfide isomerase (PDI), which is normally a soluble resident of the ER lumen, where would you expect the modified PDI to be located?

13–9 The KDEL receptor must shuttle back and forth between the ER and the Golgi apparatus to accomplish its task of ensuring that soluble ER proteins are retained in the ER lumen. In which compartment does the KDEL receptor bind its ligands more tightly? In which compartment does it bind its ligands more weakly? What is thought to be the basis for its different binding affinities in the two compartments? If you were designing the system, in which compartment would you have the highest concentration of KDEL receptor? Would you predict that the KDEL receptor, which is a transmembrane protein, would itself possess an ER retrieval signal?

13–10 How does the low pH of lysosomes protect the rest of the cell from lysosomal enzymes in case the lysosome breaks?

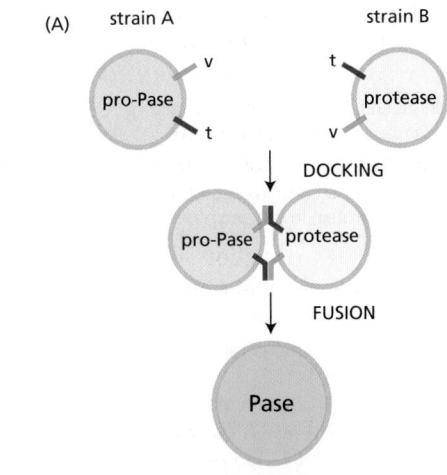

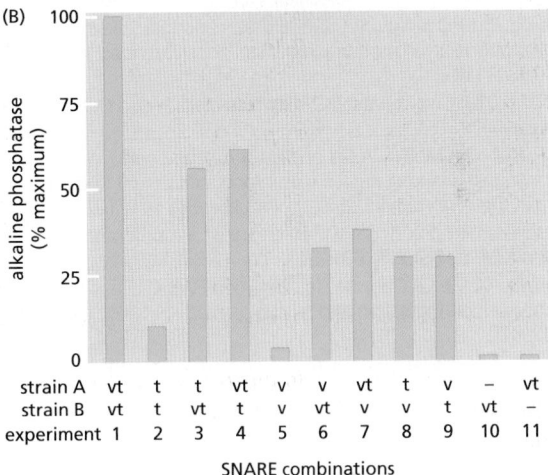

Figure Q13–2 SNARE requirements for vesicle fusion (Problem 13–7). (A) Scheme for measuring the fusion of vacuolar vesicles. (B) Results of fusions of vesicles with different combinations of v-SNAREs and t-SNAREs. The SNAREs present on the vesicles of the two strains are indicated as v (v-SNARE) and t (t-SNARE). (Adapted from B.J. Nichols et al, *Nature* 387: 199–202, 1997.)

13–11 Melanosomes are specialized lysosomes that store pigments for eventual release by exocytosis. Various cells such as skin and hair cells then take up the pigment, which accounts for their characteristic pigmentation. Mouse mutants that have defective melanosomes often have pale or unusual coat colors. One such light-colored mouse, the *Mocha* mouse (**Figure Q13–3**), has a defect in the gene for one of the subunits of the adaptor protein complex AP3, which is associated with coated vesicles budding from the *trans* Golgi network. How might the loss of AP3 cause a defect in melanosomes?

Figure Q13–3 A normal mouse and the *Mocha* mouse (Problem 13–11). In addition to its light coat color, the *Mocha* mouse has a poor sense of balance. (Courtesy of Margit Burmeister.)

REFERENCES

General

Harrison SC & Kirchhausen T (2010) Structural biology: Conservation in vesicle coats. *Nature* 466, 1048–1049.

Pfeffer SR (2013) A prize for membrane magic. *Cell* 155, 1203–1206.

Thor F, Gautschi M, Geiger R & Helenius A (2009) Bulk flow revisited: transport of a soluble protein in the secretory pathway. *Traffic* 10, 1819–1830.

The Molecular Mechanisms of Membrane Transport and the Maintenance of Compartmental Diversity

Antonny B (2011) Mechanisms of membrane curvature sensing. *Annu. Rev. Biochem.* 80, 101–123.

Ferguson SM & De Camilli P (2012) Dynamin, a membrane-remodelling GTPase. *Nat. Rev. Mol. Cell Biol.* 13, 75–88.

Frost A, Unger VM & De Camilli P (2009) The BAR domain superfamily: membrane-molding macromolecules. *Cell* 137, 191–196.

Grosshans BL, Ortiz D & Novick P (2006) Rabs and their effectors: achieving specificity in membrane traffic. *Proc. Natl Acad. Sci. USA* 103, 11821–11827.

Hughson FM (2010) Copy coats: COPI mimics clathrin and COPII. *Cell* 142, 19–21.

Jackson LP, Kelly BT, McCoy AJ et al. (2010) A large-scale conformational change couples membrane recruitment to cargo binding in the AP2 clathrin adaptor complex. *Cell* 141, 1220–1229.

Jahn R & Scheller RH (2006) SNAREs—engines for membrane fusion. *Nat. Rev. Mol. Cell Biol.* 7, 631–643.

Jean S & Kiger AA (2012) Coordination between RAB GTPase and phosphoinositide regulation and functions. *Nat. Rev. Mol. Cell Biol.* 13, 463–470.

Jin L, Pahuja KB, Wickliffe KE et al. (2012) Ubiquitin-dependent regulation of COPII coat size and function. *Nature* 482, 495–500.

Martens S & McMahon HT (2008) Mechanisms of membrane fusion: disparate players and common principles. *Nat. Rev. Mol. Cell Biol.* 9, 543–556.

McNew JA, Parlati F, Fukuda R et al. (2000) Compartmental specificity of cellular membrane fusion encoded in SNARE proteins. *Nature* 407, 153–159.

Miller EA & Schekman R (2013) COPII – a flexible vesicle formation system. *Curr. Opin. Cell Biol.* 25, 420–427.

Pfeffer SR (2013) Rab GTPase regulation of membrane identity. *Curr. Opin. Cell Biol.* 25, 414–419.

Saito K, Chen M, Bard F et al. (2009) TANGO1 facilitates cargo loading at endoplasmic reticulum exit sites. *Cell* 136, 891–902.

Seaman MN (2005) Recycle your receptors with retromer. *Trends Cell Biol.* 15, 68–75.

Transport from the ER Through the Golgi Apparatus

Ellgaard L & Helenius A (2003) Quality control in the endoplasmic reticulum. *Nat. Rev. Mol. Cell Biol.* 4, 181–191.

Emr S, Glick BS, Linstedt AD et al. (2009) Journeys through the Golgi—taking stock in a new era. *J. Cell Biol.* 187, 449–453.

Farquhar MG & Palade GE (1998) The Golgi apparatus: 100 years of progress and controversy. *Trends Cell Biol.* 8, 2–10.

Ladinsky MS, Mastronarde DN, McIntosh JR et al. (1999) Golgi structure in three dimensions: functional insights from the normal rat kidney cell. *J. Cell Biol.* 144, 1135–1149.

Munro S (2011) The golgin coiled-coil proteins of the Golgi apparatus. *Cold Spring Harb. Perspect. Biol.* 3, a005256.

Pfeffer S (2010) How the Golgi works: a cisternal progenitor model. *Proc. Natl Acad. Sci. USA* 107, 19614–19618.

Varki A (2011) Evolutionary forces shaping the Golgi glycosylation machinery: why cell surface glycans are universal to living cells. *Cold Spring Harb. Perspect. Biol.* 3, a005462.

Transport from the *Trans* Golgi Network to Lysosomes

Andrews NW (2000) Regulated secretion of conventional lysosomes. *Trends Cell Biol.* 10, 316–321.

Bonifacino JS & Rojas R (2006) Retrograde transport from endosomes to the *trans*-Golgi network. *Nat. Rev. Mol. Cell Biol.* 7, 568–579.

de Duve C (2005) The lysosome turns fifty. *Nat. Cell Biol.* 7, 847–849.

Futerman AH & van Meer G (2004) The cell biology of lysosomal storage disorders. *Nat. Rev. Mol. Cell Biol.* 5, 554–565.

Kraft C & Martens S (2012) Mechanisms and regulation of autophagosome formation. *Curr. Opin. Cell Biol.* 24, 496–501.

Mizushima N, Yoshimori T & Ohsumi Y (2011) The role of Atg proteins in autophagosome formation. *Annu. Rev. Cell Dev. Biol.* 27, 107–132.

Parzych KR & Klionsky DJ (2014) An overview of autophagy: morphology, mechanism, and regulation. *Antioxid. Redox Signal.* 20, 460–473.

Transport into the Cell from the Plasma Membrane: Endocytosis

Bonifacino JS & Traub LM (2003) Signals for sorting of transmembrane proteins to endosomes and lysosomes. *Annu. Rev. Biochem.* 72, 395–447.

Brown MS & Goldstein JL (1986) A receptor-mediated pathway for cholesterol homeostasis. *Science* 232, 34–47.

Conner SD & Schmid SL (2003) Regulated portals of entry into the cell. *Nature* 422, 37–44.

Doherty GJ & McMahon HT (2009) Mechanisms of endocytosis. *Annu. Rev. Biochem.* 78, 857–902.

Howes MT, Mayor S & Parton RG (2010) Molecules, mechanisms, and cellular roles of clathrin-independent endocytosis. *Curr. Opin. Cell Biol.* 22, 519–527.

Huotari J & Helenius A (2011) Endosome maturation. *EMBO J.* 30, 3481–3500.

Hurley JH & Hanson PI (2010) Membrane budding and scission by the ESCRT machinery: it's all in the neck. *Nat. Rev. Mol. Cell Biol.* 11, 556–566.

Kelly BT & Owen DJ (2011) Endocytic sorting of transmembrane protein cargo. *Curr. Opin. Cell Biol.* 23, 404–412.

Maxfield FR & McGraw TE (2004) Endocytic recycling. *Nat. Rev. Mol. Cell Biol.* 5, 121–132.

McMahon HT & Boucrot E (2011) Molecular mechanism and physiological functions of clathrin-mediated endocytosis. *Nat. Rev. Mol. Cell Biol.* 12, 517–533.

Mercer J & Helenius A (2012) Gulping rather than sipping: macropinocytosis as a way of virus entry. *Curr. Opin. Microbiol.* 15, 490–499.

Sorkin A & von Zastrow M (2009) Endocytosis and signalling: intertwining molecular networks. *Nat. Rev. Mol. Cell Biol.* 10, 609–622.

Tjelle TE, Lovdal T & Berg T (2000) Phagosome dynamics and function. *BioEssays* 22, 255–263.

Transport from the *Trans* Golgi Network to the Cell Exterior: Exocytosis

Burgess TL & Kelly RB (1987) Constitutive and regulated secretion of proteins. *Annu. Rev. Cell Biol.* 3, 243–293.

Li F, Pincet F, Perez E et al. (2011) Complexin activates and clamps SNAREpins by a common mechanism involving an intermediate energetic state. *Nat. Struct. Mol. Biol.* 18, 941–946.

Martin TF (1997) Stages of regulated exocytosis. *Trends Cell Biol.* 7, 271–276.

Mellman I & Nelson WJ (2008) Coordinated protein sorting, targeting and distribution in polarized cells. *Nat. Rev. Mol. Cell Biol.* 9, 833–845.

Mostov K, Su T & ter Beest M (2003) Polarized epithelial membrane traffic: conservation and plasticity. *Nat. Cell Biol.* 5, 287–293.

Pang ZP & Südhof TC (2010) Cell biology of Ca^{2+}-triggered exocytosis. *Curr. Opin. Cell Biol.* 22, 496–505.

Schuck S & Simons K (2004) Polarized sorting in epithelial cells: raft clustering and the biogenesis of the apical membrane. *J. Cell Sci.* 117, 5955–5964.

Energy Conversion: Mitochondria and Chloroplasts

To maintain their high degree of organization in a universe that is constantly drifting toward chaos, cells have a constant need for a plentiful supply of ATP, as we have explained in Chapter 2. In eukaryotic cells, most of the ATP that powers life processes is produced by specialized, membrane-enclosed, *energy-converting organelles*. These are of two types. **Mitochondria**, which occur in virtually all cells of animals, plants, and fungi, burn food molecules to produce ATP by *oxidative phosphorylation*. **Chloroplasts**, which occur only in plants and green algae, harness solar energy to produce ATP by *photosynthesis*. In electron micrographs, the most striking features of both mitochondria and chloroplasts are their extensive internal membrane systems. These internal membranes contain sets of membrane protein complexes that work together to produce most of the cell's ATP. In bacteria, simpler versions of essentially the same protein complexes produce ATP, but they are located in the cell's plasma membrane (**Figure 14–1**).

Comparisons of DNA sequences indicate that the energy-converting organelles in present-day eukaryotes originated from prokaryotic cells that were endocytosed during the evolution of eukaryotes (discussed in Chapter 1). This explains why mitochondria and chloroplasts contain their own DNA, which still encodes a subset of their proteins. Over time, these organelles have lost most of their own genomes and become heavily dependent on proteins that are encoded by genes in the nucleus, synthesized in the cytosol, and then imported into the organelle. And the eukaryotic cells now rely on these organelles not only for the ATP they need for biosynthesis, solute transport, and movement, but also for many important biosynthetic reactions that occur inside each organelle.

The common evolutionary origin of the energy-converting machinery in mitochondria, chloroplasts, and prokaryotes (archaea and bacteria) is reflected in the fundamental mechanism that they share for harnessing energy. This is known as **chemiosmotic coupling**, signifying a link between the chemical bond-forming reactions that generate ATP ("chemi") and membrane transport processes

IN THIS CHAPTER

THE MITOCHONDRION

THE PROTON PUMPS OF THE ELECTRON-TRANSPORT CHAIN

ATP PRODUCTION IN MITOCHONDRIA

CHLOROPLASTS AND PHOTOSYNTHESIS

THE GENETIC SYSTEMS OF MITOCHONDRIA AND CHLOROPLASTS

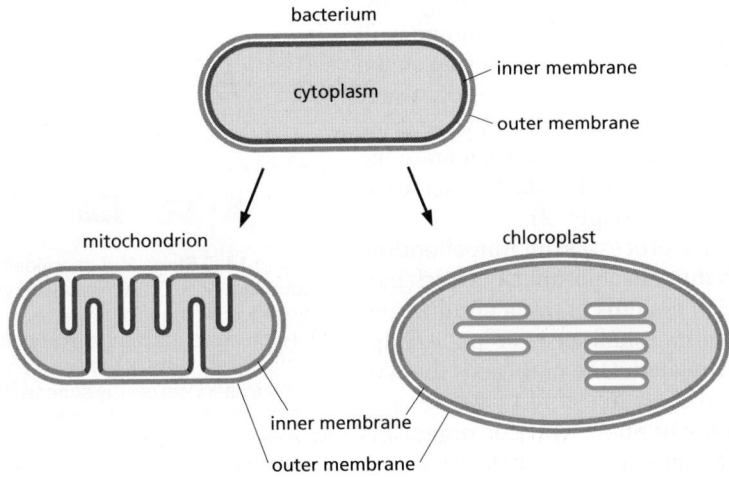

Figure 14–1 **The membrane systems of bacteria, mitochondria, and chloroplasts are related.** Mitochondria and chloroplasts are cell organelles that have originated from bacteria and have retained the bacterial energy-conversion mechanisms. Like their bacterial ancestors, mitochondria and chloroplasts have an outer and an inner membrane. Each of the membranes colored in this diagram contains energy-harvesting electron-transport chains. The deep invaginations of the mitochondrial inner membrane and the internal membrane system of the chloroplast harbor the machinery for cellular respiration and photosynthesis, respectively.

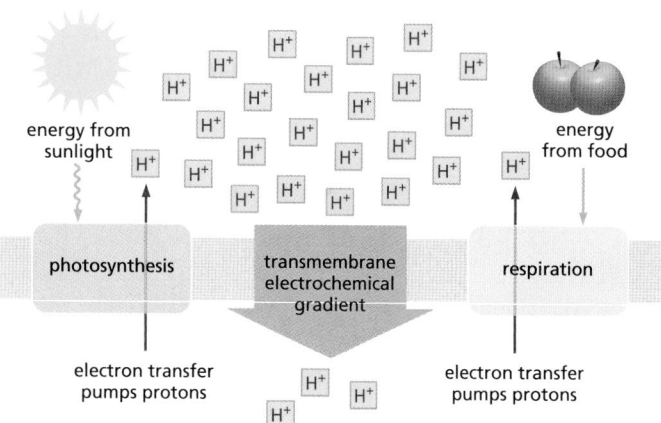

Figure 14–2 **Stage 1 of chemiosmotic coupling.** Energy from sunlight or the oxidation of food compounds is captured to generate an electrochemical proton gradient across a membrane. The electrochemical gradient serves as a versatile energy store that drives energy-requiring reactions in mitochondria, chloroplasts, and bacteria.

("osmotic"). The chemiosmotic process occurs in two linked stages, both of which are performed by protein complexes in a membrane.

Stage 1: High-energy electrons (derived from the oxidation of food molecules, from pigments excited by sunlight, or from other sources described later) are transferred along a series of electron-transport protein complexes that form an *electron-transport chain* embedded in a membrane. Each electron transfer releases a small amount of energy that is used to pump protons (H^+) and thereby generate a large *electrochemical gradient* across the membrane (**Figure 14–2**). As discussed in Chapter 11, such an electrochemical gradient provides a way of storing energy, and it can be harnessed to do useful work when ions flow back across the membrane.

Stage 2: The protons flow back down their electrochemical gradient through an elaborate membrane protein machine called *ATP synthase*, which catalyzes the production of ATP from ADP and inorganic phosphate (P_i). This ubiquitous enzyme works like a turbine in the membrane, driven by protons, to synthesize ATP (**Figure 14–3**). In this way, the energy derived from food or sunlight in stage 1 is converted into the chemical energy of a phosphate bond in ATP.

Electrons move through protein complexes in biological systems via tightly bound metal ions or other carriers that take up and release electrons easily, or by special small molecules that pick electrons up at one location and deliver them to another. For mitochondria, the first of these electron carriers is NAD^+, a water-soluble small molecule that takes up two electrons and one H^+ derived from food molecules (fats and carbohydrates) to become NADH. NADH transfers these electrons from the sites where the food molecules are degraded to the inner mitochondrial membrane. There, the electrons from the energy-rich NADH are passed from one membrane protein complex to the next, passing to a lower-energy compound at each step, until they reach a final complex in which they combine with molecular oxygen (O_2) to produce water. The energy released at each step as the electrons flow down this path from the energy-rich NADH to the low-energy water molecule drives H^+ pumps in the inner mitochondrial membrane, utilizing three different membrane protein complexes. Together, these complexes generate the proton-motive force harnessed by ATP synthase to produce the ATP that serves as the universal energy currency throughout the cell (see Chapter 2).

Figure 14–4 compares the electron-transport processes in mitochondria, which harness energy from food molecules, with those in chloroplasts, which harness energy from sunlight. The energy-conversion systems of mitochondria and chloroplasts can be described in similar terms, and we shall see later in the chapter that two of their key components are closely related. One of these is the ATP synthase, and the other is a proton pump (colored green in Figure 14–4).

Among the crucial constituents that are unique to photosynthetic organisms are the two *photosystems*. These use the green pigment chlorophyll to capture

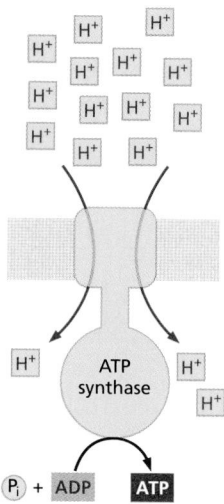

Figure 14–3 **Stage 2 of chemiosmotic coupling.** An ATP synthase *(yellow)* embedded in the lipid bilayer of a membrane harnesses the electrochemical proton gradient across the membrane, using it as a local energy store to drive ATP synthesis. The *red arrows* show the direction of proton movement through the ATP synthase.

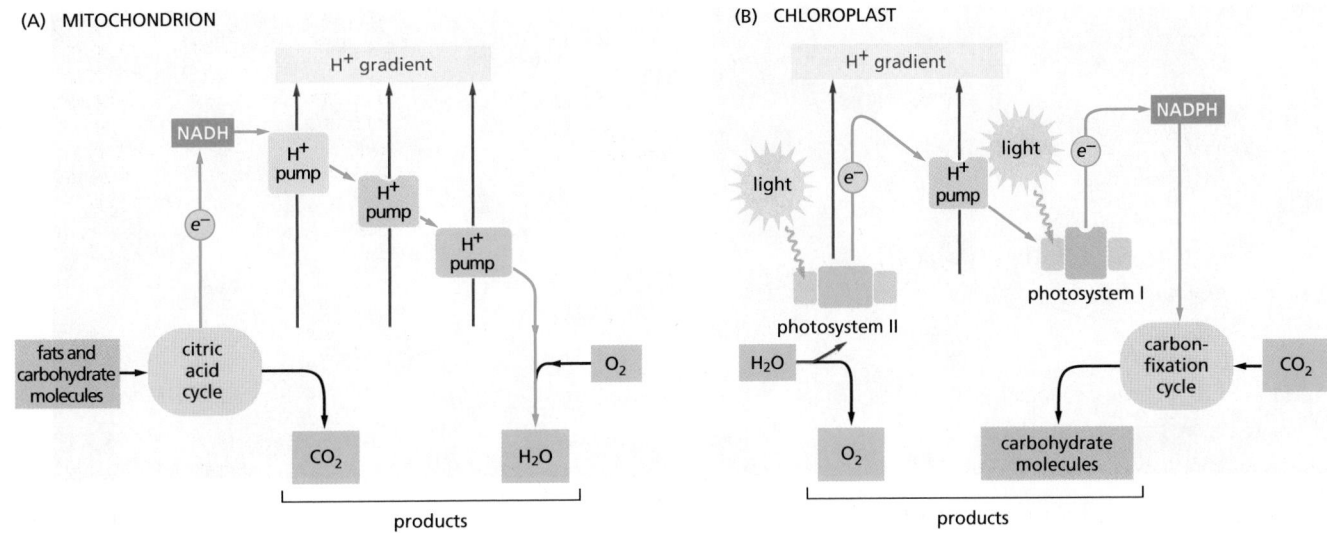

(A) MITOCHONDRION

(B) CHLOROPLAST

Figure 14–4 Electron-transport processes. (A) The mitochondrion converts energy from chemical fuels. (B) The chloroplast converts energy from sunlight. In both cases, electron flow is indicated by *blue arrows*. Each of the protein complexes *(green)* is embedded in a membrane. In the mitochondrion, fats and carbohydrates from food molecules are fed into the citric acid cycle and provide electrons to generate the energy-rich compound NADH from NAD+. These electrons then flow down an energy gradient as they pass from one complex to the next in the electron-transport chain, until they combine with molecular O_2 in the last complex to produce water. The energy released at each stage is harnessed to pump H+ across the membrane. In the chloroplast, by contrast, electrons are extracted from water through the action of light in the photosystem II complex and molecular O_2 is released. The electrons pass on to the next complex in the chain, which uses some of their energy to pump protons across the membrane, before passing to photosystem I, where sunlight generates high-energy electrons that combine with NADP+ to produce NADPH. NADPH then enters the *carbon-fixation cycle* along with CO_2 to generate carbohydrates.

light energy and power the transfer of electrons, not unlike a photocell in a solar panel. The chloroplasts drive electron transfer in the direction opposite to that in mitochondria: electrons are taken from water to produce O_2, and these electrons are used (via NADPH, a molecule closely related to the NADH used in mitochondria) to synthesize carbohydrates from CO_2 and water. These carbohydrates then serve as the source for all other compounds a plant cell needs.

Thus, both mitochondria and chloroplasts make use of an electron-transfer chain to produce an H+ gradient that powers reactions that are critical for the cell. However, chloroplasts generate O_2 and take up CO_2, whereas mitochondria consume O_2 and release CO_2 (see Figure 14–4).

THE MITOCHONDRION

Mitochondria occupy up to 20% of the cytoplasmic volume of a eukaryotic cell. Although they are often depicted as short, bacterium-like bodies with a diameter of 0.5–1 μm, they are in fact remarkably dynamic and plastic, moving about the cell, constantly changing shape, dividing, and fusing (Movie 14.1). Mitochondria are often associated with the microtubular cytoskeleton (Figure 14–5), which determines their orientation and distribution in different cell types. Thus, in highly polarized cells such as neurons, mitochondria can move long distances (up to a meter or more in the extended axons of neurons), being propelled along the tracks of the microtubular cytoskeleton. In other cells, mitochondria remain fixed at points of high energy demand; for example, in skeletal or cardiac muscle cells, they pack between myofibrils, and in sperm cells they wrap tightly around the flagellum (Figure 14–6).

Mitochondria also interact with other membrane systems in the cell, most notably the endoplasmic reticulum (ER). Contacts between mitochondria and ER define specialized domains thought to facilitate the exchange of lipids between the two membrane systems. These contacts also appear to induce mitochondrial

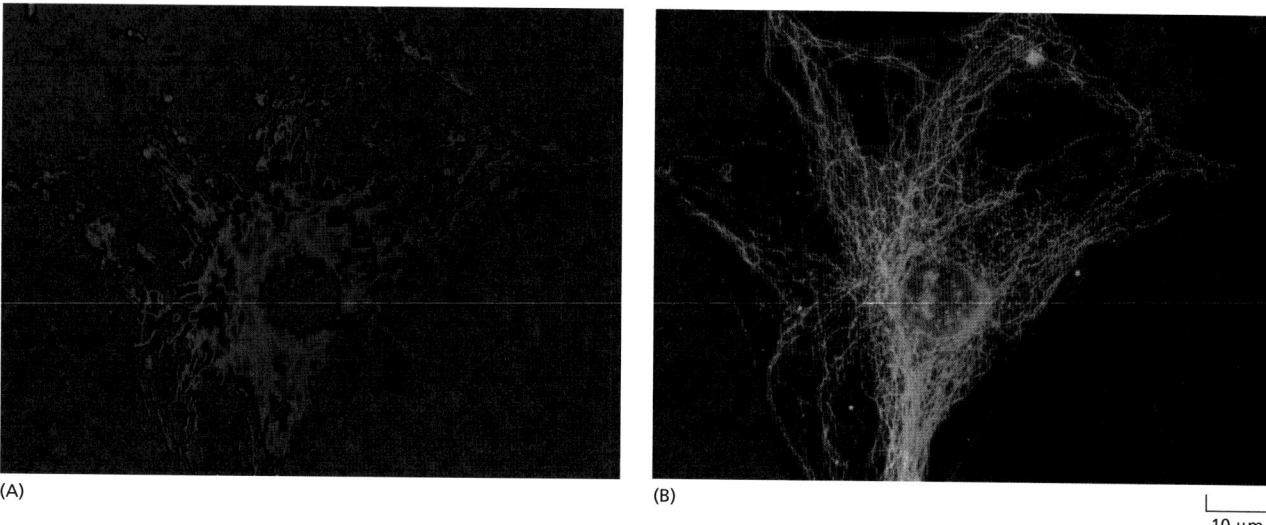

(A)

(B)

10 μm

fission, which, as we discuss later, is involved in the distribution and partitioning of mitochondria within cells (**Figure 14–7**).

The acquisition of mitochondria was a prerequisite for the evolution of complex animals. Without mitochondria, present-day animal cells would have had to generate all of their ATP through anaerobic glycolysis. When glycolysis converts glucose to pyruvate, it releases only a small fraction of the total free energy that is potentially available from glucose oxidation (see Chapter 2). In mitochondria, the metabolism of sugars is complete: pyruvate is imported into the mitochondrion and ultimately oxidized by O_2 to CO_2 and H_2O, which allows 15 times more ATP to be made from a sugar than by glycolysis alone. As explained later, this became possible only when enough molecular oxygen accumulated in the Earth's atmosphere to allow organisms to take full advantage, via respiration, of the large amounts of energy potentially available from the oxidation of organic compounds.

Mitochondria are large enough to be seen in the light microscope, and they were first identified in the nineteenth century. Real progress in understanding their internal structure and function, however, depended on biochemical procedures developed in 1948 for isolating intact mitochondria, and on electron microscopy, which was first used to look at cells at about the same time.

Figure 14–5 The relationship between mitochondria and microtubules. (A) A light micrograph of chains of elongated mitochondria in a living mammalian cell in culture. The cell was stained with a fluorescent dye (rhodamine 123) that specifically labels mitochondria in living cells. (B) An immunofluorescence micrograph of the same cell stained (after fixation) with fluorescent antibodies that bind to microtubules. Note that the mitochondria tend to be aligned along microtubules. (Courtesy of Lan Bo Chen.)

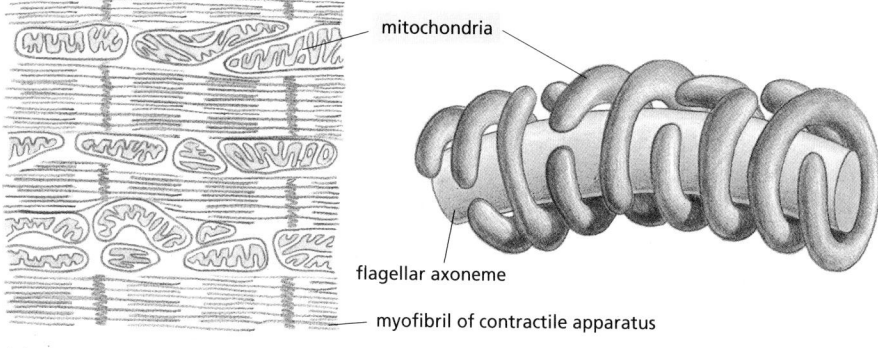

mitochondria

flagellar axoneme

myofibril of contractile apparatus

(A) CARDIAC MUSCLE

(B) SPERM TAIL

Figure 14–6 Localization of mitochondria near sites of high ATP demand in cardiac muscle and a sperm tail. (A) Cardiac muscle in the wall of the heart is the most heavily used muscle in the body, and its continual contractions require a reliable energy supply. It has limited built-in energy stores and has to depend on a steady supply of ATP from the copious mitochondria aligned close to the contractile myofibrils (see Figure 16–32). (B) During sperm development, microtubules wind helically around the flagellar axoneme, where they are thought to help localize the mitochondria in the tail to produce the structure shown.

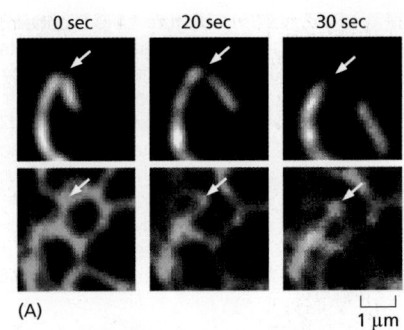

Figure 14–7 **Interaction of mitochondria with the endoplasmic reticulum.**
(A) Fluorescence light microscopy shows that tubules of the ER *(green)* wrap
around parts of the mitochondrial network *(red)* in mammalian cells. The
mitochondria then divide at the contact sites. After contact is established,
fission occurs within less than a minute, as indicated by time-lapse
microscopy. (B) Schematic drawing of an ER tubule wrapped around part of
the mitochondrial reticulum. It is thought that ER–mitochondrial contacts also
mediate the exchange of lipids between the two membrane systems.
(A, adapted from J.R. Friedman et al., *Science* 334:358–362, 2011. With
permission from AAAS.)

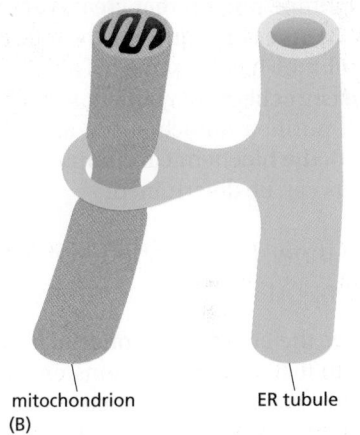

The Mitochondrion Has an Outer Membrane and an Inner Membrane

Like the bacteria from which they originated, mitochondria have an outer and
an inner membrane. The two membranes have distinct functions and properties,
and delineate separate compartments within the organelle. The inner membrane,
which surrounds the internal **mitochondrial matrix** compartment (**Figure 14–8**),
is highly folded to form invaginations known as **cristae** (the singular is crista),
which contain in their membranes the proteins of the electron-transport chain.
Where the inner membrane runs parallel to the outer membrane, between the
cristae, it is known as the *inner boundary membrane.* The narrow (20–30 nm) gap
between the inner boundary membrane and the outer membrane is known as
the **intermembrane space.** The cristae are about 20 nm-wide membrane discs
or tubules that protrude deeply into the matrix and enclose the *crista space.* The
crista membrane is continuous with the inner boundary membrane, and where
their membranes join, the membrane forms narrow membrane tubes or slits,
known as *crista junctions.*

Like the bacterial outer membrane, the **outer mitochondrial membrane** is
freely permeable to ions and to small molecules as large as 5000 daltons. This

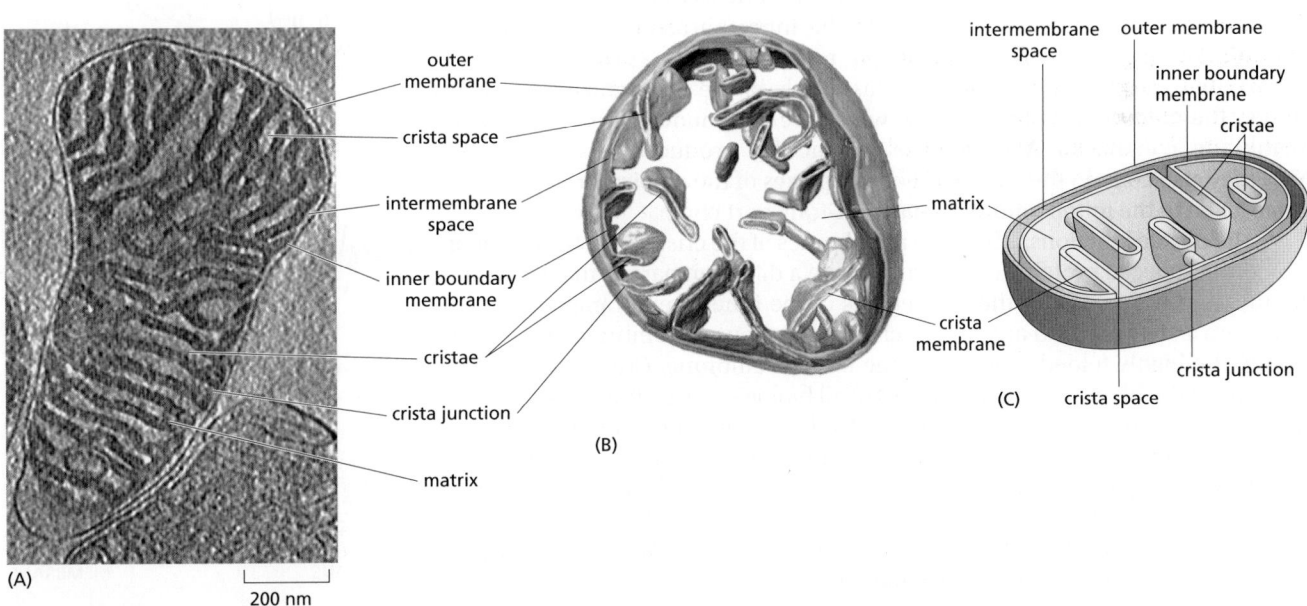

Figure 14–8 **Structure of a mitochondrion.** (A) Tomographic slice through a three-dimensional map of a mouse heart mitochondrion determined by
electron-microscope tomography. The outer membrane envelops the inner boundary membrane. The inner membrane is highly folded into tubular
or lamellar cristae, which crisscross the matrix. The dense matrix, which contains most of the mitochondrial protein, appears dark in the electron
microscope, whereas the intermembrane space and the crista space appear light due to their lower protein content. The inner boundary membrane
follows the outer membrane closely at a distance of ≈20 nm. The inner membrane turns sharply at the cristae junctions, where the cristae join the
inner boundary membrane. (B) Tomographic surface-rendered portion of a yeast mitochondrion, showing how flattened cristae project into the matrix
from the inner membrane (**Movie 14.2**). (C) Schematic drawing of a mitochondrion showing the outer membrane *(gray)*, and the inner membrane
(yellow). Note that the inner membrane is compartmentalized into the inner boundary membrane and the crista membrane. There are three distinct
spaces: the inner membrane space, the crista space, and the matrix. (A, courtesy of Tobias Brandt; B, from K. Davies et al., *Proc. Natl Acad. Sci.
USA* 109:13602–13607, 2012. With permission from the National Academy of Sciences.)

Figure 14–9 Biochemical fractionation of purified mitochondria into separate components. Large numbers of mitochondria are isolated from homogenized tissue by centrifugation and then suspended in a medium of low osmotic strength. In such a medium, water flows into mitochondria and greatly expands the matrix space *(yellow)*. While the cristae of the inner membrane unfold to accommodate the swelling, the outer membrane—which has no folds—breaks, releasing structures composed of the inner membrane surrounding the matrix. These techniques have made it possible to study the protein composition of the inner membrane (comprising a mixture of cristae, boundary membranes, and cristae junctions), the outer membrane, and the matrix.

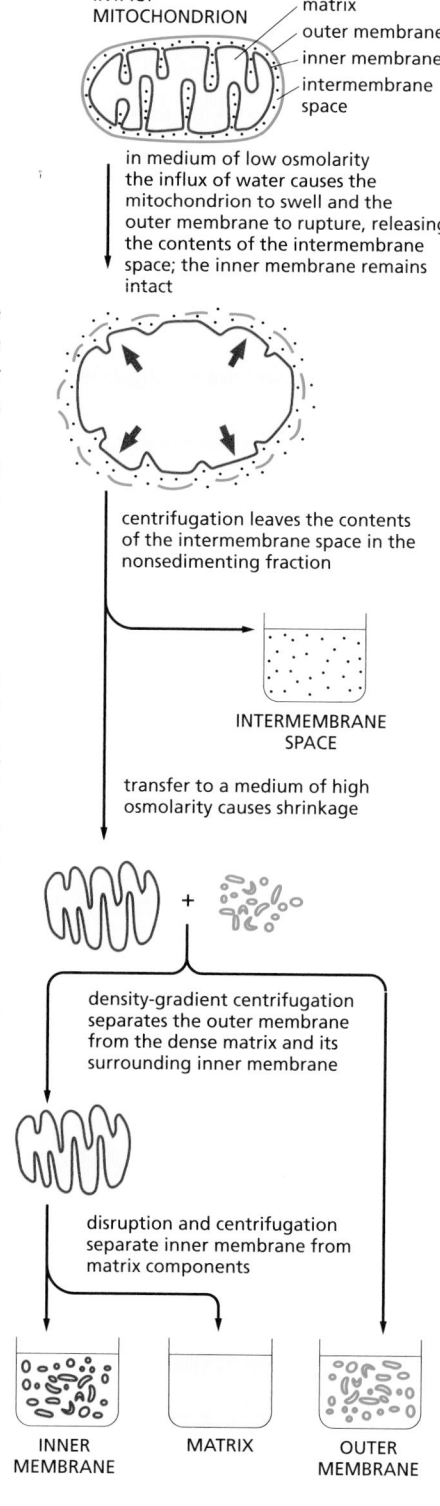

is because it contains many porin molecules, a special class of β-barrel-type membrane protein that creates aqueous pores across the membrane (see Figure 10–23). As a consequence, the intermembrane space between the outer and inner membrane has the same pH and ionic composition as the cytoplasm, and there is no electrochemical gradient across the outer membrane.

If purified mitochondria are gently disrupted and then fractionated (**Figure 14–9**), the biochemical composition of membranes and mitochondrial compartments can be determined.

The Inner Membrane Cristae Contain the Machinery for Electron Transport and ATP Synthesis

Unlike the outer mitochondrial membrane, the **inner mitochondrial membrane** is a diffusion barrier to ions and small molecules, just like the bacterial inner membrane. However, selected ions, most notably protons and phosphate, as well as essential metabolites such as ATP and ADP, can pass through it by means of special transport proteins.

The inner mitochondrial membrane is highly differentiated into functionally distinct regions with different protein compositions. As discussed in Chapter 10, the lateral segregation of membrane regions with different protein and lipid compositions is a key feature of cells. In the inner mitochondrial membrane, the boundary membrane region is thought to contain the machinery for protein import, new membrane insertion, and assembly of the respiratory-chain complexes. The membranes of the cristae, which are continuous with the boundary membrane, contain the ATP synthase enzyme that produces most of the cell's ATP; they also contain the large protein complexes of the **respiratory chain**—the name given to the mitochondrion's electron-transport chain.

At the cristae junctions, where the membranes of the cristae join the boundary membrane, special protein complexes provide a diffusion barrier that segregates the membrane proteins in the two regions of the inner membrane; these complexes are also thought to anchor the cristae to the outer membrane, thus maintaining the highly folded topology of the inner membrane. Cristae membranes have one of the highest protein densities of all biological membranes, with a lipid content of 25% and a protein content of 75% by weight. The folding of the inner membrane into cristae greatly increases the membrane area available for oxidative phosphorylation. In highly active cardiac muscle cells, for example, the total area of cristae membranes can be up to 20 times larger than the area of the cell's plasma membrane. In total, the surface area of cristae membranes in each human body adds up to roughly the size of a football field.

The Citric Acid Cycle in the Matrix Produces NADH

Together with the cristae that project into it, the matrix is the major working part of the mitochondrion. Mitochondria can use both pyruvate and fatty acids as fuel. Pyruvate is derived from glucose and other sugars, whereas fatty acids are derived from fats. Both of these fuel molecules are transported across the inner mitochondrial membrane by specialized transport proteins, and they are then converted to the crucial metabolic intermediate *acetyl CoA* by enzymes located in the mitochondrial matrix (see Chapter 2).

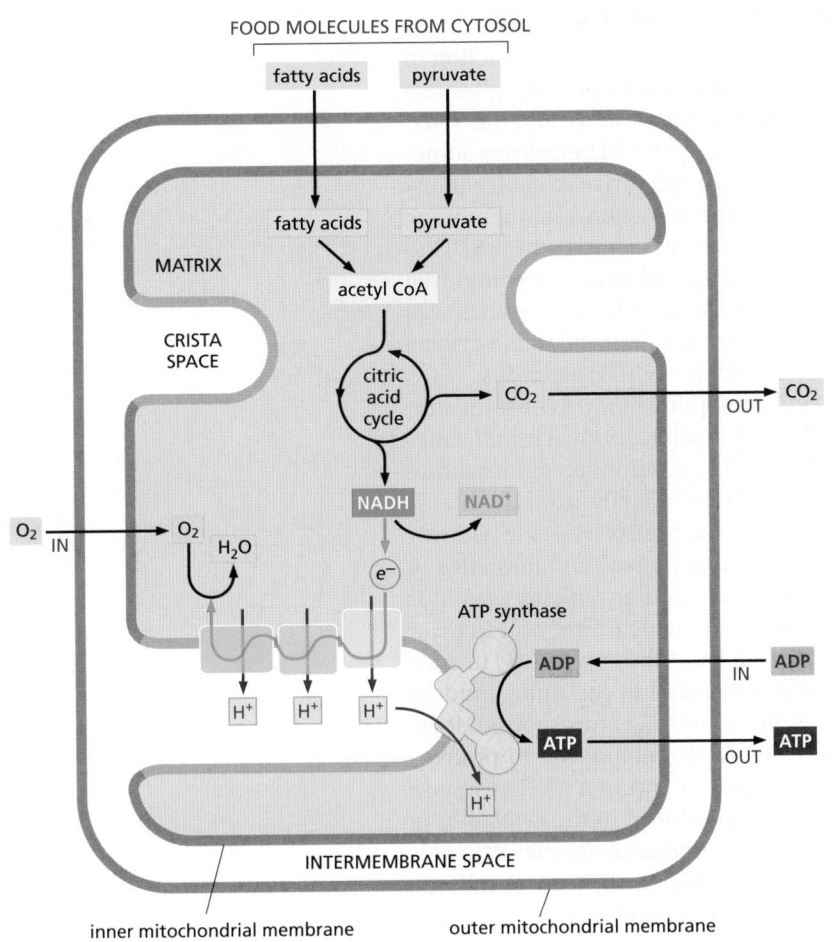

FOOD MOLECULES FROM CYTOSOL

Figure 14–10 A summary of the energy-converting metabolism in mitochondria. Pyruvate and fatty acids enter the mitochondrion *(top of the figure)* and are broken down to acetyl CoA. The acetyl CoA is metabolized by the citric acid cycle, which reduces NAD^+ to NADH, which then passes its high-energy electrons to the first complex in the electron-transport chain. In the process of oxidative phosphorylation, these electrons pass along the electron-transport chain in the inner membrane cristae to oxygen (O_2). This electron transport generates a proton gradient, which drives the production of ATP by the ATP synthase (see Figure 14–3). Electrons from the oxidation of succinate, a reaction intermediate in the citric acid cycle (see Panel 2–9, pp. 106–107), take a separate path to enter this electron-transport chain (not shown, see p. 772).

The membranes that comprise the mitochondrial inner membrane—the inner boundary membrane and the crista membrane—contain different mixtures of proteins and they are therefore shaded differently in this diagram.

The acetyl groups in acetyl CoA are oxidized in the matrix via the *citric acid cycle*, also called the Krebs cycle (see Figure 2–57 and Movie 2.6). The oxidation of these carbon atoms in acetyl CoA produces CO_2, which diffuses out of the mitochondrion to be released to the environment as a waste product. More importantly, the citric acid cycle saves a great deal of the bond energy released by this oxidation in the form of electrons carried by NADH. This NADH transfers its electrons from the matrix to the electron-transport chain in the inner mitochondrial membrane, where—through the *chemiosmotic coupling* process described previously (see Figures 14–2 and 14–3)—the energy that was carried by NADH electrons is converted into phosphate-bond energy in ATP. **Figure 14–10** outlines this sequence of reactions schematically.

The matrix contains the genetic system of the mitochondrion, including the mitochondrial DNA and the ribosomes. The mitochondrial DNA (see section on genetic systems, p. 800) is organized into compact bodies—the nucleoids—by special scaffolding proteins that also function as transcription regulatory proteins. The large number of enzymes required for the maintenance of the mitochondrial genetic system, as well as for many other essential reactions to be outlined next, accounts for the very high protein concentration in the matrix; at more than 500 mg/mL, this concentration is close to that in a protein crystal.

Mitochondria Have Many Essential Roles in Cellular Metabolism

Mitochondria not only generate most of the cell's ATP; they also provide many other essential resources for biosynthesis and cell growth. Before describing in detail the remarkable machinery of the respiratory chain, we diverge briefly to touch on some of these important roles.

Mitochondria are critical for buffering the redox potential in the cytosol. Cells need a constant supply of the electron acceptor NAD^+ for the central reaction in glycolysis that converts glyceraldehyde 3-phosphate to 1,3-bisphosphoglycerate (see Figure 2–48). This NAD^+ is converted to NADH in the process, and the NAD^+ needs to be regenerated by transferring the high-energy NADH electrons somewhere. The NADH electrons will eventually be used to help drive oxidative phosphorylation inside the mitochondrion. But the inner mitochondrial membrane is impermeable to NADH. The electrons are therefore passed from the NADH to smaller molecules in the cytosol that can move through the inner mitochondrial membrane. Once in the matrix, these smaller molecules transfer their electrons to NAD^+ to form mitochondrial NADH, after which they are returned to the cytosol for recharging—creating a so-called *shuttle system* for the NADH electrons.

In addition to ATP, biosynthesis in the cytosol requires both a constant supply of reducing power in the form of NADPH and small carbon-rich molecules to serve as building blocks (discussed in Chapter 2). Descriptions of biosynthesis often state that the needed carbon skeletons come directly from the breakdown of sugars, whereas the NADPH is produced in the cytosol by a side pathway for the breakdown of sugars (the *pentose phosphate pathway*, an alternative to glycolysis). But under conditions where nutrients abound and plenty of ATP is available, mitochondria help to generate both the reducing power and the carbon-rich building blocks (the "carbon skeletons" in Panel 2-1, pp. 90–91) needed for cell growth. For this purpose, excess citrate produced in the mitochondrial matrix by the citric acid cycle (see Panel 2-9, pp. 106–107) is transported down its electrochemical gradient to the cytosol, where it is metabolized to produce essential components of the cell. Thus, for example, as part of a cell's response to growth signals, large amounts of acetyl CoA are produced in the cytosol from citrate exported from mitochondria, accelerating the production of the fatty acids and sterols that build new membranes (described in Chapter 10). Cancer cells are frequently mutated in ways that enhance this pathway, as part of their program of abnormal growth (see Figure 20–26).

The urea cycle is a central metabolic pathway in mammals that converts the ammonia (NH_4^+) produced by the breakdown of nitrogen-containing compounds (such as amino acids) to the urea excreted in urine. Two critical steps of the urea cycle are carried out in the mitochondria of liver cells, while the remaining steps occur in the cytosol. Mitochondria also play an essential part in the metabolic adaptation of cells to different nutritional conditions. For example, under conditions of starvation, proteins in our bodies are broken down to amino acids, and the amino acids are imported into mitochondria and oxidized to produce NADH for ATP production.

The biosynthesis of *heme groups*—which, as we shall see in the next section, play a central part in electron transfer—is another critical process that is shared between the mitochondrion and the cytoplasm. Iron–sulfur clusters, which are essential not only for electron transfer in the respiratory chain (see p. 766), but also for the maintenance and stability of the nuclear genome, are produced in mitochondria (and chloroplasts). Nuclear genome instability, a hallmark of cancer, can sometimes be linked to the decreased function of cellular proteins that contain iron–sulfur clusters.

Mitochondria also have a central role in membrane biosynthesis. Cardiolipin is a two-headed phospholipid (Figure 14–11) that is confined to the inner mitochondrial membrane, where it is also produced. But mitochondria are also a major source of phospholipids for the biogenesis of other cell membranes. Phosphatidylethanolamine, phosphatidylglycerol, and phosphatidic acid are synthesized in the mitochondrion, while phosphatidylinositol, phosphatidylcholine, and phosphatidylserine are primarily synthesized in the endoplasmic reticulum (ER). As described in Chapter 12, most of the cell's membranes are assembled in the ER. The exchange of lipids between the ER and mitochondria is thought to occur at special sites of close contact (see Figure 14-7) by an as-yet unknown mechanism.

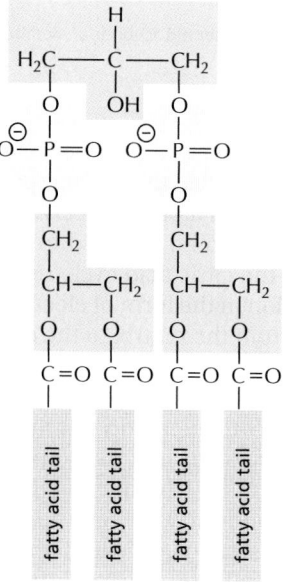

Figure 14–11 The structure of cardiolipin. Cardiolipin consists of two covalently linked phospholipid units, with a total of four rather than the usual two fatty acid chains (see Figure 10–3). Cardiolipin is only produced in the mitochondrial inner membrane, where it interacts closely with membrane proteins involved in oxidative phosphorylation and ATP transport. In cristae, its two juxtaposed phosphate groups may act as a local proton trap on the membrane surface.

Finally, mitochondria are important calcium buffers, taking up calcium from the ER and sarcoplasmic reticulum at special membrane junctions. Cellular calcium levels control muscle contraction (see Chapter 16) and alterations are implicated in neurodegeneration and apoptosis. Clearly, cells and organisms depend on mitochondria in many different ways.

We now return to the central function of the mitochondrion in respiratory ATP generation.

A Chemiosmotic Process Couples Oxidation Energy to ATP Production

Although the citric acid cycle that takes place in the mitochondrial matrix is considered to be part of aerobic metabolism, it does not itself use oxygen. Only the final step of oxidative metabolism consumes molecular oxygen (O_2) directly (see Figure 14–10). Nearly all the energy available from metabolizing carbohydrates, fats, and other foodstuffs in earlier stages is saved in the form of energy-rich compounds that feed electrons into the respiratory chain in the inner mitochondrial membrane. These electrons, most of which are carried by NADH, finally combine with O_2 at the end of the respiratory chain to form water. The energy released during the complex series of electron transfers from NADH to O_2 is harnessed in the inner membrane to generate an electrochemical gradient that drives the conversion of ADP + P_i to ATP. For this reason, the term **oxidative phosphorylation** is used to describe this final series of reactions (**Figure 14–12**).

The total amount of energy released by biological oxidation in the respiratory chain is equivalent to that released by the explosive combustion of hydrogen when it combines with oxygen in a single step to form water. But the combustion of hydrogen in a single-step chemical reaction, which has a strongly negative ΔG, releases this large amount of energy unproductively as heat. In the respiratory chain, the same energetically favorable reaction $H_2 + \frac{1}{2} O_2 \rightarrow H_2O$ is divided into small steps (**Figure 14–13**). This stepwise process allows the cell to store nearly half of the total energy that is released in a useful form. At each step, the electrons, which can be thought of as having been removed from a hydrogen molecule to

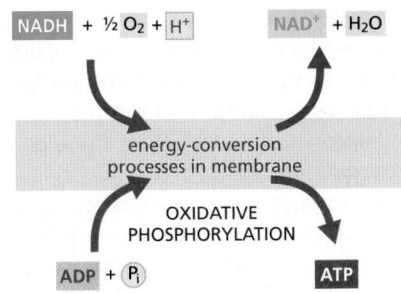

Figure 14–12 The major net energy conversion catalyzed by the mitochondrion. In the process of oxidative phosphorylation, the mitochondrial inner membrane serves as a device that changes one form of chemical-bond energy to another, converting a major part of the energy of NADH oxidation into phosphate-bond energy in ATP.

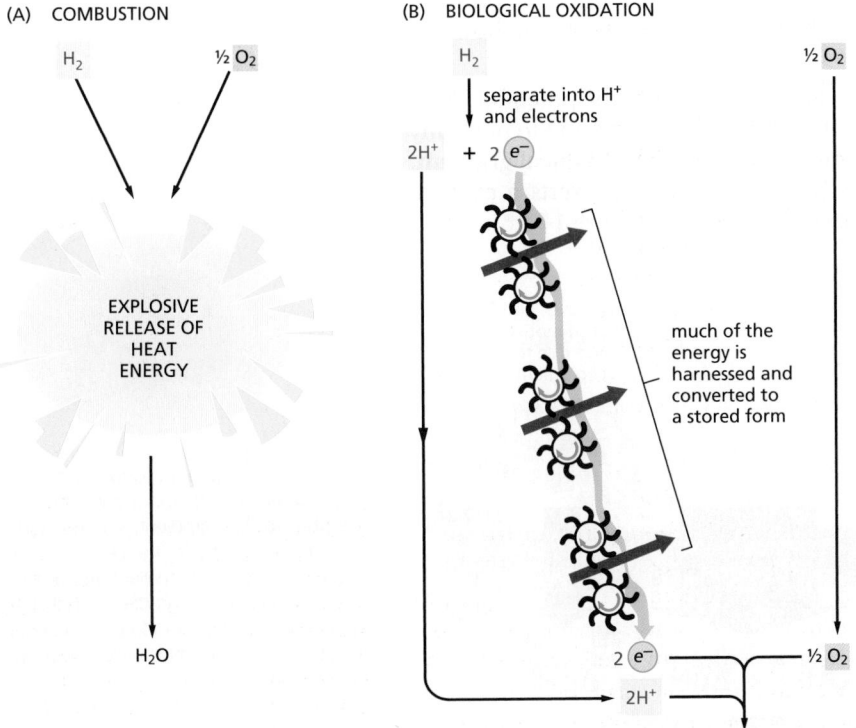

Figure 14–13 A comparison of biological oxidation with combustion. (A) If hydrogen were simply burned, nearly all of the energy would be released in the form of heat. (B) In biological oxidation reactions, about half of the released energy is stored in a form useful to the cell by means of the electron-transport chain (the respiratory chain) in the crista membrane of the mitochondrion. Only the rest of the energy is released as heat. In the cell, the protons and electrons shown here as being derived from H_2 are removed from hydrogen atoms that are covalently linked to NADH molecules.

produce two protons, pass through a series of electron carriers in the inner mitochondrial membrane. At each of three distinct steps along the way (marked by the three electron-transport complexes of the respiratory chain, see below), much of the energy is utilized for pumping protons across the membrane. At the end of the electron-transport chain, the electrons and protons recombine with molecular oxygen into water.

Water is a very low-energy molecule and is thus very stable; it can serve as an electron donor only when a large amount of energy from an external source is spent on splitting it into protons, electrons, and molecular oxygen. This is exactly what happens in oxygenic photosynthesis, where the external energy source is the sun, as we shall see later in the section on chloroplasts (p. 782).

The Energy Derived from Oxidation Is Stored as an Electrochemical Gradient

In mitochondria, the process of electron transport begins when two electrons and a proton are removed from NADH (to regenerate NAD$^+$). These electrons are passed to the first of about 20 different electron carriers in the respiratory chain. The electrons start at a large negative redox potential (see Panel 14–1, p. 765)—that is, at a high energy level—which gradually drops as they pass along the chain. The proteins involved are grouped into three large *respiratory enzyme complexes*, each composed of protein subunits that sit in the inner mitochondrial membrane. Each complex in the chain has a higher affinity for electrons than its predecessor, and electrons pass sequentially from one complex to the next until they are finally transferred to molecular oxygen, which has the highest electron affinity of all.

The net result is the pumping of H$^+$ out of the matrix across the inner membrane, driven by the energetically favorable flow of electrons. This transmembrane movement of H$^+$ has two major consequences:

1. It generates a pH gradient across the inner mitochondrial membrane, with a high pH in the matrix (close to 8) and a lower pH in the intermembrane space. Since ions and small molecules equilibrate freely across the outer mitochondrial membrane, the pH in the intermembrane space is the same as in the cytosol (generally around pH 7.4).

2. It generates a voltage gradient across the inner mitochondrial membrane, creating a *membrane potential* with the matrix side negative and the crista space side positive.

The pH gradient (ΔpH) reinforces the effect of the membrane potential (ΔV), because the latter acts to attract any positive ion into the matrix and to push any negative ion out. Together, ΔpH and ΔV make up the **electrochemical gradient**, which is measured in units of millivolts (mV). This gradient exerts a **proton-motive force**, which tends to drive H$^+$ back into the matrix (**Figure 14–14**).

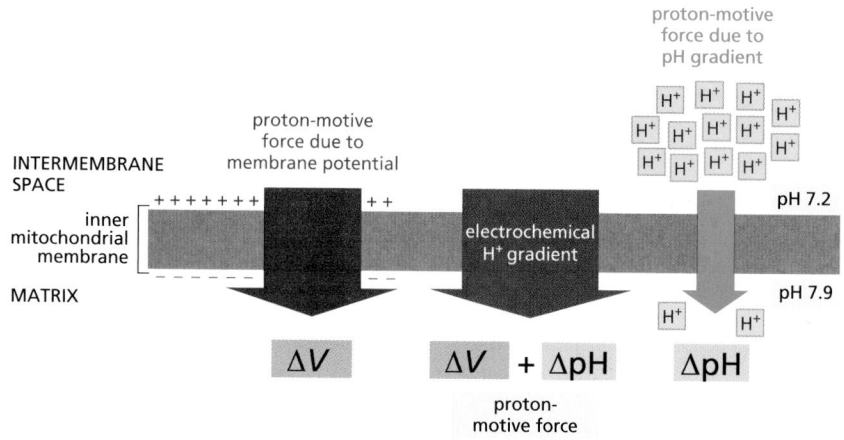

Figure 14–14 The electrochemical proton gradient across the inner mitochondrial membrane. This gradient is composed of a large force due to the membrane potential (ΔV) and a smaller force due to the H$^+$ concentration gradient—that is, the pH gradient (ΔpH). Both forces combine to generate the proton-motive force, which pulls H$^+$ back into the mitochondrial matrix. The exact relationship between these forces is expressed by the Nernst equation (see Panel 11–1, p. 616).

The electrochemical gradient across the inner membrane of a respiring mitochondrion is typically about 180 mV (inside negative), and it consists of a membrane potential of about 150 mV and a pH gradient of about 0.5 to 0.6 pH units (each ΔpH of 1 pH unit is equivalent to a membrane potential of about 60 mV). The electrochemical gradient drives not only ATP synthesis but also the transport of selected molecules across the inner mitochondrial membrane, including the import of selected proteins from the cytoplasm (discussed in Chapter 12).

Summary

The mitochondrion performs most cellular oxidations and produces the bulk of the animal cell's ATP. A mitochondrion has two separate membranes: the outer membrane and the inner membrane. The inner membrane surrounds the innermost space (the matrix) of the mitochondrion and forms the cristae, which project into the matrix. The matrix and the inner membrane cristae are the major working parts of the mitochondrion. The membranes that form cristae account for a major part of the membrane surface area in most cells, and they contain the mitochondrion's electron-transport chain (the respiratory chain).

The mitochondrial matrix contains a large variety of enzymes, including those that convert pyruvate and fatty acids to acetyl CoA and those that oxidize this acetyl CoA to CO_2 through the citric acid cycle. These oxidation reactions produce large amounts of NADH, whose high-energy electrons are passed to the respiratory chain. The respiratory chain then uses the energy derived from transporting electrons from NADH to molecular oxygen to pump H^+ out of the matrix. This produces a large electrochemical proton gradient across the inner mitochondrial membrane, composed of contributions from both a membrane potential and a pH difference. This electrochemical gradient exerts a force to drive H^+ back into the matrix. This proton-motive force is harnessed both to produce ATP and for the selective transport of metabolites across the inner mitochondrial membrane.

THE PROTON PUMPS OF THE ELECTRON-TRANSPORT CHAIN

Having considered in general terms how a mitochondrion uses electron transport to generate a proton-motive force, we now turn to the molecular mechanisms that underlie this membrane-based energy-conversion process. In describing the respiratory chain of mitochondria, we accomplish the larger purpose of explaining how an electron-transport process can pump protons across a membrane. As stated at the beginning of this chapter, mitochondria, chloroplasts, archaea, and bacteria use very similar chemiosmotic mechanisms. In fact, these mechanisms underlie the function of all living organisms—including anaerobes that derive all their energy from electron transfers between two inorganic molecules, as we shall see later.

We start with some of the basic principles on which all of these processes depend.

The Redox Potential Is a Measure of Electron Affinities

In chemical reactions, any electrons removed from one molecule are always passed to another, so that whenever one molecule is oxidized, another is reduced. As with any other chemical reaction, the tendency of such **redox reactions** to proceed spontaneously depends on the free-energy change (ΔG) for the electron transfer, which in turn depends on the relative affinities of the two molecules for electrons.

Because electron transfers provide most of the energy for life, it is worth taking the time to understand them. As discussed in Chapter 2, acids donate protons and bases accept them (see Panel 2–2, p. 93). Acids and bases exist in conjugate acid–base pairs, in which the acid is readily converted into the base by the loss of a proton. For example, acetic acid (CH_3COOH) is converted into its conjugate base, the acetate ion (CH_3COO^-), in the reaction:

$$CH_3COOH \rightleftharpoons CH_3COO^- + H^+$$

In an exactly analogous way, pairs of compounds such as NADH and NAD^+ are called **redox pairs**, since NADH is converted to NAD^+ by the loss of electrons in the reaction:

$$NADH \rightleftharpoons NAD^+ + H^+ + 2e^-$$

NADH is a strong electron donor: because two of its electrons are engaged in a covalent bond which releases energy when broken, the free-energy change for passing these electrons to many other molecules is favorable. Energy is required to form this bond from NAD^+, two electrons, and a proton (the same amount of energy that was released when the bond was broken). Therefore NAD^+, the redox partner of NADH, is of necessity a weak electron acceptor.

We can measure the tendency to transfer electrons from any redox pair experimentally. All that is required is the formation of an electrical circuit linking a 1:1 (equimolar) mixture of the redox pair to a second redox pair that has been arbitrarily selected as a reference standard, so that we can measure the voltage difference between them (**Panel 14–1**). This voltage difference is defined as the **redox potential**; electrons move spontaneously from a redox pair like NADH/NAD^+ with a lower redox potential (a lower affinity for electrons) to a redox pair like O_2/H_2O with a higher redox potential (a higher affinity for electrons). Thus, NADH is a good molecule for donating electrons to the respiratory chain, while O_2 is well suited to act as the "sink" for electrons at the end of the chain. As explained in Panel 14–1, the difference in redox potential, $\Delta E'_0$, is a direct measure of the standard free-energy change ($\Delta G°$) for the transfer of an electron from one molecule to another.

Electron Transfers Release Large Amounts of Energy

As just discussed, those pairs of compounds that have the most negative redox potentials have the weakest affinity for electrons and therefore are useful as carriers with a strong tendency to donate electrons. Conversely, those pairs that have the most positive redox potentials have the greatest affinity for electrons and therefore are useful as carriers with a strong tendency to accept electrons. A 1:1 mixture of NADH and NAD^+ has a redox potential of –320 mV, indicating that NADH has a strong tendency to donate electrons; a 1:1 mixture of H_2O and $\frac{1}{2}O_2$ has a redox potential of +820 mV, indicating that O_2 has a strong tendency to accept electrons. The difference in redox potential is 1140 mV, which means that the transfer of each electron from NADH to O_2 under these standard conditions is enormously favorable, since $\Delta G° = -109$ kJ/mole, and twice this amount of energy is gained for the two electrons transferred per NADH molecule (see Panel 14–1). If we compare this free-energy change with that for the formation of the phosphoanhydride bonds in ATP, where $\Delta G° = 30.6$ kJ/mole (see Figure 2–50), we see that, under standard conditions, the oxidation of one NADH molecule releases more than enough energy to synthesize seven molecules of ATP from ADP and P_i. (In the cell, the number of ATP molecules generated will be lower because the standard conditions are far from the physiological ones; in addition, small amounts of energy are inevitably dissipated as heat along the way.)

Transition Metal Ions and Quinones Accept and Release Electrons Readily

The electron-transport properties of the membrane protein complexes in the respiratory chain depend upon electron-carrying *cofactors*, most of which are *transition metals* such as Fe, Cu, Ni, and Mn, bound to proteins in the complexes. These metals have special properties that allow them to promote both enzyme catalysis and electron-transfer reactions. Most relevant here is the fact that their ions exist in several different oxidation states with closely spaced redox potentials, which enables them to accept or give up electrons readily; this property is

HOW REDOX POTENTIALS ARE MEASURED

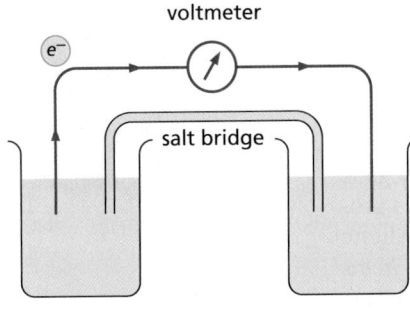

voltmeter

salt bridge

$A_{reduced}$ and $A_{oxidized}$ in equimolar amounts

1 M H⁺ and 1 atmosphere H_2 gas

One beaker *(left)* contains substance A with an equimolar mixture of the reduced ($A_{reduced}$) and oxidized ($A_{oxidized}$) members of its redox pair. The other beaker contains the hydrogen reference standard ($2H^+ + 2e^- \rightleftharpoons H_2$), whose redox potential is arbitrarily assigned as zero by international agreement. (A salt bridge formed from a concentrated KCl solution allows K^+ and Cl^- to move between the beakers, as required to neutralize the charges when electrons flow between the beakers.) The metal wire *(dark blue)* provides a resistance-free path for electrons, and a voltmeter then measures the redox potential of substance A. If electrons flow from $A_{reduced}$ to H^+, as indicated here, the redox pair formed by substance A is said to have a negative redox potential. If they instead flow from H_2 to $A_{oxidized}$, the redox pair is said to have a positive redox potential.

THE STANDARD REDOX POTENTIAL, E'_0

The standard redox potential for a redox pair, defined as E_0, is measured for a standard state where all of the reactants are at a concentration of 1 M, including H^+. Since biological reactions occur at pH 7, biologists instead define the standard state as $A_{reduced} = A_{oxidized}$ and $H^+ = 10^{-7}$ M. This standard redox potential is designated by the symbol E'_0, in place of E_0.

examples of redox reactions	standard redox potential E'_0
$NADH \rightleftharpoons NAD^+ + H^+ + 2e^-$	−320 mV
reduced ubiquinone $\rightleftharpoons$ oxidized ubiquinone $+ 2H^+ + 2e^-$	+30 mV
reduced cytochrome c $\rightleftharpoons$ oxidized cytochrome c $+ e^-$	+230 mV
$H_2O \rightleftharpoons ½O_2 + 2H^+ + 2e^-$	+820 mV

CALCULATION OF $\Delta G°$ FROM REDOX POTENTIALS

To determine the energy change for an electron transfer, the $\Delta G°$ of the reaction (kJ/mole) is calculated as follows:

$\Delta G° = -n(0.096) \Delta E'_0$, where n is the number of electrons transferred across a redox potential change of $\Delta E'_0$ millivolts (mV), and

$\Delta E'_0 = E'_0(acceptor) - E'_0(donor)$

EXAMPLE:

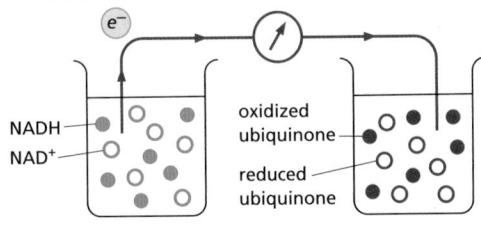

NADH
NAD⁺

oxidized ubiquinone

reduced ubiquinone

1:1 mixture of NADH and NAD⁺

1:1 mixture of oxidized and reduced ubiquinone

For the transfer of one electron from NADH to ubiquinone:

$$\Delta E'_0 = +30 - (-320) = +350 \text{ mV}$$

$\Delta G° = -n(0.096)\Delta E'_0 = -1(0.096)(350) = -34$ kJ/mole

The same calculation reveals that the transfer of one electron from ubiquinone to oxygen has an even more favorable $\Delta G°$ of −76 kJ/mole. The $\Delta G°$ value for the transfer of one electron from NADH to oxygen is the sum of these two values, −110 kJ/mole.

EFFECT OF CONCENTRATION CHANGES

As explained in Chapter 2 (see p. 60), the actual free-energy change for a reaction, ΔG, depends on the concentration of the reactants and generally will be different from the standard free-energy change, $\Delta G°$. The standard redox potentials are for a 1:1 mixture of the redox pair. For example, the standard redox potential of −320 mV is for a 1:1 mixture of NADH and NAD^+. But when there is an excess of NADH over NAD^+, electron transfer from NADH to an electron acceptor becomes more favorable. This is reflected by a more negative redox potential and a more negative ΔG for electron transfer.

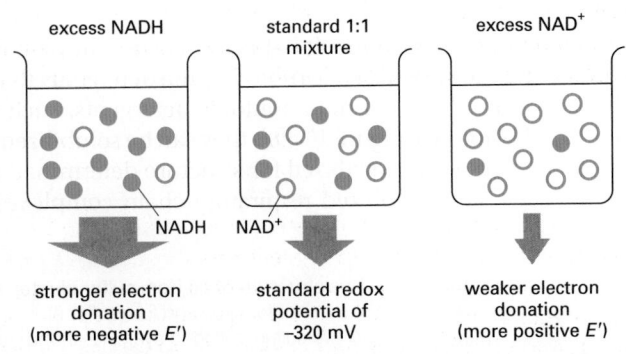

excess NADH

standard 1:1 mixture

excess NAD⁺

NADH   NAD⁺

stronger electron donation (more negative E')

standard redox potential of −320 mV

weaker electron donation (more positive E')

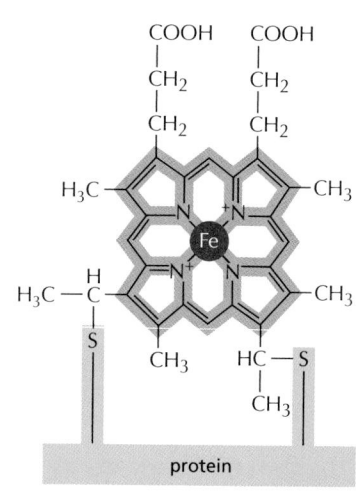

Figure 14–15 **The structure of the heme group attached covalently to cytochrome c.** The porphyrin ring of the heme is shown in *red*. There are six different cytochromes in the respiratory chain. Because the hemes in different cytochromes have slightly different structures and are kept in different local environments by their respective proteins, each has a different affinity for an electron, and a slightly different spectroscopic signature.

exploited by the membrane protein complexes in the respiratory chain to move electrons both within and between complexes.

Unlike the colorless atoms H, C, N, and O that constitute the bulk of biological molecules, transition metal ions are often brightly colored, which makes the proteins that contain them easy to study by spectroscopic methods using visible light. One family of such colored proteins, the **cytochromes**, contains a bound *heme group*, in which an iron atom is tightly held by four nitrogen atoms at the corners of a square in a *porphyrin ring* (**Figure 14–15**). Similar porphyrin rings are responsible both for the red color of blood and for the green color of leaves, binding an iron in hemoglobin or a magnesium in chlorophyll, respectively.

Iron–sulfur proteins contain a second major family of electron-transfer cofactors. In this case, either two or four iron atoms are bound to an equal number of sulfur atoms and to cysteine side chains, forming **iron–sulfur clusters** in the protein (**Figure 14–16**). Like the cytochrome hemes, these clusters carry one electron at a time.

The simplest of the electron-transfer cofactors in the respiratory chain—and the only one that is not always bound to a protein—is a quinone (called *ubiquinone*, or *coenzyme Q*). A **quinone (Q)** is a small hydrophobic molecule that is freely mobile in the lipid bilayer. This *electron carrier* can accept or donate either one or two electrons. Upon reduction (note that reduced quinones are called quinols), it picks up a proton from water along with each electron (**Figure 14–17**).

In the mitochondrial electron-transport chain, six different cytochrome hemes, eight iron–sulfur clusters, three copper atoms, a flavin mononucleotide (another electron-transfer cofactor), and ubiquinone work in a defined sequence to carry electrons from NADH to O_2. In total, this pathway involves more than 60 different polypeptides arranged in three large membrane protein complexes, each of which binds several of the above electron-carrying cofactors.

As we would expect, the electron-transfer cofactors have increasing affinities for electrons (higher redox potentials) as the electrons move along the respiratory chain. The redox potentials have been fine-tuned during evolution by the protein environment of each cofactor, which alters the cofactor's normal affinity for electrons. Because iron–sulfur clusters have a relatively low affinity for electrons, they predominate in the first half of the respiratory chain; in contrast, the heme cytochromes predominate further down the chain, where a higher electron affinity is required.

NADH Transfers Its Electrons to Oxygen Through Three Large Enzyme Complexes Embedded in the Inner Membrane

Membrane proteins are difficult to purify because they are insoluble in aqueous solutions, and they are easily disrupted by the detergents that are required to solubilize them. But by using mild nonionic detergents, such as octylglucoside or dodecyl maltoside (see Figure 10–28), they can be solubilized and purified in their native form, and even crystallized for structure determination. Each of the three different detergent-solubilized respiratory-chain complexes can be re-inserted

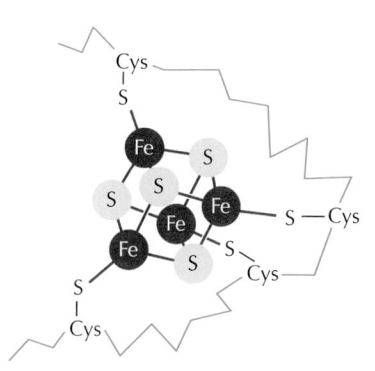

Figure 14–16 **The structure of an iron–sulfur cluster.** These dark brown clusters consist either of four iron and four sulfur atoms, as shown here, or of two irons and two sulfurs linked to cysteines in the polypeptide chain via covalent sulfur bridges, or to histidines. Although they contain several iron atoms, each iron–sulfur cluster can carry only one electron at a time. Nine different iron–sulfur clusters participate in electron transport in the respiratory chain.

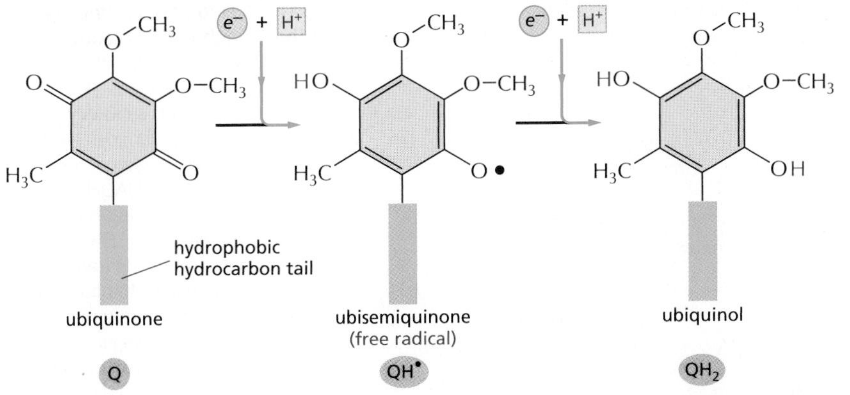

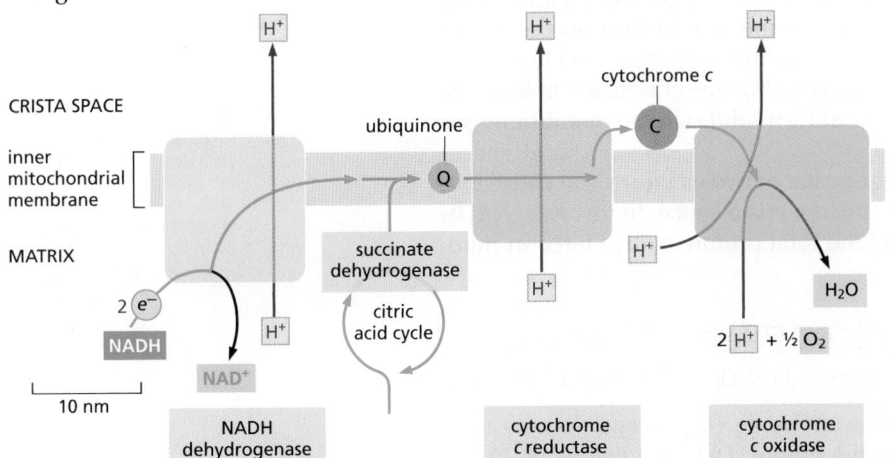

hydrophobic
hydrocarbon tail

ubiquinone   ubisemiquinone   ubiquinol
                              (free radical)

Q   QH•   QH₂

Figure 14–17 Quinone electron carriers. Ubiquinone in the lipid bilayer picks up one H⁺ *(red)* from the aqueous environment for each electron *(blue)* it accepts, in two steps, from respiratory-chain complexes. The first step involves the acquisition of a proton and an electron and converts the ubiquinone into an unstable ubisemiquinone radical. In the second step, it becomes a fully reduced ubiquinone (called ubiquinol), which is freely mobile as an electron carrier in the lipid bilayer of the membrane. When the ubiquinol donates its electrons to the next complex in the chain, the two protons are released. The long hydrophobic tail *(green)* that confines ubiquinone to the membrane consists of 6–10 five-carbon isoprene units, depending on the organism. The corresponding electron carrier in the photosynthetic membranes of chloroplasts is plastoquinone, which has almost the same structure and works in the same way. For simplicity, we refer to both ubiquinone and plastoquinone in this chapter as quinone (abbreviated as Q).

into artificial lipid bilayer vesicles and shown to pump protons across the membrane as electrons pass through them.

In the mitochondrion, the three complexes are linked in series, serving as electron-transport-driven H⁺ pumps that pump protons out of the matrix to acidify the crista space (**Figure 14–18**):

1. The **NADH dehydrogenase complex** (often referred to as *Complex I*) is the largest of these respiratory enzyme complexes. It accepts electrons from NADH and passes them through a flavin mononucleotide and eight iron–sulfur clusters to the lipid-soluble electron carrier ubiquinone. The reduced ubiquinol then transfers its electrons to cytochrome *c* reductase.

2. The **cytochrome *c* reductase** (also called the *cytochrome b-c₁ complex*) is a large membrane protein assembly that functions as a dimer. Each monomer contains three cytochrome hemes and an iron–sulfur cluster. The complex accepts electrons from ubiquinol and passes them on to the small, soluble protein cytochrome *c*, which is located in the crista space and carries electrons one at a time to cytochrome *c* oxidase.

3. The **cytochrome *c* oxidase complex** contains two cytochrome hemes and three copper atoms. The complex accepts electrons one at a time from cytochrome *c* and passes them to molecular oxygen. In total, four electrons and four protons are needed to convert one molecule of oxygen to water.

We have previously discussed how the redox potential reflects electron affinities. **Figure 14–19** presents an outline of the redox potentials measured along the respiratory chain. These potentials change in three large steps, one across each proton-translocating respiratory complex. The change in redox potential between any two electron carriers is directly proportional to the free energy released when an electron transfers between them. Each complex acts as an energy-conversion device by harnessing some of this free-energy change to pump H⁺ across the inner membrane, thereby creating an electrochemical proton gradient as electrons pass along the chain.

Figure 14–18 The path of electrons through the three respiratory-chain proton pumps. (Movie 14.3) The approximate size and shape of each complex is shown. During the transfer of electrons from NADH to oxygen *(blue arrows)*, ubiquinone and cytochrome *c* serve as mobile carriers that ferry electrons from one complex to the next. During the electron-transfer reactions, protons are pumped across the membrane by each of the respiratory enzyme complexes, as indicated *(red arrows)*.

For historical reasons, the three proton pumps in the respiratory chain are sometimes denoted as Complex I, Complex III, and Complex IV, according to the order in which electrons pass through them from NADH. Electrons from the oxidation of succinate by succinate dehydrogenase (designated as Complex II) are fed into the electron-transport chain in the form of reduced ubiquinone. Although embedded in the crista membrane, succinate dehydrogenase does not pump protons and thus does not contribute to the proton-motive force; it is therefore not considered to be an integral part of the respiratory chain.

CRISTA SPACE

inner
mitochondrial
membrane

MATRIX

2 e⁻

NADH

NAD⁺

10 nm

NADH
dehydrogenase

H⁺

ubiquinone

Q

succinate
dehydrogenase

citric
acid cycle

H⁺

H⁺

H⁺

cytochrome c

C

cytochrome
c reductase

H⁺

2 H⁺ + ½ O₂

H⁺

H₂O

cytochrome
c oxidase

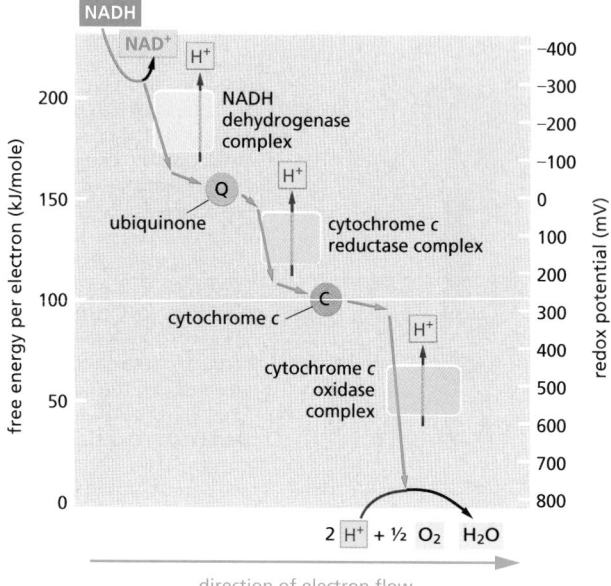

Figure 14–19 Redox potential changes along the mitochondrial electron-transport chain. The redox potential (designated E'_0) increases as electrons flow down the respiratory chain to oxygen. The standard free-energy change in kilojoules, $\Delta G°$, for the transfer of each of the two electrons donated by an NADH molecule can be obtained from the left-hand ordinate [$\Delta G° = -n(0.096) \Delta E'_0$, where n is the number of electrons transferred across a redox potential change of $\Delta E'_0$ mV]. Electrons flow through a respiratory enzyme complex by passing in sequence through the multiple electron carriers in each complex *(blue arrows)*. As indicated, part of the favorable free-energy change is harnessed by each enzyme complex to pump H^+ across the inner mitochondrial membrane *(red arrows)*. The NADH dehydrogenase pumps up to four H^+ per electron, the cytochrome c reductase complex pumps two, whereas the cytochrome c oxidase complex pumps one per electron.

Note that NADH is not the only source of electrons for the respiratory chain. The flavin $FADH_2$, which is generated by fatty acid oxidation (see Figure 2–56) and by succinate dehydrogenase in the citric acid cycle (see Figure 2–57), also contributes. Its two electrons are passed directly to ubiquinone, bypassing NADH dehydrogenase.

X-ray crystallography has elucidated the structure of each of the three respiratory-chain complexes in great detail, and we next examine each of them in turn to see how they work.

The NADH Dehydrogenase Complex Contains Separate Modules for Electron Transport and Proton Pumping

The NADH dehydrogenase complex is a massive assembly of membrane and nonmembrane proteins that receives electrons from NADH and passes them to ubiquinone. In animal mitochondria, it consists of more than 40 different protein subunits, with a molecular mass of nearly a million daltons. The x-ray structures of the NADH dehydrogenase complex from fungi and bacteria show that it is L-shaped, with both a hydrophobic membrane arm and a hydrophilic arm that projects into the mitochondrial matrix (**Figure 14–20**).

Electron transfer and proton pumping are physically separated in the NADH dehydrogenase complex, with electron transfer occurring in the matrix arm and proton pumping in the membrane arm. The NADH docks near the tip of the matrix arm, where it transfers its electrons via a bound flavin mononucleotide to a string of iron–sulfur clusters that runs down the arm, acting like a wire to carry electrons to a protein-bound molecule of ubiquinone. Electron transfer to the quinone is thought to trigger proton translocation in a set of proton pumps in the membrane arm, and for this to happen the two processes must be energetically and mechanically linked. A mechanical link is thought to be provided by a 6-nm long, amphipathic α helix that runs parallel to the membrane surface on the matrix side of the membrane arm. This helix may act like the connecting rod in a steam engine to generate a mechanical, energy-transducing power stroke that links the quinone-binding site to the proton-translocating modules in the membrane (see Figure 14–20).

The reduction of each quinone by the transfer of two electrons can cause four protons to be pumped out of the matrix into the crista space. In this way, NADH dehydrogenase generates roughly half of the total proton-motive force in mitochondria.

Cytochrome c Reductase Takes Up and Releases Protons on the Opposite Side of the Crista Membrane, Thereby Pumping Protons

As described previously, when a quinone molecule (Q) accepts its two electrons, it also takes up two protons to form a quinol (QH_2; see Figure 14–17). In

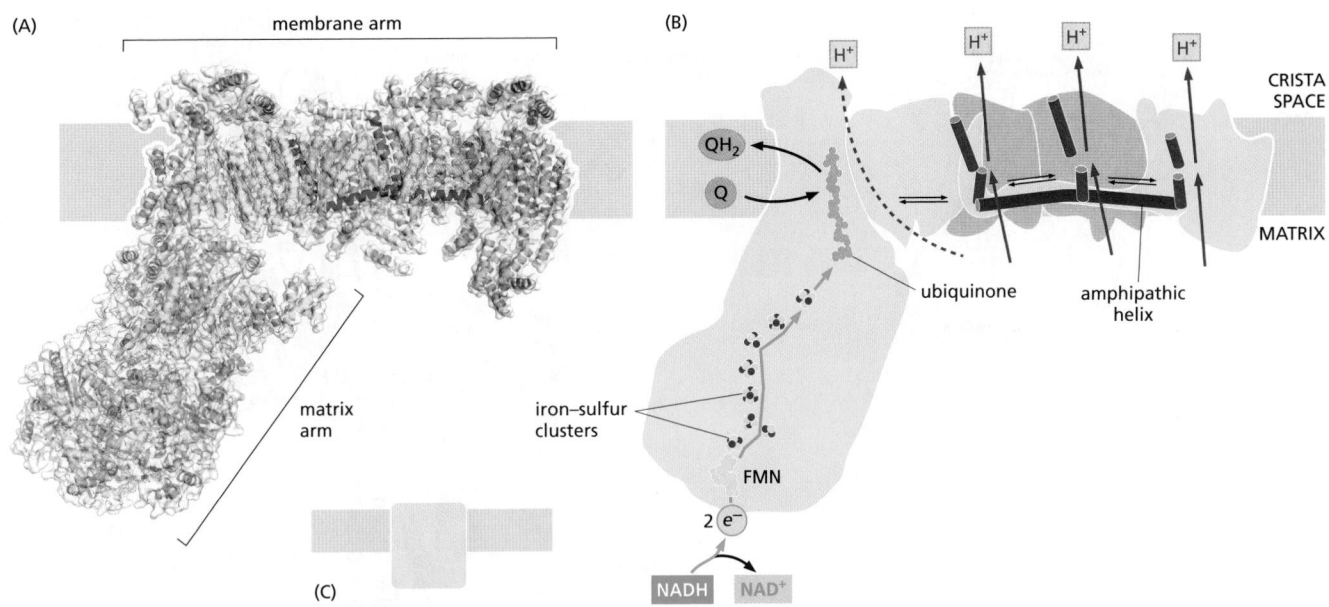

Figure 14–20 **The structure of NADH dehydrogenase.** (A) The model of the mitochondrial complex shown here is based on the x-ray structure of the smaller bacterial complex, which works in the same way. The matrix arm of NADH dehydrogenase (also known as Complex I) contains eight iron–sulfur (FeS) clusters that appear to participate in electron transport. The membrane contains more than 70 transmembrane helices, forming three distinct proton-pumping modules, while the matrix arm contains the electron-transport cofactors. (B) NADH donates two electrons, via a bound flavin mononucleotide (FMN; *yellow*), to a chain of seven iron–sulfur clusters *(red and yellow spheres)*. From the terminal iron–sulfur cluster, the electrons pass to ubiquinone *(orange)*. Electron transfer results in conformational changes *(black arrows)* that are thought to be transmitted to a long amphipathic α helix *(purple)* on the matrix side of the membrane arm, which pulls on discontinuous transmembrane helices *(red)* in three membrane subunits, each of which resembles an antiporter (see Chapter 11). This movement is thought to change the conformation of charged residues in the three proton channels, resulting in the translocation of three protons out of the matrix. A fourth proton may be translocated at the interface of the two arms *(dotted line)*. (C) This shows the symbol for NADH dehydrogenase used throughout this chapter. (Adapted from R.G. Efremov, R. Baradaran and L.A. Sazanov, *Nature* 465:441–445, 2010. PDB code: 3M9S.)

the respiratory chain, ubiquinol tranfers electrons from NADH dehydrogenase to cytochrome *c* reductase. Because the protons in this QH_2 molecule are taken up from the matrix and released on the opposite side of the crista membrane, two protons are transferred from the matrix into the crista space per pair of electrons transferred (**Figure 14–21**). This vectorial transfer of protons supplements the electrochemical proton gradient that is created by the NADH dehydrogenase proton pumping just discussed.

Cytochrome *c* reductase is a large assembly of membrane protein subunits. Three subunits form a catalytic core that passes electrons from ubiquinol to cytochrome *c*, with a structure that has been highly conserved from bacterial ancestors (**Figure 14–22**). It pumps protons by a vectorial transfer of protons that involves a binding site for a second molecule of ubiquinone; the elaborate

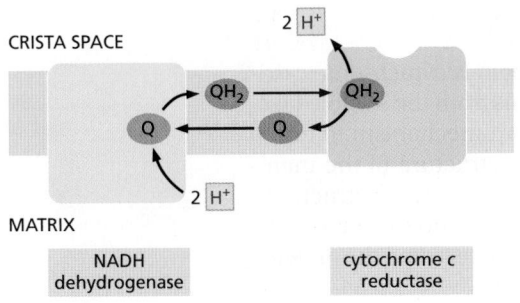

Figure 14–21 **How a directional release and uptake of protons by a quinone pumps protons across a membrane.** Two protons are picked up on the matrix side of the inner mitochondrial membrane when the reaction $Q + 2e^- + 2H^+ \rightarrow QH_2$ is catalyzed by the NADH dehydrogenase complex. This molecule of ubiquinol (QH_2) diffuses rapidly in the plane of the membrane, becoming bound to the crista side of cytochrome *c* reductase. When its oxidation by cytochrome *c* reductase generates two protons and two electrons (see Figure 14–17), the two protons are released into the crista space. The flow of electrons is not shown in this diagram.

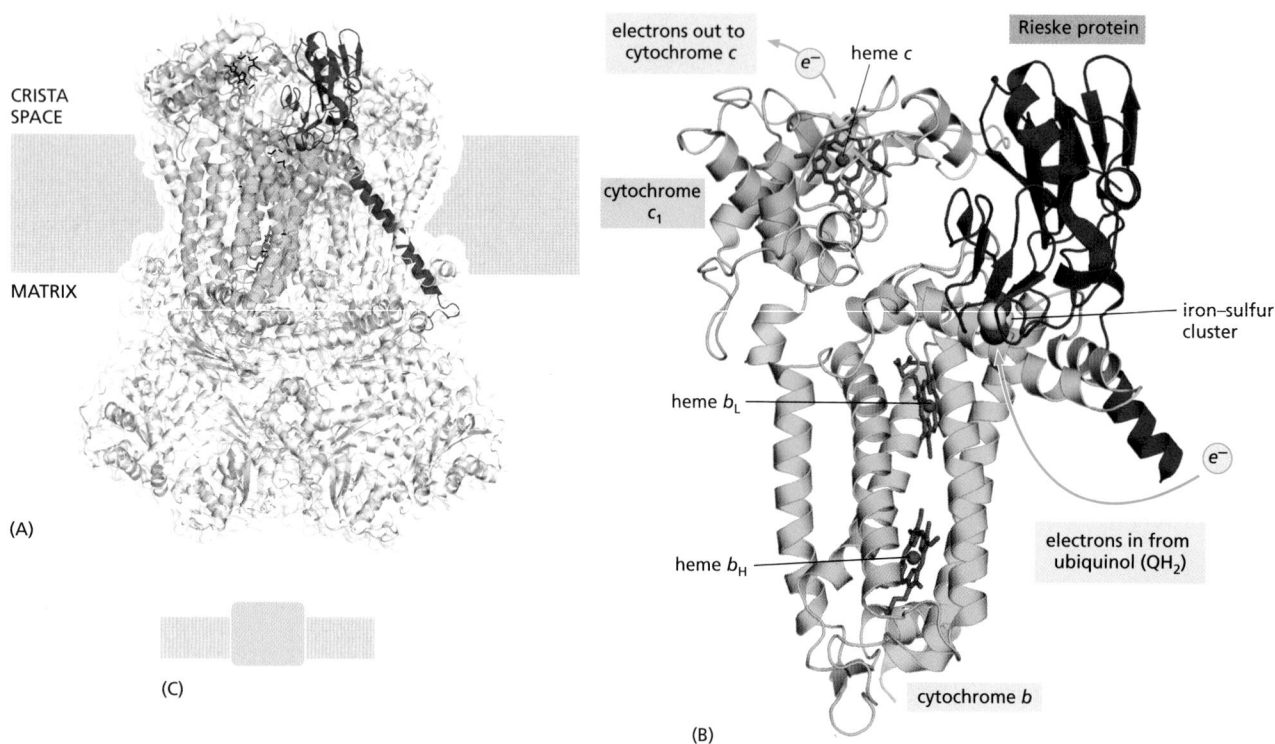

Figure 14–22 The structure of cytochrome c reductase. Cytochrome c reductase (also known as the cytochrome b-c_1 complex) is a dimer of two identical 240,000-dalton halves, each composed of 11 different protein molecules in mammals. (A) A structure graphic of the entire dimer, showing in color the three proteins that form the functional core of the enzyme complex: cytochrome b *(green)* and cytochrome c_1 *(blue)* are colored in one half, and the Rieske protein *(purple)* containing an Fe_2S_2 iron–sulfur cluster *(red and yellow)* is colored in the other. These three protein subunits interact across the two halves. (B) Transfer of electrons through cytochrome c reductase to the small, soluble carrier protein cytochrome c. Electrons entering from ubiquinol near the matrix side of the membrane are captured by the iron–sulfur cluster of the Rieske protein, which moves its iron–sulfur group back and forth to transfer these electrons to heme c *(red)*. Heme c then transfers them to the carrier molecule cytochrome c.

As detailed in Figure 14–23, only one of the two electrons from each ubiquinol is transferred through this path. To increase proton pumping, the second ubiquinol electron is passed to a molecule of ubiquinone bound to cytochrome c reductase on the opposite side of the membrane—near the matrix. (C) This shows the symbol for cytochrome c reductase used throughout this chapter. (PDB code: 1EZV.)

redox loop mechanism used is called the *Q cycle* because while one of the electrons received from each QH_2 molecule is transferred from ubiquinone through the complex to the carrier protein cytochrome c, the other electron is recycled back into the quinone pool. Through the mechanism illustrated in **Figure 14–23**, the Q cycle increases the total amount of redox energy that can be stored in the electrochemical proton gradient. As a result, for every electron that is transferred from NADH dehydrogenase to cytochrome c, two protons are pumped across the crista membrane into the crista space.

The Cytochrome c Oxidase Complex Pumps Protons and Reduces O₂ Using a Catalytic Iron–Copper Center

The final link in the mitochondrial electron-transport chain is cytochrome c oxidase. The cytochrome c oxidase complex accepts electrons from the soluble electron carrier cytochrome c, and it uses yet a different, third mechanism to pump protons across the inner mitochondrial membrane. The structure of the mammalian complex is illustrated in **Figure 14–24**. The atomic-resolution structures, combined with studies of the effect of mutations introduced into the enzyme by genetic engineering of the yeast and bacterial proteins, have revealed the detailed mechanisms of this electron-driven proton pump.

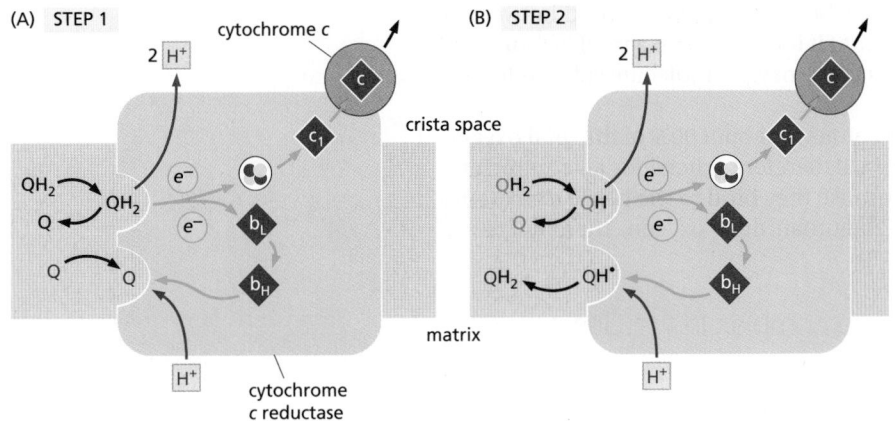

Figure 14–23 **The two-step mechanism of the cytochrome *c* reductase Q-cycle.**
(A) In step 1, ubiquinol reduced by NADH dehydrogenase docks to the cytochrome *c* reductase complex. Oxidation of the quinol produces two protons and two electrons. The protons are released into the crista space. One electron passes via an iron-sulfur cluster to heme c_1, and then to the soluble electron carrier protein cytochrome *c* on the membrane surface. The second electron passes via hemes b_L and b_H to a ubiquinone (*red* Q) bound at a separate site near the matrix side of the protein. Uptake of a proton from the matrix produces an ubisemiquinone radical (see Figure 14–17), which remains bound to this site (*red* QH• in B).

(B) In step 2, a second ubiquinol (*blue* QH_2) docks and releases two protons and two electrons, as described for step 1. One electron is passed to a second molecule of cytochrome *c*, whereas the other electron is accepted by the ubisemiquinone. The ubisemiquinone then takes up a proton from the matrix and is released into the lipid bilayer as fully reduced ubiquinol (*red* QH_2).

On balance, the oxidation of one ubiquinol in the Q cycle pumps two protons through the membrane by a directional release and uptake of protons (see Figure 14–21), while releasing another two into the crista space. In addition, in each of the two steps (A) and (B), one electron is transferred to a cytochrome *c* carrier protein (Movie 14.4).

Because oxygen has a high affinity for electrons, it can release a large amount of free energy when it is reduced to form water. Thus, the evolution of cellular respiration, in which O_2 is converted to water, enabled organisms to harness much more energy than can be derived from anaerobic metabolism. As we discuss later, the availability of the large amount of energy released by the reduction of molecular oxygen to form water is thought to have been essential to the emergence of multicellular life: this would explain why all large organisms respire. The ability of biological systems to use O_2 in this way, however, requires sophisticated chemistry. Once a molecule of O_2 has picked up one electron, it forms a superoxide radical anion ($O_2^{\bullet-}$) that is dangerously reactive and rapidly takes up an additional three electrons wherever it can get them, with destructive effects on its immediate environment. We can tolerate oxygen in the air we breathe only because the uptake of the first electron by the O_2 molecule is slow, allowing cells to use enzymes to control electron uptake by oxygen. Thus, cytochrome *c* oxidase holds

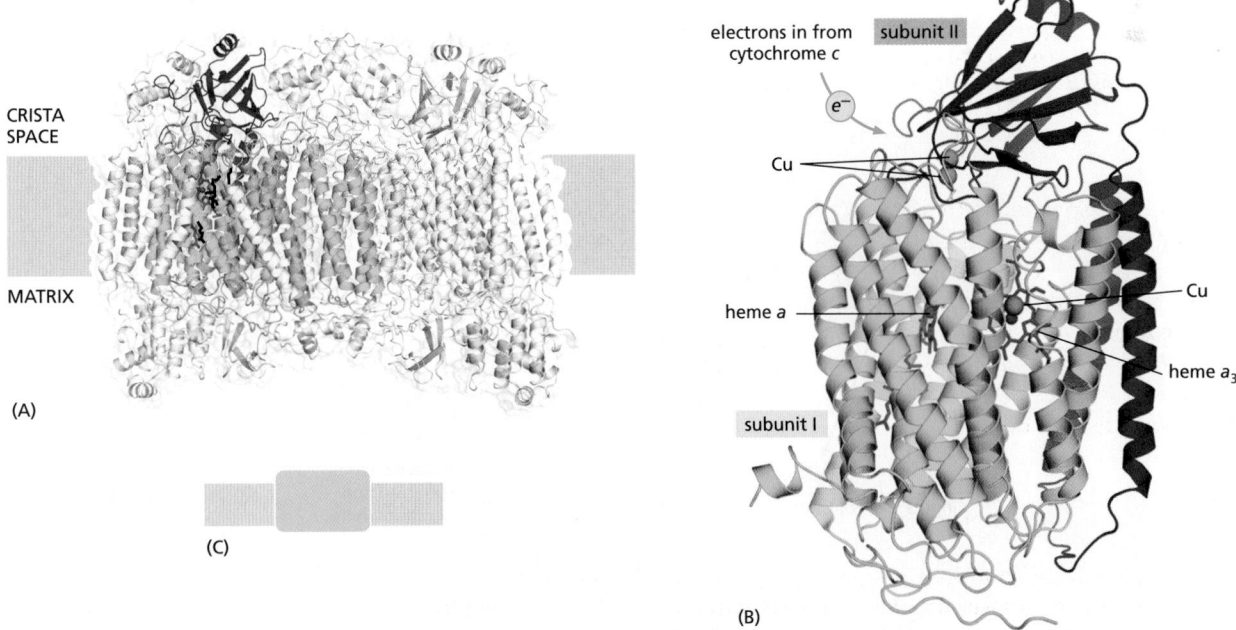

Figure 14–24 **The structure of cytochrome *c* oxidase.** The final complex in the mitochondrial electron-transfer chain consists of 13 different protein subunits, with a total mass of 204,000 daltons. (A) The entire dimeric complex is shown, positioned in the crista membrane. The highly conserved subunits I (*green*), II (*purple*), and III (*blue*) are encoded by the mitochondrial genome, and they form the functional core of the enzyme. (B) The functional core of the complex. Electrons pass through this structure from cytochrome *c* via bound copper ions (*blue spheres*) and hemes (*red*) to an O_2 molecule bound between heme a_3 and a copper ion. The four protons needed to reduce O_2 to water are taken up from the matrix; see also Figure 14–25. (C) This shows the symbol for cytochrome *c* oxidase used throughout this chapter. (PDB code: 2OCC.)

on to oxygen at a special bimetallic center, where it remains clamped between a heme-linked iron atom and a copper ion until it has picked up a total of four electrons. Only then are the two oxygen atoms of the oxygen molecule safely released as two molecules of water (**Figure 14–25**).

The cytochrome *c* oxidase reaction accounts for about 90% of the total oxygen uptake in most cells. This protein complex is therefore crucial for all aerobic life. Cyanide and azide are extremely toxic because they bind to the heme iron atoms in cytochrome *c* oxidase much more tightly than does oxygen, thereby greatly reducing ATP production.

The Respiratory Chain Forms a Supercomplex in the Crista Membrane

By using cryoelectron microscopy to examine proteins that have been very gently isolated, it can be shown that the three protein complexes that form the respiratory chain assemble into an even larger *supercomplex* in the crista membrane. As illustrated in **Figure 14–26**, this structure is thought to help the mobile electron carriers ubiquinone (in the crista membrane) and cytochrome *c* (in the crista space) transfer electrons with high efficiency. The formation of the supercomplex depends on the presence of the mitochondrial lipid cardiolipin (see Figure 14–11), which presumably works like a hydrophobic glue that holds the components together.

In addition to the three proton pumps in the supercomplex just discussed, one of the enzymes in the citric acid cycle, *succinate dehydrogenase*, is embedded in the mitochondrial crista membrane. In the course of oxidizing succinate to fumarate in the matrix, this enzyme complex captures electrons in the form of a tightly bound $FADH_2$ molecule (see Panel 2–9, pp. 106–107) and passes them to

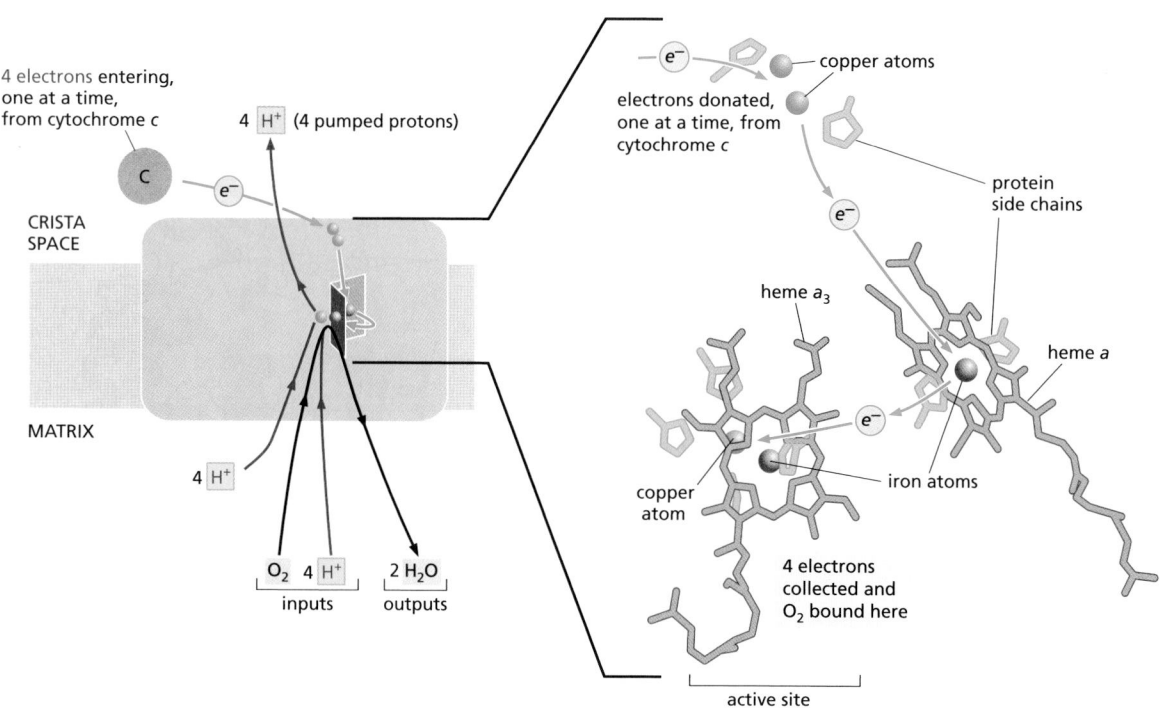

Figure 14–25 The reaction of O₂ with electrons in cytochrome *c* oxidase. Electrons from cytochrome *c* pass through the complex via bound copper ions (*blue spheres*) and hemes *(red)* to an O_2 molecule bound between heme a_3 and a copper ion. Iron ions are shown as *red spheres*. The iron atom in heme *a* serves as an electron queuing point where electrons are held so that they can be released to an O_2 molecule (not shown) that is held at the bimetallic center active site, which is formed by the central iron of the other heme (heme a_3) and a closely apposed copper atom. The four protons needed to reduce O_2 to water are removed from the matrix. For each O_2 molecule that undergoes the reaction $4e^- + 4H^+ + O_2 \rightarrow 2H_2O$, another four protons are pumped out of the matrix by mechanisms that are driven by allosteric changes in protein conformation (see Figure 14–28).

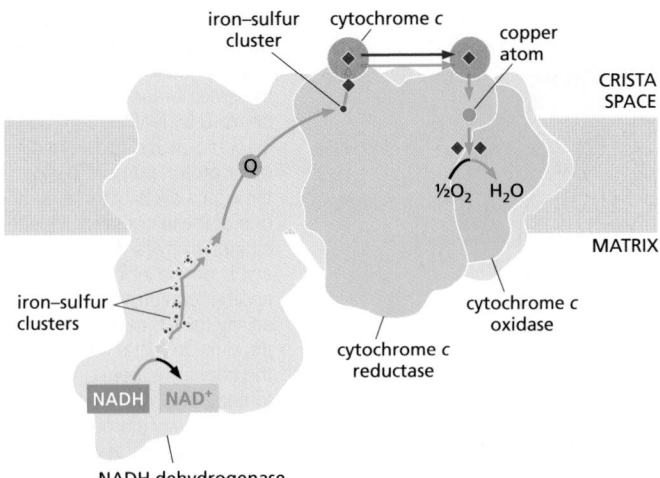

Figure 14–26 **The respiratory-chain supercomplex from bovine heart mitochondria.** The three proton-pumping complexes of the mitochondrial respiratory chain of mammalian mitochondria assemble into large supercomplexes in the crista membrane. Supercomplexes can be isolated by mild detergent treatment of mitochondria, and their structure has been deciphered by single-particle cryoelectron microscopy. The bovine heart supercomplex has a total mass of 1.7 megadaltons. Shown is a schematic of such a complex that consists of NADH dehydrogenase, cytochrome *c* reductase, and cytochrome *c* oxidase, as indicated. The facing quinol-binding sites of NADH dehydrogenase and cytochrome *c* reductase, plus the short distance between the cytochrome *c*-binding sites in cytochrome *c* reductase and cytochrome *c* oxidase, facilitate fast, efficient electron transfer. Cofactors active in electron transport are marked as a *yellow dot* (flavin mononucleotide), *red and yellow dots* (iron–sulfur clusters), Q (quinone), *red squares* (hemes), and a *blue dot* (copper atom). Only cofactors participating in the linear flow of electrons from NADH to water are shown. *Blue arrows* indicate the path of the electrons through the supercomplex. (Adapted from T. Athoff et al., *EMBO J.* 30:4652–4664, 2011.)

a molecule of ubiquinone. The reduced ubiquinol then passes its two electrons to the respiratory chain via cytochrome *c* reductase (see Figure 14–18). Succinate dehydrogenase is not a proton pump, and it does not contribute directly to the electrochemical potential utilized for ATP production in mitochondria. Thus, it is not considered to be an integral part of the respiratory chain.

Protons Can Move Rapidly Through Proteins Along Predefined Pathways

The protons in water are highly mobile: by rapidly dissociating from one water molecule and associating with its neighbor, they can rapidly flit through a hydrogen-bonded network of water molecules (see Figure 2–5). But how can a proton move through the hydrophobic interior of a protein embedded in the lipid bilayer? Proton-translocating proteins contain so-called *proton wires*, which are rows of polar or ionic side chains, or water molecules spaced at short distances, so that the protons can jump from one to the next (**Figure 14–27**). Along such predefined pathways, protons move up to 40 times faster than through bulk water. The three-dimensional structure of cytochrome *c* oxidase indicates two different proton-uptake pathways. This confirmed earlier mutagenesis studies, which had shown that replacing the side chains of particular aspartate or arginine residues, whose side chains can bind and release protons, made the cytochrome *c* oxidase less efficient as a proton pump.

But how can electron transport cause allosteric changes in protein conformations that pump protons? From the most basic point of view, if electron transport drives sequential allosteric changes in protein conformation that alter the redox state of the components, these conformational changes can be connected to protein wires that allow the protein to pump H⁺ across the crista membrane. This type of H⁺ pumping requires at least three distinct conformations for the pump protein, as schematically illustrated in **Figure 14–28**.

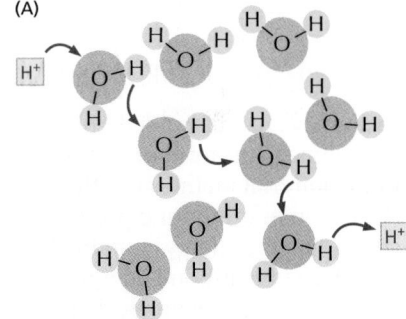

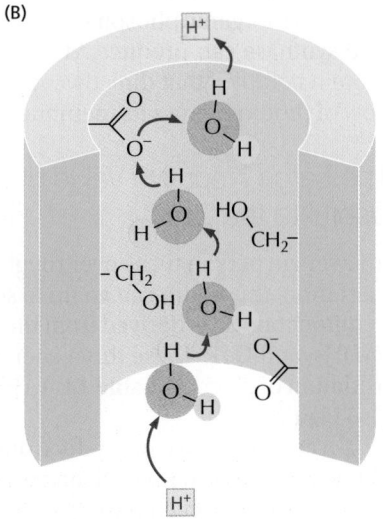

Figure 14–27 **Proton movement through water and proteins.** (A) Protons move rapidly through water, hopping from one H_2O molecule to the next by the continuous formation and dissociation of hydronium ions, H_3O^+ (see Chapter 2). In this diagram, proton jumps are indicated by *red arrows*. (B) Protons can move even more rapidly through a protein along "proton wires." These are predefined proton paths consisting of suitably spaced amino acid side chains that accept and release protons easily (Asp, Glu) or carry a waterlike hydroxyl group (Ser, Thr), along with water molecules trapped in the protein interior.

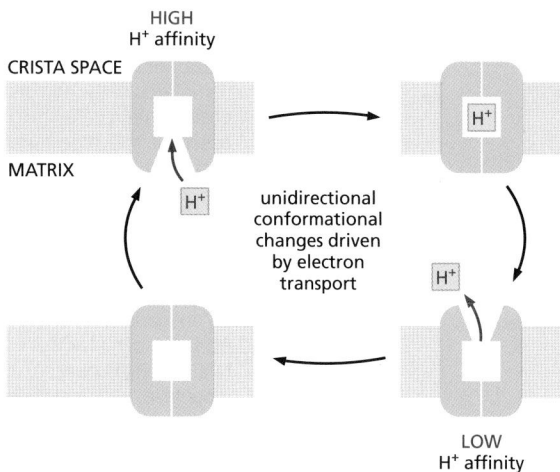

HIGH
H$^+$ affinity

CRISTA SPACE

MATRIX

H$^+$

unidirectional
conformational
changes driven
by electron
transport

H$^+$

H$^+$

LOW
H$^+$ affinity

Figure 14–28 A general model for H$^+$ pumping coupled to electron transport. This mechanism for H$^+$ pumping by a transmembrane protein is thought to be used by NADH dehydrogenase and cytochrome *c* oxidase, and by many other proton pumps. The protein is driven through a cycle of three conformations. In one of these conformations, the protein has a high affinity for H$^+$, causing it to pick up an H$^+$ on the inside of the membrane. In another conformation, the protein has a low affinity for H$^+$, causing it to release an H$^+$ on the outside of the membrane. As indicated, the transitions from one conformation to another occur only in one direction, because they are being driven by being allosterically coupled to the energetically favorable process of electron transport (discussed in Chapter 11).

Summary

The respiratory chain embedded in the inner mitochondrial membrane contains three respiratory enzyme complexes, through which electrons pass on their way from NADH to O$_2$. In these complexes, electrons are transferred along a series of protein-bound electron carriers, including hemes and iron–sulfur clusters. The energy released as the electrons move to lower and lower energy levels is used to pump protons by different mechanisms in the three respiratory enzyme complexes, each coupling lateral electron transport to vectorial proton transport across the membrane. Electrons are shuttled between enzyme complexes by the mobile electron carriers ubiquinone and cytochrome c *to complete the electron-transport chain. The path of electron flow is NADH → NADH dehydrogenase complex → ubiquinone → cytochrome* c *reductase → cytochrome* c *→ cytochrome* c *oxidase complex → molecular oxygen (O$_2$).*

ATP PRODUCTION IN MITOCHONDRIA

As we have just discussed, the three proton pumps of the respiratory chain each contribute to the formation of an electrochemical proton gradient across the inner mitochondrial membrane. This gradient drives ATP synthesis by ATP synthase, a large membrane-bound protein complex that performs the extraordinary feat of converting the energy contained in this electrochemical gradient into biologically useful, chemical-bond energy in the form of ATP (see Figure 14–10). Protons flow down their electrochemical gradient through the membrane part of this proton turbine, thereby driving the synthesis of ATP from ADP and P$_i$ in the extramembranous part of the complex. As discussed in Chapter 2, the formation of ATP from ADP and inorganic phosphate is highly unfavorable energetically. As we shall see, ATP synthase can produce ATP only because of allosteric shape changes in this protein complex that directly couple ATP synthesis to the energetically favorable flow of protons across its membrane.

The Large Negative Value of ΔG for ATP Hydrolysis Makes ATP Useful to the Cell

An average person turns over roughly 50 kg of ATP per day. In athletes running a marathon, this figure can go up to several hundred kilograms. The ATP produced in mitochondria is derived from the energy available in the intermediates NADH, FADH$_2$, and GTP. These three energy-rich compounds are produced both by the oxidation of glucose (**Table 14–1A**), and by the oxidation of fats (Table 14–1B; see also Figure 2–56).

Glycolysis alone can produce only two molecules of ATP for every molecule of glucose that is metabolized, and this is the total energy yield for the fermentation processes that occur in the absence of O$_2$ (discussed in Chapter 2). In oxidative

TABLE 14–1 Product Yields from the Oxidation of Sugars and Fats
A. Net products from oxidation of one molecule of glucose
In cytosol (glycolysis)   1 glucose → 2 pyruvate + 2 NADH + 2 ATP

In mitochondrion (pyruvate dehydrogenase and citric acid cycle)
    2 pyruvate → 2 acetyl CoA + 2 NADH
    2 acetyl CoA → 6 NADH + 2 FADH$_2$ + 2 GTP

Net result in mitochondrion   2 pyruvate → 8 NADH + 2 FADH$_2$ + 2 GTP
B. Net products from oxidation of one molecule of palmitoyl CoA (activated form of palmitate, a fatty acid)

In mitochondrion (fatty acid oxidation and citric acid cycle)
    1 palmitoyl CoA → 8 acetyl CoA + 7 NADH + 7 FADH$_2$
    8 acetyl CoA → 24 NADH + 8 FADH$_2$ + 8 GTP

Net result in mitochondrion   1 palmitoyl CoA → 31 NADH + 15 FADH$_2$ + 8 GTP

phosphorylation, each pair of electrons donated by the NADH produced in mitochondria can provide energy for the formation of about 2.5 molecules of ATP. Oxidative phosphorylation also produces 1.5 ATP molecules per electron pair from the FADH$_2$ produced by succinate dehydrogenase in the mitochondrial matrix, and from the NADH molecules produced by glycolysis in the cytosol. From the product yields of glycolysis and the citric acid cycle, we can calculate that the complete oxidation of one molecule of glucose—starting with glycolysis and ending with oxidative phosphorylation—gives a net yield of about 30 molecules of ATP. Nearly all this ATP is produced by the mitochondrial ATP synthase.

In Chapter 2, we introduced the concept of free energy (G). The free-energy change for a reaction, ΔG, determines whether that reaction will occur in a cell. We showed on pp. 60–63 that the ΔG for a given reaction can be written as the sum of two parts: the first, called the standard free-energy change, $\Delta G°$, depends only on the intrinsic characters of the reacting molecules; the second depends only on their concentrations. For the simple reaction A → B,

$$\Delta G = \Delta G° + RT \ln \frac{[B]}{[A]}$$

where [A] and [B] denote the concentrations of A and B, and ln is the natural logarithm. $\Delta G°$ is the standard reference value, which can be seen to be equal to the value of ΔG when the molar concentrations of A and B are equal (since $\ln 1 = 0$).

In Chapter 2, we discussed how the large, favorable free-energy change (large negative ΔG) for ATP hydrolysis is used, through coupled reactions, to drive many other chemical reactions in the cell that would otherwise not occur (see pp. 65–66). The ATP hydrolysis reaction produces two products, ADP and P$_i$; it is therefore of the type A → B + C, where, as demonstrated in **Figure 14–29**,

$$\Delta G = \Delta G° + RT \ln \frac{[B][C]}{[A]}$$

When ATP is hydrolyzed to ADP and P$_i$ under the conditions that normally exist in a cell, the free-energy change is roughly –46 to –54 kJ/mole (–11 to –13 kcal/mole). This extremely favorable ΔG depends on maintaining a high concentration of ATP compared with the concentrations of ADP and P$_i$. When ATP, ADP, and P$_i$ are all present at the same concentration of 1 mole/liter (so-called standard conditions), the ΔG for ATP hydrolysis drops to the standard free-energy change ($\Delta G°$), which is only –30.5 kJ/mole (–7.3 kcal/mole). At much lower concentrations of ATP relative to ADP and P$_i$, ΔG becomes zero. At this point, the rate at

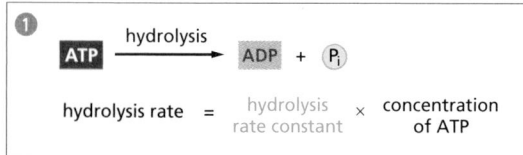

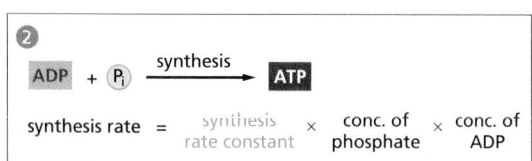

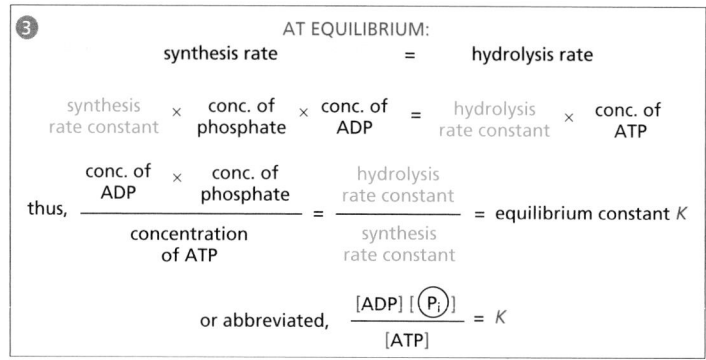

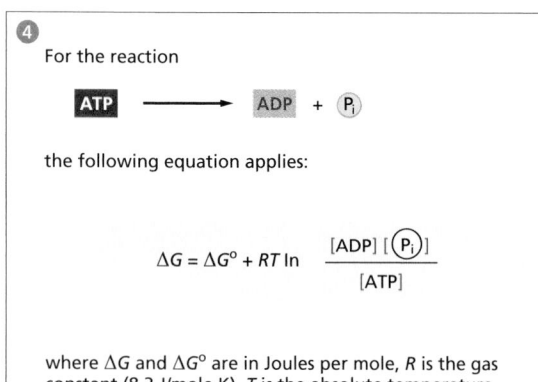

④

For the reaction

ATP ⟶ ADP + Pᵢ

the following equation applies:

$$\Delta G = \Delta G^\circ + RT \ln \frac{[ADP][P_i]}{[ATP]}$$

where ΔG and ΔG° are in Joules per mole, R is the gas constant (8.3 J/mole K), T is the absolute temperature (K), and all the concentrations are in moles per liter.
When the concentrations of all reactants are at 1 M, $\Delta G = \Delta G^\circ$ (since $RT \ln 1 = 0$). ΔG° is thus a constant defined as the standard free-energy change for the reaction.

At equilibrium the reaction has no net effect on the disorder of the universe, so $\Delta G = 0$. Therefore, at equilibrium,

$$-RT \ln \frac{[ADP][P_i]}{[ATP]} = \Delta G^\circ$$

But the concentrations of reactants at equilibrium must satisfy the equilibrium equation:

$$\frac{[ADP][P_i]}{[ATP]} = K$$

Therefore, at equilibrium,

$$\Delta G^\circ = -RT \ln K$$

We thus see that whereas ΔG° indicates the equilibrium point for a reaction, ΔG reveals *how far* the reaction is from equilibrium. ΔG is a measure of the "driving force" for any chemical reaction, just as the proton-motive force is the driving force for the translocation of protons.

which ADP and Pᵢ will join to form ATP will be equal to the rate at which ATP hydrolyzes to form ADP and Pᵢ. In other words, when $\Delta G = 0$, the reaction is at *equilibrium* (see Figure 14–29).

It is ΔG, not ΔG°, that indicates how far a reaction is from equilibrium and determines whether it can drive other reactions. Because the efficient conversion of ADP to ATP in mitochondria maintains such a high concentration of ATP relative to ADP and Pᵢ, the ATP hydrolysis reaction in cells is kept very far from equilibrium and ΔG is correspondingly very negative. Without this large disequilibrium, ATP hydrolysis could not be used to drive the reactions of the cell. At low ATP concentrations, many biosynthetic reactions would run backward and the cell would die.

Figure 14–29 The basic relationship between free-energy changes and equilibrium in the ATP hydrolysis reaction. The rate constants in boxes 1 and 2 are determined from experiments in which product accumulation is measured as a function of time (conc., concentration). The equilibrium constant shown here, K, is in units of moles per liter. (See Panel 2–7, pp. 102–103, for a discussion of free energy and see Figure 3–44 for a discussion of the equilibrium constant.)

The ATP Synthase Is a Nanomachine that Produces ATP by Rotary Catalysis

The **ATP synthase** is a finely tuned nanomachine composed of 23 or more separate protein subunits, with a total mass of about 600,000 daltons. The ATP synthase can work both in the forward direction, producing ATP from ADP and phosphate in response to an electrochemical gradient, or in reverse, generating an electrochemical gradient by ATP hydrolysis. To distinguish it from other enzymes that hydrolyze ATP, it is also called an F_1F_0 ATP synthase or F-type ATPase.

Resembling a turbine, ATP synthase is composed of both a rotor and a stator (**Figure 14–30**). To prevent the catalytic head from rotating, a stalk at the periphery of the complex (the stator stalk) connects the head to stator subunits embedded in the membrane. A second stalk in the center of the assembly (the rotor stalk) is connected to the rotor ring in the membrane that turns as protons flow through it, driven by the electrochemical gradient across the membrane. As a result, proton flow makes the rotor stalk rotate inside the stationary head, where the catalytic sites that assemble ATP from ADP and P_i are located. Three α and three β subunits of similar structure alternate to form the head. Each of the three β subunits has a catalytic nucleotide-binding site at the α/β interface. These catalytic sites are all in different conformations, depending on their interaction with the rotor stalk. This stalk acts like a camshaft, the device that opens and closes the valves in a combustion engine. As it rotates within the head, the stalk changes the conformations of the β subunits sequentially. One of the possible conformations of the catalytic sites has high affinity for ADP and P_i, and as the rotor stalk pushes the binding site into a different conformation, these two substrates are driven to form ATP. In this way, the mechanical force exerted by the central rotor stalk is directly converted into the chemical energy of the ATP phosphate bond.

Serving as a proton-driven turbine, the ATP synthase is driven by H^+ flow into the matrix to spin at about 8000 revolutions per minute, generating three molecules of ATP per turn. In this way, each ATP synthase can produce roughly 400 molecules of ATP per second.

Proton-driven Turbines Are of Ancient Origin

The membrane-embedded rotors of ATP synthases consist of a ring of identical c subunits (**Figure 14–31**). Each c subunit is a hairpin of two membrane-spanning α helices that contain a proton-binding site defined by a glutamate or aspartate in the middle of the lipid bilayer. The a subunit, which is part of the stator (see Figure 14–30), makes two narrow channels at the interface between the rotor and stator, each spanning half of the membrane and converging on the proton-binding site at the middle of the rotor subunit. Protons flow through the two half-channels down their electrochemical gradient from the crista space back into the matrix. A negatively charged side chain in the binding site accepts a proton arriving from the crista space through the first half-channel, as it rotates past the a subunit. The bound proton then rides round in the ring for a full cycle, whereupon it is thought to be displaced by a positively charged arginine in the a subunit, and escapes

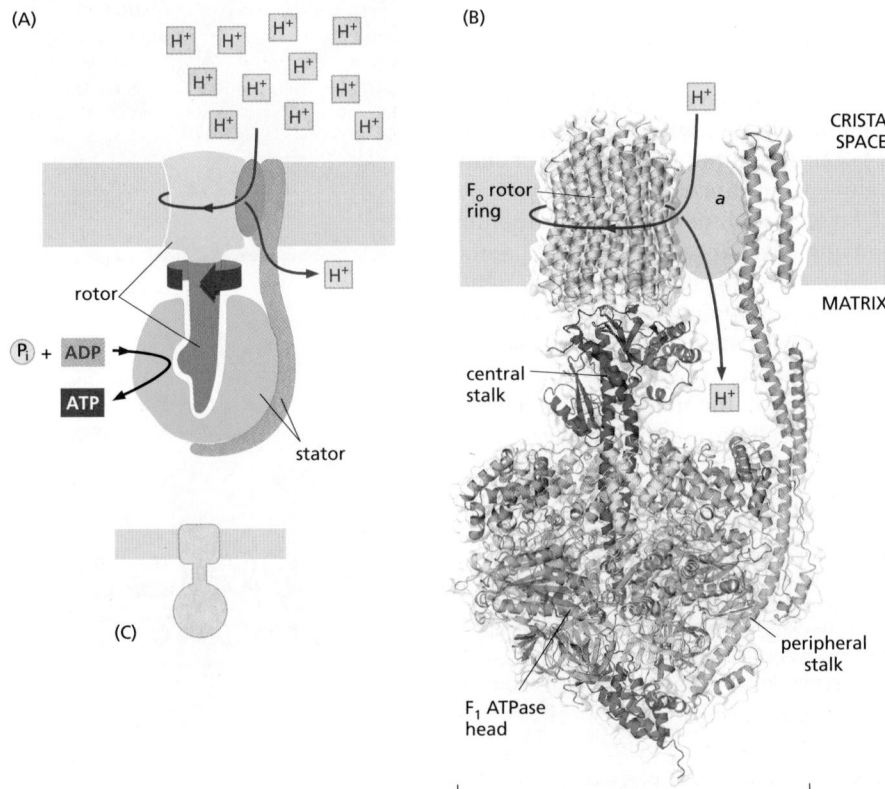

Figure 14–30 ATP synthase. The three-dimensional structure of the F_1F_o ATP synthase, determined by x-ray crystallography. Also known as an F-type ATPase, it consists of an F_o part (from "oligomycin-sensitive factor") in the membrane and the large, catalytic F_1 head in the matrix. Under mild dissociation conditions, this complex separates into its F_1 and F_o components, which can be isolated and studied individually. (A) Diagram of the enzyme complex showing how its globular head portion *(green)* is kept stationary as proton-flow across the membrane drives a rotor *(blue)* that turns inside it. (B) In bovine heart mitochondria, the F_o rotor ring in the membrane *(light blue)* has eight c subunits. It is attached to the γ subunit of the central stalk *(dark blue)* by the ε subunit *(purple)*. The catalytic F_1 head consists of a ring of three α and three β subunits *(light and dark green)*, and it directly converts mechanical energy into chemical-bond energy in ATP, as described in the text. The elongated peripheral stalk of the stator *(orange)* is connected to the F_1 head by the small δ subunit *(red)* at one end, and to the a subunit in the membrane *(pink oval)* at the other. Together with the c subunits of the ring rotating past it, the a subunit creates a path for protons through the membrane. (C) The symbol for ATP synthase used throughout this book.

The closely related ATP synthases of mitochondria, chloroplasts, and bacteria synthesize ATP by harnessing the proton-motive force across a membrane. This powers the rotation of the rotor against the stator in a counterclockwise direction, as seen from the F_1 head. The same enzyme complex can also pump protons against their electrochemical gradient by hydrolyzing ATP, which then drives the clockwise rotation of the rotor. The direction of operation depends on the net free-energy change (ΔG) for the coupled processes of H^+ translocation across the membrane and the synthesis of ATP from ADP and P_i (**Movie 14.5** and **Movie 14.6**).

Measurement of the torque that the ATP synthase can produce by ATP hydrolysis reveals that the ATP synthase is 60 times more powerful than a diesel engine of equal dimensions. PDB codes: 2WPD, 2CLY, 2WSS, 2BO5.

through the second half-channel into the matrix. Thus proton flow causes the rotor ring to spin against the stator like a proton-driven turbine.

The mitochondrial ATP synthase is of ancient origin: essentially the same enzyme occurs in plant chloroplasts and in the plasma membrane of bacteria or archaea. The main difference between them is the number of *c* subunits in the rotor ring. In mammalian mitochondria, the ring has 8 subunits. In yeast mitochondria, the number is 10; in bacteria and archaea, it ranges from 11 to 13; in plant chloroplasts, there are 14; and the rings of some cyanobacteria contain 15 *c* subunits.

The *c* subunits in the rotor ring can be thought of as cogs in the gears of a bicycle. A high gear, with a small number of cogs, is advantageous when the supply of protons is limited, as in mitochondria, but a low gear, with a large number of cogs in the wheel, is preferable when the proton gradient is high. This is the case in chloroplasts and cyanobacteria, where protons produced through the action of sunlight are plentiful. Because each rotation produces three molecules of ATP in the head, the synthesis of one ATP requires around three protons in mitochondria but up to five in photosynthetic organisms. It is the number of *c* subunits in the ring that defines how many protons need to pass through this marvelous device to make each molecule of ATP, and thereby how high a ratio of ATP to ADP can be maintained by the ATP synthase.

In principle, ATP synthase can also run in reverse as an ATP-powered proton pump that converts the energy of ATP back into a proton gradient across the membrane. In many bacteria, the rotor of the ATP synthase in the plasma membrane changes direction routinely, from ATP synthesis mode in aerobic respiration, to ATP hydrolysis mode in anaerobic metabolism. In this latter case, ATP hydrolysis serves to maintain the proton gradient across the plasma membrane, which is used to power many other essential cell functions including nutrient transport and the rotation of bacterial flagella. The V-type ATPases that acidify certain cellular organelles are architecturally similar to the F-type ATP synthases, but they normally function in reverse (see Figure 13–37).

Mitochondrial Cristae Help to Make ATP Synthesis Efficient

In the electron microscope, the mitochondrial ATP synthase complexes can be seen to project like lollipops on the matrix side of cristae membranes. Recent studies by cryoelectron microscopy and tomography have shown that this large complex is not distributed randomly in the membrane, but forms long rows of dimers along the cristae ridges (**Figure 14–32**). The dimer rows induce or stabilize these regions of high membrane curvature, which are otherwise energetically unfavorable. Indeed, the formation of ATP synthase dimers and their assembly into rows are required for cristae formation and have far-reaching consequences for cellular fitness. By contrast with bacterial or chloroplast ATP synthases, which do not form dimers, the mitochondrial complex contains additional subunits, located mostly near the membrane end of the stator stalk. Several of these subunits are found to be dimer-specific. If these subunits are mutated in yeast, the ATP synthase in the membrane remains monomeric, the mitochondria have no cristae, cellular respiration drops by half, and the cells grow more slowly.

(A)

(B)

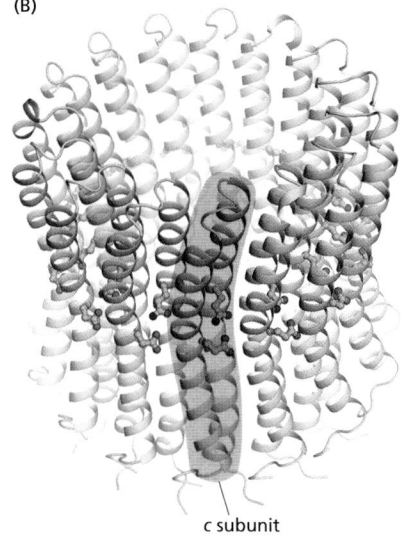

c subunit

Figure 14–31 F$_0$ ATP synthase rotor rings. (A) Atomic force microscopy image of ATP synthase rotors from the cyanobacterium *Synechococcus elongatus* in a lipid bilayer. Whereas 8 *c* subunits form the rotor in Figure 14–30, there are 13 *c* subunits in this ring. (B) The x-ray structure of the F$_0$ ring of the ATP synthase from *Spirulina platensis*, another cyanobacterium, shows that this rotor has 15 *c* subunits. In all ATP synthases, the *c* subunits are hairpins of two membrane-spanning α helices (one subunit is highlighted in *gray*). The helices are highly hydrophobic, except for two glutamine and glutamate side chains *(yellow)* that create proton-binding sites in the membrane. (A, courtesy of Thomas Meier and Denys Pogoryelov; B, PDB code: 2WIE.)

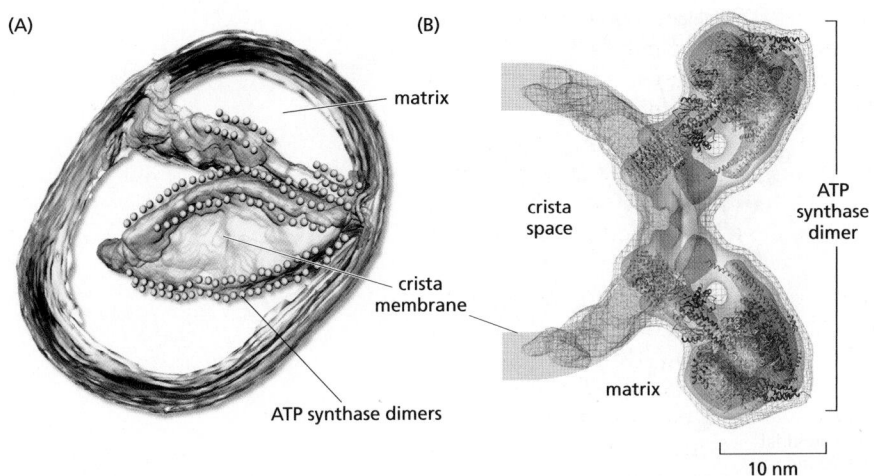

Figure 14–32 Dimers of mitochondrial ATP synthase in cristae membranes. (A) A three-dimensional map of a small mitochondrion obtained by electron microscope tomography shows that ATP synthases form long paired rows along cristae ridges. The outer membrane is *gray*, the inner membrane and cristae membranes have been colored *light blue*. Each head of an ATP synthase is indicated by a *yellow sphere*. (B) A three-dimensional map of a mitochondrial ATP synthase dimer in the crista membrane obtained by subtomogram averaging, with fitted x-ray structures (**Movie 14.7**). (A, from K. Davies et al., *Proc. Natl Acad. Sci. USA* 108:14121–14126, 2011. With permission from the National Academy of Sciences; B, from K. Davies et al., *Proc. Natl Acad. Sci. USA* 109:13602–13607, 2012. With permission from the National Academy of Sciences.)

Electron tomography suggests that the proton pumps of the respiratory chain are located in the membrane regions at either side of the dimer rows. Protons pumped into the crista space by these respiratory-chain complexes are thought to diffuse very rapidly along the membrane surface, with the ATP synthase rows creating a proton "sink" at the cristae tips (**Figure 14–33**). *In vitro* studies suggest that the ATP synthase needs a proton gradient of about 2 pH units to produce ATP at the rate required by the cell, irrespective of the membrane potential. The H^+ gradient across the inner mitochondrial membrane is only 0.5 to 0.6 pH units. The cristae thus seem to work as proton traps that enable the ATP synthase to make efficient use of the protons pumped out of the mitochondrial matrix. As we shall see in the next section, this elaborate arrangement of membrane protein complexes is absent in chloroplasts, where the H^+ gradient is much higher.

Special Transport Proteins Exchange ATP and ADP Through the Inner Membrane

Like all biological membranes, the inner mitochondrial membrane contains numerous specific *transport proteins* that allow particular substances to pass through. One of the most abundant of these is the *ADP/ATP carrier protein* (**Figure 14–34**). This carrier shuttles the ATP produced in the matrix through the inner membrane to the intermembrane space, from where it diffuses through the outer mitochondrial membrane to the cytosol. In exchange, ADP passes from the cytosol into the matrix for recycling into ATP. ATP^{4-} has one more negative charge than ADP^{3-}, and the exchange of ATP and ADP is driven by the electrochemical gradient across the inner membrane, so that the more negatively charged ATP is pushed out of the matrix, and the less negatively charged ADP is pulled in. The ADP/ATP carrier is but one member of a *mitochondrial carrier family*: the inner

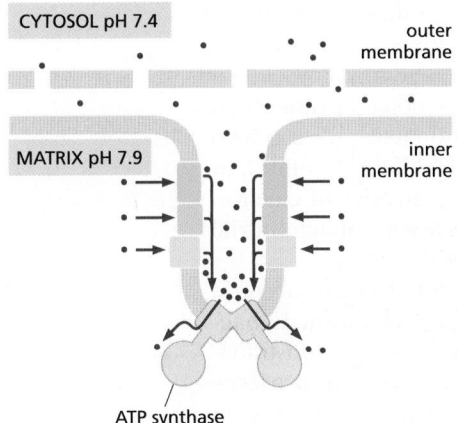

Figure 14–33 ATP synthase dimers at cristae ridges and ATP production. At the crista ridges, the ATP synthases *(yellow)* form a sink for protons *(red)*. The proton pumps of the electron-transport chain *(green)* are located in the membrane regions on either side of the crista. As illustrated, protons tend to diffuse along the membrane from their source to the proton sink created by the ATP synthase. This allows efficient ATP production despite the small H^+ gradient between the cytosol and matrix. *Red arrows* show the direction of the proton flow.

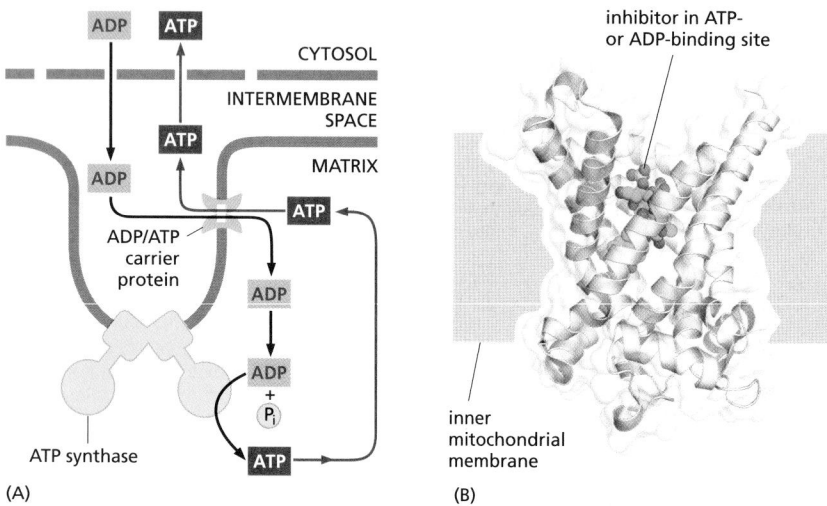

(A)   (B)

Figure 14–34 The ADP/ATP carrier protein. (A) The ADP/ATP carrier protein is a small membrane protein that carries the ATP produced on the matrix side of the inner membrane to the intermembrane space, and the ADP that is needed for ATP synthesis into the matrix. (B) In the ADP/ATP carrier, six transmembrane α helices define a cavity that binds either ADP or ATP. In this x-ray structure, the substrate is replaced by a tightly bound inhibitor instead *(colored)*. When ADP binds from outside the inner membrane, it triggers a conformational change and is released into the matrix. In exchange, a molecule of ATP quickly binds to the matrix side of the carrier and is transported to the intermembrane space. From there the ATP diffuses through the outer mitochondrial membrane to the cytoplasm, where it powers the energy-requiring processes in the cell. (B, PDB code: 1OKC.)

mitochondrial membrane contains about 20 related carrier proteins exchanging various other metabolites, including the phosphate that is required along with ADP for ATP synthesis.

In some specialized fat cells, mitochondrial respiration is uncoupled from ATP synthesis by the *uncoupling protein,* another member of the mitochondrial carrier family. In these cells, known as brown fat cells, most of the energy of oxidation is dissipated as heat rather than being converted into ATP. In the inner membranes of the large mitochondria in these cells, the uncoupling protein allows protons to move down their electrochemical gradient without passing through ATP synthase. This process is switched on when heat generation is required, causing the cells to oxidize their fat stores at a rapid rate and produce heat rather than ATP. Tissues containing brown fat serve as "heating pads," helping to revive hibernating animals and to protect newborn human babies from the cold.

Chemiosmotic Mechanisms First Arose in Bacteria

Bacteria use enormously diverse energy sources. Some, like animal cells, are aerobic; they synthesize ATP from sugars they oxidize to CO_2 and H_2O by glycolysis, the citric acid cycle, and a respiratory chain in their plasma membrane that is similar to the one in the inner mitochondrial membrane. Others are strict anaerobes, deriving their energy either from glycolysis alone (by fermentation, see Figure 2–47) or from an electron-transport chain that employs a molecule other than oxygen as the final electron acceptor. The alternative electron acceptor can be a nitrogen compound (nitrate or nitrite), a sulfur compound (sulfate or sulfite), or a carbon compound (fumarate or carbonate), for example. A series of electron carriers in the plasma membrane that are comparable to those in mitochondrial respiratory chains transfers the electrons to these acceptors.

Despite this diversity, the plasma membrane of the vast majority of bacteria contains an ATP synthase that is very similar to the one in mitochondria. In bacteria that use an electron-transport chain to harvest energy, the electron-transport chain pumps H^+ out of the cell and thereby establishes a proton-motive force across the plasma membrane that drives the ATP synthase to make ATP. In other

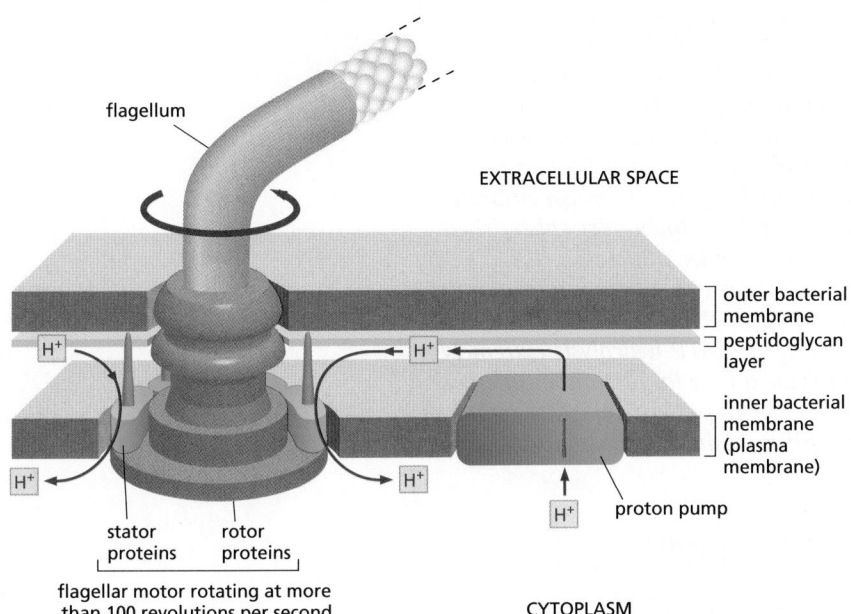

flagellum

EXTRACELLULAR SPACE

outer bacterial membrane

peptidoglycan layer

inner bacterial membrane (plasma membrane)

H⁺

proton pump

stator proteins

rotor proteins

flagellar motor rotating at more than 100 revolutions per second

CYTOPLASM

Figure 14–35 **The rotation of the bacterial flagellum driven by H⁺ flow.** The flagellum is attached to a series of protein rings (pink), which are embedded in the outer and inner membranes and rotate with the flagellum. The rotation is driven by a flow of protons through an outer ring of proteins (the stator) by mechanisms that may resemble those used by the ATP synthase. However, the flow of protons in the flagellar motor is always toward the cytosol, both during clockwise and counterclockwise rotation, whereas in ATP synthase this flow reverses with the direction of rotation (Movie 14.8).

bacteria, the ATP synthase works in reverse, using the ATP produced by glycolysis to pump H⁺ and establish a proton gradient across the plasma membrane.

Bacteria, including the strict anaerobes, maintain a proton gradient across their plasma membrane that is harnessed to drive many other processes. It can be used to drive a flagellar motor, for example (Figure 14–35). This gradient is harnessed to pump Na⁺ out of the bacterium via a Na⁺–H⁺ antiporter that takes the place of the Na⁺-K⁺ pump of eukaryotic cells. The gradient is also used for the active inward transport of nutrients, such as most amino acids and many sugars: each nutrient is dragged into the cell along with one or more protons through a specific symporter (Figure 14–36; see also Chapter 11). In animal cells, by contrast, most inward transport across the plasma membrane is driven by the Na⁺ gradient (high Na⁺ outside, low Na⁺ inside) that is established by the Na⁺-K⁺ pump (see Figure 11–15).

Some unusual bacteria have adapted to live in a very alkaline environment and yet must maintain their cytoplasm at a physiological pH. For these cells, any attempt to generate an electrochemical H⁺ gradient would be opposed by a large H⁺ concentration gradient in the wrong direction (H⁺ higher inside than outside). Presumably for this reason, some of these bacteria substitute Na⁺ for H⁺ in all of their chemiosmotic mechanisms. The respiratory chain pumps Na⁺ out of the cell, the transport systems and flagellar motor are driven by an inward flux of Na⁺, and a Na⁺-driven ATP synthase synthesizes ATP. The existence of such bacteria demonstrates a critical point: the principle of chemiosmosis is more fundamental than the proton-motive force on which it is normally based.

As we discuss next, an ATP synthase coupled to chemiosmotic processes is also a central feature of plants, where it plays critical roles in both mitochondria and chloroplasts.

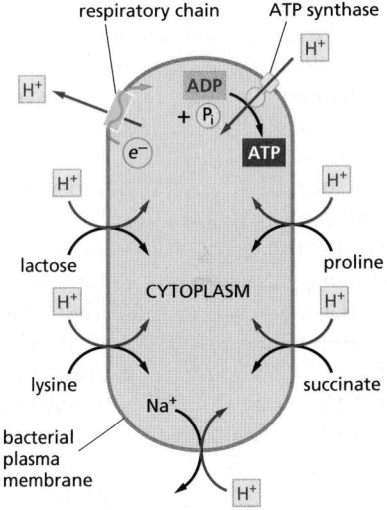

(A) **AEROBIC CONDITIONS**

respiratory chain

ATP synthase

H⁺

ADP + Pᵢ

ATP

e⁻

H⁺

H⁺

lactose

proline

H⁺

CYTOPLASM

H⁺

lysine

succinate

bacterial plasma membrane

Na⁺

H⁺

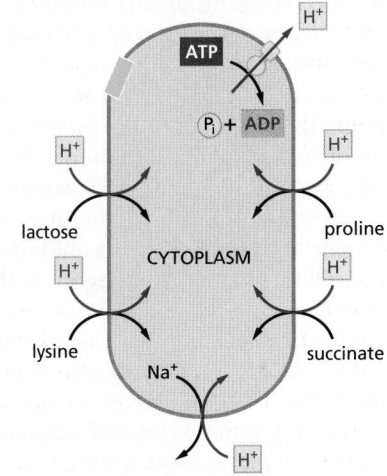

(B) **ANAEROBIC CONDITIONS**

H⁺

ATP

Pᵢ + ADP

H⁺

H⁺

lactose

proline

H⁺

CYTOPLASM

H⁺

lysine

succinate

Na⁺

H⁺

Figure 14–36 **The importance of H⁺-driven transport in bacteria.** A proton-motive force generated across the plasma membrane pumps nutrients into the cell and expels Na⁺. (A) In an aerobic bacterium, a respiratory chain fed by the oxidation of substrates produces an electrochemical proton gradient across the plasma membrane. This gradient is then harnessed to make ATP, as well as to transport nutrients (proline, succinate, lactose, and lysine) into the cell and to pump Na⁺ out of the cell. (B) When the same bacterium grows under anaerobic conditions, it derives its ATP from glycolysis. As indicated, the ATP synthase in the plasma membrane then hydrolyzes some of this ATP to establish an electrochemical proton gradient that drives the same transport processes that depend on respiratory chain proton-pumping in (A).

Summary

The large amount of free energy released when H⁺ flows back into the matrix from the cristae provides the basis for ATP production on the matrix side of mitochondrial cristae membranes by a remarkable protein machine—the ATP synthase. The ATP synthase functions like a miniature turbine, and it is a reversible device that can couple proton flow to either ATP synthesis or ATP hydrolysis. The transmembrane electrochemical gradient that drives ATP production in mitochondria also drives the active transport of selected metabolites across the inner mitochondrial membrane, including an efficient ADP/ATP exchange between the mitochondrion and the cytosol that keeps the cell's ATP pool highly charged. The resulting high cellular concentration of ATP makes the free-energy change for ATP hydrolysis extremely favorable, allowing this hydrolysis reaction to drive a very large number of energy-requiring processes throughout the cell. The universal presence of ATP synthase in bacteria, mitochondria, and chloroplasts testifies to the central importance of chemiosmotic mechanisms in cells.

CHLOROPLASTS AND PHOTOSYNTHESIS

All animals and most microorganisms rely on the continual uptake of large amounts of organic compounds from their environment. These compounds provide both the carbon-rich building blocks for biosynthesis and the metabolic energy for life. It is likely that the first organisms on the primitive Earth had access to an abundance of organic compounds produced by geochemical processes, but it is clear that these were used up billions of years ago. Since that time, virtually all of the organic materials required by living cells have been produced by *photosynthetic organisms*, including plants and photosynthetic bacteria. The core machinery that drives all photosynthesis appears to have evolved more than 3 billion years ago in the ancestors of present-day bacteria; today it provides the only major solar energy storage mechanism on Earth.

The most advanced photosynthetic bacteria are the cyanobacteria, which have minimal nutrient requirements. They use electrons from water and the energy of sunlight to convert atmospheric CO_2 into organic compounds—a process called *carbon fixation*. In the course of the overall reaction $nH_2O + nCO_2 \rightarrow$ (light) $(CH_2O)_n + nO_2$, they also liberate into the atmosphere the molecular oxygen that then powers oxidative phosphorylation. In this way, it is thought that the evolution of cyanobacteria from more primitive photosynthetic bacteria eventually made possible the development of the many different aerobic life-forms that populate the Earth today.

Chloroplasts Resemble Mitochondria But Have a Separate Thylakoid Compartment

Plants (including algae) developed much later than cyanobacteria, and their photosynthesis occurs in a specialized intracellular organelle—the **chloroplast** (**Figure 14–37**). Chloroplasts use chemiosmotic mechanisms to carry out their energy interconversions in much the same way that mitochondria do. Although much larger than mitochondria, they are organized on the same principles. They have a highly permeable outer membrane; a much less permeable inner membrane, in which membrane transport proteins are embedded; and a narrow intermembrane space in between. Together, these two membranes form the chloroplast envelope (Figure 14–37D). The inner chloroplast membrane surrounds a large space called the **stroma**, which is analogous to the mitochondrial matrix. The stroma contains many metabolic enzymes and, as for the mitochondrial matrix, it is the place where ATP is made by the head of an ATP synthase. Like the mitochondrion, the chloroplast has its own genome and genetic system. The stroma therefore also contains a special set of ribosomes, RNAs, and the chloroplast DNA.

An important difference between the organization of mitochondria and chloroplasts is highlighted in **Figure 14–38**. The inner membrane of the chloroplast is

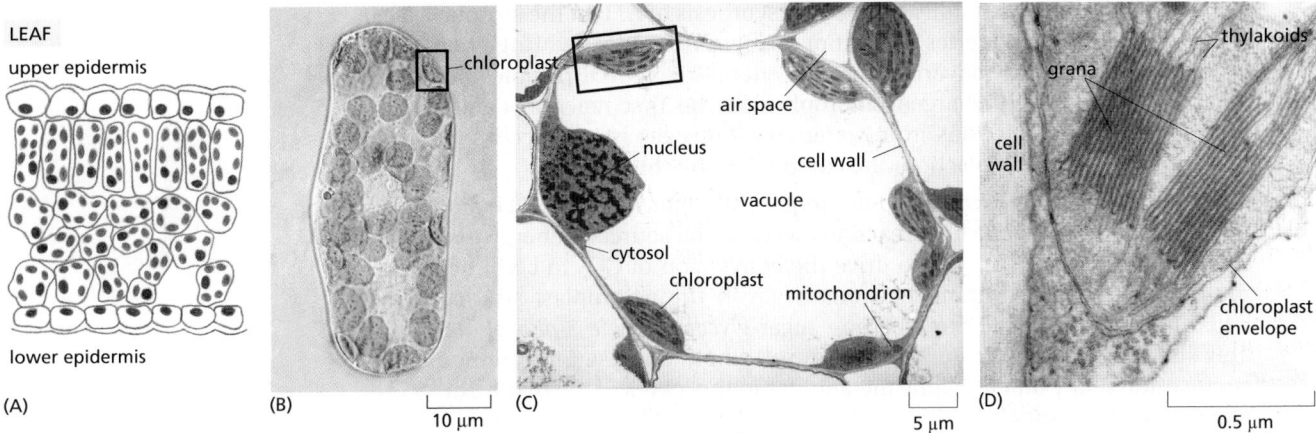

(A)

(B)

(C)

(D)

LEAF

upper epidermis

lower epidermis

chloroplast

air space

nucleus

cell wall

vacuole

cytosol

chloroplast

mitochondrion

10 μm

5 μm

thylakoids

grana

cell wall

chloroplast envelope

0.5 μm

not folded into cristae and does not contain electron-transport chains. Instead, the electron-transport chains, photosynthetic light-capturing systems, and ATP synthase are all contained in the **thylakoid membrane**, a separate, distinct membrane that forms a set of flattened, disc-like sacs, the *thylakoids*. The thylakoid membrane is highly folded into numerous local stacks of flattened vesicles called *grana*, interconnected by nonstacked thylakoids. The lumen of each thylakoid is connected with the lumen of other thylakoids, thereby defining a third internal compartment called the *thylakoid space*. This space represents a separate compartment in each chloroplast that is not connected to either the intermembrane space or the stroma.

Chloroplasts Capture Energy from Sunlight and Use It to Fix Carbon

We can group the reactions that occur during photosynthesis in chloroplasts into two broad categories:

1. The **photosynthetic electron-transfer reactions** (also called the "light reactions") occur in two large protein complexes, called *reaction centers*, embedded in the thylakoid membrane. A photon (a quantum of light) knocks an electron out of the green pigment molecule *chlorophyll* in the first reaction center, creating a positively charged chlorophyll ion. This electron then moves along an electron-transport chain and through a second reaction center in much the same way that an electron moves along the respiratory chain in mitochondria. During this electron-transport process, H$^+$ is pumped across the thylakoid membrane, and the resulting

Figure 14–37 Chloroplasts in the cell. (A) Schematic cross section through the leaf of a green plant. (B) Light microscopy of a plant leaf cell—here, a mesophyll cell from *Zinnia elegans*—shows chloroplasts as bright green bodies, measuring several micrometers across, in the transparent cell interior. (C) The electron micrograph of a thin, stained section through a wheat leaf cell shows a thin rim of cytoplasm—containing chloroplasts, the nucleus, and mitochondria—surrounding a large, water-filled vacuole. (D) At higher magnification, electron microscopy reveals the chloroplast envelope membrane and the thylakoid membrane within the chloroplast that is highly folded into grana stacks (Movie 14.9). (B, courtesy of John Innes Foundation; C and D, courtesy of K. Plaskitt.)

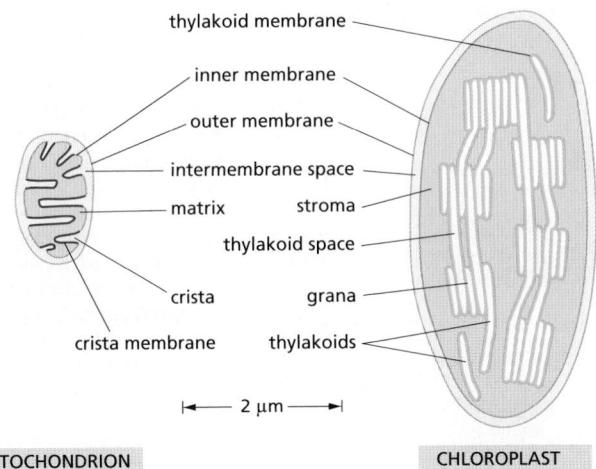

thylakoid membrane

inner membrane

outer membrane

intermembrane space

matrix

stroma

thylakoid space

crista

grana

crista membrane

thylakoids

2 μm

MITOCHONDRION

CHLOROPLAST

Figure 14–38 A mitochondrion and chloroplast compared. Chloroplasts are generally larger than mitochondria. In addition to an outer and inner envelope membrane, they contain the thylakoid membrane with its internal thylakoid space. The chloroplast thylakoid membrane, which is the site of solar energy conversion in plants and algae, corresponds to the mitochondrial cristae, which are the sites of energy conversion by cellular respiration. Unlike the crista membrane, which is continuous with the inner mitochondrial membrane at cristae junctions, the thylakoid membrane is not connected to the inner chloroplast membrane at any point.

electrochemical proton gradient drives the synthesis of ATP in the stroma. As the final step in this series of reactions, electrons are loaded (together with H$^+$) onto NADP$^+$, converting it to the energy-rich NADPH molecule. Because the positively charged chlorophyll in the first reaction center quickly regains its electrons from water (H$_2$O), O$_2$ gas is produced as a by-product. All of these reactions are confined to the chloroplast.

2. The **carbon-fixation reactions** do not require sunlight. Here the ATP and NADPH generated by the light reactions serve as the source of energy and reducing power, respectively, to drive the conversion of CO$_2$ to carbohydrate. These carbon-fixation reactions begin in the chloroplast stroma, where they generate the three-carbon sugar *glyceraldehyde 3-phosphate*. This simple sugar is exported to the cytosol, where it is used to produce sucrose and many other organic metabolites in the leaves of the plant. The sucrose is then exported to meet the metabolic needs of the nonphotosynthetic plant tissues, serving as a source of both carbon skeletons and energy for growth.

Thus, the formation of ATP, NADPH, and O$_2$ (which requires light energy directly) and the conversion of CO$_2$ to carbohydrate (which requires light energy only indirectly) are separate processes (**Figure 14–39**). However, they are linked by elaborate feedback mechanisms that allow a plant to manufacture sugars only when it is appropriate to do so. Several of the chloroplast enzymes required for carbon fixation, for example, are inactive in the dark and reactivated by light-stimulated electron-transport processes.

Carbon Fixation Uses ATP and NADPH to Convert CO$_2$ into Sugars

We have seen earlier in this chapter how animal cells produce ATP by using the large amount of free energy released when carbohydrates are oxidized to CO$_2$ and H$_2$O. The reverse reaction, in which plants make carbohydrate from CO$_2$ and H$_2$O, takes place in the chloroplast stroma. The large amounts of ATP and NADPH produced by the photosynthetic electron-transfer reactions are required to drive this energetically unfavorable reaction.

Figure 14–39 A summary of the energy-converting metabolism in chloroplasts. Chloroplasts require only water and carbon dioxide as inputs for their light-driven photosynthesis reactions, and they produce the nutrients for most other organisms on the planet. Each oxidation of two water molecules by a photochemical reaction center in the thylakoid membrane produces one molecule of oxygen, which is released into the atmosphere. At the same time, protons are concentrated in the thylakoid space. These protons create a large electrochemical gradient across the thylakoid membrane, which is utilized by the chloroplast ATP synthase to produce ATP from ADP and phosphate. The electrons withdrawn from water are transferred to a second type of photochemical reaction center to produce NADPH from NADP$^+$. As indicated, the NADPH and ATP are fed into the *carbon-fixation cycle* to reduce carbon dioxide, thereby producing the precursors for sugars, amino acids, and fatty acids. The CO$_2$ that is taken up from the atmosphere here is the source of the carbon atoms for most organic molecules on Earth.

In a plant cell, a variety of metabolites produced in the chloroplast are exported to the cytoplasm for biosyntheses. Some of the sugar produced is stored in the form of starch granules in the chloroplast, but the rest is transported throughout the plant as sucrose or converted to starch in special storage tissues. These storage tissues serve as a major food source for animals.

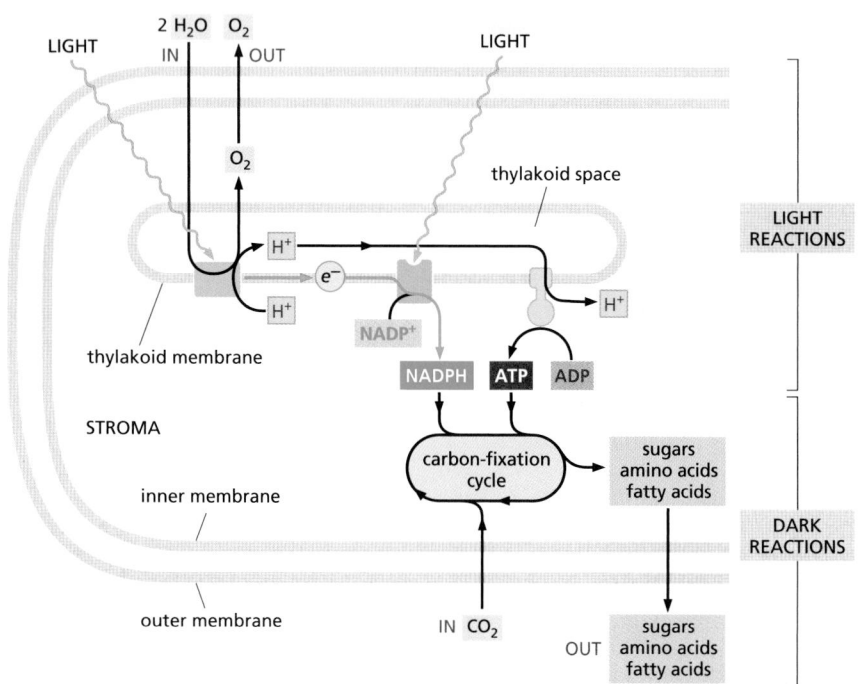

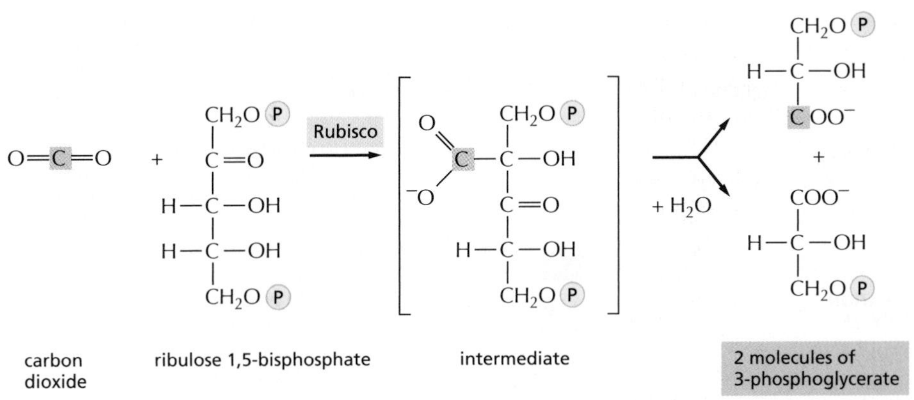

Figure 14–40 **The initial reaction in carbon fixation.** This carboxylation reaction allows one molecule each of carbon dioxide and water to be incorporated into organic carbon molecules. It is catalyzed in the chloroplast stroma by the abundant enzyme ribulose bisphosphate carboxylase, or Rubisco. As indicated, the product is two molecules of 3-phosphoglycerate.

Figure 14–40 illustrates the central reaction of **carbon fixation**, in which an atom of inorganic carbon is converted to organic carbon: CO_2 from the atmosphere combines with the five-carbon compound ribulose 1,5-bisphosphate plus water to yield two molecules of the three-carbon compound 3-phosphoglycerate. This carboxylation reaction is catalyzed in the chloroplast stroma by a large enzyme called *ribulose bisphosphate carboxylase*, or *Rubisco* for short. Because the reaction is so slow (each Rubisco molecule turns over only about 3 molecules of substrate per second, compared to 1000 molecules per second for a typical enzyme), an unusually large number of enzyme molecules are needed. Rubisco often constitutes more than 50% of the chloroplast protein mass, and it is thought to be the most abundant protein on Earth. In a global context, Rubisco also keeps the amount of the greenhouse gas CO_2 in the atmosphere at a low level.

Although the production of carbohydrates from CO_2 and H_2O is energetically unfavorable, the fixation of CO_2 catalyzed by Rubisco is an energetically favorable reaction. Carbon fixation is energetically favorable because a continuous supply of the energy-rich ribulose 1,5-bisphosphate is fed into the process. This compound is consumed by the addition of CO_2, and it must be replenished. The energy and reducing power needed to regenerate ribulose 1,5-bisphosphate come from the ATP and NADPH produced by the photosynthetic light reactions.

The elaborate series of reactions in which CO_2 combines with ribulose 1,5-bisphosphate to produce a simple sugar—a portion of which is used to regenerate ribulose 1,5-bisphosphate—forms a cycle, called the *carbon-fixation cycle*, or the Calvin cycle (**Figure 14–41**). This cycle was one of the first metabolic pathways to be worked out by applying radioisotopes as tracers in biochemistry. As indicated, each turn of the cycle converts six molecules of 3-phosphoglycerate to three molecules of ribulose 1,5-bisphosphate plus one molecule of glyceraldehyde 3-phosphate. *Glyceraldehyde 3-phosphate*, the three-carbon sugar produced by the cycle, then provides the starting material for the synthesis of many other sugars and all of the other organic molecules that form the plant.

Sugars Generated by Carbon Fixation Can Be Stored as Starch or Consumed to Produce ATP

The glyceraldehyde 3-phosphate generated by carbon fixation in the chloroplast stroma can be used in a number of ways, depending on the needs of the plant. During periods of excess photosynthetic activity, much of it is retained in the chloroplast stroma and converted to *starch*. Like glycogen in animal cells, starch is a large polymer of glucose that serves as a carbohydrate reserve, and it is stored as large granules in the chloroplast stroma. Starch forms an important part of the diet of all animals that eat plants. Other glyceraldehyde 3-phosphate molecules are converted to fat in the stroma. This material, which accumulates as fat droplets, likewise serves as an energy reserve. At night, this stored starch and fat can be broken down to sugars and fatty acids, which are exported to the cytosol to help support the metabolic needs of the plant. Some of the exported sugar enters the

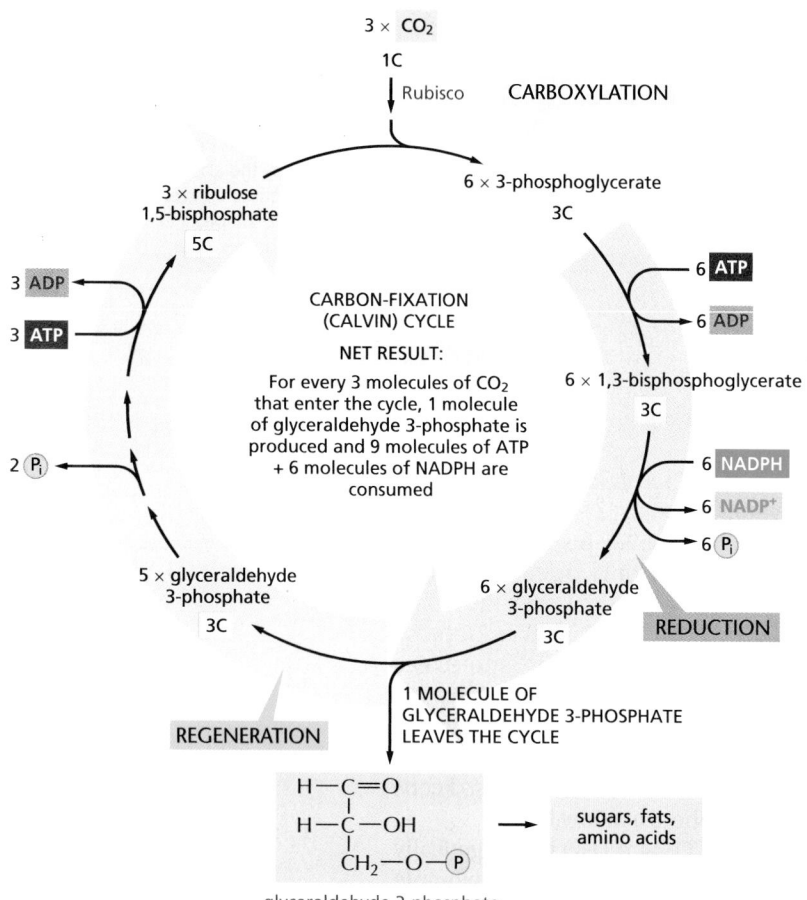

3 × CO₂

1C

Rubisco   **CARBOXYLATION**

3 × ribulose
1,5-bisphosphate
5C

6 × 3-phosphoglycerate
3C

3 ADP

3 ATP

**CARBON-FIXATION
(CALVIN) CYCLE**

NET RESULT:

For every 3 molecules of CO_2
that enter the cycle, 1 molecule
of glyceraldehyde 3-phosphate is
produced and 9 molecules of ATP
+ 6 molecules of NADPH are
consumed

6 ATP

6 ADP

6 × 1,3-bisphosphoglycerate
3C

6 NADPH

6 NADP⁺

6 Pᵢ

2 Pᵢ

5 × glyceraldehyde
3-phosphate
3C

6 × glyceraldehyde
3-phosphate
3C

REDUCTION

REGENERATION

1 MOLECULE OF
GLYCERALDEHYDE 3-PHOSPHATE
LEAVES THE CYCLE

H—C═O
|
H—C—OH   →   sugars, fats,
| amino acids
CH₂—O—Ⓟ

glyceraldehyde 3-phosphate

Figure 14–41 The carbon-fixation cycle. This central metabolic pathway allows organic molecules to be produced from CO_2 and H_2O. In the first stage of the cycle (carboxylation), CO_2 is added to ribulose 1,5-bisphosphate, as shown in Figure 14–40. In the second stage (reduction), ATP and NADPH are consumed to produce glyceraldehyde 3-phosphate molecules. In the final stage (regeneration), some of the glyceraldehyde 3-phosphate produced is used to regenerate ribulose 1,5-bisphosphate. Other glyceraldehyde 3-phosphate molecules are either converted to starch and fat in the chloroplast stroma, or transported out of the chloroplast into the cytosol. The number of carbon atoms in each type of molecule is indicated in *yellow*. There are many intermediates between glyceraldehyde 3-phosphate and ribulose 5-phosphate, but they have been omitted here for clarity. The entry of water into the cycle is also not shown (but see Figure 14–40).

glycolytic pathway (see Figure 2–46), where it is converted to pyruvate. Both that pyruvate and the fatty acids can enter the plant cell mitochondria and be fed into the citric acid cycle, ultimately leading to the production of large amounts of ATP by oxidative phosphorylation (**Figure 14–42**). Plants use this ATP in the same way that animal cells and other nonphotosynthetic organisms do to power a variety of metabolic reactions.

The glyceraldehyde 3-phosphate exported from chloroplasts into the cytosol can also be converted into many other metabolites, including the disaccharide *sucrose*. Sucrose is the major form in which sugar is transported between the cells of a plant: just as glucose is transported in the blood of animals, so sucrose is exported from the leaves to provide carbohydrate to the rest of the plant.

The Thylakoid Membranes of Chloroplasts Contain the Protein Complexes Required for Photosynthesis and ATP Generation

We next need to explain how the large amounts of ATP and NADPH required for carbon fixation are generated in the chloroplast. Chloroplasts are much larger and less dynamic than mitochondria, but they make use of chemiosmotic energy conversion in much the same way. As we saw in Figure 14–38, chloroplasts and mitochondria are organized on the same principles, although the chloroplast contains a separate thylakoid membrane system in which its chemiosmotic mechanisms occur. The thylakoid membranes contain two large membrane protein complexes, called *photosystems*, which endow plants and other photosynthetic organisms with the ability to capture and convert solar energy for their own use. Two other protein complexes in the thylakoid membrane that work together with the photosystems in photophosphorylation—the generation of ATP with sunlight—have mitochondrial equivalents. These are the heme-containing cytochrome

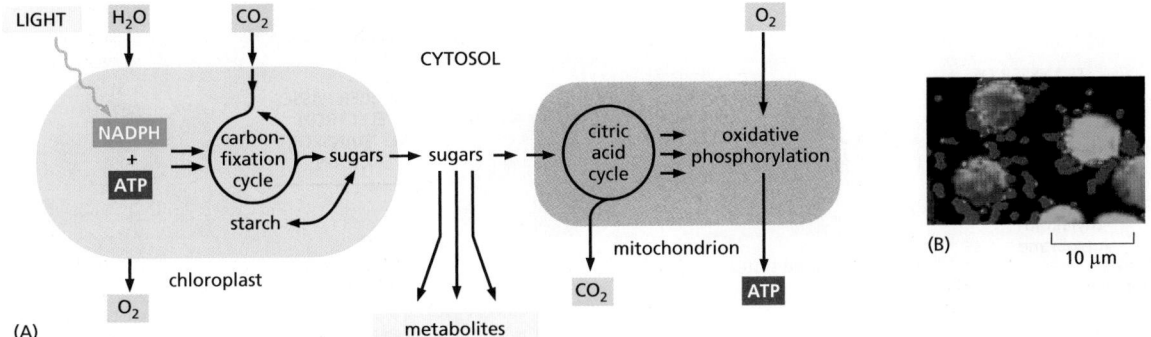

Figure 14–42 How chloroplasts and mitochondria collaborate to supply cells with both metabolites and ATP. (A)The inner chloroplast membrane is impermeable to the ATP and NADPH that are produced in the stroma during the light reactions of photosynthesis. These molecules are therefore funneled into the carbon-fixation cycle, where they are used to make sugars. The resulting sugars and their metabolites are either stored within the chloroplast—in the form of starch or fat—or exported to the rest of the plant cell. There, they can enter the energy-generating pathway that ends in ATP synthesis linked to oxidative phosphorylation inside the mitochondrion. Unlike the chloroplast, mitochondrial membranes contain a specific transporter that makes them permeable to ATP (see Figure 14–34). Note that the O_2 released to the atmosphere by photosynthesis in chloroplasts is used for oxidative phosphorylation in mitochondria; similarly, the CO_2 released by the citric acid cycle in mitochondria is used for carbon fixation in chloroplasts. (B) In a leaf, mitochondria *(red)* tend to cluster close to the chloroplasts *(green)*, as seen in this light micrograph. (B, courtesy of Olivier Grandjean.)

b_6-f complex, which both functionally and structurally resembles cytochrome c reductase in the respiratory chain; and the chloroplast ATP synthase, which closely resembles the mitochondrial ATP synthase and works in the same way.

Chlorophyll–Protein Complexes Can Transfer Either Excitation Energy or Electrons

The photosystems in the thylakoid membrane are multiprotein assemblies of a complexity comparable to that of the protein complexes in the mitochondrial electron-transport chain. They contain large numbers of specifically bound chlorophyll molecules, in addition to cofactors that will be familiar from our discussion of mitochondria (heme, iron–sulfur clusters, and quinones). **Chlorophyll**, the green pigment of photosynthetic organisms, has a long hydrophobic tail that makes it behave like a lipid, plus a porphyrin ring that has a central Mg atom and an extensive system of delocalized electrons in conjugated double bonds (**Figure 14–43**). When a chlorophyll molecule absorbs a quantum of sunlight (a photon), the energy of the photon causes one of these electrons to move from a low-energy molecular orbital to another orbital of higher energy.

The excited electron in a chlorophyll molecule tends to return quickly to its ground state, which can occur in one of three ways:

1. By converting the extra energy into heat (molecular motion) or to some combination of heat and light of a longer wavelength (fluorescence); this is what usually happens when light is absorbed by an isolated chlorophyll molecule in solution.

2. By transferring the energy—but not the electron—directly to a neighboring chlorophyll molecule by a process called *resonance energy transfer*.

3. By transferring the excited electron with its negative charge to another nearby molecule, an *electron acceptor*, after which the positively charged chlorophyll returns to its original state by taking up an electron from some other molecule, an *electron donor*.

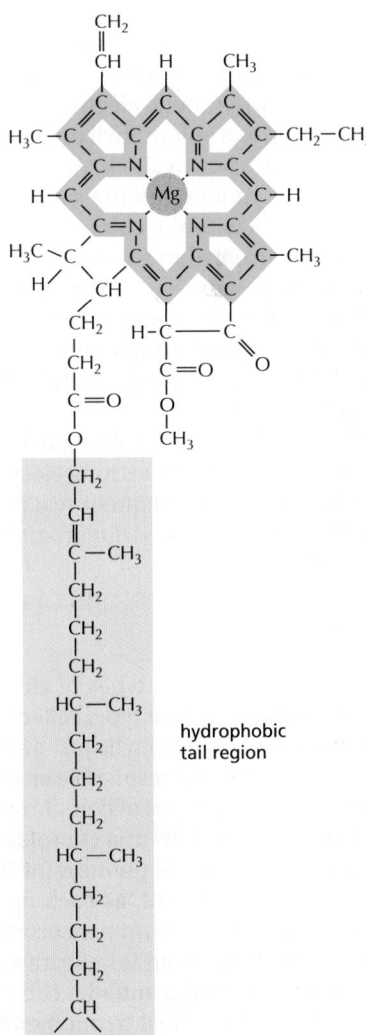

Figure 14–43 The structure of chlorophyll. A magnesium atom is held in a porphyrin ring, which is related to the porphyrin ring that binds iron in heme (see Figure 14–15). Electrons are delocalized over the bonds shaded in *blue*.

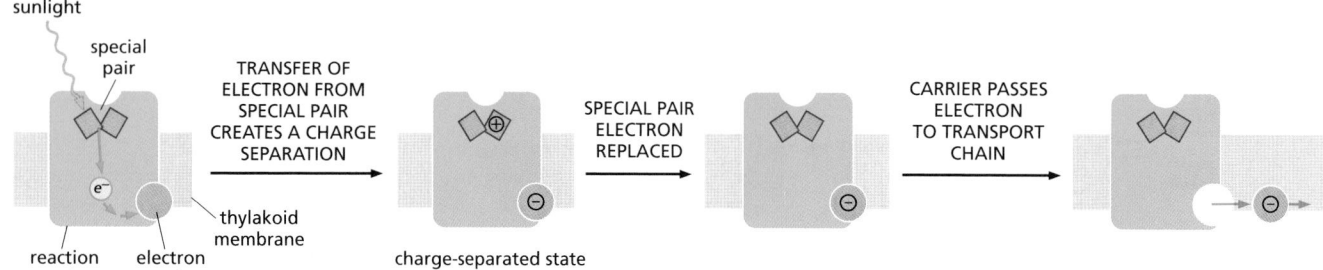

The latter two mechanisms occur when chlorophylls are attached to proteins in a *chlorophyll–protein complex*. The protein coordinates the central Mg atom in the chlorophyll porphyrin, most often through a histidine side chain located in the hydrophobic interior of a membrane, causing each of the chlorophylls in a protein complex to be held at exactly defined distances and orientations. The flow of excitation energy or electrons then depends on both the precise spatial arrangement and the local protein environment of the protein-bound chlorophylls.

When excited by a photon, most protein-bound chlorophylls simply transmit the absorbed energy to another nearby chlorophyll by the process of resonance energy transfer. However, in a few specially positioned chlorophylls, the energy difference between the ground state and the excited state is just right for the photon to trigger a light-induced chemical reaction. The special state of such chlorophyll molecules derives from their close interaction with a second chlorophyll molecule in the same chlorophyll–protein complex. Together, these two chlorophylls form a *special pair*.

The photosynthetic electron transfer process starts when a photon of suitable energy ionizes a chlorophyll molecule in such a special pair, dissociating it into an electron and a positively charged chlorophyll ion. The energized electron is passed rapidly to a quinone in the same protein complex, preventing its unproductive reassociation with the chlorophyll ion. This light-induced transfer of an electron from a chlorophyll to a mobile electron carrier is the central **charge-separation** step in photosynthesis, in which a chlorophyll becomes positively charged and an electron carrier becomes negatively charged (**Figure 14–44**). The chlorophyll ion is a very strong oxidant that is able to withdraw an electron from a low-energy substrate; in the first step of oxygenic photosynthesis, this low-energy substrate is water.

Upon transfer to a mobile carrier in the electron-transport chain, the electron is stabilized as part of a strong electron donor and made available for subsequent reactions. These subsequent reactions require more time to complete, and they result in light-generated energy-rich compounds.

Figure 14–44 **A general scheme for the charge-separation step in a photosynthetic reaction center.** In a reaction center, light energy is harnessed to generate electrons that are held at a high energy level by mobile electron carriers in a membrane. Light energy is thereby converted to chemical energy. The process starts when a photon absorbed by the special pair of chlorophylls in the reaction center knocks an electron out of one of the chlorophylls. The electron is taken up by a mobile electron carrier *(orange)* bound at the opposite membrane surface. A set of intermediary carriers embedded in the reaction center provide the path from the special pair to this carrier (not shown). The physical distance between the positively charged chlorophyll ion and the negatively charged electron carrier stabilizes the charge-separated state for a short time, during which the chlorophyll ion, a strong oxidant, withdraws an electron from a suitable compound (for example, from water, an event we will discuss in detail shortly). The electron carrier then diffuses away from the reaction center as a strong electron donor that will transfer its electron to an electron-transport chain.

A Photosystem Consists of an Antenna Complex and a Reaction Center

There are two distinct types of chlorophyll–protein complexes in the photosynthetic membrane. One type, called a *photochemical reaction center*, contains the special pair of chlorophylls just described. The other type engages exclusively in light absorption and resonance energy transfer and is called an *antenna complex*. Together, the two types of complex make up a **photosystem** (**Figure 14–45**).

The role of the **antenna complex** in the photosystem is to collect the energy of a sufficient number of photons for photosynthesis. Without it, the process would be slow and inefficient, as each reaction-center chlorophyll would absorb only about one light quantum per second, even in broad daylight, whereas hundreds per second are needed for effective photosynthesis. When light excites a chlorophyll molecule in the antenna complex, the energy passes rapidly from one protein-bound chlorophyll to another by resonance energy transfer until it reaches the special pair in the reaction center. The antenna complex is also known as a

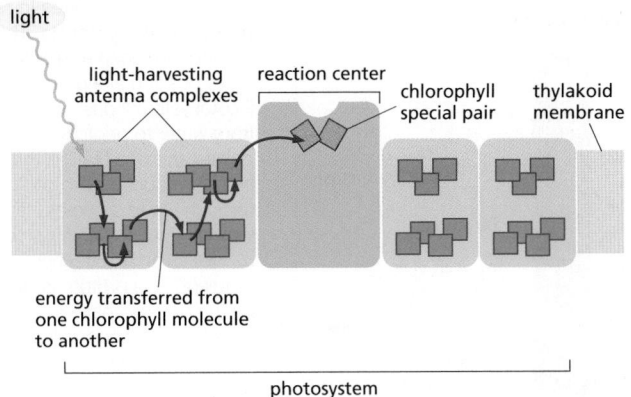

Figure 14–45 **A photosystem.** Each photosystem consists of a reaction center plus a number of light-harvesting antenna complexes. The solar energy for photosynthesis is collected by the antenna complexes, which account for most of the chlorophyll in a plant cell. The energy hops randomly by resonance energy transfer *(red arrows)* from one chlorophyll molecule to another, until it reaches the reaction center complex, where it ionizes a chlorophyll in the special pair. The chlorophyll special pair holds its electrons at a lower energy than the chlorophyll in the antenna complexes, causing the energy transferred to it from the antenna complex to become trapped there. Note that it is only energy that moves from one chlorophyll molecule to another in the antenna complex, not electrons (**Movie 14.10**).

light-harvesting complex, or LHC. In addition to many chlorophyll molecules, an LHC contains orange carotenoid pigments. The carotenoids collect light of a different wavelength from that absorbed by chlorophylls, helping to make the antenna complex more efficient. They also have an important protective role in preventing the formation of harmful oxygen radicals in the photosynthetic membrane.

The Thylakoid Membrane Contains Two Different Photosystems Working in Series

The excitation energy collected by the antenna complex is delivered to the special pair in the **photochemical reaction center**. The reaction center is a transmembrane chlorophyll–protein complex that lies at the heart of photosynthesis. It harbors the special pair of chlorophyll molecules, which acts as an irreversible trap for excitation energy (see Figure 14–45).

Chloroplasts contain two functionally different although structurally related photosystems, each of which feeds electrons generated by the action of sunlight into an electron-transfer chain. In the chloroplast thylakoid membrane, *photosystem I* is confined to the unstacked stroma thylakoids, while the stacked grana thylakoids contain *photosystem II*. The two photosystems were named in order of their discovery, not of their actions in the photosynthetic pathway, and electrons are first activated in photosystem II before being transferred to photosystem I (**Figure 14–46**). The path of the electron through the two photosystems can be described as a Z-like trajectory and is known as the *Z scheme*. In the Z scheme, the reaction center of *photosystem II* first withdraws an electron from water. The electron passes via an electron-transport chain (composed of the electron carrier plastoquinone, the cytochrome b_6-f complex, and the protein plastocyanin) to *photosystem I*, which propels the electron across the membrane in a second light-driven charge-separation reaction that leads to NADPH production.

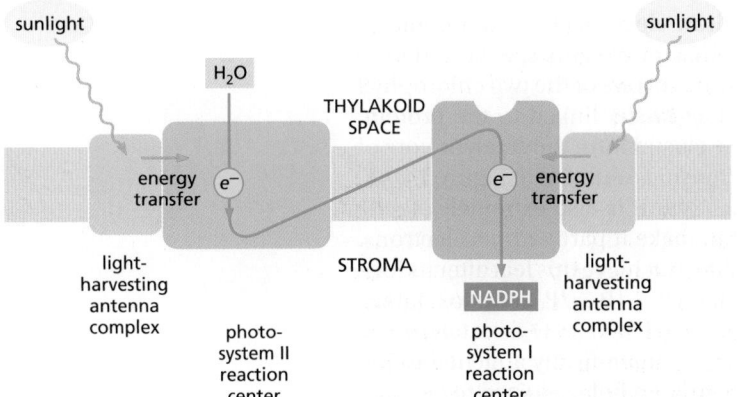

Figure 14–46 **The Z scheme for photosynthesis.** The thylakoids of plants and cyanobacteria contain two different photosystems, known as photosystem I and photosystem II, which work in series. Each of the photosystem I and II reaction centers receives excitation energy from its own set of tightly associated antenna complexes, known as LHC-I and LHC-II, by resonance energy transfer. Note that, for historical reasons, the two photosystems were named opposite to the order in which they act, with photosystem II passing its electrons to photosystem I.

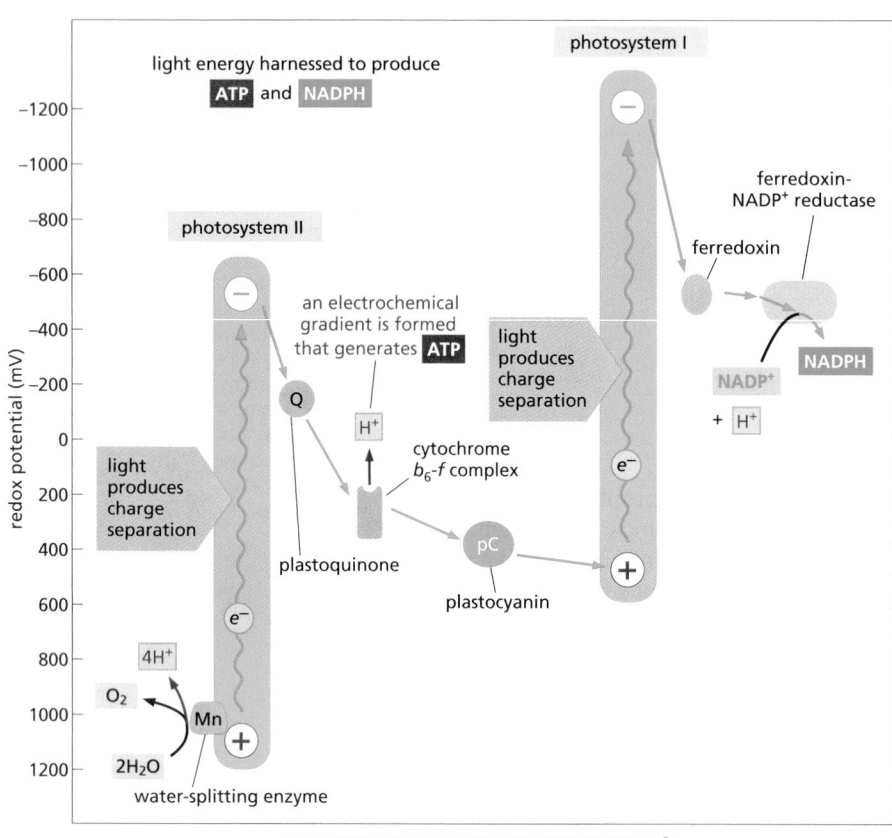

Figure 14–47 **Changes in redox potential during photosynthesis.** The redox potential for each molecule is indicated by its position along the vertical axis. Photosystem II passes electrons derived from water to photosystem I, which in turn passes them to $NADP^+$ through ferredoxin-$NADP^+$ reductase. The net electron flow through the two photosystems is from water to $NADP^+$, and it produces NADPH as well as an electrochemical proton gradient. This proton gradient is used by the ATP synthase to produce ATP. Details in this figure will be explained in the subsequent text.

The Z scheme is necessary to bridge the very large energy gap between water and NADPH (**Figure 14–47**). A single quantum of visible light does not contain enough energy both to withdraw electrons from water, which holds on to its electrons very tightly (redox potential +820 mV) and therefore is a very poor electron donor, and to force them on to $NADP^+$, which is a very poor electron acceptor (redox potential –320 mV). The Z scheme first evolved in cyanobacteria to enable them to use water as a universally available electron source. Other, simpler photosynthetic bacteria have only one photosystem. As we shall see, they cannot use water as an electron source and must rely on other, more energy-rich substrates instead, from which electrons are more readily withdrawn. The ability to extract electrons from water (and thereby to produce molecular oxygen) was acquired by plants when their ancestors took up the endosymbiotic cyanobacteria that later evolved into chloroplasts (see Figure 1–31).

Photosystem II Uses a Manganese Cluster to Withdraw Electrons From Water

In biology, only photosystem II is able to withdraw electrons from water and to generate molecular oxygen as a waste product. This remarkable specialization of photosystem II is conferred by the unique properties of one of the two chlorophyll molecules of its special pair and by a *manganese cluster* linked to the protein. These chlorophyll molecules and the manganese cluster form the catalytic core of the photosystem II reaction center, whose mechanism is outlined in **Figure 14–48**.

Water is an inexhaustible source of electrons, but it is also extremely stable; therefore a large amount of energy is required to make it part with its electrons. The only compound in living organisms that is able to achieve this feat after its ionization by light, is the chlorophyll special pair called P_{680} (P_{680}/P_{680}^+ redox potential = +1270 mV). The reaction $2H_2O + 4$ photons $\rightarrow 4H^+ + 4e^- + O_2$ is catalyzed by its adjacent manganese cluster. The intermediates remain firmly attached to the manganese cluster until two water molecules have been fully oxidized to O_2, thus

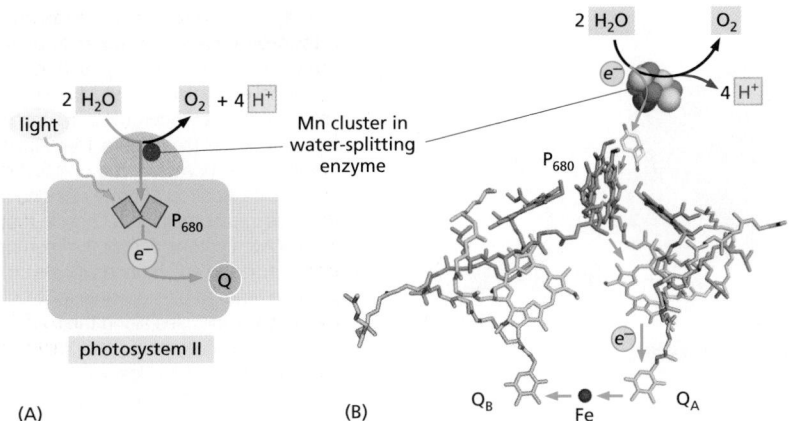

(A) (B)

Figure 14–48 The conversion of light energy to chemical energy in the photosystem II complex. (A) Schematic diagram of the photosystem II reaction center, whose special pair of chlorophyll molecules is designated as P_{680} based on the wavelength of its absorbance maximum (680 nm). (B) Cofactors and pigments at the core of the reaction center. Shown are the manganese (Mn) cluster, the tyrosine side chain that links it to the P_{680} special pair, four chlorophylls (green), two pheophytins (light blue), two plastoquinones (pink), and an iron atom (red). The path of electrons is shown by blue arrows. In the manganese cluster, four manganese atoms (light blue), one calcium atom (purple), and five oxygen atoms (red) work together to catalyze the oxidation of water. The water-splitting reaction occurs in four successive steps, each requiring the energy of one photon. Each photon turns a P_{680} reaction-center chlorophyll into a positively charged chlorophyll ion. Through an ionized tyrosine side chain (yellow), this chlorophyll ion pulls an electron away from a water molecule bound at the manganese cluster. In this way, a total of four electrons are withdrawn from two water molecules to generate molecular oxygen, which is released into the atmosphere.

ensuring that no dangerous oxygen radicals are released as the reaction proceeds. The protons released by the two water molecules are discharged to the thylakoid space, contributing to the proton gradient across the thylakoid membrane (pH lower in the thylakoid space than in the stroma). The unique protein environment that endows life with this all-important ability to oxidize water has remained essentially unchanged throughout billions of years of evolution (**Figure 14–49**).

All of the oxygen in the Earth's atmosphere has been generated in this way. Although the exact details of the water-oxidation reaction in photosystem II are still not fully understood, scientists are trying to construct an artificial system that mimics the process. If successful, this might provide a virtually endless supply of clean energy, helping to solve the world's energy crisis.

The Cytochrome b_6-f Complex Connects Photosystem II to Photosystem I

Following the path shown previously in Figure 14–48, the electrons extracted from water by photosystem II are transferred to plastoquinol, a strong electron donor similar to ubiquinol in mitochondria. This quinol, which can diffuse rapidly in the lipid bilayer of the thylakoid membrane, transfers its electrons to the *cytochrome b_6-f complex*, whose structure is homologous to the cytochrome *c* reductase in mitochondria. The cytochrome b_6-f complex pumps H$^+$ into the thylakoid space using the same Q cycle that is utilized in mitochondria (see Figure 14–21), thereby adding to the proton gradient across the thylakoid membrane.

The cytochrome b_6-f complex forms the connecting link between photosystems II and I in the chloroplast electron-transport chain. It passes its electrons

Each electron that is energized by light passes from the special pair along an electron-transfer chain inside the complex, along the indicated path to the permanently bound plastoquinone Q_A and then to plastoquinone Q_B as electron acceptors. Once Q_B has picked up two electrons (plus two protons; see Figure 14–17), it dissociates from its binding site in the complex and enters the lipid bilayer as a mobile electron carrier, being immediately replaced by a new, nonreduced molecule of plastoquinone. Note that the chlorophylls and pheophytins form two symmetrical branches of a potential electron-transport chain. Only one branch is active, thus ensuring that the plastoquinones become fully reduced in minimum time.

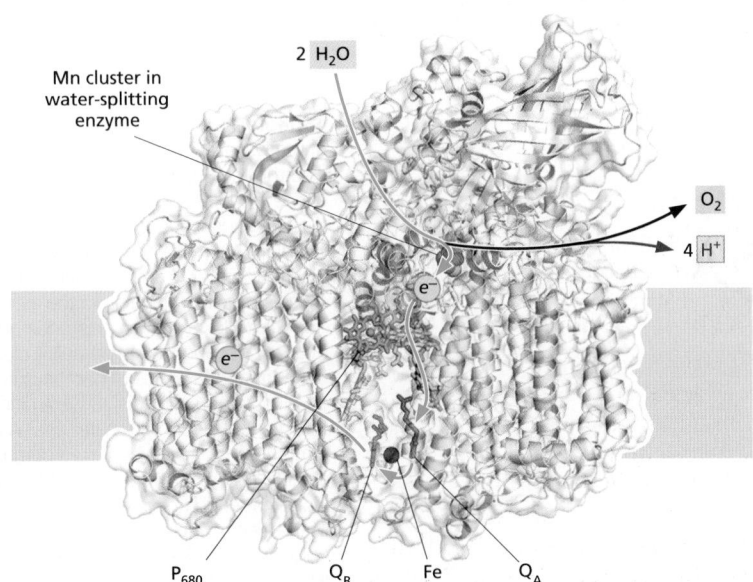

Figure 14–49 The structure of the complete photosystem II complex. This photosystem contains at least 16 protein subunits, along with 36 chlorophylls, two pheophytins, two hemes, and a number of protective carotenoids (colored). Most of these pigments and cofactors are deeply buried, tightly complexed to protein (gray). The path of electrons is indicated by the blue arrows, and is explained in Figure 14–48B. The photosystem II complex presented here is the cyanobacterial complex, which is simpler and more stable than the plant complex, which works in the same way. (PDB code: 3ARC.)

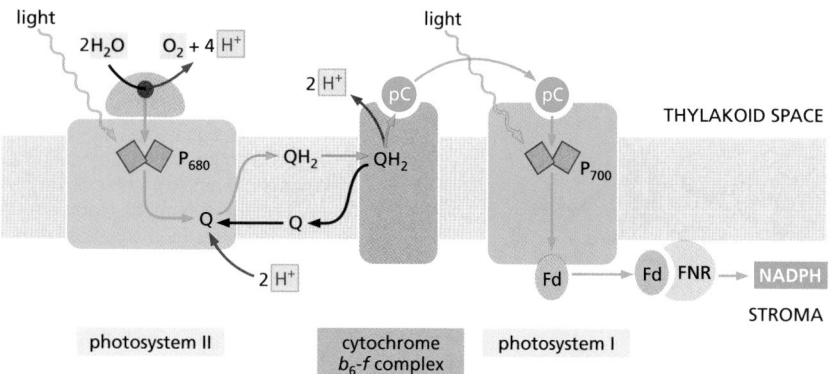

one at a time to the mobile electron carrier plastocyanin (a small copper-containing protein that takes the place of the cytochrome *c* in mitochondria), which will transfer them to photosystem I (**Figure 14–50**). As we discuss next, photosystem I then harnesses a second photon of light to further energize the electrons that it receives.

Photosystem I Carries Out the Second Charge-Separation Step in the Z Scheme

Photosystem I receives electrons from plastocyanin in the thylakoid space and transfers them, via a second charge-separation reaction, to the small protein ferredoxin on the opposite membrane surface (**Figure 14–51**). Then, in a final step, ferredoxin feeds its electrons to a membrane-associated enzyme complex, the *ferredoxin-NADP$^+$ reductase*, which uses the electrons to produce NADPH from NADP$^+$ (see Figure 14–50).

The redox potential of the NADP$^+$/NADPH pair (–320 mV) is already very low, and reduction of NADP$^+$ therefore requires a compound with an even lower redox potential. This turns out to be a chlorophyll molecule near the stromal membrane surface of photosystem I that has a redox potential of –1000 mV (chlorophyll A$_0$), making it the strongest known electron donor in biology. The reduced NADPH is released into the chloroplast stroma, where it is used for biosynthesis of glyceraldehyde 3-phosphate, amino acid precursors, and fatty acids, much of it to be exported to the cytoplasm.

Figure 14–50 Electron flow through the cytochrome b_6-f complex to NADPH. The cytochrome b_6-f complex is the functional equivalent of cytochrome *c* reductase (the cytochrome *b*-c_1 complex) in mitochondria (see Figure 14–22). Like its mitochondrial homolog, the b_6-f complex receives its electrons from a quinone and engages in a complicated Q cycle that pumps two protons across the membrane (details not shown). It hands its electrons, one at a time, to plastocyanin (pC). Plastocyanin diffuses along the membrane surface to photosystem I and transfers the electrons via ferredoxin (Fd) to the ferredoxin-NADP$^+$ reductase (FNR), where they are utilized to produce NADPH. P$_{700}$ is a special pair of chlorophylls that absorbs light of wavelength 700 nm.

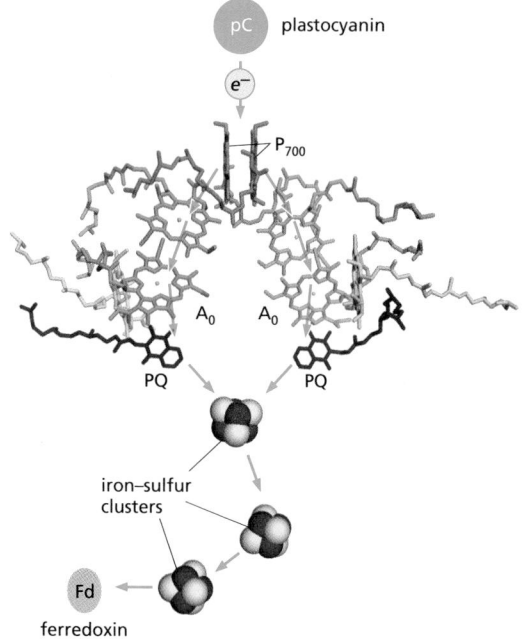

Figure 14–51 Structure and function of photosystem I. At the heart of the photosystem I complex assembly is the electron-transfer chain shown. At one end is a special pair of chlorophylls called P$_{700}$ (because it absorbs light of 700 nm wavelength), receiving electrons from plastocyanin (pC). At the other end are the A$_0$ chlorophylls, which hand the electrons on to ferredoxin via two plastoquinones (PQ; *purple*) and three iron–sulfur clusters. Even though the roles of photosystems I and II in photosynthesis are very different, their central electron-transfer chains are structurally similar, indicating a common evolutionary origin (see Figure 14–53). Note that in photosystem I both branches of the electron-transfer chain are active, unlike in photosystem II (see Figure 14–48). (PDB code: 3LW5.)

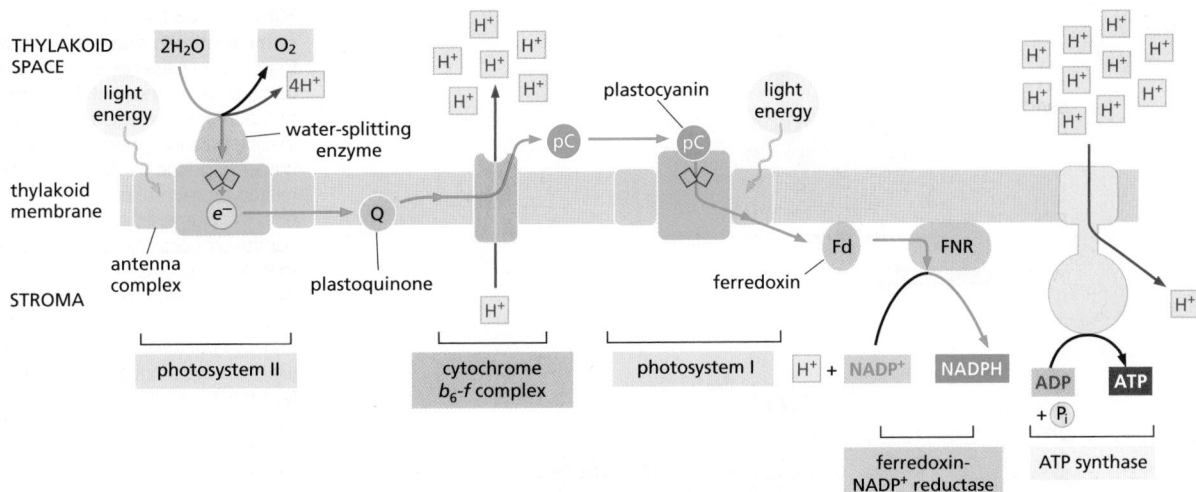

Figure 14–52 **Summary of electron and proton movements during photosynthesis in the thylakoid membrane.** Electrons are withdrawn, through the action of light energy, from a water molecule that is held by the manganese cluster in photosystem II. The electrons pass on to plastoquinone, which delivers them to the cytochrome b_6-f complex that resembles the cytochrome c reductase of mitochondria and the b-c complex of bacteria. They are then carried to photosystem I by the soluble electron carrier plastocyanin, the functional equivalent of cytochrome c in mitochondria. From photosystem I they are transferred to ferredoxin-NADP$^+$ reductase (FNR) by the soluble carrier ferredoxin (Fd; a small protein containing an iron–sulfur center). Protons are pumped into the thylakoid space by the cytochrome b_6-f complex, in the same way that protons are pumped into mitochondrial cristae by cytochrome c reductase (see Figure 14–21). In addition, the H$^+$ released into the thylakoid space by water oxidation, and the H$^+$ consumed during NADPH formation in the stroma, contribute to the generation of the electrochemical H$^+$ gradient across the thylakoid membrane. As illustrated, this gradient drives ATP synthesis by an ATP synthase that sits in the same membrane (see Figure 14–47).

The Chloroplast ATP Synthase Uses the Proton Gradient Generated by the Photosynthetic Light Reactions to Produce ATP

The sequence of events that results in light-driven production of ATP and NADPH in chloroplasts and cyanobacteria is summarized in **Figure 14–52**. Starting with the withdrawal of electrons from water, the light-driven charge-separation steps in photosystems II and I enable the energetically unfavorable (uphill) flow of electrons from water to NADPH (see Figure 14–47). Three small mobile electron carriers—plastoquinone, plastocyanin, and ferredoxin—participate in this process. Together with the electron-driven proton pump of the cytochrome b_6-f complex, the photosystems generate a large proton gradient across the thylakoid membrane. The ATP synthase molecules embedded in the thylakoid membranes then harness this proton gradient to produce large amounts of ATP in the chloroplast stroma, mimicking the synthesis of ATP in the mitochondrial matrix.

The linear Z scheme for photosynthesis thus far discussed can switch to a circular mode of electron flow through photosystem I and the b_6-f complex. Here, the reduced ferredoxin diffuses back to the b_6-f complex to reduce plastoquinone, instead of passing its electrons to the ferredoxin-NADP$^+$ reductase enzyme complex. This, in effect, turns photosystem I into a light-driven proton pump, thereby increasing the proton gradient and thus the amount of ATP made by the ATP synthase. An elaborate set of regulatory mechanisms control this switch, which enables the chloroplast to generate either more NADPH (linear mode) or more ATP (circular mode), depending on the metabolic needs of the cell.

All Photosynthetic Reaction Centers Have Evolved From a Common Ancestor

Evidence for the prokaryotic origins of mitochondria and chloroplasts abounds in their genetic systems, as we will see in the next section. But strong and direct evidence for the evolutionary origins of chloroplasts can also be found in the molecular structures of photosynthetic reaction centers revealed in recent years by crystallography. The positions of the chlorophylls in the special pair and the two branches of the electron-transfer chain are basically the same in photosystem I, photosystem II, and the photochemical reaction centers of photosynthetic bacteria (**Movie 14.11**). As a result, one can conclude that they all have evolved from a common ancestor. Evidently, the molecular architecture of the photosynthetic reaction center originated only once and has remained essentially unchanged during evolution. By contrast, the less critical antenna systems have evolved in several different ways and are correspondingly diverse in present-day photosynthetic organisms (**Figure 14–53**).

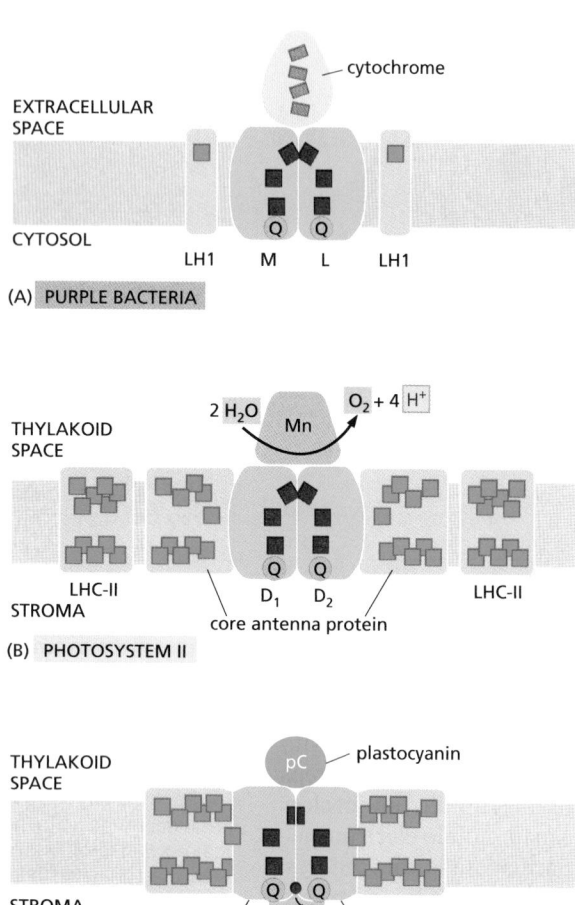

(A) **PURPLE BACTERIA**

(B) **PHOTOSYSTEM II**

(C) **PHOTOSYSTEM I**

Figure 14–53 Evolution of photosynthetic reaction centers. Pigments involved in light-harvesting are colored *green*; those involved in the central photochemical events are colored *red*. (A) The primitive photochemical reaction center of purple bacteria contains two related protein subunits, L and M, that bind the pigments involved in the central process of photosynthesis, including a special pair of chlorophyll molecules. Electrons are fed into the excited chlorophylls by a cytochrome. LH1 is a bacterial antenna complex. (B) Photosystem II contains the D_1 and D_2 proteins, which are homologous to the L and M subunits in (A). The excited P_{680} chlorophyll in the special pair withdraws electrons from water held by the manganese cluster. LHC-II is the light-harvesting complex that feeds energy into the core antenna proteins. (C) Photosystem I contains the Psa A and Psa B proteins, each of which is equivalent to a fusion of the D_1 or D_2 protein to a core antenna protein of photosystem II. The loosely bound plastocyanin (pC) feeds electrons into the excited chlorophyll pair. As indicated, in photosystem I, electrons are passed from a bound quinone (Q) through a series of three iron–sulfur centers *(red circles)*. (Adapted from K. Rhee et al., *Nature* 396:283–286, published 1998 by Macmillan Publishers Ltd. Reproduced with permission of SNCSC.)

The Proton-Motive Force for ATP Production in Mitochondria and Chloroplasts Is Essentially the Same

The proton gradient across the thylakoid membrane depends both on the proton-pumping activity of the cytochrome b_6-f complex and on the photosynthetic activity of the two photosystems, which in turn depends on light intensity. In chloroplasts exposed to light, H^+ is pumped out of the stroma (pH around 8, similar to the mitochondrial matrix) into the thylakoid space (pH 5–6), creating a gradient of 2–3 pH units across the thylakoid membrane, representing a proton-motive force of about 180 mV. This is very similar to the proton-motive force in respiring mitochondria. However, a membrane potential across the inner mitochondrial membrane makes the largest contribution to the proton-motive force that drives the mitochondrial ATP synthase to make ATP, whereas a H^+ gradient predominates for chloroplasts.

In contrast to mitochondrial ATP synthase, which forms long rows of dimers along the cristae ridges, the chloroplast ATP synthase is monomeric and located in flat membrane regions (**Figure 14–54**). Evidently, the H^+ gradient across the thylakoid membrane is high enough for ATP synthesis without the need for the elaborate arrangement of ATP synthase seen in mitochondria.

Chemiosmotic Mechanisms Evolved in Stages

The first living cells on Earth may have consumed geochemically produced organic molecules and generated their ATP by fermentation. Because oxygen was not yet present in the atmosphere, such anaerobic fermentation reactions would have dumped organic acids—such as lactic or formic acids, for example—into the

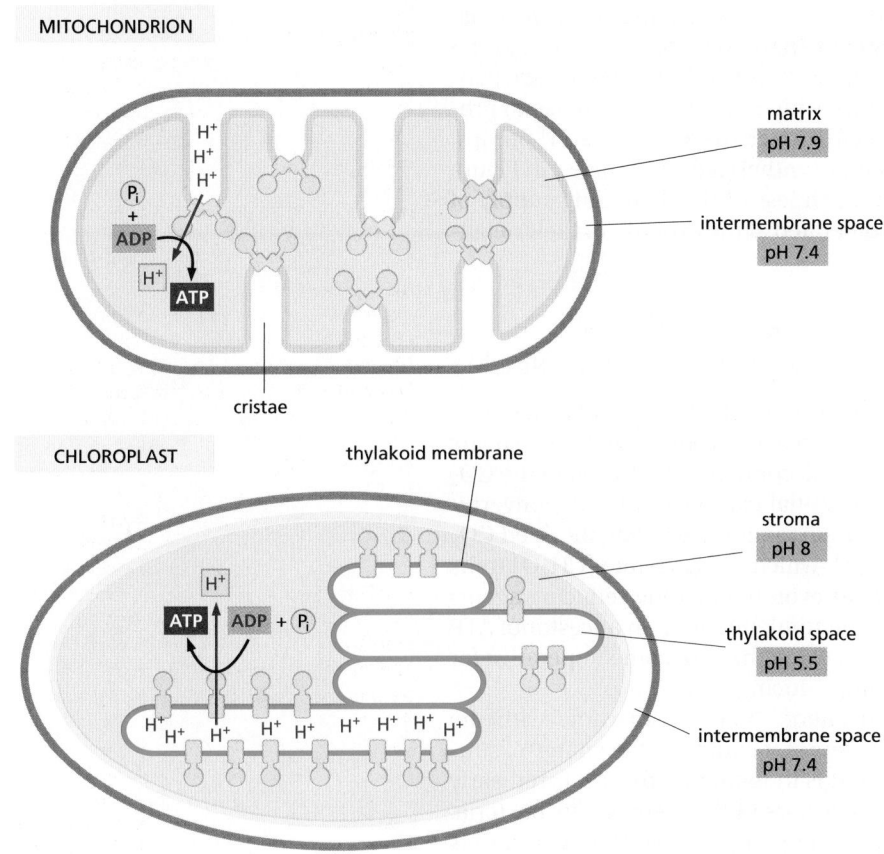

Figure 14–54 **A comparison of H⁺ concentrations and the arrangement of ATP synthase in mitochondria and chloroplasts.** In both organelles, the pH in the intermembrane space is 7.4, as in the cytoplasm. The pH of the mitochondrial matrix and the pH of the chloroplast stroma are both about 8 *(light gray)*. The pH in the thylakoid space is around 5.5, depending on photosynthetic activity. This results in a high proton-motive force across the thylakoid membrane, consisting largely of the H⁺ gradient (a high permeability of this membrane to Mg²⁺ and Cl⁻ ions allows the flow of these ions to dissipate most of the membrane potential).

In contrast to chloroplasts, the H⁺ gradient across the inner mitochondrial membrane is insufficient for ATP production, and mitochondria need a membrane potential to bring the proton-motive force to the same level as in chloroplasts. The arrangement of the mitochondrial ATP synthase in rows of dimers along the cristae ridges (see Figure 14–32) next to the respiratory-chain proton pumps may help the flow of protons along the membrane surface toward the ATP synthase, as the availability of protons is limiting for ATP production. In the chloroplast, the ATP synthase is distributed randomly in thylakoid membranes.

environment (see Figure 2–47). Perhaps such acids lowered the pH of the environment, favoring the survival of cells that evolved transmembrane proteins that could pump H⁺ out of the cytosol, thereby preventing the cell from becoming too acidic (stage 1 in **Figure 14–55**). One of these pumps may have used the energy available from ATP hydrolysis to eject H⁺ from the cell; such a proton pump could have been the ancestor of present-day ATP synthases.

As the Earth's supply of geochemically produced nutrients began to dwindle, organisms that could find a way to pump H⁺ without consuming ATP would have been at an advantage: they could save the small amounts of ATP they derived from the fermentation of increasingly scarce foodstuffs to fuel other important activities. This need to conserve resources might have led to the evolution of electron-transport proteins that allowed cells to use the movement of electrons between molecules of different redox potentials as a source of energy for pumping H⁺ across the plasma membrane (stage 2 in Figure 14–55). Some of these cells might have used the nonfermentable organic acids that neighboring cells had excreted as waste to provide the electrons needed to feed this electron-transport system. Some present-day bacteria grow on formic acid, for example, using the small amount of redox energy derived from the transfer of electrons from formic acid to fumarate to pump H⁺.

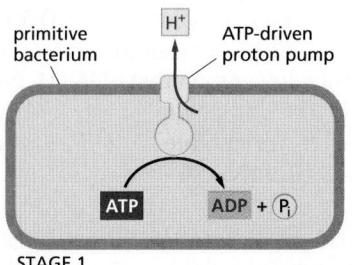

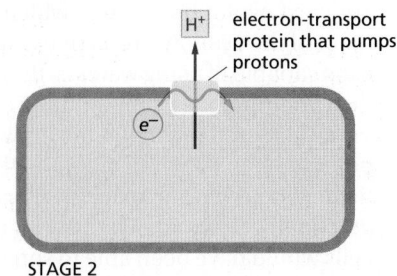

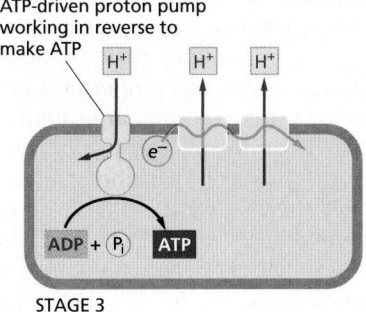

Figure 14–55 **How ATP synthesis by chemiosmosis might have evolved in stages.** The first stage could have involved the evolution of an ATPase that pumped protons out of the cell using the energy of ATP hydrolysis. Stage 2 could have involved the evolution of a different proton pump, driven by an electron-transport chain. Stage 3 would then have linked these two systems together to generate a primitive ATP synthase that used the protons pumped by the electron-transport chain to synthesize ATP. An early bacterium with this final system would have had a selective advantage over bacteria with neither of the systems or only one.

Eventually, some bacteria would have developed H^+-pumping electron-transport systems that were so efficient that they could harvest more redox energy than they needed to maintain their internal pH. Such cells would probably have generated large electrochemical proton gradients, which they could then use to produce ATP. Protons could leak back into the cell through the ATP-driven H^+ pumps, essentially running them in reverse so that they synthesized ATP (stage 3 in Figure 14–55). Because such cells would require much less of the dwindling supply of fermentable nutrients, they would have proliferated at the expense of their neighbors.

By Providing an Inexhaustible Source of Reducing Power, Photosynthetic Bacteria Overcame a Major Evolutionary Obstacle

The gradual depletion of nutrients from the environment on the early Earth meant that organisms had to find some alternative source of carbon to make the sugars that serve as the precursors for so many other cell components. Although the CO_2 in the atmosphere provides an abundant potential carbon source, to convert it into an organic molecule such as a carbohydrate requires reducing the fixed CO_2 with a strong electron donor, such as NADPH, which can generate (CH_2O) units from CO_2 (see Figure 14–41). Early in cellular evolution, strong reducing agents (electron donors) are thought to have been plentiful. But once an ancestor of ATP synthase began to generate most of the ATP, it would have become imperative for cells to evolve a new way of generating strong reducing agents.

A major evolutionary breakthrough in energy metabolism came with the development of photochemical reaction centers that could use the energy of sunlight to produce molecules such as NADPH. It is thought that this occurred early in the process of cellular evolution in the ancestors of the green sulfur bacteria. Present-day green sulfur bacteria use light energy to transfer hydrogen atoms (as an electron plus a proton) from H_2S to NADPH, thereby producing the strong reducing power required for carbon fixation. Because the redox potential of H_2S is much lower than that of H_2O (–230 mV for H_2S compared with +820 mV for H_2O), one quantum of light absorbed by the single photosystem in these bacteria is sufficient to generate NADPH via a relatively simple photosynthetic electron-transport chain.

The Photosynthetic Electron-Transport Chains of Cyanobacteria Produced Atmospheric Oxygen and Permitted New Life-Forms

The next evolutionary step, which is thought to have occurred with the development of the cyanobacteria perhaps 3 billion years ago, was the evolution of organisms capable of using water as the electron source for CO_2 reduction. This entailed the evolution of a water-splitting enzyme and also required the addition of a second photosystem, acting in series with the first, to bridge the large gap in redox potential between H_2O and NADPH. The biological consequences of this evolutionary step were far-reaching. For the first time, there would have been organisms that could survive on water, CO_2, and sunlight (plus a few trace elements). These cells would have been able to spread and evolve in ways denied to the earlier photosynthetic bacteria, which needed H_2S or organic acids as a source of electrons. Consequently, large amounts of biologically synthesized, reduced organic materials accumulated and oxygen entered the atmosphere for the first time.

Oxygen is highly toxic because the oxidation of biological molecules alters their structure and properties indiscriminately and irreversibly. Most anaerobic bacteria, for example, are rapidly killed when exposed to air. Thus, organisms on the primitive Earth would have had to evolve protective mechanisms against the rising O_2 levels in the environment. Late evolutionary arrivals, such as ourselves, have numerous detoxifying mechanisms that protect our cells from the ill effects of oxygen. Even so, an accumulation of oxidative damage to our macromolecules has been postulated to contribute to human aging, as we discuss in the next section.

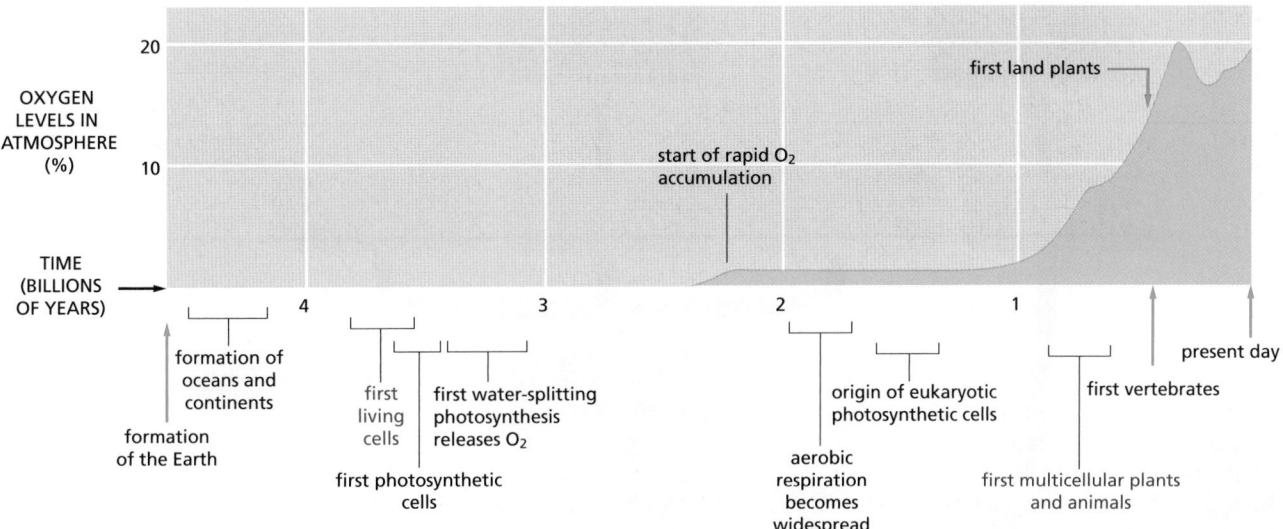

Figure 14–56 **Major events during the evolution of living organisms on Earth.** With the evolution of the membrane-based process of photosynthesis, organisms were able to make their own organic molecules from CO_2 gas. The delay of more than 10^9 years between the appearance of bacteria that split water and released O_2 during photosynthesis and the accumulation of high levels of O_2 in the atmosphere is thought to be due to the initial reaction of the oxygen with the abundant ferrous iron (Fe^{2+}) that was dissolved in the early oceans. Only when the ferrous iron was used up would oxygen have started to accumulate in the atmosphere. In response to the rising oxygen levels, nonphotosynthetic oxygen-consuming organisms evolved, and the concentration of oxygen in the atmosphere equilibrated at its present-day level.

The increase in atmospheric O_2 was very slow at first and would have allowed a gradual evolution of protective devices. For example, the early seas contained large amounts of iron in its reduced, ferrous state (Fe^{2+}), and nearly all the O_2 produced by early photosynthetic bacteria would have been used up in oxidizing Fe^{2+} to ferric Fe^{3+}. This conversion caused the precipitation of huge amounts of stable oxides, and the extensive banded iron formations in sedimentary rocks, beginning about 2.7 billion years ago, help to date the spread of the cyanobacteria. By about 2 billion years ago, the supply of Fe^{2+} was exhausted, and the deposition of further iron precipitates ceased. Geological evidence reveals how O_2 levels in the atmosphere have changed over billions of years, approximating current levels only about 0.5 billion years ago (**Figure 14–56**).

The availability of O_2 enabled the rise of bacteria that developed an aerobic metabolism to make their ATP. These organisms could harness the large amount of energy released by breaking down carbohydrates and other reduced organic molecules all the way to CO_2 and H_2O, as explained when we discussed mitochondria. Components of preexisting electron-transport complexes were modified to produce a cytochrome oxidase, so that the electrons obtained from organic or inorganic substrates could be transported to O_2 as the terminal electron acceptor. Some present-day purple photosynthetic bacteria can switch between photosynthesis and respiration depending on the availability of light and O_2, with only relatively minor reorganizations of their electron-transport chains.

In **Figure 14–57**, we relate these postulated evolutionary pathways to different types of bacteria. By necessity, evolution is always conservative, taking parts of the old and building on them to create something new. Thus, parts of the electron-transport chains that were derived to service anaerobic bacteria 3–4 billion years ago survive, in altered form, in the mitochondria and chloroplasts of today's higher eukaryotes. A good example is the overall similarity in structure and function between the cytochrome *c* reductase that pumps H⁺ in the central segment of the mitochondrial respiratory chain and the analogous cytochrome *b-f* complex in the electron-transport chains of both bacteria and chloroplasts, revealing their common evolutionary origin (**Figure 14–58**).

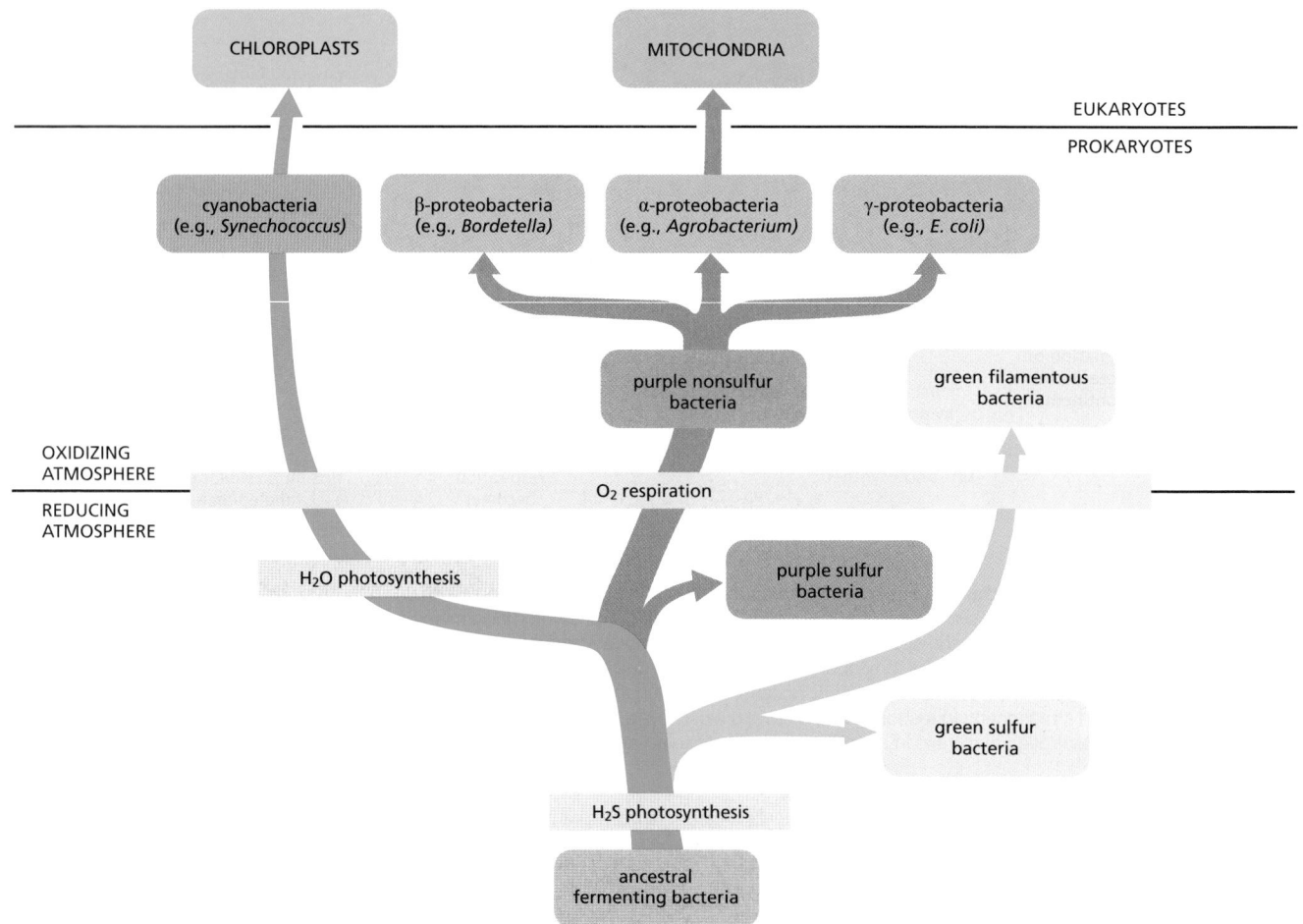

Figure 14–57 Evolutionary scheme showing the postulated origins of mitochondria and chloroplasts and their bacterial ancestors. The consumption of oxygen by respiration is thought to have first developed about 2 billion years ago. Nucleotide-sequence analyses suggest that an endosymbiotic oxygen-evolving cyanobacterium *(cyan)* gave rise to chloroplasts *(dark green)*, while mitochondria arose from an α-proteobacterium. The nearest relatives of mitochondria *(pink)* are members of three closely related groups of α-proteobacteria—the rhizobacteria, agrobacteria, and rickettsias—known to form intimate associations with present-day eukaryotic cells. Proteobacteria are *pink*, purple photosynthetic bacteria are *purple*, and other photosynthetic bacteria are *light green*.

Summary

Chloroplasts and photosynthetic bacteria have the unique ability to harness the energy of sunlight to produce energy-rich compounds. This is achieved by the photosystems, in which chlorophyll molecules attached to proteins are excited when hit by a photon. Photosystems are composed of an antenna complex that collects solar energy and a photochemical reaction center, in which the collected energy is funneled to a chlorophyll molecule held in a special position, enabling it to withdraw electrons from an electron donor. Chloroplasts and cyanobacteria contain two distinct photosystems. The two photosystems are normally linked in series in the Z scheme, and they transfer electrons from water to NADP$^+$ to form NADPH, generating a transmembrane electrochemical potential. One of the two photosystems—photosystem II—can split water by removing electrons from this ubiquitous, low-energy compound. All the molecular oxygen (O$_2$) in our atmosphere is a by-product of the water-splitting reaction in this photosystem. The three-dimensional structures of photosystems I and II are strikingly similar to the photosystems of purple photosynthetic bacteria, demonstrating a remarkable degree of conservation over billions of years of evolution.

The two photosystems and the cytochrome b$_6$-f complex reside in the thylakoid membrane, a separate membrane system in the central stroma compartment of the *chloroplast that is differentiated into stacked grana and unstacked stroma thylakoids. Electron-transport processes in the thylakoid membrane cause protons to be released into the thylakoid space. The backflow of protons through the chloroplast ATP synthase then generates ATP. This ATP is used in conjunction with the NADPH produced by photosynthesis to drive a large number of biosynthetic reactions in the chloroplast stroma, including the carbon-fixation cycle, which generates large amounts of carbohydrates from CO_2.*

In the early evolution of life, cyanobacteria overcame a major obstacle in devising a way to use solar energy to split water and fix carbon dioxide. Cyanobacteria produced both abundant organic nutrients and molecular oxygen, enabling the rise of a multitude of aerobic life-forms. The chloroplasts in plants have evolved from a cyanobacterium that was endocytosed long ago by an aerobic eukaryotic host organism.

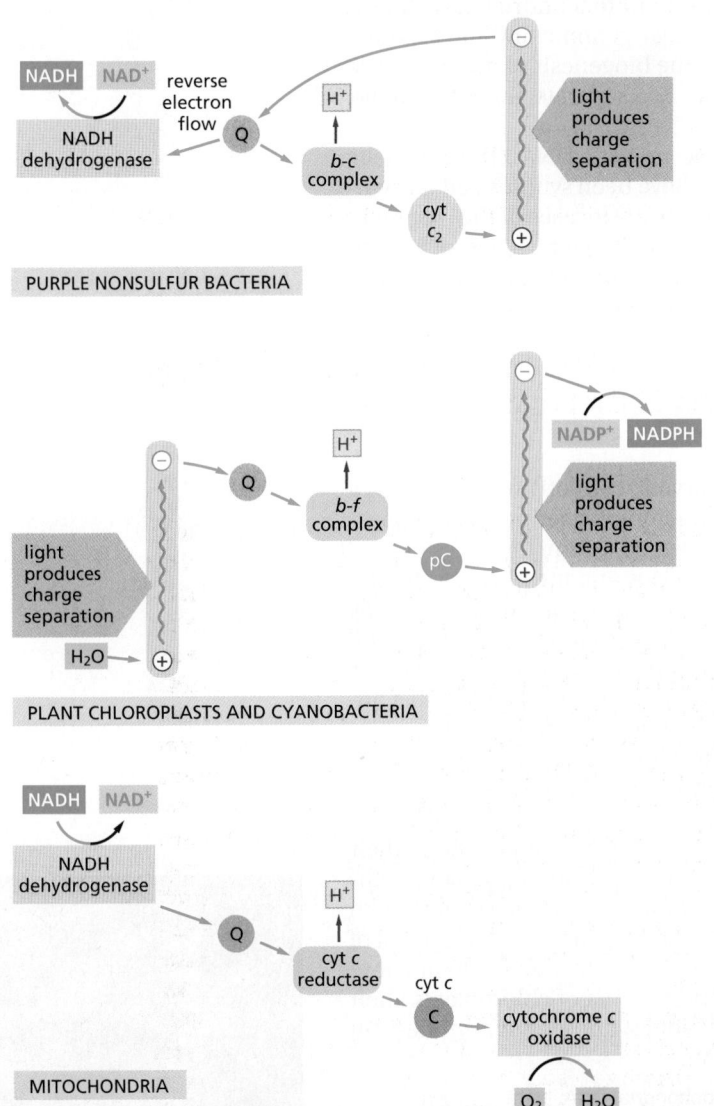

Figure 14–58 A comparison of three electron-transport chains discussed in this chapter. Bacteria, chloroplasts, and mitochondria all contain a membrane-bound enzyme complex that resembles the cytochrome *c* reductase of mitochondria. These complexes all accept electrons from a quinone carrier (Q) and pump H$^+$ across their respective membranes. Moreover, in reconstituted *in vitro* systems, the different complexes can substitute for one another, and the structures of their protein components reveal that they are evolutionarily related. Note that the purple nonsulfur bacteria use a cyclic flow of electrons to produce a large electrochemical proton gradient that drives a *reverse electron flow* through NADH dehydrogenase to produce NADH from NAD$^+$ + H$^+$ + e^-.

THE GENETIC SYSTEMS OF MITOCHONDRIA AND CHLOROPLASTS

As we discussed in Chapter 1, mitochondria and chloroplasts are thought to have evolved from endosymbiotic bacteria (see Figures 1–29 and 1–31). Both types of organelles still contain their own genomes (**Figure 14–59**). As we will discuss shortly, they also retain their own biosynthetic machinery for making RNA and organellar proteins.

Like bacteria, mitochondria and chloroplasts proliferate by growth and division of an existing organelle. In actively dividing cells, each type of organelle must double in mass in each cell generation and then be distributed into each daughter cell. In addition, nondividing cells must replenish organelles that are degraded as part of the continual process of organelle turnover, or produce additional organelles as the need arises. Organelle growth and proliferation are therefore carefully controlled. The process is complicated because mitochondrial and chloroplast proteins are encoded in two places: the nuclear genome and the separate genomes harbored in the organelles themselves. The biogenesis of mitochondria and chloroplasts thus requires contributions from two separate genetic systems, which must be closely coordinated.

Most organellar proteins are encoded by the nuclear DNA. The organelle imports these proteins from the cytosol, after they have been synthesized on cytosolic ribosomes, through the mitochondrial protein translocases of the outer and inner mitochondrial membrane—TOM and TIM. In Chapter 12, we discussed how this happens. Here, we describe the organelle genomes and genetic systems, and consider the consequences of separate organelle genomes for the cell and the organism as a whole.

The Genetic Systems of Mitochondria and Chloroplasts Resemble Those of Prokaryotes

As discussed in Chapter 12, it is thought that eukaryotic cells originated through a symbiotic relationship between an archaeon and an aerobic bacterium (a proteobacterium). The two organisms are postulated to have merged to form the ancestor of all nucleated cells, with the archeaon providing the nucleus and the proteobacterium serving as a respiring, ATP-producing endosymbiont—one that would eventually evolve into the mitochondrion (see Figure 12–3). This most likely occurred roughly 1.6 billion years ago, when oxygen had entered the atmosphere in substantial amounts (see Figure 14–56). The chloroplast was derived later, after the plant and animal lineages diverged, through endocytosis of an oxygen-producing cyanobacterium.

This *endosymbiont hypothesis* of organelle development receives strong support from the observation that the genetic systems of mitochondria and chloroplasts are similar to those of present-day bacteria. For example, chloroplast ribosomes are very similar to bacterial ribosomes, both in their structure and in their sensitivity to various antibiotics (such as chloramphenicol, streptomycin, erythromycin, and tetracyclin). In addition, protein synthesis in chloroplasts starts with *N*-formylmethionine, as in bacteria, and not with methionine as in the cytosol of eukaryotic cells. Although mitochondrial genetic systems are much less similar to those of present-day bacteria than are the genetic systems of chloroplasts, their ribosomes are also sensitive to antibacterial antibiotics, and protein synthesis in mitochondria also starts with *N*-formylmethionine.

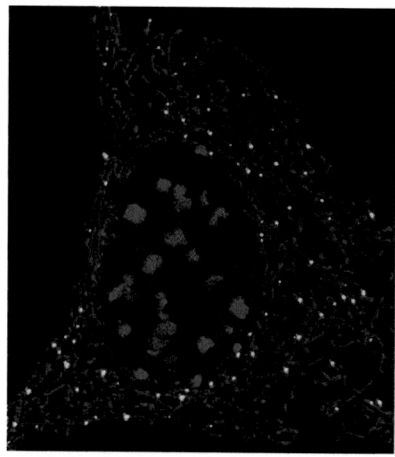

Figure 14–59 Staining of nuclear and mitochondrial DNA. In this confocal micrograph of a single fibroblast cell, the nuclear DNA is stained with a fluorescent dye *(blue)* while the mitochondrial DNA is visualized indirectly using a tagged mitochondrial transcription factor *(green)*. The mitochondrial network is stained with a fluorescent mitochondrial matrix marker *(red)*. The image was acquired using a structured illumination microscope (SIM), which yields approximately double the resolution of a confocal microscope. Numerous copies of the mitochondrial genome can be seen distributed in distinct nucleoids throughout the mitochondria that snake through the cytoplasm. (From C. Kukat et al., *Proc. Natl. Acad Sci. USA* 108(33):13534–13539, 2011. Copyright 2011 National Academy of Sciences, USA. With permission from National Academy of Sciences.)

5 µm

The processes of organelle DNA transcription, protein synthesis, and DNA replication take place where the genome is located: in the matrix of mitochondria or the stroma of chloroplasts. Although the enzymes that mediate these genetic processes are unique to the organelle, and resemble those of bacteria (or even of bacterial viruses) rather than their eukaryotic analogs, the nuclear genome encodes the vast majority of these enzymes. Indeed, most present-day mitochondrial and chloroplast proteins are encoded by genes that reside in the cell nucleus.

Over Time, Mitochondria and Chloroplasts Have Exported Most of Their Genes to the Nucleus by Gene Transfer

The nature of the organelle genes located in the nucleus of the cell demonstrates that an extensive transfer of genes from organelle to nuclear DNA has occurred in the course of eukaryotic evolution. Such successful *gene transfer* is expected to be rare, because any gene moved from the organelle needs to adapt to both nuclear transcription and cytoplasmic translation requirements. In addition, the protein needs to acquire a signal sequence that directs it to the correct organelle after its synthesis in the cytosol. By comparing the genes in the mitochondria from different organisms, we can infer that some of the gene transfers to the nucleus occurred relatively recently. The smallest and presumably most highly evolved mitochondrial genomes, for example, encode only a few hydrophobic inner-membrane proteins of the electron-transport chain, plus ribosomal RNAs (rRNAs) and transfer RNAs (tRNAs). Other mitochondrial genomes that have remained more complex tend to contain this same subset of genes along with others (**Figure 14–60**). The most complex mitochondrial genomes include genes that encode components of the mitochondrial genetic system, such as RNA polymerase subunits and ribosomal proteins; these same genes are found in the cell nucleus in yeast and all animal cells.

The proteins that are encoded by genes in the organellar DNA are synthesized on ribosomes within the organelle, using organelle-produced messenger RNA (mRNA) to specify their amino acid sequence (**Figure 14–61**). The protein traffic between the cytosol and these organelles seems to be unidirectional: proteins are normally not exported from mitochondria or chloroplasts to the cytosol. An important exception occurs when a cell is about to undergo apoptosis. As will be

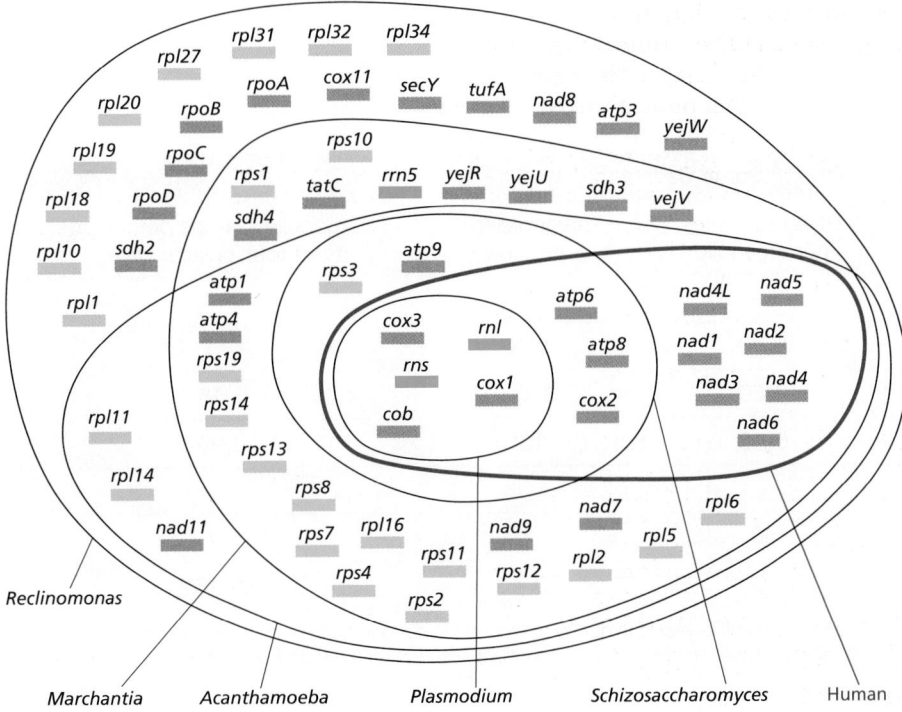

Figure 14–60 **Comparison of mitochondrial genomes.** Less complex mitochondrial genomes encode subsets of the proteins and ribosomal RNAs that are encoded by larger mitochondrial genomes. In this comparison, there are only five genes that are shared by the six mitochondrial genomes; these encode ribosomal RNAs (*rns* and *rnl*), cytochrome b (*cob*), and two cytochrome oxidase subunits (*cox1* and *cox3*). *Blue* indicates ribosomal RNAs; *green*, ribosomal proteins; and *brown*, components of the respiratory chain and other proteins. (Adapted from M.W. Gray, G. Burger and B.F. Lang, *Science* 283:1476–1481, 1999.)

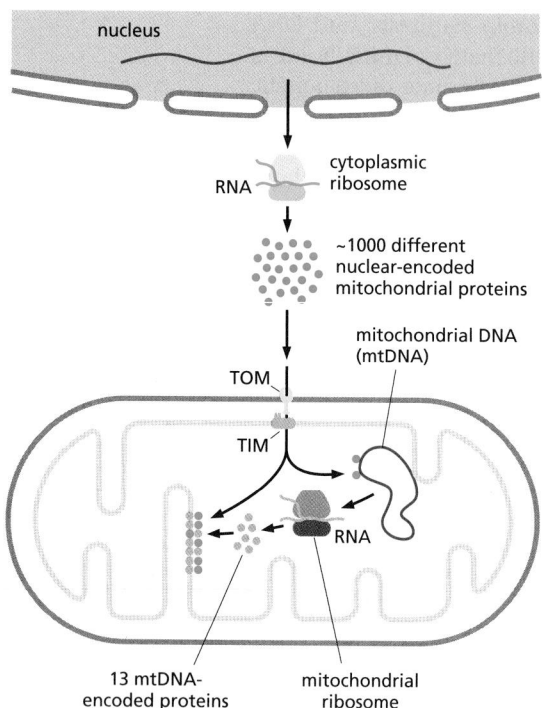

Figure 14–61 Biogenesis of the respiratory-chain proteins in human mitochondria. Most of the protein components of the mitochondrial respiratory chain are encoded by nuclear DNA, with only a small number encoded by mitochondrial DNA (mtDNA). Transcription of mtDNA produces 13 mRNAs, all of which encode subunits of the oxidative phosphorylation system, and the 24 RNAs (22 transfer RNAs and 2 ribosomal RNAs) needed for translation of these mRNAs on the mitochondrial ribosomes *(brown)*.

The mRNAs produced by transcription of nuclear genes are translated on cytoplasmic ribosomes *(green)*, which are distinct from the mitochondrial ribosomes. The nuclear-encoded mitochondrial proteins *(dark green)* are imported into mitochondria through two protein translocases called TOM and TIM, and constitute the vast majority of the approximately 1000 different protein species present in mammalian mitochondria. The nuclear-encoded mitochondrial proteins in humans include the majority of the oxidative phosphorylation system subunits, all proteins needed for expression and maintenance of mtDNA, and all proteins of the mitochondrial ribosomes.

The mtDNA-encoded subunits *(orange)* assemble together with the nuclear subunits to form a functional oxidative phosphorylation system. (Adapted from N.G. Larsson, *Annu. Rev. Biochem.* 79:683–706, 2010.)

discussed in detail in Chapter 18, during apoptosis the mitochondrion releases proteins (most notably cytochrome *c*) from the crista space through its outer mitochondrial membrane, as part of an elaborate signaling pathway that is triggered to cause cells to undergo programmed cell death.

The Fission and Fusion of Mitochondria Are Topologically Complex Processes

In mammalian cells, mitochondrial DNA makes up less than 1% of the total cellular DNA. In other cells, however, such as the leaves of higher plants or the very large egg cells of amphibians, a much larger fraction of the cellular DNA may be present in mitochondria or chloroplasts (Table 14–2), and a large fraction of the total RNA and protein synthesis takes place in the organelles.

Mitochondria and chloroplasts are large enough to be visible by light microscopy in living cells. For example, mitochondria can be visualized by expressing in cells a genetically engineered fusion of a mitochondrial protein linked to green

TABLE 14–2 Relative Amounts of Organelle DNA in Some Cells and Tissues

Organism	Tissue or cell type	DNA molecules per organelle	Organelles per cell	Organelle DNA as percentage of total cellular DNA
Mitochondrial DNA				
Rat	Liver	5–10	1000	1
Yeast*	Vegetative	2–50	1–50	15
Frog	Egg	5–10	10^7	99
Chloroplast DNA				
Chlamydomonas	Vegetative	80	1	7
Maize	Leaves	0–300**	20–40	0–15**

*The large variation in the number and size of mitochondria per cell in yeasts is due to mitochondrial fusion and fission.**In maize, the amount of chloroplast DNA drops precipitously in mature leaves, after cell division ceases: the chloroplast DNA is degraded and stable mRNAs persist to provide for protein synthesis.

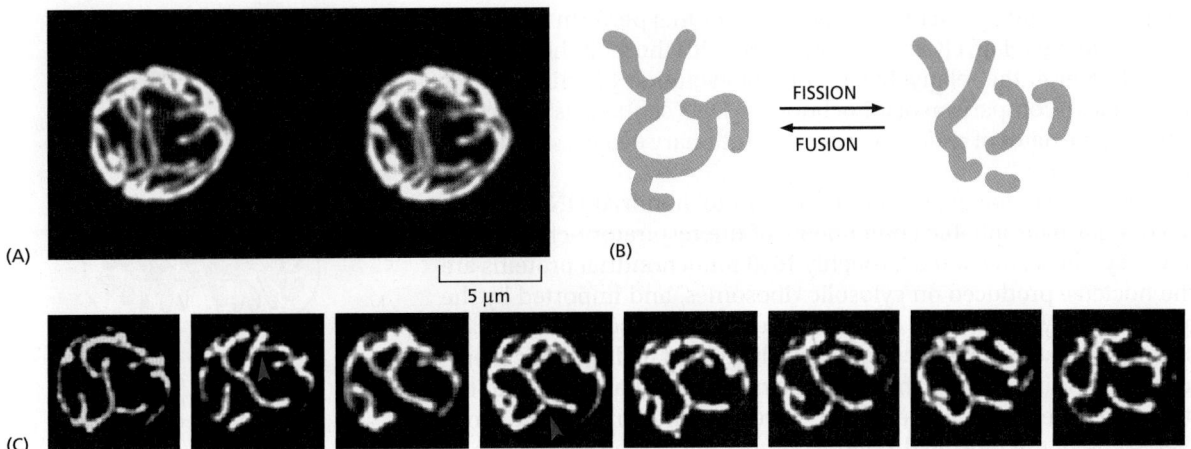

Figure 14–62 **The mitochondrial reticulum is dynamic.** (A) In yeast cells, mitochondria form a continuous reticulum on the cytoplasmic side of the plasma membrane (stereo pair). (B) A balance between fission and fusion determines the arrangement of the mitochondria in different cells. (C) Time-lapse fluorescent microscopy shows the dynamic behavior of the mitochondrial network in a yeast cell. In addition to shape changes, fission and fusion constantly remodel the network *(red arrows)*. These pictures were taken at 3-minute intervals. (A and C, from J. Nunnari et al., *Mol. Biol. Cell* 8:1233–1242, 1997. With permission from the American Society for Cell Biology.)

fluorescent protein (GFP), or cells can be incubated with a fluorescent dye that is specifically taken up by mitochondria because of their membrane potential. Such images demonstrate that the mitochondria in living cells are dynamic—frequently dividing by fission, fusing, and changing shape (**Figure 14–62** and **Movie 14.12**). The fission of mitochondria may be necessary so that small parts of the network can pinch off and reach remote regions of the cell—for example in the thin, extended axon and dendrites of a neuron.

The fission and fusion of mitochondria are topologically complex processes that must ensure the integrity of the separate mitochondrial compartments defined by the inner and outer membranes. These processes control the number and shape of mitochondria, which can vary dramatically in different cell types, ranging from multiple spherical or wormlike organelles to a highly branched, net-shaped single organelle called a *reticulum*. Each depends on its own special set of proteins. The mitochondrial fission machine works by assembling dynamin-related GTPases (discussed in Chapter 13) into helical oligomers that cause local constrictions in tubular mitochondria. GTP hydrolysis then generates the mechanical force that severs the inner and outer mitochondrial membranes in one step (**Figure 14–63**). Mitochondrial fusion requires two separate machineries, one each for the outer and the inner membrane (**Figure 14–64**). In addition to GTP hydrolysis for force generation, both mechanisms also depend on the mitochondrial proton-motive force for reasons that are still unknown.

Animal Mitochondria Contain the Simplest Genetic Systems Known

Comparisons of DNA sequences in different organisms reveal that, in vertebrates (including ourselves), the mutation rate during evolution has been roughly 100 times greater in the mitochondrial genome than in the nuclear genome. This difference is likely to be due to lower fidelity of mitochondrial DNA replication,

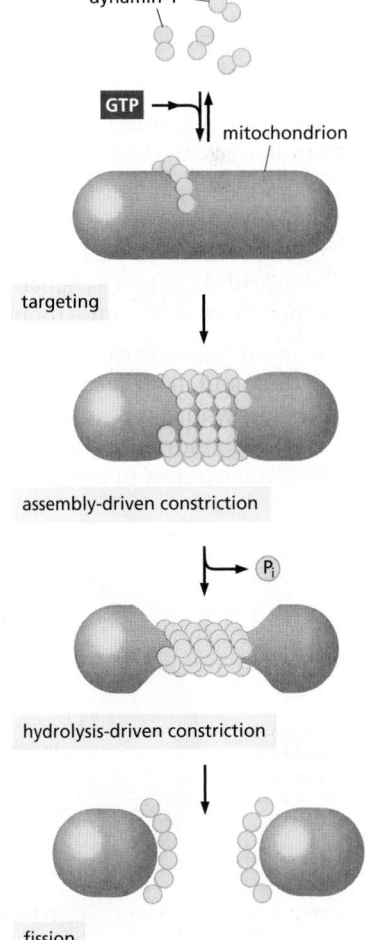

Figure 14–63 **A model for mitochondrial division.** Dynamin-1 *(yellow)* exists as dimers in the cytosol, which form larger oligomeric structures in a process that requires GTP hydrolysis. Dynamin assemblies interact with the outer mitochondrial membrane through special adaptor proteins, forming a spiral of GTP-dynamin around the mitochondrion that causes a constriction. A concerted GTP-hydrolysis event in the dynamin subunits is then thought to produce the conformational changes that result in fission. (Adapted from S. Hoppins, L. Lackner and J. Nunnari, *Annu. Rev. Biochem.* 76:751–780, 2007. With permission from Annual Reviews.)

inefficient DNA repair, or both, given that the mechanisms that perform these processes in the organelle are relatively simple compared with those in the nucleus. As discussed in Chapter 4, the relatively high rate of evolution of animal mitochondrial genes makes a comparison of mitochondrial DNA sequences especially useful for estimating the dates of relatively recent evolutionary events, such as the steps in primate evolution.

There are 13 protein-encoding genes in human mitochondrial DNA (**Figure 14–65**). These code for hydrophobic components of the respiratory-chain complexes and of ATP synthase. In contrast, roughly 1000 mitochondrial proteins are encoded in the nucleus, produced on cytosolic ribosomes, and imported by the protein import machinery in the outer and inner membrane (discussed in Chapter 12). It has been suggested that the cytosolic production of hydrophobic membrane proteins and their import into the organelle may present a problem to the cell, and that this is the reason why their genes have remained in the mitochondrion. However, some of the most hydrophobic mitochondrial proteins, such as the *c* subunit of the ATP synthase rotor ring, are imported from the cytosol in some species (though they are mitochondrially encoded in others). And the parasites *Plasmodium falciparum* and *Leishmania tarentolae*, which spend most of their life cycles inside cells of their host organisms, have retained only two or three mitochondrially encoded proteins.

The size range of mitochondrial DNAs is similar to that of viral DNAs. The mitochondrial DNA in *Plasmodium falciparum* (the human malaria parasite) has less than 6000 nucleotide pairs, whereas the mitochondrial DNAs of some land plants contain more than 300,000 nucleotide pairs (**Figure 14–66**). In animals, the mitochondrial genome is a simple DNA circle of about 16,600 nucleotide pairs (less than 0.001% of the nuclear genome), and it is nearly the same size in organisms as different from us as *Drosophila* and sea urchins.

Mitochondria Have a Relaxed Codon Usage and Can Have a Variant Genetic Code

The human mitochondrial genome has several surprising features that distinguish it from nuclear, chloroplast, and bacterial genomes:

1. *Dense gene packing.* Unlike other genomes, the human mitochondrial genome seems to contain almost no noncoding DNA: nearly every nucleotide seems to be part of a coding sequence, either for a protein or for one of the rRNAs or tRNAs. Since these coding sequences run directly into each other, there is very little room left for regulatory DNA sequences.

2. *Relaxed codon usage.* Whereas 30 or more tRNAs specify amino acids in the cytosol and in chloroplasts, only 22 tRNAs are required for mitochondrial protein synthesis. The normal codon–anticodon pairing rules are relaxed in mitochondria, so that many tRNA molecules recognize any one of the four nucleotides in the third (wobble) position. Such "2 out of 3" pairing

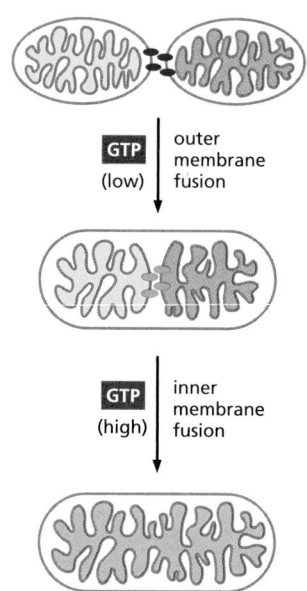

Figure 14–64 A model for mitochondrial fusion. The fusions of the outer and inner mitochondrial membranes are coordinated sequential events, each of which requires a separate set of protein factors. Outer membrane fusion is brought about by an outer-membrane GTPase *(purple)*, which forms an oligomeric complex that includes subunits anchored in the two membranes to be fused. Fusion of outer membranes requires GTP and an H+ gradient across the inner membrane. For fusion of the inner membrane, a dynamin-related protein forms an oligomeric tethering complex *(blue)* that includes subunits anchored in the two inner membranes to be fused. Fusion of the inner membranes requires GTP and the electrical component of the potential across the inner membrane. (Adapted from S. Hoppins, L. Lackner and J. Nunnari, *Annu. Rev. Biochem.* 76:751–780, 2007. With permission from Annual Reviews.)

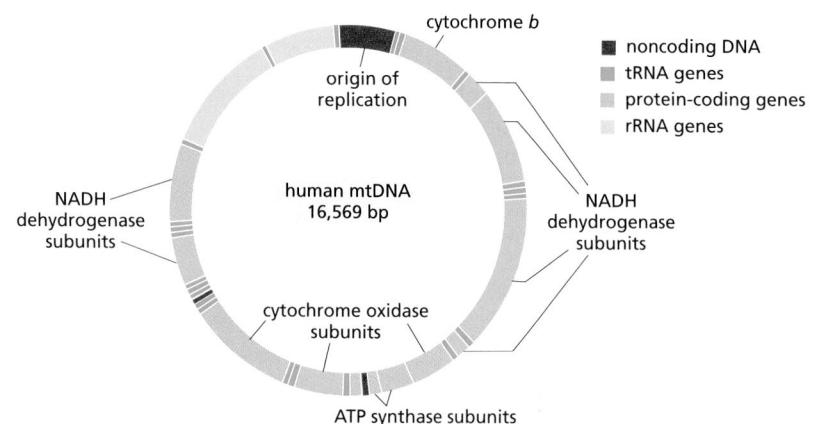

Figure 14–65 The organization of the human mitochondrial genome. The human mitochondrial genome of ≈16,600 nucleotide pairs contains 2 rRNA genes, 22 tRNA genes, and 13 protein-coding sequences. There are two transcriptional promoters, one for each strand of the mitochondrial DNA (mtDNA). The DNAs of many other animal mitochondrial genomes have been completely sequenced. Most of these animal mitochondrial DNAs encode precisely the same genes as humans, with the gene order being identical for animals ranging from fish to mammals.

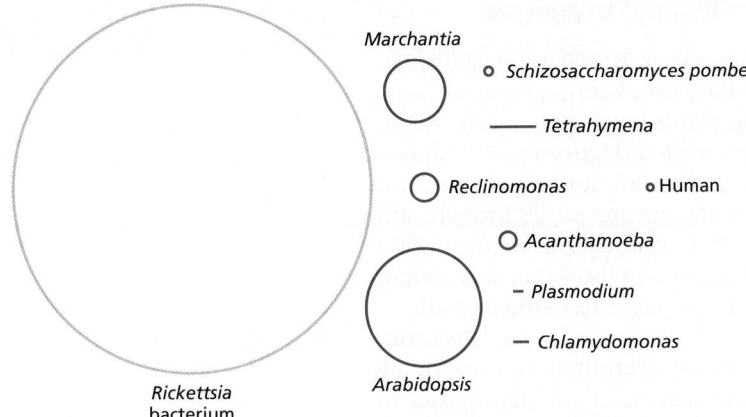

Marchantia

○ *Schizosaccharomyces pombe*

—— *Tetrahymena*

○ *Reclinomonas*   ○ Human

○ *Acanthamoeba*

– *Plasmodium*

– *Chlamydomonas*

Rickettsia
bacterium

Arabidopsis

Figure 14–66 Comparison of various sizes of mitochondrial genomes with the genome of bacterial ancestors. The complete DNA sequences for thousands of mitochondrial genomes have been determined. The lengths of a few of these mitochondrial DNAs are shown to scale— as circles for those genomes thought to be circular and lines for linear genomes. The largest circle represents the genome of *Rickettsia prowazekii*, a small pathogenic bacterium whose genome most closely resembles that of mitochondria. The size of mitochondrial genomes does not correlate well with the number of proteins encoded in them: while human mitochondrial DNA encodes 13 proteins, the 22-fold larger mitochondrial DNA of *Arabidopsis thaliana* encodes only 32 proteins—that is, about 2.5-fold as many as human mitochondrial DNA. The extra DNA that is found in *Arabidopsis*, *Marchantia*, and other plant mitochondria may be "junk DNA"—that is, noncoding DNA with no apparent function. The mitochondrial DNA of the protozoan *Reclinomonas americana* has 98 genes. (Adapted from M.W. Gray, G. Burger and B.F. Lang, *Science* 283:1476–1481, 1999.)

allows one tRNA to pair with any one of four codons and permits protein synthesis with fewer tRNA molecules.

3. *Variant genetic code.* Perhaps most surprising, comparisons of mitochondrial gene sequences and the amino acid sequences of the corresponding proteins indicate that the genetic code is different: 4 of the 64 codons have different "meanings" from those of the same codons in other genomes (Table 14–3).

The close similarity of the genetic code in all organisms provides strong evidence that they all have evolved from a common ancestor. How, then, do we explain the differences in the genetic code in many mitochondria? A hint comes from the finding that the mitochondrial genetic code in different organisms is not the same. In the mitochondrion with the largest number of genes in Figure 14–60, that of the protozoan *Reclinomonas*, the genetic code is unchanged from the standard genetic code of the cell nucleus. Yet UGA, which is a stop codon elsewhere, is read as tryptophan in the mitochondria of mammals, fungi, and invertebrates. Similarly, the codon AGG normally codes for arginine, but it codes for *stop* in the mitochondria of mammals and codes for serine in the mitochondria of *Drosophila* (see Table 14–3). Such variation suggests that a random drift can occur in the genetic code in mitochondria. Presumably, the unusually small number of proteins encoded by the mitochondrial genome makes an occasional change in the meaning of a rare codon tolerable, whereas such a change in a larger genome would alter the function of many proteins and thereby destroy the cell.

Interestingly, in many species, one or two tRNAs for mitochondrial protein synthesis are encoded in the nucleus. Some parasites, for example trypanosomes, have not retained any tRNA genes in their mitochondrial DNA. Instead, the required tRNAs are all produced in the cytosol and are thought to be imported into the mitochondrion by special tRNA translocases that are distinct from the mitochondrial protein import system.

TABLE 14–3 Some Differences Between the "Universal" Code and Mitochondrial Genetic Codes*

| Codon | "Universal" code | Mitochondrial codes | | | |
		Mammals	Invertebrates	Yeasts	Plants
UGA	STOP	*Trp*	*Trp*	*Trp*	STOP
AUA	Ile	*Met*	*Met*	*Met*	Ile
CUA	Leu	Leu	Leu	*Thr*	Leu
AGA AGG	Arg	*STOP*	*Ser*	Arg	Arg

*Red italics indicate that the code differs from the "Universal" code.

Chloroplasts and Bacteria Share Many Striking Similarities

The chloroplast genomes of land plants range in size from 70,000 to 200,000 nucleotide pairs. More than 300 chloroplast genomes have now been sequenced. Many are surprisingly similar, even in distantly related plants (such as tobacco and liverwort), and even those of green algae are closely related (**Figure 14–67**). Chloroplast genes are involved in three main processes: transcription, translation, and photosynthesis. Plant chloroplast genomes typically encode 80–90 proteins and around 45 RNAs, including 37 or more tRNAs. As in mitochondria, most of the organelle-encoded proteins are part of larger protein complexes that also contain one or more subunits encoded in the nucleus and imported from the cytosol.

The genomes of chloroplasts and bacteria have striking similarities. Basic regulatory sequences, such as transcription promoters and terminators, are virtually identical. The amino acid sequences of the proteins encoded in chloroplasts are clearly recognizable as bacterial, and several clusters of genes with related functions (such as those encoding ribosomal proteins) are organized in the same way in the genomes of chloroplasts, the bacterium *E. coli*, and cyanobacteria.

The mechanisms by which chloroplasts and bacteria divide are also similar. Both utilize *FtsZ* proteins, which are self-assembling GTPases related to tubulins (see Chapter 16). Bacterial FtsZ is a soluble protein that assembles into a dynamic ring of membrane-attached protofilaments beneath the plasma membrane in the middle of the dividing cell. The FtsZ ring acts as a scaffold for recruitment of other cell-division proteins and generates a contractile force that results in membrane constriction and eventually in cell division. Presumably, chloroplasts divide in very much the same way. Although both employ membrane-interacting GTPases, the mechanisms by which mitochondria and chloroplasts divide are fundamentally different. The machinery for chloroplast division acts from the inside, as in bacteria, while the dynamin-like GTPases divide mitochondria from the outside (see Figure 14–63). The chloroplasts have remained closer to their bacterial origins than have mitochondria, since the eukaryotic mechanisms of membrane constriction and vesicle formation have been adapted for mitochondrial fission.

The RNA editing and RNA processing that is prevalent in chloroplasts owes everything to their eukaryotic hosts. This RNA processing includes the generation of transcript 5′ and 3′ termini and the cleavage of polycistronic transcripts. In addition, an RNA editing process converts specific C residues to U and can change the amino acid specified by the edited codon. These and other RNA-based

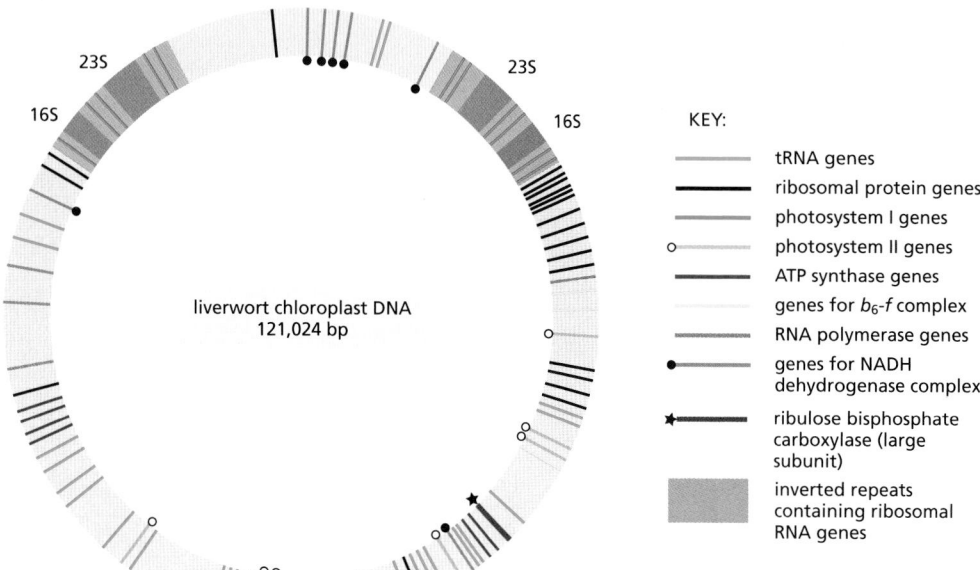

KEY:

——	tRNA genes
——	ribosomal protein genes
——	photosystem I genes
○——	photosystem II genes
——	ATP synthase genes
——	genes for b_6-f complex
——	RNA polymerase genes
●——	genes for NADH dehydrogenase complex
★——	ribulose bisphosphate carboxylase (large subunit)
▨	inverted repeats containing ribosomal RNA genes

Figure 14–67 The organization of the liverwort chloroplast genome. The chloroplast genome organization is similar in all higher plants, although the size varies from species to species—depending on how much of the DNA surrounding the genes encoding the chloroplast's 16S and 23S ribosomal RNAs is present in two copies.

processes are catalyzed by protein families that are not found in prokaryotes. One can ask why the expression of so few chloroplast genes needs to be so complex. One explanation is that the expression of chloroplast and nuclear genes must be closely coordinated. More generally, the bacterial concept of the operon as a co-regulated set of genes in a single transcription unit has been largely abandoned in chloroplasts. Polycistronic transcripts are cleaved into smaller fragments, which then require splicing or RNA editing to become functional.

Organelle Genes Are Maternally Inherited in Animals and Plants

In *Saccharomyces cerevisiae* (baker's yeast), when two haploid cells mate, they are equal in size and contribute equal amounts of mitochondrial DNA to the diploid zygote. Mitochondrial inheritance in yeasts is therefore *biparental*: both parents contribute equally to the mitochondrial gene pool of the progeny. However, during the course of the subsequent asexual, vegetative growth, the mitochondria become distributed more or less randomly to daughter cells. After a few generations, the mitochondria of any given cell contain only the DNA from one or the other parent cell, because only a small sample of the mitochondrial DNA passes from the mother cell to the bud of the daughter cell. This process is known as *mitotic segregation*, and it gives rise to a distinct form of inheritance that is called non-Mendelian, or *cytoplasmic inheritance*, in contrast to the Mendelian inheritance of nuclear genes.

The inheritance of mitochondria in animals and plants is quite different. In these organisms, the egg cell contributes much more cytoplasm to the zygote than does the male gamete (sperm in animals, pollen in plants). For example, a typical human oocyte contains about 100,000 copies of maternal mitochondrial DNA, whereas a sperm cell contains only a few. In addition, an active process ensures that the sperm mitochondria do not compete with those in the egg. As sperm mature, the DNA is degraded in their mitochondria. Sperm mitochondria are also specifically recognized then eliminated from the fertilized egg cell by autophagy in very much the same way that damaged mitochondria are removed (by ubiquitylation followed by delivery to lysosomes, as discussed in Chapter 13). Because of these two processes, the mitochondrial inheritance in both animals and plants is *uniparental*. More precisely, the mitochondrial DNA passes from one generation to the next by **maternal inheritance**.

In about two-thirds of higher plants, the chloroplast precursors from the male parent (contained in pollen grains) fail to enter the zygote, so that chloroplast as well as mitochondrial DNA is maternally inherited. In other plants, the chloroplast precursors from the pollen grains enter the zygote, making chloroplast inheritance biparental. In such plants, defective chloroplasts are a cause of variegation: a mixture of normal and defective chloroplasts in a zygote may sort out by mitotic segregation during plant growth and development, thereby producing alternating green and white patches in leaves. Leaf cells in the green patches contain normal chloroplasts, while those in the white patches contain defective chloroplasts (**Figure 14–68**).

Mutations in Mitochondrial DNA Can Cause Severe Inherited Diseases

In humans, as we have explained, all the mitochondrial DNA in a fertilized egg cell is inherited from the mother. Some mothers carry a mixed population of both mutant and normal mitochondrial genomes. Their daughters and sons will inherit this mixture of normal and mutant mitochondrial DNAs and be healthy unless the process of mitotic segregation results in a majority of defective mitochondria in a particular tissue. Muscle and the nervous system are most at risk. Because they need particularly large amounts of ATP, muscle and nerve cells are particularly dependent on fully functional mitochondria.

Numerous diseases in humans are caused by mutations in mitochondrial DNA. These diseases are recognized by their passage from affected mothers to

Figure 14–68 **A variegated leaf.** In the white patches, the plant cells have inherited a defective chloroplast. (Courtesy of John Innes Foundation.)

both their daughters and their sons, with the daughters but not the sons producing children with the disease. As expected from the random nature of mitotic segregation, the symptoms of these diseases vary greatly between different family members—including not only the severity and age of onset, but also which tissue is affected. There are also mitochondrial diseases that are caused by mutations in nuclear-encoded mitochondrial proteins; these diseases are inherited in the regular, Mendelian fashion.

The Accumulation of Mitochondrial DNA Mutations Is a Contributor to Aging

Mitochondria are marvels of efficiency in energy conversion, and they supply the cells of our body with a readily available source of energy in the form of ATP. But in highly developed, long-lived animals such as ourselves, the cells in our body age and eventually die. A factor in this inevitable process is the accumulation of deletions and point mutations in mitochondrial DNA. Oxidative damage to the cell by *reactive oxygen species* (*ROS*) such as H_2O_2, superoxide, or hydroxyl radicals also increases with age. The mitochondrial respiratory chain is the main source of ROS in animal cells, and animals in which mitochondrial superoxide dismutase—the main ROS scavenger—has been knocked out, die prematurely.

The less complex DNA replication and repair systems in mitochondria mean that accidents are corrected less efficiently. This results in a 100-fold higher occurrence of deletions and point mutations than in nuclear DNA. Mathematical modeling suggests that most of these mutations and lesions are acquired in childhood or early adult life, and then proliferate by *clonal expansion* in later life. Due to mitotic segregation, some cells will accumulate higher levels of faulty mitochondrial DNA than others. Above some threshold, serious deficiencies in respiratory-chain function will develop, producing cells that are *senescent*. In many organs of the human body, senescent cells with high levels of mitochondrial DNA damage are intermingled with normal cells, resulting in a mosaic of cells with and without respiratory-chain deficiency.

The main role of mitochondrial fusion in cellular physiology is most likely to ensure an even distribution of mitochondrial DNA throughout the mitochondrial reticulum, and to prevent the accumulation of damaged DNA in one part of the network. When the fusion machinery is defective, DNA is lost from a subset of the mitochondria in the cell. Loss of mitochondrial DNA leads to a loss of respiratory-chain function, and it can cause disease.

All of the considerations just discussed have suggested to some scientists that changes in our mitochondria are major contributors to human aging. However, there are many other processes that tend to go wrong as cells and tissues age, as one might expect given the incredible complexity of human cell biology. Despite intensive research, the issue remains unresolved.

Why Do Mitochondria and Chloroplasts Maintain a Costly Separate System for DNA Transcription and Translation?

Why do mitochondria and chloroplasts require their own separate genetic systems, when other organelles that share the same cytoplasm, such as peroxisomes and lysosomes, do not? The question is not trivial, because maintaining a separate genetic system is costly: more than 90 proteins—including many ribosomal proteins, aminoacyl-tRNA synthetases, DNA polymerase, RNA polymerase, and RNA-processing and RNA-modifying enzymes—must be encoded by nuclear genes specifically for this purpose. Moreover, as we have seen, the mitochondrial genetic system entails the risk of aging and disease.

A possible reason for maintaining this costly and potentially hazardous arrangement is the highly hydrophobic nature of the nonribosomal proteins encoded by organelle genes. This may make their production in and import from the cytoplasm simply too difficult and energy-consuming. It is also possible that the evolution (and eventual elimination) of the organellar genetic systems is still ongoing, but for now there is no alternative for the cell than to maintain separate genetic systems for its nuclear, mitochondrial, and chloroplast genes.

Summary

Mitochondria are organelles that allow eukaryotes to carry out oxidative phosphorylation, while chloroplasts are organelles that allow plants to carry out photosynthesis. Presumably as a result of their prokaryotic origins, each organelle maintains and reproduces itself in a highly coordinated process that requires the contribution of two separate genetic systems—one in the organelle and the other in the cell nucleus. The vast majority of the proteins in these organelles are encoded by nuclear DNA, synthesized in the cytosol, and then imported individually into the organelle. Other organelle proteins, as well as organelle ribosomal and transfer RNAs, are encoded by the organelle DNA; these are synthesized in the organelle itself.

The ribosomes of chloroplasts closely resemble bacterial ribosomes, while the origin of mitochondrial ribosomes is more difficult to trace. Extensive protein similarities, however, suggest that both organelles originated when a primitive eukaryotic cell entered into a stable endosymbiotic relationship with a bacterium. Although some of the genes of these former bacteria still function to make organelle proteins and RNA, most of them have been transferred into the nuclear genome, where they encode bacteria-like enzymes that are synthesized on cytosolic ribosomes and then imported into the organelle. The mitochondrial DNA replication and DNA repair processes are substantially less effective than the corresponding processes in the cell nucleus. Damage therefore accumulates in the genome of mitochondria over time; this damage may be a substantial contributor to the aging of cells and organisms, and it can cause serious diseases.

WHAT WE DON'T KNOW

• What structures are needed to form the barriers that separate and maintain the differentiated membrane domains in a single continuous membrane—as for the cristae and inner boundary membrane in mitochondria?

• How does a eukaryotic cell regulate the many functions of mitochondria, including ATP production?

• What are the origins and evolutionary history of photosynthetic complexes? Are there undiscovered types of photosynthesis present on Earth to help answer this question?

• Why is the mutation rate so much higher in mitochondria than in the nucleus (and chloroplasts)? Could this high rate have been useful to the cell?

• What mechanisms and pathways have been used during evolution to transfer genes from the mitochondrion to the nucleus?

PROBLEMS

Which statements are true? Explain why or why not.

14–1 The three respiratory enzyme complexes in the mitochondrial inner membrane tend to associate with each other in ways that facilitate the correct transfer of electrons between appropriate complexes.

14–2 The number of c subunits in the rotor ring of ATP synthase defines how many protons need to pass through the turbine to make each molecule of ATP.

14–3 Mutations that are inherited according to Mendelian rules affect nuclear genes; mutations whose inheritance violates Mendelian rules are likely to affect organelle genes.

Discuss the following problems.

14–4 In the 1860s, Louis Pasteur noticed that when he added O_2 to a culture of yeast growing anaerobically on glucose, the rate of glucose consumption declined dramatically. Explain the basis for this result, which is known as the Pasteur effect.

14–5 Heart muscle gets most of the ATP needed to power its continual contractions through oxidative phosphorylation. When oxidizing glucose to CO_2, heart muscle consumes O_2 at a rate of 10 µmol/min per g of tissue, in order to replace the ATP used in contraction and give a steady-state ATP concentration of 5 µmol/g of tissue. At this rate,

how many seconds would it take the heart to consume an amount of ATP equal to its steady-state levels? (Complete oxidation of one molecule of glucose to CO_2 yields 30 ATP, 26 of which are derived by oxidative phosphorylation using the 12 pairs of electrons captured in the electron carriers NADH and $FADH_2$.)

14–6 Both H^+ and Ca^{2+} are ions that move through the cytosol. Why is the movement of H^+ ions so much faster than that of Ca^{2+} ions? How do you suppose the speed of these two ions would be affected by freezing the solution? Would you expect them to move faster or slower? Explain your answer.

14–7 If isolated mitochondria are incubated with a source of electrons such as succinate, but without oxygen, electrons enter the respiratory chain, reducing each of the electron carriers almost completely. When oxygen is then introduced, the carriers become oxidized at different rates (**Figure Q14–1**). How does this result allow you to order

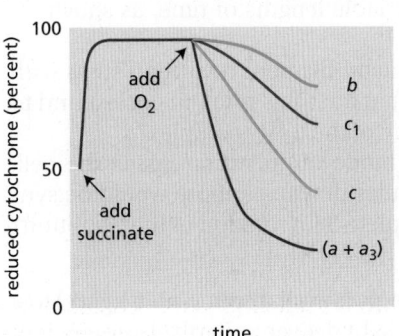

Figure Q14–1 Rapid spectrophotometric analysis of the rates of oxidation of electron carriers in the respiratory chain (Problem 14–7). Cytochromes a and a_3 cannot be distinguished and thus are listed as cytochrome $(a + a_3)$.

the electron carriers in the respiratory chain? What is their order?

14–8   Normally, the flow of electrons to O_2 is tightly linked to the production of ATP via the electrochemical gradient. If ATP synthase is inhibited, for example, electrons do not flow down the electron-transport chain and respiration ceases. Since the 1940s, several substances—such as 2,4-dinitrophenol—have been known to uncouple electron flow from ATP synthesis. Dinitrophenol was once prescribed as a diet drug to aid in weight loss. How would an uncoupler of oxidative phosphorylation promote weight loss? Why do you suppose dinitrophenol is no longer prescribed?

14–9   In actively respiring liver mitochondria, the pH in the matrix is about half a pH unit higher than it is in the cytosol. Assuming that the cytosol is at pH 7 and the matrix is a sphere with a diameter of 1 μm [$V = (4/3)\pi r^3$], calculate the total number of protons in the matrix of a respiring liver mitochondrion. If the matrix began at pH 7 (equal to that in the cytosol), how many protons would have to be pumped out to establish a matrix pH of 7.5 (a difference of 0.5 pH units)?

14–10   ATP synthase is the world's smallest rotary motor. Passage of H^+ ions through the membrane-embedded portion of ATP synthase (the F_o component) causes rotation of the single, central, axle-like γ subunit inside the head group. The tripartite head is composed of the three αβ dimers, the β subunit of which is responsible for synthesis of ATP. The rotation of the γ subunit induces conformational changes in the αβ dimers that allow ADP and Pi to be converted into ATP. A variety of indirect evidence had suggested rotary catalysis by ATP synthase, but seeing is believing.

To demonstrate rotary motion, a modified form of the $\alpha_3\beta_3\gamma$ complex was used. The β subunits were modified so they could be firmly anchored to a solid support and the γ subunit was modified (on the end that normally inserts into the F_o component in the inner membrane) so that a fluorescently tagged, readily visible filament of actin could be attached (**Figure Q14–2A**). This arrangement allows rotations of the γ subunit to be visualized as revolutions of the long actin filament. In these experiments, ATP synthase was studied in the reverse of its normal mechanism by allowing it to hydrolyze ATP. At low ATP concentrations, the actin filament was observed to revolve in steps of 120° and then pause for variable lengths of time, as shown in Figure Q14–2B.

A.   Why does the actin filament revolve in steps with pauses in between? What does this rotation correspond to in terms of the structure of the $\alpha_3\beta_3\gamma$ complex?

B.   In its normal mode of operation inside the cell, how many ATP molecules do you suppose would be synthesized for each complete 360° rotation of the γ subunit? Explain your answer.

14–11   How much energy is available in visible light? How much energy does sunlight deliver to Earth? How efficient

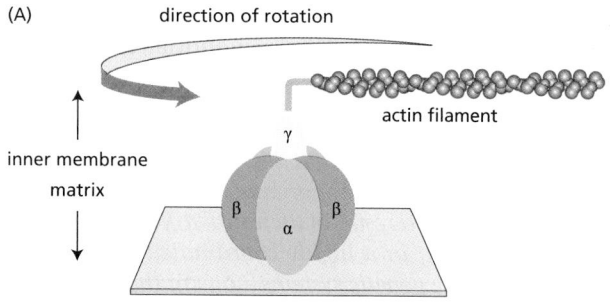

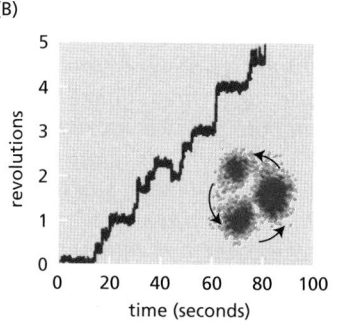

Figure Q14–2 Experimental set-up for observing rotation of the γ subunit of ATP synthase (Problem 14–10). (A) The immobilized $\alpha_3\beta_3\gamma$ complex. The β subunits are anchored to a solid support and a fluorescent actin filament is attached to the γ subunit. (B) Stepwise revolution of the actin filament. The indicated trace is a typical example from one experiment. The inset shows the positions in the revolution at which the actin filament pauses. (B, from R. Yasuda et al., *Cell* 93:1117–1124, 1998. With permission from Elsevier.)

are plants at converting light energy into chemical energy? The answers to these questions provide an important backdrop to the subject of photosynthesis.

Each quantum or photon of light has energy $h\nu$, where h is Planck's constant (6.6×10^{-37} kJ sec/photon) and ν is the frequency in sec^{-1}. The frequency of light is equal to c/λ, where c is the speed of light (3.0×10^{17} nm/ sec) and λ is the wavelength in nm. Thus, the energy (E) of a photon is

$$E = h\nu = hc/\lambda$$

A.   Calculate the energy of a mole of photons (6×10^{23} photons/mole) at 400 nm (violet light), at 680 nm (red light), and at 800 nm (near-infrared light).

B.   Bright sunlight strikes Earth at the rate of about 1.3 kJ/sec per square meter. Assuming for the sake of calculation that sunlight consists of monochromatic light of wavelength 680 nm, how many seconds would it take for a mole of photons to strike a square meter?

C.   Assuming that it takes eight photons to fix one molecule of CO_2 as carbohydrate under optimal conditions (8–10 photons is the currently accepted value), calculate how long it would take a tomato plant with a leaf area of 1 square meter to make a mole of glucose from CO_2. Assume that photons strike the leaf at the rate calculated above and, furthermore, that all the photons are absorbed and used to fix CO_2.

D. If it takes 468 kJ/mole to fix a mole of CO_2 into carbohydrate, what is the efficiency of conversion of light energy into chemical energy after photon capture? Assume again that eight photons of red light (680 nm) are required to fix one molecule of CO_2.

14–12 In chloroplasts, protons are pumped out of the stroma across the thylakoid membrane, whereas in mitochondria, they are pumped out of the matrix across the crista membrane. Explain how this arrangement allows chloroplasts to generate a larger proton gradient across the thylakoid membrane than mitochondria can generate across the inner membrane.

14–13 Examine the variegated leaf shown in **Figure Q14–3**. Yellow patches surrounded by green are common, but there are no green patches surrounded by yellow. Propose an explanation for this phenomenon.

Figure Q14–3 A variegated leaf of *Aucuba japonica* with green and yellow patches (Problem 14–13).

REFERENCES

General

Cramer WA & Knaff DB (1990) Energy Transduction in Biological Membranes: A Textbook of Bioenergetics. New York: Springer-Verlag.

Mathews CK, van Holde KE & Ahern K-G (2012) Biochemistry, 4th ed. San Francisco: Benjamin Cummings.

Nicholls DG & Ferguson SJ (2013) Bioenergetics, 4th ed. New York: Academic Press.

Schatz G (2012) The fires of life. *Annu. Rev. of Biochem.* 81, 34–59.

The Mitochondrion

Ernster L & Schatz G (1981) Mitochondria: a historical review. *J. Cell Biol.* 91, 227s–255s.

Friedman JR & Nunnari J (2014) Mitochondrial form and function. *Nature* 505, 335–43.

Mitchell P (1961) Coupling of phosphorylation to electron and hydrogen transfer by a chemi-osmotic type of mechanism. *Nature* 191, 144–148.

Pebay-Peyroula E & Brandolin G (2004) Nucleotide exchange in mitochondria: insight at a molecular level. *Curr. Opin. Struct. Biol.* 14, 420–425.

Scheffler IE (1999) Mitochondria. New York/Chichester: Wiley-Liss.

The Proton Pumps of the Electron-Transport Chain

Althoff T, Mills DJ, Popot J-L & Kühlbrandt W (2011) Arrangement of electron transport chain components in bovine mitochondrial supercomplex $I_1III_2IV_1$. *EMBO J.* 30, 4652–4664.

Baradaran R, Berrisford JM, Minhas GS & Sazanov LA (2013) Crystal structure of the entire respiratory complex I. *Nature* 494, 443–448.

Beinert H, Holm RH & Münck E (1997) Iron-sulfur clusters: nature's modular, multipurpose structures. *Science* 277, 653–659.

Berry EA, Guergova-Kuras M, Huang LS & Crofts AR (2000) Structure and function of cytochrome *bc* complexes. *Annu. Rev. Biochem.* 69, 1005–1075.

Brandt U (2006) Energy converting NADH:quinone oxidoreductase (complex I). *Annu. Rev. Biochem.* 75, 69–92.

Chance B & Williams GR (1955) A method for the localization of sites for oxidative phosphorylation. *Nature* 176, 250–254.

Cooley JW (2013) Protein conformational changes involved in the cytochrome bc_1 complex catalytic cycle *Biochim. Biophys. Acta.* 1827, 1340–45.

Gottschalk G (1997) Bacterial Metabolism, 2nd ed. New York: Springer.

Gray HB & Winkler JR (1996) Electron transfer in proteins. *Annu. Rev. Biochem.* 65, 537–561.

Hirst J (2013) Mitochondrial complex I. *Ann. Rev. Biochem.* 82, 551–75.

Hosler JP, Ferguson-Miller S & Mills DA (2006) Energy transduction: proton transfer through the respiratory complexes. *Annu. Rev. Biochem.* 75, 165–187.

Hunte C, Zickermann V & Brandt U (2010) Functional modules and structural basis of conformational coupling in mitochondrial complex I. *Science* 329, 448–451.

Keilin D (1966) The History of Cell Respiration and Cytochromes. Cambridge, UK: Cambridge University Press.

Rouault TA, Tracey A & Tong WH (2008) Iron–sulfur cluster biogenesis and human disease. *Trends Genet.* 24, 398–407.

Trumpower BL (2002) A concerted, alternating sites mechanism of ubiquinol oxidation by the dimeric cytochrome bc_1 complex. *Biochim. Biophys. Acta.* 1555, 166–173.

Tsukihara T, Aoyama H, Yamashita E et al. (1996) The whole structure of the 13-subunit oxidized cytochrome *c* oxidase at 2.8 Å. *Science* 272, 1136–1144.

ATP Production in Mitochondria

Abrahams JP, Leslie AG, Lutter R & Walker JE (1994) Structure at 2.8 Å resolution of F_1-ATPase from bovine heart mitochondria. *Nature* 370, 621–628.

Berg HC (2003) The rotary motor of bacterial flagella. *Annu. Rev. Biochem.* 72, 19–54.

Boyer PD (1997) The ATP synthase—a splendid molecular machine. *Annu. Rev. Biochem.* 66, 717–749.

Meier T, Polzer P, Diederichs K et al. (2005) Structure of the rotor ring of F-type Na^+-ATPase from *Ilyobacter tartaricus*. *Science* 308, 659–662.

Stock D, Gibbons C, Arechaga I et al. (2000) The rotary mechanism of ATP synthase. *Curr. Opin. Struct. Biol.* 10, 672–679.

von Ballmoos C, Wiedenmann A & Dimroth P (2009) Essentials for ATP synthesis by F_1F_0 ATP synthases. *Annu. Rev. Biochem.* 78, 649–672.

Chloroplasts and Photosynthesis

Barber J (2013) Photosystem II: the water-splitting enzyme of photosynthesis. *Cold Spring Harbor Symp. Quant. Biol.* 77, 295–307.

Bassham JA (1962) The path of carbon in photosynthesis. *Sci. Am.* 206, 89–100.

Blankenship RE (2002) Molecular Mechanisms of Photosynthesis. Oxford, UK: Blackwell Scientific.

Blankenship RE & Bauer CE (eds) (1995) Anoxygenic Photosynthetic Bacteria. Dordrecht: Kluwer.

Deisenhofer J & Michel H (1989) Nobel lecture. The photosynthetic reaction centre from the purple bacterium *Rhodopseudomonas viridis*. *EMBO J.* 8, 2149–2170.

De Las Rivas J, Balsera M & Barber J (2004) Evolution of oxygenic photosynthesis: genome-wide analysis of the OEC extrinsic proteins. *Trends Plant Sci.* 9:18–25.

Hohmann-Marriott MF & Blankenship RE (2011) Evolution of photosynthesis. *Annu. Rev. Plant Biol.* 62, 515–548.

Jordan P, Fromme P, Witt HT et al. (2001) Three-dimensional structure of cyanobacterial photosystem I at 2.5 Å resolution. *Nature* 411, 909–917.

Kühlbrandt W, Wang DN & Fujiyoshi Y (1994) Atomic model of plant light-harvesting complex by electron crystallography. *Nature* 367, 614–621.

Lane N & Martin WF (2012) The origin of membrane bioenergetics. *Cell* 151, 1406–1416.

Lyons TW, Reinhard CT & Planavsky NJ (2014) The rise of oxygen in earth's early ocean and atmosphere. *Nature* 506, 307–15.

Nelson N & Ben-Shem A (2004) The complex architecture of oxygenic photosynthesis. *Nat. Rev. Mol. Cell Biol.* 5, 971–982.

Orgel LE (1998) The origin of life—a review of facts and speculations. *Trends Biochem. Sci.* 23, 491–495.

Tang K-H, Tang YJ & Blankenship RE (2011) Carbon metabolic pathways in phototrophic bacteria and their broader evolutionary implications. *Front. Microbiol.* 2, 165.

Umena Y, Kawakami K, Shen J-R & Kamiya N (2011) Crystal structure of oxygen-evolving photosystem II at a resolution of 1.9 Å. *Nature* 473, 55–60.

Vinyard DJ, Ananyev GM & Dismukes GC (2013) Photosystem II: the reaction center of oxygenic photosynthesis. *Annu. Rev. Biochem.* 82, 577–606.

Yano J, Kern J, Sauer K et al. (2006) Where water is oxidized to dioxygen: structure of the photosynthetic Mn4Ca cluster. *Science* 314, 821–25.

Yoon HS (2004) A molecular timeline for the origin of photosynthetic eukaryotes. *Mol. Biol. Evol.* 21, 809–18.

The Genetic Systems of Mitochondria and Chloroplasts

Anderson S, Bankier AT, Barrell BG et al. (1981) Sequence and organization of the human mitochondrial genome. *Nature* 290, 457–465.

Bendich AJ (2004) Circular chloroplast chromosomes: the grand illusion. *Plant Cell* 16, 1661–1666.

Birky CW Jr (1995) Uniparental inheritance of mitochondrial and chloroplast genes: mechanisms and evolution. *Proc. Natl Acad. Sci. USA* 92, 11331–11338.

Bullerwell CE & Gray MW (2004) Evolution of the mitochondrial genome: protist connections to animals, fungi and plants. *Curr. Opin. Microbiol.* 7, 528–534.

Chen XJ & Butow RA (2005) The organization and inheritance of the mitochondrial genome. *Nat. Rev. Genet.* 6, 815–825.

Clayton DA (2000) Vertebrate mitochondrial DNA—a circle of surprises. *Exp. Cell Res.* 255, 4–9.

Daley DO & Whelan J (2005) Why genes persist in organelle genomes. *Genome Biol.* 6, 110.

de Duve C (2007) The origin of eukaryotes: a reappraisal. *Nat. Rev. Genet.* 8, 395–403.

Dyall SD, Brown MT & Johnson PJ (2004) Ancient invasions: from endosymbionts to organelles. *Science* 304, 253–257.

Falkenberg M, Larsson NG & Gustafsson CM (2007) DNA replication and transcription in mammalian mitochondria. *Annu. Rev. Biochem.* 76, 679–699.

Harel A, Bromberg Y, Falkowski PG & Bhattacharya D (2014) Evolutionary history of redox metal-binding domains across the tree of life. *Proc. Natl Acad. Sci. USA* 111, 7042–47.

Hoppins S, Lackner L & Nunnari J (2007) The machines that divide and fuse mitochondria. *Annu. Rev. Biochem.* 76, 751–780.

Larsson NG (2010) Somatic mitochondrial DNA mutations in mammalian aging. *Annu. Rev. Biochem.* 79, 683–706.

Ma H, Xu H & O'Farrell PH (2014) Transmission of mitochondrial mutations and action of purifying selection in *Drosophila melanogaster*. *Nat. Genet.* 46, 393–97.

Neupert W & Herrmann JM (2007) Translocation of proteins into mitochondria. *Annu. Rev. Biochem.* 76, 723–749.

Rawi A, Louvet-Vallee SS, Djeddi D et al. (2011) Postfertilization autophagy of sperm organelles prevents paternal mitochondrial DNA transmission. *Science* 334, 1144–47.

Taylor RW, Barron MJ, Borthwick GM et al. (2003) Mitochondrial DNA mutations in human colonic crypt stem cells. *J. Clin. Invest.* 112, 1351–60.

Wallace DC (1999) Mitochondrial diseases in man and mouse. *Science* 283, 1482–1488.

Williams TA, Foster PG, Cox CJ & Embley TM (2014) An archaeal origin of eukaryotes supports only two primary domains of life. *Nature* 504, 231–36.

Cell Signaling

When things change, cells respond. Every cell, from the humble bacterium to the most sophisticated eukaryotic cell, monitors its intracellular and extracellular environment, processes the information it gathers, and responds accordingly. Unicellular organisms, for example, modify their behavior in response to changes in environmental nutrients or toxins. The cells of multicellular organisms detect and respond to countless internal and extracellular signals that control their growth, division, and differentiation during development, as well as their behavior in adult tissues. At the heart of all these communication systems are regulatory proteins that produce chemical signals, which are sent from one place to another in the body or within a cell, usually being processed along the way and integrated with other signals to provide clear and effective communication.

The study of cell signaling has traditionally focused on the mechanisms by which eukaryotic cells communicate with each other using *extracellular signal molecules* such as hormones and growth factors. In this chapter, we describe the features of some of these cell–cell communication systems, and we use them to illustrate the general principles by which any regulatory system, inside or outside the cell, is able to generate, process, and respond to signals. Our main focus is on animal cells, but we end by considering the special features of cell signaling in plants.

PRINCIPLES OF CELL SIGNALING

Long before multicellular creatures roamed the Earth, unicellular organisms had developed mechanisms for responding to physical and chemical changes in their environment. These almost certainly included mechanisms for responding to the presence of other cells. Evidence comes from studies of present-day unicellular organisms such as bacteria and yeasts. Although these cells lead mostly independent lives, they can communicate and influence one another's behavior. Many bacteria, for example, respond to chemical signals that are secreted by their neighbors and accumulate at higher population density. This process, called *quorum sensing*, allows bacteria to coordinate their behavior, including their motility, antibiotic production, spore formation, and sexual conjugation. Similarly, yeast cells communicate with one another in preparation for mating. The budding yeast *Saccharomyces cerevisiae* provides a well-studied example: when a haploid individual is ready to mate, it secretes a peptide *mating factor* that signals cells of the opposite mating type to stop proliferating and prepare to mate. The subsequent fusion of two haploid cells of opposite mating type produces a diploid zygote.

Intercellular communication achieved an astonishing level of complexity during the evolution of multicellular organisms. These organisms are tight-knit societies of cells, in which the well-being of the individual cell is often set aside for the benefit of the organism as a whole. Complex systems of intercellular communication have evolved to allow the collaboration and coordination of different tissues and cell types. Bewildering arrays of signaling systems govern every conceivable feature of cell and tissue function during development and in the adult.

Communication between cells in multicellular organisms is mediated mainly by **extracellular signal molecules**. Some of these operate over long distances,

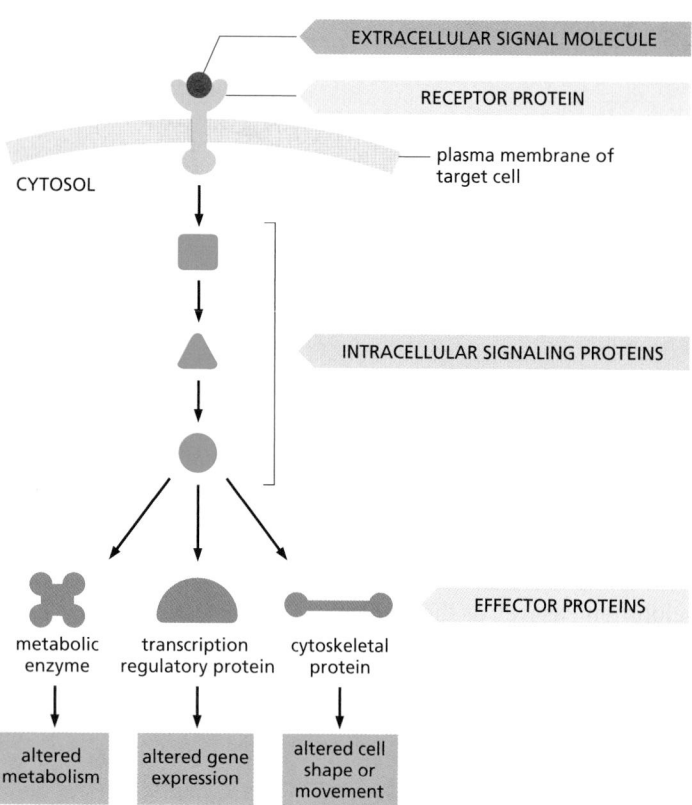

Figure 15–1 **A simple intracellular signaling pathway activated by an extracellular signal molecule.** The signal molecule usually binds to a receptor protein that is embedded in the plasma membrane of the target cell. The receptor activates one or more intracellular signaling pathways, involving a series of signaling proteins. Finally, one or more of the intracellular signaling proteins alters the activity of effector proteins and thereby the behavior of the cell.

signaling to cells far away; others signal only to immediate neighbors. Most cells in multicellular organisms both emit and receive signals. Reception of the signals depends on *receptor proteins*, usually (but not always) at the cell surface, which bind the signal molecule. The binding activates the receptor, which in turn activates one or more *intracellular signaling pathways* or *systems*. These systems depend on *intracellular signaling proteins*, which process the signal inside the receiving cell and distribute it to the appropriate intracellular targets. The targets that lie at the end of signaling pathways are generally called *effector proteins*, which are altered in some way by the incoming signal and implement the appropriate change in cell behavior. Depending on the signal and the type and state of the receiving cell, these effectors can be transcription regulators, ion channels, components of a metabolic pathway, or parts of the cytoskeleton (**Figure 15–1**).

The fundamental features of cell signaling have been conserved throughout the evolution of the eukaryotes. In budding yeast, for example, the response to mating factor depends on cell-surface receptor proteins, intracellular GTP-binding proteins, and protein kinases that are clearly related to functionally similar proteins in animal cells. Through gene duplication and divergence, however, the signaling systems in animals have become much more elaborate than those in yeasts; the human genome, for example, contains more than 1500 genes that encode receptor proteins, and the number of different receptor proteins is further increased by alternative RNA splicing and post-translational modifications.

Extracellular Signals Can Act Over Short or Long Distances

Many extracellular signal molecules remain bound to the surface of the signaling cell and influence only cells that contact it (**Figure 15–2A**). Such **contact-dependent signaling** is especially important during development and in immune responses. Contact-dependent signaling during development can sometimes operate over relatively large distances if the communicating cells extend long thin processes to make contact with one another.

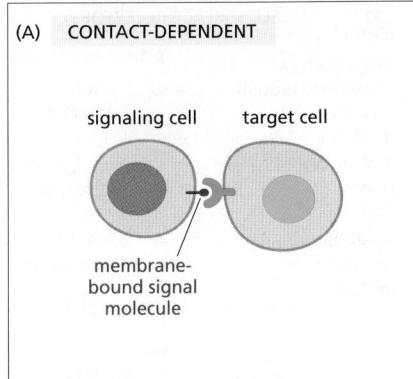

(A) CONTACT-DEPENDENT

signaling cell   target cell

membrane-
bound signal
molecule

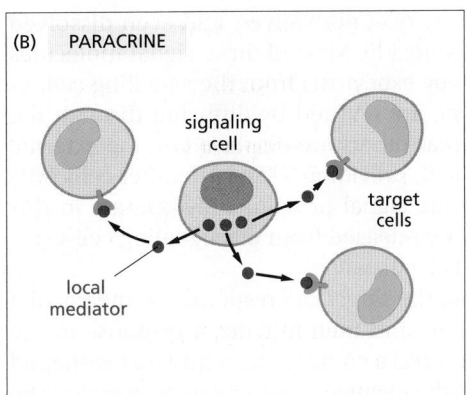

(B) PARACRINE

signaling
cell

target
cells

local
mediator

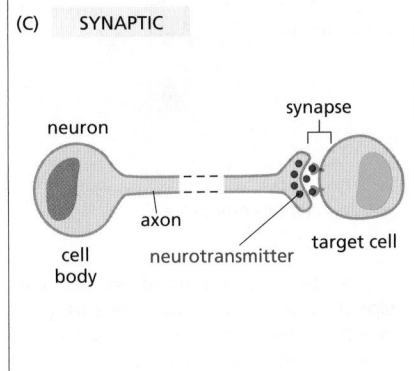

(C) SYNAPTIC

synapse

neuron

axon

cell
body

neurotransmitter

target cell

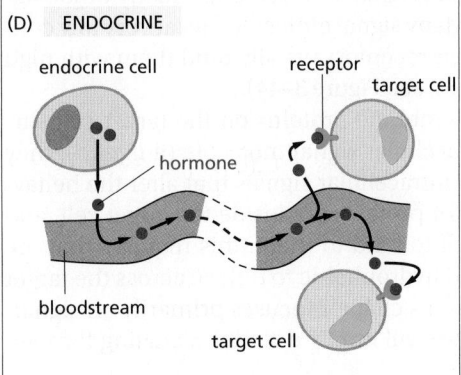

(D) ENDOCRINE

endocrine cell   receptor

target cell

hormone

bloodstream

target cell

Figure 15–2 Four forms of intercellular signaling. (A) Contact-dependent signaling requires cells to be in direct membrane–membrane contact. (B) Paracrine signaling depends on local mediators that are released into the extracellular space and act on neighboring cells. (C) Synaptic signaling is performed by neurons that transmit signals electrically along their axons and release neurotransmitters at synapses, which are often located far away from the neuronal cell body. (D) Endocrine signaling depends on endocrine cells, which secrete hormones into the bloodstream for distribution throughout the body. Many of the same types of signaling molecules are used in paracrine, synaptic, and endocrine signaling; the crucial differences lie in the speed and selectivity with which the signals are delivered to their targets.

In most cases, however, signaling cells secrete signal molecules into the extracellular fluid. Often, the secreted molecules are **local mediators**, which act only on cells in the local environment of the signaling cell. This is called **paracrine signaling** (Figure 15-2B). Usually, the signaling and target cells in paracrine signaling are of different cell types, but cells may also produce signals that they themselves respond to: this is referred to as *autocrine signaling*. Cancer cells, for example, often produce extracellular signals that stimulate their own survival and proliferation.

Large multicellular organisms like us need long-range signaling mechanisms to coordinate the behavior of cells in remote parts of the body. Thus, they have evolved cell types specialized for intercellular communication over large distances. The most sophisticated of these are nerve cells, or neurons, which typically extend long, branching processes (axons) that enable them to contact target cells far away, where the processes terminate at the specialized sites of signal transmission known as *chemical synapses*. When a neuron is activated by stimuli from other nerve cells, it sends electrical impulses (action potentials) rapidly along its axon; when the impulse reaches the synapse at the end of the axon, it triggers secretion of a chemical signal that acts as a **neurotransmitter**. The tightly organized structure of the synapse ensures that the neurotransmitter is delivered specifically to receptors on the postsynaptic target cell (Figure 15-2C). The details of this **synaptic signaling** process are discussed in Chapter 11.

A quite different strategy for signaling over long distances makes use of **endocrine cells**, which secrete their signal molecules, called **hormones**, into the bloodstream. The blood carries the molecules far and wide, allowing them to act on target cells that may lie anywhere in the body (Figure 15-2D).

Extracellular Signal Molecules Bind to Specific Receptors

Cells in multicellular animals communicate by means of hundreds of kinds of extracellular signal molecules. These include proteins, small peptides, amino acids, nucleotides, steroids, retinoids, fatty acid derivatives, and even dissolved gases such as nitric oxide and carbon monoxide. Most of these signal molecules are released into the extracellular space by exocytosis from the signaling cell, as discussed in Chapter 13. Some, however, are emitted by diffusion through the signaling cell's plasma membrane, whereas others are displayed on the external surface of the cell and remain attached to it, providing a signal to other cells only when they make contact. Transmembrane signal proteins may operate in this way, or their extracellular domains may be released from the signaling cell's surface by proteolytic cleavage and then act at a distance.

Regardless of the nature of the signal, the *target cell* responds by means of a **receptor**, which binds the signal molecule and then initiates a response in the target cell. The binding site of the receptor has a complex structure that is shaped to recognize the signal molecule with high specificity, helping to ensure that the receptor responds only to the appropriate signal and not to the many other signaling molecules surrounding the cell. Many signal molecules act at very low concentrations (typically $\leq 10^{-8}$ M), and their receptors usually bind them with high affinity (dissociation constant $K_d \leq 10^{-8}$ M; see Figure 3–44).

In most cases, receptors are transmembrane proteins on the target-cell surface. When these proteins bind an extracellular signal molecule (a *ligand*), they become activated and generate various intracellular signals that alter the behavior of the cell. In other cases, the receptor proteins are inside the target cell, and the signal molecule has to enter the cell to bind to them: this requires that the signal molecule be sufficiently small and hydrophobic to diffuse across the target cell's plasma membrane (**Figure 15–3**). This chapter focuses primarily on signaling through cell-surface receptors, but we will briefly describe signaling through intracellular receptors later in the chapter.

Each Cell Is Programmed to Respond to Specific Combinations of Extracellular Signals

A typical cell in a multicellular organism is exposed to hundreds of different signal molecules in its environment. The molecules can be soluble, bound to the extracellular matrix, or bound to the surface of a neighboring cell; they can be stimulatory or inhibitory; they can act in innumerable different combinations; and they can influence almost any aspect of cell behavior. The cell responds to this blizzard of signals selectively, in large part by expressing only those receptors and intracellular signaling systems that respond to the signals that are required for the regulation of that cell.

Most cells respond to many different signals in the environment, and some of these signals may influence the response to other signals. One of the key challenges in cell biology is to determine how a cell integrates all of this signaling information in order to make decisions—to divide, to move, to differentiate, and so on. Many cells, for example, require a specific combination of extracellular survival factors to allow the cell to continue living; when deprived of these signals, the cell activates a suicide program and kills itself—usually by *apoptosis*, a form of *programmed cell death*, as discussed in Chapter 18. Cell proliferation often depends on a combination of signals that promote both cell division and survival, as well as signals that stimulate cell growth (**Figure 15–4**). On the other hand, differentiation into a nondividing state (called *terminal differentiation*) frequently requires a different combination of survival and differentiation signals that must override any signal to divide.

In principle, the hundreds of signal molecules that an animal makes can be used in an almost unlimited number of combinations to control the diverse behaviors of its cells in highly specific ways. Relatively small numbers of types of signal molecules and receptors are sufficient. The complexity lies in the ways in which cells respond to the combinations of signals that they receive.

A signal molecule often has different effects on different types of target cells. The neurotransmitter acetylcholine (**Figure 15–5**A), for example, decreases the

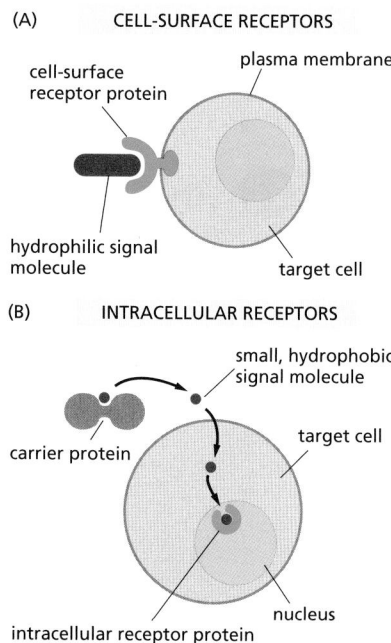

(A) CELL-SURFACE RECEPTORS

cell-surface receptor protein
plasma membrane
hydrophilic signal molecule
target cell

(B) INTRACELLULAR RECEPTORS

small, hydrophobic signal molecule
carrier protein
target cell
nucleus
intracellular receptor protein

Figure 15–3 The binding of extracellular signal molecules to either cell-surface or intracellular receptors. (A) Most signal molecules are hydrophilic and are therefore unable to cross the target cell's plasma membrane directly; instead, they bind to cell-surface receptors, which in turn generate signals inside the target cell (see Figure 15–1). (B) Some small signal molecules, by contrast, diffuse across the plasma membrane and bind to receptor proteins inside the target cell—either in the cytosol or in the nucleus (as shown here). Many of these small signal molecules are hydrophobic and poorly soluble in aqueous solutions; they are therefore transported in the bloodstream and other extracellular fluids bound to carrier proteins, from which they dissociate before entering the target cell.

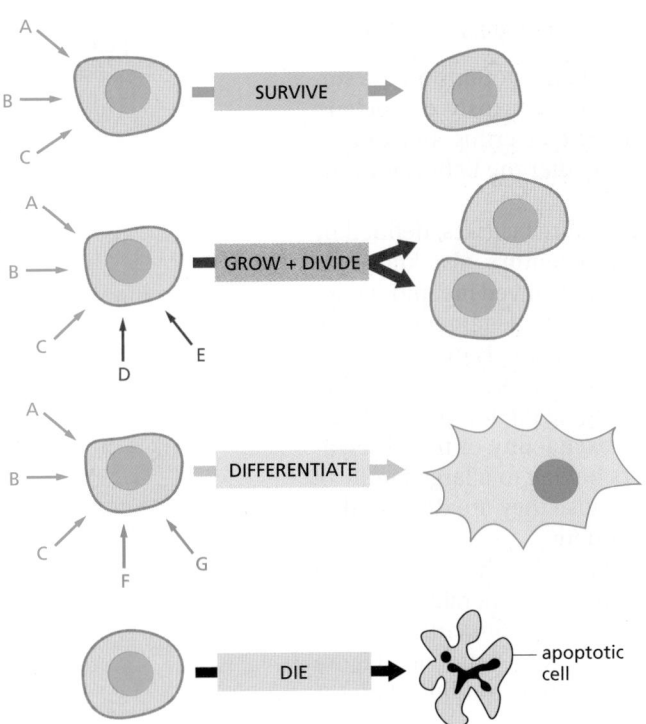

Figure 15–4 **An animal cell's dependence on multiple extracellular signal molecules.** Each cell type displays a set of receptors that enables it to respond to a corresponding set of signal molecules produced by other cells. These signal molecules work in various combinations to regulate the behavior of the cell. As shown here, an individual cell often requires multiple signals to survive *(blue arrows)* and additional signals to grow and divide *(red arrows)* or differentiate *(green arrows)*. If deprived of appropriate survival signals, a cell will undergo a form of cell suicide known as apoptosis. The actual situation is even more complex. Although not shown, some extracellular signal molecules act to inhibit these and other cell behaviors, or even to induce apoptosis.

rate of action potential firing in heart pacemaker cells (Figure 15–5B) and stimulates the production of saliva by salivary gland cells (Figure 15–5C), even though the receptors are the same on both cell types. In skeletal muscle, acetylcholine causes the cells to contract by binding to a different receptor protein (Figure 15–5D). The different effects of acetylcholine in these cell types result from differences in the intracellular signaling proteins, effector proteins, and genes that are activated. Thus, an extracellular signal itself has little information content; it simply induces the cell to respond according to its predetermined state, which depends on the cell's developmental history and the specific genes it expresses.

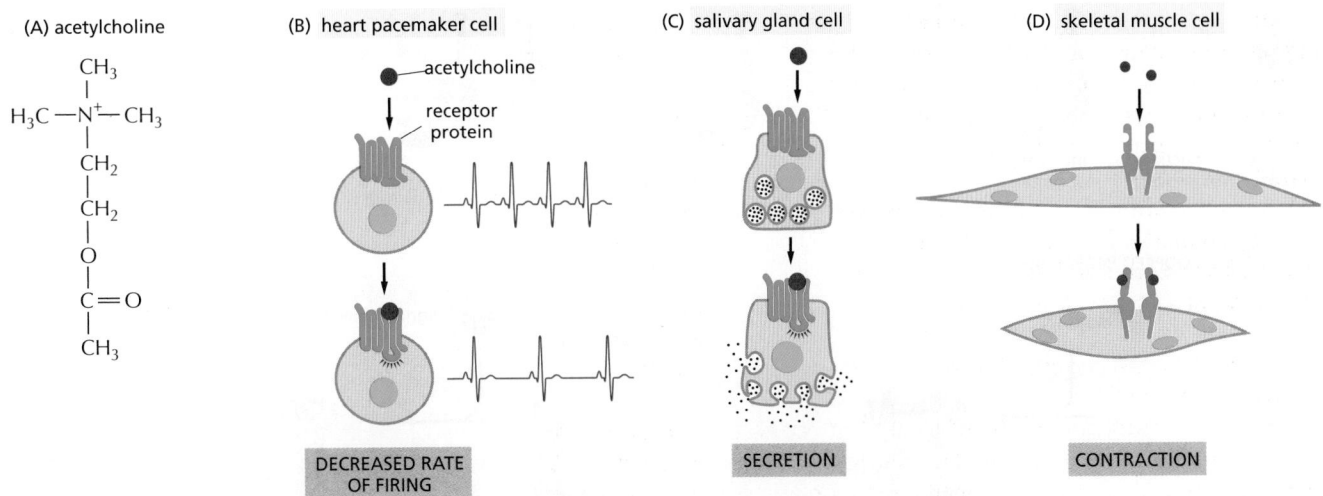

Figure 15–5 **Various responses induced by the neurotransmitter acetylcholine.** (A) The chemical structure of acetylcholine. (B–D) Different cell types are specialized to respond to acetylcholine in different ways. In some cases (B and C), acetylcholine binds to similar receptor proteins (G-protein-coupled receptors; see Figure 15–6), but the intracellular signals produced are interpreted differently in cells specialized for different functions. In other cases (D), the receptor protein is also different (here, an ion-channel-coupled receptor; see Figure 15–6).

There Are Three Major Classes of Cell-Surface Receptor Proteins

Most extracellular signal molecules bind to specific receptor proteins on the surface of the target cells they influence and do not enter the cytosol or nucleus. These cell-surface receptors act as *signal transducers* by converting an extracellular ligand-binding event into intracellular signals that alter the behavior of the target cell.

Most cell-surface receptor proteins belong to one of three classes, defined by their transduction mechanism. **Ion-channel-coupled receptors**, also known as *transmitter-gated ion channels* or *ionotropic receptors,* are involved in rapid synaptic signaling between nerve cells and other electrically excitable target cells such as nerve and muscle cells (**Figure 15–6**A). This type of signaling is mediated by a small number of neurotransmitters that transiently open or close an ion channel formed by the protein to which they bind, briefly changing the ion permeability of the plasma membrane and thereby changing the excitability of the postsynaptic target cell. Most ion-channel-coupled receptors belong to a large family of homologous, multipass transmembrane proteins. Because they are discussed in detail in Chapter 11, we will not consider them further here.

G-protein-coupled receptors act by indirectly regulating the activity of a separate plasma-membrane-bound target protein, which is generally either an enzyme or an ion channel. A *trimeric GTP-binding protein* (*G protein*) mediates the interaction between the activated receptor and this target protein (Figure 15–6B). The activation of the target protein can change the concentration of one or

(A)   ION-CHANNEL-COUPLED RECEPTORS

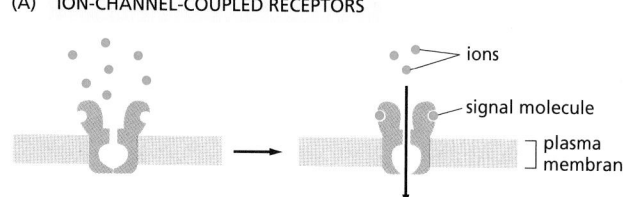

Figure 15–6 **Three classes of cell-surface receptors.** (A) Ion-channel-coupled receptors (also called transmitter-gated ion channels), (B) G-protein-coupled receptors, and (C) enzyme-coupled receptors. Although many enzyme-coupled receptors have intrinsic enzymatic activity, as shown on the left in (C), many others rely on associated enzymes, as shown on the right in (C). Ligands activate most enzyme-coupled receptors by promoting their dimerization, which results in the interaction and activation of the cytoplasmic domains.

(B)   G-PROTEIN-COUPLED RECEPTORS

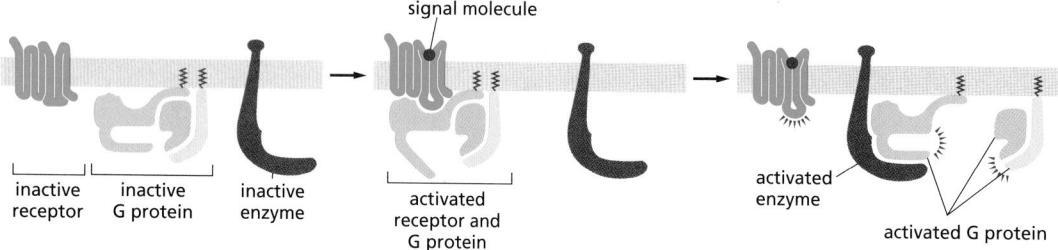

(C)   ENZYME-COUPLED RECEPTORS

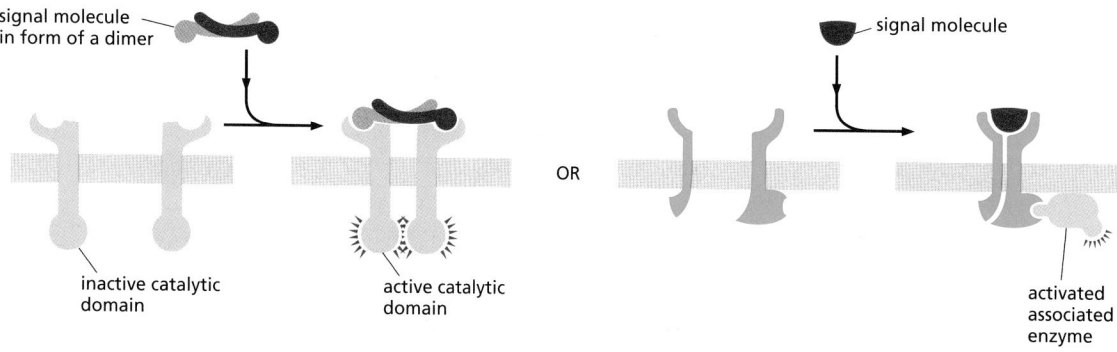

more small intracellular signaling molecules (if the target protein is an enzyme), or it can change the ion permeability of the plasma membrane (if the target protein is an ion channel). The small intracellular signaling molecules act in turn to alter the behavior of yet other signaling proteins in the cell.

Enzyme-coupled receptors either function as enzymes or associate directly with enzymes that they activate (Figure 15–6C). They are usually single-pass transmembrane proteins that have their ligand-binding site outside the cell and their catalytic or enzyme-binding site inside. Enzyme-coupled receptors are heterogeneous in structure compared with the other two classes; the great majority, however, are either protein kinases or associate with protein kinases, which phosphorylate specific sets of proteins in the target cell when activated.

There are also some types of cell-surface receptors that do not fit easily into any of these classes but have important functions in controlling the specialization of different cell types during development and in tissue renewal and repair in adults. We discuss these in a later section, after we explain how G-protein-coupled receptors and enzyme-coupled receptors operate. First, we continue our general discussion of the principles of signaling via cell-surface receptors.

Cell-Surface Receptors Relay Signals Via Intracellular Signaling Molecules

Numerous intracellular signaling molecules relay signals received by cell-surface receptors into the cell interior. The resulting chain of intracellular signaling events ultimately alters effector proteins that are responsible for modifying the behavior of the cell (see Figure 15–1).

Some intracellular signaling molecules are small chemicals, which are often called **second messengers** (the "first messengers" being the extracellular signals). They are generated in large amounts in response to receptor activation and diffuse away from their source, spreading the signal to other parts of the cell. Some, such as *cyclic AMP* and *Ca²⁺*, are water-soluble and diffuse in the cytosol, while others, such as *diacylglycerol*, are lipid-soluble and diffuse in the plane of the plasma membrane. In either case, they pass the signal on by binding to and altering the behavior of selected signaling or effector proteins.

Most intracellular signaling molecules are proteins, which help relay the signal into the cell by either generating second messengers or activating the next signaling or effector protein in the pathway. Many of these proteins behave like *molecular switches*. When they receive a signal, they switch from an inactive to an active state, until another process switches them off, returning them to their inactive state. The switching off is just as important as the switching on. If a signaling pathway is to recover after transmitting a signal so that it can be ready to transmit another, every activated molecule in the pathway must return to its original, unactivated state.

The largest class of molecular switches consists of proteins that are activated or inactivated by **phosphorylation** (discussed in Chapter 3). For these proteins, the switch is thrown in one direction by a **protein kinase**, which covalently adds one or more phosphate groups to specific amino acids on the signaling protein, and in the other direction by a **protein phosphatase**, which removes the phosphate groups (Figure 15–7A). The activity of any protein regulated by phosphorylation depends on the balance between the activities of the kinases that phosphorylate it and of the phosphatases that dephosphorylate it. About 30–50% of human proteins contain covalently attached phosphate, and the human genome encodes about 520 protein kinases and about 150 protein phosphatases. A typical mammalian cell makes use of hundreds of distinct types of protein kinases at any moment.

Protein kinases attach phosphate to the hydroxyl group of specific amino acids on the target protein. There are two main types of protein kinase. The great majority are **serine/threonine kinases**, which phosphorylate the hydroxyl groups of serines and threonines in their targets. Others are **tyrosine kinases**, which

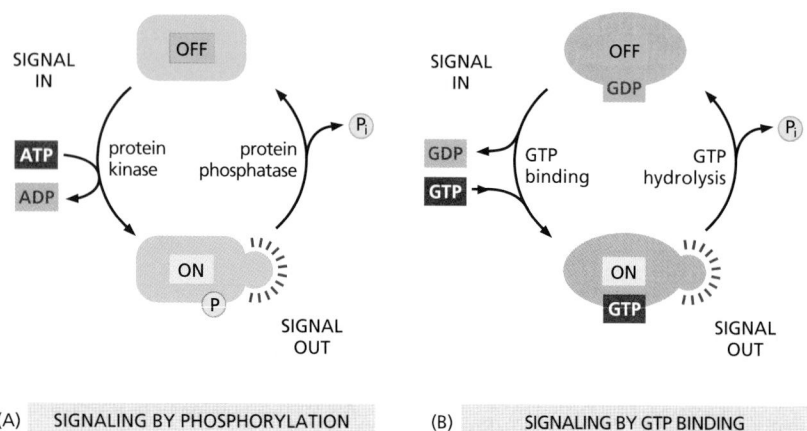

(A)   SIGNALING BY PHOSPHORYLATION   (B)   SIGNALING BY GTP BINDING

Figure 15–7 **Two types of intracellular signaling proteins that act as molecular switches.** (A) A protein kinase covalently adds a phosphate from ATP to the signaling protein, and a protein phosphatase removes the phosphate. Although not shown, many signaling proteins are activated by dephosphorylation rather than by phosphorylation. (B) A GTP-binding protein is induced to exchange its bound GDP for GTP, which activates the protein; the protein then inactivates itself by hydrolyzing its bound GTP to GDP.

phosphorylate proteins on tyrosines. The two types of protein kinase are closely related members of a large family, differing primarily in the structure of their protein substrate binding sites.

Many intracellular signaling proteins controlled by phosphorylation are themselves protein kinases, and these are often organized into **kinase cascades**. In such a cascade, one protein kinase, activated by phosphorylation, phosphorylates the next protein kinase in the sequence, and so on, relaying the signal onward and, in some cases, amplifying it or spreading it to other signaling pathways.

The other important class of molecular switches consists of **GTP-binding proteins** (discussed in Chapter 3). These proteins switch between an "on" (actively signaling) state when GTP is bound and an "off" state when GDP is bound. In the "on" state, they usually have intrinsic GTPase activity and shut themselves off by hydrolyzing their bound GTP to GDP (Figure 15–7B). There are two major types of GTP-binding proteins. Large, *trimeric GTP-binding proteins* (also called *G proteins*) help relay signals from G-protein-coupled receptors that activate them (see Figure 15–6B). Small **monomeric GTPases** (also called *monomeric GTP-binding proteins*) help relay signals from many classes of cell-surface receptors.

Specific regulatory proteins control both types of GTP-binding proteins. **GTPase-activating proteins** (**GAPs**) drive the proteins into an "off" state by increasing the rate of hydrolysis of bound GTP. Conversely, **guanine nucleotide exchange factors** (**GEFs**) activate GTP-binding proteins by promoting the release of bound GDP, which allows a new GTP to bind. In the case of trimeric G proteins, the activated receptor serves as the GEF. **Figure 15–8** illustrates the regulation of monomeric GTPases.

Not all molecular switches in signaling systems depend on phosphorylation or GTP binding. We see later that some signaling proteins are switched on or off by the binding of another signaling protein or a second messenger such as cyclic AMP or Ca^{2+}, or by covalent modifications other than phosphorylation or dephosphorylation, such as ubiquitylation (discussed in Chapter 3).

For simplicity, we often portray a signaling pathway as a series of activation steps (see Figure 15–1). It is important to note, however, that most signaling pathways contain inhibitory steps, and a sequence of two inhibitory steps can have the same effect as one activating step (**Figure 15–9**). This *double-negative* activation is very common in signaling systems, as we will see when we describe specific pathways later in this chapter.

Intracellular Signals Must Be Specific and Precise in a Noisy Cytoplasm

Ideally, an activated intracellular signaling molecule should interact only with the appropriate downstream targets, and, likewise, the targets should only be activated by the appropriate upstream signal. In reality, however, intracellular

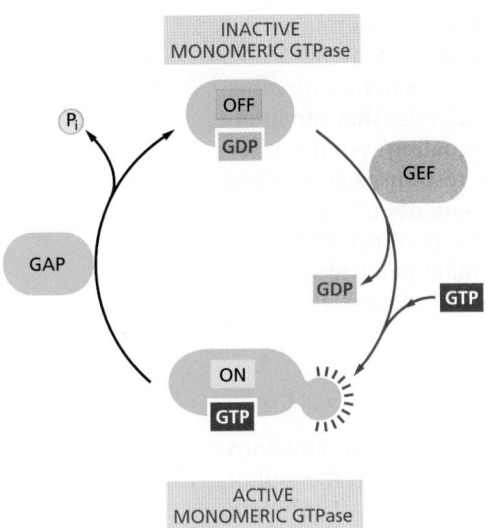

INACTIVE
MONOMERIC GTPase

ACTIVE
MONOMERIC GTPase

Figure 15–8 The regulation of a monomeric GTPase. GTPase-activating proteins (GAPs) inactivate the protein by stimulating it to hydrolyze its bound GTP to GDP, which remains tightly bound to the inactivated GTPase. Guanine nucleotide exchange factors (GEFs) activate the inactive protein by stimulating it to release its GDP; because the concentration of GTP in the cytosol is 10 times greater than the concentration of GDP, the protein rapidly binds GTP and is thereby activated.

signaling molecules share the cytoplasm with a crowd of closely related signaling molecules that control a diverse array of cellular processes. It is inevitable that an occasional signaling molecule will bind or modify the wrong partner, potentially creating unwanted cross-talk and interference between signaling systems. How does a signal remain strong, precise, and specific under these noisy conditions?

The first line of defense comes from the high affinity and specificity of the interactions between signaling molecules and their correct partners compared to the relatively low affinity of the interactions between inappropriate partners. The binding of a signaling molecule to the correct target is determined by precise and complex interactions between complementary surfaces on the two molecules. Protein kinases, for example, contain active sites that recognize a specific amino acid sequence around the phosphorylation site on the correct target protein, and they often contain additional *docking sites* that promote a specific, high-affinity interaction with the target. These and related mechanisms help provide a strong and persistent interaction between the correct partners, reducing the likelihood of inappropriate interactions with other proteins.

Another important way that cells avoid responses to unwanted background signals depends on the ability of many downstream target proteins to simply ignore such signals. These proteins respond only when the upstream signal reaches a high concentration or activity level. Consider a signaling pathway in which a protein kinase activates some downstream target protein by phosphorylation. If a response is triggered only when more than half of the target proteins are phosphorylated, then there will be little harm done if a small number of them are occasionally phosphorylated by some inappropriate protein kinase. Furthermore, constitutively active protein phosphatases will further reduce the impact of background phosphorylation by rapidly removing much of it. In these and other ways, intracellular signaling systems filter out noise, generating little or no response to low levels of stimuli.

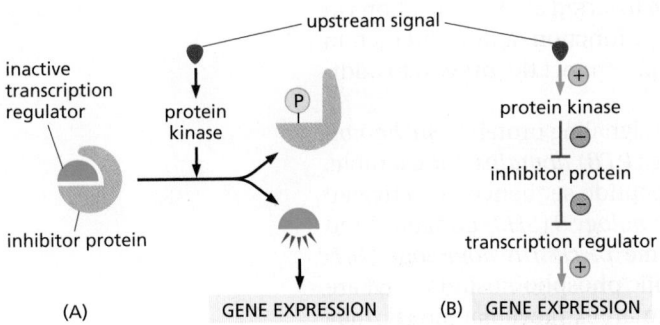

(A)

(B)

Figure 15–9 A sequence of two inhibitory signals produces a positive signal. (A) In this simple signaling system, a transcription regulator is kept in an inactive state by a bound inhibitor protein. In response to some upstream signal, a protein kinase is activated and phosphorylates the inhibitor, causing its dissociation from the transcription regulator and thereby activating gene expression. (B) This signaling pathway can be diagrammed as a sequence of four steps, including two sequential inhibitory steps that are equivalent to a single activating step.

Cells in a population often exhibit random variation in the concentration or activity of their intracellular signaling molecules. Similarly, individual molecules in a large population of molecules vary in their activity or interactions with other molecules. This *signal variability* introduces another form of noise that can interfere with the precision and efficiency of signaling. Most signaling systems, however, are built to generate remarkably robust and precise responses even when upstream signals are variable or even when some components of the system are disabled. In many cases, this *robustness* depends on the presence of backup mechanisms: for example, a signal might employ two parallel pathways to activate a single common downstream target protein, allowing the response to occur even if one pathway is crippled.

Intracellular Signaling Complexes Form at Activated Receptors

One simple and effective strategy for enhancing the specificity of interactions between signaling molecules is to localize them in the same part of the cell or even within large protein complexes, thereby ensuring that they interact only with each other and not with inappropriate partners. Such mechanisms often involve **scaffold proteins**, which bring together groups of interacting signaling proteins into *signaling complexes*, often before a signal has been received (**Figure 15–10A**). Because the scaffold holds the proteins in close proximity, they can interact at high local concentrations and be sequentially activated rapidly, efficiently, and selectively in response to an appropriate extracellular signal, avoiding unwanted cross-talk with other signaling pathways.

In other cases, such signaling complexes form only transiently in response to an extracellular signal and rapidly disassemble when the signal is gone. They often assemble around a receptor after an extracellular signal molecule has activated it. In many of these cases, the cytoplasmic tail of the activated receptor is phosphorylated during the activation process, and the phosphorylated amino acids then serve as docking sites for the assembly of other signaling proteins (Figure 15–10B). In yet other cases, receptor activation leads to the production of modified phospholipid molecules (called phosphoinositides) in the adjacent plasma membrane, which then recruit specific intracellular signaling proteins to this region of membrane, where they are activated (Figure 15–10C).

Modular Interaction Domains Mediate Interactions Between Intracellular Signaling Proteins

Simply bringing intracellular signaling proteins together into close proximity is sometimes sufficient to activate them. Thus, *induced proximity*, where a signal triggers assembly of a signaling complex, is commonly used to relay signals from protein to protein along a signaling pathway. The assembly of such signaling complexes depends on various highly conserved, small **interaction domains**, which are found in many intracellular signaling proteins. Each of these compact protein modules binds to a particular structural motif in another protein or lipid. The recognized motif in the interacting protein can be a short peptide sequence, a covalent modification (such as a phosphorylated amino acid), or another protein domain. The use of modular interaction domains presumably facilitated the evolution of new signaling pathways; because it can be inserted at many locations in a protein without disturbing the protein's folding or function, a new interaction domain added to an existing signaling protein could connect the protein to additional signaling pathways.

There are many types of interaction domains in signaling proteins. *Src homology 2* (*SH2*) *domains* and *phosphotyrosine-binding* (*PTB*) *domains*, for example, bind to phosphorylated tyrosines in a particular peptide sequence on activated receptors or intracellular signaling proteins. *Src homology 3* (*SH3*) domains bind to short, proline-rich amino acid sequences. Some *pleckstrin homology* (*PH*) domains bind to the charged head groups of specific phosphoinositides that are produced in the plasma membrane in response to an extracellular signal; they

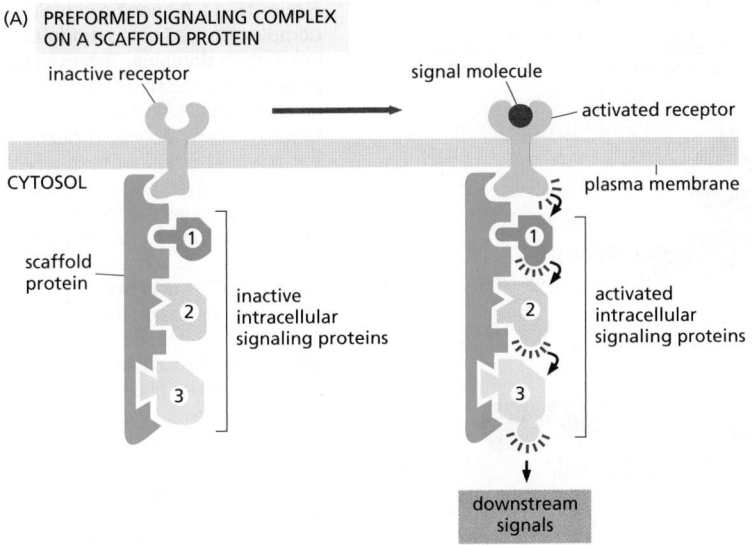

(A) PREFORMED SIGNALING COMPLEX ON A SCAFFOLD PROTEIN

inactive receptor

signal molecule

activated receptor

CYTOSOL

plasma membrane

scaffold protein

inactive intracellular signaling proteins

activated intracellular signaling proteins

downstream signals

(B) ASSEMBLY OF SIGNALING COMPLEX ON AN ACTIVATED RECEPTOR

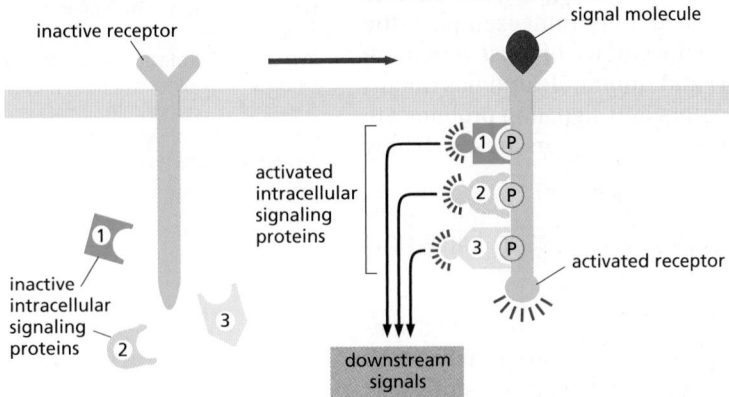

inactive receptor

signal molecule

activated intracellular signaling proteins

inactive intracellular signaling proteins

activated receptor

downstream signals

(C) ASSEMBLY OF SIGNALING COMPLEX ON PHOSPHOINOSITIDE DOCKING SITES

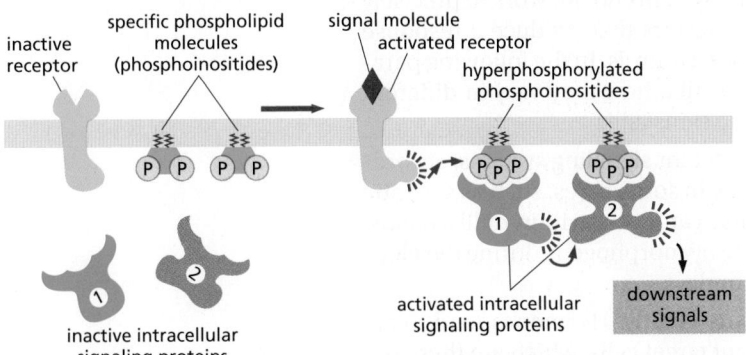

inactive receptor

specific phospholipid molecules (phosphoinositides)

signal molecule
activated receptor

hyperphosphorylated phosphoinositides

inactive intracellular signaling proteins

activated intracellular signaling proteins

downstream signals

Figure 15–10 **Three types of intracellular signaling complexes.** (A) A receptor and some of the intracellular signaling proteins it activates in sequence are preassembled into a signaling complex on the inactive receptor by a large scaffold protein. (B) A signaling complex assembles transiently on a receptor only after the binding of an extracellular signal molecule has activated the receptor; here, the activated receptor phosphorylates itself at multiple sites, which then act as docking sites for intracellular signaling proteins. (C) Activation of a receptor leads to the increased phosphorylation of specific phospholipids (phosphoinositides) in the adjacent plasma membrane; these then serve as docking sites for specific intracellular signaling proteins, which can now interact with each other.

enable the protein they are part of to dock on the membrane and interact with other similarly recruited signaling proteins (see Figure 15–10C). Some signaling proteins consist solely of two or more interaction domains and function only as **adaptors** to link two other proteins together in a signaling pathway.

Interaction domains enable signaling proteins to bind to one another in multiple specific combinations. Like Lego® bricks, the proteins can form linear or branching chains or three-dimensional networks, which determine the route followed by the signaling pathway. As an example, **Figure 15–11** illustrates how some interaction domains mediate the formation of a large signaling complex around the receptor for the hormone *insulin*.

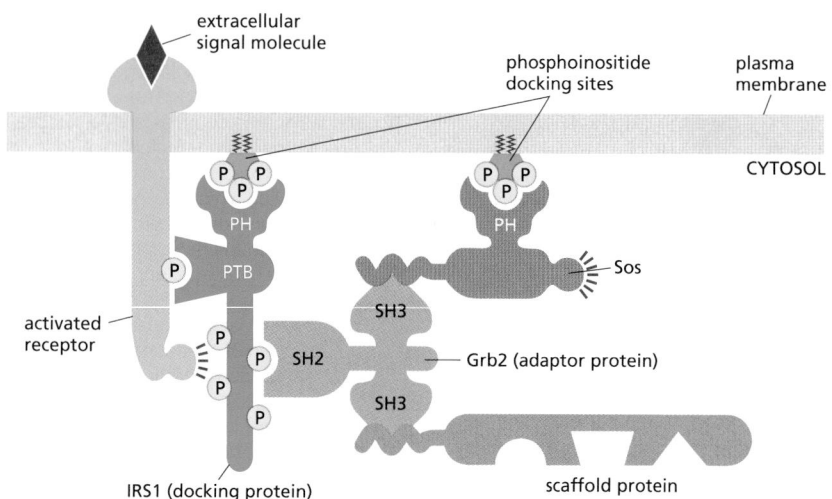

Figure 15–11 A specific signaling complex formed using modular interaction domains. This example is based on the insulin receptor, which is an enzyme-coupled receptor (a receptor tyrosine kinase, discussed later). First, the activated receptor phosphorylates itself on tyrosines, and one of the phosphotyrosines then recruits a docking protein called insulin receptor substrate-1 (IRS1) via a PTB domain of IRS1; the PH domain of IRS1 also binds to specific phosphoinositides on the inner surface of the plasma membrane. Then, the activated receptor phosphorylates IRS1 on tyrosines, and one of these phosphotyrosines binds the SH2 domain of the adaptor protein Grb2. Next, Grb2 uses one of its two SH3 domains to bind to a proline-rich region of a protein called Sos, which relays the signal downstream by acting as a GEF (see Figure 15–8) to activate a monomeric GTPase called Ras (not shown). Sos also binds to phosphoinositides in the plasma membrane via its PH domain. Grb2 uses its other SH3 domain to bind to a proline-rich sequence in a scaffold protein. The scaffold protein binds several other signaling proteins, and the other phosphorylated tyrosines on IRS1 recruit additional signaling proteins that have SH2 domains (not shown).

Another way of bringing receptors and intracellular signaling proteins together is to concentrate them in a specific region of the cell. An important example is the **primary cilium** that projects like an antenna from the surface of most vertebrate cells (discussed in Chapter 16). It is usually short and nonmotile and has microtubules in its core, and a number of surface receptors and signaling proteins are concentrated there. We shall see later that light and smell receptors are also highly concentrated in specialized cilia.

The Relationship Between Signal and Response Varies in Different Signaling Pathways

The function of an intracellular signaling system is to detect and measure a specific stimulus in one location of a cell and then generate an appropriately timed and measured response at another location. The system accomplishes this task by sending information in the form of molecular "signals" from the sensor to the target, often through a series of intermediaries that do not simply pass the signal along but process it in various ways. All signaling systems do not work in precisely the same way: each has evolved specialized behaviors that produce a response that is appropriate for the cell function that system controls. In the following paragraphs, we list some of these behaviors and describe how they vary in different systems, as a foundation for more detailed discussions later.

1. *Response timing* varies dramatically in different signaling systems, according to the speed required for the response. In some cases, such as synaptic signaling (see Figure 15–2C), the response can occur within milliseconds. In other cases, as in the control of cell fate by morphogens during development, a full response can require hours or days.

2. *Sensitivity* to extracellular signals can vary greatly. Hormones tend to act at very low concentrations on their distant target cells, which are therefore highly sensitive to low concentrations of signal. Neurotransmitters, on the other hand, operate at much higher concentrations at a synapse, reducing the need for high sensitivity in postsynaptic receptors. Sensitivity is often controlled by changes in the number or affinity of the receptors on the target cell. A particularly important mechanism for increasing the sensitivity of a signaling system is signal *amplification*, whereby a small number of activated cell-surface receptors evoke a large intracellular response either by producing large amounts of a second messenger or by activating many copies of a downstream signaling protein.

3. *Dynamic range* of a signaling system is related to its sensitivity. Some systems, like those involved in simple developmental decisions, are responsive

over a narrow range of extracellular signal concentrations. Other systems, like those controlling vision or the metabolic response to some hormones, are highly responsive over a much broader range of signal strengths. We will see that broad dynamic range is often achieved by *adaptation* mechanisms that adjust the responsiveness of the system according to the prevailing amount of signal.

4. *Persistence* of a response can vary greatly. A transient response of less than a second is appropriate in some synaptic responses, for example, while a prolonged or even permanent response is required in cell fate decisions during development. Numerous mechanisms, including positive feedback, can be used to alter the duration and reversibility of a response.

5. *Signal processing* can convert a simple signal into a complex response. In many systems, for example, a gradual increase in an extracellular signal is converted into an abrupt, switchlike response. In other cases, a simple input signal is converted into an oscillatory response, produced by a repeating series of transient intracellular signals. Feedback usually lies at the heart of biochemical switches and oscillators, as we describe later.

6. *Integration* allows a response to be governed by multiple inputs. As discussed earlier, for example, specific combinations of extracellular signals are generally required to stimulate complex cell behaviors such as cell survival and proliferation (see Figure 15–4). The cell therefore has to integrate information coming from multiple signals, which often depends on intracellular *coincidence detectors*; these proteins are equivalent to *AND gates* in the microprocessor of a computer, in that they are only activated if they receive multiple converging signals (**Figure 15–12**).

7. *Coordination* of multiple responses in one cell can be achieved by a single extracellular signal. Some extracellular signal molecules, for example, stimulate a cell to both grow and divide. This coordination generally depends on mechanisms for distributing a signal to multiple effectors, by creating branches in the signaling pathway. In some cases, the branching of signaling pathways can allow one signal to *modulate* the strength of a response to other signals.

Given the complexity that arises from behaviors like signal integration, distribution, and feedback, it is clear that signaling systems rarely depend on a simple linear sequence of steps but are often more like a network, in which information flows not just forward but in multiple directions—and sometimes even backward. A major research challenge is to understand the nature of these networks and the response behaviors they can achieve.

The Speed of a Response Depends on the Turnover of Signaling Molecules

The speed of any signaling response depends on the nature of the intracellular signaling molecules that carry out the target cell's response. When the response requires only changes in proteins already present in the cell, it can occur very rapidly: an allosteric change in a neurotransmitter-gated ion channel (discussed in Chapter 11), for example, can alter the plasma membrane electrical potential in milliseconds, and responses that depend solely on protein phosphorylation can occur within seconds. When the response involves changes in gene expression and the synthesis of new proteins, however, it usually requires many minutes or hours, regardless of the mode of signal delivery (**Figure 15–13**).

It is natural to think of intracellular signaling systems in terms of the changes produced when an extracellular signal is delivered. But it is just as important to consider what happens when the signal is withdrawn. During development, transient extracellular signals often produce lasting effects: they can trigger a change in the cell's development that persists indefinitely through cell memory mechanisms, as we discuss later (and in Chapters 7 and 22). In most cases in adult tissues, however, the response fades when a signal ceases. Often the effect is transient

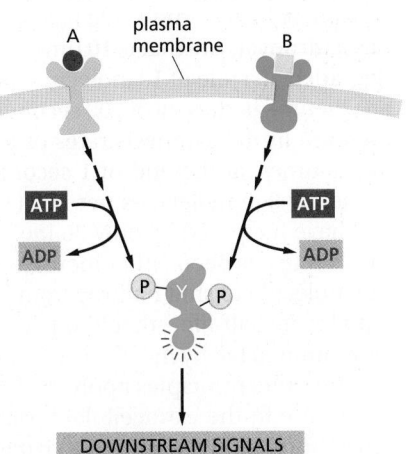

Figure 15–12 Signal integration. Extracellular signals A and B activate different intracellular signaling pathways, each of which leads to the phosphorylation of protein Y but at different sites on the protein. Protein Y is activated only when both of these sites are phosphorylated, and therefore it becomes active only when signals A and B are simultaneously present. Such proteins are often called coincidence detectors.

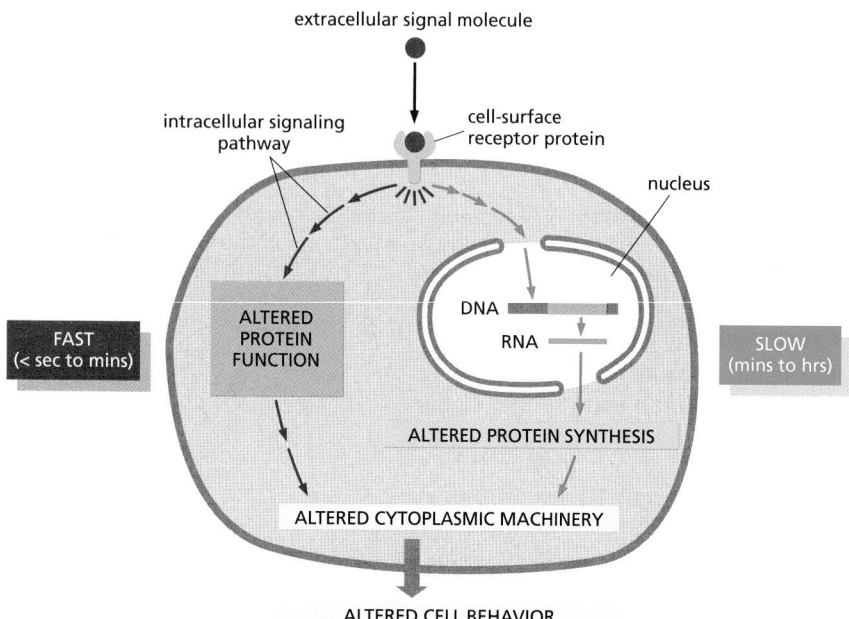

extracellular signal molecule

intracellular signaling pathway

cell-surface receptor protein

nucleus

FAST (< sec to mins)

ALTERED PROTEIN FUNCTION

DNA

RNA

SLOW (mins to hrs)

ALTERED PROTEIN SYNTHESIS

ALTERED CYTOPLASMIC MACHINERY

ALTERED CELL BEHAVIOR

Figure 15–13 **Slow and rapid responses to an extracellular signal.** Certain types of signal-induced cellular responses, such as increased cell growth and division, involve changes in gene expression and the synthesis of new proteins; they therefore occur slowly, often starting an hour or more after the signal is received. Other responses—such as changes in cell movement, secretion, or metabolism—need not involve changes in gene transcription and therefore occur much more quickly, often starting in seconds or minutes; they may involve the rapid phosphorylation of effector proteins in the cytoplasm, for example. Synaptic responses mediated by changes in membrane potential are even quicker and can occur in milliseconds (not shown). Some signaling systems generate both rapid and slow responses as shown here, allowing the cell to respond quickly to a signal while simultaneously initiating a more long-term, persistent response.

because the signal exerts its effects by altering the concentrations of intracellular molecules that are short-lived (unstable), undergoing continual turnover. Thus, once the extracellular signal is gone, the degradation of the old molecules quickly wipes out all traces of the signal's action. It follows that the speed with which a cell responds to signal removal depends on the rate of destruction, or turnover, of the intracellular molecules that the signal affects.

It is also true, although much less obvious, that this turnover rate can determine the promptness of the response when an extracellular signal arrives. Consider, for example, two intracellular signaling molecules, X and Y, both of which are normally maintained at a steady-state concentration of 1000 molecules per cell. The cell synthesizes and degrades molecule Y at a rate of 100 molecules per second, with each molecule having an average lifetime of 10 seconds. Molecule X has a turnover rate that is 10 times slower than that of Y: it is both synthesized and degraded at a rate of 10 molecules per second, so that each molecule has an average lifetime in the cell of 100 seconds. If a signal acting on the cell causes a tenfold increase in the synthesis rates of both X and Y with no change in the molecular lifetimes, at the end of 1 second the concentration of Y will have increased by nearly 900 molecules per cell ($10 \times 100 - 100$), while the concentration of X will have increased by only 90 molecules per cell. In fact, after a molecule's synthesis rate has been either increased or decreased abruptly, the time required for the molecule to shift halfway from its old to its new equilibrium concentration is equal to its half-life—that is, equal to the time that would be required for its concentration to fall by half if all synthesis were stopped (**Figure 15–14**).

The same principles apply to proteins and small molecules, whether the molecules are in the extracellular space or inside cells. Many intracellular proteins have short half-lives, some surviving for less than 10 minutes. In most cases, these are key regulatory proteins whose concentrations are rapidly controlled in the cell by changes in their rates of synthesis.

As we have seen, many cell responses to extracellular signals depend on the conversion of intracellular signaling proteins from an inactive to an active form, rather than on their synthesis or degradation. Phosphorylation or the binding of GTP, for example, commonly activates signaling proteins. Even in these cases, however, the activation must be rapidly and continuously reversed (by dephosphorylation or GTP hydrolysis to GDP, respectively, in these examples) to make rapid signaling possible. These inactivation processes play a crucial part in determining the magnitude, rapidity, and duration of the response.

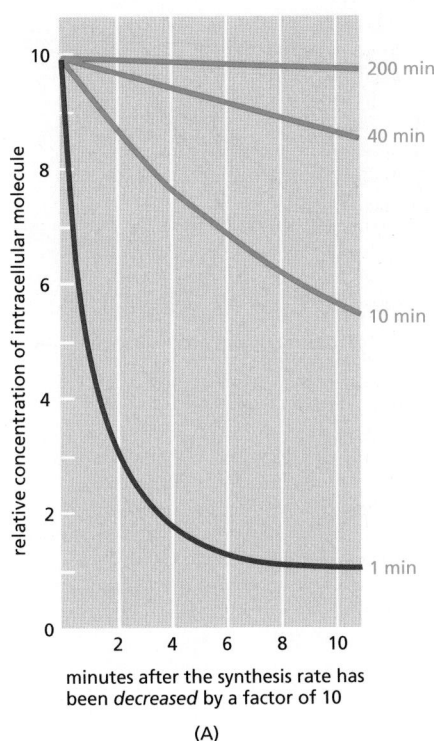

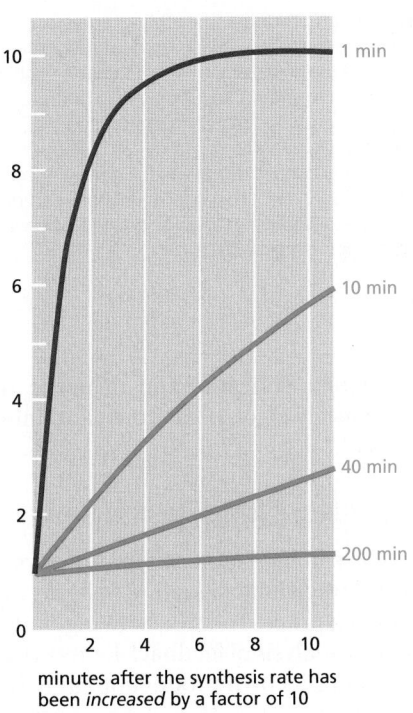

Figure 15–14 **The importance of rapid turnover.** The graphs show the predicted relative rates of change in the intracellular concentrations of molecules with differing turnover times when their synthesis rates are either (A) decreased or (B) increased suddenly by a factor of 10. In both cases, the concentrations of those molecules that are normally degraded rapidly in the cell *(red lines)* change quickly, whereas the concentrations of those that are normally degraded slowly *(green lines)* change proportionally more slowly. The numbers (in *blue*) on the right are the half-lives assumed for each of the different molecules.

Cells Can Respond Abruptly to a Gradually Increasing Signal

Some signaling systems are capable of generating a smoothly graded response over a wide range of extracellular signal concentrations (**Figure 15–15**, *blue* line); such systems are useful, for example, in the fine tuning of metabolic processes by some hormones. Other signaling systems generate significant responses only when the signal concentration rises beyond some threshold value. These abrupt responses are of two types. One is a *sigmoidal* response, in which low concentrations of stimulus do not have much effect, but then the response rises steeply and continuously at intermediate stimulus levels (Figure 15–15, *red* line). Such systems provide a filter to reduce inappropriate responses to low-level background signals but respond with high sensitivity when the stimulus falls within a small range of physiological signal concentrations. A second type of abrupt response is the *discontinuous* or *all-or-none* response, in which the response switches on completely (and often irreversibly) when the signal reaches some threshold concentration (Figure 15–15, *green* line). Such responses are particularly useful for controlling the choice between two alternative cell states, and they generally involve positive feedback, as we describe in more detail shortly.

Cells use a variety of molecular mechanisms to produce a sigmoidal response to increasing signal concentrations. In one mechanism, more than one intracellular signaling molecule must bind to its downstream target protein to induce a response. As we discuss later, for example, four molecules of the second messenger cyclic AMP must bind simultaneously to each molecule of *cyclic-AMP-dependent protein kinase* (*PKA*) to activate the kinase. A similar sharpening of response is seen when the activation of an intracellular signaling protein requires phosphorylation at more than one site. Such responses become sharper as the number of required molecules or phosphate groups increases, and if the number is large enough, responses become almost all-or-none (**Figure 15–16**).

Responses are also sharpened when an intracellular signaling molecule activates one enzyme and also inhibits another enzyme that catalyzes the opposite reaction. A well-studied example of this common type of regulation is the stimulation of glycogen breakdown in skeletal muscle cells induced by the hormone *adrenaline* (epinephrine). Adrenaline's binding to a G-protein-coupled

Figure 15–15 **Signal processing can produce smoothly graded or switchlike responses.** Some cell responses increase gradually as the concentration of extracellular signal molecule increases, eventually reaching a plateau as the signaling pathway is saturated, resulting in a *hyperbolic* response curve *(blue line)*. In other cases, the signaling system reduces the response at low signal concentrations and then produces a steeper response at some intermediate signal concentration—resulting in a *sigmoidal* response curve *(red line)*. In still other cases, the response is more abrupt and switchlike; the cell switches completely between a low and high response, without any stable intermediate response *(green line)*.

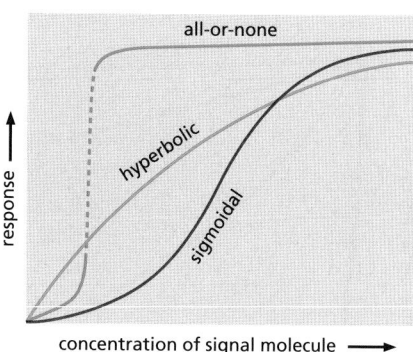

cell-surface receptor increases the intracellular concentration of cyclic AMP, which both activates an enzyme that promotes glycogen breakdown and inhibits an enzyme that promotes glycogen synthesis.

Positive Feedback Can Generate an All-or-None Response

Like intracellular metabolic pathways (discussed in Chapter 2) and the systems controlling gene activity (Chapter 7), most intracellular signaling systems incorporate feedback loops, in which the output of a process acts back to regulate that same process. We discussed the mathematical analysis of feedback loops in Chapter 8. In *positive feedback*, the output stimulates its own production; in *negative feedback*, the output inhibits its own production (**Figure 15–17**). Feedback loops are of great general importance in biology, and they regulate many chemical and physical processes in cells. Those that regulate cell signaling can either operate exclusively within the target cell or involve the secretion of extracellular signals. Here, we focus on those feedback loops that operate entirely within the target cell; even the simplest of these loops can produce complex and interesting effects.

Positive feedback in a signaling pathway can transform the behavior of the responding cell. If the positive feedback is of only moderate strength, its effect will be simply to steepen the response to the signal, generating a sigmoidal response like those described earlier; but if the feedback is strong enough, it can produce an all-or-none response (see Figure 15–15). This response goes hand in hand with a further property: once the responding system has switched to the high level of activation, this condition is often self-sustaining and can persist even after the

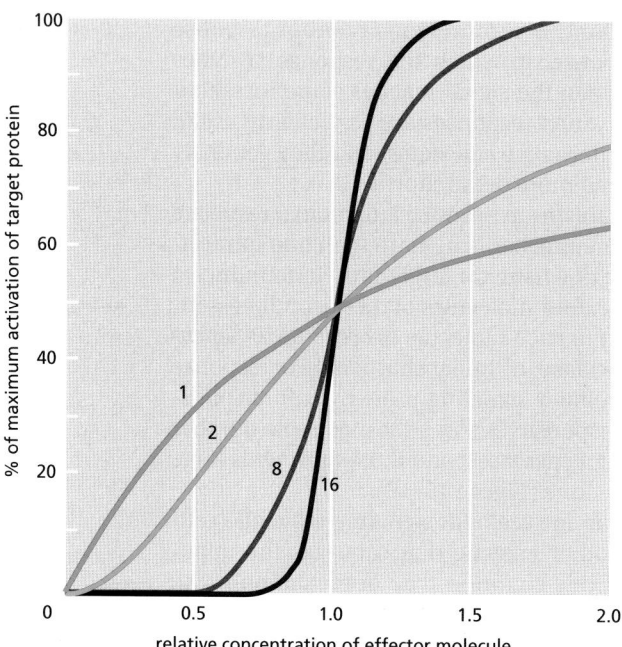

Figure 15–16 **Activation curves for an allosteric protein as a function of effector molecule concentration.** The curves show how the sharpness of the activation response increases with an increase in the number of allosteric effector molecules that must be bound simultaneously to activate the target protein. The curves shown are those expected, under certain conditions, if the activation requires the simultaneous binding of 1, 2, 8, or 16 effector molecules.

signal strength drops back below its critical value. In such a case, the system is said to be *bistable*: it can exist in either a "switched-off" or a "switched-on" state, and a transient stimulus can flip it from the one state to the other (**Figure 15–18**A and B).

Through positive feedback, a transient extracellular signal can induce long-term changes in cells and their progeny that can persist for the lifetime of the organism. The signals that trigger muscle-cell specification, for example, turn on the transcription of a series of genes that encode muscle-specific transcription regulatory proteins, which stimulate the transcription of their own genes, as well as genes encoding various other muscle-cell proteins; in this way, the decision to become a muscle cell is made permanent. This type of cell memory, which depends on positive feedback, is one of the basic ways in which a cell can undergo a lasting change of character without any alteration in its DNA sequence.

Studies of signaling responses in large populations of cells can give the false impression that a response is smoothly graded, even when strong positive feedback is causing an abrupt, discontinuous switch in the response in individual cells. Only by studying the response in single cells is it possible to see its all-or-none character (**Figure 15–19**). The misleading smooth response in a cell population is due to the random, intrinsic variability in signaling systems that we described earlier: all cells in a population do not respond identically to the same concentration of extracellular signal, especially at intermediate signal concentrations where the receptor is only partially occupied.

Negative Feedback is a Common Motif in Signaling Systems

By contrast with positive feedback, negative feedback counteracts the effect of a stimulus and thereby abbreviates and limits the level of the response, making the system less sensitive to perturbations (see Chapter 8). As with positive feedback, however, qualitatively different responses can be obtained when the feedback

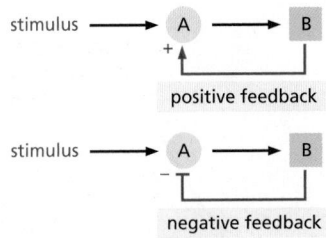

Figure 15–17 Positive and negative feedback. In these simple examples, a stimulus activates protein A, which, in turn, activates protein B. Protein B then acts back to either increase or decrease the activity of A.

Figure 15–18 Some effects of simple feedback. The graphs show the computed effects of simple positive and negative feedback loops (see Chapter 8). In each case, the input signal is an activated protein kinase (S) that phosphorylates and thereby activates another protein kinase (E); a protein phosphatase (I) dephosphorylates and inactivates the activated E kinase. In the graphs, the *red* line indicates the activity of the E kinase over time; the underlying *blue* bar indicates the time for which the input signal (activated S kinase) is present. (A) Diagram of the positive feedback loop, in which the activated E kinase acts back to promote its own phosphorylation and activation; the basal activity of the I phosphatase dephosphorylates activated E at a steady, low rate. (B) The top graph shows that, without feedback, the activity of the E kinase is simply proportional (with a short lag) to the level of stimulation by the S kinase. The bottom graph shows that, with the positive feedback loop, the transient stimulation by S kinase switches the system from an "off" state to an "on" state, which then persists after the stimulus has been removed. (C) Diagram of the negative feedback loop, in which the activated E kinase phosphorylates and activates the I phosphatase, thereby increasing the rate at which the phosphatase dephosphorylates and inactivates the phosphorylated E kinase. (D) The top graph shows, again, the response in E kinase activity without feedback. The other graphs show the effects on E kinase activity of negative feedback operating after a short or long delay. With a short delay, the system shows a strong, brief response when the signal is abruptly changed, and the feedback then drives the response back down to a lower level. With a long delay, the feedback produces sustained oscillations for as long as the stimulus is present.

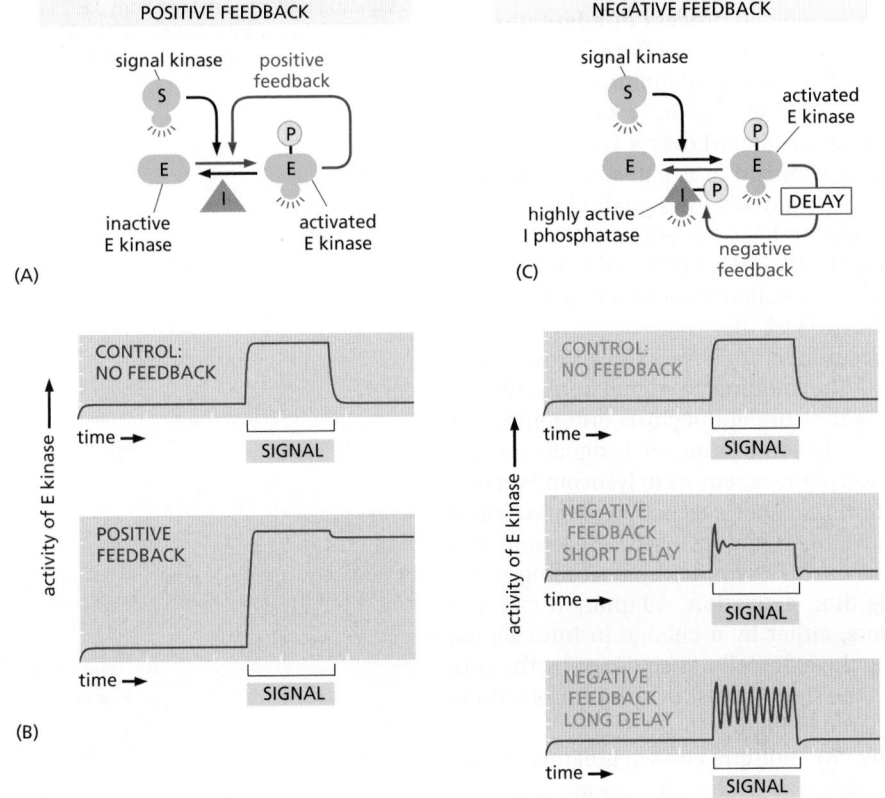

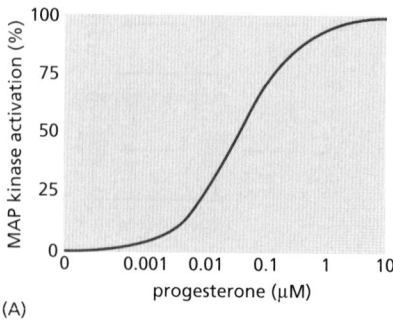

(A)

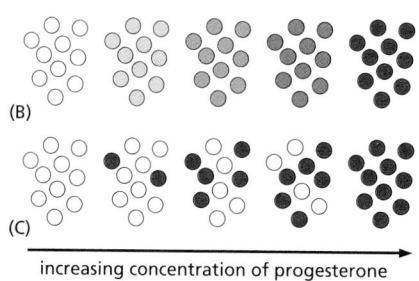

(B)

(C)

increasing concentration of progesterone

Figure 15–19 **The importance of examining individual cells to detect all-or-none responses to increasing concentrations of an extracellular signal.** In these experiments, immature frog eggs (oocytes) were stimulated with increasing concentrations of the hormone progesterone. The response was assessed by analyzing the activation of *MAP kinase* (discussed later), which is one of the protein kinases activated by phosphorylation in the response. The amount of phosphorylated (activated) MAP kinase in extracts of the oocytes was assessed biochemically. In (A), extracts of populations of stimulated oocytes were analyzed, and the activation of MAP kinase appeared to increase progressively with increasing progesterone concentration. There are two possible ways of explaining this result: (B) MAP kinase could have increased gradually in each individual cell with increasing progesterone concentration; or (C) individual cells could have responded in an all-or-none way, with the gradual increase in total MAP kinase activation reflecting the increasing number of cells responding with increasing progesterone concentration. When extracts of individual oocytes were analyzed, it was found that cells had either very low amounts or very high amounts, but not intermediate amounts, of the activated kinase, indicating that the response was essentially all-or-none at the level of individual cells, as diagrammed in (C). Subsequent studies revealed that this all-or-none response is due in part to strong positive feedback in the progesterone signaling system. (Adapted from J.E. Ferrell and E.M. Machleder, *Science* 280:895–898, 1998. With permission from AAAS.)

operates more powerfully. A delayed negative feedback with a long enough delay can produce responses that oscillate. The oscillations may persist for as long as the stimulus is present (Figure 15–18C and D) or they may even be generated spontaneously, without need of an external signal to drive them. Many such oscillators also contain positive feedback loops that generate sharper oscillations. Later in this chapter, we will encounter specific examples of oscillatory behavior in the intracellular responses to extracellular signals; all of them depend on negative feedback, generally accompanied by positive feedback.

If negative feedback operates with a short delay, the system behaves like a change detector. It gives a strong response to a stimulus, but the response rapidly decays even while the stimulus persists; if the stimulus is suddenly increased, however, the system responds strongly again, but, again, the response rapidly decays. This is the phenomenon of *adaptation*, which we now discuss.

Cells Can Adjust Their Sensitivity to a Signal

In responding to many types of stimuli, cells and organisms are able to detect the same percentage of change in a signal over a wide range of stimulus strengths. The target cells accomplish this through a reversible process of **adaptation**, or **desensitization**, whereby a prolonged exposure to a stimulus decreases the cells' response to that level of stimulus. In chemical signaling, adaptation enables cells to respond to *changes* in the concentration of an extracellular signal molecule (rather than to the absolute concentration of the signal) over a very wide range of signal concentrations. The underlying mechanism is negative feedback that operates with a short delay: a strong response modifies the signaling machinery involved, such that the machinery resets itself to become less responsive to the same level of signal (see Figure 15–18D, middle graph). Owing to the delay, however, a sudden increase in the signal is able to stimulate the cell again for a short period before the negative feedback has time to kick in.

Adaptation to a signal molecule can occur in various ways. It can result from inactivation of the receptors themselves. The binding of signal molecules to cell-surface receptors, for example, may induce the endocytosis and temporary sequestration of the receptors in endosomes. In some cases, such signal-induced receptor endocytosis leads to the destruction of the receptors in lysosomes, a process referred to as *receptor down-regulation* (in other cases, however, activated receptors continue to signal after they have been endocytosed). Receptors can also become inactivated on the cell surface—for example, by becoming phosphorylated—with a short delay following their activation. Adaptation can also occur at sites downstream of the receptors, either by a change in intracellular signaling proteins involved in transducing the extracellular signal or by the production of an inhibitor protein that blocks the signal transduction process. These various adaptation mechanisms are compared in **Figure 15–20**.

Though bewildering in their complexity, the multiple cross-regulatory signaling pathways and feedback loops that we describe in this chapter are not just a haphazard tangle, but a highly evolved system for processing and interpreting

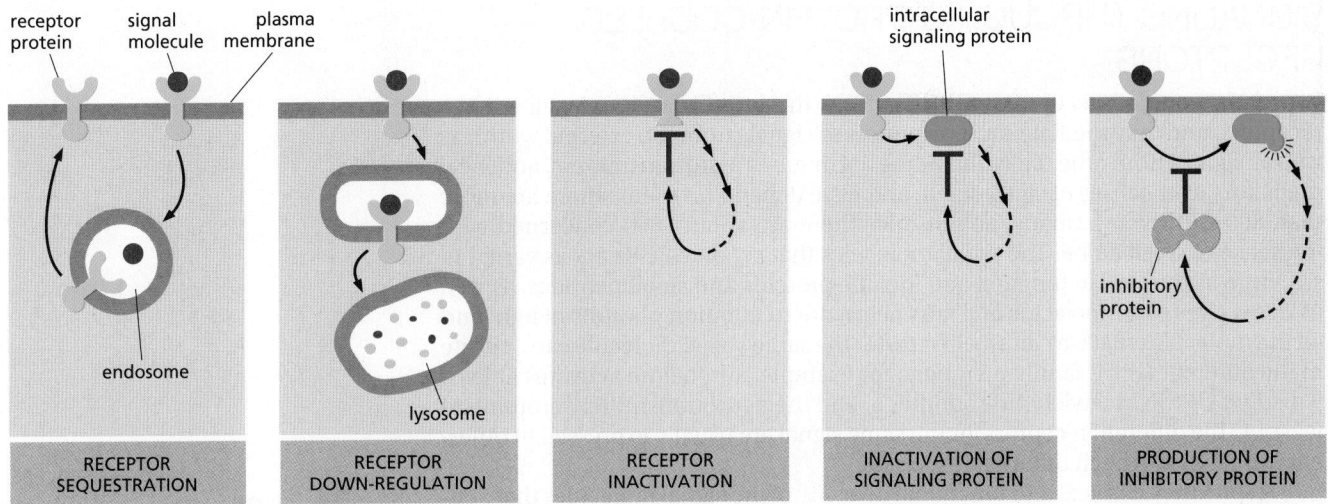

Figure 15–20 **Some ways in which target cells can become adapted (desensitized) to an extracellular signal molecule.** The mechanisms shown here that operate at the level of the receptor often involve phosphorylation or ubiquitylation of the receptor proteins.

the vast number of signals that impinge upon animal cells. The whole molecular control network, leading from the receptors at the cell surface to the genes in the nucleus, can be viewed as a computing device; and, like that other biological computing device, the brain, it presents one of the hardest problems in biology. We can identify the components and discover how they work individually. We can understand how small subsets of components work together as regulatory modules, noise filters, or adaptation mechanisms, as we have seen. However, it is a much more difficult task to understand how the system works as a whole. This is not only because the system is complex; it is also because the way in which it behaves is strongly dependent on the quantitative details of the molecular interactions, and, for most animal cells, we have only rough qualitative information. A major challenge for the future of signaling research is to develop more sophisticated quantitative and computational methods for the analysis of signaling systems, as described in Chapter 8.

Summary

Each cell in a multicellular animal is programmed to respond to a specific set of extracellular signal molecules produced by other cells. The signal molecules act by binding to a complementary set of receptor proteins expressed by the target cells. Most extracellular signal molecules activate cell-surface receptor proteins, which act as signal transducers, converting the extracellular signal into intracellular ones that alter the behavior of the target cell. Activated receptors relay the signal into the cell interior by activating intracellular signaling proteins. Some of these signaling proteins transduce, amplify, or spread the signal as they relay it, while others integrate signals from different signaling pathways. Some function as switches that are transiently activated by phosphorylation or GTP binding. Large signaling complexes form by means of modular interaction domains in the signaling proteins, which allow the proteins to form functional signaling networks.

Target cells use various mechanisms, including feedback loops, to adjust the ways in which they respond to extracellular signals. Positive feedback loops can help cells to respond in an all-or-none fashion to a gradually increasing concentration of an extracellular signal and to convert a short-lasting signal into a long-lasting, or even irreversible, response. Negative feedback allows cells to adapt to a signal molecule, which enables them to respond to small changes in the concentration of the signal molecule over a large concentration range.

SIGNALING THROUGH G-PROTEIN-COUPLED RECEPTORS

G-protein-coupled receptors (GPCRs) form the largest family of cell-surface receptors, and they mediate most responses to signals from the external world, as well as signals from other cells, including hormones, neurotransmitters, and local mediators. Our senses of sight, smell, and taste depend on them. There are more than 800 GPCRs in humans, and in mice there are about 1000 concerned with the sense of smell alone. The signal molecules that act on GPCRs are as varied in structure as they are in function and include proteins and small peptides, as well as derivatives of amino acids and fatty acids, not to mention photons of light and all the molecules that we can smell or taste. The same signal molecule can activate many different GPCR family members; for example, adrenaline activates at least 9 distinct GPCRs, acetylcholine another 5, and the neurotransmitter serotonin at least 14. The different receptors for the same signal are usually expressed in different cell types and elicit different responses.

Despite the chemical and functional diversity of the signal molecules that activate them, all GPCRs have a similar structure. They consist of a single polypeptide chain that threads back and forth across the lipid bilayer seven times, forming a cylindrical structure, often with a deep ligand-binding site at its center (**Figure 15–21**). In addition to their characteristic orientation in the plasma membrane, they all use G proteins to relay the signal into the cell interior.

The GPCR superfamily includes *rhodopsin*, the light-activated protein in the vertebrate eye, as well as the large number of olfactory receptors in the vertebrate nose. Other family members are found in unicellular organisms: the receptors in yeasts that recognize secreted mating factors are an example. It is likely that the GPCRs that mediate cell–cell signaling in multicellular organisms evolved from the sensory receptors in their unicellular eukaryotic ancestors.

It is remarkable that almost half of all known drugs work through GPCRs or the signaling pathways GPCRs activate. Of the many hundreds of genes in the human genome that encode GPCRs, about 150 encode orphan receptors, for which the ligand is unknown. Many of them are likely targets for new drugs that remain to be discovered.

Trimeric G Proteins Relay Signals From GPCRs

When an extracellular signal molecule binds to a GPCR, the receptor undergoes a conformational change that enables it to activate a **trimeric GTP-binding protein (G protein)**, which couples the receptor to enzymes or ion channels in the membrane. In some cases, the G protein is physically associated with the receptor before the receptor is activated, whereas in others it binds only after receptor activation. There are various types of G proteins, each specific for a particular set of GPCRs and for a particular set of target proteins in the plasma membrane. They all have a similar structure, however, and operate similarly.

G proteins are composed of three protein subunits—α, β, and γ. In the unstimulated state, the α subunit has GDP bound and the G protein is inactive (**Figure 15–22**). When a GPCR is activated, it acts like a guanine nucleotide exchange factor (GEF) and induces the α subunit to release its bound GDP, allowing GTP to bind in its place. GTP binding then causes an activating conformational change in the Gα subunit, releasing the G protein from the receptor and triggering dissociation of the GTP-bound Gα subunit from the G$\beta\gamma$ pair—both of which then interact with various targets, such as enzymes and ion channels in the plasma membrane, which relay the signal onward (**Figure 15–23**).

The α subunit is a GTPase and becomes inactive when it hydrolyzes its bound GTP to GDP. The time required for GTP hydrolysis is usually short because the GTPase activity is greatly enhanced by the binding of the α subunit to a second protein, which can be either the target protein or a specific **regulator of G protein signaling (RGS)**. RGS proteins act as α-subunit-specific GTPase-activating proteins (GAPs) (see Figure 15–8), and they help shut off G-protein-mediated responses in all eukaryotes. There are about 25 RGS proteins encoded in the human genome, each of which interacts with a particular set of G proteins.

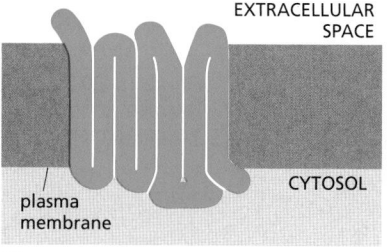

(A)

(B)

Figure 15–21 A G-protein-coupled receptor (GPCR). (A) GPCRs that bind small ligands such as adrenaline have small extracellular domains, and the ligand usually binds deep within the plane of the membrane to a site that is formed by amino acids from several transmembrane segments. GPCRs that bind protein ligands have a large extracellular domain (not shown here) that contributes to ligand binding. (B) The structure of the β_2-adrenergic receptor, a receptor for the neurotransmitter adrenaline, illustrates the typical cylindrical arrangement of the seven transmembrane helices in a GPCR. The ligand *(orange)* binds in a pocket between the helices, resulting in conformational changes on the cytoplasmic surface of the receptor that promote G-protein activation (not shown). (PDB code: 3P0G.)

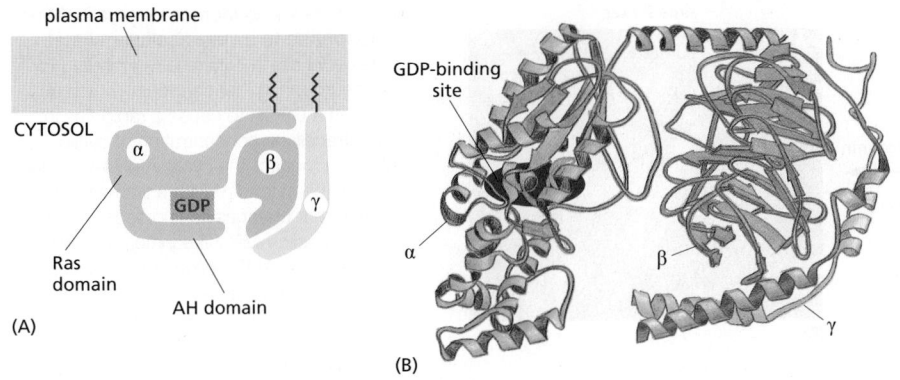

Figure 15–22 **The structure of an inactive G protein.** (A) Note that both the α and the γ subunits have covalently attached lipid molecules *(red tails)* that help bind them to the plasma membrane, and the α subunit has GDP bound. (B) The three-dimensional structure of the inactive, GDP-bound form of a G protein called G_s, which interacts with numerous GPCRs, including the $β_2$-adrenergic receptor shown in Figures 15–21 and 15–23. The α subunit contains the GTPase domain and binds to one side of the β subunit. The γ subunit binds to the opposite side of the β subunit, and the β and γ subunits together form a single functional unit. The GTPase domain of the α subunit contains two major subdomains: the "Ras" domain, which is related to other GTPases and provides one face of the nucleotide-binding pocket; and the alpha-helical or "AH" domain, which clamps the nucleotide in place. (B, based on D.G. Lombright et al., *Nature* 379:311–319, 1996.)

Some G Proteins Regulate the Production of Cyclic AMP

Cyclic AMP (cAMP) acts as a second messenger in some signaling pathways. An extracellular signal can increase cAMP concentration more than twentyfold in seconds (**Figure 15–24**). As explained earlier (see Figure 15–14), such a rapid response requires balancing a rapid synthesis of the molecule with its rapid breakdown or removal. Cyclic AMP is synthesized from ATP by an enzyme called

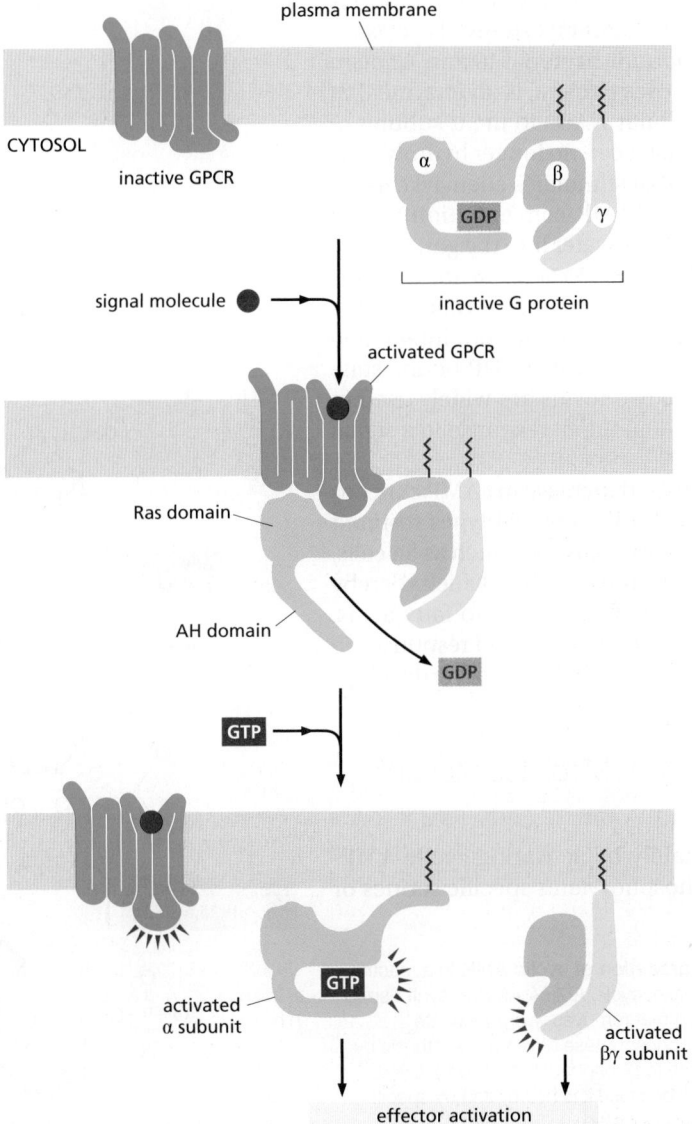

Figure 15–23 **Activation of a G protein by an activated GPCR.** Binding of an extracellular signal molecule to a GPCR changes the conformation of the receptor, which allows the receptor to bind and alter the conformation of a trimeric G protein. The AH domain of the G protein α subunit moves outward to open the nucleotide-binding site, thereby promoting dissociation of GDP. GTP binding then promotes closure of the nucleotide-binding site, triggering conformational changes that cause dissociation of the α subunit from the receptor and from the βγ complex. The GTP-bound α subunit and the βγ complex each regulate the activities of downstream signaling molecules (not shown). The receptor stays active while the extracellular signal molecule is bound to it, and it can therefore catalyze the activation of many G-protein molecules (**Movie 15.1**).

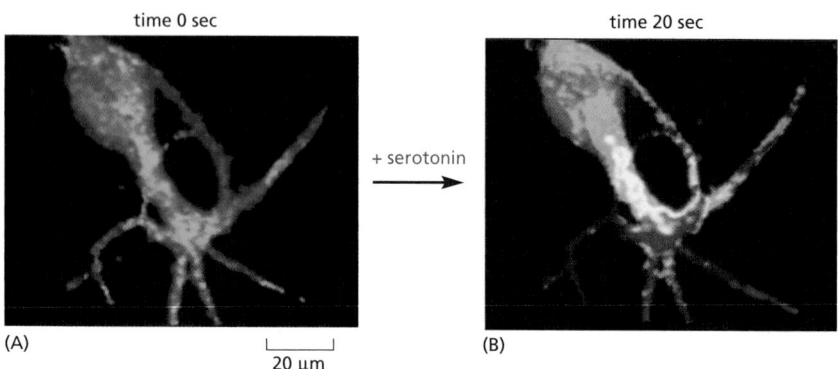

time 0 sec

+ serotonin

time 20 sec

(A)

(B)

20 μm

Figure 15–24 An increase in cyclic AMP in response to an extracellular signal. This nerve cell in culture is responding to the neurotransmitter serotonin, which acts through a GPCR to cause a rapid rise in the intracellular concentration of cyclic AMP. To monitor the cyclic AMP level, the cell has been loaded with a fluorescent protein that changes its fluorescence when it binds cyclic AMP. *Blue* indicates a low level of cyclic AMP, *yellow* an intermediate level, and *red* a high level. (A) In the resting cell, the cyclic AMP level is about 5×10^{-8} M. (B) Twenty seconds after the addition of serotonin to the culture medium, the intracellular level of cyclic AMP has increased to more than 10^{-6} M in the relevant parts of the cell, an increase of more than twentyfold. (From B.J. Bacskai et al., *Science* 260:222–226, 1993. With permission from AAAS.)

adenylyl cyclase, and it is rapidly and continuously destroyed by **cyclic AMP phosphodiesterases** (**Figure 15–25**). Adenylyl cyclase is a large, multipass transmembrane protein with its catalytic domain on the cytosolic side of the plasma membrane. There are at least eight isoforms in mammals, most of which are regulated by both G proteins and Ca^{2+}.

Many extracellular signals work by increasing cAMP concentrations inside the cell. These signals activate GPCRs that are coupled to a **stimulatory G protein** (**G_s**). The activated α subunit of G_s binds and thereby activates adenylyl cyclase. Other extracellular signals, acting through different GPCRs, reduce cAMP levels by activating an **inhibitory G protein** (**G_i**), which then inhibits adenylyl cyclase.

Both G_s and G_i are targets for medically important bacterial toxins. *Cholera toxin*, which is produced by the bacterium that causes cholera, is an enzyme that catalyzes the transfer of ADP ribose from intracellular NAD^+ to the α subunit of G_s. This ADP ribosylation alters the α subunit so that it can no longer hydrolyze its bound GTP, causing it to remain in an active state that stimulates adenylyl cyclase indefinitely. The resulting prolonged elevation in cAMP concentration within intestinal epithelial cells causes a large efflux of Cl^- and water into the gut, thereby causing the severe diarrhea that characterizes cholera. *Pertussis toxin*, which is made by the bacterium that causes pertussis (whooping cough), catalyzes the ADP ribosylation of the α subunit of G_i, preventing the protein from interacting with receptors; as a result, the G protein remains in the inactive GDP-bound state and is unable to regulate its target proteins. These two toxins are widely used in experiments to determine whether a cell's GPCR-dependent response to a signal is mediated by G_s or by G_i.

Some of the responses mediated by a G_s-stimulated increase in cAMP concentration are listed in **Table 15–1**. As the table shows, different cell types respond differently to an increase in cAMP concentration. Some cell types, such as fat cells, activate adenylyl cyclase in response to multiple hormones, all of which thereby stimulate the breakdown of triglyceride (the storage form of fat) to fatty acids. Individuals with genetic defects in the G_s α subunit show decreased responses to certain hormones, resulting in metabolic abnormalities, abnormal bone development, and mental retardation.

Cyclic-AMP-Dependent Protein Kinase (PKA) Mediates Most of the Effects of Cyclic AMP

In most animal cells, cAMP exerts its effects mainly by activating **cyclic-AMP-dependent protein kinase (PKA)**. This kinase phosphorylates specific serines or

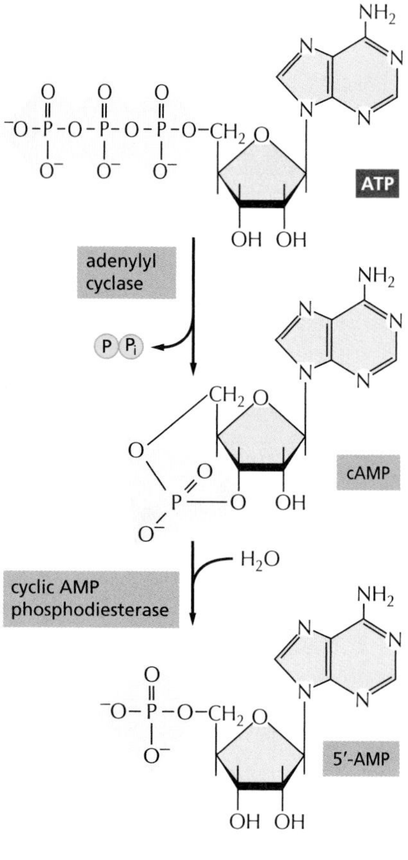

Figure 15–25 The synthesis and degradation of cyclic AMP. In a reaction catalyzed by the enzyme adenylyl cyclase, cyclic AMP (cAMP) is synthesized from ATP through a cyclization reaction that removes two phosphate groups as pyrophosphate (PP$_i$); a pyrophosphatase drives this synthesis by hydrolyzing the released pyrophosphate to phosphate (not shown). Cyclic AMP is short-lived (unstable) in the cell because it is hydrolyzed by specific phosphodiesterases to form 5′-AMP, as indicated.

ATP

adenylyl cyclase

P P$_i$

cAMP

H_2O

cyclic AMP phosphodiesterase

5′-AMP

TABLE 15–1 Some Hormone-induced Cell Responses Mediated by Cyclic AMP

Target tissue	Hormone	Major response
Thyroid gland	Thyroid-stimulating hormone (TSH)	Thyroid hormone synthesis and secretion
Adrenal cortex	Adrenocorticotrophic hormone (ACTH)	Cortisol secretion
Ovary	Luteinizing hormone (LH)	Progesterone secretion
Muscle	Adrenaline	Glycogen breakdown
Bone	Parathormone	Bone resorption
Heart	Adrenaline	Increase in heart rate and force of contraction
Liver	Glucagon	Glycogen breakdown
Kidney	Vasopressin	Water resorption
Fat	Adrenaline, ACTH, glucagon, TSH	Triglyceride breakdown

threonines on selected target proteins, including intracellular signaling proteins and effector proteins, thereby regulating their activity. The target proteins differ from one cell type to another, which explains why the effects of cAMP vary so markedly depending on the cell type (see Table 15–1).

In the inactive state, PKA consists of a complex of two catalytic subunits and two regulatory subunits. The binding of cAMP to the regulatory subunits alters their conformation, causing them to dissociate from the complex. The released catalytic subunits are thereby activated to phosphorylate specific target proteins (**Figure 15–26**). The regulatory subunits of PKA (also called A-kinase) are important for localizing the kinase inside the cell: special *A-kinase anchoring proteins (AKAPs)* bind both to the regulatory subunits and to a component of the cytoskeleton or a membrane of an organelle, thereby tethering the enzyme complex to a particular subcellular compartment. Some AKAPs also bind other signaling proteins, forming a signaling complex. An AKAP located around the nucleus of heart muscle cells, for example, binds both PKA and a phosphodiesterase that hydrolyzes cAMP. In unstimulated cells, the phosphodiesterase keeps the local cAMP concentration low, so that the bound PKA is inactive; in stimulated cells, cAMP concentration rapidly rises, overwhelming the phosphodiesterase and activating the PKA. Among the target proteins that PKA phosphorylates and activates in these cells is the adjacent phosphodiesterase, which rapidly lowers the cAMP concentration again. This negative feedback arrangement converts what might otherwise be a prolonged PKA response into a brief, local pulse of PKA activity.

Whereas some responses mediated by cAMP occur within seconds (see Figure 15–24), others depend on changes in the transcription of specific genes and take hours to develop fully. In cells that secrete the peptide hormone *somatostatin*,

Figure 15–26 The activation of cyclic-AMP-dependent protein kinase (PKA). The binding of cAMP to the regulatory subunits of the PKA tetramer induces a conformational change, causing these subunits to dissociate from the catalytic subunits, thereby activating the kinase activity of the catalytic subunits. The release of the catalytic subunits requires the binding of more than two cAMP molecules to the regulatory subunits in the tetramer. This requirement greatly sharpens the response of the kinase to changes in cAMP concentration, as discussed earlier (see Figure 15–16). Mammalian cells have at least two types of PKAs: type I is mainly in the cytosol, whereas type II is bound via its regulatory subunits and special anchoring proteins to the plasma membrane, nuclear membrane, mitochondrial outer membrane, and microtubules. In both types, once the catalytic subunits are freed and active, they can migrate into the nucleus (where they can phosphorylate transcription regulatory proteins), while the regulatory subunits remain in the cytoplasm.

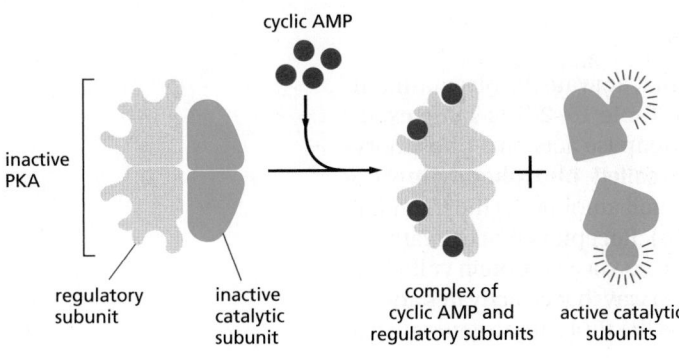

inactive PKA

regulatory subunit

inactive catalytic subunit

cyclic AMP

complex of cyclic AMP and regulatory subunits

active catalytic subunits

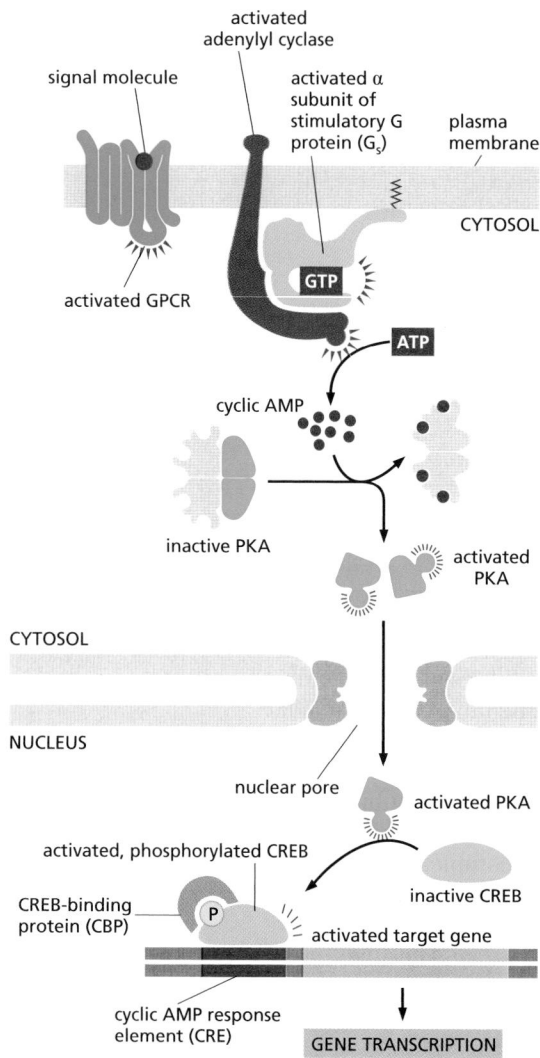

signal molecule

activated
adenylyl cyclase

activated α
subunit of
stimulatory G
protein (G$_s$)

plasma
membrane

activated GPCR

CYTOSOL

GTP

ATP

cyclic AMP

inactive PKA

activated
PKA

CYTOSOL

NUCLEUS

nuclear pore

activated PKA

activated, phosphorylated CREB

inactive CREB

CREB-binding
protein (CBP)

P

activated target gene

cyclic AMP response
element (CRE)

GENE TRANSCRIPTION

Figure 15–27 How a rise in intracellular cyclic AMP concentration can alter gene transcription. The binding of an extracellular signal molecule to its GPCR activates adenylyl cyclase via G$_s$ and thereby increases cAMP concentration in the cytosol. This rise activates PKA, and the released catalytic subunits of PKA can then enter the nucleus, where they phosphorylate the transcription regulatory protein CREB. Once phosphorylated, CREB recruits the coactivator CBP, which stimulates gene transcription. In some cases, at least, the inactive CREB protein is bound to the cyclic AMP response element (CRE) in DNA before it is phosphorylated (not shown). See Movie 15.2.

for example, cAMP activates the gene that encodes this hormone. The regulatory region of the somatostatin gene contains a short *cis*-regulatory sequence, called the *cyclic AMP response element* (*CRE*), which is also found in the regulatory region of many other genes activated by cAMP. A specific transcription regulator called **CRE-binding (CREB) protein** recognizes this sequence. When PKA is activated by cAMP, it phosphorylates CREB on a single serine; phosphorylated CREB then recruits a transcriptional coactivator called *CREB-binding protein* (*CBP*), which stimulates the transcription of the target genes (**Figure 15–27**). Thus, CREB can transform a short cAMP signal into a long-term change in a cell, a process that, in the brain, is thought to play an important part in some forms of learning and memory.

Some G Proteins Signal Via Phospholipids

Many GPCRs exert their effects through G proteins that activate the plasma-membrane-bound enzyme **phospholipase C-β (PLCβ)**. **Table 15–2** lists some examples of responses activated in this way. The phospholipase acts on a phosphorylated inositol phospholipid (a *phosphoinositide*) called **phosphatidylinositol 4,5-bisphosphate [PI(4,5)P$_2$]**, which is present in small amounts in the inner half of the plasma membrane lipid bilayer (**Figure 15–28**). Receptors that activate this **inositol phospholipid signaling pathway** mainly do so via a G protein called **G$_q$**, which activates phospholipase C-β in much the same way that G$_s$ activates adenylyl cyclase. The activated phospholipase then cleaves the PI(4,5)P$_2$ to generate two

TABLE 15–2 Some Cell Responses in Which GPCRs Activate PLCβ

Target tissue	Signal molecule	Major response
Liver	Vasopressin	Glycogen breakdown
Pancreas	Acetylcholine	Amylase secretion
Smooth muscle	Acetylcholine	Muscle contraction
Blood platelets	Thrombin	Platelet aggregation

products: **inositol 1,4,5-trisphosphate (IP$_3$)** and **diacylglycerol**. At this step, the signaling pathway splits into two branches.

IP$_3$ is a water-soluble molecule that leaves the plasma membrane and diffuses rapidly through the cytosol. When it reaches the endoplasmic reticulum (ER), it binds to and opens **IP$_3$-gated Ca^{2+}-release channels** (also called **IP$_3$ receptors**) in the ER membrane. Ca^{2+} stored in the ER is released through the open channels, quickly raising the concentration of Ca^{2+} in the cytosol (**Figure 15–29**). The increase in cytosolic Ca^{2+} propagates the signal by influencing the activity of Ca^{2+}-sensitive intracellular proteins, as we describe shortly.

At the same time that the IP$_3$ produced by the hydrolysis of PI(4,5)P$_2$ is increasing the concentration of Ca^{2+} in the cytosol, the other cleavage product of the PI(4,5)P$_2$, diacylglycerol, is exerting different effects. It also acts as a second messenger, but it remains embedded in the plasma membrane, where it has several potential signaling roles. One of its major functions is to activate a protein kinase called **protein kinase C (PKC)**, so named because it is Ca^{2+}-dependent. The initial rise in cytosolic Ca^{2+} induced by IP$_3$ alters the PKC so that it translocates from the cytosol to the cytoplasmic face of the plasma membrane. There it is activated by the combination of Ca^{2+}, diacylglycerol, and the negatively charged membrane phospholipid phosphatidylserine (see Figure 15–29). Once activated, PKC phosphorylates target proteins that vary depending on the cell type. The principles are the same as discussed earlier for PKA, although most of the target proteins are different.

Diacylglycerol can be further cleaved to release arachidonic acid, which can either act as a signal in its own right or be used in the synthesis of other small lipid signal molecules called *eicosanoids*. Most vertebrate cell types make eicosanoids, including *prostaglandins*, which have many biological activities. They participate

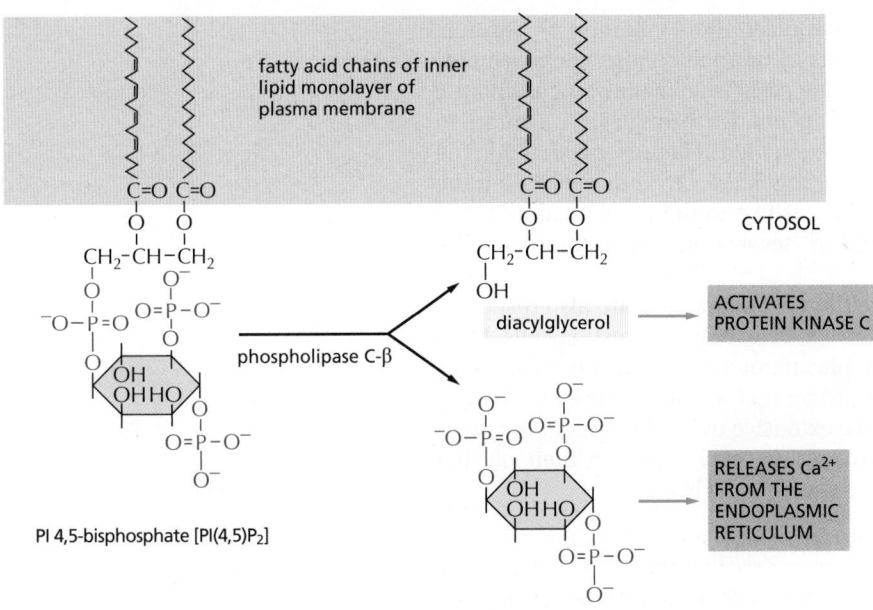

PI 4,5-bisphosphate [PI(4,5)P$_2$]

phospholipase C-β

diacylglycerol → ACTIVATES PROTEIN KINASE C

RELEASES Ca^{2+} FROM THE ENDOPLASMIC RETICULUM

inositol 1,4,5-trisphosphate (IP$_3$)

fatty acid chains of inner lipid monolayer of plasma membrane

CYTOSOL

Figure 15–28 The hydrolysis of PI(4,5) P$_2$ by phospholipase C-β. Two second messengers are produced directly from the hydrolysis of PI(4,5)P$_2$: inositol 1,4,5-trisphosphate (IP$_3$), which diffuses through the cytosol and releases Ca^{2+} from the endoplasmic reticulum, and diacylglycerol, which remains in the membrane and helps to activate protein kinase C (PKC; see Figure 15–29). There are several classes of phospholipase C: these include the β class, which is activated by GPCRs; as we see later, the γ class is activated by a class of enzyme-coupled receptors called receptor tyrosine kinases (RTKs).

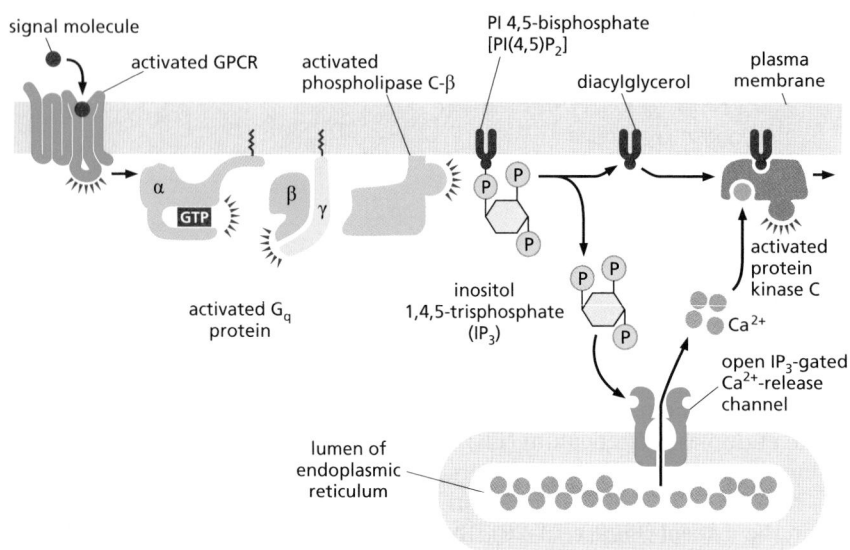

signal molecule

activated GPCR

activated phospholipase C-β

PI 4,5-bisphosphate [PI(4,5)P₂]

diacylglycerol

plasma membrane

α

GTP

β

γ

activated Gq protein

inositol 1,4,5-trisphosphate (IP₃)

activated protein kinase C

Ca²⁺

open IP₃-gated Ca²⁺-release channel

lumen of endoplasmic reticulum

Figure 15–29 How GPCRs increase cytosolic Ca²⁺ and activate protein kinase C. The activated GPCR stimulates the plasma-membrane-bound phospholipase C-β (PLCβ) via a G protein called Gq. The α subunit and βγ complex of Gq are both involved in this activation. Two second messengers are produced when PI(4,5)P₂ is hydrolyzed by activated PLCβ. Inositol 1,4,5-trisphosphate (IP₃) diffuses through the cytosol and releases Ca²⁺ from the ER by binding to and opening IP₃-gated Ca²⁺-release channels (IP₃ receptors) in the ER membrane. The large electrochemical gradient for Ca²⁺ across this membrane causes Ca²⁺ to escape into the cytosol when the release channels are opened. Diacylglycerol remains in the plasma membrane and, together with phosphatidylserine (not shown) and Ca²⁺, helps to activate protein kinase C (PKC), which is recruited from the cytosol to the cytosolic face of the plasma membrane. Of the 10 or more distinct isoforms of PKC in humans, at least 4 are activated by diacylglycerol (**Movie 15.3**).

in pain and inflammatory responses, for example, and many anti-inflammatory drugs (such as aspirin, ibuprofen, and cortisone) act in part by inhibiting their synthesis.

Ca²⁺ Functions as a Ubiquitous Intracellular Mediator

Many extracellular signals, and not just those that work via G proteins, trigger an increase in cytosolic Ca²⁺ concentration. In muscle cells, Ca²⁺ triggers contraction, and in many secretory cells, including nerve cells, it triggers secretion. Ca²⁺ has numerous other functions in a variety of cell types. Ca²⁺ is such an effective signaling mediator because its concentration in the cytosol is normally very low ($\sim 10^{-7}$ M), whereas its concentration in the extracellular fluid ($\sim 10^{-3}$ M) and in the lumen of the ER [and sarcoplasmic reticulum (SR) in muscle] is high. Thus, there is a large gradient tending to drive Ca²⁺ into the cytosol across both the plasma membrane and the ER or SR membrane. When a signal transiently opens Ca²⁺ channels in these membranes, Ca²⁺ rushes into the cytosol, and the resulting 10–20-fold increase in the local Ca²⁺ concentration activates Ca²⁺-responsive proteins in the cell.

Some stimuli, including membrane depolarization, membrane stretch, and certain extracellular signals, activate Ca²⁺ channels in the plasma membrane, resulting in Ca²⁺ influx from outside the cell. Other signals, including the GPCR-mediated signals described earlier, act primarily through IP₃ receptors to stimulate Ca²⁺ release from intracellular stores in the ER (see Figure 15–29). The ER membrane also contains a second type of regulated Ca²⁺ channel called the **ryanodine receptor** (so called because it is sensitive to the plant alkaloid ryanodine), which opens in response to rising Ca²⁺ levels and thereby amplifies the Ca²⁺ signal, as we describe shortly.

Several mechanisms rapidly terminate the Ca²⁺ signal and are also responsible for keeping the concentration of Ca²⁺ in the cytosol low in resting cells. Most importantly, there are Ca²⁺-pumps in the plasma membrane and the ER membrane that use the energy of ATP hydrolysis to pump Ca²⁺ out of the cytosol. Cells such as muscle and nerve cells, which make extensive use of Ca²⁺ signaling, have an additional Ca²⁺ transporter (a Na⁺-driven Ca²⁺ exchanger) in their plasma membrane that couples the efflux of Ca²⁺ to the influx of Na⁺.

Feedback Generates Ca²⁺ Waves and Oscillations

The IP₃ receptors and ryanodine receptors of the ER membrane have an important feature: they are both stimulated by low to moderate cytoplasmic Ca²⁺ concentrations. This *Ca²⁺-induced calcium release* (*CICR*) results in positive feedback,

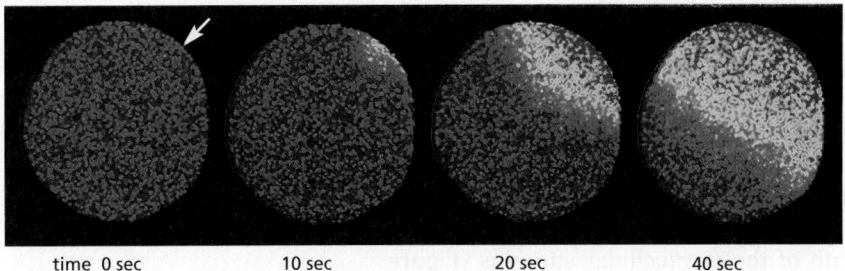

time 0 sec   10 sec   20 sec   40 sec

which has a major impact on the properties of the Ca²⁺ signal. The importance of this feedback is seen clearly in studies with Ca²⁺-sensitive fluorescent indicators, such as *aequorin* or *fura-2* (discussed in Chapter 9), which allow researchers to monitor cytosolic Ca²⁺ in individual cells under a microscope (**Figure 15–30** and **Movie 15.4**).

When cells carrying a Ca²⁺ indicator are treated with a small amount of an extracellular signal molecule that stimulates IP₃ production, tiny bursts of Ca²⁺ are seen in one or more discrete regions of the cell. These Ca²⁺ puffs or sparks reflect the local opening of small groups of IP₃-gated Ca²⁺-release channels in the ER. Because various Ca²⁺-binding proteins act as Ca²⁺ buffers and restrict the diffusion of Ca²⁺, the signal often remains localized to the site where the Ca²⁺ enters the cytosol. If the extracellular signal is sufficiently strong and persistent, however, the local Ca²⁺ concentration can reach a sufficient level to activate nearby IP₃ receptors and ryanodine receptors, resulting in a regenerative wave of Ca²⁺ release that moves through the cytosol (**Figure 15–31**), much like an action potential in an axon.

Figure 15–30 The fertilization of an egg by a sperm triggers a wave of cytosolic Ca²⁺. This starfish egg was injected with a Ca²⁺-sensitive fluorescent dye before it was fertilized. A wave of cytosolic Ca²⁺ *(red)*, released from the ER, sweeps across the egg from the site of sperm entry *(arrow)*. This Ca²⁺ wave changes the egg cell surface, preventing the entry of other sperm, and it also initiates embryonic development (**Movie 15.5**). The initial increase in Ca²⁺ is thought to be caused by a sperm-specific form of PLC (PLCζ) that the sperm brings into the egg cytoplasm when it fuses with the egg; the PLCζ cleaves PI(4,5)P₂ to produce IP₃, which releases Ca²⁺ from the egg ER. The released Ca²⁺ stimulates further Ca²⁺ release from the ER, producing the spreading wave, as we explain in Figure 15–31. (Courtesy of Stephen A. Stricker.)

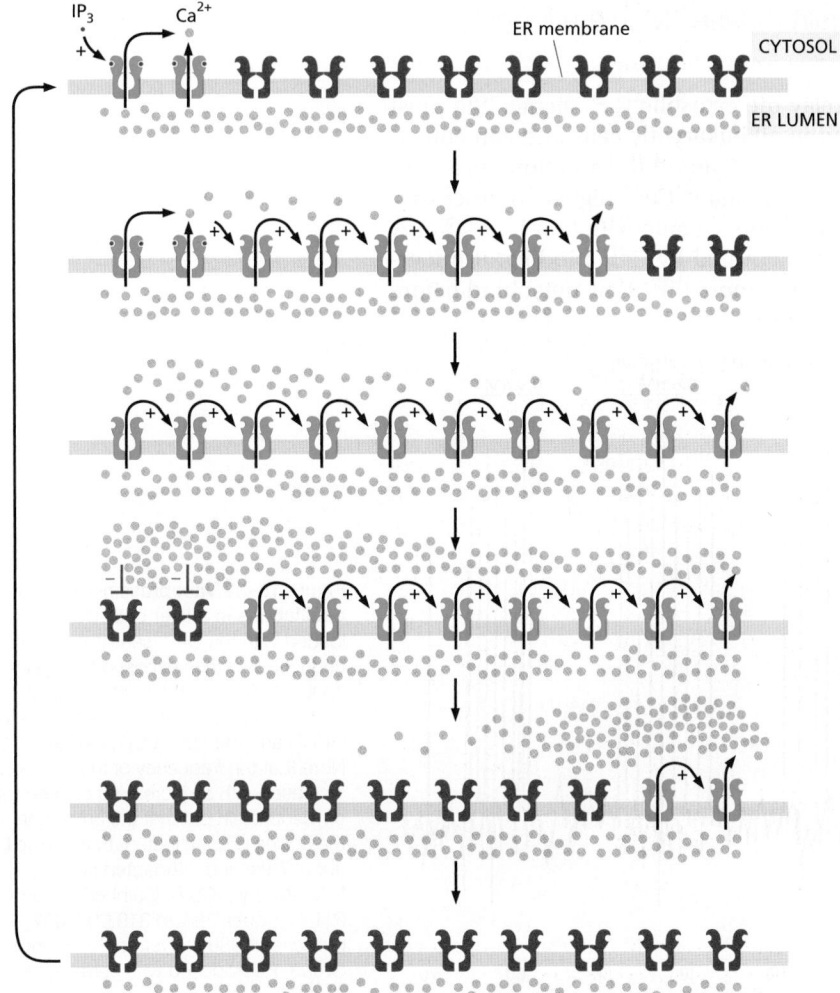

Figure 15–31 Positive and negative feedback produce Ca²⁺ waves and oscillations. This diagram shows IP₃ receptors and ryanodine receptors on a portion of the ER membrane: active receptors are in *green*; inactive receptors are in *red*. When a small amount of cytosolic IP₃ activates a cluster of IP₃ receptors at one site on the ER membrane *(top)*, the local release of Ca²⁺ promotes the opening of nearby IP₃ and ryanodine receptors, resulting in more Ca²⁺ release. This positive feedback (indicated by *positive* signs) produces a regenerative wave of Ca²⁺ release that spreads across the cell (see Figure 15–30). These waves of Ca²⁺ release move more quickly across the cell than would be possible by simple diffusion. Also, unlike a diffusing burst of Ca²⁺ ions, which will become more dilute as it spreads, the regenerative wave produces a high Ca²⁺ concentration across the entire cell. Eventually, the local Ca²⁺ concentration inactivates IP₃ receptors and ryanodine receptors *(middle;* indicated by *red* negative signs), shutting down the Ca²⁺ release. Ca²⁺-pumps reduce the local cytosolic Ca²⁺ concentration to its normal low levels. The result is a Ca²⁺ spike: positive feedback drives a rapid rise in cytosolic Ca²⁺, and negative feedback sends it back down again. The Ca²⁺ channels remain refractory to further stimulation for some period of time, delaying the generation of another Ca²⁺ spike *(bottom)*. Eventually, however, the negative feedback wears off, allowing IP₃ to trigger another Ca²⁺ wave. The end result is repeated Ca²⁺ oscillations (see Figure 15–32). Under some conditions, these oscillations can be seen as repeating narrow waves of Ca²⁺ moving across the cell.

Another important property of IP$_3$ receptors and ryanodine receptors is that they are inhibited, after some delay, by high Ca^{2+} concentrations (a form of negative feedback). Thus, the rise in Ca^{2+} in a stimulated cell leads to inhibition of Ca^{2+} release; because Ca^{2+} pumps remove the cytosolic Ca^{2+}, the Ca^{2+} concentration falls (see Figure 15–31). The decline in Ca^{2+} eventually relieves the negative feedback, allowing cytosolic Ca^{2+} to rise again. As in other cases of delayed negative feedback (see Figure 15–18), the result is an oscillation in the Ca^{2+} concentration. These oscillations persist for as long as receptors are activated at the cell surface, and their frequency reflects the strength of the extracellular stimulus (**Figure 15–32**). The frequency, amplitude, and breadth of oscillations can also be modulated by other signaling mechanisms, such as phosphorylation, which influence the Ca^{2+} sensitivity of Ca^{2+} channels or affect other components in the signaling system.

The frequency of Ca^{2+} oscillations can be translated into a frequency-dependent cell response. In some cases, the frequency-dependent response itself is also oscillatory: in hormone-secreting pituitary cells, for example, stimulation by an extracellular signal induces repeated Ca^{2+} spikes, each of which is associated with a burst of hormone secretion. In other cases, the frequency-dependent response is non-oscillatory: in some types of cells, for instance, one frequency of Ca^{2+} spikes activates the transcription of one set of genes, while a higher frequency activates the transcription of a different set. How do cells sense the frequency of Ca^{2+} spikes and change their response accordingly? The mechanism presumably depends on Ca^{2+}-sensitive proteins that change their activity as a function of Ca^{2+}-spike frequency. A protein kinase that acts as a molecular memory device seems to have this remarkable property, as we discuss next.

Ca^{2+}/Calmodulin-Dependent Protein Kinases Mediate Many Responses to Ca^{2+} Signals

Various Ca^{2+}-binding proteins help to relay the cytosolic Ca^{2+} signal. The most important is **calmodulin**, which is found in all eukaryotic cells and can constitute as much as 1% of a cell's total protein mass. Calmodulin functions as a multipurpose intracellular Ca^{2+} receptor, governing many Ca^{2+}-regulated processes. It consists of a highly conserved, single polypeptide chain with four high-affinity Ca^{2+}-binding sites (**Figure 15–33A**). When activated by Ca^{2+} binding, it undergoes a conformational change. Because two or more Ca^{2+} ions must bind before

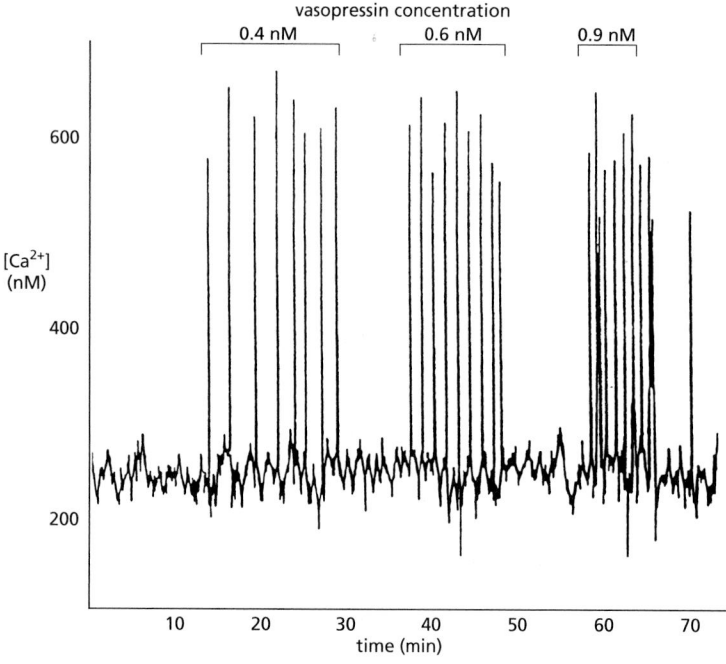

Figure 15–32 **Vasopressin-induced Ca^{2+} oscillations in a liver cell.** The cell was loaded with the Ca^{2+}-sensitive protein aequorin and then exposed to increasing concentrations of the peptide signal molecule *vasopressin*, which activates a GPCR and thereby PLCβ (see Table 15–2). Note that the frequency of the Ca^{2+} spikes increases with an increasing concentration of vasopressin but that the amplitude of the spikes is not affected. Each spike lasts about 7 seconds. (Adapted from N.M. Woods, K.S.R. Cuthbertson and P.H. Cobbold, *Nature* 319:600–602, published 1986 by Nature Publishing Group. Reproduced with permission of SNCSC.)

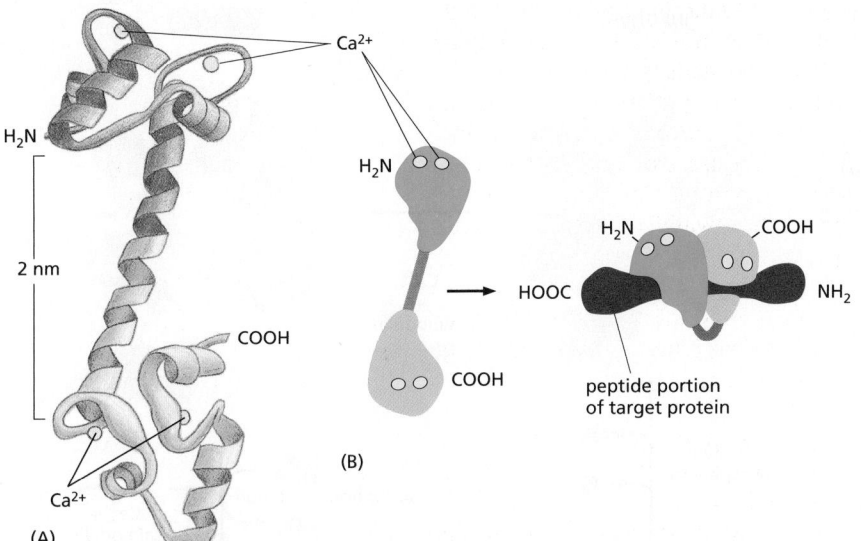

Figure 15–33 **The structure of Ca^{2+}/calmodulin.** (A) The molecule has a dumbbell shape, with two globular ends, which can bind to many target proteins. The globular ends are connected by a long, exposed α helix, which allows the protein to adopt a number of different conformations, depending on the target protein it interacts with. Each globular head has two Ca^{2+}-binding sites (**Movie 15.6**). (B) Shown is the major structural change that occurs in Ca^{2+}/calmodulin when it binds to a target protein (in this example, a peptide that consists of the Ca^{2+}/calmodulin-binding domain of a Ca^{2+}/calmodulin-dependent protein kinase). Note that the Ca^{2+}/calmodulin has "jack-knifed" to surround the peptide. When it binds to other targets, it can adopt different conformations. (A, based on x-ray crystallographic data from Y.S. Babu et al., *Nature* 315:37–40, 1985; B, based on x-ray crystallographic data from W.E. Meador, A.R. Means, and F.A. Quiocho, *Science* 257:1251–1255, 1992, and on nuclear magnetic resonance (NMR) spectroscopy data from M. Ikura et al., *Science* 256: 632–638, 1992.)

calmodulin adopts its active conformation, the protein displays a sigmoidal response to increasing concentrations of Ca^{2+} (see Figure 15–16).

The allosteric activation of calmodulin by Ca^{2+} is analogous to the activation of PKA by cyclic AMP, except that Ca^{2+}/calmodulin has no enzymatic activity itself but instead acts by binding to and activating other proteins. In some cases, calmodulin serves as a permanent regulatory subunit of an enzyme complex, but usually the binding of Ca^{2+} instead enables calmodulin to bind to various target proteins in the cell to alter their activity.

When an activated molecule of Ca^{2+}/calmodulin binds to its target protein, the calmodulin further changes its conformation, the nature of which depends on the specific target protein (Figure 15–33B). Among the many targets calmodulin regulates are enzymes and membrane transport proteins. As one example, Ca^{2+}/calmodulin binds to and activates the plasma membrane Ca^{2+}-pump that uses ATP hydrolysis to pump Ca^{2+} out of cells. Thus, whenever the concentration of Ca^{2+} in the cytosol rises, the pump is activated, which helps to return the cytosolic Ca^{2+} level to resting levels.

Many effects of Ca^{2+}, however, are more indirect and are mediated by protein phosphorylations catalyzed by a family of protein kinases called **Ca^{2+}/calmodulin-dependent kinases** (**CaM-kinases**). Some CaM-kinases phosphorylate transcription regulators, such as the CREB protein (see Figure 15–27), and in this way activate or inhibit the transcription of specific genes.

One of the best-studied CaM-kinases is **CaM-kinase II**, which is found in most animal cells but is especially enriched in the nervous system. It constitutes up to 2% of the total protein mass in some regions of the brain, and it is highly concentrated in synapses. CaM-kinase II has several remarkable properties. To begin with, it has a spectacular quaternary structure: twelve copies of the enzyme are assembled into a stacked pair of rings, with kinase domains on the outside linked to a central hub (**Figure 15–34**). This structure helps the enzyme function as a molecular memory device, switching to an active state when exposed to Ca^{2+}/calmodulin and then remaining active even after the Ca^{2+} signal has decayed. This is because adjacent kinase subunits can phosphorylate each other (a process called *autophosphorylation*) when Ca^{2+}/calmodulin activates them (Figure 15–34). Once a kinase subunit is autophosphorylated, it remains active even in the absence of Ca^{2+}, thereby prolonging the duration of the kinase activity beyond that of the initial activating Ca^{2+} signal. The enzyme maintains this activity until a protein phosphatase removes the autophosphorylation and shuts the kinase off. CaM-kinase II activation can thereby serve as a memory trace of a prior Ca^{2+} pulse, and it seems to have a role in some types of memory and learning in the vertebrate nervous system. Mutant mice that lack a brain-specific form of the enzyme have specific defects in their ability to remember where things are.

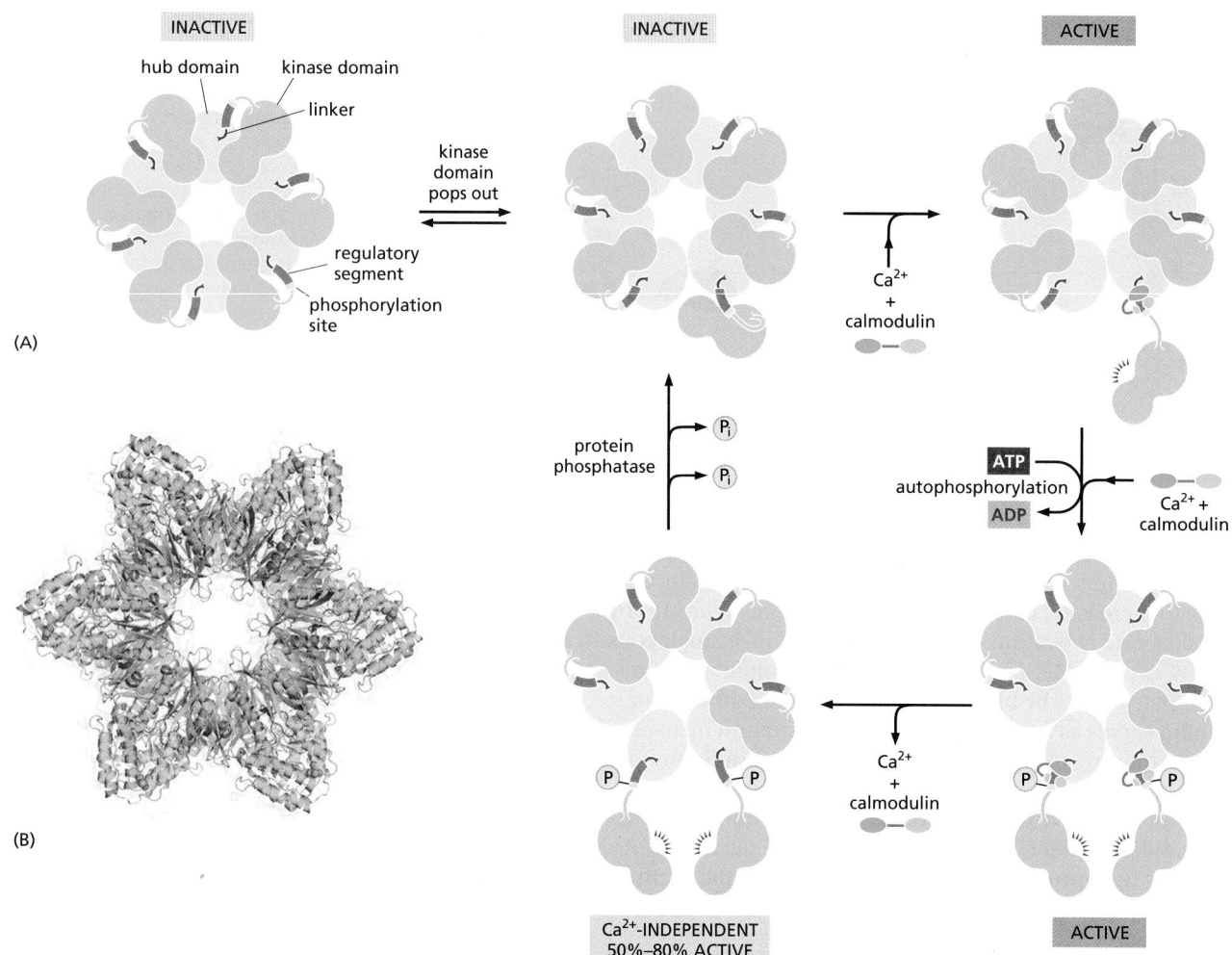

(A)

(B)

Figure 15–34 The stepwise activation of CaM-kinase II. (A) Each CaM-kinase II protein has two major domains: an amino-terminal kinase domain *(green)* and a carboxyl-terminal hub domain *(blue)*, linked by a regulatory segment. Six CaM-kinase II proteins are assembled into a giant ring in which the hub domains interact tightly to produce a central structure that is surrounded by kinase domains. The complete enzyme contains two stacked rings, for a total of 12 kinase proteins, but only one ring is shown here for clarity. When the enzyme is inactive, the ring exists in a dynamic equilibrium between two states. The first *(upper left)* is a compact state, in which the kinase domains interact with the hub, so that the regulatory segment is buried in the kinase active site and thereby blocks catalytic activity. In the second inactive state *(upper middle)*, a kinase domain has popped out and is linked to the central hub by its regulatory segment, which continues to inhibit the kinase but is now accessible to Ca^{2+}/calmodulin. If present, Ca^{2+}/calmodulin will bind the regulatory segment and prevent it from inhibiting the kinase, thereby locking the kinase in an active state *(upper right)*. If the adjacent kinase subunit also pops out from the hub, it will also be activated by Ca^{2+}/calmodulin, and the two kinases will then phosphorylate each other on their regulatory segments *(lower right)*. This autophosphorylation further activates the enzyme. It also prolongs the activity of the enzyme in two ways. First, it traps the bound Ca^{2+}/calmodulin so that it does not dissociate from the enzyme until cytosolic Ca^{2+} levels return to basal values for at least 10 seconds (not shown). Second, it converts the enzyme to a Ca^{2+}-independent form, so that the kinase remains active even after the Ca^{2+}/calmodulin dissociates from it *(lower left)*. This activity continues until the action of a protein phosphatase overrides the autophosphorylation activity of CaM-kinase II. (B) This structural model of the enzyme is based on x-ray crystallographic analysis.

The remarkable dodecameric structure of the enzyme allows it to achieve a broad range of intermediate activity states in response to different Ca^{2+} oscillation frequencies: higher frequencies tend to cause more subunits in the enzyme to reach the phosphorylated active state (see Figure 15–35). The behavior of CaM-kinase II is also controlled by the length of the linker segment between the kinase and hub domains. The linker is longer in some isoforms of the enzyme; in these isoforms, the kinase domains tend to pop out of the ring more frequently, making it more sensitive to Ca^{2+}. These and other mechanisms allow the cell to tailor the responsiveness of the enzyme to the needs of different types of neurons. (Adapted from L.H. Chao et al., *Cell* 146:732–745, 2011. PDB code: 3SOA.)

Another remarkable property of CaM-kinase II is that the enzyme can use its intrinsic memory mechanism to decode the frequency of Ca^{2+} oscillations. This property is thought to be especially important at a nerve cell synapse, where changes in intracellular Ca^{2+} levels in a postsynaptic cell as a result of neural activity can lead to long-term changes in the subsequent effectiveness of that synapse

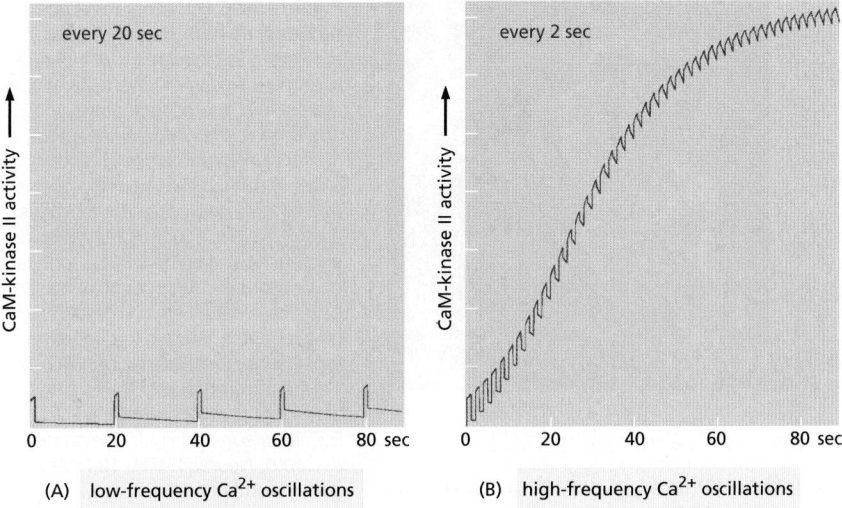

(A) low-frequency Ca²⁺ oscillations

(B) high-frequency Ca²⁺ oscillations

Figure 15–35 CaM-kinase II as a frequency decoder of Ca²⁺ oscillations. (A) At low frequencies of Ca²⁺ spikes, the enzyme becomes inactive after each spike, as the autophosphorylation induced by Ca²⁺/calmodulin binding does not maintain the enzyme's activity long enough for the enzyme to remain active until the next Ca²⁺ spike arrives. (B) At higher spike frequencies, however, the enzyme fails to inactivate completely between Ca²⁺ spikes, so its activity ratchets up with each spike. If the spike frequency is high enough, this progressive increase in enzyme activity will continue until the enzyme is autophosphorylated on all subunits and is therefore maximally activated. Although not shown, once enough of its subunits are autophosphorylated, the enzyme can be maintained in a highly active state even with a relatively low frequency of Ca²⁺ spikes (a form of cell memory). The binding of Ca²⁺/calmodulin to the enzyme is enhanced by the CaM-kinase II autophosphorylation (an additional form of positive feedback), helping to generate a more switchlike response to repeated Ca²⁺ spikes. (From P.I. Hanson, T. Meyer, L. Stryer, and H. Schulman, *Neuron* 12:943–956, 1994. With permission from Elsevier.)

(discussed in Chapter 11). When CaM-kinase II is exposed to both a protein phosphatase and repetitive pulses of Ca²⁺/calmodulin at different frequencies that mimic those observed in stimulated cells, the enzyme's activity increases steeply as a function of pulse frequency (**Figure 15–35**).

Some G Proteins Directly Regulate Ion Channels

G proteins do not act exclusively by regulating the activity of membrane-bound enzymes that alter the concentration of cyclic AMP or Ca²⁺ in the cytosol. The α subunit of one type of G protein (called G_{12}), for example, activates a guanine nucleotide exchange factor (GEF) that activates a monomeric GTPase of the *Rho family* (discussed later and in Chapter 16), which regulates the actin cytoskeleton.

In some other cases, G proteins directly activate or inactivate ion channels in the plasma membrane of the target cell, thereby altering the ion permeability—and hence the electrical excitability—of the membrane. As an example, acetylcholine released by the vagus nerve reduces the heart rate (see Figure 15–5B). This effect is mediated by a special class of acetylcholine receptors that activate the G_i protein discussed earlier. Once activated, the α subunit of G_i inhibits adenylyl cyclase (as described previously), while the βγ subunits bind to K⁺ channels in the heart muscle cell plasma membrane and open them. The opening of these K⁺ channels makes it harder to depolarize the cell and thereby contributes to the inhibitory effect of acetylcholine on the heart. (These acetylcholine receptors, which can be activated by the fungal alkaloid muscarine, are called *muscarinic acetylcholine receptors* to distinguish them from the very different *nicotinic acetylcholine receptors*, which are ion-channel-coupled receptors on skeletal muscle and nerve cells that can be activated by the binding of nicotine, as well as by acetylcholine.)

Other G proteins regulate the activity of ion channels less directly, either by stimulating channel phosphorylation (by PKA, PKC, or CaM-kinase, for example) or by causing the production or destruction of cyclic nucleotides that directly activate or inactivate ion channels. These *cyclic-nucleotide-gated ion channels* have a crucial role in both smell (olfaction) and vision, as we now discuss.

Smell and Vision Depend on GPCRs That Regulate Ion Channels

Humans can distinguish more than 10,000 distinct smells, which they detect using specialized olfactory receptor neurons in the lining of the nose. These cells use specific GPCRs called **olfactory receptors** to recognize odors; the receptors are displayed on the surface of the modified cilia that extend from each cell (**Figure 15–36**). The receptors act through cAMP. When stimulated by odorant binding,

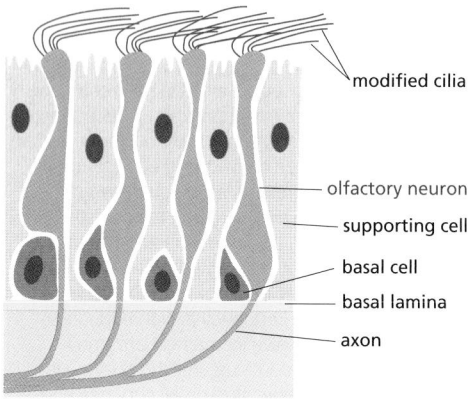

(A)

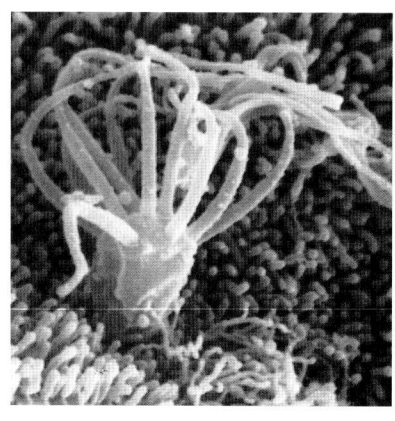

(B)

- modified cilia
- olfactory neuron
- supporting cell
- basal cell
- basal lamina
- axon

Figure 15–36 **Olfactory receptor neurons.** (A) A section of olfactory epithelium in the nose. Olfactory receptor neurons possess modified cilia, which project from the surface of the epithelium and contain the olfactory receptors, as well as the signal transduction machinery. The axon, which extends from the opposite end of the receptor neuron, conveys electrical signals to the brain when an odorant activates the cell to produce an action potential. In rodents, at least, the basal cells act as stem cells, producing new receptor neurons throughout life, to replace the neurons that die. (B) A scanning electron micrograph of the cilia on the surface of an olfactory neuron. (B, from E.E. Morrison and R.M. Costanzo, *J. Comp. Neurol.* 297:1–13, 1990. With permission from Wiley-Liss.)

they activate an olfactory-specific G protein (known as G_{olf}), which in turn activates adenylyl cyclase. The resulting increase in cAMP opens *cyclic-AMP-gated cation channels*, thereby allowing an influx of Na⁺, which depolarizes the olfactory receptor neuron and initiates a nerve impulse that travels along its axon to the brain.

There are about 1000 different olfactory receptors in a mouse and about 350 in a human, each encoded by a different gene and each recognizing a different set of odorants. Each olfactory receptor neuron produces only one of these receptors; the neuron responds to a specific set of odorants by means of the specific receptor it displays, and each odorant activates its own characteristic set of olfactory receptor neurons. The same receptor also helps direct the elongating axon of each developing olfactory neuron to the specific target neurons that it will connect to in the brain. A different set of GPCRs acts in a similar way in some vertebrates to mediate responses to *pheromones*, chemical signals detected in a different part of the nose that are used in communication between members of the same species. Humans, however, are thought to lack functional pheromone receptors.

Vertebrate vision employs a similarly elaborate, highly sensitive, signal-detection process. Cyclic-nucleotide-gated ion channels are also involved, but the crucial cyclic nucleotide is **cyclic GMP** (**Figure 15–37**) rather than cAMP. As with cAMP, a continuous rapid synthesis (by *guanylyl cyclase*) and rapid degradation (by *cyclic GMP phosphodiesterase*) controls the concentration of cyclic GMP in the cytosol.

In visual transduction responses, which are the fastest G-protein-mediated responses known in vertebrates, the receptor activation stimulated by light causes a fall rather than a rise in the level of the cyclic nucleotide. The pathway has been especially well studied in **rod photoreceptors (rods)** in the vertebrate retina. Rods are responsible for noncolor vision in dim light, whereas **cone photoreceptors (cones)** are responsible for color vision in bright light. A rod photoreceptor is a highly specialized cell with outer and inner segments, a cell body, and a synaptic region where the rod passes a chemical signal to a retinal nerve cell (**Figure 15–38**). This nerve cell relays the signal to another nerve cell in the retina, which in turn relays it to the brain.

The phototransduction apparatus is in the outer segment of the rod, which contains a stack of *discs*, each formed by a closed sac of membrane that is densely packed with photosensitive **rhodopsin** molecules. The plasma membrane surrounding the outer segment contains *cyclic-GMP-gated cation channels*. Cyclic GMP bound to these channels keeps them open in the dark. Paradoxically, light causes a hyperpolarization (which inhibits synaptic signaling) rather than a depolarization of the plasma membrane (which would stimulate synaptic signaling). Hyperpolarization (that is, the membrane potential moves to a more negative value—discussed in Chapter 11) results because the light-induced activation of rhodopsin molecules in the disc membrane decreases the cyclic GMP concentration and closes the cation channels in the surrounding plasma membrane (**Figure 15–39**).

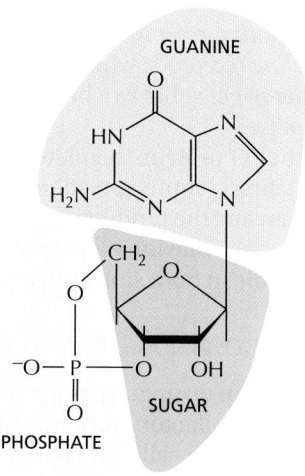

GUANINE

SUGAR

PHOSPHATE

Figure 15–37 **Cyclic GMP.**

Figure 15–38 **A rod photoreceptor cell.** There are about 1000 discs in the outer segment. The disc membranes are not connected to the plasma membrane. The inner and outer segments are specialized parts of a *primary cilium* (discussed in Chapter 16). A primary cilium extends from the surface of most vertebrate cells, where it serves as a signaling organelle.

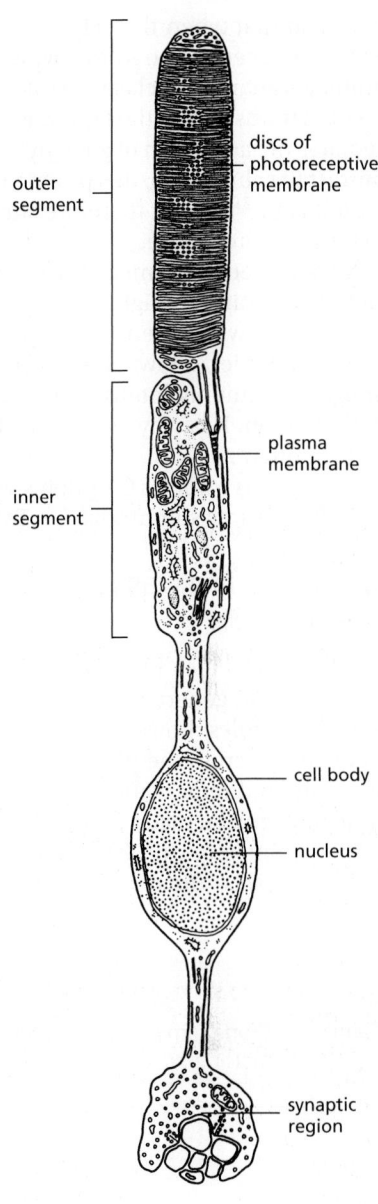

Rhodopsin is a member of the GPCR family, but the activating extracellular signal is not a molecule but a photon of light. Each rhodopsin molecule contains a covalently attached chromophore, 11-*cis* retinal, which isomerizes almost instantaneously to all-*trans* retinal when it absorbs a single photon. The isomerization alters the shape of the retinal, forcing a conformational change in the protein (opsin). The activated rhodopsin molecule then alters the conformation of the G protein *transducin* (G_t), causing the transducin α subunit to activate **cyclic GMP phosphodiesterase**. The phosphodiesterase then hydrolyzes cyclic GMP, so that cyclic GMP levels in the cytosol fall. This drop in cyclic GMP concentration decreases the amount of cyclic GMP bound to the plasma membrane cation channels, allowing more of these cyclic-GMP-sensitive channels to close. In this way, the signal quickly passes from the disc membrane to the plasma membrane, and a light signal is converted into an electrical one, through a hyperpolarization of the rod cell plasma membrane.

Rods use several negative feedback loops to allow the cells to revert quickly to a resting, dark state in the aftermath of a flash of light—a requirement for perceiving the shortness of the flash. A rhodopsin-specific protein kinase called *rhodopsin kinase* (*RK*) phosphorylates the cytosolic tail of activated rhodopsin on multiple serines, partially inhibiting the ability of the rhodopsin to activate transducin. An inhibitory protein called *arrestin* (discussed later) then binds to the phosphorylated rhodopsin, further inhibiting rhodopsin's activity. Mice or humans with a mutation that inactivates the gene encoding RK have a prolonged light response.

At the same time as arrestin shuts off rhodopsin, an RGS protein (discussed earlier) binds to activated transducin, stimulating the transducin to hydrolyze its bound GTP to GDP, which returns transducin to its inactive state. In addition, the cation channels that close in response to light are permeable to Ca^{2+}, as well as

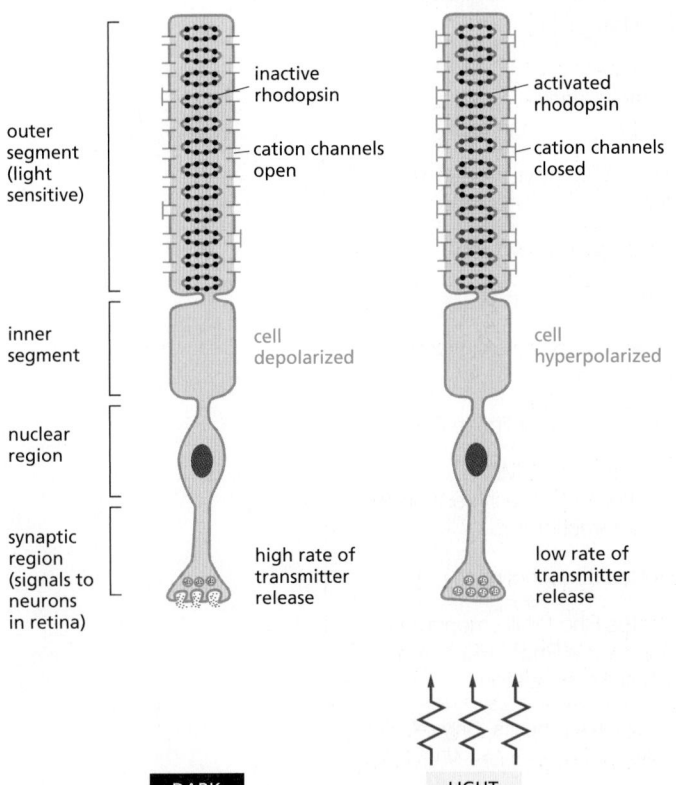

Figure 15–39 **The response of a rod photoreceptor cell to light.** Rhodopsin molecules in the outer-segment discs absorb photons. Photon absorption closes cation channels in the plasma membrane, which hyperpolarizes the membrane and reduces the rate of neurotransmitter release from the synaptic region. Because the neurotransmitter inhibits many of the postsynaptic retinal neurons, illumination serves to free the neurons from inhibition and thus, in effect, excites them. The neural connections of the retina lie between the light source and the outer segment, and so the light must pass through the synapses and rod cell nucleus to reach the light sensors.

to Na$^+$, so that when they close, the normal influx of Ca^{2+} is inhibited, causing the Ca^{2+} concentration in the cytosol to fall. The decrease in Ca^{2+} concentration stimulates guanylyl cyclase to replenish the cyclic GMP, rapidly returning its level to where it was before the light was switched on. A specific Ca^{2+}-sensitive protein mediates the activation of guanylyl cyclase in response to the fall in Ca^{2+} levels. In contrast to calmodulin, this protein is inactive when Ca^{2+} is bound to it and active when it is Ca^{2+}-free. It therefore stimulates the cyclase when Ca^{2+} levels fall following a light response.

Negative feedback mechanisms do more than just return the rod to its resting state after a transient light flash; they also help the rod to *adapt*, stepping down the response when the rod is exposed to light continuously. Adaptation, as we discussed earlier, allows the receptor cell to function as a sensitive detector of *changes* in stimulus intensity over an enormously wide range of baseline levels of stimulation. It is why we can see faint stars in a dark sky, or a camera flash in bright sunlight.

The various trimeric G proteins we have discussed in this chapter fall into four major families, as summarized in **Table 15–3**.

Nitric Oxide Is a Gaseous Signaling Mediator That Passes Between Cells

Signaling molecules like cyclic nucleotides and calcium are hydrophilic small molecules that generally act within the cell where they are produced. Some signaling molecules, however, are hydrophobic enough, small enough, or both, to pass readily across the plasma membrane and carry signals to nearby cells. An important and remarkable example is the gas **nitric oxide** (**NO**), which acts as a signal molecule in many tissues of both animals and plants.

In mammals, one of NO's many functions is to relax smooth muscle in the walls of blood vessels. The neurotransmitter acetylcholine stimulates NO synthesis by

Family	Some family members	Subunits that mediate action	Some functions
I	G_s	α	Activates adenylyl cyclase; activates Ca^{2+} channels
	G_{olf}	α	Activates adenylyl cyclase in olfactory sensory neurons
II	G_i	α	Inhibits adenylyl cyclase
		$\beta\gamma$	Activates K$^+$ channels
	G_o	$\beta\gamma$	Activates K$^+$ channels; inactivates Ca^{2+} channels
		α and $\beta\gamma$	Activates phospholipase C-β
	G_t (transducin)	α	Activates cyclic GMP phosphodiesterase in vertebrate rod photoreceptors
III	G_q	α	Activates phospholipase C-β
IV	$G_{12}/_{13}$	α	Activates Rho family monomeric GTPases (via Rho-GEF) to regulate the actin cytoskeleton

TABLE 15–3 Four Major Families of Trimeric G Proteins*

*Families are determined by amino acid sequence relatedness of the α subunits. Only selected examples are included. About 20 α subunits and at least 6 β subunits and 11 γ subunits have been described in humans.

activating a GPCR on the membranes of the endothelial cells that line the interior of the vessel. The activated receptor triggers IP$_3$ synthesis and Ca^{2+} release (see Figure 15–29), leading to stimulation of an enzyme that synthesizes NO. Because dissolved NO passes readily across membranes, it diffuses out of the cell where it is produced and into neighboring smooth muscle cells, where it causes muscle relaxation and thereby vessel dilation (**Figure 15–40**). It acts only locally because it has a short half-life—about 5–10 seconds—in the extracellular space before oxygen and water convert it to nitrates and nitrites.

The effect of NO on blood vessels provides an explanation for the mechanism of action of nitroglycerine, which has been used for about 100 years to treat patients with angina (pain resulting from inadequate blood flow to the heart muscle). The nitroglycerine is converted to NO, which relaxes blood vessels. This reduces the workload on the heart and, as a consequence, reduces the oxygen requirement of the heart muscle.

NO is made by the deamination of the amino acid arginine, catalyzed by enzymes called **NO synthases** (**NOS**) (see Figure 15–40). The NOS in endothelial cells is called *eNOS*, while that in nerve and muscle cells is called *nNOS*. Both eNOS and nNOS are stimulated by Ca^{2+}. Macrophages, by contrast, make yet another NOS, called inducible NOS (*iNOS*), that is constitutively active but synthesized only when the cells are activated, usually in response to an infection.

In some target cells, including smooth muscle cells, NO binds reversibly to iron in the active site of guanylyl cyclase, stimulating synthesis of cyclic GMP. NO can increase cyclic GMP in the cytosol within seconds, because the normal rate of turnover of cyclic GMP is high: rapid degradation to GMP by a phosphodiesterase constantly balances the production of cyclic GMP by guanylyl cyclase. The drug Viagra® and its relatives inhibit the cyclic GMP phosphodiesterase in the penis, thereby increasing the amount of time that cyclic GMP levels remain elevated in the smooth muscle cells of penile blood vessels after NO production is induced by local nerve terminals. The cyclic GMP, in turn, keeps the blood vessels relaxed and thereby the penis erect. NO can also signal cells independently of cyclic GMP. It can, for example, alter the activity of an intracellular protein by covalently nitrosylating thiol (–SH) groups on specific cysteines in the protein.

(A)

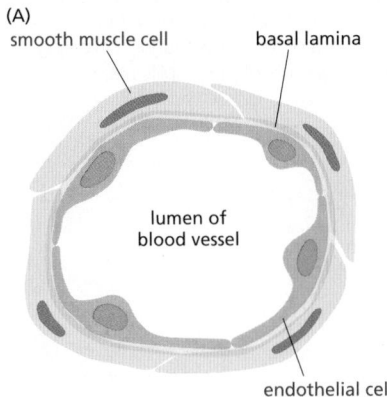

Figure 15–40 **The role of nitric oxide (NO) in smooth muscle relaxation in a blood vessel wall.** (A) Simplified cross section of a blood vessel, showing the endothelial cells lining the lumen and the smooth muscle cells around them. (B) The neurotransmitter acetylcholine stimulates blood vessel dilation by activating a GPCR—the *muscarinic acetylcholine receptor*—on the surface of endothelial cells. This receptor activates a G protein, G$_q$, thereby stimulating IP$_3$ synthesis and Ca^{2+} release by the mechanisms illustrated in Figure 15–29. Ca^{2+} activates nitric oxide synthase, causing the endothelial cells to produce NO from arginine. The NO diffuses out of the endothelial cells and into the neighboring smooth muscle cells, where it activates guanylyl cyclase to produce cyclic GMP. The cyclic GMP triggers a response that causes the smooth muscle cells to relax, increasing blood flow through the vessel.

(B)

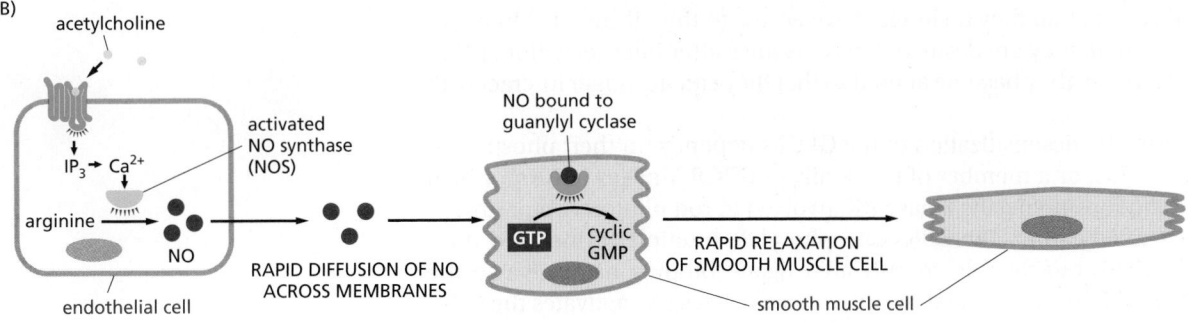

Second Messengers and Enzymatic Cascades Amplify Signals

Despite the differences in molecular details, the different intracellular signaling pathways that GPCRs trigger share certain features and obey similar general principles. They depend on relay chains of intracellular signaling proteins and second messengers. These relay chains provide numerous opportunities for amplifying the responses to extracellular signals. In the visual transduction cascade, for example, a single activated rhodopsin molecule catalyzes the activation of hundreds of molecules of transducin at a rate of about 1000 transducin molecules per second. Each activated transducin molecule activates a molecule of cyclic GMP phosphodiesterase, each of which hydrolyzes about 4000 molecules of cyclic GMP per second. This catalytic cascade lasts for about 1 second and results in the hydrolysis of more than 10^5 cyclic GMP molecules for a single quantum of light absorbed, and the resulting drop in the concentration of cyclic GMP in turn transiently closes hundreds of cation channels in the plasma membrane (**Figure 15–41**). As a result, a rod cell can respond to even a single photon of light in a way that is highly reproducible in its timing and magnitude.

Likewise, when an extracellular signal molecule binds to a receptor that indirectly activates adenylyl cyclase via G_s, each receptor protein may activate many molecules of G_s protein, each of which can activate a cyclase molecule. Each cyclase molecule, in turn, can catalyze the conversion of a large number of ATP molecules to cAMP molecules. A similar amplification operates in the IP_3 signaling pathway. In these ways, a nanomolar (10^{-9} M) change in the concentration of an extracellular signal can induce micromolar (10^{-6} M) changes in the concentration of a second messenger such as cAMP or Ca^{2+}. Because these messengers function as allosteric effectors to activate specific enzymes or ion channels, a single extracellular signal molecule can alter many thousands of protein molecules within the target cell.

Any such amplifying cascade of stimulatory signals requires counterbalancing mechanisms at every step of the cascade to restore the system to its resting state when stimulation ceases. As emphasized earlier, the response to stimulation can be rapid only if the inactivating mechanisms are also rapid. Cells therefore have efficient mechanisms for rapidly degrading (and resynthesizing) cyclic nucleotides and for buffering and removing cytosolic Ca^{2+}, as well as for inactivating the responding enzymes and ion channels once they have been activated. This is not only essential for turning a response off, but is also important for defining the resting state from which a response begins.

Each protein in the signaling relay chain can be a separate target for regulation, including the receptor itself, as we discuss next.

GPCR Desensitization Depends on Receptor Phosphorylation

As discussed earlier, when target cells are exposed to a high concentration of a stimulating ligand for a prolonged period, they can become *desensitized*, or *adapted*, in several different ways. An important class of adaptation mechanisms depends on alteration of the quantity or condition of the receptor molecules themselves.

For GPCRs, there are three general modes of adaptation (see Figure 15–20): (1) In *receptor sequestration*, they are temporarily moved to the interior of the cell (internalized) so that they no longer have access to their ligand. (2) In *receptor down-regulation*, they are destroyed in lysosomes after internalization. (3) In *receptor inactivation*, they become altered so that they can no longer interact with G proteins.

In each case, the desensitization of the GPCRs depends on their phosphorylation by PKA, PKC, or a member of the family of **GPCR kinases (GRKs)**, which includes the rhodopsin-specific kinase RK involved in rod photoreceptor desensitization discussed earlier. The GRKs phosphorylate multiple serines and threonines on a GPCR, but they do so only after ligand binding has activated the receptor, because it is the activated receptor that allosterically activates the GRK.

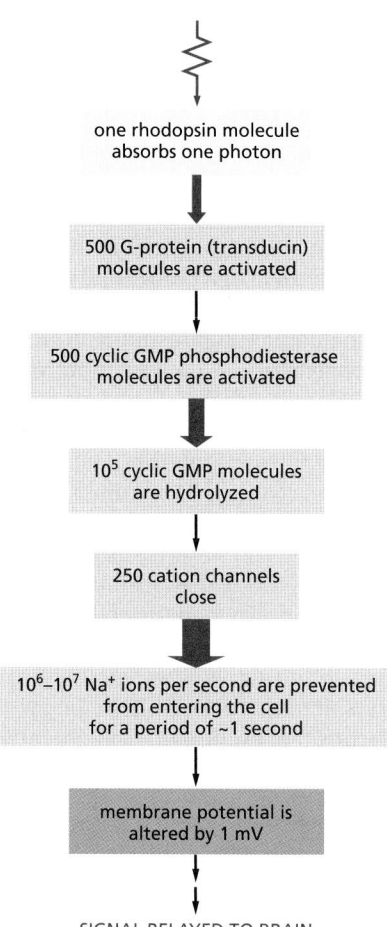

one rhodopsin molecule
absorbs one photon

↓

500 G-protein (transducin)
molecules are activated

↓

500 cyclic GMP phosphodiesterase
molecules are activated

↓

10^5 cyclic GMP molecules
are hydrolyzed

↓

250 cation channels
close

↓

10^6–10^7 Na$^+$ ions per second are prevented
from entering the cell
for a period of ~1 second

↓

membrane potential is
altered by 1 mV

↓

SIGNAL RELAYED TO BRAIN

Figure 15–41 Amplification in the light-induced catalytic cascade in vertebrate rods. The *red arrows* indicate the steps where amplification occurs, with the thickness of the arrow roughly indicating the magnitude of the amplification.

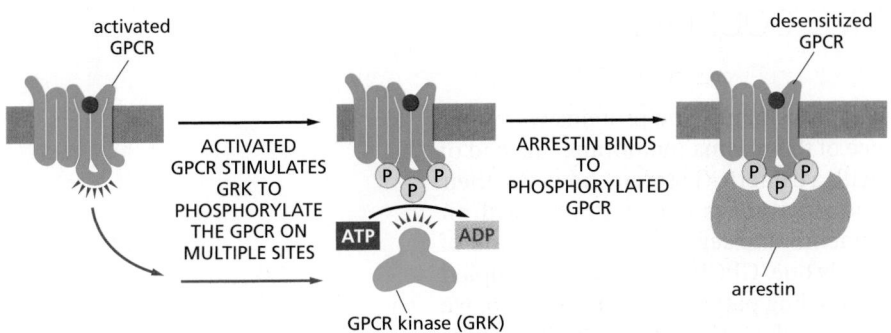

Figure 15–42 The roles of GPCR kinases (GRKs) and arrestins in GPCR desensitization. A GRK phosphorylates only activated receptors because it is the activated GPCR that activates the GRK. The binding of an arrestin to the phosphorylated receptor prevents the receptor from binding to its G protein and also directs its endocytosis (not shown). Mice that are deficient in one form of arrestin fail to desensitize in response to morphine, for example, attesting to the importance of arrestins for desensitization.

As with rhodopsin, once a receptor has been phosphorylated by a GRK, it binds with high affinity to a member of the **arrestin** family of proteins (**Figure 15–42**).

The bound arrestin can contribute to the desensitization process in at least two ways. First, it prevents the activated receptor from interacting with G proteins. Second, it serves as an adaptor protein to help couple the receptor to the clathrin-dependent endocytosis machinery (discussed in Chapter 13), inducing receptor-mediated endocytosis. The fate of the internalized GPCR–arrestin complex depends on other proteins in the complex. In some cases, the receptor is dephosphorylated and recycled back to the plasma membrane for reuse. In others, it is ubiquitylated, endocytosed, and degraded in lysosomes (discussed later).

Receptor endocytosis does not necessarily stop the receptor from signaling. In some cases, the bound arrestin recruits other signaling proteins to relay the signal onward from the internalized GPCRs along new pathways.

Summary

GPCRs can indirectly activate or inactivate either plasma-membrane-bound enzymes or ion channels via G proteins. When an activated receptor stimulates a G protein, the G protein undergoes a conformational change that activates its α subunit, thereby triggering release of a βγ complex. Either component can then directly regulate the activity of target proteins in the plasma membrane. Some GPCRs either activate or inactivate adenylyl cyclase, thereby altering the intracellular concentration of the second messenger cyclic AMP. Others activate a phosphoinositide-specific phospholipase C (PLCβ), which generates two second messengers. One is inositol 1,4,5-trisphosphate (IP₃), which releases Ca²⁺ from the ER and thereby increases the concentration of Ca²⁺ in the cytosol. The other is diacylglycerol, which remains in the plasma membrane and helps activate protein kinase C (PKC). An increase in cytosolic cyclic AMP or Ca²⁺ levels affects cells mainly by stimulating cAMP-dependent protein kinase (PKA) and Ca²⁺/calmodulin-dependent protein kinases (CaM-kinases), respectively.

PKC, PKA, and CaM-kinases phosphorylate specific target proteins and thereby alter the activity of the proteins. Each type of cell has its own characteristic set of target proteins that is regulated in these ways, enabling the cell to make its own distinctive response to the second messengers. The intracellular signaling cascades activated by GPCRs greatly amplify the responses, so that many thousands of target protein molecules are changed for each molecule of extracellular signaling ligand bound to its receptor. The responses mediated by GPCRs are rapidly turned off when the extracellular signal is removed, and activated GPCRs are inactivated by phosphorylation and association with arrestins.

SIGNALING THROUGH ENZYME-COUPLED RECEPTORS

Like GPCRs, **enzyme-coupled receptors** are transmembrane proteins with their ligand-binding domain on the outer surface of the plasma membrane. Instead of having a cytosolic domain that associates with a trimeric G protein, however, their cytosolic domain either has intrinsic enzyme activity or associates directly with an enzyme. Whereas a GPCR has seven transmembrane segments, each subunit of an enzyme-coupled receptor typically has only one. GPCRs and enzyme-coupled receptors often activate some of the same signaling pathways. In this section, we describe some of the important features of signaling by enzyme-coupled receptors, with an emphasis on the most common class of these proteins, the *receptor tyrosine kinases.*

Activated Receptor Tyrosine Kinases (RTKs) Phosphorylate Themselves

Many extracellular signal proteins act through **receptor tyrosine kinases (RTKs)**. These include many secreted and cell-surface-bound proteins that control cell behavior in developing and adult animals. Some of these signal proteins and their RTKs are listed in Table 15–4.

There are about 60 human RTKs, which can be classified into about 20 structural subfamilies, each dedicated to its complementary family of protein ligands. Figure 15–43 shows the basic structural features of a number of the families that operate in mammals. In all cases, the binding of the signal protein to the ligand-binding domain on the extracellular side of the receptor activates the tyrosine kinase domain on the cytosolic side. This leads to phosphorylation of tyrosine side chains on the cytosolic part of the receptor, creating phosphotyrosine docking sites for various intracellular signaling proteins that relay the signal.

How does the binding of an extracellular ligand activate the kinase domain on the other side of the plasma membrane? For a GPCR, ligand binding is thought to change the relative orientation of several of the transmembrane α helices, thereby shifting the position of the cytoplasmic loops relative to one another. It is unlikely, however, that a conformational change could propagate across the lipid bilayer

TABLE 15–4 Some Signal Proteins That Act Via RTKs		
Signal protein family	Receptor family	Some representative responses
Epidermal growth factor (EGF)	EGF receptors	Stimulates cell survival, growth, proliferation, or differentiation of various cell types; acts as inductive signal in development
Insulin	Insulin receptor	Stimulates carbohydrate utilization and protein synthesis
Insulin-like growth factor (IGF1)	IGF receptor-1	Stimulates cell growth and survival in many cell types
Nerve growth factor (NGF)	Trk receptors	Stimulates survival and growth of some neurons
Platelet-derived growth factor (PDGF)	PDGF receptors	Stimulates survival, growth, proliferation, and migration of various cell types
Macrophage-colony-stimulating factor (MCSF)	MCSF receptor	Stimulates monocyte/macrophage proliferation and differentiation
Fibroblast growth factor (FGF)	FGF receptors	Stimulates proliferation of various cell types; inhibits differentiation of some precursor cells; acts as inductive signal in development
Vascular endothelial growth factor (VEGF)	VEGF receptors	Stimulates angiogenesis
Ephrin	Eph receptors	Stimulates angiogenesis; guides cell and axon migration

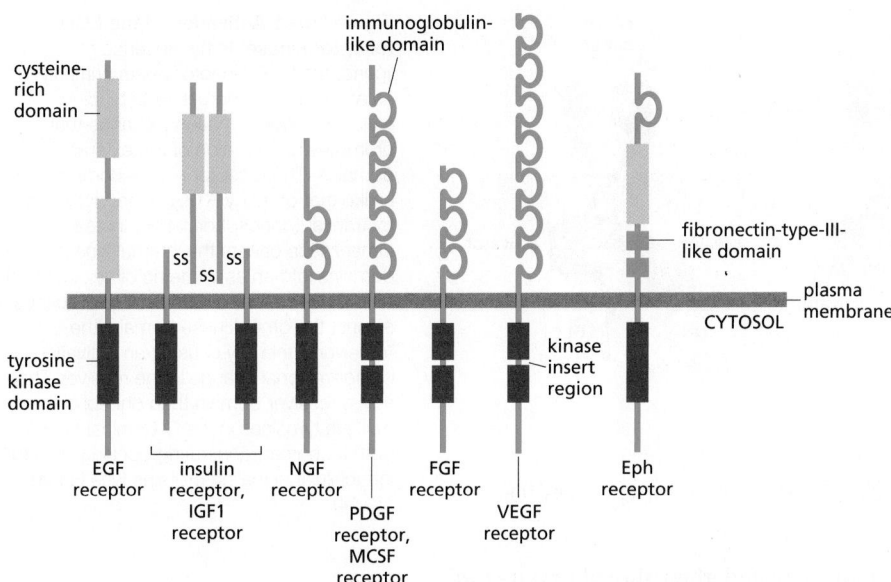

Figure 15–43 **Some subfamilies of RTKs.** Only one or two members of each subfamily are indicated. Note that in some cases, the tyrosine kinase domain is interrupted by a "kinase insert region" that is an extra segment emerging from the folded kinase domain. The functions of most of the cysteine-rich, immunoglobulin-like, and fibronectin-type-III-like domains are not known. Some of the ligands for the receptors shown are listed in Table 15–4, along with some representative responses that they mediate.

through a single transmembrane α helix. Instead, for most RTKs, ligand binding causes the receptors to dimerize, bringing the two cytoplasmic kinase domains together and thereby promoting their activation (**Figure 15–44**).

Dimerization stimulates kinase activity by a variety of mechanisms. In many cases, such as the insulin receptor, dimerization simply brings the kinase domains close to each other in an orientation that allows them to phosphorylate each other on specific tyrosines in the kinase active sites, thereby promoting conformational changes that fully activate both kinase domains. In other cases, such as the receptor for *epidermal growth factor* (*EGF*), the kinase is not activated by phosphorylation but by conformational changes brought about by interactions between the two kinase domains outside their active sites (**Figure 15–45**).

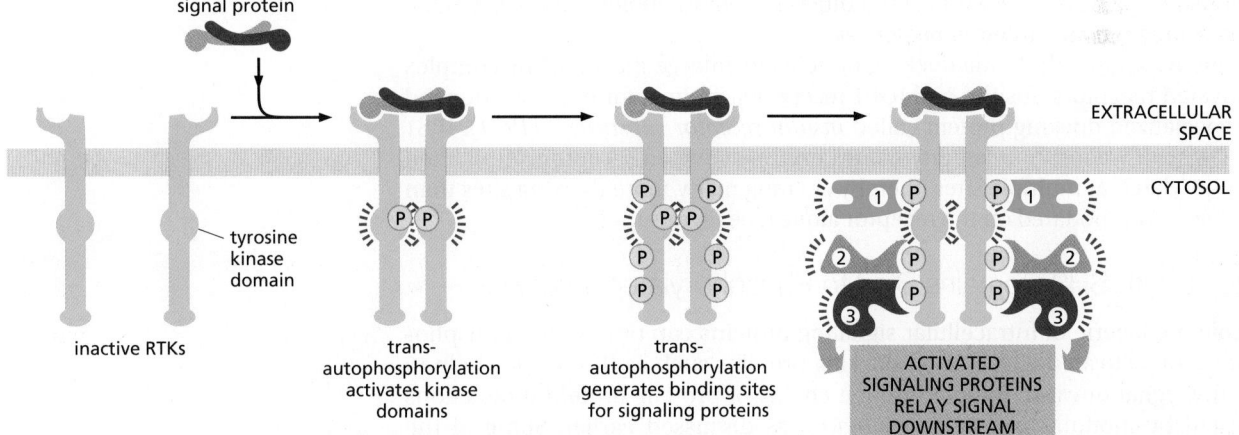

Figure 15–44 **Activation of RTKs by dimerization.** In the absence of extracellular signals, most RTKs exist as monomers in which the internal kinase domain is inactive. Binding of ligand brings two monomers together to form a dimer. In most cases, the close proximity in the dimer leads the two kinase domains to phosphorylate each other, which has two effects. First, phosphorylation at some tyrosines in the kinase domains promotes the complete activation of the domains. Second, phosphorylation at tyrosines in other parts of the receptors generates docking sites for intracellular signaling proteins, resulting in the formation of large signaling complexes that can then broadcast signals along multiple signaling pathways.

Mechanisms of dimerization vary widely among different RTK family members. In some cases, as shown here, the ligand itself is a dimer and brings two receptors together by binding them simultaneously. In other cases, a monomeric ligand can interact with two receptors simultaneously to bring them together, or two ligands can bind independently on two receptors to promote dimerization. In some RTKs—notably those in the insulin receptor family—the receptor is always a dimer (see Figure 15–43), and ligand binding causes a conformational change that brings the two internal kinase domains closer together. Although many RTKs are activated by transautophosphorylation as shown here, there are some important exceptions, including the EGF receptor illustrated in Figure 15–45.

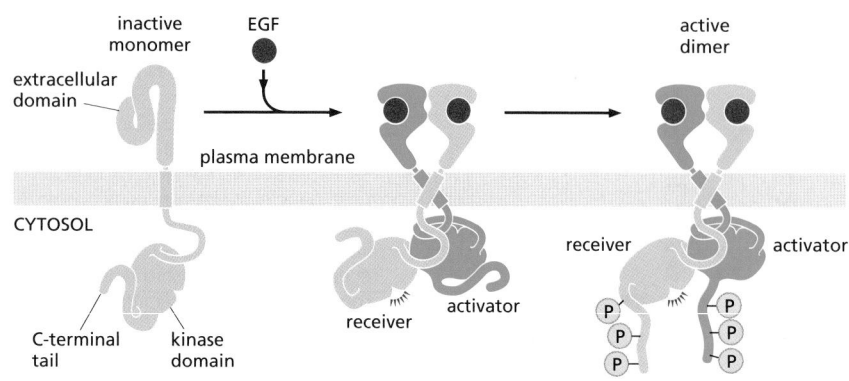

Figure 15–45 Activation of the EGF receptor kinase. In the absence of ligand, the EGF receptor exists primarily as an inactive monomer. EGF binding results in a conformational change that promotes dimerization of the external domains. The receptor kinase domain, unlike that of many RTKs, is not activated by transautophosphorylation. Instead, dimerization orients the internal kinase domains into an asymmetric dimer, in which one kinase domain (the "activator") pushes against the other kinase domain (the "receiver"), thereby causing an activating conformational change in the receiver. The active receiver domain then phosphorylates multiple tyrosines in the C-terminal tails of both receptors, generating docking sites for intracellular signaling proteins (see Figure 15–44).

Phosphorylated Tyrosines on RTKs Serve as Docking Sites for Intracellular Signaling Proteins

Once the kinase domains of an RTK dimer are activated, they phosphorylate multiple additional sites in the cytosolic parts of the receptors, typically in disordered regions outside the kinase domain (see Figure 15–44). This phosphorylation creates high-affinity docking sites for intracellular signaling proteins. Each signaling protein binds to a particular phosphorylated site on the activated receptors because it contains a specific phosphotyrosine-binding domain that recognizes surrounding features of the polypeptide chain in addition to the phosphotyrosine.

Once bound to the activated RTK, a signaling protein may become phosphorylated on tyrosines and thereby activated. In many cases, however, the binding alone may be sufficient to activate the docked signaling protein, by either inducing a conformational change in the protein or simply bringing it near the protein that is next in the signaling pathway. Thus, receptor phosphorylation serves as a switch to trigger the assembly of an intracellular signaling complex, which can then relay the signal onward, often along multiple routes, to various destinations in the cell. Because different RTKs bind different combinations of these signaling proteins, they activate different responses.

Some RTKs use additional docking proteins to enlarge the signaling complex at activated receptors. Insulin and IGF1 receptor signaling, for example, depend on a specialized docking protein called *insulin receptor substrate 1* (*IRS1*). IRS1 associates with phosphorylated tyrosines on the activated receptor and is then phosphorylated at multiple sites, thereby creating many more docking sites than could be accommodated on the receptor alone (see Figure 15–11).

Proteins with SH2 Domains Bind to Phosphorylated Tyrosines

A whole menagerie of intracellular signaling proteins can bind to the phosphotyrosines on activated RTKs (or on docking proteins such as IRS1). They help to relay the signal onward, mainly through chains of protein–protein interactions mediated by modular *interaction domains*, as discussed earlier. Some of the docked proteins are enzymes, such as **phospholipase C-γ (PLCγ)**, which functions in the same way as phospholipase C-β—activating the inositol phospholipid signaling pathway discussed earlier in connection with GPCRs (see Figures 15–28 and 15–29). Through this pathway, RTKs can increase cytosolic Ca^{2+} levels and activate PKC. Another enzyme that docks on these receptors is the cytoplasmic tyrosine kinase *Src*, which phosphorylates other signaling proteins on tyrosines. Yet another is *phosphoinositide 3-kinase* (*PI 3-kinase*), which phosphorylates lipids rather than proteins; as we discuss later, the phosphorylated lipids then serve as docking sites to attract various signaling proteins to the plasma membrane.

The intracellular signaling proteins that bind to phosphotyrosines have varied structures and functions. However, they usually share highly conserved phosphotyrosine-binding domains. These can be either **SH2 domains** (for *Src homology* *region*) or, less commonly, *PTB domains* (for *phosphotyrosine-binding*). By recognizing specific phosphorylated tyrosines, these small interaction domains enable the proteins that contain them to bind to activated RTKs, as well as to many other intracellular signaling proteins that have been transiently phosphorylated on tyrosines (**Figure 15–46**). As discussed previously, many signaling proteins also contain other interaction domains that allow them to interact specifically with other proteins as part of the signaling process. These domains include the *SH3 domain*, which binds to proline-rich motifs in intracellular proteins (see Figure 15–11).

Not all proteins that bind to activated RTKs via SH2 domains help to relay the signal onward. Some act to decrease the signaling process, providing negative feedback. One example is the *c-Cbl protein*, which can dock on some activated receptors and catalyze their ubiquitylation, covalently adding one or more ubiquitin molecules to specific sites on the receptor. This promotes the endocytosis and degradation of the receptors in lysosomes—an example of receptor down-regulation (see Figure 15–20). Endocytic proteins that contain *ubiquitin-interaction motifs (UIMs)* recognize the ubiquitylated RTKs and direct them into clathrin-coated vesicles and, ultimately, into lysosomes (discussed in Chapter 13). Mutations that inactivate c-Cbl-dependent RTK down-regulation cause prolonged RTK signaling and thereby promote the development of cancer.

As is the case for GPCRs, ligand-induced endocytosis of RTKs does not always decrease signaling. In some cases, RTKs are endocytosed with their bound signaling proteins and continue to signal from endosomes or other intracellular compartments. This mechanism, for example, allows *nerve growth factor* (*NGF*) to bind to its specific RTK (called TrkA) at the end of a long nerve cell axon and signal to the cell body of the same cell a long distance away. Here, signaling endocytic

Figure 15–46 The binding of SH2-containing intracellular signaling proteins to an activated RTK.
(A) This drawing of a receptor for *platelet-derived growth factor* (*PDGF*) shows five phosphotyrosine docking sites, three in the kinase insert region and two on the C-terminal tail, to which the three signaling proteins shown bind as indicated. The numbers on the right indicate the positions of the tyrosines in the polypeptide chain. These binding sites have been identified by using recombinant DNA technology to mutate specific tyrosines in the receptor. Mutation of tyrosines 1009 and 1021, for example, prevents the binding and activation of PLCγ, so that receptor activation no longer stimulates the inositol phospholipid signaling pathway. The locations of the SH2 *(red)* and SH3 *(blue)* domains in the three signaling proteins are indicated. (Additional phosphotyrosine docking sites on this receptor are not shown, including those that bind the cytoplasmic tyrosine kinase Src and two adaptor proteins.) It is unclear how many signaling proteins can bind simultaneously to a single RTK. (B) The three-dimensional structure of an SH2 domain, as determined by x-ray crystallography. The binding pocket for phosphotyrosine is shaded in *orange* on the right, and a pocket for binding a specific amino acid side chain (isoleucine, in this case) is shaded in *yellow* on the left. The RTK polypeptide segment that binds the SH2 domain is shown in *yellow* (see also Figure 3–40). (C) The SH2 domain is a compact, "plug-in" module, which can be inserted almost anywhere in a protein without disturbing the protein's folding or function (discussed in Chapter 3). Because each domain has distinct sites for recognizing phosphotyrosine and for recognizing a particular amino acid side chain, different SH2 domains recognize phosphotyrosine in the context of different flanking amino acid sequences. (B, based on data from G. Waksman et al., *Cell* 72:779–790, 1993. PDB code: 2SRC.)

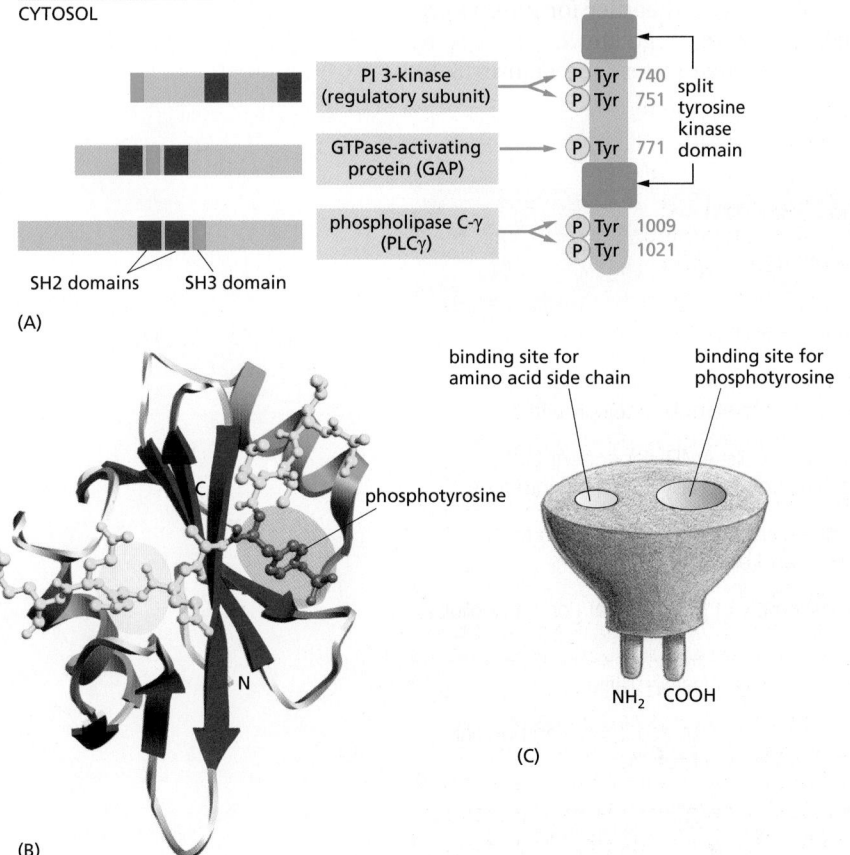

vesicles containing TrkA, with NGF bound on the lumenal side and signaling proteins docked on the cytosolic side, are transported along the axon to the cell body, where they signal the cell to survive.

Some signaling proteins are composed almost entirely of SH2 and SH3 domains and function as *adaptors* to couple tyrosine-phosphorylated proteins to other proteins that do not have their own SH2 domains (see Figure 15–11). Adaptor proteins of this type help to couple activated RTKs to the important signaling protein *Ras*, a monomeric GTPase that, in turn, can activate various downstream signaling pathways, as we now discuss.

The GTPase Ras Mediates Signaling by Most RTKs

The **Ras superfamily** consists of various families of monomeric GTPases, but only the Ras and Rho families relay signals from cell-surface receptors (Table 15–5). By interacting with different intracellular signaling proteins, a single Ras or Rho family member can coordinately spread the signal along several distinct downstream signaling pathways, thereby acting as a *signaling hub*.

There are three major, closely related Ras proteins in humans: H-, K-, and N-Ras (see Table 15–5). Although they have subtly different functions, they are thought to work in the same way, and we will refer to them simply as **Ras**. Like many monomeric GTPases, Ras contains one or more covalently attached lipid groups that help anchor the protein to the cytoplasmic face of the membrane, from where it relays signals to other parts of the cell. Ras is often required, for example, when RTKs signal to the nucleus to stimulate cell proliferation or differentiation, both of which require changes in gene expression. If Ras function is inhibited by various experimental approaches, the cell proliferation or differentiation responses normally induced by the activated RTKs do not occur. Conversely, 30% of human tumors express hyperactive mutant forms of Ras, which contribute to the uncontrolled proliferation of the cancer cells.

Like other GTP-binding proteins, Ras functions as a molecular switch, cycling between two distinct conformational states—active when GTP is bound and inactive when GDP is bound (Movie 15.7). As discussed earlier for monomeric GTPases in general, two classes of signaling proteins regulate Ras activity by influencing its transition between active and inactive states (see Figure 15–8).

TABLE 15–5 **The Ras Superfamily of Monomeric GTPases**		
Family	Some family members	Some functions
Ras	H-Ras, K-Ras, N-Ras	Relay signals from RTKs
	Rheb	Activates mTOR to stimulate cell growth
	Rap1	Activated by a cyclic-AMP-dependent GEF; influences cell adhesion by activating integrins
Rho*	Rho, Rac, Cdc42	Relay signals from surface receptors to the cytoskeleton and elsewhere
ARF*	ARF1–ARF6	Regulate assembly of protein coats on intracellular vesicles
Rab*	Rab1–60	Regulate intracellular vesicle traffic
Ran*	Ran	Regulates mitotic spindle assembly and nuclear transport of RNAs and proteins

*The Rho family is discussed in Chapter 16, the ARF and Rab proteins in Chapter 13, and Ran in Chapters 12 and 17. The three-dimensional structure of Ras is shown in Figure 3–67.

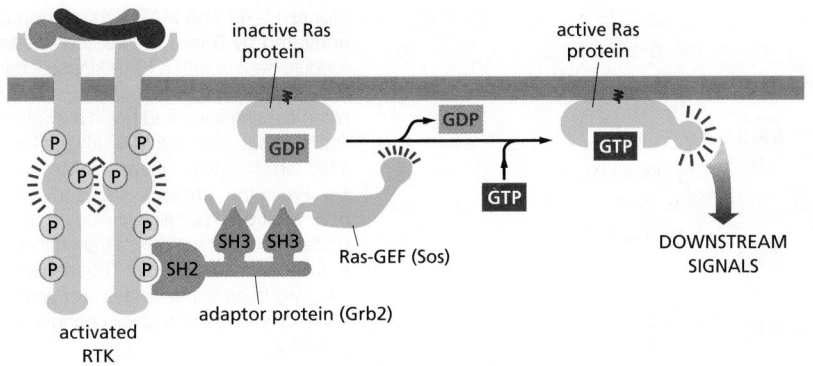

Figure 15–47 How an RTK activates Ras. Grb2 recognizes a specific phosphorylated tyrosine on the activated receptor by means of an SH2 domain and recruits Sos by means of two SH3 domains. Sos stimulates the inactive Ras protein to replace its bound GDP by GTP, which activates Ras to relay the signal downstream.

Ras guanine nucleotide exchange factors (**Ras-GEFs**) stimulate the dissociation of GDP and the subsequent uptake of GTP from the cytosol, thereby activating Ras. *Ras GTPase-activating proteins* (**Ras-GAPs**) increase the rate of hydrolysis of bound GTP by Ras, thereby inactivating Ras. Hyperactive mutant forms of Ras are resistant to GAP-mediated GTPase stimulation and are locked permanently in the GTP-bound active state, which is why they promote the development of cancer.

But how do RTKs normally activate Ras? In principle, they could either activate a Ras-GEF or inhibit a Ras-GAP. Even though some GAPs bind directly (via their SH2 domains) to activated RTKs (see Figure 15–46A), it is the indirect coupling of the receptor to a Ras-GEF that drives Ras into its active state. The loss of function of a Ras-GEF has a similar effect to the loss of function of Ras itself. Activation of the other Ras superfamily proteins, including those of the Rho family, also occurs through the activation of GEFs. The particular GEF determines in which membrane the GTPase is activated and, by acting as a scaffold, it can also determine which downstream proteins the GTPase activates.

The GEF that mediates Ras activation by RTKs was discovered by genetic studies of eye development in *Drosophila*, where an RTK called *Sevenless* (*Sev*) is required for the formation of a photoreceptor cell called R7. Genetic screens for components of this signaling pathway led to the discovery of a Ras-GEF called *Son-of-sevenless* (*Sos*). Further genetic screens uncovered another protein, now called *Grb2*, which is an adaptor protein that links the Sev receptor to the Sos protein; the SH2 domain of the Grb2 adaptor binds to the activated receptor, while its two SH3 domains bind to Sos. Sos then promotes Ras activation. Biochemical and cell biological studies have shown that Grb2 and Sos also link activated RTKs to Ras in mammalian cells, revealing that this is a highly conserved mechanism in RTK signaling (**Figure 15–47**). Once activated, Ras activates various other signaling proteins to relay the signal downstream, as we discuss next.

Ras Activates a MAP Kinase Signaling Module

Both the tyrosine phosphorylations and the activation of Ras triggered by activated RTKs are usually short-lived (**Figure 15–48**). *Tyrosine-specific protein phosphatases* quickly reverse the phosphorylations, and Ras-GAPs induce activated

Figure 15–48 Transient activation of Ras revealed by single-molecule fluorescence resonance energy transfer (FRET). (A) Schematic drawing of the experimental strategy. Cells of a human cancer cell line are genetically engineered to express a Ras protein that is covalently linked to yellow fluorescent protein (YFP). GTP that is labeled with a red fluorescent dye is microinjected into some of the cells. The cells are then stimulated with the extracellular signal protein EGF, and single fluorescent molecules of Ras-YFP at the inner surface of the plasma membrane are followed by video fluorescence microscopy in individual cells. When a fluorescent Ras-YFP molecule becomes activated, it exchanges unlabeled GDP for fluorescently labeled GTP; the energy emitted by the YFP now activates the fluorescent GTP to emit red light (called fluorescence resonance energy transfer, or FRET; see Figure 9–26). Thus, the activation of single Ras molecules can be followed by the emission of red fluorescence from a previously yellow-green fluorescent spot at the plasma membrane. As shown in (B), activated Ras molecules can be detected after about 30 seconds of EGF stimulation. The red signal peaks at about 3 minutes and then decreases to baseline by 6 minutes. As Ras-GAP is found to be recruited to the same spots at the plasma membrane as Ras, it presumably plays a major part in rapidly shutting off the Ras signal. (Modified from H. Murakoshi et al., *Proc. Natl Acad. Sci. USA* 101:7317–7322, 2004. Copyright 2004 National Academy of Sciences, USA. With permission from National Academy of Sciences.)

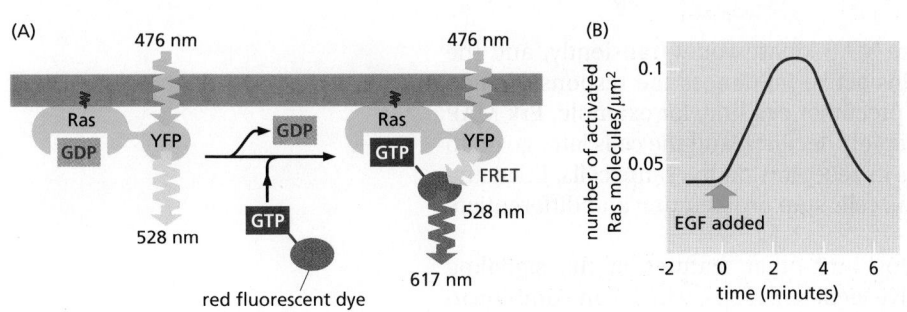

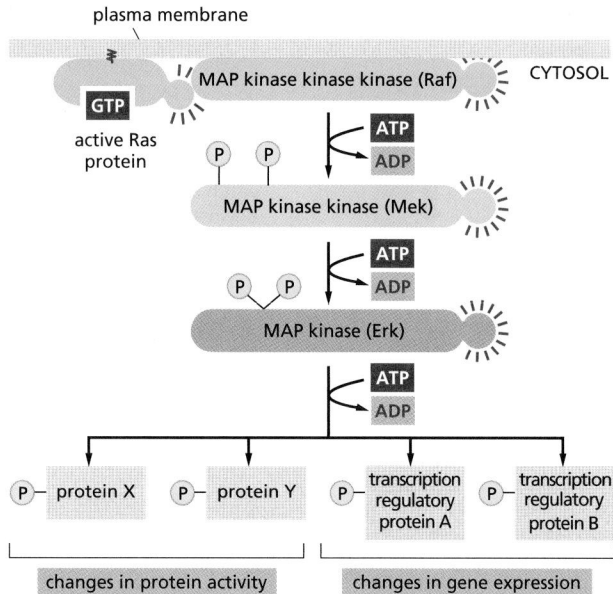

Figure 15–49 **The MAP kinase module activated by Ras.** The three-component module begins with a MAP kinase kinase kinase called *Raf.* Ras recruits Raf to the plasma membrane and helps activate it. Raf then activates the MAP kinase kinase *Mek*, which then activates the MAP kinase *Erk.* Erk in turn phosphorylates a variety of downstream proteins, including other protein kinases, as well as transcription regulators in the nucleus. The resulting changes in protein activities and gene expression cause complex changes in cell behavior.

Ras to inactivate itself by hydrolyzing its bound GTP to GDP. To stimulate cells to proliferate or differentiate, these short-lived signaling events must be converted into longer-lasting ones that can sustain the signal and relay it downstream to the nucleus to alter the pattern of gene expression. One of the key mechanisms used for this purpose is a system of proteins called the *mitogen-activated protein kinase module* (**MAP kinase module**) (**Figure 15–49**). The three components of this system form a functional signaling module that has been remarkably well conserved during evolution and is used, with variations, in many different signaling contexts.

The three components are all protein kinases. The final kinase in the series is called simply MAP kinase (MAPK). The next one upstream from this is MAP kinase kinase (MAPKK): it phosphorylates and thereby activates MAP kinase. Next above that, receiving an activating signal directly from Ras, is MAP kinase kinase kinase (MAPKKK): it phosphorylates and thereby activates MAPKK. In the mammalian **Ras–MAP-kinase signaling pathway**, these three kinases are known by shorter names: Raf (= MAPKKK), Mek (= MAPKK), and Erk (=MAPK).

Once activated, the MAP kinase relays the signal downstream by phosphorylating various proteins in the cell, including transcription regulators and other protein kinases (see Figure 15–49). Erk, for example, enters the nucleus and phosphorylates one or more components of a transcription regulatory complex. This activates the transcription of a set of *immediate early genes,* so named because they turn on within minutes after an RTK receives an extracellular signal, even if protein synthesis is experimentally blocked with drugs. Some of these genes encode other transcription regulators that turn on other genes, a process that requires both protein synthesis and more time. In this way, the Ras–MAP-kinase signaling pathway conveys signals from the cell surface to the nucleus and alters the pattern of gene expression. Among the genes activated by this pathway are some that stimulate cell proliferation, such as the genes encoding G_1 *cyclins* (discussed in Chapter 17).

Extracellular signals usually activate MAP kinases only transiently, and the period during which the kinase remains active influences the response. When EGF activates its receptors in a neural precursor cell line, for example, Erk MAP kinase activity peaks at 5 minutes and rapidly declines, and the cells later go on to divide. By contrast, when NGF activates its receptors on the same cells, Erk activity remains high for many hours, and the cells stop proliferating and differentiate into neurons.

Many factors influence the duration and other features of the signaling response, including positive and negative feedback loops, which can combine to

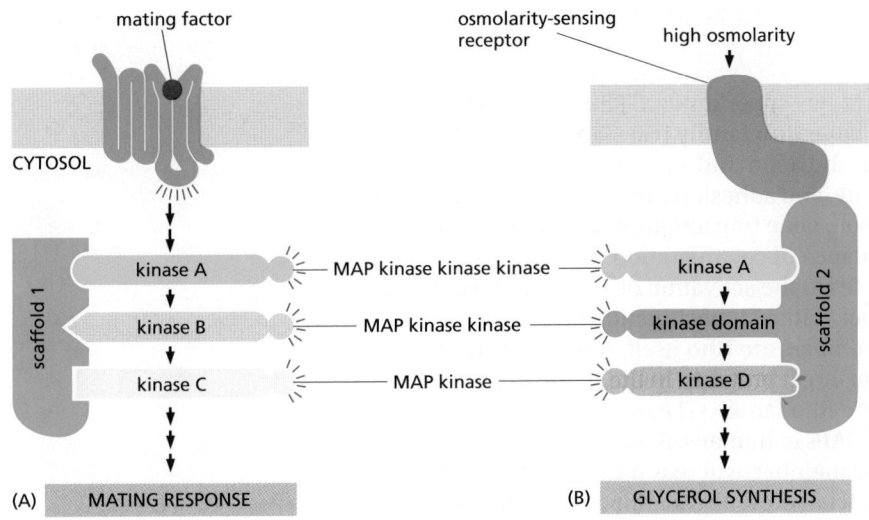

Figure 15–50 **The organization of two MAP kinase modules by scaffold proteins in budding yeast.** Budding yeast have at least six three-component MAP kinase modules involved in a variety of biological processes, including the two responses illustrated here—a mating response and the response to high osmolarity. (A) The mating response is triggered when a mating factor secreted by a yeast of opposite mating type binds to a GPCR. This activates a G protein, the βγ complex of which indirectly activates the MAPKKK (kinase A), which then relays the response onward. Once activated, the MAP kinase (kinase C) phosphorylates and thereby activates several proteins that mediate the mating response, in which the yeast cell stops dividing and prepares for fusion. The three kinases in this module are bound to scaffold protein 1. (B) In a second response, a yeast cell exposed to a high-osmolarity environment is induced to synthesize glycerol to increase its internal osmolarity. This response is mediated by an osmolarity-sensing receptor protein and a different MAP kinase module bound to a second scaffold protein. (Note that the kinase domain of scaffold 2 provides the MAPKK activity of this module.) Although both pathways use the same MAPKKK (kinase A, *green*), there is no cross-talk between them because the kinases in each module are bound to different scaffold proteins, and the osmosensor is bound to the same scaffold protein as the particular kinase it activates.

give responses that are either graded or switchlike and either brief or long lasting. In an example illustrated earlier, in Figure 15–19, MAP kinase activates a complex positive feedback loop to produce an all-or-none, irreversible response when frog oocytes are stimulated to mature by a brief exposure to the extracellular signal molecule progesterone. In many cells, MAP kinases activate a negative feedback loop by increasing the concentration of a protein phosphatase that removes the phosphate from MAP kinase. The increase in the phosphatase results from both an increase in the transcription of the phosphatase gene and the stabilization of the enzyme against degradation. In the Ras–MAP-kinase pathway shown in Figure 15–49, Erk also phosphorylates and inactivates Raf, providing another negative feedback loop that helps shut off the MAP kinase module.

Scaffold Proteins Help Prevent Cross-talk Between Parallel MAP Kinase Modules

Three-component MAP kinase signaling modules operate in all eukaryotic cells, with different modules mediating different responses in the same cell. In budding yeast, for example, one such module mediates the response to mating pheromone, another the response to starvation, and yet another the response to osmotic shock. Some of these MAP kinase modules use one or more of the same kinases and yet manage to activate different effector proteins and hence different responses. As discussed earlier, one way in which cells avoid cross-talk between the different parallel signaling pathways and ensure that each response is specific is to use scaffold proteins (see Figure 15–10A). In budding yeast cells, such scaffolds bind all or some of the kinases in each MAP kinase module to form a complex and thereby help to ensure response specificity (**Figure 15–50**).

Mammalian cells also use this scaffold strategy to prevent cross-talk between different MAP kinase modules. At least five parallel MAP kinase modules can operate in a mammalian cell. These modules make use of at least 12 MAP kinases, 7 MAPKKs, and 7 MAPKKKs. Two of these modules (terminating in MAP kinases called JNK and p38) are activated by different kinds of cell stresses, such as ultraviolet (UV) irradiation, heat shock, and osmotic stress, as well as by inflammatory cytokines; others mainly mediate responses to signals from other cells.

Although the scaffold strategy provides precision and avoids cross-talk, it reduces the opportunities for amplification and spreading of the signal to different parts of the cell, which require at least some of the components to be diffusible. It is unclear to what extent the individual components of MAP kinase modules can dissociate from the scaffold during the activation process to permit amplification.

Rho Family GTPases Functionally Couple Cell-Surface Receptors to the Cytoskeleton

Besides the Ras proteins, the other class of Ras superfamily GTPases that relays signals from cell-surface receptors is the large **Rho family** (see Table 15–5). Rho family monomeric GTPases regulate both the actin and microtubule cytoskeletons, controlling cell shape, polarity, motility, and adhesion (discussed in Chapter 16); they also regulate cell-cycle progression, gene transcription, and membrane transport. They play a key part in the guidance of cell migration and nerve axon outgrowth, mediating cytoskeletal responses to the activation of a special class of guidance receptors. We focus on this aspect of Rho family function here.

The three best-characterized family members are **Rho** itself, **Rac**, and **Cdc42**, each of which affects multiple downstream target proteins. In the same way as for Ras, GEFs activate and GAPs inactivate the Rho family GTPases; there are more than 80 Rho-GEFs and more than 70 Rho-GAPs in humans. Some of the GEFs and GAPs are specific for one particular family member, whereas others are less specific. Unlike Ras, which is membrane-associated even when inactive (with GDP bound), inactive Rho family GTPases are often bound to *guanine nucleotide dissociation inhibitors* (*GDIs*) in the cytosol, which prevent the GTPases from interacting with their Rho-GEFs at the plasma membrane.

Signaling by extracellular signaling proteins of the **ephrin** family provides an example of how RTKs can activate a Rho GTPase. Ephrins bind and thereby activate members of the *Eph* family of RTKs (see Figure 15–43). One member of the Eph family is found on the surface of motor neurons and helps guide the migrating tip of the axon (called a *growth cone*) to its muscle target. The binding of a cell-surface *ephrin* protein activates the Eph receptor, causing the growth cones to collapse, thereby repelling them from inappropriate regions and keeping them on track. The response depends on a Rho-GEF called *ephexin*, which is stably associated with the cytosolic tail of the Eph receptor. When ephrin binding activates the Eph receptor, the receptor activates a cytoplasmic tyrosine kinase that phosphorylates ephexin on a tyrosine, enhancing the ability of ephexin to activate the Rho protein RhoA. The activated RhoA (RhoA-GTP) then regulates various downstream target proteins, including some effector proteins that control the actin cytoskeleton, causing the growth cone to collapse (**Figure 15–51**).

Having considered how RTKs use GEFs and monomeric GTPases to relay signals into the cell, we now consider a second major strategy that RTKs use that depends on a quite different intracellular relay mechanism.

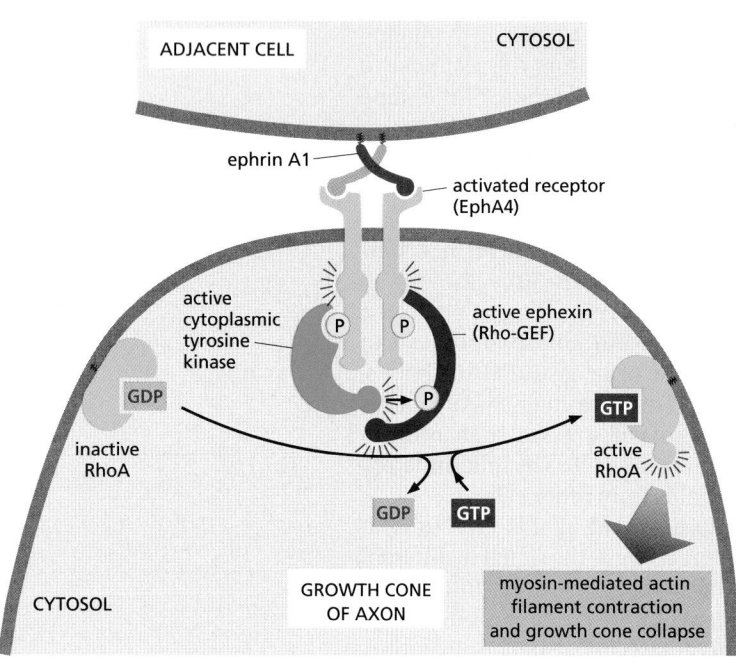

Figure 15–51 Growth cone collapse mediated by Rho family GTPases. The binding of ephrin A1 proteins on an adjacent cell activates EphA4 RTKs on the growth cone. Phosphotyrosines on the activated Eph receptors recruit and activate a cytoplasmic tyrosine kinase to phosphorylate the receptor-associated Rho-GEF ephexin on a tyrosine. This enhances the ability of the ephexin to activate RhoA. RhoA then induces the growth cone to collapse by stimulating the myosin-dependent contraction of the actin cytoskeleton.

PI 3-Kinase Produces Lipid Docking Sites in the Plasma Membrane

As mentioned earlier, one of the proteins that binds to the intracellular tail of RTK molecules is the plasma-membrane-bound enzyme **phosphoinositide 3-kinase (PI 3-kinase)**. This kinase principally phosphorylates inositol phospholipids rather than proteins, and both RTKs and GPCRs can activate it. It plays a central part in promoting cell survival and growth.

Phosphatidylinositol (PI) is unique among membrane lipids because it can undergo reversible phosphorylation at multiple sites on its inositol head group to generate a variety of phosphorylated PI lipids called **phosphoinositides**. When activated, PI 3-kinase catalyzes phosphorylation at the 3 position of the inositol ring to generate several phosphoinositides (**Figure 15–52**). The production of PI(3,4,5)P$_3$ matters most because it can serve as a docking site for various intracellular signaling proteins, which assemble into signaling complexes that relay the signal into the cell from the cytosolic face of the plasma membrane (see Figure 15–10C).

Notice the difference between this use of phosphoinositides and their use described earlier, in which PI(4,5)P$_2$ is cleaved by PLCβ (in the case of GPCRs) or PLCγ (in the case of RTKs) to generate soluble IP$_3$ and membrane-bound diacylglycerol (see Figures 15–28 and 15–29). By contrast, PI(3,4,5)P$_3$ is not cleaved by either PLC. It is made from PI(4,5)P$_2$ and then remains in the plasma membrane until specific *phosphoinositide phosphatases* dephosphorylate it. Prominent among these is the *PTEN* phosphatase, which dephosphorylates the 3 position of the inositol ring. Mutations in PTEN are found in many cancers: by prolonging signaling by PI 3-kinase, they promote uncontrolled cell growth.

There are various types of PI 3-kinases. Those activated by RTKs and GPCRs belong to class I. These are heterodimers composed of a common catalytic subunit and different regulatory subunits. RTKs activate *class Ia PI 3-kinases*, in which the regulatory subunit is an adaptor protein that binds to two phosphotyrosines on activated RTKs through its two SH2 domains (see Figure 15–46A). GPCRs activate *class Ib PI 3-kinases*, which have a regulatory subunit that binds to the βγ complex of an activated trimeric G protein (G$_q$) when GPCRs are activated by their extracellular ligand. The direct binding of activated Ras can also activate the common class I catalytic subunit.

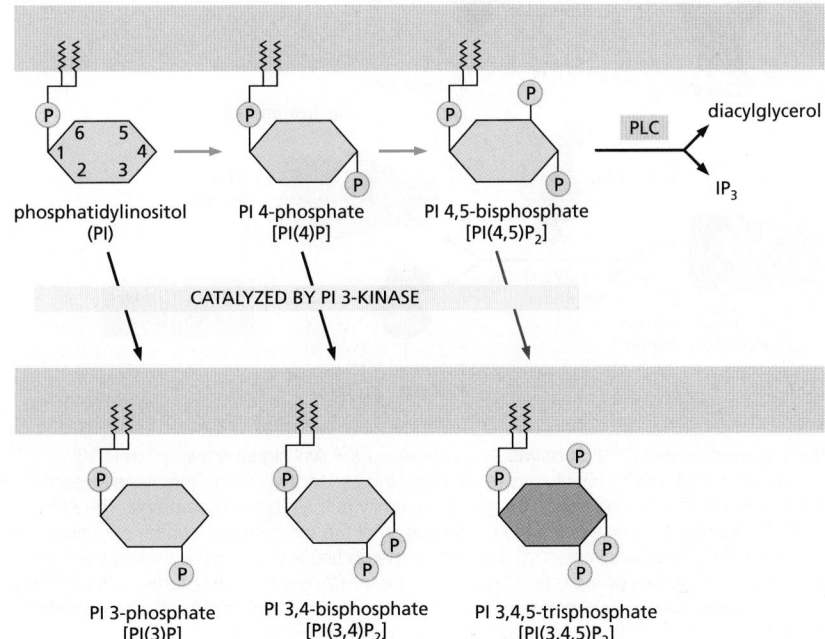

phosphatidylinositol (PI) → PI 4-phosphate [PI(4)P] → PI 4,5-bisphosphate [PI(4,5)P$_2$] → PLC → diacylglycerol, IP$_3$

CATALYZED BY PI 3-KINASE

PI 3-phosphate [PI(3)P]   PI 3,4-bisphosphate [PI(3,4)P$_2$]   PI 3,4,5-trisphosphate [PI(3,4,5)P$_3$]

Figure 15–52 **The generation of phosphoinositide docking sites by PI 3-kinase.** PI 3-kinase phosphorylates the inositol ring on carbon atom 3 to generate the phosphoinositides shown at the bottom of the figure (diverting them away from the pathway leading to IP$_3$ and diacylglycerol; see Figure 15–28). The most important phosphorylation (indicated in *red*) is of PI(4,5)P$_2$ to PI(3,4,5)P$_3$, which can serve as a docking site for signaling proteins with PI(3,4,5)P$_3$-binding PH domains. Other inositol phospholipid kinases (not shown) catalyze the phosphorylations indicated by the *green arrows*.

Intracellular signaling proteins bind to PI(3,4,5)P$_3$ produced by activated PI 3-kinase via a specific interaction domain, such as a **pleckstrin homology (PH) domain**, first identified in the platelet protein pleckstrin. PH domains function mainly as protein–protein interaction domains, and it is only a small subset of them that bind to PI(3,4,5)P$_3$; at least some of these also recognize a specific membrane-bound protein as well as the PI(3,4,5)P$_3$, which greatly increases the specificity of the binding and helps to explain why the signaling proteins with PI(3,4,5)P$_3$-binding PH domains do not all dock at all PI(3,4,5)P$_3$ sites. PH domains occur in about 200 human proteins, including the Ras-GEF Sos discussed earlier (see Figure 15–11).

One especially important PH-domain-containing protein is the serine/threonine protein kinase *Akt*. The *PI-3-kinase–Akt signaling pathway* is the major pathway activated by the hormone *insulin*. It also plays a key part in promoting the survival and growth of many cell types in both invertebrates and vertebrates, as we now discuss.

The PI-3-Kinase–Akt Signaling Pathway Stimulates Animal Cells to Survive and Grow

As discussed earlier, extracellular signals are usually required for animal cells to grow and divide, as well as to survive (see Figure 15–4). Members of the *insulin-like growth factor (IGF)* family of signal proteins, for example, stimulate many types of animal cells to survive and grow. They bind to specific RTKs (see Figure 15–43), which activate PI 3-kinase to produce PI(3,4,5)P$_3$. The PI(3,4,5)P$_3$ recruits two protein kinases to the plasma membrane via their PH domains—**Akt** (also called *protein kinase B*, or *PKB*) and *phosphoinositide-dependent protein kinase 1 (PDK1)*, and this leads to the activation of Akt (**Figure 15–53**). Once activated,

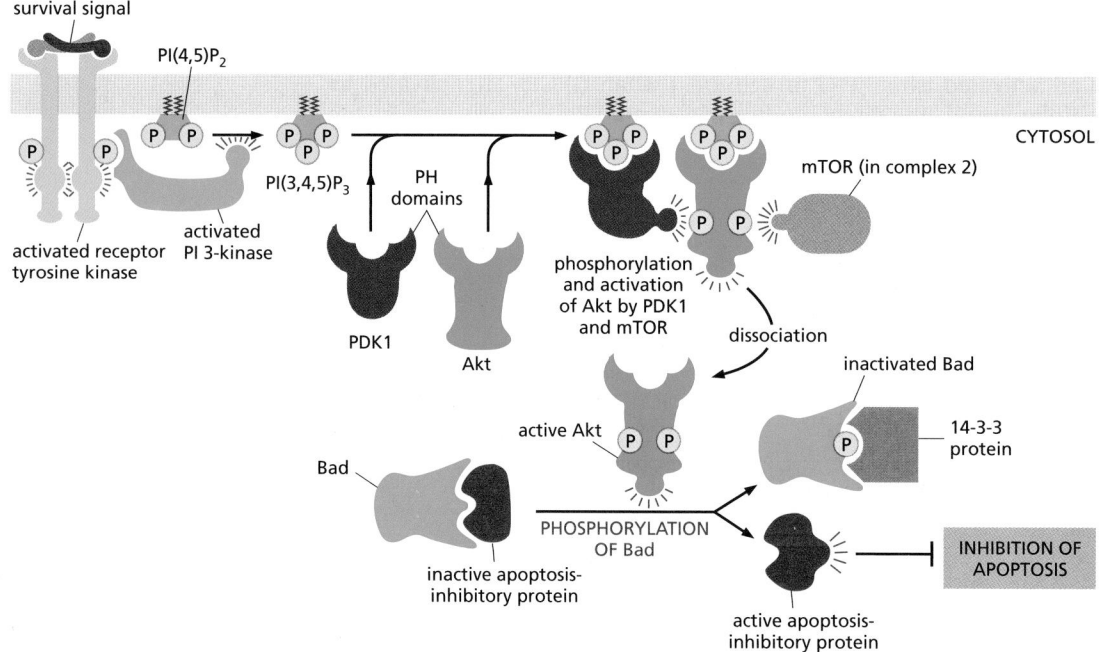

Figure 15–53 One way in which signaling through PI 3-kinase promotes cell survival. An extracellular survival signal activates an RTK, which recruits and activates PI 3-kinase. The PI 3-kinase produces PI(3,4,5)P$_3$, which serves as a docking site for two serine/threonine kinases with PH domains—Akt and the phosphoinositide-dependent kinase PDK1—and brings them into proximity at the plasma membrane. The Akt is phosphorylated on a serine by a third kinase (usually mTOR in complex 2), which alters the conformation of the Akt so that it can be phosphorylated on a threonine by PDK1, which activates the Akt. The activated Akt now dissociates from the plasma membrane and phosphorylates various target proteins, including the Bad protein. When unphosphorylated, Bad holds one or more apoptosis-inhibitory proteins (of the Bcl2 family—discussed in Chapter 18) in an inactive state. Once phosphorylated, Bad releases the inhibitory proteins, which now can block apoptosis and thereby promote cell survival. As shown, the phosphorylated Bad binds to a ubiquitous cytosolic protein called *14-3-3*, which keeps Bad out of action.

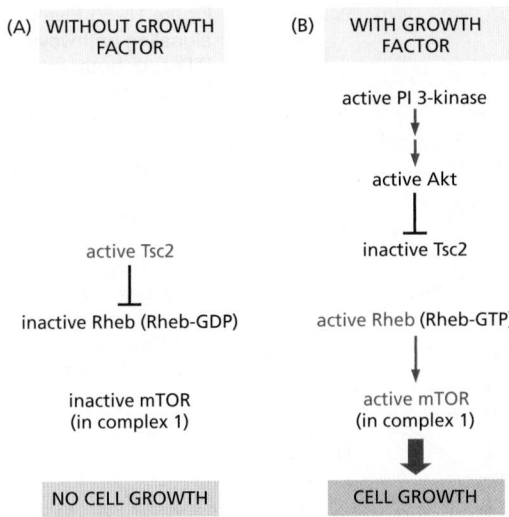

(A) WITHOUT GROWTH FACTOR	(B) WITH GROWTH FACTOR
	active PI 3-kinase
	↓
	active Akt
	⊣
active Tsc2	inactive Tsc2
⊣	
inactive Rheb (Rheb-GDP)	active Rheb (Rheb-GTP)
↓	↓
inactive mTOR (in complex 1)	active mTOR (in complex 1)
↓	⬇
NO CELL GROWTH	CELL GROWTH

Figure 15–54 Activation of mTOR by the PI-3-kinase–Akt signaling pathway. (A) In the absence of extracellular growth factors, Tsc2 (a Rheb-GAP) keeps Rheb inactive; mTOR in complex 1 is inactive, and there is no cell growth. (B) In the presence of growth factors, activated Akt phosphorylates and inhibits Tsc2, thereby promoting the activation of Rheb. Activated Rheb (Rheb-GTP) helps activate mTOR in complex 1, which in turn stimulates cell growth. Figure 15–53 shows how growth factors (or survival signals) activate Akt. The Erk MAP kinase (see Figure 15–49) can also phosphorylate and inhibit Tsc2 and thereby activate mTOR. Thus, both the PI-3-kinase–Akt and Ras–MAP-kinase signaling pathways converge on mTOR in complex 1 to stimulate cell growth.

Tsc2 is short for *tuberous sclerosis protein 2*, and it is one component of a heterodimer composed of Tsc1 and Tsc2 (not shown); these proteins are so called because mutations in either gene encoding them cause the genetic disease *tuberous sclerosis*, which is associated with benign tumors that contain abnormally large cells.

Akt phosphorylates various target proteins at the plasma membrane, as well as in the cytosol and nucleus. The effect on most of the known targets is to inactivate them; but the targets are such that these actions of Akt all conspire to enhance cell survival and growth, as illustrated for one cell survival pathway in Figure 15–53.

The control of cell growth by the **PI-3-kinase–Akt pathway** depends in part on a large protein kinase called **TOR** (named as the target of *rapamycin*, a bacterial toxin that inactivates the kinase and is used clinically as both an immunosuppressant and anticancer drug). TOR was originally identified in yeasts in genetic screens for rapamycin resistance; in mammalian cells, it is called **mTOR**, which exists in cells in two functionally distinct multiprotein complexes. *mTOR complex 1* contains the protein *raptor*; this complex is sensitive to rapamycin, and it stimulates cell growth—both by promoting ribosome production and protein synthesis and by inhibiting protein degradation. Complex 1 also promotes both cell growth and cell survival by stimulating nutrient uptake and metabolism. *mTOR complex 2* contains the protein *rictor* and is insensitive to rapamycin; it helps to activate Akt (see Figure 15–53), and it regulates the actin cytoskeleton via Rho family GTPases.

The mTOR in complex 1 integrates inputs from various sources, including extracellular signal proteins referred to as *growth factors* and nutrients such as amino acids, both of which help activate mTOR and promote cell growth. The growth factors activate mTOR mainly via the PI-3-kinase–Akt pathway. Akt activates mTOR in complex 1 indirectly by phosphorylating, and thereby inhibiting, a GAP called Tsc2. Tsc2 acts on a monomeric Ras-related GTPase called **Rheb** (see Table 15–5). Rheb in its active form (Rheb-GTP) activates mTOR in complex 1. The net result is that Akt activates mTOR and thereby promotes cell growth (**Figure 15–54**). We discuss how mTOR stimulates ribosome production and protein synthesis in Chapter 17 (see Figure 17–64).

RTKs and GPCRs Activate Overlapping Signaling Pathways

As mentioned earlier, RTKs and GPCRs activate some of the same intracellular signaling pathways. Both, for example, can activate the inositol phospholipid pathway triggered by phospholipase C. Moreover, even when they activate different pathways, the different pathways can converge on the same target proteins. **Figure 15–55** illustrates both of these types of signaling overlaps: it summarizes five parallel intracellular signaling pathways that we have discussed so far—one triggered by GPCRs, two triggered by RTKs, and two triggered by both kinds of receptors. Interactions among these pathways allow different extracellular signal molecules to modulate and coordinate each other's effects.

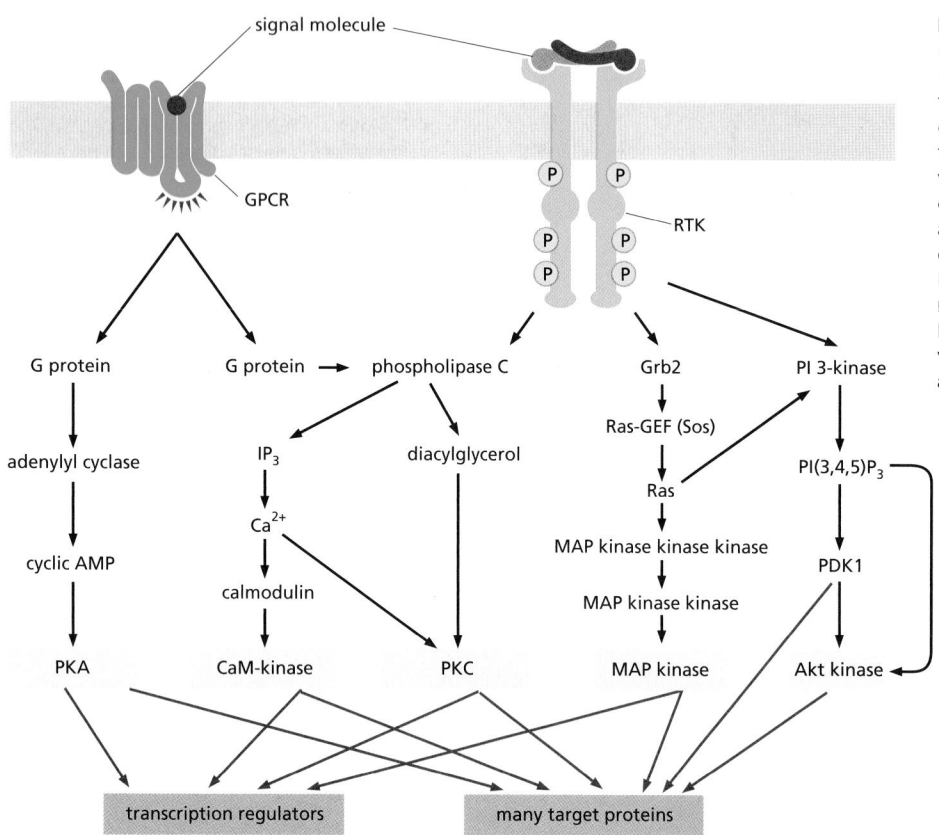

Figure 15–55 **Five parallel intracellular signaling pathways activated by GPCRs, RTKs, or both.** In this simplified example, the five kinases (shaded *yellow*) at the end of each signaling pathway phosphorylate target proteins (shaded *red*), many of which are phosphorylated by more than one of the kinases. The phospholipase C activated by the two types of receptors is different: GPCRs activate PLCβ, whereas RTKs activate PLCγ (not shown). Although not shown, some GPCRs can also activate Ras, but they do so independently of Grb2, via a Ras-GEF that is activated by Ca^{2+} and diacylglycerol.

Some Enzyme-Coupled Receptors Associate with Cytoplasmic Tyrosine Kinases

Many cell-surface receptors depend on tyrosine phosphorylation for their activity and yet lack a tyrosine kinase domain. These receptors act through **cytoplasmic tyrosine kinases**, which are associated with the receptors and phosphorylate various target proteins, often including the receptors themselves, when the receptors bind their ligand. These **tyrosine-kinase-associated receptors** thus function in much the same way as RTKs, except that their kinase domain is encoded by a separate gene and is noncovalently associated with the receptor polypeptide chain. A variety of receptor classes belong in this category, including the receptors for antigen and interleukins on lymphocytes (discussed in Chapter 24), integrins (discussed in Chapter 19), and receptors for various cytokines and some hormones. As with RTKs, many of these receptors are either preformed dimers or are cross-linked into dimers by ligand binding.

Some of these receptors depend on members of the largest family of mammalian cytoplasmic tyrosine kinases, the **Src family** (see Figures 3–10 and 3–64), which includes *Src, Yes, Fgr, Fyn, Lck, Lyn, Hck,* and *Blk*. These protein kinases all contain SH2 and SH3 domains and are located on the cytoplasmic side of the plasma membrane, held there partly by their interaction with transmembrane receptor proteins and partly by covalently attached lipid chains. Different family members are associated with different receptors and phosphorylate overlapping but distinct sets of target proteins. Lyn, Fyn, and Lck, for example, are each associated with different sets of receptors on lymphocytes. In each case, the kinase is activated when an extracellular ligand binds to the appropriate receptor protein. Src itself, as well as several other family members, can also bind to activated RTKs; in these cases, the receptor and cytoplasmic kinases mutually stimulate each other's catalytic activity, thereby strengthening and prolonging the signal (see Figure 15–51). There are even some G proteins (G_s and G_i) that can activate Src, which is one way that the activation of GPCRs can lead to tyrosine phosphorylation of intracellular signaling proteins and effector proteins.

Another type of cytoplasmic tyrosine kinase associates with *integrins*, the main receptors that cells use to bind to the extracellular matrix (discussed in Chapter 19). The binding of matrix components to integrins activates intracellular signaling pathways that influence the behavior of the cell. When integrins cluster at sites of matrix contact, they help trigger the assembly of cell–matrix junctions called *focal adhesions*. Among the many proteins recruited into these junctions is the cytoplasmic tyrosine kinase called **focal adhesion kinase** (**FAK**), which binds to the cytosolic tail of one of the integrin subunits with the assistance of other proteins. The clustered FAK molecules phosphorylate each other, creating phosphotyrosine docking sites where the Src kinase can bind. Src and FAK then phosphorylate each other and other proteins that assemble in the junction, including many of the signaling proteins used by RTKs. In this way, the two tyrosine kinases signal to the cell that it has adhered to a suitable substratum, where the cell can now survive, grow, divide, migrate, and so on.

The largest and most diverse class of receptors that rely on cytoplasmic tyrosine kinases to relay signals into the cell is the class of *cytokine receptors*, which we consider next.

Cytokine Receptors Activate the JAK–STAT Signaling Pathway

The large family of **cytokine receptors** includes receptors for many kinds of local mediators (collectively called *cytokines*), as well as receptors for some hormones, such as *growth hormone* and *prolactin* (**Movie 15.8**). These receptors are stably associated with cytoplasmic tyrosine kinases called **Janus kinases** (**JAKs**) (after the two-faced Roman god), which phosphorylate and activate transcription regulators called **STATs** (**s**ignal **t**ransducers and **a**ctivators of **t**ranscription). STAT proteins are located in the cytosol and are referred to as *latent transcription regulators* because they migrate into the nucleus and regulate gene transcription only after they are activated.

Although many intracellular signaling pathways lead from cell-surface receptors to the nucleus, where they alter gene transcription (see Figure 15–55), the **JAK–STAT signaling pathway** provides one of the more direct routes. Cytokine receptors are dimers or trimers and are stably associated with one or two of the four known JAKs (JAK1, JAK2, JAK3, and Tyk2). Cytokine binding alters the arrangement so as to bring two JAKs into close proximity so that they phosphorylate each other, thereby increasing the activity of their tyrosine kinase domains. The JAKs then phosphorylate tyrosines on the cytoplasmic tails of cytokine receptors, creating phosphotyrosine docking sites for STATs (**Figure 15–56**). Some adaptor proteins can also bind to some of these sites and couple cytokine receptors to the Ras–MAP-kinase signaling pathway discussed earlier, but these will not be discussed here.

There are at least six STATs in mammals. Each has an SH2 domain that performs two functions. First, it mediates the binding of the STAT protein to a phosphotyrosine docking site on an activated cytokine receptor. Once bound, the JAKs phosphorylate the STAT on tyrosines, causing the STAT to dissociate from the receptor. Second, the SH2 domain on the released STAT now mediates its binding to a phosphotyrosine on another STAT molecule, forming either a STAT homodimer or a heterodimer. The STAT dimer then translocates to the nucleus, where, in combination with other transcription regulatory proteins, it binds to a specific *cis*-regulatory sequence in various genes and stimulates their transcription (see Figure 15–56). In response to the hormone prolactin, for example, which stimulates breast cells to produce milk, activated STAT5 stimulates the transcription of genes that encode milk proteins. **Table 15–6** lists some of the more than 30 cytokines and hormones that activate the JAK–STAT pathway by binding to cytokine receptors.

Negative feedback regulates the responses mediated by the JAK–STAT pathway. In addition to activating genes that encode proteins mediating the cytokine-induced response, the STAT dimers can also activate genes that encode inhibitory proteins that help shut off the response. Some of these proteins bind to

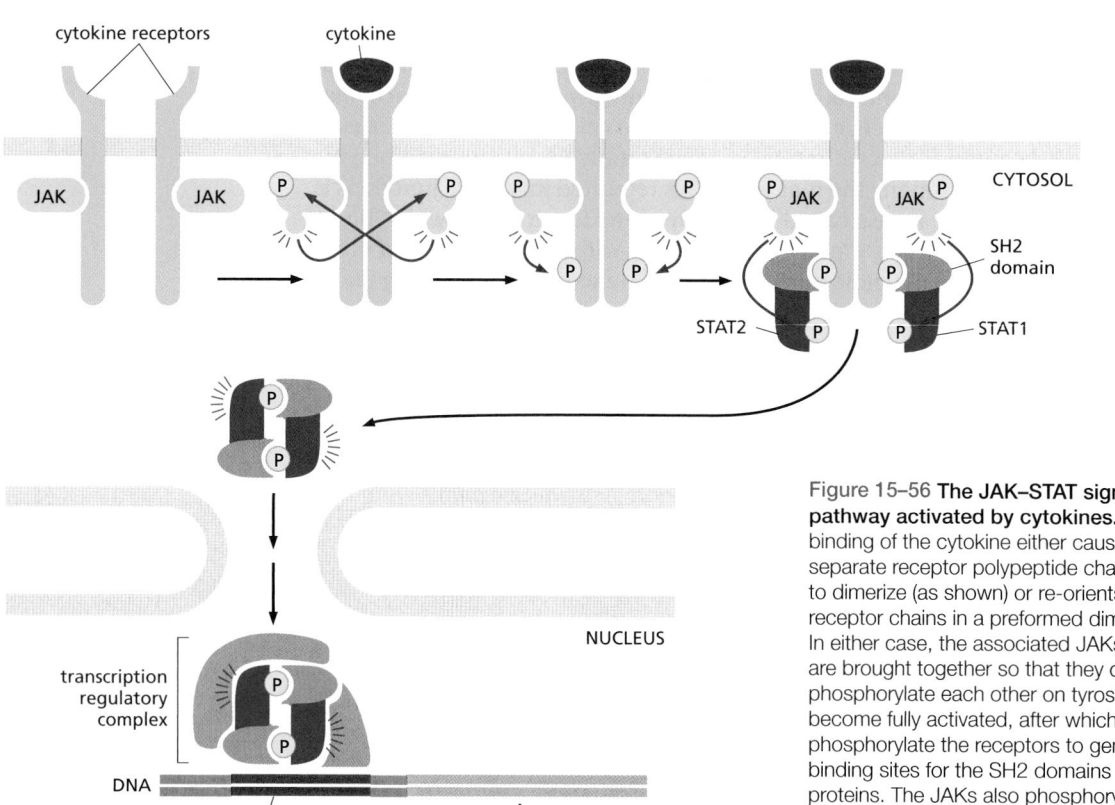

Figure 15–56 **The JAK–STAT signaling pathway activated by cytokines.** The binding of the cytokine either causes two separate receptor polypeptide chains to dimerize (as shown) or re-orients the receptor chains in a preformed dimer. In either case, the associated JAKs are brought together so that they can phosphorylate each other on tyrosines to become fully activated, after which they phosphorylate the receptors to generate binding sites for the SH2 domains of STAT proteins. The JAKs also phosphorylate the STAT proteins, which dissociate from the receptor to form dimers and enter the nucleus to control gene expression.

and inactivate phosphorylated JAKs and their associated phosphorylated receptors; others bind to phosphorylated STAT dimers and prevent them from binding to their DNA targets. Such negative feedback mechanisms, however, are not enough on their own to turn off the response. Inactivation of the activated JAKs and STATs requires dephosphorylation of their phosphotyrosines.

Protein Tyrosine Phosphatases Reverse Tyrosine Phosphorylations

In all signaling pathways that use tyrosine phosphorylation, the tyrosine phosphorylations are reversed by **protein tyrosine phosphatases**. These phosphatases are as important in the signaling process as the protein tyrosine kinases that add the phosphates. Whereas only a few types of *serine/threonine protein phosphatase*

TABLE 15–6 **Some Extracellular Signal Proteins That Act Through Cytokine Receptors and the JAK–STAT Signaling Pathway**

Signal protein	Receptor-associated JAKs	STATs activated	Some responses
Interferon-γ (IFNγ)	JAK1 and JAK2	STAT1	Activates macrophages
Interferon-α (IFNα)	Tyk2 and JAK2	STAT1 and STAT2	Increases cell resistance to viral infection
Erythropoietin	JAK2	STAT5	Stimulates production of erythrocytes
Prolactin	JAK1 and JAK2	STAT5	Stimulates milk production
Growth hormone	JAK2	STAT1 and STAT5	Stimulates growth by inducing IGF1 production
Granulocyte–Macrophage-Colony-Stimulating Factor (GMCSF)	JAK2	STAT5	Stimulates production of granulocytes and macrophages

catalytic subunits are responsible for removing phosphate groups from phosphorylated serines and threonines on proteins, there are about 100 protein tyrosine phosphatases encoded in the human genome, including some *dual-specificity phosphatases* that also dephosphorylate serines and threonines.

Like tyrosine kinases, the tyrosine phosphatases occur in both cytoplasmic and transmembrane forms. Unlike serine/threonine protein phosphatases, which generally have broad specificity, most tyrosine phosphatases display exquisite specificity for their substrates, removing phosphate groups from only selected phosphotyrosines on a subset of proteins. Together, these phosphatases ensure that tyrosine phosphorylations are short-lived and that the level of tyrosine phosphorylation in resting cells is very low. They do not, however, simply continuously reverse the effects of protein tyrosine kinases; they are often regulated to act only at the appropriate time and place.

Having discussed the crucial role of tyrosine phosphorylation and dephosphorylation in the intracellular signaling pathways activated by many enzyme-coupled receptors, we now turn to a class of enzyme-coupled receptors that rely on serine and threonine phosphorylation. These *receptor serine/threonine kinases* activate an even more direct signaling pathway to the nucleus than does the JAK–STAT pathway. They directly phosphorylate latent transcription regulators called *Smads*, which then translocate into the nucleus to control gene transcription.

Signal Proteins of the TGFβ Superfamily Act Through Receptor Serine/Threonine Kinases and Smads

The **transforming growth factor-β (TGFβ) superfamily** consists of a large number (33 in humans) of structurally related, secreted, dimeric proteins. They act either as hormones or, more commonly, as local mediators to regulate a wide range of biological functions in all animals. During development, they regulate pattern formation and influence various cell behaviors, including proliferation, specification and differentiation, extracellular matrix production, and cell death. In adults, they are involved in tissue repair and in immune regulation, as well as in many other processes. The superfamily consists of the TGFβ/*activin* family and the larger *bone morphogenetic protein* (*BMP*) family.

All of these proteins act through enzyme-coupled receptors that are single-pass transmembrane proteins with a serine/threonine kinase domain on the cytosolic side of the plasma membrane. There are two classes of these **receptor serine/threonine kinases**—*type I* and *type II*—which are structurally similar homodimers. Each member of the TGFβ superfamily binds to a characteristic combination of type-I and type-II receptor dimers, bringing the kinase domains together so that the type-II receptor can phosphorylate and activate the type-I receptor, forming an active tetrameric receptor complex.

Once activated, the receptor complex uses a strategy for rapidly relaying the signal to the nucleus that is very similar to the JAK–STAT strategy used by cytokine receptors. The activated type-I receptor directly binds and phosphorylates a latent transcription regulator of the **Smad family** (named after the first two proteins identified, Sma in *C. elegans* and Mad in *Drosophila*). Activated TGFβ/activin receptors phosphorylate Smad2 or Smad3, while activated BMP receptors phosphorylate Smad1, Smad5, or Smad8. Once one of these *receptor-activated Smads* (*R-Smads*) has been phosphorylated, it dissociates from the receptor and binds to Smad4 (called a *co-Smad*), which can form a complex with any of the five R-Smads. The Smad complex then translocates into the nucleus, where it associates with other transcription regulators and controls the transcription of specific target genes (**Figure 15–57**). Because the partner proteins in the nucleus vary depending on the cell type and state of the cell, the genes affected vary.

Activated TGFβ receptors and their bound ligand are endocytosed by two distinct routes, one leading to further activation and the other leading to inactivation. The activation route depends on clathrin-coated vesicles and leads to early endosomes (discussed in Chapter 13), where most of the Smad activation occurs. An anchoring protein called *SARA* (for *Smad anchor for receptor activation*) has

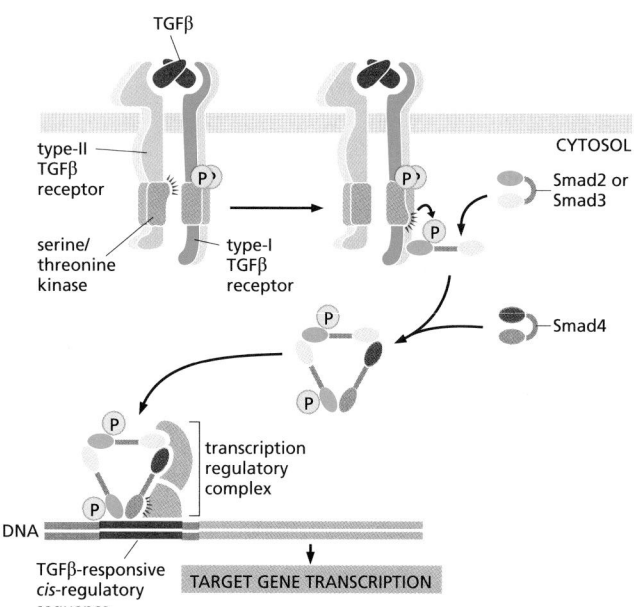

TGFβ

type-II
TGFβ
receptor

serine/
threonine
kinase

type-I
TGFβ
receptor

CYTOSOL

Smad2 or
Smad3

Smad4

transcription
regulatory
complex

DNA

TGFβ-responsive
cis-regulatory
sequence

TARGET GENE TRANSCRIPTION

Figure 15–57 The Smad-dependent signaling pathway activated by TGFβ. The TGFβ dimer promotes the assembly of a tetrameric receptor complex containing two copies each of the type-I and type-II receptors. The type-II receptors phosphorylate specific sites on the type-I receptors, thereby activating their kinase domains and leading to phosphorylation of R-Smads such as Smad2 and Smad3. Smads open up to expose a dimerization surface when they are phosphorylated, leading to the formation of a trimeric Smad complex containing two R-Smads and the co-Smad, Smad4. The phosphorylated Smad complex enters the nucleus and collaborates with other transcription regulators to control the transcription of specific target genes.

an important role in this pathway; it is concentrated in early endosomes and binds to both activated TGFβ receptors and Smads, increasing the efficiency of receptor-mediated Smad phosphorylation. The inactivation route depends on *caveolae* (discussed in Chapter 13) and leads to receptor ubiquitylation and degradation in proteasomes.

During the signaling response, the Smads shuttle continuously between the cytoplasm and the nucleus: they are dephosphorylated in the nucleus and exported to the cytoplasm, where they can be rephosphorylated by activated receptors. In this way, the effect exerted on the target genes reflects both the concentration of the extracellular signal and the time the signal continues to act on the cell-surface receptors (often several hours). Cells exposed to a morphogen at high concentration, or for a long time, or both, will switch on one set of genes, whereas cells receiving a lower or more transient exposure will switch on another set.

As in other signaling systems, negative feedback regulates the Smad pathway. Among the target genes activated by Smad complexes are those that encode *inhibitory Smads*, either Smad6 or Smad7. Smad7 (and possibly Smad6) binds to the cytosolic tail of the activated receptor and inhibits its signaling ability in at least three ways: (1) it competes with R-Smads for binding sites on the receptor, decreasing R-Smad phosphorylation; (2) it recruits a ubiquitin ligase called *Smurf*, which ubiquitylates the receptor, leading to receptor internalization and degradation (it is because Smurfs also ubiquitylate and promote the degradation of Smads that they are called *Smad ubiquitylation regulatory factors,* or Smurfs); and (3) it recruits a protein phosphatase that dephosphorylates and inactivates the receptor. In addition, the inhibitory Smads bind to the co-Smad, Smad4, and inhibit it, either by preventing its binding to R-Smads or by promoting its ubiquitylation and degradation.

Although receptor serine/threonine kinases operate mainly through the Smad pathway just described, they can also stimulate other intracellular signaling proteins such as MAP kinases and PI 3-kinase. Conversely, signaling proteins in other pathways can phosphorylate Smads and thereby influence signaling along the Smad pathway.

Summary

There are various classes of enzyme-coupled receptors, the most common of which are receptor tyrosine kinases (RTKs), tyrosine-kinase-associated receptors, and receptor serine/threonine kinases.

Ligand binding to RTKs causes their dimerization, which leads to activation of their kinase domains. These activated kinase domains phosphorylate multiple tyrosines on the receptors, producing a set of phosphotyrosines that serve as docking sites for a set of intracellular signaling proteins, which bind via their SH2 (or PTB) domains. One such signaling protein serves as an adaptor to couple some activated receptors to a Ras-GEF (Sos), which activates the monomeric GTPase Ras; Ras, in turn, activates a three-component MAP kinase signaling module, which relays the signal to the nucleus by phosphorylating transcription regulatory proteins. Another important signaling protein that can dock on activated RTKs is PI 3-kinase, which phosphorylates specific phosphoinositides to produce lipid docking sites in the plasma membrane for signaling proteins with phosphoinositide-binding PH domains, including the serine/threonine protein kinase Akt (PKB), which plays a key part in the control of cell survival and cell growth. Many receptor classes, including some RTKs, activate Rho family monomeric GTPases, which functionally couple the receptors to the cytoskeleton.

Tyrosine-kinase-associated receptors depend on various cytoplasmic tyrosine kinases for their action. These kinases include members of the Src family, which associate with many kinds of receptors, and the focal adhesion kinase (FAK), which associates with integrins at focal adhesions. The cytoplasmic tyrosine kinases then phosphorylate a variety of signaling proteins to relay the signal onward. The largest family of receptors in this class is the cytokine receptor family. When stimulated by ligand binding, these receptors activate JAK cytoplasmic tyrosine kinases, which phosphorylate STATs. The STATs then dimerize, translocate to the nucleus, and activate the transcription of specific genes. Receptor serine/threonine kinases, which are activated by signal proteins of the TGFβ superfamily, act similarly: they directly phosphorylate and activate Smads, which then oligomerize with another Smad, translocate to the nucleus, and regulate gene transcription.

ALTERNATIVE SIGNALING ROUTES IN GENE REGULATION

Major changes in the behavior of a cell tend to depend on changes in the expression of numerous genes. Thus, many extracellular signaling molecules carry out their effects, in whole or in part, by initiating signaling pathways that change the activities of transcription regulators. There are numerous examples of gene regulation in both GPCR and enzyme-coupled receptor pathways (see Figures 15–27 and 15–49). In this section, we describe some of the less common signaling mechanisms by which gene expression can be controlled. We begin with several pathways that depend on *regulated proteolysis* to control the activity and location of latent transcription regulators. We then turn to a class of extracellular signal molecules that do not employ cell-surface receptors but enter the cell and interact directly with transcription regulators to perform their functions. Finally, we briefly discuss some of the mechanisms by which gene expression is controlled by the *circadian rhythm*: the daily cycle of light and dark.

The Receptor Notch Is a Latent Transcription Regulatory Protein

Signaling through the **Notch** receptor protein is used widely in animal development. As discussed in Chapter 22, it has a general role in controlling cell fate choices and regulating pattern formation during the development of most tissues, as well as in the continual renewal of tissues such as the lining of the gut. It is best known, however, for its role in the production of *Drosophila* neural cells, which usually arise as isolated single cells within an epithelial sheet of precursor cells. During this process, when a precursor cell commits to becoming a neural cell, it signals to its immediate neighbors not to do the same; the inhibited cells develop into epidermal cells instead. This process, called *lateral inhibition*, depends on a contact-dependent signaling mechanism that is activated by a single-pass transmembrane signal protein called **Delta**, displayed on the surface of the future neural cell. By binding to the Notch receptor protein on a neighboring cell, Delta

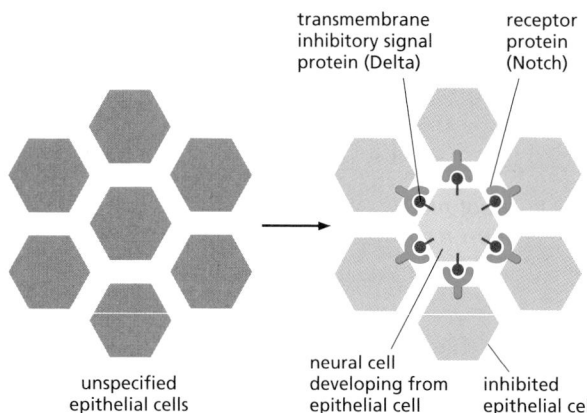

transmembrane
inhibitory signal
protein (Delta)

receptor
protein
(Notch)

unspecified
epithelial cells

neural cell
developing from
epithelial cell

inhibited
epithelial cell

Figure 15–58 Lateral inhibition mediated by Notch and Delta during neural cell development in *Drosophila*. When individual cells in the epithelium begin to develop as neural cells, they signal to their neighbors not to do the same. This inhibitory, contact-dependent signaling is mediated by the ligand Delta, which appears on the surface of the future neural cell and binds to Notch receptor proteins on the neighboring cells. In many tissues, all the cells in a cluster initially express both Delta and Notch, and a competition occurs, with one cell emerging as winner, expressing Delta strongly and inhibiting its neighbors from doing likewise. In other cases, additional factors interact with Delta or Notch to make some cells susceptible to the lateral inhibition signal and others unresponsive to it.

signals to the neighbor not to become neural (**Figure 15–58**). When this signaling process is defective, a huge excess of neural cells is produced at the expense of epidermal cells, which is lethal.

Notch is a single-pass transmembrane protein that requires proteolytic processing to function. It acts as a latent transcription regulator and provides the simplest and most direct signaling pathway known from a cell-surface receptor to the nucleus. When activated by the binding of Delta on another cell, a plasma-membrane-bound protease cleaves off the cytoplasmic tail of Notch, and the released tail translocates into the nucleus to activate the transcription of a set of Notch-response genes. The Notch tail fragment acts by binding to a DNA-binding protein, converting it from a transcriptional repressor into a transcriptional activator.

The Notch receptor undergoes three successive proteolytic cleavage steps, but only the last two depend on Delta binding. As part of its normal biosynthesis, it is cleaved in the Golgi apparatus to form a heterodimer, which is then transported to the cell surface as the mature receptor. The binding of Delta to Notch induces a second cleavage in the extracellular domain, mediated by an extracellular protease. A final cleavage quickly follows, cutting free the cytoplasmic tail of the activated receptor (**Figure 15–59**). Note that, unlike most receptors, the activation of Notch is irreversible; once activated by ligand binding, the protein cannot be used again.

This final cleavage of the Notch tail occurs just within the transmembrane segment, and it is mediated by a protease complex called γ-*secretase*, which is also responsible for the intramembrane cleavage of various other proteins. One of its essential subunits is *Presenilin*, so called because mutations in the gene encoding it are a frequent cause of early-onset familial Alzheimer's disease, a form of presenile dementia. The protease complex is thought to contribute to this and other forms of Alzheimer's disease by generating extracellular peptide fragments from a transmembrane neuronal protein; the fragments accumulate in excessive amounts and form aggregates of misfolded protein called amyloid plaques, which may injure nerve cells and contribute to their degeneration and loss.

Both Notch and Delta are glycoproteins, and their interaction is regulated by the glycosylation of Notch. The *Fringe family* of glycosyl transferases, in particular, adds extra sugars to the *O*-linked oligosaccharide (discussed in Chapter 13) on Notch, which alters the specificity of Notch for its ligands. This has provided the first example of the modulation of ligand–receptor signaling by differential receptor glycosylation.

Wnt Proteins Bind to Frizzled Receptors and Inhibit the Degradation of β-Catenin

Wnt proteins are secreted signal molecules that act as local mediators and morphogens to control many aspects of development in all animals that have been studied. They were discovered independently in flies and in mice: in *Drosophila*, the *Wingless* (*Wg*) gene originally came to light because of its role as a morphogen

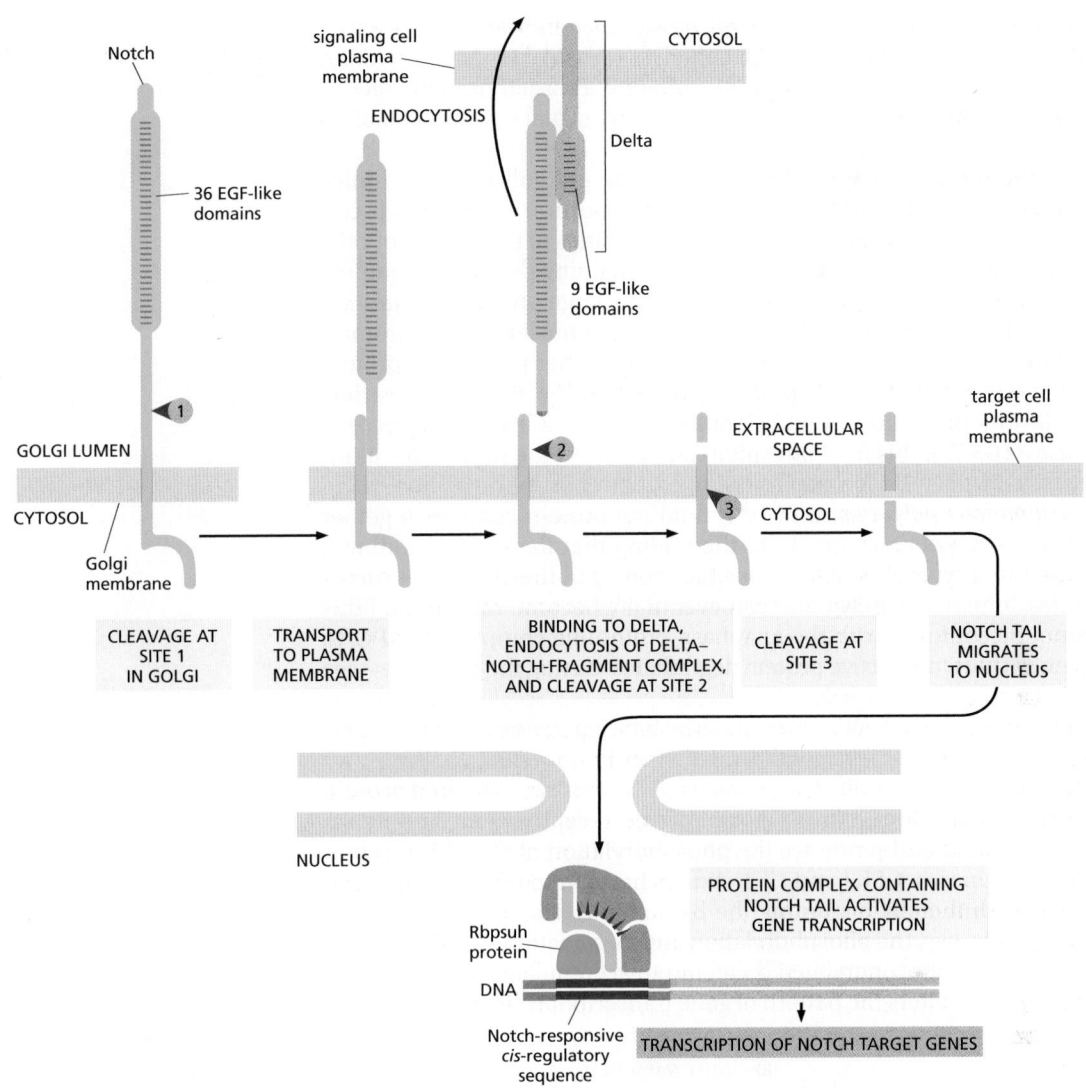

Figure 15–59 **The processing and activation of Notch by proteolytic cleavage.** The *numbered red arrowheads* indicate the sites of proteolytic cleavage. The first proteolytic processing step occurs within the *trans* Golgi network to generate the mature heterodimeric Notch receptor that is then displayed on the cell surface. The binding to Delta on a neighboring cell triggers the next two proteolytic steps: the complex of Delta and the Notch fragment to which it is bound is endocytosed by the Delta-expressing cell, exposing the extracellular cleavage site in the transmembrane Notch subunit. Note that Notch and Delta interact through their repeated EGF-like domains. The released Notch tail migrates into the nucleus, where it binds to the Rbpsuh protein, which it converts from a transcriptional repressor to a transcriptional activator.

in wing development, while in mice, the *Int1* gene was found because it promoted the development of breast tumors when activated by the integration of a virus next to it. Both of these genes encode Wnt proteins. Wnts are unusual as secreted proteins in that they have a fatty acid chain covalently attached to their N-terminus, which increases their binding to cell surfaces. There are 19 Wnts in humans, each having distinct, but often overlapping, functions.

Wnts can activate at least two types of intracellular signaling pathways. Our primary focus here is the *Wnt/β-catenin pathway* (also known as the *canonical Wnt pathway*), which is centered on the latent transcription regulator *β-catenin*. A second pathway, called the *planar polarity pathway*, coordinates the polarization of cells in the plane of a developing epithelium and depends on Rho family GTPases. Both of these pathways begin with the binding of Wnts to **Frizzled** family cell-surface receptors, which are seven-pass transmembrane proteins that resemble GPCRs in structure but do not generally work through the activation of G proteins. Instead, when activated by Wnt binding, Frizzled proteins recruit the scaffold protein **Dishevelled**, which helps relay the signal to other signaling molecules.

The **Wnt/β-catenin pathway** acts by regulating the proteolysis of the multifunctional protein **β-catenin** (or *Armadillo* in flies). A portion of the cell's β-catenin is located at cell–cell junctions and thereby contributes to the control of cell–cell adhesion (discussed in Chapter 19), while the remaining β-catenin is rapidly degraded in the cytoplasm. Degradation depends on a large protein *degradation complex*, which binds β-catenin and keeps it out of the nucleus while promoting its degradation. The complex contains at least four other proteins: a protein kinase called *casein kinase 1* (*CK1*) phosphorylates the β-catenin on a serine, priming it for further phosphorylation by another protein kinase called *glycogen synthase kinase 3* (*GSK3*); this final phosphorylation marks the protein for ubiquitylation and rapid degradation in proteasomes. Two scaffold proteins called *axin* and *Adenomatous polyposis coli* (*APC*) hold the protein complex together (**Figure 15–60A**). APC gets its name from the finding that the gene encoding it is often mutated in a type of benign tumor (adenoma) of the colon; the tumor projects into the lumen as a polyp and can eventually become malignant. (This APC should not be confused with the anaphase-promoting complex, or APC/C, that plays a central part in selective protein degradation during the cell cycle—see Figure 17–15A.)

Wnt proteins regulate β-catenin proteolysis by binding to both a Frizzled protein and a co-receptor that is related to the low-density lipoprotein (LDL) receptor (discussed in Chapter 13) and is therefore called an **LDL-receptor-related protein** (**LRP**). In a poorly understood process, the activated receptor complex recruits the Dishevelled scaffold and promotes the phosphorylation of the LRP receptor by the two protein kinases, GSK3 and CK1. Axin is brought to the receptor complex and inactivated, thereby disrupting the β-catenin degradation complex in the cytoplasm. In this way, the phosphorylation and degradation of β-catenin are prevented, enabling unphosphorylated β-catenin to accumulate and translocate to the nucleus, where it alters the pattern of gene transcription (Figure 15–60B).

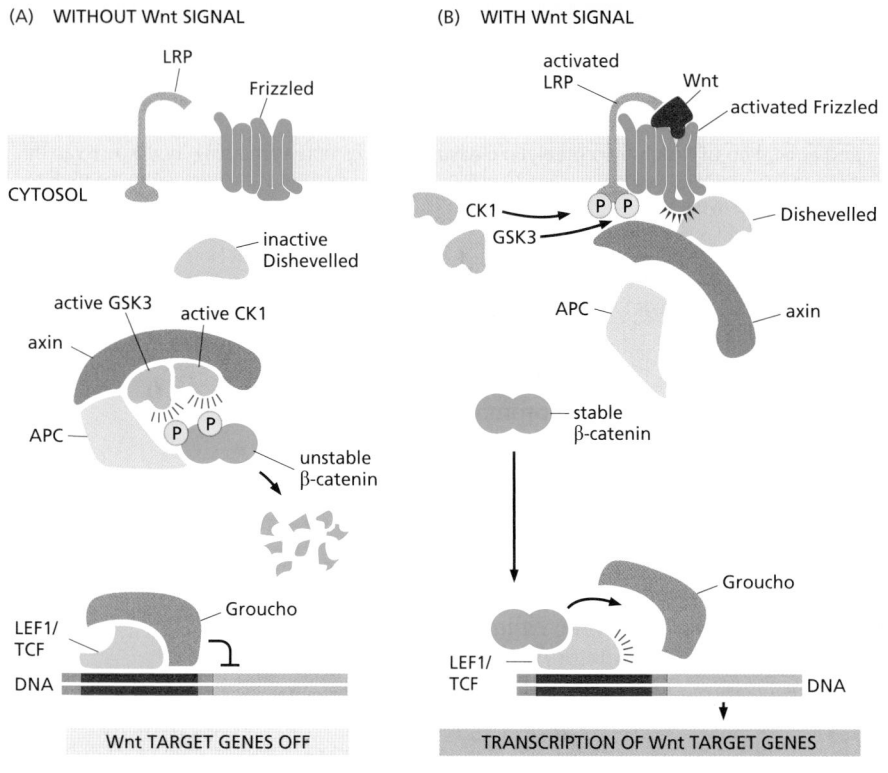

Figure 15–60 The Wnt/β-catenin signaling pathway. (A) In the absence of a Wnt signal, β-catenin that is not bound to cell–cell adherens junctions (not shown) interacts with a degradation complex containing APC, axin, GSK3, and CK1. In this complex, β-catenin is phosphorylated by CK1 and then by GSK3, triggering its ubiquitylation and degradation in proteasomes. Wnt-responsive genes are kept inactive by the Groucho co-repressor protein bound to the transcription regulator LEF1/TCF. (B) Wnt binding to Frizzled and LRP clusters the two co-receptors together, and the cytosolic tail of LRP is phosphorylated by GSK3 and then by CK1. Axin binds to the phosphorylated LRP and is inactivated and/or degraded, resulting in disassembly of the degradation complex. The phosphorylation of β-catenin is thereby prevented, and unphosphorylated β-catenin accumulates and translocates to the nucleus, where it binds to LEF1/TCF, displaces the co-repressor Groucho, and acts as a coactivator to stimulate the transcription of Wnt target genes. The scaffold protein Dishevelled is required for the signaling pathway to operate; it binds to Frizzled and becomes phosphorylated (not shown), but its precise role is unknown.

In the absence of Wnt signaling, Wnt-responsive genes are kept silent by an inhibitory complex of transcription regulatory proteins. The complex includes proteins of the *LEF1/TCF* family bound to a co-repressor protein of the *Groucho* family (see Figure 15–60A). In response to a Wnt signal, β-catenin enters the nucleus and binds to the LEF1/TCF proteins, displacing Groucho. The β-catenin now functions as a coactivator, inducing the transcription of the Wnt target genes (see Figure 15–60B). Thus, as in the case of Notch signaling, Wnt/β-catenin signaling triggers a switch from transcriptional repression to transcriptional activation.

Among the genes activated by β-catenin is *Myc*, which encodes a protein (Myc) that is an important regulator of cell growth and proliferation (discussed in Chapter 17). Mutations of the *Apc* gene occur in 80% of human colon cancers (discussed in Chapter 20). These mutations inhibit the protein's ability to bind β-catenin, so that β-catenin accumulates in the nucleus and stimulates the transcription of *c-Myc* and other Wnt target genes, even in the absence of Wnt signaling. The resulting uncontrolled cell growth and proliferation promote the development of cancer.

Various secreted inhibitory proteins regulate Wnt signaling in development. Some bind to the LRP receptors and promote their down-regulation, whereas others compete with Frizzled receptors for secreted Wnts. In *Drosophila* at least, Wnts activate negative feedback loops, in which Wnt target genes encode proteins that help shut the response off; some of these proteins inhibit Dishevelled, and others are secreted inhibitors.

Hedgehog Proteins Bind to Patched, Relieving Its Inhibition of Smoothened

Hedgehog proteins and Wnt proteins act in similar ways. Both are secreted signal molecules, which act as local mediators and morphogens in many developing invertebrate and vertebrate tissues. Both proteins are modified by covalently attached lipids, depend on secreted or cell-surface-bound heparan sulfate proteoglycans (discussed in Chapter 19) for their action, and activate latent transcription regulators by inhibiting their degradation. They both trigger a switch from transcriptional repression to transcriptional activation, and excessive signaling along either pathway in adult cells can lead to cancer. They even use some of the same intracellular signaling proteins and sometimes collaborate to mediate a response.

The **Hedgehog proteins** were discovered in *Drosophila*, where this protein family has only one member. Mutation of the *Hedgehog* gene produces a larva covered with spiky processes (denticles), like a hedgehog. At least three genes encode Hedgehog proteins in vertebrates—*Sonic, Desert,* and *Indian hedgehog*. The active forms of all Hedgehog proteins are covalently coupled to cholesterol, as well as to a fatty acid chain. The cholesterol is added during an unusual processing step, in which a precursor protein cleaves itself to produce a smaller, cholesterol-containing signal protein. Most of what we know about the Hedgehog signaling pathway came initially from genetic studies in flies, and it is the fly pathway that we summarize here.

The effects of Hedgehog are mediated by a latent transcription regulator called **Cubitus interruptus (Ci)**, the regulation of which is reminiscent of the regulation of β-catenin by Wnts. In the absence of a Hedgehog signal, Ci is ubiquitylated and proteolytically cleaved in proteasomes. Instead of being completely degraded, however, Ci is processed to form a smaller fragment, which accumulates in the nucleus, where it acts as a transcriptional repressor, helping to keep Hedgehog-responsive genes silent. The proteolytic processing of the Ci protein depends on its phosphorylation by three protein kinases—PKA and two kinases also used in the Wnt pathway, namely GSK3 and CK1. As in the Wnt pathway, the proteolytic processing occurs in a multiprotein complex. The complex includes the protein kinase *Fused* and a scaffold protein *Costal2*, which stably associates with Ci, recruits the three other kinases, and binds the complex to microtubules, thereby keeping unprocessed Ci out of the nucleus (**Figure 15–61A**).

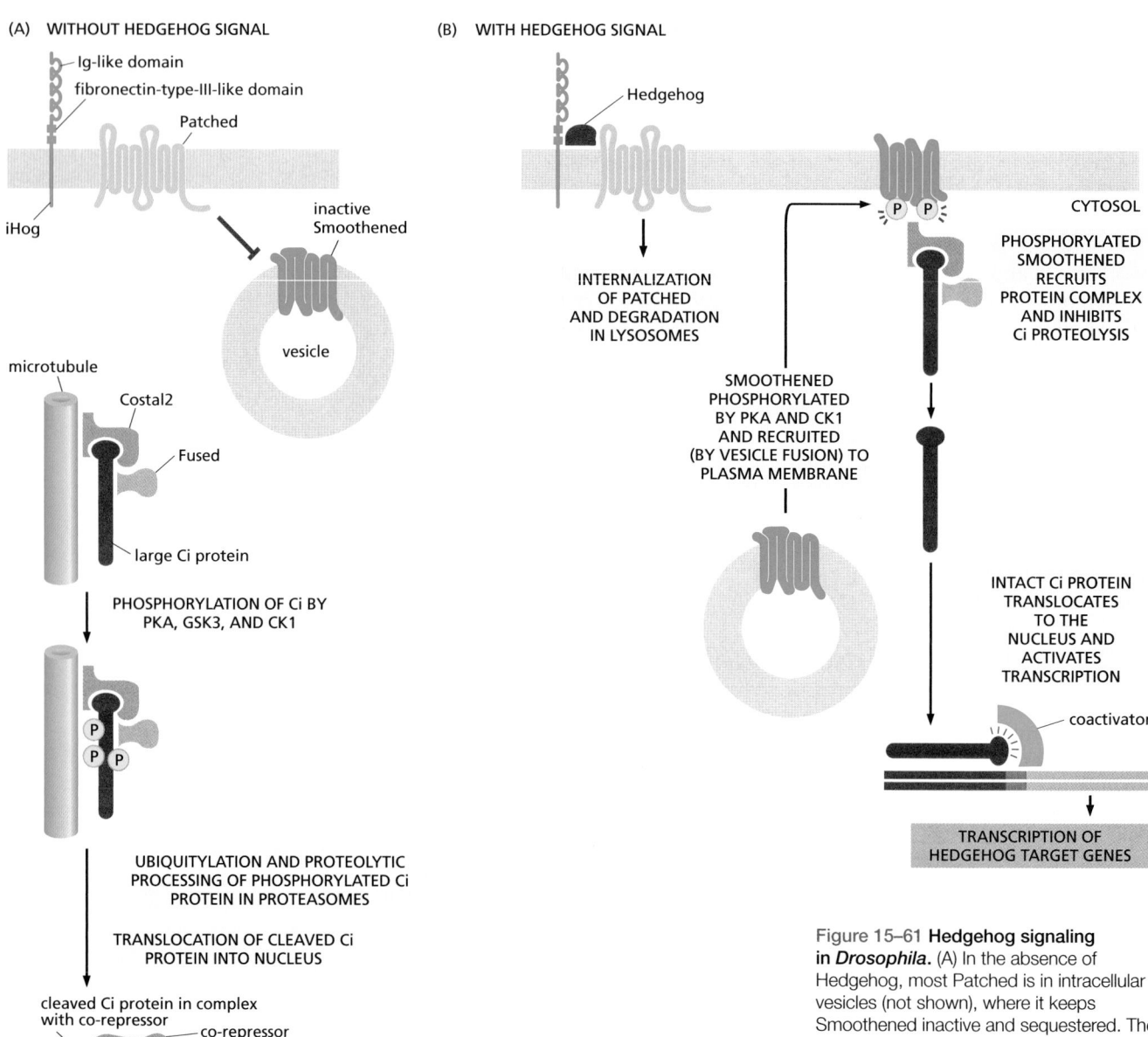

(A) WITHOUT HEDGEHOG SIGNAL

Ig-like domain
fibronectin-type-III-like domain
Patched
iHog
inactive Smoothened
vesicle

microtubule
Costal2
Fused
large Ci protein

PHOSPHORYLATION OF Ci BY PKA, GSK3, AND CK1

P P P

UBIQUITYLATION AND PROTEOLYTIC PROCESSING OF PHOSPHORYLATED Ci PROTEIN IN PROTEASOMES

TRANSLOCATION OF CLEAVED Ci PROTEIN INTO NUCLEUS

cleaved Ci protein in complex with co-repressor
co-repressor

HEDGEHOG TARGET GENES OFF

(B) WITH HEDGEHOG SIGNAL

Hedgehog

INTERNALIZATION OF PATCHED AND DEGRADATION IN LYSOSOMES

SMOOTHENED PHOSPHORYLATED BY PKA AND CK1 AND RECRUITED (BY VESICLE FUSION) TO PLASMA MEMBRANE

P P

CYTOSOL

PHOSPHORYLATED SMOOTHENED RECRUITS PROTEIN COMPLEX AND INHIBITS Ci PROTEOLYSIS

INTACT Ci PROTEIN TRANSLOCATES TO THE NUCLEUS AND ACTIVATES TRANSCRIPTION

coactivator

TRANSCRIPTION OF HEDGEHOG TARGET GENES

Figure 15–61 Hedgehog signaling in *Drosophila*. (A) In the absence of Hedgehog, most Patched is in intracellular vesicles (not shown), where it keeps Smoothened inactive and sequestered. The Ci protein is bound in a cytosolic protein degradation complex, which includes the protein kinase Fused and the scaffold protein Costal2. Costal2 recruits three other protein kinases (PKA, GSK3, and CK1; not shown), which phosphorylate Ci. Phosphorylated Ci is ubiquitylated and then cleaved in proteasomes (not shown) to form a transcriptional repressor, which accumulates in the nucleus to help keep Hedgehog target genes inactive.
(B) Hedgehog binding to iHog and Patched removes the inhibition of Smoothened by Patched. Smoothened is phosphorylated by PKA and CK1 and translocates to the plasma membrane, where it recruits the complex containing Fused, Costal2, and Ci. Costal2 releases unprocessed Ci, which accumulates in the nucleus and activates the transcription of Hedgehog target genes. Many details in the pathway are poorly understood, including the role of Fused.

Hedgehog functions by blocking the proteolytic processing of Ci, thereby changing it into a transcriptional activator. It does this by a convoluted signaling process that depends on three transmembrane proteins: Patched, iHog, and Smoothened. **Patched** is predicted to cross the plasma membrane 12 times, and, although much of it is in intracellular vesicles, some is on the cell surface where it can bind the Hedgehog protein. **iHog** is also on the cell surface and is thought to serve as a co-receptor for Hedgehog. **Smoothened** is a seven-pass transmembrane protein with a structure very similar to a GPCR, but it does not seem to act as a Hedgehog receptor or even as an activator of G proteins; it is controlled by Patched and iHog.

In the absence of a Hedgehog signal, Patched employs an unknown mechanism to keep Smoothened sequestered and inactive in intracellular vesicles (see Figure 15–61A). The binding of Hedgehog to iHog and Patched inhibits the activity of Patched and induces its endocytosis and degradation. The result is that Smoothened is liberated from inhibition and translocates to the plasma membrane, where it recruits the protein complex containing Ci, Fused, and Costal2.

Costal2 is no longer able to bind the other three kinases, and so Ci is no longer cleaved and can now enter the nucleus and activate the transcription of Hedgehog target genes (Figure 15–61B). Among the genes activated by Ci is *Patched* itself; the resulting increase in Patched protein on the cell surface inhibits further Hedgehog signaling—providing another example of negative feedback.

Many gaps remain in our understanding of the Hedgehog signaling pathway. It is not known, for example, how Patched keeps Smoothened inactive and intracellular. As the structure of Patched resembles a transmembrane transporter protein, it has been proposed that it may transport a small molecule into the cell that keeps Smoothened sequestered in vesicles.

Even less is known about the more complex Hedgehog pathway in vertebrate cells. In addition to there being at least three types of vertebrate Hedgehog proteins, there are three Ci-like transcription regulator proteins (*Gli1*, *Gli2*, and *Gli3*) downstream of Smoothened. Gli2 and Gli3 are most similar to Ci in structure and function, and Gli3 has been shown to undergo proteolytic processing like Ci and to act as either a transcriptional repressor or a transcriptional activator. Moreover, in vertebrates, Smoothened, upon activation, becomes localized to the surface of the primary cilium (discussed in Chapter 16), where the Gli proteins are also concentrated, thereby increasing the speed and efficiency of signaling.

Hedgehog signaling can promote cell proliferation, and excessive Hedgehog signaling can lead to cancer. Inactivating mutations in one of the two human *Patched* genes, for example, which lead to excessive Hedgehog signaling, occur frequently in *basal cell carcinoma* of the skin, the most common form of cancer in Caucasians. A small molecule called *cyclopamine*, made by a meadow lily, is being used to treat cancers associated with excessive Hedgehog signaling. It blocks Hedgehog signaling by binding tightly to Smoothened and inhibiting its activity. It was originally identified because it causes severe developmental defects in the progeny of sheep grazing on such lilies; these include the presence of a single central eye (a condition called *cyclopia*), which is also seen in mice that are deficient in Hedgehog signaling.

Many Stressful and Inflammatory Stimuli Act Through an NFκB-Dependent Signaling Pathway

The **NFκB proteins** are latent transcription regulators that are present in most animal cells and are central to many stressful, inflammatory, and innate immune responses. These responses occur as a reaction to infection or injury and help protect stressed multicellular organisms and their cells (discussed in Chapter 24). An excessive or inappropriate inflammatory response in animals can also damage tissue and cause severe pain, and chronic inflammation can lead to cancer; as in the case of Wnt and Hedgehog signaling, excessive NFκB signaling is found in a number of human cancers. NFκB proteins also have important roles during normal animal development: the *Drosophila* NFκB family member *Dorsal*, for example, has a crucial role in specifying the dorsal–ventral axis of the developing fly embryo (discussed in Chapter 22).

Various cell-surface receptors activate the NFκB signaling pathway in animal cells. *Toll receptors* in *Drosophila* and *Toll-like receptors* in vertebrates, for example, recognize pathogens and activate this pathway in triggering innate immune responses (discussed in Chapter 24). The receptors for *tumor necrosis factor α* (*TNFα*) and *interleukin-1* (*IL1*), which are vertebrate cytokines especially important in inducing inflammatory responses, also activate this signaling pathway. The Toll, Toll-like, and IL1 receptors belong to the same family of proteins, whereas TNF receptors belong to a different family; all of them, however, act in similar ways to activate NFκB. When activated, they trigger a multiprotein ubiquitylation and phosphorylation cascade that releases NFκB from an inhibitory protein complex, so that it can translocate to the nucleus and turn on the transcription of hundreds of genes that participate in inflammatory and innate immune responses.

There are five NFκB proteins in mammals (*RelA*, *RelB*, *c-Rel*, *NFκB1*, and *NFκB2*), and they form a variety of homodimers and heterodimers, each of which

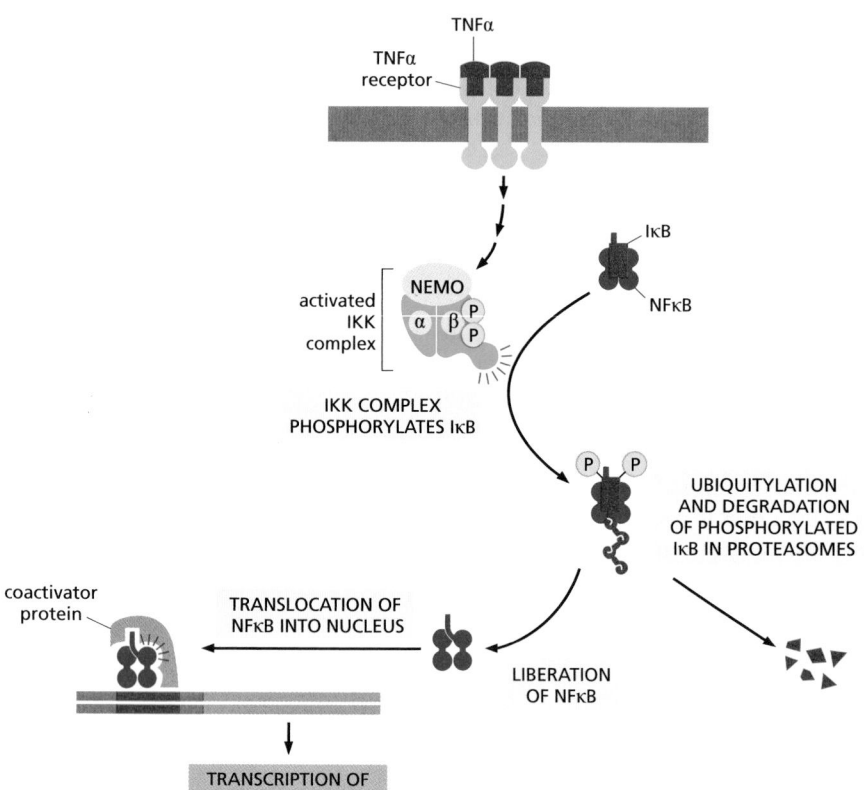

Figure 15–62 **The activation of the NFκB pathway by TNFα.** Both TNFα and its receptors are trimers. The binding of TNFα causes a rearrangement of the clustered cytosolic tails of the receptors, which now recruit various signaling proteins, resulting in the activation of a protein kinase that phosphorylates and activates IκB kinase kinase (IKK). IKK is a heterotrimer composed of two kinase subunits (IKKα and IKKβ) and a regulatory subunit called NEMO. IKKβ then phosphorylates IκB on two serines, which marks the protein for ubiquitylation and degradation in proteasomes. The released NFκB translocates into the nucleus, where, in collaboration with coactivator proteins, it stimulates the transcription of its target genes.

activates its own characteristic set of genes. Inhibitory proteins called **IκB** bind tightly to the dimers and hold them in an inactive state within the cytoplasm of unstimulated cells. There are three major IκB proteins in mammals (IκB α, β, and ε), and the signals that release NFκB dimers do so by triggering a signaling pathway that leads to the phosphorylation, ubiquitylation, and consequent degradation of the IκB proteins (**Figure 15–62**).

Among the genes activated by the released NFκB is the gene that encodes IκBα. This activation leads to increased synthesis of IκBα protein, which binds to NFκB and inactivates it, creating a negative feedback loop (**Figure 15–63A**). Experiments on TNFα-induced responses, as well as computer modeling studies of the responses, indicate that the negative feedback produces two types of NFκB responses, depending on the duration of the TNFα stimulus; importantly, the two types of responses induce different patterns of gene expression (Figure 15–63B, C, and D). The negative feedback through IκBα is required for both types of responses: in cells deficient in IκBα, even a short exposure to TNFα induces a sustained activation of NFκB, without oscillations, and all of the NFκB-responsive genes are activated.

Thus far, we have focused on the mechanisms by which extracellular signal molecules use cell-surface receptors to initiate changes in gene expression. We now turn to a class of extracellular signals that bypasses the plasma membrane entirely and controls, in the most direct way possible, transcription regulatory proteins inside the cell.

Nuclear Receptors Are Ligand-Modulated Transcription Regulators

Various small, hydrophobic signal molecules diffuse directly across the plasma membrane of target cells and bind to intracellular receptors that are transcription regulators. These signal molecules include steroid hormones, thyroid hormones, retinoids, and vitamin D. Although they differ greatly from one another in both

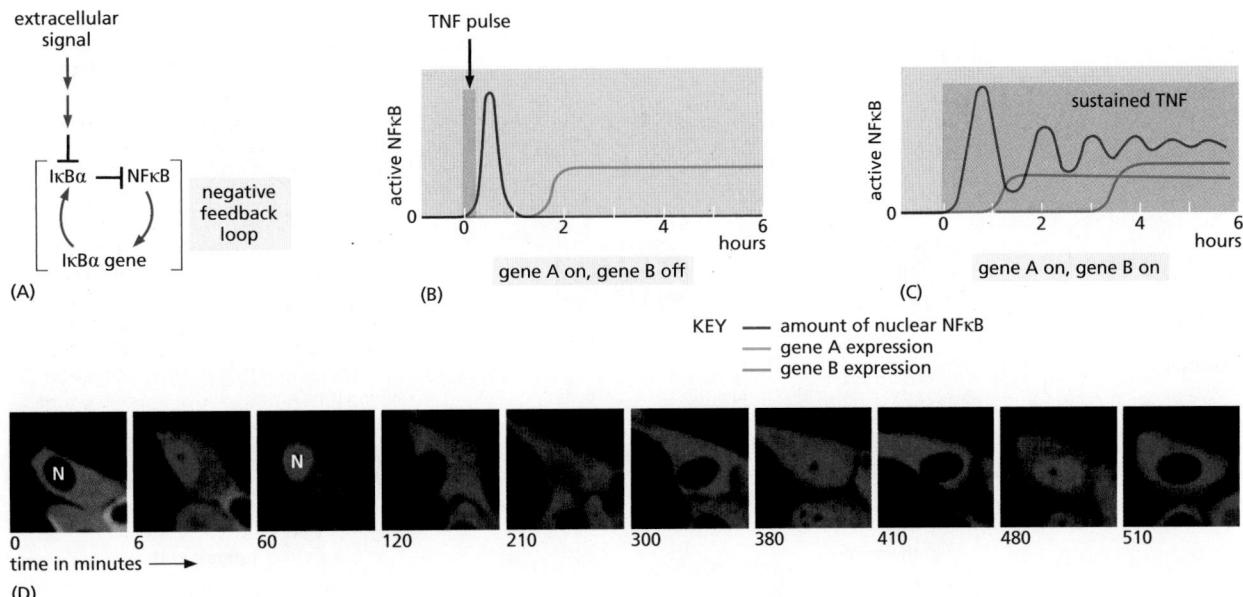

Figure 15–63 Negative feedback in the NFκB signaling pathway induces oscillations in NFκB activation. (A) Drawing showing how activated NFκB stimulates the transcription of the IκBα gene, the protein product of which acts back in the cytoplasm to sequester and inhibit NFκB there; if the stimulus is persistent, the newly made IκBα protein will then be ubiquitylated and degraded, liberating active NFκB again so that it can return to the nucleus and activate transcription (see Figure 15–62). (B) A short exposure to TNFα produces a single, short pulse of NFκB activation, beginning within minutes and ending by 1 hour. This response turns on the transcription of gene A but not gene B. (C) A sustained exposure to TNFα for the entire 6 hours of the experiment produces oscillations in NFκB activation that damp down over time. This response turns on the transcription of both genes; gene B turns on only after several hours, indicating that gene B transcription requires prolonged activation of NFκB, for reasons that are not understood. (D) These time-lapse confocal fluorescence micrographs from a different study of TNFα stimulation show the oscillations of NFκB in a cultured cell, as indicated by its periodic movement into the nucleus (N) of a fusion protein composed of NFκB fused to a red fluorescent protein. In the cell at the center of the micrographs, NFκB is active and in the nucleus at 6, 60, 210, 380, and 480 minutes, but it is exclusively in the cytoplasm at 0, 120, 300, 410, and 510 minutes. (A–C, based on data from A. Hoffmann et al., *Science* 298:1241–1245, 2002, and adapted from A.Y. Ting and D. Endy, *Science* 298:1189–1190, 2002; D, from D.E. Nelson et al., *Science* 306:704–708, 2004. All with permission from AAAS.)

chemical structure (**Figure 15–64**) and function, they all act by a similar mechanism. They bind to their respective intracellular receptor proteins and alter the ability of these proteins to control the transcription of specific genes. Thus, these proteins serve both as intracellular receptors and as intracellular effectors for the signal.

The receptors are all structurally related, being part of the very large **nuclear receptor superfamily**. Many family members have been identified by DNA sequencing only, and their ligand is not yet known; they are therefore referred to as *orphan nuclear receptors*, and they make up large fractions of the nuclear receptors encoded in the genomes of humans, *Drosophila*, and the nematode *C. elegans*. Some mammalian nuclear receptors are regulated by intracellular metabolites rather than by secreted signal molecules; the *peroxisome proliferation-activated receptors* (*PPARs*), for example, bind intracellular lipid metabolites and regulate the transcription of genes involved in lipid metabolism and fat-cell differentiation. It seems likely that the nuclear receptors for hormones evolved from such receptors for intracellular metabolites, which would help explain their intracellular location.

Steroid hormones—which include cortisol, the steroid sex hormones, vitamin D (in vertebrates), and the molting hormone *ecdysone* (in insects)—are all made from cholesterol. *Cortisol* is produced in the cortex of the adrenal glands and influences the metabolism of many types of cells. The *steroid sex hormones* are made in the testes and ovaries and are responsible for the secondary sex characteristics that distinguish males from females. *Vitamin D* is synthesized in the skin in response to sunlight; after it has been converted to its active form in

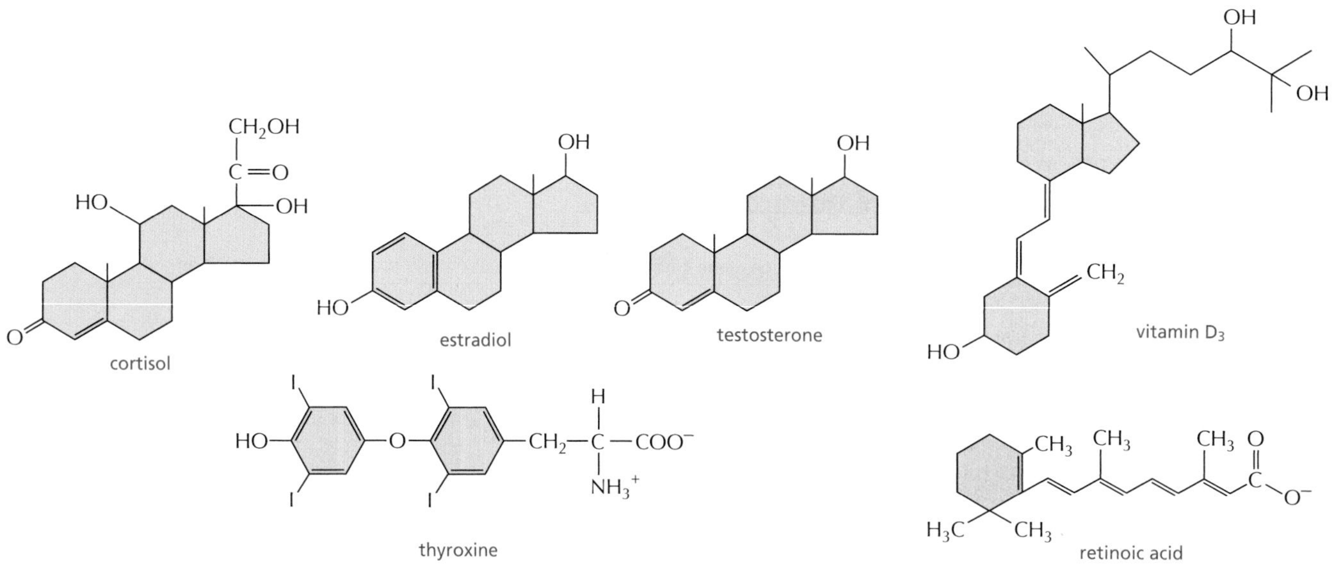

Figure 15–64 **Some signal molecules that bind to intracellular receptors.** Note that all of them are small and hydrophobic. The active, hydroxylated form of vitamin D_3 is shown. Estradiol and testosterone are steroid sex hormones.

the liver or kidneys, it regulates Ca^{2+} metabolism, promoting Ca^{2+} uptake in the gut and reducing its excretion in the kidneys. The *thyroid hormones*, which are made from the amino acid tyrosine, act to increase the metabolic rate of many cell types, while the *retinoids*, such as retinoic acid, are made from vitamin A and have important roles as local mediators in vertebrate development. Although all of these signal molecules are relatively insoluble in water, they are made soluble for transport in the bloodstream and other extracellular fluids by binding to specific carrier proteins, from which they dissociate before entering a target cell (see Figure 15–3B).

The nuclear receptors bind to specific DNA sequences adjacent to the genes that the ligand regulates. Some of the receptors, such as those for cortisol, are located primarily in the cytosol and enter the nucleus only after ligand binding; others, such as the thyroid and retinoid receptors, are bound to DNA in the nucleus even in the absence of ligand. In either case, the inactive receptors are usually bound to inhibitory protein complexes. Ligand binding alters the conformation of the receptor protein, causing the inhibitory complex to dissociate, while also causing the receptor to bind coactivator proteins that stimulate gene transcription (**Figure 15–65**). In other cases, however, ligand binding to a nuclear receptor inhibits transcription: some thyroid hormone receptors, for example, act as transcriptional activators in the absence of their hormone and become transcriptional repressors when hormone binds.

Thus far, we have focused on the control of gene expression by extracellular signal molecules produced by other cells. We now turn to gene regulation by a more global environmental signal: the cycle of light and darkness that results from the Earth's rotation.

Circadian Clocks Contain Negative Feedback Loops That Control Gene Expression

Life on Earth evolved in the presence of a daily cycle of day and night, and many present-day organisms (ranging from archaea to plants and humans) possess an internal rhythm that dictates different behaviors at different times of day. These behaviors range from the cyclical change in metabolic enzyme activities of a bacterium to the elaborate sleep–wake cycles of humans. The internal oscillators that control such diurnal rhythms are called **circadian clocks**.

Having a circadian clock enables an organism to anticipate the regular daily changes in its environment and take appropriate action in advance. Of course, the internal clock cannot be perfectly accurate, and so it must be capable of being reset by external cues such as the light of day. Thus, circadian clocks keep running

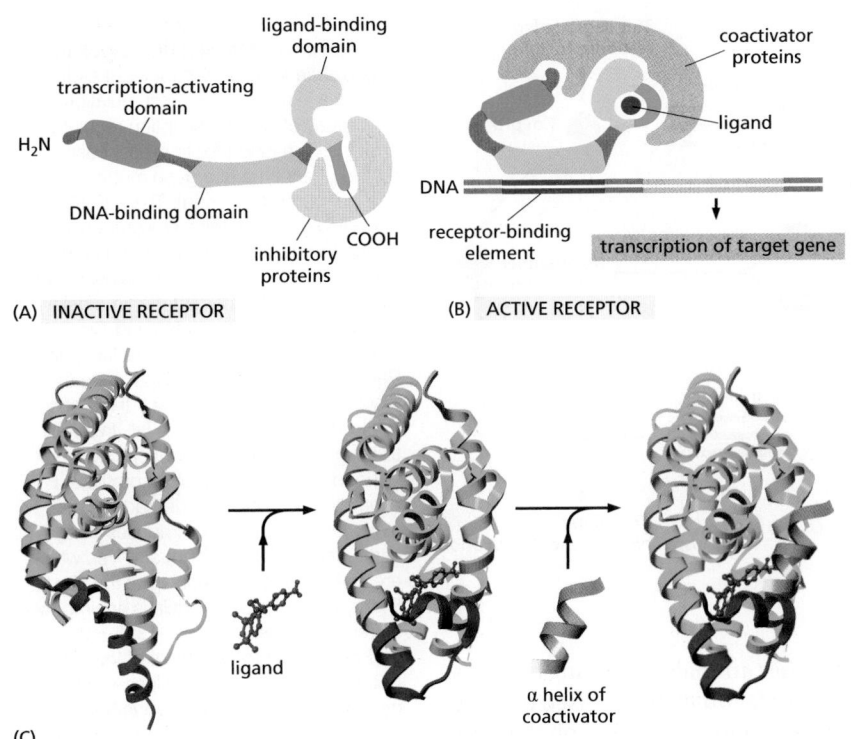

ligand-binding
domain

transcription-activating
domain

H₂N

DNA-binding domain

inhibitory
proteins

COOH

(A) INACTIVE RECEPTOR

coactivator
proteins

ligand

DNA

receptor-binding
element

transcription of target gene

(B) ACTIVE RECEPTOR

ligand

α helix of
coactivator

(C)

Figure 15–65 The activation of nuclear receptors. All nuclear receptors bind to DNA as either homodimers or heterodimers, but for simplicity we show them as monomers. (A) The receptors all have a related structure, which includes three major domains, as shown. An inactive receptor is bound to inhibitory proteins. (B) Typically, the binding of ligand to the receptor causes the ligand-binding domain of the receptor to clamp shut around the ligand, the inhibitory proteins to dissociate, and coactivator proteins to bind to the receptor's transcription-activating domain, thereby increasing gene transcription. In other cases, ligand binding has the opposite effect, causing co-repressor proteins to bind to the receptor, thereby decreasing transcription (not shown). (C) The structure of the ligand-binding domain of the retinoic acid receptor is shown in the absence (*left*) and presence (*middle*) of ligand (shown in *red*). When ligand binds, the *blue* α helix acts as a lid that snaps shut, trapping the ligand in place. The shift in the conformation of the receptor upon ligand binding also creates a binding site for a small α helix (*orange*) on the surface of coactivator proteins. (PDB codes: 1LBD, 2ZYO, and 2ZXZ.)

even when the environmental cues (changes in light and dark) are removed, but the period of this free-running rhythm is generally a little less or more than 24 hours. External signals indicating the time of day cause small adjustments in the running of the clock, so as to keep the organism in synchrony with its environment. Following more drastic shifts, circadian cycles become gradually reset (entrained) by the new cycle of light and dark, as anyone who has experienced jet lag can attest.

We might expect that the circadian clock would be a complex multicellular device, with different groups of cells responsible for different parts of the oscillation mechanism. Remarkably, however, in almost all multicellular organisms, including humans, the timekeepers are individual cells. Thus, a clock that operates in each member of a specialized group of brain cells (the SCN cells in the suprachiasmatic nucleus of the hypothalamus) controls our diurnal cycles of sleeping and waking, body temperature, and hormone release. Even if these cells are removed from the brain and dispersed in a culture dish, they will continue to oscillate individually, showing a cyclic pattern of gene expression with a period of approximately 24 hours. In the intact body, the SCN cells receive neural cues from the retina, entraining the SCN cells to the daily cycle of light and dark; they also send information about the time of day to another brain area, the pineal gland, which relays the time signal to the rest of the body by releasing the hormone melatonin in time with the clock.

Although the SCN cells have a central role as timekeepers in mammals, almost all the other cells in the mammalian body have an internal circadian rhythm, which has the ability to reset in response to light. Similarly, in *Drosophila*, many different types of cells have a similar circadian clock, which continues to cycle when they have been dissected away from the rest of the fly and can be reset by externally imposed light and dark cycles.

The working of circadian clocks, therefore, is a fundamental problem in cell biology. Although we do not yet understand all the details, studies in a wide variety of organisms have revealed the basic principles and molecular components. The key principle is that circadian clocks generally depend on *negative feedback loops*. As discussed earlier, oscillations in the activity of an intracellular signaling protein can occur if that protein inhibits its own activity with a long delay (see

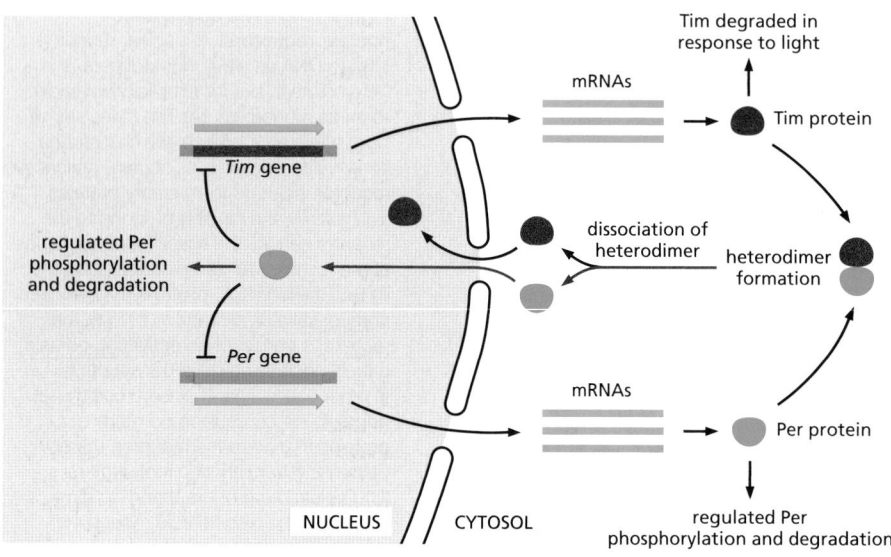

Figure 15–66 Simplified outline of the mechanism of the circadian clock in *Drosophila* cells. A central feature of the clock is the periodic accumulation and decay of two transcription regulatory proteins, Tim (short for timeless, based on the phenotype of a gene mutation) and Per (short for period). The mRNAs encoding these proteins rise gradually during the day and are translated in the cytosol, where the two proteins associate to form a heterodimer. After a time delay, the heterodimer dissociates and Tim and Per are transported into the nucleus, where Per represses the *Tim* and *Per* genes, resulting in negative feedback that causes the levels of Tim and Per to fall. In addition to this transcriptional feedback, the clock depends on numerous other proteins. For example, the controlled degradation of Per indicated in the diagram imposes delays in the accumulation of Tim and Per, which are crucial to the functioning of the clock. Steps at which specific delays are imposed are shown in *red*.

Entrainment (or resetting) of the clock occurs in response to new light–dark cycles. Although most *Drosophila* cells do not have true photoreceptors, light is sensed by intracellular flavoproteins, also called cryptochromes. In the presence of light, these proteins associate with the Tim protein and cause its degradation, thereby resetting the clock. (Adapted from J.C. Dunlap, *Science* 311:184–186, 2006.)

Figure 15–18C and D). In *Drosophila* and many other animals, including humans, the heart of the circadian clock is a delayed negative feedback loop based on transcription regulators: accumulation of certain gene products switches off the transcription of their own genes, but with a delay, so that the cell oscillates between a state in which the products are present and transcription is switched off, and one in which the products are absent and transcription is switched on (**Figure 15–66**). The negative feedback underlying circadian rhythms does not have to be based on transcription regulators. In some cell types, the circadian clock is constructed of proteins that govern their own activities through post-translational mechanisms, as we discuss next.

Three Proteins in a Test Tube Can Reconstitute a Cyanobacterial Circadian Clock

The best understood circadian clock is found in the photosynthetic cyanobacterium, *Synechococcus elongatus*. The core oscillator in this organism is remarkably simple, being composed of just three proteins—*KaiA*, *KaiB*, and *KaiC*. The central player is KaiC, a multifunctional enzyme that catalyzes its own phosphorylation and dephosphorylation in a 24-hour cycle: it gradually phosphorylates itself sequentially at two sites during the day and dephosphorylates itself during the night. This timing depends on interactions with the two other Kai proteins: KaiA binds to unphosphorylated KaiC and stimulates KaiC autophosphorylation, first at one site and then, with a delay, at the other. The second phosphorylation promotes the binding of the third protein, KaiB, which blocks the stimulatory effect of KaiA and thereby allows KaiC to dephosphorylate itself, bringing KaiC back to its dephosphorylated state. This clock depends on a negative feedback loop: KaiC drives its own phosphorylation until, after a delay, it recruits an inhibitor, KaiB, that stimulates KaiC to dephosphorylate itself. Amazingly, when the three Kai proteins are purified and incubated in a test tube with ATP, KaiC phosphorylation and dephosphorylation occur with roughly 24-hour timing over a period of several days (**Figure 15–67**).

Circadian oscillations in KaiC phosphorylation lead to parallel rhythms in the expression of large numbers of genes involved in controlling metabolic activities and cell division (see Figure 15–67). As a result, many aspects of cell behavior are synchronized with the circadian cycle.

Even in continuous darkness, cyanobacterial cells generate free-running oscillations of KaiC phosphorylation with roughly 24-hour periods. As in other circadian clocks, the cyanobacterial clock is entrained by the environmental light/dark

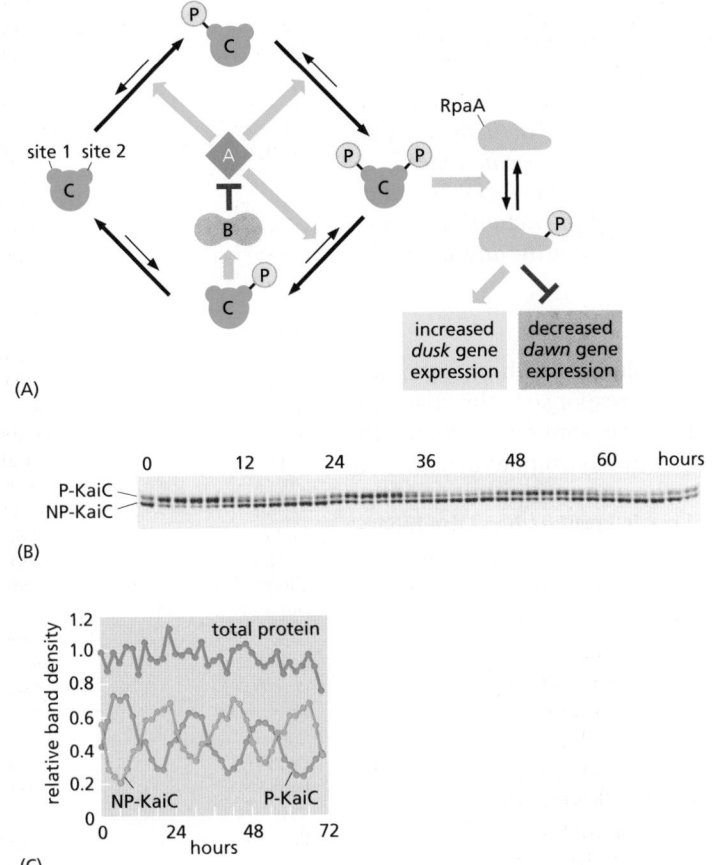

(A)

(B)

(C)

Figure 15–67 **The core circadian oscillator of cyanobacteria.** (A) KaiC is a combined kinase and phosphatase that phosphorylates and dephosphorylates itself on two adjacent sites. In the absence of other proteins, the phosphatase activity is dominant, and the protein is mostly unphosphorylated. The binding of KaiA to KaiC suppresses the phosphatase activity and promotes the kinase activity, leading to KaiC phosphorylation, first at site 1 and then at site 2, resulting in diphosphorylated KaiC. KaiC then dephosphorylates itself slowly at site 1, even in the presence of KaiA, so that KaiC is phosphorylated only at site 2. This form of KaiC interacts with KaiB, which blocks the stimulatory effects of KaiA, thereby reducing the rate of KaiC phosphorylation and allowing dephosphorylation to occur. Diphosphorylated KaiC increases in abundance during the day and peaks around dusk. It activates other proteins that phosphorylate a transcription regulator (RpaA), which then stimulates expression of some genes (the *dusk* genes that peak in early evening) and inhibits expression of other genes (the *dawn* genes that peak in the morning). When KaiC dephosphorylation gradually occurs during the night, these effects are reversed: *dusk* genes are turned off and dawn genes are turned on.

(B) In this experiment, the three Kai proteins were purified and mixed in a test tube with ATP (which is required for KaiC kinase activity). Every two hours over the next 3 days, the KaiC protein was analyzed by polyacrylamide gel electrophoresis, in which the phosphorylated form of KaiC migrates more slowly (*upper band*, P-KaiC) than the nonphosphorylated form (*lower band*, NP-KaiC). The three different phosphorylated forms of KaiC are not distinguished by this method. The phosphorylation of KaiC oscillates with a roughly 24-hour period. (C) The amount of phosphorylated and unphosphorylated KaiC in the experiment in B is plotted on this graph, along with the amount of total protein. (B and C, from M. Nakajima et al., *Science* 308:414–415, 2005. With permission from AAAS.)

cycle. Light is thought to affect the circadian clock indirectly: the activities of Kai proteins are influenced by changes in intracellular redox potential, which occur as a result of increased photosynthetic activity during the day.

Summary

Some signaling pathways that are especially important in animal development depend on proteolysis to control the activity and location of latent transcription regulatory proteins. Notch receptors are themselves such proteins, which are activated by cleavage when Delta on another cell binds to them; the cleaved cytosolic tail of Notch migrates into the nucleus, where it stimulates the transcription of Notch-responsive genes. In the Wnt/β-catenin signaling pathway, by contrast, the proteolysis of the latent transcription regulatory protein β-catenin is inhibited when a secreted Wnt protein binds to both a Frizzled and LRP receptor protein; as a result, β-catenin accumulates in the nucleus and activates the transcription of Wnt target genes.

Hedgehog signaling in flies works much like Wnt signaling. In the absence of a signal, a bifunctional, cytoplasmic transcription regulator, Ci, is proteolytically cleaved to form a transcriptional repressor that keeps Hedgehog target genes silenced. The binding of Hedgehog to its receptors (Patched and iHog) inhibits the proteolytic processing of Ci; as a result, the intact Ci protein accumulates in the nucleus and activates the transcription of Hedgehog-responsive genes. In Notch, Wnt, and Hedgehog signaling, the extracellular signal triggers a switch from transcriptional repression to transcriptional activation.

Signaling through the latent transcription regulator NFκB also depends on proteolysis. NFκB proteins are normally held in an inactive state by inhibitory IκB proteins in the cytoplasm. A variety of extracellular stimuli, including proinflammatory cytokines, trigger the phosphorylation and ubiquitylation of IκB, marking it for degradation; this enables the NFκB to translocate to the nucleus and activate the transcription of its target genes. NFκB also activates the transcription of the gene that encodes IκBα, creating a negative feedback loop, which can produce prolonged oscillations in NFκB activity with sustained extracellular signaling.

Some small, hydrophobic signal molecules, including steroid and thyroid hormones, diffuse across the plasma membrane of the target cell and activate intracellular receptor proteins that directly regulate the transcription of specific genes.

In many cell types, gene expression is governed by circadian clocks, in which delayed negative feedback produces 24-hour oscillations in the activities of transcription regulators, anticipating the cell's changing needs during the day and night.

SIGNALING IN PLANTS

In plants, as in animals, cells are in constant communication with one another. Plant cells communicate to coordinate their activities in response to the changing conditions of light, dark, and temperature, which guide the plant's cycle of growth, flowering, and fruiting. Plant cells also communicate to coordinate activities in their roots, stems, and leaves. In this final section, we consider how plant cells signal to one another and how they respond to light. Less is known about the receptors and intracellular signaling mechanisms involved in cell communication in plants than is known in animals, and we will concentrate mainly on how the receptors and intracellular signaling mechanisms differ from those used by animals.

Multicellularity and Cell Communication Evolved Independently in Plants and Animals

Although plants and animals are both eukaryotes, they have evolved separately for more than a billion years. Their last common ancestor is thought to have been a unicellular eukaryote that had mitochondria but no chloroplasts; the plant lineage acquired chloroplasts after plants and animals diverged. The earliest fossils of multicellular animals and plants date from almost 600 million years ago. Thus, it seems that plants and animals evolved multicellularity independently, each starting from a different unicellular eukaryote, some time between 1.6 and 0.6 billion years ago (**Figure 15–68**).

If multicellularity evolved independently in plants and animals, the molecules and mechanisms used for cell communication will have evolved separately and would be expected to be different. There should be some degree of resemblance, however, because the genes in both plants and animals diverged from those contained by their last common unicellular ancestor. Thus, whereas both plants and animals use nitric oxide, cyclic GMP, Ca²⁺, and Rho family GTPases for signaling, there are no homologs of the nuclear receptor family, Ras, JAK, STAT, TGFβ,

Figure 15–68 The proposed divergence of plant and animal lineages from a common unicellular eukaryotic ancestor. The plant lineage acquired chloroplasts after the two lineages diverged. Both lineages independently gave rise to multicellular organisms—plants and animals. (Paintings courtesy of John Innes Foundation.)

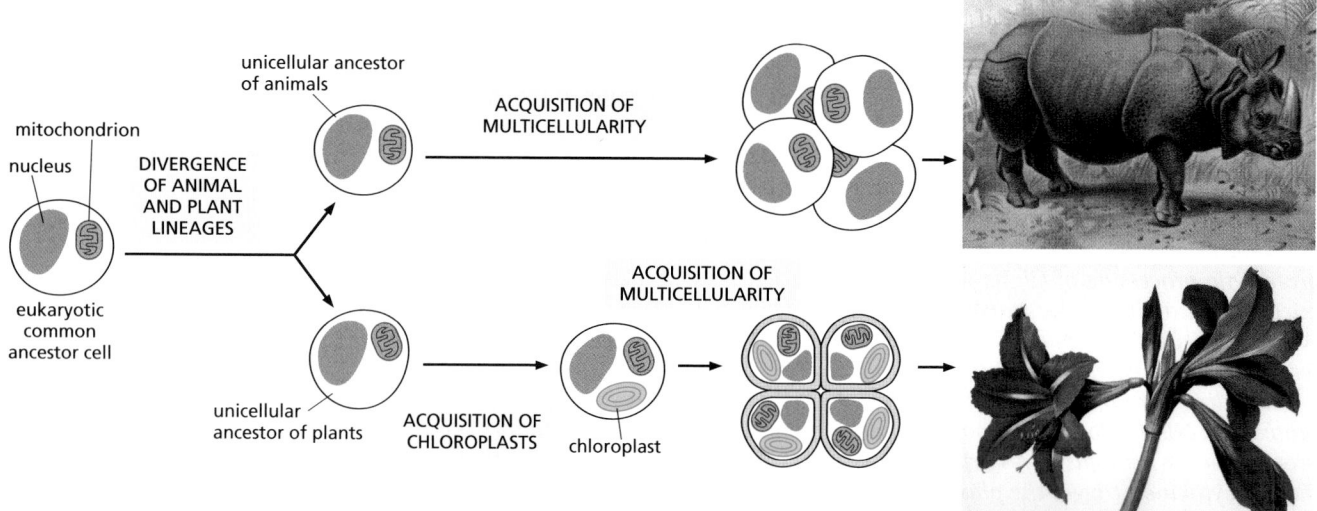

Notch, Wnt, or Hedgehog encoded by the completely sequenced genome of *Arabidopsis thaliana*, the small flowering plant. Similarly, plants do not seem to use cyclic AMP for intracellular signaling. Nevertheless, the general strategies underlying signaling are frequently very similar in plants and animals. Both, for example, use enzyme-coupled cell-surface receptors, as we now discuss.

Receptor Serine/Threonine Kinases Are the Largest Class of Cell-Surface Receptors in Plants

Most cell-surface receptors in plants are enzyme-coupled. However, whereas the largest class of enzyme-coupled receptors in animals is the receptor tyrosine kinase (RTK) class, this type of receptor is extremely rare in plants. Instead, plants rely largely on a great diversity of transmembrane *receptor serine/threonine kinases*, which have a typical serine/threonine kinase cytoplasmic domain and an extracellular ligand-binding domain. The most abundant types of these receptors have a tandem array of extracellular leucine-rich repeat structures and are therefore called **leucine-rich repeat (LRR) receptor kinases**.

There are about 175 LRR receptor kinases encoded by the *Arabidopsis* genome. These include a protein called *Bri1*, which forms part of a cell-surface steroid hormone receptor. Plants synthesize a class of steroids that are called **brassinosteroids** because they were originally identified in the mustard family *Brassicaceae*, which includes *Arabidopsis*. These signal molecules regulate the growth and differentiation of plants throughout their life cycle. Binding of a brassinosteroid to a Bri1 cell-surface receptor kinase initiates an intracellular signaling cascade that uses a GSK3 protein kinase and a protein phosphatase to regulate the phosphorylation and degradation of specific transcription regulatory proteins in the nucleus, and thereby specific gene transcription. Mutant plants that are deficient in the Bri1 receptor kinase are insensitive to brassinosteroids and are therefore dwarfs.

The LRR receptor kinases are only one of many classes of transmembrane receptor serine/threonine kinases in plants. There are at least six additional families, each with its own characteristic set of extracellular domains. The *lectin receptor kinases*, for example, have extracellular domains that bind carbohydrate signal molecules. The *Arabidopsis* genome encodes over 300 receptor serine/threonine kinases, which makes them the largest family of receptors known in plants. Many are involved in defense responses against pathogens.

Ethylene Blocks the Degradation of Specific Transcription Regulatory Proteins in the Nucleus

Various **plant growth regulators** (also called **plant hormones**) help to coordinate plant development. They include *ethylene, auxin, cytokinins, gibberellins*, and *abscisic acid*, as well as brassinosteroids. Growth regulators are all small molecules made by most plant cells. They diffuse readily through cell walls and can either act locally or be transported to influence cells further away. Each growth regulator can have multiple effects. The specific effect depends on environmental conditions, the nutritional state of the plant, the responsiveness of the target cells, and which other growth regulators are acting.

Ethylene is an important example. This small gas molecule (**Figure 15–69A**) can influence plant development in various ways; it can, for example, promote fruit ripening, leaf abscission, and plant senescence. It also functions as a stress signal in response to wounding, infection, flooding, and so on. When the shoot of a germinating seedling, for instance, encounters an obstacle, ethylene promotes a complex response that allows the seedling to safely bypass the obstacle (Figure 15–69B and C).

Plants have various ethylene receptors, which are located in the endoplasmic reticulum and are all structurally related. They are dimeric, multipass transmembrane proteins, with a copper-containing ethylene-binding domain and a domain that interacts with a cytoplasmic protein called *CTR1*, which is closely related

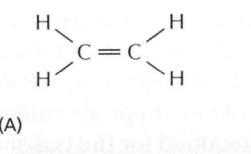

(A)

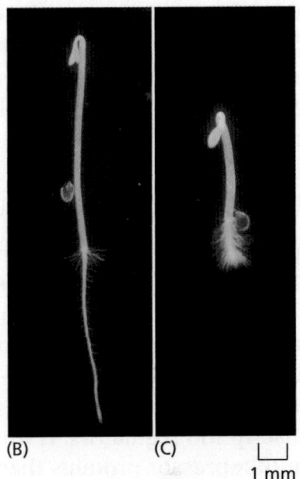

(B)   (C) |__|
                      1 mm

Figure 15–69 The ethylene-mediated triple response that occurs when the growing shoot of a germinating seedling encounters an obstacle underground. (A) The structure of ethylene. (B) In the absence of obstacles, the shoot grows upward and is long and thin. (C) If the shoot encounters an obstacle, such as a piece of gravel in the soil, the seedling responds to the encounter in three ways. First, it thickens its stem, which can then exert more force on the obstacle. Second, it shields the tip of the shoot (at *top*) by increasing the curvature of a specialized hook structure. Third, it reduces the shoot's tendency to grow away from the direction of gravity, so as to avoid the obstacle. (Courtesy of Melanie Webb.)

in sequence to the Raf MAP kinase kinase kinase discussed earlier (see Figure 15–49). Surprisingly, it is the empty receptors that are active and keep CTR1 active. By an unknown signaling mechanism, active CTR1 stimulates the ubiquitylation and degradation in proteasomes of a nuclear transcription regulator called *EIN3*, which is required for the transcription of ethylene-responsive genes. In this way, the empty but active receptors keep ethylene-response genes off. Ethylene binding inactivates the receptors, altering their conformation so that they no longer activate CTR1. The EIN3 protein is no longer ubiquitylated and degraded and can now activate the transcription of the large number of ethylene-responsive genes (**Figure 15–70**).

Regulated Positioning of Auxin Transporters Patterns Plant Growth

The plant hormone **auxin**, which is generally indole-3-acetic acid (**Figure 15–71A**), binds to receptor proteins in the nucleus. It helps plants grow toward light, grow upward rather than branch out, and grow their roots downward. It also regulates organ initiation and positioning and helps plants flower and bear fruit. Like ethylene (and like some of the animal signal molecules we have described in this chapter), auxin influences gene expression by controlling the degradation of transcription regulators. It works by stimulating the ubiquitylation and degradation of repressor proteins that block the transcription of auxin target genes in unstimulated cells (Figure 15–71B and C).

Auxin is unique in the way that it is transported. Unlike animal hormones, which are usually secreted by a specific endocrine organ and transported to target

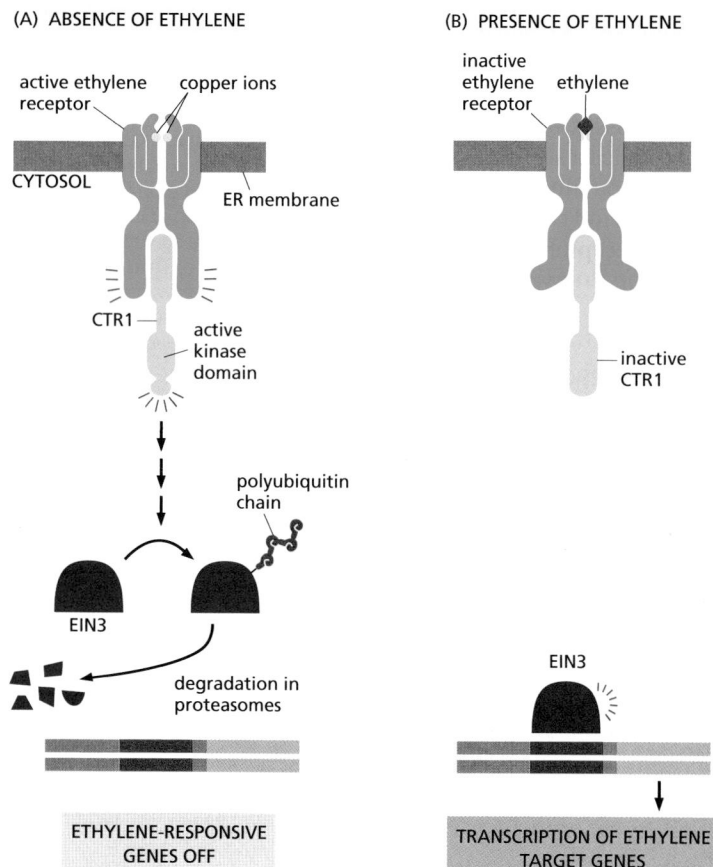

Figure 15–70 The ethylene signaling pathway. (A) In the absence of ethylene, the receptors and CTR1 are active, causing the ubiquitylation and destruction of EIN3, the transcription regulatory protein in the nucleus that is responsible for the transcription of ethylene-responsive genes. (B) The binding of ethylene inactivates the receptors and disrupts the activation of CTR1. The EIN3 protein is not degraded and can therefore activate the transcription of ethylene-responsive genes.

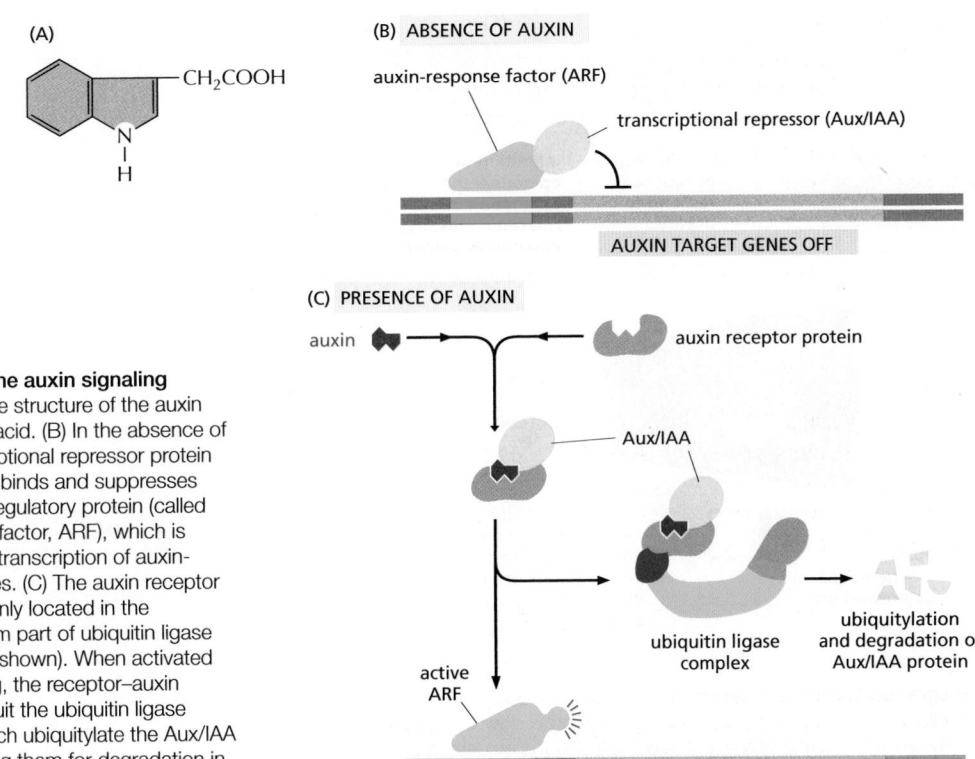

Figure 15–71 The auxin signaling pathway. (A) The structure of the auxin indole-3-acetic acid. (B) In the absence of auxin, a transcriptional repressor protein (called Aux/IAA) binds and suppresses a transcription regulatory protein (called auxin-response factor, ARF), which is required for the transcription of auxin-responsive genes. (C) The auxin receptor proteins are mainly located in the nucleus and form part of ubiquitin ligase complexes (not shown). When activated by auxin binding, the receptor–auxin complexes recruit the ubiquitin ligase complexes, which ubiquitylate the Aux/IAA proteins, marking them for degradation in proteasomes. ARF is now free to activate the transcription of auxin-responsive genes. There are many ARFs, Aux/IAA proteins, and auxin receptors that work as illustrated.

cells via the circulatory system, auxin has its own transport system. Specific plasma-membrane-bound *influx transporter proteins* and *efflux transporter proteins* move auxin into and out of plant cells, respectively. The efflux transporters can be distributed asymmetrically in the plasma membrane to make the efflux of auxin directional. A row of cells with their auxin efflux transporters confined to the basal plasma membrane, for example, will transport auxin from the top of the plant to the bottom.

In some regions of the plant, the localization of the auxin transporters, and therefore the direction of auxin flow, is highly dynamic and regulated. A cell can rapidly redistribute transporters by controlling the traffic of vesicles containing them. The auxin efflux transporters, for example, normally recycle between intracellular vesicles and the plasma membrane. A cell can redistribute these transporters on its surface by inhibiting their endocytosis in one domain of the plasma membrane, causing the transporters to accumulate there. One example occurs in the root, where gravity influences the direction of growth. The auxin efflux transporters are normally distributed symmetrically in the cap cells of the root. Within minutes of a change in the direction of the gravity vector, however, the efflux transporters redistribute to one side of the cells, so that auxin is pumped out toward the side of the root pointing downward. Because auxin inhibits root-cell elongation, this redirection of auxin transport causes the root tip to reorient, so that it grows downward again (**Figure 15–72**).

Phytochromes Detect Red Light, and Cryptochromes Detect Blue Light

Plant development is greatly influenced by environmental conditions. Unlike animals, plants cannot move when conditions become unfavorable; they have to adapt or they die. The most important environmental influence on plants is

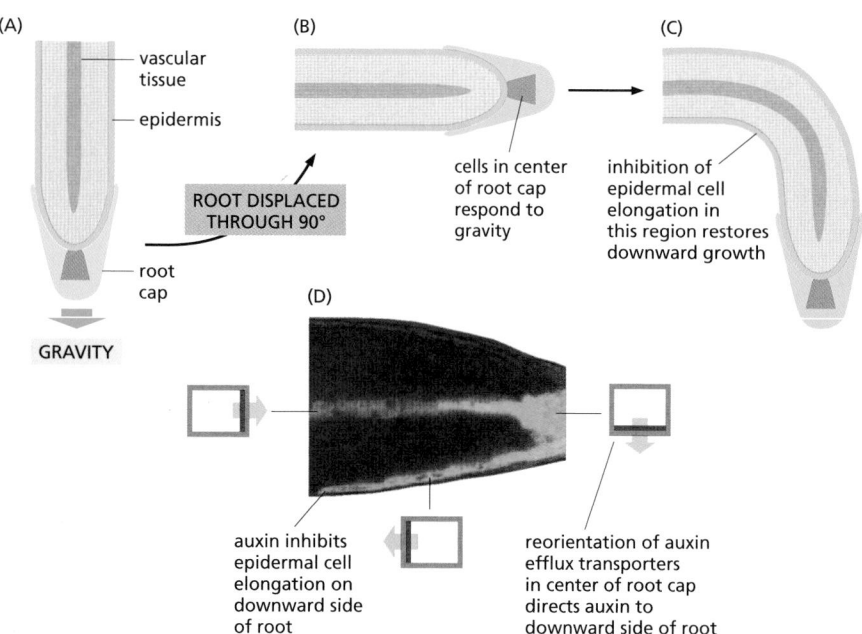

(A)

vascular tissue

epidermis

ROOT DISPLACED THROUGH 90°

root cap

GRAVITY

(B)

cells in center of root cap respond to gravity

(C)

inhibition of epidermal cell elongation in this region restores downward growth

(D)

auxin inhibits epidermal cell elongation on downward side of root

reorientation of auxin efflux transporters in center of root cap directs auxin to downward side of root

Figure 15–72 Auxin transport and root gravitropism. (A–C) Roots respond to a 90° change in the gravity vector and adjust their direction of growth so that they grow downward again. The cells that respond to gravity are in the center of the root cap, while it is the epidermal cells further back (on the lower side) that decrease their rate of elongation to restore downward growth. (D) The gravity-responsive cells in the root cap redistribute their auxin efflux transporters in response to the displacement of the root. This redirects the auxin flux mainly to the lower part of the displaced root, where it inhibits the elongation of the epidermal cells. The resulting asymmetrical distribution of auxin in the *Arabidopsis* root tip shown here is assessed indirectly, using an auxin-responsive reporter gene that encodes a protein fused to green fluorescent protein (GFP); the epidermal cells on the downward side of the root are green, whereas those on the upper side are not, reflecting the asymmetrical distribution of auxin. The distribution of auxin efflux transporters in the plasma membrane of cells in different regions of the root (shown as *gray rectangles*) is indicated in *red*, and the direction of auxin efflux is indicated by a *green arrow*. (D, photograph from T. Paciorek et al., *Nature* 435:1251–1256, published 2005 by Nature Publishing Group. Reproduced with permission of SNCSC.)

light, which is their energy source and has a major role throughout their entire life cycle—from germination, through seedling development, to flowering and senescence. Plants have thus evolved a large set of light-sensitive proteins to monitor the quantity, quality, direction, and duration of light. These are usually referred to as *photoreceptors*. However, because the term photoreceptor is also used for light-sensitive cells in the animal retina (see Figure 15–38), we shall use the term *photoprotein* instead.

All photoproteins sense light by means of a covalently attached light-absorbing chromophore, which changes its shape in response to light and then induces a change in the protein's conformation. The best-known plant photoproteins are the **phytochromes**, which are present in all plants and in some algae but are absent in animals. These are dimeric, cytoplasmic serine/threonine kinases, which respond differentially and reversibly to red and far-red light: whereas red light usually activates the kinase activity of the phytochrome, far-red light inactivates it. When activated by red light, the phytochrome is thought to phosphorylate itself and then to phosphorylate one or more other proteins in the cell. In some light responses, the activated phytochrome translocates into the nucleus, where it activates transcription regulators to alter gene transcription (**Figure 15–73**). In other cases, the activated phytochrome activates a latent transcription regulator in the cytoplasm, which then translocates into the nucleus to regulate gene transcription. In still other cases, the photoprotein triggers signaling pathways in the cytosol that alter the cell's behavior without involving the nucleus.

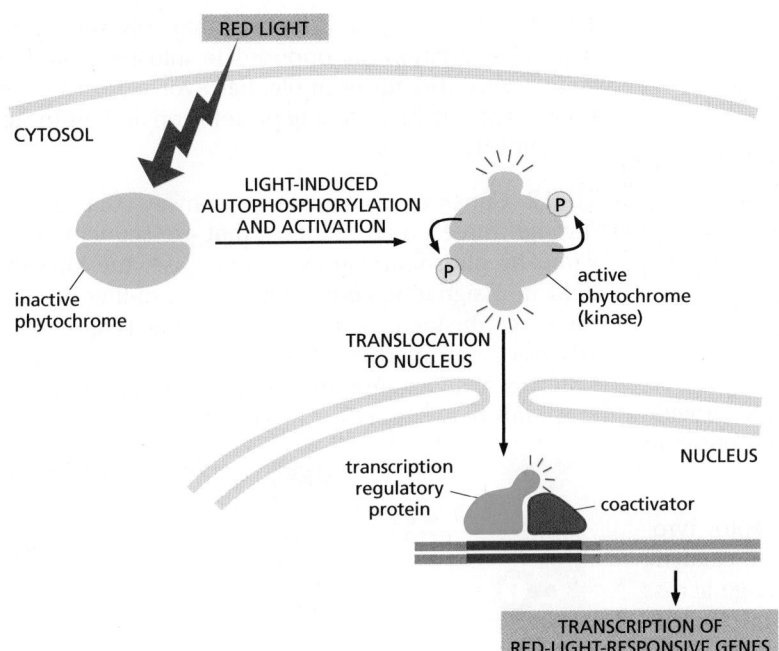

Figure 15–73 **One way in which phytochromes mediate a light response in plant cells.** When activated by red light, the phytochrome, which is a dimeric protein kinase, phosphorylates itself and then moves into the nucleus, where it activates transcription regulatory proteins to stimulate the transcription of red-light-responsive genes.

Plants sense blue light using photoproteins of two other sorts, phototropin and cryptochromes. **Phototropin** is associated with the plasma membrane and is partly responsible for *phototropism*, the tendency of plants to grow toward light. Phototropism occurs by directional cell elongation, which is stimulated by auxin, but the links between phototropin and auxin are unknown.

Cryptochromes are flavoproteins that are sensitive to blue light. They are structurally related to blue-light-sensitive enzymes called *photolyases*, which are involved in the repair of ultraviolet-induced DNA damage in all organisms, except most mammals. Unlike phytochromes, cryptochromes are also found in animals, where they have an important role in circadian clocks (see Figure 15–66). Although cryptochromes are thought to have evolved from the photolyases, they do not have a role in DNA repair.

Summary

Plants and animals are thought to have evolved multicellularity and cell communication mechanisms independently, each starting from a different unicellular eukaryote, which in turn evolved from a common unicellular eukaryotic ancestor. Not surprisingly, therefore, the mechanisms used to signal between cells in animals and in plants have both similarities and differences. Whereas animals rely heavily on GPCRs and RTKs, plants rely mainly on enzyme-coupled receptors of the receptor serine/threonine kinase type, especially ones with extracellular leucine-rich repeats. Various plant hormones, or growth regulators, including ethylene and auxin, help coordinate plant development. Ethylene acts through intracellular receptors to stop the degradation of specific nuclear transcription regulators, which can then activate the transcription of ethylene-responsive genes. The receptors for some other plant hormones, including auxin, also regulate the degradation of specific transcription regulators, although the details vary. Auxin signaling is unusual in that it has its own highly regulated transport system, in which the dynamic positioning of plasma-membrane-bound auxin transporters controls the direction of auxin flow and thereby the direction of plant growth. Light has an important role in regulating plant development. These light responses are mediated by a variety of light-sensitive photoproteins, including phytochromes, which are responsive to red light, and cryptochromes and phototropin, which are sensitive to blue light.

WHAT WE DON'T KNOW

• How does a cell integrate the information received from its many different cell-surface receptors to make all-or-none decisions?

• Much of what we know about cell signaling comes from biochemical studies of isolated proteins in test tubes. What is the precise quantitative behavior of intracellular signaling networks in an intact cell, or in an intact animal, where countless other signals and cell components might influence signaling specificity and strength?

• How do intracellular signaling circuits generate specific and dynamic signaling patterns such as oscillations and waves, and how are these patterns sensed and interpreted by the cell?

• Scaffold proteins and activated receptor tyrosine kinases nucleate the assembly of large intracellular signaling complexes. What is the dynamic behavior of these complexes, and how does this behavior influence downstream signaling?

PROBLEMS

Which statements are true? Explain why or why not.

15–1   All second messengers are water-soluble and diffuse freely through the cytosol.

15–2   In the regulation of molecular switches, protein kinases and guanine nucleotide exchange factors (GEFs) always turn proteins on, whereas protein phosphatases and GTPase-activating proteins (GAPs) always turn proteins off.

15–3   Most intracellular signaling pathways provide numerous opportunities for amplifying the responses to extracellular signals.

15–4   Binding of extracellular ligands to receptor tyrosine kinases (RTKs) activates the intracellular catalytic domain by propagating a conformational change across the lipid bilayer through a single transmembrane α helix.

15–5   Protein tyrosine phosphatases display exquisite specificity for their substrates, unlike most serine/threonine protein phosphatases, which have rather broad specificity.

15–6   Even though plants and animals independently evolved multicellularity, they use virtually all the same signaling proteins and second messengers for cell–cell communication.

Discuss the following problems.

15–7   Suppose that the circulating concentration of hormone is 10^{-10} M and the K_d for binding to its receptor is 10^{-8} M. What fraction of the receptors will have hormone bound? If a meaningful physiological response occurs when 50% of the receptors have bound a hormone molecule, how much will the concentration of hormone have to rise to elicit a response? The fraction of receptors (R) bound to hormone (H) to form a receptor–hormone complex (R–H) is [R–H]/ ([R] + [R–H]) = [R–H]/[R]$_{TOT}$ = [H]/([H] + K_d).

15–8   Cells communicate in ways that resemble human communication. Decide which of the following forms of human communication are analogous to autocrine, paracrine, endocrine, and synaptic signaling by cells.
A. A telephone conversation
B. Talking to people at a cocktail party
C. A radio announcement
D. Talking to yourself

15–9   Why do signaling responses that involve changes in proteins already present in the cell occur in milliseconds to seconds, whereas responses that require changes in gene expression require minutes to hours?

15–10   How is it that different cells can respond in different ways to exactly the same signaling molecule even when they have identical receptors?

15–11   Why do you suppose that phosphorylation/ dephosphorylation, as opposed to allosteric binding of small molecules, for example, has evolved to play such a prominent role in switching proteins on and off in signaling pathways?

15–12   Consider a signaling pathway that proceeds through three protein kinases that are sequentially activated by phosphorylation. In one case, the kinases are held in a signaling complex by a scaffolding protein; in the other, the kinases are freely diffusible (**Figure Q15–1**). Discuss the properties of these two types of organization in terms of signal amplification, speed, and potential for cross-talk between signaling pathways.

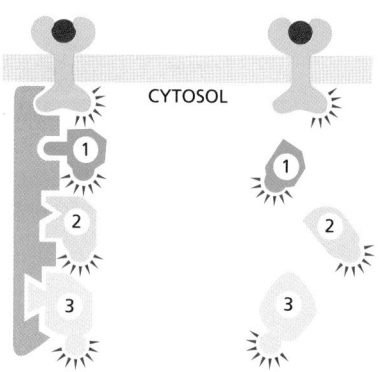

Figure Q15–1 A kinase cascade organized by a scaffolding protein or composed of freely diffusing components (Problem 15–12).

15–13   Describe three ways in which a gradual increase in an extracellular signal can be sharpened by the target cell to produce an abrupt or nearly all-or-none response.

15–14   Activation ("maturation") of frog oocytes is signaled through a MAP kinase signaling module. An increase in the hormone progesterone triggers the module by stimulating the translation of Mos mRNA, which is the frog's MAP kinase kinase kinase (**Figure Q15–2**). Maturation is easy to score visually by the presence of a white spot in the middle of the brown surface of the oocyte (see Figure Q15–2). To determine the dose–response curve for progesterone-induced activation of MAP kinase, you place 16 oocytes in each of six plastic dishes and add various concentrations of progesterone. After an overnight incubation, you crush the oocytes, prepare an extract, and determine the state of MAP kinase phosphorylation (hence, activation) by SDS polyacrylamide-gel electrophoresis (**Figure Q15–3A**). This analysis shows a graded response of MAP kinase to increasing concentrations of progesterone.

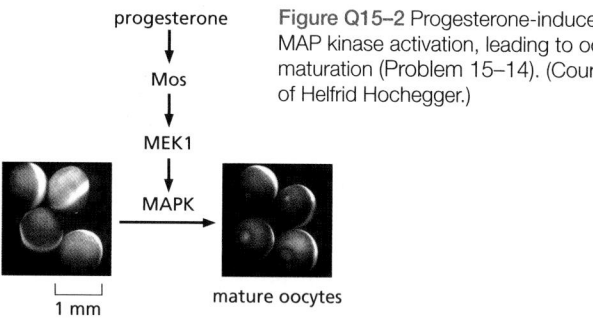

Figure Q15–2 Progesterone-induced MAP kinase activation, leading to oocyte maturation (Problem 15–14). (Courtesy of Helfrid Hochegger.)

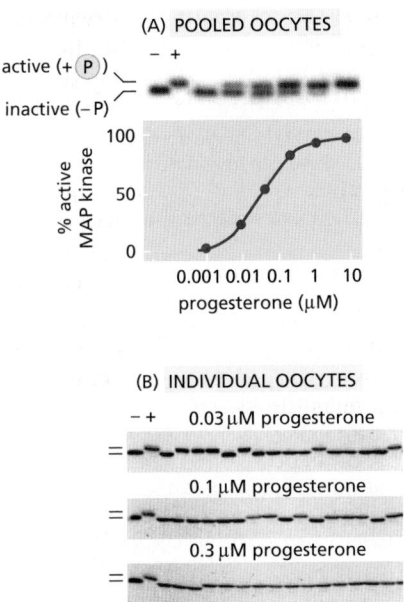

active (+ P)

inactive (– P)

(A) POOLED OOCYTES

(B) INDIVIDUAL OOCYTES

Figure Q15–3 Activation of frog oocytes (Problem 15–14). (A) Phosphorylation of MAP kinase in pooled oocytes. (B) Phosphorylation of MAP kinase in individual oocytes. MAP kinase was detected by immunoblotting using a MAP-kinase-specific antibody. The first two lanes in each gel contain nonphosphorylated, inactive MAP kinase (–) and phosphorylated, active MAP kinase (+). (From J.E. Ferrell, Jr., and E.M. Machleder, *Science* 280:895–898, 1998. With permission from AAAS.)

Before you crushed the oocytes, you noticed that not all oocytes in individual dishes had white spots. Had some oocytes undergone partial activation and not yet reached the white-spot stage? To answer this question, you repeat the experiment, but this time you analyze MAP kinase activation in individual oocytes. You are surprised to find that each oocyte has either a fully activated or a completely inactive MAP kinase (Figure Q15–3B). How can an all-or-none response in individual oocytes give rise to a graded response in the population?

15–15 Propose specific types of mutations in the gene for the regulatory subunit of cyclic-AMP-dependent protein kinase (PKA) that could lead to either a permanently active PKA or a permanently inactive PKA.

15–16 Phosphorylase kinase integrates signals from the cyclic-AMP-dependent and Ca^{2+}-dependent signaling pathways that control glycogen breakdown in liver and muscle cells (**Figure Q15–4**). Phosphorylase kinase is composed of four subunits. One is the protein kinase that catalyzes the addition of phosphate to glycogen phosphorylase to activate it for glycogen breakdown. The other three subunits are regulatory proteins that control the activity of the

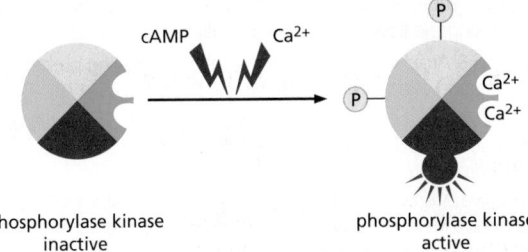

phosphorylase kinase inactive

phosphorylase kinase active

Figure Q15–4 Integration of cyclic-AMP-dependent and Ca^{2+}-dependent signaling pathways by phosphorylase kinase in liver and muscle cells (Problem 15–16).

catalytic subunit. Two contain sites for phosphorylation by PKA, which is activated by cyclic AMP. The remaining subunit is calmodulin, which binds Ca^{2+} when the cytosolic Ca^{2+} concentration rises. The regulatory subunits control the equilibrium between the active and inactive conformations of the catalytic subunit, with each phosphate and Ca^{2+} nudging the equilibrium toward the active conformation. How does this arrangement allow phosphorylase kinase to serve its role as an integrator protein for the multiple pathways that stimulate glycogen breakdown?

15–17 The Wnt planar polarity signaling pathway normally ensures that each wing cell in *Drosophila* has a single hair. Overexpression of the *Frizzled* gene from a heat-shock promoter (hs-*Fz*) causes multiple hairs to grow from many cells (**Figure Q15–5A**). This phenotype is suppressed if hs-*Fz* is combined with a heterozygous deletion (Dsh^Δ) of the *Disheveled* gene (Figure Q15–5B). Do these results allow you to order the action of Frizzled and Disheveled in the signaling pathway? If so, what is the order? Explain your reasoning.

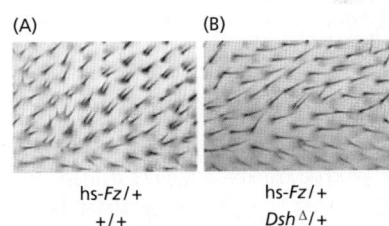

(A) hs-*Fz*/+ +/+

(B) hs-*Fz*/+ Dsh^Δ/+

Figure Q15–5 Pattern of hair growth on wing cells in genetically different *Drosophila* (Problem 15–17). (From C.G. Winter et al., *Cell* 105:81–91, 2001. With permission from Elsevier.)

REFERENCES

General

Marks F, Klingmüller U & Müller-Decker K (2009) Cellular Signal Processing: An Introduction to the Molecular Mechanisms of Signal Transduction. New York: Garland Science.

Lim W, Mayer B & Pawson T (2015) Cell Signaling: Principles and Mechanisms. New York: Garland Science.

Principles of Cell Signaling

Alon U (2007) Network motifs: theory and experimental approaches. *Nat. Rev. Genet.* 8, 450–461.

Ben-Shlomo I, Yu Hsu S, Rauch R et al (2003) Signaling receptome: a genomic and evolutionary perspective of plasma membrane receptors involved in signal transduction. *Sci. STKE* 187, RE9.

Endicott JA, Noble ME & Johnson LN (2012) The structural basis for control of eukaryotic protein kinases. *Annu. Rev. Biochem.* 81, 587–613.

Ferrell JE, Jr (2002) Self-perpetuating states in signal transduction: positive feedback, double-negative feedback and bistability. *Curr. Opin. Cell Biol.* 14, 140–148.

Good MC, Zalatan JG & Lim WA (2011) Scaffold proteins: hubs for controlling the flow of cellular information. *Science* 332, 680–686.

Ladbury JE & Arold ST (2012) Noise in cellular signaling pathways: causes and effects. *Trends Biochem. Sci.* 37, 173–178.

Mehta S & Zhang J (2011) Reporting from the field: genetically encoded fluorescent reporters uncover signaling dynamics in living biological systems. *Annu. Rev. Biochem.* 80, 375–401.

Pires-daSilva A & Sommer RJ (2003) The evolution of signalling pathways in animal development. *Nature Rev. Genet.* 4, 39–49.

Rosse C, Linch M, Kermorgant S, Cameron AJ et al. (2010) PKC and the control of localized signal dynamics. *Nat. Rev. Mol. Cell Biol.* 11, 103–112.

Scott JD & Pawson T (2009) Cell signaling in space and time: where proteins come together and when they're apart. *Science* 326, 1220–1224.

Seet BT, Dikic I, Zhou MM & Pawson T (2006) Reading protein modifications with interaction domains. *Nat. Rev. Mol. Cell Biol.* 7, 473–483.

Tyson JJ, Chen KC & Novak B (2003) Sniffers, buzzers, toggles and blinkers: dynamics of regulatory and signaling pathways in the cell. *Curr. Opin. Cell Biol.* 15, 221–231.

Ubersax JA & Ferrell JE, Jr (2007) Mechanisms of specificity in protein phosphorylation. *Nat. Rev. Mol. Cell Biol.* 8, 530–541.

Wittinghofer A & Vetter IR (2011) Structure-function relationships of the G domain, a canonical switch motif. *Annu. Rev. Biochem.* 80, 943–971.

Signaling Through G-protein-coupled Receptors

Audet M & Bouvier M (2012) Restructuring G-protein-coupled receptor activation. *Cell* 151, 14–23.

Berridge MJ, Bootman MD & Roderick HL (2003) Calcium signalling: dynamics, homeostasis and remodelling. *Nat. Rev. Mol. Cell Biol.* 4, 517–529.

Breer H (2003) Sense of smell: recognition and transduction of olfactory signals. *Biochem. Soc. Trans.* 31, 113–116.

Hoeflich KP & Ikura M (2002) Calmodulin in action: diversity in target recognition and activation mechanisms. *Cell* 108, 739–742.

Kamenetsky M, Middelhaufe S, Bank EM et al. (2006) Molecular details of cAMP generation in mammalian cells: a tale of two systems. *J. Mol. Biol.* 362, 623–639.

McConnachie G, Langeberg LK & Scott JD (2006) AKAP signaling complexes: getting to the heart of the matter. *Trends Mol. Med.* 12, 317–323.

Murad F (2006) Shattuck Lecture. Nitric oxide and cyclic GMP in cell signaling and drug development. *N. Engl. J. Med.* 355, 2003–2011.

Parker PJ (2004) The ubiquitous phosphoinositides. *Biochem. Soc. Trans.* 32, 893–898.

Rasmussen SG, DeVree BT, Zou Y et al. (2011) Crystal structure of the beta2 adrenergic receptor-Gs protein complex. *Nature* 477, 549–555.

Rhee SG (2001) Regulation of phosphoinositide-specific phospholipase C. *Annu. Rev. Biochem.* 70, 281–312.

Robishaw JD & Berlot CH (2004) Translating G protein subunit diversity into functional specificity. *Curr. Opin. Cell Biol.* 16, 206–209.

Rosenbaum DM, Rasmussen SG & Kobilka BK (2009) The structure and function of G-protein-coupled receptors. *Nature* 459, 356–363.

Shenoy SK & Lefkowitz RJ (2011) beta-Arrestin-mediated receptor trafficking and signal transduction. *Trends Pharmacol. Sci.* 32, 521–533.

Sorkin A & von Zastrow M (2009) Endocytosis and signalling: intertwining molecular networks. *Nat. Rev. Mol. Cell Biol.* 10, 609–622.

Stratton MM, Chao LH, Schulman H & Kuriyan J (2013) Structural studies on the regulation of Ca^{2+}/calmodulin dependent protein kinase II. *Curr. Opin. Struct. Biol.* 23, 292–301.

Sung CH & Chuang JZ (2010) The cell biology of vision. *J. Cell Biol.* 190, 953–963.

Willoughby D & Cooper DM (2008) Live-cell imaging of cAMP dynamics. *Nat. Methods* 5, 29–36.

Signaling Through Enzyme-coupled Receptors

Jura N, Zhang X, Endres NF et al. (2011) Catalytic control in the EGF receptor and its connection to general kinase regulatory mechanisms. *Mol. Cell* 42, 9–22.

Lemmon MA & Schlessinger J (2010) Cell signaling by receptor tyrosine kinases. *Cell* 141, 1117–1134.

Massagué J (2012) TGFβ signalling in context. *Nat. Rev. Mol. Cell Biol.* 13, 616–630.

Manning BD & Cantley LC (2007) AKT/PKB signaling: navigating downstream. *Cell* 129, 1261–1274.

Mitin N, Rossman KL & Der CJ (2005) Signaling interplay in Ras superfamily function. *Curr. Biol.* 15, R563–R574.

Pitulescu ME & Adams RH (2010) Eph/ephrin molecules—a hub for signaling and endocytosis. *Genes Dev.* 24, 2480–2492.

Qi M & Elion EA (2005) MAP kinase pathways. *J. Cell Sci.* 118, 3569–3572.

Rawlings JS, Rosler KM & Harrison DA (2004) The JAK/STAT signaling pathway. *J. Cell Sci.* 117, 1281–1283.

Roskoski R, Jr (2004) Src protein-tyrosine kinase structure and regulation. *Biochem. Biophys. Res. Commun.* 324, 1155–1164.

Schwartz MA & Madhani HD (2004) Principles of MAP kinase signaling specificity in *Saccharomyces cerevisiae*. *Annu. Rev. Genet.* 38, 725–748.

Shaw RJ & Cantley LC (2006) Ras, PI(3)K and mTOR signalling controls tumour cell growth. *Nature* 44, 424–430.

Zoncu R, Efeyan A & Sabatini DM (2011) mTOR: from growth signal integration to cancer, diabetes and ageing. *Nat. Rev. Mol. Cell Biol.* 12, 21–35.

Alternative Signaling Routes in Gene Regulation

Bray SJ (2006) Notch signalling: a simple pathway becomes complex. *Nat. Rev. Mol. Cell Biol.* 7, 678–689.

Briscoe J & Therond PP (2013) The mechanisms of Hedgehog signalling and its roles in development and disease. *Nat. Rev. Mol. Cell Biol.* 14, 416–429.

Hoffmann A & Baltimore D (2006) Circuitry of nuclear factor κB signaling. *Immunol. Rev.* 210, 171–186.

Huang P, Chandra V & Rastinejad F (2010) Structural overview of the nuclear receptor superfamily: insights into physiology and therapeutics. *Annu. Rev. Physiol.* 72, 247–272.

Markson JS & O'Shea EK (2009) The molecular clockwork of a protein-based circadian oscillator. *FEBS Lett.* 583, 3938–3947.

Mohawk JA, Green CB & Takahashi JS (2012) Central and peripheral circadian clocks in mammals. *Annu. Rev. Neurosci.* 35, 445–462.

Niehrs C (2012) The complex world of WNT receptor signalling. *Nat. Rev. Mol. Cell Biol.* 13, 767–779.

Signaling in Plants

Chen M, Chory J & Fankhauser C (2004) Light signal transduction in higher plants. *Annu. Rev. Genet.* 38, 87–117.

Dievart A & Clark SE (2004) LRR-containing receptors regulating plant development and defense. *Development* 131, 251–261.

Kim TW & Wang ZY (2010) Brassinosteroid signal transduction from receptor kinases to transcription factors. *Annu. Rev. Plant Biol.* 61, 681–704.

Mockaitis K & Estelle M (2008) Auxin receptors and plant development: a new signaling paradigm. *Annu. Rev. Cell Dev. Biol.* 24, 55–80.

Stepanova AN & Alonso JM (2009) Ethylene signaling and response: where different regulatory modules meet. *Curr. Opin. Plant Biol.* 12, 548–555.

The Cytoskeleton

For cells to function properly, they must organize themselves in space and interact mechanically with each other and with their environment. They have to be correctly shaped, physically robust, and properly structured internally. Many have to change their shape and move from place to place. All cells have to be able to rearrange their internal components as they grow, divide, and adapt to changing circumstances. These spatial and mechanical functions depend on a remarkable system of filaments called the **cytoskeleton** (Figure 16–1).

The cytoskeleton's varied functions depend on the behavior of three families of protein filaments—*actin filaments, microtubules,* and *intermediate filaments.* Each type of filament has distinct mechanical properties, dynamics, and biological roles, but all share certain fundamental features. Just as we require our ligaments, bones, and muscles to work together, so all three cytoskeletal filament systems must normally function collectively to give a cell its strength, its shape, and its ability to move.

In this chapter, we describe the function and conservation of the three main filament systems. We explain the basic principles underlying filament assembly and disassembly, and how other proteins interact with the filaments to alter their dynamics, enabling the cell to establish and maintain internal order, to shape and remodel its surface, and to move organelles in a directed manner from one place to another. Finally, we discuss how the integration and regulation of the cytoskeleton allows a cell to move to new locations.

FUNCTION AND ORIGIN OF THE CYTOSKELETON

The three major cytoskeletal filaments are responsible for different aspects of the cell's spatial organization and mechanical properties. Actin filaments determine the shape of the cell's surface and are necessary for whole-cell locomotion; they also drive the pinching of one cell into two. Microtubules determine the positions of membrane-enclosed organelles, direct intracellular transport, and form the mitotic spindle that segregates chromosomes during cell division. Intermediate filaments provide mechanical strength. All of these cytoskeletal filaments interact with hundreds of accessory proteins that regulate and link the filaments to other cell components, as well as to each other. The accessory proteins are essential for the controlled assembly of the cytoskeletal filaments in particular locations, and they include the *motor proteins*, remarkable molecular machines that convert the energy of ATP hydrolysis into mechanical force that can either move organelles along the filaments or move the filaments themselves.

In this section, we discuss the general features of the proteins that make up the filaments of the cytoskeleton. We focus on their ability to form intrinsically

IN THIS CHAPTER

FUNCTION AND ORIGIN OF THE CYTOSKELETON

ACTIN AND ACTIN-BINDING PROTEINS

MYOSIN AND ACTIN

MICROTUBULES

INTERMEDIATE FILAMENTS AND SEPTINS

CELL POLARIZATION AND MIGRATION

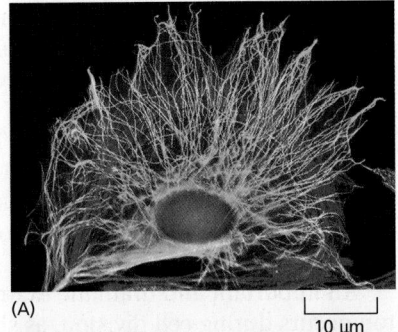

(A)   10 µm

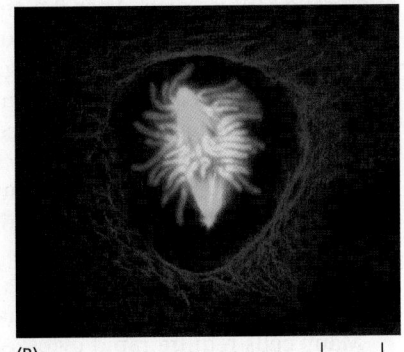

(B)   20 µm

Figure 16–1 The cytoskeleton. (A) A cell in culture has been fixed and labeled to show its cytoplasmic arrays of microtubules *(green)* and actin filaments *(red)*. (B) This dividing cell has been labeled to show its spindle microtubules *(green)* and surrounding cage of intermediate filaments *(red)*. The DNA in both cells is labeled in *blue*. (A, courtesy of Albert Tousson, High Resolution Imaging Facility, University of Alabama at Birmingham; B, courtesy of Conly Rieder.)

polarized and self-organized structures that are highly dynamic, allowing the cell to rapidly modify cytoskeletal structure and function under different conditions.

Cytoskeletal Filaments Adapt to Form Dynamic or Stable Structures

Cytoskeletal systems are dynamic and adaptable, organized more like ant trails than interstate highways. A single trail of ants may persist for many hours, extending from the ant nest to a delectable picnic site, but the individual ants within the trail are anything but static. If the ant scouts find a new and better source of food, or if the picnickers clean up and leave, the dynamic structure adapts with astonishing rapidity. In a similar way, large-scale cytoskeletal structures can change or persist, according to need, lasting for lengths of time ranging from less than a minute up to the cell's lifetime. But the individual macromolecular components that make up these structures are in a constant state of flux. Thus, like the alteration of an ant trail, a structural rearrangement in a cell requires little extra energy when conditions change.

Regulation of the dynamic behavior and assembly of cytoskeletal filaments allows eukaryotic cells to build an enormous range of structures from the three basic filament systems. The micrographs in **Panel 16–1** illustrate some of these structures. Actin filaments underlie the plasma membrane of animal cells, providing strength and shape to its thin lipid bilayer. They also form many types of cell-surface projections. Some of these are dynamic structures, such as the *lamellipodia* and *filopodia* that cells use to explore territory and move around. More stable arrays allow cells to brace themselves against an underlying substratum and enable muscle to contract. The regular bundles of *stereocilia* on the surface of hair cells in the inner ear contain stable bundles of actin filaments that tilt as rigid rods in response to sound, and similarly organized *microvilli* on the surface of intestinal epithelial cells vastly increase the apical cell-surface area to enhance nutrient absorption. In plants, actin filaments drive the rapid streaming of cytoplasm inside cells.

Microtubules, which are frequently found in a cytoplasmic array that extends to the cell periphery, can quickly rearrange themselves to form a bipolar *mitotic spindle* during cell division. They can also form *cilia*, which function as motile whips or sensory devices on the surface of the cell, or tightly aligned bundles that serve as tracks for the transport of materials down long neuronal axons. In plant cells, organized arrays of microtubules help to direct the pattern of cell wall synthesis, and in many protozoans they form the framework upon which the entire cell is built.

Intermediate filaments line the inner face of the nuclear envelope, forming a protective cage for the cell's DNA; in the cytosol, they are twisted into strong cables that can hold epithelial cell sheets together or help nerve cells to extend long and robust axons, and they allow us to form tough appendages such as hair and fingernails.

An important and dramatic example of rapid reorganization of the cytoskeleton occurs during cell division, as shown in **Figure 16–2** for a fibroblast growing in a tissue-culture dish. After the chromosomes have replicated, the interphase microtubule array that spreads throughout the cytoplasm is reconfigured into the bipolar *mitotic spindle*, which segregates the two copies of each chromosome into daughter nuclei. At the same time, the specialized actin structures that enable the fibroblast to crawl across the surface of the dish rearrange so that the cell stops moving and assumes a more spherical shape. Actin and its associated motor protein myosin then form a belt around the middle of the cell, the *contractile ring*, which constricts like a tiny muscle to pinch the cell in two. When division is complete, the cytoskeletons of the two daughter fibroblasts reassemble into their interphase structures to convert the two rounded-up daughter cells into smaller versions of the flattened, crawling mother cell.

Many cells require rapid cytoskeletal rearrangements for their normal functioning during interphase as well. For example, the *neutrophil*, a type of white

ACTIN FILAMENTS

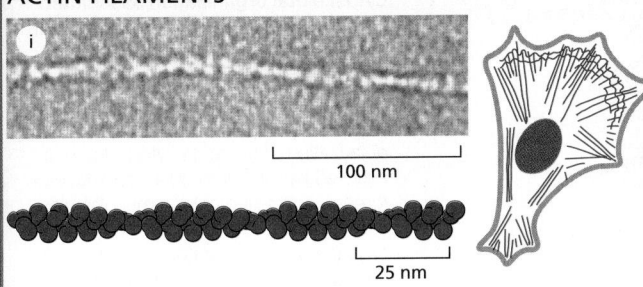

100 nm

25 nm

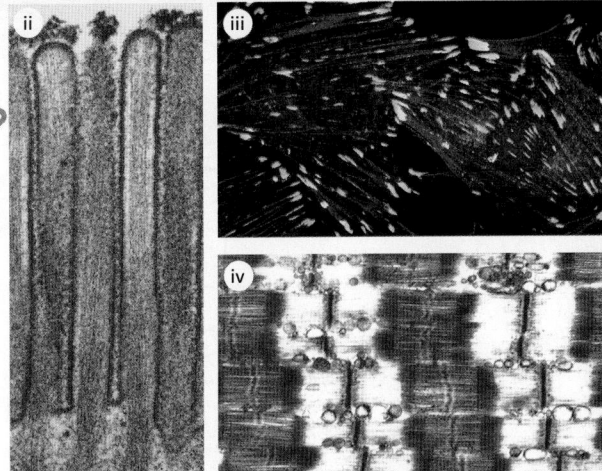

Actin filaments (also known as *microfilaments*) are helical polymers of the protein actin. They are flexible structures with a diameter of 8 nm that organize into a variety of linear bundles, two-dimensional networks, and three-dimensional gels. Although actin filaments are dispersed throughout the cell, they are most highly concentrated in the *cortex*, just beneath the plasma membrane. (i) Single actin filament; (ii) microvilli; (iii) stress fibers (*red*) terminating in focal adhesions (*green*); (iv) striated muscle.

Micrographs courtesy of R. Craig (i and iv); P.T. Matsudaira and D.R. Burgess (ii); K. Burridge (iii).

MICROTUBULES

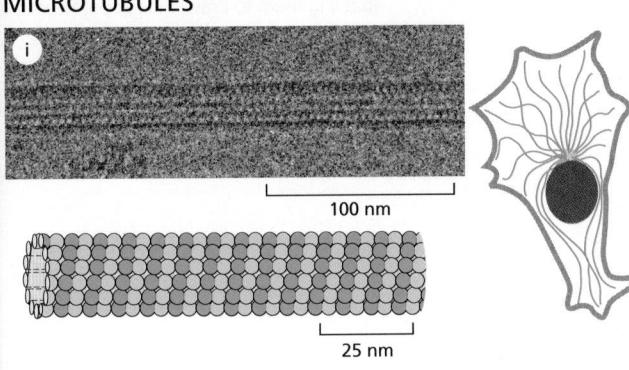

100 nm

25 nm

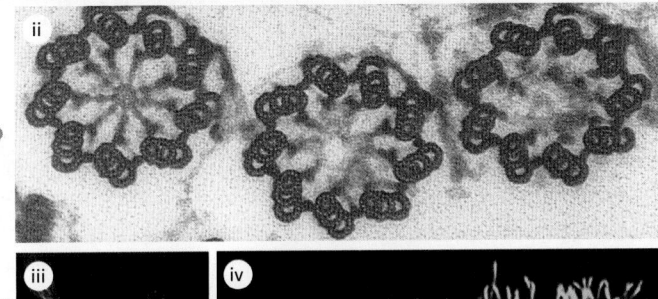

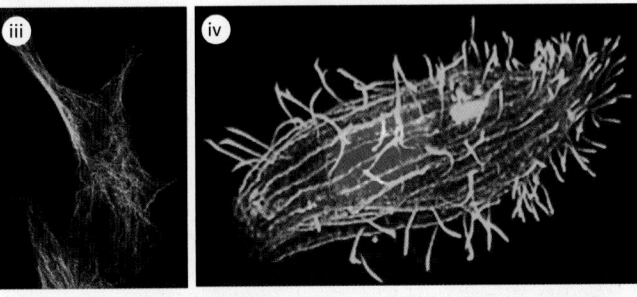

Microtubules are long, hollow cylinders made of the protein tubulin. With an outer diameter of 25 nm, they are much more rigid than actin filaments. Microtubules are long and straight and frequently have one end attached to a microtubule-organizing center (MTOC) called a *centrosome*. (i) Single microtubule; (ii) cross section at the base of three cilia showing triplet microtubules; (iii) interphase microtubule array (*green*) and organelles (*red*); (iv) ciliated protozoan.

Micrographs courtesy of R. Wade (i); D.T. Woodrow and R.W. Linck (ii); D. Shima (iii); D. Burnette (iv).

INTERMEDIATE FILAMENTS

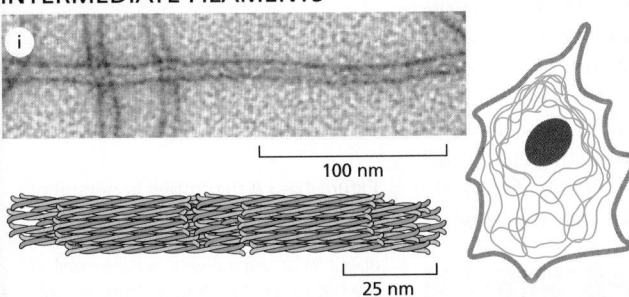

100 nm

25 nm

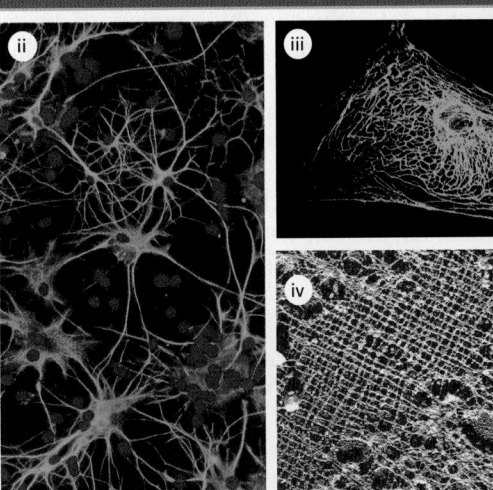

Intermediate filaments are ropelike fibers with a diameter of about 10 nm; they are made of intermediate filament proteins, which constitute a large and heterogeneous family. One type of intermediate filament forms a meshwork called the nuclear lamina just beneath the inner nuclear membrane. Other types extend across the cytoplasm, giving cells mechanical strength. In an epithelial tissue, they span the cytoplasm from one cell–cell junction to another, thereby strengthening the entire epithelium. (i) Individual intermediate filaments; (ii) Intermediate filaments (*blue*) in neurons and (iii) epithelial cell; (iv) nuclear lamina.

(i) L. Norlen et al., *Exp. Cell Res.* 313:2217-2227, 2007. With permission from Elsevier, (ii) N. Kedersha (iii) Alvin Tesler/Science Source, (iv) from U. Aebi, *Nature* 323:560-564, published 1986 by Nature Publishing Group. Reproduced with permission of SNCSC.

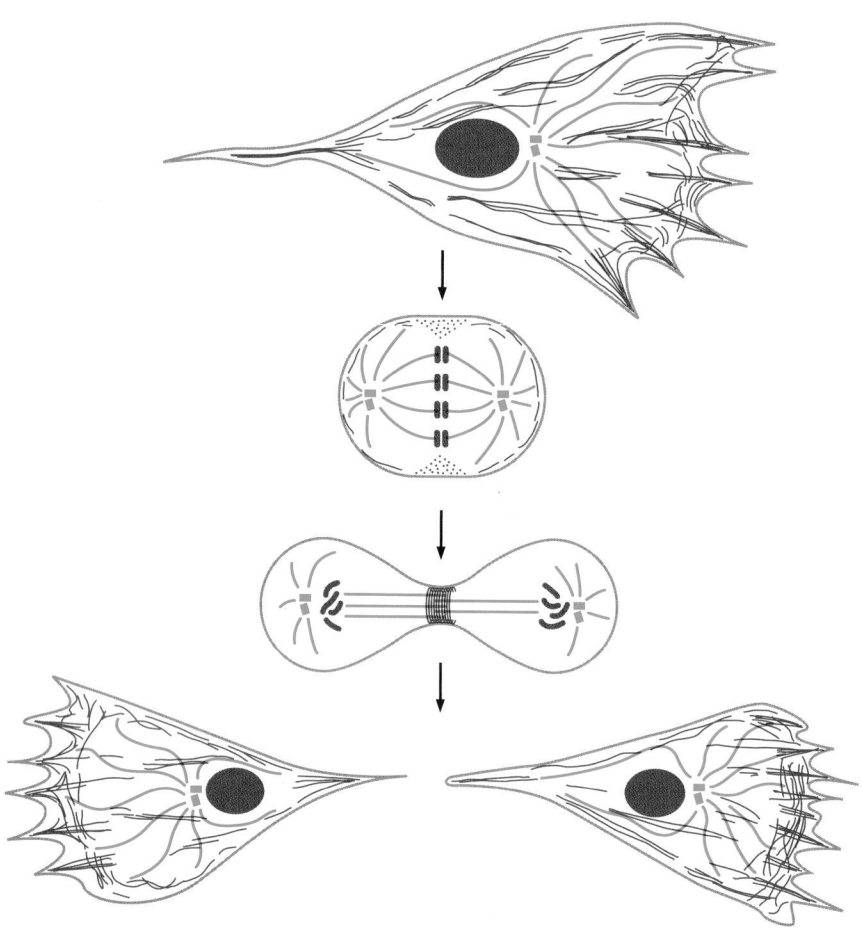

Figure 16–2 **Diagram of changes in cytoskeletal organization associated with cell division.** The crawling fibroblast drawn here has a polarized, dynamic actin cytoskeleton (shown in *red*) that assembles lamellipodia and filopodia to push its leading edge toward the right. The polarization of the actin cytoskeleton is assisted by the microtubule cytoskeleton *(green)*, consisting of long microtubules that emanate from a single microtubule-organizing center located in front of the nucleus. When the cell divides, the polarized microtubule array rearranges to form a bipolar mitotic spindle, which is responsible for aligning and then segregating the duplicated chromosomes *(brown)*. The actin filaments form a contractile ring at the center of the cell that pinches the cell in two after the chromosome segregation. After cell division is complete, the two daughter cells reorganize both the microtubule and actin cytoskeletons into smaller versions of those that were present in the mother cell, enabling them to crawl their separate ways.

blood cell, chases and engulfs bacterial and fungal cells that accidentally gain access to the normally sterile parts of the body, as through a cut in the skin. Like most crawling cells, neutrophils advance by extending a protrusive structure filled with newly polymerized actin filaments. When the elusive bacterial prey moves in a different direction, the neutrophil is poised to reorganize its polarized protrusive structures within seconds (**Figure 16–3**).

The Cytoskeleton Determines Cellular Organization and Polarity

In cells that have achieved a stable, differentiated morphology—such as mature neurons or epithelial cells—the dynamic elements of the cytoskeleton must also provide stable, large-scale structures for cellular organization. On specialized epithelial cells that line organs such as the intestine and the lung, cytoskeletal-based cell-surface protrusions including microvilli and cilia are able to maintain a constant location, length, and diameter over the entire lifetime of the cell. For the actin bundles at the cores of microvilli on intestinal epithelial cells, this is only a few days. But the actin bundles at the cores of stereocilia on the hair cells of the inner ear must maintain their stable organization for the entire lifetime of the animal, since these cells do not turn over. Nonetheless, the individual actin filaments

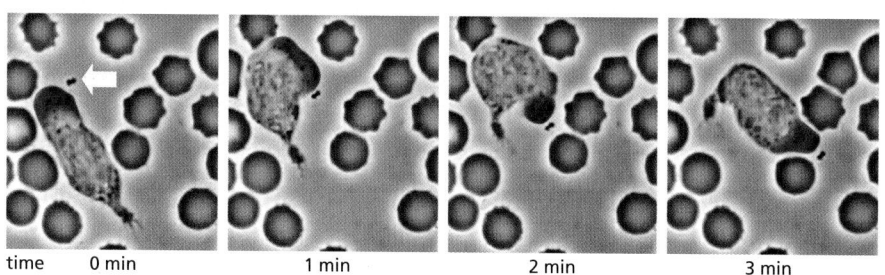

time   0 min   1 min   2 min   3 min

Figure 16–3 **A neutrophil in pursuit of bacteria.** In this preparation of human blood, a clump of bacteria *(white arrow)* is about to be captured by a neutrophil. As the bacteria move, the neutrophil quickly reassembles the dense actin network at its leading edge (highlighted in *red*) to push toward the location of the bacteria (**Movie 16.1**). Rapid disassembly and reassembly of the actin cytoskeleton in this cell enables it to change its orientation and direction of movement within a few minutes. (From a video recorded by David Rogers.)

remain strikingly dynamic and are continuously remodeled and replaced every 48 hours, even within stable cell-surface structures that persist for decades.

Besides forming stable, specialized cell-surface protrusions, the cytoskeleton is also responsible for large-scale cellular polarity, enabling cells to tell the difference between top and bottom, or front and back. The large-scale polarity information conveyed by cytoskeletal organization is often maintained over the lifetime of the cell. Polarized epithelial cells use organized arrays of microtubules, actin filaments, and intermediate filaments to maintain the critical differences between the *apical surface* and the *basolateral surface*. They also must maintain strong adhesive contacts with one another to enable this single layer of cells to serve as an effective physical barrier (**Figure 16–4**).

Filaments Assemble from Protein Subunits That Impart Specific Physical and Dynamic Properties

Cytoskeletal filaments can reach from one end of the cell to the other, spanning tens or even hundreds of micrometers. Yet the individual protein molecules that form the filaments are only a few nanometers in size. The cell builds the filaments by assembling large numbers of the small subunits, like building a skyscraper out of bricks. Because these subunits are small, they can diffuse rapidly in the cytosol, whereas the assembled filaments cannot. In this way, cells can undergo rapid structural reorganizations, disassembling filaments at one site and reassembling them at another site far away.

Actin filaments and microtubules are built from subunits that are compact and globular—*actin subunits* for actin filaments and *tubulin subunits* for microtubules—whereas intermediate filaments are made up of smaller subunits that are themselves elongated and fibrous. All three major types of cytoskeletal filaments form as helical assemblies of subunits (see Figure 3–22) that self-associate, using a combination of end-to-end and side-to-side protein contacts. Differences in the structures of the subunits and the strengths of the attractive forces between

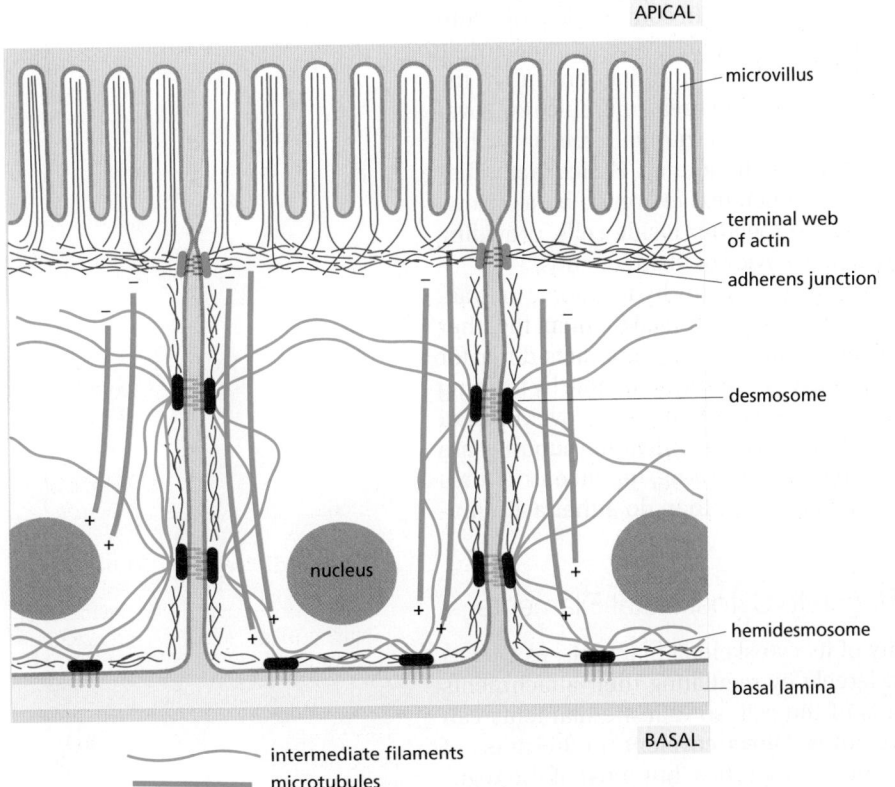

APICAL

microvillus

terminal web of actin

adherens junction

desmosome

nucleus

hemidesmosome

basal lamina

BASAL

intermediate filaments
microtubules
actin microfilaments

Figure 16–4 Organization of the cytoskeleton in polarized epithelial cells. All the components of the cytoskeleton cooperate to produce the characteristic shapes of specialized cells, including the epithelial cells that line the small intestine, diagrammed here. At the apical (upper) surface, facing the intestinal lumen, bundled actin filaments *(red)* form microvilli that increase the cell surface area available for absorbing nutrients from food. Below the microvilli, a circumferential band of actin filaments is connected to cell–cell adherens junctions that anchor the cells to each other. Intermediate filaments *(blue)* are anchored to other kinds of adhesive structures, including desmosomes and hemidesmosomes, that connect the epithelial cells into a sturdy sheet and attach them to the underlying extracellular matrix; these structures are discussed in Chapter 19. Microtubules *(green)* run vertically from the top of the cell to the bottom and provide a global coordinate system that enables the cell to direct newly synthesized components to their proper locations.

them produce important differences in the stability and mechanical properties of each type of filament. Whereas covalent linkages between their subunits hold together the backbones of many biological polymers—including DNA, RNA, and proteins—it is weak noncovalent interactions that hold together the three types of cytoskeletal polymers. Consequently, their assembly and disassembly can occur rapidly, without covalent bonds being formed or broken.

The subunits of actin filaments and microtubules are asymmetrical and bind to one another head-to-tail so that they all point in one direction. This subunit polarity gives the filaments structural polarity along their length, and makes the two ends of each polymer behave differently. In addition, actin and tubulin subunits are both enzymes that catalyze the hydrolysis of a nucleoside triphosphate—ATP and GTP, respectively. As we discuss later, the energy derived from nucleotide hydrolysis enables the filaments to remodel rapidly. By controlling when and where actin and microtubules assemble, the cell harnesses the polar and dynamic properties of these filaments to generate force in a specific direction, to move the leading edge of a migrating cell forward, for example, or to pull chromosomes apart during cell division. In contrast, the subunits of intermediate filaments are symmetrical, and thus do not form polarized filaments with two different ends. Intermediate filament subunits also do not catalyze the hydrolysis of nucleotides. Nevertheless, intermediate filaments can be disassembled rapidly when required. In mitosis , for example, kinases phosphorylate the subunits, leading to their dissociation.

Cytoskeletal filaments in living cells are not built by simply stringing subunits together in single file. A thousand tubulin subunits lined up end-to-end, for example, would span the diameter of a small eukaryotic cell, but a filament formed in this way would lack the strength to avoid breakage by ambient thermal energy, unless each subunit in the filament was bound extremely tightly to its two neighbors. Such tight binding would limit the rate at which the filaments could disassemble, making the cytoskeleton a static and less useful structure. To provide both strength and adaptability, microtubules are built of 13 **protofilaments**—linear strings of subunits joined end-to-end—that associate with one another laterally to form a hollow cylinder. The addition or loss of a subunit at the end of one protofilament makes or breaks a small number of bonds. In contrast, loss of a subunit from the middle of the filament requires breaking many more bonds, while breaking it in two requires breaking bonds in multiple protofilaments all at the same time (**Figure 16–5**). The greater energy required to break multiple noncovalent bonds simultaneously allows microtubules to resist thermal breakage, while allowing rapid subunit addition and loss at the filament ends. Helical actin filaments are much thinner and therefore require much less energy to break. However, multiple actin filaments are often bundled together inside cells, providing mechanical strength, while allowing dynamic behavior of filament ends.

As with other specific protein–protein interactions, many hydrophobic interactions and noncovalent bonds hold the subunits in a cytoskeletal filament together (see Figure 3–4). The locations and types of subunit–subunit contacts differ for the different filaments. Intermediate filaments, for example, assemble by forming strong lateral contacts between α-helical coiled-coils, which extend over most of the length of each elongated fibrous subunit. Because the individual subunits are staggered in the filament, intermediate filaments form strong, ropelike structures that tolerate stretching and bending to a greater extent than do either actin filaments or microtubules (**Figure 16–6**).

Accessory Proteins and Motors Regulate Cytoskeletal Filaments

The cell regulates the length and stability of its cytoskeletal filaments, as well as their number and geometry. It does so largely by regulating their attachments to one another and to other components of the cell, so that the filaments can form a wide variety of higher-order structures. Direct covalent modification of the filament subunits regulates some filament properties, but most of the regulation is performed by hundreds of accessory proteins that determine the spatial

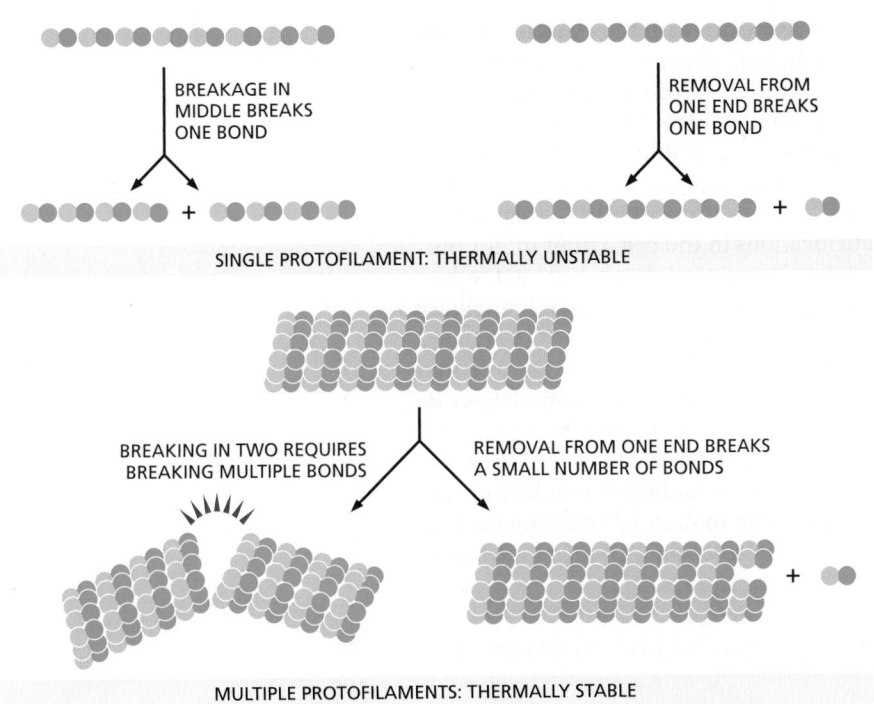

Figure 16–5 **The thermal stability of cytoskeletal filaments with dynamic ends.** A protofilament consisting of a single strand of subunits is thermally unstable, since breakage of a single bond between subunits is sufficient to break the filament. In contrast, formation of a cytoskeletal filament from more than one protofilament allows the ends to be dynamic, while enabling the filaments themselves to be resistant to thermal breakage. In a microtubule, for example, removing a single subunit dimer from the end of the filament requires breaking noncovalent bonds with a maximum of three other subunits, whereas fracturing the filament in the middle requires breaking noncovalent bonds in all thirteen protofilaments.

distribution and the dynamic behavior of the filaments, converting information received through signaling pathways into cytoskeletal action. These accessory proteins bind to the filaments or their subunits to determine the sites of assembly of new filaments, to regulate the partitioning of polymer proteins between filament and subunit forms, to change the kinetics of filament assembly and disassembly, to harness energy to generate force, and to link filaments to one another or to other cell structures such as organelles and the plasma membrane. In these processes, the accessory proteins bring cytoskeletal structure under the control of extracellular and intracellular signals, including those that trigger the dramatic transformations of the cytoskeleton that occur during each cell cycle. Acting together, the accessory proteins enable a eukaryotic cell to maintain a highly organized but flexible internal structure and, in many cases, to move.

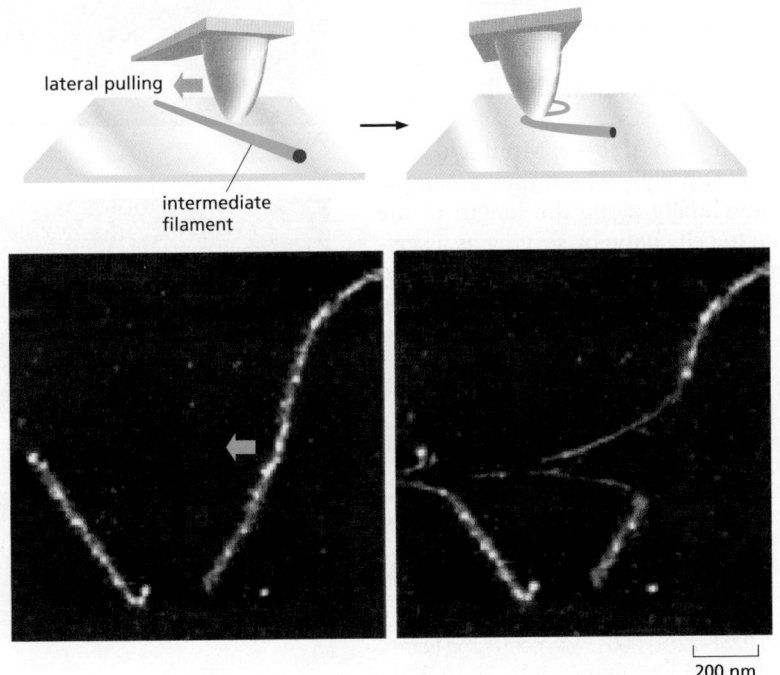

Figure 16–6 **Flexibility and stretch in an intermediate filament.** Intermediate filaments are formed from elongated fibrous subunits with strong lateral contacts, resulting in resistance to stretching forces. When a tiny mechanical probe is dragged across an intermediate filament, the filament is stretched over three times its length before it breaks, as illustrated by the fluorescently labeled filaments in the photomicrographs. This technique is termed atomic force microscopy (see Figure 9–33). (Adapted from L. Kreplak et al., *J. Mol. Biol.* 354:569–577, 2005. With permission from Elsevier.)

Among the most fascinating proteins that associate with the cytoskeleton are the **motor proteins**. These proteins bind to a polarized cytoskeletal filament and use the energy derived from repeated cycles of ATP hydrolysis to move along it. Dozens of different motor proteins coexist in every eukaryotic cell. They differ in the type of filament they bind to (either actin or microtubules), the direction in which they move along the filament, and the "cargo" they carry. Many motor proteins carry membrane-enclosed organelles—such as mitochondria, Golgi stacks, or secretory vesicles—to their appropriate locations in the cell. Other motor proteins cause cytoskeletal filaments to exert tension or to slide against each other, generating the force that drives such phenomena as muscle contraction, ciliary beating, and cell division.

Cytoskeletal motor proteins that move unidirectionally along an oriented polymer track are reminiscent of some other proteins and protein complexes discussed elsewhere in this book, such as DNA and RNA polymerases, helicases, and ribosomes. All of these proteins have the ability to use chemical energy to propel themselves along a linear track, with the direction of sliding dependent on the structural polarity of the track. All of them generate motion by coupling nucleoside triphosphate hydrolysis to a large-scale conformational change (see Figure 3–75).

Bacterial Cell Organization and Division Depend on Homologs of Eukaryotic Cytoskeletal Proteins

While eukaryotic cells are typically large and morphologically complex, bacterial cells are usually only a few micrometers long and assume simple shapes such as spheres or rods. Bacteria also lack elaborate networks of intracellular membrane-enclosed organelles. Historically, biologists assumed that a cytoskeleton was not necessary in such simple cells. We now know, however, that bacteria contain homologs of all three of the eukaryotic cytoskeletal filaments. Furthermore, bacterial actins and tubulins are more diverse than their eukaryotic versions, both in the types of assemblies they form and in the functions they carry out.

Nearly all bacteria and many archaea contain a homolog of tubulin called FtsZ, which can polymerize into filaments and assemble into a ring (called the Z-ring) at the site where the **septum** forms during cell division (**Figure 16–7**). Although the Z-ring persists for many minutes, the individual filaments within it are highly dynamic, with an average filament half-life of about thirty seconds. As the bacterium divides, the Z-ring becomes smaller until it has completely disassembled. FtsZ filaments in the Z-ring are thought to generate a bending force that drives the membrane invagination necessary to complete cell division. The Z-ring may also serve as a site for localization of enzymes required for building the septum between the two daughter cells.

Many bacteria also contain homologs of actin. Two of these, MreB and Mbl, are found primarily in rod-shaped or spiral-shaped cells where they assemble to form dynamic patches that move circumferentially along the length of the cell (**Figure 16–8A**). These proteins contribute to cell shape by serving as a scaffold to direct the synthesis of the peptidoglycan cell wall, in much the same way that microtubules help organize the synthesis of the cellulose cell wall in higher

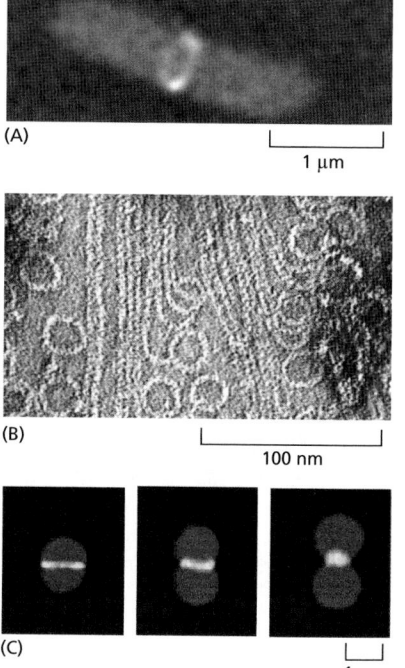

(A)   1 μm

(B)   100 nm

(C)   1 μm

Figure 16–7 **The bacterial FtsZ protein, a tubulin homolog in prokaryotes.** (A) A band of FtsZ protein forms a ring in a dividing bacterial cell. This ring has been labeled by fusing the FtsZ protein to green fluorescent protein (GFP), which allows it to be observed in living *E. coli* cells with a fluorescence microscope. (B) FtsZ filaments and circles, formed *in vitro*, as visualized using electron microscopy. (C) Dividing chloroplasts *(red)* from a red alga also cleave using a protein ring made from FtsZ *(yellow)*. (A, from X. Ma, D.W. Ehrhardt and W. Margolin, *Proc. Natl Acad. Sci. USA* 93:12998–13003, 1996; B, from H.P. Erickson et al., *Proc. Natl Acad. Sci. USA* 93:519–523, 1996. Both with permission from National Academy of Sciences; C, from S. Miyagishima et al., *Plant Cell* 13:2257–2268, 2001, with permission from American Society of Plant Biologists.)

Figure 16–8 Actin homologs in bacteria determine cell shape. (A) The MreB protein forms abundant patches made up of many short, interwoven linear or helical filaments that are seen to move circumferentially along the length of the bacterium and are associated with sites of cell wall synthesis. (B) The common soil bacterium *Bacillus subtilis* normally forms cells with a regular rodlike shape when viewed by scanning electron microscopy *(left)*. In contrast, *B. subtilis* cells lacking the actin homolog MreB or Mbl grow in distorted or twisted shapes and eventually die *(center and right)*. (A, from P. Vats and L. Rothfield, *Proc. Natl Acad. Sci. USA* 104:17795–17800, 2007. With permission from National Academy of Sciences; B, from A. Chastanet and R. Carballido-Lopez, *Front. Biosci.* 4S:1582–1606, 2012. With permission Frontiers in Bioscience.)

plant cells (see Figure 19–65). As with FtsZ, MreB and Mbl filaments are highly dynamic, with half-lives of a few minutes, and nucleotide hydrolysis accompanies the polymerization process. Mutations disrupting MreB or Mbl expression cause extreme abnormalities in cell shape and defects in chromosome segregation (Figure 16–8B).

Relatives of MreB and Mbl have more specialized roles. A particularly intriguing bacterial actin homolog is ParM, which is encoded by a gene on certain bacterial plasmids that also carry genes responsible for antibiotic resistance and cause the spread of multidrug resistance in epidemics. Bacterial plasmids typically encode all the gene products that are necessary for their own segregation, presumably as a strategy to ensure their inheritance and propagation in bacterial hosts following plasmid replication. ParM assembles into filaments that associate at each end with a copy of the plasmid, and growth of the ParM filament pushes the replicated plasmid copies apart (**Figure 16–9**). This spindle-like structure apparently arises from the selective stabilization of filaments that bind to specialized proteins recruited to the origins of replication on the plasmids. A distant relative of both tubulin and FtsZ, called TubZ, has a similar function in other bacterial species.

Thus, self-association of nucleotide-binding proteins into dynamic filaments is used in all cells, and the actin and tubulin families are very ancient, predating the split between the eukaryotic and bacterial kingdoms.

At least one bacterial species, *Caulobacter crescentus*, appears to harbor a protein with significant structural similarity to the third major class of cytoskeletal filaments found in animal cells, the intermediate filaments. A protein called crescentin forms a filamentous structure that influences the unusual crescent shape of this species; when the gene encoding crescentin is deleted, the *Caulobacter* cells grow as straight rods (**Figure 16–10**).

Figure 16–9 Role of the actin homolog ParM in plasmid segregation in bacteria.
(A) Some bacterial drug-resistance plasmids *(orange)* encode an actin homolog, ParM, that will spontaneously nucleate to form small, dynamic filaments *(green)* throughout the bacterial cytoplasm. A second plasmid-encoded protein called ParR *(blue)* binds to specific DNA sequences in the plasmid and also stabilizes the dynamic ends of the ParM filaments. When the plasmid duplicates, both ends of the ParM filaments become stabilized, and the growing ParM filaments push the duplicated plasmids to opposite ends of the cell. (B) In these bacterial cells harboring a drug-resistance plasmid, the plasmids are labeled in *red* and the ParM protein in *green*. Left, a short ParM filament bundle connects the two daughter plasmids shortly after their duplication. Right, the fully assembled ParM filament has pushed the duplicated plasmids to the cell poles. (A, adapted from E.C. Garner, C.S. Campbell and R.D. Mullins, *Science* 306:1021–1025, 2004; B, from J. Møller-Jensen et al., *Mol. Cell* 12:1477–1487, 2003. With permission from Elsevier.)

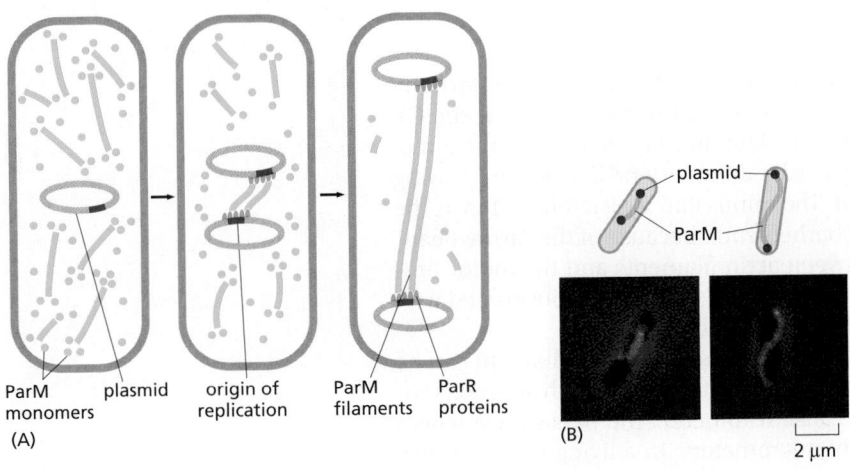

ParM monomers   plasmid   origin of replication   ParM filaments   ParR proteins

(A)

plasmid
ParM

(B)   2 µm

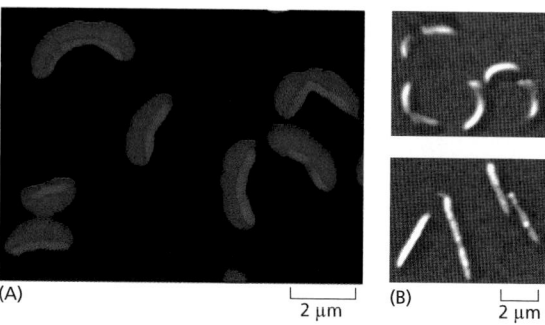

(A)   2 µm   (B)   2 µm

Figure 16–10 *Caulobacter* **and crescentin.** The sickle-shaped bacterium *Caulobacter crescentus* expresses a protein, crescentin, with a series of coiled-coil domains similar in size and organization to the domains of eukaryotic intermediate filaments. (A) The crescentin protein forms a fiber (labeled in *red*) that runs down the inner side of the curving bacterial cell wall. (B) When the gene is disrupted, the bacteria grow as straight rods *(bottom).* (From N. Ausmees, J.R. Kuhn and C. Jacobs-Wagner, *Cell* 115:705–713, 2003. With permission from Elsevier.)

Summary

The cytoplasm of eukaryotic cells is spatially organized by a network of protein filaments known as the cytoskeleton. This network contains three principal types of filaments: actin filaments, microtubules, and intermediate filaments. All three types of filaments form as helical assemblies of subunits that self-associate using a combination of end-to-end and side-to-side protein contacts. Differences in the structure of the subunits and the manner of their self-assembly give the filaments different mechanical properties. Subunit assembly and disassembly constantly remodel all three types of cytoskeletal filaments. Actin and tubulin (the subunits of actin filaments and microtubules, respectively) bind and hydrolyze nucleoside triphosphates (ATP and GTP, respectively), and assemble head-to-tail to generate polarized filaments capable of generating force. In living cells, accessory proteins modulate the dynamics and organization of cytoskeletal filaments, resulting in complex events such as cell division or migration, and generating elaborate cellular architecture to form polarized tissues such as epithelia. Bacterial cells also contain homologs of actin, tubulin, and intermediate filaments that form dynamic structures that help control cell shape and division.

ACTIN AND ACTIN-BINDING PROTEINS

The actin cytoskeleton performs a wide range of functions in diverse cell types. Each actin subunit, sometimes called globular or G-actin, is a 375-amino-acid polypeptide carrying a tightly associated molecule of ATP or ADP (**Figure 16–11**A). Actin is extraordinarily well conserved among eukaryotes. The amino acid sequences of actins from different eukaryotic species are usually about 90% identical. Small variations in actin amino acid sequence can cause significant functional differences: In vertebrates, for example, there are three isoforms of actin, termed α, β, and γ, that differ slightly in their amino acid sequences and have distinct functions. α-Actin is expressed only in muscle cells, while β- and γ-actins are found together in almost all non-muscle cells.

Actin Subunits Assemble Head-to-Tail to Create Flexible, Polar Filaments

Actin subunits assemble head-to-tail to form a tight, right-handed helix, forming a structure about 8 nm wide called filamentous or F-actin (Figure 16–11B and C). Because the asymmetrical actin subunits of a filament all point in the same direction, filaments are polar and have structurally different ends: a slower-growing *minus end* and a faster-growing *plus end*. The minus end is also referred to as the "pointed end" and the plus end as the "barbed end," because of the "arrowhead" appearance of the complex formed between actin filaments and the motor protein myosin (**Figure 16–12**). Within the filament, the subunits are positioned with their nucleotide-binding cleft directed toward the minus end.

    Individual actin filaments are quite flexible. The stiffness of a filament can be characterized by its *persistence length*, the minimum filament length at which random thermal fluctuations are likely to cause it to bend. The persistence length of an actin filament is only a few tens of micrometers. In a living cell, however,

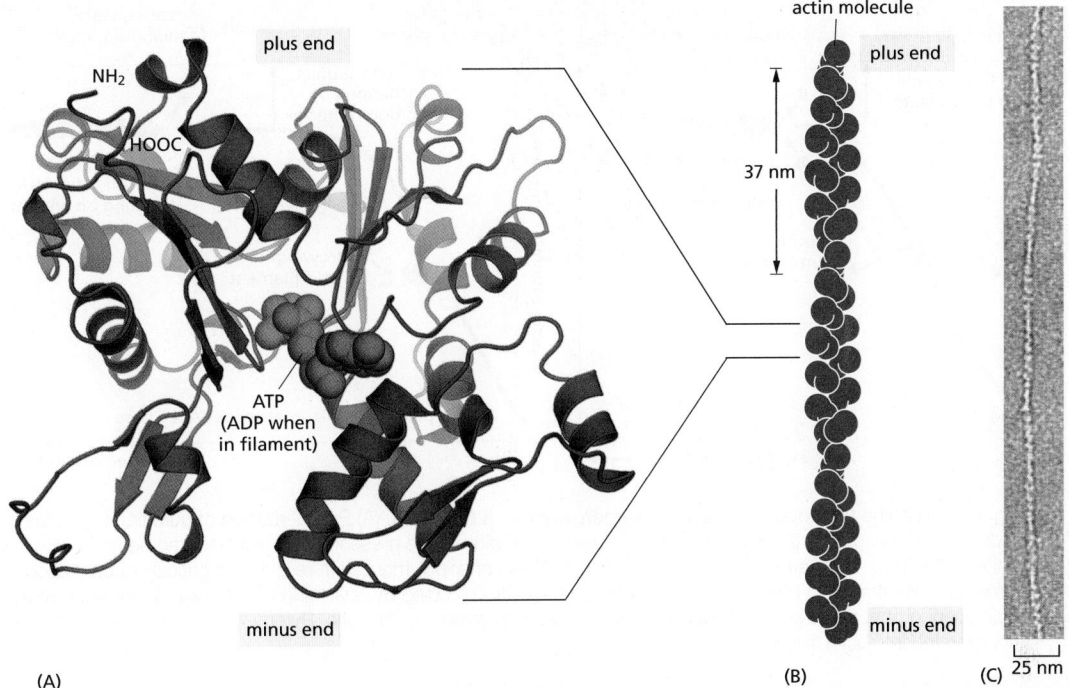

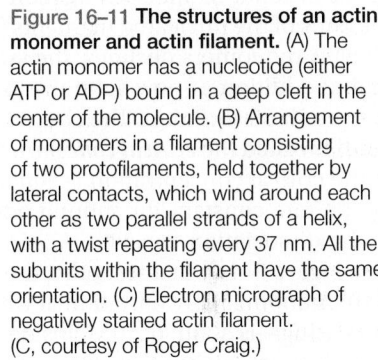

Figure 16–11 **The structures of an actin monomer and actin filament.** (A) The actin monomer has a nucleotide (either ATP or ADP) bound in a deep cleft in the center of the molecule. (B) Arrangement of monomers in a filament consisting of two protofilaments, held together by lateral contacts, which wind around each other as two parallel strands of a helix, with a twist repeating every 37 nm. All the subunits within the filament have the same orientation. (C) Electron micrograph of negatively stained actin filament. (C, courtesy of Roger Craig.)

accessory proteins cross-link and bundle the filaments together, making large-scale actin structures that are much more rigid than an individual actin filament.

Nucleation Is the Rate-Limiting Step in the Formation of Actin Filaments

The regulation of actin filament formation is an important mechanism by which cells control their shape and movement. Small oligomers of actin subunits can assemble spontaneously, but they are unstable and disassemble readily because each monomer is bound to only one or two other monomers. For a new actin filament to form, subunits must assemble into an initial aggregate, or nucleus, that is stabilized by multiple subunit–subunit contacts and can then elongate rapidly by addition of more subunits. This process is called filament *nucleation*.

Many features of actin nucleation and polymerization have been studied with purified actin in a test tube (**Figure 16–13**). The instability of smaller actin aggregates creates a kinetic barrier to nucleation. When polymerization is initiated, this barrier results in a lag phase during which no filaments are observed. During this lag phase, however, a few of the small, unstable aggregates succeed in making the transition to a more stable form that resembles an actin filament. This leads to a

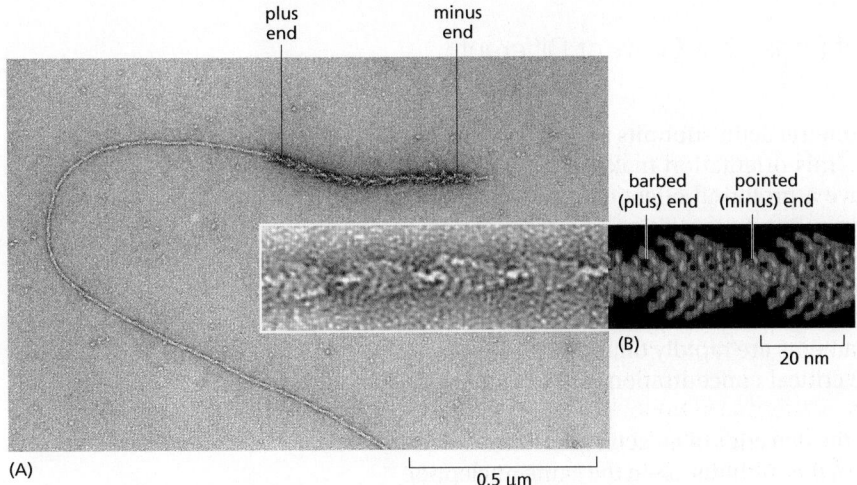

Figure 16–12 **Structural polarity of the actin filament.** (A) This electron micrograph shows an actin filament polymerized from a short actin filament seed that was decorated with myosin motor domains, resulting in an arrowhead pattern. The filament has grown much faster at the barbed (plus) end than at the pointed (minus) end. (B) Enlarged image and model showing the arrowhead pattern. (A, courtesy of Tom Pollard; B, adapted from M. Whittaker, B.O. Carragher, and R.A. Milligan, *Ultramicro.* 54:245–260, 1995. With permission from Elsevier.)

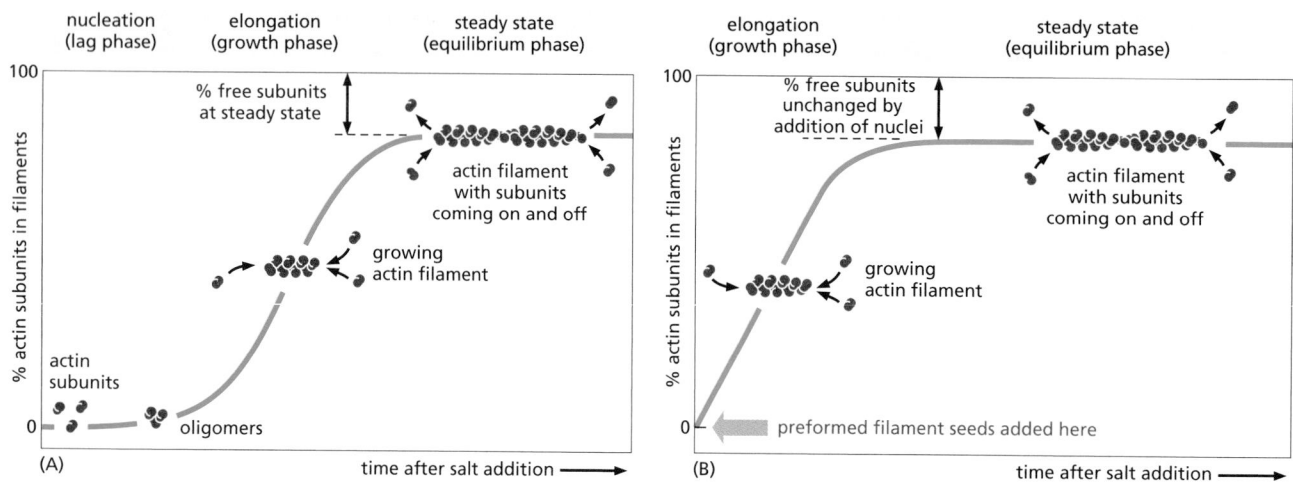

Figure 16–13 The time course of actin polymerization in a test tube. (A) Polymerization of pure actin subunits into filaments occurs after a lag phase. (B) Polymerization occurs more rapidly in the presence of preformed fragments of actin filaments, which act as nuclei for filament growth. The % free subunits after polymerization reflects the critical concentration (C_c), at which there is no net change in polymer. Actin polymerization is often studied by observing the change in the light emission from a fluorescent probe, called pyrene, that has been covalently attached to the actin. Pyrene-actin fluoresces more brightly when it is incorporated into actin filaments.

phase of rapid filament elongation during which subunits are added quickly to the ends of the nucleated filaments (Figure 16–13A). Finally, as the concentration of actin monomers declines, the system approaches a steady state at which the rate of addition of new subunits to the filament ends exactly balances the rate of subunit dissociation. The concentration of free subunits left in solution at this point is called the *critical concentration*, C_c. As explained in **Panel 16–2**, the value of the critical concentration is equal to the rate constant for subunit loss divided by the rate constant for subunit addition—that is, $C_c = k_{off}/k_{on}$, which is equal to the dissociation constant, K_d, and the inverse of the equilibrium constant, K (see Figure 3–44). In a test tube, the C_c for actin polymerization—that is, the free actin monomer concentration at which the fraction of actin in the polymer stops increasing—is about 0.2 µM. Inside the cell, the concentration of unpolymerized actin is much higher than this, and the cell has evolved mechanisms to prevent most of its monomeric actin from assembling into filaments, as we discuss later.

The lag phase in filament growth is eliminated if preexisting seeds (such as fragments of actin filaments that have been chemically cross-linked) are added to the solution at the beginning of the polymerization reaction (Figure 16–13B). The cell takes great advantage of this nucleation requirement: it uses special proteins to catalyze filament nucleation at specific sites, thereby determining the location at which new actin filaments are assembled.

Actin Filaments Have Two Distinct Ends That Grow at Different Rates

Due to the uniform orientation of asymmetric actin subunits in the filament, the structures at its two ends are different. This orientation makes the two ends of each polymer different in ways that have a profound effect on filament growth rates. The kinetic rate constants for actin subunit association and dissociation—k_{on} and k_{off}, respectively—are much greater at the plus end than the minus end. This can be seen when an excess of purified actin monomers is allowed to assemble onto polarity-marked filaments—the plus end of the filament elongates up to ten times faster (see Figure 16–12). If filaments are rapidly diluted so that the free subunit concentration drops below the critical concentration, the plus end also depolymerizes faster.

It is important to note, however, that the two ends of an actin filament have the same net affinity for actin subunits, if all of the subunits are in the same nucleotide state. Addition of a subunit to either end of a filament of n subunits results in a filament of $n + 1$ subunits. Thus, the free-energy difference, and therefore the equilibrium constant (and the critical concentration), must be the same for addition of subunits at either end of the polymer. In this case, the ratio of the rate constants, k_{off}/k_{on}, must be identical at the two ends, even though the absolute values of these rate constants are very different at each end (see Panel 16–2).

The cell takes advantage of actin filament dynamics and polarity to do mechanical work. Filament elongation proceeds spontaneously when the free-energy change (ΔG) for addition of the soluble subunit is less than zero. This is the case when the concentration of subunits in solution exceeds the critical concentration. A cell can couple an energetically unfavorable process to this spontaneous process; thus, the cell can use free energy released during spontaneous filament polymerization to move an attached load. For example, by orienting the fast-growing plus ends of actin filaments toward its leading edge, a motile cell can push its plasma membrane forward, as we discuss later.

ATP Hydrolysis Within Actin Filaments Leads to Treadmilling at Steady State

Thus far in our discussion of actin filament dynamics, we have ignored the critical fact that actin can catalyze the hydrolysis of the nucleoside triphosphate ATP. For free actin subunits, this hydrolysis proceeds very slowly; however, it is accelerated when the subunits are incorporated into filaments. Shortly after ATP hydrolysis occurs, the free phosphate group is released from each subunit, but the ADP remains trapped in the filament structure. Thus, two different types of filament structures can exist, one with the "T form" of the nucleotide bound (ATP), and one with the "D form" bound (ADP).

When the nucleotide is hydrolyzed, much of the free energy released by cleavage of the phosphate–phosphate bond is stored in the polymer. This makes the free-energy change for dissociation of a subunit from the D-form polymer more negative than the free-energy change for dissociation of a subunit from the T-form polymer. Consequently, the ratio of k_{off}/k_{on} for the D-form polymer, which is numerically equal to its critical concentration [$C_c(D)$], is larger than the corresponding ratio for the T-form polymer. Thus, $C_c(D)$ is greater than $C_c(T)$. At certain concentrations of free subunits, D-form polymers will therefore shrink while T-form polymers grow.

In living cells, most soluble actin subunits are in the T form, as the free concentration of ATP is about tenfold higher than that of ADP. However, the longer the time that subunits have been in the actin filament, the more likely they are to have hydrolyzed their ATP. Whether the subunit at each end of a filament is in the T or the D form depends on the rate of this hydrolysis compared with the rate of subunit addition. If the concentration of actin monomers is greater than the critical concentration for both the T-form and D-form polymer, then subunits will add to the polymer at both ends before the nucleotides in the previously added subunits are hydrolyzed; as a result, the tips of the actin filament will remain in the T form. On the other hand, if the subunit concentration is less than the critical concentrations for both the T-form and D-form polymer, then hydrolysis may occur before the next subunit is added and both ends of the filament will be in the D form and will shrink. At intermediate concentrations of actin subunits, it is possible for the rate of subunit addition to be faster than nucleotide hydrolysis at the plus end, but slower than nucleotide hydrolysis at the minus end. In this case, the plus end of the filament remains in the T conformation, while the minus end adopts the D conformation. The filament then undergoes a net addition of subunits at the plus end, while simultaneously losing subunits from the minus end. This leads to the remarkable property of filament **treadmilling** (**Figure 16–14**; see Panel 16–2).

At a particular intermediate subunit concentration, the filament growth at the plus end exactly balances the filament shrinkage at the minus end. Under these conditions, the subunits cycle rapidly between the free and filamentous states,

ON RATES AND OFF RATES

A linear polymer of protein molecules, such as an actin filament or a microtubule, assembles (polymerizes) and disassembles (depolymerizes) by the addition and removal of subunits at the ends of the polymer. The rate of addition of these subunits (called monomers) is given by the rate constant k_{on}, which has units of $M^{-1} sec^{-1}$. The rate of loss is given by k_{off} (units of sec^{-1}).

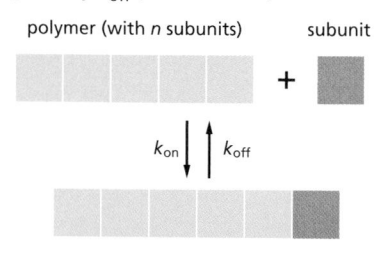

polymer (with n subunits)   subunit

k_{on} ↓ ↑ k_{off}

polymer (with $n+1$ subunits)

NUCLEATION

A helical polymer is stabilized by multiple contacts between adjacent subunits. In the case of actin, two actin molecules bind relatively weakly to each other, but addition of a third actin monomer to form a trimer makes the entire group more stable.

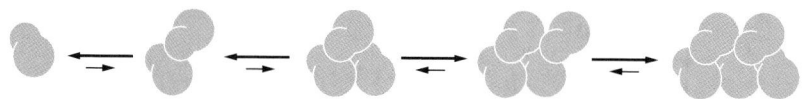

Further monomer addition can take place onto this trimer, which therefore acts as a nucleus for polymerization. For tubulin, the nucleus is larger and has a more complicated structure (possibly a ring of 13 or more tubulin molecules)—but the principle is the same.

The assembly of a nucleus is relatively slow, which explains the lag phase seen during polymerization. The lag phase can be reduced or abolished entirely by adding premade nuclei, such as fragments of already polymerized microtubules or actin filaments.

THE CRITICAL CONCENTRATION

The number of monomers that add to the polymer (actin filament or microtubule) per second will be proportional to the concentration of the free subunit ($k_{on}C$), but the subunits will leave the polymer end at a constant rate (k_{off}) that does not depend on C. As the polymer grows, subunits are used up, and C is observed to drop until it reaches a constant value, called the critical concentration (C_c). At this concentration, the rate of subunit addition equals the rate of subunit loss.

At this equilibrium,

$$k_{on} C = k_{off}$$

so that

$$C_c = \frac{k_{off}}{k_{on}} = K_d$$

(where K_d is the dissociation constant; see Figure 3–44).

TIME COURSE OF POLYMERIZATION

The assembly of a protein into a long helical polymer such as a cytoskeletal filament or a bacterial flagellum typically shows the following time course:

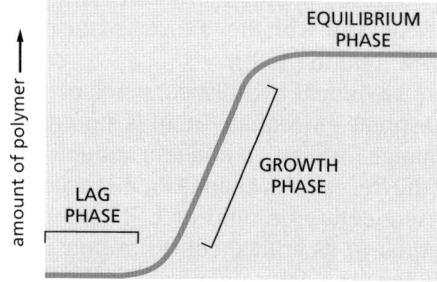

The lag phase corresponds to time taken for nucleation.

The growth phase occurs as monomers add to the exposed ends of the growing filament, causing filament elongation.

The equilibrium phase, or steady state, is reached when the growth of the polymer due to monomer addition precisely balances the shrinkage of the polymer due to disassembly back to monomers.

PLUS AND MINUS ENDS

The two ends of an actin filament or microtubule polymerize at different rates. The fast-growing end is called the plus end, whereas the slow-growing end is called the minus end. The difference in the rates of growth at the two ends is made possible by changes in the conformation of each subunit as it enters the polymer.

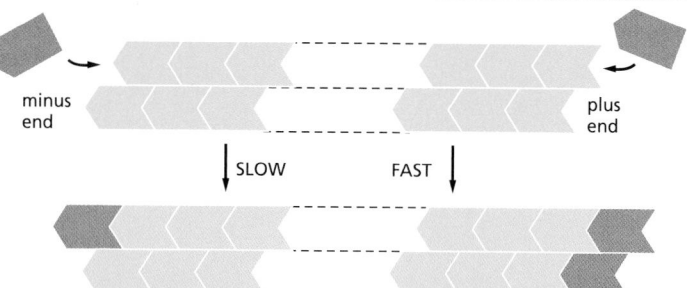

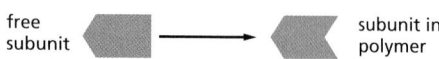

free subunit   subunit in polymer

This conformational change affects the rates at which subunits add to the two ends.

Even though k_{on} and k_{off} will have different values for the plus and minus ends of the polymer, their ratio k_{off}/k_{on}—and hence C_c—must be the same at both ends for a simple polymerization reaction (no ATP or GTP hydrolysis). This is because exactly the same subunit interactions are broken when a subunit is lost at either end, and the final state of the subunit after dissociation is identical. Therefore, the ΔG for subunit loss, which determines the equilibrium constant for its association with the end, is identical at both ends: if the plus end grows four times faster than the minus end, it must also shrink four times faster. Thus, for $C > C_c$, both ends grow; for $C < C_c$, both ends shrink.

The nucleoside triphosphate hydrolysis that accompanies actin and tubulin polymerization removes this constraint.

NUCLEOTIDE HYDROLYSIS

Each actin molecule carries a tightly bound ATP molecule that is hydrolyzed to a tightly bound ADP molecule soon after its assembly into the polymer. Similarly, each tubulin molecule carries a tightly bound GTP that is converted to a tightly bound GDP molecule soon after the molecule assembles into the polymer.

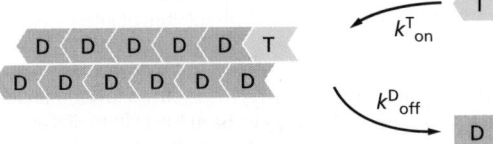

(T = monomer carrying ATP or GTP)
(D = monomer carrying ADP or GDP)

free monomer   subunit in polymer

Hydrolysis of the bound nucleotide reduces the binding affinity of the subunit for neighboring subunits and makes it more likely to dissociate from each end of the filament (see Figure 16–44 for a possible mechanism). It is usually the [T] form that adds to the filament and the [D] form that leaves.
  Considering events at the plus end only:

As before, the polymer will grow until $C = C_c$. For illustrative purposes, we can ignore k^D_{on} and k^T_{off} since they are usually very small, so that polymer growth ceases when

$$k^T_{on} C = k^D_{off} \quad \text{or} \quad C_c = \frac{k^D_{off}}{k^T_{on}}$$

This is a steady state and not a true equilibrium, because the ATP or GTP that is hydrolyzed must be replenished by a nucleotide exchange reaction of the free subunit $\left(\text{[D]} \longrightarrow \text{[T]} \right)$.

ATP CAPS AND GTP CAPS

The rate of addition of subunits to a growing actin filament or microtubule can be faster than the rate at which their bound nucleotide is hydrolyzed. Under such conditions, the end has a "cap" of subunits containing the nucleoside triphosphate—an ATP cap on an actin filament or a GTP cap on a microtubule.

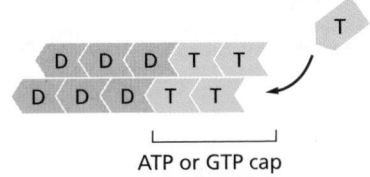

ATP or GTP cap

DYNAMIC INSTABILITY and TREADMILLING are two behaviors observed in cytoskeletal polymers. Both are associated with nucleoside triphosphate hydrolysis. Dynamic instability is believed to predominate in microtubules, whereas treadmilling may predominate in actin filaments.

TREADMILLING

One consequence of the nucleotide hydrolysis that accompanies polymer formation is to change the critical concentration at the two ends of the polymer. Since k^D_{off} and k^T_{on} refer to different reactions, their ratio k^D_{off}/k^T_{on} need not be the same at both ends of the polymer, so that:

$$C_c \text{ (minus end)} > C_c \text{ (plus end)}$$

Thus, if both ends of a polymer are exposed, polymerization proceeds until the concentration of free monomer reaches a value that is above C_c for the plus end but below C_c for the minus end. At this steady state, subunits undergo a net assembly at the plus end and a net disassembly at the minus end at an identical rate. The polymer maintains a constant length, even though there is a net flux of subunits through the polymer, known as treadmilling.

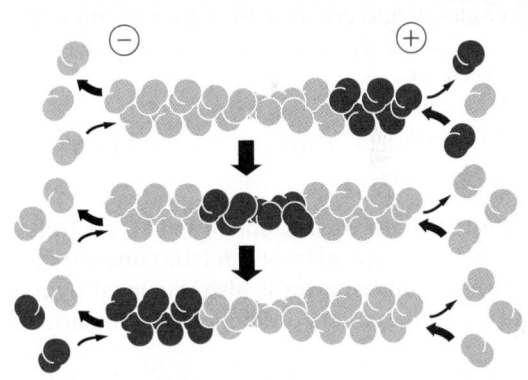

DYNAMIC INSTABILITY

Microtubules depolymerize about 100 times faster from an end containing GDP-tubulin than from one containing GTP-tubulin. A GTP cap favors growth, but if it is lost, then depolymerization ensues.

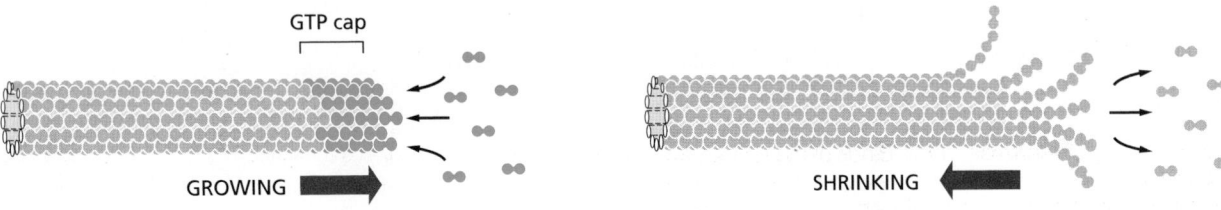

GTP cap

GROWING   SHRINKING

Individual microtubules can therefore alternate between a period of slow growth and a period of rapid disassembly, a phenomenon called dynamic instability.

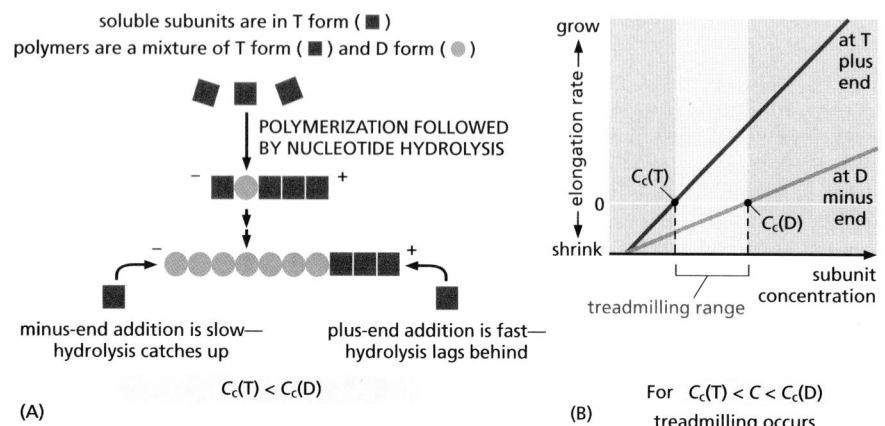

soluble subunits are in T form (■)
polymers are a mixture of T form (■) and D form (●)

POLYMERIZATION FOLLOWED
BY NUCLEOTIDE HYDROLYSIS

minus-end addition is slow—
hydrolysis catches up

plus-end addition is fast—
hydrolysis lags behind

$C_c(T) < C_c(D)$

(A)

For $C_c(T) < C < C_c(D)$
treadmilling occurs

(B)

Figure 16–14 Treadmilling of an actin filament, made possible by the ATP hydrolysis that follows subunit addition. (A) Explanation for the different critical concentrations (C_c) at the plus and minus ends. Subunits with bound ATP (T-form subunits) polymerize at both ends of a growing filament, and then undergo nucleotide hydrolysis within the filament. As the filament grows, elongation is faster than hydrolysis at the plus end in this example, and the terminal subunits at this end are therefore always in the T form. However, hydrolysis is faster than elongation at the minus end, and so terminal subunits at this end are in the D form. (B) Treadmilling occurs at intermediate concentrations of free subunits. The critical concentration for polymerization on a filament end in the T form is lower than for a filament end in the D form. If the actual subunit concentration is somewhere between these two values, the plus end grows while the minus end shrinks, resulting in treadmilling.

while the total length of the filament remains unchanged. This "steady-state treadmilling" requires a constant consumption of energy in the form of ATP hydrolysis.

The Functions of Actin Filaments Are Inhibited by Both Polymer-stabilizing and Polymer-destabilizing Chemicals

Chemical compounds that stabilize or destabilize actin filaments are important tools in studies of the filaments' dynamic behavior and function in cells. The *cytochalasins* are fungal products that prevent actin polymerization by binding to the plus end of actin filaments. *Latrunculin* prevents actin polymerization by binding to actin subunits. The *phalloidins* are toxins isolated from the *Amanita* mushroom that bind tightly all along the side of actin filaments and stabilize them against depolymerization. All of these compounds cause dramatic changes in the actin cytoskeleton and are toxic to cells, indicating that the function of actin filaments depends on a dynamic equilibrium between filaments and actin monomers (**Table 16–1**).

Actin-Binding Proteins Influence Filament Dynamics and Organization

In a test tube, polymerization of actin is controlled simply by its concentration, as described above, and by pH and the concentrations of salts and ATP. Within a cell, however, actin behavior is also regulated by numerous accessory proteins that bind actin monomers or filaments (summarized in **Panel 16–3**). At steady state

TABLE 16–1	Chemical Inhibitors of Actin and Microtubules		
Chemical	Effect on filaments	Mechanism	Original source
Actin			
Latrunculin	Depolymerizes	Binds actin subunits	Sponges
Cytochalasin B	Depolymerizes	Caps filament plus ends	Fungi
Phalloidin	Stabilizes	Binds along filaments	*Amanita* mushroom
Microtubules			
Taxol® (paclitaxel)	Stabilizes	Binds along filaments	Yew tree
Nocodazole	Depolymerizes	Binds tubulin subunits	Synthetic
Colchicine	Depolymerizes	Caps filament ends	Autumn crocus

ACTIN FILAMENTS

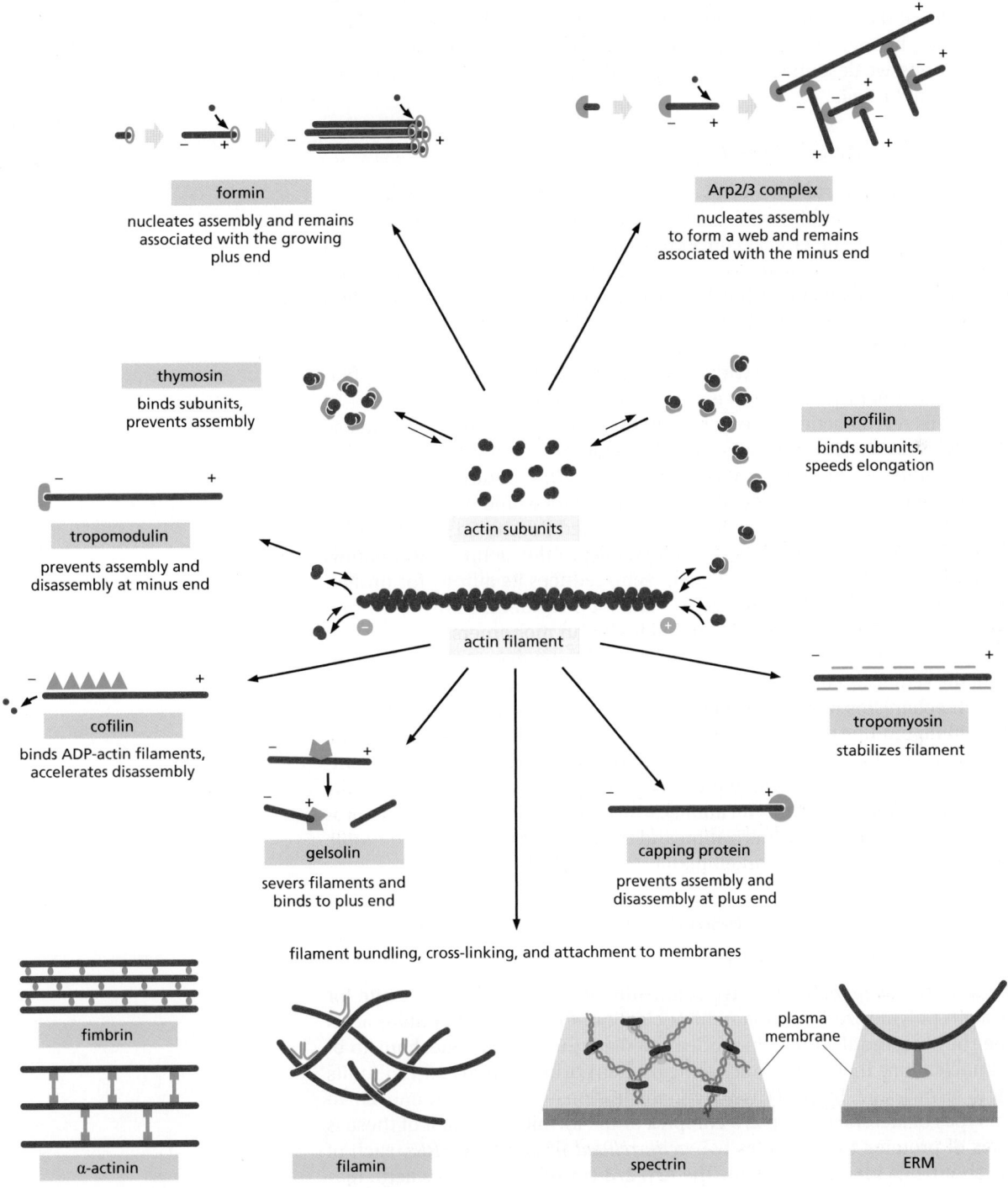

formin

nucleates assembly and remains associated with the growing plus end

Arp2/3 complex

nucleates assembly to form a web and remains associated with the minus end

thymosin

binds subunits, prevents assembly

profilin

binds subunits, speeds elongation

actin subunits

tropomodulin

prevents assembly and disassembly at minus end

actin filament

tropomyosin

stabilizes filament

cofilin

binds ADP-actin filaments, accelerates disassembly

gelsolin

severs filaments and binds to plus end

capping protein

prevents assembly and disassembly at plus end

filament bundling, cross-linking, and attachment to membranes

fimbrin

α-actinin

filamin

plasma membrane

spectrin

ERM

Some of the major accessory proteins of the actin cytoskeleton. Except for the myosin motor proteins, an example of each major type is shown. Each of these is discussed in the text. However, most cells contain more than a hundred different actin-binding proteins, and it is likely that there are important types of actin-associated proteins that are not yet recognized.

in vitro, when the monomer concentration is 0.2 μM, filament half-life, a measure of how long an individual actin monomer spends in a filament as it treadmills, is approximately 30 minutes. In a non-muscle vertebrate cell, actin half-life in filaments is only 30 seconds, demonstrating that cellular factors modify the dynamic behavior of actin filaments. Actin-binding proteins dramatically alter actin filament dynamics and organization through spatial and temporal control of monomer availability, filament nucleation, elongation, and depolymerization. In the following sections, we describe the ways in which these accessory proteins modify actin function in the cell.

Monomer Availability Controls Actin Filament Assembly

In most non-muscle vertebrate cells, approximately 50% of the actin is in filaments and 50% is soluble—and yet the soluble monomer concentration is 50–200 μM, well above the critical concentration. Why does so little of the actin polymerize into filaments? The reason is that the cell contains proteins that bind to the actin monomers and make polymerization much less favorable (an action similar to that of the drug latrunculin). A small protein called *thymosin* is the most abundant of these proteins. Actin monomers bound to thymosin are in a locked state, where they cannot associate with either the plus or minus ends of actin filaments and can neither hydrolyze nor exchange their bound nucleotide.

How do cells recruit actin monomers from this buffered storage pool and use them for polymerization? The answer depends on another monomer-binding protein called *profilin*. Profilin binds to the face of the actin monomer opposite the ATP-binding cleft, blocking the side of the monomer that would normally associate with the filament minus end, while leaving exposed the site on the monomer that binds to the plus end (**Figure 16–15**). When the profilin–actin complex binds a free plus end, a conformational change in actin reduces its affinity for profilin and the profilin falls off, leaving the actin filament one subunit longer. Profilin competes with thymosin for binding to individual actin monomers. Thus, by regulating the local activity of profilin, cells can control the movement of actin subunits from the sequestered thymosin-bound pool onto filament plus ends.

Several mechanisms regulate profilin activity, including profilin phosphorylation and profilin binding to inositol phospholipids. These mechanisms can define the sites where profilin acts. For example, profilin is required for filament assembly at the plasma membrane, where it is recruited by an interaction with acidic membrane phospholipids. At this location, extracellular signals can activate profilin to produce local actin polymerization and the extension of actin-rich motile structures such as filopodia and lamellipodia.

Actin-Nucleating Factors Accelerate Polymerization and Generate Branched or Straight Filaments

In addition to the availability of active actin subunits, a second prerequisite for cellular actin polymerization is filament nucleation. Proteins that contain actin monomer binding motifs linked in tandem mediate the simplest mechanism of filament nucleation. These actin-nucleating proteins bring several actin subunits together to form a seed. In most cases, actin nucleation is catalyzed by one of two different types of factors: the Arp 2/3 complex or the formins. The first of these is a complex of proteins that includes two *actin-related proteins*, or *ARPs*, each of which is about 45% identical to actin. The **Arp 2/3 complex** nucleates actin filament growth from the minus end, allowing rapid elongation at the plus end (**Figure 16–16**A and B). The complex can attach to the side of another actin filament while remaining bound to the minus end of the filament that it has nucleated, thereby building individual filaments into a treelike web (Figure 16–16C and D).

Formins are dimeric proteins that nucleate the growth of straight, unbranched filaments that can be cross-linked by other proteins to form parallel bundles. Each formin subunit has a binding site for monomeric actin, and the formin dimer appears to nucleate actin filament polymerization by capturing two monomers.

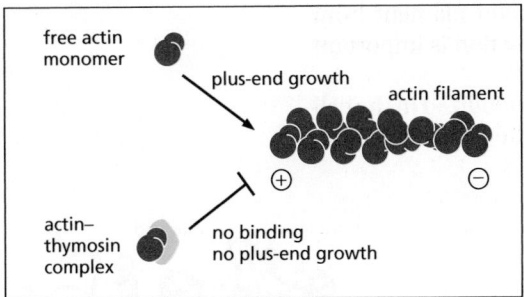

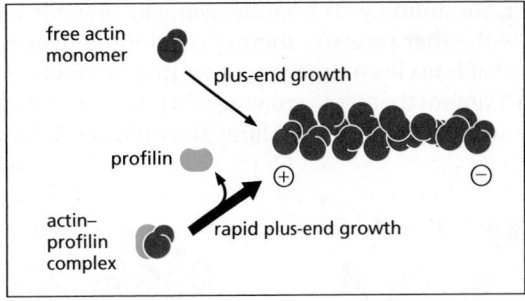

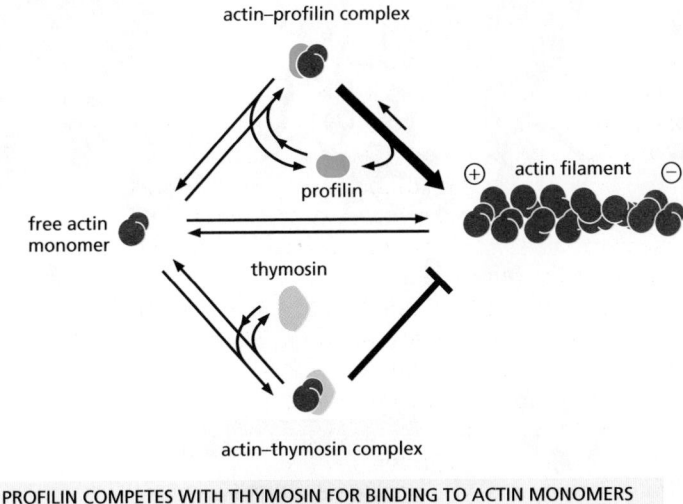

PROFILIN COMPETES WITH THYMOSIN FOR BINDING TO ACTIN MONOMERS
AND PROMOTES ASSEMBLY

Figure 16–15 **Effects of thymosin and profilin on actin polymerization.** An actin monomer bound to thymosin is sterically prevented from binding to and elongating the plus end of an actin filament (*left*). An actin monomer bound to profilin, on the other hand, is capable of elongating a filament (*right*). Thymosin and profilin cannot both bind to a single actin monomer at the same time. In a cell in which most of the actin monomer is bound to thymosin, the activation of a small amount of profilin can produce rapid filament assembly. As indicated (*bottom*), profilin binds to actin monomers that are transiently released from the thymosin-bound monomer pool, shuttles them onto the plus ends of actin filaments, and is then released and recycled for further rounds of filament elongation.

As the newly nucleated filament grows, the formin dimer remains associated with the rapidly growing plus end while still allowing the addition of new subunits at that end (**Figure 16–17**). This mechanism of filament assembly is clearly different from that used by the Arp 2/3 complex, which remains stably bound to the filament minus end, preventing subunit addition or loss at that end. Formin-dependent actin filament growth is strongly enhanced by the association of actin monomers with profilin (**Figure 16–18**).

Like profilin activation, actin filament nucleation by Arp 2/3 complexes and formins occurs primarily at the plasma membrane, and the highest density of actin filaments in most cells is at the cell periphery. The layer just beneath the plasma membrane is called the **cell cortex**, and the actin filaments in this region determine the shape and movement of the cell surface, allowing the cell to change its shape and stiffness rapidly in response to changes in its external environment.

Actin-Filament-Binding Proteins Alter Filament Dynamics

Actin filament behavior is regulated by two major classes of binding proteins: those that bind along the side of a filament and those that bind to the ends (see Panel 16–3). Side-binding proteins include *tropomyosin*, an elongated protein that binds simultaneously to six or seven adjacent actin subunits along each of the two grooves of the helical actin filament. In addition to stabilizing and stiffening the filament, the binding of tropomyosin can prevent the actin filament from interacting with other proteins; this aspect of tropomyosin function is important in the control of muscle contraction, as we discuss later.

An actin filament that stops growing and is not specifically stabilized in the cell will depolymerize rapidly, particularly at its plus end, once the actin molecules

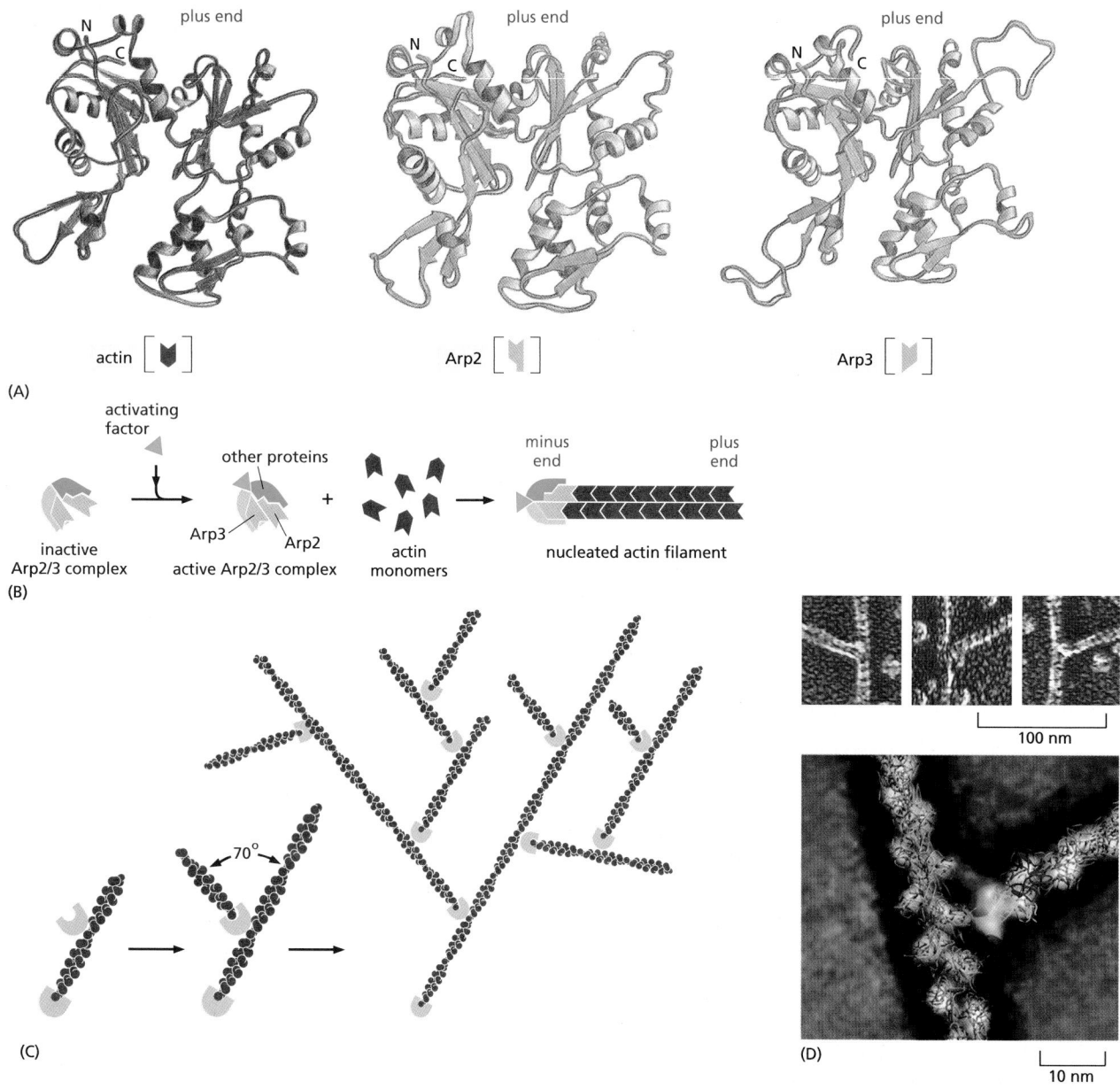

Figure 16–16 Nucleation and actin web formation by the Arp 2/3 complex. (A) The structures of Arp2 and Arp3 compared to the structure of actin. Although the face of the molecule equivalent to the plus end *(top)* in both Arp2 and Arp3 is very similar to the plus end of actin itself, differences on the sides and minus end prevent these actin-related proteins from forming filaments on their own or coassembling into filaments with actin. (B) A model for actin filament nucleation by the Arp 2/3 complex. In the absence of an activating factor, Arp2 and Arp3 are held by their accessory proteins in an orientation that prevents them from nucleating a new actin filament. When an activating factor (indicated by the *blue triangle*) binds the complex, Arp2 and Arp3 are brought together into a new configuration that resembles the plus end of an actin filament. Actin subunits can then assemble onto this structure, bypassing the rate-limiting step of filament nucleation. (C) The Arp 2/3 complex nucleates filaments most efficiently when it is bound to the side of a preexisting actin filament. The result is a filament branch that grows at a 70° angle relative to the original filament. Repeated rounds of branching nucleation result in a treelike web of actin filaments. (D) Top, electron micrographs of branched actin filaments formed by mixing purified actin subunits with purified Arp 2/3 complexes. Bottom, reconstructed image of a branch where the crystal structures of actin *(pink)* and the Arp 2/3 complex have been fitted to the electron density. The mother filament runs from top to bottom, and the daughter filament branches off to the right where the Arp 2/3 complex binds to three actin subunits in the mother filament. (D, top, from R.D. Mullins et al., *Proc. Natl Acad. Sci. USA* 95:6181–6186, 1998. Copyright 1998 National Academy of Sciences, USA. With permission from National Academy of Sciences; botttom, from N. Volkmann et al., *Science* 293:2456–2459, 2001. With permission from AAAS.)

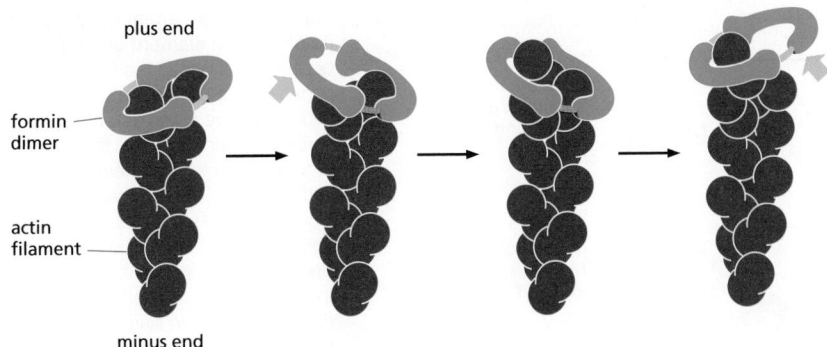

plus end

formin dimer

actin filament

minus end

Figure 16–17 **Actin elongation mediated by formins.** Formin proteins *(green)* form a dimeric complex that can nucleate the formation of a new actin filament *(red)* and remain associated with the rapidly growing plus end as it elongates. The formin protein maintains its binding to one of the two actin subunits exposed at the plus end as it allows each new subunit to assemble. Only part of the large dimeric formin molecule is shown here. Other regions regulate its activity and link it to particular structures in the cell. Many formins are indirectly connected to the cell plasma membrane and aid the insertional polymerization of the actin filament directly beneath the membrane surface.

have hydrolyzed their ATP. The binding of plus-end *capping protein* (also called *CapZ* for its location in the muscle Z band) stabilizes an actin filament at its plus end by rendering it inactive, greatly reducing the rates of filament growth and depolymerization (**Figure 16–19**). At the minus end, an actin filament may be capped by the Arp 2/3 complex that was responsible for its nucleation, although many minus ends in a typical cell are released from the Arp 2/3 complex and are uncapped.

Tropomodulin, best known for its function in the capping of exceptionally long-lived actin filaments in muscle, binds tightly to the minus ends of actin filaments that have been coated and thereby stabilized by tropomyosin. It can also transiently cap pure actin filaments and significantly reduce their elongation and depolymerization rates. A large family of tropomodulin proteins regulates actin filament length and stability in many cell types.

For maximum effect, proteins that bind the side of actin filaments coat the filament completely, and must therefore be present in high amounts. In contrast, end-binding proteins can affect filament dynamics even when they are present at very low levels. Since subunit addition and loss occur primarily at filament ends, one molecule of an end-binding protein per actin filament (roughly one molecule per 200–500 actin subunits) can be enough to transform the architecture of an actin filament network.

Severing Proteins Regulate Actin Filament Depolymerization

Another important mechanism of actin filament regulation depends on proteins that break an actin filament into many smaller filaments, thereby generating a large number of new filament ends. The fate of these new ends depends on the presence of other accessory proteins. Under some conditions, newly formed ends nucleate filament elongation, thereby accelerating the assembly of new filament structures. Under other conditions, severing promotes the depolymerization of old filaments, speeding up the depolymerization rate by tenfold or more. In addition, severing changes the physical and mechanical properties of the cytoplasm: stiff, large bundles and gels become more fluid.

One class of actin-severing proteins is the *gelsolin superfamily*. These proteins are activated by high levels of cytosolic Ca^{2+}. Gelsolin interacts with the side of the actin filament and contains subdomains that bind to two different sites: one that is exposed on the surface of the filament and one that is hidden between adjacent

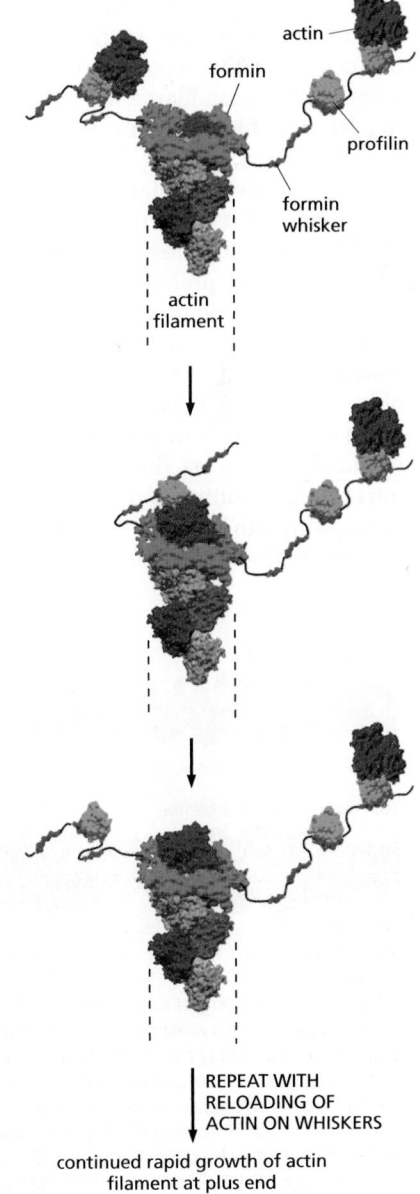

actin

formin

profilin

formin whisker

actin filament

REPEAT WITH RELOADING OF ACTIN ON WHISKERS

continued rapid growth of actin filament at plus end

Figure 16–18 **Profilin and formins.** Some members of the formin protein family have unstructured domains or "whiskers" that contain several binding sites for profilin or the profilin–actin complex. These flexible domains serve as a staging area for addition of actin to the growing plus end of the actin filament when formin is bound. Under some conditions, this can enhance the rate of actin filament elongation so that filament growth is faster than that expected for a diffusion-controlled reaction, and faster in the presence of formin and profilin than the rate for pure actin alone (see also Figure 3–78).

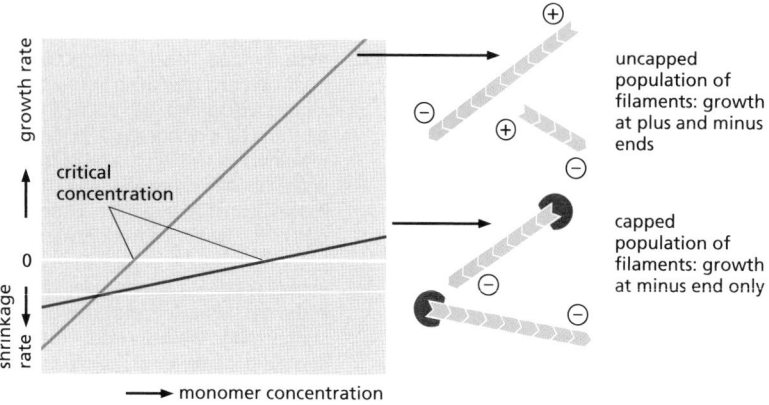

Figure 16–19 **Filament capping and its effects on filament dynamics.** A population of uncapped filaments adds and loses subunits at both the plus and minus ends, resulting in rapid growth or shrinkage, depending on the concentration of available free monomers (*green* line). In the presence of a protein that caps the plus end (*red* line), only the minus end is able to add or lose subunits; consequently, filament growth will be slower at all monomer concentrations above the critical concentration, and filament shrinkage will be slower at all monomer concentrations below the critical concentration. In addition, the critical concentration for the population shifts to that of the filament minus end.

subunits. According to one model, gelsolin binds the side of an actin filament until a thermal fluctuation creates a small gap between neighboring subunits, at which point gelsolin inserts itself into the gap to break the filament. After the severing event, gelsolin remains attached to the actin filament and caps the new plus end.

Another important actin-filament destabilizing protein, found in all eukaryotic cells, is *cofilin*. Also called *actin depolymerizing factor*, cofilin binds along the length of the actin filament, forcing the filament to twist a little more tightly (**Figure 16–20**). This mechanical stress weakens the contacts between actin subunits in the filament, making the filament brittle and more easily severed by thermal motions, generating filament ends that undergo rapid disassembly. As a result, most of the actin filaments inside cells are shorter lived than are filaments formed from pure actin in a test tube.

Cofilin binds preferentially to ADP-containing actin filaments rather than to ATP-containing filaments. Since ATP hydrolysis is usually slower than filament assembly, the newest actin filaments in the cell still contain mostly ATP and are resistant to depolymerization by cofilin. Cofilin therefore tends to dismantle the older filaments in the cell. As we will discuss later, the cofilin-mediated disassembly of old but not new actin filaments is critical for the polarized, directed growth of the actin network that is responsible for unidirectional cell crawling and the intracellular motility of pathogens. Actin filaments can be protected from cofilin by tropomyosin binding. Thus, the dynamics of actin in different subcellular locations depends on the balance of stabilizing and destabilizing accessory proteins.

(A) actin filament

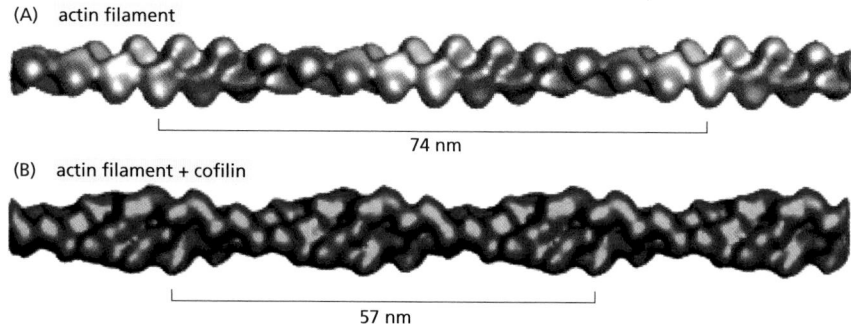

74 nm

(B) actin filament + cofilin

57 nm

Figure 16–20 **Twisting of an actin filament induced by cofilin.** (A) Three-dimensional reconstruction from cryoelectron micrographs of filaments made of pure actin. The bracket shows the span of two twists of the actin helix. (B) Reconstruction of an actin filament coated with cofilin, which binds in a 1:1 stoichiometry to actin subunits all along the filament. Cofilin is a small protein (14 kD) compared to actin (43 kD), and so the filament appears only slightly thicker. The energy of cofilin binding serves to deform the actin filament, twisting it more tightly and reducing the distance spanned by each twist of the helix. (© 1997 A. McGough et al. Originally published in *J. Cell Biol.* https://doi.org.10.1083/jcb.138.4.771. With permission from Rockefeller University Press.)

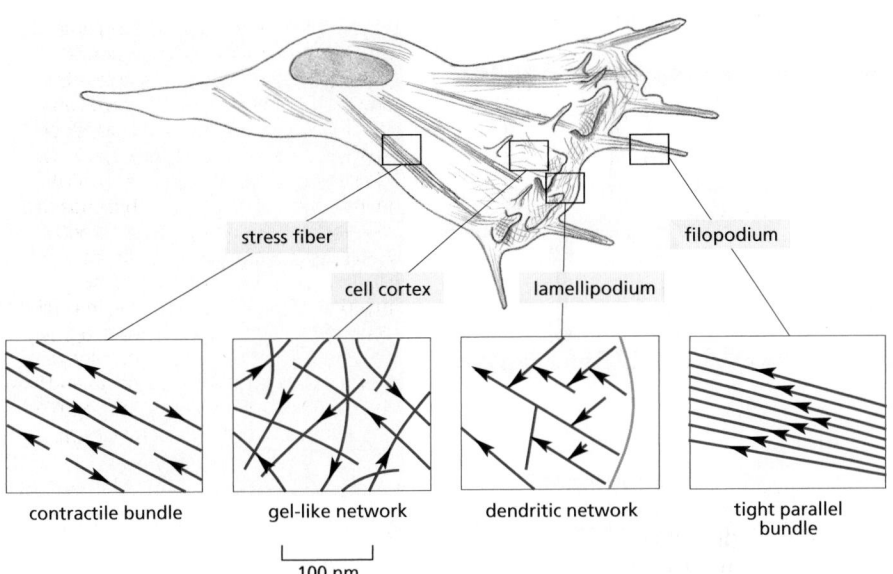

stress fiber

cell cortex

lamellipodium

filopodium

contractile bundle

gel-like network

dendritic network

tight parallel
bundle

100 nm

Figure 16–21 Actin arrays in a cell.
A fibroblast crawling in a tissue-culture
dish is shown with four areas enlarged to
show the arrangement of actin filaments.
The actin filaments are shown in *red*, with
arrowheads pointing toward the minus
end. Stress fibers are contractile and exert
tension. The actin cortex underlies the
plasma membrane and consists of gel-
like networks or dendritic actin networks
that enable membrane protrusion at
lamellopodia. Filopodia are spike-like
projections of the plasma membrane that
allow a cell to explore its environment.

Higher-Order Actin Filament Arrays Influence Cellular Mechanical Properties and Signaling

Actin filaments in animal cells are organized into several types of arrays: dendritic networks, bundles, and weblike (gel-like) networks (**Figure 16–21**). Different structures are initiated by the action of distinct nucleating proteins: the actin filaments of dendritic networks are nucleated by the Arp 2/3 complex, while bundles are made of the long, straight filaments produced by formins. The proteins nucleating the filaments in the gel-like networks are not yet well defined.

The structural organization of different actin networks depends on specialized accessory proteins. As explained earlier, Arp 2/3 organizes filaments into dendritic networks by attaching filament minus ends to the side of other filaments. Other actin filament structures are assembled and maintained by two classes of proteins: *bundling proteins,* which cross-link actin filaments into a parallel array, and *gel-forming proteins,* which hold two actin filaments together at a large angle to each other, thereby creating a looser meshwork. Both bundling and gel-forming proteins generally have two similar actin-filament-binding sites, which can either be part of a single polypeptide chain or contributed by each of two polypeptide chains held together in a dimer (**Figure 16–22**). The spacing and arrangement of these two filament-binding domains determine the type of actin structure that a given cross-linking protein forms.

Each type of bundling protein also determines which other molecules can interact with the cross-linked actin filaments. Myosin II is the motor protein that enables stress fibers and other contractile arrays to contract. The very close packing of actin filaments caused by the small monomeric bundling protein *fimbrin* apparently excludes myosin, and thus the parallel actin filaments held together by fimbrin are not contractile. On the other hand, α-actinin cross-links oppositely

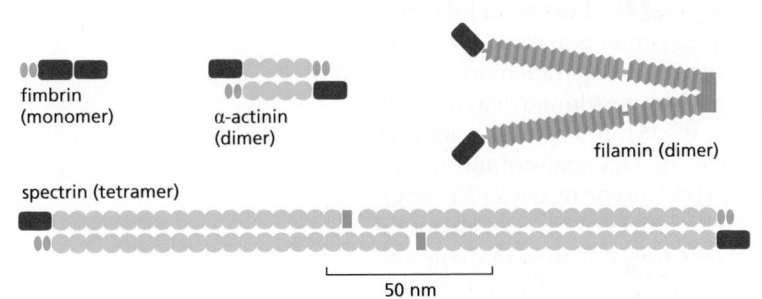

fimbrin
(monomer)

α-actinin
(dimer)

filamin (dimer)

spectrin (tetramer)

50 nm

**Figure 16–22 The modular structures of
four actin-cross-linking proteins.** Each of
the proteins shown has two actin-binding
sites *(red)* that are related in sequence.
Fimbrin has two directly adjacent actin-
binding sites, so that it holds its two actin
filaments very close together (14 nm apart),
aligned with the same polarity (see Figure
16–23A). The two actin-binding sites in
α-actinin are separated by a spacer around
30 nm long, so that it forms more loosely
packed actin bundles (see Figure 16–23A).
Filamin has two actin-binding sites with a
V-shaped linkage between them, so that it
cross-links actin filaments into a network
with the filaments oriented almost at right
angles to one another (see Figure 16–24).
Spectrin is a tetramer of two α and two
β subunits, and the tetramer has two actin-
binding sites spaced about 200 nm apart
(see Figure 10–38).

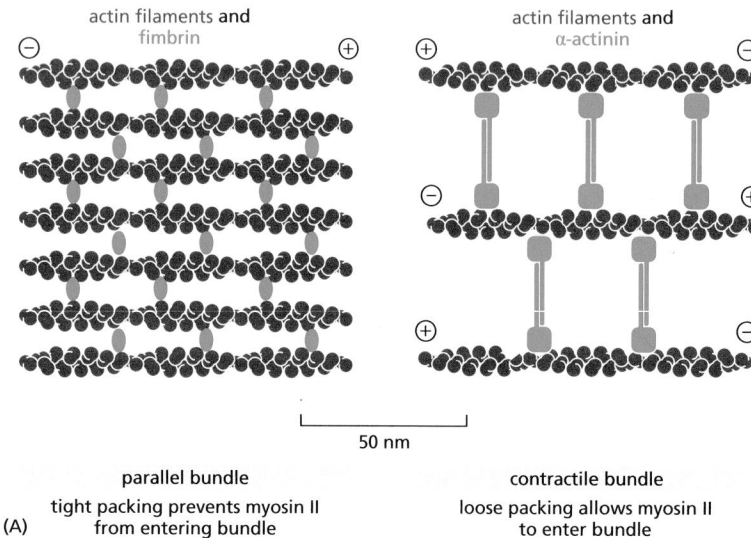

actin filaments **and** fimbrin

actin filaments **and** α-actinin

50 nm

parallel bundle
tight packing prevents myosin II
(A)   from entering bundle

contractile bundle
loose packing allows myosin II
to enter bundle

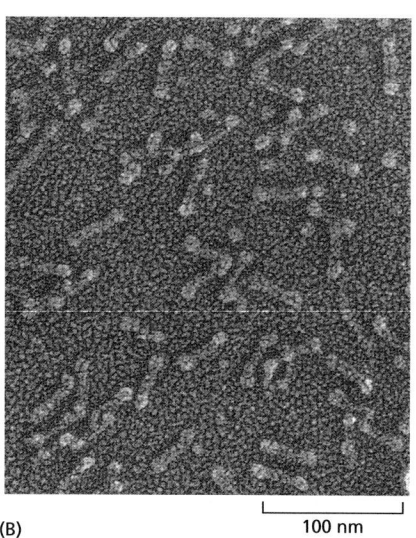

(B)   100 nm

polarized actin filaments into loose bundles, allowing the binding of myosin and formation of contractile actin bundles (**Figure 16–23**). Because of the very different spacing and orientation of the actin filaments, bundling by fimbrin automatically discourages bundling by α-actinin, and vice versa, so that the two types of bundling protein are mutually exclusive.

The bundling proteins that we have discussed so far have straight, stiff connections between their two actin-filament-binding domains. Other actin cross-linking proteins have either a flexible or a stiff, bent connection between their two binding domains, allowing them to form actin filament webs or gels, rather than actin bundles. *Filamin* (see Figure 16–22) promotes the formation of a loose and highly viscous gel by clamping together two actin filaments roughly at right angles (**Figure 16–24**A). Cells require the actin gels formed by filamin to extend the thin, sheetlike membrane projections called *lamellipodia* that help them to crawl across solid surfaces. In humans, mutations in the filamin A gene cause defects in nerve-cell migration during early embryonic development. Cells in the periventricular region of the brain fail to migrate to the cortex and instead form nodules, causing a syndrome called periventricular heterotopia (Figure 16–24B). Interestingly, in addition to binding actin, filamins have been reported to interact with a large number of cellular proteins of great functional diversity, including membrane receptors for signaling molecules, and filamin mutations can also lead to defects in development of bone, the cardiovascular system, and other organs. Thus, filamins may also function as signaling scaffolds by connecting and coordinating a wide variety of cellular processes with the actin cytoskeleton.

A very different, well-studied web-forming protein is *spectrin*, which was first identified in red blood cells. Spectrin is a long, flexible protein made out of four elongated polypeptide chains (two α subunits and two β subunits), arranged so that the two actin-filament-binding sites are about 200 nm apart (compared with 14 nm for fimbrin and about 30 nm for α-actinin; see Figure 16–23). In the red blood cell, spectrin is concentrated just beneath the plasma membrane, where it forms a two-dimensional weblike network held together by short actin filaments whose precise lengths are tightly regulated by capping proteins at each end; spectrin links this web to the plasma membrane because it has separate binding sites for peripheral membrane proteins, which are themselves positioned near the lipid bilayer by integral membrane proteins (see Figure 10–38). The resulting network creates a strong, yet flexible cell cortex that provides mechanical support for the overlying plasma membrane, allowing the red blood cell to spring back to its original shape after squeezing through a capillary. Close relatives of spectrin are found in the cortex of most other vertebrate cell types, where they also help to shape and stiffen the surface membrane. A particularly striking example of spectrin's role

Figure 16–23 The formation of two types of actin filament bundles. (A) Fimbrin cross-links actin filaments into tight bundles, which exclude the motor protein myosin II from participating in the assembly. In contrast, α-actinin, which is a homodimer, cross-links actin filaments into loose bundles, which allow myosin (not shown) to incorporate into the bundle. Fimbrin and α-actinin tend to exclude one another because of the very different spacing of the actin filament bundles that they form. (B) Electron micrograph of purified α-actinin molecules. (B, courtesy of John Heuser.)

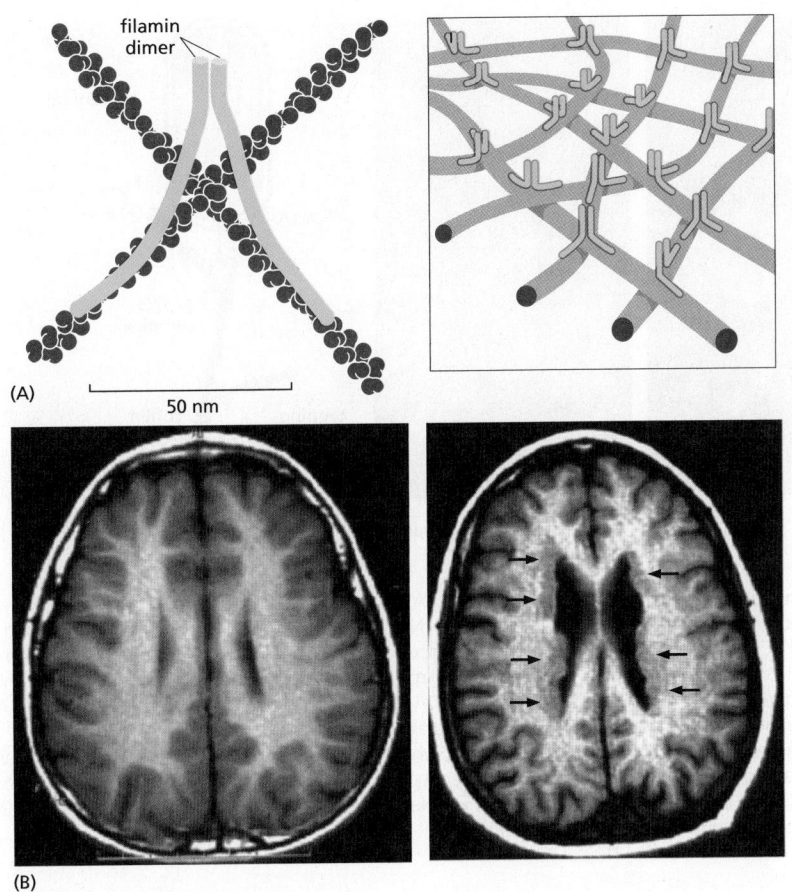

Figure 16–24 Filamin cross-links actin filaments into a three-dimensional network and is required for normal neuronal migration. (A) Each filamin homodimer is about 160 nm long when fully extended and forms a flexible, high-angle link between two adjacent actin filaments. A set of actin filaments cross-linked by filamin forms a mechanically strong web or gel. (B) Magnetic resonance imaging of a normal human brain *(left)* and of a patient with periventricular heterotopia *(right)* caused by mutation in the filamin A gene. In contrast to the smooth ventricular surface in the normal brain, a rough zone of cortical neurons (arrowheads) is seen along the lateral walls of the ventricles, representing neurons that have failed to migrate to the cortex during brain development. Remarkably, although many neurons are not in the right place, the intelligence of affected individuals is frequently normal or only mildly compromised, and the major clinical syndrome is epilepsy that often starts in the second decade of life. (B, adapted from Y. Feng and C.A. Walsh, *Nat. Cell Biol.* 6:1034–1038, published 2004 by Nature Publishing Group. Reproduced with permission of SNCSC.)

in promoting mechanical stability is the long, thin axon of neurons in the nematode worm *Caenorhabditis elegans*, where spectrin is required to keep them from breaking during the twisting motions the worms make during crawling.

The connections of the cortical actin cytoskeleton to the plasma membrane are only partially understood. Members of the *ERM* family (named for its first three members, ezrin, radixin, and moesin), help organize membrane domains through their ability to interact with transmembrane proteins and the underlying cytoskeleton. In so doing, they not only provide structural links to strengthen the cell cortex, but also regulate the activities of signal transduction pathways. Moesin also increases cortical stiffness to promote cell rounding during mitosis. Measurements by atomic force microscopy indicate that the cell cortex remains soft during mitosis when moesin is depleted. ERM proteins are thought to bind to and organize the cortical actin cytoskeleton in a variety of contexts, thereby affecting the shape and stiffness of the membrane as well as the localization and activity of signaling molecules.

Bacteria Can Hijack the Host Actin Cytoskeleton

The importance of accessory proteins in actin-based motility and force production is illustrated beautifully by studies of certain bacteria and viruses that use components of the host-cell actin cytoskeleton to move through the cytoplasm. The cytoplasm of mammalian cells is extremely viscous, containing organelles and cytoskeletal elements that inhibit diffusion of large particles like bacteria or viruses. To move around in a cell and invade neighboring cells, several pathogens, including *Listeria monocytogenes* (which causes a rare but serious form of food poisoning), overcome this problem by recruiting and activating the Arp 2/3 complex at their surface. The Arp 2/3 complex nucleates the assembly of actin filaments that generate a substantial force and push the bacterium through the

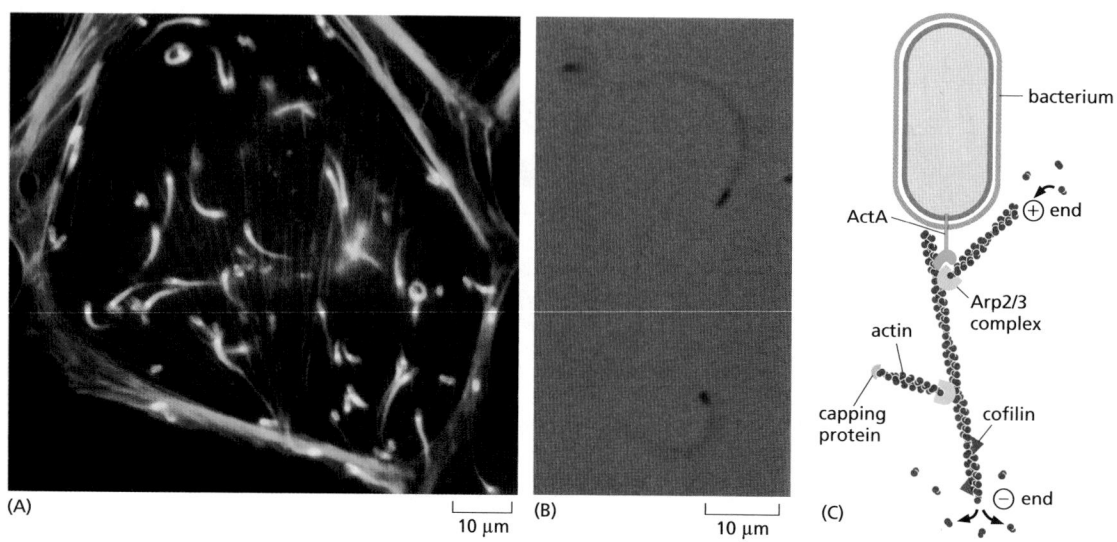

(A)   (B)   (C)

10 µm   10 µm

bacterium

ActA

⊕ end

Arp2/3 complex

actin

capping protein   cofilin

⊖ end

Figure 16–25 The actin-based movement of *Listeria monocytogenes*. (A) Fluorescence micrograph of an infected cell that has been stained to reveal bacteria in *red* and actin filaments in *green*. Note the cometlike tail of actin filaments behind each moving bacterium. Regions of overlap between red and green fluorescence appear *yellow*. (B) *Listeria* motility can be reconstituted in a test tube with ATP and just four purified proteins: actin, Arp 2/3 complex, capping protein, and cofilin. This micrograph shows the dense actin tails behind bacteria *(black)*. (C) The ActA protein on the bacterial surface activates the Arp 2/3 complex to nucleate new filament assembly along the sides of existing filaments. Filaments grow at their plus end until capped by capping protein. Actin is recycled through the action of cofilin, which enhances depolymerization at the minus ends of the filaments. By this mechanism, polymerization is focused at the rear surface of the bacterium, propelling it forward (see Movie 23.7). (A, courtesy of Julie Theriot and Tim Mitchison; B, from T.P. Loisel et al., *Nature* 401:613–616, published 1999 by Nature Publishing Group. Reproduced with permission of SNCSC.)

cytoplasm at rates of up to 1 µm/sec, leaving behind a long actin "comet tail" (**Figure 16–25**; see also Figures 23–28 and 23–29). This motility can be reconstituted in a test tube by adding the bacteria to a mixture of pure actin, Arp 2/3 complex, cofilin, and capping protein, illustrating how actin polymerization dynamics generate movement through spatial regulation of filament assembly and disassembly. As we shall see, actin-based movement of this sort also underlies membrane protrusion at the leading edge of motile cells.

Summary

Actin is a highly conserved cytoskeletal protein that is present in high concentrations in nearly all eukaryotic cells. Nucleation presents a kinetic barrier to actin polymerization, but once formed, actin filaments undergo dynamic behavior due to hydrolysis of the bound nucleotide ATP. Actin filaments are polarized and can undergo treadmilling when a filament assembles at the plus end while simultaneously depolymerizing at the minus end. In cells, actin filament dynamics are regulated at every step, and the varied forms and functions of actin depend on a versatile repertoire of accessory proteins. Approximately half of the actin is kept in a monomeric form through association with sequestering proteins such as thymosin. Nucleation factors such as the Arp 2/3 complex and formins promote formation of branched and parallel filaments, respectively. Interplay between proteins that bind or cap actin filaments and those that promote filament severing or depolymerization can slow or accelerate the kinetics of filament assembly and disassembly. Another class of accessory proteins assembles the filaments into larger ordered structures by cross-linking them to one another in geometrically defined ways. Connections between these actin arrays and the plasma membrane of cells give an animal cell mechanical strength and permit the elaboration of cortical cellular structures such as lamellipodia, filopodia, and microvilli. By inducing actin filament polymerization at their surface, intracellular pathogens can hijack the host-cell cytoskeleton and move around inside the cell.

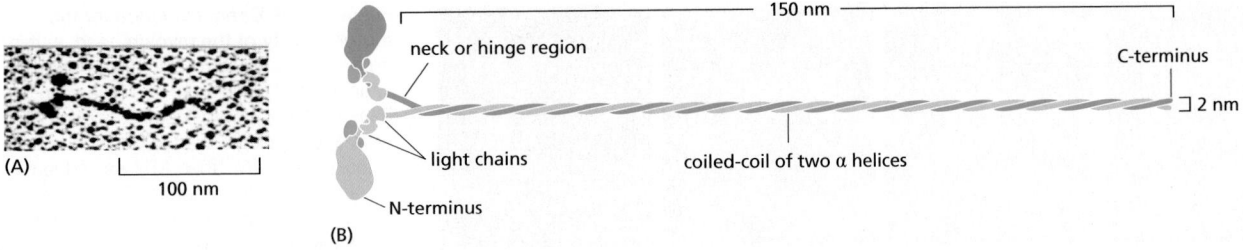

Figure 16–26 **Myosin II.** (A) The two globular heads and long tail of a myosin II molecule shadowed with platinum can be seen in this electron micrograph. (B) A myosin II molecule is composed of two heavy chains (each about 2000 amino acids long; *green*) and four light chains *(blue)*. The light chains are of two distinct types, and one copy of each type is present on each myosin head. Dimerization occurs when the two α helices of the heavy chains wrap around each other to form a coiled-coil, driven by the association of regularly spaced hydrophobic amino acids (see Figure 3–9). The coiled-coil arrangement makes an extended rod in solution, and this part of the molecule forms the tail. (A, courtesy of David Shotton.)

MYOSIN AND ACTIN

A crucial feature of the actin cytoskeleton is that it can form contractile structures that cross-link and slide actin filaments relative to one another through the action of **myosin** motor proteins. In addition to driving muscle contraction, actin–myosin assemblies perform important functions in non-muscle cells.

Actin-Based Motor Proteins Are Members of the Myosin Superfamily

The first motor protein to be identified was skeletal muscle myosin, which generates the force for muscle contraction. This protein, now called *myosin II*, is an elongated protein formed from two heavy chains and two copies of each of two light chains. Each heavy chain has a globular head domain at its N-terminus that contains the force-generating machinery, followed by a very long amino acid sequence that forms an extended coiled-coil that mediates heavy-chain dimerization (**Figure 16–26**). The two light chains bind close to the N-terminal head domain, while the long coiled-coil tail bundles itself with the tails of other myosin molecules. These tail–tail interactions form large, bipolar "thick filaments" that have several hundred myosin heads, oriented in opposite directions at the two ends of the thick filament (**Figure 16–27**).

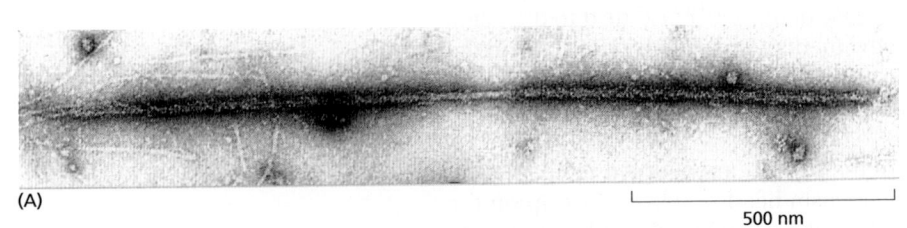

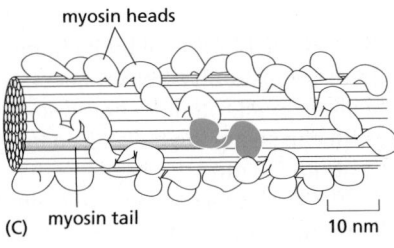

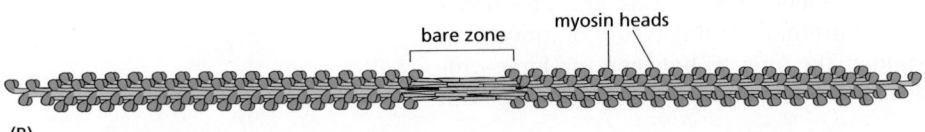

Figure 16–27 **The myosin II bipolar thick filament in muscle.** (A) Electron micrograph of a myosin II thick filament isolated from frog muscle. Note the central bare zone, which is free of head domains. (B) Schematic diagram, not drawn to scale. The myosin II molecules aggregate by means of their tail regions, with their heads projecting to the outside of the filament. The bare zone in the center of the filament consists entirely of myosin II tails. (C) A small section of a myosin II filament as reconstructed from electron micrographs. An individual myosin molecule is highlighted in *green*. The cytoplasmic myosin II filaments in non-muscle cells are much smaller, although similarly organized (see Figure 16–39). (A, from M. Stewart and R.W. Kensler, *J. Mol. Biol.* 192:831–851, 1986. With permission from Elsevier; C, based on R.A. Crowther, R. Padrón and R. Craig, *J. Mol. Biol.* 184:429–439, 1985.)

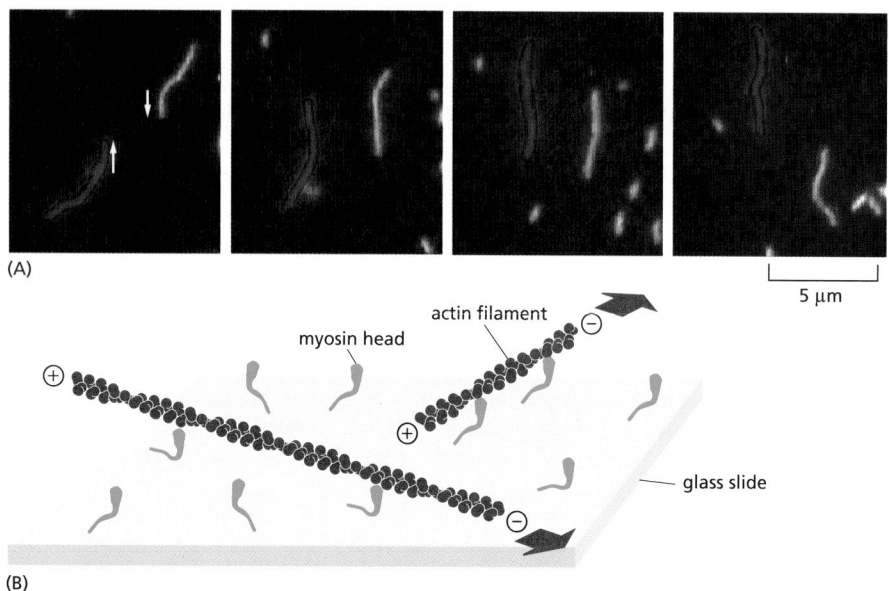

(A)

myosin head

actin filament

⊕

⊖

⊕

⊖

glass slide

(B)

Figure 16–28 **Direct evidence for the motor activity of the myosin head.** In this experiment, purified myosin heads were attached to a glass slide, and then actin filaments labeled with fluorescent phalloidin were added and allowed to bind to the myosin heads. (A) When ATP was added, the actin filaments began to glide along the surface, owing to the many individual steps taken by each of the dozens of myosin heads bound to each filament. The video frames shown in this sequence were recorded about 0.6 second apart; the two actin filaments shown (one *red* and one *green*) were moving in opposite directions at a rate of about 4 μm/sec. (B) Diagram of the experiment. The large *red* arrows indicate the direction of actin filament movement (Movie 16.2). (A, courtesy of James Spudich.)

Each myosin head binds and hydrolyzes ATP, using the energy of ATP hydrolysis to walk toward the plus end of an actin filament (**Figure 16–28**). The opposing orientation of the heads in the thick filament makes the filament efficient at sliding pairs of oppositely oriented actin filaments toward each other, shortening the muscle. In skeletal muscle, in which carefully arranged actin filaments are aligned in "thin filament" arrays surrounding the myosin thick filaments, the ATP-driven sliding of actin filaments results in a powerful contraction. Cardiac and smooth muscle contain myosin II molecules that are similarly arranged, although different genes encode them.

Myosin Generates Force by Coupling ATP Hydrolysis to Conformational Changes

Motor proteins use structural changes in their ATP-binding sites to produce cyclic interactions with a cytoskeletal filament. Each cycle of ATP binding, hydrolysis, and release propels them forward in a single direction to a new binding site along the filament. For myosin II, each step of the movement along actin is generated by the swinging of an 8.5-nm-long α helix, or *lever arm*, which is structurally stabilized by the binding of light chains. At the base of this lever arm next to the head, there is a pistonlike helix that connects movements at the ATP-binding cleft in the head to small rotations of the so-called converter domain. A small change at this point can swing the helix like a long lever, causing the far end of the helix to move by about 5.0 nm.

These changes in the conformation of the myosin are coupled to changes in its binding affinity for actin, allowing the myosin head to release its grip on the actin filament at one point and snatch hold of it again at another. The full mechanochemical cycle of nucleotide binding, nucleotide hydrolysis, and phosphate release (which causes the "power stroke") produces a single step of movement (**Figure 16–29**). At low ATP concentrations, the interval between the force-producing step and the binding of the next ATP is long enough that single steps can be observed (**Figure 16–30**).

Sliding of Myosin II Along Actin Filaments Causes Muscles to Contract

Muscle contraction is the most familiar and best-understood form of movement in animals. In vertebrates, running, walking, swimming, and flying all depend on the rapid contraction of skeletal muscle on its scaffolding of bone, while involuntary

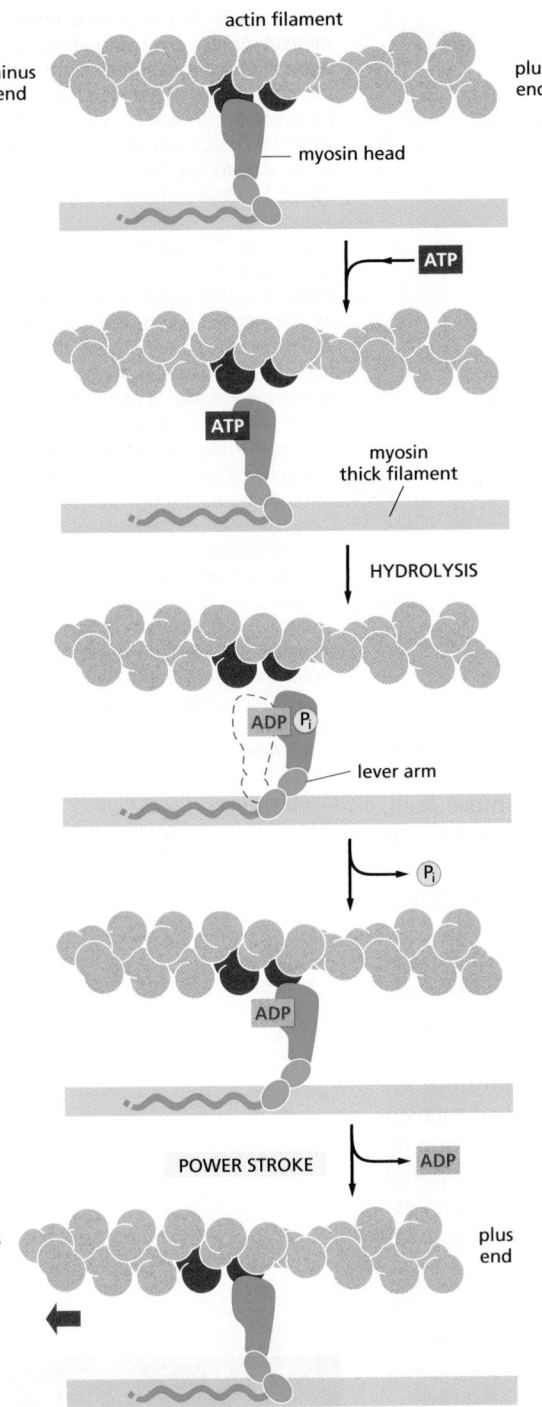

ATTACHED At the start of the cycle shown in this figure, a myosin head lacking a bound nucleotide is locked tightly onto an actin filament in a *rigor* configuration (so named because it is responsible for *rigor mortis*, the rigidity of death). In an actively contracting muscle, this state is very short-lived, being rapidly terminated by the binding of a molecule of ATP.

RELEASED A molecule of ATP binds to the large cleft on the "back" of the head (that is, on the side furthest from the actin filament) and immediately causes a slight change in the conformation of the actin-binding site, reducing the affinity of the head for actin and allowing it to move along the filament. (The space drawn here between the head and actin emphasizes this change, although in reality the head probably remains very close to the actin.)

COCKED The cleft closes like a clam shell around the ATP molecule, triggering a movement in the lever arm that causes the head to be displaced along the filament by a distance of about 5 nm. Hydrolysis of ATP occurs, but the ADP and inorganic phosphate (P_i) remain tightly bound to the protein.

FORCE-GENERATING Weak binding of the myosin head to a new site on the actin filament causes release of the inorganic phosphate produced by ATP hydrolysis, concomitantly with the tight binding of the head to actin. This release triggers the power stroke—the force-generating change in shape during which the head regains its original conformation. In the course of the power stroke, the head loses its bound ADP, thereby returning to the start of a new cycle.

ATTACHED At the end of the cycle, the myosin head is again locked tightly to the actin filament in a rigor configuration. Note that the head has moved to a new position on the actin filament.

Figure 16–29 **The cycle of structural changes used by myosin II to walk along an actin filament.** In the myosin II cycle, the head remains bound to the actin filament for only about 5% of the entire cycle time, allowing many myosins to work together to move a single actin filament (Movie 16.3). (Based on I. Rayment et al., *Science* 261:50–58, 1993.)

movements such as heart pumping and gut peristalsis depend on the contraction of cardiac muscle and smooth muscle, respectively. All these forms of muscle contraction depend on the ATP-driven sliding of highly organized arrays of actin filaments against arrays of myosin II filaments.

Skeletal muscle was a relatively late evolutionary development, and muscle cells are highly specialized for rapid and efficient contraction. The long, thin muscle fibers of skeletal muscle are actually huge single cells that form during

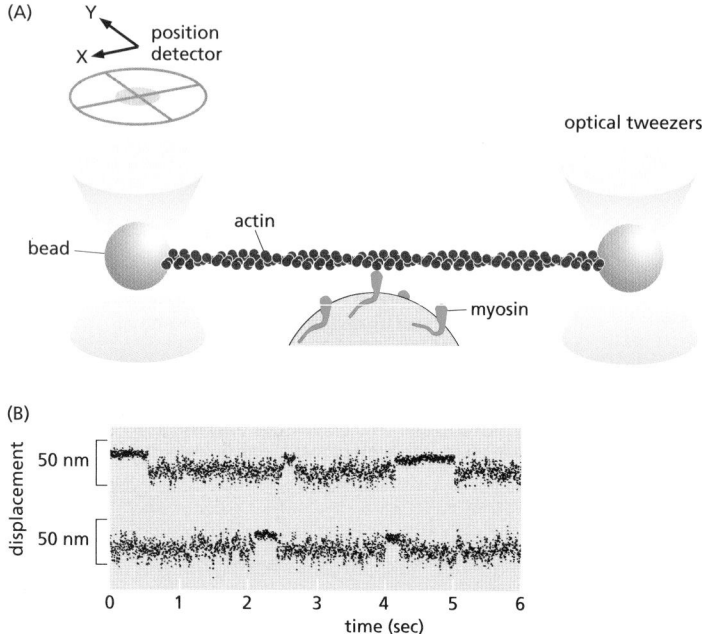

(A)

Y

position detector

X

optical tweezers

bead

actin

myosin

(B)

displacement

50 nm

50 nm

time (sec)

Figure 16–30 **The force of a single myosin molecule moving along an actin filament measured using an optical trap.** (A) Schematic of the experiment, showing an actin filament with beads attached at both ends and held in place by focused beams of light called optical tweezers (Movie 16.4). The tweezers trap and move the bead, and can also be used to measure the force exerted on the bead through the filament. In this experiment, the filament was positioned over another bead to which myosin II motors were attached, and the optical tweezers were used to determine the effects of myosin binding on movement of the actin filament. (B) These traces show filament movement in two separate experiments. Initially, when the actin filament is unattached to myosin, thermal motion of the filament produces noisy fluctuations in filament position. When a single myosin binds to the actin filament, thermal motion decreases abruptly and a roughly 10-nm displacement results from movement of the filament by the motor. The motor then releases the filament. Because the ATP concentration is very low in this experiment, the myosin remains attached to the actin filament for much longer than it would in a muscle cell. (Adapted from C. Rüegg et al., *Physiology* 17:213–218, 2002. With permission from the American Physiological Society.)

development by the fusion of many separate cells. The large muscle cell retains the many nuclei of the contributing cells. These nuclei lie just beneath the plasma membrane (**Figure 16–31**). The bulk of the cytoplasm inside is made up of myofibrils, which is the name given to the basic contractile elements of the muscle cell. A **myofibril** is a cylindrical structure 1–2 μm in diameter that is often as long as the muscle cell itself. It consists of a long, repeated chain of tiny contractile units—called *sarcomeres*, each about 2.2 μm long—which give the vertebrate myofibril its striated appearance (**Figure 16–32**).

Each sarcomere is formed from a miniature, precisely ordered array of parallel and partly overlapping thin and thick filaments. The *thin filaments* are composed of actin and associated proteins, and they are attached at their plus ends to a *Z disc* at each end of the sarcomere. The capped minus ends of the actin filaments extend in toward the middle of the sarcomere, where they overlap with *thick filaments*, the bipolar assemblies formed from specific muscle isoforms of myosin II (see Figure 16–27). When this region of overlap is examined in cross section by electron microscopy, the myosin filaments are arranged in a regular hexagonal lattice, with the actin filaments evenly spaced between them (**Figure 16–33**). Cardiac muscle and smooth muscle also contain sarcomeres, although the organization is not as regular as that in skeletal muscle.

Figure 16–31 **Skeletal muscle cells (also called muscle fibers).** (A) These huge multinucleated cells form by the fusion of many muscle cell precursors, called myoblasts. Here, a single muscle cell is depicted. In an adult human, a muscle cell is typically 50 μm in diameter and can be up to several centimeters long. (B) Fluorescence micrograph of rat muscle, showing the peripherally located nuclei *(blue)* in these giant cells. Myofibrils are stained *red*. (B, courtesy of Nancy L. Kedersha.)

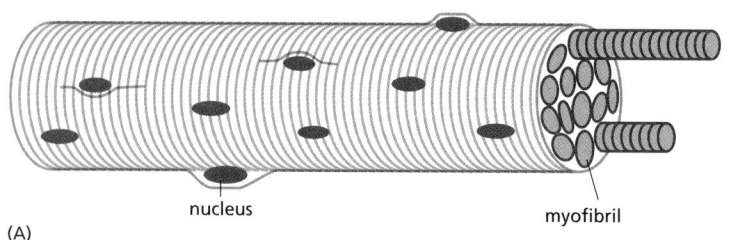

nucleus

myofibril

(A)

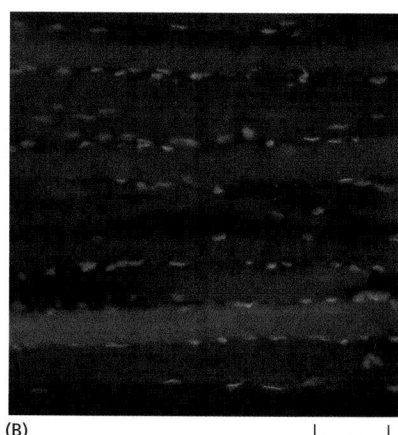

(B)

50 μm

Figure 16–32 **Skeletal muscle myofibrils.** (A) Low-magnification electron micrograph of a longitudinal section through a skeletal muscle cell of a rabbit, showing the regular pattern of cross-striations. The cell contains many myofibrils aligned in parallel (see Figure 16–31). (B) Detail of the skeletal muscle shown in (A), showing portions of two adjacent myofibrils and the definition of a sarcomere (*black* arrow). (C) Schematic diagram of a single sarcomere, showing the origin of the dark and light bands seen in the electron micrographs. The Z discs, at each end of the sarcomere, are attachment sites for the plus ends of actin filaments (thin filaments); the M line, or midline, is the location of proteins that link adjacent myosin II filaments (thick filaments) to one another. (D) When the sarcomere contracts, the actin and myosin filaments slide past one another without shortening. (A and B, courtesy of Roger Craig.)

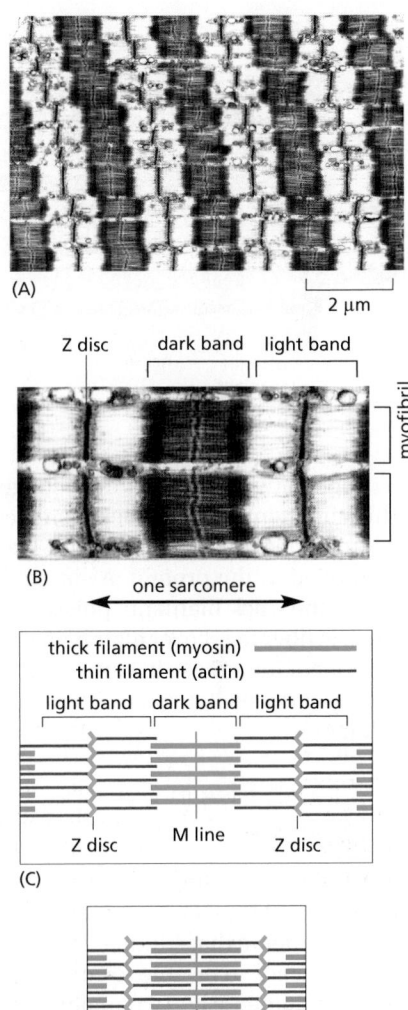

Sarcomere shortening is caused by the myosin filaments sliding past the actin thin filaments, with no change in the length of either type of filament (see Figure 16–32C and D). Bipolar thick filaments walk toward the plus ends of two sets of thin filaments of opposite orientations, driven by dozens of independent myosin heads that are positioned to interact with each thin filament. Because there is no coordination among the movements of the myosin heads, it is critical that they remain tightly bound to the actin filament for only a small fraction of each ATPase cycle so that they do not hold one another back. Each myosin thick filament has about 300 heads (294 in frog muscle), and each head cycles about five times per second in the course of a rapid contraction—sliding the myosin and actin filaments past one another at rates of up to 15 μm/sec and enabling the sarcomere to shorten by 10% of its length in less than one-fiftieth of a second. The rapid synchronized shortening of the thousands of sarcomeres lying end-to-end in each myofibril enables skeletal muscle to contract rapidly enough for running and flying, or for playing the piano.

Accessory proteins produce the remarkable uniformity in filament organization, length, and spacing in the sarcomere (**Figure 16–34**). The actin filament plus ends are anchored in the Z disc, which is built from CapZ and α-actinin; the Z disc caps the filaments (preventing depolymerization), while holding them together in a regularly spaced bundle. The precise length of each thin filament is influenced by a protein of enormous size, called *nebulin*, which consists almost entirely of a repeating 35-amino-acid actin-binding motif. Nebulin stretches from the Z disc toward the minus end of each thin filament, which is capped and stabilized by tropomodulin. Although there is some slow exchange of actin subunits at both ends of the muscle thin filament, such that the components of the thin filament turn over with a half-life of several days, the actin filaments in sarcomeres are remarkably stable compared with those found in most other cell types, whose dynamic actin filaments turn over with half-lives of a few minutes or less.

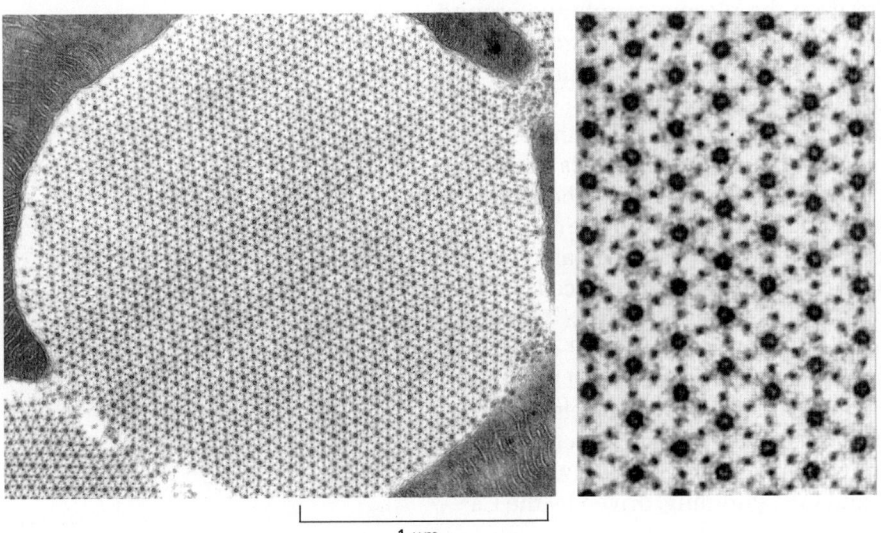

Figure 16–33 **Electron micrographs of an insect flight muscle viewed in cross section.** The myosin and actin filaments are packed together with almost crystalline regularity. Unlike their vertebrate counterparts, these myosin filaments have a hollow center, as seen in the enlargement on the right. The geometry of the hexagonal lattice is slightly different in vertebrate muscle. (From J. Auber, *J. de Microsc.* 8:197–232, 1969. With permission from Societé Française de Microscopie Électronique.)

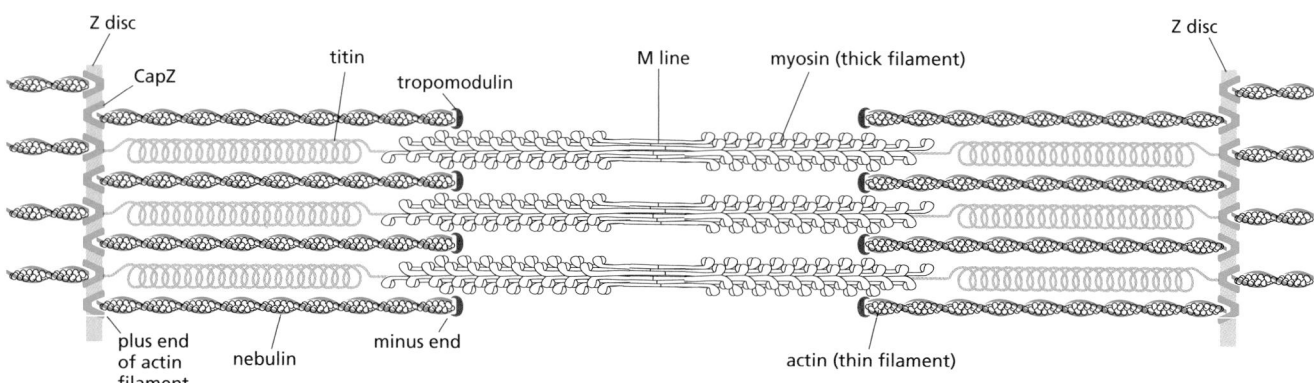

Figure 16–34 **Organization of accessory proteins in a sarcomere.** Each giant titin molecule extends from the Z disc to the M line—a distance of over 1 μm. Part of each titin molecule is closely associated with a myosin thick filament (which switches polarity at the M line); the rest of the titin molecule is elastic and changes length as the sarcomere contracts and relaxes. Each nebulin molecule is exactly the length of a thin filament. The actin filaments are also coated with tropomyosin and troponin (not shown; see Figure 16–36) and are capped at both ends. Tropomodulin caps the minus end of the actin filaments, and CapZ anchors the plus end at the Z disc, which also contains α-actinin (not shown).

Opposing pairs of an even longer template protein, called *titin*, position the thick filaments midway between the Z discs. Titin acts as a molecular spring, with a long series of immunoglobulin-like domains that can unfold one by one as stress is applied to the protein. A springlike unfolding and refolding of these domains keeps the thick filaments poised in the middle of the sarcomere and allows the muscle fiber to recover after being overstretched. In *C. elegans*, whose sarcomeres are longer than those in vertebrates, titin is longer as well, suggesting that it serves also as a molecular ruler, determining in this case the overall length of each sarcomere.

A Sudden Rise in Cytosolic Ca²⁺ Concentration Initiates Muscle Contraction

The force-generating molecular interaction between myosin thick filaments and actin thin filaments takes place only when a signal passes to the skeletal muscle from the nerve that stimulates it. Immediately upon arrival of the signal, the muscle cell needs to be able to contract very rapidly, with all the sarcomeres shortening simultaneously. Two major features of the muscle cell make extremely rapid contraction possible. First, as previously discussed, the individual myosin motor heads in each thick filament spend only a small fraction of the ATP cycle time bound to the filament and actively generating force, so many myosin heads can act in rapid succession on the same thin filament without interfering with one another. Second, a specialized membrane system relays the incoming signal rapidly throughout the entire cell. The signal from the nerve triggers an action potential in the muscle cell plasma membrane (discussed in Chapter 11), and this electrical excitation spreads swiftly into a series of membranous folds—the transverse tubules, or *T tubules*—that extend inward from the plasma membrane around each myofibril. The signal is then relayed across a small gap to the *sarcoplasmic reticulum*, an adjacent weblike sheath of modified endoplasmic reticulum that surrounds each myofibril like a net stocking (**Figure 16–35**A and B).

When the incoming action potential activates a Ca²⁺ channel in the T-tubule membrane, Ca²⁺ influx triggers the opening of Ca²⁺-release channels in the sarcoplasmic reticulum (Figure 16–35C). Ca²⁺ flooding into the cytosol then initiates the contraction of each myofibril. Because the signal from the muscle cell plasma membrane is passed within milliseconds (via the T tubules and sarcoplasmic reticulum) to every sarcomere in the cell, all of the myofibrils in the cell contract at once. The increase in Ca²⁺ concentration is transient because the Ca²⁺ is rapidly pumped back into the sarcoplasmic reticulum by an abundant, ATP-dependent Ca²⁺-pump (also called a Ca²⁺-ATPase) in its membrane (see Figure 11–13). Typically, the cytoplasmic Ca²⁺ concentration is restored to resting levels within 30 msec, allowing the myofibrils to relax. Thus, muscle contraction depends on two processes that consume enormous amounts of ATP: filament sliding, driven by the ATPase of the myosin motor domain, and Ca²⁺ pumping, driven by the Ca²⁺-pump.

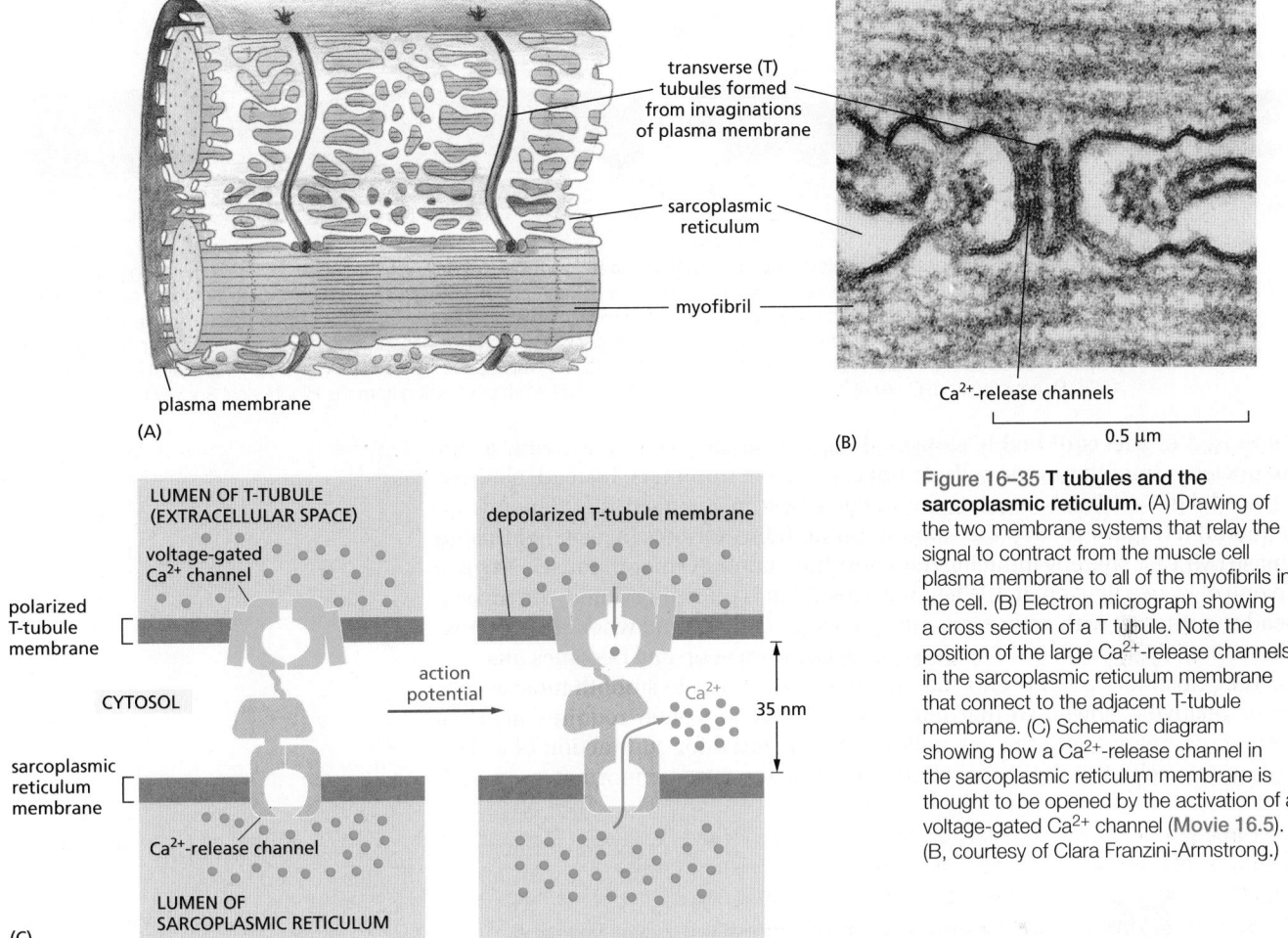

transverse (T) tubules formed from invaginations of plasma membrane

sarcoplasmic reticulum

myofibril

plasma membrane

(A)

Ca²⁺-release channels

0.5 µm

(B)

LUMEN OF T-TUBULE (EXTRACELLULAR SPACE)

voltage-gated Ca²⁺ channel

polarized T-tubule membrane

CYTOSOL

sarcoplasmic reticulum membrane

Ca²⁺-release channel

LUMEN OF SARCOPLASMIC RETICULUM

(C)

depolarized T-tubule membrane

action potential

Ca²⁺

35 nm

Figure 16–35 T tubules and the sarcoplasmic reticulum. (A) Drawing of the two membrane systems that relay the signal to contract from the muscle cell plasma membrane to all of the myofibrils in the cell. (B) Electron micrograph showing a cross section of a T tubule. Note the position of the large Ca²⁺-release channels in the sarcoplasmic reticulum membrane that connect to the adjacent T-tubule membrane. (C) Schematic diagram showing how a Ca²⁺-release channel in the sarcoplasmic reticulum membrane is thought to be opened by the activation of a voltage-gated Ca²⁺ channel (Movie 16.5). (B, courtesy of Clara Franzini-Armstrong.)

The Ca²⁺-dependence of vertebrate skeletal muscle contraction, and hence its dependence on commands transmitted via nerves, is due entirely to a set of specialized accessory proteins that are closely associated with the actin thin filaments. One of these accessory proteins is a muscle form of *tropomyosin*, the elongated protein that binds along the groove of the actin filament helix. The other is *troponin*, a complex of three polypeptides, troponins T, I, and C (named for their tropomyosin-binding, inhibitory, and Ca²⁺-binding activities, respectively). Troponin I binds to actin as well as to troponin T. In a resting muscle, the troponin I–T complex pulls the tropomyosin out of its normal binding groove into a position along the actin filament that interferes with the binding of myosin heads, thereby preventing any force-generating interaction. When the level of Ca²⁺ is raised, troponin C—which binds up to four molecules of Ca²⁺—causes troponin I to release its hold on actin. This allows the tropomyosin molecules to slip back into their normal position so that the myosin heads can walk along the actin filaments (**Figure 16–36**). Troponin C is closely related to the ubiquitous Ca²⁺-binding protein calmodulin (see Figure 15–33); it can be thought of as a specialized form of calmodulin that has acquired binding sites for troponin I and troponin T, thereby ensuring that the myofibril responds extremely rapidly to an increase in Ca²⁺ concentration.

In smooth muscle cells, so called because they lack the regular striations of skeletal muscle, contraction is also triggered by an influx of calcium ions, but the regulatory mechanism is different. Smooth muscle forms the contractile portion of the stomach, intestine, and uterus, as well as the walls of arteries and many other structures requiring slow and sustained contractions. Smooth muscle is

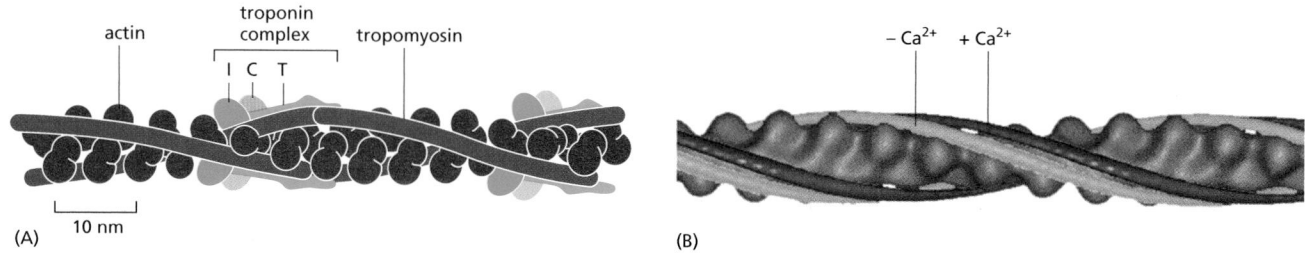

(A)

actin

troponin
complex

I C T

tropomyosin

10 nm

(B)

−Ca²⁺ +Ca²⁺

Figure 16–36 The control of skeletal muscle contraction by troponin. (A) A skeletal-muscle-cell thin filament, showing the positions of tropomyosin and troponin along the actin filament. Each tropomyosin molecule has seven evenly spaced regions with similar amino acid sequences, each of which is thought to bind to an actin subunit in the filament. (B) Reconstructed cryoelectron microscopy image of an actin filament showing the relative position of a superimposed tropomyosin strand in the presence *(dark purple)* or absence *(light purple)* of calcium. (A, adapted from G.N. Phillips, J.P. Fillers and C. Cohen, *J. Mol. Biol.* 192:111–131, 1986; B, adapted from C. Xu et al., *Biophys. J.* 77: 985–992, 1999. With permission from the Biophysical Society.)

composed of sheets of highly elongated spindle-shaped cells, each with a single nucleus. Smooth muscle cells do not express the troponins. Instead, elevated intracellular Ca^{2+} levels regulate contraction by a mechanism that depends on calmodulin (**Figure 16–37**). Ca^{2+}-bound calmodulin activates myosin light-chain kinase (MLCK), thereby inducing the phosphorylation of smooth muscle myosin on one of its two light chains. When the light chain is phosphorylated, the myosin head can interact with actin filaments and cause contraction; when it is dephosphorylated, the myosin head tends to dissociate from actin and becomes inactive.

The phosphorylation events that regulate contraction in smooth muscle cells occur relatively slowly, so that maximum contraction often requires nearly a second (compared with the few milliseconds required for contraction of a skeletal muscle cell). But rapid activation of contraction is not important in smooth

(A)

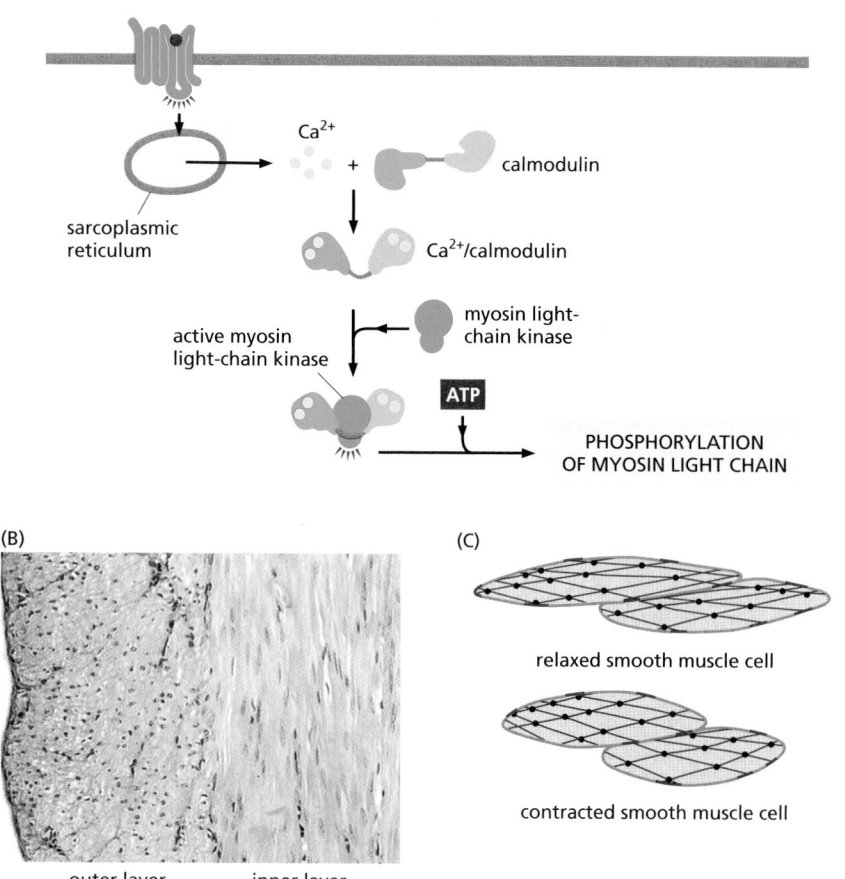

Ca^{2+}

calmodulin

sarcoplasmic
reticulum

Ca^{2+}/calmodulin

active myosin
light-chain kinase

myosin light-
chain kinase

ATP

PHOSPHORYLATION
OF MYOSIN LIGHT CHAIN

(B)

outer layer   inner layer

(C)

relaxed smooth muscle cell

contracted smooth muscle cell

Figure 16–37 Smooth muscle contraction. (A) Upon muscle stimulation by activation of cell-surface receptors, Ca^{2+} released into the cytoplasm from the sarcoplasmic reticulum (SR) binds to calmodulin (see Figure 15–29). Ca^{2+}-bound calmodulin then binds myosin light-chain kinase (MLCK), which phosphorylates myosin light chain, stimulating myosin activity. Non-muscle myosin is regulated by the same mechanism (see Figure 16–39). (B) Smooth muscle cells in a cross section of cat intestinal wall. The outer layer of smooth muscle is oriented with the long axis of its cells extending parallel along the length of the intestine, and upon contraction will shorten the intestine. The inner layer is oriented circularly around the intestine and when contracted will cause the intestine to become narrower. Contraction of both layers squeezes material through the intestine, much like squeezing toothpaste out of a tube. (C) A model for the contractile apparatus in a smooth muscle cell, with bundles of contractile filaments containing actin and myosin *(red)* oriented obliquely to the long axis of the cell. Their contraction greatly shortens the cell. In this diagram, the bundle angles are exaggerated by more than a factor of 4, to schematically illustrate the effect of contraction. In addition, only a few of the many bundles are shown. (B, courtesy of Gwen V. Childs.)

muscle: its myosin II hydrolyzes ATP about 10 times more slowly than skeletal muscle myosin, producing a slow cycle of myosin conformational changes that results in slow contraction.

Heart Muscle Is a Precisely Engineered Machine

The heart is the most heavily worked muscle in the body, contracting about 3 billion (3×10^9) times during the course of a human lifetime (Movie 16.6). Heart cells express several specific isoforms of cardiac muscle myosin and cardiac muscle actin. Even subtle changes in these cardiac-specific contractile proteins—changes that would not cause any noticeable consequences in other tissues—can cause serious heart disease (Figure 16–38).

The normal cardiac contractile apparatus is such a highly tuned machine that a tiny abnormality anywhere in the works can be enough to gradually wear it down over years of repetitive motion. *Familial hypertrophic cardiomyopathy* is a common cause of sudden death in young athletes. It is a genetically dominant inherited condition that affects about two out of every thousand people, and it is associated with heart enlargement, abnormally small coronary vessels, and disturbances in heart rhythm (cardiac arrhythmias). The cause of this condition is either any one of over 40 subtle point mutations in the genes encoding cardiac β myosin heavy chain (almost all causing changes in or near the motor domain) or one of about a dozen mutations in other genes encoding contractile proteins—including myosin light chains, cardiac troponin, and tropomyosin. Minor missense mutations in the cardiac actin gene cause another type of heart condition, called *dilated cardiomyopathy*, which can also result in early heart failure.

Actin and Myosin Perform a Variety of Functions in Non-Muscle Cells

Most non-muscle cells contain small amounts of contractile actin–myosin II bundles that form transiently under specific conditions and are much less well organized than muscle fibers. Non-muscle contractile bundles are regulated by myosin phosphorylation rather than the troponin mechanism (Figure 16–39). These contractile bundles function to provide mechanical support to cells, for example, by assembling into cortical **stress fibers** that connect the cell to the extracellular

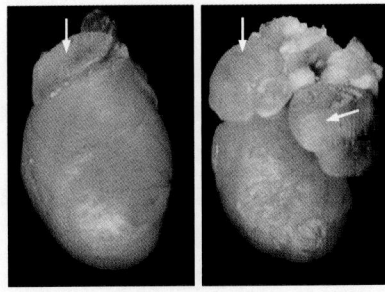

Figure 16–38 Effect on the heart of a subtle mutation in cardiac myosin. *Left,* normal heart from a 6-day-old mouse pup. *Right,* heart from a pup with a point mutation in both copies of its cardiac myosin gene, changing Arg403 to Gln. The arrows indicate the atria. In the heart from the pup with the cardiac myosin mutation, both atria are greatly enlarged (hypertrophic), and the mice die within a few weeks of birth. (From D. Fatkin et al., *J. Clin. Invest.* 103:147–153, 1999. With permission from The American Society for Clinical Investigation.)

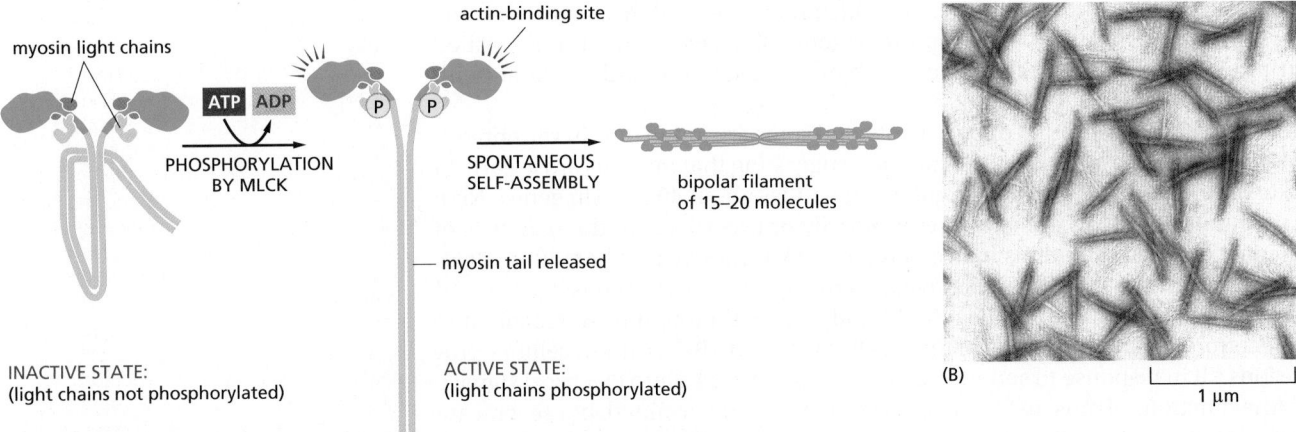

Figure 16–39 Light-chain phosphorylation and the regulation of the assembly of myosin II into thick filaments. (A) The controlled phosphorylation by the enzyme myosin light-chain kinase (MLCK) of one of the two light chains (the so-called regulatory light chain, shown in *light blue*) on non-muscle myosin II in a test tube has at least two effects: it causes a change in the conformation of the myosin head, exposing its actin-binding site, and it releases the myosin tail from a "sticky patch" on the myosin head, thereby allowing the myosin molecules to assemble into short, bipolar, thick filaments. Smooth muscle is regulated by the same mechanism (see Figure 16–37). (B) Electron micrograph of negatively stained short filaments of myosin II that have been induced to assemble in a test tube by phosphorylation of their light chains. These myosin II filaments are much smaller than those found in skeletal muscle cells (see Figure 16–27). (B, courtesy of John Kendrick-Jones.)

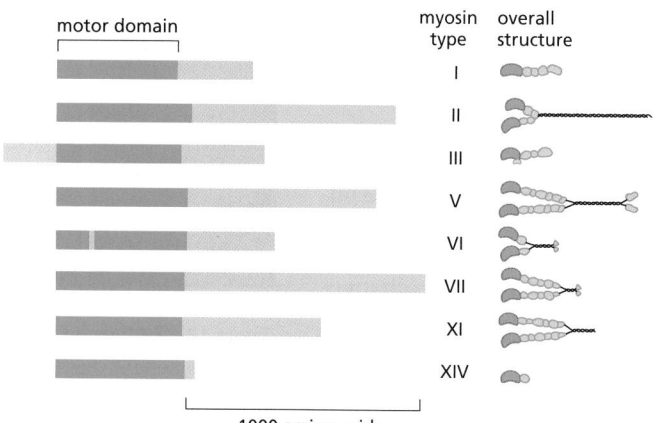

motor domain

myosin type — overall structure

I

II

III

V

VI

VII

XI

XIV

1000 amino acids

Figure 16–40 Myosin superfamily members. Comparison of the domain structure of the heavy chains of some myosin types. All myosins share similar motor domains (shown in *dark green*), but their C-terminal tails *(light green)* and N-terminal extensions *(light blue)* are very diverse. On the right are depictions of the molecular structure for these family members. Many myosins form dimers, with two motor domains per molecule, but a few (such as I, III, and XIV) seem to function as monomers, with just one motor domain. Myosin VI, despite its overall structural similarity to other family members, is unique in moving toward the minus end (instead of the plus end) of an actin filament. The small insertion within its motor head domain, not found in other myosins, is probably responsible for this change in direction.

matrix through *focal adhesions* or by forming a *circumferential belt* in an epithelial cell, connecting it to adjacent cells through *adherens junctions* (discussed in Chapter 19). As described in Chapter 17, actin and myosin II in the *contractile ring* generate the force for cytokinesis, the final stage in cell division. Finally, as discussed later, contractile bundles also contribute to the adhesion and forward motion of migrating cells.

Non-muscle cells also express a large family of other myosin proteins, which have diverse structures and functions in the cell. Following the discovery of conventional muscle myosin, a second member of the family was found in the freshwater amoeba *Acanthamoeba castellanii*. This protein had a different tail structure and seemed to function as a monomer, and so it was named *myosin I* (for one-headed). Conventional muscle myosin was renamed *myosin II* (for two-headed). Subsequently, many other myosin types were discovered. The heavy chains generally start with a recognizable myosin motor domain at the N-terminus and then diverge widely with a variety of C-terminal tail domains (**Figure 16–40**). The myosin family includes a number of one-headed and two-headed varieties that are about equally related to myosin I and myosin II, and the nomenclature now reflects their approximate order of discovery (myosin III through at least myosin XVIII). Sequence comparisons among diverse eukaryotes indicate that there are at least 37 distinct myosin families in the superfamily. All of the myosins except one move toward the plus end of an actin filament, although they do so at different speeds. The exception is myosin VI, which moves toward the minus end. The myosin tails (and the tails of motor proteins generally) have apparently diversified during evolution to permit the proteins to bind other subunits and to interact with different cargoes.

Some myosins are found only in plants, and some are found only in vertebrates. Most, however, are found in all eukaryotes, suggesting that myosins arose early in eukaryotic evolution. The human genome includes about 40 myosin genes. Nine of the human myosins are expressed primarily or exclusively in the hair cells of the inner ear, and mutations in five of them are known to cause hereditary deafness. These extremely specialized myosins are important for the construction and function of the complex and beautiful bundles of actin found in stereocilia that project from the apical surface of these cells (see Figure 9–51); these cellular protrusions tilt in response to sound and convert sound waves into electrical signals.

The functions of most of the myosins remain to be determined, but several are well characterized. The myosin I proteins often contain either a second actin-binding site or a membrane-binding site in their tails, and they are generally involved in intracellular organization—including the protrusion of actin-rich structures at the cell surface, such as microvilli (see Panel 16–1 and Figure 16–4), and endocytosis. Myosin V is a two-headed myosin with a large step size (**Figure 16–41**A) and is involved in organelle transport along actin filaments. In contrast to myosin II motors, which work in ensembles and are attached only transiently to actin filaments so as not to interfere with one another, myosin V moves continuously,

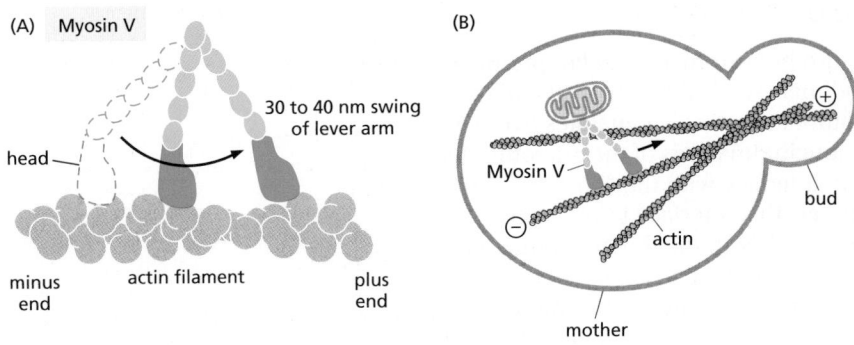

Figure 16–41 Myosin V carries cargo along actin filaments. (A) The lever arm of myosin V is long, allowing it to take a bigger step along an actin filament than myosin II (see Figure 16–29). (B) Myosin V transports cargo and organelles along actin cables, in this example moving a mitochondrion into the growing bud of a yeast cell.

or *processively*, along actin filaments without letting go. Myosin V functions are well studied in the yeast *Saccharomyces cerevisiae*, which undergoes a stereotypical pattern of growth and division called budding. Actin cables in the mother cell point toward the bud, where actin is found in patches that concentrate where cell wall growth is taking place. Myosin V motors carry a wide range of cargoes—including mRNA, endoplasmic reticulum, and secretory vesicles—along the actin cables and into the bud. In addition, myosin V mediates the correct partitioning of organelles such as peroxisomes and mitochondria between mother and daughter cells (see Figure 16–41B).

Summary

Using their neck domain as a lever arm, myosins convert ATP hydrolysis into mechanical work to move along actin filaments in a stepwise fashion. Skeletal muscle is made up of myofibrils containing thousands of sarcomeres assembled from highly ordered arrays of actin and myosin II filaments, together with many accessory proteins. Muscle contraction is stimulated by calcium, which causes the actin-filament-associated protein tropomyosin to move, uncovering myosin binding sites and allowing the filaments to slide past one another. Smooth muscle and non-muscle cells have less well-ordered contractile bundles of actin and myosin, which are regulated by myosin light-chain phosphorylation. Myosin V transports cargo by walking along actin filaments.

MICROTUBULES

Microtubules are structurally more complex than actin filaments, but they are also highly dynamic and play comparably diverse and important roles in the cell. Microtubules are polymers of the protein **tubulin**. The tubulin subunit is itself a heterodimer formed from two closely related globular proteins called *α-tubulin* and *β-tubulin*, each comprising 445–450 amino acids, which are tightly bound together by noncovalent bonds (**Figure 16–42A**). These two tubulin proteins are found only in this heterodimer, and each α or β monomer has a binding site for one molecule of GTP. The GTP that is bound to α-tubulin is physically trapped at the dimer interface and is never hydrolyzed or exchanged; it can therefore be considered to be an integral part of the tubulin heterodimer structure. The nucleotide on the β-tubulin, in contrast, may be in either the GTP or the GDP form and is exchangeable within the soluble (unpolymerized) tubulin dimer.

Tubulin is found in all eukaryotic cells, and it exists in multiple isoforms. Yeast and human tubulins are 75% identical in amino acid sequence. In mammals, there are at least six forms of α-tubulin and a similar number of β-tubulins, each encoded by a different gene. The different forms of tubulin are very similar, and they generally copolymerize into mixed microtubules in the test tube. However, they can have distinct locations in cells and tissues and perform subtly different functions. As a striking example, mutations in a particular human β-tubulin gene give rise to a paralytic eye-movement disorder due to loss of ocular nerve function. Numerous human neurological diseases have been linked to specific mutations in different tubulin genes.

Microtubules Are Hollow Tubes Made of Protofilaments

A microtubule is a hollow cylindrical structure built from 13 parallel protofilaments, each composed of αβ-tubulin heterodimers stacked head to tail and then folded into a tube (Figure 16–42B–D). Microtubule assembly generates two new types of protein–protein contacts. Along the longitudinal axis of the microtubule, the "top" of one β-tubulin molecule forms an interface with the "bottom" of the α-tubulin molecule in the adjacent heterodimer. This interface is very similar to the interface holding the α and β monomers together in the dimer subunit, and the binding energy is high. Perpendicular to these interactions, neighboring protofilaments form lateral contacts. In this dimension, the main lateral contacts are between monomers of the same type (α–α and β–β). As longitudinal and lateral contacts are repeated during assembly, a slight stagger in lateral contacts gives rise to the helical microtubule lattice. Because multiple contacts within the lattice hold most of the subunits in a microtubule in place, the addition and loss of subunits occurs almost exclusively at the microtubule ends (see Figure 16–5). These multiple contacts among subunits make microtubules stiff and difficult to bend. The persistence length of a microtubule is several millimeters, making microtubules the stiffest and straightest structural elements found in most animal cells.

The subunits in each protofilament in a microtubule all point in the same direction, and the protofilaments themselves are aligned in parallel (see Figure

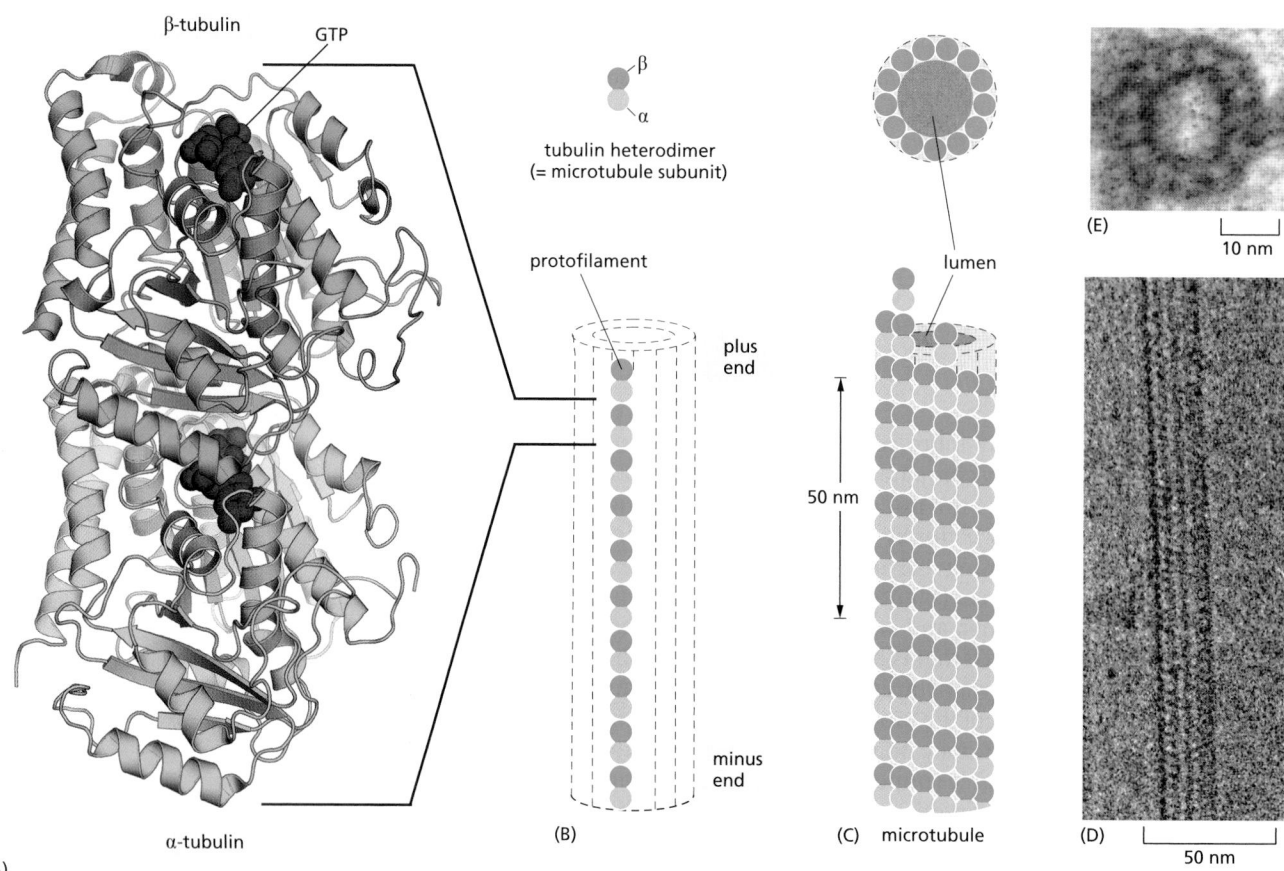

Figure 16–42 The structure of a microtubule and its subunit. (A) The subunit of each protofilament is a tubulin heterodimer, formed from a tightly linked pair of α- and β-tubulin monomers. The GTP molecule in the α-tubulin monomer is so tightly bound that it can be considered an integral part of the protein. The GTP molecule in the β-tubulin monomer, however, is less tightly bound and has an important role in filament dynamics. Both nucleotides are shown in *red*. (B) One tubulin subunit (αβ-heterodimer) and one protofilament are shown schematically. Each protofilament consists of many adjacent subunits with the same orientation. (C) The microtubule is a stiff hollow tube formed from 13 protofilaments aligned in parallel. (D) A short segment of a microtubule viewed in an electron microscope. (E) Electron micrograph of a cross section of a microtubule showing a ring of 13 distinct protofilaments. (D, courtesy of Richard Wade; E, courtesy of Richard Linck.)

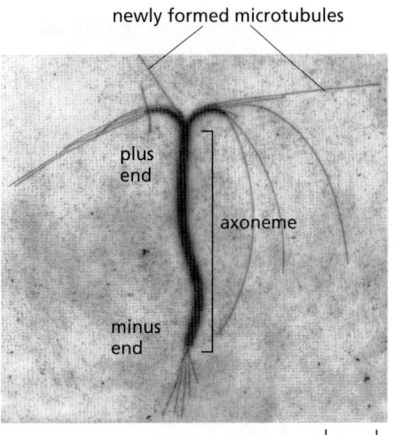

Figure 16–43 The preferential growth of microtubules at the plus end. Microtubules grow faster at one end than at the other. A stable bundle of microtubules obtained from the core of a cilium (called an axoneme) was incubated for a short time with tubulin subunits under polymerizing conditions. Microtubules grew fastest from the plus end of the microtubule bundle, the end at the *top* in this micrograph. (Courtesy of Gary Borisy.)

16–42). Therefore, the microtubule lattice itself has a distinct structural polarity, with α-tubulins exposed at the minus end and β-tubulins exposed at the plus end. As for actin filaments, the regular, parallel orientation of their subunits gives microtubules structural and dynamic polarity (**Figure 16–43**), with plus ends growing and shrinking more rapidly.

Microtubules Undergo Dynamic Instability

Microtubule dynamics, like those of actin filaments, are profoundly influenced by the binding and hydrolysis of nucleotide—GTP in this case. GTP hydrolysis occurs only within the β-tubulin subunit of the tubulin dimer. It proceeds very slowly in free tubulin subunits but is accelerated when they are incorporated into microtubules. Following GTP hydrolysis, the free phosphate group is released and the GDP remains bound to β-tubulin within the microtubule lattice. Thus, as in the case of actin filaments, two different types of microtubule structures can exist, one with the "T form" of the nucleotide bound (GTP) and one with the "D form" bound (GDP). The energy of nucleotide hydrolysis is stored as elastic strain in the polymer lattice, making the free-energy change for dissociation of a subunit from the D-form polymer more negative than the free-energy change for dissociation of a subunit from the T-form polymer. In consequence, the ratio of k_{off}/k_{on} for GDP-tubulin (its critical concentration $[C_c(D)]$) is much higher than that of GTP-tubulin. Thus, under physiological conditions, GTP-tubulin tends to polymerize and GDP-tubulin to depolymerize.

Whether the tubulin subunits at the very end of a microtubule are in the T or the D form depends on the relative rates of GTP hydrolysis and tubulin addition. If the rate of subunit addition is high—and thus the filament is growing rapidly—then it is likely that a new subunit will be added to the polymer before the nucleotide in the previously added subunit has been hydrolyzed. In this case, the tip of the polymer remains in the T form, forming a *GTP cap*. However, if the rate of subunit addition is low, hydrolysis may occur before the next subunit is added, and the tip of the filament will then be in the D form. If GTP-tubulin subunits assemble at the end of the microtubule at a rate similar to the rate of GTP hydrolysis, then hydrolysis will sometimes "catch up" with the rate of subunit addition and transform the end to a D form. This transformation is sudden and random, with a certain probability per unit time that depends on the concentration of free GTP-tubulin subunits.

Suppose that the concentration of free tubulin is intermediate between the critical concentration for a T-form end and the critical concentration for a D-form end (that is, above the concentration necessary for T-form assembly, but below that for the D form). Now, any end that happens to be in the T form will grow, whereas any end that happens to be in the D form will shrink. On a single microtubule, an end might grow for a certain length of time in a T form, but then suddenly change to the D form and begin to shrink rapidly, even while the free subunit concentration is held constant. At some later time, it might then regain a T-form end and begin to grow again. This rapid interconversion between a growing and shrinking state, at a uniform free subunit concentration, is called **dynamic instability** (**Figure 16–44**A and **Figure 16–45**; see Panel 16–2). The change from growth to shrinkage is called a *catastrophe*, while the change to growth is called a *rescue*.

In a population of microtubules, at any instant some of the ends are in the T form and some are in the D form, with the ratio depending on the hydrolysis rate and the free subunit concentration. *In vitro*, the structural difference between a T-form end and a D-form end is dramatic. Tubulin subunits with GTP bound to

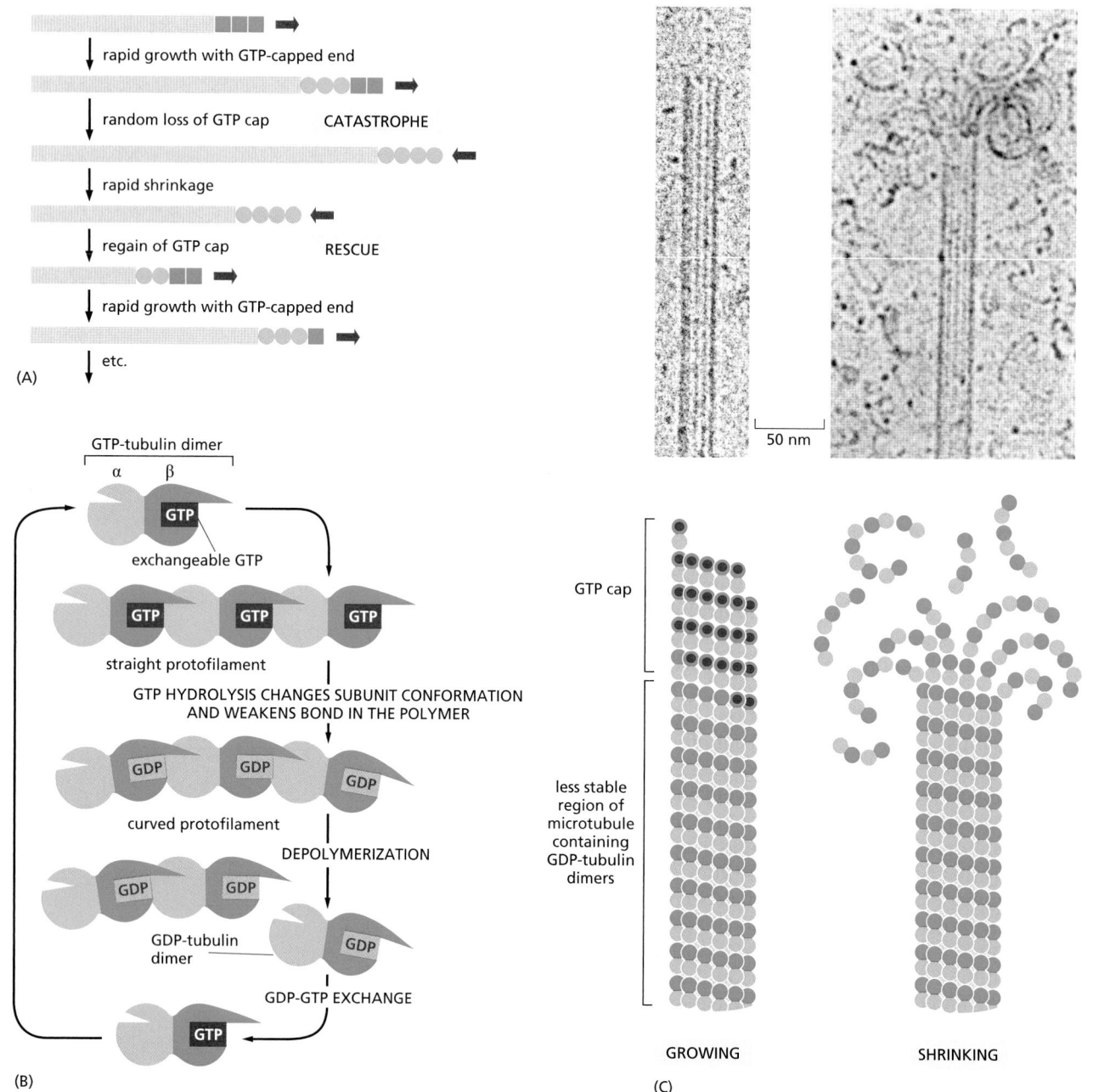

Figure 16–44 Dynamic instability due to the structural differences between a growing and a shrinking microtubule end. (A) If the free tubulin concentration in solution is between the critical concentrations of the GTP- and GDP-bound forms, a single microtubule end may undergo transitions between a growing state and a shrinking state. A growing microtubule has GTP-containing subunits at its end, forming a GTP cap. If nucleotide hydrolysis proceeds more rapidly than subunit addition, the cap is lost and the microtubule begins to shrink, an event called a "catastrophe." But GTP-containing subunits may still add to the shrinking end, and if enough add to form a new cap, then microtubule growth resumes, an event called "rescue." (B) Model for the structural consequences of GTP hydrolysis in the microtubule lattice. The addition of GTP-containing tubulin subunits to the end of a protofilament causes the end to grow in a linear conformation that can readily pack into the cylindrical wall of the microtubule. Hydrolysis of GTP after assembly changes the conformation of the subunits and tends to force the protofilament into a curved shape that is less able to pack into the microtubule wall. (C) In an intact microtubule, protofilaments made from GDP-containing subunits are forced into a linear conformation by the many lateral bonds within the microtubule wall, given a stable cap of GTP-containing subunits. Loss of the GTP cap, however, allows the GDP-containing protofilaments to relax into their more curved conformation. This leads to a progressive disruption of the microtubule. Above the drawings of a growing and a shrinking microtubule, electron micrographs show actual microtubules in each of these two states. Note particularly the curling, disintegrating GDP-containing protofilaments at the end of the shrinking microtubule. (C, © 1991 E.M. Mandelkow, E. Mandelkow and R.A. Milligan. Originally published in *J. Cell Biol.* https://doi.org/10.1083/jcb.114.5.977. With permission from Rockefeller University Press.)

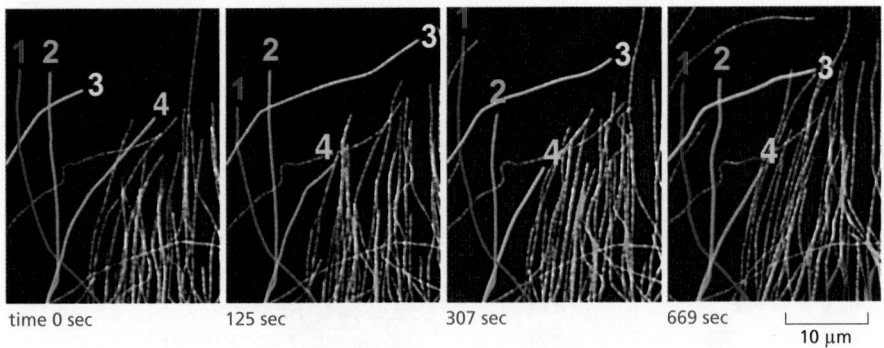

time 0 sec　　　125 sec　　　307 sec　　　669 sec

10 μm

Figure 16–45 Direct observation of the dynamic instability of microtubules in a living cell. Microtubules in a newt lung epithelial cell were observed after the cell was injected with a small amount of rhodamine-labeled tubulin. Notice the dynamic instability of microtubules at the edge of the cell. Four individual microtubules are highlighted for clarity; each of these shows alternating shrinkage and growth (Movie 16.7). (Courtesy of Wendy C. Salmon and Clare Waterman.)

the β-monomer produce straight protofilaments that make strong and regular lateral contacts with one another. But the hydrolysis of GTP to GDP is associated with a subtle conformational change in the protein, which makes the protofilaments curved (Figure 16–44B). On a rapidly growing microtubule, the GTP cap is thought to constrain the curvature of the protofilaments, and the ends appear straight. But when the terminal subunits have hydrolyzed their nucleotides, this constraint is removed, and the curved protofilaments spring apart. This cooperative release of the energy of hydrolysis stored in the microtubule lattice causes the curled protofilaments to peel off rapidly, and curved oligomers of GDP-containing tubulin are seen near the ends of depolymerizing microtubules (Figure 16–44C).

Microtubule Functions Are Inhibited by Both Polymer-stabilizing and Polymer-destabilizing Drugs

Chemical compounds that impair polymerization or depolymerization of microtubules are powerful tools for investigating the roles of these polymers in cells. Whereas *colchicine* and *nocodazole* interact with tubulin subunits and lead to microtubule depolymerization, *Taxol* binds to and stabilizes microtubules, causing a net increase in tubulin polymerization (see Table 16–1). Drugs like these have a rapid and profound effect on the organization of the microtubules in living cells. Both microtubule-depolymerizing drugs (such as nocodazole) and microtubule-polymerizing drugs (such as Taxol) preferentially kill dividing cells, since microtubule dynamics are crucial for correct function of the mitotic spindle (discussed in Chapter 17). Some of these drugs efficiently kill certain types of tumor cells in a human patient, although not without toxicity to rapidly dividing normal cells, including those in the bone marrow, intestine, and hair follicles. Taxol in particular has been widely used to treat cancers of the breast and lung, and it is frequently successful in treatment of tumors that are resistant to other chemotherapeutic agents.

A Protein Complex Containing γ-Tubulin Nucleates Microtubules

Because formation of a microtubule requires the interaction of many tubulin heterodimers, the concentration of tubulin subunits required for spontaneous nucleation of microtubules is very high. Microtubule nucleation therefore requires help from other factors. While α- and β-tubulins are the regular building blocks of microtubules, another type of tubulin, called *γ-tubulin*, is present in much smaller amounts than α- and β-tubulin and is involved in the nucleation of microtubule growth in organisms ranging from yeasts to humans. Microtubules are generally nucleated from a specific intracellular location known as a **microtubule-organizing center** (**MTOC**) where γ-tubulin is most enriched. Nucleation in many cases depends on the **γ-tubulin ring complex** (**γ-TuRC**). Within this complex, two accessory proteins bind directly to the γ-tubulin, along with several other proteins that help create a spiral ring of γ-tubulin molecules, which serves as a template that creates a microtubule with 13 protofilaments (**Figure 16–46**).

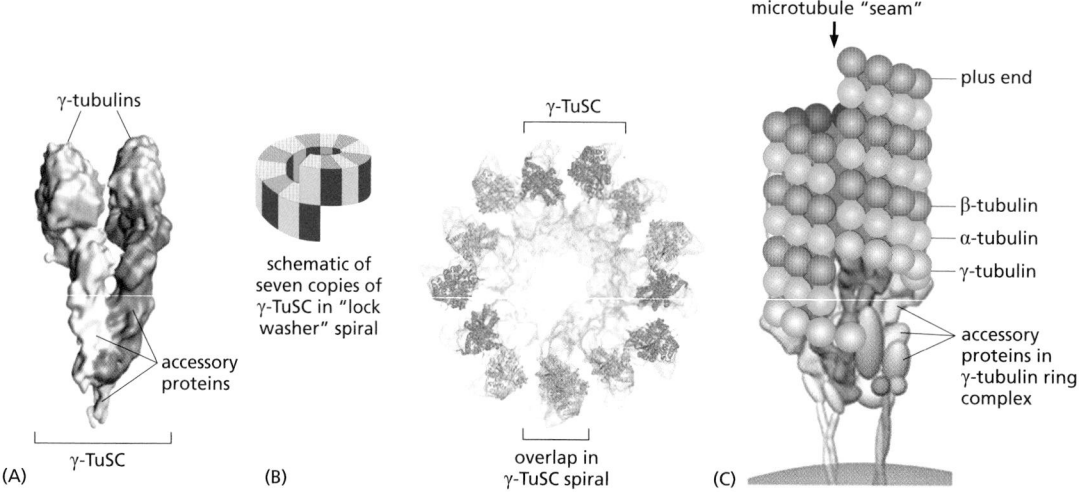

Figure 16–46 Microtubule nucleation by the γ-tubulin ring complex. (A) Two copies of γ-tubulin associate with a pair of accessory proteins to form the γ-tubulin small complex (γ-TuSC). This image was generated by high-resolution electron microscopy of individual purified complexes. (B) Seven copies of the γ-TuSC associate to form a spiral structure in which the last γ-tubulin lies beneath the first, resulting in 13 exposed γ-tubulin subunits in a circular orientation that matches the orientation of the 13 protofilaments in a microtubule. (C) In many cell types, the γ-TuSC spiral associates with additional accessory proteins to form the γ-tubulin ring complex (γ-TuRC), which is likely to nucleate the minus end of a microtubule as shown here. Note the longitudinal discontinuity between two protofilaments, which results from the spiral orientation of the γ-tubulin subunits. Microtubules often have one such "seam" breaking the otherwise uniform helical packing of the protofilaments. (A and B, from J.M. Kollman et al., *Nature* 466:879–883, published 2010 by Macmillan Publishers Ltd. Reproduced with permission of SNCSC.)

Microtubules Emanate from the Centrosome in Animal Cells

Many animal cells have a single, well-defined MTOC called the **centrosome**, which is located near the nucleus and from which microtubules are nucleated at their minus ends, so the plus ends point outward and continuously grow and shrink, probing the entire three-dimensional volume of the cell. A centrosome typically recruits more than fifty copies of γ-TuRC. In addition, γ-TuRC molecules are found in the cytoplasm, and centrosomes are not absolutely required for microtubule nucleation, since destroying them with a laser pulse does not prevent microtubule nucleation elsewhere in the cell. A variety of proteins have been identified that anchor γ-TuRC to the centrosome, but mechanisms that activate microtubule nucleation at MTOCs and at other sites in the cell are poorly understood.

Embedded in the centrosome are the **centrioles**, a pair of cylindrical structures arranged at right angles to each other in an L-shaped configuration (**Figure 16–47**). A centriole consists of a cylindrical array of short, modified microtubules arranged into a barrel shape with striking ninefold symmetry (**Figure 16–48**). Together with a large number of accessory proteins, the centrioles organize the *pericentriolar material*, where microtubule nucleation takes place. As described in Chapter 17, the centrosome duplicates and splits into two parts before mitosis, each containing a duplicated centriole pair. The two centrosomes move to opposite sides of the nucleus when mitosis begins, and they form the two poles of the mitotic spindle (see Panel 17–1).

Microtubule organization varies widely among different species and cell types. In budding yeast, microtubules are nucleated at an MTOC that is embedded in the nuclear envelope as a small, multilayered structure called the *spindle pole body*, also found in other fungi and diatoms. Higher-plant cells appear to nucleate microtubules at sites distributed all around the nuclear envelope and at the cell cortex. Neither fungi nor most plant cells contain centrioles. Despite these differences, all these cells seem to use γ-tubulin to nucleate their microtubules.

In cultured animal cells, the aster-like configuration of microtubules is robust, with dynamic plus ends pointing outward toward the cell periphery and stable minus ends collected near the nucleus. The system of microtubules radiating from

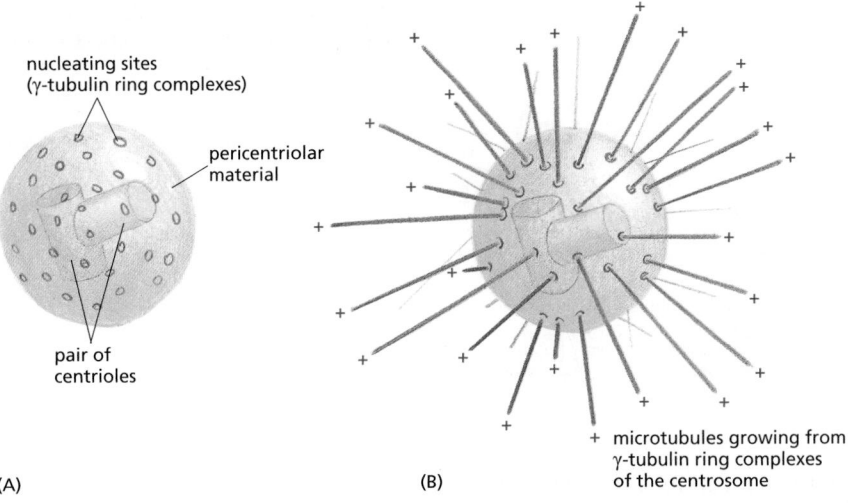

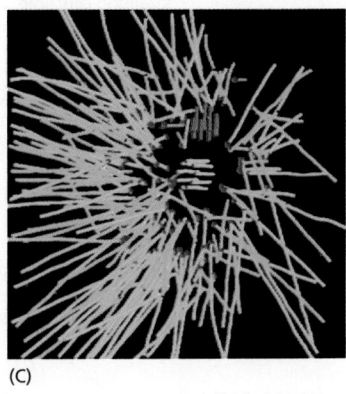

nucleating sites
(γ-tubulin ring complexes)

pericentriolar
material

pair of
centrioles

(A)

(B)

+ microtubules growing from
γ-tubulin ring complexes
of the centrosome

(C)

Figure 16–47 **The centrosome.** (A) The centrosome is the major MTOC of animal cells. Located in the cytoplasm next to the nucleus, it consists of an amorphous matrix of fibrous proteins to which the γ-tubulin ring complexes that nucleate microtubule growth are attached. This matrix is organized by a pair of centrioles, as described in the text. (B) A centrosome with attached microtubules. The minus end of each microtubule is embedded in the centrosome, having grown from a γ-tubulin ring complex, whereas the plus end of each microtubule is free in the cytoplasm. (C) In a reconstructed image of the MTOC from a *C. elegans* cell, a dense thicket of microtubules can be seen emanating from the centrosome. (C, from E.T. O'Toole et al., *J. Cell Biol.* 163:451–456, 2003. With permission from the authors.)

the centrosome acts as a device to survey the outlying regions of the cell and to position the centrosome at its center. Even in an isolated cell fragment lacking the centrosome, dynamic microtubules arrange themselves into a star-shaped array with the microtubule minus ends clustered at the center by minus-end-binding proteins (**Figure 16–49**). This ability of the microtubule cytoskeleton to find the center of the cell establishes a general coordinate system, which is then used to position many organelles within the cell. Highly differentiated cells with complex morphologies such as neurons, muscles, and epithelial cells must use additional measuring mechanisms to establish their more elaborate internal coordinate systems. Thus, for example, when an epithelial cell forms cell–cell junctions and becomes highly polarized, the microtubule minus ends move to a region near the

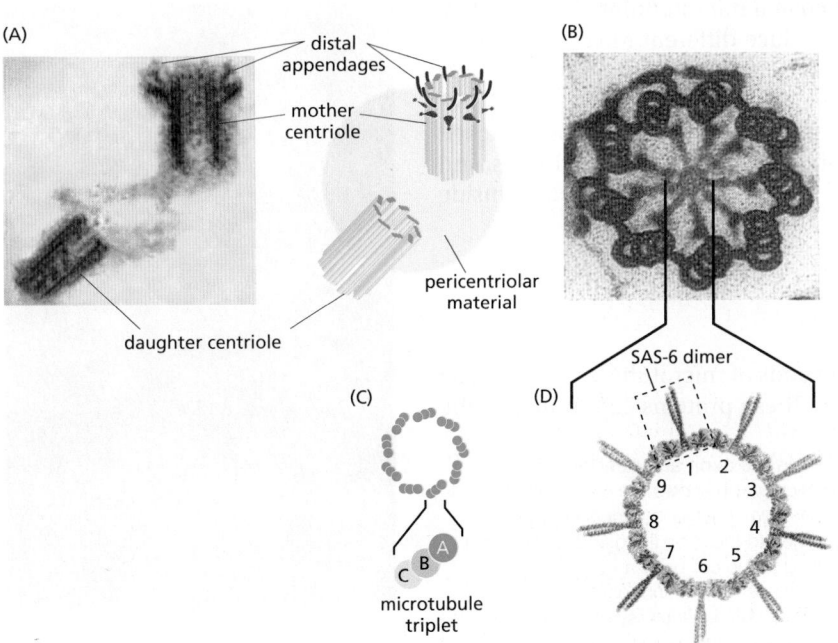

distal
appendages

mother
centriole

pericentriolar
material

daughter centriole

microtubule
triplet

SAS-6 dimer

Figure 16–48 **A pair of centrioles in the centrosome.** (A) An electron micrograph of a thin section of an isolated centrosome showing the mother centriole with its distal appendages and the adjacent daughter centriole, which formed through a duplication event during S phase (see Figure 17–26). In the centrosome, the centriole pair is surrounded by a dense matrix of pericentriolar material from which microtubules nucleate. Centrioles also function as basal bodies to nucleate the formation of ciliary axonemes (see Figure 16–68). (B) Electron micrograph of a cross section through a centriole in the cortex of a protozoan. Each centriole is composed of nine sets of triplet microtubules arranged to form a cylinder. (C) Each triplet contains one complete microtubule (the A microtubule) fused to two incomplete microtubules (the B and C microtubules). (D) The centriolar protein SAS-6 forms a coiled-coil dimer. Nine SAS-6 dimers can self-associate to form a ring. Located at the hub of the centriole cartwheel structure, the SAS-6 ring is thought to generate the ninefold symmetry of the centriole. (A, from from M. Paintrand, et al. *J. Struct. Biol.* 108:107, 1992. With permission from Elsevier; B, courtesy of Richard Linck; D, courtesy of Michel Steinmetz.)

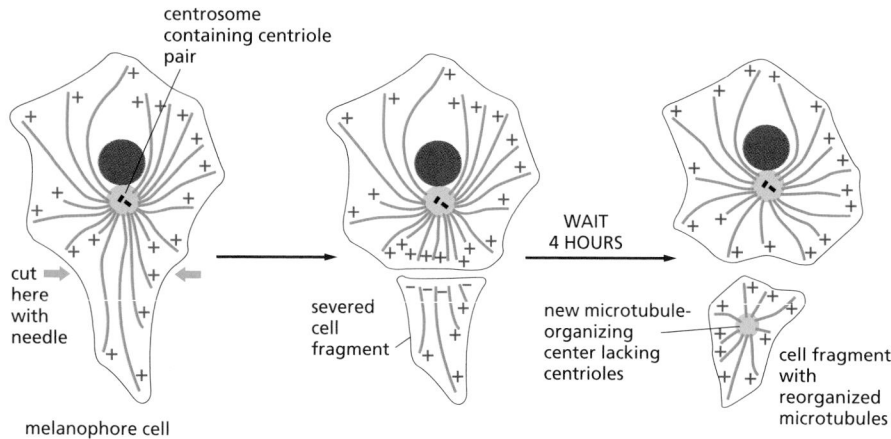

centrosome
containing centriole
pair

WAIT
4 HOURS

cut
here
with
needle

melanophore cell

severed
cell
fragment

new microtubule-
organizing
center lacking
centrioles

cell fragment
with
reorganized
microtubules

Figure 16–49 A microtubule array can find the center of a cell. After the arm of a fish pigment cell is cut off with a needle, the microtubules in the detached cell fragment reorganize so that their minus ends end up near the center of the fragment, buried in a new microtubule-organizing center.

apical plasma membrane. From this asymmetrical location, a microtubule array extends along the long axis of the cell, with plus ends directed toward the basal surface (see Figure 16–4).

Microtubule-Binding Proteins Modulate Filament Dynamics and Organization

Microtubule polymerization dynamics are very different in cells than in solutions of pure tubulin. Microtubules in cells exhibit a much higher polymerization rate (typically 10–15 µm/min, relative to about 1.5 µm/min with purified tubulin at similar concentrations), a greater catastrophe frequency, and extended pauses in microtubule growth, a dynamic behavior rarely observed in pure tubulin solutions. These and other differences arise because microtubule dynamics inside the cell are governed by a variety of proteins that bind tubulin dimers or microtubules, as summarized in **Panel 16–4**.

Proteins that bind to microtubules are collectively called **microtubule-associated proteins**, or **MAPs**. Some MAPs can stabilize microtubules against disassembly. A subset of MAPs can also mediate the interaction of microtubules with other cell components. This subset is prominent in neurons, where stabilized microtubule bundles form the core of the axons and dendrites that extend from the cell body (**Figure 16–50**). These MAPs have at least one domain that binds to the microtubule surface and another that projects outward. The length of the projecting domain can determine how closely MAP-coated microtubules pack together, as demonstrated in cells engineered to overproduce different MAPs. Cells overexpressing MAP2, which has a long projecting domain, form bundles of stable microtubules that are kept widely spaced, while cells overexpressing tau, a MAP with a much shorter projecting domain, form bundles of more closely packed microtubules (**Figure 16–51**). MAPs are the targets of several protein kinases, and phosphorylation of a MAP can control both its activity and localization inside cells.

Microtubule Plus-End-Binding Proteins Modulate Microtubule Dynamics and Attachments

Cells contain numerous proteins that bind the ends of microtubules and thereby influence microtubule stability and dynamics. These proteins can influence the

Figure 16–50 Localization of MAPs in the axon and dendrites of a neuron. This immunofluorescence micrograph shows the distribution of the proteins tau *(green)* and MAP2 *(orange)* in a hippocampal neuron in culture. Whereas tau staining is confined to the axon (long and branched in this neuron), MAP2 staining is confined to the cell body and its dendrites. The antibody used here to detect tau binds only to unphosphorylated tau; phosphorylated tau is also present in dendrites. (Courtesy of James W. Mandell and Gary A. Banker.)

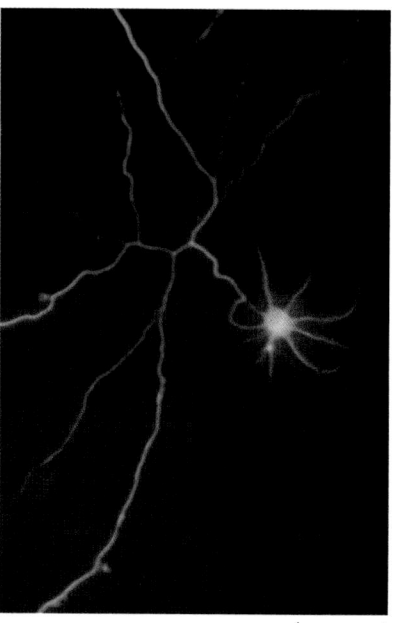

10 µm

MICROTUBULES

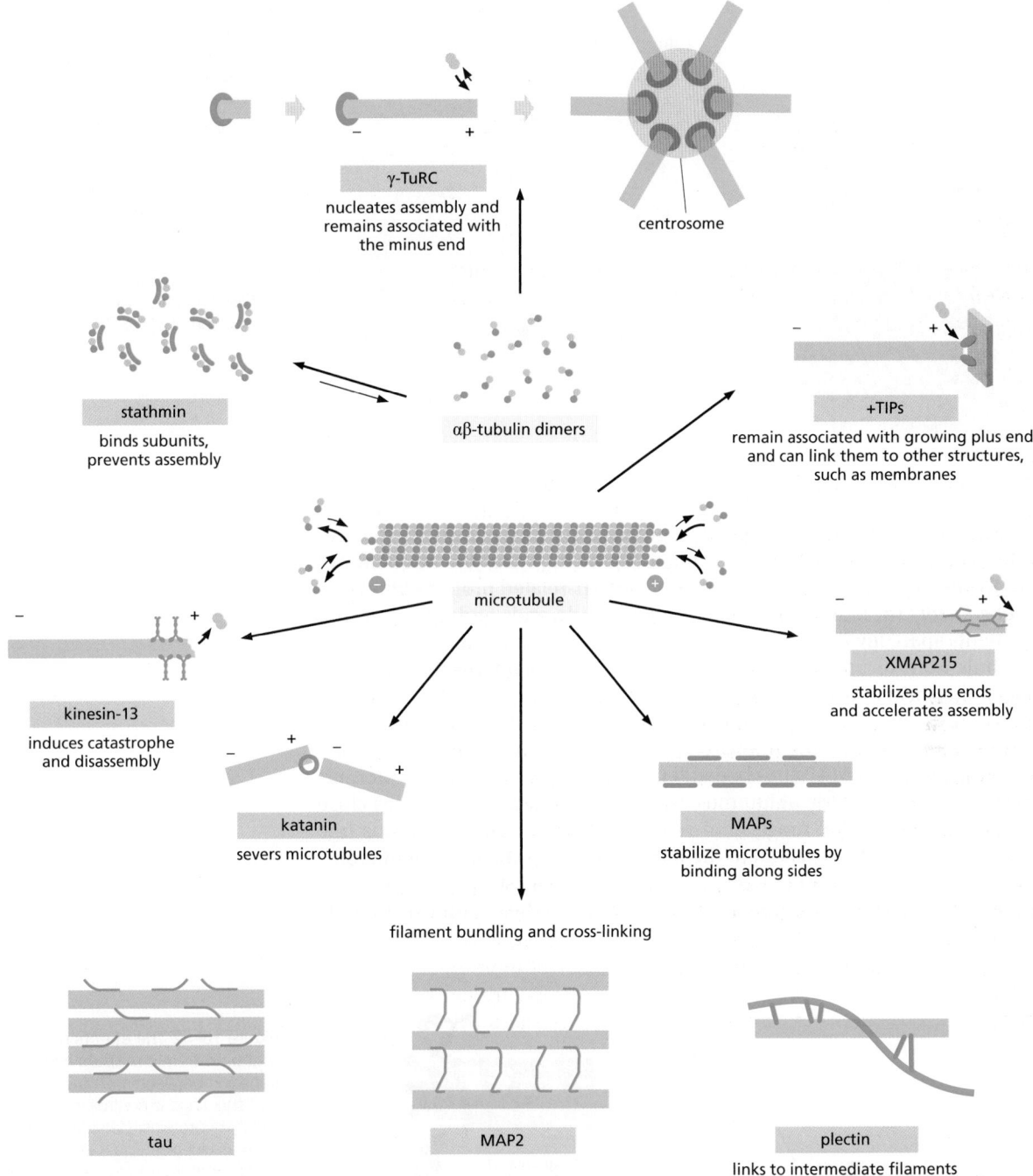

γ-TuRC
nucleates assembly and remains associated with the minus end

centrosome

αβ-tubulin dimers

stathmin
binds subunits, prevents assembly

+TIPs
remain associated with growing plus ends and can link them to other structures, such as membranes

microtubule

kinesin-13
induces catastrophe and disassembly

XMAP215
stabilizes plus ends and accelerates assembly

katanin
severs microtubules

MAPs
stabilize microtubules by binding along sides

filament bundling and cross-linking

tau

MAP2

plectin
links to intermediate filaments

Some of the major accessory proteins of the microtubule cytoskeleton. Except for two classes of motor proteins, an example of each major type is shown. Each of these is discussed in the text. However, most cells contain more than a hundred different microtubule-binding proteins, and — as for the actin-associated proteins — it is likely that there are important types of microtubule-associated proteins that are not yet recognized.

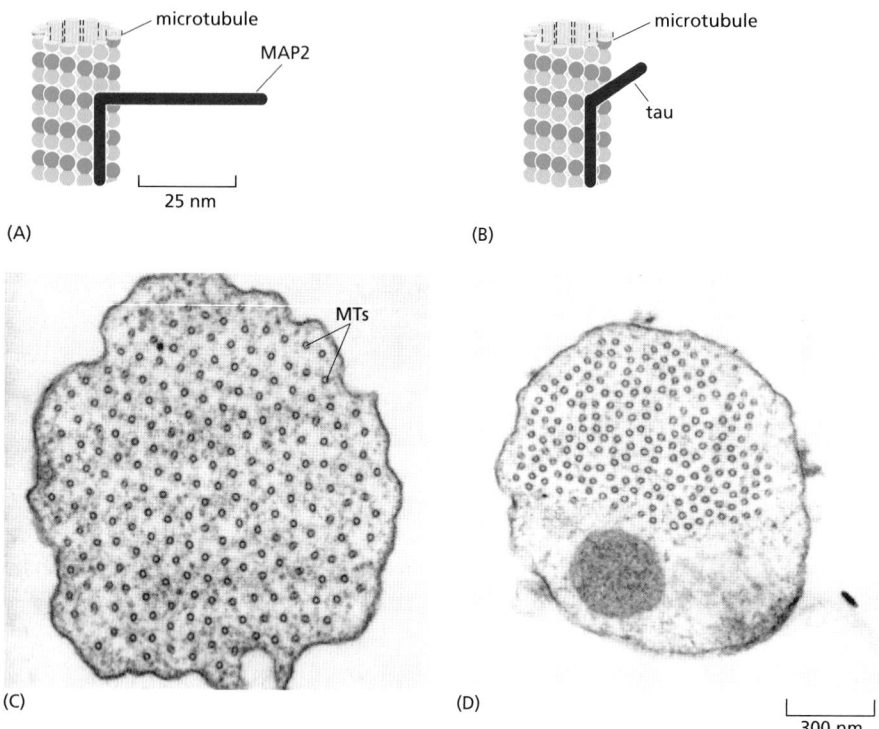

(A)

25 nm

(B)

MTs

(C)

(D)

300 nm

Figure 16–51 **Organization of microtubule bundles by MAPs.** (A) MAP2 binds along the microtubule lattice at one of its ends and extends a long projecting arm with a second microtubule-binding domain at the other end. (B) Tau possesses a shorter microtubule cross-linking domain. (C) Electron micrograph showing a cross section through a microtubule bundle in a cell overexpressing MAP2. The regular spacing of the microtubules (MTs) in this bundle results from the constant length of the projecting arms of the MAP2. (D) Similar cross section through a microtubule bundle in a cell overexpressing tau. Here the microtubules are spaced more closely together than they are in (C) because of tau's relatively short projecting arm. (C and D, from J. Chen et al., *Nature* 360:674–677, published 1992 by Nature Publishing Group. Reproduced with permission of SNCSC.)

rate at which a microtubule switches from a growing to a shrinking state (the frequency of catastrophes) or from a shrinking to a growing state (the frequency of rescues). For example, members of a family of kinesin-related proteins known as *catastrophe factors* (or kinesin-13) bind to microtubule ends and appear to pry protofilaments apart, lowering the normal activation-energy barrier that prevents a microtubule from springing apart into the curved protofilaments that are characteristic of the shrinking state (**Figure 16–52**). Another protein, called Nezha or Patronin, protects microtubule minus ends from the effects of catastrophe factors.

While very few microtubule minus-end-binding proteins have been characterized, a large subset of MAPs has been identified that are enriched at microtubule plus ends. A particularly ubiquitous example is *XMAP215*, which has close homologs in organisms that range from yeast to humans. XMAP215 binds free tubulin subunits and delivers them to the plus end, thereby promoting microtubule polymerization and simultaneously counteracting catastrophe factor activity (see Figure 16–52). The phosphorylation of XMAP215 during mitosis inhibits

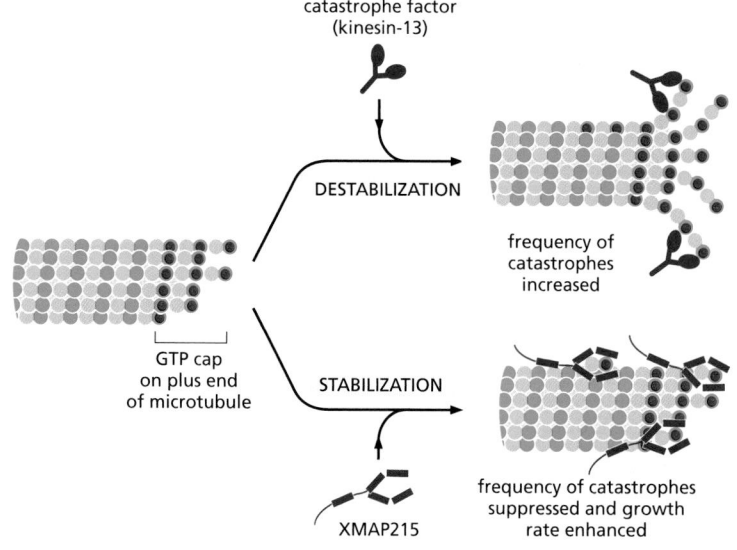

catastrophe factor
(kinesin-13)

DESTABILIZATION

frequency of
catastrophes
increased

GTP cap
on plus end
of microtubule

STABILIZATION

frequency of catastrophes
suppressed and growth
rate enhanced

XMAP215

Figure 16–52 **The effects of proteins that bind to microtubule ends.** The transition between microtubule growth and shrinkage is controlled in cells by a variety of proteins. Catastrophe factors such as kinesin-13, a member of the kinesin motor protein superfamily, bind to microtubule ends and pry them apart, thereby promoting depolymerization. On the other hand, a MAP such as XMAP215 stabilizes the end of a growing microtubule (XMAP stands for *Xenopus* microtubule-associated protein, and the number refers to its molecular mass in kilodaltons). XMAP215 binds tubulin dimers and delivers them to the microtubule plus end, thereby increasing the microtubule growth rate and suppressing catastrophes.

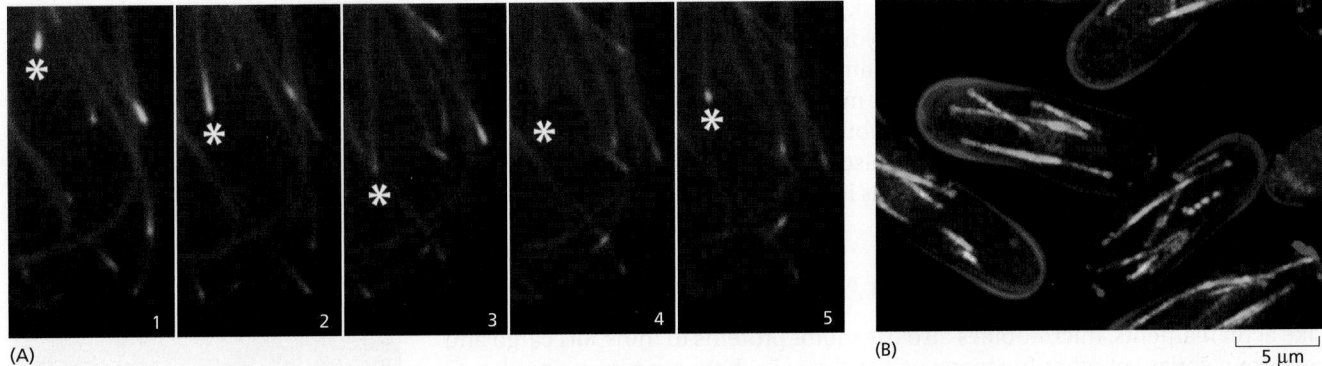

(A) (B) 5 μm

Figure 16–53 +TIP proteins found at the growing plus ends of microtubules.
(A) Frames from a fluorescence time-lapse movie of the edge of a cell expressing fluorescently labeled tubulin that incorporates into microtubules *(red)* as well as the +TIP protein EB1 tagged with a different color *(green)*. The same microtubule is marked (asterisk) in successive movie frames. When the microtubule is growing (frames 1, 2), EB1 is associated with the tip. When the microtubule undergoes a catastrophe and begins shrinking, EB1 is lost (frames 3, 4). The labeled EB1 is regained when growth of the microtubule is rescued (frame 5). See Movie 16.8. (B) In the fission yeast *Schizosaccharomyces pombe*, the plus ends of the microtubules *(green)* are associated with the homolog of EB1 *(red)* at the two poles of the rod-shaped cells. (A, courtesy of Anna Akhmanova and Ilya Grigoriev; B, courtesy of Takashi Toda.)

its activity and shifts the balance of its competition with catastrophe factors. This shift results in a tenfold increase in the dynamic instability of microtubules during mitosis, a transition that is critical for the efficient construction of the mitotic spindle (discussed in Chapter 17).

In many cells, the minus ends of microtubules are stabilized by association with a capping protein or the centrosome, or else they serve as microtubule depolymerization sites. The plus ends, in contrast, efficiently explore and probe the entire cell space. Microtubule-associated proteins called *plus-end tracking proteins (+TIPs)* accumulate at these active ends and appear to rocket around the cell as passengers at the ends of rapidly growing microtubules, dissociating from the ends when the microtubules begin to shrink (**Figure 16–53**).

The kinesin-related catastrophe factors and XMAP215 mentioned above behave as +TIPs and act to modulate the growth and shrinkage of the microtubule end to which they are attached. Other +TIPs control microtubule positioning by helping to capture and stabilize the growing microtubule end at specific cellular targets, such as the cell cortex or the kinetochore of a mitotic chromosome. EB1 and its relatives, small dimeric proteins that are highly conserved in animals, plants, and fungi, are key players in this process. EB1 proteins do not actively move toward plus ends, but rather recognize a structural feature of the growing plus end (see Figure 16–53). Several of the +TIPs depend on EB1 proteins for their plus-end accumulation and also interact with each other and with the microtubule lattice. By attaching to the plus end, these factors allow the cell to harness the energy of microtubule polymerization to generate pushing forces that can be used for positioning the spindle, chromosomes, or organelles.

Tubulin-Sequestering and Microtubule-Severing Proteins Destabilize Microtubules

As it does with actin monomers, the cell sequesters unpolymerized tubulin subunits to maintain a pool of active subunits at a level near the critical concentration. One molecule of the small protein *stathmin* (also called Op18) binds to two tubulin heterodimers and prevents their addition to the ends of microtubules (**Figure 16–54**). Stathmin thus decreases the effective concentration of tubulin subunits that are available for polymerization (an action analogous to that of the drug colchicine), and enhances the likelihood that a growing microtubule will switch to the shrinking state. Phosphorylation of stathmin inhibits its binding to tubulin, and signals that cause stathmin phosphorylation can increase the rate of microtubule elongation and suppress dynamic instability. Stathmin has been implicated in the regulation of both cell proliferation and cell death. Interestingly, mice lacking stathmin develop normally but are less fearful than wild-type mice, reflecting a role for stathmin in neurons of the amygdala, where it is normally expressed at high levels.

Severing is another mechanism employed by the cell to destabilize microtubules. To sever a microtubule, thirteen longitudinal bonds must be broken, one for each protofilament. The protein *katanin*, named after the Japanese word for

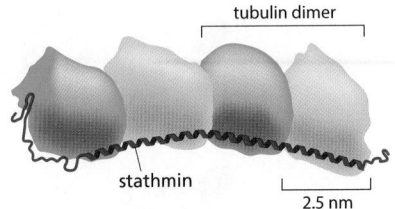

tubulin dimer

stathmin 2.5 nm

Figure 16–54 Sequestration of tubulin by stathmin. Structural studies with electron microscopy and crystallography suggest that the elongated stathmin protein binds along the side of two tubulin heterodimers. (Adapted from M.O. Steinmetz et al., *EMBO J.* 19:572–580, 2000.)

"sword," accomplishes this demanding task (**Figure 16–55**). Katanin is made up of two subunits: a smaller subunit that hydrolyzes ATP and performs the actual severing, and a larger one that directs katanin to the centrosome. Katanin releases microtubules from their attachment to a microtubule-organizing center and is thought to contribute to the rapid microtubule depolymerization observed at the poles of spindles during mitosis. It may also be involved in microtubule release and depolymerization in proliferating cells in interphase and in postmitotic cells such as neurons.

Two Types of Motor Proteins Move Along Microtubules

Like actin filaments, microtubules also use motor proteins to transport cargo and perform a variety of other functions within the cell. There are two major classes of microtubule-based motors, **kinesins** and **dyneins. Kinesin-1**, also called "conventional kinesin," was first purified from squid neurons, where it carries membrane-enclosed organelles away from the cell body toward the axon terminal by walking toward the plus end of microtubules. Kinesin-1 is similar to myosin II in having two heavy chains per active motor; these form two globular head motor domains that are held together by an elongated coiled-coil tail that is responsible for heavy-chain dimerization. One kinesin-1 light chain associates with each heavy chain through its tail domain and mediates cargo binding. Like myosin, kinesin is a member of a large protein superfamily, for which the motor domain is the common element (**Figure 16–56**). The yeast *Saccharomyces cerevisiae* has six distinct kinesins. The nematode *C. elegans* has 20 kinesins, and humans have 45.

There are at least fourteen distinct families in the kinesin superfamily. Most of them have the motor domain at the N-terminus of the heavy chain and walk toward the plus end of the microtubule. One family has the motor domain at the C-terminus and walks in the opposite direction, toward the minus end of the microtubule, while kinesin-13 has a central motor domain and does not walk at all, but uses the energy of ATP hydrolysis to depolymerize microtubule ends, as described above (see Figure 16–52). Some kinesin heavy chains are homodimers, and others are heterodimers. Most kinesins have a binding site in the tail for another microtubule; alternatively, they may link the motor to a membrane-enclosed organelle via a light chain or an adaptor protein. Many of the kinesin superfamily members have specific roles in mitotic spindle formation and in chromosome segregation during cell division.

In kinesin-1, instead of the rocking of a lever arm, small movements at the nucleotide-binding site regulate the docking and undocking of the motor head domain to a long linker region. This acts to throw the second head forward along

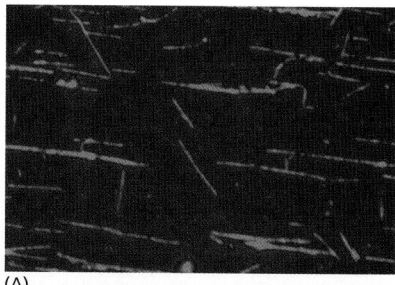

(A)

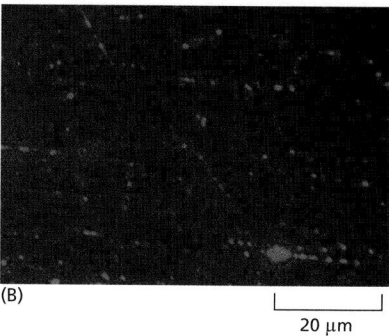

(B)

|_____|
20 µm

Figure 16–55 Microtubule severing by katanin. Taxol stabilized, rhodamine-labeled microtubules were adsorbed on the surface of a glass slide, and purified katanin was added along with ATP. (A) There are a few breaks in the microtubules 30 seconds after the addition of katanin. (B) The same field 3 minutes after the addition of katanin. The filaments have been severed in many places, leaving a series of small fragments at the previous locations of the long microtubules. (From J.J. Hartman et al., *Cell* 93:277–287, 1998. With permission from Elsevier.)

Figure 16–56 Kinesin and kinesin-related proteins. Structures of four kinesin superfamily members. As in the myosin superfamily, only the motor domains are conserved. Kinesin-1 has the motor domain at the N-terminus of the heavy chain. The middle domain forms a long coiled-coil, mediating dimerization. The C-terminal domain forms a tail that attaches to cargo, such as a membrane-enclosed organelle. Kinesin-5 forms tetramers where two dimers associate by their tails. The bipolar kinesin-5 tetramer is able to slide two microtubules past each other, analogous to the activity of the bipolar thick filaments formed by myosin II. Kinesin-13 has its motor domain located in the middle of the heavy chain. It is a member of a family of kinesins that have lost typical motor activity and instead bind to microtubule ends to promote depolymerization (see Figure 16–52). Kinesin-14 is a C-terminal kinesin that includes the *Drosophila* protein Ncd and the yeast protein Kar3. These kinesins generally travel in the opposite direction from the majority of kinesins, toward the minus end instead of the plus end of a microtubule.

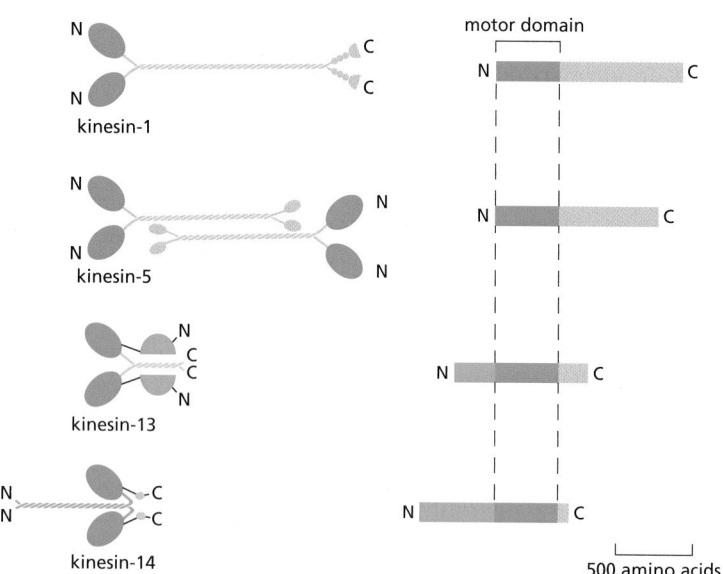

Figure 16–57 **The mechanochemical cycle of kinesin.** Kinesin-1 is a dimer of two nucleotide-binding motor domains (heads) that are connected through a long coiled-coil tail (see Figure 16–56). The two kinesin motor domains work in a coordinated manner; during a kinesin "step," the rear head detaches from its tubulin binding site, passes the partner motor domain, and then rebinds to the next available tubulin binding site. Using this "hand-over-hand" motion, the kinesin dimer can move for long distances on the microtubule without completely letting go of its track.

At the start of each step, one of the two kinesin motor domain heads, the rear or lagging head *(dark green)*, is tightly bound to the microtubule and to ATP, while the front or leading head is loosely bound to the microtubule with ADP in its binding site. The forward displacement of the rear motor domain is driven by the dissociation of ADP and binding of ATP in the leading head (between panels 2 and 3 in this drawing). The binding of ATP to this motor domain causes a small peptide called the "neck linker" to shift from a rearward-pointing to a forward-pointing conformation (the neck linker is drawn here as a *purple* connecting line between the leading motor domain and the intertwined coiled-coil). This shift pulls the rear head forward, once it has detached from the microtubule with ADP bound [detachment requires ATP hydrolysis and phosphate (P_i) release]. The kinesin molecule is now poised for the next step, which proceeds by an exact repeat of the process shown (Movie 16.9).

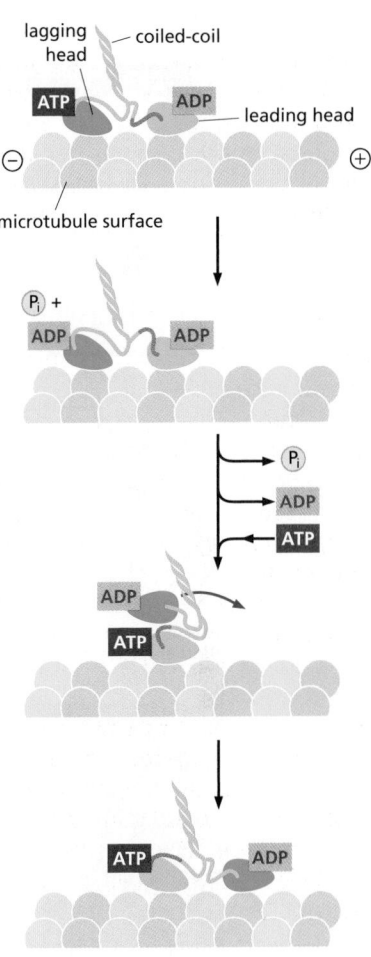

the protofilament to a binding site 8 nm closer to the microtubule plus end, which is the distance between tubulin dimers of a protofilament. The nucleotide-hydrolysis cycles in the two heads are closely coordinated, so that this cycle of linker docking and undocking allows the two-headed motor to move in a hand-over-hand (or head-over-head) stepwise manner (**Figure 16–57**).

The **dyneins** are a family of minus-end directed microtubule motors unrelated to the kinesins. They are composed of one, two, or three heavy chains (that include the motor domain) and a large and variable number of associated intermediate, light-intermediate, and light chains. The dynein family has two major branches (**Figure 16–58**). The first branch contains the *cytoplasmic dyneins*, which are homodimers of two heavy chains. Cytoplasmic dynein 1 is encoded by a single gene in almost all eukaryotic cells, but is missing from flowering plants and some algae. It is used for organelle and mRNA trafficking, for positioning the centrosome and nucleus during cell migration, and for construction of the microtubule spindle in mitosis and meiosis. Cytoplasmic dynein 2 is found only in eukaryotic organisms that have cilia and is used to transport material from the tip to the base of the cilia, a process called intraflagellar transport. *Axonemal dyneins* (also called *ciliary dyneins*) comprise the second branch and include monomers, heterodimers, and heterotrimers, with one, two, or three motor-containing heavy chains, respectively. They are highly specialized for the rapid and efficient sliding movements of microtubules that drive the beating of cilia and flagella (discussed later).

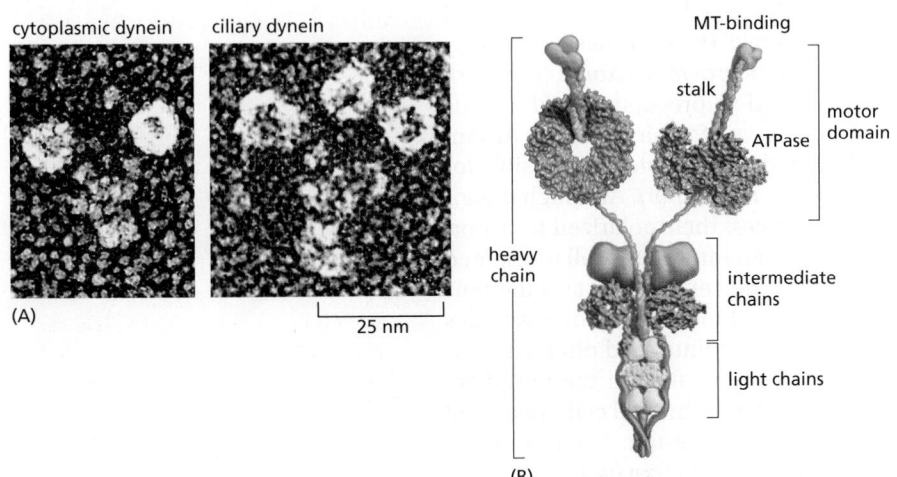

(B)

Figure 16–58 **Dyneins.** (A) Freeze-etch electron micrographs of a molecule of cytoplasmic dynein and a molecule of ciliary (axonemal) dynein. Like myosin II and kinesin-1, cytoplasmic dynein is a two-headed molecule. The ciliary dynein shown is from a protozoan and has three heads; ciliary dynein from animals has two heads. Note that the dynein head is very large compared with the head of either myosin or kinesin. (B) Schematic depiction of cytoplasmic dynein showing the two heavy chains *(blue and gray)* that contain domains for microtubule (MT) binding and ATP hydrolysis, connected by a long stalk. Bound to the heavy chain are multiple intermediate chains *(dark green)* and light chains *(light green)* that help to mediate many of dynein's functions. (A, courtesy of John Heuser; B, adapted from R. Vale, *Cell* 112:467–480, 2003. With permission from Elsevier.)

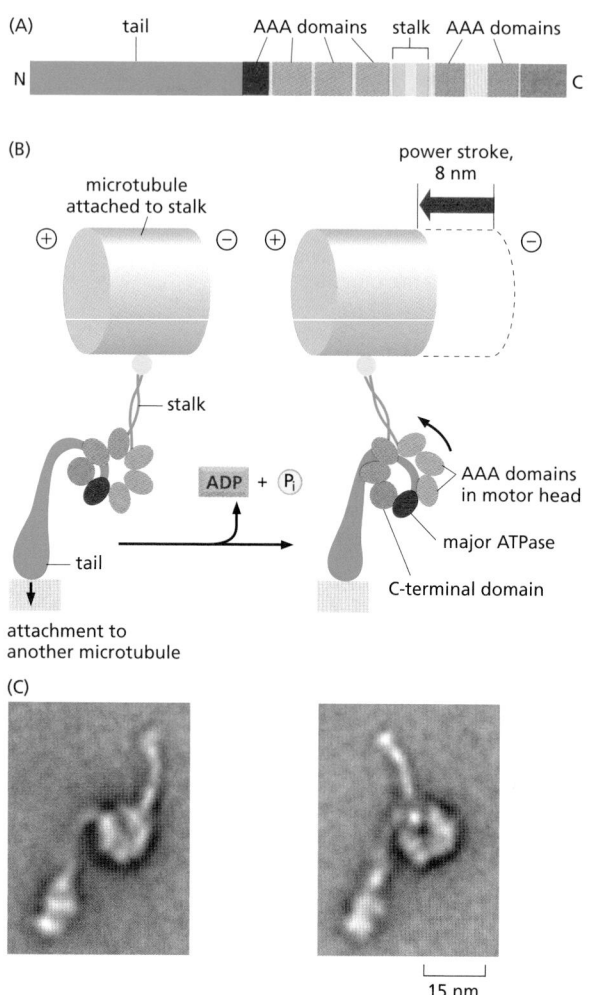

Figure 16–59 **The power stroke of dynein.** (A) The organization of the domains in each dynein heavy chain. This is a huge polypeptide, containing nearly 4000 amino acids. The number of heavy chains in a dynein is equal to its number of motor heads. (B) Illustration of dynein c, a monomeric axonemal dynein found in the unicellular green alga *Chlamydomonas reinhardtii*. The large dynein motor head is a planar ring containing a C-terminal domain *(gray)* and six AAA domains, four of which retain ATP-binding sequences, but only one of which *(dark red)* has the major ATPase activity. Extending from the head are a long, coiled-coil stalk with the microtubule-binding site at the tip, and a tail that attaches to an adjacent microtubule in the axoneme. In the ATP-bound state, the stalk is detached from the microtubule, but ATP hydrolysis causes stalk–microtubule attachment *(left)*. Subsequent release of ADP and phosphate (Pi) then leads to a large conformational "power stroke" involving rotation of the head and stalk relative to the tail *(right)*. Each cycle generates a step of about 8 nm, thereby contributing to flagellar beating (see Figure 16–65). In the case of cytoplasmic dynein, the tail is attached to a cargo such as a vesicle, and a single power stroke transports the cargo about 8-nm along the microtubule toward its minus end (see Figure 16–60). (C) Electron micrographs of purified monomeric dyneins in two different conformations representing different steps in the mechanochemical cycle. (C, from S.A. Burgess et al., *Nature* 421:715–718, published 2003 by Nature Publishing Group. Reproduced with permission of SNCSC.)

Dyneins are the largest of the known molecular motors, and they are also among the fastest: axonemal dyneins attached to a glass slide can move microtubules at the rate of 14 µm/sec. The dynein motor is structurally unrelated to myosins and kinesins, but still follows the general rule of coupling nucleotide hydrolysis to microtubule binding and unbinding as well as to a force-generating conformational change (**Figure 16–59**).

Microtubules and Motors Move Organelles and Vesicles

A major function of cytoskeletal motors in interphase cells is the transport and positioning of membrane-enclosed organelles (**Movie 16.10**). Kinesin was originally identified as the protein responsible for fast *anterograde axonal transport*, the rapid movement of mitochondria, secretory vesicle precursors, and various synapse components down the microtubule highways of the axon to the distant nerve terminals. Cytoplasmic dynein was identified as the motor responsible for transport in the opposite direction, *retrograde axonal transport*. Although organelles in most cells need not cover such long distances, their polarized transport is equally necessary. A typical microtubule array in an interphase cell is oriented with the minus ends near the center of the cell at the centrosome and the plus ends extending to the cell periphery. Thus, centripetal movements of organelles or vesicles toward the cell center require the action of minus-end directed cytoplasmic dynein motors, whereas centrifugal movements toward the periphery require plus-end directed kinesin motors. Interestingly, in animal cells, nearly all minus-end directed transport is driven by the single cytoplasmic dynein 1 motor, whereas 15 different kinesins are used for plus-end directed transport.

A clear example of the effect of microtubules and microtubule motors on the behavior of intracellular membranes is their role in organizing the endoplasmic reticulum (ER) and the Golgi apparatus. The network of ER membrane tubules aligns with microtubules and extends almost to the edge of the cell (Movie 16.11), whereas the Golgi apparatus is located near the centrosome. When cells are treated with a drug that depolymerizes microtubules, such as colchicine or nocodazole, the ER collapses to the center of the cell, while the Golgi apparatus fragments and disperses throughout the cytoplasm. *In vitro*, kinesins can tether ER-derived membranes to preformed microtubule tracks and walk toward the microtubule plus ends, dragging the ER membranes out into tubular protrusions and forming a membranous web that looks very much like the ER in cells. Conversely, dyneins are required for positioning the Golgi apparatus near the cell center of animal cells; they do this by moving Golgi vesicles along microtubule tracks toward the microtubules' minus ends at the centrosome.

The different tails and their associated light chains on specific motor proteins allow the motors to attach to their appropriate organelle cargo. Membrane-associated motor receptors that are sorted to specific membrane-enclosed compartments interact directly or indirectly with the tails of the appropriate kinesin family members. Many viruses take advantage of microtubule motor-based transport during infection and use kinesin to move from their site of replication and assembly to the plasma membrane, from which they are poised to infect neighboring cells. An outer-membrane protein of *Vaccinia* virus, for example, contains an amino acid motif that mediates binding to kinesin-1 light chain and transport along microtubules to the plasma membrane. Interestingly, this motif is present in over 450 human proteins, one-third of which are associated with human diseases. Thus, kinesin transports a diverse set of cargoes involved in a wide range of important cellular functions.

For dynein, a large macromolecular assembly often mediates attachment to membranes. Cytoplasmic dynein, itself a huge protein complex, requires association with a second large protein complex called *dynactin* to translocate organelles effectively. The dynactin complex includes a short, actin-like filament that forms from the actin-related protein Arp1 (distinct from Arp2 and Arp3, the components of the Arp 2/3 complex involved in the nucleation of conventional actin filaments) (Figure 16–60). A number of other proteins also contribute to dynein cargo binding and motor regulation, and their function is especially important in neurons, where defects in microtubule-based transport have been linked to neurological diseases. A striking example is smooth brain, or lissencephaly, a human disorder in which cells fail to migrate to the cerebral cortex of the developing brain. One type of lissencephaly is caused by defects in Lis1, a dynein-binding protein required for nuclear migration in several species. In the normal brain, migration of the nucleus directs the developing neural cell body toward its correct position in the cortex. In the absence of Lis1, however, the nuclei of migrating neurons fail to attach to dynein, resulting in nuclear-migration defects. Dynein is required continuously for neuronal function, as mutations in a dynactin subunit or in the tail region of cytoplasmic dynein lead to neuronal degeneration in humans and mice. These effects are associated with decreased retrograde axonal transport and provide strong evidence for the importance of robust axonal transport in neuronal viability.

The cell can regulate the activity of motor proteins and thereby cause either a change in the positioning of its membrane-enclosed organelles or whole-cell movements. Fish melanocytes provide one of the most dramatic examples. These giant cells, which are responsible for rapid changes in skin coloration in several species of fish, contain large pigment granules that can alter their location in response to neuronal or hormonal stimulation (Figure 16–61). The pigment granules aggregate or disperse by moving along an extensive network of microtubules that are anchored at the centrosome by their minus ends. The tracking of individual pigment granules reveals that the inward movement is rapid and smooth, while the outward movement is jerky, with frequent backward steps. Both dynein and kinesin microtubule motors are associated with the pigment granules. The

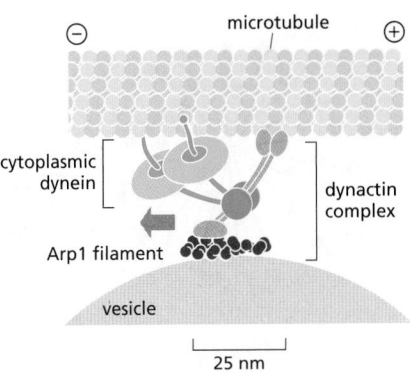

Figure 16–60 Dynactin mediates the attachment of dynein to a membrane-enclosed organelle. Dynein requires the presence of a large number of accessory proteins to associate with membrane-enclosed organelles. Dynactin is a large complex that includes components that bind weakly to microtubules, components that bind to dynein itself, and components that form a small, actin-like filament made of the actin-related protein Arp1.

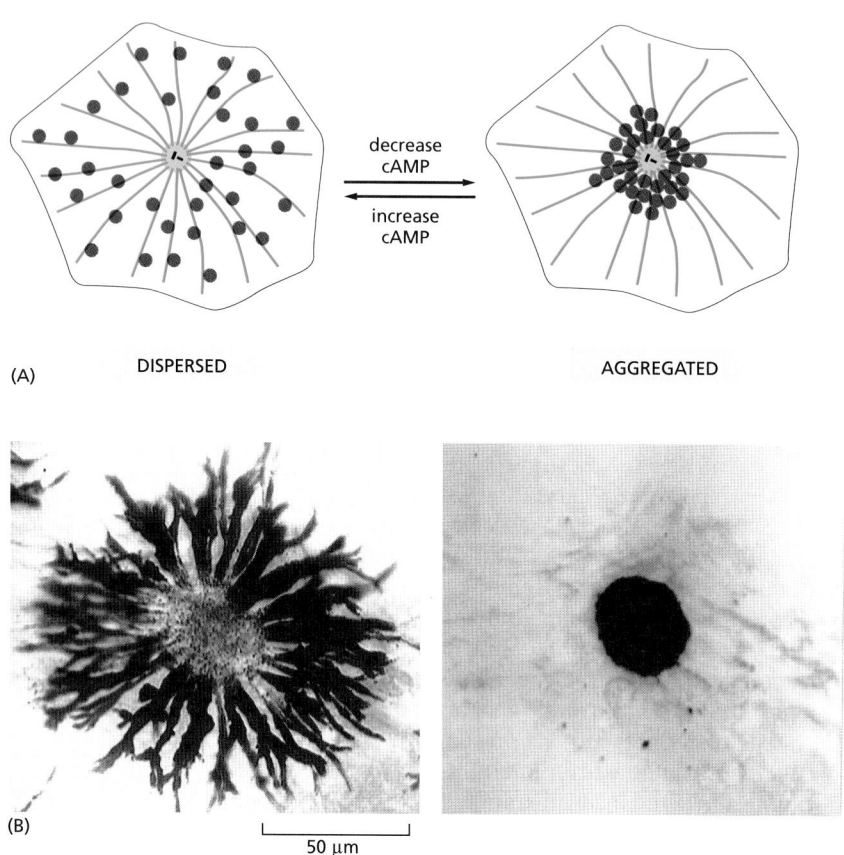

(A) DISPERSED — AGGREGATED

decrease cAMP

increase cAMP

(B)

50 μm

Figure 16–61 **Regulated melanosome movements in fish pigment cells.** These giant cells, which are responsible for changes in skin coloration in several species of fish, contain large pigment granules, or melanosomes *(brown)*. The melanosomes can change their location in the cell in response to a hormonal or neuronal stimulus. (A) Schematic view of a pigment cell, showing the dispersal and aggregation of melanosomes in response to an increase or decrease in intracellular cyclic AMP (cAMP), respectively. Both redistributions of melanosomes occur along microtubules. (B) Bright-field images of a single cell in a scale of an African cichlid fish, showing its melanosomes either dispersed throughout the cytoplasm *(left)* or aggregated in the center of the cell *(right)*. (B, courtesy of Leah Haimo.)

jerky outward movements may result from a tug-of-war between the two opposing microtubule motor proteins, with the stronger kinesin winning out overall. When intracellular cyclic AMP levels decrease, kinesin is inactivated, leaving dynein free to drag the pigment granules rapidly toward the cell center, changing the fish's color. In a similar way, the movement of other membrane organelles coated with particular motor proteins is controlled by a complex balance of competing signals that regulate both motor protein attachment and activity.

Construction of Complex Microtubule Assemblies Requires Microtubule Dynamics and Motor Proteins

The construction of the mitotic spindle and the neuronal cytoskeleton are important and fascinating examples of the power of organization by teams of motor proteins interacting with dynamic cytoskeletal filaments. As described in Chapter 17, mitotic spindle assembly depends on reorganization of the interphase array of microtubules to form a bipolar array of microtubules, with their minus ends focused at the poles and their plus ends overlapping in the center or connecting to chromosomes. Spindle assembly depends on the coordinated actions of several motor proteins and other factors that modulate polymerization dynamics (see Figures 17–23 and 17–25).

Neurons also contain complex cytoskeletal structures. As they differentiate, neurons send out specialized processes that will either receive electrical signals (*dendrites*) or transmit electrical signals (*axons*) (see Figure 16–50). The beautiful and elaborate branching morphology of axons and dendrites enables neurons to form tremendously complex signaling networks, interacting with many other cells simultaneously and making possible the complicated behavior of the higher animals. Both axons and dendrites (collectively called *neurites*) are filled with bundles of microtubules that are critical to both their structure and their function.

In axons, all the microtubules are oriented in the same direction, with their minus end pointing back toward the cell body and their plus end pointing toward the axon terminals (**Figure 16–62**). The microtubules do not reach from the cell

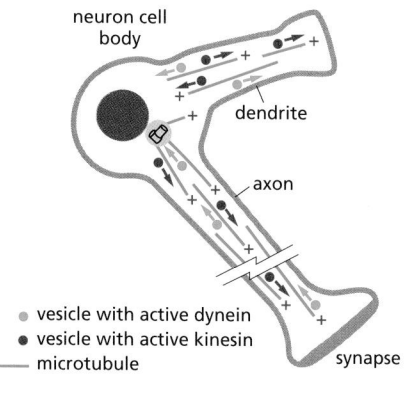

neuron cell body

dendrite

axon

● vesicle with active dynein
● vesicle with active kinesin
— microtubule

synapse

NEURON

Figure 16–62 **Microtubule organization in a neuron.** In a neuron, microtubule organization is complex. In the axon, all microtubules share the same polarity, with the plus ends pointing outward toward the axon terminus. No single microtubule stretches the entire length of the axon; instead, short overlapping segments of parallel microtubules make the tracks for fast axonal transport. In dendrites, the microtubules are of mixed polarity, with some plus ends pointing outward and some pointing inward. Vesicles can associate with both kinesin and dynein and move in either direction along the microtubules in axons and dendrites, depending on which motor is active.

body all the way to the axon terminals; each is typically only a few micrometers in length, but large numbers are staggered in an overlapping array. These aligned microtubule tracks act as a highway to transport specific proteins, protein-containing vesicles, and mRNAs to the axon terminals, where synapses are constructed and maintained. The longest axon in the human body reaches from the base of the spinal cord to the foot and is up to a meter in length. By comparison, dendrites are generally much shorter than axons. The microtubules in dendrites lie parallel to one another but their polarities are mixed, with some pointing their plus ends toward the dendrite tip, while others point back toward the cell body, reminiscent of the antiparallel microtubule array of the mitotic spindle.

Motile Cilia and Flagella Are Built from Microtubules and Dyneins

Just as myofibrils are highly specialized and efficient motility machines built from actin and myosin filaments, cilia and flagella are highly specialized and efficient motility structures built from microtubules and dynein. Both cilia and flagella are hairlike cell appendages that have a bundle of microtubules at their core. **Flagella** are found on sperm and many protozoa. By their undulating motion, they enable the cells to which they are attached to swim through liquid media. **Cilia** are organized in a similar fashion, but they beat with a whiplike motion that resembles the breaststroke in swimming. Ciliary beating can either propel single cells through a fluid (as in the swimming of the protozoan *Paramecium*) or can move fluid over the surface of a group of cells in a tissue. In the human body, huge numbers of cilia ($10^9/cm^2$ or more) line our respiratory tract, sweeping layers of mucus, trapped particles of dust, and bacteria up to the mouth where they are swallowed and ultimately eliminated. Likewise, cilia along the oviduct help to sweep eggs toward the uterus.

The movement of a cilium or a flagellum is produced by the bending of its core, which is called the **axoneme**. The axoneme is composed of microtubules and their associated proteins, arranged in a distinctive and regular pattern. Nine special doublet microtubules (comprising one complete and one partial microtubule fused together so that they share a common tubule wall) are arranged in a ring around a pair of single microtubules (**Figure 16–63**). Almost all forms of motile eukaryotic flagella and cilia (from protozoans to humans) have this characteristic arrangement. The microtubules extend continuously for the length of the axoneme, which can be 10–200 μm. At regular positions along the length of the microtubules, accessory proteins cross-link the microtubules together.

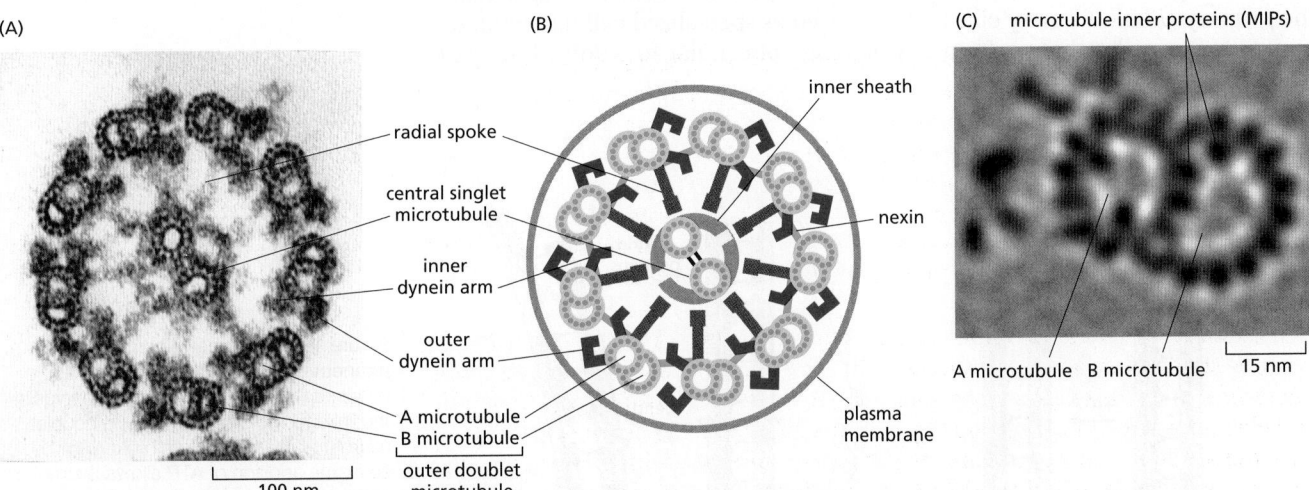

Figure 16–63 The arrangement of microtubules in a flagellum or cilium. (A) Electron micrograph of the flagellum of a green-alga cell (*Chlamydomonas*) shown in cross section, illustrating the distinctive "9 + 2" arrangement of microtubules. (B) Diagram of the parts of a flagellum or cilium. The various projections from the microtubules link the microtubules together and occur at regular intervals along the length of the axoneme. (C) High-resolution electron tomography image of an outer doublet microtubule showing structural details and features inside the microtubules called microtubule inner proteins (MIPs). (A, courtesy of Lewis Tilney; C, courtesy of Daniela Nicastro.)

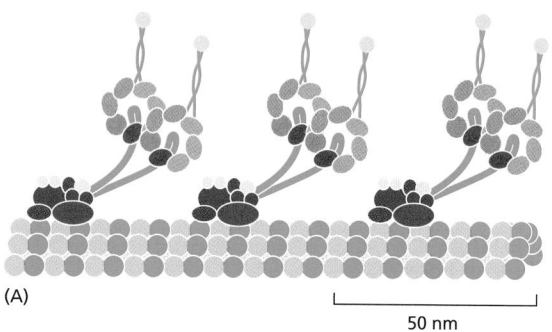

(A)

50 nm

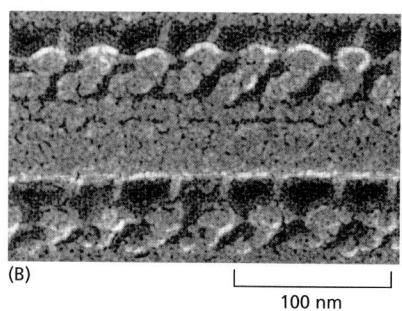

(B)

100 nm

Molecules of *axonemal dynein* form bridges between the neighboring doublet microtubules around the circumference of the axoneme (**Figure 16–64**). When the motor domain of this dynein is activated, the dynein molecules attached to one microtubule doublet (see Figure 16–59) attempt to walk along the adjacent microtubule doublet, tending to force the adjacent doublets to slide relative to one another, much as actin thin filaments slide during muscle contraction. However, the presence of other links between the microtubule doublets prevents this sliding, and the dynein force is instead converted into a bending motion (**Figure 16–65**).

In humans, hereditary defects in axonemal dynein cause a condition called primary ciliary dyskinesia or Kartagener's syndrome. This syndrome is characterized by inversion of the normal asymmetry of internal organs (sinus inversus) due to disruption of fluid flow in the developing embryo, male sterility due to immotile sperm, and a high susceptibility to lung infections due to paralyzed cilia being unable to clear the respiratory tract of debris and bacteria.

Bacteria also swim using cell-surface structures called flagella, but these do not contain microtubules or dynein and do not wave or beat. Instead, *bacterial flagella* are long, rigid helical filaments, made up of repeating subunits of the protein flagellin. The flagella rotate like propellers, driven by a special rotary motor embedded in the bacterial cell wall. The use of the same name to denote these two very different types of swimming apparatus is an unfortunate historical accident.

Primary Cilia Perform Important Signaling Functions in Animal Cells

Many cells possess a shorter, nonmotile counterpart of cilia and flagella called the *primary cilium*. Primary cilia can be viewed as specialized cellular compartments or organelles that perform a wide range of cellular functions, but share

Figure 16–64 Ciliary dynein. Ciliary (axonemal) dynein is a large protein assembly (nearly 2 million daltons) composed of 9–12 polypeptide chains, the largest of which is the heavy chain of more than 500,000 daltons. (A) The heavy chains form the major portion of the globular head and stem domains, and many of the smaller chains are clustered around the base of the stem. There are two heads in the outer dynein in metazoans (shown here), but three heads in protozoa, each formed from their own heavy chain. The tail of the molecule binds tightly to an A microtubule, while the large globular heads have an ATP-dependent binding site for a B microtubule (see Figure 16–63). When the heads hydrolyze their bound ATP, they move toward the minus end of the B microtubule, thereby producing a sliding force between the adjacent microtubule doublets in a cilium or flagellum (see Figure 16–59). (B) Freeze-etch electron micrograph of a cilium showing the dynein arms projecting at regular intervals from the doublet microtubules. (B, courtesy of John Heuser.)

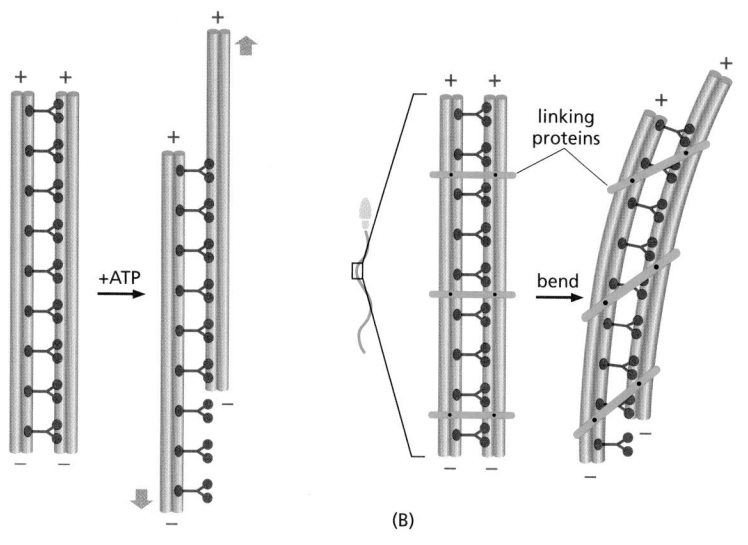

(A)   (B)

Figure 16–65 The bending of an axoneme. (A) When axonemes are exposed to the proteolytic enzyme trypsin, the linkages holding neighboring doublet microtubules together are broken. In this case, the addition of ATP allows the motor action of the dynein heads to slide one pair of doublet microtubules against the other pair. (B) In an intact axoneme (such as in a sperm), flexible protein links prevent the sliding of the doublet. The motor action therefore causes a bending motion, creating waves or beating motions.

many structural features with motile cilia. Both motile and nonmotile cilia are generated during interphase at plasma-membrane-associated structures called *basal bodies* that firmly root them at the cell surface. At the core of each basal body is a centriole, the same structure found embedded at the center of animal centrosomes, with nine groups of fused triplet microtubules arranged in a cartwheel (see Figure 16–48). Centrioles are multifunctional, contributing to assembly of the mitotic spindle in dividing cells but migrating to the plasma membrane of interphase cells to template the nucleation of the axoneme (**Figure 16–66**). Because no protein translation occurs in cilia, construction of the axoneme requires intraflagellar transport (IFT), a transport system discovered in the green algae *Chlamydomonas*. Analogous to the axon, motors move cargoes in both anterograde and retrograde directions, in this case driven by kinesin-2 and cytoplasmic dynein 2, respectively.

Primary cilia are found on the surface of almost all cell types, where they sense and respond to the exterior environment, functions best understood in the context of smell and sight. In the nasal epithelium, cilia protruding from dendrites of olfactory neurons are the site of both odorant reception and signal amplification. Similarly, the rod and cone cells of the vertebrate retina possess a primary cilium equipped with an expanded tip called the outer segment, which is specialized for converting light into a neural signal (see Figure 15–38). Maintenance of the outer segment requires continuous IFT-mediated transport of large quantities of lipids and proteins into the cilium, at rates of up to 2000 molecules per minute. The links between cilia function and the senses of sight and smell are underscored by Bardet-Biedl syndrome, a set of disorders associated with defects in IFT, the cilium, or the basal body. Patients with Bardet-Biedl syndrome cannot smell and suffer from retinal degeneration. Other characteristics of this multifaceted disorder include hearing loss, polycystic kidney disease, diabetes, obesity, and polydactyly, suggesting that primary cilia have functions in many aspects of human physiology.

Summary

Microtubules are stiff polymers of tubulin molecules. They assemble by addition of GTP-containing tubulin subunits to the free end of a microtubule, with one end (the plus end) growing faster than the other. Hydrolysis of the bound GTP takes place

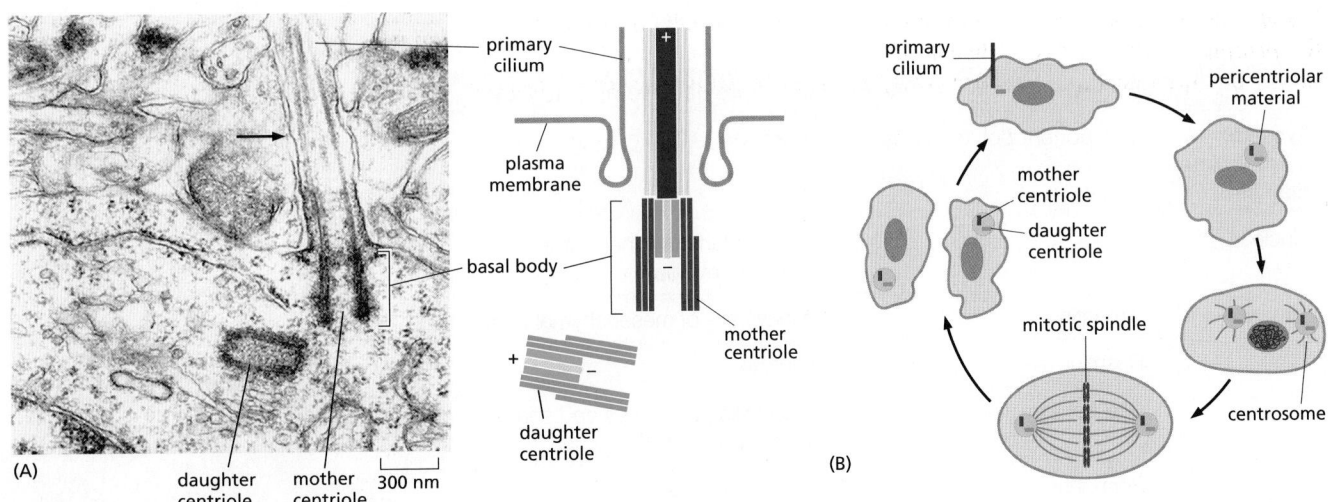

Figure 16–66 Primary cilia. (A) Electron micrograph and diagram of the basal body of a mouse neuron primary cilium. The axoneme of the primary cilium (*black arrow*) is nucleated by the mother centriole at the basal body, which localizes at the plasma membrane near the cell surface. (B) Centrioles function alternately as basal bodies and as the core of centrosomes. Before a cell enters the cell division cycle, the primary cilium is shed or resorbed. The centrioles recruit pericentriolar material and duplicate during S phase, generating two centrosomes, each of which contains a pair of centrioles. The centrosomes nucleate microtubules and localize to the poles of the mitotic spindle. Upon exit from mitosis, a primary cilium again grows from the mother centriole. (A, courtesy of Josef Spacek.)

after assembly and weakens the bonds that hold the microtubule together. Microtubules are dynamically unstable and liable to catastrophic disassembly, but they can be stabilized in cells by association with other structures. Microtubule-organizing centers such as centrosomes protect the minus ends of microtubules and continually nucleate the formation of new microtubules. Microtubule-associated proteins (MAPs) stabilize microtubules, and those that localize to the plus end (+TIPs) can alter the dynamic properties of the microtubule or mediate their interaction with other structures. Counteracting the stabilizing activity of MAPs are catastrophe factors, such as kinesin-13 proteins, that act to peel apart microtubule ends. Other kinesin family members as well as dynein use the energy of ATP hydrolysis to move unidirectionally along a microtubule. The motor dynein moves toward the minus end of microtubules, and its sliding of axonemal microtubules underlies the beating of cilia and flagella. Primary cilia are nonmotile sensory organs found on many cell types.

INTERMEDIATE FILAMENTS AND SEPTINS

All eukaryotic cells contain actin and tubulin. But the third major type of cytoskeletal protein, the *intermediate filament*, forms a cytoplasmic filament only in some metazoans—including vertebrates, nematodes, and mollusks. Intermediate filaments are particularly prominent in the cytoplasm of cells that are subject to mechanical stress and are generally not found in animals that have rigid exoskeletons, such as arthropods and echinoderms. It seems that intermediate filaments impart mechanical strength to tissues for the squishier animals.

Cytoplasmic intermediate filaments are closely related to their ancestors, the much more prevalent *nuclear lamins*, which are found in many eukaryotes but missing from unicellular organisms. The nuclear lamins form a meshwork lining the inner membrane of the nuclear envelope, where they provide anchorage sites for chromosomes and nuclear pores. Several times during metazoan evolution, lamin genes have apparently duplicated, and the duplicates have evolved to produce ropelike, cytoplasmic intermediate filaments. In contrast to the highly conserved actins and tubulin isoforms that are encoded by a handful of genes, different families of intermediate filaments are much more diverse and are encoded by 70 different human genes with distinct, cell type-specific functions (**Table 16–2**).

TABLE 16–2 Major Types of Intermediate Filament Proteins in Vertebrate Cells

Types of intermediate filament	Component polypeptides	Location
Nuclear	Lamins A, B, and C	Nuclear lamina (inner lining of nuclear envelope)
Vimentin-like	Vimentin	Many cells of mesenchymal origin
	Desmin	Muscle
	Glial fibrillary acidic protein	Glial cells (astrocytes and some Schwann cells)
	Peripherin	Some neurons
Epithelial	Type I keratins (acidic)	Epithelial cells and their derivatives (e.g., hair and nails)
	Type II keratins (neutral/basic)	
Axonal	Neurofilament proteins (NF-L, NF-M, and NF-H)	Neurons

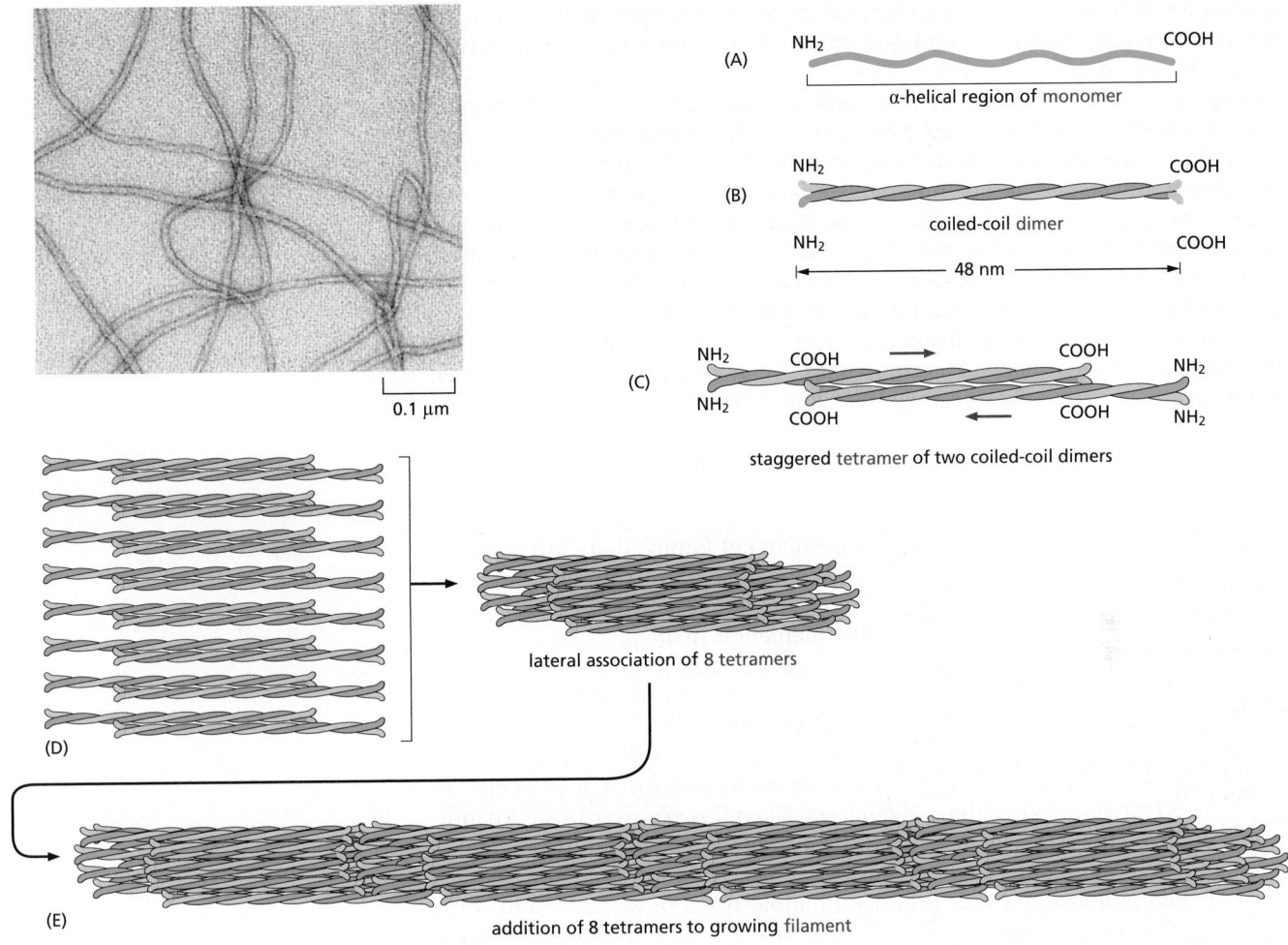

Figure 16–67 **A model of intermediate filament construction.** The monomer shown in (A) pairs with another monomer to form a dimer (B), in which the conserved central rod domains are aligned in parallel and wound together into a coiled-coil. (C) Two dimers then line up side by side to form an antiparallel tetramer of four polypeptide chains. Dimers and tetramers are the soluble subunits of intermediate filaments. (D) Within each tetramer, the two dimers are offset with respect to one another, thereby allowing it to associate with another tetramer. (E) In the final 10-nm ropelike filament, tetramers are packed together in a helical array, which has 16 dimers (32 coiled-coils) in cross section. Half of these dimers are pointing in each direction. An electron micrograph of intermediate filaments is shown on the upper left (**Movie 16.12**). (Electron micrograph from L. Norlen et al., *Exp. Cell Res.* 313:2217–2227, 2007. With permission from Elsevier.)

Intermediate Filament Structure Depends on the Lateral Bundling and Twisting of Coiled-Coils

Although their amino- and carboxy-terminal domains differ, all intermediate filament family members are elongated proteins with a conserved central α-helical domain containing 40 or so heptad repeat motifs that form an extended coiled-coil structure with another monomer (see Figure 3–9). A pair of parallel dimers then associates in an antiparallel fashion to form a staggered tetramer (**Figure 16–67**). Unlike actin or tubulin subunits, intermediate filament subunits do not contain a binding site for a nucleotide. Furthermore, since the tetrameric subunit is made up of two dimers pointing in opposite directions, its two ends are the same. The assembled intermediate filament therefore lacks the overall structural polarity that is critical for actin filaments and microtubules. The tetramers pack together laterally to form the filament, which includes eight parallel protofilaments made up of tetramers. Each individual intermediate filament therefore has a cross section of 32 individual α-helical coils. This large number of polypeptides all lined up together, with the strong lateral hydrophobic interactions typical of coiled-coil proteins, gives intermediate filaments a ropelike character. They can be easily bent, with a persistence length of less than one micrometer (compared to several millimeters for microtubules and about ten micrometers for actin), but they are extremely difficult to break and can be stretched to over three times their length (see Figure 16–6).

Less is understood about the mechanism of assembly and disassembly of intermediate filaments than of actin filaments and microtubules. In pure protein solutions, intermediate filaments are extremely stable due to tight association of subunits, but some types of intermediate filaments, including *vimentin*, form highly dynamic structures in cells such as fibroblasts. Protein phosphorylation probably regulates their disassembly, in much the same way that phosphorylation regulates the disassembly of nuclear lamins in mitosis (see Figure 12–18). As evidence for rapid turnover, labeled subunits microinjected into tissue-culture cells incorporate into intermediate filaments within a few minutes. Remodeling of the intermediate filament network accompanies events requiring dynamic cellular reorganization, such as division, migration, and differentiation.

Intermediate Filaments Impart Mechanical Stability to Animal Cells

Keratins are the most diverse intermediate filament family: there are about 20 found in different types of human epithelial cells and about 10 more that are specific to hair and nails; analysis of the human genome sequence has revealed that there are 54 distinct keratins. Every keratin filament is made up of an equal mixture of type I (acidic) and type II (neutral/basic) keratin proteins; these form a heterodimer filament subunit (see Figure 16–67). Cross-linked keratin networks held together by disulfide bonds can survive even the death of their cells, forming tough coverings for animals, as in the outer layer of skin and in hair, nails, claws, and scales. The diversity in keratins is clinically useful in the diagnosis of epithelial cancers (carcinomas), as the particular set of keratins expressed gives an indication of the epithelial tissue in which the cancer originated and thus can help to guide the choice of treatment.

A single epithelial cell may produce multiple types of keratins, and these copolymerize into a single network (**Figure 16–68**). Keratin filaments impart mechanical strength to epithelial tissues in part by anchoring the intermediate filaments at sites of cell–cell contact, called *desmosomes*, or cell–matrix contact, called *hemidesmosomes* (see Figure 16–4). We discuss these important adhesive structures in Chapter 19. Accessory proteins, such as *filaggrin*, bundle keratin filaments in differentiating cells of the epidermis to give the outermost layers of the

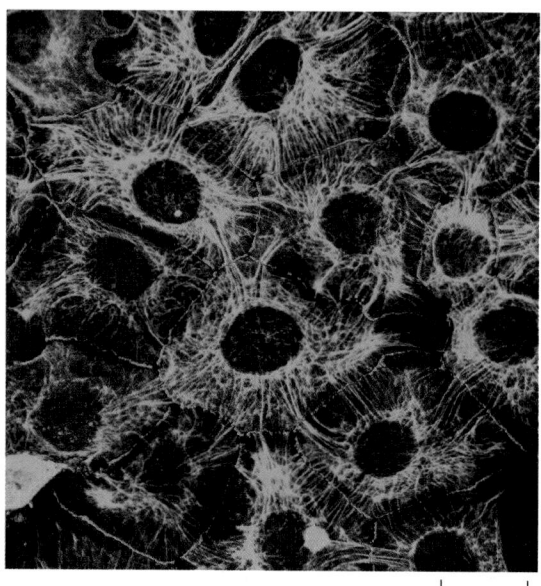

Figure 16–68 Keratin filaments in epithelial cells. Immunofluorescence micrograph of the network of keratin filaments *(blue)* in a sheet of epithelial cells in culture. The filaments in each cell are indirectly connected to those of its neighbors by desmosomes (discussed in Chapter 19). A second protein *(red)* has been stained to reveal the location of the cell boundaries. (From K.J. Green and C.A. Gaudry, *Nat. Rev. Mol. Cell Biol.* 1:208–216, published 2000 by Nature Publishing Group. Reproduced with permission of SNCSC.)

10 µm

skin their special toughness. Individuals with mutations in the gene encoding filaggrin are strongly predisposed to dry skin diseases such as eczema.

Mutations in keratin genes cause several human genetic diseases. For example, when defective keratins are expressed in the basal cell layer of the epidermis, they produce a disorder called *epidermolysis bullosa simplex*, in which the skin blisters in response to even very slight mechanical stress, which ruptures the basal cells (**Figure 16–69**). Other types of blistering diseases, including disorders of the mouth, esophageal lining, and the cornea of the eye, are caused by mutations in the different keratins whose expression is specific to those tissues. All of these maladies are typified by cell rupture as a consequence of mechanical trauma and a disorganization or clumping of the keratin filament cytoskeleton. Many of the specific mutations that cause these diseases alter the ends of the central rod domain, demonstrating the importance of this particular part of the protein for correct filament assembly.

Members of another family of intermediate filaments, called **neurofilaments**, are found in high concentrations along the axons of vertebrate neurons (**Figure 16–70**). Three types of neurofilament proteins (NF-L, NF-M, and NF-H) coassemble *in vivo*, forming heteropolymers. The NF-H and NF-M proteins have lengthy C-terminal tail domains that bind to neighboring filaments, generating aligned arrays with a uniform interfilament spacing. During axonal growth, new neurofilament subunits are incorporated all along the axon in a dynamic process that involves the addition of subunits along the filament length as well as the ends. After an axon has grown and connected with its target cell, the diameter of the axon may increase as much as fivefold. The level of neurofilament gene expression seems to directly control axonal diameter, which in turn influences how fast electrical signals travel down the axon. In addition, neurofilaments provide strength and stability to the long cell processes of neurons.

The neurodegenerative disease amyotrophic lateral sclerosis (ALS, or Lou Gehrig's disease) is associated with an accumulation and abnormal assembly of neurofilaments in motor neuron cell bodies and in the axon, aberrations that may interfere with normal axonal transport. The degeneration of the axons leads to muscle weakness and atrophy, which is usually fatal. The overexpression of human NF-L or NF-H in mice results in mice that have an ALS-like disease. However, a causative link between neurofilament pathology and ALS has not been firmly established.

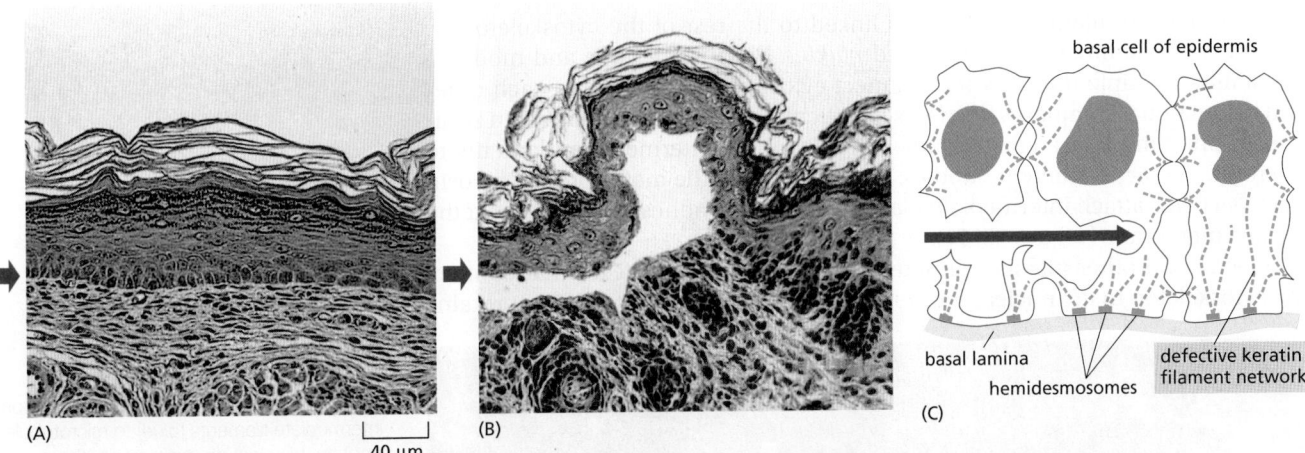

Figure 16–69 Blistering of the skin caused by a mutant keratin gene. A mutant gene encoding a truncated keratin protein (lacking both the N- and C-terminal domains) was expressed in a transgenic mouse. The defective protein assembles with the normal keratins and thereby disrupts the keratin filament network in the basal cells of the skin. Light micrographs of cross sections of (A) normal and (B) mutant skin show that the blistering results from the rupturing of cells in the basal layer of the mutant epidermis (short *red* arrows). (C) A sketch of three cells in the basal layer of the mutant epidermis, as observed by electron microscopy. As indicated by the *red* arrow, the cells rupture between the nucleus and the hemidesmosomes (discussed in Chapter 19), which connect the keratin filaments to the underlying basal lamina. (© 1991 P. A. Coulombe et al. Originally published in *J. Cell Biol*. https://doi.org/10.1083/jcb.115.6.1661. With permission from Rockefeller University Press.)

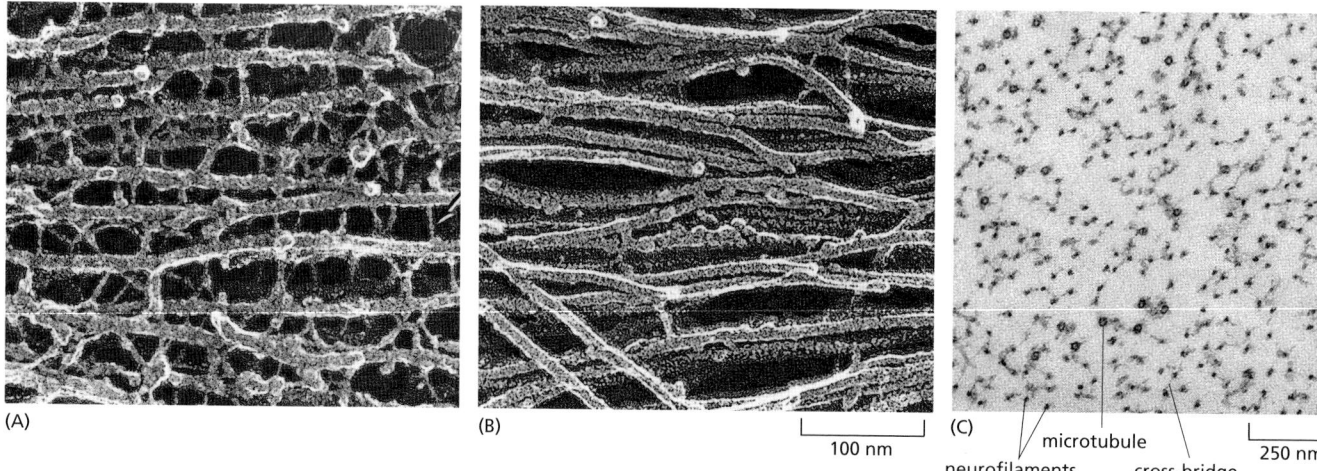

(A)

(B)

|——————| 100 nm

(C)   microtubule

neurofilaments   cross-bridge

|——————| 250 nm

Figure 16–70 **Two types of intermediate filaments in cells of the nervous system.** (A) Freeze-etch electron microscopic image of neurofilaments in a nerve cell axon, showing the extensive cross-linking through protein cross-bridges—an arrangement believed to give this long cell process great tensile strength. The cross-bridges are formed by the long, nonhelical extensions at the C-terminus of the largest neurofilament protein (NF-H). (B) Freeze-etch image of glial filaments in glial cells, showing that these intermediate filaments are smooth and have few cross-bridges. (C) Conventional transmission electron micrograph of a cross section of an axon showing the regular side-to-side spacing of the neurofilaments, which greatly outnumber the microtubules. (A and B, courtesy of Nobutaka Hirokawa; C, courtesy of Anthony Brown.)

The vimentin-like filaments are a third family of intermediate filaments. *Desmin*, a member of this family, is expressed in skeletal, cardiac, and smooth muscle, where it forms a scaffold around the Z disc of the sarcomere (see Figure 16–34). Mice lacking desmin show normal initial muscle development, but adults have various muscle-cell abnormalities, including misaligned muscle fibers. In humans, mutations in desmin are associated with various forms of muscular dystrophy and cardiac myopathy, illustrating the important role of desmin in stabilizing muscle fibers.

Besides their well-established role in maintaining the mechanical stability of the nucleus, it is becoming increasingly evident that one class of lamins, the A-type, together with many proteins of the nuclear envelope, are scaffolds for proteins that control myriad cellular processes including transcription, chromatin organization, and signal transduction. The majority of *laminopathies* are associated with mutant versions of lamin A and include tissue-specific diseases. Skeletal and cardiac abnormalities might be explained by a weakened nuclear envelope leading to cell damage and death, but laminopathies are also thought to arise from pathogenic and tissue-specific alterations in gene expression.

Linker Proteins Connect Cytoskeletal Filaments and Bridge the Nuclear Envelope

The intermediate filament network is linked to the rest of the cytoskeleton by members of a family of proteins called *plakins*. Plakins are large and modular, containing multiple domains that connect cytoskeletal filaments to each other and to junctional complexes. *Plectin* is a particularly interesting example. In addition to bundling intermediate filaments, it links the intermediate filaments to microtubules, actin filament bundles, and filaments of the motor protein myosin II; it also helps attach intermediate filament bundles to adhesive structures at the plasma membrane (**Figure 16–71**).

Plectin and other plakins can interact with protein complexes that connect the cytoskeleton to the nuclear interior. These complexes consist of SUN proteins

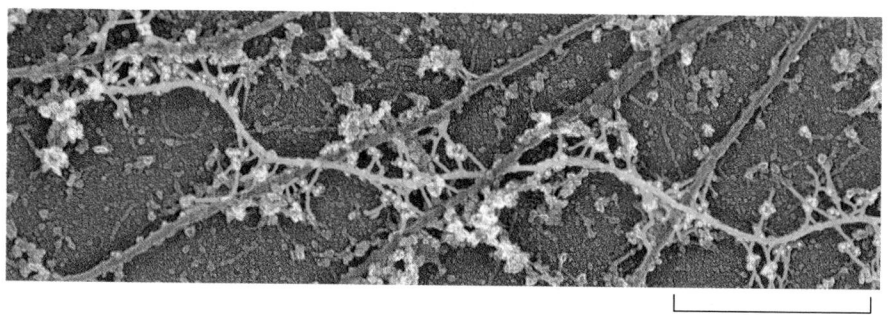

|——————| 0.5 μm

Figure 16–71 **Plectin cross-linking of diverse cytoskeletal elements.** Plectin *(green)* is seen here making cross-links from intermediate filaments *(blue)* to microtubules *(red)*. In this electron micrograph, the dots *(yellow)* are gold particles linked to anti-plectin antibodies. The entire actin filament network was removed to reveal these proteins. (© 1996 T. M. Svitkina et al. Originally published in *J. Cell Biol.* http://doi.org/10.1083/jcb.135.4.991. With permission from Rockefeller University Press.)

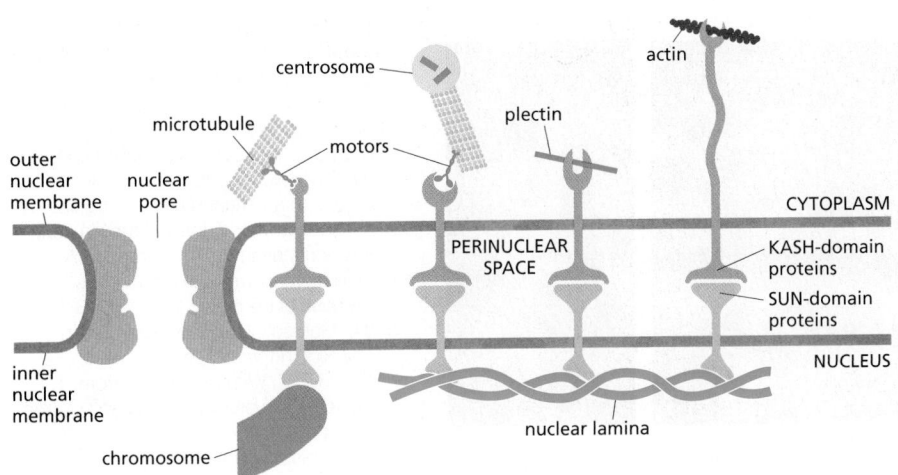

Figure 16–72 **SUN–KASH protein complexes connect the nucleus and cytoplasm through the nuclear envelope.** The cytoplasmic cytoskeleton is linked across the nuclear envelope to the nuclear lamina or chromosomes through SUN and KASH proteins (*orange* and *purple*, respectively). The SUN and KASH domains of these proteins bind within the lumen of the nuclear envelope. From the inner nuclear envelope, SUN proteins connect to the nuclear lamina or chromosomes. KASH proteins in the outer nuclear envelope connect to the cytoplasmic cytoskeleton by binding microtubule motor proteins, actin filaments, or plectin.

of the inner nuclear membrane and KASH proteins (also called nesprins) of the outer nuclear membrane (Figure 16–72). SUN and KASH proteins bind to each other within the lumen of the nuclear envelope, forming a bridge that connects the nuclear and cytoplasmic cytoskeletons. Inside the nucleus, the SUN proteins bind to the nuclear lamina or chromosomes, whereas in the cytoplasm, KASH proteins can bind directly to actin filaments and indirectly to microtubules and intermediate filaments through association with motor proteins and plakins, respectively. This linkage serves to mechanically couple the nucleus to the cytoskeleton and is involved in many cellular functions, including chromosome movements inside the nucleus during meiosis, nuclear and centrosome positioning, nuclear migration, and global cytoskeletal organization.

Mutations in the gene for plectin cause a devastating human disease that combines epidermolysis bullosa (caused by disruption of skin keratin filaments), muscular dystrophy (caused by disruption of desmin filaments), and neurodegeneration (caused by disruption of neurofilaments). Mice lacking a functional plectin gene die within a few days of birth, with blistered skin and abnormal skeletal and heart muscles. Thus, although plectin may not be necessary for the initial formation and assembly of intermediate filaments, its cross-linking action is required to provide cells with the strength they need to withstand the mechanical stresses inherent to vertebrate life.

Septins Form Filaments That Regulate Cell Polarity

GTP-binding proteins called *septins* serve as an additional filament system in all eukaryotes except terrestrial plants. Septins assemble into nonpolar filaments that form rings and cagelike structures, which act as scaffolds to compartmentalize membranes into distinct domains, or recruit and organize the actin and microtubule cytoskeletons. First identified in budding yeast, septin filaments localize to the neck between a dividing yeast mother cell and its growing bud (Figure 16–73A). At this location, septins block the movement of proteins from one side of the bud neck to the other, thereby concentrating cell growth preferentially within the bud. Septins also recruit the actin–myosin machinery that forms the contractile ring required for cytokinesis. In animal cells, septins function in cell division, migration, and vesicle trafficking. In primary cilia, for example, a ring of septin filaments assembles at the base of the cilium and serves as a diffusion barrier at the plasma membrane, restricting the movement of membrane proteins and establishing a specific composition in the ciliary membrane (Figure 16–73B and C). Reduction of septin levels impairs primary cilium formation and signaling.

There are 7 septin genes in yeast and 13 in human, and septin proteins fall into four groups on the basis of sequence relationships. In a test tube, purified septins assemble into symmetrical hetero-hexamers or hetero–octamers that

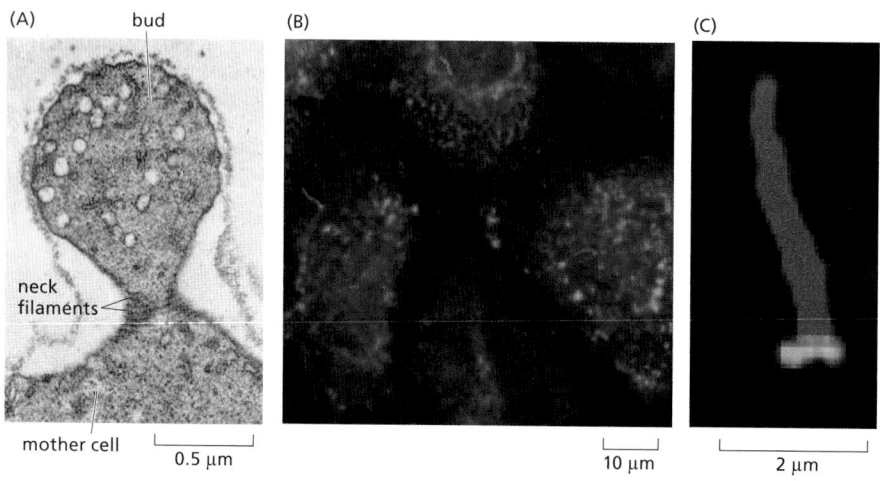

(A) bud

neck filaments

mother cell

0.5 μm

(B)

10 μm

(C)

2 μm

Figure 16–73 **Cell compartmentalization by septins.** (A) Septins form filaments in the neck region between a mother yeast cell and bud. (B) In this photomicrograph of human cultured cells, the DNA is stained *blue* and septins are labeled in *green*. The microtubules of primary cilia are labeled with an antibody that recognizes a modified (acetylated) form of tubulin *(red)* that is enriched in the axoneme. (C) A magnified image reveals a collar of septin at the base of the cilium. (A, ©1976 B. Byers and L. Goetsch. Originally published in *J. Cell Biol.* https://doi.org/10.1083 /jcb.69.3.717. With permission from Rockefeller University Press; B and C, from Q. Hu et al., *Science* 329:436–439, 2010. With permission from AAAS.)

form nonpolar paired filaments (**Figure 16–74**). GTP binding is required for the folding of septin polypeptides, but the role of GTP hydrolysis in septin function is not understood. Septin structures assemble and disassemble inside cells, but they are not as dynamic as actin filaments and microtubules.

Summary

Whereas tubulin and actin have been highly conserved in evolution, intermediate filament proteins are very diverse. There are many tissue-specific forms of interme- diate filaments in the cytoplasm of animal cells, including keratin filaments in epi- thelial cells, neurofilaments in nerve cells, and desmin filaments in muscle cells. The primary function of these filaments is to provide mechanical strength. Septins com- prise an additional system of filaments that organize compartments inside cells.

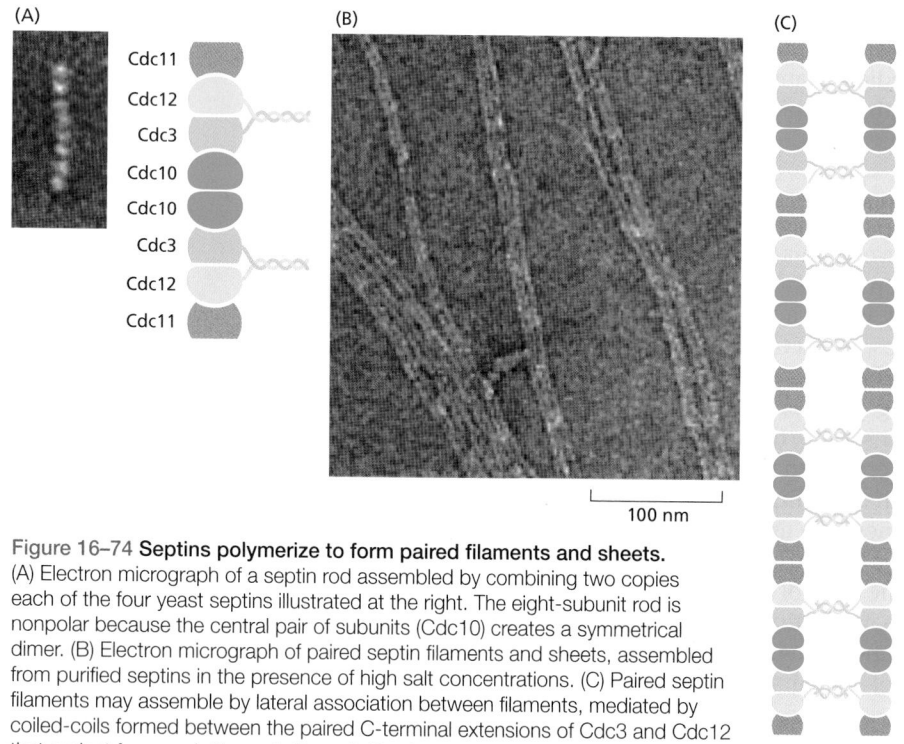

(A)

Cdc11

Cdc12

Cdc3

Cdc10

Cdc10

Cdc3

Cdc12

Cdc11

(B)

100 nm

(C)

Figure 16–74 **Septins polymerize to form paired filaments and sheets.** (A) Electron micrograph of a septin rod assembled by combining two copies each of the four yeast septins illustrated at the right. The eight-subunit rod is nonpolar because the central pair of subunits (Cdc10) creates a symmetrical dimer. (B) Electron micrograph of paired septin filaments and sheets, assembled from purified septins in the presence of high salt concentrations. (C) Paired septin filaments may assemble by lateral association between filaments, mediated by coiled-coils formed between the paired C-terminal extensions of Cdc3 and Cdc12 that project from each filament. (From A. Bertin et al., *Proc. Natl. Acad. Sci. USA* 105:8274–8279, 2008. Copyright 2008 National Academy of Sciences, USA. With permission from the National Academy of Sciences.)

CELL POLARIZATION AND MIGRATION

A central challenge in cell biology is to understand how multiple individual molecular components collaborate to produce complex cell behaviors. The process of cell migration, which we describe in this final section, relies on the coordinated deployment of the components and processes that we have explored in this chapter: the dynamic assembly and disassembly of cytoskeletal polymers, the regulation and modification of their structure by polymer-associated proteins, and the actions of motor proteins moving along the polymers or exerting tension against them. How does the cell coordinate all these activities to define its polarity and enable it to crawl?

Many Cells Can Crawl Across a Solid Substratum

Many cells move by crawling over surfaces rather than by using cilia or flagella to swim. Predatory amoebae crawl continuously in search of food, and they can easily be observed to attack and devour smaller ciliates and flagellates in a drop of pond water (see Movie 1.4). In animals, almost all cell locomotion occurs by crawling, with the notable exception of swimming sperm. During embryogenesis, the structure of an animal is created by the migrations of individual cells to specific target locations and by the coordinated movements of whole epithelial sheets (discussed in Chapter 21). In vertebrates, *neural crest cells* are remarkable for their long-distance migrations from their site of origin in the neural tube to a variety of sites throughout the embryo (see Movie 21.5). Long-distance crawling is fundamental to the construction of the entire nervous system: it is in this way that the actin-rich growth cones at the advancing tips of developing axons travel to their eventual synaptic targets, guided by combinations of soluble signals and signals bound to cell surfaces and extracellular matrix along the way.

The adult animal also seethes with crawling cells. Macrophages and neutrophils crawl to sites of infection and engulf foreign invaders as a critical part of the innate immune response. Osteoclasts tunnel into bone, forming channels that are filled in by the osteoblasts that follow after them, in a continuous process of bone remodeling and renewal. Similarly, fibroblasts migrate through connective tissues, remodeling them where necessary and helping to rebuild damaged structures at sites of injury. In an ordered procession, the cells in the epithelial lining of the intestine travel up the sides of the intestinal villi, replacing absorptive cells lost at the tip of the villus. Unfortunately, cell crawling also has a role in many cancers, when cells in a primary tumor invade neighboring tissues and crawl into blood vessels or lymph vessels and then emerge at other sites in the body to form metastases.

Cell migration is a complex process that depends on the actin-rich cortex beneath the plasma membrane. Three distinct activities are involved: *protrusion*, in which the plasma membrane is pushed out at the front of the cell; *attachment*, in which the actin cytoskeleton connects across the plasma membrane to the substratum; and *traction*, in which the bulk of the trailing cytoplasm is drawn forward (**Figure 16–75**). In some crawling cells, such as keratocytes from the fish epidermis, these activities occur simultaneously, and the cells seem to glide forward smoothly without changing shape. In other cells, such as fibroblasts, these activities are more independent, and the locomotion is jerky and irregular.

Actin Polymerization Drives Plasma Membrane Protrusion

The first step in locomotion, protrusion of a leading edge, frequently relies on forces generated by actin polymerization pushing the plasma membrane outward. Different cell types generate different types of protrusive structures, including filopodia (also known as microspikes) and lamellipodia. These are filled with dense cores of filamentous actin, which excludes membrane-enclosed organelles. The structures differ primarily in the way in which the actin is organized by actin-cross-linking proteins (see Figure 16–22).

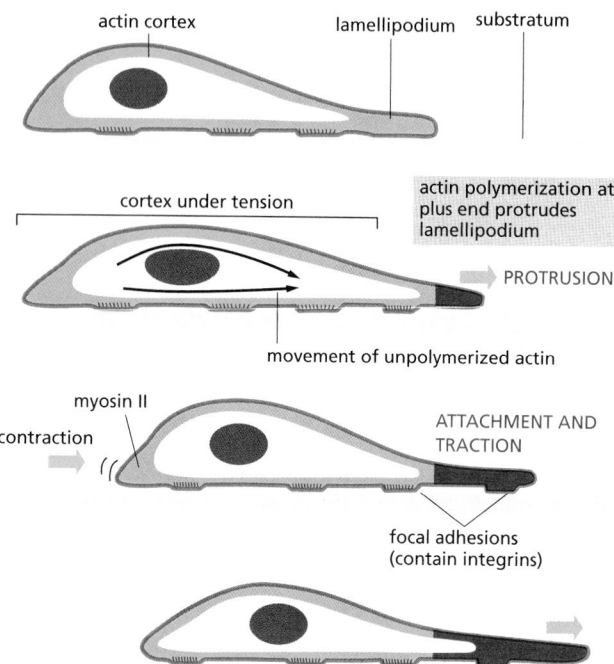

actin cortex   lamellipodium   substratum

cortex under tension

actin polymerization at
plus end protrudes
lamellipodium

→ PROTRUSION

movement of unpolymerized actin

myosin II

contraction

ATTACHMENT AND
TRACTION

focal adhesions
(contain integrins)

Figure 16–75 **A model of how forces generated in the actin-rich cortex move a cell forward.** The actin-polymerization-dependent protrusion and firm attachment of a lamellipodium at the leading edge of the cell move the edge forward (*green* arrows at front) and stretch the actin cortex. Contraction at the rear of the cell propels the body of the cell forward (*green* arrow at back) to relax some of the tension (traction). New focal contacts are made at the front, and old ones are disassembled at the back as the cell crawls forward. The same cycle can be repeated, moving the cell forward in a stepwise fashion. Alternatively, all steps can be tightly coordinated, moving the cell forward smoothly. The newly polymerized cortical actin is shown in *red*.

Filopodia, formed by migrating growth cones of neurons and some types of fibroblasts, are essentially one-dimensional. They contain a core of long, bundled actin filaments, which are reminiscent of those in microvilli but longer and thinner, as well as more dynamic. **Lamellipodia**, formed by epithelial cells and fibroblasts, as well as by some neurons, are two-dimensional, sheetlike structures. They contain a cross-linked mesh of actin filaments, most of which lie in a plane parallel to the solid substratum. **Invadopodia** and related structures known as podosomes represent a third type of actin-rich protrusion. These extend in three dimensions and are important for cells to cross tissue barriers, as when a metastatic cancer cell invades the surrounding tissue. Invadopodia contain many of the same actin-regulatory components as filopodia and lamellipodia, and they also degrade the extracellular matrix, which requires the delivery of vesicles containing matrix-degrading proteases.

A distinct form of membrane protrusion called **blebbing** is often observed *in vivo* or when cells are cultured on a pliable extracellular matrix substratum. Blebs form when the plasma membrane detaches locally from the underlying actin cortex, thereby allowing cytoplasmic flow to push the membrane outward (**Figure 16–76**). Bleb formation also depends on hydrostatic pressure within the cell, which is generated by the contraction of actin and myosin assemblies. Once blebs have extended, actin filaments reassemble on the bleb membrane to form a new

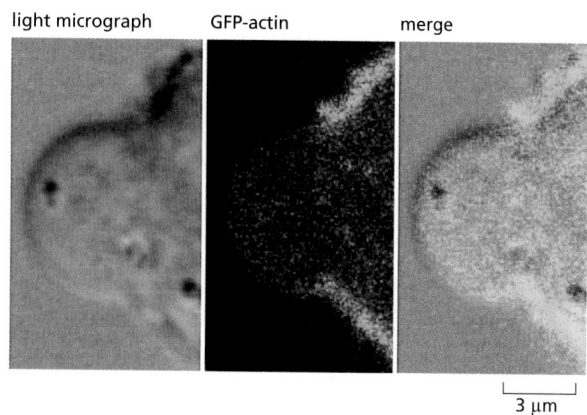

light micrograph   GFP-actin   merge

⊢ 3 μm ⊣

Figure 16–76 **Membrane bleb induced by disruption of the actin cortex.** On the left is a light micrograph showing a spherical membrane protrusion or bleb induced by laser ablation of a small region of the actin cortex. The cortex is labeled *green* in the *middle* image by expression of GFP-actin. (Courtesy of Ewa Paluch lab.)

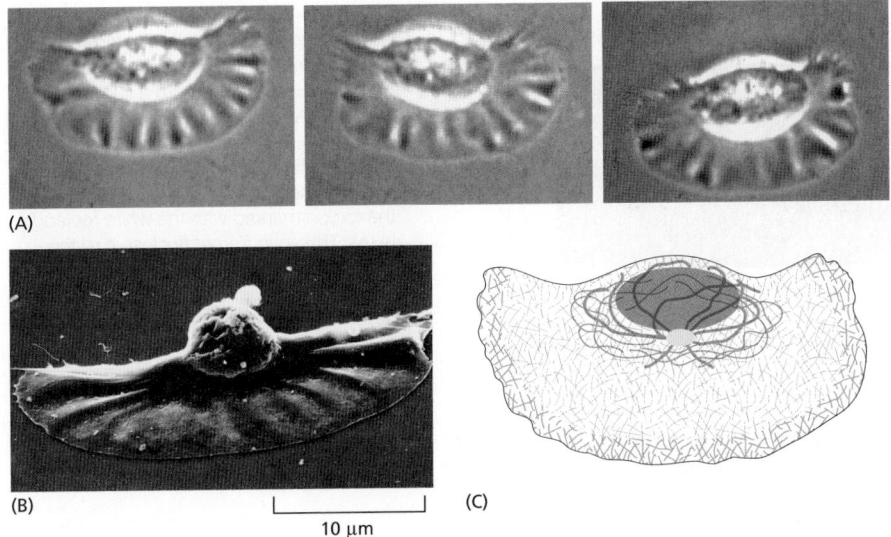

(A)

(B)

(C)

10 μm

Figure 16–77 **Migratory keratocytes from a fish epidermis.** (A) Light micrographs of a keratocyte in culture, taken about 15 seconds apart. This cell is moving at about 15 μm/min (Movie 16.13 and see Movie 1.1). (B) Keratocyte seen by scanning electron microscopy, showing its broad, flat lamellipodium and small cell body, including the nucleus, carried up above the substratum at the rear. (C) Distribution of cytoskeletal filaments in this cell. Actin filaments *(red)* fill the large lamellipodium and are responsible for the cell's rapid movement. Microtubules *(green)* and intermediate filaments *(blue)* are restricted to the regions close to the nucleus. (A and B, courtesy of Juliet Lee.)

actin cortex. Recruitment of myosin II and contraction of actin and myosin can then power retraction of membrane blebs. Alternatively, extension of new blebs from old ones can drive cell migration.

Lamellipodia Contain All of the Machinery Required for Cell Motility

Lamellipodia have been particularly well studied in the epithelial cells of the epidermis of fish and frogs; these epithelial cells are known as *keratocytes* because of their abundant keratin filaments. These cells normally cover the animal by forming an epithelial sheet, and they are specialized to close wounds very rapidly, moving at rates of up to 30 μm/min. When cultured as individual cells, keratocytes assume a distinctive shape with a very large lamellipodium and a small, trailing cell body that is not attached to the substratum (**Figure 16–77**). Fragments of this lamellipodium can be sliced off with a micropipette. Although the fragments generally lack microtubules and membrane-enclosed organelles, they continue to crawl normally, looking like tiny keratocytes.

The dynamic behavior of actin filaments in keratocyte lamellipodia can be studied by labeling a small patch of actin and examining its fate. This reveals that, while the lamellipodia crawl forward, the actin filaments remain stationary with respect to the substratum. The actin filaments in the meshwork are mostly oriented with their plus ends facing forward. The minus ends are frequently attached to the sides of other actin filaments by Arp 2/3 complexes (see Figure 16–16), helping to form the two-dimensional web (**Figure 16–78**). The web as a whole is undergoing treadmilling, assembling at the front and disassembling at the back, reminiscent of the treadmilling that occurs in individual actin filaments discussed previously (see Figure 16–14).

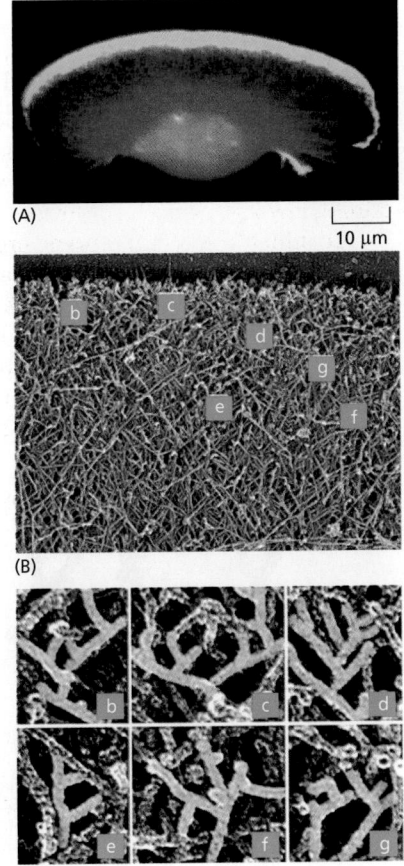

(A)

10 μm

(B)

(C)

Figure 16–78 **Actin filament nucleation and web formation by the Arp 2/3 complex in lamellipodia.** (A) A keratocyte with actin filaments labeled in *red* by fluorescent phalloidin and the Arp 2/3 complex labeled in *green* with an antibody against one of its subunits. The Arp 2/3 complex is highly concentrated near the front of the lamellipodium, where actin nucleation is most active. (B) Electron micrograph of a platinum-shadowed replica of the leading edge of a keratocyte, showing the dense actin filament meshwork. The labels denote areas enlarged in (C). (C) Close-up views of the marked regions of the actin web at the leading edge shown in (B). Numerous branched filaments can be seen, with the characteristic 70° angle formed when the Arp 2/3 complex nucleates a new actin filament off the side of a preexisting filament (see Figure 16–16). (© 1999 T. Svitkina and G. Borisy. Originally published in *J. Cell Biol.* https://doi.org/10.1083/jcb.145.5.1009. With permission from Rockefeller University Press.)

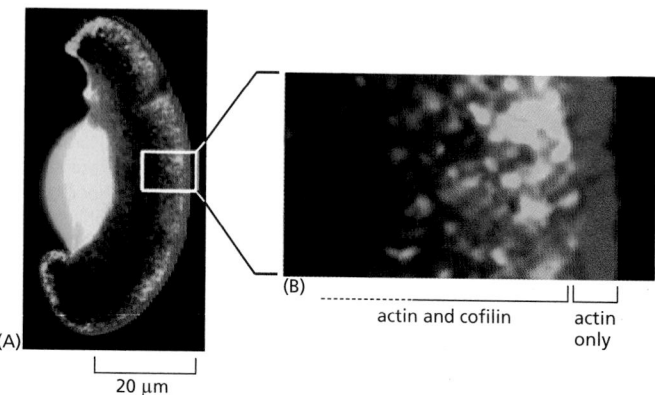

(B)

actin and cofilin   actin
only

(A)

20 μm

Figure 16–79 Cofilin in lamellipodia.
(A) A keratocyte with actin filaments labeled in *red* by fluorescent phalloidin, and cofilin labeled in *green* with a fluorescent antibody. Although the dense actin meshwork reaches all the way through the lamellipodium, cofilin is not found at the very leading edge. (B) Close-up view of the region marked with the *white* rectangle in (A). The actin filaments closest to the leading edge, which are also the ones that have formed most recently and that are most likely to contain ATP-actin (rather than ADP-actin), are generally not associated with cofilin. (© 1999 T. Svitkina and G. Borisy. Originally published in *J. Cell Biol.* https://doi.org/10.1083/jcb.145.5.1009. With permission from Rockefeller University Press.)

Maintenance of unidirectional motion by lamellipodia is thought to require the cooperation and mechanical integration of several factors. Filament nucleation is localized at the leading edge, with new actin filament growth occurring primarily in that location to push the plasma membrane forward. Most filament depolymerization occurs at sites located well behind the leading edge. Because *cofilin* (see Figure 16–20) binds cooperatively and preferentially to actin filaments containing ADP-actin (the D form), the new T-form filaments generated at the leading edge should be resistant to depolymerization by cofilin (**Figure 16–79**). As the filaments age and ATP hydrolysis proceeds, cofilin can efficiently disassemble the older filaments. Thus, the delayed ATP hydrolysis by filamentous actin is thought to provide the basis for a mechanism that maintains an efficient, unidirectional treadmilling process in the lamellipodium (**Figure 16–80**); it also explains the intracellular movement of bacterial pathogens such as *Listeria* (see Figure 16–25).

Myosin Contraction and Cell Adhesion Allow Cells to Pull Themselves Forward

Forces generated by actin filament polymerization at the front of a migrating cell are transmitted to the underlying substratum to drive cell motion. For the leading

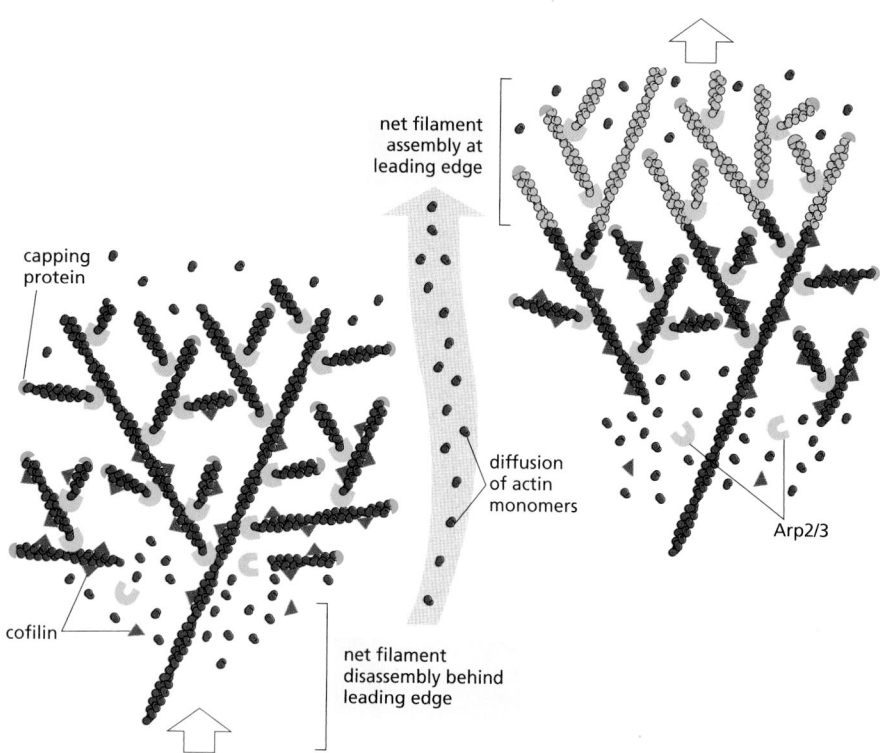

net filament
assembly at
leading edge

capping
protein

diffusion
of actin
monomers

Arp2/3

cofilin

net filament
disassembly behind
leading edge

Figure 16–80 A model for protrusion of the actin meshwork at the leading edge. Two time points during advance of the lamellipodium are illustrated, with newly assembled structures at the later time point shown in a lighter color. Nucleation is mediated by the Arp 2/3 complex at the front. Newly nucleated actin filaments are attached to the sides of preexisting filaments, primarily at a 70° angle. Filaments elongate, pushing the plasma membrane forward because of some sort of anchorage of the array behind. At a steady rate, actin filament plus ends become capped. After newly polymerized actin subunits hydrolyze their bound ATP in the filament lattice, the filaments become susceptible to depolymerization by cofilin. This cycle causes a spatial separation between net filament assembly at the front and net filament disassembly at the rear, so that the actin filament network as a whole can move forward, even though the individual filaments within it remain stationary with respect to the substratum. Not all of the actin disassembles, however, and actin at the rear of the lamellipodium contributes to subsequent steps of migration together with myosin.

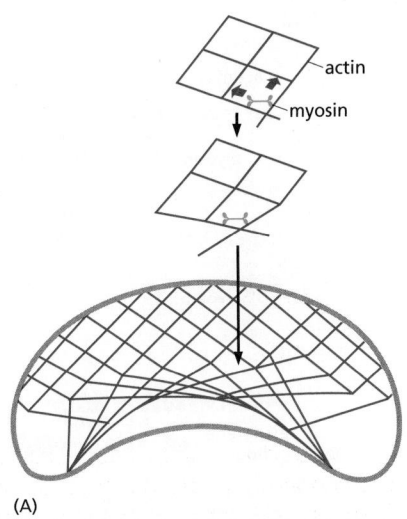

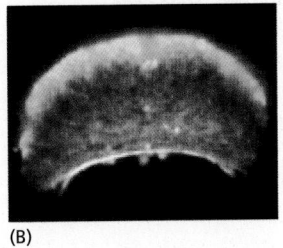

Figure 16–81 Contribution of myosin II to polarized cell motility. (A) Myosin II bipolar filaments bind to actin filaments in the lamellipodial meshwork and cause network contraction. The myosin-driven reorientation of the actin filaments forms an actin bundle that recruits more myosin II and helps generate the contractile forces required for retraction of the trailing edge of the moving cell. (B) A fragment of the large lamellipodium of a keratocyte can be separated from the main cell body either by surgery with a micropipette or by treating the cell with certain drugs. Many of these fragments continue to move rapidly, with the same overall cytoskeletal organization as the intact keratocytes. Actin *(blue)* forms a protrusive meshwork at the front of the fragment. Myosin II *(pink)* is gathered into a band at the rear. (From A. Verkhovsky et al., *Curr. Biol.* 9:11–20, 1999. With permission from Elsevier.)

edge of a migrating cell to advance, protrusion of the membrane must be followed by adhesion to the substratum at the front. Conversely, in order for the cell body to follow, contraction must be coupled with de-adhesion at the rear of the cell. The processes contributing to migration are therefore tightly regulated in space and time, with actin polymerization, dynamic adhesions, and myosin contraction being employed to coordinate movement. Myosin II operates in at least two ways to assist cell migration. The first is by helping to connect the actin cytoskeleton to the substratum through integrin-mediated adhesions. Forces generated by both actin polymerization and myosin activity create tension at attachment sites, promoting their maturation into *focal adhesions*, which are dynamic assemblies of structural and signaling proteins that link the migrating cell to the extracellular matrix (see Figure 19–59). A second mechanism involves bipolar myosin II filaments, which associate with the actin filaments at the rear of the lamellipodium and pull them into a new orientation—from nearly perpendicular to the leading edge to almost parallel to the leading edge. This sarcomere-like contraction prevents protrusion, and it pinches in the sides of the locomoting lamellipodium, helping to gather in the sides of the cell as it moves forward (Figure 16–81).

Actin-mediated protrusions can only push the leading edge of the cell forward if there are strong interactions between the actin network and the focal adhesions that link the cell to the substrate. When these interactions are disengaged, polymerization pressure at the leading edge and myosin-dependent contraction cause the actin network to slip back, resulting in a phenomenon known as retrograde flow (Figure 16–82).

The traction forces generated by locomoting cells exert a significant pull on the substratum. By growing cells on a surface coated with tiny flexible posts, the force exerted on the substratum can be calculated by measuring the deflection of each post from its vertical position (Figure 16–83). In a living animal, most crawling cells move across a semiflexible substratum made of extracellular matrix, which can be deformed and rearranged by these cell forces. Conversely, mechanical tension or stretching applied externally to a cell will cause it to assemble stress fibers and focal adhesions, and become more contractile. Although poorly understood, this two-way mechanical interaction between cells and their physical environment is thought to help vertebrate tissues organize themselves.

Cell Polarization Is Controlled by Members of the Rho Protein Family

Cell migration requires long-distance communication and coordination between one end of a cell and the other. During directed migration, it is important that the front end of the cell remain structurally and functionally distinct from the back end. In addition to driving local mechanical processes such as protrusion at the front and retraction at the rear, the cytoskeleton is responsible for coordinating cell shape, organization, and mechanical properties from one end of the cell to the other, a distance that is typically tens of micrometers for animal cells.

In many cases, including but not limited to cell migration, large-scale cytoskeletal coordination takes the form of the establishment of cell polarity, where a cell builds different structures with distinct molecular components at the front

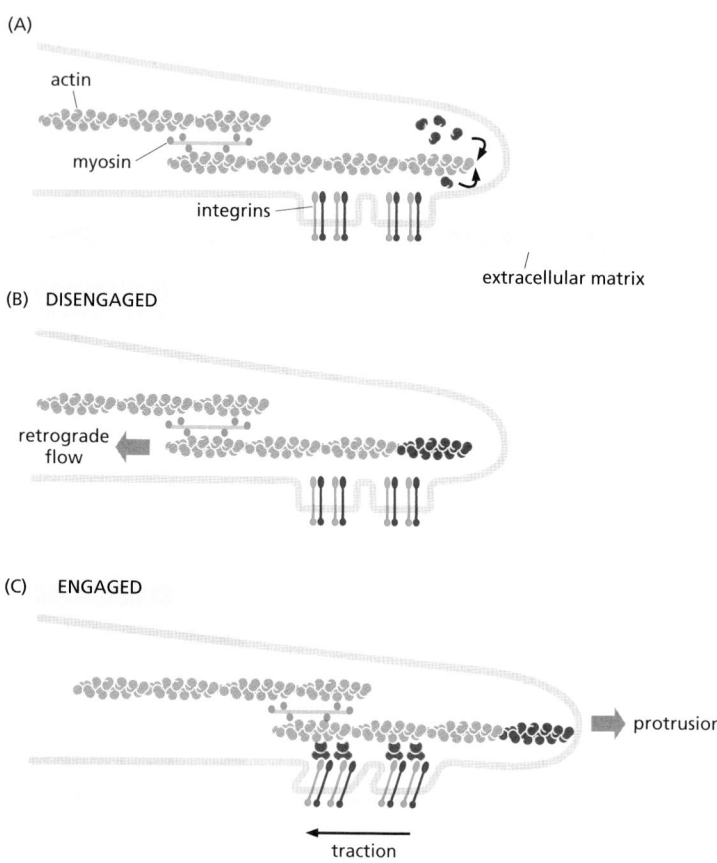

(A)

actin

myosin

integrins

extracellular matrix

(B) DISENGAGED

retrograde
flow

(C) ENGAGED

protrusion

traction

Figure 16–82 **Control of cell–
substratum adhesion at the leading
edge of a migrating cell.** (A) Actin
monomers assemble on the barbed end
of actin filaments at the leading edge.
Transmembrane integrin proteins *(blue)*
help form focal adhesions that link the
cell membrane to the substrate. (B) If
there is no interaction between the actin
filaments and focal adhesions, the actin
filament is driven rearward by newly
assembled actin. Myosin motors *(green)*
also contribute to filament movement.
(C) Interactions between actin-binding
adaptor proteins *(brown)* and integrins link
the actin cytoskeleton to the substratum.
Myosin-mediated contractile forces are
then transmitted through the focal adhesion
to generate traction on the extracellular
matrix, and new actin polymerization drives
the leading edge forward in a protrusion.

versus the back, or at the top versus the bottom. Cell locomotion requires an initial polarization of the cell to set it off in a particular direction. Carefully controlled
cell-polarization processes are also required for oriented cell divisions in tissues
and for formation of a coherent, organized multicellular structure. Genetic studies in yeast, flies, and worms have provided most of our current understanding
of the molecular basis of cell polarity. The mechanisms that generate cell polarity in vertebrates are only beginning to be explored. In all known cases, however,
the cytoskeleton has a central role, and many of the molecular components have
been evolutionarily conserved.

The establishment of many kinds of cell polarity depends on the local regulation of the actin cytoskeleton by external signals. Many of these signals seem to
converge inside the cell on a group of closely related monomeric GTPases that
are members of the **Rho protein family**—*Cdc42, Rac,* and *Rho.* Like other monomeric GTPases, the Rho proteins act as molecular switches that cycle between
an active GTP-bound state and an inactive GDP-bound state (see Figure 3–66).
Activation of Cdc42 on the inner surface of the plasma membrane triggers actin
polymerization and bundling to form filopodia. Activation of Rac promotes actin
polymerization at the cell periphery, leading to the formation of sheetlike lamellipodial extensions. Activation of Rho promotes both the bundling of actin filaments with myosin II filaments into stress fibers and the clustering of integrins

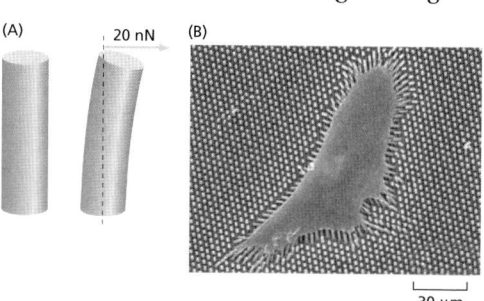

(A)   20 nN   (B)

30 μm

Figure 16–83 **Traction forces exerted
by a motile cell.** (A) Tiny flexible pillars
attached to the substratum bend in
response to traction forces. (B) Scanning
electron micrograph of a cell on a
substratum coated with pillars that are
6.1 μm in height. Pillar deflections are used
to calculate force vectors corresponding
to inward pulling forces on the underlying
substratum. (B, adapted from J. Fu et al.,
Nat. Methods 7:733–736, published 2010
by Nature America, Inc. Reproduced with
permission of SNCSC.)

actin staining   actin staining

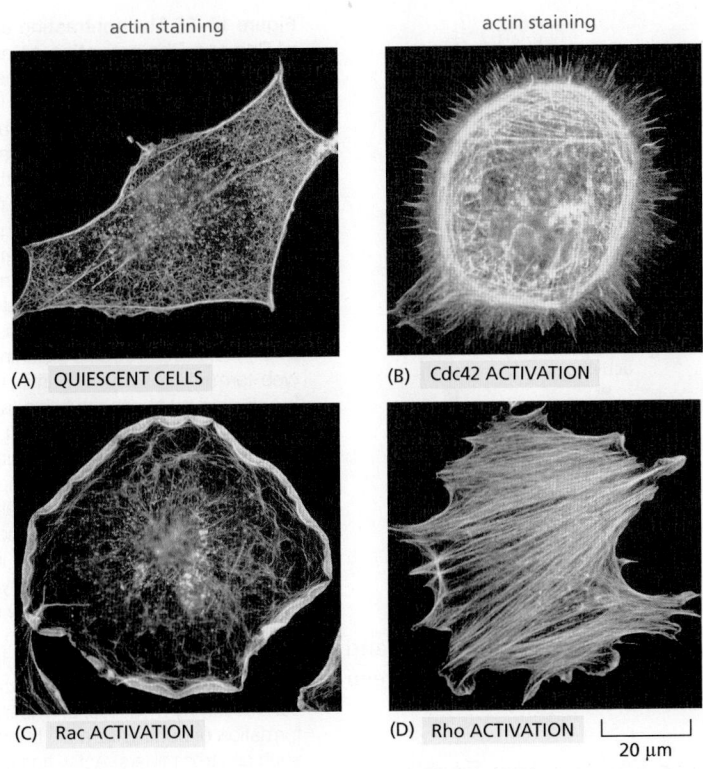

(A)   QUIESCENT CELLS

(B)   Cdc42 ACTIVATION

(C)   Rac ACTIVATION

(D)   Rho ACTIVATION

20 µm

Figure 16–84 The dramatic effects of Cdc42, Rac, and Rho on actin organization in fibroblasts. In each case, the actin filaments have been labeled with fluorescent phalloidin. (A) Serum-starved fibroblasts have actin filaments primarily in the cortex, and relatively few stress fibers. (B) Microinjection of a constitutively activated form of Cdc42 causes the protrusion of many long filopodia at the cell periphery. (C) Microinjection of a constitutively activated form of Rac, a closely related monomeric GTPase, causes the formation of an enormous lamellipodium that extends from the entire circumference of the cell. (D) Microinjection of a constitutively activated form of Rho causes the rapid assembly of many prominent stress fibers. (From A. Hall, *Science* 279:509–514, 1998. With permission from Catherine Nobes.)

and associated proteins to form focal adhesions (**Figure 16–84**). These dramatic and complex structural changes occur because each of these three molecular switches has numerous downstream target proteins that affect actin organization and dynamics.

Some key targets of activated Cdc42 are members of the **WASp protein** family. Human patients deficient in WASp suffer from Wiskott-Aldrich Syndrome, a severe form of immunodeficiency in which immune system cells have abnormal actin-based motility and platelets do not form normally. Although WASp itself is expressed only in blood cells and immune system cells, other more ubiquitous versions enable activated Cdc42 to enhance actin polymerization in many cell types. WASp proteins can exist in an inactive folded conformation and an activated open conformation. Association with Cdc42-GTP stabilizes the open form of WASp, enabling it to bind to the Arp 2/3 complex and strongly enhance its actin-nucleating activity (see Figure 16–16). In this way, activation of Cdc42 increases actin nucleation.

Rac-GTP also activates WASp family members. Additionally, it activates the cross-linking activity of the gel-forming protein filamin and inhibits the contractile activity of the motor protein myosin II. It thereby stabilizes lamellipodia and inhibits the formation of contractile stress fibers (**Figure 16–85A**).

Rho-GTP has a very different set of targets. Instead of activating the Arp 2/3 complex to build actin networks, Rho-GTP turns on formin proteins to construct parallel actin bundles. At the same time, Rho-GTP activates a protein kinase that indirectly inhibits the activity of cofilin, leading to actin filament stabilization. The same protein kinase inhibits a phosphatase acting on myosin light chains (see Figure 16–39). The consequent increase in the net amount of myosin light chain phosphorylation increases the amount of contractile myosin motor protein activity in the cell, enhancing the formation of tension-dependent structures such as stress fibers (Figure 16–85B).

In some cell types, Rac-GTP activates Rho, usually at a rate that is slow compared to Rac's activation of the Arp 2/3 complex. This enables cells to use the Rac pathway to build a new actin structure while subsequently activating the Rho pathway to generate a contractility that builds up tension in this structure. This occurs, for example, during the formation and maturation of cell–cell contacts.

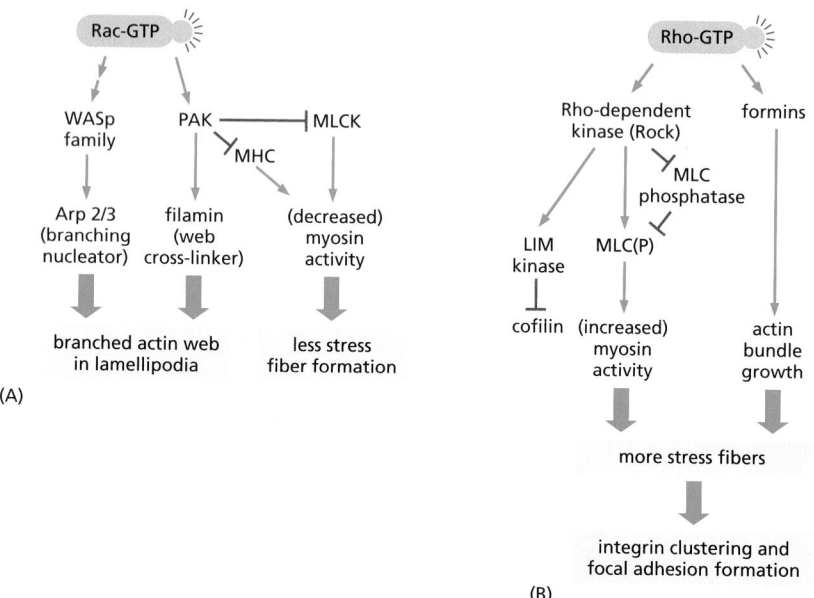

(A)

(B)

Figure 16–85 The contrasting effects of Rac and Rho activation on actin organization. (A) Activation of the small GTPase Rac leads to alterations in actin accessory proteins that tend to favor the formation of actin networks, as in lamellipodia. Several different pathways contribute independently. Rac-GTP activates members of the WASp protein family, which in turn activate actin nucleation and branched web formation by the Arp 2/3 complex. In a parallel pathway, Rac-GTP activates a protein kinase, PAK, which has several targets including the web-forming cross-linker filamin, which is activated by phosphorylation, and the myosin light chain kinase (MLCK), which is inhibited by phosphorylation. Inhibition of MLCK results in decreased phosphorylation of the myosin regulatory light chain and leads to myosin II filament disassembly and a decrease in contractile activity. In some cells, PAK also directly inhibits myosin II activity by phosphorylation of the myosin heavy chain (MHC). (B) Activation of the related GTPase Rho leads to nucleation of actin filaments by formins and increases contraction by myosin II, promoting the formation of contractile actin bundles such as stress fibers. Activation of myosin II by Rho requires a Rho-dependent protein kinase called Rock. This kinase inhibits the phosphatase that removes the activating phosphate groups from myosin II light chains (MLC); it may also directly phosphorylate the myosin light chains in some cell types. Rock also activates other protein kinases, such as LIM kinase, which in turn contributes to the formation of stable contractile actin filament bundles by inhibiting the actin depolymerizing factor cofilin. A similar signaling pathway is important for forming the contractile ring necessary for cytokinesis (see Figure 17–44).

As we will explore in more detail below, the communication between the Rac and Rho pathways also facilitates maintenance of the large-scale differences between the cell front and the cell rear during migration.

Extracellular Signals Can Activate the Three Rho Protein Family Members

The activation of the monomeric GTPases Rho, Rac, and Cdc42 occurs through an exchange of GTP for a tightly bound GDP molecule, catalyzed by guanine nucleotide exchange factors (GEFs). Of the many GEFs that have been identified in the human genome, some are specific for an individual Rho family GTPase, whereas others seem to act on multiple family members. Different GEFs are restricted to specific tissues and even specific subcellular locations, and they are sensitive to distinct kinds of regulatory inputs. GEFs can be activated by extracellular cues through cell-surface receptors, or in response to intracellular signals. GEFs may also act as scaffolds that direct GTPases to downstream effectors. Interestingly, several of the Rho family GEFs associate with the growing ends of microtubules by binding to one of the +TIPs. This provides a connection between the dynamics of the microtubule cytoskeleton and the large-scale organization of the actin cytoskeleton; such a connection is important for the overall integration of cell shape and movement.

External Signals Can Dictate the Direction of Cell Migration

Chemotaxis is the movement of a cell toward or away from a source of some diffusible chemical. These external signals act through Rho family proteins to set up large-scale cell polarity by influencing the organization of the cell motility apparatus. One well-studied example is the chemotactic movement of a class of white blood cells, called *neutrophils*, toward a source of bacterial infection. Receptor proteins on the surface of neutrophils enable them to detect very low concentrations of *N*-formylated peptides that are derived from bacterial proteins (only prokaryotes begin protein synthesis with *N*-formylmethionine). Using these receptors, neutrophils are guided to bacterial targets by their ability to detect a difference of only 1% in the concentration of these diffusible peptides on one side of the cell versus the other (**Figure 16–86**A).

In this case, and in the chemotaxis of *Dictyostelium* amoebae toward a source of cyclic AMP, binding of the chemoattractant to its G-protein-coupled receptor activates phosphoinositide 3-kinases (PI3Ks) (see Figure 15–52), which generate a signaling molecule [PI(3,4,5)P$_3$] that in turn activates the Rac GTPase. Rac

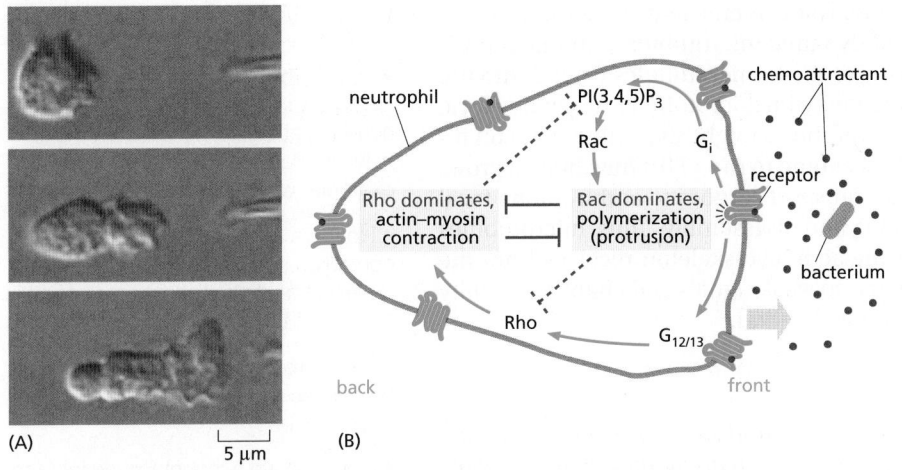

(A)

5 µm

(B)

Figure 16–86 Neutrophil polarization and chemotaxis. (A) The pipette tip at the right is leaking a small amount of the bacterial peptide formyl-Met-Leu-Phe, which is recognized by the human neutrophil as the product of a foreign invader. The neutrophil quickly extends a new lamellipodium toward the source of the chemoattractant peptide (top). It then extends this lamellipodium and polarizes its cytoskeleton so that contractile myosin II is located primarily at the rear, opposite the position of the lamellipodium (middle). Finally, the cell crawls toward the source of the peptide (bottom). If a real bacterium were the source of the peptide, rather than an investigator's pipette, the neutrophil would engulf the bacterium and destroy it (see also Figure 16–3 and Movie 16.14). (B) Binding of bacterial molecules to G-protein-coupled receptors on the neutrophil stimulates directed motility. These receptors are found all over the surface of the cell, but are more likely to be bound to the bacterial ligand at the front. Two distinct signaling pathways contribute to the cell's polarization. At the front of the cell, stimulation of the Rac pathway leads, via the trimeric G protein G_i, to growth of protrusive actin networks. Second messengers within this pathway are short-lived, so protrusion is limited to the region of the cell closest to the stimulant. The same receptor also stimulates a second signaling pathway, via the trimeric G proteins G_{12} and G_{13}, that triggers the activation of Rho. The two pathways are mutually antagonistic. Since Rac-based protrusion is active at the front of the cell, Rho is activated only at the rear of the cell, stimulating contraction of the cell rear and assisting directed movement. (A, from O.D. Weiner et al., Nat. Cell Biol. 1:75–81, published 1999 by Macmillan Magazines Ltd. Reproduced with permission of SNCSC.)

then activates the Arp 2/3 complex leading to lamellipodial protrusion. Through an unknown mechanism, accumulation of the polarized actin web at the leading edge causes further local enhancement of PI3K activity in a positive feedback loop, strengthening the induction of protrusion. The $PI(3,4,5)P_3$ that activates Rac cannot diffuse far from its site of synthesis, since it is rapidly converted back into $PI(4,5)P_2$ by a constitutively active lipid phosphatase. At the same time, binding of the chemoattractant ligand to its receptor activates another signaling pathway that turns on Rho and enhances myosin-based contractility. The two processes directly inhibit each other, such that Rac activation dominates in the front of the cell and Rho activation dominates in the rear (Figure 16–86B). This enables the cell to maintain its functional polarity with protrusion at the leading edge and contraction at the back.

Nondiffusible chemical cues attached to the extracellular matrix or to the surface of cells can also influence the direction of cell migration. When these signals activate receptors, they can cause increased cell adhesion and directed actin polymerization. Most long-distance cell migrations in animals, including neural-crest-cell migration and the travels of neuronal growth cones, depend on a combination of diffusible and nondiffusible signals to steer the locomoting cells or growth cones to their proper destinations.

Communication Among Cytoskeletal Elements Coordinates Whole-Cell Polarization and Locomotion

The interconnected cytoskeleton is crucial for cell migration. Although movement is driven primarily by actin polymerization and myosin contractility, septins and intermediate filaments also participate. For example, vimentin intermediate filament networks associate with integrins at focal adhesions, and vimentin-deficient fibroblasts display impaired mechanical stability, migration, and contractile capacity. Furthermore, disruption of linker proteins that connect different cytoskeletal elements, including several plakins and KASH proteins, leads to defects in cell polarization and migration. Thus, interactions among cytoplasmic filament systems, as well as mechanical linkage to the nucleus, are required for complex, whole-cell behaviors such as migration.

Cells also use microtubules to help organize persistent movement in a specific direction. In many locomoting cells, the position of the centrosome is influenced by the location of protrusive actin polymerization. Activation of receptors on the protruding front edge of a cell might locally activate dynein motor proteins that move the centrosome by pulling on its microtubules. Several effector proteins downstream of Rac and Rho modulate microtubule dynamics directly: for example, a protein kinase activated by Rac can phosphorylate (and thereby inhibit) the tubulin-binding protein stathmin (see Panel 16–4), thereby stabilizing microtubules.

In turn, microtubules influence actin rearrangements and cell adhesion. The centrosome nucleates a large number of dynamic microtubules, and its repositioning means that the plus ends of many of these microtubules extend into the protrusive region of the cell. Direct interactions with microtubules help guide focal adhesion dynamics in migrating cells. Microtubules might also influence actin filament formation by delivering Rac-GEFs that bind to the +TIPs traveling on growing microtubule ends. Microtubules also transport cargoes to and from the focal adhesions, thereby affecting their signaling and disassembly. Thus, microtubules reinforce the polarity information that the actin cytoskeleton receives from the outside world, allowing a sensitive response to weak signals and enabling motility to persist in the same direction for a prolonged period.

Summary

Whole-cell movements and the large-scale shaping and structuring of cells require the coordinated activities of all three basic filament systems along with a large variety of cytoskeletal accessory proteins, including motor proteins. Cell crawling—a widespread behavior important in embryonic development and also in wound healing, tissue maintenance, and immune system function in the adult animal—is a prime example of such complex, coordinated cytoskeletal action. For a cell to crawl, it must generate and maintain an overall structural polarity, which is influenced by external cues. In addition, the cell must coordinate protrusion at the leading edge (by assembly of new actin filaments), adhesion of the newly protruded part of the cell to the substratum, and forces generated by molecular motors to bring the cell body forward.

WHAT WE DON'T KNOW

- How is the cell cortex regulated locally and globally to coordinate its activities at different places on the cell surface? What determines, for example, where filopodia form?

- How are actin-regulatory proteins controlled spatially in the cytoplasm to generate multiple distinct types of actin arrays in the same cell?

- Are there biologically important processes occurring inside a microtubule?

- How can we account for the fact that there are many different kinesins and myosins in the cytoplasm but only one dynein?

- Mutations in the nuclear lamin proteins cause a large number of diseases called laminopathies. What do we not understand about the nuclear lamina that could account for this fact?

PROBLEMS

Which statements are true? Explain why or why not.

16–1   The role of ATP hydrolysis in actin polymerization is similar to the role of GTP hydrolysis in tubulin polymerization: both serve to weaken the bonds in the polymer and thereby promote depolymerization.

16–2   Motor neurons trigger action potentials in muscle cell membranes that open voltage-sensitive Ca^{2+} channels in T tubules, allowing extracellular Ca^{2+} to enter the cytosol, bind to troponin C, and initiate rapid muscle contraction.

16–3   In most animal cells, minus-end directed microtubule motors deliver their cargo to the periphery of the cell, whereas plus-end directed microtubule motors deliver their cargo to the interior of the cell.

Discuss the following problems.

16–4   The concentration of actin in cells is 50–100 times greater than the critical concentration observed for pure actin in a test tube. How is this possible? What prevents the actin subunits in cells from polymerizing into filaments? Why is it advantageous to the cell to maintain such a large pool of actin subunits?

16–5   Detailed measurements of sarcomere length and tension during isometric contraction in striated muscle provided crucial early support for the sliding-filament

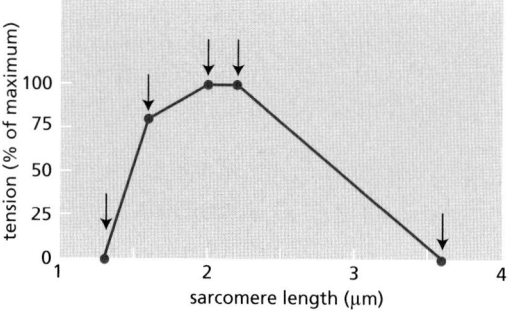

Figure Q16–1 Tension as a function of sarcomere length during isometric contraction (Problem 16–5).

model of muscle contraction. Based on your understanding of the sliding-filament model and the structure of a sarcomere, propose a molecular explanation for the relationship of tension to sarcomere length in the portions of **Figure Q16–1** marked I, II, III, and IV. (In this muscle, the length of the myosin filament is 1.6 µm, and the lengths of the actin thin filaments that project from the Z discs are 1.0 µm.)

16–6   At 1.4 mg/mL pure tubulin, microtubules grow at a rate of about 2 µm/min. At this growth rate, how many αβ-tubulin dimers (8 nm in length) are added to the ends of a microtubule each second?

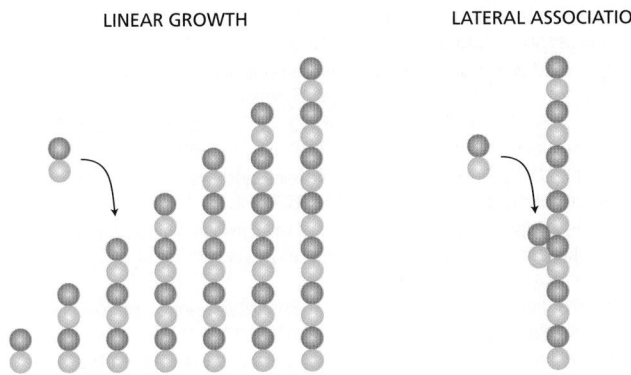

LINEAR GROWTH   LATERAL ASSOCIATION

Figure Q16–2 Model for microtubule nucleation by pure αβ-tubulin dimers (Problem 16–7).

16–7 A solution of pure αβ-tubulin dimers is thought to nucleate microtubules by forming a linear protofilament about seven dimers in length. At that point, the probabilities that the next αβ-dimer will bind laterally or to the end of the protofilament are about equal. The critical event for microtubule formation is thought to be the first lateral association (**Figure Q16–2**). How does lateral association promote the subsequent rapid formation of a microtubule?

16–8 How does a centrosome "know" when it has found the center of the cell?

16–9 The movements of single motor-protein molecules can be analyzed directly. Using polarized laser light, it is possible to create interference patterns that exert a centrally directed force, ranging from zero at the center to a few piconewtons at the periphery (about 200 nm from the center). Individual molecules that enter the interference pattern are rapidly pushed to the center, allowing them to be captured and moved at the experimenter's discretion.

Using such "optical tweezers," single kinesin molecules can be positioned on a microtubule that is fixed to a coverslip. Although a single kinesin molecule cannot be seen optically, it can be tagged with a silica bead and tracked indirectly by following the bead (**Figure Q16–3A**). In the absence of ATP, the kinesin molecule remains at the center of the interference pattern, but with ATP it moves toward the plus end of the microtubule. As kinesin moves along the microtubule, it encounters the force of the interference pattern, which simulates the load kinesin carries during its actual function in the cell. Moreover, the pressure against the silica bead counters the effects of Brownian (thermal) motion, so that the position of the bead more accurately reflects the position of the kinesin molecule on the microtubule.

A trace of the movements of a kinesin molecule along a microtubule is shown in Figure Q16–3B.

A. As shown in Figure Q16–3B, all movement of kinesin is in one direction (toward the plus end of the microtubule). What supplies the free energy needed to ensure a unidirectional movement along the microtubule?

B. What is the average rate of movement of kinesin along the microtubule?

C. What is the length of each step that a kinesin takes as it moves along a microtubule?

D. From other studies it is known that kinesin has two globular domains that can each bind to β-tubulin, and that kinesin moves along a single protofilament in a microtubule. In each protofilament, the β-tubulin subunit repeats at 8-nm intervals. Given the step length and the interval between β-tubulin subunits, how do you suppose a kinesin molecule moves along a microtubule?

E. Is there anything in the data in Figure Q16–3B that tells you how many ATP molecules are hydrolyzed per step?

16–10 A mitochondrion 1 μm long can travel the 1 meter length of the axon from the spinal cord to the big toe in a day. The Olympic men's freestyle swimming record for 200 meters is 1.75 minutes. In terms of body lengths per day, who is moving faster: the mitochondrion or the Olympic record holder? (Assume that the swimmer is 2 meters tall.)

16–11 Cofilin preferentially binds to older actin filaments and promotes their disassembly. How does cofilin distinguish old filaments from new ones?

16–12 Why is it that intermediate filaments have identical ends and lack polarity, whereas actin filaments and microtubules have two distinct ends with a defined polarity?

16–13 How is the unidirectional motion of a lamellipodium maintained?

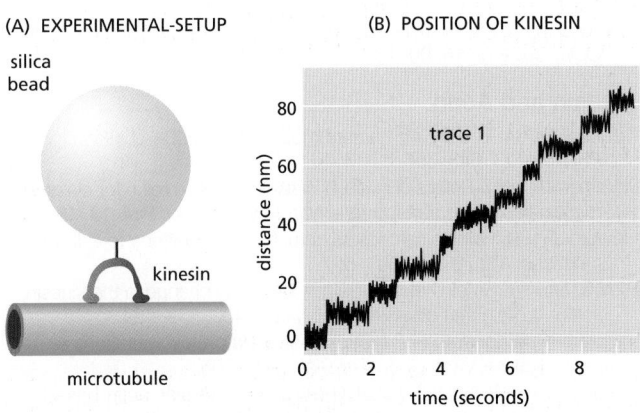

(A) EXPERIMENTAL-SETUP   (B) POSITION OF KINESIN

silica bead

kinesin

microtubule

Figure Q16–3 Movement of kinesin along a microtubule (Problem 16–9). (A) Experimental set-up, with kinesin linked to a silica bead, moving along a microtubule. (B) Position of kinesin (as visualized by the position of the silica bead) relative to the center of the interference pattern, as a function of time of movement along the microtubule. The jagged nature of the trace results from Brownian motion of the bead.

REFERENCES

General

Bray D (2001) Cell Movements: From Molecules to Motility, 2nd ed. New York: Garland Science.

Howard J (2001) Mechanics of Motor Proteins and the Cytoskeleton. Sunderland, MA: Sinauer.

Kavallaris M (2012) Cytoskeleton and Human Disease. New York: Springer, Humana Press.

Function and Origin of the Cytoskeleton

Garner EC, Bernard R, Wang W et al. (2011) Coupled, circumferential motions of the cell wall synthesis machinery and MreB filaments in *B. subtilis. Science* 333, 222–225.

Garner EC, Campbell CS & Mullins RD (2004) Dynamic instability in a DNA-segregating prokaryotic actin homolog. *Science* 306, 1021–1025.

Hill TL & Kirschner MW (1982) Bioenergetics and kinetics of microtubule and actin filament assembly-disassembly. *Int. Rev. Cytol.* 78, 1–125.

Jones LJ, Carballido-López R & Errington J (2001) Control of cell shape in bacteria: helical, actin-like filaments in *Bacillus subtilis. Cell* 104, 913–922.

Luby-Phelps K (2000) Cytoarchitecture and physical properties of cytoplasm: volume, viscosity, diffusion, intracellular surface area. *Int. Rev. Cytol.* 192, 189–221.

Oosawa F & Asakura S (1975) Thermodynamics of the Polymerization of Protein, pp. 41–55 and 90–108. New York: Academic Press.

Osawa M, Anderson DE & Erickson HP (2009) Curved FtsZ protofilaments generate bending forces on liposome membranes. *EMBO J.* 28, 3476–3484.

Pauling L (1953) Aggregation of globular proteins. *Discuss. Faraday Soc.* 13, 170–176.

Purcell EM (1977) Life at low Reynolds number. *Am. J. Phys.* 45, 3–11.

Theriot JA (2013) Why are bacteria different from eukaryotes? *BMC Biol.* 11, 119.

Actin and Actin-binding Proteins

Fehon RG, McClatchey AI & Bretscher A (2010) Organizing the cell cortex: the role of ERM proteins. *Nat. Rev. Mol. Cell Biol.* 11, 276–287.

Mullins RD, Heuser JA & Pollard TD (1998) The interaction of Arp2/3 complex with actin: nucleation, high affinity pointed end capping, and formation of branching networks of filaments. *Proc. Natl Acad. Sci. U.S.A.* 95, 6181–6186.

Zigmond SH (2004) Formin-induced nucleation of actin filaments. *Curr. Opin. Cell Biol.* 16, 99–105.

Myosin and Actin

Cooke R (2004) The sliding filament model: 1972–2004. *J. Gen. Physiol.* 123, 643–656.

Hammer JA 3rd & Sellers JR (2011) Walking to work: roles for class V myosins as cargo transporters. *Nat. Rev. Mol. Cell Biol.* 13, 13–26.

Howard J (1997) Molecular motors: structural adaptations to cellular functions. *Nature* 389, 561–567.

Rice S, Lin AW, Safer D et al. (1999) A structural change in the kinesin motor protein that drives motility. *Nature* 402, 778–784.

Vikstrom KL & Leinwand LA (1996) Contractile protein mutations and heart disease. *Curr. Opin. Cell Biol.* 8, 97–105.

Wells AL, Lin AW, Chen LQ et al. (1999) Myosin VI is an actin-based motor that moves backwards. *Nature* 401, 505–508.

Yildiz A, Forkey JN, McKinney SA et al. (2003) Myosin V walks hand-over-hand: single fluorophore imaging with 1.5-nm localization. *Science* 300, 2061–2065.

Microtubules

Aldaz H, Rice LM, Stearns T & Agard DA (2005) Insights into microtubule nucleation from the crystal structure of human gamma-tubulin. *Nature* 435, 523–527.

Brouhard GJ, Stear JH, Noetzel TL et al. (2008) XMAP215 is a processive microtubule polymerase. *Cell* 132, 79–88.

Dogterom M & Yurke B (1997) Measurement of the force-velocity relation for growing microtubules. *Science* 278, 856–860.

Doxsey S, McCollum D & Theurkauf W (2005) Centrosomes in cellular regulation. *Annu. Rev. Cell Dev. Biol.* 21, 411–434.

Galjart N (2010) Plus-end-tracking proteins and their interactions at microtubule ends. *Curr. Biol.* 20, R528–R537.

Hotani H & Horio T (1988) Dynamics of microtubules visualized by darkfield microscopy: treadmilling and dynamic instability. *Cell Motil. Cytoskeleton* 10, 229–236.

Howard J, Hudspeth AJ & Vale RD (1989) Movement of microtubules by single kinesin molecules. *Nature* 342, 154–158.

Kerssemakers JW, Munteanu EL, Laan L et al. (2006) Assembly dynamics of microtubules at molecular resolution. *Nature* 442, 709–712.

Kikkawa M (2013) Big steps toward understanding dynein. *J. Cell Biol.* 202, 15–23.

Mitchison T & Kirschner M (1984) Dynamic instability of microtubule growth. *Nature* 312, 237–242.

Reck-Peterson SL, Yildiz A, Carter AP et al. (2006) Single-molecule analysis of dynein processivity and stepping behavior. *Cell* 126, 335–348.

Sharp DJ & Ross JL (2012) Microtubule-severing enzymes at the cutting edge. *J. Cell Sci.* 125, 2561–2569.

Singla V & Reiter JF (2006) The primary cilium as the cell's antenna: signaling at a sensory organelle. *Science* 313, 629–633.

Stearns T & Kirschner M (1994) *In vitro* reconstitution of centrosome assembly and function: the central role of gamma-tubulin. *Cell* 76, 623–637.

Svoboda K, Schmidt CF, Schnapp BJ & Block SM (1993) Direct observation of kinesin stepping by optical trapping interferometry. *Nature* 365, 721–727.

Verhey KJ, Kaul N & Soppina V (2011) Kinesin assembly and movement in cells. *Annu. Rev. Biophys.* 40, 267–288.

Intermediate Filaments and Septins

Helfand BT, Chang L & Goldman RD (2003) The dynamic and motile properties of intermediate filaments. *Annu. Rev. Cell Dev. Biol.* 19, 445–467.

Isermann P & Lammerding J (2013) Nuclear mechanics and mechanotransduction in health and disease. *Curr. Biol.* 23, R1113–R1121.

Saarikangas J & Barral Y (2011) The emerging functions of septins in metazoans. *EMBO Rep.* 12, 1118–1126.

Cell Polarization and Migration

Abercrombie M (1980) The crawling movement of metazoan cells. *Proc. R. Soc. Lond. B* 207, 129–147.

Gardel ML, Schneider IC, Aratyn-Schaus Y & Waterman CM (2010) Mechanical integration of actin and adhesion dynamics in cell migration. *Annu. Rev. Cell Dev. Biol.* 26, 315–333.

Lo CM, Wang HB, Dembo M & Wang YL (2000) Cell movement is guided by the rigidity of the substrate. *Biophys. J.* 79, 144–152.

Madden K & Snyder M (1998) Cell polarity and morphogenesis in budding yeast. *Annu. Rev. Microbiol.* 52, 687–744.

Parent CA & Devreotes PN (1999) A cell's sense of direction. *Science* 284, 765–770.

Pollard TD & Borisy GG (2003) Cellular motility driven by assembly and disassembly of actin filaments. *Cell* 112, 453–465.

Rafelski SM & Theriot JA (2004) Crawling toward a unified model of cell mobility: spatial and temporal regulation of actin dynamics. *Annu. Rev. Biochem.* 73, 209–239.

Ridley A (2011) Life at the leading edge. *Cell* 145, 1012–1022.

Vitriol EA & Zheng JQ (2012) Growth cone travel in space and time: the cellular ensemble of cytoskeleton, adhesion, and membrane. *Neuron* 73, 1068–1081.

Weiner OD (2002) Regulation of cell polarity during eukaryotic chemotaxis: the chemotactic compass. *Curr. Opin. Cell Biol.* 14, 196–202.

The Cell Cycle

The only way to make a new cell is to duplicate a cell that already exists. This simple fact, first established in the middle of the nineteenth century, carries with it a profound message for the continuity of life. All living organisms, from the unicellular bacterium to the multicellular mammal, are products of repeated rounds of cell growth and division extending back in time to the beginnings of life on Earth over three billion years ago.

A cell reproduces by performing an orderly sequence of events in which it duplicates its contents and then divides in two. This cycle of duplication and division, known as the **cell cycle**, is the essential mechanism by which all living things reproduce. In unicellular species, such as bacteria and yeasts, each cell division produces a complete new organism. In multicellular species, long and complex sequences of cell divisions are required to produce a functioning organism. Even in the adult body, cell division is usually needed to replace cells that die. In fact, each of us must manufacture many millions of cells every second simply to survive: if all cell division were stopped—by exposure to a very large dose of x-rays, for example—we would die within a few days.

The details of the cell cycle vary from organism to organism and at different times in an organism's life. Certain characteristics, however, are universal. At a minimum, the cell must accomplish its most fundamental task: the passing on of its genetic information to the next generation of cells. To produce two genetically identical daughter cells, the DNA in each chromosome must first be faithfully replicated to produce two complete copies. The replicated chromosomes must then be accurately distributed (*segregated*) to the two daughter cells, so that each receives a copy of the entire genome (**Figure 17–1**). In addition to duplicating their genome, most cells also duplicate their other organelles and macromolecules; otherwise, daughter cells would get smaller with each division. To maintain their size, dividing cells must coordinate their growth (that is, their increase in cell mass) with their division.

This chapter describes the events of the cell cycle and how they are controlled and coordinated. We begin with a brief overview of the cell cycle. We then describe the *cell-cycle control system*, a complex network of regulatory proteins that triggers the different events of the cycle. We next consider in detail the major stages of the cell cycle, in which the chromosomes are duplicated and then segregated into the two daughter cells. Finally, we consider how extracellular signals govern the rates of cell growth and division and how these two processes are coordinated.

OVERVIEW OF THE CELL CYCLE

The most basic function of the cell cycle is to duplicate the vast amount of DNA in the chromosomes and then segregate the copies into two genetically identical daughter cells. These processes define the two major phases of the cell cycle. Chromosome duplication occurs during *S phase* (S for DNA *synthesis*), which requires 10–12 hours and occupies about half of the cell-cycle time in a typical mammalian cell. After S phase, chromosome segregation and cell division occur in *M phase* (M for *mitosis*), which requires much less time (less than an hour in a mammalian cell). M phase comprises two major events: nuclear division, or *mitosis*, during

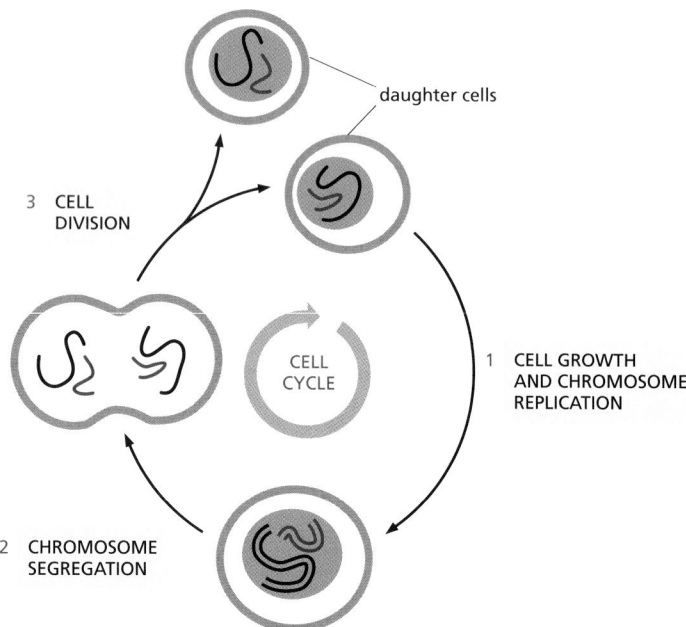

daughter cells

3 CELL
   DIVISION

CELL
CYCLE

1 CELL GROWTH
   AND CHROMOSOME
   REPLICATION

2 CHROMOSOME
   SEGREGATION

Figure 17–1 **The cell cycle.** The division of a hypothetical eukaryotic cell with two chromosomes (one *red*, and one *black*) is shown to illustrate how two genetically identical daughter cells are produced in each cycle. Each of the daughter cells will often continue to divide by going through additional cell cycles.

which the copied chromosomes are distributed into a pair of daughter nuclei; and cytoplasmic division, or *cytokinesis*, when the cell itself divides in two (**Figure 17–2**).

At the end of S phase, the DNA molecules in each pair of duplicated chromosomes are intertwined and held tightly together by specialized protein linkages. Early in mitosis at a stage called *prophase*, the two DNA molecules are gradually disentangled and condensed into pairs of rigid, compact rods called **sister chromatids**, which remain linked by *sister-chromatid cohesion*. When the nuclear envelope disassembles later in mitosis, the sister-chromatid pairs become attached to the *mitotic spindle*, a giant bipolar array of microtubules (discussed in Chapter 16). Sister chromatids are attached to opposite poles of the spindle and, eventually, align at the spindle equator in a stage called *metaphase*. The destruction of sister-chromatid cohesion at the start of *anaphase* separates the sister chromatids, which are pulled to opposite poles of the spindle. The spindle is then disassembled, and the segregated chromosomes are packaged into separate nuclei at *telophase*. Cytokinesis then cleaves the cell in two, so that each daughter cell inherits one of the two nuclei (**Figure 17–3**).

The Eukaryotic Cell Cycle Usually Consists of Four Phases

Most cells require much more time to grow and double their mass of proteins and organelles than they require to duplicate their chromosomes and divide. Partly to allow time for growth, most cell cycles have *gap phases*—a **G$_1$ phase** between M phase and S phase and a **G$_2$ phase** between S phase and mitosis. Thus, the eukaryotic cell cycle is traditionally divided into four sequential phases: G$_1$, S, G$_2$, and M. G$_1$, S, and G$_2$ together are called **interphase** (**Figure 17–4**, and see Figure 17–3). In a typical human cell proliferating in culture, interphase might occupy 23 hours of a 24-hour cycle, with 1 hour for M phase. Cell growth occurs throughout the cell cycle, except during mitosis.

The two gap phases are more than simple time delays to allow cell growth. They also provide time for the cell to monitor the internal and external environment

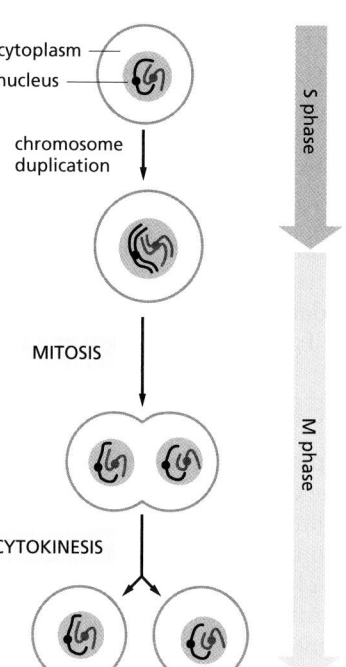

cytoplasm

nucleus

chromosome
duplication

S phase

MITOSIS

M phase

CYTOKINESIS

Figure 17–2 **The major events of the cell cycle.** The major chromosomal events of the cell cycle occur in S phase, when the chromosomes are duplicated, and M phase, when the duplicated chromosomes are segregated into a pair of daughter nuclei (in mitosis), after which the cell itself divides into two (cytokinesis).

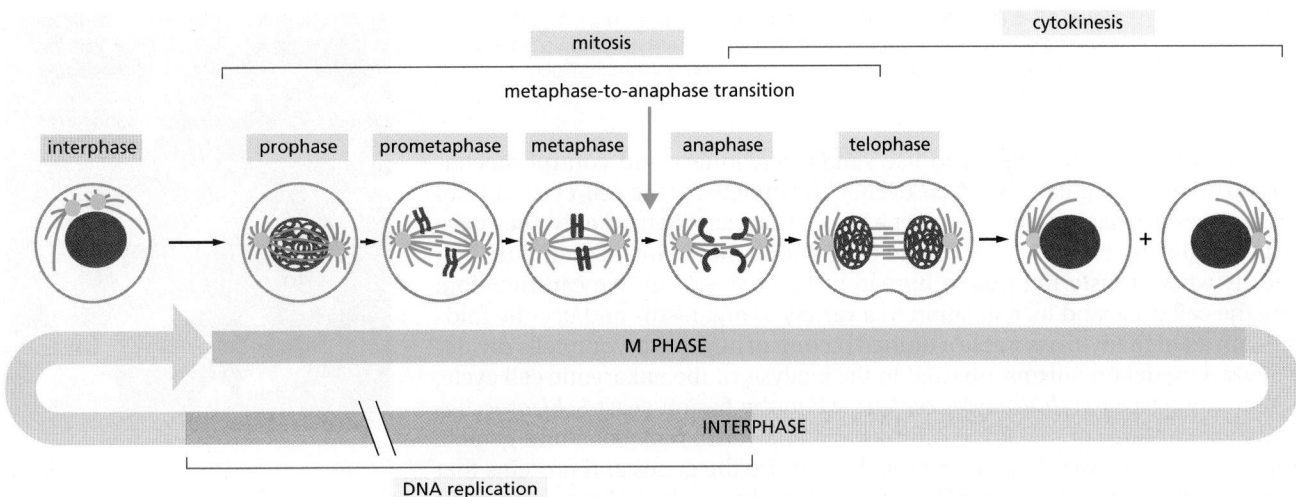

Figure 17–3 The events of eukaryotic cell division as seen under a microscope. The easily visible processes of nuclear division (mitosis) and cell division (cytokinesis), collectively called M phase, typically occupy only a small fraction of the cell cycle. The other, much longer, part of the cycle is known as interphase, which includes S phase and the gap phases (discussed in text). The five stages of mitosis are shown: an abrupt change in the biochemical state of the cell occurs at the transition from metaphase to anaphase. A cell can pause in metaphase before this transition point, but once it passes this point, the cell carries on to the end of mitosis and through cytokinesis into interphase.

to ensure that conditions are suitable and preparations are complete before the cell commits itself to the major upheavals of S phase and mitosis. The G_1 phase is especially important in this respect. Its length can vary greatly depending on external conditions and extracellular signals from other cells. If extracellular conditions are unfavorable, for example, cells delay progress through G_1 and may even enter a specialized resting state known as G_0 (G zero), in which they can remain for days, weeks, or even years before resuming proliferation. Indeed, many cells remain permanently in G_0 until they or the organism dies. If extracellular conditions are favorable and signals to grow and divide are present, cells in early G_1 or G_0 progress through a commitment point near the end of G_1 known as **Start** (in yeasts) or the **restriction point** (in mammalian cells). We will use the term Start for both yeast and animal cells. After passing this point, cells are committed to DNA replication, even if the extracellular signals that stimulate cell growth and division are removed.

Cell-Cycle Control Is Similar in All Eukaryotes

Some features of the cell cycle, including the time required to complete certain events, vary greatly from one cell type to another, even in the same organism. The basic organization of the cycle, however, is essentially the same in all eukaryotic

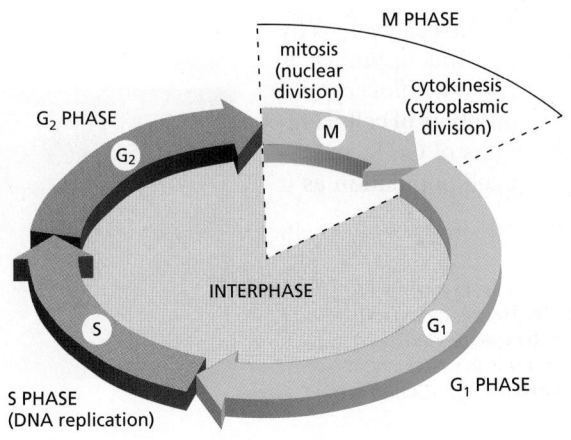

Figure 17–4 The four phases of the cell cycle. In most cells, gap phases separate the major events of S phase and M phase. G_1 is the gap between M phase and S phase, while G_2 is the gap between S phase and M phase.

Figure 17–5 **Mammalian cells proliferating in culture.** The cells in this scanning electron micrograph are rat fibroblasts. Cells at the lower left have rounded up and are in mitosis. (Courtesy of Guenter Albrecht-Buehler.)

10 μm

cells, and all eukaryotes appear to use similar machinery and control mechanisms to drive and regulate cell-cycle events. The proteins of the cell-cycle control system, for example, first appeared over a billion years ago. Remarkably, they have been so well conserved over the course of evolution that many of them function perfectly when transferred from a human cell to a yeast cell. We can therefore study the cell cycle and its regulation in a variety of organisms and use the findings from all of them to assemble a unified picture of how eukaryotic cells divide.

Several model organisms are used in the analysis of the eukaryotic cell cycle. The budding yeast *Saccharomyces cerevisiae* and the fission yeast *Schizosaccharomyces pombe* are simple eukaryotes in which powerful molecular and genetic approaches can be used to identify and characterize the genes and proteins that govern the fundamental features of cell division. The early embryos of certain animals, particularly those of the frog *Xenopus laevis*, are excellent tools for biochemical dissection of cell-cycle control mechanisms, while the fruit fly *Drosophila melanogaster* is useful for the genetic analysis of mechanisms underlying the control and coordination of cell growth and division in multicellular organisms. Cultured human cells provide an excellent system for the molecular and microscopic exploration of the complex processes by which our own cells divide.

Cell-Cycle Progression Can Be Studied in Various Ways

How can we tell what stage a cell has reached in the cell cycle? One way is simply to look at living cells with a microscope. A glance at a population of mammalian cells proliferating in culture reveals that a fraction of the cells have rounded up and are in mitosis (**Figure 17–5**). Others can be observed in the process of cytokinesis. Similarly, looking at budding yeast cells under a microscope is very useful, because the size of the bud provides an indication of cell-cycle stage (**Figure 17–6**). We can gain additional clues about cell-cycle position by staining cells with DNA-binding fluorescent dyes (which reveal the condensation of chromosomes in mitosis) or with antibodies that recognize specific cell components such as the microtubules (revealing the mitotic spindle). S-phase cells can be identified in the microscope by supplying them with visualizable molecules that are incorporated into newly synthesized DNA, such as the artificial thymidine analog bromodeoxyuridine (BrdU); cell nuclei that have incorporated BrdU are then revealed by staining with anti-BrdU antibodies (**Figure 17–7**).

Typically, in a population of cultured mammalian cells that are all proliferating rapidly but asynchronously, about 30–40% will be in S phase at any instant and become labeled by a brief pulse of BrdU. From the proportion of cells in such a population that are labeled, we can estimate the duration of S phase as a fraction of the whole cell-cycle duration. Similarly, from the proportion of cells in mitosis (the *mitotic index*), we can estimate the duration of M phase.

Another way to assess the stage that a cell has reached in the cell cycle is by measuring its DNA content, which doubles during S phase. This approach is greatly facilitated by the use of fluorescent DNA-binding dyes and a *flow cytometer*, which allows the rapid and automatic analysis of large numbers of cells (**Figure 17–8**). We can use flow cytometry to determine the lengths of G_1, S, and G_2 + M phases, by measuring DNA content in a synchronized cell population as it progresses through the cell cycle.

Figure 17–6 **The morphology of budding yeast cells.** In a normal population of proliferating yeast cells, buds vary in size according to the cell-cycle stage. Unbudded cells are in G_1. Progression through the Start transition triggers formation of a tiny bud, which grows in size during the S and M phases until it is almost the size of the mother cell. (Courtesy of Jeff Ubersax.)

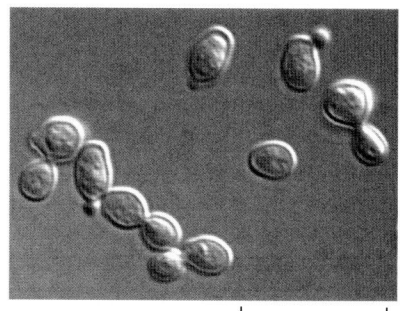

20 μm

Figure 17–7 **Labeling S-phase cells.** An immunofluorescence micrograph of BrdU-labeled epithelial cells of the zebrafish gut. The fish was exposed to BrdU, after which the tissue was fixed and prepared for labeling with fluorescent anti-BrdU antibodies *(green)*. All the cells are stained with a *red* fluorescent dye. (Courtesy of Cécile Crosnier.)

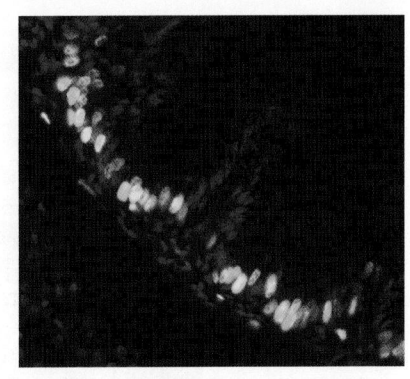

Summary

Cell division usually begins with duplication of the cell's contents, followed by distribution of those contents into two daughter cells. Chromosome duplication occurs during S phase of the cell cycle, whereas most other cell components are duplicated continuously throughout the cycle. During M phase, the replicated chromosomes are segregated into individual nuclei (mitosis), and the cell then splits in two (cytokinesis). S phase and M phase are usually separated by gap phases called G_1 and G_2, when various intracellular and extracellular signals regulate cell-cycle progression. Cell-cycle organization and control have been highly conserved during evolution, and studies in a wide range of systems have led to a unified view of eukaryotic cell-cycle control.

THE CELL-CYCLE CONTROL SYSTEM

For many years, cell biologists watched the puppet show of DNA synthesis, mitosis, and cytokinesis but had no idea of what lay behind the curtain controlling these events. It was not even clear whether there was a separate control system, or whether the processes of DNA synthesis, mitosis, and cytokinesis somehow controlled themselves. A major breakthrough came in the late 1980s with the identification of the key proteins of the control system, along with the realization that they are distinct from the proteins that perform the processes of DNA replication, chromosome segregation, and so on.

In this section, we first consider the basic principles upon which the cell-cycle control system operates. We then discuss the protein components of the system and how they work together to time and coordinate the events of the cell cycle.

The Cell-Cycle Control System Triggers the Major Events of the Cell Cycle

The **cell-cycle control system** operates much like a timer that triggers the events of the cell cycle in a set sequence (**Figure 17–9**). In its simplest form—as seen in the stripped-down cell cycles of early animal embryos, for example—the control system is rigidly programmed to provide a fixed amount of time for the completion of each cell-cycle event. The control system in these early embryonic divisions is independent of the events it controls, so that its timing mechanisms continue to operate even if those events fail. In most cells, however, the control system does respond to information received back from the processes it controls. If some malfunction prevents the successful completion of DNA synthesis, for example, signals are sent to the control system to delay progression to M phase. Such delays provide time for the machinery to be repaired and also prevent the disaster that might result if the cycle progressed prematurely to the next stage—and segregated incompletely replicated chromosomes, for example.

The cell-cycle control system is based on a connected series of biochemical switches, each of which initiates a specific cell-cycle event. This system of switches possesses many important features that increase the accuracy and reliability of cell-cycle progression. First, the switches are generally *binary* (on/off) and launch events in a complete, irreversible fashion. It would clearly be disastrous, for example, if events like chromosome condensation or nuclear-envelope breakdown were only partially initiated or started but not completed. Second, the cell-cycle control system is remarkably robust and reliable, partly because backup mechanisms and other features allow the system to operate effectively under a variety of conditions and even if some components fail. Finally, the control system is highly adaptable and can be modified to suit specific cell types or to respond to specific intracellular or extracellular signals.

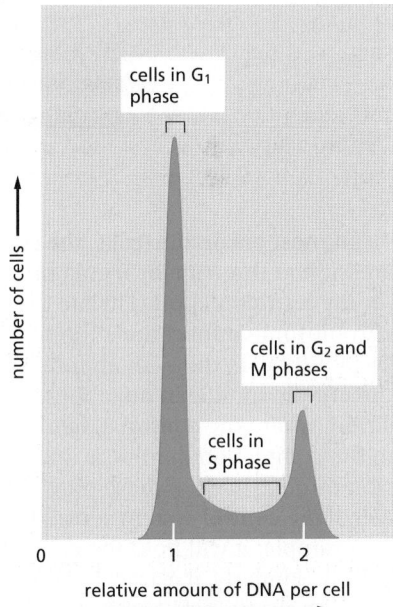

cells in G_1 phase

cells in G_2 and M phases

cells in S phase

number of cells

0   1   2

relative amount of DNA per cell

(arbitrary units)

Figure 17–8 **Analysis of DNA content with a flow cytometer.** This graph shows typical results obtained for a proliferating cell population when the DNA content of its individual cells is determined in a flow cytometer. (A flow cytometer, also called a fluorescence-activated cell sorter, or FACS, can also be used to sort cells according to their fluorescence—see Figure 8–2). The cells analyzed here were stained with a dye that becomes fluorescent when it binds to DNA, so that the amount of fluorescence is directly proportional to the amount of DNA in each cell. The cells fall into three categories: those that have an unreplicated complement of DNA and are therefore in G_1, those that have a fully replicated complement of DNA (twice the G_1 DNA content) and are in G_2 or M phase, and those that have an intermediate amount of DNA and are in S phase. The distribution of cells indicates that there are greater numbers of cells in G_1 than in G_2 + M phase, showing that G_1 is longer than G_2 + M in this population.

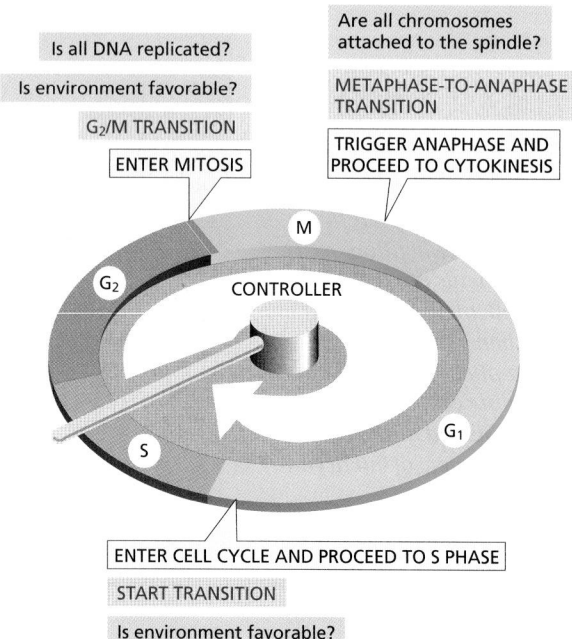

Figure 17–9 **The control of the cell cycle.** A cell-cycle control system triggers the essential processes of the cycle—such as DNA replication, mitosis, and cytokinesis. The control system is represented here as a central arm—the controller—that rotates clockwise, triggering essential processes when it reaches specific transitions on the outer dial *(yellow boxes)*. Information about the completion of cell-cycle events, as well as signals from the environment, can cause the control system to arrest the cycle at these transitions.

In most eukaryotic cells, the cell-cycle control system governs cell-cycle progression at three major regulatory transitions (see Figure 17–9). The first is Start (or the restriction point) in late G_1, where the cell commits to cell-cycle entry and chromosome duplication. The second is the **G_2/M transition**, where the control system triggers the early mitotic events that lead to chromosome alignment on the mitotic spindle in metaphase. The third is the **metaphase-to-anaphase transition**, where the control system stimulates sister-chromatid separation, leading to the completion of mitosis and cytokinesis. The control system blocks progression through each of these transitions if it detects problems inside or outside the cell. If the control system senses problems in the completion of DNA replication, for example, it will hold the cell at the G_2/M transition until those problems are solved. Similarly, if extracellular conditions are not appropriate for cell proliferation, the control system blocks progression through Start, thereby preventing cell division until conditions become favorable.

The Cell-Cycle Control System Depends on Cyclically Activated Cyclin-Dependent Protein Kinases (Cdks)

Central components of the cell-cycle control system are members of a family of protein kinases known as **cyclin-dependent kinases (Cdks)**. The activities of these kinases rise and fall as the cell progresses through the cycle, leading to cyclical changes in the phosphorylation of intracellular proteins that initiate or regulate the major events of the cell cycle. An increase in Cdk activity at the G_2/M transition, for example, increases the phosphorylation of proteins that control chromosome condensation, nuclear-envelope breakdown, spindle assembly, and other events that occur in early mitosis.

Cyclical changes in Cdk activity are controlled by a complex array of enzymes and other proteins. The most important of these Cdk regulators are proteins known as **cyclins**. Cdks, as their name implies, are dependent on cyclins for their activity: unless they are bound tightly to a cyclin, they have no protein kinase activity (**Figure 17–10**). Cyclins were originally named because they undergo a cycle of synthesis and degradation in each cell cycle. The levels of the Cdk proteins, by contrast, are constant. Cyclical changes in cyclin protein levels result in the cyclic assembly and activation of **cyclin–Cdk complexes** at specific stages of the cell cycle.

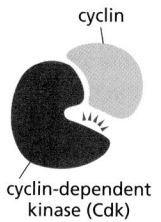

Figure 17–10 **Two key components of the cell-cycle control system.** When cyclin forms a complex with Cdk, the protein kinase is activated to trigger specific cell-cycle events. Without cyclin, Cdk is inactive.

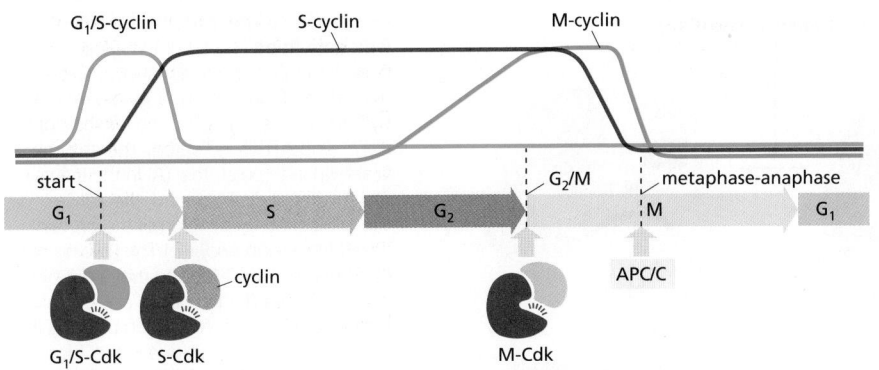

Figure 17–11 **Cyclin–Cdk complexes of the cell-cycle control system.** The concentrations of the three major cyclin types oscillate during the cell cycle, while the concentrations of Cdks (not shown) do not change and exceed cyclin amounts. In late G_1, rising G_1/S-cyclin levels lead to the formation of G_1-S-Cdk complexes that trigger progression through the Start transition. S-Cdk complexes form at the start of S phase and trigger DNA replication, as well as some early mitotic events. M-Cdk complexes form during G_2 but are held in an inactive state; they are activated at the end of G_2 and trigger entry into mitosis at the G_2/M transition. A separate regulatory protein complex, the APC/C, initiates the metaphase-to-anaphase transition, as we discuss later.

There are four classes of cyclins, each defined by the stage of the cell cycle at which they bind Cdks and function. All eukaryotic cells require three of these classes (**Figure 17–11**):

1. **G_1/S-cyclins** activate Cdks in late G_1 and thereby help trigger progression through Start, resulting in a commitment to cell-cycle entry. Their levels fall in S phase.

2. **S-cyclins** bind Cdks soon after progression through Start and help stimulate chromosome duplication. S-cyclin levels remain elevated until mitosis, and these cyclins also contribute to the control of some early mitotic events.

3. **M-cyclins** activate Cdks that stimulate entry into mitosis at the G_2/M transition. M-cyclin levels fall in mid-mitosis.

In most cells, a fourth class of cyclins, the **G_1-cyclins**, helps govern the activities of the G_1/S-cyclins, which control progression through Start in late G_1.

In yeast cells, a single Cdk protein binds all classes of cyclins and triggers different cell-cycle events by changing cyclin partners at different stages of the cycle. In vertebrate cells, by contrast, there are four Cdks. Two interact with G_1-cyclins, one with G_1/S- and S-cyclins, and one with S- and M-cyclins. In this chapter, we simply refer to the different cyclin–Cdk complexes as **G_1-Cdk, G_1/S-Cdk, S-Cdk,** and **M-Cdk. Table 17–1** lists the names of the individual Cdks and cyclins.

How do different cyclin–Cdk complexes trigger different cell-cycle events? The answer, at least in part, seems to be that the cyclin protein does not simply activate its Cdk partner but also directs it to specific target proteins. As a result, each cyclin–Cdk complex phosphorylates a different set of substrate proteins. The same cyclin–Cdk complex can also induce different effects at different times in the cycle, probably because the accessibility of some Cdk substrates changes during the cell cycle. Certain proteins that function in mitosis, for example, may become available for phosphorylation only in G_2.

TABLE 17–1 The Major Cyclins and Cdks of Vertebrates and Budding Yeast				
	Vertebrates		Budding yeast	
Cyclin–Cdk complex	Cyclin	Cdk partner	Cyclin	Cdk partner
G_1-Cdk	Cyclin D*	Cdk4, Cdk6	Cln3	Cdk1**
G_1/S-Cdk	Cyclin E	Cdk2	Cln1, 2	Cdk1
S-Cdk	Cyclin A	Cdk2, Cdk1**	Clb5, 6	Cdk1
M-Cdk	Cyclin B	Cdk1	Clb1, 2, 3, 4	Cdk1

* There are three D cyclins in mammals (cyclins D1, D2, and D3).
** The original name of Cdk1 was Cdc2 in both vertebrates and fission yeast, and Cdc28 in budding yeast.

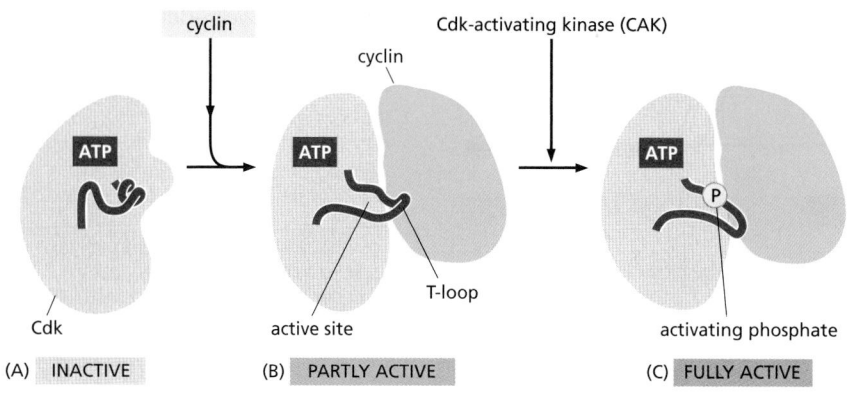

(A) INACTIVE   (B) PARTLY ACTIVE   (C) FULLY ACTIVE

Figure 17–12 **The structural basis of Cdk activation.** These drawings are based on three-dimensional structures of human Cdk2 and cyclin A, as determined by x-ray crystallography. The location of the bound ATP is indicated. The enzyme is shown in three states. (A) In the inactive state, without cyclin bound, the active site is blocked by a region of the protein called the T-loop *(red)*. (B) The binding of cyclin causes the T-loop to move out of the active site, resulting in partial activation of the Cdk2. (C) Phosphorylation of Cdk2 (by CAK) at a threonine residue in the T-loop further activates the enzyme by changing the shape of the T-loop, improving the ability of the enzyme to bind its protein substrates (**Movie 17.1**).

Studies of the three-dimensional structures of Cdk and cyclin proteins have revealed that, in the absence of cyclin, the active site in the Cdk protein is partly obscured by a protein loop, like a stone blocking the entrance to a cave (**Figure 17–12**A). Cyclin binding causes the loop to move away from the active site, resulting in partial activation of the Cdk enzyme (Figure 17–12B). Full activation of the cyclin–Cdk complex then occurs when a separate kinase, the **Cdk-activating kinase** (**CAK**), phosphorylates an amino acid near the entrance of the Cdk active site. This causes a small conformational change that further increases the activity of the Cdk, allowing the kinase to phosphorylate its target proteins effectively and thereby induce specific cell-cycle events (Figure 17–12C).

Cdk Activity Can Be Suppressed By Inhibitory Phosphorylation and Cdk Inhibitor Proteins (CKIs)

The rise and fall of cyclin levels is the primary determinant of Cdk activity during the cell cycle. Several additional mechanisms, however, help control Cdk activity at specific stages of the cycle.

Phosphorylation at a pair of amino acids in the roof of the kinase active site inhibits the activity of a cyclin–Cdk complex. Phosphorylation of these sites by a protein kinase known as **Wee1** inhibits Cdk activity, while dephosphorylation of these sites by a phosphatase known as **Cdc25** increases Cdk activity (**Figure 17–13**). We will see later that this regulatory mechanism is particularly important in the control of M-Cdk activity at the onset of mitosis.

Binding of **Cdk inhibitor proteins** (**CKIs**) inactivates cyclin–Cdk complexes. The three-dimensional structure of a cyclin–Cdk–CKI complex reveals that CKI binding stimulates a large rearrangement in the structure of the Cdk active site, rendering it inactive (**Figure 17–14**). Cells use CKIs primarily to help govern the activities of G_1/S- and S-Cdks early in the cell cycle.

Regulated Proteolysis Triggers the Metaphase-to-Anaphase Transition

Whereas activation of specific cyclin–Cdk complexes drives progression through the Start and G_2/M transitions (see Figure 17–11), progression through the metaphase-to-anaphase transition is triggered not by protein phosphorylation but by protein destruction, leading to the final stages of cell division.

The key regulator of the metaphase-to-anaphase transition is the **anaphase-promoting complex**, or **cyclosome** (**APC/C**), a member of the ubiquitin ligase family of enzymes. As discussed in Chapter 3, these enzymes are used in numerous cell processes to stimulate the proteolytic destruction of specific regulatory proteins. They polyubiquitylate specific target proteins, resulting in their destruction in proteasomes. Other ubiquitin ligases mark proteins for purposes other than destruction (discussed in Chapter 3).

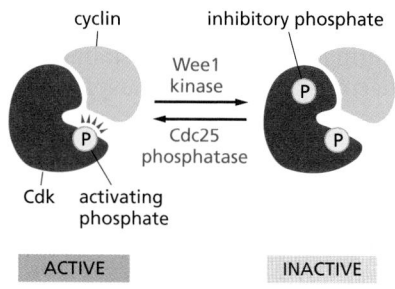

ACTIVE   INACTIVE

Figure 17–13 **The regulation of Cdk activity by phosphorylation.** The active cyclin–Cdk complex is turned off when the kinase Wee1 phosphorylates two closely spaced sites above the active site. Removal of these phosphates by the phosphatase Cdc25 activates the cyclin–Cdk complex. For simplicity, only one inhibitory phosphate is shown. CAK adds the activating phosphate, as shown in Figure 17–12.

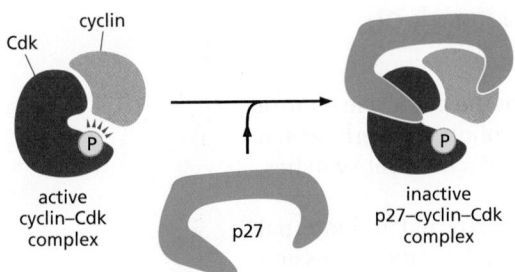

Figure 17–14 The inhibition of a cyclin–Cdk complex by a CKI. This drawing is based on the three-dimensional structure of the human cyclin A–Cdk2 complex bound to the CKI p27, as determined by x-ray crystallography. The p27 binds to both the cyclin and Cdk in the complex, distorting the active site of the Cdk. It also inserts into the ATP-binding site, further inhibiting the enzyme activity.

The APC/C catalyzes the ubiquitylation and destruction of two major types of proteins. The first is *securin*, which protects the protein linkages that hold sister-chromatid pairs together in early mitosis. Destruction of securin in metaphase activates a protease that separates the sisters and unleashes anaphase, as described later. The S- and M-cyclins are the second major targets of the APC/C. Destroying these cyclins inactivates most Cdks in the cell (see Figure 17–11). As a result, the many proteins phosphorylated by Cdks from S phase to early mitosis are dephosphorylated by various phosphatases in the anaphase cell. This dephosphorylation of Cdk targets is required for the completion of M phase, including the final steps in mitosis and then cytokinesis. Following its activation in mid-mitosis, the APC/C remains active in G_1 to provide a stable period of Cdk inactivity. When G_1/S-Cdk is activated in late G_1, the APC/C is turned off, thereby allowing cyclin accumulation in the next cell cycle.

The cell-cycle control system also uses another ubiquitin ligase called **SCF** (see Figure 3–71). It has many functions in the cell, but its major role in the cell cycle is to ubiquitylate certain CKI proteins in late G_1, thereby helping to control the activation of S-Cdks and DNA replication. SCF is also responsible for the destruction of G_1/S-cyclins in early S phase.

The APC/C and SCF are both large, multisubunit complexes with some related components (see Figure 3–71), but they are regulated differently. APC/C activity changes during the cell cycle, primarily as a result of changes in its association with an activating subunit—either **Cdc20** in mid-mitosis or **Cdh1** from late mitosis through early G_1. These subunits help the APC/C recognize its target proteins (**Figure 17–15**A). SCF activity depends on substrate-binding subunits called F-box proteins. Unlike APC/C activity, however, SCF activity is constant during the cell cycle. Ubiquitylation by SCF is controlled instead by changes in the phosphorylation state of its target proteins, as F-box subunits recognize only specifically phosphorylated proteins (Figure 17–15B).

Cell-Cycle Control Also Depends on Transcriptional Regulation

In the simple cell cycles of early animal embryos, gene transcription does not occur. Cell-cycle control depends exclusively on post-transcriptional mechanisms that involve the regulation of Cdks and ubiquitin ligases and their target proteins. In the more complex cell cycles of most cell types, however, transcriptional control provides an important additional level of regulation. Changes in cyclin gene transcription, for example, help control cyclin levels in most cells.

A variety of methods discussed in Chapter 8 have been used to analyze changes in the expression of all genes in the genome as the cell progresses through the cell cycle. The results of these studies are surprising. In budding yeast, for example, about 10% of the genes encode mRNAs whose levels oscillate during the cell cycle. Some of these genes encode proteins with known cell-cycle functions, but the functions of many others are unknown.

The Cell-Cycle Control System Functions as a Network of Biochemical Switches

Table 17–2 summarizes some of the major components of the cell-cycle control system. These proteins are functionally linked to form a robust network, which operates essentially autonomously to activate a series of biochemical switches, each of which triggers a specific cell-cycle event.

When conditions for cell proliferation are right, various external and internal signals stimulate the activation of G_1-Cdk, which in turn stimulates the expression of genes encoding G_1/S- and S-cyclins (**Figure 17–16**). The resulting activation of G_1/S-Cdk then drives progression through the Start transition. By mechanisms we discuss later, G_1/S-Cdks unleash a wave of S-Cdk activity, which initiates

(A) control of proteolysis by APC/C

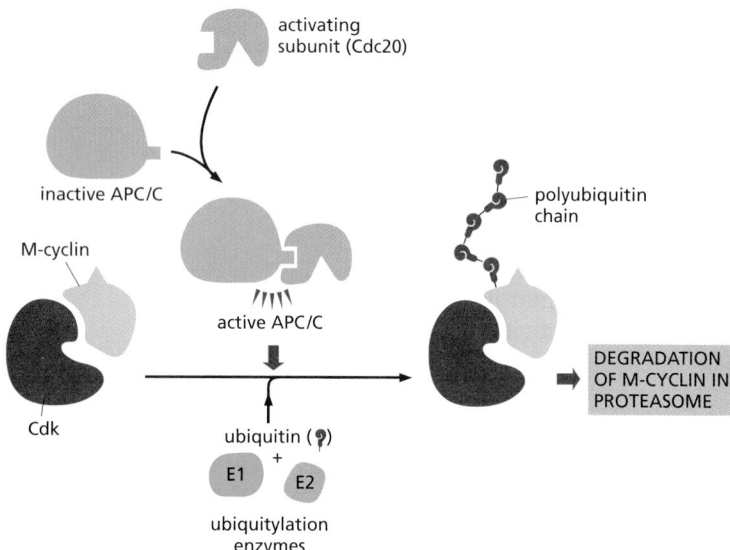

(B) control of proteolysis by SCF

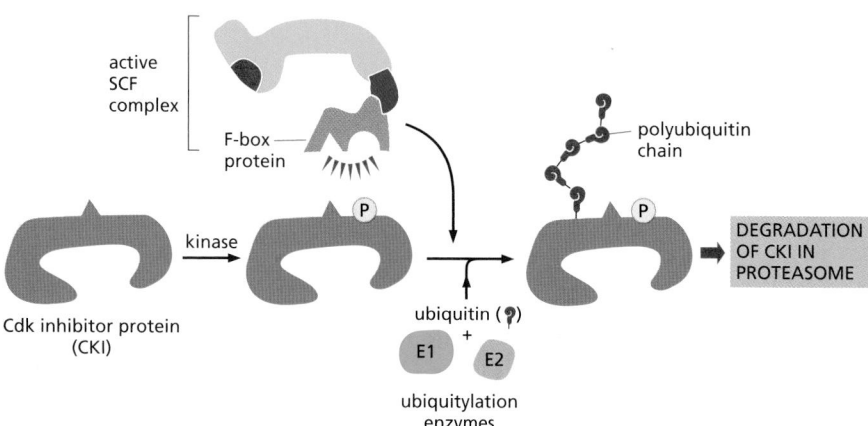

Figure 17–15 The control of proteolysis by APC/C and SCF during the cell cycle. (A) The APC/C is activated in mitosis by association with Cdc20, which recognizes specific amino acid sequences on M-cyclin and other target proteins. With the help of two additional proteins called E1 and E2, the APC/C assembles polyubiquitin chains on the target protein. The polyubiquitylated target is then recognized and degraded in a proteasome. (B) The activity of the ubiquitin ligase SCF depends on substrate-binding subunits called F-box proteins, of which there are many different types. The phosphorylation of a target protein, such as the CKI shown, allows the target to be recognized by a specific F-box subunit.

TABLE 17–2 Summary of the Major Cell Cycle Regulatory Proteins	
General name	Functions and comments
Protein kinases and protein phosphatases that modify Cdks	
Cdk-activating kinase (CAK)	Phosphorylates an activating site in Cdks
Wee1 kinase	Phosphorylates inhibitory sites in Cdks; primarily involved in suppressing Cdk1 activity before mitosis
Cdc25 phosphatase	Removes inhibitory phosphates from Cdks; three family members (Cdc25A, B, C) in mammals; primarily involved in controlling Cdk1 activation at the onset of mitosis
Cdk inhibitor proteins (CKIs)	
Sic1 (budding yeast)	Suppresses Cdk1 activity in G_1; phosphorylation by Cdk1 at the end of G_1 triggers its destruction
p27 (mammals)	Suppresses G_1/S-Cdk and S-Cdk activities in G_1; helps cells withdraw from cell cycle when they terminally differentiate; phosphorylation by Cdk2 triggers its ubiquitylation by SCF
p21 (mammals)	Suppresses G_1/S-Cdk and S-Cdk activities following DNA damage
p16 (mammals)	Suppresses G_1-Cdk activity in G_1; frequently inactivated in cancer
Ubiquitin ligases and their activators	
APC/C	Catalyzes ubiquitylation of regulatory proteins involved primarily in exit from mitosis, including securin and S- and M-cyclins; regulated by association with activating subunits Cdc20 or Cdh1
Cdc20	APC/C-activating subunit in all cells; triggers initial activation of APC/C at metaphase-to-anaphase transition; stimulated by M-Cdk activity
Cdh1	APC/C-activating subunit that maintains APC/C activity after anaphase and throughout G_1; inhibited by Cdk activity
SCF	Catalyzes ubiquitylation of regulatory proteins involved in G_1 control, including some CKIs (Sic1 in budding yeast, p27 in mammals); phosphorylation of target protein usually required for this activity

chromosome duplication in S phase and also contributes to some early events of mitosis. M-Cdk activation then triggers progression through the G_2/M transition and the events of early mitosis, leading to the alignment of sister-chromatid pairs at the equator of the mitotic spindle. Finally, the APC/C, together with its activator Cdc20, triggers the destruction of securin and cyclins, thereby unleashing sister-chromatid separation and segregation and the completion of mitosis. When mitosis is complete, multiple mechanisms collaborate to suppress Cdk activity, resulting in a stable G_1 period. We are now ready to discuss these cell-cycle stages in more detail, starting with S phase.

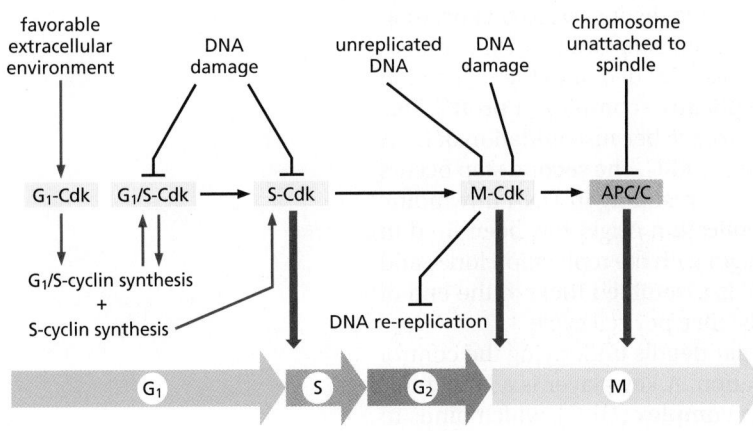

Figure 17–16 **An overview of the cell-cycle control system.** The core of the cell-cycle control system consists of a series of cyclin–Cdk complexes *(yellow)*. The activity of each complex is also influenced by various inhibitory mechanisms, which provide information about the extracellular environment, cell damage, and incomplete cell-cycle events *(top)*. These inhibitory mechanisms are not present in all cell types; many are missing in early embryonic cell cycles, for example.

Summary

The cell-cycle control system triggers the events of the cell cycle and ensures that they are properly timed and coordinated with each other. The control system responds to various intracellular and extracellular signals and arrests the cycle when the cell either fails to complete an essential cell-cycle process or encounters unfavorable environmental or intracellular conditions.

Central components of the control system are the cyclin-dependent protein kinases (Cdks), which depend on cyclin subunits for their activity. Oscillations in the activities of different cyclin–Cdk complexes control various cell-cycle events. Thus, activation of S-phase cyclin–Cdk complexes (S-Cdk) initiates S phase, whereas activation of M-phase cyclin–Cdk complexes (M-Cdk) triggers mitosis. The mechanisms that control the activities of cyclin–Cdk complexes include phosphorylation of the Cdk subunit, binding of Cdk inhibitor proteins (CKIs), proteolysis of cyclins, and changes in the transcription of genes encoding Cdk regulators. The cell-cycle control system also depends crucially on two additional enzyme complexes, the APC/C and SCF ubiquitin ligases, which catalyze the ubiquitylation and consequent destruction of specific regulatory proteins that control critical events in the cycle.

S PHASE

The linear chromosomes of eukaryotic cells are vast and dynamic assemblies of DNA and protein, and their duplication is a complex process that takes up a major fraction of the cell cycle. Not only must the long DNA molecule of each chromosome be duplicated accurately—a remarkable feat in itself—but the protein packaging surrounding each region of that DNA must also be reproduced, ensuring that the daughter cells inherit all features of chromosome structure.

The central event of chromosome duplication—DNA replication—poses two problems for the cell. First, replication must occur with extreme accuracy to minimize the risk of mutations in the next cell generation. Second, every nucleotide in the genome must be copied once, and only once, to prevent the damaging effects of gene amplification. In Chapter 5, we discuss the sophisticated protein machinery that performs DNA replication with astonishing speed and accuracy. In this section, we consider the elegant mechanisms by which the cell-cycle control system initiates the replication process and, at the same time, prevents it from happening more than once per cycle.

S-Cdk Initiates DNA Replication Once Per Cycle

DNA replication begins at *origins of replication*, which are scattered at numerous locations in every chromosome. During S phase, DNA replication is initiated at these origins when a *DNA helicase* unwinds the double helix and DNA replication enzymes are loaded onto the two single-stranded templates. This leads to the *elongation* phase of replication, when the replication machinery moves outward from the origin at two *replication forks* (discussed in Chapter 5).

To ensure that chromosome duplication occurs only once per cell cycle, the initiation phase of DNA replication is divided into two distinct steps that occur at different times in the cell cycle (**Figure 17–17**). The first step occurs in late mitosis and early G_1, when a pair of inactive DNA helicases is loaded onto the replication origin, forming a large complex called the **prereplicative complex** or **preRC**. This step is sometimes called *licensing* of replication origins because initiation of DNA synthesis is permitted only at origins containing a preRC. The second step occurs in S phase, when the DNA helicases are activated, resulting in DNA unwinding and the initiation of DNA synthesis. Once a replication origin has been fired in this way, the two helicases move out from the origin with the replication forks, and that origin cannot be reused until a new preRC is assembled there at the end of mitosis. As a result, origins can be activated only once per cell cycle.

Figure 17–18 illustrates some of the molecular details underlying the control of the two steps in the initiation of DNA replication. A key player is a large multiprotein complex called the **origin recognition complex** (**ORC**), which binds to

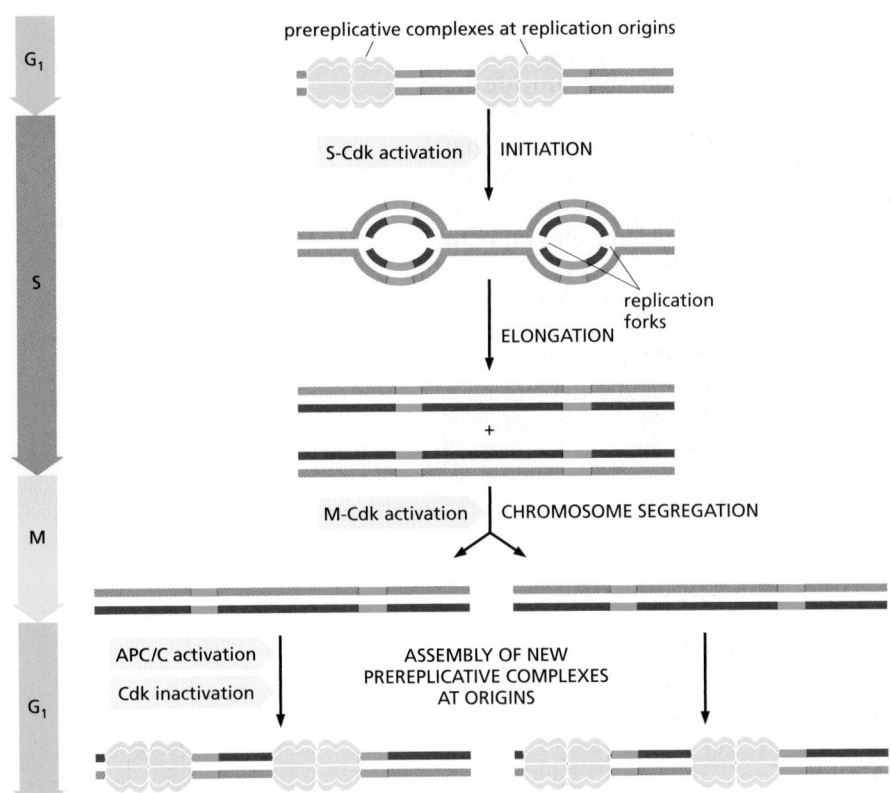

G₁

S

M

G₁

prereplicative complexes at replication origins

S-Cdk activation   INITIATION

ELONGATION

replication forks

+

M-Cdk activation   CHROMOSOME SEGREGATION

APC/C activation

Cdk inactivation

ASSEMBLY OF NEW
PREREPLICATIVE COMPLEXES
AT ORIGINS

Figure 17–17 **Control of chromosome duplication.** Preparations for DNA replication begin in late mitosis and G₁, when the DNA helicases are loaded by multiple proteins at the replication origin, forming the prereplicative complex (preRC). S-Cdk activation leads to activation of the DNA helicases, which unwind the DNA at origins to initiate DNA replication. Two replication forks move out from each origin until the entire chromosome is duplicated. Duplicated chromosomes are then segregated in M phase. S-Cdk activation in S phase also prevents assembly of new preRCs at any origin until the following G₁—thereby ensuring that each origin is activated only once in each cell cycle.

replication origins throughout the cell cycle. In late mitosis and early G₁, the proteins **Cdc6** and **Cdt1** collaborate with the ORC to load the inactive DNA helicases around the DNA next to the origin. The resulting large complex is the preRC, and the origin is now licensed for replication.

At the onset of S phase, S-Cdk triggers origin activation by phosphorylating specific initiator proteins, which then nucleate the assembly of a large protein complex that activates the DNA helicase and recruits the DNA synthesis machinery. Another protein kinase called DDK is also activated in S phase and helps drive origin activation by phosphorylating specific subunits of the DNA helicase.

At the same time as S-Cdk initiates DNA replication, several mechanisms prevent assembly of new preRCs. S-Cdk phosphorylates and thereby inhibits the ORC and Cdc6 proteins. Inactivation of the APC/C in late G₁ also helps turn off preRC assembly. In late mitosis and early G₁, the APC/C triggers the destruction of a Cdt1 inhibitor called **geminin**, thereby allowing Cdt1 to be active. When the APC/C is turned off in late G₁, geminin accumulates and inhibits the Cdt1 that is not associated with DNA. Also, the association of Cdt1 with a protein at active replication forks stimulates Cdt1 destruction. In these various ways, preRC formation is prevented from S phase to mitosis, thereby ensuring that each origin is fired only once per cell cycle. How, then, is the cell-cycle control system reset to allow replication in the next cell cycle? At the end of mitosis, APC/C activation leads to the inactivation of Cdks and the destruction of geminin. ORC and Cdc6 are dephosphorylated and Cdt1 is activated, allowing preRC assembly to prepare the cell for the next S phase.

Chromosome Duplication Requires Duplication of Chromatin Structure

The DNA of the chromosomes is extensively packaged in a variety of protein components, including histones and various regulatory proteins involved in the control of gene expression (discussed in Chapter 4). Thus, duplication of a chromosome is not simply a matter of replicating the DNA at its core but also requires the duplication of these chromatin proteins and their proper assembly on the DNA.

The production of chromatin proteins increases during S phase to provide the raw materials needed to package the newly synthesized DNA. Most importantly, S-Cdks stimulate a large increase in the synthesis of the four histone subunits that form the histone octamers at the core of each nucleosome. These subunits are assembled into nucleosomes on the DNA by nucleosome assembly factors, which typically associate with the replication fork and distribute nucleosomes on both strands of the DNA as they emerge from the DNA synthesis machinery.

Chromatin packaging helps to control gene expression. In some parts of the chromosome, the chromatin is highly condensed and is called *heterochromatin*, whereas in other regions it has a more open structure and is called *euchromatin* (discussed in Chapter 4). These differences in chromatin structure depend on a variety of mechanisms, including modification of histone tails and the presence of non-histone proteins. Because these differences are important in gene regulation,

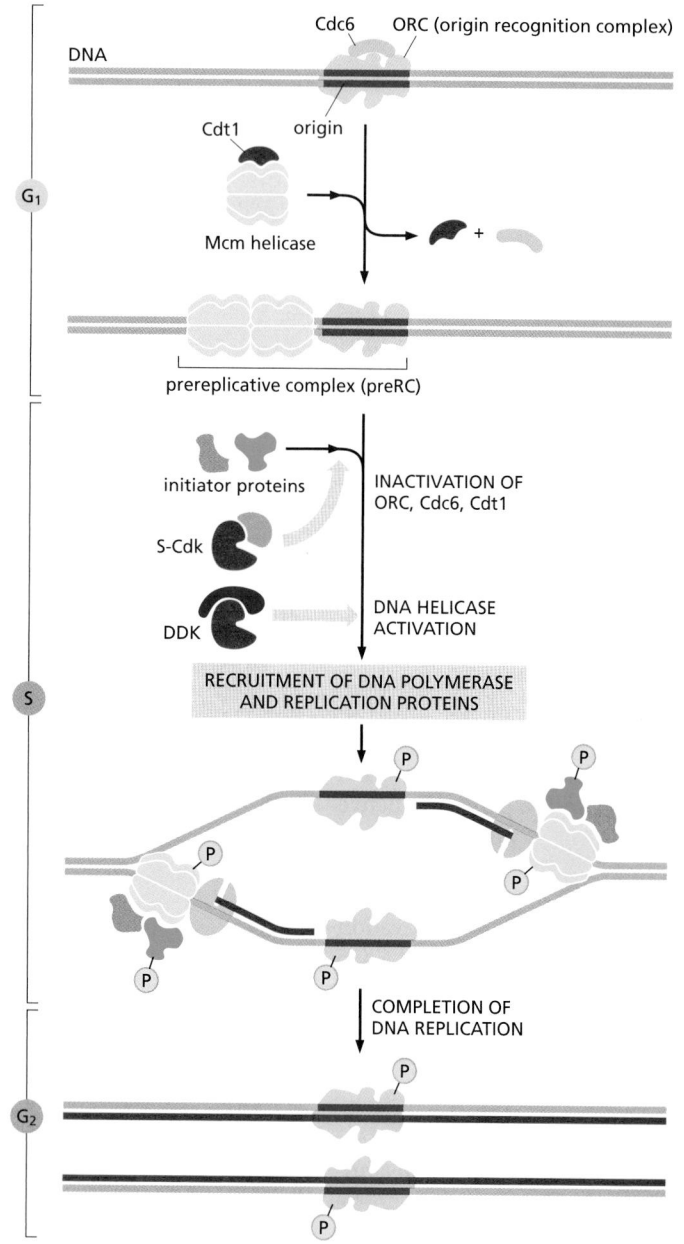

Figure 17–18 **Control of the initiation of DNA replication.** The replication origin is bound by the ORC throughout the cell cycle. In early G_1, Cdc6 associates with the ORC, and these proteins bind the DNA helicase, which contains six closely related subunits called Mcm proteins. The helicase also associates with a protein called Cdt1. Using energy provided by ATP hydrolysis, the ORC and Cdc6 proteins load two copies of the DNA helicase, in an inactive form, around the DNA next to the origin, thereby forming the prereplicative complex (preRC). At the onset of S phase, S-Cdk stimulates the assembly of several initiator proteins on each DNA helicase, while another protein kinase, DDK, phosphorylates subunits of the DNA helicase. As a result, the DNA helicases are activated and unwind the DNA. DNA polymerase and other replication proteins are recruited to the origin, and DNA replication begins. The ORC is displaced by the replication machinery and then rebinds. S-Cdk and other mechanisms also inactivate the preRC components ORC, Cdc6, and Cdt1, thereby preventing formation of new preRCs at the origins until the end of mitosis (see text).

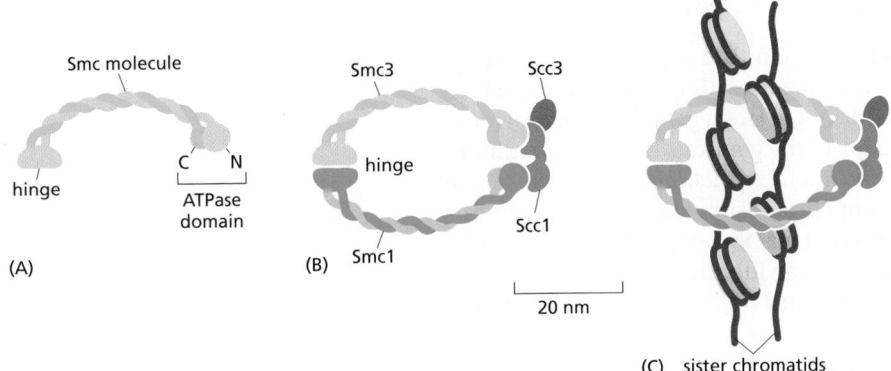

Figure 17–19 **Cohesin.** Cohesin is a protein complex with four subunits. (A) Two subunits, Smc1 and Smc3, are coiled-coil proteins with an ATPase domain at one end; (B) two additional subunits, Scc1 and Scc3, connect the ATPase head domains, forming a ring structure that may encircle the sister chromatids as shown in (C). The ATPase domains are required for cohesin loading on the DNA.

it is crucial that chromatin structure, like the DNA within, is reproduced accurately during S phase. How chromatin structure is reproduced is not well understood, however. During DNA synthesis, histone-modifying enzymes and various non-histone proteins are probably deposited onto the two new DNA strands as they emerge from the replication fork, and these proteins are thought to reproduce the local chromatin structure of the parent chromosome (see Figure 4–45).

Cohesins Hold Sister Chromatids Together

At the end of S phase, each replicated chromosome consists of a pair of identical sister chromatids glued together along their length. This sister-chromatid cohesion sets the stage for a successful mitosis because it greatly facilitates the attachment of the two sister chromatids to opposite poles of the mitotic spindle. Imagine how difficult it would be to achieve this bipolar attachment if sister chromatids were allowed to drift apart after S phase. Indeed, defects in sister-chromatid cohesion—in yeast mutants, for example—lead inevitably to major errors in chromosome segregation.

Sister-chromatid cohesion depends on a large protein complex called **cohesin**, which is deposited at many locations along the length of each sister chromatid as the DNA is replicated in S phase. Two of the subunits of cohesin are members of a large family of proteins called *SMC proteins* (for Structural Maintenance of Chromosomes). Cohesin forms giant ringlike structures, and it has been proposed that these surround the two sister chromatids (**Figure 17–19**).

Sister-chromatid cohesion also results, at least in part, from *DNA catenation*, the intertwining of sister DNA molecules that occurs when two replication forks meet during DNA synthesis. The enzyme topoisomerase II gradually disentangles the catenated sister DNAs between S phase and early mitosis by cutting one DNA molecule, passing the other through the break, and then resealing the cut DNA (see Figure 5–22). Once the catenation has been removed, sister-chromatid cohesion depends primarily on cohesin complexes. The sudden and synchronous loss of sister cohesion at the metaphase-to-anaphase transition therefore depends primarily on disruption of these complexes, as we describe later.

Summary

Duplication of the chromosomes in S phase involves the accurate replication of the entire DNA molecule in each chromosome, as well as the duplication of the chromatin proteins that associate with the DNA and govern various aspects of chromosome function. Chromosome duplication is triggered by the activation of S-Cdk, which activates proteins that unwind the DNA and initiate its replication at replication origins. Once a replication origin is activated, S-Cdk also inhibits proteins that are required to allow that origin to initiate DNA replication again. Thus, each origin is fired once and only once in each S phase and cannot be reused until the next cell cycle.

MITOSIS

Following the completion of S phase and transition through G_2, the cell undergoes the dramatic upheaval of M phase. This begins with mitosis, during which the sister chromatids are separated and distributed (*segregated*) to a pair of identical daughter nuclei, each with its own copy of the genome. Mitosis is traditionally divided into five stages—*prophase, prometaphase, metaphase, anaphase,* and *telophase*—defined primarily on the basis of chromosome behavior as seen in a microscope. As mitosis is completed, the second major event of M phase—cytokinesis—divides the cell into two halves, each with an identical nucleus. **Panel 17–1** summarizes the major events of M phase (Movie 17.2, Movie 17.3, Movie 17.4, and Movie 17.5).

From a regulatory point of view, mitosis can be divided into two major parts, each governed by distinct components of the cell-cycle control system. First, an abrupt increase in M-Cdk activity at the G_2/M transition triggers the events of early mitosis (prophase, prometaphase, and metaphase). During this period, M-Cdk and several other mitotic protein kinases phosphorylate a variety of proteins, leading to the assembly of the mitotic spindle and its attachment to the sister-chromatid pairs. The second major part of mitosis begins at the metaphase-to-anaphase transition, when the APC/C triggers the destruction of securin, liberating a protease that cleaves cohesin and thereby initiates separation of the sister chromatids. The APC/C also promotes the destruction of cyclins, which leads to Cdk inactivation and the dephosphorylation of Cdk targets, which is required for all events of late M phase, including the completion of anaphase, the disassembly of the mitotic spindle, and the division of the cell by cytokinesis.

In this section, we describe the key mechanical events of mitosis and how M-Cdk and the APC/C orchestrate them.

M-Cdk Drives Entry Into Mitosis

One of the most remarkable features of cell-cycle control is that a single protein kinase, M-Cdk, brings about all of the diverse and complex cell rearrangements that occur in the early stages of mitosis. At a minimum, M-Cdk must induce the assembly of the mitotic spindle and ensure that each sister chromatid in a pair is attached to the opposite pole of the spindle. It also triggers *chromosome condensation*, the large-scale reorganization of the intertwined sister chromatids into compact, rodlike structures. In animal cells, M-Cdk also promotes the breakdown of the nuclear envelope and rearrangements of the actin cytoskeleton and the Golgi apparatus. Each of these processes is thought to be initiated when M-Cdk phosphorylates specific proteins involved in the process, although most of these proteins have not yet been identified.

M-Cdk does not act alone to phosphorylate key proteins involved in early mitosis. Two additional families of protein kinases, the *Polo-like kinases* and the *Aurora kinases*, also make important contributions to the control of early mitotic events. The Polo-like kinase Plk, for example, is required for the normal assembly of a bipolar mitotic spindle, in part because it phosphorylates proteins involved in separation of the spindle poles early in mitosis. The Aurora kinase Aurora-A also helps control proteins that govern the assembly and stability of the spindle, whereas Aurora-B controls attachment of sister chromatids to the spindle, as we discuss later.

Dephosphorylation Activates M-Cdk at the Onset of Mitosis

M-Cdk activation begins with the accumulation of M-cyclin (cyclin B in vertebrate cells; see Table 17–1). In embryonic cell cycles, the synthesis of M-cyclin is constant throughout the cell cycle, and M-cyclin accumulation results from the high stability of the protein in interphase. In most cell types, however, M-cyclin synthesis increases during G_2 and M, owing primarily to an increase in M-cyclin gene transcription. The increase in M-cyclin protein leads to a corresponding accumulation of M-Cdk (the complex of Cdk1 and M-cyclin) as the cell approaches

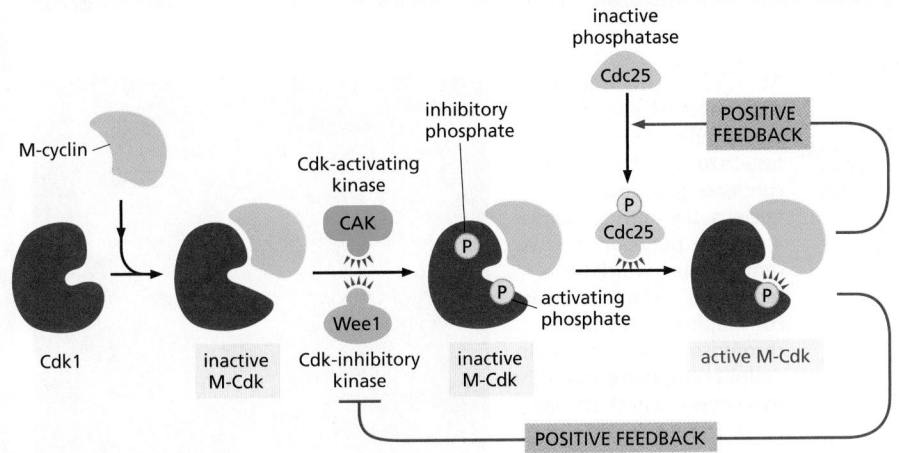

Figure 17–20 **The activation of M-Cdk.** Cdk1 associates with M-cyclin as the levels of M-cyclin gradually rise. The resulting M-Cdk complex is phosphorylated on an activating site by the Cdk-activating kinase (CAK) and on a pair of inhibitory sites by the Wee1 kinase. The resulting inactive M-Cdk complex is then activated at the end of G₂ by the phosphatase Cdc25. Cdc25 is further stimulated by active M-Cdk, resulting in positive feedback. This feedback is enhanced by the ability of M-Cdk to inhibit Wee1.

mitosis. Although the Cdk in these complexes is phosphorylated at an activating site by the Cdk-activating kinase (CAK), as discussed earlier, the protein kinase Wee1 holds it in an inactive state by inhibitory phosphorylation at two neighboring sites (see Figure 17–13). Thus, by the time the cell reaches the end of G₂, it contains an abundant stockpile of M-Cdk that is primed and ready to act but is suppressed by phosphates that block the active site of the kinase.

What, then, triggers the activation of the M-Cdk stockpile? The crucial event is the activation of the protein phosphatase Cdc25, which removes the inhibitory phosphates that restrain M-Cdk (**Figure 17–20**). At the same time, the inhibitory activity of the kinase Wee1 is suppressed, further ensuring that M-Cdk activity increases. The mechanisms that unleash Cdc25 activity in early mitosis are not well understood. One possibility is that the S-Cdks that are active in G₂ and early prophase stimulate Cdc25.

Interestingly, Cdc25 can also be activated, at least in part, by its target, M-Cdk. M-Cdk may also inhibit the inhibitory kinase Wee1. The ability of M-Cdk to activate its own activator (Cdc25) and inhibit its own inhibitor (Wee1) suggests that M-Cdk activation in mitosis involves positive feedback loops (see Figure 17–20). According to this attractive model, the partial activation of Cdc25 (perhaps by S-Cdk) leads to the partial activation of a subpopulation of M-Cdk complexes, which then phosphorylate Cdc25 and Wee1 molecules. This leads to more M-Cdk activation, and so on. Such a mechanism would quickly promote the activation of all M-Cdk complexes in the cell. As mentioned earlier, similar molecular switches operate at various points in the cell cycle to promote the abrupt and complete transition from one cell-cycle state to the next.

Condensin Helps Configure Duplicated Chromosomes for Separation

At the end of S phase, the immensely long DNA molecules of the sister chromatids are tangled in a mass of partially catenated DNA and proteins. Any attempt to pull the sisters apart in this state would undoubtedly lead to breaks in the chromosomes. To avoid this disaster, the cell devotes a great deal of energy in early mitosis to gradually reorganizing the sister chromatids into relatively short, distinct structures that can be pulled apart more easily in anaphase. These chromosomal changes involve two processes: *chromosome condensation*, in which the chromatids are dramatically compacted; and *sister-chromatid resolution*, whereby the two sisters are resolved into distinct, separable units (**Figure 17–21**). Resolution results from the decatenation of the sister DNAs, accompanied by the partial removal of cohesin molecules along the chromosome arms. As a result, when the cell reaches metaphase, the sister chromatids appear in the microscope as compact, rodlike structures that are joined tightly at their centromeric regions and only loosely along their arms.

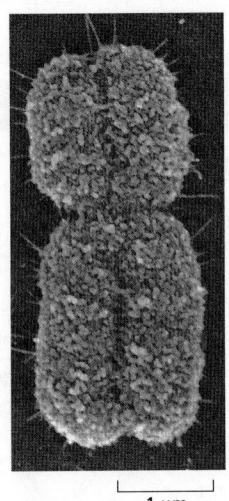

1 µm

Figure 17–21 **The mitotic chromosome.** Scanning electron micrograph of a human mitotic chromosome, consisting of two sister chromatids joined along their length. The constricted regions are the centromeres. (Courtesy of Terry D. Allen.)

1 PROPHASE

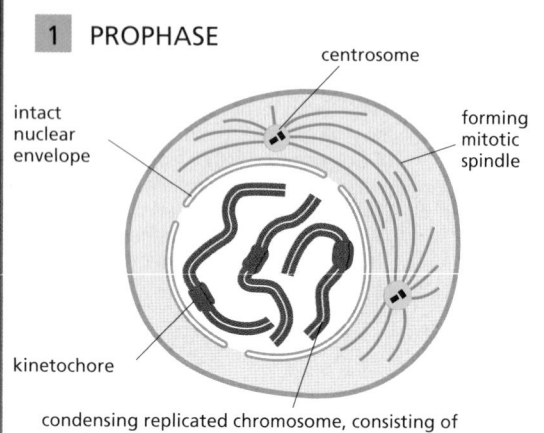

centrosome

intact nuclear envelope

forming mitotic spindle

kinetochore

condensing replicated chromosome, consisting of two sister chromatids held together along their length

At prophase, the replicated chromosomes, each consisting of two closely associated sister chromatids, condense. Outside the nucleus, the mitotic spindle assembles between the two centrosomes, which have replicated and moved apart. For simplicity, only three chromosomes are shown. In diploid cells, there would be two copies of each chromosome present. In the fluorescence micrograph, chromosomes are stained *orange* and microtubules are *green*.

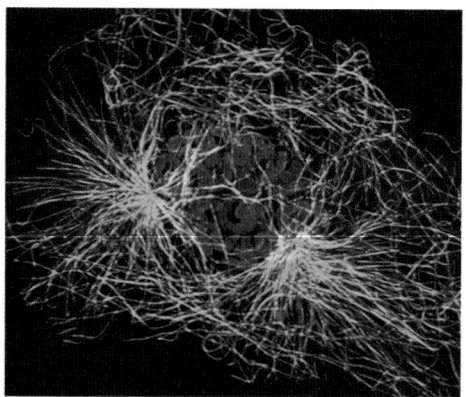

2 PROMETAPHASE

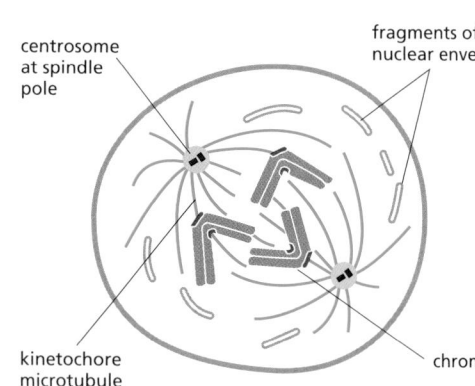

centrosome at spindle pole

fragments of nuclear envelope

kinetochore microtubule

chromosome in active motion

Prometaphase starts abruptly with the breakdown of the nuclear envelope. Chromosomes can now attach to spindle microtubules via their kinetochores and undergo active movement.

3 METAPHASE

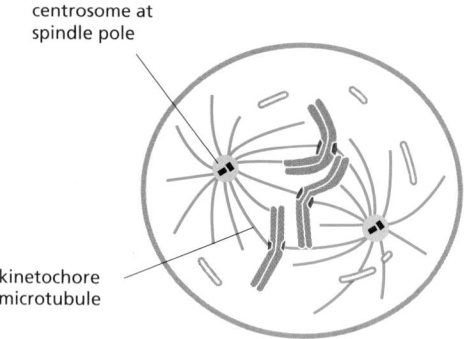

centrosome at spindle pole

kinetochore microtubule

At metaphase, the chromosomes are aligned at the equator of the spindle, midway between the spindle poles. The kinetochore microtubules attach sister chromatids to opposite poles of the spindle.

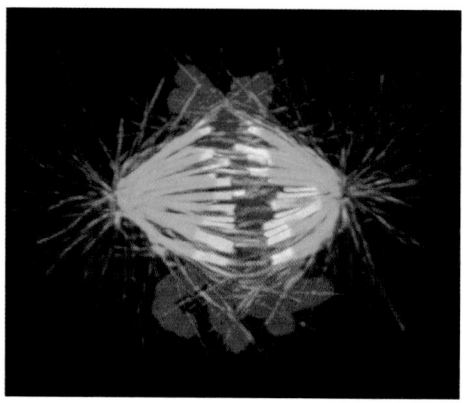

4 ANAPHASE

daughter chromosomes

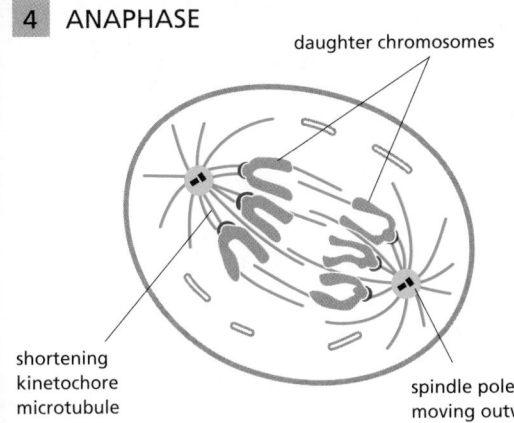

shortening
kinetochore
microtubule

spindle pole
moving outward

At anaphase, the sister
chromatids synchronously
separate to form two
daughter chromosomes,
and each is pulled slowly
toward the spindle pole it
faces. The kinetochore
microtubules get shorter,
and the spindle poles also
move apart; both
processes contribute to
chromosome segregation.

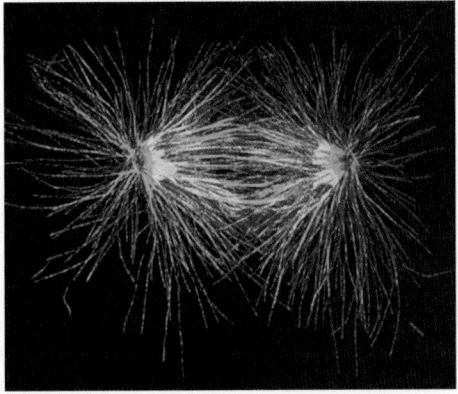

5 TELOPHASE

set of daughter chromosomes
at spindle pole

contractile ring
starting to
contract

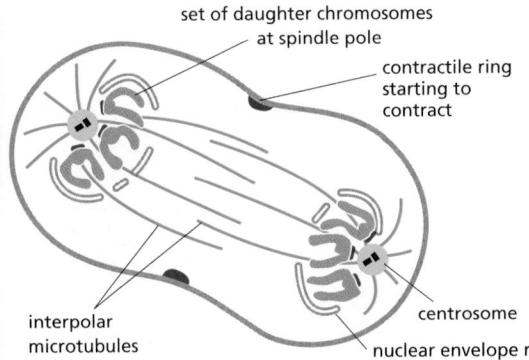

interpolar
microtubules

centrosome

nuclear envelope reassembling
around individual chromosomes

During telophase, the two
sets of daughter chromo-
somes arrive at the poles of
the spindle and decondense.
A new nuclear envelope
reassembles around each
set, completing the formation
of two nuclei and marking
the end of mitosis. The
division of the cytoplasm
begins with contraction of
the contractile ring.

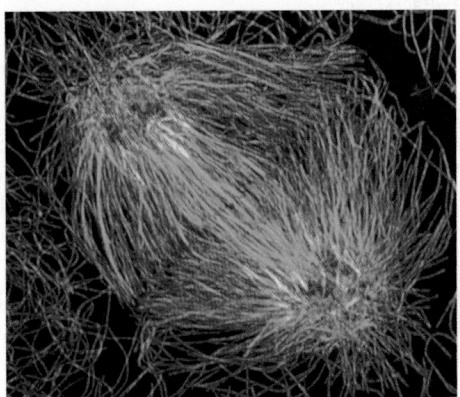

6 CYTOKINESIS

completed nuclear envelope
surrounds decondensing
chromosomes

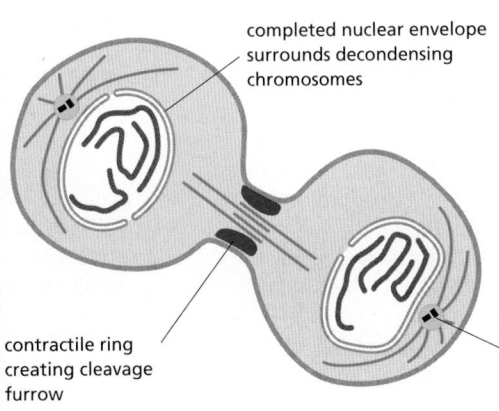

contractile ring
creating cleavage
furrow

re-formation of interphase
array of microtubules nucleated
by the centrosome

During cytokinesis, the
cytoplasm is divided in two
by a contractile ring of
actin and myosin
filaments, which pinches
the cell in two to create
two daughters, each with
one nucleus.

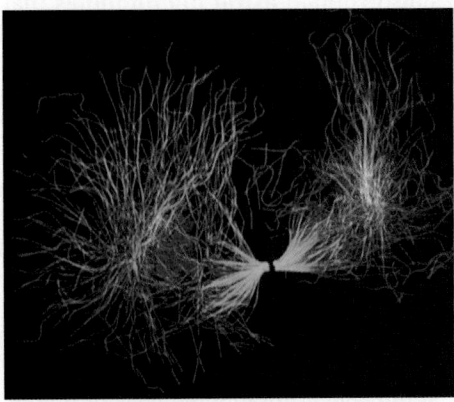

(Micrographs courtesy of Julie Canman and Ted Salmon.)

Figure 17–22 **Condensin.** (A) Condensin is a five-subunit protein complex that resembles cohesin (see Figure 17–19). The ATPase head domains of its two major subunits, Smc2 and Smc4, are held together by three additional subunits. (B) It is not clear how condensin catalyzes the restructuring and compaction of chromosome DNA, but it may form a ring structure that encircles loops of DNA within each sister chromatid.

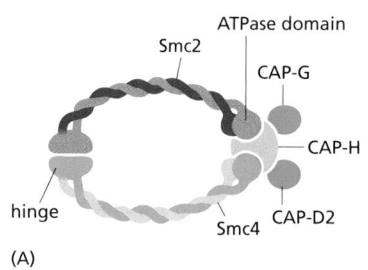

(A)

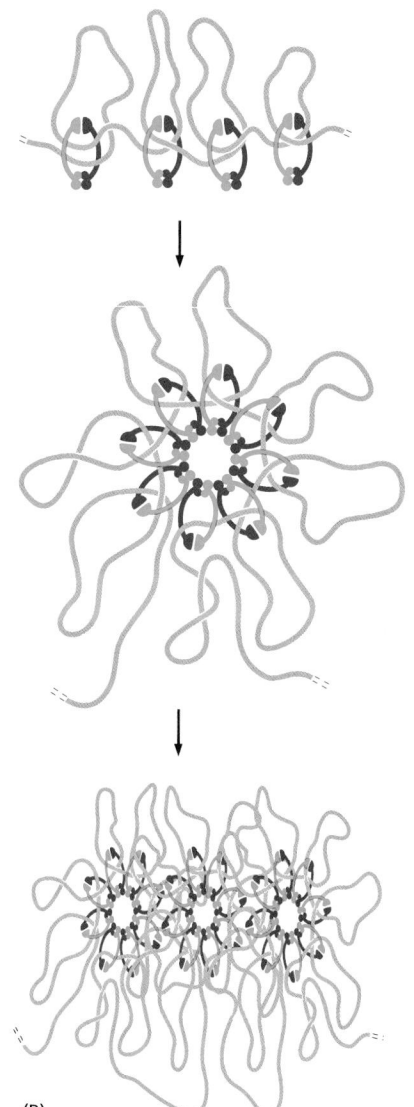

(B)

The condensation and resolution of sister chromatids depend, at least in part, on a five-subunit protein complex called **condensin**. Condensin structure is related to that of the cohesin complex that holds sister chromatids together (see Figure 17–19). It contains two SMC subunits like those of cohesin, plus three non-SMC subunits (**Figure 17–22**). Condensin may form a ringlike structure that somehow uses the energy provided by ATP hydrolysis to promote the compaction and resolution of sister chromatids. Condensin is able to change the coiling of DNA molecules in a test tube, and this coiling activity is thought to be important for chromosome condensation during mitosis. Interestingly, phosphorylation of condensin subunits by M-Cdk stimulates this coiling activity, providing one mechanism by which M-Cdk may promote chromosome restructuring in early mitosis.

The Mitotic Spindle Is a Microtubule-Based Machine

The central event of mitosis—chromosome segregation—depends in all eukaryotes on a complex and beautiful machine called the **mitotic spindle** (see Panel 17–1). The spindle is a bipolar array of microtubules, which pulls sister chromatids apart in anaphase, thereby segregating the two sets of chromosomes to opposite ends of the cell, where they are packaged into daughter nuclei (**Movie 17.6**). M-Cdk triggers the assembly of the spindle early in mitosis, in parallel with the chromosome restructuring just described. Before we consider how the spindle assembles and how its microtubules attach to sister chromatids, we briefly review the basic features of spindle structure.

The core of the mitotic spindle is a bipolar array of microtubules, the minus ends of which are focused at the two spindle poles, and the plus ends of which radiate outward from the poles (**Figure 17–23**). The plus ends of some microtubules—called the **interpolar microtubules**—overlap with the plus ends of microtubules from the other pole, resulting in an antiparallel array in the spindle midzone. The plus ends of other microtubules—the **kinetochore microtubules**—are attached to sister-chromatid pairs at large protein structures called *kinetochores*, which are located at the *centromere* of each sister chromatid. Finally, many spindles also contain **astral microtubules** that radiate outward from the poles and contact the cell cortex, helping to position the spindle in the cell.

In most somatic animal cells, each spindle pole is focused at a protein organelle called the **centrosome** (see Figures 16–47 and 16–48). Each centrosome consists of a cloud of amorphous material (called the *pericentriolar matrix)* that surrounds a pair of *centrioles* (**Figure 17–24**). The pericentriolar matrix nucleates a radial array of microtubules, with their fast-growing plus ends projecting outward and their minus ends associated with the centrosome. The matrix contains a variety of proteins, including microtubule-dependent motor proteins, coiled-coil proteins that link the motors to the centrosome, structural proteins, and components of the cell-cycle control system. Most important, it contains *γ-tubulin ring complexes*, which are the components mainly responsible for nucleating microtubules (see Figure 16–46).

Some cells—notably the cells of higher plants and the oocytes of many vertebrates—do not have centrosomes, and microtubule-dependent motor proteins and other proteins associate with microtubule minus ends to organize and focus the spindle poles.

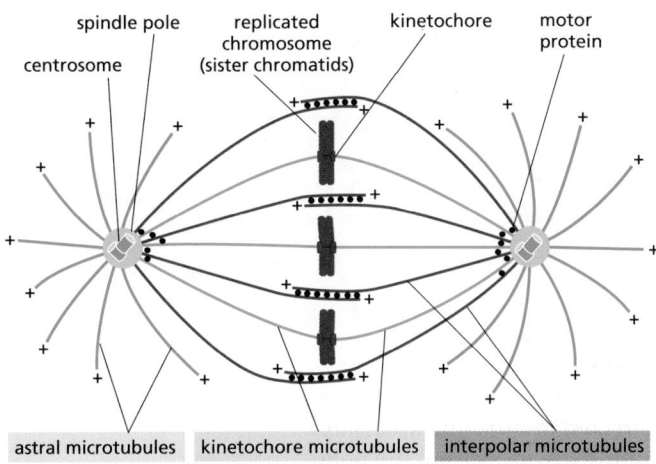

astral microtubules | kinetochore microtubules | interpolar microtubules

Figure 17–23 **The metaphase mitotic spindle in an animal cell.** The plus ends of the microtubules project away from the spindle pole, while the minus ends are anchored at the spindle poles, which in this example are organized by centrosomes. Kinetochore microtubules connect the spindle poles with the kinetochores of sister chromatids, while interpolar microtubules from the two poles interdigitate at the spindle equator. Astral microtubules radiate out from the poles into the cytoplasm.

Microtubule-Dependent Motor Proteins Govern Spindle Assembly and Function

The function of the mitotic spindle depends on numerous microtubule-dependent motor proteins. As discussed in Chapter 16, these proteins belong to two families—the kinesin-related proteins, which usually move toward the plus end of microtubules, and dyneins, which move toward the minus end. In the mitotic spindle, these motor proteins generally operate at or near the ends of the microtubules. Four major types of motor proteins—*kinesin-5, kinesin-14, kinesins-4/10,* and *dynein*—are particularly important in spindle assembly and function (**Figure 17–25**).

Kinesin-5 proteins contain two motor domains that interact with the plus ends of antiparallel microtubules in the spindle midzone. Because the two motor domains move toward the plus ends of the microtubules, they slide the two antiparallel microtubules past each other toward the spindle poles, pushing the poles apart. Kinesin-14 proteins, by contrast, are minus-end directed motors with a single motor domain and other domains that can interact with a neighboring microtubule. They can cross-link antiparallel interpolar microtubules at the spindle midzone and tend to pull the poles together. Kinesin-4 and kinesin-10

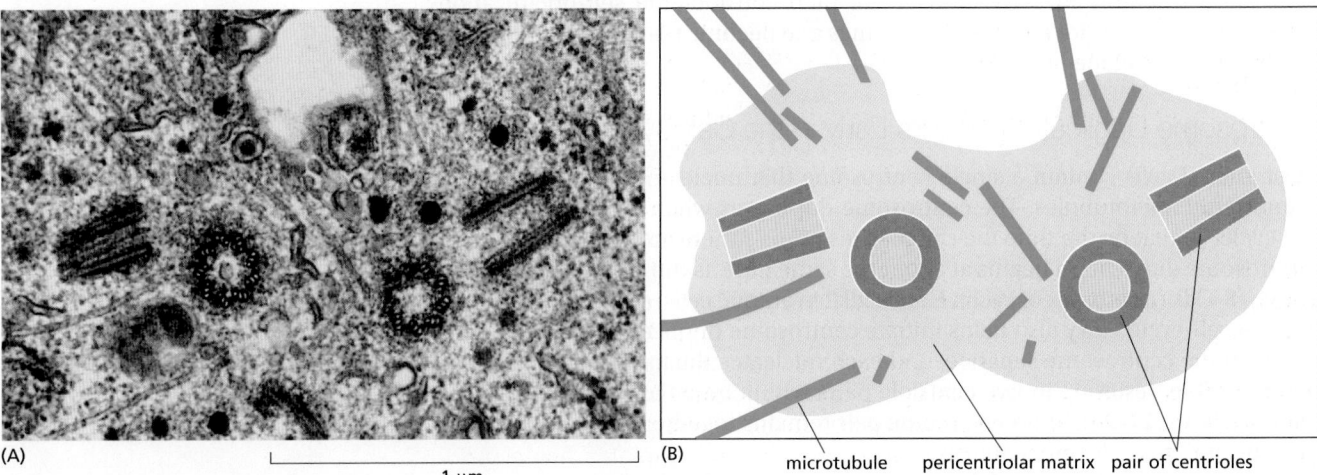

(A)

1 μm

(B) microtubule   pericentriolar matrix   pair of centrioles

Figure 17–24 **The centrosome.** (A) Electron micrograph of an S-phase mammalian cell in culture, showing a duplicated centrosome. Each centrosome contains a pair of centrioles; although the centrioles have duplicated, they remain together in a single complex, as shown in the drawing of the micrograph in (B). One centriole of each centriole pair has been cut in cross section, while the other is cut in longitudinal section, indicating that the two members of each pair are aligned at right angles to each other. The two halves of the replicated centrosome, each consisting of a centriole pair surrounded by pericentriolar matrix, will split and migrate apart to initiate the formation of the two poles of the mitotic spindle when the cell enters M phase. (A, from M. McGill, D.P. Highfield, T.M. Monahan, and B.R. Brinkley, *J. Ultrastruct. Res.* 57:43–53, 1976. With permission from Academic Press.)

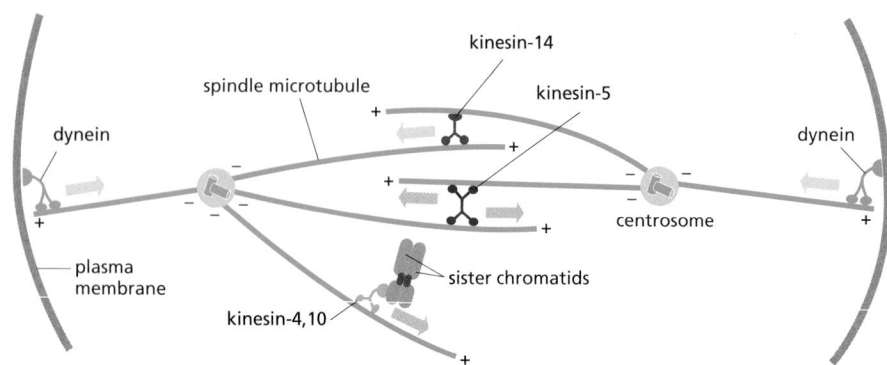

Figure 17–25 **Major motor proteins of the spindle.** Four major classes of microtubule-dependent motor proteins *(yellow boxes)* contribute to spindle assembly and function (see text). The colored arrows indicate the direction of motor protein movement along a microtubule—*blue* toward the minus end and *red* toward the plus end. (From D.O. Morgan, *The Cell Cycle: Principles of Control,* p.117. London: New Science Press, 2007. With permission from Oxford University Press.)

proteins, also called *chromokinesins,* are plus-end directed motors that associate with chromosome arms and push the attached chromosome away from the pole (or the pole away from the chromosome). Finally, dyneins are minus-end directed motors that, together with associated proteins, organize microtubules at various locations in the cell. They link the plus ends of astral microtubules to components of the actin cytoskeleton at the cell cortex, for example; by moving toward the minus end of the microtubules, the dynein motors pull the spindle poles toward the cell cortex and away from each other.

Multiple Mechanisms Collaborate in the Assembly of a Bipolar Mitotic Spindle

The mitotic spindle must have two poles if it is to pull the two sets of sister chromatids to opposite ends of the cell in anaphase. In most animal cells, several mechanisms ensure the bipolarity of the spindle. One depends on centrosomes. A typical animal cell enters mitosis with a pair of centrosomes, each of which nucleates a radial array of microtubules. The two centrosomes provide prefabricated spindle poles that greatly facilitate bipolar spindle assembly. The other mechanisms depend on the ability of mitotic chromosomes to nucleate and stabilize microtubules and on the ability of motor proteins to organize microtubules into a bipolar array. These "self-organization" mechanisms can produce a bipolar spindle even in cells lacking centrosomes.

We now describe the steps of spindle assembly, beginning with centrosome-dependent assembly in early mitosis. We then consider the self-organization mechanisms that do not require centrosomes and become particularly important after nuclear-envelope breakdown.

Centrosome Duplication Occurs Early in the Cell Cycle

Most animal cells contain a single centrosome that nucleates most of the cell's cytoplasmic microtubules. The centrosome duplicates when the cell enters the cell cycle, so that by the time the cell reaches mitosis there are two centrosomes. Centrosome duplication begins at about the same time as the cell enters S phase. The G_1/S-Cdk (a complex of cyclin E and Cdk2 in animal cells; see Table 17–1) that triggers cell-cycle entry also helps initiate centrosome duplication. The two centrioles in the centrosome separate, and each nucleates the formation of a single new centriole, resulting in two centriole pairs within an enlarged pericentriolar matrix (**Figure 17–26**). This centrosome pair remains together on one side of the nucleus until the cell enters mitosis.

There are interesting parallels between centrosome duplication and chromosome duplication. Both use a semiconservative mechanism of duplication, in which the two halves separate and serve as templates for construction of a new half. Centrosomes, like chromosomes, must replicate once and only once per cell cycle, to ensure that the cell enters mitosis with only two copies: an incorrect number of centrosomes could lead to defects in spindle assembly and thus errors in chromosome segregation.

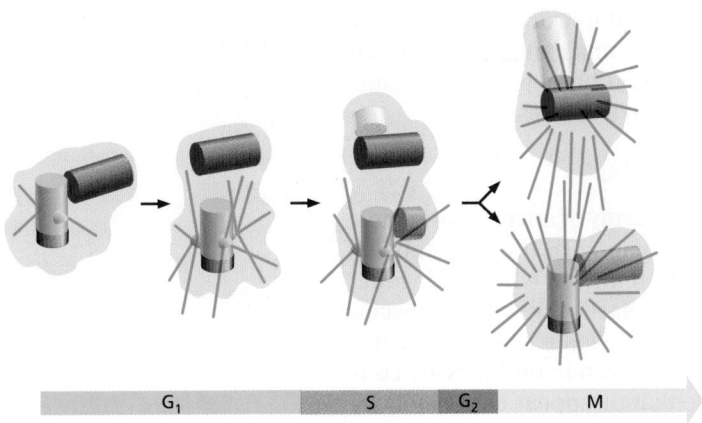

Figure 17–26 **Centriole replication.** The centrosome consists of a centriole pair and associated pericentriolar matrix *(green)*. At a certain point in G_1, the two centrioles of the pair separate by a few micrometers. During S phase, a daughter centriole begins to grow near the base of each mother centriole and at a right angle to it. The elongation of the daughter centriole is usually completed in G_2. The two centriole pairs remain close together in a single centrosomal complex until the beginning of M phase, when the complex splits in two and the two daughter centrosomes begin to separate. Each centrosome now nucleates its own radial array of microtubules (called an aster), mainly from the mother centriole.

The mechanisms that limit centrosome duplication to once per cell cycle are uncertain. In many cell types, experimental inhibition of DNA synthesis blocks centrosome duplication, providing one mechanism by which centrosome number is kept in check. Other cell types, however, including those in the early embryos of flies, sea urchins, and frogs, do not have such a mechanism and centrosome duplication continues if chromosome duplication is blocked. It is not known how such cells limit centrosome duplication to once per cell cycle.

M-Cdk Initiates Spindle Assembly in Prophase

Spindle assembly begins in early mitosis, when the two centrosomes move apart along the nuclear envelope, pulled by dynein motor proteins that link astral microtubules to the cell cortex (see Figure 17–25). The plus ends of the microtubules between the centrosomes interdigitate to form the interpolar microtubules, and kinesin-5 motor proteins associate with these microtubules and push the centrosomes apart (see Figure 17–25). Also in early mitosis, the number of γ-tubulin ring complexes in each centrosome increases greatly, increasing the ability of the centrosomes to nucleate new microtubules, a process called *centrosome maturation*.

The balance of opposing forces generated by different types of motor proteins determines the final length of the spindle. Dynein and kinesin-5 motors generally promote centrosome separation and increase spindle length. Kinesin-14 proteins do the opposite: they tend to pull the poles together (see Figure 17–25). It is not clear how the cell regulates the balance of opposing forces to generate the appropriate spindle length.

M-Cdk and other mitotic protein kinases are required for centrosome separation and maturation. M-Cdk and Aurora-A phosphorylate kinesin-5 motors and stimulate them to drive centrosome separation. Aurora-A and Plk also phosphorylate components of the centrosome and thereby promote its maturation.

The Completion of Spindle Assembly in Animal Cells Requires Nuclear-Envelope Breakdown

The centrosomes and microtubules of animal cells are located in the cytoplasm, separated from the chromosomes by the double-membrane barrier of the nuclear envelope (discussed in Chapter 12). Clearly, the attachment of sister-chromatid pairs to the spindle requires the removal of this barrier. In addition, many of the motor proteins and microtubule regulators that promote spindle assembly are associated with the chromosomes inside the nucleus, and they require nuclear-envelope breakdown to carry out their functions.

Nuclear-envelope breakdown is a complex, multistep process, which is thought to begin when M-Cdk phosphorylates several subunits of the nuclear pore complexes in the nuclear envelope. This phosphorylation initiates the disassembly of nuclear pore complexes and their dissociation from the envelope. M-Cdk also phosphorylates components of the nuclear lamina, the structural framework beneath the envelope. The phosphorylation of these lamina components and of several inner-nuclear-envelope proteins leads to disassembly of the nuclear lamina and the breakdown of the envelope membranes into small vesicles.

Microtubule Instability Increases Greatly in Mitosis

Most animal cells in interphase contain a cytoplasmic array of microtubules radiating out from the single centrosome. As discussed in Chapter 16, the microtubules of this interphase array are in a state of *dynamic instability*, in which individual microtubules are either growing or shrinking and stochastically switch between the two states. The switch from growth to shrinkage is called a *catastrophe*, and the switch from shrinkage to growth is called a *rescue*. New microtubules are continually being created to balance the loss of those that disappear completely by depolymerization.

Entry into mitosis signals an abrupt change in the cell's microtubules. The interphase array of few, long microtubules radiating from the single centrosome is converted to a larger number of shorter and more dynamic microtubules emanating from both centrosomes. During prophase, and particularly in prometaphase and metaphase (see Panel 17–1), the half-life of microtubules decreases dramatically. This increase in microtubule instability, coupled with the increased ability of centrosomes to nucleate microtubules as mentioned earlier, results in remarkably dense and dynamic arrays of spindle microtubules that are ideally suited for capturing sister chromatids.

Microtubule dynamics are controlled in the cell by a variety of regulatory proteins, including microtubule-associated proteins (MAPs) that promote stability and catastrophe factors that destabilize microtubule plus ends. Changes in the activities of these regulatory proteins are responsible for the changes in microtubule dynamics that occur during mitosis. Many of these changes result from phosphorylation of specific proteins by M-Cdk and other mitotic protein kinases.

Mitotic Chromosomes Promote Bipolar Spindle Assembly

Chromosomes are not just passive passengers in the process of spindle assembly. By creating a local environment that favors both microtubule nucleation and microtubule stabilization, they play an active part in spindle formation. The influence of the chromosomes can be demonstrated by using a fine glass needle to reposition them after the spindle has formed. For some cells in metaphase, if a single chromosome is tugged out of alignment, a mass of new spindle microtubules rapidly appears around the newly positioned chromosome, while the spindle microtubules at the chromosome's former position depolymerize. This property of the chromosomes seems to depend, at least in part, on a guanine nucleotide exchange factor (GEF) that is bound to chromatin; the GEF stimulates a small GTPase in the cytosol called *Ran* to bind GTP in place of GDP. The activated Ran-GTP, which is also involved in nuclear transport (discussed in Chapter 12), releases microtubule-stabilizing proteins from protein complexes in the cytosol, thereby stimulating the local nucleation and stabilization of microtubules around chromosomes (**Figure 17–27**). Local microtubule stabilization is also promoted by the protein kinase Aurora-B, which associates with mitotic chromosomes.

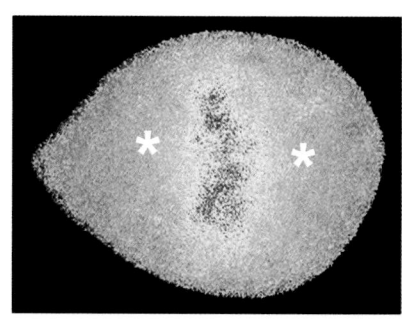

Figure 17–27 Activation of the GTPase Ran around mitotic chromosomes. The Ran protein, like other members of the small GTPase family (discussed in Chapter 15), can exist in two conformations depending on whether it is bound to GDP (inactive state) or GTP (active state). The localization of active Ran in mitosis was determined using a protein that emits fluorescence at a specific wavelength when it is activated by Ran-GTP. In the metaphase human cell shown here, Ran activity (*yellow* and *red*) is highest around the chromosomes, between the poles of the mitotic spindle (indicated by *asterisks*). (From P. Kaláb et al., *Nature* 440:697–701, published 2006 by Nature Publishing Group. Reproduced with permission from SNCSC.)

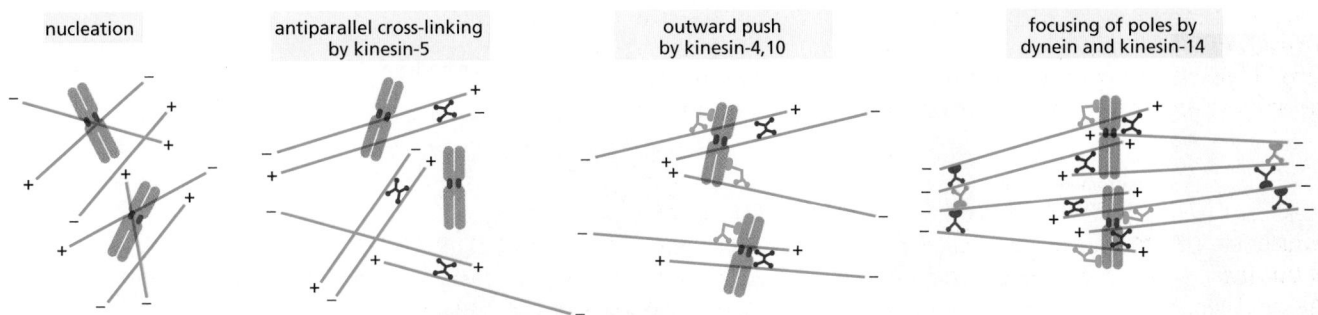

| nucleation | antiparallel cross-linking by kinesin-5 | outward push by kinesin-4,10 | focusing of poles by dynein and kinesin-14 |

Figure 17–28 Spindle self-organization by motor proteins. Mitotic chromosomes stimulate the local activation of proteins that nucleate and promote the formation of microtubules in the vicinity of the chromosomes. Kinesin-5 motor proteins (see Figure 17–25) organize these microtubules into antiparallel bundles, while plus-end directed kinesins-4 and 10 link the microtubules to chromosome arms and push minus ends away from the chromosomes. Dynein and kinesin-14 motors, together with numerous other proteins, focus these minus ends into a pair of spindle poles.

The ability of chromosomes to stabilize and organize microtubules enables cells to form bipolar spindles in the absence of centrosomes. Acentrosomal spindle assembly is thought to begin with the formation of microtubules around the chromosomes. Various motor proteins then organize the microtubules into a bipolar spindle, as illustrated in **Figure 17–28**.

Cells that normally lack centrosomes, such as those of higher plants and many animal oocytes, use this chromosome-based self-organization process to form spindles. It is also the process used to assemble spindles in certain animal embryos that have been induced to develop from eggs without fertilization (that is, *parthenogenetically*); as the sperm normally provides the centrosome when it fertilizes an egg, the mitotic spindles in these parthenogenetic embryos develop without centrosomes (**Figure 17–29**). Even in cells that normally contain centrosomes, the chromosomes help organize the spindle microtubules and, with the help of various motor proteins, can promote the assembly of a bipolar mitotic spindle if the centrosomes are removed. Although the resulting acentrosomal spindle can segregate chromosomes normally, it lacks astral microtubules, which are responsible for positioning the spindle in animal cells; as a result, the spindle is often mispositioned in the cell.

Kinetochores Attach Sister Chromatids to the Spindle

Following the assembly of a bipolar microtubule array, the second major step in spindle formation is the attachment of the array to the sister-chromatid pairs. Spindle microtubules become attached to each chromatid at its **kinetochore**, a giant, multilayered protein structure that is built at the centromeric region of the chromatid (**Figure 17–30**; also see Chapter 4). In metaphase, the plus ends of kinetochore microtubules are embedded head-on in specialized microtubule-attachment sites within the outer region of the kinetochore, furthest from the DNA. The kinetochore of an animal cell can bind 10–40 microtubules, whereas a budding yeast kinetochore can bind only one. Attachment of each microtubule depends on multiple copies of a rod-shaped protein complex called the Ndc80 complex, which is anchored in the kinetochore at one end and interacts with the sides of the microtubule at the other, thereby linking the microtubule to the kinetochore while still allowing the addition or removal of tubulin subunits at this end (**Figure 17–31**). Regulation of plus-end polymerization and depolymerization at the kinetochore is critical for the control of chromosome movement on the spindle, as we discuss later.

Kinetochore attachment to the spindle occurs by a complex sequence of events. At the end of prophase in animal cells, the centrosomes of the growing spindle generally lie on opposite sides of the nuclear envelope. Thus, when the envelope breaks down, the sister-chromatid pairs are bombarded by microtubule

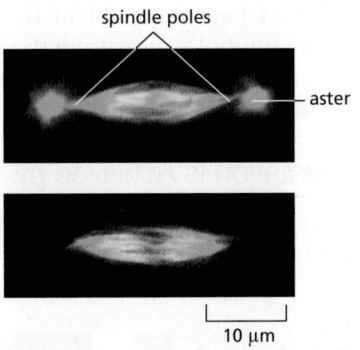

Figure 17–29 Bipolar spindle assembly without centrosomes in parthenogenetic embryos of the insect *Sciara* (or fungus gnat). The microtubules are stained *green*, the chromosomes *red*. The *top* fluorescence micrograph shows a normal spindle formed with centrosomes in a normally fertilized *Sciara* embryo. The *bottom* micrograph shows a spindle formed without centrosomes in an embryo that initiated development without fertilization. Note that the spindle with centrosomes has an aster at each pole of the spindle, whereas the spindle formed without centrosomes does not. Both types of spindles are able to segregate the replicated chromosomes. (© 1998 B. de Saint Phalle and W. Sullivan. Originally published in *J. Cell Biol.* https://doi.org/10.1083/jcb.141.6.1383. With permission from Rockefeller University Press.)

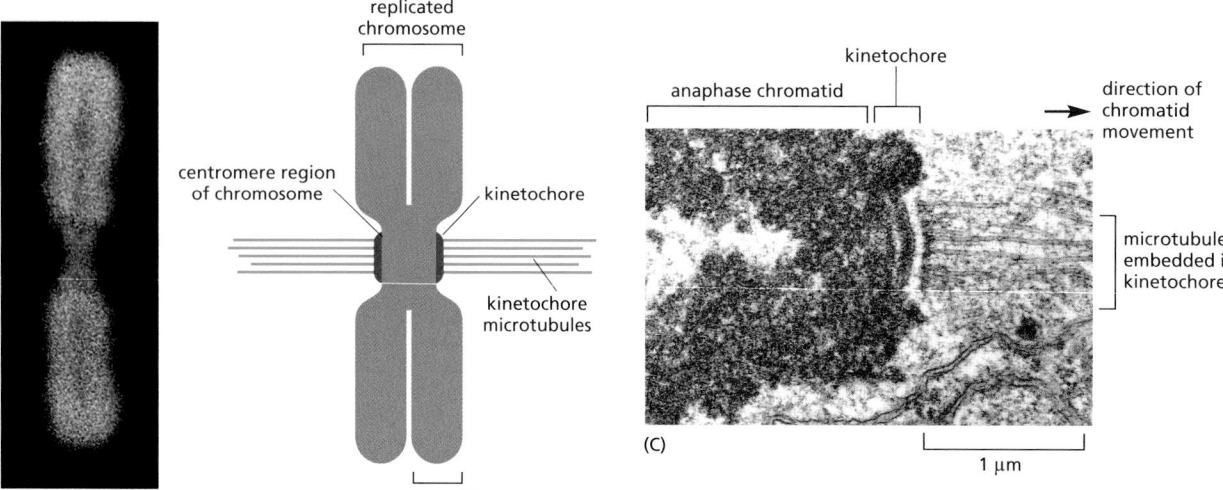

(A)

(B)

replicated chromosome

centromere region of chromosome

kinetochore

kinetochore microtubules

chromatid

anaphase chromatid

kinetochore

direction of chromatid movement

microtubules embedded in kinetochore

(C)

1 μm

Figure 17–30 **The kinetochore.**
(A) A fluorescence micrograph of a metaphase chromosome stained with a DNA-binding fluorescent dye and with human autoantibodies that react with specific kinetochore proteins. The two kinetochores, one associated with each sister chromatid, are stained *red*.
(B) A drawing of a metaphase chromosome showing its two sister chromatids attached to the plus ends of kinetochore microtubules. Each kinetochore forms a plaque on the surface of the centromere.
(C) Electron micrograph of an anaphase chromatid with microtubules attached to its kinetochore. While most kinetochores have a trilaminar structure, the one shown here (from a green alga) has an unusually complex structure with additional layers.
(A, © 1991 R.P. Zinkowski et al. Originally published in *J. Cell Biol.* https://doi .org/10.1083/jcb.113.5.1091. With permission from Rockefeller University Press; C, from J.D. Pickett-Heaps and L.C. Fowke, *Aust. J. Biol. Sci.* 23:71–92, 1970. With permission from CSIRO publishing.)

plus ends coming from two directions. However, the kinetochores do not instantly achieve the correct 'end-on' microtubule attachment to both spindle poles. Instead, detailed studies with light and electron microscopy show that most initial attachments are unstable *lateral* attachments, in which a kinetochore attaches to the side of a passing microtubule, with assistance from kinesin motor proteins in the outer kinetochore. Soon, however, the dynamic microtubule plus ends capture the kinetochores in the correct end-on orientation (**Figure 17–32**).

Another attachment mechanism also plays a part, particularly in the absence of centrosomes. Careful microscopic analysis suggests that short microtubules in the vicinity of the chromosomes become embedded in the plus-end-binding sites of the kinetochore. Polymerization at these plus ends then results in growth of the microtubules away from the kinetochore. The minus ends of these kinetochore microtubules are eventually cross-linked to other minus ends and focused by motor proteins at the spindle pole (see Figure 17–28).

Bi-orientation Is Achieved by Trial and Error

The success of mitosis demands that sister chromatids in a pair attach to opposite poles of the mitotic spindle, so that they move to opposite ends of the cell when

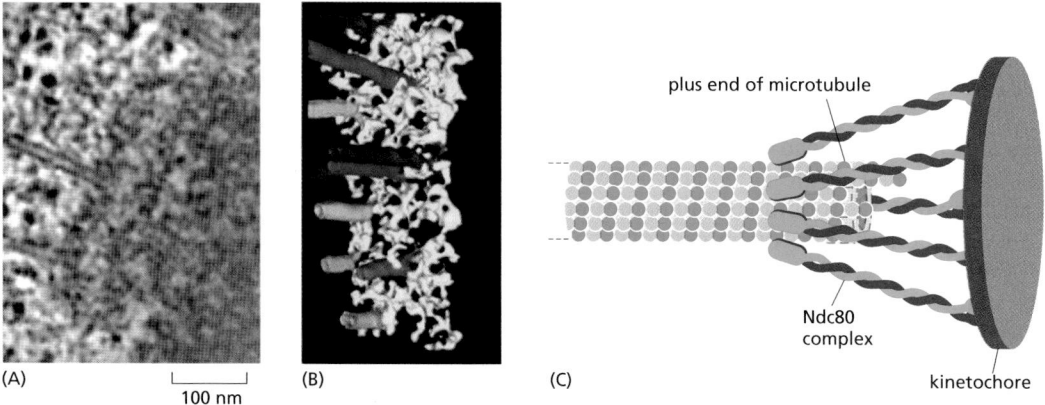

(A)

100 nm

(B)

(C)

plus end of microtubule

Ndc80 complex

kinetochore

Figure 17–31 **Microtubule attachment sites in the kinetochore.** (A) In this electron micrograph of a mammalian kinetochore, the chromosome is on the right, and the plus ends of multiple microtubules are embedded in the outer kinetochore on the left. (B) Electron tomography (discussed in Chapter 9) was used to construct a low-resolution three-dimensional image of the outer kinetochore in (A). Several microtubules (*in multiple colors*) are embedded in fibrous material of the kinetochore, which is thought to be composed of the Ndc80 complex and other proteins. (C) Each microtubule is attached to the kinetochore by interactions with multiple copies of the Ndc80 complex (*blue*). This complex binds to the sides of the microtubule near its plus end, allowing polymerization and depolymerization to occur while the microtubule remains attached to the kinetochore. (A and B, from Y. Dong et al., *Nature Cell Biol.* 9:516–522, published 2007 by Nature Publishing Group. Reproduced with permission of SNCSC.)

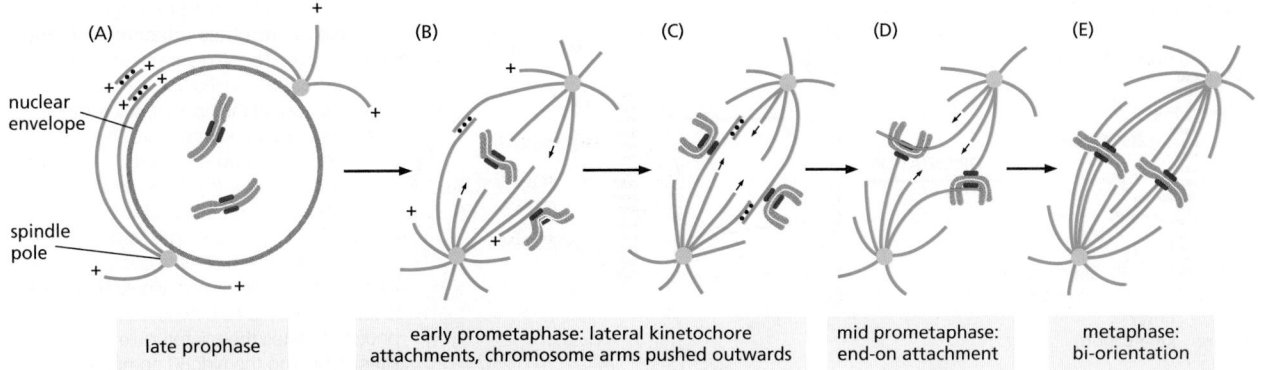

Figure 17-32 **Chromosome attachment to the mitotic spindle in animal cells.** (A) In late prophase of most animal cells, the mitotic spindle poles have moved to opposite sides of the nuclear envelope, with an array of overlapping microtubules between them. (B) Following nuclear envelope breakdown, the sister-chromatid pairs are exposed to the large number of dynamic plus ends of microtubules radiating from the spindle poles. In most cases, the kinetochores are first attached to the sides of these microtubules, while at the same time the arms of the chromosomes are pushed outward from the spindle interior, preventing the arms from blocking microtubule access to the kinetochores. (C) Eventually, the laterally-attached sister chromatids are arranged in a ring around the outside of the spindle. Most of the microtubules are concentrated in this ring, so that the spindle is relatively hollow inside. (D) Dynamic microtubule plus ends eventually encounter the kinetochores in an end-on orientation and are captured and stabilized. (E) Stable end-on attachment to both poles results in *bi-orientation*. Additional microtubules are attached to the kinetochore, resulting in a *kinetochore fiber* containing 10–40 microtubules.

they separate in anaphase. How is this mode of attachment, called **bi-orientation**, achieved? What prevents the attachment of both kinetochores to the same spindle pole or the attachment of one kinetochore to both spindle poles? Part of the answer is that sister kinetochores are constructed in a back-to-back orientation that reduces the likelihood that both kinetochores can face the same spindle pole. Nevertheless, incorrect attachments do occur, and elegant regulatory mechanisms have evolved to correct them.

Incorrect attachments are corrected by a system of trial and error that is based on a simple principle: incorrect attachments are highly unstable and do not last, whereas correct attachments become locked in place. How does the kinetochore sense a correct attachment? The answer appears to be tension (**Figure 17–33**). When a sister-chromatid pair is properly bi-oriented on the spindle, the two kinetochores are pulled in opposite directions by strong poleward forces. Sister-chromatid cohesion resists these poleward forces, creating high levels of tension within the kinetochores. When chromosomes are incorrectly attached—when both sister chromatids are attached to the same spindle pole, for example—tension is low

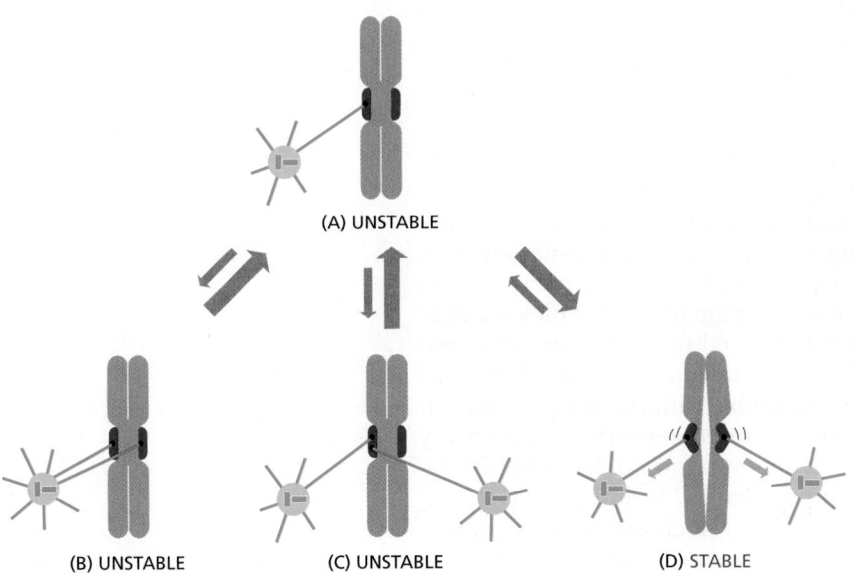

(A) UNSTABLE

(B) UNSTABLE   (C) UNSTABLE   (D) STABLE

Figure 17–33 **Alternative forms of kinetochore attachment to the spindle poles.** (A) Initially, a single microtubule from a spindle pole binds to one kinetochore in a sister-chromatid pair. Additional microtubules can then bind to the chromosome in various ways. (B) A microtubule from the same spindle pole can attach to the other sister kinetochore, or (C) microtubules from both spindle poles can attach to the same kinetochore. These incorrect attachments are unstable, however, so that one of the two microtubules tends to dissociate. (D) When a microtubule from the opposite pole binds to the second kinetochore, the sister kinetochores are thought to sense tension across their microtubule-binding sites. This triggers an increase in microtubule binding affinity, thereby locking the correct attachment in place.

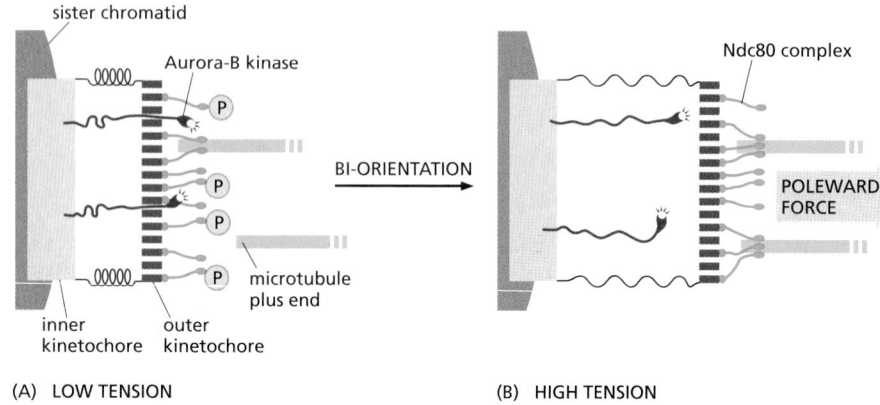

(A) LOW TENSION   (B) HIGH TENSION

Figure 17–34 How tension might increase microtubule attachment to the kinetochore. These diagrams illustrate one speculative mechanism by which bi-orientation might increase microtubule attachment to the kinetochore. A single kinetochore is shown for clarity; the spindle pole is on the right. (A) When a sister-chromatid pair is unattached to the spindle or attached to just one spindle pole, there is little tension between the outer and inner kinetochores. The protein kinase Aurora-B is tethered to the inner kinetochore and phosphorylates the microtubule attachment sites, including the Ndc80 complex *(blue)*, in the outer kinetochore as shown, thereby reducing the affinity of microtubule binding. Microtubules therefore associate and dissociate rapidly, and attachment is unstable. (B) When bi-orientation is achieved, the forces pulling the kinetochore toward the spindle pole are resisted by forces pulling the other sister kinetochore toward the opposite pole, and the resulting tension pulls the outer kinetochore away from the inner kinetochore. As a result, Aurora-B is unable to reach the outer kinetochore, and microtubule attachment sites are not phosphorylated. Microtubule binding affinity is therefore increased, resulting in the stable attachment of multiple microtubules to both kinetochores. The dephosphorylation of outer kinetochore proteins depends on a phosphatase that is not shown here.

and the kinetochore sends an inhibitory signal that loosens the grip of its microtubule attachment site, allowing detachment to occur. When bi-orientation occurs, the high tension at the kinetochore shuts off the inhibitory signal, strengthening microtubule attachment. In animal cells, tension not only increases the affinity of the attachment site but also leads to the attachment of additional microtubules to the kinetochore. This results in the formation of a thick *kinetochore fiber* composed of multiple microtubules.

The tension-sensing mechanism depends on the protein kinase Aurora-B, which is associated with the kinetochore and is thought to generate the inhibitory signal that reduces the strength of microtubule attachment in the absence of tension. It phosphorylates several components of the microtubule attachment site, including the Ndc80 complex, decreasing the site's affinity for a microtubule plus end. When bi-orientation occurs, the resulting tension somehow reduces phosphorylation by Aurora-B, thereby increasing the affinity of the attachment site (**Figure 17–34**).

Following their attachment to the two spindle poles, the chromosomes are tugged back and forth, eventually assuming a position equidistant between the two poles, a position called the **metaphase plate**. In vertebrate cells, the chromosomes then oscillate gently at the metaphase plate, awaiting the signal for the sister chromatids to separate. The signal is produced, with a predictable lag time, after the bi-oriented attachment of the last of the chromosomes.

Multiple Forces Act on Chromosomes in the Spindle

Multiple mechanisms generate the forces that move chromosomes back and forth after they are attached to the spindle, and produce the tension that is so important for the stabilization of correct attachments. In anaphase, similar forces pull the separated chromatids to opposite ends of the spindle. Three major spindle forces are particularly critical, although their strength and importance vary at different stages of mitosis.

The first major force pulls the kinetochore and its associated chromatid along the kinetochore microtubule toward the spindle pole. It is produced by proteins at the kinetochore itself. By an uncertain mechanism, depolymerization at the plus end of the microtubule generates a force that pulls the kinetochore poleward. This force pulls on chromosomes during prometaphase and metaphase but is particularly important for moving sister chromatids toward the poles after they separate in anaphase. Interestingly, this kinetochore-generated poleward force does not require ATP or motor proteins. This might seem implausible at first, but it has been shown that purified kinetochores in a test tube, with no ATP present, can remain attached to depolymerizing microtubules and thereby move. The energy that drives the movement is stored in the microtubule and is released when the microtubule depolymerizes; it ultimately comes from the hydrolysis of GTP that occurs after a tubulin subunit adds to the end of a microtubule (discussed in Chapter 16).

How does plus-end depolymerization drive the kinetochore toward the pole? As we discussed earlier (see Figure 17–31C), Ndc80 complexes in the kinetochore make multiple low-affinity attachments along the side of the microtubule. Because the attachments are constantly breaking and re-forming at new sites, the kinetochore remains attached to a microtubule even as the microtubule depolymerizes. In principle, this could move the kinetochore toward the spindle pole.

A second poleward force is provided in some cell types by **microtubule flux**, whereby the microtubules themselves are pulled toward the spindle poles and dismantled at their minus ends. The mechanism underlying this poleward movement is not clear, although it might depend on forces generated by motor proteins and minus-end depolymerization at the spindle pole. In metaphase, the addition of new tubulin at the plus end of a microtubule compensates for the loss of tubulin at the minus end, so that microtubule length remains constant despite the movement of microtubules toward the spindle pole (**Figure 17–35**). Any kinetochore that is attached to a microtubule undergoing such flux experiences a poleward force, which contributes to the generation of tension at the kinetochore in metaphase. Together with the kinetochore-based forces discussed above, flux also contributes to the poleward forces that move sister chromatids after they separate in anaphase.

A third force acting on chromosomes is the *polar ejection force*, or *polar wind*. Plus-end directed kinesin-4 and 10 motors on chromosome arms interact with interpolar microtubules and transport the chromosomes away from the spindle poles (see Figure 17–25). This force is particularly important in prometaphase and metaphase, when it helps push chromosome arms out from the spindle. This force might also help align the sister-chromatid pairs at the metaphase plate (**Figure 17–36**).

One of the most striking aspects of mitosis in vertebrate cells is the continuous oscillatory movement of the chromosomes in prometaphase and metaphase. When studied by video microscopy in newt lung cells, the movements are seen to switch between two states—a poleward state, when the chromosomes are pulled toward the pole, and an away-from-the-pole, or neutral, state, when the poleward forces are turned off and the polar ejection force pushes the chromosomes away

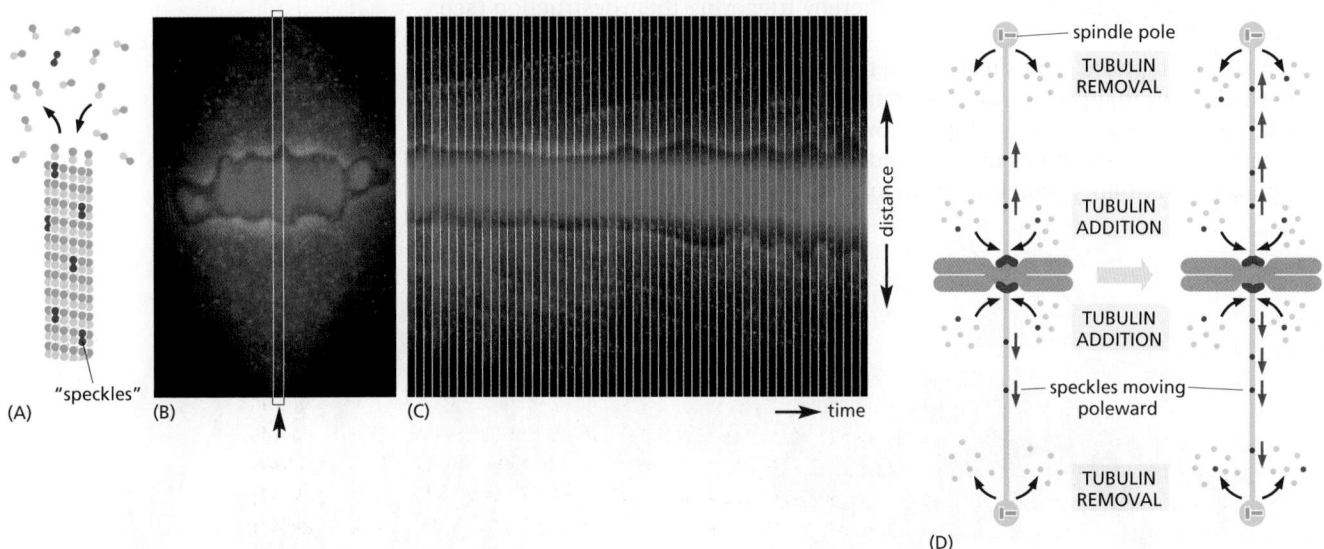

(A) (B) (C) → time (D)

"speckles"

distance

spindle pole
TUBULIN REMOVAL
TUBULIN ADDITION
TUBULIN ADDITION
speckles moving poleward
TUBULIN REMOVAL

Figure 17–35 **Microtubule flux in the metaphase spindle.** (A) To observe microtubule flux, a very small amount of fluorescent tubulin is injected into living cells so that individual microtubules form with a very small proportion of fluorescent tubulin. Such microtubules have a speckled appearance when viewed by fluorescence microscopy. (B) Fluorescence micrograph of a mitotic spindle in a living newt lung epithelial cell. The chromosomes are colored *brown*, and the tubulin speckles are *red*. (C) The movement of individual speckles can be followed by time-lapse video microscopy. Images of the thin vertical boxed region *(arrow)* in (B), taken at sequential times, show that individual speckles move toward the poles at a rate of about 0.75 μm/min, indicating that the microtubules are moving poleward. (D) Microtubule length in the metaphase spindle does not change significantly because new tubulin subunits are added at the microtubule plus end at the same rate as tubulin subunits are removed from the minus end. (B and C, from T.J. Mitchison and E.D. Salmon, *Nat. Cell Biol.* 3:E17–21, published 2001 by Nature Publishing Group. Reproduced wtih permission of SNCSC.)

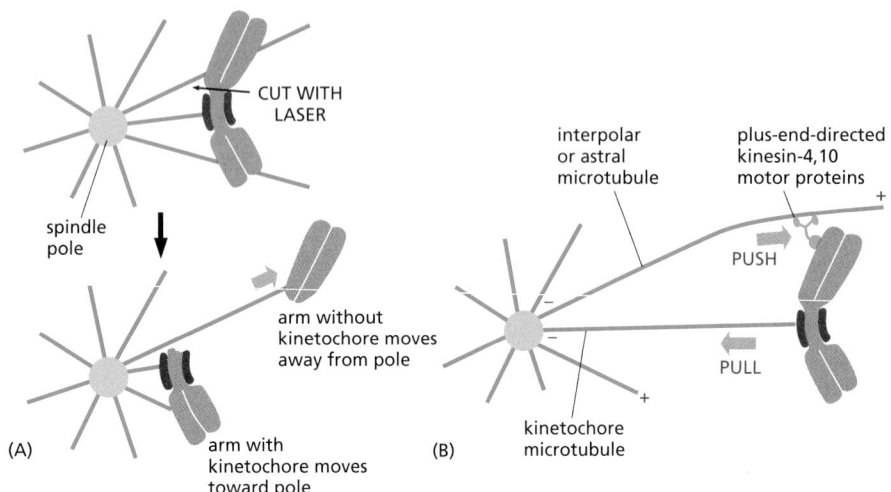

(A)

(B)

Figure 17–36 **How opposing forces may drive chromosomes to the metaphase plate.** (A) Evidence for a polar ejection force that pushes chromosomes away from the spindle poles toward the spindle equator. In this experiment, a laser beam severs a prometaphase chromosome that is attached to a single pole by a kinetochore microtubule. The part of the severed chromosome without a kinetochore is pushed rapidly away from the pole, whereas the part with the kinetochore moves toward the pole, reflecting a decreased repulsion. (B) A model of how two opposing forces may cooperate to move chromosomes to the metaphase plate. Plus-end-directed motor proteins (kinesin-4 and kinesin-10) on the chromosome arms are thought to interact with microtubules to generate the polar ejection force, which pushes chromosomes toward the spindle equator (see Figure 17–25). Poleward forces generated by depolymerization at the kinetochore, together with microtubule flux, are thought to pull chromosomes toward the pole.

from the pole. The switch between the two states may depend on the degree of tension in the kinetochore. It has been proposed, for example, that, as chromosomes move toward a spindle pole, an increasing polar ejection force generates tension in the kinetochore nearest the pole, triggering a switch to the away-from-the-pole state and gradually resulting in the accumulation of chromosomes at the equator of the spindle.

The APC/C Triggers Sister-Chromatid Separation and the Completion of Mitosis

After M-Cdk has triggered the complex processes leading up to metaphase, the cell cycle reaches its climax with the separation of the sister chromatids at the metaphase-to-anaphase transition (**Figure 17–37**). Although M-Cdk activity sets the stage for this event, the anaphase-promoting complex (APC/C) discussed earlier throws the switch that initiates sister-chromatid separation by ubiquitylating several mitotic regulatory proteins and thereby triggering their destruction (see Figure 17–15A).

During metaphase, cohesins holding the sister chromatids together resist the poleward forces that pull the sister chromatids apart. Anaphase begins with the sudden loss of sister-chromatid cohesion, which allows the sisters to separate and move to opposite poles of the spindle. The APC/C initiates the process by targeting the inhibitory protein **securin** for destruction. Before anaphase, securin

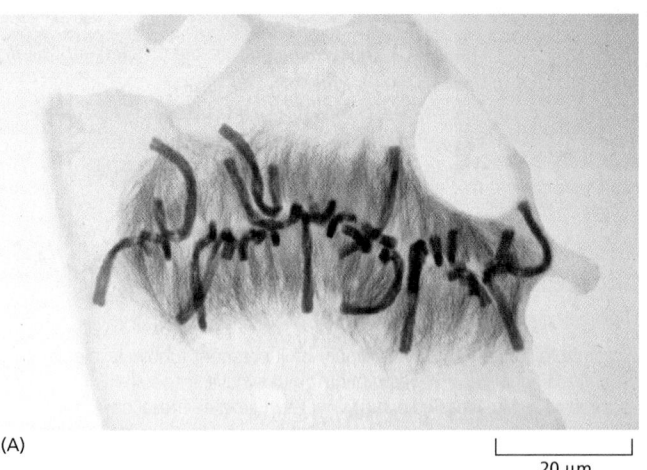

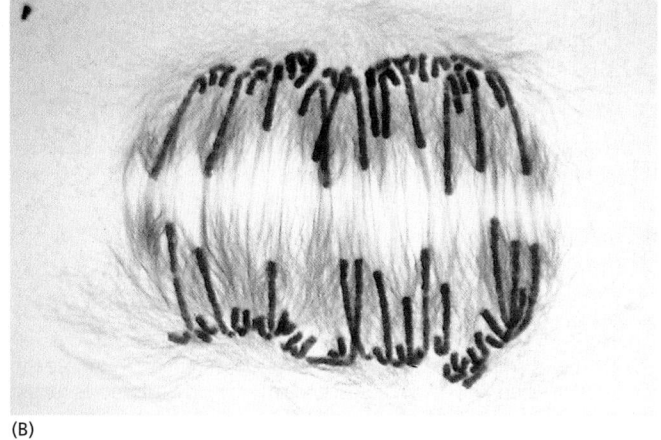

(A)

(B)

20 μm

Figure 17–37 **Sister-chromatid separation at anaphase.** In the transition from metaphase (A) to anaphase (B), sister chromatids suddenly and synchronously separate and move toward opposite poles of the mitotic spindle—as shown in these light micrographs of *Haemanthus* (lily) endosperm cells that were stained with gold-labeled antibodies against tubulin. (Courtesy of Andrew Bajer.)

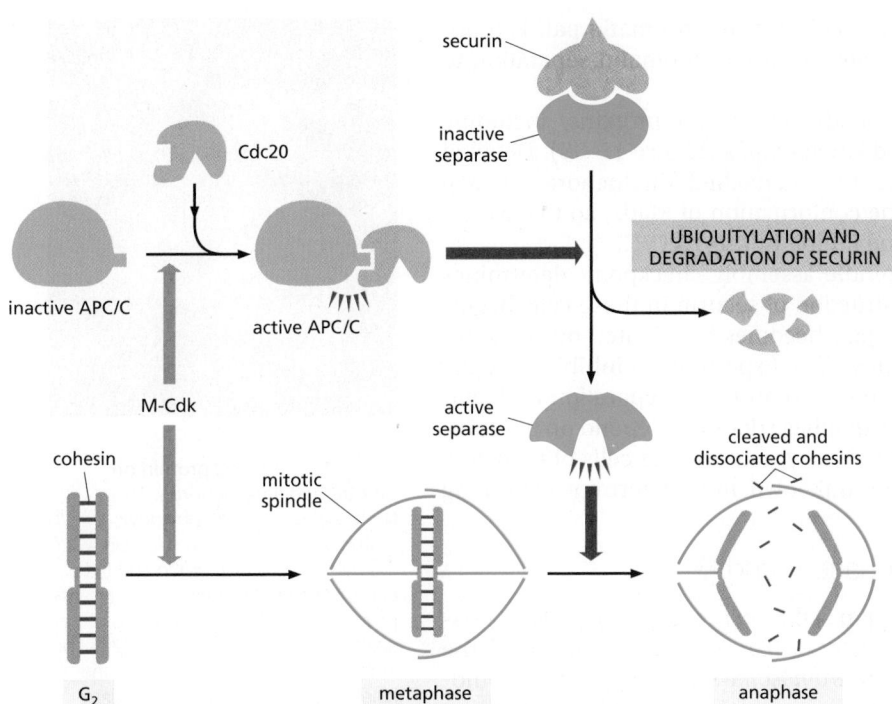

Figure 17–38 **The initiation of sister-chromatid separation by the APC/C.** The activation of APC/C by Cdc20 leads to the ubiquitylation and destruction of securin, which normally holds separase in an inactive state. The destruction of securin allows separase to cleave Scc1, a subunit of the cohesin complex holding the sister chromatids together (see Figure 17–19). The pulling forces of the mitotic spindle then pull the sister chromatids apart. In animal cells, phosphorylation by Cdks also inhibits separase (not shown). Thus, Cdk inactivation in anaphase (resulting from cyclin destruction) also promotes separase activation by allowing its dephosphorylation.

binds to and inhibits the activity of a protease called **separase**. The destruction of securin at the end of metaphase releases separase, which is then free to cleave one of the subunits of cohesin. The cohesins fall away, and the sister chromatids separate (**Figure 17–38**).

In addition to securin, the APC/C also targets the S- and M-cyclins for destruction, leading to the loss of most Cdk activity in anaphase. Cdk inactivation allows phosphatases to dephosphorylate the many Cdk target substrates in the cell, as required for the completion of mitosis and cytokinesis.

If the APC/C triggers anaphase, what activates the APC/C? The answer is only partly known. As mentioned earlier, APC/C activation requires binding to the protein Cdc20 (see Figure 17–15A). At least two processes regulate Cdc20 and its association with the APC/C. First, Cdc20 synthesis increases as the cell approaches mitosis, owing to an increase in the transcription of its gene. Second, phosphorylation of the APC/C helps Cdc20 bind to the APC/C, thereby helping to create an active complex. Among the kinases that phosphorylate and thus activate the APC/C is M-Cdk. Thus, M-Cdk not only triggers the early mitotic events leading up to metaphase, but it also sets the stage for progression into anaphase. The ability of M-Cdk to promote Cdc20–APC/C activity creates a negative feedback loop: M-Cdk sets in motion a regulatory process that leads to cyclin destruction and thus its own inactivation.

Unattached Chromosomes Block Sister-Chromatid Separation: The Spindle Assembly Checkpoint

Drugs that destabilize microtubules, such as colchicine or vinblastine (discussed in Chapter 16), arrest cells in mitosis for hours or even days. This observation led to the identification of a **spindle assembly checkpoint** mechanism that is activated by the drug treatment and blocks progression through the metaphase-to-anaphase transition. The checkpoint mechanism ensures that cells do not enter anaphase until all chromosomes are correctly bi-oriented on the mitotic spindle.

The spindle assembly checkpoint depends on a sensor mechanism that monitors the strength of microtubule attachment at the kinetochore, possibly by sensing tension as described earlier (see Figure 17–34). Any kinetochore that is not properly attached to the spindle sends out a diffusible negative signal that blocks Cdc20–APC/C activation throughout the cell and thus blocks the metaphase-to-anaphase transition. When the last sister-chromatid pair is properly bi-oriented, this block is removed, allowing sister-chromatid separation to occur.

The negative checkpoint signal depends on several proteins, including *Mad2*, which are recruited to unattached kinetochores (**Figure 17–39**). Detailed structural analyses of Mad2 suggest that the unattached kinetochore acts like an enzyme that catalyzes a change in the conformation of Mad2, so that Mad2, together with other proteins, can bind and inhibit Cdc20–APC/C.

In mammalian somatic cells, the spindle assembly checkpoint determines the normal timing of anaphase. The destruction of securin in these cells begins moments after the last sister-chromatid pair becomes bi-oriented on the spindle, and anaphase begins about 20 minutes later. Experimental inhibition of the checkpoint mechanism causes premature sister-chromatid separation and anaphase. Surprisingly, the normal timing of anaphase does not depend on the spindle assembly checkpoint in some cells, such as yeasts and the cells of early frog and fly embryos. Other mechanisms, as yet unknown, must determine the timing of anaphase in these cells.

Chromosomes Segregate in Anaphase A and B

The sudden loss of sister-chromatid cohesion at the onset of anaphase leads to sister-chromatid separation, which allows the forces of the mitotic spindle to pull the sisters to opposite poles of the cell—called *chromosome segregation*. The chromosomes move by two independent and overlapping processes. The first, **anaphase A**, is the initial poleward movement of the chromosomes, which is accompanied by shortening of the kinetochore microtubules. The second, **anaphase B**, is the separation of the spindle poles themselves, which begins after the sister chromatids have separated and the daughter chromosomes have moved some distance apart (**Figure 17–40**).

Chromosome movement in anaphase A depends on a combination of the two major poleward forces described earlier. The first is the force generated by microtubule depolymerization at the kinetochore, which results in the loss of tubulin subunits at the plus end as the kinetochore moves toward the pole. The second is provided by microtubule flux, which is the poleward movement of the microtubules toward the spindle pole, where minus-end depolymerization occurs. The relative importance of these two forces during anaphase varies in different cell types: in embryonic cells, chromosome movement depends mainly on microtubule flux, for example, whereas movement in yeast and vertebrate somatic cells results primarily from forces generated at the kinetochore.

Spindle-pole separation during anaphase B depends on motor-driven mechanisms similar to those that separate the two centrosomes in early mitosis. Plus-end directed kinesin-5 motor proteins, which cross-link the overlapping plus ends of the interpolar microtubules, push the poles apart. In addition, dynein motors that anchor astral microtubule plus ends to the cell cortex pull the poles apart (see Figure 17–25).

Although sister-chromatid separation initiates the chromosome movements of anaphase A, other mechanisms also ensure correct chromosome movements in anaphase A and spindle elongation in anaphase B. Most importantly, the completion of a normal anaphase depends on the dephosphorylation of Cdk substrates, which in most cells results from the APC/C-dependent destruction of cyclins. If M-cyclin destruction is prevented—by the production of a mutant form that is not recognized by the APC/C, for example—sister-chromatid separation generally occurs, but the chromosome movements and microtubule behavior of anaphase are abnormal.

The relative contributions of anaphase A and anaphase B to chromosome segregation vary greatly, depending on the cell type. In mammalian cells, anaphase B begins shortly after anaphase A and stops when the spindle is about twice its metaphase length; in contrast, the spindles of yeasts and certain protozoa primarily use anaphase B to separate the chromosomes at anaphase, and their spindles elongate to up to 15 times their metaphase length.

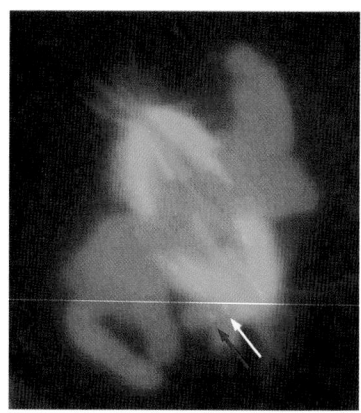

Figure 17–39 **Mad2 protein on unattached kinetochores.** This fluorescence micrograph shows a mammalian cell in prometaphase, with the mitotic spindle in *green* and the sister chromatids in *blue*. One sister-chromatid pair is attached to only one pole of the spindle. Staining with anti-Mad2 antibodies indicates that Mad2 is bound to the kinetochore of the unattached sister chromatid *(red dot, indicated by red arrow)*. A small amount of Mad2 is associated with the kinetochore of the sister chromatid that is attached to the spindle pole *(pale dot, indicated by white arrow)*. (© 1998 J.C. Waters et al. Originally published in *J. Cell Biol.* https://doi.org/10.1083/jcb.141.5.1181. With permission from Rockefeller University Press.)

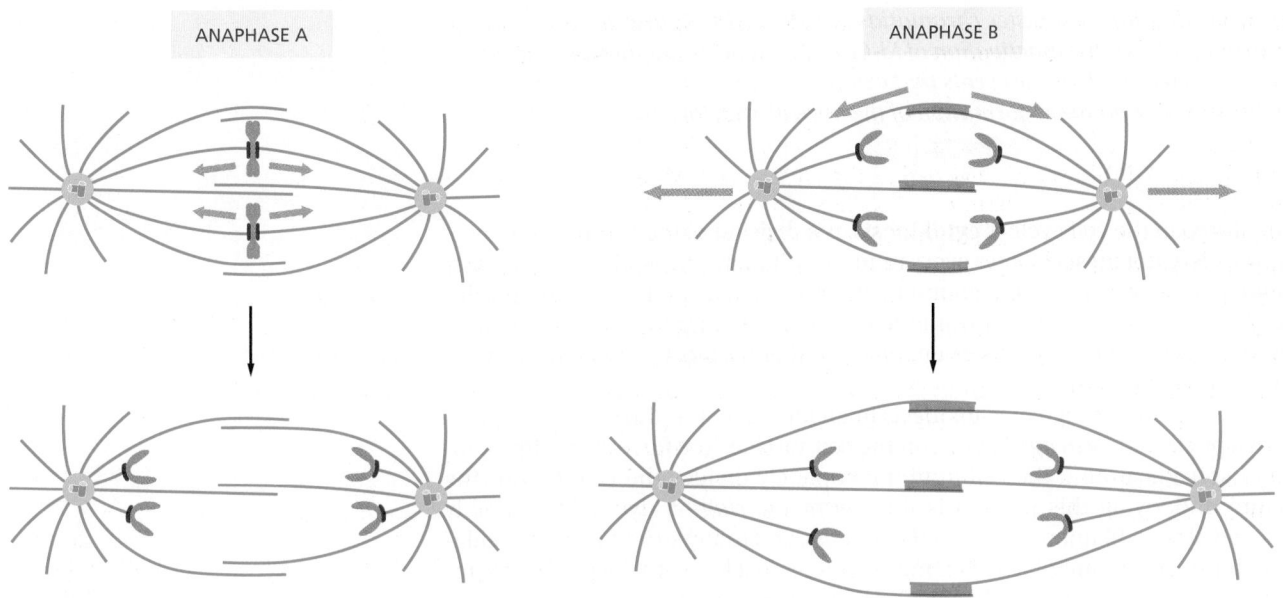

Segregated Chromosomes Are Packaged in Daughter Nuclei at Telophase

Figure 17–40 **The two processes of anaphase in mammalian cells.** Separated sister chromatids move toward the poles in anaphase A. In anaphase B, the two spindle poles move apart.

By the end of anaphase, the daughter chromosomes have segregated into two equal groups at opposite ends of the cell. In **telophase**, the final stage of mitosis, the two sets of chromosomes are packaged into a pair of daughter nuclei. The first major event of telophase is the disassembly of the mitotic spindle, followed by the re-formation of the nuclear envelope. Initially, nuclear membrane fragments associate with the surface of individual chromosomes. These membrane fragments fuse to partly enclose clusters of chromosomes and then coalesce to re-form the complete nuclear envelope. Nuclear pore complexes are incorporated into the envelope, the nuclear lamina re-forms, and the envelope once again becomes continuous with the endoplasmic reticulum. Once the nuclear envelope has re-formed, the pore complexes pump in nuclear proteins, the nucleus expands, and the mitotic chromosomes are reorganized into their interphase state, allowing gene transcription to resume. A new nucleus has been created, and mitosis is complete. All that remains is for the cell to complete its division into two.

We saw earlier that phosphorylation of various proteins by M-Cdk promotes spindle assembly, chromosome condensation, and nuclear-envelope breakdown in early mitosis. It is thus not surprising that the dephosphorylation of these same proteins is required for spindle disassembly and the re-formation of daughter nuclei in telophase. In principle, these dephosphorylations and the completion of mitosis could be triggered by the inactivation of Cdks, the activation of phosphatases, or both. Although Cdk inactivation—resulting primarily from cyclin destruction—is mainly responsible in most cells, some cells also rely on activation of phosphatases. In budding yeast, for example, the completion of mitosis depends on the activation of a phosphatase called *Cdc14*, which dephosphorylates a subset of Cdk substrates involved in anaphase and telophase.

Summary

M-Cdk triggers the events of early mitosis, including chromosome condensation, assembly of the mitotic spindle, and bipolar attachment of the sister-chromatid pairs to microtubules of the spindle. Spindle formation in animal cells depends largely on the ability of mitotic chromosomes to stimulate local microtubule nucleation and stability, as well as on the ability of motor proteins to organize microtubules into a bipolar array. Many cells also use centrosomes to facilitate spindle assembly. Anaphase is triggered by the APC/C, which stimulates the destruction of the proteins that hold the sister chromatids together. APC/C also promotes cyclin destruction and thus the inactivation of M-Cdk. The resulting dephosphorylation of Cdk targets is required for the events that complete mitosis, including the disassembly of the spindle and the re-formation of the nuclear envelope.

CYTOKINESIS

The final step in the cell cycle is **cytokinesis**, the division of the cytoplasm in two. In most cells, cytokinesis follows every mitosis, although some cells, such as early *Drosophila* embryos and some mammalian hepatocytes and heart muscle cells, undergo mitosis without cytokinesis and thereby acquire multiple nuclei. In most animal cells, cytokinesis begins in anaphase and ends shortly after the completion of mitosis in telophase.

The first visible change of cytokinesis in an animal cell is the sudden appearance of a pucker, or *cleavage furrow*, on the cell surface. The furrow rapidly deepens and spreads around the cell until it completely divides the cell in two. The structure underlying this process is the *contractile ring*—a dynamic assembly composed of actin filaments, myosin II filaments, and many structural and regulatory proteins. During anaphase, the ring assembles just beneath the plasma membrane (**Figure 17–41**; see also Panel 17–1). The ring gradually contracts, and, at the same time, fusion of intracellular vesicles with the plasma membrane inserts new membrane adjacent to the ring. This addition of membrane compensates for the increase in surface area that accompanies cytoplasmic division. When ring contraction is completed, membrane insertion and fusion seal the gap between the daughter cells.

Actin and Myosin II in the Contractile Ring Generate the Force for Cytokinesis

In interphase cells, actin and myosin II filaments form a cortical network underlying the plasma membrane. In some cells, they also form large cytoplasmic bundles called *stress fibers* (discussed in Chapter 16). As cells enter mitosis, these arrays of actin and myosin disassemble; much of the actin reorganizes, and myosin II filaments are released. As the sister chromatids separate in anaphase, actin and myosin II begin to accumulate in the rapidly assembling **contractile ring** (**Figure 17–42**), which also contains numerous other proteins that provide structural support or assist in ring assembly. Assembly of the contractile ring results in part from the local formation of new actin filaments, which depends on *formin* proteins that

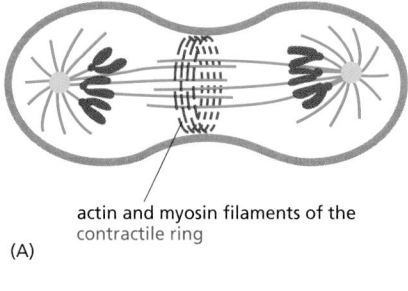

actin and myosin filaments of the
contractile ring

(A)

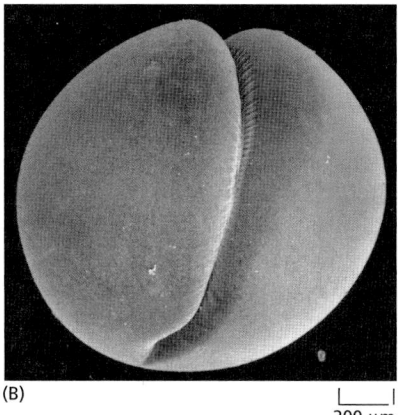

(B) |⎯⎯⎯|
                                       200 µm

(C) |⎯⎯|
                                       25 µm

Figure 17–41 Cytokinesis. (A) The actin–myosin bundles of the contractile ring are oriented as shown, so that their contraction pulls the membrane inward. (B) In this low-magnification scanning electron micrograph of a cleaving frog egg, the cleavage furrow is especially prominent, as the cell is unusually large. The furrowing of the cell membrane is caused by the activity of the contractile ring underneath it. (C) The surface of a furrow at higher magnification. (B and C, from H.W. Beams and R.G. Kessel, *Am. Sci.* 64:279–290, 1976. With permission from Sigma Xi.)

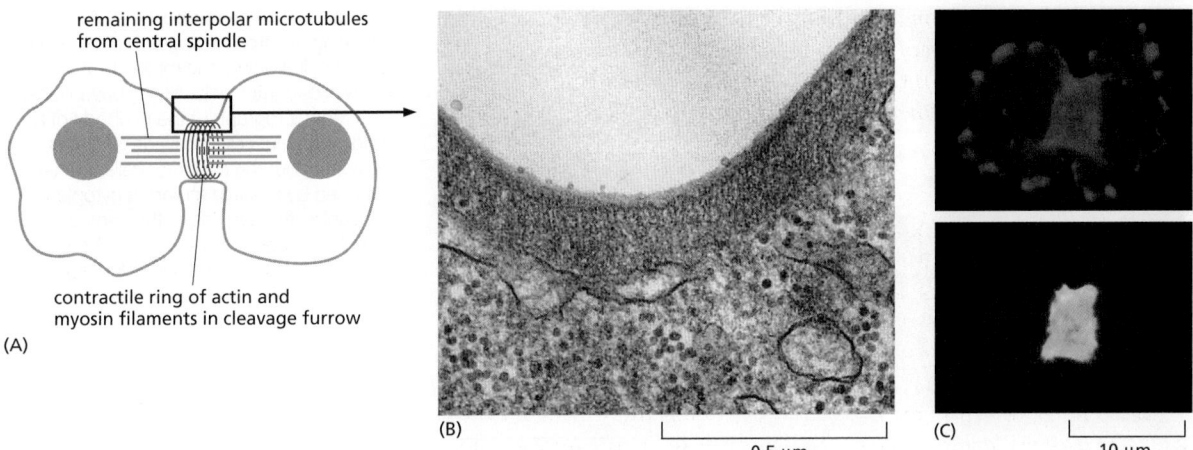

remaining interpolar microtubules
from central spindle

contractile ring of actin and
myosin filaments in cleavage furrow

(A)

(B)

0.5 μm

(C)

10 μm

Figure 17–42 The contractile ring.
(A) A drawing of the cleavage furrow in a dividing cell. (B) An electron micrograph of the ingrowing edge of a cleavage furrow of a dividing animal cell. (C) Fluorescence micrographs of a dividing slime mold amoeba stained for actin *(red)* and myosin II *(green)*. Whereas all of the visible myosin II has redistributed to the contractile ring, only some of the actin has done so; the rest remains in the cortex of the nascent daughter cells. (B, from H.W. Beams and R.G. Kessel, *Am. Sci.* 64:279–290, 1976. With permission from Sigma Xi; C, courtesy of Yoshio Fukui.)

nucleate the assembly of parallel arrays of linear, unbranched actin filaments (discussed in Chapter 16). After anaphase, the overlapping arrays of actin and myosin II filaments contract to generate the force that divides the cytoplasm in two. Once contraction begins, the ring exerts a force large enough to bend a fine glass needle that is inserted in its path. As the ring constricts, it maintains the same thickness, suggesting that its total volume and the number of filaments it contains decrease steadily. Moreover, unlike actin in muscle, the actin filaments in the ring are highly dynamic, and their arrangement changes continually during cytokinesis.

The contractile ring is finally dispensed with altogether when cleavage ends, as the plasma membrane of the cleavage furrow narrows to form the **midbody**. The midbody persists as a tether between the two daughter cells and contains the remains of the *central spindle*, a large protein structure derived from the antiparallel interpolar microtubules of the spindle midzone, packed tightly together within a dense matrix material (**Figure 17–43**). After the daughter cells separate completely, some of the components of the residual midbody often remain on the inside of the plasma membrane of each cell, where they may serve as a mark on the cortex that helps to orient the spindle in the subsequent cell division.

Local Activation of RhoA Triggers Assembly and Contraction of the Contractile Ring

RhoA, a small GTPase of the Ras superfamily (see Table 15–5), controls the assembly and function of the contractile ring at the site of cleavage. RhoA is activated at the cell cortex at the future division site, where it promotes actin filament formation, myosin II assembly, and ring contraction. It stimulates actin filament formation by activating formins, and it promotes myosin II assembly and contractions by activating multiple protein kinases, including the Rho-activated kinase Rock (**Figure 17–44**). These kinases phosphorylate the regulatory myosin light chain, a subunit of myosin II, thereby stimulating bipolar myosin II filament formation and motor activity.

RhoA is thought to be activated by a guanine nucleotide exchange factor (RhoGEF), which is found at the cell cortex at the future division site and stimulates the release of GDP and binding of GTP to RhoA (see Figure 17–44). We know little about how the RhoGEF is localized or activated at the division site, although the microtubules of the anaphase spindle seem to be involved, as we discuss next.

The Microtubules of the Mitotic Spindle Determine the Plane of Animal Cell Division

The central problem in cytokinesis is how to ensure that division occurs at the right time and in the right place. Cytokinesis must occur only after the two sets of chromosomes are fully segregated from each other, and the site of division must

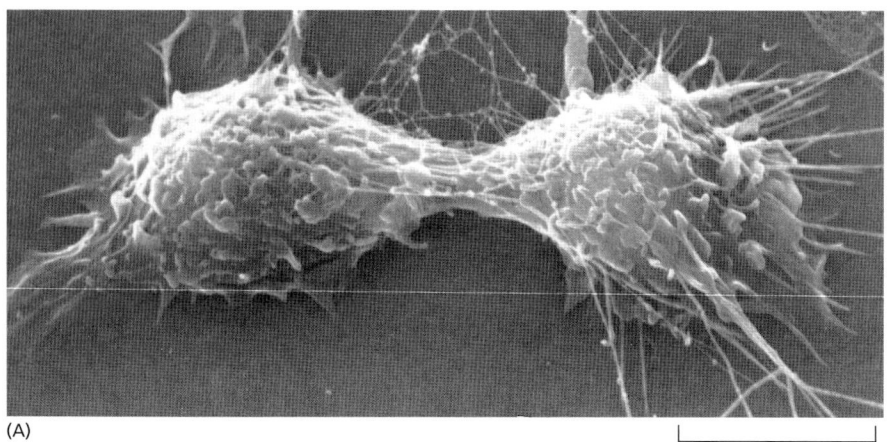

(A)

10 µm

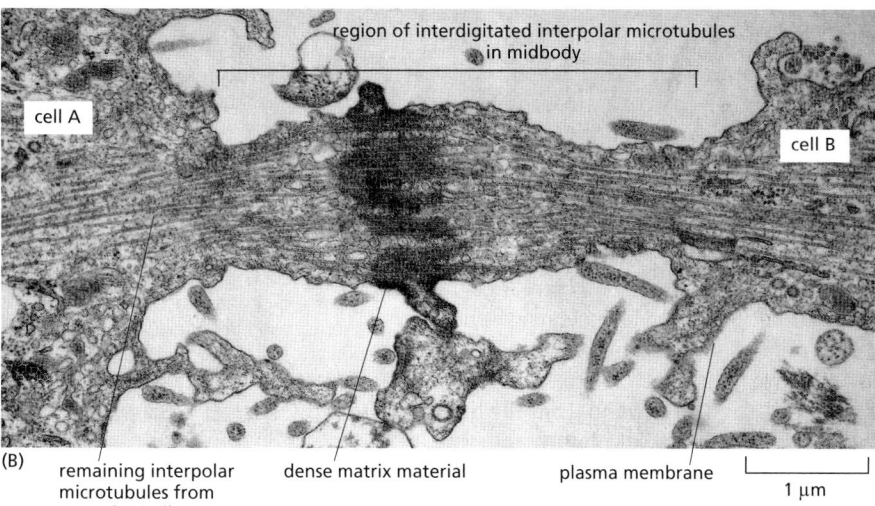

region of interdigitated interpolar microtubules in midbody

cell A

cell B

(B) remaining interpolar microtubules from central spindle   dense matrix material   plasma membrane

1 µm

Figure 17–43 The midbody. (A) A scanning electron micrograph of a cultured animal cell dividing; the midbody still joins the two daughter cells. (B) A conventional electron micrograph of the midbody of a dividing animal cell. Cleavage is almost complete, but the daughter cells remain attached by this thin strand of cytoplasm containing the remains of the central spindle. (A, courtesy of Guenter Albrecht-Buehler; B, courtesy of J.M. Mullins.)

be placed between the two sets of daughter chromosomes, thereby ensuring that each daughter cell receives a complete set. The correct timing and positioning of cytokinesis in animal cells are achieved by mechanisms that depend on the mitotic spindle. During anaphase, the spindle generates signals that initiate furrow formation at a position midway between the spindle poles, thereby ensuring that division occurs between the two sets of separated chromosomes. Because these signals originate in the anaphase spindle, this mechanism also contributes to the correct timing of cytokinesis in late mitosis. Cytokinesis also occurs at the correct time because dephosphorylation of Cdk substrates, which depends on cyclin destruction in metaphase and anaphase, initiates cytokinesis. We now describe these regulatory mechanisms in more detail, with an emphasis on cytokinesis in animal cells.

Studies of the fertilized eggs of marine invertebrates first revealed the importance of spindle microtubules in determining the placement of the contractile ring. After fertilization, these embryos cleave rapidly without intervening periods of growth. In this way, the original egg is progressively divided into smaller and smaller cells. Because the cytoplasm is clear, the spindle can be observed in real time with a microscope. If the spindle is tugged into a new position with a fine glass needle in early anaphase, the incipient cleavage furrow disappears, and a new one develops in accord with the new spindle site—supporting the idea that signals generated by the spindle induce local furrow formation.

How does the mitotic spindle specify the site of division? Three general mechanisms have been proposed, and most cells appear to employ a combination of these (Figure 17–45). The first is termed the *astral stimulation model*, in which the

Figure 17–44 Regulation of the contractile ring by the GTPase RhoA. Like other Rho family GTPases, RhoA is activated by a RhoGEF protein and inactivated by a Rho GTPase-activating protein (RhoGAP). The active GTP-bound form of RhoA is focused at the future cleavage site. By binding formins, activated RhoA promotes the assembly of actin filaments in the contractile ring. By activating Rho-activated protein kinases, such as Rock, it stimulates myosin II filament formation and activity, thereby promoting contraction of the ring.

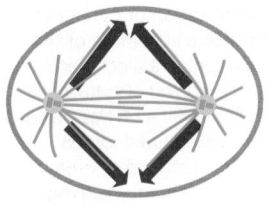
astral stimulation model

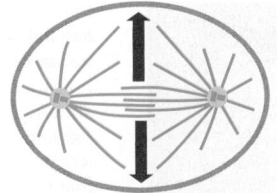
central spindle stimulation model

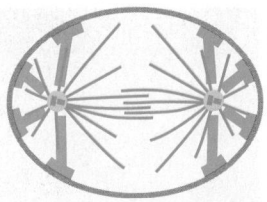
astral relaxation model

Figure 17–45 **Three current models of how the microtubules of the anaphase spindle generate signals that influence the positioning of the contractile ring.** No single model explains all the observations, and furrow positioning is probably determined by a combination of these mechanisms, with the importance of the different mechanisms varying in different organisms. See text for details.

astral microtubules carry furrow-inducing signals, which are somehow focused in a ring on the cell cortex, halfway between the spindle poles. Evidence for this model comes from ingenious experiments in large embryonic cells, which demonstrate that a cleavage furrow forms midway between two asters, even when the two centrosomes nucleating the asters are not connected to each other by a mitotic spindle (**Figure 17–46**).

A second possibility, called the *central spindle stimulation model*, is that the spindle midzone, or central spindle, generates a furrow-inducing signal that specifies the site of furrow formation at the cell cortex (see Figure 17–45). The overlapping interpolar microtubules of the central spindle associate with numerous signaling proteins, including proteins that may stimulate RhoA (**Figure 17–47**). Defects in the functions of these proteins (in *Drosophila* mutants, for example) result in failure of cytokinesis.

A third model proposes that, in some cell types, the astral microtubules promote the local relaxation of actin–myosin bundles at the cell cortex. According to this *astral relaxation model*, the cortical relaxation is minimal at the spindle equator, thus promoting cortical contraction at that site (see Figure 17–45). In the early embryos of *Caenorhabditis elegans*, for example, treatments that result in the loss of astral microtubules lead to increased contractile activity throughout the cell cortex, consistent with this model.

In some cell types, the site of ring assembly is chosen before mitosis. In budding yeasts, for example, a ring of proteins called *septins* assembles in late G$_1$ at the future division site. The septins are thought to form a scaffold onto which other components of the contractile ring, including myosin II, assemble. In plant cells, an organized band of microtubules and actin filaments, called the **preprophase band**, assembles just before mitosis and marks the site where the cell wall will assemble and divide the cell in two, as we now discuss.

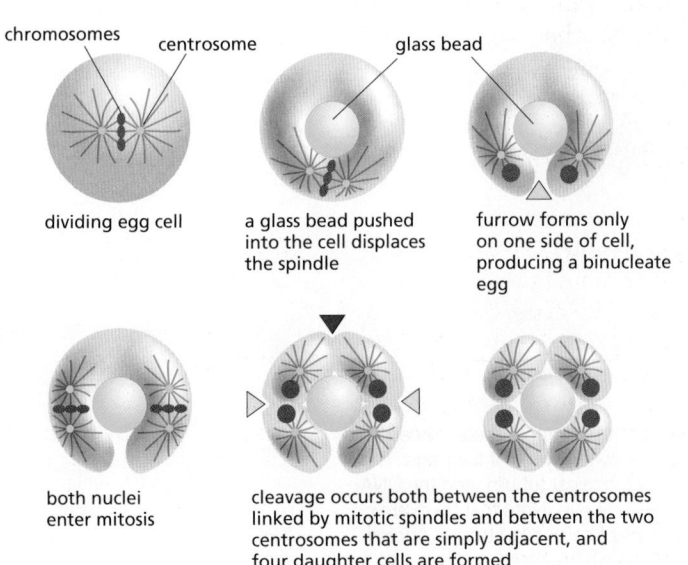

chromosomes centrosome glass bead

dividing egg cell

a glass bead pushed into the cell displaces the spindle

furrow forms only on one side of cell, producing a binucleate egg

both nuclei enter mitosis

cleavage occurs both between the centrosomes linked by mitotic spindles and between the two centrosomes that are simply adjacent, and four daughter cells are formed

Figure 17–46 **An experiment demonstrating the influence of the position of microtubule asters on the subsequent plane of cleavage in a large egg cell.** If the mitotic spindle is mechanically pushed to one side of the cell with a glass bead, the membrane furrowing is incomplete, failing to occur on the opposite side of the cell. Subsequent cleavages occur not only at the midzone of each of the two subsequent mitotic spindles *(yellow arrowheads)*, but also between the two adjacent asters that are not linked by a mitotic spindle—but in this abnormal cell share the same cytoplasm *(red arrowhead)*. Apparently, the contractile ring that produces the cleavage furrow in these cells always forms in the region midway between two asters, suggesting that the asters somehow alter the adjacent region of cell cortex to induce furrow formation between them.

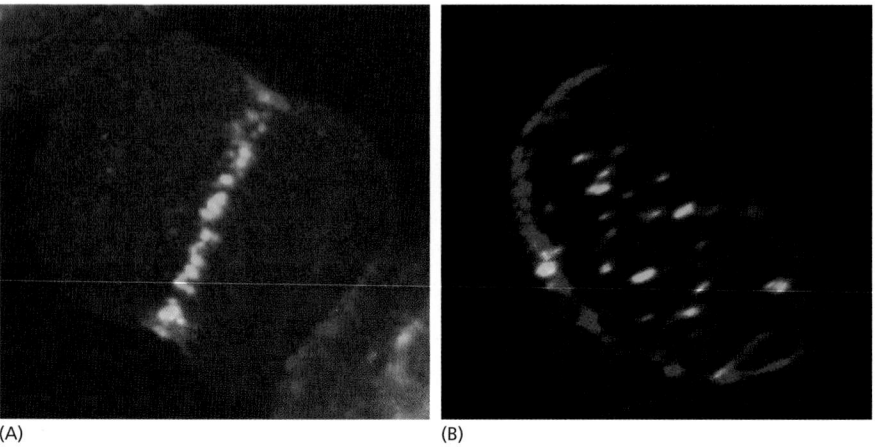

(A)   (B)

Figure 17–47 **Localization of cytokinesis regulators at the central spindle of the human cell.** (A) At center is a cultured human cell at the beginning of cytokinesis, showing the locations of the GTPase RhoA *(red)* and a protein called Cyk4 *(green)*, which is one of several regulatory proteins that form complexes at the overlapping plus ends of interpolar microtubules. These proteins are thought to generate signals that help control RhoA activity at the cell cortex. (B) When the same three-dimensional image is viewed in the plane of the contractile ring, as shown here, RhoA *(red)* is seen as a ring beneath the cell surface, while the central spindle protein Cyk4 *(green)* is associated with microtubule bundles scattered throughout the equatorial plane of the cell. (Courtesy of Alisa Piekny and Michael Glotzer.)

The Phragmoplast Guides Cytokinesis in Higher Plants

In most animal cells, the inward movement of the cleavage furrow depends on an increase in the surface area of the plasma membrane. New membrane is added at the inner edge of the cleavage furrow and is generally provided by small membrane vesicles that are transported on microtubules from the Golgi apparatus to the furrow.

Membrane deposition is particularly important for cytokinesis in higher-plant cells. These cells are enclosed by a semirigid *cell wall*. Rather than a contractile ring dividing the cytoplasm from the outside in, the cytoplasm of the plant cell is partitioned from the inside out by the construction of a new cell wall, called the **cell plate**, between the two daughter nuclei (**Figure 17–48**). The assembly of the cell plate begins in late anaphase and is guided by a structure called the **phragmoplast**, which contains microtubules derived from the mitotic spindle. Motor proteins transport small vesicles along these microtubules from the Golgi apparatus to the cell center. These vesicles, filled with polysaccharide and glycoproteins required for the synthesis of the new cell wall, fuse to form a disc-like, membrane-enclosed structure called the *early cell plate*. The plate expands outward by further

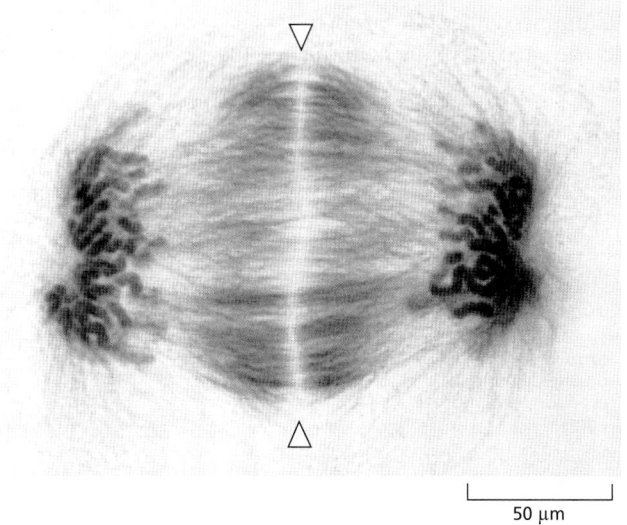

50 μm

Figure 17–48 **Cytokinesis in a plant cell in telophase.** In this light micrograph, the early cell plate (between the two *arrowheads*) has formed in a plane perpendicular to the plane of the page. The microtubules of the spindle are stained with gold-labeled antibodies against tubulin, and the DNA in the two sets of daughter chromosomes is stained with a fluorescent dye. Note that there are no astral microtubules, because there are no centrosomes in higher-plant cells. (Courtesy of Andrew Bajer.)

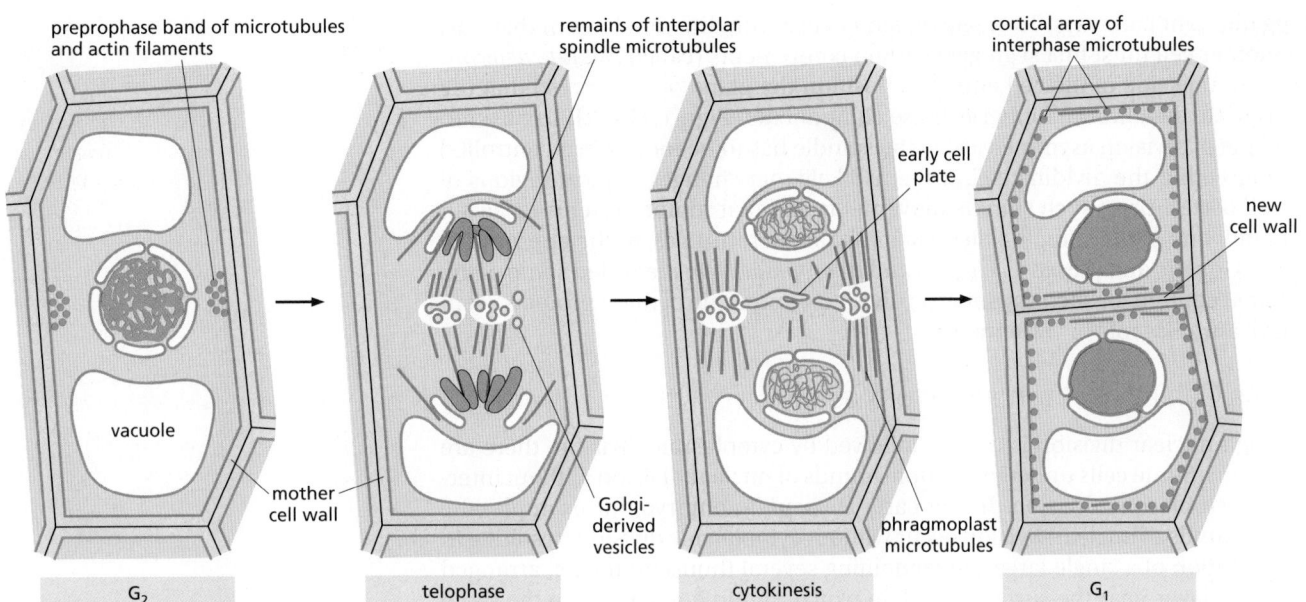

preprophase band of microtubules and actin filaments

remains of interpolar spindle microtubules

cortical array of interphase microtubules

early cell plate

new cell wall

vacuole

mother cell wall

Golgi-derived vesicles

phragmoplast microtubules

G₂ · telophase · cytokinesis · G₁

vesicle fusion until it reaches the plasma membrane and the original cell wall and divides the cell in two. Later, cellulose microfibrils are laid down within the matrix of the cell plate to complete the construction of the new cell wall (**Figure 17–49**).

Membrane-Enclosed Organelles Must Be Distributed to Daughter Cells During Cytokinesis

The process of mitosis ensures that each daughter cell receives a full complement of chromosomes. When a eukaryotic cell divides, however, each daughter cell must also inherit all of the other essential cell components, including the membrane-enclosed organelles. As discussed in Chapter 12, organelles such as mitochondria and chloroplasts cannot be assembled *de novo* from their individual components; they can arise only by the growth and division of the preexisting organelles. Similarly, cells cannot make a new endoplasmic reticulum (ER) unless some part of it is already present.

How, then, do the various membrane-enclosed organelles segregate when a cell divides? Organelles such as mitochondria and chloroplasts are usually present in large enough numbers to be safely inherited if, on average, their numbers roughly double once each cycle. The ER in interphase cells is continuous with the nuclear membrane and is organized by the microtubule cytoskeleton. Upon entry into M phase, the reorganization of the microtubules and breakdown of the nuclear envelope releases the ER. In most cells, the ER remains largely intact and is cut in two during cytokinesis. The Golgi apparatus is reorganized and fragmented during mitosis. Golgi fragments associate with the spindle poles and are thereby distributed to opposite ends of the spindle, ensuring that each daughter cell inherits the materials needed to reconstruct the Golgi in telophase.

Some Cells Reposition Their Spindle to Divide Asymmetrically

Most animal cells divide symmetrically: the contractile ring forms around the equator of the parent cell, producing two daughter cells of equal size and with the same components. This symmetry results from the placement of the mitotic spindle, which in most cases tends to center itself in the cytoplasm. Astral microtubules and motor proteins that either push or pull on these microtubules contribute to the centering process.

There are many instances in development, however, when cells divide asymmetrically to produce two cells that differ in size, in the cytoplasmic contents they inherit, or in both. Usually, the two different daughter cells are destined to develop

Figure 17–49 The special features of cytokinesis in a higher-plant cell. The division plane is established before M phase by a band of microtubules and actin filaments (the preprophase band) at the cell cortex. At the beginning of telophase, after the chromosomes have segregated, a new cell wall starts to assemble inside the cell at the equator of the old spindle. The interpolar microtubules of the mitotic spindle remaining at telophase form the phragmoplast. The plus ends of these microtubules no longer overlap but end at the cell equator. Golgi-derived vesicles, filled with cell-wall material, are transported along these microtubules and fuse to form the new cell wall, which grows outward to reach the plasma membrane and original cell wall. The plasma membrane and the membrane surrounding the new cell wall fuse, separating the two daughter cells.

along different pathways. To create daughter cells with different fates in this way, the mother cell must first segregate certain components (called *cell fate determinants*) to one side of the cell and then position the plane of division so that the appropriate daughter cell inherits these components (Figure 17–50). To position the plane of division asymmetrically, the spindle has to be moved in a controlled manner within the dividing cell. It seems likely that changes in local regions of the cell cortex direct such spindle movements and that motor proteins localized there pull one of the spindle poles, via its astral microtubules, to the appropriate region. Genetic analyses in *C. elegans* and *Drosophila* have identified some of the proteins required for such asymmetric divisions, and some of these proteins seem to have a similar role in vertebrates.

Mitosis Can Occur Without Cytokinesis

Although nuclear division is usually followed by cytoplasmic division, there are exceptions. Some cells undergo multiple rounds of nuclear division without intervening cytoplasmic division. In the early *Drosophila* embryo, for example, the first 13 rounds of nuclear division occur without cytoplasmic division, resulting in the formation of a single large cell containing several thousand nuclei, arranged in a monolayer near the surface. A cell in which multiple nuclei share the same cytoplasm is called a **syncytium**. This arrangement greatly speeds up early development, as the cells do not have to take the time to go through all the steps of cytokinesis for each division. After these rapid nuclear divisions, membranes are created around each nucleus in one round of coordinated cytokinesis called *cellularization*. The plasma membrane extends inward and, with the help of an actin–myosin ring, pinches off to enclose each nucleus (Figure 17–51).

Nuclear division without cytokinesis also occurs in some types of mammalian cells. Megakaryocytes, which produce blood platelets, and some hepatocytes and heart muscle cells, for example, become multinucleated in this way.

After cytokinesis, most cells enter G_1, in which Cdks are mostly inactive. We end this section by discussing how this state is achieved at the end of M phase.

The G_1 Phase Is a Stable State of Cdk Inactivity

A key regulatory event in late M phase is the inactivation of Cdks, which is driven primarily by APC/C-dependent cyclin destruction. As described earlier, the inactivation of Cdks in late M phase has many functions: it triggers the events of late mitosis, promotes cytokinesis, and enables the synthesis of prereplicative complexes at DNA replication origins. It also provides a mechanism for resetting the

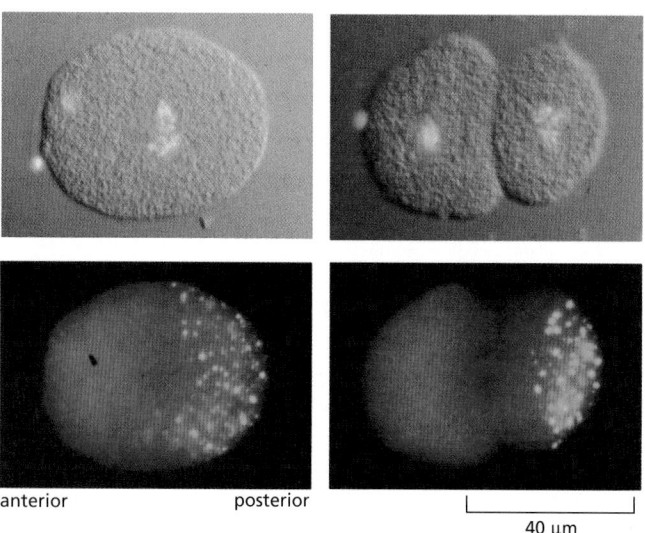

anterior   posterior

40 μm

Figure 17–50 An asymmetric cell division segregating cytoplasmic components to only one daughter cell. These light micrographs illustrate the controlled asymmetric segregation of specific cytoplasmic components to one daughter cell during the first division of a fertilized egg of the nematode *C. elegans*. The fertilized egg is shown in the *left* micrographs and the two daughter cells in the *right* micrographs. The cells *above* have been stained with a *blue*, DNA-binding, fluorescent dye to show the nucleus (and polar bodies); they are viewed by both differential-interference-contrast and fluorescence microscopy. The cells *below* are the same cells stained with an antibody against P-granules and viewed by fluorescence microscopy. These small granules are made of RNA and proteins and determine which cells become germ cells. They are distributed randomly throughout the cytoplasm of the unfertilized egg (not shown) but become segregated to the posterior pole of the fertilized egg. The cleavage plane is oriented to ensure that only the posterior daughter cell receives the P-granules when the egg divides. The same segregation process is repeated in several subsequent cell divisions, so that the P-granules end up only in cells that give rise to eggs and sperm. (Courtesy of Susan Strome.)

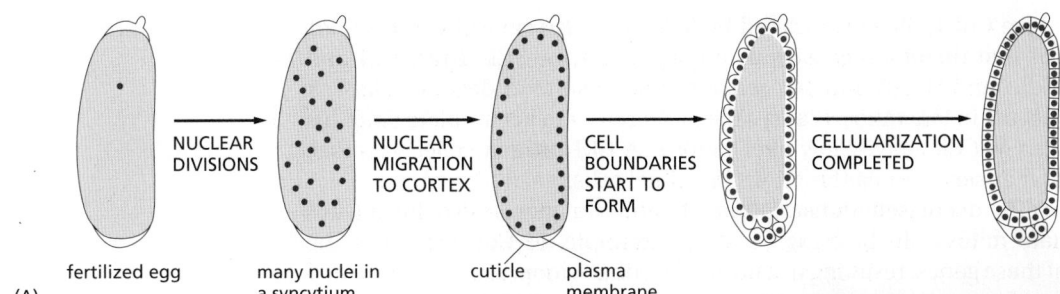

NUCLEAR DIVISIONS → NUCLEAR MIGRATION TO CORTEX → CELL BOUNDARIES START TO FORM → CELLULARIZATION COMPLETED

fertilized egg   many nuclei in a syncytium   cuticle   plasma membrane

(A)

Figure 17–51 Mitosis without cytokinesis in the early *Drosophila* embryo. (A) The first 13 nuclear divisions occur synchronously and without cytoplasmic division to create a large syncytium. Most of the nuclei migrate to the cortex, and the plasma membrane extends inward and pinches off to surround each nucleus to form individual cells in a process called cellularization. (B) Fluorescence micrograph of multiple mitotic spindles in a *Drosophila* embryo before cellularization. The microtubules are stained *green* and the centrosomes *red*. Note that all the nuclei go through the cycle synchronously; here, they are all in metaphase, with the unlabeled chromosomes seen as a dark band at the spindle equator. (B, courtesy of Kristina Yu and William Sullivan.)

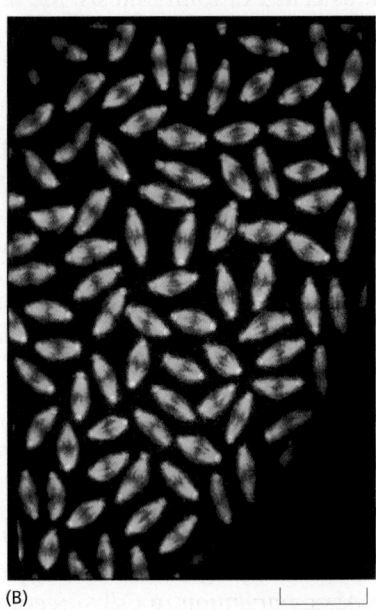

(B)

10 μm

cell-cycle control system to a state of Cdk inactivity as the cell prepares to enter a new cell cycle. In most cells, this state of Cdk inactivity generates a G_1 gap phase, during which the cell grows and monitors its environment before committing to a new cell cycle.

In early animal embryos, the inactivation of M-Cdk in late mitosis is due almost entirely to the action of Cdc20–APC/C, discussed earlier. Recall, however, that M-Cdk stimulates Cdc20–APC/C activity. Thus, the destruction of M-cyclin in late mitosis soon leads to the inactivation of all APC/C activity in an embryonic cell. This APC/C inactivation immediately after mitosis is especially useful in rapid embryonic cell cycles, as it allows the cell to quickly begin accumulating new M-cyclin for the next cycle (**Figure 17–52A**).

Rapid cyclin accumulation immediately after mitosis is not useful, however, for cells in which a G_1 phase is needed to allow control of entry into the next cell cycle. These cells employ several mechanisms to prevent Cdk reactivation after mitosis. One mechanism uses another APC/C-activating protein called Cdh1, mentioned earlier as a close relative of Cdc20 (see Table 17–2). Although both Cdh1 and Cdc20 bind to and activate the APC/C, they differ in one important respect. Whereas M-Cdk activates the Cdc20–APC/C complex, it inhibits the Cdh1–APC/C complex by directly phosphorylating Cdh1. As a result of this relationship, Cdh1–APC/C activity increases in late mitosis after the Cdc20–APC/C complex has initiated the destruction of M-cyclin. M-cyclin destruction therefore continues after mitosis: although Cdc20–APC/C activity has declined, Cdh1–APC/C activity is high (Figure 17–52B).

A second mechanism that suppresses Cdk activity in G_1 depends on the increased production of CKIs, the Cdk inhibitor proteins discussed earlier. Budding yeast cells, in which this mechanism is best understood, contain a CKI protein called *Sic1*, which binds to and inactivates M-Cdk in late mitosis and G_1 (see

Figure 17–52 The creation of a G_1 phase by stable Cdk inhibition after mitosis. (A) In early embryonic cell cycles, Cdc20–APC/C activity rises at the end of metaphase, triggering M-cyclin destruction. Because M-Cdk activity stimulates Cdc20–APC/C activity, the loss of M-cyclin leads to APC/C inactivation after mitosis, which allows M-cyclins to begin accumulating again. (B) In cells that have a G_1 phase, the drop in M-Cdk activity in late mitosis leads to the activation of Cdh1–APC/C (as well as to the accumulation of Cdk inhibitor proteins; not shown). This ensures a continued suppression of Cdk activity after mitosis, as required for a G_1 phase.

(A) embryonic cells with no G_1 phase

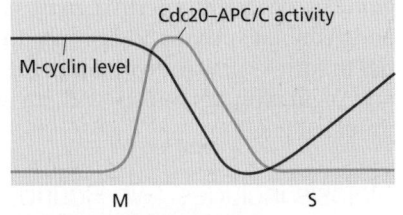

Cdc20–APC/C activity

M-cyclin level

M   S

(B) cells with G_1 phase

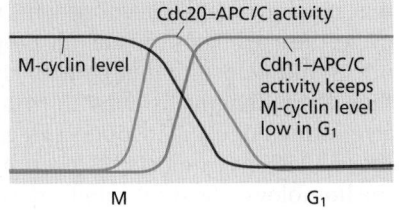

Cdc20–APC/C activity

M-cyclin level

Cdh1–APC/C activity keeps M-cyclin level low in G_1

M G_1

Table 17–2). Like Cdh1, Sic1 is inhibited by M-Cdk, which phosphorylates Sic1 during mitosis and thereby promotes its ubiquitylation by SCF. Thus, Sic1 and M-Cdk, like Cdh1 and M-Cdk, inhibit each other. As a result, the decline in M-Cdk activity that occurs in late mitosis causes the Sic1 protein to accumulate, and this CKI helps keep M-Cdk activity low after mitosis. A CKI protein called *p27* (see Figure 17–14) may serve similar functions in animal cells.

In most cells, decreased transcription of M-cyclin genes also inactivates M-Cdks in late mitosis. In budding yeast, for example, M-Cdk promotes the expression of these genes, resulting in a positive feedback loop. This loop is turned off as cells exit from mitosis: the inactivation of M-Cdk by Cdh1 and Sic1 leads to decreased M-cyclin gene transcription and thus decreased M-cyclin synthesis. Gene regulatory proteins that promote the expression of G_1/S- and S-cyclins are also inhibited during G_1.

Thus, Cdh1–APC/C activation, CKI accumulation, and decreased cyclin gene expression act together to ensure that the early G_1 phase is a time when essentially all Cdk activity is suppressed. As in many other aspects of cell-cycle control, the use of multiple regulatory mechanisms allows the system to operate with reasonable efficiency even if one mechanism fails. So how does the cell escape from this stable G_1 state to initiate a new cell cycle? The answer is that G_1/S-Cdk activity, which rises in late G_1, releases all the braking mechanisms that suppress Cdk activity, as we describe later, in the last section of this chapter.

Summary

After mitosis completes the formation of a pair of daughter nuclei, cytokinesis finishes the cell cycle by dividing the cell itself. Cytokinesis depends on a ring of actin and myosin filaments that contracts in late mitosis at a site midway between the segregated chromosomes. In animal cells, the positioning of the contractile ring is determined by signals emanating from the microtubules of the anaphase spindle. Dephosphorylation of Cdk targets, which results from Cdk inactivation in anaphase, triggers cytokinesis at the correct time after anaphase. After cytokinesis, the cell enters a stable G_1 state of low Cdk activity, where it awaits signals to enter a new cell cycle.

MEIOSIS

Most eukaryotic organisms reproduce sexually: the genomes of two parents mix to generate offspring that are genetically distinct from either parent. The cells of these organisms are generally *diploid*: that is, they contain two slightly different copies, or *homologs*, of each chromosome, one from each parent. Sexual reproduction depends on a specialized nuclear division process called *meiosis*, which produces *haploid* cells carrying only a single copy of each chromosome. In many organisms, the haploid cells differentiate into specialized reproductive cells called *gametes*—eggs and sperm in most species. In these species, the reproductive cycle ends when a sperm and egg fuse to form a diploid *zygote*, which has the potential to form a new individual. In this section, we consider the basic mechanisms and regulation of meiosis, with an emphasis on how they compare with those of mitosis.

Meiosis Includes Two Rounds of Chromosome Segregation

Meiosis reduces the chromosome number by half using many of the same molecular machines and control systems that operate in mitosis. As in the mitotic cell cycle, the cell begins the meiotic program by duplicating its chromosomes in meiotic S phase, resulting in pairs of sister chromatids that are tightly linked along their entire lengths by cohesin complexes. Unlike mitosis, however, two successive rounds of chromosome segregation then occur (**Figure 17–53**). The first of these divisions (**meiosis I**) solves the problem, unique to meiosis, of segregating the homologs. The duplicated paternal and maternal homologs pair up alongside each other and become physically linked by the process of genetic recombination. These pairs of homologs, each containing a pair of sister chromatids, then line up on the first meiotic spindle. In the first meiotic anaphase, duplicated homologs rather than sister chromatids are pulled apart and segregated into the two daughter nuclei. Only in the second division (**meiosis II**), which occurs without further DNA replication, are the sister chromatids pulled apart and segregated (as in mitosis) to produce haploid daughter nuclei. In this way, each diploid nucleus that enters meiosis produces four haploid nuclei, each of which contains either the maternal or paternal copy of each chromosome, but not both (Movie 17.7).

Figure 17–53 **Comparison of meiosis and mitosis.** For clarity, only one pair of homologous chromosomes (homologs) is shown. (A) Meiosis is a form of nuclear division in which a single round of chromosome duplication (meiotic S phase) is followed by two rounds of chromosome segregation. The duplicated homologs, each consisting of tightly bound sister chromatids, pair up and are segregated into different daughter nuclei in meiosis I; the sister chromatids are segregated in meiosis II. As indicated by the formation of chromosomes that are partly *red* and partly *gray*, homolog pairing in meiosis leads to genetic recombination (crossing-over) during meiosis I. Each diploid cell that enters meiosis therefore produces four genetically different haploid nuclei, which are distributed by cytokinesis into haploid cells that differentiate into gametes. (B) In mitosis, by contrast, homologs do not pair up, and the sister chromatids are segregated during the single division. Thus, each diploid cell that divides by mitosis produces two genetically identical diploid daughter nuclei, which are distributed by cytokinesis into a pair of daughter cells.

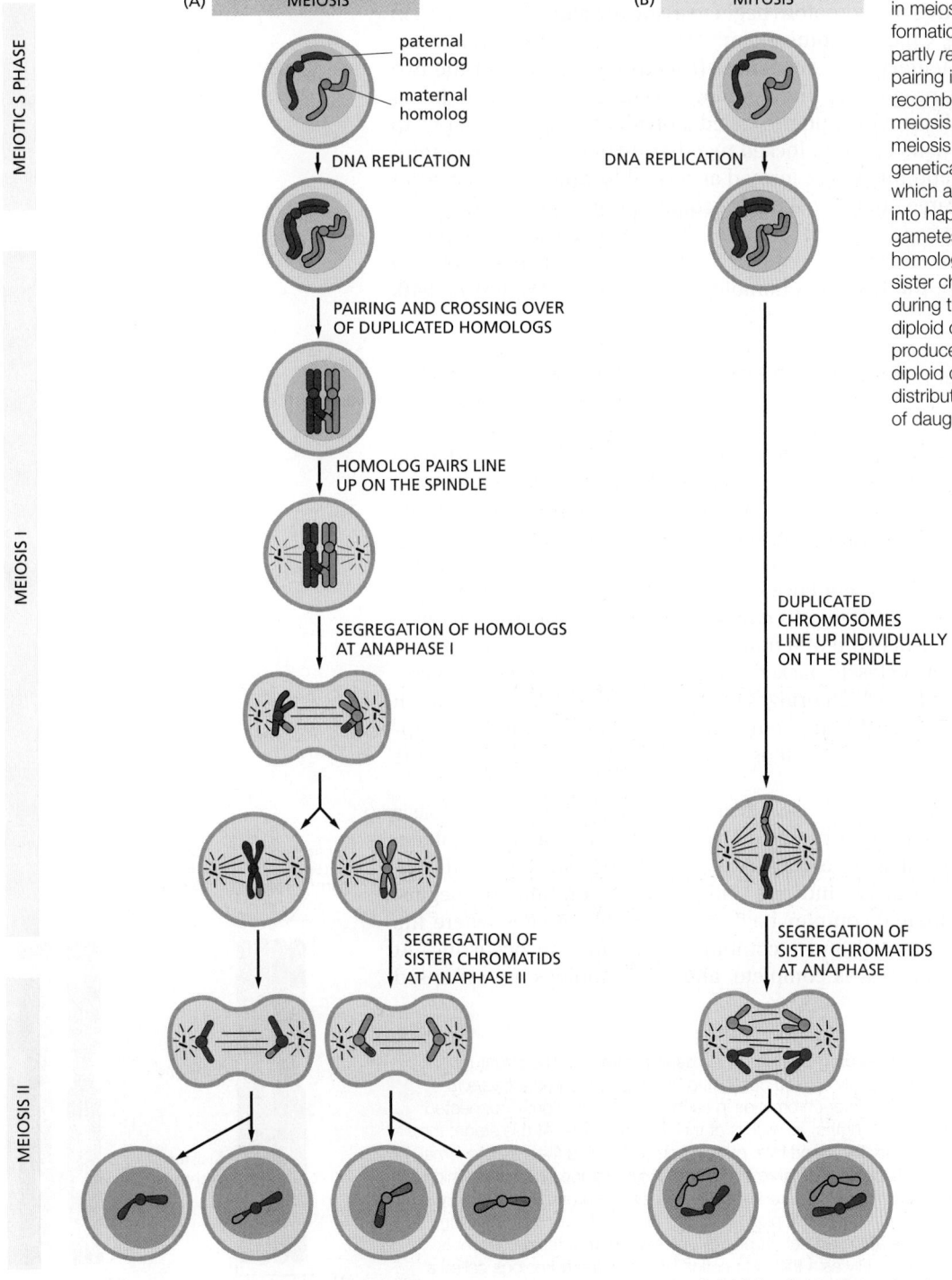

haploid daughter cells

diploid daughter cells

Duplicated Homologs Pair During Meiotic Prophase

During mitosis in most organisms, homologous chromosomes behave independently of each other. During meiosis I, however, it is crucial that homologs recognize each other and associate physically in order for the maternal and paternal homologs to be bi-oriented on the first meiotic spindle. Special mechanisms mediate these interactions.

The gradual juxtaposition of homologs occurs during a prolonged period called meiotic prophase (or prophase I), which can take hours in yeasts, days in mice, and weeks in higher plants. Like their mitotic counterparts, duplicated meiotic prophase chromosomes first appear as long threadlike structures, in which the sister chromatids are so tightly glued together that they appear as one. It is during early prophase I that the homologs begin to associate along their length in a process called **pairing**, which, in some organisms at least, begins with interactions between complementary DNA sequences (called *pairing sites*) in the two homologs. As prophase progresses, the homologs become more closely juxtaposed, forming a four-chromatid structure called a **bivalent** (**Figure 17–54A**). In most species, homolog pairs are then locked together by homologous recombination: DNA double-strand breaks are formed at several locations in each sister chromatid, resulting in large numbers of DNA recombination events between the homologs (as described in Chapter 5). Some of these events lead to reciprocal DNA exchanges called *crossovers*, where the DNA of a chromatid crosses over to become continuous with the DNA of a homologous chromatid (Figure 17–54B; also see Figure 5–54).

Homolog Pairing Culminates in the Formation of a Synaptonemal Complex

The paired homologs are brought into close juxtaposition, with their structural axes (*axial cores*) about 400 nm apart, by a mechanism that depends in most species on the double-strand DNA breaks that occur in sister chromatids. What pulls the axes together? One possibility is that the large protein machine, called a *recombination complex*, which assembles on a double-strand break in a chromatid, binds the matching DNA sequence in the nearby homolog and helps reel in this partner. This so-called *presynaptic alignment* of the homologs is followed by *synapsis*, in which the axial core of a homolog becomes tightly linked to the axial core of its partner by a closely packed array of *transverse filaments* to create a **synaptonemal complex**, which bridges the gap, now only 100 nm, between the homologs (**Figure 17–55**). Although crossing-over begins before the synaptonemal complex assembles, the final steps occur while the DNA is held in the complex.

The morphological changes that occur during homolog pairing are the basis for dividing meiotic prophase into five sequential stages—leptotene, zygotene, pachytene, diplotene, and diakinesis (**Figure 17–56**). Prophase starts with *leptotene*, when homologs condense and pair and genetic recombination begins. At *zygotene*, the synaptonemal complex begins to assemble at sites where the homologs are closely associated and recombination events are occurring. At *pachytene*, the assembly process is complete, and the homologs are synapsed

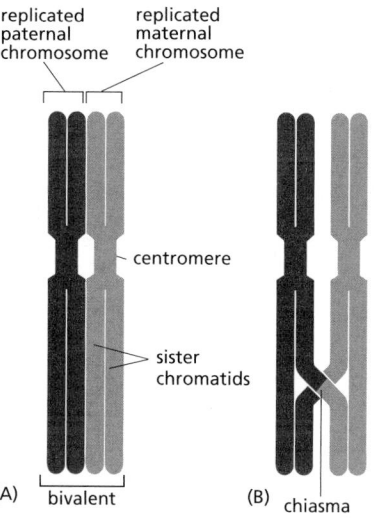

Figure 17–54 Homolog pairing and crossing-over. (A) The structure formed by two closely aligned duplicated homologs is called a *bivalent*. As in mitosis, the sister chromatids in each homolog are tightly connected along their entire lengths, as well as at their centromeres. At this stage, the homologs are usually joined by a protein complex called the *synaptonemal complex* (not shown; see Figure 17–55). (B) A later-stage bivalent in which a single crossover has occurred between nonsister chromatids. It is only when the synaptonemal complex disassembles and the paired homologs separate a little at the end of prophase I, as shown, that the crossover is seen microscopically as a thin connection between the homologs called a *chiasma*.

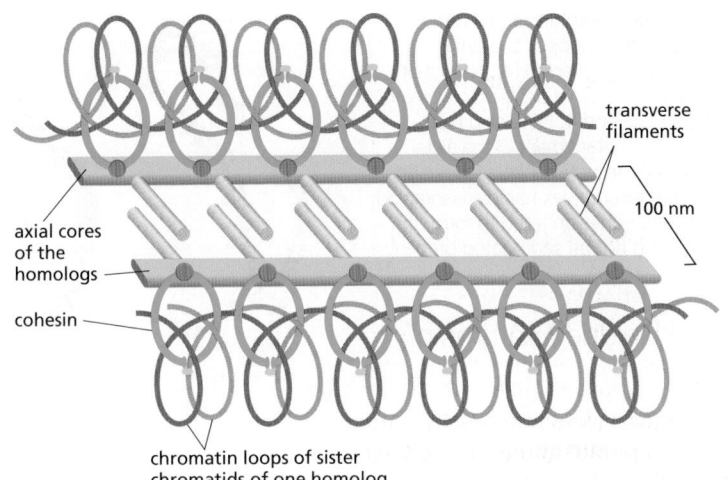

Figure 17–55 Simplified schematic drawing of a synaptonemal complex. Each homolog is organized around a protein axial core, and the synaptonemal complex forms when these homolog axes are linked by rod-shaped transverse filaments. The axial core of each homolog also interacts with the cohesin complexes that hold the sister chromatids together (see Figure 9–35). (Modified from K. Nasmyth, *Annu. Rev. Genet.* 35:673–745, 2001.)

along their entire lengths (see Figure 9–35). The pachytene stage can persist for days or longer, until desynapsis begins at *diplotene* with the disassembly of the synaptonemal complexes and the concomitant condensation and shortening of the chromosomes. It is only at this stage, after the complexes have disassembled, that the individual crossover events between nonsister chromatids can be seen as inter-homolog connections called **chiasmata** (singular **chiasma**), which now play a crucial part in holding the compact homologs together (**Figure 17–57**). The homologs are now ready to begin the process of segregation.

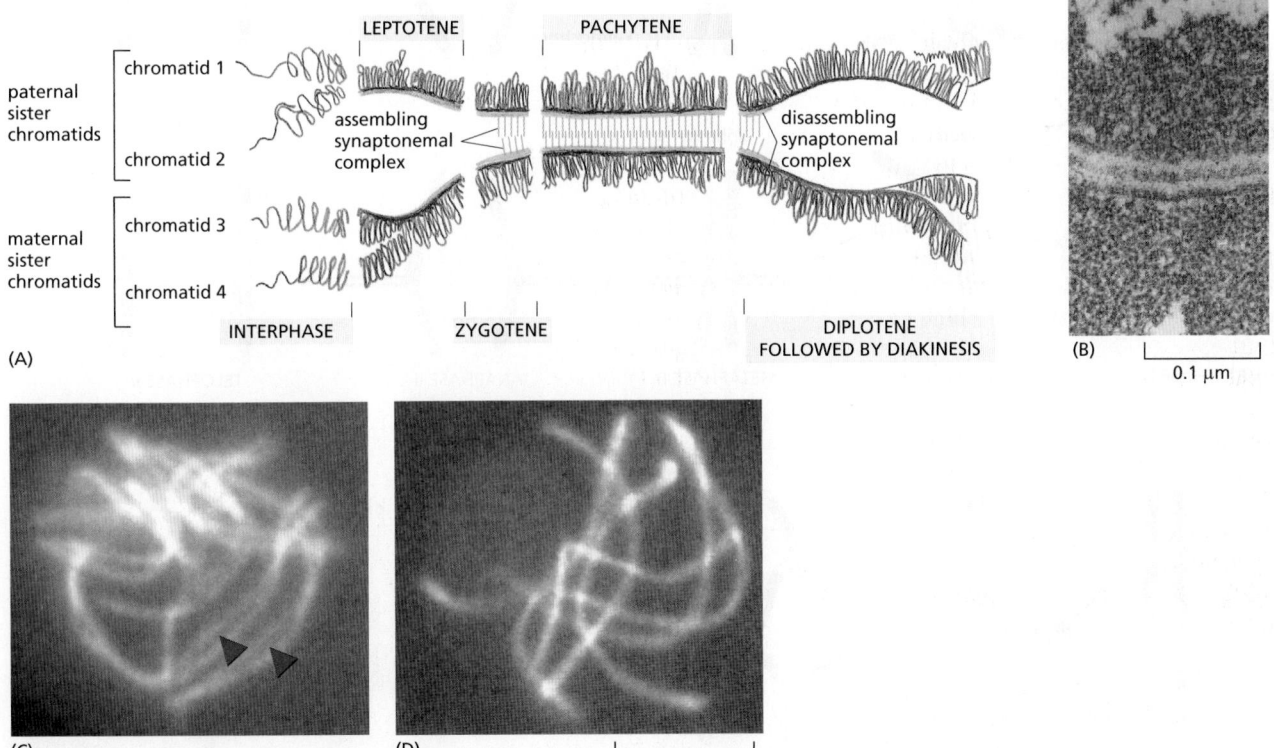

Figure 17–56 Homolog synapsis and desynapsis during the different stages of prophase I. (A) A single bivalent is shown schematically. At leptotene, the two sister chromatids coalesce, and their chromatid loops extend out from a common axial core. Assembly of the synaptonemal complex begins in early zygotene and is complete in pachytene. The complex disassembles in diplotene. (B) An electron micrograph of a synaptonemal complex from a meiotic cell at pachytene in a lily flower. (C and D) Immunofluorescence micrographs of prophase I cells of the fungus *Sordaria*. Partially synapsed bivalents at zygotene are shown in (C) and fully synapsed bivalents are shown in (D). *Red arrowheads* in (C) point to regions where synapsis is still incomplete. (B, courtesy of Brian Wells; C and D, from A. Storlazzi et al., *Genes Dev.* 17:2675–2687, 2003. With permission from Cold Spring Harbor Laboratory Press.)

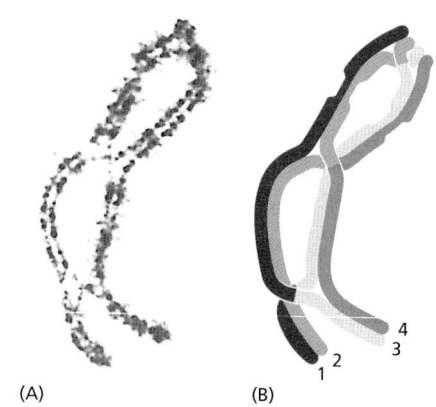

Figure 17–57 A bivalent with three chiasmata resulting from three crossover events. (A) Light micrograph of a grasshopper bivalent. (B) Drawing showing the arrangement of the crossovers in (A). Note that chromatid 1 has undergone an exchange with chromatid 3, and chromatid 2 has undergone exchanges with chromatids 3 and 4. Note also how the combination of the chiasmata and the tight attachment of the sister chromatid arms to each other (mediated by cohesin complexes) holds the two homologs together after the synaptonemal complex has disassembled; if either the chiasmata or the sister-chromatid cohesion failed to form, the homologs would come apart at this stage and not be segregated properly in meiosis I. (A, courtesy of Bernard John.)

Homolog Segregation Depends on Several Unique Features of Meiosis I

A fundamental difference between meiosis I and mitosis (and meiosis II) is that in meiosis I homologs rather than sister chromatids separate and then segregate (see Figure 17–53). This difference depends on three features of meiosis I that distinguish it from mitosis (**Figure 17–58**).

First, both sister kinetochores in a homolog must attach stably to the same spindle pole. This type of attachment is normally avoided during mitosis (see Figure 17–33). In meiosis I, however, the two sister kinetochores are fused into a single microtubule-binding unit that attaches to just one pole (see Figure 17–58A). The fusion of sister kinetochores is achieved by a complex of proteins that is

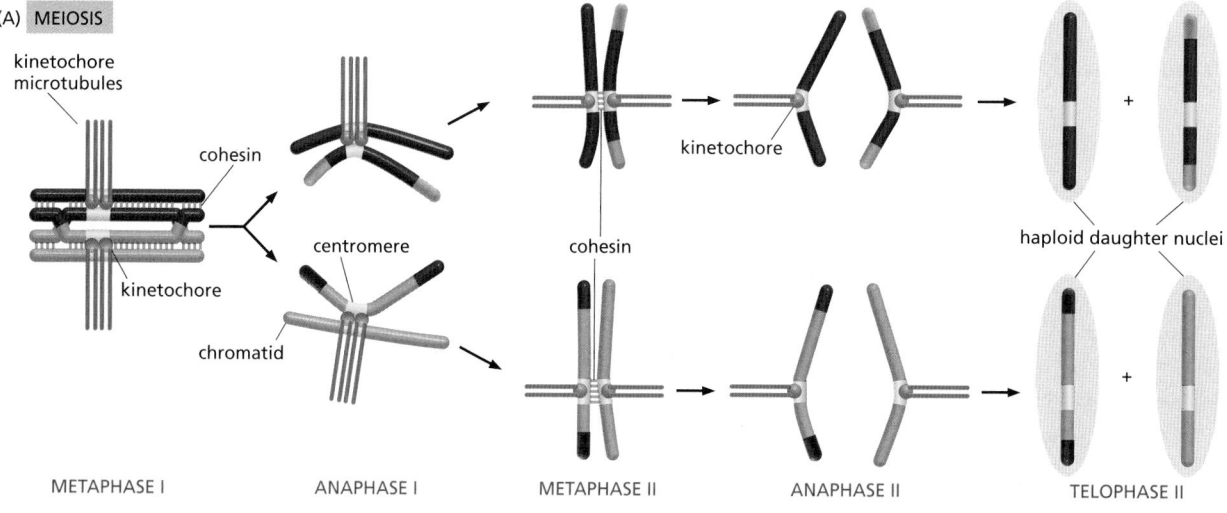

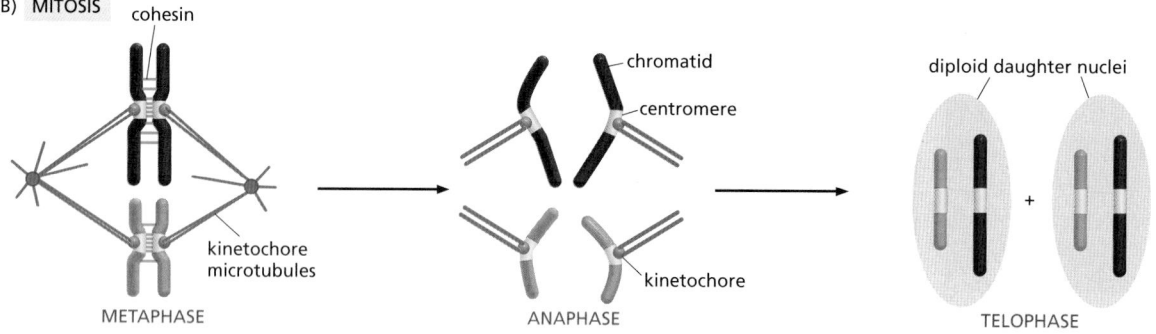

Figure 17–58 Comparison of chromosome behavior in meiosis I, meiosis II, and mitosis. Chromosomes behave similarly in mitosis and meiosis II, but they behave very differently in meiosis I. (A) In meiosis I, the two sister kinetochores are located side-by-side on each homolog and attach to microtubules from the same spindle pole. The proteolytic cleavage of cohesin along the sister-chromatid arms unglues the arms and resolves the crossovers, allowing the duplicated homologs to separate at anaphase I, while the residual cohesin at the centromeres keeps the sisters together. Cleavage of centromeric cohesin allows the sister chromatids to separate at anaphase II. (B) In mitosis, by contrast, the two sister kinetochores attach to microtubules from different spindle poles, and the two sister chromatids come apart at the start of anaphase and segregate into separate daughter nuclei.

localized at the kinetochores in meiosis I, but we do not know in any detail how these proteins work. They are removed from kinetochores after meiosis I, so that in meiosis II the sister-chromatid pairs can be bi-oriented on the spindle as they are in mitosis.

Second, crossovers generate a strong physical linkage between homologs, allowing their bi-orientation at the equator of the spindle—much like cohesion between sister chromatids is important for their bi-orientation in mitosis (and meiosis II). Crossovers hold homolog pairs together only because the arms of the sister chromatids are connected by sister-chromatid cohesion (see Figure 17–58A).

Third, cohesion is removed in anaphase I only from chromosome arms and not from the regions near the centromeres, where the kinetochores are located. The loss of arm cohesion triggers homolog separation at the onset of anaphase I. This process depends on APC/C activation, which leads to securin destruction, separase activation, and cohesin cleavage along the arms (see Figure 17–38).

Cohesins near the centromeres are protected from separase in meiosis I by a kinetochore-associated protein called *shugoshin* (from the Japanese word for "guardian spirit"). Shugoshin acts by recruiting a protein phosphatase that removes phosphates from centromeric cohesins. Cohesin phosphorylation is normally required for separase to cleave cohesin; thus, removal of this phosphorylation near the centromere prevents cohesin cleavage. Sister-chromatid pairs therefore remain linked through meiosis I, allowing their correct bi-orientation on the spindle in meiosis II. Shugoshin is inactivated after meiosis I. At the onset of anaphase II, APC/C activation triggers centromeric cohesin cleavage and sister-chromatid separation—much as it does in mitosis. Following anaphase II, nuclear envelopes form around the chromosomes to produce four haploid nuclei, after which cytokinesis and other differentiation processes lead to the production of haploid gametes.

Crossing-Over Is Highly Regulated

Crossing-over has two distinct functions in meiosis: it helps hold homologs together so that they are properly segregated to the two daughter nuclei produced by meiosis I, and it contributes to the genetic diversification of the gametes that are eventually produced. As might be expected, therefore, crossing-over is highly regulated: the number and location of double-strand breaks along each chromosome is controlled, as is the likelihood that a break will be converted into a crossover. On average, the result of this regulation is that each pair of human homologs is linked by about two or three crossovers (**Figure 17–59**).

Although the double-strand breaks that occur in meiosis I can be located almost anywhere along the chromosome, they are not distributed uniformly: they cluster at "hot spots," where the DNA is accessible, and occur only rarely in "cold spots," such as the heterochromatin regions around centromeres and telomeres.

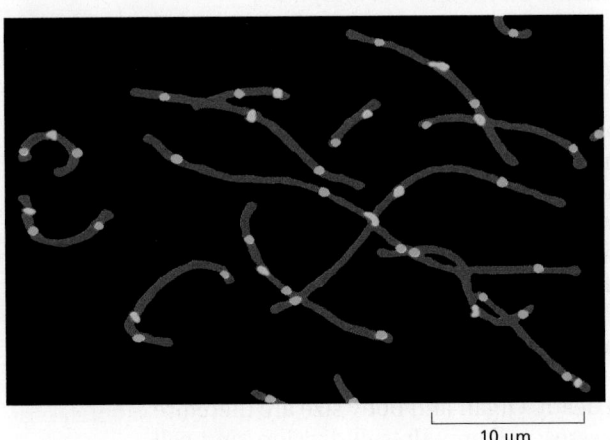

10 µm

Figure 17–59 **Crossovers between homologs in the human testis.** In these immunofluorescence micrographs, antibodies have been used to stain the synaptonemal complexes *(red)*, the centromeres *(blue)*, and the sites of crossing-over *(green)*. Note that all of the bivalents have at least one crossover and none have more than four. (Modified from A. Lynn et al., *Science* 296:2222–2225, 2002. With permission from AAAS.)

At least two kinds of regulation influence the location and number of crossovers that form, neither of which is well understood. Both operate before the synaptonemal complex assembles. One ensures that at least one crossover forms between the members of each homolog pair, as is necessary for normal homolog segregation in meiosis I. In the other, called *crossover interference*, the presence of one crossover event inhibits another from forming close by, perhaps by locally depleting proteins required for converting a double-strand DNA break into a stable crossover.

Meiosis Frequently Goes Wrong

The sorting of chromosomes that takes place during meiosis is a remarkable feat of intracellular bookkeeping. In humans, each meiosis requires that the starting cell keep track of 92 chromatids (46 chromosomes, each of which has duplicated), distributing one complete set of each type of autosome to each of the four haploid progeny. Not surprisingly, mistakes can occur in allocating the chromosomes during this elaborate process. Mistakes are especially common in human female meiosis, which arrests for years after diplotene: meiosis I is completed only at *ovulation*, and meiosis II only after the egg is fertilized. Indeed, such chromosome segregation errors during egg development are the most common cause of both spontaneous abortion (miscarriage) and mental retardation in humans.

When homologs fail to separate properly—a phenomenon called **nondisjunction**—the result is that some of the resulting haploid gametes lack a particular chromosome, while others have more than one copy of it. Upon fertilization, these gametes form abnormal embryos, most of which die. Some survive, however. *Down syndrome* in humans, for example, which is the leading cause of mental retardation, is caused by an extra copy of chromosome 21, usually resulting from nondisjunction during meiosis I in the female ovary. Segregation errors during meiosis I increase greatly with advancing maternal age.

Summary

Haploid gametes are produced by meiosis, in which a diploid nucleus undergoes two successive cell divisions after one round of DNA replication. Meiosis is dominated by a prolonged prophase. At the start of prophase, the chromosomes have replicated and consist of two tightly joined sister chromatids. Homologous chromosomes then pair up and become progressively more closely juxtaposed as prophase proceeds. The tightly aligned homologs undergo genetic recombination, forming crossovers that help hold each pair of homologs together during metaphase I. Meiosis-specific, kinetochore-associated proteins help ensure that both sister chromatids in a homolog attach to the same spindle pole; other kinetochore-associated proteins ensure that the homologs remain connected at their centromeres during anaphase I, so that homologs rather than sister chromatids are segregated in meiosis I. After meiosis I, meiosis II follows rapidly, without DNA replication, in a process that resembles mitosis, in that sister chromatids are pulled apart at anaphase.

CONTROL OF CELL DIVISION AND CELL GROWTH

A fertilized mouse egg and a fertilized human egg are similar in size, yet they produce animals of very different sizes. What factors in the control of cell behavior in humans and mice are responsible for these size differences? The same fundamental question can be asked for each organ and tissue in an animal's body. What factors determine the length of an elephant's trunk or the size of its brain or its liver? These questions are largely unanswered, but it is nevertheless possible to say what the ingredients of an answer must be.

The size of an organ or organism depends on its total cell mass, which depends on both the total number of cells and their size. Cell number, in turn, depends on the amounts of cell division and cell death. Organ and body size are therefore determined by three fundamental processes: cell growth, cell division, and cell survival. Each is tightly regulated—both by intracellular programs and by extracellular signal molecules that control these programs.

The extracellular signal molecules that regulate cell growth, division, and survival are generally soluble secreted proteins, proteins bound to the surface of cells, or components of the extracellular matrix. They can be divided operationally into three major classes:

1. *Mitogens*, which stimulate cell division, primarily by triggering a wave of G_1/S-Cdk activity that relieves intracellular negative controls that otherwise block progress through the cell cycle.

2. *Growth factors*, which stimulate cell growth (an increase in cell mass) by promoting the synthesis of proteins and other macromolecules and by inhibiting their degradation.

3. *Survival factors*, which promote cell survival by suppressing the form of programmed cell death known as *apoptosis*.

Many extracellular signal molecules promote all of these processes, while others promote one or two of them. Indeed, the term *growth factor* is often used inappropriately to describe a factor that has any of these activities. Even worse, the term *cell growth* is often used to mean an increase in cell number, or *cell proliferation*.

In addition to these three classes of stimulating signals, there are extracellular signal molecules that suppress cell proliferation, cell growth, or both; in general, less is known about them. There are also extracellular signal molecules that activate apoptosis.

In this section, we focus primarily on how mitogens and other factors, such as DNA damage, control the rate of cell division. We then turn to the important but poorly understood problem of how a proliferating cell coordinates its growth with cell division so as to maintain its appropriate size. We discuss the control of cell survival and cell death by apoptosis in Chapter 18.

Mitogens Stimulate Cell Division

Unicellular organisms tend to grow and divide as fast as they can, and their rate of proliferation depends largely on the availability of nutrients in the environment. The cells of a multicellular organism, however, divide only when the organism needs more cells. Thus, for an animal cell to proliferate, it must receive stimulatory extracellular signals, in the form of **mitogens**, from other cells, usually its neighbors. Mitogens overcome intracellular braking mechanisms that block progress through the cell cycle.

One of the first mitogens to be identified was *platelet-derived growth factor* (*PDGF*), and it is typical of many others discovered since. The path to its isolation began with the observation that fibroblasts in a culture dish proliferate when provided with *serum* but not when provided with *plasma*. Plasma is prepared by removing the cells from blood without allowing clotting to occur; serum is prepared by allowing blood to clot and taking the cell-free liquid that remains. When blood clots, platelets incorporated in the clot are stimulated to release the contents of their secretory vesicles (**Figure 17–60**). The superior ability of serum to support cell proliferation suggested that platelets contain one or more mitogens. This hypothesis was confirmed by showing that extracts of platelets could serve instead of serum to stimulate fibroblast proliferation. The crucial factor in the extracts was shown to be a protein, which was subsequently purified and named PDGF. In the body, PDGF liberated from blood clots helps stimulate cell division during wound healing.

PDGF is only one of over 50 animal proteins that are known to act as mitogens. Most of these proteins have a broad specificity. PDGF, for example, can stimulate many types of cells to divide, including fibroblasts, smooth muscle cells, and neuroglial cells. Similarly, *epidermal growth factor* (*EGF*) acts not only on epidermal cells but also on many other cell types, including both epithelial and nonepithelial cells. Some mitogens, however, have a narrow specificity; *erythropoietin*, for

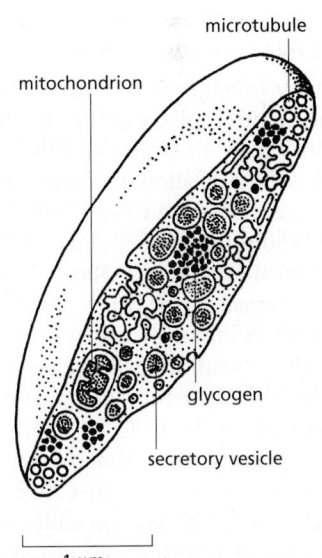

Figure 17–60 A platelet. Platelets are miniature cells without a nucleus. They circulate in the blood and help stimulate blood clotting at sites of tissue damage, thereby preventing excessive bleeding. They also release various factors that stimulate wound healing. The platelet shown here has been cut in half to show its secretory vesicles, some of which contain platelet-derived growth factor (PDGF).

example, only induces the proliferation of red blood cell precursors. Many mitogens, including PDGF, also have actions other than the stimulation of cell division: they can stimulate cell growth, survival, differentiation, or migration, depending on the circumstances and the cell type.

In some tissues, inhibitory extracellular signal proteins oppose the positive regulators and thereby inhibit organ growth. The best-understood inhibitory signal proteins are transforming growth factor-β (TGFβ) and its relatives. TGFβ inhibits the proliferation of several cell types, mainly by blocking cell-cycle progression in G_1.

Cells Can Enter a Specialized Nondividing State

In the absence of a mitogenic signal to proliferate, Cdk inhibition in G_1 is maintained by the multiple mechanisms discussed earlier, and progression into a new cell cycle is blocked. In some cases, cells partly disassemble their cell-cycle control system and withdraw from the cycle to a specialized nondividing state called **G_0**.

Most cells in our body are in G_0, but the molecular basis and reversibility of this state vary in different cell types. Most of our neurons and skeletal muscle cells, for example, are in a *terminally differentiated* G_0 state, in which their cell-cycle control system is completely dismantled: the expression of the genes encoding various Cdks and cyclins is permanently turned off, and cell division rarely occurs. Some cell types withdraw from the cell cycle only transiently and retain the ability to reassemble the cell-cycle control system quickly and re-enter the cycle. Most liver cells, for example, are in G_0, but they can be stimulated to divide if the liver is damaged. Still other types of cells, including fibroblasts and some lymphocytes, withdraw from and re-enter the cell cycle repeatedly throughout their lifetime.

Almost all the variation in cell-cycle length in the adult body occurs during the time the cell spends in G_1 or G_0. By contrast, the time a cell takes to progress from the beginning of S phase through mitosis is usually brief (typically 12–24 hours in mammals) and relatively constant, regardless of the interval from one division to the next.

Mitogens Stimulate G_1-Cdk and G_1/S-Cdk Activities

For the vast majority of animal cells, mitogens control the rate of cell division by acting in the G_1 phase of the cell cycle. As discussed earlier, multiple mechanisms act during G_1 to suppress Cdk activity. Mitogens release these brakes on Cdk activity, thereby allowing entry into a new cell cycle.

As we discuss in Chapter 15, mitogens interact with cell-surface receptors to trigger multiple intracellular signaling pathways. One major pathway acts through the monomeric GTPase **Ras**, which leads to the activation of a *mitogen-activated protein kinase (MAP kinase) cascade* (see Figure 15–49). This leads to an increase in the production of transcription regulatory proteins, including **Myc**. Myc is thought to promote cell-cycle entry by several mechanisms, one of which is to increase the expression of genes encoding G_1 cyclins (D cyclins), thereby increasing G_1-Cdk (cyclin D–Cdk4) activity. Myc also has a major role in stimulating the transcription of genes that increase cell growth.

The key function of G_1-Cdk complexes in animal cells is to activate a group of gene regulatory factors called the **E2F proteins**, which bind to specific DNA sequences in the promoters of a wide variety of genes that encode proteins required for S-phase entry, including G_1/S-cyclins, S-cyclins, and proteins involved in DNA synthesis and chromosome duplication. In the absence of mitogenic stimulation, E2F-dependent gene expression is inhibited by an interaction between E2F and members of the **retinoblastoma protein (Rb)** family. When cells are stimulated to divide by mitogens, active G_1-Cdk accumulates and phosphorylates Rb family members, reducing their binding to E2F. The liberated E2F proteins then activate expression of their target genes (**Figure 17–61**).

This transcriptional control system, like so many other control systems that regulate the cell cycle, includes feedback loops that ensure that entry into the

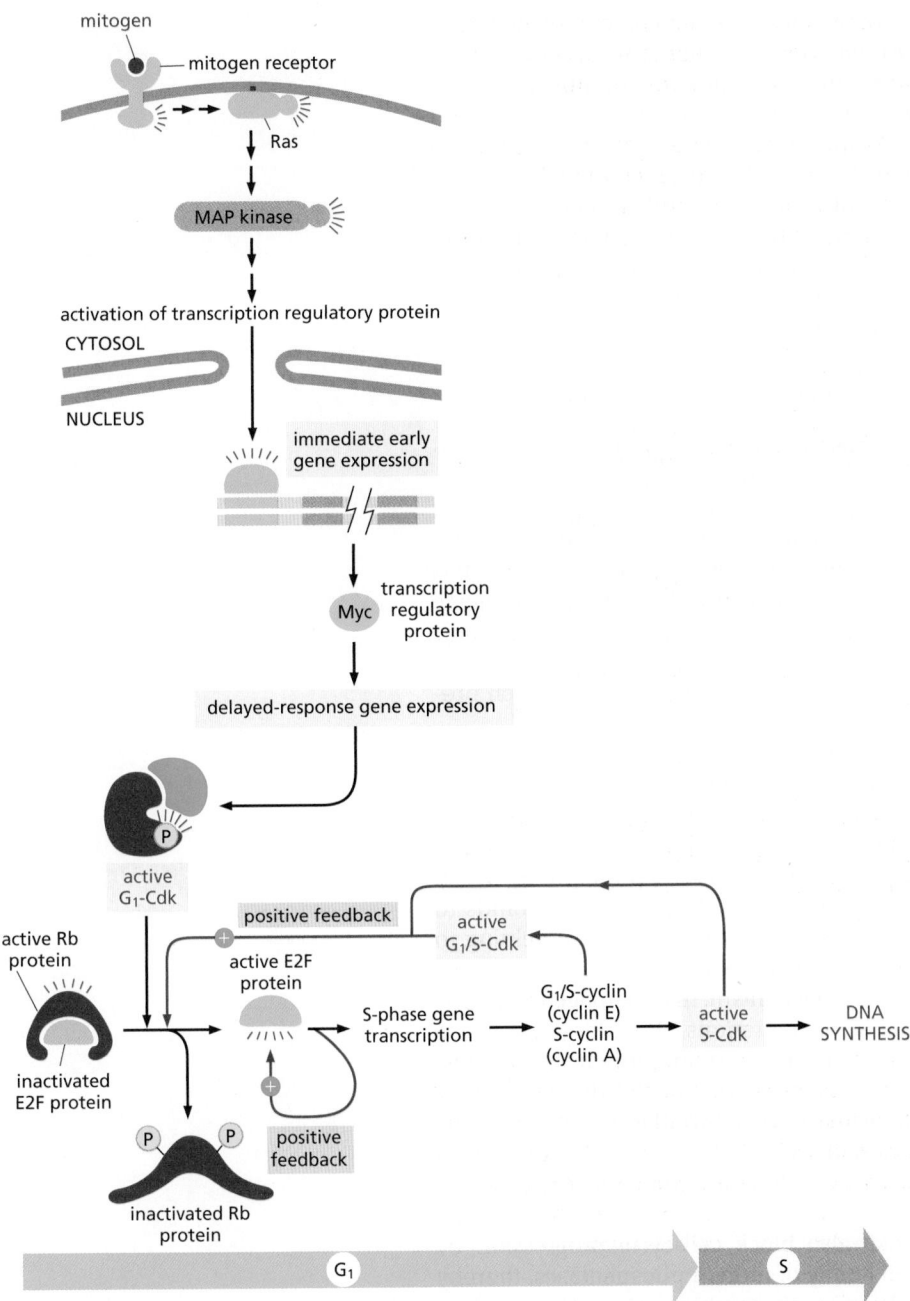

Figure 17–61 **Mitogen stimulation of cell-cycle entry.** As discussed in Chapter 15, mitogens bind to cell-surface receptors to initiate intracellular signaling pathways. One of the major pathways involves activation of the small GTPase Ras, which activates a MAP kinase cascade, leading to increased expression of numerous *immediate early* genes, including the gene encoding the transcription regulatory protein Myc. Myc increases the expression of many *delayed-response* genes, including some that lead to increased G_1-Cdk activity (cyclin D–Cdk4), which triggers the phosphorylation of members of the Rb family of proteins. This inactivates the Rb proteins, freeing the gene regulatory protein E2F to activate the transcription of G_1/S genes, including the genes for a G_1/S-cyclin (cyclin E) and S-cyclin (cyclin A). The resulting G_1/S-Cdk and S-Cdk activities further enhance Rb protein phosphorylation, forming a positive feedback loop. E2F proteins also stimulate the transcription of their own genes, forming another positive feedback loop.

cell cycle is complete and irreversible. The liberated E2F proteins, for example, increase the transcription of their own genes. In addition, E2F-dependent transcription of G_1/S-cyclin (cyclin E) and S-cyclin (cyclin A) genes leads to increased G_1/S-Cdk and S-Cdk activities, which in turn increase Rb protein phosphorylation and promote further E2F release (see Figure 17–61).

The central member of the Rb family, the Rb protein itself, was identified originally through studies of an inherited form of eye cancer in children, known as *retinoblastoma* (discussed in Chapter 20). The loss of both copies of the *Rb* gene leads to excessive proliferation of some cells in the developing retina, suggesting that the Rb protein is particularly important for restraining cell division in this tissue. The complete loss of Rb does not immediately cause increased proliferation of retinal or other types of cells, in part because Cdh1 and CKIs also help inhibit progression through G_1 and in part because other cell types contain Rb-related proteins that provide backup support in the absence of Rb. It is also likely that other proteins, unrelated to Rb, help to regulate the activity of E2F.

Additional layers of control promote an overwhelming increase in S-Cdk activity at the beginning of S phase. We mentioned earlier that the APC/C activator Cdh1 suppresses cyclin levels after mitosis. In animal cells, however, G_1- and G_1/S-cyclins are resistant to Cdh1–APC/C and can therefore act unopposed by the APC/C to promote Rb protein phosphorylation and E2F-dependent gene expression. S-cyclin, by contrast, is not resistant, and its level is initially restrained by Cdh1–APC/C activity. However, G_1/S-Cdk also phosphorylates and inactivates Cdh1–APC/C, thereby allowing the accumulation of S-cyclin, further promoting S-Cdk activation. G_1/S-Cdk also inactivates CKI proteins that suppress S-Cdk activity. The overall effect of all these interactions is the rapid and complete activation of the S-Cdk complexes required for S-phase initiation.

DNA Damage Blocks Cell Division: The DNA Damage Response

Progression through the cell cycle, and thus the rate of cell proliferation, is controlled not only by extracellular mitogens but also by other extracellular and intracellular signals. One of the most important influences is DNA damage, which can occur as a result of spontaneous chemical reactions in DNA, errors in DNA replication, or exposure to radiation or certain chemicals (discussed in Chapter 5). It is essential that damaged chromosomes are repaired before attempting to duplicate or segregate them. The cell-cycle control system can readily detect DNA damage and arrest the cycle at either of two transitions—one at Start, which prevents entry into the cell cycle and into S phase, and one at the G_2/M transition, which prevents entry into mitosis (see Figure 17–16).

DNA damage initiates a signaling pathway by activating one of a pair of related protein kinases called **ATM** and **ATR**, which associate with the site of damage and phosphorylate various target proteins, including two other protein kinases called *Chk1* and *Chk2*. These various kinases phosphorylate other target proteins that lead to cell-cycle arrest. A major target is the gene regulatory protein **p53**, which stimulates transcription of the gene encoding *p21*, a CKI protein; p21 binds to G_1/S-Cdk and S-Cdk complexes and inhibits their activities, thereby helping to block entry into the cell cycle (**Figure 17–62** and **Movie 17.8**).

DNA damage activates p53 by an indirect mechanism. In undamaged cells, p53 is highly unstable and is present at very low concentrations. This is largely because it interacts with another protein, *Mdm2*, which acts as a ubiquitin ligase that targets p53 for destruction by proteasomes. Phosphorylation of p53 after DNA damage reduces its binding to Mdm2. This decreases p53 degradation, which results in a marked increase in p53 concentration in the cell. In addition, the decreased binding to Mdm2 enhances the ability of p53 to stimulate gene transcription (see Figure 17–62).

The protein kinases Chk1 and Chk2 also block cell-cycle progression by phosphorylating members of the Cdc25 family of protein phosphatases, thereby inhibiting their function. As described earlier, these phosphatases are particularly important in the activation of M-Cdk at the beginning of mitosis (see Figure 17–20). Chk1 and Chk2 phosphorylate Cdc25 at inhibitory sites that are distinct from the phosphorylation sites that stimulate Cdc25 activity. The inhibition of Cdc25 activity by DNA damage helps block entry into mitosis (see Figure 17–16).

The DNA damage response can also be activated by problems that arise when a replication fork fails during DNA replication. When nucleotides are depleted, for example, replication forks stall during the elongation phase of DNA synthesis. To prevent the cell from attempting to segregate partially replicated chromosomes, the same mechanisms that respond to DNA damage detect the stalled replication forks and block entry into mitosis until the problems are resolved.

A low level of DNA damage occurs in the normal life of any cell, and this damage accumulates in the cell's progeny if the DNA damage response is not functioning. Over the long term, the accumulation of genetic damage in cells lacking the DNA damage response leads to an increased frequency of cancer-promoting mutations. Indeed, mutations in the *p53* gene occur in at least half of all human cancers (discussed in Chapter 20). This loss of p53 function allows the cancer cell

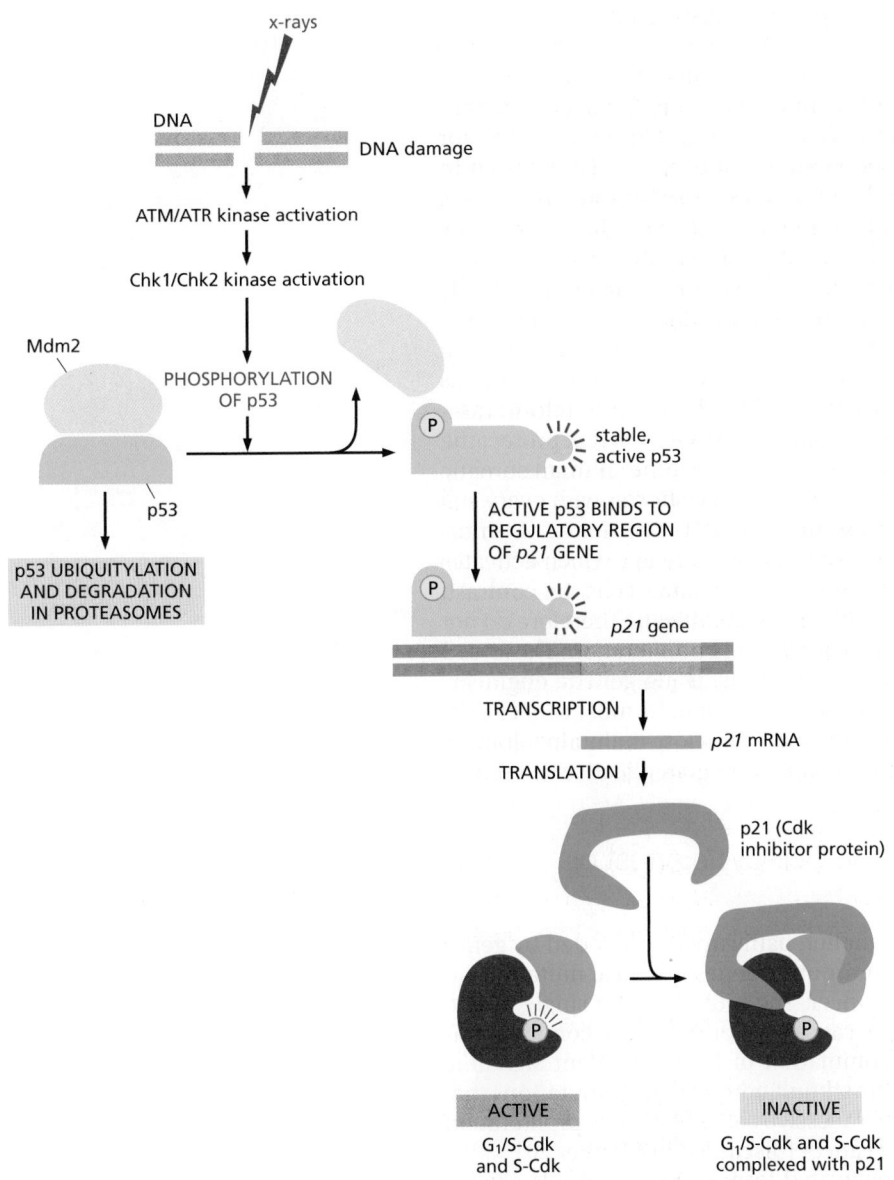

Figure 17–62 **How DNA damage arrests the cell cycle in G₁.** When DNA is damaged, various protein kinases are recruited to the site of damage and initiate a signaling pathway that causes cell-cycle arrest. The first kinase at the damage site is either ATM or ATR, depending on the type of damage. Additional protein kinases, called Chk1 and Chk2, are then recruited and activated, resulting in the phosphorylation of the transcription regulatory protein p53. Mdm2 normally binds to p53 and promotes its ubiquitylation and destruction in proteasomes. Phosphorylation of p53 blocks its binding to Mdm2; as a result, p53 accumulates to high levels and stimulates transcription of numerous genes, including the gene that encodes the CKI protein p21. The p21 binds and inactivates G_1/S-Cdk and S-Cdk complexes, arresting the cell in G_1. In some cases, DNA damage also induces either the phosphorylation of Mdm2 or a decrease in Mdm2 production, which causes a further increase in p53 (not shown).

to accumulate mutations more readily. Similarly, a rare genetic disease known as *ataxia telangiectasia* is caused by a defect in ATM, one of the protein kinases that are activated in response to x-ray-induced DNA damage; patients with this disease are very sensitive to x-rays and suffer from increased rates of cancer.

What happens if DNA damage is so severe that repair is not possible? The answer differs in different organisms. Unicellular organisms such as budding yeast arrest their cell cycle to try to repair the damage, but the cycle resumes even if the repair cannot be completed. For a single-celled organism, life with mutations is apparently better than no life at all. In multicellular organisms, however, the health of the organism takes precedence over the life of an individual cell. Cells that divide with severe DNA damage threaten the life of the organism, since genetic damage can often lead to cancer and other diseases. Thus, animal cells with severe DNA damage do not attempt to continue division, but instead commit suicide by undergoing apoptosis. Thus, unless the DNA damage is repaired, the DNA damage response can lead to either cell-cycle arrest or cell death. DNA damage-induced apoptosis often depends on the activation of p53. Indeed, it is this apoptosis-promoting function of p53 that is apparently most important in protecting us against cancer.

Many Human Cells Have a Built-In Limitation on the Number of Times They Can Divide

Many human cells divide a limited number of times before they stop and undergo a permanent cell-cycle arrest. Fibroblasts taken from normal human tissue, for example, go through only about 25–50 population doublings when cultured in a standard mitogenic medium. Toward the end of this time, proliferation slows down and finally halts, and the cells enter a nondividing state from which they never recover. This phenomenon is called **replicative cell senescence**.

Replicative cell senescence in human fibroblasts seems to be caused by changes in the structure of the **telomeres**, the repetitive DNA sequences and associated proteins at the ends of chromosomes. As discussed in Chapter 5, when a cell divides, telomeric DNA sequences are not replicated in the same manner as the rest of the genome but instead are synthesized by the enzyme **telomerase**. Telomerase also promotes the formation of protein cap structures that protect the chromosome ends. Because human fibroblasts, and many other human somatic cells, do not produce telomerase, their telomeres become shorter with every cell division, and their protective protein caps progressively deteriorate. Eventually, the exposed chromosome ends are sensed as DNA damage, which activates a p53-dependent cell-cycle arrest (see Figure 17–62). Rodent cells, by contrast, maintain telomerase activity when they proliferate in culture and therefore do not have such a telomere-dependent mechanism for limiting proliferation. The forced expression of telomerase in normal human fibroblasts, using genetic engineering techniques, blocks this form of senescence. Unfortunately, most cancer cells have regained the ability to produce telomerase and therefore maintain telomere function as they proliferate; as a result, they do not undergo replicative cell senescence.

Abnormal Proliferation Signals Cause Cell-Cycle Arrest or Apoptosis, Except in Cancer Cells

Many of the components of mitogenic signaling pathways are encoded by genes that were originally identified as cancer-promoting genes, because mutations in them contribute to the development of cancer. The mutation of a single amino acid in the small GTPase Ras, for example, causes the protein to become permanently overactive, leading to constant stimulation of Ras-dependent signaling pathways, even in the absence of mitogenic stimulation. Similarly, mutations that cause an overexpression of Myc stimulate excessive cell growth and proliferation and thereby promote the development of cancer (discussed in Chapter 20).

Surprisingly, however, when a hyperactivated form of Ras or Myc is experimentally overproduced in most normal cells, the result is not excessive proliferation but the opposite: the cells undergo either permanent cell-cycle arrest or apoptosis. The normal cell seems able to detect abnormal mitogenic stimulation, and it responds by preventing further division. Such responses help prevent the survival and proliferation of cells with various cancer-promoting mutations.

Although it is not known how a cell detects excessive mitogenic stimulation, such stimulation often leads to the production of a cell-cycle inhibitor protein called *Arf*, which binds and inhibits Mdm2. As discussed earlier, Mdm2 normally promotes p53 degradation. Activation of Arf therefore causes p53 levels to increase, inducing either cell-cycle arrest or apoptosis (**Figure 17–63**).

How do cancer cells ever arise if these mechanisms block the division or survival of mutant cells with overactive proliferation signals? The answer is that the protective system is often inactivated in cancer cells by mutations in the genes that encode essential components of the blocking mechanisms, such as Arf or p53 or the proteins that help activate them.

Cell Proliferation is Accompanied by Cell Growth

If cells proliferated without growing, they would get progressively smaller and there would be no net increase in total cell mass. In most proliferating cell populations, therefore, cell growth accompanies cell division. In single-celled organisms

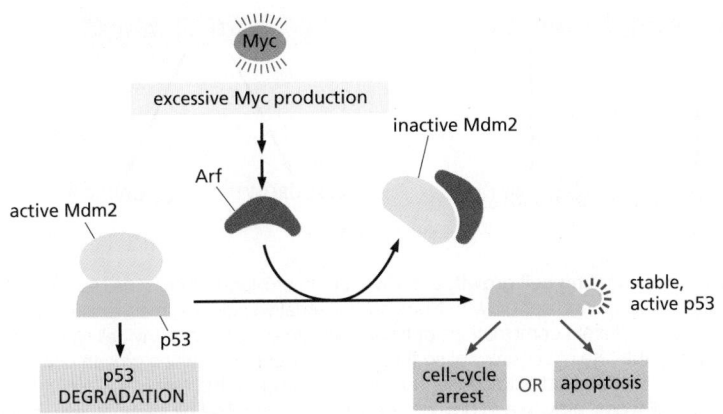

Figure 17–63 **Cell-cycle arrest or apoptosis induced by excessive stimulation of mitogenic pathways.** Abnormally high levels of Myc cause the activation of Arf, which binds and inhibits Mdm2 and thereby increases p53 levels (see Figure 17–62). Depending on the cell type and extracellular conditions, p53 then causes either cell-cycle arrest or apoptosis.

such as yeasts, both cell growth and cell division require only nutrients. In animals, by contrast, both cell growth and cell proliferation depend on extracellular signal molecules, produced by other cells, which we call **growth factors** and mitogens, respectively.

Like mitogens, the extracellular growth factors that stimulate animal cell growth bind to receptors on the cell surface and activate intracellular signaling pathways. These pathways stimulate the accumulation of proteins and other macromolecules, and they do so by both increasing their rate of synthesis and decreasing their rate of degradation. They also trigger increased uptake of nutrients and production of the ATP required to fuel the increased protein synthesis. One of the most important intracellular signaling pathways activated by growth factor receptors involves the enzyme phosphoinositide 3-kinase (*PI 3-kinase*), which adds a phosphate from ATP to the 3′ position of inositol phospholipids in the plasma membrane (discussed in Chapter 15). The activation of PI 3-kinase leads to the activation of a kinase called *TOR*, which lies at the heart of cell growth regulatory pathways in all eukaryotes. TOR activates many targets in the cell that stimulate metabolic processes , including protein synthesis. One target is a protein kinase called *S6 kinase* (*S6K*), which phosphorylates ribosomal protein S6, increasing the ability of ribosomes to translate a subset of mRNAs that mostly encode ribosomal components. TOR also indirectly activates a translation initiation factor called *eIF4E* and directly activates transcription regulators that promote the increased expression of genes encoding ribosomal subunits (**Figure 17–64**).

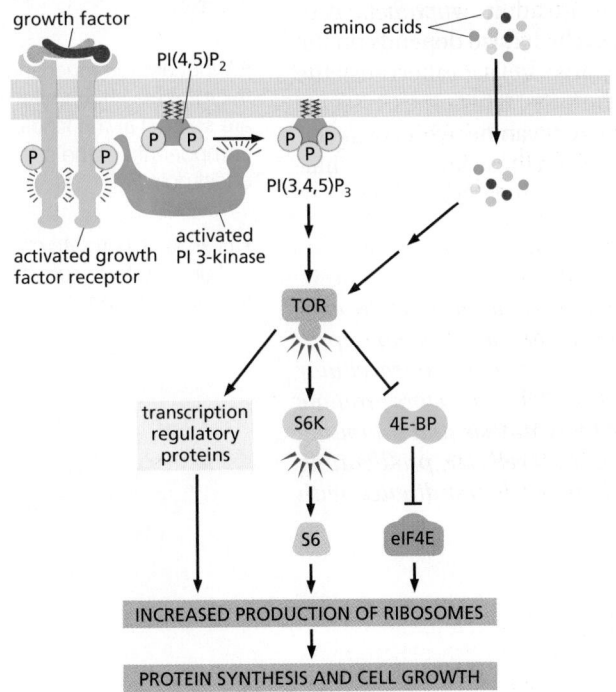

Figure 17–64 **Stimulation of cell growth by extracellular growth factors and nutrients.** The occupation of cell-surface receptors by growth factors leads to the activation of PI 3-kinase, which promotes protein synthesis through a complex signaling pathway that leads to the activation of the protein kinase TOR; extracellular nutrients such as amino acids also help activate TOR. TOR phosphorylates multiple proteins to stimulate protein synthesis, as shown; it also inhibits protein degradation (not shown). Growth factors also stimulate increased production of the transcription regulatory protein Myc (not shown), which activates the transcription of various genes that promote cell metabolism and growth. 4E-BP is an inhibitor of the translation initiation factor eIF4E. PI(4,5)P$_2$, phosphatidylinositol 4,5-bisphosphate; PI(3,4,5)P$_3$, phosphatidylinositol 3,4,5-trisphosphate.

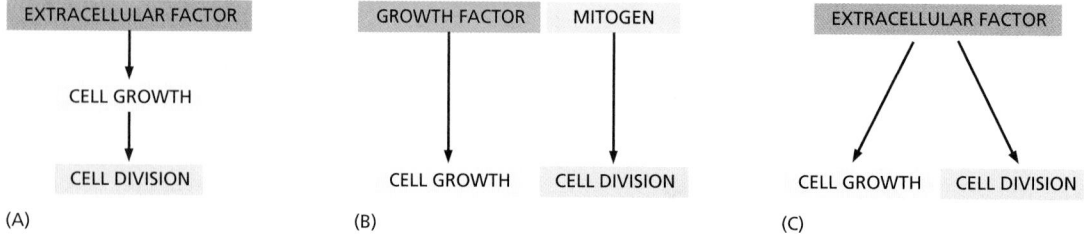

Figure 17–65 Potential mechanisms for coordinating cell growth and division. In proliferating cells, cell size is maintained by mechanisms that coordinate rates of cell division and cell growth. Numerous alternative coupling mechanisms are thought to exist, and different cell types appear to employ different combinations of these mechanisms. (A) In many cell types—particularly yeast—the rate of cell division is governed by the rate of cell growth, so that division occurs only when growth rate achieves some minimal threshold; in yeasts, it is mainly the levels of extracellular nutrients that regulate the rate of cell growth and thereby the rate of cell division. (B) In some animal cell types, growth and division can each be controlled by separate extracellular factors (growth factors and mitogens, respectively), and cell size depends on the relative levels of the two types of factors. (C) Some extracellular factors can stimulate both cell growth and cell division by simultaneously activating signaling pathways that promote growth and other pathways that promote cell-cycle progression.

Proliferating Cells Usually Coordinate Their Growth and Division

For proliferating cells to maintain a constant size, they must coordinate their growth with cell division to ensure that cell size doubles with each division: if cells grow too slowly, they will get smaller with each division, and if they grow too fast, they will get larger with each division. It is not clear how cells achieve this coordination, but it is likely to involve multiple mechanisms that vary in different organisms and even in different cell types of the same organism (Figure 17–65).

Animal cell growth and division are not always coordinated, however. In many cases, they are completely uncoupled to allow growth without division or division without growth. Muscle cells and nerve cells, for example, can grow dramatically after they have permanently withdrawn from the cell cycle. Similarly, the eggs of many animals grow to an extremely large size without dividing; after fertilization, however, this relationship is reversed, and many rounds of division occur without growth.

Compared to cell division, there has been surprisingly little study of how cell size is controlled in animals. As a result, it remains a mystery how cell size is determined and why different cell types in the same animal grow to be so different in size. One of the best-understood cases in mammals is the adult *sympathetic neuron*, which has permanently withdrawn from the cell cycle. Its size depends on the amount of *nerve growth factor* (NGF) secreted by the target cells it innervates; the greater the amount of NGF the neuron has access to, the larger it becomes. It seems likely that the genes a cell expresses set limits on the size it can be, while extracellular signal molecules and nutrients regulate the size within these limits. The challenge is to identify the relevant genes and signal molecules for each cell type.

Summary

In multicellular animals, cell size, cell division, and cell survival are carefully controlled to ensure that the organism and its organs achieve and maintain an appropriate size. Mitogens stimulate the rate of cell division by removing intracellular molecular brakes that restrain cell-cycle progression in G_1. Growth factors promote cell growth (an increase in cell mass) by stimulating the synthesis and inhibiting the degradation of macromolecules. To maintain a constant cell size, proliferating cells employ multiple mechanisms to ensure that cell growth is coordinated with cell division.

WHAT WE DON'T KNOW

• Progression through the cell cycle depends on the phosphorylation of hundreds of different proteins by cyclin–Cdk complexes. What are the molecular mechanisms ensuring that these proteins are phosphorylated at precisely the right time and place?

• During S phase, how are histones and their modifying enzymes controlled to replicate chromatin structure on the duplicated DNA?

• What is the structural basis of chromosome condensation, and how is the process stimulated during mitosis?

• What are the mechanisms by which microtubule attachment and tension are sensed at the kinetochore by the components of the spindle assembly checkpoint?

• How is cell growth coordinated with cell division to ensure that cell size remains constant?

PROBLEMS

Which statements are true? Explain why or why not.

17–1 Since there are about 10^{13} cells in an adult human, and about 10^{10} cells die and are replaced each day, we become new people every three years.

17–2 In order for proliferating cells to maintain a relatively constant size, the length of the cell cycle must match the time it takes for the cell to double in size.

17–3 While other proteins come and go during the cell cycle, the proteins of the origin recognition complex remain bound to the DNA throughout.

17–4 Chromosomes are positioned on the metaphase plate by equal and opposite forces that pull them toward the two poles of the spindle.

17–5 Meiosis segregates the paternal homologs into sperm and the maternal homologs into eggs.

17–6 If we could turn on telomerase activity in all our cells, we could prevent aging.

Discuss the following problems.

17–7 Many cell-cycle genes from human cells function perfectly well when expressed in yeast cells. Why do you suppose that is considered remarkable? After all, many human genes encoding enzymes for metabolic reactions also function in yeast, and no one thinks that is remarkable.

17–8 Hoechst 33342 is a membrane-permeant dye that fluoresces when it binds to DNA. When a population of cells is incubated briefly with Hoechst dye and then sorted in a flow cytometer, which measures the fluorescence of each cell, the cells display various levels of fluorescence as shown in Figure Q17–1.

A. Which cells in Figure Q17–1 are in the G_1, S, G_2, and M phases of the cell cycle? Explain the basis for your answer.

B. Sketch the sorting distributions you would expect for cells that were treated with inhibitors that block the cell cycle in the G_1, S, or M phase. Explain your reasoning.

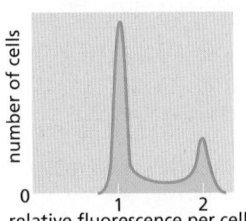

Figure Q17–1 Analysis of Hoechst 33342 fluorescence in a population of cells sorted in a flow cytometer (Problem 17–8).

17–9 The yeast cohesin subunit Scc1, which is essential for sister-chromatid cohesion, can be artificially regulated for expression at any point in the cell cycle. If expression is turned on at the beginning of S phase, all the cells divide satisfactorily and survive. By contrast, if Scc1 expression is turned on only after S phase is completed, the cells fail to divide and they die, even though Scc1 accumulates in the nucleus and interacts efficiently with chromosomes. Why do you suppose that cohesin must be present during S phase for cells to divide normally?

17–10 High doses of caffeine interfere with the DNA damage response in mammalian cells. Why then do you suppose the Surgeon General has not yet issued an appropriate warning to heavy coffee and cola drinkers? A typical cup of coffee (150 mL) contains 100 mg of caffeine (196 g/mole). How many cups of coffee would you have to drink to reach the dose (10 mM) required to interfere with the DNA damage response? (A typical adult contains about 40 liters of water.)

17–11 How many kinetochores are there in a human cell at mitosis?

17–12 A living cell from the lung epithelium of a newt is shown at different stages in M phase in Figure Q17–2. Order these light micrographs into the correct sequence and identify the stage in M phase that each represents.

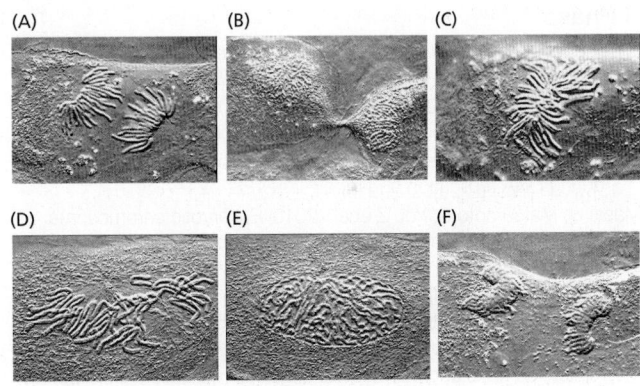

Figure Q17–2 Light micrographs of a single cell at different stages of M phase (Problem 17–12). (Courtesy of Conly L. Rieder.)

17–13 Down syndrome (trisomy 21) and Edwards syndrome (trisomy 18) are the most common autosomal trisomies seen in human infants. Does this fact mean that these chromosomes are the most difficult to segregate properly during meiosis?

17–14 The human genome consists of 23 pairs of chromosomes (22 pairs of autosomes and one pair of sex chromosomes). During meiosis, the maternal and paternal sets of homologs pair, and then are separated into gametes, so that each contains 23 chromosomes. If you assume that the chromosomes in the paired homologs are randomly assorted to daughter cells, how many potential combinations of paternal and maternal homologs can be generated during meiosis? (For the purposes of this calculation, assume that no recombination occurs.)

REFERENCES

Overview of the Cell Cycle

Morgan DO (2007) The Cell Cycle: Principles of Control. London: New Science Press.

Murray AW & Hunt T (1993) The Cell Cycle: An Introduction. New York: WH Freeman and Co.

The Cell-Cycle Control System

Evans T, Rosenthal ET, Youngblom J et al. (1983) Cyclin: a protein specified by maternal mRNA in sea urchin eggs that is destroyed at each cleavage division. *Cell* 33, 389–396.

Hartwell LH, Culotti J, Pringle JR et al (1974) Genetic control of the cell division cycle in yeast. *Science* 183, 46–51.

Holt LJ, Tuch BB, Villen J et al. (2009) Global analysis of Cdk1 substrate phosphorylation sites provides insights into evolution. *Science* 325, 1682–1686.

Nurse P, Thuriaux P & Nasmyth K (1976) Genetic control of the cell division cycle in the fission yeast *Schizosaccharomyces pombe*. *Mol. Gen. Genet.* 146, 167–178.

Pavletich NP (1999) Mechanisms of cyclin-dependent kinase regulation: structures of Cdks, their cyclin activators, and CIP and Ink4 inhibitors. *J. Mol. Biol.* 287, 821–828.

Primorac I & Musacchio A (2013) Panta rhei: The APC/C at steady state. *J. Cell Biol.* 201, 177–189.

Wittenberg C & Reed SI (2005) Cell cycle-dependent transcription in yeast: promoters, transcription factors, and transcriptomes. *Oncogene* 24, 2746–2755.

S Phase

Arias EE & Walter JC (2007) Strength in numbers: preventing rereplication via multiple mechanisms in eukaryotic cells. *Genes Dev.* 21, 497–518.

Bell SP & Kaguni JM (2013) Helicase loading at chromosomal origins of replication. *Cold Spring Harb. Perspect. Biol.* 5, a010124.

Groth A, Rocha W, Verreault A et al. (2007) Chromatin challenges during DNA replication and repair. *Cell* 128, 721–733.

Masai H, Matsumoto S, You Z et al. (2010) Eukaryotic chromosome DNA replication: where, when, and how? *Annu. Rev. Biochem.* 79, 89–130.

Siddiqui K, On KF & Diffley JF (2013) Regulating DNA replication in eukarya. *Cold Spring Harb. Perspect. Biol.* 5, a012930.

Tanaka S & Araki H (2013) Helicase activation and establishment of replication forks at chromosomal origins of replication. *Cold Spring Harb. Perspect. Biol.* 5, a010371.

Mitosis

Alushin G & Nogales E (2011) Visualizing kinetochore architecture. *Curr. Opin. Struct. Biol.* 21, 661–669.

Cuylen S & Haering CH (2011) Deciphering condensin action during chromosome segregation. *Trends Cell Biol.* 21, 552–559.

Gonczy P (2012) Towards a molecular architecture of centriole assembly. *Nat. Rev. Mol. Cell Biol.* 13, 425–435.

Hirano T (2012) Condensins: universal organizers of chromosomes with diverse functions. *Genes Dev.* 26, 1659–1678.

Joglekar AP, Bloom KS & Salmon ED (2010) Mechanisms of force generation by end-on kinetochore-microtubule attachments. *Curr. Opin. Cell Biol.* 22, 57–67.

Lampson MA & Cheeseman IM (2011) Sensing centromere tension: Aurora B and the regulation of kinetochore function. *Trends Cell Biol.* 21, 133–140.

Magidson V, O'Connell CB, Loncarek J et al. (2011) The spatial arrangement of chromosomes during prometaphase facilitates spindle assembly. *Cell* 146, 555–567.

Musacchio A & Salmon ED (2007) The spindle-assembly checkpoint in space and time. *Nat. Rev. Mol. Cell Biol.* 8, 379–393.

Nasmyth K & Haering CH (2009) Cohesin: its roles and mechanisms. *Annu. Rev. Genet.* 43, 525–558.

Nigg EA & Stearns T (2011) The centrosome cycle: Centriole biogenesis, duplication and inherent asymmetries. *Nat. Cell Biol.* 13, 1154–1160.

Rago F & Cheeseman IM (2013) The functions and consequences of force at kinetochores. *J. Cell Biol.* 200, 557–565.

Wadsworth P & Khodjakov A (2004) E pluribus unum: towards a universal mechanism for spindle assembly. *Trends Cell Biol.* 14, 413–419.

Walczak CE, Cai S & Khodjakov A (2010) Mechanisms of chromosome behaviour during mitosis. *Nat. Rev. Mol. Cell Biol.* 11, 91–102.

Cytokinesis

Fededa JP & Gerlich DW (2012) Molecular control of animal cell cytokinesis. *Nat. Cell Biol.* 14, 440–447.

Green RA, Paluch E & Oegema K (2012) Cytokinesis in animal cells. *Annu. Rev. Cell Dev. Biol.* 28, 29–58.

Jurgens G (2005) Plant cytokinesis: fission by fusion. *Trends Cell Biol.* 15, 277–283.

Oliferenko S, Chew TG & Balasubramanian MK (2009) Positioning cytokinesis. *Genes Dev.* 23, 660–674.

Pollard TD (2010) Mechanics of cytokinesis in eukaryotes. *Curr. Opin. Cell Biol.* 22, 50–56.

Rappaport R (1986) Establishment of the mechanism of cytokinesis in animal cells. *Int. Rev. Cytol.* 105, 245–281.

Schiel JA & Prekeris R (2013) Membrane dynamics during cytokinesis. *Curr. Opin. Cell Biol.* 25, 92–98.

Meiosis

Bhalla N & Dernburg AF (2008) Prelude to a division. *Annu. Rev. Cell Dev. Biol.* 24, 397–424.

Gerton JL & Hawley RS (2005) Homologous chromosome interactions in meiosis: diversity amidst conservation. *Nat. Rev. Genet.* 6, 477–487.

Hall H, Hunt P & Hassold T (2006) Meiosis and sex chromosome aneuploidy: how meiotic errors cause aneuploidy; how aneuploidy causes meiotic errors. *Curr. Opin. Genet. Dev.* 16, 323–329.

Jordan P (2006) Initiation of homologous chromosome pairing during meiosis. *Biochem. Soc. Trans.* 34, 545–549.

Lake CM & Hawley RS (2012) The molecular control of meiotic chromosomal behavior: events in early meiotic prophase in *Drosophila* oocytes. *Annu. Rev. Physiol.* 74, 425–51.

Watanabe Y (2012) Geometry and force behind kinetochore orientation: lessons from meiosis. *Nat. Rev. Mol. Cell Biol.* 13, 370–382.

Control of Cell Division and Cell Growth

Adhikary S & Eilers M (2005) Transcriptional regulation and transformation by Myc proteins. *Nat. Rev. Mol. Cell Biol.* 6, 635–645.

Bertoli C, Skotheim JM & de Bruin RA (2013) Control of cell cycle transcription during G1 and S phases. *Nat. Rev. Mol. Cell Biol.* 14, 518–528.

Dick FA & Rubin SM (2013) Molecular mechanisms underlying RB protein function. *Nat. Rev. Mol. Cell Biol.* 14, 297–306.

Jackson SP & Bartek J (2009) The DNA-damage response in human biology and disease. *Nature* 461, 1071–1078.

Jorgensen P & Tyers M (2004) How cells coordinate growth and division. *Curr. Biol.* 14, R1014–R1027.

Shimobayashi M & Hall MN (2014) Making new contacts: the mTOR network in metabolism and signalling crosstalk. *Nat. Rev. Mol. Cell Biol.* 15, 155–162.

Turner JJ, Ewald JC & Skotheim JM (2012) Cell size control in yeast. *Curr. Biol.* 22, R350–R359.

van den Heuvel S & Dyson NJ (2008) Conserved functions of the pRB and E2F families. *Nat. Rev. Mol. Cell Biol.* 9, 713–724.

Vousden KH & Lu X (2002) Live or let die: the cell's response to p53. *Nat. Rev. Cancer* 2, 594–604.

Zoncu R, Efeyan A & Sabatini DM (2011) mTOR: from growth signal integration to cancer, diabetes and ageing. *Nat. Rev. Mol. Cell Biol.* 12, 21–35.

Cell Death

The growth, development, and maintenance of multicellular organisms depend not only on the production of cells but also on mechanisms to destroy them. The maintenance of tissue size, for example, requires that cells die at the same rate as they are produced. During development, carefully orchestrated patterns of cell death help determine the size and shape of limbs and other tissues. Cells also die when they become damaged or infected, ensuring that they are removed before they threaten the health of the organism. In these and most other cases, cell death is not a random process but occurs by a programmed sequence of molecular events, in which the cell systematically destroys itself from within and is then eaten by other cells, leaving no trace. In most cases, this **programmed cell death** occurs by a process called **apoptosis**—from the Greek word meaning "falling off," as leaves from a tree.

Cells dying by apoptosis undergo characteristic morphological changes. They shrink and condense, the cytoskeleton collapses, the nuclear envelope disassembles, and the nuclear chromatin condenses and breaks up into fragments (**Figure 18-1A**). The cell surface often bulges outward and, if the cell is large, it breaks up into membrane-enclosed fragments called *apoptotic bodies*. The surface of the cell or apoptotic bodies becomes chemically altered, so that a neighboring cell or a macrophage (a specialized phagocytic cell, discussed in Chapter 22) rapidly engulfs them, before they can spill their contents (Figure 18-1B). In this way, the cell dies neatly and is rapidly cleared away, without causing a damaging inflammatory response. Because the cells are eaten and digested so quickly, there are usually few dead cells to be seen, even when large numbers of cells have died by apoptosis. This is probably why biologists overlooked apoptosis for many years and still might underestimate its extent.

In contrast to apoptosis, animal cells that die in response to an acute insult, such as trauma or a lack of blood supply, usually do so by a process called *cell necrosis*. Necrotic cells swell and burst, spilling their contents over their neighbors and eliciting an inflammatory response (Figure 18-1C). In most cases, necrosis is likely to be caused by energy depletion, which leads to metabolic defects and loss of the ionic gradients that normally exist across the cell membrane. One form of necrosis, called *necroptosis*, is a form of programmed cell death that is triggered by a specific regulatory signal from other cells, although we are only just beginning to understand the underlying mechanisms.

Some form of programmed cell death occurs in many organisms, but apoptosis is found primarily in animals. This chapter focuses on the major functions of apoptosis, its mechanism and regulation, and how excessive or insufficient apoptosis can contribute to human disease.

Apoptosis Eliminates Unwanted Cells

The amount of apoptotic cell death that occurs in developing and adult animal tissues is astonishing. In the developing vertebrate nervous system, for example, more than half of many types of nerve cells normally die soon after they are formed. It seems remarkably wasteful for so many cells to die, especially as the vast majority are perfectly healthy at the time they kill themselves. What purposes does this massive cell death serve?

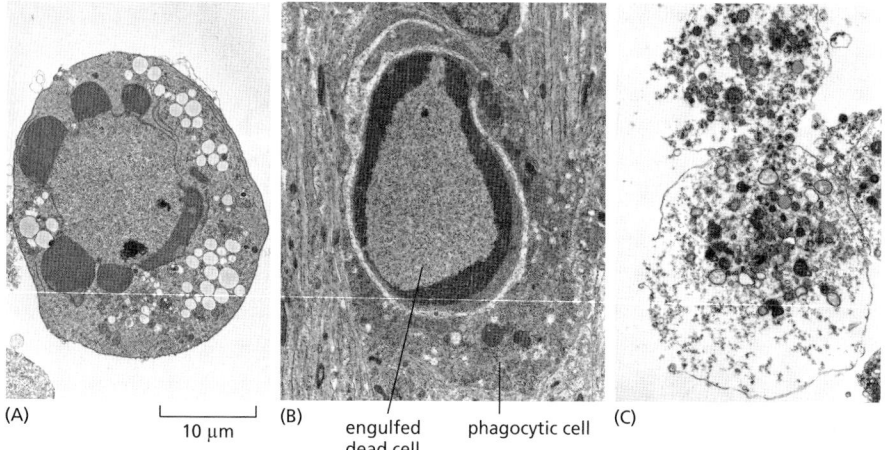

(A)   10 μm   (B)   engulfed   phagocytic cell   (C)
dead cell

Figure 18–1 Two distinct forms of cell death. These electron micrographs show cells that have died by apoptosis (A and B) or by necrosis (C). The cells in (A) and (C) died in a culture dish, whereas the cell in (B) died in a developing tissue and has been engulfed by a phagocytic cell. Note that the cells in (A) and (B) have condensed but seem relatively intact, whereas the cell in (C) seems to have exploded. The large vacuoles visible in the cytoplasm of the cell in (A) are a variable feature of apoptosis. (Courtesy of Julia Burne.)

In some cases, the answer is clear. Cell death helps sculpt hands and feet during embryonic development: they start out as spade-like structures, and the individual digits separate only as the cells between them die, as illustrated for a mouse paw in Figure 18–2. In other cases, cells die when the structure they form is no longer needed. When a tadpole changes into a frog at metamorphosis, the cells in the tail die, and the tail, which is not needed in the frog, disappears. Apoptosis also functions as a quality-control process in development, eliminating cells that are abnormal, misplaced, nonfunctional, or potentially dangerous to the animal. Striking examples occur in the vertebrate adaptive immune system, where apoptosis eliminates developing T and B lymphocytes that either fail to produce potentially useful antigen-specific receptors or produce self-reactive receptors that make the cells potentially dangerous (discussed in Chapter 24); it also eliminates most of the lymphocytes activated by an infection, after they have helped destroy the responsible microbes.

In adult tissues that are neither growing nor shrinking, cell death and cell division must be tightly regulated to ensure that they are exactly in balance. If part of the liver is removed in an adult rat, for example, liver cell proliferation increases to make up the loss. Conversely, if a rat is treated with the drug phenobarbital— which stimulates liver cell division (and thereby liver enlargement)—and then the phenobarbital treatment is stopped, apoptosis in the liver greatly increases until the liver has returned to its original size, usually within a week or so. Thus, the liver is kept at a constant size through the regulation of both the cell death rate and the cell birth rate. The control mechanisms responsible for such regulation are largely unknown.

Animal cells can recognize damage in their various organelles and, if the damage is great enough, they can kill themselves by undergoing apoptosis. An important example is DNA damage, which can produce cancer-promoting mutations if not repaired. Cells have various ways of detecting DNA damage, and undergo apoptosis if they cannot repair it.

Apoptosis Depends on an Intracellular Proteolytic Cascade That Is Mediated by Caspases

Apoptosis is triggered by members of a family of specialized intracellular proteases, which cleave specific sequences in numerous proteins inside the cell, thereby bringing about the dramatic changes that lead to cell death and engulfment. These proteases have a cysteine at their active site and cleave their target proteins at specific aspartic acids; they are therefore called **caspases** (c for cysteine and asp for aspartic acid). Caspases are synthesized in the cell as inactive precursors and are activated only during apoptosis. There are two major classes of apoptotic caspases: *initiator* caspases and *executioner* caspases.

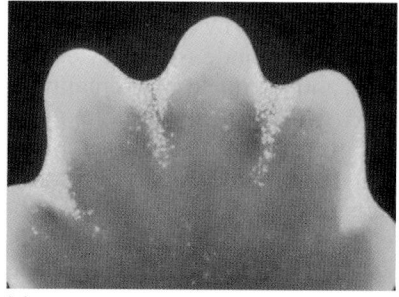

(A)

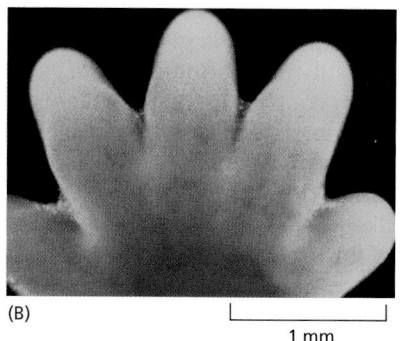

(B)   1 mm

Figure 18–2 Sculpting the digits in the developing mouse paw by apoptosis. (A) The paw in this mouse fetus has been stained with a dye that specifically labels cells that have undergone apoptosis. The apoptotic cells appear as *bright green dots* between the developing digits. (B) The interdigital cell death has eliminated the tissue between the developing digits, as seen one day later, when there are very few apoptotic cells. (From W. Wood et al., *Development* 127:5245–5252, 2000. With permission from The Company of Biologists.)

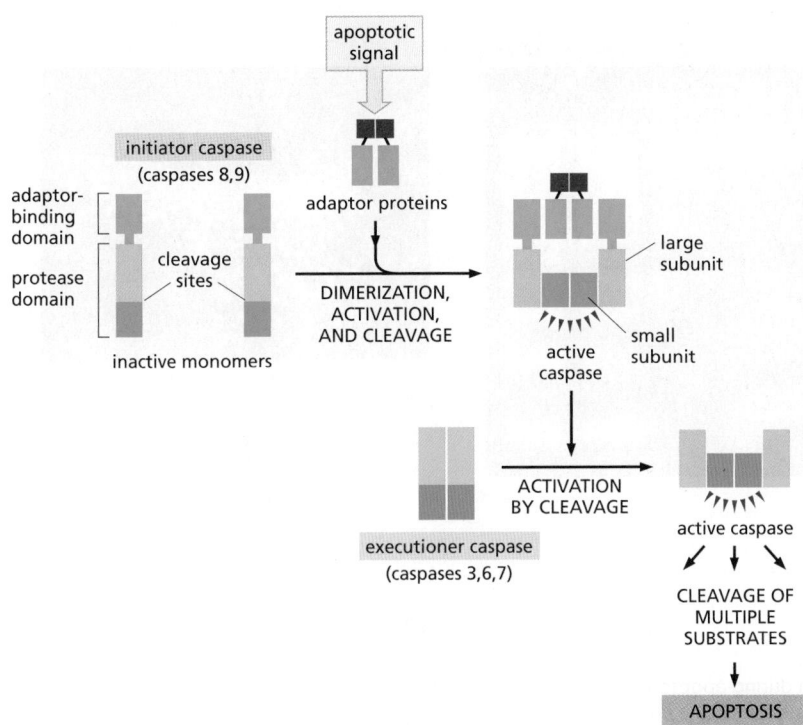

Figure 18–3 **Caspase activation during apoptosis.** An initiator caspase contains a protease domain in its carboxy-terminal region and a small protein interaction domain near its amino terminus. It is initially made in an inactive, monomeric form, sometimes called procaspase. Apoptotic signals trigger the assembly of adaptor proteins carrying multiple binding sites for the caspase amino-terminal domain. Upon binding to the adaptor proteins, the initiator caspases dimerize and are thereby activated, leading to cleavage of a specific site in their protease domains. Each protease domain is then rearranged into a large and small subunit. In some cases (not shown), the adaptor-binding domain of the initiator caspase is also cleaved (see Figure 18–5). Executioner caspases are initially formed as inactive dimers. Upon cleavage at a site in the protease domain by an initiator caspase, the executioner caspase dimer undergoes an activating conformational change. The executioner caspases then cleave a variety of key proteins, leading to the controlled death of the cell.

Initiator caspases, as their name implies, begin the apoptotic process. They normally exist as inactive, soluble monomers in the cytosol. An apoptotic signal triggers the assembly of large protein platforms that bring multiple initiator caspases together into large complexes. Within these complexes, pairs of caspases associate to form dimers, resulting in protease activation (**Figure 18–3**). Each caspase in the dimer then cleaves its partner at a specific site in the protease domain, which stabilizes the active complex and is required for the proper function of the enzyme in the cell.

The major function of the initiator caspases is to activate the **executioner caspases**. These normally exist as inactive dimers. When they are cleaved by an initiator caspase at a site in the protease domain, the active site is rearranged from an inactive to an active conformation. One initiator caspase complex can activate many executioner caspases, resulting in an amplifying proteolytic cascade. Once activated, executioner caspases catalyze the widespread protein cleavage events that kill the cell.

Various experimental approaches have led to the identification of over a thousand proteins that are cleaved by caspases during apoptosis. Only a few of these proteins have been studied in any detail. These include the nuclear lamins, the cleavage of which causes the irreversible breakdown of the nuclear lamina (discussed in Chapter 12). Another target is a protein that normally holds a DNA-degrading endonuclease in an inactive form; its cleavage frees the endonuclease to cut up the DNA in the cell nucleus (**Figure 18–4**). Other target proteins include components of the cytoskeleton and cell–cell adhesion proteins that attach cells to their neighbors; the cleavage of these proteins helps the apoptotic cell to round up and detach from its neighbors, making it easier for a neighboring cell to engulf it, or, in the case of an epithelial cell, for the neighbors to extrude the apoptotic cell from the cell sheet. The caspase cascade is not only destructive and self-amplifying but also irreversible, so that once a cell starts out along the path to destruction, it cannot turn back.

How is the initiator caspase first activated in response to an apoptotic signal? The two best-understood activation mechanisms in mammalian cells are called the *extrinsic pathway* and the *intrinsic*, or *mitochondrial, pathway*. Each uses its own initiator caspase and activation system, as we now discuss.

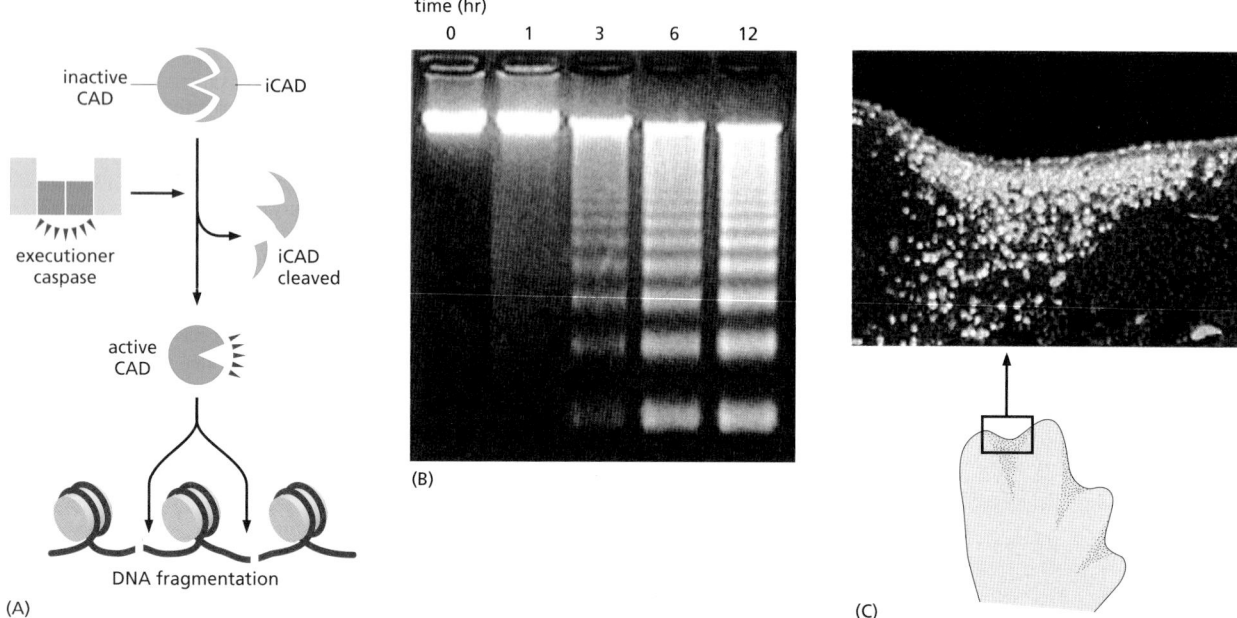

Figure 18–4 DNA fragmentation during apoptosis. (A) In healthy cells, the endonuclease CAD associates with its inhibitor, iCAD. Activation of executioner caspases in the cell leads to cleavage of iCAD, which unleashes the nuclease. Activated CAD cuts the chromosomal DNA between nucleosomes, resulting in the production of DNA fragments that form a ladder pattern (see B) upon gel electrophoresis. (B) Mouse thymus lymphocytes were treated with an antibody against the cell-surface death receptor Fas (discussed in the text), inducing the cells to undergo apoptosis. DNA was extracted at the times indicated above the figure, and the fragments were separated by size by electrophoresis in an agarose gel and stained with ethidium bromide. Because the cleavages occur in the linker regions between nucleosomes, the fragments separate into a characteristic ladder pattern on these gels. Note that in gel electrophoresis, smaller molecules are more widely separated in the lower part of the gel, so that removal of a single nucleosome has a greater apparent effect on their gel mobility. (C) Apoptotic nuclei can be detected using a technique that adds a fluorescent label to DNA ends. In the image shown here, this technique was used in a tissue section of a developing chick leg bud; this cross section through the skin and underlying tissue is from a region between two developing digits, as indicated in the underlying drawing. The procedure is called the TUNEL (TdT-mediated dUTP nick end labeling) technique because the enzyme terminal deoxynucleotidyl transferase (TdT) adds chains of labeled deoxynucleotide (dUTP) to the 3′-OH ends of DNA fragments. The presence of large numbers of DNA fragments therefore results in bright fluorescent dots in apoptotic cells. (B, from D. McIlroy et al., *Genes Dev.* 14:549–558, 2000. With permission from Cold Spring Harbor Laboratory Press; C, from V. Zuzarte-Luís and J.M. Hurlé, *Int. J. Dev. Biol.* 46:871–876, 2002. With permission from UPV/EHU Press.)

Cell-Surface Death Receptors Activate the Extrinsic Pathway of Apoptosis

Extracellular signal proteins binding to cell-surface **death receptors** trigger the **extrinsic pathway** of apoptosis. Death receptors are transmembrane proteins containing an extracellular ligand-binding domain, a single transmembrane domain, and an intracellular *death domain*, which is required for the receptors to activate the apoptotic program. The receptors are homotrimers and belong to the *tumor necrosis factor (TNF) receptor* family, which includes a receptor for TNF itself and the *Fas* death receptor. The ligands that activate the death receptors are also homotrimers; they are structurally related to one another and belong to the *TNF family* of signal proteins.

A well-understood example of how death receptors trigger the extrinsic pathway of apoptosis is the activation of **Fas** on the surface of a target cell by **Fas ligand** on the surface of a killer (cytotoxic) lymphocyte. When activated by the binding of Fas ligand, the death domains on the cytosolic tails of the Fas death receptors bind intracellular adaptor proteins, which in turn bind initiator caspases (primarily caspase-8), forming a **death-inducing signaling complex (DISC)**. Once dimerized and activated in the DISC, the initiator caspases cleave their partners and then activate downstream executioner caspases to induce apoptosis (**Figure 18–5**). In some cells, the extrinsic pathway recruits the intrinsic apoptotic pathway to amplify the caspase cascade and kill the cell.

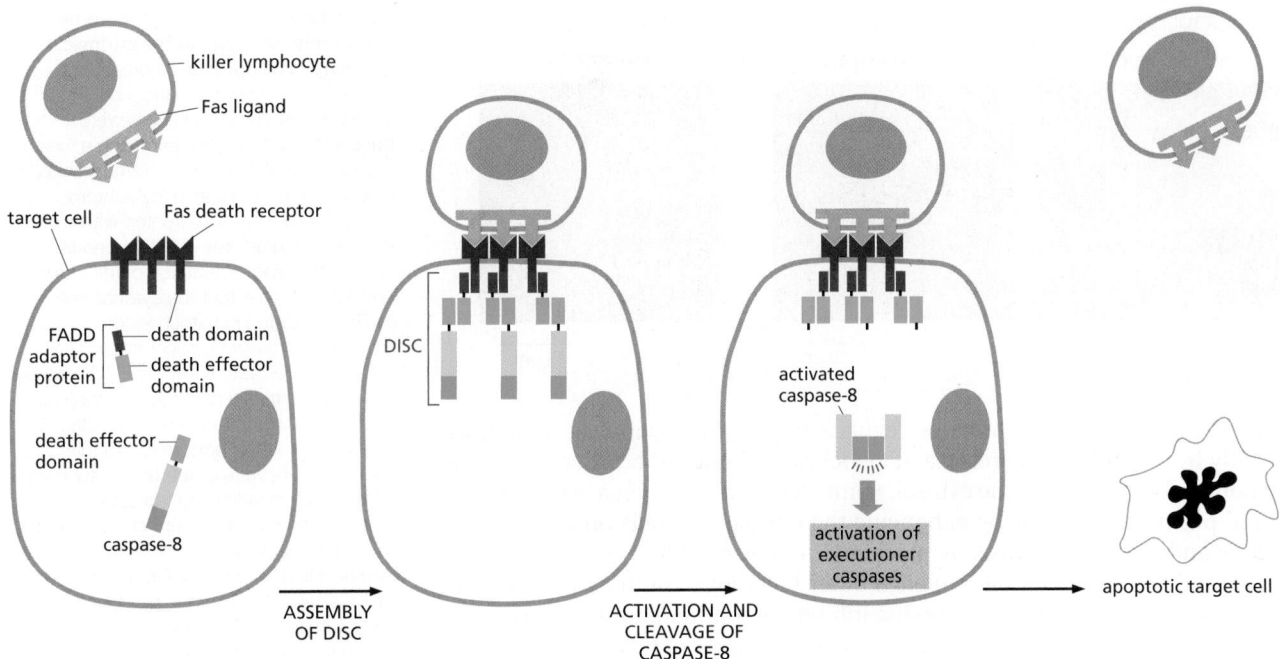

Figure 18–5 **The extrinsic pathway of apoptosis activated through Fas death receptors.** Trimeric Fas ligands on the surface of a killer lymphocyte interact with trimeric Fas receptors on the surface of the target cell, leading to clustering of several ligand-bound receptor trimers (only one trimer is shown here for clarity). Receptor clustering activates death domains on the receptor tails, which interact with similar domains on the adaptor protein FADD (FADD stands for Fas-associated death domain). Each FADD protein then recruits an initiator caspase (caspase-8) via a death effector domain on both FADD and the caspase, forming a death-inducing signaling complex (DISC). Within the DISC, two adjacent initiator caspases interact and cleave one another to form an activated protease dimer, which then cleaves itself in the region linking the protease to the death effector domain. This stabilizes and releases the active caspase dimer into the cytosol, where it activates executioner caspases by cleaving them.

Many cells produce inhibitory proteins that act to restrain the extrinsic pathway. For example, some cells produce the protein *FLIP*, which resembles an initiator caspase but has no protease activity because it lacks the key cysteine in its active site. FLIP dimerizes with caspase-8 in the DISC; although caspase-8 appears to be active in these heterodimers, it is not cleaved at the site required for its stable activation, and the apoptotic signal is blocked. Such inhibitory mechanisms help prevent the inappropriate activation of the extrinsic pathway of apoptosis.

The Intrinsic Pathway of Apoptosis Depends on Mitochondria

Cells can also activate their apoptosis program from inside the cell, often in response to stresses, such as DNA damage, or in response to developmental signals. In vertebrate cells, these responses are governed by the **intrinsic**, or **mitochondrial**, **pathway** of apoptosis, which depends on the release into the cytosol of mitochondrial proteins that normally reside in the intermembrane space of these organelles (see Figure 12–19). Some of the released proteins activate a caspase proteolytic cascade in the cytoplasm, leading to apoptosis.

A key protein in the intrinsic pathway is **cytochrome *c***, a water-soluble component of the mitochondrial electron-transport chain. When released into the cytosol (**Figure 18–6**), it takes on a new function: it binds to an adaptor protein called **Apaf1** (*apoptotic protease activating factor-1*), causing the Apaf1 to oligomerize into a wheel-like heptamer called an **apoptosome**. The Apaf1 proteins in the apoptosome then recruit initiator caspase-9 proteins, which are thought to be activated by proximity in the apoptosome, just as caspase-8 is activated in the DISC. The activated caspase-9 molecules then activate downstream executioner caspases to induce apoptosis (**Figure 18–7**).

Bcl2 Proteins Regulate the Intrinsic Pathway of Apoptosis

The intrinsic pathway of apoptosis is tightly regulated to ensure that cells kill themselves only when it is appropriate. A major class of intracellular regulators of the intrinsic pathway is the **Bcl2 family** of proteins, which, like the caspase family, has been conserved in evolution from worms to humans; a human Bcl2 protein, for example, can suppress apoptosis when expressed in the worm *Caenorhabditis elegans*.

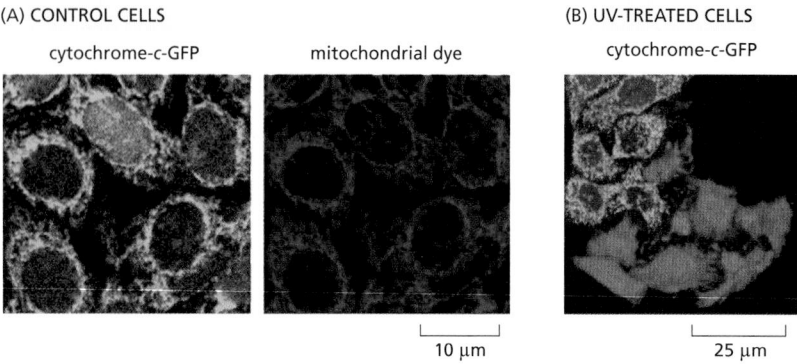

(A) CONTROL CELLS

cytochrome-*c*-GFP   mitochondrial dye

(B) UV-TREATED CELLS

cytochrome-*c*-GFP

10 μm   25 μm

Figure 18–6 Release of cytochrome *c* from mitochondria in the intrinsic pathway of apoptosis. Fluorescence micrographs of human cancer cells in culture. (A) The control cells were transfected with a gene encoding a fusion protein consisting of cytochrome *c* linked to green fluorescent protein (cytochrome-*c*-GFP); they were also treated with a red dye that accumulates in mitochondria. The overlapping distribution of the *green* and *red* indicates that the cytochrome-*c*-GFP is located in mitochondria. (B) Cells expressing cytochrome-*c*-GFP were irradiated with ultraviolet (UV) light to induce the intrinsic pathway of apoptosis and were photographed 5 hours later. The six cells in the bottom half of this micrograph have released their cytochrome *c* from mitochondria into the cytosol, whereas the cells in the upper half of the micrograph have not yet done so (Movie 18.1). (From J.C. Goldstein et al., *Nat. Cell Biol.* 2:156–162, published 2000 by Nature Publishing Group. Reproduced with permission of SNCSC.)

Mammalian Bcl2 family proteins regulate the intrinsic pathway of apoptosis mainly by controlling the release of cytochrome *c* and other intermembrane mitochondrial proteins into the cytosol. Some Bcl2 family proteins are *pro-apoptotic* and promote apoptosis by enhancing the release, whereas others are *anti-apoptotic* and inhibit apoptosis by blocking the release. The pro-apoptotic and anti-apoptotic proteins can bind to each other in various combinations to form heterodimers in which the two proteins inhibit each other's function. The balance between the activities of these two functional classes of Bcl2 family proteins largely determines whether a mammalian cell lives or dies by the intrinsic pathway of apoptosis.

As illustrated in **Figure 18–8**, the anti-apoptotic Bcl2 family proteins, including *Bcl2* itself (the founding member of the Bcl2 family) and *BclX*$_L$, share four distinctive *Bcl2 homology (BH) domains* (BH1–4). The pro-apoptotic Bcl2 family proteins

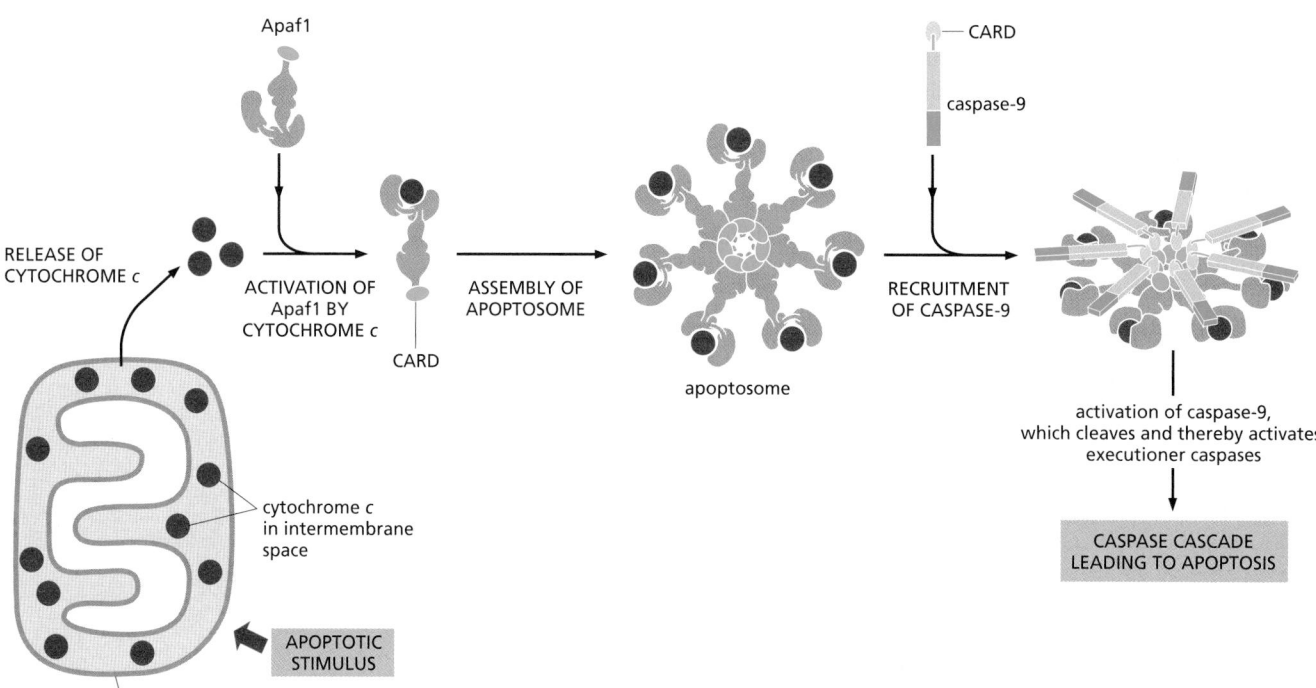

Figure 18–7 The intrinsic pathway of apoptosis. Intracellular apoptotic stimuli cause mitochondria to release cytochrome *c*, which interacts with Apaf1. The binding of cytochrome *c* causes Apaf1 to unfold partly, exposing a domain that interacts with the same domain in other activated Apaf1 molecules. Seven activated Apaf1 proteins form a large ring complex called the apoptosome. Each Apaf1 protein contains a caspase recruitment domain (CARD), and these are clustered above the central hub of the apoptosome. The CARDs bind similar domains in multiple caspase-9 molecules, which are thereby recruited into the apoptosome and activated. The mechanism of caspase-9 activation is not clear: it probably results from dimerization and cleavage of adjacent caspase-9 proteins, but it might also depend on interactions between caspase-9 and Apaf1. Once activated, caspase-9 cleaves and thereby activates downstream executioner caspases. Note that the CARD is related in structure and function to the death effector domain of caspase-8 (see Figure 18–5). Some scientists use the term "apoptosome" to refer to the complex containing caspase-9.

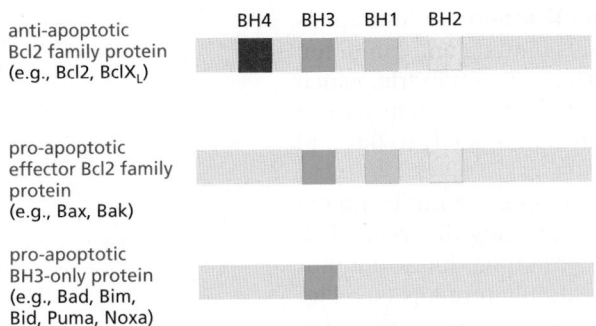

anti-apoptotic
Bcl2 family protein
(e.g., Bcl2, BclX$_L$)

pro-apoptotic
effector Bcl2 family
protein
(e.g., Bax, Bak)

pro-apoptotic
BH3-only protein
(e.g., Bad, Bim,
Bid, Puma, Noxa)

Figure 18–8 The three classes of Bcl2 family proteins. Note that the BH3 domain is the only BH domain shared by all Bcl2 family members; it mediates the direct interactions between pro-apoptotic and anti-apoptotic family members.

consist of two subfamilies—the *effector Bcl2 family proteins* and the *BH3-only proteins*. The main effector proteins are *Bax* and *Bak*, which are structurally similar to Bcl2 but lack the BH4 domain. The BH3-only proteins share sequence homology with Bcl2 in only the BH3 domain.

When an apoptotic stimulus triggers the intrinsic pathway, the pro-apoptotic **effector Bcl2 family proteins** become activated and aggregate to form oligomers in the mitochondrial outer membrane, inducing the release of cytochrome *c* and other intermembrane proteins by an unknown mechanism (**Figure 18–9**). In mammalian cells, **Bax** and **Bak** are the main effector Bcl2 family proteins, and at least one of them is required for the intrinsic pathway of apoptosis to operate: mutant mouse cells that lack both proteins are resistant to all pro-apoptotic signals that normally activate this pathway. Whereas Bak is bound to the mitochondrial outer membrane even in the absence of an apoptotic signal, Bax is mainly located in the cytosol and translocates to the mitochondria only after an apoptotic signal activates it. As we discuss below, the activation of Bax and Bak usually depends on activated pro-apoptotic BH3-only proteins.

The **anti-apoptotic Bcl2 family proteins** such as **Bcl2** itself and **BclX$_L$** are also located on the cytosolic surface of the outer mitochondrial membrane, where they help prevent inappropriate release of intermembrane proteins. The anti-apoptotic Bcl2 family proteins inhibit apoptosis mainly by binding to and inhibiting pro-apoptotic Bcl2 family proteins—either on the mitochondrial membrane or in the cytosol. On the outer mitochondrial membrane, for example, they bind to Bak and prevent it from oligomerizing, thereby inhibiting the release of cytochrome *c* and other intermembrane proteins. There are at least five mammalian anti-apoptotic Bcl2 family proteins, and every mammalian cell requires at least one to survive. Moreover, a number of these proteins must be inhibited for the intrinsic pathway to induce apoptosis; the BH3-only proteins mediate the inhibition.

The **BH3-only proteins** are the largest subclass of Bcl2 family proteins. The cell either produces or activates them in response to an apoptotic stimulus, and they are thought to promote apoptosis mainly by inhibiting anti-apoptotic Bcl2 family

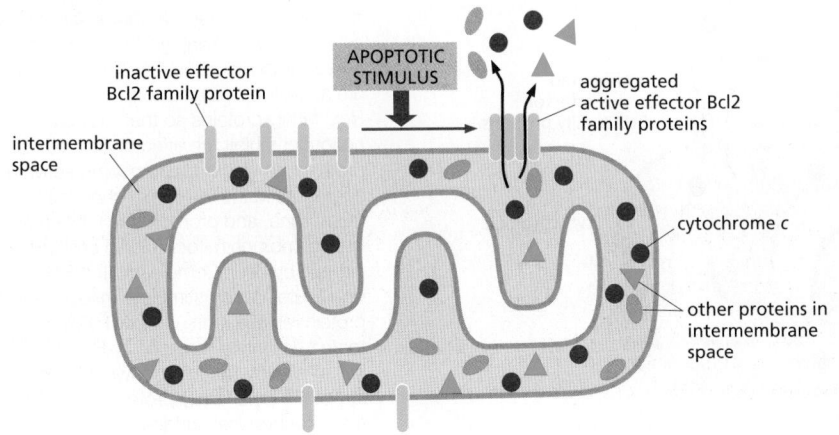

Figure 18–9 The role of pro-apoptotic effector Bcl2 family proteins (mainly Bax and Bak) in the release of mitochondrial intermembrane proteins in the intrinsic pathway of apoptosis. When activated by an apoptotic stimulus, the effector Bcl2 family proteins aggregate on the outer mitochondrial membrane and release cytochrome *c* and other proteins from the intermembrane space into the cytosol by an unknown mechanism.

proteins. Their BH3 domain binds to a long hydrophobic groove on anti-apop-totic Bcl2 family proteins, neutralizing their activity. This binding and inhibition enables the aggregation of Bax and Bak on the surface of mitochondria, which triggers the release of the intermembrane mitochondrial proteins that induce apoptosis (**Figure 18–10**). Some BH3-only proteins may bind directly to Bax and Bak to help stimulate their aggregation.

BH3-only proteins provide the crucial link between apoptotic stimuli and the intrinsic pathway of apoptosis, with different stimuli activating different BH3-only proteins. Some extracellular survival signals, for example, block apopto-sis by inhibiting the synthesis or activity of certain BH3-only proteins (see Fig-ure 18–12B). Similarly, in response to DNA damage that cannot be repaired, the tumor suppressor protein **p53** accumulates (discussed in Chapters 17 and 20) and activates the transcription of genes that encode the BH3-only proteins *Puma* and *Noxa*. These BH3-only proteins then trigger the intrinsic pathway, thereby elimi-nating a potentially dangerous cell that could otherwise become cancerous.

As mentioned earlier, in some cells the extrinsic apoptotic pathway recruits the intrinsic pathway to amplify the caspase cascade to kill the cell. The BH3-only protein *Bid* is the link between the two pathways. Bid is normally inactive. How-ever, when death receptors activate the extrinsic pathway in some cells, the initia-tor caspase, caspase-8, cleaves Bid, producing an active form of Bid that trans-locates to the outer mitochondrial membrane and inhibits anti-apoptotic Bcl2 family proteins, thereby amplifying the death signal.

(A) INACTIVE INTRINSIC PATHWAY

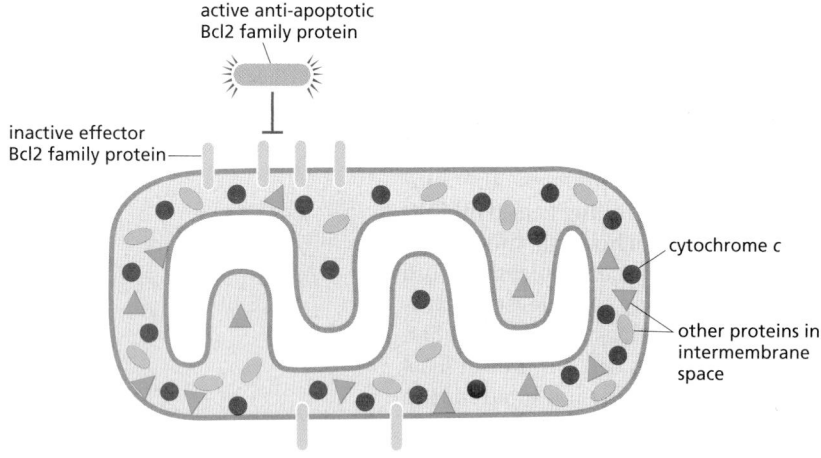

(B) ACTIVATION OF INTRINSIC PATHWAY

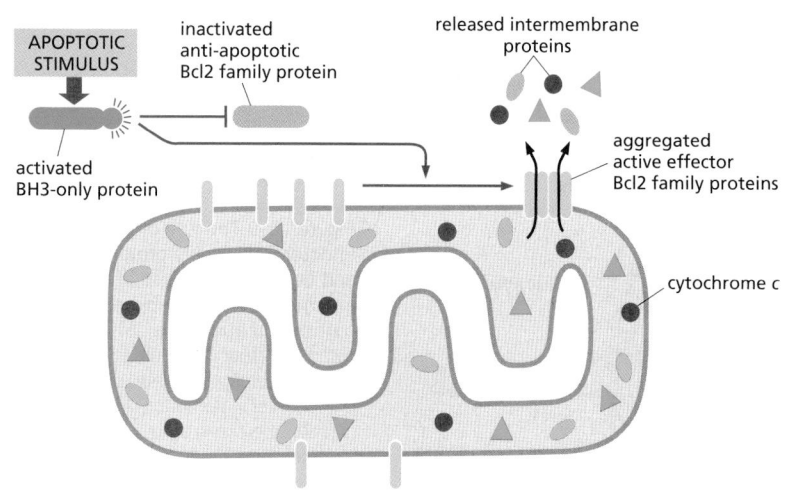

Figure 18–10 **How pro-apoptotic BH3-only and anti-apoptotic Bcl2 family proteins regulate the intrinsic pathway of apoptosis.** (A) In the absence of an apoptotic stimulus, anti-apoptotic Bcl2 family proteins bind to and inhibit the effector Bcl2 family proteins on the mitochondrial outer membrane (and in the cytosol—not shown). (B) In the presence of an apoptotic stimulus, BH3-only proteins are activated and bind to the anti-apoptotic Bcl2 family proteins so that they can no longer inhibit the effector Bcl2 family proteins; the latter then become activated, aggregate in the outer mitochondrial membrane, and promote the release of intermembrane mitochondrial proteins into the cytosol. Some activated BH3-only proteins may stimulate mitochondrial protein release more directly by binding to and activating the effector Bcl2 family proteins. Although not shown, the anti-apoptotic Bcl2 family proteins are bound to the mitochondrial surface.

IAPs Help Control Caspases

Because activation of a caspase cascade leads to certain death, the cell employs multiple robust mechanisms to ensure that these proteases are activated only when appropriate. One line of defense is provided by a family of proteins called **inhibitors of apoptosis** (**IAPs**). These proteins were first identified in certain insect viruses (baculoviruses), which encode IAP proteins to prevent a host cell that is infected by the virus from killing itself by apoptosis. It is now known that most animal cells also make IAP proteins.

All IAPs have one or more BIR (baculovirus IAP repeat) domains, which enable them to bind to and inhibit activated caspases. Some IAPs also polyubiquitylate caspases, marking the caspases for destruction by proteasomes. In this way, the IAPs set an inhibitory threshold that caspases must overcome to trigger apoptosis.

In *Drosophila* at least, the inhibitory barrier provided by IAPs can be neutralized by **anti-IAP** proteins, which are produced in response to various apoptotic stimuli. There are numerous anti-IAPs in flies, including *Reaper*, *Grim*, and *Hid*, and their only structural similarity is their short, N-terminal, IAP-binding motif, which binds to the BIR domain of IAPs, preventing the domain from binding to a caspase. Deletion of the three genes encoding Reaper, Grim, and Hid blocks apoptosis in flies. Conversely, inactivation of one of the two genes that encode IAPs in *Drosophila* causes all of the cells in the developing fly embryo to undergo apoptosis. Clearly, the balance between IAPs and anti-IAPs is tightly regulated and is crucial for controlling apoptosis in the fly.

The role of mammalian IAP and anti-IAP proteins in apoptosis is less clear. Anti-IAPs are released from the mitochondrial intermembrane space when the intrinsic pathway of apoptosis is activated, blocking IAPs in the cytosol and thereby promoting apoptosis. However, mice appear to develop normally if they are missing either the major mammalian IAP (called XIAP) or the two known mammalian anti-IAPs (called Smac/Diablo and Omi). Worms do not even contain a caspase-inhibiting IAP protein. Apparently, the tight control of caspase activity is achieved by different mechanisms in different animals.

Extracellular Survival Factors Inhibit Apoptosis in Various Ways

Intercellular signals regulate most activities of animal cells, including apoptosis. These extracellular signals are part of the normal "social" controls that ensure that individual cells behave for the good of the organism as a whole—in this case, by surviving when they are needed and killing themselves when they are not. Some extracellular signal molecules stimulate apoptosis, whereas others inhibit it. We have discussed signal proteins such as Fas ligand that activate death receptors and thereby trigger the extrinsic pathway of apoptosis. Other extracellular signal molecules that stimulate apoptosis are especially important during vertebrate development: a surge of thyroid hormone in the bloodstream, for example, signals cells in the tadpole tail to undergo apoptosis at metamorphosis. In mice, locally produced signal proteins stimulate cells between developing fingers and toes to kill themselves (see Figure 18–2). Here, however, we focus on extracellular signal molecules that inhibit apoptosis, which are collectively called **survival factors**.

Most animal cells require continuous signaling from other cells to avoid apoptosis. This surprising arrangement apparently helps ensure that cells survive only when and where they are needed. Nerve cells, for example, are produced in excess in the developing nervous system and then compete for limited amounts of survival factors that are secreted by the target cells that they normally connect to (see Figure 21–81). Nerve cells that receive enough survival signals live, while the others die. In this way, the number of surviving neurons is automatically adjusted so that it is appropriate for the number of target cells they connect with (**Figure 18–11**). A similar competition for limited amounts of survival factors produced by neighboring cells is thought to control cell numbers in other tissues, both during development and in adulthood.

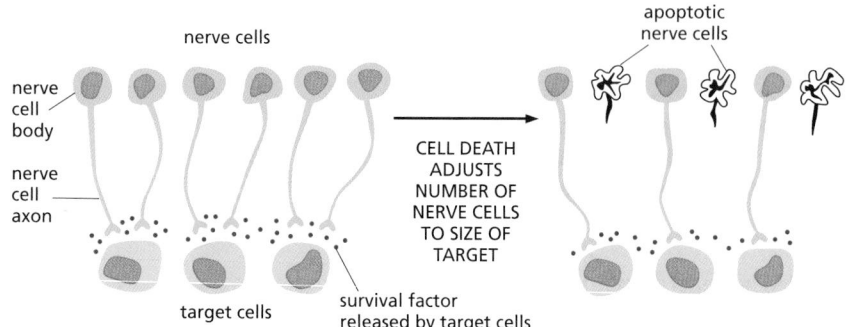

Figure 18–11 The role of survival factors and cell death in adjusting the number of developing nerve cells to the amount of target tissue. More nerve cells are produced than can be supported by the limited amount of survival factors released by the target cells. Therefore, some nerve cells receive an insufficient amount of survival factors to avoid apoptosis. This strategy of overproduction followed by culling helps ensure that all target cells are contacted by nerve cells and that the extra nerve cells are automatically eliminated.

Survival factors usually bind to cell-surface receptors, which activate intracellular signaling pathways that suppress the apoptotic program, often by regulating members of the Bcl2 family of proteins. Some survival factors, for example, stimulate the synthesis of anti-apoptotic Bcl2 family proteins such as Bcl2 itself or BclX$_L$ (**Figure 18–12A**). Others act by inhibiting the function of pro-apoptotic BH3-only proteins such as *Bad* (Figure M18–12B). In *Drosophila*, some survival factors act by phosphorylating and inactivating anti-IAP proteins such as Hid, thereby enabling IAP proteins to suppress apoptosis (Figure 18–12C). Some developing neurons, like those illustrated in Figure 18–11, use an ingenious alternative approach: survival-factor receptors stimulate apoptosis—by an unknown mechanism—when they are not occupied, and then stop promoting death when survival factor binds. The end result in all these cases is the same: cell survival depends on survival factor binding.

Phagocytes Remove the Apoptotic Cell

Apoptotic cell death is a remarkably tidy process: the apoptotic cell and its fragments do not break open and release their contents, but instead remain intact as they are efficiently eaten—or *phagocytosed*—by neighboring cells, leaving no trace and therefore triggering no inflammatory response (see Figure 18–1B and Movie 13.5). This engulfment process depends on chemical changes on the surface of the apoptotic cell, which displays signals that recruit phagocytic cells. An especially important change occurs in the distribution of the negatively charged phospholipid *phosphatidylserine* on the cell surface. This phospholipid is normally located exclusively in the inner leaflet of the lipid bilayer of the plasma membrane (see Figure 10–15), but it flips to the outer leaflet in apoptotic cells. The

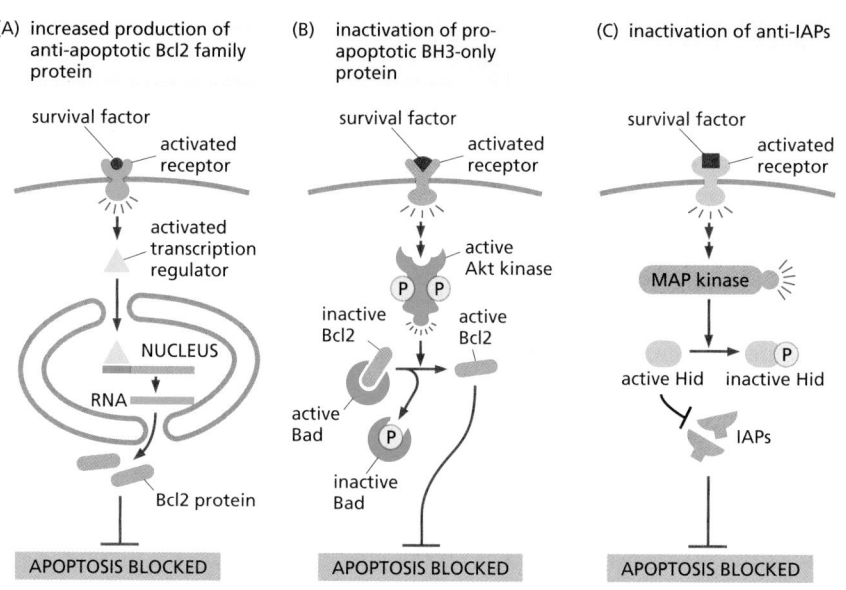

Figure 18–12 Three ways that extracellular survival factors can inhibit apoptosis. (A) Some survival factors suppress apoptosis by stimulating the transcription of genes that encode anti-apoptotic Bcl2 family proteins such as Bcl2 itself or BclX$_L$. (B) Many others activate the serine/threonine protein kinase Akt, which, among many other targets, phosphorylates and inactivates the pro-apoptotic BH3-only protein Bad (see Figure 15–53). When not phosphorylated, Bad promotes apoptosis by binding to and inhibiting Bcl2; once phosphorylated, Bad dissociates, freeing Bcl2 to suppress apoptosis. Akt also suppresses apoptosis by phosphorylating and inactivating transcription regulatory proteins that stimulate the transcription of genes encoding proteins that promote apoptosis (not shown). (C) In *Drosophila*, some survival factors inhibit apoptosis by stimulating the phosphorylation of the anti-IAP protein Hid. When not phosphorylated, Hid promotes cell death by inhibiting IAPs. Once phosphorylated, Hid no longer inhibits IAPs, which become active and block apoptosis. MAP kinase, mitogen-activated protein kinase.

underlying mechanism is poorly understood, but the external exposure of phosphatidylserine is likely to depend on caspase cleavage of some protein involved in phospholipid distribution in the membrane. A variety of soluble "bridging" proteins interact with the exposed phosphatidylserine on the apoptotic cell. These bridging proteins also interact with specific receptors on the surface of a neighboring cell or macrophage, triggering cytoskeletal and other changes that initiate the engulfment process.

Macrophages do not phagocytose healthy cells in the animal—despite the fact that healthy cells normally expose some phosphatidylserine on their surfaces. Healthy cells express signal proteins on their surface that interact with inhibitory receptors on macrophages that block phagocytosis. Thus, in addition to expressing cell-surface signals such as phosphatidylserine that stimulate phagocytosis, apoptotic cells must lose or inactivate these "don't eat me" signals that block phagocytosis.

Either Excessive or Insufficient Apoptosis Can Contribute to Disease

There are many human disorders in which excessive numbers of cells undergo apoptosis and thereby contribute to tissue damage. Among the most dramatic examples are heart attacks and strokes. In these acute conditions, many cells die by necrosis as a result of ischemia (inadequate blood supply), but some of the less affected cells die by apoptosis. It is hoped that, in the future, drugs that block apoptosis—such as specific caspase inhibitors—will prove useful in saving such cells.

There are other conditions where too few cells die by apoptosis. Mutations in mice and humans, for example, that inactivate the genes that encode the Fas death receptor or the Fas ligand prevent the normal death of some lymphocytes, causing these cells to accumulate in excessive numbers in the spleen and lymph glands. In many cases, this leads to autoimmune disease, in which the lymphocytes react against the individual's own tissues.

Decreased apoptosis also makes an important contribution to many tumors, as cancer cells often regulate their apoptotic program abnormally. The *Bcl2* gene, for example, was first identified in a common form of lymphocyte cancer in humans, where a chromosome translocation causes excessive production of the Bcl2 protein; indeed, Bcl2 gets its name from this *B cell lymphoma*. The high level of Bcl2 protein in the lymphocytes that carry the translocation promotes the development of cancer by inhibiting apoptosis, thereby prolonging lymphocyte survival and increasing their number; it also decreases the cells' sensitivity to anticancer drugs, which commonly work by causing cancer cells to undergo apoptosis.

Similarly, the gene encoding the tumor suppressor protein p53 is mutated in about 50% of human cancers so that it no longer promotes apoptosis or cell-cycle arrest in response to DNA damage. The lack of p53 function therefore enables the cancer cells to survive and proliferate even when their DNA is damaged; in this way, the cells accumulate more mutations, some of which make the cancer more malignant (discussed in Chapter 20). As many anticancer drugs induce apoptosis (and cell-cycle arrest) by a p53-dependent mechanism (discussed in Chapters 17 and 20), the loss of p53 function also makes cancer cells less sensitive to these drugs.

If decreased apoptosis contributes to many cancers, then we might be able to treat those cancers with drugs that stimulate apoptosis. This line of thinking has recently led to the development of small chemicals that interfere with the function of anti-apoptotic Bcl2 family proteins such as Bcl2 and $BclX_L$. These chemicals bind with high affinity to the hydrophobic groove on anti-apoptotic Bcl2 family proteins, blocking their function in essentially the same way that BH3-only proteins do (**Figure 18–13**). The intrinsic pathway of apoptosis is thereby stimulated, which in certain tumors increases the amount of cell death.

Most human cancers arise in epithelial tissues such as those in the lung, intestinal tract, breast, and prostate. Such cancer cells display many abnormalities in

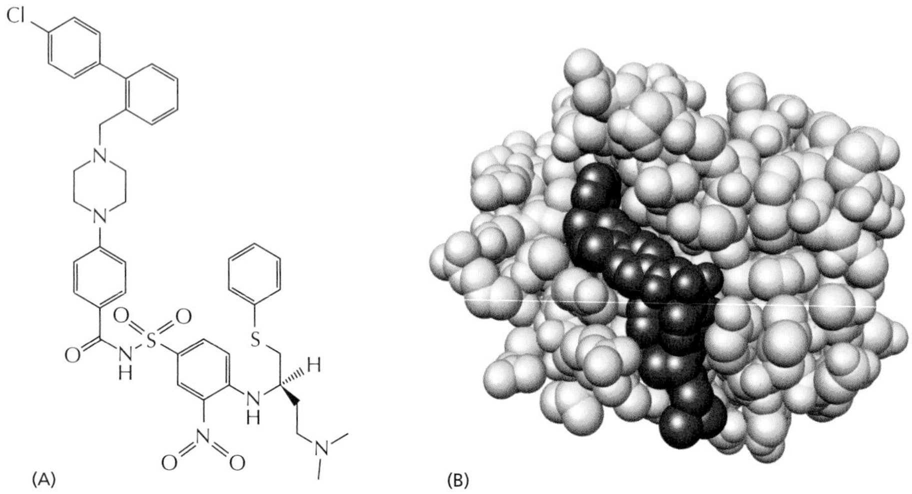

(A)   (B)

Figure 18–13 **How the chemical ABT-737 inhibits anti-apoptotic Bcl2 family proteins.** As shown in Figure 18–10B, an apoptotic signal results in activation of BH3-only proteins, which interact with a long hydrophobic groove in anti-apoptotic Bcl2 family proteins, thereby preventing them from blocking apoptosis. Using the crystal structure of the groove, the drug shown in (A), called ABT-737, was designed and synthesized to bind tightly in the groove, as shown for the anti-apoptotic Bcl2 family protein, BclX$_L$, in (B). By inhibiting the activity of these proteins, the drug promotes apoptosis in any cell that depends on them for survival. (PDB code: 2YXJ.)

their behavior, including a decreased ability to adhere to the extracellular matrix and to one another at specialized cell–cell junctions. In the next chapter, we discuss the remarkable structures and functions of the extracellular matrix and cell junctions.

Summary

Animal cells can activate an intracellular death program and kill themselves in a controlled way when they are irreversibly damaged, no longer needed, or are a threat to the organism. In most cases, these deaths occur by apoptosis: the cells shrink, condense, and frequently fragment, and neighboring cells or macrophages rapidly phagocytose the cells or fragments before there is any leakage of cytoplasmic contents. Apoptosis is mediated by proteolytic enzymes called caspases, which cleave specific intracellular proteins to help kill the cell. Caspases are present in all nucleated animal cells as inactive precursors. Initiator caspases are activated when brought into proximity in activation complexes: once activated, they cleave and thereby activate downstream executioner caspases, which then cleave various target proteins in the cell, producing an amplifying, irreversible proteolytic cascade.

Cells use at least two distinct pathways to activate initiator caspases and trigger a caspase cascade leading to apoptosis: the extrinsic pathway is activated by extracellular ligands binding to cell-surface death receptors; the intrinsic pathway is activated by intracellular signals generated when cells are stressed. Each pathway uses its own initiator caspases, which are activated in distinct activation complexes: in the extrinsic pathway, the death receptors recruit caspase-8 via adaptor proteins to form the DISC; in the intrinsic pathway, cytochrome c released from the intermembrane space of mitochondria activates Apaf1, which assembles into an apoptosome and recruits and activates caspase-9.

Intracellular Bcl2 family proteins and IAP proteins tightly regulate the apoptotic program to ensure that cells kill themselves only when it benefits the animal. Both anti-apoptotic and pro-apoptotic Bcl2 family proteins regulate the intrinsic pathway by controlling the release of mitochondrial intermembrane proteins, while IAP proteins inhibit activated caspases and promote their degradation.

WHAT WE DON'T KNOW

• How many forms of programmed cell death exist? What are the underlying mechanisms and benefits of each?

• Thousands of caspase substrates have been identified. Which ones are the critical proteins that must be cleaved to trigger the major cell remodeling events underlying apoptosis?

• How did the intrinsic pathway of apoptosis evolve, and what is the advantage of having mitochondria play such a central role in regulating apoptosis?

• How are "don't eat me" signals eliminated or inactivated during apoptosis to allow the cells to be phagocytosed?

PROBLEMS

Which statements are true? Explain why or why not.

18–1   In normal adult tissues, cell death usually balances cell division.

18–2   Mammalian cells that do not have cytochrome c should be resistant to apoptosis induced by DNA damage.

Discuss the following problems.

18–3   One important role of Fas and Fas ligand is to mediate the elimination of tumor cells by killer lymphocytes. In a study of 35 primary lung and colon tumors, half the tumors were found to have amplified and overexpressed a gene for a secreted protein that binds to Fas ligand. How do you suppose that overexpression of this protein might contribute to the survival of these tumor cells? Explain your reasoning.

18–4   Development of the nematode *Caenorhabditis elegans* generates exactly 959 somatic cells; it also produces an additional 131 cells that are later eliminated by apoptosis. Classical genetic experiments in *C. elegans* isolated mutants that led to the identification of the first genes involved in apoptosis. Of the many mutations affecting apoptosis in the nematode, none have ever been found in the gene for cytochrome c. Why do you suppose that such a central effector molecule in apoptosis was not found in the many genetic screens for "death" genes that have been carried out in *C. elegans*?

18–5   Imagine that you could microinject cytochrome c into the cytosol of wild-type mammalian cells and of cells that were doubly defective for Bax and Bak. Would you expect one, both, or neither type of cell to undergo apoptosis? Explain your reasoning.

18–6   In contrast to their similar brain abnormalities, newborn mice deficient in Apaf1 or caspase-9 have distinctive abnormalities in their paws. Apaf1-deficient mice fail to eliminate the webs between their developing digits, whereas caspase-9-deficient mice have normally formed digits (**Figure Q18–1**). If Apaf1 and caspase-9 function in the same apoptotic pathway, how is it possible for these deficient mice to differ in web-cell apoptosis?

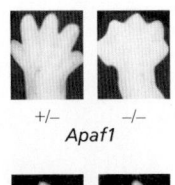

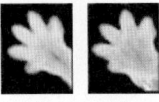

Figure Q18–1 Appearance of paws in *Apaf1*$^{-/-}$ and *Casp9*$^{-/-}$ newborn mice relative to normal newborn mice (Problem 18–6). (From H. Yoshida et al., *Cell* 94:739–750, 1998. With permission from Elsevier.)

18–7   When human cancer cells are exposed to ultraviolet (UV) light at 90 mJ/cm^2, most of the cells undergo apoptosis within 24 hours. Release of cytochrome c from mitochondria can be detected as early as 6 hours after exposure of a population of cells to UV light, and it continues to increase for more than 10 hours thereafter. Does this mean that individual cells slowly release their cytochrome c over this time period? Or, alternatively, do individual cells release their cytochrome c rapidly but with different cells being triggered over the longer time period?

To answer this fundamental question, you have fused the gene for green fluorescent protein (GFP) to the gene for cytochrome c, so that you can observe the behavior of individual cells by confocal fluorescence microscopy. In cells that are expressing the cytochrome c–GFP fusion, fluorescence shows the punctate pattern typical of mitochondrial proteins. You then irradiate these cells with UV light and observe individual cells for changes in the punctate pattern. Two such cells (outlined in white) are shown in **Figure Q18–2**A and B. Release of cytochrome c–GFP is detected as a change from a punctate to a diffuse pattern of fluorescence. Times after UV exposure are indicated as hours:minutes below the individual panels.

Which model for cytochrome c release do these observations support? Explain your reasoning.

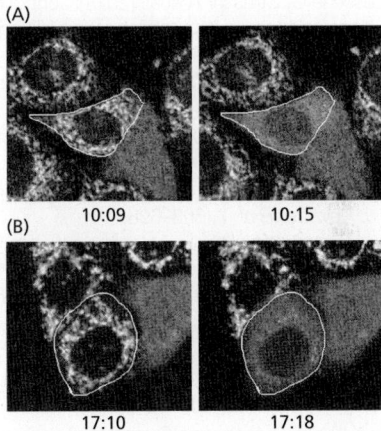

Figure Q18–2 Time-lapse video fluorescence microscopic analysis of cytochrome c–GFP release from mitochondria of individual cells (Problem 18–7). (A) Cells observed for 6 minutes, 10 hours after UV irradiation. (B) Cells observed for 8 minutes, 17 hours after UV irradiation. One cell in (A) and one in (B), each *outlined in white*, have released their cytochrome c–GFP during the time frame of the observation, which is shown as hours:minutes below each panel. (From J.C. Goldstein et al., *Nat. Cell Biol.* 2:156–162, published 2000 by Nature Publishing Group. Reproduced with permission of SNCSC.)

18–8   Fas ligand is a trimeric, extracellular protein that binds to its receptor, Fas, which is composed of three identical transmembrane subunits (**Figure Q18–3**). The binding of Fas ligand alters the conformation of Fas so that it binds an adaptor protein, which then recruits and activates caspase-8, triggering a caspase cascade that leads to cell death. In humans, the autoimmune lymphoproliferative syndrome (ALPS) is associated with dominant mutations in Fas that include point mutations and C-terminal truncations. In individuals that are heterozygous for such mutations, lymphocytes do not die at their normal rate and accumulate in abnormally large numbers, causing a variety of clinical problems. In contrast to these patients, individuals that are heterozygous for mutations that eliminate Fas expression entirely have no clinical symptoms.

A.   Assuming that the normal and dominant forms of Fas are expressed to the same level and bind Fas ligand equally, what fraction of Fas–Fas ligand complexes on a lymphocyte from a heterozygous ALPS patient would be expected to be composed entirely of normal Fas subunits?

B.   In an individual heterozygous for a mutation that eliminates Fas expression, what fraction of Fas–Fas ligand complexes would be expected to be composed entirely of normal Fas subunits?

C.   Why are the Fas mutations that are associated with ALPS dominant, while those that eliminate expression of Fas are recessive?

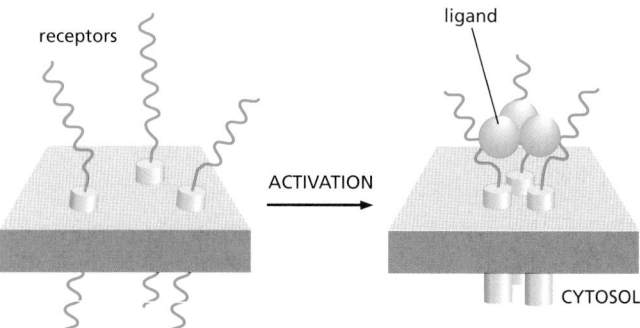

Figure Q18–3 The binding of trimeric Fas ligand to Fas (Problem 18–8).

REFERENCES

Crawford ED & Wells JA (2011) Caspase substrates and cellular remodeling. *Annu. Rev. Biochem.* 80, 1055–1087.

Czabotar PE, Lessene G, Strasser A et al. (2014) Control of apoptosis by the BCL-2 protein family: implications for physiology and therapy. *Nat. Rev. Mol. Cell Biol.* 15, 49–63.

Danial NN & Korsmeyer SJ (2004) Cell death: critical control points. *Cell* 116, 205–219.

Elliott MR & Ravichandran KS (2010) Clearance of apoptotic cells: implications in health and disease. *J. Cell Biol.* 189, 1059–1070.

Ellis RE, Yuan JY & Horvitz RA (1991) Mechanisms and functions of cell death. *Annu. Rev. Cell Biol.* 7, 663–698.

Fadok VA & Henson PM (2003) Apoptosis: giving phosphatidylserine recognition an assist—with a twist. *Curr. Biol.* 13, R655–R657.

Green DR (2011) Means to an End: Apoptosis and Other Cell Death Mechanisms. Cold Spring Harbor, New York: Cold Spring Harbor Laboratory Press.

Jacobson MD, Weil M & Raff MC (1997) Programmed cell death in animal development. *Cell* 88, 347–354.

Jiang X & Wang X (2004) Cytochrome C-mediated apoptosis. *Annu. Rev. Biochem.* 73, 87–106.

Kerr JF, Wyllie AH & Currie AR (1972) Apoptosis: a basic biological phenomenon with wide-ranging implications in tissue kinetics. *Brit. J. Cancer* 26, 239–257.

Kumar S (2007) Caspase function in programmed cell death. *Cell Death Differ.* 14, 32–43.

Lavrik I, Golks A & Krammer PH (2005) Death receptor signaling. *J. Cell Sci.* 118, 265–267.

Lessene G, Czabotar PE & Colman PM (2008) BCL-2 family antagonists for cancer therapy. *Nat. Rev. Drug Discov.* 7, 989–1000.

Mace PD & Riedl SJ (2010) Molecular cell death platforms and assemblies. *Curr. Opin. Cell Biol.* 22, 828–836.

Nagata S (2005) DNA degradation in development and programmed cell death. *Annu. Rev. Immunol.* 23, 853–875.

Raff MC (1999) Cell suicide for beginners. *Nature* 396, 119–122.

Tait SW & Green DR (2013) Mitochondrial regulation of cell death. *Cold Spring Harb. Perspect. Biol.* 5, a008706.

Vanden Berghe T, Linkermann A, Jouan-Lanhouet S et al. (2014) Regulated necrosis: the expanding network of non-apoptotic cell death pathways. *Nat. Rev. Mol. Cell Biol.* 15, 135–147.

Vousden KH (2005) Apoptosis. p53 and PUMA: a deadly duo. *Science* 309, 1685–1686.

Willis SN & Adams JM (2005) Life in the balance: how BH3-only proteins induce apoptosis. *Curr. Opin. Cell Biol.* 17, 617–625.

Yuan S & Akey CW (2013) Apoptosome structure, assembly, and procaspase activation. *Structure* 21, 501–515.

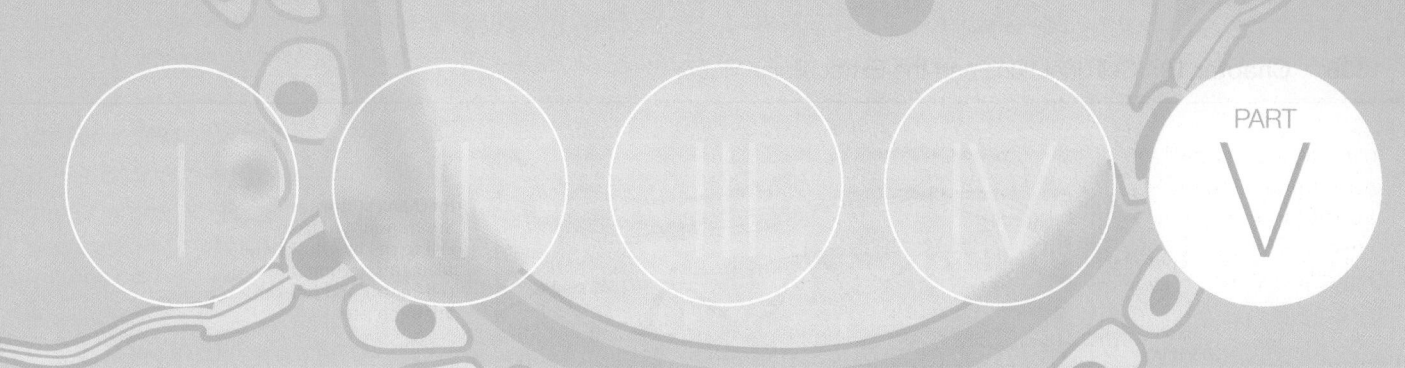

CELLS IN THEIR SOCIAL CONTEXT

Cell Junctions and the Extracellular Matrix

Of all the social interactions between cells in a multicellular organism, the most fundamental are those that hold the cells together. Cells may be linked by direct interactions, or they may be held together within the *extracellular matrix*, a complex network of proteins and polysaccharide chains that the cells secrete. By one means or another, cells must cohere if they are to form an organized multicellular structure that can withstand and respond to the various external forces that try to pull it apart.

The mechanisms of cohesion govern the architecture of the body—its shape, its strength, and the arrangement of its different cell types. The making and breaking of the attachments between cells and the modeling of the extracellular matrix govern the way cells move within the organism, guiding them as the body grows, develops, and repairs itself. Attachments to other cells and to extracellular matrix control the orientation and behavior of the cell's cytoskeleton, thereby allowing cells to sense and respond to changes in the mechanical features of their environment. Thus, the apparatus of cell junctions and the extracellular matrix is critical for every aspect of the organization, function, and dynamics of multicellular structures. Defects in this apparatus underlie an enormous variety of diseases.

The key features of cell junctions and the extracellular matrix are best illustrated by considering two broad categories of tissues that are found in all animals (**Figure 19–1**). **Connective tissues**, such as bone or tendon, are formed from an extracellular matrix produced by cells that are distributed sparsely in the matrix. It is the matrix—rather than the cells—that bears most of the mechanical stress to which the tissue is subjected. Direct attachments between one cell and another are relatively rare, but the cells have important attachments to the matrix. These *cell–matrix junctions* link the cytoskeleton to the matrix, allowing the cells to move through the matrix and monitor changes in its mechanical properties.

In **epithelial tissues**, such as the lining of the gut or the epidermal covering of the skin, cells are tightly bound together into sheets called **epithelia**. The extracellular matrix is less pronounced, consisting mainly of a thin mat called the *basal lamina* (or *basement membrane*) underlying the sheet. Within the epithelium, cells are attached to each other directly by *cell–cell junctions*, where cytoskeletal filaments are anchored, transmitting stresses across the interiors of the cells, from

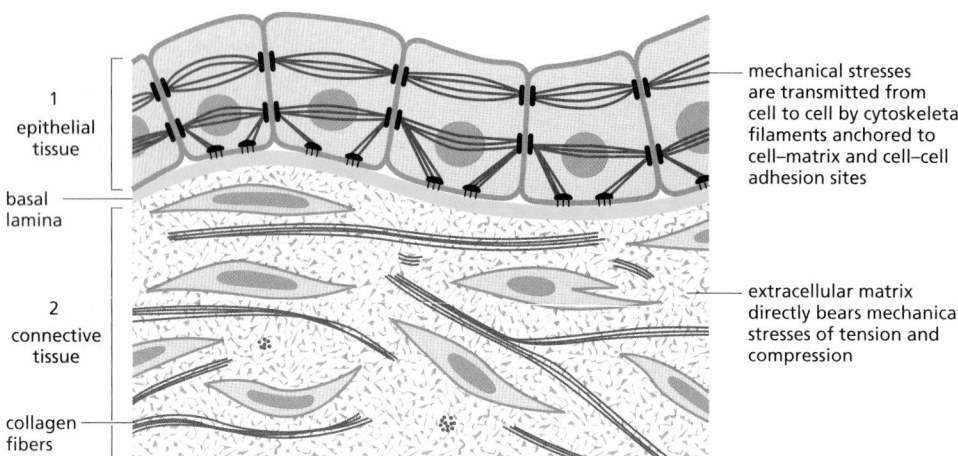

1
epithelial
tissue

basal
lamina

2
connective
tissue

collagen
fibers

mechanical stresses
are transmitted from
cell to cell by cytoskeletal
filaments anchored to
cell–matrix and cell–cell
adhesion sites

extracellular matrix
directly bears mechanical
stresses of tension and
compression

adhesion site to adhesion site. The cytoskeleton of epithelial cells is also linked to the basal lamina through cell–matrix junctions.

Figure 19–2 provides a closer view of epithelial cells to illustrate the major types of cell–cell and cell–matrix junctions that we will discuss in this chapter. The diagram shows the typical arrangement of junctions in a *simple columnar* epithelium such as the lining of the small intestine of a vertebrate. Here, a single layer of tall cells stands on a basal lamina, with the cells' uppermost surface, or *apex*, free and exposed to the extracellular medium. On their sides, or *lateral* surfaces, the cells make junctions with one another. Two types of **anchoring junctions** link the cytoskeletons of adjacent cells: **adherens junctions** are anchorage sites for actin filaments; **desmosomes** are anchorage sites for intermediate filaments. Two additional types of anchoring junctions link the cytoskeleton of the epithelial cells to the basal lamina: *actin-linked cell–matrix junctions* anchor actin filaments to the matrix, while *hemidesmosomes* anchor intermediate filaments to it.

Figure 19–1 Two main ways in which animal cells are bound together. In connective tissue, the main stress-bearing component is the extracellular matrix. In epithelial tissue, it is the cytoskeletons of the cells themselves, linked from cell to cell by adhesive junctions. Cell–matrix attachments bond epithelial tissue to the connective tissue beneath it.

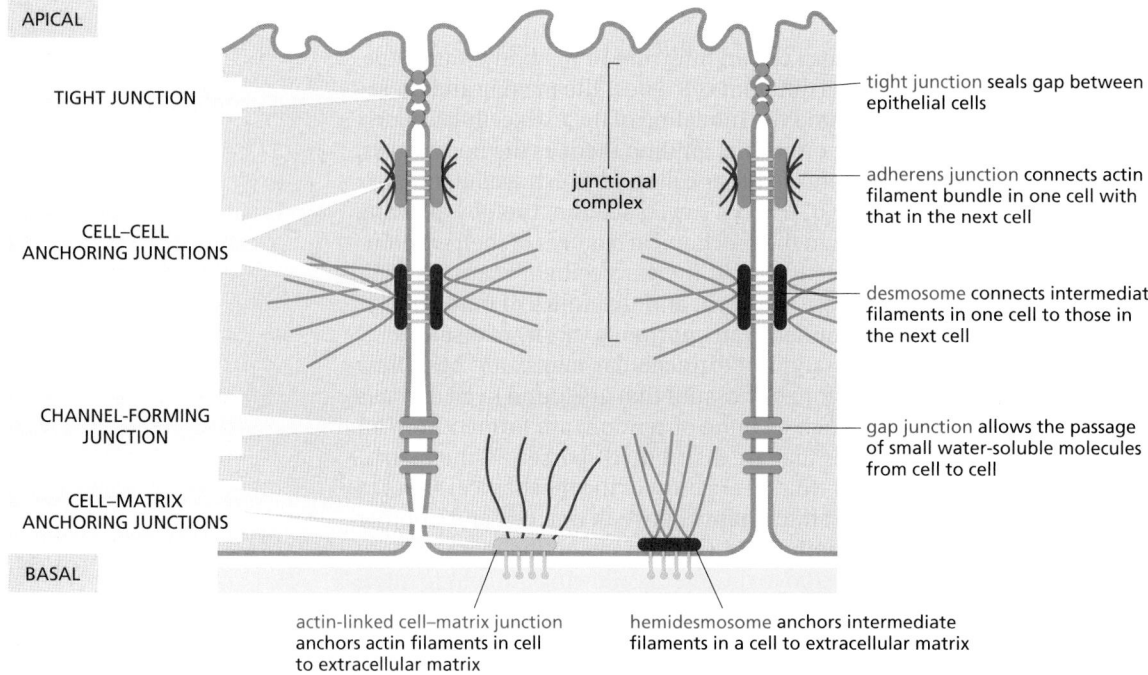

APICAL

TIGHT JUNCTION

CELL–CELL
ANCHORING JUNCTIONS

CHANNEL-FORMING
JUNCTION

CELL–MATRIX
ANCHORING JUNCTIONS

BASAL

junctional
complex

tight junction seals gap between
epithelial cells

adherens junction connects actin
filament bundle in one cell with
that in the next cell

desmosome connects intermediate
filaments in one cell to those in
the next cell

gap junction allows the passage
of small water-soluble molecules
from cell to cell

actin-linked cell–matrix junction
anchors actin filaments in cell
to extracellular matrix

hemidesmosome anchors intermediate
filaments in a cell to extracellular matrix

Figure 19–2 A summary of the various cell junctions found in a vertebrate epithelial cell, classified according to their primary functions.
In the most apical portion of the cell, the relative positions of the junctions are the same in nearly all vertebrate epithelia. The tight junction occupies the most apical position, followed by the adherens junction (adhesion belt) and then by a special parallel row of desmosomes; together these form a structure called a junctional complex. Gap junctions and additional desmosomes are less regularly organized. Two types of cell-matrix anchoring junctions tether the basal surface of the cell to the basal lamina. The drawing is based on epithelial cells of the small intestine.

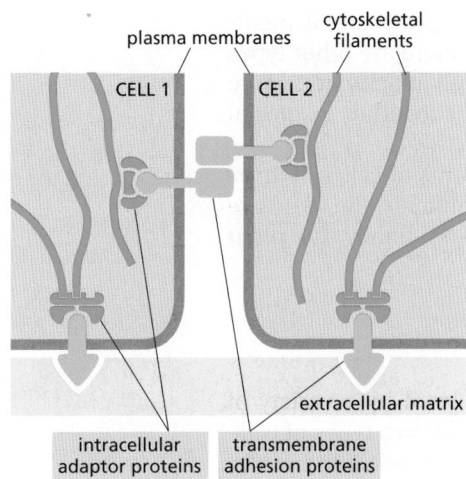

plasma membranes

cytoskeletal filaments

CELL 1   CELL 2

extracellular matrix

intracellular adaptor proteins

transmembrane adhesion proteins

Figure 19–3 Transmembrane adhesion proteins link the cytoskeleton to extracellular structures. The external linkage may be either to other cells (cell–cell junctions, mediated typically by cadherins) or to extracellular matrix (cell–matrix junctions, mediated typically by integrins). The internal linkage to the cytoskeleton is generally indirect, via intracellular adaptor proteins, to be discussed later.

Two other types of cell–cell junction are shown in Figure 19–2. *Tight junctions* hold the cells closely together near the apex, sealing the gap between the cells and thereby preventing molecules from leaking across the epithelium. Near the basal end of the cells are channel-forming junctions, called *gap junctions,* that create passageways linking the cytoplasms of adjacent cells.

Each of the four major anchoring junction types depends on **transmembrane adhesion proteins** that span the plasma membrane, with one end linking to the cytoskeleton inside the cell and the other end linking to other structures outside it (**Figure 19–3**). These cytoskeleton-linked transmembrane proteins fall neatly into two superfamilies, corresponding to the two basic kinds of external attachment. Proteins of the **cadherin** superfamily chiefly mediate attachment of cell to cell (**Movie 19.1**). Proteins of the **integrin** superfamily chiefly mediate attachment of cells to matrix. There is specialization within each family: some cadherins link to actin and form adherens junctions, while others link to intermediate filaments and form desmosomes; likewise, some integrins link to actin and form actin-linked cell–matrix junctions, while others link to intermediate filaments and form hemidesmosomes (**Table 19–1**).

TABLE 19–1 Anchoring Junctions

Junction	Transmembrane adhesion protein	Extracellular ligand	Intracellular cytoskeletal attachment	Intracellular adaptor proteins
Cell–Cell				
Adherens junction	Classical cadherins	Classical cadherin on neighboring cell	Actin filaments	α-Catenin, β-catenin, plakoglobin (γ-catenin), p120-catenin, vinculin
Desmosome	Nonclassical cadherins (desmoglein, desmocollin)	Desmoglein and desmocollin on neighboring cell	Intermediate filaments	Plakoglobin (γ-catenin), plakophilin, desmoplakin
Cell–Matrix				
Actin-linked cell–matrix junction	Integrin	Extracellular matrix proteins	Actin filaments	Talin, kindlin, vinculin, paxillin, focal adhesion kinase (FAK), numerous others
Hemidesmosome	$\alpha_6\beta_4$ Integrin, type XVII collagen	Extracellular matrix proteins	Intermediate filaments	Plectin, BP230

There are some exceptions to these rules. Some integrins, for example, mediate cell–cell rather than cell–matrix attachment. Moreover, there are other types of cell adhesion molecules that can provide transient cell–cell attachments more flimsy than anchoring junctions, but sufficient to stick cells together in special circumstances.

We begin the chapter with a discussion of the major forms of cell–cell junctions. We then consider in turn the extracellular matrix of animals, the structure and function of integrin-mediated cell–matrix junctions, and, finally, the plant cell wall, a special form of extracellular matrix.

CELL–CELL JUNCTIONS

Cell–cell junctions come in many forms and can be regulated by a variety of mechanisms. The best understood and most common are the two types of cell–cell anchoring junctions, which employ cadherins to link the cytoskeleton of one cell with that of its neighbor. Their primary function is to resist the external forces that pull cells apart. The epithelial cells of your skin, for example, must remain tightly linked when they are stretched, pinched, or poked. Cell–cell anchoring junctions must also be dynamic and adaptable, so that they can be altered or rearranged when tissues are remodeled or repaired, or when there are changes in the forces acting on them.

In this section, we focus primarily on the cadherin-based anchoring junctions. We then briefly describe tight junctions and gap junctions. Finally, we consider the more transient cell–cell adhesion mechanisms employed by some cells in the bloodstream.

Cadherins Form a Diverse Family of Adhesion Molecules

Cadherins are present in all multicellular animals whose genomes have been analyzed. They are also present in the choanoflagellates, which can exist either as free-living unicellular organisms or as multicellular colonies and are thought to be representatives of the group of protists from which all animals evolved. Other eukaryotes, including fungi and plants, lack cadherins, and they are also absent from bacteria and archaea. Cadherins therefore seem to be part of the essence of what it is to be an animal.

The cadherins take their name from their dependence on Ca^{2+} ions: removing Ca^{2+} from the extracellular medium causes adhesions mediated by cadherins to come apart. The first three cadherins to be discovered were named according to the main tissues in which they were found: *E-cadherin* is present on many types of epithelial cells; *N-cadherin* on nerve, muscle, and lens cells; and *P-cadherin* on cells in the placenta and epidermis. All are also found in other tissues. These and other **classical cadherins** are closely related in sequence throughout their extracellular and intracellular domains.

There are also a large number of **nonclassical cadherins** that are more distantly related in sequence, with more than 50 expressed in the brain alone. The nonclassical cadherins include proteins with known adhesive function, such as the diverse *protocadherins* found in the brain, and the *desmocollins* and *desmogleins* that form desmosomes (see Table 19–1). Other family members are involved primarily in signaling. Together, the classical and nonclassical cadherin proteins constitute the **cadherin superfamily** (**Figure 19–4**), with more than 180 members in humans.

Cadherins Mediate Homophilic Adhesion

Anchoring junctions between cells are usually symmetrical: if the linkage is to actin in the cell on one side of the junction, it will be to actin in the cell on the other side. In fact, the binding between cadherins is generally **homophilic** (like-to-like, **Figure 19–5**): cadherin molecules of a specific subtype on one cell bind to cadherin molecules of the same or closely related subtype on adjacent cells.

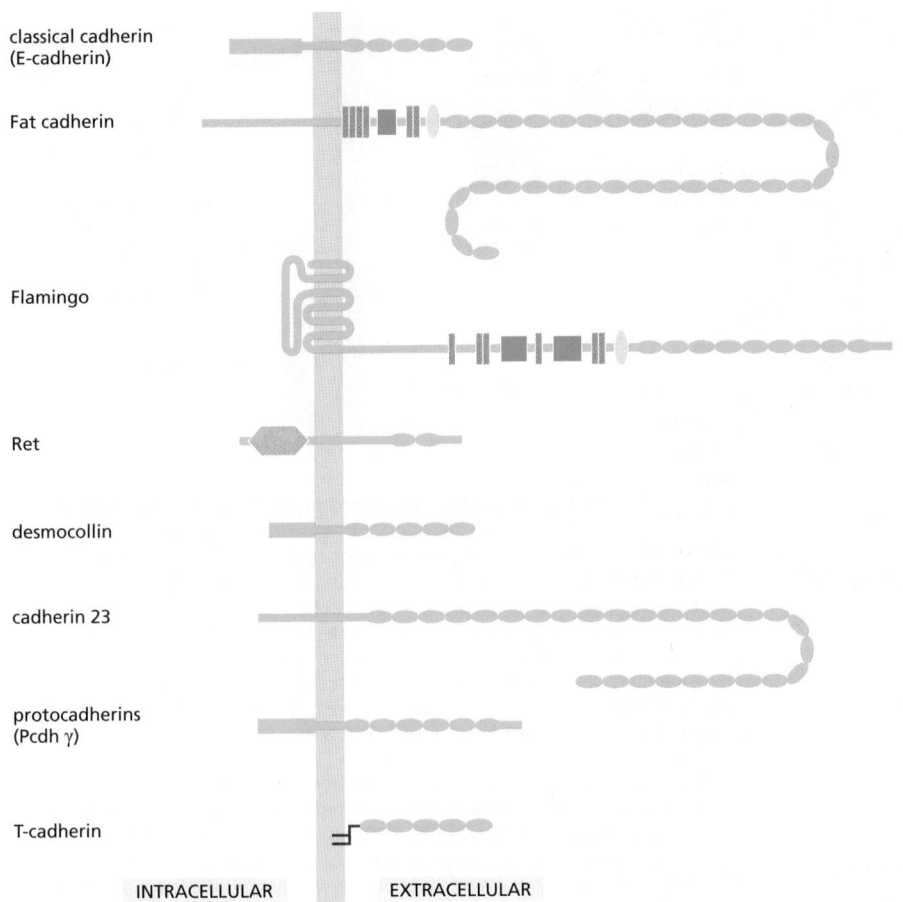

classical cadherin
(E-cadherin)

Fat cadherin

Flamingo

Ret

desmocollin

cadherin 23

protocadherins
(Pcdh γ)

T-cadherin

INTRACELLULAR   EXTRACELLULAR

Figure 19–4 The cadherin superfamily.
The diagram shows some of the diversity among cadherin superfamily members. These proteins all have extracellular portions containing multiple copies of the extracellular cadherin domain *(green ovals)*. In the classical cadherins of vertebrates there are 5 of these domains, and in desmogleins and desmocollins there are 4 or 5, but some nonclassical cadherins have more than 30. The intracellular portions are more varied, reflecting interactions with a wide variety of intracellular ligands, including signaling molecules and adaptor proteins that connect the cadherin to the cytoskeleton. In some cases, such as T-cadherin, a transmembrane domain is not present and the protein is attached to the plasma membrane by a glycosylphosphatidylinositol (GPI) anchor. The differently colored motifs in Fat, Flamingo, and Ret represent conserved domains that are also found in other protein families.

The spacing between the cell membranes at an anchoring junction is precisely defined and depends on the structure of the participating cadherin molecules. All the members of the superfamily, by definition, have an extracellular portion consisting of several copies of the *extracellular cadherin (EC) domain*. Homophilic binding occurs at the N-terminal tips of the cadherin molecules—the cadherin domains that lie furthest from the membrane. These terminal domains each form a knob and a nearby pocket, and the cadherin molecules protruding from opposite cell membranes bind by insertion of the knob of one domain into the pocket of the other (**Figure 19–6A**).

Each cadherin domain forms a more-or-less rigid unit, joined to the next cadherin domain by a hinge. Ca^{2+} ions bind to sites near each hinge and prevent it from flexing, so that the whole string of cadherin domains behaves as a rigid and slightly curved rod. When Ca^{2+} is removed, the hinges can flex, and the structure becomes floppy (Figure 19–6B). At the same time, the conformation at the N-terminus is thought to change slightly, weakening the binding affinity for the matching cadherin molecule on the opposite cell.

Unlike receptors for soluble signal molecules, which bind their specific ligand with high affinity, cadherins (and most other cell–cell adhesion proteins) typically bind to their partners with relatively low affinity. Strong attachments result from the formation of many such weak bonds in parallel. When binding to oppositely oriented partners on another cell, cadherin molecules are often clustered side-to-side with many other cadherin molecules on the same cell (Figure 19–6C). The strength of this junction is far greater than that of any individual intermolecular bond, and yet regulatory mechanisms can easily disassemble the junction by separating the molecules sequentially, just as two pieces of fabric can be joined strongly by Velcro and yet easily peeled apart from the sides. A similar "Velcro principle" also operates at cell–cell and cell–matrix adhesions formed by other types of transmembrane adhesion proteins.

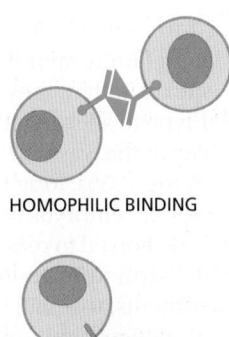

HOMOPHILIC BINDING

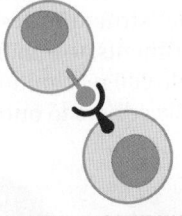

HETEROPHILIC BINDING

Figure 19–5 Homophilic versus heterophilic binding. Cadherins in general bind homophilically; some other cell adhesion molecules, discussed later, bind heterophilically.

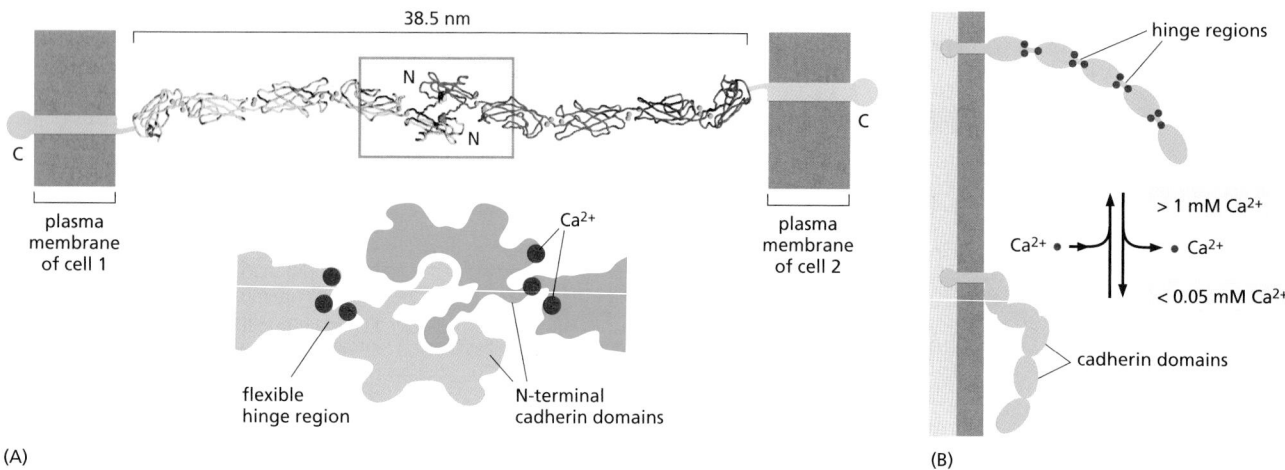

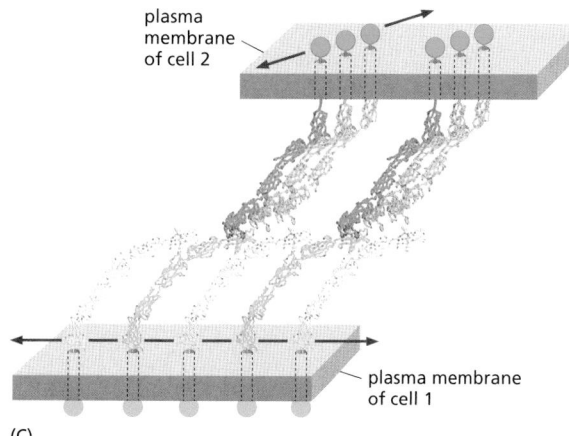

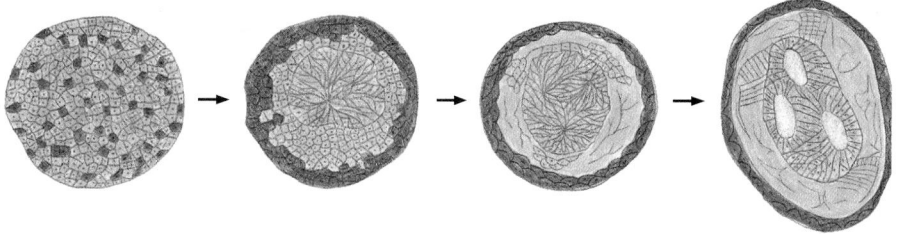

Figure 19–6 **Cadherin structure and function.** (A) The extracellular region of a classical cadherin contains five copies of the extracellular cadherin domain (see Figure 19–4) separated by flexible hinge regions. Ca^{2+} ions *(red dots)* bind in the neighborhood of each hinge, preventing it from flexing. As a result, the extracellular region forms a rigid, curved structure as shown here. To generate cell–cell adhesion, the cadherin domain at the N-terminal tip of one cadherin molecule binds the N-terminal domain from a cadherin molecule on another cell. The structure was determined by x-ray diffraction of the crystallized C-cadherin extracellular region. (B) In the absence of Ca^{2+}, increased flexibility in the hinge regions results in a floppier molecule that is no longer oriented correctly to interact with a cadherin on another cell—and adhesion fails. (C) At a typical cell–cell junction, an organized array of cadherin molecules functions like Velcro to hold cells together. Cadherins on the same cell are thought to be coupled by side-to-side interactions between their N-terminal head regions, resulting in a linear array like the alternating *green* and *light green* cadherins on the lower cell shown here. These arrays are thought to interact with similar linear arrays on an adjacent cell (*blue* cadherin molecules, top cell). The linear arrays on one cell are perpendicular to those on the other cell, as indicated by the *red arrows.* Multiple perpendicular arrays on both cells interact to form a tight-knit mat of cadherin proteins. (A, based on T.J. Boggon et al., *Science* 296:1308–1313, 2002; C, based on O.J. Harrison et al. *Structure* 19:244–256, 2011.)

Cadherin-Dependent Cell–Cell Adhesion Guides the Organization of Developing Tissues

Cadherins form specific homophilic attachments, explaining why there are so many different family members. Cadherins are not like glue, making cell surfaces generally sticky. Rather, they mediate highly selective recognition, enabling cells of a similar type to stick together and to stay segregated from other types of cells.

Selectivity in the way that animal cells consort with one another was first demonstrated in the 1950s, long before the discovery of cadherins, in experiments in which amphibian embryos were dissociated into single cells. These cells were then mixed up and allowed to reassociate. Remarkably, the dissociated cells often reassembled into structures resembling those of the original embryo (**Figure 19–7**). These experiments, together with numerous more recent experiments, reveal that selective cell–cell recognition systems make cells of the same differentiated tissue preferentially adhere to one another.

Figure 19–7 **Sorting out.** Cells from different layers of an early amphibian embryo will sort out according to their origins. In the classical experiment shown here, mesoderm cells *(green)*, neural plate cells *(blue)*, and epidermal cells *(red)* have been disaggregated and then reaggregated in a random mixture. They sort out into an arrangement reminiscent of a normal embryo, with a "neural tube" internally, epidermis externally, and mesoderm in between. (Modified from P.L. Townes and J. Holtfreter, *J. Exp. Zool.* 128:53–120, 1955. With permission from Wiley-Liss.)

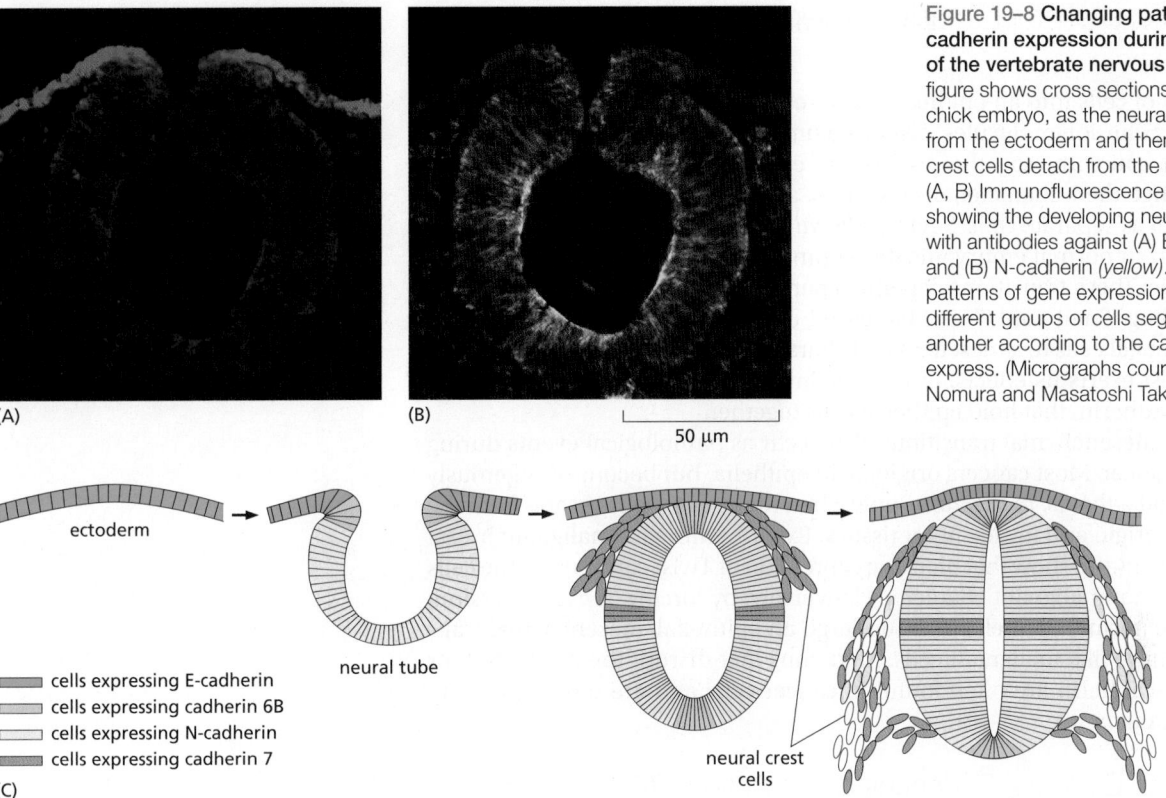

(A) (B) 50 μm

Figure 19–8 **Changing patterns of cadherin expression during construction of the vertebrate nervous system.** The figure shows cross sections of the early chick embryo, as the neural tube detaches from the ectoderm and then as neural crest cells detach from the neural tube. (A, B) Immunofluorescence micrographs showing the developing neural tube labeled with antibodies against (A) E-cadherin *(blue)* and (B) N-cadherin *(yellow)*. (C) As the patterns of gene expression change, the different groups of cells segregate from one another according to the cadherins they express. (Micrographs courtesy of Miwako Nomura and Masatoshi Takeichi.)

ectoderm

neural tube

⬛ cells expressing E-cadherin
⬜ cells expressing cadherin 6B
⬜ cells expressing N-cadherin
⬛ cells expressing cadherin 7

(C)

neural crest cells

Cadherins play a crucial part in these cell-sorting processes during development. The appearance and disappearance of specific cadherins correlate with steps in embryonic development where cells regroup and change their contacts to create new tissue structures. In the vertebrate embryo, for example, changes in cadherin expression are seen when the neural tube forms and pinches off from the overlying ectoderm: neural tube cells lose E-cadherin and acquire other cadherins, including N-cadherin, while the cells in the overlying ectoderm continue to express E-cadherin (Figure 19–8A and B). Then, when the neural crest cells migrate away from the neural tube, these cadherins become scarcely detectable, and another cadherin (cadherin 7) appears that helps hold the migrating cells together as loosely associated cell groups (Figure 19–8C). Finally, when the cells aggregate to form a ganglion, they switch on expression of N-cadherin again. If N-cadherin is artificially overexpressed in the emerging neural crest cells, the cells fail to escape from the neural tube.

Studies with cultured cells further support the idea that the homophilic binding of cadherins controls these processes of tissue segregation. In a line of cultured fibroblasts called *L cells*, for example, cadherins are not expressed and the cells do not adhere to one another. When these cells are transfected with DNA encoding E-cadherin, E-cadherins on one cell bind to E-cadherins on another, resulting in cell–cell adhesion. If L cells expressing different cadherins are mixed together, they sort out and aggregate separately, indicating that different cadherins preferentially bind to their own type (Figure 19–9A), mimicking what happens when cells derived from tissues that express different cadherins are mixed together. A similar segregation of cells occurs if L cells expressing different amounts of the same cadherin are mixed together (Figure 19–9B). It therefore seems likely that both qualitative and quantitative differences in the expression of cadherins have a role in organizing tissues.

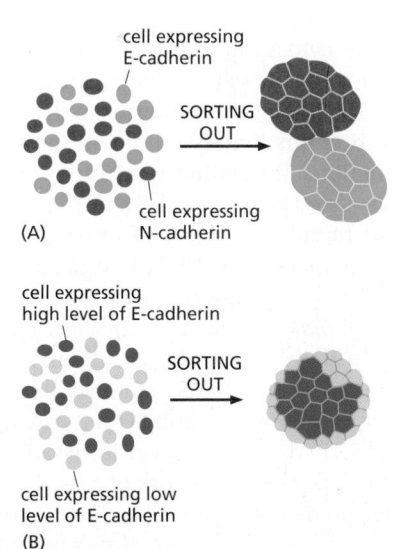

cell expressing E-cadherin

SORTING OUT

(A) cell expressing N-cadherin

cell expressing high level of E-cadherin

SORTING OUT

cell expressing low level of E-cadherin

(B)

Figure 19–9 **Cadherin-dependent cell sorting.** Cells in culture can sort themselves out according to the type and level of cadherins they express. This can be visualized by labeling different populations of cells with dyes of different colors. (A) Cells expressing N-cadherin sort out from cells expressing E-cadherin. (B) Cells expressing high levels of E-cadherin sort out from cells expressing low levels of E-cadherin. The cells expressing high levels adhere more strongly and end up internally.

Epithelial–Mesenchymal Transitions Depend on Control of Cadherins

The assembly of cells into an epithelium is a reversible process. By switching on expression of adhesion molecules, dispersed unattached *mesenchymal cells*, such as fibroblasts, can come together to form an epithelium. Conversely, epithelial cells can change their character, disassemble, and migrate away from their parent epithelium as separate cells. Such *epithelial–mesenchymal transitions* play an important part in normal embryonic development; the origin of the neural crest is one example. These transitions depend in part on transcription regulatory proteins called Slug, Snail, and Twist. Increased expression of Twist, for example, converts epithelial cells to a mesenchymal character, and switching it off does the opposite. Twist exerts its effects, in part, by inhibiting expression of cadherins, including E-cadherin, that hold epithelial cells together.

Epithelial–mesenchymal transitions also occur as pathological events during adult life, in cancer. Most cancers originate in epithelia, but become dangerously prone to spread—that is, *malignant*—only when the cancer cells escape from the epithelium of origin and invade other tissues. Experiments with malignant breast cancer cells in culture show that blocking expression of Twist can convert the cells back toward a nonmalignant character. Conversely, by forcing Twist expression, one can make normal epithelial cells undergo an epithelial–mesenchymal transition and behave like malignant cells. Mutations that disrupt the production or function of E-cadherin are often found in cancer cells and are thought to help make them malignant.

Catenins Link Classical Cadherins to the Actin Cytoskeleton

The extracellular domains of cadherins mediate homophilic binding at adherens junctions. The intracellular domains of typical cadherins, including all classical and some nonclassical ones, interact with filaments of the cytoskeleton: actin at adherens junctions and intermediate filaments at desmosomes (see Table 19–1). These cytoskeletal linkages are essential for efficient cell–cell adhesion, as cadherins that lack their cytoplasmic domains cannot stably hold cells together.

The linkage of cadherins to the cytoskeleton is indirect and depends on adaptor proteins that assemble on the cytoplasmic tail of the cadherin. At adherens junctions, the cadherin tail binds two such proteins: *β-catenin* and a distant relative called *p120-catenin*; a third protein called *α-catenin* interacts with β-catenin and recruits a variety of other proteins to provide a dynamic linkage to actin filaments (**Figure 19–10**). At desmosomes, cadherins are linked to intermediate filaments through other adaptor proteins, including a β-catenin-related protein called *plakoglobin*, as we discuss later.

In their mature form, adherens junctions are enormous protein complexes containing hundreds to thousands of cadherin molecules, packed into dense, regular arrays that are linked on the extracellular side by lateral interactions between cadherin domains, as we discussed earlier (see Figure 19–6C). On the cytoplasmic side, a complex network of catenins, actin regulators, and contractile actin bundles holds the cluster of cadherins together and links it to the actin cytoskeleton. Assembling a structure of this complexity is not a simple task, and it involves a complex sequence of events controlled by the actin-regulatory proteins discussed in Chapter 16. The general features of the assembly process are summarized in **Figure 19–11**.

Adherens Junctions Respond to Forces Generated by the Actin Cytoskeleton

Most adherens junctions are linked to contractile bundles of actin filaments and non-muscle myosin II. These junctions are therefore subjected to pulling forces generated by the attached actin. The pulling forces are important for junction assembly and maintenance: disruption of myosin activity, for example, results in

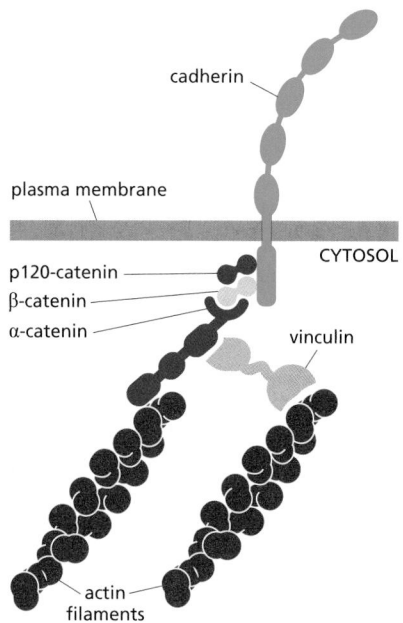

Figure 19–10 The linkage of classical cadherins to actin filaments. The cadherins are coupled indirectly to actin filaments through an adaptor protein complex containing p120-catenin, β-catenin, and α-catenin. Other proteins, including vinculin, associate with α-catenin and help provide the linkage to actin. β-Catenin has a second, and very important, function in intracellular signaling, as we discuss in Chapter 15 (see Figure 15–60). For clarity, this diagram does not show the cadherin of the adjacent cell in the junction.

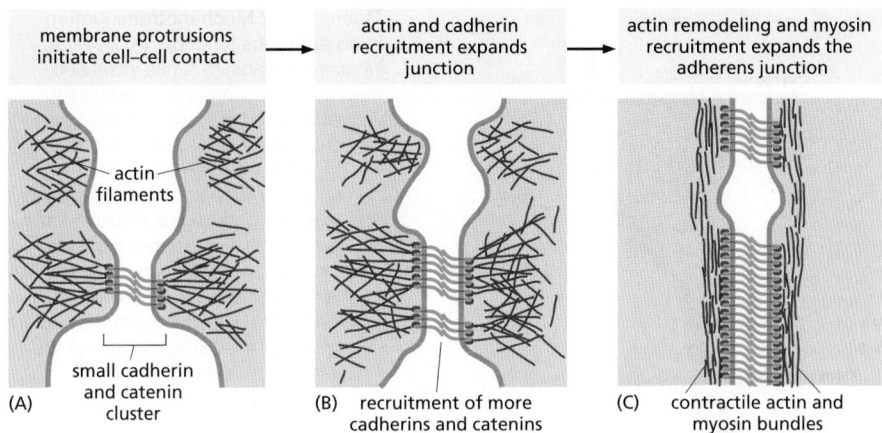

| membrane protrusions initiate cell–cell contact | actin and cadherin recruitment expands junction | actin remodeling and myosin recruitment expands the adherens junction |

(A) actin filaments / small cadherin and catenin cluster

(B) recruitment of more cadherins and catenins

(C) contractile actin and myosin bundles

Figure 19–11 **Assembly of an adherens junction.** (A) Assembly begins when two unattached epithelial cell precursors explore their surroundings with membrane protrusions, generated by local nucleation of actin networks. When the cells make contact, small cadherin and catenin clusters take shape at the contact sites and associate with actin, leading to activation of the small monomeric GTPase Rac (not shown), an important actin regulator (see Figure 16–85). (B) Rac promotes additional actin protrusions in the vicinity, expanding the size of the contact zone and thereby promoting further recruitment of cadherins and their associated catenin proteins. (C) Eventually, Rac is inactivated and replaced by the related GTPase Rho (not shown), which shifts actin remodeling toward the assembly of linear, contractile filament bundles. Rho also promotes the assembly of myosin II filaments that associate with bundles of actin filaments to generate contractile activity. This contractile activity generates tension that stimulates further actin recruitment and expansion of the junction, in part through the mechanisms illustrated in Figure 19–12.

the disassembly of many adherens junctions. Furthermore, the contractile forces acting on a junction in one cell are balanced by contractile forces at the junction of the opposite cell, so that no cell pulls others toward it and thereby disrupts the uniform distribution of cells in the tissue.

We do not understand the mechanisms responsible for maintaining this balance. Adherens junctions seem to sense the forces acting on them and modify local actin and myosin behavior to balance the forces on both sides of the junction. Evidence for these mechanisms comes from studies of pairs of cultured mammalian cells connected by adherens junctions. If contractile activity in one cell is increased experimentally, the adherens junctions linking the two cells increase in size, and the contractile activity of the second cell increases to match that of the first—resulting in a balance of forces across the junction. These and other experiments suggest that adherens junctions are not simply passive sites of protein-protein binding but are dynamic tension sensors that regulate their behavior in response to changing mechanical conditions. This ability to transduce a mechanical signal into a change in junctional behavior is an example of *mechanotransduction*. We will see later that it is also important at cell–matrix junctions.

The mechanotransduction at cell–cell junctions is thought to depend, at least in part, on proteins in the cadherin complex that alter their shape when stretched by tension. The protein α-catenin, for example, is stretched from a folded to an extended conformation when contractile activity increases at the junction. The unfolding exposes a cryptic binding site for another protein, vinculin, which promotes the recruitment of more actin to the junction (**Figure 19–12**). By mechanisms such as this, pulling on a junction makes it stronger. Furthermore, as noted above, pulling on a junction in one cell will increase the contractile force generated in the attached cell.

In some cell types, actin contractility reduces cell–cell adhesion, particularly if large forces are involved. Large actin-based contractile forces might, in some tissues, pull sufficiently hard on the edges of cell–cell adhesions to peel them apart, particularly if contraction is coupled to additional regulatory mechanisms that weaken the adhesion. This mechanism might be important in certain forms of tissue remodeling during development, as we describe next.

Tissue Remodeling Depends on the Coordination of Actin-Mediated Contraction With Cell–Cell Adhesion

Adherens junctions are an essential part of the machinery for modeling the shapes of multicellular structures in the animal body. By indirectly linking the actin filaments in one cell to those in its neighbors, they enable the cells in the tissue to use their actin cytoskeletons in a coordinated way.

Adherens junctions occur in various forms. In many nonepithelial tissues, they appear as small punctate or linear attachments that connect the cortical

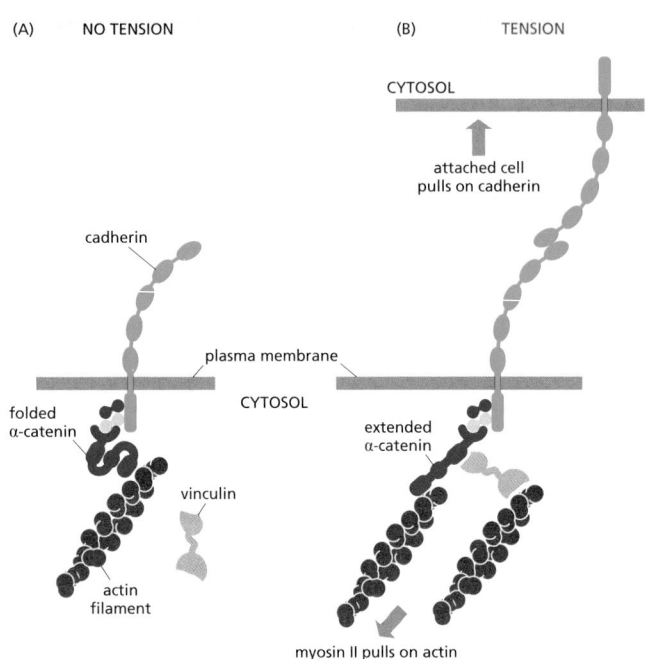

(A)   NO TENSION

(B)   TENSION

CYTOSOL

attached cell
pulls on cadherin

cadherin

plasma membrane

CYTOSOL

folded
α-catenin

extended
α-catenin

vinculin

actin
filament

myosin II pulls on actin

Figure 19–12 Mechanotransduction in an adherens junction. (A) Cell–cell junctions are able to sense increased tension and respond by strengthening their actin linkages. Tension sensing is thought to depend in part on α-catenin (see Figure 19–10). (B) When actin filaments are pulled from within the cell by non-muscle myosin II, the resulting force unfolds a domain in α-catenin, thereby exposing an otherwise hidden binding site for the adaptor protein vinculin. Vinculin then promotes additional actin recruitment, strengthening the linkages between the junction and the cytoskeleton.

actin filaments beneath the plasma membranes of two interacting cells. In heart muscle, they anchor the actin bundles of the contractile apparatus and act in parallel with desmosomes to link the contractile cells end-to-end. But the prototypical examples of adherens junctions occur in epithelia, where they often form a continuous **adhesion belt** (or *zonula adherens*) just beneath the apical face of the epithelium, encircling each of the interacting cells in the sheet (**Figure 19-13**). Within each cell, a contractile bundle of actin filaments and myosin II lies adjacent to the adhesion belt, oriented parallel to the plasma membrane and tethered to

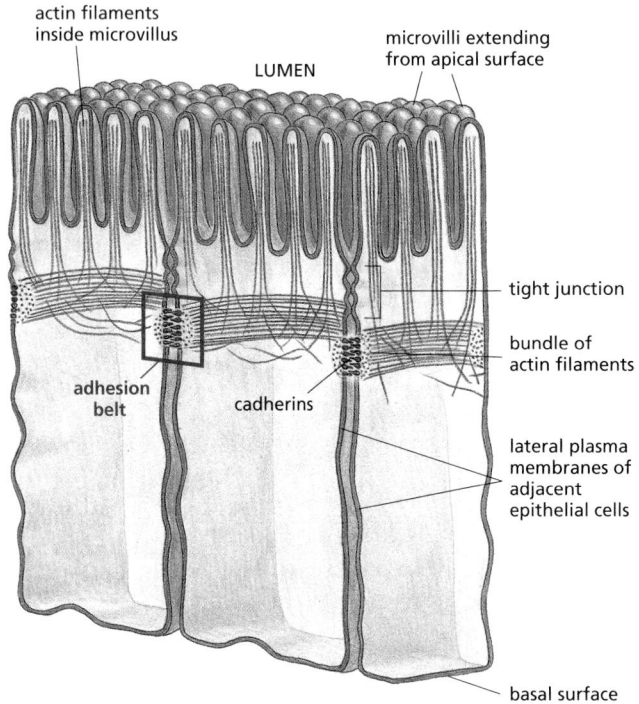

actin filaments
inside microvillus

microvilli extending
from apical surface

LUMEN

tight junction

bundle of
actin filaments

adhesion
belt

cadherins

lateral plasma
membranes of
adjacent
epithelial cells

basal surface

Figure 19–13 Adherens junctions between epithelial cells in the small intestine. These cells are specialized for absorption of nutrients; at their apex, facing the lumen of the gut, they have many microvilli (protrusions that increase the absorptive surface area). The adherens junction takes the form of an *adhesion belt*, encircling each of the interacting cells. Its most obvious feature is a contractile bundle of actin filaments running along the cytoplasmic surface of the junctional plasma membrane. The actin filament bundles are tethered by intracellular proteins to cadherins, which bind to cadherins on the adjacent cell. In this way, the actin filament bundles in adjacent cells are tied together. For clarity, this drawing does not show most of the other cell–cell and cell–matrix junctions of epithelial cells (see Figure 19–2).

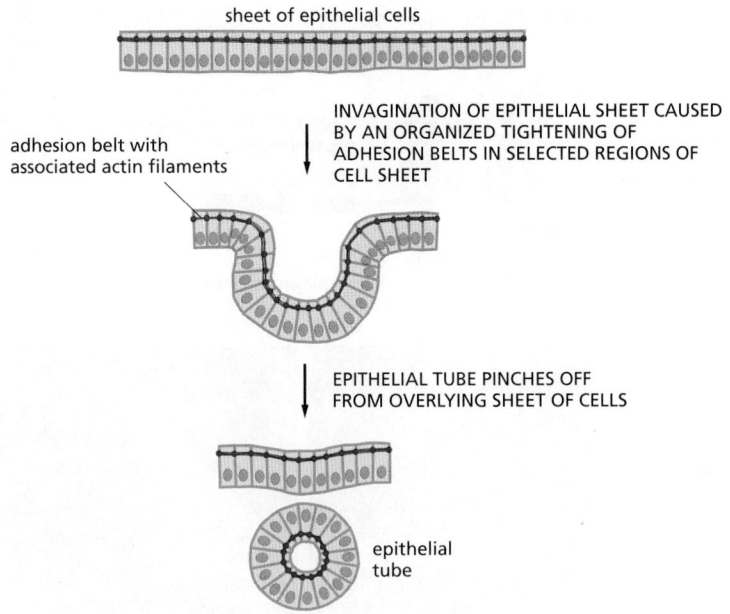

sheet of epithelial cells

adhesion belt with associated actin filaments

INVAGINATION OF EPITHELIAL SHEET CAUSED BY AN ORGANIZED TIGHTENING OF ADHESION BELTS IN SELECTED REGIONS OF CELL SHEET

EPITHELIAL TUBE PINCHES OFF FROM OVERLYING SHEET OF CELLS

epithelial tube

Figure 19–14 **The folding of an epithelial sheet to form an epithelial tube.** The oriented contraction of the bundles of actin and myosin filaments running along adhesion belts causes the epithelial cells to narrow at their apex and helps the epithelial sheet to roll up into a tube. An example is the formation of the neural tube in early vertebrate development (see Figure 19–8).

it by the cadherins and their associated intracellular adaptor proteins. The actin–myosin bundles are thus linked, via the cadherins, into an extensive transcellular network. Coordinated contraction of this network provides the motile force for a fundamental process in animal morphogenesis—the folding of epithelial cell sheets into tubes, vesicles, and other related structures (**Figure 19–14**).

The coordination of cell–cell adhesion and actin contractility is beautifully illustrated by cellular rearrangements that occur early in the development of the fruit fly *Drosophila melanogaster*. Soon after gastrulation, the outer epithelium of the embryo is elongated by a process called *germ-band extension*, in which the cells converge inward toward the dorsal–ventral axis and extend along the anterior–posterior axis (**Figure 19–15**). Actin-dependent contraction along specific cell boundaries is coordinated with a loss of specific adherens junctions to allow cells to insert themselves between other cells (a process called *intercalation*), resulting in a longer and narrower epithelium. The mechanisms underlying the loss of adhesion along specific cell boundaries are not clear, but they depend in part on increased degradation of β-catenin, due to its phosphorylation by a protein kinase that is localized specifically at those boundaries.

Desmosomes Give Epithelia Mechanical Strength

Desmosomes are structurally similar to adherens junctions but contain specialized cadherins that link to intermediate filaments instead of actin filaments. Their main function is to provide mechanical strength. Desmosomes are important

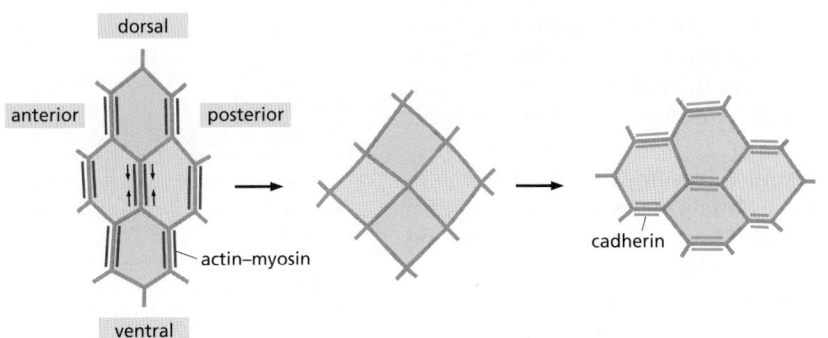

dorsal

anterior · posterior

actin–myosin

ventral

cadherin

Figure 19–15 **Remodeling of cell–cell adhesions in embryonic *Drosophila* epithelium.** Depicted at *left* is a group of cells in the outer epithelium of a *Drosophila* embryo. During germ-band extension, cells converge toward each other (*middle*) on the dorsal–ventral axis and then extend (*right*) along the anterior–posterior axis. The result is intercalation: cells that were originally far apart along the dorsal–ventral axis (*dark green*) are inserted between the cells (*light green*) that separated them. These rearrangements depend on the spatial regulation of actin–myosin contractile bundles, which are localized primarily at the vertical cell boundaries (*red, left*). Contraction of these bundles is accompanied by removal of E-cadherin (not shown) at the same cell boundaries, resulting in shrinkage and loss of adhesion along the vertical axis (*middle*). New cadherin-based adhesions (*blue, right*) then form and expand along horizontal boundaries, resulting in extension of the cells in the anterior–posterior dimension.

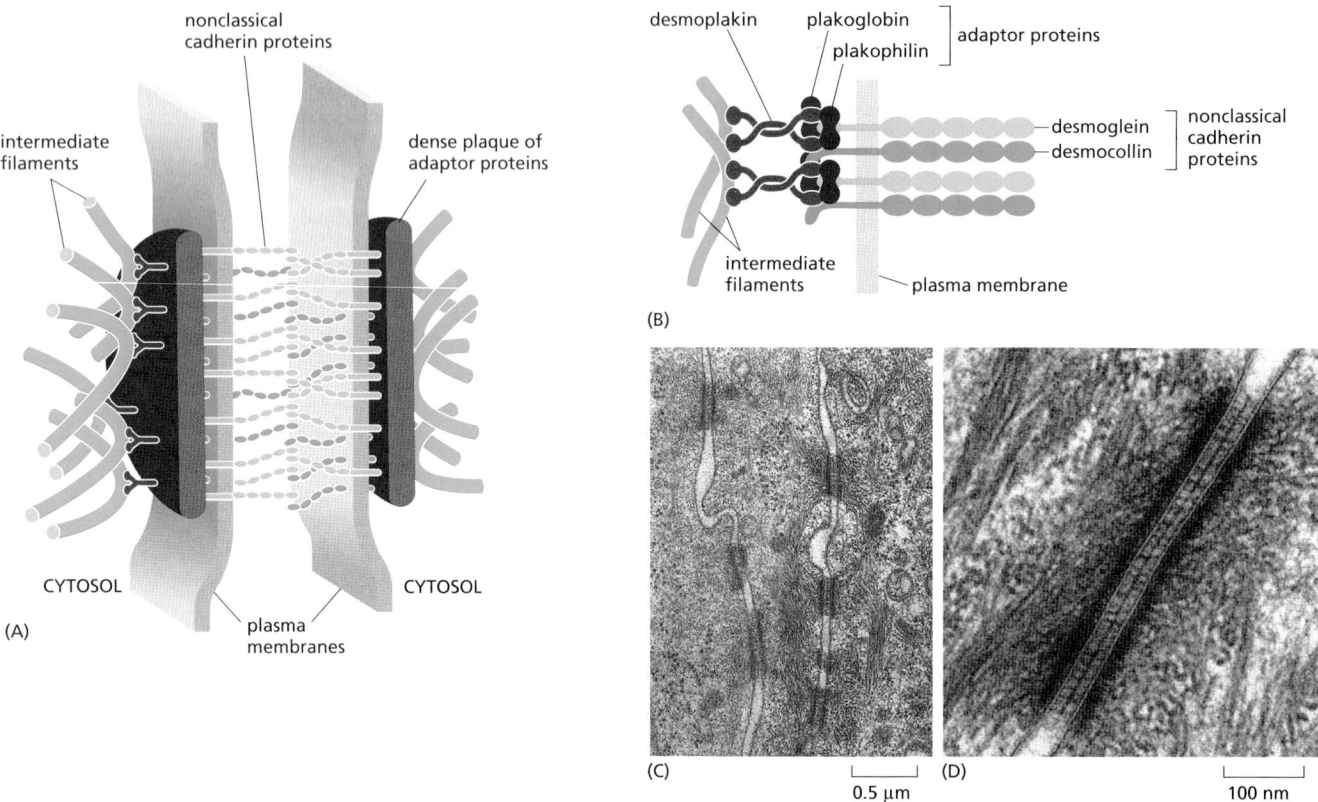

Figure 19–16 Desmosomes. (A) The structural components of a desmosome. On the cytoplasmic surface of each interacting plasma membrane is a dense plaque composed of a mixture of intracellular adaptor proteins. A bundle of keratin intermediate filaments is attached to the surface of each plaque. Transmembrane nonclassical cadherins bind to the plaques and interact through their extracellular domains to hold the adjacent membranes together. (B) Some of the molecular components of a desmosome. Desmoglein and desmocollin are nonclassical cadherins. Their cytoplasmic tails bind *plakoglobin* (γ-catenin) and *plakophilin* (a distant relative of p120-catenin), which in turn bind to *desmoplakin*. Desmoplakin binds to the sides of intermediate filaments, thereby tying the desmosome to these filaments. (C) An electron micrograph of desmosome junctions between three epidermal cells in the skin of a baby mouse. (D) Part of the same tissue at higher magnification, showing a single desmosome, with intermediate filaments attached to it. (C and D, from W. He, P. Cowin and D.L. Stokes, *Science* 302:109–113, 2003. With permission from AAAS.)

in vertebrates but are not found, for example, in *Drosophila*. They are present in most mature vertebrate epithelia and are particularly plentiful in tissues that are subject to high levels of mechanical stress, such as heart muscle and the epidermis, the epithelium that forms the outer layer of the skin.

Figure 19–16A shows the general structure of a desmosome, and Figure 19–16B shows some of the proteins that form it. Desmosomes typically appear as buttonlike spots of adhesion, riveting the cells together (Figure 19–16C). Inside the cell, the bundles of ropelike intermediate filaments that are anchored to the desmosomes form a structural framework of great tensile strength (Figure 19–16D), with linkage to similar bundles in adjacent cells, creating a network that extends throughout the tissue (Figure 19–17). The particular type of intermediate filaments attached to the desmosomes depends on the cell type: they are *keratin filaments* in most epithelial cells, for example, and *desmin filaments* in heart muscle cells.

The importance of desmosomes is demonstrated by some forms of the potentially fatal skin disease *pemphigus*. Affected individuals make antibodies against one of their own desmosomal cadherin proteins. These antibodies bind to and disrupt the desmosomes that hold their epidermal cells (keratinocytes) together. This results in a severe blistering of the skin, with leakage of body fluids into the loosened epithelium.

Tight Junctions Form a Seal Between Cells and a Fence Between Plasma Membrane Domains

Sheets of epithelial cells enclose and partition the animal body, lining all its surfaces and cavities, and creating internal compartments where specialized processes occur. The epithelial sheet seems to be one of the inventions that lie at the origin of animal evolution, diversifying in a huge variety of ways but retaining an organization based on a set of conserved molecular mechanisms.

Essentially all epithelia are anchored to other tissue on one side—the **basal** side—and free of such attachment on their opposite side—the **apical** side. A basal lamina lies at the interface with the underlying tissue, mediating the attachment, while the apical surface of the epithelium is generally bathed by extracellular fluid. Thus, all epithelia are structurally **polarized**, and so are their individual cells: the basal end of a cell, adherent to the basal lamina below, differs from the apical end, exposed to the medium above.

Correspondingly, all epithelia have at least one function in common: they serve as selective permeability barriers, separating the fluid that permeates the tissue on their basal side from fluid with a different chemical composition on their apical side. This barrier function requires that the adjacent cells be sealed together by **tight junctions**, so that molecules cannot leak freely across the cell sheet.

The epithelium of the small intestine provides a good illustration of tight-junction structure and function (see Figure 19–2). This epithelium has a *simple columnar* structure; that is, it consists of a single layer of tall (columnar) cells. These are of several differentiated types, but the majority are absorptive cells, specialized for uptake of nutrients from the internal cavity, or *lumen*, of the gut. The absorptive cells have to transport selected nutrients across the epithelium from the lumen into the extracellular fluid on the other side. From there, these nutrients diffuse into small blood vessels to provide nourishment to the organism. This *transcellular transport* depends on two sets of transport proteins in the plasma membrane of the absorptive cell. One set is confined to the apical surface of the cell (facing the lumen) and actively transports selected molecules into the cell from the gut. The other set is confined to the *basolateral* (basal and lateral) surfaces of the cell, and it allows the same molecules to leave the cell by passive transport into the extracellular fluid on the other side of the epithelium. For this transport activity to be effective, the spaces between the epithelial cells must be tightly sealed, so that the transported molecules cannot leak back into the gut lumen through these spaces (**Figure 19–18**). Moreover, the transport proteins must be correctly distributed in the plasma membranes: the apical transporters must be delivered to the cell apex and must not be allowed to drift to the basolateral membrane, and the basolateral transporters must be delivered to and remain in the basolateral membrane. Tight junctions, besides sealing the gaps between the cells, also function as "fences" that help prevent apical or basolateral proteins from diffusing into the wrong region.

The sealing function of tight junctions is easy to demonstrate experimentally: a low-molecular-weight tracer added to one side of an epithelium will generally not pass beyond the tight junction (**Figure 19–19**). This seal is not absolute, however. Although all tight junctions are impermeable to macromolecules, their permeability to ions and other small molecules varies. Tight junctions in the epithelium lining the small intestine, for example, are 10,000 times more permeable to inorganic ions, such as Na^+, than the tight junctions in the epithelium lining the urinary bladder. The movement of ions and other molecules between epithelial cells is called *paracellular transport*, and tissue-specific differences in transport rates generally result from differences in the proteins that form tight junctions.

Tight Junctions Contain Strands of Transmembrane Adhesion Proteins

When tight junctions are visualized by freeze-fracture electron microscopy, they are seen as a branching network of *sealing strands* that completely encircles the

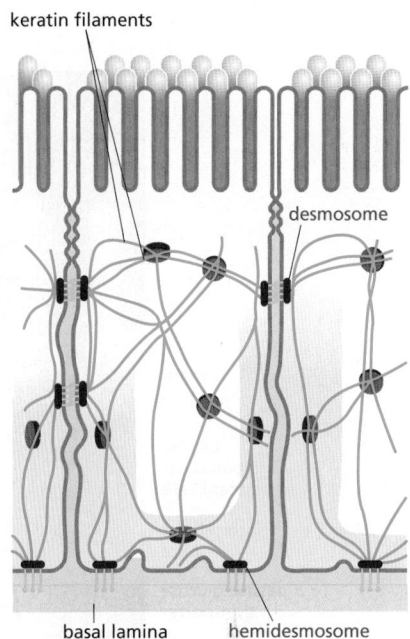

Figure 19–17 Desmosomes, hemidesmosomes, and the intermediate filament network. The keratin intermediate filament networks of adjacent cells—in this example, epithelial cells of the small intestine—are indirectly connected to one another through desmosomes, and to the basal lamina through hemidesmosomes.

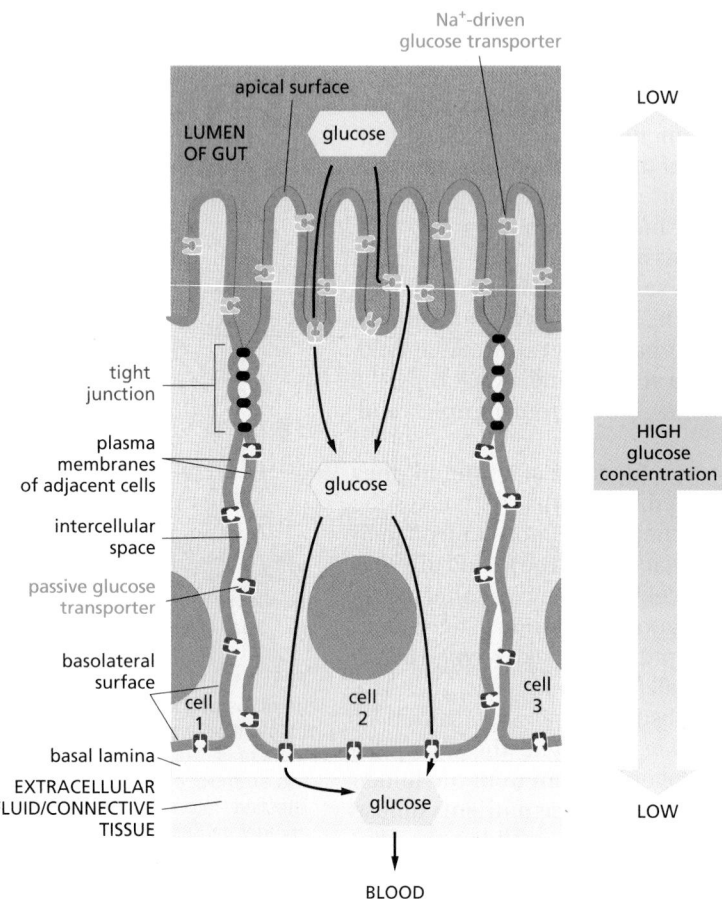

Figure 19–18 **The role of tight junctions in transcellular transport.** For clarity, only the tight junctions are shown. Transport proteins are confined to different regions of the plasma membrane in epithelial cells of the small intestine. This segregation permits a vectorial transfer of nutrients across the epithelium from the gut lumen to the blood. In the example shown, glucose is actively transported into the cell by Na^+-driven glucose transporters at its apical surface, and it leaves the cell through passive glucose transporters in its basolateral membrane. Tight junctions are thought to confine the transport proteins to their appropriate membrane domains by acting as diffusion barriers, or "fences," within the lipid bilayer of the plasma membrane; these junctions also block the backflow of glucose from the basal side of the epithelium into the gut lumen (see Movie 11.2).

apical end of each cell in the epithelial sheet (**Figure 19–20**A and B). In conventional electron micrographs, the outer leaflets of the two interacting plasma membranes are tightly apposed where sealing strands are present (Figure 19–20C). Each sealing strand is composed of a long row of transmembrane homophilic adhesion proteins embedded in each of the two interacting plasma membranes. The extracellular domains of these proteins adhere directly to one another to occlude the intercellular space (**Figure 19–21**).

The main transmembrane proteins forming these strands are the *claudins*, which are essential for tight-junction formation and function. Mice that lack the *claudin-1* gene, for example, fail to make tight junctions between the cells in the epidermal layer of the skin; as a result, the baby mice lose water rapidly by evaporation through the skin and die within a day after birth. Conversely, if nonepithelial cells such as fibroblasts are artificially caused to express claudin genes, they

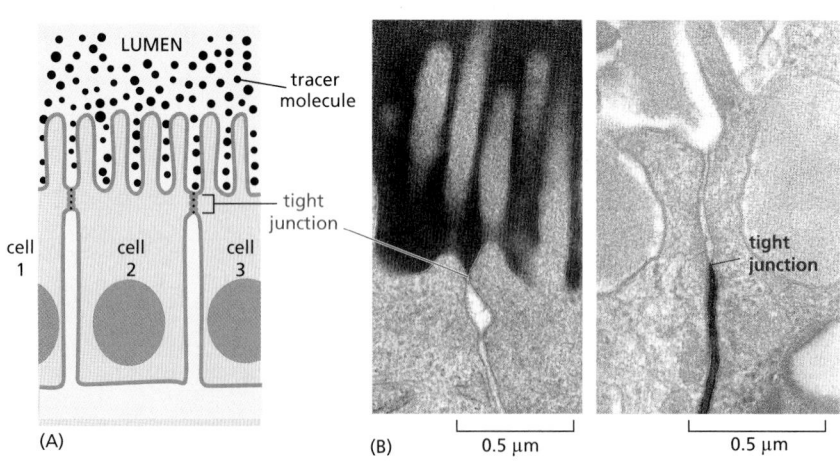

Figure 19–19 **The role of tight junctions in allowing epithelia to serve as barriers to solute diffusion.** (A) The drawing shows how a small extracellular tracer molecule added on one side of an epithelium is prevented from crossing the epithelium by the tight junctions that seal adjacent cells together. Adherens junctions and other cell junctions are not shown for clarity. (B) Electron micrographs of cells in an epithelium in which a small, extracellular, electron-dense tracer molecule has been added to either the apical side (on the *left*) or the basolateral side (on the *right*). The tight junction blocks passage of the tracer in both directions. (B, courtesy of Daniel Friend, by permission of E.L. Bearer.)

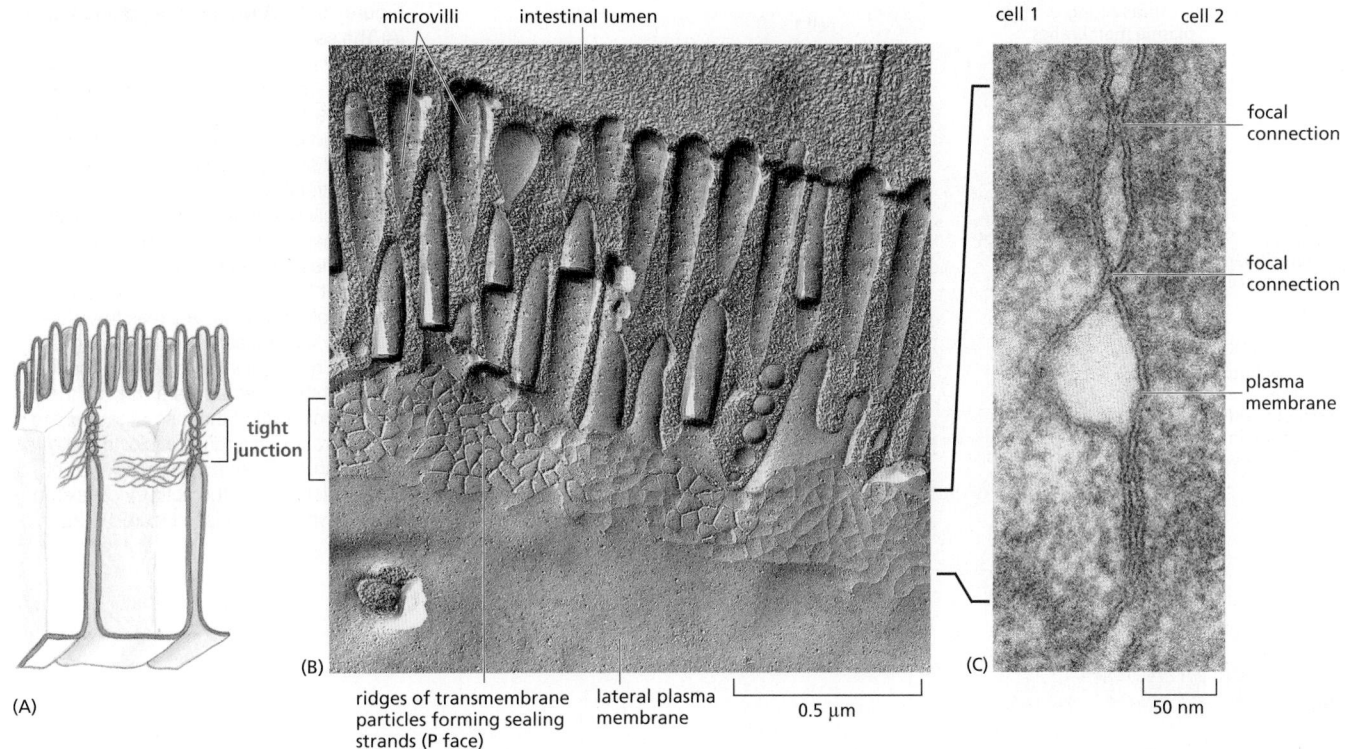

Figure 19–20 **The structure of a tight junction between epithelial cells of the small intestine.** The junctions are shown (A) schematically, (B) in a freeze-fracture electron micrograph, and (C) in a conventional electron micrograph. In (B), the plane of the micrograph is parallel to the plane of the membrane, and the tight junction appears as a band of branching sealing strands that encircle each cell in the epithelium (see Figure 19–21A). In (C), the junction is seen in cross section as a series of focal connections between the outer leaflets of the two interacting plasma membranes, each connection corresponding to a sealing strand in cross section. (B and C, from N.B. Gilula, in Cell Communication [R.P. Cox, ed.], pp. 1–29. New York: Wiley, 1974. With permission from John Wiley & Sons.)

will form tight-junctional connections with one another. Normal tight junctions also contain a second major transmembrane protein called *occludin*, which is not essential for the assembly or structure of the tight junction but is important for limiting junctional permeability. A third transmembrane protein, *tricellulin*, is required to seal cell membranes together and prevent transepithelial leakage at the points where three cells meet.

The claudin protein family has many members (24 in humans), and these are expressed in different combinations in different epithelia to confer particular permeability properties on the epithelial sheet. They are thought to form *paracellular pores*—selective channels allowing specific ions to cross the tight-junctional barrier, from one extracellular space to another. A specific claudin found in kidney epithelial cells, for example, is needed to let Mg^{2+} pass between the cells of the kidney tubules so that this ion can be resorbed from the urine into the blood. A mutation in the gene encoding this claudin results in excessive loss of Mg^{2+} in the urine.

Scaffold Proteins Organize Junctional Protein Complexes

Like the cadherin molecules of an adherens junction, the claudins and occludins of a tight junction interact with each other on their extracellular sides to promote junction assembly. Also as in adherens junctions, the organization of adhesion proteins in a tight junction depends on additional proteins that bind the cytoplasmic side of the adhesion proteins. The key organizational proteins at tight junctions are the *zonula occludens* (*ZO*) proteins. The three major members of the ZO family—ZO-1, ZO-2, and ZO-3—are large **scaffold proteins** that provide a structural support on which the tight junction is built. These intracellular molecules

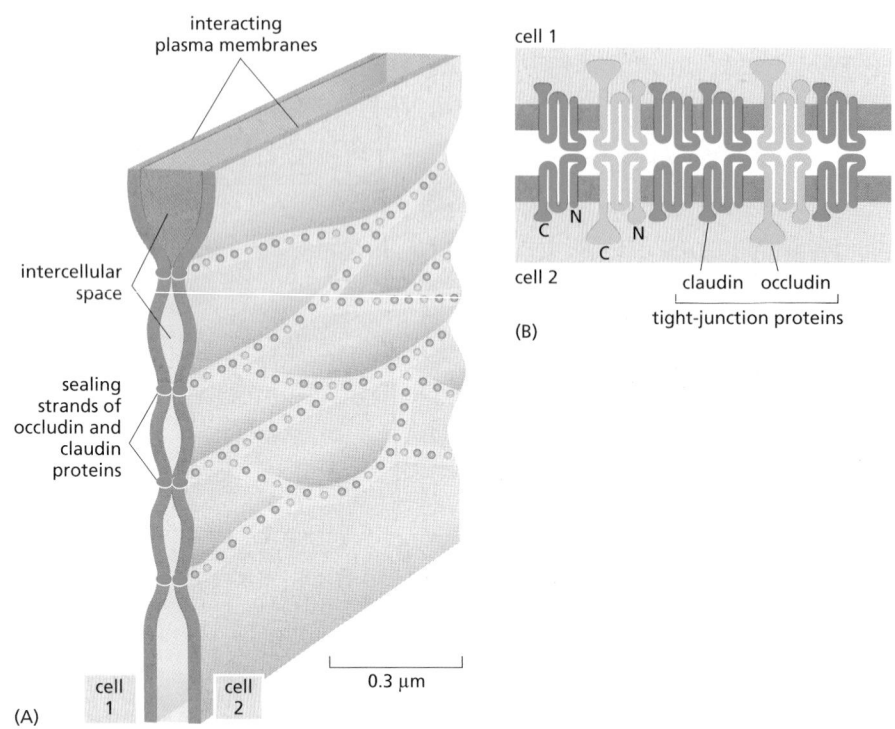

interacting
plasma membranes

intercellular
space

sealing
strands of
occludin and
claudin
proteins

cell
1

cell
2

(A)

0.3 µm

cell 1

cell 2

(B)

claudin   occludin

tight-junction proteins

Figure 19–21 A model of a tight junction.
(A) The sealing strands hold adjacent plasma membranes together. The strands are composed of transmembrane proteins that make contact across the intercellular space and create a seal. (B) The molecular composition of a sealing strand. The major extracellular components of the tight junction are members of a family of proteins with four transmembrane domains. One of these proteins, claudin, is the most important for the assembly and structure of the sealing strands, whereas the related protein occludin has the less critical role of determining junction permeability. The two termini of these proteins are both on the cytoplasmic side of the membrane, where they interact with large scaffolding proteins that organize the sealing strands and link the tight junction to the actin cytoskeleton (not shown here, but see Figure 19–22).

consist of strings of protein-binding domains, typically including several **PDZ domains**—segments about 80 amino acids long that can recognize and bind the C-terminal tails of specific partner proteins (**Figure 19–22**). One domain of these scaffold proteins can attach to a claudin protein, while others can attach to occludin or the actin cytoskeleton. Moreover, one molecule of scaffold protein can bind to another. In this way, the cell can assemble a mat of intracellular proteins that organizes and positions the sealing strands of the tight junction.

The tight-junctional network of sealing strands usually lies just apical to adherens and desmosome junctions that bond the cells together mechanically; the whole assembly is called a *junctional complex* (see Figure 19-2). The parts of this junctional complex depend on each other for their formation. For example, anti-cadherin antibodies that block the formation of adherens junctions also block the formation of tight junctions.

Gap Junctions Couple Cells Both Electrically and Metabolically

Tight junctions block the passageways through the gaps between epithelial cells, preventing extracellular molecules from leaking from one side of an epithelium to the other. Another type of junctional structure has a radically different function: it bridges gaps between adjacent cells so as to create direct channels from the

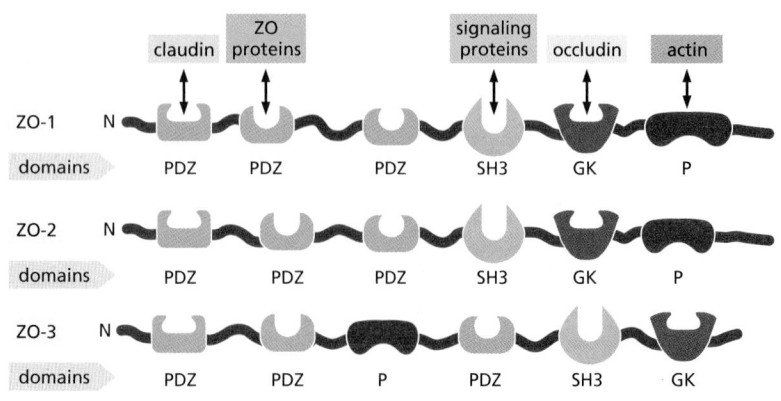

claudin   ZO proteins   signaling proteins   occludin   actin

ZO-1   N

domains   PDZ   PDZ   PDZ   SH3   GK   P

ZO-2   N

domains   PDZ   PDZ   PDZ   SH3   GK   P

ZO-3   N

domains   PDZ   PDZ   P   PDZ   SH3   GK

Figure 19–22 Scaffold proteins at the tight junction. The scaffold proteins ZO-1, ZO-2, and ZO-3 are concentrated beneath the plasma membrane at tight junctions. Each of the proteins contains multiple protein-binding domains, including three PDZ domains, an SH3 domain, and a GK domain, linked together like beads on a flexible string. These domains enable the proteins to interact with each other and with numerous other partners, as indicated here, to generate a tightly woven protein network that organizes the sealing strands of the tight junction and links them to the actin cytoskeleton. Scaffold proteins with similar structure help organize other junctional complexes, including those at neural synapses.

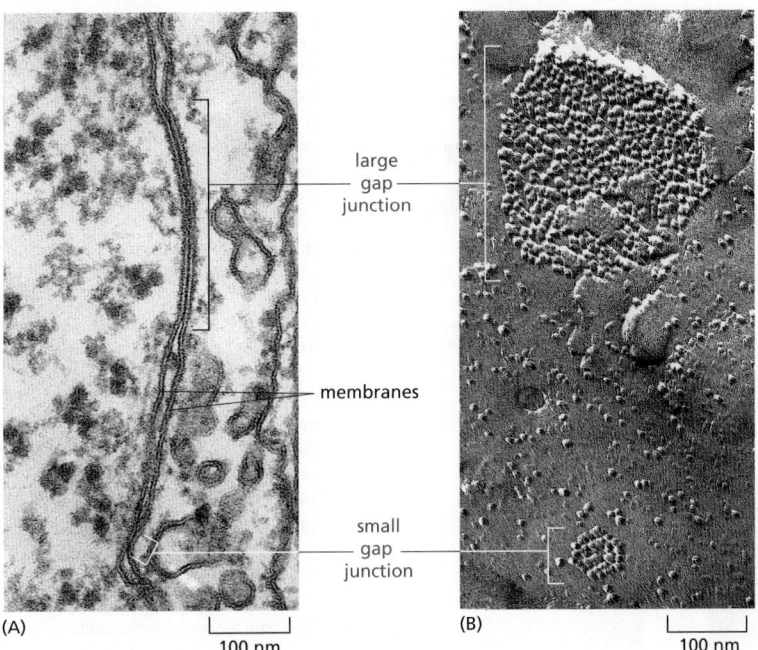

Figure 19–23 **Gap junctions as seen in the electron microscope.** (A) Thin-section and (B) freeze-fracture electron micrographs of a large and a small gap-junction plaque between fibroblasts in culture. In (B), each gap junction is seen as a cluster of homogeneous intramembrane particles. Each intramembrane particle corresponds to a connexon (see Figure 19–25). (From N.B. Gilula, in Cell Communication [R.P. Cox, ed.], pp. 1–29. New York: Wiley, 1974. With permission from John Wiley & Sons.)

cytoplasm of one to that of the other. These channels are called **gap junctions**.

Gap junctions are present in most animal tissues, including connective tissues as well as epithelia and heart muscle. Each gap junction appears in conventional electron micrographs as a patch where the membranes of two adjacent cells are separated by a uniform narrow gap of about 2–4 nm (**Figure 19–23**). The gap is spanned by channel-forming proteins, of which there are two distinct families, called the *connexins* and the *innexins*. Connexins are the predominant gap-junction proteins in vertebrates, with 21 isoforms in humans. Innexins are found in the gap junctions of invertebrates.

Gap junctions have a pore size of about 1.4 nm, which allows the exchange of inorganic ions and other small water-soluble molecules, but not of macromolecules such as proteins or nucleic acids (**Figure 19–24**). An electric current injected into one cell through a microelectrode causes an electrical disturbance in the neighboring cell, due to the flow of ions carrying electric charge through gap junctions. This electrical coupling via gap junctions serves an obvious purpose in tissues containing electrically excitable cells: action potentials can spread rapidly from cell to cell, without the delay that occurs at chemical synapses. In vertebrates, for example, electrical coupling through gap junctions synchronizes the contractions of heart muscle cells as well as those of the smooth muscle cells responsible for the peristaltic movements of the intestine. Gap junctions also occur in many tissues whose cells are not electrically excitable. In principle, the sharing of small metabolites and ions provides a mechanism for coordinating the activities of individual cells in such tissues and for smoothing out random fluctuations in small-molecule concentrations in different cells.

A Gap-Junction Connexon Is Made of Six Transmembrane Connexin Subunits

Connexins are four-pass transmembrane proteins, six of which assemble to form a *hemichannel*, or **connexon**. When the connexons in the plasma membranes of two cells in contact are aligned, they form a continuous aqueous channel that connects the two cell interiors (**Figure 19–25**). A gap junction consists of many such connexon pairs in parallel, forming a sort of molecular sieve. Not only does this sieve provide a communication channel between cells, but it also provides a form of cell–cell adhesion that supplements the cadherin- and claudin-mediated adhesions we discussed earlier.

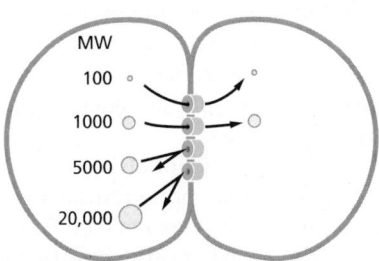

Figure 19–24 **Determining the size of a gap-junction channel.** When fluorescent molecules of various sizes are injected into one of two cells coupled by gap junctions, molecules with a molecular weight (MW) of less than about 1000 daltons can pass into the other cell, but larger molecules cannot. Thus, the coupled cells share their small molecules (such as inorganic ions, sugars, amino acids, nucleotides, vitamins, and the intracellular signaling molecules cyclic AMP and inositol trisphosphate) but not their macromolecules (proteins, nucleic acids, and polysaccharides).

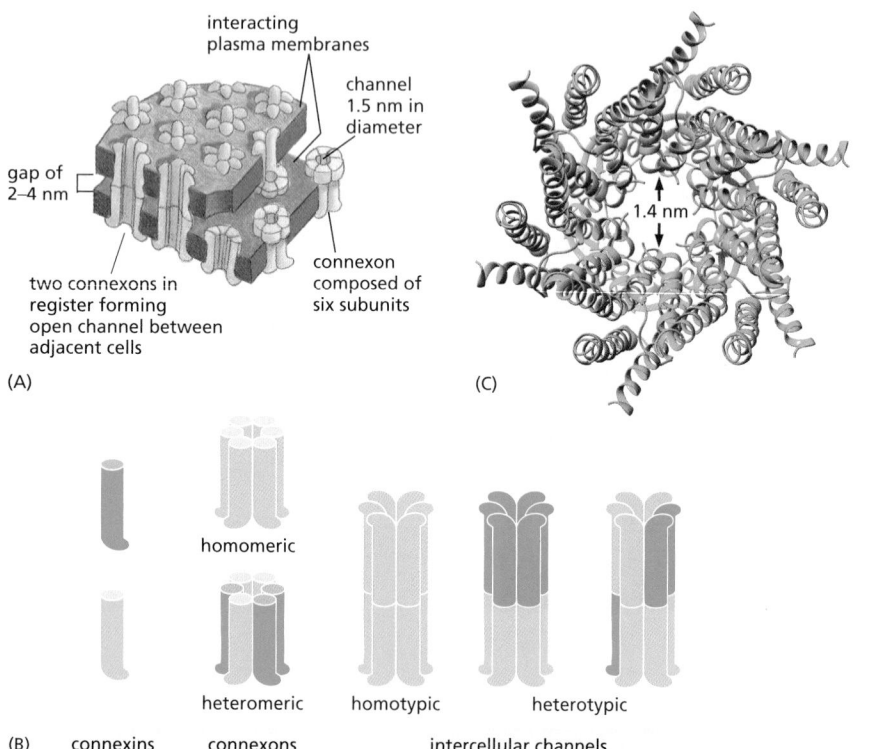

(A)

(B) connexins   connexons   intercellular channels

(C)

Figure 19–25 Gap junctions. (A) A drawing of the interacting plasma membranes of two adjacent cells connected by gap junctions. Each lipid bilayer is shown as a pair of *red* sheets. Protein assemblies called connexons *(green)*, each of which is formed by six connexin subunits, penetrate the apposed lipid bilayers. Two connexons join across the intercellular gap to form a continuous aqueous channel connecting the two cells. (B) The organization of connexins into connexons, and connexons into intercellular channels. The connexons can be homomeric or heteromeric, and the intercellular channels can be homotypic or heterotypic. (C) The high-resolution structure of a homomeric gap-junction channel, determined by x-ray crystallography of human connexin 26. In this view, we are looking down on the pore, formed from six connexin subunits. The structure illustrates the general features of the channel and suggests a pore size of about 1.4 nm, as predicted from studies of gap-junction permeability with molecules of various sizes (see Figure 19–24). (PDB code: 2ZW3.)

Gap junctions in different tissues can have different properties because they are formed from different combinations of connexins, creating channels that differ in permeability and regulation. Most cell types express more than one type of connexin, and two different connexin proteins can assemble into a heteromeric connexon, with its own distinct properties. Moreover, adjacent cells expressing different connexins can form intercellular channels in which the two aligned half-channels are different (see Figure 19–25B).

Like conventional ion channels (discussed in Chapter 11), individual gap-junction channels do not remain open all the time; instead, they flip between open and closed states. These changes are triggered by a variety of stimuli, including the voltage difference between the two connected cells, the membrane potential of each cell, and various chemical properties of the cytoplasm, including the pH and concentration of free Ca^{2+}. Some subtypes of gap junctions can also be regulated by extracellular signals such as neurotransmitters. We are only just beginning to understand the physiological functions and structural basis of these various gating mechanisms.

Each gap-junctional plaque is a dynamic structure that can readily assemble, disassemble, or be remodeled, and it can contain a cluster of a few to many thousands of connexons (see Figure 19–23B). Studies with fluorescently labeled connexins in living cells show that new connexons are continually added around the periphery of an existing junctional plaque, while old connexons are removed from the middle of it and destroyed (**Figure 19–26**). This turnover is rapid: the connexin molecules have a half-life of only a few hours.

The mechanism of removal of old connexons from the middle of the plaque is not known, but the route of delivery of new connexons to its periphery seems clear: they are inserted into the plasma membrane by exocytosis, like other integral membrane proteins, and then diffuse in the plane of the membrane until they bump into the periphery of a connexon plaque and become trapped. This has a corollary: the plasma membrane away from the gap junction should contain connexons—hemichannels—that have not yet paired with their counterparts on another cell. It is thought that these unpaired hemichannels are normally held in a closed conformation, preventing the cell from losing its small molecules by

Figure 19–26 **Connexin turnover at a gap junction.** Cells were transfected with a slightly modified connexin gene, coding for a connexin with a short amino acid tag containing four cysteines in the sequence Cys-Cys-X-X-Cys-Cys (where X denotes an arbitrary amino acid). This *tetracysteine tag* can bind strongly to certain small fluorescent dye molecules, which can be added to the culture medium and will readily enter cells by diffusing across the plasma membrane. In the experiment shown, a green dye was added first to label all the connexin molecules in the cells, and the cells were then washed and incubated for 4 or 8 hours. At the end of this time, a red dye was added to the medium and the cells were washed again and fixed. Connexin molecules already present at the beginning of the experiment are labeled green (and take up no red dye because their tetracysteine tags are already saturated with green dye), while connexins synthesized subsequently, during the 4- or 8-hour incubation, are labeled red. The fluorescence images show gap junctions between pairs of cells treated in this way. The central part of the gap-junction plaque is *green*, indicating that it consists of old connexin molecules, while the periphery is *red*, indicating that it consists of connexins synthesized during the previous 4 or 8 hours. The longer the time of incubation, the smaller the green central patch of old molecules, and the larger the peripheral ring of new molecules that have been recruited to replace the old ones. (From G. Gaietta et al., *Science* 296:503–507, 2002. With permission from AAAS.)

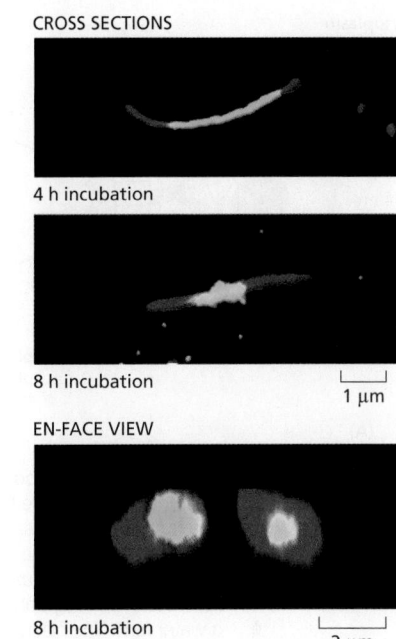

CROSS SECTIONS

4 h incubation

8 h incubation

1 µm

EN-FACE VIEW

8 h incubation

2 µm

leakage through them. But there is also evidence that in some circumstances they can open and serve as channels for the release of small signal molecules.

In Plants, Plasmodesmata Perform Many of the Same Functions as Gap Junctions

The tissues of a plant are organized on different principles from those of an animal. This is because plant cells are imprisoned within tough *cell walls* composed of an extracellular matrix rich in cellulose and other polysaccharides, as we discuss later. The cell walls of adjacent cells are firmly cemented to those of their neighbors, which eliminates the need for anchoring junctions to hold the cells in place. But a need for direct cell–cell communication remains. Thus, plant cells have only one class of intercellular junctions, **plasmodesmata**. Like gap junctions, they directly connect the cytoplasms of adjacent cells.

In plants, the cell wall between a typical pair of adjacent cells is at least 0.1 µm thick, and so a structure very different from a gap junction is required to mediate communication across it. Plasmodesmata solve the problem. With a few specialized exceptions, every living cell in a higher plant is connected to its living neighbors by these structures, which form fine cytoplasmic channels through the intervening cell walls. As shown in **Figure 19–27**A, the plasma membrane of one cell is continuous with that of its neighbor at each plasmodesma, which connects the cytoplasms of the two cells by a roughly cylindrical channel with a diameter of 20–40 nm.

Running through the center of the channel in most plasmodesmata is a narrower cylindrical structure, the *desmotubule*, which is continuous with elements of the smooth endoplasmic reticulum (ER) in each of the connected cells (Figure 19–27B–D). Between the outside of the desmotubule and the inner face of the cylindrical channel formed by plasma membrane is an annulus of cytosol through which small molecules can pass from cell to cell. As each new cell wall is assembled during the cytokinesis phase of cell division, plasmodesmata are created within it. They form around elements of smooth ER that become trapped across the developing cell plate (discussed in Chapter 17). They can also be inserted *de novo* through preexisting cell walls, where they are commonly found in dense clusters called *pit fields*. When no longer required, plasmodesmata can be removed.

In spite of the radical difference in structure between plasmodesmata and gap junctions, they seem to function in remarkably similar ways. Evidence obtained by injecting tracer molecules of different sizes suggests that plasmodesmata allow the passage of molecules with a molecular weight of less than about 800, which

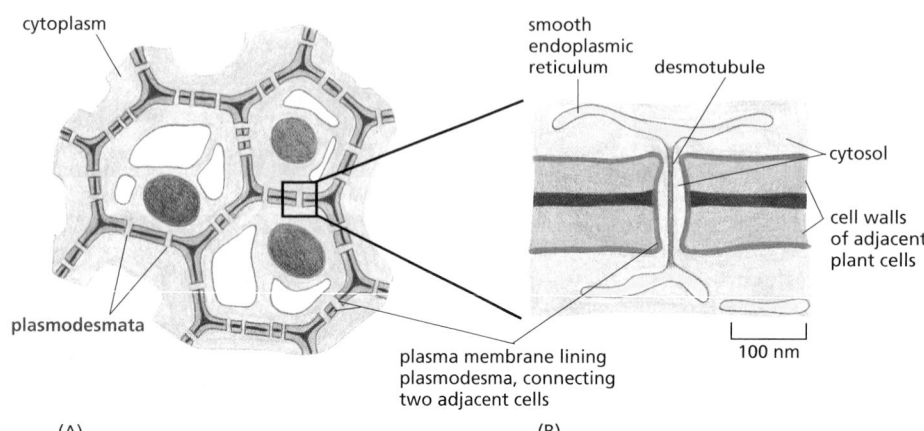

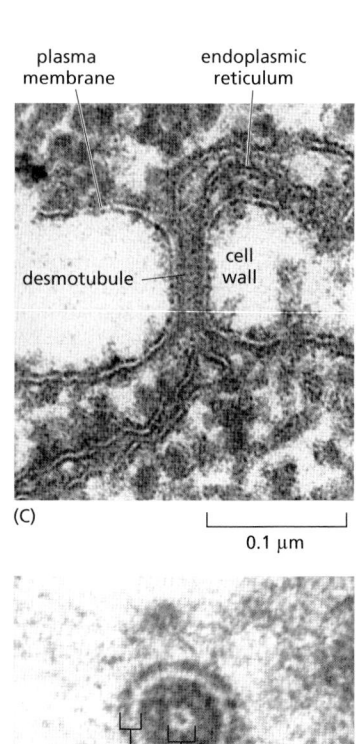

Figure 19–27 **Plasmodesmata.** (A) The cytoplasmic channels of plasmodesmata pierce the plant cell wall and connect cells in a plant together. (B) Each plasmodesma is lined with plasma membrane that is common to the two connected cells. It usually also contains a fine tubular structure, the desmotubule, derived from smooth endoplasmic reticulum. (C) Electron micrograph of a longitudinal section of a plasmodesma from a water fern. The plasma membrane lines the pore and is continuous from one cell to the next. Endoplasmic reticulum and its association with the central desmotubule can also be seen. (D) A similar plasmodesma seen in cross section. (C and D, from R.L. Overall, J. Wolfe, and B.E.S. Gunning, in *Protoplasma* 111(2)134–150, published 1982 by Springer-Verlag. Reproduced with permission of SNCSC.)

is similar to the molecular-weight cutoff for gap junctions. As with gap junctions, transport through plasmodesmata is regulated. Dye-injection experiments, for example, show that there can be barriers to the movement of even low-molecular-weight molecules between certain cells, or groups of cells, that are connected by apparently normal plasmodesmata; the mechanisms that restrict communication in these cases are not understood.

Selectins Mediate Transient Cell–Cell Adhesions in the Bloodstream

We now complete our overview of cell–cell junctions and adhesion by briefly describing some of the more specialized adhesion mechanisms used in some tissues. In addition to those we have already discussed, at least three other superfamilies of cell–cell adhesion proteins are important: the *integrins*, the *selectins*, and the adhesive *immunoglobulin (Ig) superfamily* members. We shall discuss integrins in more detail later: their main function is in cell–matrix adhesion, but a few of them mediate cell–cell adhesion in specialized circumstances. Ca^{2+} dependence provides one simple way to distinguish among these classes of adhesion proteins experimentally. Selectins, like cadherins and integrins, require Ca^{2+} for their adhesive function; Ig superfamily members do not.

Selectins are cell-surface carbohydrate-binding proteins (*lectins*) that mediate a variety of transient cell–cell adhesion interactions in the bloodstream. Their main role, in vertebrates at least, is in governing the traffic of white blood cells into normal lymphoid organs and any inflamed tissues. White blood cells lead a nomadic life, roving between the bloodstream and the tissues, and this necessitates special adhesive behavior. The selectins control the binding of white blood cells to the endothelial cells lining blood vessels, thereby enabling the blood cells to migrate out of the bloodstream into a tissue.

Each selectin is a transmembrane protein with a conserved lectin domain that binds to a specific oligosaccharide on another cell (**Figure 19–28**A). There are at least three types: *L-selectin* on white blood cells, *P-selectin* on blood platelets and on endothelial cells that have been locally activated by an inflammatory response, and *E-selectin* on activated endothelial cells. In a lymphoid organ, such as a lymph

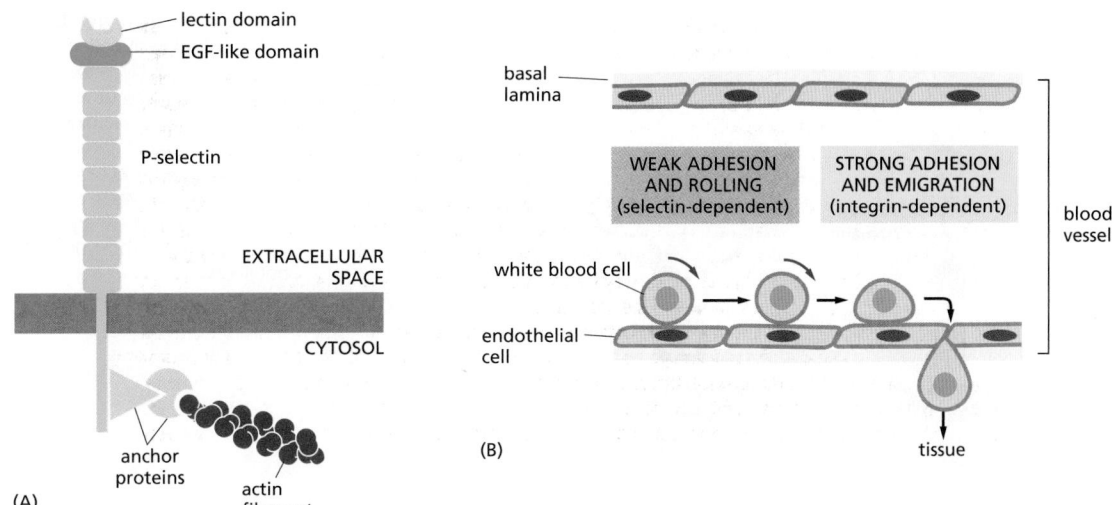

Figure 19–28 **The structure and
function of selectins.** (A) The structure
of P-selectin. The selectin attaches to the
actin cytoskeleton through adaptor proteins
that are still poorly characterized. (B) How
selectins and integrins mediate the cell–cell
adhesions required for a white blood cell
to migrate out of the bloodstream into a
tissue. First, selectins on endothelial cells
bind to oligosaccharides on the white
blood cell, so that it becomes loosely
attached and rolls along the vessel wall.
Then the white blood cell activates a cell-
surface integrin called LFA1, which binds
to a protein called ICAM1 (belonging to
the Ig superfamily) on the membrane of
the endothelial cell. The white blood cell
adheres to the vessel wall and then crawls
out of the vessel by a process that requires
another immunoglobulin superfamily
member called PECAM1 (or CD31), not
shown (Movie 19.2). EGF, epidermal
growth factor.

node or the spleen, the endothelial cells express oligosaccharides that are recognized by L-selectin on lymphocytes, causing the lymphocytes to loiter and become trapped. At sites of inflammation, the roles are reversed: the endothelial cells switch on expression of selectins that recognize the oligosaccharides on white blood cells and platelets, flagging the cells down to help deal with the local emergency. Selectins do not act alone, however; they collaborate with integrins, which strengthen the binding of the blood cells to the endothelium. The cell–cell adhesions mediated by both selectins and integrins are *heterophilic*—that is, the binding is to a molecule of a different type: selectins bind to specific oligosaccharides on glycoproteins and glycolipids, while integrins bind to specific Ig-family proteins.

Selectins and integrins act in sequence to let white blood cells leave the bloodstream and enter tissues (Figure 19–28B). The selectins mediate a weak adhesion because the binding of the lectin domain of the selectin to its carbohydrate ligand is of low affinity. This allows the white blood cell to adhere weakly and reversibly to the endothelium, rolling along the surface of the blood vessel, propelled by the flow of blood. The rolling continues until the blood cell activates its integrins. As we discuss later, these transmembrane molecules can be switched into an adhesive conformation that enables them to latch onto specific macromolecules external to the cell—in the present case, proteins on the surfaces of the endothelial cells. Once it has attached in this way, the white blood cell escapes from the bloodstream into the tissue by crawling out of the blood vessel between adjacent endothelial cells.

Members of the Immunoglobulin Superfamily Mediate Ca²⁺-Independent Cell–Cell Adhesion

The chief endothelial cell proteins that are recognized by the white blood cell integrins are called *ICAMs* (*intercellular cell adhesion molecules*) or *VCAMs* (*vascular cell adhesion molecules*). They are members of another large and ancient family of cell-surface molecules—the **immunoglobulin (Ig) superfamily**. These contain one or more extracellular Ig-like domains that are characteristic of antibody molecules. They have many functions outside the immune system that are unrelated to immune defenses.

While ICAMs and VCAMs on endothelial cells both mediate heterophilic binding to integrins, many other Ig superfamily members appear to mediate homophilic binding. An example is the *neural cell adhesion molecule* (*NCAM*), which is expressed by various cell types, including most nerve cells, and can take different forms, generated by alternative splicing of an RNA transcript produced from a single gene (**Figure 19–29**). Some forms of NCAM carry an unusually large

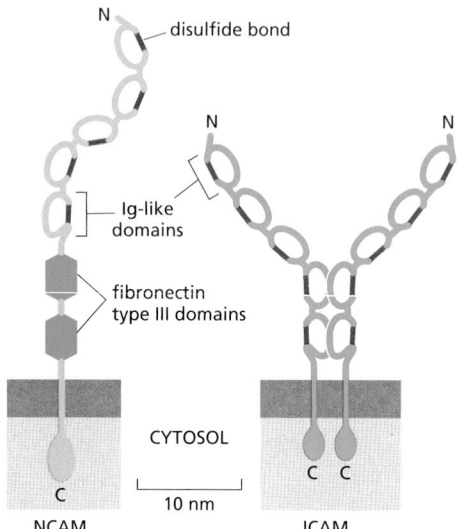

disulfide bond

Ig-like domains

fibronectin type III domains

CYTOSOL

10 nm

NCAM

ICAM

Figure 19–29 **Two members of the Ig superfamily of cell–cell adhesion molecules.** NCAM is expressed on neurons and many other cell types, and mediates homophilic binding. ICAM is expressed on endothelial cells and some other cell types and binds heterophilically to an integrin on white blood cells. Both NCAM and ICAM are glycoproteins, but their attached carbohydrate chains are not shown.

quantity of sialic acid (with chains containing hundreds of repeating sialic acid units). By virtue of their negative charge, the long polysialic acid chains can interfere with cell adhesion (because like charges repel one another); thus, these forms of NCAM can serve to inhibit adhesion, rather than cause it.

A cell of a given type generally uses an assortment of different adhesion proteins to interact with other cells, just as each cell uses an assortment of different receptors to respond to the many soluble extracellular signal molecules in its environment. Although cadherins and Ig superfamily members are frequently expressed on the same cells, the adhesions mediated by cadherins are much stronger, and they are largely responsible for holding cells together, segregating cell collectives into discrete tissues, and maintaining tissue integrity. Molecules such as NCAM seem to contribute more to the fine-tuning of these adhesive interactions during development and regeneration, playing a part in various specialized adhesive phenomena, such as that discussed for blood cells and endothelial cells. Thus, while mutant mice that lack N-cadherin die early in development, those that lack NCAM develop relatively normally but show some mild abnormalities in the development of certain specific tissues, including parts of the nervous system.

Summary

In epithelia, as well as in some other types of tissue, cells are directly attached to one another through strong cell–cell adhesions, mediated by transmembrane proteins called cadherins, which are anchored intracellularly to the cytoskeleton. Cadherins generally bind to one another homophilically: the head of one cadherin molecule binds to the head of a similar cadherin on an opposite cell. This selectivity enables mixed populations of cells of different types to sort out from one another according to the specific cadherins they express, and it helps to control cell rearrangements during development.

The "classical" cadherins at adherens junctions are linked to the actin cytoskeleton by intracellular adaptor proteins called catenins. These form an anchoring complex on the intracellular tail of the cadherin molecule, and are involved not only in physical anchorage but also in the detection of and response to tension and other regulatory signals at the junction.

Tight junctions seal the gaps between cells in epithelia, creating a barrier to the diffusion of molecules across the cell sheet and also helping to separate the populations of proteins in the apical and basolateral plasma membrane domains of the epithelial cell. Claudins are the major transmembrane proteins forming tight junctions. Intracellular scaffold proteins organize the claudins and other junctional proteins into a complex protein network that is linked to the actin cytoskeleton.

The cells of many animal tissues are coupled by gap junctions, which take the form of plaques of clustered connexons, which usually allow molecules smaller than about 1000 daltons to pass directly from the inside of one cell to the inside of the next. Cells connected by gap junctions share many of their inorganic ions and other small molecules and are therefore chemically and electrically coupled.

Three additional classes of transmembrane adhesion proteins mediate more transient cell–cell adhesion: selectins, immunoglobulin (Ig) superfamily members, and integrins. Selectins are expressed on white blood cells, blood platelets, and endothelial cells; they bind heterophilically to carbohydrate groups on cell surfaces, helping to mediate the adhesive interactions between these cells. Ig superfamily proteins also play a part in these interactions, as well as in many other adhesive processes; some of them bind homophilically, some heterophilically. Integrins, though they mainly serve to attach cells to the extracellular matrix, can also mediate cell–cell adhesion by binding to specific Ig superfamily proteins.

THE EXTRACELLULAR MATRIX OF ANIMALS

Tissues are not made up solely of cells. They also contain a remarkably complex and intricate network of macromolecules constituting the *extracellular matrix.* This matrix is composed of many different proteins and polysaccharides that are secreted locally and assembled into an organized meshwork in close association with the surfaces of the cells that produce them.

The classes of macromolecules constituting the extracellular matrix in different animal tissues are broadly similar, but variations in the relative amounts of these different classes of molecules and in the ways in which they are organized give rise to an amazing diversity of materials. The matrix can become calcified to form the rock-hard structures of bone or teeth, or it can form the transparent substance of the cornea, or it can adopt the ropelike organization that gives tendons their enormous tensile strength. It forms the jelly in a jellyfish. Covering the body of a beetle or a lobster, it forms a rigid carapace. Moreover, the extracellular matrix is more than a passive scaffold to provide physical support. It has an active and complex role in regulating the behavior of the cells that touch it, inhabit it, or crawl through its meshes, influencing their survival, development, migration, proliferation, shape, and function.

In this section, we describe the major features of the extracellular matrix in animal tissues, with an emphasis on vertebrates. We begin with an overview of the major classes of macromolecules in the matrix, after which we turn to the structure and function of the *basal lamina,* the thin layer of specialized extracellular matrix that lies beneath all epithelial cells. In the sections that follow, we then describe the varied types of cell–matrix junctions through which cells are connected to the matrix.

The Extracellular Matrix Is Made and Oriented by the Cells Within It

The macromolecules that constitute the extracellular matrix are mainly produced locally by cells in the matrix. As we discuss later, these cells also help to organize the matrix: the orientation of the cytoskeleton inside the cell can control the orientation of the matrix produced outside. In most connective tissues, the matrix macromolecules are secreted by cells called **fibroblasts** (**Figure 19–30**). In certain specialized types of connective tissues, such as cartilage and bone, however, they are secreted by cells of the fibroblast family that have more specific names: *chondroblasts,* for example, form cartilage, and *osteoblasts* form bone.

The extracellular matrix is constructed from three major classes of macromolecules: (1) glycosaminoglycans (*GAGs*), which are large and highly charged polysaccharides that are usually covalently linked to protein in the form of *proteoglycans*; (2) fibrous proteins, which are primarily members of the *collagen* family; and (3) a large class of noncollagen *glycoproteins*, which carry conventional asparagine-linked oligosaccharides (described in Chapter 12). All three classes of macromolecule have many members and come in a great variety of shapes and

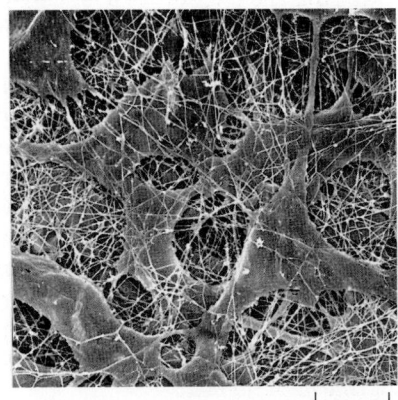

10 µm

Figure 19–30 Fibroblasts in connective tissue. This scanning electron micrograph shows tissue from the cornea of a rat. The extracellular matrix surrounding the fibroblasts is here composed largely of collagen fibrils. The glycoproteins, hyaluronan, and proteoglycans, which normally form a hydrated gel filling the interstices of the fibrous network, have been removed by enzyme and acid treatment. (Courtesy of T. Nishida.)

sizes (**Figure 19–31**). Mammals are thought to have almost 300 matrix proteins, including about 36 proteoglycans, about 40 collagens, and over 200 glycoproteins, which usually contain multiple subdomains and self-associate to form multimers. Add to this the large number of matrix-associated proteins and enzymes that can modify matrix behavior by cross-linking, degradation, or other mechanisms, and one begins to see that the matrix is an almost infinitely variable material. Each tissue contains its own unique blend of matrix components, resulting in an extracellular matrix that is specialized for the needs of that tissue.

The proteoglycan molecules in connective tissue typically form a highly hydrated, gel-like "ground substance" in which collagens and glycoproteins are embedded. The polysaccharide gel resists compressive forces on the matrix while permitting the rapid diffusion of nutrients, metabolites, and hormones between the blood and the tissue cells. The collagen fibers strengthen and help organize the matrix, while other fibrous proteins, such as the rubberlike *elastin*, give it resilience. Finally, the many matrix glycoproteins help cells migrate, settle, and differentiate in the appropriate locations.

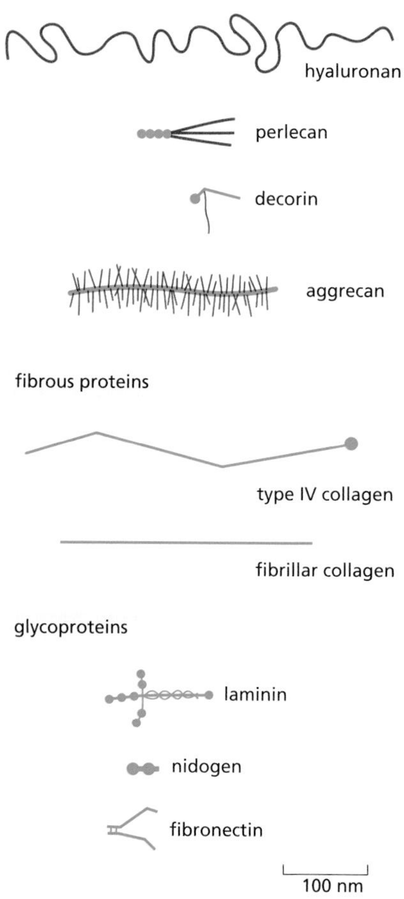

Glycosaminoglycan (GAG) Chains Occupy Large Amounts of Space and Form Hydrated Gels

Glycosaminoglycans (**GAGs**) are unbranched polysaccharide chains composed of repeating disaccharide units. One of the two sugars in the repeating disaccharide is always an amino sugar (*N*-acetylglucosamine or *N*-acetylgalactosamine), which in most cases is sulfated. The second sugar is usually a uronic acid (glucuronic or iduronic). Because there are sulfate or carboxyl groups on most of their sugars, GAGs are highly negatively charged (**Figure 19–32**). Indeed, they are the most anionic molecules produced by animal cells. Four main groups of GAGs are distinguished by their sugars, the type of linkage between the sugars, and the number and location of sulfate groups: (1) *hyaluronan*, (2) *chondroitin sulfate* and *dermatan sulfate*, (3) *heparan sulfate*, and (4) *keratan sulfate*.

Polysaccharide chains are too stiff to fold into compact globular structures, and they are strongly hydrophilic. Thus, GAGs tend to adopt highly extended conformations that occupy a huge volume relative to their mass (**Figure 19–33**), and they form hydrated gels even at very low concentrations. The weight of GAGs in connective tissue is usually less than 10% of the weight of proteins, but GAG chains fill most of the extracellular space. Their high density of negative charges attracts a cloud of cations, especially Na^+, that are osmotically active, causing large amounts of water to be sucked into the matrix. This creates a swelling pressure, or turgor, that enables the matrix to withstand compressive forces (in contrast to collagen fibrils, which resist stretching forces). The cartilage matrix that lines the knee joint, for example, can support pressures of hundreds of atmospheres in this way.

Defects in the production of GAGs can affect many different body systems. In one rare human genetic disease, for example, there is a severe deficiency in the synthesis of dermatan sulfate disaccharide. The affected individuals have a short stature, a prematurely aged appearance, and generalized defects in their skin, joints, muscles, and bones.

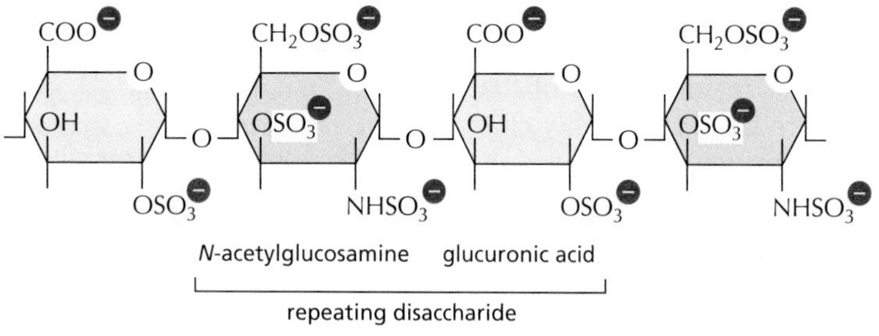

proteoglycans and GAGs

hyaluronan

perlecan

decorin

aggrecan

fibrous proteins

type IV collagen

fibrillar collagen

glycoproteins

laminin

nidogen

fibronectin

100 nm

Figure 19–31 The comparative shapes and sizes of some of the major extracellular matrix macromolecules. Protein is shown in *green*, and glycosaminoglycan (GAG) in *red*.

COO$^-$ CH$_2$OSO$_3^-$ COO$^-$ CH$_2$OSO$_3^-$

OH   OSO$_3^-$ OH   OSO$_3^-$

OSO$_3^-$ NHSO$_3^-$ OSO$_3^-$ NHSO$_3^-$

N-acetylglucosamine   glucuronic acid

repeating disaccharide

Figure 19–32 The repeating disaccharide sequence of a heparan sulfate glycosaminoglycan (GAG) chain. These chains can consist of as many as 200 disaccharide units, but are typically less than half that size. There is a high density of negative charges along the chain due to the presence of both carboxyl and sulfate groups. The molecule is shown here with its maximal number of sulfate groups. *In vivo*, the proportion of sulfated and nonsulfated groups is variable. Heparin typically has >70% sulfation, while heparan sulfate has <50%.

Hyaluronan Acts as a Space Filler During Tissue Morphogenesis and Repair

Hyaluronan (also called *hyaluronic acid* or *hyaluronate*) is the simplest of the GAGs (**Figure 19–34**). It consists of a regular repeating sequence of up to 25,000 disaccharide units, is found in variable amounts in all tissues and fluids in adult animals, and is especially abundant in early embryos. Hyaluronan is not a typical GAG because it contains no sulfated sugars, all its disaccharide units are identical, its chain length is enormous, and it is not generally linked covalently to any core protein. Moreover, whereas other GAGs are synthesized inside the cell and released by exocytosis, hyaluronan is spun out directly from the cell surface by an enzyme complex embedded in the plasma membrane.

Hyaluronan is thought to have a role in resisting compressive forces in tissues and joints. It is also important as a space filler during embryonic development, where it can be used to force a change in the shape of a structure, as a small quantity expands with water to occupy a large volume. Hyaluronan synthesized locally from the basal side of an epithelium can deform the epithelium by creating a cell-free space beneath it, into which cells subsequently migrate. In the developing heart, for example, hyaluronan synthesis helps in this way to drive formation of the valves and septa that separate the heart's chambers. Similar processes occur in several other organs. When cell migration ends, the excess hyaluronan is generally degraded by the enzyme *hyaluronidase*. Hyaluronan is also produced in large quantities during wound healing, and it is an important constituent of joint fluid, in which it serves as a lubricant.

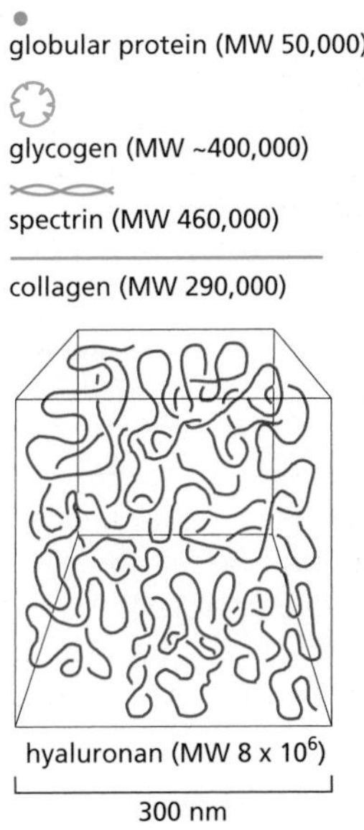

Proteoglycans Are Composed of GAG Chains Covalently Linked to a Core Protein

Except for hyaluronan, all GAGs are covalently attached to protein as **proteoglycans**, which are produced by most animal cells. Membrane-bound ribosomes make the polypeptide chain, or *core protein*, of a proteoglycan, which is then threaded into the lumen of the endoplasmic reticulum. The polysaccharide chains are mainly assembled on this core protein in the Golgi apparatus before delivery to the exterior of the cell by exocytosis. First, a special *linkage tetrasaccharide* is attached to a serine side chain on the core protein to serve as a primer for polysaccharide growth; then, one sugar at a time is added by specific glycosyl transferases (**Figure 19–35**). While still in the Golgi apparatus, many of the polymerized sugars are covalently modified by a sequential and coordinated series of reactions. Epimerizations alter the configuration of the substituents around individual carbon atoms in the sugar molecule; sulfations increase the negative charge.

Proteoglycans are clearly distinguished from other glycoproteins by the nature, quantity, and arrangement of their sugar side chains. By definition, at least one of the sugar side chains of a proteoglycan must be a GAG. Whereas glycoproteins generally contain relatively short, branched oligosaccharide chains that contribute only a small fraction of their weight, proteoglycans can contain as much as

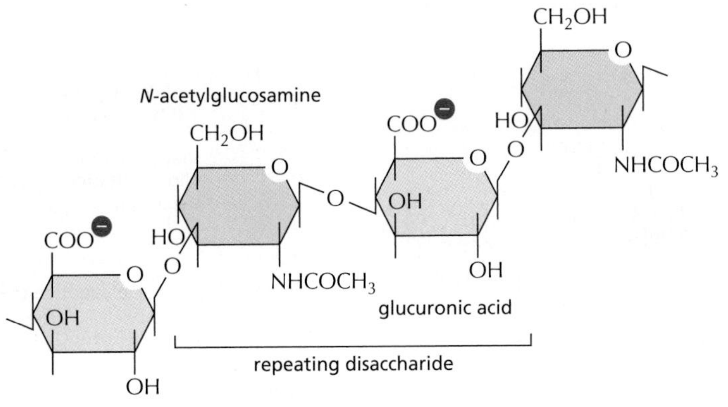

globular protein (MW 50,000)

glycogen (MW ~400,000)

spectrin (MW 460,000)

collagen (MW 290,000)

hyaluronan (MW 8 x 10⁶)

300 nm

Figure 19–33 **The relative dimensions and volumes occupied by various macromolecules.** Several proteins, a glycogen granule, and a single hydrated molecule of hyaluronan are shown.

N-acetylglucosamine

CH₂OH

COO⁻

CH₂OH

COO⁻ HO

HO

NHCOCH₃

glucuronic acid

NHCOCH₃

OH

OH

NHCOCH₃

OH

repeating disaccharide

Figure 19–34 **The repeating disaccharide sequence in hyaluronan, a relatively simple GAG.** This ubiquitous molecule in vertebrates consists of a single long chain of up to 25,000 sugar monomers. Note the absence of sulfate groups.

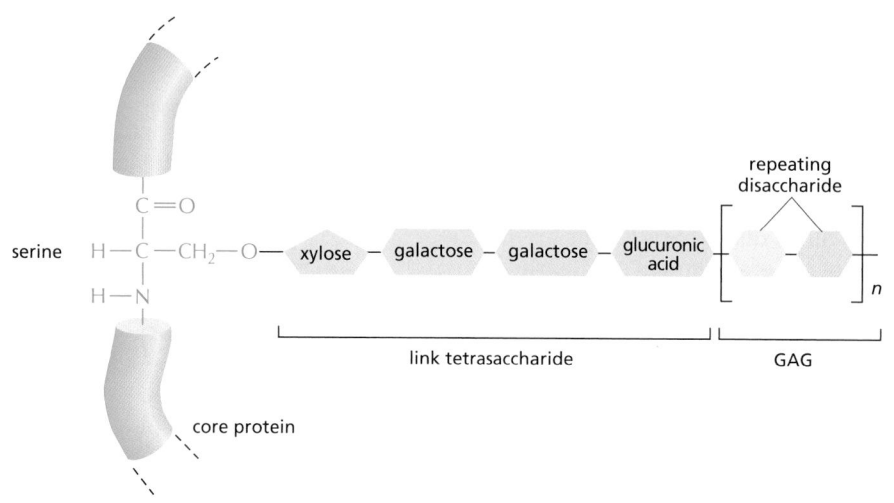

serine

link tetrasaccharide

GAG

repeating disaccharide

core protein

Figure 19–35 **The linkage between a GAG chain and its core protein in a proteoglycan molecule.** A specific link tetrasaccharide is first assembled on a serine side chain. The rest of the GAG chain, consisting mainly of a repeating disaccharide unit, is then synthesized, with one sugar being added at a time. In chondroitin sulfate, the disaccharide is composed of D-glucuronic acid and N-acetyl-D-galactosamine; in heparan sulfate, it is either D-glucuronic acid or L-iduronic acid and N-acetyl-D-glucosamine; in keratan sulfate, it is D-galactose and N-acetyl-D-glucosamine.

95% carbohydrate by weight, mostly in the form of long, unbranched GAG chains, each typically about 80 sugars long.

In principle, proteoglycans have the potential for almost limitless heterogeneity. Even a single type of core protein can carry highly variable numbers and types of attached GAG chains. Moreover, the underlying repeating sequence of disaccharides in each GAG can be modified by a complex pattern of sulfate groups. The core proteins, too, are diverse, though many of them share some characteristic domains such as the LINK domain, involved in binding to GAGs.

Proteoglycans can be huge. The proteoglycan *aggrecan*, for example, which is a major component of cartilage, has a mass of about 3×10^6 daltons with over 100 GAG chains. Other proteoglycans are much smaller and have only 1–10 GAG chains; an example is *decorin*, which is secreted by fibroblasts and has a single GAG chain (**Figure 19–36**). Decorin binds to collagen fibrils and regulates fibril assembly and fibril diameter; mice that cannot make decorin have fragile skin that has reduced tensile strength. The GAGs and proteoglycans of these various types can associate to form even larger polymeric complexes in the extracellular matrix. Molecules of aggrecan, for example, assemble with hyaluronan in cartilage matrix to form aggregates that are as big as a bacterium (**Figure 19–37**). Moreover, besides associating with one another, GAGs and proteoglycans associate with fibrous matrix proteins such as collagen and with protein meshworks such as the basal lamina, creating extremely complex composites (**Figure 19–38**).

Not all proteoglycans are secreted components of the extracellular matrix. Some are integral components of plasma membranes and have their core protein either inserted across the lipid bilayer or attached to the lipid bilayer by a glycosylphosphatidylinositol (GPI) anchor. Among the best-characterized plasma membrane proteoglycans are the *syndecans*, which have a membrane-spanning core protein whose intracellular domain is thought to interact with the actin cytoskeleton and with signaling molecules in the cell cortex. Syndecans are located on the surface of many types of cells, including fibroblasts and epithelial cells. In

Figure 19–36 **Examples of a small (decorin) and a large (aggrecan) proteoglycan found in the extracellular matrix.** The figure compares these two proteoglycans with a typical secreted glycoprotein molecule, pancreatic ribonuclease B. All three are drawn to scale. The core proteins of both aggrecan and decorin contain oligosaccharide chains as well as the GAG chains, but these are not shown. Aggrecan typically consists of about 100 chondroitin sulfate chains and about 30 keratan sulfate chains linked to a serine-rich core protein of almost 3000 amino acids. Decorin "decorates" the surface of collagen fibrils, hence its name.

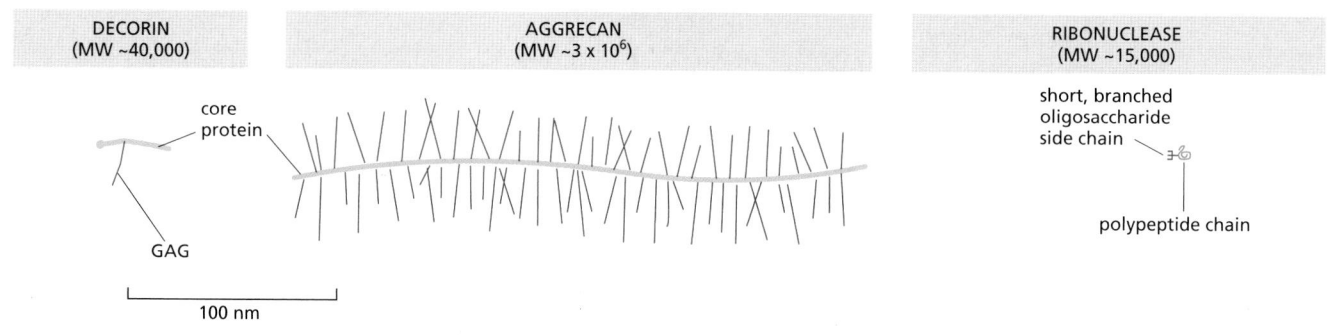

DECORIN
(MW ~40,000)

core protein

GAG

AGGRECAN
(MW ~3 x 10⁶)

RIBONUCLEASE
(MW ~15,000)

short, branched oligosaccharide side chain

polypeptide chain

100 nm

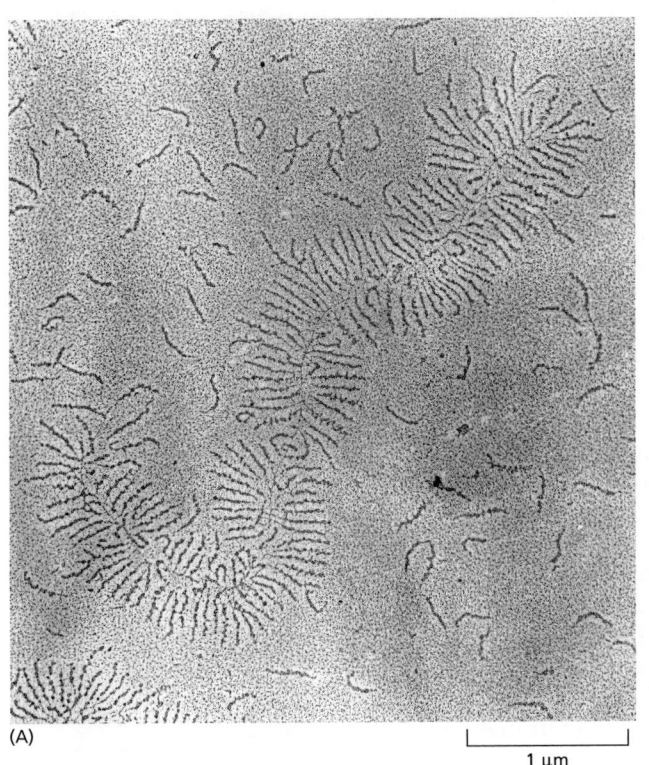

(A)

1 µm

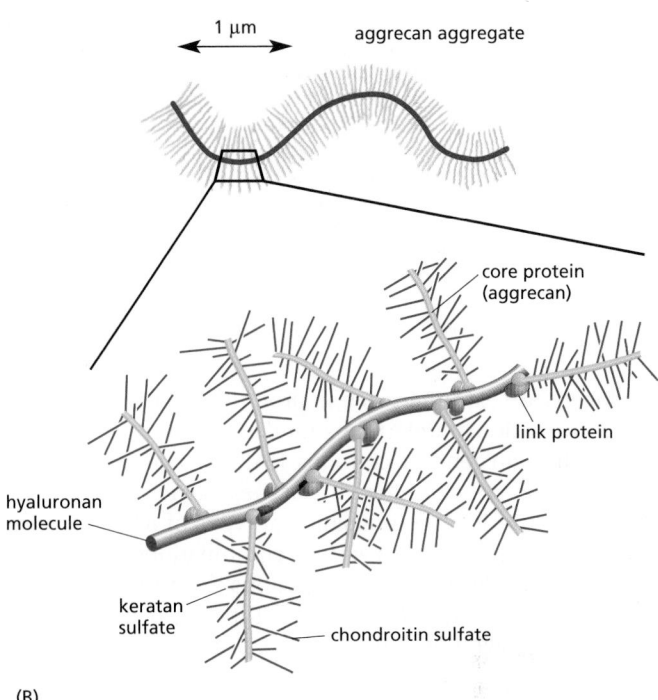

1 µm

aggrecan aggregate

core protein
(aggrecan)

link protein

hyaluronan
molecule

keratan
sulfate

chondroitin sulfate

(B)

Figure 19–37 An aggrecan aggregate from fetal bovine cartilage. (A) An electron micrograph of an aggrecan aggregate shadowed with platinum. Many free aggrecan molecules are also visible. (B) A drawing of the giant aggrecan aggregate shown in (A). It consists of about 100 aggrecan monomers (each like the one shown in Figure 19–36) noncovalently bound through the N-terminal domain of the core protein to a single hyaluronan chain. A link protein binds both to the core protein of the proteoglycan and to the hyaluronan chain, thereby stabilizing the aggregate. The link proteins are members of a family of hyaluronan-binding proteins, some of which are cell-surface proteins. The molecular mass of such a complex can be 10^8 daltons or more, and it occupies a volume equivalent to that of a bacterium, which is about 2 × 10^{-12} cm³. (A, from J.A. Buchwalter, P.J. Roughley and L.C. Rosenberg, *Microsc. Res. Tech.* 28(5):398–408, 1994. With permission from John Wiley & Sons.)

fibroblasts, syndecans can be found in cell–matrix adhesions, where they modulate integrin function by interacting with fibronectin on the cell surface and with cytoskeletal and signaling proteins inside the cell. As we discuss later, syndecan and other proteoglycans also interact with soluble peptide growth factors, influencing their effects on cell growth and proliferation.

Collagens Are the Major Proteins of the Extracellular Matrix

The **collagens** are a family of fibrous proteins found in all multicellular animals. They are secreted in large quantities by connective-tissue cells, and in smaller quantities by many other cell types. As a major component of skin and bone, collagens are the most abundant proteins in mammals, where they constitute 25% of the total protein mass.

The primary feature of a typical collagen molecule is its long, stiff, triple-stranded helical structure, in which three collagen polypeptide chains, called *α chains*, are wound around one another in a ropelike superhelix (**Figure 19–39**).

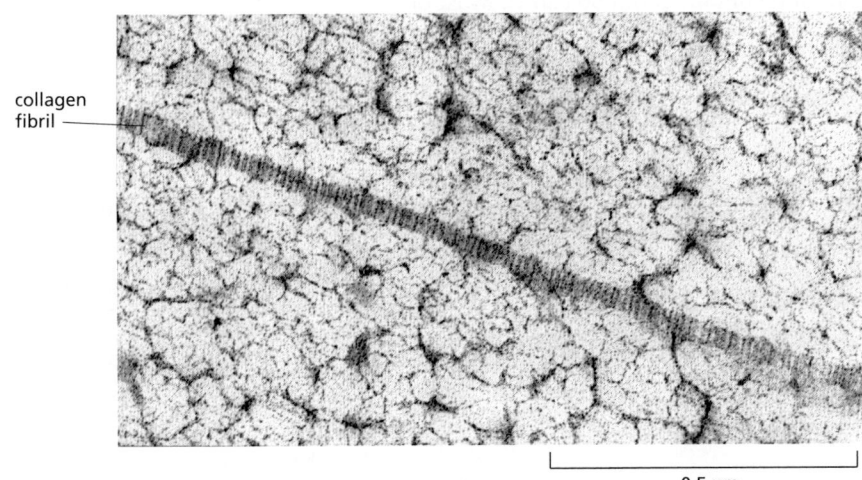

collagen
fibril

0.5 µm

Figure 19–38 Proteoglycans in the extracellular matrix of rat cartilage. The tissue was rapidly frozen at –196°C, and fixed and stained while still frozen (a process called freeze substitution) to prevent the GAG chains from collapsing. In this electron micrograph, the proteoglycan molecules are seen to form a fine filamentous network in which a single striated collagen fibril is embedded. The more darkly stained parts of the proteoglycan molecules are the core proteins; the faintly stained threads are the GAG chains. (© 1984 E.B. Hunziker and R.K. Schenk. Originally published in *J. Cell Biol.* https://doi.org/10.1083/jcb.98.1.277. With permission from Rockefeller University Press.)

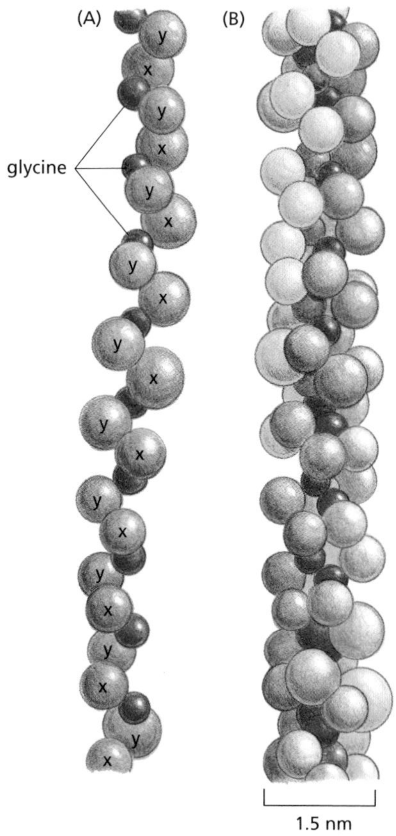

Figure 19–39 **The structure of a typical collagen molecule.** (A) A model of part of a single collagen α chain, in which each amino acid is represented by a sphere. The chain is about 1000 amino acids long. It is arranged as a left-handed helix, with three amino acids per turn and with glycine as every third amino acid. Therefore, an α chain is composed of a series of triplet Gly-X-Y sequences, in which X and Y can be any amino acid (although X is commonly proline and Y is commonly hydroxyproline, a form of proline that is chemically modified during collagen synthesis in the cell). (B) A model of part of a collagen molecule, in which three α chains, each shown in a different color, are wrapped around one another to form a triple-stranded helical rod. Glycine is the only amino acid small enough to occupy the crowded interior of the triple helix. Only a short length of the molecule is shown; the entire molecule is 300 nm long. (From a model by B.L. Trus.)

Collagens are extremely rich in proline and glycine, both of which are important in the formation of the triple-stranded helix.

The human genome contains 42 distinct genes coding for different collagen α chains. Different combinations of these genes are expressed in different tissues. Although in principle thousands of types of triple-stranded collagen molecules could be assembled from various combinations of the 42 α chains, only a limited number of triple-helical combinations are possible, and roughly 40 types of collagen molecules have been found. Type I is by far the most common, being the principal collagen of skin and bone. It belongs to the class of **fibrillar collagens**, or fibril-forming collagens: after being secreted into the extracellular space, they assemble into higher-order polymers called **collagen fibrils**, which are thin structures (10–300 nm in diameter) many hundreds of micrometers long in mature tissues, where they are clearly visible in electron micrographs (**Figure 19–40**; see also Figure 19–38). Collagen fibrils often aggregate into larger, cablelike bundles, several micrometers in diameter, that are visible in the light microscope as *collagen fibers*.

Collagen types IX and XII are called *fibril-associated collagens* because they decorate the surface of collagen fibrils. They are thought to link these fibrils to one another and to other components in the extracellular matrix. Type IV is a *network-forming collagen*, forming a major part of basal laminae, while type VII molecules form dimers that assemble into specialized structures called *anchoring fibrils*. Anchoring fibrils help attach the basal lamina of multilayered epithelia to the underlying connective tissue and therefore are especially abundant in the skin. There are also a number of "collagen-like" proteins containing short collagen-like segments. These include collagen type XVII, which has a transmembrane domain and is found in hemidesmosomes, and type XVIII, the core protein of a proteoglycan in basal laminae.

Many proteins appear to have evolved by repeated duplications of an original DNA sequence, giving rise to a repetitive pattern of amino acids. The genes that encode the α chains of most of the fibrillar collagens provide a good example: they are very large (up to 44 kilobases in length) and contain about 50 exons. Most of

(A)   (B)

glycine

1.5 nm

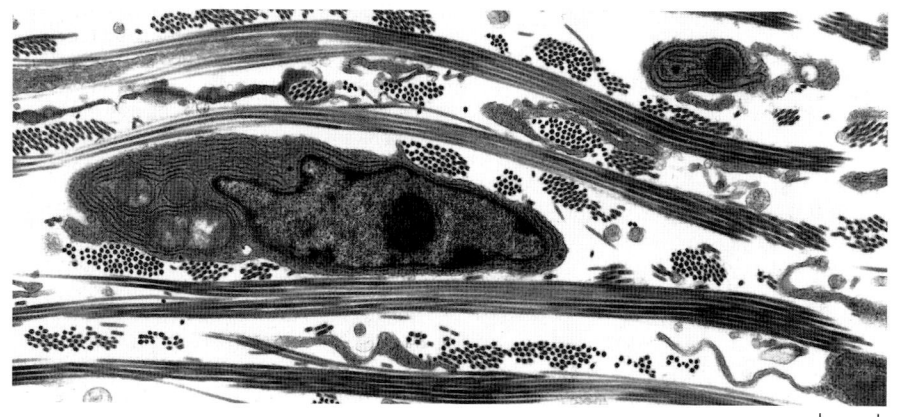

1 µm

Figure 19–40 **A fibroblast surrounded by collagen fibrils in the connective tissue of embryonic chick skin.** In this electron micrograph, the fibrils are organized into bundles that run approximately at right angles to one another. Therefore, some bundles are oriented longitudinally, whereas others are seen in cross section. The collagen fibrils are produced by fibroblasts. (From C. Ploetz, E.I. Zycband and D.E. Birk, *J. Struct. Biol.* 106:73–81, 1991. With permission from Elsevier.)

TABLE 19–2 Some Types of Collagen and Their Properties

	Type	Polymerized form	Tissue distribution	Mutant phenotype
Fibril-forming (fibrillar)	I	Fibril	Bone, skin, tendons, ligaments, cornea, internal organs (accounts for 90% of body collagen)	Severe bone defects, fractures (*osteogenesis imperfecta*)
	II	Fibril	Cartilage, intervertebral disc, notochord, vitreous humor of the eye	Cartilage deficiency, dwarfism (*chondrodysplasia*)
	III	Fibril	Skin, blood vessels, internal organs	Fragile skin, loose joints, blood vessels prone to rupture (*Ehlers–Danlos syndrome*)
	V	Fibril (with type I)	As for type I	Fragile skin, loose joints, blood vessels prone to rupture
	XI	Fibril (with type II)	As for type II	Myopia, blindness
Fibril-associated	IX	Lateral association with type II fibrils	Cartilage	Osteoarthritis
Network-forming	IV	Sheetlike network	Basal lamina	Kidney disease (glomerulonephritis), deafness
	VII	Anchoring fibrils	Beneath stratified squamous epithelia	Skin blistering
Transmembrane	XVII	Nonfibrillar	Hemidesmosomes	Skin blistering
Proteoglycan core protein	XVIII	Nonfibrillar	Basal lamina	Myopia, detached retina, hydrocephalus

Note that types I, IV, V, IX, and XI are each composed of two or three types of α chains (distinct, nonoverlapping sets in each case), whereas types II, III, VII, XVII, and XVIII are composed of only one type of α chain each.

the exons are 54, or multiples of 54, nucleotides long, suggesting that these collagens originated through multiple duplications of a primordial gene containing 54 nucleotides and encoding exactly six Gly-X-Y repeats (see Figure 19–39).

Table 19–2 provides additional details for some of the collagen types discussed in this chapter.

Secreted Fibril-Associated Collagens Help Organize the Fibrils

In contrast to GAGs, which resist compressive forces, collagen fibrils form structures that resist tensile forces. The fibrils have various diameters and are organized in different ways in different tissues. In mammalian skin, for example, they are woven in a wickerwork pattern so that they resist tensile stress in multiple directions; leather consists of this material, suitably preserved. In tendons, collagen fibrils are organized in parallel bundles aligned along the major axis of tension. In mature bone and in the cornea, they are arranged in orderly plywoodlike layers, with the fibrils in each layer lying parallel to one another but nearly at right angles to the fibrils in the layers on either side. The same arrangement occurs in tadpole skin (Figure 19–41).

The connective-tissue cells themselves determine the size and arrangement of the collagen fibrils. The cells can express one or more genes for the different types of fibrillar collagen molecules. But even fibrils composed of the same mixture of collagens have different arrangements in different tissues. How is this achieved? Part of the answer is that cells can regulate the disposition of the collagen

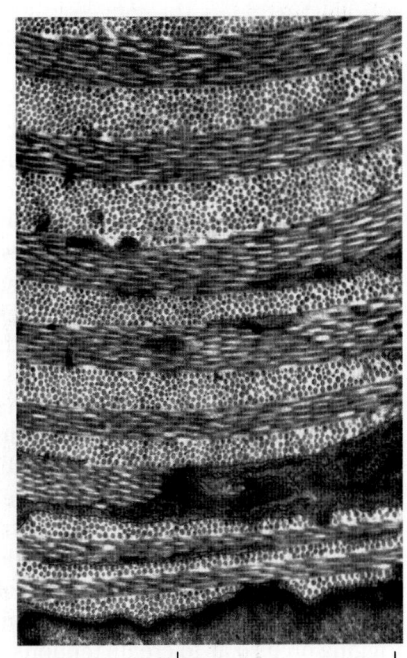

Figure 19–41 Collagen fibrils in the tadpole skin. This electron micrograph shows the plywoodlike arrangement of the fibrils: successive layers of fibrils are laid down nearly at right angles to each other. This organization is also found in mature bone and in the cornea. (Courtesy of Jerome Gross.)

5 μm

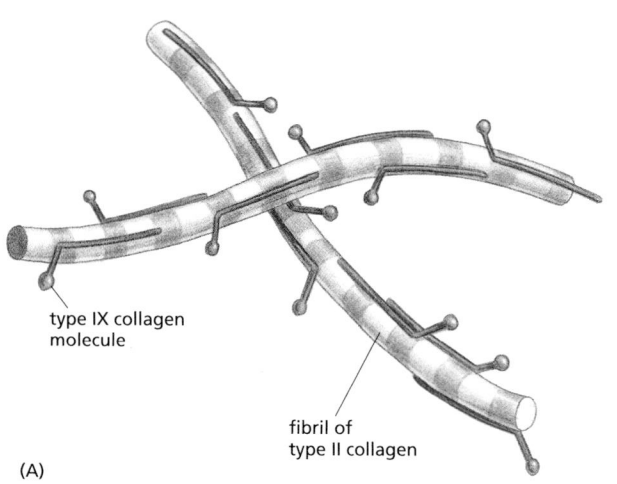

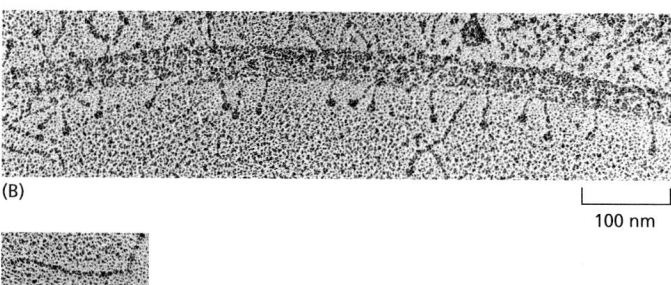

100 nm

(B)

(C)

Figure 19–42 **Type IX collagen.** (A) Type IX collagen molecules binding in a periodic pattern to the surface of a fibril containing type II collagen. (B) Electron micrograph of a rotary-shadowed type-II-collagen-containing fibril in cartilage, decorated by type IX collagen molecules. (C) An individual type IX collagen molecule. (B and C, © 1998 L. Vaughan et al. Originally published in *J. Cell Biol.* https://doi.org/10.1083/jcb.106.3.991. With permission from Rockefeller University Press.)

molecules after secretion by guiding collagen fibril formation near the plasma membrane. In addition, cells can influence this organization by secreting, along with their fibrillar collagens, different kinds and amounts of other matrix macromolecules. In particular, they secrete the fibrous protein *fibronectin*, as we discuss later, and this precedes the formation of collagen fibrils and helps guide their organization.

Fibril-associated collagens, such as types IX and XII collagens, are thought to be especially important in organizing collagen fibrils. They differ from fibrillar collagens in the following ways. First, their triple-stranded helical structure is interrupted by one or two short nonhelical domains, which makes the molecules more flexible than fibrillar collagen molecules. Second, they do not aggregate with one another to form fibrils in the extracellular space. Instead, they bind in a periodic manner to the surface of fibrils formed by the fibrillar collagens. Type IX molecules bind to type-II-collagen-containing fibrils in cartilage, the cornea, and the vitreous of the eye (**Figure 19-42**), whereas type XII molecules bind to type-I-collagen-containing fibrils in tendons and various other tissues.

Fibril-associated collagens are thought to mediate the interactions of collagen fibrils with one another and with other matrix macromolecules to help determine the organization of the fibrils in the matrix.

Cells Help Organize the Collagen Fibrils They Secrete by Exerting Tension on the Matrix

Cells interact with the extracellular matrix mechanically as well as chemically, and studies in culture suggest that the mechanical interaction can have dramatic effects on the architecture of connective tissue. Thus, when fibroblasts are mixed with a meshwork of randomly oriented collagen fibrils that form a gel in a culture dish, the fibroblasts tug on the meshwork, drawing in collagen from their surroundings and thereby causing the gel to contract to a small fraction of its initial volume. By similar activities, a cluster of fibroblasts surrounds itself with a capsule of densely packed and circumferentially oriented collagen fibers.

If two small pieces of embryonic tissue containing fibroblasts are placed far apart on a collagen gel, the intervening collagen becomes organized into a compact band of aligned fibers that connect the two explants (**Figure 19-43**). The fibroblasts subsequently migrate out from the explants along the aligned collagen fibers. Thus, the fibroblasts influence the alignment of the collagen fibers, and the collagen fibers in turn affect the distribution of the fibroblasts.

Fibroblasts may have a similar role in organizing the extracellular matrix inside the body. First they synthesize the collagen fibrils and deposit them in the correct orientation. Then they work on the matrix they have secreted, crawling over it and tugging on it so as to create tendons and ligaments and the tough, dense layers of connective tissue that surround and bind together most organs.

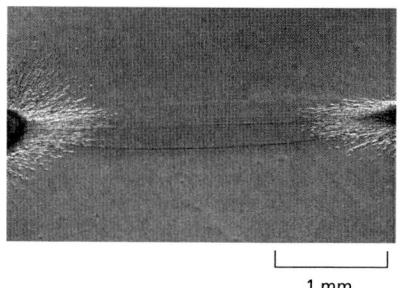

1 mm

Figure 19–43 **The shaping of the extracellular matrix by cells.** This micrograph shows a region between two pieces of embryonic chick heart (rich in fibroblasts as well as heart muscle cells) that were cultured on a collagen gel for 4 days. A dense tract of aligned collagen fibers has formed between the explants, presumably as a result of the fibroblasts in the explants tugging on the collagen. (From D. Stopak and A.K. Harris, *Dev. Biol.* 90:383–398, 1982. With permission from Academic Press.)

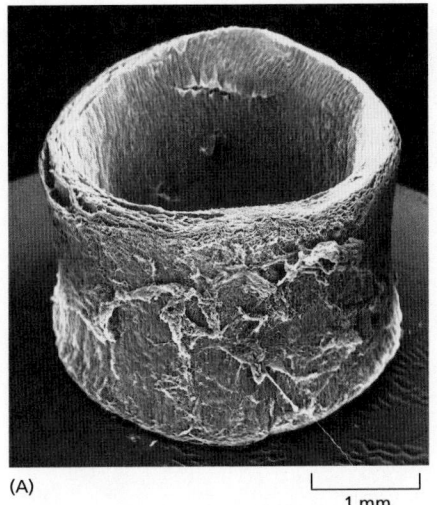 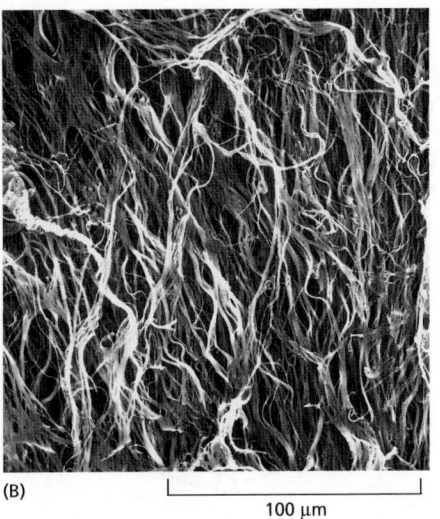

(A)

1 mm

(B)

100 μm

Figure 19–44 **Elastic fibers.** These scanning electron micrographs show (A) a low-power view of a segment of a dog's aorta and (B) a high-power view of the dense network of longitudinally oriented elastic fibers in the outer layer of the same blood vessel. All the other components have been digested away with enzymes and formic acid. (From K.S. Haas et al., *Anat. Rec.* 230:86–96, 1991. With permission from Wiley-Liss.)

Elastin Gives Tissues Their Elasticity

Many vertebrate tissues, such as skin, blood vessels, and lungs, need to be both strong and elastic in order to function. A network of **elastic fibers** in the extracellular matrix of these tissues gives them the resilience to recoil after transient stretch (**Figure 19–44**). Elastic fibers are at least five times more extensible than a rubber band of the same cross-sectional area. Long, inelastic collagen fibrils are interwoven with the elastic fibers to limit the extent of stretching and prevent the tissue from tearing.

The main component of elastic fibers is **elastin**, a highly hydrophobic protein (about 750 amino acids long), which, like collagen, is unusually rich in proline and glycine but, unlike collagen, is not glycosylated. Soluble *tropoelastin* (the biosynthetic precursor of elastin) is secreted into the extracellular space and assembled into elastic fibers close to the plasma membrane, generally in cell-surface infoldings. After secretion, the tropoelastin molecules become highly cross-linked to one another, generating an extensive network of elastin fibers and sheets.

The elastin protein is composed largely of two types of short segments that alternate along the polypeptide chain: hydrophobic segments, which are responsible for the elastic properties of the molecule; and alanine- and lysine-rich α-helical segments, which are cross-linked to adjacent molecules by covalent attachment of lysine residues. Each segment is encoded by a separate exon. There is still uncertainty concerning the conformation of elastin molecules in elastic fibers and how the structure of these fibers accounts for their rubberlike properties. However, it seems that parts of the elastin polypeptide chain, like the polymer chains in ordinary rubber, adopt a loose "random coil" conformation, and it is the random coil nature of the component molecules cross-linked into the elastic fiber network that allows the network to stretch and recoil like a rubber band (**Figure 19–45**).

Elastin is the dominant extracellular matrix protein in arteries, comprising 50% of the dry weight of the largest artery—the aorta (see Figure 19–44). Mutations in the elastin gene causing a deficiency of the protein in mice or humans result in narrowing of the aorta and other arteries and excessive proliferation of smooth muscle cells in the arterial wall. Apparently, the normal elasticity of an artery is required to restrain the proliferation of these cells.

Elastic fibers do not consist solely of elastin. The elastin core is covered with a sheath of *microfibrils*, each of which has a diameter of about 10 nm. The microfibrils appear before elastin in developing tissues and seem to provide scaffolding to guide elastin deposition. Arrays of microfibrils are elastic in their own right, and in some places they persist in the absence of elastin: they help to hold the lens in its place in the eye, for example. Microfibrils are composed of a number of distinct glycoproteins, including the large glycoprotein *fibrillin*, which binds to

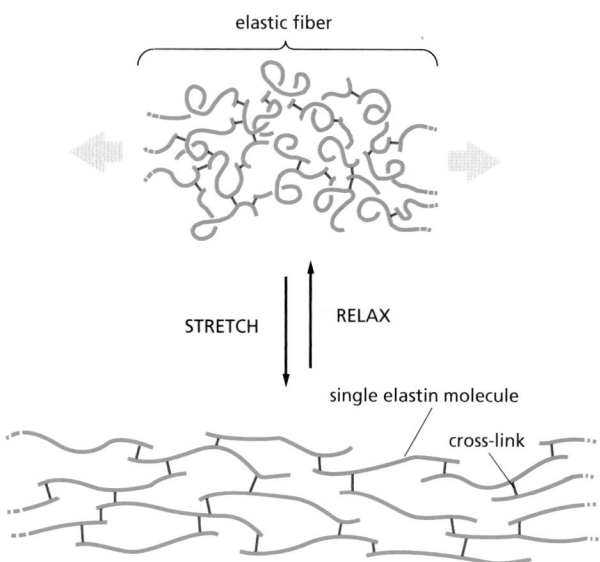

Figure 19–45 Stretching a network of elastin molecules. The molecules are joined together by covalent bonds *(red)* to generate a cross-linked network. In this model, each elastin molecule in the network can extend and contract in a manner resembling a random coil, so that the entire assembly can stretch and recoil like a rubber band.

elastin and is essential for the integrity of elastic fibers. Mutations in the fibrillin gene result in *Marfan's syndrome*, a relatively common human disorder. In the most severely affected individuals, the aorta is prone to rupture; other common effects include displacement of the lens and abnormalities of the skeleton and joints. Affected individuals are often unusually tall and lanky: Abraham Lincoln is suspected to have had the condition.

Fibronectin and Other Multidomain Glycoproteins Help Organize the Matrix

In addition to proteoglycans, collagens, and elastic fibers, the extracellular matrix contains a large and varied assortment of glycoproteins that typically have multiple domains, each with specific binding sites for other matrix macromolecules and for receptors on the surface of cells (**Figure 19–46**). These proteins therefore contribute to both organizing the matrix and helping cells attach to it. Like the proteoglycans, they also guide cell movements in developing tissues, by serving

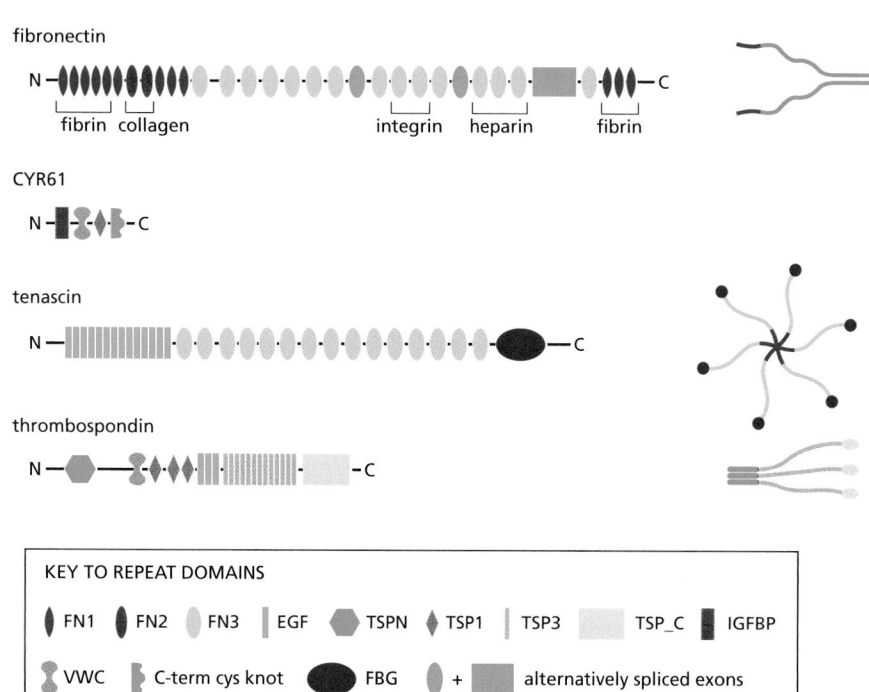

Figure 19–46 Complex glycoproteins of the extracellular matrix. Many matrix glycoproteins are large scaffold proteins containing multiple copies of specific protein-interaction domains. Each domain is folded into a discrete globular structure, and many such domains are arrayed along the protein like beads on a string. This diagram shows four representative proteins among the roughly 200 matrix glycoproteins that are found in mammals. Each protein contains multiple repeat domains, with the names listed in the key at the bottom. Fibronectin, for example, contains numerous copies of three different *fibronectin repeats* (types I–III, labeled here as FN1, FN2, and FN3). Two type III repeats near the C-terminus contain important binding sites for cell-surface integrins, whereas other FN repeats are involved in binding fibrin, collagen, and heparin, as indicated (see Figure 19–47). Other matrix proteins contain repeated sequences resembling those of epidermal growth factor (EGF), a major regulator of cell growth and proliferation; these repeats might serve a similar signaling function in matrix proteins. Other proteins contain domains, such as the insulin-like growth factor-binding protein (IGFBP) repeat, that bind and regulate the function of soluble growth factors. To add more structural diversity, many of these proteins are encoded by RNA transcripts that can be spliced in different ways, adding or removing exons, such as those in fibronectin. Finally, the scaffolding and regulatory functions of many matrix proteins are further expanded by assembly into multimeric forms, as shown at the right: fibronectin forms dimers linked at the C-termini, whereas tenascin and thrombospondin form N-terminally linked hexamers and trimers, respectively. Other domains include four repeats from thrombospondin (TSPN, TSP1, TSP3, TSP_C). VWC, von Willebrand type C; FBG, fibrinogen-like. (Adapted from R.O. Hynes and A. Naba, *Cold Spring Harb. Perspect. Biol.* 4:a004903, 2012. With permission from Cold Spring Harbor Laboratory Press.)

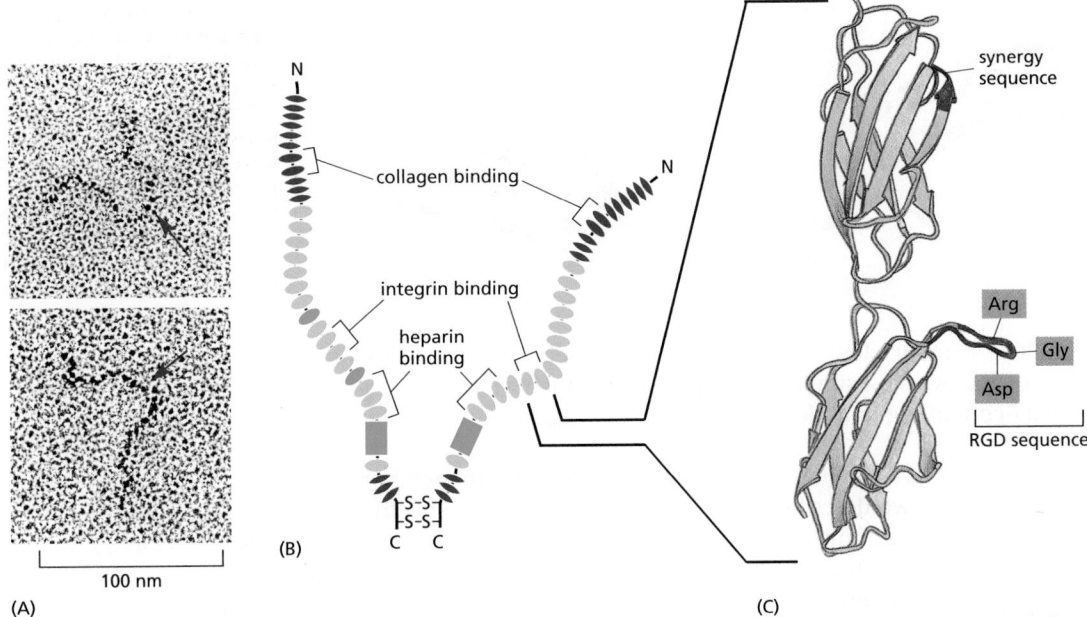

(B)

100 nm

(A)

(C)

as tracks along which cells can migrate or as repellents that keep cells out of forbidden areas. They can also bind and thereby influence the function of peptide growth factors and other small molecules produced by nearby cells.

The best-understood member of this class of matrix proteins is **fibronectin**, a large glycoprotein found in all vertebrates and important for many cell-matrix interactions. Mutant mice that are unable to make fibronectin die early in embryogenesis because their endothelial cells fail to form proper blood vessels. The defect is thought to result from abnormalities in the interactions of these cells with the surrounding extracellular matrix, which normally contains fibronectin.

Fibronectin is a dimer composed of two very large subunits joined by disulfide bonds at their C-terminal ends. Each subunit contains a series of small repeated domains, or modules, separated by short stretches of flexible polypeptide chain (**Figure 19–47**). Each domain is usually encoded by a separate exon, suggesting that the fibronectin gene, like the genes encoding many matrix proteins, evolved by multiple exon duplications. In the human genome, there is only one fibronectin gene, containing about 50 exons of similar size, but the transcripts can be spliced in different ways to produce multiple fibronectin isoforms (see Figure 19–46). The major repeat domain in fibronectin is called the **type III fibronectin repeat**, which is about 90 amino acids long and occurs at least 15 times in each subunit. This repeat is among the most common of all protein domains in vertebrates.

Fibronectin Binds to Integrins

One way to analyze a complex multifunctional protein molecule such as fibronectin is to synthesize individual regions of the protein and test their ability to bind other proteins. By these and other methods, it was possible to show that one region of fibronectin binds to collagen, another to proteoglycans, and another to specific integrins on the surface of various types of cells (see Figure 19–47B). Synthetic peptides corresponding to different segments of the integrin-binding domain were then used to show that binding depends on a specific tripeptide sequence (*Arg-Gly-Asp*, or *RGD*) that is found in one of the type III repeats (see Figure 19–47C). Even very short peptides containing this **RGD sequence** can compete with fibronectin for the binding site on cells, thereby inhibiting the attachment of the cells to a fibronectin matrix.

Several extracellular proteins besides fibronectin also have an RGD sequence that mediates cell-surface binding. Many of these proteins are components of the extracellular matrix, while others are involved in blood clotting. Peptides

Figure 19–47 **The structure of a fibronectin dimer.** (A) Electron micrographs of individual fibronectin dimer molecules shadowed with platinum; *red arrows* mark the joined C-termini. (B) The two polypeptide chains are similar but generally not identical (being made from the same gene but from differently spliced mRNAs). They are joined by two disulfide bonds near the C-termini. Each chain is almost 2500 amino acids long and is folded into multiple domains (see Figure 19–46). As indicated, some domains are specialized for binding to a particular molecule. For simplicity, not all of the known binding sites are shown. (C) The three-dimensional structure of the ninth and tenth type III fibronectin repeats, as determined by x-ray crystallography. Both the Arg-Gly-Asp (RGD) and the "synergy" sequences shown in *red* are important for binding to integrins on cell surfaces. (A, from J. Engel et al., *J. Mol. Biol.* 150:97–120, 1981. With permission from Academic Press; C, from Daniel J. Leahy, *Annu. Rev. Cell Dev. Biol.* 13:363–393, 1997. With permission from Annual Reviews.)

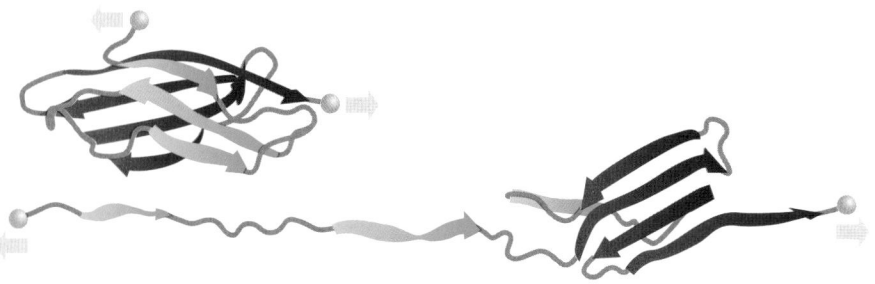

Figure 19–48 Tension-sensing by fibronectin. Some type III fibronectin repeats are thought to unfold when fibronectin is stretched. The unfolding exposes cryptic binding sites that interact with other fibronectin molecules resulting in the formation of fibronectin filaments like those shown in Figure 19–49. (From V. Vogel and M. Sheetz, *Nat. Rev. Mol. Cell Biol.* 7:265–275, published 2006 by Nature Publishing Group. Reproduced with permission of SNCSC.)

containing the RGD sequence have been useful in the development of anti-clotting drugs. Some snakes use a similar strategy to cause their victims to bleed: they secrete RGD-containing anti-clotting proteins called *disintegrins* into their venom.

The cell-surface receptors that bind RGD-containing proteins are members of the integrin family, which we describe in detail later. Each integrin specifically recognizes its own small set of matrix molecules, indicating that tight binding requires more than just the RGD sequence. Moreover, RGD sequences are not the only sequence motifs used for binding to integrins: many integrins recognize and bind to other motifs instead.

Tension Exerted by Cells Regulates the Assembly of Fibronectin Fibrils

Fibronectin can exist both in a soluble form, circulating in the blood and other body fluids, and as insoluble *fibronectin fibrils*, in which fibronectin dimers are cross-linked to one another by additional disulfide bonds and form part of the extracellular matrix. Unlike fibrillar collagen molecules, however, which can self-assemble into fibrils in a test tube, fibronectin molecules assemble into fibrils only on the surface of cells, and only where those cells possess appropriate fibronectin-binding proteins—in particular, integrins. The integrins provide a linkage from the fibronectin outside the cell to the actin cytoskeleton inside it. The linkage transmits tension to the fibronectin molecules—provided that they also have an attachment to some other structure—and stretches them, exposing cryptic binding sites in the fibronectin molecules (**Figure 19–48**). This allows them to bind directly to one another and to recruit additional fibronectin molecules to form a fibril (**Figure 19–49**). This dependence on tension and interaction with cell surfaces ensures that fibronectin fibrils assemble where there is a mechanical need for them and not in inappropriate locations such as the bloodstream.

Many other extracellular matrix proteins contain multiple copies of the type III fibronectin repeat (see Figure 19–46), and it is possible that tension exerted on these proteins also uncovers cryptic binding sites and thereby influences their behavior.

The Basal Lamina Is a Specialized Form of Extracellular Matrix

Thus far in this section we have reviewed the general principles underlying the structure and function of the major classes of extracellular matrix components. We now describe how some of these components are assembled into a specialized type of extracellular matrix called the **basal lamina** (also known as the **basement membrane**). This exceedingly thin, tough, flexible sheet of matrix molecules is an essential underpinning of all epithelia. Although small in volume, it has a critical role in the architecture of the body. Like the cadherins, it seems to be one of the defining features common to all multicellular animals, and it seems to have appeared very early in their evolution. The major molecular components of the basal lamina are among the most ancient extracellular matrix macromolecules.

Basal laminae are typically 40–120 nm thick. A sheet of basal lamina not only lies beneath epithelial cells but also surrounds individual muscle cells, fat cells,

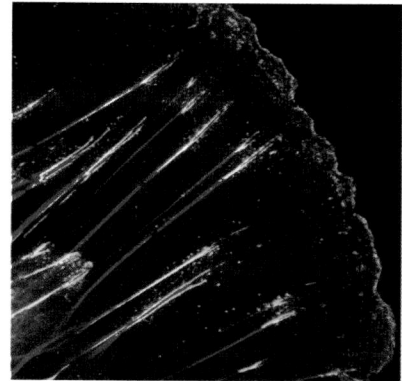

Figure 19–49 Organization of fibronectin into fibrils at the cell surface. This fluorescence micrograph shows the front end of a migrating mouse fibroblast. Extracellular fibronectin is stained *green* and intracellular actin filaments are stained *red*. The fibronectin is initially present as small dotlike aggregates near the leading edge of the cell. It accumulates at focal adhesions (sites of anchorage of actin filaments, discussed later) and becomes organized into fibrils parallel to the actin filaments. Integrin molecules spanning the cell membrane link the fibronectin outside the cell to the actin filaments inside it (see Figure 19–55). Tension exerted on the fibronectin molecules through this linkage is thought to stretch them, exposing binding sites that promote fibril formation. (Courtesy of Roumen Pankov and Kenneth Yamada.)

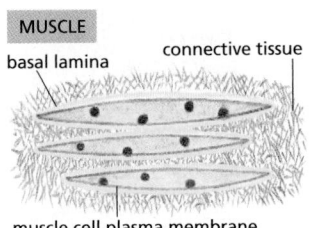

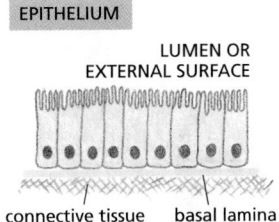

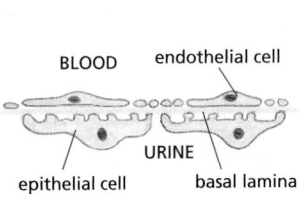

and Schwann cells (which wrap around peripheral nerve cell axons to form myelin). The basal lamina thus separates these cells and epithelia from the underlying or surrounding connective tissue and forms the mechanical connection between them. In other locations, such as the kidney glomerulus, a basal lamina lies between two cell sheets and functions as a selective filter (Figure 19–50). Basal laminae have more than simple structural and filtering roles, however. They are able to determine cell polarity; influence cell metabolism; organize the proteins in adjacent plasma membranes; promote cell survival, proliferation, or differentiation; and serve as highways for cell migration.

The mechanical role is nevertheless essential. In the skin, for example, the epithelial outer layer—the epidermis—depends on the strength of the basal lamina to keep it attached to the underlying connective tissue—the dermis. In people with genetic defects in certain basal lamina proteins or in a special type of collagen that anchors the basal lamina to the underlying connective tissue, the epidermis becomes detached from the dermis. This causes a blistering disease called *junctional epidermolysis bullosa*, a severe and sometimes lethal condition.

Laminin and Type IV Collagen Are Major Components of the Basal Lamina

The basal lamina is synthesized by the cells on each side of it: the epithelial cells contribute one set of basal lamina components, while cells of the underlying bed of connective tissue (called the *stroma*, Greek for "bedding") contribute another set (Figure 19–51). Although the precise composition of the mature basal lamina varies from tissue to tissue and even from region to region in the same lamina, it

Figure 19–50 **Three ways in which basal laminae are organized.** Basal laminae *(yellow)* surround certain cells (such as skeletal muscle cells), underlie epithelia, and are interposed between two cell sheets (as in the kidney glomerulus). Note that, in the kidney glomerulus, both cell sheets have gaps in them, and the basal lamina has a filtering as well as a supportive function, helping to determine which molecules will pass into the urine from the blood. The filtration also depends on other protein-based structures, called *slit diaphragms*, that span the intercellular gaps in the epithelial sheet.

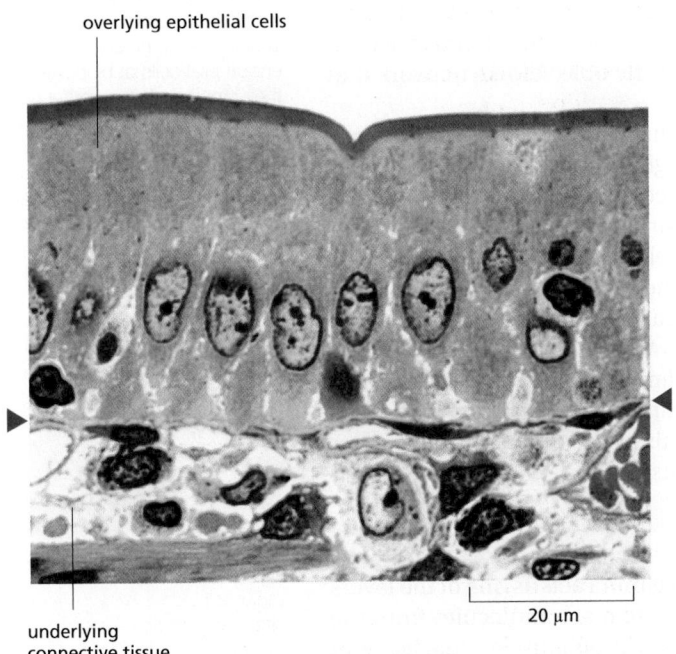

Figure 19–51 **The basal lamina supports a sheet of epithelial cells.** Light micrograph of the epithelial sheet that lines the small intestine. The sheet of columnar cells sits on a thin mat-like structure, the basal lamina *(red* arrowheads), which is woven from type IV collagen and laminin proteins. A network of other collagen fibrils and fibers in the underlying connective tissue interacts with the lower face of the lamina. (Jose Luis Calvo/Shutterstock)

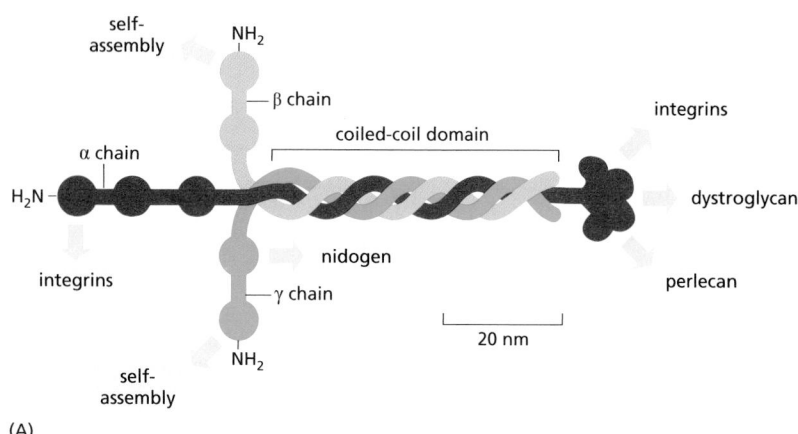

(A)

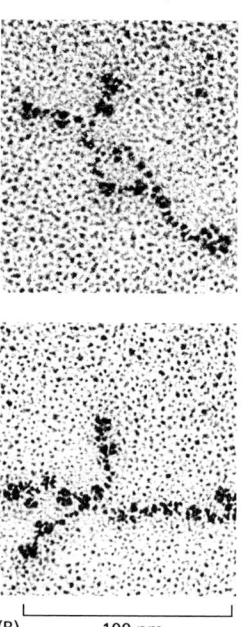

(B)   100 nm

Figure 19–52 The structure of laminin.
(A) The best-understood family member
is laminin-111, shown here with some
of its binding sites for other molecules
(yellow boxes). Laminins are multidomain
glycoproteins composed of three
polypeptides (α, β, and γ) that are disulfide-
bonded into an asymmetric crosslike
structure. Each of the polypeptide chains
is more than 1500 amino acids long. Five
types of α chains, four types of β chains,
and three types of γ chains are known, and
various combinations of these subunits can
assemble to form a large variety of different
laminins, which are named according
to numbers assigned to each of their
three subunits: laminin-111, for example,
contains α1, β1, and γ1 subunits. Each
isoform tends to have a specific tissue
distribution: laminin-332 is found in skin,
laminin-211 in muscle, and laminin-411 in
endothelial cells of blood vessels. Through
their binding sites for other proteins, laminin
molecules play a central part in organizing
basal laminae and anchoring them to
cells. (B) Electron micrographs of laminin
molecules shadowed with platinum.
(B, from J. Engel et al., *J. Mol. Biol.*
150:97–120, 1981. With permission
from Academic Press.)

typically contains the glycoproteins *laminin, type IV collagen,* and *nidogen,* along with the proteoglycan *perlecan.* Other common basal lamina components are fibronectin and *type XVIII collagen* (an atypical member of the collagen family, forming the core protein of a proteoglycan).

Laminin is the primary organizer of the sheet structure, and, early in development, basal laminae consist mainly of laminin molecules. Laminins comprise a large family of proteins, each composed of three long polypeptide chains (α, β, and γ) held together by disulfide bonds and arranged in the shape of an asymmetric bouquet, like a bunch of three flowers whose stems are twisted together at the foot but whose heads remain separate (**Figure 19–52**). These heterotrimers can self-assemble *in vitro* into a network, largely through interactions between their heads, although interaction with cells is needed to organize the network into an orderly sheet. Since there are several isoforms of each type of chain, and these can associate in different combinations, many different laminins can be produced, creating basal laminae with distinctive properties. The laminin γ1 chain is, however, a component of most laminin heterotrimers; mice lacking it die during embryogenesis because they are unable to make basal laminae.

Type IV collagen is a second essential component of mature basal laminae, and it, too, exists in several isoforms. Like the *fibrillar collagens* that constitute the bulk of the protein in connective tissues such as bone or tendon, type IV collagen molecules consist of three separately synthesized long protein chains that twist together to form a ropelike superhelix; however, they differ from the fibrillar collagens in that the triple-stranded helical structure is interrupted in more than 20 regions, allowing multiple bends. Type IV collagen molecules interact via their terminal domains to assemble extracellularly into a flexible, feltlike network that gives the basal lamina tensile strength.

Laminin and type IV collagen interact with other basal lamina components, such as the glycoprotein nidogen and the proteoglycan perlecan, resulting in a highly cross-linked network of proteins and proteoglycans (**Figure 19–53**). The laminin molecules that generate the initial sheet structure first join to each other while bound to receptors on the surface of the cells that produce laminin. The cell-surface receptors are primarily members of the integrin family, but another important type of laminin receptor is *dystroglycan,* a proteoglycan with a core protein that spans the cell membrane, dangling its GAG chains in the extracellular space. Together, these receptors organize basal lamina assembly: they hold the laminin molecules by their feet, leaving the laminin heads positioned to interact so as to form a two-dimensional network. This laminin network then coordinates the assembly of the other basal lamina components.

Basal Laminae Have Diverse Functions

In the kidney glomerulus, an unusually thick basal lamina acts as one of the layers of a molecular filter, helping to prevent the passage of macromolecules from the blood into the urine as urine is formed (see Figure 19–50). The proteoglycan in

(A)

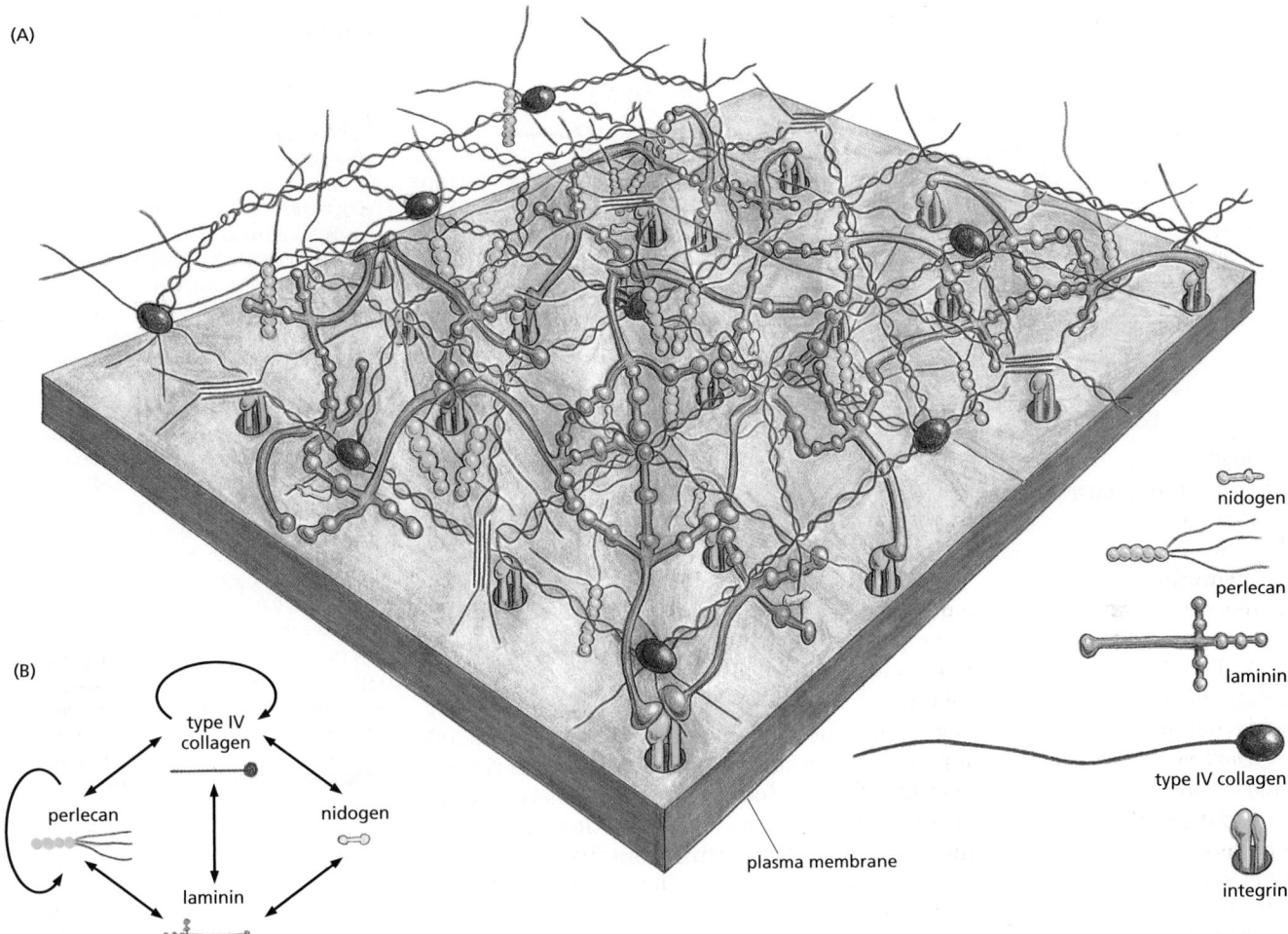

nidogen

perlecan

laminin

type IV collagen

integrin

plasma membrane

(B)

type IV
collagen

perlecan

nidogen

laminin

Figure 19–53 **A model of the molecular structure of a basal lamina.** (A) The basal lamina is formed by specific interactions (B) between the proteins laminin, type IV collagen, and nidogen, and the proteoglycan perlecan. *Arrows* in (B) connect molecules that can bind directly to each other. There are various isoforms of type IV collagen and laminin, each with a distinctive tissue distribution. Transmembrane laminin receptors (integrins and dystroglycan) in the plasma membrane are thought to organize the assembly of the basal lamina; only the integrins are shown. (Based on H. Colognato and P.D. Yurchenco, *Dev. Dyn.* 218:213–234, 2000. With permission from Wiley-Liss.)

the basal lamina is important for this function: when its GAG chains are removed by specific enzymes, the filtering properties of the lamina are destroyed. Type IV collagen also has a role: in a human hereditary kidney disorder (*Alport syndrome*), mutations in a type IV collagen gene result in an irregularly thickened and dysfunctional glomerular filter. Laminin mutations, too, can disrupt the function of the kidney filter, but in a different way—by interfering with the differentiation of the cells that contact it and support it.

The basal lamina can act as a selective barrier to the movement of cells, as well as a filter for molecules. The lamina beneath an epithelium, for example, usually prevents fibroblasts in the underlying connective tissue from making contact with the epithelial cells. It does not, however, stop macrophages, lymphocytes, or nerve processes from passing through it, using specialized protease enzymes to cut a hole for their transit. The basal lamina is also important in tissue regeneration after injury. When cells in tissues such as muscles, nerves, and epithelia are damaged or killed, the basal lamina often survives and provides a scaffold along which regenerating cells can migrate. In this way, the original tissue architecture is readily reconstructed.

A particularly striking example of the role of the basal lamina in regeneration comes from studies of the *neuromuscular junction*, the site where the nerve terminals of a motor neuron form a chemical synapse with a skeletal muscle cell (discussed in Chapter 11). In vertebrates, the basal lamina that surrounds the muscle cell separates the nerve and muscle cell plasma membranes at the synapse, and the synaptic region of the lamina has a distinctive chemical character,

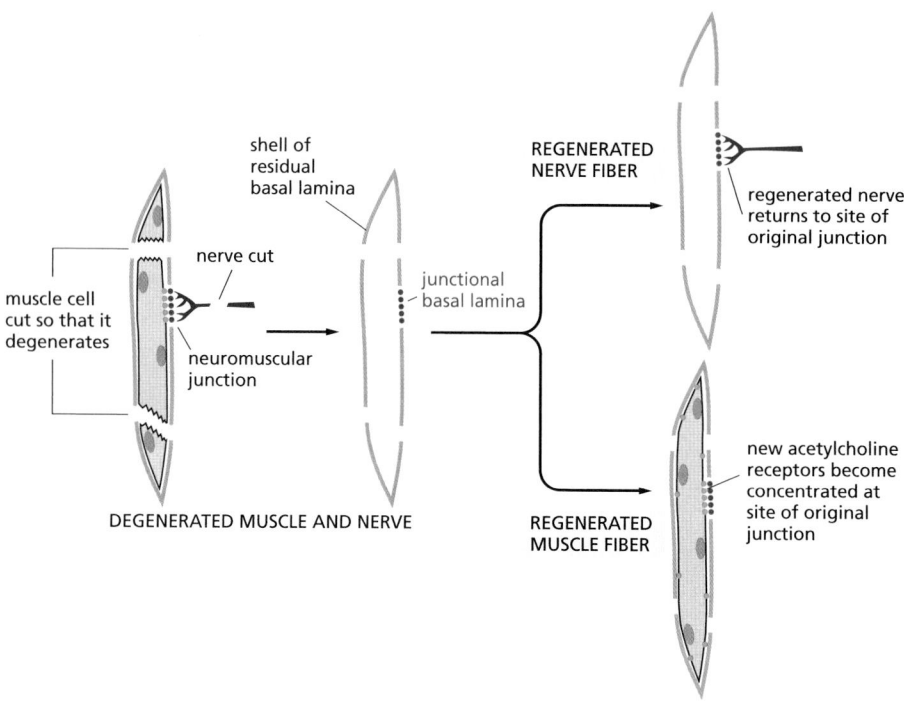

Figure 19–54 Regeneration experiments demonstrating the special character of the junctional basal lamina at a neuromuscular junction. If a frog muscle and its motor nerve are destroyed, the basal lamina around each muscle cell remains intact and the sites of the old neuromuscular junctions are still recognizable. When the nerve, but not the muscle, is allowed to regenerate *(upper right)*, the junctional basal lamina directs the regenerating nerve to the original synaptic site. When the muscle, but not the nerve, is allowed to regenerate *(lower right)*, the junctional basal lamina causes newly made acetylcholine receptors *(blue)* to accumulate at the original synaptic site. These experiments show that the junctional basal lamina controls the localization of synaptic components on both sides of the lamina. Some of the molecules responsible for these effects have been identified. Motor neuron axons, for example, deposit agrin in the junctional basal lamina, where it regulates the assembly of acetylcholine receptors and other proteins in the junctional plasma membrane of the muscle cell. Reciprocally, muscle cells deposit a particular isoform of laminin in the junctional basal lamina, and this molecule is likely to interact with specific ion channels on the presynaptic membrane of the neuron.

with special isoforms of type IV collagen and laminin and a proteoglycan called *agrin*. After a nerve or muscle injury, the basal lamina at the synapse has a central role in reconstructing the synapse at the correct location (**Figure 19–54**). Defects in components of the basal lamina at the synapse are responsible for some forms of muscular dystrophy, in which muscles develop normally but then degenerate later in life.

Cells Have to Be Able to Degrade Matrix, as Well as Make It

The ability of cells to degrade and destroy extracellular matrix is as important as their ability to make it and bind to it. Rapid matrix degradation is required in processes such as tissue repair, and even in the seemingly static extracellular matrix of adult animals there is a slow, continuous turnover, with matrix macromolecules being degraded and resynthesized. This allows bone, for example, to be remodeled so as to adapt to changes in the stresses on it.

From the point of view of individual cells, the ability to cut through matrix is crucial in two ways: it enables them to divide while embedded in matrix, and it enables them to travel through it. Cells in connective tissues generally need to be able to stretch out in order to divide. If a cell lacks the enzymes needed to degrade the surrounding matrix, it is strongly inhibited from dividing, as well as being hindered from migrating.

Localized degradation of matrix components is also required wherever cells have to escape from confinement by a basal lamina. It is needed during normal branching growth of epithelial structures such as glands, for example, to allow the population of epithelial cells to increase, and needed also when white blood cells migrate across the basal lamina of a blood vessel into tissues in response to infection or injury. Matrix degradation is important both for the spread of cancer cells through the body and for their ability to proliferate in the tissues that they invade (discussed in Chapter 20).

In general, matrix components are degraded by extracellular proteolytic enzymes (proteases) that act close to the cells that produce them. Many of these proteases belong to one of two general classes. The largest group, with about 50 members in vertebrates, is the **matrix metalloproteases**, which depend on bound Ca^{2+} or Zn^{2+} for activity. The second group is the **serine proteases**, which have a highly reactive serine in their active site. Together, metalloproteases and serine proteases cooperate to degrade matrix proteins such as collagen, laminin, and fibronectin. Some metalloproteases, such as the *collagenases*, are highly specific, cleaving particular proteins at a small number of sites. In this way, the structural integrity of the matrix is largely retained, while the limited amount of proteolysis that occurs is sufficient for cell migration. Other metalloproteases may be less specific, but, because they are anchored to the plasma membrane, they can act just where they are needed; it is this type of matrix metalloprotease that is crucial for a cell's ability to divide when embedded in matrix.

Clearly, the activities of the proteases that degrade the matrix must be tightly controlled, if the fabric of the body is not to collapse in a heap. Numerous mechanisms are therefore employed to ensure that matrix proteases are activated only at the correct time and place. Protease activity is generally confined to the cell surface by specific anchoring proteins, by membrane-associated activators, and by the production of specific protease inhibitors in regions where protease activity is not needed.

Matrix Proteoglycans and Glycoproteins Regulate the Activities of Secreted Proteins

The physical properties of extracellular matrix are important for its fundamental roles as a scaffold for tissue structure and as a substrate for cell anchorage and migration. The matrix also has an important impact on cell signaling. Cells communicate with each other by secreting signal molecules that diffuse through the extracellular fluid to influence other cells (discussed in Chapter 15). En route to their targets, the signal molecules encounter the tightly woven meshwork of the extracellular matrix, which contains a high density of negative charges and protein-interaction domains that can interact with the signal molecules, thereby altering their function in a variety of ways.

The highly charged heparan sulfate chains of proteoglycans, for example, interact with numerous secreted signal molecules, including *fibroblast growth factors* (FGFs) and *vascular endothelial growth factor* (VEGF), which (among other effects) stimulate a variety of cell types to proliferate. By providing a dense array of growth factor binding sites, proteoglycans are thought to generate large local reservoirs of these factors, limiting their diffusion and focusing their actions on nearby cells. Similarly, proteoglycans might help generate steep growth factor gradients in an embryo, which can be important in the patterning of tissues during development. FGF activity can also be enhanced by proteoglycans, which oligomerize the FGF molecules, enabling them to cross-link and activate their cell-surface receptors more effectively.

The importance of proteoglycans as regulators of the distribution and activity of signal molecules is illustrated by the severe developmental defects that can occur when specific proteoglycans are inactivated by mutation. In *Drosophila*, for example, the function of several signal proteins during development is governed by interactions with the membrane-associated proteoglycans *Dally* and *Dally-like*. These members of the *glypican* family are thought to concentrate signal proteins in specific locations and act as co-receptors that collaborate with the conventional cell-surface receptor proteins; as a result, they promote signaling in the correct location and prevent it in the wrong locations. In the *Drosophila* ovary, for example, Dally is partly responsible for the restricted localization and function of a signaling protein called Dpp, which blocks differentiation of the germline stem cells: when the gene encoding Dally is mutated, Dpp activity is greatly reduced and oocyte development is abnormal.

Several matrix proteins also interact with signal proteins. The type IV collagen of basal laminae interacts with Dpp in *Drosophila*, for example. Fibronectin contains a type III fibronectin repeat that interacts with VEGF, and another domain that interacts with another growth factor called hepatocyte growth factor (HGF), thereby promoting the activities of these factors. As discussed earlier, many matrix glycoproteins contain extensive arrays of binding domains, and the arrangement of these domains is likely to influence the presentation of signal proteins to their target cells (see Figure 19–46).

Finally, many matrix glycoproteins contain domains that bind directly to specific cell-surface receptors, thereby generating signals that influence the behavior of the cells, as we describe in the next section.

Summary

Cells are embedded in an intricate extracellular matrix, which not only binds the cells together but also influences their survival, development, shape, polarity, and migratory behavior. The matrix contains various protein fibers interwoven in a network of glycosaminoglycan (GAG) chains. GAGs are negatively charged polysaccharide chains that (except for hyaluronan) are covalently linked to protein to form proteoglycan molecules. GAGs attract water and occupy a large volume of extracellular space. Proteoglycans are also found on the surface of cells, where they often function as co-receptors to help cells respond to secreted signal proteins. Fiber-forming proteins give the matrix strength and resilience. The fibrillar collagens (types I, II, III, V, and XI) are ropelike, triple-stranded helical molecules that aggregate into long fibrils in the extracellular space, thereby providing tensile strength. They also form structures to which cells can be anchored, often via large multidomain glycoproteins, such as laminin and fibronectin, that bind to integrins on the cell surface. Elasticity is provided by elastin molecules, which form an extensive cross-linked network of fibers and sheets that can stretch and recoil.

The basal lamina is a specialized form of extracellular matrix that underlies epithelial cells or is wrapped around certain other cell types, such as muscle cells. Basal laminae are organized on a framework of laminin molecules, which are linked together by their side-arms and bind to integrins and other receptors in the basal plasma membrane of overlying epithelial cells. Type IV collagen molecules, together with the protein nidogen and the large heparan sulfate proteoglycan perlecan, assemble into a sheetlike mesh that is an essential component of all mature basal laminae. Basal laminae provide mechanical support for epithelia; they form the interface and attachment between epithelia and connective tissue; they serve as filters in the kidney; they act as barriers to keep cells in their proper compartments; they influence cell polarity and cell differentiation; and they guide cell migration during development and tissue regeneration.

CELL–MATRIX JUNCTIONS

Cells make extracellular matrix, organize it, and degrade it. The matrix in its turn exerts powerful influences on the cells. The influences are exerted chiefly through transmembrane cell adhesion proteins that act as *matrix receptors.* These proteins tie the matrix outside the cell to the cytoskeleton inside it, but their role goes far beyond simple passive mechanical attachment. Through them, components of the matrix can affect almost any aspect of a cell's behavior. The matrix receptors have a crucial role in epithelial cells, mediating their interactions with the basal lamina beneath them. They are no less important in connective-tissue cells, mediating the cells' interactions with the matrix that surrounds them.

Several types of molecules can function as matrix receptors or co-receptors, including the transmembrane proteoglycans. But the principal receptors on animal cells for binding most extracellular matrix proteins are the integrins. Like the cadherins and the key components of the basal lamina, integrins are part of the fundamental architectural toolkit that is characteristic of multicellular animals. The members of this large family of homologous transmembrane adhesion molecules have a remarkable ability to transmit signals in both directions across the plasma membrane. The binding of a matrix component to an integrin can send a message into the interior of the cell, and conditions in the cell interior can send a signal outward to control binding of the integrin to the matrix. Tension applied to an integrin can cause it to tighten its grip on intracellular and extracellular structures, and loss of tension can loosen its hold, so that molecular signaling complexes fall apart on either side of the membrane. In this way, integrins can serve not only to transmit mechanical and molecular signals, but also to convert one type of signal into the other.

Integrins Are Transmembrane Heterodimers That Link the Extracellular Matrix to the Cytoskeleton

There are many varieties of integrins, but they all conform to a common plan. An integrin molecule is composed of two noncovalently associated glycoprotein subunits called α and β. Both subunits span the cell membrane, with short intracellular C-terminal tails and large N-terminal extracellular domains (**Figure 19–55**). The extracellular domains bind to specific amino acid sequence motifs in extracellular matrix proteins or, in some cases, in proteins on the surfaces of other cells. The best-understood binding site for integrins is the RGD sequence mentioned earlier (see Figure 19–47), which is found in fibronectin and other extracellular matrix proteins. Some integrins bind a Leu-Asp-Val (LDV) sequence in fibronectin and other proteins. Additional integrin-binding sequences, as yet poorly defined, exist in laminins and collagens.

Humans contain 24 types of integrins, formed from the products of 8 different β-chain genes and 18 different α-chain genes, dimerized in different combinations. Each integrin dimer has distinctive properties and functions. Moreover, because the same integrin molecule in different cell types can have different ligand-binding specificities, it seems that additional cell-type-specific factors can interact with integrins to modulate their binding activity. The binding of integrins to their matrix ligands is also affected by the concentration of Ca^{2+} and Mg^{2+} in the extracellular medium, reflecting the presence of divalent cation-binding domains in the α and β subunits. The divalent cations can influence both the affinity and the specificity of the binding of an integrin to its extracellular ligands.

The intracellular portion of an integrin dimer binds to a complex of several different proteins, which together form a linkage to the cytoskeleton. For all but one of the 24 varieties of human integrins, this intracellular linkage is to actin filaments. These linkages depend on proteins that assemble at the short cytoplasmic tails of the integrin subunits (see Figure 19–55). A large adaptor protein called *talin* is a component of the linkage in many cases, but numerous additional proteins are also involved. Like the actin-linked cell–cell junctions formed by cadherins, the actin-linked cell–matrix junctions formed by integrins may be small, inconspicuous, and transient, or large, prominent, and durable. Examples of the latter are the *focal adhesions* that form when fibroblasts have sufficient time to establish strong attachments to the rigid surface of a culture dish, and the *myotendinous junctions* that attach muscle cells to their tendons.

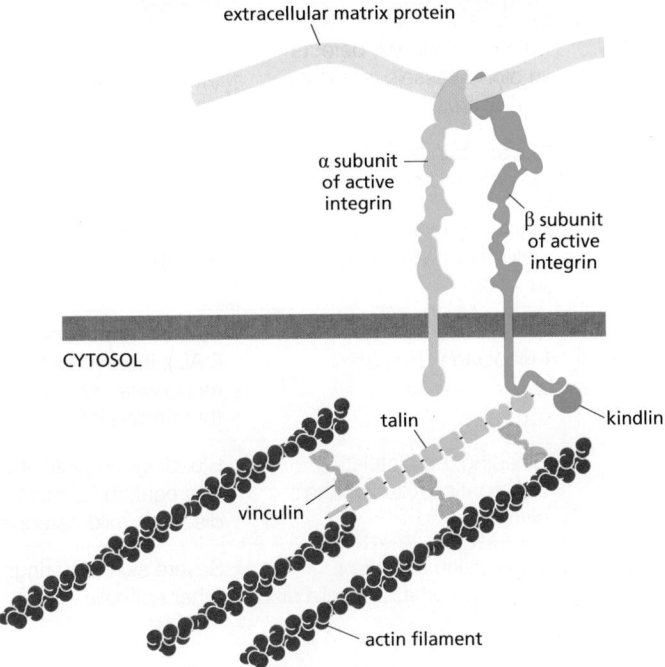

Figure 19–55 **The subunit structure of an active integrin molecule, linking extracellular matrix to the actin cytoskeleton.** The N-terminal heads of the integrin chains attach directly to an extracellular protein such as fibronectin; the C-terminal intracellular tail of the integrin β subunit binds to adaptor proteins that interact with filamentous actin. The best-understood adaptor is a giant protein called talin, which contains a string of multiple domains for binding actin and other proteins, such as vinculin, that help reinforce and regulate the linkage to actin filaments. One end of talin binds to a specific site on the integrin β subunit cytoplasmic tail; other regulatory proteins, such as kindlin, bind at another site on the tail.

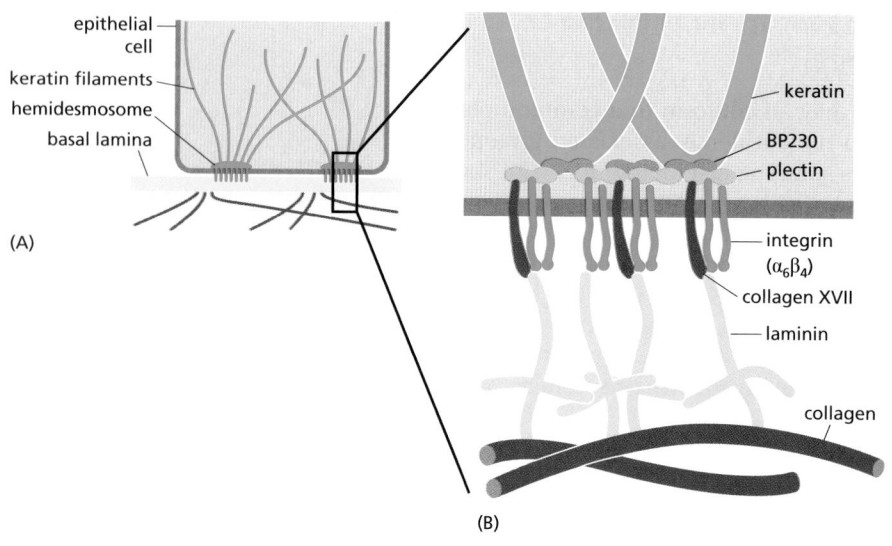

epithelial cell
keratin filaments
hemidesmosome
basal lamina

(A)

keratin
BP230
plectin
integrin (α₆β₄)
collagen XVII
laminin

collagen

(B)

Figure 19–56 Hemidesmosomes.
(A) Hemidesmosomes spot-weld epithelial cells to the basal lamina, linking laminin outside the cell to keratin filaments inside it. (B) Molecular components of a hemidesmosome. A specialized integrin ($\alpha_6\beta_4$ integrin) spans the membrane, attaching to keratin filaments intracellularly via adaptor proteins called plectin and BP230, and to laminin extracellularly. The adhesive complex also contains, in parallel with the integrin, an unusual collagen family member known as collagen type XVII; this has a membrane-spanning domain attached to its extracellular collagenous portion. Defects in any of these components can give rise to a blistering disease of the skin. One such disease, called *bullous pemphigoid*, is an autoimmune disease in which the immune system develops antibodies against collagen XVII or BP230.

In epithelia, the most prominent cell–matrix attachment sites are the hemidesmosomes, where a specific type of integrin anchors the cells to laminin in the basal lamina. Here, uniquely, the intracellular attachment is to keratin intermediate filaments, via the intracellular adaptor proteins plectin and BP230 (**Figure 19–56**).

Integrin Defects Are Responsible for Many Genetic Diseases

Although there is some overlap in the activities of the different integrins—at least five bind laminin, for example—it is the diversity of integrin functions that is more remarkable. **Table 19–3** lists some varieties of integrins and the problems that result when individual integrin α or β chains are defective.

The β_1 subunit forms dimers with at least 12 distinct α subunits and is found on almost all vertebrate cells: $\alpha_5\beta_1$ is a fibronectin receptor and $\alpha_6\beta_1$ is a laminin

TABLE 19–3 Some Types of Integrins

Integrin	Ligand*	Distribution	Phenotype when α subunit is mutated	Phenotype when β subunit is mutated
$\alpha_5\beta_1$	Fibronectin	Ubiquitous	Death of embryo; defects in blood vessels, somites, neural crest	Early death of embryo (at implantation)
$\alpha_6\beta_1$	Laminin	Ubiquitous	Severe skin blistering; defects in other epithelia also	Early death of embryo (at implantation)
$\alpha_7\beta_1$	Laminin	Muscle	Muscular dystrophy; defective myotendinous junctions	Early death of embryo (at implantation)
$\alpha_L\beta_2$ (LFA1)	Ig superfamily counterreceptors (ICAM1)	White blood cells	Impaired recruitment of leucocytes	Leukocyte adhesion deficiency (LAD); impaired inflammatory responses; recurrent life-threatening infections
$\alpha_{IIb}\beta_3$	Fibrinogen	Platelets	Bleeding; no platelet aggregation (Glanzmann's disease)	Bleeding; no platelet aggregation (Glanzmann's disease); mild osteopetrosis
$\alpha_6\beta_4$	Laminin	Hemidesmosomes in epithelia	Severe skin blistering; defects in other epithelia also	Severe skin blistering; defects in other epithelia also

*Not all ligands are listed.

receptor on many types of cells. Mutant mice that cannot make any β_1 integrins die early in embryonic development. Mice that are only unable to make the α_7 subunit (the partner for β_1 in muscle) survive but develop muscular dystrophy (as do mice that cannot make the laminin ligand for the $\alpha_7\beta_1$ integrin).

The β_2 subunit forms dimers with at least four types of α subunit and is expressed exclusively on the surface of white blood cells, where it has an essential role in enabling these cells to fight infection. The β_2 integrins mainly mediate cell–cell rather than cell–matrix interactions, binding to specific ligands on another cell, such as an endothelial cell. The ligands are members of the Ig superfamily of cell–cell adhesion molecules. We have already described an example earlier in the chapter: an integrin of this class ($\alpha_L\beta_2$, also known as LFA1) on white blood cells enables them to attach firmly to the Ig family protein ICAM1 on vascular endothelial cells at sites of infection (see Figure 19–28B). People with the genetic disease called *leukocyte adhesion deficiency* fail to synthesize functional β_2 subunits. As a consequence, their white blood cells lack the entire family of β_2 receptors, and they suffer repeated bacterial infections.

The β_3 integrins are found on blood platelets (as well as various other cells), and they bind several matrix proteins, including the blood clotting factor *fibrinogen*. Platelets have to interact with fibrinogen to mediate normal blood clotting, and humans with *Glanzmann's disease*, who are genetically deficient in β_3 integrins, suffer from defective clotting and bleed excessively.

Integrins Can Switch Between an Active and an Inactive Conformation

A cell crawling through a tissue—a fibroblast or a macrophage, for example, or an epithelial cell migrating along a basal lamina—has to be able both to make and to break attachments to the matrix, and to do so rapidly if it is to travel quickly. Similarly, a circulating white blood cell has to be able to switch on or off its tendency to bind to endothelial cells in order to crawl out of a blood vessel at a site of inflammation. Furthermore, if force is to be applied where it is needed, the making and breaking of the extracellular attachments in all these cases has to be coupled to the prompt assembly and disassembly of cytoskeletal attachments inside the cell. The integrin molecules that span the membrane and mediate the attachments cannot simply be passive, rigid objects with sticky patches at their two ends. They must be able to switch between an active state, where they readily form attachments, and an inactive state, where they do not.

Structural studies, using a combination of electron microscopy and x-ray crystallography, suggest that integrins exist in multiple structural conformations that reflect different states of activity (**Figure 19–57**). In the inactive state, the external segments of the integrin dimer are folded together into a compact structure that cannot bind matrix proteins. In this state, the cytoplasmic tails of the dimer are

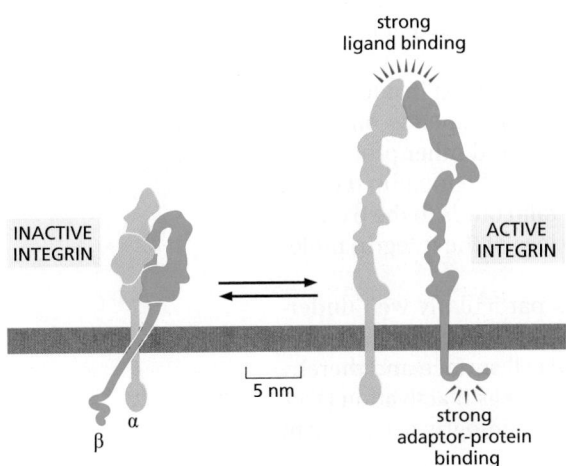

Figure 19–57 **Integrins exist in two major activity states.** Inactive (folded) and active (extended) structures of an integrin molecule, based on data from x-ray crystallography and other methods.

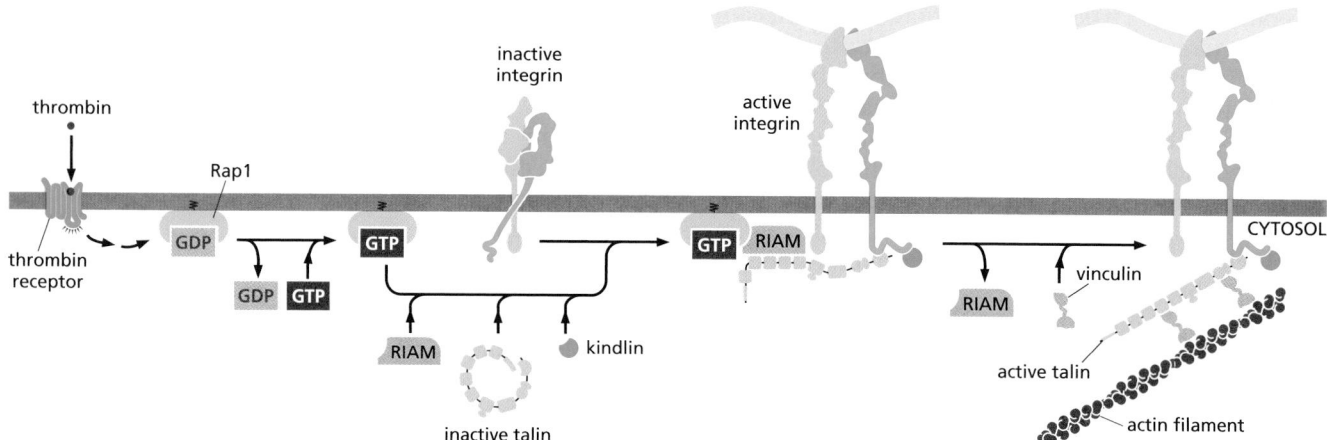

Figure 19–58 Activation of integrins by intracellular signaling. Signals received from outside the cell can act through various intracellular mechanisms to stimulate integrin activation. In platelets, as illustrated here, the extracellular signal protein thrombin activates a G-protein-coupled receptor on the cell surface, thereby initiating a signaling pathway that leads to activation of Rap1, a member of the monomeric GTPase family. Activated Rap1 interacts with the protein RIAM, which then recruits talin to the plasma membrane. Together with another protein called kindlin, talin interacts with the integrin β chain to trigger integrin activation. Talin then interacts with adaptor proteins such as vinculin, resulting in the formation of an actin linkage (see Figure 19–55).

Talin regulation depends in part on an interaction between its flexible C-terminal rod domain and the N-terminal head domain that contains the integrin-binding site. This interaction is thought to maintain talin in an inactive state when it is free in the cytoplasm. When talin is recruited by RIAM to the plasma membrane, the talin head domain interacts with a phosphoinositide called $PI(4,5)P_2$ (not shown here, but see Figure 15–28), resulting in dissociation of the rod domain. Talin unfolds to expose its binding sites for integrin and other proteins.

hooked together, preventing their interaction with cytoskeletal linker proteins. In the active state, the two integrin subunits are unhooked at the membrane to expose the intracellular binding sites for cytoplasmic adaptor proteins, and the external domains unfold and extend, like a pair of legs, to expose a high-affinity matrix-binding site at the tips of the subunits. Thus, the switch from inactive to active states depends on a major conformational change that simultaneously exposes the external and internal ligand-binding sites at the ends of the integrin molecule. External matrix binding and internal cytoskeleton linkages are thereby coupled.

Switching between the inactive and active states is regulated by a variety of mechanisms that vary, depending on the needs of the cell. In some cases, activation occurs by an "outside-in" mechanism: the binding of an external matrix protein, such as the RGD sequence of fibronectin, can drive some integrins to switch from the low-affinity inactive state to the high-affinity active state. As a result, binding sites for talin and other cytoplasmic adaptor proteins are exposed on the tail of the β chain. The binding of these adaptor proteins then leads to attachment of actin filaments to the intracellular end of the integrin molecule (see Figure 19–55). In this way, when the integrin catches hold of its ligand outside the cell, the cell reacts by tying the integrin molecule to the cytoskeleton, so that force can be applied at the point of cell attachment.

The chain of cause and effect can also operate in reverse, from inside to outside. This "inside-out" integrin-activation process generally depends on intracellular regulatory signals that stimulate the ability of talin and other proteins to interact with the β chain of the integrin. Talin competes with the integrin α chain for its binding site on the tail of the β chain. Thus, when talin binds to the β chain, it blocks the intracellular α–β linkage, allowing the two legs of the integrin molecule to spring apart.

The regulation of "inside-out" integrin activation is particularly well understood in platelets, where an extracellular signal protein called thrombin binds to a specific G-protein-coupled receptor (GPCR) on the cell surface and thereby activates an intracellular signaling pathway that leads to integrin activation (**Figure 19–58**). It is likely that similar signaling pathways govern integrin activation in numerous other cell types.

Integrins Cluster to Form Strong Adhesions

Integrins, like other cell adhesion molecules, differ from cell-surface receptors for hormones and for other extracellular soluble signal molecules in that they usually bind their ligand with lower affinity and are present at a 10–100-fold higher concentration on the cell surface. The Velcro principle, mentioned earlier in the context of cadherin adhesion (see Figure 19–6C), operates here too. Following their activation, integrins cluster together to create a dense plaque in which many integrin molecules are anchored to cytoskeletal filaments. The resulting protein structure can be remarkably large and complex, as seen in the focal adhesion made by a fibroblast on a fibronectin-coated surface culture dish.

The assembly of mature cell–matrix junctional complexes depends on the recruitment of dozens of different scaffolding and signaling proteins. Talin is a major component of many cell–matrix complexes, but numerous other proteins also make important contributions. These include the *integrin-linked kinase* (*ILK*) and its binding partners *pinch* and *parvin*, which together form a trimeric complex that serves as an organizing hub at many junctions. Cell–matrix junctions also employ several actin-binding proteins, such as vinculin, *zyxin*, *VASP*, and *α-actinin*, to promote the assembly and organization of actin filaments. Another critical component of many cell–matrix junctions is the *focal adhesion kinase* (*FAK*), which interacts with multiple components in the junction and serves an important function in signaling, as we describe next.

Extracellular Matrix Attachments Act Through Integrins to Control Cell Proliferation and Survival

Like other transmembrane cell adhesion proteins, integrins do more than just create attachments. They also activate intracellular signaling pathways and thereby allow control of almost any aspect of the cell's behavior according to the nature of the surrounding matrix and the state of the cell's attachments to it.

Many cells will not grow or proliferate in culture unless they are attached to extracellular matrix; nutrients and soluble growth factors in the culture medium are not enough. For some cell types, including epithelial, endothelial, and muscle cells, even cell survival depends on such attachments. When these cells lose contact with the extracellular matrix, they undergo apoptosis. This dependence of cell growth, proliferation, and survival on attachment to a substratum is known as **anchorage dependence**, and it is mediated mainly by integrins and the intracellular signals they generate. Mutations that disrupt or override this form of control, allowing cells to escape from anchorage dependence, occur in cancer cells and play a major part in their invasive behavior.

Our understanding of anchorage dependence has come mainly from studies of cells living on the surface of matrix-coated culture dishes. For connective-tissue cells that are normally surrounded by matrix on all sides, this is a far cry from the natural environment. Walking over a two-dimensional plain is very different from clambering through a three-dimensional jungle. The types of contacts that cells make with a rigid substratum are not the same as those, much less well studied, that they make with the deformable web of fibers of the extracellular matrix, and there are substantial differences in cell behavior in the two contexts. Nevertheless, it is likely that the same basic principles apply. Both *in vitro* and *in vivo*, intracellular signals generated at cell–matrix adhesion sites are crucial for cell proliferation and survival.

Integrins Recruit Intracellular Signaling Proteins at Sites of Cell–Matrix Adhesion

The mechanisms by which integrins signal into the cell interior are complex, involving several pathways, and integrins and conventional signaling receptors often influence one another and work together to regulate cell behavior, as we have already emphasized. The Ras/MAP kinase pathway (see Figure 15–49), for

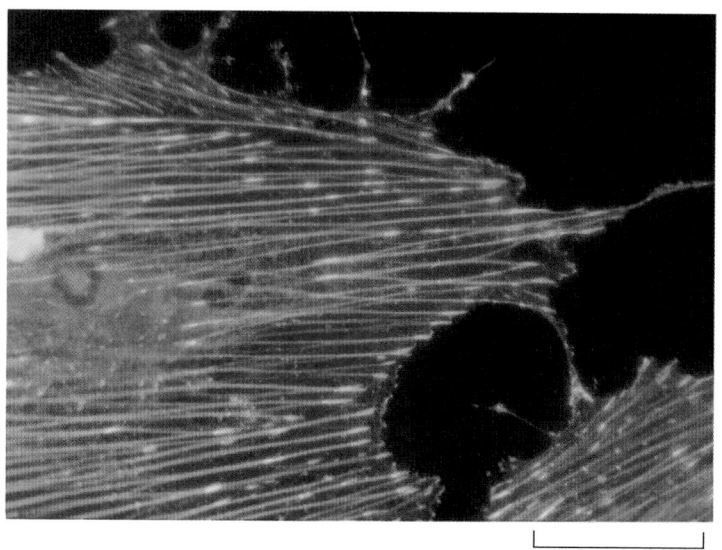

10 µm

Figure 19–59 **Tyrosine phosphorylation at focal adhesions.** A fibroblast cultured on a fibronectin-coated substratum and stained with fluorescent antibodies: actin filaments are stained *green* and activated proteins that contain phosphotyrosine are *red*, giving *orange* where the two components overlap. The actin filaments terminate at focal adhesions, where the cell attaches to the substratum by means of integrins. Proteins containing phosphotyrosine are also concentrated at these sites, reflecting the local activation of FAK and other protein kinases. Signals generated at such adhesion sites help regulate cell division, growth, and survival. (Courtesy of Keith Burridge.)

example, can be activated both by conventional signaling receptors and by integrins, but cells often need both kinds of stimulation of this pathway at the same time to give sufficient activation to induce cell proliferation. Integrins and conventional signaling receptors also cooperate to promote cell survival (discussed in Chapters 15 and 18).

One of the best-studied modes of integrin signaling depends on a cytoplasmic protein tyrosine kinase called **focal adhesion kinase (FAK)**. In studies of cells cultured on plastic dishes, focal adhesions are often prominent sites of tyrosine phosphorylation (**Figure 19–59**), and FAK is one of the major tyrosine-phosphorylated proteins found at these sites. When integrins cluster at cell–matrix contacts, FAK is recruited to the integrin β subunit by intracellular adaptor proteins such as talin or *paxillin* (which binds to one type of integrin α subunit). The clustered FAK molecules phosphorylate each other on a specific tyrosine, creating a phosphotyrosine docking site for members of the Src family of cytoplasmic tyrosine kinases. In addition to phosphorylating other proteins at the adhesion sites, these kinases then phosphorylate FAK on additional tyrosines, creating docking sites for a variety of additional intracellular signaling proteins. In this way, outside-in signaling from integrins, via FAK and Src family kinases, is relayed into the cell in much the same way as receptor tyrosine kinases generate signals (as discussed in Chapter 15).

Cell–Matrix Adhesions Respond to Mechanical Forces

Like the cell–cell junctions we described earlier, cell–matrix junctions can sense and respond to the mechanical forces that act on them. Most cell–matrix junctions, for example, are connected to a contractile actin network that tends to pull the junctions inward, away from the matrix. When cells are attached to a rigid matrix that strongly resists such pulling forces, the cell–matrix junction is able to sense the resulting high tension and trigger a response in which it recruits additional integrins and other proteins to increase the junction's ability to withstand that tension. Cell attachment to a relatively soft matrix generates less tension and therefore a less robust response. These mechanisms allow cells to sense and respond to differences in the rigidity of extracellular matrices in different tissues.

We saw earlier that mechanotransduction at cadherin-based cell–cell junctions likely depends on junctional proteins that change their structure when the junction is stretched by tension (see Figure 19–12). The same is true for cell–matrix junctions. The long C-terminal tail domain of talin, for example, includes a large number of binding sites for the actin-regulatory protein vinculin. Many of these sites are hidden inside folded protein domains but are exposed when those

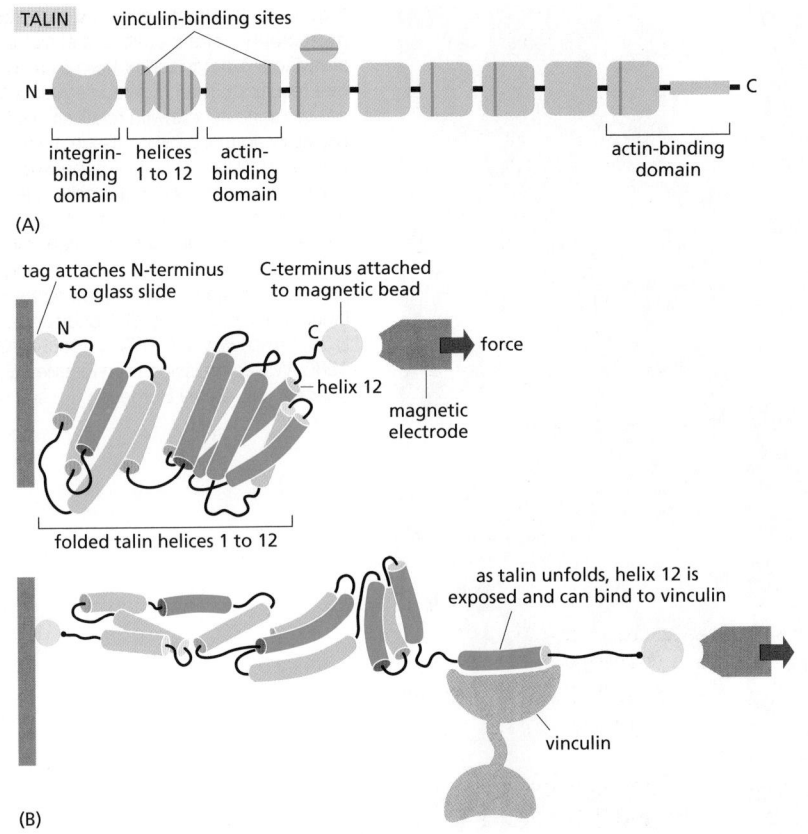

Figure 19–60 Talin is a tension sensor at cell–matrix junctions. Tension across cell–matrix junctions stimulates the local recruitment of vinculin and other actin-regulatory proteins, thereby strengthening the junction's attachment to the cytoskeleton. The experiments presented here tested the hypothesis that tension is sensed by the talin adaptor protein that links integrins to actin filaments (see Figure 19–55). (A) The long, flexible, C-terminal region of talin is divided into a series of folded domains, some of which contain vinculin-binding sites (*dark green lines*) that are thought to be hidden and therefore inaccessible. One domain near the N-terminus, for example, comprises a folded bundle of 12 α helices containing five vinculin-binding sites. (B) This experiment tested the hypothesis that tension stretches the 12-helix domain, thereby exposing vinculin-binding sites. A fragment of talin containing this domain was attached to an apparatus in which the domain could be stretched, as shown here. The fragment was labeled at its N-terminus with a tag that sticks to the surface of a glass slide on a microscope stage. The C-terminal end of the fragment was bound to a tiny magnetic bead, so the talin fragment could be stretched using a small magnetic electrode. The solution around the protein contained fluorescently tagged vinculin proteins. After the talin protein was stretched, excess vinculin solution was washed away, and the microscope was used to determine if any fluorescent vinculin proteins were bound to the talin protein. In the absence of stretching (*top*), most talin molecules did not bind vinculin. When the protein was stretched (*bottom*), two or three vinculin molecules were bound (only one is shown here for clarity). (Adapted from A. del Rio et al., *Science* 323:638–641, 2009.)

domains are unfolded by stretching the protein (**Figure 19–60**). The N-terminal end of talin binds integrin and the C-terminal end binds actin (see Figure 19–55); thus, when actin filaments are pulled by myosin motors inside the cell, the resulting tension stretches the talin rod, thereby exposing vinculin-binding sites. The vinculin molecules then recruit and organize additional actin filaments. Tension thereby increases the strength of the junction.

Summary

Integrins are the principal cell-surface receptors used by animal cells to bind to the extracellular matrix: they function as transmembrane linkers between the extracellular matrix and the cytoskeleton. Most integrins connect to actin filaments, while those at hemidesmosomes bind to intermediate filaments. Integrin molecules are heterodimers, and the binding of extracellular matrix ligands or intracellular activator proteins such as talin results in a dramatic conformational switch from an inactive to an active state. This creates an allosteric coupling between binding to matrix outside the cell and binding to the cytoskeleton inside it, allowing the integrin to convey signals in both directions across the plasma membrane. Complex assemblies of proteins become organized around the intracellular tails of activated integrins, producing intracellular signals that can influence almost any aspect of cell behavior, from proliferation and survival, as in the phenomenon of anchorage dependence, to cell polarity and guidance of migration. Integrin-based cell–matrix junctions are also capable of mechanotransduction: they can sense and respond to mechanical forces acting across the junction.

THE PLANT CELL WALL

Each cell in a plant deposits, and is in turn completely enclosed by, an elaborate extracellular matrix called the *plant cell wall*. It was the thick cell walls of cork, visible in a primitive microscope, that in 1663 enabled Robert Hooke to distinguish and name cells for the first time. The walls of neighboring plant cells, cemented

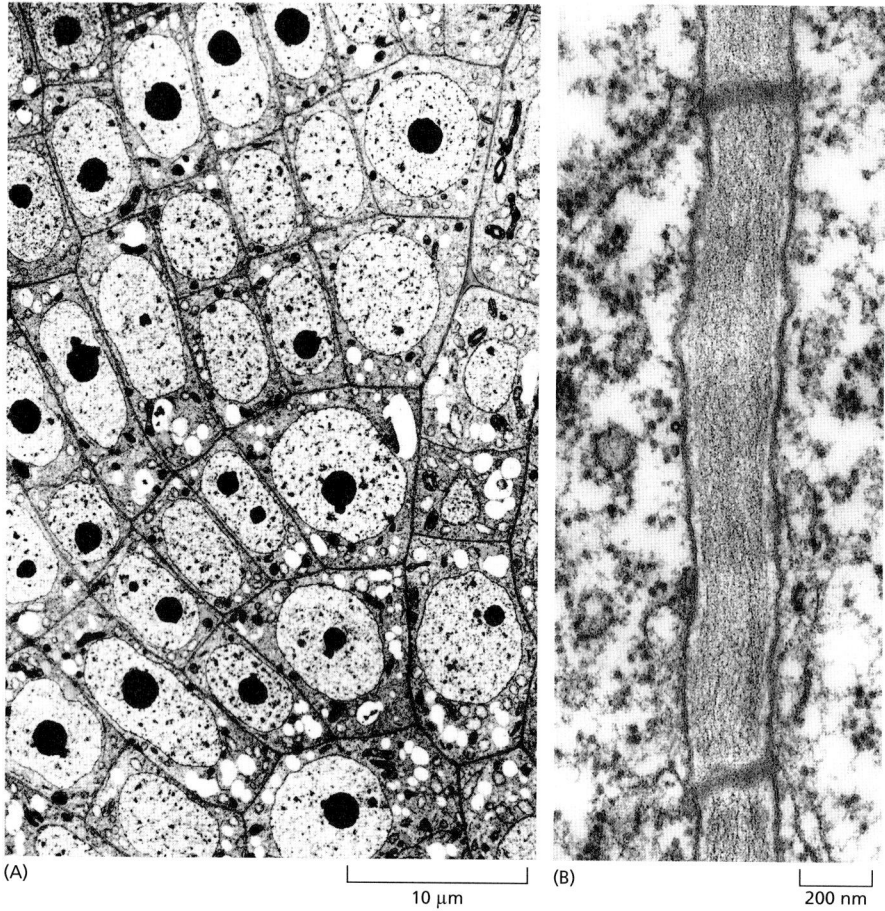

(A)

10 μm

(B)

200 nm

Figure 19–61 Plant cell walls. (A) Electron micrograph of the root tip of a rush, showing the organized pattern of cells that results from an ordered sequence of cell divisions in cells with relatively rigid cell walls. In this growing tissue, the cell walls are still relatively thin, appearing as fine black lines between the cells in the micrograph. (B) Section of a typical cell wall separating two adjacent plant cells. The two dark transverse bands correspond to plasmodesmata that span the wall (see Figure 19–27). (A, from C. Busby and B. Gunning, *Eur. J. Cell Biol.* 21:214–223, 1980. With permission from Elsevier; B, courtesy of Jeremy Burgess.)

together to form the intact plant (**Figure 19-61**), are generally thicker, stronger, and, most important of all, more rigid than the extracellular matrix produced by animal cells. In evolving relatively rigid walls, which can be up to many micrometers thick, early plant cells forfeited the ability to crawl about and adopted a sedentary lifestyle that has persisted in all present-day plants.

The Composition of the Cell Wall Depends on the Cell Type

All cell walls in plants have their origin in dividing cells, as the cell plate forms during cytokinesis to create a new partition wall between the daughter cells (discussed in Chapter 17). The new cells are usually produced in special regions called *meristems*, and they are generally small in comparison with their final size. To accommodate subsequent cell growth, the walls of the newborn cells, called **primary cell walls**, are thin and extensible, although tough. Once cell growth stops, the wall no longer needs to be extensible: sometimes the primary wall is retained without major modification, but, more commonly, a rigid **secondary cell wall** is produced by depositing new layers of matrix inside the old ones. These new layers generally have a composition that is significantly different from that of the primary wall. The most common additional polymer in secondary walls is **lignin**, a complex network of covalently linked phenolic compounds found in the walls of the xylem vessels and fiber cells of woody tissues.

Although the cell walls of higher plants vary in both composition and organization, they are all constructed, like animal extracellular matrices, using a structural principle common to all fiber-composites, including fiberglass and reinforced concrete. One component provides tensile strength, while another, in which the first is embedded, provides resistance to compression. While the principle is the same in plants and animals, the chemistry is different. Unlike the animal extracellular matrix, which is rich in protein and other nitrogen-containing polymers, the plant cell wall is made almost entirely of polymers that contain no nitrogen, including *cellulose* and lignin. For a sedentary organism that depends on CO_2, H_2O, and sunlight, these two abundant biopolymers represent "cheap," carbon-based structural materials, helping to conserve the scarce fixed nitrogen available in the soil that generally limits plant growth. Thus trees, for example, make a huge investment in the cellulose and lignin that comprise the bulk of their biomass.

In the cell walls of higher plants, the tensile fibers are made from the polysaccharide cellulose, the most abundant organic macromolecule on Earth, tightly linked into a network by *cross-linking glycans*. In primary cell walls, the matrix in which the cross-linked cellulose network is embedded is composed of *pectin*, a highly hydrated network of polysaccharides rich in galacturonic acid. Secondary cell walls contain additional molecules to make them rigid and permanent; lignin, in particular, forms a hard, waterproof filler in the interstices between the other components. All of these molecules are held together by a combination of covalent and noncovalent bonds to form a highly complex structure, whose composition, thickness, and architecture depend on the cell type.

The plant cell wall thus has a "skeletal" role in supporting the structure of the plant as a whole, a protective role as an enclosure for each cell individually, and a transport role, helping to form channels for the movement of fluid in the plant. When plant cells become specialized, they generally adopt a specific shape and produce specially adapted types of walls, according to which the different types of cells in a plant can be recognized and classified. We focus here, however, on the primary cell wall and the molecular architecture that underlies its remarkable combination of strength, resilience, and plasticity, as seen in the growing parts of a plant.

The Tensile Strength of the Cell Wall Allows Plant Cells to Develop Turgor Pressure

The aqueous extracellular environment of a plant cell consists of the fluid contained in the walls that surround the cell. Although the fluid in the plant cell wall contains more solutes than does the water in the plant's external milieu (for example, soil), it is still hypotonic in comparison with the cell interior. This osmotic imbalance causes the cell to develop a large internal hydrostatic pressure, or **turgor pressure**, which pushes outward on the cell wall, just as an inner tube pushes outward on a tire. The turgor pressure increases just to the point where the cell is in osmotic equilibrium, with no net influx of water despite the salt imbalance. The turgor pressure generated in this way may reach 10 or more atmospheres, about five times that in the average car tire. This pressure is vital to plants because it is the main driving force for cell expansion during growth, and it provides much of the mechanical rigidity of living plant tissues. Compare the wilted leaf of a dehydrated plant, for example, with the turgid leaf of a well-watered one. It is the mechanical strength of the cell wall that allows plant cells to sustain this internal pressure.

The Primary Cell Wall Is Built from Cellulose Microfibrils Interwoven with a Network of Pectic Polysaccharides

Cellulose gives the primary cell wall tensile strength. Each cellulose molecule consists of a linear chain of at least 500 glucose residues that are covalently linked to one another to form a ribbonlike structure, which is stabilized by hydrogen bonds within the chain (**Figure 19–62**). In addition, hydrogen bonds between adjacent cellulose molecules cause them to stick together in overlapping parallel arrays, forming bundles of about 40 cellulose chains, all of which have the same polarity. These highly ordered crystalline aggregates, many micrometers long, are called **cellulose microfibrils**, and they have a tensile strength comparable to that of steel. Sets of microfibrils are arranged in layers, or lamellae, with each microfibril about 20–40 nm from its neighbors and connected to them by long cross-linking

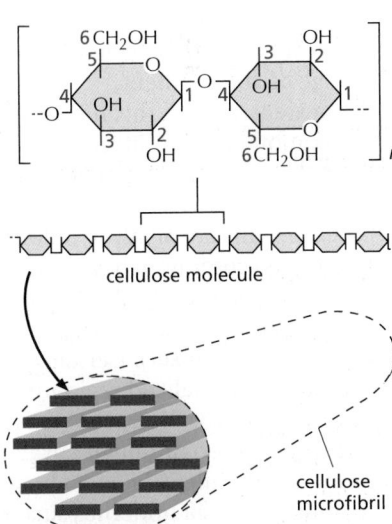

Figure 19–62 **Cellulose.** Cellulose molecules are long, unbranched chains of β1,4-linked glucose units. Each glucose residue is inverted with respect to its neighbors, and the resulting disaccharide repeat occurs hundreds of times in a single cellulose molecule. About 16 individual cellulose molecules assemble to form a strong, hydrogen-bonded cellulose microfibril.

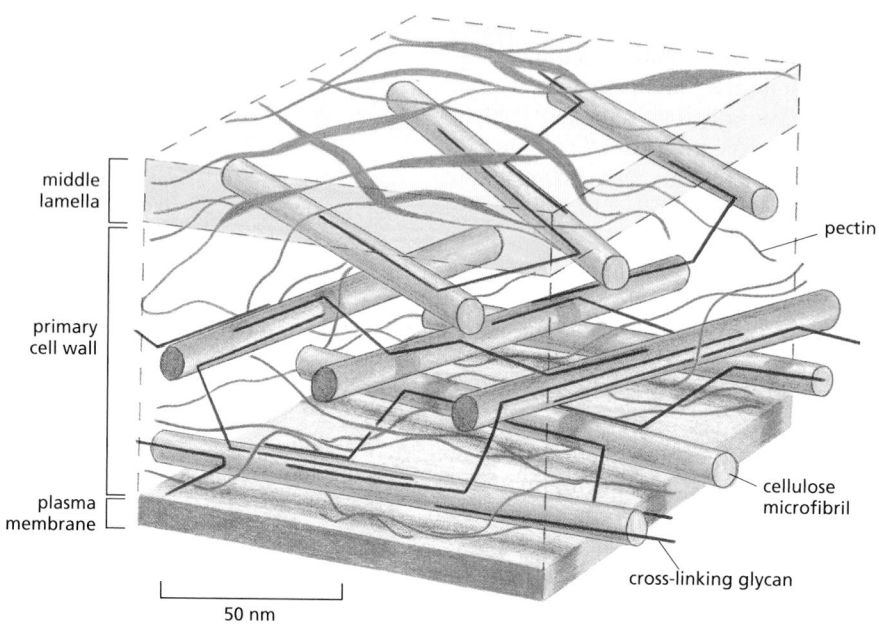

middle
lamella

primary
cell wall

plasma
membrane

pectin

cellulose
microfibril

cross-linking glycan

50 nm

Figure 19–63 **Scale model of a portion of a primary plant cell wall showing the two major polysaccharide networks.** The orthogonally arranged layers of cellulose microfibrils *(green)* are tied into a network by the cross-linking glycans *(red)* that form hydrogen bonds with the microfibrils. This network is coextensive with a network of pectin polysaccharides *(blue)*. The network of cellulose and cross-linking glycans provides tensile strength, while the pectin network resists compression. Cellulose, cross-linking glycans, and pectin are typically present in roughly equal amounts in a primary cell wall. The middle lamella is especially rich in pectin, and it cements adjacent cells together.

glycan molecules, which are attached by hydrogen bonds to the surface of the microfibrils. The primary cell wall consists of several such lamellae arranged in a plywoodlike network (**Figure 19–63**).

The **cross-linking glycans** are a heterogeneous group of branched polysaccharides that bind tightly to the surface of each cellulose microfibril and thereby help to cross-link the microfibrils into a complex network. There are many classes of cross-linking glycans, but they all have a long linear backbone composed of one type of sugar (glucose, xylose, or mannose) from which short side chains of other sugars protrude. It is the backbone sugar molecules that form hydrogen bonds with the surface of cellulose microfibrils, cross-linking them in the process. Both the backbone and the side-chain sugars vary according to the plant species and its stage of development.

Coextensive with this network of cellulose microfibrils and cross-linking glycans is another cross-linked polysaccharide network based on **pectins** (see Figure 19–63). Pectins are a heterogeneous group of branched polysaccharides that contain many negatively charged galacturonic acid units. Because of their negative charge, pectins are highly hydrated and associated with a cloud of cations, resembling the glycosaminoglycans of animal cells in the large amount of space they occupy (see Figure 19–33). When Ca^{2+} is added to a solution of pectin molecules, it cross-links them to produce a semirigid gel (it is pectin that is added to fruit juice to make jam set). Certain pectins are particularly abundant in the *middle lamella*, the specialized region that cements together the walls of adjacent cells (see Figure 19–63); here, Ca^{2+} cross-links are thought to help hold cell wall components together. Although covalent bonds also play a part in linking the components, very little is known about their nature. Regulated separation of cells at the middle lamella underlies such processes as the ripening of tomatoes and the abscission (detachment) of leaves in the fall.

In addition to the two polysaccharide-based networks that form the bulk of all plant primary cell walls, proteins are present, contributing up to about 5% of the wall's dry mass. Many of these proteins are enzymes, responsible for wall turnover and remodeling, particularly during growth. Another class of wall proteins, like collagen, contains high levels of hydroxyproline. These proteins are thought to strengthen the wall, and they are produced in greatly increased amounts as a local response to attack by pathogens. From the genome sequence of *Arabidopsis*, it has been estimated that more than 700 genes are required to synthesize, assemble, and remodel the plant cell wall.

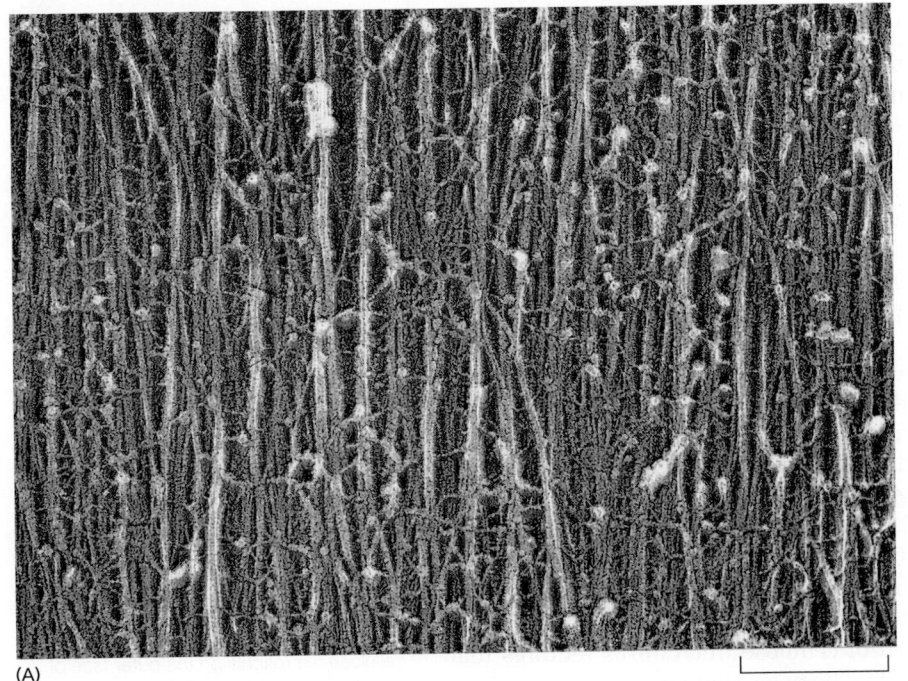

(A)

200 nm

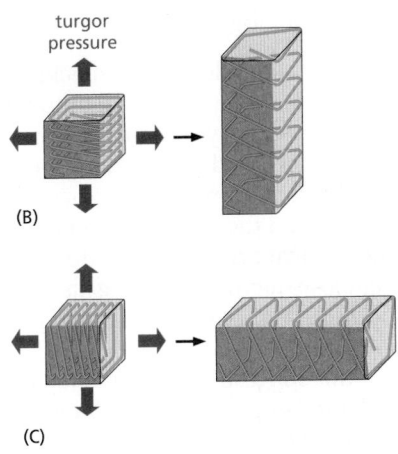

turgor pressure

(B)

(C)

Figure 19–64 **Cellulose microfibrils influence the direction of cell elongation.** (A) The orientation of cellulose microfibrils in the primary cell wall of an elongating carrot cell is shown in this electron micrograph of a shadowed replica from a rapidly frozen and deep-etched cell wall. The cellulose microfibrils are aligned parallel to one another and perpendicular to the axis of cell elongation. The microfibrils are cross-linked by, and interwoven with, a complex web of matrix molecules (compare with Figure 19–63). (B, C) The cells in (B) and (C) start off with identical shapes (shown here as cubes) but with different net orientations of cellulose microfibrils in their walls. Although turgor pressure is uniform in all directions, cell wall loosening allows each cell to elongate only in a direction perpendicular to the orientation of the innermost layer of microfibrils, which have great tensile strength. Cell expansion occurs in concert with the insertion of new wall material. The final shape of an organ, such as a shoot, is determined in part by the direction in which its component cells can expand. (A, courtesy of Brian Wells and Keith Roberts.)

Oriented Cell Wall Deposition Controls Plant Cell Growth

Once a plant cell has left the meristem where it is generated, it can grow dramatically, commonly by more than a thousand times in volume. The manner of this expansion determines the final shape of each cell, and hence the final form of the plant as a whole. Turgor pressure inside the cell drives the expansion, but it is the behavior of the cell wall that governs its direction and extent. Complex wall-remodeling activities are required, as well as the deposition of new wall materials. Because of their crystalline structure, the individual cellulose microfibrils in the wall are unable to stretch, and this gives them a crucial role in the process. For the cell wall to stretch or deform, the microfibrils must either slide past one another or become more widely separated, or both. The orientation of the microfibrils in the innermost layers of the wall governs the direction in which the cell expands. Cells in plants therefore anticipate their future morphology by controlling the orientation of the cellulose microfibrils that they deposit in the wall (**Figure 19–64**).

Unlike most other matrix macromolecules, which are made in the endoplasmic reticulum and Golgi apparatus and are secreted, cellulose is spun out from the surface of the cell by a plasma-membrane-bound enzyme complex (*cellulose synthase*), which uses as its substrate the sugar nucleotide UDP-glucose supplied from the cytosol. Each enzyme complex, or *rosette*, has a sixfold symmetry (see Figure 19–65) and contains the protein products of three separate cellulose synthase (*CESA*) genes. Each CESA protein is essential for the production of a cellulose microfibril. Three *CESA* genes are required for primary cell wall synthesis and a different three for secondary cell wall synthesis.

As they are being synthesized, the nascent cellulose chains assemble into microfibrils. These are spun out on the extracellular surface of the plasma membrane, forming a layer, or lamella, in which all the microfibrils have more or less the same alignment (see Figure 19–63). Each new lamella is deposited internally to the previous one, so that the wall consists of concentrically arranged lamellae, with the oldest on the outside. The most recently deposited microfibrils in elongating cells commonly lie perpendicular to the axis of cell elongation, although the orientation of the microfibrils in the outer lamellae that were laid down earlier may be different (see Figure 19–64B and C).

Microtubules Orient Cell Wall Deposition

An important clue to the mechanism that dictates microfibril orientation came from observations of the microtubules in plant cells. These are frequently arranged in the cortical cytoplasm with the same orientation as the cellulose microfibrils that are currently being deposited in the cell wall in that region. These cortical microtubules form a *cortical array* close to the cytosolic face of the plasma membrane, held there by poorly characterized proteins. The congruent orientation of the cortical array of microtubules (lying just inside the plasma membrane) and cellulose microfibrils (lying just outside) is seen in many types and shapes of plant cells and is present during both primary and secondary cell wall deposition, suggesting a causal relationship.

This suggestion can be tested by treating a plant tissue with a microtubule-depolymerizing drug so as to disassemble the entire system of cortical microtubules. The consequences for subsequent cellulose deposition, however, are not as straightforward as might be expected. The drug treatment does not disrupt the production of new cellulose microfibrils, and in some cases cells can continue to deposit new microfibrils in the preexisting orientation. Any developmental switch in the orientation of the microfibril pattern that would normally occur between successive lamellae, however, is invariably blocked. It seems that a preexisting orientation of microfibrils can be propagated even in the absence of microtubules, but any change in the deposition of cellulose microfibrils requires that intact microtubules be present to determine the new orientation.

These observations are consistent with the following model. The cellulose-synthesizing rosettes embedded in the plasma membrane spin out long cellulose molecules. As the synthesis of cellulose molecules and their self-assembly into microfibrils proceeds, the distal end of each microfibril presumably forms indirect cross-links to the previous layer of wall material, orienting the new microfibril in parallel with the old ones as it becomes integrated into the texture of the wall. Since the microfibril is stiff, the rosette at its growing, proximal end has to move as it deposits the new material. Traveling in the plane of the membrane, the rosette moves in the direction defined by the way in which the far end of the microfibril is anchored in the existing wall. In this way, each layer of microfibrils would tend to be spun out from the membrane in the same orientation as the layer laid down previously, with the rosettes following the direction of the preexisting oriented microfibrils outside the cell. Oriented microtubules inside the cell, however, can force a change in the direction in which the rosettes move: they can create boundaries in the plasma membrane that act like the banks of a canal to constrain rosette movement (**Figure 19–65**). In this view, cellulose synthesis can occur independently of microtubules; but it is constrained spatially when cortical microtubules are present to define membrane microdomains within which the enzyme complex can move.

Figure 19–65 One model of how the orientation of newly deposited cellulose microfibrils might be determined by the orientation of cortical microtubules. (A) The large cellulose synthase complexes, or *rosettes*, are integral membrane proteins that continuously synthesize cellulose microfibrils on the outer face of the plasma membrane. The distal ends of the stiff microfibrils become integrated into the texture of the wall, and their elongation at the proximal end pushes the synthase complex along in the plane of the membrane. Because the cortical array of microtubules is attached to the plasma membrane in a way that confines this complex to defined membrane channels, the orientation of these microtubules—when they are present—determines the axis along which the new microfibrils are laid down. (B, C) Two electron micrographs show the tight association of the cortical microtubules with the plasma membrane. One shows the microtubules in cross section while the other shows a microtubule in longitudinal section. Both emphasize the constant gap of about 20 nm between membrane and microtubule; the connecting molecules responsible remain obscure. (B and C, courtesy of Andrew Staehelin.)

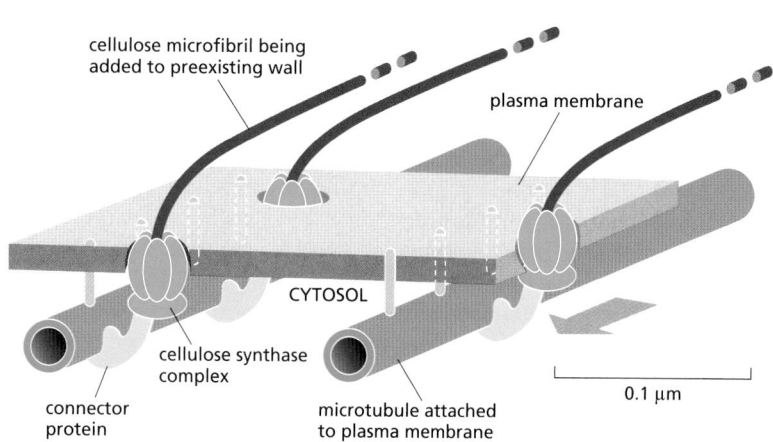

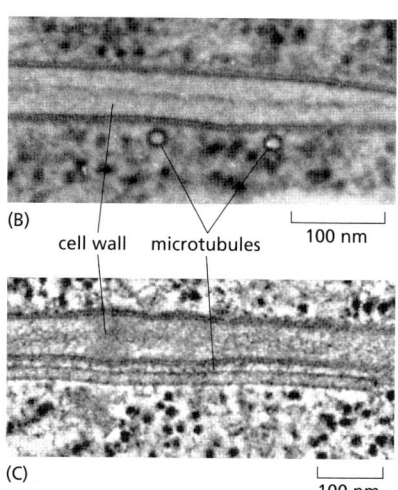

(A)

(B)   cell wall   microtubules   100 nm

(C)   100 nm

cellulose microfibril being added to preexisting wall

plasma membrane

CYTOSOL

connector protein

cellulose synthase complex

microtubule attached to plasma membrane

0.1 µm

In this way, plant cells can change their direction of expansion by a sudden change in the orientation of their cortical array of microtubules. Because plant cells cannot move (being constrained by their walls), the entire morphology of a multicellular plant presumably depends on a coordinated, highly patterned deployment of cortical microtubule orientations during plant development. It is not known how these orientations are controlled, although it has been shown that the microtubules can reorient rapidly in response to extracellular stimuli, including plant growth regulators such as ethylene and auxins (discussed in Chapter 15).

Microtubules are not, however, the only cytoskeletal elements that influence wall deposition. Local foci of cortical actin filaments can also direct the deposition of new wall material at specific sites on the cell surface, contributing to the elaborate final shaping of many differentiated plant cells.

Summary

Plant cells are surrounded by a tough extracellular matrix, or cell wall, which is responsible for many of the unique features of a plant's lifestyle. The wall is composed of a network of cellulose microfibrils and cross-linking glycans, embedded in a highly cross-linked matrix of pectin polysaccharides. In secondary cell walls, lignin may be deposited to make them waterproof, hard, and woody. A cortical array of microtubules can control the orientation of newly deposited cellulose microfibrils, which in turn determine the direction of cell expansion and therefore the final shape of the cell and, ultimately, of the plant as a whole.

PROBLEMS

Which statements are true? Explain why or why not.

19–1 Given the numerous processes inside cells that are regulated by changes in Ca^{2+} concentration, it seems likely that Ca^{2+}-dependent cell–cell adhesions are also regulated by changes in Ca^{2+} concentration.

19–2 Tight junctions perform two distinct functions: they seal the space between cells to restrict paracellular flow and they fence off plasma membrane domains to prevent the mixing of apical and basolateral membrane proteins.

19–3 The elasticity of elastin derives from its high content of α helices, which act as molecular springs.

19–4 Integrins can convert mechanical signals into intracellular molecular signals.

Discuss the following problems.

19–5 Comment on the following (1922) quote from Warren Lewis, who was one of the pioneers of cell biology. "Were the various types of cells to lose their stickiness for one another and for the supporting extracellular matrix, our bodies would at once disintegrate and flow off into the ground in a mixed stream of cells."

19–6 Cell adhesion molecules were originally identified using antibodies raised against cell-surface components to block cell aggregation. In the adhesion-blocking assays, the researchers found it necessary to use antibody fragments, each with a single binding site (so-called Fab fragments), rather than intact IgG antibodies, which are Y-shaped molecules with two identical binding sites. The

WHAT WE DON'T KNOW

• What are the regulatory mechanisms that control the rearrangement of cell–cell junctions in epithelia during early development? What roles do mechanical force and tension play in these rearrangements?

• How do extracellular matrix proteins and carbohydrates influence the localization and actions of extracellular signal molecules or their cell-surface receptors?

• How do intracellular adaptor proteins coordinate the activation of integrin proteins and their interactions with cytoskeletal components and their response to changes in mechanical force acting on cell–matrix junctions?

• Given that extracellular matrix molecules have the ability to present ordered arrays of signals to cells, might the exact spatial relationships between such signals carry a message beyond that of the individual signals themselves?

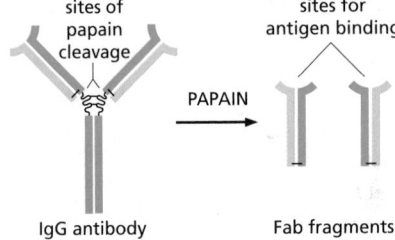

Figure Q19–1 Production of Fab fragments from IgG antibodies by digestion with papain (Problem 19–6).

Fab fragments were generated by digesting the IgG antibodies with papain, a protease, to separate the two binding sites (**Figure Q19–1**). Why do you suppose it was necessary to use Fab fragments to block cell aggregation?

19–7 The food-poisoning bacterium *Clostridium perfringens* makes a toxin that binds to members of the claudin family of proteins, which are the main constituents of tight junctions. When the C-terminus of the toxin is bound to a claudin, the N-terminus can insert into the adjacent cell membrane, forming holes that kill the cell. The portion of the toxin that binds to the claudins has proven to be a valuable reagent for investigating the properties of tight junctions. MDCK cells are a common choice for studies of tight junctions because they can form an intact epithelial sheet with high transepithelial resistance. MDCK cells express two claudins: claudin-1, which is not bound by the toxin, and claudin-4, which is.

When an intact MDCK epithelial sheet is incubated with the C-terminal toxin fragment, claudin-4 disappears, becoming undetectable within 24 hours. In

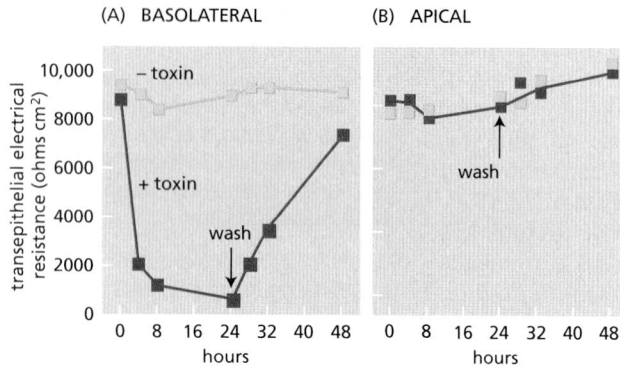

Figure Q19–2 Effects of *Clostridium* toxin on the barrier function of MDCK cells (Problem 19–7). (A) Addition of toxin from the basolateral side of the epithelial sheet. (B) Addition of toxin from the apical side of the epithelial sheet. For a given voltage, a higher resistance (ohms cm²) gives less paracellular current.

TABLE Q19–1 Phenotypes of mice with genetic defects in components of the basal lamina (Problem 19–8).

Protein	Genetic defect	Phenotype
Nidogen-1	Gene knockout (–/–)	None
Nidogen-2	Gene knockout (–/–)	None
Laminin γ-1	Nidogen binding-site deletion (+/–)	None
Laminin γ-1	Nidogen binding-site deletion (–/–)	Dead at birth

+/– stands for heterozygous, –/– stands for homozygous.

the absence of claudin-4, the cells remain healthy and the epithelial sheet appears intact. The mean number of strands in the tight junctions that link the cells also decreases over 24 hours from about four to about two, and they are less highly branched. A functional assay for the integrity of the tight junctions shows that transepithelial resistance decreases dramatically in the presence of the toxin, but the resistance can be restored by washing out the toxin (**Figure Q19–2A**). Curiously, the toxin produces these effects only when it is added to the basolateral side of the sheet; it has no effect when added to the apical surface (Figure Q19–2B).

A.   How can it be that two tight-junction strands remain, even though all of the claudin-4 has disappeared?

B.   Why do you suppose the toxin works when it is added to the basolateral side of the epithelial sheet, but not when added to the apical side?

19–8 It is not an easy matter to assign particular functions to specific components of the basal lamina, since the overall structure is a complicated composite material with both mechanical and signaling properties. Nidogen, for example, cross-links two central components of the basal lamina by binding to the laminin γ-1 chain and to type IV collagen. Given such a key role, it was surprising that mice with a homozygous knockout of the gene for nidogen-1 were entirely healthy, with no abnormal phenotype. Similarly, mice homozygous for a knockout of the gene for nidogen-2 also appeared completely normal. By contrast, mice that were homozygous for a defined mutation in the gene for laminin γ-1, which eliminated just the binding site for nidogen, died at birth with severe defects in lung and kidney formation. The mutant portion of the laminin γ-1 chain is thought to have no other function than to bind nidogen, and does not affect laminin structure or its ability to assemble into the basal lamina. How would you explain these genetic observations, which are summarized in **Table Q19–1**? What would you predict would be the phenotype of a mouse that was homozygous for knockouts of both nidogen genes?

19–9 Discuss the following statement: "The basal lamina of muscle fibers serves as a molecular bulletin board, in which adjoining cells can post messages that direct the differentiation and function of the underlying cells."

19–10 The affinity of integrins for matrix components can be modulated by changes to their cytoplasmic domains: a process known as inside-out signaling. You have identified a key region in the cytoplasmic domains of αIIbβ3 integrin that seems to be required for inside-out signaling (**Figure Q19–3**). Substitution of alanine for either D723 in the β chain or R995 in the α chain leads to a high level of spontaneous activation, under conditions where the wild-type chains are inactive. Your advisor suggests that you convert the aspartate in the β chain to an arginine (D723R) and the arginine in the α chain to an aspartate (R995D). You compare all three α chains (R995, R995A, and R995D) against all three β chains (D723, D723A, and D723R). You find that all pairs have a high level of spontaneous activation, except D723 vs R995 (the wild type) and D723R vs R995D, which have low levels. Based on these results, how do you think the αIIbβ3 integrin is held in its inactive state?

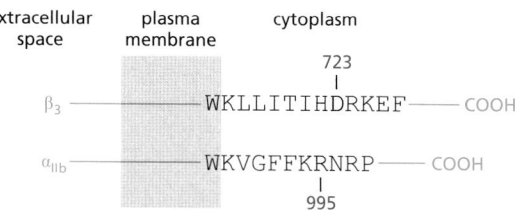

Figure Q19–3 Schematic representation of αIIbβ3 integrin (Problem 19–10). The D723 and R995 residues are indicated. (From P.E. Hughes et al., *J. Biol. Chem.* 271:6571–6574, 1996. With permission from American Society for Biochemistry and Molecular Biology.)

19–11 The glycosaminoglycan polysaccharide chains that are linked to specific core proteins to form the proteoglycan components of the extracellular space are highly negatively charged. How do you suppose these negatively charged polysaccharide chains help to establish a hydrated gel-like environment around the cell? How would the properties of these molecules differ if the polysaccharide chains were uncharged?

19–12 At body temperature, L-aspartate in proteins racemizes to D-aspartate at an appreciable rate. Most proteins in the body have a very low level of D-aspartate, if it can be detected at all. Elastin, however, has a fairly high level of D-aspartate. Moreover, the amount of D-aspartate increases in direct proportion to the age of the person from whom the sample was taken. Why do you suppose that most proteins have little if any D-aspartate, while elastin has levels of D-aspartate that increase steadily with age?

19–13 Your boss is coming to dinner! All you have for a salad is some wilted, day-old lettuce. You vaguely recall that there is a trick to rejuvenating wilted lettuce, but you cannot remember what it is. Should you soak the lettuce in salt water, soak it in tap water, or soak it in sugar water, or maybe just shine a bright light on it and hope that photosynthesis will perk it up?

19–14 A plant must be able to respond to changes in the water status of its surroundings. It does so by the flow of water molecules through water channels called aquaporins. The hydraulic conductivity of a single aquaporin is 4.4×10^{-22} m^3 per second per MPa (megapascal) of pressure. What does this correspond to in terms of water molecules per second at atmospheric pressure? [Atmospheric pressure is 0.1 MPa (1 bar) and the concentration of water is 55.5 M.]

REFERENCES

General

Beckerle M ed. (2002) Cell Adhesion. Oxford: Oxford University Press.

Hynes RO & Yamada KM (eds) (2011) Extracellular Matrix Biology (Cold Spring Harbor Perspectives in Biology). Cold Spring Harbor: Cold Spring Harbor Laboratory Press.

Cell–Cell Junctions

Brasch J, Harrison OJ, Honig B & Shapiro L (2012) Thinking outside the cell: how cadherins drive adhesion. *Trends Cell Biol.* 22, 299–310.

Gomez GA, McLachlan RW & Yap AS (2011) Productive tension: force-sensing and homeostasis of cell-cell junctions. *Trends Cell Biol.* 21, 499–505.

Goodenough DA & Paul DL (2003) Beyond the gap: functions of unpaired connexon channels. *Nat. Rev. Mol. Cell Biol.* 4, 285–294.

Gumbiner BM (2005) Regulation of cadherin-mediated adhesion in morphogenesis. *Nat. Rev. Mol. Cell Biol.* 6, 622–634.

Harris TJ & Tepass U (2010) Adherens junctions: from molecules to morphogenesis. *Nat. Rev. Mol. Cell Biol.* 11, 502–514.

King N, Hittinger CT & Carroll SB (2003) Evolution of key cell signaling and adhesion protein families predates animal origins. *Science* 301, 361–363.

Leckband DE, le Duc Q, Wang N & de Rooij J (2011) Mechanotransduction at cadherin-mediated adhesions. *Curr. Opin. Cell Biol.* 23, 523–530.

Lecuit T, Lenne PF & Munro E (2011) Force generation, transmission, and integration during cell and tissue morphogenesis. *Annu. Rev. Cell Dev. Biol.* 27, 157–184.

Litjens SH, de Pereda JM & Sonnenberg A (2006) Current insights into the formation and breakdown of hemidesmosomes. *Trends Cell Biol.* 16, 376–383.

Maule AJ, Benitez-Alfonso Y & Faulkner C (2011) Plasmodesmata—membrane tunnels with attitude. *Curr. Opin. Plant Biol.* 14, 683–690.

McEver RP & Zhu C (2010) Rolling cell adhesion. *Annu. Rev. Cell Dev. Biol.* 26, 363–396.

Nakagawa S, Maeda S & Tsukihara T (2010) Structural and functional studies of gap junction channels. *Curr. Opin. Struct. Biol.* 20, 423–430.

Shin K, Fogg VC & Margolis B (2006) Tight junctions and cell polarity. *Annu. Rev. Cell Dev. Biol.* 22, 207–236.

Takeichi M (2007) The cadherin superfamily in neuronal connections and interactions. *Nat. Rev. Neurosci.* 8, 11–20.

Thomason HA, Scothern A, McHarg S & Garrod DR (2010) Desmosomes: adhesive strength and signalling in health and disease. *Biochem. J.* 429, 419–433.

The Extracellular Matrix of Animals

Aszodi A, Legate KR, Nakchbandi I & Fassler R (2006) What mouse mutants teach us about extracellular matrix function. *Annu. Rev. Cell Dev. Biol.* 22, 591–621.

Bulow HE & Hobert O (2006) The molecular diversity of glycosaminoglycans shapes animal development. *Annu. Rev. Cell Dev. Biol.* 22, 375–407.

Couchman JR (2010) Transmembrane signaling proteoglycans. *Annu. Rev. Cell Dev. Biol.* 26, 89–114.

Domogatskaya A, Rodin S & Tryggvason K (2012) Functional diversity of laminins. *Annu. Rev. Cell Dev. Biol.* 28, 523–553.

Hynes RO (2009) The extracellular matrix: not just pretty fibrils. *Science* 326, 1216–1219.

Hynes RO & Naba A (2012) Overview of the matrisome—an inventory of extracellular matrix constituents and functions. *Cold Spring Harb. Perspect. Biol.* 4, a004903.

Kielty CM, Sherratt MJ & Shuttleworth CA (2002) Elastic fibres. *J. Cell Sci.* 115, 2817–2828.

Larsen M, Artym VV, Green JA & Yamada KM (2006) The matrix reorganized: extracellular matrix remodeling and integrin signaling. *Curr. Opin. Cell Biol.* 18, 463–471.

Lu P, Takai K, Weaver VM & Werb Z (2011) Extracellular matrix degradation and remodeling in development and disease. *Cold Spring Harb. Perspect. Biol.* 3, a005058.

Ricard-Blum S (2011) The collagen family. *Cold Spring Harb. Perspect. Biol.* 3, a004978.

Sasaki T, Fässler R & Hohenester E (2004) Laminin: the crux of basement membrane assembly. *J. Cell Biol.* 164, 959–963.

Toole BP (2001) Hyaluronan in morphogenesis. *Semin. Cell Dev. Biol.* 12, 79–87.

Yurchenco PD (2011) Basement membranes: cell scaffoldings and signaling platforms. *Cold Spring Harb. Perspect. Biol.* 3, a004911.

Cell–Matrix Junctions

Calderwood DA, Campbell ID & Critchley DR (2013) Talins and kindlins: partners in integrin-mediated adhesion. *Nat. Rev. Mol. Cell Biol.* 14, 503–517.

Campbell ID & Humphries MJ (2011) Integrin structure, activation, and interactions. *Cold Spring Harb. Perspect. Biol.* 3, a004994.

Hoffman BD, Grashoff C & Schwartz MA (2011) Dynamic molecular processes mediate cellular mechanotransduction. *Nature* 475, 316–323.

Hogg N, Patzak I & Willenbrock F (2011) The insider's guide to leukocyte integrin signalling and function. *Nat. Rev. Immunol.* 11, 416–426.

Kanchanawong P, Shtengel G, Pasapera AM et al. (2010) Nanoscale architecture of integrin-based cell adhesions. *Nature* 468, 580–584.

Luo BH & Springer TA (2006) Integrin structures and conformational signaling. *Curr. Opin. Cell Biol.* 18, 579–586.

Moser M, Legate KR, Zent R & Fässler R (2009) The tail of integrins, talin, and kindlins. *Science* 324, 895–899.

Ross TD, Coon BG, Yun S et al. (2013) Integrins in mechanotransduction. *Curr. Opin. Cell Biol.* 25, 613–618.

Shattil SJ, Kim C & Ginsberg MH (2010) The final steps of integrin activation: the end game. *Nat. Rev. Mol. Cell Biol.* 11, 288–300.

The Plant Cell Wall

Albersheim P, Darvill A, Roberts K et al. (2011) Plant Cell Walls: From Chemistry to Biology. New York: Garland Science.

Braidwood L, Breuer C & Sugimoto K (2013) My body is a cage: mechanisms and modulation of plant cell growth. *New Phyto.* 210, 388–402.

Keegstra K (2010) Plant cell walls. *Plant Physiol.* 154, 483–486.

Li S, Lei L, Somerville C et al. (2011) Cellulose synthase interactive protein 1 (CSI1) links microtubules and cellulose synthase complexes. *Proc. Natl. Acad. Sci. USA* 109, 189–190.

Lloyd C (2011) Dynamic microtubules and the texture of plant cell walls. *Int. Rev. Cell Mol. Biol.* 287, 287–329.

McFarlane HE, Döring A & Perrson S (2014) The cell biology of cellulose synthesis. *Annu. Rev. Plant Biol.* 65, 69–94.

Somerville C (2006) Cellulose synthesis in higher plants. *Annu. Rev. Cell Dev. Biol.* 22, 53–78.

Szymanski DB & Cosgrove DJ (2009) Dynamic Coordination of cytoskeletal and cell wall systems during cell wall biogenesis. *Curr. Biol.* 19, R800–R811.

Wightman R & Turner SR (2008) The roles of the cytoskeleton during cellulose deposition at the secondary cell wall. *Plant J.* 54, 794–805.

Wolf S, Hématy K & Höfte H (2012) Growth control and cell wall signaling in plants. *Annu. Rev. Plant Biol.* 63, 381–407.

Cancer

About one in five of us will die of cancer, but that is not why we devote a chapter to this disease. Cancer cells break the most basic rules of cell behavior by which multicellular organisms are built and maintained, and they exploit every kind of opportunity to do so. These transgressions help to reveal what the normal rules are and how they are enforced. As a result, cancer research helps to illuminate the fundamentals of cell biology—especially cell signaling (Chapter 15), the cell cycle and cell growth (Chapter 17), programmed cell death (apoptosis, Chapter 18), and the control of tissue architecture (Chapters 19 and 22). Of course, with a deeper understanding of these normal processes, we also gain a deeper understanding of the disease and better tools to treat it.

In this chapter, we first consider what cancer is and describe the natural history of the disease from a cellular standpoint. We then discuss the molecular changes that make a cell cancerous. And we end the chapter by considering how our enhanced understanding of the molecular basis of cancer is leading to improved methods for its prevention and treatment.

CANCER AS A MICROEVOLUTIONARY PROCESS

The body of an animal operates as a society or ecosystem, whose individual members are cells that reproduce by cell division and organize themselves into collaborative assemblies called *tissues*. This ecosystem is very peculiar, however, because self-sacrifice—as opposed to survival of the fittest—is the rule. Ultimately, all of the somatic cell lineages in animals are committed to die: they leave no progeny and instead dedicate their existence to the support of the germ cells, which alone have a chance of continued survival (discussed in Chapter 21). There is no mystery in this, for the body is a clone derived from a fertilized egg, and the genome of the somatic cells is the same as that of the germ-cell lineage that gives rise to sperm or eggs. By their self-sacrifice for the sake of the germ cells, the somatic cells help to propagate copies of their own genes.

Thus, unlike free-living cells such as bacteria, which compete to survive, the cells of a multicellular organism are committed to collaboration. To coordinate their behavior, the cells send, receive, and interpret an elaborate set of extracellular signals that serve as *social controls*, directing cells how to act (discussed in Chapter 15). As a result, each cell behaves in a socially responsible manner—resting, growing, dividing, differentiating, or dying—as needed for the good of the organism.

Molecular disturbances that upset this harmony mean trouble for a multicellular society. In a human body with more than 10^{14} cells, billions of cells experience mutations every day, potentially disrupting the social controls. Most dangerously, a mutation may give one cell a selective advantage, allowing it to grow and divide slightly more vigorously and survive more readily than its neighbors and in this way to become a founder of a growing mutant clone. A mutation that promotes such selfish behavior by individual members of the cooperative can jeopardize the future of the whole enterprise. Over time, repeated rounds of mutation, competition, and natural selection operating within the population of somatic cells can cause matters to go from bad to worse. These are the basic ingredients of cancer: it is a disease in which an individual mutant clone of cells begins by prospering at the expense of its neighbors. In the end—as the clone grows, evolves, and spreads—it can destroy the entire cellular society (Movie 20.1).

In this section, we discuss the development of cancer as a microevolutionary process that takes place within the course of a human life-span in a subpopulation of cells in the body. But the process depends on the same principles of mutation and natural selection that have driven the evolution of living organisms on Earth for billions of years.

Cancer Cells Bypass Normal Proliferation Controls and Colonize Other Tissues

Cancer cells are defined by two heritable properties: (1) they reproduce in defiance of the normal restraints on cell growth and division, and (2) they invade and colonize territories normally reserved for other cells. It is the combination of these properties that makes cancers particularly dangerous. An abnormal cell that grows (increases in mass) and proliferates (divides) out of control will give rise to a tumor, or *neoplasm*—literally, a new growth. As long as the neoplastic cells have not yet become invasive, however, the tumor is said to be **benign**. For most types of such neoplasms, removing or destroying the mass locally usually achieves a complete cure. A tumor is considered a true cancer if it is **malignant**; that is, when its cells have acquired the ability to invade surrounding tissue. Invasiveness is an essential characteristic of cancer cells. It allows them to break loose, enter blood or lymphatic vessels, and form secondary tumors called **metastases** at other sites in the body (Figure 20–1). In general, the more widely a cancer spreads, the harder it becomes to eradicate. It is generally metastases that kill the cancer patient.

Cancers are traditionally classified according to the tissue and cell type from which they arise. **Carcinomas** are cancers arising from epithelial cells, and they are by far the most common cancers in humans. They account for about 80% of cases, perhaps because most of the cell proliferation in adults occurs in epithelia. In addition, epithelial tissues are the most likely to be exposed to the various forms of physical and chemical damage that favor the development of cancer. **Sarcomas** arise from connective tissue or muscle cells. Cancers that do not fit in either of these two broad categories include the various **leukemias** and **lymphomas**, derived from white blood cells and their precursors (hemopoietic cells), as well as cancers derived from cells of the nervous system. Figure 20–2 shows the types of cancers that are common in the United States, together with their incidence and death rates. Each broad category has many subdivisions according to the specific cell type, the location in the body, and the microscopic appearance of the tumor.

In parallel with the set of names for malignant tumors, there is a related set of names for benign tumors: an *adenoma*, for example, is a benign epithelial tumor with a glandular organization; the corresponding type of malignant tumor is an *adenocarcinoma* (Figure 20–3). Similarly, a *chondroma* and a *chondrosarcoma* are, respectively, benign and malignant tumors of cartilage.

Most cancers have characteristics that reflect their origin. Thus, for example, the cells of a *basal-cell carcinoma*, derived from a keratinocyte stem cell in the skin, generally continue to synthesize cytokeratin intermediate filaments, whereas the cells of a *melanoma*, derived from a pigment cell in the skin, will often (but not always) continue to make pigment granules. Cancers originating from different cell types are, in general, very different diseases. Basal-cell carcinomas of the skin, for example, are only locally invasive and rarely metastasize, whereas melanomas can become much more malignant and often form metastases. Basal-cell carcinomas are readily cured by surgery or local irradiation, whereas malignant melanomas, once they have metastasized widely, are usually fatal.

Later, we shall see that there is also a different way to classify cancers, one that cuts across the traditional classification by site of origin: we can classify them in terms of the mutations that make the tumor cells cancerous. The final section of the chapter will show how this information can be crucial to the design and choice of treatments.

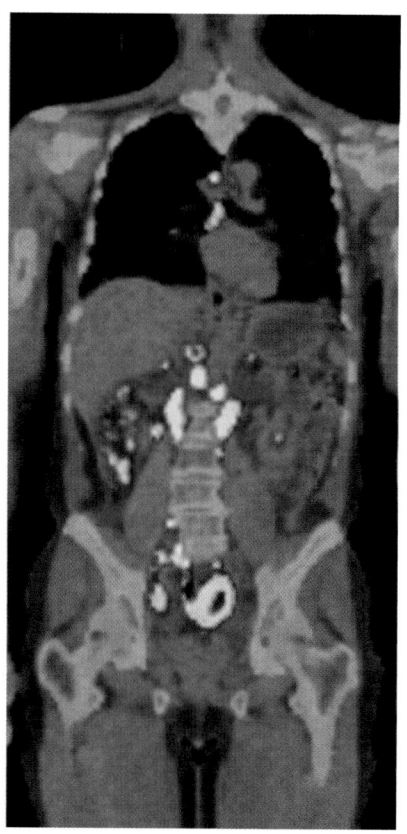

Figure 20–1 Metastasis. Malignant tumors typically give rise to metastases, making the cancer hard to eradicate. Shown in this fusion image is a whole-body scan of a patient with metastatic non-Hodgkin's lymphoma (NHL). The background image of the body's tissues was obtained by CT (computed x-ray tomography) scanning. Overlaid on this image, a PET (positron emission tomography) scan reveals the tumor tissue *(yellow)*, detected by its unusually high uptake of radioactively labeled fluorodeoxyglucose (FDG). High FDG uptake occurs in cells with unusually active glucose uptake and metabolism, which is a characteristic of cancer cells (see Figure 20–12). The yellow spots in the abdominal region reveal multiple metastases. (Courtesy of S.S. Gambhir.)

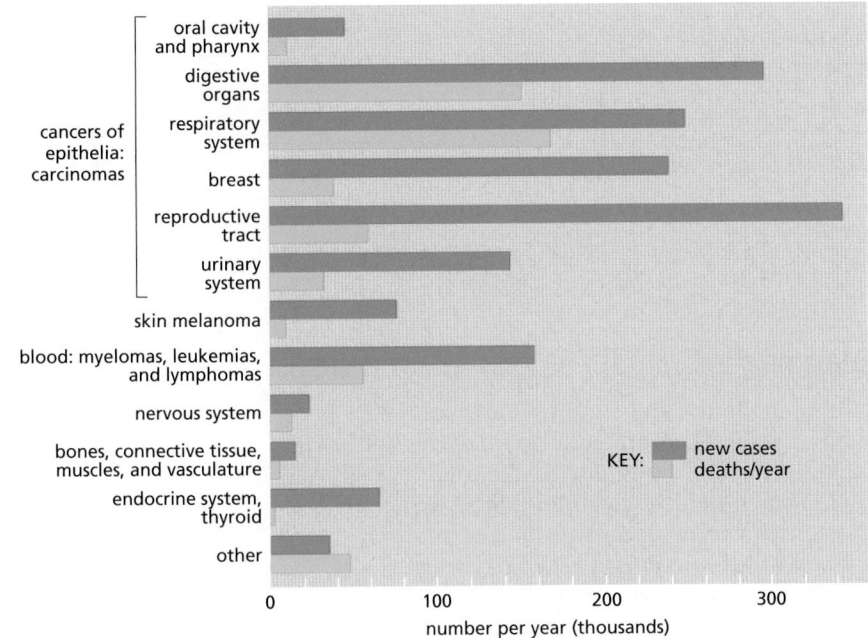

Figure 20–2 **Cancer incidence and mortality in the United States.** The total number of new cases diagnosed in 2012 in the United States was 1,665,540, and total cancer deaths were 585,720. Note that deaths reflect cases diagnosed at many different times and that somewhat less than half of the people who develop cancer die of it. In the world as a whole, the five most common cancers are those of the lung, stomach, breast, colon/rectum, and uterine cervix (included in the figure under the heading of reproductive tract), and the total number of new cancer cases recorded per year is just over 6 million. Skin cancers other than melanomas are not included in these figures, since almost all are cured easily and many are unrecorded.

The data for the United Kingdom are similar. However, incidences are different in some other parts of the world, reflecting widespread exposures to different infectious agents and environmental toxins. (Data from American Cancer Society, Cancer Facts and Figures, 2014.)

Most Cancers Derive from a Single Abnormal Cell

Even when a cancer has metastasized, we can usually trace its origins to a single **primary tumor**, arising in a specific organ. The primary tumor is thought to derive by cell division from a single cell that initially experienced some heritable change. Subsequently, additional changes accumulate in some of the descendants of this cell, allowing them to outgrow, out-divide, and often outlive their neighbors. By the time it is first detected, a typical human cancer will have been developing for many years and will already contain a billion cancer cells or more (**Figure 20–4**). Tumors will usually also contain a variety of other cell types; for example, fibroblasts will be present in the supporting connective tissue associated with a carcinoma, in addition to inflammatory and vascular endothelial cells. How can we be sure that the cancer cells are the clonal descendants of a single abnormal cell?

One way of proving clonal origin is through molecular analysis of the chromosomes in tumor cells. In almost all patients with *chronic myelogenous leukemia* (*CML*), for example, we can distinguish the leukemic white blood cells from the patient's normal cells by a specific chromosomal abnormality: the so-called *Philadelphia chromosome*, created by a translocation between the long arms of chromosomes 9 and 22 (**Figure 20–5**). When the DNA at the site of translocation is cloned and sequenced, it is found that the site of breakage and rejoining of the translocated fragments is identical in all the leukemic cells in any given patient, but that this site differs slightly (by a few hundred or thousand base pairs) from one patient to another. This is the expected result if, and only if, the cancer in each patient arises from a unique accident occurring in a single cell. We will see later

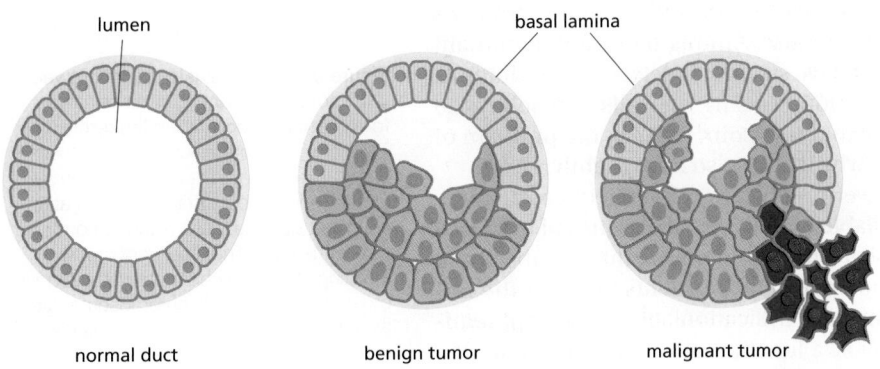

Figure 20–3 **Benign versus malignant tumors.** A benign glandular tumor (*pink* cells; an adenoma) remains inside the basal lamina (*yellow*) that marks the boundary of the normal structure (a duct, in this example). In contrast, a malignant glandular tumor (*red* cells; an adenocarcinoma) can develop from a benign tumor cell, and it destroys the integrity of the tissue, as shown. There are many different forms that such tumors may take.

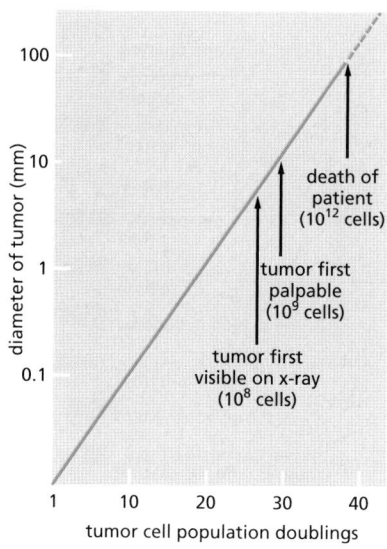

Figure 20–4 **The growth of a typical human tumor, such as a tumor of the breast.** The diameter of the tumor is plotted on a logarithmic scale. Years may elapse before the tumor becomes noticeable. The doubling time of a typical breast tumor, for example, is about 100 days. However, particularly virulent tumors may grow much more rapidly.

how this particular translocation promotes the development of CML by creating a novel hybrid gene encoding a protein that promotes cell proliferation.

Many other lines of evidence, from a variety of cancers, point to the same conclusion: most cancers originate from a single aberrant cell.

Cancer Cells Contain Somatic Mutations

If a single abnormal cell is to give rise to a tumor, it must pass on its abnormality to its progeny: the aberration has to be heritable. Thus, the development of a clone of cancer cells depends on genetic changes. The tumor cells contain **somatic mutations**: they have one or more shared detectable abnormalities in their DNA sequence that distinguish them from the normal cells surrounding the tumor, as in the example of CML just described. (The mutations are called *somatic* because they occur in the soma, or body cells, not in the germ line). Cancers are also driven by *epigenetic changes*—persistent, heritable changes in gene expression that result from modifications of chromatin structure without alteration of the cell's DNA sequence. But somatic mutations that alter DNA sequence appear to be a fundamental and universal feature, and cancer is in this sense a genetic disease.

Factors that cause genetic changes tend to provoke the development of cancer. Thus, **carcinogenesis** (the generation of cancer) can be linked to *mutagenesis* (the production of a change in the DNA sequence). This correlation is particularly clear for two classes of external agents: (1) *chemical carcinogens* (which typically cause simple local changes in the nucleotide sequence), and (2) *radiation* such as x-rays (which typically cause chromosome breaks and translocations) or ultraviolet (UV) light (which causes specific DNA base alterations).

As would be expected, people who have inherited a genetic defect in one of several DNA repair mechanisms, causing their cells to accumulate mutations at an elevated rate, run a heightened risk of cancer. Those with the disease *xeroderma pigmentosum*, for example, have defects in the system that repairs DNA damage induced by UV light, and they have a greatly increased incidence of skin cancers.

A Single Mutation Is Not Enough to Change a Normal Cell into a Cancer Cell

An estimated 10^{16} cell divisions occur in a normal human body in the course of a typical lifetime; in a mouse, with its smaller number of cells and its shorter lifespan, the number is about 10^{12}. Even in an environment that is free of mutagens, mutations would occur spontaneously at an estimated rate of about 10^{-6} mutations per gene per cell division—a value set by fundamental limitations on the accuracy of DNA replication and repair (see pp. 237–238). Thus, in a typical lifetime, every single gene is likely to have undergone mutation on about 10^{10} separate occasions in a human, or on about 10^6 occasions in a mouse. Among the resulting mutant cells, we might expect a large number that have sustained deleterious mutations in genes that regulate cell growth and division, causing the cells to disobey the normal restrictions on cell proliferation. From this point of view, the problem of cancer seems to be not why it occurs, but why it occurs so infrequently.

Clearly, if a mutation in a single gene were enough to convert a typical healthy cell into a cancer cell, we would not be viable organisms. Many lines of evidence indicate that the development of a cancer typically requires that a substantial number of independent, rare genetic and epigenetic accidents occur in the lineage that emanates from a single cell. One such indication comes from epidemiological studies of the incidence of cancer as a function of age (**Figure 20–6**). If a

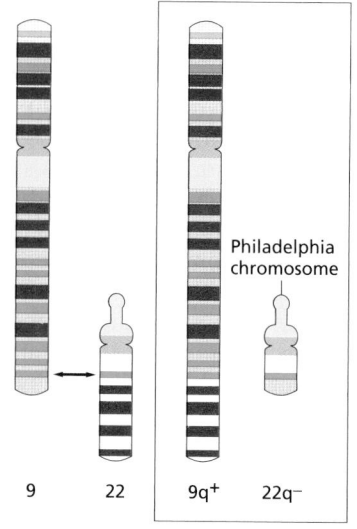

Figure 20–5 **The translocation between chromosomes 9 and 22 responsible for chronic myelogenous leukemia.** The normal structures of chromosomes 9 and 22 are shown at the left. When a translocation occurs between them at the indicated site, the result is the abnormal pair at the right. The smaller of the two resulting abnormal chromosomes (22q⁻) is called the Philadelphia chromosome, after the city where the abnormality was first recorded.

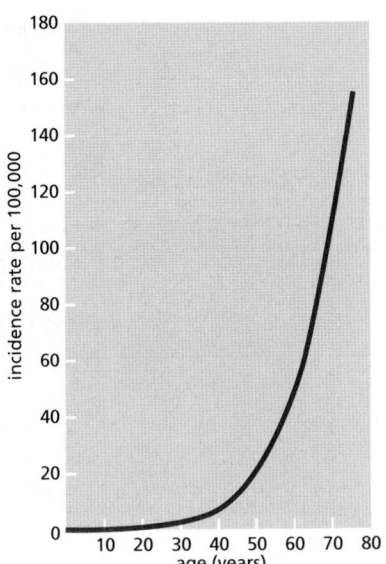

Figure 20–6 **Cancer incidence as a function of age.** The number of newly diagnosed cases of colon cancer in women in England and Wales in 1 year is plotted as a function of age at diagnosis, relative to the total number of individuals in each age group. The incidence of cancer rises steeply as a function of age. If only a single mutation were required to trigger the cancer and this mutation had an equal chance of occurring at any time, the incidence of this cancer would be the same at all ages. Analyses of this type suggest that the development of a solid tumor instead requires five to eight independent accidents ("hits") that occur randomly over time. This calculation assumes that the mutation rate remains constant as a cancer evolves, where in fact it often increases (see p. 1097). (Data from C. Muir et al., Cancer Incidence in Five Continents, Vol. V. Lyon: International Agency for Research on Cancer, 1987.)

single mutation were responsible for cancer, occurring with a fixed probability per year, the chance of developing cancer in any given year of life should be independent of age. In fact, for most types of cancer, the incidence rises steeply with age—as would be expected if cancer is caused by a progressive, random accumulation of a set of mutations in a single lineage of cells.

As discussed later, these indirect arguments have now been confirmed by systematically sequencing the genomes of the tumor cells from individual cancer patients and cataloging the mutations that they contain.

Cancers Develop Gradually from Increasingly Aberrant Cells

For those cancers known to have a specific external cause, the disease does not usually become apparent until long after exposure to the causal agent. The incidence of lung cancer, for example, does not begin to rise steeply until after decades of heavy smoking (**Figure 20–7**). Similarly, the incidence of leukemias in Hiroshima and Nagasaki did not show a marked rise until about 5 years after the explosion of the atomic bombs, and industrial workers exposed for a limited period to chemical carcinogens do not usually develop the cancers characteristic of their occupation until 10, 20, or even more years after the exposure. During this long incubation period, the prospective cancer cells undergo a succession of changes, and the same presumably applies to cancers where the initial genetic lesion has no such obvious external cause.

The concept that the development of a cancer requires a gradual accumulation of mutations in a number of different genes helps to explain the well-known phenomenon of **tumor progression**, whereby an initial mild disorder of cell behavior evolves gradually into a full-blown cancer. Chronic myelogenous leukemia again provides a clear example. It begins as a disorder characterized by a nonlethal overproduction of white blood cells and continues in this form for several years before changing into a much more rapidly progressing illness that usually ends in death within a few months. In the early chronic phase, the leukemic cells are distinguished mainly by the chromosomal translocation (the Philadelphia chromosome) mentioned previously, although there may well be other, less visible

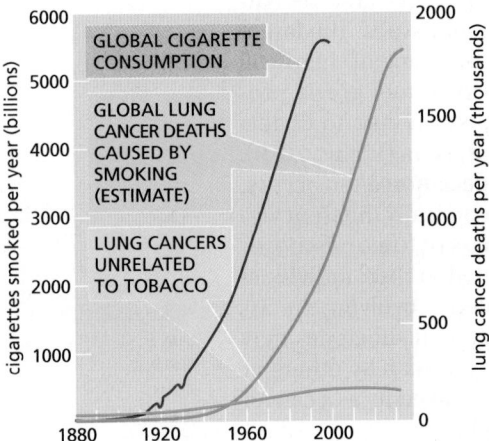

Figure 20–7 **Smoking and the onset of lung cancer.** A major increase in cigarette smoking *(red line)* has caused a dramatic rise in lung cancer deaths *(green line)*, with a lag time of about 35 years. Because global cigarette smoking peaked in 1990, global lung cancer deaths are expected to decline after a similar lag. (Data from R.N. Proctor, *Nat. Rev. Cancer* 1:82–86, 2001).

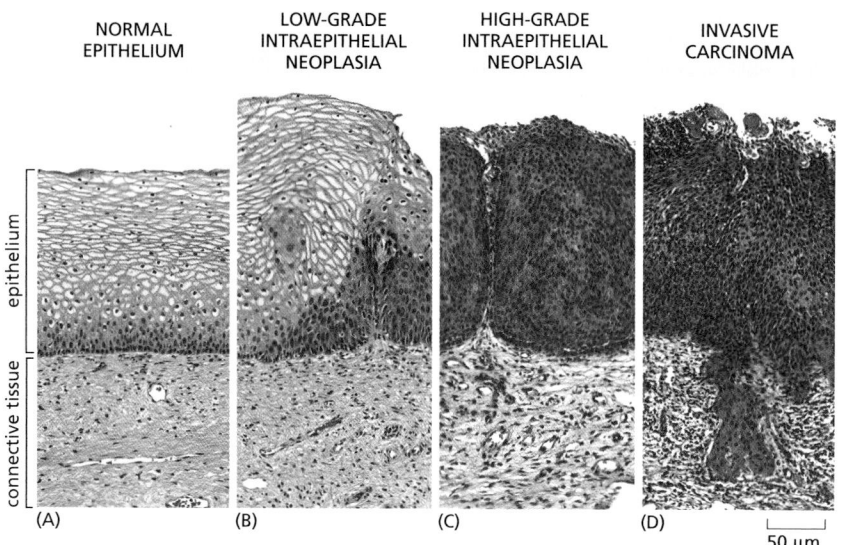

NORMAL
EPITHELIUM

LOW-GRADE
INTRAEPITHELIAL
NEOPLASIA

HIGH-GRADE
INTRAEPITHELIAL
NEOPLASIA

INVASIVE
CARCINOMA

epithelium

connective tissue

(A)   (B)   (C)   (D)

50 μm

Figure 20–8 Stages of progression in the development of cancer of the epithelium of the uterine cervix. Pathologists use standardized terminology to classify the types of disorders they see, so as to guide the choice of treatment. (A) In a stratified squamous epithelium, dividing cells are confined to the basal layer. (B) In this low-grade intraepithelial neoplasia (right half of image), dividing cells can be found throughout the lower third of the epithelium; the superficial cells are still flattened and show signs of differentiation, but this is incomplete. (C) In high-grade intraepithelial neoplasia, cells in all the epithelial layers are proliferating and exhibit defective differentiation. (D) True malignancy begins when the cells move through or destroy the basal lamina that underlies the basal layer of epithelium and invade the underlying connective tissue. (Photographs courtesy of Andrew J. Connolly.)

genetic or epigenetic changes. In the subsequent acute phase, cells that show not only the translocation but also several other chromosomal abnormalities overrun the hemopoietic (blood-forming) system. It appears that cells from the initial mutant clone have undergone further mutations that make them proliferate even more vigorously, so that they come to outnumber both the normal blood cells and their ancestors with the primary chromosomal translocation.

Carcinomas and other solid tumors evolve in a similar way (**Figure 20–8**). Although many such cancers in humans are not diagnosed until a relatively late stage, in some cases it is possible to observe the earlier steps and, as we shall see later, to relate them to specific genetic changes

Tumor Progression Involves Successive Rounds of Random Inherited Change Followed by Natural Selection

From all the evidence, therefore, it seems that cancers arise by a process in which an initial population of slightly abnormal cells—descendants of a single abnormal ancestor—evolve from bad to worse through successive cycles of random inherited change followed by natural selection. Correspondingly, tumors grow in fits and starts, as additional advantageous inherited changes arise and the cells bearing them flourish. Tumor progression involves a large element of chance and usually takes many years, which may be why the majority of us will die of causes other than cancer.

At each stage of progression, some individual cell acquires an additional mutation or epigenetic change that gives it a selective advantage over its neighbors, making it better able to thrive in its environment—an environment that, inside a tumor, may be harsh, with low levels of oxygen, scarce nutrients, and the natural barriers to growth presented by the surrounding normal tissues. The larger the number of tumor cells, the higher the chance that at least one of them will undergo a change that favors it over its neighbors. Thus, as the tumor grows, progression accelerates. The offspring of the best-adapted cells continue to divide, eventually producing the dominant clones in the developing lesion (**Figure 20–9**).

Just as in the evolution of plants and animals, a kind of speciation often occurs: the original cancer cell lineage can diversify to give many genetically different vigorous subclones of cells. These may coexist in the same mass of tumor tissue; or they may migrate and colonize separate environments suited to their individual quirks, where they settle, thrive, and progress as independently evolving metastases. As new mutations arise within each tumor mass, different subclones may gain an advantage and come to predominate, only to be overtaken by others or outgrown by their own sub-subclones. The increasing genetic diversity as a cancer progresses is one of the chief factors that make cures difficult.

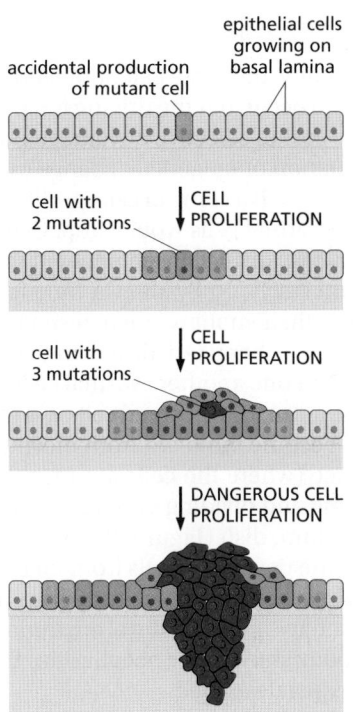

Figure 20–9 **Clonal evolution.** In this schematic diagram, a tumor develops through repeated rounds of mutation and proliferation, giving rise eventually to a clone of fully malignant cancer cells. At each step, a single cell undergoes a mutation that either enhances cell proliferation or decreases cell death, so that its progeny become the dominant clone in the tumor. Proliferation of each clone hastens the occurrence of the next step of tumor progression by increasing the size of the cell population that is at risk of undergoing an additional mutation. The final step depicted here is invasion through the basement membrane, an initial step in metastasis. In reality, there are more than the three steps shown here, and a combination of genetic and epigenetic changes are involved. Not shown here is the fact that, over time, a variety of competing subclones will often arise in a tumor. As we will discuss later, this heterogeneity complicates cancer therapies (see Figure 20–30).

Human Cancer Cells Are Genetically Unstable

Most human cancer cells accumulate genetic changes at an abnormally rapid rate and are said to be **genetically unstable**. The extent of this instability and its molecular origins differ from cancer to cancer and from patient to patient, as we shall discuss in a later section. The basic phenomenon was evident even before modern molecular analyses. For example, the cells of many cancers show grossly abnormal sets of chromosomes, with duplications, deletions, and translocations that are visible at mitosis (**Figure 20–10**). When the cells are maintained in culture, these patterns of chromosomal disruption can often be seen to evolve rapidly and in a seemingly haphazard way. And for many years, pathologists have used an abnormal appearance of the cell nucleus to identify and classify cancer cells in tumor biopsies; in particular, cancer cells can contain an unusually large amount of heterochromatin—a condensed form of interphase chromatin that silences genes (see pp. 194–195). This suggested that epigenetic changes of chromatin structure can also contribute to the cancer cell phenotype, as recently confirmed by molecular analysis.

The genetic instability observed in cancer cells can arise from defects in the ability to repair DNA damage or to correct replication errors of various kinds. These alterations lead to changes in DNA sequence and produce rearrangements such as DNA translocations and duplications. Also common are defects in chromosome segregation during mitosis, which provide another possible source of chromosome instability and changes in karyotype.

From an evolutionary perspective, none of this should be a surprise: anything that increases the probability of random changes in gene function heritable from one cell generation to the next—and that is not too deleterious—is likely to speed the evolution of a clone of cells toward malignancy, thereby causing this property to be selected for during tumor progression.

(A)

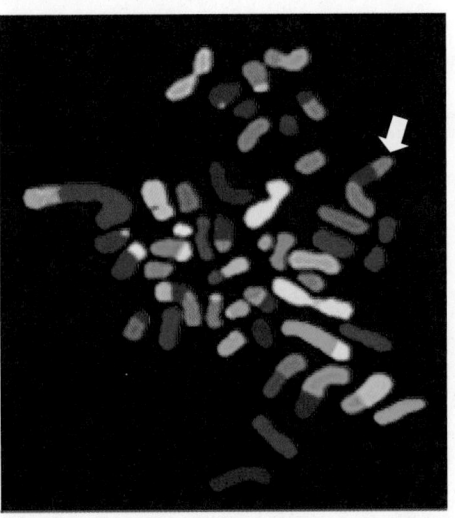

(B)

Figure 20–10 **Chromosomes from a breast tumor displaying abnormalities in structure and number.** Chromosomes were prepared from a breast tumor cell in metaphase, spread on a glass slide, and stained with (A) a general DNA stain or (B) a combination of fluorescently labeled DNA molecules that color each normal human chromosome differently (see Figure 4–10). The staining (displayed in false color) shows multiple translocations, including a doubly translocated chromosome *(white arrow)* that is made up of two pieces of chromosome 8 *(green-brown)* and a piece of chromosome 17 *(purple)*. The karyotype also contains 48 chromosomes, instead of the normal 46. (Courtesy of Joanne Davidson and Paul Edwards.)

Cancer Cells Display an Altered Control of Growth

Mutability and large cell population numbers create the opportunities for mutations to occur, but the driving force for development of a cancer has to come from some sort of selective advantage possessed by the mutant cells. Most obviously, a mutation or epigenetic change can confer such an advantage by increasing the rate at which a clone of cells proliferates or by enabling it to continue proliferating when normal cells would stop. Cancer cells that can be grown in culture, or cultured cells artificially engineered to contain the types of mutations encountered in cancers, typically show a **transformed** phenotype. They are abnormal in their shape, their motility, their responses to growth factors in the culture medium, and, most characteristically, in the way they react to contact with the substratum and with one another. Normal cells will not divide unless they are attached to the substratum; transformed cells will often divide even if held in suspension. Normal cells become inhibited from moving and dividing when the culture reaches confluence (where the cells are touching one another); transformed cells continue moving and dividing even after confluence, and so pile up in layer upon layer in the culture dish (**Figure 20–11**). In addition, transformed cells no longer require all of the positive signals from their surroundings that normal cells require.

Their behavior in culture gives a hint of the ways in which cancer cells may misbehave in their natural environment, embedded in a tissue. But cancer cells in the body show other peculiarities that mark them out from normal cells, beyond those just described.

Cancer Cells Have an Altered Sugar Metabolism

Given sufficient oxygen, normal adult tissue cells will generally fully oxidize almost all the carbon in the glucose they take up to CO_2, which is lost from the body as a waste product. A growing tumor needs nutrients in abundance to provide the building blocks to make new macromolecules. Correspondingly, most tumors have a metabolism more similar to that of a growing embryo than to that of normal adult tissue. Tumor cells consume glucose avidly, importing it from the blood at a rate that can be as much as 100 times higher than neighboring normal cells. Moreover, only a small fraction of this imported glucose is used for production of ATP by oxidative phosphorylation. Instead, a great deal of lactate is produced, and many of the remaining carbon atoms derived from glucose are diverted for use as raw materials for synthesis of the proteins, nucleic acids, and lipids required for tumor growth (**Figure 20–12**).

This tendency of tumor cells to de-emphasize oxidative phosphorylation even when oxygen is plentiful, while at the same time taking up large quantities of glucose, can be shown to promote cancer cell growth and is called the *Warburg*

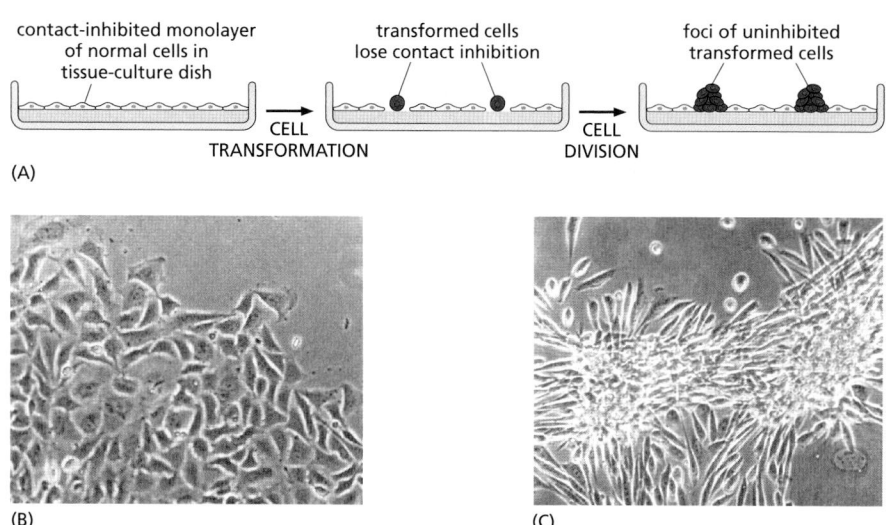

(A)

(B)   (C)

Figure 20–11 **Loss of contact inhibition by cancer cells in cell culture.** Most normal cells stop proliferating once they have carpeted the dish with a single layer of cells: proliferation seems to depend on contact with the dish, and to be inhibited by contacts with other cells—a phenomenon known as "contact inhibition." Cancer cells, in contrast, usually disregard these restraints and continue to grow, so that they pile up on top of one another, as shown (Movie 20.2). (A) Schematic drawing. (B and C) Light micrographs of normal (B) and transformed (C) fibroblasts. (B and C, courtesy of Lan Bo Chen.)

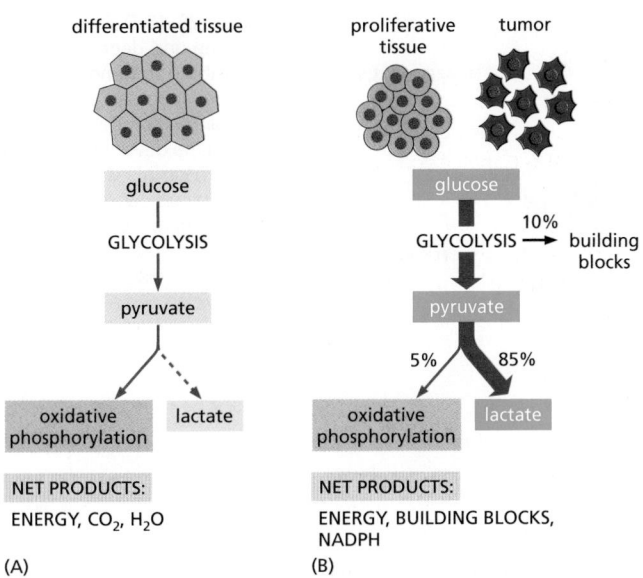

Figure 20–12 **The Warburg effect in tumor cells reflects a dramatic change in glucose uptake and sugar metabolism.** (A) Cells that are not proliferating will normally oxidize nearly all of the glucose that they import from the blood to produce ATP through the oxidative phosphorylation that takes place in their mitochondria. Only when deprived of oxygen will these cells generate most of their ATP from glycolysis, converting the pyruvate produced to lactate in order to regenerate the NAD$^+$ that they need to keep glycolysis going (see Figure 2–47). (B) Tumor cells, by contrast, will generally produce abundant lactate even in the presence of oxygen. This results from a greatly increased rate of glycolysis that is fed by a very large increase in the rate of glucose import. In this way, tumor cells resemble the rapidly proliferating cells in embryos (and during tissue repair), which likewise require for biosynthesis a large supply of the small-molecule building blocks that can be produced from imported glucose (see also Figure 20–26).

effect—so named because Otto Warburg first noticed the phenomenon in the early twentieth century. It is this abnormally high glucose uptake that allows tumors to be selectively imaged in whole-body scans (see Figure 20–1), thereby providing a way to monitor cancer progression and responses to treatment.

Cancer Cells Have an Abnormal Ability to Survive Stress and DNA Damage

In a large multicellular organism, there are powerful safety mechanisms that guard against the trouble that can be caused by damaged and deranged cells. For example, internal disorder gives rise to danger signals in the faulty cell, activating protective devices that can eventually lead to apoptosis (see Chapter 18). To survive, cancer cells require additional mutations to elude or break through these defenses against cellular misbehavior.

Cancer cells are found to contain mutations that drive the cell into an abnormal state, where metabolic processes may be unbalanced and essential cell components may be produced in ill-matched proportions. States of this type, where the cell's homeostatic mechanisms are inadequate to cope with an imposed disturbance, are loosely referred to as states of *cell stress*. As one example, chromosome breakage and other forms of DNA damage are commonly observed during the development of cancer, reflecting the genetic instability that cancer cells display. Thus, to survive and divide without limit, a prospective cancer cell must accumulate mutations that disable the normal safety mechanisms that would otherwise induce a cell that is stressed, in this or in other ways, to commit suicide. In fact, one of the most important properties of many types of cancer cells is that they fail to undergo apoptosis when a normal cell would do so (**Figure 20–13**).

While cancer cells tend to avoid apoptosis, this does not mean that they rarely die. On the contrary, in the interior of a large solid tumor, cell death often occurs on a massive scale: living conditions are difficult, with severe competition among the cancer cells for oxygen and nutrients. Many die, but typically much more by necrosis than by apoptosis (**Figure 20–14**). The tumor grows because the cell birth rate outpaces the cell death rate, but often by only a small margin. For this reason, the time that a tumor takes to double in size can be far longer than the cell-cycle time of the tumor cells.

Human Cancer Cells Escape a Built-in Limit to Cell Proliferation

Many normal human cells have a built-in limit to the number of times they can divide when stimulated to proliferate in culture: they permanently stop dividing

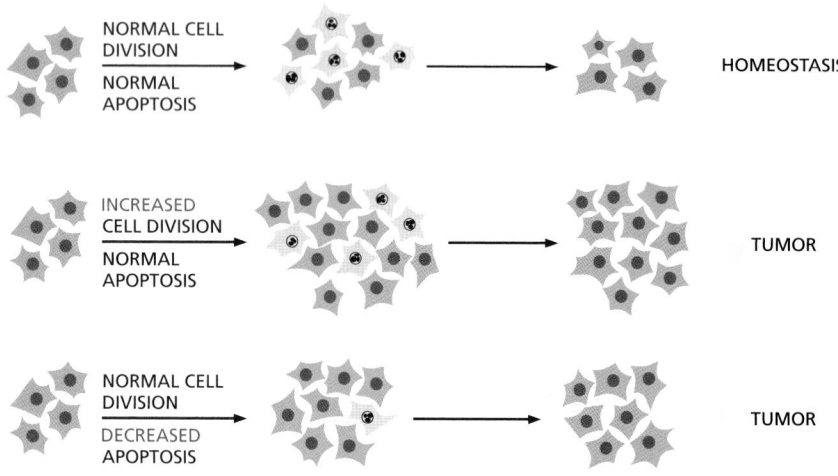

Figure 20–13 **Both increased cell division and decreased apoptosis can contribute to tumorigenesis.** In normal tissues, apoptosis balances cell division to maintain homeostasis (see Movie 18.1). During the development of cancer, either an increase in cell division or an inhibition of apoptosis can lead to the increased cell numbers important for tumorigenesis. The cells fated to undergo apoptosis are *gray* in this diagram. Both an increase in cell division and a decrease in apoptosis normally contribute to tumor growth.

after a certain number of population doublings (25–50 for human fibroblasts, for example). This cell-division-counting mechanism is termed **replicative cell senescence**, and it generally depends on the progressive shortening of the telomeres at the ends of chromosomes, a process that eventually changes their structure (discussed in Chapter 17). As discussed in Chapter 5, the replication of telomere DNA during S phase depends on the enzyme *telomerase*, which maintains a special telomeric DNA sequence that promotes the formation of protein cap structures to protect chromosome ends. Because many proliferating human cells (stem cells being an exception) are deficient in telomerase, their telomeres shorten with every division, and their protective caps deteriorate, creating a DNA damage signal. Eventually, the altered chromosome ends can trigger a permanent cell-cycle arrest, causing a normal cell to die.

Human cancer cells avoid replicative cell senescence in one of two ways. They can maintain the activity of telomerase as they proliferate, so that their telomeres do not shorten or become uncapped, or they can evolve an alternate mechanism based on homologous recombination (called ALT) for elongating their chromosome ends. Regardless of the strategy used, the result is that the cancer cells continue to proliferate under conditions when normal cells would stop.

The Tumor Microenvironment Influences Cancer Development

While the cancer cells in a tumor are the bearers of dangerous mutations and are often grossly abnormal, the other cells in the tumor—especially those of the supporting connective tissue, or **stroma**—are far from passive bystanders. The

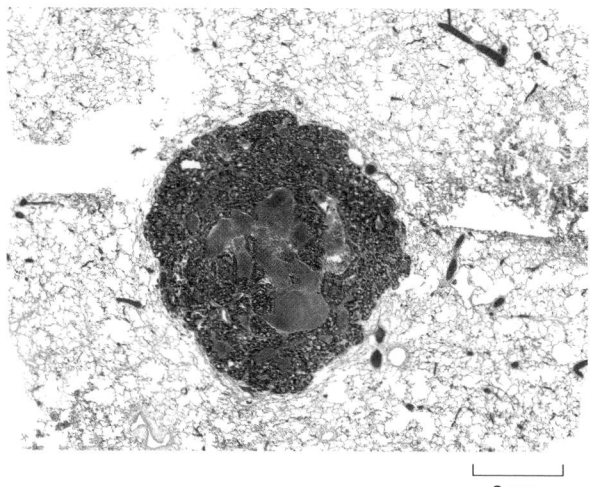

2 mm

Figure 20–14 **Cross-section of a colon adenocarcinoma that has metastasized to the lung.** This tissue slice shows well-differentiated colorectal cancer cells forming cohesive glands in the lung. The metastasis has central pink areas of necrosis where dying cancer cells have outgrown their blood supply. Such anoxic regions are common in the interior of large tumors. (Courtesy of Andrew J. Connolly.)

development of a tumor relies on a two-way communication between the tumor cells and the tumor stroma, just as the normal development of epithelial organs relies on communication between epithelial cells and mesenchymal cells (discussed in Chapter 22).

The stroma provides a framework for the tumor. It is composed of normal connective tissue containing fibroblasts and inflammatory white blood cells, as well as the endothelial cells that form blood and lymphatic vessels with their attendant pericytes and smooth muscle cells (Figure 20–15). As a carcinoma progresses, the cancer cells induce changes in the stroma by secreting signal proteins that alter the behavior of the stromal cells, as well as proteolytic enzymes that modify the extracellular matrix. The stromal cells in turn act back on the tumor cells, secreting signal proteins that stimulate cancer cell growth and division as well as proteases that further remodel the extracellular matrix. In these ways, the tumor and its stroma evolve together, like weeds and the ecosystem that they invade, and the tumor becomes dependent on its particular stromal cells. Experiments using mice indicate that the growth of some transplanted carcinomas depends on the tumor-associated fibroblasts and normal fibroblasts will not do. Such environmental requirements help to protect us from cancer, as we discuss next in considering the critical phenomenon called metastasis.

Cancer Cells Must Survive and Proliferate in a Foreign Environment

Cancer cells generally need to spread and multiply at new sites in the body in order to kill us, through a process called metastasis. This is the most deadly—and least understood—aspect of cancer, being responsible for 90% of cancer-associated deaths. By spreading through the body, a cancer becomes almost impossible to eradicate by either surgery or local irradiation. **Metastasis** is itself a multistep process: the cancer cells first have to invade local tissues and vessels, move through the circulation, leave the vessels, and then establish new cellular colonies at distant sites (Figure 20–16). Each of these events is complex, and most of the molecular mechanisms involved are not yet clear.

For a cancer cell to become dangerous, it must break free of constraints that keep normal cells in their proper places and prevent them from invading neighboring tissues. Invasiveness is thus one of the defining properties of malignant tumors, which show a disorganized pattern of growth and ragged borders, with extensions into the surrounding tissue (see, for example, Figure 20–8). Although the underlying molecular changes are not well understood, invasiveness almost certainly requires a disruption of the adhesive mechanisms that normally keep cells tethered to their proper neighbors and to the extracellular matrix. For carcinomas, this change resembles the *epithelial–mesenchymal transition* (*EMT*) that occurs in some epithelial tissues during normal development (see p. 1042).

The next step in metastasis—the establishment of colonies in distant organs—begins with entry into the circulation: the invasive cancer cells must penetrate the

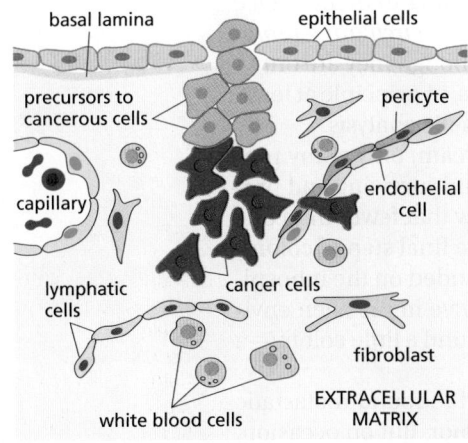

Figure 20–15 **The tumor microenvironment plays a role in tumorigenesis.** Tumors consist of many cell types, including cancer cells, endothelial cells, pericytes (vascular smooth muscle cells), fibroblasts, and inflammatory white blood cells. Communication among these and other cell types plays an important part in tumor development. Note, however, that only the cancer cells are thought to be genetically abnormal in a tumor.

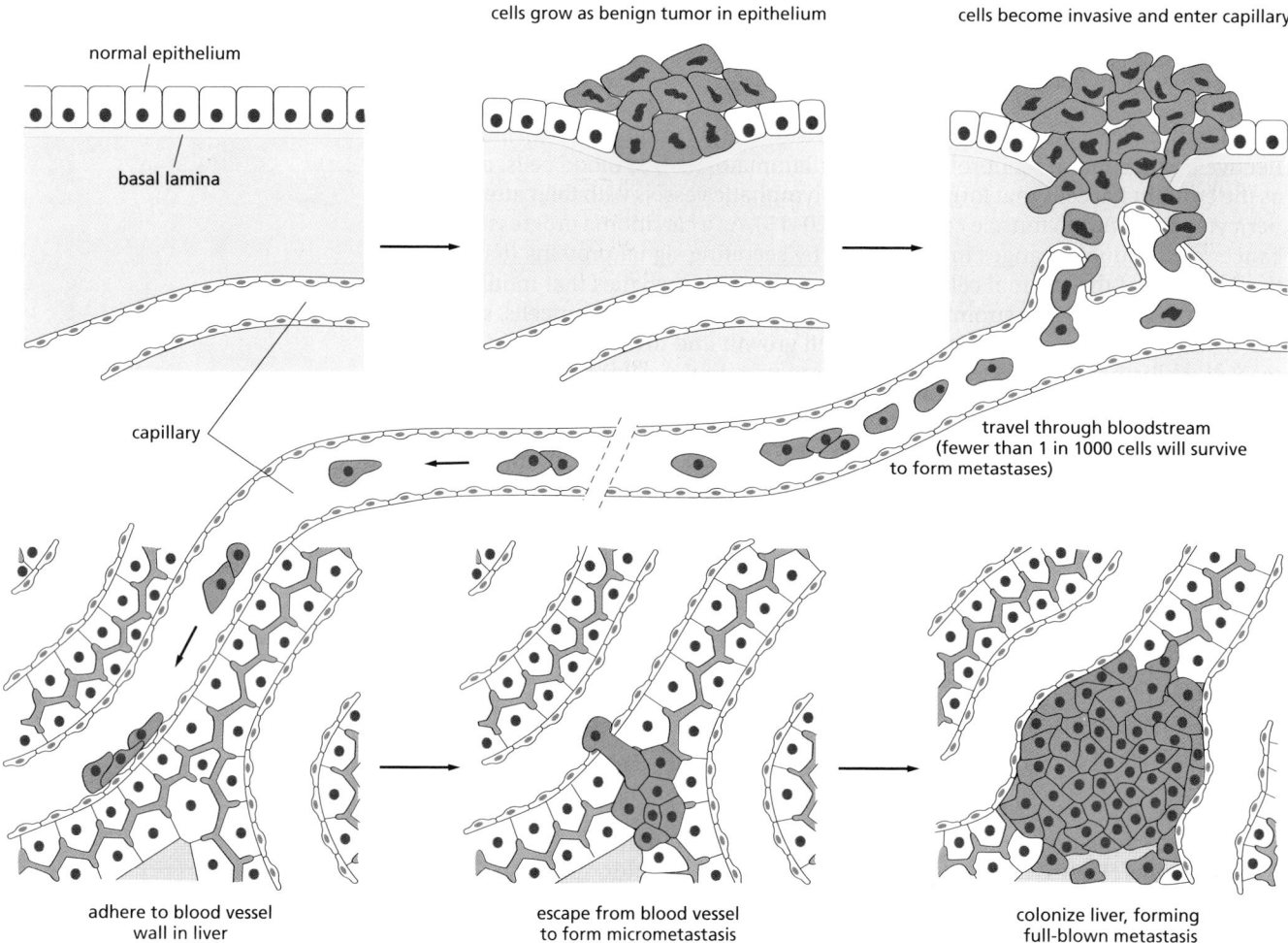

normal epithelium

basal lamina

cells grow as benign tumor in epithelium

cells become invasive and enter capillary

capillary

travel through bloodstream
(fewer than 1 in 1000 cells will survive
to form metastases)

adhere to blood vessel
wall in liver

escape from blood vessel
to form micrometastasis

colonize liver, forming
full-blown metastasis

Figure 20–16 Steps in the process of metastasis. This example illustrates the spread of a tumor from an organ such as the bladder to the liver. Tumor cells may enter the bloodstream directly by crossing the wall of a blood vessel, as diagrammed here, or, more commonly perhaps, by crossing the wall of a lymphatic vessel that ultimately discharges its contents (lymph) into the bloodstream. Tumor cells that have entered a lymphatic vessel often become trapped in lymph nodes along the way, giving rise to lymph-node metastases.

Studies in animals show that typically far fewer than one in every thousand malignant tumor cells that enter the bloodstream will colonize a new tissue so as to produce a detectable tumor at a new site.

wall of a blood or lymphatic vessel. Lymphatic vessels, being larger and having more flimsy walls than blood vessels, allow cancer cells to enter in small clumps; such clumps may then become trapped in lymph nodes, giving rise to lymph-node metastases. The cancer cells that enter blood vessels, in contrast, seem to do so singly. With modern techniques for sorting cells according to their surface properties, it has become possible in some cases to detect these *circulating tumor cells* (*CTCs*) in samples of blood from cancer patients, even though they are only a minute fraction of the total blood-cell population. These cells, in principle at least, provide a useful sample of the tumor-cell population for genetic analysis.

Of the cancer cells that enter the lymphatics or bloodstream, only a tiny proportion succeed in making their exit, settling in new sites, and surviving and proliferating there as founders of metastases. Experiments show that fewer than one in thousands, perhaps one in millions, manage this feat. The final step of colonization seems to be the most difficult: like the Vikings who landed on the inhospitable shores of Greenland, the migrant cells may fail to survive in the alien environment; or they may only thrive there for a short while to found a little colony—a *micrometastasis*—that then dies out (**Movie 20.3**).

Many cancers are discovered before they have managed to found metastatic colonies and can be cured by destruction of the primary tumor. But on occasion, an undetected micrometastasis will remain dormant for many years, only to reveal its presence by erupting into growth to form a large secondary tumor long after the primary tumor has been removed.

Many Properties Typically Contribute to Cancerous Growth

Clearly, to produce a cancer, a cell must acquire a range of aberrant properties—a collection of subversive new skills—as it evolves. Different cancers require different combinations of these properties. Nevertheless, cancers all share some common features. By definition, they all ignore or misinterpret normal social controls so as to proliferate and spread where normal cells would not. These defining properties are commonly combined with other features that help the miscreants to arise and thrive. A list of the key attributes of cancer cells in general would include the following, all of which we have just discussed:

1. They grow (biosynthesize) when they should not, aided by a metabolism shifted from oxidative phosphorylation toward aerobic glycolysis.
2. They go through the cell-division cycle when they should not.
3. They escape from their home tissues (that is, they are invasive) and survive and proliferate in foreign sites (that is, they metastasize).
4. They have abnormal stress responses, enabling them to survive and continue dividing in conditions of stress that would arrest or kill normal cells, and they are less prone than normal cells to commit suicide by apoptosis.
5. They are genetically and epigenetically unstable.
6. They escape replicative cell senescence, either by producing telomerase or by acquiring another way of stabilizing their telomeres.

In the next section of the chapter, we examine the mutations and molecular mechanisms that underlie these and other properties of cancer cells.

Summary

Cancer cells, by definition, grow and proliferate in defiance of normal controls (that is, they are neoplastic) and are able to invade surrounding tissues and colonize distant organs (that is, they are malignant). By giving rise to secondary tumors, or metastases, they become difficult to eradicate by surgery or local irradiation. Cancers are thought to originate from a single cell that has experienced an initial mutation, but the progeny of this cell must undergo many further changes, requiring additional mutations and epigenetic events, to become cancerous. Tumor progression usually takes many years and reflects the operation of a Darwinian-like process of evolution, in which somatic cells undergo mutation and epigenetic changes accompanied by natural selection.

Cancer cells acquire a variety of special properties as they evolve, multiply, and spread. Their mutant genomes enable them to grow and divide in defiance of the signals that normally keep cell proliferation under tight control. As part of the evolutionary process of tumor progression, cancer cells acquire a collection of additional abnormalities, including defects in the controls that permanently stop cell division or induce apoptosis in response to cell stress or DNA damage, and in the mechanisms that normally keep cells from straying from their proper place. All of these changes increase the ability of cancer cells to survive, grow, and divide in their original tissue and then to metastasize, founding new colonies in foreign environments. The evolution of a tumor also depends on other cells present in the tumor microenvironment, collectively called stromal cells, that the cancer attracts and manipulates.

Since many changes are needed to confer this collection of asocial behaviors, it is not surprising that most cancer cells are genetically and/or epigenetically unstable. This instability is thought to be selected for in the clones of aberrant cells that are able to produce tumors, because it greatly accelerates the accumulation of the further genetic and epigenetic changes that are required for tumor progression.

CANCER-CRITICAL GENES: HOW THEY ARE FOUND AND WHAT THEY DO

As we have seen, cancer depends on the accumulation of inherited changes in somatic cells. To understand it at a molecular level we need to identify the mutations and epigenetic changes involved and to discover how they give rise to cancerous cell behavior. Finding the relevant cells is often easy; they are favored by natural selection and call attention to themselves by giving rise to tumors. But how do we identify those genes with the cancer-promoting changes among all the other genes in the cancerous cells? A typical cancer depends on a whole set of mutations and epigenetic changes—usually a somewhat different set in each individual patient. In addition, a given cancer cell will also contain a large number of somatic mutations that are accidental by-products—so-called *passengers* rather than *drivers*—of its genetic instability, and it can be difficult to distinguish these meaningless changes from those changes that have a causative role in the disease. Despite these difficulties, many of the genes that are repeatedly altered in human cancers have been identified over the past 40 years. We will call such genes, for want of a better term, **cancer-critical genes**, meaning all genes whose alteration contributes to the causation or evolution of cancer by driving tumorigenesis.

In this section, we shall first discuss how cancer-critical genes are identified. We shall then examine their functions and the parts they play in conferring on cancer cells the properties outlined in the first part of the chapter. We shall end the section by discussing colon cancer as an extended example, showing how a succession of changes in cancer-critical genes enables a tumor to evolve from one pattern of bad behavior to another that is worse.

The Identification of Gain-of-Function and Loss-of-Function Cancer Mutations Has Traditionally Required Different Methods

Cancer-critical genes are grouped into two broad classes, according to whether the cancer risk arises from too much activity of the gene product or too little. Genes of the first class, in which a gain-of-function mutation can drive a cell toward cancer, are called **proto-oncogenes**; their mutant, overactive or overexpressed forms are called **oncogenes**. Genes of the second class, in which a loss-of-function mutation can contribute to cancer, are called **tumor suppressor genes**. In either case, the mutation may lead toward cancer directly (by causing cells to proliferate when they should not) or indirectly—for example, by causing genetic or epigenetic instability and so hastening the occurrence of other inherited changes that directly stimulate tumor growth. Those genes whose alteration results in genomic instability represent a subclass of cancer-critical genes that are sometimes called *genome maintenance genes*.

As we shall see, mutations in oncogenes and tumor suppressor genes can have similar effects in promoting the development of cancer; overproduction of a signal for cell proliferation, for example, can result from either kind of mutation. Thus, from the point of view of a cancer cell, oncogenes and tumor suppressor genes—and the mutations that affect them—are flip sides of the same coin. The techniques that led to the discovery of these two categories of genes, however, are quite different.

The mutation of a single copy of a proto-oncogene that converts it to an oncogene has a dominant, growth-promoting effect on a cell (**Figure 20–17**A). Thus, we can identify the oncogene by its effect when it is *added*—by DNA transfection, for example, or through infection with a viral vector—to the genome of a suitable type of tester cell or experimental animal. In the case of the tumor suppressor gene, on the other hand, the cancer-causing alleles produced by the change are generally recessive: often (but not always) both copies of the normal gene must be removed or inactivated in the diploid somatic cell before an effect is seen (Figure 20–17B). This calls for a different experimental approach, one focusing on discovering what is *missing* in the cancer cell.

(A) overactivity mutation (gain of function)

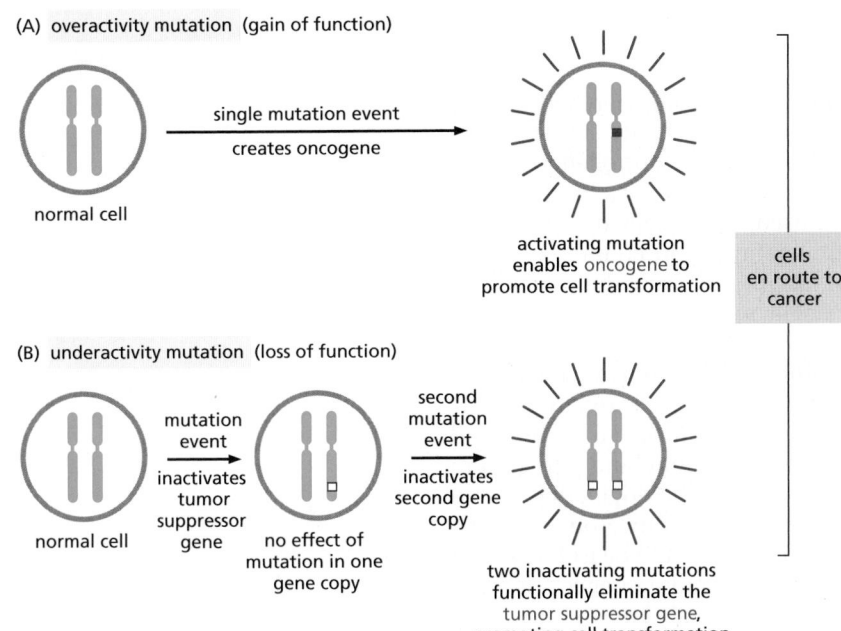

normal cell

single mutation event
creates oncogene

activating mutation
enables oncogene to
promote cell transformation

cells
en route to
cancer

(B) underactivity mutation (loss of function)

normal cell

mutation
event

inactivates
tumor
suppressor
gene

no effect of
mutation in one
gene copy

second
mutation
event

inactivates
second gene
copy

two inactivating mutations
functionally eliminate the
tumor suppressor gene,
promoting cell transformation

Figure 20–17 Cancer-critical mutations fall into two readily distinguishable categories, dominant and recessive. In this diagram, activating mutations are represented by *solid red boxes*, inactivating mutations by *hollow red boxes*. (A) Oncogenes act in a dominant manner: a gain-of-function mutation in a single copy of the cancer-critical gene can drive a cell toward cancer. (B) Mutations in tumor suppressor genes, on the other hand, generally act in a recessive manner: the function of both alleles of the cancer-critical gene must be lost to drive a cell toward cancer. Although in this diagram the second allele of the tumor suppressor gene is inactivated by mutation, it is often inactivated instead by loss of the second chromosome. Not shown is the fact that mutation of some tumor suppressor genes can have an effect even when only one of the two gene copies is damaged.

We begin by discussing some examples of each class of cancer-critical genes to illustrate basic principles. These examples are chosen also for their historical importance: the experiments that led to their discovery—at different times and by different methods—marked turning points in the understanding of cancer.

Retroviruses Can Act as Vectors for Oncogenes That Alter Cell Behavior

The search for the genetic causes of human cancer took a devious route, beginning with clues that came from the study of **tumor viruses**. Although viruses are involved only in a minority of human cancers, a set of viruses that infect animals provided critical early tools for studying cancer.

One of the first animal viruses to be implicated in cancer was discovered over 100 years ago in chickens, when an infectious agent that causes connective-tissue tumors, or sarcomas, was characterized as a virus—the *Rous sarcoma virus*. Like all the other *RNA tumor viruses* discovered since, it is a **retrovirus**. When it infects a cell, its RNA genome is copied into DNA by reverse transcription, and the DNA is inserted into the host genome, where it can persist and be inherited by subsequent generations of cells. Something in the DNA inserted by the Rous sarcoma virus made the host cells cancerous, but what was it? The answer was a surprise. It turned out to be a piece of DNA that was unnecessary for the virus's own survival or reproduction; instead, it was a passenger, a gene called *v-Src*, that the virus had picked up on its travels. *v-Src* was unmistakably similar, but not identical, to a gene—*c-Src*—that was discovered in the normal vertebrate genome. *c-Src* had evidently been caught up accidentally by the retrovirus from the genome of a previously infected host cell, and it had undergone mutation in the process to become an oncogene (*v-Src*).

This Nobel Prize-winning finding was followed by a flood of discoveries of other viral oncogenes carried by retroviruses that cause cancer in nonhuman animals. Each such oncogene turned out to have a counterpart proto-oncogene in the normal vertebrate genome. As was the case for *Src*, these other oncogenes generally differed from their normal counterparts, either in structure or in level of expression. But how did this relate to typical human cancers, most of which are not infectious and in which retroviruses play no part?

Different Searches for Oncogenes Converged on the Same Gene—*Ras*

In an attempt to answer the above question, other researchers searched directly for oncogenes in the genomes of human cancer cells. They did this by searching for DNA fragments from cancer cells that could provoke uncontrolled proliferation when introduced into noncancerous cell lines. As tester cells for the assay, cell lines derived from mouse fibroblasts were used. These cells had been previously selected for their ability to proliferate indefinitely in culture, and they are thought to already contain alterations that take them part of the way toward malignancy. For this reason, the addition of a single oncogene can sometimes be enough to produce a dramatic effect.

When DNA was extracted from the human tumor cells, broken into fragments, and introduced into the cultured cells, occasional colonies of abnormally proliferating cells began to appear in the culture dish. These cells showed a transformed phenotype, outgrowing the untransformed cells in the culture and piling up in layer upon layer (see Figure 20–11). Each colony was a clone originating from a single cell that had incorporated a DNA fragment that drove cancerous behavior. This fragment, which carried markers of its human origin, could be isolated from the transformed cultured mouse cells. And once isolated and sequenced, it could be recognized: it contained a human version of a gene already known from study of a retrovirus that caused tumors in rats—an oncogene called *v-Ras*.

The newly discovered oncogene was clearly derived by mutation from a normal human gene, one of a small family of proto-oncogenes called **Ras**. This discovery in the early 1980s of the same oncogene in human tumor cells and in an animal tumor virus was electrifying. The implication that cancers are caused by mutations in a limited number of cancer-critical genes transformed our understanding of the molecular biology of cancer.

As discussed in Chapter 15, normal Ras proteins are monomeric GTPases that help transmit signals from cell-surface receptors to the cell interior (see Movie 15.7). The *Ras* oncogenes isolated from human tumors contain point mutations that create a hyperactive Ras protein that cannot shut itself off by hydrolyzing its bound GTP to GDP. Because this makes the protein hyperactive, its effect is dominant—that is, only one of the cell's two gene copies needs to change to have an effect. One or another of the three human *Ras* family members is mutated in perhaps 30% of all human cancers. *Ras* genes are thus among the most important of all cancer-critical genes.

Genes Mutated in Cancer Can Be Made Overactive in Many Ways

Figure 20–18 summarizes the types of accidents that can convert a proto-oncogene into an oncogene. (1) A small change in DNA sequence such as a point

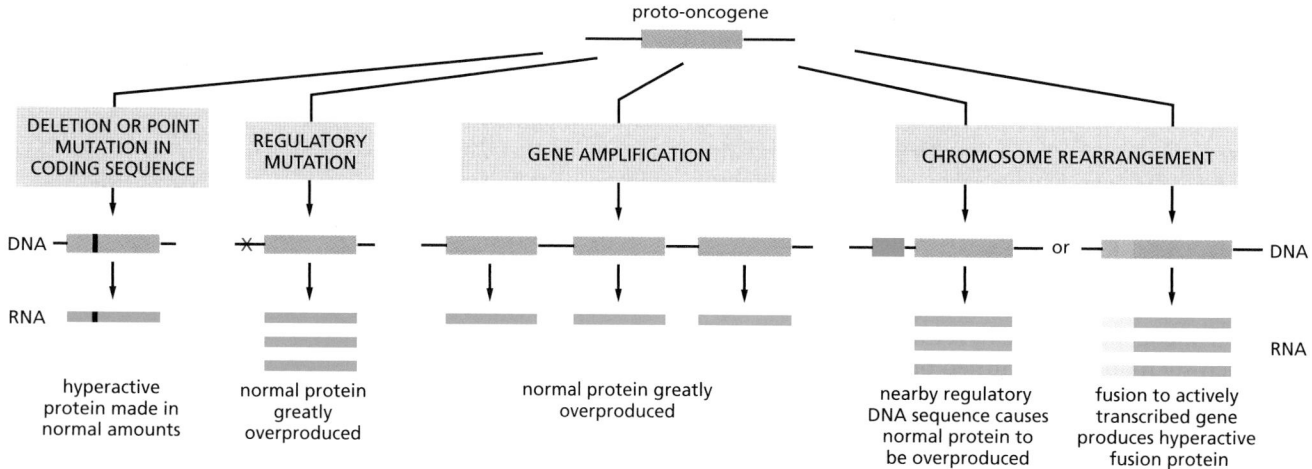

Figure 20–18 The types of accidents that can convert a proto-oncogene into an oncogene.

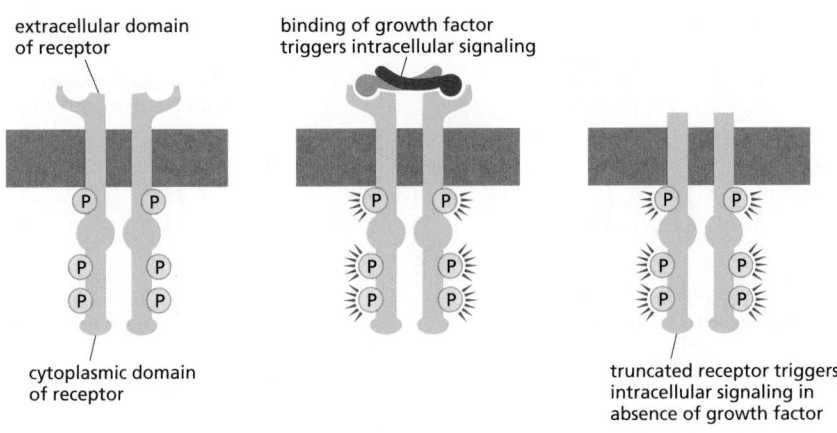

extracellular domain of receptor

binding of growth factor triggers intracellular signaling

cytoplasmic domain of receptor

truncated receptor triggers intracellular signaling in absence of growth factor

Figure 20–19 **Mutation of the epidermal growth factor (EGF) receptor can make it active even in the absence of EGF, and consequently oncogenic.** Only one of the possible types of activating mutations is illustrated here.

mutation or deletion may produce a hyperactive protein when it occurs within a protein-coding sequence, or lead to protein overproduction when it occurs within a regulatory region for that gene. (2) Gene amplification events, such as those that can be caused by errors in DNA replication, may produce extra gene copies; this can lead to overproduction of the protein. (3) A chromosomal rearrangement—involving the breakage and rejoining of the DNA helix—may either change the protein-coding region, resulting in a hyperactive fusion protein, or alter the control regions for a gene so that a normal protein is overproduced.

As one example, the receptor for the extracellular signal protein *epidermal growth factor (EGF)* can be activated by a deletion that removes part of its extracellular domain, causing it to be active even in the absence of EGF (**Figure 20–19**). It thus produces an inappropriate stimulatory signal, like a faulty doorbell that rings even when nobody is pressing the button. Mutations of this type are frequently found in the most common type of human brain tumor, called glioblastoma.

As another example, the *Myc protein*, which acts in the nucleus to stimulate cell growth and division (see Chapter 17), generally contributes to cancer by being overproduced in its normal form. In some cases, the gene is amplified—that is, errors of DNA replication lead to the creation of large numbers of gene copies in a single cell. Or a point mutation can stabilize the protein, which normally turns over very rapidly. More commonly, the overproduction appears to be due to a change in a regulatory element that acts on the gene. For example, a chromosomal translocation can inappropriately bring powerful gene regulatory sequences next to the *Myc* protein-coding sequence, so as to produce unusually large amounts of *Myc* mRNA. Thus, in Burkitt's lymphoma, a translocation brings the *Myc* gene under the control of sequences that normally drive the expression of antibody genes in B lymphocytes. As a result, the mutant B cells tend to proliferate excessively and form a tumor. Different specific chromosome translocations are common in other cancers.

Studies of Rare Hereditary Cancer Syndromes First Identified Tumor Suppressor Genes

Identifying a gene that has been inactivated in the genome of a cancer cell requires a different strategy from finding a gene that has become hyperactive: one cannot, for example, use a cell transformation assay to identify something that simply is not there. The key insight that led to the discovery of the first tumor suppressor gene came from studies of a rare type of human cancer, **retinoblastoma**, which arises from cells in the retina of the eye that are converted to a cancerous state by an unusually small number of mutations. As often happens in biology, the discovery arose from examination of a special case, but it turned out to reveal a gene of widespread importance.

Retinoblastoma occurs in childhood, and tumors develop from neural precursor cells in the immature retina. About one child in 20,000 is afflicted. One form of the disease is hereditary, and the other is not. In the hereditary form,

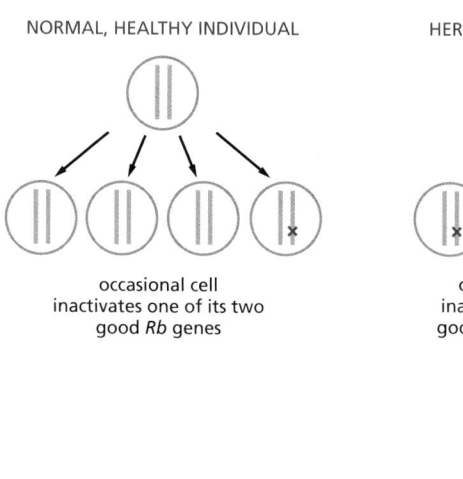

NORMAL, HEALTHY INDIVIDUAL

occasional cell
inactivates one of its two
good *Rb* genes

RESULT: NO TUMOR

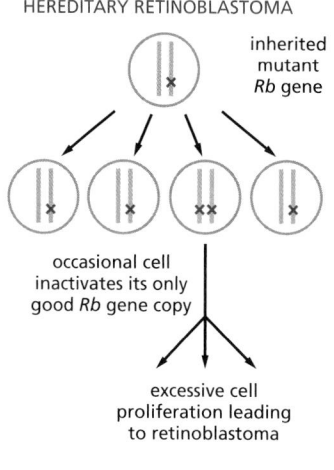

HEREDITARY RETINOBLASTOMA

inherited
mutant
Rb gene

occasional cell
inactivates its only
good *Rb* gene copy

excessive cell
proliferation leading
to retinoblastoma

RESULT: MOST PEOPLE WITH INHERITED
MUTATION DEVELOP MULTIPLE TUMORS
IN BOTH EYES

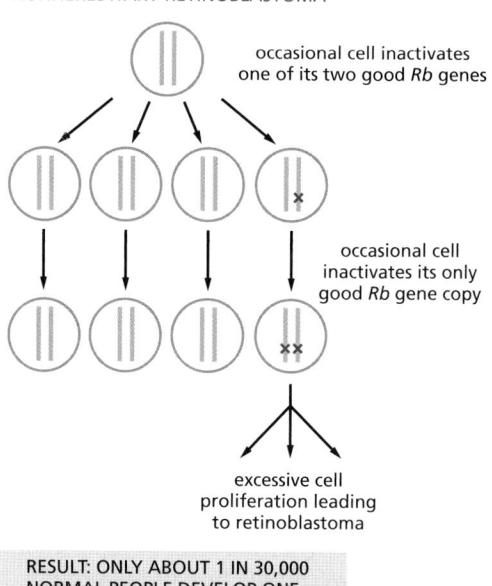

NONHEREDITARY RETINOBLASTOMA

occasional cell inactivates
one of its two good *Rb* genes

occasional cell
inactivates its only
good *Rb* gene copy

excessive cell
proliferation leading
to retinoblastoma

RESULT: ONLY ABOUT 1 IN 30,000
NORMAL PEOPLE DEVELOP ONE
TUMOR IN ONE EYE

Figure 20–20 **The genetic mechanisms that cause retinoblastoma.** In the hereditary form, all cells in the body lack one of the normal two functional copies of the *Rb* tumor suppressor gene, and tumors occur where the remaining copy is lost or inactivated by a somatic event (either mutation or epigenetic silencing). In the nonhereditary form, all cells initially contain two functional copies of the gene, and the tumor arises because both copies are lost or inactivated through the coincidence of two somatic events in a single line of cells.

multiple tumors usually arise independently, affecting both eyes; in the nonhereditary form, only one eye is affected, and by only one tumor. A few individuals with retinoblastoma have a visibly abnormal karyotype, with a deletion of a specific band on chromosome 13 that, if inherited, predisposes an individual to the disease. Deletions of this same region are also encountered in tumor cells from some patients with the nonhereditary disease, which suggested that the cancer was caused by loss of a critical gene in that location.

Using the location of this chromosomal deletion, it was possible to clone and sequence the **Rb gene**. It was then discovered that those who suffer from the hereditary form of the disease have a deletion or loss-of-function mutation present in one copy of the *Rb* gene in every somatic cell. These cells are predisposed to becoming cancerous, but do not do so if they retain one good copy of the gene. The retinal cells that are cancerous are defective in both copies of *Rb* because of a somatic event that has eliminated the function of the previously good copy.

In patients with the nonhereditary form of the disease, by contrast, the noncancerous cells show no defect in either copy of *Rb*, while the cancerous cells have become defective in both copies. These nonhereditary retinoblastomas are very rare because they require two independent events that inactivate the same gene on two chromosomes in a single retinal cell lineage (**Figure 20–20**).

The *Rb* gene is also missing in several common types of sporadic cancer, including carcinomas of lung, breast, and bladder. These more common cancers arise by a more complex series of genetic changes than does retinoblastoma, and they make their appearance much later in life. But in all of them, it seems, loss of *Rb* function is frequently a major step in the progression toward malignancy.

The *Rb* gene encodes the **Rb protein**, which is a universal regulator of the cell cycle present in almost all cells of the body (see Figure 17–61). It acts as one of the main brakes on progress through the cell-division cycle, and its loss can allow cells to enter the cell cycle inappropriately, as we discuss later.

Both Genetic and Epigenetic Mechanisms Can Inactivate Tumor Suppressor Genes

For tumor suppressor genes, it is their inactivation that is dangerous. This inactivation can occur in many ways, with different combinations of mishaps serving to eliminate or cripple both gene copies. The first copy may, for example, be lost by a small chromosomal deletion or inactivated by a point mutation. The second copy is commonly eliminated by a less specific and more probable mechanism:

HEALTHY CELL WITH ONLY ONE NORMAL *Rb* GENE COPY

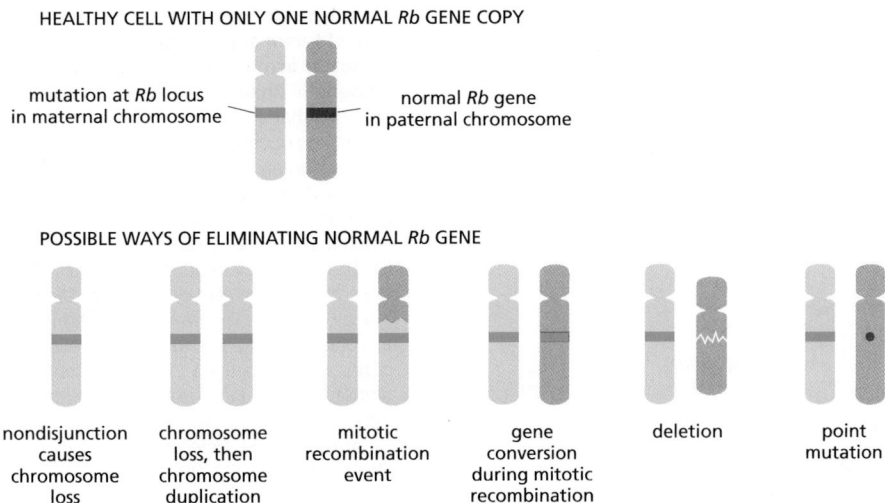

POSSIBLE WAYS OF ELIMINATING NORMAL *Rb* GENE

Figure 20–21 **Six ways of losing the remaining good copy of a tumor suppressor gene through a change in DNA sequences.** A cell that is defective in only one of its two copies of a tumor suppressor gene—for example, the *Rb* gene—usually behaves as a normal, healthy cell; the diagrams below show how this cell may lose the function of the other gene copy as well and thereby progress toward cancer. A seventh possibility, frequently encountered with some tumor suppressors, is that the gene may be silenced by an epigenetic change, without alteration of the DNA sequence, as illustrated in Figure 20–22. (After W.K. Cavenee et al., *Nature* 305:779–784, 1983.)

the chromosome carrying the remaining normal copy may be lost from the cell through errors in chromosome segregation; or the normal gene, along with neighboring genetic material, may be replaced by a mutant version through either a *mitotic recombination* event or a *gene conversion* that accompanies it (see p. 286).

Figure 20–21 summarizes the range of ways in which the remaining good copy of a tumor suppressor gene can be lost through a DNA sequence change, using the *Rb* gene as an example. It is important to note that, except for the point mutation mechanism illustrated at the far right, these pathways all produce cells that carry only a single type of DNA sequence in the chromosomal region containing their *Rb* genes—a sequence that is identical to the sequence in the original mutant chromosome.

Epigenetic changes provide another important way to permanently inactivate a tumor suppressor gene. Most commonly, the gene may become packaged into heterochromatin and/or the C nucleotides in CG sequences in its promoter may become methylated in a heritable manner (see pp. 404–405). These mechanisms can irreversibly silence the gene in a cell and in all of its progeny. Analysis of methylation patterns in cancer genomes shows that epigenetic gene silencing is a frequent event in tumor progression, and epigenetic mechanisms are now thought to help inactivate several different tumor suppressor genes in most human cancers (Figure 20–22).

Systematic Sequencing of Cancer Cell Genomes Has Transformed Our Understanding of the Disease

Methods such as those we have described above shone a spotlight on a set of cancer-critical genes that were identified in a piecemeal fashion. Meanwhile, the rest of the cancer cell genome remained in darkness: it was a mystery how many other mutations might lurk there, of what types, in which varieties of cancer, at what frequencies, with what variations from patient to patient, and with what consequences. With the sequencing of the human genome and the dramatic advances in DNA sequencing technology (see Panel 8–1, pp. 478–481), it has become possible to see the whole picture—to view cancer cell genomes in their entirety. This transforms our understanding of the disease.

Cancer cell genomes can be scanned systematically in several different ways. At one extreme—the most costly, but no longer prohibitively so—one can determine a tumor's complete genome sequence. More cheaply, one can focus just on the 21,000 or so genes in the human genome that code for protein (the so-called *exome*), looking for mutations in the cancer cell DNA that alter the amino acid sequence of the product or prevent its synthesis (Figure 20–23). There are also efficient techniques to survey the genome for regions that have undergone

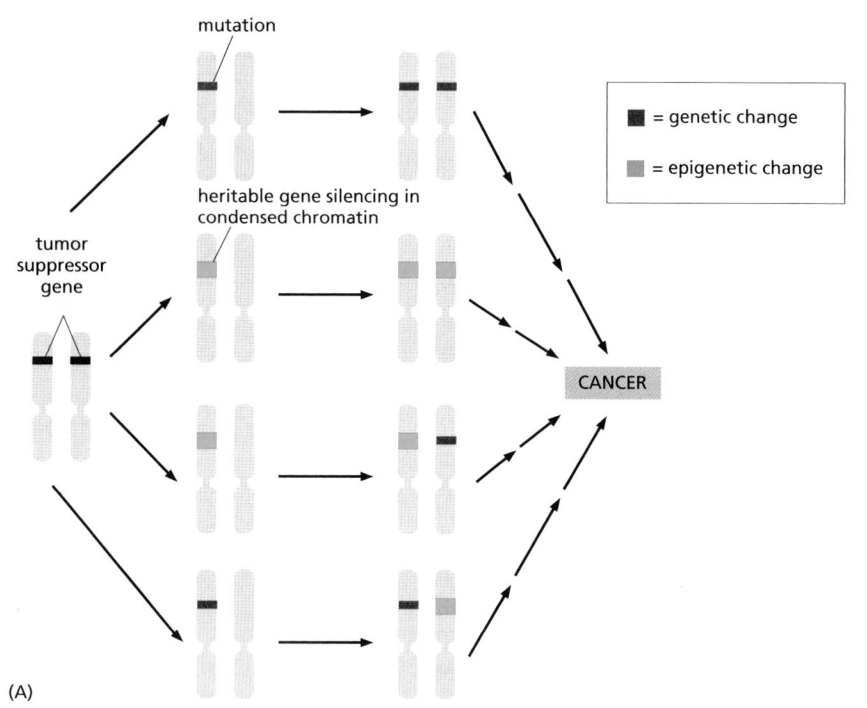

(A)

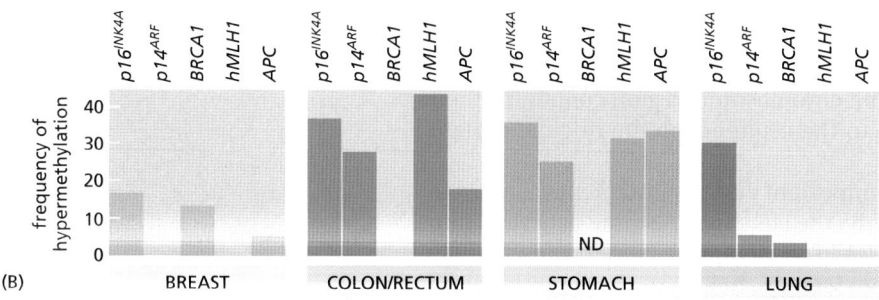

(B)

Figure 20–22 The pathways leading to loss of tumor suppressor gene function in cancer involve both genetic and epigenetic changes. (A) As indicated, the changes that silence tumor suppressor genes can occur in any order. Both DNA methylation and the packaging of a gene into condensed chromatin can prevent its expression in a way that is inherited when a cell divides (see Figure 4–44). (B) The frequency of gene silencing by hypermethylation observed in four different types of cancer. The five genes listed at the top can all function as tumor suppressor genes; *BRCA1* and *hMLH1* affect genome stability and are in the subclass known as genome maintenance genes. ND, no data. (Adapted from M. Esteller et al., *Cancer Res.* 61:3225–3229, 2001.)

deletion or duplication, without the need for complete sequence information. The genome can be scanned for epigenetic changes. And finally, alterations in levels of gene expression can be systematically determined by analysis of mRNAs (see Figure 7–3). These approaches generally involve comparing cancer cells with normal controls—ideally, noncancerous cells originating in the same tissue and from the same patient.

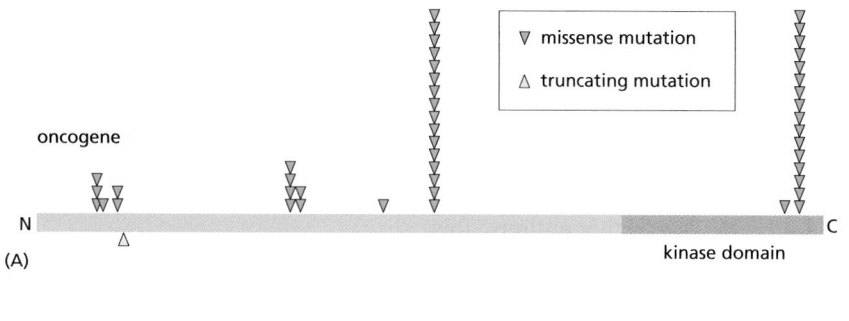

(A)

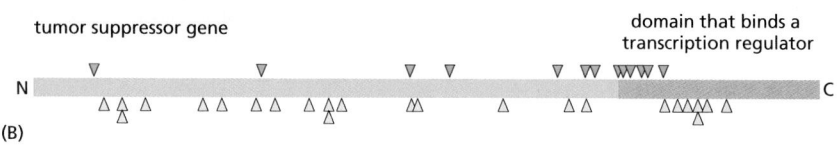

(B)

Figure 20–23 The distinct types of DNA sequence changes found in oncogenes compared to tumor suppressor genes. In this diagram, mutations that change an amino acid are denoted by *blue* arrowheads, whereas mutations that truncate the polypeptide chain are marked by *yellow* arrowheads. (A) As in this example, oncogene mutations can be detected by the fact that the same nucleotide change is repeatedly found among the missense mutations in a gene. (B) For tumor suppressor genes, by contrast, missense mutations that abort protein synthesis by creating stop codons predominate. (Adapted from B. Vogelstein et al., *Science* 339:1546–1558, 2013.)

Many Cancers Have an Extraordinarily Disrupted Genome

Cancer genome analysis reveals, first of all, the scale of gross genetic disruption in cancer cells. This varies greatly from one type of cancer and one cancer patient to another, both in severity and in character. In some cases, the karyotype—the set of chromosomes as they appear at mitosis—is normal or nearly so, but many point mutations are detected in individual genes, suggesting a failure of the repair mechanisms that normally correct local errors in the replication or maintenance of DNA sequences. Often, however, the karyotype is severely disordered, with many chromosome breaks and rearrangements. In some breast cancers, for example, genome sequencing reveals an astonishing scene of genetic chaos (**Figure 20–24**), with hundreds of chromosome breaks and translocations, resulting in many deletions, duplications, and amplifications of parts of the genome. In such cells, the normal machinery for avoidance or repair of DNA double-strand breaks is evidently somehow defective, destabilizing the genome by giving rise to broken chromosomes whose fragments then rejoin in random combinations. From the pattern of changes, one can infer that this disruptive process has occurred repeatedly during the evolution of the tumor, with a progressive increase of genetic disorder. Breast cancers showing the most extreme chromosome disorder are usually hard to treat and have a gloomy prognosis.

One survey of more than 3000 individual cancer specimens showed that on average 24 separate blocks of genetic material were duplicated in each tumor, amounting to 17% of the normal genome, and 18 blocks were deleted, amounting to 16% of the normal genome. Many of these changes were found repeatedly, suggesting that they contain cancer-critical genes whose loss (tumor suppressor genes) or gain (oncogenes) confers a selective advantage.

Whole-genome analysis also helps to explain some cancers that seem, at first sight, to be exceptions to the general rules. An example is retinoblastoma, with its early onset during childhood. If cancers in general require an accumulation of many genetic changes and are thus diseases of old age, what makes retinoblastoma different? Whole-genome sequencing confirms that in retinoblastoma, the tumor cells contain loss-of-function mutations in the *Rb* gene; but, astonishingly, they contain practically no mutations or genome rearrangements that affect any other oncogene or tumor suppressor gene. Instead, they contain many epigenetic modifications, which alter the level of expression of many known cancer-critical genes—as many as 15 in one well-analyzed case.

Many Mutations in Tumor Cells are Merely Passengers

Cancer cells generally contain many mutations in addition to gross chromosome abnormalities: point mutations can be scattered over the genome as a whole at a rate of about one per million nucleotide pairs, in addition to the abnormalities

Figure 20–24 The chromosomal rearrangements in breast cancer cells. The results of an extensive DNA sequencing analysis performed on two different primary tumors are displayed as "Circos plots." In each plot, the reference DNA sequences of the 22 autosomes and single sex chromosome (X) of a normal human female (3.2 billion nucleotide pairs) are aligned end-to-end to form a circle. Colored lines within the circle are then used to indicate the chromosome alterations found in the particular primary tumor. As indicated, *purple* lines connect sites at which two different chromosomes have become joined to create an interchromosomal rearrangment, while *green* lines connect the sites of rearrangements found within a single chromosome. The intrachromosomal rearrangements can be seen to predominate, and most join neighboring sections of DNA that were originally located within 2 million nucleotide pairs of each other. The increases in copy number, shown in *blue*, reveal the amplified DNA sequences (see the highly amplified regions indicated). (Adapted from P.J. Stephens et al., *Nature* 462:1005–1010, 2009.)

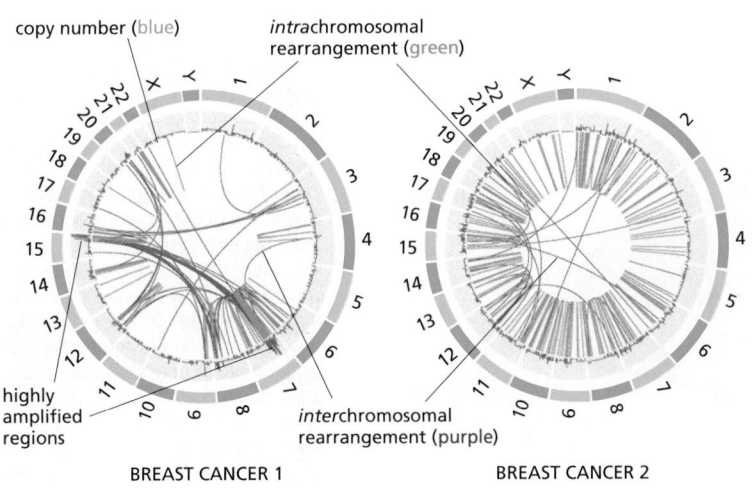

copy number (blue)

*intra*chromosomal rearrangement (green)

highly amplified regions

*inter*chromosomal rearrangement (purple)

BREAST CANCER 1

BREAST CANCER 2

attributed to chromosome breakage and rejoining. Systematic surveys of the protein-coding genes in common solid tumors—such as those of the breast, colon, brain, or pancreas—have revealed that an average of 33 to 66 genes have undergone somatic mutation affecting the sequence of their protein product. Mutations in noncoding regions of the genome are much more numerous, as one would expect from the much larger fraction of the genome that noncoding DNA represents. But they are considerably more difficult to interpret.

The high frequency of mutations testifies to the genetic instability of many cancer cells, but it leaves us with a difficult problem. How can we discover which of the mutations are **drivers** of cancer—that is, causal factors in the development of the disease—and which are merely **passengers**—mutations that happen to have occurred in the same cell as the driver mutations, thanks to genetic instability, but are irrelevant to the development of the disease? A simple criterion is based on frequency of occurrence. Driver mutations affecting a gene that plays a part in the disease will be seen repeatedly, in many different patients. In contrast, passenger mutations, occurring at more-or-less random locations in the genome and conferring no selective advantage on the cancer cell, are unlikely to be found in the same genes in different patients.

Figure 20–25 shows the results of an analysis of this sort for a large sample of colorectal cancers. The different sites in the genome are laid out on a two-dimensional array, with chromosome serial number along one axis and position within each chromosome along the other. The frequency with which mutations are encountered is shown by height above this plane, creating a mutation "landscape" with mountains (sites where mutations are found in a large proportion of the tumors in the sample), hills (where mutations are found less frequently but still more often than would be expected for a random scattering over the genome), and hillocks (sites of occasional mutations, occurring at a frequency no higher than would be expected for mutations scattered at random in each individual tumor). The mountains and the hills are strong candidates to be the sites of driver mutations—in other words, sites of cancer-critical genes; the hillocks are likely to correspond to passengers. Indeed, many of the mountains and hills turn out to be sites of known oncogenes or tumor suppressor genes, whereas the hillocks mostly correspond to genes that have no known or probable role in causation of cancer. Of course, some hillocks may correspond to genes that are mutated in only a few rare patients but are nevertheless cancer-critical for them.

About One Percent of the Genes in the Human Genome Are Cancer-Critical

From studies such as the one just described, it is estimated that the number of driver mutations for an individual case of cancer (the sum of meaningful epigenetic and genetic changes in both coding sequences and regulatory regions) is typically on the order of 10, explaining why cancer progression generally involves an increase in genetic and/or epigenetic instability that enhances the rate of such changes.

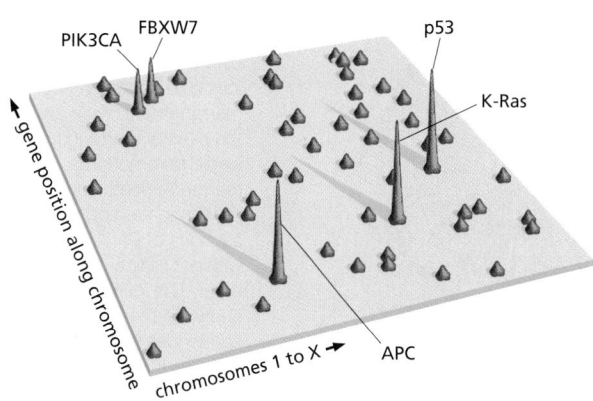

Figure 20–25 **The mutation landscape in colorectal cancer.** In this two-dimensional representation of the human genome, the *green* surface depicts the 22 human autosomes plus the X sex chromosome as being laid out side-by-side in numerical order from left to right, with the DNA sequence of each chromosome running from back to front. The mountains represent the locations of genes mutated with high frequency in different, independent tumors. As indicated, these are suspected driver mutations in the adenomatous polyposis coli (APC), K-Ras, p53, phosphoinositide 3-kinase (PIK3CA), and ubiquitin ligase (FBXW7) proteins. (Adapted from L.D. Wood et al., *Science* 318:1108–1113, 2007.)

By compiling the data for different types of cancer, each with its own range of identified driver mutations, we can develop a comprehensive catalog of genes that are strongly suspected to be cancer-critical. Current estimates put the total number of such genes at about 300, about 1% of the genes in the human genome. These cancer-critical genes are amazingly diverse. Their products include secreted signal proteins, transmembrane receptors, GTP-binding proteins, protein kinases, transcription regulators, chromatin modifiers, DNA repair enzymes, cell–cell adhesion molecules, cell-cycle controllers, apoptosis regulators, scaffold proteins, metabolic enzymes, components of the RNA splicing machinery, and more besides. All these are susceptible to mutations that can contribute, in one way or another, in one tissue or another, to the evolution of cells with the cancerous properties that we listed earlier on page 1103.

Clearly, the molecular changes that cause cancer are complex. As we now explain, however, the complexity is not quite as daunting as it may initially seem.

Disruptions in a Handful of Key Pathways Are Common to Many Cancers

Some genes, like *Rb* and *Ras*, are mutated in many cases of cancer and in cancers of many different types. The involvement of genes such as *Rb* and *Ras* in cancer is no surprise, now that we understand their normal functions: they control fundamental processes of cell division and growth. But even these common culprits feature in considerably less than half of individual cases. What is happening to the control of these processes in the many cases of cancer where, for example, *Rb* is intact or *Ras* is not mutated? What part do mutations in the hundreds of other cancer-critical genes play in the development of the disease? With our increasing knowledge of the normal functions of the genes in the human genome, it is becoming easier to see patterns in the cataloged driver mutations and to give some simplifying answers to these questions.

Glioblastoma—the commonest type of human brain tumor—provides a good example. Analysis of the genomes of tumor cells from 91 patients identified a total of at least 79 genes that were mutated in more than one individual. The normal functions of most of these genes were known or could be guessed, allowing them to be assigned to specific biochemical or regulatory pathways. Three functional groupings stood out, accounting for a total of 21 of the recurrently mutated genes. One of these groupings consisted of genes in the *Rb pathway* (that is, *Rb* itself, along with genes that directly regulate *Rb*); this pathway governs initiation of the cell-division cycle. Another consisted of genes in the same regulatory subnetwork as *Ras*—a more loosely defined system of genes referred to as the *RTK/Ras/PI3K pathway*, after three of its core components; this pathway serves to transmit signals for cell growth and cell division from the cell exterior into the heart of the cell. The third grouping consisted of genes in a pathway regulating responses to stress and DNA damage—the *p53 pathway*. We shall have more to say about each of these pathways below.

Out of all tumors, 74% had identifiable mutations in all three pathways. If one were to trace these three pathways further upstream and include all the components, known and unknown, on which they depend, this percentage would almost certainly be even higher. In other words, in almost every case of glioblastoma, there are mutations that disrupt each of three fundamental controls: the control of cell growth, the control of cell division, and the control of responses to stress and DNA damage.

Strikingly, in any given tumor-cell clone, there is a strong tendency for no more than one gene to be mutated in each pathway. Evidently, what matters for tumor evolution is the disruption of the control mechanism, and not the genetic means by which that is achieved. Thus, for example, in a patient whose tumor cells have no mutation in *Rb* itself, there is generally a mutation in some other component of the Rb pathway, producing a similar biological effect.

Similar patterns are seen in other types of cancers. A survey of many specimens of the major variety of ovarian cancer, for example, identified 67% of patients as having mutations in the Rb pathway, 45% in the Ras/PI3K pathway (defined more narrowly than in the glioblastoma study), and more than 96% in the p53 pathway. Allowing for additional pathway components not included in the analysis, it seems that most cases of this type of cancer, too, have mutations disrupting the same three controls, leading to misregulated cell growth, misregulated cell proliferation, and abnormal disregard of stress and DNA damage. It seems that these three fundamental controls are subverted in one way or another in virtually every type of cancer.

We have devoted an entire chapter to the cell cycle and growth controls (Chapter 17). Some important details of the other two control pathways are reviewed next.

Mutations in the PI3K/Akt/mTOR Pathway Drive Cancer Cells to Grow

Cell proliferation is not simply a matter of progression through the cell cycle; it also requires cell growth, which involves complex anabolic processes through which the cell synthesizes all the necessary macromolecules from small-molecule precursors. If a cell divides inappropriately without growing first, it will get smaller at each division and will ultimately die or become too small to divide. Cells appear to require two separate signals to grow and divide (**Figure 20–26**). Cancer depends, therefore, not only on a loss of restraints on cell-cycle progression, but also on disrupted control of cell growth.

The phosphoinositide 3-kinase (PI 3-kinase)/Akt/mTOR intracellular signaling pathway is critical for cell growth control. As described in Chapter 15, various extracellular signal proteins, including insulin and insulin-like growth factors,

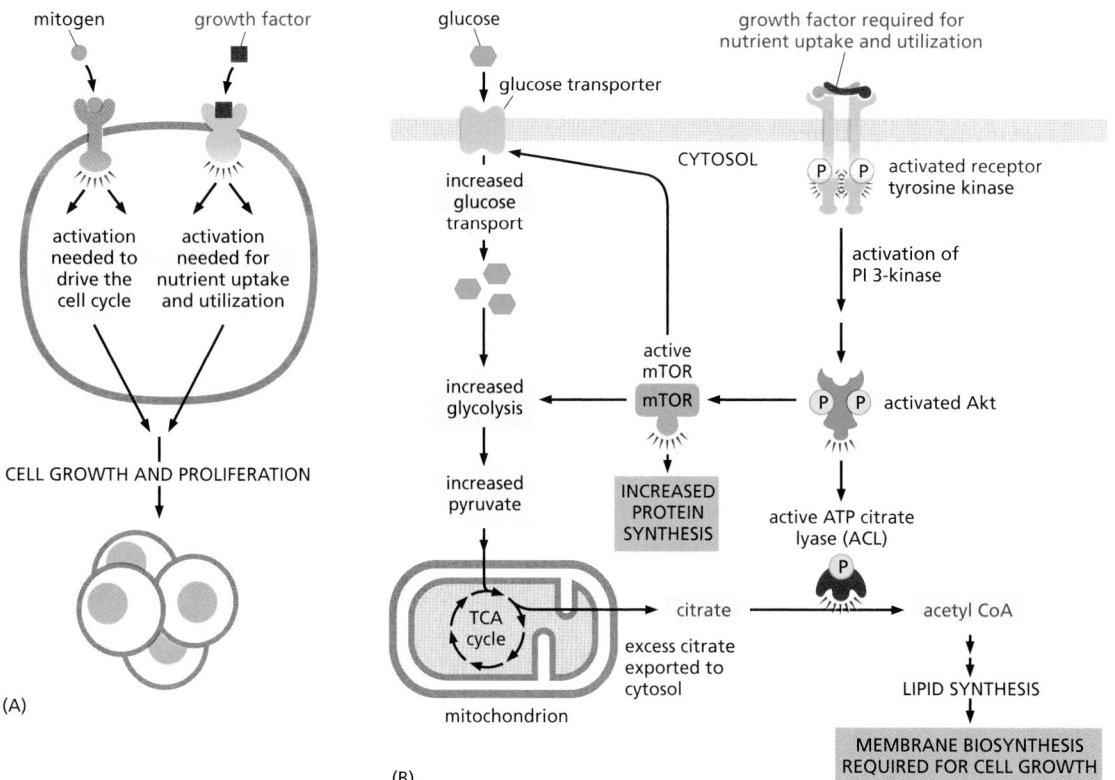

Figure 20–26 Cells seem to require two types of signals to proliferate. (A) In order to multiply successfully, most normal cells are suspected to require both extracellular signals that drive cell-cycle progression (shown here as *blue* mitogen) and extracellular signals that drive cell growth (shown here as *red* growth factor). How mitogens activate signaling through the Rb pathway to drive entry into the cell cycle is described in Figure 17–61. (B) Diagram of the signaling system containing Akt that drives cell growth through greatly stimulating glucose uptake and utilization, including a conversion of the excess citric acid produced from sugar intermediates in mitochondria into the acetyl CoA that is needed in the cytosol for lipid synthesis and new membrane production. As indicated, protein synthesis is also increased. This system becomes abnormally activated early in tumor progression. TCA cycle indicates the tricarboxylic acid cycle (citric acid cycle).

normally activate this pathway. In cancer cells, however, the pathway is activated by mutation so that the cell can grow in the absence of such signals. The resulting abnormal activation of the protein kinases Akt and mTOR not only stimulates protein synthesis (see Figure 17–64), but also greatly increases both glucose uptake and the production of the acetyl CoA in the cytosol required for cell lipid synthesis, as outlined in Figure 20–26B.

The abnormal activation of the PI 3-kinase/Akt/mTOR pathway, which normally occurs early in the process of tumor progression, helps to explain the excessive rate of glycolysis that is observed in tumor cells, known as the Warburg effect, as discussed earlier (see Figure 20–12). As expected from our previous discussion, cancers can activate this pathway in many different ways. Thus, for example, a growth factor receptor can become abnormally activated, as in Figure 20–19. Also very common in cancers is the loss of the PTEN phosphatase, an enzyme that normally suppresses the PI 3-kinase/Akt/mTOR pathway by dephosphorylating the PI $(3,4,5)$ P_3 molecules that the PI 3-kinase forms (see pp. 859–861). *PTEN* is thus a common tumor-suppressor gene.

Of course, mutation is not the only way to overactivate the pathway: high levels of insulin in the circulation can have a similar effect. This may explain why the risk of cancer is significantly increased, by a factor of two or more, in people who are obese or have type 2 diabetes. Their insulin levels are abnormally high, driving cancer cell growth without need of mutation in the PI 3-kinase/Akt/mTOR pathway.

Mutations in the p53 Pathway Enable Cancer Cells to Survive and Proliferate Despite Stress and DNA Damage

That cancer cells must break the normal rules governing cell growth and cell division is obvious: that is part of the definition of cancer. It is not so obvious why cancer cells should also be abnormal in their response to stress and DNA damage, and yet this too is an almost universal feature. The gene that lies at the center of this response, the *p53* gene, is mutated in about 50% of all cases of cancer—a higher proportion than for any other known cancer-critical gene. When we include with *p53* the other genes that are closely involved in its function, we find that most cases of cancer harbor mutations in the p53 pathway. Why should this be? To answer, we must first consider the normal function of this pathway.

In contrast to Rb, most cells in the body have very little p53 protein under normal conditions: although the protein is synthesized, it is rapidly degraded. Moreover, p53 is not essential for normal development. Mice in which both copies of the gene have been deleted or inactivated typically appear normal in all respects except one—they universally develop cancer before 10 months of age. These observations suggest that p53 has a function that is required only in special circumstances. In fact, cells raise their concentration of p53 protein in response to a whole range of conditions that have only one obvious thing in common: they are, from the cell's point of view, pathological, putting the cell in danger of death or serious injury. These conditions include DNA damage, putting the cell at risk from a faulty genome; telomere loss or shortening (see p. 1016), also dangerous to the integrity of the genome; hypoxia, depriving the cell of the oxygen it needs to keep its metabolism going; osmotic stress, causing the cell to swell or shrivel; and oxidative stress, generating dangerous levels of highly reactive free radicals.

Yet another form of stress that can activate the p53 pathway arises, it seems, when regulatory signals are so intense or uncoordinated as to drive the cell beyond its normal limits and into a danger zone where its mechanisms of control and coordination break down, as in an engine driven badly or too fast. The p53 concentration rises, for example, when *Myc* is overexpressed to oncogenic levels.

All these circumstances call for desperate action, which may take either of two forms: the cell can block any further progress through the division cycle in order to take time out to repair or recover from the pathological condition; or it can accept that it must die, and do so in a way that minimizes damage to the organism. A good death, from this point of view, is a death by apoptosis. In apoptosis,

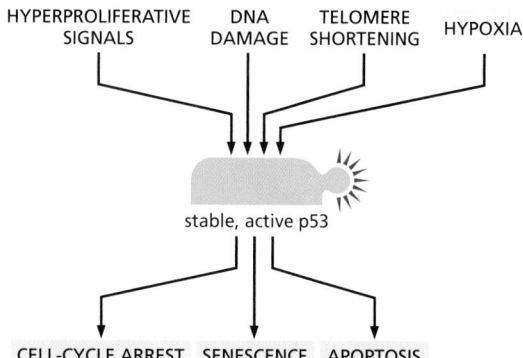

HYPERPROLIFERATIVE   DNA   TELOMERE   HYPOXIA
SIGNALS   DAMAGE   SHORTENING

stable, active p53

CELL-CYCLE ARREST   SENESCENCE   APOPTOSIS

Figure 20–27 **Modes of action of the p53 tumor suppressor.** The p53 protein is a cellular stress sensor. In response to hyperproliferative signals, DNA damage, hypoxia, telomere shortening, and various other stresses, the p53 levels in the cell rise. As indicated, this may either arrest cell cycling in a way that allows the cell to adjust and survive, trigger cell suicide by apoptosis, or cause cell "senescence"—an irreversible cell-cycle arrest that stops damaged cells from dividing.

the cell is phagocytosed by its neighbors and its contents are efficiently recycled. A bad death is a death by necrosis. In necrosis, the cell bursts or disintegrates and its contents are spilled into the extracellular space, inducing inflammation.

The p53 pathway, therefore, behaves as a sort of antenna, sensing the presence of a wide range of dangerous conditions, and when any are detected, triggering appropriate action—either a temporary or permanent arrest of cell cycling (senescence), or suicide by apoptosis (**Figure 20–27**). These responses serve to prevent deranged cells from proliferating. Cancer cells are indeed generally deranged, and their survival and proliferation thus depend on inactivation of the p53 pathway. If the p53 pathway were active in them, they would be halted in their tracks or die (**Movie 20.4**).

The p53 protein performs its job mainly by acting as a transcription regulator (see Movie 17.8). Indeed, the most common mutations observed in p53 in human tumors are in its DNA-binding domain, where they cripple the ability of p53 to bind to its DNA target sequences. Because p53 binds to DNA as a tetramer, a single mutant subunit within a tetrameric complex can be enough to block its function. Thus, mutations in *p53* can have a dominant negative effect, causing loss of p53 function even when the cell also contains a wild-type version of the gene. For this reason, in contrast with other tumor suppressor genes such as *Rb*, the development of cancer does not always require that both copies of *p53* be knocked out.

As discussed in Chapter 17, the p53 protein exerts its inhibitory effects on the cell cycle, in part at least, by inducing the transcription of *p21*, which encodes a protein that binds to and inhibits the cyclin-dependent kinase (Cdk) complexes required for progression through the cell cycle. By blocking the kinase activity of these Cdk complexes, the p21 protein prevents the cell from progressing through S phase and replicating its DNA.

The mechanism by which p53 induces apoptosis includes stimulation of the expression of many pro-apoptotic genes, as described in Chapter 18.

Genome Instability Takes Different Forms in Different Cancers

If the p53 pathway is functional, a cell with unrepaired DNA damage will stop dividing or die; it cannot proliferate. Mutations in the p53 pathway are, therefore, generally present in cancer cells showing genome instability—which is to say, the majority. But how does this genome instability originate? Here too, cancer genome studies are illuminating.

In ovarian cancers, for example, chromosome breaks, translocations, and deletions are very common, and these aberrations correlate with a high frequency of mutations and epigenetic silencing in the genes needed for repair of DNA double-strand breaks by homologous recombination, especially *Brca1* and *Brca2* (see pp. 281–282). In a subset of colorectal cancers with DNA mismatch repair defects, on the other hand, one instead finds many point mutations scattered throughout the genome (see pp. 250–251). In both kinds of cancer, the genome is commonly destabilized, but different types of mutations can bring this about.

Cancers of Specialized Tissues Use Many Different Routes to Target the Common Core Pathways of Cancer

Mutations in core components of the machinery that regulates cell growth, division, and survival, such as Rb, Ras, PTEN, or p53, are not the only way to pervert the control of these processes. Specialized tissues depend on a variety of pathways, as discussed in Chapter 15, to relay environmental signals to the core control machinery, and each pathway lays the cells open to subversion in a different set of ways. Thus, in different cancers, we can find examples of driver mutations in practically all the major signaling pathways through which cells communicate during development and tissue maintenance (discussed in Chapters 21 and 22).

In glioblastoma, for example, most patients have mutations in one or other of a set of cell-surface receptor tyrosine kinases, especially the EGF receptor mentioned earlier (linking into the Ras/PI3K pathway), suggesting that the cells from which the cancer originates are normally controlled by this route. The cells of the prostate gland, on the other hand, respond to the androgen hormone testosterone, and in prostate cancer, components of the androgen receptor signaling pathway (a variety of nuclear hormone receptor signaling; see Chapter 15) are often mutated. In the normal gut lining, Wnt signaling is critical, and Wnt pathway mutations are present in most colorectal cancers. Pancreatic cancers generally have mutations in the transforming growth factor-β (TGFβ) signaling pathway. Activating mutations in the Notch pathway are present in more than 50% of T cell acute lymphocytic leukemias, and so on.

Cells are generally regulated by several different types of external signals that must act in combination, representing a "fail-safe" control mechanism that protects the organism as a whole from cancer. These signals are different in different tissues. As expected, therefore, the corresponding cancers often have mutations in several signaling pathways concurrently. This is true of the examples we have just listed, which commonly have mutations in other signaling pathways in addition to the ones that we have singled out.

Studies Using Mice Help to Define the Functions of Cancer-Critical Genes

The ultimate test of a gene's role in cancer has to come from investigations in the intact, mature organism. The most favored organism for experimental studies is the mouse. To explore the function of a candidate oncogene or tumor suppressor gene, one can make a transgenic mouse that overexpresses it or a knockout mouse that lacks it. Using the techniques described in Chapter 8, one can engineer mice in which the misexpression or deletion of the gene is restricted to a specific set of cells, or in which expression of the gene can be switched on at will at a chosen point in time, or both, to see whether and how tumors develop. Moreover, to follow the growth of tumors from day to day in the living mouse, the cells of interest can be genetically marked and made visible by expression of a fluorescent or luminescent reporter (**Figure 20–28**). In these ways, one can begin to clarify the part that each cancer-critical gene plays in cancer initiation or progression.

Figure 20–28 Monitoring tumor growth and metastasis in a mouse with a luminescent reporter. A mouse was genetically engineered in a way that allows both copies of its *PTEN* tumor suppressor gene to be inactivated in the prostate gland, simultaneously with the prostate-specific activation of a gene engineered to produce the enzyme luciferase (derived from fireflies). After an injection of luciferin (the substrate molecule for luciferase) into the mouse's bloodstream, the cells in the prostate emit light and can be detected by their bioluminescense in a live mouse, as seen in the 67-day-old animal at the left. Cells lacking the PTEN phosphatase enzyme contain elevated amounts of the Akt activator, PI(3,4,5)P$_3$, and this causes the prostate cells to proliferate abnormally, progressing over time to form a cancer. In this way, the process of metastasis could be followed in the same animal over the course of a year. The light intensity in these experiments is proportional to the number of prostate-cell descendants, increasing from *light blue* to *green*, to *yellow*, to *red* in this representation. (Adapted from C.-P. Liao et al., *Cancer Res.* 67:7525–7533, 2007. With permission from American Association for Cancer Research.)

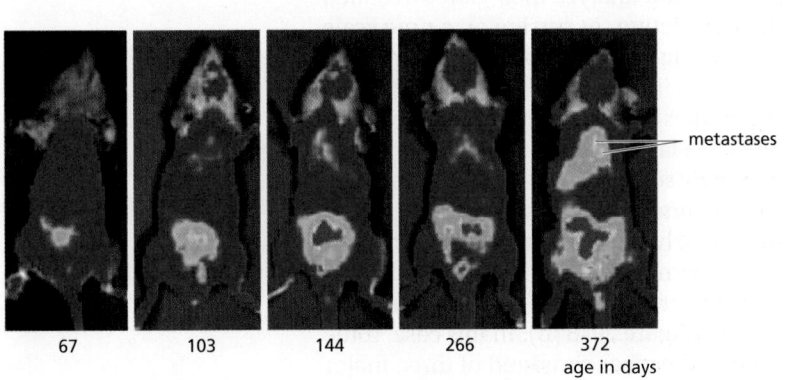

metastases

| 67 | 103 | 144 | 266 | 372 |

age in days

Figure 20–29 **Oncogene collaboration in transgenic mice.** The graphs show the incidence of tumors in three types of transgenic mouse strains, one carrying a *Myc* oncogene, one carrying a *Ras* oncogene, and one carrying both oncogenes. For these experiments, two lines of transgenic mice were first generated. One carries an inserted copy of an oncogene created by fusing the proto-oncogene *Myc* with the mouse mammary tumor virus regulatory DNA (which then drives *Myc* overexpression in the mammary gland). The other line carries an inserted copy of the *Ras* oncogene under control of the same regulatory element. Both strains of mice develop tumors much more frequently than normal, most often in the mammary or salivary glands. Mice that carry both oncogenes together are obtained by crossing the two strains. These hybrids develop tumors at a far higher rate still, much greater than the sum of the rates for the two oncogenes separately. Nevertheless, the tumors arise only after a delay and only from a small proportion of the cells in the tissues where the two genes are expressed. Further accidental changes, in addition to the two oncogenes, are apparently required for the development of cancer. (After E. Sinn et al., *Cell* 49:465–475, 1987.)

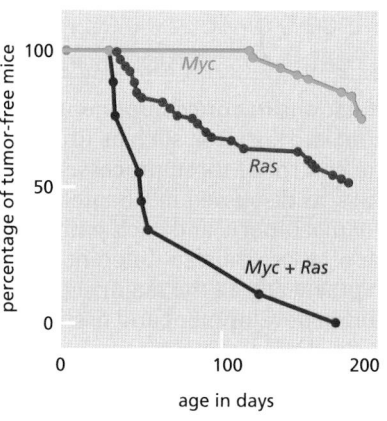

Transgenic mouse studies confirm, for example, that a single oncogene is generally not enough to turn a normal cell into a cancer cell. Thus, in mice engineered to express a *Myc* or *Ras* oncogenic transgene, some of the tissues that express the oncogene may show enhanced cell proliferation, and, over time, occasional cells will undergo further changes to give rise to cancers. Most cells expressing the oncogene, however, do not give rise to cancers. Nevertheless, from the point of view of the whole animal, the inherited oncogene is a serious menace because it creates a high risk that a cancer will arise somewhere in the body. Mice that express both *Myc* and *Ras* oncogenes (bred by mating a transgenic mouse carrying a *Myc* oncogene with one carrying a *Ras* oncogene) develop cancers earlier and at a much higher rate than either parental strain (**Figure 20–29**); but, again, the cancers originate as scattered, isolated tumors among noncancerous cells. Thus, even cells expressing these two oncogenes must undergo further, randomly generated changes to become cancerous. This strongly suggests that multiple mutations are required for tumorigenesis, as supported by a great deal of other evidence discussed earlier. Experiments using mice with deletions of tumor suppressor genes lead to similar conclusions.

Cancers Become More and More Heterogeneous as They Progress

From simple histology, looking at stained tissue sections, it is clear that some tumors contain distinct sectors, all clearly cancerous, but differing in appearance because they differ genetically: the cancer cell population is heterogeneous. Evidently, within the initial clone of cancerous cells, additional mutations have arisen and thrived, creating diverse subclones. Today, the ability to analyze cancer genomes lets us look much deeper into the process.

One approach involves taking samples from different regions of a primary tumor and from the metastases that it has spawned. With modern methods, it is even possible to take representative single cells and analyze their genomes. Such studies reveal a classic picture of Darwinian evolution, occurring on a time scale of months or years rather than millions of years, but governed by the same rules of natural selection (**Figure 20–30**).

One such investigation compared the genomes of 100 individual cells from different regions of a primary tumor of the breast. A large fraction—just over half—of the chosen cells was genetically normal or nearly so: these were connective-tissue cells and other cell types, such as those of the immune system, that were mixed up with the cancer cells. The cancer cells themselves were distinguished by their severely disrupted genomes. The detailed pattern of gene deletions and amplifications in each such cell revealed how closely it was related to the others, and from this data one could draw up a family tree (Figure 20–30B). In this case, three main branches of the tree were seen; that is, the cancer consisted of three major

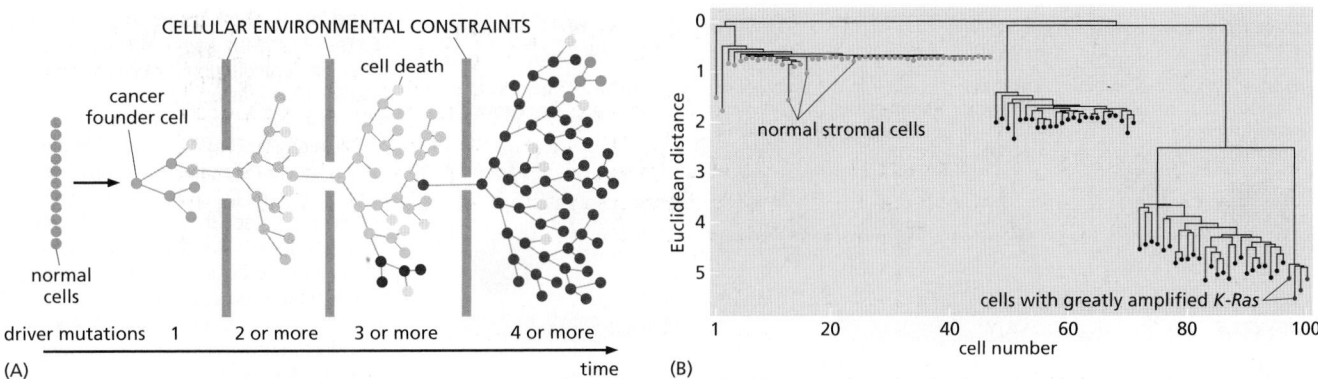

(A)

(B)

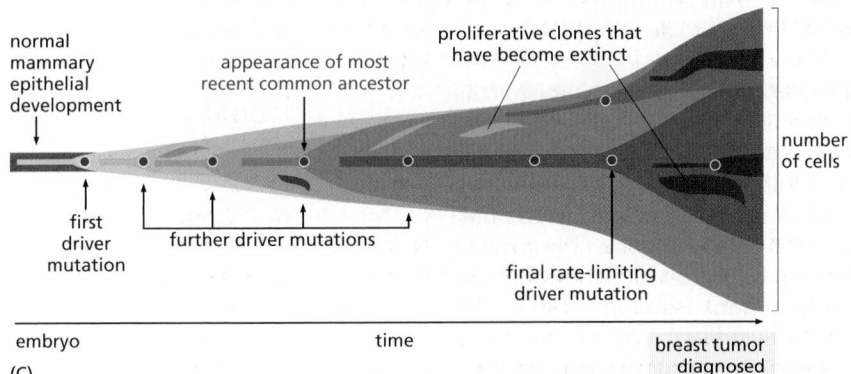

(C)

Figure 20–30 How cancers progress as a series of subclones. (A) Schematic illustration of the pattern of mutation and natural selection in a clone of tumor cells. (B) A family tree of cancer cells sampled from different regions of a single breast tumor, showing how the cells have evolved and diversified from a common ancestor, the cancer founder cell. The genome of each of the indicated 100 cells from a human breast tumor was sequenced to produce an evolutionary tree. About half of these cells were normal cells from the stroma (*blue* cells). The *red* cells have greatly amplified their *K-Ras* gene. Note that many subclones appear to have died out, including the one that contained the founder cells for the three subclones that survive.

(C) A depiction of how driver mutations are thought to cause cancer progression over long periods of time, before producing a large enough clone of proliferating cells to be detected as a tumor. The data indicate that driver mutations occur only rarely in a background of long-lived subclones of cells that continually accumulate passenger mutations without gaining a growth advantage. (A, adapted from M. Greaves, *Semin. Cancer Biol.* 20:65–70, 2010. With permission from Elsevier; B, adapted from N. Navin et al., *Nature* 472:90–94, 2011; C, adapted from S. Nik-Zainal et al., *Cell* 149:994–1007, 2012. With permission from Elsevier.)

subclones. From the shared abnormalities, one could deduce that their last common ancestor—the presumed founder of the cancer—was already very different from a normal cell, but that the first split between branches occurred early, when the tumor was small. This was followed by a large amount of additional change within each branch. A hint of the future could be seen in the smallest of the three major subclones: its cells were distinguished by a massive amplification of a *Ras* oncogene. Given more time, perhaps they would have out-competed the other cancer cells and taken over the whole tumor.

Similar results have been obtained with other cancers. Clearly, cancer cells are constantly mutating, multiplying, competing, evolving, and diversifying as they exploit new ecological niches and react to the treatments that are used against them (Figure 20–30C). Diversification accelerates as they metastasize and colonize new territories, where they encounter new selection pressures. The longer the evolutionary process continues, the harder it becomes to catch them all in the same net and kill them.

The Changes in Tumor Cells That Lead to Metastasis Are Still Largely a Mystery

Perhaps the most significant gap in our understanding of cancer concerns invasiveness and metastasis. For a start, it is not clear exactly what new properties a cancer cell must acquire to become metastatic. In some cases, it is possible that invasion and metastasis require no further genetic changes beyond those needed to violate the normal controls on cell growth, cell division, and cell death. On the other hand, it may be that, for some cancers, metastasis requires a large number of additional mutations and epigenetic changes. Clues are coming from comparisons of the genomes of cells of primary tumors with the cells of metastases that they have spawned. The results appear complex and variable from one cancer to another. Nevertheless, some general principles have emerged.

As we discussed earlier, it is helpful to distinguish three phases of tumor progression required for a carcinoma to metastasize (see Figure 20–16). First, the cells

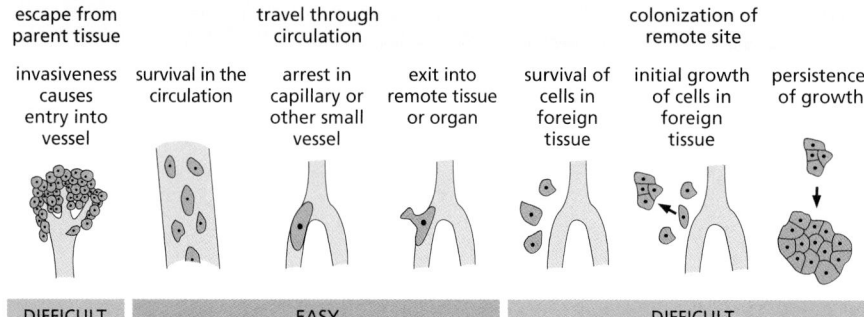

escape from parent tissue		travel through circulation		colonization of remote site		
invasiveness causes entry into vessel	survival in the circulation	arrest in capillary or other small vessel	exit into remote tissue or organ	survival of cells in foreign tissue	initial growth of cells in foreign tissue	persistence of growth

DIFFICULT	EASY	DIFFICULT

Figure 20–31 **The barriers to metastasis.** Studies of labeled tumor cells leaving a tumor site, entering the circulation, and establishing metastases show which steps in the metastatic process, outlined in Figure 20–16, are difficult or "inefficient," in the sense that they are steps in which large numbers of cells fail and are lost. It is in these difficult steps that cells from highly metastatic tumors are observed to have much greater success than cells from a nonmetastatic source. It seems that the ability to escape from the parent tissue, and an ability to survive and grow in the foreign tissue, are key properties that cells must acquire to become metastatic. (Adapted from A.F. Chambers et al., *Breast Cancer Res.* 2:400–407, 2000.)

must escape the normal confines of their parent epithelium and begin to invade the tissue immediately beneath. Second, they must travel via the blood or lymph to lodge in distant sites. Third, they must survive there and multiply. It is the first and last steps in this sequence that are the most difficult to accomplish for most cancers (**Figure 20–31**).

The first step, local invasiveness, requires a relaxation of the mechanisms that normally hold epithelial cells together. As mentioned earlier, this step resembles the normal developmental process known as the *epithelial–mesenchymal transition* (*EMT*), in which epithelial cells undergo a shift in character, becoming less adhesive and more migratory (discussed in Chapter 19). A key part of the EMT process involves switching off expression of the *E-cadherin* gene. The primary function of the transmembrane E-cadherin protein is in cell–cell adhesion, binding epithelial cells together through adherens junctions (see Figure 19–13). In some carcinomas of the stomach and of the breast, *E-cadherin* has been identified as a tumor suppressor gene, and a loss of *E-cadherin* may promote cancer development by facilitating local invasiveness.

The initial entry of tumor cells into the circulation is helped by the presence of a dense supply of blood vessels and sometimes lymphatic vessels, which tumors attract to themselves as they grow larger and become hypoxic in their interior. This process, called *angiogenesis*, is caused by the secretion of angiogenic factors that promote the growth of blood vessels, such as vascular endothelial growth factor (VEGF; see Figure 22–26). An abnormal fragility and leakiness of the new vessels that form may help the cells that have become invasive to enter and then move through the circulation with relative ease.

The remaining steps in metastasis, involving exit from a blood or lymphatic vessel and the effective colonization of remote sites, are much harder to study. To discover which of the later steps in metastasis present cancer cells with the greatest difficulties, one can label the cells with a fluorescent dye or green fluorescent protein (GFP), inject them into the bloodstream of a mouse, and then monitor their fate (**Movie 20.5**). In such experiments, one observes that many cells survive in the circulation, lodge in small vessels, and exit into the surrounding tissue, regardless of whether they come from a tumor that metastasizes or one that does not. Some cells die immediately after they enter foreign tissue; others survive entry into the foreign tissue but fail to proliferate. Still others divide a few times and then stop, forming micrometastases containing ten to several thousand cells. Very few establish full-blown metastases.

What, if anything, distinguishes the survivors from the failures? A clue may come from the fact that in many types of tumors, the cancer cells show a kind of heterogeneity that resembles the heterogeneity seen among the cells of those normal tissues that renew themselves continually by a stem-cell strategy, as we discuss next.

A Small Population of Cancer Stem Cells May Maintain Many Tumors

Self-renewing tissues, where cell division continues throughout life, are the breeding ground for the great majority of human cancers. They include the epidermis

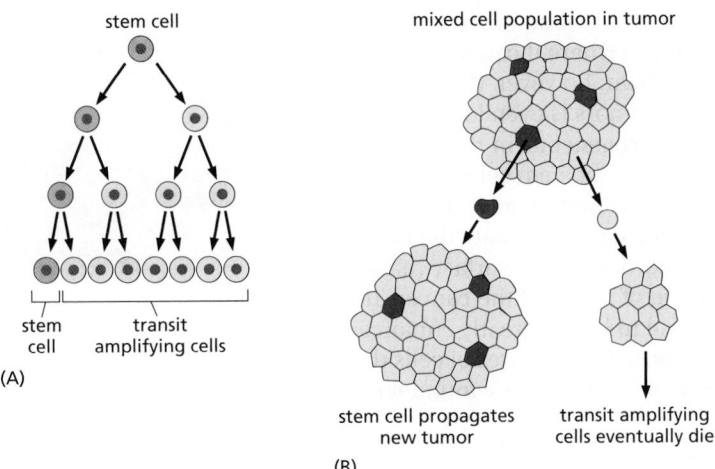

(A)

(B)

stem cell propagates
new tumor

transit amplifying
cells eventually die

stem cell

mixed cell population in tumor

stem
cell

transit
amplifying cells

Figure 20–32 **Cancer stem cells can be responsible for tumor growth and yet remain only a small part of the tumor-cell population.** (A) How stem cells produce transit amplifying cells. (B) How a small proportion of cancer stem cells can maintain a tumor. Suppose, for example, that each daughter of a cancer stem cell has a probability slightly greater than 50% of retaining stem-cell potential and a probability slightly less than 50% of becoming a transit amplifying cell that is committed to a program of cell divisions that stops after 10 division cycles. While the number of cancer stem cells will increase slowly but steadily to give a growing tumor, the non-stem cells that they give rise to will always outnumber the stem cells by a large factor—in this example, by a factor of about 1000. (If the cell-division-cycle and survival times for the two classes of cells are equal.)

(the outer epithelial layer of the skin), the lining of the digestive and reproductive tracts, and the bone marrow, where blood cells are generated (see Chapter 22). In almost all these tissues, renewal depends on the presence of stem cells, which divide to give rise to terminally differentiated cells, which do not divide. This creates a mixture of cells that are genetically identical and closely related by lineage, but are in different states of differentiation. Many tumors seem likewise to consist of cells in varied states of differentiation, with different capacities for cell division and self-renewal.

To see the implications, it is helpful to consider how normal stem-cell systems operate. When a normal stem cell divides, each daughter cell has a choice—it can remain a stem cell, or it can commit to a pathway leading to differentiation. A stem-cell daughter remains in place to generate more cells in the future. A committed daughter typically undergoes some rounds of cell proliferation (as a so-called *transit amplifying cell*) but then stops dividing, terminally differentiates, and eventually is discarded and replaced (it may die by apoptosis, with recycling of its materials, or be shed from the body). On average, the two fates—stem cell or differentiating cell—normally occur with equal probability, so that half the daughters of stem-cell divisions take the one path and half take the other. In a healthy body, feedback controls regulate the process, adjusting this balance of cell-fate choices to correct for any departure from the proper cell population numbers. Thus, the number of stem cells remains approximately constant, and the terminally differentiated cells are continually replaced at a steady rate. Because of the divisions undergone by the transit amplifying cells, the stem cells may be vastly outnumbered by the cells that are committed to terminal differentiation and have lost the capacity for self-renewal. But the stem cells, though few and far between and often relatively slowly dividing, carry the whole responsibility for maintenance of the tissue in the long term.

Some cancers seem to be organized in a similar way: they consist of rare **cancer stem cells** capable of dividing indefinitely, together with much larger numbers of dividing transit amplifying cells that are derived from the cancer stem cells but have a limited capacity for self-renewal (**Figure 20–32**). These non-stem cells appear to constitute the great majority of the cell population in some tumors.

The Cancer Stem-Cell Phenomenon Adds to the Difficulty of Curing Cancer

Evidence for the cancer stem-cell phenomenon comes chiefly from experiments in which individual cells from a cancer are tested for their ability to give rise to fresh tumors: a standard assay is to implant the cells into an immunodeficient mouse (**Figure 20–33**). It has been known for half a century that there is usually only a small chance—typically much less than 1%—that a tumor cell chosen at random and tested in this way will generate a new tumor. This by itself does not prove that the tumor cells are heterogeneous: like seeds scattered on difficult ground, each of them may have only a small chance of finding a spot where it can survive and grow. Modern technologies for sorting cells have shown, however, that in some cancers at least, the rate of success in founding new tumors is even lower than it would otherwise be because the cancer cells are heterogeneous in their state of differentiation, and only a small subset of them—the cancer stem cells—have the special properties needed for tumor propagation. For example, in several types of cancer, including breast cancers and leukemias, one can fractionate the tumor cells using monoclonal antibodies that recognize a particular cell-surface marker that is present on the normal stem cells in the tissue of origin of the cancer. The purified cancer cells expressing this marker are found to have a greatly enhanced ability to found new tumors. And the new tumors consist of mixtures of cells that express the marker and cells that do not, all generated from the same founder cell that expressed the marker.

Figure 20–33 **An immunodeficient mouse, as used in transplantation assays to test human cancer cells for their ability to found new tumors.** This *nude* mouse has a mutation that blocks development of the thymus and, as a side effect, robs it of hair. Because it has practically no T cells, it tolerates grafts of cells even from other species. (Courtesy of Envigo.)

Experiments with breast cancer cells have revealed that, instead of following a rigid program from stem cell to transit amplifying cell to terminally differentiated cell, these cancer cells can randomly switch to and fro—with a certain low transition probability—between different states of differentiation that express different molecular markers. In one state, they behave like stem cells, dividing slowly but capable of founding new tumors; in other states, they behave like transit amplifying cells, dividing rapidly but unable to found new tumors in a standard transplant assay. But a single cell in any of these states—given time in culture, or a congenial environment in the body—will give rise to a mixed population that includes all the other states as well.

The cancer stem-cell phenomenon, whatever its basis, implies that even when the tumor cells are genetically similar, they are phenotypically diverse. A treatment that wipes out those in one state is likely to allow survival of others that remain a danger. Radiotherapy or a cytotoxic drug, for example, may selectively kill off the rapidly dividing cells, reducing the tumor volume to almost nothing, and yet spare a few slowly dividing cells that go on to resurrect the disease. This greatly adds to the difficulty of cancer therapy, and it is part of the reason why treatments that seem at first to succeed often end in relapse and disappointment.

Colorectal Cancers Evolve Slowly Via a Succession of Visible Changes

At the beginning of this chapter, we saw that most cancers develop gradually from a single aberrant cell, progressing from benign to malignant tumors by the accumulation of a number of independent genetic and epigenetic changes. We have discussed what some of these changes are in molecular terms and seen how they contribute to cancerous behavior. We now examine one of the common human cancers more closely, using it to illustrate and enlarge upon some of the general principles and molecular mechanisms we have introduced. We take **colorectal cancer** as our example.

Colorectal cancers arise from the epithelium lining the colon (the large intestine) and rectum (the terminal segment of the gut). The organization of this tissue is broadly similar to that of the small intestine, discussed in detail in Chapter 22 (pp. 1217–1221). For both the small and large intestine, the epithelium is renewed at an extraordinarily rapid rate, taking about a week to completely replace most of the epithelial sheet. In both regions, the renewal depends on stem cells that lie in deep pockets of the epithelium, called intestinal crypts. The signals that maintain the stem cells and control the normal organization and renewal of the epithelium are beginning to be quite well understood, as explained in Chapter 22. Mutations that disrupt these signals begin the process of tumor progression for most colorectal cancers (**Movie 20.6**).

Colorectal cancers are common, currently causing nearly 60,000 deaths a year in the United States, or about 10% of total deaths from cancer. Like most cancers, they are not usually diagnosed until late in life (90% occur after the age of 55). However, routine examination of normal adults with a colonoscope (a fiber optic device for viewing the interior of the colon and rectum) often reveals a small benign tumor, or adenoma, of the gut epithelium in the form of a protruding mass of tissue called a *polyp* (see Figure 22–4). These adenomatous polyps are believed to be the precursors of a large proportion of colorectal cancers. Because the progression of the disease is usually very slow, there is typically a period of about 10 years in which the slowly growing tumor is detectable but has not yet turned malignant. Thus, when people are screened by colonoscopy in their fifties and the polyps are removed through the colonoscope—a quick and easy surgical procedure—the subsequent incidence of colorectal cancer is much lower: according to some studies, less than a quarter of what it would be otherwise.

In microscopic sections of polyps smaller than 1 cm in diameter, the cells and their arrangement in the epithelium usually appear almost normal. The larger the polyp, the more likely it is to contain cells that look abnormally undifferentiated and form abnormally organized structures. Sometimes, two or more distinct areas can be distinguished within a single polyp, with the cells in one area appearing relatively normal and those in the other appearing clearly cancerous, as though they have arisen as a mutant subclone within the original clone of adenomatous cells. At later stages in the disease, some tumor cells become invasive in a small fraction of the polyps, first breaking through the epithelial basal lamina, then spreading through the layer of muscle that surrounds the gut, and finally metastasizing to lymph nodes via lymphatic vessels and to liver, lung, and other organs via blood vessels.

A Few Key Genetic Lesions Are Common to a Large Fraction of Colorectal Cancers

What are the mutations that accumulate with time to produce this chain of events? Of those genes so far discovered to be involved in colorectal cancer, three stand out as most frequently mutated: the proto-oncogene *K-Ras* (a member of the *Ras* gene family), in about 40% of cases; *p53*, in about 60% of cases; and the tumor suppressor gene *Apc* (discussed below), in more than 80% of cases. Others are involved in smaller numbers of colon cancers, and some of these are listed in **Table 20–1**.

The role of *Apc* first came to light through study of certain families showing a rare type of hereditary predisposition to colorectal cancer, called *familial*

TABLE 20–1 Some Genetic Abnormalities Detected in Colorectal Cancer Cells

Gene	Class	Pathway affected	Human colon cancers (%)
K-Ras	Oncogene	Receptor tyrosine kinase signaling	40
β-Catenin[1]	Oncogene	Wnt signaling	5–10
Apc[1]	Tumor suppressor	Wnt signaling	>80
p53	Tumor suppressor	Response to stress and DNA damage	60
TGFβ receptor II[2]	Tumor suppressor	TGFβ signaling	10
Smad4[2]	Tumor suppressor	TGFβ signaling	30
MLH1 and other DNA mismatch repair genes (often silenced by DNA methylation)	Tumor suppressor (genetic stability)	DNA mismatch repair	15

[1,2]The genes with the same superscript numeral act in the same pathway, and therefore only one of the components is mutated in an individual cancer.

Figure 20–34 **Colon of familial adenomatous polyposis coli patient compared with normal colon.** (A) The normal colon wall is a gently undulating but smooth surface. (B) The polyposis colon is completely covered by hundreds of projecting polyps, each resembling a tiny cauliflower when viewed with the naked eye. (Courtesy of Mark Arends.)

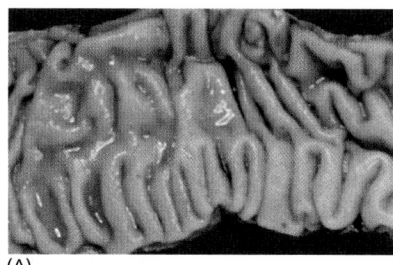

(A)

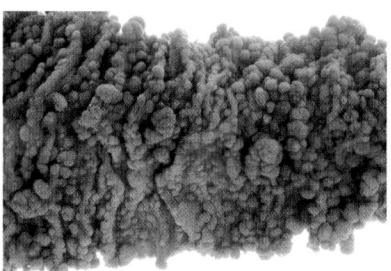

(B)

adenomatous polyposis coli (*FAP*). In this syndrome, hundreds or thousands of polyps develop along the length of the colon (**Figure 20–34**). These polyps start to appear in early adult life, and if they are not removed, one or more will almost always progress to become malignant; the average time from the first detection of polyps to the diagnosis of cancer is 12 years. The disease can be traced to a deletion or inactivation of the tumor suppressor gene *Apc*, named after the syndrome. Individuals with FAP have inactivating mutations or deletions of one copy of the *Apc* gene in all their cells and show loss of heterozygosity in tumors, even in the benign polyps. Most patients with colorectal cancer do not have the hereditary condition. Nevertheless, in more than 80% of the cases, their cancer cells (but not their normal cells) have inactivated both copies of the *Apc* gene through mutations acquired during the patient's lifetime. Thus, by a route similar to that which we discussed for retinoblastoma, mutation of the *Apc* gene was identified as one of the central ingredients of colorectal cancer.

The Apc protein, as we now know, is an inhibitory component of the *Wnt signaling pathway* (discussed in Chapter 15). It binds to the *β-catenin* protein, another component of the Wnt pathway, and helps to induce the protein's degradation. By inhibiting β-catenin in this way, Apc prevents the β-catenin from migrating to the nucleus, where it would act as a transcriptional regulator to drive cell proliferation and maintain the stem-cell state (see Figure 15–60). Loss of Apc results in an excess of free β-catenin and thus leads to an uncontrolled expansion of the stem-cell population. This causes massive increase in the number and size of the intestinal crypts (see Figure 22–4).

When the *β-catenin* gene was sequenced in a collection of colorectal tumors, it was discovered that, many of the tumors that did not have *Apc* mutations had activating mutations in the *β-catenin* protein instead. Thus, it is excessive activity in the Wnt signaling pathway that is critical for the initiation of this cancer, rather than any single oncogene or tumor suppressor gene that the pathway contains.

This being so, why is the *Apc* gene in particular so often the most common culprit in colorectal cancer? The Apc protein is large and it interacts not only with β-catenin but also with various other cell components, including microtubules. Loss of Apc appears to increase the frequency of mitotic spindle defects, leading to chromosome abnormalities when cells divide. This additional, independent cancer-promoting effect could explain why *Apc* mutations feature so prominently in the causation of colorectal cancer.

Some Colorectal Cancers Have Defects in DNA Mismatch Repair

In addition to the hereditary disease (FAP) associated with *Apc* mutations, there is a second, more common kind of hereditary predisposition to colon carcinoma in which the course of events differs from the one we have described for FAP. In this more common condition, called *hereditary nonpolyposis colorectal cancer* (*HNPCC*), the probability of colon cancer is increased without any increase in the number of colorectal polyps (adenomas). Moreover, the cancer cells are unusual, in that they have a normal (or almost normal) karyotype. The majority of colorectal tumors in non-HNPCC patients, in contrast, have gross chromosomal abnormalities, with multiple translocations, deletions, and other aberrations, as well as having many more chromosomes than normal (**Figure 20–35**).

The mutations that predispose HNPCC individuals to colorectal cancer occur in one of several genes that code for central components of the DNA *mismatch repair system*. These genes are homologous in structure and function to the *MutL* and *MutS* genes in bacteria and yeast (see Figure 5–19). Only one of the two copies of the involved gene is defective, so the repair system is still able to remove

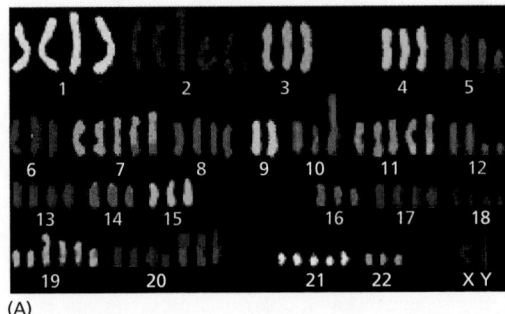 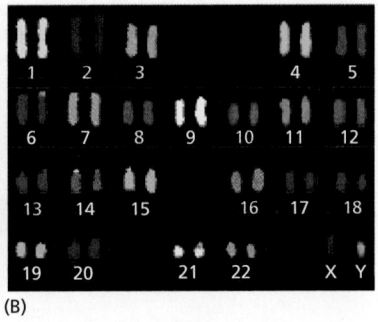

(A)　　　　　　　　　　　　　(B)

Figure 20–35 **Chromosome complements (karyotypes) of colon cancers showing different kinds of genetic instability.** (A) The karyotype of a typical cancer shows many gross abnormalities in chromosome number and structure. Considerable variation can also exist from cell to cell (not shown). (B) The karyotype of a tumor that has a stable chromosome complement with few chromosomal anomalies; the genetic abnormalities are mostly invisible, having been created by defects in DNA mismatch repair. All of the chromosomes in this figure were stained as in Figure 4–10, the DNA of each human chromosome being marked with a different combination of fluorescent dyes. (Courtesy Wael Abdel-Rahman and Paul Edwards.)

the inevitable DNA replication errors that occur in the patient's cells. However, as discussed previously, these individuals are at risk, because the accidental loss or inactivation of the remaining good gene copy will immediately elevate the spontaneous mutation rate by a hundredfold or more (discussed in Chapter 5). These genetically unstable cells then can presumably speed through the standard processes of mutation and natural selection that allow clones of cells to progress to malignancy.

This particular type of genetic instability produces invisible changes in the chromosomes—most notably changes in individual nucleotides and short expansions and contractions of mono- and dinucleotide repeats such as AAAA... or CACACA.... Once the defect in HNPCC patients was recognized, the epigenetic silencing or mutation of mismatch repair genes was found in about 15% of the colorectal cancers occurring in people with no inherited predisposing mutation.

Thus, the genetic instability found in many colorectal cancers can be acquired in at least two ways. The majority of the cancers display a form of chromosomal instability that leads to visibly altered chromosomes, whereas in the others the instability occurs on a much smaller scale and reflects a defect in DNA mismatch repair. Indeed, many carcinomas show either chromosomal instability or defective mismatch repair—but rarely both. These findings clearly demonstrate that genetic instability is not an accidental by-product of malignant behavior but a contributory cause—and that cancer cells can acquire this instability in multiple ways.

The Steps of Tumor Progression Can Often Be Correlated with Specific Mutations

In what order do *K-Ras*, *p53*, *Apc*, and the other identified colorectal cancer-critical genes mutate, and what contribution does each of them make to the asocial behavior of the cancer cell? There is no single answer, because colorectal cancer can arise by more than one route: thus, we know that in some cases, the first mutation can be in a DNA mismatch repair gene; in others, it can be in a gene regulating cell proliferation. Moreover, as previously discussed, a general feature such as genetic instability or a tendency to proliferate abnormally can arise in a variety of ways, through mutations in different genes.

Nevertheless, certain sets of mutations are particularly common in colorectal cancer, and they occur in a characteristic order. Thus, in most cases, mutations inactivating the *Apc* gene appear to be the first, or at least a very early step, as they are detected at the same high frequency in small benign polyps as in large malignant tumors. Changes that lead to genetic and epigenetic instability are likely also to arise early in tumor progression, since they are needed to drive the later steps.

Activating mutations in the *K-Ras* gene occur later, as they are rare in small polyps but common in larger ones that show disturbances in cell differentiation and histological pattern.

Inactivating mutations in *p53* are thought to come later still, as they are rare in polyps but common in carcinomas (**Figure 20–36**). We have seen that loss of *p53* function allows cancer cells to endure stress and to avoid apoptosis and cell-cycle arrest. Additionally, loss of *p53* is related to the heightened activation

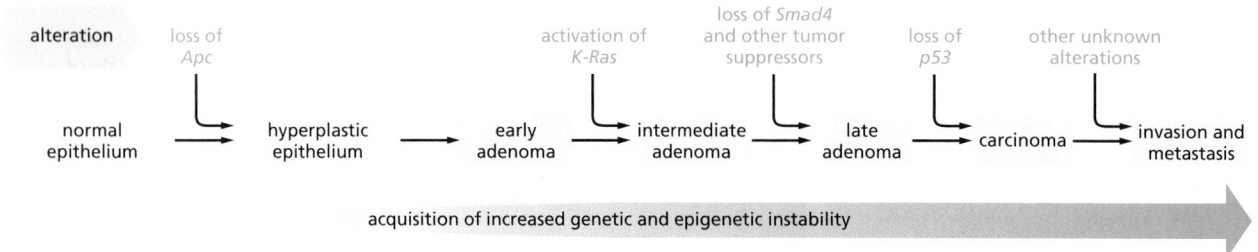

of oncogenes such as *Ras*. Experiments in mice show that an initial low level of oncogene activation can give rise to a slowly growing tumor even while *p53* is functional: genes such as *Ras* are, after all, part of the normal machinery of growth control, and moderate activation is not stressful for a cell and does not call the p53 protein into play. Progression of a tumor from slow to rapid, malignant growth, however, involves activation of oncogenes beyond normal physiological limits to a higher, stressful level. If the p53 protein is present and functional, this should lead to cell-cycle arrest or death. Only by losing p53 function can the cancer cells with hyperactive oncogenes survive and progress.

The steps we have just described are only part of the picture. It is important to emphasize that each case of colorectal cancer is different, with its own detailed combination of mutations, and that even for the mutations that are commonly shared, the sequence of occurrence may vary. The same is true for cancers in general.

Advances in molecular biology have recently provided the tools to find out precisely which genes are amplified, deleted, mutated, or misregulated by epigenetic mechanisms in the tumor cells of any given patient. As we discuss in the next section, such information promises to become as important for the diagnosis and treatment of cancer as was the breakthrough of being able to identify microorganisms for the treatment of infectious diseases.

Figure 20–36 Suggested typical sequence of genetic changes underlying the development of a colorectal carcinoma. This oversimplified diagram provides a general idea of the way mutation and tumor development are related. But many other mutations are generally involved, and different colon cancers can progress through different sequences of mutations (and/or epigenetic changes).

Summary

The molecular analysis of cancer cells reveals two classes of cancer-critical genes: oncogenes and tumor suppressor genes. A set of these genes becomes altered by a combination of genetic and epigenetic accidents to drive tumor progression. Many cancer-critical genes code for components of the social control pathways that regulate when cells grow, divide, differentiate, or die. In addition, a subclass of tumor suppressors can be categorized as "genome maintenance genes," because their normal role is to help maintain genome integrity.

The inactivation of the p53 pathway, which occurs in nearly all human cancers, allows genetically damaged cells to escape apoptosis and continue to proliferate. Inactivation of the Rb pathway also occurs in most human cancers, illustrating how fundamental each of these pathways is for protecting us against cancer.

The sequencing of cancer cell genomes reveals that—except for the cancers of childhood—many cancers acquire 10 or so driver mutations over the long course of tumor progression, along with a considerably larger number of passenger mutations of no consequence. The same methods reveal how subclones of cells arise and die out as a tumor ages. Tumors thus contain a heterogeneous mixture of cells, some—the so-called cancer stem cells—being much more dangerous than others.

We can often correlate the steps of tumor progression with mutations that activate specific oncogenes and inactivate specific tumor suppressor genes, with colon cancer providing a good example. But different combinations of mutations and epigenetic changes are found in different types of cancer, and even in different patients with the same type of cancer, reflecting the random way in which these inherited changes arise. Nevertheless, many of the same changes are encountered repeatedly, suggesting that there are a limited number of ways to breach our defenses against cancer.

CANCER PREVENTION AND TREATMENT: PRESENT AND FUTURE

We can apply the growing understanding of the molecular biology of cancer to sharpen our attack on the disease at three levels: prevention, diagnosis, and treatment. Prevention is always better than cure, and indeed many cancers can be prevented, especially by avoiding smoking. Highly sensitive molecular assays promise new opportunities for earlier and more precise diagnosis, with the aim of detecting primary tumors while they are still small and have not yet metastasized. Cancers caught at these early stages can often be nipped in the bud by surgery or radiotherapy, as we saw for colorectal polyps. Nevertheless, full-blown malignant disease will continue to be common for many years to come, and cancer treatments will continue to be needed.

In this section, we first examine the preventable causes of cancer and then consider how advances in our understanding at a molecular level are beginning to transform the treatment of the disease.

Epidemiology Reveals That Many Cases of Cancer Are Preventable

A certain irreducible background incidence of cancer is to be expected regardless of circumstances. As discussed in Chapter 5, mutations can never be absolutely avoided because they are an inescapable consequence of fundamental limitations on the accuracy of DNA replication and repair. If a person could live long enough, it is inevitable that at least one of his or her cells would eventually accumulate a set of mutations sufficient for cancer to develop.

Nevertheless, environmental factors seem to play a large part in determining the risk for cancer. This is demonstrated most clearly by a comparison of cancer incidence in different countries: for almost every cancer that is common in one country, there is another country where the incidence is much lower. Because migrant populations tend to adopt the pattern of cancer incidence typical of their new host country, the differences are thought to be due mostly to environmental, not genetic, factors. From such findings, it has been suggested that 80–90% of cancers should be avoidable, or at least postponable (**Figure 20–37**).

Unfortunately, different cancers have different environmental risk factors, and a population that escapes one such danger is usually exposed to another. This is not, however, inevitable. There are some human subgroups whose way of life substantially reduces the total cancer death rate among individuals of a given age. Under the current conditions in the United States and Europe, approximately one in five people will die of cancer. But the incidence of cancer among strict Mormons in Utah—who avoid alcohol, coffee, cigarettes, drugs, and casual sex—is only about half the incidence for non-practicing members of the same family or for Americans in general. Cancer incidence is also low in certain relatively affluent populations in Africa.

Although such observations on human populations indicate that cancer can often be avoided, it has been difficult in most cases—with tobacco as a striking exception—to pinpoint the specific environmental factors responsible for these large population differences or to establish how they act. Nevertheless, several important classes of environmental cancer risk factors have been identified (Figure 20–37B). One thinks first of mutagens. But there are also many other influences—including the amount of food we eat, the hormones that circulate in our bodies, and the irritations, infections, and damage to which we expose our tissues—that are no less important and favor development of the disease in other ways.

Sensitive Assays Can Detect Those Cancer-Causing Agents that Damage DNA

Many quite disparate chemicals are carcinogenic when they are fed to experimental animals or painted repeatedly on their skin. Examples include a range

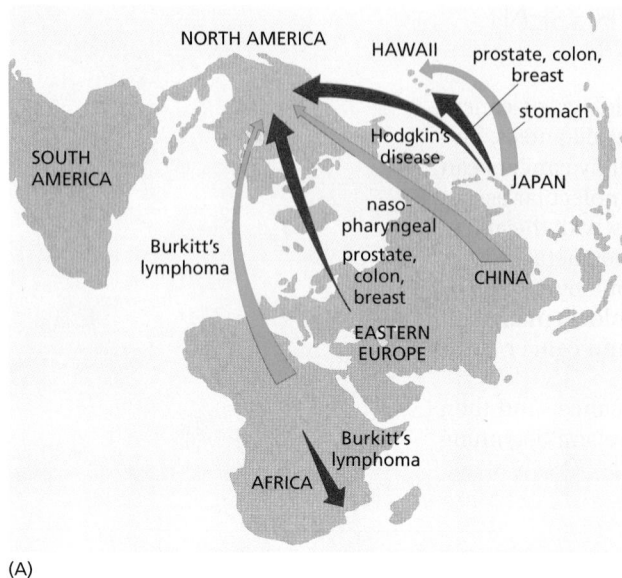

(A)

cause	cancers caused (percent of total)	number of deaths in US (annual)	magnitude of reduction possible (percent)
smoking	33	189,000	75
diet, overweight, and obesity	25	143,000	50
lack of exercise	5	28,600	85
viruses	5	28,600	100
alcohol	3	17,200	50
UV and ionizing radiation	2	11,400	50
occupational carcinogens	5	28,600	50

(B)

Figure 20–37 **Cancer incidence is related to environmental influences.** (A) This map of the world shows the rates of cancer increasing *(red arrows)* or decreasing *(blue arrows)* when specific populations move from one location to another. Such observations suggest the importance of environmental factors, including diet, in dictating cancer risk. (B) Some estimated effects of environment and lifestyle on cancer in the United States (US). The table shows both the yearly deaths in the US attributable to each cancer and the estimated percentage of that cancer that could be eliminated through prevention. (B, data from G.A. Colditz, K.Y. Wolin and S. Gehlert, *Sci. Transl. Med.* 4:127rv4, 2012.)

of aromatic hydrocarbons and derivatives of them such as aromatic amines, nitrosamines, and alkylating agents such as mustard gas. Although these **chemical carcinogens** are diverse in structure, a large proportion of them have at least one shared property—they cause mutations. In one common test for mutagenicity (the *Ames test*), the carcinogen is mixed with an activating extract prepared from rat liver cells (to mimic the biochemical processing that occurs in an intact animal). The mixture is then added to a culture of specially designed test bacteria and the bacterial mutation rate measured. Most of the compounds scored as mutagenic by this rapid and convenient assay in bacteria also cause mutations or chromosome aberrations when tested on mammalian cells.

A few of these carcinogens act directly on DNA. But generally the more potent ones are relatively inert chemically; these chemicals become damaging only after they have been converted to a more reactive molecule by metabolic processes in the liver, catalyzed by a set of intracellular enzymes known as the *cytochrome P-450 oxidases.* These enzymes normally help to convert ingested toxins into harmless and easily excreted compounds. Unhappily, their activity on certain chemicals generates products that are highly mutagenic. Examples of carcinogens activated in this way include *benzo[a]pyrene*, a cancer-causing chemical present in coal tar and tobacco smoke and the fungal toxin *aflatoxin B1* (**Figure 20–38**).

Fifty Percent of Cancers Could Be Prevented by Changes in Lifestyle

Tobacco smoke is the most important carcinogen in the world today. Even though many other chemical carcinogens have been identified, none of these appear to be responsible for anything like the same numbers of human cancer deaths. It is sometimes thought that the main environmental causes of cancer are the products of a highly industrialized way of life—the rise in pollution, the enhanced use of food additives, and so on—but there is little evidence to support this view. The idea may have come in part from the identification of some highly carcinogenic materials used in industry, such as 2-naphthylamine and asbestos. Except for the increase in cancers caused by smoking, however, age-adjusted death rates for most common human cancers have stayed much the same over the past half-century, or, in some cases, have declined significantly (**Figure 20–39**). Survival rates, moreover, have improved. Thirty years ago, less than 50% of patients lived more than five years from the time of diagnosis; now, more than two-thirds do so.

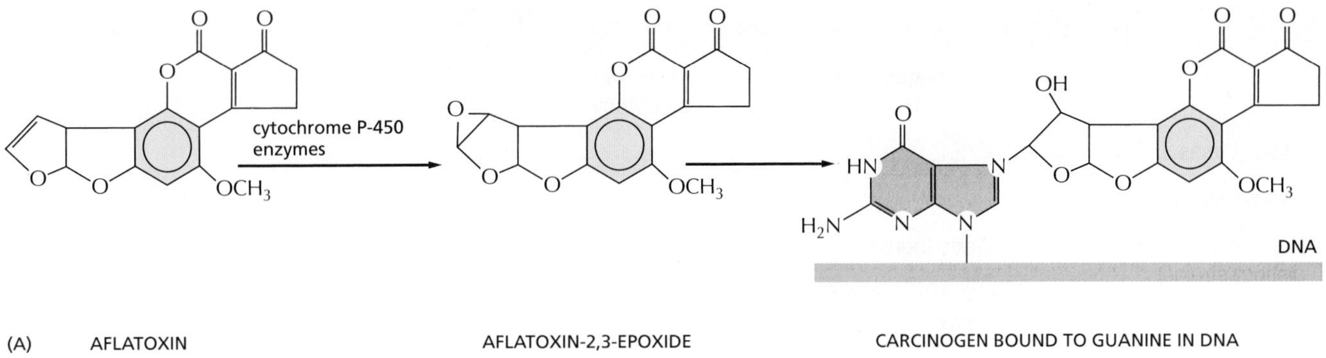

(A)   AFLATOXIN   AFLATOXIN-2,3-EPOXIDE   CARCINOGEN BOUND TO GUANINE IN DNA

cytochrome P-450 enzymes

Figure 20–38 **Some known carcinogens.** (A) Carcinogen activation. A metabolic transformation must activate many chemical carcinogens before they will cause mutations by reacting with DNA. The compound illustrated here is *aflatoxin B1*, a toxin from a mold (*Aspergillus flavus oryzae*) that grows on grain and peanuts when they are stored under humid tropical conditions. Aflatoxin is an important cause of liver cancer in the tropics. (B) Different carcinogens cause different types of cancer. (B, data from Cancer and the Environment: Gene Environment Interactions, National Academies Press, 2002.)

- **VINYL CHLORIDE:**
  liver angiosarcoma

- **BENZENE:**
  acute leukemias

- **ARSENIC:**
  skin carcinomas, bladder cancer

- **ASBESTOS:**
  mesothelioma

- **RADIUM:**
  osteosarcoma

(B)

Most of the carcinogenic factors that are known to be significant are by no means specific to the modern world. The most potent known carcinogen, by certain assays at least, is aflatoxin B1 (see Figure 20–38). It is produced by fungi that naturally contaminate foods such as tropical peanuts and is an important cause of liver cancer in Africa and Asia.

Except for tobacco, chemical toxins and mutagens are of lesser importance as contributory causes of cancer than other factors that are more a matter of personal choice. One important factor is the quantity of food we eat: as mentioned earlier, the risk of cancer is greatly increased in people who are obese. In fact, it is estimated that as many as 50% of all cancers could be avoided by simple, identifiable changes in lifestyle (see Figure 20–37B).

Viruses and Other Infections Contribute to a Significant Proportion of Human Cancers

Cancer in humans is not an infectious disease, and most human cancers do not have any infectious cause. However, a small but significant proportion of human cancers, perhaps 15% in the world as a whole, are thought to arise by mechanisms that involve viruses, bacteria, or parasites. Evidence for their involvement comes partly from the detection of viruses in cancer patients and partly from epidemiology. Thus, cancer of the uterine cervix is associated with infection with a papillomavirus, while liver cancer is very common in parts of the world (Africa and Southeast Asia) where hepatitis-B viral infections are common. Chronic infection

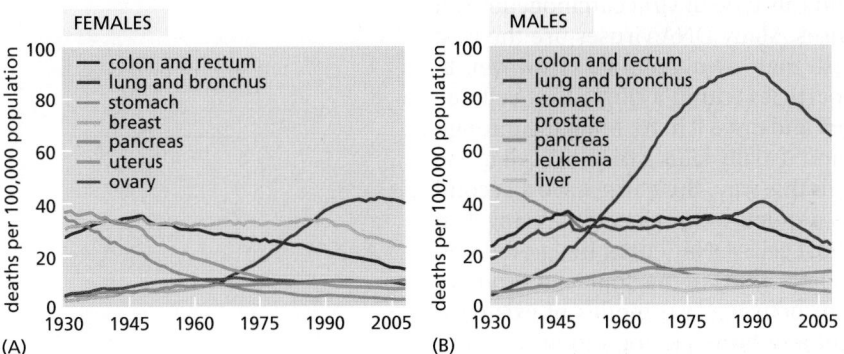

(A)

Figure 20–39 **Age-adjusted cancer death rates, United States, 1930–2008.** Selected death rates, adjusted to the age distribution of the US population, are plotted for (A) females and (B) males. Note the dramatic rise in lung cancer for both sexes, following the pattern of tobacco smoking, and the fall in deaths from stomach cancer, thought to be related to a fall in rates of infection with *Helicobacter pylori*. Recent reductions in other cancer death rates may correspond to improvements in detection and treatment. Age-adjusted data like these are needed to compensate for the inevitable increase in cancer as people live longer, on average. (American Cancer Society. Cancer Facts and Figures, 2012. Atlanta: American Cancer Society, Inc. With permission from American Cancer Society.)

TABLE 20–2 Viruses Associated with Human Cancers		
Virus	Associated cancer	Areas of high incidence
DNA viruses		
Papovavirus family		
Papillomavirus (many distinct strains)	Warts (benign)	Worldwide
	Carcinoma of the uterine cervix	Worldwide
Hepadnavirus family		
Hepatitis-B virus	Liver cancer (hepatocellular carcinoma)	Southeast Asia, tropical Africa
Herpesvirus family		
Epstein–Barr virus	Burkitt's lymphoma (cancer of B lymphocytes)	West Africa, Papua New Guinea
	Nasopharyngeal carcinoma	Southern China, Greenland
Human herpesvirus 8	Kaposi's sarcoma	Central and Southern Africa
RNA viruses		
Retrovirus family		
Human T-cell leukemia virus type I (HTLV-1)	Adult T-cell leukemia/lymphoma	Japan, West Indies
Human immunodeficiency virus (HIV, the AIDS virus)	Kaposi's sarcoma (via human herpesvirus 8)	Central and Southern Africa
Flavivirus family		
Hepatitis-C virus	Liver cancer (hepatocellular carcinoma)	Worldwide
For all these viruses, the number of people infected is much larger than the number who develop cancer: the viruses must act in conjunction with other factors. As described in the text, different viruses contribute to cancer in different ways.		

with hepatitis-C virus, which has infected 170 million people worldwide, is also clearly associated with the development of liver cancer.

The main culprits, as shown in Table 20–2, are the DNA viruses. The **DNA tumor viruses** cause cancer by the most direct route—by interfering with controls of the cell cycle and apoptosis. To understand this type of viral carcinogenesis, it is important to review the life history of viruses. Many DNA viruses use the host cell's DNA replication machinery to replicate their own genomes. However, to produce a large number of infectious virus particles within a single host cell, the DNA virus has to commandeer this machinery and drive it hard, breaking through the normal constraints on DNA replication and usually killing the host cell in the process. Many DNA viruses reproduce only in this way. But some have a second option: they can propagate their genome as a quiet, well-behaved passenger in the host cell, replicating in parallel with the host cell's DNA (either integrated into the host genome, or as an extrachromosomal plasmid) in the course of ordinary cell-division cycles. These viruses will switch between two modes of existence according to circumstances, remaining latent and harmless for a long time, but then proliferating in occasional cells in a process that kills the host cell and generates large numbers of infectious particles.

Neither of these conditions converts the host cell to a cancerous character, nor is it in the interest of the virus to do so. But for viruses with a latent phase, accidents can occur that prematurely activate some of the viral proteins that the virus would normally use in its replicative phase to allow the viral DNA to replicate independently of the cell cycle. As described in the example below, this type of accident can switch on the persistent proliferation of the host cell itself, leading to cancer.

Cancers of the Uterine Cervix Can Be Prevented by Vaccination Against Human Papillomavirus

The **papillomaviruses** are a prime example of DNA tumor viruses. They are responsible for human warts and are especially important as a cause of carcinoma of the uterine cervix: this is the second commonest cancer of women in the world as a whole, representing about 6% of all human cancers. Human papillomaviruses (**HPV**) infect the cervical epithelium and maintain themselves in a latent phase in the basal layer of cells as extrachromosomal plasmids, which replicate in step with the chromosomes. Infectious virus particles are generated through a switch to a replicative phase in the outer epithelial layers, as progeny of these cells begin to differentiate before being sloughed from the surface. Here, cell division should normally stop, but the virus interferes with this cell-cycle arrest so as to allow replication of its own genome. Usually, the effect is restricted to the outer layers of cells and is relatively harmless, as in a wart. Occasionally, however, a genetic accident causes the viral genes that encode the proteins that prevent cell-cycle arrest to integrate into the host chromosome and become active in the basal layer, where the stem cells of the epithelium reside (see Figure 22–10). This can lead to cancer, with the viral genes acting as oncogenes (**Figure 20–40**).

The whole process, from initial infection to invasive cancer, is slow, taking many years. It involves a long intermediate stage when the affected patch of cervical epithelium is visibly disordered but the cells have not yet begun to invade the underlying connective tissue—a phenomenon called *intraepithelial neoplasia*. Many such lesions regress spontaneously. Moreover, at this stage, it is still easy to cure the condition by destroying or surgically removing the abnormal tissue. Fortunately, the presence of such lesions can be detected by scraping off a sample of cells from the surface of the cervix and viewing it under the microscope (the "Pap smear" technique).

Better still, a vaccine has now been developed that protects against infection with the relevant strains of human papillomavirus. This vaccine, given to girls before puberty and thus before they become sexually active, has been shown to greatly reduce their risk of ever developing cervical cancer. Because the virus spreads through sexual activity, it is now recommended that both young males and young females be routinely vaccinated. Mass immunization programs have begun in several countries.

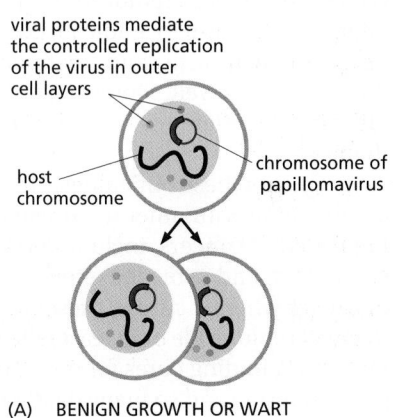

(A)   BENIGN GROWTH OR WART

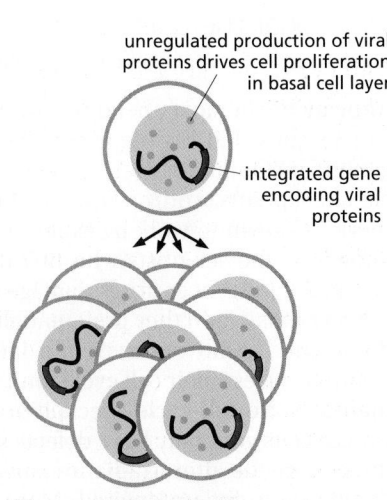

(B)   EVOLUTION OF MALIGNANT TUMOR

Figure 20–40 **How certain papillomaviruses are thought to give rise to cancer of the uterine cervix.** Papillomaviruses have double-stranded circular DNA chromosomes of about 8000 nucleotide pairs. These chromosomes are normally stably maintained in the basal cells of the epithelium as plasmids *(red circles)*, whose replication is regulated so as to keep step with the chromosomes of the host. (A) Normally, the virus perturbs the host cell cycle only when the virus is programmed to produce infectious progeny, in the outer layers of an epithelium. This is relatively harmless. (B) Rare accidents can cause the integration of a fragment of such a plasmid into a chromosome of the host, altering the environment of the viral genes in the basal cells of an epithelium. This can disrupt the normal control of viral gene expression. The unregulated production of certain viral proteins (E6 and E7) interferes with the control of cell division in the basal cells, thereby helping to generate a cancer *(bottom)*.

Infectious Agents Can Cause Cancer in a Variety of Ways

In papillomaviruses, the viral genes that are mainly to blame are called *E6* and *E7*. The protein products of these viral oncogenes interact with many host-cell proteins, but, in particular, they bind to two key tumor suppressor proteins of the host cell, putting them both out of action and so permitting the cell to replicate its DNA and divide in an uncontrolled way. One of these host proteins is Rb; the other is p53. Other DNA tumor viruses use similar mechanisms to inhibit Rb and p53, underlining the central importance of inactivating both of these tumor suppressor pathways if a cell is to escape the normal constraints on proliferation.

In other cancers, viruses have indirect tumor-promoting actions. The hepatitis-B and C viruses, for example, favor the development of liver cancer by causing chronic inflammation (hepatitis), which stimulates an extensive cell division in the liver that promotes the eventual evolution of tumor cells. In AIDS, the human immunodeficiency virus (HIV) promotes development of an otherwise rare cancer called Kaposi's sarcoma by destroying the immune system, thereby permitting a secondary infection with a human herpesvirus (HHV-8) that has a direct carcinogenic action. By causing severe inflammation, chronic infection with parasites and bacteria can also promote the development of some cancers. For example, chronic infection of the stomach with the bacterium *Helicobacter pylori*, which causes ulcers, appears to be a major cause of stomach cancer; dramatic falls in the incidence of stomach cancer over the last half-century (see Figure 20–39) correlate with a decline in the incidence of *Helicobacter* infections.

The Search for Cancer Cures Is Difficult but Not Hopeless

The difficulty of curing a cancer is similar to the difficulty of getting rid of weeds. Cancer cells can be removed surgically or destroyed with toxic chemicals or radiation, but it is hard to eradicate every single one of them. Surgery can rarely ferret out every metastasis, and treatments that kill cancer cells are generally toxic to normal cells as well. Moreover, unlike normal cells, cancer cells can mutate rapidly and will often evolve resistance to the poisons and irradiation used against them.

In spite of these difficulties, effective cures using anticancer drugs (alone or in combination with other treatments) have already been found for some formerly highly lethal cancers, including Hodgkin's lymphoma, testicular cancer, choriocarcinoma, and some leukemias and other cancers of childhood. Even for types of cancer where a cure at present seems beyond our reach, there are treatments that will prolong life or at least relieve distress. But what prospect is there of doing better and finding cures for the most common forms of cancer, which still cause great suffering and so many deaths?

Traditional Therapies Exploit the Genetic Instability and Loss of Cell-Cycle Checkpoint Responses in Cancer Cells

Anticancer therapies need to take advantage of some molecular peculiarity of cancer cells that distinguishes them from normal cells. One such property is genetic instability, reflecting deficiencies in chromosome maintenance, cell-cycle checkpoints, and/or DNA repair. Remarkably, the most widely used cancer therapies seem to work by exploiting these abnormalities, although this was not known by the scientists who first developed the treatments. Ionizing radiation and most anticancer drugs damage DNA or interfere with chromosome segregation at mitosis, and they preferentially kill cancer cells because cancer cells have a diminished ability to survive the damage. Normal cells treated with radiation, for example, arrest their cell cycle until they have repaired the damage to their DNA, thanks to the cell-cycle checkpoint responses discussed in Chapter 17. Because cancer cells generally have defects in their checkpoint responses, they may continue to divide after irradiation, only to die after a few days because the genetic damage remains unrepaired. More generally, most cancer cells are physiologically deranged to a stressful degree: they live dangerously. Even though the cells in a tumor have evolved to be unusually tolerant of minor DNA damage, they are hypersensitive to the much greater amount of damage that can be created by radiation and by DNA-damaging drugs. A small increase of genetic damage can be enough to tip the balance between proliferation and death.

Unfortunately, while the molecular defects present in cancer cells often enhance their sensitivity to cytotoxic agents, they can also increase their resistance. For example, where a normal cell might die by apoptosis in response to DNA damage, thanks to the stress response mediated by p53, a cancer cell may escape apoptosis because its p53 is lacking. Cancers vary widely in their sensitivity to cytotoxic treatments, some responding to one drug, some to another, probably reflecting the particular kinds of defects that a particular cancer has in DNA repair, cell-cycle checkpoints, and the control of apoptosis.

New Drugs Can Kill Cancer Cells Selectively by Targeting Specific Mutations

Radiotherapy and traditional cytotoxic drugs are rather weakly selective: they hurt normal cells as well as the cancer cells, and the safety margin is narrow. The dose often cannot be raised high enough to kill all the cancer cells, because this would kill the patient, and curative treatments, where achievable, generally require a combination of several cytotoxic agents. The side effects can be harsh and hard to endure. How can we do better?

An ideal treatment is one that is cell-lethal in combination with some lesion that is present in the cancer cells, but harmless to cells where this lesion is absent. Such a treatment is said to be *synthetic-lethal* (from the original sense of the word *synthesis*, meaning "putting together"): it kills only in partnership with the cancer-specific mutation. As we become increasingly able to pinpoint the specific alterations in cancer cells that make them different from their normal neighbors, new opportunities for such precisely targeted treatments are coming into view. We end this chapter with some examples of new treatments of this type that are already being put into practice.

PARP Inhibitors Kill Cancer Cells That Have Defects in *Brca1* or *Brca2* Genes

As we have emphasized, the genetic instability of cancer cells makes the cells both dangerous and vulnerable—dangerous because of the enhancement in their ability to evolve and proliferate, and vulnerable because treatment that leads to still more extreme genetic disruption can take them over the brink and kill them. In some cancers, genetic instability results from an identified fault in one of the many devices on which normal cells depend for DNA repair and maintenance. In this case, a drug is tailored to block a complementary part of the DNA repair machinery can lead to such severe genetic damage that the cancer cells die.

Detailed studies of the mechanisms for DNA maintenance discussed in Chapter 5 reveal a surprising amount of apparent redundancy. Thus, knocking out a particular pathway for DNA repair is generally less disastrous than one might expect, because alternate repair pathways exist. For example, stalled DNA replication forks can arise when the fork encounters a single-strand break in a template strand, but cells can avoid the disaster that would otherwise result either by directly repairing these single-strand breaks, or, if that fails, repairing the broken fork that results by homologous recombination (see Figure 5–50). Suppose that the cells in a particular cancer have become genetically unstable by acquiring a mutation that reduces their ability to repair broken replication forks by homologous recombination. Might it be possible to eradicate that cancer by treating it with a drug that inhibits the repair of single-strand breaks, thereby greatly increasing the number of forks that break? The consequences of such drug treatment might be expected to be relatively harmless for normal cells, but lethal for the cancer.

This strategy appears to work to kill the cells in at least one class of cancers—those that have inactivated both copies of either their *Brca1* or their *Brca2* tumor

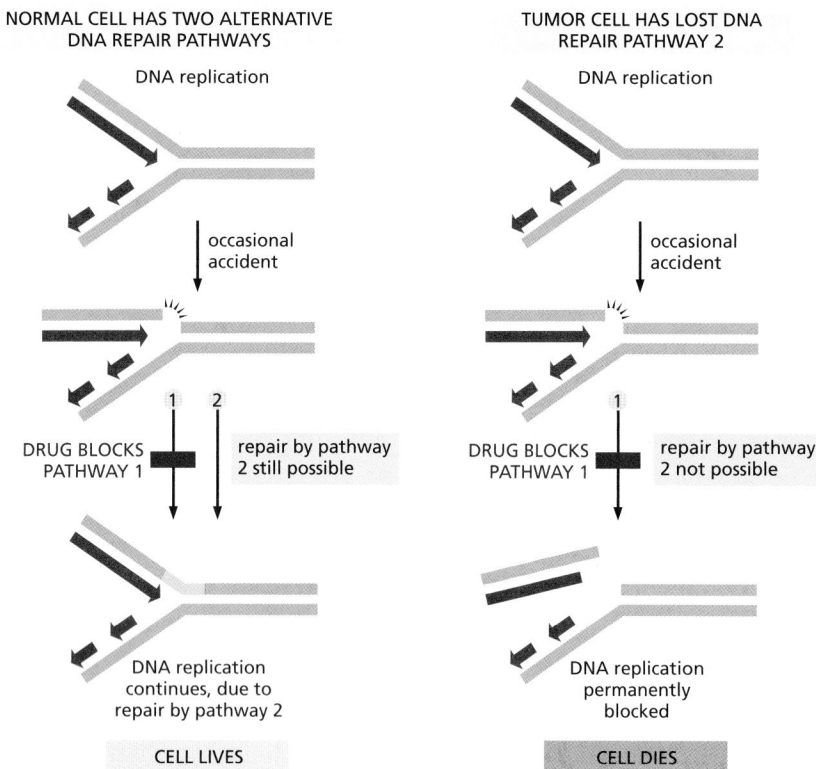

NORMAL CELL HAS TWO ALTERNATIVE
DNA REPAIR PATHWAYS

DNA replication

occasional
accident

DRUG BLOCKS
PATHWAY 1

repair by pathway
2 still possible

DNA replication
continues, due to
repair by pathway 2

CELL LIVES

TUMOR CELL HAS LOST DNA
REPAIR PATHWAY 2

DNA replication

occasional
accident

DRUG BLOCKS
PATHWAY 1

repair by pathway
2 not possible

DNA replication
permanently
blocked

CELL DIES

Figure 20–41 **How a tumor's genetic instability can be exploited for cancer therapy.** As explained in Chapter 5, the maintenance of DNA sequences is so critical for life that cells have evolved multiple pathways for repairing DNA damage and reducing DNA replication errors. As illustrated, a DNA replication fork will stall whenever it encounters a break in a DNA template strand. In this example, normal cells have two different repair pathways that help them to avoid the problem, pathways 1 and 2. They are therefore not harmed by treatment with a drug that blocks repair pathway 1. But, because the inactivation of repair pathway 2 was selected for during the evolution of the tumor cell, the tumor cells are killed by the same drug treatment.

In the actual case that underlies this example, the function of repair pathway 1 (requiring the PARP protein discussed in the text) is to remove persistent, accidental breaks in a DNA single strand before they are encountered by a moving replication fork. Pathway 2 is the recombination-dependent process (requiring the Brca2 and Brca1 proteins) for repairing stalled replication forks illustrated in Figure 5–50. PARP inhibitors have promise for treating cancers with defective *Brca2* or *Brca1* tumor suppressor genes.

suppressor genes. As described in Chapter 5, Brca2 is an accessory protein that interacts with the Rad51 protein (the RecA analog in humans) in the repair of DNA double-strand breaks by homologous recombination. Brca1 is another protein that is also required for this repair process. Like *Rb*, the *Brca1* and *Brca2* genes were discovered as mutations that predispose humans to cancer—in this case, chiefly cancers of the breast and ovaries (though unlike *Rb*, they seem to be involved in only a small proportion of such cancers). Individuals who inherit one mutant copy of *Brca1* or *Brca2* develop tumors that have inactivated the second copy of the same gene, presumably because this change makes the cells genetically unstable and speeds tumor progression.

While Brca1 and Brca2 are needed for the repair of DNA double-strand breaks, single-strand breaks are repaired by other machinery, involving an enzyme called PARP (polyADP-ribose polymerase). This understanding of the basic mechanisms of DNA repair led to a striking discovery: drugs that block PARP activity kill *Brca*-deficient cells with extraordinary selectivity. At the same time, PARP inhibition has very little effect on normal cells; in fact, mice that have been engineered to lack PARP1—the major PARP family member involved in DNA repair—remain healthy under laboratory conditions. This result suggests that, while the repair pathway requiring PARP provides a first line of defense against persistent breaks in a DNA strand, these breaks can be repaired efficiently by a genetic recombination pathway in normal cells. In contrast, tumor cells that have acquired their genetic instability by the loss of Brca1 or Brca2 have lost this second line of defense, and they are therefore uniquely sensitive to PARP inhibitors (**Figure 20–41**).

PARP inhibitors are still under clinical trial, but they have produced some striking results, causing tumors to regress in many Brca-deficient patients and delaying progression of their disease, with relatively few disagreeable side effects. These drugs also appear to be applicable to cancers with other mutations that cause defects in the cell's homologous recombination machinery—a small, though significant, proportion of cancer cases.

PARP inhibition provides an example of the type of rational, highly selective approach to cancer therapy that is beginning to be possible. Along with other new treatments to be discussed below, it raises high hopes for treating many other cancers.

Small Molecules Can Be Designed to Inhibit Specific Oncogenic Proteins

An obvious tactic for treating cancer is to attack a tumor expressing an oncogene with a drug designed to specifically block the function of the protein that the oncogene produces. But how can such a treatment avoid hurting the normal cells that depend on the function of the proto-oncogene from which the oncogene has evolved, and why should the drug kill the cancer cells, rather than simply calm them down? One answer may lie in the phenomenon of *oncogene dependence*. Once a cancer cell has undergone an oncogenic mutation, it will often undergo further mutations, epigenetic changes, or physiological adaptations that make it reliant on the hyperactivity of the initial oncogene, just as drug addicts become reliant on high doses of their drug. Blocking the activity of the oncogenic protein may then kill the cancer cell without significantly harming its normal neighbors. Some remarkable successes have been achieved in this way.

As we saw earlier, chronic myelogenous leukemia (CML) is usually associated with a particular chromosomal translocation, visible as the Philadelphia chromosome (see Figure 20–5). This results from chromosome breakage and rejoining at the sites of two specific genes, *Abl* and *Bcr*. The fusion of these genes creates a hybrid gene, called *Bcr-Abl*, that codes for a chimeric protein consisting of the N-terminal fragment of Bcr fused to the C-terminal portion of Abl (**Figure 20–42**). Abl is a tyrosine kinase involved in cell signaling. The substitution of the Bcr fragment for the normal N-terminus of Abl makes it hyperactive, so that it stimulates inappropriate proliferation of the hemopoietic precursor cells that contain it and prevents these cells from dying by apoptosis—which many of them would normally do. As a result, excessive numbers of white blood cells accumulate in the bloodstream, producing CML.

The chimeric Bcr-Abl protein is an obvious target for therapeutic attack. Searches for synthetic drug molecules that can inhibit the activity of tyrosine kinases discovered one, called *imatinib* (trade name Gleevec®), that blocks Bcr-Abl (**Figure 20–43**). When the drug was first given to patients with CML, nearly all of them showed a dramatic response, with an apparent disappearance of the cells carrying the Philadelphia chromosome in over 80% of patients. The response appears relatively durable: after years of continuous treatment, many patients have not progressed to later stages of the disease—although imatinib-resistant cancers emerge with a probability of about 5% per year during the early years.

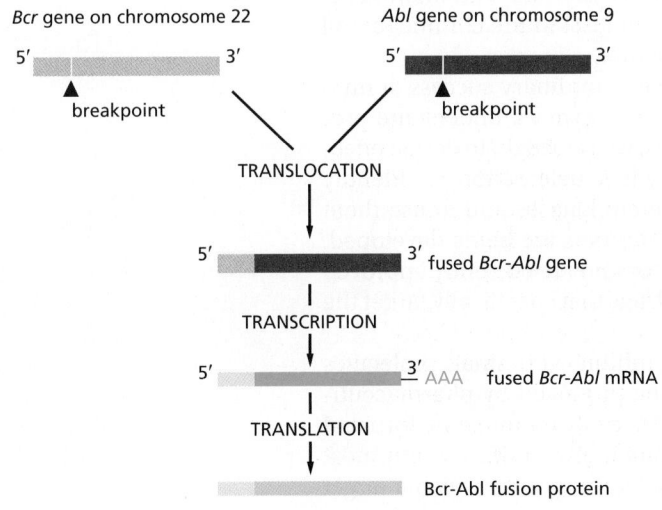

Figure 20–42 **The conversion of the** *Abl* **proto-oncogene into an oncogene in patients with chronic myelogenous leukemia.** The chromosome translocation responsible joins the *Bcr* gene on chromosome 22 to the *Abl* gene from chromosome 9, thereby generating a Philadelphia chromosome (see Figure 20–5). The resulting fusion protein has the N-terminus of the Bcr protein joined to the C-terminus of the Abl tyrosine protein kinase; in consequence, the Abl kinase domain becomes inappropriately active, driving excessive proliferation of a clone of hemopoietic cells in the bone marrow.

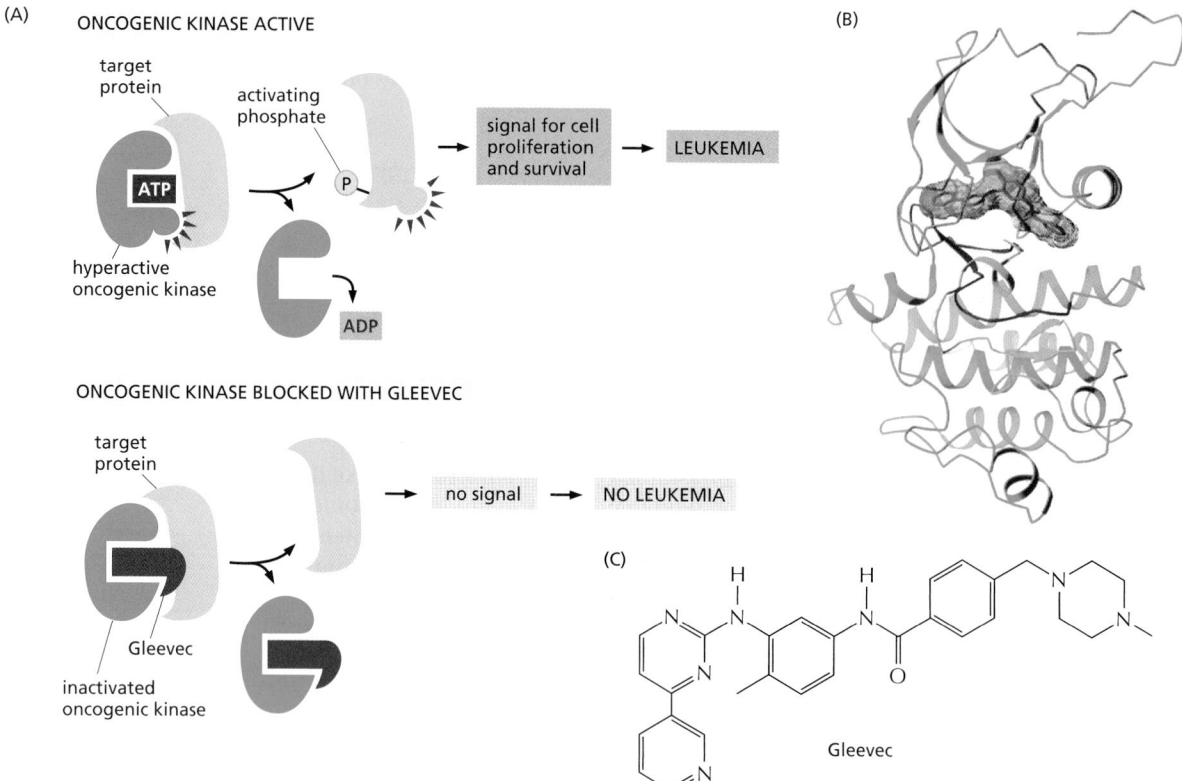

Figure 20–43 How imatinib (Gleevec) blocks the activity of Bcr-Abl protein and halts chronic myelogenous leukemia.
(A) Imatinib sits in the ATP-binding pocket of the tyrosine kinase domain of Bcr-Abl and thereby prevents Bcr-Abl from transferring a phosphate group from ATP onto a tyrosine residue in a substrate protein. This blocks transmission of a signal for cell proliferation and survival. (B) The structure of the complex of imatinib (solid *blue* object) with the tyrosine kinase domain of the Abl protein (ribbon diagram), as determined by x-ray crystallography. (C) The chemical structure of the drug. It can be given by mouth; it has side effects, but they are usually quite tolerable. (B, from T. Schindler et al., *Science* 289:1938–1942, 2000. With permission from AAAS.)

Results are not so good for those patients who have already progressed to the more acute phase of myeloid leukemia, known as blast crisis, where genetic instability has set in and the march of the disease is far more rapid. These patients show a response at first and then relapse because the cancer cells develop a resistance to imatinib. This resistance is usually associated with secondary mutations in the part of the *Bcr-Abl* gene that encodes the kinase domain, disrupting the ability of imatinib to bind to Bcr-Abl kinase. Second-generation inhibitors that function effectively against a whole range of imatinib-resistant mutants have now been developed. By combining one or more of these new inhibitors with imatinib as the initial therapy (see below), it seems that CML—at least in the chronic (early) stage—may be on its way to becoming a curable disease.

Despite the complications with resistance, the extraordinary success of imatinib is enough to drive home an important principle: once we understand precisely what genetic lesions have occurred in a cancer, we can begin to design effective rational methods to treat it. This success story has fueled efforts to identify small-molecule inhibitors for other oncogenic protein kinases and to use them to attack the appropriate cancer cells. Increasing numbers are being developed. These include molecules that target the EGF receptor and are currently approved for the treatment of some lung cancers, as well as drugs that specifically target the B-Raf oncoprotein in melanomas.

Protein kinases have been relatively easy to inhibit with small molecules like imatinib, and many kinase inhibitors are being produced by pharmaceutical companies in the hope that they can be effective as drugs for some forms of cancer. Many cancers lack an oncogenic mutation in a protein kinase. But most tumors contain inappropriately activated signaling pathways, for which a target

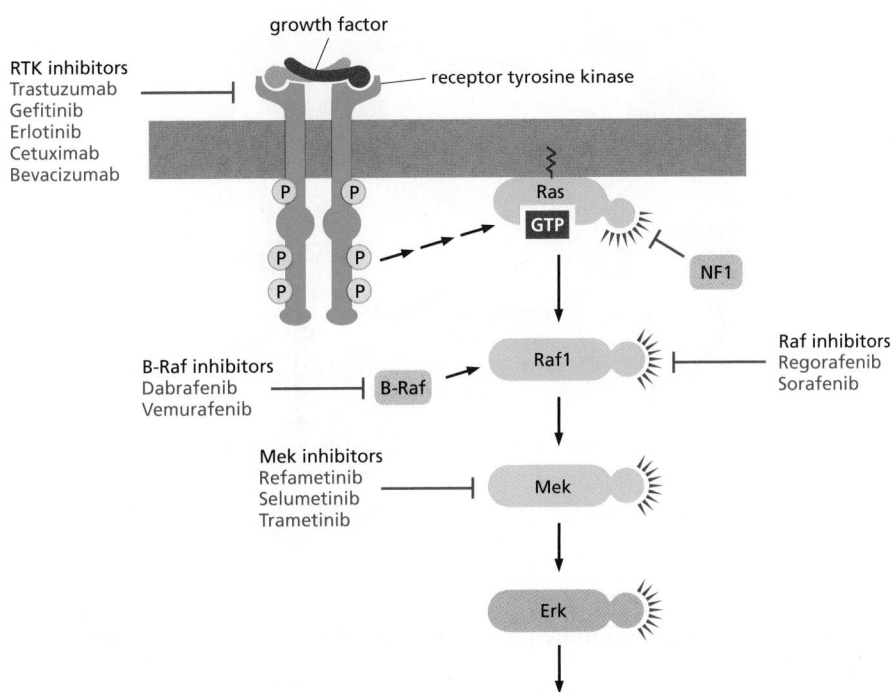

Figure 20–44 **Some anticancer drugs and drug targets in the Ras–MAP-kinase signaling pathway.** Each of the signaling proteins in this diagram has been identified as a product of a cancer-critical gene, with the exception of Raf1 and Erk. This Ras–MAP-kinase signaling pathway is triggered by a variety of receptor tyrosine kinases (RTKs), including the EGF receptor (see Figures 15–47 and 15–49). Those drugs that are antibodies end in "mab," while those that are small molecules end in "nib." (Adapted from B. Vogelstein et al, *Science* 339:1546–1558, 2013.)

somewhere in the pathway can hopefully be found (**Movie 20.7**). As an example, **Figure 20–44** displays some of the anticancer drugs and drug targets that are currently being tested for a pathway frequently activated in cancers.

Many Cancers May Be Treatable by Enhancing the Immune Response Against the Specific Tumor

Cancers have complex interactions with the immune system, and its various components may sometimes help as well as hinder tumor progression. But for more than a century it has been a dream of cancer researchers to somehow harness the immune system in a controlled and efficient way to exterminate cancer cells, just as it exterminates infectious organisms. There are finally signs that this dream may one day be realized, at least for some forms of cancer.

The simplest type of immunological therapy, conceptually at least, is to inject the patient with antibodies that target the cancer cells. This approach has had some successes. About 25% of breast cancers, for example, express unusually high levels of the Her2 protein (human epidermal growth factor receptor 2), a receptor tyrosine kinase related to the EGF receptor that plays a part in the normal development of mammary epithelium. A monoclonal antibody called *trastuzumab* (trade name *Herceptin*®) that binds to Her2 and inhibits its function slows the growth of breast tumors in humans that overexpress Her2, and it is now an approved therapy for these cancers (see Figure 20–44). A related approach uses antibodies to deliver poisons to the cancer cells. Antibodies against proteins that are abundant on the surface of a particular type of cancer cell but rare on normal cells can be armed with a toxin that kills those cells that bind the antibody molecule.

A great deal of current excitement centers around a different type of approach, based on the relatively recent recognition that the microenvironment in a tumor is highly immunosuppressive. As a result, the cancer victim's immune system is prevented from destroying the tumor cells. Recall that, from the thousands of tumor genome sequences thus far determined, we know that a typical cancer cell will contain on the order of 50 proteins with a mutation that alters an amino acid sequence, most of these being "passenger" mutations, as previously explained (see p. 1104). Many of these mutant proteins will be recognized by the patient's immune system as foreign, but—to allow the cancer cells to survive throughout the course of tumor progression—the cancer cells have evolved a set of anti-immune defenses. These

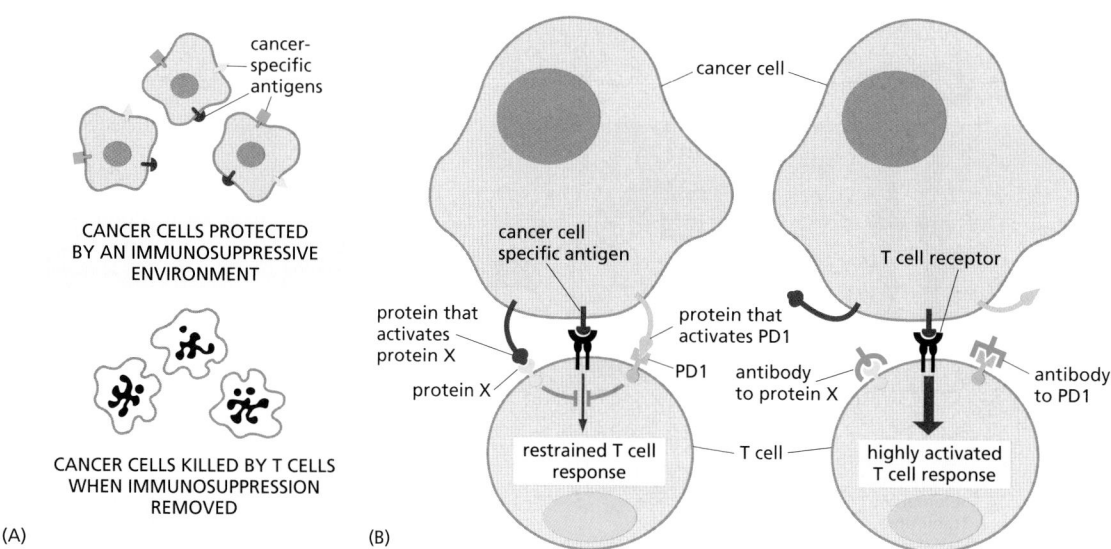

defenses include the expression on the cancer cell surface of one or more proteins that bind to inhibitory receptors on activated T cells.

The normal immune system is subject to complex controls that keep its activity within safe bounds and prevent autoimmunity from developing. The inhibitory receptors that are expressed on the surface of activated T cells have an important normal function: they control the immune response by down-regulating the T cell response under appropriate circumstances. But in the context of a tumor, the down-regulation is inappropriate, because it prevents the organism from killing the cancer cells that are threatening its survival.

In its attack on infectious organisms, the natural immune system usually eliminates every last trace of infection and maintains this immunity in the long term. The challenge is to find ways of recruiting the immune system to attack cancers with similar efficiency and specificity, hunting the cancer cells down by virtue of the tumor-specific antigens that they express. With this aim, a new type of anti-cancer therapy focuses on overcoming the immunosuppressive environment in a tumor through the use of specific antibodies that prevent the tumor cells from engaging with the inhibitory receptors on T cells. As illustrated in **Figure 20–45**A, blocking the action of the immune suppressors with such treatments should unleash an immune attack on the cancer cells. Importantly, multiple antigens are recognized as foreign; thus, the cancer cells cannot escape through the mutational loss of a single antigen, making it difficult for the tumor to escape from the T cell attack.

This is a potentially dangerous strategy. If one provokes the immune system to recognize the cancer cells as targets for destruction, there is a risk of autoimmune side effects with dire consequences for normal tissues of the body, since the cancer cells and the normal cells are close cousins and share most of their molecular features. Nevertheless, several recent successes seem to hold great promise for the future.

One of the many molecules involved in keeping the activity of the normal immune system within safe bounds is a protein called CTLA4 (cytotoxic T-lymphocyte-associated protein 4), which functions as an inhibitory receptor on the surface of T cells. If the function of CTLA4 is blocked, the T cells become more reactive and may mount an attack on cells that they would otherwise leave in peace. In particular, the T cells may attack tumor cells that are recognizably abnormal but whose presence was previously tolerated. With this in mind, cancer immunologists developed a monoclonal antibody, called *ipilimumab*, that binds to CTLA4 and blocks its action. Injected repeatedly into patients with metastatic melanoma, this antibody increases their median lifespan by several months and, in one large trial, enabled as many as a quarter of them to survive for five years

Figure 20–45 Therapies designed to remove the immunosuppressive microenvironment in tumors. (A) The cells in tumors will produce many mutant proteins. As described in Chapter 24, peptides from these proteins will be displayed on MHC complexes on the tumor-cell surface and would normally activate a T cell response that destroys the tumor (see Figure 24–42). However, as schematically illustrated, during the course of tumor progression, the cancer cells have evolved immunosuppressive mechanisms that protect them from such killing. (B) The cells in tumors often protect themselves from immune attack by expressing proteins on their surface that bind to and thereby activate the inhibitory receptors on T cells. As indicated, this makes the tumor susceptible to specific antibody therapies. In this diagram, two such inhibitory receptors are shown, PD1 and a hypothetical protein X. Different tumors are thought to protect themselves by activating different members of a large set of T cell inhibitory receptors, some of which are not yet well characterized.

or more—far beyond expectations for comparable patients without this treatment. Even more promising are recent clinical trials using a combination of two antibodies, one against CTLA4 and the other against PD1, a second cell-surface receptor on T cells that normally restrains their activity.

In clinical trials using such techniques, a substantial fraction of the patients can respond in a dramatic way, with their cancer being driven into remission for years, while the treatment fails to help others with the same type of cancer. One possible explanation is that, while most tumors express proteins that protect them from T-cell attack, these proteins are different for different tumors. Thus, while some tumors will respond dramatically when treated with an antibody that blocks a particular immunosuppressive agent, many others will not. If true, one can foresee an era of personalized immunotherapy, in which each patient's tumor is molecularly analyzed to determine its particular mechanisms of immunosuppression. The patient would then be treated with a specific cocktail of antibodies designed to remove these blocks (see Figure 20–45).

Cancers Evolve Resistance to Therapies

High hopes have to be tempered with sobering realities. We have seen that genetic instability can provide an Achilles heel that cancer therapies can exploit, but at the same time it can make eradicating the disease more difficult by allowing the cancer cells to evolve resistance to therapeutic drugs, often at an alarming rate. This applies even to the drugs that target genetic instability itself. Thus, PARP inhibitors give valuable remission of illness, but in the long term the disease generally comes back. For example, *Brca*-deficient cancers can sometimes develop resistance to PARP inhibitors by undergoing a second mutation in an affected *Brca* gene that restores its function. By then, the cancer is already out of control and it may be too late to affect the course of the disease with additional treatments.

There are many different strategies by which cancers can evolve resistance to anticancer drugs. Often, a cancer will be dramatically reduced in size by an initial drug treatment, with all of the detectable tumor cells seeming to disappear. But months or years later the cancer will reappear in an altered form that is resistant to the drug that was at first so successful. In such cases, the initial drug treatment has evidently failed to destroy some tiny fraction of cells in the original tumor-cell population. These cells have escaped death because they carry a protective mutation or epigenetic change, or perhaps simply because they were lurking in a protected environment. They eventually regenerate the cancer by continuing to proliferate, mutating and evolving still further as they do so.

In some cases, cells that are exposed to one anticancer drug evolve a resistance not only to that drug but also to other drugs to which they have never been exposed. This phenomenon of **multidrug resistance** frequently correlates with amplification of a part of the genome that contains a gene called *Mdr1* or *Abcb1*. This gene encodes a plasma-membrane-bound transport ATPase of the ABC transporter superfamily (discussed in Chapter 11), which pumps lipophilic drugs out of the cell (see Movie 11.5). The overproduction of this protein (or some of its other family members) by a cancer cell can prevent the intracellular accumulation of many cytotoxic drugs, making the cell insensitive to them.

In the to-and-fro struggle between advanced metastatic cancer and the therapist, as current practice stands, the cancer usually wins in the end. Does it have to be so? As we discuss below, there is reason to think that by attacking a cancer with many weapons at once—instead of using them one after another, each until it fails—it may be possible to do much better.

Combination Therapies May Succeed Where Treatments with One Drug at a Time Fail

Nowadays, cancers caught at an early stage can often be cured, by surgery, radiation, or drugs. For most cancers that have progressed and metastasized widely, however, cure is still beyond us. Treatments such as those described above can

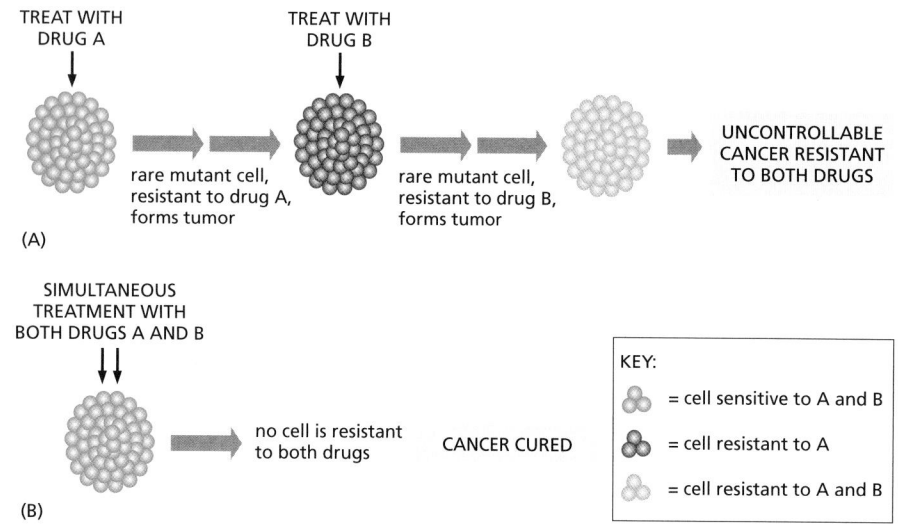

TREAT WITH
DRUG A

TREAT WITH
DRUG B

rare mutant cell,
resistant to drug A,
forms tumor

rare mutant cell,
resistant to drug B,
forms tumor

UNCONTROLLABLE
CANCER RESISTANT
TO BOTH DRUGS

(A)

SIMULTANEOUS
TREATMENT WITH
BOTH DRUGS A AND B

no cell is resistant
to both drugs

CANCER CURED

(B)

KEY:

= cell sensitive to A and B

= cell resistant to A

= cell resistant to A and B

Figure 20–46 Why multidrug treatments can be more effective than sequential treatments for cancer therapy. (A) Because tumor cells are hypermutable, two single-drug treatments that are given sequentially often allow for the selection of mutant cell clones that are resistant to both drugs. (B) Simultaneous treatment with both drugs can be more effective.

give valuable remissions, but sooner or later these are typically followed by relapse.

Nevertheless, for some relatively rare forms of advanced cancer, curative therapies have been developed. These generally involve a cocktail of several different anticancer agents: by trial and error, certain combinations of cytotoxic drugs have been found to wipe out the cancer completely. Discovering such combinations has hitherto involved a long, hard search. But now, armed with our new tools for identifying the specific genetic lesions that cancer cells contain, the prospects are better.

The logic of combination therapies is the same as that behind the current treatment of HIV-AIDS with a cocktail of three different protease inhibitors: whereas there may always be some cells in the initial population carrying the rare mutations that confer resistance to any one drug treatment, there should be no cell carrying the whole set of rare mutations that would confer resistance to several different drugs delivered simultaneously. In contrast, sequential drug treatments will allow the few cells resistant to the first drug to multiply to large numbers. Within this large population of cells resistant to the first drug, a small number of cells are likely to have arisen that are resistant to the next drug also; and so on (Figure 20–46).

We Now Have the Tools to Devise Combination Therapies Tailored to the Individual Patient

Efficient, rational combination drug therapy requires three things. First, we have to identify multiple peculiarities of cancer cells that make them vulnerable in ways that normal cells are not. Second, we have to devise drugs (or other treatments) that target each of these vulnerabilities. Third, we have to match the combination of drugs to the specific set of peculiarities present in the cancer cells of the individual patient.

The first requirement is already partially met: we now have large catalogs of cancer-critical genes that are commonly mutated in cancer cells. The second requirement is harder, but attainable: we have described some remarkable recent successes, and for cancer researchers there is excitement in the air. It is becoming increasingly possible to use our growing knowledge of cell and molecular biology to design new drugs against designated targets. At the same time, efficient, high-throughput automated methods are available to screen large libraries of chemicals for any that may be effective against cells with a given cancer-related defect. In such searches, the goal is synthetic lethality: a cell death that occurs when and only when a particular drug is put together with a particular cancer cell abnormality. Through these and other approaches, the repertoire of precisely targeted anticancer drugs is rapidly increasing.

This brings us to the third requirement: the therapy—the choice of drugs to be given in combination—must be tailored to the individual patient. Here, too, the prospects are bright. Cancers evolve by a fundamentally random process, and each patient is different; but modern methods of genome analysis now let us characterize the cells from a tumor biopsy in exhaustive detail so as to discover which cancer-critical genes are affected in a particular case. Admittedly, this is not straightforward: the tumor cells in an individual patient are heterogeneous and do not all contain the same genetic lesions. With increased understandings of the pathways of cancer evolution, however, and with the experience gained from many different cases, it should become possible to make good guesses at the optimal therapies to use.

From the perspective of the patient, the pace of advance in cancer research can seem frustratingly slow. Each new drug has to be tested in the clinic, first for safety and then for efficacy, before it can be released for general use. And if the drug is to be used in combination with others, the combination therapy must then go through the same long process. Strict ethical rules constrain the conduct of trials, which means that they take time—typically several years. But slow and cautious steps, taken systematically in the right direction, can lead to great advances. There is still far to go, but the examples that we have discussed provide proof of principle and grounds for optimism.

From the cancer research effort, we have learned a great deal of what we know about the molecular biology of the normal cell. Now, more and more, we are discovering how to put that knowledge to use in the battle with cancer itself.

Summary

Our growing understanding of the cell biology of cancers has already begun to lead to better ways of preventing, diagnosing, and treating these diseases. Anticancer therapies can be designed to destroy cancer cells preferentially by exploiting the properties that distinguish cancer cells from normal cells, including the cancer cells' dependence on oncogenic proteins and the defects they harbor in their DNA repair mechanisms. We now have good evidence that, by increasing our understanding of normal cell control mechanisms and exactly how they are subverted in specific cancers, we can eventually devise drugs to kill cancers precisely by attacking specific molecules critical for the growth and survival of the cancer cells. In addition, great progress has recently been made through sophisticated immunological approaches to cancer therapy. And, as we become better able to determine which genes are altered in the cells of any given tumor, we can begin to tailor treatments more accurately to each individual patient.

WHAT WE DON'T KNOW

• What is required to enable a cancer cell to metastasize?

• How can the molecular analysis of an individual tumor be more effectively used to design effective therapies to kill it?

• Can we identify general features common to all cancer cells—such as their production of misfolded, mutated proteins—that can be used for the targeted destruction of many different types of cancers?

• Can sensitive and reliable blood tests be devised to detect cancers very early, before they have grown to a size where treatment with a single drug will generally be defeated by the survival of a preexisting resistant variant?

• How can the observed environmental effects on cancer rates be exploited to reduce avoidable cancers?

• Can new technologies be devised to reveal exactly how a quiescent micrometastasis converts to a full-blown metastatic tumor?

PROBLEMS

Which statements are true? Explain why or why not.

20–1 The chemical carcinogen dimethylbenz[a]anthracene (DMBA) must be an extraordinarily specific mutagen since 90% of the skin tumors it causes have an A-to-T alteration at exactly the same site in the mutant *Ras* gene.

20–2 In the cellular regulatory pathways that control cell growth and proliferation, the products of oncogenes are stimulatory components and the products of tumor suppressor genes are inhibitory components.

20–3 Cancer therapies directed solely at killing the rapidly dividing cells that make up the bulk of a tumor are unlikely to eliminate the cancer from many patients.

20–4 The main environmental causes of cancer are the products of our highly industrialized way of life such as pollution and food additives.

Discuss the following problems.

20–5 In contrast to colon cancer, whose incidence increases dramatically with age, incidence of osteosarcoma—a tumor that occurs most commonly in the long bones—peaks during adolescence. Osteosarcomas are relatively rare in young children (up to age 9) and in adults (over 20). Why do you suppose that the incidence of osteosarcoma does not show the same sort of age-dependence as colon cancer?

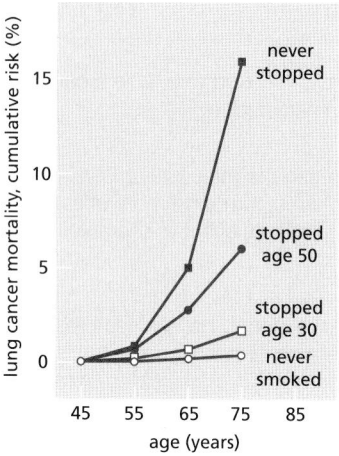

Figure Q20–1 Cumulative risk of lung cancer mortality for nonsmokers, smokers, and former smokers (Problem 20–6). Cumulative risk is the running total of deaths, as a percentage, for each group. Thus, for continuing smokers, 1% died of lung cancer between ages 45 and 55; an additional 4% died between 55 and 65 (giving a cumulative risk of 5%); and 11% more died between 65 and 75 (for a cumulative risk of 16%).

20–8   Virtually all cancer treatments are designed to kill cancer cells, usually by inducing apoptosis. However, one particular cancer—acute promyelocytic leukemia (APL)—has been successfully treated with all-*trans*-retinoic acid, which causes the promyelocytes to differentiate into neutrophils. How might a change in the state of differentiation of APL cancer cells help the patient?

20–9   One major goal of modern cancer therapy is to identify small molecules—anticancer drugs—that can be used to inhibit the products of specific cancer-critical genes. If you were searching for such molecules, would you design inhibitors for the products of oncogenes or the products of tumor suppressor genes? Explain why you would (or would not) select each type of gene.

20–6   Mortality due to lung cancer was followed in groups of males in the United Kingdom for 50 years. **Figure Q20–1** shows the cumulative risk of dying from lung cancer as a function of age and smoking habits for four groups of males: those who never smoked, those who stopped at age 30, those who stopped at age 50, and those who continued to smoke. These data show clearly that individuals can substantially reduce their cumulative risk of dying from lung cancer by stopping smoking. What do you suppose is the biological basis for this observation?

20–7   A small fraction—2 to 3%—of all cancers, across many subtypes, displays a quite remarkable phenomenon: tens to hundreds of rearrangements that primarily involve a single chromosome, or chromosomal region. The breakpoints can be tightly clustered, with several in a few kilobases; the junctions of the rearrangements often involve segments of DNA that were not originally close together on the chromosome. The copy number of various segments within the rearranged chromosome was found to be either zero, indicating deletion, or one, indicating retention.

You can imagine two ways in which such multiple, localized rearrangements might happen: a progressive rearrangements model with ongoing inversions, deletions, and duplications involving a localized area, or a catastrophic model in which the chromosome is shattered into fragments that are stitched back together in random order by nonhomologous end joining (**Figure Q20–2**).
A.   Which of the two models in Figure Q20–2 accounts more readily for the features of these highly rearranged chromosomes? Explain your reasoning.
B.   For whichever model you choose, suggest how such multiple rearrangements might arise. (The true mechanism is not known.)
C.   Do you suppose such rearrangements are likely to be causative events in the cancers in which they are found, or are they probably just passenger events that are unrelated to the cancer? If you think they could be driver events, suggest how such rearrangements might activate an oncogene or inactivate a tumor suppressor gene.

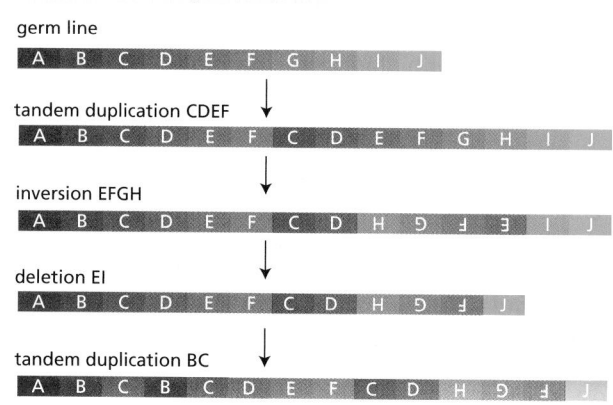

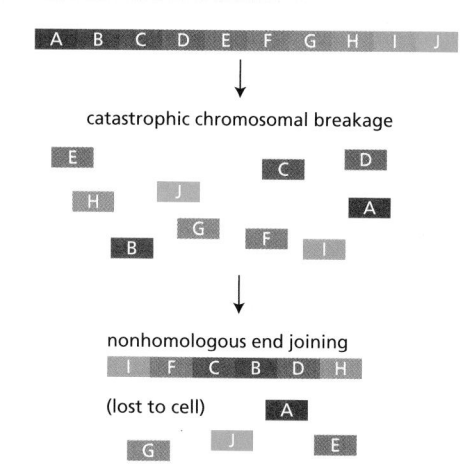

Figure Q20–2 Two models to explain the multiple, localized chromosome rearrangements found in some cancers (Problem 20–7). The progressive rearrangements model shows a sequence of rearrangements that disrupts the chromosome, generating increasingly complex chromosomal configurations. The chromosome catastrophe model shows the chromosome being fragmented and then reassembled randomly, with some pieces left out.

20–10 PolyADP-ribose polymerase (PARP) plays a key role in the repair of DNA single-strand breaks. In the presence of the PARP inhibitor olaparib, single-strand breaks accumulate. When a replication fork encounters a single-strand break, it converts it to a double-strand break, which in normal cells is then repaired by homologous recombination. In cells defective for homologous recombination, however, inhibition of PARP triggers cell death.

Patients who have only one functional copy of the *Brca1* gene, which is required for homologous recombination, are at much higher risk for cancer of the breast and ovary. Cancers that arise in these tissues in these patients can be treated successfully with olaparib. Explain how it is that treatment with olaparib kills the cancer cells in these patients, but does not harm their normal cells.

20–11 The Tasmanian devil, a carnivorous Australian marsupial, is threatened with extinction by the spread of a fatal disease in which a malignant oral–facial tumor interferes with the animal's ability to feed. You have been called in to analyze the source of this unusual cancer. It seems clear to you that the cancer is somehow spread from devil to devil, very likely by their frequent fighting, which is accompanied by biting around the face and mouth. To uncover the source of the cancer, you isolate tumors from 11 devils captured in widely separated regions and examine them. As might be expected, the karyotypes of the tumor cells are highly rearranged relative to that of the wild-type devil (**Figure Q20–3**). Surprisingly, you find that the karyotypes from all 11 tumor samples are very similar. Moreover, one of the Tasmanian devils has an inversion on chromosome 5 that is not present in its facial tumor. How do you suppose this cancer is transmitted from devil to devil? Is it likely to arise as a consequence of an infection by a virus or microorganism? Explain your reasoning.

(A)

Tasmanian devil (*Sarcophilus harrisii*)

(B)

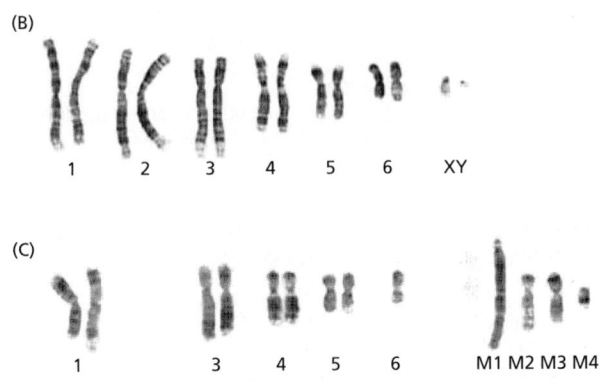

Figure Q20–3 Karyotypes of cells from Tasmanian devils (Problem 20–11). (A) A Tasmanian devil. (B) Normal karyotype for a male Tasmanian devil. The karyotype has 14 chromosomes, including XY. (C) Karyotype of cancer cells found in each of the 11 facial tumors studied. The karyotype has 13 chromosomes, no sex chromosomes, no chromosome 2 pair, one chromosome 6, two chromosome 1 with deleted long arms, and four highly rearranged marker chromosomes (M1–M4). (A, from John Gould, *The Mammals of Australia, Vol.1* (1863); B and C, from A.M. Pearse and K. Swift, *Nature* 439:549, published 2006 by Nature Publishing Group. Reproduced with permission of SNCSC.)

REFERENCES

General

Bishop JM (2004) How to Win the Nobel Prize: An Unexpected Life in Science. Cambridge, MA: Harvard University Press.
Hanahan D & Weinberg RA (2011) Hallmarks of cancer: the next generation. *Cell* 144, 646–674.
Vogelstein B, Papadopoulos N, Velculescu VE et al. (2013) Cancer genome landscapes. *Science* 339, 1546–1558.
Weinberg RA (2013) The Biology of Cancer, 2nd ed. Garland Science: New York.

Cancer as a Microevolutionary Process

Brown JM & Attardi LD (2005) The role of apoptosis in cancer development and treatment response. *Nat. Rev. Cancer* 5, 231–237.
Chambers AF, Naumov GN, Vantyghem S & Tuck AB (2000) Molecular biology of breast cancer metastasis. Clinical implications of experimental studies on metastatic inefficiency. *Breast Cancer Res.* 2, 400–407.
Chi P, Allis CD & Wang GG (2010) Covalent histone modifications—miswritten, misinterpreted and mis-erased in human cancers. *Nat. Rev. Cancer* 10, 457–469.

Fidler IJ (2003) The pathogenesis of cancer metastasis: the 'seed and soil' hypothesis revisited. *Nat. Rev. Cancer* 3, 453–458.
Hoeijmakers JHJ (2001) Genome maintenance mechanisms for preventing cancer. *Nature* 411, 366–374.
Joyce JA & Pollard JW (2009) Microenvironmental regulation of metastasis. *Nat. Rev. Cancer* 9, 239–252.
Lowe SW, Cepero E & Evan G (2004) Intrinsic tumour suppression. *Nature* 432, 307–315.
Nowell PC (1976) The clonal evolution of tumor cell populations. *Science* 194, 23–28.
Stephens PJ, McBride DJ, Lin M-L et al. (2009) Complex landscapes of somatic rearrangement in human breast cancer genomes. *Nature* 462, 1005–1010.
Thiery JP (2002) Epithelial-mesenchymal transitions in tumour progression. *Nat. Rev. Cancer* 2, 442–454.
Vander Heiden MG, Cantley LC & Thompson CB (2009) Understanding the Warburg effect: the metabolic requirements of cell proliferation. *Science* 324, 1029–1033.
Zink D, Fischer AH & Nickerson JA (2004) Nuclear structure in cancer cells. *Nat. Rev. Cancer* 4, 677–687.

Cancer-Critical Genes: How They Are Found and What They Do

Berdasco M & Esteller M (2010) Aberrant epigenetic landscape in cancer: how cellular identity goes awry. *Dev. Cell* 19, 698–711.

Brognard J & Hunter T (2011) Protein kinase signaling networks in cancer. *Curr. Opin. Genet. Dev.* 21, 4–11.

Eilers M & Eisenman R (2008) Myc's broad reach. *Genes Dev.* 22, 2755–2766.

Feinberg AP (2007) Phenotypic plasticity and the epigenetics of human disease. *Nature* 447, 433–440.

Garraway LA & Lander ES (2013) Lessons from the cancer genome. *Cell* 153, 17–37.

Greaves M & Maley CC (2012) Clonal evolution in cancer. *Nature* 481, 306–313.

Junttila MR & Evan GI (2009) p53—a Jack of all trades but master of none. *Nat. Rev. Cancer* 9, 821–829.

Levine AJ (2009) The common mechanisms of transformation by the small DNA tumor viruses: the inactivation of tumor suppressor gene products: p53. *Virology* 384, 285–293.

Lu P, Weaver VM & Werb Z (2012) The extracellular matrix: a dynamic niche in cancer progression. *J. Cell Biol.* 196, 395–406.

Mitelman F, Johansson B & Mertens F (2007) The impact of translocations and gene fusions on cancer causation. *Nat. Rev. Cancer* 7, 233–245.

Negrini S, Gorgoulis VG & Halazonetis TD (2010) Genomic instability—an evolving hallmark of cancer. *Nat. Rev. Mol. Cell Biol.* 11, 220–228.

Nguyen DX, Bos PD & Massagué J (2009) Metastasis: from dissemination to organ-specific colonization. *Nat. Rev. Cancer* 9, 274–284.

Radtke F & Clevers H (2005) Self-renewal and cancer of the gut: two sides of a coin. *Science* 307, 1904–1909.

Rowley JD (2001) Chromosome translocations: dangerous liaisons revisited. *Nat. Rev. Cancer* 1, 245–250.

Shaw RJ & Cantley LC (2006) Ras, PI(3)K and mTOR signalling controls tumour cell growth. *Nature* 441, 424–430.

Suvà ML, Riggi N & Bernstein BE (2013) Epigenetic reprogramming in cancer. *Science* 339, 1567–1570.

Weinberg RA (1995) The retinoblastoma protein and cell cycle control. *Cell* 81, 323–330.

Cancer Prevention and Treatment: Present and Future

Al-Hajj M, Becker MW, Wicha M et al. (2004) Therapeutic implications of cancer stem cells. *Curr. Opin. Genet. Dev.* 14, 43–47.

Ames B, Durston WE, Yamasaki E & Lee FD (1973) Carcinogens are mutagens: a simple test system combining liver homogenates for activation and bacteria for detection. *Proc. Natl Acad. Sci. USA* 70, 2281–2285.

Bozic I, Reiter JG, Allen B et al. (2013) Evolutionary dynamics of cancer in response to targeted combination therapy. *eLife* 2, e00747.

Doll R & Peto R (1981) The causes of cancer: quantitative estimates of avoidable risks of cancer in the United States today. *J. Natl Cancer Inst.* 66, 1191–1308.

Druker BJ & Lydon NB (2000) Lessons learned from the development of an Abl tyrosine kinase inhibitor for chronic myelogenous leukemia. *J. Clin. Invest.* 105, 3–7.

Huang P & Oliff A (2001) Signaling pathways in apoptosis as potential targets for cancer therapy. *Trends Cell Biol.* 11, 343–348.

Jain RK (2005) Normalization of tumor vasculature: an emerging concept in antiangiogenic therapy. *Science* 307, 58–62.

Jonkers J & Berns A (2004) Oncogene addiction: sometimes a temporary slavery. *Cancer Cell* 6, 535–538.

Kalos M & June CH (2013) Adoptive T cell transfer for cancer immunotherapy in the era of synthetic biology. *Immunity* 39, 49–60.

Loeb LA (2011) Human cancers express mutator phenotypes: origin, consequences and targeting. *Nat. Rev. Cancer* 11, 450–457.

Lord CJ & Ashworth A (2012) The DNA damage response and cancer therapy. *Nature* 481, 287–294.

Pardoll DM (2012) The blockade of immune checkpoints in cancer immunotherapy. *Nat. Rev. Cancer* 12, 252–264.

Peto J (2001) Cancer epidemiology in the last century and the next decade. *Nature* 411, 390–395.

Sawyers C (2004) Targeted cancer therapy. *Nature* 432, 294–297.

Schreiber RD, Old LJ & Smyth MJ (2011) Cancer immunoediting: integrating immunity's roles in cancer suppression and promotion. *Science* 331, 1565–1570.

Sliwkowski MX & Mellman I (2013) Antibody therapeutics in cancer. *Science* 341, 1192–1198.

Varmus H, Pao W, Politi K et al. (2005) Oncogenes come of age. *Cold Spring Harb. Symp. Quant. Biol.* 70, 1–9.

Ward RJ & Dirks PB (2007) Cancer stem cells: at the headwaters of tumor development. *Annu. Rev. Pathol.* 2, 175–189.

Development of Multicellular Organisms

An animal or plant starts its life as a single cell—a fertilized egg, or **zygote**. During development, this cell divides repeatedly to produce many different kinds of cells, arranged in a final pattern of spectacular complexity and precision. The goal of developmental cell biology is to understand the cellular and molecular mechanisms that direct this amazing transformation (Movie 21.1).

Plants and animals have very different ways of life, and they use different developmental strategies; in this chapter, we focus mainly on animals. Four processes are fundamental to animal development: (1) cell proliferation, which produces many cells from one; (2) cell–cell interactions, which coordinate the behavior of each cell with that of its neighbors; (3) cell specialization, or **differentiation**, which creates cells with different characteristics at different positions; and (4) cell movement, which rearranges the cells to form structured tissues and organs (Figure 21–1). It is on the fourth point that plant development differs radically: plant cells are unable to migrate or move independently through the embryo because each one is contained within a cell wall, through which it is cemented to its neighbors, as discussed in Chapter 19.

In a developing animal embryo, the four fundamental processes are happening in a kaleidoscopic variety of ways, as they give rise to different parts of the organism. Like the members of an orchestra, the cells in the embryo have to play their individual parts in a highly coordinated manner. In the embryo, however, there is no conductor—no central authority—to direct the performance. Instead, development is a self-assembly process in which the cells, as they grow and proliferate, organize themselves into increasingly complex structures. Each of the millions of cells has to choose for itself how to behave, selectively utilizing the genetic instructions in its chromosomes.

At each stage in its development, the cell is presented with a limited set of options, so that its developmental pathway branches repeatedly, reflecting a large set of sequential choices. Like the decisions we make in our own lives, the choices made by the cell are based on its internal state—which largely reflects its history—and on current influences from other cells, especially its close neighbors. To understand development, we need to know how each choice is controlled and how it depends on previous choices. Beyond that, we need to understand how the choices, once made, influence the cell's chemistry and behavior, and how cell behaviors act synergistically to determine the structure and function of the body.

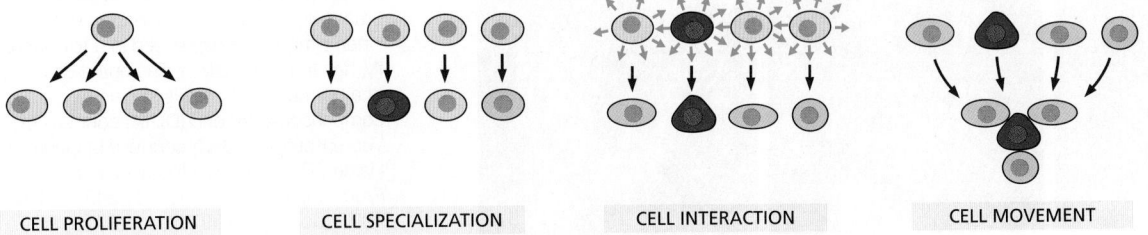

CELL PROLIFERATION   CELL SPECIALIZATION   CELL INTERACTION   CELL MOVEMENT

Figure 21–1 **The four essential cell processes that allow a multicellular organism to be made.**

As cells become specialized they change not only their chemistry but also their shape and their attachments to other cells and to the extracellular matrix. They move and rearrange themselves to create the complex architecture of the body, with all its tissues and organs, each structured precisely and defined in size. To understand this process of form generation, or *morphogenesis*, we will need to take account of the mechanical, as well as the biochemical, interactions between the cells.

At first glance, one would no more expect the worm, the flea, the eagle, and the giant squid all to be generated by the same developmental mechanisms than one would suppose that the same methods were used to make a shoe and an airplane. Remarkably, however, research in the past 30 years has revealed that much of the basic machinery of development is essentially the same in all animals—not just in all vertebrates, but in all the major phyla of invertebrates too. Recognizably similar, evolutionarily related molecules define the specialized animal cell types, mark the differences between body regions, and help create the animal body pattern. Homologous proteins are often functionally interchangeable between very different species. Thus, a human protein produced artificially in a fly, for example, can perform the same function as the fly's own version of that protein (**Figure 21–2**). Thanks to an underlying unity of mechanisms, developmental biologists have been making great strides toward a coherent understanding of animal development.

We begin this chapter with an overview of some of the basic mechanisms that operate in animal development. We then discuss, in sequence, how cells in the embryo diversify to form patterns in space, how the timing of developmental events is controlled, how cell movements contribute to morphogenesis, and how the size of an animal is regulated. We end by considering the most challenging aspect of development—the mechanisms that enable a highly complex nervous system to form.

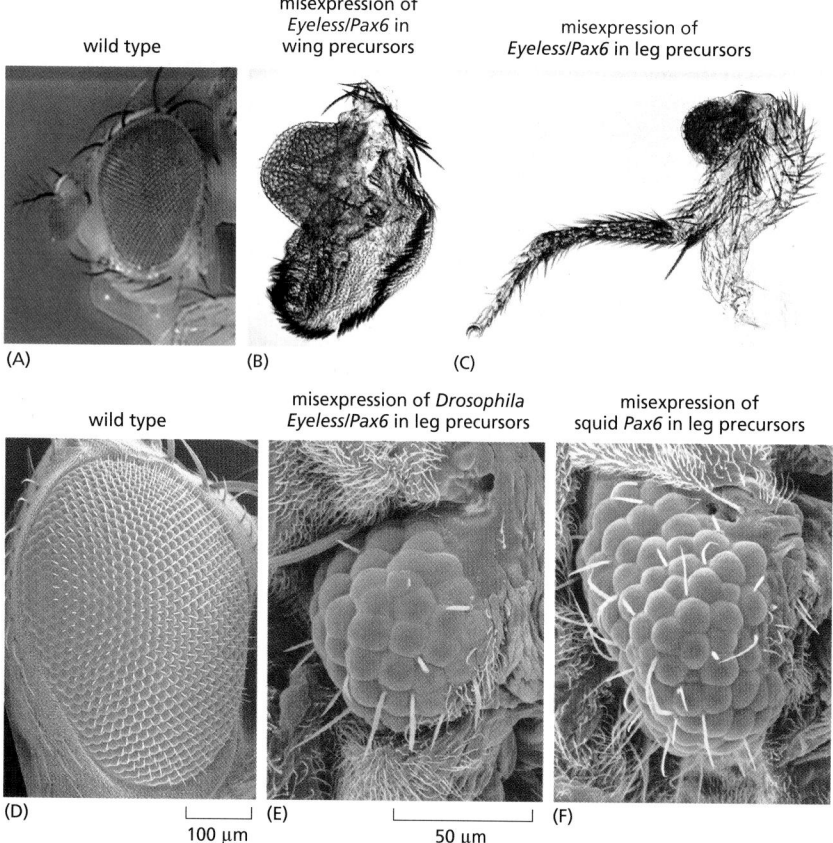

Figure 21–2 Homologous proteins can function interchangeably. (A–C) The Eyeless protein (also called Pax6) controls eye development in *Drosophila* and, when misexpressed during development, can cause an eye to form in an abnormal site, such as a wing (B) or a leg (C). The scanning electron micrographs show a patch of eye tissue on the leg of a fly resulting from misexpression of *Drosophila Eyeless* (E) and of squid *Pax6* (F). The homologous protein from a human or practically any animal possessing eyes, when similarly misexpressed in a transgenic fly, has the same effect. The entire eye of a normal *Drosophila* is shown for comparison in (A) and (D). (A, courtesy of Thom Kaufman; B–C, courtesy of Georg Halder; D–F, from S.I. Tomarev, et al. *Proc. Natl Acad. Sci. USA* 94:2421–2426, 1997. Copyright 1997 National Academy of Sciences, USA. With permission from National Academy of Sciences.)

OVERVIEW OF DEVELOPMENT

Animals live by eating other organisms. Thus, despite their remarkable diversity, animals as different as worms, mollusks, insects, and vertebrates share anatomical features that are fundamental to this way of life. Epidermal cells form a protective outer layer; gut cells absorb nutrients from ingested food; muscle cells allow movement; and neurons and sensory cells control behavior. These diverse cell types are organized into tissues and organs, forming a sheet of skin covering the exterior, a mouth for feeding, and an internal gut tube to digest food—with muscles, nerves, and other tissues arranged in the space between the skin and the gut tube. Many animals have clearly defined axes—an anteroposterior axis, with mouth and brain anterior and anus posterior; a dorsoventral axis, with back dorsal and belly ventral; and a left-right axis. In this section, we discuss some fundamental mechanisms underlying animal development, beginning with how the basic animal body plan is established.

Conserved Mechanisms Establish the Basic Animal Body Plan

The shared anatomical features of animals develop through conserved mechanisms. After fertilization, the zygote usually divides rapidly, or **cleaves**, to form many smaller cells; during this cleavage, the embryo, which cannot yet feed, does not grow. This phase of development is initially driven and controlled entirely by the material deposited in the egg by the mother. The embryonic genome remains inactive until a point is reached when maternal mRNAs and proteins rather abruptly begin to be degraded. The embryo's genome is activated, and the cells cohere to form a **blastula**—typically a solid or a hollow fluid-filled ball of cells. Complex cell rearrangements called **gastrulation** (from the Greek "gaster," meaning "belly") then transform the blastula into a multilayered structure containing a rudimentary internal gut (**Figure 21–3**). Some cells of the blastula remain external, constituting the **ectoderm**, which will give rise to the epidermis and the nervous system; other cells invaginate, forming the **endoderm**, which will give rise to the gut tube and its appendages, such as lung, pancreas, and liver. Another group of cells moves into the space between ectoderm and endoderm and forms the **mesoderm**, which will give rise to muscles, connective tissues, blood, kidney, and various other components. Further cell movements and accompanying cell differentiations create and refine the embryo's architecture.

Figure 21–3 The early stages of development, as exemplified by a frog. (A) A fertilized egg divides to produce a blastula—a sheet of epithelial cells often surrounding a cavity. During gastrulation, some of the cells tuck into the interior to form the mesoderm *(green)* and endoderm *(yellow)*. Ectodermal cells *(blue)* remain on the outside. (B) A cross section through the trunk of an amphibian embryo shows the basic animal body plan, with a sheet of ectoderm on the outside, a tube of endoderm on the inside, and mesoderm sandwiched between them. The endoderm forms the epithelial lining of the gut, from the mouth to the anus. It gives rise not only to the pharynx, esophagus, stomach, and intestines, but also to many associated structures. The salivary glands, liver, pancreas, trachea, and lungs, for example, all develop from the wall of the digestive tract and grow to become systems of branching tubes that open into the gut or pharynx. The endoderm forms only the epithelial components of these structures—the lining of the gut and the secretory cells of the pancreas, for example. The supporting muscular and fibrous elements arise from the mesoderm.

The mesoderm gives rise to the connective tissues—at first, to the loose mesh of cells in the embryo known as mesenchyme, and ultimately to cartilage, bone, and fibrous tissue, including the dermis (the inner layer of the skin). The mesoderm also forms the muscles, the entire vascular system—including the heart, blood vessels, and blood cells—and the tubules, ducts, and supporting tissues of the kidneys and gonads. The notochord forms from the mesoderm and serves as the core of the future backbone and the source of signals that coordinate the development of surrounding tissues.

The ectoderm will form the epidermis (the outer, epithelial layer of the skin) and epidermal appendages such as hair, sweat glands, and mammary glands. It will also give rise to the whole of the nervous system, central and peripheral, including not only neurons and glia but also the sensory cells of the nose, the ear, the eye, and other sense organs. (B, after T. Mohun et al., *Cell* 22:9–15, 1980.)

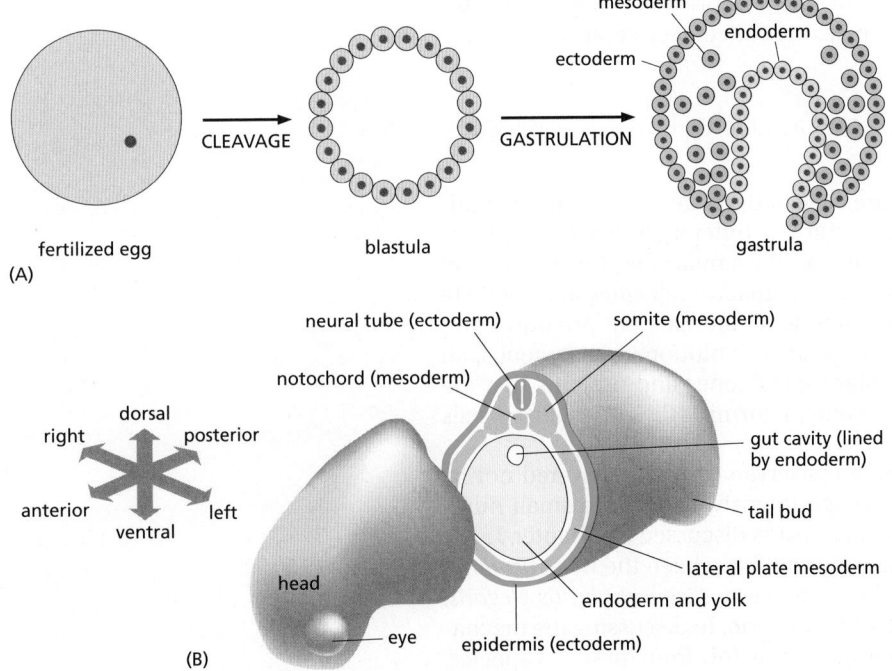

The ectoderm, mesoderm, and endoderm formed during gastrulation constitute the three **germ layers** of the early embryo. Many later developmental transformations will produce the elaborately structured organs. But the basic body plan and axes set up in miniature during gastrulation are preserved into adult life, when the organism may be billions of times larger (Movie 21.2).

The Developmental Potential of Cells Becomes Progressively Restricted

Concomitant with the refinement of the body plan, the individual cells become more and more restricted in their developmental potential. During the blastula stages, cells are often **totipotent** or **pluripotent**—they have the potential to give rise to all or almost all of the cell types of the adult body. The pluripotency is lost as gastrulation proceeds: a cell located in the endodermal germ layer, for example, can give rise to the cell types that will line the gut or form gut-derived organs such as the liver or pancreas, but it no longer has the potential to form mesoderm-derived structures such as skeleton, heart, or kidney. Such a cell is said to be *determined* for an endodermal fate. Thus, **cell determination** starts early and progressively narrows the options as the cell steps through a programmed series of intermediate states—guided at each step by its genome, its history, and its interactions with neighbors. The process reaches its limit when a cell undergoes **terminal differentiation** to form one of the highly specialized cell types of the adult body (Figure 21–4). Although there are cell types in the adult that retain some degree of pluripotency, their range of options is generally narrow (discussed in Chapter 22).

Cell Memory Underlies Cell Decision-Making

Underlying the richness and astonishingly complex outcomes of development is **cell memory** (see p. 404). Both the genes a cell expresses and the way it behaves depend on the cell's past, as well as on its present circumstances. The cells of our body—the muscle cells, the neurons, the skin cells, the gut cells, and so on—maintain their specialized characters largely because they retain a record of the extracellular signals their ancestors received during development, rather than because they continually receive such instructions from their surroundings. Despite their radically different phenotypes, they retain the same complete genome that was present in the zygote; their differences arise instead from differential gene expression. We have discussed the molecular mechanisms of gene regulation, cell memory, cell division, cell signaling, and cell movement in previous chapters. In this chapter, we shall see how these basic processes are collectively deployed to create an animal.

Several Model Organisms Have Been Crucial for Understanding Development

The anatomical features that animals share have undergone many extreme modifications in the course of evolution. As a result, the differences between species are usually more striking to our human eye than the similarities. But at the level of the underlying molecular mechanisms and the macromolecules that mediate them, the reverse is true: the similarities among all animals are profound and extensive. Through more than half a billion years of evolutionary divergence, all animals have retained unmistakably similar sets of genes and proteins that are responsible for generating their body plans and for forming their specialized cells and organs.

This astonishing degree of evolutionary conservation was discovered not by broad surveys of animal diversity, but through intensive study of a small number of representative species—the model organisms discussed in Chapter 1. For animal developmental biology, the most important have been the fly *Drosophila melanogaster*, the frog *Xenopus laevis*, the roundworm *Caenorhabditis elegans*, the mouse *Mus musculus*, and the zebrafish *Danio rerio*. In discussing the mechanisms of development, we shall draw our examples mainly from these few species.

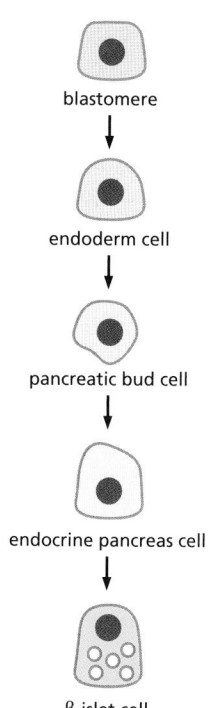

blastomere

endoderm cell

pancreatic bud cell

endocrine pancreas cell

β-islet cell

Figure 21–4 The lineage from blastomere to differentiated cell type. As development proceeds, cells become more and more specialized. Blastomeres have the potential to give rise to most or all cell types. Under the influence of signaling molecules and gene regulatory factors, cells acquire more restricted fates until they differentiate into highly specialized cell types, such as the pancreatic β-islet cells that secrete the hormone insulin.

Genes Involved in Cell–Cell Communication and Transcriptional Control Are Especially Important for Animal Development

What are the genes that animals share with one another but not with other kingdoms of life? These would be expected to include genes required specifically for animal development but not needed for unicellular existence. Comparison of animal genomes with the genome of budding yeast—a unicellular eukaryote—suggests that three classes of genes are especially important for multicellular organization. The first class includes genes that encode proteins used for cell–cell adhesion and cell signaling; hundreds of human genes encode signal proteins, cell-surface receptors, cell adhesion proteins, or ion channels that are either not present in yeast or present in much smaller numbers. The second class includes genes encoding proteins that regulate transcription and chromatin structure: more than 1000 human genes encode transcription regulators, but only about 250 yeast genes do so. As we shall see, the development of animals is dominated by cell–cell interactions and by differential gene expression. The third class of noncoding RNAs has a more uncertain status: it includes genes that encode microRNAs (miRNAs); there are at least 500 of these in humans. Along with the regulatory proteins, they play a significant part in controlling gene expression during animal development, but the full extent of their importance is still unclear. The loss of individual miRNA genes in *C. elegans*, where their functions have been well studied, rarely leads to obvious phenotypes, suggesting that the roles of miRNAs during animal development are often subtle, serving to fine-tune the developmental machinery rather than to form its core structures.

Regulatory DNA Seems Largely Responsible for the Differences Between Animal Species

As discussed in Chapter 7, each gene in a multicellular organism is associated with many thousands of nucleotides of noncoding DNA that contains regulatory elements. These regulatory elements determine when, where, and how strongly the gene is to be expressed, according to the transcription regulators and chromatin structures that are present in the particular cell (**Figure 21–5**). Consequently, a change in the regulatory DNA, even without any change in the coding DNA, can alter the logic of the gene-regulatory network and change the outcome of development.

As discussed in Chapter 4, when we compare the genomes of different animal species, we find that evolution has altered the coding and regulatory DNA to different extents. The coding DNA, for the most part, has been highly conserved, the noncoding regulatory DNA much less so. It seems that changes in regulatory DNA are largely responsible for the dramatic differences between one class of animals and another (see p. 227). We can view the protein products of the coding sequences as a conserved kit of common molecular parts, and the regulatory DNA as instructions for assembly: with different instructions, the same kit of parts can be used to make a whole variety of different body structures. We will return to this important concept later.

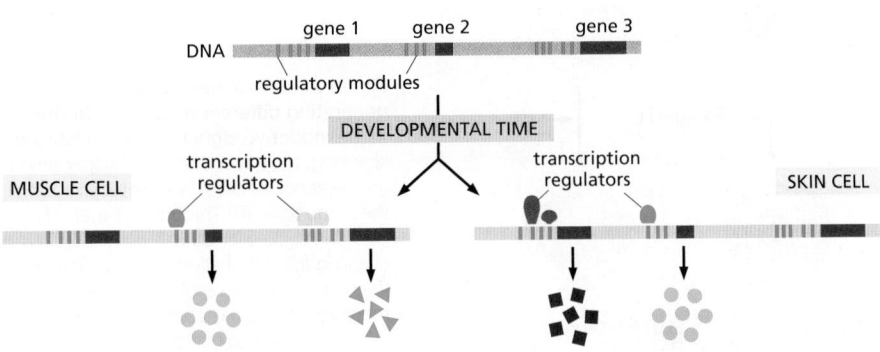

Figure 21–5 **Regulatory DNA defines the gene expression patterns in development.** The genome is the same in a muscle cell as in a skin cell, but different genes are active because these cells express different transcription regulators that bind to gene regulatory elements. For example, transcription regulators in skin cells recognize a regulatory element in gene 1, leading to its activation, whereas a different set of regulators is present in muscle cells, binding to and activating gene 3. Transcriptional regulators that activate the expression of gene 2 are present in both cell types.

Small Numbers of Conserved Cell–Cell Signaling Pathways Coordinate Spatial Patterning

Spatial patterning of a developing animal requires that cells become different according to their positions in the embryo, which means that cells must respond to extracellular signals produced by other cells, especially their neighbors. In what is probably the commonest mode of spatial patterning, a group of cells starts out with the same developmental potential, and a signal from cells outside the group then induces one or more members of the group to change their character. This process is called *inductive signaling*. Generally, the inductive signal is limited in time and space so that only a subset of the cells capable of responding—the cells close to the source of the signal—take on the induced character (**Figure 21–6**). Some inductive signals depend on cell–cell contact; others act over a longer range and are mediated by molecules that diffuse through the extracellular medium or are transported in the bloodstream (see Figure 15–2).

Most of the known inductive events in animal development are governed by a small number of highly conserved signaling pathways, including transforming growth factor-β (TGFβ), Wnt, Hedgehog, Notch, and receptor tyrosine kinase (RTK) pathways (discussed in Chapter 15). The discovery of the limited vocabulary that developing cells use for intercellular communication has emerged over the past 25 years as one of the great simplifying features of developmental biology.

Through Combinatorial Control and Cell Memory, Simple Signals Can Generate Complex Patterns

But how can this small number of signaling pathways generate the huge diversity of cells and patterns? Three kinds of mechanisms are responsible. First, through gene duplication, the basic components of a pathway often come to be encoded by small families of closely related homologous genes. This allows for diversity in the operation of the pathway, according to which family member is employed in a given situation. Notch signaling, for example, may be mediated by Notch1 in one tissue, but by its homolog Notch4 in another. Second, the response of a cell to a given signal protein depends on the other signals that the cell is receiving concurrently (**Figure 21–7A**). As a result, different combinations of signals can generate a large variety of different responses. Third, and most fundamental, the effect of activating a signaling pathway depends on the previous experiences of the responding cell: past influences leave a lasting mark, registered in the state of the cell's chromatin and the selection of transcription regulatory proteins and RNA molecules that the cell contains. This cell memory enables cells with different

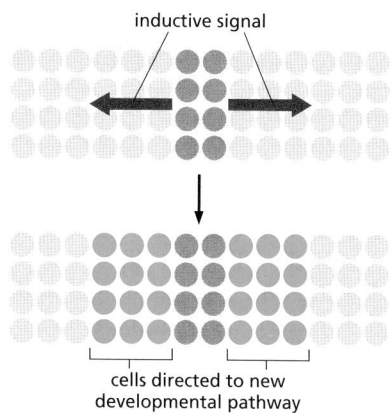

Figure 21–6 **Inductive signaling.**

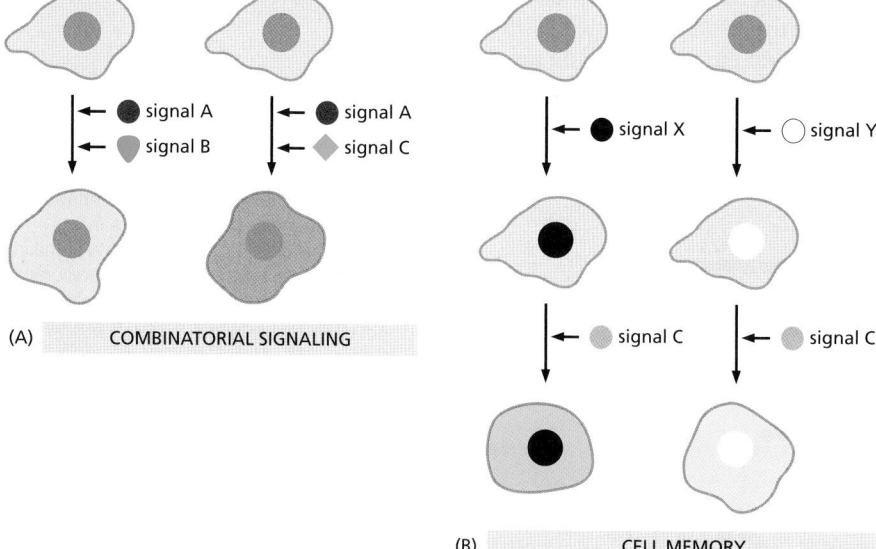

(A)   COMBINATORIAL SIGNALING

(B)   CELL MEMORY

Figure 21–7 **Two mechanisms for generating different responses to the same inductive signal.** (A) In combinatorial signaling, the effect of a signal depends on the presence of other signals received at the same time. (B) Through cell memory, previous signals (or other events) can leave a lasting trace that alters the response to the current signal (see Figure 7–54). The memory trace is represented here in the coloring of the cell nucleus.

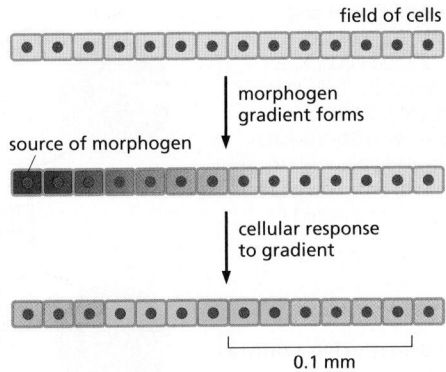

field of cells

morphogen gradient forms

source of morphogen

cellular response to gradient

0.1 mm

Figure 21–8 Gradient formation and interpretation. A gradient forms by localized production of an inducer—a morphogen—that diffuses away from its source. Different concentrations of morphogen (or different durations of exposure) induce different gene expression patterns and cell fates in responding cells. Diffusive transport can generate gradients only over short distances, and morphogens generally act over distances of 1 mm or less.

histories to respond to the same signals differently (Figure 21–7B). Thus, the same few signaling pathways can be used repeatedly at different times and places with different outcomes, so as to generate patterns of unlimited complexity.

Morphogens Are Long-Range Inductive Signals That Exert Graded Effects

Signal molecules often govern simple yes–no choices—one outcome when their concentration is high, another when it is low. In many cases, however, the responses are more finely graded: a high concentration of a signal molecule may, for example, direct cells into one developmental pathway, an intermediate concentration into another, and a low concentration into yet another.

One common way to generate such different concentrations of a signal molecule is for the molecule to diffuse out from a localized signaling source, creating a concentration gradient. Cells at different distances from the source are driven to behave in a variety of different ways, according to the signal concentration that they experience (**Figure 21–8**). A signal molecule that imposes a pattern on a whole field of cells in this way is called a **morphogen**. In the simplest case, a specialized group of cells produces a morphogen at a steady rate, and the morphogen is then degraded as it diffuses away from this source. The speed of diffusion and the half-life of the morphogen will together determine the range and steepness of its resulting gradient (**Figure 21–9**).

This simple mechanism can be modified in various ways. Receptors on the surface of cells along the way, for example, may trap the diffusing morphogen and cause it to be endocytosed and degraded, shortening its effective half-life. Alternatively, the morphogen may bind to molecules in the extracellular matrix such as heparan sulfate proteoglycan (discussed in Chapter 19), thereby greatly reducing its diffusion rate.

Lateral Inhibition Can Generate Patterns of Different Cell Types

Morphogen gradients, and other kinds of inductive signal, exploit an existing asymmetry in the embryo to create further asymmetries and differences between cells: already, at the outset, some cells are specialized to produce the morphogen and thereby impose a pattern on another class of cells that are sensitive to it. But

Figure 21–9 Setting up a signal gradient by diffusion. (A–C) Each graph shows six successive stages in the buildup of the concentration of a signal molecule that is produced at a steady rate at the origin, with production starting at time 0. In all cases, the molecule undergoes degradation as it diffuses away from the source, and the graphs are calculated on the assumption that diffusion is occurring along two axes in space (for example, radially from a source in an epithelial sheet). (A) The pattern of the morphogen assuming that the molecule has a half-life of 170 minutes, and that it diffuses with an effective diffusion constant of D = 1 μm^2 sec^{-1}, typical of a small protein molecule in extracellular tissues. Note that the gradient is already close to its steady-state form within an hour and that the concentration at steady state falls off exponentially with distance. (B) A threefold increase in the diffusion constant of the morphogen extends its range but lowers its concentration next to the source, whereas (C) a threefold increase in morphogen half-life increases its concentration throughout the tissue. Effects of the morphogen will depend not just on its concentration at some critical moment, but also on how each target cell integrates its response over time. (Courtesy of Patrick Müller.)

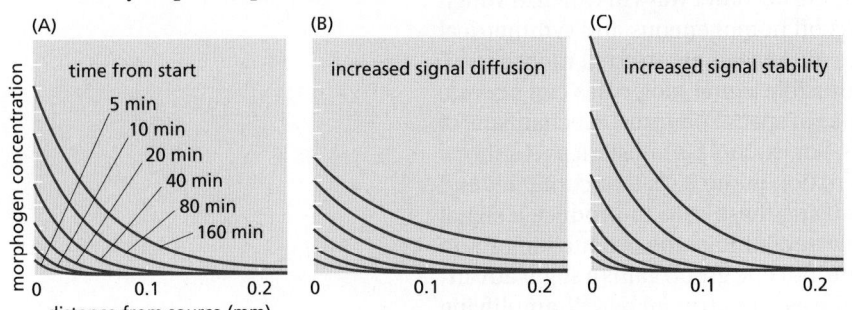

(A) time from start
5 min
10 min
20 min
40 min
80 min
160 min

morphogen concentration

0   0.1   0.2
distance from source (mm)

(B) increased signal diffusion

0   0.1   0.2

(C) increased signal stability

0   0.1   0.2

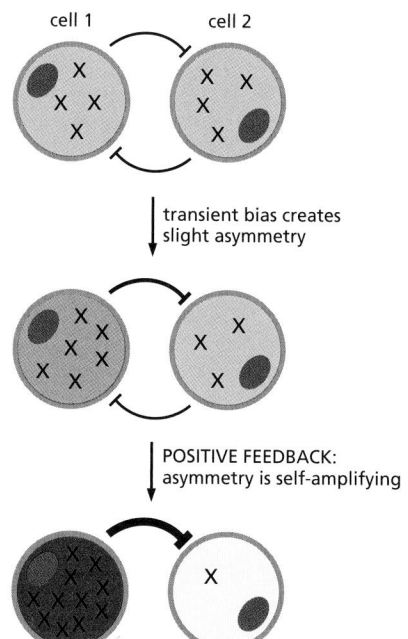

Figure 21–10 Genesis of asymmetry through lateral inhibition and positive feedback. In this example, two cells interact, each producing a substance X that acts on the other cell to inhibit its production of X, an effect known as lateral inhibition. An increase of X in one of the cells leads to a positive feedback that tends to increase X in that cell still further, while decreasing X in its neighbor. This can create a runaway instability, making the two cells become radically different. Ultimately, the system comes to rest in one or the other of two opposite stable states. The final choice of state represents a form of memory: the small influence that initially directed the choice is no longer required to maintain it.

what if there is no clear initial asymmetry? Can a regular pattern arise spontaneously within a set of cells that are initially all alike?

The answer is yes. The fundamental principle underlying such *de novo* pattern formation is positive feedback: cells can exchange signals in such a way that any small initial discrepancy between cells at different sites becomes self-amplifying, driving the cells toward different fates. This is most clearly illustrated in the phenomenon of *lateral inhibition*, a form of cell–cell interaction that forces close neighbors to become different and thereby generates fine-grained patterns of different cell types.

Consider a pair of adjacent cells that start off in a similar state. Each of these cells can both produce and respond to a certain signal molecule X, with the added rule that the stronger the signal a cell receives, the weaker the signal it generates (Figure 21–10). If one cell produces more X, the other is forced to produce less. This gives rise to a positive feedback loop that tends to amplify any initial difference between the two adjacent cells. Such a difference may arise from a bias imposed by some present or past external factor, or it may simply originate from spontaneous random fluctuations, or "noise"—an inevitable feature of the genetic control circuitry in cells (discussed in Chapter 7). In either case, lateral inhibition means that if cell 1 makes a little more of X, it will thereby cause cell 2 to make less; and because cell 2 makes less X, it delivers less inhibition to cell 1 and so allows the production of X in cell 1 to rise higher still; and so on, until a steady state is reached where cell 1 produces a lot of X and cell 2 produces very little. In the standard case, the signal molecule X acts in the receiving cell by regulating gene transcription, and the result is that the two cells are driven along different pathways of differentiation.

In almost all tissues, a balanced mixture of different cell types is required. Lateral inhibition provides a common way to generate the mixture. As we shall see, lateral inhibition is very often mediated by exchange of signals at cell–cell contacts via the Notch signaling pathway, driving cell diversification by enabling individual cells that express one set of genes to direct their immediate neighbors to express a different set, in exactly the way we have described (see also Figure 15–58).

Short-Range Activation and Long-Range Inhibition Can Generate Complex Cellular Patterns

Lateral inhibition mediated by the Notch pathway is not the only example of pattern generation through **positive feedback**: there are other ways in which, through the same basic principle, a system that starts off homogeneous and symmetrical can pattern itself spontaneously, even in the absence of an external morphogen. Positive feedback processes mediated by diffusible signal molecules can operate over broad arrays of cells to create many types of spatial patterns. Mechanisms of this sort are called *reaction-diffusion systems*. For example, a substance A (a short-range activator) may stimulate its own production in the cells that contain it and in their immediate neighbors, while also causing these cells to produce a signal I (a long-range inhibitor) that diffuses widely and inhibits the production of A in cells farther away. If the cells all start the same, but one group gains a slight advantage by making a little more A than the rest, the asymmetry can be self-amplifying

Figure 21–11 **Pattern generation by a reaction-diffusion system.** From (A) a uniform field of cells, (B) local positive feedback and (C) long-range inhibition can (D) generate patterns within the initially uniform field. The patterns can be complex, resembling the spots of a leopard (as shown) or the stripes of a zebra; or they can be simple, with creation of a single cluster of specialized cells that can, for example, go on to serve as the source of a morphogen gradient.

(**Figure 21–11**). Such short-range activation combined with long-range inhibition can account for the formation of clusters of cells within an initially homogeneous tissue that become specialized as localized **signaling centers**.

Asymmetric Cell Division Can Also Generate Diversity

Cell diversification does not always depend on extracellular signals: in some cases, daughter cells are born different as a result of an **asymmetric cell division**, in which some important molecule or molecules are distributed unequally between the two daughters. This asymmetric inheritance ensures that the two daughter cells develop differently (**Figure 21–12**). Asymmetric division is a common feature of early development, where the fertilized egg already has an internal pattern and cleavage of this large cell segregates different determinants into separate blastomeres. We shall see that asymmetric division also plays a part in some later developmental processes.

Initial Patterns Are Established in Small Fields of Cells and Refined by Sequential Induction as the Embryo Grows

The signals that organize the spatial pattern of cells in an embryo generally act over short distances and govern relatively simple choices. A morphogen, for example, typically acts over a distance of less than 1 mm—an effective range for diffusion —and directs choices between several developmental options for the cells on which it acts. Yet the organs that eventually develop are much larger and more complex than this.

The cell proliferation that follows the initial specification accounts for the size increase, while the refinement of the initial pattern is explained by a series of local inductions plus other interactions that add successive levels of detail on an initially simple sketch. For example, as soon as two types of cells are present in a developing tissue, one of them can produce a signal that induces a subset of the neighboring cells to specialize in a third way. The third cell type can in turn signal

(A) uniform field of cells

(B) short-range activator (*green*) in one cell stimulates its own production

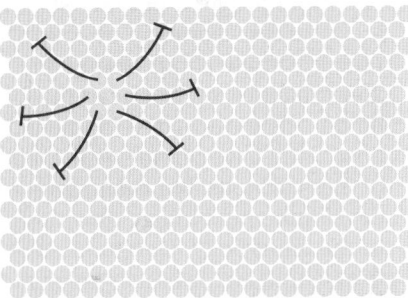

(C) long-range inhibitor (*red*) blocks production of activator by other cells in the neighborhood

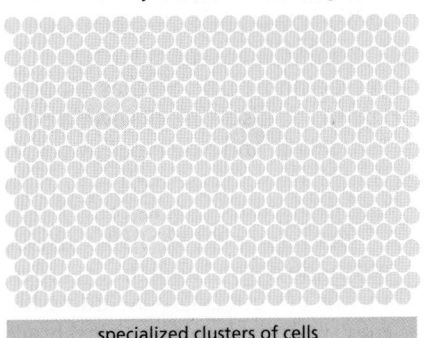

specialized clusters of cells

(D)

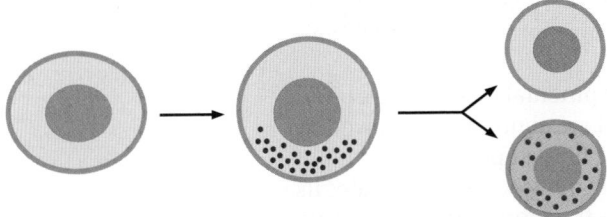

1. asymmetric division: sister cells born different

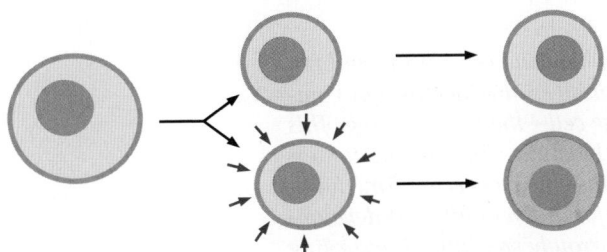

2. symmetric division: sister cells become different as a result of influences acting on them after their birth

Figure 21–12 **Two ways of making sister cells different.**

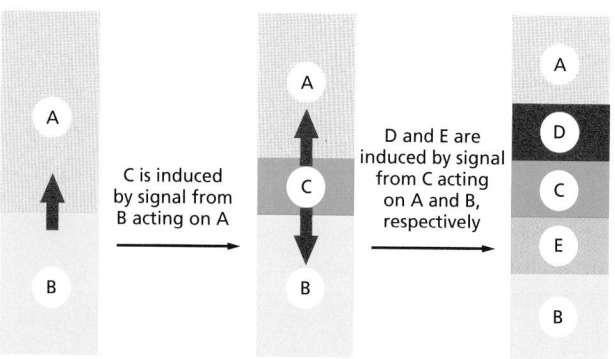

back to the other two cell types nearby, generating a fourth and a fifth cell type, and so on (**Figure 21–13**).

This strategy for generating a progressively more complicated pattern is called **sequential induction**. It is chiefly through sequential inductions that the body plan of a developing animal, after being first roughed out in miniature, becomes elaborated with finer and finer details as development proceeds.

Developmental Biology Provides Insights into Disease and Tissue Maintenance

The rapid progress in understanding animal development has been one of the great success stories in biology over the last few decades, and it has important practical implications. Some 2 to 5% of all human babies are born with anatomical abnormalities, such as heart malformations, truncated limbs, cleft palate, or spina bifida. Advances in developmental biology help us understand how these defects arise, even if we cannot yet prevent or cure most of them.

Less obvious, but even more important from a practical point of view, is that developmental biology provides insights into the workings of cells and tissues in the adult body. Developmental processes do not halt at birth; they continue throughout life, as tissues are maintained and repaired. The fundamental mechanisms of cell growth and division, cell–cell signaling, cell memory, cell adhesion, and cell movement are involved in adult tissue maintenance and repair—just as they are in embryo development.

Embryos are simpler than adults, and they allow us to analyze such basic processes more easily. Studies of the early *Drosophila* embryo, for example, were crucial to the discovery of several conserved signaling pathways, including the Wnt, Hedgehog, and Notch pathways. They also provided the key to understanding the central role of these pathways in the maintenance of normal adult human tissues and in diseases such as cancer.

In Chapter 22, we shall consider how other developmental mechanisms operate in the normal adult body, especially in tissues that are continually renewed by means of stem cells—including the gut, skin, and the hematopoietic system. But now, we must look more closely at the way in which an early embryo generates its spatial pattern of specialized cells, beginning with the transformations that create the adult body plan.

Summary

Animal development is a self-assembly process, in which the cells of the embryo become different from one another and organize themselves into increasingly complex structures. The process begins with a single large cell—the fertilized egg. This cell cleaves to form many smaller cells, producing a blastula. The blastula undergoes gastrulation to generate the three germ layers of the embryo—ectoderm, mesoderm, and endoderm—consisting of cells determined for different fates. As development continues, the cells become more and more narrowly specialized according to their locations and their interactions with one another. Through cell memory, these cell–cell interactions, even though transient, can have lasting effects on each cell's internal state. Thus, a succession of simple cues that a cell receives at different times can direct it along a complex developmental pathway. At each step, the cell becomes further restricted in the range of final states open to it. The process reaches its limit when the cell differentiates to form one of the specialized cell types of the adult body.

Differences between developing cells arise in various ways and have to be properly coordinated in space. In one common strategy, initially similar cells within a group become different by exposure to different levels of an inductive signal or morphogen emanating from a source outside the group. Neighboring cells can also become different by lateral inhibition, in which a cell signals to its neighbors not to follow the same fate. These cell–cell interactions are mediated by a small number of highly conserved signaling pathways, which are used repeatedly in different organisms and at different times during development. Not all cell diversification arises by cell–cell interactions, however: daughter cells can be born different as a result of asymmetric cell division.

Regulators of transcription and chromatin structure bind to regulatory DNA and determine the fate of each cell. Differences of body plan seem to arise to a large extent from differences in the regulatory DNA associated with each gene. This DNA has a central role in defining the sequential program of development, calling genes into action at specific times and places according to the pattern of gene expression that was present in each cell at the previous developmental stage.

Development has been most thoroughly studied in a handful of model organisms. But most of the genes and mechanisms thereby identified are used in all animals and repeatedly at different stages of development. Thus, insights from worms, flies, fish, frogs, and mice deeply inform our understanding of embryology, birth defects, and adult tissue maintenance in humans.

MECHANISMS OF PATTERN FORMATION

A developing multicellular organism has to create a pattern in fields of cells where there was little or none before. Some of the early microscopists imagined the entire shape and structure of the human body to be already present in the sperm as a "homunculus," a miniature human; after fertilization, the homunculus would simply grow and generate a full-sized human. We now know that this view is incorrect and that development is a progression from simple to complex, through a gradual refinement of an animal's anatomy. To see how the whole sequence of events of spatial patterning and cell determination is set in train, we must return to the egg and the early embryo.

Different Animals Use Different Mechanisms to Establish Their Primary Axes of Polarization

Surprisingly, the earliest steps of animal development are among the most variable, even within a phylum. A frog, a chicken, and a mammal, for example, even though they develop in similar ways later, make eggs that differ radically in size and structure, and they begin their development with different sequences of cell divisions and cell specializations. Gastrulation occurs in all animal embryos, but the details of its timing, of the associated pattern of cell movements, and of the shape and size of the embryo as gastrulation proceeds are highly variable. Likewise, there is great variation in the time and manner in which the primary axes of the body become marked out. However, this *polarization* of the embryo usually becomes discernible very early, before gastrulation begins: it is the first step of spatial patterning.

Three axes generally have to be established. The *animal–vegetal (A-V)* axis, in most species, defines which parts are to become internal (through the movements of gastrulation) and which are to remain external. (The bizarre name dates from a century ago and has nothing to do with vegetables.) The *anteroposterior (A-P)* axis specifies the locations of future head and tail. The *dorsoventral (D-V)* axis specifies the future back and belly.

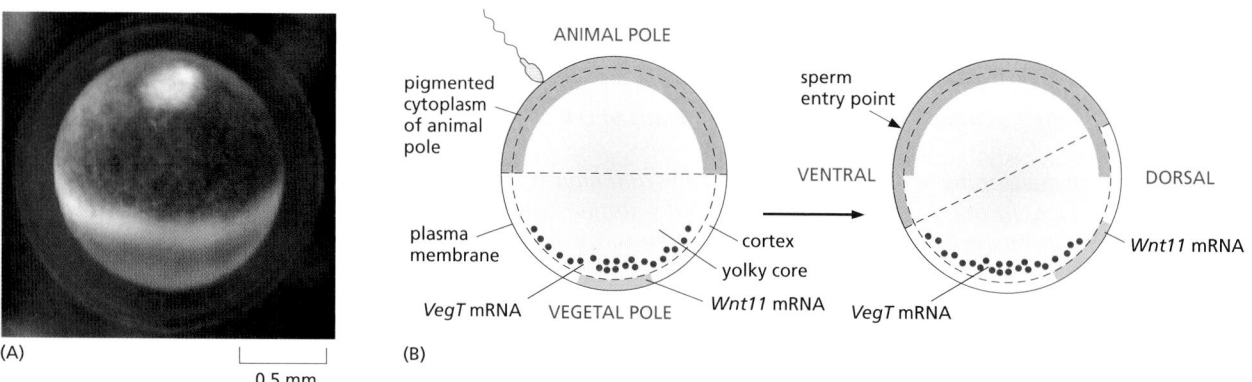

(A)

0.5 mm

(B)

Figure 21–14 The frog egg and its asymmetries. (A) Side view of a *Xenopus* egg photographed just before fertilization. (B) The asymmetric distribution of molecules inside the egg, and how this changes following fertilization so as to define a dorsoventral as well as an animal-vegetal asymmetry. Fertilization, through a reorganization of the microtubule cytoskeleton, triggers a rotation of the egg cortex (a layer a few μm deep) through about 30° relative to the core of the egg; the direction of rotation determined by the site of sperm entry. Some components are carried still further to the future dorsal side by active transport along microtubules. The resulting dorsal concentration of *Wnt11* mRNA leads to dorsal production of the Wnt11 signal protein and defines the dorsoventral polarity of the future embryo. Vegetally localized VegT defines the vegetal source of signals that will induce endoderm and mesoderm. (A, courtesy of Tony Mills.)

At one extreme, the egg is spherically symmetrical, and the axes only become defined during embryogenesis. The mouse comes close to being an example, with little obvious sign of polarity in the egg. Correspondingly, the **blastomeres** produced by the first few cell divisions seem to be all alike and are remarkably adaptable. If the early mouse embryo is split in two, a pair of identical twins can be produced—two complete, normal individuals from a single cell. Similarly, if one of the cells in a two-cell mouse embryo is destroyed by pricking it with a needle and the resulting "half-embryo" is placed in the uterus of a foster mother to develop, in many cases a perfectly normal mouse will emerge.

At the opposite extreme, the structure of the egg defines the future axes of the body. This is the case for most species, including insects such as *Drosophila*, as we shall see shortly. Many other organisms lie between the two extremes. The egg of the frog *Xenopus*, for example, has a clearly defined A-V axis even before fertilization: the nucleus near the top defines the animal pole, while the mass of yolk (the embryo's food supply, destined to be incorporated in the gut) toward the bottom defines the vegetal pole. Several types of mRNA molecules are already localized in the vegetal cytoplasm of the egg, where they produce their protein products. After fertilization, these mRNAs and proteins act in and on the cells in the lower and middle part of the embryo, giving the cells there specialized characters, both by direct effects and by stimulating the production of secreted signal proteins. For example, mRNA encoding the transcription regulator VegT is deposited at the vegetal pole during oogenesis. After fertilization, this mRNA is translated, and the resulting VegT protein activates a set of genes that code for signal proteins that induce mesoderm and endoderm, as discussed later.

The D-V axis of the *Xenopus* embryo, by contrast, is defined through the act of fertilization. Following entry of the sperm, the outer cortex of the egg cytoplasm rotates relative to the central core of the egg, so that the animal pole of the cortex becomes slightly shifted to one side (**Figure 21–14**). Treatments that block the rotation allow cleavage to occur normally but produce an embryo with a central gut and no dorsal structures or D-V asymmetry. Thus, this cortical rotation is required to define the D-V axis of the future body by creating the D-V axis of the egg.

The site of sperm entry that biases the direction of the cortical rotation in *Xenopus*, perhaps through the centrosome that the sperm brings into the egg—inasmuch as the rotation is associated with a reorganization of the microtubules nucleated from the centrosome in the egg cytoplasm. The reorganization leads to a microtubule-based transport of several cytoplasmic components, including the mRNA coding for Wnt11, a member of the Wnt family of signal proteins, moving it toward the future dorsal side (see Figure 21–14). This mRNA is soon translated and the Wnt11 protein secreted from cells that form in that region of the embryo activates the Wnt signaling pathway (see Figure 15–60). This activation is crucial for triggering the cascade of subsequent events that will organize the dorsoventral axis of the body. (The A-P axis of the embryo will only become clear later, in the process of gastrulation.)

Although different animal species use a variety of different mechanisms to specify their axes, the outcome has been relatively well conserved in evolution: head is distinguished from tail, back from belly, and gut from skin. It seems that it does not much matter what tricks the embryo uses to break the initial symmetry and set up this basic body plan.

Studies in *Drosophila* Have Revealed the Genetic Control Mechanisms Underlying Development

It is the fly *Drosophila*, more than any other organism, that has provided the key to our present understanding of how genes govern development. Decades of genetic study culminated in a large-scale genetic screen, focusing especially on the early embryo and searching for mutations that disrupt its pattern. This revealed that the key developmental genes fall into a relatively small set of functional classes defined by their mutant phenotypes. The discovery of these genes and the subsequent analysis of their functions was a famous *tour de force* and had a revolutionary impact on all of developmental biology, earning its discoverers a Nobel Prize. Some parts of the developmental machinery revealed in this way are conserved between flies and vertebrates, some parts not. But the logic of the experimental approach and the general strategies of genetic control that it revealed have transformed our understanding of multicellular development in general.

To understand how the early developmental machinery operates in *Drosophila*, it is important to note a peculiarity of fly development. Like the eggs of other insects, but unlike most vertebrates, the *Drosophila* egg—shaped like a cucumber—begins its development with an extraordinarily rapid series of nuclear divisions without cell division, producing multiple nuclei in a common cytoplasm—a **syncytium**. The nuclei then migrate to the cell cortex, forming a structure called the *syncytial blastoderm*. After about 6000 nuclei have been produced, the plasma membrane folds inward between them and partitions them into separate cells, converting the syncytial blastoderm into the *cellular blastoderm* (**Figure 21–15**).

We shall see that the initial patterning of the *Drosophila* embryo depends on signals that diffuse through the cytoplasm at the syncytial stage and exert their actions on genes in the rapidly dividing nuclei, before the partitioning of the egg into separate cells. Here, there is no need for the usual forms of cell–cell signaling; neighboring regions of the syncytial blastoderm can communicate by means of transcription regulatory proteins that move through the cytoplasm of the giant multinuclear cell.

Egg-Polarity Genes Encode Macromolecules Deposited in the Egg to Organize the Axes of the Early *Drosophila* Embryo

As in most insects, the main axes of the future body of *Drosophila* are defined before fertilization by a complex exchange of signals between the developing egg,

Figure 21–15 Development of the *Drosophila* egg from fertilization to the cellular blastoderm stage.

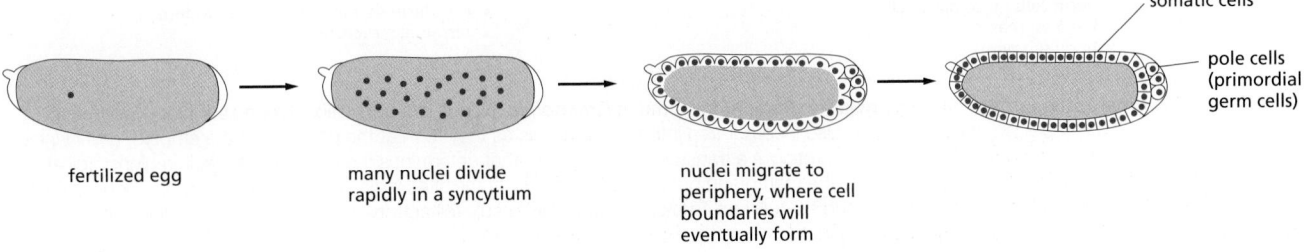

fertilized egg

many nuclei divide rapidly in a syncytium

nuclei migrate to periphery, where cell boundaries will eventually form

somatic cells

pole cells (primordial germ cells)

or *oocyte*, and the *follicle cells* that surround it in the ovary. In the stages before fertilization, the anteroposterior and dorsoventral axes of the future embryo become defined by four systems of **egg-polarity genes** that create landmarks—either mRNA or protein—in the developing oocyte. Following fertilization, each landmark serves as a beacon, providing a signal that organizes the developmental process in its neighborhood.

The nature of the genes emerged from studies of mutants in which the patterning of the embryo was altered. One class of mutations gave embryos with disrupted polarity—for example, tail-end structures at both ends of the body, with no head-end structures. This class of mutations identified the set of egg-polarity genes. The egg-polarity gene responsible for the signal that organizes the anterior end of the embryo is called **Bicoid**. A deposit of *Bicoid* mRNA molecules is localized, before fertilization, at the anterior end of the egg. Upon fertilization, the mRNA is translated to produce Bicoid protein. This protein is an intracellular morphogen and transcription regulator that diffuses away from its source to form a concentration gradient within the syncytial cytoplasm, with its maximum at the head end of the embryo (**Figure 21–16**). The different concentrations of Bicoid along the A-P axis help determine different cell fates by regulating the transcription of genes in the nuclei of the syncytial blastoderm (discussed in Chapter 7).

Of the three other egg-polarity gene systems, two contribute to patterning the syncytial nuclei along the A-P axis and one to patterning them along the D-V axis. Together with the *Bicoid* group of genes, and acting in a broadly similar way, their gene products mark out three fundamental partitions of body regions—head versus rear, dorsal versus ventral, and endoderm versus mesoderm and ectoderm—as well as a fourth partition, no less fundamental to the body plan of animals: the distinction between germ cells and somatic cells (**Figure 21–17**).

The egg-polarity genes have a further special feature: they are all **maternal-effect genes**, in that it is the mother's genome rather than the zygote's genome that is critical. For example, a fly whose chromosomes are mutant in both copies of the *Bicoid* gene but who is born from a mother carrying one normal copy of Bicoid develops perfectly normally, without any defects in the head pattern. However, if that offspring is a female, she cannot deposit any functional *Bicoid* mRNA into her own eggs, which will therefore develop into headless embryos, regardless of the father's genotype.

The egg-polarity genes act first in a hierarchy of gene systems that define a progressively more detailed pattern of body parts. In the next few pages, we begin with the molecular mechanisms that pattern the developing *Drosophila* embryo and larva along the A-P axis, before considering the patterning along the D-V axis.

(A) *Bicoid* mRNA

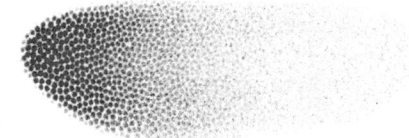

(B) Bicoid protein

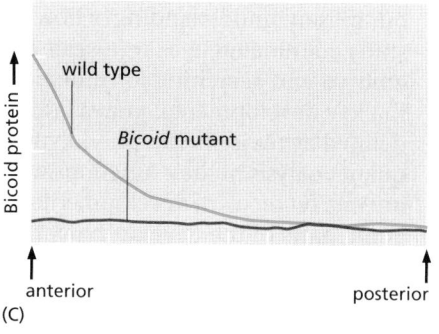

Figure 21–16 The Bicoid protein gradient. (A) *Bicoid* mRNA is deposited at the anterior pole during oogenesis. (B) Local translation followed by diffusion generates the Bicoid protein gradient. (C) Absence of the Bicoid protein gradient in embryos from *Bicoid* homozygous mutant mothers. (A and B, courtesy of Stephen Small.)

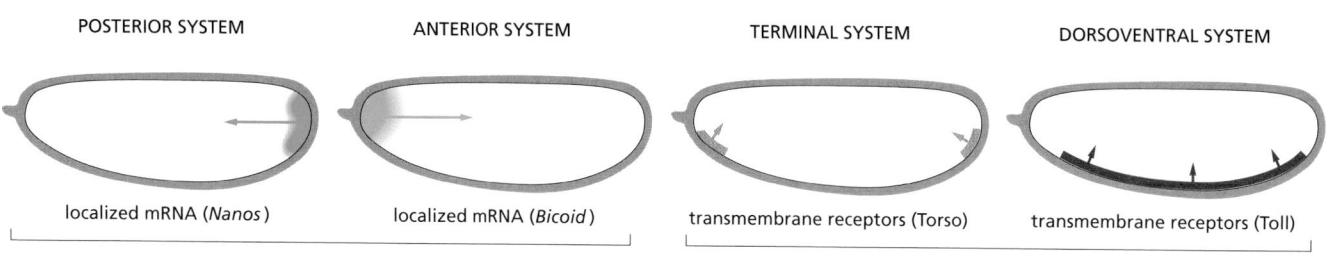

Figure 21–17 The organization of the four egg-polarity gradient systems in *Drosophila.* Nanos is a translational repressor that governs the formation of the abdomen. Localized *Nanos* mRNA is also incorporated into the germ cells as they form at the posterior of the embryo, and Nanos protein is necessary for germ-line development. Bicoid protein is a transcriptional activator that determines the head and thoracic regions. Toll and Torso are receptor proteins that are distributed all over the membrane but are activated only at the sites indicated by the coloring, through localized exposure to the extracellular ligands Spaetzle (the ligand for Toll) and Trunk (the ligand for Torso). Toll activity determines the mesoderm and Torso activity determines the formation of terminal structures.

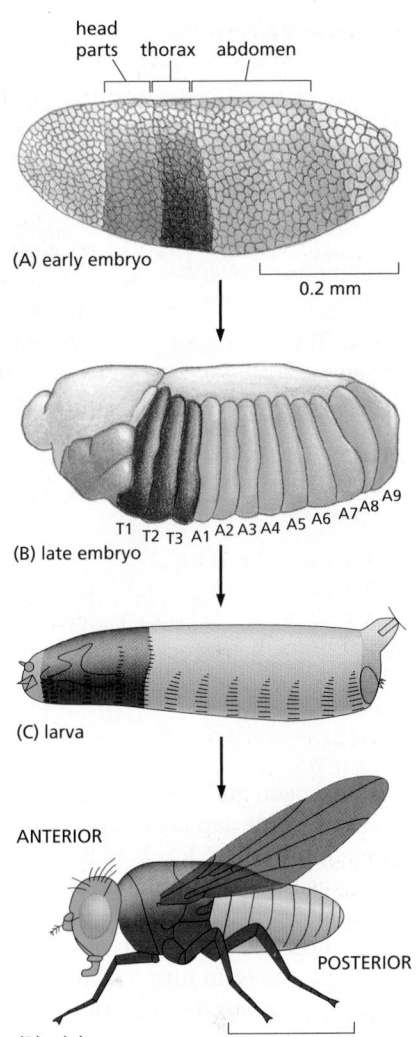

Figure 21–18 **The origins of the *Drosophila* body segments.** (A) At 3 hours, the embryo (shown in side view) is at the blastoderm stage and no segmentation is visible, although a fate map can be drawn showing the future segmented regions (color). (B) At 10 hours, all the segments are clearly defined (T1: first thoracic segment; A1: first abdominal segment). See Movie 21.3. (C) The segments of the *Drosophila* larva and their correspondence with regions in the embryo. (D) The segments of the *Drosophila* adult and their correspondence with regions in the embryo.

Three Groups of Genes Control *Drosophila* Segmentation Along the A-P Axis

The body of an insect is divided along its A-P axis into a series of **segments**. The segments are repetitions of a theme with variations: each segment forms highly specialized structures, but all built according to a similar fundamental plan (**Figure 21–18**). The gradients of transcription regulators set up along the A-P axis in the early embryo by the egg-polarity genes are the prelude to creation of the segments. These regulators initiate the orderly transcription of *segmentation genes*, which refine the pattern of gene expression to define the boundaries and ground plan of the individual segments. Segmentation genes are expressed by subsets of cells in the embryo, and their products are the first components that the embryo's own genome contributes to embryonic development; they are therefore called *zygotic-effect genes*, to distinguish them from the earlier-acting maternal-effect genes. Mutations in segmentation genes can alter either the number of segments or their basic internal organization.

The **segmentation genes** fall into three groups according to their mutant phenotypes (**Figure 21–19**). It is convenient to think of the three groups as acting in sequence, although in reality their functions overlap in time. First to be expressed is a set of at least six **gap genes**, whose products mark out coarse A-P subdivisions of the embryo. Mutations in a gap gene eliminate one or more groups of adjacent segments: in the mutant *Krüppel*, for example, the larva lacks eight segments. Next comes a set of eight **pair-rule genes**. Mutations in these genes cause a series of deletions affecting alternate segments, leaving the embryo with only half as many segments as usual; although all the mutants display this two-segment periodicity, they differ in the precise pattern. Finally, there are at least 10 **segment-polarity genes**, in which mutations produce a normal number of segments but with a part of each segment deleted and replaced by a mirror-image duplicate of all or part of the rest of the segment.

In parallel with the segmentation process, a further set of genes—the *homeotic selector*, or *Hox*, genes—serves to define and preserve the differences between one segment and the next, as we describe shortly.

The phenotypes of the various segmentation mutants suggest that the segmentation genes form a coordinated system that subdivides the embryo progressively into smaller and smaller domains along the A-P axis, each distinguished by a different pattern of gene expression. Molecular genetics has helped to reveal how this system works.

A Hierarchy of Gene Regulatory Interactions Subdivides the *Drosophila* Embryo

Like *Bicoid*, most of the segmentation genes encode transcription regulator proteins. Their control by the egg-polarity genes and their actions on one another and on still other genes can be deciphered by comparing gene expression in normal and mutant embryos. By using appropriate probes to detect RNA transcripts or their protein products, one can observe genes switch on and off in changing patterns. By comparing these patterns in different mutants, one can begin to discern the logic of the entire gene control system.

The products of the egg-polarity genes provide the global positional signals in the early embryo (see Figure 21–17). The Bicoid protein, as we have seen, acts as

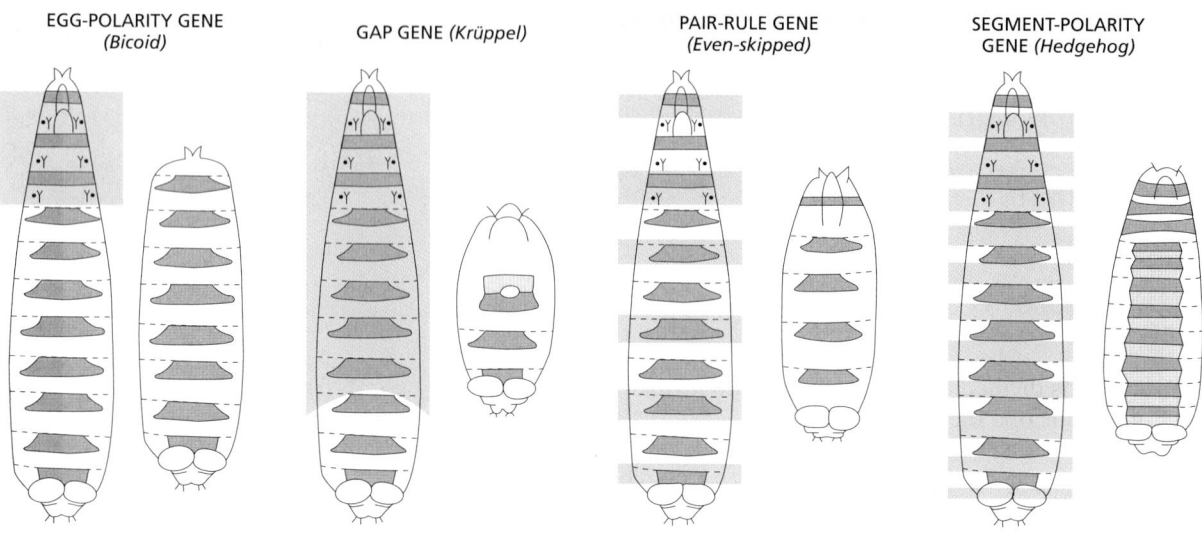

Figure 21–19 Examples of the phenotypes of mutations affecting egg-polarity genes and the three types of segmentation genes. In each case, the areas shaded in *green* on the normal larva (left) are deleted in the mutant or are replaced by mirror-image duplicates of the unaffected regions. (Modified from C. Nüsslein-Volhard and E. Wieschaus, *Nature* 287:795–801, 1980.)

a morphogen and activates different sets of genes at different positions along the A-P axis: some gap genes are only activated in regions with high levels of Bicoid, others only where levels of Bicoid are lower. After the gap gene products refine their positions by mutual repression, they provide a second tier of positional signals that act more locally to regulate finer details of patterning. Gap genes act by controlling the expression of yet other genes, including the pair-rule genes. The pair-rule genes, in turn, collaborate with one another and with the gap genes to set up a regular, periodic pattern of expression of the segment-polarity genes, which collaborate with one another to define the internal pattern of each individual segment (**Figure 21–20**).

The initial steps in creation of the segmental pattern occur before cellularization of the syncytial blastoderm and are governed by the combinatorial effects of transcription regulators, as discussed in detail in Chapter 7 for the regulation of the expression of the pair-rule gene *Even-skipped* (see pp. 394–396). After cellularization, the segment-polarity genes further subdivide each segment into smaller domains. A large subset of the segment-polarity genes codes for components of two signaling pathways—the Wnt pathway and the Hedgehog pathway, including the secreted signal proteins Wingless (the first-named member of the Wnt family) and Hedgehog. (The Hedgehog pathway was first discovered through study of *Drosophila* segmentation, and it takes its name from the prickly appearance of the surface of the *Hedgehog* mutant embryo.) Wingless and Hedgehog are synthesized in different bands of cells that serve as signaling centers within each segment. The two proteins mutually maintain each other's expression, while regulating the expression of genes such as *Engrailed* in neighboring cells (**Figure 21–21**). In such a manner, a series of sequential inductions creates a fine-grained pattern of gene expression within each segment.

Egg-Polarity, Gap, and Pair-Rule Genes Create a Transient Pattern That Is Remembered by Segment-Polarity and *Hox* Genes

The gap genes and pair-rule genes are activated within the first few hours after fertilization. Their mRNA products initially appear in patterns that only approximate the final picture; then, within a short time, this fuzzy initial pattern resolves itself into a regular, crisply defined system of stripes. But this pattern itself is unstable

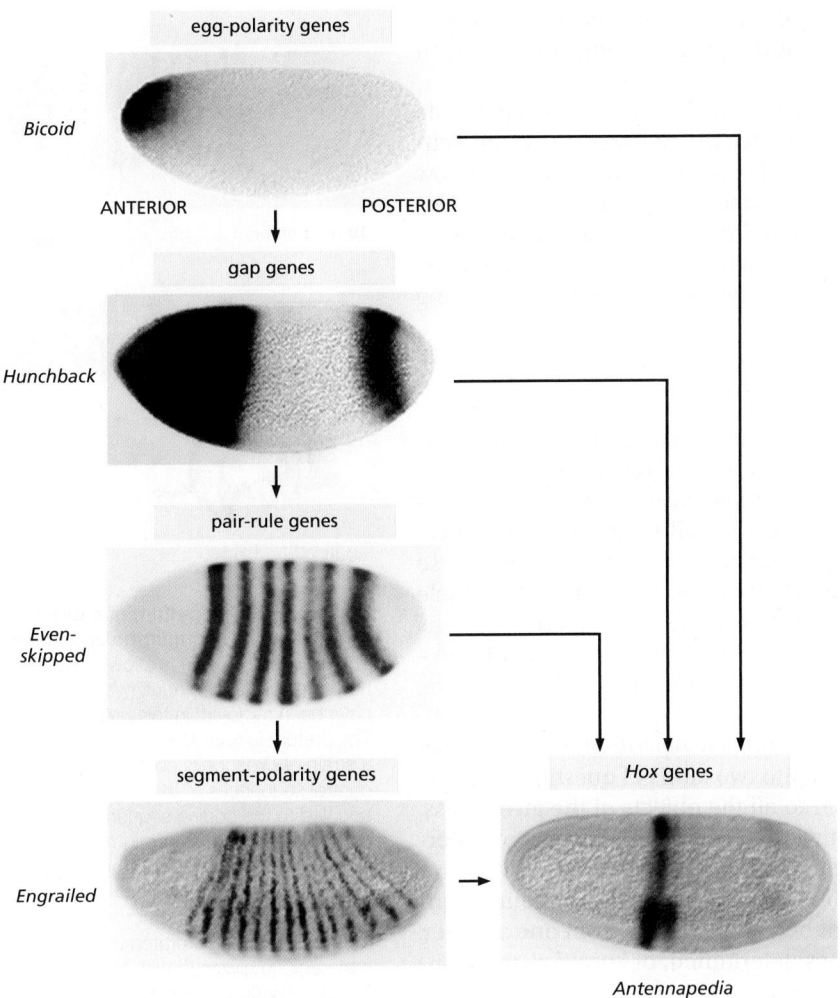

Figure 21–20 **An example of the regulatory hierarchy of egg-polarity, segmentation, and *Hox* genes.** As discussed in the text, there are three groups of segmentation genes. The photographs show mRNA expression patterns of representative examples of genes of each type. (Courtesy of Stephen Small.)

and transient: as the embryo proceeds through gastrulation and beyond, the pattern disintegrates. The genes' actions, however, have passed on an enduring memory of their patterns of expression by inducing the expression of certain segment polarity genes along with *Hox* genes (discussed shortly). After a period of pattern refinement mediated by cell–cell interactions, the expression patterns of these new groups of patterning genes is stabilized to provide *positional labels* that serve to maintain the segmental organization of the larva and adult fly.

The segment-polarity gene *Engrailed* provides a good example. Its RNA transcripts form a series of 14 bands in the cellular blastoderm, each approximately one-cell wide. These stripes lie immediately anterior to similar stripes of expression of another segment polarity gene, *Wingless*. As the cells in the developing embryo continue to grow, divide, and move, a mutually reinforcing signal between the Wingless expressing cells and the Engrailed expressing cells maintains narrow stripes of their expression (see Figure 21–21). After three cell cycles, newly expressed regulators stabilize an Engrailed expression pattern that will last

Figure 21–21 **Mutual maintenance of *Hedgehog* and *Wingless* expression.** Engrailed is a transcription regulator *(blue)* that drives the expression of *Hedgehog*. *Hedgehog* encodes a secreted protein *(red)* that activates its signaling pathway in neighboring cells and thereby drives them to express the *Wingless* gene. In turn, *Wingless* encodes a secreted protein *(green)* that acts back on neighbors of the *Wingless*-expressing cell to maintain their expression of *Engrailed* and *Hedgehog*. As indicated, the same control loop repeats along the A-P axis of the fly. (Based on S. Dinardo et al., *Curr. Opin. Genet. Dev.* 4:529–534, 1994.)

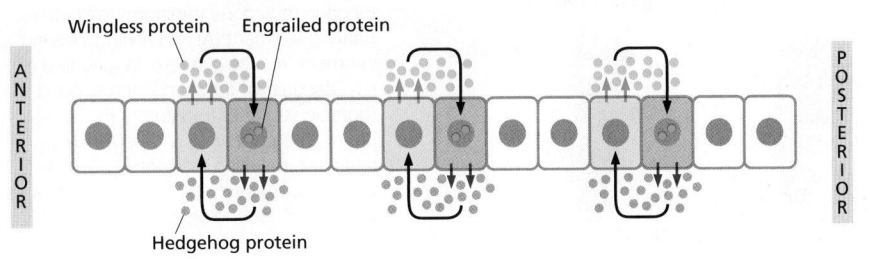

throughout the life of the fly, long after the signals that induced and refined it have disappeared. The segment borders will form at the posterior edge of each such Engrailed stripe (**Figure 21–22**).

In addition to regulating the segment-polarity genes, the products of pair-rule genes collaborate with those of gap genes to induce the precisely localized activation of a further set of genes—originally called *homeotic selector genes* and now often called *Hox genes*, for reasons that will become clear shortly. It is the *Hox* genes that permanently distinguish one segment from another. In the next section, we examine these important genes in detail and consider their role in cell memory; we shall see that this role is critical in a wide range of animals, including ourselves.

Hox Genes Permanently Pattern the A-P Axis

As animal development proceeds, the body becomes more and more complex. But again and again, in every species and at every level of organization, we find that complex structures are made by repeating a few basic themes, with variations. Thus, a limited number of basic differentiated cell types, such as muscle cells or fibroblasts, recur with subtle individual variations in different sites. These cell types are organized into a limited variety of tissue types, such as muscle or tendon, which again are repeated with subtle variations in different regions of the body. From the various tissues, organs such as teeth or digits are built—molars and incisors, fingers and thumbs and toes—a few basic kinds of structure, repeated with variations.

Wherever we find this phenomenon of *modulated repetition*, we can break down the developmental biologist's problem into two kinds of questions: what is the basic construction mechanism common to all the objects of the given class, and how is this mechanism modified to give the observed variations in different animals? The segments of the insect body provide a good example. We have thus far sketched the way in which the rudiment of a single body segment is constructed and how cells within each segment become different from one another. We now consider how one segment becomes determined, or *specified*, to be different from another.

The first glimpse of the answer to this problem came over 80 years ago, with the discovery of a set of mutations in *Drosophila* that cause bizarre disturbances in the organization of the adult fly. In the *Antennapedia* mutant, for example, legs sprout from the head in place of antennae, whereas in the *Bithorax* mutant, portions of an extra pair of wings appear where normally there should be the much smaller appendages called halteres (**Figure 21–23**). These mutations transform parts of the body into structures appropriate to other positions, and they are called *homeotic* mutations (from the Greek "homoios," meaning similar) because the transformation is between structures of a recognizably similar general type, changing one kind of limb, or one kind of segment, into another. It was eventually discovered that a whole set of genes, the **homeotic selector genes**, or *Hox* **genes**, serve to permanently specify the A-P characters of the whole set of animal segments. These genes are all related to one another as members of a multigene family.

There are eight *Hox* genes in the fly, and they all lie in one or the other of two gene clusters known as the ***Bithorax* complex** and the ***Antennapedia* complex**.

10-hour embryo   ⊢——⊣ 100 μm

adult   ⊢——⊣ 500 μm

Figure 21–22 The pattern of expression of *Engrailed*, a segment-polarity gene. The *Engrailed* pattern is shown in a 10-hour embryo and an adult (whose wings have been removed in this preparation). The pattern is revealed by constructing a strain of *Drosophila* containing the control sequences of the *Engrailed* gene coupled to the coding sequence of the reporter *LacZ*, whose product is detected histochemically through the brown product generated by immunohistochemistry against LacZ (10-hour embryo) or through the blue product generated by a reaction that LacZ catalyzes (adult). Note that the *Engrailed* pattern, once established, is preserved throughout the animal's life. (Courtesy of Tom Kornberg.)

Figure 21–23 Homeotic mutations. *Ultrabithorax*, or *Ubx*, is one of three genes in the *Bithorax* gene complex (a *Hox* gene cluster). *Ubx* is responsible for all of the differences between the second and third thoracic segments. (A, B) *Ubx* loss-of-function mutations transform the haltere-bearing segment (A) into a wing-bearing segment, resulting in four-winged flies (B). (C) *Ubx* gain-of-function in the second thoracic segment transforms this wing-bearing segment into a haltere-bearing segment, resulting in wingless flies. (A and B, courtesy of the Archives, California Institute of Technology; C, courtesy of L.S. Shashidhara.)

wild type   loss of *Ubx*   gain of *Ubx*

(A)   haltere   (B)   (C)

The genes in the *Bithorax* complex control the differences among the abdominal and thoracic segments of the body, while those in the *Antennapedia* complex control the differences among thoracic and head segments. Comparisons with other species show that the same genes are present in essentially all animals, including humans. These comparisons also reveal that the *Antennapedia* and *Bithorax* complexes are the two halves of a single entity, called the **Hox complex**, that has become split in the course of the fly's evolution, and whose members operate in a coordinated way to exert their control over the head-to-tail pattern of the body.

The products of the *Hox* genes, the **Hox proteins**, are transcription regulators, all of which possess a highly conserved, 60-amino-acid-long DNA-binding *homeodomain* (see p. 376). The corresponding motif in the DNA sequence is called a "homeobox," from which, by abbreviation, the *Hox* complex takes its name. There are many homeobox-containing genes, but only those located in a *Hox* complex are *Hox* genes.

Hox Proteins Give Each Segment Its Individuality

The Hox proteins can be viewed as molecular address labels possessed by the cells of each segment: these labels give the cells in each region a **positional value**—that is, an intrinsic character that differs according to a cell's location. If the address labels in a developing *Drosophila* segment are changed, the segment behaves as though it were located somewhere else; if all the *Hox* genes in an embryo are deleted, the body segments in the larva will all be alike.

To a first approximation, each *Hox* gene is normally expressed in those regions that develop abnormally when that gene is mutated or absent. How does each Hox protein give a segment its permanent identity? All the Hox proteins are similar in their DNA-binding regions, but they are very different in the regions that interact with the other proteins with which the Hox proteins form transcriptional regulatory complexes. The different protein partners act together with the Hox proteins to dictate which DNA binding sites will be recognized, as well as whether the effect on transcription at those sites will be activation or repression. Acting in this way, the Hox proteins modulate the actions of many other transcription regulators. Hundreds of genes are under this type of Hox-modulated control, including genes for cell–cell signaling, transcriptional regulation, cell polarity, cell adhesion, cytoskeletal function, cell growth, and cell death, all conspiring (in ways that are not yet understood) to give each segment its distinctive Hox-dependent character.

Hox Genes Are Expressed According to Their Order in the Hox Complex

How, then, is the expression of the *Hox* genes themselves regulated? The coding sequences of the eight *Hox* genes in the *Antennapedia* and *Bithorax* complexes in *Drosophila* are interspersed amid a much larger quantity of regulatory DNA. This DNA includes binding sites for the products of the egg-polarity and segmentation genes, thereby serving as an interpreter of the multiple items of spatial information supplied to it by all these transcription regulators. The net result is that the particular set of *Hox* genes transcribed is appropriate for each location along the A-P body axis.

The pattern of *Hox* gene expression exhibits a remarkable regularity that suggests an additional form of control. The sequence in which the genes are ordered along the chromosome, in both the *Antennapedia* and the *Bithorax* complexes, corresponds almost exactly to the order in which they are expressed along the A-P axis of the body (**Figure 21–24**). This hints at some process of gene activation, perhaps dependent on chromatin structures that propagate along the *Hox* complexes, switching on one *Hox* gene after another according to their order along the chromosome. The most "posterior" of the *Hox* genes that are expressed in a cell generally dominates, driving down expression and activity of the "anterior" genes and dictating the character of the segment. The gene regulatory mechanisms

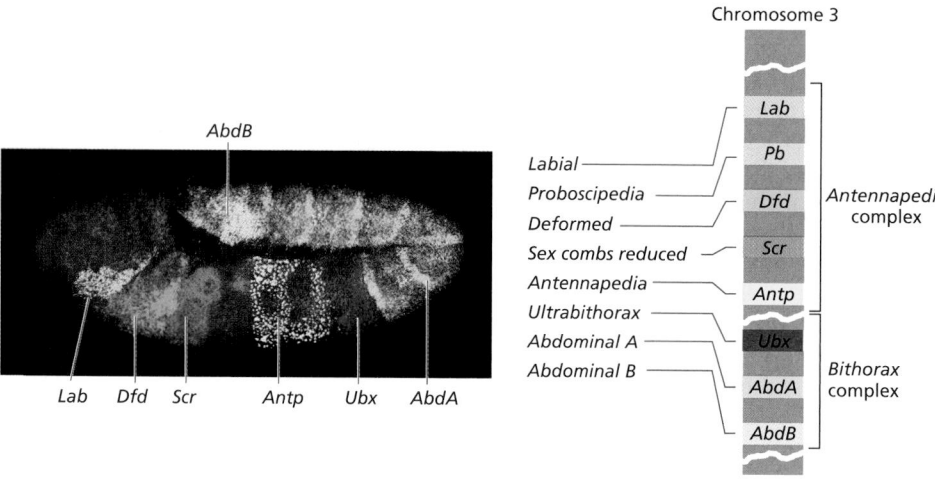

Figure 21–24 **The patterns of expression compared to the chromosomal locations of the genes of the Hox complex.** The diagram shows the sequence of genes in each of the two subdivisions of the chromosomal complex. This corresponds, with minor deviations, to the spatial sequence in which the genes are expressed, shown in the photograph of a *Drosophila* embryo at the so-called germ band retraction stage, about 10 hours after fertilization. The embryo has been stained by *in situ* hybridization with differently labeled probes to detect the mRNA products of different *Hox* genes in different colors. (Photograph courtesy of William McGinnis, adapted from D. Kosman et al., *Science* 305:846, 2004. With permission from AAAS.)

underlying these phenomena are still not well understood, but their consequences are profound. We shall see that the serial organization of gene expression in the *Hox* complex is a fundamental feature that has been highly conserved in the course of animal evolution.

Trithorax and Polycomb Group Proteins Enable the *Hox* Complexes to Maintain a Permanent Record of Positional Information

The spatial pattern of expression of the genes in the *Hox* complex is set up by signals acting early in development, but the consequences are long lasting. Although the pattern of expression undergoes complex adjustments as development proceeds, the *Hox* complexes serve to stamp each cell and its progeny with a permanent record of the A-P position that the cell occupied in the early embryo. In this way, the cells of each segment are equipped with a long-term memory of their location along the A-P axis of the body. This memory trace is somehow imprinted on the *Hox* complexes, and it governs the segment-specific identity not only of the larval segments, but also of the structures of the adult fly.

The molecular mechanism of this memory of positional information relies on two types of regulation. One is from the *Hox* genes themselves: many of the Hox proteins autoactivate the transcription of their own genes, thereby helping to keep the genes on indefinitely. Another crucial input is from two large, complementary sets of proteins, called the **Trithorax group** and the **Polycomb group**, which stamp the chromatin of the *Hox* complex with a heritable record of its embryonic state of activation or repression. These are key general regulators of chromatin structure that can be shown to be critical for cell memory: if genes of the *Trithorax* or *Polycomb* group are defective, the pattern of expression of the *Hox* genes is set up correctly at first, but it is not correctly maintained as the embryo grows older.

The two sets of regulators act in opposite ways. Trithorax group proteins are needed to maintain the transcription of *Hox* genes in cells where transcription has already been switched on. In contrast, Polycomb group proteins form stable complexes that bind to the chromatin of the *Hox* complex and maintain the repressed state in cells where *Hox* genes have not been activated at the critical time (**Figure 21–25**). How such changes in chromatin can store developmental cell memory is discussed in Chapters 4 and 7.

The D-V Signaling Genes Create a Gradient of the Transcription Regulator Dorsal

As with the patterning along the *Drosophila* A-P axis just discussed, the patterning along the dorsoventral (D-V) axis begins with maternal gene products that define

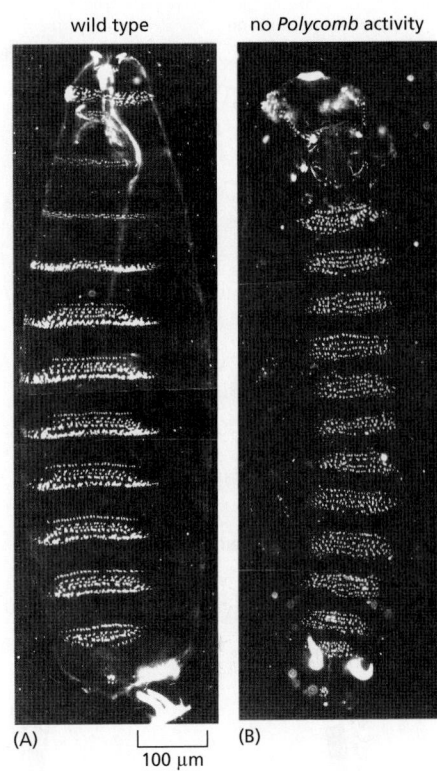

Figure 21–25 **The role of genes of the *Polycomb* group.** (A) Photograph of a wild-type *Drosophila* embryo. (B) Photograph of a mutant embryo defective for the gene *Extra sex combs* (*Esc*) and derived from a mother also lacking this gene. The gene belongs to the *Polycomb* group. Essentially all segments have been transformed to resemble the most posterior abdominal segment. In the mutant, the pattern of expression of the homeotic selector genes, which is roughly normal initially, is unstable in such a way that all these genes soon become switched on all along the body axis. (From G. Struhl, *Nature* 293:36–41, published 1981 by Nature Publishing Group. Reproduced with permission of SNCSC.)

this axis in the egg (see Figure 21–17), and it then progresses through zygotic gene products that further subdivide the D–V axis in the embryo.

Initially, a protein that is produced by follicle cells underneath the future ventral region of the embryo leads to the localized activation of a transmembrane receptor, called **Toll**, on the ventral side of the egg membrane. The various maternal genes required for this process are called *D–V egg-polarity genes*. (Curiously, *Drosophila* Toll and vertebrate Toll-like proteins also operate in innate immune responses, as discussed in Chapter 24). The localized activation of Toll controls the distribution of **Dorsal**, a transcription regulator of the NFκB family discussed in Chapter 15. The Toll-regulated activity of Dorsal, like that of NFκB, depends on the translocation of Dorsal from the cytosol, where it is held in an inactive form, to the nucleus, where it regulates gene expression (see Figure 15–62). In the newly laid egg, both Dorsal mRNA and protein are distributed uniformly in the cytosol. After the nuclei in the syncytial blastoderm have migrated to the surface of the embryo, but before cellularization (see Figure 21–15), Toll receptor activation on the ventral side induces a remarkable redistribution of the Dorsal protein. On the dorsal side, the protein remains in the cytosol, but ventrally it becomes concentrated in the nuclei, with a smooth gradient of nuclear localization between these two extremes (**Figure 21–26**).

Once inside the nucleus, the Dorsal protein acts as a morphogen and turns on or off the expression of different sets of genes depending on Dorsal's concentration. The expression of each responding gene depends on its regulatory DNA—specifically, on the number and affinity of the binding sites that this DNA contains for Dorsal and other transcription regulators. In this way, the regulatory DNA interprets the positional signal provided by the nuclear Dorsal protein gradient, so as to define a D–V series of territories—distinctive bands of cells that run the length of the embryo. Most ventrally—where the nuclear concentration of Dorsal protein is highest—it switches on, for example, the expression of a gene called *Twist*, which is specific for mesoderm. Most dorsally, where the nuclear concentration of Dorsal protein is lowest, the cells switch on a gene called *Decapentaplegic* (*Dpp*). And in an intermediate region, where the nuclear concentration of Dorsal protein is high enough to repress *Dpp* but too low to activate *Twist*, the cells switch on another set of genes, including one called *Short gastrulation* (*Sog*) (**Figure 21–27A**).

Products of the genes directly regulated by the Dorsal protein generate in turn more local signals, which define finer subdivisions along the D–V axis. These signals act during cellularization and take the form of conventional extracellular

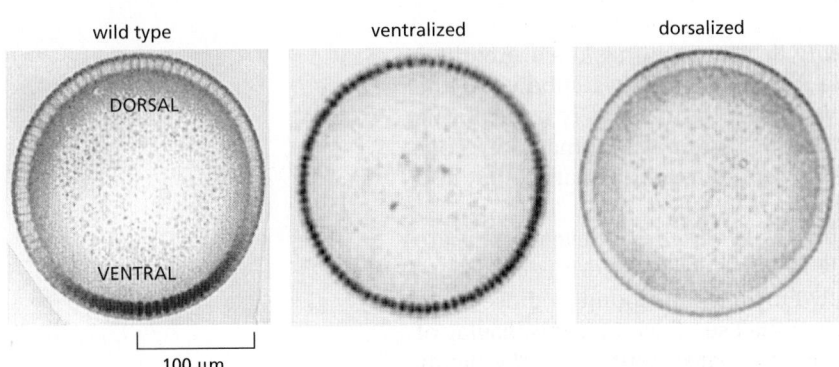

Figure 21–26 **The concentration gradient of Dorsal protein in the nuclei of the blastoderm.** In wild-type *Drosophila* embryos, the protein is present in the dorsal cytoplasm and absent from the dorsal nuclei; ventrally, it is depleted in the cytoplasm and concentrated in the nuclei. In a mutant in which the Toll pathway is activated everywhere and not just ventrally, Dorsal protein is everywhere concentrated in the nuclei; the result is a ventralized embryo. Conversely, in a mutant in which the Toll signaling pathway is inactivated, Dorsal protein everywhere remains in the cytoplasm and is absent from the nuclei; the result is a dorsalized embryo. (From S. Roth, D. Stein and C. Nüsslein-Volhard, *Cell* 59:1189–1202, 1989. With permission from Elsevier.)

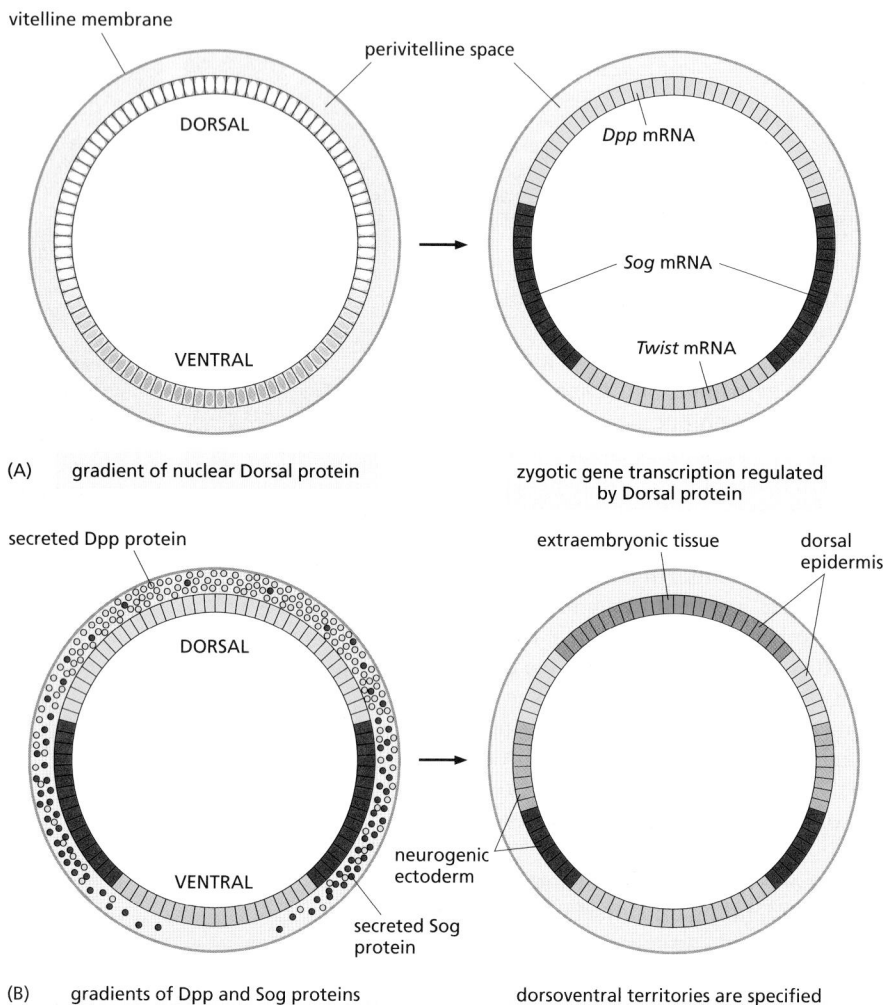

(A) gradient of nuclear Dorsal protein

zygotic gene transcription regulated by Dorsal protein

(B) gradients of Dpp and Sog proteins

dorsoventral territories are specified

Figure 21–27 How morphogen gradients guide a patterning process along the dorsoventral axis of the *Drosophila* embryo. (A) Initially, a gradient of Dorsal protein defines three broad territories of gene expression, marked here by the expression of three representative genes—*Dpp*, *Sog*, and *Twist*. (B) Slightly later, the cells expressing *Dpp* and *Sog* secrete, respectively, the signal proteins Dpp (a TGFβ family member) and Sog (an antagonist of Dpp). These two proteins then diffuse and interact with one another (and with certain other factors) to create the dorsoventral (D–V) territories shown.

signal proteins. In particular, *Dpp* codes for a secreted TGFβ-family protein, which forms a local morphogen gradient in the dorsal part of the embryo. *Sog* encodes another secreted protein that is produced by the *neurogenic ectoderm* (which gives rise to the nervous system) and acts as an antagonist of Dpp protein. The opposing diffusion gradients of these two signal proteins create a steep gradient of Dpp activity: the highest Dpp activity levels, in combination with certain other factors, cause development of the most dorsal tissue of all—an extraembryonic membrane. Intermediate levels cause development of dorsal epidermis; and the absence of Dpp activity allows the development of neurogenic ectoderm (Figure 21–27B).

A Hierarchy of Inductive Interactions Subdivides the Vertebrate Embryo

The molecular genetic analysis of *Drosophila* development has uncovered how a cascade of transcription regulators and signaling pathways subdivides the embryo. The same principle of progressive pattern refinement is used during the development of all animal embryos, including vertebrates. Remarkably, conservation is not restricted to the general strategy of pattern formation, but also extends to many of the molecules involved.

As mentioned previously, the earliest phases of vertebrate development are surprisingly variable, even between closely related species, and it is even hard to say precisely how the axes of an early fly embryo correspond to those of an early frog or mouse embryo. Nevertheless, we shall see that amid this display of evolutionary plasticity, some features of early development turn out to be highly conserved. The same is true of later developmental stages also, often to an astonishing degree. From our own anatomy, it is obvious that we are cousins to birds and fish. But looking at molecular mechanisms, we see that we are cousins to flies and worms too.

In the following pages, we discuss how vertebrate embryos are patterned by the interplay of signaling molecules and transcription regulators. We begin by discussing the formation and patterning of the embryonic axes in amphibians, taking the frog *Xenopus* as our example. We have already broached this topic earlier in the chapter. Here, we pick up the thread and draw comparisons with the fly.

As noted earlier, the origins of the embryonic axes and the three germ layers in the frog can be traced back to the blastula (see Figure 21–3A). By labeling individual blastomeres, we can track cells through all their divisions, transformations, and migrations and see what they become and where they come from. The precursors of ectoderm, mesoderm, and endoderm are arranged in order along the animal-vegetal axis of the blastula: the endoderm derives from the most vegetal blastomeres, the ectoderm from the most animal, and the mesoderm from a middle set. Within each of these territories, the cells have diverse fates according to their positions along the D-V axis of the later embryo. For ectoderm, epidermal precursors are located ventrally, and future neurons are found dorsally; for mesoderm, precursors for notochord, muscle, kidney, and blood are arranged from dorsal to ventral. All this can be represented by a **fate map** that shows which cell types derive from which regions of the blastula (**Figure 21–28**). The fate map confronts us with the central question: how are the cells in different positions driven toward their different fates? We have already explained how maternal factors deposited in the developing frog egg define its animal-vegetal axis, and how cortical rotation triggered by fertilization defines the orientation of the dorsoventral axis (see Figure 21–14). But how does the establishment of axes lead on to the subdivision of the embryo into the future body parts?

The maternal gene products lead to the formation of signaling centers on the vegetal and dorsal sides of the embryo. The dorsal signaling center in particular has a special place in the history of developmental biology. Experiments in the early twentieth century identified it as a small cluster of cells, located on the dorsal side of the amphibian embryo, with an extraordinary property: when the cells were transplanted to an opposite site, they could trigger a radical reorganization of the neighboring tissue, causing it to form a second whole-body axis (**Figure 21–29**). The discovery of this signaling center, called the **Organizer**, led the way to a pioneering analysis of the chain of inductive interactions that establish the framework of the vertebrate body.

In contrast to the *Drosophila* syncytial embryo, the fertilized frog egg undergoes rapid cleavage divisions that result in an embryo consisting of thousands of cells. Patterning must therefore be mediated by extracellular signal molecules

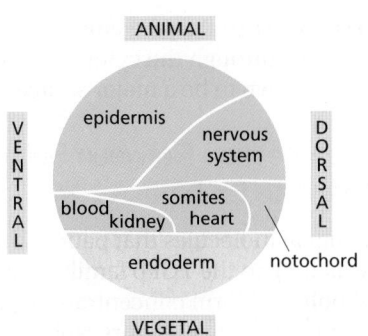

Figure 21–28 Blastula fate map in a frog embryo. The endoderm derives from the most vegetal blastomeres *(yellow)*, the ectoderm from the most animal *(blue)*, and the mesoderm from a middle set *(green)* that contributes also to endoderm and ectoderm. Different cell types derive from different positions along the dorsoventral axis.

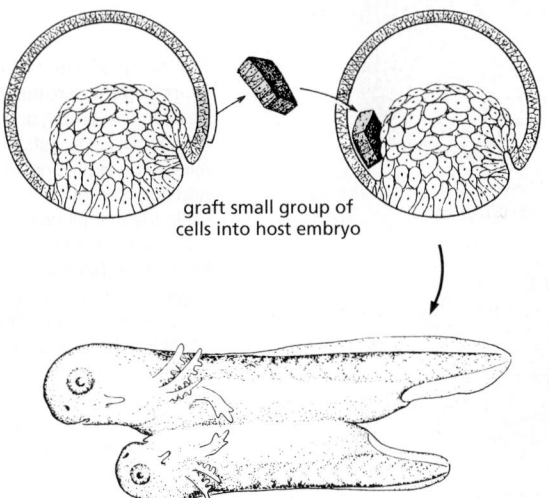

graft small group of cells into host embryo

Figure 21–29 Induction of a secondary axis by the Organizer. An amphibian embryo receives a graft of a small cluster of cells taken from a specific site, called the Organizer region, on the dorsal side of another embryo at the same stage. Signals from the graft organize the behavior of neighboring cells of the host embryo, causing development of a pair of conjoined (Siamese) twins. See **Movie 21.4**. [After J. Holtfreter and V. Hamburger, in Analysis of Development (B.H. Willier, P.A. Weiss and V. Hamburger, eds), pp. 230–296. Philadelphia: Saunders, 1955.]

that diffuse through the embryo from cell to cell, not by transcription regulators that move through the cytoplasm of a syncytium. Not surprisingly, the Organizer is now known to be a major source of secreted protein signals.

A Competition Between Secreted Signaling Proteins Patterns the Vertebrate Embryo

The signal molecules that pattern the frog embryo along the animal-vegetal (A-V) axis belong to the TGFβ family: they are secreted by a signaling center at the vegetal pole and form concentration gradients along the A-V axis. The *Nodal* protein acts over a relatively short range: cells near the vegetal pole are exposed to high levels of it and respond by switching on genes that promote the development of endoderm; cells further away are exposed to lower levels and activate genes that promote the formation of mesoderm. The cells at the vegetal pole that produce Nodal also produce a more rapidly diffusing TGFβ-like protein called *Lefty*, which antagonizes Nodal. The result is a high ratio of Lefty to Nodal at the animal pole, where Lefty predominates and Nodal signaling is blocked; this causes the cells there to develop as ectoderm (**Figure 21–30**A). Thus, a mid-range activation by Nodal, combined with a long-range inhibition by Lefty, sets up the pattern of progenitors along the A-V axis for the three germ layers—endoderm, mesoderm, and ectoderm.

The frog's dorsal signaling system uses a different set of secreted signals from that of the vegetal signaling system to subdivide the germ-layer territories according to location along the D-V axis of the embryo. It exerts its influence by secreting two inhibitory signal proteins, called *Chordin* and *Noggin*. These antagonize the action of *bone morphogenetic proteins* (*BMPs*; members of yet another subclass of the TGFβ family), which themselves are secreted throughout the embryo. In this way, Chordin and Noggin form a dorsal-to-ventral gradient that blocks BMP signaling on the dorsal side but allows it to remain high on the ventral side (Figure 21–30B). Ectodermal cells that experience high levels of BMP signaling are driven to epidermal fates, whereas cells that experience little or no BMP signaling remain neural.

Knowing the signals that specify the three germ layers and various tissue types of the vertebrate body, one can reproduce this specification in a culture dish. Frog cells taken from the animal-pole region of the embryo, for example,

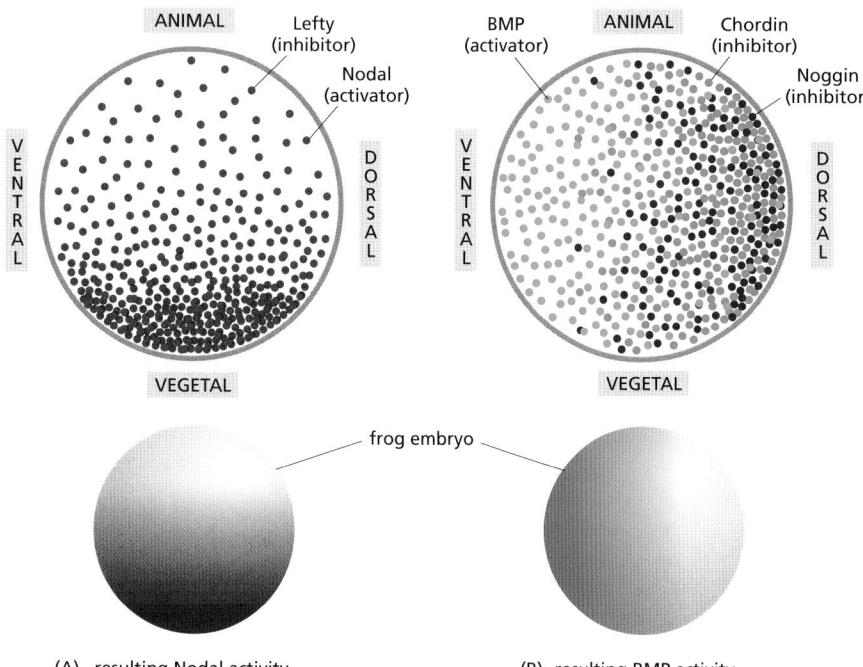

(A)   resulting Nodal activity   (B)   resulting BMP activity

Figure 21–30 How Nodal and bone morphogenic protein (BMP) signaling pattern the embryonic axes. Nodal and its antagonist Lefty pattern the animal-vegetal axis, while BMP and its antagonists Chordin and Noggin pattern the dorsoventral axis. (A) In the animal pole region, where Nodal levels are low relative to Lefty, Lefty blocks Nodal from binding to its receptors. In the vegetal region, there is an excess of Nodal, resulting in Nodal pathway activation. (B) Along the dorsoventral axis, BMP is widely present but Chordin and Noggin are concentrated at the dorsal side: there, they bind to BMP and block its binding to receptors. The resulting patterns of Nodal and BMP activity are illustrated at the bottom of the figure.

will differentiate into blood (a ventral mesodermal tissue) when diverted from their original fate by exposure to intermediate concentrations of Nodal and high concentrations of BMP. Similarly, mouse or human embryonic stem cells can be coaxed to generate specific cell types by exposing them in culture to appropriate combinations of signal molecules. In this way, the insights gained through studies of animal development can be used to generate the cell types needed for regenerative medicine, as we discuss in the next chapter.

The Insect Dorsoventral Axis Corresponds to the Vertebrate Ventral-Dorsal Axis

The signaling systems that pattern the D-V axis in *Drosophila* and in vertebrates are similar. In *Drosophila,* as we saw, Dpp and its inhibitor Sog are responsible, whereas in vertebrates, BMP and its inhibitors Chordin and Noggin do the job. Dpp is a member of the BMP family, while Sog is a homolog of Chordin. Both in flies and frogs, high levels of the inhibitors define the region that is neurogenic, and high levels of BMP/Dpp activity define the region that is not. These and other molecular parallels strongly suggest that this aspect of body patterning has been conserved in evolution from insects to vertebrates. Curiously, however, the axis is inverted: dorsal in the fly corresponds to ventral in the vertebrate (**Figure 21–31**). At some point in evolution, it seems that the ancestor of one of these classes of animals took to living life upside-down.

Hox Genes Control the Vertebrate A-P Axis

The conservation of developmental mechanisms between *Drosophila* and vertebrates extends beyond the D-V signaling system. *Hox* genes are found in almost every animal species studied, where they are often grouped in complexes similar to the insect *Hox* complex. In mice and humans, for example, there are four such complexes—called the *HoxA, HoxB, HoxC,* and *HoxD* complexes—each on a different chromosome. Individual genes in each complex can be recognized by their sequences as counterparts of specific members of the *Drosophila* set. Indeed, mammalian *Hox* genes can function in *Drosophila* as partial replacements for the corresponding *Drosophila Hox* genes. It appears that each of the four mammalian *Hox* complexes is, roughly speaking, the equivalent of one complete insect *Hox* complex (that is, an *Antennapedia* complex plus a *Bithorax* complex) (**Figure 21–32**).

The ordering of the genes within each vertebrate *Hox* complex is essentially the same as in the insect *Hox* complex, suggesting that all four vertebrate complexes originated by duplications of a single primordial complex and have preserved its basic organization. Most tellingly, when the expression patterns of the *Hox* genes are examined in the vertebrate embryo, it turns out that the members of each complex are expressed in a head-to-tail series along the axis of the body, just as they are in *Drosophila*. As in *Drosophila,* vertebrate *Hox* gene expression patterns are often aligned with vertebrate segments. This alignment is especially clear in the hindbrain (see Figure 21–32), where the segments are called *rhombomeres*.

The products of the vertebrate *Hox* genes, the Hox proteins, specify positional values that control the A-P pattern of parts in the hindbrain, neck, and trunk (as well as some other parts of the body). As in *Drosophila,* when a posterior *Hox* gene is artificially expressed in an anterior region, it can convert the anterior tissue to

Figure 21–31 The vertebrate body plan as a dorsoventral inversion of the insect body plan. Note the correspondence with regard to the circulatory system as well as the gut and nervous system. In insects, the circulatory system is represented by a tubular heart and a main dorsal blood vessel, which pumps blood out into the tissue spaces through one set of apertures and receives blood back from the tissues through another set. Unlike vertebrates, insects have no system of capillary vessels to contain the blood as it percolates through the tissues. Nevertheless, heart development depends on homologous genes in vertebrates and insects, reinforcing the relationship between the two body plans. (After E.L. Ferguson, *Curr. Opin. Genet. Dev.* 6:424–431, 1996.)

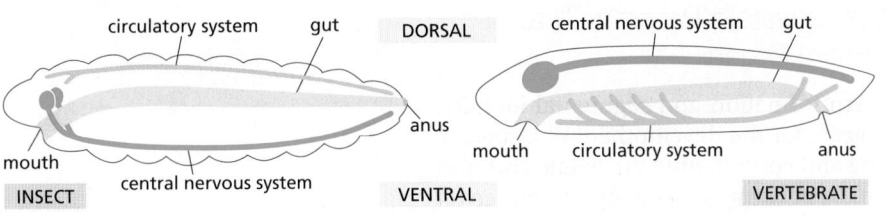

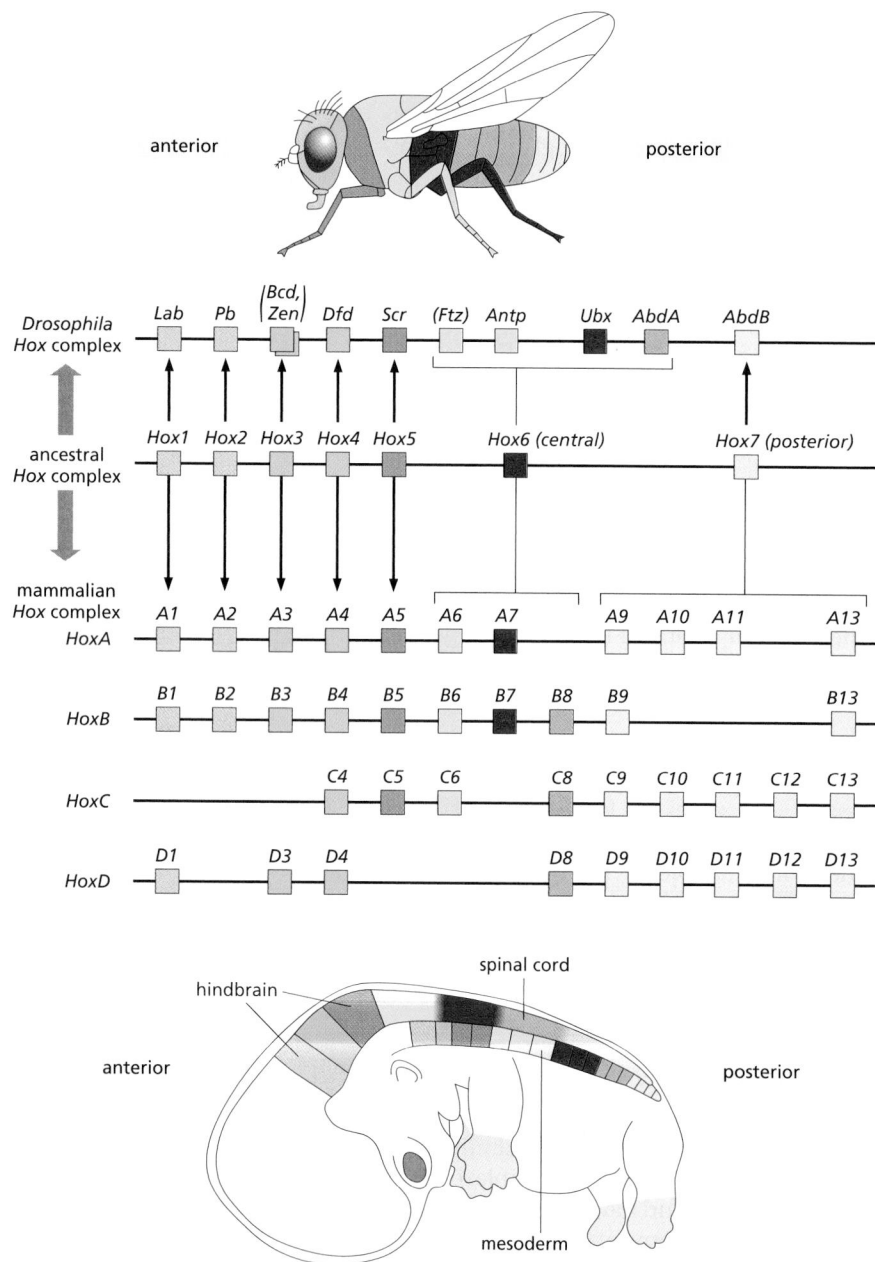

Figure 21–32 The *Hox* complexes of an insect and a mammal, compared and related to body regions. The genes of the *Antennapedia* and *Bithorax* complexes of *Drosophila* are shown in their chromosomal order in the top line. The corresponding genes of the four mammalian *Hox* complexes are shown below, also in chromosomal order. The gene expression domains in fly and mammal are indicated in a simplified form by color in the cartoons of animals above and below. There is a remarkable parallelism. However, the details of the patterns depend on developmental stage and vary somewhat from one mammalian *Hox* complex to another. Also, in many cases, genes shown here as expressed in an anterior domain are also expressed more posteriorly, overlapping the domains of more posterior *Hox* genes.

The complexes are thought to have evolved as follows: first, in some common ancestor of worms, flies, and vertebrates, a single primordial homeotic selector gene underwent repeated duplication to form a series of such genes in tandem—the ancestral *Hox* complex. In the *Drosophila* sublineage, this single complex became split into separate *Antennapedia* and *Bithorax* complexes. Meanwhile, in the lineage leading to the mammals, the whole complex was repeatedly duplicated to give four *Hox* complexes. The parallelism is not perfect because apparently some individual genes have been duplicated and others lost. Still others have been co-opted for different purposes (genes in parentheses in the top line) over the time that has elapsed since the complexes diverged. (Based on a diagram courtesy of William McGinnis.)

a posterior character. Conversely, loss of posterior *Hox* genes allows the posterior tissue where they are normally expressed to adopt an anterior character (**Figure 21–33**). Because of a redundancy between genes in the four *Hox* gene clusters, the transformations observed in mouse *Hox* mutants are not always so straightforward as those in the fly, and they are often incomplete. Nonetheless, it seems clear that the fly and the mouse use essentially the same molecular machinery to impart individual characteristics to successive regions along at least a part of the A-P axis.

Some Transcription Regulators Can Activate a Program That Defines a Cell Type or Creates an Entire Organ

Just as there are genes that regulate pattern formation and segmental identity, there are genes whose products act as triggers for the development of a specific cell type or even a specific organ, initiating and coordinating the whole complex program of gene expression that is required. An example is the *MyoD/myogenin*

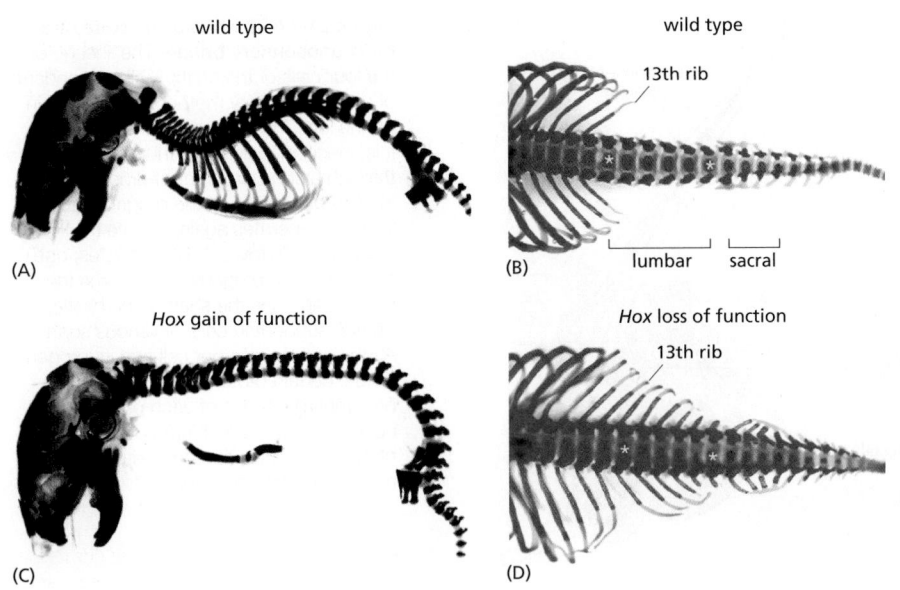

wild type (A)

wild type (B)
13th rib
lumbar   sacral

Hox gain of function (C)

Hox loss of function (D)
13th rib

Figure 21–33 Control of anteroposterior pattern by *Hox* genes in the mouse. (A,B) A normal mouse (wild type) has about 65 vertebrae, differing in structure according to their position along the body axis: 7 cervical (neck), 13 thoracic (with ribs), 6 lumbar [bracketed by *yellow* asterisks in (B)], 4 sacral [bracketed by *red* asterisks in (B)], and about 35 caudal (tail). (A) shows a side view and (B) shows a dorsal view; for clarity, the limbs have been removed in each picture.

(C) The *HoxA10* gene is normally expressed in the lumbar region (together with its paralogs *HoxC10* and *HoxD10*); here it has been artificially expressed in the developing vertebral tissue all along the body axis. As a result, the cervical and thoracic vertebrae are all converted to a lumbar character. (D) Conversely, when *HoxA10* is removed along with *HoxC10* and *HoxD10*, vertebrae that should normally have a lumbar or sacral character take on a thoracic character instead. (A and C, from M. Carapuço et al., *Genes Dev.* 19:2116–2121, 2005. With permission from Cold Spring Harbor Laboratory Press; B and D, from D.M. Wellik and M.R. Capecchi, *Science* 301:363–367, 2003. With permission of AAAS.)

family of transcription regulators that we encountered in Chapter 7. These proteins drive cells to differentiate into muscle, expressing muscle-specific actins and myosins and all the other specialized cytoskeletal, metabolic, and membrane proteins that a muscle cell needs. Analogously, members of the Achaete/Scute family of transcription regulators drive cells to become neural progenitors. In both these examples, the proteins belong to the basic helix–loop–helix (bHLH) class of transcription regulators (see p. 377), and the same is true for many of the other proteins that induce the differentiation of particular cell types. These *master transcription regulators* exert their powerful differentiation-inducing activity by binding to many different regulatory sites in the genome and thereby controlling the expression of large numbers of downstream target genes. In one well-studied case, that of an Achaete/Scute family member called Atonal homolog 1 (Atoh1), the number of direct target genes in the mouse genome is more than 600. It is important to note, however, that even such powerful drivers of cell differentiation can have radically different effects according to the context and history of the cells in which they act: Atoh1, for example, drives differentiation of certain classes of neurons in the brain, of sensory hair cells in the inner ear, and of secretory cells in the lining of the gut.

Other genes encoding transcription regulators can drive the formation and assembly of the multiple cell types that constitute an entire organ. A famous example is the transcription regulator Eyeless. When it is artificially expressed in a patch of cells in the leg precursors of *Drosophila*, a well-organized eye-like organ develops on the leg, with the various eye cell types correctly arranged (see Figure 7–35B); conversely, loss of the *Eyeless* gene results in flies that lack eyes. Moreover, loss of the Eyeless homolog Pax6 in vertebrates likewise leads to loss of eye structures. Similar organ-selector proteins are known for foregut, heart, pancreas, and other organs. They are all master transcription regulators that directly regulate hundreds of target genes, the products of which then specify and construct the different elements of the appropriate organ. However, as in the example of Atoh1, they usually exert their specific effect only in combination with the right partners, which are only expressed in cells that were appropriately primed during their earlier development.

Notch-Mediated Lateral Inhibition Refines Cellular Spacing Patterns

After the establishment of the basic body plan and the generation of organ precursors, many further steps of pattern refinement are required to achieve the adult pattern of terminally differentiated cells in tissues and organs. As we discussed

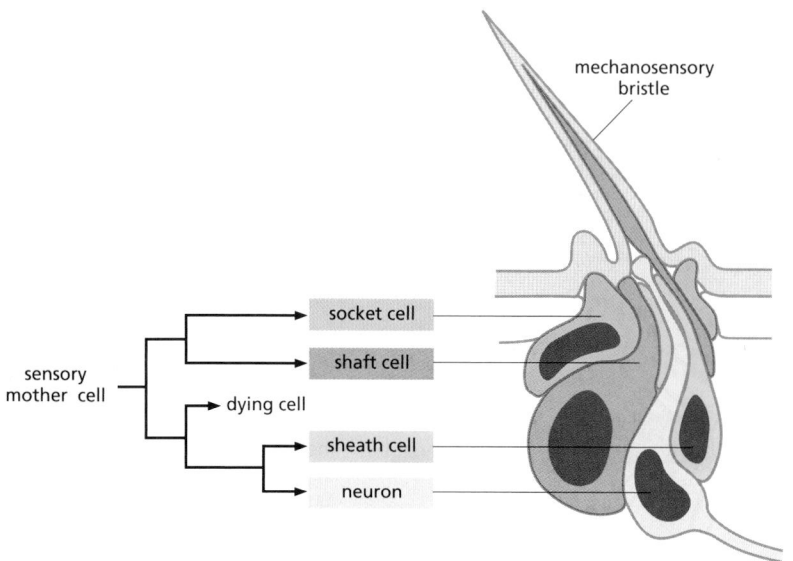

Figure 21–34 The basic structure of a mechanosensory bristle. The lineage of the four cells of the bristle—all descendants of a single sensory mother cell—is shown on the left. The sensory mother cell, once it is specified, generates this set of cells through a short program of division cycles. In each generation of the progeny, lateral inhibition operates again to drive the newborn cells toward different fates: one of the ultimate progeny will become the neuron; another, the shaft of the bristle; others, supporting cells of various sorts. As the sensory mother cell and its progeny divide, certain proteins are allocated preferentially to one of each pair of newborn sister cells, biasing the outcome of the lateral-inhibition competition mediated by Notch signaling.

earlier, lateral inhibition mediated by Notch signaling is crucial for both cell diversification and fine-grained patterning in an enormous variety of tissues in all animals.

One example is the development of **sensory bristles** in *Drosophila*, most easily seen on the fly's back, but also present on most of its other exposed surfaces. Each of these is a miniature sense organ, consisting of a sensory neuron and a small set of supporting cells. Some bristles respond to chemical stimuli, others to mechanical stimuli, but they are all constructed in a similar way (**Figure 21–34**). The proneural genes *Achaete* and *Scute* mentioned earlier mark the patches of epidermis within which bristles will form. Mutations that eliminate the expression of these genes at some of their usual sites block development of bristles at just those sites, and mutations that cause expression in abnormal sites cause bristles to develop there.

The initial cells expressing the proneural genes are called proneural cells, and they are primed to take the neurosensory pathway of differentiation, but which of the cells will actually do so depends on competitive interactions among them. In the first round of these interactions, a single cell within each small group of proneural cells is picked to serve as the progenitor of the bristle. This single cell is called the *sensory mother cell*. It becomes distinct from the other cells of the cluster through lateral inhibition mediated by the Notch signaling pathway. This operates in the way we discussed earlier. The cells in the proneural cluster initially all express both the transmembrane receptor Notch and its transmembrane ligand *Delta*, along with proteins that regulate the signaling activity of Delta. Wherever Delta activates Notch, an inhibitory signal is transmitted that diminishes the tendency of the Notch-activated cell to specialize as a sensory mother cell. At first, all the cells in the cluster inhibit one another. However, receipt of the signal in a given cell diminishes that cell's ability to fight back by delivering the inhibitory Delta signal in return. This creates a competitive situation, from which a single cell in each cluster—the future sensory mother cell—eventually emerges as winner, sending a strong inhibitory signal to its immediate neighbors but receiving no such signal in return (**Figure 21–35**). If a cell that would normally become a sensory mother cell is genetically disabled from doing so, a neighboring proneural cell, freed from lateral inhibition, will become a sensory mother cell instead.

The sensory mother cell goes through a short program of further divisions to generate the set of cells that form the final bristle. Notch signaling acts repeatedly at successive stages in this program to drive the descendants of the sensory mother cell along different pathways and assign them to their various specialized fates. However, it does so in conjunction with additional mechanisms that bias the outcome of the competition mediated by lateral inhibition. Determinants that

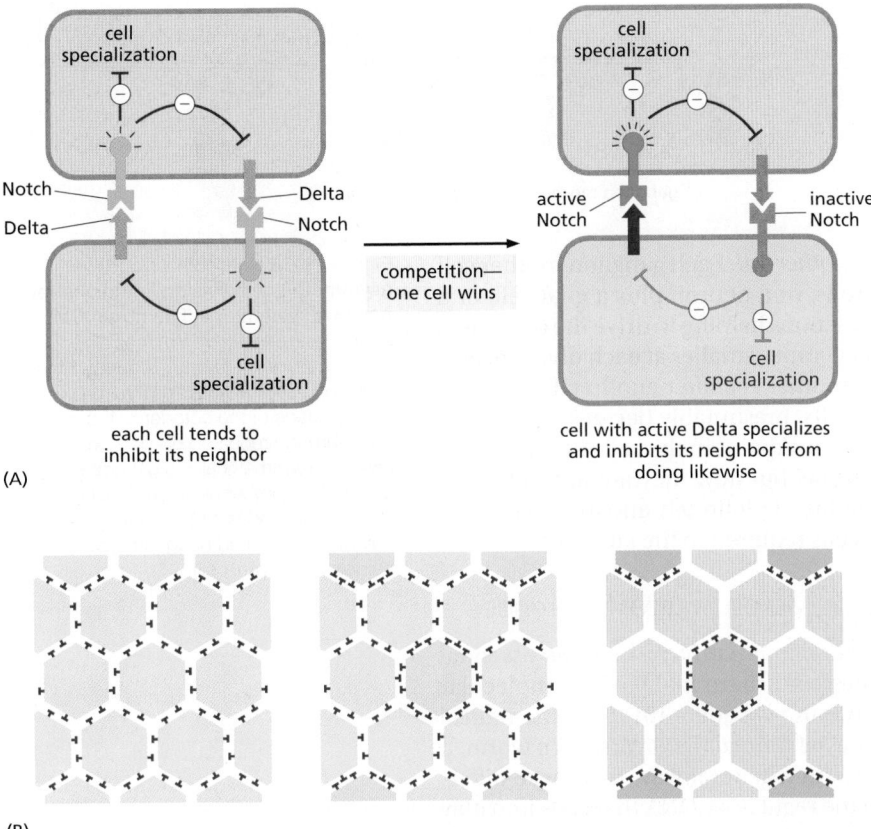

(A) each cell tends to inhibit its neighbor

cell with active Delta specializes and inhibits its neighbor from doing likewise

(B)

Figure 21–35 **Lateral inhibition.** (A) The basic mechanism of Notch-mediated competitive lateral inhibition, illustrated for just two interacting cells. In this diagram, the absence of color on proteins or effector lines indicates inactivity. (B) The outcome of the same process operating in a larger patch of cells. At first, all cells in the patch are equivalent, expressing both the transmembrane receptor Notch and its transmembrane ligand Delta. Each cell has a tendency to specialize (as a sensory mother cell), and each sends an inhibitory signal to its neighbors to discourage them from also specializing in that way. This creates a competitive situation. As soon as an individual cell gains any advantage in the competition, that advantage becomes magnified. The winning cell, as it becomes more strongly committed to differentiating as a sensory mother cell, also inhibits its neighbors more strongly. Conversely, as these neighbors lose their capacity to differentiate as sensory mothers, they also lose their capacity to inhibit other cells from doing so. Lateral inhibition thus makes adjacent cells follow different fates.

Although the interaction is thought to be normally dependent on cell–cell contacts, the future sensory mother cell may be able to deliver an inhibitory signal to cells that are more than one cell diameter away—for example, by sending out long protrusions to touch them.

are asymmetrically localized inside the dividing cells have this role in sensory bristle development. They are also important in other contexts, as we now discuss.

Asymmetric Cell Divisions Make Sister Cells Different

Cell diversification does not always have to depend on extracellular signals: in some cases, sister cells are born different as a result of an asymmetric cell division, during which some significant set of molecules is divided unequally between them. This asymmetrically segregated molecule (or set of molecules) then acts as a determinant for one of the cell fates by directly or indirectly altering the pattern of gene expression within the daughter cell that receives it (see Figure 21–12). We have already encountered the asymmetric segregation of molecules in the context of the early frog embryo: *VegT* RNA is localized in the vegetal region of the fertilized egg. Following cell division, only vegetal daughter cells will inherit *VegT* RNA.

Asymmetric divisions often occur at the beginning of development, but they are also encountered at some later stages. As mentioned for the sensory bristle, they can set the scene for an exchange of Notch signals between the daughter cells, with the signaling occurring after the cells have become separate and reinforcing the differences between them. In the central nervous system, asymmetric divisions have a key role in generating the very large numbers of neurons and glial cells that are needed. A special class of cells becomes committed as neural precursors, but instead of differentiating directly as neurons or glial cells, these undergo a long series of asymmetric divisions through which a succession of additional neurons and glial cells are added to the population. The process is best understood in *Drosophila*, although there are many hints that something similar occurs also in vertebrate neurogenesis.

In the embryonic central nervous system of *Drosophila*, the nerve-cell precursors, or *neuroblasts*, are initially singled out from the neurogenic ectoderm by a typical lateral-inhibition mechanism that depends on Notch. Each neuroblast then divides repeatedly in an asymmetric fashion (**Figure 21–36**). At each division, one daughter remains as a neuroblast, while the other, which is much

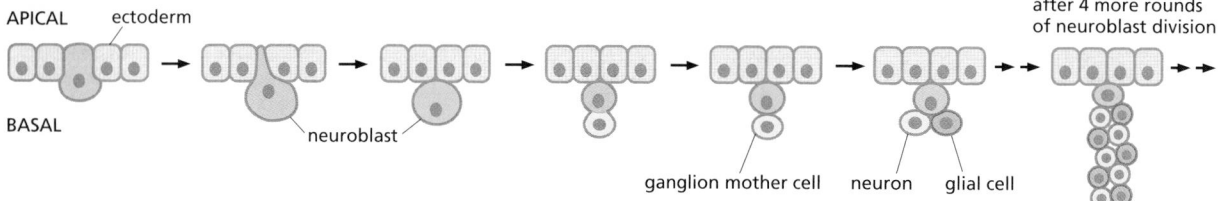

APICAL

ectoderm

BASAL

neuroblast

ganglion mother cell   neuron   glial cell

after 4 more rounds
of neuroblast division

smaller, becomes specialized as a *ganglion mother cell*. Each ganglion mother cell will divide only once, giving a pair of neurons, or a neuron plus a glial cell, or a pair of glial cells, with Notch-mediated interactions helping to drive the daughters along different paths. The neuroblast itself becomes smaller at each division, as it parcels out its substance into one ganglion mother cell after another. Eventually, typically after about 12 cycles, the process halts, presumably because the neuroblast becomes too small to pass the cell-size checkpoint in the cell-division cycle. Later, in the larva, neuroblast divisions resume, but now they are accompanied by cell growth, permitting the process to continue indefinitely and to generate the much larger numbers of neurons and glial cells required in the adult fly.

Figure 21–36 Neuroblasts and asymmetric cell division in the central nervous system of a fly embryo. The neuroblast originates as a specialized ectodermal cell. It is singled out by lateral inhibition and emerges from the basal (internal) face of the ectoderm. It then goes through repeated division cycles, dividing asymmetrically to generate a series of ganglion mother cells. Each ganglion mother cell divides just once to give a pair of differentiated daughters (typically a neuron plus a glial cell).

Differences in Regulatory DNA Explain Morphological Differences

In the preceding sections, we have seen that animals contain the same essential cell types, have a similar collection of genes, and share many of the molecular mechanisms of pattern formation. But how can we square this with the radical differences that we see in the body structures of animals as diverse as a worm, a fly, a frog, and a mouse? We asserted earlier, in a general way, that these differences usually seem to reflect differences in the regulatory DNA that calls into play the components of the conserved basic kit of parts. We must now examine the evidence a little more closely.

When we compare animal species with similar basic body plans—different vertebrates, for example, such as fish, birds, and mammals—we find that corresponding genes usually have similar sets of regulatory elements: the regulatory DNA sequences have been well conserved and are recognizably homologous in the different animals. The same is true if we compare different species of nematode worms or insects. But, when we compare vertebrate regulatory regions with those of worms or flies, it is hard to see any such resemblance. The protein-coding sequences are unmistakably similar, but the corresponding regulatory DNA sequences appear mostly very different, suggesting that the differences in body plans mainly reflect differences in regulatory DNA. Although variations in the proteins themselves also contribute, differences in regulatory DNA would be enough to generate radically different tissues and body structures even if the proteins were the same.

It is not yet possible to trace the genetic steps that have led to all the spectacular diversity of animals. Their lineages have diverged over hundreds of millions of years, and in most cases too many changes have occurred for us to be able to say that this or that feature results from this or that mutation. The picture is clearer, however, for more recent evolutionary events. Studies of both closely related animal populations and plant populations whose members have different morphologies have revealed that dramatic developmental effects can result from subtle changes in regulatory DNA.

A well-studied example is the morphological diversity found in stickleback fish. After the last ice age ended about 10,000 years ago, marine sticklebacks colonized many newly formed freshwater streams and lakes. Marine sticklebacks extend sharp spines from their pelvic skeleton. These spines are thought to help protect the fish from soft-mouthed fish predators. In contrast, several populations of freshwater sticklebacks have lost these spines, usually in lakes that lack such predators. The different morphologies reflect differences in control of the expression of a transcription regulator called Pitx1. Whereas marine sticklebacks express the *Pitx1* gene in the pelvic bone precursor cells that will form the spikes,

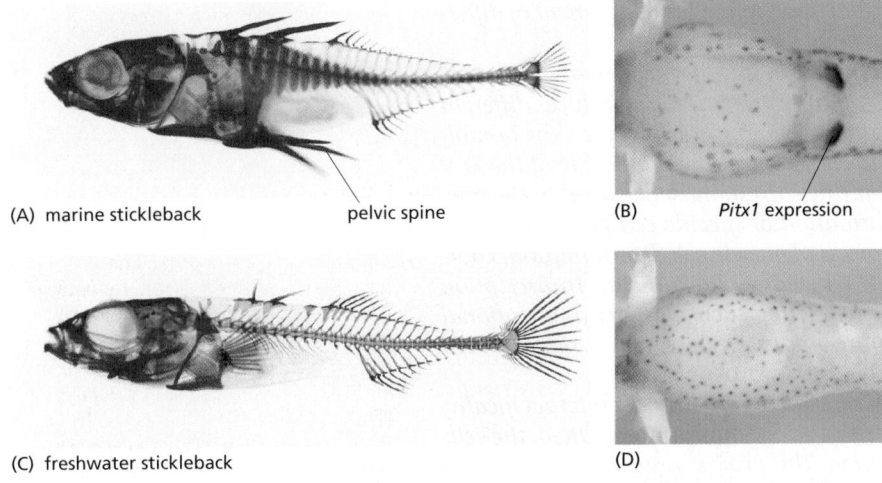

(A) marine stickleback   pelvic spine

(B)   *Pitx1* expression

(C) freshwater stickleback

(D)

Figure 21–37 Morphological diversity in stickleback fish is caused by changes in regulatory elements. (A–D) Pelvic spines are present in marine (A) but not in freshwater (C) populations. Correspondingly, *Pitx1* is expressed in the pelvic area in marine (B) but not in freshwater (D) fish. The lack of expression in the pelvic area of freshwater populations is caused by mutations in an enhancer element. Other enhancers and sites of expression for *Pitx1* are the same in marine and freshwater sticklebacks. (Courtesy of Michael D. Shapiro.)

freshwater sticklebacks have lost this expression as a result of a change at the *Pitx1* locus. These changes do not lie in the coding sequence. Instead, each is a small deletion of a block of adjacent regulatory DNA that controls *Pitx1* expression specifically in the pelvic cells (**Figure 21–37**).

The Pitx1 protein has important functions elsewhere in the body, so that the DNA sequences that encode this protein must be retained. The regulatory DNA responsible for *Pitx1* expression at these other sites is also unchanged in the two populations of sticklebacks. The evolution of pelvis development in sticklebacks shows how the modular nature of regulatory DNA elements that we encountered in Chapter 7 (see Figure 7–29) allows independent modification of the different parts of the body, even when formation of those body parts depends on the same proteins.

In the recent evolution of plants, changes of body structure can be traced in a similar way to changes in regulatory DNA. For example, these account for a large part of the dramatic difference between the wild teosinte plant and its modern descendant, maize, through some 10,000 years of mutation and selection by Native Americans.

Summary

Drosophila has been the foremost model organism for the study of the genetics of animal development. Its embryonic pattern is initiated by the products of maternal-effect genes called egg-polarity genes, which operate by setting up graded distributions of transcription regulators in the egg and early embryo. The gradient of Bicoid protein along the A-P axis, for example, helps initiate the orderly expression of gap genes, pair-rule genes, and segment-polarity genes. These three classes of segmentation genes, through a hierarchy of interactions, become expressed in some regions of the embryo and not others, progressively subdividing the embryo along the A-P axis into a regular series of repeating modular units called segments.

Superimposed on the pattern of gene expression that repeats itself in every segment, there is a serial pattern of expression of Hox genes that confer on each segment a different identity. These genes are grouped in complexes and are arranged in a sequence that matches their sequence of expression along the A-P axis of the body.

Although Hox gene expression is initiated in the embryo, it is subsequently maintained by the action of chromatin-binding proteins of the Polycomb and Trithorax group, which stamp the chromatin of the Hox complex with a heritable record of its embryonic state of repression or activation, respectively. Hox complexes homologous to that of Drosophila are found in virtually every type of animal, where they help pattern the A-P axis of the body.

Signaling gradients are also set up along the dorsoventral (D-V) axis. Initially, Toll signaling generates a nuclear gradient of Dorsal protein, which induces an extracellular signaling gradient of the TGFβ-family protein Dpp and its antagonist,

Sog. This creates a gradient of Dpp activity that helps refine the assignment of different characters to cells at different positions along the D-V axis.

In Xenopus, *the polarity of the egg and the site of sperm entry set up the embryonic axes. A gradient generated by the TGFβ-family protein Nodal induces different fates along the animal-vegetal axis, whereas BMP and Chordin—proteins homologous to* Drosophila *Dpp and Sog, respectively—control the patterning of the D-V axis. This axis is inverted, so that dorsal in the fly corresponds to ventral in the frog.*

Transcription regulators control the formation of specific cell types. Members of the MyoD/myogenin family drive the process of muscle cell determination, coordinating the many components required, whereas Achaete/Scute transcription regulators control neural fate. Other genes encoding such master transcriptional regulators can regulate the formation of entire organs. Eyeless, for example, is both necessary and sufficient to generate eye structures in Drosophila.

To refine the anatomical pattern within such an organ, the cells interact locally, both by diffusible inductive signals and by short-range mechanisms. Often, the cells compete with one another by lateral inhibition. This process results in activation of the Notch signaling pathway in one cell and inhibition in its neighbors, generating two different cell types. Asymmetric cell divisions, in which daughter cells inherit different molecular determinants from the mother cell, provide an additional way to organize a fine-grained diversity of cell types.

Evidence from recent evolutionary events indicates that anatomical changes are mostly driven by changes in regulatory DNA sequences that determine when and where developmental genes are expressed. How the striking diversity in body structures has evolved over longer times remains largely unknown, although it seems likely that similar principles apply.

DEVELOPMENTAL TIMING

Developmental events unfold over minutes, hours, days, weeks, months, or even years, with each organism following its own strict timetable. The cascades of inductive interactions and transcriptional regulatory events described earlier take time, as signals are transmitted and transcription regulators are synthesized and then bind to DNA to activate or repress their target genes. At the beginning of this chapter, we compared development with an orchestral performance. There are many players, and each must do the right thing at the right time; yet there is no leader or conductor to set the tempo and coordinate the timing of all the different events. Each developmental process must thus occur at an appropriate rate, tuned by evolution to fit with the timing of other processes in the embryo or in the environment. The control of timing is one of the most important problems in developmental biology, but also one of the least understood.

Molecular Lifetimes Play a Critical Part in Developmental Timing

Developmental processes are complex, but they are built up from simple steps. A first challenge is to understand the timing of these steps. How long does it take, for example, to switch the expression of a gene on or off? This is not like throwing a light switch: it involves delays. First, it takes time to make an mRNA molecule: the RNA polymerase must travel the length of the gene, the primary RNA transcript must be spliced and otherwise processed, and the resulting mRNA must be exported from the nucleus and delivered to the site where it will be translated. This adds up to what one might call the *gestation time* of the individual molecule. Second, it takes time for the individual mRNA molecules to accumulate to their fully effective concentration; as explained in Chapter 15, this *accumulation time* is dictated by the average lifetime of the molecules—the longer they last, the higher their ultimate concentration, and the longer the time taken to attain it. Similar delays occur at the next step, where the mRNA is translated into protein: synthesis of each individual protein molecule involves a gestation delay, and attainment of an effective concentration of protein molecules involves an accumulation delay that depends on the protein's lifetime. The time for the whole gene switching process is just the sum of the gestation delays and the accumulation delays (basically, the molecular lifetimes) for both the mRNA and the protein molecules. Somewhat counterintuitively, it is the combined length of these delays, rather than the rate of molecular synthesis (the number of molecules synthesized per second), that chiefly determines the switching time.

The same additive principle applies to long cascades of gene switching, where gene A activates gene B, and gene B activates gene C, and so on. It also applies in other circumstances, such as in signaling pathways where one protein directly regulates the activation of the next. In all these cases, molecular lifetimes, along with gestation delays, play a key part in determining the pace of development. The lifetimes of mRNA and protein molecules are enormously variable, from a few minutes or hours to days or more, explaining much of the variation we see in the tempo of developmental events.

Gene switching delays, however, are not the be-all and end-all of developmental timing. Development involves many other kinds of delay that contribute to timing. Chromatin structure takes time to remodel. Inductive signals take time to diffuse across a field of cells (see Figure 21–9). Cells take time to move and rearrange themselves in space. Nevertheless, the timing of gene switching plays a fundamental part in developmental timing, as illustrated in an especially clear and striking way by a gene-expression oscillator that controls the segmentation of the vertebrate body axis, as we now explain.

A Gene-Expression Oscillator Acts as a Clock to Control Vertebrate Segmentation

The main body axis of all vertebrates has a repetitive, periodic structure, seen in the series of vertebrae, ribs, and segmental muscles of the neck, trunk, and tail. These segmental structures originate from the mesoderm that lies as a long slab on either side of the embryonic midline. This slab becomes broken up into a regular repetitive series of separate blocks, or **somites**—cohesive groups of cells, separated by clefts (**Figure 21–38**A). The somites form (as bilateral pairs) one after another, in a regular rhythm, starting in the region of the head and ending in the tail. Depending on the species, the final number of somites ranges from less than 40 (in a frog or a zebrafish) to more than 300 (in a snake).

The posterior, most immature part of the mesodermal slab, called the *presomitic mesoderm*, supplies the required cells: as the cells proliferate, this mesoderm

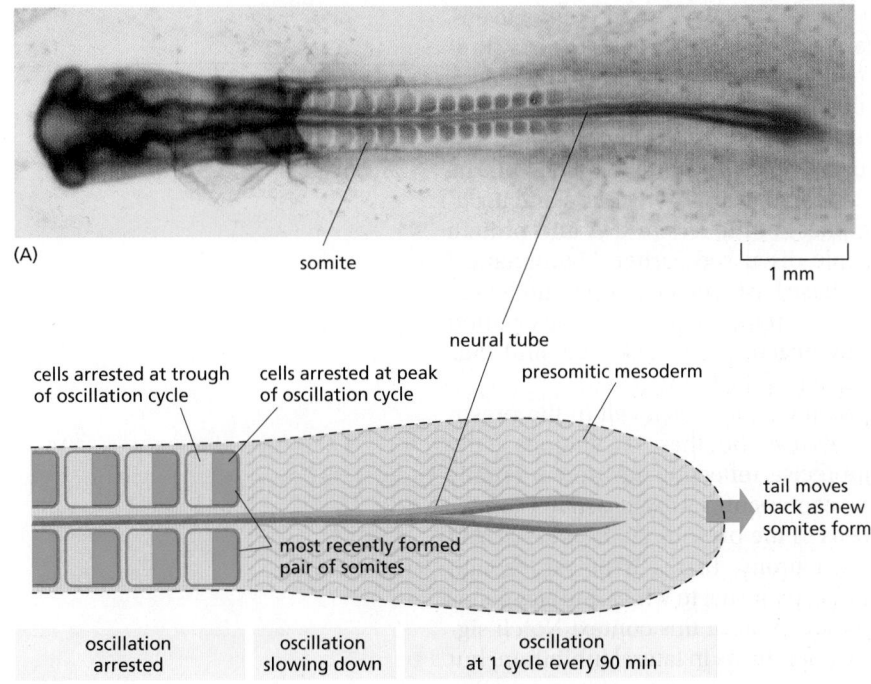

Figure 21–38 Somite formation in the chick embryo. (A) A chick embryo at 40 hours of incubation. (B) How the temporal oscillation of gene expression in the presomitic mesoderm becomes converted into a spatial alternating pattern of gene expression in the formed somites. In the posterior part of the presomitic mesoderm, each cell oscillates with a cycle time of 90 minutes. As cells mature and emerge from the presomitic region, their oscillation is gradually slowed down and finally brought to a halt, leaving them in a state that depends on the phase of the cycle they happen to be in at the critical moment. In this way, a temporal oscillation of gene expression traces out an alternating spatial pattern. (A, from Y.J. Jiang, L. Smithers and J. Lewis, *Curr. Biol.* 8:R868–R871, 1998. With permission from Elsevier.)

retreats tailward, extending the embryo (Figure 21–38B). In the process, it deposits a trail of somites formed from cells that group together into blocks as they emerge from the anterior end of the presomitic region. The special character of the pre-somitic mesoderm is maintained by a combination of fibroblast growth fac-tor (FGF) and Wnt signals, produced by a signaling center at the tail end of the embryo, and the range of these signals seems to define the length of the preso-mitic mesoderm. The somites emerge with clocklike timing, but what determines the rhythm of the process?

In the posterior part of the presomitic mesoderm, the expression of certain genes oscillates in time. Snapshots of gene expression taken by fixing embryos for analysis at different times in the oscillation cycle reveal what is happening, and the oscillations can now also be observed in time-lapse movies of embryos containing fluorescent reporters of individual oscillating genes. One new somite pair is formed in each oscillation cycle, and, in mutants where the oscillations fail to occur, somite segmentation is disrupted: the cells may still break up, belatedly, into separate clusters, but they do so in a haphazard, irregular way. The gene-ex-pression oscillator controlling regular segmentation is called the **segmentation clock**. The length of one complete oscillation cycle depends on the species: it is 30 minutes in a zebrafish, 90 minutes in a chick, 120 minutes in a mouse.

As cells emerge from the presomitic mesoderm to form somites—in other words, as they escape from the influence of the FGF and Wnt signals—their oscil-lation stops. Some become arrested in one state, some in another, according to the phase of the oscillation cycle at the time they leave the presomitic region. In this way, the temporal oscillation of gene expression in the presomitic mesoderm leaves its trace in a spatially periodic pattern of gene expression in the matur-ing mesoderm; this in turn dictates how the tissue will break up into physically separate blocks, through effects on the pattern of cell–cell adhesion (see Figure 21–38B).

How does the segmentation clock work? The first somite oscillator genes to be discovered were *Hes* genes, which are key components of the Notch signaling pathway. They are directly regulated by the activated form of Notch, and they code for inhibitory transcription regulators that inhibit the expression of other genes, including *Delta*. As well as regulating other genes, the products of *Hes* genes can directly regulate their own expression, creating a remarkably simple negative feedback loop. Autoregulation of certain specific *Hes* genes (depending on spe-cies) is thought to be the basic generator of the oscillations of the somite clock. Although the machinery has been modified in various ways in different species, the underlying principle seems to be conserved. When the key *Hes* gene is tran-scribed, the amount of Hes protein product builds up until it is sufficient to block *Hes* gene transcription; synthesis of the protein ceases; the protein then decays, permitting transcription to begin again; and so on, cyclically (**Figure 21–39**). The period of oscillation, which determines the size of each somite, depends on the delay in the feedback loop. This equals the sum of the gestation delays and accu-mulation delays (that is, the molecular lifetimes) of the *Hes* mRNA and protein molecules, according to the additive principle discussed earlier. Mathematical modeling (see Chapter 8) allows us to relate these basic molecular parameters to the cycle time of the segmentation clock: to a first approximation, the cycle period is simply equal to twice the total delay in the negative feedback loop, and thus twice the sum of the delays occurring at each step of the loop.

The feedback loop just described is intracellular, and each cell in the preso-mitic mesoderm can generate oscillations on its own. But these oscillations at the single-cell level are somewhat erratic and imprecise, reflecting the fundamentally noisy, stochastic nature of the control of gene expression, as discussed in Chapter 7. A mechanism is needed to keep all the cells in the presomitic mesoderm that will form a particular somite oscillating in synchrony. This is achieved through cell–cell communication via the Notch signaling pathway, to which the *Hes* genes are coupled. The gene regulatory circuitry is such that in this context Notch sig-naling does not drive neighboring cells to be different, as in lateral inhibition, but does just the opposite: it keeps them in unison. In mutants where Notch signaling

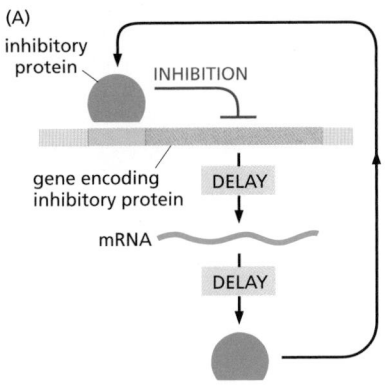

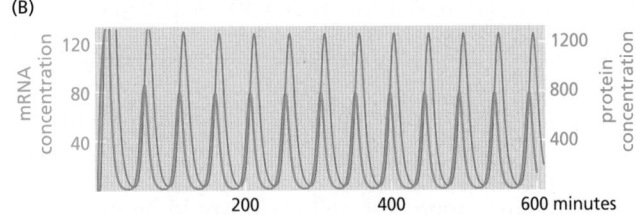

Figure 21–39 Delayed negative feedback giving rise to oscillating gene expression.
(A) A single gene, coding for a transcription regulator that inhibits its own expression, can behave as an oscillator. For oscillation to occur, there must be a delay (or several delays) in the feedback circuit, and the lifetimes of the mRNA and protein (which contribute to the delay) must be short compared with the total delay. The total delay determines the period of oscillation. It is thought that a feedback circuit like this, based on a pair of redundantly acting genes called *Her1* and *Her7* in the zebrafish—or their counterpart, *Hes7*, in the mouse—is the pacemaker of the segmentation clock governing somite formation. (B) The predicted oscillation of *Her1* and *Her7* mRNA and protein, computed using rough estimates of the feedback circuit parameters appropriate to this gene in the zebrafish. Concentrations are measured as numbers of molecules per cell. The predicted period is close to the observed period, which is 30 minutes per somite in the zebrafish (depending on temperature).

fails, including mutants defective in Delta or Notch itself, the cells drift out of synchrony and somite segmentation is again disrupted. This leads to gross deformity of the vertebral column—an extraordinary display of the consequences of the noisy temporal control of gene expression at the single-cell level, writ large in the structure of the vertebrate body as a whole.

Intracellular Developmental Programs Can Help Determine the Time-Course of a Cell's Development

Although signaling between cells plays an essential part in driving the progress of development, this does not mean that cells always need signals from other cells to prod them into changing their character as development proceeds. Some of these changes are intrinsic to the cell (like the ticking of the segmentation clock) and depend on *intracellular developmental programs* that can operate even when the cell is removed from its normal environment.

The best-understood example is in the development of neural precursor cells, or neuroblasts, in the embryonic *Drosophila* central nervous system. These cells, as we saw, are initially singled out from the neurogenic ectoderm of the embryo by a typical lateral-inhibition mechanism that depends on Notch, and they then proceed through an entirely predictable series of asymmetric cell divisions to generate ganglion mother cells that divide to form neurons and glial cells (see Figure 21–36). The neuroblast changes its internal state as it goes through its set program of divisions, generating different cell types with a reproducible sequence and timing. These successive changes in neuroblast specification occur through the sequential expression of specific transcription regulators. For example, most embryonic neuroblasts sequentially express the transcription regulators Hunchback, Krüppel, Pdm, and Cas in a fixed order (**Figure 21–40**). When a neuroblast divides, the set of transcription regulators expressed at that time is inherited by the ganglion mother cell and its neural progeny; thus, the differentiated neural cells are endowed with different characters according to their time of birth.

Remarkably, when neuroblasts are taken from an embryo and maintained in culture, isolated from their normal surroundings, they step through much the same stereotyped developmental program as if they had been left in the embryo. Moreover, many of the neuroblast transitions occur even when cell division is blocked. The neuroblasts seem to have a built-in timer that determines when each of the transcription regulators is expressed, and this timer can continue to run in the absence of cell division. The molecular basis of the timing is largely unknown; in part, at least, it must depend on the time taken for gene switching, as described above; but it may well also depend on slow progressive changes in chromatin structure. These too can serve to measure the passage of time in the embryo.

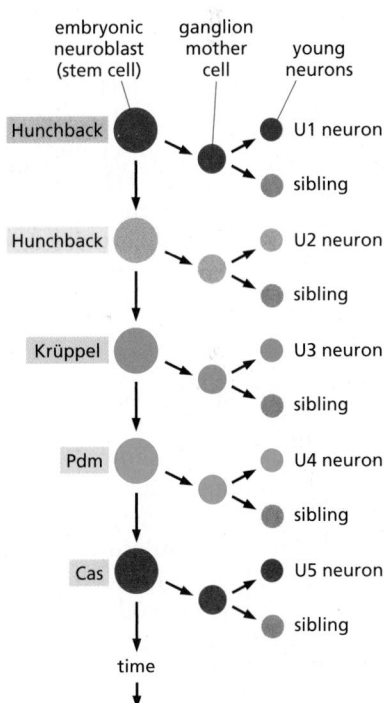

Figure 21–40 Temporal patterning of neuroblast fate in *Drosophila*. Hunchback, Krüppel, Pdm, and Cas are transcription regulators that are expressed consecutively in the cell lineage of neuroblasts during development of the *Drosophila* nervous system. At successive time steps, correlated with cell division, the neuroblast switches its pattern of gene expression. Each neuroblast division produces one daughter that remains a neuroblast and expresses the updated set of genes, and one ganglion mother cell that maintains the expression of this gene set and differentiates into specific cell types accordingly. (After B.J. Pearson and C.Q. Doe, *Nature* 425:624–628, 2003.)

Cells Rarely Count Cell Divisions to Time Their Development

Many specialized cells in animals develop from proliferating progenitor cells that stop dividing and terminally differentiate after a limited number of cell divisions. In these cases, differentiation is coordinated with withdrawal from the cell cycle, but it is usually not known how the coordination is achieved. It has often been suggested that the cell-division cycle might serve as an intracellular timer to control the timing of cell differentiation. The cell cycle would be the ticking clock that sets the tempo of other developmental processes, with maturational changes in gene expression being dependent on cell-cycle progression. Most of the evidence, however, indicates that this tempting idea is wrong. Although there are examples where cells change their maturation state with each division and the change depends on cell division, this is not the general rule. As we just saw for neuroblasts in the *Drosophila* embryo, cells in developing animals often carry on with their normal timetable of maturation and differentiation even when cell division is artificially blocked; necessarily, some abnormalities occur, if only because a single undivided cell cannot differentiate in two ways at once. But it seems that most developing cells can change their state without a requirement for cell division. Developmental control genes can switch the cell-division-cycle machinery on or off, and it is the dynamics of these genes, rather than the cell cycle, that sets the tempo of development.

MicroRNAs Often Regulate Developmental Transitions

Genetic screens are useful for tracking down the genes involved in almost any biological process, and they have been used to search for mutations that alter developmental timing. Such screens were performed in the nematode *Caenorhabditis elegans* (**Figure 21–41**). This worm is small, relatively simple, and precisely structured. The anatomy of its development is highly predictable and has been described in extraordinary detail, so that one can map out the exact lineage of every cell in the body and see exactly how the developmental program is altered in a mutant. Genetic screens in *C. elegans* revealed mutations that disrupt developmental timing in a particularly striking way: in these so-called **heterochronic** mutants, certain cells in a larva at one stage of development behave as though they were in a larva at a different stage of development, or cells in the adult carry on dividing as though they belonged to a larva (**Figure 21–42**).

Genetic analyses showed that the products of the heterochronic genes act in series, forming regulatory cascades. Unexpectedly, two genes at the top of their respective cascades, called *Lin4* and *Let7*, were found to code not for protein but instead for **microRNAs (miRNAs)**—short, untranslated, regulatory RNA molecules, 21 or 22 nucleotides long. These act by binding to complementary sequences in the noncoding regions of mRNA molecules transcribed from other heterochronic genes, thereby repressing their translation and promoting their degradation, as discussed in Chapter 7. Increasing levels of *Lin4* miRNA govern the progression from first-stage larva cell behaviors to third-stage larva cell behaviors.

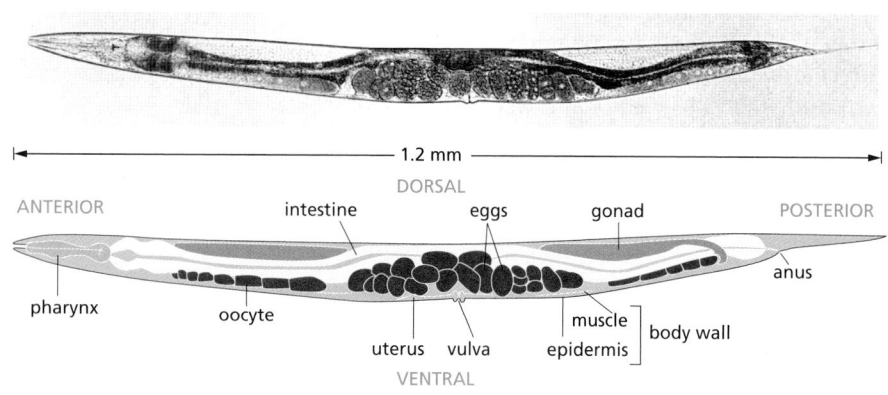

Figure 21–41 *Caenorhabditis elegans.* A side view of an adult hermaphrodite is shown. (From J.E. Sulston and H.R. Horvitz, *Dev. Biol.* 56:110–156, 1977. With permission from Academic Press.)

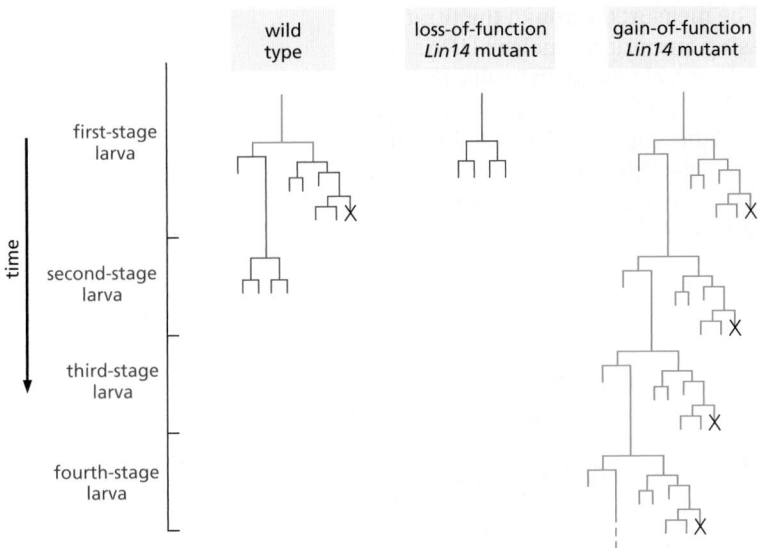

Figure 21–42 **Heterochronic mutations in the *Lin14* gene of *C. elegans*.** Only the effects on one of the many altered lineages are shown. A loss-of-function (recessive) mutation in *Lin14* causes premature occurrence of the pattern of cell division and differentiation characteristic of a late larva, so that the animal reaches its final state prematurely and with an abnormally small number of cells. The gain-of-function (dominant) mutation has the opposite effect, causing cells to reiterate patterns of cell divisions characteristic of the first larval stage, continuing through as many as five or six molt cycles. The cross denotes a programmed cell death. *Green lines* represent cells that contain Lin14 protein (which binds to DNA), *red lines* those that do not. (Adapted from V. Ambros and H.R. Horvitz, *Science* 226:409–416, 1984; and P. Arasu, B. Wightman and G. Ruvkun, *Genes Dev.* 5:1825–1833, 1991.)

Increasing levels of *Let7* miRNA govern the progression from late larva to adult. In fact, *Lin4* and *Let7* were the first miRNAs to be described in any animal: it was through developmental genetic studies in *C. elegans* that the importance of this whole class of molecules for gene regulation in animals was discovered.

More generally, in many animals, miRNAs help regulate the transitions between different stages of development. For example, in flies, fish, and frogs, the maternal mRNAs that are loaded into the egg in the mother are removed during early development when the genome of the embryo begins to be transcribed; at this stage, the embryo begins to express specific miRNAs that target many maternal mRNAs for translational repression and degradation.

Thus, miRNAs can sharpen developmental transitions by blocking and removing mRNAs that define an earlier developmental stage. But how is the timing of miRNA expression itself controlled? In the case of the miRNAs that disable maternal mRNAs in frogs and fish, expression is activated at the end of the series of rapid, synchronous divisions that cleave the fertilized egg into many smaller cells. As the division rate of these blastomeres slows, widespread transcription of the embryo's genome begins (**Figure 21–43**). This event, where the embryo's own genome largely takes over control of development from maternal macromolecules, is called the **maternal-zygotic transition (MZT)**, and it occurs with roughly similar timing in most animal species, with the exception of mammals.

One trigger for the MZT appears to be the nuclear-to-cytoplasmic ratio. During cleavage, the total amount of cytoplasm in the embryo remains constant, but the number of cell nuclei increases exponentially. As a critical threshold is reached in the ratio of cytoplasm to DNA, the cell cycles lengthen and transcription is initiated. Thus, haploid embryos undergo the MZT one cell cycle later than diploid embryos, which contain twice as much DNA per cell. According to one

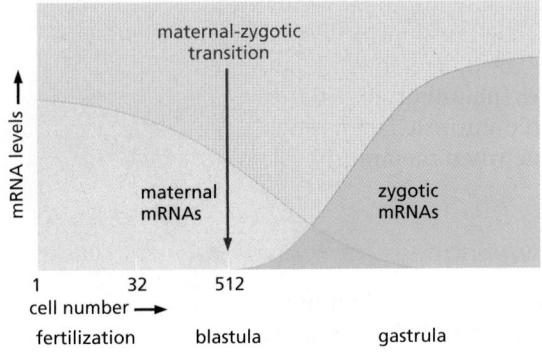

Figure 21–43 **The maternal-zygotic transition in a zebrafish embryo.** Maternal mRNAs are deposited by the mother into the egg and drive early development. These mRNAs are degraded during different stages of embryogenesis, including blastula and gastrula stages, but a relatively abrupt change occurs at the maternal-zygotic transition (MZT). Before this, the embryonic (zygotic) genome is transcriptionally inactive; afterward, zygotic genes start to be transcribed. In zebrafish embryos, the zygotic genome begins to be activated at the 512-cell stage.

model, the nuclear-to-cytoplasmic ratio might be measured through the titration of a transcription repressor against the increasing amount of nuclear DNA. The total amount of repressor would stay constant during cleavage divisions, but the amount of repressor per genome would decrease, falling by a half with each round of DNA synthesis, until loss of repression allowed the zygotic genome to become transcriptionally active. The newly synthesized transcripts include the miRNAs that recognize many of the transcripts deposited in the egg by the mother, directing their translational repression and rapid degradation.

Hormonal Signals Coordinate the Timing of Developmental Transitions

We have so far emphasized timing mechanisms that operate locally and separately in the different parts of the embryo, or in specific subsystems of the molecular control machinery. Evolution has tuned each of these largely independent processes to run at an appropriate rate, matched to the needs of the organism as a whole. For some purposes, however, this is not enough: a global coordinating signal is required. This is especially true where changes have to occur throughout the body in response to a cue that depends on the environment. For example, when an insect or amphibian undergoes *metamorphosis*—the transition from larva to adult—almost every part of the body is transformed. The timing of metamorphosis depends on external factors such as the supply of food, which determines when the animal reaches an appropriate size. All the bodily changes have to be triggered together at the right time, even though they are occurring in widely separated sites. The coordination in such cases is provided by **hormones**—signal molecules that spread throughout the body.

The metamorphosis of amphibians provides a spectacular example. During this developmental transition, amphibians switch from an aquatic to a terrestrial life. Larva-specific organs such as gills and tail disappear, and adult-specific organs such as legs form. This dramatic transformation is triggered by thyroid hormone, produced in the thyroid gland. If the gland is removed or if thyroid hormone action is blocked, metamorphosis does not occur, although growth continues, producing a giant tadpole. Conversely, a dose of thyroid hormone given to a tadpole by an experimenter can trigger metamorphosis prematurely.

The thyroid hormone is distributed through the vascular system and induces changes throughout the animal by binding to intracellular nuclear hormone receptors, which regulate hundreds of genes. This does not mean, however, that target tissues all respond in the same way to the hormone: organs differ not only in their levels of thyroid hormone receptors and levels of extracellular proteins that locally regulate the amount of active hormone, but also in the sets of genes that respond. Thyroid hormone induces muscle in the limbs to grow and muscle in the tail to die. The timing of the responses also differs: for example, the legs form early in response to a very low concentration of circulating hormone, but it requires a high level of the hormone to induce resorption of the tail.

A surge of thyroid hormone triggers metamorphosis, but how is the timing of the surge controlled? One mechanism depends on coupling hormone synthesis to the size of the thyroid gland, which reflects the size of the tadpole. Only when the gland attains a certain size does it produce enough thyroid hormone to initiate metamorphosis. However, environmental cues other than nutrition also play a part: conditions such as temperature and light are sensed by the nervous system, which regulates the secretion of another tier of hormones (neurohormones) that stimulate the secretion of thyroid hormone. Thus, tadpole-intrinsic factors such as size combine with environmental factors to determine when metamorphosis begins.

Environmental Cues Determine the Time of Flowering

Another striking example of environmentally controlled developmental timing is the flowering of plants. Flowering involves a transformation of the behavior of the cells at the growing apex of the plant shoot—the *apical meristem*. During ordinary vegetative growth, these cells behave as stem cells, generating a steady succession of new leaves and new segments of stalk. In flowering, the meristem cells switch to making the components of a flower, with its sepals and petals, its stamens carrying pollen, and its ovary containing the female gametes.

To time the switch correctly, the plant has to take account of both past and present conditions. One important cue, for many plants, is day length. To sense this, the plant uses its circadian clock—an endogenous 24-hour rhythm of gene expression—to generate a signal for flowering only when there is light for the appropriate part of the day. The clock itself is influenced by light, and the plant in effect uses the clock to compare past to present lighting conditions. Important parts of the genetic circuitry underlying these phenomena have been identified, including the phytochromes and cryptochromes that act as light receptors (discussed in Chapter 15). The flowering signal that is carried from the leaves to the stem cells via the vasculature depends on the product of *Flowering locus T (Ft)*.

But this signal will trigger flowering only if the plant is in a receptive condition from prior long-term cold exposure. Many plants need winter before they will flower—a process called *vernalization*. Cold over a period of weeks or months progressively reduces the level of expression of a remarkable gene called *Flowering locus C (Flc)*. *Flc* encodes a transcriptional repressor that suppresses expression of the *Ft* flowering promoter.

How does vernalization shut down *Flc* so as to lift the block to flowering? The effect involves a noncoding RNA called *Coolair* that overlaps with the *Flc* gene and is produced when the temperature is low (**Figure 21–44**). Together with cold-induced chromatin modifiers, including Polycomb-group proteins, *Coolair* coordinates the switching of *Flc* chromatin to a silent state (discussed in Chapters 4 and 7). The degree of silencing depends on the length of cold exposure enabling the plants to distinguish the odd chilly night from the whole of winter.

The effect on the chromatin is long lasting, persisting through many rounds of cell division even as the weather grows warmer. Thus vernalization creates a persistent block in production of Flc, enabling the Ft signal to be generated when day length is sufficiently long.

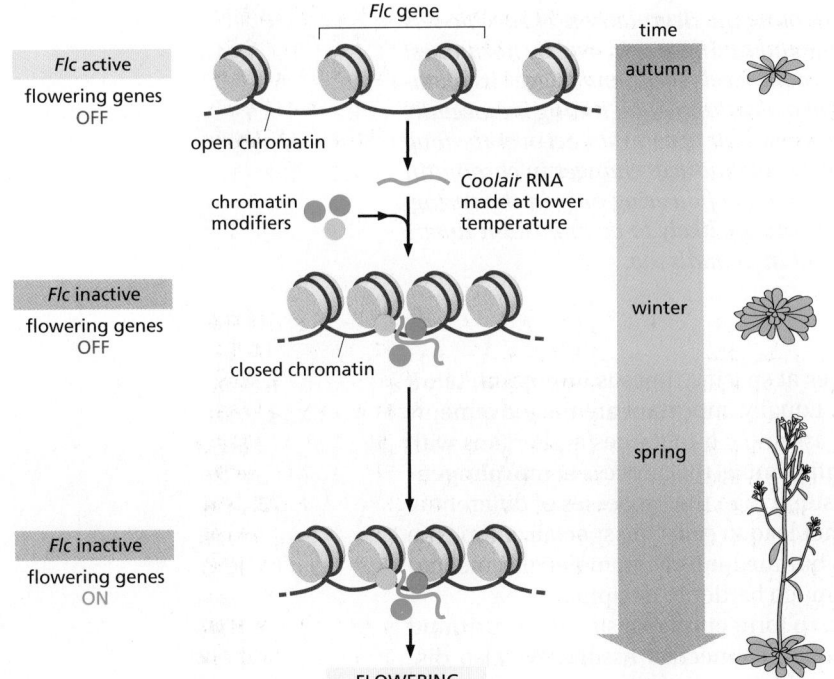

Figure 21–44 Temporal control of flowering in *Arabidopsis*. The *Flc* gene is active and blocks flowering when plants have been grown without exposure to winterlike temperatures. Exposure to a prolonged period of cold leads to the production of the noncoding RNA *Coolair*, which overlaps with the *Flc* gene. Coolair induces long-term chromatin changes that turn off *Flc*. These changes persist after the end of the cold period and allow the plant to flower when other environmental conditions are favorable for flowering.

Mutations affecting the regulation of *Flc* expression alter the time of flowering and thus the ability of a plant to flourish in a given climate. The whole control system governing the switch to flowering is thus of vital importance for agriculture, especially in an era of rapid climate change.

The example of vernalization suggests a general point about the role of chromatin modification in developmental timing. The plant uses changes in chromatin to record its experience of prolonged cold. It may be that in other organisms—animals as well as plants—slow, progressive changes in chromatin structure provide long-term timers for those mysterious developmental processes that unfold slowly, over a period of days, weeks, months, or years. Such chromatin timers may be among the most important clocks in the embryo, but as yet we understand very little about them.

Summary

Developmental timing is controlled at many levels. It takes time to switch a gene on or off, and this time delay depends on the lifetimes of the molecules involved, which can vary widely. Cascades of gene regulation involve cascades of delays. Feedback loops can give rise to temporal oscillations in gene expression, and these may serve to generate spatially periodic structures. During vertebrate segmentation, for example, expression of the Hes *genes oscillates, and one new pair of somites is formed during each oscillation cycle.* Hes *genes encode transcription repressor proteins that can act back on expression of the* Hes *genes themselves. This negative feedback generates oscillations with a period that reflects the delay in the autoregulatory gene switching loop. The period of oscillation of this "segmentation clock" controls the sizes of the somites. Notch signaling between neighboring cells synchronizes their oscillations: when Notch signaling fails, the cells drift out of synchrony because of genetic noise in their individual clocks, and the segmental organization of the vertebral column is disrupted.*

Timing does not always depend on cell–cell interactions; many developing animal cells have intrinsic developmental programs that play out even in isolated cells in culture. Neuroblasts in Drosophila *embryos, for example, go through set programs of asymmetric divisions, generating different neural cell types at each division with a predictable sequence and timing, through a cascade of gene switching events. Studies in both vertebrates and invertebrates show that such programs are rarely governed by the timing of cell division and can unfold even when cell division is blocked. MicroRNAs produced at critical moments sharpen developmental transitions by blocking the translation and promoting the degradation of specific sets of mRNAs. Global coordination of developmental timing is achieved by hormones: as a tadpole grows, for example, thyroid hormone levels surge and trigger its metamorphosis into a frog. Environmental control of developmental timing is especially striking in plants and reveals the presence of molecular timers that act over the long term. In vernalization, for example, prolonged cold induces changes in chromatin that chart the passage through winter so as to allow flowering only in the spring. Slow, progressive changes in chromatin structure are likely to be important timers in the long-term programming of development in animals too.*

MORPHOGENESIS

The specialization of cells into distinct types at specific times is important, but it is only one aspect of animal development. Equally important are the movements and deformations that cells go through to assemble into tissues and organs with specific shapes and sizes. Like developmental timing, this process of **morphogenesis** ("form generation") is less well understood than the processes of differential gene expression and inductive signaling that lead to cell-type specialization. The cell movements can be readily described, but the underlying molecular mechanisms that coordinate the movements are much harder to decipher.

In Chapter 19, we saw how cells cohere to form epithelial sheets or surround themselves with extracellular matrix to create connective tissues. We also discussed how the basic features of tissues, such as the polarity of epithelia, arise from the properties of individual cells. In this section, we consider how the rearrangements of cells during animal development give shape to the embryo and to all the individual organs and appendages of the body.

A small number of cell processes are basic to morphogenesis. Individual cells can migrate through the embryo along defined tracks. They can crawl over one another in a coordinated way to elongate, constrict, or thicken a tissue. They can segregate from their neighbors and form physically separate groups. They can change their shape so as to deform an epithelial sheet into a tube or a vesicle. By stretching out while holding on to their companions, specialized sets of cells can form growing tubular networks such as the system of blood or lymph vessels. Mass migrations, as occur in gastrulation, can transform the entire topology of the embryo. Underlying all these processes are changes in cell shape and changes in cell contacts—either with other cells or with extracellular matrix. We begin by considering the migration of individual cells.

Cell Migration Is Guided by Cues in the Cell's Environment

The birthplace of cells is often far from their ultimate location in the body. Our skeletal muscles, for example, derive from muscle cell precursors, or *myoblasts*, in somites, from which they migrate into the limbs and other regions. The routes that the migrant cells follow and the selection of sites that they colonize determine the eventual pattern of muscles in the body. The embryonic connective tissues form the framework through which the myoblasts travel, and these tissues provide the cues that guide myoblast distribution. No matter which somite they come from, the myoblasts that migrate into a forelimb bud will form the pattern of muscles appropriate to a forelimb, and those that migrate into a hindlimb bud will form the pattern appropriate to a hindlimb. It is the connective tissue that provides the patterning information.

As a migrant cell travels through the embryonic tissues, it repeatedly extends surface projections that probe its immediate surroundings, testing for cues to which it is particularly sensitive by virtue of its specific assortment of cell-surface receptor proteins. Inside the cell, these receptors are connected to the cortical actin and myosin cytoskeleton, which moves the cell along. Some extracellular matrix molecules, such as the protein fibronectin, provide adhesive sites that help the cell advance; others, such as chondroitin sulfate proteoglycan, inhibit locomotion and repel immigration. The nonmigrant cells along the migration pathway may likewise have inviting or repellent macromolecules on their surface; some may even extend filopodia to make their presence known.

Among the many guiding influences, a few stand out as especially important. In particular, many types of migrating cells are guided by chemotaxis that depends on a G-protein-coupled receptor (called CXCR4), which is activated by an extracellular ligand called CXCL12. Cells expressing this receptor can snuffle their way along tracks marked out by CXCL12 (**Figure 21–45**). Chemotaxis toward sources of CXCL12 plays a major part in guiding the migrations of lymphocytes

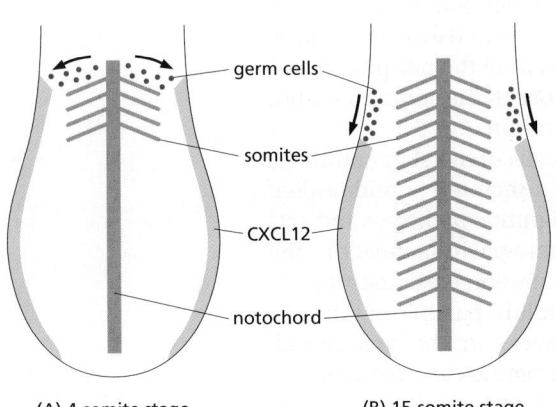

(A) 4-somite stage   (B) 15-somite stage

Figure 21–45 CXCL12 guides migrating germ cells. Zebrafish germ cells migrate to domains that express CXCL12. As the sites of CXCL12 expression change, cells follow the CXCL12 track and are guided to the region where the gonad develops at a later developmental stage. (A) At the 4-somite stage, germ cells move from a position that is close to the midline to more lateral regions where CXCL12 is expressed. (B) As the CXCL12 expression retracts, germ cells are guided to more posterior positions.

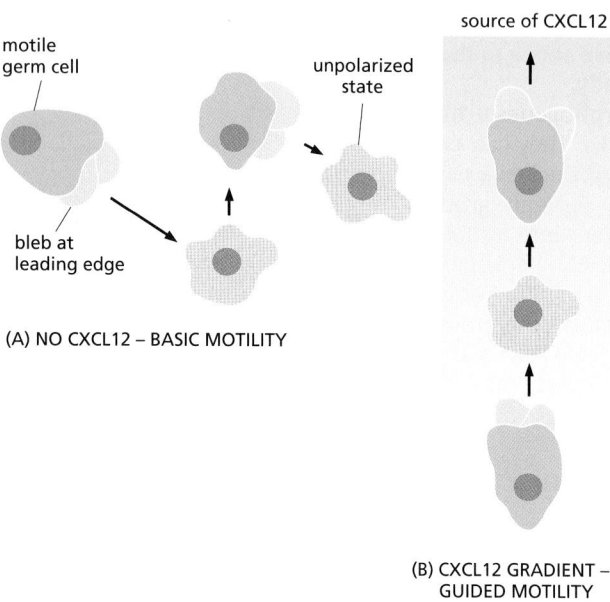

(A) NO CXCL12 – BASIC MOTILITY

(B) CXCL12 GRADIENT – GUIDED MOTILITY

Figure 21–46 Directional migration by local blebbing. Germ cells migrate via protrusions that define the leading edge of the cell. The persistence and site of the protrusions are biased toward higher levels of CXCL12. Thus, germ cells migrate up the CXCL12 gradient.

and various other white blood cells; of neurons in the developing brain; of myoblasts entering limb buds; of primordial germ cells as they travel toward the developing gonads; and of cancer cells when they metastasize.

Detailed studies of primordial-germ-cell migration have shown that CXCL12 signaling does not induce cell migration per se but rather serves to control its direction. In the absence of CXCL12 signaling, germ cells still display the membrane blebbing associated with cell migration, but the position of the cell front where blebs form is randomly chosen (Figure 21–46); if CXCL12 signaling is intact, blebbing is more frequent on the side of the cell that faces the source of CXCL12, resulting in directional migration.

The Distribution of Migrant Cells Depends on Survival Factors

The final distribution of migrant cells depends not only on the routes they take, but also on whether they survive the journey and thrive in the environment they find at the journey's end. Specific sites provide survival factors needed for specific types of migrant cells to survive.

Among the most important sets of migrant cells in the vertebrate embryo are those of the **neural crest**. They arise from the border region between the part of the ectoderm that will form epidermis and the part that will form the central nervous system. As the neural ectoderm rolls up to form the neural tube, the neural crest cells break loose from the epithelial sheet along this border region and set out on their long migrations (see Figure 19–8 and Movie 21.5). They settle ultimately in many sites and give rise to a surprising diversity of cell types. Some lodge in the skin and specialize as pigment cells; still others form skeletal tissue in the face. Still others will differentiate into the neurons and glial cells of the peripheral nervous system—not only in the sensory ganglia that lie close to the spinal cord, but also, following a much longer migration, in the wall of the gut.

The neural crest cells that give rise to the pigment cells of the skin and those that develop into the nerve cells of the gut depend on a secreted peptide called endothelin-3, which is produced by tissues along the migration pathways and acts as a survival factor for the migrating crest cells. In mutants with a defect in the gene for endothelin-3 or its receptor, many of these migrating crest cells die. As a result, the mutant individuals have nonpigmented (albino) patches of skin and a deficit of nerve cells in the intestine, especially its lower end, the large bowel, which becomes abnormally distended for lack of proper neural control—a potentially lethal condition called megacolon.

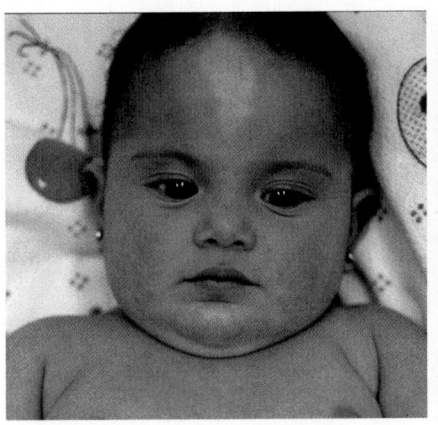

Figure 21–47 **Effect of mutations in the** *Kit* **gene.** Both the baby and the mouse are heterozygous for a loss-of-function mutation that leaves them with only half the normal quantity of *Kit* gene product. In both cases, pigmentation is defective because pigment cells depend on the gene product as a receptor for a survival factor. (Courtesy of R.A. Fleischman, from R.A. Fleischman et al., *Proc. Natl Acad. Sci. USA* 88:10885–10889, 1991.)

Another important survival signal for many types of migratory cells, including primordial germ cells, blood cell precursors, and neural-crest-derived pigment cells, depends on a receptor tyrosine kinase called *Kit*. This is expressed on the surface of the migrant cells, and a protein ligand, called Steel factor, is produced by the cells of the tissue through which the cells migrate and/or in which they come to settle. Individuals with mutations in the genes for either of these proteins have deficits in pigmentation, blood cells, and germ cells (**Figure 21–47**).

Changing Patterns of Cell Adhesion Molecules Force Cells Into New Arrangements

Patterns of gene expression govern embryonic cell movements in many ways. They regulate cell motility, cell shape, and the production of proteins that guide migration. Importantly, they also determine the sets of adhesion molecules that the cells display on their surface. Through changes in its surface molecules, a cell can break old attachments and make new ones. Cells in one region may develop surface properties that make them cohere with one another and become segregated from a neighboring group of cells with different surface chemistry.

Experiments done half a century ago on early amphibian embryos showed that the effects of selective cell–cell adhesion can be so powerful that they can bring about an approximate reconstruction of the normal structure of an early postgastrulation embryo after the cells have been artificially dissociated and mixed up. When these cells are reaggregated into a random mixture, the cells spontaneously sort themselves out according to their original germ-layer origins (**Figure 21–48**). As discussed in Chapter 19, cadherin proteins have a central role in the sorting process (see Figure 19–9). Cadherins belong to a large and varied family of Ca^{2+}-dependent cell–cell adhesion proteins, and they and other cell–cell adhesion proteins are differentially expressed in the various tissues of the early embryo. Antibodies against these proteins interfere with the normal selective adhesion between cells of a similar type.

Changes in the patterns of expression of the various cadherins correlate closely with the changing patterns of association among cells during various developmental processes, including gastrulation, neural tube formation, and somite formation. These cell rearrangements are likely to be regulated and driven in part by

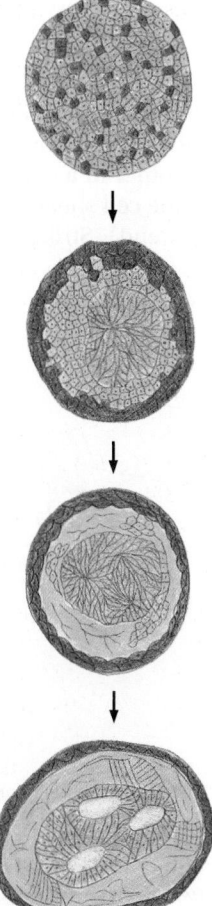

Figure 21–48 **Sorting out by adhesion.** Cells from different parts of an early amphibian embryo will sort out according to their origins. In the classical experiment shown here, mesoderm cells *(green)*, neural plate cells *(blue)*, and epidermal cells *(red)* have been disaggregated and then reaggregated in a random mixture. They sort out into an arrangement reminiscent of a normal embryo, with a "neural tube" internally, epidermis externally, and mesoderm in between. (Modified from P.L. Townes and J. Holtfreter, *J. Exp. Zool.* 128:53–120, 1955. With permission from Wiley-Liss.)

the cadherin pattern. In particular, cadherins appear to have a major role in controlling the formation and dissolution of epithelial sheets and clusters of cells (see Movie 19.1). They not only glue one cell to another but also provide anchorage for intracellular actin filaments at the sites of cell–cell adhesion. In this way, the pattern of stresses and movements in the developing tissue is regulated according to the pattern of cell adhesions.

Repulsive Interactions Help Maintain Tissue Boundaries

The different types of cadherins enable different types of cells to cohere selectively: cells expressing one type of cadherin will maximize their contact with cells expressing the same cadherin and thereby segregate from other cells, creating specific tissue boundaries. Cell mixing can be inhibited and boundaries created and maintained in another way as well: cells of different types can sometimes actively repel one another. The bidirectional activation of Eph receptors and ephrins discussed in Chapter 15 often mediates such repulsion, acting at interfaces between different groups of cells to keep the groups from mixing, and repelling invasion by inappropriate visitors. Ephrin–Eph signaling operates, for example, at the boundaries of the rhombomeres discussed earlier. Neighboring rhombomeres express complementary combinations of ephrins and Eph receptors, and this keeps the cells in adjacent rhombomeres strictly segregated, with a boundary between them that is sharply defined (**Figure 21–49**).

Groups of Similar Cells Can Perform Dramatic Collective Rearrangements

Cadherin-mediated cell sorting and ephrin–Eph-mediated repulsion exemplify how differences in cell-surface properties can drive tissue arrangements, causing cells that express different sets of genes to separate from one another. However, groups of cells that are all similar can also undergo dramatic rearrangements. During frog gastrulation, for example, cells in one region of the surface epithelium invaginate and migrate as a sheet into the interior of the embryo and converge toward the embryonic midline. The movement is driven mainly by an active rearrangement of the migrating cells, called **convergent extension**. Here the cells crawl over one another in a coordinated way, displacing their neighbors as they migrate, causing the cell sheet to narrow along one axis (converge) and elongate along another (extend). Strikingly, small, square fragments of tissue from the

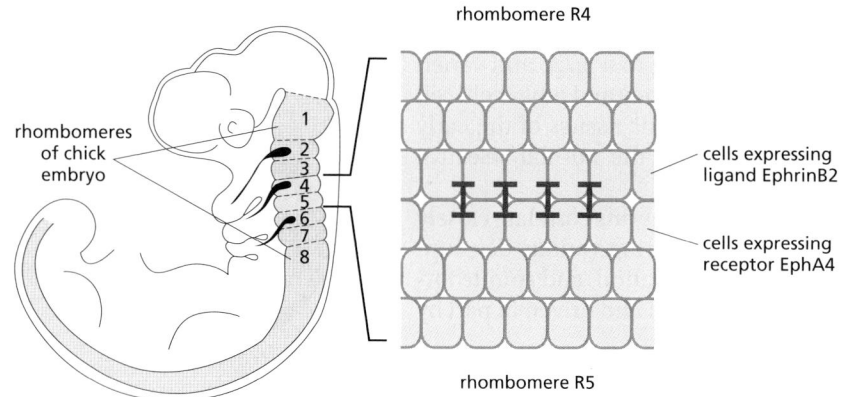

Figure 21–49 Sorting out by repulsion. Ephrin–Eph signaling in hindbrain segmentation in a chick embryo. Each pair of rhombomeres (segments in the hindbrain) is associated with a branchial arch (a modified gill rudiment) to which it sends innervation. Rhombomeres are distinguished from one another by expression of different *Hox* genes (see Figure 21–32). Mutual repulsion *(red bars)* between cells that express EphrinB2 in rhombomere 4 and EphA4 in rhombomere 5 creates a sharp boundary.

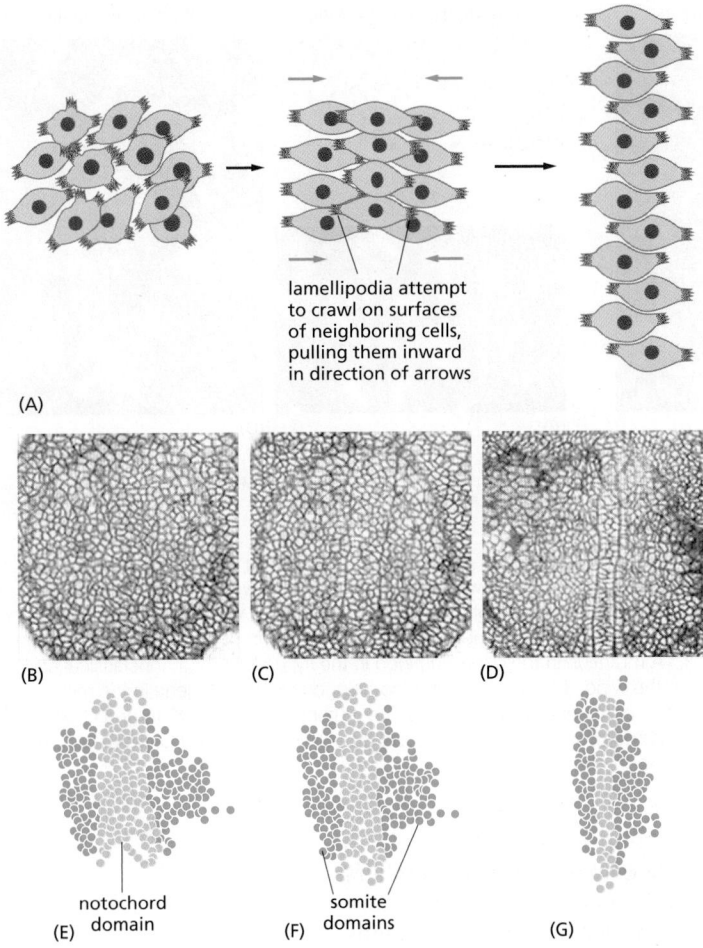

(A)

lamellipodia attempt
to crawl on surfaces
of neighboring cells,
pulling them inward
in direction of arrows

(B)

(C)

(D)

notochord
(E) domain

somite
(F) domains

(G)

Figure 21–50 Convergent extension and its cellular basis. (A) Schematic diagram of cell behaviors that underlie convergent extension. The cells form lamellipodia, with which they attempt to crawl over one another. Alignment of the lamellipodial movements along a common axis leads to convergent extension. The process depends on the Wnt–Frizzled/ planar-cell-polarity signaling pathway and is cooperative, presumably because cells that are already aligned exert forces that tend to align their neighbors in the same way.

(B–G) The pattern of convergent extension of dorsal mesoderm during zebrafish gastrulation at 8.8 (B, E), 9.3 (C, F), and 11.3 (D, G) hours after fertilization. Cells that will give rise to the notochord are labeled in *green*, and cells that will give rise to somites and muscle are labeled in *blue*. The notochord and somite domains are spatially separate from the start of the recording (B, E), but their boundaries are at first barely visible and only a little later become obvious. Convergence narrows the notochord domain to a width of about two cells at the last time point (D, G). (A, after J. Shih and R. Keller, *Development* 116:901–914, 1992; B–G, after N.S. Glickman et al., *Development* 130:873–887, 2003. With permission from The Company of Biologists.)

appropriate region of the embryo, isolated in culture, will spontaneously narrow and elongate, just as they would in the embryo (**Figure 21–50**). The alignment of the cell movements depends on the same signaling pathway that is involved in generating *planar cell polarity* within developing epithelia, as we discuss next.

Planar Cell Polarity Helps Orient Cell Structure and Movement in Developing Epithelia

Cells within an epithelium always have an apical–basal polarity (discussed in Chapter 19), but the cells of many epithelia show an additional polarity at right angles to this axis: the cells are all arranged as if they had an arrow written on them, pointing in a specific direction in the plane of the epithelium. This type of polarity is called **planar cell polarity**. In the wing of a fly, for example, each epithelial cell has a tiny asymmetrical projection, called a wing hair, on its surface, and the hairs all point toward the tip of the wing. Similarly, in the inner ear of a vertebrate, each mechanosensory hair cell has a precisely oriented asymmetric bundle of actin-filled, rodlike protrusions called stereocilia sticking up from its apical plasma membrane as a detector of sound and of forces such as gravity. Tilting the bundle in one direction causes ion channels in the membrane to open, electrically activating the cell; tilting in the opposite direction has the opposite effect. For the ear to function properly, the hair cells must be oriented correctly. Planar cell polarity is also important in the respiratory tract, where every ciliated cell must orient the beating of its cilia so as to sweep mucus upward, away from the lungs.

Screens for mutants with misoriented wing hairs in *Drosophila* have identified a set of genes that is critical for planar cell polarity. Some of these genes code for

epidermal cells in fly wing   sensory hair cells in mouse ear   epidermal cells in fly wing   sensory hair cells in mouse ear

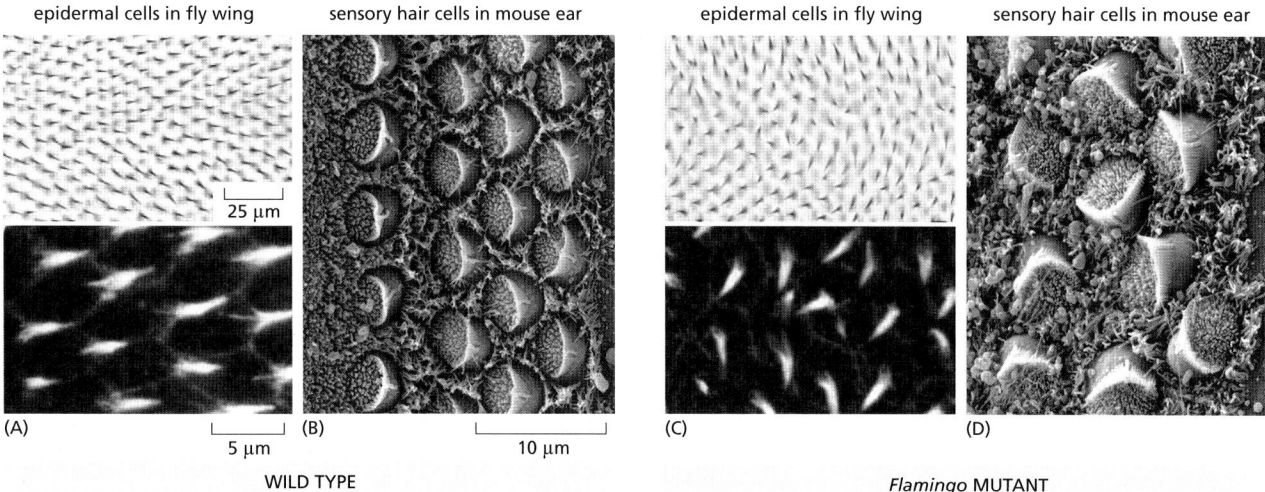

25 μm

(A)   5 μm   (B)   10 μm   (C)   (D)

WILD TYPE   *Flamingo* **MUTANT**

Figure 21–51 Planar cell polarity. (A) Wing hairs on the wing of a fly. Each cell in the wing epithelium forms a small, spiky protrusion or "hair" at its apex, and all the hairs point the same way, toward the tip of the wing. This reflects a planar polarity in the structure of each cell. (B) Sensory hair cells in the inner ear of a mouse similarly have a well-defined planar polarity, manifest in the oriented pattern of stereocilia (actin-filled protrusions) on their surface. The detection of sound depends on the correct, coordinated orientation of the hair cells. (C) A mutation in the gene *Flamingo* in the fly, coding for a nonclassical cadherin, disrupts the pattern of planar cell polarity in the wing. (D) A mutation in a homologous *Flamingo* gene in the mouse randomizes the orientation of the planar cell polarity vector of the hair cells in the ear. The mutant mice are deaf. (A and C, from J. Chae et al., *Development* 126:5421–5429, 1999. With permission from The Company of Biologists; B and D, from J.A. Curtin et al., *Curr. Biol.* 13:1129–1133, 2003. With permission from Elsevier.)

components of the Wnt signaling pathway, others code for specialized members of the cadherin superfamily, while the functions of others are uncertain. These components of planar-cell-polarity signaling are assembled at cell–cell junctions in the epithelium in such a way as to exert a polarizing influence that can propagate from cell to cell. Essentially the same system of proteins controls planar cell polarity in vertebrates; mice deficient in homologs of the *Drosophila* planar polarity genes have a variety of defects, including incorrectly oriented hair cells in the inner ear, making them deaf (**Figure 21–51**).

Interactions Between an Epithelium and Mesenchyme Generate Branching Tubular Structures

Animals require specialized types of epithelial surfaces for many functions, including excretion, absorption of nutrients, and gas exchange. Where large surfaces are required, they are often organized as branching tubular structures. The lung is an example. It originates from epithelial buds that grow out from the floor of the foregut and invade neighboring mesenchyme to form the bronchial tree, a system of tubes that branch repeatedly as they extend. Endothelial cells that form the lining of blood vessels invade the same mesenchyme, thereby creating a system of closely apposed airways and blood vessels, as required for gas exchange in the lung (**Figure 21–52**). This whole process of *branching morphogenesis* depends on signals that pass in both directions between the growing epithelial buds and the mesenchyme. Genetic studies in mice indicate that FGF proteins and their receptor tyrosine kinases play a central part in these signaling processes. FGF signaling has various roles in development, but it is especially important in the many interactions that occur between a developing epithelium and mesenchyme.

In the case of lung development, FGF10 is expressed in clusters of mesenchyme cells that lie near the tips of the growing epithelial tubes, and its receptor is expressed in the invading epithelial cells. In FGF10-deficient mutant mice, a primary bud of lung epithelium is formed but fails to grow out into the mesenchyme to create a branching bronchial tree. Conversely, a microscopic bead soaked in

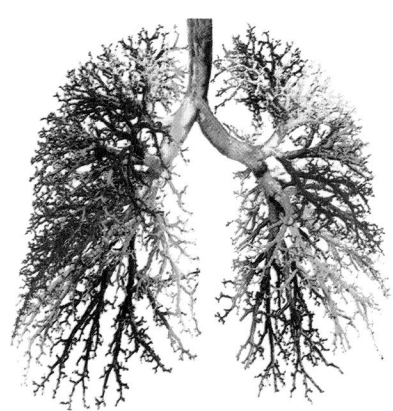

Figure 21–52 The airways of the lung, shown in a cast of the adult human bronchial tree. Resins of different colors have been injected into different branches of the tree of airways. (James Cavallini/ Science Source.)

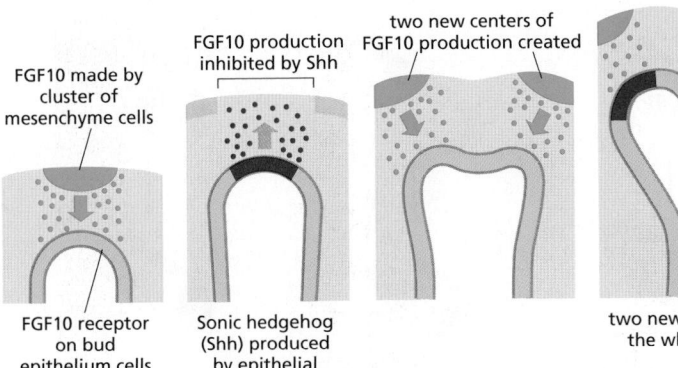

FGF10 made by cluster of mesenchyme cells

FGF10 production inhibited by Shh

two new centers of FGF10 production created

FGF10 receptor on bud epithelium cells

Sonic hedgehog (Shh) produced by epithelial cells at tip of growing bud

two new buds are formed and the whole process repeats

Figure 21–53 **Branching morphogenesis of the lung.** How FGF10 and Sonic hedgehog are thought to induce the growth and branching of the buds of the bronchial tree. Many other signal molecules, such as BMP4, are also expressed in this system, and the suggested branching mechanism is only one of several possibilities.

As indicated, FGF10 protein is expressed in clusters of mesenchyme cells near the tips of the growing epithelial tubes, and its receptor is expressed in the epithelial cells themselves. The Sonic hedgehog signal is sent in the opposite direction, from the epithelial cells at the tips of the buds back to the mesenchyme. The patterns of gene expression and their timing suggest that the Sonic hedgehog signal may serve to shut off FGF10 expression in the mesenchyme cells closest to the growing tip of a bud, splitting the FGF10-secreting cluster into two separate clusters, which in turn cause the bud to branch into two.

FGF10 and placed near embryonic lung epithelium in culture will induce a bud to form and grow out from the epithelium toward the bead. Evidently, the epithelium invades the mesenchyme only by invitation, in response to FGF10.

But what makes the growing epithelial tubes of the lung branch repeatedly as they invade the mesenchyme? This depends on a Sonic hedgehog signal that is sent in the opposite direction, from the epithelial cells at the tips of the buds back to the mesenchyme, as shown in **Figure 21–53**. In mice lacking Sonic hedgehog, the lung epithelium grows and differentiates, but it forms a sac instead of a branching tree of tubules.

FGF signaling acts in a remarkably similar way in the formation of the air-exchange system of insects, which consists of a pattern of fine, air-filled channels called *tracheae* and *tracheoles*. These originate from the epidermis covering the surface of the body and extend inward to invade the underlying tissues, branching and narrowing as they go (**Figure 21–54**). The FGF acts on cells at the tips of the advancing tracheae, causing them to extend filopodia and migrate toward the source of the FGF signal. Because the tip cells remain connected to the remainder of the tracheal epithelium, the pulling force that they generate elongates the tracheal tube.

Initially, the pattern of FGF production in fly embryos is defined by the D-V and A-P patterning systems discussed earlier. In later stages of development, however, FGF expression is induced by transcription regulators called *hypoxia-inducible factors* (*HIFs*) that are activated by hypoxia (low oxygen levels). In this way, hypoxia stimulates the formation of finer and finer and more extensively branched trachea, until the oxygen supply is sufficient to stop the process. Hypoxia and HIFs have similar roles in vertebrates, especially in the development of blood vessels, as we shall see in the next chapter.

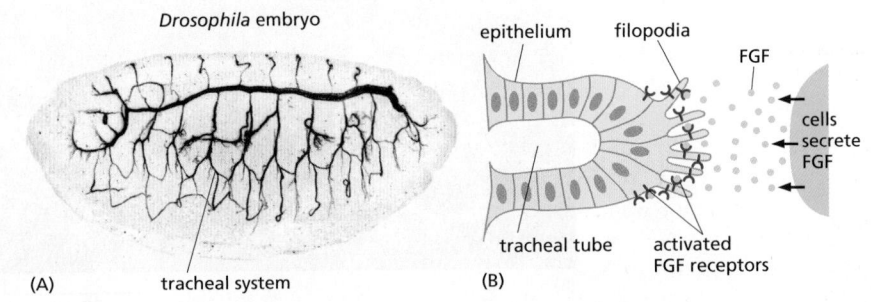

Drosophila embryo

epithelium filopodia

FGF

cells secrete FGF

tracheal tube activated FGF receptors

(A) tracheal system

(B)

Figure 21–54 **Branching morphogenesis of airways in a fly.** (A) *Drosophila* embryonic tracheal system. (B) FGF (produced in *Drosophila* by the *Branchless* gene) signals from surrounding cells to the tracheal epithelium and activates its FGF receptors, leading to filopodia formation and tube elongation. [A, from G. Manning and M.A. Krasnow, in The Development of *Drosophila* (A. Martinez-Arias and M. Bate, eds), Vol. 1, pp. 609–685. New York: Cold Spring Harbor Laboratory Press, 1993. With permission from Cold Spring Harbor Laboratory Press.]

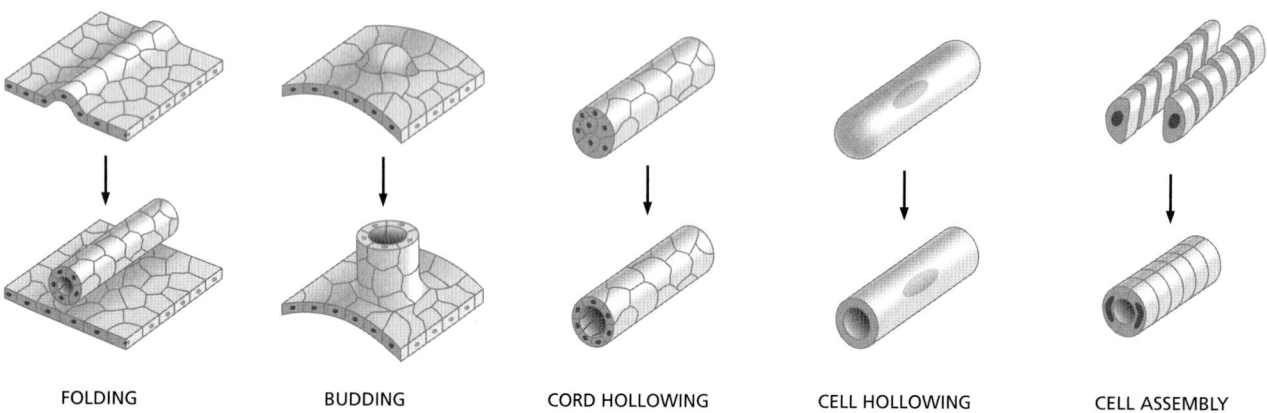

FOLDING   BUDDING   CORD HOLLOWING   CELL HOLLOWING   CELL ASSEMBLY

An Epithelium Can Bend During Development to Form a Tube or Vesicle

The creation of systems of tubes such as blood vessels and airways is a complex process, and it can involve various additional forms of cell behavior, as sketched in Figure 21–55.

As explained in Chapter 19, the process that converts an epithelial sheet into a tube depends on contraction of specific bundles of actin filaments. With the help of myosin motor proteins, actin filament bundles can shorten, causing the epithelial cells to narrow at their apex. These actin bundles are connected from cell to cell by adherens junctions, and if their contraction is coordinated along a specific axis, the result will be that the sheet bends and rolls up into a tube (Figure 21–56). The vertebrate neural tube, which we discuss in the last section of this chapter, originates in this way.

Figure 21–55 The forms of cell behavior involved in tube formation. Folding generates the neural tube, budding underlies the formation of lungs and trachea, cord hollowing occurs during the formation of mammalian salivary glands, cell hollowing is involved in the formation of tracheal terminal cell tubes, and cell assembly generates the heart tube that forms at the earliest stage of heart development.

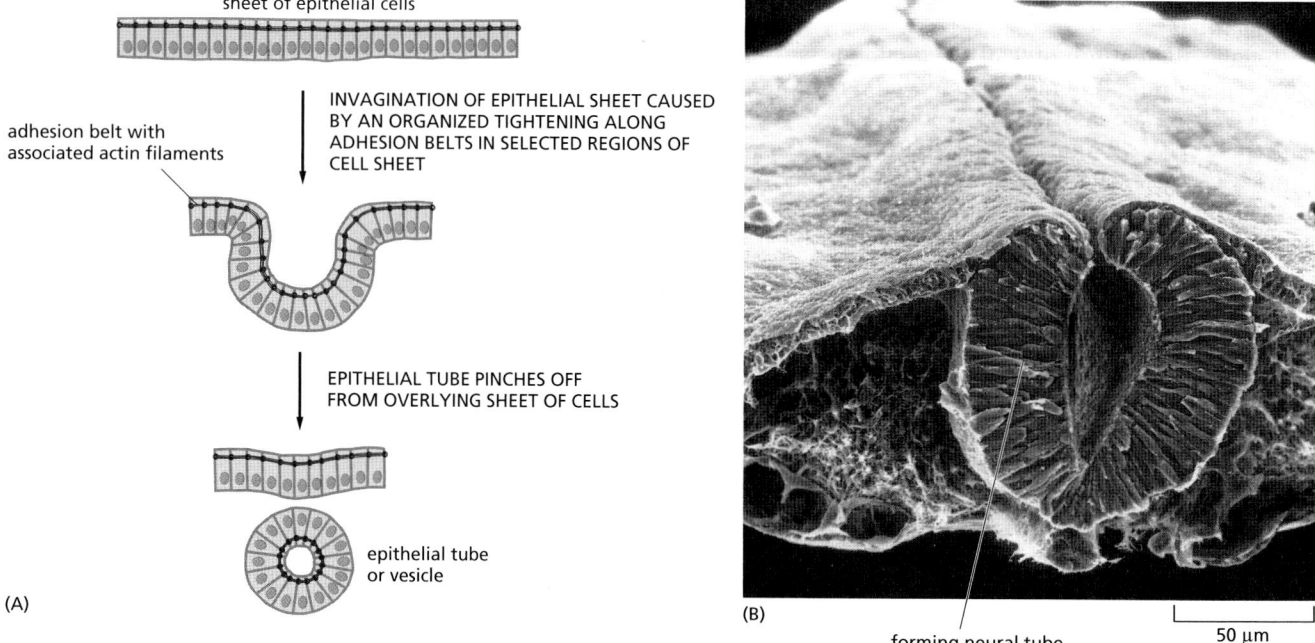

sheet of epithelial cells

INVAGINATION OF EPITHELIAL SHEET CAUSED BY AN ORGANIZED TIGHTENING ALONG ADHESION BELTS IN SELECTED REGIONS OF CELL SHEET

adhesion belt with associated actin filaments

EPITHELIAL TUBE PINCHES OFF FROM OVERLYING SHEET OF CELLS

epithelial tube or vesicle

(A)

(B)

forming neural tube   50 μm

Figure 21–56 Bending of an epithelial sheet to form a tube. Contraction of apical bundles of actin filaments linked from cell to cell via adherens junctions causes the epithelial cells to narrow at their apex. Depending on whether the contraction is oriented along one axis of the sheet or is equal in all directions, the epithelium will either roll up into a tube or invaginate to form a vesicle. (A) Diagram showing how an apical contraction along one axis of an epithelial sheet can cause the sheet to form a tube. (B) Scanning electron micrograph of a cross section through the trunk of a two-day chick embryo, showing the formation of the neural tube by the process diagrammed in (A). (B, courtesy of Jean-Paul Revel.)

Summary

Animal development involves dramatic cell movements, including the guided migration of individual cells, the adhesion and repulsion of groups of cells, and the complex extension, branching, or rolling up of epithelial tissues. Migrant cells, such as those of the neural crest, break loose from their original neighbors and travel through the embryo to colonize new sites. Many migrant cells, including primordial germ cells, are guided by chemotaxis dependent on the receptor CXCR4 and its ligand CXCL12. In general, cells that have similar adhesion molecules on their surfaces cohere and tend to segregate from other cell groups with different surface properties. Selective cell–cell adhesion is often mediated by cadherins; repulsion is often driven by ephrin–Eph signaling. Within an epithelial sheet, cells can rearrange themselves to drive epithelial convergence and extension, as in gastrulation. Many movements are coordinated through a Wnt-dependent planar-polarity signaling pathway that is also responsible for orienting cells correctly in various types of epithelium. Elaborate branched tubular structures, such as the airways of the lung, are generated through bidirectional signaling between an epithelial bud and the mesenchyme that it invades, in a process called branching morphogenesis. Epithelial tubes and vesicles can originate in various ways, most simply by the rolling up and pinching off of a segment of epithelium, as in the formation of the neural tube.

GROWTH

One of the most fundamental aspects of animal development is one we know surprisingly little about—how the size of an animal or an organ is determined. Why, for example, do we grow to be so much larger than a mouse? Even within a species, size can vary greatly; a Great Dane, for instance, can weigh over 40 times more than a Chihuahua (**Figure 21–57**).

Three variables define the size of an organ or organism: the number of cells, the size of the cells, and the quantity of extracellular material per cell. Size differences can arise from changes in any of these factors (**Figure 21–58**). If we compare a mouse with a human, for example, we find that the difference lies chiefly in the number of cells, there being roughly 3000 times more cells in a human, corresponding to a body that is roughly 3000 times more massive. Wild and cultivated species of food plants, on the other hand, often differ in body size chiefly because of differences of cell size.

The challenge, therefore, is to understand how cell numbers, cell size, and extracellular matrix production are regulated. First of all, we need to identify the signals that drive or inhibit growth. Then we need to discover how the signals themselves are regulated. In many cases, the size of an organ or of the body as a whole seems to be controlled homeostatically, so that the correct size is reached and maintained even in the face of drastic disturbances. This suggests that the developing structure somehow senses its own size and uses this information to regulate the signals for its own growth or shrinkage. In most cases, the nature of this feedback control remains a profound mystery.

In other cases, the duration of growth and the final size seem to be dictated by intracellular programs that take no cognizance of the size the structure has attained. These intracellular programs, too, present many mysteries, as we saw in our discussion of developmental timing. Very often, it seems, the sizes and proportions of body parts must depend on combinations of size-measuring feedback controls and intracellular programs, as well as on environmental influences such as nutrition.

The variation in control strategies is nicely illustrated by some classic transplantation experiments. If several fetal thymus glands are transplanted into a developing mouse, each grows to its characteristic adult size. In contrast, if multiple fetal spleens are transplanted, each ends up smaller than normal, but collectively they grow to the size of one adult spleen. Thus, thymus growth is regulated by local mechanisms intrinsic to the individual organ, whereas spleen growth is controlled by a feedback mechanism that senses the quantity of spleen tissue in the body as a whole. In neither case is the mechanism known.

Figure 21–57 **Members of the same species can have dramatically different sizes.** The Chihuahua weighs 2–5 kilograms, whereas a Great Dane weighs 45–90 kilograms. (Courtesy of Deanne Fitzmaurice.)

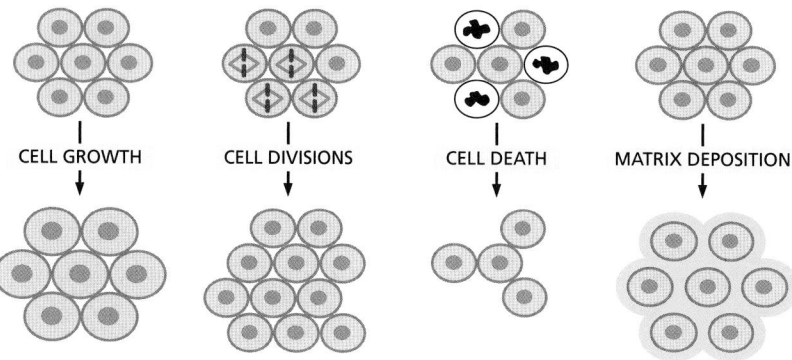

Figure 21–58 **Determinants of organ size.**

CELL GROWTH   CELL DIVISIONS   CELL DEATH   MATRIX DEPOSITION

The Proliferation, Death, and Size of Cells Determine Organism Size

The nematode worm *C. elegans* illustrates the different ways in which size differences can arise. This creature follows an astonishingly precise and predictable developmental program. Each individual of a given sex is generated by almost exactly the same sequences of cell divisions and cell deaths, and consequently has precisely the same number of somatic cells—959 in the adult hermaphrodite (the sex of the majority of these animals)—although the number of germ cells is more variable from worm to worm. The stereotyped development makes it possible to trace somatic cell lineages in exhaustive detail. More than 1000 cell divisions generate 1090 somatic cells during hermaphrodite development, but 131 of these cells undergo apoptotic cell death. Thus, precise regulation of both cell division and cell death determines the final numbers of somatic cells in the worm. In fact, genetic screens in *C. elegans* identified the first genes responsible for apoptosis and its regulation—thereby revolutionizing our molecular understanding of this form of programmed cell death (discussed in Chapter 18).

The final number of somatic cells in the adult worm is already present at sexual maturity (around three days after fertilization), after which no more somatic cells are generated. Yet the worm continues to grow, doubling in size between sexual maturity and death 2–3 weeks later. This doubling results from somatic cell growth: although the cells no longer divide, they continue to go through rounds of DNA synthesis; this *endoreplication* of the genome makes the cells *polyploid*. As in all organisms, the size of a cell is proportional to its ploidy—that is, the number of genome copies that it contains: a doubling of ploidy roughly doubles cell volume. By artificial manipulation of somatic cell ploidy, and thereby somatic cell size, the size of the worm as a whole can be increased or decreased. Thus the worm's final size is set by a combination of programmed cell divisions and cell deaths, along with regulation of the sizes of individual cells through changes in ploidy.

In plants, as in animals, cell size increases as ploidy increases (**Figure 21–59**). This effect has been exploited in the agricultural breeding of plants for large size: most of the major fruits and vegetables that we consume are polyploid.

Animals and Organs Can Assess and Regulate Total Cell Mass

The size of an animal or organ depends on both cell number and cell size—that is, on total cell mass. Remarkably, many animals and organs can somehow assess their total cell mass and regulate it, providing evidence for feedback controls of the sort highlighted earlier in our introductory account of general principles of growth control. In contrast with *C. elegans*, if cell size is artificially increased or decreased in these cases, cell numbers adjust to maintain a normal total cell mass. This has been beautifully illustrated by experiments done long ago in salamanders, where cell size can be manipulated by altering the animal's ploidy. As shown in Figure 21–59E, salamanders of different ploidies end up being the same size with very different numbers of cells. The individual cells in a pentaploid salamander,

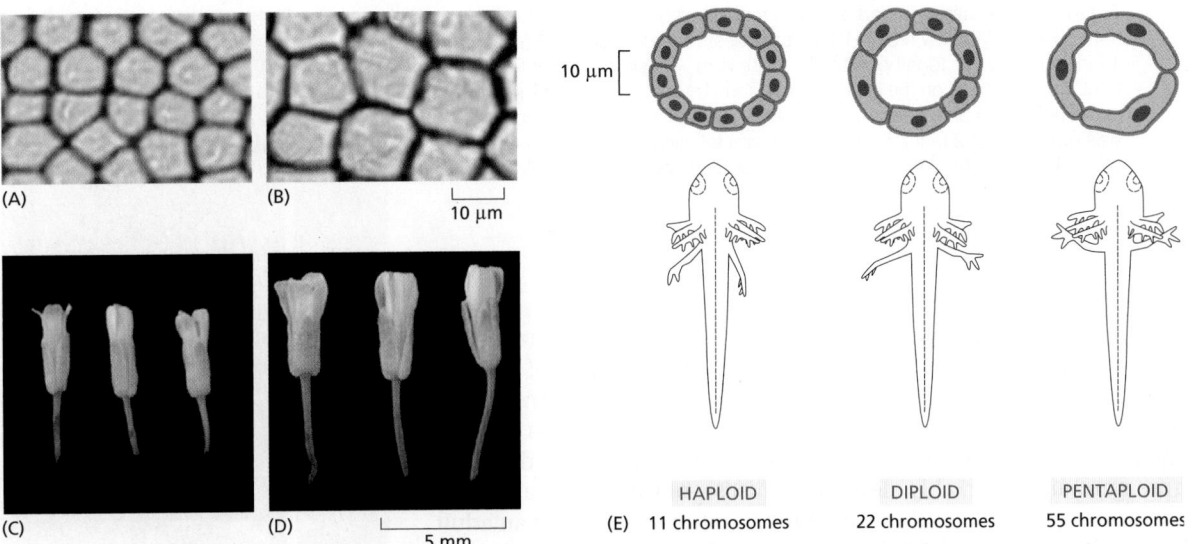

(A)

(B)

10 μm

(C)

(D)

5 mm

10 μm

(E)

HAPLOID
11 chromosomes

DIPLOID
22 chromosomes

PENTAPLOID
55 chromosomes

Figure 21–59 **Effects of ploidy on cell size and organ size.** In all organisms, from bacteria to humans, cell size is proportional to ploidy—the number of copies of the genome per cell. This is illustrated for (A–D) *Arabidopsis* flowers and (E) for salamanders. In each case, the upper panels show cells in a specific tissue [a petal for *Arabidopsis*, a pronephric (kidney) tubule for the salamander]; the lower panels show the gross anatomy—flowers for *Arabidopsis*, the whole body for the salamander. In the case of *Arabidopsis* flowers, increase in cell size increases organ size. By contrast, the salamander and its individual organs attain their normal standard size regardless of ploidy, because large cell size is compensated for by fewer cells. This indicates that the size of an organism or organ in this species is not controlled simply by counting cell divisions or cell numbers; size must somehow be regulated at the level of total cell mass. [A–D, from C. Breuer et al., *Plant Cell* 19:3655–3668, 2007. With permission from the American Society of Plant Biologists; E, adapted from G. Fankhauser, in Analysis of Development (B.H. Willier, P.A. Weiss and V. Hamburger, eds), pp. 126–150. Philadelphia: Saunders, 1955.]

for example, are about five times the size of those in a haploid salamander, but there are only one-fifth as many cells. This scaling operates not only in the body as a whole, but in its individual organs.

The **imaginal discs** of *Drosophila* provide another striking example of homeostatic size control. These are epithelial pouches that grow by cell proliferation during the larval period and, during the pupal stage, form the organs and extremities of the adult fly (**Figure 21–60**). Experiments have been chiefly done on the wing imaginal disc. Mutations in components of the cell-cycle control machinery can be used to speed up or slow down the rate of cell division in the disc. Remarkably, such mutations can result in an excessive number of abnormally small cells

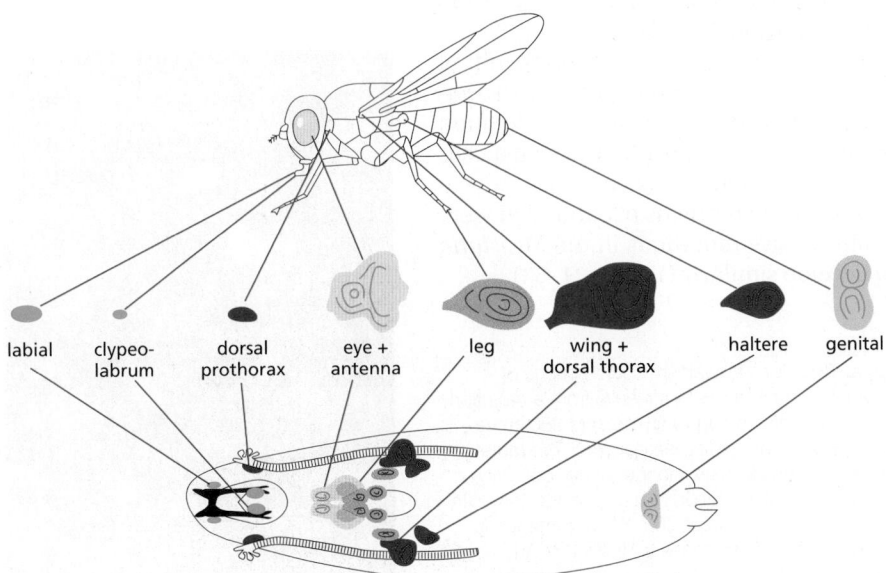

labial   clypeo-labrum   dorsal prothorax   eye + antenna   leg   wing + dorsal thorax   haltere   genital

Figure 21–60 **The imaginal discs in the *Drosophila* larva (below) and the structures in the adult (above) that they give rise to.** [After J.W. Fristrom et al., in Problems in Biology: RNA in Development (E.W. Hanley, ed.), p. 382. Salt Lake City: University of Utah Press, 1969.]

Figure 21–61 **Pituitary dwarf and pituitary giant.** The "giant" on the right is Robert Wadlow (1914–1940), the tallest recorded man at 8 feet 11 inches (2.72 m), together with his father, who was almost 6 feet tall (1.82 m). The dwarf on the left is General Tom Thumb, which was the stage name of Charles Sherwood Stratton (1838–1883). On his 18th birthday, he was measured at 2 feet 8.5 inches (82.6 cm) tall, and at his death, he was 3 feet 4 inches (102 cm). (Left, Bettmann/Getty; right, History and Art Collection/Alamy Stock Photo.)

1 meter

or a reduced number of abnormally large cells, respectively, leaving the size (area) and patterning of the adult wing practically unchanged. Thus, the size of the disc is not regulated so as to contain a set number of cells. Instead, there must be a regulatory mechanism that halts growth when the disc's total cell mass reaches the appropriate value, so that the size and pattern of the adult wing that develops from the disc are normal. Remarkably, developing discs—or even disc fragments, taken out of their normal context and transplanted into the abdomen of an adult female—will grow until they reach their normal size. Clearly, the mechanisms that regulate disc size are intrinsic to the disc.

We still have very little idea how organisms or organs assess their total cell mass or monitor their own growth. Nevertheless, we are beginning to understand some of the signal molecules that drive or halt growth in response to the mysterious cues that convey information about the size attained.

Extracellular Signals Stimulate or Inhibit Growth

We have already seen how some signals act systemically as hormones to regulate the development of the animal as a whole. Some of these serve to regulate growth. In mammals, for instance, **growth hormone (GH)** is secreted by the pituitary gland into the bloodstream and stimulates growth throughout the body: excessive production of growth hormone leads to gigantism, and too little leads to dwarfism (**Figure 21–61**). Pituitary dwarfs have bodies and organs that are proportionately small, unlike achondroplastic dwarfs, for example, whose limbs are disproportionately short, usually because of a mutation in a gene encoding an FGF receptor that disrupts normal cartilage development (**Figure 21–62**).

Growth hormone stimulates growth largely by inducing the liver and other organs to produce insulin-like growth factor 1 (IGF1), which acts mainly as a local signal within many tissues to increase cell survival, cell growth, cell proliferation, or some combination of these, depending on the cell type. Large breeds of dogs such as Great Danes owe their great size to high levels of IGF1, while miniature breeds such as Chihuahuas have low levels (see Figure 21–57).

Not all growth-regulating extracellular signals stimulate growth; some inhibit it, by promoting cell death or inhibiting cell growth, cell division, or both. *Myostatin* is a TGFβ family member that specifically inhibits the growth and proliferation of myoblasts—the precursor cells that fuse to form the huge, multinucleated cells of skeletal muscle. When the *Myostatin* gene is deleted in mice, muscles grow to be several times larger than normal. Remarkably, two breeds of cattle that were bred for large muscles have both turned out to have mutations in the *Myostatin* gene; whippet dogs mutant for *Myostatin* develop similarly (**Figure 21–63**).

Figure 21–62 **Achondroplasia.** This type of dwarfism occurs in one of 10,000–100,000 births; in more than 99% of cases it results from a mutation at an identical site in the genome, corresponding to amino acid 380 in the FGF receptor FGFR3 (a glycine in the transmembrane domain). The mutation is dominant, and almost all cases are due to new, independently occurring mutations, implying an extraordinarily high mutation rate at this particular site in the genome. The defect in FGF signaling causes dwarfism by interfering with the growth of cartilage in developing long bones. (The Picture Art Collection/Alamy Stock Photo.)

Figure 21–63 Myostatin limits muscle growth. A mutation in the *myostatin* gene leads to a great increase in the mass of muscle tissue, as illustrated in this Belgian Blue bull. This animal, produced by cattle breeders, was found to have a mutation in the gene encoding myostatin. (Yann Arthus-Bertrand/Getty Images.)

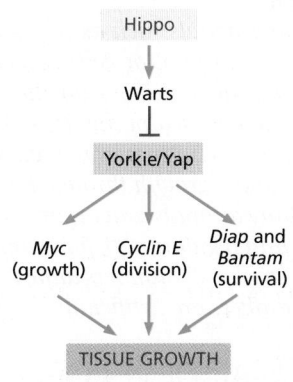

Figure 21–64 Hippo pathway. Hippo, a protein kinase, limits growth by phosphorylation and activation of the kinase Warts, which in turn phosphorylates and inactivates the transcriptional coactivator Yorkie (called Yap in vertebrates). When unphosphorylated, Yorkie/Yap drives tissue growth: it activates the transcription of the growth-promoting gene *Myc*, the cell-cycle progression gene *Cyclin E*, the anti-apoptotic gene *Diap*, and the microRNA *Bantam*. Hippo-induced phosphorylation of Yorkie/Yap blocks this effect.

Like TGFβ itself, myostatin acts through the Smad intracellular signaling pathway (see Figure 15–57) to inhibit muscle growth specifically. Another intracellular signaling pathway, called the *Hippo pathway*, inhibits organ and organism growth more generally. It was discovered in *Drosophila*, but it operates in vertebrates as well. It inhibits growth both by promoting cell death (by blocking an apoptosis inhibitor) and by inhibiting cell-cycle progression (by inhibiting the expression of the cell-cycle gene *Cyclin E*). Some components of the pathway in *Drosophila* are shown in **Figure 21–64**. The organs of animals that are abnormally resistant to Hippo repression can grow to a monstrous size (**Figure 21–65**).

It is important to note that in all species nutritional conditions also play a fundamental part in regulating the pace and extent of growth, and in animals they do so through hormonal signal networks that are highly conserved between vertebrates and invertebrates. Although we do not have space for details here, genetic experiments, especially in *Drosophila*, have begun to unravel the logic of these controls, and to indicate how they may operate alongside other machinery, such as the Hippo pathway, to determine final size.

Summary

The sizes of animals and their organs vary widely and largely depend on total cell mass. This in turn depends on the size and number of cells, which are increased through cell growth and cell division, respectively. Cell numbers are reduced by programmed cell death. Each of these processes depends on both intracellular and

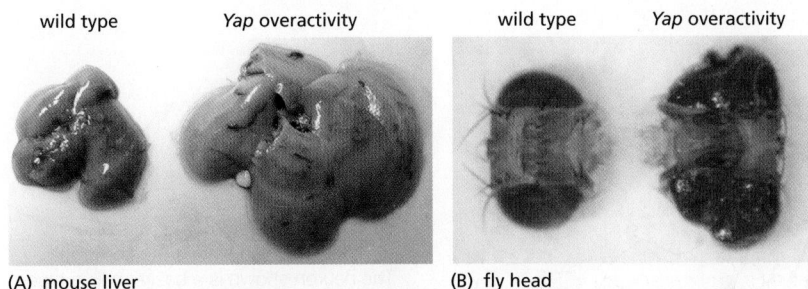

Figure 21–65 Overcoming Hippo repression increases organ size. (A) Livers from control and *Yap*-overexpressing mice. In these mice, Hippo signaling is insufficient to block Yap. (B) Adult heads from control and *Yap*-overexpressing flies. In the mutant flies, Hippo signaling is unable to block Yap. (From J. Dong et al., *Cell* 130:1120–1133, 2007. With permission from Elsevier.)

extracellular signals. The mystery is how these processes are regulated and coordinated to produce and maintain the characteristic final size of the adult organ or animal.

Some signals such as survival factors, growth factors, and mitogens stimulate growth by promoting cell survival, cell growth, and cell division, respectively, while other signal molecules do the opposite. Although most of these signals operate locally to help sculpt the size and shape of the animal, its organs, and appendages, others act as hormones to regulate the growth of the animal as a whole. Nutrients can regulate growth through hormonal signals in the entire body.

Many animals and organs can, by unknown mechanisms, assess their total cell mass and regulate it. If, for example, cell size is artificially increased or decreased in these cases, cell numbers adjust to maintain a normal total cell mass. Conversely, if cell numbers are artificially increased or decreased, cell size adjusts to compensate.

NEURAL DEVELOPMENT

The development of the nervous system poses problems that have little parallel in other tissues. A typical nerve cell, or neuron, has a structure unlike that of any other class of cells, with a long axon and branching dendrites, both of which make many synaptic connections to other cells (**Figure 21–66**). The central challenge of neural development is to explain how the axons and dendrites grow out, find their right partners, and synapse with them selectively to create a neural network—an electrical signaling system—that functions correctly to guide behavior (**Figure 21–67**). The problem is formidable: the human brain contains more than 10^{11} neurons, each of which, on average, has to make connections with a thousand others, according to a regular and predictable wiring plan. The precision required is not so great as in a man-made computer, because the brain performs its computations in a different way and is more tolerant of vagaries in individual components. But the human brain nevertheless outstrips all other biological structures in its organized complexity.

The components of a typical nervous system—the various classes of neurons, glial cells, sensory cells, and muscles—originate in a number of widely separate locations in the embryo. Thus, in the first phase of neural development, the different parts of the nervous system develop according to their own local programs: neurons are born and assigned specific characters according to the place and time of their birth, under the control of inductive signals and transcription regulators, by mechanisms of the types we have already discussed. In the next phase, newborn neurons extend axons and dendrites along specific routes toward their target cells, guided by extracellular signals that attract or repel them. In the third phase, neurons form synapses with other neurons or muscle cells, setting up a provisional but orderly network of connections. In the final phase, which continues into adult life, the synaptic connections are adjusted and refined through mechanisms that usually depend on synaptic signaling between the cells involved

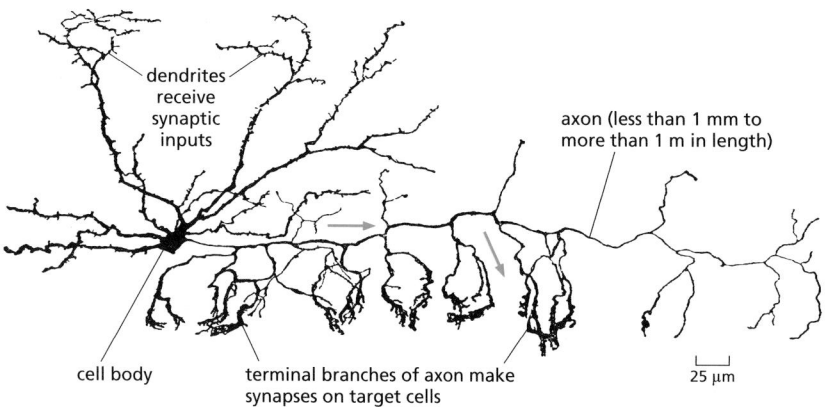

dendrites receive synaptic inputs

axon (less than 1 mm to more than 1 m in length)

cell body

terminal branches of axon make synapses on target cells

25 μm

Figure 21–66 A typical neuron of a vertebrate. The arrows indicate the direction in which signals are conveyed. The neuron shown is a basket cell, a type of neuron in the cerebellum. (Adapted from S. Ramón y Cajal, Histologie du Système Nerveux de l'Homme et des Vertébrés, 1909–1911. Paris: Maloine; reprinted, Madrid: C.S.I.C., 1972.)

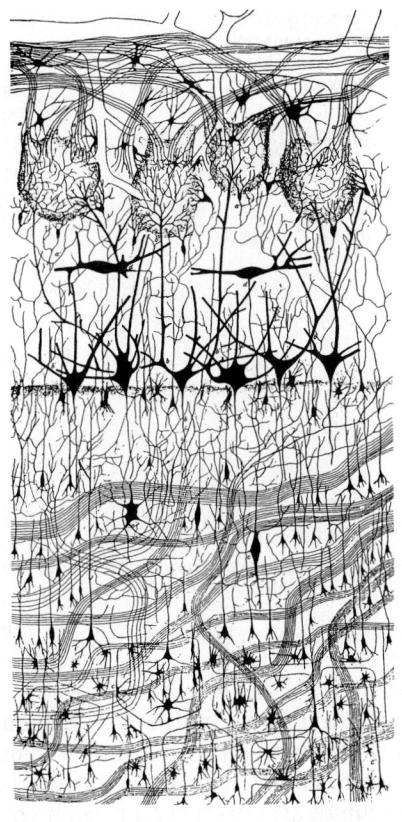

Figure 21–67 **The complex organization of nerve cell connections.** This drawing depicts a section through a small part of a mammalian brain—the olfactory bulb of a dog—stained by the Golgi technique. The black objects are neurons; the thin lines are axons and dendrites, through which the various sets of neurons are interconnected according to precise rules. (From C. Golgi, *Riv. sper. freniat. Reggio-Emilia* 1:405–425, 1875.)

(**Figure 21–68**). At all stages, neurons are in intimate contact with various types of non-neuronal supporting cells—the **glial cells**.

Neurons Are Assigned Different Characters According to the Time and Place of Their Birth

We start our account here with the first phase of neural development: the generation of neural progenitors and their differentiation into hundreds of different neuronal subtypes, along with a much smaller number of glial types. Although the nervous system is exceptional in the extent of cell diversity, the process depends on the same principles that generate different cell types in other organs. We have already discussed some of the underlying machinery in the developing *Drosophila* nervous system. We turn now to vertebrates.

The vertebrate spinal cord, the brain, and the retina of the eye together constitute the central nervous system (CNS). They all originate as parts of the **neural tube**, whose formation was described earlier (see Figure 21–56). The brain and eyes develop from the anterior neural tube and the spinal cord from the posterior.

The developmental anatomy is seen at its simplest in the **spinal cord**. As it develops, the epithelium forming the walls of the posterior neural tube becomes enormously thickened as the cells proliferate and differentiate, creating a highly organized structure of neurons and glial cells, surrounding a small central channel. Bands of neurons with different future functions—and expressing different genes—are laid out along the dorsoventral axis of the tube. Motor neurons (those that control the muscles) are located ventrally, whereas neurons that process sensory information are found dorsally. This pattern is established by opposing gradients of morphogens. These are secreted by specialized groups of cells that run the length of the ventral and dorsal midlines of the neural tube (**Figure 21–69**). The two morphogen gradients—consisting of Sonic hedgehog protein from the ventral source and BMP and Wnt from the dorsal source—help induce different groups of proliferating neural progenitor cells and differentiating neurons to express different combinations of transcription regulators. These regulators in turn drive the production of different combinations of neurotransmitters, receptors, cell–cell adhesion proteins, and other molecules, creating terminally differentiated neurons that will form synaptic connections selectively with the right partners and exchange appropriate signals with them.

Figure 21–68 **The four phases of neural development.**

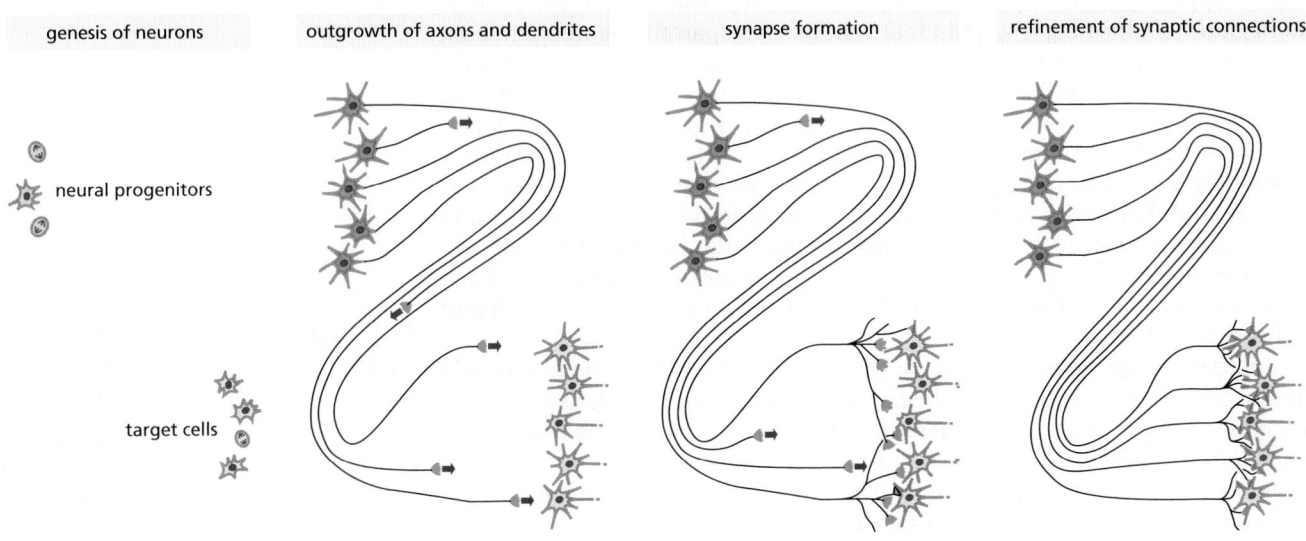

| genesis of neurons | outgrowth of axons and dendrites | synapse formation | refinement of synaptic connections |

neural progenitors

target cells

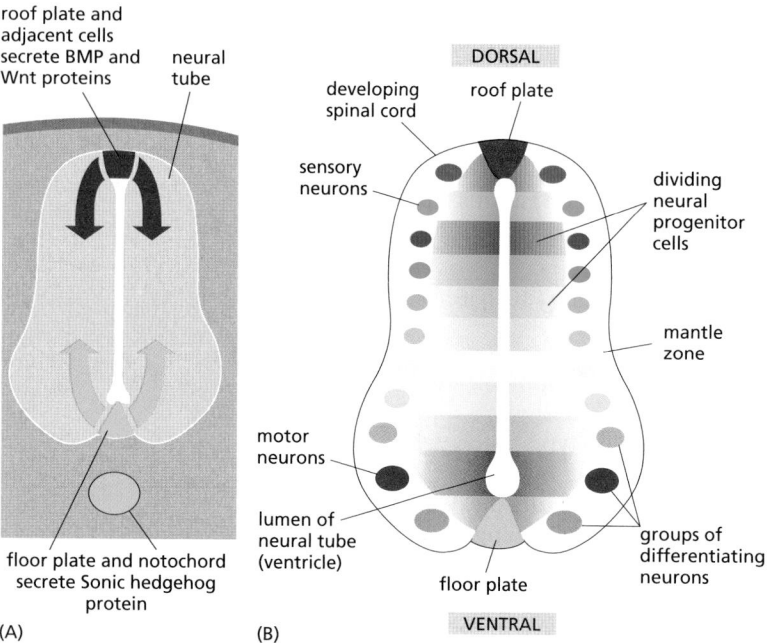

Figure 21–69 **A schematic cross section of the spinal cord of a chick embryo, showing how cells at different levels along the dorsoventral axis acquire different characters.** (A) Signals that direct the dorsoventral pattern. Sonic hedgehog protein from the notochord and the floor plate (the ventral midline of the neural tube) and BMP and Wnt proteins from the roof plate (the dorsal midline) act as morphogens to control gene expression. (B) The resulting patterns of cell fates in the developing spinal cord. Different groups of proliferating neural progenitor cells (in the ventricular zone, close to the lumen of the neural tube) and of differentiating neurons (in the mantle zone, further out) express different combinations of transcription regulators. Neurons expressing different transcription regulators will form connections with different partners and may make different combinations of neurotransmitters and receptors. Colors represent different cell types and combinations of regulatory proteins.

Extracellular morphogen gradients, however, are not the only way to generate cell diversity. As we saw earlier in our discussion of *Drosophila* neuroblasts (see Figure 21–36), different cell types can also be generated by temporal patterning, in which an intracellular program changes the character of a progenitor cell over time, giving rise to different cell types as development progresses. This mechanism also seems to operate in vertebrate neurogenesis. The most striking illustration comes from study of another part of the CNS—the mammalian cerebral cortex.

Although the **cerebral cortex** is the most complex structure in the human body, it has a simple beginning—from the anterior neural tube. As in the spinal cord, the cells that form the walls of the tube proliferate, and the neuroepithelium thickens and expands as they divide. On a predictable schedule, the divisions of the neuroepithelial cells begin to produce a succession of cells committed to terminal differentiation as neurons. These future neurons are born close to the lumen (the central cavity) of the tube. From here, they migrate outward, losing attachment to the lumenal surface and crawling outward along neighboring cells that continue to span the full thickness of the neuroepithelium. These latter neuroepithelial cells do double duty, functioning as progenitors of neurons and glia, and as supporters of the epithelial architecture. They become stretched out as *radial glial cells*, forming a scaffold that continues to span the neuroepithelium even as this grows to an enormous thickness (**Figure 21–70**). At the same time, the radial glial cells continue to divide as neural precursors, giving rise to both neurons and glial cells—new radial glial cells as well as glial cells of other types. The newborn neurons, migrating along the radial glial cells, find their appropriate resting places in the developing cortex, where they mature, and from these sites they send out their axons and dendrites. The first-born neurons settle closest to their birthplace near the lumen, while neurons born later crawl past them to settle farther out (**Figure 21–71**). The successive generations of neurons thus build up as a series of cortical layers, ordered by birthdate and endowed with different intrinsic characters.

Strikingly, single cortical progenitor cells isolated in culture generate distinct types of cortical neurons and glial cells, with the timing and characteristics appropriate to specific cortical layers. These observations suggest that the neural progenitors in the developing mammalian cortex, much like the *Drosophila* neuroblasts, step through an intracellular developmental program that generates the ordered succession of different nerve cell types.

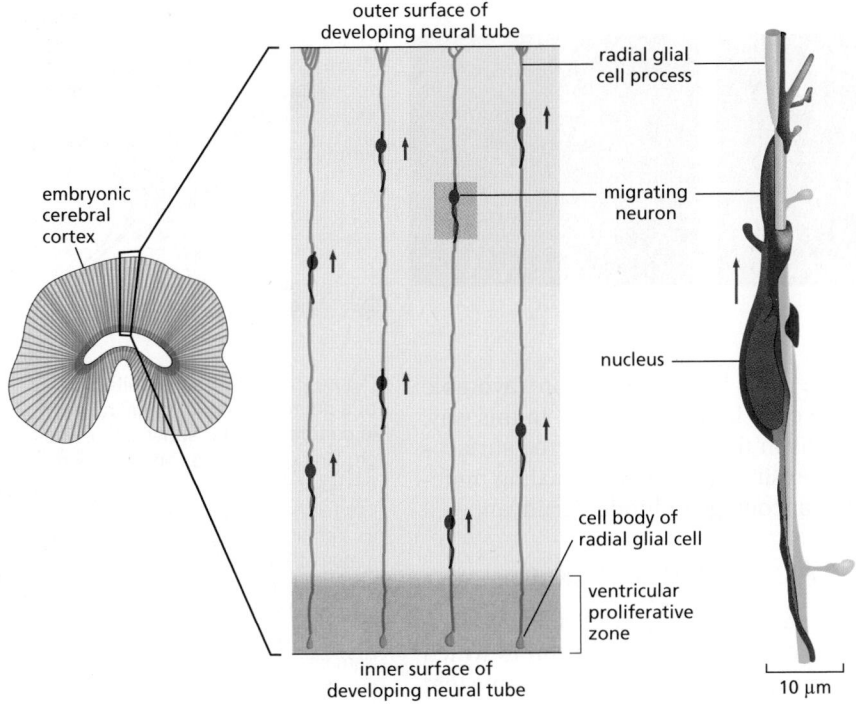

outer surface of
developing neural tube

radial glial
cell process

migrating
neuron

embryonic
cerebral
cortex

nucleus

cell body of
radial glial cell

ventricular
proliferative
zone

inner surface of
developing neural tube

10 μm

Figure 21–70 **Migration of immature neurons.** Before sending out axons and dendrites, newborn neurons often migrate from their birthplace and settle in another location. The diagrams are based on reconstructions from sections of the cerebral cortex (part of the neural tube) of a monkey and rely on a staining technique that picks out at random a small subset of the whole dense mass of neuroepithelial cells. The neurons go through their final cell division close to the inner, lumenal face of the neural tube (in the ventricular proliferative zone) and then migrate outward by crawling along radial glial cells that form a scaffold. Each of these latter cells extends from the inner to the outer surface of the tube, a distance that may be as long as 2 cm in the cerebral cortex of the developing brain of a primate.

The radial glial cells can be considered as persisting cells of the original columnar epithelium of the neural tube that become extraordinarily stretched as the wall of the tube thickens. They also serve as neural stem cells: depending on stage and region, the newborn neurons can be generated from radial glial cells that undergo mitosis while their nuclei are close to the inner surface of the tube, or they can be generated from a nearby class of specialized progenitors in the ventricular proliferative zone. (After P. Rakic, *J. Comp. Neurol.* 145:61–84, 1972.)

The Growth Cone Pilots Axons Along Specific Routes Toward Their Targets

According to the character assigned to it during its early development, a neuron will proceed to make connections with specific partners. This phase of neural development involves a type of morphogenesis unique to the nervous system, in which axons and dendrites extend along specific routes toward their target cells. A typical neuron sends out one long axon and many dendrites, which are usually shorter. The axon projects to distant target cells to which the neuron will eventually send signals. The dendrites will receive incoming signals from axon terminals of other neurons. Axons and dendrites extend by growth at their tip, where one sees an irregular, spiky enlargement called a **growth cone** (**Figure 21–72** and **Movie 21.6**). The growth cone is both the engine that produces the crawling movement and the steering apparatus that directs the tip along the proper path. Cytoskeletal machinery in the growth cone creates active protrusions, in the form of filopodia and lamellipodia (see Chapter 16 for details): when such a protrusion

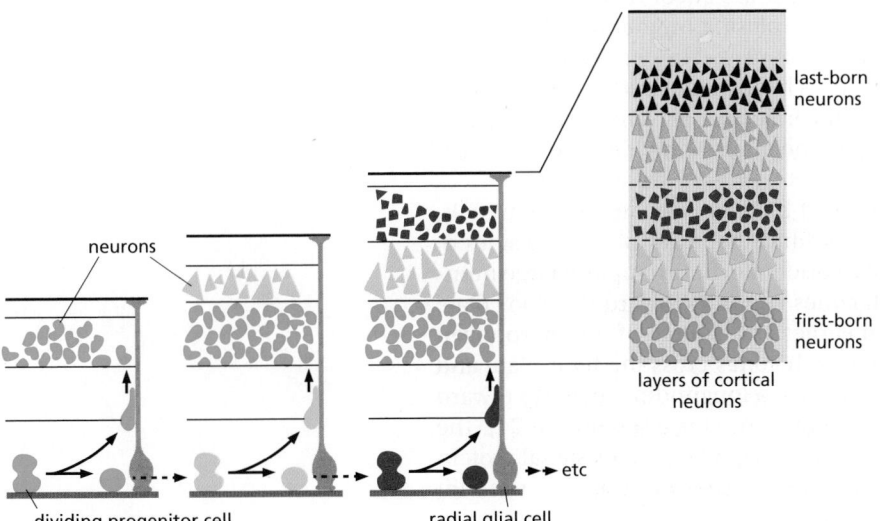

neurons

last-born
neurons

first-born
neurons

layers of cortical
neurons

dividing progenitor cell

radial glial cell

etc

Figure 21–71 **Programmed production of different types of neurons at different times from dividing progenitors in the cerebral cortex of the brain of a mammal.** Close to one face of the cortical neuroepithelium, progenitor cells divide, in stem-cell fashion, to produce successive generations of neurons (colored here *blue, green, red, orange,* and *black*). The neurons migrate out toward the opposite face of the epithelium by crawling along the surfaces of radial glial cells, as shown in Figure 21–70. The first-born neurons settle closest to their birthplace, while neurons born later crawl past them to settle farther out. Successive generations of neurons thus occupy different layers in the cortex and have different intrinsic characters according to their birth dates.

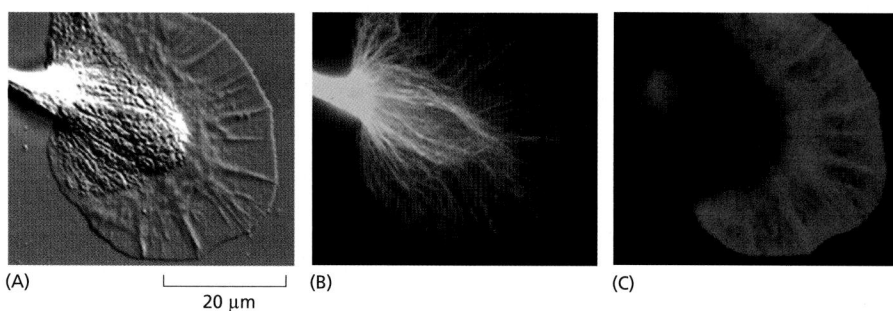

(A)

20 μm

(B)

(C)

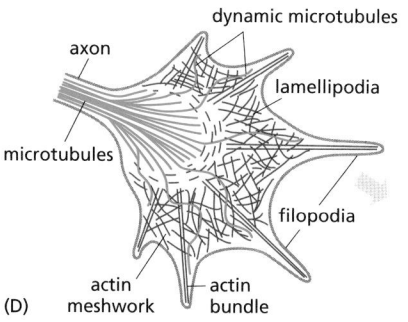

(D)

Figure 21–72 **Internal architecture of a neuronal growth cone, as seen in culture on a flat substratum.** The growth cone forms as an expansion of the tip of the growing axon. (A) Image by interference-contrast microscopy. (B) Immunostaining to show microtubules *(green)*. (C) Immunostaining to show actin filaments *(red)*. (D) Diagram of the cytoskeletal machinery. Filopodia form and push forward by assembly of actin filaments at the leading edge of the growth cone. Microtubules stabilize the directional decisions made by the actin-rich protrusions. Filopodia adhering to the flat substratum contract and pull the growth cone forward. (Images by Chi-Hung Lin, Paul Forscher Laboratory, Yale University, New Haven, CT.)

contacts an unfavorable surface, it withdraws; when it contacts a more favorable surface, it persists longer, steering the growth cone in that direction. In this way, the growth cone is guided by subtle variations in the properties of the surfaces over which it moves. At the same time, it is sensitive to specific signaling molecules, which—as we discuss next—can either encourage or hinder its advance.

A Variety of Extracellular Cues Guide Axons to their Targets

Growth cones generally travel toward their targets along predictable routes, according to programs stored in the memory of the particular neuron to which they belong (Movie 21.7). In the simplest case, a growth cone can take a route that has been pioneered by other neurites, which they follow by contact guidance. As a result, nerve fibers in a mature animal are usually found grouped together in tight parallel bundles (called fascicles or fiber tracts). Such crawling of growth cones along axons is partly mediated by homophilic cell–cell adhesion molecules—membrane glycoproteins that help a cell displaying them to stick to any other cell that displays the same molecules. As discussed in Chapter 19, many homophilic adhesion molecules fall into one of two main classes: they are members of either the immunoglobulin superfamily, such as *N-CAM*, or the Ca^{2+}-dependent cadherin family, such as *N-cadherin*. Members of both families are generally present on the surfaces of growth cones, of axons, and of various other cell types that growth cones crawl over, including glial cells in the central nervous system and muscle cells in the periphery of the body. Growth cones also migrate over components of the extracellular matrix. When tested with neurons growing in a culture dish, some of the matrix molecules, such as laminin, favor axon outgrowth, while others, such as chondroitin sulfate proteoglycans, discourage it. But exactly how the matrix functions to guide axons in intact animals remains to be discovered.

Growth cones are generally guided by a succession of different cues at different stages of their journey, as summarized in Figure 21–73. Many of these cues involve specific signaling molecules. Some of these are encountered in the extracellular matrix, while others are attached to the plasma membrane of cells that the growth cones touch. Another important part is played by chemotactic factors; these are proteins secreted from cells that act as beacons at strategic points along the path—some attracting, others repelling. The trajectory of *commissural axons*—axons that cross from one side of the body to the other—provides a well-studied example.

Commissural axons are a general feature of bilaterally symmetrical animals, such as us, because they are required to coordinate behavior of the two sides of the body. In the developing spinal cord of a vertebrate, for example, a large number of neurons send their axonal growth cones ventrally toward the floor plate (the same structure that we encountered earlier as a source of the morphogen Sonic hedgehog—see Figure 21–69). The growth cones cross the floor plate and then turn abruptly through a right angle to follow a longitudinal path up toward the brain, parallel to the floor plate but never again crossing it (Figure 21–74). The first stage of the journey depends on a concentration gradient of the signal protein **Netrin**, secreted by the cells of the floor plate: the commissural growth cones sniff their way toward its source.

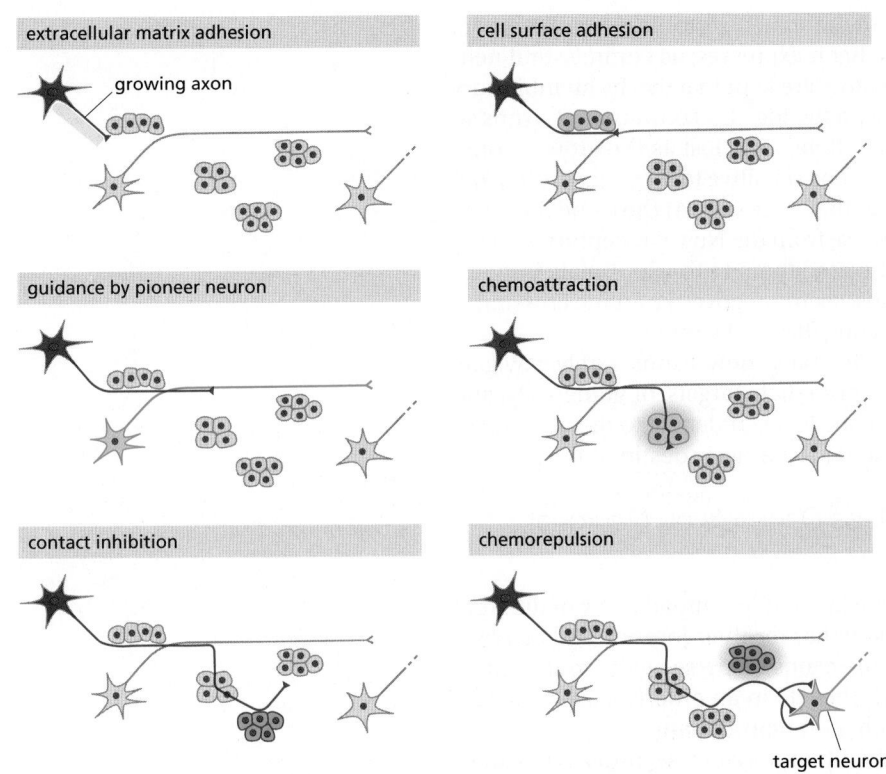

Figure 21–73 **Mechanisms of growth-cone guidance.** Growth cones use a variety of extracellular cues to navigate to distant targets. They can adhere to the extracellular matrix or to the surfaces of other cells, or they can be repelled by them; they can crawl, for example, by homophilic adhesion along the axons of pioneer neurons; and they can be attracted or repelled by soluble guidance signals. (After E. Kandel et al., Principles of Neural Science, 5th ed., New York: McGraw Hill Medical, 2012.)

If commissural growth cones are attracted to the floor plate, why do they cross it and emerge on the other side, instead of staying in the attractive territory? And having crossed it, why do they never cross back again? The answers lie in a change in the responsiveness of the growth cones during their journey. As the growth cones cross the midline, they lose sensitivity to Netrin and become sensitive instead to a signal protein called **Slit** (see Figure 21–74). Slit is also produced by the floor plate, but it has the opposite effect to that of Netrin: it repels the growth

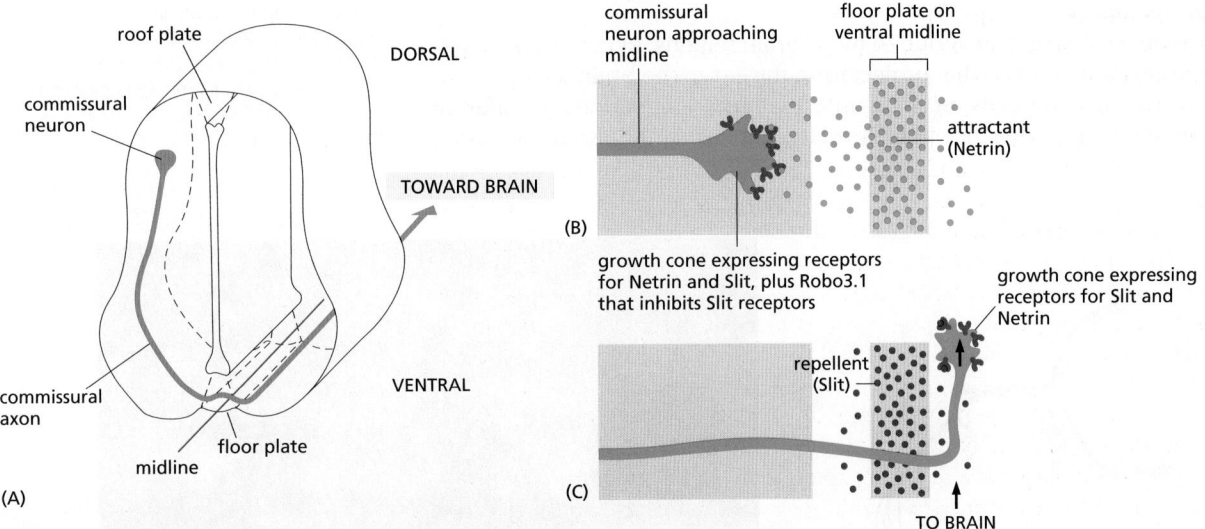

Figure 21–74 **The guidance of commissural axons.** (A) The pathway taken by commissural axons in the embryonic spinal cord of a vertebrate. (B) Attraction to the midline. The growth cone is first attracted to the floor plate by Netrin, which is secreted by the floor-plate cells and acts on the receptor DCC in the axonal membrane. (C) Repulsion from the midline after crossing it. As the growth cone crosses the floor plate, Slit comes into play: it binds to its receptors Robo1 and Robo2 and acts as a repellent to keep the growth cone from re-entering the floor plate. In addition, it blocks responsiveness to the attractant Netrin. Before crossing the midline, the commissural neurons express Robo3.1, an alternative splice form of Robo3 that is related to Robo proteins but blocks Slit signaling. As neurites cross the midline, Robo3.1 is lost and growth cones become responsive to Slit and are repelled from the midline.

cones, preventing them from re-entering the midline territory. The responses of the growth cone depend on the receptors that it expresses: as commissural neurons approach the floor plate, the Slit receptors are kept inactive by an inhibitory protein (Robo3.1) in the same membrane, allowing the commissural axons to grow to the midline without being repelled. Robo3.1 is lost as the growth cones cross the midline; now the growth cones become sensitive to repulsion by Slit and are thereby prevented from crossing back to the other side. At the same time, signals from the Slit receptors interfere with those from the Netrin receptors, making the growth cones deaf to the signal that attracted them to the floor plate initially. A similar mechanism, using similar proteins, seems to govern midline crossing of commissural axons in other animals, including flies and worms.

The guidance of commissural axons illustrates how axons rarely navigate directly to their targets. Instead, they use intermediate targets, or guideposts, and switch their sensitivities as they move from one local guidepost to the next, steering their way through a complex environment to a far-away destination.

The Formation of Orderly Neural Maps Depends on Neuronal Specificity

In many cases, neurons of a similar type are laid out in a broad array of different positions, but send out axons that come together for their journey and arrive at the target region in a tight bundle. There the axons disperse again, to terminate at different sites in the target territory. This they do in an orderly way, creating a regular mapping from one territory to another—a **neural map**.

The axon projection from the eye to the brain provides an important example. The neurons in the retina that convey visual information back to the brain are called *retinal ganglion cells* (*RGCs*). There are more than a million of them in humans, each one reporting on a different part of the visual field. Their axons converge on the optic nerve head at the back of the eye and travel together along the developing optic nerve toward the brain. Their main site of termination, in most vertebrates other than mammals, is the *optic tectum*—a broad expanse of cells in the midbrain. In connecting with tectal neurons, the RGC axons distribute themselves in a predictable pattern according to the arrangement of their cell bodies in the retina: RGCs that are neighbors in the retina connect with target cells that are neighbors in the tectum. The orderly projection creates a *retinotopic map* of visual space on the tectum (**Figure 21-75**).

Orderly maps of this sort are found in many brain regions. In the auditory system, for example, the neurons that project from the ear to the brain form a tonotopic map in which brain cells receiving information about sounds of different pitch are ordered along a line, like the keys of a piano. And in the somatosensory

Figure 21–75 The neural map from eye to brain in a young zebrafish. (A) Diagrammatic view, looking down on the top of the head. (B) Fluorescence micrograph. Fluorescent tracer dyes have been injected into each eye—*red* into the anterior part, *green* into the posterior part. The tracer molecules have been taken up by the neurons in the retina and carried along their axons, revealing the paths they take to the optic tectum in the brain and the map that they form there. (Courtesy of Chi-Bin Chien, from D.H. Sanes, T.A. Reh and W.A. Harris, Development of the Nervous System. San Diego, CA: Academic Press, 2000.)

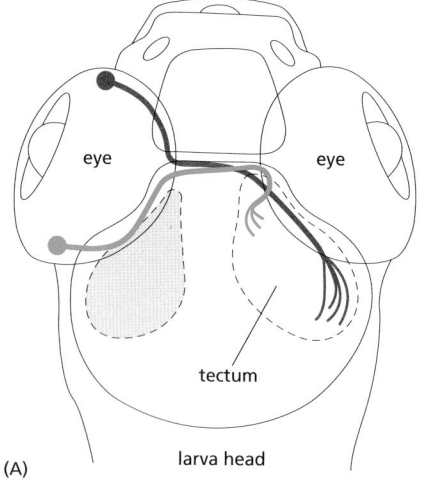

(A) eye   eye   tectum   larva head

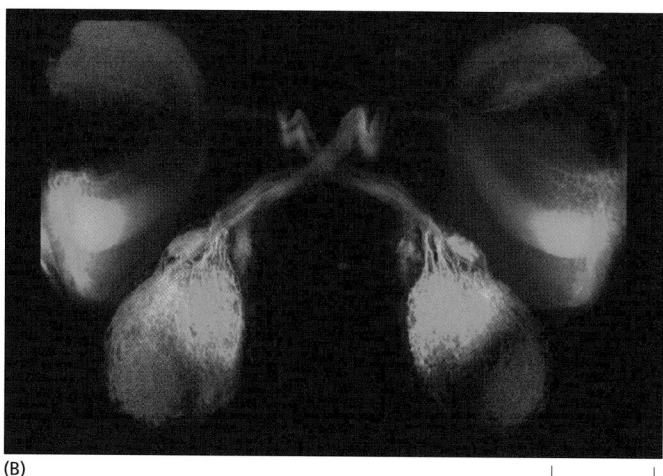

(B)

100 μm

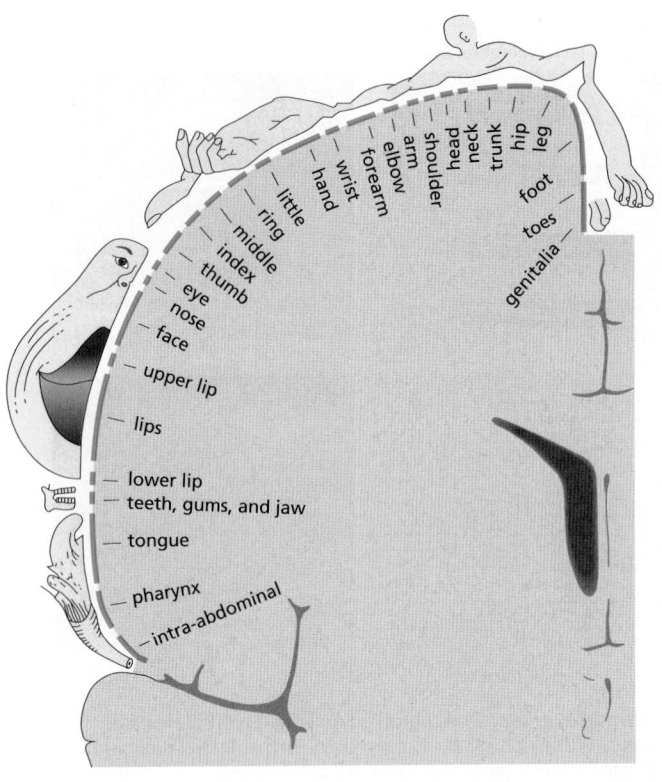

Figure 21–76 **A map of the body surface in the human brain.** The surface of the body is mapped onto the somatosensory region of the cerebral cortex by using an orderly system of nerve cell connections to pair body sites with the brain sites that receive their sensory information. This means that the map in the brain is largely faithful to the topology of the body surface, even though different body regions are represented at different magnifications according to their density of innervation. The homunculus (the "little man" in the brain) has big lips, for example, because the lips are a particularly large and important source of sensory information. The map was determined by stimulating different points in the cortex of conscious patients during brain surgery and recording what they said they felt. (Interphoto/Alamy Stock Photo.)

system, neurons conveying information about touch map onto the cerebral cortex so as to mark out a "homunculus"—a small, distorted, two-dimensional image of the body surface (Figure 21–76).

The retinotopic map of visual space in the optic tectum is the best characterized of all these maps. How does it arise? A famous experiment in the 1940s on frogs provided an important clue. If the optic nerve of a frog is cut, it will regenerate. The retinal axons grow back to the optic tectum, restoring normal vision. If, however, the eye is in addition rotated in its socket at the time of cutting of the nerve, so as to put originally ventral retinal cells in the position of dorsal retinal cells, vision is still restored, but with an awkward flaw: the animal behaves as though it sees the world upside down and left–right inverted (Figure 21–77). If food is dangled in front of it, for example, it will lunge perversely backward. This is because the misplaced retinal cells make the connections appropriate to their original, not their actual, positions. It seems that the retinal ganglion cells (RGCs) have positional values—position-specific biochemical properties representing records of their original location in the retina, assigned perhaps by earlier morphogen gradients, and making RGCs on opposite sides of the retina intrinsically different.

Such nonequivalence among neurons is referred to as **neuronal specificity**. It is this intrinsic characteristic that guides the retinal axons to their appropriate target sites in the tectum. Those target sites themselves are distinguishable by the

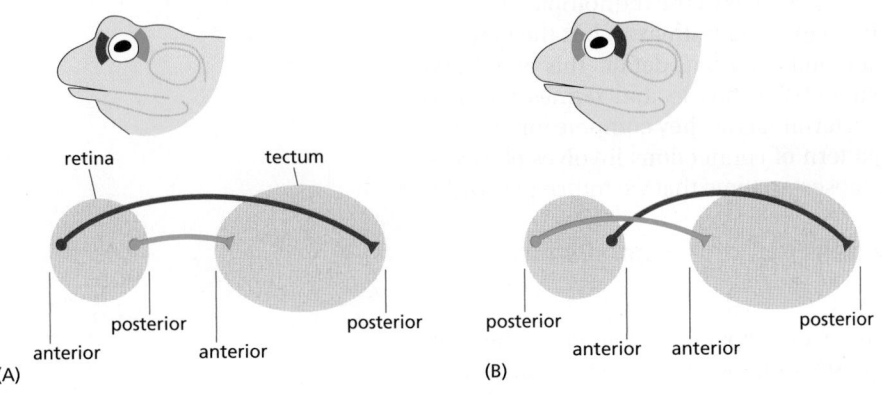

Figure 21–77 **Neurons in different regions of the retina project axons to different regions in the tectum.** (A) Neurons (RGCs) in the anterior retina project axons to the posterior tectum (as shown in Figure 21–75 for zebrafish). (B) Regeneration experiments show that retinal neurons have an intrinsic preference for the part of the tectum they normally connect to. If the eye is surgically rotated when the optic nerve is cut, the regenerating retinal axons connect to their original targets, creating an inverted map. (After E. Kandel et al., Principles of Neural Science, 5th ed., New York: McGraw Hill Medical, 2012.)

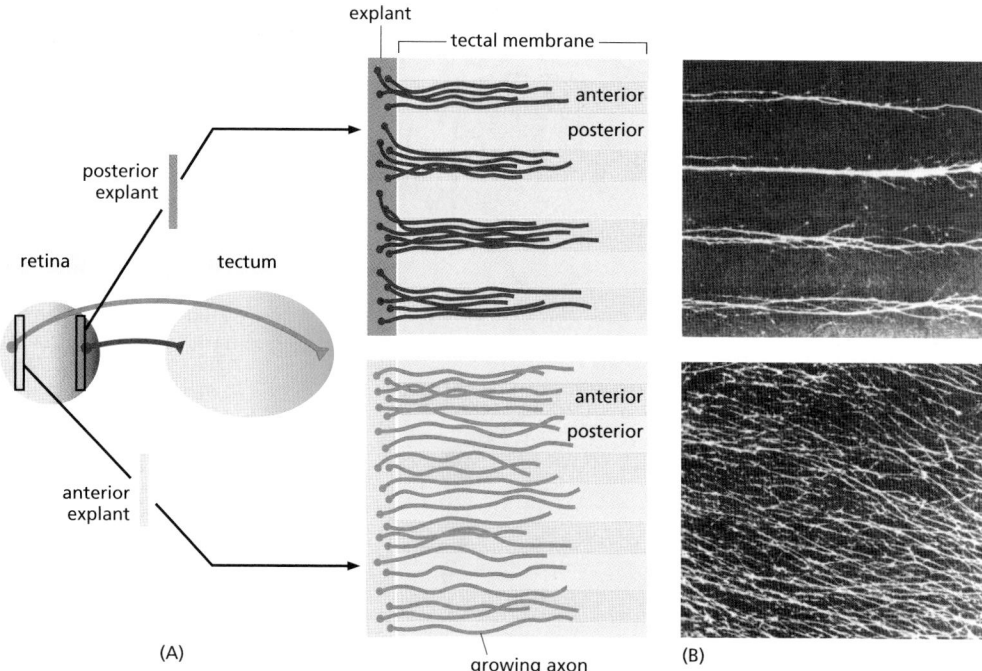

Figure 21–78 Selectivity of retinal axons growing over tectal membranes. (A) Diagram of an experiment performed with cells from a chick embryo. The culture substratum is coated with alternating stripes of membrane prepared either from posterior tectum or from anterior tectum. Axons from posterior retina grow on anterior tectal membrane but are repelled by posterior tectal membrane. Axons from anterior retina show different (less selective) behavior. (B) Photograph of results. The retinal axons, growing out from the left, are made visible by staining them with a fluorescent marker. The selective pattern of outgrowth shows that anterior tectum differs from posterior tectum, and anterior retina correspondingly differs from posterior retina. In the intact organism, this serves to orient a retinotopic map; the map is refined by subsequent competitive interactions among the anterior and posterior retinal axons, which push the anterior retinal cells off anterior tectal territory. (From J. Walter et al., *Development* 101:685–696, 1987.With permission from the Company of Biologists.)

retinal axons because the tectal cells also carry positional labels. Thus, the neural map depends on a correspondence between two systems of positional markers, one in the retina and the other in the tectum.

How are these markers used to make the map? When posterior axons are allowed to grow out over a carpet of anterior or posterior tectal membranes in a culture dish, they show selectivity. Posterior axons strongly prefer the anterior tectal membranes, as *in vivo*, whereas anterior axons show no preference or prefer posterior tectal membranes (Figure 21–78). The key difference between anterior and posterior tectum is not an attractive factor on the anterior tectum but a repulsive factor on the posterior tectum, to which posterior retinal axons are sensitive but anterior retinal axons are not. If a posterior retinal growth cone touches posterior tectal membrane, it collapses its filopodia and withdraws.

In this system, as in others that we have mentioned, the repulsive interactions are mediated by ephrin–Eph signaling—specifically, EphrinA–EphA signaling for the anteroposterior axis (Figure 21–79). An analogous mechanism based on EphB–EphrinB signaling orients the dorsoventral axis of the retinotopic map.

These mechanisms serve to orient the map along both axes, but they are not enough by themselves to ensure accurate point-to-point detail. This is brought about through a long process of adjustment that fills in and refines the map through interactions among the RGC axon terminals as they compete for territory on the tectum. This refinement of the pattern of connections involves electrical signaling in the system of developing synapses—a topic that we return to shortly.

Both Dendrites and Axonal Branches From the Same Neuron Avoid One Another

Axons and dendrites from different neurons can repel one another, or they can cohere; they can collaborate to form synapses, or they can compete. Remarkably,

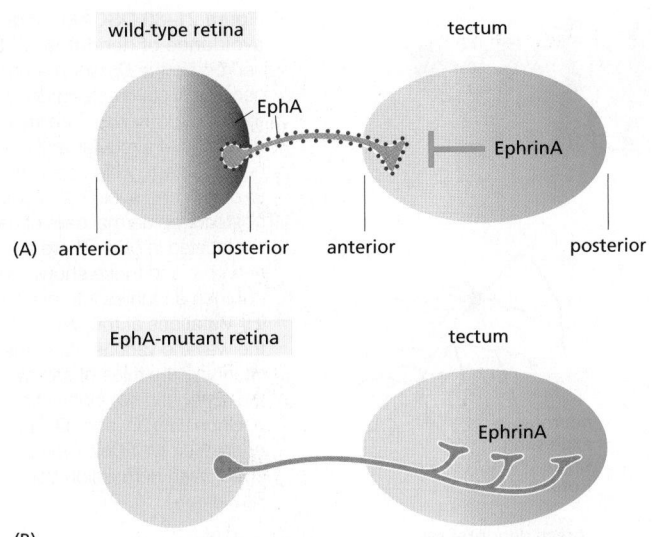

wild-type retina

tectum

EphA

EphrinA

(A) anterior   posterior   anterior   posterior

EphA-mutant retina   tectum

EphrinA

(B)

Figure 21–79 **Ephrin signaling orients the retinotopic map.** (A) Neurons in the posterior retina express EphA. As their axons reach the tectum, they are repelled by high levels of EphrinA protein in the posterior tectum and project preferentially to the anterior tectum. (B) In EphA-mutant mice, posterior retinal axons feel no such repulsion and project more widely within the tectum. (After E. Kandel et al., Principles of Neural Science, 5th ed., New York: McGraw Hill Medical, 2012.)

axons or dendrites can also repel each other when they arise from a single neuron. Such self-avoidance prevents the neuron from making purposeless synapses with itself; it also helps the cell spread out its processes widely so as to innervate a broad territory.

Self-avoidance poses a problem. If the same self-recognition molecule were used in every neuron, all neurons in the brain would repel each other. Some classes of neurons do show this sort of mutual repulsion, creating solitary territories—a phenomenon called *tiling*; but in most cases, axons and dendrites from different neurons can overlap with one another. How then can the processes put out by a single neuron distinguish between self and non-self? This conundrum has been partially resolved by the discovery of a remarkable set of proteins that endow each neuron with a label unlike that of its neighbors. These are the *DSCAM* proteins in *Drosophila* and the *protocadherins* in vertebrates. As described in Chapter 7, DSCAM proteins are extraordinary for the number of isoforms that can be generated by alternative RNA splicing—more than 30,000 variants for DSCAM1 (see Figure 7–57). Diversity arises from alternative exons that code for three highly variable extracellular immunoglobulin domains. Each DSCAM1 isoform engages in homophilic binding (see Figure 19–5), but remarkably, all the variable domains need to be identical for this to occur. Thus, one cell surface will bind to another via DSCAM only when the two cell surfaces express identical isoforms. The result of binding is repulsion, although the detailed mechanisms are poorly understood.

If alternative splicing occurs in a random fashion in each cell, neighboring processes from different neurons are unlikely to express the same DSCAM1 variant, so only the processes of the same cell will repel one another. Neurons that lack all DSCAM1 variants have severe defects in neuronal self-avoidance. Engineering *Drosophila* so that all of its neurons produce a single isoform restores self-avoidance; but now the processes of neighboring neurons express the same isoform and repel each other, resulting in the phenomenon of tiling (**Figure 21–80**).

Vertebrate neurons use a similar self-avoidance strategy to pattern their axons and dendrites, but instead of DSCAMs, they use protocadherins for self/non-self discrimination. The *Protocadherin* locus encodes 58 related cadherin-like transmembrane proteins that are expressed in different combinations in single neurons. Homophilic recognition results in self-avoidance of dendrites emanating from the same neuron; neighboring dendrites of different neurons express different protocadherins and thus evade repulsion. Thus, although insect DSCAM and vertebrate protocadherin proteins share no sequence homology, they mediate similar self-avoidance strategies.

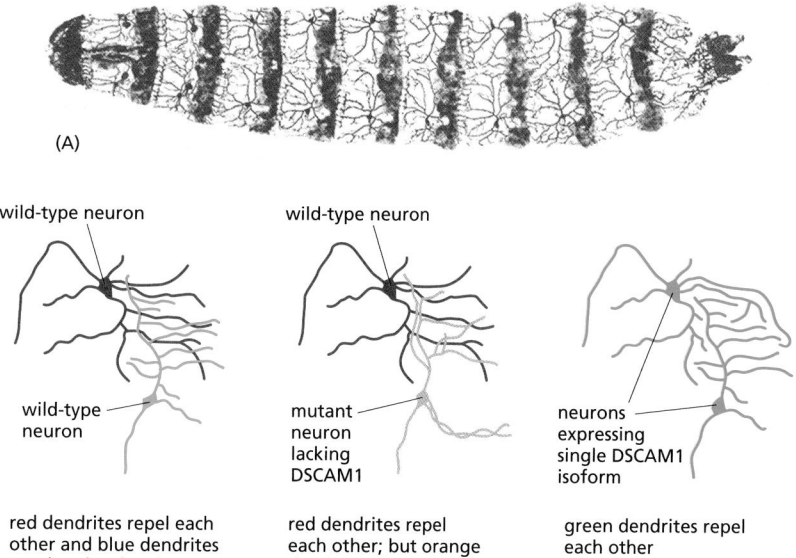

(A)

wild-type neuron

wild-type neuron

wild-type neuron

wild-type neuron

mutant neuron lacking DSCAM1

neurons expressing single DSCAM1 isoform

red dendrites repel each other and blue dendrites repel each other; but blue and red do not repel each other

red dendrites repel each other; but orange dendrites repel neither themselves nor red dendrites

green dendrites repel each other

(B)

Figure 21–80 **DSCAM mediates self-avoidance of dendrites.** (A) Sensory neurons in the *Drosophila* peripheral nervous system extend dendrites along the larval body wall. The image shows the dendrites of a regular array of photosensing neurons *(red)*, which allow the larva to detect and avoid harmful light. The posterior epidermal cells of each segment are labeled in *blue*. There are many neurons, and those shown here spread out their dendrites into overlapping fields. (B) Mutations at the *Dscam* locus upset the way the various dendrites interact, changing the rules of self-avoidance and the distribution of innervation. (A, courtesy of Chun Han; B, after D. Hattori et al., *Annu. Rev. Cell Dev. Biol.* 24:597–620, 2008. With permission from Annual Reviews.)

Target Tissues Release Neurotrophic Factors That Control Nerve Cell Growth and Survival

Eventually, axonal growth cones reach the target region where they must halt and make synapses. These synapses, as a rule, are destined to transmit neural signals in one direction, from axon to target cell. The development of synapses, however, depends on signaling in both directions: signals from the target tissue not only help control which growth cones synapse where (as we discuss shortly), but can also regulate how many of the innervating neurons survive.

Many types of vertebrate neurons are produced in excess; up to 50% or more of some of them die soon after they reach their target, even though they appear perfectly normal and healthy up to the time of their death. About half of all the motor neurons that send axons to skeletal muscle, for example, die within a few days after making contact with their target muscle cells. A similar proportion of the sensory neurons that innervate the skin die after their growth cones have arrived there.

This large-scale *normal neuronal death* often seems to reflect the outcome of a competition, in which the target tissue releases a limited amount of a specific **neurotrophic factor** that the neurons innervating the tissue require to survive; those that do not get enough die by programmed cell death. If the amount of target tissue is increased—for example, by grafting an extra limb bud onto the side of the embryo—more limb-innervating neurons survive; conversely, if the limb bud is cut off, the same neurons all die (**Figure 21–81**). In this way, although individuals may vary in their bodily proportions, they always retain the right number of motor neurons to innervate all their muscles and the right number of sensory neurons to innervate their body surface. The strategy of overproduction followed by death of surplus cells may seem wasteful, but it provides a simple and effective means to adjust the number of innervating neurons according to the amount of tissue requiring innervation.

The first neurotrophic factor to be identified, and still the best characterized, is called *nerve growth factor* (*NGF*)—the founding member of the **neurotrophin** family of signal proteins. It promotes the survival and growth of specific classes of sensory neurons and of sympathetic neurons (a subclass of peripheral neurons that control contractions of smooth muscle and secretion from exocrine glands).

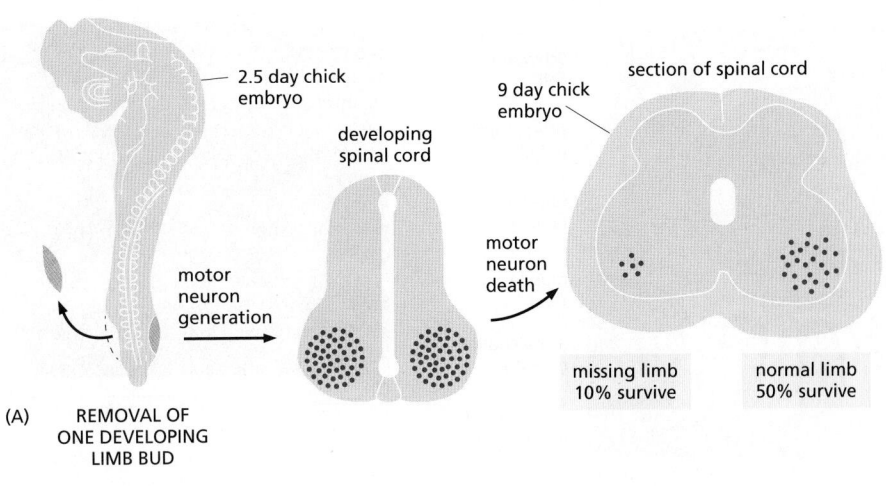

(A) REMOVAL OF ONE DEVELOPING LIMB BUD

2.5 day chick embryo

developing spinal cord

motor neuron generation

9 day chick embryo

section of spinal cord

motor neuron death

missing limb 10% survive

normal limb 50% survive

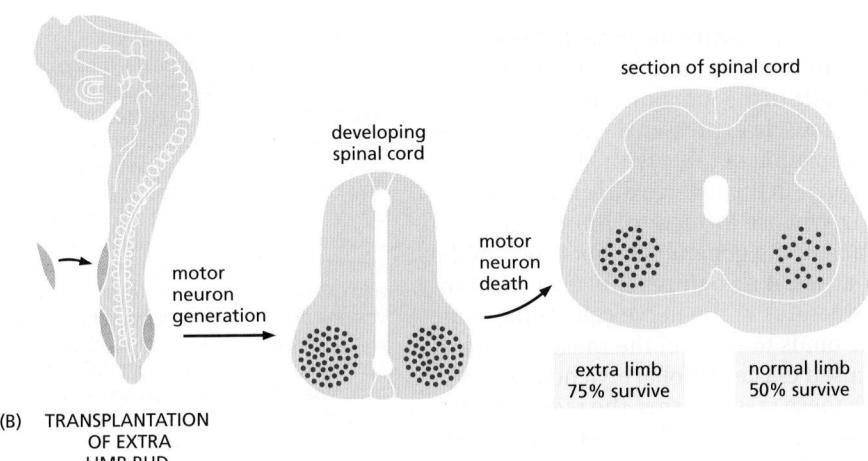

(B) TRANSPLANTATION OF EXTRA LIMB BUD

developing spinal cord

motor neuron generation

motor neuron death

section of spinal cord

extra limb 75% survive

normal limb 50% survive

Figure 21–81 The survival of motor neurons depends on signals provided by the target muscles. (A) Removal of the limb bud shortly after arrival of motor axons results in the death of motor neurons in the spinal cord on the amputated side. (B) Transplantation of an extra limb bud increases the survival of motor neurons. (After E. Kandel et al., Principles of Neural Science, 5th ed., New York: McGraw Hill Medical, 2012.)

NGF is produced by the tissues that these neurons innervate. When extra NGF is provided, extra sensory and sympathetic neurons survive, just as if extra target tissue were present. Conversely, in a mouse with a mutation that inactivates the gene for NGF or for its receptor (a receptor tyrosine kinase called TrkA), almost all sympathetic neurons and the NGF-dependent sensory neurons are lost. There are many neurotrophic factors, only a few of which belong to the neurotrophin family, and they act in different combinations to promote the survival and growth of different classes of neurons.

Formation of Synapses Depends on Two-Way Communication Between Neurons and Their Target Cells

At journey's end, the task of a growth cone is to halt its travels and make synapses with specific target cells. Synapses were introduced in Chapter 11, where we discussed channels and the electrical properties of membranes. Two main classes of synapses are found in vertebrates; those made with muscle cells and those made with other neurons. Synapse formation is best understood in the case of the highly specialized connections between motor neurons and skeletal muscle cells—so-called **neuromuscular junctions** (see Figure 11–38). During synapse formation, the axonal growth cone differentiates into a *nerve terminal* that contains synaptic vesicles filled with the neurotransmitter acetylcholine, while acetylcholine receptors become clustered in the muscle cell plasma membrane at the site of synapse formation. A synaptic cleft separates the pre- and postsynaptic plasma membranes, and a thin sheet of basal lamina lies in this space between them (**Figure 21–82**).

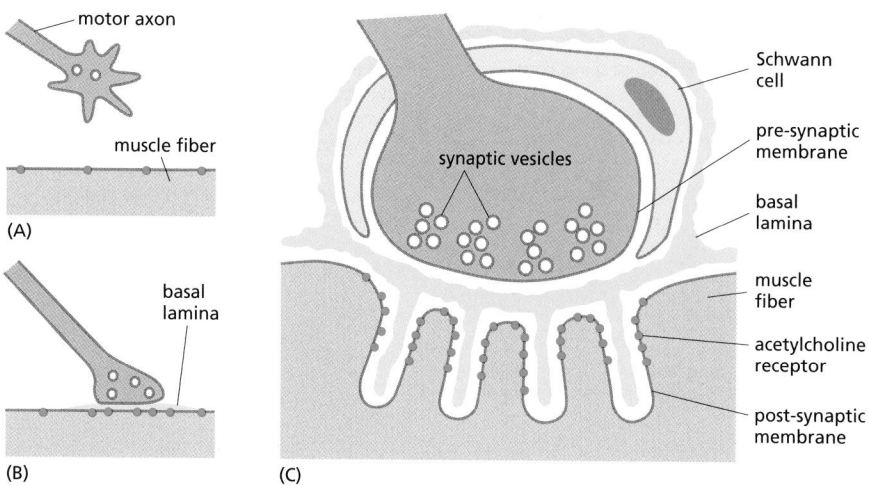

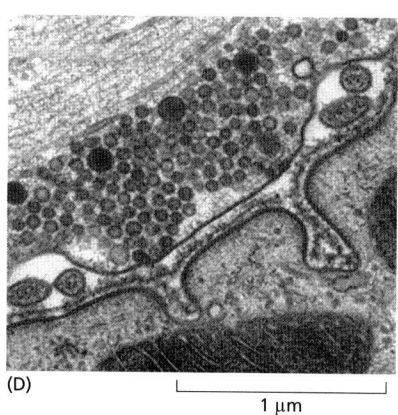

Formation of the synapse involves two-way communication between the muscle cell and axonal growth cone: each of them, under the influence of the other, must reorganize the molecules on its side of the junction. The growth cone releases the signal protein **Agrin**, while the muscle expresses the Agrin receptor LRP4. Agrin binding to LRP4 stimulates association of LRP4 with MuSK, a receptor tyrosine kinase. LRP4 also serves as a signal in the reverse direction, from the muscle to the axon (**Figure 21–83**). During synapse formation, MuSK and LRP4 cluster in the muscle cell plasma membrane in the general neighborhood of the future synapse. As the growth cone approaches, it recognizes LRP4, which stimulates the differentiation of presynaptic structures in the nerve cell. At the same time, Agrin released from the growth cone binds to LRP4 in the muscle cell; this activates MuSK, and promotes a more focused clustering of acetylcholine receptors in the muscle cell membrane. Through these mechanisms, the reciprocal signaling of LRP4 from muscle to growth cone—and of Agrin from growth cone to muscle—induces the coordinated, localized differentiation of pre- and postsynaptic structures.

Synapse formation between neurons in the CNS is far more challenging, both for the neurons and for the scientists trying to understand the molecular basis of its specificity, and it remains poorly understood.

Figure 21–82 Formation of the neuromuscular junction. (A) The growth cone of a motor axon approaches the muscle fiber. (B) Initial synapse formation is characterized by the accumulation of synaptic vesicles at the axon terminal and the formation of a specialized basal lamina in the synaptic cleft. (C) As the neuromuscular junction matures, the synaptic cleft accumulates basal lamina and extracellular matrix proteins, synaptic vesicles cluster at presynaptic release sites, and neurotransmitter receptors cluster at postsynaptic sites. Schwann (glial) cells accompany the motor axon and wrap around its terminus outside the region of synaptic contact. (D) Transmission electron micrograph of the region of synaptic contact. (D, micrograph by John Heuser. From *J. Electron Microsc.* 60 (Suppl 1), 2011. With permission from Oxford University Press.)

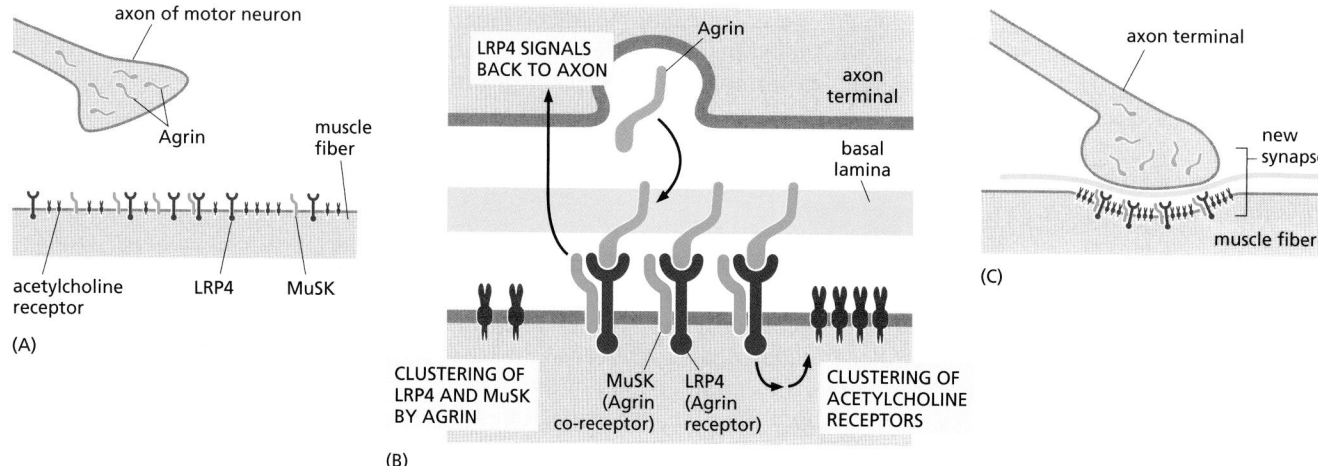

Figure 21–83 Reciprocal signaling during neuromuscular synapse differentiation. (A) The Agrin receptor LRP4 and its co-receptor MuSK cluster in the muscle cell membrane in the general neighborhood of the future synapse. (B) As the growth cone approaches, it recognizes LRP4, which stimulates differentiation of presynaptic structures. Reciprocally, Agrin is released from the nerve terminal, binds to a complex of LRP4 and MuSK in the muscle, and (C) promotes the further and more focused clustering of the LRP4 and acetylcholine receptors in the muscle cell. Although the Agrin/MuSK/LRP4 machinery organizes the synapse, the process also depends on electrical signaling via the acetylcholine receptors. It is not yet known how LRP4 signals to the motor axon.

Synaptic Pruning Depends on Electrical Activity and Synaptic Signaling

The two-way exchange of signals between axon growth cones and muscle cells controls the initial formation of neuromuscular junctions, but it is only the first step in the establishment of the final pattern of the synaptic connections. Each muscle cell at first receives synapses from several motor neurons, but in the end it is left innervated by only one. This process of **synapse elimination** depends on active synaptic communication and electrical activity. If synaptic transmission is blocked by a toxin that binds to the acetylcholine receptors in the muscle cell membrane, or if axonal electrical activity is blocked by a toxin that binds to sodium channels in the axon plasma membrane, the muscle cell retains its multiple innervation beyond the normal time of elimination.

The phenomenon of *activity-dependent synapse elimination* is encountered in almost every part of the developing vertebrate nervous system (**Figure 21–84**). It has a key role, for example, in the refinement of the retinotopic map discussed earlier. Synapses are first formed in abundance and distributed over a broad target field; then the system of connections is pruned back and remodeled by competitive processes that depend on electrical activity and synaptic signaling. The elimination of synapses in this way is distinct from the elimination of surplus neurons by cell death, and it occurs after the period of normal neuronal death is over. Synapse remodeling during neural development, however, involves more than just synapse elimination; it also involves synapse reinforcement, as we discuss next.

Neurons That Fire Together Wire Together

Throughout the nervous system, and throughout life, activity-dependent elimination and reinforcement of synapses plays a fundamental part in adjusting the detailed anatomy of the neural network according to functional requirements. The importance of these processes, and their underlying rules, emerged half a century ago from a groundbreaking series of experiments on the developing visual system of young mammals.

In the brain of most mammals, axons relaying visual inputs from the two eyes are brought together in a specific neuronal layer in the visual region of the cerebral cortex. Here, they form two overlapping maps of the external visual field, one as perceived through the right eye, the other as perceived through the left. Although there may be a tendency for right- and left-eye inputs to be segregated even before synaptic communication begins, a large proportion of the axons carrying information from the two eyes at early stages form synapses together on shared target neurons in the visual cortex. A period of early electrical signaling activity,

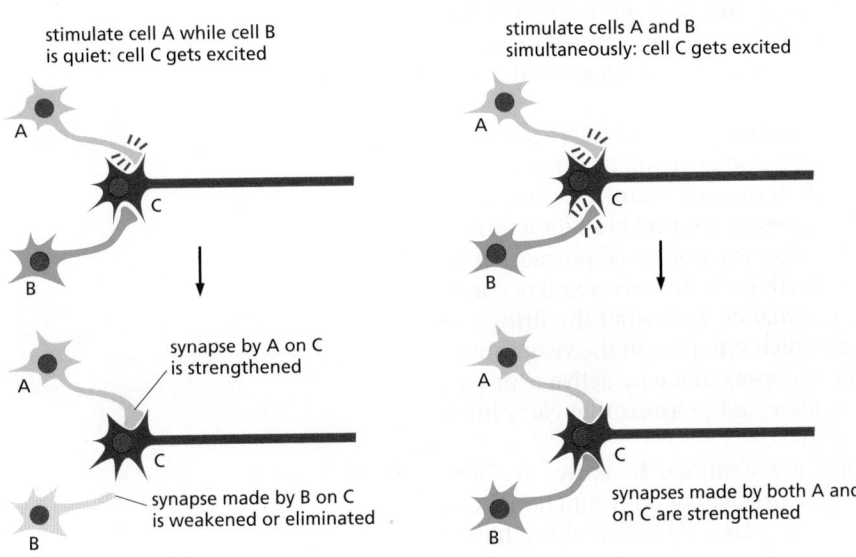

stimulate cell A while cell B is quiet: cell C gets excited

stimulate cells A and B simultaneously: cell C gets excited

synapse by A on C is strengthened

synapse made by B on C is weakened or eliminated

synapses made by both A and B on C are strengthened

Figure 21–84 **Synapse modification and its dependence on electrical activity.** Experiments in several systems indicate that synapses are strengthened or weakened by electrical activity according to the rule shown in the diagram. The underlying principle appears to be that each excitation of a target cell tends to weaken any synapse where the presynaptic axon terminal has been quiet, but to strengthen any synapse where the presynaptic axon terminal has just been active. As a result, any synapse that is repeatedly weakened and rarely strengthened is eventually eliminated altogether.

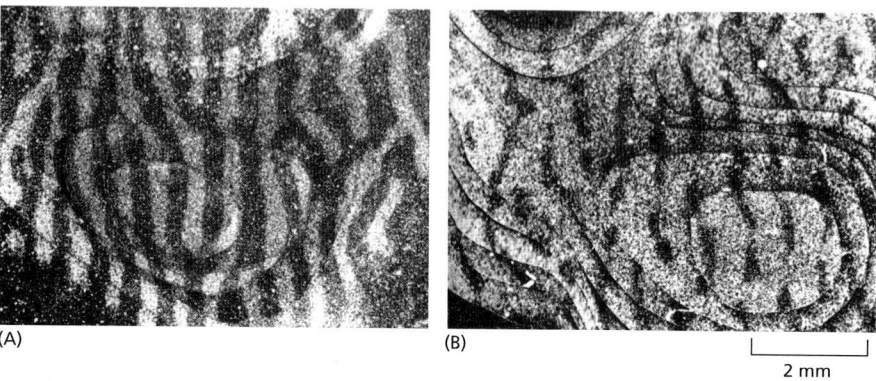

Figure 21–85 **Ocular dominance columns in the visual cortex of a monkey's brain, and their sensitivity to visual experience.** (A) Normally, stripes of cortical cells driven by the right eye alternate with stripes, of equal width, driven by the left eye. The stripes, set up before birth, are revealed here by injecting a radioactive tracer molecule into one eye, allowing time for this tracer to be transported to the visual cortex, and detecting radioactivity there by autoradiography, in sections cut parallel to the cortical surface. (B) If one eye is kept covered after birth, during the sensitive period of development, and thus deprived of visual experience, its stripes shrink and those of the active eye expand. In this way, the deprived eye may lose the power of vision almost entirely. (From D.H. Hubel, T.N. Wiesel and S. LeVay, *Philos. Trans. R. Soc. Lond. B Biol. Sci.* 278:377–409, 1977. With permission from The Royal Society.)

however, occurring spontaneously and independently in each retina before birth, leads to a remarkable pattern of *ocular dominance columns* in the visual cortex: stripes of cells driven by inputs from the right eye alternating with stripes driven by inputs from the left eye (**Figure 21–85**).

The basis for these phenomena became clear from ingenious experiments interfering artificially with visual experience and altering the coordination of electrical signaling in the two eyes. These studies, and many others subsequently, have highlighted a simple but profoundly important principle that seems to govern synapse reinforcement and elimination throughout the nervous system. When two (or more) neurons synapsing on the same target cell fire at the same time, they reinforce their connections to that cell; when they fire at different times, they compete, so that all but one of them tend to be eliminated. This **firing rule** is expressed in the catchphrase "neurons that fire together wire together."

The firing rule provides a simple interpretation of the developmental phenomenon we have just described in the mammalian visual system. A pair of axons bringing information from neighboring sites in the left eye will frequently fire together, and therefore wire together, as will a pair of axons from neighboring sites in the right eye; but a right-eye axon and a left-eye axon will rarely fire together, and will instead compete. Indeed, if activity from both eyes is silenced using toxins that block axonal electrical activity or synaptic signaling, as described above, the inputs fail to segregate correctly.

The segregation of inputs from the two eyes is only the first of a series of activity-dependent adjustments of visual connections, whose maintenance is extraordinarily sensitive to experience early in life. If, during a certain *sensitive period* (ending at about 5 years of age in humans), one eye is kept covered for a time so as to deprive it of visual stimulation, while the other eye is allowed normal stimulation, the deprived eye loses its synaptic connections to the cortex and becomes almost entirely, and irreversibly, blind. In accordance with what the firing rule would predict, a competition has occurred in which synapses in the visual cortex made by inactive axons are eliminated while synapses made by active axons are consolidated. In this way, cortical territory is allocated to axons that carry information and is not wasted on those that are silent.

Activity-dependent synaptic changes are not confined to early life. They also occur in the adult brain, where many synapses show both functional and morphological alterations with use. This *synaptic plasticity* is thought to have a fundamental role in learning and memory. Clearly, for the nervous system as for other parts of the body, developmental processes do not end at birth, as we discuss in the next chapter.

Summary

The development of the nervous system proceeds in four phases. First, neurons and glial cells are generated from dividing neural progenitor cells. Then, the newborn neurons send out axons and dendrites toward their targets. Next, they make synaptic connections with appropriate target cells so that communication can begin. Finally, excessive neurons are eliminated by normal neuronal cell death, after which the system of synaptic connections is refined and remodeled according to the pattern of electrical and synaptic activity in the neural network.

Neurons born at different times and places are specialized to express different sets of genes, and they have a cell memory that plays a major role in determining the connections they will form. Their specialization depends not only on spatial patterning by morphogens but also on intrinsic developmental programs that unfold as the neural progenitors proliferate. Axons and dendrites grow out from the neurons by means of growth cones, which follow specific pathways delineated by attractive and repellant signals along the way, including cell-surface and extracellular matrix molecules and soluble signal proteins to which growth cones from different classes of neurons respond differently. In many parts of the nervous system, neural maps are set up—orderly projections of one array of neurons onto another. In the retinotopic system, the map is based on the matching of complementary systems of position-specific cell-surface markers—ephrins and Eph receptors—possessed by the two sets of cells. Other cell-surface molecules such as DSCAM proteins in Drosophila *and protocadherins in vertebrates mediate self-avoidance between the branches arising from a single neuron, helping the cell spread out its processes.*

The formation of synapses involves back-and-forth signaling between target cells and the growth cone. After the growth cones have reached their targets and initial connections have formed, individual synapses are eliminated in some places and reinforced in others by mechanisms that depend on synaptic and electrical activity. These mechanisms adjust the architecture of the neural network according to the way in which it is used.

WHAT WE DON'T KNOW

- What regulates the pace of development? Why does a mouse embryo develop faster than a human embryo, for example?

- What are the mechanisms that allow cell memory to be stored during development, explaining how each cell's history determines its future behavior?

- How do signals move through tissues? What are the roles of the extracellular matrix and of elongated cell projections?

- How does a cell know exactly where it is in a multicellular organism? How does it know that its neighbors are the correct ones and that, if not, it should move or kill itself?

- How do cells respond to tiny gradients of molecules in their environment, as required for knowing their positions? How are morphogen gradients reliably interpreted?

- What are the genetic changes that allow the repurposing of existing body parts during evolution? For example, how did bat wings evolve from arms?

- How do cells use genetic instructions to form the shape of something as complex as the human nose?

PROBLEMS

Which statements are true? Explain why or why not.

21–1   In the early cleavage stages, when the embryo cannot yet feed, the developmental program is driven and controlled entirely by the material deposited in the egg by the mother.

21–2   Because of the many later developmental transformations that produce the elaborately structured organs, the body plan set up during gastrulation bears little resemblance to the body plan in the adult.

21–3   As development progresses, individual cells become more and more restricted in the range of cell types they can give rise to.

21–4   At different stages of embryonic development, the same signals are used over and over again by different cells, but with different biological outcomes.

21–5   Changes in the coding regions of genes involved in development are primarily responsible for the differences between species.

21–6   The cell cycle is the ticking clock that sets the tempo of developmental processes, with maturational changes in gene expression being dependent on cell-cycle progression.

Discuss the following problems.

21–7   Name the four processes that are fundamental to animal development, and describe each of them in a single sentence.

21–8   What are the three germ layers formed during gastrulation, and what are the principal structures each gives rise to in the adult?

21–9 In the early *Drosophila* embryo, there seems to be no requirement for the usual forms of cell–cell signaling; instead, transcriptional regulators and mRNA molecules move freely between nuclei. How can that be?

21–10 Morphogens play a key role in development, creating concentration gradients that inform cells of where they are and how to behave. Examine the simple patterns represented by the flags in Figure Q21–1. Which do you suppose could be created by a gradient of a single morphogen? Which would require gradients of two morphogens? Assuming that such patterns were present in a sheet of cells, explain how they could be created by morphogens.

Figure Q21–1 National flags from three countries (Problem 21–10). (Left, railway fx/Shutterstock; center, Creative Photo Corner/ Shutterstock; right, Derek Brumby/Shutterstock.)

21–11 Two adjacent cells in the nematode worm normally differentiate into an anchor cell (AC) and a ventral uterine precursor (VU) cell, but which of the two becomes the AC and which becomes the VU cell is completely random: the cells have an equal chance of adopting either fate, but they always adopt different fates. Mutations of *Lin12* alter these fates. In hyperactive *Lin12* mutants, both cells become VU cells, while in inactive *Lin12* mutants, both cells become ACs. Thus, Lin12 is central to the decision-making process. In genetic mosaics in which one precursor cell has the hyperactive Lin12 and the other precursor has the inactive Lin12, the cell with the hyperactive Lin12 always becomes the VU cell and the cell with inactive Lin12 always becomes the AC. Assuming that one cell sends a signal and the other cell receives it, explain how these results suggest that *Lin12* encodes a protein required to receive the signal. Offer a suggestion for how the fates of these two precursor cells are normally decided in wild-type worms.

21–12 It was clear from the early days of studying development that certain "morphogenetic" substances were present in the egg and segregated asymmetrically into cells of the developing embryo. One such investigation in ascidian (sea squirt) embryos examined endodermal alkaline phosphatase, which could be visualized by a histochemical stain. Treatment of embryos with cytochalasin B stopped cell division, but did not block expression of alkaline phosphatase at the appropriate time. Treatment with actinomycin D, which blocks transcription, did not interfere with expression of alkaline phosphatase. Treatment with puromycin, which blocks translation, eliminated expression of alkaline phosphatase. What is the likely nature of the morphogenetic substance that gives rise to alkaline phosphatase?

21–13 The mouse *HoxA3* and *HoxD3* genes are paralogs that occupy equivalent positions in their respective *Hox* gene clusters and share roughly 50% identity in their protein-coding sequences. Mice with defects in *HoxA3* have deficiencies in pharyngeal tissues, whereas mice with defects in *HoxD3* have deficiencies in the axial skeleton, suggesting quite different functions for the paralogs. Thus, it came as a surprise when it was found that replacing a defective *HoxD3* gene with the normal *HoxA3* gene corrected the deficiency, as did the reciprocal experiment of replacing a mutant *HoxA3* gene with a normal *HoxD3* gene. Neither transplaced gene, however, could supply its normal function; that is, a normal *HoxA3* gene at the *HoxD3* locus could not correct the deficiency caused by a mutant *HoxA3* gene at the *HoxA3* locus. The same was true for the *HoxD3* gene. If the *HoxA3* and *HoxD3* genes are equivalent, how do you suppose they can play such distinct roles in development? Why do you suppose they cannot perform their normal function in a new location?

21–14 The segmentation of somites in vertebrate embryos is thought to depend on oscillations in the expression of the *Hes7* gene. Mathematical modeling explains these oscillations in terms of the delays in production of the unstable Hes7 protein, which acts as a transcription regulator to shut off its own expression. Once Hes7 decays, with a half-life of about 20 minutes, its transcription resumes. To test this model, you decide to reduce the total delay by removing one, two, or all three of the introns from the *Hes7* gene in mice. Why do you expect that intron removal would reduce the delay? What would you predict would happen to the oscillation time, and somite formation, if the model were correct?

21–15 The oscillatory clock that drives somite formation in vertebrates involves three essential components Her7 (an unstable repressor of its own synthesis), Delta (a transmembrane signaling molecule), and Notch (a transmembrane receptor for Delta). Notch is bound by Delta on neighboring cells, activating the Notch signaling pathway, which then activates *Her7* transcription. Normally, this system works flawlessly to create sharply defined somites (Figure Q21–2A). In the absence of Delta, however, only the first five somites form normally, and the rest are poorly defined (Figure Q21–2B). If a pulse of Delta is supplied later, somite formation returns to normal in the regions where Delta was present (Figure Q21–2C). A diagram of the connections between the components of the clock and how they interact in adjacent cells is shown in Figure Q21–2D. In the absence of Delta, why do the cells become unsynchronized? What is it about the presence of Delta that keeps adjacent cells oscillating in synchrony?

21–16 The extracellular protein factor Decapentaplegic (Dpp) is critical for proper wing development in *Drosophila* (Figure Q21–3A). It is normally expressed in a narrow stripe in the middle of the wing, along the anterior–posterior boundary. Flies that are defective for Dpp form stunted "wings" (Figure Q21–3B). If an additional copy of the gene is placed under control of a promoter that is active in the anterior part of the wing, or in the posterior

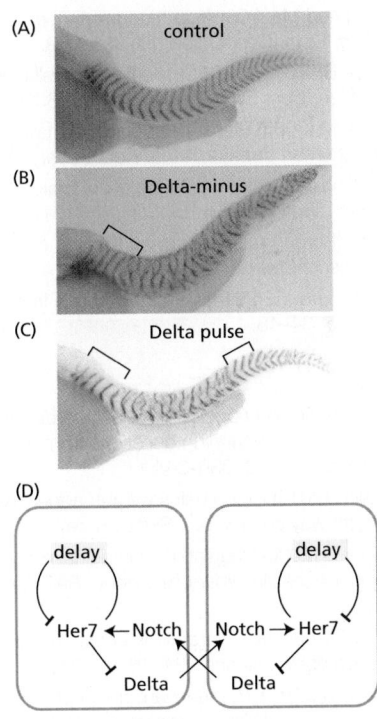

Figure Q21–2 Somite formation in zebrafish embryos (Problem 21–15). (A) Wild-type embryos with normal somites. (B) Somite formation in embryos lacking Delta. The bracket indicates normal-looking somites where they initially form. (C) Somite formation in embryos lacking Delta, but receiving a pulse of Delta expression at the time indicated by the right-hand bracket. (D) Interactions among components of the oscillatory clock in adjacent cells. (Adapted from C. Soza-Ried et al., *Development* 141:1780–1788, 2014. With permission from The Company of Biologists.)

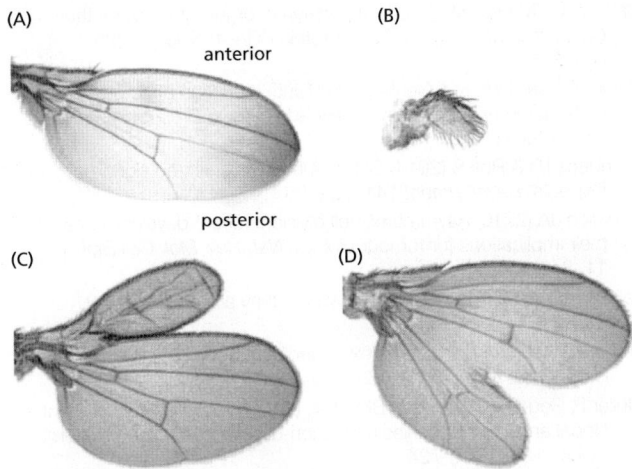

Figure Q21–3 Effects of Dpp expression on wing development in *Drosophila* (Problem 21–16). (A) Normal Dpp expression. (B) Absence of Dpp expression. (C) Additional anterior Dpp expression. (D) Additional posterior Dpp expression. (From M. Zecca, K. Basler and G. Struhl, *Development* 121:2265–2278, 1995. With permission from The Company of Biologists.)

part of the wing, a large mass of wing tissue composed of normal-looking cells is produced at the site of Dpp expression (Figure Q21–3C and D). Does Dpp stimulate cell division, cell growth, or both? How can you tell?

21–17 The highly branched structures of neurons would seem to make it almost inevitable that they should make unproductive synapses with themselves, yet they manage to avoid this outcome very effectively. How is this accomplished in vertebrates?

REFERENCES

General

Carroll SB (2006) Endless Forms Most Beautiful: The New Science of Evo Devo. New York: W.W. Norton & Co., Inc.

Gilbert SF (2013) Developmental Biology, 10th ed. Sunderland, MA: Sinauer Associates, Inc.

Wolpert L & Tickle C (2010) Principles of Development, 3rd ed. Oxford, UK: Oxford University Press.

Overview of Development

Gurdon JB (2013) The egg and the nucleus: a battle for supremacy (Nobel Lecture). *Angew. Chem. Int. Ed. Engl.* 52, 13890–13899.

Istrail S & Davidson EH (2005) Logic functions of the genomic cis-regulatory code. *Proc. Natl Acad. Sci. USA* 102, 4954–4959.

Levine M (2010) Transcriptional enhancers in animal development and evolution. *Curr. Biol.* 20, R754–R763.

Lewis J (2008) From signals to patterns: space, time, and mathematics in developmental biology. *Science* 322, 399–403.

Meinhardt H & Gierer A (2000) Pattern formation by local self-activation and lateral inhibition. *Bioessays* 22, 753–760.

Rogers KW & Schier AF (2011) Morphogen gradients: from generation to interpretation. *Annu. Rev. Cell Dev. Biol.* 27, 377–407.

Shubin N, Tabin C & Carroll S (2009) Deep homology and the origins of evolutionary novelty. *Nature* 457, 818–823.

Mechanisms of Pattern Formation

Andrey G & Duboule D (2014) SnapShot: Hox gene regulation. *Cell* 156, 856–856.e1.

Baker NE (2011) Proximodistal patterning in the *Drosophila* leg: models and mutations. *Genetics* 187, 1003–1010.

Chan YF, Marks ME, Jones FC et al. (2010) Adaptive evolution of pelvic reduction in sticklebacks by recurrent deletion of a Pitx1 enhancer. *Science* 327, 302–305.

Davis RL, Weintraub H & Lassar AB (1987) Expression of a single transfected cDNA converts fibroblasts to myoblasts. *Cell* 51, 987–1000.

De Robertis EM (2006) Spemann's organizer and self-regulation in amphibian embryos. *Nat. Rev. Mol. Cell Biol.* 4, 296–302.

DiNardo S, Heemskerk J, Dougan S & O'Farrrell PH (1994) The making of a maggot: patterning the *Drosophila* embryonic epidermis. *Curr. Opin. Genet. Dev.* 4, 529–534.

Driever W & Nüsslein-Volhard C (1988) A gradient of bicoid protein in *Drosophila* embryos. *Cell* 54, 83–93.

Fowlkes CC, Luengo CL, Keränen VE et al. (2008) A quantitative spatiotemporal atlas of gene expression in the *Drosophila* blastoderm. *Cell* 133, 364–74.

Furman DP & Bukharina TA (2008) How *Drosophila Melanogaster* forms its mechanoreceptors. *Curr. Genomics* 9, 312–323.

Gaudet J & Mango SE (2002) Regulation of organogenesis by the *Caenorhabditis elegans* FoxA protein PHA-4. *Science* 295, 821–825.

Halder G, Callaerts P & Gehring WJ (1995) Induction of ectopic eyes by targeted expression of the eyeless gene in *Drosophila. Science* 267, 1788–1792.

Kornberg TB & Roy S (2014) Cytonemes as specialized signaling filopodia. *Development* 141, 729–36.

Knoblich JA (2010) Asymmetric cell division: recent developments and their implications for tumour biology. *Nat. Rev. Mol. Cell Biol.* 11, 849–860.

Lander AD (2013) How cells know where they are. *Science* 339, 923–27.

Lewis EB (1978) A gene complex controlling segmentation in *Drosophila. Nature* 276, 565–570.

Müller P, Rogers KW, Jordan BM et al. (2012) Differential diffusivity of Nodal and Lefty underlies a reaction-diffusion patterning system. *Science* 336, 721–724.

Nüsslein-Volhard C & Wieschaus E (1980) Mutations affecting segment number and polarity in *Drosophila. Nature* 287, 795–801.

Ringrose L & Paro R (2007) Polycomb/Trithorax response elements and epigenetic memory of cell identity. *Development* 134, 223–232.

Shulman JM & St Johnston D (1999) Pattern formation in single cells. *Trends Cell Biol.* 9, M60–64.

von Dassow G, Meir E, Munro EM & Odell GM (2000) The segment polarity network is a robust developmental module. *Nature* 406, 188–192.

Developmental Timing

Brown DD & Cai L (2007) Amphibian metamorphosis. *Dev. Biol.* 306, 20–33.

Giraldez AJ, Mishima Y, Rihel J et al. (2006) Zebrafish MiR-430 promotes deadenylation and clearance of maternal mRNAs. *Science* 312, 75–79.

Isshiki T, Pearson B, Holbrook S & Doe CQ (2001) *Drosophila* neuroblasts sequentially express transcription factors which specify the temporal identity of their neuronal progeny. *Cell* 106, 511–521.

Lee RC, Feinbaum RL & Ambros V (1993) The *C. elegans* heterochronic gene lin-4 encodes small RNAs with antisense complementarity to lin-14. *Cell* 75, 843–854.

Lewis J (2003) Autoinhibition with transcriptional delay: a simple mechanism for the zebrafish somitogenesis oscillator. *Curr. Biol.* 13, 1398–1408.

Pourquié O (2011) Vertebrate segmentation: from cyclic gene networks to scoliosis. *Cell* 145, 650–663.

Song J, Irwin J & Dean C (2013) Remembering the prolonged cold of winter. *Curr. Biol.* 23, R807–R811.

Wightman B, Ha I & Ruvkun G (1993) Posttranscriptional regulation of the heterochronic gene lin-14 by lin-4 mediates temporal pattern formation in *C. elegans. Cell* 75, 855–862.

Morphogenesis

Green AA, Kennaway JR, Hanna AI et al. (2010) Genetic control of organ shape and tissue polarity. *PLoS Biol.* 8, e1000537.

Le Douarin NM & Kalcheim C (1999) The Neural Crest, 2nd ed. Cambridge, UK: Cambridge University Press.

Matis M & Axelrod JD (2013) Regulation of PCP by the fat signaling pathway. *Genes Dev.* 27, 2207–20.

Ochoa-Espinosa A & Affolter M (2012) Branching morphogenesis: from cells to organs and back. *Cold Spring Harb. Perspect. Biol.* 4, pii: a008243.

Raz E & Reichman-Fried M (2006) Attraction rules: germ cell migration in zebrafish. *Curr. Opin. Genet. Dev.* 16, 355–359.

Revenu C & Gilmour DE (2009) MT2.0: shaping epithelia through collective migration. *Curr. Opin. Genet. Dev.* 19, 338–342.

Simons M & Mlodzik M (2008) Planar cell polarity signaling: from fly development to human disease. *Annu. Rev. Genet.* 42, 517–540.

Solnica-Krezel L & Sepich DS (2012) Gastrulation: making and shaping germ layers. *Annu. Rev. Cell Dev. Biol.* 28, 687–717.

Takeichi M (2011) Self-organization of animal tissues: cadherin-mediated processes. *Dev. Cell* 21, 24–26.

Walck-Shannon E & Hardin J (2014) Cell intercalation from top to bottom. *Nature* 15, 34–48.

Growth

Andersen DS, Colombani J & Léopold P (2013) Coordination of organ growth: principles and outstanding questions from the world of insects. *Trends Cell Biol.* 23, 336–344.

Enderle L & McNeill H (2013) Hippo gains weight: added insights and complexity to pathway control. *Sci. Signal.* 6, re7.

Hariharan IK & Bilder D (2006) Regulation of imaginal disc growth by tumor-suppressor genes in *Drosophila. Annu. Rev. Genetics* 40, 335–61.

Johnston LA (2009) Competitive interactions between cells: death, growth, and geography. *Science* 324, 1679–1682.

Lawrence PA & Casal J (2013) The mechanisms of planar cell polarity, growth and the hippo pathway: some known unknowns. *Dev. Biol.* 377, 1–8.

Pan D (2010) The hippo signaling pathway in development and cancer. *Dev. Cell* 19, 491–505.

Restrepo S, Zartman JJ & Basler K (2014) Coordination of patterning and growth by the morphogen DPP. *Curr. Biol.* 24, R245–R255.

Neural Development

Burden SJ, Yumoto N & Zhang W (2013) The role of MuSK in synapse formation and neuromuscular disease. *Cold Spring Harb. Perspect. Biol.* 5, a009167.

Dessaud E, McMahon AP & Briscoe J (2008) Pattern formation in the vertebrate neural tube: a sonic hedgehog morphogen-regulated transcriptional network. *Development* 135, 2489–2503.

Hubel DH & Wiesel TN (1965) Binocular interaction in striate cortex of kittens reared with artificial squint. *J. Neurophysiol.* 28, 1041–1059.

Kolodkin AL & Tessier-Lavigne M (2011) Mechanisms and molecules of neuronal wiring: a primer. *Cold Spring Harb. Perspect. Biol.* 3, pii: a001727.

Luo L & Flanagan JG (2007) Development of continuous and discrete neural maps. *Neuron* 56, 284–300.

Rakic P (1988) Specification of cerebral cortical areas. *Science* 241, 170–176.

Reichardt LF (2006) Neurotrophin-regulated signalling pathways. *Philos. Trans. R. Soc. Lond. B Biol. Sci.* 361, 1545–1564.

Sanes DH, Reh TA & Harris WA (2011) Development of the Nervous System, 3rd ed. San Diego, CA: Academic Press.

Sperry RW (1963) Chemoaffinity in the orderly growth of nerve fiber patterns and connections. *Proc. Natl Acad. Sci. USA* 50, 703–710.

Zipursky SL & Sanes JR (2010) Chemoaffinity revisited: dscams, protocadherins, and neural circuit assembly. *Cell* 143, 343–353.

Stem Cells and Tissue Renewal

Cells evolved originally as free-living individuals, and such cells still dominate the Earth and its oceans. But the cells that matter most to us, as human beings, are specialized members of a multicellular community. These cells have lost features needed for independent survival and acquired peculiarities that serve the needs of the body as a whole. Although they share the same genome, they are spectacularly diverse in structure, chemistry, and behavior. There are more than 200 different named cell types in the human body that collaborate with one another to form many different tissues, arranged into organs performing widely varied functions. To understand them, it is not enough to analyze cells in a culture dish: we need also to know how they live, work, and die in their natural habitat, the intact body.

In Chapters 7 and 21, we saw how the various cell types become different in the embryo and how cell memory and signals from their neighbors enable them to remain different thereafter. In Chapter 19, we discussed the technology used to build multicellular tissues—the devices that bind cells together and the extracellular materials that give them support. But the adult body is not static: it is a structure in dynamic equilibrium, where new cells are continually being born, differentiating, and dying. Homeostatic mechanisms maintain a proper balance, so that the tissue architecture is preserved despite the constant replacement of old cells by new. In this chapter, we focus on these developmental processes that continue throughout life. In doing so, we shall illustrate some of the diversity of specialized cell types and see how they work together to perform their tasks.

We shall examine in particular the role played in many tissues by *stem cells*—cells that are specialized to provide a fresh supply of differentiated cells where these need to be continually replaced, or when they are required in great number for purposes of repair and regeneration. We shall see that while many tissues renew and repair themselves, some others do not; there, lost cells are lost forever, causing deafness, blindness, dementia, and other ills.

In the final section of the chapter, we discuss how stem cells can be generated and manipulated artificially, and we confront the practical question that underlies the current storm of interest in stem-cell technology: How can we use our understanding of the processes of cell differentiation and tissue renewal to improve upon nature, and make good those injuries and failings of the human body that have hitherto seemed to be beyond repair?

STEM CELLS AND RENEWAL IN EPITHELIAL TISSUES

Among all the self-renewing tissues in a mammal, the champion—for speed at least—is the lining of the small intestine: the long, convoluted portion of the gut tube that is chiefly responsible for absorption of nutrients from the gut lumen. To introduce stem cells, we take the small intestine as our starting point—not only because it renews itself at a greater rate than any other tissue in the body, but also because the molecular mechanisms that control its organization are particularly well understood. It thereby provides a beautiful illustration of the principles of stem-cell systems that have broad applicability.

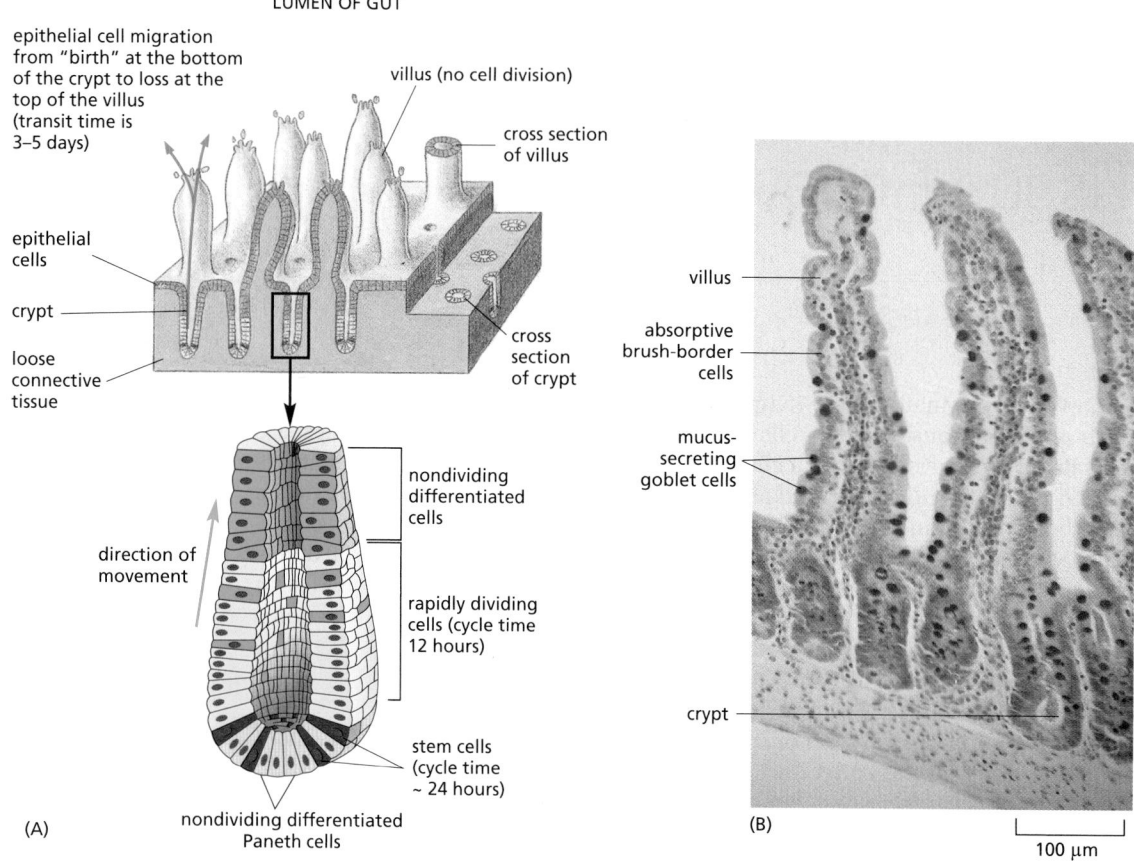

LUMEN OF GUT

epithelial cell migration from "birth" at the bottom of the crypt to loss at the top of the villus (transit time is 3–5 days)

villus (no cell division)

cross section of villus

epithelial cells

crypt

loose connective tissue

cross section of crypt

nondividing differentiated cells

direction of movement

rapidly dividing cells (cycle time 12 hours)

stem cells (cycle time ~ 24 hours)

nondividing differentiated Paneth cells

(A)

villus

absorptive brush-border cells

mucus-secreting goblet cells

crypt

(B)

100 μm

The Lining of the Small Intestine Is Continually Renewed Through Cell Proliferation in the Crypts

The lining of the small intestine (and of most other regions of the gut) is a single-layered epithelium, only one cell thick. This epithelium covers the surfaces of the *villi* that project into the lumen, and it lines the *crypts* that descend into the underlying connective tissue (**Figure 22–1**). Dividing cells are restricted to the crypts, and differentiated cells, no longer dividing, pour out of the crypts in a steady stream onto the villi. There are four main types of nondividing differentiated cells—one absorptive and three secretory (**Figure 22–2**):

1. *Absorptive cells* (also called *brush-border cells* or *enterocytes*) have densely packed microvilli on their exposed surfaces. Their job is to take up nutrients from the gut lumen. To this end, they also produce hydrolytic enzymes that perform some of the final steps of extracellular digestion. They are the majority cell type in the epithelium.

2. *Goblet cells* secrete mucus into the gut lumen that covers the epithelium with a protective coat.

3. *Paneth cells* form part of the innate immune defense system (discussed in Chapter 24) and secrete proteins that kill bacteria.

4. *Enteroendocrine cells,* of more than 15 different subtypes, secrete serotonin and peptide hormones that act on neurons and other cell types in the gut wall and regulate the growth, proliferation, and digestive activities of cells of the gut and other tissues.

As if on a conveyor belt, the absorptive, goblet, and enteroendocrine cells travel mainly upward from their site of birth in the crypt, by a sliding movement in the plane of the epithelial sheet, to cover the surfaces of the villi. Within 3–5 days (in the mouse) after emerging from the crypts, the cells reach the tips of the villi, where they undergo apoptosis and are finally discarded into the gut lumen (see

Figure 22–1 Renewal of the gut lining. (A) The pattern of cell turnover and proliferation in the epithelium that forms the lining of the small intestine. Stem cells *(red)* lie at the crypt base, interspersed among nondividing differentiated cells (Paneth cells). Progeny of the stem cells move mainly upward from the crypts onto the villi; after a few quick divisions, they cease dividing and differentiate—some of them while still in the crypt, most of them as they emerge from the crypt. The Paneth cells, like the other nondividing differentiated cells, are continually replaced by progeny of the stem cells, but they migrate downward to the crypt base and survive there for many weeks. (B) Photograph of a section of part of the lining of the small intestine, showing the crypts and villi. Note the mixture of differentiated cell types, all generated from the stem cells; these are primarily absorptive cells, with mucus-secreting goblet cells (stained *red*) interspersed among them. Enteroendocrine cells (not labeled) are less numerous and less easy to identify without special stains.

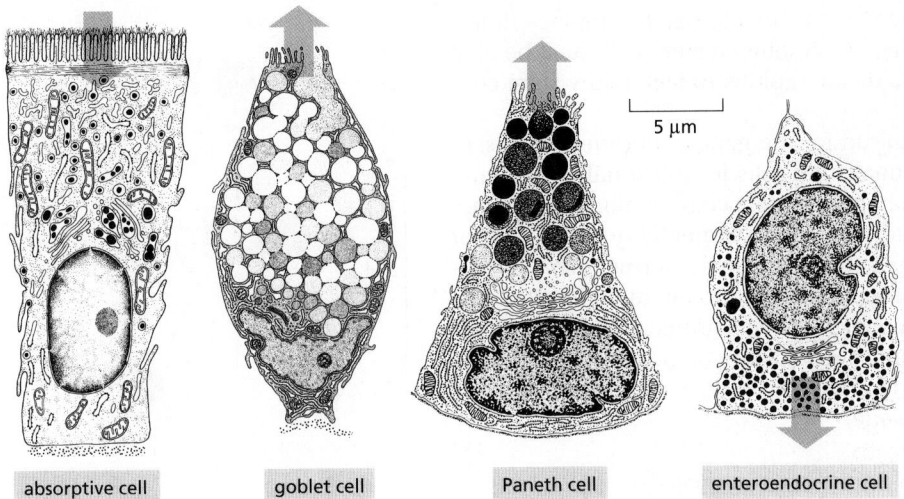

absorptive cell   goblet cell   Paneth cell   enteroendocrine cell

5 µm

Figure 22–2 The four main differentiated cell types found in the epithelial lining of the small intestine. All cells are oriented with the gut lumen at *top*. Broad *orange arrows* indicate direction of secretion or uptake of materials for each type of cell. All of these cells are generated from undifferentiated multipotent stem cells living near the bottoms of the crypts (see Figure 22–1). *Absorptive (brush-border) cells* outnumber the other cell types in the epithelium by about 10:1 or more. The microvilli on their apical surface provide a 30-fold increase of surface area, not only for the import of nutrients but also for the anchorage of enzymes that perform the final stages of extracellular digestion, breaking down small peptides and disaccharides into monomers that can be transported across the cell membrane. *Goblet cells* secrete mucus; these are the commonest of the secretory cell types. *Paneth cells* secrete (along with some growth factors) *cryptdins*—proteins of the defensin family that kill bacteria. Different subtypes of *enteroendocrine cells* secrete serotonin and peptide hormones into the gut wall (and thence the blood). Cholecystokinin is a hormone released from enteroendocrine cells in response to the presence of nutrients in the gut. It binds to receptors on nearby sensory nerve endings, which relay a signal to the brain to stop the feeling of hunger once one has eaten enough. (Absorptive and goblet cells, Don W. Fawcett/Science Source; Paneth and enteroendocrine cells, from R.V. Krstić, *Illustrated Encyclopedia of Human Histology*. Berlin: Springer-Verlag, 1984. With permission from Springer Nature.)

Movie 20.6). The Paneth cells in the crypts are produced in much smaller numbers and have a different migration pattern. They live at the bottom of the crypts, where they too are continually replaced, although not so rapidly, persisting for several weeks (in the mouse) before undergoing apoptosis and being phagocytosed by their neighbors.

The central problem is to understand the processes in the crypt that generate a continual supply of all these nondividing, terminally differentiated cell types.

Stem Cells of the Small Intestine Lie at or Near the Base of Each Crypt

The general pattern of cell proliferation and migration in the gut lining is revealed by a simple labeling method that uses injected pulses of tritiated (radioactive) thymidine or of a thymidine analog that can be detected in tissue sections. Cells that are in S phase of the division cycle incorporate the marker molecule into their DNA, and their fate can then be followed over subsequent hours and days. If a cell divides after incorporation of the label, the label becomes diluted, halving with each cell cycle. This can be quantified. Experiments based on this labeling method confirm, first of all, that dividing cells are confined to the crypts and that the differentiated cell types listed above do not divide. Second, the most rapidly dividing cells, with a cycle time of about 12 hours in the mouse, are shown to lie in the middle and upper parts of the crypt, and these cells are all fated to differentiate and stop dividing (see Figure 22–1A). Just above the base of the crypt, interspersed among the Paneth cells, lie cells that divide more slowly. These are the **stem cells**, which feed some of their progeny into the higher levels of the crypt destined for differentiation, while other progeny remain at the crypt base to continue the whole process. The rapidly dividing cells above these stem cells are derived from them, but already committed to differentiation. These cells are called **committed precursors** or **transit amplifying cells**, since their divisions serve to amplify the number of differentiated cells that ultimately result from each stem-cell division.

The Two Daughters of a Stem Cell Face a Choice

Stem cells have a critical role in a variety of tissues, and it is useful to list their defining properties:

1. A stem cell is not itself **terminally differentiated**: that is, it is not at the end of a pathway of differentiation.

2. It can divide without limit (or at least for the lifetime of the animal).

3. When it divides, each daughter has a choice: it can either remain a stem cell, or it can embark on a course that commits it to terminal differentiation (**Figure 22–3**).

Stem cells are required wherever there is a recurring need to replace differentiated cells that cannot themselves divide. Although a stem cell must be able to divide, it does not necessarily have to divide rapidly; in fact, many stem cells divide at a relatively slow rate.

Stem cells are of many types, specialized for the genesis of different classes of terminally differentiated cells—intestinal stem cells for intestinal epithelium, epidermal stem cells for epidermis, hematopoietic stem cells for blood, and so on. Each stem-cell system nevertheless raises similar fundamental questions. What are the distinguishing features of the stem cell in molecular terms? What conditions serve to keep the stem cell in its proper place and to maintain its stem-cell character? What decides whether a given daughter cell commits to differentiation or remains a stem cell? In a tissue where several distinct types of differentiated cells must be produced, are they all derived from a single type of stem cell, or is there a distinct type of stem cell for each one?

Wnt Signaling Maintains the Gut Stem-Cell Compartment

For the gut, the beginnings of an answer to these questions came from studies of cancer of the colon and rectum (the lower end of the gut, also known as the large intestine). Some people have a hereditary predisposition to colorectal cancer and, in advance of the invasive disease, develop large numbers of small precancerous tumors (adenomas) in the lining of this part of the gut (**Figure 22–4**). The appearance of these tumors suggests that they have arisen from intestinal crypt cells that have failed to halt their proliferation in the normal way. As discussed in Chapter 20, the cause has been traced to mutations in the *Apc* (*adenomatous polyposis coli*) gene: the tumors arise from cells that have lost both gene copies. Because *Apc* codes for a protein that prevents inappropriate activation of the Wnt signaling pathway (see Figure 15–60), this loss of Apc is presumed to mimic the effect of continual exposure to a Wnt signal. The suggestion, therefore, is that Wnt signaling normally keeps crypt cells in a proliferative state, and that a cessation of exposure to Wnt signaling normally makes them stop dividing as they leave the crypt.

Stem Cells at the Crypt Base Are Multipotent, Giving Rise to the Full Range of Differentiated Intestinal Cell Types

It has long been suspected that all the differentiated cell types in the lining of the intestine derive from a single type of stem cell. But firm proof was lacking, and the precise nature and location of the stem cells were disputed.

To solve the problem, and indeed to understand the organization of any stem-cell system, we need to discover how its cells are related to one another—who

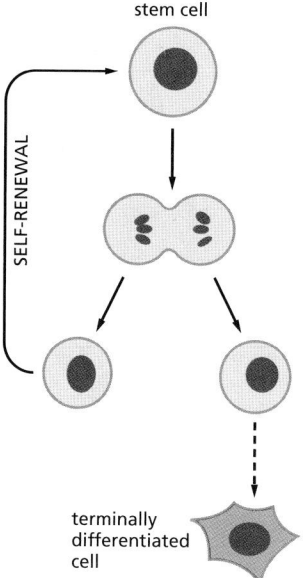

Figure 22–3 The definition of a stem cell. Each daughter produced when a stem cell divides can either remain a stem cell or go on to become terminally differentiated. In many cases, the daughter that opts for terminal differentiation undergoes additional cell divisions before terminal differentiation is completed; such cells are called transit amplifying cells.

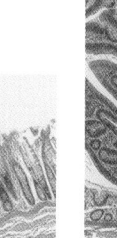
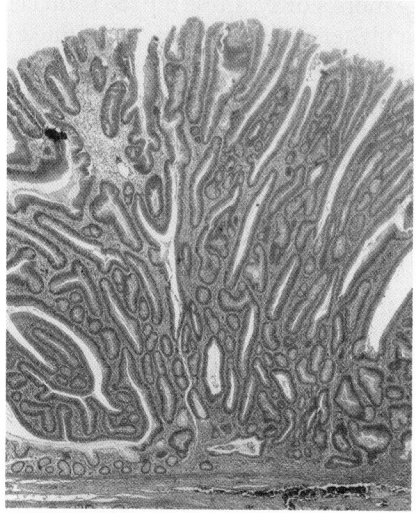

NORMAL COLON

ADENOMA

200 µm

Figure 22–4 An adenoma in the human colon, compared with normal tissue from an adjacent region of the same person's colon. The specimen is from a patient with an inherited mutation in one of his two copies of the *Apc* gene. A mutation in the other *Apc* gene copy, occurring in a colon epithelial cell during adult life, has given rise to a clone of cells that behave as though the Wnt signaling pathway is permanently activated. As a result, the cells of this clone form an adenoma—an enormous, steadily expanding mass of giant cryptlike structures.

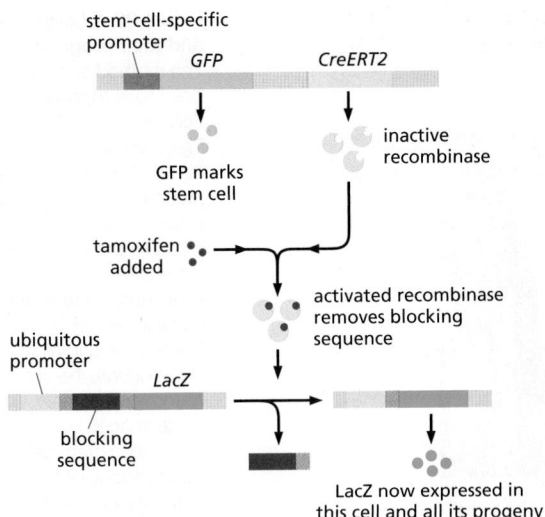

Figure 22–5 Clonal analysis using a genetic marker. A modern method for tracking cell lineage uses transgenic animals containing two transgenes, which together drive expression of a readily detected and heritable marker protein in a small subset of stem cells. The first transgene *(top)* carries two adjacent protein-coding sequences, *GFP* and *CreERT2*, both expressed under the control of the *Lgr* promoter that is active only in stem cells and not in their differentiated progeny. *GFP* encodes green fluorescent protein (see Chapter 9), which is used here simply to confirm expression in the entire stem-cell population. The *CreERT2* gene encodes a chimeric form of the Cre recombinase called CreERT, which consists of Cre recombinase linked to the estrogen receptor protein; this enzyme becomes active as a recombinase only when it binds the artificial estrogen analog tamoxifen.

The second transgene *(bottom)* carries a marker gene, *LacZ*, under the control of a promoter that is active in all cells. The *LacZ* gene encodes β-galactosidase, an enzyme that can be detected histochemically in tissues (see Figure 7–28). However, *LacZ* expression in the transgene shown here is prevented by a blocking sequence *(red)* that is flanked by *LoxP* sites *(pink;* see Figure 5–66). When tamoxifen is provided, CreERT becomes active—leading to a recombination event that removes the blocking DNA sequence (and leaves one *LoxP* site behind). As a result, the LacZ marker is expressed. Because this change is heritable, the marker continues to be expressed in all cells descended from those in which a recombination event has occurred. With a low dose of the inducer molecule tamoxifen, it is possible to activate the marker at random in just a few widely spaced cells, which, in the course of time, give rise to widely separated and easily distinguished clones of progeny (see Figure 22–6).

is descended from whom, or, equivalently, what progeny will be produced from any given cell. This can best be done using a heritable marker that can be activated in an individual cell, thus allowing the identification of the clone of progeny descended from that cell. A modern method uses transgenic animals to create a visible genetic mark in just a few widely spaced cells, which, in the course of time, give rise to widely separated and easily distinguished clones of progeny, as explained in **Figure 22–5**.

A search among genes that are strongly upregulated in response to Wnt signaling revealed one, called *Lgr5*, that is expressed in gut stem cells specifically. The technique described in Figure 22–5 can be used to create a genetic mark in a random subset of *Lgr5*-expressing cells—a mark that is inherited by the progeny of each cell. These *Lgr5* cells divide with a cycle time of about 24 hours, and within a few days marked clones are seen extending from the crypt bases up along the sides of the villi. After as long as 60 days or more, many of these clones still persist, retaining one or more members at the crypt base and extending all the way up to the tips of the villi (**Figure 22–6**). Moreover, each single clone typically contains all the major differentiated gut cell types—absorptive, goblet, Paneth, and enteroendocrine—in their normal proportions. The *Lgr5*-expressing cells, therefore, are true stem cells that are *multipotent*—that is, able to generate a diverse set of differentiated cell types.

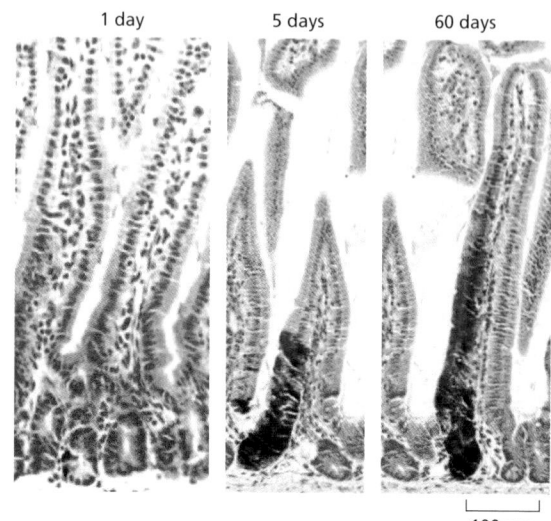

1 day   5 days   60 days

100 µm

Figure 22–6 *Lgr5*-expressing stem cells and their progeny in the small intestine. The method shown in Figure 22–5 was used here to mark single intestinal stem cells and trace the fates of their progeny. The *Lgr5* gene encodes a member of the family of G-protein-linked transmembrane receptors, and it is expressed specifically in stem cells near the crypt base. Because the *Lgr5* promoter was used to drive expression of *CreERT2*, treatment with a low dose of tamoxifen resulted in occasional stem cells expressing *LacZ*. These cells and all of their progeny could subsequently be detected with a *blue* histochemical stain. All of the *blue* cells in these images derive from a single *Lgr5*-expressing stem cell. After 60 days, the *blue* progeny of this cell are seen to extend all the way up a villus. These progeny can be shown to include all types of differentiated cells, as well as persistent *Lgr5*-expressing cells at the crypt base. This proves that *Lgr5*-expressing cells are multipotent stem cells. (From N. Barker et al., *Nature* 449:1003–1007, published 2007 by Nature Publishing Group. Reproduced with permission of SNCSC.)

The Two Daughters of a Stem Cell Do Not Always Have to Become Different

If the number of stem cells in a crypt is to remain stable, each stem-cell division must on average generate one daughter that remains a stem cell and one that becomes committed to differentiation. In principle, this could be achieved in at least two ways (**Figure 22–7**).

One mechanism—the simplest at first sight—would be through asymmetric division: processes internal to the dividing stem cell could distribute regulatory factors asymmetrically to its two daughters, as occurs in *Drosophila* neuroblast divisions (see Figure 21–36). The factors inherited by one daughter would cause it to remain a stem cell, while those inherited by the other would drive it toward differentiation. This strategy would guarantee that the original stem cell would give rise to precisely one stem cell in every subsequent cell generation.

An alternative strategy would be based on a choice that each daughter makes independently of its sister: in normal circumstances, each would have a 50% probability of remaining as a stem cell and a 50% probability of commitment to differentiation. Sometimes the two daughters of a stem cell would thus have opposite fates, sometimes the same. The choice that each cell makes might either be stochastic, like the flip of a coin, or governed by the environment in which the cell finds itself. A strategy of independent choices is more flexible than that of strict asymmetric division. In particular, environmental factors can control the balance of probabilities, adjusting them in favor of the stem-cell option where more stem cells are needed, as they often are, either for growth or for damage repair.

Clonal analysis gives a way to distinguish between the two strategies, since they give quite different predictions as to the expected number of clones of different sizes produced from individual stem cells (see Figure 22–7). For the gut, the findings seem clear: the independent-choice theory fits the observations, and the asymmetric-division theory does not.

Paneth Cells Create the Stem-Cell Niche

There are about 15 *Lgr5*-expressing stem cells in each crypt. They are slim and columnar, and they sit at the crypt base interspersed among the Paneth cells (see Figure 22–6). This is the intestinal **stem-cell niche**: the Paneth cells generate signals, including a strong Wnt signal, that act over a short range to maintain the stem-cell state. Signal proteins from the connective tissue surrounding the crypt base help to reinforce the localizing signal from the Paneth cells; Lgr5 itself is a receptor for one of these proteins, called R-spondin.

In the intestine, it seems that the niche created by the Paneth cells has space for only a limited number of stem cells, and when these divide, it is a random

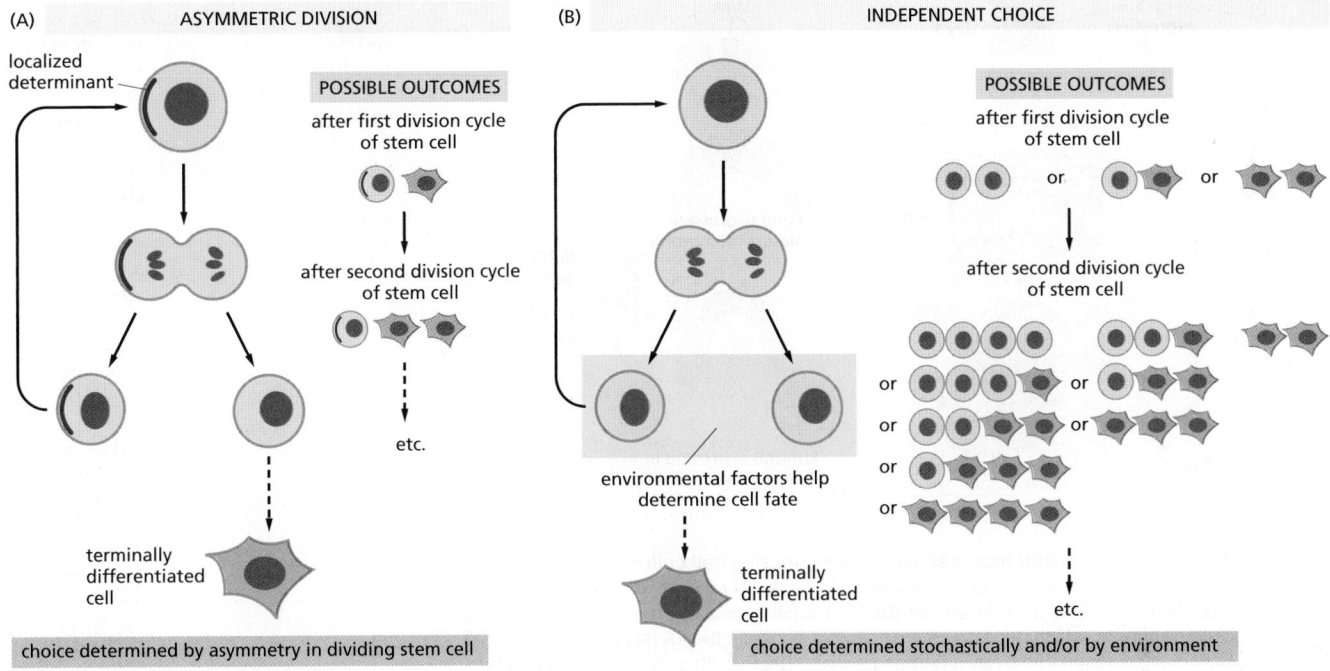

Figure 22–7 Two ways for a stem cell to produce daughters with different fates: asymmetric division and independent choice. (A) The asymmetric-division strategy gives a clone consisting of precisely one stem cell plus a steadily increasing number of differentiating cells, in proportion to the number of cell divisions. (B) The independent-choice strategy is more variable in its outcome. With a choice made at random by each daughter and with a 50% probability for each one to remain a stem cell or differentiate, there is, for example, a 25% chance at the first division that both daughters will differentiate, so that the clone eventually goes extinct. Or, at this division or later, a preponderance of daughters may chance to retain stem-cell character, creating a clone that persists and increases in size. With the help of some mathematics, the probability distribution of clone sizes generated from a single stem cell at any given time can be predicted on this stochastic assumption. The observations in the gut and elsewhere fit the stochastic independent-choice strategy, but not the asymmetric-division strategy.

matter which of them are pushed out of the nest and condemned to differentiation and which stay in place as stem cells for the future. In most other stem-cell systems where the question has been examined, it appears that the fates of the daughters of a stem cell are assigned in a similar way, independently and subject to influence from the cells' environment.

A Single *Lgr5*-expressing Cell in Culture Can Generate an Entire Organized Crypt-Villus System

The Paneth cells themselves are progeny of the stem cells, suggesting that the intestinal stem-cell system is in some way self-maintaining and self-organizing. This is demonstrated in a striking way by taking single dissociated *Lgr5*-expressing cells and allowing them to proliferate in culture, embedded in a cell-free matrix rich in the basal-lamina component laminin (mimicking basal lamina). The cells proliferate, forming at first small, round epithelial vesicles. Within a few days, however, one or another of the cells in the vesicle, at random, begins to differentiate as a Paneth cell. This induces its neighbors to behave as stem cells and initiates transformation of the simple vesicle into an organized structure, or *organoid* (**Figure 22–8A,B**). Protrusions resembling crypts grow out into the surrounding matrix and contain Paneth cells, *Lgr5*-expressing stem cells, and the transit amplifying cells derived from them; these cell types are confined to the cryptlike structures. Terminally differentiated, nondividing absorptive cells line the other parts of the organoid epithelium, with their microvilli facing the lumen. Goblet and enteroendocrine cells are also present, scattered through the epithelium, and the whole "minigut" structure, with all its cell types, grows and renews itself in much the same way as the lining of the normal intestine.

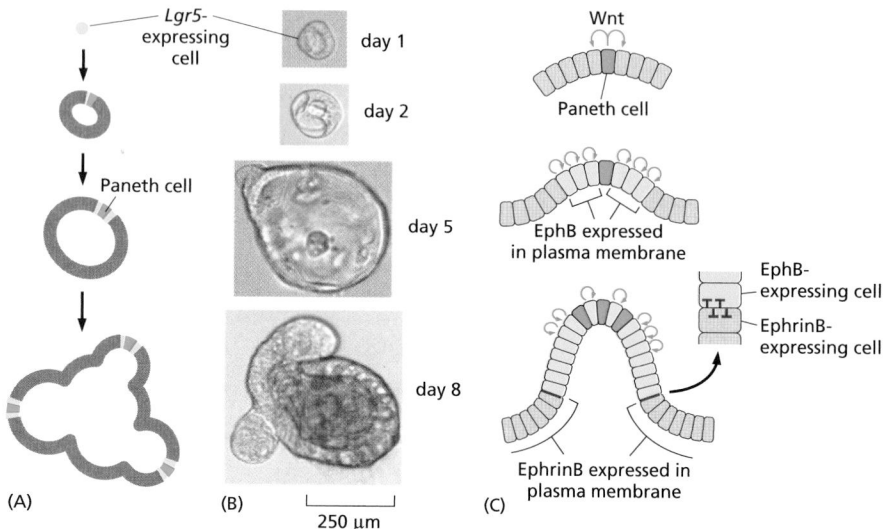

Figure 22–8 Genesis of a minigut from a single *Lgr5*-expressing cell cultured in a cell-free matrix. (A,B) The founder cell first divides to form a small vesicle. At random, one or more of the cells in this vesicle differentiates as a Paneth cell *(blue)*. This cell maintains *Lgr5* expression *(yellow)* in its immediate neighbors, which persist as stem cells that generate the full range of intestinal cell types. (C) Schematic diagram of the key organizing signals. The Paneth cells organize crypts by producing a Wnt signal that acts on neighboring cells and keeps them proliferating in the stem-cell state. A repulsive interaction based on ephrin–Eph binding causes the crypt cell types (which express EphB, induced by Wnt) to segregate from the nondividing differentiated villus cell types (which express EphrinB). Both ephrin and Eph are cell-surface proteins attached to the plasma membrane; in many tissues, two cells that contain a different member of this pair repel each other when they touch (see Figure 21–49). (Adapted from T. Sato and H. Clevers, *Science* 340:1190–1194, 2013. With permission from AAAS.)

Ephrin–Eph Signaling Drives Segregation of the Different Gut Cell Types

The remarkable self-organizing behavior of the cultured organoids suggests that some interaction among the different epithelial cells drives them to segregate from one another. The ephrin–Eph signaling pathway (discussed in Chapter 15) appears to be responsible. The cells that live in the crypts express EphB receptor proteins, while absorptive, goblet, and enteroendocrine cells, as they begin to differentiate, switch off expression of this receptor and instead switch on expression of its ligands, cell-surface proteins of the EphrinB family (Figure 22–8C). In various other tissues, cells expressing Eph proteins are repelled by contacts with cells expressing ephrins on their surface (see Figures 21–49 and 21–79). It seems that the same is true in the gut lining, and that this mechanism serves to keep the cells segregated and in their proper places. In EphB knockout mutants, the populations become mixed, so that, for example, Paneth cells wander out onto the villi.

Notch Signaling Controls Gut Cell Diversification and Helps Maintain the Stem-Cell State

If a single type of stem cell generates all the differentiated cell types in the gut lining, what causes the progeny of this stem cell to diversify? Notch signaling has this role in many other systems, where it mediates lateral inhibition—a competitive interaction that drives neighboring cells toward different fates (see Figure 15–58 and Figure 21–35). All the essential components of the Notch pathway are expressed in the crypts; it seems that Wnt signaling maintains them there. If Notch signaling is abruptly blocked, within a few days all the cells in the crypts differentiate as goblet cells, and absorptive cells cease to be produced; conversely, if Notch signaling is artificially activated in all the cells, absorptive cells continue

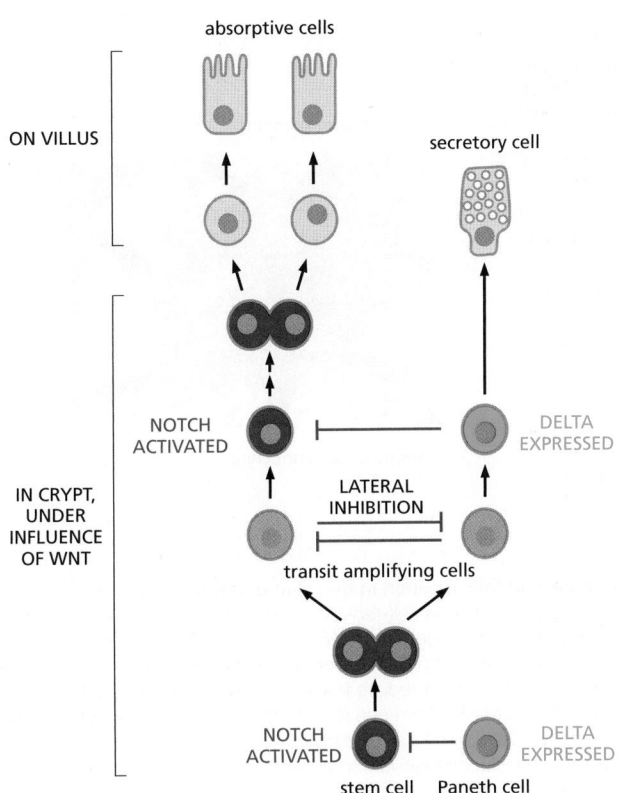

Figure 22–9 How Notch signaling, in combination with Wnt, maintains stem cells and drives cell diversification in the intestine. Wnt signaling leads to expression of Notch and Delta in the cells of the crypt, and Delta–Notch signaling in the crypt mediates lateral inhibition between adjacent cells. Cells expressing higher levels of Delta eventually activate Notch in their neighbors, adopt a secretory fate, and stop dividing; their neighbors, with activated Notch, are prevented from differentiating and keep on dividing. Essentially the same process operates at the crypt base, where the Paneth cells express higher levels of Delta to prevent stem cells from differentiating, and in the transit amplifying population, where nascent secretory cells express higher levels of Delta. Division continues in the Notch-activated cells as they move up the crypt, until they escape from the influence of Wnt and emerge onto the villi to become absorptive cells.

to be generated but no goblet cells are produced. This reflects the lateral inhibition mechanism operating in normal animals: the nascent goblet (and other secretory) cells express the Notch ligand Delta and thereby activate Notch in their neighbors, inhibiting them from differentiating as secretory (**Figure 22–9**).

Delta–Notch signaling is crucial not only in the transit amplifying population, but also at the crypt base: the Paneth cells express Delta and this activates Notch in the stem cells, inhibiting differentiation. Without this influence, the stem cells lose their special character and differentiate as secretory cells. Thus maintenance of the intestinal stem-cell state requires a combination of signals, with both Wnt and Notch acting as central players.

The Epidermal Stem-Cell System Maintains a Self-Renewing Waterproof Barrier

Stem-cell systems are organized in many different ways, but they share some underlying principles. Consider the **epidermis**, for example—the outer, epithelial covering of the body. The epidermis undergoes continual renewal, but, unlike the lining of the gut, it is multilayered or *stratified*. Stem cells are located in the basal layer, and their progeny move outward toward the exposed surface, differentiating as they go. They end up as lifeless scales or *squames*, which are eventually shed from the surface of the skin (**Figure 22–10**). Even though the architecture of this tissue is very different from that of the intestine, many of the same basic principles apply. The stem cells depend for their existence on signals from a specific niche, in this case the basal lamina and underlying connective tissue. The daughters of stem cells that are committed to differentiation undergo several divisions as transit amplifying cells (while still in the basal layer) before differentiating. Finally, a stochastic independent-choice mechanism dictates the fates of the daughters of a stem-cell division, allowing for increase in the number of stem cells when needed for growth or wound healing. Most of the same signaling pathways that organize the intestinal stem-cell system are also involved in regulating the epidermal stem-cell system, although with different individual roles.

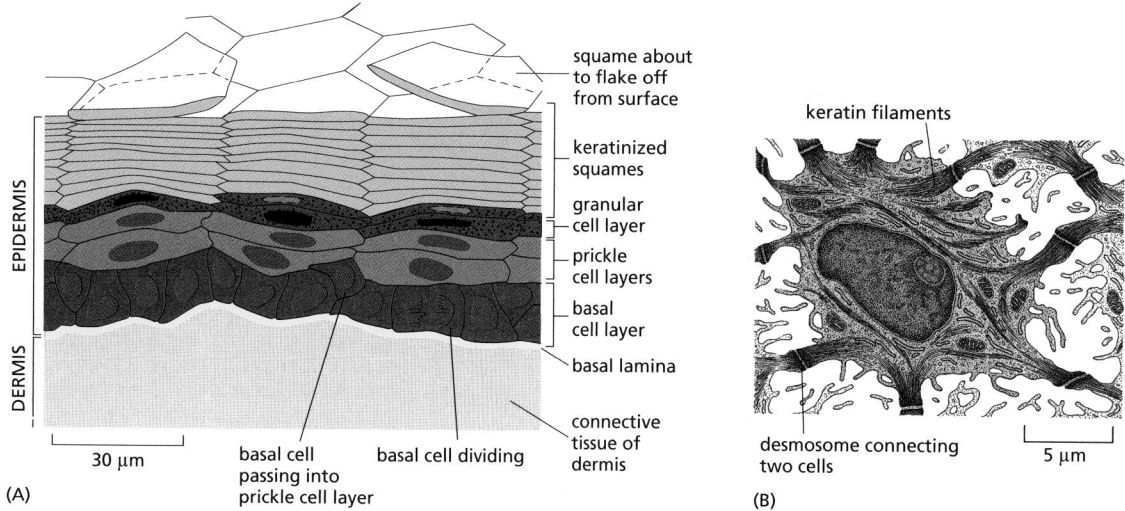

(A)

(B)

squame about to flake off from surface

keratinized squames

granular cell layer

prickle cell layers

basal cell layer

basal lamina

connective tissue of dermis

EPIDERMIS

DERMIS

30 μm

basal cell passing into prickle cell layer

basal cell dividing

keratin filaments

desmosome connecting two cells

5 μm

Figure 22–10 The multilayered structure of the epidermis, as seen in thin skin of a mouse. (A) The epidermis forms the outer covering of the skin, creating a waterproof barrier that is self-repairing and continually renewed. Beneath this lies a relatively thick layer of connective tissue, which includes the tough, collagen-rich dermis (from which leather is made) and the underlying fatty subcutaneous layer or hypodermis. The cells of the epidermis are called keratinocytes, because their characteristic differentiated activity is the synthesis of keratin intermediate filament proteins, which give the epidermis its toughness. These cells change their appearance and properties from one layer to the next, progressing through a regular program of differentiation. Those in the innermost layer, attached to an underlying basal lamina, are termed basal cells, and it is usually only these that divide: the basal cell population includes relatively small numbers of stem cells along with larger numbers of transit amplifying cells derived from them. Above the basal cells are several layers of larger prickle cells, shown in top view in (B), whose numerous desmosomes—each a site of anchorage for thick tufts of keratin filaments—are just visible in the light microscope as tiny prickles around the cell surface. Beyond the prickle cells lies the thin, darkly staining granular cell layer, where the cells are sealed together to form a waterproof barrier; this marks the boundary between the inner, metabolically active strata and the outermost layer of the epidermis, consisting of dead cells whose intracellular organelles have disappeared. These outermost cells are reduced to flattened scales, or squames, filled with densely packed keratin, which are eventually shed from the surface of the skin. The time from exit of a cell from the basal layer to its loss by shedding at the surface is a week or two, depending on body region and species.

In addition to the cells destined for keratinization, the deep layers of the epidermis include small numbers of cells (not shown) that invade this tissue and have quite different origins and functions. These immigrants include dendritic cells, called Langerhans cells, derived from bone marrow and belonging to the immune system; melanocytes (pigment cells) derived from the neural crest; and Merkel cells, which are associated with nerve endings in the epidermis. (B, from R.V. Krstić, *Ultrastructure of the Mammalian Cell: An Atlas*. Berlin: Springer-Verlag, 1979. With permission from Springer Nature.)

Tissue Renewal That Does Not Depend on Stem Cells: Insulin-Secreting Cells in the Pancreas and Hepatocytes in the Liver

Some types of cells can divide even though fully differentiated, allowing for renewal and regeneration without the use of stem cells. The *insulin-secreting cells* (β *cells*) of the pancreas are one example. Their mode of renewal has a special importance, because it is the loss of these cells (through autoimmune attack) that is responsible for type 1 (juvenile-onset) diabetes; they are also a significant factor in the type 2 (adult-onset) form of the disease. The β cells are normally sequestered in cell clusters called *islets of Langerhans*. These islets contain no obvious subset of cells specialized to act as stem cells, yet fresh β cells are continually generated within them. Lineage tracing studies, similar to those described above for the gut, show that the renewal of this population normally occurs by simple duplication of the existing insulin-expressing cells, and not by means of stem cells.

Another tissue that can renew by simple duplication of fully differentiated cells is the liver. The main cell type in the liver is the *hepatocyte*, a large cell that performs the liver's metabolic functions. Hepatocytes normally live for a year or more and renew themselves through cell division at a very slow rate. Powerful homeostatic mechanisms operate to adjust the rate of cell proliferation or the rate of cell death, or both, so as to keep the organ at its normal size or restore it to that size in case of damage. A dramatic effect is seen if large numbers of hepatocytes are removed surgically or are killed by poisoning with carbon tetrachloride. Within a day or so after either sort of damage, a surge of cell division occurs among the surviving hepatocytes, quickly replacing the lost tissue. If two-thirds of a rat's liver is removed, for example, a liver of nearly normal size can regenerate from the remainder by hepatocyte proliferation within about two weeks.

Both the pancreas and the liver contain small populations of stem cells that can be called into play as a backup mechanism for production of the differentiated cell types in more extreme circumstances. This imparts resilience to the mechanisms of renewal and repair.

Some Tissues Lack Stem Cells and Are Not Renewable

The variety among tissues in the capacity for self-renewal is illustrated in a striking way by comparing the olfactory epithelium in the nose, the auditory epithelium of the inner ear, and the photoreceptive epithelium of the retina. These three sensory structures, which like the epidermis develop from the ectodermal layer of the early embryo, differ radically in their self-renewal capabilities. The olfactory epithelium contains a population of stem cells that give rise to differentiated cells that have a limited life-span and are continually replaced. But unlike the epidermis, these differentiated cells (the olfactory receptor cells) are neurons, with cell bodies lying in the olfactory epithelium and axons that extend back to the olfactory lobes in the brain. The continual renewal of this epithelium therefore involves continual production of fresh axons, which have to navigate back to the appropriate sites in the brain.

In contrast, in mammals at least, the auditory epithelium and the retinal epithelium lack stem cells, and their sensory receptor cells—the sensory hair cells in the ear, the photoreceptors in the retina—are irreplaceable. If they are destroyed—whether by too much exposure to loud noise, by looking into the beam of a laser, or through degenerative processes in old age—the loss is permanent.

Summary

Many tissues in the adult mammalian body are continually renewed by stem cells. Stem cells, by definition, are not terminally differentiated and have the ability to divide throughout the organism's lifetime, yielding some progeny that differentiate and others that remain stem cells. The lining of the gut renews itself more rapidly than any other tissue in the mammalian body and provides a paradigm for the workings of stem-cell systems. In the small intestine, there is a continual upward flow from crypts, where new cells are generated by cell division, onto villi that are composed of nondividing differentiated cells. Wnt signaling maintains cell proliferation in the crypts, and overactivation of the Wnt pathway gives rise to tumors. Stem cells lie at each crypt base and are distinguished by expression of Lgr5 and certain other genes. The Lgr5+ stem cells are multipotent, each capable of generating several different types of differentiated cells as well as new stem cells. The balance of fate choices is adjusted according to need, allowing increase in the number of stem cells where more are needed for growth or repair. In a suitable cell-free culture medium, a single Lgr5+ stem cell can generate a self-organizing "minigut," containing all the standard intestinal epithelial cell types.

Other self-renewing epithelia, such as the epidermis with its multilayered (stratified) architecture, have stem cells and their differentiating progeny arranged in different ways but are governed by similar basic principles. However, tissue renewal and repair does not always have to depend on stem cells. Thus, the population of insulin-producing cells in the pancreas is enlarged and renewed by simple duplication of existing insulin-producing cells. Similarly, in the liver, differentiated hepatocytes remain able to divide throughout life and can dramatically increase their division rate when the need arises. At an opposite extreme, some tissues, such as the sensory epithelia of the ear and the eye, do not undergo any turnover and are not renewable: their cells, once lost, are lost forever.

FIBROBLASTS AND THEIR TRANSFORMATIONS: THE CONNECTIVE-TISSUE CELL FAMILY

From epithelia, with their varied patterns of renewal and their enormous variety of protective, absorptive, secretory, sensory, and biosynthetic functions, we turn now to **connective tissues**. Connective tissues typically consist of cells dispersed in extracellular matrix that they themselves secrete, as discussed in Chapter 19. They originate from the mesodermal (middle) layer of the early embryo, sandwiched between ectoderm and endoderm (see Chapter 21, Figure 21–3).

In the adult body, virtually all epithelia are supported by a connective-tissue bed, or *stroma*; and specialized types of connective tissue, such as bone, cartilage, and tendon, form the supporting framework of the body as a whole. No less important than its mechanical role, connective tissue also contains the blood vessels that bring the oxygen and nourishment on which all cells depend. Cells of the immune system roam through connective tissue, passing in and out of blood vessels and lymphatics, and providing defence against infection; and through the meshes of connective tissue run peripheral nerves. Also embedded in connective tissue are the muscles that enable us to move. In these many ways, the cells that form connective tissue and synthesize its various types of extracellular matrix contribute to the support and repair of almost every tissue and organ.

Connective-tissue cells belong to a family of cell types that are related by origin, and they are often remarkably interconvertible. The family includes *fibroblasts*, *cartilage cells*, and *bone cells*, all of which are specialized for the secretion of collagenous extracellular matrix and are jointly responsible for the architectural framework of the body. The connective-tissue family also includes *fat cells* (*adipocytes*) and *smooth muscle cells*. **Figure 22–11** illustrates these cell types and the interconversions that are thought to occur between them. The adaptability of the differentiated character of connective-tissue cells is an important feature of responses to many types of damage.

Fibroblasts Change Their Character in Response to Chemical and Physical Signals

Fibroblasts seem to be the least specialized cells in the connective-tissue family. They are dispersed in connective tissue throughout the body, where they secrete a nonrigid extracellular matrix that is rich in type I or type III collagen, or both, as discussed in Chapter 19. When a tissue is injured, the fibroblasts nearby proliferate, migrate into the wound (**Movie 22.1**), and produce large amounts of collagenous matrix that helps to isolate and repair the damaged tissue. Their ability to thrive in the face of injury, together with their solitary lifestyle, may explain why fibroblasts are the easiest of cells to grow in culture—a feature that has made them a favorite subject for cell biological studies.

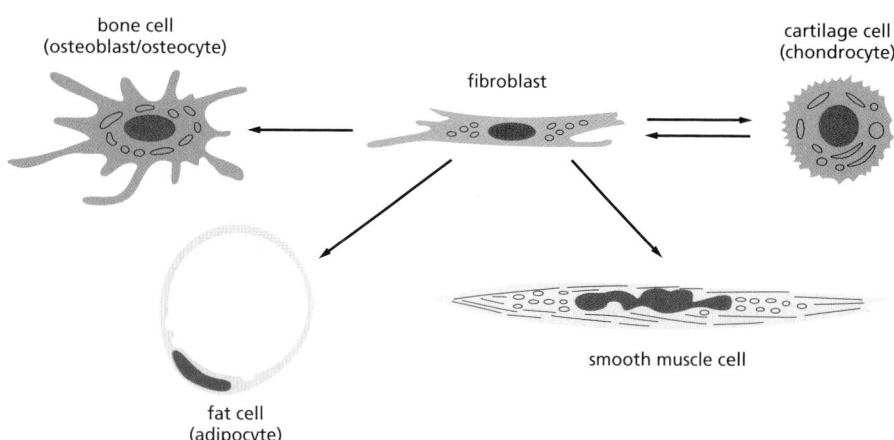

Figure 22–11 The family of connective-tissue cells. Arrows show the interconversions that are thought to occur within the family. For simplicity, the fibroblast is shown as a single cell type, but it is uncertain how many types of fibroblasts exist and whether the differentiation potential of different types is restricted in different ways.

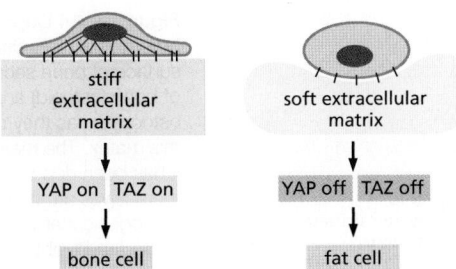

Figure 22–12 **Control of fibroblast differentiation by the physical properties of the extracellular matrix.** On a stiff matrix, the cells form strong adhesions, spread out, and tend to turn into bone cells. On a soft matrix, where the cells are unable to form strong anchorages, they fail to spread and tend to differentiate as fat cells. These effects depend on transcription regulators (YAP and TAZ proteins) that move into the cell nucleus in response to tension developed in actin–myosin bundles in the cytoplasm. (Based on S. Dupont et al., *Nature* 474:179–183, 2011.)

A class of connective-tissue cells in the bone marrow, called *bone marrow stromal* cells, provides an example of radical connective-tissue versatility. These cells, which can be regarded as a kind of fibroblast, can be isolated from the bone marrow and propagated in culture. Large clones of progeny can be generated in this way from single ancestral stromal cells. Depending on the culture conditions, the members of such a clone either can continue proliferating to produce more cells of the same type, or can differentiate as fat cells, cartilage cells, or bone cells. The fate of the cells depends on physical as well as chemical signals: embedded in a stiff, unyielding matrix, they tend to turn into bone cells, whereas in a softer, more elastic matrix, they tend to turn into fat cells. This effect is mediated by an intracellular pathway that responds to tension in actin–myosin bundles and relays a signal to specific transcription regulators in the nucleus (**Figure 22–12**). Because of their self-renewing, multipotent character, the bone marrow stromal cells, and other cells with similar properties, are referred to as *mesenchymal stem cells*.

Osteoblasts Make Bone Matrix

Cartilage and bone are tissues of very different character; but they are closely related in origin, and the formation of the skeleton depends on an intimate partnership between them.

Cartilage tissue is structurally simple, consisting of cells of a single type—chondrocytes—embedded in a more or less uniform, highly hydrated matrix consisting of proteoglycans and type II collagen (discussed in Chapter 19). The cartilage matrix is deformable, and the tissue grows by expanding as the chondrocytes divide and secrete more matrix (**Figure 22–13**). **Bone**, by contrast, is dense and rigid; it grows by apposition—that is, by deposition of additional matrix on free surfaces. Like reinforced concrete, the bone matrix is predominantly a mixture of tough fibers (type I collagen fibrils), which resist pulling forces, and solid particles (calcium phosphate as *hydroxylapatite* crystals), which resist compression. The bone matrix is secreted by **osteoblasts** that lie at the surface of the existing matrix and deposit fresh layers of bone onto it. Some of the osteoblasts remain free at the surface, while others gradually become embedded in their own secretion. This freshly formed material (consisting chiefly of type I collagen) is rapidly converted into hard bone matrix by the deposition of calcium phosphate crystals in it.

Once imprisoned in hard matrix, the original bone-forming cell, now called an **osteocyte**, has no opportunity to divide, although it continues to secrete additional matrix in small quantities around itself. The osteocyte, like the chondrocyte, occupies a small cavity, or *lacuna*, in the matrix, but unlike the chondrocyte

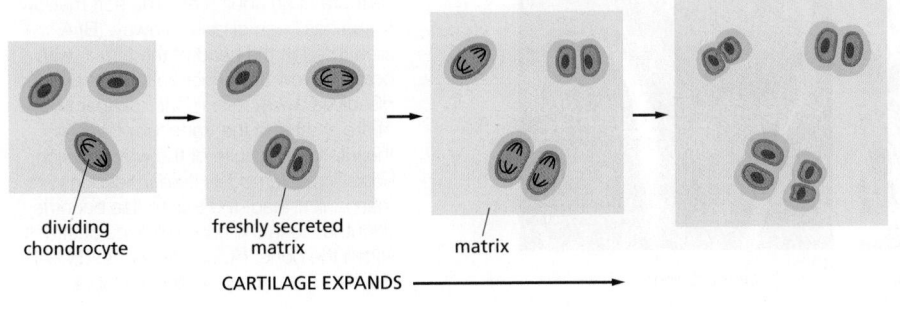

dividing chondrocyte

freshly secreted matrix

matrix

CARTILAGE EXPANDS ⟶

Figure 22–13 **The growth of cartilage.** The tissue expands as the chondrocytes divide and make more matrix. The freshly synthesized matrix with which each cell surrounds itself is shaded *dark green*. Cartilage may also grow by recruiting fibroblasts from the surrounding tissue and converting them into chondrocytes.

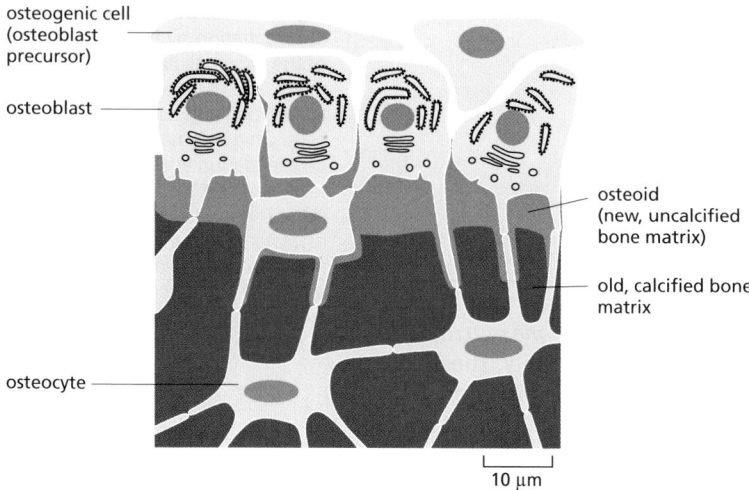

osteogenic cell (osteoblast precursor)

osteoblast

osteoid (new, uncalcified bone matrix)

old, calcified bone matrix

osteocyte

10 μm

Figure 22–14 Deposition of bone matrix by osteoblasts. Osteoblasts lining the surface of bone secrete the organic matrix of bone (osteoid) and are converted into osteocytes as they become embedded in this matrix. The matrix calcifies soon after it has been deposited. The osteoblasts themselves are thought to derive from osteogenic stem cells that are closely related to fibroblasts.

it is not isolated from its fellows. Tiny channels, or *canaliculi*, radiate from each lacuna and contain cell processes from the resident osteocyte, enabling it to form gap junctions with adjacent osteocytes (**Figure 22–14**). Blood vessels and nerves run through the tissue, keeping the bone cells alive and reacting when the bone is damaged.

A mature bone has a complex and beautiful architecture, in which dense plates of *compact bone* tissue enclose spaces spanned by light frameworks of *trabecular bone*—a filigree of delicate shafts and flying buttresses of bone tissue, with soft marrow in the interstices (**Figure 22–15**). The creation, maintenance, and repair of this structure depend not only on the cells of the connective-tissue family that synthesize matrix, but also on a separate class of cells called *osteoclasts* that degrade it, as we explain below.

Bone Is Continually Remodeled by the Cells Within It

For all its rigidity, bone is by no means a permanent and immutable tissue. Running through the hard extracellular matrix are channels and cavities occupied by living cells, which account for about 15% of the weight of compact bone. These cells are engaged in an unceasing process of remodeling: while osteoblasts deposit new bone matrix, osteoclasts demolish old bone matrix. This mechanism provides for continuous turnover and replacement of the matrix in the interior of the bone.

Osteoclasts (**Figure 22–16**) are large, multinucleated cells that originate, like macrophages, from hematopoietic stem cells in the bone marrow (discussed later

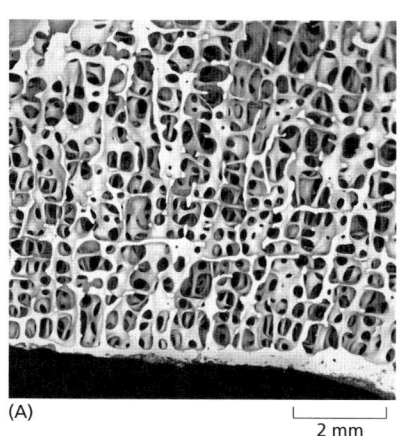

(A)

2 mm

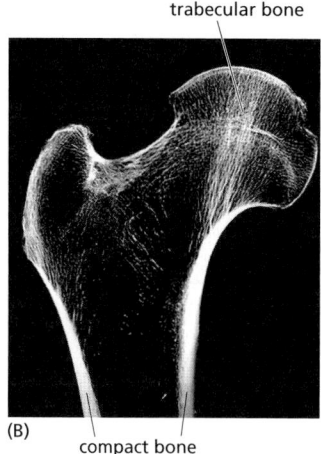

trabecular bone

(B)

compact bone

Figure 22–15 Trabecular and compact bone. (A) Low-magnification scanning electron micrograph of trabecular bone in a vertebra of an adult man. The soft marrow tissue has been dissolved away. (B) A slice through the head of the femur, with bone marrow and other soft tissue likewise dissolved away, reveals the compact bone of the shaft and the trabecular bone in the interior. Because of the way in which bone tissue remodels itself in response to mechanical load, the trabeculae become oriented along the principle axes of stress within the bone. (A, courtesy of and © Alan Boyde; B, courtesy of Clinton Rubin.)

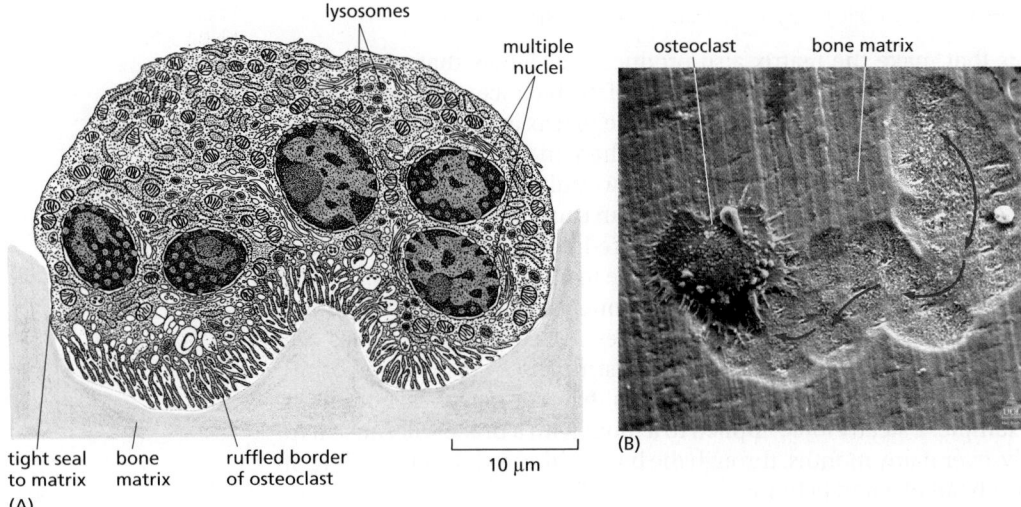

Figure 22–16 Osteoclasts. (A) Drawing of an osteoclast in cross section. This giant, multinucleated cell erodes bone matrix. The "ruffled border" is a site of secretion of acids (to dissolve the bone minerals) and hydrolases (to digest the organic components of the matrix). Osteoclasts vary in shape, are motile, and often send out processes to resorb bone at multiple sites. They develop from monocytes and can be viewed as specialized macrophages. (B) An osteoclast on bone matrix, seen by scanning electron microscopy. The osteoclast has been crawling over the matrix, eating it away, and leaving a trail of pits where it has done so. (A, from R.V. Krstić, *Ultrastructure of the Mammalian Cell: An Atlas*. Berlin: Springer-Verlag, 1979. With permission from Springer Nature; B, courtesy of and © Alan Boyde.)

in this chapter). The precursor cells are released into the bloodstream and collect at sites of bone resorption, where they fuse to form the multinucleated osteoclasts, which cling to surfaces of the bone matrix and eat it away. Osteoclasts are capable of tunneling deep into the substance of compact bone, forming cavities that are then invaded by other cells. A blood capillary grows down the center of such a tunnel, and the walls of the tunnel become lined with a layer of osteoblasts (**Figure 22–17**). These osteoblasts lay down concentric layers of new matrix, which gradually fill the cavity, leaving only a narrow canal surrounding the new blood vessel. At the same time as some tunnels are filling up with bone, others are being bored by osteoclasts, cutting through older concentric systems.

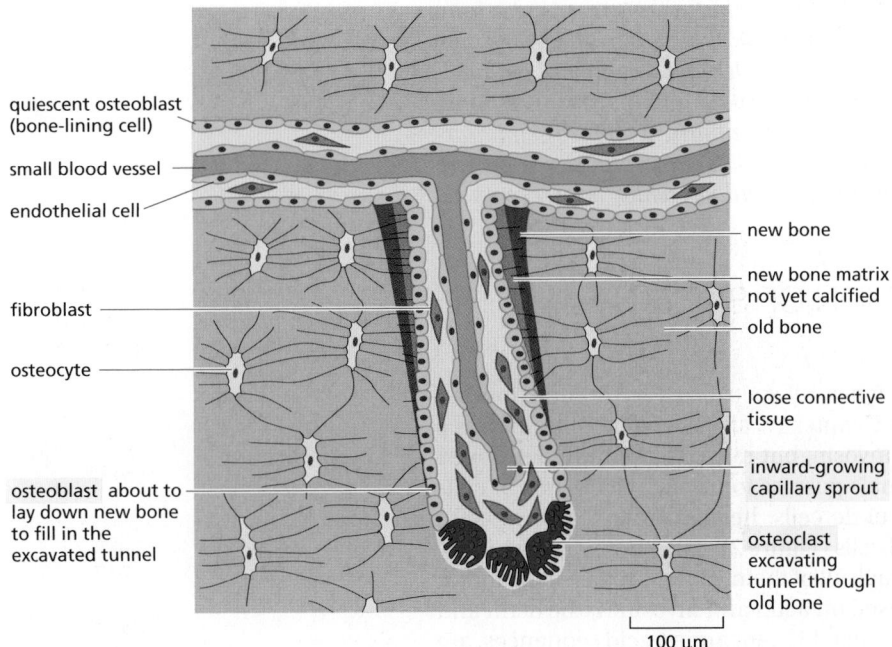

Figure 22–17 The remodeling of compact bone. Osteoclasts acting together in a small group excavate a tunnel through the old bone, advancing at a rate of about 50 µm per day. Osteoblasts enter the tunnel behind them, line its walls, and begin to form new bone, depositing layers of matrix at a rate of 1–2 µm per day. At the same time, a capillary sprouts down the center of the tunnel. The tunnel eventually becomes filled with concentric layers of new bone, with only a narrow central canal remaining. Each such canal, besides providing a route of access for osteoclasts and osteoblasts, contains one or more blood vessels that transport the nutrients the bone cells require for survival. Typically, about 5–10% of the bone in a healthy adult mammal is replaced in this way each year. (After Z.F.G. Jaworski, B. Duck and G. Sekaly, *J. Anat.* 133:397–405, 1981.)

Osteoclasts Are Controlled by Signals From Osteoblasts

The osteoblasts that make the matrix also produce the signals that recruit and activate the osteoclasts to degrade it. Disturbance of the balance can lead to *osteoporosis*, where there is excessive erosion of the bone matrix and weakening of the bone, or to the opposite condition, *osteopetrosis*, where the bone becomes excessively thick and dense. Hormonal signals have powerful effects on this balance. Chronic use of corticosteroid drugs, for example, can cause osteoporosis as a side effect; but this can be treated by other drugs that redress the balance, including agents that block the factors that osteoblasts secrete to recruit osteoclasts.

Local controls allow bone to be deposited in one place while it is resorbed in another. Through such controls over the process of remodeling, bones are endowed with a remarkable ability to adjust their structure in response to long-term variations in the load imposed on them. It is this that makes orthodontics possible, for example: a steady force applied to a tooth with a brace will cause it to move gradually, over many months, through the bone of the jaw, by remodeling of the bone tissue ahead of it and behind it.

Bone can also undergo much more rapid and dramatic reconstruction when the need arises. Some cells capable of forming new cartilage persist in the connective tissue that surrounds a bone. If the bone is broken, the cells in the neighborhood of the fracture repair it by a process that resembles the way bones develop in the embryo: cartilage is first laid down to bridge the gap and is then replaced by bone. The capacity for self-repair, so strikingly illustrated by the tissues of the skeleton, is a property of living structures that has no parallel among present-day man-made objects.

Summary

The family of connective-tissue cells includes fibroblasts, cartilage cells, bone cells, fat cells, and smooth muscle cells. Some classes of fibroblasts, such as the mesenchymal stem cells of bone marrow, seem to be able to transform into any of the other members of the family. These transformations of connective-tissue cell type are regulated by the composition of the surrounding extracellular matrix, by cell shape, and by hormones and growth factors. Cartilage and bone both consist of cells and solid matrix that the cells secrete around themselves—chondrocytes in cartilage, osteoblasts in bone (osteocytes being osteoblasts that have become trapped within the bone matrix). The matrix of cartilage is deformable so that the tissue can grow by swelling, whereas bone is rigid and can grow only by apposition. While osteoblasts secrete bone matrix, they also produce signals that recruit monocytes from the circulation to become osteoclasts, which degrade bone matrix. Through the activities of these antagonistic classes of cells, bone undergoes a perpetual remodeling through which it can adapt to the load it bears and alter its density in response to hormonal signals. Moreover, adult bone retains an ability to repair itself if fractured, by reactivation of the mechanisms that governed its embryonic development: cells in the neighborhood of the break convert into cartilage, which is later replaced by bone.

GENESIS AND REGENERATION OF SKELETAL MUSCLE

The term "muscle" includes many cell types, all specialized for contraction but in other respects dissimilar. As noted in Chapter 16, all eukaryotic cells possess a contractile system involving actin and myosin, but muscle cells have developed this apparatus to a high degree. Mammals possess four main categories of cells specialized for contraction: skeletal muscle cells, heart (cardiac) muscle cells, smooth muscle cells, and myoepithelial cells (**Figure 22–18**). These differ in function, structure, and development. Although all of them generate contractile forces by using organized filament systems based on actin and myosin II, the actin and myosin molecules employed have somewhat different amino acid sequences, are

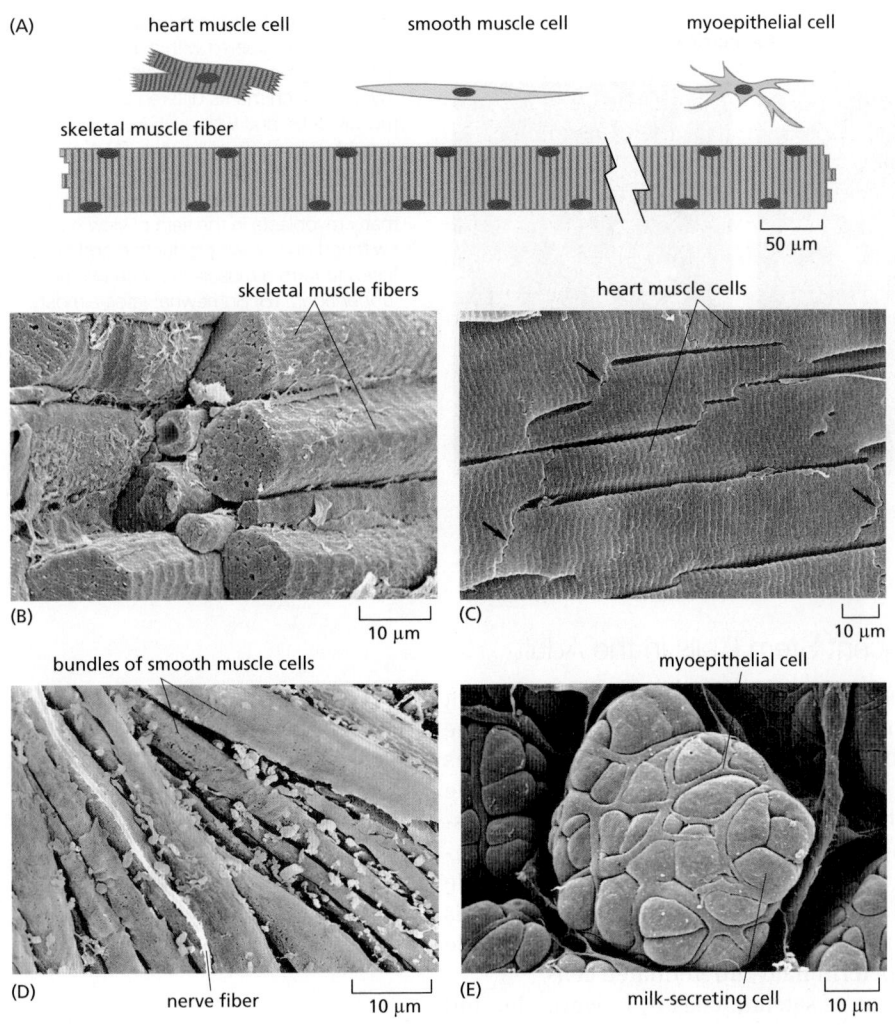

Figure 22–18 **The four classes of muscle cells of a mammal.** (A) Schematic drawings (to scale). (B–E) Scanning electron micrographs. Skeletal muscle fibers from a human (B) are giant cells with many nuclei and are formed by cell fusion. The other types of muscle cells are more conventional, generally having only a single nucleus. Heart muscle cells (C, from a rat) resemble skeletal muscle fibers in that their actin and myosin filaments are aligned in very orderly arrays to form a series of contractile units called sarcomeres, so that the cells have a striated (striped) appearance. The arrows in (C) point to intercalated discs—end-to-end junctions between the heart muscle cells; skeletal muscle cells in long muscles are joined end-to-end in a similar way. Smooth muscle cells (D) from the human oviduct help propel an egg down the oviduct from the ovary to the uterus. They are so named because they do not appear striated; they belong to the connective-tissue family and are closely related to fibroblasts. The functions of smooth muscle vary greatly, from propelling food along the digestive tract to erecting hairs in response to cold or fear. Myoepithelial cells (E, from a secretory alveolus of a lactating rat mammary gland) also have no striations, but unlike all other muscle cells they lie in epithelia and are derived from the ectoderm. They form the dilator muscle of the eye's iris and serve to expel saliva, sweat, and milk from the corresponding glands. (B, Steve Gschmeissner/Science Source; C, from T. Fujiwara, "Cardiac Muscle" in E.D. Canal, *Handbook of Microscopic Anatomy,* published 1986 by Springer-Verlag. Reproduced with permission of SNCSC; D, Professor Pietro M. Motta & Enrico Vizza/Science Source; E, from T. Nagato et al., *Cell and Tissue Research* 209(1);1–10, published 1980 by Springer-Verlag. Reproduced with permission of SNCSC.)

differently arranged in the cell, and are associated with different sets of proteins that control contraction.

We focus in this section on skeletal muscle cells, which are responsible for practically all movements that are under voluntary control. These cells can be very large (2–3 cm long and 100 μm in diameter in an adult human) and are often called *muscle fibers* because of their highly elongated shape. Each one is a syncytium, containing many nuclei within a common cytoplasm. In an intact muscle, they are bundled tightly together, with fibroblasts (and some fat cells) in the interstices between them and blood vessels and nerve fibers running through the tissue. The mechanisms of muscle contraction were discussed in Chapter 16. Here we consider the unusual strategy by which the multinucleate skeletal muscle cells are generated and maintained.

Myoblasts Fuse to Form New Skeletal Muscle Fibers

During development, certain cells, originating from the somites of a vertebrate embryo at a very early stage, become determined as **myoblasts**, the precursors of skeletal muscle fibers. After a period of proliferation, the myoblasts undergo a dramatic change of state: they stop dividing, switch on the expression of a whole battery of muscle-specific genes required for terminal differentiation, and fuse with one another to form multinucleate skeletal muscle fibers (**Figure 22–19**). Once differentiation and cell fusion have occurred, the cells do not divide and the nuclei never again replicate their DNA.

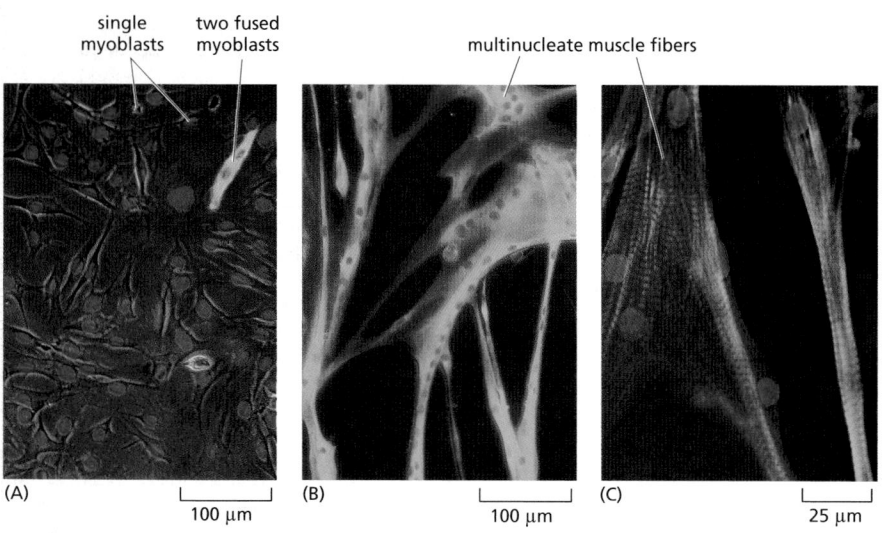

single myoblasts   two fused myoblasts   multinucleate muscle fibers

(A)   ⌊___100 μm___⌋   (B)   ⌊___100 μm___⌋   (C)   ⌊___25 μm___⌋

Figure 22–19 Myoblast fusion in culture. The culture is stained with a fluorescent antibody *(green)* against skeletal muscle myosin, which marks differentiated muscle cells, and with a DNA-specific dye *(blue)* to show cell nuclei. (A) A short time after a change to a culture medium that favors differentiation, just two of the many myoblasts in the field of view have switched on myosin production and have fused to form a muscle cell with two nuclei *(upper right)*. (B) Somewhat later, almost all the cells have differentiated and fused. (C) High-magnification view, showing characteristic striations (fine transverse stripes) in two of the multinucleate muscle cells. (Courtesy of Jacqueline Gross and Terence Partridge.)

Some Myoblasts Persist as Quiescent Stem Cells in the Adult

Even though humans do not normally generate new skeletal muscle fibers in adult life, they still have the capacity to do so, and existing muscle fibers can resume growth when the need arises. Cells capable of serving as myoblasts are retained as small, flattened, and inactive cells lying in close contact with the mature muscle cell and contained within its sheath of basal lamina (**Figure 22–20**). If the muscle is damaged or stimulated to grow, these *satellite cells* are activated to proliferate, and their progeny can fuse to repair the damaged muscle or to allow muscle growth. Satellite cells, or some subset of the satellite cells, are thus the stem cells of adult skeletal muscle, normally held in reserve in a quiescent state but available when needed as a self-renewing source of terminally differentiated cells.

The process of muscle repair by means of satellite cells is, however, limited in what it can achieve. In one form of *muscular dystrophy*, for example, a genetic defect in the cytoskeletal protein dystrophin damages differentiated skeletal muscle cells. As a result, satellite cells proliferate to repair the damaged muscle fibers. This regenerative response is, however, unable to keep pace with the damage, and connective tissue eventually replaces the muscle cells, blocking any further possibility of regeneration. A decline of capacity for repair likewise contributes to the weakening of muscle in the elderly.

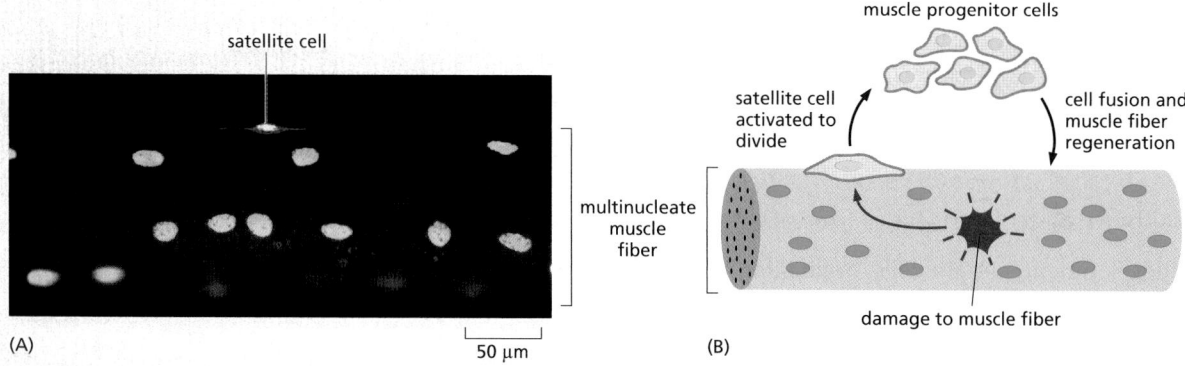

satellite cell

multinucleate muscle fiber

muscle progenitor cells

satellite cell activated to divide

cell fusion and muscle fiber regeneration

damage to muscle fiber

(A)   ⌊___50 μm___⌋   (B)

Figure 22–20 Satellite cells repair skeletal muscle fibers. (A) The specimen is stained with an antibody *(red)* against a muscle cadherin, M-cadherin, which is present on both the satellite cell and the muscle fiber and is concentrated at the site where their membranes are in contact. The nuclei of the muscle fiber are stained *green*, and the nucleus of the satellite cell is stained *blue*. (B) Schematic of the repair of a damaged muscle fiber by proliferation and fusion of satellite cells. (A, courtesy of Terence Partridge.)

Summary

Skeletal muscle fibers are one of four main categories of vertebrate cells specialized for contraction, and they are responsible for all voluntary movement. Each skeletal muscle fiber is a syncytium and develops by the fusion of many myoblasts. Myoblasts proliferate extensively, but once they have fused, they can no longer divide. Fusion generally follows the onset of myoblast differentiation, in which many genes encoding muscle-specific proteins are switched on coordinately. Some myoblasts persist in a quiescent state as satellite cells in adult muscle; when a muscle is damaged, these cells are reactivated to proliferate and to fuse in order to replace the muscle cells that have been lost. They are the stem cells of skeletal muscle, and exhaustion of their regenerative capacity is responsible for some forms of muscular dystrophy as well as for the decline of muscle mass in old age.

BLOOD VESSELS, LYMPHATICS, AND ENDOTHELIAL CELLS

Almost all tissues depend on a blood supply, and the blood supply depends on **endothelial cells**, which form the linings of the blood vessels. Endothelial cells have a remarkable capacity to adjust their number and arrangement to suit local requirements. They create an adaptable life-support system, extending by cell migration into almost every region of the body. If it were not for endothelial cells extending and remodeling the network of blood vessels, tissue growth and repair would be impossible. Cancerous tissue is as dependent on a blood supply as is normal tissue, and this has led to a surge of interest in endothelial cell biology, in the hope that it may be possible to block the growth of tumors by attacking the endothelial cells that bring them nourishment.

Endothelial Cells Line All Blood Vessels and Lymphatics

The largest blood vessels are arteries and veins, which have a thick, tough wall of connective tissue and many layers of smooth muscle cells (**Figure 22–21**). The inner wall is lined by an exceedingly thin single sheet of endothelial cells, the *endothelium*, separated from the surrounding outer layers by a basal lamina. The amounts of connective tissue and smooth muscle in the vessel wall vary according to the vessel's diameter and function, but the endothelial lining is always present. In the finest branches of the vascular tree—the capillaries and sinusoids—the walls consist of nothing but endothelial cells and a basal lamina (**Figure 22–22**), together with a few scattered *pericytes*. Related to vascular smooth muscle cells, pericytes wrap themselves around the small vessels and strengthen them (**Figure 22–23**).

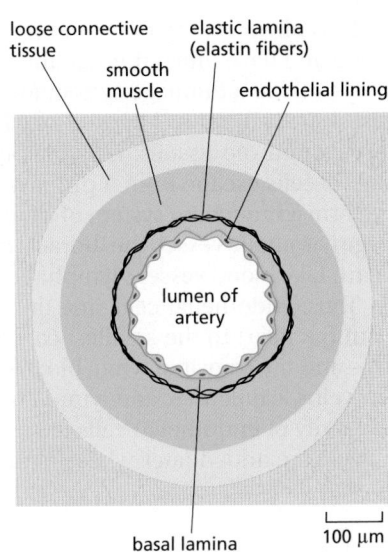

Figure 22–21 Diagram of a small artery in cross section. The endothelial cells form the endothelial lining, which although inconspicuous, is the fundamental component. Compare with the capillary in Figure 22–22.

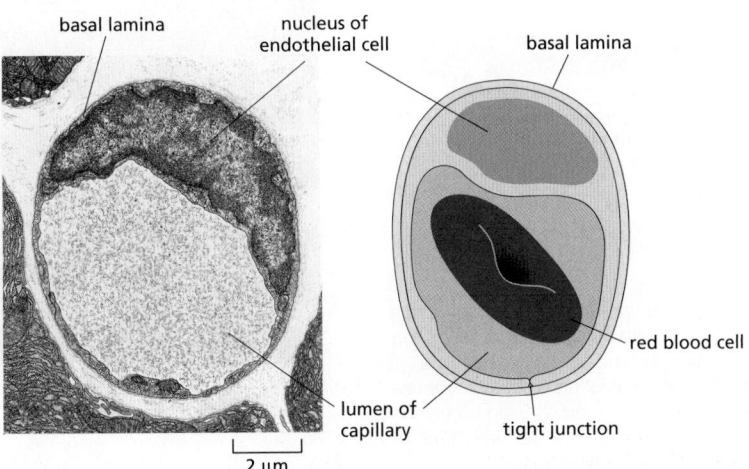

Figure 22–22 Capillaries. Electron micrograph *(left)* of a cross section of a small capillary in the pancreas. The wall is formed by a single endothelial cell surrounded by a basal lamina, as seen most clearly in the drawing to the *right*. (© 1974 R.P. Bolender. Originally published in *J. Cell Biol.* https://doi.org/10.1083/jcb.61.2.269. With permission from Rockefeller University Press.)

Less obvious than the blood vessels are the lymphatic vessels. These carry no blood and have much thinner and more permeable walls than the blood vessels. They provide a drainage system for the fluid (lymph) that seeps out of the blood vessels, as well as an exit route for white blood cells that have migrated from blood vessels into the tissues. Less happily, they can also provide the path by which cancer cells escape from a primary tumor to invade other tissues. The lymphatics form a branching system of tributaries, all ultimately discharging into a single large lymphatic vessel, the thoracic duct, which opens into a large vein close to the heart. Like blood vessels, lymphatics are lined with endothelial cells.

Thus, endothelial cells line the entire blood and lymphatic vascular system, from the heart to the smallest capillary, and they control the passage of materials—and the transit of white blood cells—into and out of the bloodstream. Arteries, veins, capillaries, and lymphatics all develop from small vessels constructed primarily of endothelial cells and a basal lamina: connective tissue and smooth muscle are added later where required, under the influence of signals from the endothelial cells.

Endothelial Tip Cells Pioneer Angiogenesis

To understand how the vascular system comes into being and how it adapts to the changing needs of tissues, we have to understand endothelial cells. How do they become so widely distributed, and how do they form channels that connect in just the right way for blood to circulate through the tissues and for lymph to drain back to the bloodstream?

Endothelial cells originate at specific sites in the early embryo from precursors that also give rise to blood cells. From these sites, the early embryonic endothelial cells migrate, proliferate, and differentiate to form the first rudiments of blood vessels—a process called *vasculogenesis*. Subsequent growth and branching of the vessels throughout the body occurs mainly by proliferation and movement of the endothelial cells of these first vessels, in a process called **angiogenesis**.

Angiogenesis occurs in a broadly similar way in the young organism as it grows and in the adult during tissue repair and remodeling. We can watch the behavior of the cells in naturally transparent structures, such as the cornea of the eye or the fin of a tadpole, or in tissue culture, or in the embryo. The embryonic retina, which blood vessels invade according to a predictable timetable, provides a convenient example for experimental study. Each new vessel originates as a capillary sprout from the side of an existing capillary or small venule (**Figure 22–24**). At the tip of the sprout, leading the way, is an endothelial cell with a distinctive character. This *tip cell* has a pattern of gene expression somewhat different from that of the endothelial stalk cells following behind it, and while they divide, it does not. The tip cell's most striking feature is that it puts out many long filopodia, resembling

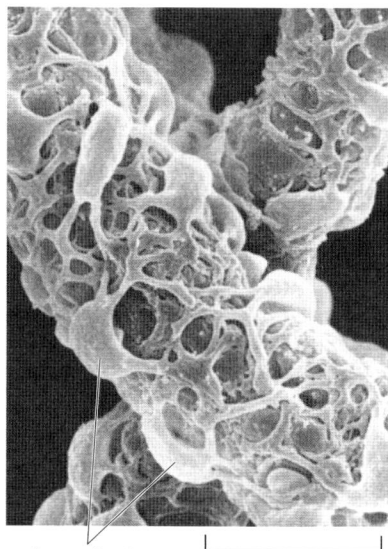

pericytes clinging
to outer face of
small blood vessel

Figure 22–23 Pericytes. The scanning electron micrograph shows pericytes wrapping their processes around a small blood vessel (a post-capillary venule) in the mammary gland of a cat. Pericytes are also present around capillaries, but are much more sparsely distributed there. (From T. Fujiwara and Y. Uehara, *Am. J. Anat.* 170:39–54, 1984. With permission from Wiley-Liss.)

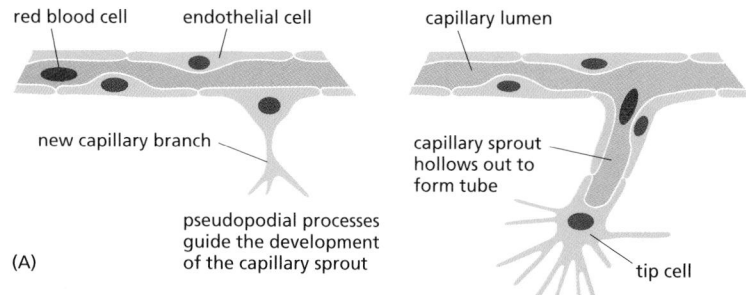

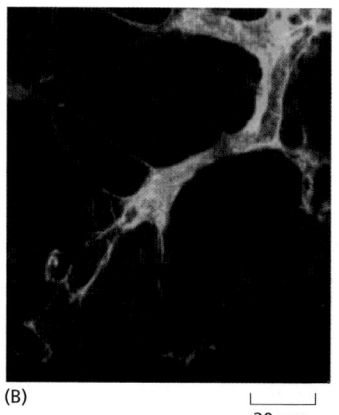

Figure 22–24 Angiogenesis. (A) A new blood capillary forms by the sprouting of an endothelial cell from the wall of an existing small vessel. An endothelial tip cell, with many filopodia, leads the advance of each capillary sprout. The endothelial stalk cells trailing behind the tip cell become hollowed out to form a lumen. (B) Blood capillaries sprouting in the retina of an embryonic mouse that had a red dye injected into the bloodstream, revealing the capillary lumen opening up behind the tip cell (**Movie 22.2**). (B, © 2003 H. Gerhardt et al. Originally published in *J. Cell Biol.* https://doi.org/10.1083/jcb.200302047. With permission from Rockefeller University Press.)

those of a neuronal growth cone. The column of stalk cells behind it, meanwhile, becomes hollowed out to form a lumen.

The endothelial tip cells that pioneer the growth of normal capillaries not only look like neuronal growth cones, but also respond similarly to signals in the environment. In fact, many of the same guidance molecules are involved, including the netrins, slits, and ephrins mentioned in our account of neural development in the previous chapter. The corresponding receptors are expressed in the tip cells and guide the vascular sprouts along specific pathways in the embryo, often in parallel with nerves. Perhaps the most important of the guidance molecules for endothelial cells, however, is one that is chiefly dedicated to the control of vascular development: *vascular endothelial growth factor*, or *VEGF*.

Tissues Requiring a Blood Supply Release VEGF

Almost every cell, in almost every tissue of a vertebrate, is located within 50–100 μm of a blood capillary. What mechanism ensures that the system of blood vessels branches into every nook and cranny? How is it adjusted so perfectly to the local needs of the tissues, not only during normal development but also in pathological circumstances? Wounding, for example, induces a burst of capillary growth in the neighborhood of the damage, to satisfy the high metabolic requirements of the repair process (**Figure 22–25**). Local irritants and infections also cause a proliferation of new capillaries, most of which regress and disappear when the inflammation subsides. Less benignly, a small sample of tumor tissue implanted in the cornea, which normally lacks blood vessels, causes blood vessels to grow quickly toward the implant from the vascular margin of the cornea; the growth rate of the tumor increases abruptly as soon as the vessels reach it.

In all these cases, the invading endothelial cells respond to signals produced by the tissue that they invade. The signals are complex, but a key part is played by **vascular endothelial growth factor** (**VEGF**). The regulation of blood vessel growth to match the needs of the tissue depends on the control of VEGF production, through changes in the stability of its mRNA and in its rate of transcription. The latter control is relatively well understood. A shortage of oxygen, in practically any type of cell, causes an increase in the intracellular level of a transcription factor called **hypoxia-inducible factor 1α** (**HIF1α**). HIF1α stimulates transcription of *Vegf* (and of other genes whose products are needed when oxygen is in short supply). The VEGF protein is secreted, diffuses through the tissue, and acts on nearby endothelial cells, stimulating them to proliferate, to produce proteases to help them digest their way through the basal lamina of the parent capillary or venule, and to form sprouts. The tip cells of the sprouts detect the VEGF gradient and move toward its source. As the new vessels form, bringing blood to the tissue, the oxygen concentration rises. The HIF1α activity then declines, VEGF production is shut off, and angiogenesis comes to a halt (**Figure 22–26**).

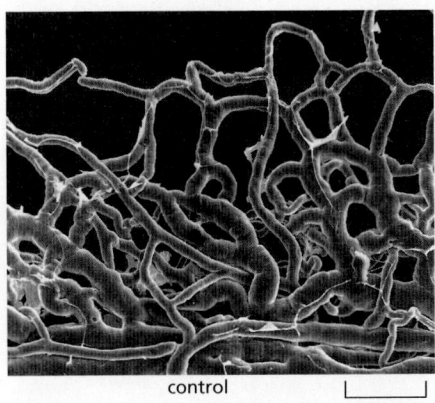

control   100 μm

60 hours after wounding   100 μm

Figure 22–25 **New capillary formation in response to wounding.** Scanning electron micrographs of casts of the system of blood vessels surrounding the margin of the cornea show the reaction to wounding. The casts are made by injecting a resin into the vessels and letting the resin set; this reveals the shape of the lumen, as opposed to the shape of the cells. Sixty hours after wounding, many new capillaries have begun to sprout toward the site of injury, which is just above the top of the picture. Their oriented outgrowth reflects a chemotactic response of the endothelial cells to an angiogenic factor released at the wound. (Courtesy of Peter C. Burger.)

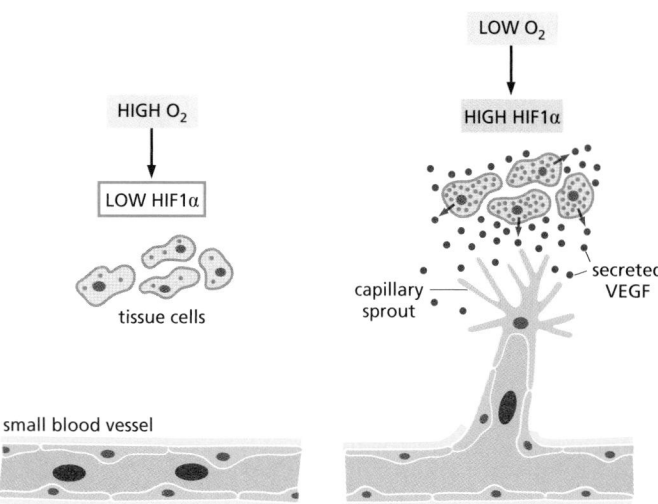

Figure 22–26 **The regulatory mechanism controlling blood vessel growth according to a tissue's need for oxygen.** Lack of oxygen triggers the secretion of VEGF, which stimulates angiogenesis.

Signals from Endothelial Cells Control Recruitment of Pericytes and Smooth Muscle Cells to Form the Vessel Wall

The vascular network is continually remodeled as it grows and adapts. A newly formed vessel may enlarge; or it may sprout side branches; or it may regress. Smooth muscle and other connective-tissue cells that pack themselves around the endothelium (see Figure 22–23) help to stabilize vessels as they enlarge. This process of vessel wall formation begins with recruitment of pericytes. Small numbers of these cells travel outward in company with the stalk cells of each endothelial sprout. The recruitment and proliferation of pericytes and smooth muscle cells to form a vessel wall depend on platelet-derived growth factor-B (PDGF-B) secreted by the endothelial cells and on PDGF receptors in the pericytes and smooth muscle cells. In mutants lacking this signal protein or its receptor, these vessel wall cells are missing in many regions. As a result, the embryonic blood vessels develop microaneurysms—microscopic pathological dilatations—that eventually rupture, as well as other abnormalities, reflecting the importance of signals exchanged in both directions between the exterior cells of the wall and the endothelial cells.

Summary

Endothelial cells are the fundamental elements of the vascular system. They form a single cell layer that lines all blood vessels and lymphatics and regulates exchanges between the bloodstream and the surrounding tissues. New vessels originate as endothelial sprouts from the walls of existing small vessels. A specialized motile endothelial tip cell at the leading edge of each sprout puts out filopodia that respond to gradients of guidance molecules in the environment, leading the growth of the sprout in much the same way as the growth cone of a neuron is led. The endothelial stalk cells following behind become hollowed out to form a capillary tube. Signals from endothelial cells organize the growth and development of the connective-tissue cells that form the surrounding layers of the vessel wall.

A homeostatic mechanism ensures that blood vessels permeate every region of the body. Cells that are short of oxygen increase their concentration of hypoxia-inducible factor 1α (HIF1α), which stimulates the production of vascular endothelial growth factor (VEGF). VEGF acts on endothelial cells, causing them to proliferate and invade the hypoxic tissue to supply it with new blood vessels. As new vessels enlarge, they recruit increasing numbers of pericytes—cells that cling to the outside of the endothelial tube and mature into the smooth muscle coat that is needed to give the vessel strength.

A HIERARCHICAL STEM-CELL SYSTEM: BLOOD CELL FORMATION

The function of blood vessels is to carry blood, and it is to blood itself that we now turn. Blood contains many types of cells, with functions that range from the transport of oxygen to the production of antibodies. Some of these cells stay within the vascular system, while others use the vascular system only as a means of transport and perform their function elsewhere. All blood cells, however, have certain similarities in their life history. They all have limited life-spans and are produced throughout the life of the animal. Most remarkably, they are all generated ultimately from a common stem cell, located (in adult humans) in the bone marrow. This *hematopoietic* (blood-making) *stem cell* is thus multipotent, giving rise to all the types of terminally differentiated blood cells as well as some other types of cells, such as the osteoclasts in bone, as mentioned earlier. The hematopoietic system is the most complex of the stem-cell systems in the mammalian body, and it is exceptionally important in medical practice.

Red Blood Cells Are All Alike; White Blood Cells Can Be Grouped in Three Main Classes

Blood cells can be classified as red or white. The **red blood cells**, or **erythrocytes**, remain within the blood vessels and transport O_2 and CO_2 bound to hemoglobin. The **white blood cells**, or **leukocytes**, combat infection and in some cases phagocytose and digest debris. Leukocytes, unlike erythrocytes, must make their way across the walls of small blood vessels and migrate into tissues to perform their tasks. In addition, the blood contains large numbers of **platelets**, which are not entire cells but small, detached cell fragments or "minicells" derived from the cortical cytoplasm of large cells called *megakaryocytes*. Platelets adhere specifically to the endothelial cell lining of damaged blood vessels, where they help to repair breaches and aid in blood clotting.

All red blood cells belong in a single class, following the same developmental trajectory as they mature, and the same is true of platelets; but there are many distinct types of white blood cells. White blood cells are traditionally grouped into three major categories—granulocytes, monocytes, and lymphocytes—based on their appearance in the light microscope.

Granulocytes contain numerous lysosomes and secretory vesicles (or granules) and are subdivided into three classes according to the morphology and staining properties of these organelles (**Figure 22–27**). The differences in staining reflect major differences of chemistry and function. *Neutrophils* (also called *polymorphonuclear leukocytes* because of their multilobed nucleus) are the most common type of granulocyte; they phagocytose and destroy microorganisms, especially bacteria, and thus have a key role in innate immunity to bacterial infection, as discussed in Chapter 24 (see Movie 16.1). *Basophils* secrete histamine (and, in some species, serotonin) to help mediate inflammatory reactions; they are closely related to *mast cells*, which reside in connective tissues but are also generated from the hematopoietic stem cells. *Eosinophils* help to destroy parasites and modulate allergic inflammatory responses.

Once they leave the bloodstream, **monocytes** (see Figure 22–27D) mature into **macrophages**, which, together with neutrophils, are the main "professional phagocytes" in the body. As discussed in Chapter 13, both types of phagocytic cells contain specialized lysosomes that fuse with newly formed phagocytic vesicles (phagosomes), exposing phagocytosed microorganisms to a barrage of enzymatically produced, highly reactive molecules of superoxide (O_2^-) and hypochlorite (ClO^-, the active ingredient in bleach), as well as to attack by a concentrated mixture of lysosomal hydrolase enzymes that become activated in the phagosome. Macrophages, however, are much larger and longer-lived than neutrophils. They recognize and remove senescent, dead, and damaged cells in many tissues, and they are unique in being able to ingest large microorganisms such as protozoa.

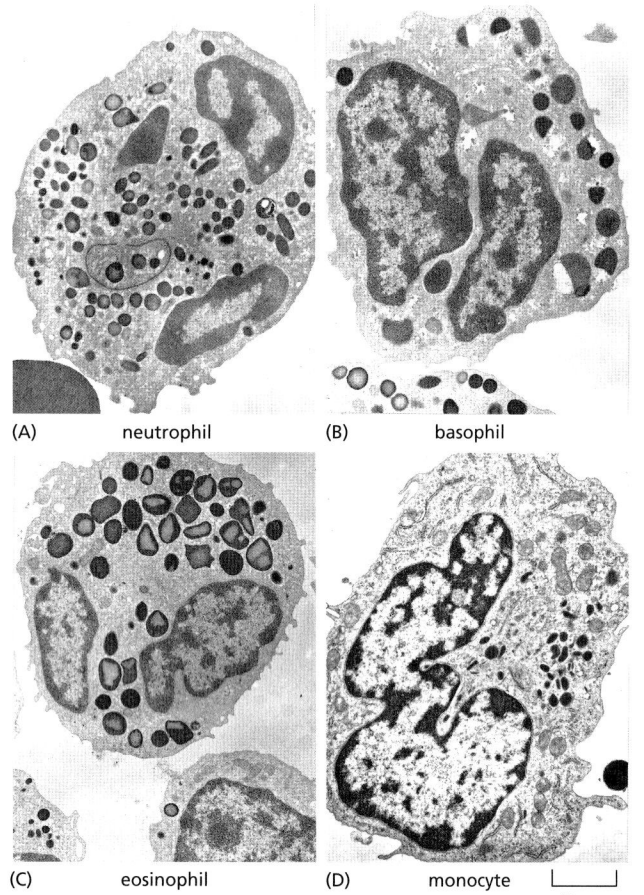

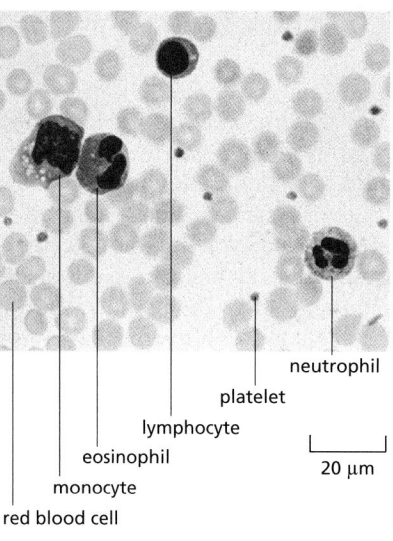

(A)   neutrophil   (B)   basophil

(C)   eosinophil   (D)   monocyte

2 µm

red blood cell

monocyte

eosinophil

lymphocyte

platelet

neutrophil

20 µm

(E)

Figure 22–27 White blood cells. (A–D) These electron micrographs show (A) a neutrophil, (B) a basophil, (C) an eosinophil, and (D) a monocyte. Electron micrographs of lymphocytes are shown in Figure 24–14. Each of the cell types shown here has a different function, which is reflected in the distinctive types of secretory granules and lysosomes it contains. There is only one nucleus per cell, but it has an irregular lobed shape, and in (A), (B), and (C) the connections between the lobes are out of the plane of section. (E) A light micrograph of a blood smear stained with the Romanowsky stain, which colors the white blood cells strongly. (A–D, courtesy of Dorothy Bainton; E, courtesy of David Mason.)

Monocytes also give rise to *dendritic cells.* Like macrophages, dendritic cells are migratory cells that can ingest foreign substances and organisms, but they do not have as active an appetite for phagocytosis and instead have a crucial role as presenters of foreign antigens to lymphocytes to trigger an immune response. Dendritic cells in the epidermis (called *Langerhans cells*), for example, ingest foreign antigens and carry these trophies back from the skin to present to lymphocytes in lymph nodes.

There are two main classes of **lymphocytes**, both involved in immune responses: *B lymphocytes* make antibodies, while *T lymphocytes* kill virus-infected cells and regulate the activities of other white blood cells. In addition, there are lymphocyte-like cells called *natural killer (NK) cells*, which kill some types of tumor cells and virus-infected cells. The production of lymphocytes is a specialized topic discussed in detail in Chapter 24. Here we concentrate mainly on the development of the other blood cells, often referred to collectively as **myeloid cells**.

Table 22–1 summarizes the various types of blood cells and their functions.

The Production of Each Type of Blood Cell in the Bone Marrow Is Individually Controlled

Most white blood cells function in tissues other than the blood; blood simply transports them to where they are needed. A local infection or injury in any tissue rapidly attracts white blood cells into the affected region as part of the inflammatory response, which helps fight the infection or heal the wound (Movie 22.3).

The inflammatory response is complex and is governed by many different signal molecules produced locally by mast cells, nerve endings, platelets, and white

TABLE 22–1 Blood Cells

Type of cell	Main functions	Typical concentration in human blood (cells/liter)
Red blood cells (erythrocytes)	Transport O_2 and CO_2	5×10^{12}
White blood cells (leukocytes)		
Granulocytes		
Neutrophils (polymorphonuclear leukocytes)	Phagocytose and destroy invading bacteria	5×10^9
Eosinophils	Destroy larger parasites and modulate allergic inflammatory responses	2×10^8
Basophils	Release histamine (and in some species serotonin) in certain immune reactions	4×10^7
Monocytes	Become tissue macrophages, which phagocytose and digest invading microorganisms and foreign bodies as well as damaged senescent cells	4×10^8
Lymphocytes		
B cells	Make antibodies	$\sim 0.3 \times 10^9$
T cells	Kill virus-infected cells and regulate activities of other leukocytes	$\sim 2 \times 10^9$
Natural killer (NK) cells	Kill virus-infected cells and some tumor cells	1×10^8
Platelets (cell fragments arising from megakaryocytes in bone marrow)	Initiate blood clotting	3×10^{11}

Humans contain about 5 liters of blood, accounting for 7% of body weight. Red blood cells constitute about 45% of this volume and white blood cells about 1%, the rest being the liquid blood plasma.

blood cells, as well as by the activation of complement (discussed in Chapter 24). Some of these signal molecules act on the endothelial lining of nearby capillaries, helping white blood cells to first stick and then make an exit from the bloodstream into the tissue where they are needed, as described in Chapter 19 (see Figure 19–28 and Movie 19.2). Damaged or inflamed tissues and local endothelial cells secrete other molecules called *chemokines*, which act as chemoattractants for specific types of white blood cells, causing them to become polarized and crawl toward the source of the attractant. As a result, large numbers of white blood cells enter the affected tissue (**Figure 22–28**).

Other signal molecules produced during an inflammatory response escape into the blood and stimulate the bone marrow to produce more leukocytes and release them into the bloodstream. The regulation tends to be cell-type specific: some bacterial infections, for example, cause a selective increase in neutrophils, while infections with some protozoa and other parasites cause a selective increase in eosinophils. (For this reason, physicians routinely use differential white blood cell counts to aid in the diagnosis of infectious and other inflammatory diseases.)

In other circumstances, erythrocyte production is selectively increased—for example, in response to anemia (lack of hemoglobin) due to blood loss, and in the process of acclimatization when one goes to live at high altitude, where oxygen is scarce. Thus, blood cell formation, or *hematopoiesis*, necessarily involves complex controls, which regulate the production of each type of blood cell individually to meet changing needs.

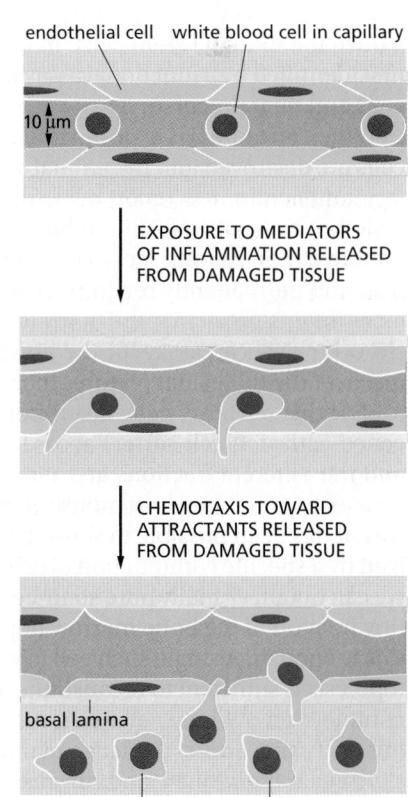

Figure 22–28 Chemotaxis of white blood cells to damaged tissue. A chemoattractive signal released from a site of damage, which is toward the bottom of the page, causes white blood cells to exit from the capillary by crawling between adjacent endothelial cells, as shown.

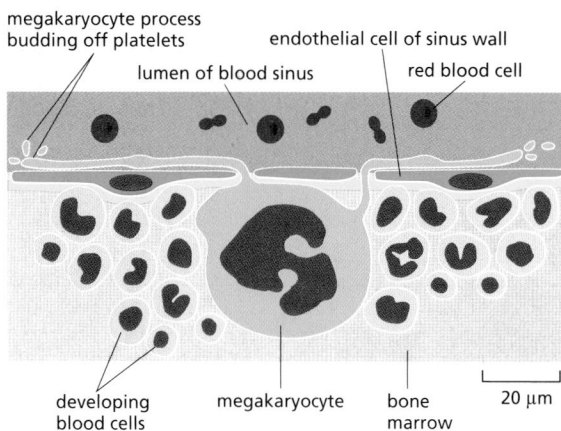

Figure 22–29 **A megakaryocyte among other developing blood cells in the bone marrow.** The megakaryocyte's enormous size results from its having a highly polyploid nucleus. One megakaryocyte produces about 10,000 platelets, which split off from long processes that extend through holes in the walls of an adjacent blood sinus.

Bone Marrow Contains Multipotent Hematopoietic Stem Cells, Able to Give Rise to All Classes of Blood Cells

In the bone marrow, the developing blood cells and their precursors, including the stem cells, are intermingled with one another, as well as with fat cells and other stromal cells (connective-tissue cells), which produce a delicate supporting meshwork of collagen fibers and other extracellular matrix components. In addition, the whole tissue is richly supplied with thin-walled blood vessels, called *blood sinuses*, into which the new blood cells are discharged. **Megakaryocytes** are also present; these, unlike other blood cells, remain in the bone marrow when mature and are one of its most striking features, being extraordinarily large (diameter up to 60 μm) with a highly polyploid nucleus. They normally lie close beside blood sinuses, and they extend processes through holes in the endothelial lining of these vessels; platelets pinch off from the processes and are swept away into the blood (**Figure 22–29** and **Movie 22.4**).

Because of the complex arrangement of the cells in bone marrow, it is difficult to identify in ordinary tissue sections any but the immediate precursors of the mature blood cells. There is no obvious visible characteristic by which we can recognize the ultimate stem cells. In the case of hematopoiesis, the stem cells were first identified by a functional assay that exploited the wandering lifestyle of blood cells and their precursors.

When an animal is exposed to a large dose of x-rays, most of the hematopoietic cells are destroyed and the animal dies within a few days as a result of its inability to manufacture new blood cells. The animal can be saved, however, by a transfusion of cells taken from the bone marrow of a healthy, immunologically compatible donor. Among these cells there are some that can colonize the irradiated host and permanently reequip it with hematopoietic tissue (**Figure 22–30**). Such experiments prove that the marrow contains hematopoietic stem cells. They also show how we can assay for the presence of hematopoietic stem cells and hence discover the molecular features that distinguish them from other cells.

For this purpose, cells taken from bone marrow are sorted (using a fluorescence-activated cell sorter) according to the surface antigens that they display, and the different fractions are transfused back into irradiated mice. If a fraction rescues an irradiated host mouse, it must contain hematopoietic stem cells. In this way, it has been possible to show that the hematopoietic stem cells are characterized by a specific combination of cell-surface proteins, and by appropriate sorting we can obtain virtually pure stem-cell preparations. The stem cells turn out to be a tiny fraction of the bone marrow population—about 1 cell in 50,000–100,000; but this is enough. A single such cell injected into a host mouse with defective hematopoiesis is sufficient to reconstitute its entire hematopoietic system, generating a complete set of blood cell types, as well as fresh stem cells. This and other experiments (using artificial lineage markers) show that the individual hematopoietic stem cell is *multipotent* and can give rise to the complete range of blood cell types, both myeloid and lymphoid, as well as to new stem cells like itself (**Figure 22–31**).

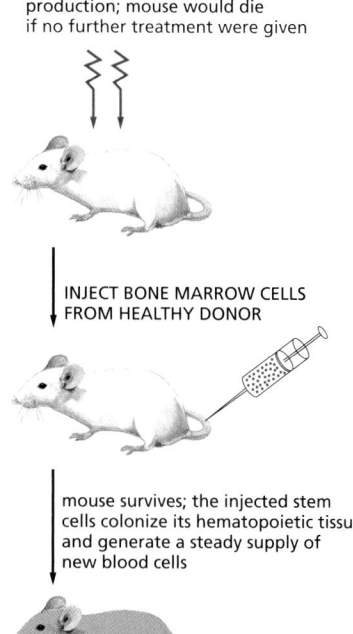

Figure 22–30 **Rescue of an irradiated mouse by a transfusion of bone marrow cells.** An essentially similar procedure is used in the treatment of leukemia in human patients by bone marrow transplantation.

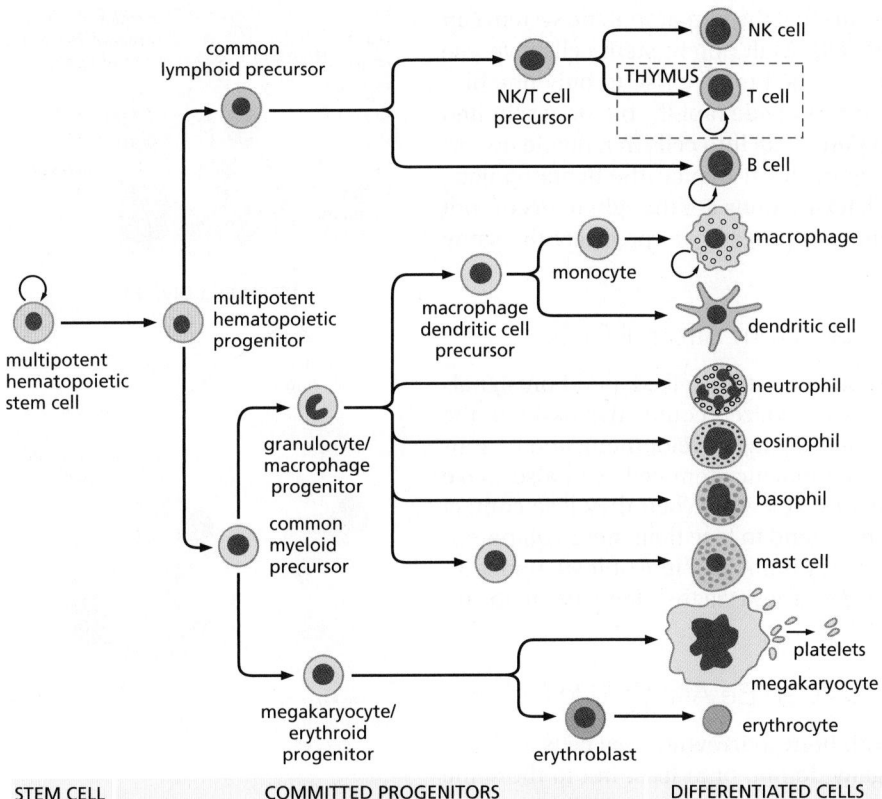

STEM CELL   **COMMITTED PROGENITORS**   **DIFFERENTIATED CELLS**

Figure 22–31 A tentative scheme of hematopoiesis. The multipotent stem cell normally divides infrequently to generate either more multipotent stem cells, which are self-renewing, or committed progenitor cells, which are limited in the number of times that they can divide before differentiating to form mature blood cells. As they go through their divisions, the progenitors become progressively more specialized in the range of cell types that they can give rise to, as indicated by the branching of this cell-lineage diagram. In adult mammals, all of the cells shown develop mainly in the bone marrow—except for T lymphocytes, which as indicated develop in the thymus, and macrophages and osteoclasts, which develop from blood monocytes. Some dendritic cells may also derive from monocytes.

Commitment Is a Stepwise Process

Hematopoietic stem cells do not jump directly from a multipotent state into a commitment to just one pathway of differentiation; instead, they go through a series of progressive restrictions. The first step, usually, is commitment to either a myeloid or a lymphoid fate. This is thought to give rise to two kinds of progenitor cells, one capable of generating large numbers of all the different types of myeloid cells, and the other giving rise to large numbers of all the different types of lymphoid cells. Further steps give rise to progenitors committed to the production of just one cell type. The steps of commitment correlate with changes in the expression of specific transcription regulators, needed for the production of different subsets of blood cells.

Divisions of Committed Progenitor Cells Amplify the Number of Specialized Blood Cells

Hematopoietic progenitor cells generally become committed to a particular pathway of differentiation long before they cease proliferating and terminally differentiate. The committed progenitors go through many rounds of cell division to amplify the ultimate number of cells of the given specialized type. In this way, a single stem-cell division can lead to the production of thousands of differentiated progeny, which explains why the number of stem cells is such a small fraction of the total population of hematopoietic cells. For the same reason, a high rate of blood cell production can be maintained even though the stem-cell division rate is low. The smaller the number of division cycles that the stem cells themselves have to undergo in the course of a lifetime, the lower the risk of generating stem-cell mutations, which would give rise to persistent mutant clones of cells in the body—a particular danger in the hematopoietic system where, as discussed in Chapter 20, a relatively small accumulation of mutations can be sufficient to cause cancer. A low rate of stem-cell division also slows the process of replicative cell senescence (discussed in Chapter 17).

The stepwise nature of commitment means that the hematopoietic system can be viewed as a hierarchical family tree of cells. Multipotent stem cells give rise to committed progenitor cells, which are specified to give rise to only one or a few blood cell types. The committed progenitors divide rapidly, but only a limited number of times, before they terminally differentiate into cells that divide no further and die after several days or weeks. Figure 22–31 depicts the hematopoietic family tree. It should be noted, however, that variations are thought to occur: not all stem cells generate the identical patterns of progeny via precisely the same sequence of steps.

Stem Cells Depend on Contact Signals From Stromal Cells

Like the stem cells of other tissues, hematopoietic stem cells depend on signals from their niche, in this case created by the specialized connective tissue of the bone marrow. (This is the site in adult humans; during development, and in non-human mammals such as the mouse, hematopoietic stem cells can also make their home in other tissues—notably liver and spleen.) When they lose contact with their niche, the hematopoietic stem cells tend to lose their stem-cell potential (**Figure 22–32**). Evidently the loss of potency is not absolute or instantaneous, however, since the stem cells can still survive journeys via the bloodstream to colonize other sites in the body.

Factors That Regulate Hematopoiesis Can Be Analyzed in Culture

While the stem cells depend on contact with bone marrow stromal cells for long-term maintenance, their committed progeny do not, or at least not to the same degree. These cells can thus be dispersed and cultured in a semisolid matrix of dilute agar or methylcellulose, and factors derived from other cells can be added artificially to the medium. The semisolid matrix inhibits migration, so that the progeny of each isolated precursor cell remain together as an easily distinguishable colony. A single committed neutrophil progenitor, for example, may give rise to a clone of thousands of neutrophils. Such culture systems have provided a way to assay for the factors that support hematopoiesis and hence to purify them and explore their actions. These substances are glycoproteins and are usually called **colony-stimulating factors** (**CSFs**). Some of these factors circulate in the blood and act as hormones, while others act in the bone marrow as secreted local mediators; still others take the form of membrane-bound signals that act through cell–cell contact.

An important example of the latter is a protein called *Steel* or *Stem Cell Factor* (*SCF*). This is expressed both in the bone marrow stroma (where it helps to define the stem-cell niche) and along pathways of migration, and it occurs both in a membrane-bound and a soluble form. It binds to a receptor tyrosine kinase called Kit, and it is required during development for guidance and survival not only of hematopoietic cells but also of other migratory cell types—specifically, germ cells and pigment cells.

Erythropoiesis Depends on the Hormone Erythropoietin

The best understood of the CSFs that act as hormones is the glycoprotein erythropoietin, which is produced in the kidneys and regulates *erythropoiesis*, the formation of red blood cells, to which we now turn.

The erythrocyte is by far the most common type of cell in the blood (see Table 22–1). When mature, it is packed full of hemoglobin and contains hardly any of the usual cell organelles. In an erythrocyte of an adult mammal, even the nucleus, endoplasmic reticulum, mitochondria, and ribosomes are absent, having been extruded from the cell in the course of its development (**Figure 22–33**). The erythrocyte therefore cannot grow or divide, and it has a limited life-span—about 120 days in humans or 55 days in mice. Worn-out erythrocytes are phagocytosed and digested by macrophages in the liver and spleen, which remove more than 10^{11}

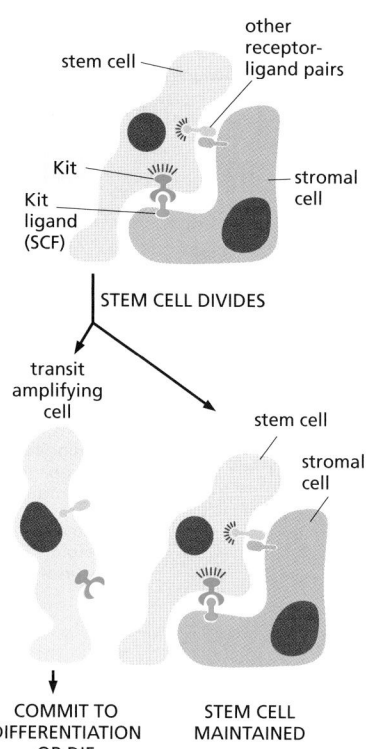

Figure 22–32 Dependence of hematopoietic stem cells on contact with stromal cells. The contact-dependent interaction between the Kit receptor and its ligand is one of several signaling mechanisms thought to be involved in hematopoietic stem-cell maintenance. The real system is certainly more complex. Moreover, the dependence of hematopoietic cells on contact with stromal cells cannot be absolute, since small numbers of the functional stem cells can be found free in the circulation. SCF, stem-cell factor.

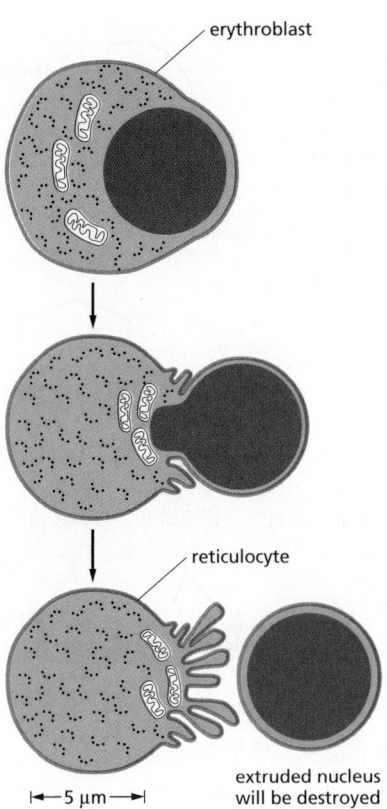

Figure 22–33 **A developing red blood cell (erythroblast).** The cell is shown extruding its nucleus to become an immature erythrocyte (a reticulocyte), which then leaves the bone marrow and passes into the bloodstream. The reticulocyte will lose its mitochondria and ribosomes within a day or two to become a mature erythrocyte. Erythrocyte clones develop in the bone marrow on the surface of a macrophage, which phagocytoses and digests the nuclei discarded by the erythroblasts.

erythroblast

reticulocyte

5 μm

extruded nucleus will be destroyed

senescent erythrocytes in each of us each day. Young erythrocytes actively protect themselves from this fate: they have a protein on their surface that binds to an inhibitory receptor on macrophages and thereby prevents their phagocytosis.

A lack of oxygen or a shortage of erythrocytes stimulates specialized cells in the kidney to synthesize and secrete increased amounts of **erythropoietin** into the bloodstream. The erythropoietin, in turn, boosts the production of erythrocytes. The effect is rapid: the rate of release of new erythrocytes into the bloodstream rises steeply 1–2 days after an increase in erythropoietin levels in the bloodstream. Clearly, the hormone must act on cells that are close precursors of the mature erythrocytes.

The cells that respond to erythropoietin can be identified by culturing bone marrow cells in a semisolid matrix in the presence of erythropoietin. In a few days, colonies of about 60 erythrocytes appear, each founded by a single committed erythroid progenitor cell. This progenitor depends on erythropoietin for its survival as well as its proliferation. It does not yet contain hemoglobin, and it is derived from an earlier type of committed erythroid progenitor whose survival and proliferation are governed by other factors.

Multiple CSFs Influence Neutrophil and Macrophage Production

The two classes of cells dedicated to phagocytosis, neutrophils and macrophages, develop from a common progenitor cell called a **granulocyte/macrophage (GM) progenitor cell**. Like the other granulocytes (eosinophils and basophils), neutrophils circulate in the blood for only a few hours before migrating out of capillaries into the connective tissues or other specific sites, where they survive for only a few days. They then die by apoptosis and are phagocytosed by macrophages. Macrophages, in contrast, can persist for months or perhaps even years outside the bloodstream, where they can be activated by local signals to resume proliferation.

At least seven distinct CSFs that stimulate neutrophil and macrophage colony formation in culture have been defined, and some or all of these are thought to act in different combinations to regulate the selective production of these cells *in vivo*. These CSFs are synthesized by various cell types—including endothelial cells, fibroblasts, macrophages, and lymphocytes—and their concentration in the blood typically increases rapidly in response to bacterial infection in a tissue, thereby increasing the number of phagocytic cells released from the bone marrow into the bloodstream.

The CSFs not only operate on the precursor cells to promote the production of differentiated progeny, they also activate the specialized functions (such as phagocytosis and target-cell killing) of the terminally differentiated cells. CSFs can be synthesized artificially and are now widely used in human patients to stimulate the regeneration of hematopoietic tissue and to boost resistance to infection.

The Behavior of a Hematopoietic Cell Depends Partly on Chance

CSFs are defined as factors that promote the production of colonies of differentiated blood cells. But precisely what effect does a CSF have on an individual hematopoietic cell? The factor might control the rate of cell division or the number of division cycles that the progenitor cell undergoes before differentiating; it might act late in the hematopoietic lineage to facilitate differentiation; it might act early to influence commitment; or it might simply increase the probability of cell survival (**Figure 22–34**). By monitoring the fate of isolated individual hematopoietic cells in culture, it has been possible to show that a single CSF, such as granulocyte/

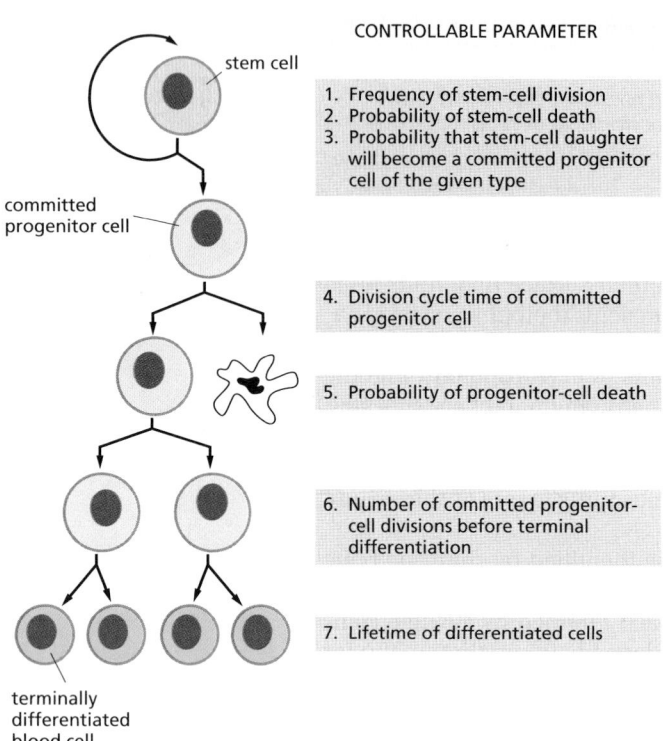

CONTROLLABLE PARAMETER

stem cell

1. Frequency of stem-cell division
2. Probability of stem-cell death
3. Probability that stem-cell daughter will become a committed progenitor cell of the given type

committed progenitor cell

4. Division cycle time of committed progenitor cell

5. Probability of progenitor-cell death

6. Number of committed progenitor-cell divisions before terminal differentiation

7. Lifetime of differentiated cells

terminally differentiated blood cell

Figure 22–34 Some of the parameters through which the production of blood cells of a specific type might be regulated. Studies in culture suggest that various colony-stimulating factors (CSFs) can affect all of these aspects of hematopoiesis.

macrophage CSF, can exert all these effects, although it is still not clear which are most important *in vivo*.

Studies *in vitro* indicate, moreover, that there is a large element of chance in the way a hematopoietic cell behaves—a reflection, presumably, of "noise" in the genetic control system, as discussed in Chapters 7 and 8. If two sister cells are taken immediately after a cell division and cultured apart under identical conditions, they frequently give rise to colonies that contain different types of blood cells or the same types of blood cells in different numbers. Thus, both the programming of cell division and the process of commitment to a particular path of differentiation seem to involve random events at the level of the individual cell, even though the behavior of the multicellular system as a whole is regulated in a reliable way. The sequence of cell fate restrictions shown earlier, in Figure 22–31, conveys the impression of a program executed with computer-like logic and precision. Individual cells may be more varied, quirky, and erratic, and may sometimes progress by other decision pathways from the stem-cell state toward terminal differentiation.

Regulation of Cell Survival Is as Important as Regulation of Cell Proliferation

The default behavior of hematopoietic cells in the absence of CSFs is death by apoptosis (discussed in Chapter 18), and the control of cell survival plays a central part in regulating the numbers of blood cells. The amount of apoptosis in the vertebrate hematopoietic system is enormous: billions of neutrophils die in this way each day in an adult human, for example. In fact, most neutrophils produced in the bone marrow die there without ever functioning. This futile cycle of production and destruction presumably serves to maintain a reserve supply of cells that can be promptly mobilized to fight infection whenever it flares up, or phagocytosed and digested for recycling when all is quiet. Compared with the life of the organism, the lives of cells are cheap.

Too little cell death can be as dangerous to the health of a multicellular organism as too much proliferation. As noted in Chapter 18, mutations that inhibit cell death by causing excessive production of the intracellular apoptosis inhibitor

Bcl2 promote the development of cancer in B lymphocytes. Indeed, the capacity for unlimited self-renewal is a dangerous property for any cell to possess. Many cases of leukemia arise through mutations that confer this capacity on committed hematopoietic precursor cells that would normally be fated to differentiate and die after a limited number of division cycles.

Summary

The many types of blood cells, including erythrocytes, lymphocytes, granulocytes, and macrophages, all derive from a common multipotent stem cell. In the adult, hematopoietic stem cells are found mainly in bone marrow, and they depend on signals from the marrow stromal (connective-tissue) cells to maintain their stem-cell character. The stem cells are few and far between, and they normally divide infrequently to produce more stem cells (self-renewal) and various committed progenitor cells (transit amplifying cells), each able to give rise to only one or a few types of blood cells. The committed progenitor cells divide extensively under the influence of various protein signal molecules (colony-stimulating factors, or CSFs) and then terminally differentiate into mature blood cells, which usually die after several days or weeks.

Studies of hematopoiesis have been greatly aided by in vitro *assays in which stem cells or committed progenitor cells form clonal colonies when cultured in a semisolid matrix. The progeny of stem cells seem to make their choices between alternative developmental pathways in a partly random manner. Cell death by apoptosis, controlled by the availability of CSFs, also plays a central part in regulating the numbers of mature differentiated blood cells.*

REGENERATION AND REPAIR

As we have seen, many of the tissues of the body are not only self-renewing but also self-repairing, and this is largely thanks to stem cells and the feedback controls that regulate their behavior and maintain homeostasis. There are, however, limits to what these natural repair mechanisms can achieve. In most parts of the human brain, for example, nerve cells that die, as in Alzheimer's disease, are not replaced. Likewise, when heart muscle dies for lack of oxygen, as in a heart attack, it is replaced by scar tissue rather than new heart muscle.

Some animals do far better than humans and can regenerate entire organs, such as whole limbs, after amputation. Among the invertebrates, there are some species that can even regenerate all the tissues of the body from a single somatic cell. These phenomena encourage the hope that human cells might be coaxed by artificial measures into similar feats of repair and regeneration, so as to replace the skeletal muscle fibers that degenerate in victims of muscular dystrophy, the nerve cells that die in patients with Parkinson's disease, the insulin-secreting cells that are lacking in type 1 diabetics, the heart muscle cells that die in a heart attack, and so on. As we learn more about the basic cell biology, these goals, once only a dream, are beginning to seem attainable.

In this section, we start with some examples of the remarkable regenerative abilities of some animal species, as an indication of what is possible in principle. We shall then discuss how we can improve upon the natural repair processes of the human body and treat disease by exploiting the properties of the various types of stem cells found in human tissues. In the final section of the chapter, we shall see how a deeper understanding of the molecular biology of cell differentiation and of stem cells has revealed ways to convert one type of cell into another, opening up radically new possibilities.

Planarian Worms Contain Stem Cells That Can Regenerate a Whole New Body

Schmidtea mediterranea is a small freshwater flatworm, or *planarian*, just under a centimeter long when grown to full size (**Figure 22–35**). It has an epidermis, a gut,

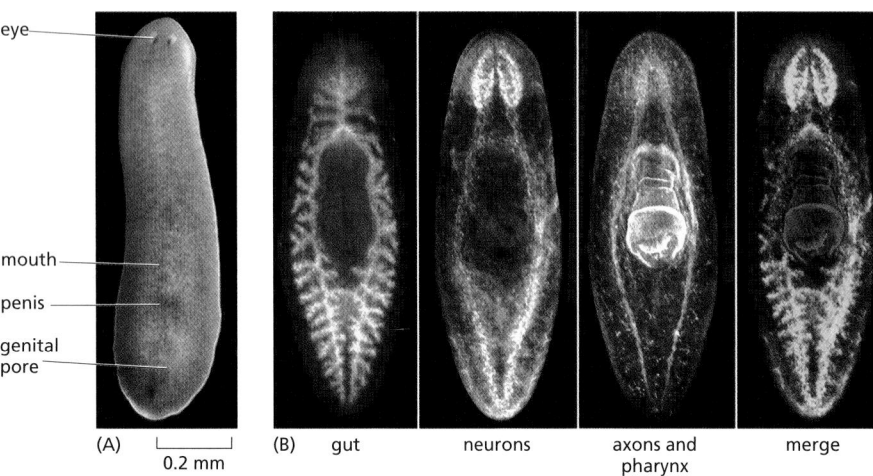

Figure 22–35 **The planarian worm, *Schmidtea mediterranea.*** (A) External view. (B) Immunostaining with three different antibodies, revealing the internal anatomy. (A, courtesy of A. Sánchez Alvarado; B, from A. Sánchez Alvarado, *BMC Biol.* 10:88, 2012. With permission from the author.)

a brain, a pair of primitive eyes, a peripheral nervous system, musculature, and excretory and reproductive organs—most of the basic body parts familiar in other animals, although all relatively simple by vertebrate standards and built from about 20–25 distinct differentiated cell types. For more than a century, planarians such as *Schmidtea* have intrigued biologists because of their extraordinary capacity for regeneration: a small tissue fragment taken from almost any part of the body will reorganize itself and grow to form a complete new animal. This property goes with another: when the animal is starved, it gets smaller and smaller, by reducing its cell numbers while maintaining essentially normal body proportions. This behavior is called *degrowth*, and it can continue until the animal is as little as one-twentieth or even a smaller fraction of its full size. Supplied with food, it will grow back to full size again. Cycles of degrowth and growth can be repeated indefinitely, without impairing survival or fertility.

Underlying this behavior is a process of continual cell turnover. Along with the differentiated cells, which do not divide, there is a population of small, apparently undifferentiated dividing cells called neoblasts. The neoblasts constitute about 20% of the cells in the body and are widely distributed within it; by cell division, they serve as stem cells for the production of new differentiated cells. Differentiated cells, meanwhile, are continually dying by apoptosis, allowing their corpses to be phagocytosed and digested by neighboring cells. Through this cell cannibalism, the constituents of the dying cells can be efficiently recycled. Cell birth continues in a dynamic balance with cell death and cell cannibalism, no matter whether the animal is fed or starved. In conditions of starvation, the balance is evidently tilted toward cell cannibalism, and in conditions of plenty, toward cell birth.

A high dose of x-rays halts all cell division, puts a stop to cell turnover, and destroys the capacity for regeneration. The result is death after a delay of several weeks. The animal can be rescued, however, by injecting into it a single neoblast isolated from an unirradiated donor (**Figure 22–36**). In a certain proportion of cases, the injected cell divides to form a clone of progeny that eventually repopulate the entire body, creating a healthy regenerative individual with an apparently complete set of differentiated cell types as well as dividing neoblasts. Genetic markers prove that these are all derived from the single neoblast that was injected. It follows that at least some neoblasts are *totipotent* (or at least highly *pluripotent*) stem cells; that is, cells able to give rise to all (or at least almost all) of the cell types that make up the body of a flatworm, including more neoblasts like themselves.

Some Vertebrates Can Regenerate Entire Organs

One might think that such powers of regeneration would be a prerogative of small, simple, primitive animals. But some vertebrates, too, especially fish and amphibians, show remarkable regenerative abilities. A newt, for example, can regenerate

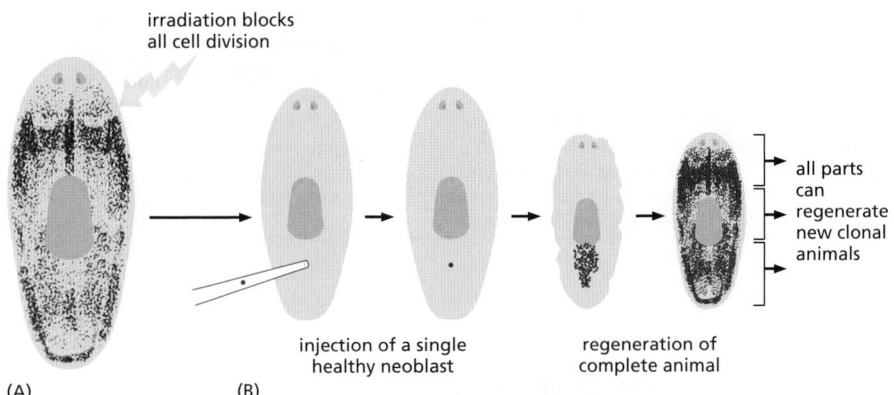

Figure 22–36 **Regeneration of a planarian from a single somatic cell.**
(A) The distribution of dividing cells (neoblasts, *blue*) in the adult body. Irradiation blocks all cell division and prevents regeneration, but (B) a single unirradiated neoblast cell injected into the irradiated animal is able to reconstitute all tissues. This eventually produces a complete animal that consists entirely of the progeny of this one cell and can regenerate. (Adapted from E.M. Tanaka and P.W. Reddien, *Dev. Cell* 21:172–185, 2011.)

a whole amputated limb. In this process, differentiated cells seem to revert to an embryonic character by first forming on the amputation stump a *blastema*—a small bud resembling an embryonic limb bud. The blastema then grows and its cells differentiate to form a correctly patterned replacement for the limb that has been lost, in what looks like a recapitulation of embryonic limb development (**Figure 22–37**). A large contribution to the blastema comes from the skeletal muscle cells in the limb stump. These multinucleate cells re-enter the cell cycle, dedifferentiate, and break up into mononucleated cells, which then proliferate within the blastema, before eventually redifferentiating. But do they redifferentiate only into muscle, or do they behave like neoblasts in the planarian and give rise to the full range of cell types needed to reconstruct the missing part of the limb? Careful lineage tracing, using genetic markers, shows (contrary to previous belief) that the cells are restricted according to their origins: muscle-derived cells give rise only to muscle, connective-tissue cells only to connective tissues, epidermal cells only to epidermal cells. The cells in the adult vertebrate body are, after all, less adaptable than the cells of the flatworm: by working in concert, they can replace the lost structure, but each cell type is far from totipotent.

Why a newt can regenerate a whole limb—as well as many other body parts—but a mammal cannot remains a profound mystery.

Stem Cells Can Be Used Artificially to Replace Cells That Are Diseased or Lost: Therapy for Blood and Epidermis

Earlier in this chapter, we saw how mice can be irradiated to kill off their hematopoietic cells, and then rescued by a transfusion of new stem cells, which repopulate the bone marrow and restore blood cell production (see Figure 22–30). In the same way, patients with some forms of leukemia or lymphoma can be irradiated or chemically treated to destroy their cancerous cells along with the rest of their hematopoietic tissue, and then can be rescued by a transfusion of healthy, noncancerous hematopoietic stem cells. In favorable cases, these can be sorted out from samples of the patient's own hematopoietic tissue before it is ablated. They are then transfused back afterward, avoiding problems of immune rejection.

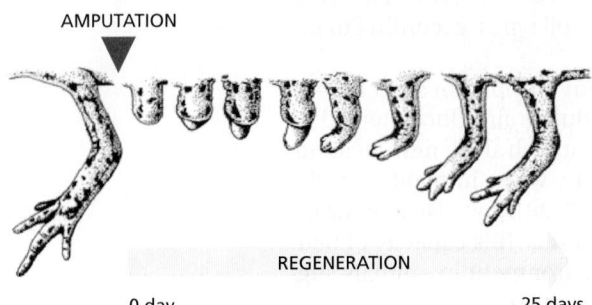

Figure 22–37 **Newt limb regeneration.** The time-lapse sequence shows the stages of regeneration after amputation at the level of the humerus. The sequence spans the events of wound healing, dedifferentiation of stump tissues, blastema formation, and redifferentiation. (Courtesy of Susan Bryant and David Gardiner.)

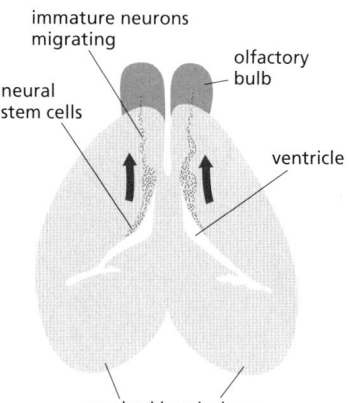

Figure 22–38 **The continuing production of neurons in an adult mouse brain.** The brain is viewed from above, in a cut-away section, to show the region lining the ventricles of the forebrain where neural stem cells are found. These cells continually produce progeny that migrate to the olfactory bulb, where they differentiate as neurons. The constant turnover of neurons in the olfactory bulb is presumably linked in some way to the turnover of the olfactory receptor neurons that project to it from the olfactory epithelium, as mentioned earlier. In adult humans, there is a continuing turnover of neurons in the hippocampus, a region specially concerned with learning and memory. (Adapted from B. Barres, *Cell* 97:667–670, 1999.)

Another example of the use of stem cells is in the repair of the skin after extensive burns. By culturing cells from undamaged regions of the burned patient's skin, it is possible to obtain epidermal stem cells quite rapidly in large numbers. These can then be used (through rather long and complicated procedures) to repopulate the damaged body surface.

Neural Stem Cells Can Be Manipulated in Culture and Used to Repopulate the Central Nervous System

The central nervous system (the CNS) is the most complex tissue in the body, at an opposite extreme from the epidermis. And yet fish and amphibians can regenerate large parts of the brain, spinal cord, and eyes after they have been cut away. In adult mammals, however, these tissues have very little capacity for self-repair, and stem cells capable of generating new neurons are hard to find—so hard to find, indeed, that for many years they were thought to be absent.

We now know, however, that neural stem cells that generate both neurons and glial cells do persist in certain parts of the adult mammalian brain (**Figure 22–38**). Neuronal turnover occurs on a dramatic scale in certain songbirds' brains, where large numbers of neurons die each year and are replaced by newborn neurons as part of a process by which the birds refine their song for each new breeding season. In the adult human brain, there is a continuing turnover of neurons in the hippocampus, a region specially concerned with learning and memory. Here, plasticity of adult function is associated with turnover of a specific subset of neurons. About 1400 fresh neurons in this class are generated every day, giving a turnover of 1.75% of the population per year.

Fragments taken from self-renewing regions of the adult brain, or from the brain of a fetus, can be dissociated and used to establish cell cultures, where they give rise to floating "neurospheres"—clusters consisting of a mixture of neural stem cells with neurons and glial cells derived from the stem cells. These neurospheres can be propagated through many cell generations, or their cells can be taken at any time and implanted back into the brain of an intact animal. Here they will produce differentiated progeny, in the form of neurons and glial cells.

Using slightly different culture conditions, with the right combination of growth factors in the medium, the neural stem cells can be grown as a monolayer and induced to proliferate as an almost pure stem-cell population without attendant differentiated progeny. By a further change in the culture conditions, these cells can be induced at any time to differentiate to give either a mixture of neurons and glial cells (**Figure 22–39**), or just one of these two cell types, according to the composition of the culture medium.

Neural stem cells, whether derived as above or from pluripotent stem cells as described in the next section, can be grafted into an adult brain. Once there, they show a remarkable ability to adjust their behavior to match their new location. Stem cells from the mouse hippocampus, for example, when implanted in the mouse olfactory-bulb-precursor pathway (see Figure 22–38), give rise to neurons that become correctly incorporated into the olfactory bulb. This capacity of neural stem cells and their progeny to adapt to a new environment in animals suggests applications in the treatment for diseases where neurons degenerate, and

fetal brain or ES cells ⟶ neurospheres (A) ⟶ pure culture of neural stem cells (B) ⟶ mixture (C) of differentiated neurons *(red)* and glial cells *(green)*; cell nuclei are *blue*

dissociate cells and culture in suspension in medium 1

dissociate and culture as monolayer in medium 2

switch to medium 3

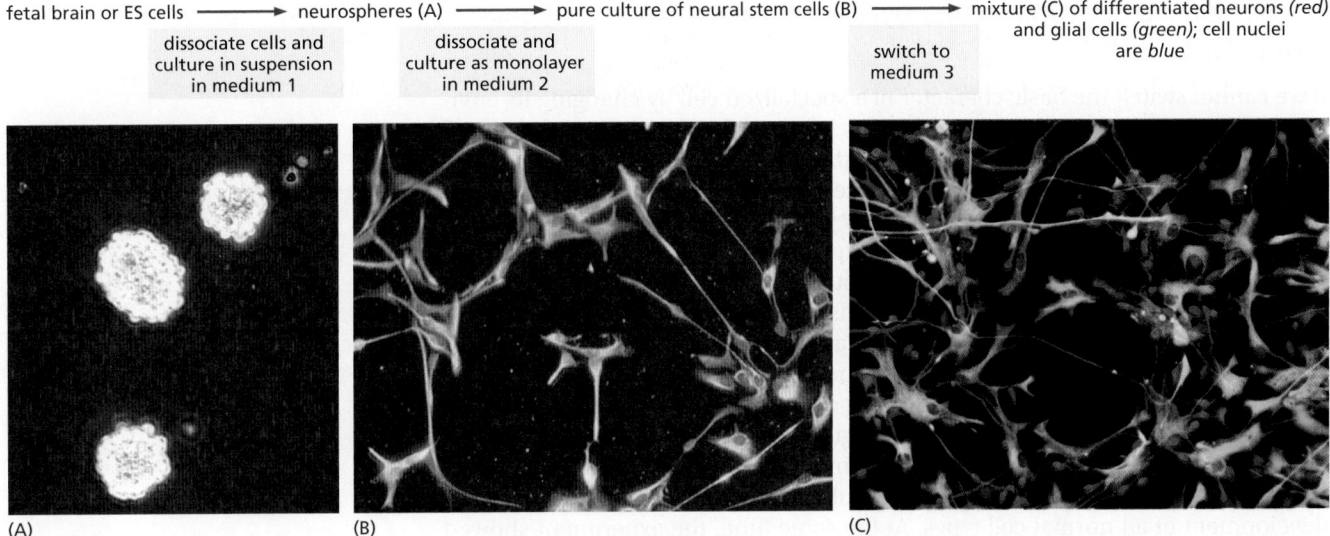

(A)　　　　　　　　　　(B)　　　　　　　　　　(C)

for injuries of the central nervous system. For example, might it be possible to use injected neural stem cells to replace the neurons that die in Parkinson's disease or to repair accidents that sever the spinal cord?

Figure 22–39 Neural stem cells. Shown are the steps leading from fetal brain tissue, via neurospheres (A), to a pure culture of neural stem cells (B). These stem cells can be kept proliferating as such indefinitely, or, through a change of medium, can be caused to differentiate (C) into neurons *(red)* and glial cells *(green)*. Neural stem cells with the same properties can also be derived, via a similar series of steps, from embryonic stem (ES) or induced pluripotent stem (iPS) cells (discussed later in this chapter). (Micrographs from L. Conti et al., *PLoS Biol.* 3:1594–1606, 2005. With permission from the authors.)

Summary

Animals vary in their capacity for regeneration. At one extreme, planarian worms contain stem cells (neoblasts) that support continual turnover of all cell types, and an entire worm can be regenerated from practically any small body fragment or even from a single neoblast cell. Newts can regenerate limbs and other large body parts after amputation, but the cells remain restricted according to their origins: muscle cells in the regenerate derive from muscle, epidermis from epidermis, and so on. In mammals, regeneration is more limited. Nevertheless, it is becoming possible to go beyond the natural limits of wound healing by exploiting stem-cell biology. Thus, certain regions of the nervous system contain stem cells that support production of neurons in these sites throughout life. Neural stem cells can be obtained from these sites or from fetal brains, grown in culture, and then grafted back into other sites in the brain, where they are able to generate neurons appropriate to the new location.

CELL REPROGRAMMING AND PLURIPOTENT STEM CELLS

When cells are transplanted from one site in the mammalian body to another or are removed from the body and maintained in culture, they remain largely faithful to their origins. Each type of specialized cell has a memory of its developmental history and seems fixed in its specialized fate. Some limited transformations can certainly occur, as we saw in our account of the connective-tissue cell family, and some stem cells can generate a variety of differentiated cell types, but the possibilities are restricted. Each type of stem cell serves for the renewal of one particular type of tissue, and the whole pattern of self-renewing and differentiated cells in the adult body is amazingly stable. What, at a fundamental molecular level, is the nature of these stable differences between cell types? Is there any way to override the cell memory mechanisms and force a switch from one state to another that is radically different?

We have already discussed these fundamental questions from a general standpoint in Chapter 7. Here we consider them more closely in the context of stem-cell biology, where there has been a recent revolution in our understanding and in our ability to manipulate states of cell differentiation. With further research, these advances would seem to have important practical consequences.

Nuclei Can Be Reprogrammed by Transplantation into Foreign Cytoplasm

If we cannot switch the basic character of a specialized cell by changing its environment, can we do so by interfering with its inner workings in a more direct and drastic way? An extreme treatment of this sort is to take the nucleus of the cell and transplant it into the cytoplasm of a large cell of a different type. If the specialized character is defined and maintained by cytoplasmic factors, the transplanted nucleus should switch its pattern of gene expression to conform with that of the host cell. In Chapter 7, we described a famous experiment of this sort, using the frog *Xenopus*. In this experiment, the nucleus of a differentiated cell (a cell from the lining of a tadpole's gut) was used to replace the nucleus of an oocyte (an egg-cell precursor arrested in prophase of the first meiotic division, in readiness for fertilization). The resulting hybrid cell went on, in a certain fraction of cases, to develop into a complete normal frog (see Figure 7–2A). This was crucial evidence for what is now a central principle of developmental biology: the cell nucleus, even that of a differentiated cell, contains a complete genome, capable of supporting development of all normal cell types. At the same time, the experiment showed that cytoplasmic factors can indeed reprogram a nucleus: the oocyte cytoplasm can drive the gut cell nucleus back to an early embryonic state, from which it can then step through the changing patterns of gene expression that lead all the way to a complete adult organism.

The full story, however, is not quite so simple. First, the reprogramming in such experiments is not perfect. When the transplanted nucleus is taken from a gut cell, for example, a gene that is normally specific to the gut is found to be expressed persistently, even in the muscle cells of the final animal. Second, the experiment succeeds in only a limited proportion of cases, and this success rate becomes lower and lower, the more mature the animal from which the transplanted nucleus is taken: very large numbers of transplantations must be done to score a single success if the nucleus comes from a differentiated cell of an adult frog.

Nuclear transplantation can be done in mammals too, with basically similar results. Thus, a nucleus taken from a differentiated cell in the mammary gland of an adult sheep and transplanted into an enucleated sheep's egg was able to support development of an apparently normal sheep—the famous Dolly. Again, the success rate is low: many transplantations have to be done to obtain one such individual.

Reprogramming of a Transplanted Nucleus Involves Drastic Epigenetic Changes

In a typical fully differentiated cell, there seem to be mechanisms maintaining the pattern of gene expression that cytoplasmic factors cannot easily override. An obvious possibility is that the stability of the pattern of gene expression in an adult cell may depend, in part at least, on self-perpetuating modifications of chromatin, as discussed in Chapter 4. As explained in Chapter 7, the phenomenon of X-inactivation in mammals provides a clear example of such epigenetic control. Two X chromosomes exist side by side in each female cell, exposed to the same chemical environment, but while one remains active, the other persists from one cell generation to the next in a condensed inactive state; cytoplasmic factors cannot be responsible for the difference, which must instead reflect mechanisms intrinsic to the individual chromosome. Elsewhere in the genome also, controls at the level of chromatin act in combination with other forms of regulation to govern the expression of each gene. Genes can be shut down completely, or switched on constitutively, or maintained in a labile state where they can be readily switched on or off according to changing circumstances.

The reprogramming of a nucleus transplanted into an oocyte involves dramatic changes in chromatin. The nucleus swells, increasing its volume 50-fold as the chromosomes decondense; there is a wholesale alteration in patterns of methylation of DNA and histones; the standard histone H1 (the histone that links adjacent nucleosomes) is replaced by a variant form that is peculiar to the oocyte and early embryo; and the preexisting type of histone H3 is also replaced at many sites by a distinct isoform. Evidently, the egg contains factors that reset the state of the chromatin in the nucleus, wiping out old histone modifications on chromatin and imposing new ones. Reprogrammed in this way, the genome becomes competent once again to initiate embryonic development and to give rise to the full range of differentiated cell types.

Embryonic Stem (ES) Cells Can Generate Any Part of the Body

A fertilized egg, or an equivalent cell produced by nuclear transplantation, is a remarkable thing: it can generate a whole new multicellular individual, and that means that it can give rise to every normal type of specialized cell, including even egg or sperm cells for production of the next generation. A cell in such a state is said to be **totipotent**, and it is said to be **pluripotent** if it can give rise to most cell types but not absolutely all. Nevertheless, such a progenitor is not a stem cell: it is not self-renewing, but is instead dedicated to a program of progressive differentiation. If it were the only available starting point for study and exploitation of pluripotent cells, the enterprise would require a continual supply of fresh fertilized eggs or fresh nuclear transplantation procedures—an awkward requirement for studies in experimental animals, and unacceptable for practical applications in humans.

Here, however, nature has been unexpectedly kind to scientists. It is possible to take an early mouse embryo, at the blastocyst stage, and through cell culture to derive from it a class of stem cells called **embryonic stem cells**, or **ES cells**. These originate from the inner cell mass of the early embryo (the cluster of cells that give rise to the body of the embryo proper, as opposed to extraembryonic structures), and they have an extraordinary property: given suitable culture conditions, they will continue proliferating indefinitely and yet retain an unrestricted developmental potential. Their only limitation is that they do not give rise to extraembryonic tissues such as those of the placenta. Thus they are classified as pluripotent, rather than totipotent. But this is a minor restriction. If ES cells are put back into a blastocyst, they become incorporated into the embryo and can give rise to all the tissues and cell types in the body, integrating perfectly into whatever site they may come to occupy, and adopting the character and behavior that normal cells would show at that site. They can even give rise to germ cells, from which a new generation of animals can be derived (**Figure 22–40**).

ES cells let us move between cell culture, where we can use powerful techniques for genetic transformation and selection, and the intact organism, where we can discover how such genetic manipulations affect development and physiology. Thus, ES cells have opened the way to efficient genetic engineering in mammals, leading to a revolution in our understanding of mammalian molecular biology.

Cells with properties similar to those of mouse ES cells can now be derived from early human embryos and from human fetal germ cells, and even, as we shall explain below, from differentiated cells taken from adult mammalian tissues. In this way, one can obtain a potentially inexhaustible supply of pluripotent

Figure 22–40 Production and pluripotency of ES cells. ES cells are derived from the inner cell mass (ICM) of the early embryo. The ICM cells are transferred to a culture dish containing an appropriate medium, where they become converted to ES cells and can be kept proliferating indefinitely without differentiating. The ES cells can be taken at any time—after genetic manipulation, if desired—and injected back into the inner cell mass of another early embryo. There they take part in formation of a well-formed chimeric animal that is a mixture of ordinary and ES-derived cells. The ES-derived cells can differentiate into any of the cell types in the body, including germ cells from which a new generation of mice can be produced. These next-generation progeny are no longer chimeric, but consist of cells that all inherit half their genes from the cultured ES cell line.

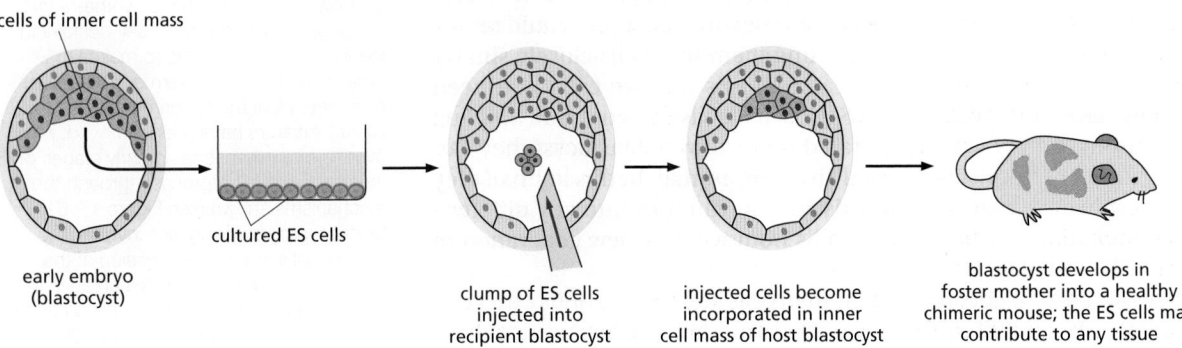

cells of inner cell mass

early embryo (blastocyst)

cultured ES cells

clump of ES cells injected into recipient blastocyst

injected cells become incorporated in inner cell mass of host blastocyst

blastocyst develops in foster mother into a healthy chimeric mouse; the ES cells may contribute to any tissue

cells. Grown in culture, these can be manipulated, by suitable choice of culture conditions, to give rise to large quantities of almost any type of differentiated cell, opening the way to important practical applications. Before discussing them, however, we consider the underlying biology.

A Core Set of Transcription Regulators Defines and Maintains the ES Cell State

What is it that gives ES cells and related types of pluripotent stem cells their extraordinary capabilities? What can they tell us about the fundamental mechanisms underlying stemness, cell differentiation, and the stability of the differentiated state?

For some attributes, the answer is simple. For example, an essential feature of ES cells is that they must avoid senescence. As discussed in Chapter 17, this is the fate of fibroblasts and many other types of somatic cells: they are limited in the number of times they will divide, in part at least because they lack telomerase activity, with the result that their telomeres become progressively eroded in each division cycle, leading eventually to cell-cycle arrest. ES cells, by contrast, express high levels of active telomerase, allowing them to escape senescence and continue dividing indefinitely. This is a property shared with other, more specialized types of stem cells, such as those of the adult intestine, which similarly can carry on dividing for hundreds or thousands of cycles.

The deeper problem is to explain how the whole complex pattern of gene expression in an ES cell is organized and maintained. As a first step, one can look for genes expressed specifically in ES cells or in the corresponding pluripotent cells of the early embryo. This approach identifies a relatively small number of candidate ES-critical genes; that is, genes that seem to be essential in one way or another for the peculiar character of ES cells. A gene called *Oct4*, for example, is exclusively expressed in ES cells and in related classes of cells in the intact organism—specifically, in the germ-cell lineage and in the inner cell mass and its precursors. *Oct4* codes for a transcription regulator. When it is lost from ES cells, they lose their ES cell character; and when it is missing in an embryo, the cells that should specialize as inner cell mass are diverted into an extraembryonic pathway of differentiation and their development is aborted.

Fibroblasts Can Be Reprogrammed to Create Induced Pluripotent Stem Cells (iPS Cells)

In Chapter 7, we saw that fibroblasts and some other cell types can be driven to switch their character and differentiate as muscle cells if the master muscle-specific transcription regulator MyoD is artificially expressed in them. Could the same technique be used to convert adult cell types into ES cells, through forced expression of factors such as Oct4? This question was tackled by transfecting fibroblasts with retroviral vectors carrying genes that one might hope to have such an effect. A total of 24 candidate ES-critical genes were tested in this way. None of them was able by itself to cause the conversion; but in certain combinations they could do so. In 2006, the first breakthrough experiments whittled down the requirement to a core set of four factors, all of them transcription regulators: Oct4, Sox2, Klf4, and Myc, known as the OSKM factors for short. When coexpressed, these could reprogram mouse fibroblasts, permanently converting them into cells closely similar to ES cells (**Figure 22–41**). ES-like cells created in this way are called **induced pluripotent stem cells**, or **iPS cells**. Like ES cells, iPS cells can continue dividing indefinitely in culture, and when incorporated into a mouse blastocyst they can participate in creation of a perfectly formed chimeric animal. In this animal, they can contribute to the development of any tissue and can turn into any differentiated cell type, including functional germ cells from which a new generation of mice can be raised (see Figure 22–40).

iPS cells can now be derived from adult human cells and from various other differentiated cell types besides fibroblasts. Numerous methods can be used to drive

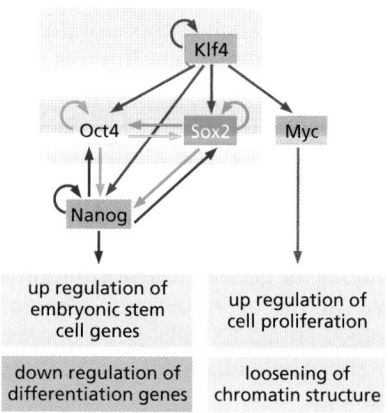

up regulation of embryonic stem cell genes

up regulation of cell proliferation

down regulation of differentiation genes

loosening of chromatin structure

Figure 22–41 Reprogramming fibroblasts to iPS cells with the OSKM factors. As indicated, the master gene regulator proteins Oct4, Sox2, and Klf4 (OSK) induce both their own and each other's synthesis (*gray shading*). This generates a self-sustaining feedback loop that helps to maintain cells in an embryonic stem cell-like state, even after all of the experimentally added OSKM initiators have been removed. Myc overexpression speeds up early stages of the reprogramming process through the mechanisms shown (see Figure 17–61). Stable reprogramming also involves the permanently induced expression of the *Nanog* gene, which produces an additional master transcription regulator. (Adapted from J. Kim et al., *Cell* 132:1049–1061, 2008.)

expression of the transforming OSKM factors, including methods that leave no trace of foreign DNA in the reprogrammed cell. Variations of the original cocktail of transcription regulators can drive the conversion, with different specialized cell types having somewhat different requirements. Myc overexpression, for example, turns out not to be absolutely necessary, although it enhances the efficiency of the process. And differentiated cell types may express some of the required factors as part of their normal phenotype. For example, cells of the dermal papilla of hair follicles already express Sox2, Klf4, and Myc; to convert them into iPS cells, it is enough to force them artificially to express Oct4. Oct4, indeed, seems to have a central role and to be generally indispensable for the creation of iPS cells.

Reprogramming Involves a Massive Upheaval of the Gene Control System

Converting a differentiated cell into an iPS cell is not like flicking a switch on some predictable, precisely engineered piece of machinery. Only a few of the cells that receive the OSKM factors will actually become iPS cells—one in several thousand in the original experiments, and still only a small minority with more recent, improved techniques. In fact, the success of the original experiments depended on clever selection to pick out those few cells where the conversion had occurred (**Figure 22–42**).

Conversion to an iPS character by the OSKM factors is not only inefficient but also slow: fibroblasts take ten days or more from introduction of the conversion factors before they begin to express markers of the iPS state. This suggests that the transformation involves a long cascade of changes. These changes are being extensively studied, and they affect both the expression of individual genes and the state of the chromatin. The results of one such study are outlined in **Figure 22–43**. The process begins with a Myc-induced cell proliferation and loosening of chromatin structure that promotes the binding of the other three master regulators to many hundreds of different sites in the genome. At a large proportion of these sites, Oct4, Sox2, and Klf4 all bind in concert. The binding sites include the endogenous *Oct4*, *Sox2*, and *Klf4* genes themselves, which eventually creates the types of positive feedback loops just described that makes expression of these genes self-sustaining (see Figure 22–41). But self-induction of *Oct4*, *Sox2*, and *Klf4* is only a small part of the transformation that occurs. The three core factors activate some target genes and repress others, producing a cascade of effects that reorganize the gene control system globally and at every level, changing the patterns of histone modification, DNA methylation, and chromatin compaction, as well as the expression of innumerable proteins and noncoding RNAs. By the end of this complex process, the resulting iPS cell is no longer dependent on the artificially generated factors that triggered the change: it has settled into a stable, self-sustaining state of coordinated gene expression, making its own Oct4, Sox2, Klf4, and Myc (and all the other essential ingredients of a pluripotent stem cell) from its own endogenous copies of the genes.

Figure 22–42 **A strategy used to select cells that have converted to an iPS character.** The experiment makes use of a gene (*Fbx15*) that is present in all cells but is normally expressed only in ES and early embryonic cells (although not required for their survival). A fibroblast cell line is genetically engineered to contain a gene that produces an enzyme that degrades G418 under the control of the *Fbx15* regulatory sequence. G418 is an aminoglycoside antibiotic that blocks protein synthesis in both bacteria and eukaryotic cells. When the OSKM factors are artificially expressed in this cell line, a small proportion of the cells undergo a change of state and activate the *Fbx15* regulatory sequence, driving expression of the G418-resistance gene. When G418 is added to the culture medium, these are the only cells that survive and proliferate. When tested, they turn out to have an iPS character.

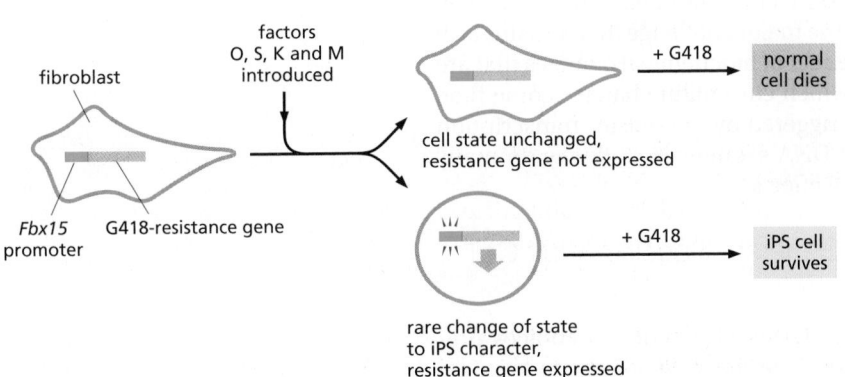

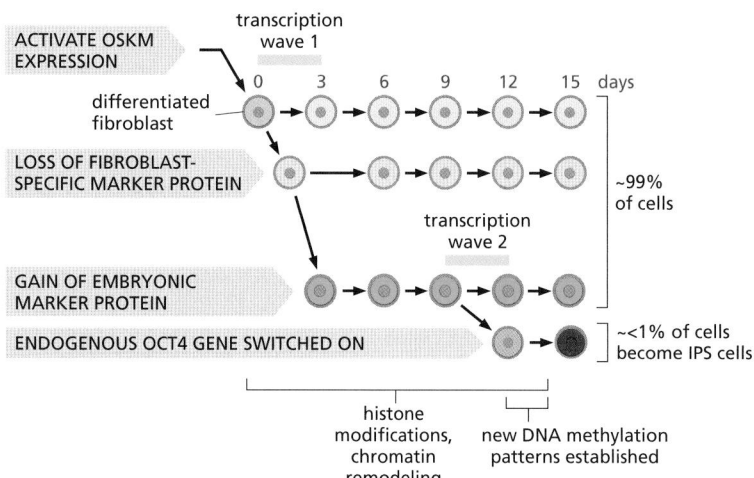

Figure 22–43 **A summary of some of the major events that accompany the reprogramming of mouse fibroblasts to iPS cells.** By sorting cells at various times after the OSKM induction shown, one can carry out detailed biochemical analyses on the different cell populations shown. This led to the discovery that two major waves of new gene transcription are induced, but that the second wave occurs only in the subset of cells expressing an embryonic marker protein. Some 1500 genes are found to be differentially expressed in these cells, compared to the large majority of cells that fail to progress toward iPS cells. As indicated, major DNA methylation changes are observed only after the alteration of chromatin structures.

In the first transcription wave, among the genes prominently induced are those for cell proliferation, metabolism, and cytoskeletal organization; in contrast, genes associated with fibroblast development are repressed. In the second transcription wave, genes required for embryonic development and for stem cell maintenance are induced. (Adapted from J.M. Polo et al., *Cell* 151:1617–1632, 2012.)

An Experimental Manipulation of Factors that Modify Chromatin Can Increase Reprogramming Efficiencies

The low efficiency and slow rate of conversion suggest that there is some barrier blocking the switch from the differentiated state to the iPS state in these experiments, and that overcoming this barrier is a difficult process that involves a large element of chance. Likewise, the outcome is variable, with significant differences between the individual lines of transformed cells that are generated, even when the initial differentiated cells are genetically and phenotypically identical. Only some of the candidate iPS lines pass all the tests of pluripotency. At a molecular level, there are differences even among the fully validated iPS cells: although they share many features, they vary in details of their gene expression patterns and, for example, in their patterns of DNA methylation.

Overcoming these difficulties will be critical for improving our understanding of how cell specialization is controlled and organized in multicellular organisms; it should also facilitate many medical advances. Thus, intensive research is being carried out on the reprogramming process. One approach aims at obtaining a much clearer picture of the role that chromatin structures play in gene regulation in eukaryotes.

From our discussion of nuclear transplantation, one might expect that any reprogramming of a differentiated cell would require a radical and widespread change in the chromatin structure of selected genes. Not only are such changes observed, but a large number of different experiments reveal that the efficiency of the reprogramming process can be substantially increased by altering the activity of proteins that affect chromatin structure. **Figure 22–44** categorizes some of the factors that have been shown to enhance the transformation of fibroblasts to iPS cells; those in the top three rows—chromatin remodelers, histone modifying enzymes, and histone variants—are especially well known to have profound effects on the organization of nucleosomes in chromatin (discussed in Chapter 4).

We can only touch briefly here on the massive amounts of data that have been accumulating in this exciting research area. The major challenge that remains is to obtain a systems-level model for the complex set of biochemical changes that are involved in reprogramming. For example, which chromatin changes come first, and which then follow? How can these be triggered by the master transcription regulators through their binding to specific DNA sequences, and why do many cells in a population appear resistant to these effects?

ES and iPS Cells Can Be Guided to Generate Specific Adult Cell Types and Even Whole Organs

We can think of embryonic development in terms of a series of choices presented to cells as they follow a road that leads from the fertilized egg to terminal

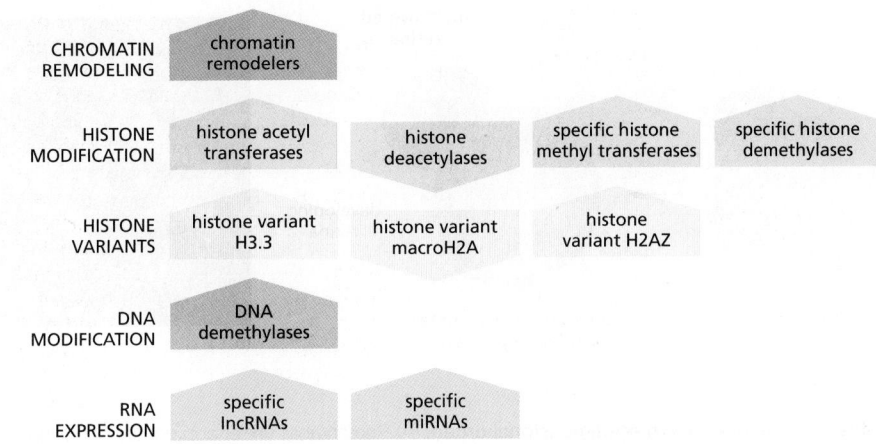

Figure 22–44 **Factors that have been observed to enhance reprogramming efficiency.** Emphasized here are those factors that can alter chromatin states, with those in the top three rows having the most direct effects. An *up arrow* indicates that reprogramming is increased when the activity of the indicated factor is increased; a *down arrow* indicates that reprogramming is increased when the activity of the indicated factor is decreased. Thus, for example, increased activity of histone acetyl transferases and increased activity of histone deacetylases have opposite effects, as expected from their biochemical activities (see p. 196).

differentiation. After their long sojourn in culture, the ES cells or iPS cells and their progeny can still read the signs at each branch in the highway and respond as normal embryonic cells would. If ES or iPS cells are implanted directly into an embryo at a later stage or into an adult tissue, however, they fail to receive the appropriate sequence of cues; their differentiation then is not properly controlled, and they will often give rise to a tumor of the type known as a *teratoma*, containing a mixture of cell types inappropriate to the site in the body.

In culture, by exposing the ES or iPS cell to an appropriate sequence of signal proteins and growth factors, delivered with the right timing, it is possible to guide the cell along a pathway that approximates a normal developmental pathway, so as to convert it into one of the standard specialized adult cell types (**Figure 22–45** and **Movie 22.5**). Success requires trial and error, but has now been achieved for many different final specialized states, including neuronal, muscular, and intestinal cell types. In a few cases, it has even been possible, by careful manipulation of the culture conditions, to get ES or iPS cells to interact with one another so as to construct an entire organ, albeit on a small scale (**Figure 22–46**).

Figure 22–45 **Production of differentiated cells from ES or iPS cells in culture.** These cells can be cultured indefinitely as pluripotent cells when attached as a monolayer to a dish. Alternatively they can be detached and allowed to form aggregates called embryoid bodies, which causes the cells to begin to specialize. Cells from embryoid bodies, cultured in media with different factors added, can then be driven to differentiate in various ways. (Based on E. Fuchs and J.A. Segre, *Cell* 100:143–155, 2000.)

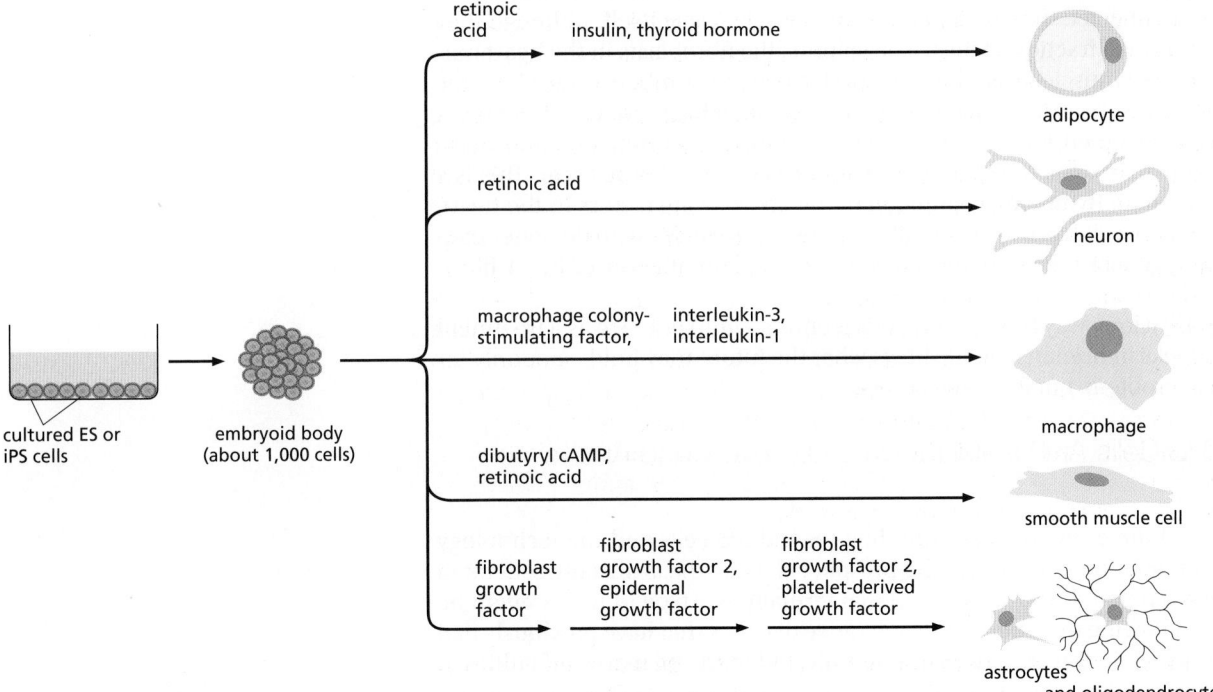

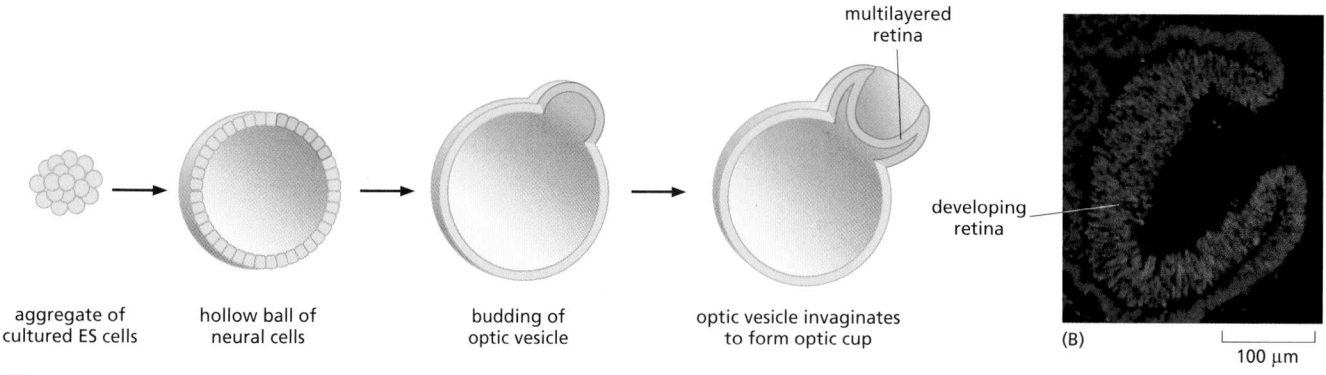

(A)

(B)

100 μm

Figure 22–46 Cultured ES cells can give rise to a three-dimensional organ. (A) Remarkably, under appropriate conditions, mouse ES cells in culture can proliferate, differentiate, and interact to form a three-dimensional, eye-like structure, which includes a multilayered retina similar in organization to the one that forms *in vivo*. (B) Fluorescent micrograph of an optic cup formed by ES cells in culture. The structure includes a developing retina, containing multiple layers of neural cells, which produce a protein *(pink)* that serves as a marker for retinal tissue. (B, from M. Eiraku et al., *Nature* 472:51–56, published 2011 by Macmillan Publishers Ltd. Reproduced with permission of SNCSC.)

Cells of One Specialized Type Can Be Forced to Transdifferentiate Directly Into Another

The route we have just described, from one mode of differentiation to another via conversion to an iPS cell, seems needlessly roundabout. Could we not convert cell type A into cell type B directly, without backtracking to the embryonic-like iPS state? For many years, it has been known that such *transdifferentiation* can be achieved in a few special cases, such as the conversion of fibroblasts into skeletal muscle cells by forced expression of MyoD (see p. 396). But now, with the insights that have come from the study of ES and iPS cells, ways are being found to bring about such interconversions in a much wider range of cases.

An elegant example comes from studies of the heart. By forcing expression of an appropriate combination of factors—not Oct4, Sox2, Klf4, and Myc, but Gata4, Mef2c, and Tbx5—it is possible to convert heart fibroblasts directly into heart muscle cells. This has been done in the living mouse, using retroviral vectors, and the transformation occurs with high efficiency when the vectors carrying the transgenes are injected directly into the heart muscle tissue itself. Although they occupy only a small fraction of the tissue volume, the fibroblasts in the heart outnumber the heart muscle cells, and they survive in large numbers even where the heart muscle cells have died. Thus, in a typical nonfatal heart attack, where heart muscle cells have died for lack of oxygen, the fibroblasts proliferate and make collagenous matrix so as to replace the lost muscle with a fibrous scar. This is a poor sort of repair. By forcing expression of the appropriate factors in the heart, as described above, it has proved possible, in the mouse at least, to do better than nature and regenerate lost heart muscle by transdifferentiation of heart fibroblasts.

We are still a long way from putting this technique into practice as a treatment for heart attacks in humans, but it shows what the future may hold—not only for this medical problem, but for many others.

ES and iPS Cells Are Useful for Drug Discovery and Analysis of Disease

A large part of the excitement surrounding ES and iPS cells and the technology of transdifferentiation comes from the prospect of using the artificially generated cells for tissue repair. It begins to seem that virtually any type of tissue might be replaceable, allowing treatment of degenerative diseases that have previously had no cure. Research in this area is moving rapidly, but there are many difficulties to be overcome.

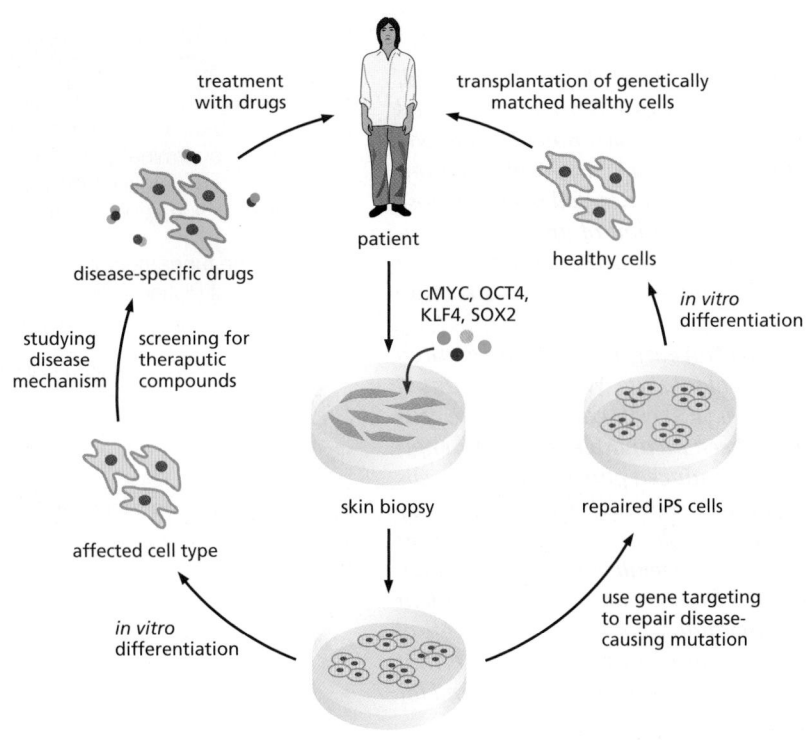

treatment with drugs

transplantation of genetically matched healthy cells

disease-specific drugs

patient

healthy cells

studying disease mechanism

screening for theraputic compounds

cMYC, OCT4, KLF4, SOX2

in vitro differentiation

affected cell type

skin biopsy

repaired iPS cells

in vitro differentiation

use gene targeting to repair disease-causing mutation

patient-specific iPS cells

Figure 22–47 **Use of iPS cells for drug discovery and for analysis and treatment of genetic disease.** The left side of the diagram shows how differentiated cells that are generated from iPS cells derived from a patient with a genetic disease can be used for analysis of the disease mechanism and for discovery of therapeutic drugs. The right side of the diagram shows how the genetic defect might be repaired in the iPS cells, which could then be induced to differentiate in an appropriate way and grafted back into the patient without danger of immune rejection. (Adapted from D.A. Robinton and G.Q. Daley, *Nature* 481:295–305, published 2012 by Macmillan Publishers Ltd. Reproduced with permission of SNCSC.)

With the advent of iPS cells and direct transdifferentiation, at least one major hurdle has been surmounted, in principle at least: the problem of immune rejection. ES cells, because they are created from early embryos that generally come from unrelated donors, will never be genetically identical to the cells of the patient receiving the transplant. The transplanted cells and their progeny are therefore liable to rejection by the immune system. Both iPS and transdifferentiated cells, in contrast, can be generated from a small sample of the patient's own tissue and so should escape immune attack when transplanted back into the same individual.

Tissue repair by transplantation, however, is not the only application for which ES, iPS, and transdifferentiated cells can be used: there are other ways in which they promise to be more immediately valuable. In particular, they can be used to generate large, homogeneous populations of specialized cells of any chosen type in culture; and these can serve for investigation of disease mechanisms and in the search for new drugs acting on a specific cell type (**Figure 22–47**).

Where a disease has a genetic cause, we can derive iPS cells from sufferers and use these cells to produce the specific cell types that malfunction, to investigate how the malfunction occurs, and to screen for drugs that might help to put it right. *Timothy syndrome* provides an example. In this rare genetic condition, there is a severe, life-threatening disorder in the rhythm of the heart beat (as well as several other abnormalities), as a result of a mutation in a specific type of Ca^{2+} channel. To study the underlying pathology, researchers took skin fibroblasts from patients with the disorder, generated iPS cells from the fibroblasts, and drove the iPS cells to differentiate into heart muscle cells. These cells, when compared with heart muscle cells prepared similarly from normal control individuals, showed irregular contractions and abnormal patterns of Ca^{2+} influx and electrical activity that could be characterized in detail. From this finding, it is a small step to development of an *in vitro* assay for drugs that might correct the misbehavior of the heart muscle cells.

This approach to drug discovery—where iPS cells are prepared from the individual patient, differentiated into the relevant cell type, and used to test candidate drugs *in vitro*—would seem to represent a huge advance on the slow, costly traditional methods that involve administration of test compounds to large numbers of people.

Summary

In the adult mammalian body, the various types of stem cells are highly specialized, each giving rise to a limited range of differentiated cell types. Cells become restricted to specific pathways of differentiation during embryonic development. One way to force a return to a pluripotent or totipotent state is by nuclear transplantation: the nucleus of a differentiated cell can be injected into an enucleated oocyte, whose cytoplasm reprograms the genome back to an approximation of an early embryonic state. This allows production of an entire new individual. The reversion of the genome to this state involves radical, genome-wide changes in chromatin structure and DNA methylation.

Remarkably, cells taken from the inner cell mass of an early mammalian embryo can be propagated in culture indefinitely in a pluripotent state. When transplanted back into a host early embryo, these embryonic stem (ES) cells can contribute cells to any tissue, including the germ line. ES cells have been invaluable for genetic engineering in mice. Cells with similar properties, called induced pluripotent stem cells (iPS cells), can be generated from adult differentiated cells such as fibroblasts by forced expression of a cocktail of key transcription regulators. A similar method can be used to reprogram adult cells directly from one specialized state to another. In principle, iPS cells generated from cells biopsied from an adult human patient could be used for tissue repair in that same individual, avoiding the problem of immune rejection. More immediately, they provide a source of specialized cells that can be used to analyze in vitro the effects of mutations affecting human cells and for screening for drugs for treatment of genetic diseases.

WHAT WE DON'T KNOW

- What determines tissue and organ size? How do the cells in each tissue know when to terminate their growth and division, so as to limit the size of an organ or tissue appropriately?

- What is the fundamental molecular difference that distinguishes a stem cell?

- How is the correct balance between stem cells, progenitor cells, and differentiated cells maintained in a tissue or organ?

- What role does chromatin structure play in cell memory and in cell reprogramming?

- How are molecules inherited asymmetrically during cell division?

- How do germ cells avoid aging?

PROBLEMS

Which statements are true? Explain why or why not.

22–1   In the small intestine, stem cells in the crypts divide asymmetrically to maintain the population of cells that make up the villi; after each division, one daughter remains a stem cell and the other begins to divide rapidly to produce differentiated progeny.

22–2   Stem cells, being stem cells, are by definition the same in all tissues.

22–3   Every tissue that can be renewed is renewed from a tissue-specific population of stem cells.

22–4   Disturbance of the balance in the activities of osteoblasts and osteoclasts in favor of osteoclasts can give rise to the condition known as osteoporosis, the brittle-bone syndrome of the elderly.

Discuss the following problems.

22–5   In the 1950s, scientists fed ^{3}H-thymidine to rats to label cells that were synthesizing DNA, and then followed the fates of labeled cells for periods of up to a year. They found three patterns of cell labeling in different tissues. Cells in some tissues such as neurons in the central nervous system and the retina did not get labeled. Muscle, kidney, and liver, by contrast, each showed a small number of labeled cells that retained their label, apparently without further division or loss. Finally, cells such as those in the squamous epithelia of the tongue and esophagus were labeled in fairly large numbers, with radioactive pairs of nuclei visible in 12 hours; however, the labeled cells disappeared over time. Which of these three patterns of labeling would you expect to see if the labeled cells were generated by stem cells? Explain your answer.

22–6   At any given time, intestinal crypts of mice comprise about 15 stem cells and 10 Paneth cells. After cell division, which occurs about once a day, the daughter cells remain stem cells only if they maintain contact with a Paneth cell. This constant competition for Paneth-cell contact raises the possibility that crypts might become monoclonal over time; that is, the crypt cells at one point in time might derive from only 1 of the 15 stem cells that existed at some earlier time. To test this possibility, you use the so-called confetti marker that upon activation expresses any one of three fluorescent proteins in the stem cells of the crypt. You then examine crypts at various times to determine whether they contain cells with multiple colors or only one color (**Figure Q22–1**). Do the crypts become monoclonal over time or not? How can you tell?

22–7   The origin of new β cells of the pancreas—from stem cells or from preexisting β cells—was not resolved until a decade ago, when the technique of lineage tracing was used to decide the issue. Using transgenic mice that expressed a tamoxifen-activated form of Cre recombinase under the control of the insulin promoter, which is active only in β cells, investigators could remove an inhibitory segment of DNA and thereby allow expression of human placental alkaline phosphatase (HPAP), which can be detected by histochemical staining. After a pulse of tamoxifen that converted about 30% of β cells in young mice to

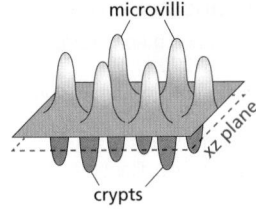

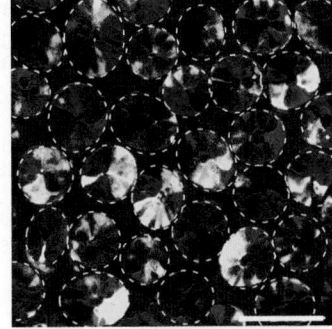

4 days

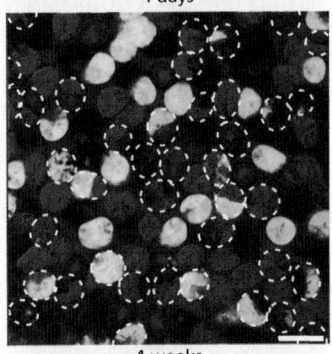

4 weeks

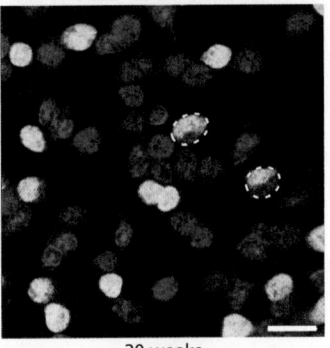

30 weeks

Figure Q22–1 Fluorescent cells in crypts in mouse intestines at various times after activation of expression of fluorescent proteins (Problem 22–6). The images are taken in the xz plane, which cuts through multiple crypts, as indicated in the schematic drawing. Roughly 50 crypts are visible in each section. *Dotted white circles* identify some individual crypts. Scale bars are 100 μm. (Adapted from H.J. Snippert et al., *Cell* 143:134–144, 2010. With permission from Elsevier.)

cells that express HPAP, the investigators followed the percentage of labeled β cells for a year, during which time the total number of β cells in the pancreas increased by 6.5-fold. How do you suppose the percentage of β cells would change over time if new β cells were derived from stem cells? What if new β cells were derived from preexisting β cells? Which hypothesis do the results in **Figure Q22–2** support?

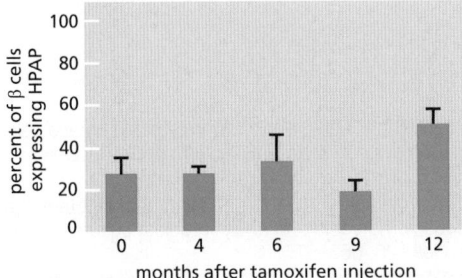

Figure Q22–2 Percentage of labeled β cells in pancreatic islets of mice at different ages (Problem 22–7). All mice were injected with a pulse of tamoxifen at 6 to 8 weeks of age and then stained for human placental alkaline phosphatase (HPAP) at various times afterward. Error bars represent standard deviations.

22–8 One of the earliest assays for hematopoietic stem cells made use of their ability to form colonies in the spleens of heavily irradiated mice. By varying the amounts of transplanted bone marrow cells, investigators showed that the number of spleen colonies varied linearly with dose and that the curve passed through the origin, suggesting that single cells were capable of forming individual colonies. However, because colony formation was rare relative to the numbers of transplanted cells, it was possible that undispersed clumps of two or more cells were the actual initiators.

A classic paper resolved this issue by exploiting rare, cytologically visible genome rearrangements generated by irradiation. Recipient mice were first irradiated to deplete bone marrow cells, and then they were irradiated a second time after transplantation to generate rare genome rearrangements in the transplanted cell population. Spleen colonies were then screened to find ones that carried genome rearrangements. How do you suppose this experiment distinguishes between colonization by single cells versus cellular aggregates?

22–9 It is possible to purify hematopoietic stem cells using a combination of antibodies directed against cell-surface targets. By removing cells that expressed surface markers characteristic of specific lineages such as B cells, granulocytes, myelomonocytic cells, and T cells, investigators generated a population of cells enriched for stem cells. They further enriched this population for putative stem cells by positively selecting for cells that expressed suspected stem-cell surface markers. Spleen colony formation in irradiated mice by these putative stem cells and the unfractionated bone marrow cells is shown in **Figure Q22–3**. Given that only about 1 in 10 cells lodges in the spleen, do these results support the idea that the enriched population consists mostly of hematopoietic stem cells? What additional information would you need to have to feel confident that the enriched cells are true stem cells? What proportion of bone marrow cells are hematopoietic stem cells?

22–10 Generation of induced pluripotent stem (iPS) cells was first accomplished using retroviral vectors to carry the OSKM (Oct4, Sox2, Klf4, and Myc) set of transcription regulators into cells. The efficiency of fibroblast reprogramming was typically low (0.01%), in part because large numbers of retroviruses must integrate to bring about reprogramming and each integration event carries with it the risk of inappropriately disrupting or activating a critical gene. In what other ways, or other forms, do you suppose you might deliver the OSKM transcription regulators so as to avoid these problems?

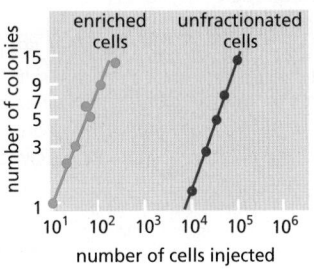

Figure Q22–3 Spleen colony formation by cells enriched for stem cells and by unfractionated bone marrow cells (Problem 22–9).

REFERENCES

General

Fawcett DW & Jensh R (2002) Bloom and Fawcett's Concise Histology, 2nd ed. New York/London: Arnold.

Gurdon JB & Melton DA (2008) Nuclear reprogramming in cells. *Science* 322, 1811–1815.

Li L & Xie T (2005) Stem cell niche: structure and function. *Annu. Rev. Cell Dev. Biol.* 21, 605–631.

Losick VP, Morris LX, Fox DT & Spradling A (2011) *Drosophila* stem cell niches: a decade of discovery suggests a unified view of stem cell regulation. *Dev. Cell* 21, 159–171.

Young B, Woodford P & O'Dowd G (2014) Wheater's Functional Histology: A Text and Colour Atlas, 6th ed. Edinburgh: Churchill Livingstone/Elsevier.

Stem Cells and Renewal in Epithelial Tissues

Barker N, van Es JH, Kuipers J et al. (2007) Identification of stem cells in small intestine and colon by marker gene *Lgr5*. *Nature* 449, 1003–1007.

Blanpain C & Fuchs E (2014) Plasticity of epithelial stem cells in tissue regeneration. *Science* 344, 1242281.

Crosnier C, Stamataki D & Lewis J (2006) Organizing cell renewal in the intestine: stem cells, signals and combinatorial control. *Nat. Rev. Genet.* 7, 349–359.

Sato T, van Es JH, Snippert HJ et al. (2011) Paneth cells constitute the niche for Lgr5 stem cells in intestinal crypts. *Nature* 469, 415–418.

Sato T & Clevers H (2013) Growing self-organizing mini-guts from a single intestinal stem cell: mechanism and applications. *Science* 340, 1190–1194.

Stanger BZ, Tanaka AJ & Melton DA (2007) Organ size is limited by the number of embryonic progenitor cells in the pancreas but not the liver. *Nature* 445, 886–891.

Taub R (2004) Liver regeneration: from myth to mechanism. *Nat. Rev. Mol. Cell Biol.* 5, 836–847.

Watt FM & Huck WTS (2013) Role of the extracellular matrix in regulating stem cell fate. *Nat. Rev. Mol. Cell Biol.* 14, 467–473.

Fibroblasts and Their Transformations: The Connective-Tissue Cell Family

Cooper KL, Oh S, Sung Y et al. (2013) Multiple phases of chondrocyte enlargement underlie differences in skeletal proportions. *Nature* 495, 375–378.

Karsenty G & Wagner EF (2002) Reaching a genetic and molecular understanding of skeletal development. *Dev. Cell* 2, 389–406.

Rinn JL, Bondre C, Gladstone HB et al. (2006) Anatomic demarcation by positional variation in fibroblast gene expression programs. *PLoS Genet.* 2, e119.

Seeman E & Delmas PD (2006) Bone quality—the material and structural basis of bone strength and fragility. *N. Engl. J. Med.* 354, 2250–2261.

Zelzer E & Olsen BR (2003) The genetic basis for skeletal diseases. *Nature* 423, 343–348.

Genesis and Regeneration of Skeletal Muscle

Bassel-Duby R & Olson EN (2006) Signaling pathways in skeletal muscle remodeling. *Annu. Rev. Biochem.* 75, 19–37.

Buckingham M (2006) Myogenic progenitor cells and skeletal myogenesis in vertebrates. *Curr. Opin. Genet. Dev.* 16, 525–532.

Collins CA, Olsen I, Zammit PS et al. (2005) Stem cell function, self-renewal, and behavioral heterogeneity of cells from the adult muscle satellite cell niche. *Cell* 122, 289–301.

Lee SJ (2004) Regulation of muscle mass by myostatin. *Annu. Rev. Cell Dev. Biol.* 20, 61–86.

Weintraub H, Davis R, Tapscott S et al. (1991) The myoD gene family: nodal point during specification of the muscle cell lineage. *Science* 251, 761–766.

Blood Vessels, Lymphatics, and Endothelial Cells

Carmeliet P & Tessier-Lavigne M (2005) Common mechanisms of nerve and blood vessel wiring. *Nature* 436, 193–200.

Folkman J & Haudenschild C (1980) Angiogenesis *in vitro*. *Nature* 288, 551–556.

Gerhardt H, Golding M, Fruttiger M et al. (2003) VEGF guides angiogenic sprouting utilizing endothelial tip cell filopodia. *J. Cell Biol.* 161, 1163–1177.

Lawson ND & Weinstein BM (2002) *In vivo* imaging of embryonic vascular development using transgenic zebrafish. *Dev. Biol.* 248, 307–318.

Pugh CW & Ratcliffe PJ (2003) Regulation of angiogenesis by hypoxia: role of the HIF system. *Nat. Med.* 9, 677–684.

Tammela T & Alitalo K (2010) Lymphangiogenesis: molecular mechanisms and future promise. *Cell* 140, 460–476.

A Hierarchical Stem-Cell System: Blood Cell Formation

Orkin SH & Zon LI (2008) Hematopoiesis: an evolving paradigm for stem cell biology. *Cell* 132, 631–644.

Shizuru JA, Negrin RS & Weissman IL (2005) Hematopoietic stem and progenitor cells: clinical and preclinical regeneration of the hematolymphoid system. *Annu. Rev. Med.* 56, 509–538.

Regeneration and Repair

Brockes JP & Kumar A (2008) Comparative aspects of animal regeneration. *Annu. Rev. Cell Dev. Biol.* 24, 525–549.

Tanaka EM & Reddien PW (2011) The cellular basis for animal regeneration. *Dev. Cell* 21, 172–185.

Wagner DE, Wang IE & Reddien PW (2011) Clonogenic neoblasts are pluripotent adult stem cells that underlie planarian regeneration. *Science* 332, 811–816.

Cell Reprogramming and Pluripotent Stem Cells

Apostolou E & Hochedlinger K (2013) Chromatin dynamics during cellular reprogramming. *Nature* 502, 462–471.

Egawa N, Kitaoka S, Tsukita K et al. (2012) Drug screening for ALS using patient-specific induced pluripotent stem cells. *Sci. Transl. Med.* 4, 145ra104.

Eggan K, Baldwin K, Tackett M et al. (2004) Mice cloned from olfactory sensory neurons. *Nature* 428, 44–49.

Fox IJ, Daley GQ, Goldman SA et al. (2014) Use of differentiated pluripotent stem cells as replacement therapy for treating disease. *Science* 345, 1247391.

Inoue H, Nagata N, Kurokawa H & Yamanaka S (2014) iPS cells: a game changer for future medicine. *EMBO J.* 33, 409–417.

Kim J, Chu J, Shen X et al. (2008) An extended transcriptional network for pluripotency of embryonic stem cells. *Cell* 132, 1049–1061.

Orkin SH & Hochedlinger K (2011) Chromatin connections to pluripotency and cellular reprogramming. *Cell* 145, 835–850.

Polo JM, Anderssen E, Walsh RM et al. (2012) A molecular roadmap of reprogramming somatic cells into iPS cells. *Cell* 151, 1617–1632.

Radzisheuskaya A & Silver JCR (2014) Do all roads lead to Oct4? The emerging concepts of induced pluripotency. *Trends Cell Biol.* 24, 275–284.

Sasai Y, Eiraku M & Suga H (2012) *In vitro* organogenesis in three dimensions: self-organising stem cells. *Development* 139, 4111–4121.

Soza-Ried J & Fisher AG (2012) Reprogramming somatic cells towards pluripotency by cellular fusion. *Curr. Opin. Genet. Dev.* 22, 459–465.

Takahashi K & Yamanaka S (2006) Induction of pluripotent stem cells from mouse embryonic and adult fibroblast cultures by defined factors. *Cell* 126, 663–676.

Theunissen TW & Jaenisch R (2014) Molecular control of induced pluripotency. *Cell Stem Cell* 14, 720–734.

Watanabe A, Yamada Y & Yamanaka S (2013) Epigenetic regulation in pluripotent stem cells: a key to breaking the epigenetic barrier. *Philos. Trans. R. Soc. Lond. B Biol. Sci.* 368, 20120292.

Yamanaka S (2013) The winding road to pluripotency (Nobel Lecture). *Angew. Chem. Int. Ed. Engl.* 52, 13900–13909.

Pathogens and Infection

Infectious diseases currently cause about one-quarter of all human deaths worldwide, more than all forms of cancer combined and second only to cardiovascular diseases. In addition to the continuing heavy burden of ancient diseases such as tuberculosis and malaria, newer infectious diseases continually emerge. The current pandemic (worldwide epidemic) of *AIDS* (*acquired immune deficiency syndrome*), was first clinically observed in 1981 and has since caused more than 35 million deaths worldwide. Moreover, some diseases long thought to result from other causes are now recognized to be associated with infections. Most gastric ulcers, for example, are caused not by stress or spicy food, but by infection of the stomach lining by the bacterium *Helicobacter pylori*.

The burden of infectious diseases is not spread equally across the planet. Poorer countries and communities suffer disproportionately, often due to poor public sanitation and health systems. Some infectious diseases, however, occur primarily or exclusively in industrialized communities: Legionnaire's disease, for example, a bacterial infection of the lungs, commonly spreads through air-conditioning systems.

Since the mid-1800s, physicians and scientists have struggled to identify the agents—collectively called **pathogens**—that are capable of causing infectious diseases. More recently, the advent of microbial genetics and molecular cell biology has greatly enhanced our understanding of the causes and mechanisms of infectious diseases. We now know that pathogens frequently exploit the attributes of their host's cells in order to infect them. This understanding can give us new insights into normal cell biology, as well as strategies for treating and preventing infectious diseases.

Although pathogens are understandably a focus of attention, only a relatively small fraction of the microbial species we encounter are pathogens. Much of the biomass of the Earth is made up of microbes, and they produce everything from the oxygen we breathe to the soil nutrients we use to grow food. Even those species of microbes that colonize the human body do not generally cause disease. The collective of microorganisms that reside in or on an organism is called the **microbiota**. Many of these microbes have a beneficial effect on the health of the organism, assisting its normal development and physiology.

In this chapter, we give an overview of the different kinds of pathogens, as well as those microorganisms that colonize our body without causing trouble. We then discuss the cell biology of infection—the molecular interactions between pathogens and their host. In Chapter 24, we consider how our innate and adaptive immune systems collaborate to defend us against pathogens.

INTRODUCTION TO PATHOGENS AND THE HUMAN MICROBIOTA

We normally think of pathogens as hostile invaders, but a pathogen, like any other organism, is simply exploiting an available niche in which to live and procreate. Living on or in a host organism is a very effective strategy, and it is possible that every organism on Earth is subject to some type of infection (**Figure 23–1**). A human host is a nutrient-rich, warm, and moist environment, which remains at a

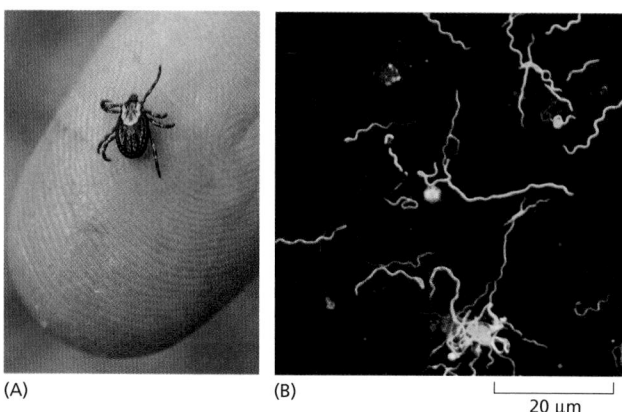

(A)   (B)

20 µm

Figure 23–1 **Parasitism at many levels.**
(A) Most animals harbor parasites, an example being the blacklegged tick or deer tick (*Ixodes scapularis*), shown here on a human finger. Although ticks of this species thrive on white-tailed deer and other wild mammals, they can also live on humans. (B) Ticks themselves harbor their own parasites including the bacterium *Borrelia burgdorferi*, stained here with a vital dye that labels living bacteria *green* and dead bacteria *red*. These spiral-shaped bacteria live in deer ticks and can be transmitted to humans during a tick's blood meal. *Borrelia burgdorferi* causes Lyme disease, which is characterized by a bull's-eye-shaped skin rash and fever; if the infection is left untreated, various complications can result, including arthritis and neurological abnormalities. The idea that parasites have their own parasites was noted by Jonathan Swift in 1733:
"So, naturalists observe, a flea
Has smaller fleas that on him prey;
And these have smaller still to bite 'em;
And so proceed ad infinitum."
(A, National Geographic Image Collection/Alamy Stock Photo; B, courtesy of M. Embers.)

uniform temperature and constantly renews itself. It is not surprising that many microorganisms have evolved the ability to survive and reproduce in this desirable niche. In this section, we discuss some of the common features that microorganisms must have in order to colonize the human body or cause disease, and we explore the wide variety of organisms that are known to cause disease.

The Human Microbiota Is a Complex Ecological System That Is Important for Our Development and Health

The human body contains about 10^{13} human cells, as well as a microbiota consisting of approximately 10^{14} bacterial, fungal, and protozoan cells, which represent thousands of microbial species—the so-called **normal flora**. The combined genomes of the various species of the human microbiota, called the **microbiome**, contain more than 5×10^6 genes—more than 100 times greater than the number of genes in the human genome itself. A consequence of this genomic diversity is that the microbiota expands the range of biochemical and metabolic activities available to the humans.

The microbiota is usually confined to the skin, mouth, digestive tract, and vagina. With the exception of microbes colonizing the skin, it consists primarily of anaerobic bacteria, with distinct communities of species inhabiting each body part. These communities vary considerably between individual humans, even between close relatives or identical twins. Although the microbiota of an individual is generally consistent over time, it is influenced by a variety of factors, including age, diet, health status, and antibiotic use.

There are various ecological relationships that these microbes have with their host. In **mutualism**, both the microbe and host benefit. The anaerobic bacteria that inhabit our intestines, for example, gain shelter and a nutrient supply but also contribute to the digestion of our food, produce important nutrients for us, and are essential for the normal development of our gastrointestinal tract and innate and adaptive immune systems. In **commensalism**, the microbe benefits but offers no benefit and causes no harm: for example, we are infected with many viruses that have no noticeable effect on our health. In **parasitism**, the microbe benefits to the detriment of the host, as is often the case for pathogens.

Many infectious diseases are caused by a single pathogen. There is increasing evidence, however, that an imbalance in the community of microbes that constitute the microbiota can contribute to some diseases, including autoimmune and allergic diseases, obesity, inflammatory bowel disease, and diabetes. Remarkably, in such cases of microbiota imbalance (referred to as *dysbiosis*), the transfer of the microbiota from a healthy individual to someone suffering from the disease can be beneficial and sometimes curative, as in the case of *Clostridium difficile* colitis caused by overgrowth of the bacterium.

Pathogens Interact with Their Hosts in Different Ways

If it is normal for us to live with a community of microbes, why are some of them capable of causing us illness or death? Although the ability of a particular microorganism to cause disease depends on many factors, it requires that the pathogen possess specialized pathogenic characteristics that allow it to live in humans.

Primary pathogens can cause overt disease in most healthy people. Some primary pathogens cause acute, life-threatening epidemic infections and spread rapidly from one sick or dying host to another; historically important examples include the bacterium *Vibrio cholerae*, which causes cholera, and the variola and influenza viruses, which cause smallpox and flu, respectively. Others may persistently infect a single individual for years without causing overt disease; examples include the bacterium *Mycobacterium tuberculosis* (which can cause the life-threatening lung infection tuberculosis) and the intestinal worm *Ascaris*. Although these potential primary pathogens can make some people critically ill, billions of people carry these foreign organisms in an asymptomatic way, often unaware that they are infected. It is sometimes difficult to draw a line between the asymptomatic presence of such pathogens and the normal microbiota. Some microbes of the normal flora can act as **opportunistic pathogens**, in that they cause disease only if our immune systems are weakened or if they gain access to a normally sterile part of the body.

In order to survive and multiply, a successful pathogen must be able to: (1) enter the host (usually by breaking an epithelial barrier); (2) find a nutritionally compatible niche in the host's body; (3) avoid, subvert, or circumvent the host's innate and adaptive immune responses; (4) replicate, using host resources; and (5) exit one host and spread to another. Pathogens have evolved various mechanisms that maximally exploit the biology of their host organisms to help accomplish these tasks. For some pathogens, these mechanisms are adapted to a unique host species, whereas for others the mechanisms are sufficiently general to permit invasion, survival, and replication in a wide variety of hosts. Because pathogens have evolved the ability to interface directly with the molecular machinery of host cells, we have learned a great deal about cell biological principles by studying them.

Our constant exposure to pathogens has strongly influenced human evolution. In modern times, humans have learned how to limit the ability of pathogens to infect us through improvements in public health measures and childhood nutrition, vaccines, antimicrobial drugs, and routine testing of blood used for transfusions. As we learn more about the mechanisms by which pathogens cause disease (called *pathogenesis*), our creativity and resourcefullness will continue to serve as an important addition to our immune systems in fighting infectious diseases.

Pathogens Can Contribute to Cancer, Cardiovascular Disease, and Other Chronic Illnesses

Some viral and bacterial pathogens can cause or contribute to chronic, life-threatening illnesses that are not normally classified as infectious diseases. An important example is cancer. As discussed in Chapter 20, the oncogene concept—that certain altered genes can trigger cell transformation and tumor development—came initially from studies of the *Rous sarcoma virus*, which causes a form of cancer (sarcomas) in chickens. One of the viral genes encodes an overactive homolog of the host tyrosine kinase Src (see Figure 3–63), which has been implicated in many kinds of cancer. Several human cancers are also known to have a viral origin. *Human papillomavirus*, for example, which causes genital warts, is responsible for more than 90% of cervical cancers (see Figure 20–40). The recent development of a vaccine against the most abundant cancer-associated strains of human papillomavirus promises to prevent many of these cancers in the future. In other cases, chronic tissue damage caused by infection can increase the likelihood of cancer. Inflammation caused by the stomach-dwelling bacterium *H. pylori* can be a major contributor to stomach cancer, as well as to gastric ulcers.

The major causes of death in wealthy industrialized nations are cardiovascular diseases. They frequently result from *atherosclerosis*, the accumulation in blood

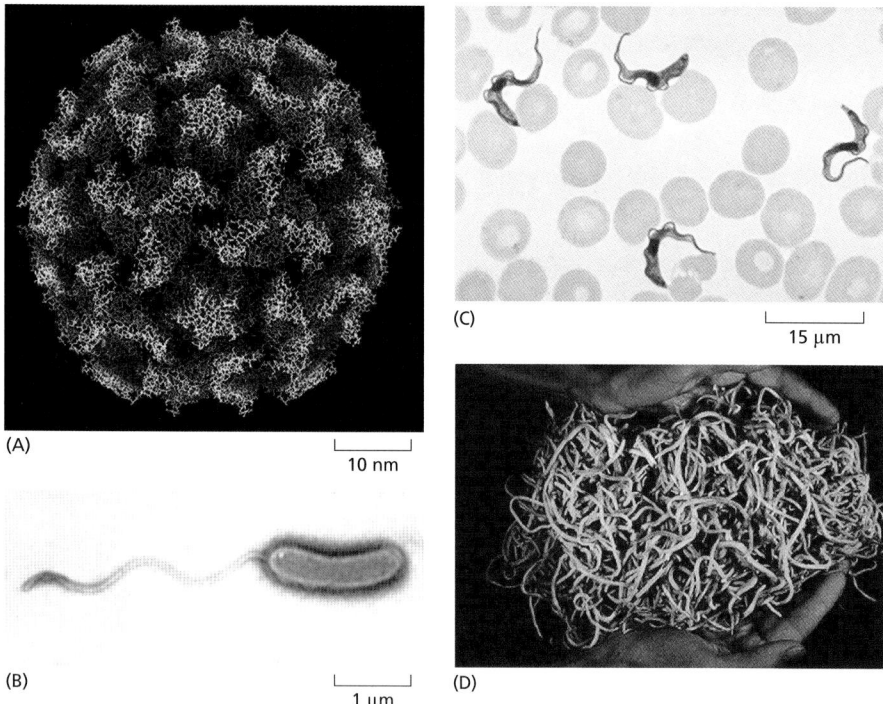

(A)

10 nm

(B)

1 μm

(C)

15 μm

(D)

Figure 23–2 Pathogens in many forms.
(A) The structure of the protein coat, or *capsid*, of poliovirus. This virus was once a common cause of paralysis, but the disease (poliomyelitis) has been greatly reduced by widespread vaccination. (B) The bacterium *Vibrio cholerae*, the causative agent of the epidemic, diarrheal disease cholera. (C) The protozoan parasite *Trypanosoma brucei (purple)* in a field of erythrocytes (red blood cells; *pink*). This parasite causes African sleeping sickness, a potentially fatal disease of the central nervous system. (D) This clump of *Ascaris* nematodes was removed from the obstructed intestine of a two-year-old boy. (A, courtesy of Robert Grant, Stephan Crainic, and James M. Hogle; B, photo kindly provided by John Mekalanos; C, CDC, Department of Health and Human Services; D, from J.K. Baird et al., *Am. J. Trop. Med. Hyg.* 35:314–318, 1986. Photograph by Daniel H. Connor. With permission from American Society of Tropical Medicine and Hygiene.)

vessel walls of fatty deposits that can block blood flow and cause heart attacks and strokes. A hallmark of early atherosclerosis is the appearance in blood vessel walls of clumps of macrophages called foam cells, which recruit other white blood cells into the forming *atherosclerotic plaque*. Foam cells in atherosclerotic plaques often contain the bacterial pathogen *Chlamydia pneumoniae*, which commonly causes pneumonia in humans and is a significant risk factor for atherosclerosis in humans and animal models. Other bacterial species are also implicated in atherosclerosis, including bacteria usually associated with teeth and gums, such as *Porphyromonas gingivalis*. As we learn more about the interactions between pathogens and the human body, it seems likely that more chronic conditions will be found to have a link to an infectious agent.

Pathogens Can Be Viruses, Bacteria, or Eukaryotes

Many types of pathogens cause disease in humans. The most familiar are viruses and bacteria. Viruses cause diseases ranging from AIDS and smallpox to the common cold. Viruses are essentially fragments of nucleic acid (DNA or RNA) that generally encode a relatively small number of gene products, wrapped in a protective shell of proteins (**Figure 23–2**A) and (in some cases) an outer membrane envelope (see Figure 5–62). Much larger and more complex than viruses, bacteria are prokaryotic cells, which perform most of their basic metabolic functions themselves, relying on the host primarily for nutrition (Figure 23–2B).

Some other infectious agents are eukaryotic organisms. These range from single-celled fungi and protozoa (Figure 23–2C) to large, complex metazoa such as parasitic worms. One of the most common human parasites, shared by about a billion people at present, is the nematode worm *Ascaris lumbricoides*, which infects the gut (Figure 23–2D). It closely resembles its harmless nematode cousin *Caenorhabditis elegans*, which is used as a model organism for genetic and developmental biological research (see Figure 1–39). *C. elegans*, however, is only about 1 mm in length, whereas *Ascaris* can reach 30 cm.

We now introduce the basic features of each of the major types of pathogens, before we examine the mechanisms that pathogens use to infect their hosts.

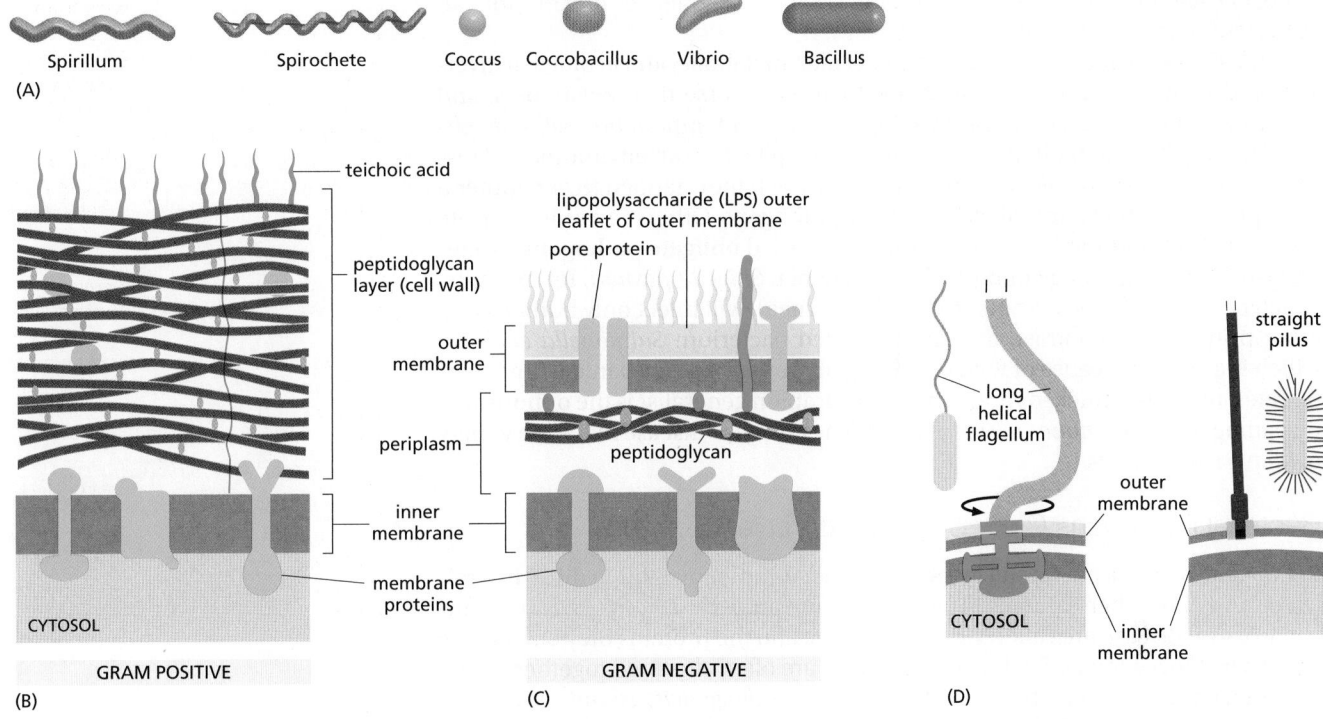

Figure 23–3 Bacterial shapes and cell-surface structures. (A) Bacteria are traditionally classified by shape. (B and C) They are also classified as *Gram positive* or *Gram negative*. (B) Gram-positive bacteria such as *Streptococcus* and *Staphylococcus* have a single membrane and a thick cell wall made of cross-linked *peptidoglycan*. They are called Gram positive because they retain the violet dye used in the Gram-staining procedure. (C) Gram-negative bacteria such as *Escherichia coli* (*E. coli*) and *Salmonella* have two membranes, separated by the *periplasm* (see Figure 11–17). The peptidoglycan cell wall of these organisms is located in the periplasm and is thinner than in Gram-positive bacteria; they therefore fail to retain the dye in the Gram-staining procedure. The inner membrane of both Gram-positive and Gram-negative bacteria is a phospholipid bilayer. The inner leaflet of the outer membrane of Gram-negative bacteria is also made primarily of phospholipids, whereas the outer leaflet of the outer membrane is composed of a unique glycosylated lipid called *lipopolysaccharide* (*LPS*). (D) Cell-surface appendages are important for bacterial behavior. Many bacteria swim using the rotation of helical flagella. The bacterium illustrated has only a single flagellum at one pole; however, many have multiple flagella. Straight *pili* (also called *fimbriae*) are used to adhere to various surfaces in the host, as well as to facilitate genetic exchange between bacteria. Some kinds of pili can retract to generate force and thereby help bacteria move along surfaces.

Bacteria Are Diverse and Occupy a Remarkable Variety of Ecological Niches

Although **bacteria** generally lack internal membranes, they are highly sophisticated cells whose organization and behaviors have attracted the attention of many scientists. Bacteria are classified broadly by their shape—as rods, spheres (*cocci*), or spirals (**Figure 23–3**A)— as well as by their so-called **Gram-staining** properties, which reflect differences in the structure of the bacterial cell wall. **Gram-positive** bacteria have a thick layer of peptidoglycan cell wall outside their inner (plasma) membrane (Figure 23–3B), whereas **Gram-negative** bacteria have a thinner peptidoglycan cell wall. In both cases, the cell wall protects against lysis by osmotic swelling, and it is a target of host antibacterial proteins such as lysozyme and antibiotics such as penicillin. Gram-negative bacteria are also covered outside the cell wall by an outer membrane containing *lipopolysaccharide* (*LPS*) (Figure 23–3C). Both peptidoglycan and LPS are unique to bacteria and are recognized as *pathogen-associated molecular patterns* (*PAMPs*) by the host innate immune system, as discussed in Chapter 24. The surface of bacterial cells can also display an array of appendages, including flagella and pili, which enable bacteria to swim or adhere to desirable surfaces, respectively (Figure 23–3D). Apart from cell shape and structure, differences in ribosomal RNA and genomic DNA sequence are also used for phylogenetic classification. Because bacterial genomes are small—typically between 1,000,000 and 5,000,000 nucleotide pairs (compared to more than

3,000,000,000 for humans)—they are now simple to sequence, making this an important new classification tool.

Bacteria also exhibit extraordinary molecular, metabolic, and ecological diversity. At the molecular level, bacteria are far more diverse than eukaryotes, and they can occupy ecological niches having extremes of temperature, salt concentrations, and nutrient limitation. Some bacteria replicate in an environmental reservoir such as water or soil and only cause disease if they happen to encounter a susceptible host; these are called **facultative pathogens**. Others can only replicate inside the body of their host and are therefore called **obligate pathogens**. Bacteria also differ in the range of hosts they will infect. *Shigella flexneri*, for example, which causes epidemic dysentery (bloody diarrhea), will infect only humans and other primates. By contrast, the closely related bacterium *Salmonella enterica*, which is a common cause of food poisoning in humans, can also infect other vertebrates, including chickens and turtles. A champion generalist is the opportunistic pathogen *Pseudomonas aeruginosa*, which can cause disease in a wide variety of plants and animals.

Bacterial Pathogens Carry Specialized Virulence Genes

Pathogenic bacteria and their closest nonpathogenic relatives often differ in a relatively small number of genes. Genes that contribute to the ability of an organism to cause disease are called **virulence genes**, and the proteins they encode are called **virulence factors**. Such virulence genes are often clustered together on the bacterial chromosome; large clusters are called *pathogenicity islands*. Virulence genes can also be carried on *bacteriophages* (bacterial viruses) or *transposons* (see Table 5–4), both of which integrate into the bacterial chromosome, or on extrachromosomal *virulence plasmids* (**Figure 23–4**A).

Pathogenic bacteria are thought to emerge when groups of virulence genes are transferred together into a previously avirulent bacterium by a process called **horizontal gene transfer** (to distinguish it from vertical gene transfer from parent to offspring). Horizontal transfer can occur by one of three mechanisms: natural *transformation* by released naked DNA, *transduction* by bacteriophages, or sexual exchange by *conjugation* (Figure 23–4B and Movie 23.1). Sequencing the genomes of large numbers of pathogenic and nonpathogenic bacteria has indicated that horizontal gene transfer has made important contributions to bacterial evolution, enabling species to inhabit new ecological and nutritional niches, as well as to cause disease. Even within a single bacterial species, the amount of chromosomal

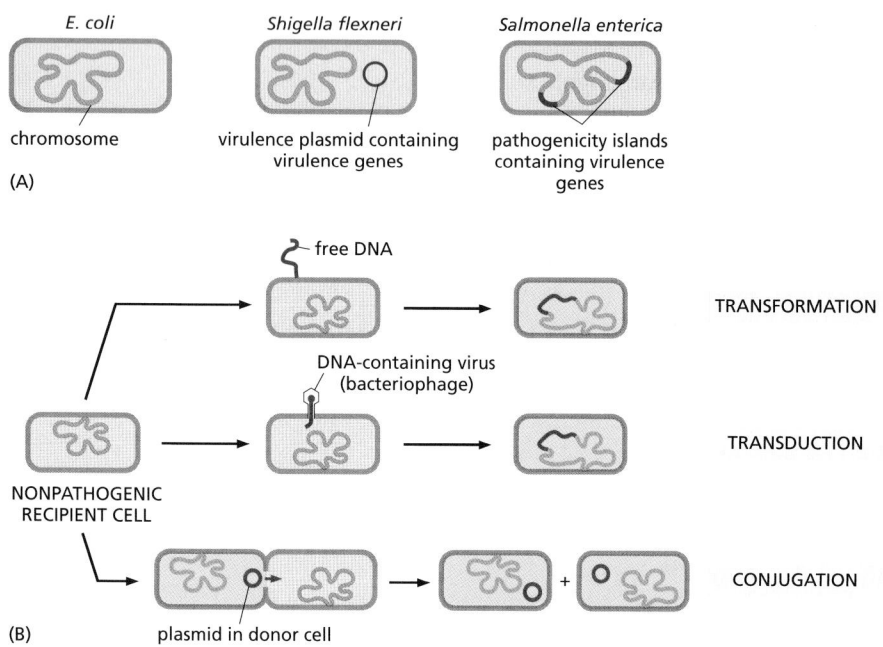

(A)

(B)

Figure 23–4 Genetic differences between pathogenic and nonpathogenic bacteria. (A) Genetic differences between nonpathogenic *E. coli* and two closely related food-borne pathogens—*Shigella flexneri*, which causes dysentery, and *Salmonella enterica*, a common cause of food poisoning. Nonpathogenic *E. coli* has a single circular chromosome. The chromosome of *S. flexneri* differs from that of *E. coli* in a limited number of locations; most of the genes required for pathogenesis (virulence genes) are carried on an extrachromosomal virulence plasmid. The chromosome of *S. enterica* carries two large inserts (pathogenicity islands) not found in the *E. coli* chromosome; these inserts each contain many virulence genes. (B) Bacterial pathogens evolve by horizontal gene transfer. This can occur by three mechanisms: natural *transformation*, in which naked DNA is taken in by competent bacteria; *transduction*, in which bacterial viruses (*bacteriophages*) transfer DNA from one bacterium into another; and *conjugation*, during which plasmid DNA, and even chromosomal DNA, is transferred from a donor to a recipient bacterium.

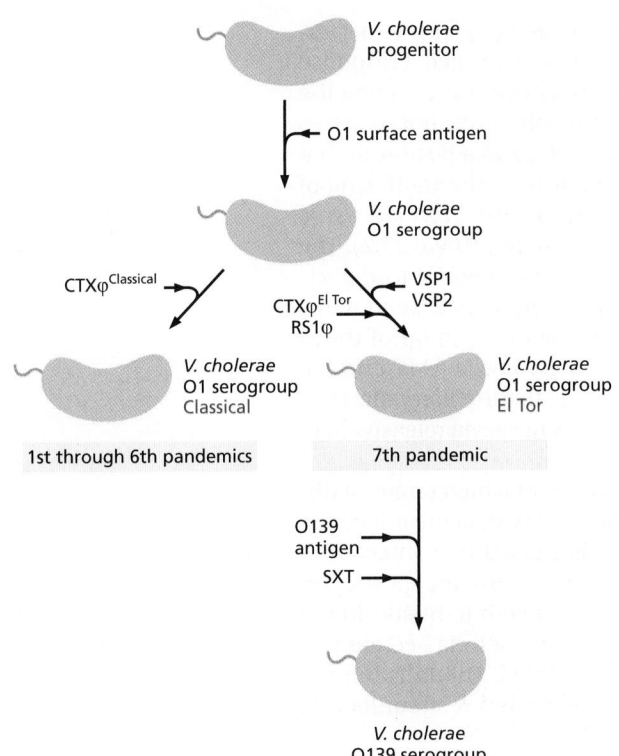

Figure 23–5 Comparative-genomics-based model for the evolution of pathogenic *Vibrio cholerae* strains. Progenitor strains in the wild first acquired the biosynthetic pathway necessary to make the O1 antigen type of carbohydrate chain on the outer-membrane lipopolysaccharide (see Figure 23–3C). Incorporation of the CTXφ bacteriophage created the Classical pathogenic strains responsible for the first six worldwide epidemics of cholera between 1817 and 1923. Sometime in the twentieth century, an O1 strain in the environment picked up the CTXφ bacteriophage again, along with an associated bacteriophage RS1φ and two pathogenicity islands (VSP1 and VSP2), creating the El Tor strain that emerged as the seventh worldwide pandemic in 1961. In 1992, an El Tor strain was isolated that had picked up a new DNA cassette, enabling it to produce the O139 antigen type of carbohydrate chain rather than the O1 type. This altered the bacterium's interaction with the human immune system, without diminishing its virulence; this bacterium also picked up a new pathogenicity island (SXT). An electron micrograph of *Vibrio cholerae* (*V. cholerae*) is shown in Figure 23–2B.

variation is astonishing; the genomes of different strains of *Escherichia coli* can differ by as much as 25%. Such variation has led to the concept that a bacterial species has both a *core genome* common to all isolates within the species and a larger *pan-genome* consisting of all genes present in the full spectrum of isolates.

Acquisition of genes and gene clusters can drive the rapid evolution of pathogens and turn nonpathogens into pathogens. Consider, for example, *Vibrio cholerae*—the Gram-negative bacterium that causes the epidemic diarrheal disease cholera. Of the hundreds of strains of *Vibrio cholerae*, the only ones that cause pandemic human disease are those infected with a mobile bacteriophage (CTXφ) containing genes encoding the two subunits of the toxin that causes the diarrhea. As summarized in **Figure 23–5**, seven pandemics of *V. cholerae* have arisen since 1817. The first six were caused by the periodic reemergence of so-called Classical strains. In addition to the toxin-encoding bacteriophage, these Classical strains shared a similar O1 surface antigen, part of the LPS in the outer membrane (see Figure 23–3C). In 1961, the seventh pandemic began, caused by a new strain named "El Tor," which arose when an O1-expressing strain acquired two bacteriophages and at least two new pathogenicity islands. El Tor eventually displaced the Classical strains. In 1992, a new strain emerged in which O1 was replaced with another O-antigen variant called O139, which was not recognized by antibodies present in the blood of survivors of previous cholera epidemics. The O139 strain also contains a transposon-like element that encodes antibiotic resistance. As this example makes clear, the rapid evolution of bacterial pathogens can be likened to an arms race which pits the survival of a bacterium against our immune systems and the tools of modern medicine. Similar struggles for survival take place between all pathogens and humans, and understanding these conflicts provides key insights into the evolution of pathogens and greatly informs us how we treat new outbreaks of infectious diseases.

Bacterial Virulence Genes Encode Effector Proteins and Secretion Systems to Deliver Effector Proteins to Host Cells

What are the gene products that enable a bacterium to cause disease in a healthy host? For pathogenic bacteria that live outside of host cells, called *extracellular bacterial pathogens,* virulence genes often encode secreted toxic proteins (*toxins*) that interact with host cell structural or signaling proteins to elicit a response that is beneficial to the pathogen. Several of these bacterial toxins are among the most potent of known human poisons. Bacterial toxins are often composed of two protein components—an A subunit with enzymatic activity, and a B subunit that binds to specific receptors on the host cell surface and directs the trafficking of the A subunit to the cytosol by various routes (**Figure 23–6**). The *Vibrio cholerae* phage, for example, encodes the two subunits of **cholera toxin** (Movie 23.2). The A subunit catalyzes the transfer of an ADP-ribose moiety from NAD$^+$ to the trimeric G protein G$_s$ (see Figure 15–23), which activates adenylyl cyclase to make cyclic AMP (see Figure 15–25). ADP-ribosylation prevents inactivation of the G protein and results in the overaccumulation of intracellular cyclic AMP and the release of ions and water into the intestinal lumen, leading to the watery diarrhea associated with cholera. The infection then spreads to new hosts via released bacteria, which can contaminate food and water.

Some pathogenic bacteria secrete multiple toxins, each of which targets a different signaling pathway in host cells. Anthrax, for example, is an acute infectious disease of sheep, cattle, and occasionally humans. It is caused by contact with spores of the Gram-positive bacterium *Bacillus anthracis.* Dormant spores can survive in soil for long periods. If inhaled, ingested, or rubbed into breaks in the skin, spores can germinate and the bacteria replicate. The bacteria secrete two toxins with identical B subunits but different A subunits. The B subunits bind to a host cell-surface receptor protein to transfer the two different A subunits into

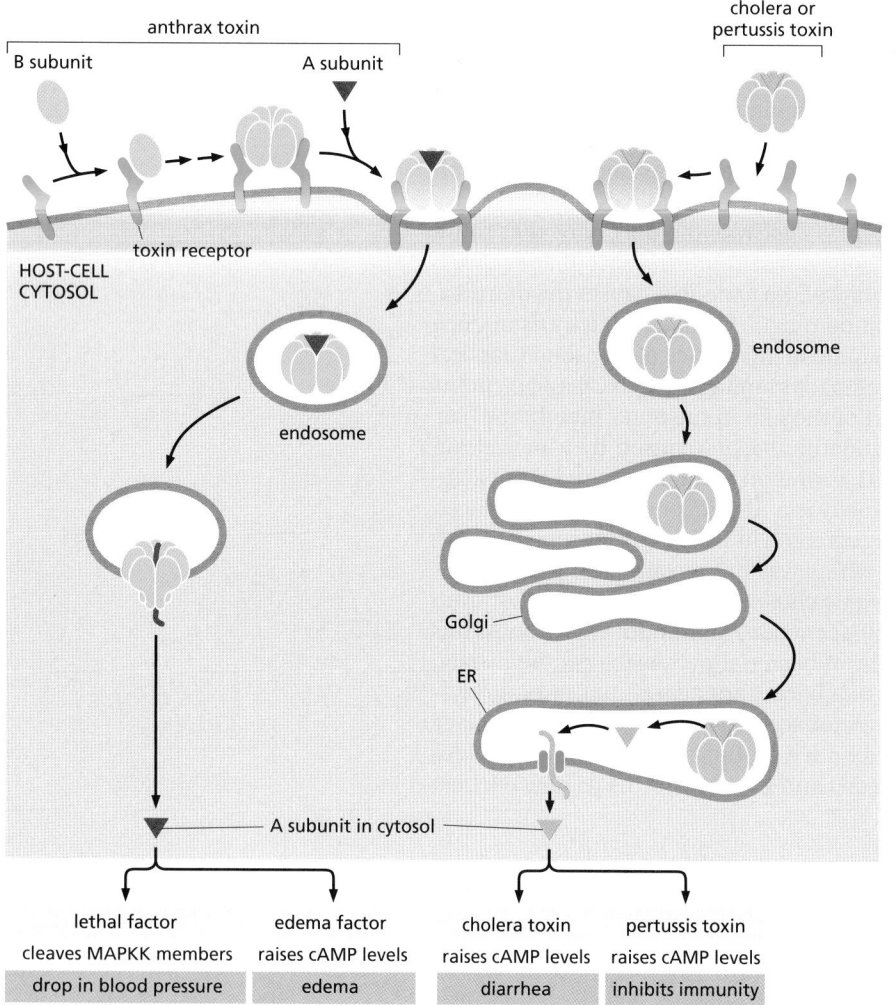

Figure 23–6 **Bacterial toxin entry into host cells.** Bacterial toxins are often composed of A and B protein subunits. The B (binding) subunit of the toxin interacts with host-cell toxin receptors, enabling endocytosis and intracellular trafficking of B subunit as well as its associated and enzymatically active A subunit(s). In the case of *Bacillus anthracis*, the B subunit changes conformation in the low pH environment of the endosome to form a pore through which two different A subunits, lethal factor and edema factor, are transported across the membrane of the endosome in an unfolded conformation. In the cases of *Vibrio cholerae* toxin and *Bordetella pertussis* toxin, the B and A subunits are transported to the Golgi apparatus and then to the endoplasmic reticulum (ER), where the A subunits are then translocated into the cytosol in an unfolded conformation through a protein-translocation channel.

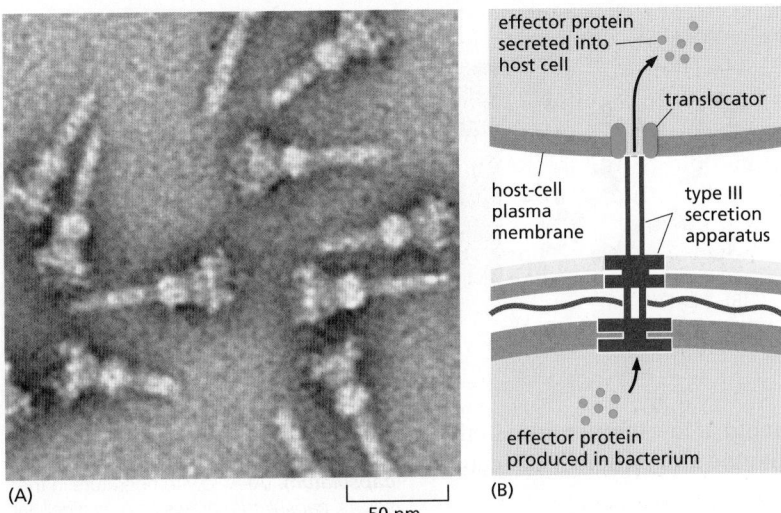

(A)

50 nm

(B)

effector protein
secreted into
host cell

translocator

host-cell
plasma
membrane

type III
secretion
apparatus

effector protein
produced in bacterium

Figure 23–7 **Type III secretion systems
that can deliver effector proteins into
the cytosol of a host cell.** (A) Electron
micrograph of purified type III secretion
systems, each of which consists of
over two dozen proteins. (B) The large
lower ring is embedded in the bacterial
inner membrane, and the smaller upper
ring is embedded in the bacterial outer
membrane. During infection, docking of
the tip of the hollow needle at a host-cell
plasma membrane results in the secretion
of bacterial translocator proteins *(green)*,
which form a pore in the host membrane,
through which bacterial effector proteins
are then secreted into the host cell.
(A, courtesy of Thomas C. Marlovits and
Oliver Schraidt.)

host cells (see Figure 23–6). The A subunits are called **lethal factor** and **edema factor**. The A subunit of edema toxin is an adenylyl cyclase that catalyzes the production of cyclic AMP (see Figure 15–25), leading to an ion imbalance that can cause an accumulation of extracellular fluid (*edema*) in the skin or lung. The A subunit of lethal toxin is a protease that cleaves several activated members of the mitogen-activated protein kinase kinase (MAP kinase kinase) family (see Figure 15–49), disrupting intracellular signaling and leading to immune cell dysfunction and cell death. Injection of lethal toxin into the bloodstream of an animal causes shock (a large fall in blood pressure) and death.

Apart from toxins, bacteria use specialized **secretion systems** to secrete many other *effector proteins* that interact with host cells. Gram-negative bacteria have a *general secretion system* and several classes of *accessory secretion systems* (types I–VI). A subset of these accessory secretion systems, called *contact-dependent secretion systems*, is present in many bacteria that contact or live inside host cells. The **type III secretion system** (Figure 23–7), for example, injects into the host-cell cytoplasm *effector proteins* that can elicit a variety of host cell responses that enable the bacterium to invade or survive. There is a remarkable degree of structural similarity between the type III syringe and the base of a bacterial flagellum. Because flagella are found in a wider range of bacteria than are type III secretion systems, and the secretion systems appear to be adaptations specific for pathogenesis, it seems likely that the type III secretion systems evolved from flagella. Other types of delivery systems used by bacterial pathogens appear to have evolved independently. For example, *type IV secretion systems* are closely related to the conjugation apparatus that many bacteria use to exchange genetic material.

Fungal and Protozoan Parasites Have Complex Life Cycles Involving Multiple Forms

Pathogenic fungi and protozoan parasites are eukaryotes, as are their hosts. Consequently, antifungal and antiparasitic drugs are often less effective and more toxic to the host than are antibiotics that target bacteria. A second characteristic of fungal and parasitic infections that makes them difficult to treat is the tendency of the pathogens to switch among several different forms during their life cycles. A drug that is effective at killing one form can be ineffective at killing another form; therefore the population can survive the treatment.

Fungi include both unicellular *yeasts* (such as *Saccharomyces cerevisiae* and *Schizosaccharomyces pombe*, which are used to bake bread and brew beer, and as model organisms for cell biology research) and filamentous, multicellular *molds* (like those found on moldy fruit or bread). Most of the important pathogenic fungi exhibit *dimorphism*—the ability to grow in either yeast or mold form. The yeast-to-mold or mold-to-yeast transition is frequently associated with infection.

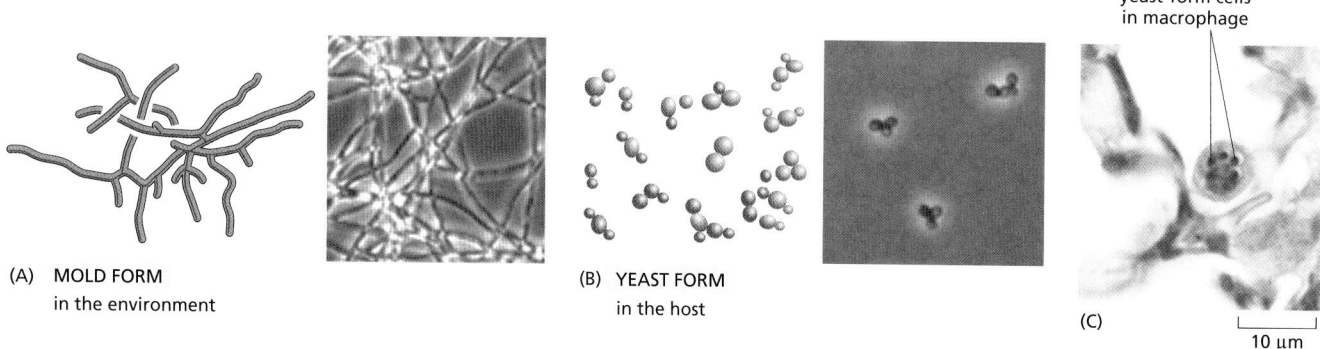

(A)   MOLD FORM
      in the environment

(B)   YEAST FORM
      in the host

yeast-form cells
in macrophage

(C)   10 μm

Figure 23–8 Dimorphism in the pathogenic fungus *Histoplasma capsulatum*. (A) At low temperature in the soil, *H. capsulatum* grows as a multicellular filamentous mold consisting of many individual cells connected together. (B) After it is inhaled into the lung of a mammal, the increase in temperature causes a switch to a yeast form consisting of small clumps of round cells. (C) A stained histologic section of a mouse lung infected with *H. capsulatum*, showing a macrophage containing yeast forms of the pathogen. (A and B, courtesy of Sinem Beyhan and Anita Sil; C, courtesy of Davina Hocking Murray and Anita Sil.)

Histoplasma capsulatum, for example, grows as a mold at low temperature in the soil, but it switches to a yeast form when inhaled into the lung, where it can cause the disease histoplasmosis (**Figure 23–8**).

Protozoan parasites are single-celled eukaryotes with more elaborate life cycles than fungi, and they frequently require more than one host. **Malaria** is the most devastating protozoal disease, infecting more than 200 million people every year and killing upward of 500,000. It is caused by four species of *Plasmodium*, which are transmitted to humans by the bite of the female *Anopheles* mosquito.

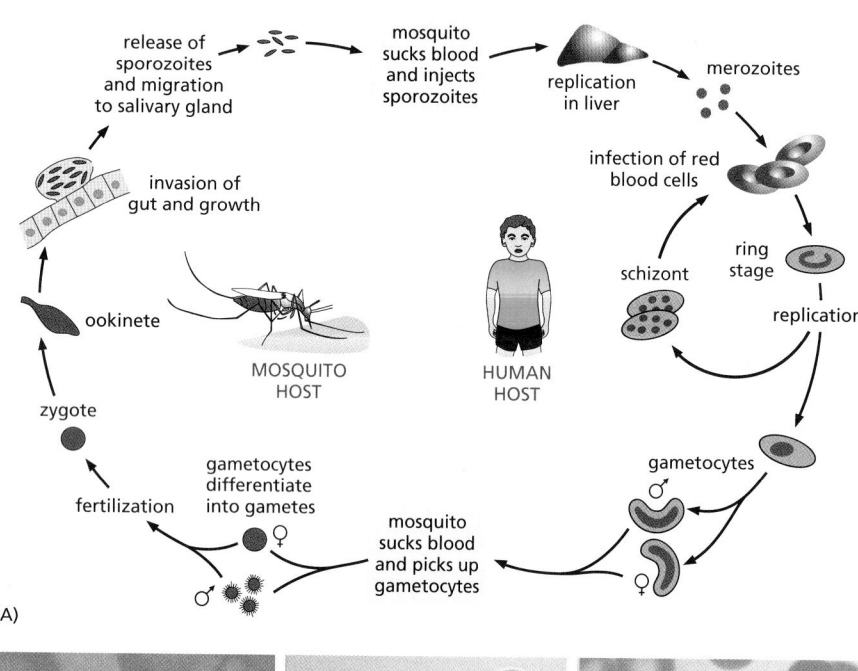

release of sporozoites and migration to salivary gland

mosquito sucks blood and injects sporozoites

replication in liver

merozoites

infection of red blood cells

invasion of gut and growth

schizont

ring stage

replication

ookinete

MOSQUITO HOST

HUMAN HOST

gametocytes

zygote

fertilization

gametocytes differentiate into gametes

mosquito sucks blood and picks up gametocytes

(A)

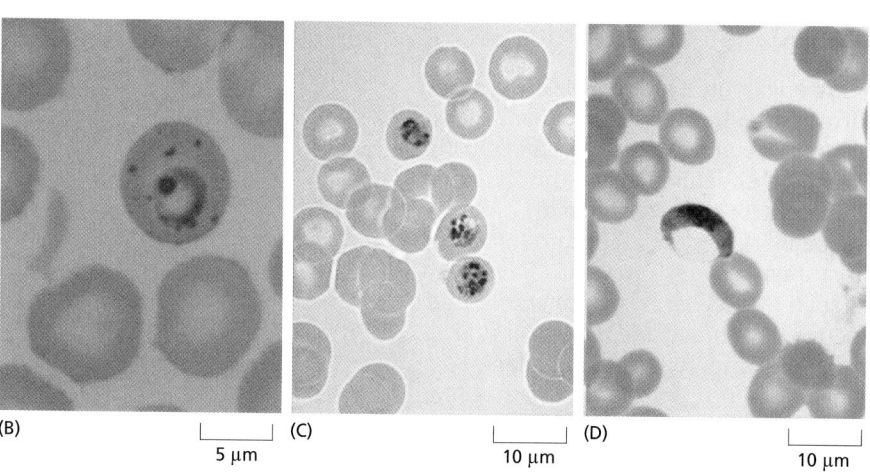

(B)   5 μm   (C)   10 μm   (D)   10 μm

Figure 23–9 The complex life cycle of malaria parasites. (A) The sexual cycle of *Plasmodium falciparum* requires passage between a human host and an insect host (Movie 23.3). (B)–(D) Blood smears from people with malaria, showing three different forms of the parasite that appear in red blood cells: (B) ring stage; (C) schizont; and (D) gametocyte. (B–D, courtesy of the Centers for Disease Control, Division of Parasitic Diseases, DPDx.)

Plasmodium falciparum causes the most serious form of malaria and is the most intensively studied of the malaria-causing parasites. It exists in many distinct forms, and it requires both the human and mosquito hosts to complete its sexual cycle (Figure 23–9). Several of these forms are highly specialized to invade and replicate in specific tissues—the lining of the insect gut, the human liver, and the human red blood cell. Even within a single host cell type, the red blood cell, the *Plasmodium* parasite undergoes a complex sequence of developmental events, reflected in striking morphological changes (Figure 23–9B–D).

All Aspects of Viral Propagation Depend on Host Cell Machinery

Bacteria, fungal, and protozoan pathogens are living cells themselves. They use their own machinery for DNA replication, transcription, and translation, and, for the most part, they provide their own sources of metabolic energy. **Viruses**, by contrast, are the ultimate hitchhikers, carrying little more than information in the form of nucleic acid. Most clinically important human viruses have small genomes consisting of double-stranded DNA or single-stranded RNA (Table 23–1), and we now have complete genome sequences of almost all of them.

Viral genomes typically encode three types of protein: proteins for replicating the genome, proteins for packaging the genome and delivering it to more host cells, and proteins for modifying the structure or function of the host cell to enhance the replication of the virus (see Figure 7–62). In general, viral replication involves (1) entry into the host cell, (2) disassembly of the infectious virus particle, (3) replication of the viral genome, (4) transcription of viral genes and synthesis of viral proteins, (5) assembly of these viral components into progeny virus particles,

TABLE 23–1 Viruses That Cause Human Disease

Virus	Genome type	Disease
Herpes simplex virus 1	Double-stranded DNA	Recurrent cold sores
Epstein–Barr virus (EBV)	Double-stranded DNA	Infectious mononucleosis
Varicella-zoster virus	Double-stranded DNA	Chickenpox and shingles
Smallpox virus (Variola)	Double-stranded DNA	Smallpox
Human papillomavirus	Double-stranded DNA	Warts, cancer
Adenovirus	Double-stranded DNA	Respiratory disease
Hepatitis-B virus	Part single-, part double-stranded DNA	Hepatitis B
Human immunodeficiency virus (HIV-1)	Single-stranded RNA [+] strand	Acquired immune deficiency syndrome (AIDS)
Poliovirus	Single-stranded RNA [+] strand	Poliomyelitis
Rhinovirus	Single-stranded RNA [+] strand	Common cold
Hepatitis-A virus	Single-stranded RNA [+] strand	Hepatitis A
Hepatitis-C virus	Single-stranded RNA [+] strand	Hepatitis C
Yellow fever virus	Single-stranded RNA [+] strand	Yellow fever
Coronavirus	Single-stranded RNA [+] strand	Common cold, respiratory disease
Rabies virus	Single-stranded RNA [−] strand	Rabies
Mumps virus	Single-stranded RNA [−] strand	Mumps
Measles virus	Single-stranded RNA [−] strand	Measles
Influenza virus type A	Single-stranded RNA [−] strand	Respiratory disease (flu)

Figure 23–10 **A simple viral life cycle.** The hypothetical simple virus shown here consists of a small double-stranded DNA molecule that codes for only a single viral capsid protein. To reproduce, the viral genome must first enter a host cell, where it is replicated to produce multiple copies, which are transcribed and translated to produce the viral coat protein. The viral genomes can then assemble spontaneously with the coat protein to form a new virus particle, which escapes from the host cell. No known virus is this simple.

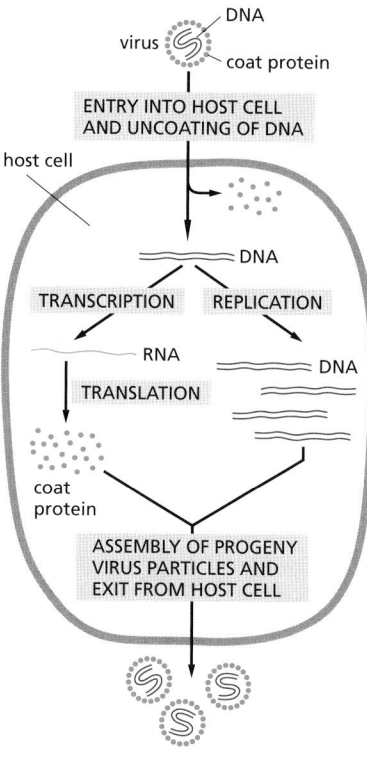

and (6) release of progeny virions (**Figure 23–10**). A single virus particle (a *virion*) that infects a single host cell can produce thousands of progeny.

Virions come in a wide variety of shapes and sizes (**Figure 23–11**), and although most have relatively small genomes, genome size can vary considerably. The recently discovered giant viruses of amoebae, called *pandoraviruses*, are the largest known viruses, with 700 nm particles and double-stranded DNA genomes of over 2,000,000 nucleotide pairs. The virions of *poxvirus* are also large: they are 250–350 nm long and enclose a genome of double-stranded DNA of about 270,000 nucleotide pairs. At the other end of the size scale are the virions of *parvovirus*, which are less than 30 nm in diameter and have a single-stranded DNA genome of fewer than 5000 nucleotides.

Viral genomes are packaged in a protein coat, called a **capsid**, which in some viruses is further enclosed by a lipid bilayer membrane, or envelope. The capsid is made of one or several proteins, arranged in regular arrays that generally produce structures with either helical symmetry, which results in a cylindrical structure (for example, influenza, measles, and bunyavirus), or icosahedral symmetry (for example, poliovirus and herpesvirus; see Figure 23–11). Some viruses instead produce capsids with more complicated structures (for example, poxviruses). When the capsid is packaged with the viral genome, the structure is called a *nucleocapsid*. The nucleocapsids of *nonenveloped viruses* usually leave an infected cell by lysing it. For *enveloped viruses*, by contrast, the nucleocapsid is enclosed within a lipid bilayer membrane that the virus acquires in the process of budding from the host-cell plasma membrane, which it does without disrupting the membrane or killing the cell (**Figure 23–12**). Enveloped viruses can cause persistent infections that may last for years, often without noticeable deleterious effects on the host.

Because the host cell performs most of the critical steps in viral replication, the identification of effective antiviral drugs that do not harm the host can be difficult. Probably the most effective strategy for containing viral diseases is through vaccinating of potential hosts. Highly successful vaccination programs have effectively

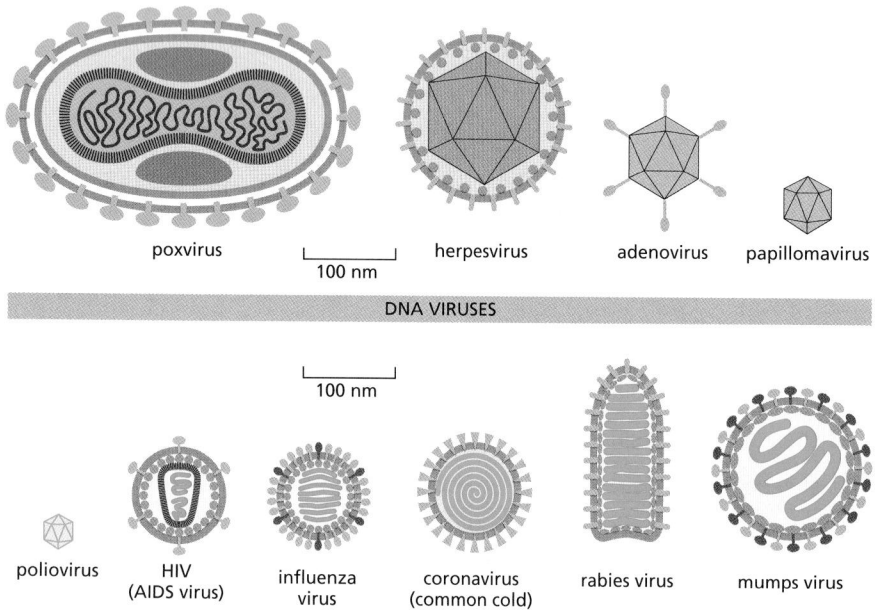

Figure 23–11 **Examples of viral morphology.** As shown, both DNA and RNA viruses vary greatly in both size and shape.

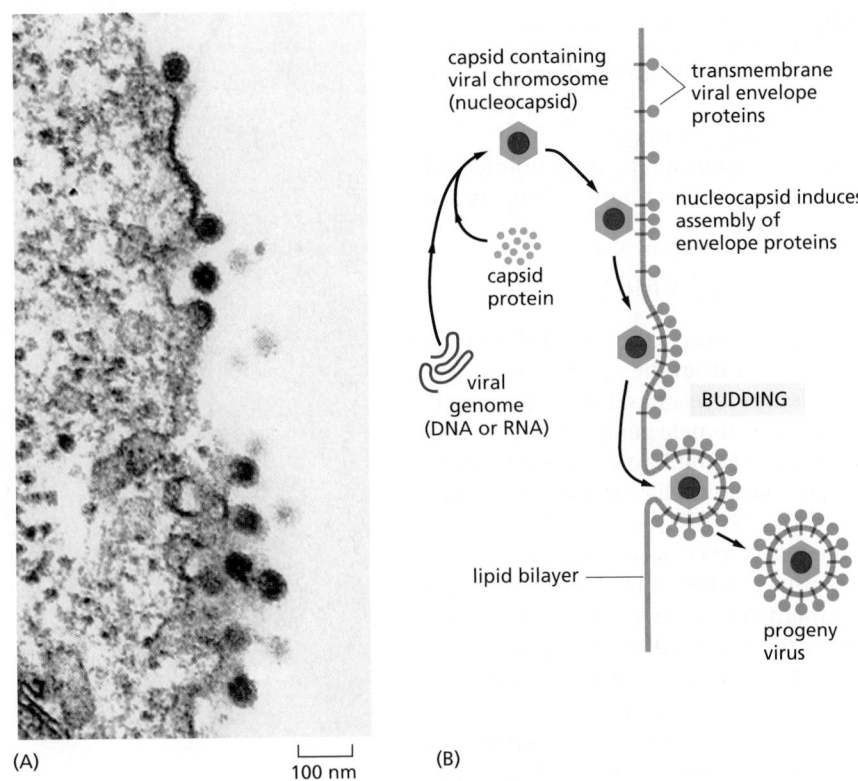

capsid containing
viral chromosome
(nucleocapsid)

transmembrane
viral envelope
proteins

nucleocapsid induces
assembly of
envelope proteins

capsid
protein

viral
genome
(DNA or RNA)

BUDDING

lipid bilayer

progeny
virus

(A)   100 nm   (B)

Figure 23–12 **Acquisition of a viral envelope.** (A) Electron micrograph of an animal cell from which six copies of an enveloped virus (*Semliki forest virus*) are budding. (B) Schematic drawing of the envelope assembly and budding processes. The lipid bilayer that surrounds the viral capsid is derived directly from the plasma membrane of the host cell. In contrast, the proteins in this lipid bilayer (shown in *green*) are encoded by the viral genome. (A, from Loewy, A, et al., *J. Virol.* 69:469–75, 1995. Reproduced with permission from the American Society for Microbiology.)

eliminated smallpox infection from the planet, and the eradication of poliomyelitis is approaching completion (**Figure 23–13**).

Summary

Infectious diseases are caused by pathogens, which include viruses, bacteria, and fungi, as well as protozoan and metazoan parasites. All pathogens must have mechanisms for entering their host and for evading immediate destruction by the host. The great majority of bacteria are not pathogenic to humans. Those that are pathogenic produce specific virulence factors that mediate the bacteria's interactions with the host; these proteins change the behavior of host cells in ways that promote the replication and spread of the bacteria. Eukaryotic pathogens such as fungi and protozoan parasites typically pass through several different forms during the course of infection; the ability to switch among these forms is usually required for these pathogens to survive in a host and cause disease. In some cases, such as malaria, parasites must pass sequentially through several host species to complete their life cycles. Unlike bacteria and eukaryotic parasites, viruses have no metabolism of their own and no intrinsic ability to produce the proteins encoded by their DNA or RNA genomes; they rely on subverting the machinery of the host cell.

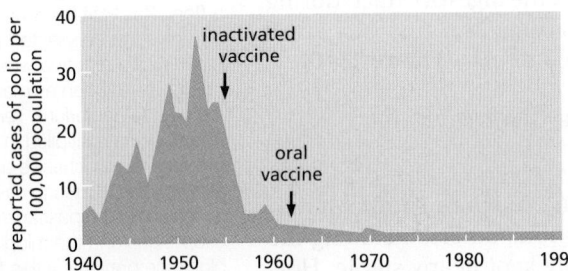

Figure 23–13 **Effective control of a viral disease through vaccination.** The graph shows the number of cases of poliomyelitis reported per year in the United States. The *arrows* indicate the timing of the introduction of the Salk vaccine (inactivated virus given by injection) and the Sabin vaccine (live attenuated virus given orally).

CELL BIOLOGY OF INFECTION

The mechanisms through which pathogens cause disease are as diverse as the pathogens themselves. Nonetheless, all pathogens must carry out certain common tasks: they must gain access to the host, reach an appropriate niche, avoid host defenses, replicate, and exit from the infected host to spread to an uninfected one. In this section, we examine the common strategies that many pathogens use to accomplish these tasks.

Pathogens Overcome Epithelial Barriers to Infect the Host

The first step in infection is for the pathogen to gain access to the host. A thick covering of skin protects most parts of the human body from the environment. The protective boundaries of some other human tissues (eyes, nasal passages, respiratory tract, mouth, digestive tract, urinary tract, and female genital tract) are less robust. In the lungs and small intestine, for example, the barrier is just a single monolayer of epithelial cells. Nonetheless, all these epithelia serve as barriers to infection.

Wounds in barrier epithelia allow pathogens direct access to unoccupied niches within otherwise sterile host tissues. This avenue of entry requires little in the way of pathogen specialization, and many members of the normal flora can cause serious illness if they enter through such wounds. *Staphylococci* from the skin and nose, or *Streptococci* from the throat and mouth, are two examples of opportunistic bacterial pathogens that are responsible for many serious infections resulting from breaches in epithelial barriers. The recent emergence of bacterial strains of *Staphylococcus* that are resistant to the antibiotics commonly used for treatment (for example, methicillin-resistant *Staphylococcus aureus*, or MRSA, which infects up to 50,000,000 people worldwide) is of particular concern. Papillomaviruses, which cause warts and cervical cancer, also take advantage of breaches in epithelial barriers.

Primary pathogens, however, need not wait for a wound to gain access to their host. One efficient way for such a pathogen to cross the skin is to catch a ride in the saliva of a biting arthropod. A diverse group of bacteria, viruses, and protozoa has developed the ability to survive in insects and then use them as *vectors* to spread from one mammalian host to another. As discussed earlier, the *Plasmodium* protozoan that causes malaria develops through several forms in its life cycle, including some that are specialized for survival in a human and others that are specialized for survival in a mosquito (see Figure 23–9). Viruses that are spread by insect bites cause yellow fever and Dengue fever, as well as many kinds of viral encephalitis (inflammation of the brain). These viruses replicate in both insect cells and mammalian cells, as required for their transmission by an insect vector.

The efficient spread of a pathogen via an insect vector requires that an individual insect consumes a blood meal from an infected host and transfers the pathogen to a naive host. In a few striking cases, the pathogen alters the behavior of the insect so that its transmission to a new host is more likely. An example is the bacterium *Yersinia pestis*, which causes bubonic plague. It multiplies in the flea's foregut to form aggregated masses that physically block the digestive tract; during each repeated, but futile, attempt at feeding, some of the bacteria in the foregut are flushed into the bite site, thus transmitting plague to a new host (**Figure 23–14**).

Pathogens That Colonize an Epithelium Must Overcome Its Protective Mechanisms

Whereas many epithelial barriers such as the skin and the lining of the mouth and large intestine are densely populated by normal flora, others, including the lining of the lower lung and the bladder, are normally kept nearly sterile. How do these epithelia avoid bacterial colonization? A layer of protective mucus covers the respiratory epithelium, and the coordinated beating of cilia sweeps the mucus and trapped bacteria up and out of the lung. The epithelial lining of the bladder and the upper gastrointestinal tract also has a thick layer of mucus, and

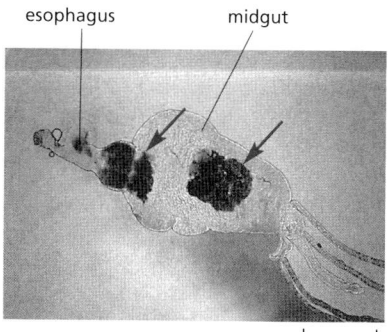

esophagus   midgut

⊢——————⊣
100 µm

Figure 23–14 **Plague bacteria within a flea.** This light micrograph shows the digestive tract dissected from a flea that had dined about two weeks previously on the blood of an animal infected with the plague bacterium, *Yersinia pestis*. The bacteria multiplied in the flea gut to produce large cohesive aggregates *(red arrows)*; the bacterial mass on the left is occluding the passage between the esophagus and the midgut. This type of blockage prevents a flea from digesting its blood meals, so that hunger causes it to bite repeatedly, disseminating the infection. (From B.J. Hinnebusch, E.R. Fischer and T.G. Schwan, *J. Infect. Dis.* 178:1406–1415, 1998.)

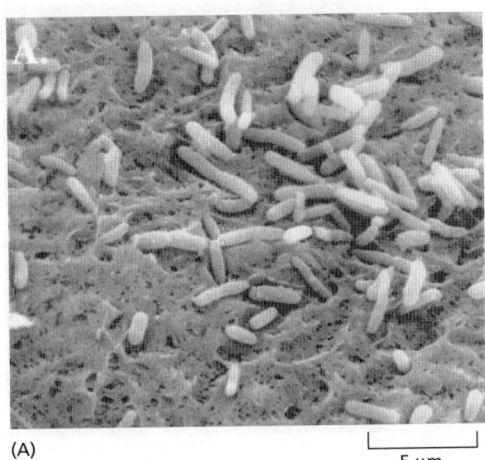

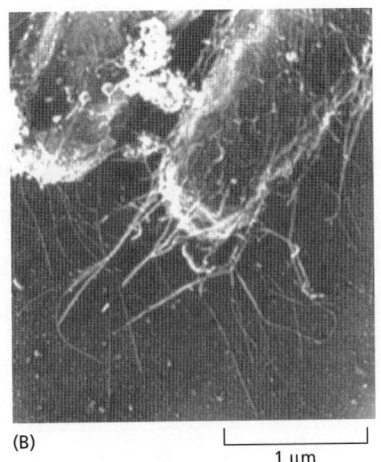

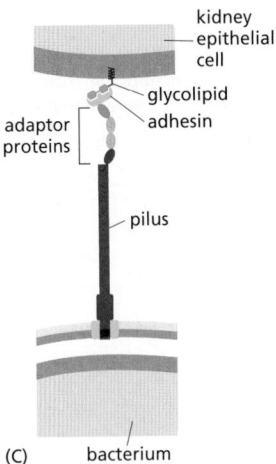

(A)

5 μm

(B)

1 μm

(C)

these organs are periodically flushed by urination and by peristalsis, respectively, which washes away most microbes.

Pathogenic bacteria and eukaryotic parasites that infect these epithelial surfaces have evolved specific mechanisms for overcoming these protective mechanisms. Those that infect the urinary tract, for example, adhere tightly to the epithelial lining via specific **adhesins**, which are proteins or protein complexes that recognize and bind to cell-surface molecules on the epithelium. An important group of adhesins in *E. coli* strains that infect the kidney are components of the *pili*—surface projections that can be several micrometers long and thus able to span the thickness of the protective mucus layer; at the tip of each pilus is an adhesin protein that binds tightly to the D-galactose–D-galactose disaccharide on glycolipids on the surface of kidney cells (**Figure 23–15**). Strains of *E. coli* that infect the bladder rather than the kidney express a second kind of pilus with a different adhesin protein that binds to bladder epithelial cells. It is the specificity of the adhesin proteins on the tips of the two types of pili that is responsible for the bacteria's colonizing of the different parts of the urinary tract.

The stomach is an especially hostile environment for pathogens. Besides the thick layer of mucus and peristaltic washing, it is filled with acid (average pH ≈2), which is lethal to almost all bacteria ingested in food. Yet, it is home to a microbiota of hundreds of resident species, including the bacterium *H. pylori*, which, as we discussed earlier, is the major cause of stomach ulcers and some stomach cancers. The hypothesis that a persistent bacterial infection could cause stomach ulcers was initially met with skepticism. The young Australian doctor who made the initial discovery finally proved the point: he drank a pure culture of *H. pylori* and developed inflammation of the stomach, which often precedes the development of ulcers. A short course of antibiotics can now effectively cure a patient of recurrent stomach ulcers. Remarkably, *H. pylori* is able to persist for life as a commensal in most humans. One way in which it survives in the stomach is by producing the enzyme *urease*, which converts urea to ammonia that neutralizes the acid in its immediate vicinity. The bacterium also uses its flagellum for chemotactic motility, allowing it to seek out the more neutral pH near the surface of gastric epithelial cells. *H. pylori* virulence proteins that target both epithelial and immune cells help *H. pylori* persist in the stomach, but they can also induce chronic inflammation, alteration in host gene expression, changes in cell proliferation and apoptosis, and disruption of cell–cell junctions, all of which are predisposing factors for stomach cancer.

Extracellular Pathogens Disturb Host Cells Without Entering Them

Extracellular pathogens can cause serious disease without entering host cells. *Bordetella pertussis*, the bacterium that causes whooping cough, for example, colonizes the respiratory epithelium and circumvents the normal mechanism that

Figure 23–15 **Pathogenic *E. coli* in the infected bladder of a mouse.** (A) Scanning electron micrograph of uropathogenic *E. coli*, a common cause of bladder and kidney infections. The bacteria are attached to the surface of epithelial cells lining the infected bladder. (B) A close-up view of one of the bacteria showing the pili on its surface. (C) An *E. coli* pilus has adaptor proteins on its tip that bind to glycolipids on the surface of kidney cells. (A, from G.E. Soto and S.J. Hultgren, *J. Bacteriol.* 181:1059–1071, 1999. With permission from the American Society for Microbiology; B, from D.G. Thanassi and S.J. Hultgren, *Methods* 20:111–126, 2000. With permission from Elsevier.)

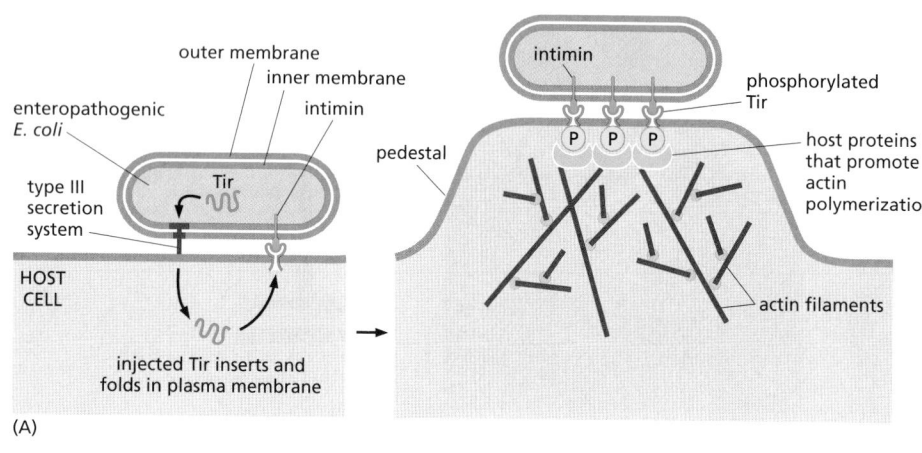

(A)

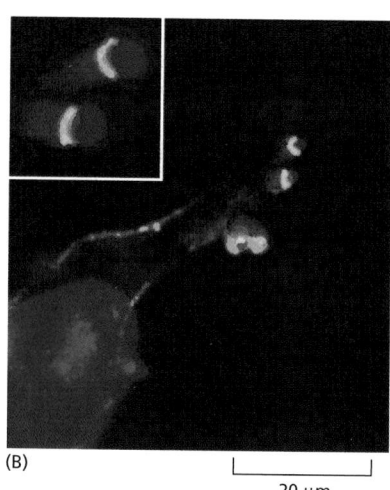

(B)

20 µm

clears the respiratory tract by expressing adhesins that bind ciliated epithelial cells. The adherent bacteria produce toxins that eventually kill the ciliated cells, compromising the host's ability to clear the infection. The most familiar of these is *pertussis toxin*, which, like the cholera toxin discused above, has an A subunit that ADP-ribosylates the α subunit of the G protein G_i, inhibiting the G protein from suppressing the activity of the host cell's adenylyl cyclase, thereby increasing the production of cyclic AMP (see Figure 23–6). This toxin also interferes with the chemotactic pathway that neutrophils use to seek out and destroy invading bacteria (see Figures 16–3 and 16–86). *B. pertussis* colonization of the respiratory tract causes severe coughing, which helps spread the infection.

Not all extracellular pathogens that colonize an epithelium exert their effect through toxins. Enteropathogenic *E. coli* (EPEC), which causes diarrhea in young children, uses a type III secretion system (see Figure 23–7) to deliver its own special receptor protein (called *Tir*) into the plasma membrane of a host intestinal epithelial cell (**Figure 23–16**). The extracellular domain of Tir binds to the bacterial surface protein *intimin*, triggering actin polymerization in the host cell that results in the formation of a unique cell-surface protrusion called a *pedestal*; this pushes the tightly adherent bacteria up about 10 µm from the host-cell membrane, thereby promoting bacterial movement along the cell surface. A similar strategy is used by vaccinia virus (the virus that was used as a vaccine to eradicate smallpox) to form mobile pedestals, which promote spread of the virus from cell to cell. The study of how EPEC and vaccinia virus promote actin polymerization has been of major importance in understanding how intracellular signaling pathways regulate the cytoskeleton in normal, uninfected cells (discussed in Chapter 16). Although pedestal formation promotes the spread of these pathogens, the sympoms of EPEC infection (severe diarrhea) are caused by the loss of absorptive microvilli and disruption of signaling pathways in epithelial cells, which are triggered by Tir and other effector proteins.

Intracellular Pathogens Have Mechanisms for Both Entering and Leaving Host Cells

Many pathogens have to enter host cells to cause disease. These **intracellular pathogens** include all viruses and many bacteria and protozoa. Each of these has a preferred niche for replication and survival within host cells. Bacteria and protozoa replicate either in the cytosol or within a membrane-enclosed compartment. While most RNA viruses replicate within the cytosol, most DNA viruses replicate in the nucleus. Life inside a host cell has several advantages. The pathogens are not accessible to *antibodies*, nor are they easy targets for phagocytic cells (discussed in Chapter 24); furthermore, intracellular bacteria and protozoa are bathed in a rich source of nutrients, and viruses have access to the host cell's biosynthetic

Figure 23–16 Interaction of enteropathogenic *E. coli* (EPEC) with host intestinal epithelial cells. (A) When EPEC contacts an epithelial cell in the lining of the human gut, it delivers a bacterial protein called Tir into the host cell through a type III secretion system. Tir then inserts into the plasma membrane of the host cell, where it functions as a receptor for the bacterial adhesin protein intimin. Next, a host-cell protein tyrosine kinase phosphorylates the intracellular domain of Tir on tyrosines. Phosphorylated Tir recruits host-cell proteins (including an adaptor protein, a WASp protein, and the Arp 2/3 complex) that trigger actin polymerization (see Figure 16–16). Consequently, a branched network of actin filaments assembles underneath the bacterium, forming an actin pedestal (Movie 23.4). (B) EPEC on a pedestal. In this fluorescence micrograph, the DNA of the EPEC and host cell is labeled in *blue*, Tir protein is labeled in *green*, and host-cell actin filaments are labeled in *red*. The inset shows a close-up view of the two upper bacteria on pedestals. (B, from D. Goosney et al., *Annu. Rev. Cell Dev. Biol.* 16:173–189, 2000. With permission from Annual Reviews.)

machinery for their reproduction. This lifestyle, however, requires that the pathogen have mechanisms for entering host cells, for finding a suitable subcellular niche where it can replicate, and for exiting from the infected cell to spread the infection. Below we consider some of the myriad ways that individual intracellular pathogens exploit and modify host cell biology to satisfy these requirements.

Viruses Bind to Virus Receptors at the Host Cell Surface

The first step for any intracellular pathogen is to bind to the surface of the host target cell. Viruses accomplish this by the binding of viral surface proteins to **virus receptors** displayed on the host cell. The first virus receptor identified was an *E. coli* surface protein that is recognized by the bacteriophage lambda; the protein normally functions to transport the sugar maltose from outside the bacterium to the inside where it is used as an energy source. Receptors need not be proteins, however: an envelope protein of herpes simplex virus, for example, binds to heparan sulfate proteoglycans (discussed in Chapter 19) on the surface of certain vertebrate host cells, and simian virus 40 (SV40) binds to a glycolipid. The specificity of virus-receptor interactions often serves as a barrier preventing the spread of a virus from one species to another. Acquiring the ability to bind to a new receptor often requires multiple changes in a virus, but it can be crucial in allowing the cross-species transmission that can result in new disease outbreaks.

Viruses that infect animal cells generally exploit cell-surface receptor molecules that are either ubiquitous (such as the sialic-acid-containing oligosaccharides used by the influenza virus) or found uniquely on those cell types in which the virus replicates (such as the neuron-specific proteins used by rabies virus). Although a virus usually uses a single type of host-cell receptor, some viruses use more than one type. An important example is HIV-1, which requires two types of receptors to enter a host cell. Its primary receptor is CD4, a cell-surface protein on helper T cells and macrophages that is involved in immune recognition (discussed in Chapter 24). It also requires a co-receptor, which is either CCR5 (a receptor for β-chemokines) or CXCR4 (a receptor for α-chemokines), depending on the particular variant of the virus; macrophages are susceptible only to HIV variants that use CCR5 for entry, whereas helper T cells are most efficiently infected by variants that use CXCR4 (**Figure 23–17**). The viruses that are found within the first few months after HIV infection almost invariably use CCR5, which explains why individuals that carry a defective *CCR5* gene are less susceptible to HIV infection. In the later stages of infection, viruses often either switch to use CXCR4 or adapt to use both co-receptors through the accumulation of mutations; in this way, the virus can change the cell types it infects as the disease progresses. It may seem paradoxical that viruses would infect immune cells, as we might expect that virus binding would trigger an immune response; but invasion of an immune cell can be a useful way for a virus to weaken the immune response and travel around the body to infect other immune cells.

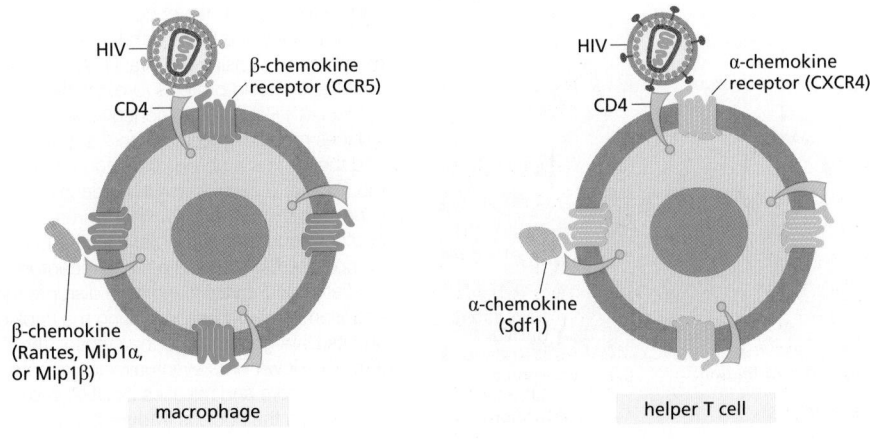

Figure 23–17 Receptor and co-receptors for HIV. All strains of HIV require the CD4 protein as a primary receptor. Early in an infection, most of the viruses use CCR5 as a co-receptor, allowing them to infect macrophages and their precursors, monocytes. As the infection progresses, mutant variants arise that now use CXCR4 as a co-receptor, enabling them to infect helper T cells efficiently. The natural ligand for the chemokine receptors (Sdf1 for CXCR4; Rantes, Mip1α, or Mip1β for CCR5) blocks co-receptor function and prevents viral invasion.

Viruses Enter Host Cells by Membrane Fusion, Pore Formation, or Membrane Disruption

After recognition and attachment to the host cell surface, the virus must enter the cell to replicate. Some **enveloped viruses** enter the host cell by fusing their envelope membrane with the plasma membrane. Most viruses, whether enveloped or nonenveloped, activate signaling pathways in the cell that induce endocytosis, commonly via clathrin-coated pits (see Figure 13–7), leading to internalization into endosomes. Large viruses that do not fit into clathrin-coated vesicles, such as poxviruses, often enter cells by *macropinocytosis*, a process by which membrane ruffles fold over and entrap fluid into macropinosomes (see Figure 13–50). Once inside endosomes, fusion of the viral envelope occurs from the lumenal side of the endosome membrane. The mechanism of membrane fusion mediated by viral spike glycoproteins has similarities with SNARE-mediated membrane fusion during normal vesicular trafficking (discussed in Chapter 13).

Enveloped viruses regulate fusion both to ensure that they fuse only with the appropriate host cell membrane and to prevent fusion with one another. For viruses such as HIV-1 that fuse at neutral pH with the plasma membrane (**Figure 23–18**A), binding to receptors or co-receptors usually triggers a conformational change in a viral envelope protein that exposes a normally buried fusion peptide (see Figure 13–21). Other enveloped viruses, such as influenza A virus, only fuse with a host cell membrane after endocytosis (Figure 23–18B); in this case, it is frequently the acid environment in the late endosome that triggers the conformational change in a viral surface protein that exposes the fusion peptide. The H$^+$

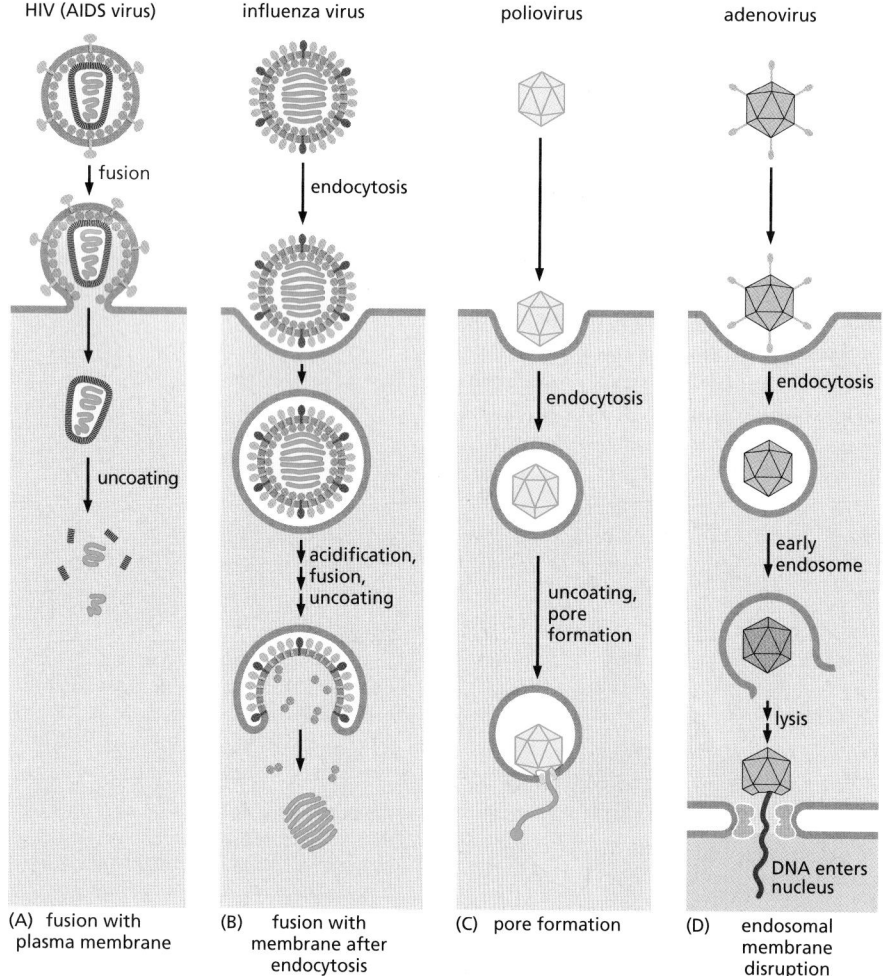

Figure 23–18 Four virus entry strategies.
(A) Some enveloped viruses, such as HIV, fuse directly with the host-cell plasma membrane to release their RNA genome *(blue)* and capsid proteins *(brown)* into the cytosol. (B) Other enveloped viruses, such as influenza virus, first bind to cell-surface receptors, triggering receptor-mediated endocytosis; when the endosome acidifies, the virus envelope fuses with the endosomal membrane, releasing the viral RNA genome *(blue)* and capsid proteins *(brown)* into the cytosol. (C) Poliovirus, a nonenveloped virus, induces receptor-mediated endocytosis, and then forms a pore in the endosomal membrane to extrude its RNA genome *(blue)* into the cytosol. (D) Adenovirus, another nonenveloped virus, uses a more complicated strategy: it induces receptor-mediated endocytosis and then disrupts the endosomal membrane, releasing the capsid and its DNA genome into the cytosol; the trimmed-down virus eventually docks onto a nuclear pore and releases its DNA *(red)* directly into the nucleus (Movie 23.5).

HIV (AIDS virus)   influenza virus   poliovirus   adenovirus

(A) fusion with plasma membrane

(B) fusion with membrane after endocytosis

(C) pore formation

(D) endosomal membrane disruption

pumped into the early endosome also has another effect; it enters the influenza virion through an ion channel in the viral envelope and triggers changes in the viral capsid. These priming steps allow the capsids to disassemble once released into the cytosol after virus fusion with the late endosomal membrane.

Nonenveloped viruses use different strategies to enter host cells—strategies that do not rely on membrane fusion. *Poliovirus*, which causes poliomyelitis, binds to a cell-surface receptor, triggering both receptor-mediated endocytosis (see Figure 13–52) and a conformational change in the viral particle. The conformational change exposes a hydrophobic projection on one of the capsid proteins, which inserts into the endosomal membrane to form a pore. The viral RNA genome then enters the cytosol through the pore, leaving the capsid in the endosome (Figure 23–18C). Other nonenveloped viruses such as *adenovirus* disrupt the endosomal membrane after they are taken up by receptor-mediated endocytosis. One of the proteins released from the capsid lyses the endosomal membrane, releasing the remainder of the virus into the cytosol. During endosomal trafficking and subsequent transport within the cytosol, adenoviruses undergo multiple uncoating steps, which sequentially remove structural proteins and ready the virus particles to release their DNA into the nucleus through nuclear pore complexes (Figure 23–18D).

Bacteria Enter Host Cells by Phagocytosis

Bacteria are much larger than viruses—too large to be taken up either through pores or by receptor-mediated endocytosis. Instead, they enter host cells by phagocytosis, which is a normal function of phagocytes such as neutrophils, macrophages, and dendritic cells (discussed in Chapter 24). These phagocytes patrol the tissues of the body and ingest and destroy microbes; however, some intracellular bacterial pathogens such as *M. tuberculosis* use this to their advantage and have evolved to survive and multiply inside macrophages.

Some bacterial pathogens can invade host cells that are normally nonphagocytic. One way they do so is by expressing an invasion protein that binds with high affinity to a host-cell receptor, which is often a cell–cell or cell–matrix adhesion protein (discussed in Chapter 19). For example, *Yersinia pseudotuberculosis* (a bacterium that causes diarrhea and is a close relative of the plague bacterium *Y. pestis*) expresses a protein called invasin that has an RGD motif that is similar to fibronectin's and likewise is recognized by host-cell β_1 integrins (see Figure 19–55). *Listeria monocytogenes*, which causes a rare but serious form of food poisoning, invades host cells by expressing a protein that binds to the cell–cell adhesion protein E-cadherin (see Figure 19–6). For both these bacterial species, binding of the bacterial invasion proteins to the host cell adhesion proteins stimulates signaling through members of the Rho family of small GTPases (discussed in Chapter 16). This in turn activates proteins in the WASp family and the Arp 2/3 complex, leading to actin polymerization at the site of bacterial attachment. Actin polymerization, together with the assembly of a clathrin coat (see Figure 13–6), drives the advancement of the host cell's plasma membrane over the adhesive surface of the microbe, resulting in the phagocytosis of the bacterium—a process known as the *zipper mechanism* of invasion (**Figure 23–19A**).

A second pathway by which bacteria can invade nonphagocytic cells is known as the *trigger mechanism* (Figure 23–19B). It is used by various pathogens that cause food poisoning, including *Salmonella enterica*, and it is initiated when the bacterium injects a set of effector molecules into the host-cell cytosol through a type III secretion system (see Figure 23–7). Some of these effector molecules activate Rho family proteins, which in turn stimulate actin polymerization, as just discussed. Other bacterial effector proteins interact with host-cell cytoskeletal elements more directly, nucleating and stabilizing actin filaments and causing the rearrangement of actin cross-linking proteins. The overall effect is to cause the formation of localized ruffles on the surface of the host cell (Figure 23–19C and D), which fold over and engulf the bacteria by a process that resembles macropinocytosis. The appearance of cells being invaded by use of the trigger mechanism

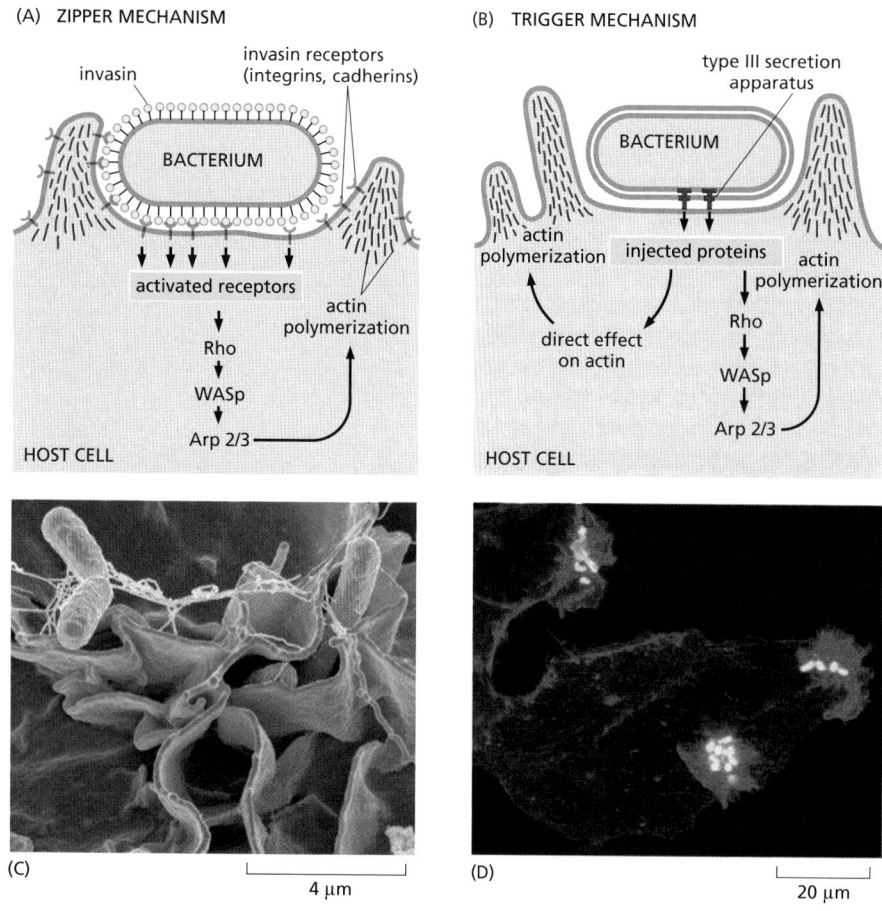

(A) ZIPPER MECHANISM

(B) TRIGGER MECHANISM

Figure 23–19 Mechanisms used by bacteria to induce phagocytosis by host cells that are normally nonphagocytic. (A) In the zipper mechanism, bacteria express an invasion protein that binds with high affinity to a host-cell receptor, which is often a cell–cell or cell–matrix adhesion protein. (B) In the trigger mechanism, bacteria inject a set of effector molecules into the host-cell cytosol through a type III secretion system called SPI1 (*Salmonella* pathogenicity island 1), inducing membrane ruffling. Both the zipper and trigger mechanisms cause the polymerization of actin at the site of bacterial attachment by activating Rho family small GTPases and the Arp 2/3 complex. (C) A scanning electron micrograph showing a very early stage of *Salmonella enterica* invasion by the trigger mechanism. Bacteria (pseudocolored *yellow*) are shown surrounded by a small membrane ruffle. (D) Fluorescence micrograph showing that the large ruffles that engulf the *Salmonella* bacteria are actin-rich. The bacteria are labeled in *green* and actin filaments in *red*; because of the color overlap, the bacteria appear *yellow*. (C, from Rocky Mountain Laboratories, NIAID, NIH; D, from J.E. Galán, *Annu. Rev. Cell Dev. Biol.* 17:53–86, 2001. With permission from Annual Reviews.)

(C)   4 µm

(D)   20 µm

is similar to the ruffling induced by some extracellular growth factors, suggesting that the bacteria exploit normal intracellular signaling pathways.

Intracellular Eukaryotic Parasites Actively Invade Host Cells

The uptake of viruses and bacteria into host cells is carried out largely by the host, with the pathogen being a relatively passive participant. In contrast, intracellular eukaryotic parasites, which are typically much larger than other types of intracellular pathogens, invade host cells through a variety of complex pathways that usually require energy expenditure by the parasite.

Toxoplasma gondii, a cat parasite that also causes occasional serious human infections, is an example. When this protozoan contacts a host cell, it protrudes an unusual microtubule-based structure called a *conoid*, which facilitates entry into the host cell (**Figure 23–20**). The energy for invasion seems to come from actin polymerization in the parasite rather than host cytoskeleton, and invasion also requires at least one unusual parasite myosin motor protein (Class XIV; see Figure 16–40). At the point of contact, the parasite discharges effector proteins from secretory organelles into the host cell, and these proteins target various host pathways to enable invasion, to block an innate immune response, and promote survival. As the parasite moves into the host cell, a membrane derived from the host-cell plasma membrane surrounds it. Remarkably, the parasite removes host transmembrane proteins from the surrounding membrane as it forms, so that the parasite is protected in a membrane-enclosed compartment that does not fuse with lysosomes and does not participate in host-cell membrane trafficking processes (see Figure 23–20). The specialized membrane is selectively porous: it allows the parasite to take up small metabolic intermediates and nutrients from the host cell's cytosol but excludes macromolecules. Malaria parasites invade human red blood cells using a similar mechanism.

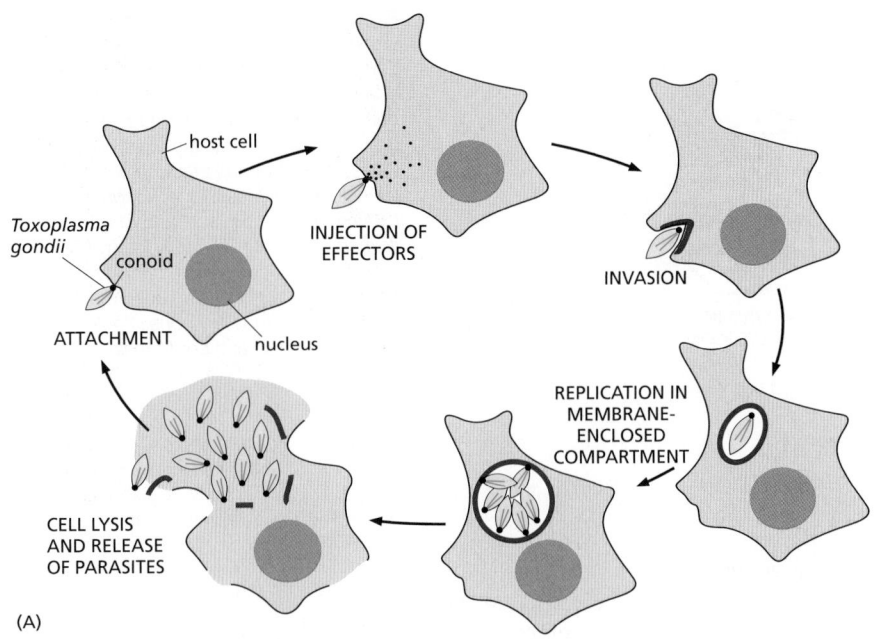

(A)

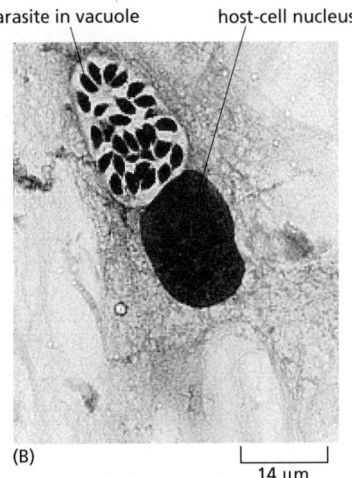

(B)

Figure 23–20 **The life cycle of the intracellular parasite *Toxoplasma gondii*.** (A) After attachment to a host cell, *T. gondii* uses its conoid to inject effector proteins that facilitate invasion. As the host cell's plasma membrane invaginates to surround the parasite, it somehow removes the normal host-cell membrane proteins, so that the compartment (shown in *red*) does not fuse with lysosomes. After several rounds of replication, the parasite causes the compartment to break down and the host cell to lyse, releasing the progeny parasites to infect other host cells (**Movie 23.6**). (B) Light micrograph of *T. gondii* replicating within a membrane-enclosed compartment (a vacuole) in a cultured cell. (B, courtesy of Manuel Camps and John Boothroyd.)

The protozoan *Trypanosoma cruzi*, which causes Chagas disease in Mexico and Central and South America, uses two alternative invasion strategies. In a *lysosome-dependent pathway*, the parasite attaches to host cell-surface receptors, inducing a local increase in Ca^{2+} in the host cell's cytosol. The Ca^{2+} signal recruits lysosomes to the site of parasite attachment, and the lysosomes fuse with the host cell's plasma membrane, allowing the parasites rapid access to the lysosomal compartment (**Figure 23–21**). In a *lysosome-independent pathway*, the parasite penetrates the host-cell plasma membrane by inducing the membrane to invaginate, without the recruitment of lysosomes.

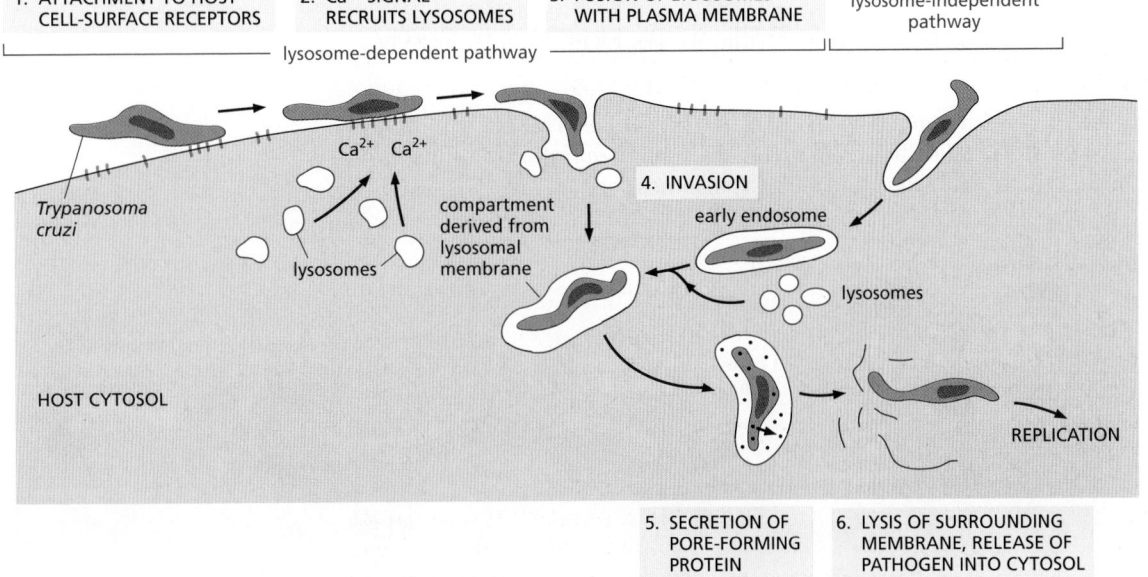

Figure 23–21 **The two alternative strategies that *Trypanosoma cruzi* uses to invade host cells.** In the lysosome-dependent pathway *(left)*, *T. cruzi* recruits host-cell lysosomes to its site of attachment to the host cell. The lysosomes fuse with the invaginating plasma membrane to create an intracellular compartment constructed almost entirely of lysosomal membrane. After a brief stay in the compartment, the parasite secretes a pore-forming protein that disrupts the surrounding membrane, thereby allowing the parasite to escape into the host-cell cytosol and proliferate. In the lysosome-independent pathway *(right)*, the parasite induces the host plasma membrane to invaginate and pinch off without recruiting lysosomes; then lysosomes fuse with the endosome prior to the parasite's escape into the cytosol.

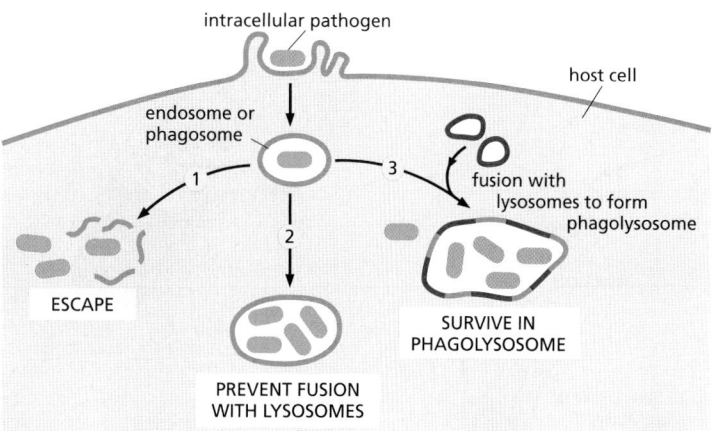

Figure 23–22 **Choices that an intracellular pathogen faces.** After entry into a host cell, generally through phagocytosis into a membrane-enclosed compartment, intracellular pathogens can use one of three strategies to survive and replicate. Pathogens that follow strategy (1) include all viruses, *Trypanosoma cruzi*, *Listeria monocytogenes*, and *Shigella flexneri*. Those that follow strategy (2) include *Mycobacterium tuberculosis* and *Legionella pneumophila*. Those that follow strategy (3) include *Salmonella enterica*, *Coxiella burnetii*, and *Leishmania*.

Some Intracellular Pathogens Escape from the Phagosome into the Cytosol

The intracellular parasites just discussed raise a general problem that faces all intracellular pathogens, including viruses, bacteria, and eukaryotic parasites: they must find a cell compartment in which they can replicate. After their endocytosis by a host cell, they usually find themselves in an endosomal compartment, which normally would fuse with lysosomes to form a *phagolysosome*—a dangerous place for pathogens. To survive, pathogens use a variety of strategies. Some escape from the endosomal compartment before such fusion. Others remain in the endosomal compartments but modify it so that it no longer fuses with lysosomes. Still others have evolved to weather the harsh conditions in the phagolysosome (**Figure 23–22**).

Trypanosoma cruzi uses the escape route by secreting a pore-forming toxin that lyses the lysosome membrane, releasing the parasite into the host cell's cytosol (see Figure 23–21). The bacterium *Listeria monocytogenes* uses a similar strategy. Following phagocytosis by the zipper mechanism, it secretes a protein called *listeriolysin O*, which disrupts the phagosomal membrane, releasing the bacteria into the cytosol (**Figure 23–23**).

Many Pathogens Alter Membrane Traffic in the Host Cell to Survive and Replicate

The survival and reproduction of many intracellular pathogens requires that they modify membrane (vesicular) traffic in the host cell. They may, for example, prevent the normal fusing of endosomes with lysosomes, or adapt themselves to

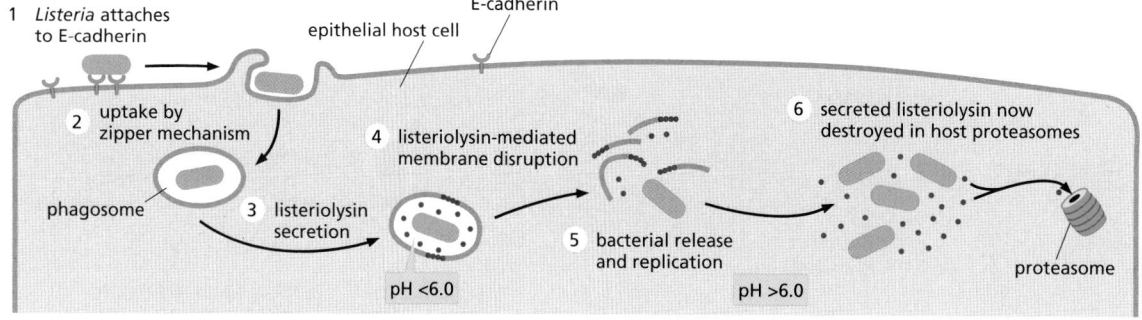

Figure 23–23 **Escape of *Listeria monocytogenes* by selective destruction of the phagosomal membrane.** The bacterium attaches to E-cadherin on the surface of host epithelial cells and induces its own uptake by the zipper mechanism (see Figure 23–19A). Within the phagosome, the bacterium secretes the protein listeriolysin O, which is activated at pH <6 and forms oligomers in the phagosome membrane, thereby creating large pores and eventually disrupting the membrane. Once in the host-cell cytosol, the bacteria begin to replicate and continue to secrete listeriolysin O; because the pH in the cytosol is >6, however, the listeriolysin O there is inactive and is also rapidly degraded by proteasomes. Thus, the host cell's plasma membrane remains intact.

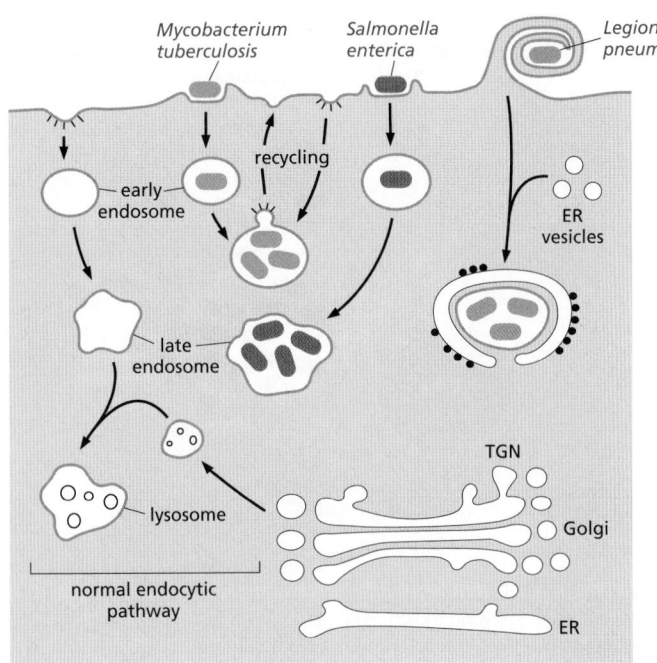

Figure 23–24 Modifications of membrane traffic in host cells by bacterial pathogens. Intracellular bacterial pathogens, including *Mycobacterium tuberculosis*, *Salmonella enterica*, and *Legionella pneumophila*, all replicate in membrane-enclosed compartments, but the compartments differ. *M. tuberculosis* remains in a compartment that has early endosomal markers and continues to communicate with the plasma membrane via transport vesicles. *S. enterica* replicates in a compartment that has late endosomal markers and does not communicate with the plasma membrane. *L. pneumophila* replicates in an unusual compartment that is wrapped in rough endoplasmic reticulum (ER) membrane and communicates with the ER via transport vesicles. TGN, *trans* Golgi network.

resist the lysosome's antimicrobial armaments. Intracellular pathogens must also provide a pathway for importing nutrients from the host cytosol into their compartment of choice.

Different pathogens have distinct strategies for altering membrane traffic in the host cell (**Figure 23–24**). *M. tuberculosis* prevents the early endosome that contains the bacteria from maturing, so the endosome never acidifies or acquires the other characteristics of a late endosome or lysosome. This strategy requires the activity of its type VII secretion system, as well as mycobacterial lipid products that mimic host lipids and influence vesicular traffic. Phagosomes containing *Salmonella enterica*, in contrast, acidify and acquire markers of late endosomes and lysosomes, but the bacteria slow the process of phagosomal maturation. They do so by injecting effector proteins through a second type III secretion system. These effectors activate host kinesin motor proteins to pull membrane tubules outward from the phagosome along cytoplasmic microtubules, forming a specialized compartment called the *Salmonella*-containing vacuole (**Figure 23–25**).

Other bacteria seem to find shelter in intracellular compartments that are distinct from those of the usual endocytic system. One example is *Legionella pneumophila*, which was first recognized as a human pathogen in 1976, when it was found to be the cause of a type of pneumonia known as **Legionnaire's disease**. *L. pneumophila* is normally a parasite of freshwater amoebae, but it is commonly

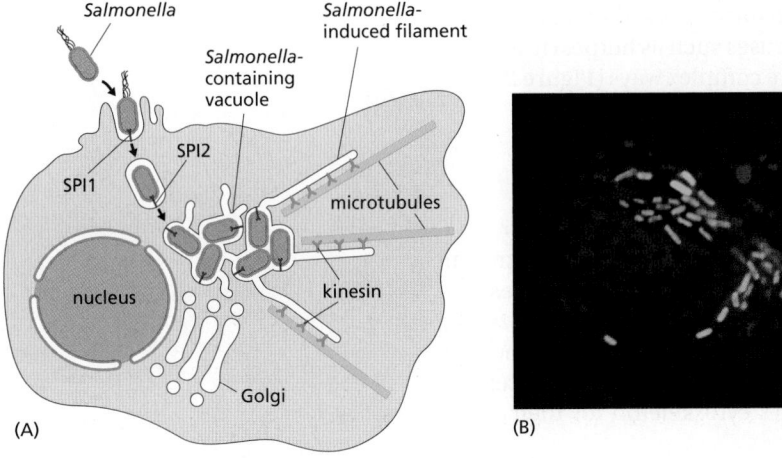

Figure 23–25 *Salmonella enterica* residing in a modified phagosomal compartment called the *Salmonella*-containing vacuole. These bacteria invade the host cell using an SPI1 type III secretion system to inject effector proteins that induce the trigger mechanism of microbe entry illustrated in Figure 23–19B. (A) Following its engulfment into a phagosome, the bacterium inactivates its SPI1 type III secretion system and activates its SPI2 type III secretion system to inject different effector proteins, which remodel the phagosome into the specialized *Salmonella*-containing vacuole. One of the injected effector proteins activates host kinesin motor proteins to pull membrane tubules outward toward the plus ends of the microtubules (see Figure 16–42). (B) Fluorescence micrograph showing *S. enterica* in a *Salmonella*-containing vacuole. The bacteria are stained *green*, the microtubules *red*, and the nucleus *blue*. (B, courtesy of Stéphane Méresse.)

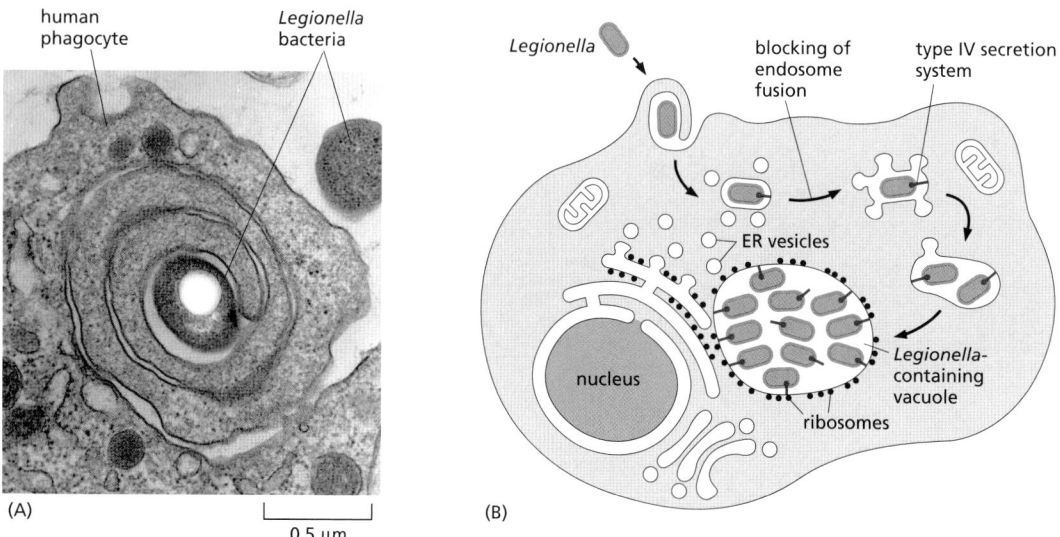

Figure 23–26 *Legionella pneumophila* **residing in a compartment with characteristics similar to those of the rough endoplasmic reticulum (ER).** (A) Electron micrograph showing the unusual coiled structure that the *Legionella pneumophila* bacterium induces on the surface of a phagocyte during the invasion process. Some other pathogens, including the bacterium *Borrelia burgdorferi*, which causes Lyme disease, the eukaryotic pathogen *Leishmania*, and the yeast *Candida albicans*, can also invade cells using this type of coiling phagocytosis. (B) Following invasion, *L. pneumophila* uses its type IV secretion system to secrete effector proteins that block phagosome–endosome fusion and phagosome maturation. It also secretes effector proteins that promote the fusion of the phagosome with ER-derived vesicles, thereby creating a *Legionella*-containing vacuole with characteristics similar to the rough ER. (A, from M.A. Horwitz, *Cell* 36:27–33, 1984. With permission from Elsevier.)

spread to humans by central air-conditioning systems, which harbor infected amoebae and produce microdroplets of water that are easily inhaled. Once in the lung, the bacteria are engulfed by macrophages by an unusual process called coiling phagocytosis (**Figure 23–26**A). *L. pneumophila* uses a type IV secretion system to inject effector proteins into the phagocyte that modulate the activity of proteins that regulate vesicular traffic, including SNARE proteins and Rab and Arf family small GTPases (discussed in Chapter 13). The effector proteins thereby prevent the phagosome from fusing with endosomes and promote its fusion with vesicles derived from the endoplasmic reticulum, converting the phagosome into a compartment that resembles the rough endoplasmic reticulum (Figure 23–26B).

Viruses can also alter membrane traffic in the host cell. Enveloped viruses make use of host cell membranes to acquire their own envelope membrane. In the simplest cases, virally encoded glycoproteins are inserted into the endoplasmic reticulum membrane and follow the secretory pathway through the Golgi apparatus to the plasma membrane; the viral capsid proteins and genome assemble into nucleocapsids, which acquire their envelope as they bud off from the plasma membrane (see Figure 23–12). This mechanism is used by many enveloped viruses including HIV-1. Other enveloped viruses such as herpesviruses and vaccinia virus acquire their lipid envelopes in more complex ways (**Figure 23–27**).

Viruses and Bacteria Use the Host-Cell Cytoskeleton for Intracellular Movement

As mentioned earlier, many pathogens escape into the cytosol rather than remaining in a membrane-enclosed compartment. The cytosol of mammalian cells is extremely viscous, as it is crowded with protein complexes, organelles, and cytoskeletal filaments, all of which inhibit the diffusion of particles the size of a bacterium or a viral nucleocapsid. Thus, to reach a particular region of the host cell a pathogen must be actively moved there. As with transport of intracellular organelles, pathogens generally use the host cell's cytoskeleton for their active movement.

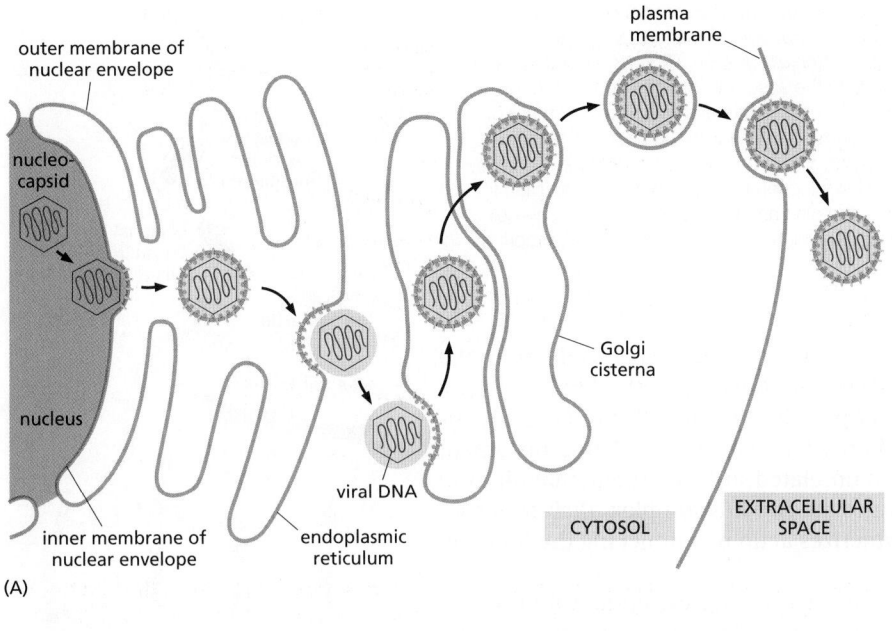

(A)

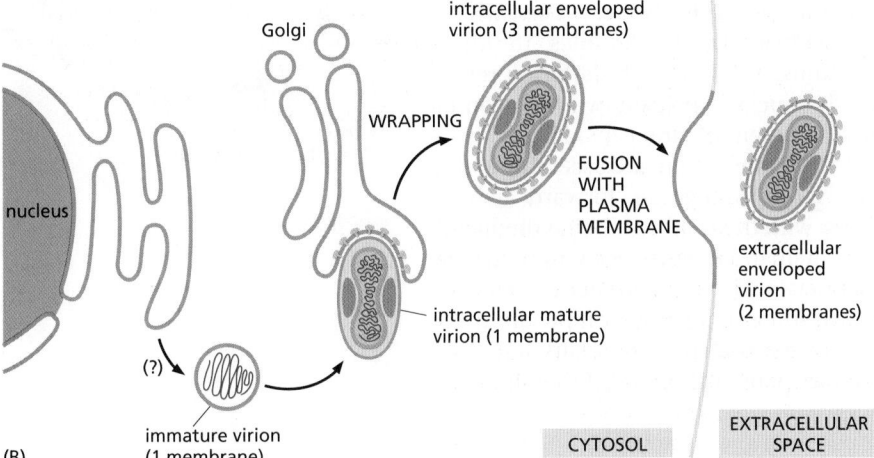

(B)

Figure 23–27 **Complex strategies for viral envelope acquisition.** (A) Herpesvirus nucleocapsids assemble in the nucleus and then bud through the inner nuclear membrane into the space between the inner and outer nuclear membranes, acquiring a lipid bilayer membrane coat. The virus particles then apparently lose this coat when they fuse with the endoplasmic reticulum membrane to escape into the cytosol. Subsequently, the nucleocapsids bud into the Golgi apparatus and bud out again on the other side, thereby acquiring two new membrane coats in the process. The virus then buds from the cell surface with a single membrane when its outer membrane fuses with the plasma membrane. (B) Vaccinia virus (which is closely related to the virus that causes smallpox and is used to vaccinate against smallpox) assembles in "replication factories" in the cytosol, far away from the plasma membrane. The immature virion, with one membrane, is then surrounded by two additional membranes, both acquired from the Golgi apparatus by a poorly understood wrapping mechanism, to form the intracellular enveloped virion. After fusion of the outermost membrane with the host-cell plasma membrane, the extracellular enveloped virion is released from the host cell.

Several bacteria that replicate in the host cell's cytosol have adopted a remarkable mechanism that depends on actin polymerization for movement. These bacteria include the human pathogens *Listeria monocytogenes*, *Shigella flexneri*, *Rickettsia rickettsii* (which causes Rocky Mountain spotted fever), and *Burkholderia pseudomallei* (which causes melioidosis, a disease characterized by severe respiratory symptoms). Baculovirus, an insect virus, also uses this mechanism for intracellular movement. All of these pathogens induce the nucleation and assembly of host-cell actin filaments at one pole of the bacterium or virus. The growing filaments generate force and push the pathogens through the cytosol at rates of up to 1 μm/sec (**Figure 23–28**). New filaments form at the rear of each pathogen and are left behind like a rocket trail as the microbe advances; the filaments depolymerize within a minute or so as they encounter depolymerizing factors in the cytosol. For *L. monocytogenes* and *S. flexneri*, the moving bacteria collide with the plasma membrane and move outward, inducing the formation of long, thin, host-cell protrusions with the bacteria at their tip. As shown in Figure 23–28, a neighboring cell often engulfs these projections, allowing the bacteria to enter the neighbor's cytoplasm without exposure to the extracellular environment, thereby avoiding antibodies produced by the host's adaptive immune system. For *B. pseudomallei*, movement and collision of the bacteria with the plasma membrane promotes cell–cell fusion, which serves a similar purpose of immune avoidance while allowing continued bacterial replication.

Figure 23–28 The actin-based movement of bacterial pathogens within and between host cells. (A) Following invasion, bacterial pathogens such as *Listeria monocytogenes*, *Shigella flexneri*, *Rickettsia rickettsii*, and *Burkholderia pseudomallei* induce the assembly of actin-rich tails in the host-cell cytoplasm, which drives rapid bacterial movement. For most of these pathogens, the moving bacteria collide with the host-cell plasma membrane to form membrane-covered protrusions, which are engulfed by neighboring cells—spreading the infection from cell to cell. In contrast, for *B. pseudomallei*, collision with the plasma membrane promotes cell–cell fusion, creating a conduit through which bacteria can invade neighboring cells (Movie 23.7).

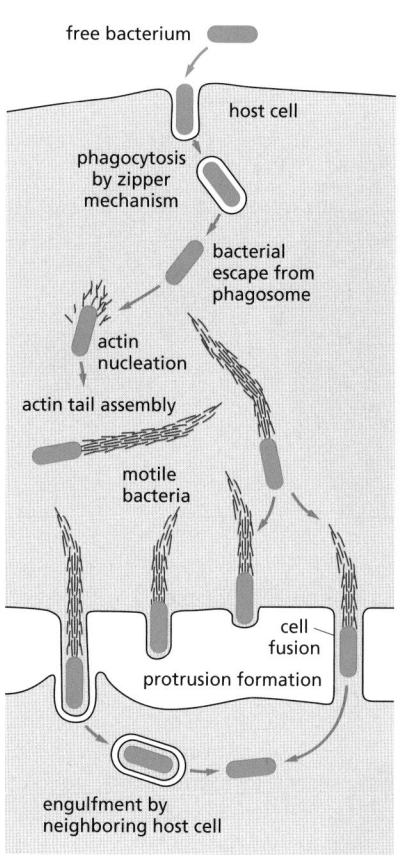

The molecular mechanisms of pathogen-induced actin assembly differ for the different pathogens, suggesting that they evolved independently (**Figure 23–29**). *L. monocytogenes* and baculovirus produce proteins that directly bind to and activate the Arp 2/3 complex to initiate the formation of an actin tail and movement (see Figure 16–16). *S. flexneri* produces an unrelated surface protein that binds to and activates N-WASp, which then activates the Arp 2/3 complex. *Rickettsia* species produce a protein that directly polymerizes actin by mimicking the function of host formin proteins (see Figure 16–17).

Many viral pathogens rely primarily on microtubule-dependent motor proteins rather than actin polymerization to move within the host-cell cytosol. Viruses that infect neurons, such as the *neurotropic alpha herpesviruses*, which include the virus that causes chickenpox, provide important examples. The virus enters sensory neurons at the tips of their axons, and microtubule-based retrograde "backward" axonal transport carries the nucleocapsids down the axon to the nucleus. The transport is mediated by attachment of viral capsid proteins to the motor protein dynein (see Figure 16–58). After replication and assembly in the nucleus, the enveloped virions are then carried by antegrade "forward" axonal transport along microtubules to the axon tips, with the transport being mediated by the attachment of a different viral capsid protein to a kinesin motor protein (see Figure 16–56). A large number of viruses associate with either dynein or kinesin motor proteins to move along microtubules at some stage in their replication. As microtubules serve as oriented tracks for vesicular transport in eukaryotic cells, it is not surprising that many viruses have independently evolved the ability to exploit them for their own transport.

Viruses Can Take Over the Metabolism of the Host Cell

Viruses use basic host cell machinery for most aspects of their reproduction: they depend on host-cell ribosomes to produce their proteins, and most use host-cell DNA and RNA polymerases for their own replication and transcription. Many viruses encode proteins that modify the host transcription or translation apparatus to favor the synthesis of viral RNAs and proteins over those of the host cell, shifting the synthetic capacity of the cell toward the production of new virus particles. Poliovirus, for example, encodes a protease that specifically cleaves the TATA-binding component of TFIID (see Figure 6–17), shutting off transcription of most of the host cell's protein-coding genes. Influenza virus produces a protein that blocks both the splicing and the polyadenylation of host-cell RNA transcripts, preventing their export into the cytosol (see Figure 6–38).

Viruses also alter translation by the host. Translation initiation for most host-cell mRNAs depends on recognition of their 5′ cap by translation initiation factors (see Figure 6–70). This initiation process is often inhibited during viral infection, so that the host-cell ribosomes can be used more efficiently for the synthesis of viral proteins. Some viral genomes encode endonucleases that cleave off the 5′ cap from host-cell mRNAs; some go even further by using the liberated 5′ caps as primers to synthesize viral mRNAs, a process called *cap snatching*. Several other viral RNA genomes encode proteases that cleave certain translation initiation factors; these viruses rely on 5′ cap-independent translation of their own RNA, using internal ribosome entry sites (IRESs) (see Figure 7–68).

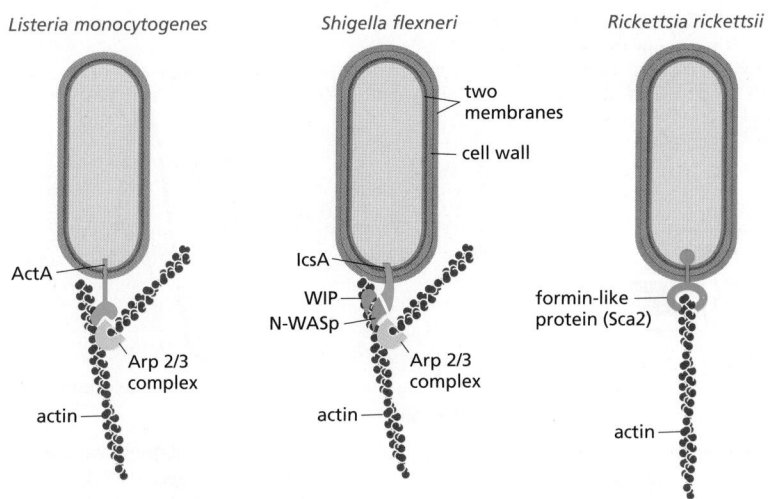

Figure 23–29 **Molecular mechanisms for actin nucleation by various bacterial pathogens.** *Listeria monocytogenes* and *Shigella flexneri* induce actin nucleation by recruiting and activating the host Arp 2/3 complex (see Figure 16–16), although each uses a different recruitment strategy: *L. monocytogenes* expresses a surface protein, ActA, that directly binds to and activates the Arp 2/3 complex; *S. flexneri* expresses a surface protein, IcsA (unrelated to ActA), that recruits the host protein N-WASp, which in turn recruits the Arp 2/3 complex, along with other host proteins, including WIP (WASp-interacting protein). *Rickettsia rickettsii* uses an entirely different strategy; it expresses a surface protein, Sca2, that directly nucleates actin polymerization by mimicking the activity of host formin proteins.

A few DNA viruses use host-cell DNA polymerase to replicate their genome. Unfortunately for these viruses, DNA polymerase is expressed at high levels only during S phase of the cell cycle, and most cells that these viruses infect spend most of their time in G_1 phase. Adenovirus has evolved a mechanism to drive the host cell into S phase, so that the cell produces large amounts of active DNA polymerase, which then replicates the viral genome; to accomplish this, the adenovirus genome also encodes proteins that inactivate both Rb (see Figure 17–61) and p53 (see Figure 17–62), two key suppressors of cell-cycle progression. As might be expected for any mechanism that encourages unregulated DNA replication, these viruses can promote, under some circumstances, the development of cancer. Other DNA viruses, including poxviruses and mimivirus, encode their own DNA and RNA polymerases, as well as some transcription regulators, allowing them to bypass usual host pathways and replicate outside the nucleus.

RNA viruses must always encode their own replication proteins because host cells lack polymerase enzymes that use RNA as a template. For RNA viruses with a single-stranded genome, the replication strategy depends on whether the RNA is a positive [+] strand, which contains translatable information like mRNA, or a complementary negative [–] strand. When the RNA is a positive [+] strand, the incoming viral genome is used to produce the viral RNA polymerase and viral proteins; the viral polymerase is then used to replicate the viral RNA and to generate mRNAs for the production of more viral proteins. For viruses with a negative [–] strand RNA genome (such as influenza and measles virus), an RNA polymerase enzyme is packaged as a structural protein of the incoming viral capsids.

Retroviruses such as HIV-1, which have a positive [+] strand RNA genome, are a special class of RNA virus because they carry with them a viral *reverse transcriptase* enzyme. After entry to the host cell, the reverse transcriptase uses the viral RNA genome as a template to synthesize a double-stranded DNA copy of the viral genome, which enters into the nucleus and integrates into the host cell's chromosomes (see Figure 5–62). It is later transcribed by the cell's DNA-dependent RNA polymerase to produce viral genomes and proteins.

Pathogens Can Evolve Rapidly by Antigenic Variation

The complexity and specificity of the interplay between pathogens and their host cells might suggest that virulence would be difficult to acquire by random mutation. Yet, new pathogens are constantly emerging, and old pathogens are constantly changing in ways that make familiar infections more difficult to prevent or treat. Pathogens have two advantages that enable them to evolve rapidly. First, they replicate very quickly, providing a great deal of material for natural selection to work with. Whereas humans and chimpanzees have acquired a 2% difference in genome sequences over about 8 million years of divergent evolution, poliovirus

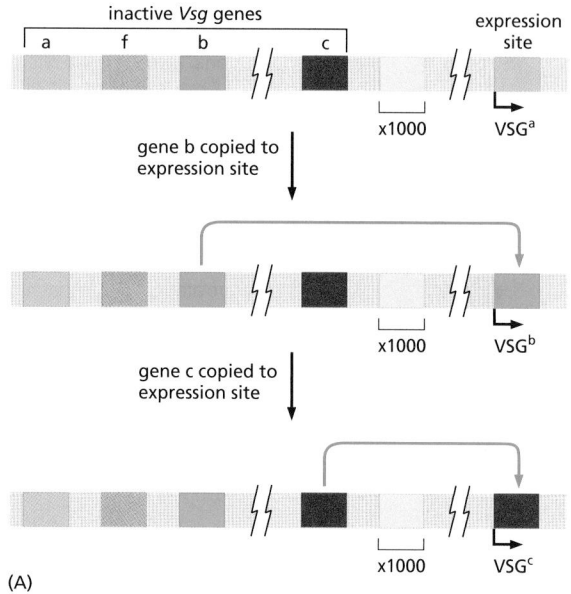

(A)

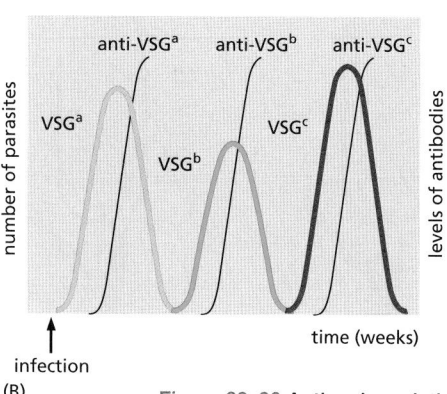

(B)

infection

Figure 23–30 Antigenic variation in trypanosomes. (A) There are about 1000 distinct *Vsg* genes in *Trypanosoma brucei*, and they are expressed one at a time from approximately 20 expression sites in the genome. To be expressed, an inactive gene is copied and the copy is moved into an expression site through DNA recombination. Each *Vsg* gene encodes a different surface protein (antigen). These switching events allow the trypanosome to repeatedly change the surface antigen it expresses. (B) A person infected with trypanosomes expressing VSGᵃ mounts a protective antibody response, which clears most of the parasites expressing this antigen. However, a few of the trypanosomes will have switched to expression of VSGᵇ, which can now proliferate until anti-VSGᵇ antibodies clear them. By that time, however, some parasites will have switched to VSGᶜ, and so the cycle continues.

manages a 2% change in its genome in 5 days—about the time it takes the virus to pass from the human mouth to the gut. Second, selective pressures act rapidly on this genetic variation. The host's adaptive immune system and modern microbicidal drugs, both of which destroy pathogens that fail to change, are the main sources of these selective pressures.

An example of an adaptation to the selective pressure imposed by the adaptive immune system is the phenomenon of **antigenic variation**. An important adaptive immune response against many pathogens is the host's production of antibodies that recognize specific molecules (*antigens*) on the pathogen's surface (discussed in Chapter 24). Many pathogens have evolved mecanisms that deliberately change these antigens during the course of an infection, enabling them to evade antibodies. Some eukaryotic parasites, for example, undergo programmed rearrangements of the genes encoding their surface antigens. A striking example occurs in *Trypanosoma brucei*, a protozoan parasite that causes African sleeping sickness and is spread by tsetse flies. (*T. brucei* is a relative of *T. cruzi*—see Figure 23–21—but it replicates extracellularly rather than intracellularly.) *T. brucei* is covered with a single type of glycoprotein, called *variant-specific glycoprotein* (*VSG*), which elicits in the host a protective antibody response that rapidly clears most of the parasites. The trypanosome genome, however, contains about 1000 different *Vsg* genes or pseudogenes, each encoding a VSG with a distinct amino acid sequence. Only one of these genes is expressed at any one time, from one of approximately 20 possible expression sites in the genome. Gene rearrangements that copy different *Vsg* genes into expression sites repeatedly change the VSG protein displayed on the surface of the pathogen. In this way, a few trypanosomes with an altered VSG escape the initial antibody-mediated clearance, replicate, and cause the disease to recur, leading to a chronic cyclic infection (**Figure 23–30**).

Bacterial pathogens can also rapidly change their surface antigens. As discussed in Chapter 5, *Salmonella enterica* bacteria switch between expressing either of two versions of the protein flagellin, the structural component of the bacterial flagellum (see Figure 23–3D), in a process called **phase variation** (see Figure 5–65). Species of the genus *Neisseria* are also champions at this. These Gram-negative cocci can cause meningitis and sexually transmitted diseases. They undergo genetic recombination very similar to that just described for eukaryotic pathogens, which enables them to vary the pilin protein they use to attach to host cells. By inserting one of the multiple silent copies of variant *pilin* genes into a single expression locus, they can express many slightly different versions of the protein and repeatedly change the amino acid sequence over time. *Neisseria* bacteria are also extremely adept at taking up DNA from their environment by natural transformation and incorporating it into their genomes, further contributing to their extraordinary variability. The end result of this considerable variation is a plethora of different surface compositions with which to bewilder the host adaptive immune system. It is therefore not surprising that it has been difficult to develop an effective vaccine against *Neisseria* infections, although there are now several that protect against *Neisseria meningitidis*, a common cause of fatal meningitis.

Error-Prone Replication Dominates Viral Evolution

In contrast to the DNA rearrangements in bacteria and parasites, viruses rely on an error-prone replication mechanism for antigenic variation. Retroviral genomes, for example, acquire on average one point mutation every replication cycle, because the viral reverse transcriptase (see Figure 5–62) needed to produce DNA from the viral RNA genome lacks the proofreading activity of DNA polymerases. A typical, untreated HIV infection may eventually produce HIV genomes with every possible point mutation. By a process of mutation and selection within each host, most viruses change over time—from a form that is most efficient at infecting macrophages to one more efficient at infecting T cells, as described earlier (see Figure 23–17). Similarly, once a patient is treated with an antiviral drug, the viral genome can quickly mutate and be selected for its resistance to the drug. Remarkably, only about one-third of the nucleotide positions in the coding sequence of the viral genome are invariant, and nucleotide sequences in some parts of the genome, such as the *Env* gene (see Figure 7–62), can differ by as much as 30% from one HIV isolate to another. This extraordinary genomic plasticity greatly complicates attempts to develop vaccines against HIV. It has also led to the rapid emergence of new HIV strains. Nucleotide sequence comparisons between various strains of HIV and the very similar simian immunodeficiency virus (SIV) isolated from a variety of monkey species suggest that the most virulent type of HIV, HIV-1, may have jumped from primates to humans multiple independent times, starting as long ago as 1908 (**Figure 23–31**).

Influenza viruses are an important exception to the rule that error-prone replication dominates viral evolution. They are unusual in that their genome consists of several (usually eight) strands of RNA. When two strains of influenza infect the same host, the RNA strands of the two strains can reassort to form a new type of influenza virus. In normal years, influenza is a mild disease in healthy adults, although it can be life-threatening in the very young and very old. Different influenza strains infect fowl such as ducks and chickens, but only a subset of these strains can infect humans, and transmission from fowl to humans is rare. In 1918, however, a particularly virulent variant of avian influenza crossed the species barrier to infect humans, triggering the catastrophic pandemic of 1918 called the Spanish flu, which killed 20–50 million people worldwide. Subsequent influenza pandemics have been triggered by genome reassortment, in which a new RNA segment from an avian form of the virus replaced one or more of the viral RNA segments from the human form (**Figure 23–32**). In 2009, a new H1N1 swine virus emerged that derived genes from pig, avian, and human influenza viruses. Such recombination events allowed the new virus to replicate rapidly and spread through an immunologically naive human population. Generally, within two or three years, the human population develops immunity to a new recombinant strain of virus, and the infection rate drops to a steady-state level. Because the recombination events are unpredictable, it is not possible to know when the next influenza pandemic will occur or how severe it might be.

Drug-Resistant Pathogens Are a Growing Problem

The development of drugs that cure rather than prevent infections has had a major impact on human health. **Antibiotics**, which are either bactericidal (they kill bacteria) or bacteriostatic (they inhibit bacterial growth without killing), are the most successful class of such drugs. Penicillin was one of the first antibiotics

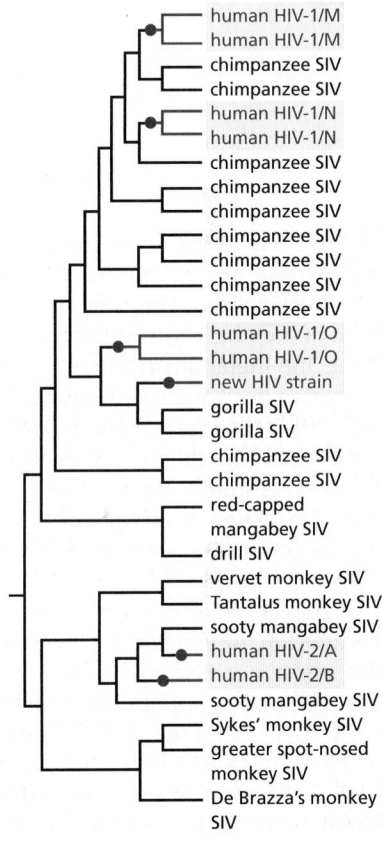

● = jumps from monkey and ape to human

Figure 23–31 Diversification of HIV-1, HIV-2, and related strains of SIV. HIV comprises different viral families, all descended from SIV (simian immunodeficiency virus). On three separate occasions, SIV was passed from a chimpanzee to a human, resulting in three HIV-1 groups: major (M), outlier (O), and non-M non-O (N). The HIV-1 M group is the most common and is primarily responsible for the global AIDS epidemic. On two separate occasions, SIV was passed from a sooty mangabey monkey to a human, resulting in the two HIV-2 groups. In 2009, a new strain of HIV was discovered that appears to have resulted from SIV passage from a gorilla to a human.

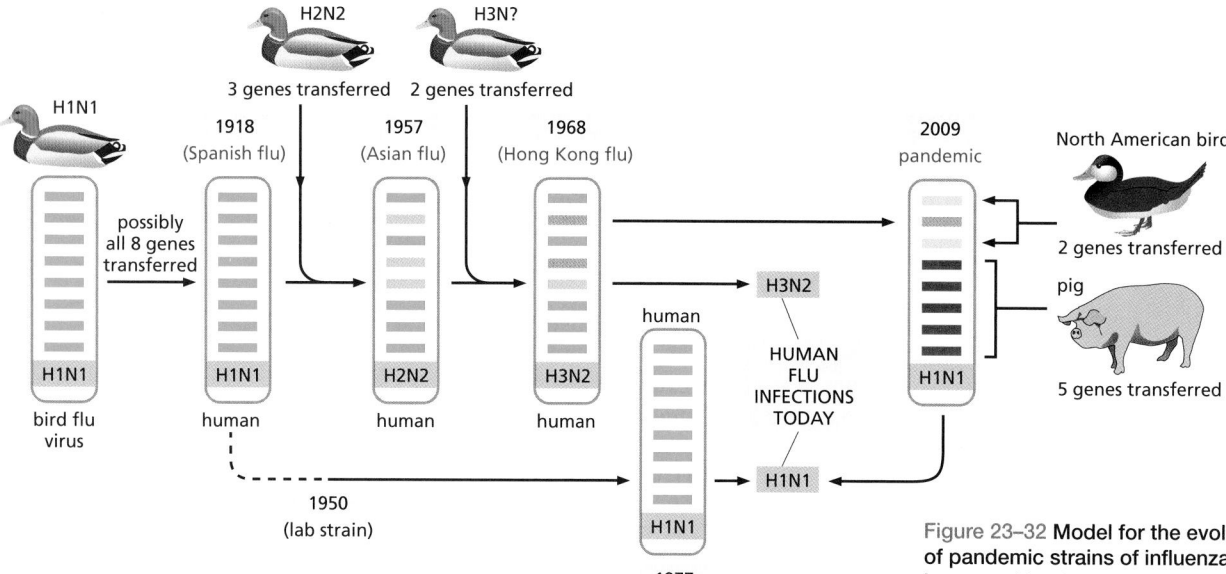

Figure 23–32 **Model for the evolution of pandemic strains of influenza virus by recombination.** Influenza A virus is a natural pathogen of birds, particularly waterfowl, and it is always present in wild bird populations. In 1918, a particularly virulent form of the virus crossed the species barrier from birds to humans and caused a devastating worldwide epidemic. This strain was designated H1N1, referring to the specific forms of its main antigens, hemagglutinin (H) and neuraminidase (N). Changes in the virus, rendering it less virulent, and the rise of adaptive immunity in the human population, prevented the pandemic from continuing in subsequent seasons, although H1N1 influenza strains continued to cause serious disease every year in very young and very old people. In 1957, a new pandemic arose when three genes were replaced by equivalent genes from an avian virus *(green bars)*; the new strain (designated H2N2) was not effectively cleared by antibodies in people who had previously contracted only H1N1 forms of influenza. In 1968, another pandemic was triggered when two genes were replaced from another avian virus; the new virus was designated H3N2. In 1977, there was a resurgence of H1N1 influenza, which had previously been almost completely replaced by the N2 strains. Molecular sequence information suggests that this minor pandemic may have been caused by an accidental release of an influenza strain that had been held in a laboratory since about 1950. In 2009, a new H1N1 swine virus emerged that had derived five genes from pig influenza viruses, two from avian influenza viruses, and one from a human influenza virus. As indicated, most human influenza today is caused by H1N1 and H3N2 strains.

used to treat infections in humans, just in time to prevent tens of thousands of deaths from infected battlefield wounds in World War II. Because bacteria (see Figure 1–17) are not closely related evolutionarily to the eukaryotes they infect, much of their basic machinery for DNA replication and transcription, RNA translation, and metabolism differs from that of their host. These differences enable us to develop antibacterial drugs that exhibit *selective toxicity*, in that they specifically inhibit these processes in bacteria without disrupting them in the host. Most of the antibiotics that we use to treat bacterial infections are small molecules that inhibit macromolecular synthesis in bacteria by targeting bacterial enzymes that either are distinct from their eukaryotic counterparts or are involved in pathways such as cell wall biosynthesis that are absent in animals (Figure 23–33 and see Table 6–4).

However, bacteria continuously evolve and strains resistant to antibiotics rapidly develop, often within a few years of the introduction of a new drug. Similar drug resistance also arises rapidly when treating viral infections with antiviral drugs. The virus population in an HIV-infected person treated with the reverse transcriptase inhibitor AZT, for example, will acquire complete resistance to the drug within a few months. The current protocol for treatment of HIV infections involves the simultaneous use of three drugs, which helps to minimize the acquisition of resistance for any one of them.

There are three general strategies by which a pathogen can develop drug resistance: (1) it can alter the molecular target of the drug so that it is no longer sensitive to the drug; (2) it can produce an enzyme that modifies or destroys the drug; or (3) it can prevent the drug's access to the drug target by, for example, actively pumping the drug out of the pathogen (Figure 23–34).

Once a pathogen has chanced upon an effective drug-resistance strategy, the newly acquired or mutated genes that confer the resistance are frequently spread throughout the pathogen population by horizontal gene transfer. They may even spread between pathogens of different species. The highly effective but expensive antibiotic *vancomycin*, for example, is used as a treatment of last resort for many severe, hospital-acquired, Gram-positive bacterial infections that are resistant to most other known antibiotics. Vancomycin prevents one step in bacterial cell wall synthesis—the cross-linking of peptidoglycan chains in the bacterial cell wall (see Figure 23–3B). Resistance can arise if the bacterium synthesizes a cell wall using different subunits that do not bind vancomycin. The most effective form of vancomycin resistance depends on the acquisition of a transposon (see Figure 5–60) containing seven genes, the products of which work together to sense the

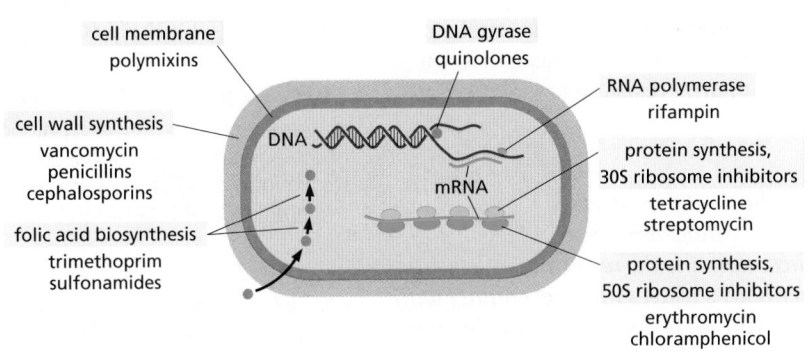

Figure 23–33 **Antibiotic targets.** Although there are many antibiotics in clinical use, they have a narrow range of targets, which are highlighted in *yellow*. A few representative antibiotics in each class are listed. Nearly all antibiotics used to treat human infections fall into one of these categories. The vast majority inhibit either bacterial protein synthesis or bacterial cell wall synthesis.

presence of vancomycin, shut down the normal pathway for bacterial cell wall synthesis, and produce a different type of cell wall.

Drug-resistance genes acquired by horizontal transfer frequently come from environmental microbial reservoirs. Nearly all antibiotics used to treat bacterial infections today are based on natural products produced by fungi or bacteria. Penicillin, for example, is made by the mold *Penicillium*, and more than 50% of the antibiotics currently used in the clinic are made by Gram-positive bacteria of the genus *Streptomyces*, which reside in the soil. It is believed that microorganisms produce antimicrobial compounds, many of which have probably existed on Earth for hundreds of millions of years, as weapons in their competition with other microorganisms in the environment. Surveys of bacteria taken from soil samples that have never been exposed to antibiotic drugs used in modern medicine reveal that the bacteria are typically already resistant to about seven or eight of the antibiotics widely used in clinical practice. When pathogenic microorganisms are faced with the selective pressure provided by antibiotic treatments, they can apparently draw upon this immense source of genetic material to acquire resistance.

Like most other aspects of infectious disease, human behavior has exacerbated the problem of drug resistance. Many patients take antibiotics for symptoms that are typically caused by viruses (flu-like illnesses, colds, sore throats, and earaches) and these drugs have no effects. Persistent and chronic misuse of antibiotics can eventually result in antibiotic-resistant normal flora, which can then transfer the resistance to pathogens. Antibiotics are also misused in agriculture, where they are commonly employed as food additives to promote the growth and health of farm animals. An antibiotic closely related to vancomycin was commonly added to cattle feed in Europe; the resulting resistance in the normal flora of these animals is widely believed to be one of the original sources for vancomycin-resistant bacteria that now threaten the lives of hospitalized patients.

Figure 23–34 **Three general mechanisms of antibiotic resistance.** (A) A nonresistant wild-type bacterial cell bathed in a drug *(red triangles)* that binds to and inhibits an essential enzyme *(light green)* will be killed due to enzyme inhibition. (B) A bacterium that has altered the drug's target enzyme so that the drug no longer binds to the enzyme will survive and proliferate. In many cases, a single point mutation in the gene encoding the target protein can generate resistance. (C) A bacterium that expresses an enzyme *(dark green)* that either degrades or covalently modifies the drug will survive and proliferate. Some resistant bacteria, for example, make β-lactamase enzymes, which cleave penicillin and similar molecules. (D) A bacterium that expresses or up-regulates an efflux pump that ejects the drug from the bacterial cytoplasm (using energy derived from either ATP hydrolysis or the electrochemical gradient across the bacterial plasma membrane) will survive and proliferate. Some efflux pumps, such as the TetR efflux pump, are specific for a single drug (in this case, tetracycline), whereas others, called multidrug resistance (MDR) efflux pumps, are capable of exporting a wide variety of structurally dissimilar drugs. Upregulation of an MDR pump can render a bacterium resistant to a very large number of different antibiotics in a single step.

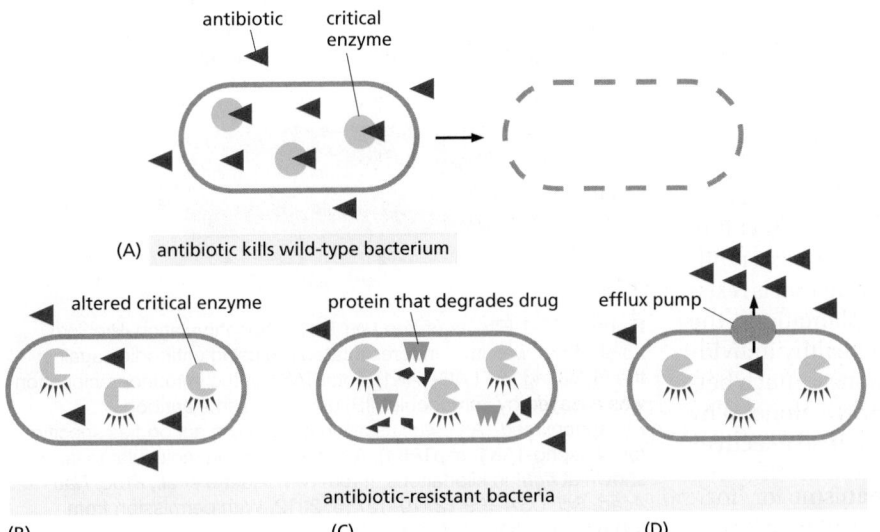

Because the acquisition of drug resistance is almost inevitable, it is crucial that we continue to develop innovative treatments for infectious diseases. We must also take additional measures to delay the onset of drug resistance.

Summary

All pathogens share the ability to interact with host cells in diverse ways that promote the replication and spread of the pathogen. Pathogens often colonize the host by adhering to or invading the epithelial surfaces that line the respiratory, gastrointestinal, and urinary tracts, as well as the other body surfaces in direct contact with the environment. Intracellular pathogens, including all viruses and many bacteria and protozoa, invade host cells by one of several mechanisms. Viruses rely largely on receptor-mediated endocytosis, whereas bacteria exploit cell adhesion and phagocytic pathways; in both cases, the host cell provides the machinery and energy for the invasion. Protozoa, by contrast, employ unique invasion strategies that usually require significant metabolic expense on the part of the invader. Once inside, intracellular pathogens seek out a cell compartment that is favorable for their survival and replication, frequently altering host membrane traffic and exploiting the host-cell cytoskeleton for intracellular movement. Pathogens evolve rapidly, so that new infectious diseases frequently emerge, and old pathogens acquire new ways to evade our attempts at treatment, prevention, and eradication.

WHAT WE DON'T KNOW

• What are the genetic and molecular features that differ between pathogens and members of the normal human microbiota? How can our immune system distinguish between the two?

• To what extent are common host-cell biological pathways and molecules hijacked by diverse microbes?

• Can host-cell defense molecules be mobilized by drugs to fight infection?

PROBLEMS

Which statements are true? Explain why or why not.

23–1 Our adult bodies harbor about 10 times more microbial cells than human cells.

23–2 The microbiomes from healthy humans are all very similar.

23–3 Pathogens must enter host cells to cause disease.

23–4 Viruses replicate their genomes in the nucleus of the host cell.

23–5 You should not take antibiotics for diseases caused by viruses.

Discuss the following problems.

23–6 In order to survive and multiply, a successful pathogen must accomplish five tasks. Name them.

23–7 *Clostridium difficile* infection is the leading cause of hospital-associated gastrointestinal illness. It is typically treated with a course of antibiotics, but the infection recurs in about 20% of cases. *C. difficile* infections are difficult to eradicate because the bacteria exist in two forms: a replicating, toxin-producing form and a spore form that is resistant to antibiotics. Fecal microbiota transplantation—the transfer of normal gut microbiota from a healthy individual—can resolve >90% of recurrent infections, a much better cure rate than further antibiotic treatment alone. Why do you suppose microbiota transplantation is so effective?

23–8 What are the three general mechanisms for horizontal gene transfer?

23–9 The Gram-negative bacterium *Yersinia pestis*, the causative agent of the plague, is extremely virulent. Upon infection, *Y. pestis* injects a set of effector proteins into macrophages that suppresses their phagocytic behavior and also interferes with their innate immune responses. One of the effector proteins, YopJ, acetylates serines and threonines on various MAP kinases, including the MAP kinase kinase kinase TAK1, which controls a key signaling step in the innate immune response pathway. To determine how YopJ interferes with TAK1, you transfect human cells with active YopJ (YopJ^WT) or inactive YopJ (YopJ^CA) and with FLAG-tagged active TAK1 (TAK1^WT) or inactive TAK1 (TAK1^K63W), and assay for total TAK1 and for phosphorylated TAK1, using antibodies against the FLAG tag or against phosphorylated TAK1 (**Figure Q23–1**). How does YopJ block the TAK1 signaling pathway? How do you suppose the serine/threonine acetylase activity of YopJ might interfere with TAK1 activation?

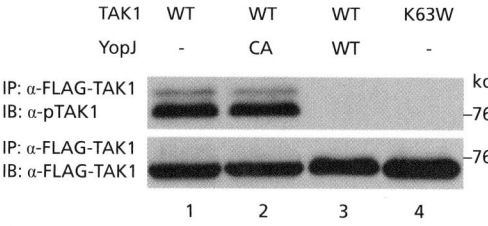

Figure Q23–1 Effects of YopJ on TAK1 phosphorylation (Problem 23–9). TAK1 was immunoprecipitated (IP) using antibodies against the FLAG tag (α-FLAG-TAK1). Total TAK1 in the immunoprecipitation was assayed by immunoblot (IB) using the same antibody. Phosphorylated TAK1 was assayed by IB using antibodies specific for phospho-TAK1 (α-pTAK1). A scale of protein molecular mass is shown at *right* in kilodaltons. (From N. Paquette et al., *Proc. Natl Acad. Sci. USA* 109:12710–12715, 2012. With permission from National Academy of Sciences.)

23–10 The intracellular bacterial pathogen *Salmonella typhimurium*, which causes gastroenteritis, injects effector proteins to promote its invasion into nonphagocytic host cells by the trigger mechanism. *S. typhimurium* first stimulates membrane ruffling to promote invasion, and then suppresses membrane ruffling once invasion is complete. This behavior is mediated in part by injection of two effector proteins: SopE, which promotes membrane ruffling and invasion, and SptP, which blocks the effects of SopE. Both effector proteins target the monomeric GTPase, Rac, which in its active form promotes membrane ruffling. How do you suppose SopE and SptP affect Rac activity? How do you suppose the effects of SopE and SptP are staggered in time if they are injected simultaneously?

23–11 John Snow is widely regarded as the father of modern epidemiology. Most famously, he investigated an outbreak of cholera in London in 1854 that killed more than 600 victims before it was finished. Snow recorded where the victims lived, and plotted the data on a map, along with the locations of the water pumps that served as the source of water for the public (**Figure Q23–2**). He concluded that the disease was most likely spread in the water, although he could find nothing suspicious-looking in it. His conclusion ran counter to the then-current belief that cholera was from "miasmas" in bad air. Very few believed his theory during the next 50 years, with the "bad air" theory persisting until at least 1901. What do you suppose Snow saw in the data that led him to his conclusion? Why do you think most scientists remained skeptical for so long?

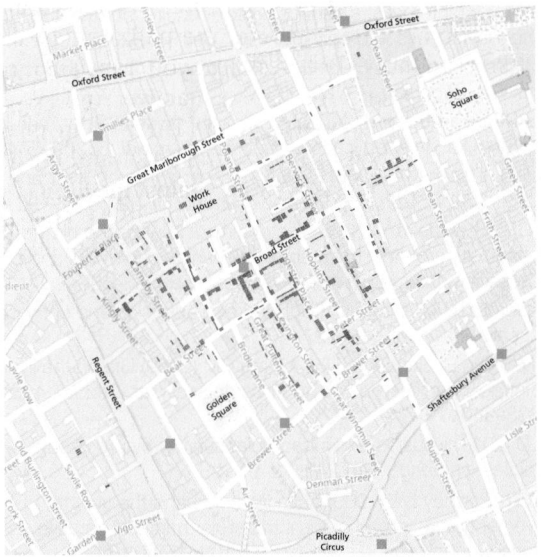

Figure Q23–2 A map of where the victims of the 1854 cholera outbreak lived, superimposed on a modern map of the area (Problem 23–11). The locations of the victims' houses are indicated by the small *red rectangles*. Stacks of rectangles indicate multiple cases occurring in the same house. Public water pumps are shown as *blue squares*. (Adapted from Wellcome Library, London/Google Maps.)

23–12 Influenza epidemics account for 250,000 to 500,000 deaths globally each year. These epidemics are markedly seasonal, occurring in temperate climates in the northern and southern hemispheres during their respective

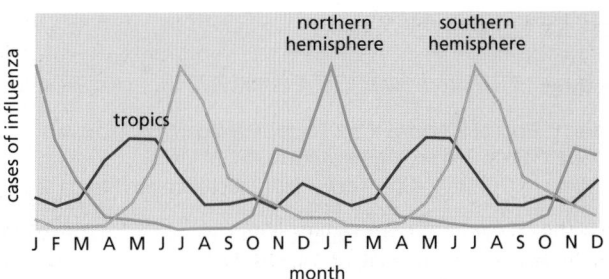

Figure Q23–3 Seasonal patterns of influenza epidemics (Problem 23–12). Cases of influenza at different times of the year are shown for the northern hemisphere *(blue)*, the southern hemisphere *(orange)*, and the tropics *(red)*.

winters. By contrast, in the tropics, there is significant influenza activity year round, with a peak in the rainy season (**Figure Q23–3**). Can you suggest some possible explanations for the patterns of influenza epidemics in temperate zones and the tropics?

23–13 Several negative-strand viruses carry their genome as a set of discrete RNA segments. Examples include influenza virus (eight segments), Rift Valley fever virus (three segments), Hantavirus (three segments), and Lassa virus (two segments), to name a few. Why does segmentation of the genome provide a strong evolutionary advantage for these viruses?

23–14 Avian influenza viruses readily infect birds, but are transmitted to humans very rarely. Similarly, human influenza viruses spread readily to other humans, but have never been detected in birds. The key to this specificity lies in the viral capsid protein, hemagglutinin, which binds to sialic acid residues on cell-surface glycoproteins, triggering virus entry into the cell (**Movie 23.8**). Hemagglutinin on human viruses recognizes sialic acid in a 2-6 linkage with galactose, whereas avian hemagglutinin recognizes sialic acid in a 2-3 linkage with galactose. Humans make carbohydrate chains that have only the 2-6 linkage between sialic acid and galactose; birds make only the 2-3 linkage; but pigs make carbohydrate chains with both linkages. How does this situation make pigs ideal hosts for generating new strains of human influenza viruses?

23–15 The majority of antibiotics used in the clinic are made as natural products by bacteria. Why do you suppose bacteria make the very agents we use to kill them?

23–16 In the early days of penicillin research, it was discovered that bacteria in the air could destroy the penicillin, a big problem for large-scale production of the drug. How do you suppose this occurs?

23–17 When the Oxford team of Ernst Chain and Norman Heatley had laboriously collected their first two grams of penicillin (probably no more than 2% pure!), Chain injected two normal mice with 1 g each of this preparation, and waited to see what would happen. The mice survived with no apparent ill effects. Their boss, Howard Florey, was furious at what he saw as a waste of good antibiotic. Why was this experiment important?

REFERENCES

General

Cossart P, Boquet P, Normark S & Rappuoli R (eds) (2005) Cellular Microbiology, 2nd ed. Washington, DC: ASM Press.

Engleberg NC, DiRita V & Dermody T (2012) Schaechter's Mechanisms of Microbial Disease, 5th ed. Philadelphia, PA: Lippincott, Williams & Wilkins.Norkin LA (2010) Virology: Molecular Biology and Pathogenesis. Washington, DC: ASM Press.

Wilson BA, Salyers AA, Whitt DD & Winkler ME (2011) Bacterial Pathogenesis: A Molecular Approach, 3rd ed. Washington, DC: ASM Press.

Introduction to Pathogens and the Human Microbiota

Aly AS, Vaughan AM & Kappe SH (2009) Malaria parasite development in the mosquito and infection of the mammalian host. *Annu. Rev. Microbiol.* 63, 195–221.

Baltimore D (1971) Expression of animal virus genomes. *Bacteriol. Rev.* 35, 235–241.

Clemente JC, Ursell LK, Parfrey LW & Knight R (2012) The impact of the gut microbiota on human health: an integrative view. *Cell* 148, 1258–1270.

Crick FH & Watson JD (1956) Structure of small viruses. *Nature* 177, 473–475.

Fauci A & Morens DM (2012) The perpetual challenge of infectious diseases. *N. Engl. J. Med.* 366, 454–461.

Frost LS, Leplae R, Summers AO & Toussaint A (2005) Mobile genetic elements: the agents of open source evolution. *Nat. Rev. Microbiol.* 3, 722–732.

Galán JE & Wolf-Watz H (2006) Protein delivery into eukaryotic cells by type III secretion machines. *Nature* 444, 567–573.

Hacker J & Kaper JB (2000) Pathogenicity islands and the evolution of microbes. *Annu. Rev. Microbiol.* 54, 641–679.

Nelson EJ, Harris JB, Morris JG Jr et al. (2009) Cholera transmission: the host, pathogen and bacteriophage dynamic. *Nat. Rev. Microbiol.* 7, 693–702.

Pflughoeft KJ & Versalovic J (2012) Human microbiome in health and disease. *Annu. Rev. Pathol.* 7, 99–122.

Polk DB & Peek RM Jr (2010) *Helicobacter pylori*: gastric cancer and beyond. *Nat. Rev. Cancer* 10, 403–414.

Poulin R & Morand S (2000) The diversity of parasites. *Q. Rev. Biol.* 75, 277–293.

Rappleye CA & Goldman WE (2006) Defining virulence genes in the dimorphic fungi. *Annu. Rev. Microbiol.* 60, 281–303.

Thomas CM & Nielsen KM (2005) Mechanisms of, and barriers to, horizontal gene transfer between bacteria. *Nat. Rev. Microbiol.* 3, 711–721.

Votteler J & Sundquist WI (2013) Virus budding and the ESCRT pathway. *Cell Host Microbe* 14, 232–241.

Young JAT & Collier RJ (2007) Anthrax toxin: receptor binding, internalization, pore formation, and translocation. *Annu. Rev. Biochem.* 76, 243–265.

Cell Biology of Infection

Alix E, Mukherjee S & Roy CR (2011) Subversion of membrane transport pathways by vacuolar pathogens. *J. Cell Biol.* 195, 943–952.

Beiting DP & Roos DS (2011) A systems biological view of intracellular pathogens. *Immunol. Rev.* 240, 117–128.

Brandenburg B & Zhuang X (2007) Virus trafficking – learning from single-virus tracking. *Nat. Rev. Microbiol.* 5, 197–208.

Cossart P & Sansonetti PJ (2004) Bacterial invasion: the paradigms of enteroinvasive pathogens. *Science* 304, 242–248.

Daugherty MD & Malik HS (2012) Rules of engagement: molecular insights from host-virus arms races. *Annu. Rev. Genet.* 46, 677–700.

Davies J & Davies D (2010) Origins and evolution of antibiotic resistance. *Microbiol. Mol. Biol. Rev.* 74, 417–433.

Dimitrov DS (2004) Virus entry: molecular mechanisms and biomedical applications. *Nat. Rev. Microbiol.* 2, 109–122.

Duffy S, Shackelton LA & Holmes EC (2008) Rates of evolutionary change in viruses: patterns and determinants. *Nat. Rev. Genet.* 9, 267–276.

Forsberg KJ, Reyes A, Wang B et al. (2012) The shared antibiotic resistome of soil bacteria and human pathogens. *Science* 337, 1107–1111.

Ghedin E, Sengamalay NA, Shumway M et al. (2005) Large-scale sequencing of human influenza reveals the dynamic nature of viral genome evolution. *Nature* 437, 1162–1166.

Goldberg DE, Siliciano RF & Jacobs WR Jr (2012) Outwitting evolution: fighting drug-resistant TB, malaria, and HIV. *Cell* 148, 1271–1283.

Haglund CM & Welch MD (2011) Pathogens and polymers: microbe-host interactions illuminate the cytoskeleton. *J. Cell Biol.* 195, 7–17.

Ham H, Sreelatha A & Orth K (2011) Manipulation of host membranes by bacterial effectors. *Nat. Rev. Microbiol.* 9, 635–646.

Hayward RD, Leong JM, Koronakis V & Campellone KG (2006) Exploiting pathogenic *Escherichia coli* to model transmembrane receptor signalling. *Nat. Rev. Microbiol.* 4, 358–370.

Kenny B, DeVinney R, Stein M et al. (1997) Enteropathogenic *E. coli* (EPEC) transfers its receptor for intimate adherence into mammalian cells. *Cell* 91, 511–520.

Lusso P (2006) HIV and the chemokine system: 10 years later. *EMBO J.* 25, 447–456.

Medina RA & García-Sastre A (2011) Influenza A viruses: new research developments. *Nat. Rev. Microbiol.* 9, 590–603.

Mengaud J, Ohayon H, Gounon P et al. (1996) E-cadherin is the receptor for internalin, a surface protein required for entry of *L. monocytogenes* into epithelial cells. *Cell* 84, 923–932.

Mercer J, Schelhaas M & Helenius A (2010) Virus entry by endocytosis. *Annu. Rev. Biochem.* 79, 803–833.

Miller S & Krijnse-Locker J (2008) Modification of intracellular membrane structures for virus replication. *Nat. Rev. Microbiol.* 6, 363–374.

Mullins JI & Jensen MA (2006) Evolutionary dynamics of HIV-1 and the control of AIDS. *Curr. Top. Microbiol. Immunol.* 299, 171–192.

Parrish CR & Kawaoka Y (2005) The origins of new pandemic viruses: the acquisition of new host ranges by canine parvovirus and influenza A viruses. *Annu. Rev. Microbiol.* 59, 553–586.

Pizarro-Cerdá J & Cossart P (2006) Bacterial adhesion and entry into host cells. *Cell* 124, 715–727.

Ray K, Marteyn B, Sansonetti PJ & Tang CM (2009) Life on the inside: the intracellular lifestyle of cytosolic bacteria. *Nat. Rev. Microbiol.* 7, 333–340.

Sibley LD (2011) Invasion and intracellular survival by protozoan parasites. *Immunol. Rev.* 240, 72–91.

Tilney LG & Portnoy DA (1989) Actin filaments and the growth, movement, and spread of the intracellular bacterial parasite, *Listeria monocytogenes. J. Cell Biol.* 109, 1597–1608.

Vink C, Rudenko G & Seifert HS (2012) Microbial antigenic variation mediated by homologous DNA recombination. *FEMS Microbiol. Rev.* 36, 917–948.

Walsh D & Mohr I (2011) Viral subversion of the host protein synthesis machinery. *Nat. Rev. Microbiol.* 9, 860–875.

Welch MD & Way M (2013) Arp2/3-mediated actin-based motility: a tail of pathogen abuse. *Cell Host Microbe* 14, 242–255.

The Innate and Adaptive Immune Systems

As we discussed in Chapter 23, all living organisms serve as hosts for other species, usually in relationships that are benign or even mutually helpful. But all organisms, and all cells in a multicellular organism, need to defend themselves against infection by harmful invaders, collectively called **pathogens**, which can be microbes (bacteria, viruses, or fungi), or larger parasites. Even bacteria defend themselves against viruses, using intracellular proteins called *restriction factors*, which block viral propagation. Invertebrates use a variety of defense strategies, including protective barriers, toxic molecules, restriction factors, and phagocytic cells that ingest and destroy invading pathogens. Vertebrates, too, depend on such *innate immune responses*, but they can also harness more sophisticated and specific mechanisms, called *adaptive immune responses*. The innate responses occur first, calling the adaptive immune responses into play if required, in which case, both types of responses work together to eliminate the pathogen (**Figure 24–1**).

Whereas innate immune responses are general defense reactions that can involve almost any cell type in an organism, the adaptive immune responses are highly specific to the particular pathogen that induced them and depend on a class of white blood cells (leukocytes) called *lymphocytes*. There are two major classes of lymphocytes that mount adaptive immune responses—B lymphocytes (*B cells*), which secrete *antibodies* that bind specifically to the pathogen, and T lymphocytes (*T cells*), which can either directly kill cells infected with the pathogen or produce secreted or cell-surface signal proteins that stimulate other host cells to help eliminate the pathogen (**Figure 24–2**). Unlike innate immune responses, which are generally short-lasting, the adaptive responses provide long-lasting protection: a person who recovers from measles or is vaccinated against it, for example, is protected for life against measles by the adaptive immune system, although not against other common viruses, such as those that cause mumps or chickenpox.

Both the innate and adaptive immune systems have evolved sensing mechanisms that enable them to recognize harmful invaders (pathogens) and distinguish them from both the host's own cells and molecules and harmless or beneficial foreign organisms and their molecules. The innate system relies on sensor proteins that recognize particular types or patterns of molecules that are common to pathogens but are absent or sequestered in the host. The adaptive system, by contrast, uses unique genetic mechanisms to produce a virtually limitless diversity of related proteins—receptors on T and B cells and secreted antibodies—that, between them, can bind almost any foreign molecule. This remarkable strategy enables the adaptive immune system to react specifically against any pathogen, even if the animal never encountered it before. But, it also requires that the system learn not to react against self molecules or harmless foreign ones; if these learning mechanisms fail, harmful autoimmune or allergic responses result.

In this chapter, we focus on vertebrate immune responses and the features that distinguish them from other kinds of cell responses. We begin with innate immune defenses and then discuss the highly specialized properties of the adaptive immune system.

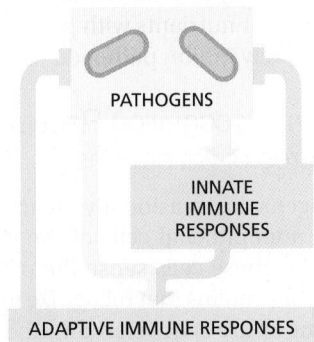

Figure 24–1 Innate and adaptive immune responses. Innate immune responses are activated directly by pathogens and defend all multicellular organisms against infection. In vertebrates, pathogens, together with the innate immune responses they activate, also stimulate adaptive immune responses, which then work together with innate immune responses to help fight the infection.

THE INNATE IMMUNE SYSTEM

Adaptive immune responses are slow to develop when a vertebrate first encounters a new pathogen. This is because the specific B cells and T cells that can respond to a particular pathogen are initially few in number and must be stimulated to proliferate and differentiate before they can mount effective adaptive immune responses, which can take days. By contrast, a single bacterium that divides every hour can generate almost twenty million progeny in a single day, producing a full-blown infection. Vertebrates, therefore, rely on their **innate immune system** to defend them against infection during the first critical hours and days of exposure to a new pathogen. Plants and invertebrates lack adaptive immune systems and therefore rely entirely on innate immunity for protection against pathogens.

In this section, we consider some of the strategies the innate immune system uses to recognize pathogens and to provide a first line of defense against them.

Epithelial Surfaces Serve as Barriers to Infection

In vertebrates, the first encounters with infectious organisms are typically at the epithelial surfaces that form the skin and line the respiratory, digestive, urinary, and reproductive tracts. These epithelia provide both physical and chemical barriers to invasion by pathogens: tight junctions between epithelial cells bar entry between the cells, and a variety of substances secreted by the cells discourage the attachment and entry of pathogens. The keratinized epithelial cells of the skin, for example, form a thick physical barrier, and the sebaceous glands in the skin secrete fatty acids and lactic acid, which inhibit bacterial growth. In addition, epithelial cells in all tissues, including those in plants and invertebrates, secrete antimicrobial molecules called **defensins**. Defensins are positively charged, amphipathic peptides that bind to and disrupt the membranes of many pathogens, including enveloped viruses, bacteria, fungi, and parasites.

The epithelial cells that line internal organs such as the respiratory and digestive tracts also secrete slimy mucus, which sticks to the epithelial surface and makes it difficult for pathogens to adhere. The beating of cilia on the surface of the epithelial cells lining the respiratory tract and the peristaltic action of the intestine also discourage the adherence of pathogens. Moreover, as we discuss in Chapter 23, healthy skin and gut are populated by enormous numbers of harmless (and often helpful) *commensal* microbes, collectively called the *normal flora*, which compete for nutrients with pathogens; some also produce antimicrobial peptides that actively inhibit pathogen proliferation.

Pattern Recognition Receptors (PRRs) Recognize Conserved Features of Pathogens

Pathogens do occasionally breach the epithelial barricades, in which case underlying, nonepithelial cells of the innate immune system provide the next line of defense. These cells sense the presence of pathogens largely through the use of receptor proteins that recognize microbe-associated molecules that either are not present or are sequestered in the host organism. Because these microbial molecules often occur in repeating patterns, they are called **pathogen-associated molecular patterns** (**PAMPs**), even though they are not unique to microbes that can cause disease. PAMPs are present in various microbial molecules, including nucleic acids, lipids, polysaccharides, and proteins.

The special receptor proteins that recognize PAMPs are called **pattern recognition receptors** (**PRRs**). Some PRRs are transmembrane proteins on the surface of many types of host cells, where they recognize extracellular pathogens; on professional phagocytic cells (phagocytes) such as *macrophages* and *neutrophils* (discussed in Chapter 22), they can mediate the uptake of the pathogens into phagosomes, which then fuse with lysosomes, where the pathogens are destroyed. Other PRRs are located intracellularly, where they can detect intracellular pathogens such as viruses; these PRRs are either free in the cytosol or associated with

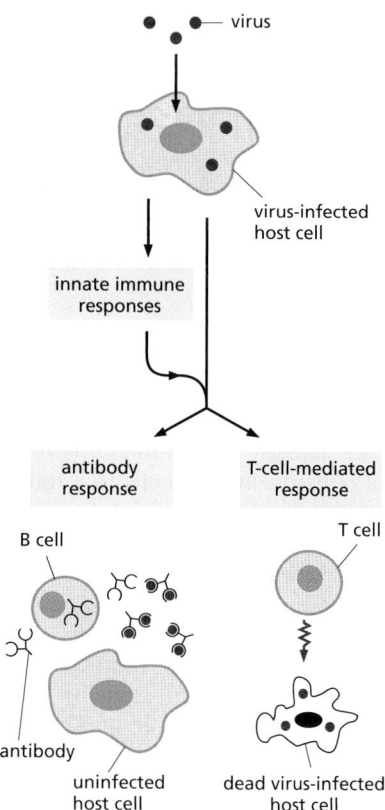

Figure 24–2 The two main classes of adaptive immune responses. Lymphocytes carry out both classes of adaptive responses—shown here as responses to a viral infection. In one class, B cells secrete antibodies that specifically bind to and neutralize extracellular viruses, by preventing the viruses from infecting host cells. In the other, T cells mediate the response; in this example, they kill the virus-infected host cells. In both cases, innate immune responses help activate the adaptive immune responses through pathways that are not shown.

Figure 24–3 **A scanning electron micrograph of a mutant fruit fly that died from a fungal infection.** The fly is covered with fungal hyphae, as it lacked a Toll receptor, which helps protect *Drosophila* from fungal infections. (From B. Lemaitre et al., *Cell* 86:973–983, 1996. With permission from Elsevier.)

the membranes of the endolysosomal system (discussed in Chapter 13). Still other PRRs are secreted and bind to the surface of extracellular pathogens, marking them for destruction by either phagocytes or blood proteins that are part of the *complement system* (discussed later).

There Are Multiple Classes of PRRs

The first PRR identified was the *Toll receptor* in *Drosophila*, which was well-known for its role in fly development (see Figure 21–17). It was later discovered to be also required for the production of antimicrobial peptides that protect the fly against fungal infections (**Figure 24–3**). Toll is a transmembrane glycoprotein with a large extracellular domain that contains a series of leucine-rich repeats. Soon it was discovered that both plants and animals have a variety of **Toll-like receptors** (**TLRs**) that function as PRRs in innate immune responses against various pathogens. Mammals make at least 10 different TLRs, each recognizing distinct ligands: TLR3, for example, recognizes double-stranded viral RNA in the endosomal lumen (**Figure 24–4**); TLR4 recognizes lipopolysaccharide (LPS) on the outer

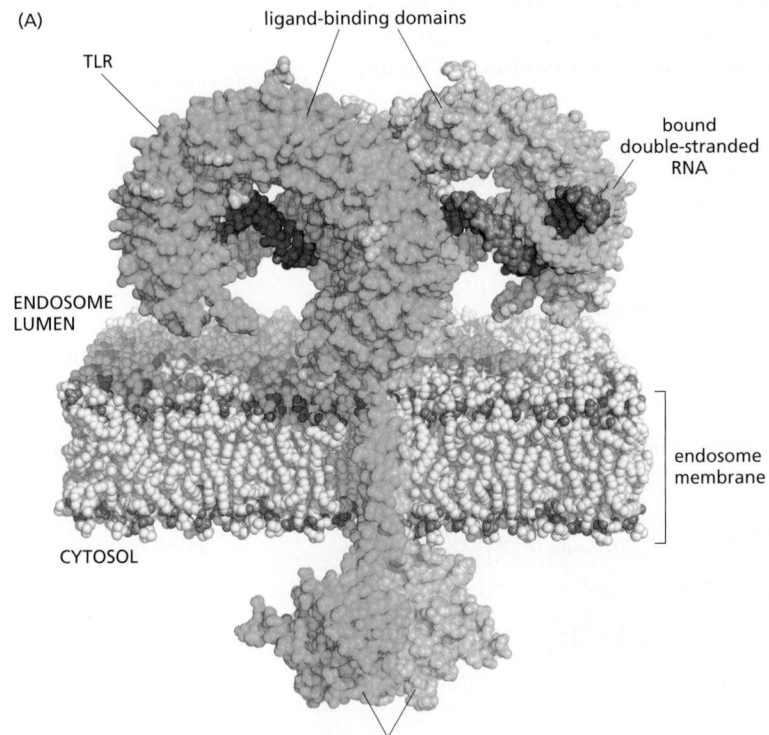

(A)

TLR

ligand-binding domains

bound double-stranded RNA

ENDOSOME LUMEN

endosome membrane

CYTOSOL

cytosolic domains

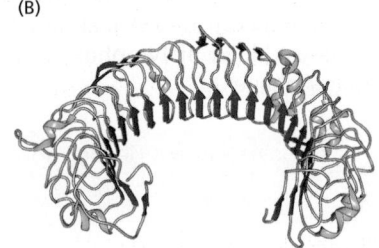

(B)

Figure 24–4 **A Toll-like receptor.** The structure of human TLR3 is shown (*green*), with a double-stranded RNA molecule (dsRNA, *blue*) bound to it. The receptor is a homodimer in the membrane of endosomes. The binding of dsRNA to the two horseshoe-shaped domains on the lumenal side of the endosome brings the two cytosolic domains together, allowing adaptor proteins in the cytosol to assemble into a large signaling complex (not shown). (B) The crystal structure of a lumenal domain of the receptor, which contains 23 conventional leucine-rich repeats, each of which contributes a β strand to the continuous β sheet (*red*) that lines the concave surface of the structure. (A, adapted from L. Liu et al., *Science* 320:379–381, 2008; B, adapted from J. Choe, M.S. Kelker and I.A. Wilson, *Science* 309: 581–585, 2006. Both with permission from AAAS. PDB: 1ZIW.)

membrane of Gram-negative bacteria; TLR5 recognizes the protein that forms the bacterial flagellum; and TLR9 recognizes short, unmethylated sequences of bacterial, viral, or protozoan DNA, called CpG motifs, which are uncommon in vertebrate DNA.

In addition to TLRs, vertebrates use several other families of PRRs to detect pathogens. One is the large family of **NOD-like receptors** (**NLRs**). Like TLRs, NLRs have leucine-rich repeat motifs, but they are exclusively cytoplasmic and recognize a distinct set of bacterial molecules. Individuals who are homozygous for a particular mutant allele of the NLR gene *NOD2* have a greatly increased risk of developing Crohn's disease, a chronic inflammatory disease of the small intestine, possibly triggered by a bacterial infection. Another class of PRRs consists of *RIG-like receptors* (*RLRs*), which are members of the RNA helicase family of proteins. They are also exclusively cytoplasmic and detect viral pathogens. A fourth class of PRRs consists of *C-type lectin receptors* (*CLRs*), which are transmembrane cell-surface proteins that recognize carbohydrates (which is why they are called lectins) on various microbes. Table 24–1 summarizes some PRRs and their ligands and locations in cells. Collectively, these and other PRRs act as an alarm system to alert the innate and adaptive immune systems that an infection is brewing (Movie 24.1).

When a cell-surface or intracellular PRR binds a PAMP, it stimulates the cell to secrete a variety of cytokines and other extracellular signal molecules. Some of these inhibit viral replication, but most induce a local inflammatory response that helps eliminate the pathogen, as we now discuss.

Activated PRRs Trigger an Inflammatory Response at Sites of Infection

When a pathogen invades a tissue, it activates PRRs on or in various cells of the innate immune system, resulting in an **inflammatory response** at the site of infection. The inflammatory response depends on changes in local blood vessels and is characterized clinically by local pain, redness, heat, and swelling. The blood vessels dilate and become permeable to fluid and proteins, leading to local swelling and an accumulation of blood proteins that aid in defense. At the same time, the endothelial cells lining the local blood vessels are stimulated to express cell adhesion proteins, which promote the attachment and escape of white blood cells or *leukocytes* (see Figure 19–29B), adding to the local swelling; initially neutrophils escape, followed later by lymphocytes and monocytes (the blood-borne precursors of macrophages).

TABLE 24–1 Some Pattern Recognition Receptors (PRRs)

Receptor	Location	Ligand	Origin of ligand
Toll-like receptors (TLRs)			
TLR3	Endolysosomal system	Double-stranded RNA	Viruses
TLR4	Plasma membrane	Bacterial lipopolysaccharide (LPS); viral coat proteins	Bacteria; viruses
TLR5	Plasma membrane	Flagellin	Bacteria
TLR9	Endolysosomal system	Unmethylated CpG DNA	Bacteria, viruses, protozoa
NOD-like receptors (NLRs)			
NOD2	Cytoplasm	Degradation products of peptidoglycans	Bacteria
Retinoic acid-inducible gene 1-like receptors (RLRs)			
RIG1	Cytoplasm	Double-stranded RNA	Viruses
C-type lectin receptors (CLRs)			
Dectin1	Plasma membrane	β-Glucan	Fungi

The activation of PRRs results in the production of a large variety of extracellular signal molecules that mediate the inflammatory response at the site of an infection. These include both lipid signal molecules, such as prostaglandins, and protein (or peptide) signal molecules called **cytokines**. Some of the most important **pro-inflammatory cytokines** are *tumor necrosis factor-α* (*TNFα*), *interferon-γ* (*IFNγ*), a variety of *chemokines* (which recruit leukocytes), and various *interleukins* (*ILs*) that we discuss later, including IL1, IL6, IL12, and IL17. In addition, a secreted PRR (mannose-binding lectin) activates the complement system when the PRR binds to a pathogen; fragments of complement proteins released during complement activation stimulate an inflammatory response (discussed shortly; see Figure 24–7).

When activated by PAMPs, most cell-surface and intracellular PRRs stimulate the production of multiple pro-inflammatory cytokines by activating intracellular signaling pathways that switch on transcription regulators, including NFκB, to induce the transcription of the relevant cytokine genes (see Figure 15–62). Some PRRs, however, can also stimulate pro-inflammatory cytokine production by a different mechanism: when activated, several cytoplasmic NLRs assemble with adaptor proteins and specific protease precursors of the caspase family (discussed in Chapter 18) to form **inflammasomes**, in which pro-inflammatory cytokines such as IL1 are cleaved from their inactive precursor proteins by activated caspases. These cytokines are then released from the cell by a poorly understood, unconventional secretion pathway. Inflammasomes closely resemble apoptosomes in their assembly and structure, but, in apoptosomes, procaspases are activated to initiate a proteolytic caspase cascade that leads to apoptosis (see Figure 18–7).

NLR-dependent inflammasome assembly can also be triggered in the absence of infection if cells are damaged or stressed. Such cells produce "danger signals," such as altered or misplaced self molecules, which can activate the relevant NLRs: the arthritis caused by uric acid crystals formed in the joints of individuals with gout, who have abnormally high uric acid levels in their blood, is a painful example.

Phagocytic Cells Seek, Engulf, and Destroy Pathogens

In all animals, the recognition of a microbial invader is usually quickly followed by its engulfment by a phagocytic cell. Macrophages are long-lived phagocytes that reside in most vertebrate tissues; they are among the first cells to encounter invading microbes, whose PAMPs activate the macrophages to secrete pro-inflammatory signal molecules. Neutrophils are short-lived phagocytes that are abundant in blood but are not present in healthy tissues; they are rapidly recruited to sites of infection by various attractive molecules, including formylmethionine-containing peptides (which are released by microbes but are not made by mammalian cells), chemokines secreted by activated macrophages, and peptide fragments produced from cleaved, activated complement proteins. The recruited neutrophils contribute their own pro-inflammatory cytokines.

In addition to their PRRs, macrophages and neutrophils display a variety of cell-surface receptors that recognize fragments of complement proteins or antibodies bound to the surface of a pathogen. The binding of such a pathogen to these receptors leads to its phagocytosis (**Figure 24–5**) and an attack on the ingested pathogen once inside a phagolysosome. The phagocytes possess an impressive armory of weapons to kill the invader, including enzymes such as lysozyme and acid hydrolases that can degrade the pathogen's cell wall. The cells assemble *NADPH oxidase complexes* on the phagolysosomal membrane, where the complexes catalyze the production of highly toxic oxygen-derived compounds, including superoxide (O_2^-), hydrogen peroxide, and hydroxyl radicals. A transient increase in oxygen consumption by the phagocytic cells, called the *respiratory burst*, accompanies the production of these toxic compounds. Whereas macrophages generally survive this killing frenzy and live to kill again, neutrophils do not. Dead and dying neutrophils are a major component of the pus that forms in acute bacterially infected wounds; their half-life in the human bloodstream is only a few hours.

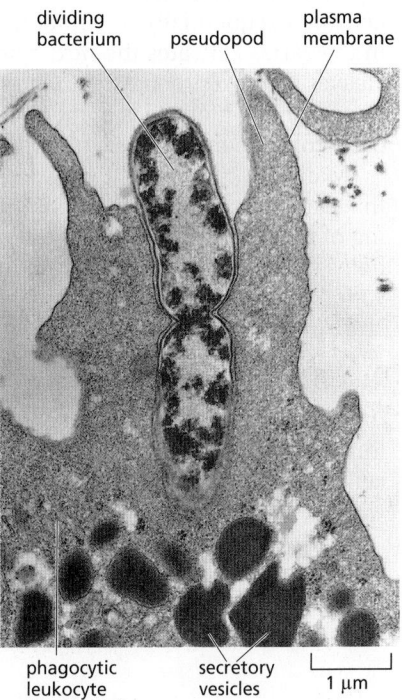

dividing bacterium   pseudopod   plasma membrane

phagocytic leukocyte (neutrophil)   secretory vesicles   1 μm

Figure 24–5 **Antibody-activated phagocytosis.** Electron micrograph of a neutrophil phagocytosing an antibody-coated bacterium, which is in the process of dividing. The process in which antibody (or complement) coating of a pathogen increases the efficiency with which the pathogen is phagocytosed is called *opsonization*. (Courtesy of Dorothy F. Bainton, from R.C. Williams, Jr. and H.H. Fudenberg, Phagocytic Mechanisms in Health and Disease. New York: Intercontinental Medical Book Corporation, 1971.)

If a pathogen is too large to be successfully phagocytosed (if it is a large para-site such as a worm, for example), a group of macrophages, neutrophils, or eosin-ophils (another type of leukocyte) will gather around the invader. They secrete defensins and other damaging agents and release the toxic products of the respi-ratory burst. This barrage is often sufficient to destroy the pathogen (**Figure 24–6**).

Complement Activation Targets Pathogens for Phagocytosis or Lysis

The blood and other extracellular fluids contain numerous proteins with antimi-crobial activity, some of which are produced in response to an infection, while others are produced constitutively. The most important of these are components of the **complement system**, which consists of about thirty interacting soluble proteins that are mainly made continuously by the liver and are inactive until an infection or another trigger activates them. They were originally identified by their ability to amplify and "complement" the action of antibodies made by B cells, but some are also secreted PRRs, which directly recognize PAMPs on microbes.

The *early complement components* consist of three sets of proteins, belong-ing to three distinct pathways of complement activation—the *classical pathway*, the *lectin pathway*, and the *alternative pathway*. The early components of all three pathways act locally to cleave and activate **C3**, which is the pivotal comple-ment component (**Figure 24–7**); individuals with a C3 deficiency are subject to repeated bacterial infections. The early components are proenzymes, which are activated sequentially by proteolytic cleavage. The cleavage of each proenzyme in the series activates the next component to generate a serine protease, which cleaves the next proenzyme in the series, and so on. Since each activated enzyme cleaves many molecules of the next proenzyme in the chain, the activation of the early components consists of an amplifying *proteolytic cascade*.

Many of these protein cleavages liberate a biologically active small fragment that can attract neutrophils, plus a membrane-binding larger fragment. The bind-ing of the large fragment to a cell membrane, usually the surface of a pathogen, helps stimulate the next reaction in the sequence. In this way, complement acti-vation is largely kept confined to the cell surface where it began. In particular, the large fragment of C3, called C3b, binds covalently to the surface of the pathogen. Here, it recruits protein fragments produced by cleavage of other early complement components to form proteolytic complexes that catalyze the subsequent steps in the complement cascade. The early events in complement activation have diverse functions: C3b-binding receptors on phagocytic cells enhance the ability of these cells to phagocytose the pathogen, and similar receptors on B cells enhance the

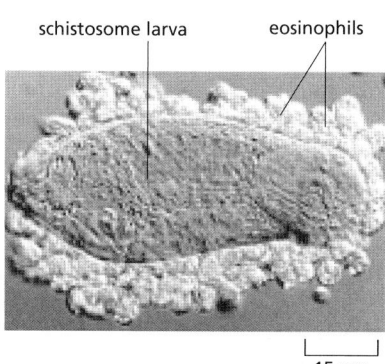

schistosome larva   eosinophils

15 μm

Figure 24–6 Eosinophils attacking a parasite. Phagocytes cannot ingest large parasites such as the schistosome larva shown here. When the larva is coated with antibody or complement components, however, eosinophils (and other leukocytes) can recognize it and collectively kill it by secreting a large variety of toxic molecules. (Courtesy of Anthony Butterworth.)

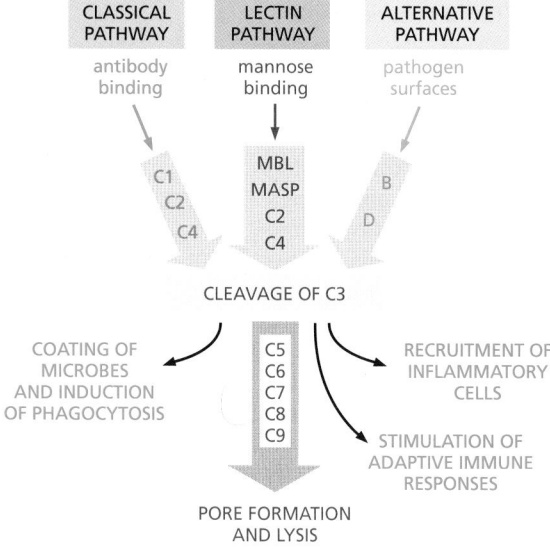

Figure 24–7 The principal stages in complement activation by the classical, lectin, and alternative pathways. In all three pathways, the reactions of complement activation usually take place on the surface of an invading microbe, such as a bacterium, and lead to the cleavage of C3 and the various consequences shown. As indicated, the complement proteins C1 to C9, mannose-binding lectin (MBL), MBL-associated serine protease (MASP), and factors B and D are the central components of the complement system. The early components are shown within *gray arrows*, while the late components are shown within a *brown arrow*. The functions of the protein fragments produced during complement activation are indicated by the *black arrows*. The various complement proteins that regulate the system are omitted.

ability of these cells to make antibodies against various microbial molecules on C3b-coated pathogens; the smaller fragment of C3 (called C3a), as well as small fragments of C4 and C5, act independently as diffusible signals to promote an inflammatory response by recruiting leukocytes to the site of infection.

As indicated in Figure 24–7, antibodies bound to the surface of a pathogen activate the *classical pathway*. *Mannose-binding lectin*, mentioned earlier, is a secreted PRR that initiates the *lectin pathway* of complement activation when it recognizes bacterial or fungal glycolipids and glycoproteins bearing terminal mannose and fucose sugars in a particular spatial conformation. These initial binding events in the classical and lectin pathways cause the recruitment and activation of the early complement components. Finally, molecules on the surface of pathogens will often directly activate the *alternative pathway*.

Host cells produce various plasma membrane molecules that prevent complement reactions from proceeding on their cell surface. The most important of these is the carbohydrate moiety sialic acid, a common constituent of cell-surface glycoproteins and glycolipids (see Figure 10–16). Because pathogens generally lack sialic acid, they are singled out for complement-mediated destruction, while host cells are spared. Some pathogens, including the bacterium *Neisseria gonorrhoeae* that causes the sexually transmitted disease gonorrhea, coat themselves with a layer of sialic acid to effectively hide from the complement system.

Membrane-immobilized C3b, produced by any of the three pathways, triggers a further cascade of reactions that leads to the assembly of the *late complement components* to form *membrane attack complexes*. These protein complexes assemble in the pathogen membrane near the site of C3 activation, forming aqueous pores through the membrane (**Figure 24–8**). For this reason, and because they perturb the structure of the lipid bilayer in their vicinity, they make the membrane leaky and can, in some cases, cause the microbe to lyse.

The self-amplifying, inflammatory, and destructive properties of the complement cascade make it essential that key activated components be rapidly inactivated after they are generated, ensuring that the attack does not spread to nearby host cells. Inactivation is achieved in at least two ways. First, specific inhibitor proteins in the blood or on the surface of host cells terminate the cascade, by either binding or cleaving certain complement components once the components have been activated by proteolytic cleavage. Second, many of the activated components in the cascade are unstable; unless they bind immediately to either the next component in the complement cascade or to a nearby membrane, they rapidly inactivate.

Virus-Infected Cells Take Drastic Measures to Prevent Viral Replication

Because host-cell ribosomes make a virus's proteins and host-cell lipids form the membranes of enveloped viruses, PAMPs are generally not present on the surface of viruses. Therefore, the only general way that a host cell PRR can recognize the presence of a virus is to detect unusual elements of the viral genome, such as the

Figure 24–8 Assembly of the late complement components to form a membrane attack complex. The cleavage of the early complement components (shown within *gray arrows* in Figure 24–7) results in the formation of C3b-containing proteolytic complexes called C5 convertases (not shown). These then cleave the first of the late components, C5, to produce C5a and C5b. As illustrated, C5b rapidly assembles with C6 and C7 to form C567, which then binds firmly via C7 to the membrane. One molecule of C8 binds to the complex to form C5678. The binding of a molecule of C9 to C5678 induces a conformational change in C9 that exposes a hydrophobic region and causes C9 to insert into the lipid bilayer of the target membrane. This starts a chain reaction in which the altered C9 binds a second molecule of C9, which can then bind another molecule of C9, and so on. In this way, a ring of C9 molecules forms a large transmembrane channel in the membrane.

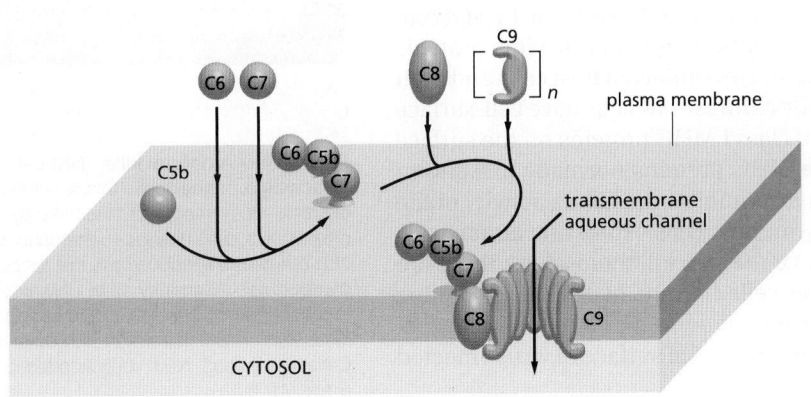

double-stranded RNA (dsRNA) that is an intermediate in the life cycle of many viruses and is recognized by several PRRs including the Toll-like receptor TLR3; in addition, DNA virus genomes frequently contain significant amounts of the CpG motifs discussed earlier, which can be recognized by TLR9 (see Table 24–1, p. 1300).

Mammalian cells are particularly adept at recognizing the presence of dsRNA, which activates intracellular PRRs that induce the host cell to produce and secrete two antiviral cytokines—**interferon-α (IFNα)** and **interferon-β (IFNβ)**. These interferons are referred to as *type I interferons* to distinguish them from IFNγ, which is a type II interferon and has different functions, as we discuss later. Type I interferons act in both an autocrine fashion on the infected cells that produced it and a paracrine fashion on uninfected neighbors. They bind to a common cell-surface receptor, which activates the JAK–STAT intracellular signaling pathway (see Figure 15–56) to stimulate specific gene transcription and thereby the production of more than 300 proteins, including many cytokines, reflecting the complexity of the cell's acute response to a viral infection.

The production of type I interferons appears to be a general response of mammalian cells to a viral infection, and viral components other than dsRNA can trigger it. The type I interferons help block viral replication in multiple ways. They activate a latent ribonuclease that nonspecifically degrades single-stranded RNA. They also indirectly activate a protein kinase that phosphorylates and inactivates the protein synthesis initiation factor eIF2 (discussed in Chapter 6), thereby shutting down most protein synthesis in the infected host cell. Apparently, by destroying most of its own RNA and transiently halting most of its protein synthesis, the host cell inhibits viral replication without killing itself. If these measures fail, the cell takes an even more extreme step to prevent the virus from replicating: it kills itself by undergoing apoptosis, often with the help of immune killer cells, as we discuss next.

Natural Killer Cells Induce Virus-Infected Cells to Kill Themselves

Type I interferons also have less direct ways of blocking viral replication. One of these is to enhance the activity of **natural killer cells (NK cells)**, which are leukocytes related to T and B cells but are part of the innate immune system and are recruited early to sites of inflammation. Like *cytotoxic T cells* of the adaptive immune system (discussed later), NK cells destroy virus-infected cells by inducing the infected cells to kill themselves by undergoing apoptosis (discussed in Chapter 18). We consider how killer cells induce apoptosis later, when we discuss how cytotoxic T cells do it (see Figure 24–43). Although they kill in the same way, the means by which cytotoxic T cells and NK cells distinguish the surface of virus-infected cells from that of uninfected cells are different (**Movie 24.2**).

Both cytotoxic T cells and NK cells recognize the same special class of cell-surface proteins on a host cell to help determine if the cell is virus-infected, but they use distinct receptors to do so. The special cell-surface proteins recognized are called *class I MHC proteins*, because they are encoded by genes in the *major histocompatibility complex*; almost all nucleated cells in vertebrates express these genes, and we discuss them in detail later. Cytotoxic T cells use both *T cell receptors (TCRs)* and *co-receptors* to recognize peptide fragments of viral proteins bound to class I MHC proteins on the surface of virus-infected host cells and then induce the infected cells to kill themselves. By contrast, NK cells have cell-surface *inhibitory receptors* that monitor the level of class I MHC proteins on the surface of other host cells: the high levels of these MHC proteins normally present on healthy cells engage these receptors and thereby inhibit the killing activity of the NK cells. The NK cells thus focus primarily on host cells expressing abnormally low levels of class I MHC proteins and induce them to kill themselves; these are mainly virus-infected cells and some cancer cells (**Figure 24–9**). NK cell killing activity is stimulated when various *activating receptors* on the NK cell surface recognize specific proteins that are greatly increased on the surface of virus-infected cells and some cancer cells.

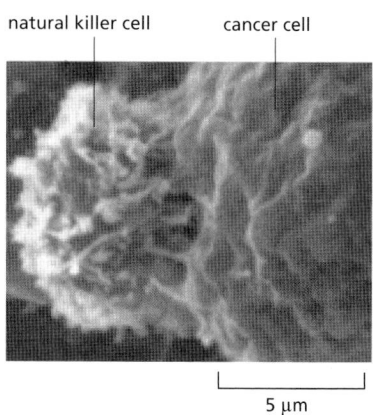

natural killer cell   cancer cell

5 μm

Figure 24–9 **A natural killer (NK) cell attacking a cancer cell.** This scanning electron micrograph was taken shortly after the NK cell attached to the cancer cell, but before it induced the cell to die by apoptosis. (Courtesy of J.C. Hiserodt, in Mechanisms of Cytotoxicity by Natural Killer Cells [R.B. Herberman and D. Callewaert, eds.]. New York: Academic Press, 1995.)

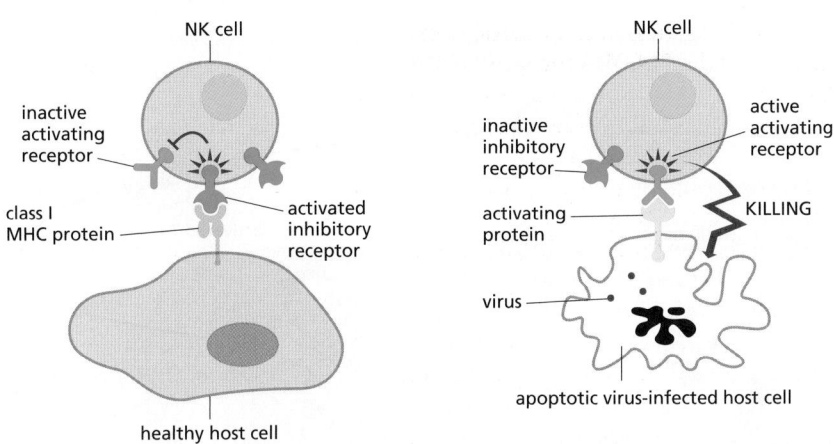

Figure 24–10 **How an NK cell recognizes its target.** An NK cell preferentially attacks infected host cells and cancer cells because these cells have on their surface both activating proteins and, in some cases, abnormally low levels of class I MHC proteins. (A) The high levels of class I MHC proteins found on normal host cells activate inhibitory receptors on the NK cell that suppress the killing activity of the NK cell. (B) In contrast, the activating proteins on infected cells and cancer cells bind to activating receptors on the NK cell and stimulate the killing activity of the cell.

The reason that class I MHC protein levels are often low on virus-infected cells is that many viruses have developed a variety of mechanisms to inhibit the expression of these proteins on the surface of the host cells they infect, in order to avoid detection by cytotoxic T cells: some viruses encode proteins that block class I MHC gene transcription; others block the intracellular assembly of peptide–MHC complexes; still others block the transport of these complexes to the cell surface. By evading recognition by cytotoxic T cells in these ways, however, a virus incurs the wrath of NK cells, which recognize the infected cells as being different—both because the infected cells express little class I MHC protein and because they express large amounts of other surface proteins that are recognized by the activating receptors on the NK cells (**Figure 24–10**).

Dendritic Cells Provide the Link Between the Innate and Adaptive Immune Systems

Dendritic cells are crucially important components of the innate immune system. They are a heterogeneous class of cells that are widely distributed in the tissues and organs of vertebrates. They express a large variety of PRRs, which enable dendritic cells to recognize and phagocytose invading pathogens and their products and to become activated in the process. The activated dendritic cells cleave the proteins of the pathogen into peptide fragments, which bind to newly synthesized MHC proteins, which then carry the fragments to the dendritic cell surface. The activated cells then migrate to a nearby lymphoid organ such as a lymph node (also called a lymph gland), where they present the peptide–MHC complexes to T cells of the adaptive immune system, activating the T cells to join in the battle against the specific pathogen (**Figure 24–11**).

In addition to the complexes of MHC proteins and microbial peptides displayed on their cell surface, activated dendritic cells also display cell-surface *co-stimulatory proteins* that help activate T cells (see Figure 24–11). As we discuss later, the activated dendritic cells also secrete a variety of cytokines that influence the type of response that the T cells make, ensuring that it is appropriate to fight the particular pathogen. In these ways, dendritic cells serve as crucial links between the innate immune system, which provides a rapid first line of defense against invading pathogens, and the adaptive immune system, which mounts slower but more powerful and highly specific responses to attack an invader, as we now discuss.

Summary

All multicellular organisms possess innate immune defenses against invading pathogens; these defenses include physical and chemical barriers and various defensive cell responses. In vertebrates, these innate defense responses can also

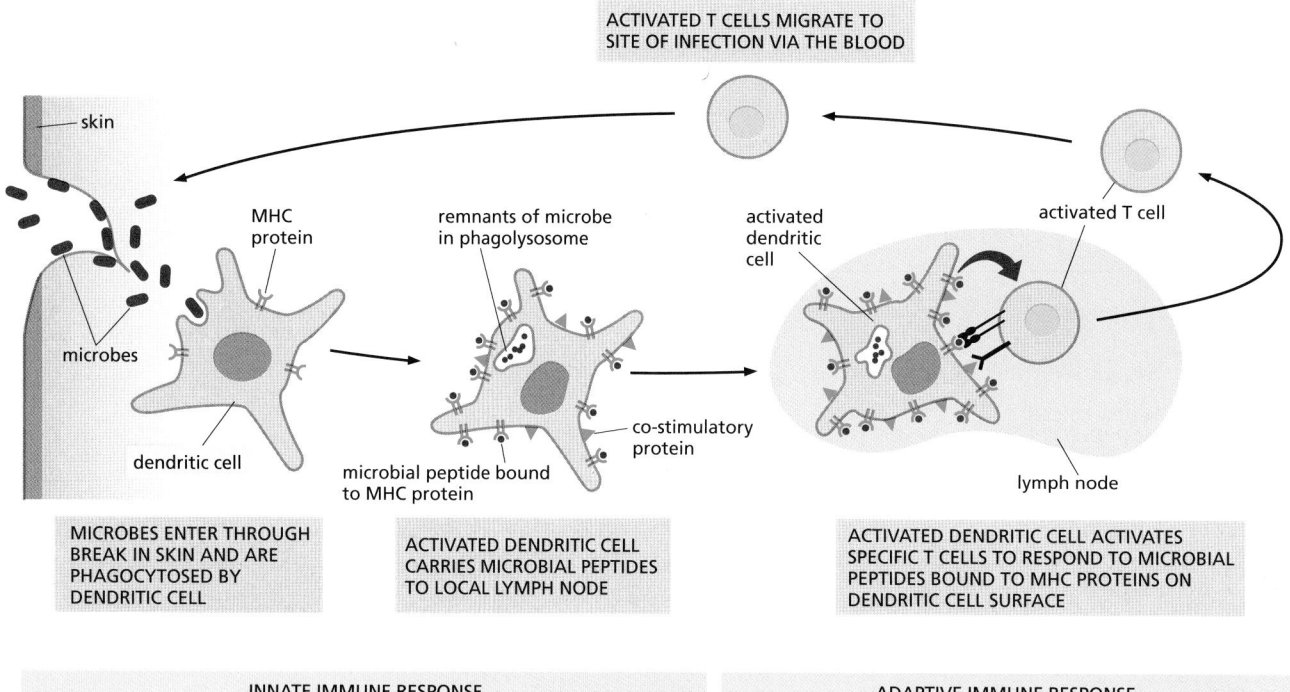

Figure 24–11 **Dendritic cells as functional links between the innate and adaptive immune systems.** Dendritic cells pick up invading microbes or their products at the site of an infection. The microbial PAMPs activate the dendritic cells to express *co-stimulatory proteins* and increased amounts of MHC proteins on their surface and to migrate via lymphatic vessels to a nearby lymph node. In the lymph node, the activated dendritic cells activate T cells that express appropriate receptors for the co-stimulatory proteins and the microbial peptides bound to MHC proteins on the dendritic cell surface. The activated T cells proliferate, and some of their progeny migrate to the original site of infection, where they help eliminate the microbes, either by activating local macrophages or by killing infected host cells (not shown). In addition, some of the activated T cells help stimulate specific B cells in the lymph node to secrete antibodies against the microbe (not shown).

A crucial feature of dendritic cell activation is that the pathogen provides an individual dendritic cell with both the peptides for presentation to T cells and the PAMP signals that activate the dendritic cell to express co-stimulatory proteins. In this way, the individual dendritic cell has all it needs to activate specific T cells that recognize the peptide–MHC complexes on its surface (Movie 24.3).

recruit specific and more powerful adaptive immune responses to help fight the infection. Innate immune responses rely on the ability of host cells to recognize characteristic features of microbial molecules called pathogen-associated molecular patterns, or PAMPs, which can be associated with a pathogen's proteins, lipids, sugars, or nucleic acids. PAMPs are mainly recognized by pattern recognition receptors (PRRs), including the Toll-like receptors (TLRs) found on or in both plant and animal cells. In vertebrates, some PRRs are secreted and can activate complement when they bind microbial PAMPs. The complement system, which can also be activated by antimicrobial antibodies bound to pathogens, consists of a group of blood proteins that are activated in sequence to help fight infections, by disrupting the pathogen's membrane, stimulating an inflammatory response, or targeting the microbe for phagocytosis—mainly by macrophages and neutrophils. The phagocytes use a combination of degradative enzymes, antimicrobial peptides, and oxygen-derived toxic molecules to kill invading pathogens; in addition, they secrete various signal molecules that help trigger an inflammatory response.

Cells infected by a virus produce and secrete type I interferons (IFNα and IFNβ), which induce a complex set of host-cell responses that inhibit viral replication. The interferons also enhance the killing activity of natural killer (NK) cells. An NK cell kills infected host cells because they express large amounts of surface proteins that activate the NK cell; the killing is especially efficient when infected cells express reduced amounts of class I MHC proteins, which, when present in normal amounts on a host cell surface inhibit the killing activity of NK cells.

Dendritic cells of the innate immune system functionally link innate immune responses to adaptive immune responses. The cells become activated when their PRRs pick up microbes and their products at sites of infection and phagocytose them. The activated cells cleave the microbial proteins into peptide fragments, which bind to newly made MHC proteins, which transport the fragments to the cell surface. The activated dendritic cells then carry the peptide–MHC complexes to a lymph organ, where they activate appropriate T cells to make specific adaptive immune responses against the microbes.

OVERVIEW OF THE ADAPTIVE IMMUNE SYSTEM

A dramatic "big bang" in immune defense mechanisms occurred when jawed vertebrates evolved and acquired an **adaptive immune system**. This sophisticated defense system depends on B and T lymphocytes (B and T cells), which, during their development, rearrange particular DNA sequences in various combinations so that, together, the cells can produce an almost limitless variety of B and T cell receptors and antibodies. Collectively, these proteins can bind to essentially any molecule, including small chemicals, carbohydrates, lipids, and proteins; individually, they can distinguish between molecules that are very similar—such as between two proteins that differ in only a single amino acid, or between two optical isomers of the same small molecule. By this strategy, the adaptive immune system can recognize and respond specifically to any pathogen, including new mutant forms. However, because the genetic rearrangement process produces both receptors that can bind to self molecules as well as receptors that can bind to foreign molecules, vertebrates have had to evolve special mechanisms to ensure that B and T cells do not react against the host's own molecules and cells—a process called *immunological self-tolerance.*

Moreover, many harmless foreign substances enter the body, for example, as food or inhaled material, and it would be pointless and potentially dangerous to mount adaptive immune responses against them. Such inappropriate responses are normally avoided because innate immune responses are required to call adaptive immune responses into play and do so only when the innate cells' PRRs recognize microbial PAMPs, as we discussed earlier. One can trick the adaptive immune system into responding to a harmless foreign molecule, such as a foreign protein, by co-injecting a molecule (often of microbial origin) called an *adjuvant*, which activates PRRs. This trick is called **immunization** and is the basis of vaccination. Any substance capable of stimulating B or T cells to make a specific adaptive immune response against it is referred to as an **antigen** (*anti*body *gen*erator).

There are two broad classes of adaptive immune responses—*antibody responses* and *T-cell-mediated immune responses*, and most pathogens induce both classes of responses. In **antibody responses**, B cells are activated to secrete antibodies, which are proteins that circulate in the bloodstream and permeate the other body fluids, where they can bind specifically to the foreign antigen that stimulated their production (see Figure 24–2). Binding of antibody neutralizes extracellular viruses and microbial toxins (such as tetanus toxin or cholera toxin) by blocking their ability to bind to receptors on host cells. Antibody binding also marks invading pathogens for destruction, both by making it easier for phagocytes of the innate immune system to ingest and destroy them and by activating the complement system.

In **T-cell-mediated immune responses**, T cells recognize foreign antigens that are bound to MHC proteins on the surface of host cells such as dendritic cells, which are specialized for presenting antigen to T cells and are therefore referred to as *professional antigen-presenting cells* (*APCs*). Because MHC proteins carry fragments of pathogen proteins from inside a host cell to the cell surface, T cells can detect pathogens hiding inside a host cell and either kill the infected cell (see Figure 24–2) or stimulate phagocytes or B cells to help eliminate the pathogens.

In this section, we discuss the origins and general properties of B and T cells. In later sections, we consider the specific properties and functions of these cells.

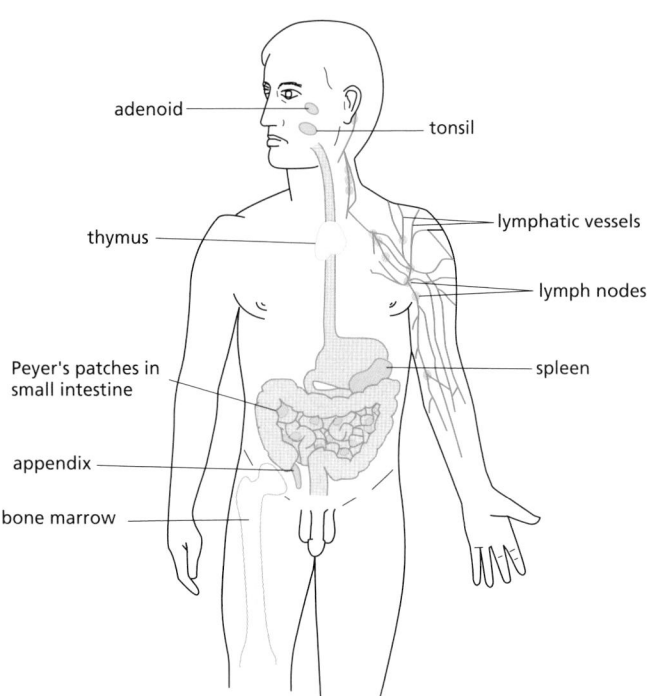

Figure 24–12 Human lymphoid organs. Lymphocytes develop from lymphoid progenitor cells in the thymus and bone marrow *(yellow)*, which are therefore called *central* (or *primary*) *lymphoid organs*. The newly formed lymphocytes migrate from these primary organs to *peripheral* (or *secondary*) *lymphoid organs*, where they can react with foreign antigen. Only some of the peripheral lymphoid organs *(blue)* and lymphatic vessels *(green)* are shown; many lymphocytes, for example, are found in the skin and respiratory tract. As we discuss later, the lymphatic vessels ultimately empty into the bloodstream (not shown).

B Cells Develop in the Bone Marrow, T Cells in the Thymus

There are about 2×10^{12} lymphocytes in the human body, making the immune system comparable in cell mass to the liver or the brain. They occur in large numbers in the blood and lymph (the colorless fluid in the lymphatic vessels, which connect the lymph nodes in the body to each other and to the bloodstream). They are also concentrated in **lymphoid organs**, such as the thymus, lymph nodes, and spleen (**Figure 24–12**), and many are also found in other organs, including skin, lung, and gut.

T cells and B cells derive their names from the organs in which they develop: T cells develop in the *thymus*, and B cells, in adult mammals, develop in the *bone marrow*. Both types of cells develop from lymphoid progenitor cells that are produced from multipotent *hematopoietic stem cells*, which are found mainly in the bone marrow (**Figure 24–13**). The hematopoietic stem cells give rise to more than just lymphocytes: as discussed in Chapter 22, they produce all of the cells of the

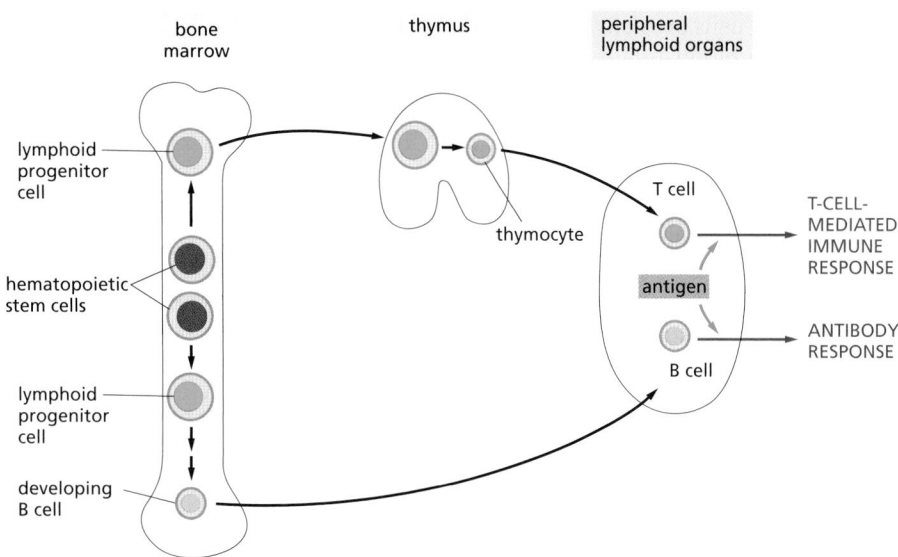

Figure 24–13 The development of B and T cells. The central lymphoid organs, where lymphocytes develop from lymphoid progenitor cells, are labeled in *yellow boxes*. The lymphoid progenitor cells develop from multipotent hematopoietic stem cells in the bone marrow. Some lymphoid progenitor cells develop locally in the bone marrow into immature B cells, while others migrate to the thymus (via the bloodstream) where they develop into thymocytes (developing T cells). Foreign antigens activate B cells and T cells mainly in peripheral lymphoid organs, such as lymph nodes or the spleen.

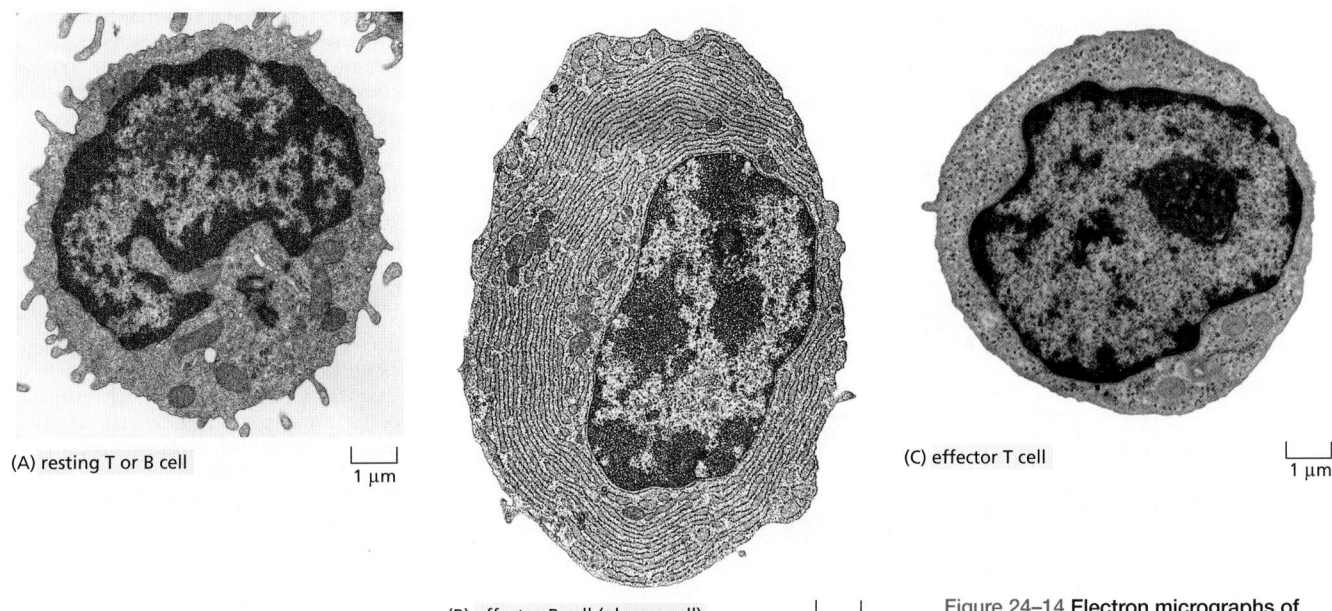

(A) resting T or B cell

1 μm

(B) effector B cell (plasma cell)

1 μm

(C) effector T cell

1 μm

Figure 24–14 Electron micrographs of resting and effector lymphocytes.
(A) This resting lymphocyte could be either a B cell or a T cell, as these cells are difficult to distinguish morphologically until antigen activates them to become effector cells. (B) An effector B cell (a plasma cell). It is filled with an extensive rough endoplasmic reticulum (ER), which is distended with antibody molecules that are secreted in large amounts. (C) An effector T cell, which has relatively little rough ER but is filled with free ribosomes; it secretes cytokines, but in relatively small amounts. The three cells are shown at the same magnification. (A and B, from D. Zucker-Franklin et al., *Atlas of Blood Cells: Function and Pathology*, 2nd ed., 1988. Reprinted with permission of Wolters Kluwer; C, David M. Phillips/Science Source.)

hematopoietic system, including erythrocytes, leukocytes, and platelets (see Figure 22–32).

Because they are sites where lymphocytes develop from lymphoid progenitor cells, the thymus and bone marrow are referred to as **central (primary) lymphoid organs** (see Figure 24–12). As we discuss later, most B and T cells die in the central lymphoid organs soon after they develop, without ever functioning. Others, however, mature and migrate via the blood to the **peripheral (secondary) lymphoid organs**—mainly the lymph nodes, spleen, and epithelium-associated lymphoid tissues in the gastrointestinal tract, respiratory tract, and skin. It is in these peripheral lymphoid organs that foreign antigens activate B and T cells (see Figure 24–13).

B and T cells become morphologically distinguishable from each other only after antigen has activated them: resting B and T cells look very similar, even in an electron microscope (**Figure 24–14**A). After activation by an antigen, both proliferate and mature into *effector cells*. Effector B cells secrete antibodies; in their most mature form, called *plasma cells*, they are filled with an extensive rough endoplasmic reticulum that is busily making antibodies (Figure 24–14B). In contrast, effector T cells (Figure 24–14C) contain very little endoplasmic reticulum and secrete a variety of cytokines rather than antibodies. Whereas B-cell-derived antibodies are widely distributed by the bloodstream, T-cell-derived cytokines mainly act locally on neighboring cells, although some are carried via the blood and act on distant host cells.

Immunological Memory Depends On Both Clonal Expansion and Lymphocyte Differentiation

The most remarkable feature of the adaptive immune system is that it can respond to millions of different foreign antigens in a highly specific way. Human B cells, for example, collectively, can make more than 10^{12} different antibody molecules that react specifically with the antigen that induced their production. How can B cells and T cells respond specifically to such an enormous diversity of foreign antigens? The answer for both B and T cells is the same. As each lymphocyte develops in a central lymphoid organ, it becomes committed to react with a particular antigen before ever being exposed to the antigen. It expresses this commitment in the form of cell-surface receptors that specifically bind the antigen. When a lymphocyte encounters its antigen in a peripheral lymphoid organ, the binding of the antigen to the receptors (with help from co-stimulatory signals, discussed later)

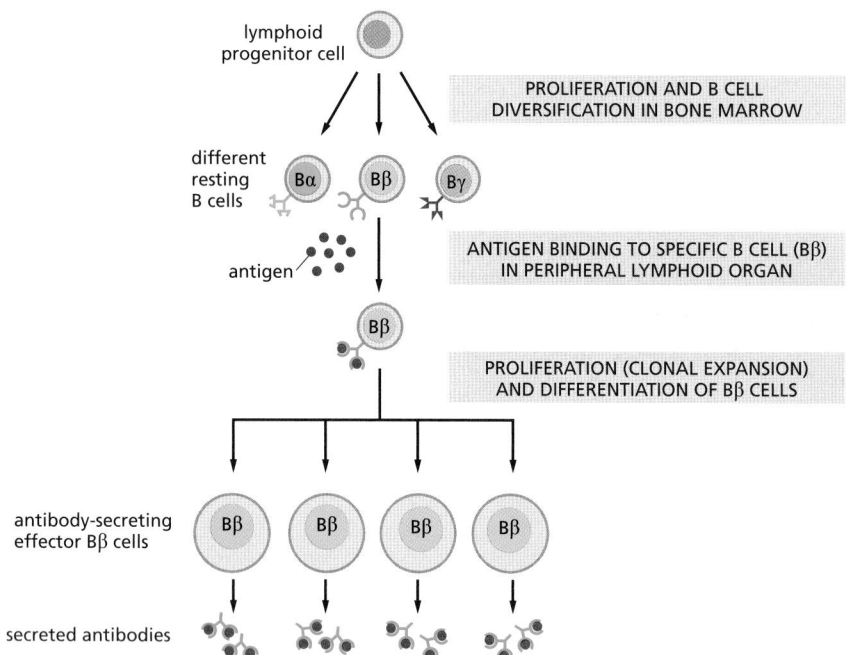

lymphoid
progenitor cell

PROLIFERATION AND B CELL
DIVERSIFICATION IN BONE MARROW

different
resting
B cells   Bα   Bβ   Bγ

antigen

ANTIGEN BINDING TO SPECIFIC B CELL (Bβ)
IN PERIPHERAL LYMPHOID ORGAN

Bβ

PROLIFERATION (CLONAL EXPANSION)
AND DIFFERENTIATION OF Bβ CELLS

antibody-secreting
effector Bβ cells   Bβ   Bβ   Bβ   Bβ

secreted antibodies

Figure 24–15 Clonal selection. An antigen activates only those lymphocytes that are already committed to respond to it. The committed cell expresses cell-surface receptors that specifically recognize the antigen. The human adaptive immune system consists of many millions of different T and B lymphocyte clones, with cells within a clone expressing the same unique antigen receptor. Before its first encounter with antigen, a clone would usually contain only one or a small number of cells. A particular antigen may activate hundreds of different clones, each expressing a different antigen receptor that binds either a different part of the antigen or the same part with a different binding affinity. Although only B cells are shown here, T cells are selected in a similar way. Note that the antigen receptors on the B cells labeled β in this diagram have the same antigen-binding site as the antibodies secreted by the effector Bβ cells. As we discuss later, B cells require co-stimulatory signals from T cells to become activated by antigen to proliferate and differentiate into antibody-secreting cells (not shown).

activates the lymphocyte; this causes the lymphocyte to proliferate, thereby producing many more cells with the same receptor—a process called *clonal expansion*. The encounter with antigen also causes some of the cells to differentiate into *effector cells*. An antigen therefore selectively stimulates those cells that express complementary antigen-specific receptors and are thus already committed to respond to it (**Figure 24–15**). This arrangement, called **clonal selection**, provides an explanation for **immunological memory**, whereby we develop lifelong immunity to many common infectious diseases after our initial exposure to the pathogen—either through natural infection or vaccination.

It is easy to demonstrate such immunological memory in experimental animals. If an animal is immunized once with antigen A, an immune response (antibody, T-cell-mediated, or both) can be detected after several days; the response rises rapidly and exponentially, and then, more gradually, declines. This is the characteristic course of a **primary immune response**, occurring on an animal's first exposure to an antigen. If, after some weeks, months, or even years have elapsed, the animal is immunized again with antigen A, it will usually produce a **secondary immune response** that differs from the primary response: the lag period is shorter, because there are now many more preexisting B or T cells (or both) with specificity for antigen A, and the response is greater and more efficient. These differences indicate that the animal has "remembered" its first exposure to antigen A. If the animal is given a different antigen (for example, antigen B) instead of a second immunization with antigen A, the response is typical of a primary, and not a secondary, immune response. The secondary response therefore reflects antigen-specific immunological memory for antigen A (**Figure 24–16**).

Immunological memory depends on both lymphocyte proliferation and differentiation. In an adult animal, the peripheral lymphoid organs contain a mixture of lymphocytes in at least three stages of maturation: *naïve cells*, *effector cells*, and *memory cells*. When **naïve cells** encounter their specific foreign antigen for the first time, the antigen stimulates some of them to proliferate and differentiate into **effector cells**, which then carry out an immune response (effector B cells secrete antibody, whereas effector T cells either kill infected cells or influence the response of other immune cells—by secreting cytokines, for example). Some of the antigen-stimulated naïve cells multiply and differentiate into **memory cells**, which are more easily and more quickly induced to become effector cells by a later encounter with the same antigen: like naïve cells, when memory cells encounter their antigen, they give rise to either effector cells or more memory cells (**Figure 24–17**).

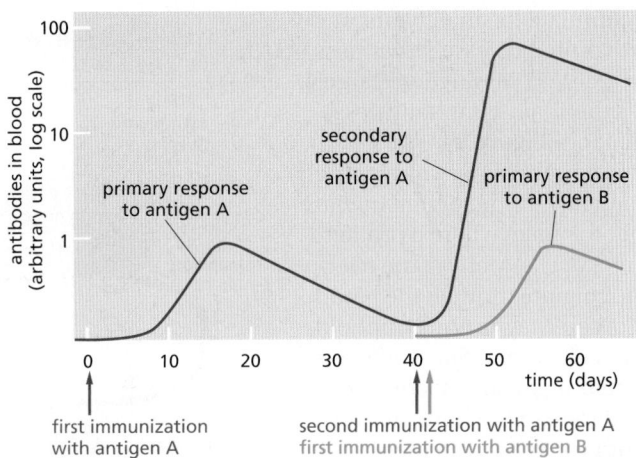

Figure 24–16 Immunological memory: primary and secondary antibody responses. The secondary response induced by a second exposure to antigen A is faster and greater than the primary response and is specific for A, indicating that the adaptive immune system has specifically remembered its previous encounter with antigen A. The same type of immunological memory is observed in T-cell-mediated responses (not shown). As we discuss later, the types of antibodies produced in the secondary response are different from those produced in the primary response, and these antibodies bind the antigen more tightly.

Thus, during the primary response, clonal expansion and differentiation of antigen-stimulated naïve cells creates many memory cells, which are able to respond to the same antigen more sensitively, rapidly, and effectively. And, unlike most effector cells, which die within days or weeks, memory cells can persist for the lifetime of the animal, even in the absence of their specific antigen, thereby providing lifelong immunological memory. Although most effector B and T cells die after an immune response is over, some survive as effector cells and help provide long-term protection against the pathogen. A small proportion of the plasma cells produced in a primary B cell response, for example, can survive for many months or years in the bone marrow, where they continue to secrete their specific antibodies into the bloodstream.

Lymphocytes Continuously Recirculate Through Peripheral Lymphoid Organs

Pathogens generally enter the body through an epithelial surface, usually through the skin, gut, or respiratory tract. To induce an adaptive immune response, microbes or their products must travel from these entry points to a peripheral lymphoid organ, such as a lymph node or the spleen, which are the sites where lymphocytes are activated (see Figure 24–11). The route and destination depend on the site of entry. Lymphatic vessels carry antigens that enter through the skin or respiratory tract to local lymph nodes; antigens that enter through the gut end up in gut-associated peripheral lymphoid organs such as Peyer's patches; and the spleen filters out antigens that enter the blood (see Figure 24–12). As discussed earlier (see Figure 24–11), in many cases, activated dendritic cells will carry the antigen from the site of infection to the peripheral lymphoid organ, where they play a crucial part in activating T cells, as we discuss later.

But only a tiny fraction of naïve B and T cells can recognize a particular microbial antigen in a peripheral lymphoid organ, a reasonable estimate being between 1/10,000 and 1/1,000,000 of each class of lymphocyte, depending on the antigen. How do these rare cells find an antigen-presenting cell displaying their specific antigen? The answer is that the lymphocytes continuously recirculate between one peripheral lymphoid organ and another via the lymph and blood. In a lymph node, for example, lymphocytes continually leave the bloodstream by squeezing out between specialized endothelial cells lining small veins called *postcapillary*

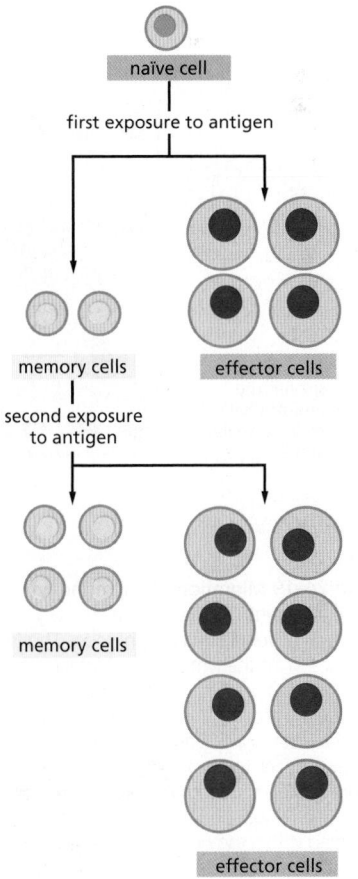

Figure 24–17 A model for the cellular basis of immunological memory. When stimulated by their specific antigen and co-stimulatory signals, naïve lymphocytes proliferate and differentiate. Most become effector cells, which function and then usually die, while others become memory cells. During a subsequent exposure to the same antigen, the memory cells respond more readily, rapidly, and efficiently than did the naïve cells: they proliferate and give rise to effector cells and to more memory cells. Some memory T cells also develop from a minority of effector T cells (not shown). It is not known how the decision to become an effector cell versus a memory cell is made.

venules. After percolating through the node, they accumulate in small lymphatic vessels that leave the node and connect with other lymphatic vessels that pass through other lymph nodes downstream (see Figure 24–12). Passing into larger and larger vessels, the lymphocytes eventually enter the main lymphatic vessel (the *thoracic duct*), which carries them back into the blood (**Figure 24–18**).

The continuous recirculation of a lymphocyte between the blood and lymph ends only if its specific antigen activates it in a peripheral lymphoid organ. In that case, the lymphocyte remains in the peripheral lymphoid organ, where it proliferates and differentiates into either effector cells or memory cells. Many of the effector T cells leave the lymphoid organ via the lymph and migrate through the blood to the site of infection (see Figure 24–11), whereas others stay in the lymphoid organ and help activate (or suppress) other immune cells there. Some effector B cells (plasma cells) remain in the peripheral lymphoid organ and secrete antibodies into the blood for days until they die; others migrate to the bone marrow, where they secrete antibodies into the blood for months or years. The memory T and B cells produced join the recirculating pool of lymphocytes.

Lymphocyte recirculation depends on specific interactions between the lymphocyte cell surface and the surface of the endothelial cells lining the blood vessels in the peripheral lymphoid organs. Lymphocytes that enter a lymph node via the blood, for example, adhere weakly to specialized endothelial cells lining the postcapillary venules via *homing receptors* that belong to the *selectin* family of cell-surface lectins that bind to specific sugar groups on the endothelial cell surface (see Figure 19–28). The lymphocytes roll slowly along the surface of the endothelial cells until another, much stronger adhesion system, dependent on an integrin protein, is called into play by chemokines secreted by the endothelial cells. Now, the lymphocytes stop rolling, and they crawl out of the blood vessel into the lymph node by using yet another cell adhesion protein called CD31 (**Figure 24–19**). Although B and T cells initially enter the same region of a lymph node, different chemokines guide them to separate regions of the node—B cells to *lymphoid follicles* and T cells to the *paracortex* (**Figure 24–20**).

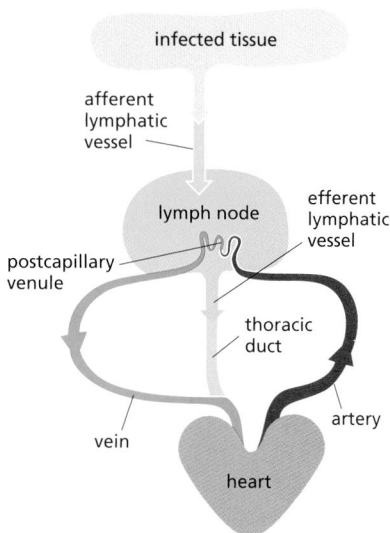

Figure 24–18 The path followed by lymphocytes as they continuously recirculate between the lymph and blood. The circulation through a lymph node *(yellow)* is shown here. Microbial antigens are usually carried into the lymph node by activated dendritic cells (not shown), which enter the node via afferent lymphatic vessels draining an infected tissue *(green)*. B and T cells, by contrast, enter via the blood, migrating out of the bloodstream into the lymph node through postcapillary venules. Unless they encounter their antigen, the B and T cells leave the lymph node via efferent lymphatic vessels, which eventually join the thoracic duct. The thoracic duct empties into a large vein carrying blood to the heart, completing the circulation cycle for T and B cells. A typical circulation cycle for these lymphocytes takes about 12–24 hours.

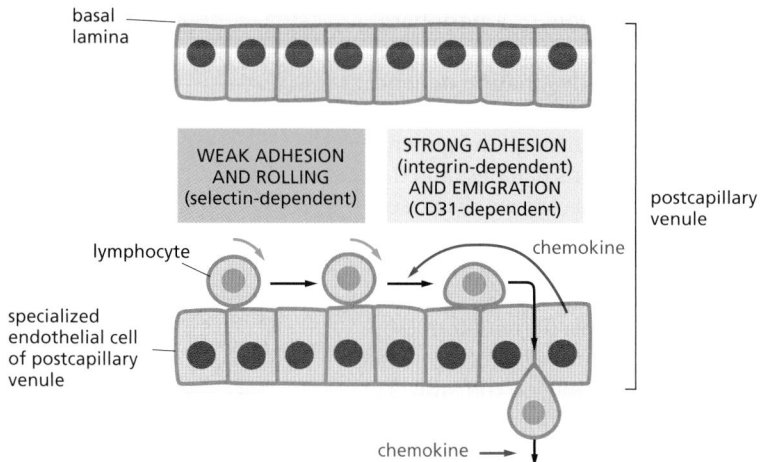

Figure 24–19 Migration of a lymphocyte out of the bloodstream into a lymph node.
A circulating lymphocyte adheres weakly to the surface of the specialized endothelial cells lining a postcapillary venule in a lymph node. This initial adhesion is mediated by L-selectin (discussed in Chapter 19) on the lymphocyte surface. The adhesion is sufficiently weak to enable the lymphocyte, pushed by the flow of blood, to roll along the surface of the endothelial cells. Stimulated by chemokines secreted by specialized endothelial cells in the node *(curved red arrow)*, the lymphocyte rapidly activates a stronger adhesion system, mediated by an integrin. This strong adhesion enables the cell to stop rolling. The lymphocyte then uses an immunoglobulin-like cell adhesion protein (CD31) to bind to the junctions between adjacent endothelial cells and migrate out of the venule. The subsequent migration of the lymphocyte in the lymph node depends on chemokines produced within the node *(straight red arrow)*. The migration of other types of leukocytes out of the bloodstream into sites of infection occurs in a similar way (see Figure 19–28 and Movie 19.2).

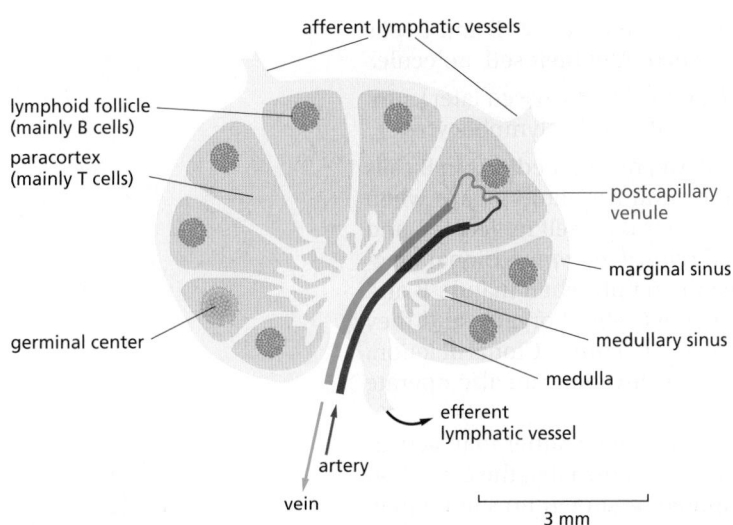

Figure 24–20 **A simplified drawing of a human lymph node.** B cells are primarily clustered in structures called lymphoid follicles, whereas T cells are found mainly in the paracortex. Chemokines attract both types of lymphocytes into the lymph node from the blood via postcapillary venules (see Figure 24–19). B and T cells then migrate to their respective areas, attracted by different chemokines. If they do not encounter their specific antigen, both B cells and T cells then enter the medullary sinuses and leave the node via the efferent lymphatic vessel. This vessel ultimately empties into the bloodstream, allowing the lymphocytes to begin another cycle of circulation through a peripheral lymphoid organ (see Figure 24–18). During an infection, proliferation of pathogen-specific B cells produces a germinal center in some lymphoid follicles.

Unless they encounter their antigen, both B and T cells soon leave the lymph node via efferent lymphatic vessels. If they encounter their antigen, however, they are stimulated to display adhesion receptors that trap the cells in the node; the cells accumulate at the junction between the B cell and T cell areas, where the rare antigen-specific B and T cells can interact, leading to their proliferation and differentiation into either effector cells or memory cells. Many of the effector cells leave the node, expressing different chemokine receptors that help guide them to their new destinations—effector plasma B cells to the bone marrow and effector T cells to sites of infection.

Immunological Self-Tolerance Ensures That B and T Cells Do Not Attack Normal Host Cells and Molecules

As discussed earlier, cells of the innate immune system use PRRs to distinguish microbial molecules from self molecules made by the host. The adaptive immune system has the far more difficult recognition task of responding specifically to an almost unlimited number of foreign molecules while not responding to the large number of self molecules. How does it accomplish this feat? It helps that self molecules normally do not induce the innate immune reactions required to activate adaptive immune responses. But even when an infection or tissue injury triggers innate reactions, the vast majority of self molecules normally still fail to induce an adaptive immune response. Why?

One important reason is that the adaptive immune system "learns" not to respond to self molecules. Normal mice, for example, cannot mount an immune response against one of their own protein components of the complement system called C5 (see Figure 24–7). However, mutant mice that lack the gene encoding C5 but are otherwise genetically identical to normal mice of the same strain can make a strong immune response to this blood protein when immunized with it. The **immunological self-tolerance** exhibited by normal mice persists only for as long as the self molecule remains in the body: if a self molecule such as C5 is experimentally removed from an adult mouse, the animal gains the ability to respond to it after a few weeks or months, as new B and T cells develop in the absence of C5. Thus, the adaptive immune system is genetically capable of responding to self molecules, but it learns not to do so.

Self-tolerance depends on a number of distinct mechanisms, including the following (**Figure 24–21**):

1. In *receptor editing*, developing B cells that recognize self molecules change their antigen receptors so that the cells no longer do so.

2. In *clonal deletion*, potentially self-reactive B and T cells die by apoptosis when they encounter their particular self molecule.

3. In *clonal inactivation* (also called clonal anergy), self-reactive B and T cells become functionally inactivated when they encounter their self molecule.

4. In *clonal suppression*, self-reactive *regulatory T cells* (discussed later) suppress the activity of other types of potentially self-reactive lymphocytes.

Some of these mechanisms—especially the first two, receptor editing in B cells and clonal deletion of B and T cells—operate in central lymphoid organs when newly formed self-reactive B and T cells first encounter their self molecules, and they are largely responsible for the process called *central tolerance*. Clonal inactivation and clonal suppression, by contrast, operate mainly when mature B and T cells encounter their self molecules in peripheral lymphoid organs, and they are largely responsible for the process called *peripheral tolerance*. Clonal deletion, however, can also operate peripherally, and clonal inactivation can also operate centrally.

Why does the binding of a self molecule lead to tolerance rather than activation? The answer is still not completely known. As we discuss later, the activation of a B or T cell by its antigen in a peripheral lymphoid organ requires more than just antigen binding: it requires co-stimulatory signals, which are provided by a *helper T cell* (discussed later) in the case of a B cell and by an activated dendritic cell in the case of a naïve T cell. The production of such signals is usually triggered by exposure to a pathogen, but a self-reactive lymphocyte normally encounters its self antigen in the absence of such signals. Under these conditions, the lymphocyte will not only fail to be activated, it will often be rendered tolerant—being either killed or inactivated, or actively suppressed by a regulatory T cell (see Figure 24–21). In peripheral lymphoid organs, both T cell tolerance and activation usually occur on the surface of a dendritic cell.

For reasons that are usually unknown, self-tolerance mechanisms sometimes fail, causing T or B cells (or both) to react against the animal's own molecules.

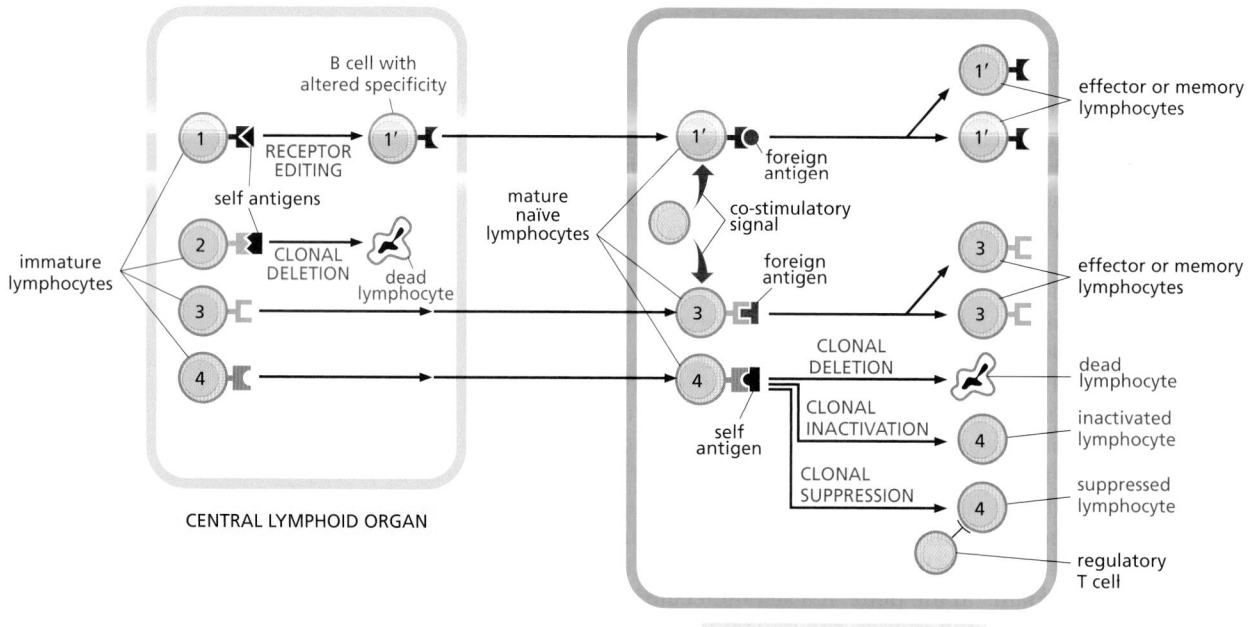

Figure 24–21 Mechanisms of immunological self-tolerance. When a self-reactive immature B cell binds its self molecule in the central lymphoid organ where the cell is produced, it may alter its antigen receptor so that it is no longer self-reactive (cell 1); this process is called receptor editing. Alternatively, when either an immature B or T cell binds its self molecule in a central lymphoid organ, it may die by apoptosis, a process called clonal deletion (cell 2). Because these two forms of tolerance (shown on the left) occur in central lymphoid organs, they are called *central tolerance*.

When a self-reactive naïve B or T cell escapes tolerance in the central lymphoid organ and binds its self molecule in a peripheral lymphoid organ (cell 4), or in another peripheral tissue, it will generally not be activated, because the binding usually occurs in the absence of sufficient co-stimulatory signals; instead, the cell may die by apoptosis (often after a period of proliferation), be inactivated, or be suppressed by a regulatory T cell. These forms of tolerance (shown on the right) are called *peripheral tolerance*. As discussed later, the cells providing the co-stimulatory signals are T lymphocytes for B cells and usually dendritic cells for T cells (not shown). For T cells at least, both activation and tolerance in a peripheral lymphoid organ usually occurs on the surface of a dendritic cell, although the dendritic cells are different in the two cases.

Myasthenia gravis is an example of such an **autoimmune disease**. Most of the affected individuals make antibodies against the acetylcholine receptors on their own skeletal muscle cells; these receptors are required for the muscle to contract normally in response to nerve stimulation, which releases acetylcholine (see Figure 11–39). The antibodies interfere with the normal functioning of the receptors so that the patients become weak and may die because they cannot breathe. Similarly, in *juvenile (type 1) diabetes*, adaptive immune reactions against insulin-secreting β cells in the pancreas kill these cells, leading to severe insulin deficiency.

Summary

Innate immune responses triggered by pathogens at sites of infection help activate adaptive immune responses in peripheral lymphoid organs. The adaptive immune system is composed of many millions of B and T cell clones, with the cells in each clone sharing a unique cell-surface receptor that enables them to bind a particular pathogen antigen. The binding of antigen to these receptors, with the help of co-stimulatory signals, stimulates the lymphocyte to proliferate and differentiate into an effector cell that can help eliminate the pathogen. Effector B cells secrete antibodies, which can act over long distances to help eliminate extracellular pathogens and their toxins. Effector T cells, by contrast, produce cell-surface and secreted co-stimulatory molecules, which mainly act locally to help other immune cells eliminate the pathogen; in addition, some T cells can induce infected host cells to kill themselves.

During a primary adaptive immune response to an antigen, lymphocytes that recognize the antigen proliferate so that there are more of them to respond the next time, during a secondary response to the same antigen; moreover, during a primary response, some lymphocytes differentiate into memory cells, which can respond faster and more efficiently the next time the same pathogen invades. These two mechanisms are largely responsible for immunological memory. Both B and T cells circulate continuously between one peripheral lymphoid organ and another via the blood and lymph; only if they encounter their specific foreign antigen in a peripheral lymphoid organ do they stop migrating, proliferate, and differentiate into effector cells or memory cells. Lymphocytes that would react against self molecules either alter their receptors (in the case of B cells) or are eliminated or inactivated; they can also be suppressed by regulatory T cells. These mechanisms collectively are responsible for immunological self-tolerance, which ensures that the adaptive immune system normally avoids attacking the molecules and cells of the host.

B CELLS AND IMMUNOGLOBULINS

Vertebrates inevitably die of infection if they are unable to make antibodies. **Antibodies** are secreted proteins that defend us against extracellular pathogens in several ways. They bind to viruses and microbial toxins, thereby preventing them from binding to and damaging host cells (see Figure 24–2). When bound to an extracellular pathogen or its products, antibodies also recruit some of the components of the innate immune system, including various types of leukocytes and components of the complement system, which work together to inactivate or eliminate the invaders.

Synthesized exclusively by B cells, antibodies are produced in billions of forms, each with a different amino acid sequence. They belong to the class of proteins called **immunoglobulins** (abbreviated as **Igs**) and are among the most abundant protein components in the blood. In this section, we discuss the structure and function of immunoglobulins and how they are made in so many different forms.

B Cells Make Immunoglobulins (Igs) as Both Cell-Surface Antigen Receptors and Secreted Antibodies

The first Igs made by a newly formed B cell are not secreted but are instead inserted into the plasma membrane, where they serve as receptors for antigen. They are called **B cell receptors** (**BCRs**), and each B cell has approximately 10^5 of them in its plasma membrane. Each BCR is stably associated with invariant transmembrane proteins that activate intracellular signaling pathways when antigen binds to the BCR; we discuss these invariant proteins later, when we consider how B cells are activated with the assistance of *helper T cells.*

Each B cell clone produces a single species of BCR, with a unique antigen-binding site. When an antigen and a helper T cell activate a naïve or a memory B cell, the B cell proliferates and differentiates into an effector cell, which then produces and secretes large amounts of soluble (rather than membrane-bound) Ig. The secreted Ig is now called an antibody, and it has the same unique antigen-binding site as the BCR (**Figure 24–22**).

A typical Ig molecule is bivalent, with two identical antigen-binding sites. It consists of four polypeptide chains—two identical *light chains* and two identical *heavy chains.* The N-terminal parts of both light and heavy chains usually cooperate to form the antigen-binding surface, while the more C-terminal parts of the heavy chains form the tail of the Y-shaped protein (**Figure 24–23**). The tail mediates many of the activities of antibodies, and antibodies with the same antigen-binding sites can have any one of a number of different tail regions, each of which gives the antibody different functional properties, such as the ability to activate complement or to bind to receptor proteins on phagocytic cells that bind a specific type of antibody tail.

Mammals Make Five Classes of Igs

In mammals, there are five major *classes* of Igs, each of which mediates a characteristic biological response following antigen binding to an antibody: IgA, IgD, IgE, IgG, and IgM, each with its own class of heavy chain—α, δ, ε, γ, and μ, respectively. IgA molecules have α chains, IgG molecules have γ chains, and so on. Moreover, there are four human IgG subclasses (IgG1, IgG2, IgG3, and IgG4), with γ_1, γ_2, γ_3, and γ_4 heavy chains, respectively. There are also two IgA subclasses in humans. In addition to the various classes and subclasses of heavy chains, higher vertebrates have two types of light chains, κ and λ, which seem to be functionally indistinguishable. Either type of light chain may be associated with any of the heavy chains, but an individual Ig molecule always contains identical light chains and identical heavy chains: an IgG molecule, for instance, may have either κ or λ light chains, but not one of each. As a result, an Ig's antigen-binding sites are always identical (see Figure 24–22).

The various heavy chains give a distinctive conformation to the tail region of antibodies, so that each class (and subclass) has characteristic properties of its own. **IgM** is always the first class of Ig that a developing B cell in the bone marrow makes. It forms the BCRs on the surface of *immature naïve B cells.* After these cells

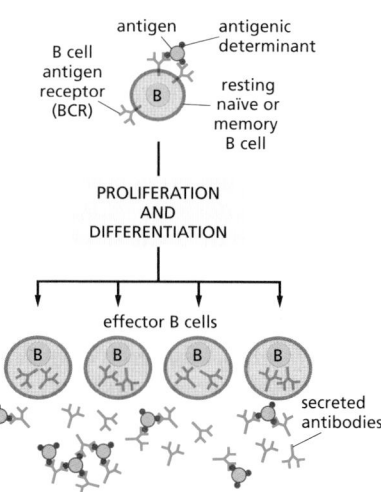

Figure 24–22 The B cell receptors (BCRs) and secreted antibodies made by a B cell clone. The binding of an antigen to BCRs on either a naïve or memory B cell (together with co-stimulatory signals provided by helper T cells—not shown) activates the cell to proliferate and differentiate into effector B cells. The effector cells produce and secrete antibodies with a unique antigen-binding site, which is the same as that of the cell-surface BCRs. Because antibodies have two identical antigen-binding sites, they can cross-link antigens, as shown for an antigen with multiple identical *antigenic determinants.*

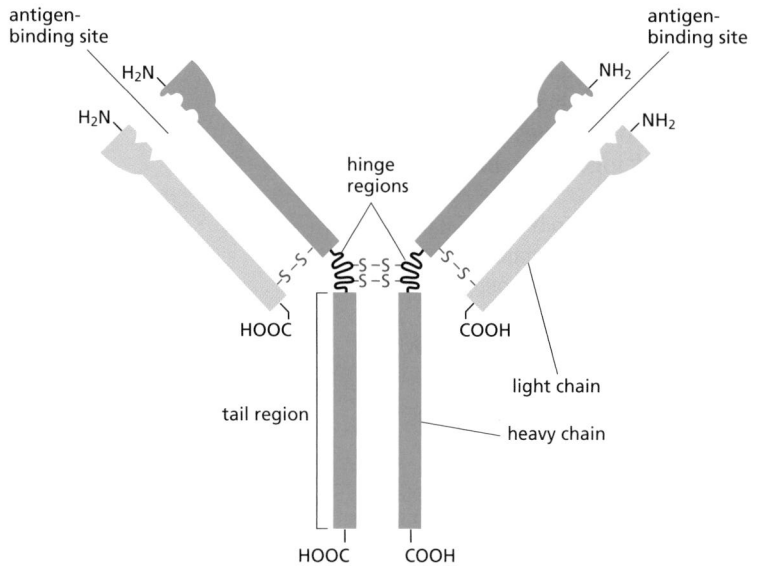

Figure 24–23 A schematic drawing of a bivalent antibody molecule. The two heavy chains each have a hinge region, which, because of its flexibility, improves the efficiency with which the antibody can cross-link antigens (see Figure 24–22). The two heavy chains also form the tail of the antibody, which determines its functional properties. The heavy and light chains are held together by both covalent S–S bonds *(red)* and noncovalent bonds (not shown).

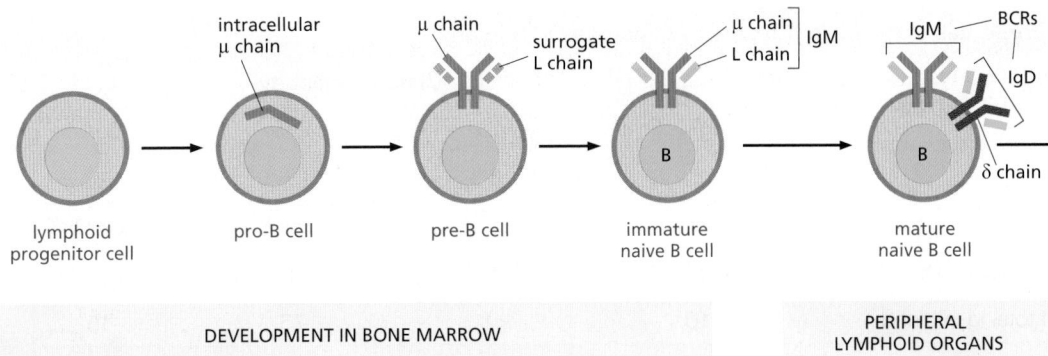

DEVELOPMENT IN BONE MARROW

PERIPHERAL
LYMPHOID ORGANS

leave the bone marrow, they start to produce **IgD** BCRs as well, with the same antigen-binding site as the IgM BCRs. These cells are now called *mature naïve B cells*, as they can now respond to their specific foreign antigen in peripheral lymphoid organs (**Figure 24–24**). IgM is also the major class of antibody secreted into the blood in the early stages of a primary antibody response on first exposure to an antigen. In its secreted form, IgM is a wheel-like pentamer composed of five four-chain units, giving it a total of 10 antigen-binding sites that allow it to bind strongly to pathogens; in its antigen-bound form, IgM is highly efficient at activating complement, which is important in early antibody responses to pathogens.

The major antibody class in the blood is **IgG**. These antibodies are four-chain monomers (see Figure 24–23), and they are produced in especially large quantities during secondary antibody responses. The tail region of some subclasses of IgG antibodies that are bound to antigen can activate complement and also bind to specific receptors on macrophages and neutrophils. Largely by means of such **Fc receptors** (so-named because antibody tails are called Fc regions), these phagocytic cells bind, ingest, and destroy infecting microorganisms that have become coated with the IgG antibodies produced in response to the infection; the activated Fc receptors also signal the phagocyte to secrete pro-inflammatory cytokines (**Movie 24.4**).

The tail region of **IgE** antibodies binds to another class of Fc receptors on the surface of *mast cells* in tissues and of *basophils* in the blood. Because antigen-free IgE antibodies bind with high affinity to such Fc receptors, the antibodies act as antigen receptors on these cells. Antigen binding to the bound antibodies activates the Fc receptors and stimulates the cells to secrete a variety of cytokines and biologically active amines, especially *histamine*, which causes blood vessels to dilate and become leaky; this helps leukocytes, antibodies, and complement components to enter sites where mast cells have been activated. The release of amines from mast cells and basophils is largely responsible for the symptoms of such *allergic reactions* as hay fever, asthma, and hives. In addition, mast cells secrete factors that attract and activate leukocytes called *eosinophils*, which also have Fc receptors that bind IgE molecules and can kill extracellular parasitic worms, especially if the worms are coated with IgE antibodies (see Figure 24–6).

IgA is the principal antibody class in secretions, including saliva, tears, milk, and respiratory and intestinal secretions. Yet another class of Fc receptors, located on the relevant epithelial cells, guides the secretion by binding antigen-free IgA dimers and transporting them across the epithelium. The properties of the various classes of antibodies in humans are summarized in **Table 24–2**.

All classes of Ig can be made in a membrane-bound form, as well as in a soluble, secreted form. The two forms differ only in the C-terminus of their heavy chain. The heavy chains of membrane-bound Ig molecules (BCRs) have a transmembrane hydrophobic C-terminus, which anchors them in the lipid bilayer of the B cell's plasma membrane. The heavy chains of secreted antibody molecules, by contrast, have instead a hydrophilic C-terminus, which allows them to escape from the cell. The switch in the character of the Ig molecules made occurs because the activation of B cells by antigen and helper T cells induces a change in the way in which the heavy-chain RNA transcripts are made and processed in the nucleus (see Figure 7–59).

Figure 24–24 Stages of B cell development. All of the stages shown occur before the cells bind their specific antigen. The first cells in the B cell lineage that make Ig are called *pro-B cells*; they make μ heavy chains, which remain in the endoplasmic reticulum until a special type of light chain is made called a surrogate light chain. The surrogate light chains substitute for genuine light chains and assemble with μ chains to form a receptor molecule that inserts into the plasma membrane. The cells are now called *pre-B cells*. Signaling from this pre-B cell receptor allows the cells to make bona fide light chains, which combine with μ chains to form four-chain IgM molecules that serve as cell-surface BCRs on *immature naïve B cells*. After these cells leave the bone marrow, they start to express IgD BCRs as well, which have the same antigen-binding sites as the IgM BCRs; it is this *mature naïve B cell* that reacts with its specific foreign antigen in peripheral lymphoid organs.

TABLE 24–2 Properties of the Major Classes of Antibodies in Humans

Properties	Class of antibody				
	IgM	IgD	IgG	IgA	IgE
Heavy chains	μ	δ	γ	α	ε
Light chains	κ or λ	κ or λ	κ or λ	κ or λ	κ or λ
Number of four-chain units	5	1	1	1 or 2	1
Percentage of total Ig in blood	10	<1	75	15	<1
Activates classical complement pathway	+	–	+ (some subclasses)	–	–
Crosses from mother to fetus	–	–	+ (some subclasses)	–	–
Binds to macrophages and neutrophils	+ (macrophages only)	–	+ (some subclasses)	+	–
Binds to mast cells and basophils	–	–	+ (some subclasses)	–	+

Ig Light and Heavy Chains Consist of Constant and Variable Regions

Both light and heavy chains have a variable amino acid sequence at their N-terminal ends but a constant sequence at their C-terminal ends. Whereas the **constant region** and **variable region** of a light chain are the same size, the constant region of a heavy chain is about three or four times longer, depending on the class (**Figure 24–25**).

The variable regions of the light and heavy chains come together to form the antigen-binding sites, and the variability of their amino acid sequences provides the structural basis for the diversity of these binding sites. The greatest diversity occurs in three small **hypervariable regions** in the variable regions of both light and heavy chains. Only about 5–10 amino acids in each hypervariable region form the actual antigen-binding site (**Figure 24–26**). As a result, the size of the **antigenic determinant** that an Ig molecule recognizes is generally comparably small: it can consist of fewer than 10 amino acids on the surface of a globular protein, for example (see Figure 24–22).

Both light and heavy chains are made up of repeating segments—each about 110 amino acids long and each containing one intrachain disulfide bond. Each repeating segment folds independently to form a compact functional unit called an **immunoglobulin (Ig) domain**. As shown in **Figure 24–27**A, a light chain consists of one variable (V_L) and one constant (C_L) domain, whereas a heavy chain has one variable and three or four constant domains: the variable domains of the light and heavy chains pair to form the antigen-binding region. Each Ig domain has a very similar three-dimensional structure, consisting of a sandwich of two β

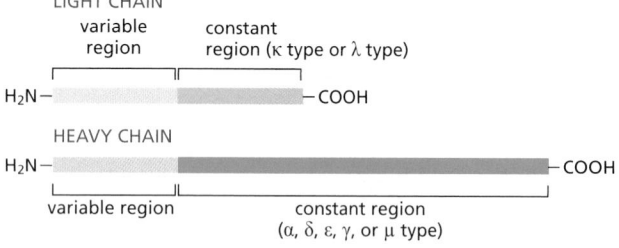

Figure 24–25 Constant and variable regions of Ig chains. The variable regions of the light and heavy chains form the antigen-binding sites, while the constant regions of the heavy chains determine the other biological properties of an Ig protein. The different subclasses of IgG antibodies have different γ-chain constant regions.

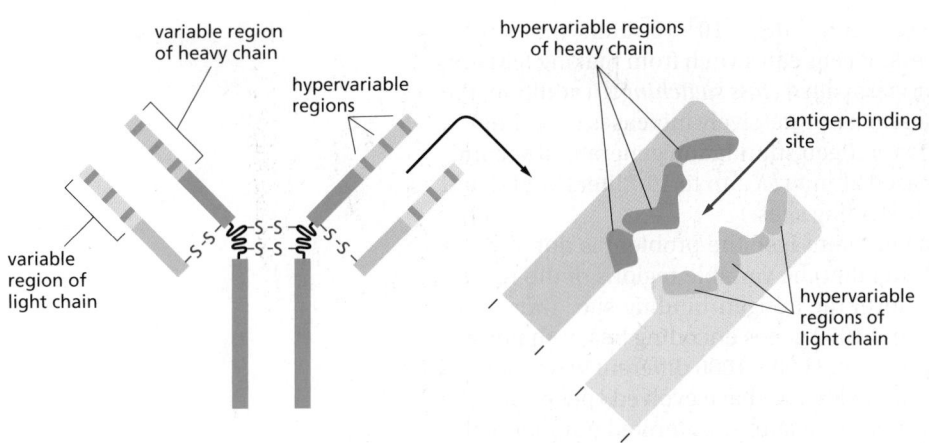

Figure 24–26 **Ig hypervariable regions.** Highly schematized drawing of how the three hypervariable regions in each light and heavy chain together form each antigen-binding site of an Ig protein.

sheets held together by a disulfide bond; the variable domains are unique in that each has its particular set of hypervariable regions, which are arranged in three *hypervariable loops* that cluster together at the ends of the variable domains to form the antigen-binding site (Figure 24–27B).

Ig Genes Are Assembled From Separate Gene Segments During B Cell Development

Even in the absence of antigen stimulation, a human can probably make more than 10^{12} different Ig molecules—its preimmune, **primary Ig repertoire**. The primary repertoire consists of IgM and IgD proteins and is apparently large enough to ensure that there will be an antigen-binding site to fit almost any potential

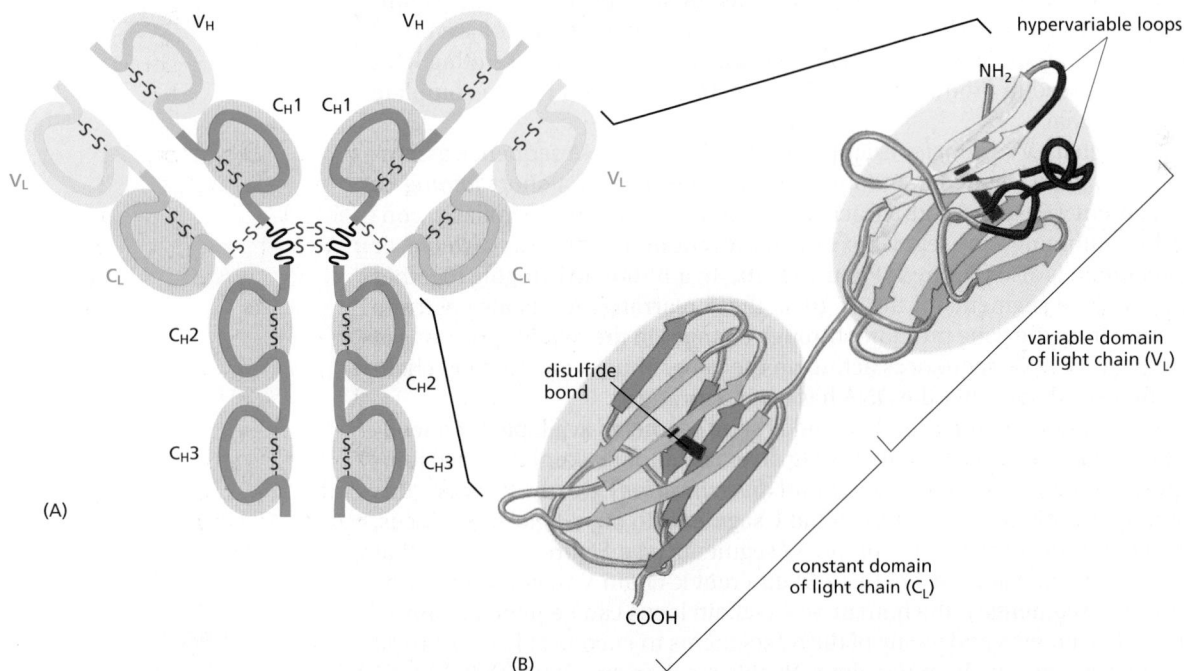

Figure 24–27 **Ig domains.** (A) The light and heavy chains in an Ig protein are each folded into similar repeating domains. The variable domains (shaded in *blue*) of the light and heavy chains (V_L and V_H) make up the antigen-binding sites, while the constant domains (shaded in *gray*) of the heavy chains (mainly C_H2 and C_H3) determine the other biological properties of the protein. The heavy chains of IgM and IgE do not have a hinge region and have an extra constant domain (C_H4). Hydrophobic interactions between domains on adjacent chains help hold the chains together in the Ig molecule: V_L binds to V_H, C_L binds to C_H1, and so on. (B) X-ray crystallography-based structures of the Ig domains of a light chain (**Movie 24.5**). Both the variable and constant domains have a similar overall structure, consisting of two β sheets joined by a disulfide bond *(red)*. Note that all the hypervariable regions *(black)* form loops at the far end of the variable domain, where they come together to form part of the antigen-binding site. All Igs are glycosylated on their C_H2 domains (not shown); the attached oligosaccharide chains vary from Ig to Ig and can greatly influence the biological properties of the protein, largely by affecting its binding to Fc receptors on immune cells.

antigenic determinant, albeit with low affinity—$K_a \approx 10^5$-10^7 liters/mole. After stimulation by antigen and helper T cells, B cells can switch from making IgM and IgD to making other classes of Ig—a process called *class switching*. In addition, the binding affinity of these Igs for their antigen progressively increases over time—a process called *affinity maturation*. Thus, antigen stimulation generates a **secondary Ig repertoire**, with a greatly increased affinity (K_a up to 10^{11} liters/mole) and diversity of both Ig classes and antigen-binding sites.

How can each of us make so many different Igs? The problem is not quite as formidable as it might first appear. Recall that the variable regions of the Ig light and heavy chains usually combine to form the antigen-binding site. Thus, if we had 1000 genes encoding light chains and 1000 genes encoding heavy chains, we could, in principle, combine their products in 1000 × 1000 different ways to make 10^6 different antigen-binding sites. Nonetheless, we have evolved special genetic mechanisms to enable our B cells to generate an almost unlimited number of different light and heavy chains in a remarkably economical way. We do so in two steps. First, before antigen stimulation, developing B cells join together separate *gene segments* in DNA to create the genes that encode the primary repertoire of low-affinity IgM and IgD proteins. Second, after antigen stimulation, the assembled *Ig genes* can undergo two further changes—mutations that can increase the affinity of their antigen-binding site and DNA rearrangements that switch the class of Ig made. Together, these changes produce the secondary repertoire of high-affinity IgG, IgE, and IgA proteins.

We produce our primary Ig repertoire by joining separate Ig **gene segments** together during B cell development. Each type of Ig chain—κ light chains, λ light chains, and heavy chains—is encoded by a separate locus on a separate chromosome. Each locus contains a large number of gene segments encoding the V region of an Ig chain, and one or more gene segments encoding the C region. During the development of a B cell in the bone marrow, a complete coding sequence for each of the two Ig chains to be synthesized is assembled by site-specific genetic recombination (discussed in Chapter 5). Once a V-region coding sequence is assembled next to a C-region sequence, it can then be co-transcribed and the resulting RNA transcript processed to produce an mRNA molecule that codes for the complete Ig polypeptide chain.

Each light-chain V region, for example, is encoded by a DNA sequence assembled from two gene segments—a long **V gene segment** and a short *joining* or **J gene segment** (**Figure 24–28**). Each heavy-chain V region is similarly constructed by combining gene segments, but here an additional *diversity segment*, or **D gene segment**, is also required (**Figure 24–29**). In addition to bringing together the separate gene segments of the Ig gene, these rearrangements also activate transcription from the gene promoter through changes in the relative positions of the *cis*-regulatory DNA sequences acting on the gene. Thus, a complete Ig chain can be synthesized only after the DNA has been rearranged.

The large number of inherited *V*, *J*, and *D* gene segments available for encoding Ig chains contributes substantially to Ig diversity, and the combinatorial joining of these segments (called *combinatorial diversification*) greatly increases this contribution. Any of the 35 or so functional *V* segments in our κ light-chain locus, for example, can be joined to any of the 5 *J* segments (see Figure 24–28), so that this locus can encode at least 175 (35 × 5) different κ-chain V regions. Similarly, any of the 40 *V* segments in the human heavy-chain locus can be joined to any of the 23 or so *D* segments and to any of the 6 *J* segments to encode at least 5520 (40 × 23 × 6) different heavy-chain V regions. By this mechanism alone, called **V(D)J recombination**, a human can produce 295 different V_L regions (175 κ and 120 λ) and 5520 different V_H regions. In principle, these could then be combined to make over 1.5×10^6 (295 × 5520) different antigen-binding sites.

V(D)J recombination is mediated by an enzyme complex called *V(D)J recombinase*, which recognizes recombination signal sequences in the DNA that flanks each gene segment to be joined. Although the process ensures that only appropriate gene segments recombine, a variable number of nucleotides are often lost from the ends of the recombining gene segments, and one or more randomly

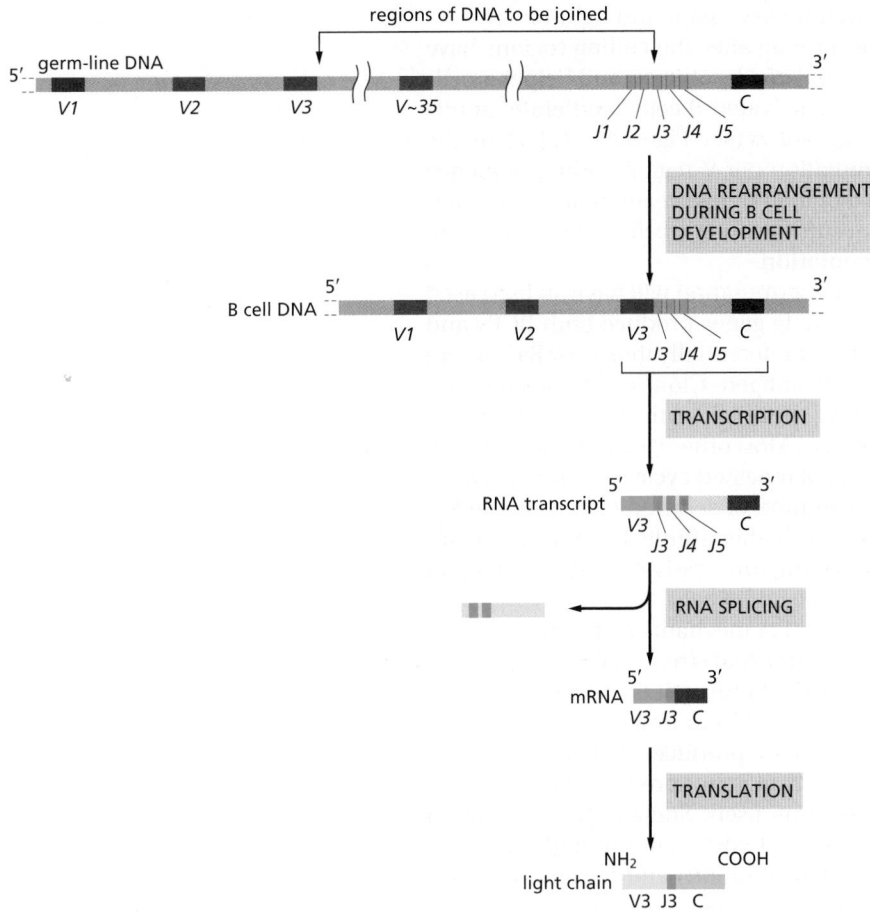

Figure 24–28 **The *V–J* joining process involved in making a human κ light chain.** In the "germ-line" DNA (where the Ig gene segments are not rearranged and are therefore not being expressed), the cluster of five *J* gene segments is separated from the C-region coding sequence by a short intron and from the 35 or so functional *V* gene segments by thousands of nucleotide pairs. During the development of a B cell, a randomly chosen *V* gene segment (*V3* in this case) is moved to lie precisely next to one of the *J* gene segments (*J3* in this case). The "extra" *J* gene segments (*J4* and *J5*) and the intron sequence are transcribed (along with the joined *V3* and *J3* gene segments and the C-region coding sequence) and then removed by RNA splicing to generate mRNA molecules with contiguous *V3*, *J3*, and *C* sequences, as shown. These mRNAs are then translated into κ light chains. A *J* gene segment encodes the 15 or so C-terminal amino acids of the V region, and a short sequence containing the *V–J* segment junction encodes the third hypervariable region, which is the most variable part of the light-chain V region.

chosen nucleotides are also inserted. This random loss and gain of nucleotides at joining sites is called **junctional diversification**, and it enormously increases the diversity of V-region coding sequences created by V(D)J recombination (up to about 10^8-fold), specifically in the third hypervariable region. This increased diversification comes at a price, however. In many cases, it shifts the reading frame to produce a nonfunctional gene, in which case the developing B cell fails to make a functional Ig molecule and consequently dies in the bone marrow. Once a B cell makes a functional heavy chain and light chain that form an antigen-binding site, it turns off the V(D)J recombination process, thereby ensuring that the cell makes Ig of only one antigen-binding specificity.

B cells making BCRs that bind strongly to self antigens in the bone marrow would be dangerous. Such B cells maintain expression of an active V(D)J recombinase and are activated by such self-binding to undergo a second round of V(D)J recombination in a light-chain locus, thereby changing the specificity of its BCR—the process of **receptor editing** discussed earlier; self-reactive B cells that fail to change their specificity die by apoptosis, in the process of clonal deletion (see Figure 24–21).

Antigen-Driven Somatic Hypermutation Fine-Tunes Antibody Responses

As mentioned earlier, with the passage of time following an infection or vaccination, there is usually a progressive increase in the affinity of the antibodies produced against the pathogen. This phenomenon of **affinity maturation** is due to

Figure 24–29 **The human heavy-chain locus.** There are 40 *V* segments, about 23 *D* segments, 6 *J* segments, and an ordered cluster of C-region coding sequences, each cluster encoding a different class of heavy chain. The *D* segment (and part of the *J* segment) encodes amino acids in the third hypervariable region, which is the most variable part of the heavy-chain V region. The genetic mechanisms involved in producing a heavy chain are the same as those shown in Figure 24–28 for light chains, except that two DNA rearrangement steps are required instead of one: first a *D* segment joins to a *J* segment, and then a *V* segment joins to the rearranged *DJ* segment. The rearrangements lead to the production of a VDJC mRNA that encodes a complete Ig heavy chain. The figure is not drawn to scale: the total length of the heavy-chain locus is over two megabases. Moreover, a number of details are omitted: for example, the exons encoding each C-region Ig domain and the hinge region (see Figure 24–27) and the different subclasses of C_γ-coding segments are not shown.

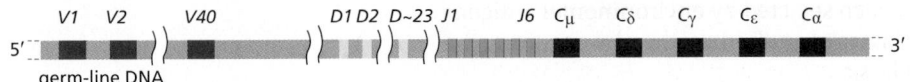

the accumulation of point mutations in both heavy-chain and light-chain V-region coding sequences. The mutations occur long after the coding regions have been assembled. After B cells have been stimulated by antigen and helper T cells in a peripheral lymphoid organ, some of the activated B cells proliferate rapidly in the lymphoid follicles and form *germinal centers* (see Figure 24–20). Here, the B cells mutate at the rate of about one mutation per V-region coding sequence per cell generation. Because this is about a million times greater than the spontaneous mutation rate in other genes and occurs in somatic cells rather than germ cells, the process is called **somatic hypermutation**.

Very few of the altered Igs generated by hypermutation will have an increased affinity for the antigen. But, because the same Ig genes produce both BCRs and secreted antibodies, the antigen will stimulate preferentially those few B cells that do make BCRs with increased affinity for the antigen. Clones of these altered B cells will preferentially survive and proliferate, especially as the amount of antigen decreases to very low levels late in the response. Most other B cells in the germinal center will die by apoptosis. Thus, as a result of repeated cycles of somatic hypermutation followed by antigen-driven proliferation of selected clones of effector and memory B cells, antibodies of increasingly higher affinity become abundant during an adaptive immune response, providing progressively better protection against the pathogen (**Movie 24.6**).

A breakthrough in understanding the molecular mechanism of somatic hypermutation came with the identification of an enzyme that is required for the process. It is called **activation-induced deaminase** (**AID**) because it is expressed specifically in activated B cells and deaminates cytosine (C) to uracil (U) during transcription of V-region coding DNA. The deamination produces U:G mismatches in the DNA double helix, and the repair of these mismatches produces various types of mutations, depending on the repair pathway used. Somatic hypermutation affects only actively transcribed DNA, because AID works only on single-stranded DNA (which is transiently exposed during transcription) and because proteins involved in the transcription of V-region coding sequences are required to recruit the AID enzyme. AID is also required for activated B cells to switch from IgM and IgD production to the production of the other classes of Ig, as we now discuss.

B Cells Can Switch the Class of Ig They Make

After a developing B cell leaves the bone marrow, before it interacts with antigen, it expresses both IgM and IgD BCRs on its surface, both with the same antigen-binding sites (see Figure 24–24). Stimulation by antigen and helper T cells activates many of these mature naïve B cells to become IgM-secreting effector cells, so that IgM antibodies dominate the primary antibody response. Later in the immune response, however, when activated B cells are undergoing somatic hypermutation, the combination of antigen and helper-T-cell-derived cytokines (discussed later) stimulates many of the B cells to switch from making membrane-bound IgM and IgD to making IgG, IgE, or IgA, in the process of **class switching**. Some of these cells become memory cells that express the corresponding class of Ig as BCRs on their surface, while others become effector cells that secrete the Ig molecules as antibodies. The IgG, IgE, and IgA molecules retain their original antigen-binding site and are collectively referred to as *secondary classes* of Igs, because they are produced only after antigen stimulation, dominate secondary antibody responses, and make up the secondary Ig repertoire.

As discussed earlier, the constant region of an Ig heavy chain determines the class of the Ig. Thus, the ability of B cells to switch the class of antibody they make without changing the antigen-binding site implies that the same assembled V_H-region coding sequence (which specifies the antigen-binding part of the heavy chain) can sequentially associate with different C_H-coding sequences. This has important functional implications. It means that, in an individual animal, a particular antigen-binding site that has been selected by environmental antigens can be distributed among the various classes of antibodies, thereby acquiring the different biological properties of each class.

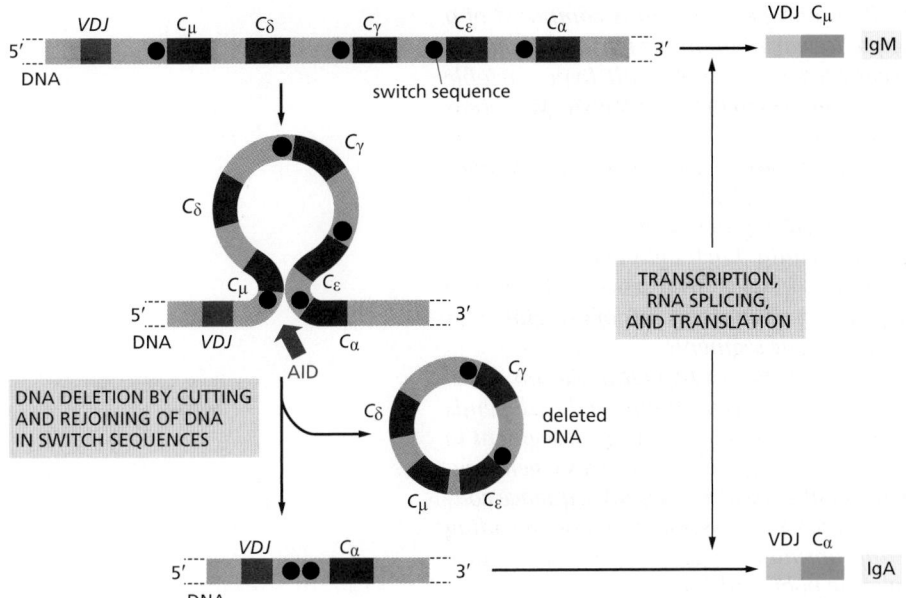

Figure 24–30 **An example of the DNA rearrangement that occurs in class switch recombination.** A B cell making IgM molecules with a V region encoded by a particular assembled *VDJ* DNA sequence is stimulated to switch to making IgA molecules with the same V region. In the process, it deletes the DNA between the *VDJ* sequence and the C_α-coding sequence. Specific DNA sequences (*switch sequences*) located upstream of each C_H-coding sequence (except C_δ, as B cells don't switch from C_μ to C_δ) can recombine with one another, with the deletion of the intervening DNA, as shown here. As discussed in the text, the recombination process depends on AID, the same enzyme that is involved in somatic hypermutation. When switching from IgM to IgG or IgE, the C-region coding sequences downstream of C_γ or C_δ, which remain after the DNA deletion, are removed during RNA splicing.

When a B cell switches from making IgM and IgD to one of the secondary classes of Ig, an irreversible change occurs in the DNA—a process called **class switch recombination**. It entails the deletion of all the C_H-coding sequences between the assembled VDJ-coding sequence and the particular C_H-coding sequence that the cell is destined to express. Class switch recombination differs from V(D)J recombination in several ways. (1) It happens after antigen stimulation, mainly in germinal centers, and depends on helper T cells. (2) It uses different recombination signal sequences, called *switch sequences*, which flank the different C_H-coding segments. (3) It involves cutting and joining the switch sequences, which are noncoding sequences, and leaves the assembled V_H-region coding sequence unchanged (**Figure 24–30**). (4) Most importantly, the molecular mechanism is different. It depends on AID, which is also involved in somatic hypermutation, rather than on the V(D)J recombinase. The cytokines that activate class switching induce the production of transcription regulators that activate transcription from the relevant switch sequences, allowing the recruitment of AID to these sites.

Once bound, AID initiates switch recombination by deaminating some cytosines to uracil in the vicinity of these switch sequences. Excision of these uracils is thought to lead to double-strand breaks in the participating switch regions, which are then joined by a form of nonhomologous end joining (discussed in Chapter 5).

Thus, whereas the primary Ig repertoire in humans (and mice) is generated by V(D)J joining mediated by V(D)J recombinase, the secondary antibody repertoire is generated by somatic hypermutation and class switch recombination, both of which are mediated by AID. **Figure 24–31** lists the main mechanisms that we have discussed in this chapter that diversify Igs.

Summary

Each B cell clone makes Ig molecules with a unique antigen-binding site. Initially, the Ig molecules are inserted into the plasma membrane and serve as B cell receptors (BCRs) for antigen. Antigen binding to the BCRs, together with co-stimulatory signals from helper T cells, activates the B cells to proliferate and differentiate into either memory cells or antibody-secreting effector cells. The effector cells secrete large amounts of antibodies with the same antigen-binding site as the BCRs.

A typical Ig molecule is composed of four polypeptide chains—two identical heavy chains and two identical light chains. Parts of both the heavy and light chains form the two identical antigen-binding sites. There are multiple classes of Ig (IgA, IgD, IgE, IgG, and IgM), each with a distinctive heavy chain, which determines the

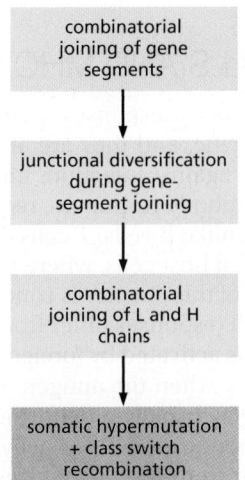

Figure 24–31 **The main mechanisms of Ig diversification in mice and humans.** Those shaded in *green* occur during B cell development in the bone marrow, whereas the two mechanisms shaded in *red* occur when B cells are stimulated by foreign antigen and helper T cells in germinal centers in peripheral lymphoid organs, either late in a primary response or in a secondary response.

biological properties of the Ig class. Each light and heavy chain is composed of a number of Ig domains. The amino acid sequence variation in the variable domains of both light and heavy chains is concentrated in several small hypervariable regions, which form loops at one end of these domains to produce the antigen-binding site.

Igs are encoded by loci on three different chromosomes, each of which is responsible for producing a different polypeptide chain—a κ light chain, a λ light chain, or a heavy chain. Each locus contains separate gene segments that encode different parts of the variable region of the particular Ig chain. Each light-chain locus contains one or more constant- (C-) region coding sequences and sets of variable (V) and joining (J) gene segments. The heavy-chain locus contains sets of C-region coding sequences and sets of V, diversity (D), and J gene segments.

During B cell development in the bone marrow, before antigen stimulation, separate gene segments are brought together by site-specific recombination that depends on a V(D)J recombinase. A V_L gene segment recombines with a J_L gene segment to produce a DNA sequence coding for the V region of a light chain, and a V_H gene segment recombines with a D and a J_H gene segment to produce a DNA sequence coding for the V region of a heavy chain. Each of the newly assembled V-region coding sequences is then co-transcribed with the appropriate C-region sequence to produce an RNA molecule that codes for the complete Ig polypeptide chain.

By randomly combining inherited gene segments that code for the variable regions during B cell development, humans can make hundreds of different light chains and thousands of different heavy chains. Because the antigen-binding site is formed where the hypervariable loops of the V_L and V_H domains come together in the final Ig molecule, the heavy and light chains can potentially pair to form Igs with millions of different antigen-binding sites. A loss or gain of nucleotides at the site of gene-segment joining increases this number enormously. The Igs made by such V(D)J recombination before antigen stimulation are IgMs and IgDs with low affinity for binding antigen, and they constitute the primary Ig repertoire.

Igs are further diversified following antigen stimulation in peripheral lymphoid organs by the AID- and helper-T-cell-dependent processes of somatic hypermutation and class switch recombination, which together produce the high-affinity IgG, IgE, and IgA Igs that constitute the secondary Ig repertoire. The process of class switching allows the same antigen-binding site to be incorporated into antibodies that have different tails and therefore different biological properties.

T CELLS AND MHC PROTEINS

Like antibody responses, T-cell-mediated immune responses are exquisitely antigen-specific, and they are at least as important as antibodies in defending vertebrates against infection. Indeed, most adaptive immune responses, including most antibody responses, require helper T cells for their initiation. Most importantly, unlike B cells, T cells can help eliminate pathogens that have entered the interior of host cells, where they are invisible to B cells and antibodies. Much of the rest of this chapter is concerned with how T cells accomplish this feat.

T cell responses differ from B cell responses in at least two crucial ways. First, a T cell is activated by foreign antigen to proliferate and differentiate into effector cells only when the antigen is displayed on the surface of an *antigen-presenting cell* (*APC*), usually a dendritic cell in a peripheral lymphoid organ. One reason T cells require APCs for activation is that the form of antigen they recognize is different from that recognized by the Igs produced by B cells. Whereas Igs can recognize antigenic determinants on the surface of pathogens and soluble folded proteins, for example, T cells can only recognize fragments of protein antigens that have been produced by partial proteolysis inside a host cell. As mentioned earlier, newly formed *MHC proteins* capture these peptide fragments and carry them to the surface of the host cell, where T cells can recognize them.

The second difference is that, once activated, effector T cells act mainly at short range, either within a secondary lymphoid organ or after they have migrated to a site of infection. Effector B cells, by contrast, secrete antibodies that can act far away. Effector T cells interact directly with another host cell in the body, which they either kill (if it is an infected host cell, for example) or signal in some way (if it is a B cell or macrophage, for example). We will refer to such host cells as *target cells*. As is the case with APCs, target cells must display an antigen bound to an MHC protein on their surface for a T cell to recognize them.

There are three main classes of T cells—cytotoxic T cells, helper T cells, and regulatory T cells. When activated, they function as effector cells (see Figure 24–17), each with their own distinct activities. Effector *cytotoxic T cells* directly kill cells that are infected with a virus or some other intracellular pathogen. Effector *helper T cells* help stimulate the responses of other immune cells—mainly macrophages, dendritic cells, B cells, and cytotoxic T cells; as we will see, there are a variety of functionally distinct subtypes of helper T cells. Effector *regulatory T cells* suppress the activity of other immune cells.

In this section, we describe these classes and subclasses of T cells and their respective functions. We discuss how they recognize foreign antigens on the surface of APCs or target cells and the crucial part played by MHC proteins in the recognition process. We begin by considering the cell-surface receptors that T cells use to recognize antigen.

T Cell Receptors (TCRs) Are Ig-like Heterodimers

T cell receptors (**TCRs**), unlike Igs made by B cells, exist only in membrane-bound form. They are composed of two transmembrane, disulfide-linked polypeptide chains, each of which contains two Ig-like domains—one variable and one constant. On most T cells, the TCRs have one α chain and one β chain (**Figure 24–32**).

The genetic loci that encode the α and β chains are located on different chromosomes. Like an Ig heavy-chain locus (see Figure 24–29), the TCR loci contain separate *V*, *D*, and *J* gene segments (or just *V* and *J* gene segments in the case of the α-chain locus), which are brought together by site-specific recombination during T cell development in the thymus. With one exception, T cells use the same mechanisms to generate antigen-binding site diversity of their TCRs as B cells use to generate antigen-binding site diversity of their Igs, and they use the same V(D)J recombinase; thus, humans or mice deficient in this recombinase cannot make functional B or T cells. The mechanism that does not operate in TCR diversification is antigen-driven somatic hypermutation. Thus, the affinities of TCRs tend to be low ($K_a \approx 10^5$–10^7 liters/mole). Various co-receptors and cell–cell adhesion proteins, however, greatly strengthen the binding of a T cell to an APC or target cell.

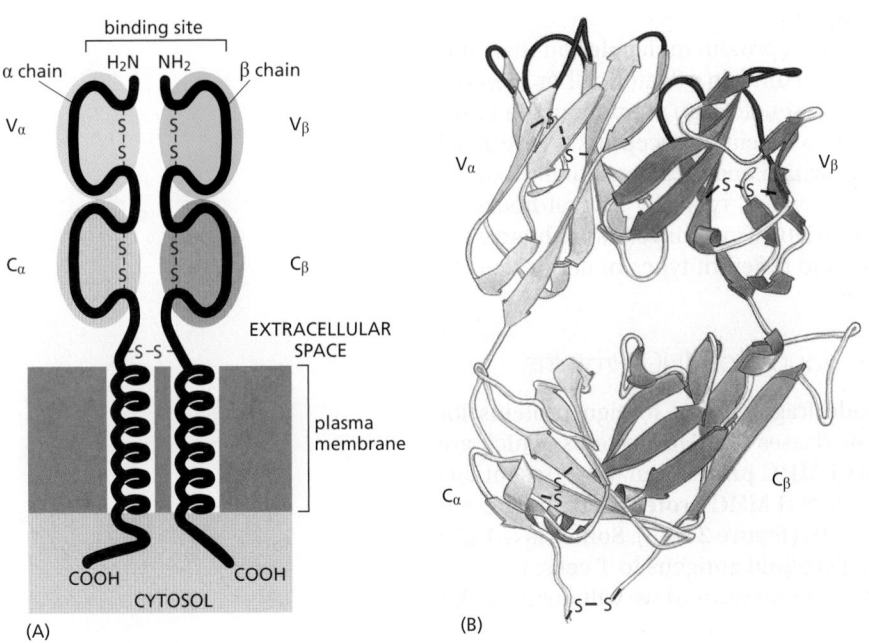

(A)   (B)

Figure 24–32 A T cell receptor (TCR) heterodimer. (A) Schematic drawing showing that the receptor is composed of an α and a β polypeptide chain. Each chain has a large extracellular part that is folded into two Ig-like domains—one variable (V) and one constant (C). A V_α and a V_β domain (shaded in *blue*) form the antigen-binding site. Unlike Igs, which have two binding sites for antigen, TCRs have only one. The αβ-heterodimer is noncovalently associated with a large set of invariant membrane-bound proteins (not shown), which help activate the T cell when the TCRs bind their specific antigen (see Figure 24–45B). A typical T cell has about 30,000 TCRs on its surface. (B) The three-dimensional structure of the extracellular part of a TCR. The antigen-binding site is formed by the hypervariable loops of both the V_α and V_β domains *(black)*, and it is similar in its overall dimensions and geometry to the antigen-binding site of an Ig molecule. (B, based on K.C. Garcia et al., *Science* 274:209–219, 1996.)

Instead of making α and β chains, a minority of T cells makes a different but related type of TCR heterodimer, composed of γ chains and δ chains. Although these *γ/δ T cells* normally make up only 5–10% of the T cells in human blood, they can be the dominant T cell population in epithelia (in the skin and gut, for example). They have some properties in common with natural killer (NK) cells and with an enlarging category of T-like cells that have features of both innate and adaptive immune cells, which are sometimes collectively referred to as *innate lymphoid cells.* The cells in all these categories tend to be enriched in mucosal tissues, respond early to infection, display little immunological memory, and, compared with B and T cells, have surface receptors of restricted diversity. We will not discuss them further.

As with BCRs, TCRs are tightly associated in the plasma membrane with a number of invariant membrane-bound proteins that are involved in passing the signal from an antigen-activated receptor to the cell interior. We will discuss these proteins in more detail later, when we consider some of the molecular events involved in T and B cell activation. First, we consider the special ways in which T cells recognize foreign antigen on the surface of an APC or target cell.

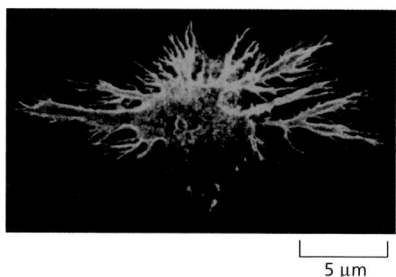

Figure 24–33 Immunofluorescence micrograph of a dendritic cell in culture. These APCs derive their name from their long processes, or "dendrites." The cell has been labeled with a monoclonal antibody that recognizes a surface antigen on these cells. (Courtesy of David Katz.)

Activated Dendritic Cells Activate Naïve T Cells

Generally, naïve T cells, including naïve helper and cytotoxic T cells, proliferate and differentiate into effector cells and memory cells only when they see their specific antigen on the surface of an activated **dendritic cell** in a peripheral lymphoid organ (**Figure 24–33**). The activated dendritic cell displays the antigen in a complex with MHC proteins on its surface, along with co-stimulatory proteins (see Figure 24–11). The memory T cells that develop, however, can be activated by the same antigen–MHC complex on the surface of other types of APCs (target cells), including macrophages and B cells—as well as by dendritic cells.

Immature dendritic cells are located in most tissues—underlying epithelial layers of the skin and gut, for example—where they are constantly sampling and processing proteins in their environment. They become activated to mature when their pattern recognition receptors (PRRs) encounter pathogen associated molecular patterns (PAMPs) on an invading pathogen or its products. The pathogen or products are ingested, and the microbial proteins are cleaved into peptide fragments, which are loaded onto MHC proteins, as we discuss later. The activated dendritic cells then migrate via the lymph from the site of infection to local lymph nodes or gut-associated lymphoid organs, where they present the foreign antigens, displayed as peptide–MHC complexes on the dendritic cell surface, for recognition by the relevant T cells (see Figure 24–11).

Activated dendritic cells display three types of protein molecules on their surface that have a role in activating a T cell to become an effector cell or memory cell (**Figure 24–34**): (1) *MHC proteins*, which present foreign peptides to the TCRs; (2) *co-stimulatory proteins*, which bind to complementary receptors on the T cell surface; and (3) *cell–cell adhesion molecules*, which enable a T cell to bind to the dendritic cell for long enough to become activated, typically several hours. In addition, activated dendritic cells secrete a variety of cytokines that influence the type of effector helper T cell that develops, and different types of dendritic cells promote different outcomes (discussed later).

T Cells Recognize Foreign Peptides Bound to MHC Proteins

MHC proteins capture and display peptide fragments of foreign proteins for presentation to T cells. There are two main classes of MHC proteins, which are structurally and functionally distinct. **Class I MHC proteins** mainly present foreign peptides to cytotoxic T cells, whereas **class II MHC proteins** mainly present foreign peptides to helper and regulatory T cells (**Figure 24–35**). Some class-I-like MHC proteins present microbial lipid and glycolipid antigens to T cells, but they are not encoded within the MHC region of the genome, and we will not consider them further.

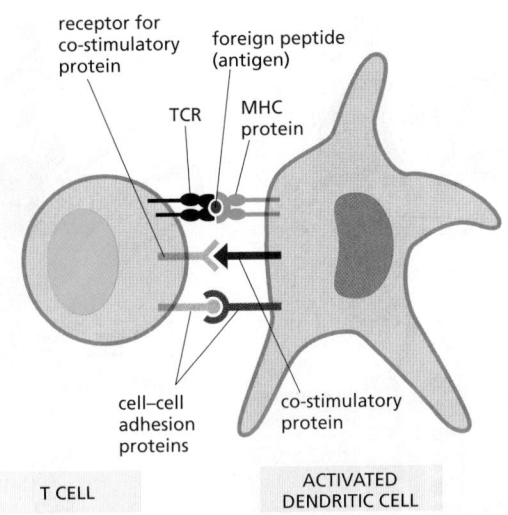

Figure 24–34 The three general types of proteins on the surface of an activated dendritic cell involved in activating a T cell. Although only membrane-bound co-stimulatory molecules are shown, activated dendritic cells also secrete soluble co-stimulatory molecules. The invariant polypeptide chains that are always stably associated with the TCR are not shown; they are illustrated in Figure 24–45B.

Both class I and class II MHC proteins are heterodimers, in which two extracellular domains form a *peptide-binding groove*, which always has a variable small peptide bound in it. In class I MHC proteins, the two domains that form the peptide-binding groove are provided by the transmembrane α chain, which is noncovalently associated with a small subunit called $β_2$-microglobulin; in class II MHC proteins, a different α chain and a large noncovalently associated β chain each contribute an extracellular domain to form the peptide-binding groove (**Figure 24–36**). A TCR binds to both the peptide and the ridges of the binding groove. Humans have three major class I proteins, called *HLA-A*, *HLA-B*, and *HLA-C*, and three class II proteins, called *HLA-DR*, *HLA-DP*, and *HLA-DQ* (HLA stands for *h*uman-*l*eukocyte-*a*ssociated, as these proteins were first demonstrated on human leukocytes). **Figure 24–37** shows how the genes that encode these proteins are arranged on human chromosome 6.

There are important differences between the class I and class II MHC proteins with regard to the cell types that express them and the origin of the peptides in their peptide-binding grooves. Almost all of our nucleated cells express class I proteins. Their peptide-binding groove displays one of a diverse collection of peptides (typically 8–10 amino acids in length). In a healthy cell, the peptides originate from the cell's own cytosolic and nuclear proteins that have undergone partial degradation in proteasomes in the processes of normal protein turnover and quality control mechanisms. Some of the peptide fragments produced in this way are actively transported into the lumen of the endoplasmic reticulum (ER), through a specialized transporter in the ER membrane, where they are loaded onto newly synthesized class I MHC α chains; once a peptide binds, the α chain can assemble with its partner chain. The resulting self-peptide–MHC complex is then transported through the Golgi apparatus to the cell surface. Such complexes are not dangerous, however, because the cytotoxic T cells that could recognize

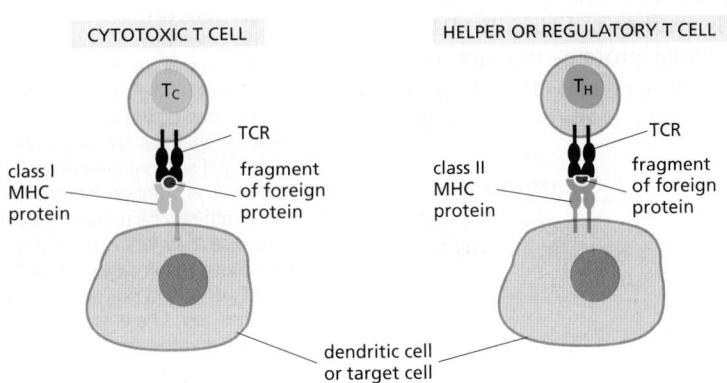

Figure 24–35 Recognition by T cells of foreign peptides bound to MHC proteins. Cytotoxic T cells recognize foreign peptides in association with class I MHC proteins, whereas helper T cells and regulatory T cells recognize foreign peptides in association with class II MHC proteins. In both cases, the T cell recognizes the peptide–MHC complexes on the surface of an APC—either a dendritic cell or a target cell. Some regulatory T cells recognize self peptides in association with class II MHC proteins (not shown).

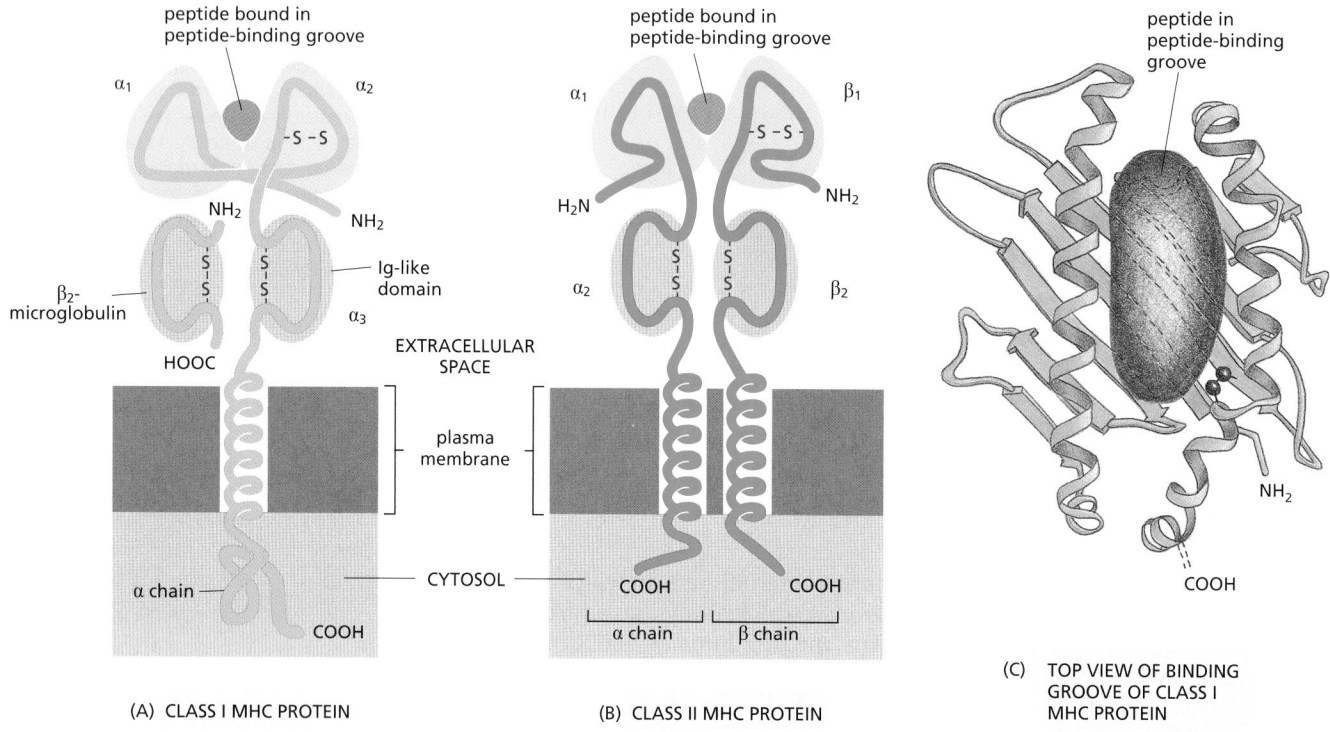

(A) CLASS I MHC PROTEIN

(B) CLASS II MHC PROTEIN

(C) TOP VIEW OF BINDING GROOVE OF CLASS I MHC PROTEIN

Figure 24–36 Class I and class II MHC proteins. (A) The α chain of the class I molecule has three extracellular domains, α_1, α_2, and α_3, each encoded by a separate exon. The α chain is noncovalently associated with a smaller polypeptide chain, β_2-microglobulin, which is not encoded within the MHC region of the genome. The α_3 domain and β_2-microglobulin are Ig-like. While β_2-microglobulin is invariant, the α chain is extremely polymorphic, mainly in the α_1 and α_2 domains. (B) In class II MHC proteins, both the α chain and the β chain are encoded within the MHC and are polymorphic, mainly in the α_1 and β_1 domains; the α_2 and β_2 domains are Ig-like. Thus, there are striking similarities between class I and class II MHC proteins. In both, the two outermost domains (shaded in *blue*) are polymorphic and interact to form a groove that binds peptide fragments. (C) The three-dimensional structure of the peptide-binding groove of a human class I MHC protein is viewed from above, with bound peptide shown schematically; a peptide must be bound in the groove for the MHC protein to assemble and be transported to the cell surface. The sides of the groove are formed by two α helices, and the floor is formed by a β pleated sheet. The S–S disulfide bond is shown in *red* (**Movie 24.8** and **Movie 24.9**). (C, adapted from P.J. Bjorkman et al., *Nature* 329:506–512, 1987.)

them have been either eliminated or inactivated, or suppressed by regulatory T cells in the process of self-tolerance. By contrast, in a cell infected by a pathogen such as a virus, the pathogen proteins will be processed in the same way, and peptides derived from them will be displayed on the infected cell surface bound to class I MHC proteins; there, they are recognized by cytotoxic T cells expressing the appropriate TCRs, thereby targeting the infected cell for destruction (**Figure 24–38**).

In general, only **antigen-presenting cells** (**APCs**) express class II MHC proteins. Dendritic cells are referred to as *professional APCs*, as they are specialized for this function and only they can activate naïve T cells. Other immune cells that are targets of effector T cell regulation, including B cells and macrophages, are *nonprofessional APCs*. All APCs load their newly synthesized class II MHC proteins with peptides derived mainly from extracellular proteins that are endocytosed and delivered to endosomes. The newly synthesized class II MHC proteins initially contain an *invariant chain*, which occupies the peptide-binding groove

Figure 24–37 Human MHC genes. This simplified schematic drawing shows the location of the genes that encode the transmembrane subunits of class I *(light green)* and class II *(dark green)* MHC proteins. The genes shown encode three types of class I MHC proteins (HLA-A, HLA-B, and HLA-C) and three types of class II MHC proteins (HLA-DP, HLA-DQ, and HLA-DR). An individual can therefore make six types of class I MHC proteins (three encoded by maternal genes and three by paternal genes) and more than six types of class II MHC proteins. Because of the extreme polymorphism of the MHC genes, the chances are very low that the maternal and paternal alleles will be the same. The number of class II MHC proteins that can be made is greater than six because there are two *DR* β genes and because maternally encoded and paternally encoded polypeptide chains can sometimes pair. The entire region shown spans about seven million base pairs and contains other genes that are not shown.

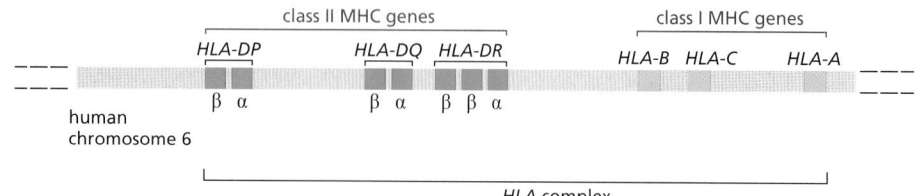

and prevents it from prematurely binding a peptide until the class II MHC protein reaches specialized vesicles, which fuse with endosomes. Here, the invariant chain is removed and peptide fragments (typically 12–20 amino acids long) produced from endocytosed proteins can bind to the groove of the class II MHC proteins, which are then transported to the plasma membrane for display on the surface of the APC. In a healthy host cell, class II MHC protein grooves are loaded with self-peptides derived from normal proteins and will be ignored by T cells because of self-tolerance mechanisms. During an infection, however, pathogen proteins are also endocytosed and processed in the same way, enabling APCs to present pathogen peptides bound to class II MHC proteins to T cells expressing an appropriate TCR (**Figure 24–39**).

The distinction just discussed between the antigen-processing pathways for loading peptides onto class I and class II MHC proteins is not absolute. Dendritic cells, for example, need to be able to activate cytotoxic T cells to kill virus-infected cells even when the virus does not infect dendritic cells themselves. To do so, specialized subsets of dendritic cells use a process called **cross-presentation**, which begins when these noninfected dendritic cells phagocytose virus-infected host cells or their fragments. The ingested viral proteins are then released by an unknown mechanism from phagolysosomes into the cytosol, where they are degraded in proteasomes; the resulting protein fragments are then transported into the ER lumen, where they load onto assembling class I MHC proteins. Cross-presentation in dendritic cells is not confined to endocytosed pathogens and their products: it also operates to activate cytotoxic T cells against tumor antigens of cancer cells and the MHC proteins of foreign organ grafts.

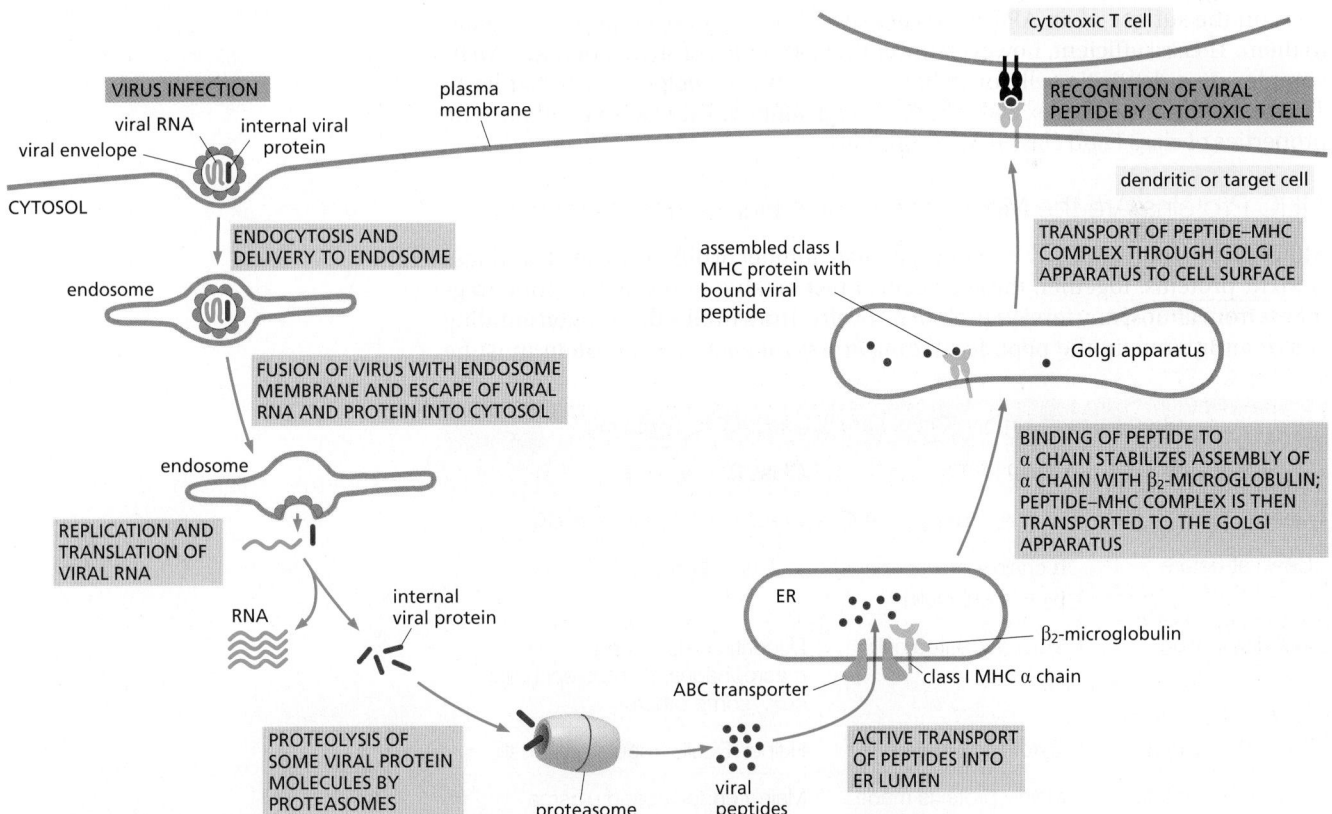

Figure 24–38 The processing of an extracellular foreign protein for presentation to cytotoxic T cells. An effector cytotoxic T cell kills a virus-infected cell when it recognizes fragments of an internal viral protein bound to class I MHC proteins on the surface of the infected cell. Not all viruses enter the cell in the way that this enveloped RNA virus does, but fragments of internal viral proteins always follow the pathway shown. Only a small proportion of the viral proteins synthesized in the cytosol are degraded and transported to the cell surface, but this is sufficient to attract an attack by a cytotoxic T cell. Several chaperone proteins in the ER lumen aid the folding and assembly of class I MHC proteins (not shown). The assembly of class I MHC proteins and their transport to the cell surface require the binding of either a self or foreign peptide (**Movie 24.10**).

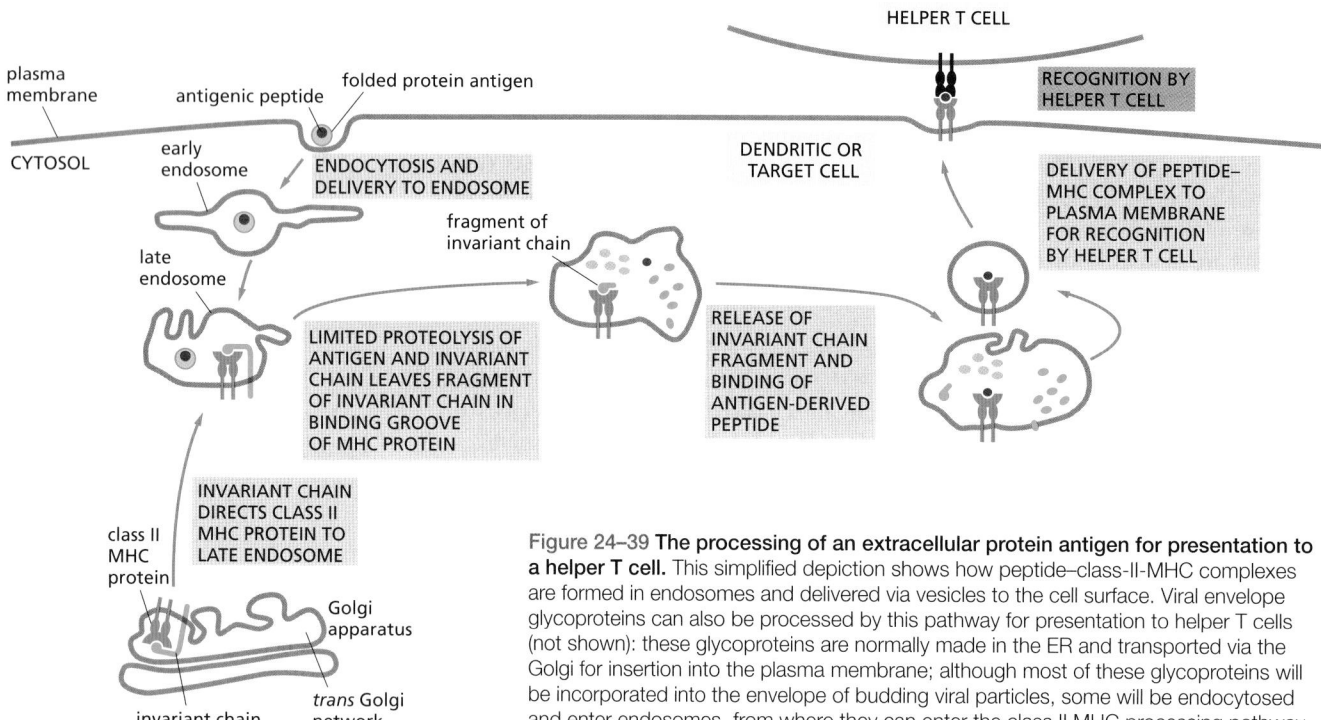

Figure 24–39 The processing of an extracellular protein antigen for presentation to a helper T cell. This simplified depiction shows how peptide–class-II-MHC complexes are formed in endosomes and delivered via vesicles to the cell surface. Viral envelope glycoproteins can also be processed by this pathway for presentation to helper T cells (not shown): these glycoproteins are normally made in the ER and transported via the Golgi for insertion into the plasma membrane; although most of these glycoproteins will be incorporated into the envelope of budding viral particles, some will be endocytosed and enter endosomes, from where they can enter the class II MHC processing pathway.

During an infection, only a small fraction of the many thousands of MHC proteins on the surface of an APC or target cell will have pathogen peptides bound to them. This is sufficient, however: fewer than 50 copies of such a peptide–MHC complex on a dendritic cell, for example, can activate a helper T cell that has a TCR that binds the complex with a high-enough affinity. **Table 24–3** compares the properties of class I and class II MHC proteins.

MHC Proteins Are the Most Polymorphic Human Proteins Known

Although any individual can make only a small number of different class I and class II MHC proteins, together, these proteins must be able to present peptide fragments from almost any foreign protein to T cells. Thus, unlike the antigen-binding site of an Ig protein, the peptide-binding groove of each MHC protein must be

TABLE 24–3 Properties of Human Class I and Class II MHC Proteins

	Class I	Class II
Genetic loci	HLA-A, HLA-B, HLA-C	HLA-DP, HLA-DQ, HLA-DR
Chain structure	α chain + β_2-microglobulin	α chain + β chain
Cell distribution	Most nucleated cells	Dendritic cells, B cells, macrophages, thymus epithelial cells, some others
Presents antigen to	Cytotoxic T cells	Helper T cells, regulatory T cells
Source of peptide fragments	Mainly proteins made in cytoplasm	Mainly endocytosed plasma membrane and extracellular proteins
Polymorphic domains	$\alpha_1 + \alpha_2$	$\alpha_1 + \beta_1$
Recognition by co-receptor	CD8	CD4

able to bind a very large number of different peptides. The genes encoding class I and class II MHC proteins (see Figure 24–37) are the most *polymorphic* known in higher vertebrates: in the human population, for example, there are more than 2000 allelic variants of these genes. The corresponding variations in the MHC proteins are concentrated in the floor and walls of the peptide-binding grooves and allow MHC molecules in different individuals to bind different arrays of peptides.

It is thought that infectious diseases have been an important driving force for generating this remarkable MHC polymorphism. In the evolutionary war between pathogens and the adaptive immune system, pathogens will tend to change their proteins through mutation so that the peptides derived from them will not fit in the MHC peptide-binding grooves. When a pathogen succeeds, it can sweep through a population as an epidemic. In such circumstances, the few individuals who produce a new allelic form of MHC protein that can bind peptides derived from the altered pathogen will have a large selective advantage. This type of selection will tend to promote and maintain a large diversity of MHC proteins in the population. In West Africa, for example, individuals with a specific MHC allele (HLA-B53) have a reduced susceptibility to a severe form of malaria that is endemic there; although this allele is rare elsewhere, it is found in 25% of the West African population.

The extensive diversity of human MHC proteins is the main reason that individuals who receive a foreign organ transplant must be treated with strong immunosuppressive drugs to prevent the immunological rejection of the grafted organ. Of all the foreign proteins that the graft expresses, the MHC proteins are by far the most powerful stimulators of the recipient's T cells, which would rapidly destroy the graft if they were not prevented from doing so by such drugs. Foreign MHC proteins are powerful T cell stimulants because T cells respond to them in the same way they respond to self MHC proteins that have foreign peptides bound to them; for this reason, the proportion of a person's T cells that can specifically recognize any foreign MHC protein is relatively high.

CD4 and CD8 Co-receptors on T Cells Bind to Invariant Parts of MHC Proteins

The affinity of TCRs for peptide–MHC complexes on an APC is usually too low by itself to mediate a functional interaction between the two cells. T cells normally require *accessory receptors* to help stabilize the interaction by increasing the overall strength of the cell–cell adhesion. Unlike TCRs or MHC proteins, the accessory receptors are invariant and do not bind to foreign peptides. Once bound to the surface of a dendritic cell, for example, a T cell increases the strength of the binding by activating an integrin adhesion protein (discussed in Chapter 19), which then binds more strongly to an Ig-like protein on the surface of the dendritic cell. This increased adhesion enables the T cell to remain bound long enough to become activated.

When an accessory receptor has a direct role in activating the T cell by generating its own intracellular signals, it is called a **co-receptor**. The most important and best understood of the co-receptors on T cells are the *CD4* and *CD8* proteins, both of which are single-pass transmembrane proteins with extracellular Ig-like domains. Like TCRs, they recognize MHC proteins, but, unlike TCRs, they bind to invariant parts of the MHC protein, far away from the peptide-binding groove. **CD4** is expressed on both helper T cells and regulatory T cells and binds to class II MHC proteins, whereas **CD8** is expressed on cytotoxic T cells and binds to class I MHC proteins (**Figure 24–40**).

CD4 and CD8 contribute to T cell recognition by helping the T cell to focus on particular MHC proteins, and thereby on particular types of target cells. Thus, the recognition of class I MHC proteins by CD8 allows cytotoxic T cells to focus on any type of infected host cell, while the recognition of class II MHC proteins by CD4 allows helper and regulatory T cells to focus on the target immune cells that they help or suppress, respectively. The cytoplasmic tail of the CD4 and CD8 proteins is associated with a member of the Src family of cytoplasmic tyrosine kinases

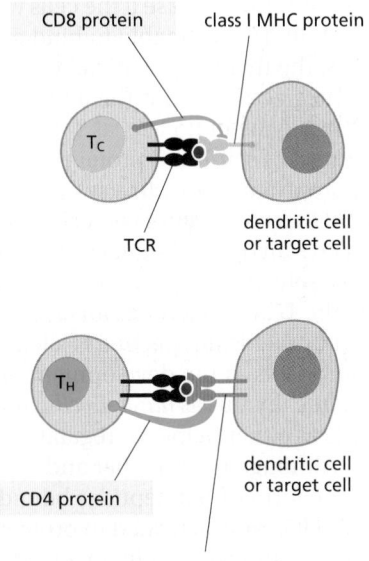

Figure 24–40 CD4 and CD8 co-receptors on the surface of T cells. Cytotoxic T cells (T$_C$) express CD8, which recognizes class I MHC proteins, whereas helper T cells (T$_H$) and regulatory T cells (not shown) express CD4, which recognizes class II MHC proteins. Note that the co-receptors bind to the same MHC protein that the TCR has engaged, so that they are brought together with TCRs during the antigen-recognition process. Whereas the TCR binds to the variable (polymorphic) parts of the MHC protein that form the peptide-binding groove, the co-receptor binds to the invariant part, well away from the binding groove.

(discussed in Chapter 15) called *Lck*, which phosphorylates various intracellular proteins on tyrosines and thereby participates in the activation of the T cell (discussed later).

The AIDS virus (HIV) uses CD4 molecules (as well as chemokine receptors) to enter helper T cells (see Figure 23–17). AIDS patients are susceptible to infection by microbes that are not normally dangerous because HIV depletes helper T cells. As a result, most AIDS patients die of infection within several years of the onset of symptoms, unless they are treated with a combination of anti-HIV drugs. HIV also uses CD4 and chemokine receptors to enter macrophages, which also have both types of receptors on their surface.

Developing Thymocytes Undergo Negative and Positive Selection

T cell development begins when bone-marrow-derived lymphoid progenitor cells enter the thymus from the bloodstream. There, the cells receive a variety of signals from thymus stromal cells, epithelial cells, macrophages, and dendritic cells, which promote their stepwise development into mature **thymocytes**. At one step, the progenitor cells are induced to express V(D)J recombinase and begin to rearrange their TCR gene segments. Soon thereafter, the cells express both CD4 and CD8 co-receptors, and these so-called *double-positive thymocytes* migrate inward and interact with thymus dendritic cells or epithelial cells expressing self peptides bound to class I and class II MHC proteins. If the TCR on the thymocyte binds with high affinity to these complexes, a strong signal will be transmitted, causing the cell to undergo apoptosis. This process, called **negative selection**, is an example of clonal deletion (see Figure 24–21), and it eliminates thymocytes that could potentially attack normal host cells and tissues and thereby cause an autoimmune disease if the cells were to continue to mature and leave the thymus.

If its TCR is unable to bind at all to a self-peptide–MHC complex in the thymus, the thymocyte will fail to receive the signals it needs to survive and will die of "neglect;" without the ability to recognize self-MHC proteins, a T cell would generally be of no use, as T cells can only see pathogen-derived peptides in the context of self-MHC proteins. Thymocytes that express a TCR that binds with an appropriate affinity to a self peptide bound to either a class I MHC protein (using CD8 as a co-receptor) or a class II MHC protein (using CD4 as a co-receptor) will receive an optimal signal to survive and continue to mature, a process called **positive selection** (**Figure 24–41**). As part of this maturation process, and depending on the TCR's preference for class I or class II MHC proteins, the CD4 or CD8 co-receptor that is not needed is silenced by DNA methylation of the respective gene; this results in the development of CD4 or CD8 *single-positive thymocytes*, which exit the thymus as *naïve T cells* and enter the recirculating pool of T cells—the CD4 cells as either helper or regulatory T cells and the CD8 cells as cytotoxic T cells.

Although naïve helper and cytotoxic T cells constantly receive survival signals in the form of self peptides bound to MHC proteins that the T cells bind weakly, a T cell is only activated to proliferate and mount an immune response if its TCR binds with high affinity to a peptide–MHC complex and receives co-stimulatory signals at the same time. Generally, this happens only when the T cell encounters an activated dendritic cell (in a peripheral lymphoid organ) that expresses an MHC protein with a foreign peptide derived from a pathogen in its binding groove. Only then will the naïve T cell proliferate and differentiate into an effector or memory T cell.

Negative selection in the thymus is a major mechanism for ensuring that peripheral T cells do not react with host cells expressing MHC proteins with peptides derived from self proteins in their peptide-binding grooves. This mechanism, however, requires that the APCs in the thymus display an array of peptides on their MHC molecules that will reflect the self proteins in peripheral tissues, as well as in the thymus. The thymus, however, would not be expected to produce many of the proteins that are specifically expressed in other organs. As an example, it would not be expected to produce insulin, and yet it is crucial to delete thymocytes with TCRs that could recognize insulin-derived peptides bound to MHC

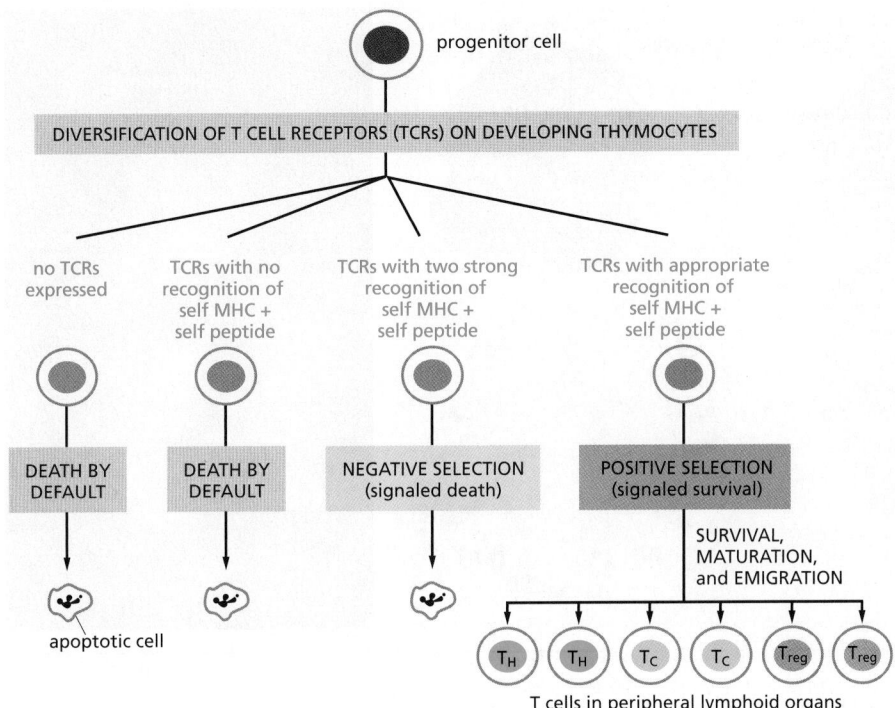

Figure 24–41 Positive and negative selection in the thymus. Developing thymocytes with TCRs that would potentially enable them to respond to peptides in association with self MHC proteins after they leave the thymus are positively selected: the binding of their TCRs to self peptides bound to self MHC proteins in the thymus signals such cells to survive, mature, and migrate to peripheral lymphoid organs. All of the other thymocytes undergo apoptosis—either because they do not express TCRs that recognize self MHC proteins with self peptides bound or because they recognize such complexes too well and undergo negative selection.

The regulatory T cells (T$_{reg}$ cells) that are positively selected in the thymus are called *natural T$_{reg}$ cells* to distinguish them from *induced T$_{reg}$ cells*, which develop in peripheral lymphoid organs from naive helper T cells (T$_H$ cells), as we discuss shortly.

proteins on the surface of insulin-secreting β cells in the pancreas. Any failure to do so would result in the T-cell-dependent destruction of the β cells and, as a consequence, cause *type 1 (or juvenile) diabetes*.

The mechanism that enables the deletion of all such cells in the thymus depends on a subpopulation of epithelial cells in the thymus that express a transcriptional regulator called **AIRE** (autoimmune regulator). By a poorly understood mechanism, the AIRE protein promotes the production of small amounts of mRNA from many genes that encode such "organ-specific" proteins, including the insulin gene. When the peptides derived from the proteins encoded by these genes are bound by MHC proteins and displayed on the surface of the epithelial cells in the thymus medulla, this is sufficient to provoke the deletion of the potentially self-reactive thymocytes. Mutations that inactivate the *AIRE* gene cause a severe multiorgan autoimmune disease in both mice and humans, indicating the importance of AIRE in self-tolerance.

Cytotoxic T Cells Induce Infected Target Cells to Kill Themselves

Cytotoxic T cells (T$_C$ cells), like the NK cells discussed earlier, protect us against intracellular pathogens, including viruses, bacteria, and parasites, that multiply in the cytoplasm of a host cell. T$_C$ cells kill infected host cells before the pathogen can escape to infect neighboring host cells. Before it can kill, however, a naïve T$_C$ cell has to become an effector cell by activation on an APC, usually an activated dendritic cell that has pathogen-derived peptides bound to class I MHC proteins—a process that depends on helper T cells. The effector T$_C$ cell can then recognize any target cell harboring the same pathogen and expressing some of the same peptide–MHC complexes on its surface: its TCRs cluster, along with CD8 co-receptors, adhesion molecules, and intracellular signaling proteins (discussed later), at the interface between the two cells, forming an **immunological synapse**. In this process, the effector T$_C$ cell reorganizes its cytoskeleton to focus its killing apparatus on the target cell, secreting its toxic proteins into a confined space (**Figure 24–42**); in this way, it avoids killing neighboring cells. A similar synapse forms when an effector helper T cell interacts with its target cell, except that the co-receptor is CD4 (**Movie 24.11**).

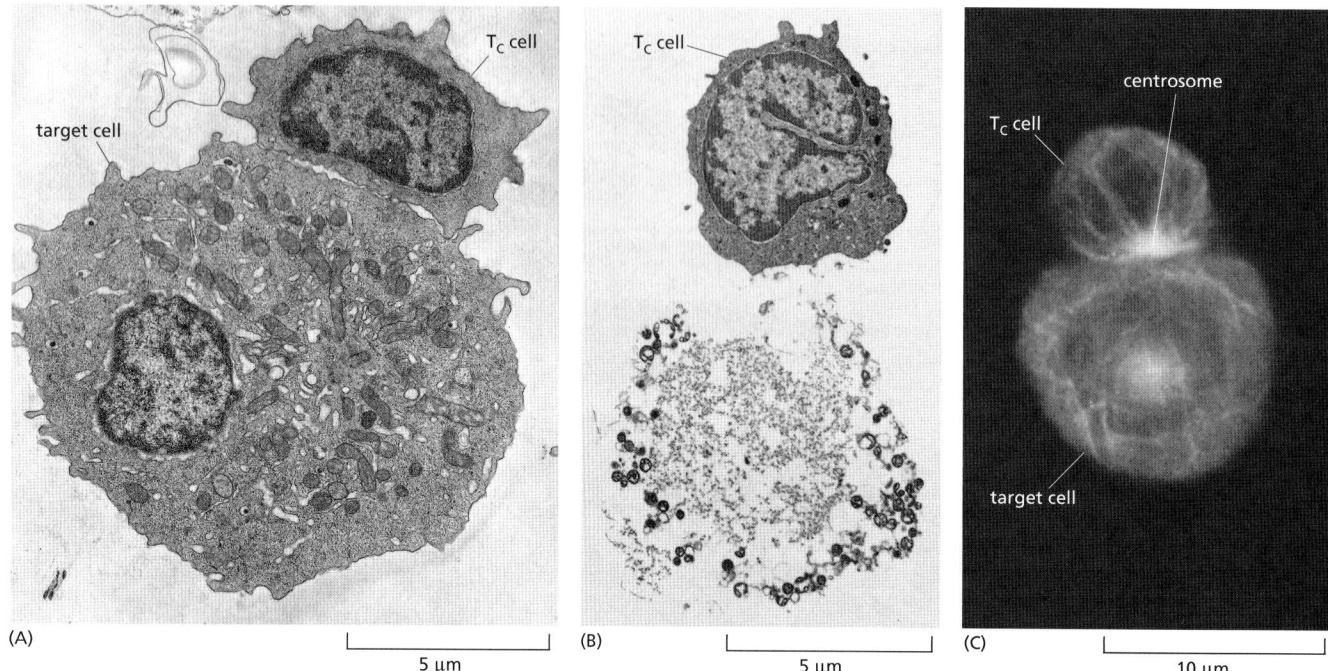

(A)   5 μm   (B)   5 μm   (C)   10 μm

Figure 24–42 Effector cytotoxic T cells killing target cells in culture. (A) Electron micrograph showing an effector cytotoxic T cell (T_C cell) binding to a target cell. The T_C cells were obtained from mice immunized with the target cells, which are foreign tumor cells. (B) Electron micrograph showing a T_C cell and a tumor cell that the T_C cell has killed. In an animal, as opposed to a culture dish, the killed target cell would be phagocytosed by neighboring cells (especially macrophages) long before it disintegrated in the way that it has here. (C) Immunofluorescence micrograph of a T_C cell and tumor cell after immunofluorescence staining with anti-tubulin antibodies. Note that the centrosome in the T_C cell is located at the point of cell–cell contact with the target cell—an immunological synapse. The secretory granules (not visible) in the T_C cell are initially transported along microtubules to the centrosome, which then moves to the synapse, delivering the granules to where they can release their contents. (A and B, from D. Zagury et al., *Eur. J. Immunol.* 5:818–822, 1975. With permission from John Wiley & Sons; C, © 1982 B. Geiger et al. Originally published in *J. Cell Biol.* https://doi.org/10.1083/jcb.95.1.137. With permission from Rockefeller University Press.)

An effector T_C cell (or an NK cell) can employ one of two strategies to kill the target, both of which operate by inducing the target cell to activate caspases and kill itself by undergoing apoptosis. One mechanism uses a protein called *Fas ligand* on the killer-cell surface, which binds to a transmembrane receptor protein called *Fas* on the target cell; this mechanism is discussed in Chapter 18 (see Figure 18–5). The other mechanism is the main one used by both NK cells and T_C cells to kill an infected target cell. The killer cell stores various toxic proteins within secretory vesicles in its cytoplasm that it releases into the synaptic space by exocytosis. The toxic proteins include *perforin* and proteases called *granzymes*. The perforin is homologous to complement component C9 and polymerizes in the target-cell plasma membrane (see Figure 24–8), forming a transmembrane pore that disrupts the membrane and allows the granzymes to enter the target cell. Once in the cytosol, the granzymes help activate caspases, thereby inducing apoptosis (**Figure 24–43**).

Figure 24–43 The main way that an effector T_C cell (or NK cell) kills an infected target cell. This simplified drawing shows how the killer cell releases perforin and granzymes onto the surface of an infected target cell by localized exocytosis at an immunological synapse. The high concentration of Ca^{2+} in the extracellular fluid causes the perforin to assemble into transmembrane channels in the target-cell plasma membrane, allowing the granzymes to enter the target-cell cytosol. The granzymes cleave and activate procaspases to initiate a caspase cascade, leading to apoptosis (see Figure 18–3). A single cytotoxic cell can kill multiple target cells in sequence. It remains a mystery why the released perforins do not form pores in the plasma membrane of the killer cell itself (Movie 24.12 and Movie 24.13).

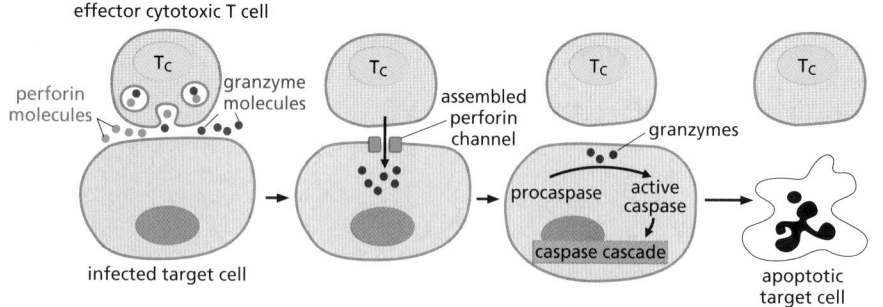

effector cytotoxic T cell

perforin molecules   granzyme molecules   assembled perforin channel   granzymes   active caspase   procaspase   caspase cascade

infected target cell   apoptotic target cell

Effector Helper T Cells Help Activate Other Cells of the Innate and Adaptive Immune Systems

In contrast to T_C cells, **helper T cells** (T_H **cells**) are crucial for defense against both extracellular and intracellular pathogens, and they express CD4 rather than CD8 co-receptors and recognize foreign peptides bound to class II rather than class I MHC proteins. Once naïve T_H cells are induced on activated dendritic cells to become effector cells, they can help activate other cells: they help activate B cells to become antibody-secreting cells and later to undergo Ig class switching and somatic hypermutation; they help activate macrophages to destroy any intracellular pathogens multiplying within the macrophage's phagosomes; they help induce naïve T_C cells to become effector cells that can kill infected target cells; and they stimulate the activated dendritic cell that activated them to maintain the dendritic cell in an activated state. In each case, the effector T_H cell recognizes the same complex of foreign peptide and class II MHC protein on the target-cell surface that it initially recognized on the activated dendritic cell. As discussed later, the T_H cell stimulates the target cell both by secreting a variety of cytokines and by displaying co-stimulatory proteins on its surface.

Naïve Helper T Cells Can Differentiate Into Different Types of Effector T Cells

When activated by binding to a foreign peptide bound to a class II MHC protein on an activated dendritic cell, a naïve T_H cell can differentiate into several distinct types of effector T cells, depending on the nature of the pathogen and the cytokines they encounter. These cells include four subtypes of helper cells—T_H1, T_H2, T_{FH}, and T_H17 cells—and regulatory (suppressor) T cells. **Figure 24–44** summarizes both the cytokines that induce these effector T cells and some of the cytokines the effector cells secrete, as well as the master transcription regulators that control the effector cell's development.

Naïve T_H cells activated by dendritic cells secreting the cytokine interleukin-12 (*IL12*) develop into **T_H1 cells**. These effector cells produce *interferon-γ* (*IFNγ*),

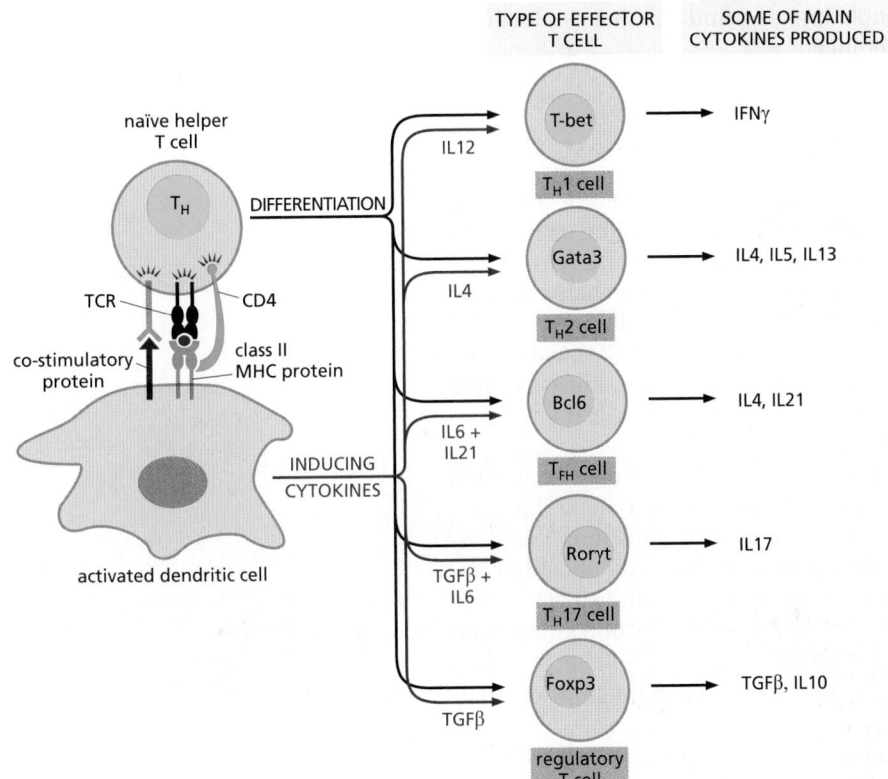

Figure 24–44 **Differentiation of naïve helper T cells into different types of effector helper cells or regulatory T cells in a peripheral lymphoid organ.** The cytokines produced by the activating dendritic cell (and by other cells in the environment) mainly determine which type of effector T cell develops, as indicated. Some of the main cytokines produced by each type of effector cell are also shown, and the master transcription regulator for each subset is indicated in the nucleus. There is increasing evidence that some of the effector cells are plastic and can change the cytokines they produce in response to changes in their environment (not shown).

which is critical for the activation of macrophages to destroy pathogens that either invaded the macrophage or were ingested by it; the IFNγ can also induce B cells to switch the class of Ig they are making. Naïve T$_H$ cells activated in the presence of *IL4* develop into **T$_H$2 cells**. These effector cells are important for the control of extracellular pathogens, including parasites. They stimulate B cells to undergo somatic hypermutation and to switch the class of Ig they produce: for example, the T$_H$2 cells themselves produce IL4, which can induce B cells to switch from making IgM and IgD to making IgE antibodies, which can bind to mast cells, as discussed earlier. Naïve T$_H$ cells activated in the presence of *IL6* and *IL21* develop into **follicular helper T cells** (**T$_{FH}$**), which are located in lymphoid follicles and secrete a variety of cytokines, including IL4 and IL21; these cells are especially important for stimulating B cells to undergo Ig class switching and somatic hypermutation. Naïve T$_H$ cells activated in the presence of *IL6* and *TGFβ* develop into **T$_H$17 cells**. These effector cells secrete *IL17*, which recruits neutrophils and stimulates epithelial cells and fibroblasts in the skin and gut to produce pro-inflammatory cytokines. T$_H$17 cells are important in controlling extracellular bacterial and fungal infections and in wound healing, but they can also have a major role in autoimmune diseases and allergy.

In some cases, naïve T$_H$ cells that encounter their antigen in a peripheral lymphoid organ in the presence of TGFβ and the absence of IL6 develop into **induced regulatory T cells** (**T$_{reg}$ cells**), which suppress rather than help immune cells; as mentioned earlier, **natural T$_{reg}$ cells** develop in the thymus during thymocyte development (see Figure 24–41). In either case, the T$_{reg}$ cells suppress the development, activation, or function of most other types of immune cells, by means of both secreted suppressive cytokines such as *IL10* and TGFβ and inhibitory proteins on the T$_{reg}$ cell surface. Induced T$_{reg}$ cells seem mainly to suppress immune responses to foreign antigens—preventing responses to harmless ingested or inhaled antigens and limiting responses against pathogens to avoid excessive responses that cause unwanted pathology; natural T$_{reg}$ cells are needed to prevent immune responses to self molecules (see Figure 24–21). T$_{reg}$ cells express the transcription regulator *FoxP3*, which serves as both a marker of these cells and a master controller of their development: if the gene encoding this protein is inactivated in mice or humans, the individuals fail to produce T$_{reg}$ cells and develop a fatal autoimmune disease involving multiple organs—findings that establish the crucial importance of T$_{reg}$ cells in self-tolerance.

Both T and B Cells Require Multiple Extracellular Signals For Activation

Foreign antigen binding to BCRs or TCRs initiates the process whereby the T and B cells are stimulated to proliferate and differentiate into effector or memory cells. As mentioned earlier, these antigen receptors do not act on their own: they are stably associated with invariant transmembrane polypeptide chains that are required to relay the signal into the cell. In B cells, these are called *Igα* and *Igβ* (**Figure 24–45**A), while in T cells they exist in a complex called *CD3*, composed of four types of polypeptide chains (Figure 24–45B). In both cases, the associated proteins help convert extracellular antigen binding to the TCR or BCR into intracellular signals, and they do so in similar ways.

Antigen binding to BCRs or TCRs clusters these receptors and their associated invariant chains (and CD4 or CD8 co-receptors in the case of TCRs). This clustering activates a Src family cytoplasmic tyrosine kinase to phosphorylate tyrosines on the cytoplasmic tails of some of the invariant chains. The phosphotyrosines then serve as docking sites for a second cytoplasmic tyrosine kinase, which becomes phosphorylated and activated by the first kinase; the second kinase then relays the signal downstream by phosphorylating other intracellular signaling proteins on tyrosines. Some of these early events in the signaling pathway activated by BCRs are shown in **Figure 24–46**.

Signaling through BCRs or TCRs and their associated proteins alone is not sufficient to activate a lymphocyte to proliferate and differentiate. Extracellular

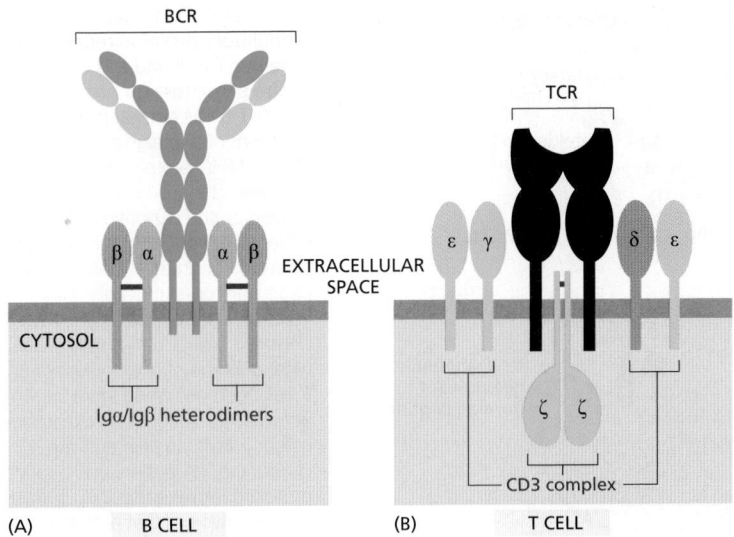

Figure 24–45 The invariant chains associated with BCRs and TCRs. (A) Each BCR is associated with two invariant heterodimers, each composed of an Igα and an Igβ polypeptide chain linked by a sulfide bond (*red*). (B) Each TCR is associated with an invariant CD3 complex composed of two disulfide-bonded ζ chains, two ε chains, and one δ and one γ chain; these chains form homodimers or heterodimers, as shown.

co-stimulatory signals produced by another cell are also required, and they are provided by membrane-bound proteins (see Figure 24–34) and secreted cytokines. Indeed, signaling through the BCR or TCR with insufficient co-stimulation can either eliminate the lymphocyte (clonal deletion) or inactivate it, with both of these mechanisms contributing to self-tolerance (see Figure 24–21). For a naïve T cell, an activated dendritic cell provides the co-stimulatory signals; these include the transmembrane *B7 proteins*, which are recognized by the co-receptor protein *CD28* on the surface of the T cell (**Figure 24–47**A). For a B cell, an effector T$_H$ cell provides the co-stimulatory signals; these include the transmembrane *CD40 ligand*, which binds to *CD40 receptors* on the B cell (Figure 24–47B). The CD40 ligand on effector T$_H$ cells acts in two other situations: (1) it acts back on CD40 receptors on the dendritic cell surface to increase and sustain the activation of the dendritic cell, creating a positive feedback loop; and (2) it acts as a co-stimulatory signal on the surface of an effector T$_H$1 cell, allowing the T cell to help activate an infected macrophage to destroy the pathogens it harbors.

In addition to receptors for co-stimulatory proteins, both B and T cells have inhibitory proteins on their surface that help regulate the cell's activity, preventing excessive or inappropriate responses. Two such proteins expressed by T cells have attracted great attention because of their roles in suppressing the ability of T cells to inhibit cancer progression: CTLA4 and PD1 proteins inhibit T cell activity in different ways, and monoclonal antibodies against either or especially both can relieve the inhibition and allow T cells to dramatically destroy the tumors in some patients with metastatic cancer (see Figure 20–45).

Figure 24–46 Early signaling events in a B cell activated by the binding of specific foreign antigen to its BCRs. If the antigen is on the surface of a pathogen or is a soluble macromolecule with two or more identical antigenic determinants (as shown), it cross-links adjacent BCRs, causing them and their associated invariant chains to cluster, as shown. A Src-like cytoplasmic tyrosine kinase (which can be *Fyn* or *Lyn*) is associated with the cytosolic tail of Igβ; it joins the cluster and phosphorylates both the Igα and Igβ invariant chains (for simplicity, only the phosphorylation on Igβ is shown). A transmembrane protein tyrosine phosphatase called *CD45* is also required to remove inactivating phosphates from these Src-like kinases (not shown). The resulting phosphotyrosines on Igα and Igβ serve as docking sites for another Src-like tyrosine kinase called *Syk*, which becomes phosphorylated and thereby activated to relay the signal downstream.

The pathway from TCRs is similar (including a requirement for CD45), except that the first Src-like kinase is *Lck*, which is associated with a CD4 or CD8 co-receptor and phosphorylates tyrosines on all the CD3 polypeptide chains shown in Figure 24–45B; the second Src-like kinase is *ZAP70*, which is homologous to the Syk kinase in B cells (**Movie 24.14**).

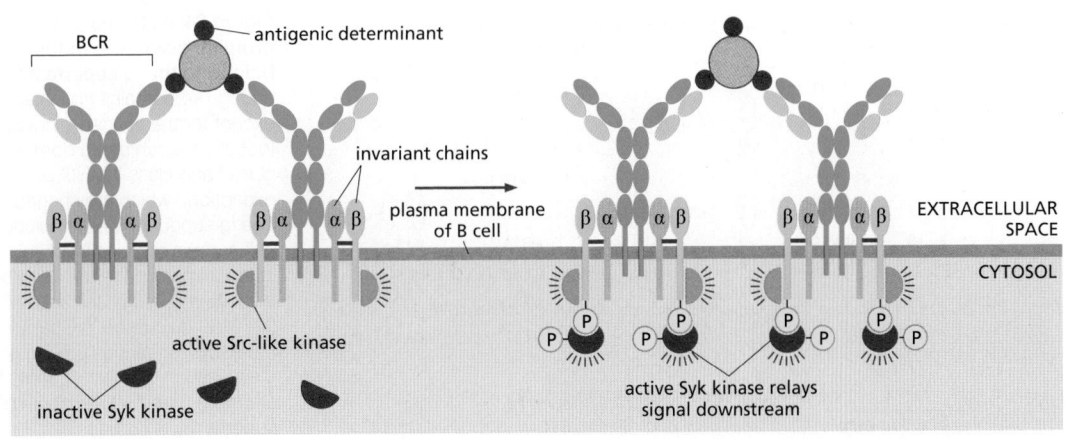

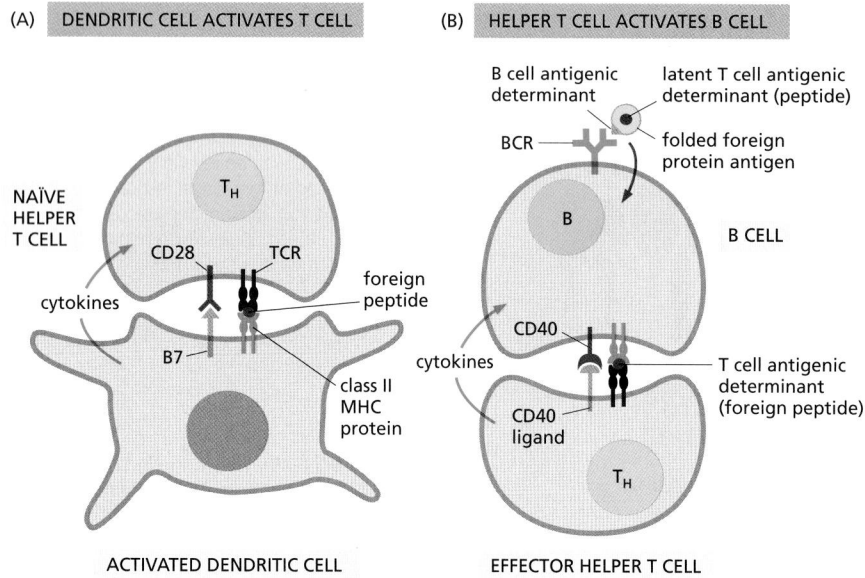

(A) DENDRITIC CELL ACTIVATES T CELL

(B) HELPER T CELL ACTIVATES B CELL

Figure 24–47 Comparison of the co-stimulatory proteins required to activate a helper T cell and a B cell in response to the same foreign protein. (A) A naïve helper T cell is activated by a peptide fragment of a foreign protein bound to a class II MHC protein on the surface of an activated dendritic cell. The co-stimulatory protein on the dendritic cell (a B7 protein—either CD80 or CD86) binds to the CD28 co-receptor on the T cell, providing a necessary co-stimulatory signal to the T cell; in addition, cytokines secreted by the dendritic cell (or other nearby cells) influence what subtype of effector helper cell the T cell becomes (see Figure 24–44). (B) Once activated to become an effector cell, the helper T cell can help activate B cells that have the same peptide–MHC protein complexes on their surface as the dendritic cell that activated the T cell. These B cells have BCRs that bind an antigenic determinant on the surface of a folded foreign protein and endocytose the protein *(red arrow)*; the protein is then cleaved into peptides, which are carried to the B cell surface by class II MHC proteins, where some of them can be recognized by the TCRs on the helper T cell (see Figure 24–39). Note that the BCRs and TCRs recognize different antigenic determinants of the protein. As indicated, the co-stimulatory protein used by the effector helper T cell is CD40 ligand, which binds to the CD40 co-receptor on the B cell; the T cell also secretes cytokines such as IL4 to help stimulate the B cell to undergo somatic hypermutation and class switching (not shown). The CD4 co-receptor on T_H cells is omitted in both (A) and (B) for simplicity.

Many Cell-Surface Proteins Belong to the Ig Superfamily

Most of the proteins that mediate antigen recognition and cell–cell recognition in the immune system contain one or more Ig or Ig-like domains, suggesting that the proteins have a common evolutionary history. Included in this very large **Ig superfamily** are antibodies, TCRs, MHC proteins, the CD4, CD8, and CD28 co-receptors, the B7 co-stimulatory proteins, and most of the invariant polypeptide chains associated with TCRs and BCRs, as well as the various Fc receptors on lymphocytes and other leukocytes. Many of these proteins are dimers or higher oligomers, in which Ig or Ig-like domains of one chain interact with those in another (**Figure 24–48**).

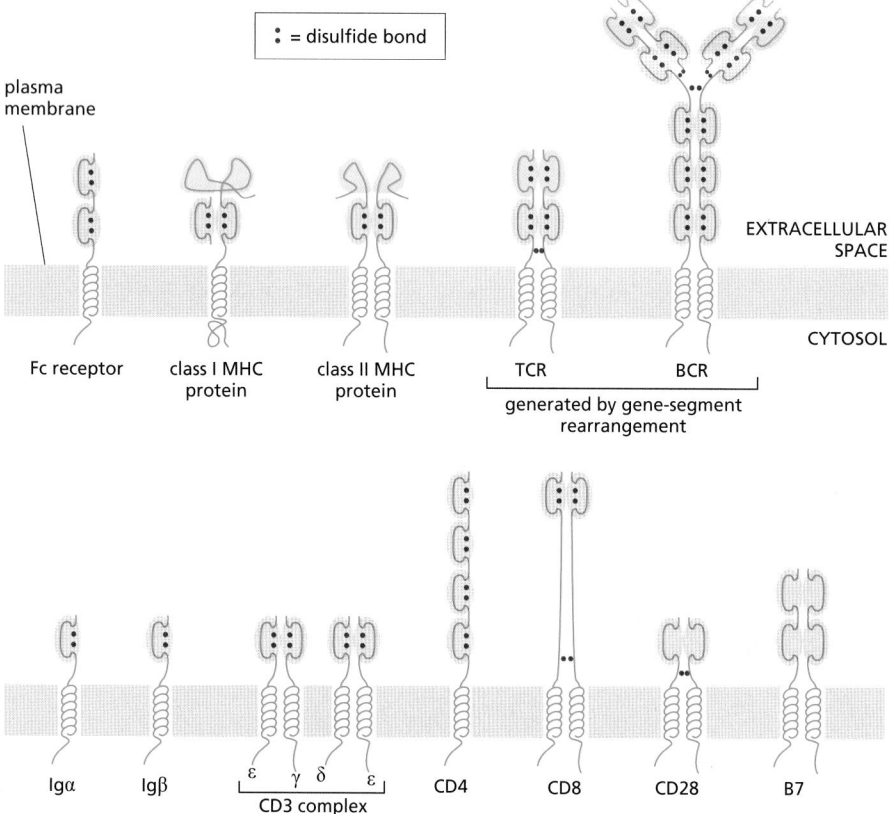

Figure 24–48 Some of the cell-surface proteins discussed in this chapter that belong to the Ig superfamily. The Ig and Ig-like domains are shaded in *gray,* except for the antigen-binding domains (not all of which are Ig domains—the class I and class II MHC proteins are the exception), which are shaded in *blue.* The Ig superfamily also includes many cell-surface proteins involved in cell–cell interactions outside the immune system, such as the neural cell adhesion molecule (N-CAM) discussed in Chapter 19 and the receptors for various protein growth factors discussed in Chapter 15 (not shown). There are more than 750 members of the Ig superfamily in humans.

In both vertebrates and invertebrates, many proteins in the Ig superfamily are also found outside immune systems, where they often function in cell–cell recognition and adhesion processes, both during development and in adult tissues. It seems likely that the entire gene superfamily evolved from a primordial gene coding for a single Ig-like domain, similar to that encoding β_2-microglobulin (see Figure 24–36). In present-day family members, a separate exon usually encodes the amino acids in each Ig-like domain, consistent with the likelihood that new family members arose during evolution by exon and gene duplications.

Summary

There are three main functionally distinct classes of T cells. Cytotoxic T cells (T_C cells) directly kill infected cells by secreting perforins and granzymes that induce the infected cells to undergo apoptosis. Helper T cells (T_H cells) help activate cytotoxic T cells to kill their target cells, B cells to make antibody responses, macrophages to destroy the microorganisms they harbor, and dendritic cells to activate T cells. Regulatory T cells (T_{reg} cells) produce suppressive proteins (such as the cytokines IL10 and TGFβ) to inhibit other immune cells.

All T cells express cell-surface antigen receptors (TCRs), which are encoded by genes that are assembled from multiple gene segments during T cell development in the thymus. TCRs recognize peptide fragments of foreign proteins that are displayed in association with MHC proteins on the surface of antigen-presenting cells (APCs) and target cells. Naïve T cells are activated in peripheral lymphoid organs by activated dendritic cells, which secrete cytokines and express peptide–MHC complexes, co-stimulatory proteins, and various cell–cell adhesion molecules on their cell surface.

Class I MHC proteins present foreign peptides to T_C cells, whereas class II MHC proteins present foreign peptides to T_H cells and T_{reg} cells. Whereas class I MHC proteins are expressed on almost all nucleated vertebrate cells, class II MHC proteins are normally restricted to APCs, including dendritic cells, macrophages, and B lymphocytes. Both classes of MHC proteins have a single peptide-binding groove, which binds a large set of small peptide fragments produced intracellularly by normal protein-degradation processes: class I MHC proteins mainly bind fragments produced in the cytosol, whereas class II MHC proteins mainly bind fragments produced in endocytic compartments. The peptide–MHC complexes are transported to the cell surface, where complexes that contain a peptide derived from a foreign protein are recognized by TCRs, which interact with both the peptide and the walls of the peptide-binding groove. T cells also express CD4 or CD8 co-receptors, which recognize invariant regions of MHC proteins: T_H cells and T_{reg} cells express CD4, which recognizes class II MHC proteins; T_C cells express CD8, which recognizes class I MHC proteins.

A combination of positive and negative selection operates during T cell development in the thymus to help ensure that only T cells with potentially useful TCRs survive, mature, and emigrate, while all of the others die by apoptosis. The naïve T_H and T_C cells that leave the thymus constantly receive survival signals when their TCRs recognize self-peptide–MHC complexes, but they can only be activated when their TCRs encounter foreign peptides in the grooves of MHC proteins on an activated dendritic cell. The natural T_{reg} cells that leave the thymus suppress self-reactive lymphocytes to help maintain self-tolerance.

The production of an effector T cell from a naïve T cell requires multiple signals from an activated dendritic cell. MHC–peptide complexes on the dendritic cell surface provide one signal, by binding to both TCRs and a CD4 co-receptor on a T_H or T_{reg} cell. Co-stimulatory proteins on the dendritic cell surface and secreted cytokines are the other signals. When naïve T_H cells are initially activated on a dendritic cell, they differentiate into T_H1, T_H2, T_{FH}, or T_H17 effector helper cells or into induced T_{reg} cells, depending mainly on the cytokines in their environment. T_H1 cells secrete interferon-γ (IFNγ) to activate macrophages and to induce B cells to switch the class of Ig they make; T_H2 and T_{FH} cells secrete other cytokines that also induce B cells to switch Ig class; and T_H17 cells secrete IL17 to promote inflammatory responses

WHAT WE DON'T KNOW

• What initiates an autoimmune disease such as type 1 diabetes or multiple sclerosis?

• When a naïve or memory T or B cell is activated by antigen and co-stimulatory signals, how does it decide whether to become an effector cell or memory cell? Are there cells that are pre-committed to becoming either effector or memory cells, for example, or is the decision determined solely by extracellular signals?

• Why do some of us make IgE antibodies against harmless antigens and thereby develop hay fever and allergic asthma, while most of us do not, and why is the proportion of such allergic individuals increasing?

• How does a cytotoxic T cell (or NK cell) avoid being killed by the perforin and granzymes that it secretes to kill a target cell?

and wound healing. The effector helper T_H cells recognize the same complex of foreign peptide and class II MHC protein on the target-cell surface as they initially recognized on the dendritic cell that activated them. They activate their target cells by producing a combination of membrane-bound and secreted co-stimulatory proteins. T_{reg} cells suppress immune cells using cell-surface and secreted inhibitory proteins.

Both T cells and B cells require multiple signals for activation. Antigen binding to the TCRs or BCRs provides one signal, while co-stimulatory proteins binding to co-receptors and cytokines binding to their complementary receptors provide the others. Effector T_H cells provide the co-stimulatory signals for B cells, whereas APCs provide them for T cells.

PROBLEMS

Which statements are true? Explain why or why not.

24–1   T cells whose receptors strongly bind a self-peptide–MHC complex are killed off in peripheral lymphoid organs when they encounter the self peptide on an antigen-presenting dendritic cell.

24–2   To guarantee that the antigen-presenting cells in the thymus will display a complete repertoire of self peptides to allow elimination of self-reactive T cells, the thymus recruits dendritic cells from all over the body.

24–3   The antibody diversity created by the combinatorial joining of V, D, and J segments by V(D)J recombination pales in comparison to the enormous diversity created by the random gain and loss of nucleotides at V, D, and J joining sites.

Discuss the following problems.

24–4   Why do living trees not rot? Redwood trees, for example, can live for centuries, but once they die they decay fairly quickly. What might this suggest?

24–5   It would be disastrous if a complement attack were not confined to the surface of the pathogen that is the target of the attack. Yet, the proteolytic cascade involved in the attack liberates biologically active molecules at several steps: one that diffuses away and one that remains bound to the target surface. How does the complement reaction remain localized when active products leave the surface?

24–6   Based on its sequence similarity to Apobec1, which deaminates Cs to Us in RNA, activation-induced deaminase (AID) was originally proposed to work on RNA. But definitive experiments in *E. coli* demonstrated that AID deaminates Cs to Us in DNA. The authors of the paper expressed AID in bacteria and followed mutations in a selectable gene. They found that AID expression increased mutations about fivefold above the background level in the absence of AID expression. More importantly, they found that 80% of the induced mutations were G→A or C→T. Does this fit with your expectation if AID-induced mutations arose by deamination of C to U in the DNA?

[Hint: imagine what would happen if the G:U mismatch created by AID was replicated several times; how would the sequences of the final mutations relate to the original G-C base pair?]

24–7   For many years it was a complete mystery how cytotoxic T cells could see a viral protein that seemed to be present only in the nucleus of the virus-infected cell. The answer was revealed in a classic paper that took advantage of a clone of T cells whose T cell receptor was directed against an antigen assoicated with the nuclear protein of the 1968 strain of influenza virus. The authors of the paper found that when they incubated high concentrations of certain peptides derived from the viral nuclear protein, the cells became sensitive to lysis by subsequent incubation with the cytotoxic T cells. Using various peptides from the 1968 strain and the 1934 strain (with which the cytotoxic T cells did not react), the authors defined the particular peptide responsible for the T cell response (**Figure Q24–1**).

A.   Which part of the viral protein gives rise to the peptide that is recognized by the clone of cytotoxic T cells?

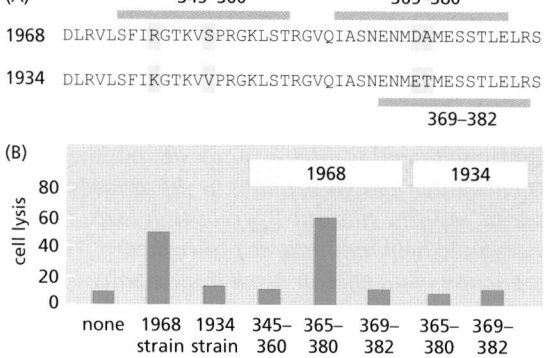

Figure Q24–1 Viral nuclear protein recognition by cytotoxic T cells (Problem 24–7). (A) Sequences of a segment of the nuclear protein from the 1968 and 1934 strains of influenza virus. Peptides used in the experiments in (B) are highlighted by *pink bars*. The amino acid differences between the viral proteins are highlighted in *blue*. (B) Cytotoxic T-cell-mediated lysis of target cells. The target cells were untreated (none), infected with virus (1968 or 1934 strain), or preincubated with high concentrations of the indicated viral peptide.

Why do not all viral peptides sensitize the target cells for lysis by the cytotoxic T cells?

B. It is thought the MHC molecules come to the cell surface with peptides already bound. If that is so, how do you imagine that these experiments worked?

24–8 Working out the rules by which T cells interact with their target cells was complicated. Some of the key observations came from studying the way cytotoxic T cells killed cells infected with choriomeningitis virus (LCMV). Cytotoxic T cells derived from mice expressing "k-type" class I MHC proteins lysed LCMV-infected cells expressing the same k-type MHC protein, but they did not lyse infected cells from mice expressing "d-type" class I MHC proteins (**Figure Q24–2**). Similarly, cytotoxic T cells from d-type mice lysed infected d-type cells, but not infected k-type cells. LCMV can kill both k-type and d-type mice.

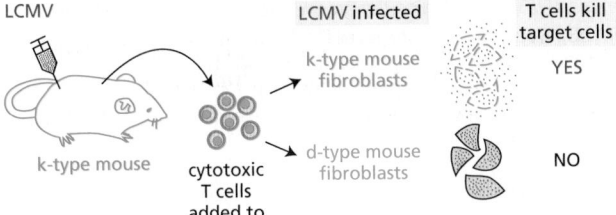

Figure Q24–2 Pattern of killing of LCMV-infected fibroblasts by cytotoxic T cells from an LCMV-infected k-type mouse (Problem 24–8).

A. If homozygous d-type mice were bred to homozygous k-type mice to generate d-type/k-type heterozygous progeny, would you expect that cytotoxic T cells from these heterozygotes, when infected with LCMV, to be able to lyse infected d-type cells? How about infected k-type cells? Explain your answers.

B. Oddly enough, LCMV infection does not kill mice that lack a thymus—such as "nude" mice, so called because they also lack hair. If a thymus is transplanted back into a nude mouse, it will die when infected with LCMV. Suppose that a d-type/k-type heterozygous nude mouse was given a thymus from an d-type donor. Would you expect its cytotoxic T cells to be able to lyse infected d-type cells? How about infected k-type cells? Explain your answers.

24–9 Before exposure to a foreign antigen, T cells with receptors specific for the antigen are a tiny fraction of the T cells—on the order of 1 in 10^5 or 1 in 10^6 T cells. After exposure to the antigen, only a small number of dendritic cells typically display the antigen on their surface. How long does it take for such antigen-presenting dendritic cells to interact with the antigen-specific T cells, which is the key first step in T cell activation and clonal expansion? The dynamics of the search process were examined by labeling dendritic cells red and T cells green, so that contacts in an intact lymph node could be scored visually using two-photon fluorescence microscopy (**Figure Q24–3A**). The frequency of contacts between dendritic cells and T cells from such experiments is given in Figure 24–3B. Assuming that 100 dendritic cells present the specific antigen, how long would it take them to scan 10^5 T cells? How long for 10^6 T cells?

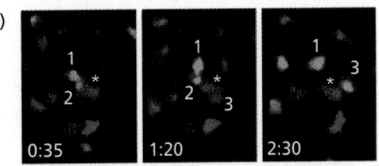

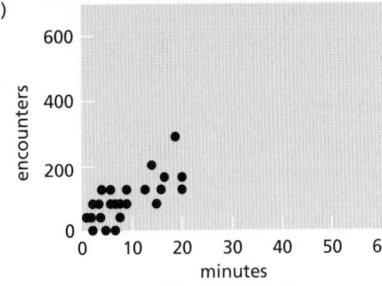

Figure Q24–3 Scanning of the T cell repertoire by dendritic cells (Problem 24–9). (A) Contacts between different T cells and one dendritic cell. T cells are *green* and dendritic cells are *red*. The dendritic cell labeled with an asterisk contacts a total of three T cells (numbered) over time in this sequence of images. Times are shown as hours: minutes. (B) Plot of T cell contacts for individual dendritic cells over time. (A, from P. Bousso and E. Robey, *Nat. Immunol.* 4:579–585, published 2003 by Nature Publishing Group. Reproduced with permission of SNCSC.)

24–10 At first glance, it would seem a dangerous strategy for the thymus to actively promote the survival, maturation, and emigration of developing T cells that bind weakly to self peptides bound to self MHC molecules. Would it not be safer to get rid of these T cells, along with those that bind strongly to such self-peptide–MHC complexes, as this would seem a more secure way to avoid autoimmune reactions?

24–11 CD4 proteins on helper and regulatory T cells serve as co-receptors that bind to invariant parts of class II MHC proteins. CD4 is thought to increase the adhesion between T cells and antigen-presenting cells (APCs) that are initially connected only weakly by the T cell receptor bound to its specific peptide–MHC complex. To test this possibility, you label cell-surface MHC molecules with a fluorescently labeled peptide so that you can detect individual peptide–MHC complexes at the interface between the APCs and the T cells in a culture dish. To detect T cell responses—the sign of a productive contact—you load them with a Ca^{2+} indicator dye, as cytosolic Ca^{2+} increases when lymphocytes are active. You now count the peptide–MHC complexes at a large number of interfaces (immunological synapses) and measure the resulting uptake of Ca^{2+} in the adherent T cells (**Figure Q24–4**, *red circles*). When you repeat the experiment in the presence of blocking antibodies against CD4, you get a different result (*blue circles*). Do these results support or refute the notion that CD4 augments T cell receptor binding? Explain your answer.

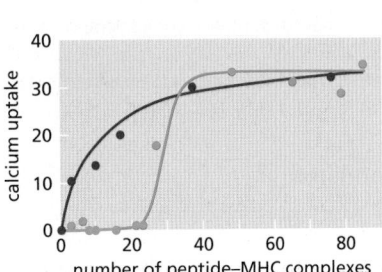

Figure Q24–4 Role of CD4 in the T cell response (Problem 24–11). The uptake of Ca^{2+} in cells with different numbers of fluorescently labeled peptide–MHC complexes at the interface between the T cells and the antigen-presenting cells. The results in the absence of CD4-blocking antibodies are shown by the *red curve*; results in the presence of CD4 antibodies are shown by the *blue curve*.

REFERENCES

General

Murphy K (2011) Janeway's Immunobiology, 8th ed. New York: Garland Science.

Abbas AK, Lichtman AH & Pillai S (2014) Cellular and Molecular Immunology, 8th ed. Philadelphia: WB Saunders.

Parham P (2015) The Immune System, 4th ed. New York: Garland Science.

The Innate Immune System

Beutler B & Rietschel ET (2003) Innate immune sensing and its roots: the story of endotoxin. *Nat. Rev. Immunol.* 3, 169–176.

Davis BK, Wen H & Ting JP (2011) The inflammasome NLRs in immunity, inflammation, and associated diseases. *Annu. Rev. Immunol.* 29, 707–735.

Gallo RL & Hooper LV (2012) Epithelial antimicrobial defence of the skin and intestine. *Nat. Rev. Immunol.* 12, 503–516.

Hoffmann J & Akira S (2013) Innate immunity. *Curr. Opin. Immunol.* 25, 1–3.

Holers VM (2014) Complement and its receptors: new insights into human disease. *Annu. Rev. Immunol.* 32, 433–459.

Ivashkiv LB & Donlin LT (2014) Regulation of type 1 interferon responses. *Nat. Rev. Immunol.* 14, 36–49.

Janeway CA Jr & Medzhitov R (2002) Innate immune recognition. *Annu. Rev. Immunol.* 20, 197–216.

Kumar H, Kawai T & Akira S (2011) Pathogen recognition by the innate immune system. *Int. Rev. Immunol.* 30, 16–34.

Lannier LL (2008) Up on the tightrope: natural killer cell activation and inhibition. *Nat. Immunol.* 9, 495–502.

Murray PJ & Wynn TA (2011) Protective and pathogenic functions of macrophages. *Nat. Rev. Immunol.* 11, 723–737.

Pluddemann A, Mukhopadhyay S & Gordon S (2011) Innate immunity to intracellular pathogens: macrophage receptors and responses to microbial entry. *Immunol. Rev.* 240, 11–24.

Schenten D & Medzhitov R (2011) The control of adaptive immune responses by the innate immune system. *Adv. Immunol.* 109, 87–124.

Overview of the Adaptive Immune System

Denucci CC, Mitchell JS & Shimizu Y (2009) Integrin function in T-cell homing to lymphoid and nonlymphoid sites: getting there and staying there. *Crit. Rev. Immunol.* 29, 87–109.

Girard J-P, Moussion C & Förster R (2012) HEVs, lymphatics and homeostatic immune cell trafficking in lymph nodes. *Nat. Rev. Immunol.* 12, 762–773.

Klein L, Yudanin NA & Restifo NP (2013) Human memory T cells: generation, compartmentalization and homeostasis. *Nat. Rev. Immunol.* 14, 24–35.

MacKeod MK, Kappler JW & Marrack P (2010) CD4 memory T cells: generation, reactivation, and reassignment. *Immunol.* 130, 10–15.

McHeyzer-Williams M, Okitsu S, Wang N & McHeyzer-Williams L (2011) Molecular programming of B cell memory. *Nat. Rev. Immunol.* 12, 24–34.

Rothenburg EV (2014) Transcriptional control of early T and B cell developmental choices. *Annu. Rev. Immunol.* 32, 283–321.

Schwartz RH (2012) Historical overview of immunological tolerance. *Cold Spr. Harb. Perspect. Biol.* 4, 1–14.

Sprent J & Surh CD (2011) Normal T cell homeostasis: the conversion of naïve cells into memory-phenotype cells. *Nat. Rev. Immunol.* 12, 478–484.

B Cells and Immunoglobulins

Eibel H, Kraus H, Sic H et al. (2014) B cell biology: an overview. *Curr. Allergy Asthma Rep.* 14, 434–444.

Ganesh K & Neuberger MS (2011) The relationship between hypothesis and experiment in unveiling the mechanisms of antibody gene diversification. *FASEB J.* 25, 1123–1132.

Kurosaki T, Shinohara H, & Baba Y (2010) B cell signalling and fate decision. *Annu. Rev. Immunol.* 29, 21–55.

Nimmerjahn F & Ravetch JV (2007) Fc-receptors as regulators of immunity. *Adv. Immunol.* 96, 179–204.

Schatz DG & Yanhong J (2010) Recombination centers and the orchestration of V(D)J recombination. *Nat. Rev. Immunol.* 11, 251–263.

Schomchik MJ & Weisel F (2012) Germinal center selection and development of memory B and plasma cells. *Immunol. Rev.* 247, 52–63.

Schroeder HW & Cavacini L (2010) Structure and function of immunoglobulins. *J. Aller. Clin. Immunol.* 125, S41–S52.

Xu Z, Zan H, Pone EJ et al. (2012) Immunoglobulin class-switch DNA recombination: induction, targeting and beyond. *Nat. Rev. Immunol.* 12, 517–531.

T Cells and the MHC

Bjorkman PJ (2006) Finding the groove. *Nat. Immunol.* 7, 787–789.

Blum JS, Wearsch PA & Cresswell P (2013) Pathways of antigen processing. *Annu. Rev. Immunol.* 31, 443–473.

Clambey ET, Davenport B, Kappler JW et al. (2014) Molecules in medicine mini review: the αβ T cell receptor. *J. Mol. Med.* 92, 735–741.

Dustin ML (2014) What counts in the immunological synapse. *Mol. Cell* 54, 255–262.

Ewen CL, Kane KP & Bleackley RC (2012) A quarter century of granzyme. *Cell Death Differ.* 19, 28–35.

Hniffa M, Collin M & Ginhoux F (2013) Ontogeny and functional specialization of dendritic cells in human and mouse. *Adv. Immunol.* 120, 1–49.

Harwood NE & Batista FD (2010) Early events in B cell activation. *Annu. Rev. Immunol.* 28, 185–210.

Hsieh CS, Lee HM & Lio CW (2012) Selection of regulatory T cells in the thymus. *Nat. Rev. Immunol.* 10, 157–167.

Huppa JB & Davis MM (2013) The interdisciplinary science of T-cell recognition. *Adv. Immunol.* 119, 1–50.

Joffre OP, Segura E, Savina A & Amigorena S (2012) Cross-presentation by dendritic cells. *Nat. Rev. Immunol.* 12, 557–569.

Josefowicz SZ, Lu L-F & Rudensky AY (2012) Regulatory T cells: mechanisms of differentiation and function. *Annu. Rev. Immunol.* 30, 531–564.

Klein L, Kyewski B, Allen PM & Hogquist KA (2014) Positive and negative selection of the T cell repertoire: what the thymocytes see (and don't see). *Nat. Rev. Immunol.* 14, 377–391.

Liu K & Nussenzweig MC (2010) Origin and development of dendritic cells. *Immunol. Rev.* 234, 45–54.

Mathis D & Benoist C (2009) AIRE. *Annu. Rev. Immunol.* 27, 287–312.

McDevitt HO (2000) Discovering the role of the major histocompatibility complex in the immune response. *Annu. Rev. Immunol.* 18, 1–17.

Ohkura N, Kitagawa Y & Sakagucci S (2013) Development and maintenance of regulatory T cells. *Semin. Immunol.* 23, 424–430.

Ramsdell F & Ziegler SF (2014) Fox3P and scurfy: how it all began. *Nat. Rev. Immunol.* 14, 343–349.

Rudolph MG, Stanfield RL & Wilson IA (2006) How TCRs bind MHCs, peptides, and coreceptors. *Annu. Rev. Immunol.* 24, 419–466.

Schuette V & Burgdorf S (2014) The ins-and-outs of endosomal antigens for cross presentation. *Curr. Opin. Immunol.* 26, 63–68.

Shevach EM (2011) Biological functions of regulatory T cells. *Adv. Immunol.* 112, 137–176.

Steinmar RM (2012) Decisions about dendritic cells: past, present, and future. *Annu. Rev. Immunol.* 30, 1–22.

Trombetta ES & Mellman I (2005) Cell biology of antigen processing *in vitro* and *in vivo*. *Annu. Rev. Immunol.* 23, 975–1028.

Zhu J, Yamane H & Paul WE (2010) Peripheral CD4+ T-cell differentiation regulated by networks of cytokines and transcription factors. *Annu. Rev. Immunol.* 28, 445–489.

Glossary

ABC transporters   A large family of membrane transport proteins that use the energy of ATP hydrolysis to transfer peptides or small molecules across membranes. (Figure 11–16)

acetyl CoA   Small water-soluble activated carrier molecule. Consists of an acetyl group linked to coenzyme A (CoA) by an easily hydrolyzable thioester bond. (Figure 2–38)

acetylcholine receptor (AChR)   Membrane protein that responds to binding of acetylcholine (ACh). The nicotinic AChR is a transmitter-gated ion channel that opens in response to ACh. The muscarinic AChR is not an ion channel, but a G-protein-coupled cell-surface receptor.

acid   A proton donor. Substance that releases protons (H^+) when dissolved in water, forming hydronium ions (H_3O^+) and lowering the pH. (Panel 2–2, pp. 92–93)

acid hydrolases   Hydrolytic enzymes—including proteases, nucleases, glycosidases, lipases, phospholipases, phosphatases, and sulfatases—that work best at acidic pH; these enzymes are found within the lysosome.

action potential   Rapid, transient, self-propagating electrical excitation in the plasma membrane of a cell such as a neuron or muscle cell. Action potentials, or nerve impulses, make possible long-distance signaling in the nervous system. (Figure 11–31)

activated carrier   Small diffusible molecule that stores easily exchangeable energy in the form of one or more energy-rich covalent bonds. Examples are ATP, acetyl CoA, $FADH_2$, NADH, and NADPH. (Figure 2–31)

activation energy   The extra energy that must be acquired by atoms or molecules in addition to their ground-state energy in order to reach the transition state required for them to undergo a particular chemical reaction. (Figure 2–21)

activation-induced deaminase (AID)   The enzyme catalyzing the processes of somatic hypermutation and immunoglobulin class switching in activated B cells.

active site   Region of an enzyme surface to which a substrate molecule binds in order to undergo a catalyzed reaction. (Figure 1–7)

active transport   Movement of a molecule across a membrane or other barrier driven by energy other than that stored in the electrochemical or concentration gradient of the transported molecule.

adaptation   (1) Adaptation (desensitization): adjustment of sensitivity following repeated stimulation. The mechanism that allows a cell to react to small changes in stimuli even against a high background level of stimulation. (2) Evolutionary adaptation: an evolved trait.

adaptive immune system   System of lymphocytes providing highly specific and long-lasting defense against pathogens in vertebrates. It consists of two major classes of lymphocytes: B lymphocytes (B cells), which secrete antibodies that bind specifically to the pathogen or its products, and T lymphocytes (T cells), which can either directly kill cells infected with the pathogen or produce secreted or cell-surface signal proteins that stimulate other host cells to help eliminate the pathogen. (Figure 24–2)

adaptor protein, adaptor   General term for a protein that functions solely to link two or more different proteins together in an intracellular signaling pathway or protein complex. (Figure 15–11)

adenylyl cyclase (adenylate cyclase)   Membrane-bound enzyme that catalyzes the formation of cyclic AMP from ATP. An important component of some intracellular signaling pathways.

adherens junction   Cell junction in which the cytoplasmic face of the plasma membrane is attached to actin filaments. Examples include adhesion belts linking adjacent epithelial cells and focal contacts on the lower surface of cultured fibroblasts.

adhesins   Specific proteins or protein complexes of pathogenic bacteria that recognize and bind cell-surface molecules on the host cells to enable tight adhesion and colonization of tissues.

adhesion belt   Adherens junctions in epithelia that form a continuous belt (zonula adherens) just beneath the apical face of the epithelium, encircling each of the interacting cells in the sheet.

ADP (adenosine 5′-diphosphate)   Nucleotide produced by hydrolysis of the terminal phosphate of ATP. Regenerates ATP when phosphorylated by an energy-generating process such as oxidative phosphorylation. (Figure 2–33)

aerobic respiration   Process by which a cell obtains energy from sugars or other organic molecules by allowing their carbon and hydrogen atoms to combine with the oxygen in air to produce CO_2 and H_2O, respectively.

affinity maturation   Progressive increase in the affinity of antibodies for the immunizing antigen with the passage of time after immunization.

Agrin   Signal protein released by an axonal growth cone during formation of the synapse between it and a muscle cell.

AIRE (autoimmune regulator)   A protein expressed by a subpopulation of epithelial cells in the thymus that stimulates the production of small amounts of self proteins characteristic of other organs, exposing developing thymocytes to these proteins for the purpose of self-tolerance.

Akt   Serine/threonine protein kinase that acts in the PI-3-kinase/Akt intracellular signaling pathway involved especially in signaling cells to grow and survive. Also called protein kinase B (PKB).

allele   One of several alternative forms of a gene. In a diploid cell, each gene will typically have two alleles, occupying the corresponding position (locus) on homologous chromosomes.

allosteric protein   A protein that can adopt at least two distinct conformations, and for which the binding of a ligand at one site causes a conformational change that alters the activity of the protein at a second site; this allows one type of molecule in a cell to alter the fate of a molecule of another type, a feature widely exploited in enzyme regulation.

allostery (adjective **allosteric**)   Change in a protein's conformation brought about by the binding of a regulatory ligand (at a site other than the protein's catalytic site), or by covalent modification. The change in conformation alters the activity of the protein and can form the basis of directed movement. (Figures 3–57 and 16–29)

alpha helix (**α helix**)   Common folding pattern in proteins, in which a linear sequence of amino acids folds into a right-handed helix stabilized by internal hydrogen-bonding between backbone atoms. (Figure 3–7)

alternative RNA splicing   Production of different RNAs from the same gene by splicing the transcript in different ways. (Figure 7–57)

amino acid   Organic molecule containing both an amino group and a carboxyl group. Those that serve as building blocks of proteins are alpha amino acids, having both the amino and carboxyl groups linked to the same carbon atom. ($NH_2CHRCOOH$, Panel 3–1, pp. 112–113)

aminoacyl-tRNA synthetase   Enzyme that attaches the correct amino acid to a tRNA molecule to form an aminoacyl-tRNA. (Figure 6–54)

AMPA receptor   Glutamate-gated ion channel in the mammalian central nervous system that carries most of the depolarizing current responsible for excitatory postsynaptic potentials.

amphiphilic   Having both hydrophobic and hydrophilic regions, as in a phospholipid or a detergent molecule.

amyloid fibrils   Self-propagating, stable β-sheet aggregates built from hundreds of identical polypeptide chains that become layered one over the other to create a continuous stack of β sheets. The unbranched fibrous structure can contribute to human diseases when not controlled.

anaphase   (1) Stage of mitosis during which sister chromatids separate and move away from each other. (2) Anaphase I and II: stages of meiosis during which chromosome homolog pairs separate (I), and then sister chromatids separate (II). (Panel 17–1, pp. 980–981)

anaphase A   Stage of mitosis during which chromosome segregation occurs as chromosomes move toward the two spindle poles.

anaphase B   Stage of mitosis during which chromosome segregation occurs as spindle poles separate and move apart.

anaphase-promoting complex (**APC/C; cyclosome**)   Ubiquitin ligase that catalyzes the ubiquitylation and destruction of securin and M- and S-cyclins, initiating the separation of sister chromatids in the metaphase-to-anaphase transition during mitosis.

anchorage dependence   Dependence of cell growth, proliferation, and survival on attachment to a substratum.

anchoring junction   Cell junction that attaches cells to neighboring cells or to the extracellular matrix. (Table 19–1, p. 1037)

angiogenesis   Growth of new blood vessels by sprouting from existing ones.

antenna complex   Part of a photosystem that captures light energy and channels it into the photochemical reaction center. It consists of protein complexes that bind large numbers of chlorophyll molecules and other pigments.

***Antennapedia* complex**   One of two gene clusters in *Drosophila* that contain *Hox* genes; genes in the *Antennapedia* complex control the differences among the thoracic and head segments of the body.

anti-apoptotic Bcl2 family proteins   Proteins (e.g., Bcl2, BclX$_L$) on the cytosolic surface of the outer mitochondrial membrane that bind and inhibit pro-apoptotic Bcl2 family proteins and thereby help prevent inappropriate activation of the intrinsic pathway of apoptosis.

anti-IAP   Produced in response to various apoptotic stimuli and, by binding to IAPs and preventing their binding to a caspase, neutralize the inhibition of apoptosis provided by IAPs.

antibiotic   Substance such as penicillin or streptomycin that is toxic to microorganisms. Often a natural product of a particular microorganism or plant.

antibody   Protein secreted by activated B cells in response to a pathogen or foreign molecule. Binds tightly to the pathogen or foreign molecule, inactivating it or marking it for destruction by phagocytosis or complement-induced lysis. (Figure 24–23)

antibody response   Adaptive immune response in which B cells are activated to secrete antibodies that circulate in the bloodstream or enter other body fluids, where they can bind specifically to the foreign antigen that stimulated their production.

anticodon   Sequence of three nucleotides in a transfer RNA (tRNA) molecule that is complementary to a three-nucleotide codon in a messenger RNA (mRNA) molecule.

antigen   A molecule that can induce an adaptive immune response or that can bind to an antibody or T cell receptor.

antigen-presenting cell   Cell that displays foreign antigen complexed with an MHC protein on its surface for presentation to T lymphocytes.

antigenic determinant   Specific region of an antigen that binds to an antibody or a complementary receptor on the surface of a B cell (BCR) or T cell (TCR).

antigenic variation   Ability to change the antigens displayed on the cell surface; a property of some pathogenic microorganisms that enables them to evade attack by the adaptive immune system.

antiparallel   Describes the relative orientation of the two strands in a DNA double helix or two paired regions of a polypeptide chain; the polarity of one strand is opposite to that of the other.

antiporter   Carrier protein that transports two different ions or small molecules across a membrane in opposite directions, either simultaneously or in sequence. (Figure 11–8)

Apaf1   Adaptor protein of the intrinsic apoptotic pathway; on binding cytochrome *c*, oligomerizes to form an apoptosome.

apical   Referring to the tip of a cell, a structure, or an organ. The apical surface of an epithelial cell is the exposed free surface, opposite to the basal surface. The basal surface rests on the basal lamina that separates the epithelium from other tissue.

apoptosis Form of programmed cell death, in which a "suicide" program is activated within an animal cell, leading to rapid cell death mediated by intracellular proteolytic enzymes called caspases.

apoptosome Heptamer of Apaf1 proteins that forms on activation of the intrinsic apoptotic pathway; it recruits and activates initiator caspases that subsequently activate downstream executioner caspases to induce apoptosis.

aquaporin (water channel) Channel protein embedded in the plasma membrane that greatly increases the cell's permeability to water, allowing transport of water, but not ions, at a high rate across the membrane.

archaeon (plural arch[a]ea) (archaebacterium) Single-celled organism without a nucleus, superficially similar to bacteria. At a molecular level, more closely related to eukaryotes in genetic machinery than are bacteria. Archaea and bacteria together make up the prokaryotes. (Figure 1–17)

ARF proteins Monomeric GTPase in the Ras superfamily responsible for regulating both COPI coat assembly and clathrin coat assembly. (Table 15–5, p. 854)

ARP (actin-related protein) complex (Arp 2/3 complex) Complex of proteins that nucleates actin filament growth from the minus end.

arrestin Member of a family of proteins that contributes to GPCR desensitization by preventing the activated receptor from interacting with G proteins and serving as an adaptor to couple the receptor to clathrin-dependent endocytosis. (Figure 15–42)

astral microtubule In the mitotic spindle, any of the microtubules radiating from the aster which are not attached to a kinetochore of a chromosome.

asymmetric cell division Cell division in which some important molecule or molecules are distributed unequally between the two daughter cells, causing these cells to become different from each other.

ATM (ataxia telangiectasia mutated protein) Protein kinase activated by double-strand DNA breaks. If breaks are not repaired, ATM initiates a signal cascade that culminates in cell cycle arrest. Related to ATR.

ATP (adenosine 5′-triphosphate) Nucleoside triphosphate composed of adenine, ribose, and three phosphate groups. The principal carrier of chemical energy in cells. The terminal phosphate groups are highly reactive in the sense that their hydrolysis, or transfer to another molecule, takes place with the release of a large amount of free energy. (Figure 2–33)

ATP synthase (F_1F_o ATPase) Transmembrane enzyme complex in the inner membrane of mitochondria and the thylakoid membrane of chloroplasts. Catalyzes the formation of ATP from ADP and inorganic phosphate during oxidative phosphorylation and photosynthesis, respectively. Also present in the plasma membrane of bacteria.

ATR (ataxia telangiectasia and Rad3 related protein) Protein kinase activated by DNA damage. If damage remains unrepaired, ATR helps initiate a signal cascade that culminates in cell cycle arrest. Related to ATM.

autoimmune disease Pathological state in which the body mounts a disabling adaptive immune response against one or more of its own molecules.

autophagosome Organelle surrounded by a double membrane contains engulfed cytoplasmic cargo in the initial stages of autophagy.

autophagy Digestion of cytoplasm and worn-out organelles by the cell's own lysosomes.

auxin Plant hormone, commonly indole-3-acetic acid, with numerous roles in plant growth and development.

axon Long nerve cell projection that can rapidly conduct nerve impulses over long distances so as to deliver signals to other cells.

axoneme Bundle of microtubules and associated proteins that forms the core of a cilium or a flagellum in eukaryotic cells and is responsible for their movements.

bacterial artificial chromosome (BAC) Cloning vector that can accommodate large pieces of DNA, typically up to 1 million base pairs.

bacteriorhodopsin Pigmented protein found in the plasma membrane of a salt-loving archaeon, *Halobacterium salinarium* (*Halobacterium halobium*). Pumps protons out of the cell in response to light.

bacterium (plural bacteria) (eubacterium) Member of the domain bacteria, one of the three main branches of the tree of life (archaea, bacteria, and eukaryotes). Bacteria and archaea both lack a distinct nuclear compartment, and together comprise the prokaryotes. (Figure 1–17)

Bak A main effector Bcl2 family protein of the intrinsic pathway of apoptosis in mammalian cells that is bound to the mitochondrial outer membrane even in the absence of an apoptotic signal; activation is usually by activated pro-apoptotic BH3-only proteins.

basal Situated near the base. Opposite the apical surface.

basal lamina (plural basal laminae) Thin mat of extracellular matrix that separates epithelial sheets, and many other types of cells such as muscle or fat cells, from connective tissue. Sometimes called basement membrane. (Figure 19–51)

base (1) A substance that can reduce the number of protons in solution, either by accepting H^+ ions directly, or by releasing OH^- ions, which then combine with H^+ to form H_2O. (2) The purines and pyrimidines in DNA and RNA are organic nitrogenous bases and are often referred to simply as bases. (Panel 2–2, pp. 92–93)

base excision repair DNA repair pathway in which single faulty bases are removed from the DNA helix and replaced. *Compare* **nucleotide excision repair**. (Figure 5–41)

base pair Two nucleotides in an RNA or DNA molecule that are held together by hydrogen bonds—for example, G paired with C, and A paired with T or U.

basement membrane Thin mat of extracellular matrix that separates epithelial sheets, and many other types of cells such as muscle or fat cells, from connective tissue. Also called basal lamina. (Figure 19–51)

Bax A main effector Bcl2 family protein of the intrinsic pathway of apoptosis in mammalian cells; located mainly in the cytosol and translocates to the mitochondria only after activation, usually by activated pro-apoptotic BH3-only proteins.

B cell receptor (BCR) The transmembrane immunoglobulin protein on the surface of a B cell that serves as its receptor for antigen.

Bcl2 Anti-apoptotic Bcl2 family protein of the outer mitochondrial membrane that binds and inhibits pro-apoptotic Bcl2 family proteins and prevents inappropriate activation of the intrinsic pathway of apoptosis.

Bcl2 family   Family of intracellular proteins that either promote or inhibit apoptosis by regulating the release of cytochrome *c* and other mitochondrial proteins from the intermembrane space into the cytosol.

BclX$_L$ Anti-apoptotic Bcl2 family protein of the outer mitochondrial membrane that binds and inhibits pro-apoptotic Bcl2 family proteins and prevents inappropriate activation of the intrinsic pathway of apoptosis.

benign   Of tumors: self-limiting in growth, and noninvasive.

beta sheet (β sheet)   Common structural motif in proteins in which different sections of the polypeptide chain run alongside each other, joined together by hydrogen-bonding between atoms of the polypeptide backbone. Also known as a β pleated sheet. (Figure 3–7)

beta-catenin (β-catenin)   Multifunctional cytoplasmic protein involved in cadherin-mediated cell–cell adhesion, linking cadherins to the actin cytoskeleton. Can also act independently as a transcription regulatory protein. Has an important role in animal development as part of a Wnt signaling pathway.

BH3-only proteins   The largest subclass of Bcl2 family proteins. Produced or activated in response to an apoptotic stimulus and promote apoptosis mainly by inhibiting anti-apoptotic Bcl2 family proteins.

bi-orientation   The attachment of sister chromatids to opposite poles of the mitotic spindle, so that they move to opposite ends of the cell when they separate in anaphase.

binding site   Region on the surface of one molecule (usually a protein or nucleic acid) that can interact with another molecule through noncovalent bonding.

BiP   Endoplasmic reticulum (ER)-resident chaperone protein member of the family of hsp70-type chaperone proteins.

***Bithorax* complex**   One of two gene clusters in *Drosophila* that contain *Hox* genes; genes in the *Bithorax* complex control the differences among the abdominal and thoracic segments of the body.

bivalent   A four-chromatid structure formed during meiosis, consisting of a duplicated chromosome tightly paired with its homologous duplicated chromosome.

blastomere   One of the many cells formed by the cleavage of a fertilized egg.

blastula   Early stage of an animal embryo, usually consisting of a hollow ball of epithelial cells surrounding a fluid-filled cavity, before gastrulation begins.

blebbing   Membrane protrusion formed when the plasma membrane detaches locally from the underlying actin cortex, allowing cytoplasmic flow and hydrostatic pressure within the cell to push the membrane outward.

bone   Dense and rigid connective tissue comprising a mixture of tough fibers (type I collagen fibrils), which resist pulling forces, and solid particles (calcium phosphate as hydroxylapatite crystals), which resist compression.

brassinosteroids   Class of steroid signal molecules in plants that regulate the growth and differentiation of plants throughout their life cycle via binding to a cell-surface receptor kinase to initiate a signaling cascade.

bright-field microscope   Normal light microscope in which the image is obtained by simple transmission of light through the object being viewed.

buffer   Solution of weak acid or weak base that resists the pH change that would otherwise occur when small quantities of acid or base are added.

C3   The pivotal complement protein that is activated by the early components of all three complement pathways (the classical pathway, the lectin pathway, and the alternative pathway). (Figure 24–7)

Ca^{2+} pump (calcium pump, Ca^{2+} ATPase)   Transport protein in the membrane of sarcoplasmic reticulum of muscle cells (and elsewhere). Pumps Ca^{2+} out of the cytoplasm into the sarcoplasmic reticulum using the energy of ATP hydrolysis.

Ca^{2+}-activated K$^+$ channel   Opens in response to the raised concentration of Ca^{2+} in nerve cells that occurs in response to an action potential. Increased K$^+$ permeability makes the membrane harder to depolarize, increasing the delay between action potentials and decreasing the response of the cell to constant, prolonged stimulation (adaptation).

Ca^{2+}/calmodulin-dependent kinase (CaM-kinase)   Serine/threonine protein kinase that is activated by Ca^{2+}/calmodulin. Indirectly mediates the effects of an increase in cytosolic Ca^{2+} by phosphorylating specific target proteins. (Figure 15–33)

cadherin   Member of the large cadherin superfamily of transmembrane adhesion proteins. Mediates homophilic Ca^{2+}-dependent cell–cell adhesion in animal tissues. (Figure 19–3 and Table 19–1, p. 1037)

cadherin superfamily   Family of classical and nonclassical cadherin proteins with more than 180 members in humans.

calmodulin   Ubiquitous intracellular Ca^{2+}-binding protein that undergoes a large conformation change when it binds Ca^{2+}, allowing it to regulate the activity of many target proteins. In its activated (Ca^{2+}-bound) form, it is called Ca^{2+}/calmodulin. (Figure 15–33)

calnexin   Carbohydrate-binding chaperone protein in the endoplasmic reticulum (ER) membrane that binds to oligosaccharides on incompletely folded proteins and retains them in the ER.

calreticulin   Carbohydrate-binding chaperone protein in the endoplasmic reticulum (ER) lumen that binds to oligosaccharides on incompletely folded proteins and retains them in the ER.

CaM-kinase II   Multifunctional Ca^{2+}/calmodulin-dependent protein kinase that phosphorylates itself and various target proteins when activated. Found in most animal cells but is especially abundant at synapses in the brain, and is involved in some forms of synaptic plasticity in vertebrates. (Figure 15–34)

cancer stem cells   Rare cancer cells capable of dividing indefinitely.

cancer-critical genes   Genes whose alteration contributes to the causation or evolution of cancer by driving tumorigenesis.

capsid   Protein coat of a virus, formed by the self-assembly of one or more types of protein subunit into a geometrically regular structure. (Figure 3–27)

carbohydrate layer   The carbohydrate-rich zone on the eukaryotic cell surface attributable to glycoproteins, glycolipids, and proteoglycans of the plasma membrane.

carbon-fixation reaction   Process by which inorganic carbon (as atmospheric CO$_2$) is incorporated into organic molecules. The second stage of photosynthesis. (Figure 14–40)

carcinogenesis   The generation of cancer.

carcinoma   Cancer of epithelial cells. The most common form of human cancer.

cargo   The membrane components and soluble molecules carried by transport vesicles.

cartilage   Form of connective tissue composed of cells (chondrocytes) embedded in a matrix rich in type II collagen and chondroitin sulfate proteoglycan.

caspase   Intracellular protease that is involved in mediating the intracellular events of apoptosis.

catalyst   Substance that can lower the activation energy of a reaction (thus increasing its rate), without itself being consumed by the reaction.

caveola (plural **caveolae**)   Invaginations at the cell surface that bud off internally to form pinocytic vesicles. Thought to form from lipid rafts, regions of membrane rich in certain lipids.

caveolins   Family of unusual integral membrane proteins that are the major structural proteins in caveolae.

CD4   Co-receptor protein on helper T cells and regulatory T cells that binds to a nonvariable part of class II MHC proteins (on antigen-presenting cells) outside the peptide-binding groove. (Figure 24–40)

CD8   Co-receptor protein on cytotoxic T cells that binds to a nonvariable part of class I MHC proteins (on antigen-presenting cells and infected target cells) outside the peptide-binding groove. (Figure 24–40)

Cdc20   Activating subunit of the anaphase-promoting complex (APC/C).

Cdc25   Protein phosphatase that dephosphorylates Cdks and increases their activity.

Cdc42   Member of the Rho family of monomeric GTPases that regulate the actin and microtubule cytoskeletons, cell-cycle progression, gene transcription, and membrane transport.

Cdc6   Protein essential in the preparation of DNA for replication. With Cdt1 it binds to an origin recognition complex on chromosomal DNA and helps load the Mcm proteins onto the complex to form the prereplicative complex.

Cdh1   Activating subunit of the anaphase-promoting complex (APC/C).

Cdk inhibitor protein (CKI)   Protein that binds to and inhibits cyclin–Cdk complexes, primarily involved in the control of G_1 and S phases.

Cdk-activating kinase (CAK)   Protein kinase that phosphorylates Cdks in cyclin–Cdk complexes, activating the Cdk.

cDNA clone   Clone containing double-stranded cDNA molecules derived from the protein-coding mRNA molecules present in a cell.

cDNA library   Collection of cloned DNA molecules representing complementary DNA copies of the mRNA produced by a cell.

Cdt1   Protein essential in the preparation of DNA for replication. With Cdc6 it binds to origin recognition complexes on chromosomes and helps load the Mcm proteins on to the complex, forming the prereplicative complex.

cell cortex   Specialized layer of cytoplasm on the inner face of the plasma membrane. In animal cells it is an actin-rich layer responsible for movements of the cell surface.

cell cycle (**cell-division cycle**)   Reproductive cycle of a cell: the orderly sequence of events by which a cell duplicates its chromosomes and, usually, the other cell contents, and divides into two. (Figure 17–4)

cell determination   Process whereby a cell progressively loses the potential to form other cell types, as development proceeds.

cell doctrine   The proposal that all living organisms are composed of one or more cells and that all cells arise from the division of other living cells.

cell memory   Retention by cells and their descendants of persistently altered patterns of gene expression, without any change in DNA sequence. *See also* **epigenetic inheritance**.

cell plate   Flattened membrane-bounded structure that forms by fusing vesicles in the cytoplasm of a dividing plant cell and is the precursor of the new cell wall.

cell-cycle control system   Network of regulatory proteins that governs progression of a eukaryotic cell through the cell cycle.

cellulose   Long, unbranched chains of glucose; major constituent of plant cell walls.

cellulose microfibril   Highly ordered crystalline aggregate formed from bundles of about 40 cellulose chains, arranged with the same polarity and stuck together in overlapping parallel arrays by hydrogen bonds between adjacent cellulose molecules.

central (primary) lymphoid organ   Organ in which T or B lymphocytes are produced from precursor cells. In adult mammals, these are the thymus and bone marrow, respectively. (Figure 24–12)

centriole   Short cylindrical array of microtubules, closely similar in structure to a basal body. A pair of centrioles is usually found at the center of a centrosome in animal cells. (Figure 16–48)

centromere   Constricted region of a mitotic chromosome that holds sister chromatids together. This is also the site on the DNA where the kinetochore forms so as to capture microtubules from the mitotic spindle. (Figure 4–43)

centrosome   Centrally located organelle of animal cells that is the primary microtubule-organizing center (MTOC) and acts as the spindle pole during mitosis. In most animal cells it contains a pair of centrioles. (Figures 16–47 and 17–24)

cerebral cortex   Outermost layer of the hemispheres of the brain; the most complex structure in the human body.

CG island   Region of DNA in vertebrate genomes with a greater than average density of CG sequences; these regions generally remain unmethylated.

channel (membrane channel)   Transmembrane protein complex that allows inorganic ions or other small molecules to diffuse passively across the lipid bilayer. (Figure 11–3)

channelrhodopsin   Photosensitive protein forming a cation channel across the membrane that opens in response to light.

charge separation   In photosynthesis, the light-induced transfer of a high-energy electron from chlorophyll to an acceptor molecule resulting in the formation of a positive charge on the chlorophyll and a negative charge on a mobile electron carrier.

chemical biology   Name given to a strategy that uses large-scale screening of hundreds of thousands of small molecules in biological assays to identify chemicals that affect a particular biological process and that can then be used to study it.

chemical carcinogens   Disparate chemicals that are carcinogenic—due to the ability to cause mutations—when fed to experimental animals or painted repeatedly on their skin.

chemical group   Certain combinations of atoms—such as methyl ($-CH_3$), hydroxyl ($-OH$), carboxyl ($-COOH$), carbonyl ($-C=O$), phosphate ($-PO_3^{2-}$), sulfhydryl ($-SH$), and amino ($-NH_2$) groups—that have distinct chemical and physical properties and influence the behavior of the molecule in which the group occurs.

chemiosmotic coupling (chemiosmosis)   Mechanism in which an electrochemical proton gradient across a membrane (composed of a pH gradient plus a membrane potential) is used to drive an energy-requiring process, such as ATP production or the rotation of bacterial flagella.

chemotaxis   Movement of a cell toward or away from some diffusible chemical.

chiasma (plural chiasmata)   X-shaped connection visible between paired homologous chromosomes during meiosis. Represents a site of chromosomal crossing-over, a form of genetic recombination.

chlorophyll   Light-absorbing green pigment that plays a central part in photosynthesis in bacteria, plants, and algae.

chloroplast   Organelle in green algae and plants that contains chlorophyll and carries out photosynthesis.

cholera toxin   Secreted toxic protein of *Vibrio cholerae* responsible for causing the watery diarrhea associated with cholera. Comprises an A subunit with enzymatic activity and a B subunit that binds to host-cell receptors to direct subunit A to the host-cell cytosol.

cholesterol   An abundant lipid molecule with a characteristic four-ring steroid structure. An important component of the plasma membranes of animal cells. (Figure 10–4)

chromatin   Complex of DNA, histones, and non-histone proteins found in the nucleus of a eukaryotic cell. The material of which chromosomes are made.

chromatin immunoprecipitation   Technique by which chromosomal DNA bound by a particular protein can be isolated and identified by precipitating it by means of an antibody against the protein. (Figures 8–66 and 8–67)

chromosome   Structure composed of a very long DNA molecule and associated proteins that carries part (or all) of the hereditary information of an organism. Especially evident in plant and animal cells undergoing mitosis or meiosis, during which each chromosome becomes condensed into a compact rodlike structure visible in the light microscope.

cilium (plural cilia)   Hairlike extension of a eukaryotic cell containing a core bundle of microtubules. Many cells contain a single nonmotile cilium, while others contain large numbers that perform repeated beating movements. *Compare* **flagellum**.

circadian clock   Internal cyclical process that produces a particular change in a cell or organism with a period of around 24 hours, for example the sleep-wakefulness cycle in humans.

***cis* face**   Face on the same or near side.

***cis* Golgi network (CGN)**   Network of fused vesicular tubular clusters that is closely associated with the *cis* face of the Golgi apparatus and is the compartment at which proteins and lipids enter the Golgi.

***cis*-regulatory sequences**   DNA sequences to which transcription regulators bind to control the rate of gene transcription. In nearly all cases, these sequences must be on the same chromosome (that is, *in cis*) to the genes they control. (Figure 7–18)

cisternal maturation model   One hypothesis for how the Golgi apparatus achieves and maintains its polarized structure and how molecules move from one cisterna to another. This model views the cisternae as dynamic structures that mature from early to late by acquiring and then losing specific Golgi-resident proteins as they move through the Golgi stack with cargo.

citric acid cycle [tricarboxylic acid (TCA) cycle, Krebs cycle]   Central metabolic pathway found in aerobic organisms. Oxidizes acetyl groups derived from food molecules, generating the activated carriers NADH and $FADH_2$, some GTP, and waste CO_2. In eukaryotic cells, it occurs in the mitochondria. (Panel 2–9, pp. 106–107)

clamp loader   Protein complex that utilizes ATP hydrolysis to load the sliding clamp on to a primer–template junction in the process of DNA replication.

class I MHC protein   One of two classes of major histocompatibility complex (MHC) protein. Found on the surface of almost all vertebrate cell types, where it can present foreign peptides derived from a pathogen such as a virus to cytotoxic T cells. (Figures 24–35 and 24–36A)

class II MHC protein   One of two classes of major histocompatibility complex (MHC) protein. Found on the surface of various antigen-presenting cells, where it presents peptides to helper and regulatory T cells. (Figures 24–35 and 24–36B)

class switching   Change from making one class of immunoglobulin (for example, IgM) to making another class (for example, IgG) that many B cells undergo during the course of an adaptive immune response. Involves DNA rearrangements called class-switch recombination. (Figure 24–30)

class-switch recombination   An irreversible change at the DNA level when a B cell switches from making IgM and IgD to making one of the secondary classes of immunoglobulin.

classical cadherins   Family of cadherin proteins, including E-cadherin, N-cadherin, and P-cadherin, that are closely related in sequence throughout their extracellular and intracellular domains.

clathrin   Protein that assembles into a polyhedral cage on the cytosolic side of a membrane so as to form a clathrin-coated pit, which buds off by endocytosis to form an intracellular clathrin-coated vesicle. (Figure 13–6)

clathrin-coated pits   Specialized regions typically occupying about 2% of the total plasma membrane area at which the endocytic pathway often begins.

clathrin-coated vesicles   Coated vesicles that transport material from the plasma membrane and between endosomal and Golgi compartments.

cleavage   (1) Physical splitting of a cell into two.
(2) Specialized type of cell division seen in many early embryos whereby a large cell becomes subdivided into many smaller cells without growth.

clonal selection   From a population of T and B lymphocytes with a vast repertoire of randomly generated antigen-specific receptors, a given foreign antigen activates (selects) only those lymphocyte clones that display a receptor that fits the antigen. Explains how the adaptive immune system can respond to millions of different antigens in a highly specific way. (Figure 24–15)

co-receptor In immunology: an accessory receptor on B cells or T cells that does not bind antigen but binds to a co-stimulatory signal and helps activate the lymphocyte, by helping to activate an intracellular signaling pathway.

co-stimulatory signal In immunology: a secreted or membrane-bound signal protein that helps activate an antigen-responding B cell or T cell.

co-translational Occurring as translation proceeds. Examples include the import of a protein into the endoplasmic reticulum before the polypeptide chain is completely synthesized (co-translational translocation, Figure 12–32), and the folding of a nascent protein into its secondary and tertiary structure as it emerges from a ribosome. (Figure 6–79)

coat-recruitment GTPases Members of a family of monomeric GTPases that have important roles in vesicle transport, being responsible for coat assembly at the membrane.

coated vesicle Small membrane-enclosed organelle with a cage of proteins (the coat) on its cytosolic surface. Formed by the pinching off of a coated region of membrane (coated pit). Some coats are made of clathrin, others are made from other proteins.

codon Sequence of three nucleotides in a DNA or mRNA molecule that represents the instruction for incorporation of a specific amino acid into a growing polypeptide chain.

coenzyme Small molecule tightly associated with an enzyme that participates in the reaction that the enzyme catalyzes, often by forming a covalent bond to the substrate. Examples include biotin, NAD^+, and coenzyme A.

cohesin, cohesin complex Complex of proteins that holds sister chromatids together along their length before their separation. (Figure 17–19)

coiled-coil Especially stable rodlike protein structure formed by two or more α helices coiled around each other. (Figure 3–9)

collagen Fibrous protein rich in glycine and proline that is a major component of the extracellular matrix in animals, conferring tensile strength. Exists in many forms: type I, the most common, is found in skin, tendon, and bone; type II is found in cartilage; type IV is present in basal laminae. (Figures 3–23 and 19–40)

collagen fibril A higher-order collagen polymer of fibrillar collagens that assemble into thin structures (10–300 nm in diameter) many hundreds of micrometers long in mature tissues.

colony-stimulating factor (CSF) General name for numerous signal molecules that control differentiation of blood cells.

colorectal cancer Cancer arising from the epithelium lining the colon (the large intestine) and rectum (the terminal segment of the gut).

column chromatography Technique for separation of a mixture of substances in solution by passage through a column containing a porous solid matrix. Substances are retarded to different extents by their interaction with the matrix and can be collected separately from the column. Depending on the matrix, separation can be according to charge, hydrophobicity, size, or the ability to bind to other molecules.

commensalism Ecological relationship between microbes and their host in which the microbe benefits but offers no benefit and causes no harm.

committed precursor Cell derived from a stem cell that divides for a limited number of times before terminally differentiating; also known as a transit amplifying cell.

complement system System of blood proteins that can be activated by antibody–antigen complexes or pathogens to help eliminate the pathogens, by directly causing their lysis, by promoting their phagocytosis, or activating an inflammatory response. (Figure 24–7)

complementary (1) Of nucleic acid sequences: capable of forming a perfect base-paired duplex with each other. (Figure 4–4) (2) Of other interacting molecules, such as an enzyme and its substrate: having biochemical or structural features that marry up, so that noncovalent bonding is facilitated. (Figure 2–3)

complementation test Test to determine whether two mutations that produce similar phenotypes are in the same or different genes. (Panel 8–2, pp. 487)

complex oligosaccharides Broad class of N-linked oligosaccharides, attached to mammalian glycoproteins in the endoplasmic reticulum and modified in the Golgi apparatus, containing N-acetylglucosamine, galactose, sialic acid, and fucose residues.

condensin, condensin complex Complex of proteins involved in chromosome condensation prior to mitosis. Target for M-Cdk. (Figure 17–22)

conditional mutation Mutation that changes a protein or RNA molecule so that its function is altered only under some conditions, such as at an unusually high or unusually low temperature.

cone photoreceptor (cone) Photoreceptor cell in the vertebrate retina that is responsible for color vision in bright light.

confocal microscope Type of light microscope that produces a clear image of a given plane within a solid object. It uses a laser beam as a pinpoint source of illumination and scans across the plane to produce a two-dimensional "optical section." (Figure 9–19)

conformation The folded, three-dimensional structure of a polypeptide chain.

connective tissue Any supporting tissue that lies between other tissues and consists of cells embedded in a relatively large amount of extracellular matrix. Includes bone, cartilage, and loose connective tissue.

connexin Protein component of gap junctions, a four-pass transmembrane protein. Six connexins assemble in the plasma membrane to form a connexon, or "hemichannel." (Figure 19–25)

connexon Water-filled pore in the plasma membrane formed by a ring of six connexin protein subunits. Half of a gap junction: connexons from two adjoining cells join to form a continuous channel through which ions and small molecules can pass. (Figure 19–25)

consensus nucleotide sequence A summary or "average" of a large number of individual nucleotide sequences derived by comparing many sequences with the same basic function and tallying up the most common nucleotides found at each position. (Figure 6–12)

consensus sequence Average or most typical form of a sequence that is reproduced with minor variations in a group of related DNA, RNA, or protein sequences. Indicates the nucleotide or amino acid most often found at each position. Preservation of a sequence implies that it is functionally important.

conservative site-specifc recombination A type of DNA recombination that takes place between short, specific sequences of DNA and occurs without the gain or loss of nucleotides. It does not require extensive homology between the recombining DNA molecules.

constant region In immunology: region of an immunoglobulin or T cell receptor chain that has a constant amino acid sequence.

constitutive secretory pathway Pathway present in all cells by which molecules such as plasma membrane proteins are continually delivered to the plasma membrane from the Golgi apparatus in vesicles that fuse with the plasma membrane. The default route to the plasma membrane if no other sorting signals are present. (Figure 13–63)

contact-dependent signaling Form of intercellular signaling in which signal molecules remain bound to the surface of the signaling cell and influence only cells that contact it.

contractile ring Ring containing actin and myosin that forms under the surface of animal cells undergoing cell division. Contracts to pinch the two daughter cells apart. (Figure 17–42)

convergent extension Rearrangement of cells within a tissue that causes it to extend in one dimension and shrink in another. (Figure 21–50)

COPI-coated vesicles Coated vesicles that transport material early in the secretory pathway, budding from Golgi compartments.

COPII-coated vesicles Coated vesicles that transport material early in the secretory pathway, budding from the endoplasmic reticulum.

copy number variations (CNVs) A difference between two individuals in the same population in the number of copies of a particular block of DNA sequence. This variation arises from occasional duplications and deletions of these sequences.

cortex The cytoskeletal network in the cortical region of the cytosol just beneath the plasma membrane.

coupled reaction Linked pair of chemical reactions in which the free energy released by one serves to drive the other. (Figure 2–29)

covalent bond Stable chemical link between two atoms produced by sharing one or more pairs of electrons. (Panel 2–1, pp. 90–91)

CRE-binding (CREB) protein Transcription regulator that recognizes the cyclic AMP response element (CRE) in the regulatory region of genes activated by cAMP. On activation by PKA, phosphorylated CREB recruits a transcriptional coactivator (CREB-binding protein; CBP) to stimulate transcription of target genes.

CRISPR A defense mechanism in bacteria using small noncoding RNA molecules (crRNAs) to seek out and destroy invading viral genomes through complementary base-pairing and targeted nuclease digestion.

crista (plural **cristae**) A specialized invagination of the inner mitochondrial membrane.

cross-linking glycan One of a heterogeneous group of branched polysaccharides that help to cross-link cellulose microfibrils into a complex network. Has a long linear backbone

of one sugar type (glucose, xylose, or mannose) with short side chains of other sugars.

cross-presentation A process in which extracellular proteins taken up by specialized dendritic cells can give rise to peptides that can be presented by class I MHC proteins to cytotoxic T cells.

crRNAs Small noncoding RNAs (≈30 nucleotides) that are the effectors of CRISPR-mediated immunity in bacteria.

cryoelectron microscopy Technique for examining a thin film of an aqueous suspension of biological material that has been frozen rapidly enough to create vitreous ice. The specimen is then kept frozen and transferred to the electron microscope. Image contrast is low, but is generated solely by the macromolecular structures present.

cryptochrome Plant flavoprotein sensitive to blue light. Structurally related to blue-light-sensitive enzymes called photolyases (involved in the repair of ultraviolet-induced DNA damage) but do not have a role in DNA repair. Also found in animals, where they have an important role in circadian clocks.

Cubitus interruptus (Ci) Latent transcription regulator that mediates the effects of Hedgehog.

cyclic AMP (cAMP) Nucleotide that is generated from ATP by adenylyl cyclase in response to various extracellular signals. It acts as a small intracellular signaling molecule, mainly by activating cAMP-dependent protein kinase (PKA). It is hydrolyzed to AMP by a phosphodiesterase. (Figure 15–25)

cyclic AMP phosphodiesterase Specific enzyme that rapidly and continuously destroys cyclic AMP, forming 5′-AMP. (Figure 15–25).

cyclic AMP-dependent protein kinase (protein kinase A, PKA) Enzyme that phosphorylates target proteins in response to a rise in intracellular cyclic AMP. (Figure 15–26)

cyclic GMP (cGMP) Nucleotide that is generated from GTP by guanylyl cyclase in response to various extracellular signals.

cyclic GMP phosphodiesterase Specific enzyme that rapidly hydrolyzes and degrades cyclic GMP.

cyclin Protein that periodically rises and falls in concentration in step with the eukaryotic cell cycle. Cyclins activate crucial protein kinases (called cyclin-dependent protein kinases, or Cdks) and thereby help control progression from one stage of the cell cycle to the next.

cyclin-dependent kinase (Cdk) Protein kinase that has to be complexed with a cyclin protein in order to act. Different cyclin–Cdk complexes trigger different steps in the cell-division cycle by phosphorylating specific target proteins. (Figure 17–10)

cyclin–Cdk complex Protein complex formed periodically during the eukaryotic cell cycle as the level of a particular cyclin increases. A cyclin-dependent kinase (Cdk) then becomes partially activated. (Figures 17–10 and 17–11, and Table 17–1, p. 969)

cyclosome *see* **anaphase-promoting complex**

cytochrome Colored heme-containing protein that transfers electrons during respiration and photosynthesis.

cytochrome c Soluble component of the mitochondrial electron-transport chain. Its release into the cytosol from the mitochondrial intermembrane space also initiates apoptosis. (Figure 14–26)

cytochrome c oxidase complex Third of the three electron-driven proton pumps in the respiratory chain. It accepts electrons from cytochrome *c* and generates water using molecular oxygen as an electron acceptor. (Figure 14–18)

cytochrome *c* reductase   Second of the three electron-driven proton pumps in the respiratory chain. Accepts electrons from ubiquinone and passes them to cytochrome *c*. (Figure 14–18)

cytokine   Extracellular signal protein or peptide that acts as a local mediator in cell–cell communication.

cytokine receptor   Cell-surface receptor that binds a specific cytokine or hormone and acts through the JAK–STAT signaling pathway. (Figure 15–56)

cytokinesis   Division of the cytoplasm of a plant or animal cell into two, as distinct from the associated division of its nucleus (which is mitosis). Part of M phase. (Panel 17–1, pp. 980–981)

cytoplasm   Contents of a cell that are contained within its plasma membrane but, in the case of eukaryotic cells, outside the nucleus.

cytoplasmic tyrosine kinase   Enzyme activated by certain cell-surface receptors (tyrosine-kinase-associated receptors) that transmits the receptor signal onward by phosphorylating target cytoplasmic proteins on tyrosine side chains.

cytoskeleton   System of protein filaments in the cytoplasm of a eukaryotic cell that gives the cell shape and the capacity for directed movement. Its most abundant components are actin filaments, microtubules, and intermediate filaments.

cytosol   Contents of the main compartment of the cytoplasm, excluding membrane-bounded organelles such as endoplasmic reticulum and mitochondria.

cytotoxic T cell (T$_C$ cell)   Type of T cell responsible for killing host cells infected with a virus or another type of intracellular pathogen. (Figure 24–42)

dark-field microscopy   Type of light microscopy in which oblique rays of light focused on the specimen do not enter the objective lens, but light that is scattered by components in the living cell can be collected to produce a bright image on a dark background. (Figure 9–7)

death receptor   Transmembrane receptor protein that can signal the cell to undergo apoptosis when it binds its extracellular ligand. (Figure 18–5)

death-inducing signaling complex (DISC)   Activation complex in which initiator caspases interact and are activated following binding of extracellular ligands to cell-surface death receptors in the extrinsic pathway of apoptosis.

deep RNA sequencing   *see* **RNA-seq**

default pathway   The transport pathway of proteins directly to the cell surface via the nonselective constitutive secretory pathway, entry into which does not require a particular signal.

defensin   Positively charged, amphipathic, antimicrobial peptide—secreted by epithelial cells—that binds to and disrupts the membranes of many pathogens.

delayed K$^+$ channel   Neuronal voltage-gated K$^+$ channel that opens following membrane depolarization but during the falling phase of an action potential due to slower activation kinetics than Na$^+$ channels; opening permits K$^+$ efflux, driving the membrane potential back toward its original negative value, ready to transmit a second impulse.

Delta   Single-pass transmembrane signal protein displayed on the surface of cells that binds to the Notch receptor protein on a neighboring cell, activating a contact-dependent signaling mechanism.

dendrite   Extension of a nerve cell, often elaborately branched, that receives stimuli from other nerve cells.

dendritic cell   The most potent type of antigen-presenting cell, which takes up antigen and processes it for presentation to T cells. It is required for activating naïve T cells. (Figure 24–11)

deoxyribonucleic acid (DNA)   Polynucleotide formed from covalently linked deoxyribonucleotide units. The store of hereditary information within a cell and the carrier of this information from generation to generation. (Figure 4–3 and Panel 2–6, pp. 100–101)

depolarization   Deviation in the electric potential across the plasma membrane towards a positive value. A depolarized cell has a potential that is positive outside and negative inside.

desensitization   *see* **adaptation**

desmosome   Anchoring cell–cell junction, usually formed between two epithelial cells. Characterized by dense plaques of protein into which intermediate filaments in the two adjoining cells insert. (Figure 19–2)

detergent   Small amphiphilic molecule, more soluble in water than lipids, that disrupts hydrophobic associations and destroys the lipid bilayer thereby solubilizing membrane proteins.

D gene segment   A short DNA sequence that encodes a part of the variable region of an immunoglobulin heavy chain or the β chain of a T cell receptor (TCR).

diacylglycerol (DAG)   Lipid produced by the cleavage of inositol phospholipids in response to extracellular signals. Composed of two fatty acid chains linked to glycerol, it serves as a small signaling molecule to help activate protein kinase C (PKC). (Figure 15–28)

dideoxy sequencing   The standard enzymatic method of DNA sequencing. (Panel 8–1, p. 478)

differential-interference-contrast microscope   Type of light microscope that exploits the interference effects that occur when light passes through parts of a cell of different refractive indices. Used to view unstained living cells.

differentiation   Process by which a cell undergoes a change to an overtly specialized cell type.

diffusion   The net drift of molecules through space due to random thermal movements.

Dishevelled   Scaffold protein recruited to the Frizzled family of cell-surface receptors upon their activation by Wnt binding that helps relay the signal to other signaling molecules.

DNA cloning   (1) The act of making many identical copies (typically billions) of a DNA molecule—the amplification of a particular DNA sequence. (2) Also, the isolation of a particular stretch of DNA (often a particular gene) from the rest of the cell's genome.

DNA helicase   Enzyme that is involved in opening the DNA helix into its single strands for DNA replication.

DNA library   Collection of cloned DNA molecules, representing either an entire genome (genomic library) or complementary DNA copies of the mRNA produced by a cell (cDNA library).

DNA ligase   Enzyme that joins the ends of two strands of DNA together with a covalent bond to make a continuous DNA strand.

DNA methylation   Addition of methyl groups to DNA.

Extensive methylation of the cytosine base in CG sequences is used in plants and animals to help keep genes in an inactive state.

DNA microarray   A large array of short DNA molecules (each of known sequence) bound to a glass microscope slide or other suitable support. Used to monitor expression of thousands of genes simultaneously: mRNA isolated from test cells is converted to cDNA, which in turn is hybridized to the microarray. (Figure 8–64)

DNA polymerase   Enzyme that synthesizes DNA by joining nucleotides together using a DNA template as a guide.

DNA primase   Enzyme that synthesizes a short strand of RNA on a DNA template, producing a primer for DNA synthesis. (Figure 5–10)

DNA repair   A set of processes for repairing the many accidental lesions that occur continually in DNA.

DNA replication   Process by which a copy of a DNA molecule is made.

DNA supercoiling   A conformation with loops or coils that DNA adopts in response to superhelical tension; conversely, creating various loops or coils in the helix can create such tension.

DNA topoisomerase (topoisomerase)   Enzyme that binds to DNA and reversibly breaks a phosphodiester bond in one or both strands. Topoisomerase I creates transient single-strand breaks, allowing the double helix to swivel and relieving superhelical tension. Topoisomerase II creates transient double-strand breaks, allowing one double helix to pass through another and thus resolving tangles. (Figures 5–21 and 5–22)

DNA tumor virus   General term for a variety of different DNA viruses that can cause tumors.

DNA-only transposon   Transposable element that exists as DNA throughout its life cycle. Many move by cut-and-paste transposition. *See also* **transposon**.

dolichol   Isoprenoid lipid molecule that anchors the precursor oligosaccharide in the endoplasmic reticulum membrane during protein glycosylation.

domain (protein domain)   Portion of a protein that has a tertiary structure of its own. Larger proteins are generally composed of several domains, each connected to the next by short flexible regions of polypeptide chain. Homologous domains are recognized in many different proteins.

Dorsal protein   Transcription regulator of the NFκB family regulating gene expression and involved in establishing the dorsoventral axis in the embryo.

double helix   The three-dimensional structure of DNA, in which two antiparallel DNA chains, held together by hydrogen-bonding between the bases, are wound into a helix. (Figure 4–5)

drivers   Mutations that are causal factors in the development of cancer.

dynamic instability   Sudden conversion from growth to shrinkage, and vice versa, in a protein filament such as a microtubule or actin filament. (Panel 16–2, pp. 902–903)

dynamin   Cytosolic GTPase that binds to the neck of a clathrin-coated vesicle in the process of budding from the membrane, and which is involved in completing vesicle formation.

dynein   Large motor protein that undergoes ATP-dependent movement along microtubules.

E2F protein   Transcription regulatory protein that switches on many genes that encode proteins required for entry into the S phase of the cell cycle.

early endosome   Common receiving compartment with which most endocytic vesicles fuse and where internalized cargo is sorted either for return to the plasma membrane or for degradation by inclusion in a late endosome.

ectoderm   Embryonic epithelial tissue that is the precursor of the epidermis and nervous system.

edema factor   One of the two A subunits of anthrax toxin; an adenylyl cyclase that catalyzes production of cAMP, leading to ion imbalance and consequent edema in the skin or lung.

effector Bcl2 family proteins   Pro-apoptotic proteins of the intrinsic pathway of apoptosis that in response to an apoptotic stimulus become activated and aggregate to form oligomers in the mitochondrial outer membrane, inducing the release of cytochrome *c* and other intermembrane proteins. Bax and Bak are the main effector Bcl2 family proteins in mammalian cells.

effector cell   Cell that carries out the final response or function in a particular process. The main effector cells of the immune system, for example, are activated lymphocytes and phagocytes that help eliminate pathogens.

egg-polarity genes   Genes in the *Drosophila* egg that define the anteroposterior and dorsoventral axes of the future embryo through the creation of landmarks (mRNA or protein) in the egg that provide signals organizing the developmental process.

elastic fiber   Extensible fiber formed by the protein elastin in many animal connective tissues, such as in skin, blood vessels, and lungs, which gives them their stretchability and resilience.

elastin   Extracellular protein that forms extensible fibers (elastic fibers) in connective tissues.

electrochemical gradient   Combined influence of a difference in the concentration of an ion on two sides of a membrane and the electrical charge difference across the membrane (membrane potential). Ions or charged molecules can move passively only down their electrochemical gradient.

electron microscope   Microscope that uses a beam of electrons to create the image.

electron microscope (EM) tomography   Technique for viewing three-dimensional specimens in the electron microscope in which multiple views are taken from different directions by tilting the specimen holder. The views are combined computationally to give a three-dimensional image.

electron-transport chain   Series of reactions in which electron carrier molecules pass electrons "down the chain" from higher to successively lower energy levels. The energy released during such electron movement can be used to power various processes. Electron-transport chains present in the inner mitochondrial membrane (called the respiratory chain) and in the thylakoid membrane of chloroplasts generate a proton gradient across the membrane that is used to drive ATP synthesis. See especially Figures 14–18 and 14–52.

electrostatic attraction   A noncovalent, ionic bond between two molecules carrying groups of opposite charge. (Panel 2–3, pp. 94–95)

embryonic stem cells (ES cells)   Cells derived from the inner cell mass of the early mammalian embryo. Capable of giving rise to all the cells in the body. Can be grown in culture, genetically modified, and inserted into a blastocyst to develop a transgenic animal.

endocrine cell   Specialized animal cell that secretes a hormone into the blood. Usually part of a gland, such as the thyroid or pituitary gland.

endocytic vesicle   Vesicle formed as material ingested by the cell during endocytosis is progressively enclosed by a small portion of the plasma membrane, which first invaginates and then pinches off to form the vesicle.

endocytosis   Uptake of material into a cell by an invagination of the plasma membrane and its internalization in a membrane-enclosed vesicle. *See also* **pinocytosis** and **phagocytosis**.

endoderm   Embryonic tissue that is the precursor of the gut and associated organs.

endoplasmic reticulum (ER)   Labyrinthine membrane-bounded compartment in the cytoplasm of eukaryotic cells, where lipids are synthesized and membrane-bound proteins and secretory proteins are made. (Figure 12–33)

endosome maturation   Process by which early endosomes mature to late endosomes and endolysosomes; in the conversion process, the endosome membrane protein composition changes, the endosome moves from the cell periphery to close to the nucleus, and the endosome ceases to recycle material to the plasma membrane and irreversibly commits its remaining contents to degradation.

endothelial cell   Flattened cell type that forms a sheet (the endothelium) lining all blood and lymphatic vessels.

entropy (S)   Thermodynamic quantity that measures the degree of disorder or randomness in a system; the higher the entropy, the greater the disorder. (Panel 2–7, pp. 102–103)

enveloped virus   Virus with a capsid surrounded by a lipid bilayer membrane (the envelope), which is often derived from the host-cell plasma membrane when the virus buds from the cell. (Figure 23–12)

enzyme   Protein that catalyzes a specific chemical reaction.

enzyme-coupled receptor   A major type of cell-surface receptor that has a cytoplasmic domain that either has enzymatic activity or is associated with an intracellular enzyme. In either case, the enzymatic activity is stimulated by an extracellular ligand binding to the receptor. (Figure 15–6)

ephrin   One of a family of membrane-bound protein ligands for the Eph receptor tyrosine kinases (RTKs) that, among many other functions, stimulate repulsion or attraction responses that guide the migration of cells and nerve cell axons during animal development.

epidermis   Epithelial layer covering the outer surface of the body. Has different structures in different animal groups. The outer layer of plant tissue is also called the epidermis.

epigenetic inheritance   Inheritance of phenotypic changes in a cell or organism that do not result from changes in the nucleotide sequence of DNA. Can be due to positive feedback loops of transcription regulators or to heritable modifications in chromatin such as DNA methylation or histone modifications. (Figure 7–53)

epistasis analysis   Analysis to discover the order in which the genes act, by investigating if a mutation in one gene can mask the effect of a mutation in another gene when both mutations are present in the same organism or cell.

epithelial tissues   Tissues, such as the lining of the gut or the epidermal covering of the skin, in which cells are closely bound together into sheets called epithelia.

epithelium (plural **epithelia**)   Sheet of cells covering the outer surface of a structure or lining a cavity.

equilibrium   State in a chemical reaction where there is no net change in free energy to drive the reaction in either direction. The ratio of product to substrate reaches a constant value at chemical equilibrium. (Figure 2–30)

equilibrium constant (K)   The ratio of forward and reverse rate constants for a reaction. Equal to the association or affinity constant (K_a) for a simple binding reaction (A + B $\rightleftharpoons$ AB). *See also* **affinity constant**, **dissociation constant**. (See page 62)

ER lumen   Space enclosed by the membrane of the endoplasmic reticulum (ER).

ER resident protein   Protein that remains in the endoplasmic reticulum (ER) or its membranes and carries out its function there, as opposed to proteins that are present in the ER only in transit.

ER retention signal   Short amino acid sequence on a protein that prevents it from moving out of the endoplasmic reticulum (ER). Found on proteins that are resident in the ER and function there.

ER signal sequence   N-terminal signal sequence that directs proteins to enter the endoplasmic reticulum (ER). Cleaved off by signal peptidase after entry.

ER tail-anchored proteins   Membrane proteins anchored in the endoplasmic reticulum (ER) membrane by a single transmembrane α helix contained at their C-terminus.

erythrocyte   Small hemoglobin-containing blood cell of vertebrates that transports oxygen to, and carbon dioxide from, tissues. Also called a red blood cell.

erythropoietin   A hormone produced by the kidney that stimulates the production of red blood cells in bone marrow.

ESCRT protein complexes   Four protein complexes (ESCRT-0, ESCRT-1, ESCRT-2, and ESCRT-3) that act sequentially to shepherd mono-ubiquitylated membrane proteins on endosomal membranes into intralumenal vesicles. ESCRT-3 complex catalyzes the pinching-off reaction.

ethylene   Small gas molecule that is a plant growth regulator influencing plant development in various ways including promoting fruit ripening, leaf abscission, and plant senescence and functioning as a stress signal in response to wounding, infection, and flooding.

euchromatin   Region of an interphase chromosome that stains diffusely; "normal" chromatin, as opposed to the more condensed heterochromatin.

eukaryote   Organism composed of one or more cells that have a distinct nucleus. Member of one of the three main divisions of the living world, the other two being bacteria and archaea. (Figure 1–17)

eukaryotic initiation factor (eIF)   Protein that helps load initiator tRNA on to the ribosome, thus initiating translation.

excitatory neurotransmitter   Neurotransmitter that opens cation channels in the postsynaptic membrane, causing an influx of Na^+, and in many cases Ca^{2+}, that depolarizes the postsynaptic membrane toward the threshold potential for firing an action potential.

executioner caspases   Apoptotic caspases that catalyze the widespread cleavage events during apoptosis that kill the cell.

exocytosis   Excretion of material from the cell by vesicle fusion with the plasma membrane; can occur constitutively or be regulated.

exon   Segment of a eukaryotic gene that consists of a sequence of nucleotides that will be represented in mRNA or in a final transfer, ribosomal, or other mature RNA molecule. In protein-coding genes, exons encode the amino acids in the protein. An exon is usually adjacent to a noncoding DNA segment called an intron. (Figure 4–15)

exosome   Large protein complex with an interior rich in $3'$-to-$5'$ RNA exonucleases; degrades RNA molecules to produce ribonucleotides.

extracellular pathogens   Pathogens that disturb host cells and can cause serious disease without replicating in host cells.

extracellular signal molecule   Any secreted or cell-surface chemical signal that binds to receptors and regulates the activity of the cell expressing the receptor.

extrinsic pathway   Pathway of apoptosis triggered by extracellular signal proteins binding to cell-surface death receptors.

facultative pathogens   Bacteria that replicate in an environmental reservoir such as water or soil and only cause disease if they happen to encounter a susceptible host.

FAD/FADH2 (flavin adenine dinucleotide/reduced flavin adenine dinucleotide)   Electron carrier system that functions in the citric acid cycle and fatty acid oxidation. One molecule of FAD gains two electrons plus two protons in becoming the activated carrier $FADH_2$. (Figure 2–39)

Fas (Fas protein, Fas death receptor)   Transmembrane death receptor that initiates apoptosis when it binds its extracellular ligand (Fas ligand). (Figure 18–5)

Fas ligand   Ligand that activates the cell-surface death receptor, Fas, triggering the extrinsic pathway of apoptosis.

fat   Energy-storage lipid in cells. Composed of triglycerides—fatty acids esterified with glycerol.

fate map   Representation showing which cell types will later derive from which regions of a tissue; e.g. from the blastula. (Figure 21–28)

Fc receptor   One of a family of cell-surface receptors that bind the tail region (Fc region) of an antibody molecule. Different Fc receptors are specific for different classes of antibodies, such as IgG, IgA, or IgE.

feedback inhibition   The process in which a product of a reaction feeds back to inhibit a previous reaction in the same pathway. (Figures 3–55 and 3–56)

fermentation   Anaerobic energy-yielding metabolic pathway involving the oxidation of organic molecules. Anaerobic glycolysis refers to the process whereby pyruvate is converted into lactate or ethanol, with the conversion of NADH to NAD^+. (Figure 2–47)

fibril-associated collagen   Mediates the interactions of collagen fibrils with one another and with other matrix macromolecules to help determine the organization of the fibrils in the matrix. This collagen (including types IX and XII) has a flexible triple-stranded helical structure and binds to the surface of the fibrils rather than forming aggregates.

fibrillar collagen   Class of fibril-forming collagens (including type I collagen, the most common type and the principal collagen of skin and bone) that have long ropelike structures with few or no interruptions and which assemble into collagen fibrils.

fibroblast   Common cell type found in connective tissue. Secretes an extracellular matrix rich in collagen and other extracellular matrix macromolecules. Migrates and proliferates readily in wounded tissue and in tissue culture.

fibronectin   Extracellular matrix protein involved in adhesion of cells to the matrix and guidance of migrating cells during embryogenesis. Integrins on the cell surface are receptors for fibronectin.

filopodium (plural filopodia) (microspike)   Thin, spike-like protrusion with an actin filament core, generated on the leading edge of a crawling animal cell. (Figure 16–21)

firing rule   Important principle governing synapse reinforcement and elimination during development of the nervous system: when two (or more) neurons synapsing on the same target cell fire at the same time, they reinforce their connections to that cell; when they fire at different times, they compete, so that all but one of them tend to be eliminated.

flagellum (plural flagella)   Long, whiplike protrusion whose undulations drive a cell through a fluid medium. Eukaryotic flagella are longer versions of cilia. Bacterial flagella are smaller and completely different in construction and mechanism of action. *Compare* **cilium**.

fluorescence microscope   Microscope designed to view material stained with fluorescent dyes or proteins. Similar to a light microscope but the illuminating light is passed through one set of filters before the specimen, to select those wavelengths that excite the dye, and through another set of filters before it reaches the eye, to select only those wavelengths emitted when the dye fluoresces. (Figure 9–12)

fluorescence recovery after photobleaching (FRAP)   Technique for monitoring the kinetic parameters of a protein by analyzing how fluorescent protein molecules move into an area of the cell bleached by a beam of laser light. (Figure 9–29)

fluorescence resonance energy transfer (FRET)   Technique for monitoring the closeness of two fluorescently labeled molecules (and thus their interaction) in cells. Also known as Förster resonance energy transfer. (Figure 9–26)

focal adhesion kinase (FAK)   Cytoplasmic tyrosine kinase present at cell–matrix junctions (focal adhesions) in association with the cytoplasmic tails of integrins.

focal adhesion kinase (FAK)   Cytoplasmic tyrosine kinase present at cell–matrix junctions (focal adhesions) in association with the cytoplasmic tails of integrins.

follicular helper T cell (TFH)   Type of T cell located in lymphoid follicles that secretes various cytokines to stimulate B cells to undergo antibody class switching and somatic hypermutation.

formin   Dimeric protein that nucleates the growth of straight, unbranched actin filaments that can be cross-linked by other proteins to form parallel bundles.

Förster resonance energy transfer   *see* **fluorescence resonance energy transfer (FRET)**

FRAP   *see* **fluorescence recovery after photobleaching**

free energy (G) (Gibbs free energy)   The energy that can be extracted from a system to drive reactions. Takes into account changes in both energy and entropy. (Panel 2–7, pp. 102–103)

free ribosome   Ribosome that is free in the cytosol, unattached to any membrane.

free-energy change (ΔG) see ΔG.

FRET see **fluorescence resonance energy transfer**

Frizzled Family of cell-surface receptors that are seven-pass transmembrane proteins that resemble GPCRs in structure but do not generally work through the activation of G proteins. Activated by Wnt binding to recruit the scaffold protein Dishevelled, which helps relay the signal to other signaling molecules.

fungus (plural **fungi**) Kingdom of eukaryotic organisms that includes the yeasts, molds, and mushrooms. Many plant diseases and a relatively small number of animal diseases are caused by fungi.

fusion protein Engineered protein that combines two or more normally separate polypeptides. Produced from a recombinant gene.

ΔG Change in the free energy during a reaction: the free energy of the product molecules minus the free energy of the starting molecules. A large negative value of ΔG indicates that the reaction has a strong tendency to occur. (Panel 2–7, pp. 102–103)

G_0 State of withdrawal from the eukaroytic cell-division cycle by entry into a quiescent digression from the G_1 phase. A common, sometimes permanent, state for differentiated cells.

G_1 phase Gap 1 phase of the eukaryotic cell-division cycle, between the end of mitosis and the start of DNA synthesis. (Figure 17–4)

G_1-Cdk Cyclin–Cdk complex formed in vertebrate cells by a G_1-cyclin and the corresponding cyclin-dependent kinase (Cdk). (Table 17–1, p. 969)

G_1-cyclin Cyclin present in the G_1 phase of the eukaryotic cell cycle. Forms complexes with Cdks that help govern the activity of the G_1/S-cyclins, which control progression to S phase.

G_1/S-Cdk Cyclin–Cdk complex formed in vertebrate cells by a G_1/S-cyclin and the corresponding cyclin-dependent kinase (Cdk). (Figure 17–11 and Table 17–1, p. 969)

G_1/S-cyclin Cyclin that activates Cdks in late G_1 of the eukaryotic cell cycle and thereby helps trigger progression through Start, resulting in a commitment to cell-cycle entry. Its level falls at the start of S phase. (Figure 17–11)

G_2 phase Gap 2 phase of the eukaryotic cell-division cycle, between the end of DNA synthesis and the beginning of mitosis. (Figure 17–4)

G_2/M transition Point in the eukaryotic cell cycle at which the cell checks for completion of DNA replication before triggering the early mitotic events that lead to chromosome alignment on the spindle. (Figure 17–9)

ganglioside Any glycolipid having one or more sialic acid residues in its structure. Found in the plasma membrane of eukaryotic cells and especially abundant in nerve cells. (Figure 10–16)

gap gene In *Drosophila* development, a gene that is expressed in specific broad regions along the anteroposterior axis of the early embryo, and which helps designate the main divisions of the insect body. (Figure 21–20)

gap junction Communicating channel-forming cell–cell junction present in most animal tissues that allows ions and small molecules to pass from the cytoplasm of one cell to the cytoplasm of the next.

gastrulation Important stage in animal embryogenesis during which the embryo is transformed from a ball of cells to a structure with a gut (a gastrula).

gated transport Movement of proteins between the cytosol and the nucleus through nuclear pore complexes in the nuclear envelope that function as selective gates.

geminin Protein that prevents the formation of new prereplicative complexes during S phase and mitosis, thus ensuring that the chromosomes are replicated only once in each cell cycle.

gene Region of DNA that is transcribed as a single unit and carries information for a discrete hereditary characteristic, usually corresponding to (1) a single protein (or set of related proteins generated by variant post-transcriptional processing), or (2) a single RNA (or set of closely related RNAs).

gene control region The set of linked DNA sequences regulating expression of a particular gene. Includes promoter and *cis*-regulatory sequences required to initiate transcription of the gene and control the rate of transcription. (Figure 7–17)

gene conversion Process by which DNA sequence information can be transferred from one DNA helix (which remains unchanged) to another DNA helix whose sequence is altered. It often accompanies general recombination events. (Figure 5–59)

gene family The set of genes in an organism related in DNA sequence due to their derivation from the same ancestor.

gene segments In immunology: short DNA sequences that are joined together during B cell and T cell development to produce the coding sequences for immunoglobulins and T cell receptors, respectively. (Figure 24–28)

general transcription factor Any of the proteins whose assembly at all promoters of a given type is required for the binding and activation of RNA polymerase and the initiation of transcription. (Table 6–3, p. 311)

genetic code The set of rules specifying the correspondence between nucleotide triplets (codons) in DNA or RNA and amino acids in proteins. (Figure 6–48)

genetic instability Abnormally increased spontaneous mutation rate, such as occurs in cancer cells.

genetic screen Procedure for discovery of genes affecting a specific phenotype by surveying large numbers of mutagenized individuals.

genetics The study of the genes of an organism on the basis of heredity and variation.

genome The totality of genetic information belonging to a cell or an organism; in particular, the DNA that carries this information.

genome annotation Process attempting to mark out all the genes (protein-coding and noncoding) in a genome and ascribing functions to each.

genomic imprinting Phenomenon in which a gene is either expressed or not expressed in the offspring depending on which parent it is inherited from. (Figure 7–49)

genomic library Collection of cloned DNA molecules representing an entire genome.

genotype Genetic constitution of an individual cell or organism. The particular combination of alleles found in a specific individual. (Panel 8–2, p. 486)

germ cell A cell in the germ line of an organism, which includes the haploid gametes and their specified diploid precursor cells. Germ cells contribute to the formation of a new generation of organisms and are distinct from somatic cells, which form the body and leave no descendants.

germ layer One of the three primary tissue layers (endoderm, mesoderm, and ectoderm) of an animal embryo. (Figure 21–3)

glial cell Supporting non-neural cell of the nervous system. Includes oligodendrocytes and astrocytes in the vertebrate central nervous system and Schwann cells in the peripheral nervous system.

glycogen Polysaccharide composed exclusively of glucose units. Used to store energy in animal cells. Large granules of glycogen are especially abundant in liver and muscle cells. (Figure 2–51 and Panel 2–4, pp. 96–97)

glycolipid Lipid molecule with a sugar residue or oligosaccharide attached. (Panel 2–5, pp. 98–99)

glycolysis Ubiquitous metabolic pathway in the cytosol in which sugars are incompletely degraded with production of ATP. Literally, "sugar splitting." (Figure 2–46 and Panel 2–8, pp. 104–105)

glycoprotein Any protein with one or more saccharide or oligosaccharide chains covalently linked to amino acid side chains. Most secreted proteins and most proteins exposed on the outer surface of the plasma membrane are glycoproteins.

glycosaminoglycan (GAG) Long, linear, highly charged polysaccharide composed of a repeating pair of sugars, one of which is always an amino sugar. Mainly found covalently linked to a protein core in extracellular matrix proteoglycans. Examples include chondroitin sulfate, hyaluronan, and heparin. (Figure 19–32)

glycosylphosphatidylinositol anchor (GPI anchor) Lipid linkage by which some membrane proteins are bound to the membrane. The protein is joined, via an oligosaccharide linker, to a phosphatidylinositol anchor during its travel through the endoplasmic reticulum. (Figure 12–52)

Golgi apparatus (Golgi complex) Complex organelle in eukaryotic cells, centered on a stack of flattened, membrane-enclosed spaces, in which proteins and lipids transferred from the endoplasmic reticulum are modified and sorted. It is the site of synthesis of many cell wall polysaccharides in plants and extracellular matrix glycosaminoglycans in animal cells. (Figure 13–26)

GPCR kinase (GRK) Member of a family of enzymes that phosphorylates multiple serines and threonines on a GPCR to produce receptor desensitization. (Figure 15–42)

G protein (trimeric GTP-binding protein) A trimeric GTP-binding protein with intrinsic GTPase activity that couples GPCRs to enzymes or ion channels in the plasma membrane. (Table 15–3, p. 846)

G-protein-coupled receptor (GPCR) A seven-pass cell-surface receptor that, when activated by its extracellular ligand, activates a G protein, which in turn activates either an enzyme or ion channel in the plasma membrane. (Figures 15–6 and 15–21)

G_q Class of G protein that couples GPCRs to phospholipase C-β to activate the inositol phospholipid signaling pathway.

Gram negative Description for bacteria that do not stain with Gram stain as a result of having a thinner peptidoglycan cell wall outside their inner (plasma) membrane, and on an additional outer membrane.

Gram positive Description for bacteria that stain positive with Gram stain due to a thick layer of peptidoglycan cell wall outside their inner (plasma) membrane.

Gram staining A technique for classifying bacteria based on differences in the structure of the bacterial cell wall and outer surface.

granulocyte Category of white blood cell distinguished by conspicuous cytoplasmic granules. Includes neutrophils, basophils, and eosinophils. Arises from a granulocyte/macrophage (GM) progenitor cell. (Figure 22–27)

granulocyte/macrophage (GM) progenitor cell Committed progenitor cell in the bone marrow that gives rise to neutrophils and macrophages. (Figure 22–31)

green fluorescent protein (GFP) Fluorescent protein isolated from a jellyfish. Widely used as a marker in cell biology. (Figure 9–24)

growth cone Migrating motile tip of a growing nerve cell axon or dendrite. (Figure 21–72)

growth factor Extracellular signal protein that can stimulate a cell to grow. They often have other functions as well, including stimulating cells to survive or proliferate. Examples include epidermal growth factor (EGF) and platelet-derived growth factor (PDGF).

growth hormone (GH) Mammalian hormone secreted by the pituitary gland into the bloodstream that stimulates growth throughout the body.

GTP (guanosine 5′-triphosphate) Nucleoside triphosphate produced by the phosphorylation of GDP (guanosine diphosphate). Like ATP, it releases a large amount of free energy on hydrolysis of its terminal phosphate group. Has a special role in microtubule assembly, protein synthesis, and cell signaling. (Figure 2–58)

GTP-binding protein Also called GTPase; an enzyme that converts GTP to GDP.

GTPase An enzyme that converts GTP to GDP. GTPases fall into two large families. Large *trimeric G proteins* are composed of three different subunits and mainly couple GPCRs to enzymes or ion channels in the plasma membrane. Small monomeric *GTP-binding proteins* (also called *monomeric GTPases*) consist of a single subunit and help relay signals from many types of cell-surface receptors and have roles in intracellular signaling pathways, regulating intracellular vesicle trafficking, and signaling to the cytoskeleton. Both trimeric G proteins and monomeric GTPases cycle between an active GTP-bound form and an inactive GDP-bound form and frequently act as molecular switches in intracellular signaling pathways. See page 820.

GTPase-activating protein (GAP) Protein that binds to a GTPase and inhibits it by stimulating its GTPase activity, causing the enzyme to hydrolyze its bound GTP to GDP. (Figure 15–8)

guanine nucleotide exchange factor (GEF) Protein that binds to a GTPase and activates it by stimulating it to release its tightly bound GDP, thereby allowing it to bind GTP in its place. (Figure 15–8)

haplotype block Combination of alleles and DNA markers that has been inherited in a large, linked block on one chromosome of a homologous pair—undisturbed by genetic recombination—across many generations.

Hedgehog protein   Secreted extracellular signal molecule that has many different roles controlling cell differentiation and gene expression in animal embryos and adult tissues. Excessive Hedgehog signaling can lead to cancer.

helper T cell (T$_H$ cell)   Type of T cell that helps activate B cells to make antibodies, cytotoxic T cells to become effector cells, and macrophages to kill ingested pathogens. They can also help activate dendritic cells.

heterochromatin   Chromatin that is highly condensed even in interphase; generally transcriptionally inactive. (*Compare* with **euchromatin**.)

heterochronic   Describes genes involved in developmental timing; mutation results in cells of a specific fate behaving as cells at a different stage of development.

high-mannose oligosaccharides   Broad class of *N*-linked oligosaccharides, attached to mammalian glycoproteins in the endoplasmic reticulum, containing two *N*-acetylglucosamine residues and many mannose residues.

high-performance liquid chromatography (HPLC)   Type of chromatography that uses columns packed with tiny beads of matrix; the solution to be separated is pushed through under high pressure.

histone   One of a group of small abundant proteins, rich in arginine and lysine, that combine to form the nucleosome cores around which DNA is wrapped in eukaryotic chromosomes. (Figure 4–24)

histone chaperone (chromatin assembly factor)   Protein that binds free histones, releasing them once they have been incorporated into newly replicated chromatin. (Figure 4–27)

histone H1   "Linker" (as opposed to "core") histone protein that binds to DNA where it exits from a nucleosome and helps package nucleosomes into the 30-nm chromatin fiber. (Figure 4–30)

Holliday junction (cross-strand exchange)   X-shaped structure observed in DNA undergoing recombination, in which the two DNA molecules are held together at the site of crossing-over, also called a cross-strand exchange. (Figure 5–55)

homeotic selector gene   In *Drosophila* development, a gene that defines and preserves the differences between body segments.

homolog   One of two or more genes that are similar in sequence as a result of derivation from the same ancestral gene. The term covers both orthologs and paralogs. (Figure 1–21) *See* **homologous chromosomes**.

homologous   Genes, proteins, or body structures that are similar as a result of a shared evolutionary origin.

homologous chromosomes (homologs)   The maternal and paternal copies of a particular chromosome in a diploid cell.

homologous recombination (general recombination)   Genetic exchange between a pair of identical or very similar DNA sequences, typically those located on two copies of the same chromosome. Also a DNA repair mechanism for double-strand breaks. (Figures 5–48, 5–50, and 5–54)

homophilic   Binding between molecules of the same kind, especially those involved in cell–cell adhesion. (Figure 19–5)

horizontal gene transfer   Gene transfer between bacteria via natural transformation by released naked DNA, transduction by bacteriophages, or sexual exchange by conjugation.

hormone   Signal molecule secreted by an endocrine cell into the bloodstream, which can then carry the signal to distant target cells.

***Hox* complex**   A gene complex consisting of a series of *Hox* genes

***Hox* genes**   Genes coding for transcription regulators, each gene containing a homeodomain, and specifying body-region differences. *Hox* mutations typically cause homeotic transformations.

Hox proteins   Transcription regulator proteins encoded by *Hox* genes; possess a highly conserved, 60-amino-acid-long DNA-binding homeodomain.

HPV   Human papillomavirus; infects the cervical epithelium and is important as a cause of carcinoma of the uterine cervix.

hyaluronan (hyaluronic acid)   Type of nonsulfated glycosaminoglycan with a regular repeating sequence of up to 25,000 identical disaccharide units, not linked to a core protein. Found in the fluid lubricating joints and in many other tissues. (Figures 19–33 and 19–34)

hybridization   In molecular biology, the process whereby two complementary nucleic acid strands form a base-paired duplex DNA-DNA, DNA-RNA, or RNA-RNA molecule. Forms the basis of a powerful technique for detecting specific nucleotide sequences. (Figures 5–47 and 8–33)

hybridoma   Hybrid cell line generated by fusion of a tumor cell and another cell type. Monoclonal antibodies are produced by hybridoma lines obtained by fusing antibody-secreting B cells with cells of a B lymphocyte tumor. (Figure 8–4)

hydrogen bond   Noncovalent bond in which an electropositive hydrogen atom is partially shared by two electronegative atoms. (Panel 2–3, pp. 94–95)

hydronium ion (H$_3$O$^+$)   Water molecule associated with an additional proton. The form generally taken by protons in aqueous solution.

hydrophilic   Dissolving readily in water. Literally, "water loving."

hydrophobic (lipophilic)   Not dissolving readily in water. Literally, "water-fearing."

hydrophobic force   Force exerted by the hydrogen-bonded network of water molecules that brings two nonpolar surfaces together by excluding water between them. (Panel 2–3, pp. 94–95)

hypervariable region   In immunology: any of the three small parts of the variable region of an immunoglobulin or T cell receptor chain that show the highest variability from molecule to molecule and contribute to the antigen-binding site. (Figure 24–26)

hypoxia-inducible factor 1α (HIF1α).   Transcription regulator, the intracellular levels of which increase in response to a shortage of oxygen, that stimulates transcription of the *VEGF* gene to promote angiogenesis.

IκB   Inhibitory proteins that bind tightly to NFκB dimers and hold them in an inactive state within the cytoplasm of unstimulated cells.

Ig superfamily   Large and diverse family of proteins that contain immunoglobulin or immunoglobulin-like domains. Most are involved in cell–cell interactions or antigen recognition. (Figure 24–48)

IgA   Immunoglobulin A; the principal class of antibody in secretions, including saliva, tears, milk, and respiratory and intestinal secretions.

IgD   Immunoglobulin D; produced by immature naïve B cells after leaving the bone marrow. Transmembrane IgD and IgM proteins, with the same antigen-binding site, form the B cell receptors (BCRs) on these cells.

IgE   Immunoglobulin E; binds with high affinity via its tail region to a class of Fc receptors on the surface of mast cells (tissues) or basophils (blood), where it acts as an antigen receptor; antigen binding stimulates the secretion of cytokines and biologically active amines, which help attract white blood cells, antibodies, and complement proteins to the site of activation.

IgG   Immunoglobulin G; the major antibody class in the blood, produced in especially large quantities during secondary antibody responses. The tail region of some IgG subclasses can bind to specific Fc receptors on macrophages and neutrophils. Antigen–IgG complexes can activate complement.

IgM   Immunoglobulin M; the first class of immunoglobulin that a developing B cell in the bone marrow makes, forming B-cell receptors on its surface. IgM antibodies are the major class of antibody secreted into the blood in the early stages of a primary antibody response on first exposure to an antigen, where their pentameric structure (with 10 antigen-binding sites) allows strong binding to pathogens. When bound to antigen, it is highly efficient activation of complement.

iHog   Protein with four or five immunoglobulin-like domains and two or three fibronectin-type-III-like domains; located on the cell surface and thought to serve as co-receptors for Hedgehog proteins.

image processing   Computer based techniques in microscopy that process digital images in order to extract latent information. Enables compensation for some optical faults in microscopes, enhanced contrast to improve detection of small differences in light intensity, and subtraction of background irregularities in the optical system.

imaginal disc   Group of cells that are set aside, apparently undifferentiated, in the *Drosophila* embryo and which will develop into an adult structure, e.g., eye, leg, wing. Overt differentiation occurs at metamorphosis. (Figure 21–60)

immunization   Method of inducing adaptive immune responses to pathogens or foreign molecules, usually involving the co-injection of an adjuvant, a molecule (often of microbial origin) that helps activate innate immune responses required for the adaptive responses.

immunoblotting   *see* **Western blotting**

immunoglobulin (Ig) superfamily   Large and diverse family of proteins that contain immunoglobulin domains or immunoglobulin-like domains. Most are involved in cell–cell interactions or antigen recognition. (Figure 24–48)

immunoglobulin domain (Ig domain)   Characteristic protein domain of about 100 amino acids that is found in immunoglobulin light and heavy chains. Similar domains, known as immunoglobulin-like (Ig-like) domains, are present in many other proteins, which, together with Igs, constitute the Ig superfamily. (Figure 24–27)

immunogold electron microscopy   Method to localize specific macromolecules using a primary antibody that binds to the molecule of interest and is then detected with a secondary antibody to which a colloidal gold particle has been attached. The gold particle is electron-dense and can be seen as a black dot in the electron microscope. (Figure 9–45)

immunological memory   Long-lived property of the adaptive immune system that follows a primary immune response to many antigens, such that a subsequent encounter with the same antigen will provoke a more rapid and stronger secondary immune response. (Figure 24–16)

immunological self-tolerance   The lack of response of the adaptive immune system to an antigen. Tolerance to self molecules is crucial to avoid autoimmune diseases. (Figure 24–21)

immunological synapse   The highly organized interface that develops between a T cell and an antigen-presenting cell (APC) or target cell it is in contact with, formed by T-cell receptors binding to antigen–MHC complexes on the APC and cell-adhesion proteins binding to their counterparts on the APCs.

induced fit   A principle for increasing the specificity of substrate recognition by proteins and RNAs. In protein synthesis, a ribosome, or enzyme folds around a codon–anticodon interaction and only when the match is correct is the subsequent reaction allowed to proceed.

induced pluripotent stem cells (iPS cells)   Cells that are induced by artificial expression of specific transcription regulators to look and behave like the pluripotent embryonic stem cells that are derived from embryos.

induced regulatory T cell   A regulatory T cell (T_{reg} cell) that develops from naive helper T cells when they are activated in the presence of TGFβ in the absence of IL6.

inflammasome   Intracellular protein complex formed after activation of cytoplasmic NOD-like receptors with adaptor proteins. It contains a caspase enzyme that cleaves pro-inflammatory cytokines from their precursor proteins.

inflammatory response   Local response of a tissue to injury or infection—characterized clinically by redness, swelling, heat, and pain. Caused by invasion of white blood cells, which are attracted by and secrete various cytokines.

inhibitors of apoptosis (IAPs)   Intracellular protein inhibitors of apoptosis.

inhibitory G protein (G_i)   Trimeric G protein that can regulate ion channels and inhibit the enzyme adenylyl cyclase in the plasma membrane. *See also* **G protein**. (Table 15–3, p. 846)

inhibitory neurotransmitter   Neurotransmitter that opens transmitter-gated Cl^- or K^+ channels in the postsynaptic membrane of a nerve or muscle cell and thus tends to inhibit the generation of an action potential.

initial segment   Specialized membrane region at the base of a nerve axon (adjacent to the cell body) that is rich in voltage-gated Na^+ channels plus other classes of ion channels that all contribute to the encoding of membrane depolarization into action potential frequency.

initiator caspases   Apoptotic caspases that begin the apoptotic process, activating the executioner caspases.

initiator tRNA   Special tRNA that intiates translation. It always carries the amino acid methionine, forming the complex Met-tRNAi. (Figure 6–70)

innate immune response   An early immune response in all organisms to a pathogen, which includes the production of antimicrobial molecules and the activation of phagocytic cells. Such a response is not specific for the pathogen, in contrast to an adaptive immune response.

inner membrane   Mitochondrial membrane that encloses the matrix space and forms extensive invaginations called cristae.

inner mitochondrial membrane   Mitochondrial membrane that encloses the matrix space and forms extensive invaginations called cristae.

inner nuclear membrane One of two concentric membranes comprising the nuclear envelope; continuous with the outer nuclear membrane; contains specific proteins as anchoring sites for chromatin and the nuclear lamina.

inositol 1,4,5-trisphosphate (IP$_3$) Small intracellular signaling molecule produced during activation of the inositol phospholipid signaling pathway. Acts to release Ca^{2+} from the endoplasmic reticulum. (Figures 15–28 and 15–29)

inositol phospholipid signaling pathway Intracellular signaling pathway that starts with the activation of phospholipase C and the generation of IP$_3$ and diacylglycerol (DAG) from inositol phospholipids in the plasma membrane. The DAG helps to activate protein kinase C. (Figures 15–28 and 15–29)

integrin Transmembrane adhesion protein that is involved in the attachment of cells to the extracellular matrix and to each other. (Figure 19–3 and Table 19–1, p. 1037)

interaction domain Compact protein module, found in many intracellular signaling proteins, that binds to a particular structural motif (e.g., a short peptide sequence, a covalent modification, or another protein domain) in another protein or lipid.

interferon-α (IFNα) and **interferon-β (IFNβ)** Cytokines (type I interferons) produced by mammalian cells as a general response to a viral infection.

intermembrane space Compartment of mitochondrion between by the outer and inner mitochondrial membranes.

internal ribosome entry site (IRES) Specific site in a eukaryotic mRNA, other than at the 5′ end, at which translation can be initiated. (Figure 7–68)

interphase Long period of the cell cycle between one mitosis and the next. Includes G_1 phase, S phase, and G_2 phase. (Figure 17–4)

interpolar microtubule In the mitotic or meiotic spindle, a microtubule interdigitating at the equator with the microtubules emanating from the other pole. (Figure 17–23)

intracellular pathogens Pathogens, including all viruses and many bacteria and protozoa, that enter and replicate inside host cells to cause disease.

intrinsic pathway (mitochondrial pathway) Pathway of apoptosis activated from inside the cell in response to stress or developmental signals; depends on the release into the cytosol of mitochondrial proteins normally resident in the mitochondrial intermembrane space.

intron Noncoding region of a eukaryotic gene that is transcribed into an RNA molecule but is then excised by RNA splicing during production of the mRNA or other functional RNA. (Figure 4–15)

invadopodia Actin-rich protrusions extending in three-dimensions that are important for cells to cross tissue barriers by degrading the extracellular matrix.

ion channel Transmembrane protein complex that forms a water-filled channel across the lipid bilayer through which specific inorganic ions can diffuse down their electrochemical gradients. (Figure 11–22)

ion-channel-coupled receptor (transmitter-gated ion channel, ionotropic receptor) Ion channel found at chemical synapses in the postsynaptic plasma membranes of nerve and muscle cells. Opens only in response to the binding of a specific extracellular neurotransmitter. The resulting inflow of ions leads to the generation of a local electrical signal in the postsynaptic cell. (Figures 15–6 and 11–35)

ion-sensitive indicators Molecules whose light emission reflects the local concentration of a particular ion; some are luminescent (emitting light spontaneously) while others are fluorescent (emitting light on exposure to light).

IP$_3$-gated Ca^{2+}-release channel (IP$_3$ receptor) Gated Ca^{2+} channel in the ER membrane that opens on binding cytosolic IP$_3$, releasing stored Ca^{2+} into the cytosol. (Figure 15–29)

iron–sulfur cluster Electron-transporting group consisting of either two or four iron atoms bound to an equal number of sulfur atoms, found in a class of electron-transport proteins. (Figure 14–16)

***J* gene segment** Short DNA sequences that encodes part of the variable region of light and heavy immunoglobulin chains and of α and β chains of T cell receptors. (Figures 24–28 and 24–29)

JAK–STAT signaling pathway Signaling pathway activated by cytokines and some hormones, providing a rapid route from the plasma membrane to the nucleus to alter gene transcription. Involves cytoplasmic Janus kinases (JAKs), and signal transducers and activators of transcription (STATs).

Janus kinases (JAKs) Cytoplasmic tyrosine kinases associated with cytokine receptors, which phosphorylate and activate transcription regulators called STATs.

junctional diversification The random loss and gain of nucleotides at joining sites during V(D)J recombination that occurs during B and T cell development when the cells are assembling the gene segments that encode their antigen receptors. It enormously increases the diversity of V-region coding sequences.

K$^+$ leak channel K^+-transporting ion channel in the plasma membrane of animal cells that remains open even in a "resting" cell.

karyotype Display of the full set of chromosomes of a cell, arranged with respect to size, shape, and number.

keratin Type of intermediate filament, commonly produced by epithelial cells.

kinase cascade Intracellular signaling pathway in which one protein kinase, activated by phosphorylation, phosphorylates the next protein kinase in the sequence, and so on, relaying the signal onward.

kinesin Member of one of the two main classes of motor proteins that use the energy of ATP hydrolysis to move along microtubules. (Figure 16–56)

kinesin-1 Motor protein associated with microtubules that transports cargo within the cell; also called "conventional kinesin."

kinetic proofreading A principle for increasing the specificity of catalysis. In the synthesis of DNA, RNA, and proteins, it refers to a time delay that begins with an irreversible step (such as ATP or GTP hydrolysis) and during which incorrect base pairs are more likely to dissociate than correct pairs.

kinetochore Large protein complex that connects the centromere of a chromosome to microtubules of the mitotic spindle. (Figure 17–30)

kinetochore microtubule In the mitotic or meiotic spindle, a microtubule that connects the spindle pole to the kinetochore of a chromosome.

lagging strand One of the two newly synthesized strands of DNA found at a replication fork. The lagging strand is made in discontinuous lengths that are later joined covalently. (Figure 5–7)

lamellipodium (plural **lamellipodia**) Flattened, sheetlike protrusion supported by a meshwork of actin filaments, which is extended at the leading edge of a crawling animal cell. (Figures 16–77 and 16–79)

laminin Extracellular matrix fibrous protein found in basal laminae, where it forms a sheetlike network. (Figures 19–52 and 19–53)

lampbrush chromosome Huge chromosome paired in preparation for meiosis, found in immature amphibian eggs; consisting of large loops of chromatin extending out from a linear central axis. (Figure 4–47)

late endosome Compartment formed from a bulbous, vacuolar portion of early endosomes by a process called endosome maturation; late endosomes fuse with one another and with lysosomes to form endolysosomes that degrade their contents.

LDL-receptor-related protein (**LRP**) Co-receptor bound by Wnt proteins in the regulation of β-catenin proteolysis.

leading strand One of the two newly synthesized strands of DNA found at a replication fork. The leading strand is made by continuous synthesis in the 5′-to-3′ direction. (Figure 5–7)

lectin Protein that binds tightly to a specific sugar. Abundant lectins from plant seeds are used as affinity reagents to purify glycoproteins or to detect them on the surface of cells.

Legionnaire's disease Type of pneumonia resulting from infection with *Legionella pneumophila*, a parasite of freshwater amoebae that is spread to humans by air-conditioning systems that harbor infected amoebae and produce microdroplets of water that are easily inhaled.

lethal factor One of the two A subunits of anthrax toxin; a protease that cleaves several activated members of the MAP kinase kinase family and causes a large fall in blood pressure and death on entry into the bloodstream of an animal.

leucine-rich repeat (**LRR**) **receptor kinases** Common type of receptor serine/threonine kinase in plants that contains a tandem array of leucine-rich repeat sequences in its extracellular portion.

leukemia Cancer of white blood cells.

leukocyte General name for all the nucleated blood cells lacking hemoglobin. Also called white blood cells. Includes lymphocytes, granulocytes, and monocytes. (Figure 22–27)

ligand Any molecule that binds to a specific site on a protein or other molecule. From Latin *ligare*, to bind.

light microscope One of a class of microscopes that uses visible light to create the image.

lignin Network of cross-linked phenolic compounds that forms a supporting network throughout the cell walls of xylem and woody tissue in plants.

limit of resolution In microscopy, the smallest distance apart at which two point objects can be resolved as separate. Just under 0.2 μm for conventional light microscopy, a limit determined by the wavelength of light.

linkage In ligand binding, the conformational coupling between two separate ligand-binding sites on a protein, such that a conformational change in the protein induced by binding of one ligand affects the binding of a second ligand.

lipid bilayer (**phospholipid bilayer**) Thin double sheet of phospholipid molecules that forms the core structure of all cell membranes. The two layers of lipid molecules are packed with their hydrophobic tails pointing inward and their hydrophilic heads outward, exposed to water. (Figure 10–1 and Panel 2–5, pp. 98–99)

lipid droplets Storage form in cells for excess lipids; comprised of a single monolayer of phospholipids and proteins that surrounds neutral lipids that can be retrieved from droplets as required by the cell.

lipid raft Small region of a membrane enriched in sphingolipids and cholesterol. (Figure 10–13)

liposome Artificial phospholipid bilayer vesicle formed from an aqueous suspension of phospholipid molecules. (Figure 10–9)

local mediator Extracellular signal molecule that acts on neighboring cells.

long noncoding RNA (**lncRNA**) One of a large group (≈8000 in humans) of RNAs longer than 200 nucleotides and not coding for protein. The functions, if any, of most lncRNAs is unknown but individual lncRNA are known to play important roles in the cell, for example, in telomerase function and genomic imprinting. In a general sense, lncRNAs are believed to act as scaffolds, holding together proteins and nucleic acids to speed up a wide variety of reactions in the cell.

long-term depression (**LTD**) A long-lasting (hours or more) decrease in the sensitivity of certain synapses in the brain triggered by NMDA receptor activation. As the opposing process to long-term potentiation, it is thought to be involved in learning and memory.

long-term potentiation (**LTP**) Long-lasting increase (days to weeks) in the sensitivity of certain synapses in the brain, induced by a short burst of repetitive firing in the presynaptic neurons. (Figure 11–44)

loss of heterozygosity The result of errant homologous recombination that uses the homolog from the other parent instead of the sister chromatid as the template, converting the sequence of the repaired DNA to that of the other homolog.

low-density lipoprotein (**LDL**) Large complex composed of a single protein molecule and many esterified cholesterol molecules, together with other lipids. The form in which cholesterol is transported in the blood and taken up into cells. (Figure 13–51)

lumen The space inside a hollow structure. In cells: the cavity enclosed by an organelle membrane. In tissues: the cavity enclosed by a sheet of cells.

lymphocyte White blood cell responsible for the specificity of adaptive immune responses. Two main types: B cells, which produce antibody, and T cells, which interact directly with other effector cells of the immune system and with infected cells. T cells develop in the thymus and are responsible for cell-mediated immunity. B cells develop in the bone marrow in mammals and are responsible for the production of circulating antibodies.

lymphoid organ An organ containing large numbers of lymphocytes. Lymphocytes are produced in primary lymphoid organs and respond to antigen in peripheral lymphoid organs. (Figure 24–12)

lymphoma Cancer of lymphocytes, in which the cancer cells are mainly found in lymphoid organs (rather than in the blood, as in leukemias).

lysosomal storage diseases   Genetic diseases resulting from defects in or a lack of one or more functional hydrolases in lysosomes of some cells, leading to accumulation of undigested substrates in lysosomes and consequent cell pathology.

lysosome   Membrane-enclosed organelle in eukaryotic cells containing digestive enzymes, which are typically most active at the acid pH found in the lumen of lysosomes. (Figure 13–37)

lysozyme   Enzyme that catalyzes the cutting of polysaccharide chains in the cell walls of bacteria.

M-Cdk (M-phase Cdk)   Cyclin-Cdk complex formed in vertebrate cells by an M-cyclin and the corresponding cyclin-dependent kinase (Cdk). (Figure 17–11 and Table 17–1, p. 969)

M-cyclin   A cyclin found in all eukaryotic cells that promotes the events of mitosis. (Figure 17–11)

M6P receptor proteins   Transmembrane receptor proteins present in the *trans* Golgi network that recognize the mannose 6-phosphate (M6P) groups added exclusively to lysosomal enzymes, marking the enzymes for packaging and delivery to early endosomes.

macromolecule   Polymers constructed of long chains of covalently linked, small organic (carbon-containing) molecules. The principal building blocks from which a cell is constructed and the components that confer the most distinctive properties of living things.

macrophage   Phagocytic cell derived from blood monocytes, resident in most tissues but able to roam. It has both scavenger and antigen-presenting functions in immune responses.

macropinocytosis   Clathrin-independent, dedicated degradative endocytic pathway induced in most cell types by cell-surface receptor activation by specific cargoes.

malaria   Protozoal disease caused by four species of *Plasmodium*, which are transmitted to humans by the bite of the female *Anopheles* mosquito.

malignant   Of tumors and tumor cells: invasive and/or able to undergo metastasis. A malignant tumor is a cancer. (Figure 20–3)

MAP kinase module (mitogen-activated protein kinase module)   An intracellular signaling module composed of three protein kinases, acting in sequence, with MAP kinase as the third. Typically activated by a Ras protein in response to extracellular signals. (Figure 15–49)

master transcription regulator   A transcription regulator specifically required for formation of a particular cell type. Artificial expression of master transcription regulators (alone or in combination with others) will often convert one cell type into another.

maternal inheritance   A form of inheritance observed when following mitochondria in animals and plants, where mitochondrial DNA is inherited only through the female germ line.

maternal-effect gene   Gene that acts in the mother to specify maternal mRNAs and proteins in the egg. Maternal-effect mutations affect the development of the embryo even if the embryo itself has not inherited the mutated gene.

maternal-zygotic transition (MZT)   Event in animal development where the embryo's own genome largely takes over control of development from maternally deposited macromolecules.

matrix metalloprotease Ca^{2+}- or Zn^{2+}-dependent proteolytic enzyme present in the extracellular matrix that degrades matrix proteins. Includes the collagenases.

matrix space   Large internal compartment of the mitochondrion.

mechanosensitive channels   Transmembrane ion channels that open in response to a mechanical stress on the lipid bilayer in which they are embedded.

megakaryocyte   Large myeloid cell with a multilobed nucleus that remains in the bone marrow when mature. Buds off platelets from long cytoplasmic processes. (Figures 22–29)

meiosis I   The first of two rounds of chromosome segregation following meiotic chromosome duplication; segregates the homologs, each composed of a tightly linked pair of sister chromatids.

meiosis II   The second of two rounds of chromosome segregation following meiotic chromosome duplication; segregates the sister chromatids of each homolog.

membrane potential   Voltage difference across a membrane due to a slight excess of positive ions on one side and of negative ions on the other. A typical membrane potential for an animal cell plasma membrane is –60 mV (inside negative relative to the surrounding fluid). (Figure 11–23)

membrane protein   Amphiphilic protein of diverse structure and function that associates with the lipid bilayer of cell membranes. (Figure 10–17)

membrane transport protein   Membrane protein that mediates the passage of ions or molecules across a membrane. The two main classes are transporters (also called carriers or permeases) and channels. (Figure 11–4)

membrane-associated protein   Membrane protein not extending into the hydrophobic interior of the lipid bilayer but bound to either face of the membrane by noncovalent interactions with other membrane proteins. (Figure 10–17)

membrane-bending proteins   Attach to specific membrane regions as needed and act to control local membrane curvature and thus confer on membranes their characteristic three-dimensional shapes.

membrane-bound ribosome   Ribosome attached to the cytosolic face of the endoplasmic reticulum. The site of synthesis of proteins that enter the endoplasmic reticulum. (Figure 12–38)

memory cell   In immunology: a T or B lymphocyte generated following antigen stimulation that is more easily and more quickly induced to become an effector cell or another memory cell by a later encounter with the same antigen. (Figure 24–17)

mesoderm   Embryonic tissue that is the precursor to muscle, connective tissue, skeleton, and many of the internal organs. (Figure 21–3)

messenger RNA (mRNA)   RNA molecule that specifies the amino acid sequence of a protein. Produced in eukaryotes by processing of an RNA molecule made by RNA polymerase as a complementary copy of DNA. It is translated into protein in a process catalyzed by ribosomes. (Figure 6–20)

metabolism   The sum total of the chemical processes that take place in living cells. All of catabolism plus anabolism. (Figure 2–14)

metabotropic receptors   Neurotransmitter receptors that regulate ion channels indirectly through the activation of second-messenger molecules.

metaphase plate   Imaginary plane at right angles to the mitotic spindle and midway between the spindle poles; the plane in which chromosomes are positioned at metaphase. (Panel 17–1, pp. 980–981)

metaphase-to-anaphase transition   Transition in the eukaryotic cell cycle preceding sister-chromatid separation at anaphase. If the cell is not ready to proceed to anaphase, the cell cycle is halted at this point. (Figure 17–9, and Panel 17–1, pp. 980–981)

metastases   Secondary tumors, at sites in the body additional to that of the primary tumor, resulting from cancer cells breaking loose, entering blood or lymphatic vessels, and colonizing separate environments.

metastasis   The spread of cancer cells from their site of origin to other sites in the body. (Figures 20–1 and 20–16)

MHC complex (major histocompatibility complex)   Cluster of genes in one vertebrate chromosome (chromosome 6 in humans) that code for a set of highly polymorphic cell-surface glycoproteins (MHC proteins). (Figure 24–37)

microbiome   The combined genomes of the various species of a defined microbiota.

microbiota   The collective of microorganisms that reside in or on an organism.

microelectrode   A piece of fine glass tubing, pulled to an even finer tip, that is used to inject electric current into cells or to study the intracellular concentrations of common inorganic ions (such as H^+, Na^+, K^+, Cl^-, and Ca^{2+}) in a single living cell by insertion of its tip directly into the cell interior through the plasma membrane.

microRNAs (miRNAs)   Short (~21 nucleotide) eukaryotic RNAs, produced by the processing of specialized RNA transcripts coded in the genome, that regulate gene expression through base-pairing with mRNA. (Figure 7–75)

microsome   Small vesicle derived from endoplasmic reticulum that is produced by fragmentation when cells are homogenized. (Figure 12–34)

microtubule flux   Movement of individual tubulin molecules in the microtubules of the spindle toward the poles by loss of tubulin at their minus ends. Helps to generate the poleward movement of sister chromatids after they separate in anaphase. (Figure 17–35)

microtubule-associated protein (MAP)   Any protein that binds to microtubules and modifies their properties. Many different kinds have been found, including structural proteins, such as MAP2, and motor proteins, such as dynein. [Not to be confused with the "MAP" (mitogen-activated protein kinase) of "MAP kinase."]

microtubule-organizing center (MTOC)   Region in a cell, such as a centrosome or a basal body, from which microtubules grow.

midbody   Structure formed at the end of cleavage that can persist for some time as a tether between the two daughter cells in animals. (Figure 17–43)

mitochondrial hsp70   Part of a multisubunit protein assembly bound to the matrix side of the TIM23 complex that acts as a motor to pull mitochondrial precursor proteins into the matrix space.

mitochondrial matrix   Large internal compartment of the mitochondrion. The corresponding compartment in a chloroplast is known as the stroma.

mitochondrial precursor proteins   Proteins first fully synthesized in the cytosol and then translocated into mitochondrial subcompartments as directed by one or more signal sequences.

mitochondrion (plural mitochondria)   Membrane-bounded organelle, about the size of a bacterium, that carries out oxidative phosphorylation and produces most of the ATP in eukaryotic cells. (Figure 1–28)

mitogen   Extracellular signal molecule that stimulates cells to proliferate.

mitotic chromosome   Highly condensed duplicated chromosome as seen at mitosis, consisting of two sister chromatids held together at the centromere.

mitotic spindle   Bipolar array of microtubules and associated molecules that forms in a eukaryotic cell during mitosis and serves to move the duplicated chromosomes apart. (Figure 17–23 and Panel 17–1, pp. 980–981)

model organism   A species that has been studied intensively over a long period and thus serves as a "model" for deriving fundamental biological principles.

molecular chaperone (chaperone)   Protein that helps guide the proper folding of other proteins, or helps them avoid misfolding. Includes heat-shock proteins (hsp).

monoallelic gene expression   Expression of only one of the two copies of a gene in a diploid genome, occurring, for example, as a result of imprinting or X-chromosome inactivation.

monoclonal antibody   Antibody secreted by a hybridoma cell line. Because the hybridoma is generated by the fusion of a single B cell with a single tumor cell, each hybridoma produces antibodies that are all identical. (Page 444)

monocyte   Type of white blood cell that leaves the bloodstream and matures into a macrophage in tissues. (Figure 22–27)

monomeric GTPase   A single-subunit enzyme that converts GTP to GDP (also called small monomeric GTP-binding proteins). Cycles between an active GTP-bound form and an inactive GDP-bound form and frequently acts as a molecular switch in intracellular signaling pathways.

morphogen   Diffusible signal molecule that can impose a pattern on a field of cells by causing cells in different places to adopt different fates. (Figure 21–8)

morphogenesis   Developmental process in which cells undergo movements and deformations in order to assemble into tissues and organs with specific shapes and sizes.

motor protein   Protein that uses energy derived from nucleoside triphosphate hydrolysis to propel itself along a linear track (protein filament or other polymeric molecule).

mRNA degradation control   Regulation by a cell of gene expression by selectively preserving or destroying certain mRNA molecules in the cytoplasm.

mTOR   The mammalian version of the large protein kinase called TOR, involved in cell signaling; mTOR exists in two functionally distinct multiprotein complexes.

multidrug resistance   An observed phenomenon in which cells exposed to one anticancer drug evolve a resistance not only to that drug, but also to other drugs to which they have never been exposed.

multidrug resistance (MDR) protein   Type of ABC transporter protein that can pump hydrophobic drugs (such as some anticancer drugs) out of the cytoplasm of eukaryotic cells.

multipass transmembrane protein   Membrane protein in which the polypeptide chain crosses the lipid bilayer more than once. (Figure 10–17)

multivesicular bodies   Intermediates in the endosome maturation process; early endosomes that are on their way to becoming late endosomes.

mutation   Heritable change in the nucleotide sequence of a chromosome. (Panel 8–2, pp. 486–487)

mutation rate   The rate at which changes (mutations) occur in DNA sequences.

mutualism   Ecological relationship between microbes and their host in which both the microbe and host benefit.

Myc   Transcription regulatory protein that is activated when a cell is stimulated to grow and divide by extracellular signals. It activates the transcription of many genes, including those that stimulate cell growth. (Figure 17–61)

myelin sheath   Insulating layer of specialized cell membrane wrapped around vertebrate axons. Produced by oligodendrocytes in the central nervous system and by Schwann cells in the peripheral nervous system. (Figure 11–33)

myeloid cell   Any white blood cell other than a lymphocyte. (Figure 22–31)

myoblast   Mononucleated, undifferentiated muscle precursor cell. A skeletal muscle cell is formed by the fusion of multiple myoblasts. (Figure 22–19)

myofibril   Long, highly organized bundle of actin, myosin, and other proteins in the cytoplasm of muscle cells that contracts by a sliding filament mechanism.

myosin   Type of motor protein that uses the energy of ATP hydrolysis to move along actin filaments.

Na$^+$-K$^+$ pump (Na$^+$-K$^+$ ATPase)   Transmembrane carrier protein found in the plasma membrane of most animal cells that pumps Na$^+$ out of and K$^+$ into the cell, using energy derived from ATP hydrolysis. (Figure 11–15)

NAD$^+$/NADH (nicotinamide adenine dinucleotide/reduced nicotinamide adenine dinucleotide)   Electron carrier system that participates in oxidation–reduction reactions, such as the oxidation of food molecules. NAD$^+$ accepts the equivalent of a hydride ion (H$^-$, a proton plus two electrons) to become the activated carrier NADH. The NADH formed donates its high-energy electrons to the ATP-generating process of oxidative phosphorylation. (Figure 2–36)

NADH dehydrogenase complex   First of the three electron-driven proton pumps in the mitochondrial respiratory chain, also known as Complex I. It accepts electrons from NADH and passes them to a quinone. (Figure 14–18)

NADP$^+$/NADPH (nicotinamide adenine dinucleotide phosphate/reduced nicotinamide adenine dinucleotide phosphate)   Electron carrier system closely related to NAD$^+$/NADH, but used almost exclusively in reductive biosynthetic, rather than catabolic, pathways. (Figure 2–36)

naïve cell   In immunology: a T or B lymphocyte that proliferates and differentiates into an effector cell or memory cell when it encounters its specific foreign antigen for the first time. (Figure 24–17)

natural killer cell (NK cell)   Cytotoxic cell of the innate immune system that can kill virus-infected cells and some cancer cells.

natural regulatory T cell   A regulatory T cell (T$_{reg}$ cell) that develops in the thymus and helps maintain self-tolerance.

negative selection   Process by which thymocytes expressing a T cell receptor with high affinity for a self peptide bound to a self-MHC protein are eliminated by undergoing apoptosis.

negative staining   A technique in electron microscopy enabling fine detail of isolated macromolecules to be seen. Samples are prepared such that a very thin film of heavy-metal salt covers everywhere except where excluded by the presence of macromolecules, which allow electrons to pass through, creating a reverse or negative image of the molecule.

Nernst equation   Equation that computes relates the electrical potential (voltage) generated by differences in ion concentrations across a membrane.

Netrin   Signal protein, secreted by cells of the neural tube floor plate, responsible for attracting growth cones of commissural axons toward and across the midline.

neural crest   Collection of cells located along the line where the neural tube pinches off from the surrounding epidermis in the vertebrate embryo. Neural crest cells migrate to give rise to a variety of tissues, including neurons and glia of the peripheral nervous system, pigment cells of the skin, and the bones of the face and jaws. (Figure 19–8)

neural map   Regular mapping of neurons of a similar type from one territory to another, such that there are orderly projections of one array of neurons onto another.

neural tube   Tube of ectoderm that will form the brain and spinal cord in a vertebrate embryo. (Figure 21–56)

neurofilament   Type of intermediate filament found in nerve cells. (Figure 16–72)

neuromuscular junction   Specialized chemical synapse between an axon terminal of a motor neuron and a skeletal muscle cell. (Figure 11–37)

neuron (nerve cell)   Impulse-conducting cell of the nervous system, with extensive processes specialized to receive, conduct, and transmit signals. (Figures 11–28 and 21–66)

neuronal specificity   Nonequivalence among neurons; an intrinsic characteristic that guides axons to their appropriate target sites.

neurotransmitter   Small signal molecule secreted by the presynaptic nerve cell at a chemical synapse to relay the signal to the postsynaptic cell. Examples include acetylcholine, glutamate, GABA, glycine, and many neuropeptides.

neurotrophic factor   Factor released in limited amounts by a target tissue that the neurons innervating that tissue require to survive.

neurotrophin   Family of signal proteins that promote the survival and growth of specific classes of neurons.

neutrophil   White blood cell that is specialized for the uptake of particulate material by phagocytosis. Enters tissues that become infected or inflamed. (Figure 24–5)

NFκB protein   Latent transcription regulator that is activated by various intracellular signaling pathways when cells are stimulated during immune, inflammatory, or stress responses. Also has important roles in animal development. (Figure 15–62)

nitric oxide (NO)   Gaseous signal molecule that is widely used in cell–cell communication in both animals and plants. (Figure 15–40)

nitrogen fixation   Biochemical process carried out by certain bacteria that reduces atmospheric nitrogen (N$_2$) to ammonia, leading eventually to various nitrogen-containing metabolites.

NMDA receptor   Subclass of glutamate-gated ion channel in the mammalian central nervous system critical for long-term potentiation and long-term depression. NMDA-receptor channels are doubly gated, opening only when glutamate is bound to the receptor and, simultaneously, the membrane is strongly depolarized.

NO synthase (NOS)   Enzyme that synthesizes nitric oxide (NO) by the deamination of arginine. (Figure 15–40B)

NOD-like receptors (NLRs)   Large family of pattern recognition receptors (PRRs) with leucine-rich repeat motifs; they are exclusively cytoplasmic and recognize a distinct set of microbial molecules.

nonclassical cadherins   Large family of cadherins that are more distantly related in sequence than classical cadherins and include proteins involved in adhesion (including protocadherins, desmocollins, and desmogleins) and signaling.

noncoding RNA   An RNA molecule that is the final product of a gene and does not code for protein. These RNAs serve as enzymatic, structural, and regulatory components for a wide variety of processes in the cell.

nondisjunction   Event occurring occasionally during meiosis in which a pair of homologous chromosomes fails to separate so that the resulting germ cell has either too many or too few chromosomes.

nonenveloped virus   Virus consisting of a nucleic acid core and a protein capsid only. (Figure 23–18C,D)

nonhomologous end joining   A DNA repair mechanism for double-strand breaks in which the broken ends of DNA are brought together and rejoined by DNA ligation, generally with the loss of one or more nucleotides at the site of joining.

nonretroviral retrotransposons   Type of transposable element that moves by being first transcribed into an RNA copy that is converted to DNA by reverse transcriptase then inserted elsewhere in the genome. The mechanism of insertion differs from that of the retroviral-like transposons. (Table 5–4, p. 288)

nonsense-mediated mRNA decay   Mechanism for degrading aberrant mRNAs containing in-frame internal stop codons before they can be translated into protein. (Figure 6–76)

normal flora   The human microbiota consisting of approximately 10^{14} bacterial, fungal, and protozoan cells, representing thousands of microbial species.

Notch   Transmembrane receptor protein (and latent transcription regulator) involved in many cell-fate choices in animal development, for example in the specification of nerve cells from ectodermal epithelium. Its ligands are cell-surface proteins such as Delta and Serrate. (Figure 15–59)

NSF   Hexameric ATPase that disassembles a complex of a v-SNARE and a t-SNARE. (Figure 13–20)

nuclear envelope   Double membrane (two bilayers) surrounding the nucleus. Consists of an outer and inner membrane and is perforated by nuclear pores. The outer membrane is continuous with the endoplasmic reticulum. (Figures 4–9 and 12–7)

nuclear export receptors   Bind to both the export signal and nuclear pore complex proteins to guide their cargo through the nuclear pore complex to the cytosol.

nuclear export signal   Sorting signal contained in the structure of molecules and complexes, such as nuclear RNPs and new ribosomal subunits, that are transported from the nucleus to the cytosol through nuclear pore complexes. (Figure 12–13)

nuclear import receptors   Recognize nuclear localization signals to initiate nuclear import of proteins containing the appropriate nuclear localization signal.

nuclear lamin   Protein subunit of the intermediate filaments that form the nuclear lamina.

nuclear lamina   Fibrous meshwork of proteins on the inner surface of the inner nuclear membrane. It is made up of a network of intermediate filaments formed from nuclear lamins.

nuclear localization signal (NLS)   Signal sequence or signal patch found in proteins destined for the nucleus that enables their selective transport into the nucleus from the cytosol through the nuclear pore complexes. (Figures 12–9 and 12–13)

nuclear magnetic resonance (NMR) spectroscopy   NMR is the resonant absorption of electromagnetic radiation at a specific frequency by atomic nuclei in a magnetic field, due to flipping of the orientation of their magnetic dipole moments. The NMR spectrum provides information about the chemical environment of the nuclei. NMR is used widely to determine the three-dimensional structure of small proteins and other small molecules. The principles of NMR are also used for medical diagnostic purposes in magnetic resonance imaging (MRI). (Figure 8–22)

nuclear pore complex (NPC)   Large multiprotein structure forming an aqueous channel (the nuclear pore) through the nuclear envelope that allows selected molecules to move between nucleus and cytoplasm. (Figure 12–8)

nuclear receptor superfamily   Intracellular receptors for hydrophobic signal molecules such as steroid and thyroid hormones and retinoic acid. The receptor-ligand complex acts as a transcription factor in the nucleus. (Figure 15–65)

nuclear transport receptor (karyopherin)   Protein that escorts macromolecules either into or out of the nucleus: nuclear import receptor or nuclear export receptor. (Figure 12–13)

nucleolus   A prominent structure in the nucleus where rRNA is transcribed and ribosomal subunits are assembled. (Figure 4–9)

nucleoporin   Any of a number of different proteins that make up nuclear pore complexes.

nucleosome   Beadlike structure in eukaryotic chromatin, composed of a short length of DNA wrapped around an octameric core of histone proteins. The fundamental structural unit of chromatin. (Figures 4–22 and 4–23)

nucleotide   Nucleoside with one or more phosphate groups joined in ester linkages to the sugar moiety. DNA and RNA are polymers of nucleotides. (Panel 2–6, pp. 100–101)

nucleotide excision repair   Type of DNA repair that corrects damage of the DNA double helix, such as that caused by chemicals or UV light, by cutting out the damaged region on one strand and resynthesizing it using the undamaged strand as template. *Compare* **base excision repair**. (Figure 5–41)

O-linked glycosylation   Addition of one or more sugars to a hydroxyl group on a protein.

obligate pathogens   Bacteria that can only replicate inside their host.

olfactory receptors   G-protein-coupled receptors on the modified cilia of olfactory receptor neurons that recognize odors. The receptors activate adenylyl cyclase via an olfactory-specific G protein (Golf) and resultant increases in cAMP open cyclic-AMP-gated cation channels, allowing Na+ influx and depolarization and initiation of a nerve impulse.

oligodendrocyte Glial cell in the vertebrate central nervous system that forms a myelin sheath around axons. *Compare* **Schwann cell**.

oncogene An altered gene whose product can act in a dominant fashion to help make a cell cancerous. Typically, an oncogene is a mutant form of a normal gene (proto-oncogene) involved in the control of cell growth or division. (Figure 20–17)

open reading frame (ORF) A continuous nucleotide sequence free from stop codons in at least one of the three reading frames (and thus with the potential to code for protein).

opportunistic pathogens Microbes of the normal flora that can cause disease only if the immune systems are weakened or if they gain access to a normally sterile part of the body.

optogenetics Use of genetically engineered channelrhodopsin and other light-responsive ion channels and transporters to modulate neuron function and hence analyze the neurons and circuits underlying complex functions, including behaviors in whole animals. (Figure 11–32)

organelle Subcellular compartment or large macromolecular complex, often membrane-enclosed, that has a distinct structure, composition, and function. Examples are nucleus, nucleolus, mitochondrion, Golgi apparatus, and centrosomes. (Figure 1–25)

Organizer Specialized tissue at the dorsal lip of the blastopore in an amphibian embryo; a source of signals that help to orchestrate formation of the embryonic body axis.

origin recognition complex (ORC) Large protein complex that is bound to the DNA at origins of replication in eukaryotic chromosomes throughout the cell cycle. (Figure 5–31)

orthologs Genes or proteins from different species that are similar in sequence because they are descendants of the same gene in the last common ancestor of those species. *Compare* **paralogs**. (Figure 1–21)

osteoblast Cell that secretes matrix of bone. (Figure 22–14)

osteoclast Macrophage-like cell that erodes bone, enabling it to be remodeled during growth and in response to stresses throughout life. (Figure 22–16)

osteocyte Nondividing cell in bone that develops from an osteoblast and is embedded in bone matrix. (Figure 22–14)

outer membrane Mitochondrial membrane that is in contact with the cytosol.

outer mitochondrial membrane Membrane that separates the organelle from the cytosol.

outer nuclear membrane One of two concentric membranes comprising the nuclear envelope; surrounds the inner nuclear membrane and is continuous with the inner nuclear membrane and the membrane of the endoplasmic reticulum.

OXA complex Protein translocator in the inner mitochondrial membrane that mediates insertion of inner membrane proteins.

oxidation (verb **oxidize**) Loss of electrons from an atom, as occurs during the addition of oxygen to a molecule or when a hydrogen is removed. Opposite of reduction. (Figure 2–20)

oxidative phosphorylation Process in bacteria and mitochondria in which ATP formation is driven by the transfer of electrons through the electron transport chain to molecular oxygen. Involves the intermediate generation of a proton gradient (pH gradient) across a membrane and a chemiosmotic coupling of that gradient to the ATP synthase. (Figures 14–12)

oxidative phosphorylation Process in bacteria and mitochondria in which ATP formation is driven by the transfer of electrons through the electron-transport chain to molecular oxygen. Involves the intermediate generation of an electrochemical proton gradient across a membrane and a chemiosmotic coupling of that gradient to the ATP synthase. (Figure 14–10)

P-type pumps A class of ATP-driven pumps comprising structurally and functionally related multipass transmembrane proteins that phosphorylate themselves during the pumping cycle. The class includes many of the ion pumps responsible for setting up and maintaining gradients of Na+, K+, H+, and Ca2+ across cell membranes. (Figure 11–12)

p53 A transcription regulatory protein that is activated by damage to DNA and is involved in blocking further progression through the cell cycle. (Figures 20–37 and 20–40)

p53 Tumor suppressor gene that is mutated in about half of human cancers. Encodes a transcription regulator that is activated by damage to DNA and is involved in blocking further progression through the cell cycle. (Figure 20–27)

pair-rule gene In *Drosophila* development, a gene expressed in a series of regular transverse stripes along the body of the embryo and which helps to determine its segments. (Figure 21–19)

pairing In meiosis, the lining up of the two homologous chromosomes along their length. (Figure 17–54)

papillomaviruses Class of viruses responsible for human warts and a prime example of DNA tumor viruses, being a cause of cancer of the uterine cervix.

paracrine signaling Short-range cell–cell communication via secreted signal molecules that act on neighboring cells. (Figure 15–2)

paralogs Genes or proteins that are similar in sequence because they are the result of a gene duplication event occurring in an ancestral organism. Those in two different organisms are less likely to have the same function than are orthologs. *Compare* **orthologs**. (Figure 1–21)

parasitism Ecological relationship between microbes and their host in which the microbe benefits to the detriment of the host, as is often the case for pathogens.

passengers Mutations that have occurred in the same cell as driver mutations, but which are irrelevant to the development of the cancer.

passive transport (facilitated diffusion) Transport of a solute across a membrane down its concentration gradient or its electrochemical gradient, using only the energy stored in the gradient. (Figure 11–4)

patch-clamp recording Electrophysiological technique in which a tiny electrode tip is sealed onto a patch of cell membrane, thereby making it possible to record the flow of current through individual ion channels in the patch. (Figure 11–34)

Patched Transmembrane protein predicted to cross the plasma membrane 12 times; much is in intracellular vesicles and some is on the cell surface where it binds the Hedgehog protein.

oxidative phosphorylation Process in bacteria and mitochondria in which ATP formation is driven by the transfer

pathogen (adjective **pathogenic**)   An organism, cell, virus, or prion that causes disease.

pathogen-associated molecular patterns (**PAMPs**)   Microbe-associated molecules, either not present or sequestered in the host organism, that often occur in repeating patterns that are recognized by pattern recognition receptors (PRRs) in or on cells of the innate immune system. PAMPs are present in various microbial molecules, including nucleic acids, lipids, polysaccharides, and proteins.

pattern recognition receptor (**PRR**)   Receptor present on or in cells of the innate immune system that recognizes and is activated by microbial pathogen-associated molecular patterns (PAMPs).

PDZ domain   Protein-binding domain present in many scaffold proteins, and often used as a docking site for intracellular tails of transmembrane proteins. (Figure 19–22)

pectin   Mixture of polysaccharides rich in galacturonic acid which forms a highly hydrated matrix in which cellulose is embedded in plant cell walls. (Figure 19–63)

peripheral (**secondary**) **lymphoid organ**   Lymphoid organ in which T cells and B cells interact and respond to foreign antigens. Examples are spleen, lymph nodes, and mucosal-associated lymphoid organs. (Figure 24–12)

peroxins   Form a protein translocator that participates in the import of proteins into peroxisomes.

peroxisome   Small membrane-bounded organelle that uses molecular oxygen to oxidize organic molecules. Contains some enzymes that produce and others that degrade hydrogen peroxide (H_2O_2). (Figure 12–27)

pH scale   Common measure of the acidity of a solution: "p" refers to power of 10, "H" to hydrogen. Defined as the negative logarithm of the hydrogen ion concentration in moles per liter (M). pH = $-\log$ [H^+]. Thus a solution of pH 3 will contain 10^{-3} M hydrogen ions. pH less than 7 is acidic and pH greater than 7 is alkaline.

phagocytosis   Process by which unwanted cells, debris, and other bulky particulate material is endocytosed ("eaten") by a cell. Prominent in carnivorous cells, such as *Amoeba proteus*, and in vertebrate macrophages and neutrophils. From Greek *phagein*, to eat.

phagosome   Large intracellular membrane-enclosed vesicle that is formed as a result of phagocytosis. Contains ingested extracellular material. (Figure 13–61)

phase variation   The random switching of phenotype and expression of proteins involved in infection at frequencies much higher than mutation rates.

phase-contrast microscope   Type of light microscope that exploits the interference effects that occur when light passes through material of different refractive indices. Used to view living cells. (Figure 9–7)

phenotype   The observable character (including both physical appearance and behavior) of a cell or organism. (Panel 8–2, p. 486)

phosphatidylinositol 4,5-bisphosphate [**PI(4,5)P$_2$, PIP$_2$**]   Membrane inositol phospholipid (a phosphoinositide) that is cleaved by phospholipase C into IP$_3$ and diacylglycerol at the beginning of the inositol phospholipid signaling pathway. It can also be phosphorylated by PI 3-kinase to produce PIP$_3$ docking sites for signaling proteins in the PI-3-kinase–Akt signaling pathway. (Figures 15–28 and 15–53)

phosphoglyceride   Phospholipid derived from glycerol, abundant in biomembranes. (Figures 10–2 and 10–3)

phosphoinositide   A lipid containing a phosphorylated inositol derivative. Minor component of the plasma membrane, but important in demarking different membranes and for intracellular signal transduction in eukaryotic cells. (Figure 15–52)

phosphoinositide 3-kinase (**PI 3-kinase**)   Membrane-bound enzyme that is a component of the PI-3-kinase–Akt intracellular signaling pathway. It phosphorylates phosphatidylinositol 4,5-bisphosphate at the 3 position on the inositol ring to produce PIP$_3$ docking sites in the membrane for other intracellular signaling proteins. (Figure 15–53)

phosphoinositides (**PIPs; phosphatidylinositol phosphates**)   A lipid containing a phosphorylated inositol derivative. Minor component of the plasma membrane, but important in demarking different membranes and for intracellular signal transduction in eukaryotic cells. (Figure 13–10)

phospholipase C (**PLC**)   Membrane-bound enzyme that cleaves inositol phospholipids to produce IP$_3$ and diacylglycerol in the inositol phospholipid signaling pathway. PLCβ is activated by GPCRs via specific G proteins, while PLCγ is activated by RTKs. (Figure 15–55)

phospholipid   The main category of lipids used to construct biomembranes. Generally composed of two fatty acids linked through glycerol (or sphingosine) phosphate to one of a variety of polar groups. (Figure 10–3, and Panel 2–5, pp. 98–99)

phosphorylation   Reaction in which a phosphate group is covalently coupled to another molecule.

photoactivation   Technique for studying intracellular processes in which an inactive form of a molecule of interest is introduced into the cell, and is then activated by a focused beam of light at a precise spot in the cell. (Figure 9–28)

photochemical reaction center   The part of a photosystem that converts light energy into chemical energy in photosynthesis. (Figure 14–44)

photosynthetic electron-transfer reactions   Light-driven reactions in photosynthesis in which electrons move along an electron-transport chain in a membrane, generating ATP and NADPH.

photosystem   Multiprotein complex involved in photosynthesis that captures the energy of sunlight and converts it to useful forms of energy: a reaction center plus an antenna (Figure 14–45)

phototropin   Photoprotein associated with the plant plasma membrane that senses blue light and is partly responsible for phototropism.

phragmoplast   Structure made of microtubules and actin filaments that forms in the prospective plane of division of a plant cell and guides formation of the cell plate. (Figure 17–49)

phytochrome   Plant photoprotein that senses light via a covalently attached light-absorbing chromophore, which changes its shape in response to light and then induces a change in the protein's conformation. Plant phytochromes are dimeric, cytoplasmic serine/threonine kinases, which respond differentially and reversibly to red and far-red light to alter cell behavior.

PI-3-kinase–Akt pathway   Intracellular signaling pathway that stimulates animal cells to survive and grow. (Figure 15–53)

pinocytosis   Literally, "cell drinking." Type of endocytosis in which soluble materials are continually taken up from the environment in small vesicles and moved into endosomes along with the membrane-bound molecules. *Compare* **phagocytosis**. (Figure 13–48)

piRNAs (piwi-interacting RNAs) A class of small noncoding RNAs made in the germ line that, in complex with Piwi proteins, keep in check the movement of transposable elements by transcriptionally silencing transposon genes and destroying RNAs produced by them.

planar cell polarity Type of cellular asymmetry seen in some epithelia, such that each cell has a polarity vector oriented in the plane of the epithelium. (Figure 21–51)

plant growth regulator (plant hormone) Signal molecule that helps coordinate growth and development. Examples are ethylene, auxins, gibberellins, cytokinins, abscisic acid, and the brassinosteroids.

plasma membrane The membrane that surrounds a living cell. (Figure 10–1)

plasmid vector Small, circular molecules of double-stranded DNA derived from plasmids that occur naturally in bacterial cells; widely used for gene cloning.

plasmodesma (plural plasmodesmata) Plant equivalent of a gap junction. Communicating cell–cell junction in plants in which a channel of cytoplasm lined by plasma membrane connects two adjacent cells through a small pore in their cell walls.

platelet Cell fragment, lacking a nucleus, that breaks off from a megakaryocyte in the bone marrow and is found in large numbers in the bloodstream. Helps initiate blood clotting when blood vessels are injured. (Figure 22–29)

pleckstrin homology domain (PH domain) Protein domain found in some intracellular signaling proteins. Some PH domains in intracellular signaling proteins bind to phosphatidylinositol 3,4,5-trisphosphate produced by PI 3-kinase, bringing the signaling protein to the plasma membrane when PI 3-kinase is activated.

pluripotent Describes a cell that has the potential to give rise to all or almost all of the cell types of the adult body.

polarized In epithelia, that the basal end of a cell, adherent to the basal lamina below, differs from the apical end, exposed to the medium above; thus, all epithelia and their individual cells are structurally polarized.

Polycomb group Set of proteins critical for cell memory for some genes. They form complexes as part of the chromatin of the *Hox* complex, where they maintain a repressed state in cells where *Hox* genes have not been activated.

polymerase chain reaction (PCR) Technique for amplifying specific regions of DNA by the use of sequence-specific primers and multiple cycles of DNA synthesis, each cycle being followed by a brief heat treatment to separate complementary strands. (Figure 8–36)

polymorphisms Describes genome sequences that coexist as two or more sequence variants at high frequency in a population.

polypeptide Linear polymer of amino acids. Proteins are large polypeptides, and the two terms can be used interchangeably. (Panel 3–1, pp. 112–113)

polypeptide backbone Repeating sequence of atoms along the core of the polypeptide chain.

polyribosome mRNA engaged with multiple ribosomes in the act of translation.

polytene chromosome Giant chromosome in which the DNA has undergone repeated replication and the many copies have stayed together in precise alignment. (Figures 4–50 and 4–51)

porin Channel-forming proteins of the outer membranes of bacteria, mitochondria, and chloroplasts.

position effect variegation Alteration in gene expression resulting from change in the position of the gene in relation to other chromosomal domains, especially heterochromatic domains. When an active gene is placed next to heterochromatin, the inactivating influence of the heterochromatin can spread to affect the gene to a variable degree, giving rise to position effect variegation. (Figure 4–31)

positional value A cell's internal record of its positional information in a multicellular organism; an intrinsic character that differs according to a cell's location.

positive feedback Control mechanism whereby the end product of a reaction or pathway stimulates its own production or activation.

positive selection In immunology: process of thymocyte maturation in which thymocytes expressing a T cell receptor with appropriate affinity for a self peptide bound to a self MHC protein is signaled to survive and continue development.

post-transcriptional controls Any control on gene expression that is exerted at a stage after transcription has begun. (Figure 7–54)

post-translational Occurring after completion of translation.

preprophase band Circumferential band of microtubules and actin filaments that forms around a plant cell under the plasma membrane prior to mitosis and cell division. (Figure 17–49)

prereplicative complex (preRC) Multiprotein complex that is assembled at origins of replication during late mitosis and early G_1 phases of the cell cycle; a prerequisite to license the assembly of a preinitiation complex, and the subsequent initiation of DNA replication. (Figures 17–17 and 17–18)

primary cell wall The first cell wall produced by a developing plant cell; it is thin and flexible, allowing room for cell growth. (Figure 19–63)

primary cilium Short, single, nonmotile cilium lacking dynein that arises from a centriole and projects from the surface of many animal cell types. Some signaling proteins are concentrated in the primary cilium. (Figure 15–38)

primary Ig repertoire The billions of IgM and IgD immunoglobulin molecules made by the B cells of an adaptive immune system in the absence of antigen stimulation.

primary immune response Adaptive immune response to an antigen that is made on first encounter with that antigen. (Figure 24–16)

primary pathogens Pathogens that can cause overt disease in most healthy people. Some cause acute, life-threatening epidemic infections and spread rapidly between hosts; other potential primary pathogens may persistently infect a single individual for years without causing overt disease, the host often being unaware that they are infected.

primary structure Linear sequence of monomer units in a polymer, such as the amino acid sequence of a protein.

primary tumor Tumor at the original site at which a cancer first arose. Secondary tumors develop elsewhere by metastasis.

prion disease Transmissible spongiform encephalopathy—such as Kuru and Creutzfeldt–Jakob disease (CJD) in humans, scrapie in sheep, and bovine spongiform encephalopathy (BSE, or "mad cow disease") in cows—that is caused and transmitted by an infectious, abnormally folded protein (prion). (Figure 3–33)

pro-inflammatory cytokine Any cytokine that stimulates an inflammatory response.

programmed cell death A form of cell death in which a cell kills itself by activating an intracellular death program.

prokaryote Single-celled microorganism whose cells lack a well-defined, membrane-enclosed nucleus. Either a bacterium or an archaeon. (Figure 1–17)

promoter Nucleotide sequence in DNA to which RNA polymerase binds to begin transcription. (Figure 7–17)

proteasome Large protein complex in the cytosol with proteolytic activity that is responsible for degrading proteins that have been marked for destruction by ubiquitylation or by some other means. (Figures 6–83 and 6–84)

protein The major macromolecular constituent of cells. A linear polymer of amino acids linked together by peptide bonds in a specific sequence. (Figure 3–1)

protein activity control The selective activation, inactivation, degradation, or compartmentalization of specific proteins after they have been made. One of the means by which a cell controls which proteins are active at a given time or location in the cell.

protein domain *see* **domain**

protein glycosylation Process of transferring a single saccharide or preformed precursor oligosaccharide to proteins.

protein kinase Enzyme that transfers the terminal phosphate group of ATP to one or more specific amino acids (serine, threonine, or tyrosine) of a target protein.

protein kinase C (PKC) Ca^{2+}-dependent protein kinase that, when activated by diacylglycerol and an increase in the concentration of cytosolic Ca^{2+}, phosphorylates target proteins on specific serine and threonine residues. (Figure 15–29)

protein phosphatase Enzyme that catalyzes phosphate removal from amino acids of a target protein.

protein subunit An individual protein chain in a protein composed of more than one chain.

protein translocation Process of moving a protein across a membrane.

protein translocator Membrane-bound protein that mediates the transport of another protein across a membrane. (Figure 12–21)

protein tyrosine phosphatase Enzyme that removes phosphate groups from phosphorylated tyrosine residues on proteins.

proteoglycan Molecule consisting of one or more glycosaminoglycan chains attached to a core protein. (Figure 19–38)

proteomics Study of all the proteins, including all the covalently modified forms of each, produced by a cell, tissue, or organism. Proteomics often investigates changes in this larger set of proteins—in "the proteome"—caused by changes in the environment or by extracellular signals.

proto-oncogene Normal gene, usually concerned with the regulation of cell proliferation, that can be converted into a cancer-promoting oncogene by mutation.

protofilament Linear string of microtubule subunits joined end to end; multiple protofilaments associate with one another laterally to construct and provide strength and adaptability to microtubules.

proton (H^+) Positively charged subatomic particle that forms part of an atomic nucleus. Hydrogen has a nucleus composed of a single proton (H^+).

proton-motive force The force exerted by the electrochemical proton gradient that moves protons across a membrane.

protozoan parasite Parasitic, nonphotosynthetic, single-celled, motile eukaryotic organism, for example *Plasmodium*.

pseudogene Nucleotide sequence of DNA that has accumulated multiple mutations that have rendered an ancestral gene inactive and nonfunctional.

purified cell-free system Fractionated cell homogenate that retains a particular biological function of the intact cell, and in which biochemical reactions and cell processes can be more easily studied.

purifying selection Natural selection operating in a population to slow genome changes and reduce divergence by eliminating individuals carrying deleterious mutations.

quantitative RT-PCR (reverse transcription–polymerase chain reaction) Technique in which a population of mRNAs is converted into cDNAs via reverse transcription, and the cDNAs are then amplified by PCR. The quantitative part relies on a direct relationship between the rate at which the PCR product is generated and the original concentration of the mRNA species of interest.

quaternary structure Three-dimensional relationship of the different polypeptide chains in a multisubunit protein or protein complex.

quinone (Q) Small, lipid-soluble, mobile electron carrier molecule found in the respiratory and photosynthetic electron-transport chains. (Figure 14–17)

Rab cascade An ordered recruitment of sequentially acting Rab proteins into Rab domains on membranes, which changes the identity of an organelle and reassigns membrane dynamics.

Rab effectors Molecules that bind activated, membrane-bound Rab proteins and act as downstream mediators of vesicle transport, membrane tethering, and fusion.

Rab proteins Monomeric GTPase in the Ras superfamily present in plasma and organelle membranes in its GTP-bound state, and as a soluble cytosolic protein in its GDP-bound state. Involved in conferring specificity on vesicle docking. (Table 15–5, p. 854)

Rac Member of the Rho family of monomeric GTPases that regulate the actin and microtubule cytoskeletons, cell-cycle progression, gene transcription, and membrane transport.

Rad51 Eukaryotic protein that catalyzes synapsis of DNA strands during genetic recombination. Called RecA in *E. coli*.

Ran (Ran protein) Monomeric GTPase of the Ras superfamily present in both cytosol and nucleus. Required for the active transport of macromolecules into and out of the nucleus through nuclear pore complexes. (Table 15–5, p. 854)

rapidly inactivating K^+ channel Neuronal voltage-gated K^+ channel, open when the membrane is depolarized, with a specific voltage sensitivity and kinetics of inactivation that induce a reduced rate of action potential firing at levels of stimulation only just above the threshold required, thereby resulting in a firing rate proportional to the strength of the depolarizing stimulus.

Ras   A small family of proto-oncogenes that are frequently mutated in cancers, each of which produces a Ras monomeric GTPase.

Ras (Ras protein)   Monomeric GTPase of the Ras superfamily that helps to relay signals from cell-surface receptor tyrosine kinase receptors to the nucleus, frequently in response to signals that stimulate cell division. Named for the *ras* gene, first identified in viruses that cause rat sarcomas. (Figure 3–67)

Ras superfamily   Large superfamily of monomeric GTPases (also called small GTP-binding proteins) of which Ras is the prototypical member. (Table 15–5, p. 854)

Ras-GAPs   Ras GTPase-activating proteins; increase the rate of hydrolysis of bound GTP by Ras, thereby inactivating Ras.

Ras-GEFs   Ras guanine nucleotide exchange factors; stimulate the dissociation of GDP and the subsequent uptake of GTP from the cytosol, thereby activating Ras.

Ras–MAP-kinase signaling pathway   Intracellular signaling pathway that relays signals from activated receptor tyrosine kinases to effector proteins in the cell including transcription regulators in the nucleus.

Rb gene   The gene that is defective in both copies in individuals with retinoblastoma; its protein product plays a central role in cell-cycle control.

reading frame   The phase in which nucleotides are read in sets of three to encode a protein. An mRNA molecule can be read in any one of three reading frames, only one of which will give the required protein. (Figure 6–49)

RecA (RecA protein)   Prototype for a class of DNA-binding proteins that catalyze synapsis of DNA strands during genetic recombination. (Figure 5–49)

receptor   Any protein that binds a specific signal molecule (ligand) and initiates a response in the cell. Some are on the cell surface, while others are inside the cell. (Figure 15–3)

receptor editing   Process by which a developing B cell that recognizes a self molecule changes its antigen receptors so that the cell no longer does so.

receptor serine/threonine kinase   Cell-surface receptor with an extracellular ligand-binding domain and an intracellular kinase domain that phosphorylates signaling proteins on serine or threonine residues in response to ligand binding. The TGFβ receptor is an example. (Figure 15–57)

receptor tyrosine kinase (RTK)   Cell-surface receptor with an extracellular ligand-binding domain and an intracellular kinase domain that phosphorylates signaling proteins on tyrosine residues in response to ligand binding. (Figure 15–43 and Table 15–4, p. 850)

receptor-mediated endocytosis   Internalization of receptor–ligand complexes from the plasma membrane by endocytosis. (Figure 13–52)

recombinant DNA technology   Collection of techniques by which DNA segments from different sources are combined to make a new DNA, often called a recombinant DNA. Recombinant DNAs are widely used in the cloning of genes, in the genetic modification of organisms, and in the production of large amounts of rare proteins.

recycling endosome   Organelle that provides an intermediate stage on the passage of recycled receptors back to the cell membrane. Regulates plasma membrane insertion of some proteins. (Figure 13–58)

red blood cell   Small hemoglobin-containing blood cell of vertebrates that transports oxygen to, and carbon dioxide from, tissues. Also called an erythrocyte.

redox pair   Pair of molecules in which one acts as an electron donor and one as an electron acceptor in an oxidation–reduction reaction: for example, NADH (electron donor) and NAD^+ (electron acceptor). (Panel 14–1, p. 765)

redox potential   The affinity of a redox pair for electrons, generally measured as the voltage difference between an equimolar mixture of the pair and a standard reference. NADH/NAD^+ has a low redox potential and O_2/H_2 has a high redox potential (high affinity for electrons). (Panel 14–1, p. 765)

redox reaction   Reaction in which one component becomes oxidized and the other reduced; an oxidation–reduction reaction. (Panel 14–1, p. 765)

reduction (verb **reduce**)   Addition of electrons to an atom, as occurs during the addition of hydrogen to a biological molecule or the removal of oxygen from it. Opposite of oxidation. (Figure 2–20)

regulated nuclear transport   Mechanisms controlling export of mRNAs from the nucleus to the cytosol that can be used to regulate gene expression. Also includes the selective import of proteins and RNA molecules into the nucleus.

regulated secretory pathway   A second secretory pathway found mainly in cells specialized for secreting products rapidly on demand—such as hormones, neurotransmitters, or digestive enzymes—in which soluble proteins and other substances are initially stored in secretory vesicles for later release. (Figure 13–62)

regulator of G protein signaling (RGS)   A GAP protein that binds to a trimeric G protein and enhances its GTPase activity, thus helping to limit G-protein-mediated signaling. (Figure 15–8)

regulatory site   Region of an enzyme surface to which a regulatory molecule binds and thereby influences the catalytic events at the separate active site.

regulatory T cell (T_{reg})   A type of T cell that suppresses the development, activation, or function of other immune cells via secreted cytokines or cell-surface inhibitory proteins.

replication fork   Y-shaped region of a replicating DNA molecule at which the two strands of the DNA are being separated and the daughter strands are being formed. (Figures 5–7 and 5–18)

replication origin   Location on a DNA molecule at which duplication of the DNA begins. (Figures 4–19 and 5–23)

replicative cell senescence   Phenomenon observed in primary cell cultures in which cell proliferation slows down and finally irreversibly halts.

respiratory chain (electron-transport chain)   Electron-transport chain present in the inner mitochondrial membrane that generates an electrochemical gradient across the membrane that is used to drive ATP synthesis. (Figures 14–4 and 14–10)

resting membrane potential   Electrical potential across the plasma membrane of a cell at rest, i.e. a cell that has not been stimulated to open additional ion channels than those that are normally open.

restriction nuclease One of a large number of nucleases that can cleave a DNA molecule at any site where a specific short sequence of nucleotides occurs. Extensively used in recombinant DNA technology. (Figure 8–24)

restriction point Important transition at the end of G_1 in the eukaryotic cell cycle; commits the cell to enter S phase. The term was originally used for this transition in the mammalian cell cycle; in this book we use the term Start. (Figure 17–9)

retinoblastoma A rare type of human cancer arising from cells in the retina of the eye that are converted to a cancerous state by an unusually small number of mutations. Studies of retinoblastoma led to the discovery of the first tumor suppressor gene.

retinoblastoma protein (Rb protein) Tumor suppressor protein involved in the regulation of cell division. Mutated in the cancer retinoblastoma, as well as in many other tumors. Its normal activity is to regulate the eukaryotic cell cycle by binding to and inhibiting the E2F proteins, thus blocking progression to DNA replication and cell division. (Figure 17–61)

retroviral-like retrotransposons A large family of transposons that move themselves in and out of chromosomes by a mechanism similar to that used by retroviruses, being first transcribed into an RNA copy that is converted to DNA by reverse transcriptase then inserted elsewhere in the genome. (Table 5–4, p. 288)

retrovirus RNA-containing virus that replicates in a cell by first making an RNA–DNA intermediate and then a double-strand DNA molecule that becomes integrated into the cell's DNA. (Figure 5–62)

reverse genetics Approach to discovering gene function that starts from the DNA (gene) and its protein product and then creates mutants to analyze the gene's function.

reverse transcriptase Enzyme first discovered in retroviruses that makes a double-strand DNA copy from a single-strand RNA template molecule.

RGD sequence Tripeptide sequence of arginine-glycine-aspartic acid that forms a binding site for integrins; present in fibronectin and some other extracellular proteins. (Figure 19–47C)

Rheb A monomeric Ras-related GTPase that in its active form (Rheb-GTP) activates mTOR, which promotes cell growth.

Rho Member of the Rho family of monomeric GTPases that regulate the actin and microtubule cytoskeletons, cell-cycle progression, gene transcription, and membrane transport.

Rho family Family of monomeric GTPases within the Ras superfamily involved in signaling the rearrangement of the cytoskeleton. Includes Rho, Rac, and Cdc42. (Table 15–5, p. 854)

rhodopsin Seven-span membrane protein of the GPCR family that acts as a light sensor in rod photoreceptor cells in the vertebrate retina. Contains the light-sensitive prosthetic group retinol. (Figure 15–39)

ribosomal RNA (rRNA) Any one of a number of specific RNA molecules that form part of the structure of a ribosome and participate in the synthesis of proteins. Often distinguished by their sedimentation coefficient (e.g., 28S rRNA, 5S rRNA).

ribosome Particle composed of rRNAs and ribosomal proteins that catalyzes the synthesis of protein using information provided by mRNA. (Figure 6–64)

ribozyme An RNA molecule with catalytic activity.

RNA (ribonucleic acid) Polymer formed from covalently linked ribonucleotide monomers. *See also* **messenger RNA**, **ribosomal RNA**, **transfer RNA**. (Figure 6–4)

RNA editing Type of RNA processing that alters the nucleotide sequence of an RNA transcript after it is synthesized by inserting, deleting, or altering individual nucleotides.

RNA interference (RNAi) As originally described, mechanism by which an experimentally introduced double-stranded RNA induces sequence-specific destruction of complementary mRNAs. The term RNAi is often used to include the inhibition of gene expression by microRNAs (miRNAs) and piwi RNAs (piRNAs), which are encoded in the cell's own genome.

RNA polymerase Enzyme that catalyzes the synthesis of an RNA molecule on a DNA template from ribonucleoside triphosphate precursors. (Figure 6–9)

RNA primer Short stretch of RNA synthesized on a DNA template. It is required by DNA polymerases to start their DNA synthesis.

RNA processing control Regulation by a cell of gene expression by controlling the processing of RNA transcripts, which includes their splicing.

RNA splicing Process in which intron sequences are excised from RNA transcripts. A major process in the nucleus of eukaryotic cells leading to formation of messenger RNAs (mRNAs).

RNA transport and localization control Regulation by a cell of gene expression by selecting which completed mRNAs are exported from the nucleus to the cytosol and determining where in the cytosol they are localized.

RNA world Hypothesis that early life on Earth was based primarily on RNA molecules that both stored genetic information and catalyzed biochemical reactions.

RNA-seq Sequencing the entire repertoire of RNA from a cell or tissue; also known as deep RNA sequencing.

robustness The ability of biological regulatory systems to function normally in the face of perturbations such as exposure to frequent and/or extreme variations in external conditions or the concentrations or activities of key components.

rod photoreceptor (rod) Photoreceptor cell in the vertebrate retina that is responsible for noncolor vision in dim light.

rough endoplasmic reticulum (rough ER) Endoplasmic reticulum with ribosomes on its cytosolic surface. Involved in the synthesis of secreted and membrane-bound proteins.

rRNA gene Gene that specifies a ribosomal RNA (rRNA).

ryanodine receptor A regulated Ca^{2+} channel in the ER membrane that opens in response to rising Ca^{2+} levels and thus amplifies the Ca^{2+} signal.

SAM complex Protein translocator that helps β-barrel proteins to fold properly in the outer mitochondrial membrane.

Sanger sequencing *see* **dideoxy sequencing**

Sar1 protein Monomeric GTPase responsible for regulating COPII coat assembly at the endoplasmic reticulum membrane.

sarcoma Cancer of connective tissue.

scaffold protein Protein that binds groups of intracellular signaling proteins into a signaling complex, often anchoring the complex at a specific location in the cell. (Figure 15–10)

scanning electron microscope Type of electron microscope that produces an image of the surface of an object. (Figure 9–50)

S-Cdk Cyclin–Cdk complex formed in vertebrate cells by an S-cyclin and the corresponding cyclin-dependent kinase (Cdk). (Figure 17–11 and Table 17–1, p. 969)

SCF Family of ubiquitin ligases formed as a complex of several different proteins. One is involved in regulating the eukaryotic cell cycle, directing the destruction of inhibitors of S-Cdks in late G_1 and thus promoting the activation of S-Cdks and DNA replication. (Figures 3–71 and 17–15)

Schwann cell Glial cell responsible for forming myelin sheaths in the peripheral nervous system. *Compare* **oligodendrocyte**. (Figure 11–33)

S-cyclin Member of a class of cyclins that accumulate during late G_1 phase and bind Cdks soon after progression through Start; they help stimulate DNA replication and chromosome duplication. Levels remain high until late mitosis, after which these cyclins are destroyed. (Figure 17–11)

Sec61 complex Three-subunit core of the protein translocator that transfers polypeptide chains across the endoplasmic reticulum membrane.

second messenger (small intracellular mediator) Small intracellular signaling molecule that is formed or released for action in response to an extracellular signal and helps to relay the signal within the cell. Examples include cyclic AMP, cyclic GMP, IP_3, Ca^{2+}, and diacylglycerol.

secondary cell wall Permanent rigid cell wall that is laid down underneath the thin primary cell wall in certain plant cells that have completed their growth.

secondary Ig repertoire Immunoglobulins produced by B cells after antigen- and helper-T-cell-induced somatic hypermutation and class switching. Compared to the primary Ig repertoire, these Igs have a greatly increased diversity of both Ig classes and antigen-binding sites and have increased affinity for antigen.

secondary immune response The adaptive immune response that occurs in response to a second or subsequent exposure to an antigen. The response is more rapid in onset and stronger than the primary immune response. (Figure 24–16)

secondary structure Regular local folding pattern of a polymeric molecule; in proteins, α helices and β sheets.

secretion system Specialized bacterial systems that secrete effector proteins that interact with host cells.

secretory vesicle Membrane-enclosed organelle in which molecules destined for secretion are stored prior to release. Sometimes called secretory granule because darkly staining contents make the organelle visible as a small solid object. (Figures 13–65)

securin Protein that binds to the protease separase and thereby prevents its cleavage of the protein linkages that hold sister chromatids together in early mitosis. Securin is destroyed at the metaphase-to-anaphase transition. (Figure 17–38)

segment Divisions of an insect body along its anteroposterior axis, each forming highly specialized structures, but all built according to a similar fundamental plan.

segment-polarity gene In *Drosophila* development, a gene involved in specifying the anteroposterior organization of each body segment. (Figure 21–19)

segmentation clock The gene-expression oscillator controlling regular segmentation during vertebrate embryonic development.

segmentation genes Genes expressed by subsets of cells in the embryo that refine the pattern of gene expression so as to define the boundaries and ground plan of the individual body segments.

selectin Member of a family of cell-surface carbohydrate-binding proteins that mediate transient, Ca^{2+}-dependent cell–cell adhesion in the bloodstream—for example between white blood cells and the endothelium of the blood vessel wall. (Figure 19–28)

selectivity filter The part of an ion channel structure that determines which ions it can transport. (Figures 11–24 and 11–25)

sensory bristles Miniature sense organs present on most exposed surfaces of *Drosophila*, consisting of a sensory neuron and supporting cells and responding to chemical or mechanical stimuli.

separase Protease that cleaves the cohesin protein linkages that hold sister chromatids together. Acts at anaphase, enabling chromatid separation and segregation. (Figure 17–38)

septum Structure formed during bacterial cell division by the inward growth of the cell wall and plasma membrane and that divides the cell into two.

sequential induction Development process that generates a progressively more complicated pattern. A series of local inductions whereby one of two cell types present in a developing tissue can produce a signal to induce neighboring cells to specialize in a third way; the third cell type can then signal back to the other two cell types nearby to generate a fourth and a fifth cell type, and so on.

serine protease Type of protease that has a reactive serine in the active site. (Figures 3–12 and 3–39)

serine/threonine kinase Enzyme that phosphorylates specific proteins on serine or threonines.

SH2 domain Src homology region 2, a protein domain present in many signaling proteins. Binds a short amino acid sequence containing a phosphotyrosine. (Panel 3–2, pp. 142–143)

side chain The part of an amino acid that differs between amino acid types. The side chains give each type of amino acid its unique physical and chemical properties. (Panel 3–1, pp. 112–113)

signal patch Protein-sorting signal that consists of a specific three-dimensional arrangement of atoms on the folded protein's surface. (Figure 13–46)

signal peptidase Enzyme that removes a terminal signal sequence from a protein once the sorting process is complete. (Figure 12–35)

signal sequence Short continuous sequence of amino acids that determines the eventual location of a protein in the cell. An example is the N-terminal sequence of 20 or so amino acids that directs nascent secretory and transmembrane proteins to the endoplasmic reticulum. (Table 12–3, p. 648)

signal-recognition particle (SRP) Ribonucleoprotein particle that binds an ER signal sequence on a partially synthesized polypeptide chain and directs the polypeptide and its attached ribosome to the endoplasmic reticulum. (Figure 12–36)

signaling center   Cluster of specialized cells in developing tissues that serves as a source of developmental signals—for example, the generation of a morphogen gradient.

single-nucleotide polymorphism (SNP)   A variation between individuals in a population due to a relatively common difference in a specific nucleotide at a defined point in the DNA sequence.

single-particle reconstruction   Computational procedure in electron microscopy in which images of many identical molecules are obtained and digitally combined to produce an averaged three-dimensional image, thereby revealing structural details that are hidden by noise in the original images. (Figures 9–54 and 9–55)

single-pass transmembrane protein   Membrane protein in which the polypeptide chain crosses the lipid bilayer only once. (Figure 10–24)

single-strand DNA-binding (SSB) protein   Protein that binds to the single strands of the opened-up DNA double helix, preventing helical structures from reforming while the DNA is being replicated. (Figure 5–15)

sister chromatids   Tightly linked pair of chromosomes that arise from chromosome duplication during S phase. They separate during M phase and segregate into different daughter cells. (Figure 17–21)

sliding clamp   Protein complex that holds the DNA polymerase on DNA during DNA replication. (Figure 5–17)

Slit   Signal protein, secreted by cells of the neural tube floor plate, responsible for repelling the growth cones of commissural axons after they have crossed the midline, thereby ensuring these neurons do not re-cross the midline.

Smad family   Latent transcription regulators that are phosphorylated and activated by receptor serine/threonine kinases and carry the signal from the cell surface to the nucleus. (Figure 15–57)

small interfering RNAs (siRNAs)   Short (21–26 nucleotide) double-stranded RNAs that inhibit gene expression by directing destruction of complementary mRNAs. Production of siRNAs is usually triggered by exogenously introduced double-stranded RNA. (Figure 7–77)

small nuclear RNA (snRNA)   Small RNA molecules that are complexed with proteins to form the ribonucleoprotein particles (snRNPs) involved in RNA splicing. (Figures 6–28 and 6–29)

small nucleolar RNA (snoRNA)   Small RNAs found in the nucleolus, with various functions, including guiding the modifications of precursor rRNA. (Table 6–1, p. 305, and Figure 6–41)

smooth endoplasmic reticulum (smooth ER)   Region of the endoplasmic reticulum not associated with ribosomes. Involved in detoxification reactions, Ca^{2+} storage, and lipid synthesis. (Figure 12–33)

Smoothened   Seven-pass transmembrane protein with a structure very similar to a GPCR but does not seem to act as a Hedgehog receptor or as an activator of G proteins; it is controlled by the Patched and iHog proteins.

SNARE proteins (SNAREs)   Members of a large family of transmembrane proteins present in organelle membranes and the vesicles derived from them. SNAREs catalyze the many membrane fusion events in cells. They exist in pairs—a v-SNARE in the vesicle membrane that binds specifically to a complementary t-SNARE in the target membrane.

sodium dodecyl sulfate polyacrylamide-gel electrophoresis (SDS-PAGE)   Type of electrophoresis used to separate proteins by size. The protein mixture to be separated is first treated with a powerful negatively charged detergent (SDS) and with a reducing agent (β-mercaptoethanol), before being run through a polyacrylamide gel. The detergent and reducing agent unfold the proteins, free them from association with other molecules, and separate the polypeptide subunits.

somatic cell   Any cell of a plant or animal other than cells of the germ line. From Greek *soma,* body.

somatic hypermutation   In immunology: accumulation of point mutations in the assembled variable-region-coding sequences of immunoglobulin genes that occurs when B cells are activated to form memory cells. Results in the production of antibodies with altered antigen-binding sites, some of which bind antigen with increased affinity; it is responsible for affinity maturation in antibody responses.

somatic mutations   In cancer, one or more detectable abnormalities in the DNA sequence of tumor cells that distinguish them from the normal somatic cells surrounding the tumor.

somite   One of a series of paired blocks of mesoderm that form during early development and lie on either side of the notochord in a vertebrate embryo. They give rise to the segments of the body axis, including the vertebrae, muscles, and associated connective tissue. (Figure 21–38)

sorting signal   Signal sequence or signal patch that directs the delivery of a protein to a specific location, such as a particular intracellular compartment.

spectrin   Abundant protein associated with the cytosolic side of the plasma membrane in red blood cells, forming a network that supports the membrane. Also present in other cells. (Figure 10–38)

S phase   Period of a eukaryotic cell cycle in which DNA is synthesized. (Figure 17–4)

spinal cord   Bundle of neurons and support cells that extends from the brain.

spindle assembly checkpoint   Regulatory system that operates during mitosis to ensure that all chromosomes are properly attached to the spindle before sister-chromatid separation starts. (Figure 17–9 and Panel 17–1, pp. 980–981)

spliceosome   Large assembly of RNA and protein molecules that performs pre-mRNA splicing in eukaryotic cells.

Src (Src protein family)   Family of cytoplasmic tyrosine kinases (pronounced "sark") that associate with the cytoplasmic domains of some enzyme-linked cell-surface receptors (for example, the T cell antigen receptor) that lack intrinsic tyrosine kinase activity. They transmit a signal onward by phosphorylating the receptor itself and specific intracellular signaling proteins on tyrosines. (Figure 3–10)

SRP (signal-recognition particle) receptor   Component in the endoplasmic reticulum (ER) membrane that guides the signal recognition particle to the ER membrane.

starch   Polysaccharide composed exclusively of glucose units, used as an energy-storage material in plant cells. (Figure 2–51)

Start (restriction point)   Important transition at the end of G_1 in the eukaryotic cell cycle. Passage through Start commits the cell to enter S phase. The term was originally used for this point in the yeast cell cycle only; the equivalent point in the mammalian cell cycle was called the restriction point. In this book we use Start for both. (Figure 17–9)

start-transfer signal   Short amino acid sequence that enables a polypeptide chain to start being translocated across the endoplasmic reticulum membrane through a protein translocator. Multipass membrane proteins sometimes have both N-terminal (signal sequence) and internal start-transfer signals. (Figure 12–42)

STAT (signal transducer and activator of transcription)   Latent transcription regulator that is activated by phosphorylation by Janus kinases (JAKs) and enters the nucleus in response to signaling from receptors of the cytokine receptor family. (Figure 15–56)

stem cell   Undifferentiated cell that can continue dividing indefinitely, throwing off daughter cells that can either commit to differentiation or remain a stem cell (in the process of self-renewal). (Figure 22–3)

stem-cell niche   The specialized microenvironment in a tissue in which self-renewing stem cells can be maintained.

steroid hormones   Hormones, including cortisol, estrogen, and testosterone, that are hydrophobic lipid molecules derived from cholesterol that activate intracellular nuclear receptors.

stimulatory G protein (Gₛ)   G protein that, when activated, activates the enzyme adenylyl cyclase and thus stimulates the production of cyclic AMP. *See also* **G protein**. (Table 15–3, p. 846)

stochastic   Random. Involving chance, probability, or random variables.

stop-transfer signal   Hydrophobic amino acid sequence that halts translocation of a polypeptide chain through the endoplasmic reticulum membrane, thus anchoring the protein chain in the membrane. (Figure 12–42)

strand exchange   Reaction in which one of the single-strand 3′ ends from one duplex DNA molecule penetrates another duplex and searches it for homologous sequences through base-pairing. Also called strand invasion.

strand-directed mismatch repair   A proofreading system that removes DNA replication errors missed by the DNA polymerase proofreading exonuclease. Detects the potential for DNA helix distortion from noncomplementary base pairs then recognizes and excises the mismatch in the newly synthesized strand and resynthesizes the excised segment using the old strand as a template.

stress fibers   Cortical fibers of contractile actin-myosin II bundles that connect the cell to the extracellular matrix or adjacent cells through focal adhesions or a circumferential belt and adherens junctions.

stroma   (1) "Bedding": the connective tissue in which a glandular or other epithelium is embedded. Stromal cells provide the environment necessary for the development of other cells within the tissue. (2) The large interior space of a chloroplast, containing enzymes that incorporate CO_2 into sugars. (Figure 14–38)

substrate   Molecule on which an enzyme acts.

superresolution   Describes several approaches in light microscopy that bypass the limit imposed by the diffraction of light and successfully allow objects as small as 20 nm to be imaged and clearly resolved.

survival factor   Extracellular signal that promotes cell survival by inhibiting apoptosis. (Figure 18–12)

symporter   Carrier protein that transports two types of solute across the membrane in the same direction. (Figure 11–8)

synapse   Communicating cell–cell junction that allows signals to pass from a nerve cell to another cell. In a chemical synapse, the signal is carried by a diffusible neurotransmitter. (Figure 19–22) In an electrical synapse, a direct connection is made between the cytoplasms of the two cells via gap junctions. (Figure 11–34 and 19–23)

synapse elimination   Process by which each muscle cell at first receives synapses from several motor neurons, but is ultimately left innervated by only one.

synaptic plasticity   Changes in the strength with which a chemical synapse transmits a signal. It is thought to be important in memory formation, where concentrations of postsynaptic AMPA receptor are modulated in response to a synapse's activity.

synaptic signaling   Intercellular signaling performed by neurons that transmit signals electrically along their axons and release neurotransmitters at synapses, which are often located far away from the neuronal cell body.

synaptic vesicle   Small neurotransmitter-filled secretory vesicle found at the axon terminals of nerve cells. Its contents are released into the synaptic cleft by exocytosis when an action potential reaches the axon terminal.

synaptonemal complex   Structure that holds paired homologous chromosomes tightly together in pachytene of prophase I in meiosis and promotes the final steps of crossing-over. (Figures 17–55 and 17–56)

syncytium   Mass of cytoplasm containing many nuclei enclosed by a single plasma membrane. Typically the result either of cell fusion or of a series of incomplete division cycles in which the nuclei divide but the cell does not.

TATA box   Sequence in the promoter region of many eukaryotic genes that binds a general transcription factor (TFIID) and hence specifies the position at which transcription is initiated. (Figure 6–14)

T cell receptor (TCR)   Transmembrane receptor for antigen on the surface of T lymphocytes, consisting of an immunoglobulin-like heterodimer. (Figure 24–32)

T-cell-mediated immune response   Any adaptive immune response mediated by antigen-specific T cells.

telomerase   Enzyme that elongates telomere sequences in DNA, which occur at the ends of eukaryotic chromosomes.

telomere   End of a chromosome, associated with a characteristic DNA sequence that is replicated in a special way. Counteracts the tendency of the chromosome otherwise to shorten with each round of replication. From Greek *telos*, end, and *meros,* portion.

telomere   End of a chromosome, associated with a characteristic DNA sequence that is replicated in a special way. Counteracts the tendency of the chromosome otherwise to shorten with each round of replication. From Greek *telos*, end.

telophase   Final stage of mitosis in which the two sets of separated chromosomes decondense and become enclosed by nuclear envelopes. (Panel 17–1, pp. 980–981)

template   Single strand of DNA or RNA whose nucleotide sequence acts as a guide for the synthesis of a complementary strand. (Figure 1–3)

terminal differentiation   The limit of cell determination when a cell forms one of the highly specialized cell types of the adult body.

terminally differentiated   A cell at the limit of cell determination, being one of the highly specialized cell types of the adult body.

terminator   Signal in bacterial DNA that halts transcription; in eukaryotes, transcription terminates after cleavage and polyadenylation of the newly synthesized RNA.

tertiary structure   Complex three-dimensional form of a folded polymer chain, especially a protein or RNA molecule.

T$_{FH}$ cell   *see* **follicular helper T cell**

T$_H$1 cell   A type of effector helper T cell that secretes interferon-γ to help activate macrophages and induces B cells to switch the class of antibody they make. (Figure 24–44)

T$_H$17 cell   A type of effector helper T cell that secretes IL17, which recruits neutrophils and stimulates an inflammatory response. (Figure 24–44)

T$_H$2 cell   A type of effector helper T cell that helps activate B cells to produce antibodies, to undergo somatic hypermutation, and switch the class of immunoglobulin produced. (Figure 24–44)

thylakoid   Flattened sac of membrane in a chloroplast that contains chlorophyll and other pigments and carries out the light-trapping reactions of photosynthesis. Stacks of thylakoids form the grana of chloroplasts. (Figures 14–35 and 14–36)

thylakoid membrane   Chloroplast membrane system that contains the large membrane protein complexes for photosynthesis and photophosphorylation.

thymocytes   Developing T cells in the thymus.

tight junction   Cell–cell junction that seals adjacent epithelial cells together, preventing the passage of most dissolved molecules from one side of the epithelial sheet to the other. (Figures 19–2 and 19–21)

TIM complexes   Protein translocators in the mitochondrial inner membrane. The TIM23 complex mediates the transport of proteins into the matrix and the insertion of some proteins into the inner membrane; the TIM22 complex mediates the insertion of a subgroup of proteins into the inner membrane. (Figure 12–21)

Toll   A transmembrane receptor protein. On the ventral side of the *Drosophila* egg membrane, its activation controls the distribution of Dorsal, a transcription regulator of the NFκB family.

Toll-like receptors (TLRs)   Family of pattern recognition receptors (PRRs) on or in cells of the innate immune system. They recognize pathogen-associated immunostimulants (PAMPs) associated with microbes. (Figure 24–4)

TOM complex   Multisubunit protein complex that transports proteins across the mitochondrial outer membrane. (Figure 12–21)

TOR   Large, serine/threonine protein kinase that is activated by the PI-3-kinase–Akt signaling pathway and promotes cell growth.

totipotent   Describes a cell that is able to give rise to all the different cell types in an organism.

trans **face**   Face on the other (far) side.

trans **Golgi network (TGN)**   Network of interconnected tubular and cisternal structures closely associated with the *trans* face of the Golgi apparatus and the compartment from which proteins and lipids exit the Golgi, bound for the cell surface or another compartment.

transcellular transport   Transport of solutes, such as nutrients, across an epithelium, by means of membrane transport proteins in the apical and basal faces of the epithelial cells. (Figure 11–11)

transcription (DNA transcription)   Copying of one strand of DNA into a complementary RNA sequence by the enzyme RNA polymerase. (Figures 6–1 and 6–8)

transcription regulators   General name for any protein that binds to a specific DNA sequence (known as a *cis*-regulatory sequence) to influence the transcription of a gene.

transcriptional control   Regulation by a cell of gene expression by controlling when and how often a given gene is transcribed.

transcytosis   Uptake of material at one face of a cell by endocytosis, its transfer across a cell in vesicles, and discharge from another face by exocytosis. (Figure 13–58)

transfer RNA (tRNA)   Set of small RNA molecules used in protein synthesis as an interface (adaptor) between mRNA and amino acids. Each type of tRNA molecule is covalently linked to a particular amino acid. (Figure 6–50)

transferrin receptor   Cell-surface receptor for transferrin (a soluble protein that carries iron) that delivers iron to the cell interior via receptor-mediated endocytosis and recycling of the receptor–transferrin complex.

transformed   A cell with an altered phenotype that behaves in many ways like a cancer cell (i.e., unregulated proliferation, anchorage-independent growth in culture).

transforming growth factor-β superfamily (TGFβ superfamily)   Large family of structurally related secreted proteins that act as hormones and local mediators to control a wide range of functions in animals, including during development. It includes the TGFβ/activin and bone morphogenetic protein (BMP) subfamilies. (Figure 15–57)

transgene   The foreign or modified gene that has been added to create a transgenic organism.

transgenic organism   Plant or animal that has stably incorporated one or more genes from another cell or organism (through insertion, deletion, and/or replacement) and can pass them on to successive generations. (Figures 8–53 and 8–70)

transit amplifying cell   Cell derived from a stem cell that divides a limited number of times before terminally differentiating.

transition state   Structure that forms transiently in the course of a chemical reaction and has the highest free energy of any reaction intermediate. Its formation is a rate-limiting step in the reaction. (Figure 3–47)

translation (RNA translation)   Process by which the sequence of nucleotides in an mRNA molecule directs the incorporation of amino acids into protein. Occurs on a ribosome. (Figures 6–1 and 6–64)

translational control   Regulation by a cell of gene expression by selecting which mRNAs in the cytoplasm are translated by ribosomes.

translocon   The assembly of a translocator associated with other membrane complexes, such as enzymes that modify the growing polypeptide chain.

transmembrane adhesion proteins Cytoskeleton-linked transmembrane molecules with one end linking to the cytoskeleton inside the cell and the other end linking to other structures outside it.

transmembrane protein Membrane protein that extends through the lipid bilayer, with part of its mass on either side of the membrane. (Figure 10–17)

transmitter-gated ion channel (ion-channel-coupled receptor, ionotropic receptor) Ion channel found at chemical synapses in the postsynaptic plasma membranes of nerve and muscle cells. Opens only in response to the binding of a specific extracellular neurotransmitter. The resulting inflow of ions leads to the generation of a local electrical signal in the postsynaptic cell. (Figures 11–36 and 15–6)

transport vesicle Membrane-enclosed transport containers that bud from specialized coated regions of donor membrane and pass from one cell compartment to another as part of the cell's membrane transport processes; vesicles can be spherical, tubular, or irregularly shaped.

transporter (carrier protein, permease) Membrane transport protein that binds to a solute and transports it across the membrane by undergoing a series of conformational changes. Transporters can transport ions or molecules passively down an electrochemical gradient or can link the conformational changes to a source of metabolic energy such as ATP hydrolysis to drive active transport. *Compare* **channel protein**. *See also* **membrane transport protein**. (Figure 11–3)

transposable element (transposon) Segment of DNA that can move from one genome position to another by transposition. (Table 5–4, p. 288)

transposition (transpositional recombination) Movement of a DNA sequence from one genome site to another. (Table 5–4, p. 288)

transposon *see* **transposable element**

treadmilling Process by which a polymeric protein filament is maintained at constant length by addition of protein subunits at one end and loss of subunits at the other. (Panel 16–2, pp. 902–903)

trimeric GTP-binding protein *see* **G protein**

Trithorax group Set of proteins critical for cell memory that maintains the transcription of *Hox* genes in cells where transcription has already been switched on.

t-SNAREs Transmembrane SNARE protein, usually composed of three proteins and found on target membranes where it interacts with v-SNAREs on vesicle membranes.

tubulin The protein subunit of microtubules. (Panel 16–1, p. 891, and Figure 16–42)

γ-tubulin ring complex (γ-TuRC) Protein complex containing γ-tubulin and other proteins that is an efficient nucleator of microtubules and caps their minus ends.

tumor progression Process by which an initial mildly disordered cell behavior gradually evolves into a full-blown cancer. (Figures 20–8 and 20–9)

tumor suppressor gene Gene that appears to help prevent formation of a cancer. Loss-of-function mutations in such genes favor the development of cancer. (Figure 20–17)

tumor virus Virus that can help make the cell it infects cancerous.

turgor pressure Large hydrostatic pressure developed inside a plant cell as the result of the intake of water by osmosis; it is the force driving cell expansion in plant growth and it maintains the rigidity of plant stems and leaves.

two-dimensional gel electrophoresis Technique combining two different separation procedures—separation by charge (isoelectric focusing) in the first dimension, then separation by size in a direction at a right angle to that of the first step—to resolve up to 2000 proteins in the form of a two-dimensional protein map.

type III fibronectin repeat The major repeat domain in fibronectin, it is about 90 amino acids long and occurs at least 15 times in each subunit. The repeat is among the most common of all protein domains in vertebrates.

type III secretion system One of several secretion systems in Gram negative bacteria; delivers effector proteins into host cells in a contact-dependent manner. (Figure 23–7)

type IV collagen An essential component of mature basal laminae consisting of three long protein chains twisted into a ropelike superhelix with multiple bends. Separate molecules assemble into a flexible, felt-like network that gives the basal lamina tensile strength.

tyrosine kinase Enzyme that phosphorylates specific proteins on tyrosines. *See also* **cytoplasmic tyrosine kinase**.

tyrosine-kinase-associated receptor Cell-surface receptor that functions similarly to RTKs, except that the kinase domain is encoded by a separate gene and is noncovalently associated with the receptor polypeptide chain.

ubiquitin Small, highly conserved protein present in all eukaryotic cells that becomes covalently attached to lysines of other proteins. Attachment of a short chain of ubiquitins to such a lysine can tag a protein for intracellular proteolytic destruction by a proteasome. (Figure 3–69)

ubiquitin ligase Any one of a large number of enzymes that attach ubiquitin to a protein, often marking it for destruction in a proteasome. The process catalyzed by a ubiquitin ligase is called ubiquitylation. (Figure 3–71)

unfolded protein response Cellular response triggered by an accumulation of misfolded proteins in the endoplasmic reticulum. Involves expansion of the ER and increased transcription of genes that code for endoplasmic reticulum chaperones and degradative enzymes. (Figure 12–51)

uniporter Carrier protein that transports a single solute from one side of the membrane to the other. (Figure 11–8)

V(D)J recombination Somatic recombination process by which gene segments are brought together to form a functional gene for a polypeptide chain of an immunoglobulin or T cell receptor. (Figure 24–28)

vacuole Large fluid-filled compartment found in most plant and fungal cells, typically occupying more than a third of the cell volume. (Figure 13–41)

van der Waals attraction Type of (individually weak) noncovalent bond that is formed at close range between nonpolar atoms. (Table 2–1, p. 45 and Panel 2–3, pp. 94–95)

variable region Region of an immunoglobulin or T cell receptor polypeptide chain that is the most variable and contributes to the antigen-binding site. (Figures 24–25 and 24–32)

vascular endothelial growth factor (VEGF) Secreted protein that stimulates the growth of blood vessels. (Table 15–4, p. 850, and Figure 22–26)

vesicle transport model   One hypothesis for how the Golgi apparatus achieves and maintains its polarized structure and how molecules move from one cisterna to another. This model holds that Golgi cisternae are long-lived structures that retain their characteristic set of Golgi-resident proteins firmly in place, and cargo proteins are transported from one cisterna to the next by transport vesicles.

vesicular transport   Transport of proteins from one cell compartment to another by means of membrane-bounded intermediaries such as vesicles or organelle fragments.

***V* gene segment**   A DNA sequence encoding most of the variable region of an immunoglobulin or T cell receptor polypeptide chain. There are many different *V* gene segments, one of which becomes joined to a *D* or *J* gene segment by somatic recombination when an individual lymphoid progenitor cell begins to differentiate into a B or T lymphocyte. (Figure 24–28)

virulence factor   Protein, encoded by a virulence gene, that contributes to an organism's ability to cause disease.

virulence gene   Gene that contributes to an organism's ability to cause disease.

virus   Particle consisting of nucleic acid (RNA or DNA) enclosed in a protein coat and capable of replicating within a host cell and spreading from cell to cell. (Figure 23–11)

virus receptor   Molecule on the host cell surface to which virus surface proteins bind to enable binding of virus to the cell surface.

voltage-gated cation channel   Type of ion channel found in the membranes of electrically excitable cells (such as nerve, endocrine, egg, and muscle cells). Opens in response to a shift in membrane potential past a threshold value.

voltage-gated K⁺ channel   Ion channel in the membrane of nerve cells that opens in response to membrane depolarization, enabling K⁺ efflux and rapid restoration of the negative membrane potential.

voltage-gated Na⁺ channel   Ion channel in the membrane of nerve and skeletal muscle cells that opens in response to a stimulus causing sufficient depolarization, allowing Na⁺ to enter the cell down its electrochemical gradient

v-SNAREs   Transmembrane SNARE protein, comprising a single polypeptide chain, usually found in vesicle membranes where it interacts with t-SNAREs in target membranes.

V-type pumps   Turbine-like protein machines constructed from multiple different subunits that use the energy of ATP hydrolysis to drive transport across a membrane. The V-type proton pump transfers H⁺ into organelles such as lysosomes to acidify their interior. (Figure 11–12)

WASp protein   Key target of activated Cdc42. Exists in an inactive folded conformation and an activated open conformation; association with Cdc42 stabilizes the open form, enabling binding to the Arp 2/3 complex and enhancing actin-nucleating activity.

Wee1   Protein kinase that inhibits Cdk activity by phosphorylating amino acids in the Cdk active site. Important in regulating entry into M phase of the cell cycle.

Western blotting   Technique by which proteins are separated by electrophoresis and immobilized on a paper sheet and then analyzed, usually by means of a labeled antibody. Also called immunoblotting.

white blood cell   General name for all the nucleated blood cells lacking hemoglobin. Also called leukocytes. Includes lymphocytes, granulocytes, and monocytes. (Figure 22–27)

Wnt protein   Member of a family of secreted signal proteins that have many different roles in controlling cell differentiation, proliferation, and gene expression in animal embryos and adult tissues.

Wnt/β-catenin pathway   Signaling pathway activated by binding of a Wnt protein to its cell-surface receptors. The pathway has several branches. In the major (canonical) branch, activation causes increased amounts of β-catenin to enter the nucleus, where it regulates the transcription of genes controlling cell differentiation and proliferation. Overactivation of the Wnt/β-catenin pathway can lead to cancer. (Figure 15–60)

X-inactivation   Inactivation of one copy of the X chromosome in the somatic cells of female mammals.

X-inactivation center (XIC)   Site in an X chromosome at which inactivation is initiated and spreads outward.

x-ray crystallography   Technique for determining the three-dimensional arrangement of atoms in a molecule based on the diffraction pattern of x-rays passing through a crystal of the molecule. (Figure 8–21)

zygote   Diploid cell produced by fusion of a male and female gamete. A fertilized egg.

Index

Note
The index covers the main text but not the problems or summaries. F indicates coverage in a Figure only, on a page separated from the indexed text treatment and T indicates inclusion in a Table. Figures on the same page as relevant text are not indexed separately.

Acronyms have generally been preferred to their expansions. Common names have been preferred for organs, so "stomach cancer" not "gastric cancer."

Sorting is word-by-word, ignoring spaces and punctuation, so that "ATP synthesis" will appear before "ATPases."

Page numbers with an F refer to a figure; page numbers with a T refer to a table.

Page numbers with an F refer to a figure; page numbers with a T refer to a table.

Page numbers with an F refer to a figure; page numbers with a T refer to a table.

Page numbers with an F refer to a figure; page numbers with a T refer to a table.

Page numbers with an F refer to a figure; page numbers with a T refer to a table.

Page numbers with an F refer to a figure; page numbers with a T refer to a table.

Page numbers with an F refer to a figure; page numbers with a T refer to a table.

Page numbers with an F refer to a figure; page numbers with a T refer to a table.

Page numbers with an F refer to a figure; page numbers with a T refer to a table.

Page numbers with an F refer to a figure; page numbers with a T refer to a table.

Page numbers with an F refer to a figure; page numbers with a T refer to a table.

Page numbers with an F refer to a figure; page numbers with a T refer to a table.

Page numbers with an F refer to a figure; page numbers with a T refer to a table.

Page numbers with an F refer to a figure; page numbers with a T refer to a table.

Page numbers with an F refer to a figure; page numbers with a T refer to a table.

Page numbers with an F refer to a figure; page numbers with a T refer to a table.

Page numbers with an F refer to a figure; page numbers with a T refer to a table.

Page numbers with an F refer to a figure; page numbers with a T refer to a table.

The Genetic Code

1st Position (5′ end) ↓	2nd Position				3rd Position (3′ end) ↓
	U	C	A	G	
U	Phe	Ser	Tyr	Cys	U
	Phe	Ser	Tyr	Cys	C
	Leu	Ser	STOP	STOP	A
	Leu	Ser	STOP	Trp	G
C	Leu	Pro	His	Arg	U
	Leu	Pro	His	Arg	C
	Leu	Pro	Gln	Arg	A
	Leu	Pro	Gln	Arg	G
A	Ile	Thr	Asn	Ser	U
	Ile	Thr	Asn	Ser	C
	Ile	Thr	Lys	Arg	A
	Met	Thr	Lys	Arg	G
G	Val	Ala	Asp	Gly	U
	Val	Ala	Asp	Gly	C
	Val	Ala	Glu	Gly	A
	Val	Ala	Glu	Gly	G

Amino Acids and Their Symbols			Codons
A	Ala	Alanine	GCA GCC GCG GCU
C	Cys	Cysteine	UGC UGU
D	Asp	Aspartic acid	GAC GAU
E	Glu	Glutamic acid	GAA GAG
F	Phe	Phenylalanine	UUC UUU
G	Gly	Glycine	GGA GGC GGG GGU
H	His	Histidine	CAC CAU
I	Ile	Isoleucine	AUA AUC AUU
K	Lys	Lysine	AAA AAG
L	Leu	Leucine	UUA UUG CUA CUC CUG CUU
M	Met	Methionine	AUG
N	Asn	Asparagine	AAC AAU
P	Pro	Proline	CCA CCC CCG CCU
Q	Gln	Glutamine	CAA CAG
R	Arg	Arginine	AGA AGG CGA CGC CGG CGU
S	Ser	Serine	AGC AGU UCA UCC UCG UCU
T	Thr	Threonine	ACA ACC ACG ACU
V	Val	Valine	GUA GUC GUG GUU
W	Trp	Tryptophan	UGG
Y	Tyr	Tyrosine	UAC UAU